36th Annual USSD Conference 2016

Celebrating the Value of Dams and Levees – Yesterday, Today and Tomorrow

Denver, Colorado, USA
11 - 15 April 2016

Volume 1 of 2

ISBN: 978-1-5108-2942-8

Printed from e-media with permission by:

Curran Associates, Inc.
57 Morehouse Lane
Red Hook, NY 12571

proceedings.com

Some format issues inherent in the e-media version may also appear in this print version.

Printed by Curran Associates, Inc. (2016)

United States Society on Dams (USSD)
1616 Seventeenth Street #483
Denver, CO 80202
USA

Phone: 303-628-5430
Fax: 303-628-5431

gene@ussdams.org

Additional copies of this publication are available from:

Curran Associates, Inc.
57 Morehouse Lane
Red Hook, NY 12571 USA
Phone: 845-758-0400
Fax: 845-758-2633
Email: curran@proceedings.com
Web: www.proceedings.com

United States Society on Dams

USSD

Celebrating the Value of Dams and Levees — Yesterday, Today and Tomorrow

36th Annual USSD Conference
Denver, Colorado, April 11-15, 2016

Gross Dam is a concrete gravity arch dam in Boulder County, owned by Denver Water. A planned modification will raise the dam 131 feet with a downstream roller compacted concrete buttress, to a final height of 471 feet. The new dam and reservoir will provide an additional 77,000 acre-feet of municipal water storage. Denver Water expects to start final design engineering in late 2016, with construction set to begin in 2018.

U.S. Society on Dams

Vision

A world class organization dedicated to advancing the role of dam and levee systems and building the community of practice.

Mission

USSD, as the United States member of the International Commission on Large Dams, is dedicated to:

ADVOCATE: Champion the role of dam and levee systems in society.

EDUCATE: Be the premier source for technical information about dam and levee systems.

COLLABORATE: Build networks and relationships to strengthen the community of practice.

CULTIVATE: Nurture the growth of the community of practice.

FOREWORD

Included in these Proceedings are papers presented during the **36th USSD Annual Meeting and Conference**, held April 11-15, 2016, in Denver, Colorado. A separate book includes abstracts for each paper, as well as abstracts for presentations not included in the Proceedings.

The theme of the 36th USSD Conference was ***Celebrating the Value of Dams and Levees — Yesterday, Today and Tomorrow.*** The Conference technical program was organized by several USSD Committees under the leadership of **Rodney Eisenbraun**, RJH Consultants, Inc. Dams and levees have provided and will continue to provide significant benefits to our society. Many of our existing dams and levees provide greater benefits today than they did when they were first constructed and they will be needed to provide even greater benefits to address the challenges of the future. USSD has been and continues to be a source for balance and reason — neither ignoring the negative impacts nor their benefits. USSD and its members have worked and continue to work to identify and advance real solutions that minimize impacts and maximize the benefits of dams and levees. The papers included in these Proceedings address the Conference Theme, and other critical technical and management topics for the dam professional.

Technical Sessions during the Conference focused on best practices and new technologies related to concrete dams, embankment dams and levees; dam safety; public safety and security; construction and rehabilitation; hydraulics; foundations, monitoring; and seismic and environmental issues.

The papers in the Proceedings were selected from abstracts submitted in response to a Call for Papers, and include both oral and poster presentations. Authors are specialists with broad experience from government agencies, utilities, academia, water districts, consulting firms and private industry.

The Conference Organizing Committee extends thanks and appreciation to the **Bureau of Reclamation**, Host of the 36th Annual Meeting and Conference.

Special thanks are also extended to the Committee Members who selected the abstracts and reviewed the technical papers, and to the authors who prepared the papers included in the Proceedings.

CONTENTS

Conference Theme

Is This Community Worth Saving? 1
Robert Beduhn, HDR

Unintended Consequences of Dam Removal 11
Melinda Dirdal, **Greg Paxson** and **Dave Campbell**, Schnabel Engineering

The Kentucky River: Managing 19th Century Infrastructure to Meet 21st Century Needs 29
Daniel A. Gilbert, **Jeffrey S. Dingrando** and **April L. Welshans**,fMGS Stantec Consulting Services, Inc.; and **David Hamilton**, Kentucky River Authority

Historical Perspective — Dams, Hydro, and Multi-Purpose Projects 45
David Kleiner, MWH

Hydraulics

Use of Labyrinth Piano Key Weir (PKW) Spillway to Improve Dam Safety for a Large Set of Dams 61
Frédéric Laugier, **Julien Vermeulen** and **Thomas Pinchard**, EDF-CIH

Ensuring Water Supply Reliability with Innovative PK Weir Spillway Design 79
G. Robblee, **S. Kees** and **B.M. Crookston**, Schnabel Engineering; and **K. Keel**, Town of Hillsborough

Importance of Site Considerations for Labyrinth Spillway Hydraulic Design — Upper Brushy Creek Dam 7 Modernization 97
Tina Stanard and **Victor Vasquez**, Freese and Nichols, Inc.; **Ruth Haberman**, Upper Brushy Creek Water Control and Improvement District; **Blake Tullis**, Utah State University; and **Bruce Savage**, Idaho State University

Automated Incremental Hazard and Sensitivity Analysis of a Hypothetical Dam Breach Model 107
Mikell Warms, **Dan Christensen** and **Christopher Goodell**, WEST Consultants, Inc.; and **Cynthia Oestreich**, Aista Utilities

GSSA Modeling for Analysis of Flood Design Features at the Picayune Strand Restoration Project 123
Charles W. Downer, **Jaime A. Graulau-Santiago**, **Brian E. Skahill**, **David M. Weston**, **Nawa Raj Pradhan** and **Aaron R. Byrd**, Corps of Engineers

Using the Variable Infiltration Capacity Model in a Semi-Stochastic Format for Hydrologic Hazard Estimates for Dam Safety Risk Analysis 139
Frank Dworak, Bureau of Reclamation

Dynamic Desert Floods at a Dry Dam 149
Robert A. Eichinger, **Zachary Whitten** and **Michael Gerlach**, Stantec Consulting Services, Inc.

Cultivating Ingenuity and Safety In Alabama: The Taming of Lake Ogletree Reservoir 165
James R. Crowder, **Brian M. Crookston**, **Bradley T. Boyer** and **J. Tyler Coats**, Schnabel Engineering

Replacing Ruby Dam Outlet Uses Computational Fluid Dynamics to Model Energy Dissipation 179
Frank Lan, **Robert Waddell** and **Michael Zusi**, AECOM; and **Brian Grant**, Montana DNRC

Reducing Scour with Flow Splitters, Extension Fins and Computational Fluid Dynamics.......................................199
Duane McClelland and **Daniel Morse**, CH2M; and **Nathan M. Gullo** and **Jason A. Eichler**, LCRA

Sediment Transport Analysis in Support of Prado Basin Sediment Management Project.......................................221
Craig P. Baxter, Golder Associates Inc.; **Nathan Scheevel**, Scheevel Engineering, LLC; and **George W. Annandale**, Golder Associates Inc.

Verification of Sediment Yield Calculations in the Arid Southwest.......................................237
Brian Wahlin and **Jesse Piotrowski**, WEST Consultants, Inc.; and **Chuck Davis**, Beaver Creek Hydrology

Post-Removal Reservoir Erosion: Elwha Dam and Glines Canyon Dam Removals, Washington 251
Jennifer A. Bountry, Bureau of Reclamation; **Andy Ritchie**, National Park Service; and **Timothy J. Randle**, Bureau of Reclamation

Unit Hydrograph Peaking Factors and their Application to PMF Simulations.............269
Michael Bartles and **Matthew Fleming**, Corps of Engineers

Areal Reduction Factors — Colorado Front Range and Analysis of the 2013 Storm.......................................287
Doug Hultstrand and **Bill Kappel**, Applied Weather Associates, LLC

Evaluating Effort Levels of Inflow Design Flood Study Methods for Dams...................303
S. Samuel Lin, Federal Energy Regulatory Commission; and **Jason E. Hedien**, MWH Americas, Inc.

Monitoring

2015 USACE Instrumentation and Monitoring Program: Internal Peer Review Results 315
Georgette Hlepas and **Travis Tutka**, Corps of Engineers

The Hydrogeological Model of Boone Dam 335
Scottie Barrentine, **Andrew Dodson** and **Lindsay McMichen**, Tennessee Valley Authority

Relearning How to Look at Piezometric Data for Seepage Evaluation 351
Richard S. Olsen and **Isaac J. Stephens**, Corps of Engineers

Evaluation and Monitoring Response to Upstream Slope Failure at an Embankment Dam 379
Jay N. Stateler and **Jeff Wormer**, Bureau of Reclamation

On Calero Pond: Evaluation of Downstream Wetland and Potential Dam Seepage 395
Matthew Powers, GEI Consultants Inc.; **Bal Ganjoo**, Santa Clara Valley Water District; and **Mark Freitas**, GEI Consultants Inc.

Maintaining Performance — Rehabilitation of Critical Drains at Crab Orchard Dam 415
David Snyder and **Robert Saber**, Gannett Fleming, Inc.; and **David Hibbs**, U.S. Fish and Wildlife Service

Foundation Protection during Post Tensioned Anchor Construction 435
William B. Stiller and **Y. Erwin Supranata**, HDR; and **Stephen J. Mockler**, Brookfield Renewable Energy Group

Environment

Buchanan Dam — Keeping Current for Eight Decades 445
Victor M. Vasquez, Freese and Nichols, Inc.; **Doug Witkowski,** Lower Colorado River Authority; **M. Leslie Boyd**, Freese and Nichols, Inc.; **Monica Masters**, Lower Colorado River Authority; and **Layne Bukhair**, Freese and Nichols, Inc.

Risk Management for Dam Construction Projects — An Owner's Perspective 469
Susan S. Hou, **Tedman Lee**, **Alan Johanson**, and **Daniel L. Wade**, San Francisco Public Utilities Commission

Seismic

Clover and Junction Dams: Risk-Based Seismic Evaluation of Two Slab and Buttress Dams 485
Scott Jones, **Steve O'Brien** and **David Hughes**, AECOM

Nonlinear Seismic Evaluation and Acceptance Criteria for Arch Dams 497
Yusof Ghanaat, Quest Structures, Inc.

Developing Time Histories with Acceptable Record Parameters for Dillon Dam 515
Dina Bourliea Hunt, MWH Global; **Justin Beutel**, Pacific Gas and Electric Company; and **Christine Weber**, MWH Global

Seismic Deformation Analyses of Embankment Dams: A Reviewer's Checklist 535
Ross W. Boulanger, University of California, Davis; and **Michael H. Beaty**, Beaty Engineering LLC

Lessons from Earthquake Lake in Preparing for Seismically Induced Landslides 547
Roger L. Kay, Corps of Engineers

Concrete

Non-Destructive Testing of Post-Tensioned Anchorages at JT Myers and Markland Dams 563
Robert F. Lindyberg and **Afshin Karshenas**, FDH Velocitel; **Jeff Lillycrop**, **Charles E. Wiggins**, **Larry L. Dalton II** and **Matthew Watts**, Corps of Engineers; and **Brandon Compton,** FDH Velocitel

USACE Portland District Spillway Gate Rehabilitation Program 581
Matthew Craig, **Laurie Ebner**, **William Fortuny**, **David Hamernik**, **Matt Hess** and **Mark Sawka**, Corps of Engineers

Lessons Learned from the Spillway Gate Failure at Folsom Dam 599
Jerzy Salamon, Bureau of Reclamation

A Finite Element Approach to Predict the Stability of Gravity Dams Laying on Rough Rock Interfaces 611
Sylvain Renaud and **Najib Bouaanani,** Polytechnique Montréal; and **Benjamin Miquel**, Hydro-Québec

Applications of FLOW-3D for Stability Analyses of Concrete Spillways at FERC Projects 627
Elise Dombeck, Federal Energy Regulatory Commission

The Stability of Masonry Walls and Dams: A Review of the Historical Design of Concrete Dams........643
Kevin Griebenow, Federal Energy Regulatory Commission

A New and Advanced Dam Performance and Response Integration Program........657
Ziyad Duron, Harvey Mudd College; and **John Yen**, Southern California Edison

Practical Analysis and Assessment of AAR-Affected Dams and Hydroelectric Plants........675
Dan D. Curtis, **Lingmim (Frank) Feng**, **Gurinderbir S. Sooch** and **Jiqin (Tom) Zheng**, Hatch, Ltd; and **John Fletcher,** NB Power

Using the Trial Load Method to Optimize Feasibility Design of Watana Dam........695
Aled Hughes, **Glenn Tarbox** and **Brian Sadden**, MWH Americas, Inc.; and **Bryan Carey**, Alaska Energy Authority

Linear and Nonlinear Seismic Analysis for the Post Tensioning System Design of the Wanapum Dam Spillway Monoliths........711
Bashar S. Sudah, **Juan E. Quiroz**, **Bryan J. Brumley**, **David P. Thompson** and **Richard E. Barrie**, MWH

Static and Dynamic Mass Concrete Material Properties of a Concrete Gravity Dam........727
Soheil Razavi Darbar, **David Queen** and **Chris Hatton**, BC Hydro; **Timothy Dolen**, Dolen and Associates LLC; and **Katie Bartojay**, Bureau of Reclamation

Levees

Using Risk Assessment to Inform Levee System Evaluation Decisions for the National Flood Insurance Program........749
David A. Margo, **Lea G. Adams**, **Scott E. Shewbridge** and **Richard Varuso**, Corps of Engineers

Using Tolerable Risk Guidelines in the California Delta........765
Lawrence H. Roth and **Jessica Ludy**, Arcadis U.S.

Geotechnical Evaluation of the Levees in the Central Valley, California........781
Richard Millet, AECOM; **Steve Mahnke**, California Department of Water Resources; **Mary Perlea**, Corps of Engineers; and **Ray Costa**, Geotechnical Consultant

Phased Investigation Approach for a Partial Slope Failure on the Rio Grande........799
Isaac J. Stephens, **Lucas A. Walshire** and **Joseph B. Dunbar**, Corps of Engineers

Embankment Dams

Reducing Uncertainties when Assessing the Stability of Rockfill Dams: Characterizing Analysis Parameters 813
Daniel B. Rogers, **Kyle R. Blakley** and **Jeffrey S. Dingrando**, Stantec Consulting Services Inc.; and **Jeff W. Munsey**, Tennessee Valley Authority

Don't Judge a Rockfill Dam by its Name Alone 829
Brian Benson and **Peter Amos**, Damwatch Engineering Ltd

Shear Strength of Compacted Beaumont Clay for Consolidated-Undrained Conditions 849
Mark J. Thompson and **David J. Bentler**, CH2M; **Thomas L. Brandon**, Virginia Polytechnic Institute and State University; and **Daniel R. VandenBerge**, Tennessee Technological University

Estimating the Peak Friction Angle of Sandy Soils In Situ with State-Based Overburden Normalized CPT Tip Resistance 865
Robert A. Jaeger, GEI Consultants, Inc.; and **Ian P. Maki**, California Department of Water Resources

Decades of Foundation Deterioration Threatening Public Safety 883
George L. Sills, George Sills Geotechnical Engineering Consultant, LLC; and **Leslie F. Harder**, HDR

An Empirical Model for Estimating Internal Erosion Critical Gradients for Inclined Exit Face Conditions 909
Richard A. Keizer, GEI Consultants, Inc., **John D. Rice**, Utah State University; and **Robert A. Jaeger**, GEI Consultants, Inc.

Laboratory Measurements of Critical Gradients of Cohesionless Soils 927
Bryant A. Robbins, **Axel Montalvo-Bartolomei**, **Jamie López-Soto** and **Isaac J. Stephens**, Corps of Engineers

Reliability Slope Stability Analysis for Rapid Drawdown 939
Mark J. Thompson and **David J. Bentler**, CH2M; and **Thomas L. Brandon**, Virginia Polytechnic Institute and State University

Embankment Dam Analysis Parameters Estimated from 100 Years of Laboratory Tests 955
David W. Harris, David Harris Engineering and Assistance, LLC/Allegiance Consulting, LLC; and **Garrett Sutley**, Bureau of Reclamation

Calaveras Dam Replacement Project — Providing a Seismically-Reliable Water Source 969
John Roadifer and **Michael Forrest**, AECOM; and **Daniel Wade**, **Susan Hou**, **Tedman Lee** and **Carman Ng**, San Francisco Public Utilities Commission

Dam Decommissioning

Non-Linear Modeling of Sediment Retention Structures for Dam Removal Project.......985
Kristen M. Martin and **Erik J. Malvick**, California Department of Water Resources

Recent Progress Towards Removal of Matilija Dam 1001
Seth Gentzler and **John Roadifer**, AECOM; **Derek Booth** and **Yantao Cui**, Stillwater Sciences; and **Peter Sheydayi**, Ventura County Water Protection District

Dam Safety

Risk Analysis Work for Dams — "Big Picture" Considerations...................................1017
Jay N. Stateler, Bureau of Reclamation; **Kim de Rubertis**, Consulting Engineer; and **Doug Boyer**, Federal Energy Regulatory Commission

Implementing a Risk-Informed Approach to Dam Safety in a Regulatory Environment..1035
Douglas D. Boyer and **William Allerton**, Federal Energy Regulatory Commission

Use of Quantitative Risk Results to Inform Dam Safety Governance: Practical Cases in Europe...1053
Ignacio Escuder-Bueno, Universitat Politècnica de València; **Adrián Morales-Torres**, iPresas; and **Jesica Tamara Castillo-Rodríguez**, Universitat Politècnica de València

The Role of Risk Assessment in the Bureau of Reclamation's Dam Safety Program...1069
William R. Fiedler and **William Engemoen**, Bureau of Reclamation

Semi-Quantitative Portfolio Risk Assessment for Prioritizing Dam Safety Activities...1083
John Yen, **Matthew Muto** and **Nicolas Von Gersdorff,** Southern California Edison Company; and **Gregg A. Scott**, Scott Consulting, LLC

Bureau of Reclamation Risk-Informed Approach to Selecting the Inflow Design Flood...1097
William R. Fiedler, Bureau of Reclamation; and **John H. LaBoon**, Consulting Civil Engineer

The Wanapum Spillway Monolith Cracking Incident Dam Safety Management Implications...1119
Robin G. Charlwood, Robin Charlwood & Associates; **Adam N. Jones** and **Kevin Snyder**, HDR; **Kevin Marshall** and **David Mishalanie**, Grant County Public Utility District No. 2

Panama Canal Borinquen Dams and Third Set of Locks Operations and Maintenance Strategy 1133
Rachael V. Bisnett, **Wonnie Kim** and **Donald J. Montgomery**, MWH

Safety in Numbers? 1145
Kim de Rubertis, Consulting Engineer

Public Safety and Security at Dams Are Not Mutually Exclusive 1161
William F. Foos, **Frank Calcagno** and **Paul G. Schweiger**, Gannett Fleming, Inc.

Emergency Action Planning — Canadian Practice and Lessons Learned at Ontario Power Generation 1173
Tali Serota and **Tony Bennett**, Ontario Power Generation

The Criticality of Engagement Before, During and After a Dam Safety Emergency 1191
Bernie Auld, **Benjamin Heath** and **Keil Neff**, Tennessee Valley Authority

Construction

Stagecoach Spillway Replacement and Performance at Electra Lake 1201
Danielle Tripp Hannes and **Stephen L. Jamieson**, W. W. Wheeler and Associates, Inc.

High Capacity Anchors on Existing Concrete Dams — Application to Two TVA Dams 1217
Husein Hasan and **Aaron Nottis**, Tennessee Valley Authority; **Brian Krohmer**, Barnard Construction Company; and **Dennis Triplett** and **Boris Caro Vargas**, Nicholson Construction Company

Lessons Learned on Post-Tensioned Trunnion Anchorage — Folsom Auxiliary Spillway 1235
Cecily M. Nolan and **Cheuk Y. Wan**, Corps of Engineers

Uncovering and Addressing a History of Significant Incidents After an Annual Dam Safety Inspection 1251
Steven Davidheiser and **Robert A. Kline, Jr.**, Gannett Fleming, Inc.; and **Amanda Lothes,** Newport News Waterworks

Emergency Repairs to Prevent a Piping Failure in the Lake Manatee Dam 1267
Jeffrey A. Beriswill and **Nickolas J. Depin**, Amec Foster Wheeler Environment & Infrastructure, Inc.; and **Mark Simpson** and **Bruce MacLeod**, Manatee County Utilities Department

Jet Grouting for Echo Dam Seismic Remediation 1283
Tara Schenk-McFarland and **Mark Bliss**, Bureau of Reclamation

Improvements to Seepage Cutoff Wall Specifications Considering Lessons Learned from Construction 1297
Michael Kynett, Corps of Engineers; **Khaled Chowdhury** and **Richard Millet**, AECOM; **Mary Perlea**, Corps of Engineers; and **Francke Walberg**, AECOM

Foundations

Levee Assessment Tool: Using Limited Data to Inspect/Evaluate Levees for Various Failure Modes Against Established Criteria 1321
A. Wohletz and **S. Wiesner**, Kleinfelder

Fault Investigations During Borinquen Dam 1E Foundation Panama Canal Expansion 1341
David Schug, **Paul Salter**, **Dan Meier**, **Robert Urban**, **Chris Goetz**, **Lelio Mejia** and **James Toose**, AECOM; and **Derek Irving**, **Pastora Franceschi**, **Maximiliano De Puy**, **Jaime Arrocha** and **Roberto Miranda**, Autoridad del Canal de Panamá

3D Modeling at Boundary Dam, WA: Visualizing a Geologically Complex Site 1365
Robert Cannon and **Frederic Snider**, Schnabel Engineering; **Kim Pate** and **Andre Ball**, Seattle City Light; and **Gary Rogers** and **J. Hawkins Gagnon**, Schnabel Engineering

Underseepage and Heave Mitigation at Douglas Saddle Dam No. 1 1379
Daniel A. Gilbert, **Alan F. Rauch** and **Paul J. Cooper**, Stantec; and **Scottie L. Barrentine**, Tennessee Valley Authority

A Deep Seepage Cutoff for the Thornton Composite Reservoir 1395
Joseph R. Kovacich, MWH; **Louis Storino**, Metropolitan Water Reclamation District of Greater Chicago; **Raymond J. Franz** and **Jonathon R. Bulger**, Hayward Baker Inc.; and **Brian J. Kazyak**, MWH

Designing a Hybrid Dam on a Geologic Profile with Highly Variable Engineering Properties 1419
Richard E. Riker and **Jacob Esterhuizen**, CH2M; **Don Deere**, Deere and Ault Consultants, Inc.; and **Gabriel Fernandez-Delgado**, University of Illinois

Lunch Presentation and Poster Session

Things You Wanted to Know About Individual Consulting, but Were Afraid to Ask 1447
Larry K. Nuss, Nuss Engineering; and **Kenneth D. Hansen**, Consulting Engineer

Construction of La Romaine Complex in Northern Quebec, Canada: Behaviour of Asphalt Core Rockfill Dams 1463
Vlad Alicescu, **Jean Pierre Tournier**, **Redha Kara** and **Dan Rosculet**, Hydro-Québec

Selecting an Intensity Measure for Fragility Curves Derived from Performance-Based Testing 1483
Spenser Anderson, **Kunal Menda**, **Mo Zhao** and **Ziyad Duron**, Harvey Mudd College

Retrofit and Closure of CCR Impoundments to Meet New Regulations 1497
Matt Bishop, **Brian Owens** and **Luke Williams**, Amec Foster Wheeler

Antecedent Snow Covered Area Analysis for Inflow Design Floods 1507
Steven F. Daly and **Angela Duren**, Corps of Engineers

Evaluation of New Storage with Optimal Hydropower: Weiser-River Basin, Idaho 1521
Jeremy Giovando, Corps of Engineers

Field Tests of Placed Riprap as Erosion Protection Against Accidental Overtopping and Leakage 1531
Priska Helene Hiller, **Fredrikke Kjosavik**, **Leif Lia** and **Jochen Aberle**, Norwegian University of Science and Technology

Design-Build Construction of the Panama Canal Borinquen Dams 1W, 2W, and 2E 1541
Gabriel Martinez, **Wonnie Kim** and **Carlos Zambrano**, MWH

A Performance-Based Approach to Developing Fragility Curves for Structures Subject to Seismic Loading Conditions 1559
Kunal Menda, **Mo Zhao**, **Spenser Anderson** and **Ziyad Duron**, Harvey Mudd College

Implementing Lessons Learned from Wheeler Lock Collapse on New Lock Construction Projects 1575
Michael A. Morrison, Tennesee Valley Authority; and **Bernard Schulte**, Corps of Engineers

Assessing and Evaluating the Operation of Pressure Cells in Embankment Dams: A Case Study of Five Iranian Embankment Dams 1591
Bahareh Shoghli, University of North Dakota; **Ali Asghar Mirghasemi**, University of Tehran; and **Yeo Howe Lim**, University of North Dakota

Stability of Slopes in Embankment Dams 1603
Gopi Siddappa and **Santhosh G. Gowda**, P.E.S. College of Engineering

Ritschard Dam Monitoring: Awareness Leads to Learning which Leads to Action 1621
Ray D. Tenney and **Don B. Meyer**, Colorado River Water Conservation District; and **Michael R. May**, AECOM

Boundary Dam Spillway No. 2 Modifications 1639
John Werner and **James H. Rutherford**, Hatch Associates Consultants Inc.; **Kim Pate** and **Daniel Kirschbaum**, Seattle City Light; **Joe Groeneveld**, Hatch Associates Consultants Inc.; **Keith Moen,** HDR; and **Justin Arnold**, Alden Laboratories Inc.

Selecting A Damage Measure for Fragility Curves Derived from Performance-Based Testing.....1655
Mo Zhao, **Spenser Anderson**, **Kunal Menda** and **Ziyad Duron**, Harvey Mudd College

IS THIS COMMUNITY WORTH SAVING?

Robert Beduhn, PE[1]

ABSTRACT

Across the country, communities face tough decisions. In an era of decreased federal funding, climate change, sea level rise, and increasing disaster costs, many major flood control programs are caught in a financing quandary that is jeopardizing long-term community resiliency and viability. Resiliency is the ability for social, economic, and infrastructure systems to experience an extreme climatic, seismic, or human-made event (shock) and either perform through such an event or return to normal function rapidly following the event. Designing for resiliency is a fundamentally different design paradigm than designing for a specified return interval, such as using statistical analysis of weather or earthquake data. In the United States, entire communities and societies have been built on the premise of a static climate and environment and the predictability of future floods and droughts. Now is the time to begin discussing what our country will look like given that the original assumptions—those that formed the basis of our water-based infrastructure such as dams, levees, and waterways—are not static; they are changing and make our infrastructure prone to failure. Bold leadership is needed to reformulate our infrastructure planning paradigms into a more adaptable and resilient approach that anticipates a greater range of loads and incorporates the ability to adapt and change. This leadership needs to consider the impacts of a global economy, that business activity may not stay in the United States, and that much of our water-based infrastructure was designed on a false premise of climate stationarity.

INTRODUCTION

Across the country, communities face tough decisions. In an era of decreased federal funding, climate change, sea level rise, and increasing disaster costs, many major flood control programs are caught in a financing quandary that is jeopardizing long-term community resiliency and viability. Business risk metrics are prompting business leaders to ponder their long-term commitment to any given region or community depending on the hazard risks that threaten business continuity.

Under the National Economic Development plan, federal water resource projects must consider regional economic development (RED) and national economic development (NED) when considering the benefits versus the cost of projects. The benefit to cost ratio approach has also been used to consider many locally financed projects as well. In the last several decades, it has been a challenge to finance projects with federal funds with benefit to cost ratios below 2.5. Projects with marginal benefit to cost ratios are left in limbo, stuck between an identified community need but lacking a clear federal cost share partner to leverage and advance the project.

[1] Director, Civil Works, HDR, 8404 Indian Hills Drive, Omaha, NE 68114, bob.beduhn@hdrinc.com

In addition, the business community is increasingly more sophisticated in making risk decisions. The increasing availability of global climatic and risk models is providing greater ability to evaluate supply chain, business continuity, and economic risks in a manner that monetizes such factors. A paradigm is emerging in which businesses can determine which assets are exposed to disaster risks. In areas were major disaster mitigation spending is in limbo, anecdotal evidence is mounting that business decisions will look to maximize revenue from those properties in the short term by minimizing short-term investments. Burden is placed on the local community to solve the greater disaster risk with the threat that the business will just close the facility after the next disaster and transfer those business functions to another, more resilient location. The RED or NED is no help to the community trying to preserve its business environment because the economic activity is assumed to relocate to another area of the country or region. Where does that leave the local community?

The paper will explore how the community resiliency movement, economics, and increased disaster risk are changing the approach to major dam, levee, and flood control programs. The paper will discuss the addition of community lifeline systems into the disaster management conversation. In some communities, it may be necessary to move, abandon, or relocate portions of the community to ensure its continued existence.

RESILIENCY

Resiliency can be defined as the ability for social, economic, and infrastructure systems to experience an extreme climatic, seismic, or human-made event (shock) and either perform through such an event or return to normal function rapidly following the event. Designing for resiliency is a fundamentally different design paradigm than designing for a specified return interval, such as using statistical analysis of weather or earthquake data.

The concept of community resiliency did not just happen. It was born out of concerns regarding human-induced climate change and the broad-reaching concern regarding social consequences of natural disasters. The modern resiliency movement can trace its roots to December 1999, when the United Nations General Assembly adopted the International Strategy for Disaster Reduction and formed the UNISDR Secretariat to ensure its implementation (UNISDR 2015a). UNISDR is the United Nations Office for Disaster Risk Reduction that was formed to implement the "Hyogo Framework for Action 2005–2015: Building the Resiliency of Nations and Communities to Disasters" (UNISDR 2015b).

At the center of the Hyogo Framework are three main goals:

- Integration of disaster risk reduction into sustainable development policies and planning
- Development and strengthening of institutions, mechanisms, and capacities to build hazard resilience
- Systematic incorporation of risk reduction approaches into the implementation of emergency preparedness and response and recovery programs

In March 2015, the mandate of the UNISDR was updated by passage of the Sendai Framework for Disaster Risk Reduction 2015–2030 (United Nations 2015). The Sendai Framework is built on lessons learned from the Hyogo Framework. While there were many lessons learned, several social needs were identified that rely on infrastructure resiliency to resolve long-term community needs. Most notably is the recognition that short-term sustainability gains of a community—in terms of reductions in poverty, improvements in public health, and reliability of food, energy, education systems, and jobs—can be rolled back by a lack of infrastructure resiliency to a disaster shock. The underpinning of a stable society is a resilient infrastructure system.

A more recent international trend, in addition to the UNISDR, is the increasing interest of nongovernmental foundations and investment and insurance interests in promoting and supporting the resiliency movement. These include, among others, the Rockefeller Foundation's 100 Resilient Cities initiative, the R!SE Disaster Risk Sensitive Investments initiative, Global Infrastructure Basel (GIB) Foundation, and the U.S. Business Council for Sustainable Development. Brief summaries of each organization's mission and goals follow:

100 Resilient Cities (100RC) – This initiative "is dedicated to helping cities around the world become more resilient to the physical, social and economic challenges that are a growing part of the 21st century. 100RC supports the adoption and incorporation of a view of resilience that includes not just the shocks—earthquakes, fires, floods, etc.—but also the stresses that weaken the fabric of a city on a day to day or cyclical basis" (Rockefeller Foundation 2015).

R!SE Disaster Risk Sensitive Investments – "The R!SE Initiative is an ambitious global response to a daunting global challenge, a new way of collaborating, to unlock the potential for public and private sector actors who are ready and willing to make a step forward and take leadership on disaster risk reduction" (UNISDR 2015c).

Global Infrastructure Basel (GIB) Foundation – This organization "is a Swiss foundation based in Basel working to promote sustainable and resilient infrastructure through sustainable infrastructure design and financing on a global scale" (GIB 2015).

U.S. Business Council for Sustainable Development – The council "is an action oriented and member-led nonprofit business association that harnesses the power of collaborative projects, platforms and partnerships to develop, deploy and scale solutions to ecosystems, energy, materials and water challenges" (U.S. Business Council for Sustainable Development 2015).

These organizations, among others, demonstrate the global nature of the resiliency movement. It crosses geopolitical boundaries by being coordinated at the United Nations to improve the lives of affected people and involves an international finance, business, and insurance system that focus on business continuity and reductions in catastrophic losses resulting from disasters. A recent report by PricewaterhouseCoopers predicts global economic growth at around 3 percent annually (2015). The report highlights that the global economy will double by 2037 and nearly triple by 2050. That growth will occur in combination with ongoing global urbanization, marked by an increase from 54 percent of people living in urban areas today to 66 percent living in urban areas by 2050 (United Nations 2014).

This combination of economic growth and increasing urbanization will likely mean that over 50 percent of the infrastructure needed to meet the global population and economic growth in 2050 has not yet been designed. The world is looking at U.S. disasters such as Hurricanes Katrina and Sandy and realizing it must do better to compete in an increasingly global economy. Economic and investment decisions will be made based on a community's or a country's overall resiliency and sustainability ratings. The winners in the global economy will be those communities and countries that demonstrate shock resilience to business disruptions, maintain continuity of services, and overcome chronic social stresses.

Fear - Fight, Retreat, or Both

In many respects, the resiliency design approach is based on meeting the basic physical and safety needs of the community before, during, and after the disaster. Traditionally, most of the country's water and safety infrastructure is based on life safety criteria or a predetermined return period interval based on the cost to benefit ratio. Earthquake building codes, for example, are intended to provide only sufficient protection to allow occupants to escape the building before collapse. In most circumstances, these buildings will not be safe for use following the event. Maslow's Hierarchy of Human Needs identifies the basic needs of humans (Figure 1).

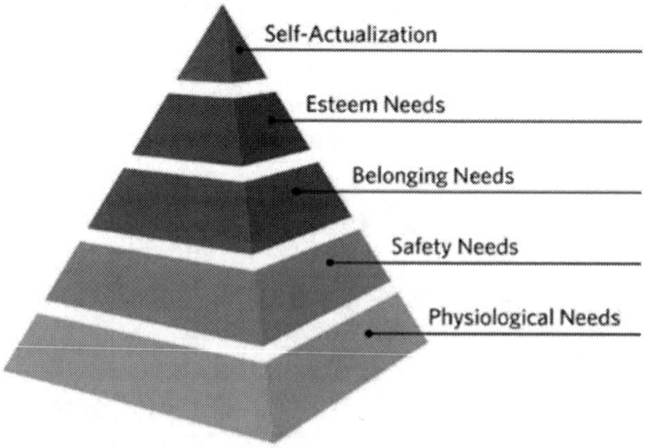

Figure 1. Maslow's Hierarchy of Human Needs

Most dam, levee, and flood control systems work to meet the basic physiological need of survival and safety—but, only up to a point based on economic considerations of RED or NED, and what if the return interval for the flood event is incorrect? The rarity of the event drives down the design level. However, there is little confidence in our ability to predict extreme events, know their real frequency, especially based on historic data. In the United States, the period of record for establishing our design criteria encompasses a relatively short time span. The Blue Hill Meteorological Observatory, in Milton, Massachusetts, is the longest continuous recording weather observatory in the United States. It began weather forecasts in 1895. A total of 120 years of weather observations are available at this site. Coastline surveys began sooner, in 1807, with Thomas Jefferson ordering a systematic survey of the U.S. coastline to facilitate trade and commerce. Consider that most of our hydrological, sea level, and seismic data are derived from a much shorter period of record. Planners and engineers have come to rely on these short periods of time to plan infrastructure and communities. It has become the standard of care to then extrapolate those datasets to represent less frequent, more extreme events. The estimated frequency of these events is based on a limited data set demonstrated to be non-stationary in its distribution. The poor confidence in the frequency estimate of these major events is a fundamental error in benefit to cost ratios.

In many locations, the failure of a large dam, levee, or perimeter flood control system results in complete collapse of the community's ability to provide essential life safety and/or postdisaster needs. However, most members of the public do not spend their day thinking about whether their building is safe, whether they are downstream of a dam, or whether they live behind a levee. However, as innumerable events have shown, it is not until the event is occurring that fear really kicks in. Even then, there is a false confidence that our infrastructure will sustain a community through and after the event. In almost all cases, the fear of an event, however remote, does not prompt significant infrastructure or community investment in disaster preparedness, mitigation, or resilient infrastructure upgrades.

Did Hurricane Katrina prompt the New York City/Newark area to invest ahead of Hurricane Sandy? Has the Fukushima earthquake in Japan prompted West Coast states and communities to take bold action? Did flash floods across the country washing out small dams in numerous communities prompt South Carolina officials to address their backlog of aging, unmaintained private dams? The answer is no—the fear of a disaster, despite analog proofs of their occurrence, does little to drive elected officials to undertake proactive investments to prevent human, ecological, and economic disaster.

The common mantra is to make bold vows such as "Never Again" or "We Will Rebuild" and put forth bold visions to rebuild the community. However, achieving those realities is often difficult. What if the impending disaster gives the community hints that it is coming? These hints are evident along our coastlines. Take Boston, Massachusetts; Alexandria, Virginia; Miami, Florida; or Jean Lafitte, Louisiana. Sea level rise and higher high tides, combined with relatively frequent storm surges, are beginning to routinely disrupt transportation and commerce and to degrade public and private property. Mitigation plans to mitigate are modest: higher sea walls, living shorelines, more pump stations, and elevated roadways. Cost estimates are in the order of $15 million to $40 million for each project to alleviate nuisance flooding. However, are these smaller systems building a false sense of security and promoting investment in vulnerable coastal flooding areas? Because of sea level rise and associated higher high tides, the coastal areas' land use decisions are beginning to resemble the riverine floodplains of the late 1960s through 1980s, when, under the encouragement of the National Flood

Insurance Program, a rash of "100-year" levees were constructed that promoted development in floodplain areas. Will the slow mitigation of nuisance coastal flooding continue to promote coastal investment and development? In his book, *High Tide on Main Street*, author John Englander lays out a worst-case sea level rise scenario based on published scientific data. His conclusion is that the United States will experience significant financial losses in coastal real estate investments when the insurance and mortgage industry realizes the inherent risks and loss of value the inevitable sea level rise and disasters will bring. A recent article in the *Louisville Magazine* validates this claim; realtors routinely shun many of the 12,000 properties within the 100-year floodplain (Austin 2016), leaving homeowners to wait for disaster and take government buyouts only after losing their homes and possessions. Time will tell whether the stay and fight scenario will pan out. The Dutch have made it work, but only after massive national investment in levees, pumps, and canals to fight off the North Sea.

After a disaster, some communities have successfully rebuilt more resilient, while others have struggled with finding the funding and making tough community choices. Other communities have simply ceased to exist in their original locations. The 1997 Red River of the North Flood—which devastated the communities of Grand Forks, North Dakota, and East Grand Forks, Minnesota—is a good example (City of Grand Forks 2016). These communities had fought flood after flood. In the Red River Valley, there is a long history of flood fighting. A series of temporary and permanent levees had been built that severely restricted the floodplain of the river. In April 1997, a combination of rain and snow pushed the river over its levees and prompted the largest peacetime evacuation in U.S. history (until Hurricane Katrina), where 50,000 people got up and left town. The recovery plan included a massive buyout of the river floodplain neighborhoods, diversion of major creeks and waterways, and the construction of a higher, stouter levee system farther back from the river. A significant portion of the river floodplain was abandoned from development and restored to green space. Retreat and rebuild was the only answer. Grand Forks and East Grand Forks were fortunate to receive substantial funding from state and federal sources. Many communities are not so fortunate and sleep with one eye open, waiting for the next disaster. Communities such as Reno, Nevada; Napa, California; and Cedar Rapids, Iowa have partial or stalled projects attributable to low benefit to cost ratios and stalled local funding. For example, the 2008 flood event on the Cedar River is a statistical outlier within recorded history, but is it actually that rare? Given the non-stationarity of the climate, is it the start of a rising trend? The traditional NED approach assumes the stationarity of the weather and leads communities to then enact only the improvements that they can afford or live with a potentially increasing risk over a multiyear implementation schedule. Given the increased globalization of the world's economy, it may be prudent to revisit the RED and NED assumptions related to climate stationarity and that business activity remains in the United States or the region. For example, in Cedar Rapids, the 2008 floods damaged a number of major employers in town. The power plant was closed for good. Other major employers such as Quaker Oats remained, but only after the City of Cedar Rapids committed to a long-term flood control plan that includes a system of buyouts, levees, and floodwalls (City of Cedar Rapids 2016).

Some communities have taken more drastic measures. Following the 1993 flood on the Mississippi River and when the waters receded, the people of Valmeyer, Illinois, signed on for a bold experiment to move their entire town to higher ground. Over 2 years, Valmeyer was rebuilt only 2 miles to the east, but on land hundreds of feet higher. The entire community was relocated out of the Mississippi River floodplain. The ongoing cost to fight the river would have meant the end of the community without this bold move. As of 2013, the relocated town of Valmeyer has continued to grow (Horng 2008). At the time of the flood, approximately 900 people lived in the small town, and the population quickly dropped to fewer than 600 immediately after the flood. In 2013, the town had more than 1,200 residents and was growing. The decision to retreat saved this community and allowed it to thrive (Stewart 2011).

The coastline of Louisiana is a good example of the slow steady retreat, rebuild and fight, and then repeat the process for a broader coastal community. It is a case study in changing economies, major shocks, and slow changes that continue to transform the economy, ecology, and communities. The land loss is well-documented (Couvillion et al. 2011). A small example from coastal Louisiana is the Biloxi-Chitimacha tribe. The tribe was forced to develop plans to leave its ancestral home and rebuild behind levees because of the ongoing land loss in coastal Louisiana (Sturgis 2009).

The current rate of change in coastal Louisiana is so alarming that the State and its federal partners have embarked on a 50-year plan to slow the rate of coastal loss and change it to a more manageable value (Coastal Protection and Restoration Authority of Louisiana 2012).

The analog examples are clear. In the face of more extreme weather, climate change, and greater natural disasters and a global economy, decisional criteria are changing. The economic activity assumptions of the RED and NED must be challenged. Communities must first ask a fundamental question: Do we retreat or fight? If we stay and fight, what does that look like? Is it a stout perimeter system that keeps the floodwaters at bay? Or is it a combination of a more moderate perimeter system and adaptable/resilient lifeline community infrastructure that can protect public safety and return a community to normal quickly?

CONCLUSION

Jared Diamond, in his 2005 book, *Collapse: How Societies Choose to Fail or Succeed,* outlines a number of factors that caused human societies and cities to collapse. Major factors include environmental changes, climate change, hostile neighbors, and trade disruptions. What is most interesting is that the community collapse happens very quickly in a historical context. Most often, the collapse occurs within a generation if not less. Some appear quite abrupt, like entire neighborhoods in Grand Forks/East Grand Forks or even portions of New Orleans being abandoned. The failure to adapt to change, retreat to a defendable and sustainable position, and recognize the fragility of the communities' built and social infrastructure resulted in their demise. There is a growing trend of communities with recognized and likely hazards that are just waiting—treading water until someone throws them a lifeline. There is growing evidence that, given federal financial imbalances, significant public funding will not be available to rebuild. These communities will likely shrink and experience significant economic loss following the disaster.

In the United States, entire communities and societies have been built on the premise of a static climate, environment, and the predictability of future floods and droughts. Now is the time to begin a discussion of what our country will look like given that the original assumptions—those that formed the basis of our water-based infrastructure such as dams, levees, and waterways—are not static; they are changing and make our infrastructure prone to failure. Dams and levees provide a means to adapt to climate extremes, manage the adaptation, and reduce the occurrence of loss. Bold leadership is needed to reformulate our infrastructure planning paradigms into a more adaptable and resilient approach that anticipates a greater range of loads and incorporates the ability to adapt, change, and, if necessary, recognize the need to retreat. This leadership needs to consider the impacts of a global economy, that business activity may not stay in the United States, and that much of our water-based infrastructure was designed on a false premise of climate stationarity. Therefore, it may be more productive to evaluate a range of climate scenarios and disasters to make fully informed investment decisions regarding our infrastructure.

REFERENCES

Austin, Mary Chellis. 2016. "Flood Assurance." *Louisville Magazine*, January.

City of Cedar Rapids. 2016. "Cedar Rapids City of Five Seasons." Accessed January 27, 2016. http://www.cedar-rapids.org/Pages/home.aspx

City of Grand Forks. 2016. "Grand Forks." Accessed January 27, 2016. http://grandforksgov.com/.

Coastal Protection and Restoration Authority of Louisiana. 2012. "Louisiana's Comprehensive Master Plan for a Sustainable Coast." Accessed January 27, 2016. http://coastal.la.gov/a-common-vision/2012-coastal-master-plan/

Couvillion, Brady R., John A. Barras, Gregory D. Steyer, William Sleavin, Michelle Fischer, Holly Beck, Nadine Trahan, Brad Griffin, and David Heckman. 2011. "Land Area Change in Coastal Louisiana from 1932 to 2010." U.S. Geological Survey Scientific Investigations Map 3164, scale 1:265,000. http://pubs.usgs.gov/sim/3164/downloads/SIM3164_Pamphlet.pdf.

Diamond, Jared M. 2005. *Collapse, How Societies Choose to Fail or Succeed*. New York: Viking.

Englander, John. 2012. *High Tide on Main Street*. Boca Raton, FL: The Science Bookshelf.

Global Infrastructure Basel (GIB). 2015. "The 6th GIB Sustainable Infrastructure Summit, 26–27 May 2016, Basel, Switzerland. Accessed June 5, 2015. www.gib-foundation.org.

Horng, Tony. 2008. "How a Flood Town Became a City on a Hill." abc News. June 20. Accessed January 27, 2016. http://abcnews.go.com/US/story?id=5214038&page=1.

PricewaterhouseCoopers. 2015. "The World in 2015: Will the Shift in Global Economic Power Continue?" February 2015. Accessed June 5, 2015. www.pwc.com/gx/en/issues/the-economy/assets/world-in-2050-february-2015.pdf.

Rockefeller Foundation. 2015. "100 Resilient Cities." Accessed June 5, 2015. www.100resilientcities.org.

Stewart, Brent. 2011. "From '93 to Now: Valmeyer Shows Relocation is Possible." *The Southern Illinoisan*. August 21. Accessed January 27, 2016. http://thesouthern.com/news/local/from-to-now-valmeyer-shows-relocation-is-possible/article_fe886176-cbac-11e0-bdc4-001cc4c002e0.html.

Sturgis, Sue. 2009. "As the Land Disappears, an Indian Tribe Plans to Abandon Its Ancestral Louisiana Home." The Institute for Southern Studies. October 9. Accessed January 27, 2016. http://www.southernstudies.org/2009/10/as-the-land-disappears-an-indian-tribe-plans-to-abandon-its-ancestral-louisiana-home.html.

UNISDR. 2015a. "Our Mandate." Accessed June 5, 2015. www.unisdr.org/who-we-are/mandate.

———. 2015b. "Hyogo Framework for Action (HFA)." Accessed June 5, 2015. www.unisdr.org/we/coordinate/hfa.

———. 2015c. "R!SE Disaster Risk-Sensitive Investments. Accessed June 5, 2015. www.preventionweb.net/rise/home.

United Nations. 2014. "World Urbanization Prospects: Highlights." 2014 Revision. Accessed June 5, 2015. esa.un.org/unpd/wup/Highlights/WUP2014-Highlights.pdf.

———. 2015. "UN World Conference on Disaster Risk Reduction: 2015 Sendai, Japan." Accessed June 5, 2015. www.wcdrr.org.

U.S. Business Council for Sustainable Development. 2015. "United States Business Council for Sustainable Development." Accessed June 5, 2015. usbcsd.org.

UNINTENDED CONSEQUENCES OF DAM REMOVAL

Melinda Dirdal, PE[1]
Greg Paxson, PE[2]
Dave Campbell, PE[3]

ABSTRACT

Removal of dams judged to have outlived their usefulness, or where their usefulness is judged to be outweighed by impacts posed by their presence, has increased significantly over the past 20 years. Dam removal case histories typically emphasize the benefits associated with the removal; however, there appears to be less attention given to the lost benefits and unrealized potential benefits. There are likely cases where all of the benefits of the dam and lake are not considered in the evaluation of repair versus removal.

When compared to dam repair and repurposing benefits, a larger emphasis on dam removal benefits, both tangible and intangible, appears to exist. To provide needed balance in consideration of both alternatives, this paper addresses benefits commonly provided by dams, including those the dam was specifically designed to provide (hydropower, water supply, etc.) and those that are often inherent to dams. This paper also discusses the value consideration of alternative facility uses, either separate from or integrated with existing uses. This essentially entails integrated resource planning to evaluate all viable project use options in order to maximize the overall value.

Examples of benefits, both quantitative and/or qualitative, that may be overlooked include:

- Flood risk reduction
- Increased property values
- Recreational use
- Water Supply
- Hydropower

These benefits are considered in the context of the economic, public safety and quality of life consequences that result from removal of dams that need to be accounted for in deciding the fate of an existing dam.

[1] Project Engineer, Schnabel Engineering, 1380 Wilmington Pike, Suite 100, West Chester, PA 19380, mdirdal@schnabel-eng.com.
[2] Principal, Schnabel Engineering, 1380 Wilmington Pike, Suite 100, West Chester, PA 19380, gpaxson@schnabel-eng.com.
[3] Senior Consultant, Schnabel Engineering, 1380 Wilmington Pike, Suite 100, West Chester, PA 19380, davec@schnabel-eng.com

INTRODUCTION

Water is an essential resource. The ongoing California drought has painfully exposed the dangers of insufficient stored water, creating an exaggerated reliance on unreliable streamflows. Although dams provide significant stored water and are critical to our infrastructure, dam removal has gained increased popularity in the past 20 years. American Rivers reports that in the past century, about 1185 dams have been removed in the United States; data for the past 15 years is presented in Figure *1* (from American Rivers, 2014). The growing trend of dam removals is evident, especially considering over 60% of dam removals of the past century have occurred in the last 15 years.

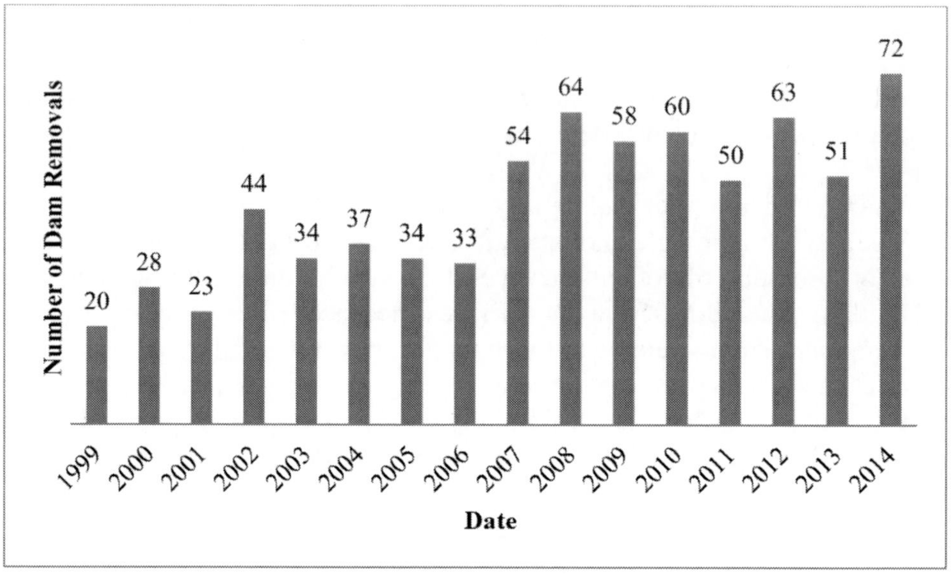

Figure 1. Dam Removals in the United States since 1999

While dam removal is appropriate in many cases, careful consideration of impacts should be given prior to removing a dam. Dams sometimes have detrimental environmental impacts that exceed their utility, and small mill dams are many times silted in and no longer provide benefit beyond a waterfall setting. But for many dams, the decision to remove or rehabilitate the dam is complex and, if not comprehensively assessed, there is the potential for unintended consequences and/or the lost opportunity cost associated with removal. Dams are often removed when the cost of maintenance and/or upgrading is greater than the direct benefits gained from the primary function of the dam, but dams commonly have benefits that extend well beyond their primary function. While it is commonly understood that the primary function would be lost following a dam removal, decision makers may not always consider the totality of a dam's benefits and/or the potential for repurposing.

Dam removal can trigger a multitude of issues related to community interest, safety, and quality of life. Removal can also instigate changes to a basin that has adjusted both aquatically and terrestrially to the presence of the lake. Therefore, public involvement in decision making is critical, and local residents must be educated on the impacts of the proposed actions. Decision-makers must give weight to residents' voices in their consideration of options that are beneficial to the community's interests and well-being, whether supportive of repair or removal. Dam removal results in a societal and economic loss of the primary function of the dam, as well as loss of ancillary benefits and possibly not-yet-realized potential.

FLOOD RISK REDUCTION

While only 17 percent of dams in the United States were constructed with the primary purpose of providing flood protection (USACE, 2015), some level of flood protection is intrinsically attributable to most dams. Downstream residents benefit from flood attenuation and rely upon the protection provided. The probability of flood damages increases due to a loss of reservoir attenuation, triggering greater outflows. If a dam is removed, the dam owner may not directly pay for the increased downstream flood risk, but there could be significant costs to society as a whole. Costs may be shifted from the dam owner to the individual home owners, business owners, insurance companies, or government agencies providing emergency relief; or costs may, through litigation, accrue to the dam owner or agency responsible for triggering the dam removal.

The societal impacts of flood damage can be far-reaching. Merz et al (2010) discuss four types of flood damage:

1. "Direct, tangible: damage to private buildings and contents; destruction of infrastructure such as roads, railroads; erosion of agricultural soil; destruction of harvest; damage to livestock; evacuation and rescue measures; business interruption inside the flooded area; clean-up costs.
2. Direct, intangible: loss of life; injuries; loss of memorabilia; psychological distress, damage to cultural heritage; negative effects on ecosystems.
3. Indirect, tangible: disruption of public services outside the flooded area; induced production losses to companies outside the flooded area (e.g. suppliers of flooded companies); cost of traffic disruption; loss of tax revenue due to migration of companies in the aftermath of floods.
4. Indirect, intangible: trauma; loss of trust in authorities."

Tangible flood damages are most commonly considered in flood risk studies because it is easier to monetize. While intangible flood damages are difficult to quantify, they are important to consider and should be included as an additional negative impact of flood damages for community consideration.

The United States Army Corps of Engineers (USACE) and the United States Department of Agriculture, Natural Resources Conservation Service (USDA NRCS) both provide significant benefits throughout the United States through their flood mitigation programs. USACE estimates that their Flood Risk Management program saved the U.S. Treasury $7.95 billion in 2010[4]. This was based on the costs that would have occurred due to business and non-business casualty losses and disaster assistance payments if the flood protection had not been in place (USACE, 2013). Similarly, the NRCS (2004) estimated

[4] Throughout this paper, dollar amounts have been adjusted to 2015 USD, based on somewhat conservative Social Security COLA data.

that their watershed program protected $15 million in property in 2003. In both studies, the value of the flood protection was based on estimates related to specific flood events during the year of study, and included only direct tangible cost savings.

Flood Risk Reduction of Flood Control and. Recreational Dams Examples

While it is understood that flood control dams have a significant impact on flooding, it is important to recognize how flood levels specifically translate to homes that would be affected and to the resulting damages from a flood. Eight NRCS PL-534 flood control dams were assessed for a 100-year flood with and without the dam in place. For the purposes of this assessment, the following approach was applied:

- The peak inflow was modeled through the downstream channel for the "no dam" condition, and the peak outflow was modeled for the "with dam" condition.
- The downstream channel and floodplain cross-sectional data within the watershed were developed in ArcGIS using a three-meter Digital Elevation Model (3M DEM).
- A steady flow HEC-RAS model was used to estimate downstream flood elevations.

The results of this analysis were used to develop downstream inundation limits resulting from the flood, and affected houses were tabulated using GIS. Figure 2 shows the aggregate number of houses to be affected by flooding with and without the flood control dams. Overall, the eight dams reduced the number of affected houses by more than 75%. Flooding depths at the houses that were affected with the dams in place were reduced by 66% to 83%.

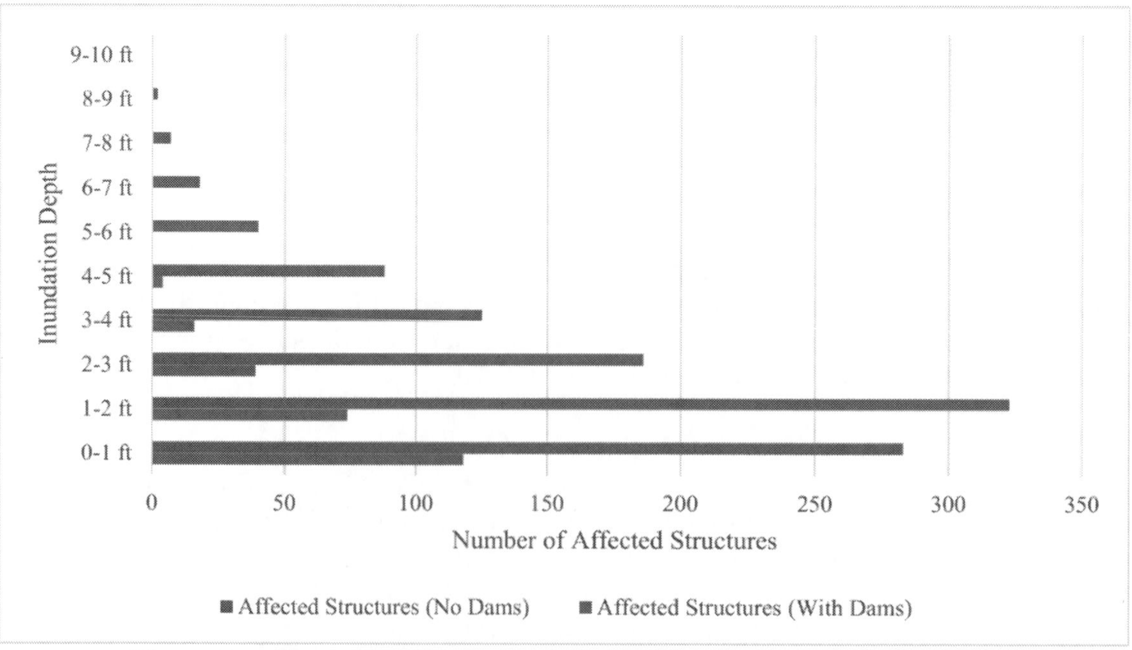

Figure 2. Affected Structures With and Without Flood Control Dams

Thorough analyses of downstream flooding would typically be considered prior to the removal of a flood control dam. However, many non-flood control dams provide meaningful flood risk reduction. Similar to **Error! Reference source not found.**, Figure 2 shows the aggregate number of houses that would be affected by flooding with and without two selected non-flood control, recreational dams. Using similar procedures as presented above, except that LiDAR data was used to develop the cross-sectional data, these two dams reduced the number of affected structures by more than 40%. Flooding depths at structures that were affected with the dams in place were reduced by 3% to 99%. This type of analysis is applicable to dams with many different primary purposes.

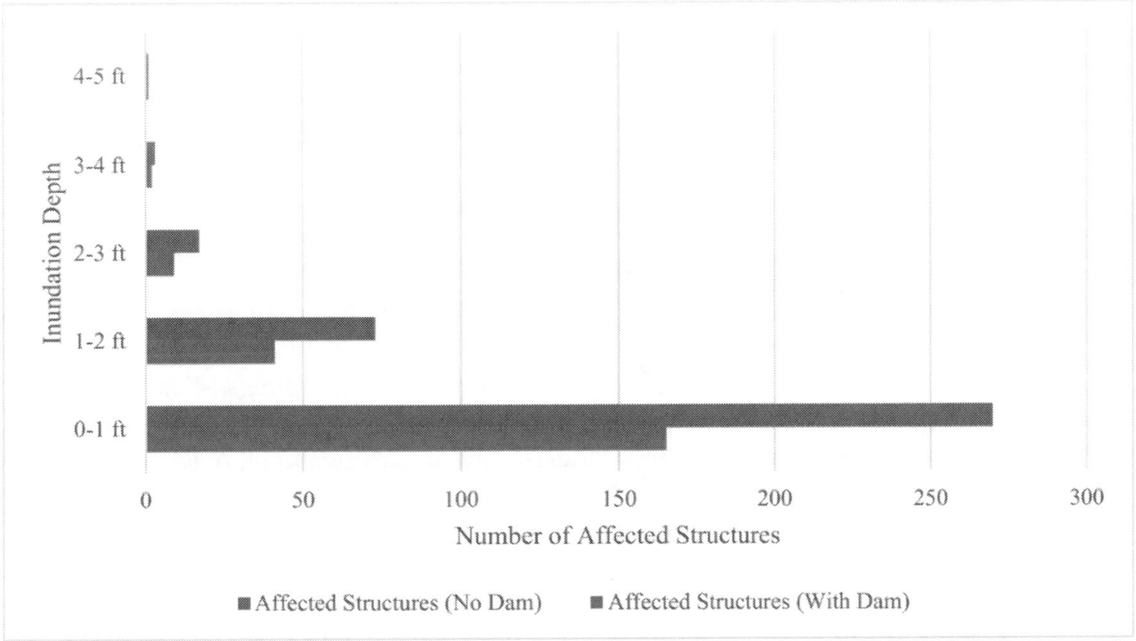

Figure 3. Affected Structures With and Without Non-Flood Control Dams

<u>Flood Risk Reduction from Water Supply and Hydropower Dams (Lakes with Fluctuating Pool Levels)</u>

Most water supply reservoirs are seasonally cyclic, with higher summer demands and lower streamflows depleting storage levels from mid-summer to mid-fall. This time period, of course, corresponds to thunderstorm and hurricane season. Although water supply planners rarely lower reservoirs with intent, summer reservoir lowering is part of the natural seasonal variation, and the flood attenuation benefits may be substantial. In September 1999, Hurricane Floyd caused major flooding across much of the eastern United States. However, flood damages may have been worse had the flood not come on the heels of a major drought, particularly in the Northeast. NOAA (2000) notes that despite heavy rainfall, the available storage in low reservoirs helped lessen flooding.

The Delaware River has a history of flooding, with significant community impacts. In August 1955, Hurricanes Connie and Diane hit the northeastern United States within a week of each other, causing extensive property damage and loss of life. The USACE estimates that a similar event with today's landscape would cause about $2.8 billion in damages (USACE, n.d., para 4). Storms in September 2004, April 2005, and June 2006 produced floods of a 100-year magnitude in three consecutive years. USACE estimated that these three floods caused close to $745 million in damages in New York, New Jersey and Pennsylvania (USACE, n.d., para 5). New York City owns three large water supply reservoirs within the Delaware River Basin and historically did not operate the lakes to provide flood risk reduction, prompting criticism from residents that the reservoirs actually exacerbate flooding (Berrett, 2006). In 2007, Delaware, New Jersey, New York, and Pennsylvania agreed to the Flexible Flow Management Program (FFMP), which is intended "to assist in mitigating the impacts of flooding" (USACE, 2016). The FFMP includes provisions for drawdown of the three reservoirs to create a "void" that reduces downstream flows.

Campbell and Harrison (personal communication, August 4, 2008) analyzed effect of seasonal variation in reservoir pool levels compared to frequency storms for a recently constructed water supply reservoir in northern Virginia. They calculated the probability of a lowered pool elevation during the summer months based on the safe yield analysis. Then a range of return period flood events were studied to compare peak outflows for the following two conditions:

- An assumed full reservoir condition
- Lowered reservoir levels reflective of water supply diversions at the project's safe yield (1-2 ft below normal pool)

The model considering reservoir levels 2 ft below normal pool indicated that a 500-year flood outflow was approximately equal to the full reservoir 100-year flood outflow. Water supply operations for this reservoir inherently reduce an annual discharge probability of 1% to a probability of 0.2%.

Similar to water supply reservoirs, many hydropower reservoirs commonly consume a portion of reservoir storage to provide needed peaking power during the lower-flow summer months, thus providing additional flood protection to downstream residents when the pool level is low. While a reduced reservoir level cannot be depended upon, operations that consume reservoir storage can decrease the real probability for a range of flood events. Many dams are also designed using a principal and auxiliary spillway combination to reduce overall construction costs. These designs inherently provide significant flood detention storage prior to activation of the auxiliary spillway, leading to significant attenuation of peak discharges.

Summary – Flood Risk Reduction

Because both flood control and non-flood control dams can have a large impact on downstream flooding, local communities need to be informed at the beginning of an evaluation process that considers dam removal. Many times, downstream residents are unaware that an upstream dam is providing a level of flood protection. Community education and involvement in decision-making processes can be critical to consideration of dam removal as an alternative to repair or repurposing.

PROPERTY VALUES

Not only do property values decrease if properties are prone to flooding, but they also increase based on proximity to water. Society places high value on scenic views and access to both land and water based recreational opportunities that are available in areas adjoining lakes. Therefore, property values, and resulting tax revenues, are higher in areas surrounding lakes.

The additional property value attributed to a lake can be due to scenic views and/or recreational access. Lansford and Jones (1995) studied property values surrounding lakes, and found that the marginal value of lakefront property is about $96,000, regardless of house size and characteristics. Not only do lakefront properties hold additional value, but also proximity to a lake adds value. The value based on proximity to the lake declines rapidly within the first 150-ft from the lake, and more slowly up to about a 3000-ft distance from the lake. Figure 4 uses data provided by Lansford and Jones (1995) with the adjustment for inflation to show declining property value as distance from a lake increases.

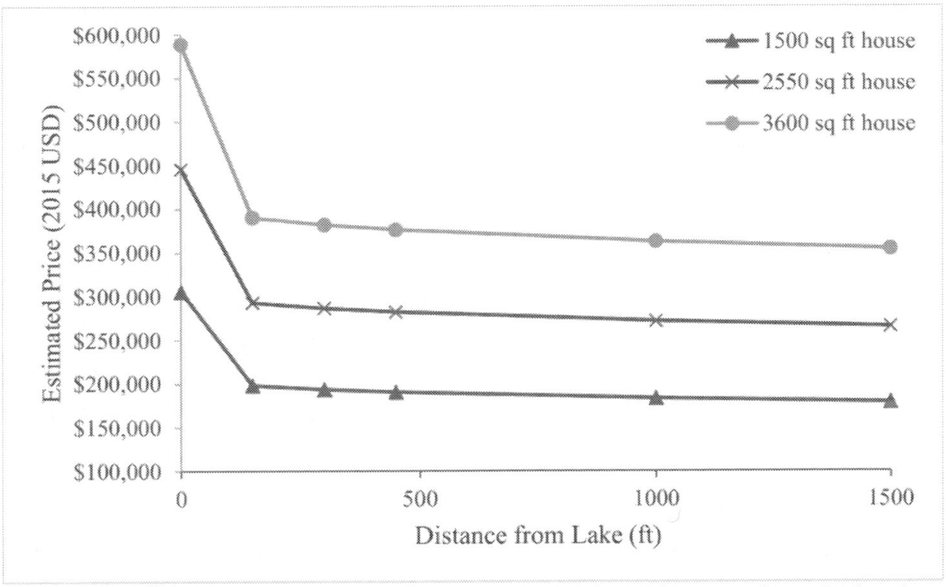

Figure 4. Predicted Sales Price of Houses at Varying Distance from Lake Austin.

Higher property values due to lakes also translate to higher tax revenues for the community. Kosciusko County, Indiana contains over 100 lakes, bordered by primary and secondary residences. Bosch et al (2013) studied the impact of the 44 largest lakes on property value and property tax in Kosciusko County. They analyzed lake-front, channel-front, and lake-view residential properties. The tax revenue credited to lakes was a percentage of tax revenues from primary residences, as shown in Figure 5, because the residences would likely exist without the presence of lakes. All of the tax revenues from secondary residences were considered because they would not likely exist without the lakes. The total credited to the lakes was $15 million in additional property tax revenue, or 25% of the total county property tax revenue for 2012. The estimated tax revenue due to lakes is considered to be conservative because it does not consider commerce, businesses and industries that exist because of the lakes.

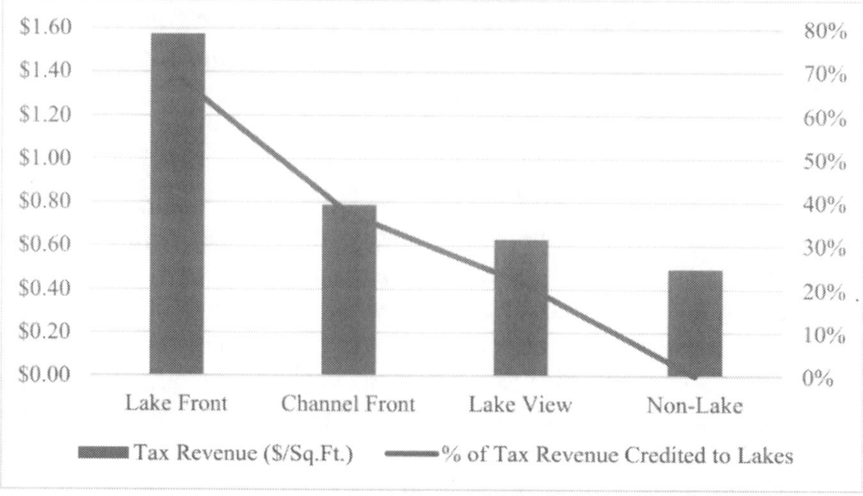

Figure 5. Tax Revenue Based on Property Type.

Although there are many factors that contribute to property values, lakes have a significant impact on property values and tax revenues. Increased property values that result from a man-made lake are a secondary benefit.

RECREATION AND QUALITY OF LIFE

Communities develop around lakes because people place high value on recreation and quality of life. Recreational lakes are impounded by over 30% of the dams in the United States, which is a higher percentage than any other primary purpose (USACE, 2015). Kakoyannis and Stankey (2002) highlight the importance of water-based recreation, stating, "The presence of water surfaces is essential to or enhances the satisfaction of recreationists engaging in most outdoor recreation activities" (p. 2). They point out that some activities are dependent on water, while others, such as hiking, camping, and studying nature, are enhanced by water. In their 2013 study, the USACE estimated that their lakes generated $3.6 billion in 2010. The revenue was generated from both day visits to the facilities and overnight visits, such as camping. Recreational dams provide enjoyment for individuals, provide a foundation for the recreation industry, and promote tourism to the region.

The value of recreational dams is particularly evident when a community faces the loss of a dam for a period of time. The PA Fish and Boat Commission had to drain several of their dams during the past 10 years, and the community response was inspiring. A conservancy group formed in Butler, PA, a small community north of Pittsburgh, invested their time, energy, and money to help rehabilitate their dam. State Senator Randy Vulakovich said, "Every now and then, you'll see a project in the community that people really get behind and they're willing to do big things and small things. This community and the surrounding communities, too, seem to be really getting behind [the dam rehabilitation] project and they want their lake back" (Weigand, 2013). Lakes bring people together through their recreational value, and the lakes become an integral part of the community.

While people generally value recreational lakes, many do not have access to sizeable natural lakes. In the United States, there are about 250 large natural lakes (>10 square miles); however, about 100 of these lakes are located in Alaska, 100 are split between 5 states, and the remainder are split among 17 states (Smardon, 1988). Over half of the states do not have larger natural lakes; therefore, people rely on dams to impound lakes for recreational activities.

The use of reservoirs for recreational purposes is beneficial because it is non-consumptive and, when well managed, does not impair other water uses. Many dams create significant recreational opportunities as a secondary purpose. For example, Buford Dam in northern Georgia has primary purposes of providing hydroelectricity, navigation, flood control, and water supply; but it also impounds the 59-sq. mi. Lake Lanier that has 692 miles of shoreline at normal pool. The lake adds significant recreational benefits to the community. It is the home of many resort hotels and marinas, and it was a venue for part of the 1996 Olympic Games.

Similarly, the Glen Canyon Dam was constructed with the primary purpose of providing hydropower and flow regulation; however, the area has become well-known for its recreational benefits. Activities include boating, fishing, swimming, jet skiing, and whitewater rafting downstream of the dam. Douglas and Harpman (1994) studied the dam's effect on employment in the area. They found that non-residents spent more than $55 million in 1990 and over $46 million in 1985 due to the impoundment and downstream releases, including activity and travel fees. Jobs are created directly from recreation and tourism, and also from the entire community that develops in the region, including real estate, car dealers, hospitals, and food stores. They found that for every job directly related to the impoundment, there is an additional 0.306 jobs created indirectly to support the community. The economic boost to the entire region, directly and indirectly related to recreational activities, is another potentially important benefit from a dam.

WATER SUPPLY

Meeting water supply needs has become increasingly complicated due to climatic stressors along with increased difficulty in obtaining permits to construct new dams. It is not unusual for planning and permitting to take fifteen to twenty years, with an additional five or more years for design, construction and reservoir filling. Therefore, the period for forecasting future water supply needs should begin at least 20 years before a new reservoir water supply is likely available for use. The forecasting process is compounded by significant uncertainties in both water demand and water supply.

The uncertainties of water demand derive from inherent difficulties in making reliable determinations of long-term population projections, per capita use rates, and allocations for industrial demand. Water supply uncertainties are related to limited streamflow records and climatic changes. Existing streamflow records have a very limited timeframe, especially in geologic terms. Campbell (2015) states:

> "Forensic techniques provide clear evidence of droughts of far greater intensity over many centuries compared to those reflected in modern records. Even without human influence, the past 50 to 70 years of stream gage data masks the potential for very severe multiyear droughts…Studies by Columbia University's Earth Institute indicate, 'It was a lot drier in the 19th century than it has been recently, but there were so few people around, it didn't harm anyone. Now, we are building big urban centers that make us vulnerable to even slight downturns.'…At this point, regulatory agencies have opposed building in additional factors of safety to address any future increased severity in drought conditions that would mitigate the loss of safe yield on new projects, terming these factors as 'speculative' and resulting in increased impacts to the natural environment."

The current California drought has painfully exposed the dangers of insufficient stored water, highlighting the exaggerated reliance on unreliable streamflows when confronted with multi-year drought conditions. "It is imperative for water providers to recognize that droughts do not announce their arrival, their intensity, or their duration with sufficient forewarning to allow for development of additional sources in response to that drought event… [Therefore,] it is imperative to plan well ahead" (Campbell, 2012). Existing dams slated for potential removal could provide additional water storage for a very uncertain future. Unless an area is blessed with an abundance of water supply, it would be prudent to set aside dams that are suited for water supply to meet future needs, even if they not necessarily needed at the time.

Water supply planners also need to address increasing costs in developing new sources. Mitigating the impacts of new sources on streams, wetlands, and habitat, and maintaining minimum in-stream flow standards are ecologically important, but they have become very costly. Over time, higher environmental flow requirements have led to the need for increases in reservoir storage, higher capacity diversion pumps, and larger diversion pipes to provide a defined yield (unmet demand). Ironically, larger reservoirs, pumps, and pipes, increase the magnitude impacts that need to be mitigated. In brief, environmental considerations for water supply projects have become more complex and more demanding, which has significantly increased the scale and cost of the projects.

The above factors have contributed to the cost of water supply projects escalating at a rate significantly exceeding the general economy. With long term escalation of water supply projects exceeding overall market escalation by factors of about 3 to 5 for a given safe yield, there is a significant price to pay in deferring the process (Campbell, 2015). If the water supply reservoir cost trends of the past 30 years continue, there could also be an incredibly high price to pay for replacing a decommissioned dam that would have been suitable for use as a water supply source. For example, if a new water supply reservoir is needed in 20 years, past trends indicate that an even more intensified permitting and mitigation process will escalate costs three to five times (or more) compared to current cost bases. If a dam could be repurposed to a water supply dam at that time, the cost savings could be very significant. Much of the cost of existing dams has already been incurred, and environmental impacts from constructed facilities and the impoundment would be minimal, with the exception of added diversion facilities.

Due to competing water demands, water supply planners must utilize water resources wisely. Harou et al (2009) claim that "future water management will shift from building new water supply systems to better operating existing ones". This shift could occur to a large extent through the repurposing of existing dams to provide water supply. Repurposing dams could provide significant resource conservation and economic value, given the uncertainties of escalating permit-related costs and climate dynamics on water supplies.

HYDROPOWER

Hydropower is a clean and renewable source of energy that also benefits from the use of existing dams due to costs and timeframes associated with permit requirements. It has become extremely difficult to permit new dams for hydropower, with the exception of small watershed upper reservoirs for adding pumped storage. Wind and solar facilities have gained considerable favor, but these power production sources cannot be synchronized with need. Pump storage hydropower systems can supplement these sources, providing more consistent and uninterrupted power (Missouri Department of Natural Resources, pp 22-23).

Given the difficulty of permitting new projects, the future of hydropower must rely on maintaining and optimizing existing hydropower plants, or retrofitting existing dams to provide hydropower. The use of existing dams is critical to the hydropower industry because "many of the monetary costs and environmental impacts of dam construction have already been incurred at [non-powered dams (NPDs)], so adding power to the existing dam structure can often be achieved at lower cost, with less risk, and in a shorter timeframe than development requiring new dam construction" (Hadjerioua et al, 2012, vii). There is great potential in the United States to retrofit existing dams to utilize hydropower.

The U.S. Department of Energy has conducted research on NPDs to evaluate their potential as hydropower dams. They assessed over 54,000 NPDs in the continental United States, and found that the top 597 NPDs each have a potential capacity for 1MW or more, as shown in Figure 6. They used the following equation to calculate the potential power generation, where the capacity factor (C_f) was computed from NHAAP regional data from 2001-2008. The total potential power generation computed for the continental United States was about 45 TWh/yr (Hadjerioua et al, 2012). That converts to 45,000,000 MWh/yr, which would power approximately 4 million houses per year based on data from the U.S. Energy Information Administration.

$$\text{Potential Generation (MWh)} = \text{Potential Capacity (MW)} * C_f * 365 * 24 \qquad (1)$$

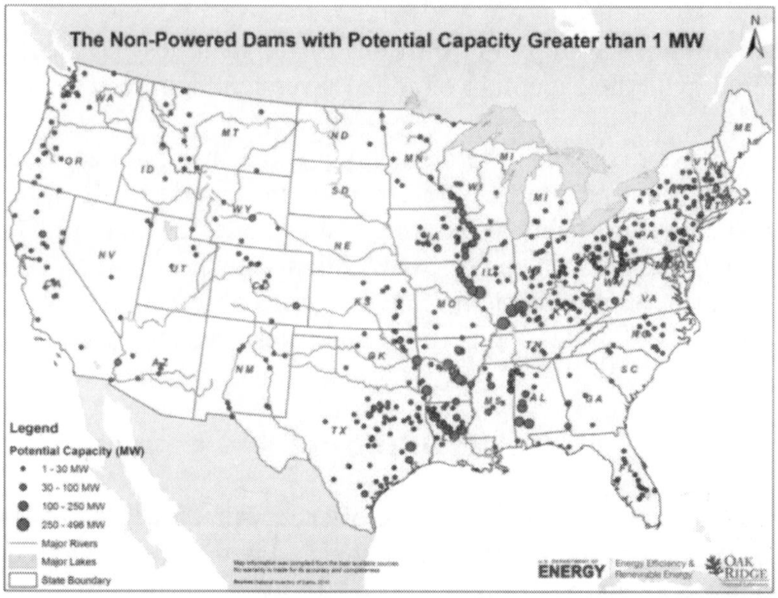

Figure 6. Locations of Non-Powered Dams with a Potential Capacity Greater than 1MW (Hadjerioua et al, 2012).

Costs associated with retrofitting dams and hydropower operating costs would vary significantly, both geographically and based on the available head, so retrofitting of some sites may not be economically feasible. However, there is potential for retrofitting existing dams to produce hydropower. Dam decommissioning would reduce e opportunities to retrofit some of these existing dams to provide hydropower.

CONSIDERATIONS FOR DAM REMOVAL

Dam removal has increased in recent years, and there are several factors to consider for a successful dam removal. Public safety and well-being should always be a top priority. Environmental factors, including river restoration and aquatic migration, and economic factors are typically a significant consideration in the decision to remove a dam. A dam removal should also reflect public input (especially from impacted parties).

There are many successful dam removal projects that have had a positive impact on the communities. Several of the large dam removals have significantly improved the stream ecology and should not be overlooked. There are also many small dams that no longer have significant value, and their removal can make a positive impact on the health of the ecosystem, and in many cases improve public safety.

Case Study 1: Elwha River Restoration Project

The Elwha River Restoration Project, located near Port Angeles, Washington, is the largest dam removal to date. Elwha Dam and Glines Canyon Dam had blocked fish passage and trapped sediments, which was harmful to the river's ecology and caused degradation of the river delta (USSD Committee on Dam Decommissioning, 2015, pp.

136-148). Since the removal of the dams, the river restoration process has begun and significant improvements have already been observed.

Although there are long-term benefits of the Elwha dam removal, there were some short-term issues that should be considered in future projects. Sediment amounts were larger than anticipated and clogged the Port Angeles Water Treatment Facility. Predicted sediment was 24 million cy, but actual volume was about 34 million cy. The sediment also contained lead and PCB contaminants that needed to be remediated. There was initially a decrease in the total number of fish due to increased turbidity (Nicole, 2012). In the end, the Elwha River Restoration project was costly, with a total budget of $325 million, but the region appears to be rebounding and returning to its natural state.

It is important to recognize that some of the factors that contributed to this successful dam removal are site specific. Although sediments at Elwha contained contaminants, the sediments were likely far cleaner than other areas due to the Olympic National Park being located immediately upstream of the reservoir. Transport and dispersal of contaminated soils are both harmful and costly to remedy. Therefore, both short-term and long-term impacts should be considered, especially in areas where upstream land disturbance activities would lead to release and dispersal of contaminants.

Another important aspect of the Elwha dam removal was the scale. Large dam removals have a large impact on the community and attract a lot of attention. Small to mid-sized dams tend to attract less attention; therefore, removal of smaller dams may not receive sufficient public attention. Projects must be assessed from many perspectives and involve the community that would be impacted. Each dam site is unique and must be thoroughly evaluated on an individual basis.

Case Study 2: Battle Creek Restoration Project

Another noteworthy dam removal is in progress at the Battle Creek Restoration Project in Manton, California. The project is multi-faceted, and has a goal of restoring the environment along the river while preserving some of the existing hydroelectric generation. The project includes the removal of 5 small dams ranging in height from 5-20 feet, modifications to 3 dams, and other modifications that will restore 48 miles of fish habitat.

Although the project results in the loss of 1/3 of the hydroelectric generation, the ecological benefits are significant. The decision to remove some of the dams was based on costs to install and maintain suitable fish ladders and intake screens, and the cost analysis considered lost hydroelectric power (USSD Committee on Dam Decommissioning, 2015, p. 116). The project shows a good example of weighing costs and benefits. While the project still accomplishes its purpose to produce hydropower, although at a reduced rate, it will greatly improve the ecology of the stream.

CONCLUSION

The decision to remove a dam is complex and should be carefully considered so that negative societal and economic consequences of the dam removal may be minimized, mitigated or accepted as providing a net benefit. Furthermore, the loss of the primary benefits is commonly recognized in dam removal assessments. However, many dams provide multiple secondary benefits that can be of considerable significance to reservoir area residents, recreation users and those living downstream of the project.

There are many cases where decommissioning dams is appropriate. Like Elwha Dam, there are dams that are outdated and create significant environmental impacts. Removing these dams allows the river to redevelop biodiversity and return to its natural state. There are also many small dams whose purposes are obsolete. Mill dams were once critical to industry in the United States. Many mill dams do not currently have a purpose, and many have not been maintained. In these cases, there are few benefits that would be lost due to decommissioning, and aquatic habitats would be improved as a result. In addition, many mill dams present public safety issues.

However, dams are a necessary part of the infrastructure in the United States. In many cases if dams were assessed more holistically to consider secondary benefits and beneficial repurposing, the focus would shift from decommissioning to optimization. The consequences of dam removal extend beyond the immediate loss of the primary purpose. Planners, regulators, designers, owners, and the public need to recognize the full range of implications before decommissioning a dam. If a water supply dam is to be removed, what are the long term implications? Will new projects be needed in the future to meet increased water supply, hydropower, recreation and other needs? What would those future costs and impacts be? Are there other beneficial purposes that the dam currently serves or has the potential to serve?

Such foresight may require input across a range of disciplines and perspectives. Education and engagement of the local citizenry is critical to the discussion of dam removal. Many dams play a large role in their communities, and the decision to remove or repair a dam should not be taken lightly.

ACKNOWLEDGEMENTS

The authors thank Jay Halligan for his assistance with GIS used in preparing this paper.

REFERENCES

American Rivers. (2014). Dams Removed to Restore Rivers. Retrieved from http://www.americanrivers.org/wp-content/uploads/2015/01/DamsRemoved_1998-2014.pdf?3ec0c7.

Berrett, Dan. (2006, August 17). NYC: We'll lower reservoirs' levels to prevent flooding. *Pocono Record*, Retrieved from http://www.poconorecord.com/article/20060817/News/608170324.

Bosch, N., Burke, A., & Verma, N. (2013). Lake impacts on property taxes and values in Kosciusko County. Retrieved September 26, 2015, from http://www.in.gov/dnr/fishwild/files/fw-LARE_Lake_Impacts_Property_Taxes_Kosciusko_County_Indiana_Grace_College_Sep_25_2013.pdf

Campbell, D. (2012). Reservoir Water Supply Planning for an Uncertain Future. Presented at the NC AWWA WEA 2012 Annual Conference, Raleigh Convention Center, Raleigh, NC.

Campbell, D. (Aug-Sept 2012). Georgia Water Supply Planning – Watching the Horizon. The Georgia Engineer, 32-33.

Campbell, D. (June-July 2015). Georgia Water Supply Planning – Protecting Your Horizon. The Georgia Engineer, 8-12.

Davis, S. A., Johnson, N. B., Hansen, W. J., Reynolds Jr, F. R., Warren, J., Foley, C. O., & Fulton, R. L. (1988). National Economic Development Procedures Manual. Urban Flood Damage (No. IWR-88-R-2). ARMY ENGINEER INST FOR WATER RESOURCES ALEXANDRIA VA.

Douglas, A. J., & Harpman, D. A. (1995). Estimating recreation employment effects with IMPLAN for the Glen Canyon Dam region. Journal of Environmental Management, 44(3), 233-247.

Duffield, J. W., & NEIIER, C. J. (1992). Recreation Benef ts of Instream Flow: Application to Montana's Big Hole and Bitterroot Rivers. Water Resources Research, 28(9), 2169-2181.

Fahey, G. L. (1989). Evaluation of the recreational benefits of major dams in Queensland, Australia. Water Science & Technology, 21(2), 229-235.

Hadjerioua, B., Wei, Y., & Kao, S. C. (2012). An assessment of energy potential at non-powered dams in the United States. Prepared for the US Department of Energy, Wind and Water Power Program, Budget Activity Number ED, 19(07), 04.

Harou, J. J., Pulido-Velazquez, M., Rosenberg, D. E., Medellín-Azuara, J., Lund, J. R., & Howitt, R. E. (2009). Hydro-economic models: Concepts, design, applications, and future prospects. Journal of Hydrology, 375(3), 627-643.

Kakoyannis, C., & Stankey, G. H. (2002). Assessing and evaluating recreational uses of water resources: Implications for an integrated management framework.

Lansford, N. H., & Jones, L. L. (1995). Marginal price of lake recreation and aesthetics: an hedonic approach. Journal of Agricultural and Applied Economics, 27(01), 212-223. doi: 10.1017/S107407080001974X.

Mapes, L. (2013). Elwha dam-removal project held back as silt estimate too low. Seattle Times, Retrieved September 27, 2015, from http://www.seattletimes.com/seattle-news/elwha-dam-removal-project-held-back-as-silt-estimate-too-low/.

Merz, B., Kreibich, H., Schwarze, R., & Thieken, A. (2010). Review article "Assessment of economic flood damage". Natural Hazards and Earth System Science, 10(8), 1697-1724. doi: 10.5194/nhess-10-1697-2010, 2010.

Missouri Department of Natural Resources (n.d.) "Energy For Missouri: Today and Tomorrow - Educator's Guide." Retrieved from http://dnr.mo.gov/education/energy/docs/hydropower.pdf.

Nicole, W. (2012). Lessons of the Elwha River: managing health hazards during dam removal. Environmental health perspectives, 120(11), a430.

Nijhuis, M. (August 27, 2014). World's Largest Dam Removal Unleashes U.S. River After Century of Electric Production. National Geographic, Retrieved September 27, 2015, from http://news.nationalgeographic.com/news/2014/08/140826-elwha-river-dam-removal-salmon-science-olympic/.

NOAA (2000). Hurricane Floyd Floods of September 1999, Service Assessment, U.S. Department of Commerce, National Oceanic and Atmospheric Administration, National Weather Service, Silver Spring, MD, June 2000.

Smardon, R. C. (1988). Water recreation in North America. Landscape and Urban Planning, 16(1), 127-143.

Smith, D. I. (1994). Flood damage estimation- A review of urban stage-damage curves and loss functions. Water S. A., 20(3), 231-238.

United States Army Corps of Engineers (USACE). (n.d). History of Delaware River Flooding. Retrieved from http://www.nap.usace.army.mil/Missions/CivilWorks/DelawareRiverBasinComprehensiveStudy/HistoryofDelawareRiverFlooding.aspx

United States Army Corps of Engineers (USACE). (2015). National Inventory on Dams. Retrieved from http://nid.usace.army.mil/cm_apex/f?p=838:1:0::NO.

United States Army Corps of Engineers (USACE), Institute for Water Resources (IWR). (2013). Value to the Nation of the U.S. Army Corps of Engineers' Civil Works Programs. Retrieved from http://www.iwr.usace.army.mil/Portals/70/docs/iwrreports/2013-R-09_vtn.pdf.

United State Army Corps of Engineers (USACE), (2016). History of Relevant Flood Studies and Related Actions. Retrieved from http://www.nap.usace.army.mil/Missions/CivilWorks/DelawareRiverBasinComprehensiveStudy/HistoryofDelawareRiverFlooding.aspx

United States Department of Agriculture Natural Resources Conservation Service (USDA NRCS). (2004). Benefit/Cost Analysis for the Emergency Watershed Protection Program Final Rule. Retrieved from http://www.nrcs.usda.gov/Internet/FSE_DOCUMENTS/nrcs143_007328.pdf.

USSD Committee on Dam Decommissioning. (2015). *Guidelines for Dam Decommissioning Projects.* Retrieved from http://www.ussdams.org/15Decommissioning.PDF.

Weigand, Jodi. (2013). Grassroots group wants to restore lake. *Trib Total Media,* Retrieved from http://triblive.com/neighborhoods/yourallekiskivalley/yourallekiskivalleymore/3767513-74/lake-conservancy-glade#axzz3y1ikr6AT.

THE KENTUCKY RIVER: MANAGING 19TH CENTURY INFRASTRUCTURE TO MEET 21ST CENTURY NEEDS

Daniel A. Gilbert, PE[1]
Jeffrey S. Dingrando, PE, PG[2]
April L. Welshans, PE[3]
David Hamilton, PE[4]

ABSTRACT

The Kentucky River navigation system consists of fourteen locks and dams, originally constructed between 1836 and 1917. Earlier structures on the lower portion of the river were built of timber cribbing, rockfill, and stone masonry, while later structures on the upper river were first generation concrete structures. The Kentucky River Authority (KRA), a state agency, received the structures in 1986 in varying states of disrepair, due in large part to their age and lack of routine maintenance in recent decades.

Beginning in the 1990s, the KRA embarked on a long-term strategy to prioritize and renovate or repair many of the structures. Near-term repairs were implemented at several structures, such as sealing of lock chambers, installation of valves to transfer water during drought, and bank stabilization to guard against flanking failures. A thorough condition assessment of KRA's portfolio of locks and dams was performed, with a focus on potential failure modes that could result in loss of pool. Based on this assessment, a capital improvement program of over $90M was developed and initiated to address preservation of the Kentucky River Basin's primary water supply, as well as navigation and recreation. Four replacement dams (two complete, one in construction and one in design) secure key water supplies for several communities and industrial users. Lock and dam renovations have implemented in-the-wet construction technologies to reduce costs, schedule and construction risk. The projects include mass underwater concrete placements, in-the-wet technologies, extensive use of divers, foundation improvements (grouting and cutoff walls), and interaction with historic structures.

This paper will provide a brief history of the system, conditions of the system prior to renovation, the process used to prioritize repairs, and description of the major dam renovation and lock rehabilitation projects and discussion of project challenges (design and construction) on these historic projects.

---

[1] Stantec Consulting Services Inc., 1409 N. Forbes Rd. Lexington, KY 40511, 859-422-3000, daniel.gilbert@stantec.com
[2] Stantec Consulting Services Inc., 1409 N. Forbes Rd. Lexington, KY 40511, 859-422-3000, jeff.dingrando@stantec.com
[3] Stantec Consulting Services Inc., 1409 N. Forbes Rd. Lexington, KY 40511, 859-422-3000, april.welshans@stantec.com
[4] Kentucky River Authority, 627 Wilkinson Boulevard, Frankfort, KY 40601, 502-564-3773, david.hamilton@ky.gov

INTRODUCTION

The Kentucky River navigation system (Figure 1) consists of fourteen locks and dams. When completed in 1917, the system provided a 6-foot deep, 255-mile long slackwater navigation pool from the Ohio River at Carrollton, Kentucky, to the "Three Forks", where the South Fork, Middle Fork, and North Fork of the Kentucky River converge at Beattyville, Kentucky. Lock and Dam Nos. 1 through 5 were built by the Commonwealth of Kentucky during 1836 to 1842, then ceded to the U.S. government in 1880 (Johnson 1984; Johnson and Parrish 1999). The U.S. Army Corps of Engineers (USACE) built Lock and Dam Nos. 6 through 14 from 1888 to 1917. Reflecting the evolution in engineering and construction over this time frame, structures on the lower portion of the river were built of timber cribbing, rockfill, and stone masonry, while the later structures on the upper reaches were built of concrete and steel.

The system was conceived to promote economic development in central Kentucky, by linking timber, salt, iron, and coal producing regions in eastern Kentucky with markets reached via the Ohio River. The lower river was an important trade route in the 1800s, but the completed system failed to attract significant traffic in the 1900s. There has been no documented commercial traffic on the upper reaches of the river since the 1970s. Today, the pools impounded by the system of dams on the Kentucky River are highly valued sources for municipal and industrial raw water across the region. Concrete walls have been built across some of the lock chambers, to protect this valuable water supply resource against the potential failure of a miter gate.

Since 1986, the fourteen locks and dams on the Kentucky River have been operated and maintained by the Kentucky River Authority (KRA), an agency of the Commonwealth of Kentucky. The USACE retains ownership of the Lock and Dam Nos. 1 through 4, having transferred ownership of Nos. 5 through 14 to the KRA in 2006.

As structures that protect valued water supplies across the region, assessments and repairs have been undertaken since the late 1990s to ensure that this resource is protected both now and in the future. This document will discuss the history of the Kentucky River, a 2008 condition assessment of the facilities, and repairs that have been undertaken since 2006 to improve these structures and protect their water supplies.

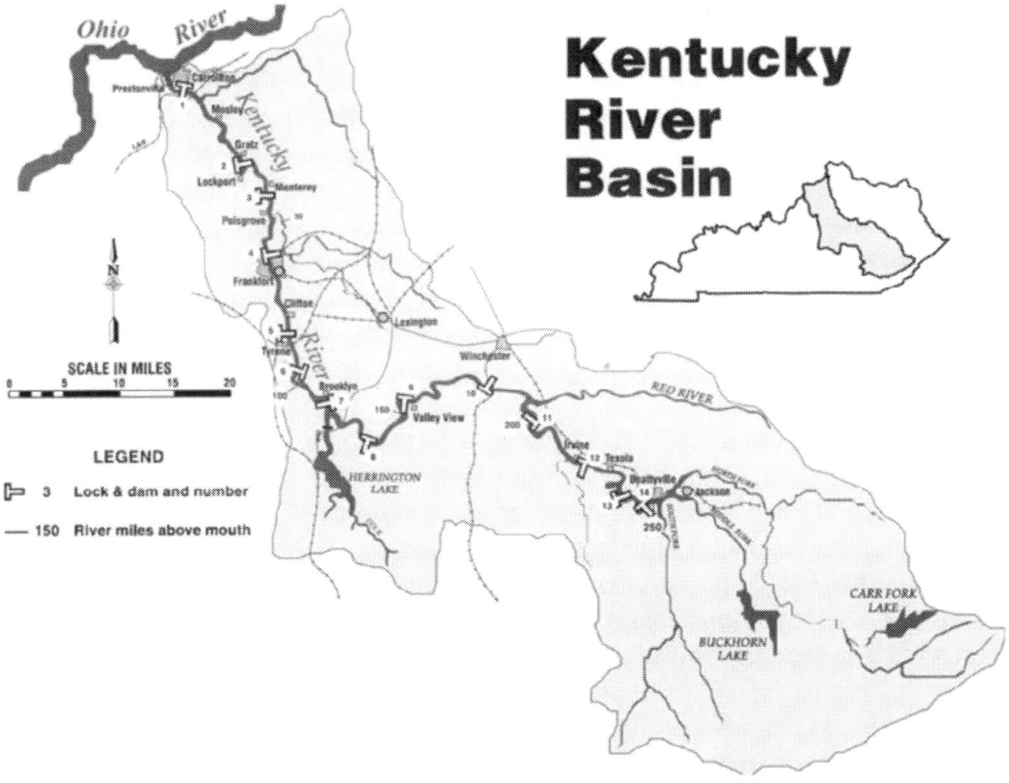

Figure 1. The Kentucky River navigation system (from Johnson and Parrish 1999)

HISTORY

Navigational improvements on the Kentucky River began in 1818, when the State appropriated funds to clear obstructions and deepen the channel at shoals. Histories of the engineering improvements and commercial navigation on the Kentucky River over the past 200 years are provided by Johnson (1984) and Johnson and Parrish (1999).

In 1828 and 1829, the State engaged the USACE to survey the river for navigation. In 1835, the Kentucky Board of Internal Improvement performed a study which developed initial plans and selected the sites for Lock and Dam Nos. 1 through 4. The following year drawings and specifications were developed for Lock and Dam Nos. 1 through 5. The first five locks and dams were not completed until 1842 and had a total construction cost of approximately $1.1 million. The stone-masonry locks were solidly built, but the dams were "cheaply built, stone-filled timber crib dams that leaked badly from the start" (Johnson 1984).

In the first decade of operation, the project supported significant river traffic and generated revenues in excess of operating and maintenance costs. River commerce and revenues suffered during the Civil War, and Lock No. 1 was sabotaged by Confederate raiders. The Kentucky River project fell into disrepair in the years after the war. By 1873, floods were undermining the dams and most had been breached in their centers. In 1877 and 1878, a survey of the Kentucky River found Lock Nos. 1 through 5 to be in a serviceable condition, but the timber crib dams were rotted or breached.

In 1880, the State gave the Kentucky River navigation project to the federal government, where it was placed under the control of the USACE. It was proposed to close the breaches in Dam Nos. 1 through 3 by demolishing the old dams down to the waterline and building new timber crib structures upon the remnants. However, discovering that the old dams were in such poor condition, it was concluded that it would be cheaper to demolish the existing dams and build entirely new structures. One of the old dams was soaked in oil and torched, but the waterlogged timbers would not burn. The dams were thus ripped apart piece by piece and replaced. Reconstruction of Lock and Dam Nos. 4 and 5 was completed in 1883.

In 1887, the USACE developed plans and started quarrying stone for Lock and Dam No. 6. Completed in 1891, No. 6 was the first of the fourteen locks and dams to be built by the USACE. Lock and Dam Nos. 7 and 8 were completed in 1897 and 1900, respectively. All three of these facilities had timber crib dams and stone masonry locks. Lock Nos. 1 through 8 included wooden miter gates. Photographs of the construction of Dam No. 7 and a masonry lock chamber are presented in **Error! Reference source not found.** and **Error! Reference source not found.**, respectively.

Starting with Lock and Dam No. 9, the USACE built the remaining projects on the Kentucky River using concrete and steel, and outfitted the locks with steel miter gates. Lock and Dam Nos. 9 through 12 were completed from 1901 to 1910. During the same time period, the older timber crib dams (Nos. 1 through 8) were repaired by removing the upper sections and capping the remaining timber and rock fill with several feet of concrete. The present river system was completed in 1917, when Lock and Dam No. 14 was opened in Beattyville.

Figure 2. Construction of a timber crib dam at Kentucky River No. 7, 1896-1899 (from Johnson and Parrish 1999)

Figure 3. Construction of a masonry lock on the Kentucky River (from Johnson and Parrish 1999)

In 1905, the breakup of a large ice and log jam upriver caused a flash flood that flanked Lock Nos. 9 and 10. At both sites, the rushing water eroded a new channel through the river bank, around the lock land walls. The new channels at each site were closed by constructing timber crib "auxiliary dams" across each breach. Dam Nos. 11 through 14 were built with moveable crests, which allowed the release of water as the river rose, a design modification intended to prevent the damaging floods that had flanked Lock Nos. 9 and 10.

SYSTEM CONDITIONS

20th Century Repairs

As noted, numerous repairs were made to Dam Nos. 1 through 8 over the years. Typical repairs to the original rock-filled timber crib dam structures included one or more of the following: addition of a timber crib apron at the downstream toe, placement of a concrete cap atop the existing timber cribbing and apron, placement of derrick stone (large riprap) at the toe, and installation of sheet piling along the upstream face of the dam. The histories of Dam No. 9 through 14 note fewer repairs, which is expected as the dams (except for the auxiliary dams) were constructed of concrete monoliths rather than rock-filled timber cribbing.

The USACE undertook numerous repair operations between the 1950s and 1990s in an effort to repair the deteriorating structures. Between 1956 and 1999, steel sheet piles were driven along the upstream face of several of the dams, with the exception of Dam Nos. 9 and 10, to reduce leakage through the aging structures. From the 1970s through the 1990s, the deteriorated concrete caps at Dam Nos. 1, 2, 4, 5, 6 (partial), 7 (partial), and 8 were removed and replaced, and additional derrick stone was placed at the toe of the structures for scour protection. During the same time period, Dam Nos. 9 through 14 underwent minor repairs to address defects in the concrete dam faces.

In the 1950s and 1960s, the wooden miter gates at Lock Nos. 1 through 8 were replaced with steel gates. Additional repairs to the miter, quoin and sill timbers were also performed over the years. Between 1994 and 2007, concrete cut-off walls were built across Lock Nos. 5 through 9 and 11 through 14 to protect against the loss of pool should the miter gates fail. The construction of these walls resulted in the loss of navigation on the Kentucky River above Lock No. 5.

2008 Assessment Study

In 2008, an Assessment Study of Kentucky River Dam Nos. 1 through 8 and 11 through 14 was performed to support the KRA's goal of prioritizing repairs for elements that retain pools. Lock and Dam Nos. 9 and 10 had been studied extensively prior to the assessment and plans for repair and/or replacement were already in place, thus they were excluded from the assessment. Facility assessments consisted of a review of historic documents and both above and below water observations of structures. The assessment primarily focused upon pool retaining elements and did not include structural analysis of elements. Significant observations from this assessment effort for investigated sites are summarized as follows:

- Dam – missing derrick stone, undermining at the dam/foundation interface, section loss at the concrete apron/timber cribbing interface, holes or gaps in sheet piling and associated leakage, tilted or missing apron slabs near the river wall connection, settling of concrete infill between the original dam and upstream sheet piling.
- Dam Far Abutment – monoliths tilted toward river channel, erosion of backfill, lack of armoring (riprap), cracking and section loss in concrete or masonry walls, seepage through cracks in concrete, leakage through karst features in rock abutments, rock highwall instability, oversteepened backfill or sediment deposits behind walls, excessive tree growth or vegetation behind walls.
- Dam Downstream Training Wall – erosion of backfill, lack of armoring, excessive tree growth or vegetation behind walls, slackwater behind wall due to scour or partial flanking, failure of wall due to shifting or collapse of concrete monoliths.
- Upstream River Lock Wall – deteriorated stones in upper courses along length of wall and in all courses near upper gates, concrete section loss near upper gates, vegetation growth on masonry and concrete surfaces.
- Upper Guide Wall – monoliths tilted toward river channel, oversteepened backfill or sediment deposits behind walls, widened construction joints, concrete section loss, vegetation growth on concrete surface, seepage through face of wall.
- Upper Gates – leakage through or around timber seals and valves, rotting timbers, cracks in gate steel including cracked welds, loss of tension in diagonals, impact damage to horizontal members, suspected distress of gate anchorages and/or pintles.

A quasi risk-based analysis method was developed not only to consider the condition of an element, but also to examine factors that contribute to the risk of a loss of pool in the event of an element failure. The contributing factors were as follows: Condition Index (i.e., physical condition on a 0 to 100 scale, lower numbers meaning worse condition); probability of pool loss due to failure of element; speed of progression of pool loss (fragility); ability to intervene for emergency repairs; and cost to intervene for emergency repairs. Through the use of a weighted averaging formula, a "composite score" (1 to 5 scale, lower numbers meaning a worse score) was computed to identify the elements with the worst condition and/or the highest risk. This process provided a relative ranking to be developed for those elements that, from a visual standpoint, were found to have defects. The KRA used the ranking system to prioritize its use of limited resources to reduce the greatest amount of risk per dollar spent on repairs or replacement structures.

The highest priority elements noted during the study included the Dam Far Abutments, Dams, Upper Gates (where there is no Concrete Bulkhead Wall), and Dam Downstream Training Walls. Dams consisting of timber cribbing with a concrete cap (Dam Nos. 1 through 8) were generally found to be in much worse condition and pose a greater risk than the younger monolithic concrete dams (Dam Nos. 11 through14). The assessment noted that the Upper Gates in their current condition are the most fragile element of the water barrier, meaning that a sudden catastrophic failure leading to a significant loss of pool was the most probable failure mode.

Repair schemes were developed to suggest a reasonable concept to provide a short term improvement in condition and/or reduction in risk and to associate a budget level cost with each repair. The short term repairs had a design life, in general terms, of 10 to 25 years. The study evaluated risks and repairs at the element level, allowing the KRA to use their fiscal resources in smaller increments to target areas identified as higher risk to dam safety and preservation of water supply. In addition, the assessment provided the KRA with recommendations and budgetary estimates to replace or renovate entire lock and dam facilities to meet current design criteria and provide a longer service life with greater reliability.

Given the age of the facilities, KRA may wish to implement a long term program where one site is overhauled every few years, such that the entire system of dams is eventually replaced. However, such decisions regarding major expenditures may also consider other factors such as long term goals for recreation and risks associated with consequential damage to water supply or other infrastructure in certain pools. When considering overall risks, inclusion of consequential damages could greatly affect how sites and/or elements are prioritized for repair or replacement.

SELECTION OF PROJECTS

The KRA is charged with developing comprehensive plans for the management of the Kentucky River Basin, including long range water supply, drought response and ground water protection plans. In alignment with this mission, the KRA developed a long-term, capital improvement program to secure the water supply of the Kentucky River basin. This included the renovation and/or replacement of numerous lock and dam facilities as well as addressing high dam safety risk concerns. Development of the capital improvement program also considered long-term goals for navigation and recreation.

In general, the KRA projects over the past 10 years have focused on large-scale renovations of entire locks and/or dams, as opposed to smaller projects focused on specified elements at a site. Recent major projects, in chronological order, are as follows:

- Lock and Dam No. 9 – A new dam was constructed across the entire width of the river, including across the front of the lock. Pool No. 9 is by far the most used water supply on the Kentucky River.
- Lock and Dam No. 3 – A new dam was constructed across the entire width of the river, except that it tied into the existing lock. The original dam was one of the highest risk structures identified by the 2008 Assessment Study. Pool No. 3 is used for a significant new water treatment plant constructed soon after the new dam was built.
- Lock and Dam No. 8 - A new dam was constructed across the entire width of the river, except that it tied into the existing lock. Pool No. 8 is a significant water supply source, and leakage through and beneath the dam had historically led to water quality problems during times of drought.
- Lock Nos. 1 through 4 – This work was accomplished in two projects: Lock Nos. 3 and 4, followed by Lock Nos. 1 and 2. The upper and lower gates were rehabilitated to allow recreational navigation to resume. This also addressed the pool security issues with the upper gates; these were identified as high risk elements in the 2008 Assessment Study.
- Lock and Dam No. 10 – This project is currently in the design phase, and is expected to include a new dam across the entire width of the river, plus major renovation of the lock for both pool security and recreational navigation. Pool No. 10 provides significant water supply and reserve capacity that can be transferred downstream during times of drought.

LOCK AND DAM RENOVATION APPROACH

The renovation design must consider what type of construction approaches will be allowed by the Owner. Allowable approaches must consider construction means and methods, materials, Contractor risk, Owner risk, and other factors. A broad spectrum of possible approaches should be considered early in the design process. Conventional approaches should be weighed against innovative techniques that may be adapted from other fields. Approaches should be judged with respect to schedule, economic, environmental, and other criteria.

The Kentucky River is volatile in nature compared to larger, mainstem rivers such as the Ohio or Mississippi Rivers. The Kentucky River can quickly rise or fall 5 to 10 feet or more, several times per year. The volatility of this river system is a major construction risk driver and an important consideration in evaluating a lock and dam renovation approach. A typical hydrograph for the Kentucky River is presented in Figure 4 and demonstrates the response of a river system with little flood control capacity. The Kentucky River can rise as much as 2 feet per hour during significant precipitation events.

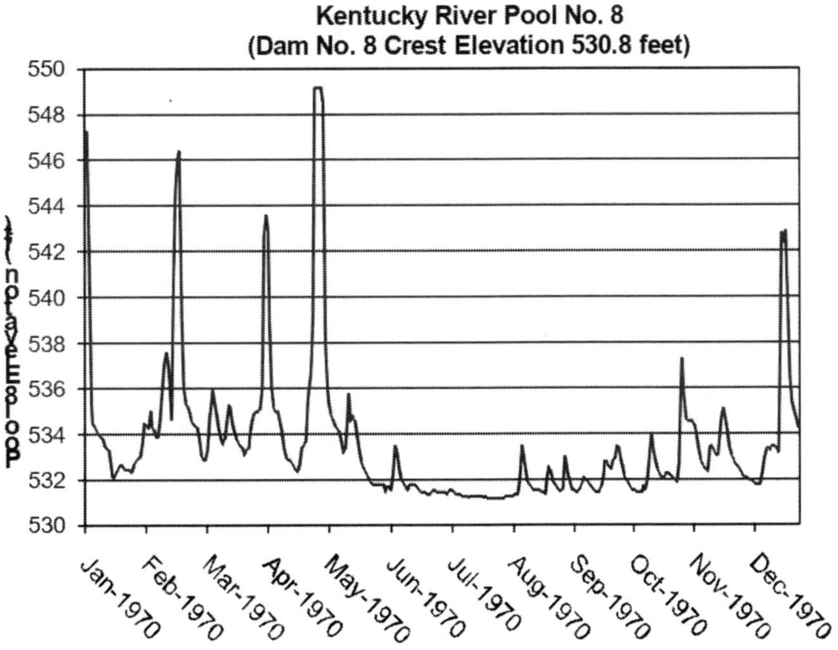

Figure 4. Typical Hydrograph for Kentucky River Lock and Dam No. 8 (1970)

The selected approach also has to consider the geologic setting. To date, the major renovation projects have generally been in areas of Kentucky with limestone bedrock, some of which is karstic (i.e., presence of voids, solution features, soil-filled crevices, etc.). The bedding planes in rock are generally horizontal, and shallow weak seams have been identified that can control the sliding stability of the dam. Along the profile of the dam, the rockline is generally flat beneath most of the channel, except at one of the abutments where it is steeply sloped. The sloping rockline at the abutment controls the design of the dam for this connection.

In-the-Dry vs. In-the-Wet Construction

A major decision on project approach is whether to perform lock and dam construction inside a dewatered cofferdam (in-the-dry construction) or without a cofferdam (in-the-wet construction). Both methods have been used successfully, and each has advantages and disadvantages depending on the project setting and the work to be performed.

A conventional fixed cofferdam approach can facilitate numerous types of dam construction including roller compacted concrete (RCC) or a more traditional cast-in-place monolithic dam. This approach also allows additional options for major rehabilitation of the existing structures (e.g., concrete encapsulation). Regardless of the renovation approach, in-the-dry construction methods require the use of cofferdams to dewater the construction area when river diversion is not a feasible option. For the Kentucky River system, multiple stages of cofferdams would be required to allow the river to bypass the active, dewatered portion of the construction footprint. Any cofferdam system requires an in-depth consideration of design factors including the effective height of the cofferdam, sequence of construction, scour, passage of flows, potential for overtopping, dam safety, economic and construction risks associated with overtopping, sediment transport, and stream flow characteristics. Due to the volatility of the Kentucky River system, selection of an effective height of a temporary cofferdam system and allocation of construction risk is a difficult task. Geologic conditions also require special consideration when evaluating in-the-dry construction approaches. Structural geology and karst features identified along the Kentucky River system can transmit large flows of water and present additional challenges and risks when dewatering. The site specific geology may require foundation pre-treatment (grouting, etc.) beneath the temporary coffer dam structures. However, even with this level of foundation treatment, dewatering at these sites may not be feasible.

Despite risks associated with river volatility and dewatering, there are several benefits to in-the-dry construction. These include improved quality control and quality assurance during construction, flexibility in dam geometry, improved connection between new and existing structures, rapid construction (after temporary cofferdams are in place) and the ability to accommodate variations in the rock profile along the alignment of the dam.

In recent years there has been momentum from dam owners to consider alternate means and methods to construct dams (i.e., without conventional cofferdams). Compared to cofferdam approaches, in-the-wet approaches to dam construction may include substantial construction cost savings, shorter construction durations, reduced environmental impacts, reduced construction risks to the Owner and Contractor, and increased safety during construction. In-the-wet approaches include concrete-filled cellular sheet pile dams, concrete-filled float-in or lift-in pre-cast segments, and other innovative technologies. An in-the-wet approach mitigates the risks associated with dewatering in karst geology or foundations with significant structural geology features.

While river volatility is still a major concern, constructing a dam in smaller segments afforded by in-the-wet approaches can limit the exposure of construction elements and reduce construction risk. By eliminating the need for temporary cofferdam systems and significant dewatering efforts, the construction footprint and corresponding environmental impacts are reduced. Disadvantages of in-the-wet approaches include significant underwater work utilizing divers and more complicated quality assurance and quality control.

Selected Dam Renovation Alternative

While an in-the-dry approach on the Kentucky River can provide many benefits, the risks and cost impacts associated with cofferdams and dewatering on a volatile river system in a karst geologic setting outweighs the benefits of this approach. The selected dam renovation approach involved an innovative in-the-wet approach using concrete-filled cellular sheet pile cells. Interconnected cellular sheet pile structures, similar to temporary coffer dam structures, were used. However, the structures are permanent and are backfilled with tremie concrete, as opposed to temporary cofferdams that are typically backfilled with sand or dredged material. The result is an efficient, massive, self-sustaining structure that can be constructed in a variety of sequences.

Pre-dredging of the site is usually performed as the first step in construction of a cell. Individual cells are constructed by placing a reusable template in the desired position and then driving interconnecting, flat sheet piles to rock. The arrangement of sheet piles conforms to the shape of the template. Multiple templates can be used to construct cells concurrently, which can reduce construction duration. Once the sheet piles have been driven to rock, the cell interior is excavated and the rock foundation is cleaned and prepared for concrete. Final cell and bedrock preparations are performed by divers after dredging operations have been performed. Diver preparations include the removal of rock overhangs, debris, sediment, loose rocks and jetting/airlifting of the foundation surface. In special circumstances, it may be necessary for divers to perform treatments of open joints or karst features. Once the cell foundation is prepared, tremie concrete is continuously placed into the cell to within a few feet of the desired crest elevation. The upper portion of the cell is then dewatered (sheet piles stick up above the pool) and the concrete surface is prepared for in-the-dry installation of a reinforced, concrete cell cap. After the cell cap is complete, the sheet piling is cut to grade along the cell cap.

Circular main cells are connected by smaller arc cells, which are constructed in similar fashion. Depending on the foundation conditions, this approach can be used to construct gravity structures as tall as 50 feet. Main cells can usually be constructed in about a four to six week period, limiting the amount of time work elements are vulnerable to river conditions. Once the cell infill concrete is placed within a cell, the structure is considered stable and risks associated with a riverine environment are mitigated.

In several instances on the Kentucky River, foundation improvements have also been required to treat significant geologic features. Treatments have included grouting and concrete secant pile walls. The foundation improvements are typically performed from the top of the cell infill concrete, prior to pouring the cell cap. Scour protection for the replacement dams has consisted of interlocking, precast toe blocks installed along the toe of the dam and tremie concreted in place. Figure 5 presents an isometric view of the typical dam renovation approach, although the foundation treatment is not always necessary.

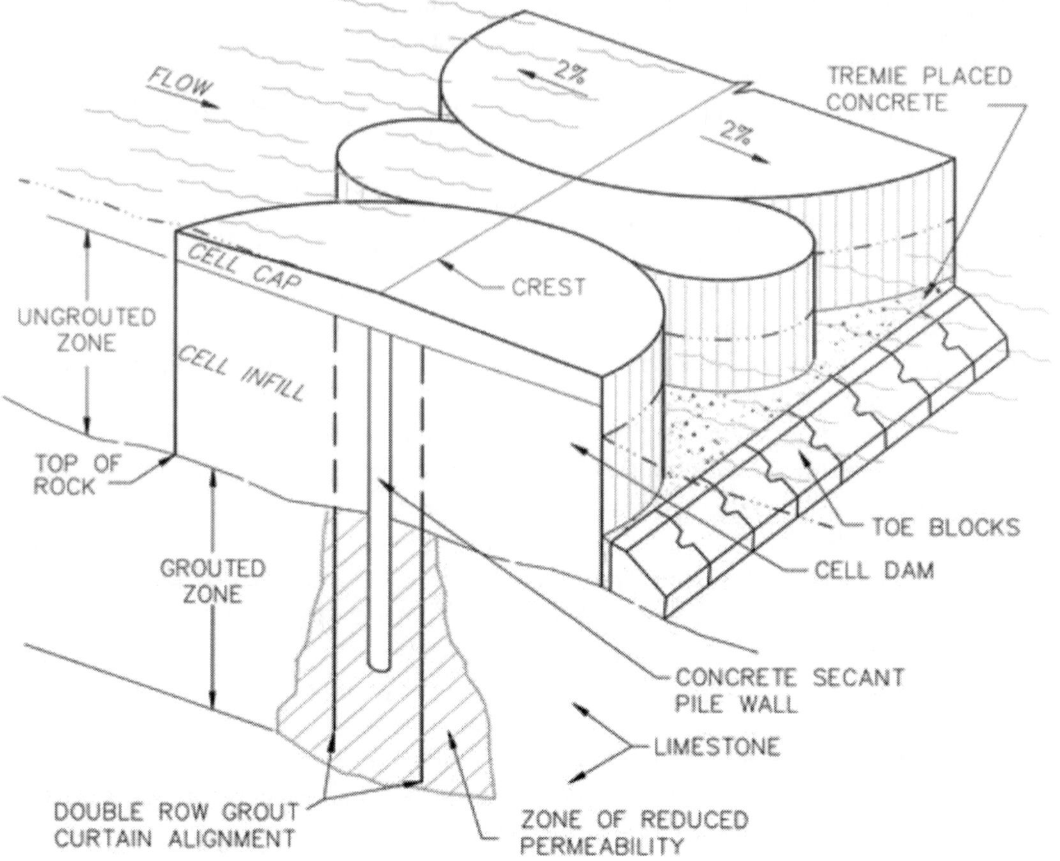

Figure 5. Dam No. 8 Isometric View of Renovation Approach with Foundation Improvements and Scour Protection

This approach provides several advantages over conventional cofferdam construction methods. As with temporary cofferdams, the sheet piling easily conforms to an irregular rockline. Additionally, the use of specialty sheet piles and configurations provide a variety of options when interfacing with existing structures or steep sloping rocklines on the abutments. The simple approach allows average size marine contractors to perform the work with standard marine construction equipment, which results in a more competitive bidding process. Materials used for construction are readily available and do not require significant lead time compared to more exotic approaches. In summary, this method of construction reduces risk to both the Owner and the Contractor.

The renovated dam structures must tie into soil or rock abutment conditions and must interface with the century old infrastructure. The Kentucky River lock and dam sites typically have a rock abutment consisting a steep sloping rockline on one side of the dam and a soil abutment usually composed of alluvial and residual soil deposits. The tie-in to the rock abutments have used closure structures consisting of internally and externally braced PZ sheet piling. The sloping rock surface requires extensive underwater preparations by divers. Tie-ins to existing infrastructure have been addressed with small connection cells creating a watertight connection, but not relying on the existing structure for stability. Soil abutments have consisted of massive abutment cells serving as an extension of the renovated dam with sheet pile cutoff walls extended well into the abutments. Figure 6 presents the completed renovation of Kentucky River Lock and Dam No. 9.

Figure 6. Renovated Kentucky River Lock and Dam No. 9

RENOVATION CHALLENGES

Complexities and challenges presented themselves in each of these projects during both the design and construction phases.

Design Challenges

A major design challenge associated with these projects, consistent with other infrastructure of this vintage, has been the lack of information related to as-built geometry and condition of existing structures. The lock and dam structures have undergone a continual process of new construction, demolition, rehabilitation, repair, and modification throughout their life cycles. Unfortunately, the documentation of these efforts is widely varying in quantity and quality. Original construction documentation for the older sites (1830 to 1860) often does not exist, and historical accounts often indicate these sites may have been designed by Contractors relying heavily on experience instead of analysis. As the state of the practice matured, design and construction documentation became more common and more detailed.

Even when historical documents are available it is sometimes unclear what was constructed. Some drawings show proposed structures that were designed but were not constructed, or were constructed and demolished at a later date or replaced with a newer structure. Significant repair efforts performed by USACE in the 1990s were considered maintenance work and were based on minimal design sketches and field adjustments. In some cases, only construction photos are available to document the work. Finally, some features that can be seen in the field are simply not documented, or else the documentation was not available from USACE or KRA.

This incomplete knowledge of repairs coupled with limited information of the condition of existing infrastructure has steered the dam renovation approach towards replacement structures as opposed to renovating existing dam infrastructure. The risk of concealed conditions and differing conditions related to interaction with existing infrastructure has required additional planning, to accommodate whatever existing conditions are found during construction.

Construction Challenges

The underwater portions of the work performed by divers are a significant aspect of these projects. This includes cleaning and preparation of limestone foundations for the dams and existing lock infrastructure for tie-ins of the new structures. As anticipated, quality assurance and quality control has been a challenge on these projects, as it is with any in-the-wet construction approach. This challenge was addressed through the design process and by providing multiple levels of quality assurance during construction. A reduction in the concrete to rock interface strength was utilized in sizing the cell dam to account for the possibility of small isolated areas of sediment left on the rock foundation. In addition, a robust foundation inspection program was performed by the Engineer (including divers) to observe and accept the conditions prior to tremie concrete. After each tremie placement, verification cores were performed to capture samples of the hardened concrete and the concrete-rock interface. A similar approach to quality assurance and verification was implemented on tie-in structures to the existing infrastructure. This approach has been successful in meeting the strict foundation preparation requirements and achieving or exceeding the required concrete to rock interface strengths necessary to meet stability criteria.

Working in a riverine environment such as the Kentucky River has been a significant construction challenge. The river system is "flashy" (Figure 4) with little upstream flood storage capacity. As a result, even moderate precipitation events can have adverse impacts to construction activities and put partially constructed elements at risk. Even with good planning and diligent monitoring of weather conditions and river forecasts, delays and efforts associated with preparing the work site for elevated river conditions should be anticipated. On two different occasions, sudden rises in pool led to collapse of sheet pile cells that were driven but not yet filled with concrete. These collapses led to significant financial and schedule impacts for the projects.

CONCLUSION

The Kentucky River navigation system consists of fourteen locks and dams, originally constructed between 1836 and 1917. The Kentucky River Authority (KRA), a state agency created in 1986, received the structures (in 1986) in varying states of disrepair, due in large part to their age and lack of routine maintenance in recent decades.

Beginning in the late 1990s, the KRA embarked on a long-term strategy to prioritize and renovate or repair many of the structures. Near-term repairs were implemented at several structures, such as sealing of lock chambers, installation of valves to transfer water during drought, and bank stabilization to guard against flanking failures. A thorough condition assessment of KRA's portfolio of locks and dams was performed, with a focus on potential failure modes that could result in loss of pool. Based on this assessment, a capital improvement program of over $90M was developed and initiated to address water supply, navigation and recreation. Based on the major projects completed to date, the following conclusions are offered:

- Given the overall poor condition of the existing structures and the uncertainties associated with submerged/concealed portions, emphasis has been placed on building new dams instead of attempting to repair existing dams.
- Designs have focused on using in-the-wet approaches, to reduce construction risk associated with a highly volatile river system and uncertainty regarding dewatering during construction.
- Design and construction challenges related to the in-the-wet approach include difficult quality control and quality assurance for submerged work and difficult foundation preparation work performed by divers.
- Even with an in-the-wet approach, construction risks relating to rapidly changing river elevations are still present. Careful planning by the Contractor is essential to limiting exposure of partially constructed elements to the riverine environment.

ACKNOWLEDGEMENTS

Stantec would like to acknowledge the Kentucky River Authority for its responsible management of the Kentucky River lock and dam system.

REFERENCES

Johnson, L. R. (1984). *The Falls City Engineers; A History of the Louisville District, Corps of Engineers, United States Army, 1970-1983.* United States Army Engineer District, Louisville, Kentucky.

Johnson, L. R., and Parrish, C. E. (1999). *Engineering the Kentucky River: The Commonwealth's Waterway.* United States Army Engineer District, Louisville, Kentucky.

Stantec Consulting Services Inc. (2008). "Final Report, Assessment Study for Kentucky River Dam Nos. 1-8 and 11-14, Various Counties, Kentucky." Prepared for Kentucky River Authority, October.

U.S. Army Corps of Engineers, Louisville District. (1898-1994). Construction and repair/rehabilitation plans/drawings/specifications.

HISTORICAL PERSPECTIVE — DAMS, HYDRO, AND MULTI-PURPOSE PROJECTS

David Kleiner, PE[1]

ABSTRACT

Many old projects dating back to the 1920s and 30s have been maintained and are still operating. Dams, reservoirs, and hydro plants that are 40 to 60 and 70 years old, whether domestic or international, are still operating. Specific projects survived wars, bombing, earthquakes, but were later restored to operating condition.

Hydro-Quebec began its dam and hydro construction in the 1940s and continues today with design and construction of four projects on the Romaine River, one of which is operating, two are under construction, and one is starting detail design.

This paper summarizes a few case histories of dam and hydro projects, both domestic and international, that have been maintained and upgraded over the years and are currently operating, thus sustaining their useful life indefinitely. These include:

- St. Mary's Falls Hydro Plant, Michigan, constructed 1902, underseepage problems 1902 to 1926.
- Dix River Dam, Kentucky, concrete-face rockfill dam and hydro, constructed 1924, substantial leakage and repairs over the years.
- Loup River dam and hydro, Nebraska, constructed 1936. In 2012, bonds were issued for upgrades to the powerplants.
- Ambuklao, Philippines, multi-purpose, constructed 1954, heavily damaged by earthquake, reconstructed intake, headrace, and penstocks, back in service 2011.
- Derbendi Khan, Iraq, multi-purpose project, constructed 1962, survived Iran-Iraq war.
- El Salvador, multiple dam and hydro projects, constructed 1954 to 1984, survived civil war, equipment upgrades in mid-2000s.
- Hydro-Quebec, James Bay System plus multiple other dam and hydro projects, 1945 to present. Hydro-Quebec acquired projects from private companies built in the 1920s and earlier, all are still operating.

INTRODUCTION

During the preparation of a presentation at MWH concerning the history of Harza Engineering Company, I realized that many very old and some not so old projects, Harza's among them, dating back to the early 1900s were still operating. Some have faced severe challenges and survived. The following examples are but a few of the many hundreds and, probably, thousands of dam and hydro projects that have similar histories.

---

[1] Consultant, 1026 S Moorings Drive, Arlington Heights, IL 60005, david.e.kleiner@mwhglobal.com

ST. MARY'S FALLS HYDRO PLANT, MICHIGAN — 1902

One of the oldest operating hydro plants in the United States is the St. Mary's Falls Hydro Plant. The St. Mary's River connects Lake Superior to the basin of Lakes Huron and Michigan. The drop in water level is about 20 feet. A 2-mile long, 200-ft wide, canal connects the river above the rapids to the powerhouse. The layout of the project is shown in Figure 1.

During testing of the plant prior to its grand opening in October 1902, there were signs of problems. Under only 7 feet of head difference, the powerhouse in the marked area of Figure 2 had deflected. The upstream headworks were closed, the canal was drained, repairs were made, but under 16 feet of head difference, deflections increased to about 2 inches. It soon became apparent that leakage below the powerhouse had eroded the soils from around the pile foundations. Figure 2 illustrates where leakage occurred and the cavity below the powerhouse.

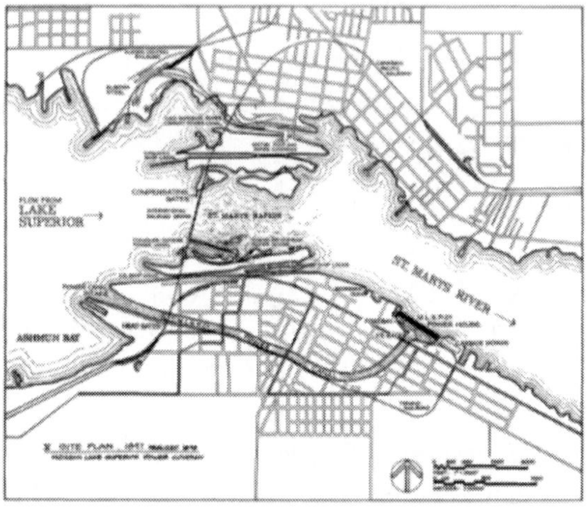

Figure 1. Layout of St. Mary's Hydro Plant

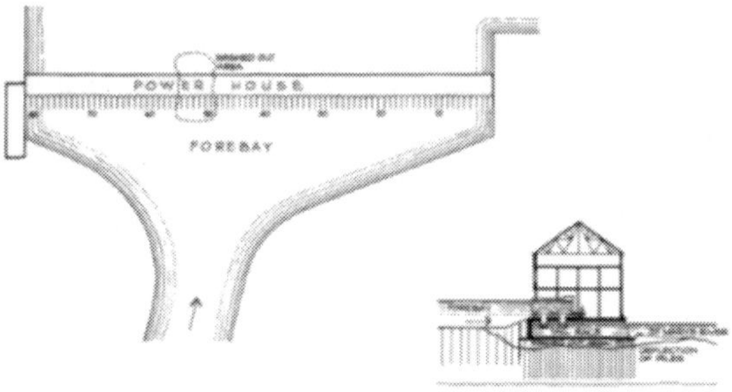

Figure 2. Location and Size of Washed Out Area, 1903

The foundation soils consist of 15 feet of silt, a gravel layer, followed by hard clay over bedrock. Borings indicated that the top elevation of the hard clay did not vary across the length of the powerhouse. The design included constructing cofferdams to isolate the powerhouse, excavation through the silt and gravel, and driving hardened wood piles through the clays. A small creek had eroded the clays in a local area with the result that this section of the foundation had not been properly excavated and backfilled. In 1903, further deflections had occurred. The canal was drained again, and upon inspection, "a massive cavern had been washed out from beneath the foundations, approximately 100 feet long, 120 feet wide, and an average of 10 feet deep" (Reynolds, 1982). Repairs were made, but the project was shut down a number of times with leakage problems from 1903 to 1910. In 1926, the plant was again shut down, the canal drained, and leaks repaired. No further repairs were reported.

The following summary of the plant is taken from the website of the current owner, Cloverland Electric Cooperative:

> "The plant is a quarter mile long, 80 feet wide and has seventy-four horizontal shaft turbines located on the generation floor level. Each turbine has four runners (blades) that drive the 60-cycle generators.
>
> Edison Sault Electric Company purchased the hydro plant and canal in 1963 from the Union Carbide Company for \$1.5 million. An additional \$1 million was spent to convert the plant from 25-cycle electricity to 60-cycle electricity. In 1992, the company completed an \$8 million modernization and automation project in the plant which enhanced both safety and efficiency. On May 4, 2010, Cloverland Electric Cooperative purchased Edison Sault Electric Company from Wisconsin Energy Corporation.
>
> • Three manufacturers (no longer in existence) built the turbines. Many replacement parts are now machined in-house.
>
> • The hydro plant generates between 25 and 30 megawatts of electricity on average. It produces one-fifth of the power needs of the eastern Upper Peninsula (of Michigan).
>
> • Half of the original wood bearings for the turbines are still in use today. The bearings were made from a rare dense wood found in Central and South America."

Figures 3 and 4 illustrate the hydro plant as it appears on Cloverland's website.

Figure 3. St. Mary's Falls as it looks today

Figure 4. Interior view, now a tourist attraction

DIX RIVER DAM AND HYDRO STATION, KENTUCKY — 1924

In 1920, Mr. Leroy Francis Harza founded the Harza Engineering Company. Among his projects during the early 1920s were the Dix River Dam and Hydro Station for Kentucky Utilities. In 1923, Mr. Harza toured the West to inspect existing dams and became convinced that a rockfill dam with a concrete face slab was technically feasible and could be justified economically. The dam was designed, constructed, and ready for reservoir filling by early 1925.

A 1936 memo by Mr. Harza summarizes his visit to the project and comments on the leakage that was occurring through the dam and its abutments and whether there was a dam safety concern. Mr. Harza did not believe that dam safety was an issue, but recommended some repairs to reduce leakage.

Much of the following is taken from a paper prepared by L. A. Schmidt and presented at the ASCE June 1958 symposium on Rockfill Dams.

Figure 5 illustrates the cross-section of the dam. The body of the dam consists of dumped rockfill using rock excavated from the spillway, mainly limestone. The first lift was about 120 ft high and was placed by 16-yd^3 side dump rail cars hauled on a system of switchback railroad tracks that were moved and extended as the rockfill progressed across the canyon. River diversion was through a tunnel. The second lift was 70-feet high, constructed from a trestle at the axis of the dam, again using side dump rail cars. Upon reaching the level of the trestle, the railroad tracks were shifted upstream and downstream to complete the platform. The third lift was also 70-feet high, constructed again from a trestle. This lift brought the rockfill to within 15 ft of the crest of the dam. Five feet of camber, 1.8% of the height, was included to accommodate settlement during reservoir impoundment and project operation.

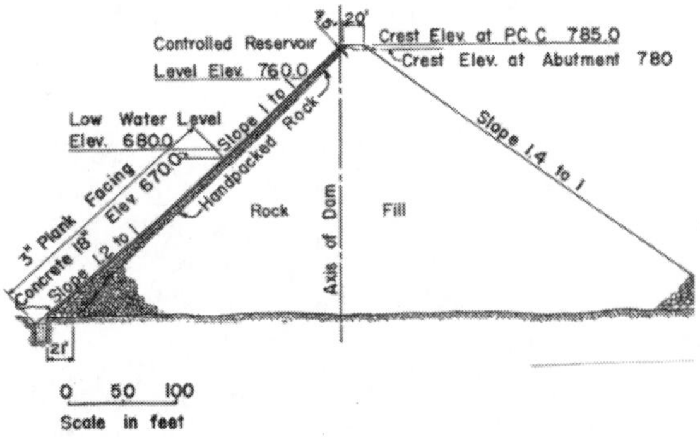

Figure 5. Dix River Dam, cross-section at maximum height, 280 feet, taken from Schmidt, 1958

Figure 6. Derrick stone placement, taken from Schmidt, 1958

Figure 6 illustrates treatment of the upstream face using derrick stone plus hand-placed smaller rock to form a relatively smooth planar surface to support the concrete slab. The derrick stone was selected from the dumped rockfill as construction progressed. These large stone blocks were moved to the upstream face using cables and chains from a derrick set at the face. Halfway through construction, it became apparent that the spillway would not yield enough derrick stone. An off-site quarry was selected to provide the required additional volume of material.

The face slab thickness varied from 18 inches at the base to 8 inches at the top of the dam. Vertical expansion joints were spaced 48 feet on center and supported by 36-inch by 18-inch deep recesses in the hand-placed stone and filled with heavily reinforced concrete. 0.5% steel reinforcement was placed each way in the center of the slab. Reinforcement did not pass through the joints. Each vertical joint contained a copper waterstop and 1 inch thick corkboard. Horizontal joints were similarly constructed spaced at about 70 feet as measured along the slope. No perimeter joint at the contact between the face slab and foundation cutoff was included in the design.

The foundation and abutments consist of thin bedded limestone with interbeds of calcareous shale. The limestone was known to contain solution features. Foundation treatment consisted of excavating an 8-foot wide trench into "solid rock" and backfilling with reinforced concrete. Foundation grouting was also performed, but open solution features were not sufficiently treated.

At the time of Mr. Harza's visit in 1936, total leakage through the dam, abutments, and foundation was close to 100 cfs. Cinders were dumped on the face in an effort to reduce leakage in 1937 and 1938. These efforts resulted in leakage reductions to about 80 cfs at full pool.

In 1944, under low flow conditions, the reservoir dropped to about 64 feet below the normal maximum pool. Inspections of the face slab indicated numerous breaks in the vertical and horizontal joints. The face slab contained numerous hairline cracks, more at the right abutment where vertical and horizontal movements had occurred. Repairs were made to many of the joints. The abutments were grouted with "predominantly asphalt" over a 3-year period starting in 1944 and another 3-year period starting in 1950. As of 1953, total leakage was about 50 cfs, but by 1955, leakage had increased to about 70 cfs. Throughout the 33 years of operation to 1957, leakage remained clear with no sign of internal erosion within the dam or the abutments and foundation.

Maximum crest settlement as of 1957 was 4.2 feet as compared to the constructed camber of 5 feet. Settlement continued, but at a small rate.

Performance

An article appeared in a local newspaper in January 2010 concerning Dix Dam safety. The writer of the article interviewed several board members of the Kentucky Council for Dix Dam Safety, a community organization that has expressed their concerns. Jeff Fraley, General Manager, E. W. Generation Station and Dix Dam and Powerhouse responded to questions. In summary, he stated the following:

- "Through time, the leakage rates increased, and we have an agreement with the state that once it gets up to a certain point, we make plans to address the leakage."

- "If the face of the dam was in distress or if there was a significant change, we would be able to see it in monitoring due to leakage....It's normal for this dam to leak because of the construction."
- In 1988 and 1989, the leakage rate had increased, "and we studied it and removed some of the debris," Fraley said. "Even without removing it we noticed we had an issue with a joint and it was repaired. We were able to see immediate results in the leakage rate after the repair was done."
- "It is a clear flow so there is not a migration of materials. The rock is staying in place." Fraley said the leakage through Dix Dam "does not concern us because of the nature of the structure".

Originally a principal source of electric energy, the hydro plant is now used mainly when heavy rainfall results in above normal lake elevations. Today, with all three units running, the plant produces 24 megawatts of power.

Figure 7. Dix River Dam and spillway, powerhouse in lower right, E. W. Brown Generating Station in background

LOUP POWER DISTRICT, NEBRASKA — 1937

During the 1920s, Harza Engineering Company grew to a staff of about 50. Then the Great Depression starting in 1929 reduced the company to a staff of two, Leroy F. Harza and a secretary. His big break was winning the design for the Loup Project in the early 1930s. The New Deal under President Franklin D. Roosevelt had created the Public Works Administration. The Loup Power District secured a $7.3 million loan and grant for the Loup Project, consisting of two hydro plants and irrigation works. Construction began in 1934.

The project consists of the following:

- Headworks structure to divert the Loup River into the canal, the first two miles of which are a de-silting basin,
- Monroe hydro plant, 8 MW, located 13 miles downstream from the headworks,
- Thirteen miles of canal terminating at two regulating reservoirs,
- Columbus hydro plant, 40 MW, and
- Five miles of tailrace canal that conveys water to the Platte River.

The total system is 33 miles long with head loss of about 165 feet. Monroe Powerhouse began commercial operation March 5, 1937. Loup Power District currently serves a population of about 50,000 and has operated continuously since 1937, 78 years. In 2012, Loup Power issued bonds totaling $15,000,000 to "fund constructing, equipping and furnishing various portions of its electric system, including, without limitation, improvements to its generation, transmission and distribution facilities".

,

Figure 8. Columbus Powerhouse, Loup Power District

AMBUKLAO, PHILIPPINES — 1955

Project Description

Because of World War II, Harza Engineering Company was again reduced to a staff of two: Leroy F. Harza and a secretary. At the conclusion of the war, veterans returned and the company began actively seeking and obtaining work. In the early 1950s, Harza Engineering Company designed and monitored the construction of Ambuklao Dam, a hydro and flood control project in the Philippines.

A description of the design and construction of the dam appeared in the December 1958 issue of the ASCE Journal of the Soil Mechanics and Foundation Division. The maximum dam section is shown in Figure 9. The central core rockfill dam was 430 feet high with three graded filter zones at the downstream face of the core and two filter zones at the upstream face. The fine filter, adjacent to the downstream face of the core, consisted of fine concrete aggregate, ranging from 3/8 inch to the #200 sieve. The second downstream filter ranged from about 2 inches to 0.25 inches. The third filter zone ranged from about 8 inches to 3 inches and formed a transition to the rockfill shell zones. Because of the finer-graded Zone C rockfill, only a single fine filter was used adjacent to the upstream face of the core. During construction, the fine filter was reduced to a width of 3.5 feet in the upper part of the dam because there was the possibility of a shortage of good quality fine aggregate.

The dam had originally been designed with dumped rockfill shells both upstream and downstream, but the first quarry test blast indicated that much of the rockfill would be considerably finer graded and that portions of the rockfill would contain weathered as well as fresh rock. Consequently, the outer upstream shell of the dam consisted of finer graded rockfill and was compacted in 24 inch layers with a 50-ton rubber tired roller. The downstream shell consisted of boulders grading down to gravels. The downstream shell was end dumped in 30 to 90-foot layers. Both the upstream and downstream rockfill shells were sluiced, but the effectiveness of this effort is unknown.

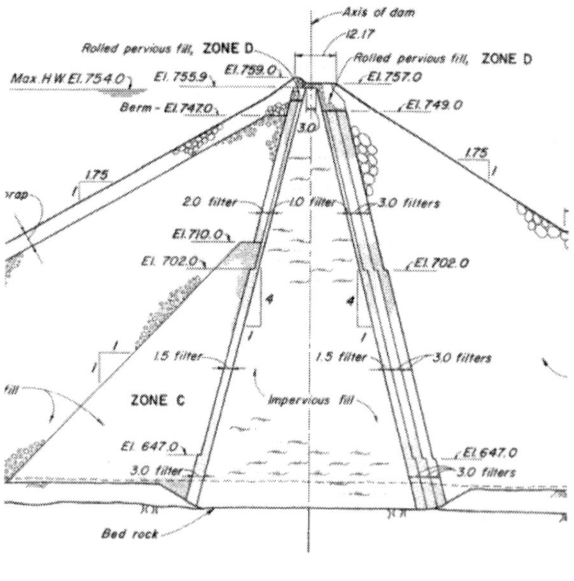

Figure 9. Ambuklao maximum section, from Harza project examples, dimensions and elevations are in meters

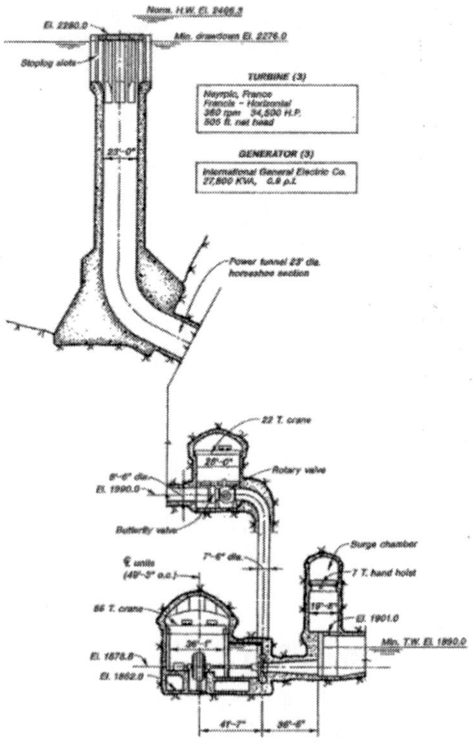

Figure 10. Ambuklao Intake and Powerhouse, from Harza project examples, dimensions are in feet and inches, elevations in tenths of a foot

I

After 4.5 years of project operation, the maximum crest settlement was 3.1 feet, 0.77% of the height at that section as compared to the camber of 3.25 feet.

The power features at Ambuklao include the following, see Figure 10:

- Free standing power intake within the reservoir,
- Sloping headrace tunnel leading to three penstocks,
- Valve chamber,
- Vertical penstocks to the turbines, and
- Tailrace tunnel with separate chamber for the draft tube gates.

The three generators were rated at 25 MW each for a total of 75 MW.

July 16, 1990 Earthquake, Richter M7.7

This earthquake caused the loss of more than 1,700 lives and major damage to the northern part of the Island of Luzon. Estimated peak ground acceleration at the dam site was on the order of 0.60 to 0.65g. At the time of the earthquake, the reservoir was at elevation 2466 (in feet) and was subsequently drawn down to elevation 2433. Crest settlement of about 3.5 feet occurred toward the left abutment close to the contact with the right approach wall of the spillway. The right approach wall of the spillway was partially founded on fill and rotated upstream about 2 feet along with dam. The deformations that occurred did not impair the reservoir holding capability of the dam, but repairs to the dam and spillway would be required prior to raising the reservoir to normal operating levels. Repairs to the dam and spillway were later undertaken.

The powerhouse continued to operate until logs and debris clogged the scroll case of the turbines. Sediment from underwater landslides filled much of the intake and penstocks. During removal the logs and debris, the powerhouse flooded because of loosening and leakage through the draft tube bulkhead seal of Unit 3.

Nine years later, in 1999, the project was officially decommissioned.

Reconstruction of Ambuklao, 2007 to 2011

In late 2007, SN Aboitiz Power-Benguet, Inc. (SNAP-Benguet) won the public bid for Ambuklao and the downstream power facility Binga, which were sold under the power sector privatization program of the Philippine government. SNAP-Benguet is a joint venture between SN Power of Norway and Aboitiz Power of the Philippines.

SNAP-Benguet began a massive rehabilitation project that restored Ambuklao to operating status and increased its capacity from 75 MW to 105 MW. The project required the construction of a new intake, headrace and penstock, elevating the tailrace tunnel outlet, and de-silting of the tailrace tunnel. All electrical and mechanical equipment were replaced including new turbines, main inlet valves, generators, transformers, control systems, circuit breakers and hydraulic structures.

It was not possible to isolate the existing intake with a cofferdam so a new intake, drop shaft and headrace tunnel were required, see Figure 11. The typhoon of 2009 forced additional changes with connections directly to the penstocks.

In June 2011, Unit 3 became the first turbine unit to go on-line, followed by the other two units. The rehabilitated and upgraded Ambuklao hydro project was formally inaugurated in October 2011.

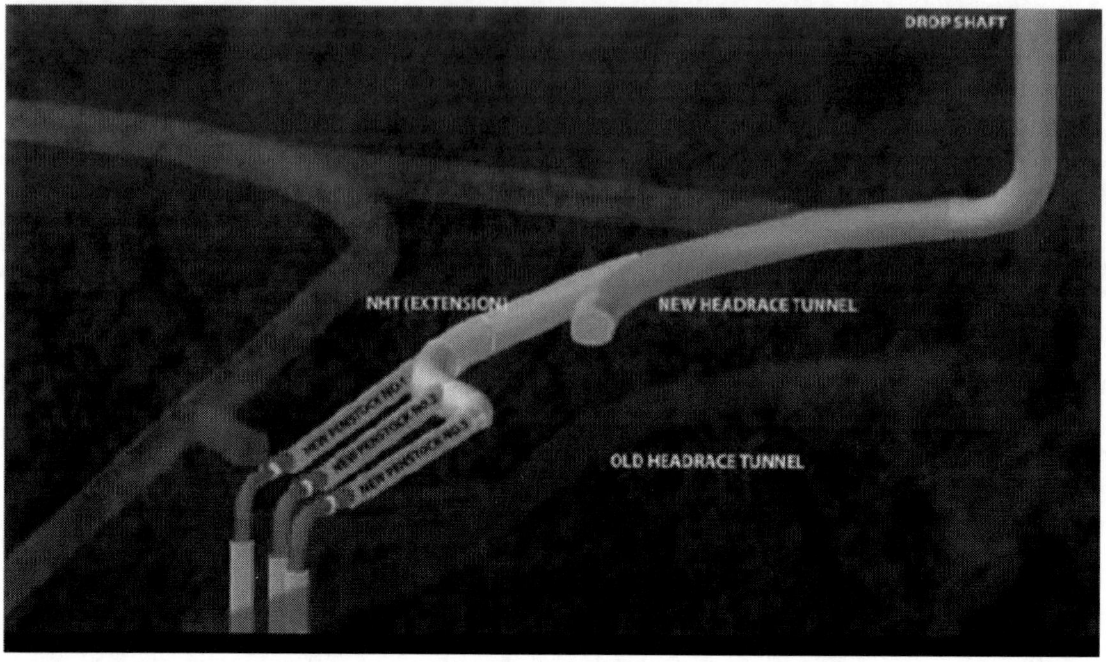

Figure 11. screen shot from a YouTube video of the revised headrace tunnel, manifold, and penstocks.

DERBENDI KHAN DAM, IRAQ--1962

The project was designed during the 1950s to provide hydroelectric energy, flood control, and irrigation regulation. Construction was completed in 1962. Derbendi Khan is located northeast of Baghdad on the Diyala River, a tributary to the Tigris River and close to the border with Iran.

The dam is a central-core rockfill structure, 450 feet high, similar in design to Ambuklao. Figure 12 illustrates the central portion of the dam. The use of multiple filters consisting of graded materials was common in the 1950s and 1960s. Later, the profession moved away from this conservative design to “transition zones” that often consisted of unprocessed material that were broadly graded and subject to segregation. This change in design concept plus improperly designed and constructed “filters” has led to many of the internal erosion issues that we face today.

Both the upstream and downstream shells consisted of end-dumped rockfill in 33-foot lifts (10 m) and sluiced with hydraulic monitors.

The power station, with three 83MW units, is located at the toe of the dam.

The following excerpts from Wikipedia summarize the history of the project to the present:

> "After the reservoir filled, several problems occurred. In 1967, there was a major slope failure about 100 m (328 ft) upstream of the dam. This and other slope failures are continually under repair. The bedrock beneath the dam has to be re-grouted and the crest of the dam settled too much, required it to be repaired. (Author note: this is the direct result of the end-dumping of the rockfill shells in 33 ft layers without proper compaction). The rip-rap on the upstream face of the dam was also repaired in 1999 and 2000…
>
> New spillway gates were installed between 1989 and 1990 after they were removed in 1988 because of the Iran-Iraq War. During the war, the spillway and substation were damaged from bombing. The power station was damaged from bombing in 1990 as well…
>
> In 2007, the World Bank began a US$40 million project to repair the Darbandikhan and Dokan Dams. Repairs to the Dam were expected to cost over $20 million and be complete in late 2012."

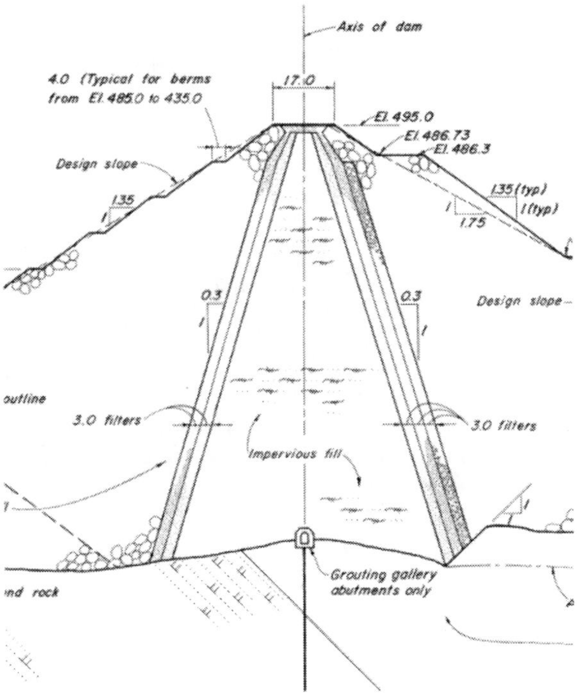

Figure 12, Derbendi Khan, taken from Harza Project Examples, dimensions and elevations are in meters

Figure 13. Derbendi Khan Dam

RIO LEMPA HYDRO, EL SALVADOR — 1950 TO PRESENT

Over about a 35 year period, Harza designed and monitored the construction of four hydro plants on the Rio Lempa in El Salvador:

- Guajabo (5 de Noviembre) 45 MW 1954
- Guajoyo 15 MW 1964
- Cerron Grande 135 MW 1977 See Figure 14
- San Lorenzo (15 de Septiembre) 180 MW 1984 See Figure 15

The two units at Cerron Grande were rated at 67.7 MW each, but were capable of producing 86 MW. In the early 2000s, technical and economic feasibility studies led to upgrading each unit to 99 MW by installing new runners, a rewound generator and ventilation modifications to optimize air circulation through the generator stator and rotor. According to Wikipedia, the two units at San Lorenzo are currently being upgraded to increase their capacity by 24MW, bringing the Rio Lempa hydro projects to more than 400MW.

From Wikipedia:

> "The Salvadoran Civil War (1979–1992) was a conflict between the military-led government of El Salvador and the Farabundo Martí National Liberation Front (FMLN), an "umbrella organization" of five left-wing guerrilla groups. The full-fledged civil war saw extreme violence from both sides."

During the conflict, the military protected the hydro projects, but the guerrillas downed many transmission towers during the conflict. It became increasingly unsafe to travel by road to the San Lorenzo project. My last visit in 1984 was via military helicopter from the San Salvador international airport.

Figure 14. Cerron Grande Hydro Project

Figure 15. San Lorenzo Hydro Project

HYDRO-QUEBEC, CANADA — 1945 TO PRESENT

From small beginnings in1944, Hydro-Quebec (owned solely by the province of Quebec) has grown into a giant hydroelectric utility with more than 36,000 MW capacity as of 2014 when Romaine 2 came on line. The completion of the other three hydro projects on the Romaine River, two of which are under construction, will add another 1000 MW. Upon its formation and through the early 1960s, Hydro-Quebec expanded by acquiring hydro resources from private companies, some of which were constructed in the early 1900s, and by developing new hydro projects.

Phase 1 of the James Bay hydro complex was started in the 1970s and ultimately completed in 1996. This added 16,000 MW of hydro capacity. Phase 2 consisted of:

- Constructing the Eastmain 1 dams, dikes, spillway, and powerhouse in 2005,
- Diverting the Rupert River northward to the Phase 1 James Bay system,
- Maintaining the bays and pre-diversion Rupert River regime using groins and weirs and minimum flow rate criteria throughout the year, and
- Constructing the Eastmain 1A powerhouse in 2012, and the Sarcelle powerhouse in 2013.

These facilities added 1400 MW capacity and increased the energy generated at two of the Phase 1 generating stations on the La Grande River, see Figure 16.

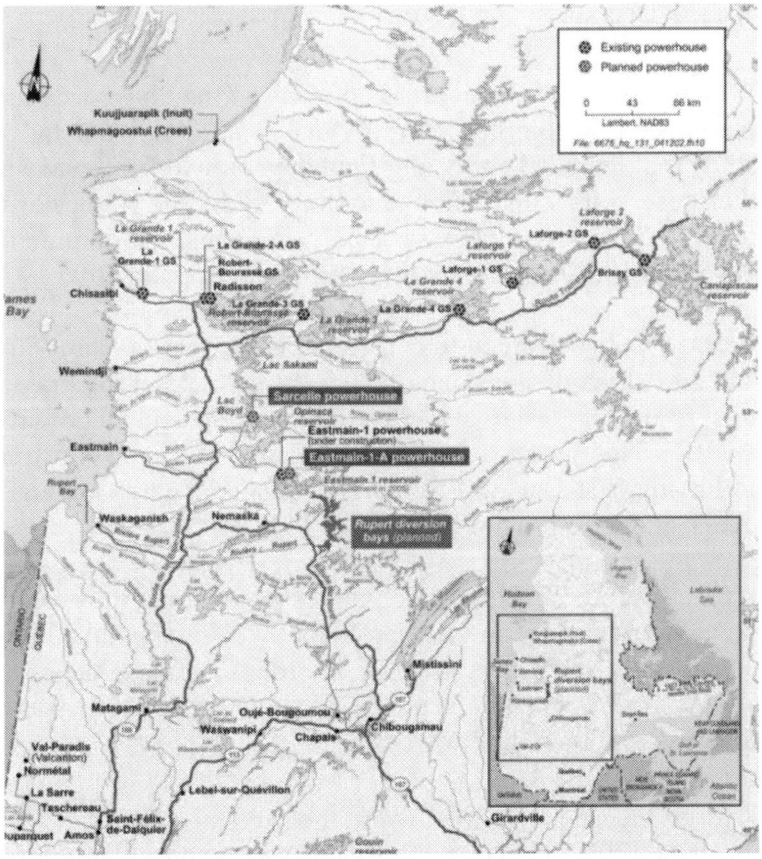

Figure 16. James Bay Complex

During the construction of the first part of Phase 1, completed in 1985, the agreement with the Cree Nation was not fully honored. The protests by the Cree Nation caused certain power-purchase agreements to be cancelled and stalled the second part of Phase 1 for several years. Phase 1 was eventually completed in 1996. An "Agreement Respecting a New Relationship between the Cree Nation and the Government of Quebec", signed in 2001, allowed Phase 2 to be completed peacefully.

My personal activity with Hydro-Quebec is as a member of their Boards of Engineering Experts on projects constructed from 2001 to the present. Some personal observations follow:

- Hydro-Quebec acts as the general contractor on new hydro construction. Major civil works are separated into smaller contracts such that contractors within the province can compete for the work. A large Hydro-Quebec field staff is placed on site to provide quality control and to manage all civil, mechanical, and electrical contractors.

- Maintaining the schedule is all-important. Remote field camps are expensive to operate and maintain. Technical and contractual issues are handled promptly such that the start-up dates are fulfilled.

- Hydro-Quebec is an innovator:

    - When riprap erosion occurred on many of the Phase 1 dams and dikes of the James Bay complex, Hydro-Quebec undertook a detailed research study that included collecting field data on wind and wave action on their reservoirs, conducting laboratory hydraulic studies of wave action on various rock particle sizes, analyzing construction practices, and developing a "Practical Guide—Riprap Sizing" in 1997.

    - Hydro-Quebec decided to construct the Nemiscau Dam (50 feet high and one of many dams and dikes for Rupert River Diversion) in order to understand the design and construction of an asphalt core rockfill dam, ACRD. This successful effort led to the design and construction of the following ACRDs: 360-foot high Romaine 2 dam (see Figure 17), 5 dikes, and the Romaine 1 dam.

    - Hydro-Quebec developed techniques to construct their new powerhouses faster and more efficiently using pre-cast concrete wall panels and a steel superstructure in order to enclose the powerhouse for winter work as quickly as possible. Construction time was reduced to three years for a major powerhouse.

    - Hydro-Quebec is moving toward the exclusive use of 3D drawings. I understand that Romaine 4 drawings will be provided to prospective contractors in 3D only.

- Most of Hydro-Quebec's rockfill dams consist of a central core of silty sand, Zone 1, with an absolute minimum of 15% passing the #200 sieve. Typically, the material is glacial till with little or no plasticity. The adjacent filters consist of about 70% clean sands and gravels, materials that are commonly found in natural glacial deposits, such as eskers. Segregation in these filters does not occur. The outer shoulders of the dams consist of fine rockfill grading to coarse rockfill and riprap at the outer slopes, see Figure 18.

Figure 17. Romaine 2, asphalt core rockfill dam

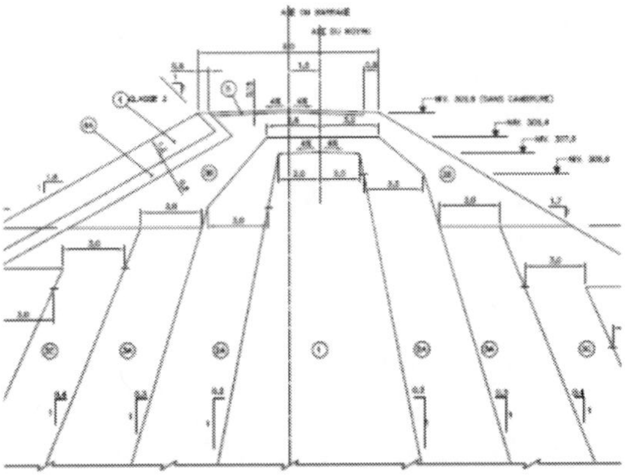

Figure 18. Rupert Dam, typical Hydro-Quebec central core rockfill dam

The following is a summary Hydro-Quebec hydro electricity generation as compared to the United States hydro generation:

- Hydro-Quebec: Total electricity generation during 2013, 210 Tera Watt hrs, TWhr, of which 97% is hydro, 204 TWhr.
- United States: Total electricity generation during 2014, 4090 TWhr, of which 6% is hydro, 245 TWhr.

Hydro-Quebec is one of the largest, if not the largest, hydro-electric utility in the world.

CONCLUSION

The examples summarized above illustrate that with proper maintenance, upgrades, and in some cases reconstruction, dams and hydro projects survive and thrive even through the hardship of old age, war, devastating earthquake, and typhoons.

ACKNOWLEDGEMENT

Kim De Rubertis provided private correspondence and photos of Ambuklao reservoir at the site of the free-standing original intake. His advice helped persuade SNAP-Benguet to abandon the original intake and construct a new intake at the abutment where constructing a cofferdam was feasible.

REFERENCES

Davis, C. V., "The Derbendi Khan Dam", Journal of the Power Division, ASCE, Paper 1741, August 1958.

Fucik, E. M., Edbrooke, R. F., "Ambuklao Rock-Fill Dam, Design and Construction" Journal of the Soil Mechanics and Foundations Division, ASCE, Proceedings Paper 1864, December 1958.

Harza Engineering Company, "Harza-65 years", 1985.

Harza Engineering Company, Design Guides, "Project Design Examples", 1994.

Harza, R. D., "Harza Engineering Company, Developing Water Resources Worldwide, 1920-1984", text of speech to The Newcomen Society of the United States, 1984.

Reynolds, T. S., "Sault Ste. Marie, A Project Report", U. S. Department of the Interior, National Park Service, 1982.

Schmidt, L. A., "Performance and Maintenance of Dix River Dam", Journal of the Power Division, ASCE, Paper 1683, June 1958.

USSD, "Observed Performance of Dams during Earthquakes", Volume II, October, 2000.

Valiente, B. S., Roddy, R. "Cerron Grande: El Salvador's Largest Hydro Project", Water Power & Dam Construction, July, 1978.

Website, Cloverland Electric Cooperative, current status of St. Mary's Falls Hydroelectic Station.

Website, Hydro-Quebec, history and status of projects within the Province of Quebec.

Website, Louisville Gas & Electric, current status of Dix Dam Generating Station.

Website, Loup Power District, current status of Loup River hydro projects.

Website, SNAP Benguet, data concerning rehabilitation and upgrading of Ambuklao hydro project.

Wikipedia, Derbendi Khan Dam, history and current conditions.

Wikipedia, Civil War, El Salvador, 1979-1980.

USE OF LABYRINTH PIANO KEY WEIR (PKW) SPILLWAY TO IMPROVE DAM SAFETY FOR A LARGE SET OF DAMS

Frédéric Laugier[1]
Thomas Pinchard[2]
Julien Vermeulen[3]

ABSTRACT

Piano Key Weir (PKW) labyrinth spillway have been a quickly developing innovation during the past decade. First publications suitable for engineering purposes were issued in 2003 by Lemperiere and Ouamane. Immediately after this publication, the first worldwide PKW was studied from 2003 to 2005 and constructed in 2006 at Goulours Dam (France). During the last ten years, EDF designed and built about 11 PKW spillways with design hydraulic head between 1 and 1,5 m. They were part of existing dam rehabilitation projects in France with regard to dam safety improvement issues. Specific discharge capacity can reach 10 m3/s/ml for 1 m hydraulic head and exceed 20 m3/s/ml for 2 m head. PKW have very simple geometrical features, basically triangles and rectangles. Thus they are easy to design and build.

PKW concept is suitable either for existing dam rehabilitation or for new dams. In the latter case, a combination "gates + PKW" has revealed many advantages by optimizing cost, reliability and other technical issues such as reservoir level control or sediment passage.

In parallel to this quick engineering development, many research actions have been carried out in by various universities and institutions in various countries (France, Switzerland, Belgium, Algeria, India, Vietnam, USA etc..). An important collaboration between all parties provided a better understanding of different PKW fundamental parameters such as hydraulics or structural design.

INTRODUCTION

Piano Key Weir (PKW) labyrinth spillways are an innovative technical-economical optimization of traditional labyrinth weirs. They were introduced by Lemperiere and Ouamane [2003] ten years ago and they are now well known in the sphere of free-flow spillways. PKW appear to be an interesting option to increase spillway discharge capacity by providing high specific flow rates. They can also be used to increase the storage of reservoirs while keeping the maximum water level constant. Unlikely traditional labyrinth weirs, PKW can be installed on the top of most existing concrete dams. As free-flow spillways they also provide an excellent reliability and would generally require less maintenance than active controle spillways.

[1] EDF-CIH, Savoie Technolac 73370 Le Bourget du Lac, France, frederic.laugier@edf.fr
[2] EDF-CIH, Savoie Technolac 73370 Le Bourget du Lac, France, thomas.pinchard@edf.fr
[3] EDF-CIH, Savoie Technolac 73370 Le Bourget du Lac, France, julien.vermeulen@edf.fr

After building the first worldwide PKW spillway at Goulours, France [Laugier, 2006], EDF actually decided to launch other PKW projects to rehabilitate dams having an insufficient discharge. 8 other PKW units were built from 2008 to 2015. Two additional PKW units are under construction by EDF and will be completed in 2016 and 2017. A couple of other units are planned in the near future.

PKW MAIN FEATURES

PKW are based on the traditional labyrinth spillway concept consisting in increasing the overflowing length by "folding" the crest of the spillway. PKW are an answer to the main drawback of traditional labyrinth spillway : They can be installed at the top of the narrow crest of most existing concrete gravity or arch dams with a small footprint on foundation. Although PKW geometry might appear rather simple at first sight, its geometry is potentially governed by more than 20 independent parameters. The international scientific interest about PKW required a unity in terminology. A workgroup involving the Laboratory of Hydraulic Constructions of Lausanne University, the hydraulic laboratory of the University of Liège and EDF – Hydro Engineering Center developed a specific nomenclature [Pralong, 2011]. This naming convention matches as much as possible to the existing labyrinth convention which previously existed [Tullis, 1995]. It was presented during the first workshop on PKW and labyrinth spillways held in Liege in 2011.

The main geometrical parameters presented in figure 1 are :

- The PKW height P;
- The PKW total width W;
- The PKW upstream-downstream length B;
- The keys widths (Wi for inlet key and Wo for outlet key);
- The overhang lengths (Bi for downstream overhang and Bo for upstream overhang);
- The sidewall thickness Ts

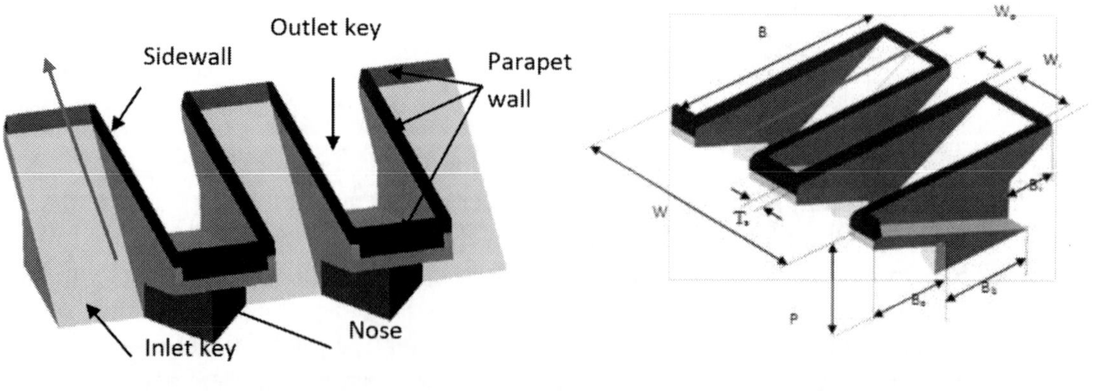

Figure 1. Main geometrical parameters of PKW naming convention.

The main geometrical parameters governing PKW discharge capacity are :

- Ratio "L/W". This "labyrinth" ratio represents the developed length. It is a major factor for small heads.

- PKW height "P". High P value improves the flow approach and distribution within PKWs inlet keys. It also reduces submergence effects in outlet keys.
- Key width ratio "Wi/Wo". High Wi/Wo value improves flow approach and distribution within PKWs inlet keys. On the other hand, high Wi/Wo value increases submergence effects in outlet keys.
- Sidewall thickness Ts. Thicker sidewalls significantly reduce the available space on the dam crest. This structural date can affect up to 10-15% PKW specific discharge capacity (for example the reduction of Malarce PKW sidewall thickness from 40 cm to 20 cm on top of wall increased PKW specific discharge capacity by 10%).

SUMMARY OF ACHIEVED PKW

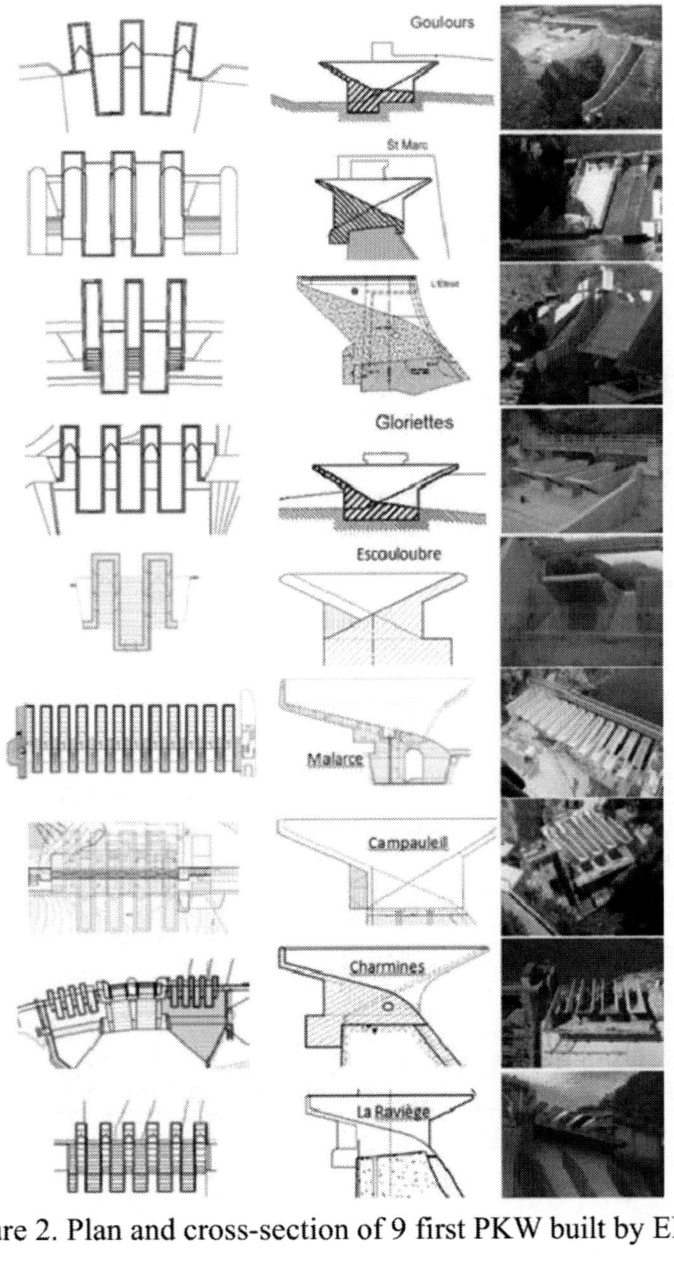

Figure 2. Plan and cross-section of 9 first PKW built by EDF.

Nine PKW were built by EDF from 2006 to 2015, about one unit per year (figure 2 and table 1). They were constructed on various types of dams and for very different topographical configurations. This highlights the flexibility of the concept with regard to existing dams. Eight PKW units were built for flood passage issues. One PKW unit (Escouloubre) was built for an operation optimization purpose on an expansion basin in case of sudden powerplant breakdown.

	Goulours	St Marc	Etroit	Gloriettes	Escoul.	Malarce	Campauleil	Charmines	Raviège	Gage	Record
Dam type	Concrete gravity	Concrete gravity	Concrete gravity	Slim arched	Concrete gravity	Concrete gravity	Concrete gravity	Cc gravity and earthf.	Concrete gravity	Slim arched	Concrete gravity
H on ground level (m)	20	40	25	43	13.4	28.4	15.8	17	37	39.6	16
Dam's date of construction	1946	1932	1933	1951	1971	1968	1939	1948	1958	1954	1956
PKW's date of construction	2006	2008	2009	2010	2011	2012	2014	2015	2014-15	2015-17	2015-16
PKW localization	Bank	Dam crest	Dam crest	Bank	Bank	Dam crest	Dam crest	Dam crest	Dam crest	Bank	Dam crest
Dam design flowed (m^3/s)	162	660	500	150	13	4600	192	685	1700	675	1700
Q PKW at MWL(m^3/s)	68	138	82	90	13	570	120	300	300	455	1300
q_{sW} at H = 1 m ($m^3/s/m$)	5	6.5	9	6.8	~ 4	9.4	8.3	8	8.5	11	4.4
Total project cost (k€)	400	1800	1500	1600	615	4500	2800	6500	4000~5000	15000~20000	2200~2800
PKW Cost (k€)	300	400	400	400	230	800	450	1800	1000~1200	900~1300	2200~2800
Head on PKW at MWL (3) (m)	0.95	1.35	0.95	0.8	0.65	1.5	0.9	1	1.4	1.75	4.4
Cconstruction (month)	3	6	6	5+5	6	12	6	6	8+8	30	8+8
B (m)	9.3	12.7	12.2	10	5.1	13.46	13.1	13.24	13.24	13	9,45
B_o (m)	3.35	4	3.2	3.5	1.2	6.63	4.9	3.97	4	4	4,5
B_i (m)	1.5	4	2	2.6	1.2	2.03	2.8	4.41	3.33	3	2,5
B_b (m)	4.4	4.7	7	3.6	2.7	4.8	3.5	4.87	5.91	6	4,5
P (m)	3.1	4.2	5.3	3	1.8	4.4	5.35	4.38	4.67	6	3
W (m)	12	18	18.7	18.5	4.8	42.5	16.55	2x23	25.8	26.6	4x12.5
W_i (m)	2.7	3.1	2.45~2.75	2.3~2.5	1.3	1.25~1.65	1.55	2.4	2.4	1.6	1.52
W_o (m)	1.5~1.8	2.2	1.5	1.5	0.9	1.58	1.4	1.6	1.65	1.3	1.52
T_s (m)	0.2	0.25	0.35	0.25~0.35	0.3	0.2~0.4	0.35	0.35	0.25~0.4	0.25~0.4	0.3~0.4
L (m)	59	77	78	86.8	22	350	115	2*120	177	208	258
P_{px} (m)	0	0	0.5	0	0	1.65	0.7	1	1	0.8	0
B_n (m)	1	1.5	1	1.5	1	1.5	1	1.15	2	2	1
Nose form	triang.	triang.	triang.	round	triang. extruded	triang.	triang.	round	triang.	triang.	triang.

Figure 3. 9 Achieved and 2 under construction PKW (characteristics).

Most of the built PKWs have a small number of alveoli (One to three inlet keys) and relatively balanced overhangs. The PKW of Malarce is by far the largest one ever with 12 units and an upstream overhang three times longer than the downstream one.

The maximum hydraulic head on these spillways is in the range between 0.5 m and 1.5 m with a specific discharge capacity up to 10 m3/s/ml for a one meter head, thus from 2 to 4 times better than classic linear weirs. Record PKW project is an exception to that with a maximum hydraulic head of 4,4 m.

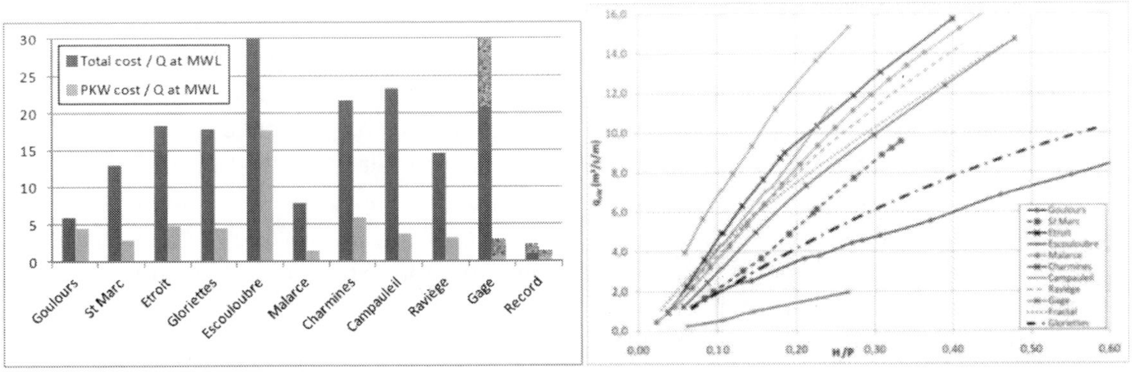

Figure 4.a) Economical ratios and b) Specific discharge versus H/P for PKW

Figure 4a shows a summary of PKW projects costs. We tried to distinguish the overall project cost from the PKW of the sole PKW structure. This effort is not so easy to achieve as it depends on the contractor strategy in the way it splits the units prices of the bill quantities. However, this figure provides general orders. In most cases, the PKW cost correspond to a minor percentage of the overall project cost. Other costs such as installation fees or downstream dissipation structures can weigh as much or more than the PKW structure.

For example, for the case of Gage project, the downstream works include the construction of an expensive 9 m diameter tunnel. On the other hand, for the case of Record project, the PKW structure cost is the main part of the project cost.
Because Record PKW has by far the highest maximum hydraulic head, it also has the lowest PKW and project cost ratio. Indeed, PKW or project cost would be very similar with only 2 m hydraulic head for instance, apart from a couple of stabilizing anchors.
In terms of pure hydraulic performance, Gage PKW is the most efficient one with a specific discharge larger than 11 m3/s/ml for a 1 m head. It is rather more efficient that Malarce PKW although it is slightly simpler to build (larger upstream overhang). This efficiency partly comes from the PKW height “P” which is a major factor for PKW discharge capacity.

PKW GENERAL DESIGN

PKW for existing dams

Design constraints and considerations will be quite different whether the project concerns a new dam or an existing one. This point was emphasized by Machiel [2012] who distinguished purely hydraulic optimization and technico-economical optimization.

In the case of existing dams, many parameters are imposed by the existing structures: for example the total available width on a dam crest, the difference between maximum water level (MWL) and normal water level (NWL), the base length (Bb). PKW height parameter P is also quite dependant on the capacity to keep a lowered reservoir level during works and the corresponding cost in term of operation loss. All these crucial parameters change for each existing dam and will lead each time to a specific design. However, in most cases, the overall cost of the project will be governed by the specific PKW discharge capacity. In most cases, it will be more interesting to optimize the hydraulic PKW specific discharge capacity. This point is illustrated by figure 4 below. As we already mentioned, for most PKW projects, the cost of the PKW represents less that 25% of the total project cost. This has to be seriously considered when designing a new spillway.

Most EDF dams are equipped with gates in order to optimize the available reservoir volume for hydroelectric purposes. Technico-economical comparisons are performed for new spillway projets. They often show that the installation of free flow spillway is more suitable for cost, maintainance and operation reasons rather than the installation of an additional mechanical gate. However, a minimum water head between NWL and MWL is required so that free flow spillway can properly operate. The analysis of EDF major dams characteristic shows that the available water head for free flow spillway (Maximum Water Level minus Normal Water Level) is less than 1 m for 40% of them. This small water head points out the relevancy of labyrinth or PKW for these applications.

Figure 4. a) Campauleil PKW and b) Raviege PKW achieved in 2014 and 2015

PKW for new dams

In new projects, the requirements are very different from those encountered on the rehabilitation ones. Labyrinth configurations or alternative PKW design can be relevant in the case where the PKW foundation extent is less limited. For example:

- No need to maintain a high reservoir level for operating purposes allowing to have higher structures.
- Larger available space on top of RCC dam for instance with a possibility to have structures without overhangs.
- Available space on the banks for spillway foundation.
  In this case, Labyrinth weirs should be closely compared to PKW on a technico-economical point of view. Both structure were already compared in recent papers by Anderson [2013] or Paxson [2013, 2014].

After 10 years of design experience, it is still difficult to propose a typical set of parameters which would fit for every project. The optimal configuration has to be derived considering all project data, which include economical data, construction planning, flood passage during works, energy dissipation structures downstream, geological rock quality, etc.

However, two typical situations can be distinguished and might lead to two different patterns:

- For rehabilitation issues: The technical-economical analysis might often lead to focus on hydraulic performances. Typical geometrical ratios could be such as follows : L/W ~ 6 to 7, P ~ 4 to 5 m, Wi/Wo > 1.3, Ts ~ 35 to 40 cm with specific discharge capacity qsw close 8 m3/s/ml for H = 1 m.
- For new schemes, the technico-economical analysis might lead to slightly different PKW shapes with following typical geometrical ratios: L/W ~ 4 to 6, P ~ 4 to 10 m, Wi/Wo > 1 and smaller specific discharge capacity qsw close to 4-6 m3/s/ml for H = 1 m.

This principle was applied for Record PKW. For this rehabilitation project, things appear such a per a new dam project. The overall existing dam equipment was removed and totally replaced by 4 PKW units as it can be seen in figure 5. The maximum hydraulic head is 4,4 m. The ratio H/P is larger than 1 and the ratio wi/wo is 1.These ratios are very different from the one used for usual rehabilitation projects whereby existing gates are kept. In addition Record specific discharge for H=1 m is much smaller than for other PKW. However, PKW record is the cheapest one by far in terms of €/m3/s as it can discharge more than 25 m3/s under a maximum head of 4,4 m.

Figure 5.a) Record dam before PKW works in 2014 and b) in 2016

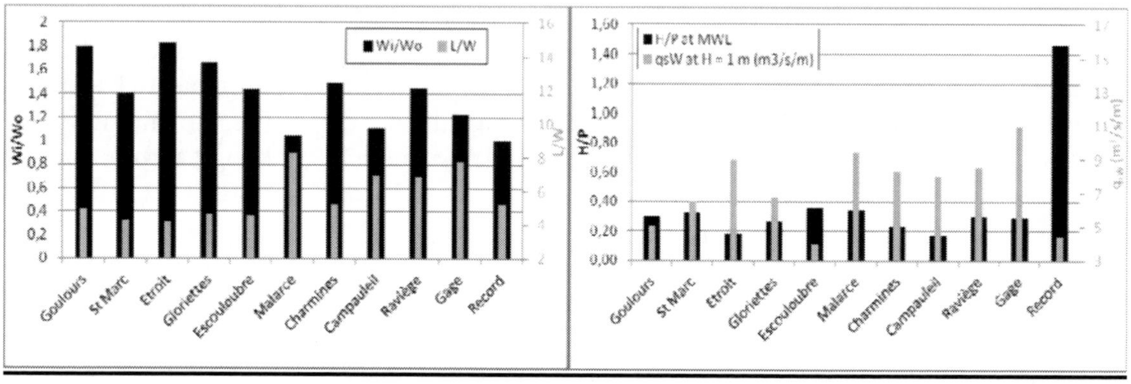

Figure 6.a) Plan parametrical ratio of PKW – b) Main hydraulic features of PKW

<u>Other projects requirements</u>

Finally, optimizing a flood passage issue project is not only about hydraulic performance of the spillway. It should include all project aspects. Malarce PKW gives a good example of a project which provided a general improvement in terms of dam safety. First of all, the dam discharge capacity was increased from 4000 to 4600 m^3/s. In addition the new PKW also doubled the minimum time that the Operator is given to reach the dam and operate other mechanical gates. During a flood event, the PKW operates first. Gates start when the PKW reaches half its discharge capacity. Considering that flood events in this area are extremely quick in terms of flood gradient, this is a vital help for Operators.

OTHER PKW HYDRAULIC FEATURES

The experience reveals some interesting points concerning the hydraulic efficiency of different devices and configurations.

- The nose : It is built under the upstream overhangs to improve the inlet key flow similarly to profiled piers upstream usual spillways. Various shapes can be designed:

triangular, parabolic, rounded, or triangular extruded in the base under the outlet key. A sensitivity analysis has been performed on CFD software and gains in specific discharge are a couple of percents. In addition, this system also provides more PKW stability with a supplementary mass of concrete which is favorably localized upstream the structure.

- Parapets walls : They have been installed on some PKW built by EDF, all along the crest or only on the outlet key. The idea was to increase the outlet key volume for a better evacuation. Simulations reveal that the global gain is limited to 5% for a parapet wall localized on the outlet key crest. For some configuration, this gain might be quite small or even negative. The gain will be very small if the outlet key slope is important, but more interesting for flatter outlet key slopes as the upper part of they might be easily submerged for higher flows. This optimization should be kept for very constrained projects and should be precisely studied with a physical or numerical model.
- Crest shape : Parametric tests were also performed to compare different patterns. Such tests were already carried out by Falvey [2003] for traditional vertical labyrinth weirs. Crest shape is also linked to the way it is actually realized on site. Quarter rounded, half rounded, flat and chamfered crest have been realized on EDF PKW and a numerical analysis confirmed Falvey results so that the quarter rounded and half rounded are the most efficient However, the influence is almost insignificant. Construction and durability constraints are more critical. Considering the difficulty of formwork assembly and long-lasting considerations, the quarter or half-rounded shapes should be preferred to respect both hydraulic and building imperatives.
- Sidewall thickness : It has been shown that the sidewall thickness has a major influence on the discharge capacity of the PKW, especially for narrow keys. Yet, it is rather hard to built thin concrete walls respecting the structural constraints of the project. In order to improve the flow conditions and to keep a good internal resistance, battered or inclined sidewalls have been imagined, large at the PKW base and getting thinner close to the top. This kind of geometry is used on Malarce PKW (40 cm to 20 cm) and gains about 5% to 10% are expected compared to thick sidewalls. On a practical point of view, it is easier to place the inclination of the wall only on one side to keep the other face of the formwork vertical.

Figure 7.a) Nose and upstream beam at Saint-Marc, b) parapet wall at l'Etroit and c) inclined sidewalls at La Raviège

PKW AERATION

In order to aerate the flow on the PKW, air intakes (>300 mm diameter) have been implemented for all PKW projects. Aeration pipes start below downstream alveoli and go inside the PKW structure. They come out to air intakes located either on the dam crest or in the drainage gallery.

For reliability reasons, two air intakes on either sides of PKW are installed. It will ensure a redundancy in case of partial air blockage in one pipe.

It is not clear whether this aeration systems are really necessary for PKW. Indeed physical models seem to show that flow naturally aerated and rather stable. However it is difficult to conclude because the scale effects are not similar for air and water actions. Vibrations could appear for a specific range of discharge flood.

Because the additional cost of aeration devices were was small compared to the PKW project, it was decided to install them by default. This is one advantage of PKW compare to traditional vertical labyrinth whereby the installation of aeration pipes might be complex.

Figure 8. Aeration device for Malarce PKW

However, following difficulties can be encountered :

- Getting a clear and reasonable design criteria for the aeration device. Up to know very basic design criteria were used, typically the air discharge to be a ratio of the water discharge
- In addition, for some projects, the installation of big aeration pipes can be a real and costly issue. That was the case of Record with very high specific discharge rates and hydraulic head (4,4 m – 25 m3/s/ml) but a small PKW (3 m high).

Anenometers have been installed in 2012 in Malarce PKW. Since then, Malarce PKW experienced more than 15 floods. Only preliminary data are available and a deep analysis of the air monitoring has still to be carried out. Figure 9 show an example of chart derived from one of the flood events in September 2014 (220 m3/s for H = 64 cm) These preliminary results seems to show:

- A linear relationship for aeration flow between air discharge and hydraulic head.
- 3.1 m3/s of air for 220 m3/s (much lower than what was designed : 10 m3/s) equivalent to 1.4 % of the water discharge rate.

These results still need to be confirmed. Engineering processing of available data is still on going. But they might lead to a real optimization of aeration systems for weirs.

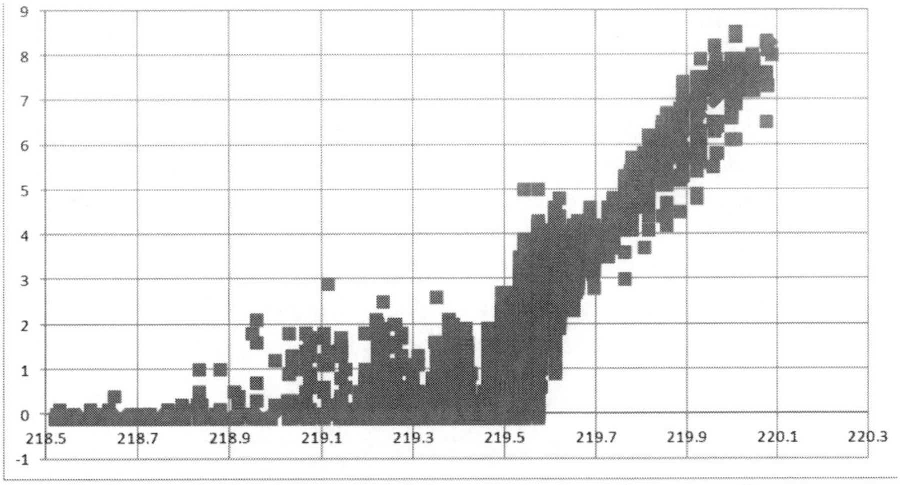

Figure 9.) Preliminary air discharge in Malarce aeration pipe according to reservoir level

PKW STRUCTURAL DESIGN

Demolition

Demolition aspects are also an important part of rehabilitation project to install PKW in existing dam structures. It is necessary to ensure all structures close to the new location of the PKW are not damaged by any explosion or vibrations. To achieve this goal, a two phases approach is recommended :

- Firstly : Diamond-cable saws can be used to physically separate structures.
- Secondly : Demolition is finalized by the use of explosives.

This method was applied on most concerned dams. It allowed a proper control of vibration and effects of explosives shots.

Overall stability

PKW overall stability can be calculated according to usual gravity dams rules. At first sight, PKW could appear to be unbalanced structures, hard to stabilize without anchoring. But, calculations generally show that the PKW center of mass at NWL is located above the base ensuring the self stability of the PKW structure for small heads. At low water level, some PKW configurations are self stable and would not require any additional anchor provided:

- a sufficient friction coefficient at the interface between the existing dam concrete and the PKW concrete

- a waterproofing / drainage system at the interface which can be proved efficient and reliable.

Some specific load cases (earthquake with empty reservoir, particular construction phasing,…) might nevertheless require to fix anchor bar. They bring additional safety margin for a limited overcost.

For some projects, an additional upstream beam was added to increase the total PKW mass and to avoid the installation of anchor bars linking the old and new structures. This was carried out for Saint-Marc and l'Etroit projects. The purpose was to take into account alkali-aggregate reaction affecting dam concrete. Swelling concrete could impact new structures if anchored by bringing imposed displacements and additional stresses. In addition to the upstream additional mass, the upstream beam provides an additional safety margin of sliding factor by creating a upstream shear key for the PKW.

Another simple measure can also be taken to improve external PKW stability during the demolition phase. The demolished section can be carried out with a 5 dg angle compare to the horizontal. This measure does not cost much but slightly improve the PKW stability.

Tendons

Project sometimes require to install prestressed anchors to improve the stability of the PKW structures, or more often the stability of the dam block.

Thus it is recommended to install anchor heads in outlets keys. It provides a permanent access to anchor heads when PKW does not discharge. In addition anchor heads remain out of water most of times.

Hydrodynamic Forces

Taking into account hydrodynamic forces remains a challenging technical issue. It is hard to address with physical models because measuring high speed turbulent flows is difficult.

On the other hand, numerical models give an access to a unlimited range of out data. If the numerical model can be validated with a physical one, then it becomes a very interesting tool.

For PKW, hydrodynamic forces in key outlets are a matter of interest for overall stability as well as per internal reinforcement design.

A couple of tests were performed on physical models and numerical ones. According to the current results, following recommendations can be proposed :

- Do not take into account hydrodynamic forces for transient load cases (100 to 1000 years return period flood)
- For extreme floods (more than 1000 years), it is possible to consider a linear equivalent hydrodynamic force stating half the hydraulic head upstream and 0 downstream.

These measures are probably conservatives. We are expecting additional physical and numerical tests to improve these recommendations.

Dam joints

It is common that PKW might be constructed on several dam blocks and overlap dam block joints. In this case, there are at least three options with regard to the position of dam block joints versus PKW : In outlet key, in inlet key, or along a thickened sidewall separating inlet and outlet key.

3D structural analysis shows that it seems preferable to set dam block joint in an outlet key. This configuration enables to limit tensile stresses in lateral walls compare to other configurations.

Waterproofing and drainage

Being calculated as a gravity structure, waterproofing and drainage system should be carefully designed and be very reliable on site.

It is therefore recommended to install waterstop systems between the existing dam and the new PKW. Interfaces with dam block joints are very difficult issues corresponding to triple point which are difficult to treat.

In any cases, the existing dam block joint should be cleared with caution to allow an efficient link between the new waterstop and the existing joint.

For old dams, existing dam block joints are made with copper sheets. Linking copper and waterstop rubber is not easy and there is few feedback available for this specific topic. A solution consists of tighting strongly the copper sheet to the rubber with mechanical means.

It is also strongly advised to get a redundancy of the waterproofing system. For instance hydroblowing joints can be set upstream the waterstop. As an ultimate redundant measure, it is possible to install injection pipes (Fuko type) between the waterstop and the hydroblowing joint. In case of unacceptable leakage, it is possible to carry out additional injection at low cost.

Drainage system. It is recommended to install a drainage system right downstream the waterstop at the interface dam / PKW. It can be achieved by creating draining voids filled with gravels and geotextile sheets. This network might be subsequently linked to the drainage gallery or to any place where the drainage flow can be measured as well as the drainage system efficiency.

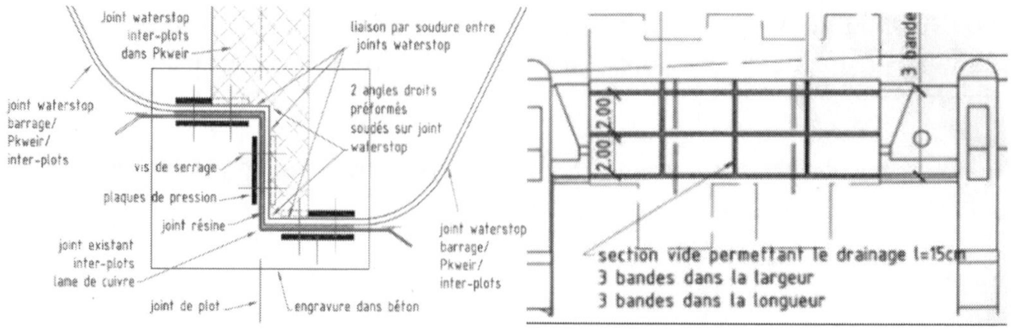

Figure 10. a) Principe of link between copper sheet joint and waterstop , .b) Typical drainage system at PKW / dam interface

Reinforcement

PKW are very stiff and hyperstatic structures. Modal analysis show that the frequency of the first modal mode is very high. Calculated stresses and required reinforcement appear to be reasonable according to the thinness of PKW sidewalls. A special attention shall be paid to thermal loading and to the potentially important temperature contract which can occur between wall faces which are submerged and the other ones.

Ice loading. In mountain or cold areas, ice loading can be very important, several tons per meter. A special care should be paid to avoid a dramatic increase of reinforcement quantity in the thin PKW sidewalls.

Several options can be considered :

- The first option consists in a deep analysis of the reservoir level across the year in cold times. It is sometimes possible to demonstrate that cold temperatures correspond to lower reservoir levels.
- Second options The ice layer is very sensitive to the variation of the reservoir level. If it can be showed that the reservoir level frequently varies by a couple of meters, then the ice layer can subsequently be fragilized or broken. A couple of papers were published during the last 20 years on this topic.
- Third option : The PKW structural analysis should consider ice loading as an evanescent thermal loading with structural displacement. An iterative calculation can be performed to take into account the actual sidewall displacement. When moving the thermal load applied to the sidewall will decrease.

Earthquake loading. Earthquake loading is not a simple issue when the PKW is installed on the top of high dams. In this cas, it is necessary to take the modal amplification caused by seismic dynamic response of the dam. This amplification can be high and be larger than 3 or 4 for 30 to 40 m high standard concrete gravity dams. For higher dams or arch dams, the amplification could be much larger. The earthquake loading might subsequently become the most demanding case.

Architectural feature

As per all labyrinth structures, PKW structures have a strong impact on the visual environment of people. They change the traditional way people look and feel with regard to dams, especially for dams that are located in inhabited or touristic place.

Therefore, a special emphasis could be laid on the architectural aspect to improve the way PKW look and the way people would feel about it and accept.

With regard to these considerations, special efforts made for Charmines PKW should be mentioned. The external shape of the PKW was slightly modified : The angles between inlet keys and PKW base were substituted by parabolic smooth lines. The visual effect is quite clear. Many people can appreciate it as the dam crest is equipped for public passage with pedestrian footbridge over the PKW.

Figure 11. Charmines PKW achieved in 2014

FLOATING DEBRIS

The floating debris issue is a crucial one with regard to the actual and safe operation of spillways. Following comments can be made:

- For many rehabilitation projects, the existing dam is often equipped with high discharge rate mechanical device such as radial gates or flap gates. Their pecific discharge capacity is generally much higher than PKW one. For example, a 4 m high gate will have a specific discharge capacity larger than 20 m3/s/ml. Therefore during the firt hours of the flood and when the PKW barely starts to work, the specific discharge of the PKW will be much smaller. Thus floating debris are clearly attracted by existing gates.

- PKW are free flow weirs. Naturally when the reservoir level increases, debris will be flushed away bu the PKW.
- In addition, trial tests were performed on most physical models realized for PKW projects. Research projects were also performed by laboratories [Pfister, 2013] and [Crockston, 2015]. These tests allowed to measure :
  - The rate of debris passing through
  - The residual discharge capacity of a PKW partially blocked by debris.

  Goulours PKW physical model was tested with reduced trees equivalent to 4 to 10 m high trees. Experimental results show that :
  - The blockage of the PKW only affects a small part of the developed discharging crest (5 to 10%) which corresponds to the front upstream width of outlet keys. This part is physically more impacted by debris.
  - Tests carried out with colored water and 3D numerical simulations show the flow pattern and the current lines. Most part of the flow go through the bottom of the PKW under the debris. The flow thus gets inside the inlet keys with few head loss.
- Finally even if the PKW seems visually blocked by debris, the residual discharge capacity is close to 80% of the nominal one without debris.

As a conclusion, we can say that PKW are very resilient toward debris.

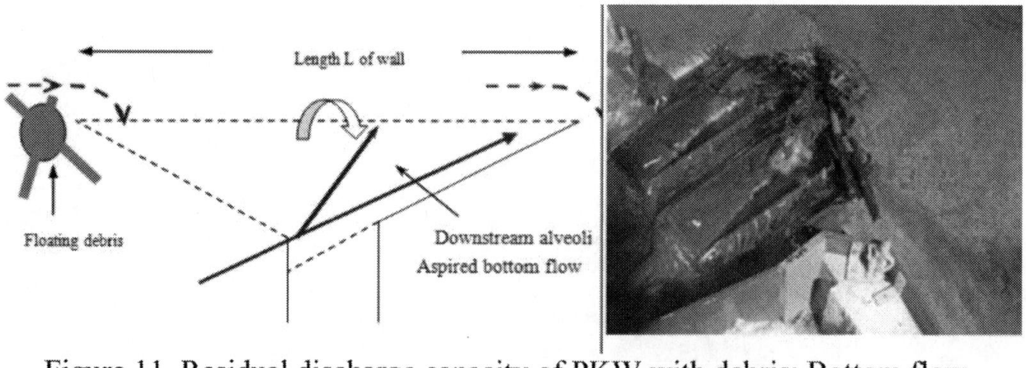

Figure 11. Residual discharge capacity of PKW with debris: Bottom flow

CONCLUSIONS

Piano Key Weir (PKW) labyrinth spillway have been a quickly developing innovation innovation in the world of dam engineering used to rather long time project developments. The first worldwide PKW was studied from 2003 to 2005 and constructed in 2006 at Goulours Dam.

PKW are based on the traditional labyrinth spillway concept consisting in increasing the overflowing length by "folding" the crest of the spillway. PKW are an answer to the main drawback of traditional labyrinth spillway : They can be installed at the top of the narrow crest of most existing concrete gravity or arch dams with a small footprint on foundation. They also have two essential qualities :

- They have large specific discharge capacities which allow to significantly increase existing dam overall discharge capacity.
- They are free flow spillway. Thus they have an excellent reliability because they operate without any electro-mechanical component nor human operator which might fail. They also do not require any energy source or remote control. In addition, they require little maintainance.

During the last ten years, EDF designed and built about 10 PKW spillways with design hydraulic head mostly between 1 and 1,5 m. They were part of existing dam rehabilitation projects in France. Other PKW are under development for new dams abroad. Specific discharge capacity can reach 10 m3/s/ml for 1 m hydraulic head and exceed 20 m3/s/ml for 2 m head.

PKW have very simple geometrical features, basically triangles and rectangles. Thus they are easy to design and build. EDF projects were kept under control in terms of cost and delays. EDF projects also confirmed PKW main quality: They can easily be build on the top of the narrow crest of existing concrete dam with usual means because of the cantilevered geometry.

PKW concept is suitable either for existing dam rehabilitation or for new dams. In the latter case, a combination "gates + PKW" has revealed many advantages by optimizing cost, reliability and other technical issues such as reservoir level control or sediment passage.

REFERENCES

Anderson, R.M. and Tullis, B.P. (2013). "Piano Key Hydraulics and Labyrinth Weir Comparisons." J. Irrig. Drain., 139(3), 246-253.

Crokston, Mortensen, Stanart, Tullis, Vasquez. 2015. Debris and maintenance of labyrinth spillway – USSD Annual Conference.

Erpicum, S., Laugier, F., Pirotton, M., Boillat, J., Reverchon, B., and Schleiss, A. (2011). Labyrinth and Piano Key Weirs: PKW 2011. CRC Press, London, UK.

Erpicum, S., Laugier, F., Pfister, M., Pirotton, M., Cicéro, G., and Schleiss, A. (2013). Labyrinth and Piano Key Weirs II: PKW 2013. CRC Press, London, UK.

Laugier, F. Design and construction of the first Piano Key Weir spillway at Goulours dam. 2007. Hydropower & Dams 14(5): 94-100.

Lemperiere, F., Ouamane, A. The Piano Key weir: a new cost-effective solution for spillways. 2003. Hydropower & Dams 10(5): 144-149.

Machiels, O. 2012. Experimental study of the hydraulic behaviour of Piano Key Weirs. Phd Thesis, University of Liège, Belgium.

Paxson G.S., Tullis B.P., Hertel D.J., 2013b. Comparison of Piano Key Weirs with labyrinth and gated spillways: hydraulics, cost, constructability and operations. Labyrinth and Piano Key Weirs II – PKW 2013, CRC Press

Paxson G.S., Laugier F.,. Labyrinth and Piano Key Weirs, Perspectives and Case Histories from the USA and France – ASDSO 2014 Annual conference – San Diego.

M. Pfister, B. Tullis, A. Schleiss2013. Effect of driftwood on hydraulic head of Piano Key weirs. Labyrinth and Piano Key Weirs II – PKW 2013, CRC Press

Pralong J., Vermeulen J., Blancher B., Laugier F., Erpicum S., Machiels O., Pirotton M., Boillat J.-L., Leite RibeiroM., Schleiss A. A naming convention for the Piano Key Weirs geometrical parameters. Labyrinth and Piano Key Weirs – PKW 2011, 2011, CRC Press

Pinchard T., Farges J.L., Boutet J.M., Lochu A., Laugier F., 2013. Design and construction of a Piano Key Weir spillway on Malarce dam. Labyrinth and Piano Key Weirs II – PKW 2013, CRC Press

Tullis, P., Amanian, N., and Waldron, D. (1995). "Design of labyrinth weir spillways." J. of Hydr. Eng., 121(3), 247-255.

ENSURING WATER SUPPLY RELIABILITY WITH INNOVATIVE PK WEIR SPILLWAY DESIGN

G. Robblee, P.E.[1]
S. Kees, P.E.[2]
B.M. Crookston, Ph.D., P.E.[3]
K. Keel, P.E.[4]

ABSTRACT

The West Fork Eno River Reservoir is located approximately eight miles northwest of the Town of Hillsborough, North Carolina and is their primary water supply. Construction was started in 1998 with initial reservoir filling performed in 2000. The project was originally designed and permitted to allow future expansion of the reservoir by raising the normal pool elevation by 10 feet. Engineering for the reservoir expansion is currently underway; the Town retained Schnabel Engineering to design the spillway modifications to support raising the reservoir.

The existing spillway configuration is similar to a side channel spillway and posed several unique challenges. For example, downstream of the control section, the spillway chute bends 70 degrees while the width of the chute simultaneously transitions from 215 feet to 80 feet. Spillway wall heights are also insufficient to contain the spillway design flood. Due to the hydraulic complexities associated with this spillway, a three-dimensional computational fluid dynamics model (CFD model) was developed to estimate spillway performance under various flood conditions. The CFD model facilitated the selection of the chute wall heights and provided valuable insights during the evaluation of spillway alternatives for the control section.

The alternatives analysis for the new spillway control section considered an ogee weir, a labyrinth weir, and a piano key (PK) weir. The labyrinth alternative was eliminated due to size constraints. Both the ogee and the PK weir alternative were viable, but the PK weir allowed the control section to be narrowed by about 45 percent while still passing the spillway design flood. The significantly smaller and compact PK weir proved to be the most cost-effective alternative. As far as we know, this will be the first PK weir designed and built in the US.

INTRODUCTION

The existing West Fork Eno River Reservoir Dam is a 2,500-foot long, 64-foot high earthen embankment with a crest width of 20 feet at El 653.7 feet. Construction of the dam started in 1998 with completion and initial reservoir filling in 2000. The dam

[1] Schnabel Engineering, Greensboro, NC, (336) 274-9456., grobblee@schnabel-eng.com
[2] Schnabel Engineering, Greensboro, NC, (336) 274-9456., skees@schnabel-eng.com
[3] Schnabel Engineering, West Chester, PA, (610) 696-6066, bcrookston@gmail.com
[4] Town of Hillsborough, Hillsborough NC, (919) 732-1270, kenny.keel@hillsboroughnc.org

impounds a 227-acre lake at normal pool condition (El. 632.3 feet) and discharges into the West Fork Eno River. The dam's intake tower and conduit serve as the principal spillway and are used to release water for minimum stream flow requirements and for water supply when the reservoir is at or below normal pool. The intake tower height is 62.5 feet, with the top of the tower at El 648.0 feet.

The dam's auxiliary spillway is a concrete chute located on the right abutment. The control section of the auxiliary spillway is 215 feet wide with the crest at El 632.3 feet (normal pool). The auxiliary spillway chute has a 70 degree bend located immediately downstream of the control section, where it transitions to an 80-foot wide chute (63 percent width reduction). The 80-foot wide section of the auxiliary spillway chute is 670 feet long and terminates with a small hydraulic jump stilling basin. The design storm for the dam is based on 3/4 of the probable maximum precipitation (3/4 PMP). A plan view of the existing auxiliary spillway is shown on Figure 1.

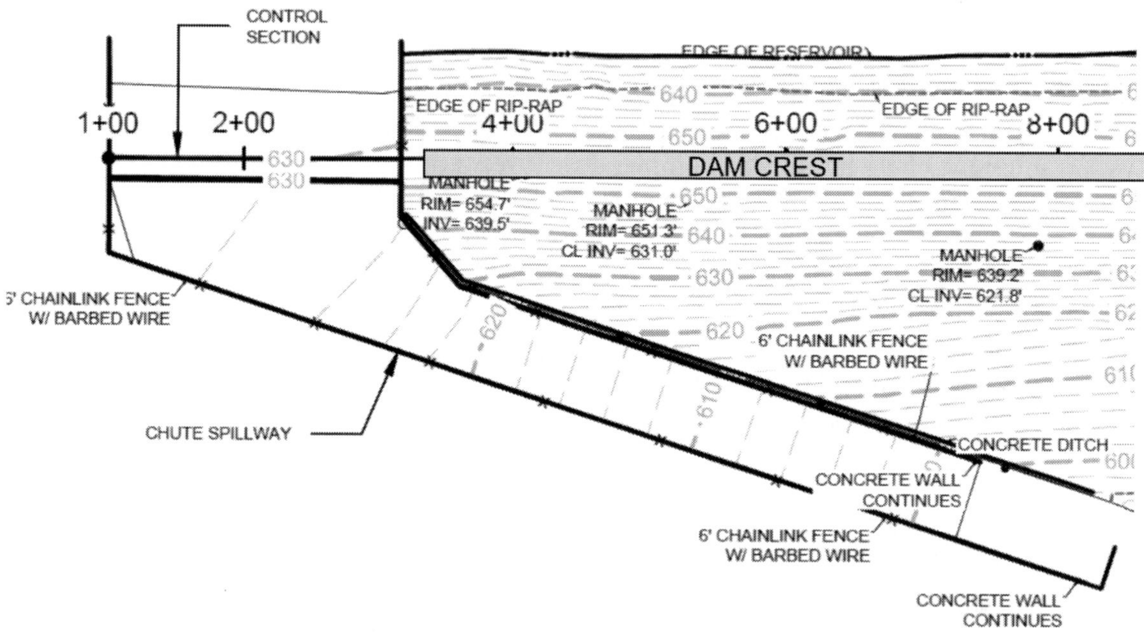

Figure 1. Plan View of the Existing Auxiliary Spillway

The original design of the dam and reservoir included a two-phased development plan. Land acquisition, constructing the existing dam and spillway structures, and clearing the reservoir to the Phase 1 normal pool elevation was performed in Phase 1. Phase 2 included constructing a 10-foot-high ogee weir in the auxiliary spillway, clearing the reservoir rim between the Phase 1 and Phase 2 normal pool elevations, and raising roadway embankments to provide adequate freeboard between the reservoir surface and adjacent roadways. The original dam and spillways were designed and permitted with the expectation that Phase 2 of the project would be completed by the end of 2018.

As part of the preliminary engineering for Phase 2 of the project, the design of the existing dam and spillways were reviewed to identify project components that may need to be modified to safely raise the normal reservoir level and pass the design storm. The project requirements, as discussed below, are based on the review of the existing design details and site reconnaissance. The review concluded that the earth embankment dam and intake structure were likely adequate and would not require modification prior to raising the normal operation pool 10 feet. However, several design details for the existing auxiliary spillway were identified that required detailed evaluations.

PROJECT REQUIREMENTS

Background

Figure 2 shows a section through the spillway control section with the original proposed Phase 2 ogee-shaped weir cast on top of the existing control section. A number of design details shown on the Phase 1 Record Drawings could result in potential distress of the auxiliary spillway during flood loading or less than desirable performance over the life of the structure. The design details are being addressed as part of the design of auxiliary spillway modifications include:

- Control section layout and weir type
- Adequacy of the seepage cut-offs beneath the control section
- Sliding stability of the control section
- Chute wall height
- Chute slab details

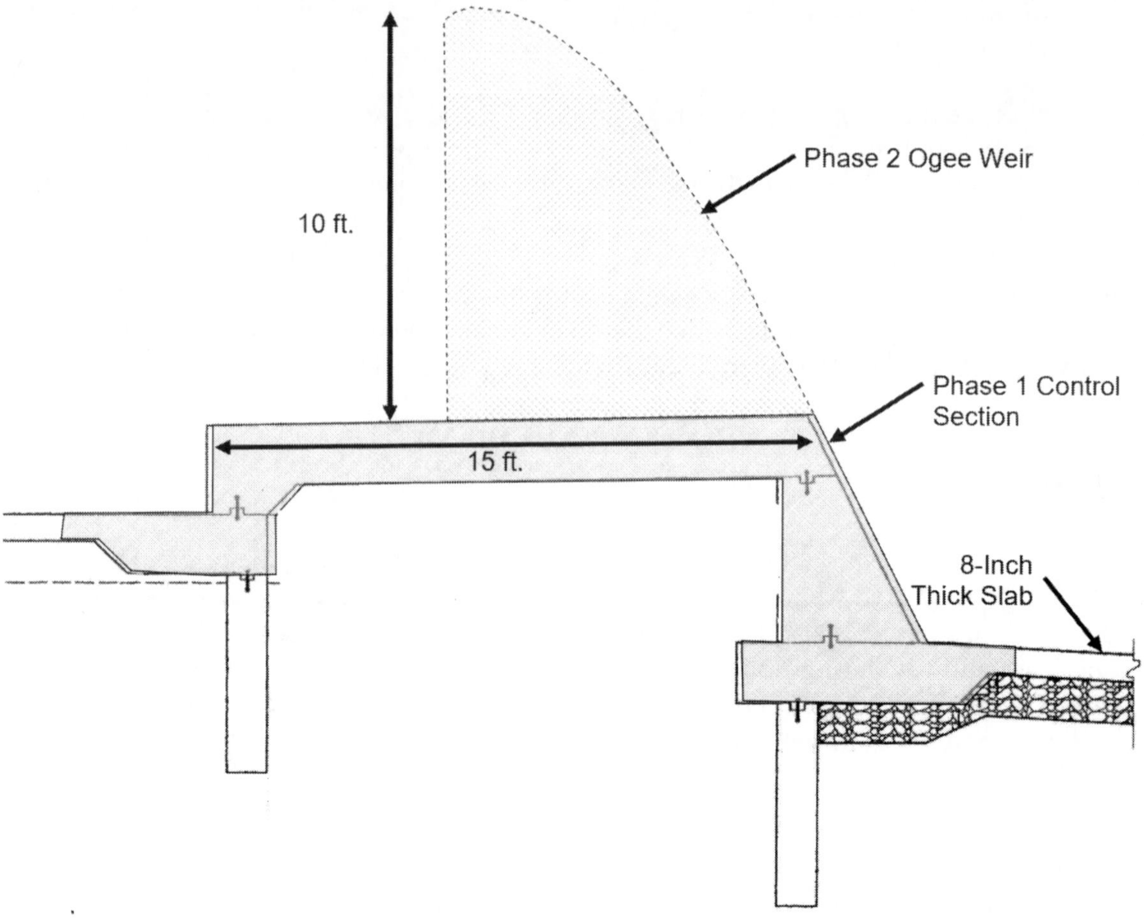

Figure 2. Section through Existing Auxiliary Spillway Control Section with Originally Proposed Ogee Weir (Phase 2)

Control Section Layout and Weir Type

Schnabel Engineering initially considered an ogee weir, labyrinth weir, and a piano key (PK) weir as alternatives for the control section of the modified auxiliary spillway. These three types of weirs were evaluated to determine weir configurations and sizes needed to pass the spillway design flood (SDF). A labyrinth weir was eliminated early in the evaluation process due to the relatively large space requirements needed to accommodate cycle depths (upstream to downstream direction) in comparison to the footprint of the originally proposed ogee weir. Hydraulic calculations were performed to detail the ogee weir option and the PK weir option. Based upon these calculations, the proposed ogee weir would be 10 feet tall and 215 feet wide. The proposed PK weir would be 10 feet tall and 115 feet wide. The hydraulic analyses for the preliminary design of the PK weir are discussed in greater detail below. Both the ogee weir alternative and the PK weir alternative would fit on the revised 17-foot wide base slab configuration that was developed to address sliding stability concerns. The proposed Phase 2 geometry of the ogee weir and PK weir alternatives are shown on Figure 3 and 4 below.

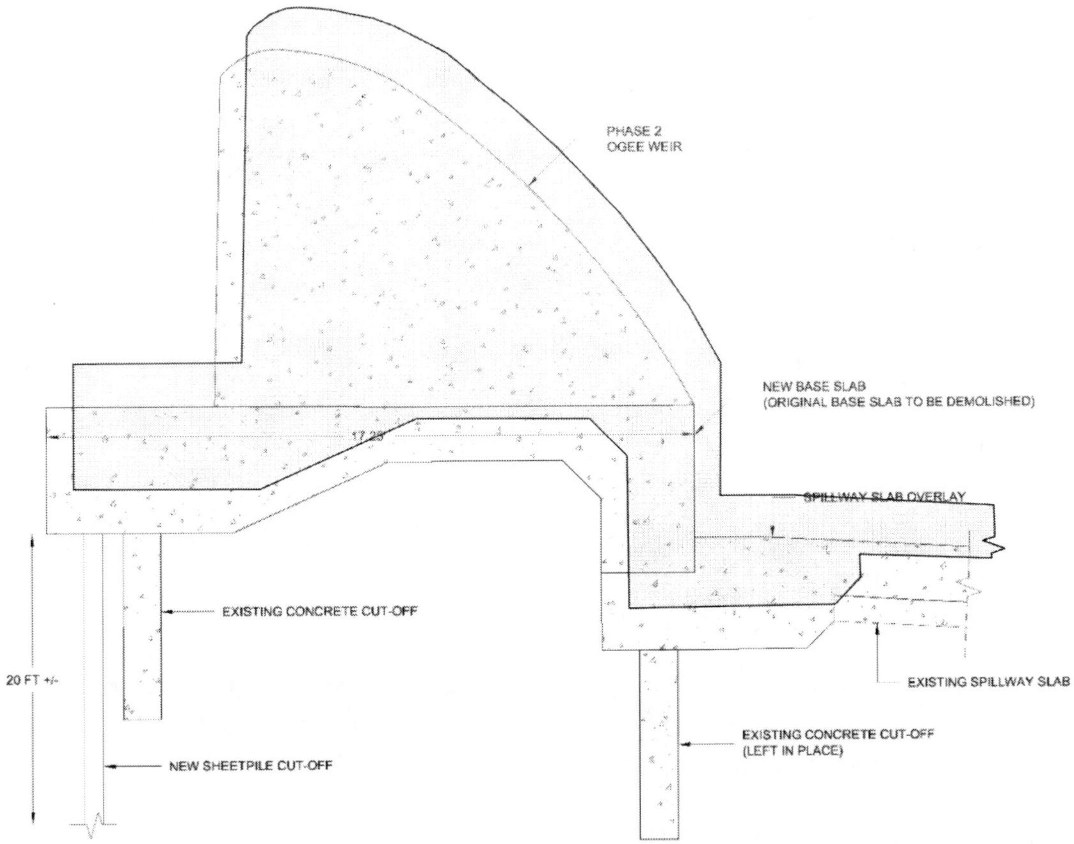

Figure 3. Section through Proposed Ogee Weir and New Control Section

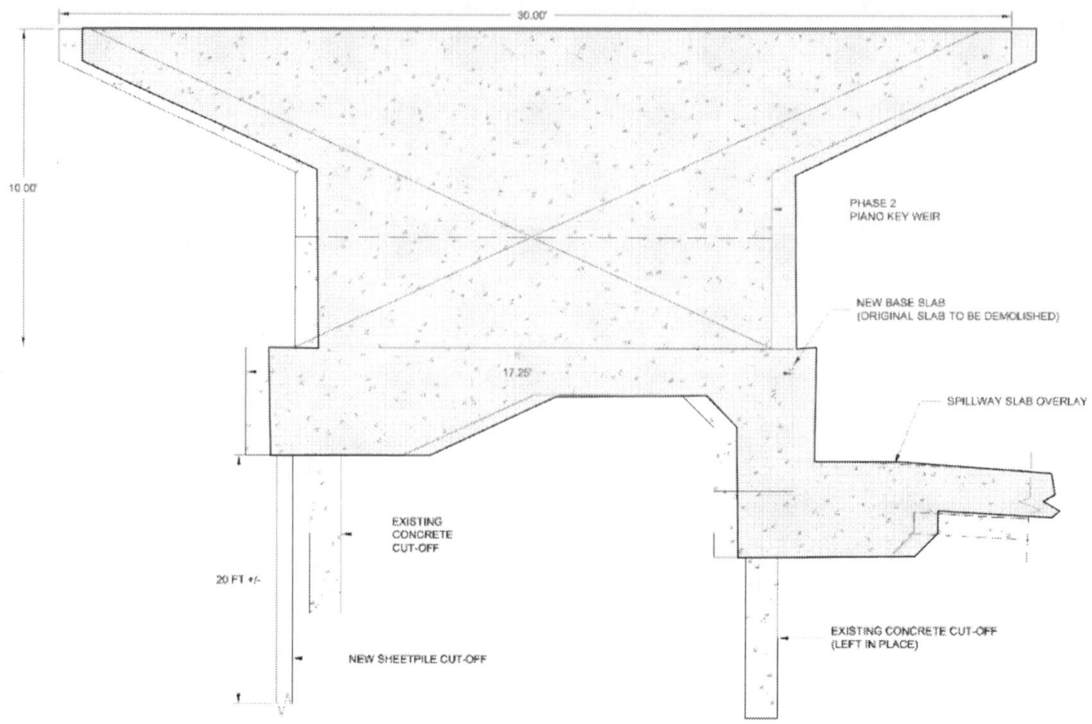

Figure 4. Section through Proposed PK Weir and New Control Section

Adequacy of the Existing Seepage Cut-off

The existing spillway section includes two concrete turn-downs or seepage cut-offs below the control section. A concrete slab extends about 88 feet upstream of the existing auxiliary spillway control section. The concrete joints between the concrete slab and the spillway control section and between the concrete slab and the spillway wing walls included water stops, which suggest that the purpose of the slab was to reduce seepage. However, there were no seepage control measures to prevent lateral seepage below the auxiliary spillway structure where it terminates (left and right). Borings performed through the control section during the Phase 2 preliminary engineering phase of the project encountered water levels below the control section that were equal to the reservoir level. This suggests that the upstream concrete slab and upstream concrete turn-down were not successful in creating head loss or reducing uplift below the control section. Both spillway modification alternatives will include a steel sheet pile seepage cut-off that will extend vertically to rock and will extend laterally into each earth section on the left and right sides of the auxiliary spillway.

Chute Wall Height

As previously noted, the auxiliary spillway includes a complex transition immediately downstream of the control section. Reservoir routings of the existing reservoir system for the 3/4 PMP results in 18,500 cubic feet per second (cfs) passing through the auxiliary spillway with a flow depth of 6.0 feet at the control section.

Hydraulic performance of the existing auxiliary spillway, with a focus on the approach conditions to the control section, hydraulic conditions immediately downstream, and the flow field as it transitions into the narrow lateral chute, was evaluated via a numerical model developed using a commercially available CFD solver. The existing auxiliary spillway's downstream transition area and chute were found to have inadequate capacity for the 3/4 PMP event. The CFD model predicts that the existing downstream walls will overtop by more than 4 feet in select areas, as shown in Figure 5. Therefore, the sidewalls would need to be raised to avoid potential technical issues for proposed alternatives (i.e., ogee weir and PK weir).

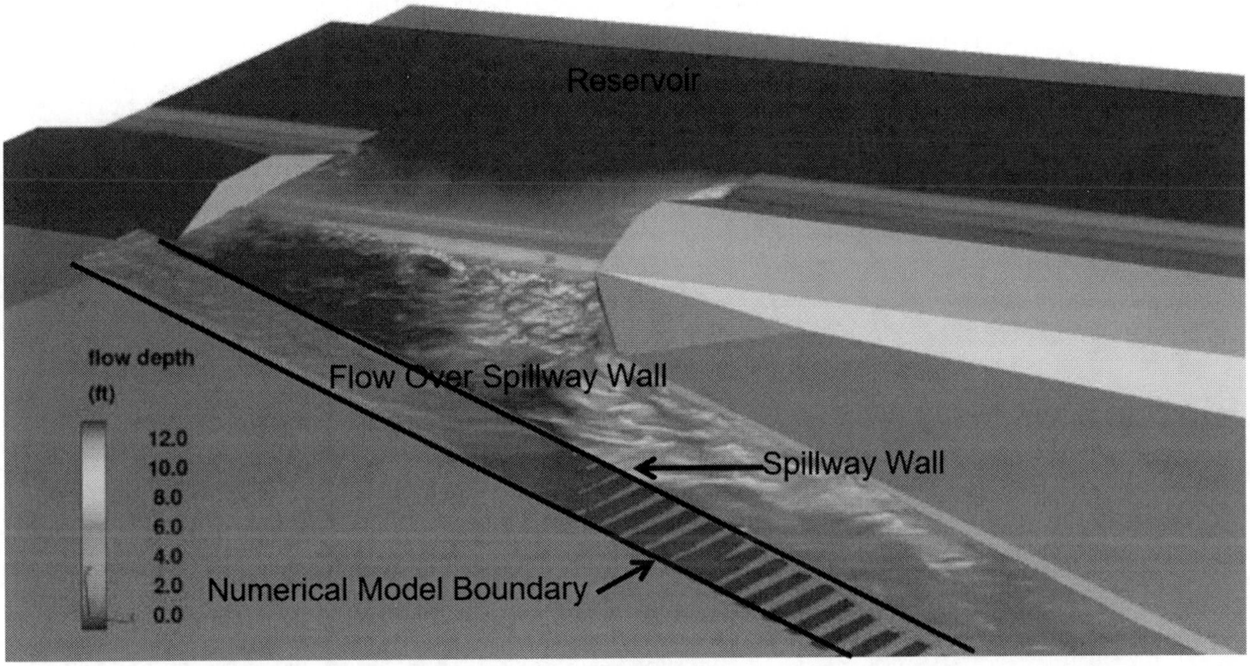

Figure 5. CFD Model Results for Existing Conditions

Chute Slab Thickness

The existing auxiliary spillway chute slab is nominally 8-inches thick, except at expansion joints where the slab is about 14-inches thick. The slab thickness was considered insufficient for all flow conditions. Site reconnaissance identified portions of the slab within 40 feet of the control section exhibiting signs of distress such as cracking spalling, and joint offsets. The Phase 2 conceptual modification for the spillway included constructing chute slab overlays to thicken the slabs in the upper portion of the spillway chute and at the stilling basin.

HYDRAULIC DESIGN OF PK WEIR

Background on Piano Key Weirs

The Piano Key Weir, also commonly identified as PK weirs (Figure 6a), was originally presented by Hydrocoop (Blanc and Lempérièr 2001) as an economic optimization of the labyrinth weir [(Figure 6b) Crookston et al. 2013]. International interest began in earnest about six years ago, with the completion of many research studies (Erpicum et al. 2012, 2014), four hydraulic design methods (Anderson and Tullis 2013, Liet Ribeiro et al 2014, Machiels et al 2012, and Kabiri Samani and Javaheri 2012), and numerous completed projects, the first being Goulours Dam in France in 2006 (Laugier et al. 2014). For example, PK weirs have been constructed on gravity dams (primarily in France), in channels (Australia), at hydropower facilities (Switzerland), and as part of massive run-of-river dams in Vietnam (see Figure 7). Thus, there are a variety of applications where a PK weir has been selected as the preferred conveyance structure when compared to other more-common weirs and gates.

Figure 6. PK weir (a) and labyrinth weir (b)

Figure 7. Van Phong Dam, Vietnam (photo courtesy Michel Hotakhanh)

A schematic of a Type-A PK weir (W_i/W_o = 1.25) with relevant geometric parameters is presented in Figure 8, with variables defined herein. Note that the crest thickness, T_s, may be constant or may vary between the sidewall crest, inlet key crest, and outlet key crest.

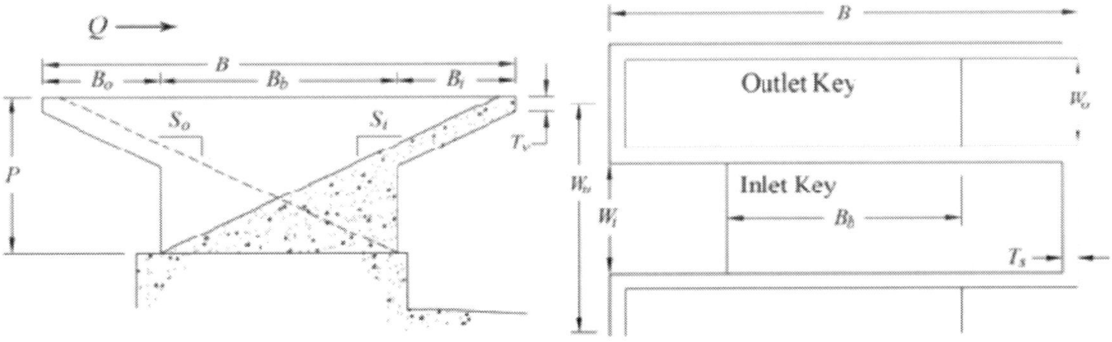

Figure 8. PK Weir Notation per Pralong et al. (2011)

PK Weir Hydraulics

Two PK weir hydraulic design methods were selected (for comparison) to estimate the head-discharge rating curve for the proposed Type-A PK weir. The hydraulic design method by Anderson and Tullis (2013) uses a traditional form of the weir equations (Eq. 1) with dimensionless experimentally determined discharge coefficients, $C_{d(1.25)}$.

$$Q_P = \frac{2}{3} C_{d(1.25)} L_c \sqrt{2g} H_T^{3/2} \tag{1}$$

In Eq. 1, Q_P is the flow rate conveyed over the PK weir, L_c is the centerline length of the crest, g is the acceleration constant of gravity, and H_T is the total energy or driving head referencing the weir crest, comprised of the piezometric head, h, and the velocity head, $v^2/2g$. Discharge coefficients can be determined graphically from the literature or from Anderson (personal communication 2010, 2015) using unpublished curve-fit equations that are a function of the headwater ratio, H_T/P where P is the height of the weir. No modification to the length of the crest (i.e., reduction) is merited.

The second PK weir hydraulic design method selected is by Leite Ribeiro et al (2014). Their approach uses a discharge enhancement ratio, r (Eq. 2) and references the discharge over a linear sharp-crested weir, Q_S (see Eq. 3). This approach is similar to the flow magnification ratio presented by Hay and Taylor (1970) for labyrinth spillways.

$$r = \frac{Q_P}{Q_S} \tag{2}$$

$$Q_s = C_s W \sqrt{2g} H_T^{3/2} \quad (3)$$

In Eq. 3, C_S is a discharge coefficient for a sharp-crested weir estimated by Hager and Schleiss (2009) to be 0.42 and W is the length of the linear weir, also equivalent to the width of the spillway or channel. To estimate the PK weir discharge, Eq. 2 is rearranged as $Q_S r = Q_P$ and r is computed via Eq. 4, which has been simplified from Leite Ribeiro et al (2014) for the PK weir geometry proposed herein.

$$r = 1 + 0.24\left(\frac{(L_c - W)P}{WH_T}\right)^{0.9}\left(\frac{W_i}{W_o}\right)^{0.05}\left(0.3 + \frac{B_o + B_i}{B}\right)^{-0.50} \quad (4)$$

In Eq. 4, W_i and W_o are the inlet and outlet key widths, respectively. Also, B_i and B_o the upstream and downstream overhang lengths, with the remaining base equivalent to B_b. The total length or depth (upstream-downstream direction) of the PK weir, B, is computed as $B_i+B_o+B_b=B$.

It was estimated that the PK weir would be required to pass approximately 17,800 cfs for the spillway design flood. The proposed PK weir geometry for the alternatives analysis is summarized in Table 1, with a weir height of 10 ft and 8 keys. The resulting head-discharge rating curves are presented in Figure 9. As shown, excellent agreement exists between the methodology developed at École Politechnique Fédéral de Lausanne in Switzerland and the methodology developed at the Utah Water Research Laboratory in the USA. Additional analyses to estimate weir hydraulics were also performed using CFD; a portion of the results is shown in Figure 9 (for details, see Crookston et al. - in press).

Table 1. Preliminary PK Weir Geometry for Spillway Design Flood

Q_P	17,840 cfs	H_T	7.4 ft	P	10.0 ft
S_i	0.47	W_i	7.0 ft	B_i	7.5 ft
S_o	0.47	W_o	5.6 ft	B_o	7.5 ft
N	8.0	W_i/W_o	1.25	B_b	15.0 ft
t_w	1.0 ft	Wu	14.6 ft	B	30.0 ft
Crest	flat	W	116.8 ft	B_i/B	0.25

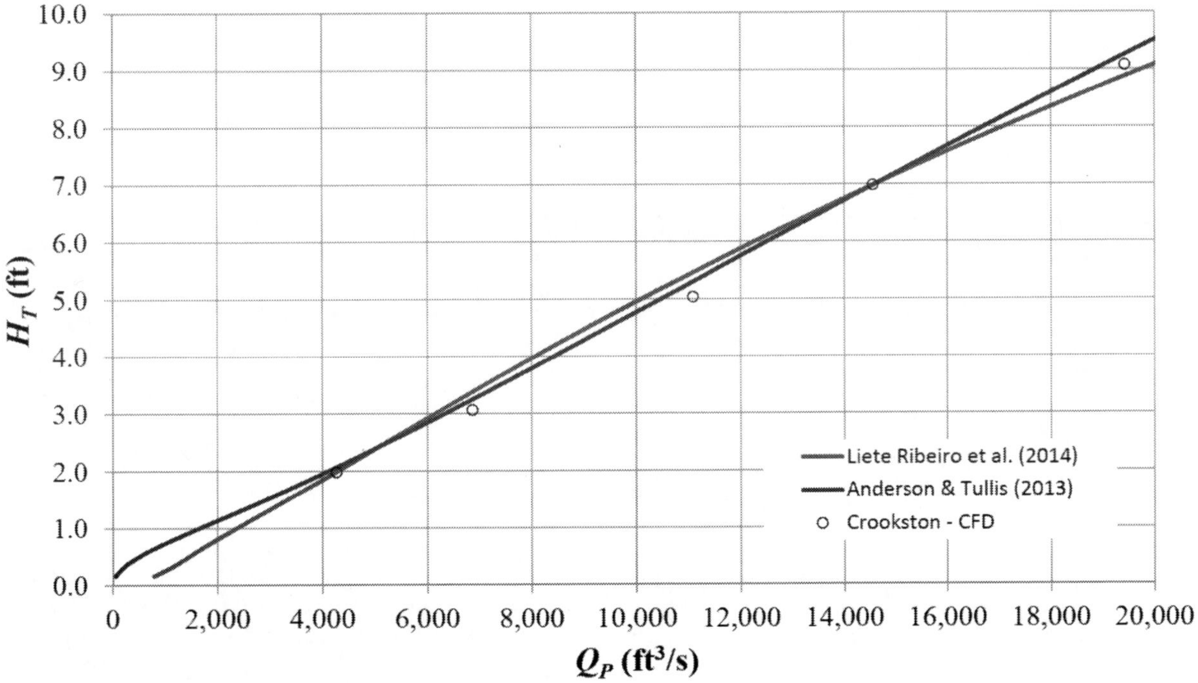

Figure 9. PK weir head-discharge rating curves

Numerical Modeling of Modified Spillway Alternatives - CFD

The Renormalized-Group (RNG) turbulence model for turbulent kinetic energy and dissipation was selected based upon model applicability and the numerical analyses of PK weirs by Crookston et al. (in press), where multiple turbulence models were compared to experimental head-discharge data by Anderson (2011). The CFD software solves the Reynolds-Averaged Navier-Stokes (RANS) equations using a finite-volume method; Eqs. 5 and 6 are implemented for conservation of mass and momentum.

$$V_F \frac{\partial \rho}{\partial t} + \frac{\partial}{\partial x}(\rho u A_x) + \frac{\partial}{\partial y}(\rho v A_y) + \frac{\partial}{\partial z}(\rho w A_z) + \xi \frac{\rho u A_x}{s} = R_{DIF} \tag{5}$$

$$\frac{\partial U_i}{\partial t} + \frac{1}{V_F}\left(U_j A_j \frac{\partial u_i}{\partial x_j} \right) = -\frac{1}{\rho}\frac{\partial P'}{\partial x_i} + g_i + f_i \tag{6}$$

The CFD model included a portion of the reservoir, the spillway, and downstream chute and basin that were drafted as solids using the 3D modeling workspace in AutoCAD Civil 3D. Figure 10 shows a schematic of the model for the PK weir alternative. The domain was discretized using hexahedral cells; dual mesh blocks were implemented for existing conditions and the ogee option. A single mesh block was used for the PK weir option. To increase computation efficiency, a domain-removing solid that enclosed areas outside the embankment and spillway was included (not shown). A solid was also prepared that represented water in the reservoir for the respective scenarios. Multiple simulations were performed of different cell sizes to establish a mesh-independent solution. Computations were performed using a customized work station with dual Xeon CPU's at 3.4 GHz (32 cores) with 64 GB of RAM. For example, the finalized PK weir simulation for the SDF (3/4 PMP) resulted in approximately 180 million cells and computation time was 70 hours, 55 min, 16 seconds.

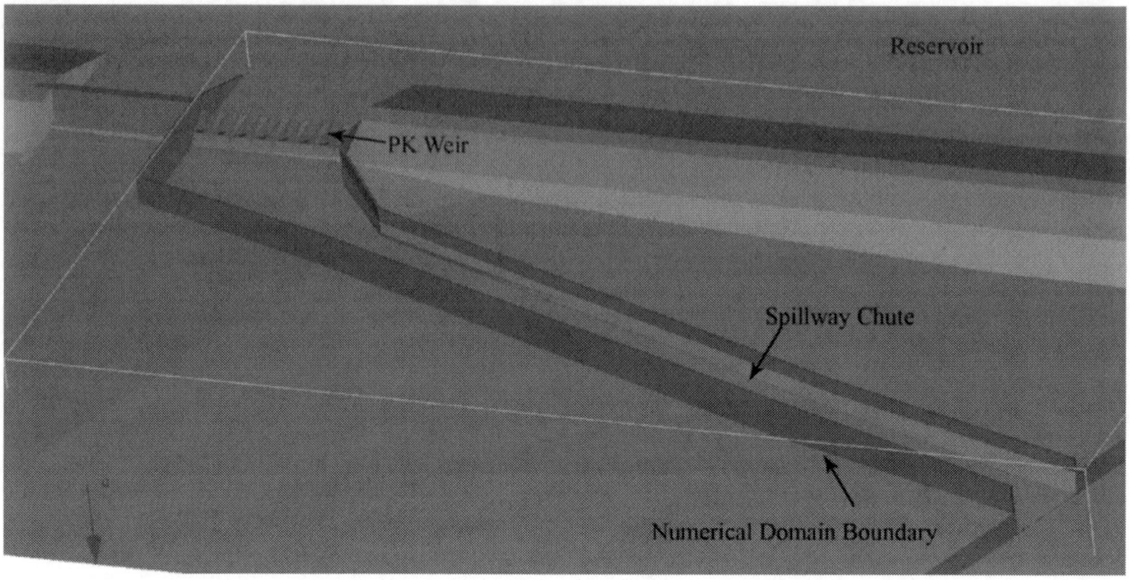

Figure 10. CFD Model of PK Weir Alternative

For the proposed ogee alternative, the SDF flows pass through critical depth near the weir crest and an irregular hydraulic jump forms at the base. The CFD model showed that the maximum flow depth in the chute is about 16 feet compared to an existing wall height of 8 feet (Figure 11). In addition to the downstream right sidewall, the left chute sidewall nearest the embankment also would need to be raised to contain the flow.

Figure 11. CFD Model Output for Ogee Alternative

CFD results for the PK weir alternative indicated that an irregular hydraulic jump would also form downstream of the weir (Figure 12). The flow depth in the chute for the PK weir alterative is similar to the flow depth for the ogee weir alternative. Additional analyses are being performed using published literature to further examine the complex flow field at the base of the PK weir.

Figure 12. CFD predicted hydraulic conditions for the 3/4 PMP, PK weir option

Laugier et al (2013) indicated that PK weirs are generally stable when experiencing hydraulic loadings; however, maximum headwater ratios for structures built by Electricite de France have been about $0.15 \leq H_T/P \leq 0.35$. In some instances, PK weirs designed in France have incorporated additional mass concrete at the upstream base. The maximum headwater ratio for Hillsborough is H_T/P=0.74 which is within the limits of H/P values tested in laboratories. Estimated pressures from the CFD simulations were incorporated into the stability analysis of the weir. Example pressure gradients at select cross-sections are presented in Figure 13.

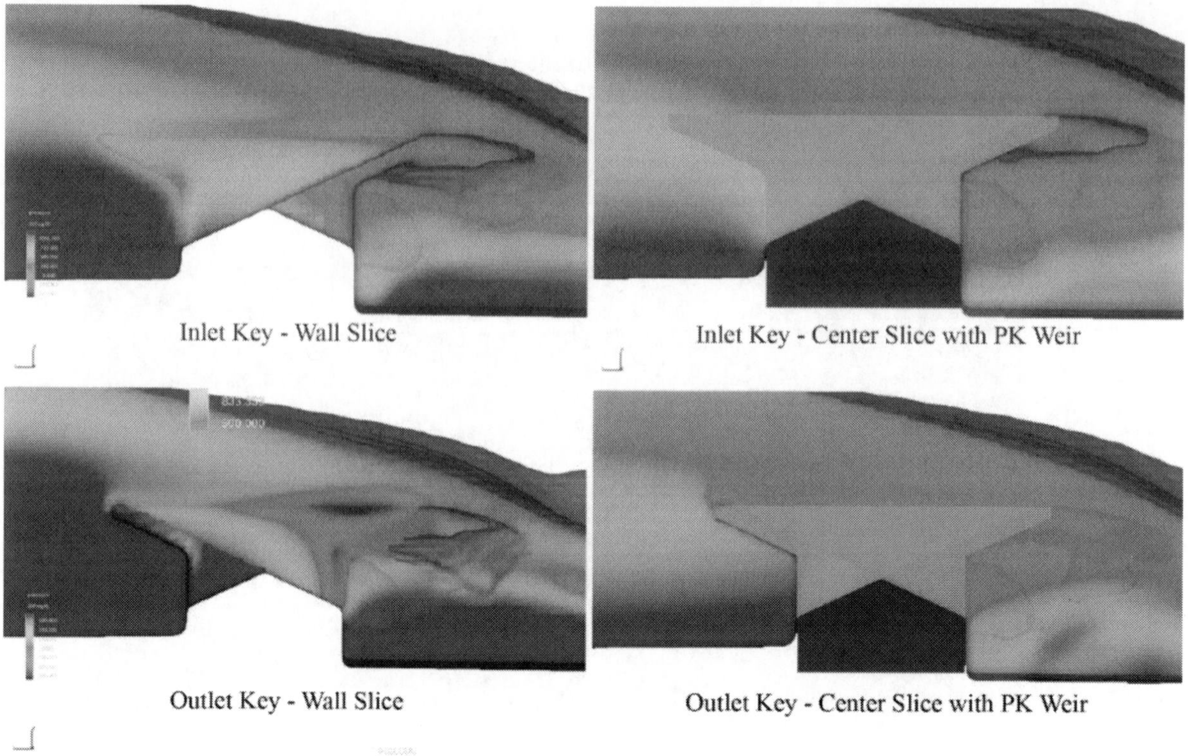

Figure 13. Predicted pressure fields for the 3/4 PMP

A portion of the stability analysis is presented in Figure 14. As shown, pressures (in terms of feet of water) from dynamic hydraulic forces at the PK weir sidewall, five cross-sections across the inlet key, and four cross-sections across the outlet key were obtained. Moment arms and resultant forces (point of reference is 7.5 ft downstream of the leading PK weir edge) were computed. To ensure stability, the PK weir will be structurally connected to the new control section base slab.

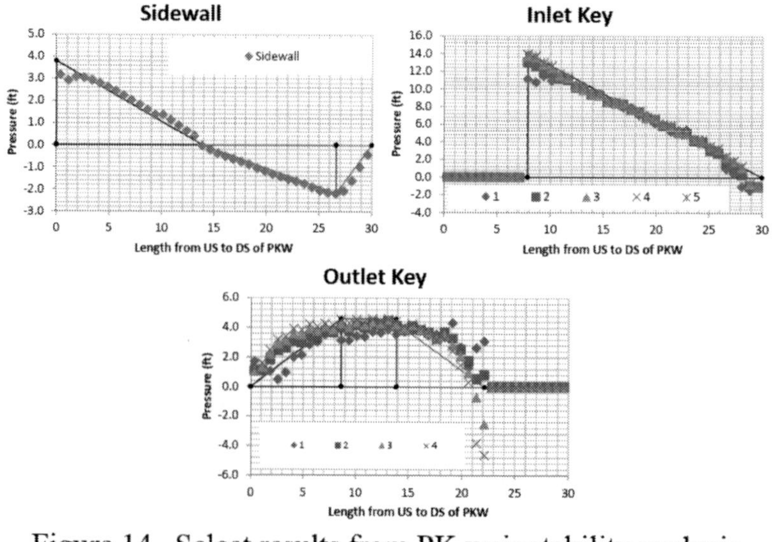

Figure 14. Select results from PK weir stability analysis

A more detailed discussion on the hydraulic design and numerical modeling of the proposed PK weir for this project can be found in Crookston et al. (in press).

COST ANALYSIS

The ogee weir alternative is the same width as the existing structure. The PK weir alternative is 45% narrower than the ogee weir alternative and will require an earth embankment closure section to fill the existing spillway. The reduction in spillway width for the PK weir alternative also affects other project components as listed below:

- The right spillway sidewall will be demolished.
- The need to replace versus raise spillway sidewalls.
- The length of raised spillway sidewalls.
- The length of the proposed sheet pile seepage cut-off is reduced.
- The need to repair a damaged section of the existing spillway slab is eliminated due its location.
- The area of spillway chute slab to receive a concrete overlay is reduced.

Schnabel Engineering performed quantity takeoffs and estimated construction costs for each alternative. The reduction in spillway width and the change from a mass concrete weir structure to a structural concrete weir structure greatly reduced the concrete volume needed for the project. However, the structural concrete is expected to be more costly on a per cubic yard basis.

Differences in material quantities and unit prices for each spillway alternative are shown in Table 2 below. Table 2 does not show project components that are the same for each alternative.

Table 2. Cost Comparison between Ogee Weir and PK Weir

ITEM	*Unit*	*Ogee Weir Alternative*			*PK Weir Alternative*		
		Quantity	*Unit Price*	*Cost*	*Quantity*	*Unit Price*	*Cost*
Demolition	LS	1	$120,000	$120,000	1	$180,000	$180,000
Concrete – Slab Overlay	CY	940	$780	$733,200	470	$780	$366,600
Concrete – Raise Chute Walls	CY	204	$1,020	$208,080	80	$1,020	$81,600
Concrete PK Weir Walls	CY	-	-	-	260	$1,200	$312,000
Concrete Ogee Mass Concrete	CY	960	$780	$748,800	-	-	-
Concrete Control Section Slab	CY	600	$780	$468,000	330	$780	$257,400
Concrete - New Wall Footings	CY	-	-	-	385	$780	$300,300
Concrete – New Wall Stems	CY	-	-	-	490	$1,020	$499,800
Sheet Pile Cutoff	SY	4,300	$48	$206,400	2400	$48	$115,200
Earthfill	CY	1,100	$18	$19,800	8,400	$18	$151,200
SUB-TOTALS				**$2,504,280**			**$2,264,100**

The results of the cost analysis suggest that the construction cost for the PK weir alternative is about $240,000 less than the ogee weir alternative. However, the design of the PK weir is expected to be more complex and require more engineering. Due to increased engineering, the potential project savings was decreased to about $200,000.

CONCLUSIONS

Several original design details were identified during the preliminary engineering stage of the Phase 2 project that potentially could result in adverse spillway performance especially for the higher Phase 2 reservoir level. This project shows the importance of examining all components of a structure prior to performing a major upgrade.

Both the ogee weir and the PK weir alternatives were considered technically feasible options for the required auxiliary spillway modifications needed to raise the West Fork of Eno Reservoir by 10 feet. The potential savings if the PK weir alternative was adopted was not a large percentage of the overall dam modification budget (< 7 %) and a smaller percentage of the overall project budget (<3%) if considering reservoir clearing costs and roadway improvement costs. However, $200,000 is a considerable savings that the Town of Hillsborough could use for other projects in the community and therefore the PK weir alternative was selected.

ACKNOWLEDGEMENTS

The authors wish to thank Dr. Michael Pfister of École Polytechnique Fédéral de Lausanne and Frederic Laugier and colleagues at Electricite de France.

REFERENCES

Anderson R.M. (2011). "Piano Key Weir Head Discharge Relationships." M.S. *Thesis*, Utah State University, Logan, Utah.

Anderson, R.M. Tullis, B.P. (2013). Piano Key Hydraulics and Labyrinth Weir Comparisons. *J. Irrig. Drain. Engng.*, 139(3):246-253.

Blanc, P. and Lempériѐr, F. (2001). Labyrinth spillways have a promising future, *International Jouirnal of Hydropower and Dams,* 8(4), 129-131.

Crookston, B.M., Crowley, L. and Pfister, M. (in press). "Piano Key Weir for Enlargement of the West Fork Eno River Reservoir." Proc. of the 6th IAHR International Symposium on Hydraulic Structures (ISHS2016), 27-30 June 2016, Portland, Oregon, B. Crookston and B. Tullis Editors, Utah State University.

Erpicum, S., Laugier, F. Boilatt, J., Pirotton, M. Reverchon, B. (2012). Labyrinth and piano key weirs – PKW 2011. CRC Press, London, UK.

Erpicum, S., Laugier, F., Pfister, M., Pirotton, M., Cicero, G., & Schleiss, A. (2014). Labyrinth and piano key weirs II – PKW 2013. CRC Press, London, UK.

Hay, N. and Taylor, G. (1970). Performance and design of labyrinth weirs. *J. of Hydr. Engrg., ASCE,* 96(11) ,2337-2357.

Kabiri-Samani, A. Javaheri, A. (2012). Discharge coefficient for free and submerged flow over Piano Key weirs. *J. Hydraulic Res.,* 50(1), 114-120.

Laugier, F. Vermeulen, J. and Lefebvre, V. (2014). Overview of Piano Key Weirs experience developed at EDF during the past few years. *Labyrinth and piano key weirs-PKW 2013,* CRC press, London, 213-226.

Leite Ribeiro, M., Pfister. M., Schleiss, A., Boillat, J.-L. (2012) Hydraulic design of A-type Piano Key weirs. *J. Hydraulic Res.* 50(4): 400-408.

Machiels, O., Pirotton, M., Archambeau, P., Dewals, B., Erpicum, S. (2014). Experimental parametric study and design of Piano Key weirs. *J. Hydraulic Res.* 52(3) 326-335.

IMPORTANCE OF SITE CONSIDERATIONS FOR LABYRINTH SPILLWAY HYDRAULIC DESIGN — UPPER BRUSHY CREEK DAM 7 MODERNIZATION

Tina Stanard, P.E.[1]
Victor Vasquez, P.E.[2]
Ruth Haberman, P.E.[3]
Blake Tullis, PhD[4]
Bruce Savage, PhD[5]

ABSTRACT

The Upper Brushy Creek Water Control and Improvement District owns and maintains 23 earthen flood control dams in Central Texas that were built by the U.S. Soil Conservation Service (SCS) over fifty years ago. This paper focuses on the rehabilitation of one of the District's largest dams, Dam 7, to meet Texas dam safety hydrologic requirements. Dam 7 is a high-hazard dam located on the border of the City of Austin within Williamson County. The dam is a popular recreational destination due to an adjacent city park as well as a hike and bike trail that crosses the dam. An innovative design was required to increase the auxiliary spillway capacity while accommodating the trail and keeping the project footprint within the limited easement area. A labyrinth weir was selected as a practical solution due to its ability to maximize discharge within a limited footprint.

Like many SCS flood control structures, the existing auxiliary spillway at Dam 7 is located on an abutment; however, the approach channel creates a substantial flow angle (an approximate 70 degree bend). Initial assessments used published data for labyrinth weir design; however, additional analyses were required to evaluate the effect of the approach angle. Computational fluid dynamics (CFD) and physical hydraulic models were performed as part of the labyrinth spillway design. The modeling results showed that that there was an approximate 35% decrease in hydraulic capacity as compared to the same labyrinth weir designed using published data (which assume the weir is perpendicular to the flow). The models were used to refine the weir layout to maximize discharge, minimize required excavation of the approach channel, and evaluate options for downstream energy dissipation.

This paper presents the design approach using CFD and physical models. It also presents unique architectural and landscaping features used to incorporate the concrete structure into the recreational and natural surroundings of the site.

---

[1] Freese and Nichols, Inc., 10431 Morado Circle, Suite 300, Austin, Texas, 78759. (512) 617-3120, Tina.Stanard@freese.com
[2] Freese and Nichols, Inc. 10431 Morado Circle, Suite 300, Austin, Texas 78759. (512) 617-3142, VMV@freese.com
[3] Upper Brushy Creek Water Control and Improvement District, 1850 Round Rock Avenue, Suite 100, Round Rock, Texas, 78681. (512) 284-7685, Ruth.Haberman@upperbrushycreekwcid.org
[4] Utah State University, 1415 Old Main Hill, Logan, Utah, 84322. (435) 797-3194 Blake.Tullis@usu.edu
[5] Idaho State University, 921 South 8th Ave, Pocatello, Idaho, 83202. (208) 282-3131 savabruc@isu.edu

INTRODUCTION

The U.S. Soil Conservation Service, now Natural Resources Conservation Service (NRCS), constructed numerous earthen flood control dams in rural Central Texas in the 1950's and 1960's. The Upper Brushy Creek Water Control and Improvement District (District) was created to maintain 23 of these flood control dams in the Brazos River basin in Williamson County, Texas. Williamson County is just north of Austin, Texas, and has experienced significant population growth since these dams were built. Most of these dams are now high-hazard dams due to substantial development downstream.

Upper Brushy Creek Floodwater Retarding Dam No. 7 (Dam 7) is one of the largest and most visible of the District's dams. The dam is adjacent to Brushy Creek Lake Park and is a popular recreational destination due to a hike and bike trail that crosses the dam. Dam 7 did not meet state dam safety hydrologic requirements as regulated by the Texas Commission on Environmental Quality (TCEQ). Freese and Nichols, Inc. (FNI) was retained by the District to design the modifications to meet dam safety requirements.

DAM OVERVIEW

Completed in 1965, Dam 7 is an earthen embankment dam with a total length of 2,350 feet and a maximum height of 56 feet. The dam is located on South Brushy Creek with a drainage area of approximately 16 square-miles. The dam creates Brushy Creek Lake which has an impoundment of 70 acre-feet at normal pool and a maximum impoundment of 5,600 acre-feet at top of dam. An overview of the site is shown on Figure 1.

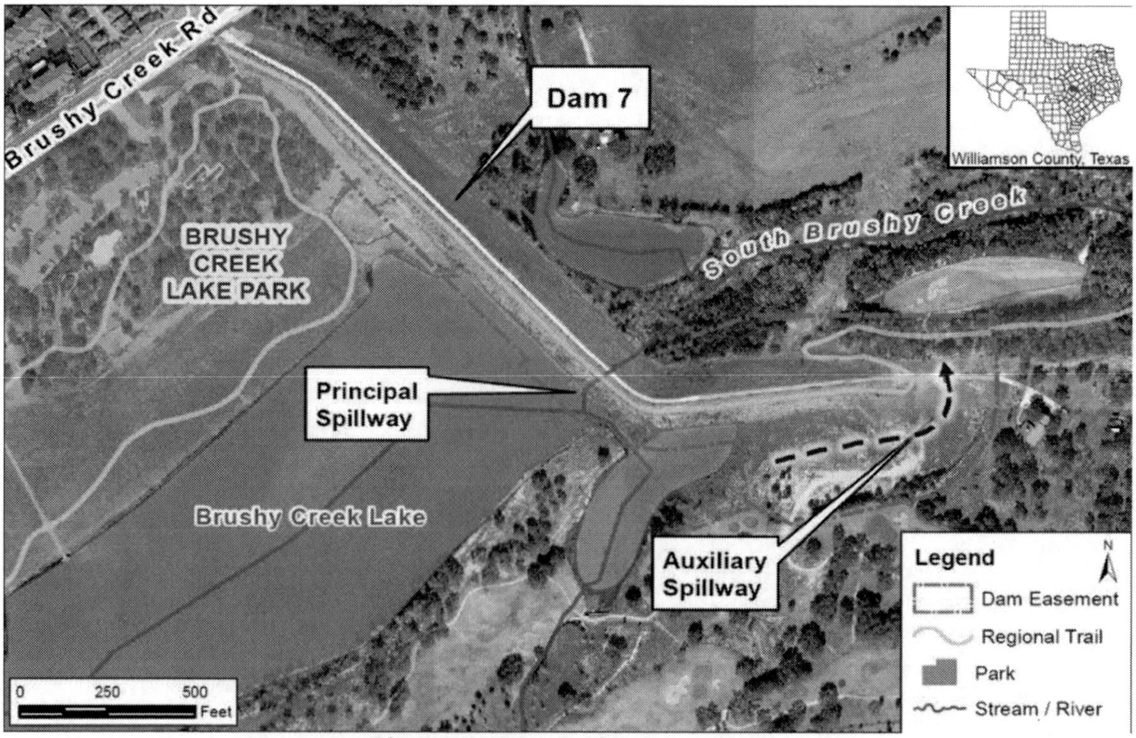

Figure 1. Dam 7 Site Overview

The principal spillway consists of a concrete inlet riser with a crest elevation of 805 feet and a 30-inch diameter concrete pipe through the dam that discharges into a plunge pool. Like many NRCS flood control dams, the auxiliary spillway is located on an abutment and consists of a rock-cut channel. The embankment has a dog-leg shape as shown on Figure 1, and the auxiliary spillway approach channel requires the flow to take a sharp turn before passing over the control section. Immediately downstream of the control section is a naturally-occurring steep bluff where the elevation drops approximately thirty feet. The flow over this natural bluff causes erosion downstream when the spillway engages as shown in Figure 2.

Figure 2. Auxiliary Spillway Engaged in September 2010

Dam 7 is currently classified as a high-hazard dam due to several homes downstream within the breach impact zone. Based on its size and hazard classification, Dam 7 is required to pass 82 percent of the Probable Maximum Flood (PMF). Dam 7 was evaluated and determined to be hydraulically inadequate because its existing spillways can only safely pass 47 percent of the PMF before the dam is overtopped.

CONCEPTUAL DESIGN

Upgrading Dam 7 to meet hydrologic requirements had several challenges, including the following site considerations:

- Limited easement and adjacent property restrictions (the west end of the dam is bordered by a major road, the east end is a home on private property, and north of the spillway is delineated Waters of the United States)
- High visibility to the public due to the adjacent Brushy Creek Lake Park

- Impact to the Brushy Creek Regional Trail which runs along the dam and down the spillway and is heavily trafficked by walkers, runners, and bikers at all times of the year

FNI evaluated several alternatives to increase the auxiliary spillway capacity including replacing the existing auxiliary spillway with a labyrinth weir, adding a second auxiliary spillway, and widening the existing broad-crested weir. The District selected to install a labyrinth weir in the existing auxiliary spillway based on minimizing the following: modifications to the embankment, impacts to Brushy Creek Lake Park and Regional Trail users, and overall permitting requirements. The labyrinth weir design could also incorporate the trail and aesthetic enhancements.

The preliminary design of the labyrinth weir used published empirical design guides for labyrinth weirs which are based on labyrinth weirs installed perpendicular to the flow (Crookston and Tullis, 2013). The preliminary design estimated that a 300-foot wide, 10-cycle labyrinth with the crest elevation of 828.2 feet would pass 55,000 cubic feet per second (cfs) with headwater at the top of dam elevation 836 feet. However, based on previous experience of the project team, it was anticipated the non-standard angle of the approach flow to the labyrinth weir would reduce the efficiency. The angle of the approach flow relative to the labyrinth weir cycle axes is approximately 70 degrees. Therefore, additional analyses were required to evaluate the effect of the approach angle and determine the site-specific discharge of the labyrinth weir spillway.

FNI developed a design approach utilizing a Computational Fluid Dynamics (CFD) numerical model and a physical hydraulic model to develop and refine a site-specific design for the labyrinth weir. The CFD modeling was performed by Bruce Savage, PhD, and the physical model was constructed and tested at Utah Water Research Laboratory (UWRL) under the direction of Blake Tullis, PhD. The CFD model was used to evaluate multiple labyrinth configurations and determine the number of cycles necessary to pass the required flow. Multiple configurations were simulated and evaluated in a cost-effective manner using numerical modeling. Once the labyrinth layout was selected, the physical model was used to refine the design and establish the spillway discharge rating curve. The physical model was also utilized to design a stepped spillway chute and impact basin to transition flow from the labyrinth weir to the downstream channel. The goal of the stepped spillway was to protect the downstream area from erosion to the extent possible within the existing easement and limits due to existing waters of the United States.

HYDRAULIC MODELING

CFD Model

CFD modeling is the simulation of fluid flow using numerical methods and algorithms. FNI provided the 3D geometry of the existing terrain and proposed structure developed in preliminary design. Dr. Bruce Savage developed the numerical model utilizing the CFD program Flow-3D® by Flow Science, Inc. The purpose of the CFD modeling was to evaluate the flow capacity of the labyrinth weir and approach angle as well as associated water surface elevations (WSE) in the reservoir. The modeled domain extended past the auxiliary spillway channel into the main portion of the reservoir. The initial geometry was a 10-cycle labyrinth weir, and it was simulated using a PMF flow rate of 49,000 cfs. The initial simulation showed that the embankment would be overtopped in the main portion of the reservoir where the maximum WSE occurred as shown in Figure 3. The WSE decreased near the labyrinth due to the drawdown effect as flow accelerates toward the spillway.

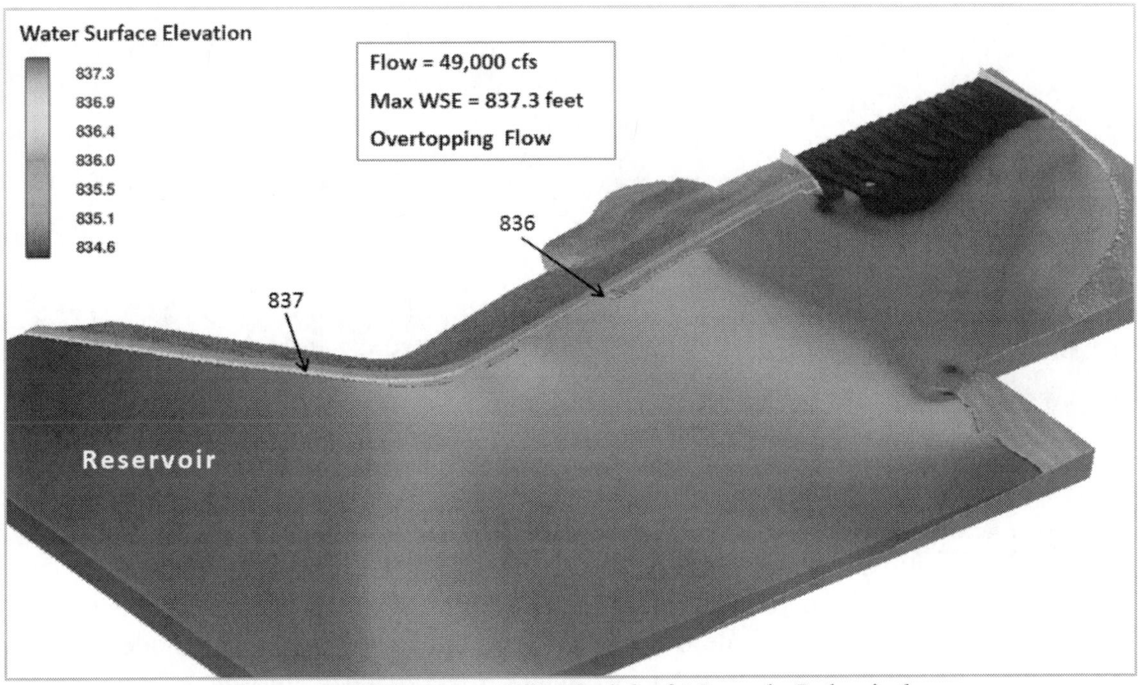

Figure 3. WSE from CFD Model of 10-cycle Labyrinth

To improve the hydraulic efficiency, the design team developed several modified spillway configurations to the 10-cycle linear labyrinth weir including revised training walls, addition of flow vanes, and revised labyrinth alignments. The CFD model was used to perform comparative studies on the modified configurations and found none of the configurations could pass the necessary flow. Therefore, an additional cycle was added, and the final configuration for evaluation was an 11-cycle linear labyrinth weir. The CFD model estimated that a 330-foot wide, 11-cycle labyrinth would pass 50,500 cfs at elevation 836.9 feet which was more than the required flow and a maximum WSE below the maximum top of dam design elevation. The design team evaluated the impact of an additional cycle to the embankment and project construction costs before proceeding with construction of the physical model of the 11-cycle labyrinth.

Upon comparison to an 11-cycle labyrinth weir designed using empirical data (installed perpendicular to the flow), the CFD results indicated an approximate 35% decrease in hydraulic capacity. The CFD model also depicted flow separation occurring at the left training wall and labyrinth weir cycles due to the approach angle as shown in Figure 4. The flow separation decreased on the right as the right bank of the auxiliary spillway channel trained the flow around the turn.

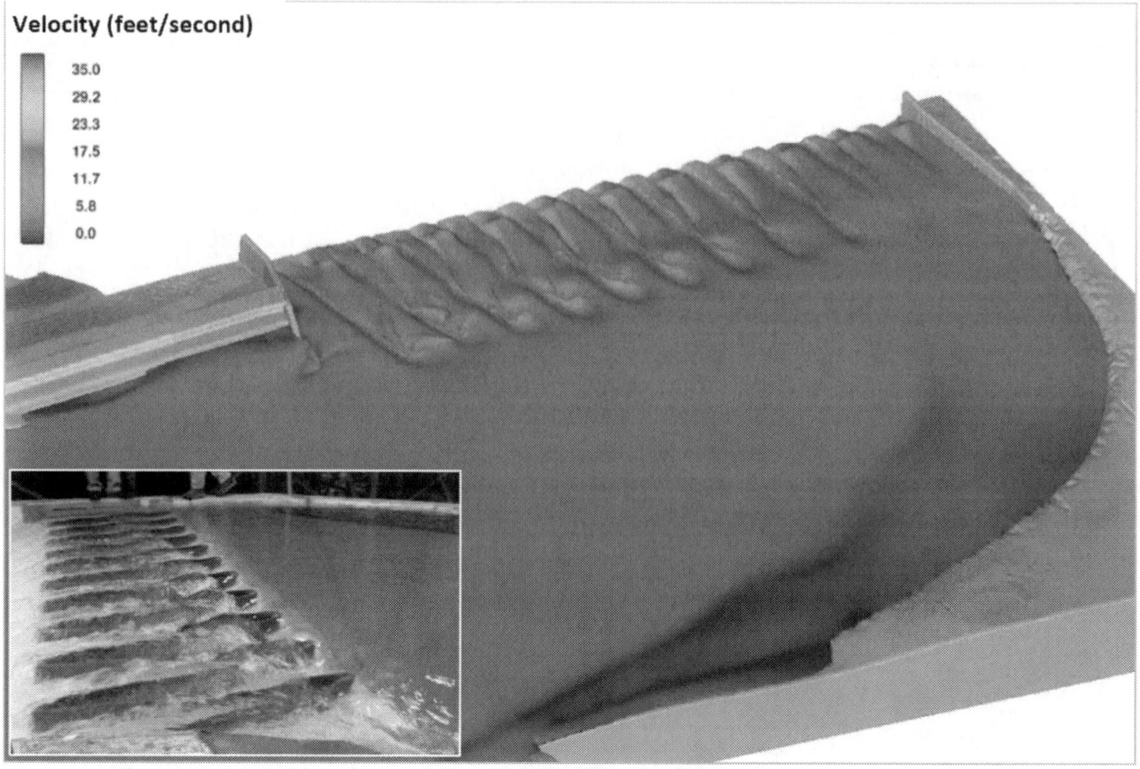

Figure 4. Velocity from CFD Model (Inset: Physical Model Comparison)

Physical Model

A 1:24-scale physical model of the 11-cycle labyrinth spillway was constructed and tested at the UWRL at Utah State University led by Dr. Blake Tullis. The model is shown in Figure 5. FNI provided the proposed geometry of approach channel and the labyrinth structure. The physical model was first tested to verify that the 11-cycle labyrinth could pass the required flow. Based on the CFD model results, the maximum reservoir WSE was expected to occur in the main section of the reservoir beyond the extents of the physical model. The extents of the CFD and physical models are compared in Figure 6.

Figure 5. Physical Model at UWRL

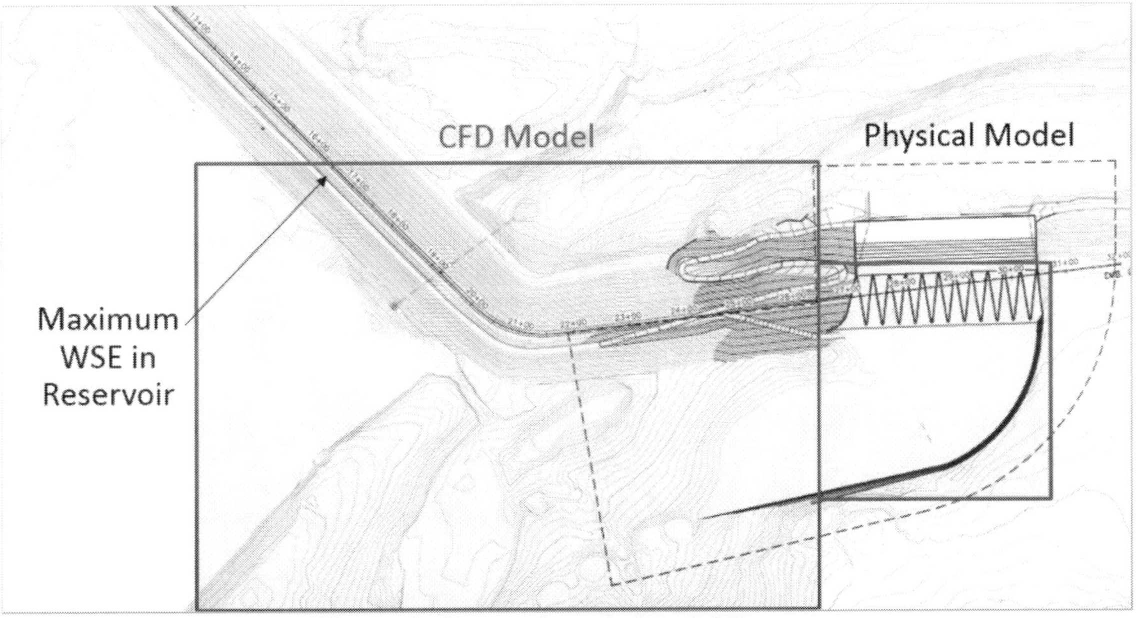

Figure 6. Comparison on Model Extents

Therefore, the spillway discharge rating curve provided by the physical model had to be correlated to the maximum reservoir WSE location along the embankment to predict the maximum lake level. FNI developed a correlation for the physical model results with the CFD model results to develop a rating curve based on the maximum WSE along the embankment. The physical model indicated that a 330-foot wide, 11-cycle labyrinth could pass 50,500 cfs at elevation 836.5 feet which passed the required flow and was slightly improved from the CFD model results.

The physical model was next used to refine the design. The design team modified the upstream left training wall to improve the efficiency by lowering the top of the training wall to be almost flush with the embankment and widening the angle of the wall toward the approach flow. The physical model verified the training walls could be converged (15 degrees on the left training wall, 10 degrees on the right training wall) to reduce construction costs. The training wall heights were selected using a water surface profile provided from the physical model.

For the stepped spillway chute design, multiple configurations of chute step sizes, chute blocks, and baffle blocks were evaluated with the physical model to determine the configuration that provided the greatest energy dissipation. The configurations were evaluated based on the height and location of the hydraulic jump. The selected stepped chute configuration is shown in Figure 7.

Figure 7. Stepped Spillway with Chute and Baffle Blocks

Lastly, multiple configurations of reduced excavation in the approach channel were evaluated to minimize the project footprint and save construction costs. The selected excavation configuration is shown in Figure 8.

Figure 8. Reduced Excavation

AESTHETIC ENHANCEMENTS

The design includes relocation of a section of the trail to accommodate the new labyrinth spillway structure. The design team incorporated architectural and landscaping features into the structure to embrace the recreational and natural surroundings of the site. Features include the following which are shown in the project rendering presented as Figure 9:

- A scenic overlook area at the crest of the embankment which faces the labyrinth spillway with benches shaped like the spillway baffle blocks.
- Backfill for plantings and crushed granite for the trail across the impact basin of the spillway.
- Concrete formliners and stain to mimic natural limestone installed on the vertical faces of the stepped spillway adjacent to the trail.
- Concrete formliners with architectural features on the training wall and fossil stamps in the baffle blocks adjacent to the trail.

Figure 9.Dam 7 Project Rendering

CONCLUSION

Empirical data is useful for preliminary design and standard spillway designs; however, CFD and physical models are valuable tools for the design of site-specific or non-standard spillways. The models found the approach flow angle at Dam 7 reduced the discharge of the labyrinth weir by 35 percent as compared to a labyrinth weir designed using empirical data which assumes the weir is perpendicular to the flow. While it was understood there would be an impact, this reduction was more than anticipated by the design team. However, even with the 35-percent reduction, the labyrinth weir substantially increased the hydraulic capacity from the existing broad-crested weir and passed the required flow. The site-specific design through modeling was critical to the labyrinth spillway hydraulic design at Dam 7.

The models provided a better understanding of the design and confidence that the project goals were met. The CFD was a cost-effective approach to determine the labyrinth spillway configuration and size that would pass the required flow. The physical model was used to refine the design to maximize discharge, minimize required excavation of the approach channel, and evaluate options for downstream energy dissipation. The total cost of the modeling constituted approximately 10 percent of the design fee and 1 percent of the anticipated construction cost. The physical modeling was estimated to reduce construction costs by over 2 percent of the anticipated cost. The physical model also provided a visual understanding and hands-on design approach for the owner and design team.

ACKNOWLEDGEMENTS

FNI would like to thank the Upper Brushy Creek Water Control and Improvement District for permitting discussion of the project and collaborating on this paper. The modernization of Dam 7 will improve flood protection to the downstream residents, and completion of the project could not have been possible without support from District staff and board of directors, as well as the cooperation and timely responses from the regulatory agencies involved in the project.

REFERENCES

Crookston and Tullis. (2013). "Hydraulic Design and Analysis of Labyrinth Weirs. I: Discharge Relationships." *ASCE Journal of Irrigation and Drainage Engineering*, May 2013.

FNI. (2015). *Dam 7 Modernization TX01365 Design Report*, dated October 2015.

FNI. (2013). "Dam 7 Modernization Alternatives Study Technical Memorandum," dated March 29, 2013.

Savage, Bruce, Ph.D., P.E. (2013). *Labyrinth Model Study of Dam 7 Using Computational Fluid Dynamics Report*, dated October 2013.

Texas Administrative Code, Title 30, Part 1, Chapter 299 Dams and Reservoirs.

UWRL. (2014). *Dam 7 Spillway Hydraulic Model Study Report*, dated March 6, 2014.

AUTOMATED INCREMENTAL HAZARD AND SENSITIVITY ANALYSIS OF A HYPOTHETICAL DAM BREACH MODEL

Mikell Warms, E.I.T.[1]
Dan Christensen, P.E.[2]
Christopher Goodell, P.E., D.WRE[3]
Cynthia Oestreich, P.E.[4]

ABSTRACT

An HEC-RAS model was developed for a dam breach analysis of a concrete dam along the Spokane River. The model is a combination of one and two-dimensional elements: A large 2-D area including the upstream reservoir, the downstream floodplain, a tributary channel, and the dam itself was developed inline of a 1-D reach, a method which is now available using HEC-RAS Version 5.0 Beta. Three dam breach scenarios (high, medium, and low flow) were tested to determine the sensitivity of the model to changes in breach parameters. In addition, three different erosion geometries of possible scour scenarios (including modifications to cross sections and terrain upstream of the dam) were tested to determine the effect and sensitivity that reservoir bed scour has on the breach hydrograph. An incremental hazard analysis was completed to determine the critical flood flows that cause flooding downstream near populated areas. This process was automated using Python scripts to run the model numerous times with different flood parameters and to extract and analyze pertinent data from the HEC-RAS hierarchical data format (HDF) output files on the fly. 2-D area cells in the flood plain downstream of the dam, along the thalweg of the river, and near sensitive developed areas downstream of the dam were chosen to determine the greatest change in flood depths due to a dam breach during all possible antecedent conditions. The flood flows tested consisted of a sunny day breach to the probable maximum flood (PMF), in 10,000 cfs increments.

INTRODUCTION

As part of the Federal Energy Regulatory Commission's (FERC) Emergency Action Plan (EAP) recommendations, independent hydroelectric utilities are asked to provide dam breach analyses to better understand the potential consequences of a breach and to determine emergency response if the dams were to fail. An incremental hazard analysis is required to determine the antecedent conditions posing the greatest incremental consequences near populated/sensitive areas. Often this analysis is completed visually, by comparing inundation extents from breach and no-breach scenarios. This paper aims to suggest an additional analysis that could assist and automate the required incremental hazard analysis.

[1] Mikell P. Warms, E.I.T., WEST Consultants, Inc., mwarms@westconsultants.com, 503.946.8536
[2] Daniel Christensen, P.E., WEST Consultants, Inc., dchristensen@westconsultants.com, 503.946.8536
[3] Christopher Goodell, P.E., D.WRE, WEST Consultants, Inc., cgoodell@westconsultants.com, 503.946.8536
[4] Cynthia Oestreich, P.E., Avista Utilities, cynthia.oestreich@avistacorp.com, 509.489.0500

An HEC-RAS model was developed for a dam breach analysis of Nine Mile Dam, a concrete dam along the Spokane River. The model is a combination of one and two-dimensional elements: A large 2-D area including the upstream reservoir, the downstream floodplain, a tributary channel, and the dam itself were developed, a technique which is now available using HEC-RAS Version 5.0 Beta. Three dam breach scenarios (high, medium, and low flow) were tested to determine the sensitivity of the model to changes in breach parameters. In addition, three different erosion geometries depicting possible scour scenarios (modifications to cross sections and terrain upstream of the dam) were tested to determine the effect that reservoir bed scour has on the breach hydrograph.

In addition to the three breach scenarios and erosion geometries, the annual peak flow (25,000 cfs), the 100-year flow (52,000 cfs), the probable maximum flood (PMF) (256,800 cfs), and the intermediate flows from the sunny day flow (10,000 cfs) to the PMF, in 10,000 cfs increments were evaluated. Rather than create an HEC-RAS plan for each possible combination of the aforementioned geometries, breach scenarios, and flood flows (a total of 348 possible combinations including "no-breach" scenarios), this process was automated using Python scripts to run the model for each possible combination and to extract and analyze pertinent data from the HEC-RAS hierarchical data format (HDF) output files on the fly. 2-D area cells in the flood plain downstream of the dam, cells along the thalweg of the river, and cells near sensitive developed areas downstream of the dam were emphasized to determine the greatest change in flood depths due to a dam breach during flood events.

PROJECT AREA

Nine Mile Dam is located in Spokane County, Nine Mile Falls, Washington. It impounds Nine Mile Reservoir on the Spokane River approximately 53 miles downstream from its source at Coeur d'Alene Lake. The drainage basin above Nine Mile Dam is approximately 5,000 square miles, and characterized by lightly forested mountain ranges separated by valleys.

Nine Mile Hydroelectric Development (HED) is a run-of-river operation. Outflow from Coeur d'Alene Lake is regulated by Post Falls HED. Three additional projects are located on the Spokane River within the Spokane city limits. Upriver Dam, Upper Falls HED, and Monroe Street HED are located about 21.9, 16.1, and 15.9 river miles upstream from Nine Mile Dam, respectively.

Initial construction of the existing Nine Mile Dam was completed in 1908. The concrete gravity structure totals a length of 466 ft, which includes a 225 ft long gated concrete ogee spillway, a 139 ft long integral dam/powerhouse section with concrete abutments, and a 102 ft long cutoff wall on the west end. In 2010, a series of Obermeyer crest gates were installed on the spillway crest and placed into service (shown in Figure 1). The Obermeyer gates were constructed in three separate, continuous zones with lengths of 31.5 feet, 157.0 feet, and 31.5 feet. The Obermeyer gates are pneumatically actuated, bottom-hinged, steel crest gates; which are supported and operated by inflatable air bladders that raise and lower the crest gates. At full capacity, the Obermeyer crest gates can raise the pool elevation by 10 ft. The normal operating level of the reservoir is maintained at elevation 1608.16 ft. There are no penstocks at Nine Mile Dam. The powerhouse includes four horizontal turbines located in steel and concrete bulkhead chambers (turbine pits). Each turbine is regulated by a vertical roller gate (20 ft high and 15 ft wide). In 1997, a 60" diameter low level tunnel was constructed to divert sediment laden water away from the intakes, and to mitigate additional sediment accumulation near the entrance of the intakes. Nine Mile Dam impounds a riverine reservoir with a storage capacity of approximately 4,600 acre-feet.

Figure 1. Downstream Face of Nine Mile Dam (Approximate Breach Section Width Shown in Red)

The Little Spokane River discharges into the Spokane River approximately 1 mile downstream of Nine Mile Dam. Backwater effects from high stages in the Spokane River extend up the Little Spokane River following a breach event.
Long Lake reservoir is formed by the Long Lake and Crescent Dams and extends 24 miles upstream to the tailwater of Nine Mile Dam. The normal pool elevation of Long Lake Reservoir is 1538.36 ft (Long Lake EAP, 2011).

Little Falls Hydroelectric Development (HED) is located approximately 5 miles downstream of Long Lake Dam and is included in this study. Little Falls Buttress Dam is a concrete dam with a crest elevation of 1366.36 ft (as-built drawings). The normal pool elevation is 1364.36 ft (Little Falls EAP, 2010).

The downstream boundary of the project is approximately 3 river miles downstream of Little Falls Dam on the Spokane River at River Mile 28.636. The upstream boundary extends approximately 7 miles past Nine Mile Dam on the Spokane River at River Mile 64.95. A map of the Spokane River is shown in Figure 2.

Downstream of the Nine Mile Reservoir, the Spokane River flows approximately 58 miles before reaching the confluence with the Columbia River. The bed slope in this part of the river averages 0.001 ft/ft. Aerial photos show the land cover to be moderately wooded with sporadic farmland and clearings on the upper floodplains. Between Long Lake Dam and Little Falls Dam, two tributaries, Little Chamokane Creek and Chamokane Creek empty into the Spokane River. Lake Roosevelt pool elevation affects the water level in the Spokane River up to the tailwater below Little Falls Dam. The normal pool level for Lake Roosevelt is 1292.9 ft (Little Falls EAP, 2010).

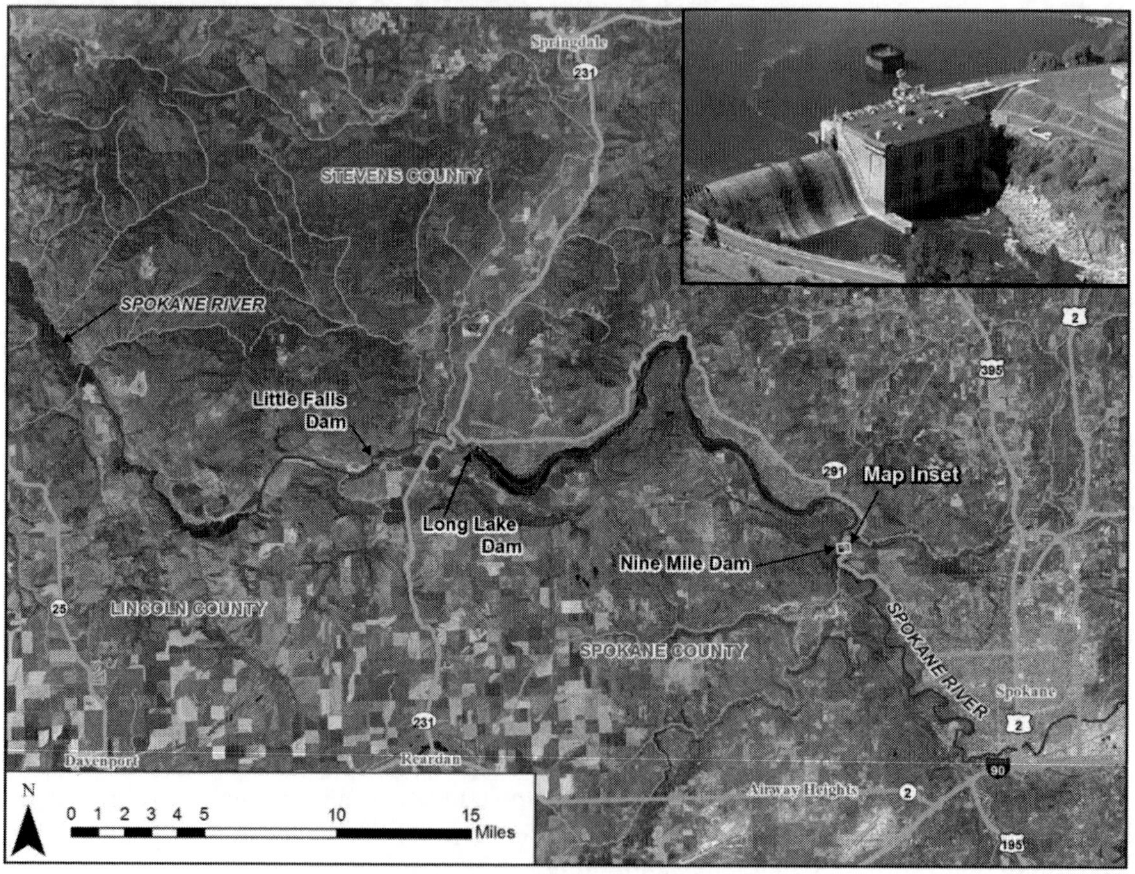

Figure 2. Project Area.

MODEL GEOMETRY

HEC-RAS Version 5.0 Beta (August 21, 2015 Build) was used for the Nine Mile Dam breach analysis. HEC-RAS has traditionally been a 1-D unsteady flow routing software package capable of integrating complex channels and structures under dynamic hydraulic conditions. However, with the release of HEC-RAS Version 5.0 Beta, 2-D areas are now able to be modeled in conjunction with the traditional 1-D geometry. This capability helps to properly simulate movement of water in areas where flow is highly irregular (floodplains, storage areas, etc.). HEC-RAS also has the capability of modeling dam breach events under a wide range of scenarios. The river schematic and geometry data for the model were extracted from GIS using ArcGIS Version 10 and HEC-GeoRAS Version 10. This includes the stream centerline, cross-sections, lateral structures, land cover, and other geometric features of the stream.

The HEC-RAS model geometry begins 7 miles upstream of Nine Mile Dam at River Mile 64.95 and terminates approximately 3 miles downstream of Little Falls Dam at River Mile 28.636. The entire study area includes approximately 36.5 river miles of the Spokane River.

The model contains two inline structures representing Long Lake Dam and Little Falls HED's intake structures, and two lateral structures representing Long Lake/Crescent Dam and the Little Falls spillway and buttress dam wall that runs parallel to the river upstream of Little Falls Dam. A large 2-D area was added encompassing the Spokane River from Nine Mile Reservoir to Long Lake, including the overflow area northwest of Nine Mile Dam, the downstream floodplain, and the Little Spokane River tributary and storage area. A storage area connection (essentially a hydraulic/inline structure inside a 2-D area) was created to represent Nine Mile Dam. Inside the 2-D area there are four 2-D Storage Area Break Lines (lines that enforce the 2-D area mesh generation to align the computational cell faces along flow boundaries and features) representing Nine Mile Dam, S.R. 291 (which serves as a boundary between the Little Spokane River and the Spokane River) and 2 for a small ridge in the floodplain downstream of Nine Mile Dam.

Reservoirs

Nine Mile Dam impounds the Spokane River in a long narrow canyon, which maintains riverine characteristics throughout the entire reservoir. Cross sections are used to define reservoir geometries for the Long Lake and Little Falls reservoir, as well as the upstream section of Nine Mile reservoir so that the dynamic reservoir routing method could be applied in the HEC-RAS model. The dynamic routing method best captures the true elevation-volume relationships of reservoirs during fast drawdowns and provides accurate routing of the flood waves during hypothetical breaches of long, narrow reservoirs. The approximately 1-mile section of Nine Mile reservoir closest to Nine Mile Dam was included in the 2-D area.

Channel geometry within Nine Mile reservoir was modified to better represent the streambed profile during a breach event. Currently, Nine Mile reservoir is partially filled with sediment from a persistent inflowing sediment load over the lifetime of the project. During a breach event, a portion of the built-up sediment would be washed downstream. The terrain and cross sections upstream of Nine Mile Dam were modified by adding an erosion channel, which would form during a dam breach event. Three erosion channels (three HEC-RAS geometries and terrains) were included with varying lengths of upstream erosion (Figure 4). The erosion channel defined in each geometry is a static feature in the model, and does not form as a result of the dam breach and hydraulic properties in the river.

Cross Sections

Cross sections are used to define the shape of the stream and its characteristics, such as Manning's n values, expansion and contraction, and ineffective flow areas. Cross sections were developed using ArcGIS Version 10 and HEC-GeoRAS Version 10. LiDAR and bathymetric terrain data was used to define the cross-sections.

A total of 243 Spokane River cross sections and eight Little Falls cross sections are defined in the model geometry (Figure 3). Three erosion channels (High, Medium and Low) were included with varying lengths of upstream erosion (Figure 4). For the high erosion geometry, 35 cross sections were modified (approximately 3 miles upstream of Nine Mile Dam and the 2-D area) by lowering the channel bed elevations to represent the eroded reservoir channel. This eroded channel bed is continued in the terrain for the 2-D area until it reaches Nine Mile Dam. It was assumed that the erosion channel would take the approximate shape of the embankment breach. The dimensions of the erosion channel at Nine Mile Dam are based on the maximum breach scenario dimensions (See dam breach section). The width of the erosion channel is twice the width of the dam breach (180 ft) and equal to the depth of the dam breach (90 ft). The channel then tapers to various lengths and depths upstream depending on the high, medium, or low scour scenario (Figure 4). The upstream bed channel slope and proposed breach shape were used to produce a smooth bed channel profile in the reservoir. The Medium erosion channel consisted of the same cross sectional shape as the High erosion channel but only extended upstream to the 2-D area boundary (0.75 miles upstream). The Low erosion channel extended upstream 1000 ft to represent localized scour without assuming the formation of an erosion channel upstream.

Inconsistencies in the terrain data immediately upstream of the Little Falls and Long Lake Dams were identified. The bathymetry data for these sections contained incorrect elevations (the bathymetry recorded the normal pool elevation instead of the channel bottom). Cross sections and terrain in this area were modified to better represent the Spokane River channel as observed in a site visit.

2-D Flow Area

The 2-D Flow areas are used to perform two-dimensional hydrodynamic flow routing within the unsteady flow analysis portion of HEC-RAS. 2-D areas in HEC-RAS are used in a number of ways, including detailed 2-D channel modeling, detailed 2-D floodplain modeling, and combined 1-D channels with 2-D flow and floodplain areas. 2-D flow modeling requires a detailed digital terrain model (DTM) as well as a shapefile of Manning's n-values that span the entire 2-D area. Elevation data in the form of LiDAR for the modeled length of the Spokane River was available in certain sections, and supplemented by the USGS NED for the Little Spokane storage area. The combined DTM was developed using ArcGIS Version 10 and converted to a GeoTIFF using HEC's

RAS Mapper tool. A shapefile of Manning's n-values was created using ArcGIS Version 10, supplemented by aerial imagery, observations during the site visit, and engineering judgment and experience. Once the terrain and land cover were added to HEC-RAS, a 2-D computational mesh was created.

The 2-D area created for this analysis begins approximately 1 mile upstream of Nine Mile Dam and ends approximately 4 miles downstream, encompassing the Spokane River from Nine Mile Reservoir to Long Lake, including the overflow area northwest of Nine Mile Dam, the downstream floodplain, and the Little Spokane River storage area. Originally, only the Little Spokane tributary was converted to a 2-D area. However a larger area was deemed necessary upon inspecting the likelihood of flood waters overtopping the left bank adjacent to Nine Mile Dam and flowing down historical channels on left bank floodplain downstream of Nine Mile Dam.

For this analysis, a 150x150 foot 2-D cell size was chosen based on engineering judgment and experience. While many 2-D models require a smaller grid size (often the grid size is set equal to terrain grid size), HEC-RAS employs a subgrid bathymetry scheme for 2-D modeling, so therefore a smaller grid size is unnecessary. In addition, four 2-D Area Break Lines (lines that enforce the 2-D area mesh generation to align the computational cell faces along barriers to flow) representing Nine Mile Dam, S.R. 291 (an elevated flow barrier between the Little Spokane River and the Spokane River), and a small ridge in the floodplain downstream of Nine Mile Dam were implemented. These break lines force the mesh to align such that the flow into and out of cells at natural flow barriers is correctly modeled (Figure 3). In total, there are 8187 cells in the 2-D mesh.

Structures

The HEC-RAS model includes one bridge, two inline structures, two lateral structures, and one 2-D area connection. Nine Mile Dam is modeled as a 2-D area connection inside the 2-D area and is the upstream-most structure. Storage area and 2-D area connections can contain all of the properties of inline and lateral structures. All of Nine Mile Dam's components are included in the 2-D area connection (concrete ogee spillway section, intake and powerhouse unit, concrete abutments, and cutoff wall). As discussed in the dam breach section below, the concrete spillway is the primary focus of the hypothetical dam failure analysis. For reference, the spillway crest at Nine Mile Dam is defined at elevation 1608.16 ft (includes 10 ft high Obermeyer crest gates).

Long Lake Dam spillway and intake structure are defined in the HEC-RAS model by Inline Structure 33.941. Eight spillway lift gates (29 feet high and 25 feet wide) regulate discharge over the spillway and four 16 ft diameter (tapering to 14') penstocks route flow to its hydroelectric generating units. The top of dam elevation is 1542.36 ft (Long Lake EAP, 2011) and the spillway lift gate inverts are at elevation 1510.36 ft.

Crescent Dam is a concrete arch dam located on the left bank just upstream of Long Lake Dam. For modeling purposes, it is defined as a lateral structure. Top of dam elevation is 1539.36 ft.

Little Falls Dam is a concrete gravity dam with a crest elevation of 1366.36 ft (as-built drawings). In plan view, Little Falls Dam and spillway structure are in the shape of an 'S'. A 604 ft spillway is located at the upstream section that runs partly inline and parallel with the river. The spillway is composed of a free overflow and gated sections. The gated section is comprised of two tainter gates. The free overflow section has a crest elevation of 1358.36. The pool level is regulated by 6 ft high flashboards that are placed on-top of the free overflow section. Cables running the length of the free overflow spillway are attached to the back support of the flashboards. The cables can be used to quickly rip off the flashboards during a high flood event. In the model, the flash boards are assumed to completely break up and breach when they are overtopped by 2 feet.

Seven Mile Bridge is included in the hydraulic model at River Station 62.021. Three other bridges (West Charles Rd, SR 291 over the Little Spokane River, and Hwy 25) are located within the project area, but are omitted in the hydraulic model. The Charles Road Bridge was omitted because the flood wave would not reach the road deck and piers. The Hwy 25 bridge is downstream far enough such that a breach would get close to reaching the low chord elevation, and the Little Spokane River Bridge (S.R. 291) is modeled as a 2-D area break line and is discussed further in the model results below. The downstream bridges are not expected to have an impact on the water surface elevations because of the attenuation that occurs on the floodwave before reaching the bridges.

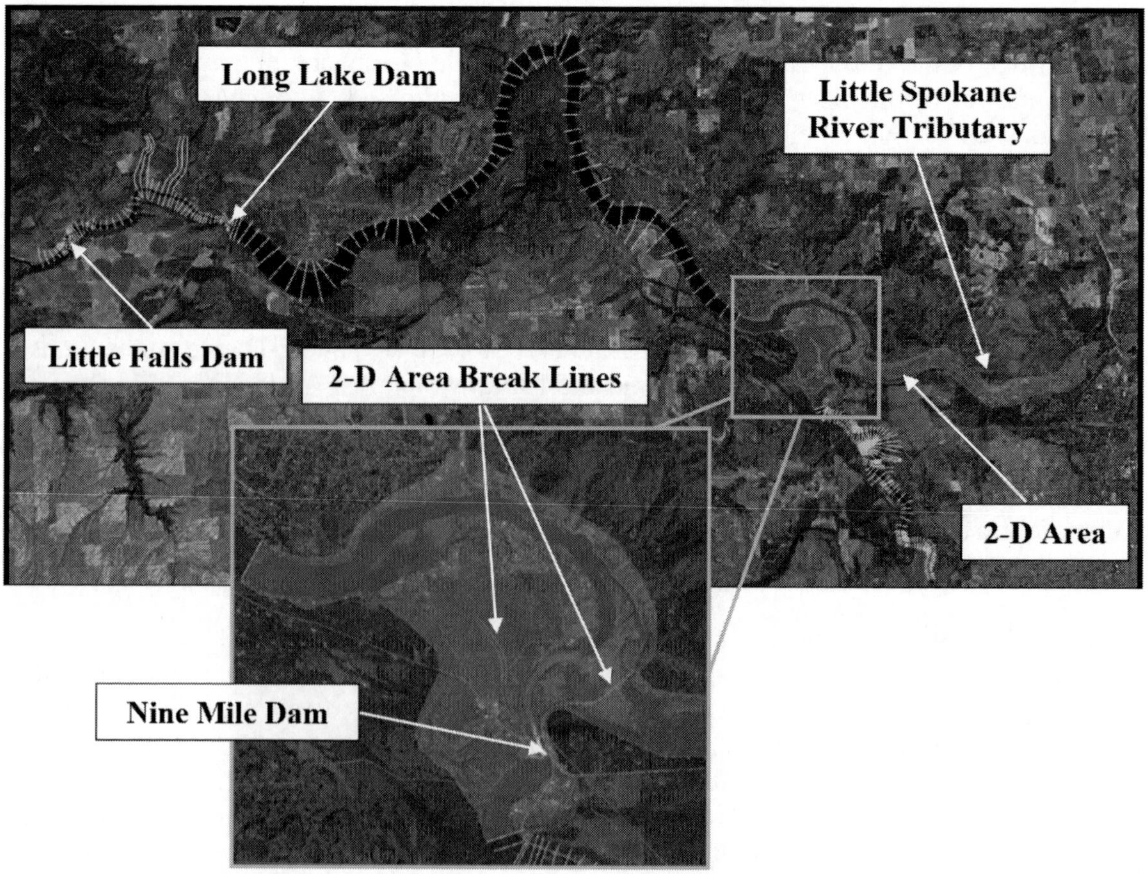

Figure 3. Plan View of HEC-RAS model.

DAM BREACH ANALYSIS

The purpose of this study is to analyze the downstream effects of dam failures during all possible antecedent flows at Nine Mile Dam. Three separate dam breach scenarios for each flood flow and erosion scenarios were analyzed and compared to 'no breach' scenarios: 1) a low section failure of Nine Mile Dam (including Obermeyer crest gates) which erodes to an elevation of 1590 feet, and 2) a medium section failure of Nine Mile Dam (including Obermeyer crest gates) which erodes to an elevation of 1573 feet, and 3) a large section failure of Nine Mile Dam (including Obermeyer crest gates) which erodes to the foundation (and elevation of 1538.16 feet). The width of all breaches is the same width as the monolith next to the power house (90 feet). Because these are hypothetical events, the actual breach size, location, and timing are unknown and must be estimated. The estimation of the breach parameters provides a range of sizes and formation times and is discussed further below.

Breach Parameter Selection

The main breach parameters used in the HEC-RAS dam breach model are breach width, breach formation time, reservoir pool elevation at time of failure, breach side slopes, breach discharge coefficients, and breach rate of growth progression. For earthen dams, breach parameters can be estimated using standard breach parameter equations based on empirical data. Nine Mile Dam is a concrete gravity dam. Unlike earthen embankment dam failures, widely published data does not exist for the failure of concrete gravity dams. The estimated dam breach parameters are based on all available project data, input from dam project engineers, and engineering judgment.

Dam Section Failure

For the dam section failure scenario, a sensitivity analysis was performed to evaluate the range of possible results given the uncertainty of the dam breach parameter inputs. The parameters for the three scenarios, the high, medium, and low flow sets are shown in Table 1 below.

Table 1. Breach Parameters for each Breach Plan: Dam Section Failure Scenario.

	Nine Mile Dam		
	HIGH	**MED**	**LOW**
Final Bottom Width (ft)	90	90	90
Final Bottom Elevation (ft)	1538.16	1573	1590
Left Side Slope (H:1)	0	0	0
Right Side Slope (H:1)	0	0	0
Breach Weir Coeff.	2.6	2.6	2.6
Full Formation Time (min)	10	10	10

The breach dimensions for the high flow plan were developed by considering the physical limitations of the site as observed from the site visit, a previous dam stability review (Raytheon, 1999), as-built drawings, and conversations with dam project engineers. A breach invert of 1538.16 ft was recommended based on project drawings and findings in the 1999 dam stability review (Raytheon). For this breach scenario, it is assumed that a horizontal shear plane would develop at the bedrock interface and would shear through the concrete along any base asperities.

Nine Mile Dam was constructed with cyclopean boulders and poured in horizontal lifts. According to construction drawings, the concrete gravity dam was built in three sections (approximately 55 ft, 80 ft, and 90 ft lengths). The longest section of 90 feet was conservatively chosen as the breach width, assuming that the vertical shear keys and tendons fail at the construction joints. The breach width, side slopes, breach weir coefficient, and formation time are constant for the high, medium and low sensitivity runs.

The lateral extents and side slopes of the breach opening were estimated to match the construction joints as found in the as-built drawings. The medium flow scenario is based on using an average breach depth value. The low flow breach scenario uses the breach parameter set that produces the lowest, yet still realistic peak discharge from the breach. The failure of concrete dam structures tend to be catastrophic in nature. FERC suggests using a breach formation of 6 to 18 minutes for masonry or concrete dams (LuBean, 1987). Therefore, a breach formation time of 10 minutes was selected for Nine Mile Dam.

Incremental Hazard and Erosion Analysis

An incremental hazard analysis was completed to determine the antecedent flow conditions which yield the greatest consequences (referred to as the Inflow Design Flood (IDF) for the purposes of this study) near populated areas downstream of Nine Mile Dam (near the Little Spokane River confluence) during a breach. The maximum inundation extent and depths of flood flows downstream of Nine Mile dam were compared for each breach scenario (high, medium, low, none), each erosion geometry (high, medium, low), and each flow (from 10,000 cfs to 260,000 cfs in 10,000 cfs increments, as well as the typical, 100-year, and probable maximum floods) as discussed previously—a total of 348 HEC-RAS plans.

A python script was written to create the text files for all (348) possible plans and flow profiles (.pXX and .uXX files). HEC-RAS was then run for all 348 plans (using the "Run Multiple Plans" option in HEC-RAS) and the runs were completed in approximately 3-4 days of computation time.

Once completed, a separate python script was written to extract data from the output HDF files inside the 2-D area cells downstream of the breach. All 2-D area cells downstream of Nine Mile Dam were analyzed, but special attention was given to the cells along the thalweg of the Spokane River and cells near heavily populated and developed areas.

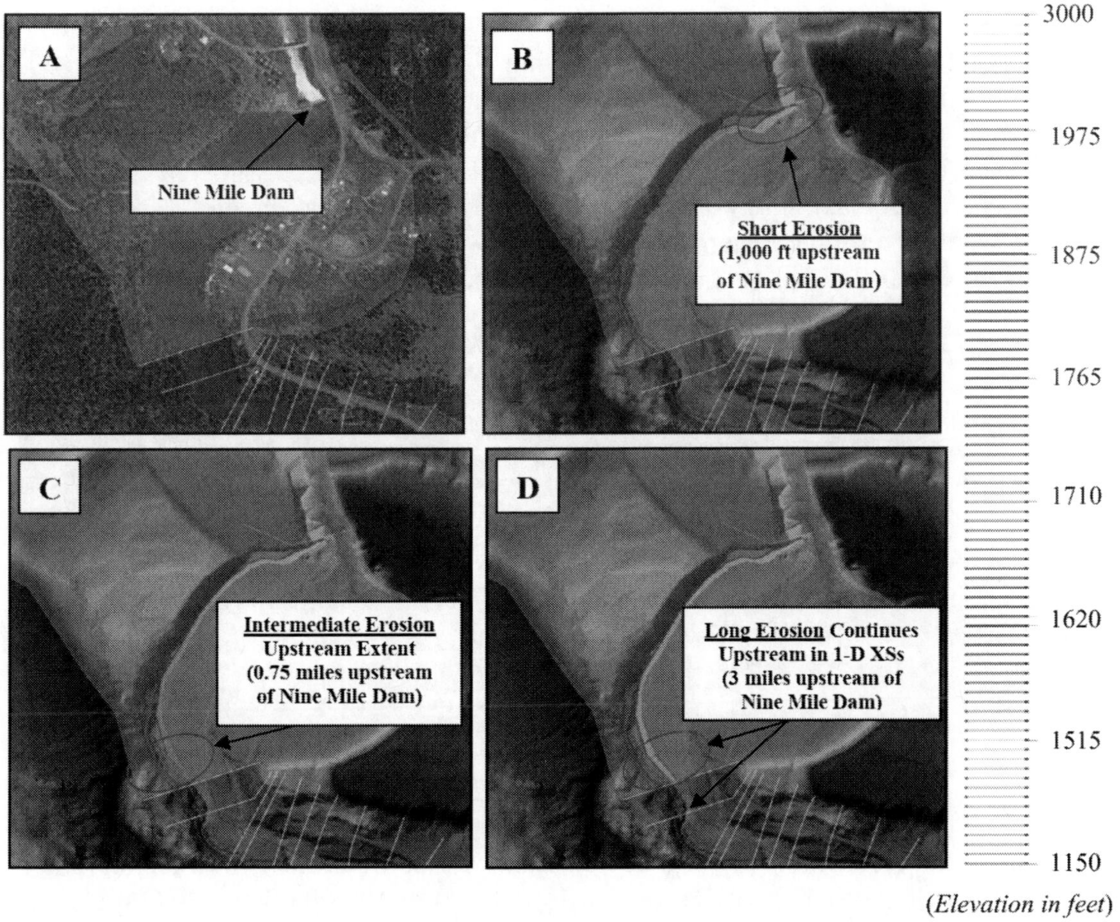

Figure 4 Description	
A	Aerial Imagery of Nine Mile Reservoir.
B	Terrain for short length erosion scenario in Nine Mile Reservoir. (Represents minimal localized scour without formation of erosion channel upstream of Nine Mile Reservoir. No change to cross sections.)
C	Terrain for intermediate length erosion scenario in Nine Mile Reservoir. (Represents intermediate scour without formation of erosion channel upstream of Nine Mile Reservoir. No change to cross sections)
D	Terrain for long length erosion scenario in Nine Mile Reservoir. Scour continued into cross section geometry. (Modified cross sections; width = 180', height = 90')

Figure 4. Aerial Imagery and Terrain for Differing Erosion Scenarios

MODEL RESULTS

To get a first look at the results without viewing all 348 possible plans, the 1-YR, 100-YR, and PMF floods were compared for each breach scenario to determine the incremental differences of the flood flows along the thalweg resulting from the different breach sizes (Figure 5). As expected, the high flow set of breach parameters produced the greatest change in maximum water surface elevation. In addition, greater differences near the dam were seen for the lower (1-YR) flow. However, the flow stabilizes fairly quickly, whereas the larger flows show greater changes much further downstream of the dam.

		Upstream								Difference in Maximum WSE (Breach vs. No Breach) of 2D Flow Area Cells																*Downstream*	
Low Breach	1-YR	1.1	0.9	0.9	0.8	0.6	0.4	0.2	0.2	0.1	0.0	0.0	0.0	0.0	0.0	0.0	0.0	0.0	0.0	0.0	0.0	0.0	0.0	0.0	0.0	0.0	5.1
Low Breach	100-YR	0.1	0.2	0.2	0.2	0.1	0.1	0.1	0.1	0.1	0.1	0.1	0.1	0.1	0.1	0.1	0.1	0.1	0.1	0.1	0.1	0.1	0.1	0.1	0.1	0.1	2.6
Low Breach	PMF	0.0	0.0	0.0	0.0	0.0	0.0	0.0	0.0	0.0	0.0	0.0	0.0	0.0	0.0	0.0	0.0	0.0	0.0	0.0	0.0	0.0	0.0	0.0	0.0	0.0	0.0
Med Breach	1-YR	4.0	3.6	3.3	3.0	2.5	1.9	1.5	1.3	1.1	1.0	0.9	0.8	0.6	0.5	0.4	0.3	0.2	0.2	0.1	0.1	0.1	0.1	0.0	0.0	0.0	27.5
Med Breach	100-YR	1.6	2.1	1.9	1.7	1.5	1.3	1.2	1.2	1.2	1.1	1.1	1.0	1.0	0.9	0.9	0.9	0.8	0.8	0.8	0.8	0.8	0.8	0.7	0.7	0.7	27.7
Med Breach	PMF	0.5	0.2	0.1	0.1	0.1	0.1	0.1	0.1	0.1	0.1	0.1	0.1	0.1	0.1	0.0	0.0	0.0	0.0	0.0	0.1	0.1	0.1	0.1	0.1	0.1	1.9
High Breach	1-YR	8.2	7.8	7.1	6.4	5.3	4.0	3.0	2.6	2.1	1.9	1.7	1.5	1.2	1.0	0.9	0.7	0.6	0.5	0.5	0.4	0.4	0.4	0.3	0.3	0.3	59.0
High Breach	100-YR	5.1	5.2	4.6	4.2	3.5	3.1	2.9	2.8	2.7	2.6	2.5	2.4	2.2	2.0	2.0	1.9	1.9	1.9	1.8	1.8	1.8	1.7	1.7	1.6	1.6	65.6
High Breach	PMF	2.5	1.6	1.1	0.7	0.3	0.3	0.3	0.3	0.3	0.3	0.2	0.2	0.2	0.2	0.2	0.2	0.2	0.2	0.2	0.2	0.2	0.2	0.2	0.2	0.2	10.7

Figure 5. Initial Results from Incremental Hazard Analysis
(values shown in differential ft).

From the initial results, it is possible to narrow down the flood flows of interest, rather than viewing all plans. The simulation for flood flows of 10,000 cfs to 120,000 cfs, in increments of 10,000 cfs for the high breach and high erosion scenarios were compared to the no breach scenarios for those flood flows.

		Upstream								Difference in Maximum WSE (Breach vs. No Breach) of 2D Flow Area Cells																*Downstream*	
High Breach	120000 cfs	4.5	4.5	4.0	3.5	2.8	2.5	2.4	2.3	2.2	2.1	1.9	1.8	1.7	1.5	1.5	1.4	1.3	1.3	1.3	1.2	1.2	1.2	1.1	1.0	1.0	51.3
High Breach	110000 cfs	4.8	4.9	4.4	3.9	3.3	2.9	2.8	2.8	2.6	2.5	2.4	2.3	2.1	2.0	2.0	1.9	1.9	1.8	1.8	1.8	1.8	1.7	1.7	1.6	1.6	63.5
High Breach	100-YR (102700)	5.1	5.2	4.6	4.2	3.5	3.1	2.9	2.8	2.7	2.6	2.5	2.4	2.2	2.0	2.0	1.9	1.9	1.9	1.8	1.8	1.8	1.7	1.7	1.6	1.6	65.6
High Breach	90000 cfs	5.5	5.7	5.1	4.6	4.0	3.4	3.1	3.0	2.9	2.7	2.6	2.4	2.2	2.1	2.0	1.9	1.9	1.8	1.8	1.8	1.7	1.7	1.6	1.6	1.5	68.5
High Breach	80000 cfs	5.9	6.1	5.5	5.0	4.3	3.6	3.3	3.1	3.0	2.8	2.6	2.5	2.3	2.1	2.0	1.9	1.8	1.8	1.8	1.4	1.7	1.6	1.6	1.5	1.5	70.8
Med Breach	70000 cfs	6.2	6.4	5.9	5.3	4.6	3.8	3.4	3.2	3.0	2.8	2.7	2.5	2.2	2.0	1.9	1.8	1.7	1.7	1.6	1.6	1.6	1.5	1.5	1.4	1.4	71.8
High Breach	60000 cfs	6.5	6.8	6.2	5.7	4.9	4.0	3.4	3.2	3.0	2.8	2.6	2.4	2.2	1.9	1.8	1.7	1.6	1.5	1.5	1.4	1.4	1.4	1.3	1.3	1.2	71.7
High Breach	50000 cfs	6.7	7.1	6.5	6.0	5.2	4.1	3.5	3.2	2.9	2.7	2.5	2.3	2.0	1.8	1.7	1.5	1.4	1.3	1.3	1.2	1.2	1.2	1.1	1.1	1.0	70.6
High Breach	40000 cfs	7.2	7.4	6.8	6.3	5.4	4.2	3.5	3.1	2.8	2.6	2.4	2.2	1.9	1.6	1.5	1.3	1.2	1.1	1.1	1.1	1.0	1.0	1.0	0.9	0.9	69.4
High Breach	30000 cfs	7.8	7.7	7.0	6.4	5.4	4.2	3.3	2.9	2.5	2.2	2.1	1.8	1.6	1.4	1.2	1.1	1.0	0.9	0.8	0.8	0.8	0.7	0.7	0.6	0.6	65.4
High Breach	1-YR (25000)	8.2	7.8	7.1	6.4	5.3	4.0	3.0	2.6	2.1	1.9	1.7	1.5	1.2	1.0	0.9	0.7	0.6	0.5	0.5	0.4	0.4	0.4	0.3	0.3	0.3	59.0
High Breach	20000 cfs	8.6	7.9	7.1	6.3	5.1	3.7	2.6	2.2	1.7	1.5	1.4	1.2	0.9	0.8	0.6	0.5	0.4	0.3	0.3	0.2	0.2	0.2	0.1	0.1	0.1	54.0
High Breach	10000 cfs	9.2	7.8	6.8	5.9	4.5	3.1	2.2	1.9	1.6	1.4	1.3	1.2	1.0	0.9	0.8	0.8	0.7	0.5	0.5	0.5	0.4	0.4	0.2	0.1	0.0	53.8

Figure 6. Incremental Hazard Analysis (values shown in differential ft).

From the above heatmap (Figure 6), we can deduce that the highest incremental change along the thalweg occurs for the flood flows in the 60,000-70,000 cfs range. This "sweet spot" seems to be the peak of incremental hazard of flood flows for this specific location. However, we are mostly concerned with possible damage to property in the floodplain, not just what is happening along the thalweg. Figures 7-8 below shows the difference in maximum water surface between breach and no breach scenarios for the 70,000 and 80,000 cfs floods, respectively. Empty cells in the floodplain of the below images are due automated raster creation near irregular 2-D cell geometry along the 2-D area break lines.

Figure 7. Incremental Hazard Analysis for all 2-D area cells. 70,000 cfs flood. (values shown in differential ft between breach and no-breach scenario)

What we are concerned with in Figures 7-8 is the level of flooding outside the main channel area on the left and right overbanks, in addition to the level of backwater effects in the Little Spokane River tributary. The 70,000 cfs flood and breach scenario (Figure 7) shows major differences on the left overbank, including areas with many residential buildings. In addition, there are mild differences on the right overbank which floods a state park, a few buildings, as well as SR 291.

Figure 8. Incremental Hazard Analysis for all 2-D area cells. 60,000 cfs flood. (values shown in differential ft between breach and no-breach scenario)

The 60,000 cfs flood and breach scenario (Figure 6) shows some differences on the left overbank, including areas with many residential buildings. In addition, there are mild differences on the right overbank which floods a state park, a few buildings, as well as SR 291. The backwater effects are not as severe in Figure 8 (60,000 cfs) than in Figure 7 (70,000 cfs). Visually, it seems that the 70,000 cfs has a higher incremental risk to downstream life and property than 60,000 cfs. While the process to run each plan is automated, the final decision should also include visual interpretation of the results. These maps can be easily created in a similarly automated fashion using ArcPy or gdal scripts in python.

As a final test of the 2-D flow areas, the mean change in feet for all cells was calculated and compared in Table 2 below:

Table 2. Average change in depth between WSE of Breach and No-Breach Scenarios

Average change in depth of all 2-D area cells *(values in differential ft between breach and no breach scenario)*			
50,000 cfs	**60,000 cfs**	**70,000 cfs**	**80,000 cfs**
0.48	0.50	0.56	0.49

SUMMARY

A dam breach analysis for Nine Mile Dam was performed using a 1-D/2-D unsteady HEC-RAS hydraulic model of the Spokane River. The modeled section of the Spokane River extends from River Mile 62 above Nine Mile Reservoir downstream to River Mile 26 below the Little Falls Dam. The model provided peak stages and discharges in the Spokane River for dam breach events at all possible antecedent flows from 10,000 cfs to 260,000 cfs (in 10,000 cfs increments).

An incremental hazard analysis was completed to determine the critical flows that cause flooding downstream of Nine Mile Dam near populated areas (near the Little Spokane River confluence). The maximum inundation extent and flood depths near populated areas downstream of Nine Mile were compared. The inflow design flood was determined to be 70,000 cfs. It was also determined that the high breach scenario provided the most conservative scenario. The sensitivity analysis showed that there could be a large range of possible discharge hydrographs for the dam failure event at Nine Mile Dam.

REFERENCES

Avista Corp, 1999. Nine Mile Hydroelectric Development, *Report For Dam Stability Review For Sediment Loading*, prepared by Raytheon Infrastructure Inc., April 1999.

Barnes, H. H. Jr., 1987. *Roughness Characteristics of Natural Channels*, U.S. Geological Survey Water-Supply Paper 1849.

BOSS International, 2003. *BOSS DAMBRK User's Manual*, Chapter 6.

Bureau of Reclamation. (2012). Franklin D. Roosevelt Lake – Grand Coulee Dam 2010-11 Survey. Technical Report No. SRH-2012-06.

Chow, VT, 1959. *Open Channel Hydraulics*, McGraw-Hill Book Company, New York, NY.

EBASCO Services, Inc., 1990. *Post-Tensioned Anchors Plans and Sections*, Nine Mile HED Dam Stabilization, WAS-1455-002, May, 1990.

Jarrett, R. D., 1984. *Hydraulics of High-Gradient Streams*, Journal of Hydraulic Engineering, Vol. 110, No. 11.

Long Lake Emergency Action Plan, 2011. Provided to WEST by Avista Corp.

Little Falls Emergency Action Plan, 2010. Provided to WEST by Avista Corp.

LuBean, H. E., 1987. *Dam Failure and Inundation Study for Nine Mile Hydroelectric Development*, FERC Project No. 2545, The Washington Water Power Co., April 1987.

Obermeyer Hydro, Inc., 2009. *Plan Layout & Construction Notes* , Nine Mile Dam, 05-116-101-1. Provided to WEST by Avista.

U.S. Army Corps of Engineers, 2000. *Hydraulic River Analysis HEC-RAS: User's Manual Version* Hydrologic Engineering Center, March 2000.

U.S. Army Corps of Engineers, 2001, Hydrologic Engineering Center. *HEC-RAS Hydraulic Reference Manual*, January 2001.

U.S. Army Corps of Engineers, 2007. Hydrologic Engineering Center, *Dam Breach Analysis using the Hydrologic Engineering Center's River Analysis System HEC-RAS*, Training Manual, 2007.

GSSHA MODELING FOR ANALYSIS OF FLOOD DESIGN FEATURES AT THE PICAYUNE STRAND RESTORATION PROJECT

Charles W. Downer[1]
Jaime A. Graulau-Santiago[2]
Brian E. Skahill[3]
David M. Weston[4]
Nawa Raj Pradhan[5]
Aaron R. Byrd[6]

ABSTRACT

This paper describes the application of the USACE physically-based Gridded Sub-Surface Hydrologic Analysis (GSSHA) model for the assessment of flooding potential within the influence area of the Picayune Strand Restoration Project (PSRP). GSSHA was selected due to its ability to simulate fully coupled rainfall distribution, extraction, retention, overland flow, and one dimensional channel flow and the capability to develop inset sub-models within the boundaries of a larger model domain (telescopic refinement). This capability allowed for high resolution analysis and refinement of the proposed flood protection features.

Hydraulic models were developed from existing and design data, populated with parameter values from previous hydrodynamic modeling efforts. Parameters were then tuned to observed stage and flow data using the Secant Levenberg-Marquardt (SLM) method, a nonlinear least squares minimization computer-based local search method. The calibrated model reproduced canal flows and stages with a Nash-Sutcliffe Forecast Efficiencies, generally 0.9 or higher. Subsequent uncertainty analysis allowed water stages to be estimated with 95% certainty. These hydrologic models and analysis demonstrated that some of the proposed features were either unnecessary or over-designed with respect to the original proposed plan. The refinement of these features resulted in construction cost savings of $40M approximately.

INTRODUCTION

The Picayune Strand Restoration Project (PSRP) will functionally restore more than 700 hectares of land located in a largely undeveloped portion of southwestern Collier County, Florida to its pre-development condition. Formerly known as the Southern Golden Gate Estates (Figure 1), the project area was originally planned as a residential subdivision in

[1] Research Hydraulic Engineer, Hydrologic Systems Branch, Coastal and Hydraulics Laboratory, Engineer Research and Development Center, 3909 Halls Ferry Road, Vicksburg, MS 39180, Charles.W.Downer@erdc.dren.mil
[2] Chief, Interagency Modeling Section, Water Resources Engineering Branch, Jacksonville District, US Army Corps of Engineers, 701 San Marco Blvd, Jacksonville, FL 332207
[3] Research Hydraulic Engineer, Hydrologic Systems Branch, Coastal and Hydraulics Laboratory, Engineer Research and Development Center, 3909 Halls Ferry Road, Vicksburg, MS 39180
[4] Senior Engineer, Water Management Section, Operations Division, Jacksonville District, US Army Corps of Engineers, 701 San Marco Blvd, Jacksonville, FL 332207-8175
[5] Research Hydraulic Engineer, Hydrologic Systems Branch, Coastal and Hydraulics Laboratory, Engineer Research and Development Center, 3909 Halls Ferry Road, Vicksburg, MS 39180
[6] Branch Chief, Hydrologic Systems Branch, Coastal and Hydraulics Laboratory, Engineer Research and Development Center, 3909 Halls Ferry Road, Vicksburg, MS 39180

the 1950s; roads and drainage canals were constructed in the 1960s and early 1970s for that purpose. The planned residential development failed and the roads and four canals have since over-drained the area. This over drainage has resulted in a reduction of aquifer recharge, increased freshwater shock load discharges to downstream estuaries, invasion by upland vegetation, loss of ecological connectivity and associated habitat, and increased frequency of forest fires. The recommended plan for the PSRP (USACE 2004) is to remove the existing infrastructure of the subdivision and restore its pre-drainage hydrology and ecological function for beneficial effects on flora and fauna to the PSRP area and surrounding public lands.

The structural hydraulic features of the PSRP are the construction of a series of pump stations, levees, spreader berms, and canal plugs to slow water flowing through existing canals and redistribute it across the landscape (Figure 2). The U.S. Army Corps of Engineers (USACE) Jacksonville District (SAJ) in partnership with South Florida Water Management District (SFWMD) is constructing these features. With the PSRP, the USACE is required to maintain the existing level of flood protection to existing landowners after the project is built. In addition, USACE projects of this magnitude are designed with a 100-year return period level of flood protection.

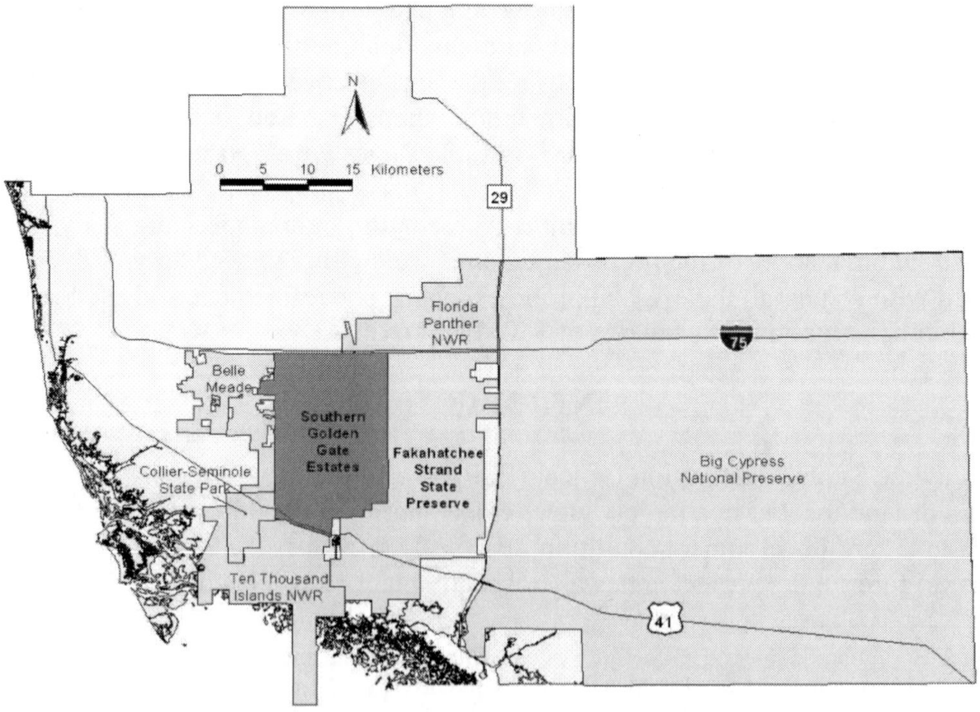

Figure 1. Southern Golden Glades Estates location

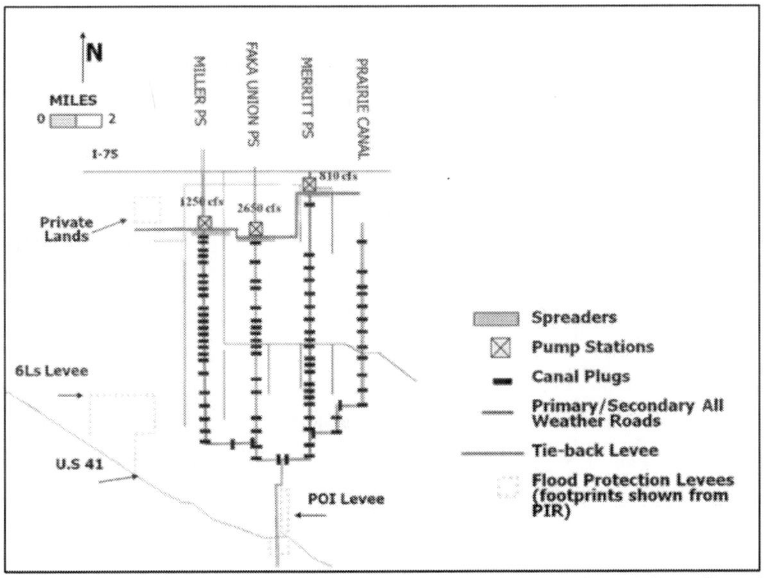

Figure 2. PSRP Preferred Alternative features and areas requiring flood protection.

To adequately refine the hydraulic design parameters for the flood mitigation features, a tiered model design-analysis approach was used in this study. A large scale, relatively low resolution (Tier1) model enabled analysis of the system at a regional level and provided the ability to estimate the response of the system in terms of flood stages and flows relative to historical and synthetic storm events. This model was then used to provide boundary conditions for more detailed, higher resolution inset models (Tier 2). Together, these models enabled detailed analysis of flow and stage behavior in the vicinity of the flood mitigation features in the recommended plan. Tier2 models purpose was to identify potential adverse impacts in and around privately owned lands for different storm frequencies (i.e., the 5, 10, 25, and 100-yr events), and to determine if the proposed features comply with the flood protection requirements. Whenever a protection feature was identified to be necessary, the design-criteria was further refined, including the evaluation of risk and uncertainty.

The GSSHA spatially distributed hydrologic model (Downer et al., 2005) developed and supported at the Engineer Research and Development Center Coastal and Hydraulics Laboratory Hydrologic Systems Branch (ERDC-CHL-HSB) was selected to represent the PSRP existing and proposed project conditions at the Tier1 and Tier2 scales. The GSSHA hydrologic model has the capacity to simulate, on a continuous basis, numerous processes relevant to the hydrologic response of a watershed system subject to meteorological forcing, including rainfall distribution, plant interception, surface retention, evapotranspiration, vertical infiltration, two-dimensional overland flow, one-dimensional channel flow, two-dimensional groundwater flow for the unconfined aquifer system and related surface water-groundwater interaction, lake/reservoir levels, and snow accumulation and melt. Additional information about the modeling tool can be found in the GSSHA wiki (http://www.gssahwiki.com).

MODEL DEVELOPMENT

Project Condition Descriptions. The baseline for the modeling is the as-is, or Existing Condition (Condition 1) depicted in Figure 3. The area is drained from north to south with a series of canals Miller, Faka Union, Merritt and Prairie. The existing condition also includes numerous roads that crisscross the area. To achieve the project goals of more natural (sheetflow) across the region, the canals are largely removed and the flood protection of surrounding areas is provided with a combination of spreader canals, berms, and levees. The Proposed Project Condition (Conditions 2) is shown schematically in Figures 2, and contains the following features:

1. Existing canals are plugged except Merritt, Faka Union and Miller Canals are open from I-75 to pump stations in the northern (upper) part of the project.
2. A tie-back levee that runs from Merritt Canal to 1829 m west of the Miller Pump Station contains the water to be pumped.
3. Water from the Private Lands area north of the Tieback levee is routed to the Miller pump station with a new interior drainage canal.
4. The pump stations discharge into bays with spreader berms where water gravity flows thru a series of weirs in the berm.
5. Most existing roads are degraded.
6. Levees provide protection for the private areas within the project footprint.
7. A remnant of Faka Union Canal provides drainage at the southern boundary of the project.

Modeling Approach. The approach to conduct the modeling was to develop the Existing Condition (Conditions 1) Tier1 model, calibrate and verify against observed stage and discharge data (resulting from two historical tropical storm events), include the features for the preferred alternative (Conditions 2) and then develop the Tier2 models (Conditions 1 and 2) using boundary conditions from the larger Tier 1 models. Then use this suite of models to simulate the series of return period storms events. To simplify the following discussion, the models will be referred by Tier level first (T1 or T2) followed by the condition (C1) or (C2), e.g. T1C2 to reference Tier 1 Conditions 2 model.

Tier 1 model. The T1 model provides the basis for the more refined T2 models. It includes sufficient detail to provide large scale impacts and boundary conditions for the T2 models. Important features of the T1 Model are:

Domain and boundary conditions. Selected based on available data, the need for coherent boundary conditions, and inclusion of important areas. At the northern boundary, the model coincides with Interstate 75 (Figure 3). The interstate has an embankment and a drainage canal with structures that controls flow into the domain. No overland flow boundary condition was prescribed at this boundary. Flow into the domain from the area upstream of the model domain is only through the canals in the existing conditions, and into the remnant canals immediately upstream the pump stations in C2. These boundary flows were extracted from a previous hydrodynamic model used for the selection of the preferred alternative (SFWMD, 2006). Flows into the canals were input as inflow hydrographs into the canals.

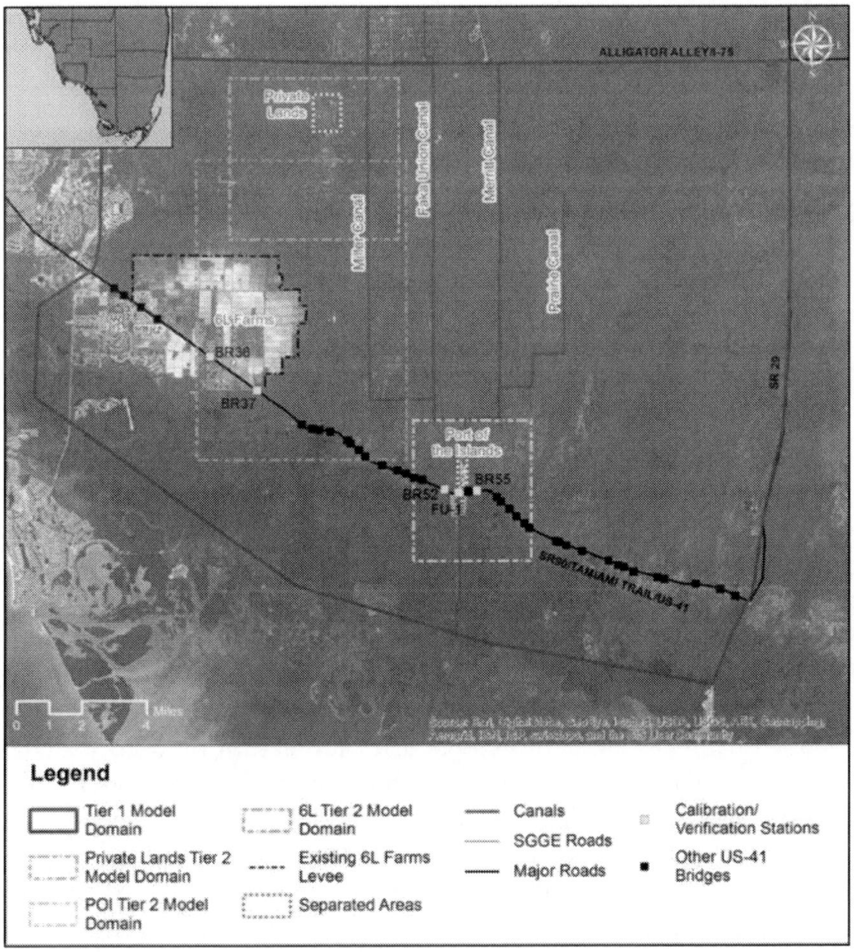

Figure 3. Model Domains and features for existing conditions (Condition 1)

At the southern boundary, a head boundary representative of the tidal stage fluctuations over the period of simulation was specified. This was also derived from the previous hydrodynamic model (SFWMD, 2006). Eastern and western boundaries are no-overland flow conditions, supported by local topography. Topographic data were derived from a number of sources and composited into a series of topographic maps at various grid scales (3, 5, 10, 50, 100m) that was further used to assign elevations to the model grid cells. Total area in the computational domain is about 928km^2.

Grid size and time step. A grid size analysis was performed to select the optimal cell size for the model in order to capture all the relevant spatial data, without undue computational burden. The final grid size of the Tier1 model was 120m. Simulation of the model with this grid size was facilitated by the inclusion of the overland flow structures that represent the bridges/culverts under US 41, which are on the order of 30m in length. Total cell number in the domain was 64,920. With the overland flow structures, this grid size was more than adequate to capture all salient spatial variability in the model domain. With this grid size, the model runs comfortably with time steps on the order of 10s.

Hydraulic structures and controls. The PSRP Existing Condition models contain numerous hydraulic structures and other overland flow control features, as described above. In GSSHA the varying structures are simulated in the following ways:

- Canals: simulated explicitly in the model as one dimensional channel flow. Cross section and bottom elevation data were extracted from "as built" drawings. The same time varying stage used for the southern overland flow boundary is used to specify the stage at the downstream end of the model, the Faka Union Canal. At the upstream end of the canals, a flow hydrograph boundary condition was specified. Weirs in the canals were included in the model. Relevant physical information was obtained from the previous hydrodynamic model (SFWMD, 2006). Discharge coefficients were initially prescribed from this model and then calibrated to observed data. Canals that were either plugged or physically removed as part of the preferred alternative were removed from the stream network.
- Roads, Levees, Spreader Canals, and other Embankments: Overland flow impediments are included in the GSSHA model as embankments. Embankments block water across faces on the overland flow grid. Flow over the embankment is computed using a broad crested weir equation. Water surface elevations on both sides of the embankment are analyzed for backwater effects during these computations. Hydraulic structures through embankments were represented as overland flow structures with specified rating curves. The embankment representing the US 41 road contained 37 of these overland flow structures. Elevations for existing structures were obtained from "as built" drawings. Elevations for proposed structures were obtained from design drawings. As a result of this modeling some of the structure size, location, and elevation were modified, and reflected back into the modeling efforts.
- Overland Roughness and Retention: Overland roughness is specified based on the 1995 land use and vegetation map. The model contains 18 land use/vegetation types. These land uses are used to assign overland hydraulic roughness coefficients (Mannings' coefficient) for the model. The primary land use/vegetation types in the model are: hydric flatland (26%), swamp forest (15%), cypress (12%), mangrove (12%), wet prairie (9%), and unknown (10%). Roughness coefficients were originally assigned based on the previous hydrodynamic model used during the alternative selection process, and then calibrated to historic events. The PSRP domain is largely a swamp and the area has depression features at both the grid and sub-grid scale. These depression features retain water at low overland flow stages (10^{-2} m magnitude) but have little effect on flood flows, which may be several meters in depth. During model development it was decided to minimize the explicit modeling of the fine scale depression features by smoothing out the overland flow grid on a macroscale level and then including depression storage for lost depression storage during the grid smoothing process. A detailed analysis was conducted to calculate the difference in storage between the smoothed 120m grid used in the model and the original 3m topographic digital elevation model used to develop the model grid. Forty two separate depression storage areas were identified and added to the

GSSHA simulation. Flow on the overland flow plane in GSSHA will not occur until the retention depth has been exceeded; then flow above the retention depth is calculated using the Manning formula driven by the water surface slope.

- Soils and Infiltration: While the event based simulations of the suite of return period storms are simulated with a high water table assumption, i.e. no infiltration, the need to establish initial conditions for calibration and verification required that soils and infiltration be considered in the model. Soils were derived from the 1990 Natural Resources Conservation Service survey for Collier County, Florida as provided by the South Florida Water Management District (SFWMD). Four different soil classifications were used in the model: Immokalee, Boca, Riviera, and Sanibel.

Tier 2 models. T2 models were developed to refine and represent with higher fidelity some of the relevant hydraulic features in the proposed alternative. They were also used to refine the hydraulic design criteria for the proposed flood protection features and analyze potential adverse effects in areas where protection is needed. Three T2 inset models were developed for the existing (C1) and proposed project conditions (C2): Private lands, 6Ls, and the Port of Islands (POI). These model domains and relevant features are shown in Figure 3.

Many data sets in the T2 models were imported from the T1 model. Among these were soil properties, land use, roughness coefficients, retention depths, and any other parameter whose representation in these two domains was not dependant on model resolution. In the C1 models the grid resolution allowed features such as roads to be explicitly included in the overland flow model. In C2 models, these features were degraded per the project alternative. If a canal traversed the model domain, that canal was included in the T2 model. In all such cases the upstream boundary condition was an input hydrograph and the downstream boundary condition was a time varying specified head.

The T2 models share many common properties. The domains and grid sizes for the models were selected to allow the features in the respective models to be simulated accurately and efficiently. Specified head overland flow boundary conditions derived from the T1 models were prescribed along the boundaries for all the models, unless otherwise noted below. All three models were constructed with a 20m grid size and timesteps on the order of 10s. Model specific features are detailed below.

Private Lands Model

1. Overland flow boundary condition on the southern boundary in C2 model, was modified to a no-change in flow boundary condition to remove water from the domain to avoid artificial boundary backwater effects.
2. In C2, the Miller Canal pump system was implicitly simulated by a time series of outflows assigned to model cells. This flow time series was representative of how the pump system would operate during the storm events simulated.
3. In C2, the proposed alternative included the tieback embankment and the spreader berm (Figure 2). The spreader berm includes a series of 10 weirs designed to convey water from inside the internal spreader berm area into the landscape in a sheet flow

regime. T2 model resolution allowed for the inclusion of these hydraulic details in the model. The "as built" drawings for the spreader berm, embankments and weirs were used to define these features.

6Ls Model

1. US41 culverts as in C1 and C2. An additional culvert included in the C2 Model in the southern most segment of the USACE's proposed protective levee.
2. Embankments in C1 include the US41 road and the 6Ls levee.
3. In C2, the proposed alternative includes a protective levee offset about 60m from the 6Ls farm boundary.

Port of Islands (POI) Model

1. In C2, a no flow boundary condition at the upstream end of the Faka Union canal since in the preferred alternative all canals segments south of the pump stations and north of this junction will be plugged.
2. Hydraulic structures in C1 and C2 are the US Hwy 41 culverts and an existing weir in the Faka Union Canal located just north of US Hwy 41.
3. Portion of the Faka Union Canal that will remain open in the preferred alternative is the only channel within the model domain. This applies to both conditions since the T2 model domain dissects the Faka Union canal at the junction of the three main canals. The difference in how the channel is treated in C1 and C2 resides in the boundary conditions applied at this junction as described previously
4. The preferred alternative does not alter the existing conditions topography in the vicinity of the POI. Therefore, only one source of data was used to represent the topography in C1 and C2.
5. The only embankment in C1 and C2 is the US Hwy 41. In the Project Implementation Report (USACE, 2004), a protective embankment feature was being considered in the area around POI. T2C2 model simulations showed that this area was not adversely impacted relative to the T2C1 conditions thus, the proposed feature did not significantly add flood protection beyond the existing conditions.

RESULTS

Model calibration

Table 1 lists the thirteen adjustable parameters in the T1C1 model, as well as the lower and upper limits imposed on them during the calibration process. Four uniform values, applied to all canals, were adjusted: Manning's roughness, *ch_n*, a discharge coefficient for weirs in the canal network, *dcf*, an adjustment factor to the canal thalweg elevations, *adj*, and another adjustment factor for the bankfull depth of the canal, *adj2*. On the overland flow plane initial values for roughness and retention depth were assigned based on land use. These values were adjusted with a factor to the base values for roughness (*mult1*) and retention (*mult2*). Stage-discharge rating curves of structures through US41 obtained from previous modeling efforts were also modified with a structure specific value: *"rtc_mult_36", "rtc_mult_37", "rtc_mult_52", "rtc_mult_55", "rtc_mult_w", and "rtc_mult_e".*

Estimation of the thirteen specified adjustable parameters for the PSRP GSSHA T1C1 model was undertaken by matching observed and simulated flows and stages at Faka Union 1 (FU-1), and also by matching observed and simulated stages at four ditch gaging locations along the US Hwy 41 (Bridges 36, 37, 52, and 55) for a single storm event (Tropical Storm Jerry; 08/20/1995 14:00 – 08/26/1995 11:00), resulting in a total of 830 observations for use in the calibration process. The objective function, which is the quantitative measure of model to measurement misfit, is defined in the weighted least squares sense. Weight assignment for each observation group was adjusted such that at the beginning of the computer-based calibration each observation group was seen to be of equal importance.

The Secant Levenberg-Marquardt (SLM) method, a nonlinear least squares minimization computer-based local search method (Skahill et al. 2009), as implemented by Skahill et al. (2012), was employed to efficiently calibrate the PSRP GSSHA T1C1 model. With the exception of the two parameters, *adj* and *adj2*, which can take on negative values, the remaining eleven adjustable model parameters were log transformed in attempts to better accommodate scaling issues resulting from the use of different units for different parameters, and to decrease the degree of nonlinearity of the parameter estimation problem; past experience has demonstrated that greater efficiency and stability of the parameter estimation process can often be achieved through this means (Skahill et al. 2009).

Table 1. Specified adjustable parameters for the PSRP GSSHA T1C1 model.

Parameter	Parameter description	lower bound	upper bound
ch_n	Channel Manning's n value	0.025	0.075
dcf	Forward discharge coefficient for weirs in channel network	1	5
adj	Adjustment to all node elevation values in the channel input file	-0.2	0.2
adj2	Adjustment to bankfull depth values in the channel input file	-1	1
mult1	Multiplier for original overland roughness values	0.2	5
mult2	Multiplier for original retention depth values	1.00E-01	1.00E+01
rain_mult	Multiplier for original input rainfall values	1	1.1
rtc_mult_36	Multiplier for original rating curve values for 36	1.00E-02	1.00E+02
rtc_mult_37	Multiplier for original rating curve values for 37	1.00E-02	1.00E+02
rtc_mult_52	Multiplier for original rating curve values for 52	1.00E-02	1.00E+02
rtc_mult_55	Multiplier for original rating curve values for 55	1.00E-02	1.00E+02
rtc_mult_w	Multiplier for remaining original rating curve values west of FU	1.00E-02	1.00E+02
rtc_mult_e	Multiplier for remaining original rating curve values east of FU	1.00E-02	1.00E+02

The SLM local search with the PSRP GSSHA T1C1 model was configured for efficiency. In particular, no full update of the model sensitivity matrix was employed together with single column cyclic updating (Skahill et al. 2009, and Skahill et al. 2012). The SLM local search completed in 33 forward model calls, resulting in a 40% reduction in the objective function value. The final estimated model parameter set is listed in Table 2 along with an estimate of its uncertainty and sensitivity. It should be noted that confidence limits provide only an indication of parameter uncertainty. They rely on a linearity assumption which may not extend as far in parameter space as the confidence limits themselves. Clearly, the rating curve multipliers (*rtc_mult_36*, *rtc_mult_37*, *rtc_mult_52*, *rtc_mult_55*, *rtc_mult_w*, and *rtc_mult_e*) were identified with the most uncertainty relative to the remaining model parameters. This is not surprising given that rating curves were derived from previous modeling efforts (SFWMD, 2006). Sensitivity testing indicated that parameters *adj* and *adj2* were most sensitive relative to the remaining specified adjustable model parameters. Measurement to model misfit (Table 3) indicates that the calibrated model is of predictive value.

Table 2. Final estimated parameter set for the PSRP GSSHA T1C1 model.

Parameter	Estimated value	95% confidence limits lower limit	upper limit	Sensitivity
ch_n	0.045844	0.044531	0.047195	1.409143
dcf	1.61301	1.521033	1.71055	2.01872
adj	0.001531	-0.004223	0.007285	47.02873
adj2	0.000042	0.00002	0.000063	624.9205
mult1	1.100165	0.988629	1.224284	0.741138
mult2	0.856535	0.83239	0.881381	1.830177
rain_mult	1.065394	1.043216	1.088043	3.064292
rtc_mult_36	0.93399	0.682793	1.277602	0.62694
rtc_mult_37	0.952339	0.608473	1.490534	0.87159
rtc_mult_52	0.901433	0.698453	1.163402	1.406658
rtc_mult_55	0.965863	0.763623	1.221665	1.438024
rtc_mult_w	0.999786	0.684533	1.460226	1.302043
rtc_mult_e	0.96641	0.699074	1.335978	0.537431

Table 3. Summary statistics of model to measurement misfit.

Data	R^2	Nash-Sutcliffe Efficiency	Mean Error	Mean Absolute Error	RMSE
All data	0.99	0.99	0.18	0.72	1.99
All FU-1 data	0.99	0.99	0.56	1.97	3.41
FU-1 discharge data	0.98	0.97	1.12	3.93	4.83
FU-1 stage data	0.97	0.97	0.00	0.03	0.04
All bridge stage data	0.90	0.87	-0.02	0.07	0.11
Bridge 36 stages	0.86	0.77	-0.07	0.10	0.14

Bridge 37 stages	0.86	0.77	-0.02	0.12	0.15
Bridge 52 stages	0.98	0.97	0.02	0.04	0.05
Bridge 55 stages	0.99	0.99	0.00	0.02	0.03

Model Predictions

The calibrated model parameter set developed for the T1C1 model were used to simulate the T1C2 conditions and further in the T2C2 models. All T1C2 and T2C2 were used to simulate 5 day storm events with 5, 10, 25 and 100 yr return periods. Two types of model predictions were used to assess the effects of the project on the private lands: point elevations from grid cells along flood protective features and spatial maps of maximum flood depth.

Detailed water surface profiles were analyzed at critical points in the T1C2 model and in each of the three T2C2 models. Water surface profiles were extracted from selected grid cells within the GSSHA model. Locations of these cells are shown in Figure 4. Water surface profiles at each of these locations were adjusted to account for uncertainty in the parameter values and determine the water levels at the 95% confidence limit (see U.S. Army Corps of Engineers, Jacksonville District, 2012, and references cited therein; i.e. EM 1110-2-1619, ER 1105-2-101, EC 1102-2-6067). For the T1 model a nonlinear calibration constrained predictive analysis was used (Cooley and Vecchia, 1987; Vecchia and Cooley, 1987) to determine the 95% confidence levels at two points near the POI (Figure 4). Nonlinear predictive analysis is nontrivial in terms of its model simulation time requirements and due to computational constraints, an additional 6 points analyzed at the T1 level and all points evaluated at the T2 level only permitted a less model run intensive linear uncertainty analysis (Doherty, 2004). Levee superiority analysis was performed to assess whether the levee crest was at least 0.6 m greater than the 95% confidence level for the 100-yr event.

Maximum water surface profiles of selected cells near critical areas in each of the T2 models are shown in Figure 5. In general, the water levels at the critical points in the project were lower than expected, and generally indicated minimal impact of the project on flooding in the private lands.

To further quantify the residual flooding caused by the proposed PSRP restoration features the T2 values of gridded maximum flood stage in C1 were subtracted from the T2C2 flood stages (Figures not shown). This analysis indicated either no change in water level or a lower water level for the C2 model relative to the C1 model. Results for all other storm events showed similar trends as the 100yr frequency storm. These results indicated that the proposed protection feature that was conceptualized as a protective levee just east of the developed areas north of US41 was not necessary or does not provided any additional flood protection relative to the existing conditions. Moreover, the proposed restoration features for the PSRP will most likely alleviate flooding conditions in this vicinity. This is consistent with the water surface profile comparison shown in Figure 5(a).

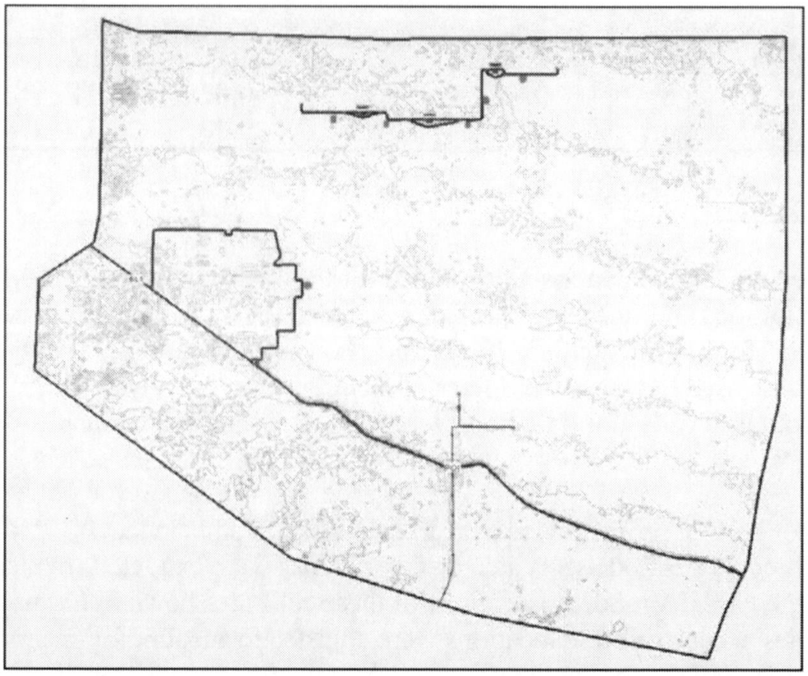

Figure 4. Location of points (magenta colored dots) selected for detailed maximum stage and uncertainty analysis.

A similar analysis for the Private Lands T2C2 model showed that the tie-back levee protective feature that was originally conceptualized in the PIR (Figure 2) extending about 3,000m west of the Miller pump station to the south of the Private Lands could be reduced to about 1,800m in length without compromising the level of flood protection as originally intended for this feature. Maximum stage difference on a cell by cell basis (results not show) indicates that increases in water depths in C2 relative to C1 do not extend to the private lands areas.

The water surface profile just south, along the alignment of the western portion of the tie-back levee system (Figure 5b), shows that one of the reasons why the water levels do not increase in C2 relative to C1 near the private lands areas is due to the existence of a spoil mound located about 540m east from the western end of the system. This spoil mound runs about 1000m south and prevents water from the spreader berm from flanking around the proposed levee where it could reach the Private Lands area. The resolution selected for these T2 models captured all these important details allowing for a refinement of the protective feature design relative to what was conceptualized in earlier studies.

Due to the way in which water is spread over the landscape by the proposed restoration features (i.e. the Miller pump station and spreader berm system), flooding extent in areas just outside the 6L farm increase in C2 relative to C1 (results not show). The residual flooding east of the proposed 6Ls protective feature varies from 0.15m east of the farms to approximately 1.0m in areas just south of the restoration features. A water surface profile (Figure 5c) along the proposed feature for the 10 and 100yr storm event frequencies confirmed this. Based on these findings, the Private Lands levee was modified as noted and the Port of the Isles levee was eliminated from consideration.

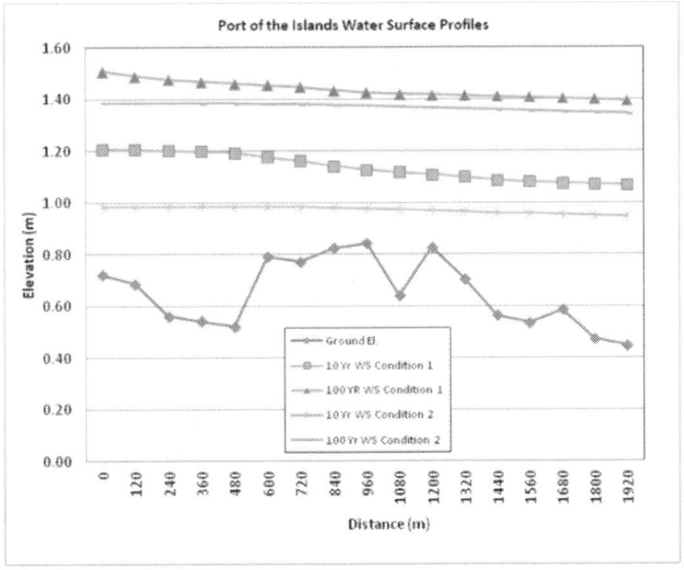

(a)

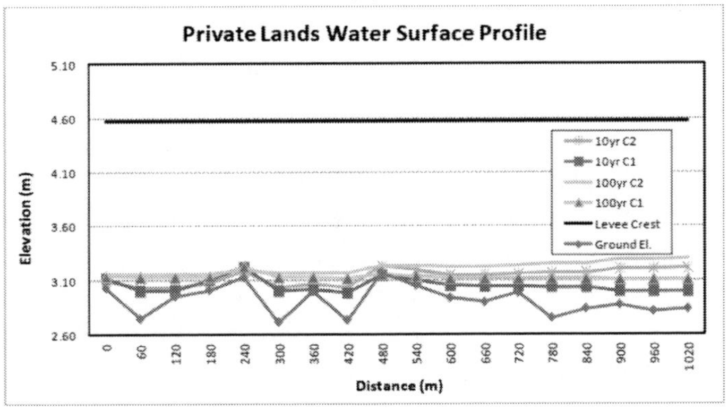

(b)

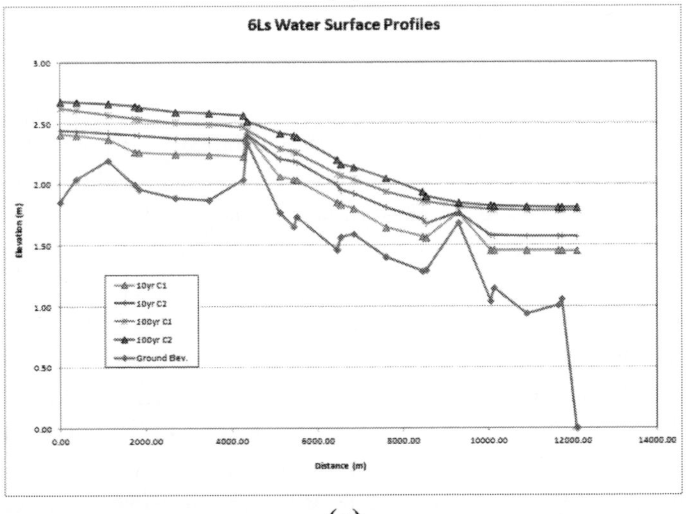

(c)

Figure 5. Water surface profiles of selected cells for T2C2 models for (a) Port of the Islands, (b) Private Lands, and (c) 6L farms

DISCUSSION

The results, slightly counter intuitive, raise two questions, that is, how did the removal of the canals specifically designed for flood control not significantly increase flooding, and why are the results significantly different from previous modeling results. In the first case, it is relevant that the existing canal system was designed to contain only the 10 yr return period flow. This is evident in Figure 5 where in all cases the C1 maximum stage for the 10 year event is lower than the C2 maximum stage, so the canals were providing flood protection for storms of this frequency. However, for storms with lower return frequencies i.e., larger magnitude events, the canals are inadequate for conveying the flood waters out of the system and overbank flooding occurs throughout the network. For these events, the canals are exacerbating the flooding near the canals by concentrating all the water into one area. Being specifically developed to provide 100-yr flood level prediction, the project condition seeks to spread the flood water over a larger area, as opposed to concentrating the flow. The spreading of the water over the northeast portion of the project accomplishes two things. First, as is the intent of the project, the water has a larger flow area to spread. Second, and perhaps more importantly, placing most of the flow on the overland toward the eastern side of the project results in most of the water following the natural drainage, toward the south and east, and away from the areas of concern, which are located on the western side of the project.

Differences from previous modeling results can largely be attributed to the level of detail in this modeling exercise. Although the PIR modeling was conducted with tools with similar capabilities, the PIR model encompassed a much larger area, at a much coarser scale for other modeling purposes not necessarily similar to the purpose of this study. In addition, elevation data available at the time did not support the development of fine scale models in the area. When combined with high resolution elevation data, at 3m, the fine resolution GSSHA model was able to clarify the smaller fine scale features. Use of the tiered modeling approach allowed the finer resolution modeling to be conducted in a reasonable computational timeframe.

CONCLUSIONS

The GSSHA hydrologic model was used to execute a multi-tiered, level of service flood study analysis on the existing and the proposed alternatives for the PSRP. Initial model parameters were derived from previous large scale coarser hydrodynamic models used to screen out the preferred alternative. The models were calibrated to existing flow and stage data in the canals and overland flow domains. Parameters specified as adjustable as part of the T1C1 computer-based model calibration work activity not only included model process related parameters, but also parameters related to model configuration, input precipitation forcing, and adjustment of the base rating curve values which controlled the flow through the numerous bridge crossings along US Hwy 41. A post-calibration model predictive uncertainty analysis included these elements in the quantification of the uncertainty associated with specific stage predictions selected to be analyzed with each project alternative.

Employment of two tractable methodologies for uncertainty analysis enabled quantification; at the 95% confidence level, of the maximum simulated stage at predetermined locations within the computational domains of the models representing the proposed project condition (C2). All of the GSSHA models developed to characterize project alternatives were forced with the 5 day 100 year design storm. The results from these analyses supported risk-based hydrologic design and refinement of PSRP protection features. The higher resolution models used in the current analysis indicated that some of these features were either unnecessary or over-designed with respect to the selected plan.

REFERENCES

Cooley, R. L., and A. V. Vecchia. 1987. Calculation of nonlinear confidence and prediction intervals for ground-water flow models. *Water Resour. Bull.* 23 (4), 581-599.

Doherty, J. 2004. PEST: Model Independent Parameter Estimation. Fifth edition of user manual. Watermark Numerical Computing, Brisbane, Australia.

Downer, C. W., F. L. Ogden, J. Neidzialek, and S. Liu, 2005. GSSHA: A model for simulating diverse streamflow generating processes. Chapter 6 in: Singh, V.P. and D. Frevert, *Watershed Models*, CRC Press.

Downer, C. W., and F. L. Ogden, 2006. *GSSHA Users' Manual*, ERDC/CHL SR-06-1.

Downer, C. W., and F. L. Ogden. 2003. Prediction of runoff and soil moistures at the watershed scale: Effects of model complexity and parameter assignment. *Water Resources Research*, 39(3): 1045.

Skahill, B., J. Baggett, S. Frankenstein, and C. W. Downer. 2009. "More efficient PEST compatible model independent model calibration." Environmental Modelling & Software (24), 517–529.

Skahill, B. E., C. W. Downer, and J. S. Baggett. 2012. "A Practical Guide to Calibration of a GSSHA Hydrologic Model using ERDC Automated Model Calibration Software – Efficient Local Search." ERDC/CHL TR-12-3 Technical Report, U.S. Army Engineer Research and Development Center, Vicksburg, MS.

U.S. Army Corps of Engineers, Jacksonville District. 2012. SCOPE OF WORK, ARCHITECT-ENGINEER SERVICES FOR, Water Resource Engineering, For Picayune Strand Restoration Project, Collier County, Florida, Modification 1. January

U.S. Army Corps of Engineers, Jacksonville District. 2004. Picayune Strand Restoration Project Final Integrated Project Implementation Report and Environmental Impact Statement, Collier County, FL

Vecchia, A.V. and Cooley, R.L., 1987. Simultaneous confidence and prediction intervals for nonlinear regression models with application to a ground water flow model. *Water Resour. Res.*, 23 (7), 1237-1250.

USING THE VARIABLE INFILTRATION CAPACITY MODEL IN A SEMI-STOCHASTIC FORMAT FOR HYDROLOGIC HAZARD ESTIMATES FOR DAM SAFETY RISK ANALYSIS

Frank Dworak[1]

ABSTRACT

The Bureau of Reclamation (Reclamation) uses multiple methods to develop hydrologic loadings for dam safety risk analysis, which typically consist of peak flow flood hydrographs for a full range of Annual Exceedance Probabilities (AEPs). Reclamation has a standard set of tools for conducting Hydrologic Hazard Analyses (HHAs) however they often rely on custom approaches for unique projects. Reclamation is currently conducting an HHA for Grand Coulee Dam, Washington, to develop probabilistic flood frequency estimates for the drainage basin that covers approximately 74,100 square miles, of which 39,500 square miles are situated in the province of British Columbia, Canada. This challenging project caused Reclamation to think outside of the box to use a custom approach of modeling the hydrology semi-stochastically with the Variable Infiltration Capacity (VIC) model.

The VIC model is a continuous, semi-distributed macroscale hydrological model that balances both water and surface energy within each grid cell. Specifically, Reclamation is using a VIC model with $1/16^{th}$ degree latitude/longitude spatial resolution that was previously developed for the Columbia River Basin by the University of Washington in collaboration with multiple local and federal government agencies, as well as the British Columbia Ministry of the Environment, in support of the Columbia Basin Climate Change Scenarios Project. Reclamation will be using the model in a semi-stochastic format with frequency precipitation, custom temporal and spatial storm distributions, and historic datasets to develop hydrologic hazard estimates.

Reclamation will discuss the advantages and challenges of using an existing calibrated hydrologic model for the Grand Coulee Dam project.

INTRODUCTION

The Bureau of Reclamation is conducting a Hydrologic Hazard Analysis (HHA) for Grand Coulee Dam to develop probabilistic flood frequency estimates for use in a dam safety risk analysis. This study is ongoing at the time of this paper and is scheduled to be completed in 2017. Reclamation uses multiple methods to develop hydrologic loadings for dam safety risk analysis, which typically consist of peak flow flood hydrographs for a full range of Annual Exceedance Probabilities (AEPs). Reclamation has a standard set of tools for conducting HHAs, however they often rely on custom approaches for unique projects. Grand Coulee Dam has one of the largest, highly regulated drainage basins in Reclamation's inventory of dams and required a custom approach of modeling the

[1] Hydrologic Engineer, Flood Hydrology and Meteorology Group, Bureau of Reclamation, PO Box 25007 86-68250, Denver, CO 80225-0007, 303-445-2547, fdworak@usbr.gov

hydrology semi-stochastically using the Variable Infiltration Capacity (VIC) model to complete the HHA.

Dam and Drainage Basin

Grand Coulee Dam is a concrete gravity structure located on the Columbia River in north-central Washington State, located approximately 75 miles west-northwest from Spokane, Washington. The towns of Grand Coulee and Coulee Dam are located immediately upstream and downstream, respectively, of the dam. The drainage basin is approximately 74,100 mi^2, of which approximately 39,500 mi^2 are situated in the province of British Columbia. The drainage basin is composed of mountainous terrain, as well as large expanses of high desert plains in eastern Washington, eastern Oregon, and western Idaho. Elevations throughout the drainage basin range from approximately 1,300 feet near the dam to over 13,000 feet in the mountains. The Columbia River Basin (CRB) has large variations in mean annual precipitation. West of the Cascade Mountains, many regions receive over 30 inches of precipitation annually, while east of the Cascades precipitation is less than 20 inches, with some areas receiving as little as 7 inches (Climate Impacts Group, 2014). Annual precipitation in the Cascade Mountains can be as much as 100 inches. By several accounts, more than 70% of streamflow in the region originates as mountain snowpack (Hamlet et al. 2005; Elsner et al. 2010). Most of the drainage basin accumulates a snowpack throughout the winter followed by rainy periods in the spring and early summer. The high flow period for the Columbia River is driven by snowmelt during the spring and early summer and rain.

Modeling hydrology for Grand Coulee Dam is complicated by the extremely large and diverse international drainage basin, and by the Columbia River being operated as a system to minimize floods, maximize power generation, and to meet the multitude of other objectives of the system of dams. There are 76 major dams located within the Columbia River Basin (CRB) that are owned and operated by a mix of the U.S. federal, state, provincial, or local governments; public utilities; and private entities. Grand Coulee Dam is one of the most important flood control projects on the Columbia River, which needs to be taken into account when estimating the hydrology as part of a risk analysis. Grand Coulee Dam flood operations can highly affect the safety of downstream dams and power generation projects. Therefore, the flood risk for Grand Coulee Dam is not only about the project itself, but more about minimizing the flood risk for the system of dams, power projects, and population at risk downstream of Grand Coulee Dam.

Scoping Study

Reclamation completes multiple Issue Evaluation HHAs each year; and all of these studies are unique and require some degree of customization to the modeling approach. Due to the magnitude of the Grand Coulee Dam HHA, Reclamation's Dam Safety Office funded a scoping study to gain a better understanding of the historic hydrology and hydraulic (H&H) work completed for Grand Coulee Dam and the CRB, and to develop a scope that will address all of the H&H concerns for Grand Coulee Dam. Some of the main objectives of the scoping study were to:

- complete a comprehensive literature review of past hydrology studies;
- develop a list and understanding of vested parties for the project (dam owners, researchers, contacts);
- look into past and current H&H studies and modeling being completed for CRB;
- develop a scope and budget for the HHA study that will address any dam safety concerns.

The scoping study resulted in an enormous amount of information about Grand Coulee Dam, the Columbia River Basin, and a large number of H&H studies already being completed that could be helpful for our project.

The scoping study resulted in a greater understanding of exactly how complicated the Grand Coulee Dam and Columbia River basins are. However, it also resulted in a realization that there has been some great H&H modeling completed within the CRB for various projects that would be helpful for the HHA. There are two important large scale studies in the CRB that helped to carve out the scope of the Grand Coulee Dam HHA; the 2014/2024 revision to the Columbia River Treaty between the U.S. and Canada (USACE, 2012) and the Columbia Basin Climate Change Scenarios Project (Hamlet et al. 2013).

The USACE has made a comprehensive effort to review the river operations, hydraulics, and hydrology involved in updating the 2014/2024 revision to the Columbia River Treaty between the U.S. and Canada (USACE, 2012). This effort has involved developing a detailed model of the Columbia River System, called the Hydrologic Engineering Center's Watershed Assessment Tool (HEC-WAT), which includes updated terrain and bathymetry, river operations, routing scenarios, climate change, and potential updates to the Columbia River Treaty. The model is capable of estimating daily and flood control operations for the Columbia River system. The model is also capable of computing Monte Carlos simulations of river routings incorporating multiple forecast, operations and hydrology scenarios. For our study, the HEC-WAT model will play an important role in allowing us to model the Columbia River system as a whole to determine how hydrology at Grand Coulee Dam will impact downstream infrastructure.

The Columbia Basin Climate Change Scenarios Project (Hamlet et al. 2013) came about through a collaboration of the Washington Department of Ecology, the Bonneville Power Administration, the Northwest Power and Conservation Council, the Oregon Water Resources Department and the British Columbia Ministry of the Environment. Through the study, 77 future climate and hydrologic scenarios were developed with respect to three future time horizons (2020s, 2040s, and 2080s) to fill an identified gap in accessible climate change information for regional watershed planning studies. The project built upon hydrologic model development from the Washington State Climate Change Impacts Assessment (Miles et al. 2010). Specifically, the Variable Infiltration Capacity (VIC) model and associated input datasets were developed at $1/16^{th}$ degree latitude/longitude spatial resolution (approximately $30km^2$ or 7400 acres per cell) over the Columbia Basin and coastal drainages within Washington and Oregon. The model was calibrated on a monthly timestep over 12 subbasins of the Columbia Basin. Hydrologic variables, such as snowpack, runoff, soil moisture, and evapotranspiration were produced at the grid resolution of the VIC model; and simulated natural streamflow (absent any management effects) was generated at approximately 300 locations throughout the basin.

An important quality of both of these modeling efforts is that they were both developed around many similar common computation points. Therefore, they can be used together to combine large scale rainfall runoff modeling with detailed hydraulic routing through the CRB.

Variable Infiltration Capacity Model

The Variable Infiltration Capacity (VIC) model (Liang et al., 1994, 1996) is a large-scale, semi-distributed hydrologic model. The model has been extensively used in hydrologic and climate change studies in the United States and internationally. The model balances both the water and surface energy budgets within the grid cell; and its sub-grid variations are captured statistically (**Error! Reference source not found.**). The VIC model is generally driven by daily inputs of precipitation, maximum and minimum air temperature, and windspeed. Additional model forcings that drive the water balance, such as solar (short-wave) and long-wave radiation, relative humidity, vapor pressure and vapor pressure deficit, are calculated in a preprocessing step within the model. The user may define the VIC model outputs; however, typical outputs on a grid cell level include runoff, baseflow, snow water equivalent, soil moisture, and evapotranspiration.

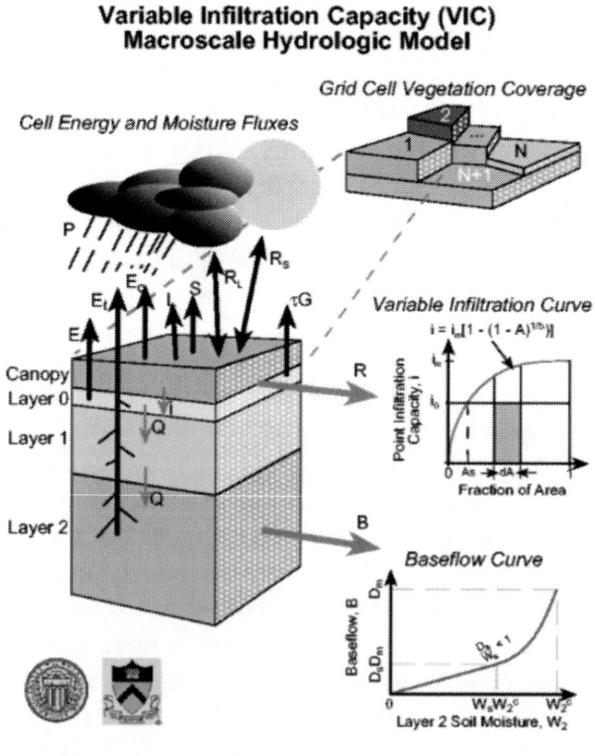

Figure1. Overview schematic of Variable Infiltration Capacity Model.

Distinguishing characteristics of the VIC model include: subgrid variability in land surface vegetation classes, sub grid variability in the soil moisture storage capacity, drainage from the lower soil moisture zone (base flow) as a nonlinear recession, and inclusion of topography that allows for orographic precipitation and temperature lapse rates resulting in more realistic hydrology in mountainous regions.

The model can be implemented in full energy balance mode or water balance mode; which has additional assumptions, but is more computationally efficient. For climate change studies like the Washington State Climate Change Impacts Assessment and the Columbia Basin Climate Change Scenarios Project, in which numerous simulations are made corresponding with climate change scenarios, the VIC model was implemented in water balance mode. To simulate streamflow, VIC results are typically post-processed with a separate routing model (Lohmann, et al., 1996; 1998a; b) based on a linear transfer function to simulate the streamflow.

Like any model, the VIC model has strengths and weaknesses, which need to be taken into consideration when applying to a specific application. Strengths include its physical representation of physical processes, such as snowpack evolution, evapotranspiration, and runoff infiltration and its computational efficiency when applied in water balance mode, among other things. One identified weakness is its limited representation of groundwater dynamics, which can be significant in watersheds with large surface water – groundwater interactions. However, the model has been shown to simulate reconstructed observed natural streamflow and other water balance parameters successfully over the Columbia River Basin.

Why Use the VIC model?

The 1/16th degree resolution VIC model configuration developed as part of the Washington State Climate Change Impacts Assessment and the Columbia Basin Climate Change Scenarios Project will be used in this study to simulate hydrology in the project area under various meteorological scenarios. There are several reasons why we chose the VIC model for this project;

- It is already calibrated for the Grand Coulee Dam and Columbia River drainage basins.
- It shares many similar common computation points with the HEC-WAT model, therefore the output from the VIC model can be used to route flows through a network of dams using the HEC-WAT model.
- Its input format allows for easy use of custom meteorology formats.
- The model has 96 years of historic forcing data from 1915-2011.
- Reclamation can develop an interface to run the model semi-stochastically for the HHA.

The model calibration was completed at the University of Washington in 2010 at the 1/16th degree scale and included 12 large watersheds in the CRB (see **Error! Reference source not found.**). Elsner et al. (2010) primarily used model parameters calibrated at the 1/8th degree scale and applied them at 1/16th degree. Further model evaluation at smaller watershed scales was conducted for streamflow locations for which naturalized flow data was available. Model calibration typically involves calibration of soil parameters, as along with parameters that define the variable infiltration and baseflow curves. Calibration may also involve parameters within the routing model (described below), as well as parameters of the snow model that operates within the VIC model. The available 1/16th degree resolution model was calibrated using the typical soil and infiltration parameters.

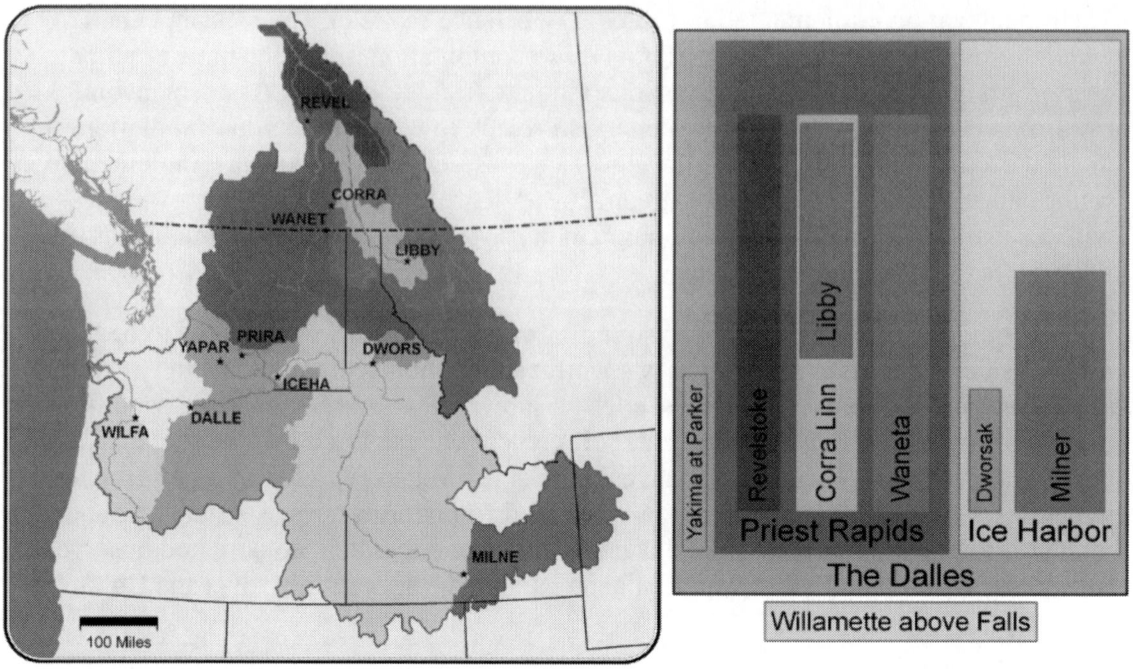

Figure 1. Overview map of Columbia River watershed and coastal drainages to Washington and Oregon, with subbasins used in calibrating the VIC model.

As previously mentioned, simulated streamflow from the VIC model is typically computed using a separate routing model. The model routes the runoff and baseflow from each VIC model grid cell to estimate daily streamflow at select sites. There are approximately 300 streamflow sites that were selected as part of the Columbia Basin Climate Change Scenarios Project. Many of these sites correspond with streamflow gage locations and major management points, including reservoirs. Similarly, many of the streamflow sites correspond to a naturalized flow dataset computed by BPA and the USACE (2015) that represents no regulation and no irrigation for a number of sites within the CRB. This dataset has been very helpful to confirm model calibration at key locations for the HHA.

One of Reclamation's main concerns is if the current calibration of the VIC model will give an accurate estimate of large floods, specifically peak discharge and volume. Currently, the model performs well estimating monthly volumes for the period of record. The model accurately captures many of the annual peaks represented in the historic record, however it also over and under estimates many others. The May, 1948 flood was off by 0.2 percent, while the June, 1974 flood was overestimated by approximately 8 percent. Further investigation is needed to determine if the current model calibration will adequately capture extremely large floods or if further calibration is needed. Similarly, there is a need to determine if the confidence limits estimated from stochastic modeling fully capture the confidence in the model calibration for annual peak flows.

Model Application

Reclamation plans to use the VIC model in a semi-stochastic manner with frequency precipitation, custom spatial and temporal distributions, and historic antecedent conditions. Reclamation is developing a stochastic interface using the Python programming language that will drive the stochastic portion of the model. The interface will be programmed to use the historic record as the range of possible antecedent conditions for the Grand Coulee Dam basin. A time period will be randomly selected based on a seasonal flood period and selected within the 96-year historic period to serve as the basis of the hydrologic event. Meanwhile, a separate random forcing will be selected and used to drive the storm event. The stochastic interface will randomly select a frequency precipitation, a temporal distribution, and a spatial distribution for a storm to be placed within the selected date in the historic record. The model will be allowed two years to run prior to this selected date to establish antecedent conditions. This will be done several thousand times to develop an adequate population to statistically estimate frequency floods for Grand Coulee Dam.

The greatest challenge for this approach is handling the large magnitude of data that will be produced during the runs. The input for each run will need to be recorded and the overall modeling will need to be repeatable. However, the post processing of the data will need to extract only the key information and delete unneeded information for each run to minimize data storage. Because each run may take up to an hour, several machines may be used to complete the modeling. The output from the VIC model will need to be formatted as input to the HEC-WAT model and methodology to use both models together is still being developed. If an automated method to run both models cannot be developed, individual HEC-WAT runs for specific VIC model output may be used.

SUMMARY

Reclamation is performing a HHA for Grand Coulee Dam for use in a Dam Safety risk analysis. Due to the extraordinarily large size and the highly regulated nature of the Grand Coulee Dam and Columbia River drainage basins, Reclamation is using a custom approach with the VIC model to model the hydrology for the project and keep the project to a manageable scope, budget, and schedule. This custom approach required additional time researching the methodology during a scoping study to ensure that the modeling approach would work. However, in addition to modeling the hydrology with the best available technology, using a model that has already been developed and calibrated should save funding and allow for a reasonable schedule. Leveraging the past and current work being completed in the CRB will allow Reclamation to complete an Issue Evaluation (IE) level HHA study for Grand Coulee Dam and help guide dam safety decisions for extreme floods.

REFERENCES

Deems, J. and Hamlet, A.F. (2010) Historical Meteorological Driving Data Set, Chapter 3 in Final Report for the Columbia Basin Climate Change Scenarios Project. Climate Impact Group, University of Washington, http://warm.atmos.washington.edu/2860/report/

Dworak, F., England, J., Elsner, M., and Klinger, R., (2015) Modelling Extreme Flood Hydrology for Grand Coulee Dam. Proceedings of papers of the 5th Federal Interagency Hydrologic Modeling Conference and the 10th Federal Interagency Sedimentation Conference, Reno, NV, April, 2015.

England, J.F. Jr., (2010) Hydrologic Hazard Analysis. Section 3 in Dam Safety Risk Analysis Best Practices Training Manual, Bureau of Reclamation, Denver, CO, 12 p.

Hamlet, A.F., P. Carrasco, J. Deems, M.M. Elsner, T. Kamstra, C. Lee, S-Y Lee, G.S. Mauger, E.P. Salathé, I. Tohver, and L.W. Binder (2010) Final Report for the Columbia Basin Climate Change Scenarios Project. Climate Impact Group, University of Washington, http://warm.atmos.washington.edu/2860/report/

Hamlet, A.F., M.M. Elsner, G.S. Mauger, S-Y. Lee, I. Tohver, and R.A. Norheim. 2013. An overview of the Columbia Basin Climate Change Scenarios Project: Approach, methods, and summary of key results. Atmosphere-Ocean 51(4):392-415, doi: 10.1080/07055900.2013.819555.

Hansen, E.M., Fenn, D.D., Corrigan, P., Vogel, J.L., Schreiner, L.C. and Stodt, R.W. (1994) Probable Maximum Precipitation-Pacific Northwest States, Columbia River (including portions of Canada), Snake River and Pacific Coastal Drainages. Hydrometeorological Report No. 57, National Weather Service, National Oceanic and Atmospheric Administration, U.S. Department of Commerce, Silver Spring, MD, 338 p.

Hosking, J.R.M. and Wallis, J.R. (1997) Regional Frequency Analysis – An Approach Based on L-Moments. Cambridge University Press, 224 p.

Liang X, Lettenmaier DP,Wood EF, Burges SJ (1994) A simple hydrologically based model of land surface water and energy fluxes for GSMs. J Geophys Res 99(D7):14,415–14,428

Liang X, Wood EF, Lettenmaier DP (1996) Surface soil moisture parameterization of the VIC-2L model: evaluation and modifications. Glob Planet Change 13:195–206

Livneh, Ben, Eric A. Rosenberg, Chiyu Lin, Bart Nijssen, Vimal Mishra, Kostas M. Andreadis, Edwin P. Maurer, Dennis P. Lettenmaier, 2013: A Long-Term Hydrologically Based Dataset of Land Surface Fluxes and States for the Conterminous United States: Update and Extensions*,#. J. Climate, 26, 9384–9392.

Lohmann, D., R. Nolte-Holube, and E. Raschke. 1996. A large-scale horizontal routing model to be coupled to land surface parameterization schemes. Tellus, 48 A, pp. 708–72.

Maurer E.P., Wood A.W., Adam J.C., Lettenmaier D.P., and B. Nijssen. 2002. A long-term hydrologically based dataset of land surface fluxes and states for the conterminous United States. Journal of Climate, 15:3237–3251.

Miles, E.L., M.M. Elsner, J.S. Littell, L.C. Whitely Binder, and D.P. Lettenmaier. 2010. Assessing regional impacts and adaptation strategies for climate change: The Washington Climate Change Impacts Assessment. Climatic Change 102(1-2): 9-27, doi: 10.1007/s10584-010-9853-2.

Snover, A.K., P.W. Mote, S.M. Capalbo, S. Eigenbrode, P. Glick, J.S. Littell, R. Raymondi, and S. Reeder. (In press). The Northwest. To appear in the Third U.S. National Climate Assessment, scheduled for release in early 2014.

Swain, R.E., England, J.F. Jr., Bullard, K.L. and Raff, D.A. (2006) Guidelines for Evaluating Hydrologic Hazards, Bureau of Reclamation, Denver, CO, 83 p.

U.S. Army Corps of Engineers (USACE) (1991) Review of Flood Control, Columbia River Basin, Columbia River and Tributaries Study, CRT-63, Final Report, June 1991, Department of the Army, U.S. Army Corps of Engineers, North Pacific Division, Portland, Oregon, 132 p. and appendix.

U.S. Army Corps of Engineers (USACE) (1997) Columbia River Basin System Flood Control Review - Preliminary Analysis Report, February, 1997, Department of the Army, U.S. Army Corps of Engineers, North Pacific Division, Portland, Oregon.

U.S. Army Corps of Engineers (USACE) (2003) Columbia River Treaty Flood Control Operating Plan, May, 2003, Department of the Army, U.S. Army Corps of Engineers, Northwestern Division, Portland, Oregon, 42 p., tables, charts, and appendices.

U.S. Army Corps of Engineers (USACE) (2009) Historic Development of 1894 Flood Hydrographs for the Columbia River Basin – Draft. U.S. Army Corps of Engineers, Seattle District. July 30, 2009.

U.S. Army Corps of Engineers (USACE) (2012). "DRAFT Columbia River Treaty 2014/2024 Review – Datasets and Models for Flood Risk Assessment Report," U.S. Army Corps of Engineers and Bonneville Power Administration, June, 2012.

U.S. Geological Survey (USGS) (1949) Floods of May-June 1948 in Columbia River basin, with a section on magnitude and frequency of floods, U.S. Geological Survey Water-Supply Paper 1080, 476 p.

U.S. Weather Bureau (USWB) (1945) Tentative Estimates of Maximum Possible Flood-Producing Meteorological Conditions in the Columbia River Basin. Hydrometeorological Report No. 18, U.S. Department of Commerce, Weather Bureau, Washington, D.C.

U.S. Weather Bureau (USWB) (1966) Probable Maximum Precipitation, Northwest States. Hydrometeorological Report No. 43, U.S. Department of Commerce, Weather Bureau, Washington, D.C., 228 p.

DYNAMIC DESERT FLOODS AT A DRY DAM

Robert A. Eichinger, P.E.[1]
Zachary Whitten, P.E.[2]
Michael Gerlach, P.E.[3]

ABSTRACT

The Inflow Design Flood (IDF) of a dam may be defined as the reservoir flood inflow magnitude selected on the basis of dam size and hazard potential classification. The IDF is used in the design or evaluation of a dam, particularly for determining the maximum crest height of the dam, auxiliary spillway capacity, freeboard, flood storage requirements, and other dam design requirements such as foundation lateral extents. Under normal every-day dam design, one IDF is usually selected for the dam and used to establish the previous design elements. However, sometimes one IDF is not always the case.

This paper presents the approach of determining the IDF for the rehabilitation of the Vineyard Road Flood Retarding Structure (FRS) project; a 5-mile long earthen dam with two auxiliary spillways and a 100 square mile watershed. Following the Arizona Department of Water Resources' (ADWR) IDF methodology of using the GIS-based Precipitation Evaluation Tool (PET), a PMP precipitation depth was derived for a 72-hour storm over the entire 100 square mile watershed. Since the ADWR PET/PMP guidelines recommend not applying a 6-hour PMP storm over watersheds greater than 50 square miles, the tool was applied to determine precipitation depths of critically centered 6-hour PMP's throughout the watershed on areas of 50 square miles or less. During initial level-pool hydrologic routing, the 72-hour PMP over the entire watershed resulted in the highest water surface elevations and therefore the probable IDF for dam design. However, using subsequent two-dimensional hydrodynamic routing with FLO-2D, separate critically centered 6-hr PMP storms, one at the north and one at the south end of the dam, resulted in more conservative water surface elevations at the ends of the dam. What resulted was a composite IDF derived from the maximum water surface elevations of the three individual PMFs. The composite IDF was used to establish the maximum dam crest elevation and the two 6-hr PMFs were used to evaluate and determine auxiliary spillway hydraulics and capacities.

INTRODUCTION

The Powerline, Vineyard Road, and Rittenhouse (PVR) Flood Retarding Structures (FRS) are earthen flood control embankment dams constructed by the Natural Resources Conservation Service (NRCS). The structures were built from 1967 to 1969 with federal assistance under the PL 83-566 Watershed Protection and Flood Protection Act of 1954.

---

[1] Senior Associate, Stantec Consulting Services, Inc. Phoenix, Arizona, bob.eichinger@stantec.com
[2] Water Resources Engineer, Stantec Consulting Services, Inc. Phoenix, Arizona, zach.whitten@stantec.com
[3] Project Manager, Stantec Consulting Services, Inc. Phoenix, Arizona, mike.gerlach@stantec.com

The Flood Control District of Maricopa County (District) operates and maintains the dams and associated facilities under an operating agreement with the NRCS. The dams are located in Pinal County, south of the City of Apache Junction, Arizona (see Figure 1).

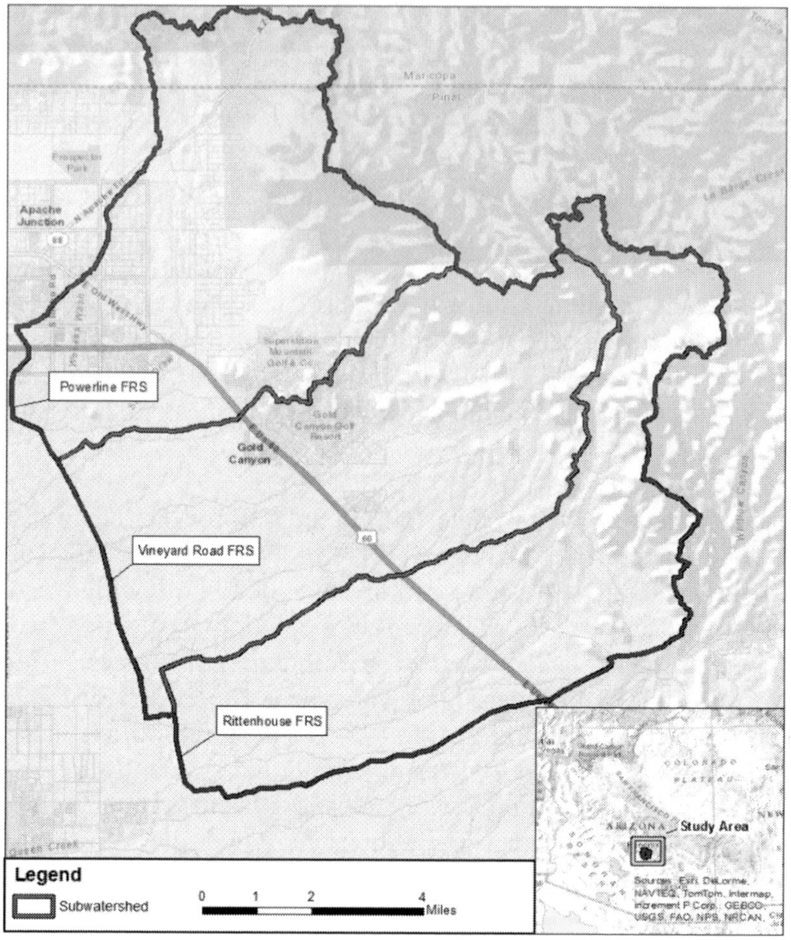

Figure 1. Project Location Map.

The three structures were designed for a 50-year life span. Each FRS has a principal spillway and an earthen open-channel auxiliary spillway, with the exception of Vineyard Road FRS, which has two auxiliary spillways. The principal spillways for both Powerline and Vineyard Road FRS discharge into the Powerline Floodway. The Rittenhouse FRS principal spillway discharges into the flood pool behind Vineyard Road FRS, which then outfalls to the Powerline Floodway.

The NRCS classifies the PVR structures as high hazard class, which means failure may cause loss of life and serious damage to homes, commercial building, utilities, and transportation infrastructure. The Arizona Department of Water Resources (ADWR), the state's dam safety jurisdictional agency, classifies the dams as intermediate size and high hazard potential. ADWR assigned an "unsafe, non-emergency, elevated risk' condition to the Powerline FRS due to the dam's close proximity to an earth fissure. Although there are no safety deficiencies assigned to Vineyard Road FRS and Rittenhouse FRS a number of potential failure modes have been identified for all three dams which must be eliminated through design and construction or adequately mitigated.

The District has identified a need to rehabilitate or replace the three dams. The District conducted a planning study to develop alternatives for the overall rehabilitation or replacement of the PVR structures and to maintain or improve upon the level of flood protection currently provided by the dams for an extended project life of 100 years. The selected alternative was subsequently modified and now consists of replacing the Powerline FRS with a constructed channel designed to the 100-year frequency storm event that discharges into the Vineyard Road FRS flood pool and raising and rehabilitating both the Vineyard Road and Rittenhouse FRS. Figure 2 (Project Overview Map) gives a visual representation of the preferred alternative.

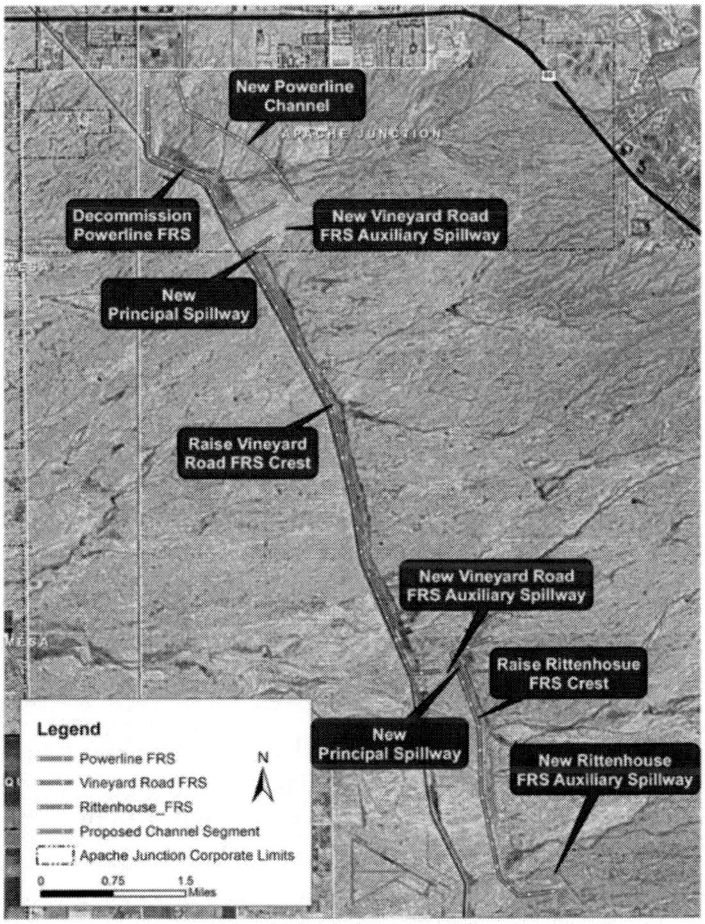

Figure 2. Project Overview Map.

EXISTING VINEYARD ROAD FLOOD RETARDING STRUCTURE

The headwaters of the Vineyard Road FRS watershed are in the Superstition Mountains to the east and the watershed drains through a series of alluvial surfaces before reaching the Vineyard Road FRS. The total contributing watershed area is approximately 52.7 square miles. Construction of the dam was completed in July 1968.

The existing Vineyard Road FRS is a homogeneous earthen embankment 28,829 feet (5.46 miles) in length, with a maximum height of 16.5 feet. The embankment crest elevation was designed at 1,581.4 feet, and the two auxiliary spillway crest elevations were designed at 1,576.7 feet to provide 4.7 feet of freeboard. The reservoir capacity is 4,310 acre-feet which provides for 178 acre-feet of sediment and 4,132 acre-feet of flood water storage. At full capacity, the sediment pool area and flood pool area would be 150 acres and 840 acres, respectively. The maximum recorded impoundment for the Vineyard Road FRS was 897 acre-feet with a flood stage of 5.9 feet on January 11, 1993. Other historical storms of note include an October 1972 storm, where the reservoir stage may have reached the auxiliary spillway crest and floodwaters may have flowed through the auxiliary spillway.

The principal spillway is an ungated, 100-foot long, 54-inch diameter RCP located at the right abutment of the FRS. The pipe is situated on a concrete cradle and three seepage cutoff collars are spaced at 20-foot intervals along the pipe. At the upstream inlet of the spillway, there is a trash rack to screen debris and the spillway discharges into a earth lined channel through an outlet structure. The downstream outlet of the principal spillway features an energy dissipater, and the principal spillway discharge capacity is 368 cubic feet per second (cfs) at the crest elevation of the auxiliary spillway.

The Vineyard Road FRS auxiliary spillways are located at the right and left abutments of the FRS. Both spillways are 300-foot wide excavated earthen open channels with a combined capacity of 12,800 cfs. The control sections of the auxiliary spillways are compacted, benched sills, located one foot above level upstream approach channels. The approach channels are on constructed bends that directs flows around the abutments of the FRS.

REGULATORY FRAMEWORK

The Arizona Revised Statutes (A.R.S.) §45-1201 assigns the responsibility for supervision of the safety of dams to the Arizona Department of Water Resources (ADWR). ADWR's dam safety procedures are promulgated in the Arizona Administrative Code – Title 12, Chapter 15, Article 12. Article 12 of the Code defines the IDF as "the reservoir flood inflow magnitude selected on the basis of size and hazard potential classification for emergency spillway design requirements of a dam". The IDF requirements for the design of a high hazard potential dam for determining the spillway minimum capacity (for all dam size classifications) is defined as "for a high hazard potential dam, the applicant shall design the dam to withstand an inflow design flood that varies from 0.5 PMF to the full PMF, with size increasing based on the persons at risk and potential for downstream damage". The District has selected the full PMF as the IDF for the rehabilitation of the Vineyard Road FRS.

ARIZONA STATEWIDE PMP STUDY

ADWR completed a statewide Probable Maximum Precipitation Study for Arizona (PMP Study) (Applied Weather, 2013). This study provides PMP values for any drainage basin within Arizona. The PMP values are used in the computation of the Probable Maximum Flood (PMF). In addition, the temporal distribution of that PMP rainfall by each storm type is provided. Results of the statewide PMP study supersede the Hydrometeorological Report (HMR) 49 for application in Arizona.

APPLYING ARIZONA STATEWIDE PMP STUDY

Determining the IDF for a watershed can be broken down into three basic steps: determine the amount of rainfall that can realistically be expected on the watershed, analyze the hydrology of the rainfall to determine how much runoff the rainfall produces then perform hydraulic analysis to determine effect of the excess runoff. For the first step for determining the amount of rainfall is accomplished by using the Arizona Statewide PMP Study. PMP estimates for the watershed were derived from data published in the PMP Study. That data and procedures for calculating the PMP depths are implemented in the GIS based application, Precipitation Evaluation Tool (PET). The PET provides depth-area reduced PMP estimates for the 6-hour Local Storm, 72-hour Winter Storm and 72-hour Tropical Storm.

The tool will give a result for all three storm types regardless of the size of the watershed. This is important to note because the ADWR PMP Study states that the maximum storm size of a local 6-hour PMP cannot exceed 50 square miles. The preferred alternative replaces the Powerline FRS with a 100-year channel that discharges into the Vineyard Road FRS impoundment area. This increases the total contributing watershed area for the Vineyard Road FRS from 52.7 square miles to approximately 100 square miles. Therefore, the PET tool will calculate a PMP estimate for all three storms but the Local 6-hour PMP over the entire 100 square miles was not considered.

Output for this tool is provided in a gridded format with a cell size of 90 arc-seconds (approximately 1.5 miles). From the grid data, a PMP depth is estimated by averaging the depth of each grid within the watershed. Also provided by the tool is the time distribution for each of the storms is also provided by the PET tool. Figure 3 illustrates an example of the output of the PET for the 72-hour Tropical Storm.

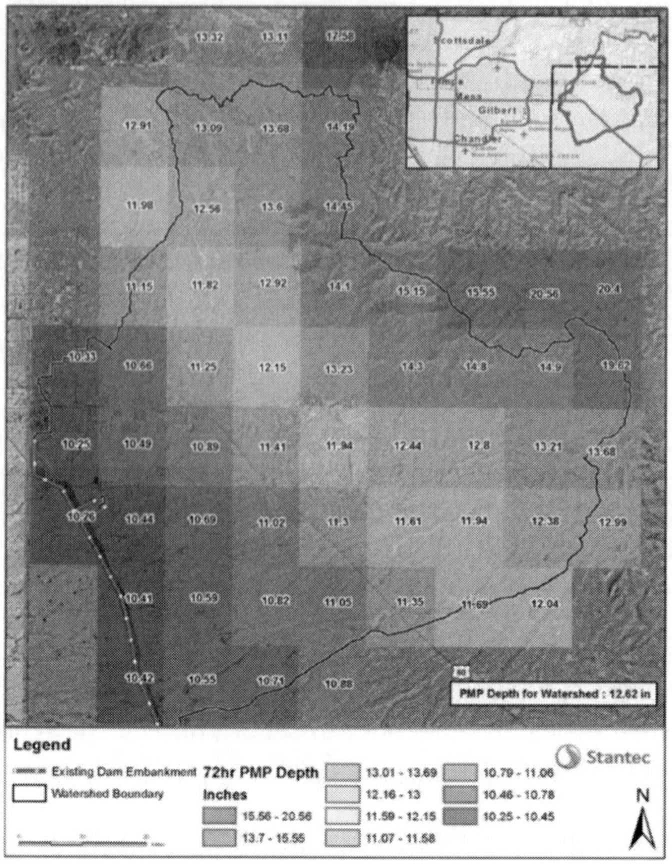

Figure 3. PET 72-hr Tropical Storm Output.

Though the 6-hour PMP over the entire 100 square mile watershed would not ordinarily be considered, this does not preclude the possibility of applying the 6-hour PMP to smaller sub watersheds (with an upper limit of 50 square miles) within the 100 square mile area. This is important to consider given the size of the total watershed, the non-uniform dynamic inflow characteristics into the Vineyard FRS impoundment area and the presence of the auxiliary spillways located 5 miles apart at each end of the dam. To investigate this possibility, multiple critically centered 6-hour PMP storms centered over portions of the watershed totally 50 square miles were evaluated. After initial calculations, two critically centered 6-hour PMP storms were selected to be fully investigated. The two sub watersheds, one centered over a northern portion of the watershed (North 6-hour PMF Watershed) and the other centered over the original Vineyard Road Watershed (South 6-hour PMF Watershed), were selected to determine critical conditions at both ends of the dam. Table 1 lists the results of the PET tool applies to the Vineyard Road FRS watershed of 100 square miles. Figure 4 shows the watershed map for the two 6-hour PMP storm areas.

Table 1. Vineyard Road FRS PMPs.

PMP Storm Type	PMP Depth (in)
6-Hour Local Storm	N/A
72-Hour Tropical Storm	12.62
72-Hour Winter Storm	9.13
North 6-Hour Local Storm	8.45
South 6-Hour Local Storm	8.27

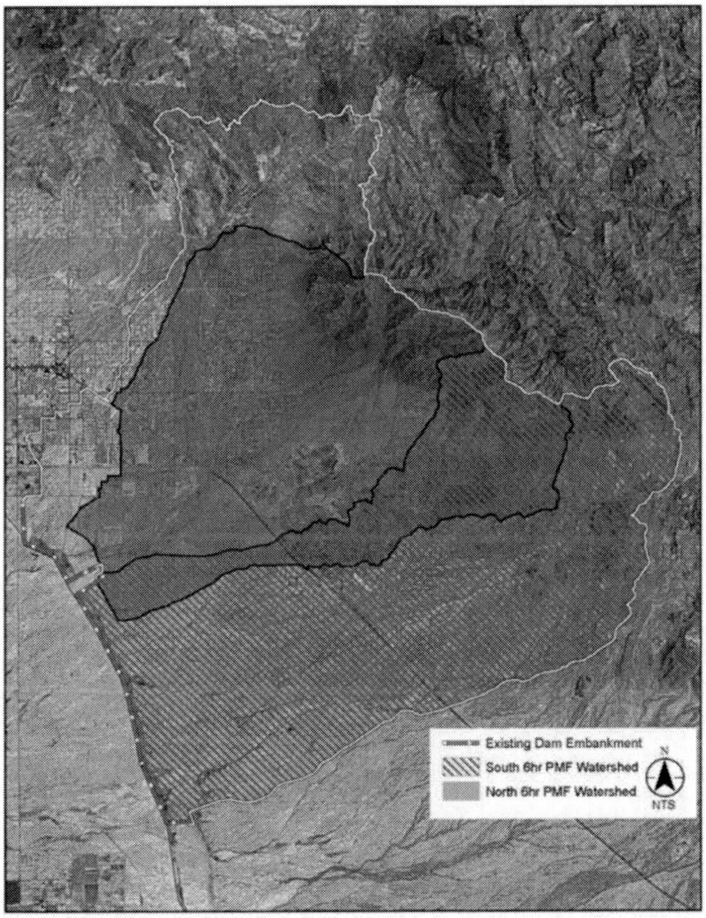

Figure 4. Critically Centered 6-hour Storm Watershed Map.

ROUTING PMF THROUGH IMPOUNDMENT

The PMP depths and time distributions derived from the PET tool together with the District's hydrology methodology were coded into the HEC-1 program (preferred hydrology model for the District) to develop rainfall-runoff models for the combined watershed for both existing and future land use conditions. The HEC-1 model results include runoff hydrographs for each PMP storm type. As noted previously the Vineyard Road FRS is over five miles in length and with two auxiliary spillways, one each at each end of the dam, and thus it was determined that a single inflow location would not adequately model the hydrodynamics of the watershed characteristics. Therefore, the watershed was subdivided into multiple subareas to produce inflow hydrographs at multiple locations along the length of the dam. Figure 5 shows the multiple inflow locations. The inflow hydrographs were dynamically routed using the FLO-2D model, an objective of which was to determine the maximum water surface elevations and determine the PMF water surface profiles along the length of the dam.

Table 2 shows the peak discharge and runoff volumes for each watershed from the HEC-1 model results for the Tropical 72-hour PMP. Tables 3 and 4 show the peak discharges and runoff volumes from each watershed for the North and South 6-hour PMP, respectively. Each table is set up to show how much runoff the original two watersheds (Powerline and Vineyard Road) produce to help illustrate the dynamics of the watershed. It should be noted that that the critically centered North 6-hour storm covers both watersheds (approximately 72% over Powerline and 28% over Vineyard Road) and the South 6-hour storm is over just Vineyard road).

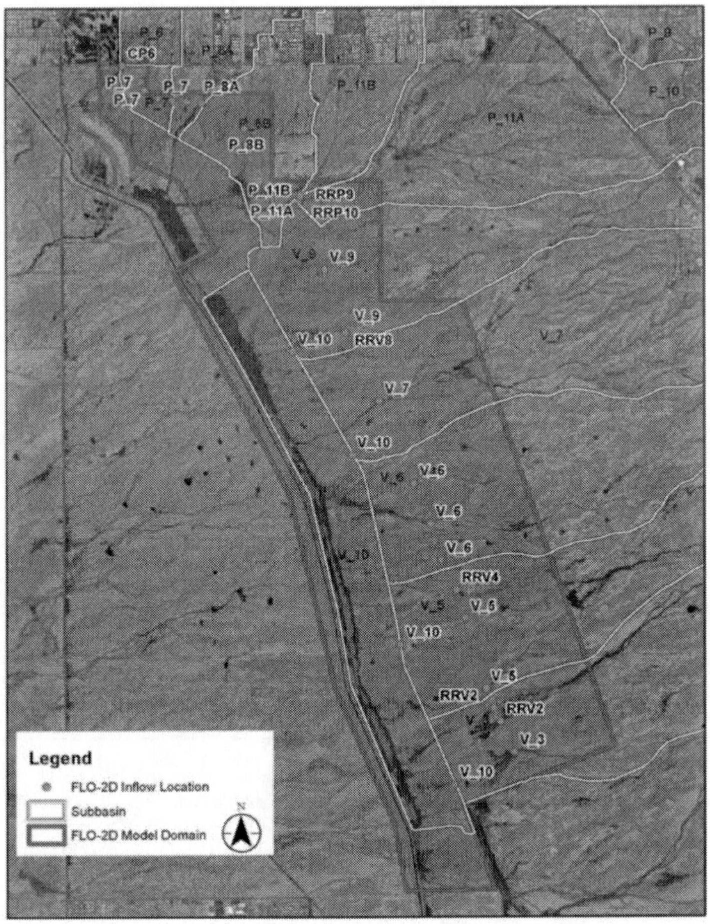

Figure 5. FLO-2D Model Inflow Locations.

Table 2. HEC-1 Model Results for Tropical 72 hour Storm.

Original Watershed	**Watershed Area sq mi**	**HEC -1 Results**	
		Peak Discharge cfs	**Runoff Volume acre-ft**
Powerline	45.0	16,910	14,050
Vineyard Road	53.7	21,740	16,730
Total into Vineyard Road Impoundment	98.7	38,650	30,780

Table 3. HEC-1 Model Results For Critically Centered North 6-hr Storm.

Original Watershed	**Total Watershed Area sq mi**	**Area of Storm within Original Watershed sq mi**	**HEC -1 Results**	
			Peak Discharge cfs	**Runoff Volume acre-ft**
Powerline	45.0 33.0		63,570	11,350
Vineyard Road	53.7	13.0	21,880	4,670
Total into Vineyard Road Impoundment	-	46.0	84,240	16,020

Table 4. HEC-1 Model Results For Critically Centered South 6-hr Storm.

Original Watershed	**Total Watershed Area sq mi**	**Area of Storm within Original Watershed sq mi**	**HEC -1 Results**	
			Peak Discharge cfs	**Runoff Volume acre-ft**
Powerline	45.0	0.0	N/A	N/A
Vineyard Road	53.7	53.7	74,330	18,980
Total into Vineyard Road Impoundment	-	53.7	74,330	18,980

The results show that the peak discharges for the critically centered 6-hour PMP storms are almost double that of the 72-hour tropical storm (2.3 and 1.9 times for the North and South, respectively). However, the runoff volume for the 72-hour PMP was 1.95 times and 1.65 times higher compared to the North 6-hour and the South 6-hour, respectively.

The IDF for Vineyard Road cannot be determined based soley on the hydrology results summarized in table 2 through 4. The maximum water surface elevations at the dam are a combination of peak discharges, runoff volumes and hydrodynamic characteristics. Each storm was then dynamically routed along the dam.

HYDRODYNAMIC FLO-2D RESULTS

The storm specific PMF inflow hydrographs were routed through the Vineyard Road FRS flood pool using the hydrodynamic flood routing program FLO-2D. The initial routing conditions started with an empty storage pool and a full sediment pool. Figure 6 shows the 72-hour tropical PMF maximum water surface elevations from FLO-2D within the study area and Figure 7 shows the maximum water surface profiles along the face of the dam for the three PMP storms.

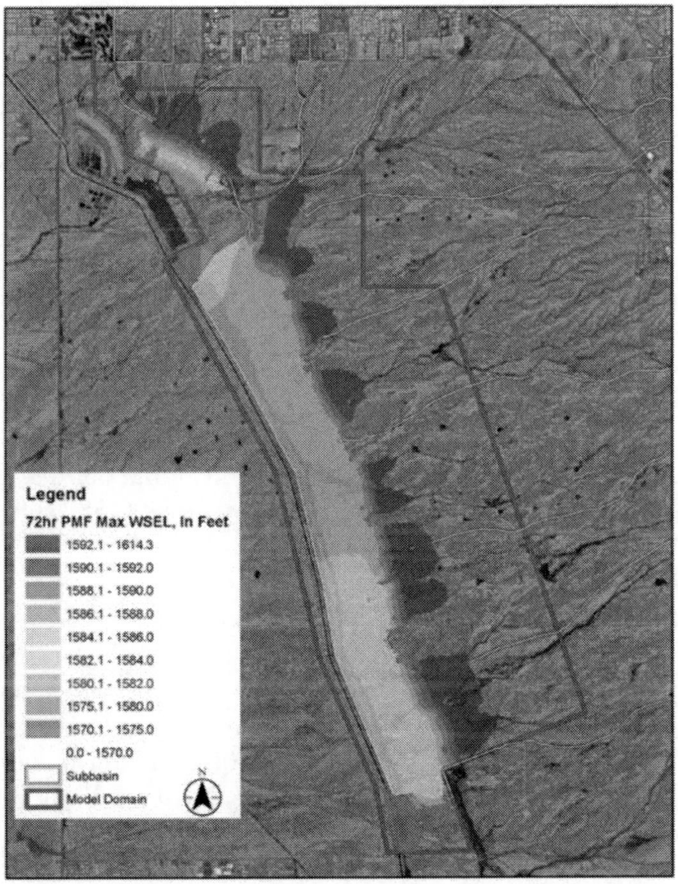

Figure 6. 72-hour Tropical PMF Maximum Water Surface Elevations

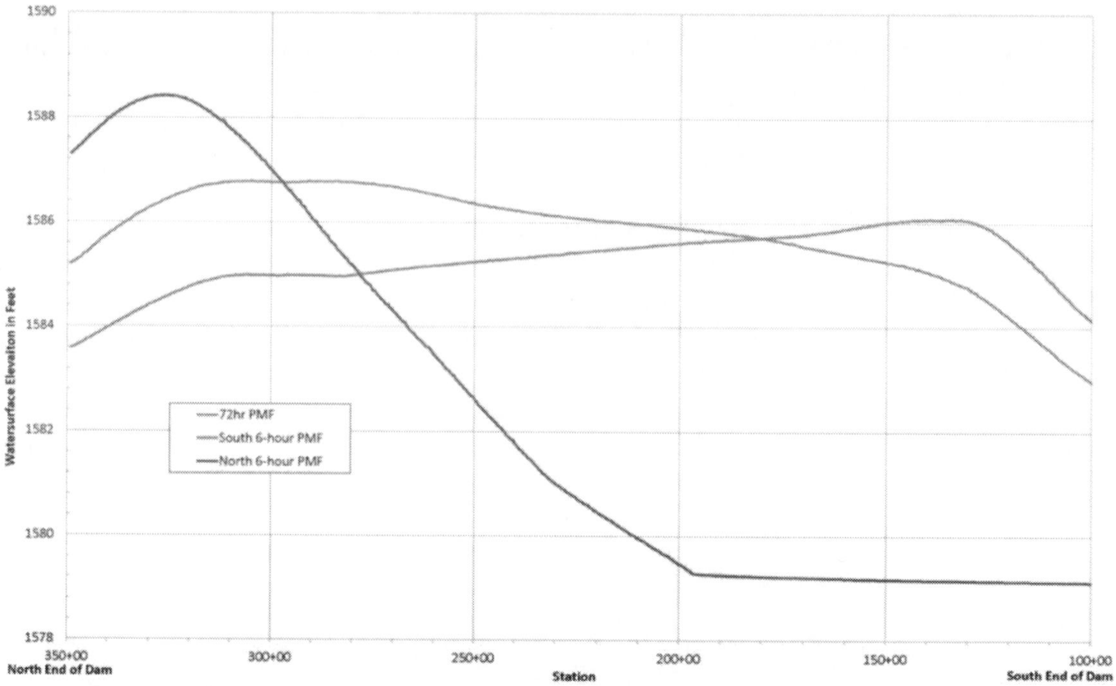

Figure 7. PMF Maximum Water Surface Profiles

The maximum PMF water surface elevations vary along the length of the dam as shown in Figure 7. This is due to a combination of several factors: (1) concentrated inflow from the 45 square mile Powerline Watershed, (2) the non-uniform distribution of inflow along the 5 miles length of dam from the Vineyard Road watershed, (3) the five mile length of the Vineyard Road FRS, and (4) the presence of two auxiliary spillways at opposite ends of the dam with different spillway crest lengths (900 feet and 600 feet for the north and south auxiliary spillways, respectively) and crest elevations (1582.0 feet and 1580.0 feet for the north and south auxiliary spillways, respectively).

The design of the auxiliary spillway crest elevations are governed by the more conservative criteria from NRCS and District criteria. The NRCS criteria states that the auxiliary spillway crest elevation is to be set to the maximum water surface elevation of the principal spillway hydrograph (PSH) while the District criteria states that flood retarding structure must provide protection for the 100-year storm event. The PSH was developed using the NRCS SITES model which uses level pool routing. The 24-hour 100-year storm was routed using FLO-2D. Using SITES, the resulting PSH water surface elevation was 1579.8 ft. Using FLO-2D for the 24-hour, 100-year storm event, at the north end of the dam, the maximum water surface elevation was 1581.5 feet and at the south the water surface elevation was approximately 1579.2 feet. Therefore, the north auxiliary spillway crest is governed by the 24-hour, 100-year storm while the south auxiliary spillway was governed by the PSH. Figure 8 below shows the results of the 24-hour, 100-year storm and the SITES PSH water surface elevations.

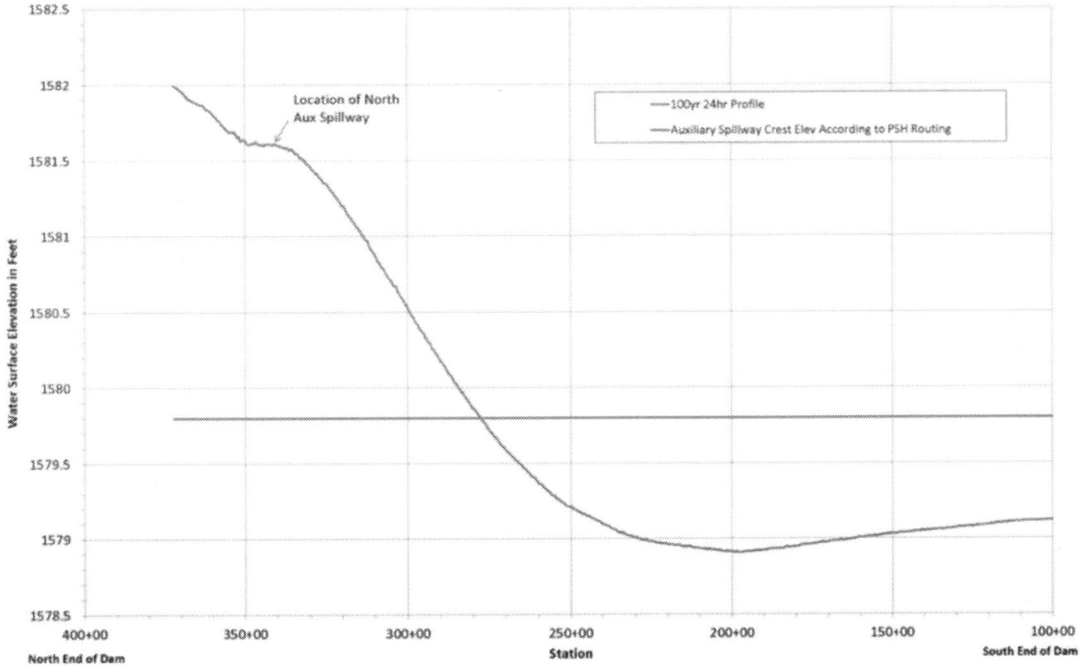

Figure 8 . Water Surface Profiles for 100-year, 24 Hour and PSH.

For the three PMF profiles, each PMF has a segment along the length of the dam in which that PMF would be considered the maximum water surfaces. As shown in Figure 7, the South 6-hour PMF has higher water surface elevations at the south end of the dam, the North 6-hour PMF has higher water surface elevations at the north end and the 72-hour PMF tropical storm in the middle segment. Therefore, none of the three PMFs could be considered a single IDF for the Vineyard Road FRS. Consequently, all three PMFs would be considered the IDF and a composite PMF profile would be used to establish the crest elevation for the rehabilitation of the Vineyard Road FRS. Figure 9 shows the design composite PMF profile which is used to determine the dam crest elevations along the length of the dam.

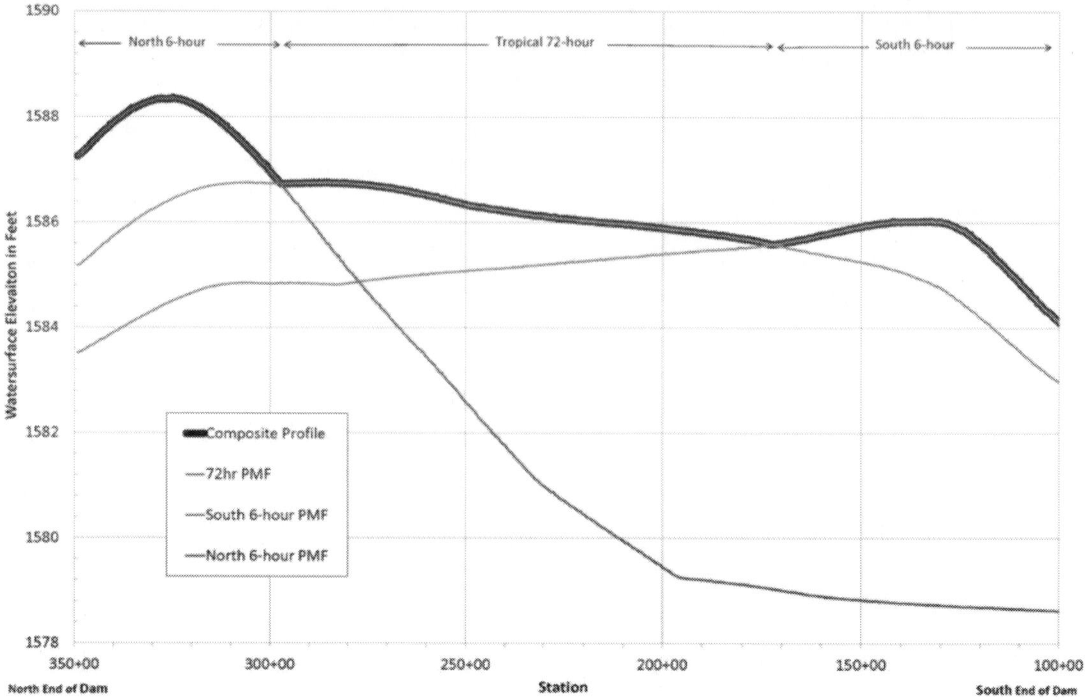

Figure 9. PMF Composite Maximum Water Surface Profile.

While the composite PMF was used to determine the dam crest elevations, the individual PMFs were examined to determine which governed the design of the auxiliary spillways. As shown in Table 5 below, the critically centered 6-hour storms produced higher peak outflow discharges than the tropical 72-hour. The North 6-hour produced a peak discharge nearly 2 times higher than the Tropical 72-hour on the north auxiliary spillway and the South 6-hour produced a peak discharge 1.7 times larger on the south auxiliary spillway. Figures 10 and 11 illustrate the outflow hydrographs for the north and south auxiliary spillways, respectively.

Table 5. Vineyard Road FRS Auxiliary Spillway Peak Discharges.

PMF Storm	Peak Outflow, cfs	
	North Auxiliary Spillway	South Auxiliary Spillway
72-hour Tropical	18,420	7,780
North 6-hour	36,390	No Spill
South 6-hour	5,660	12,910

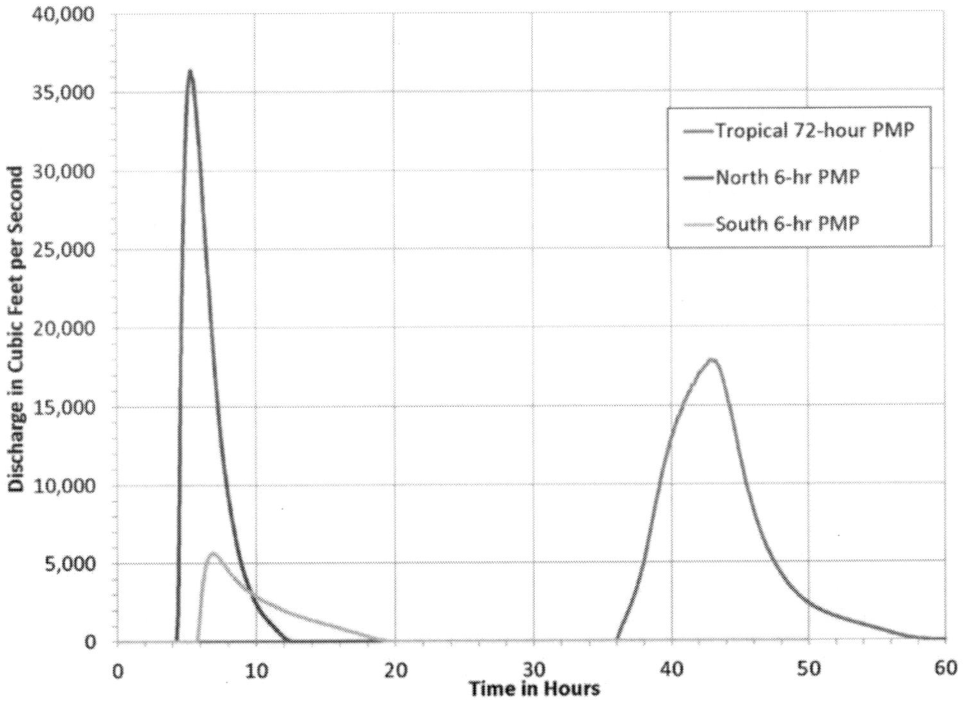

Figure 10. North Auxiliary Spillway Hydrographs.

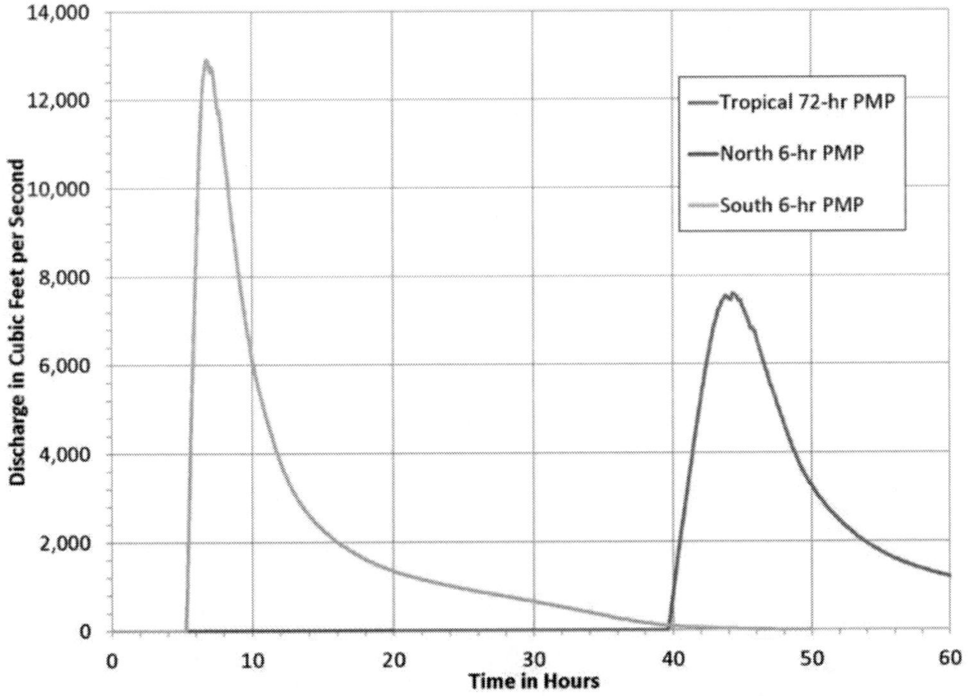

Figure 11. South Auxiliary Spillway Hydrographs.

CONCLUSION

Under every-day considerations, one would anticipate a single controlling inflow design flood for the design of a dam. However, as this paper demonstrates, the Vineyard Road Flood Retarding Structure has three PMF storms with controlling water surface elevation segments along the length of the structure. At the north end, the North 6-hour produces the highest peak water surface elevations. In the middle, the 72-hour PMF produces the highest and at the south it is the South 6-hour PMF. Due to this dynamic situation, the maximum water surface elevations that are used to determine the crest elevations of the dam is a composite of the three PMF storms.

The selection of the design PMF for each of the two auxiliary spillways was determined by dynamic routing. The north auxiliary spillway design was governed by the North 6-hour PMF and the south auxiliary spillway by the South 6-hour PMF.

In the future, if a similar project within Arizona occurs on a watershed that is greater than 50 square miles, it is important that multiple storms be investigated. As this paper demonstrates, due to the dynamic flooding in deserts, it is likely that such a watershed will have more than a single controlling storm event.

CULTIVATING INGENUITY AND SAFETY IN ALABAMA: THE TAMING OF LAKE OGLETREE RESERVOIR

James R. Crowder, P.E.[1]
Brian M. Crookston, PhD, P.E.[2]
Bradley T. Boyer, P.E.[3]
J. Tyler Coats, P.E.[4]

ABSTRACT

Labyrinth spillways are commonly implemented in dam rehabilitation projects to satisfy design flood criteria due to their hydraulic performance, flexible design, and cost effectiveness. However, their various complex hydraulic behaviors in some circumstances can present unique challenges to designers. This paper presents a unique project located in the State of Alabama, showcasing the various challenges and the approach selected for the rehabilitation of the Lake Ogletree spillway. With a myriad of modifications to the dam and spillway since 1941, and potential flooding to surrounding infrastructure from spillway releases, an innovative approach to the layout of the spillway was required. The rehabilitation includes a large, multi-stage labyrinth weir (19-ft maximum weir height) and saddle areas designed for overtopping. Hydraulic challenges included generally matching the existing outflow hydrograph for more frequent storm events while still passing the spillway design storm. The design also considered nappe vibration, a phenomena that can produce acoustic pressure waves that affect nearby structures. This paper also presents results from a detailed CFD model of the spillway and basin.

INTRODUCTION

The Water Works Board of the City of Auburn (AWWB) plans to upgrade the spillway systems for an existing reservoir dam to improve the performance of the dam and facilitate more convenient operations and maintenance of the dam and reservoir. The existing dam and spillway are located in the City of Auburn, Alabama and approximately 2 miles southeast of the intersection of Ogletree Road and Wrights Mill Road. The subject dam impounds Chewacla Creek, creating an approximately 300-acre reservoir known as Lake Ogletree. The reservoir serves as the primary source of drinking water for the AWWB. According to information obtained from the AWWB, the subject lake and dam are approximately 75 years old. Since its original construction, the dam and spillway have been modified to increase storage capacity.

Lake Ogletree Dam is an earthen embankment approximately 50 feet in height and approximately 250 feet in length across the crest. The dam has two spillways, one of

[1]Schnabel Engineering, Alpharetta, GA, (770) 781-8008, jcrowder@schnabel-eng.com
[2]Schnabel Engineering, West Chester, PA, (610) 696-6066, bcrookston@gmail.com
[3]Schnabel Engineering, Alpharetta, GA, (770) 781-8008, bboyer@schnabel-eng.com
[4]Schnabel Engineering, Alpharetta, GA, (770) 781-8008, tcoats@schnabel-eng.com

which is no longer in service. A reinforced-concrete vertical riser overflow tower with a nearly horizontal outlet conduit exists in the dam embankment. This spillway is no longer in service. The other spillway is a concrete overflow section founded on residual soil to the west of the dam proper. Approximately 525 feet of wooden flashboards have been constructed across the crest of the spillway. These flashboards can be used to raise and lower the lake level approximately 4 to 5 feet when needed. During normal operating conditions, the subject dam impounds a reservoir with a surface area of approximately 300 acres.

Dam Rehabilitation Site Challenges

Existing embankment dams may have a variety of challenges depending on location, site orientation, local environmental factors, and operation and maintenance concerns. The following paragraphs describe distinct features of the project that presented challenges to the design of the modifications to the Lake Ogletree Spillway.

Water Supply. Lake Ogletree serves as the primary water source for the city of Auburn, Alabama. Seasonal demand from the reservoir ranges from 5 million gallons of water per day (MGD) to 10 MGD. This demand includes water for agricultural purposes, as well as to support Auburn University and its 27,000 students. Further, an endangered freshwater mussel population has been identified in Chewacla Creek immediately downstream of the existing embankment. The AWWB is required to maintain a consistent flow of 2 MGD from the reservoir to support the mussel population.

Non-linear Weirs. Spillways that feature a crest length that exceeds the width of the spillway are known as non-linear weirs (Fig. 1) and are commonly implemented in dam rehabilitation projects as they can effectively meet project requirements, including hydraulic and spillway construction cost objectives (Crookston et al. 2013). The additional crest length can increase discharge for a given upstream pool elevation, relative to a linear weir of equivalent crest shape. The crest can be placed in an arc (e.g., horseshoe weirs) or folded (e.g. box-drop inlets, labyrinth weirs, piano key weirs) in plan view to accommodate site constraints. Site conditions at Lake Ogletree Reservoir are particularly challenging (limitations to raising the dam and overtopping protection) favoring an upgrade to a non-linear weir.

Figure 1. Arced weir (a), box-inlet drop spillway (b),
labyrinth weir (c) and piano key weir (d)

Labyrinth and piano-key weirs have the potential to increase unit discharges by over six times, relative to linear weirs. However, more efficient spillways should be viewed as part of the reservoir system, which naturally includes the reservoir, adjacent areas, and potential impacts due to increased outflows from non-linear weirs. Dabling et al. (2013, 2015) conducted experiments where the outflow hydrograph of a labyrinth weir was modified by incorporating multiple crest elevations. It was demonstrated that a superposition technique to develop the spillway rating curve was acceptable for design; a highly efficient weir could be designed using empirical published methods to match higher frequency storms (e.g., 5-yr, 25-yr) while still providing adequate conveyance for the spillway design flood (e.g., PMF).

Free-overfall spillways such as overflow gates and weirs can display nappe vibration (Naudascher and Rockwell 2005), a phenomena that occurs at relatively low flows (flow depths less than about 0.5 ft) and is characterized by an oscillating nappe (appears as horizontal banding) and low-frequency acoustic pressure waves (Crookston et al. 2014). These pressure waves do not negatively affect the spillway and are not harmonic to structural frequencies. However, the oscillations can reach decibel levels that are unpleasant and the acoustic pressure waves can rattle windows and walls of nearby structures. This behavior has been observed in fountains, linear weirs, overflow gates, and labyrinth spillways, to name a few; it is not limited to non-linear weirs. Fundamental understanding and design guidance regarding nappe vibration for labyrinth spillways and other non-linear weirs is limited (Lodomez et al. in press).

REGULATORY CLIMATE IN ALABAMA

At the present, the State of Alabama is the only state that has no regulations pertaining to the design and construction of dams, beyond those required for environmental permitting related to the construction work. No state entity exists that assures proper planning, design, and construction review of dams to protect the safety and welfare of Alabama citizens and their property. No entity requires action from dam owners when safety concerns arise.

With no state dam safety regulations governing the design and construction of dams, there is no local standard of practice and, therefore, no prevailing standard of care for the designer. Criteria used for analyses and design had to be carefully and collaboratively considered by both the owner and engineer in terms of risks, consequences, and project costs. However, it should be recognized that regulations are many times enacted to induce people and organizations to do what is right, and are therefore directed at those that wouldn't otherwise do so. The City of Auburn should be commended for their focus (in an environment without dam safety regulations) on providing a safe and dependable water supply for their customers, while minimizing the potential for their project to have an adverse impact on others.

Auburn took action in an admirable way. The City internally embraced high standards of safety by adopting the regulations from its close neighbor, Georgia, and going a step further by committing to a spillway design that can safely pass the full PMP. As the owner of Lake Ogletree Dam, the City of Auburn is setting an example for others to follow, both within and outside of Alabama. It is hoped that their maturity, sense of duty, and prudent judgement add to the efforts of others working diligently to establish a dam safety program in the State of Alabama.

SITE AND PROJECT ORIENTATION

Site Description

The origination of Chewacla Creek, located southeast of Opelika, AL near Old Columbus Road, is about 7.7 miles northeast/upstream of the dam site. The creek flows generally from northeast to southwest through the project site. The stream, in the vicinity of the dam site, averages 15 feet in width and flows from the east to the west. The area around the existing reservoir and dam site are primarily forested with mature pine and hardwood trees and some undergrowth. Chewacla Creek, as it makes its way through Lake Ogletree, generally flows from the north to the south, but makes an abrupt turn to the east at the location of the existing auxiliary spillway located on a natural saddle.

The existing embankment (Figure 2) is well-maintained with good ground cover, and its centerline runs in a north-south direction. The aforementioned saddle is situated at a 90 degree angle to the earth embankment between a knob of ground that forms the right contact point of the embankment fill and the right abutment. The centerline of the saddle runs in an east-west direction, parallel to the flow of Chewacla Creek. The majority of the crest and downstream portion of the saddle is paved with concrete, and a series of removable, wooden flashboards are situated along the centerline of the crest of the saddle. These flashboards set the normal pool elevation of the reservoir.

Figure 2. Lake Ogletree Reservoir – spillways and embankments (existing)

Portions of the pavement are in poor condition as indicated by cracks and spalls in the concrete. There is evidence of the existence of seepage between the top of residual ground and the bottom of the pavement. Broken concrete pipe, rubble, and backfill concrete have been randomly placed at various locations along the toe of the saddle between the paved spillway and Chewacla Creek in an attempt to reduce erosion. The area between the paved spillway and Chewacla Creek is forested with medium to large trees and some undergrowth. Ground elevations at the dam site range from 446 ft-MSL in the creek channel to 499 ft-MSL and higher at the crest of the dam and its abutments.

Project Description

Rehabilitation of Existing Principal Spillway Riser and Conduit

The existing principal spillway riser and conduit will be rehabilitated to act as a lake drain in the event the reservoir water level needs to be lowered, as well as to provide a means to supply water to Chewacla Creek immediately downstream of the dam for environmental purposes.

During periods of high water elevations, the current riser structure becomes completely inundated. An access platform will be added to the top of the structure to facilitate access to the riser when the water in the reservoir is elevated. Currently, the riser is effectively sealed and empty, and water only enters the structure by leaking through an existing access hatch in the top of the structure.

The riser structure has two existing gates, one near the middle of the structure and one near the base of the structure. There are plans to rehabilitate the higher of the two gates, decommission the existing low-level gate, and add a new low-level gate at the upstream end of the principal spillway conduit on the downstream, inside face of the structure. After the proposed renovations are completed, the existing riser will be allowed to fill with water thus removing hydraulic loading on the structure during normal operating conditions.

In addition to the gate rehabilitation, the existing conduit will be slip-lined with a smaller diameter ductile iron pipe and the annular space between the new and existing conduits will be grouted. A control valve will be installed on the downstream end of the proposed conduit for emergency purposes, should access to the existing riser be interrupted or other problems with the existing riser be encountered.

Construction of Proposed Labyrinth-Crested Chute Spillway

The majority of the construction effort for the project will focus on rehabilitating the spillway system located on the saddle. The current spillway system has several deficiencies and is difficult to operate and maintain. The existing 525-foot wide flashboard system and concrete-pavement spillway will be demolished and replaced by a proposed 218-foot wide reinforced-concrete, labyrinth-crested chute spillway. Construction of the structural spillway will involve the excavation of residual materials located in the right end and abutment of the saddle. In addition to the placement of reinforced concrete, construction of the spillway will include the installation of a lower permeability soil-bentonite blanket over the spillway inlet channel, a drainage system beneath the spillway to collect seepage and reduce the migration of fine-grained soils downstream, and outlet channel protection in the form of large diameter riprap.

Additional design features associated with the structural spillway include a bridge over the spillway. The bridge will be a highway-rated structure supported by abutments within the spillway sidewalls and two intermediate piers immediately upstream of the labyrinth wall. The bridge will facilitate access from the northern (right) abutment to the southern (left) portions of the dam and saddle, primarily for maintenance purposes. Another design feature includes two, 30-inch diameter plug valves that will be installed with associated piping and trash guard to facilitate control of water during construction and to provide a way to lower the water surface level in the reservoir during normal operations, if necessary. The valves will be located on either side of the structural spillway in the vicinity of the downstream tie-out of the labyrinth walls. Vaults will be constructed around the valves to facilitate access for maintenance purposes, and the vault covers will provide safe footing for operating the valves. The pipe fittings necessary to convey flows from the upstream side of the labyrinth wall to the downstream side will be encased in concrete after their installation.

REHABILITATION CHALLENGES AND APPROACHES

Summary of Challenges

Maintain reservoir above minimum operational elevation during construction

Due to its importance as the primary water source for the city of Auburn, the reservoir must remain at an acceptable operational level to provide water to the city's customers. After consulting with city officials, an operating pool level schedule during construction was developed and incorporated into the construction specifications for the project.

The current normal pool elevation for Lake Ogletree is 486 ft-MSL. Historically, the lake level fluctuates throughout the course of the year. During the fall and winter months (September through January), the lake level falls to an average elevation of 476 ft-MSL. During the spring and summer, the lake is operated at its normal pool elevation. As discussed previously, wooden flash boards are installed on the existing concrete spillway after the winter and removed at the end of the summer to manage the operating pool elevation of the reservoir.

Based on this historic information, the operating pool level schedule involves the addition and removal of the flash boards at similar points in time. During construction in the fall and winter months, the maximum water surface in the reservoir will be 481.1 ft-MSL (the approximate elevation of the top of the existing concrete spillway with all flash boards removed). After the winter when demand increases, the flash boards will be installed and the lake must be maintained at elevation 484 ft-MSL from May 1st until August 31st.

Provide minimum outflow from reservoir during construction

The presence of endangered freshwater mussels immediately downstream of the existing dam in Chewacla Creek requires a constant minimum flow be released from the reservoir, including during construction activities. Construction specifications have been written such that an uninterrupted flow of water at the downstream toe of the existing dam embankment is maintained at all times during construction. This uninterrupted flow must be 2 MGD (3.1 cubic feet per second).

Control water through the project site

Due to the need to maintain the reservoir pool at a relatively high elevation during construction, control of water was critically considered during the development of the plans for the proposed spillway rehabilitation. The watershed size at the existing dam is approximately 33 square miles. In developing the plans and determining the location of the proposed spillway, the design team decided to push the proposed spillway into the right abutment of the saddle area to take advantage of the lowest stage of the existing concrete spillway during construction. The low stage has a width of 225 feet that can be utilized to control water during construction.

Control of water is typically written into the construction specifications as the sole responsibility of the contractor. On this project, due to the critical nature of the construction work and the need to maintain the operational capacity of the reservoir, the owner elected to require the contractor to design and implement diversion measures that would store and discharge stormwater runoff resulting from the 100-year, 24-hour storm event without water from the reservoir entering the working areas. Furthermore, prospective bidders had to demonstrate previous successful experience in controlling surface water from large drainage basins through dam construction sites. If such experience could not be provided, the prospective bidder was not allowed to bid on the project.

Rehabilitating the Dam

Embankment Dams

Although no modifications to the existing dam embankment are to be implemented under the current scope of engineering work, a new embankment will be constructed in the vicinity of the existing auxiliary spillway, adjacent to the proposed chute spillway, in order to remove the remaining portions of the existing auxiliary spillway from service. The primary saddle embankment will be approximately 11 feet in height and constructed to an elevation of 497.2 ft-MSL. The saddle embankment will have side slopes ranging from 2H:1V to 4H:1V, wave protection riprap on the upstream slope, a concrete cutoff wall located in the foundation of the embankment, a stormwater drainage system in the downstream slope, and a subsurface/internal seepage control drainage system within the downstream slope. The concrete cutoff wall will be located along the approximate centerline of embankment and will extend vertically from the approximate 100-year reservoir flood elevation down into existing residual soils.

Rehabilitating the Spillway

Active-control spillways, such as gates, allow releases from the dam to be regulated. However, a mechanical system adds levels of complexity associated with an operator, operation procedures, monitoring, and system maintenance. As a result, an active-control spillway can increase risk to the owner and the public. When possible, a passive-control spillway is preferred due to its reliability and simplicity.

Lake Ogletree presented a significant challenge: it was undesirable to increase peak outflows for more frequently occurring storms due to potential flooding impacts. Therefore, a traditional labyrinth weir capable of passing the design flood was not feasible. To avoid a more costly or gated spillway, an innovative approach was taken by restricting flows from smaller floods to portions of the labyrinth via segments with different crest elevations. This staged approach allowed for the necessary modification of the spillway outflow hydrograph while still providing an economical hydraulic structure that meets discharge capacity requirements.

The hydraulic design method by Crookston and Tullis (2013a,b) was used to estimate labyrinth discharge capacity and size each stage. This method uses a traditional form of the weir equation, presented in Eq. 1:

$$Q = \frac{2}{3} C_{d(\alpha^\circ)} L_c \sqrt{2g} H_T^{3/2} \tag{1}$$

where Q is discharge over the weir, $C_{d(\alpha^\circ)}$ is an experimentally determined weir coefficient, L_c is the centerline crest length, g is the acceleration constant of gravity, and H_T is the total energy upstream of the weir measured relative to the spillway crest.

Figure 3. Rendering of proposed multi-stage labyrinth weir

The proposed labyrinth weir geometry features five stages with an overview presented in Figure 3. As shown, the labyrinth spillway has a maximum wall height of 19 feet and is approximately 220-ft wide and 140-ft deep (upstream to downstream distance). The spillway rating curve was developed to pass 75 percent of the PMP without increasing peak outflows for the 5 year, 25 year, and 100 year floods for existing conditions, with two notches for base flows. Note that the reservoir and spillway are capable of routing the full PMF without overtopping the dam. Apex size was governed by structural design and constructability; they were set at minimum allowable lengths. Several characteristics of the proposed labyrinth geometry stages are summarized in Table 1.

Table 1. Lake Ogletree Staged Labyrinth Spillway

Segment	Crest Elevation	# Cycles	Storm Event
Notches	486.18	-	base flows
Stage 1	486.5	1.5	$\leq$ 5 yr
Stage 2	488.0	1	$\leq$ 25 yr
Stage 3	489.0	0.5	$\leq$ 100 yr
Stage 4	490.0	2.5	> 100 yr

The vertical distance from the labyrinth crest at elevation 490 to the stilling basin floor is 44 feet. A SAF stilling basin was selected as an effective means for energy dissipation and was sized per Blaisdell (1959), noting that this methodology utilizes the kinetic flow factor, not Froude number, for dimensioning the basin. The unit discharge exiting a labyrinth weir is not uniform; this is even more so with a staged labyrinth. Therefore, a smooth chute length was included to improve the unit discharge spatial uniformity entering a basin for lesser floods. For more extreme events, tailwater was estimated to have a significant influence on chute and basin hydraulics.

Numerical modeling of the spillway, chute, and basin was performed using a commercially available computational fluid dynamics (CFD) solver. The primary objectives of the CFD modelling efforts were to examine upstream approach conditions to the labyrinth spillway (floor slab elevations/minimize excavation, grading, wing walls) and impacts of tailwater to the hydraulic performance of the chute and stilling basin. It was also used to examine general flow fields and velocity columns for sizing riprap downstream of the basin. Following this modeling effort, the owner elected to incorporate a bridge over the spillway that would require the addition of two piers. A sectional numerical model was implemented to examine the interaction of the flow with the piers. An example of the results is presented graphically in Figs. 4 and 5.

CFD modeling utilized the Re-Normalized Group (RNG) turbulence model (Yakhot et al. 1992), with a modification to include the influence of fractional volumes and a generalized turbulence production for buoyancy. The RNG turbulence model has wide application to open channel flows and is commonly used to numerically simulate flows in spillways. Furthermore, the performance of several turbulence models in estimating discharge for labyrinth spillways were investigated by Crookston et al. (2012); was found capable of providing results within 3 percent of experimental data obtained in controlled laboratory environments.

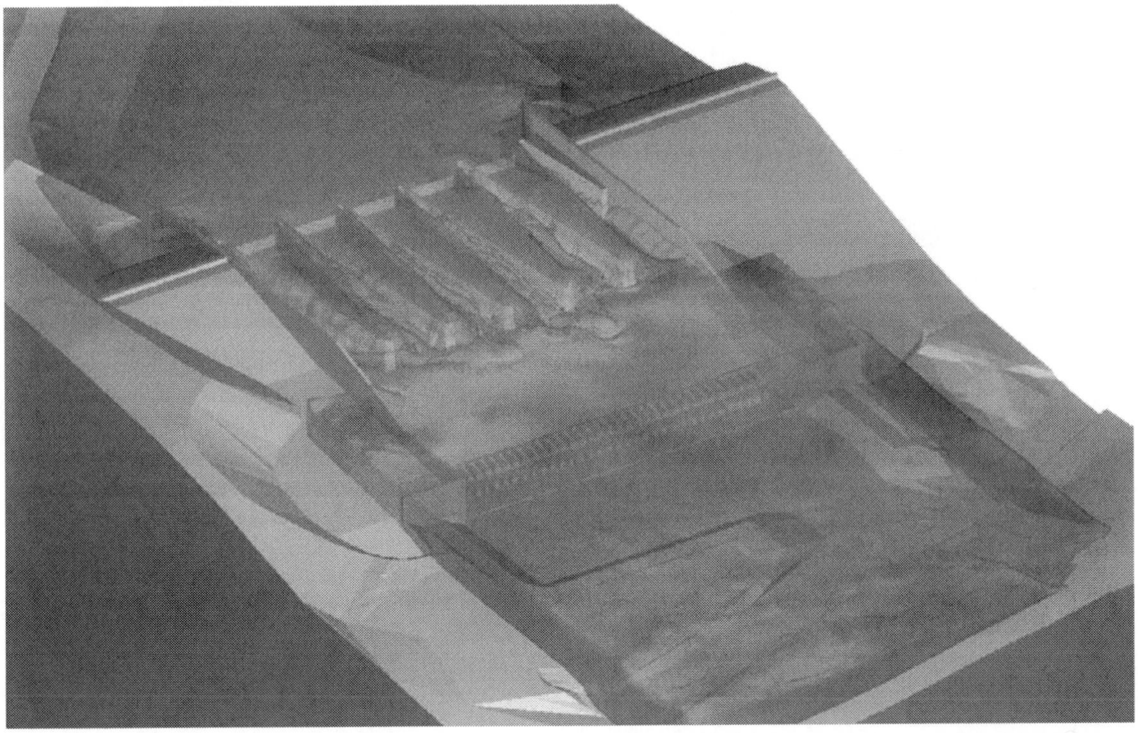

Figure 4. CFD numerical simulation results example, 3/4 PMP

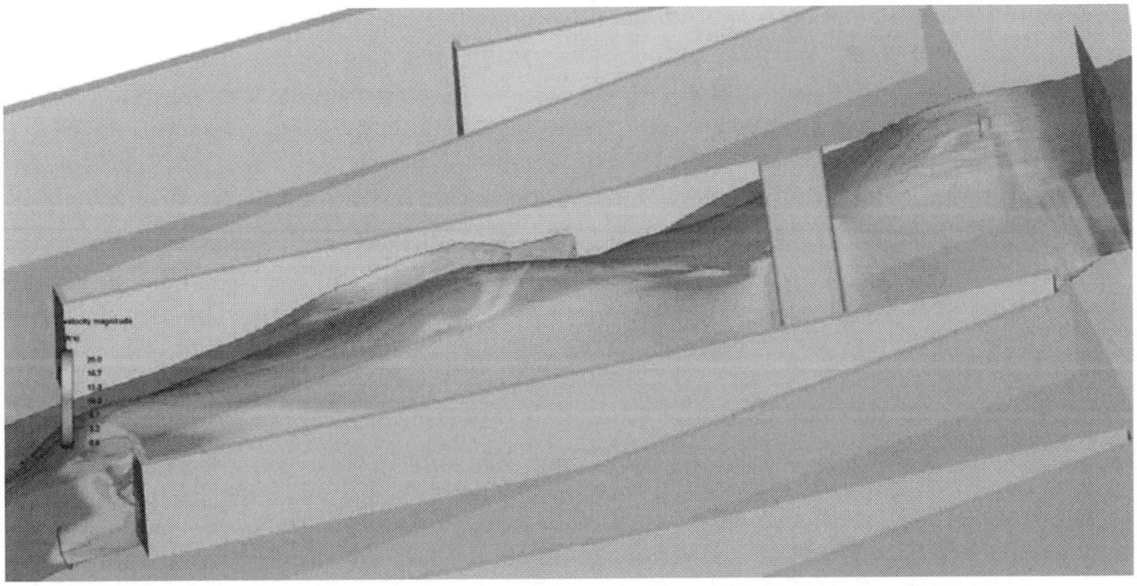

Figure 5. Sectional numerical model that includes a proposed bridge pier

A variety of free-overflow hydraulic structures (e.g., overflow gates, weirs, fountains), when operating at low heads, have displayed an acoustic phenomenon commonly termed nappe vibration. It is characterized by a thin nappe with oscillations (appear as horizontal banding) that can produce significant noise levels and acoustic pressure waves (see Fig.

6). The acoustic pressure waves have been witnessed to affect nearby people and structures. Examples include Avon Dam in Australia, Dunedin Fountain in New Zealand, Black Canyon Dam in Idaho, North Fork Smith River Dam in Montana, and Coon Rapids Dam in Minnesota, among others.

Figure 6. Example of nappe vibrations in a free-overfall structure

Limited information is available regarding the physical processes of nappe vibration and quantification of effective mitigation techniques (Naudascher and Rockwell (2005). Therefore, two collaborative research studies have been undertaken. The first collaborative effort was between Schnabel Engineering and Utah State University, where various crest treatments were tested, including roughness elements and flow splitters (Anderson 2014, Crookston et al 2014). Efforts focused on prototype-scale hydraulic modeling in a controlled laboratory environment, and gathering field observations in the USA. The second collaborative effort between the Université de Liège and Schnabel Engineering has recently completed. Experiments were performed with a prototype-scale hydraulic model for two crest shapes. Research efforts focused on quantifying the mitigation effectiveness of 10 crest modifications. The experimental data includes sound measurements, oscillation frequencies, and discharge efficiency. Fourier transforms were applied to the recorded acoustic signals to estimate the amplitude and frequency of the nappe vibrations for comparison. Details of this study, including tested crest modifications, are presented in Lodomez et al. (in press).

The most effective crest modification is not a flow splitter or roughness element. It is a step strategically placed, and can be applied to new or existing structures, is applicable to any crest shape, is simple and economical, and can be formed of concrete or steel. It also has a minimal impact on discharge (0.3% to 4% reduction in C_d) and is anticipated to be durable and not collect floating debris. Peak sound levels were reduced up to 40 percent with no vibrations observed. Due to its favorable performance, this crest modification will be recommended for use on the labyrinth weir crest at the Lake Ogletree Reservoir.

CONCLUSIONS

The City of Auburn internally embraced high standards of safety by adopting the regulations from its close neighbor, Georgia, and going a step further by committing to a spillway design that can safely pass the full PMP. As the owner of Lake Ogletree Dam, the City of Auburn is setting an example for others to follow, both within and outside of Alabama. With a myriad of modifications to the dam and spillway since 1941, and potential flooding to surrounding infrastructure from spillway releases, an innovative approach to the layout of the spillway was required. The rehabilitation includes a large, multi-stage labyrinth weir (19-ft maximum weir height) and saddle areas designed for overtopping. Hydraulic challenges included generally matching the existing outflow hydrograph for more frequent storm events while still passing the spillway design storm. The design team utilized a detailed CFD model of the spillway and basin in these efforts. The labyrinth weir will also include a crest modification to mitigate potential nappe vibrations, a phenomena that can produce acoustic pressure waves that affect nearby structures.

ACKNOWLEDGEMENTS

The authors wish to acknowledge Schnabel, Utah State University, and the Université de Liège.

REFERENCES

Anderson, A. (2014). "Causes and Countermeasures for Nappe Oscillation: An Experimental Approach." M.S. Thesis, Utah State University, Logan, Utah.

Blaisdell, F. (1959). "The SAF Stilling Basin. A Structure to Dissipate the Destructive Energy in High-velocity Flow from Spillways." U.S. Dept. Agriculture, Agr. Handbook 156.

Crookston, B.M., Anderson, A., Shearin-Feimster, L., and Tullis, B.P. (2014). "Mitigation Investigation of Flow-induced Vibrations at a Rehabilitated Spillway." in "Hydraulic Structures and Society - Engineering Challenges and Extremes", Proc. of the 5th IAHR International Symposium on Hydraulic Structures (ISHS2014), 25-27 June 2014, Brisbane, Australia, H. Chanson and L. Toombes Editors, The University of Queensland, Brisbane, Australia, 9 pages (DOI: 10.14264/uql.2014.9) (ISBN 978-1-74272-115-6).

Crookston, B.M., Paxson, G.S., and Campbell, D.B. (2013). "Effective Spillways: Harmonizing labyrinth weir hydraulic efficiency and project requirements." in Labyrinth and Piano Key Weirs II – PKW 2013. S. Erpicum, F. Laugier, M. Pfister, M. Pirotton, G Cicéro, A. Schleiss Editors. CRC Press, London, UK

Crookston, B.M., Paxson, G.S., and Savage, B.M. (2012). "Hydraulic Performance of Labyrinth Weirs for High Headwater Ratios." 4th International Symposium on Hydraulic Structures. Porto, Portugal.

Crookston, B.M. and Tullis, B.P. (2013). "Hydraulic design and analysis of labyrinth weirs. Part 1: Discharge relationships." *J. Irrig. Drain. Eng.,* 139(5) 363-370.

Crookston, B.M. and Tullis, B.P. (2013). "Hydraulic design and analysis of labyrinth weirs. Part 2: Nappe Aeration, Instability, and Vibration." *J. Irrig. Drain. Eng.* 139(5) 371-377.

Dabling, M. R., Tullis, B.P., and Crookston, B.M. (2013). "Staged labyrinth weir hydraulics." *J. Irrig. Drain. Eng.,* 139(11), 955-960.

Dabling, M.R., Tullis, B.P, and Crookston, B.M. (2015). "Closure to 'Staged labyrinth weir hydraulics' by M.R. Dabling, B.P. Tullis and B.M. Crookston." *J. Irrig. Drain. Eng.,* 10.1061/(ASCE)IR. 1943-4774.0000900 , 07015018.

Lodomez, M. Crookston. B.M., Tullis, B.P., Pirotton, M., and Erpicum, S. (in press). "Nappe Vibration Mitigation Techniques for Free-overfall Structures." Proc. of the 6th IAHR International Symposium on Hydraulic Structures (ISHS2016).

Naudascher, E. and Rockwell, D. (205). *Flow Induced Vibrations. An Engineering Guide.* Dover Publications, Inc. Mineola, N.Y.

Yakhot, V., Orszag, S., Thangam, S., Gatski, T. and Speziale, C. (1992). "Development of Turbulence Models for Shear Flows by a Double Expansion Technique." Physics of Fluids A, Vol. 4., No. 7. pp1510-1520.

REPLACING RUBY DAM OUTLET USES COMPUTATIONAL FLUID DYNAMICS TO MODEL ENERGY DISSIPATION

Frank Lan, Ph.D., P.E.[1]
Robert Waddell, P.E.[2]
Michael Zusi, P.E.[3]
Brian Grant, P.E.[4]

ABSTRACT

After 75 years of hard work, the old outlet works at Ruby Dam was ready for rehabilitation and new jet flow gates selected to regulate discharges needed an energy dissipation structure for high-velocity releases. Three-dimensional Computational Fluid Dynamics (CFD) modeling was utilized to help design the new energy dissipation structure. Gate selection criteria, project design, and post-construction performance are presented, and its operation is compared to the designer's expectations.

Ruby Dam was constructed in the 1930's in Madison County, Montana on the Ruby River, a tributary to the Jefferson River, one of the three rivers that form the Missouri River at Three Forks, Montana. Ruby Reservoir is primarily used for agricultural irrigation, making it a vital economic asset to the area, which underscores the importance of a reliable outlet works. Ruby Dam is a high-hazard earth embankment with multiple appurtenant structures including a low-level outlet and principal spillway. The original low-level outlet included a 6-foot-diameter gate valve (guard valve) and a 6-foot-diameter flow-regulating butterfly valve in a vertical gate shaft from the dam crest. The gate valve was abandoned in place and the butterfly valve was replaced with a new knife gate valve. During the recent rehabilitation, a new 6-foot diameter steel-lined outlet conduit was constructed through the downstream discharge tunnel, grouted in place, and extended to a new valve house located at the downstream toe of the dam, where two new jet flow gates were installed to regulate flow. These jet flow gates discharge water at relatively high velocity into a Type III stilling basin that was designed using CFD modeling.

INTRODUCTION

Ruby Dam is located on the Ruby River about seven miles south of Alder, MT and was constructed between 1936 and 1939, with the first filling to full pool occurring in 1939. The dam is owned and managed by the Montana Department of Natural Resources and Conservation (DNRC) and is operated by the Ruby River Water Users Association (WUA). Ruby Reservoir is primarily used for agricultural irrigation; however, additional uses include recreation and regulation of stream flow for downstream environmental enhancement.

[1] AECOM, 8181 East Tufts Avenue, Denver, CO 80237, frank.lan@aecom.com
[2] AECOM, 8181 East Tufts Avenue, Denver, CO 80237, bob.waddell@aecom.com
[3] AECOM, 8181 East Tufts Avenue, Denver, CO 80237, michael.zusi@aecom.com
[4] Montana DNRC, 1424 9th Avenue, Helena, MT 59620, bgrant@mt.gov

Ruby Dam is a high-hazard, 111-foot (ft)-high, 846-ft-long embankment structure with multiple appurtenant structures including a low-level outlet, the rehabilitation of which in 2014 is the topic of this paper. The reservoir has a normal storage of 38,850 acre-ft at Elevation (EL.) 5,393 ft, the spillway crest elevation. A 9.5-ft diameter vertical concrete valve shaft is located just upstream of the dam axis. This valve shaft penetrates the embankment dam, passing into the underlying rock foundation just upstream of the existing tunnel. Access to the valve shaft was through a wood-framed structure on the dam crest. The gate chamber located at the bottom of the shaft at the tunnel level housed a 72-inch bonneted knife gate valve (guard gate) and a 72-inch butterfly valve (operating valve). Both valves were controlled by manual operators located on the operating deck just below the dam crest.

Downstream of these valves, the original low-level outlet consisted of a 90-inch diameter horseshoe shaped tunnel cut into bedrock in the left abutment. During normal operation, the old tunnel portal was mostly submerged by water impounded behind a weir across the discharge channel, which empties into the river below the weir. The old butterfly valve controlled reservoir releases into the tunnel and the old gate valve was used to isolate the reservoir from the tunnel, mainly for inspections of the tunnel, butterfly valve, and associated piping.

The butterfly valve was inadequate and had experienced damage from cavitation at high flow rates, requiring periodic maintenance. Maximum releases from the outlet works were limited by the original valve configuration and were unable to meet dam safety criteria for reservoir evacuation. There was also concern about the age and long-term operational capability of the upstream gate valve.

DESIGN OBJECTIVES

The purpose of the rehabilitation of Ruby Dam Outlet Works was ultimately to address deficiencies of the existing outlet works, to increase reliability, reduce future long-term maintenance, and increase the reliable hydraulic capacity of the outlet works while considering the potential for future hydropower generation. One important design consideration was the scheduling of the work. Work had to be completed without lowering the reservoir. In fact, the work had to occur during winter months while the reservoir was filling, and had to be completed in time to release irrigation water through the newly completed outlet in late spring.

Conceptual analyses and design of the new outlet works were developed by others prior to AECOM's involvement. A feasibility study was completed by HKM Engineering Inc. of Billings, MT in January 2008 that included an impact basin for energy dissipation from a jet flow gate. AECOM expanded upon those designs during an alternatives evaluation phase, and selected a hydraulic jump stilling basin for energy dissipation. The design presented in this paper is based on the preferred alternative selected for final design that was developed by AECOM. Outlet Works improvements met the design objectives, and included the following features:

- Line the existing horseshoe-shaped tunnel located beneath the embankment in the left abutment foundation downstream of the existing gate chamber with a new 72-inch diameter steel outlet conduit. The annulus between the new liner and tunnel was grouted from the new knife gate to the downstream tunnel portal.
- Downstream of the tunnel, the new outlet conduit would be encased in reinforced concrete and terminate in a new concrete valve house.
- The original upstream gate valve located in the valve shaft would be abandoned in place and the downstream butterfly valve would be removed. The original gate valve was not removed since it needed to remain closed during the rehabilitation as it as there was no means to install a bulkhead in the upstream tunnel. This guard valve is the only thing between the work and a full reservoir, and it was abandoned in the up (opened) position. The original gate valve body was encased in concrete and a shear pin and redundant clamping device was installed on the gate shaft to hold it in the open position, with the gate leaf out of the flow path.
- A new knife gate would be installed in the gate shaft to replace the operational function of old gate valve, allowing it to be abandoned in place in the open position. A new concrete structure would replace the existing framed structure over the gate shaft.
- Reservoir releases would be regulated by a downstream 60-inch jet-flow gate for large releases and an 18-inch jet-flow gate valve for small releases. Both of these valves would be located in the new valve house.
- The flow would discharge into a new concrete terminal structure for energy dissipation before flowing into the Ruby River.
- The design modifications would include provisions to connect a bypass for future hydropower. A flanged bifurcation was constructed in the outlet conduit, and fitted in a dished head which was then encased in lean concrete for a future tie-in.
- Improvements to the spillway were designed and constructed in 2010 and 2011 prior to constructing the new Outlet Works under a separate contract. These spillway improvements included a gate structure in the new labyrinth weir to provide flows to the Ruby River to reduce the duration and expense of bypass pumping that was required to maintain flows in the river during closure of the Outlet Works.

EXISTING VALVE AND TUNNEL DEFICIENCIES

The original outlet works had two main deficiencies that were interrelated. First, at the maximum release rate for irrigation, 500 cubic feet per second (cfs), cavitation damage to the butterfly valve accelerated, which required periodic repair. Second, at flows of 600 cfs, popping and banging from the butterfly valve was observed and was thought indicative of potential damage to the valve if flows were maintained for a significant period of time. MT dam safety regulations required an emergency drawdown flow of 1,100 cfs for high hazard dams for dams of this hazard classification and size. This discharge was not considered possible with the existing outlet works and operating at flows greater than 600 cfs was considered imprudent. Modifications to the butterfly valve were considered (e.g. increasing air supply to the valve, or removing and rehabbing the valve) but were not considered to be a reliable long-term remedy.

The existing gate valve had only a few minor deficiencies. It leaked a few gpm through the stem packing and at times did not seal tight upon closing which limited the ability to inspect the gate chamber and downstream tunnel. Tightening the gland to reduce the leakage was considered problematic since the stem packing gland was missing a bolt, and one of the remaining bolts was a different size.

GATE SELECTION PROCESS

During the final design process, the preferred arrangement for the regulating valves was decided early on, but the type of valves and means to dissipate energy remained undecided. Placing the regulating valve(s) at the toe of the dam was considered an improvement over the existing arrangement given that the valves would be easily accessible, air demand would not be an issue, and substantial modifications to the gate chamber would not be required. Placing the valves at the toe of the dam required installing several hundred feet of steel outlet conduit through the existing tunnel and grouting it into place. In the final analysis, the competing alternatives for determining which regulating valves to select consisted of the following:

Table 1. Summary of Gate Selection Criteria

Alternative	**Advantages**	**Disadvantages**
No. 1 - Large JFG for Irrigation Releases and Max Release up to 1,100 cfs. Small JFG for winter flows of 20-30 cfs.	A) Lower maintenance B) Less spray at discharge C) Easier to protect from winter conditions at site	A) Energy dissipation structure needed
No. 2 - Two similarly sized Fixed Cone Valves with discharge of 600 cfs each for Irrigation Releases and Max Release. A third small JFG for winter flows.	A) Design for energy dissipation structures for these valve are well established	A) Higher maintenance B) Harder to protect from winter conditions

For the Owner, reliability and low maintenance were of primary importance. Therefore Alternative No. 1 was the preferred alternative, but the means to dissipate the energy had to be resolved.

ENERGY DISSIPATION OPTIONS

With the original outlet works arrangement, there was no defined energy dissipation structure. Flow discharged into the concrete lined tunnel immediately downstream of the 72-inch butterfly valve. Flow was conveyed through the tunnel to the concrete terminal structure (weir) located about 220-ft downstream of the butterfly valve. Over this length, the tunnel sloped about 0.005ft/ft. To some extent, energy was dissipated through friction losses over the length of the tunnel and by tailwater at the tunnel portal. During the initial development of the outlet works design concept, several basin concepts were investigated. The basins had to be capable of handling free discharge from the jet-flow gate for a wide range of discharges. During the initial development, four concepts were considered; free jet (plunge pool); concrete chute basin, hydraulic jump chute basin and a Type VI impact basin.

Through the initial screening process, two viable stilling basin alternatives emerged; a free jet (plunge pool) option and a hydraulic jump chute basin (stilling basin) option, based on site conditions and the location of the proposed valve house relative to the original outlet channel.

Further research was performed to develop and refine basin concepts that could be developed, sized and model studied using CFD. Design references from the U.S. Bureau of Reclamation (USBR) (Engineering Monograph No. 25) and U.S. Army Corp of Engineers (USACE) (USACE, 1980) were reviewed. In addition, physical model studies performed by Reclamation for the outlet works at Wickiup (Oregon), Cherry Creek (Colorado), and Norton Dam (Kansas), and other dams were reviewed and studied.

Wickiup Dam is located on the Deschutes River in northern Oregon. The dam is about 87 feet high and the outlet works includes two 96-inch diameter valves with a maximum discharge of 3,950 cfs. The basin consists of two side by side concrete discharge channels.

Cherry Creek Dam is a 147-foot high flood control dam built by the USACE on Cherry Creek, south of Denver, Colorado. The outlet works consists of two 96-inch conduits and one 84-inch conduit which discharges into a concrete chute stilling basin. The maximum discharge for all three conduits is 6,000cfs.

Norton Dam, built by the USBR on Prairie Dog Creek as part of the Missouri River Basin Project, in north-western Kansas near Norton, Kansas is 100-ft high and the outlet works includes two 3-foot x 7-foot conduits, discharging into an 8-foot-wide, 55-foot-long vertically curved chute. Flows then report to a 55-foot-long Type II hydraulic jump stilling basin (Type III includes baffle blocks). The design outflow from the outlet works is about 385 cfs.

The hydraulic design of the Ruby Dam stilling basin was completed with a combination of conventional basin calculations and state-of-the-art, three-dimensional CFD modeling. Through the CFD modeling, it was decided that the free jet (plunge pool) option could potentially cause waves and undulating flow in the plunge pool, and more importantly, it could lead to excessive high flow velocity at the bottom of the plunge pool that would cause significant scour in the downstream channel area, which could impact side slope stability. As a result, the hydraulic jump chute basin was selected as the final design for the Outlet Works for Ruby Dam.

Figure 1. Existing Channel downstream of Ruby Dam Outlet Works tunnel.

The stilling basin was sized for the 60-inch-diameter jet flow gate under the following hydraulic conditions:

- Maximum discharge of approximately 1,133 cfs at the Stage 2 design reservoir water level of 5,400-ft. Exit flow velocity from this jet-flow gate would be approximately 58 ft/s for this discharge.
- Tailwater elevation in the downstream channel is at approximately elevation 5,293 ft for 1,133 cfs discharge. The new outlet arrangement will shift the discharge in the channel downstream and at a much higher velocity compared to existing conditions. The following presents the results of the hydraulic analysis for the selected stilling basin for the Outlet Works.

<u>Selected Chute Basin Characteristics</u>

The chute basin was sized according to the USACE guidelines for supported jets, the Hydraulic Design of Stilling Basins and Energy Dissipators, Engineering Monograph No. 25 (Peterka, 1984) for Type III basins, and geotechnical assumptions for excavation and slope stability. The selected basin has a width of 18 ft with a floor elevation at 5,279 ft. The 18-ft-wide basin is also used to accommodate the exit flow from the 18-inch-diameter jet flow gate that is adjacent to the larger 60-inch-diameter valve.

The chute floor profile from the outlet exit to the stilling basin, which was necessary to support the jet and spread the flow, was designed according to the parabolic profile described by the following equation (USACE, 1980):

$$y = -x \tan\theta - \frac{gx^2}{2(1.25V_{sm})^2 \cos^2\theta} \quad (1)$$

where,

x and y = horizontal and vertical coordinates measured from the beginning of the curve (ft)

θ = angle with the horizontal of the valve invert at the beginning of the vertical curve (degree)

g = acceleration due to gravity (32.2-ft/sec^2)

V_{sm} = average velocity for smooth pipe flow at the beginning of the curve (ft/s)

The calculations to size the hydraulic jump basin followed the procedures outlined for Type III basins in USBR Engineering Monograph No. 25 – Hydraulic Design of Stilling Basins and Energy Dissipators (Peterka, 1984). The calculations assumed that the jet flow would spread uniformly before it enters the bottom of the chute (i.e., the unit discharge would be generally uniform across the section). Due to the uncertainty of flow uniformity at the toe of the chute, the final basin sizes were confirmed by three-dimensional CFD models. Table 2 presents a summary of the initial and final selected design parameters for the chute basin.

An isometric three-dimensional view of the proposed chute stilling basin alternative is shown in Figure 2. The basin plan and profile is shown in Figure 3.

Table 2. Summary of Design Input Parameters for Selected Chute Type Basin

Invert Design Parameters	**18-ft-Wide Basin**	
	Initially Calculated	**Final Selection**
Basin Elevation, ft	5,279	**5,279**
Chute Length, ft	81.3	**81.3**
Flow Depth at end of Chute, ft	0.91	**1.20***
Basin Length, ft	45	**45**
Chute Block width, ft	0.9	**1.0**
Chute Block Height, ft	0.9	**1.0**
Baffle Piers Height, ft	2.5	**2.5**
Baffle Piers width, inches	22.5	**21**
Baffle Piers End Gap, ft	11.0	**13.5**
End Sill height, inches	19.2	**20**

* Flow depth and velocity at toe of chute were based on FLOW-3D model results

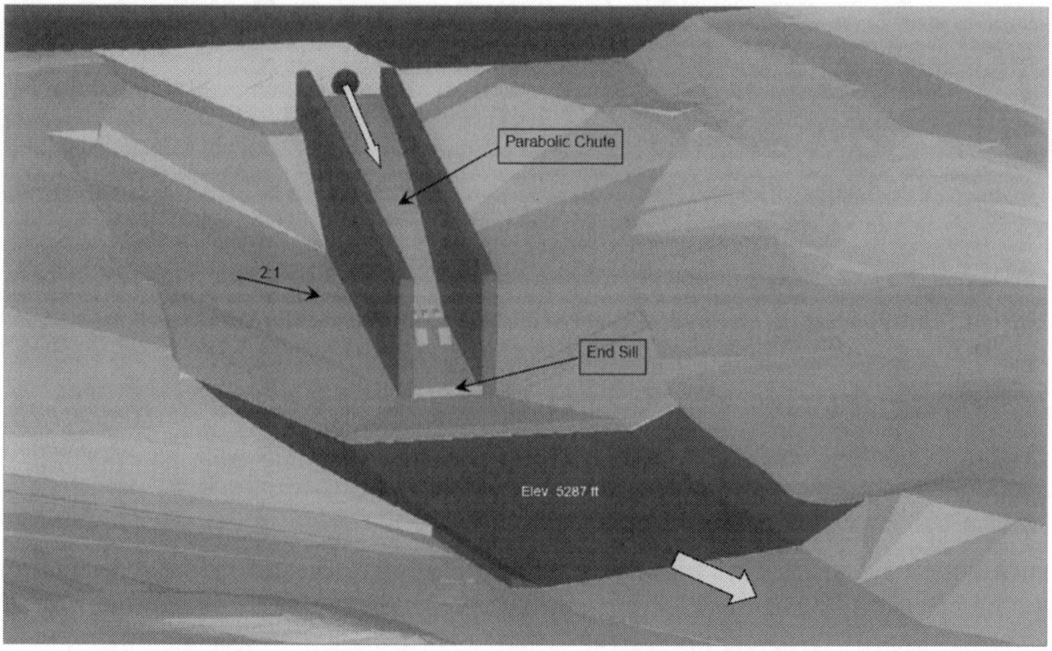

Figure 2. Isometric View of Selected Chute Basin Alternative

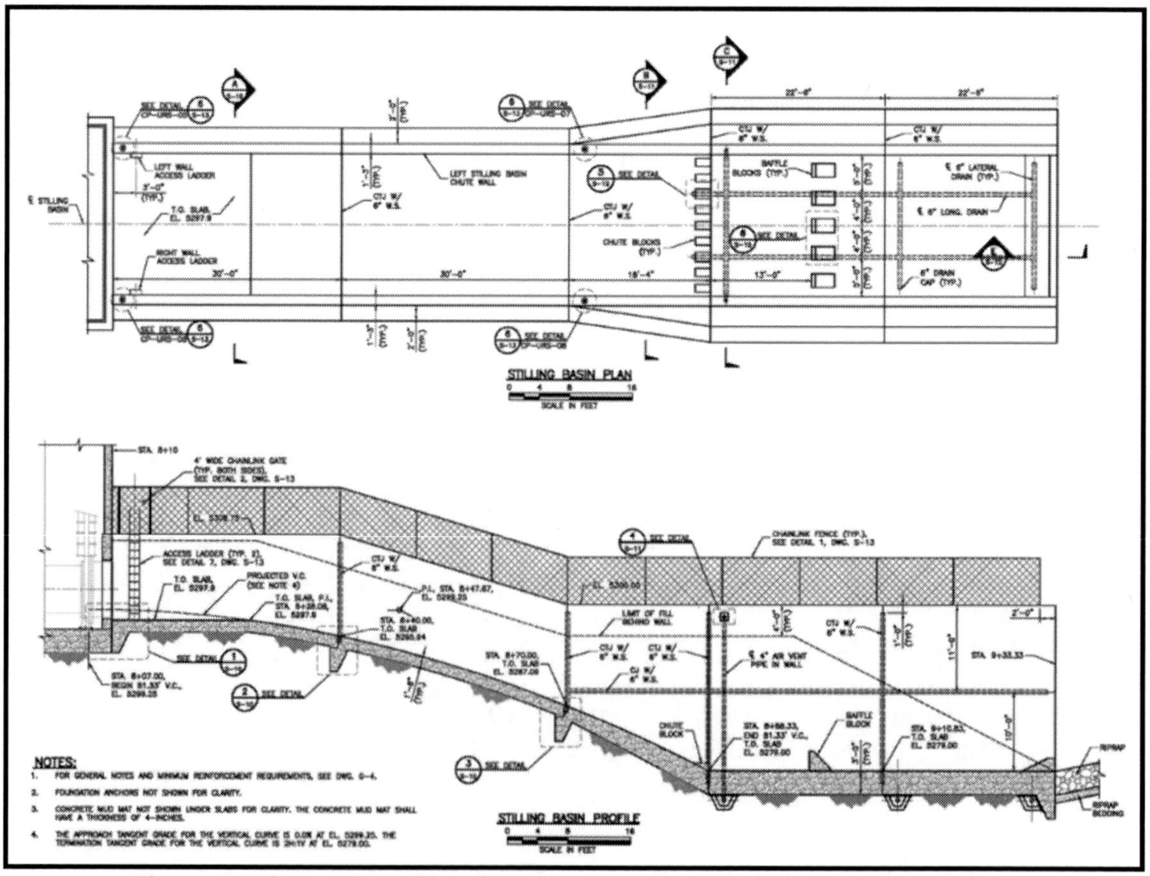

Figure 3. Plan and Profile of Selected 18-ft-wide Chute Stilling Basin

COMPUTATIONAL FLUID DYNAMICS MODELING

Since the proposed stilling basin designs presented above are marginally different from example prototypes that could be found in the U.S., a CFD modeling effort was conducted with the following purposes:

1. Confirm hydraulic calculations for the sizing of the selected chute basin for the design discharge of 1,133 cfs.
2. Estimate maximum water surface profile for selecting wall heights and for concrete wall sizing.
3. Determine if any adverse hydraulic conditions could occur for either of the valves for a range of operational flows up to the design discharge.
4. Estimate near-bed velocities to identify where riprap can be used as compared to concrete lining. The velocities could then be used for sizing the riprap lining.

The CFD modeling software used for analyzing the stilling basin hydraulics for Ruby Dam Outlet Works was the FLOW-3D computer program developed by Flow Science, Inc. of Santa Fe, New Mexico, USA. FLOW-3D is a well-tested, reliable CFD computer program that has been used successfully in many water resources projects, particularly for spillway hydraulic simulations. The FLOW-3D program was previously used by the authors in the spillway design phase for the Ruby Dam Spillway Modification Project. The software is capable of analyzing time-dependent (transient) problems in one, two and three dimensions, and is based on a finite difference solution of the complete Navier-Stokes equations. Because the program is based on the fundamental laws of mass, momentum, and energy conservation, it can be applied to many types of flow process. One of the major strengths of FLOW-3D for hydraulic analysis is its ability to accurately model problems involving free surface flow, by using the Volume of Fluid (VOF) technique. The VOF method consists of three components: a scheme to locate the surface, an algorithm to track the surface as a sharp interface moving through a computational grid, and a means of applying boundary conditions at the surface. It handles transitions between sub-critical and super-critical flow within a single model set up. These capabilities make the model well-suited for simulating the varied and complex flow conditions that will occur in a stilling basin. Version 9.4.1 of FLOW-3D was used for the modeling of the Ruby Dam outlet structure.

Model Setup

The FLOW-3D model domain for the Ruby Dam outlet stilling basin is illustrated in Figure 4. A single or multiple mesh blocks with various mesh sizes was used for the computation domain, with fine meshes around the chute and the outlet while coarser meshes away from the chute.

Boundary conditions for the model include the upstream inflow to the basin from the 60-inch jet flow gate. A total discharge was given at the upstream boundary, which would internally translate into uniform flow velocity exiting the outlet. The gate was assumed to open at three positions (fully, half, and about a quarter open) in the current analysis. The downstream boundary was specified as "Outflow" boundary, which allows the model to determine the downstream water level according to the topography.

The chute basin models started with an initial water level in the pool downstream of the outlet at El. 5,290, or about 3 ft below the final tailwater elevation that is expected for the design flow of 1,133 cfs. The lower tailwater elevation was used since it is likely that the final estimated tailwater would take time to establish as it is controlled by the downstream channel (not structure controlled). Additionally, this would provide insight into the performance of the outlet from a lower tailwater to a higher tailwater, and reveal any adverse conditions that could develop. The models were run for a specified time period until a relatively steady state solution was obtained.

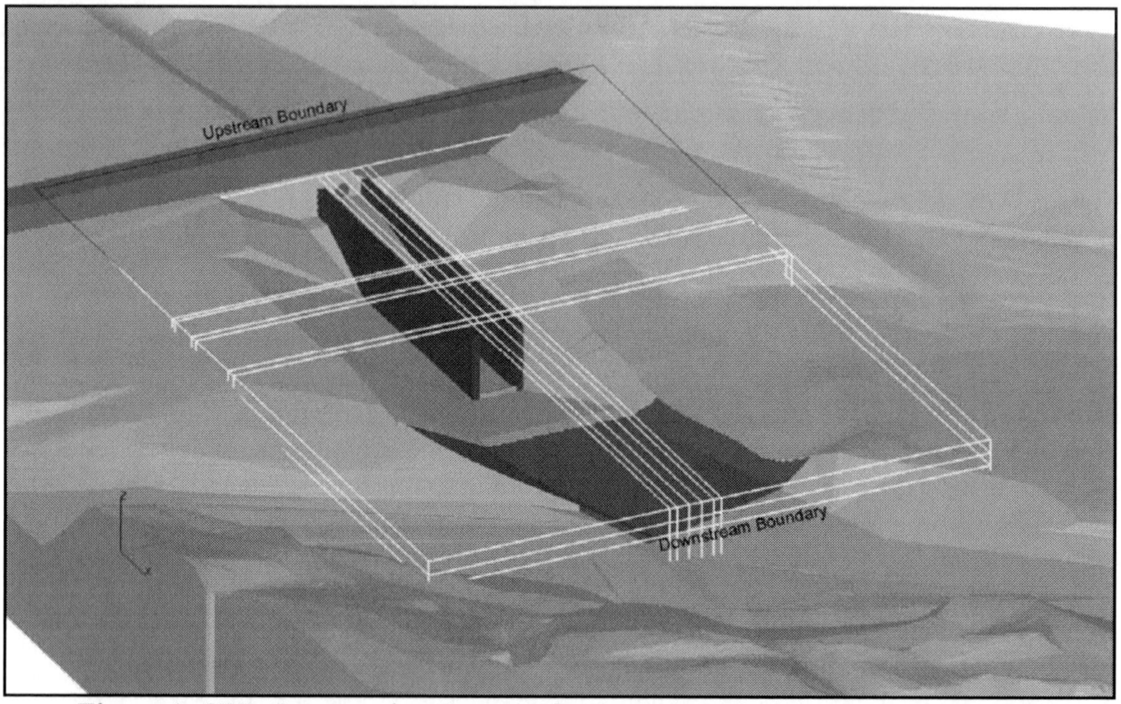

Figure 4. CFD Computation Domain for the Selected Chute Basin Alternative

CFD Modeling Results

The FLOW-3D model was run for three different openings at the maximum normal reservoir water level of 5,400 ft fully open, half open, and an opening that would produce 300 cfs outflow. The outflow with the 60-inch-diameter valve at the fully open position is 1,133 cfs, half open generates about 605 cfs, and the 300 cfs requires about 29% opening. Figures 5 and 6 show isometric views of the simulated flow velocity for the fully open scenario. The results show a well formed hydraulic jump within the basin at a tailwater level of 5,292.3 ft, about 1 foot below the HEC-RAS calculated tailwater level at a flow discharge of 1,133 cfs. The hydraulic jump is submerged at this tailwater elevation. The results indicate that the jet spreads over the width of the chute rather quickly. When the spreading reaches the walls, a relatively high fin forms on the inside of both walls, as illustrated by a cross-sectional view of the spreading on Figure 7. The fin height diminishes as the flow approaches the pool downstream as shown on Figure 8. This phenomenon was similar to the flow characteristics observed in the Wickiup physical model study.

Figure 9 is a cross-sectional view of the flow depth and velocity across the basin at the end of the curved chute, where the jump is submerged.

The simulated near-bottom flow velocity contours for the 18-ft wide basin option with an initial tailwater level of 5,292.3 ft, slightly lower than the terminal tailwater level of 5,293.0 ft at 1,133 cfs, are shown in Figure 10, which indicates near-bed velocities of less than 5 ft/s exiting the basin. Near the baffle blocks, the maximum flow velocity is about 40 ft/s, as shown in Figure 11, which is not high enough to cause a cavitation concern for the baffle blocks.

A cross-sectional view of the pressure distribution along the centerline of the valve opening immediately downstream of the 60-inch-diameter jet flow gate indicates a zone of negative pressure immediately downstream of the valve, as shown in Figure 12. The minimum pressure (relative to atmospheric pressure) reaches about -500 lbf/ft^2, or -3.5 lbs/in^2. Negative pressure is also present where the leveled chute section meets the curved chute section.

Further FLOW-3D simulations were completed for the gate half-open scenario, as well as the 300 cfs scenario (29% open). Figure 13 presents a three-dimensional view of the simulated flow velocity for the gate half-open scenario and Figure 14 presents an isometric view of the simulated flow velocity for the 300 cfs scenario (29% open). With outflow lower than the design flow, the hydraulic jump inside the basin is fairly submerged, even with a low tailwater elevation of 5,290 ft.

Figure 5. Simulated Flow Velocity for the Selected 18-ft-Wide Chute Basin at Q = 1,133 cfs with Tailwater Elev. 5292.3-ft.

Figure 6. Profile View of Simulated Flow Velocity for the Selected 18-ft-Wide Chute Basin at Q = 1,133 cfs with Tailwater Elev. 5,292.3-ft.

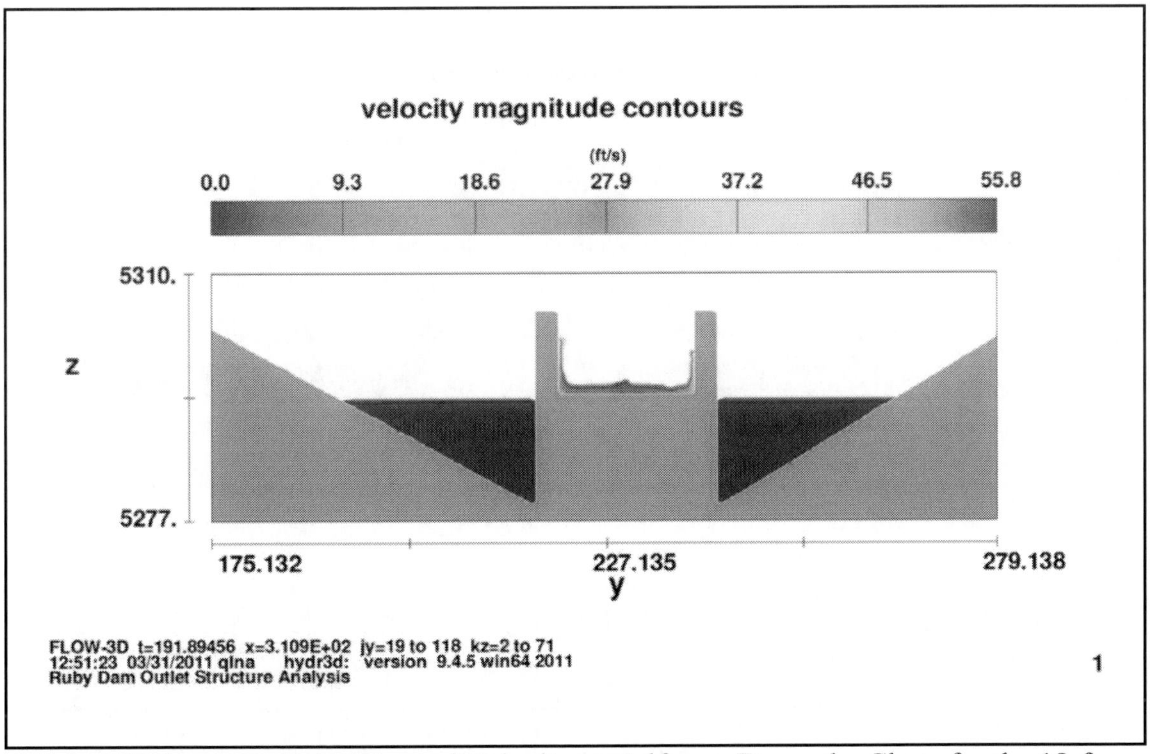

Figure 7. Cross-sectional View of Velocity at Half-way Down the Chute for the 18-ft-wide Basin Option (Looking Upstream)

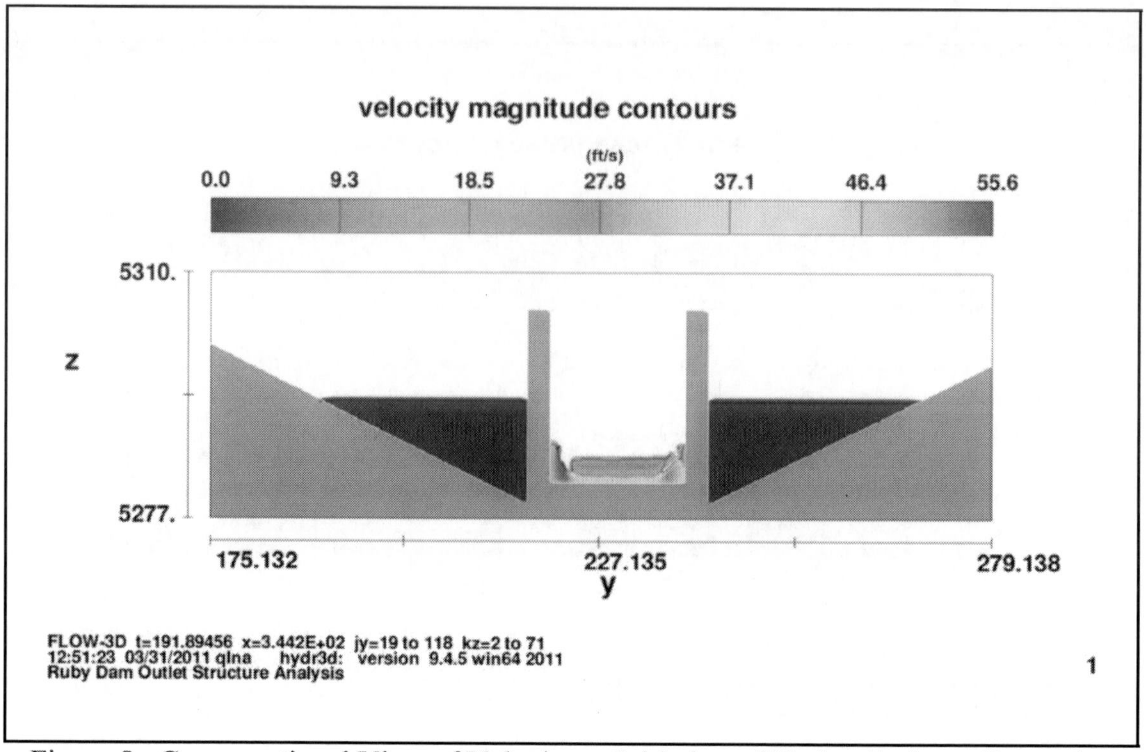

Figure 8. Cross-sectional View of Velocity at Initiation of Hydraulic Jump Down the Chute for the 18-ft-wide Basin Option (Looking Upstream)

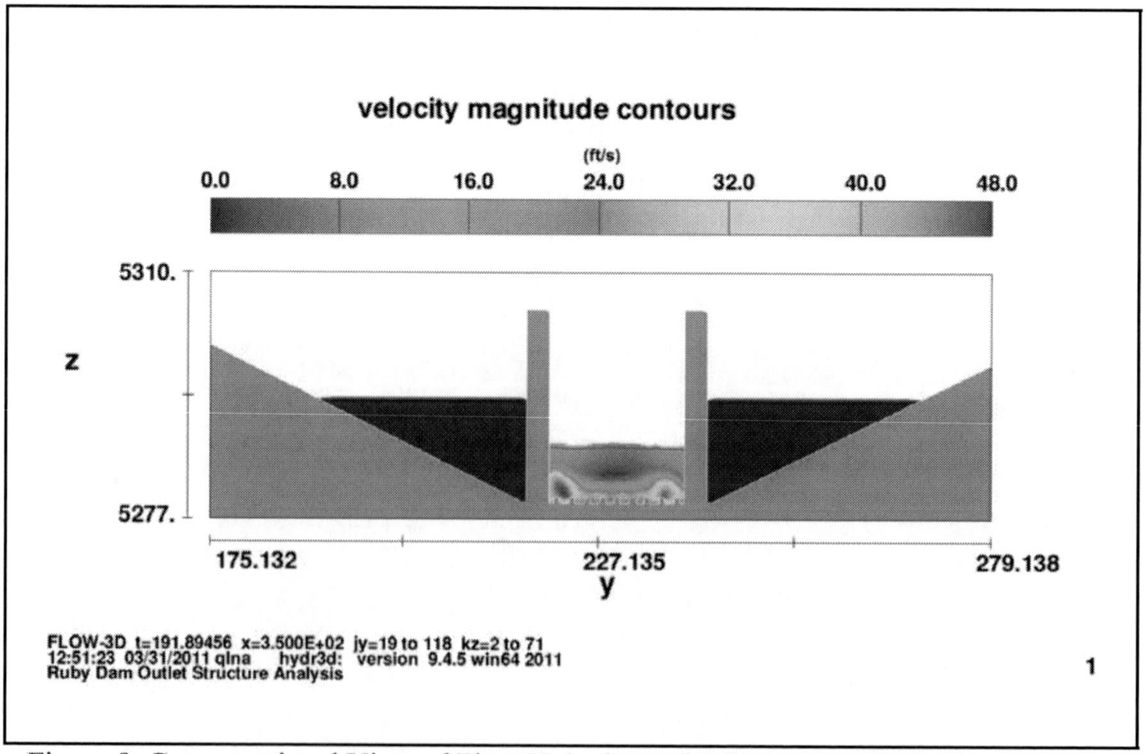

Figure 9. Cross-sectional View of Flow Velocity at the End of the Chute for the 18-ft-wide Basin Option (Looking Upstream)

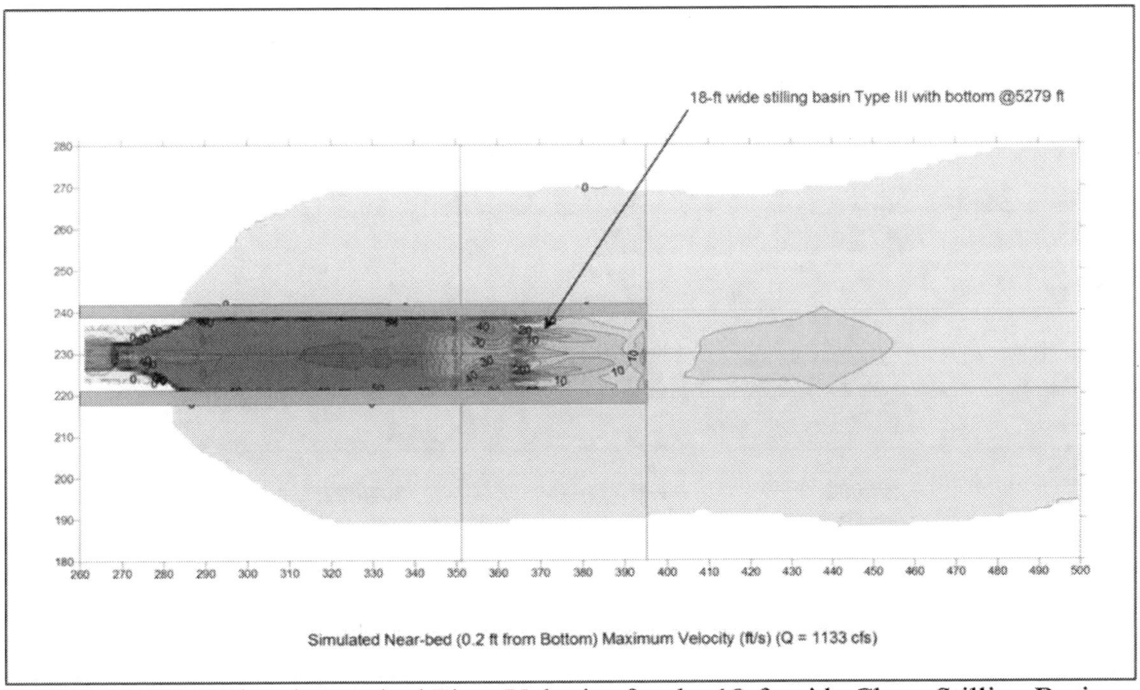

Figure 10. Simulated Near-bed Flow Velocity for the 18-ft-wide Chute Stilling Basin Alternative (Q = 1,133 cfs, Final Plan)

Figure 11. Simulated Maximum Flow Velocity near the Baffle Blocks at Q = 1,133 cfs (Final Design)

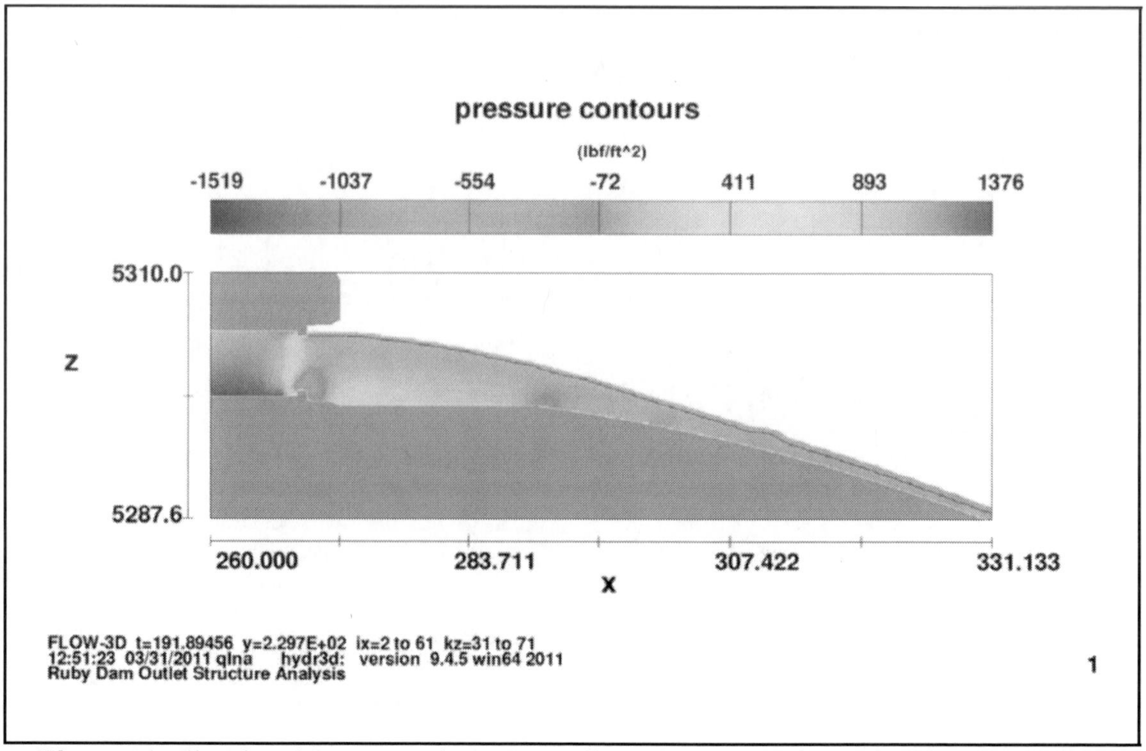

Figure 12. Simulated Pressure along Centerline of Opening for the 18-ft-wide Chute Basin at Q = 1,133 cfs.

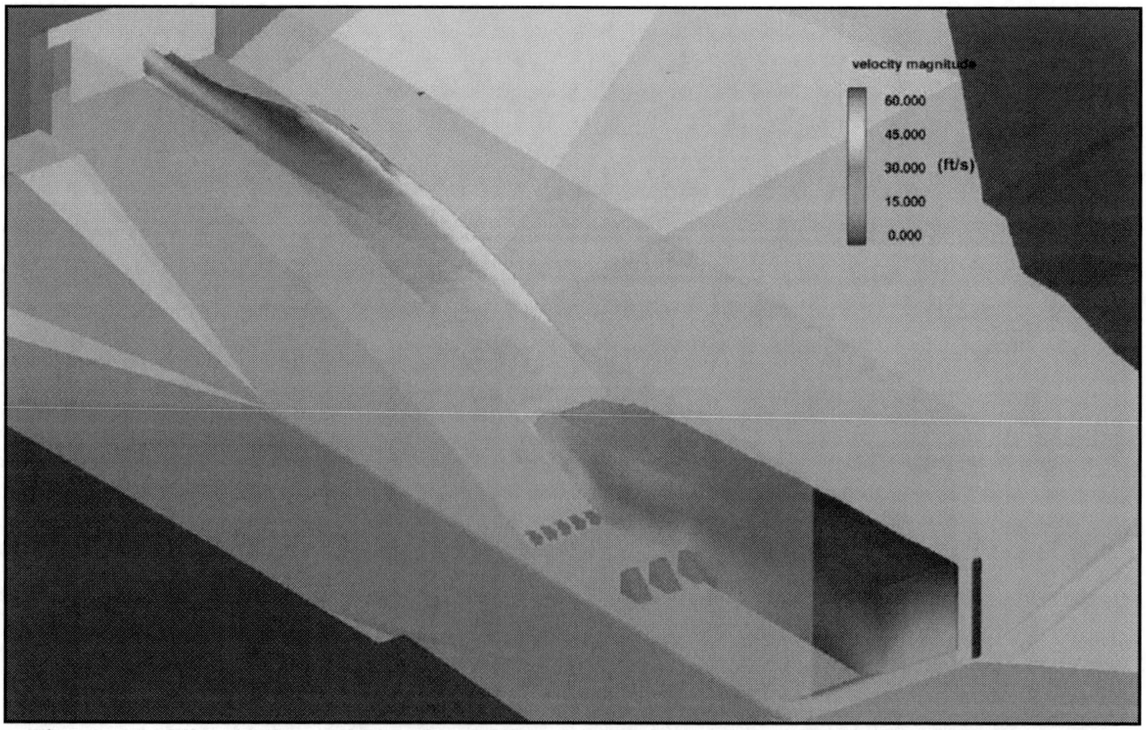

Figure 13. Sliced View of Simulated Flow Velocity for 18-ft-wide Chute Basin at Q = 600 cfs with Tailwater Elev. 5290-ft.

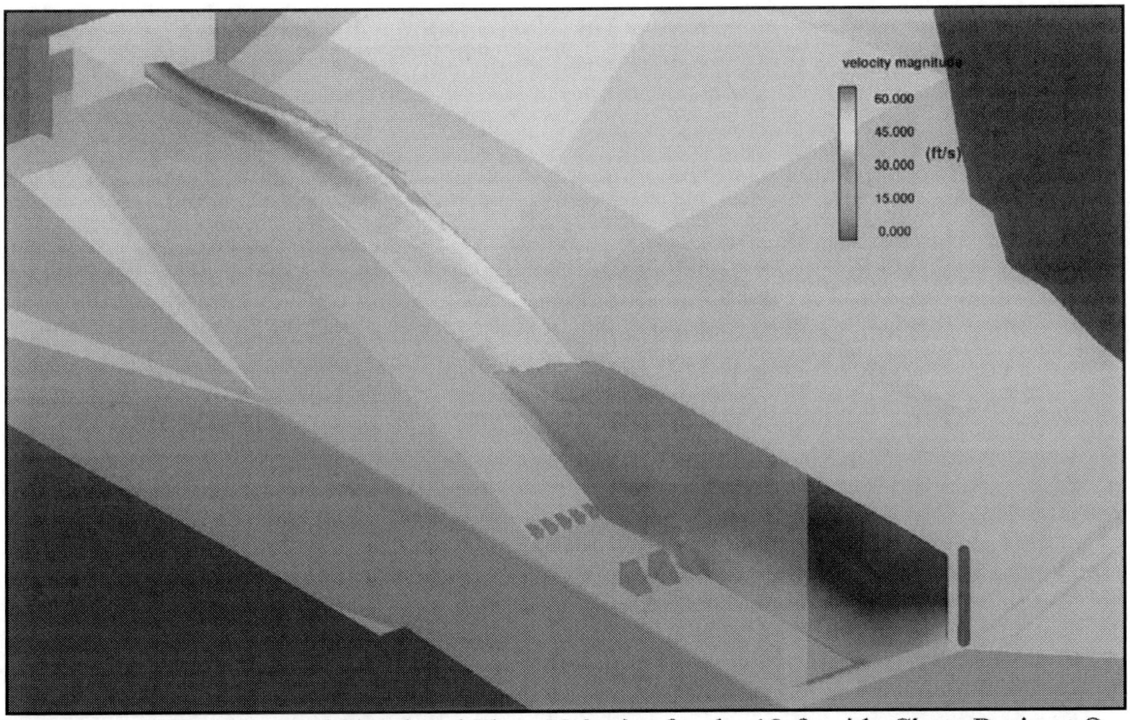

Figure 14. Sliced View of Simulated Flow Velocity for the 18-ft-wide Chute Basin at Q = 300 cfs with Tailwater Elev. 5,290-ft.

Validation of CFD Model Results

The validation of any numerical model to prototype structures is critical to the design. A literature search of published physical model and prototype studies for chute basin stilling basins downstream of dam Outlet Works with similar head, discharge, and gate type, yielded several examples.

Chute type basins are the typical stilling basins used by the USBR. Several examples of these types of basins, with similar head (100 ft) and unit discharge (220 cfs/ft) to the Ruby Dam Outlet Works are presented below:

- Wickiup Dam, Oregon. The dam is 87-ft-high with an outlet designed to discharge 3,950 cfs at the normal full reservoir water level. The outlet consists of two, 90-inch jet valves discharging into two, wall-separated Type III basins, with a unit discharge of approximately 263 cfs/ft. Each basin is approximately 15-ft wide and 50-ft long. The physical model study can be found in USBR's Hydraulic Laboratory Report No. 56 (USBR, 1939).
- Cherry Creek Dam, Denver, Colorado. The dam is about 145 ft above the stream bed and has a maximum capacity of 250,000 acre-ft. Three outlet conduits discharge a total of about 6,000 cfs at the maximum reservoir water level, or approximately 257 cfs/ft. The stilling basin consists of a parabola shaped chute followed by a 90-ft-long modified Type III basin with two rows of baffles/blocks and an end sill. The USBR model study can be found in Hydraulic Laboratory Report Hyd. 241 (USBR, 1948).

- Norton Dam, Kansas. The dam is about 100 ft above the stream bed. The outlet works exits from two 3-ft x 7-ft conduits, discharges into an 8-foot-wide, 55-ft-long vertically curved chute, and then reports to a 55-ft-long hydraulic jump stilling basin of Type II (Type III includes baffle blocks). The design outflow from the Outlet Works is about 385 cfs (128 cfs/ft) at the top of the flood control level. The physical model report can be found in USBR's Hydraulic Branch Report No. Hyd-497 (USBR, 1963).

Each of the stilling basins presented above resulted in similar depth, height and width ratios as that presented for the Ruby Dam basin. The results were generally similar in longitudinal profile and cross-section, in particular with the fins riding up on the walls as water is conveyed down the parabolic chute, spreading of flow prior to entering the basin, and the formation of a hydraulic jump within the basin.

Figure 15. View of chute and stilling basin operating at almost full discharge

PREDICTED VERSUS ACTUAL OPERATION OF BASIN

The basin at its full design flow (fully open at maximum normal reservoir level) appears to perform well with the jet contained inside the basin and no significant waves propagated to the downstream channel. The pool downstream of the basin was slightly under-built on the left side and was raised one foot to contain the full design flow. The topography of the pool was such that the right side was fully contained by an existing slope, whereas the left side was built up with fill, which was raised to contain the full design flow.

Air entrainment appears to be stronger than predicted in the numerical model, creating some splashing and mist in the air, most likely due to the incapability of CFD to predict splash under such a high pressure exit conditions. The intermittent splashes do not appear to cause any safety concern or added potential for scour around the basin, however, it was decided that wall at the downstream portion of the chute on the left side be increased to contain the expected splashes inside the walls.

No bed erosion downstream of the basin was observed.

CONCLUSION

Design of the stilling basin using CFD modeling predicted the hydraulic performance of the basin under actual operating conditions reasonably well, although actual air entrainment and misting were more than predicted due to the limitation of the CFD. CFD modeling was found to be a cost effective alternative to performing a physical model study under some circumstances, however, physical model studies may still be a critical component of the design especially when the hydraulics are complicated.

Construction of the project went well, with few Construction Change Orders. The cold weather encountered through most of the construction period was manageable using proper cold-weather concreting procedures and diligent monitoring of protection and concrete curing conditions. One of the most challenging issues arose when an existing 10-inch valve mounted on the old 72-inch guard gate body froze on a particularly cold night in January after the tunnel was opened up with the removal of the old butterfly valve. The 10-inch valve body partially cracked and spewed water under full-reservoir pressure into the bottom of the shaft. The cracked valve was unable to be removed because doing so would have been dangerous with it under full-reservoir pressure, and would have emptied the reservoir. The contractor was able to temporarily stop the leaks to proceed with the work and meet the project schedule. Final repair of the cracked valve had to wait until the reservoir level reached its minimum for the yearly cycle since the valve shaft had to be flooded to the reservoir level to equalize pressure on both sides of the valve, and the shaft flooding level had to be minimized for structural reasons. Divers were sent into the flooded shaft to remove the cracked valve and install a blind flange in its place.

ACKNOWLEDGEMENTS

AECOM would like to thank our client, Montana DNRC for the opportunity to design and oversee construction of this challenging and rewarding project. The Contractor, Johnson-Wilson Constructors, Inc. of Helena, Montana completed the project on time and on budget with excellent quality workmanship and construction project management and the Ruby Valley irrigators received their water that year as planned.

REFERENCES

Flow Science, Inc., Santa Fe, New Mexico, USA. *Flow-3D Users Manual.*

Peterka, A. J., 1984. Hydraulic Design of Stilling Basins and Energy Dissipators, A Water Resources Technical Publication, Engineering Monograph No. 25, U. S. Department of the Interior, Bureau of Reclamation.

USACE, 1980. Engineer Manual 1110-2-1602, Engineering and Design, Hydraulic Design and Reservoir Outlet Works, October 15, 1980, Department of the Army, Corps of Engineers, U.S.A.

USBR, 1987, Design of Small Dams, U.S. Department of the Interior, Bureau of Reclamation

REDUCING SCOUR WITH FLOW SPLITTERS, EXTENSION FINS AND COMPUTATIONAL FLUID DYNAMICS

Duane M. McClelland, P.E.,[1]
Daniel Morse, Ph.D.[2]
Nathan M. Gullo, P.E.[3]
Jason A. Eichler, P.E., PMP[4]

ABSTRACT

Computational Fluid Dynamics (CFD) provided key insights to reduce scour potential at Lane City Dam, a diversion dam located on the lower Colorado River in Texas. The dam breached soon after it was built in 1984 and has since experienced progressive bank and channel erosion, requiring several repairs to address scour, undermining, headcutting and structural damage. During rehabilitation design, CH2M used CFD simulations to evaluate hydraulic attack, spillway geometries, and gate operations to design scour reductions.

Over the range of flows, there are both low and high tailwater conditions outside the standard stilling basin design range. The dam has multiple crest elevations, a bascule gate, and sloped abutments that contribute to complex three-dimensional flows. CFD modeling demonstrated significant differences in the location, magnitude and direction of near-bed velocities at different flow rates, and confirmed high velocities at locations of observed scour. High velocities from lateral sweeping flow, plunging flow and vertical and horizontal back eddies were particularly important. It was also found that the benefits of opening or closing the bascule gate depended on the flow rate.

CFD modeling demonstrated hydraulic benefits from the rehabilitation design, which includes concrete dividing walls between tiered stilling basins, two sheet-pile walls (fins) that extend downstream of the stilling basin to interrupt sweeping and recirculating flow patterns, and selective use of bank revetments and channel armoring. Examples also illustrate the use of CFD to validate design performance for a spillway, conveyance system and river outfall for the proposed off-channel Lane City Reservoir.

INTRODUCTION

Lane City Dam (LCD) is a run-of-river diversion dam that is owned and operated by the Lower Colorado River Authority (LCRA) in Texas. LCD is located approximately 950 feet downstream of LCRA's Gulf Coast Irrigation Division Plant No. 2, which is a pump station that lifts Colorado River water from the LCD impoundment to a canal for irrigation deliveries. In 2015, CH2M completed design of Lane City Reservoir, which

[1] Water Resources Senior Technologist, CH2M, 1100 112th Ave NE, Ste 500, Bellevue, WA 98004; (425) 233-3432; Duane.McClelland@ch2m.com
[2] Hydraulic Engineer, CH2M, 1100 NE Circle Blvd, Site 300, Corvallis, OR 97330; (541) 768-3212; Daniel.Morse@ch2m.com
[3] Senior Engineer, Safety of Dams Lead , LCRA, Box 220, L300, Austin, TX 78767-0220; (512) 578-3326; Nathan.Gullo@lcra.org
[4] Senior Engineer, LCRA, 3700 Lake Austin Blvd, Austin TX 78703; (512) 578-7782; Jason.Eichler@lcra.org

will be a new 40,000 af off-channel reservoir impounded by a ring dam located on the Colorado River floodplain. A new relift pump station will deliver water from the canal to the reservoir (see Figure 1). By cycling the reservoir and improving overall Colorado River water delivery efficiencies, LCRA will achieve both firm delivery objectives and additional average-year deliveries far beyond the reservoir capacity. Construction began in early 2016 and the reservoir is expected to be operational in 2018.

Figure 1. Location of LCD relative to new off-channel reservoir project.

LCD (Figure 2) was built in 1984 to provide an adequate pumping pool during low river flows. It has gained importance to ensure reliable diversions to the new off-channel reservoir. The diversion dam has a history of failure and near-failure, and remains in disrepair due to progressive scour of the channel and its banks. CH2M assessed the condition of the dam and developed alternatives to reduce the risk of dam failure, loss of diversion pool, and associated channel migration and erosion at the pump intakes.

Figure 2. Lane City Dam

The selected alternative was the installation of a continuous cantilever sheet-pile cutoff at the end of the LCD stilling basin, extending across adjacent banks and floodplains. In this way, primary failure modes would be addressed and repetitive repair and armoring of the downstream banks could be reduced. To ensure adequate sheet-pile embedment, it was important to limit scour potential along the piles. In light of the persistence and complexity of historical scour, computational fluid dynamic (CFD) models of the channel and its range of flow regimes were developed to assess means to reduce or manage scour.

WHAT IS CFD?

The governing equations of fluid motion describing momentum are highly non-linear. These are solved together with mass conservation (continuity equation) to form the Navier-Stokes equations. Most commonly a turbulence closure model is also included (Ferziger and Peric, 2002). Analytical solutions of fluid flow exist only for simple geometries and boundary conditions. CFD is a computational, iterative approach to solving flow within a given domain. This is done by subdividing the domain of interest into discrete sub-volumes and applying numerical schemes to solve the fundamental fluid equations simultaneously within each sub-volume across the entire domain. CFD can then be applied to any 3D volume, regardless of geometry.

The resulting power of CFD is that inputs and solved variables are available anywhere within the 3D fluid domain, supporting evaluation of complex 3D hydraulics for various geometries, flow rates and boundary conditions. Graphical and numerical CFD output is flexible, supporting both informative evaluations and satisfying data presentations. CFD is often used as a more efficient and economical surrogate for physical models, or as a means of pretesting a proposed physical configuration before it is built.

Like any model, CFD results must be interpreted and tested when uncertain. The validity of outputs are commensurate with the accuracy and detail of the inputs; model setup and interpretation of results requires considerable experience, skill and judgment. The accuracy of additional models which handle physics such as turbulence, solids transport, or chemical and biological processes can add to the uncertainty of a given model.

Sensitivity testing of CFD inputs is one way to test CFD model results. Analysis of physical modelling results is another, and can provide additional information for use in altering the model to explore multiple model configurations. In the case of LCD, the fine sediment in its mobile-bed channel does not scale well, so CFD modeling offered advantages over physical modeling by providing a lower-cost alternative and flexibility to explore numerous alternative scenarios while potentially providing more meaningful results. Key to confidence in the LCD results was validation against the historical performance of the LCD channel itself as a full-size physical model for comparison.

CFD can be used to explore a wide range of complex hydraulic questions. Although not described in this paper, examples of CFD applications that led to significant design improvements for the larger Lane City Reservoir project are identified on Figure 3. One challenge of CFD modeling is that compelling insights can spawn new questions that beg for additional CFD runs. Wise managers will carefully frame initial questions and target follow-up runs to efficiently answer questions and manage costs.

LANE CITY DAM – A HISTORY OF SCOUR

The original LCD had a hydraulic height of 9 feet and consisted of a two-stage concrete weir mounted on a steel PZ 27 sheet-pile cutoff wall. A 25-feet-long SAF-style drop-structure stilling basin[5] spanned the normal-flow channel. The weir's second stage was two feet higher and spanned a sandbar on the right side[6] and the channel bank on the left side. Riprap provided its only stilling protection. Sheet-pile tie-back wingwalls were embedded in the river banks at each end of the dam. A 25-foot-wide bascule gate was located at the thalweg to lower the impoundment each winter for water rights.

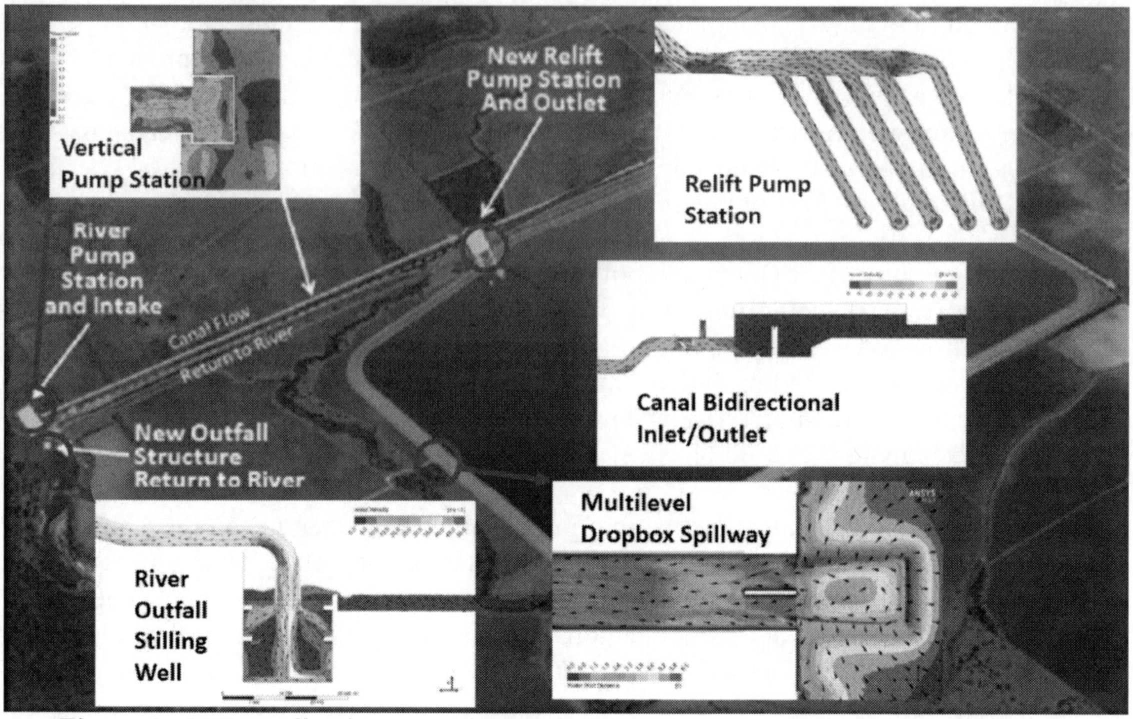

Figure 3. CFD applications to enhance design of the Lane City Reservoir Project

Figure 4 shows the general channel geometry before (upstream) and after (downstream) construction of LCD: Hydraulic attack caused by flow concentration, rollers, turbulence, and eddies have scoured away the downstream sandbar and the channel bed, banks and overbanks, resulting in channel enlargement, degradation, and undermining.

The diversion dam also became vulnerable to flanking failures. As illustrated on Figure 5, a 9,000 cfs flood occurred in 1984 and eroded portions of the right bank. A 51,600 cfs flood occurred a few years later and rerouted the channel through the west overbank. The impoundment was restored by filling the overbank area with a more erosion-resistant cap. As shown on Figure 6, in 1991 a 61,900 cfs flood nearly flanked the taller left abutment, migrating the bank 20 feet eastward and encroaching beyond the left wingwall.

[5] SAF structures were modeled at the St. Anthony Falls (SAF) Hydraulics Lab at the University of Minnesota. The stilling basin is a flat concrete apron with a row of baffle piers near a rectangular end sill.
[6] Left and right are looking downstream throughout this paper.

Figure 4. Left upstream channel resembles pre-construction; scour pool right.

Figure 5. Preliminary flanking scour through right overbank (circa 1984).

Figure 6. Near flanking scour of right bank (left downstream, right upstream).

A series of structural repairs and patches were constructed to sustain the diversion. The squared wingwalls were trimmed to the bank slopes and a concrete cap was extended to top of bank, concrete aprons were added downstream of the second-tier weir, sheet piles were installed along each bank to prevent flanking, cellular concrete mats (CCMs) were installed to stabilize the bank toe, and mounded riprap stabilized the left upstream bank. Subsequently, a 73,000 cfs flood in 1998 damaged the CCM bank protection and bank piles. During repairs in the 2000s, concrete mattresses and grout were overlaid on the damaged left-bank CCMs and portions of the right-bank CCMs were grouted. Palisades (pilings with netting between) were added at the downstream end of the CCMs to arrest headcutting bank erosion that was working its way behind the bank piles.

Despite these measures, progressive photographs show that degradation has continued. As shown on Figure 7, the stilling apron has cracked and settled where it was undermined in front of the bascule gate and subsequently backfilled with grouted riprap in 1994. Underwater photography revealed additional voids under the stilling basin toe key in 2014.

The left and right extension aprons are also undermined. Figure 8 shows the right extension apron, where flooding in 2013 first caused riprap settlement followed by sediment re-aggradation. Figure 9 shows progressive undermining settlement of the left revetment at the left apron and wingwall. As the revetment settled downstream, the left bank piles were also exposed, cantilevered over 10 feet, with some failures. Revetment and bank pile failure has also begun at the downstream end on the right bank (Figure 10). Figure 11 shows that headcutting erosion has begun to flank each bank.

Figure 7. Stilling basin (left 1994 settlement repairs; right 2014 photo of scour cavity).

Figure 8. Right apron (left 2013, right 2014 after settlement and re-aggradation).

Figure 9. Left apron (left 2000, right 2013 settlement scarps).

Figure 10. Bank revetment and pile failures on left and right banks.

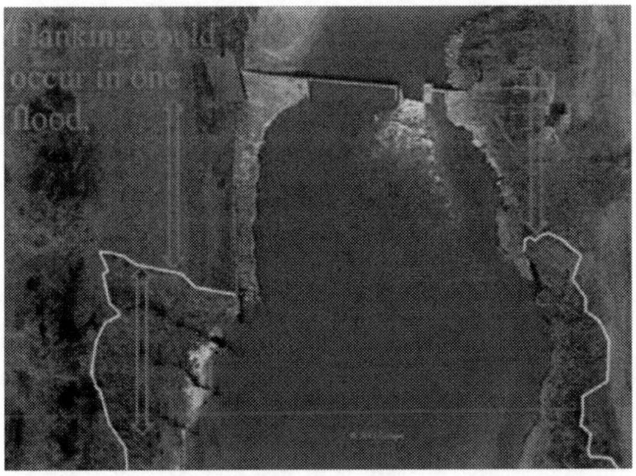

Figure 11. Headcuts flank the bank piles.

LCD is subject to extreme variations in flow. Low flows are typically a few hundred cfs. Floods around 30,000 cfs in Figure 12 submerge the dam two or three times a decade. Channel submerging flows of about 85,000 cfs 30 feet above the dam crest have a 100-year recurrence interval. The LCD rehabilitation project is being undertaken to reduce the risk of LCD failure over this full range of flows to maintain an adequate diversion pool.

Figure 12. Floods left to right: March 15, 2007; November 3, 2013; May 29, 2015

A study of historical aerial photographs and channel geomorphology indicated that the bend in the river occupied by LCD and the upstream pump station was naturally trying to move south (away from the pump station intake and into the downstream left bank). In the mid-1990s, the channel did shift south, leaving the pump intakes buried in a sandbar. The channel was retrained toward the pump intakes via upstream Palisades and channel excavation. Dam failure or flanking could potentially promote rapid channel migration, either away from the pump intakes or toward the pump station itself.

LANE CITY DAM DESIGN OBJECTIVES

The design concept to harden LCD against scour-induced failure was to install a row of cantilever sheet piles along the toe of the three existing aprons, repair or replace aprons where necessary, and extend the pile cutoffs up each bank and across the overbank to protect against a repeat of previous flanking failures. To reduce costs, this concept was scaled back from alternatives involving tie-back piles, circular gravity structures, a sloping grouted riprap apron extension, or dam replacement.

Originally constructed across a sandbar, the sandbar has since been scoured away, leaving an asymmetrical dam in a destabilized channel. The key design concept was to isolate repairs near the dam to reduce the need to maintain the failing bank protection. If much of the existing bank revetments were to be considered sacrificial, then it was imperative that the new cantilever piles be designed for any scour depth and pattern that might reasonably occur over the next 50-year life cycle. Bathymetry indicated deep scour in the channel, with a plunge pool as much as 25 feet lower than the original channel invert, and deeper scour depths possible prior to re-aggradation.

The primary objectives of CFD modeling were to evaluate the following questions:

1. Does the dam asymmetry promote scour?
2. How does bascule gate operation affect scour?
3. What is a reasonable design scour depth?
4. Optimize design configurations to limit required revetment preservation.

LANE CITY DAM CFD MODEL

To analyze spillway and river flow at LCD, CH2M used ANSYS CFX version 15 CFD software. The fluid domain was modeled using an inhomogeneous, multiphase approach, including separate momentum and velocity fields for both water and air. Turbulence was modeled using the Shear Stress Transport (SST) model. The SST model, developed by Menter (1993, 1994), allows the model to effectively utilize the best of two types of turbulence models. The near wall turbulence is handled using the more robust turbulent energy, turbulent frequency k-w model. The free stream portions of the flow can be handled using the turbulent energy, turbulent dissipation rate k-e model, which is more suited to the physics away from the wall. For additional details, the reader is pointed to Menter (1993, 1994).

Existing topography, bathymetry and design drawings were used as a basis to develop a 3-D surface of LCD and its stilling basin, aprons, channel and banks, as shown on Figure 13. This surface was later refined to more accurately capture each end of the dam, test design modifications, and evaluate effects of channel scour on near-bed velocities.

Headwater and tailwater rating tables from a 1-D HEC-RAS model were selected for upstream and downstream boundary conditions. CH2MHILL modified the existing model in the vicinity of the dam to reflect surveyed geometries. The bounding cross sections were selected far from the dam to minimize error propagation. Table 1 shows a limited number of flow rates that were tested to represent the range of anticipated design flows.

Since the goal of the study was to understand and reduce scour potential, the primary interest was visualizing 3-D flow patterns within the channel and high-velocity hot-spots near the riverbed, especially those near the dam that could expose cantilever cutoff piles.

The first step was to validate that the existing-conditions model was reasonable by explaining and illustrating both observed surface flow patterns and near-bed velocities that have caused or sustained the existing riverbed scour pattern. In this way, the site itself was used as a full-scale model for CFD comparison.

Figure 13. Existing LCD design surface

Table 1. Boundary Conditions

Approximate Scenario	Total Flow Rate (cfs)	Tailwater Elevation (feet)
Max gate-only flow	1,450	46.24
Max flow without operating the second-tier weir and aprons when gate open	4,000	50.69
Submerged basin flow	7,000	53.52
2-year storm	15,000	58.72
Recent bankfull flow rate	30,000	65.98
Maximum flood experienced	75,000	78.86

In this mobile-bed environment in which upstream sediment is constantly replacing downstream sediment, the focus was not on a threshold velocity that could mobilize sediment, but rather relative velocities that might preferentially promote scour or deposition relative to nearby areas. Near-bed velocities were taken six inches above each surface to avoid the zero-velocity boundary conditions at the surfaces themselves.

LCD CFD RESULTS - EXISTING CONDITIONS

CFD modeling indicated that normal flows through the open gate may help sustain the deepest scour hole, which is located left of the gate and 90 to 120 feet downstream. As shown by surface turbulence on Figure 14 (also Figures 2, 4 and 11), water passes asymmetrically between the gate walls and trends left toward the scour hole, parallel to the scoured bank. This asymmetry is caused first by asymmetry at the gate entrance (the gate vault extends upstream), and then is reinforced by asymmetry in the downstream grouted riprap repair (Figure 7) and the bathymetry. White water at only the right side of the gate walls (Figures 2, 11 and 14) indicates flow separation as water hugs left.

This same flow separation was made visible at the surface by the CFD model at 1,450 cfs. The first three images on Figure 14 show the flow separation, which disappears on the bottom right graphic of Figure 14 when a gate-wall extension is added to create a symmetric gate entrance. The slower blue arrows on either side of the gate jet generally point upstream as water eddies both horizontally and vertically. These eddies are visible during gate operations. Eddies reach farther upstream on the bottom right graphic of Figure 14 because flow through the gate has been straightened, opening up the sides.

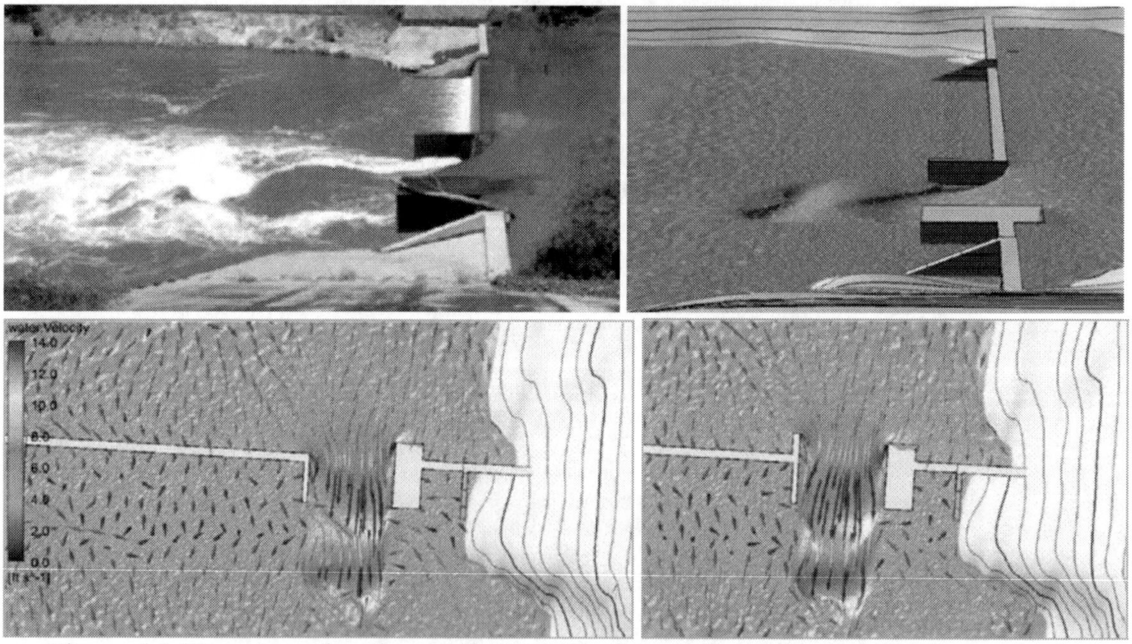

Figure 14. Top: Actual and modelled 1,450 cfs gate flow. Bottom: surface velocity vectors without (left) and with (right) gate-wall extension

Streamlines on Figure 15 show that high-velocity near-bed flows curve across the center of the scour hole and may resist low-flow aggradation when the gate is open. Subsequent analyses indicated that high-velocity sweeping flows approach the deepest scour hole from the left extension wall at 7,000 cfs and the left wingwall at 15,000 cfs; as a result, scour on the left side of the channel appears to originate from different locations at different flow rates, originating farther up the bank at higher flows.

Figure 15. Streamlines without gate-wall extension

Figures 14 and 15 show a double hump in gate discharges that can be seen at the surface in Figure 2 as flow rises over the baffle piers and riprap to fall toward the channel bottom. High-velocity red flow lines beyond the end of stilling basin indicate that the stilling basin is ineffective at forcing a hydraulic jump for lower gate flows. Analyses confirmed that the low-flow tailwater is too shallow, allowing basin blowout.

Figure 16 shows that the central stilling basin performs well at 4,000 cfs with the gate up. However, high-velocity flows from the sides of the channel cross the higher wingwalls and aprons without being stilled, as indicated by the red arrows in Figure 16. At this flow rate, the side flows are forced up and over the diving central flows, creating horizontal and vertical back eddies at the surface across most of the channel. Figure 17 shows that the eddies form into three cells mid-depth, and these cells confine the highest velocities to the center of the channel near the riverbed.

By 15,000 cfs, the flow pattern in Figures 16 and 17 is nearly reversed. Figure 18 shows that dominant flows down the center of the channel plunge less as the dam is submerged, rising higher in the channel and compressing side eddies. This opens up space for horizontal and vertical eddies to develop near the channel bottom, carrying plunging flow from each side across the bank revetments, up the center of the channel and into the stilling basin, to upwell in an eddy into the surface flow as indicated by the red arrows.

Figure 19 illustrates that once the dam is submerged, overtopping flow is buoyed by excessive tailwater beyond the stilling basin and recirculating undercurrents persist upstream to the dam face, contrary to what is commonly assumed for an end sill. This backflow into the stilling basin has been documented in the laboratory (Hanna 2014) and explains why there is a perpetual matt of sediment within the LCD stilling basin: it has likely been deposited primarily from downstream. Depending on local velocities and the portion of the hydrograph, the upstream flow could either scour or deposit at the dam toe.

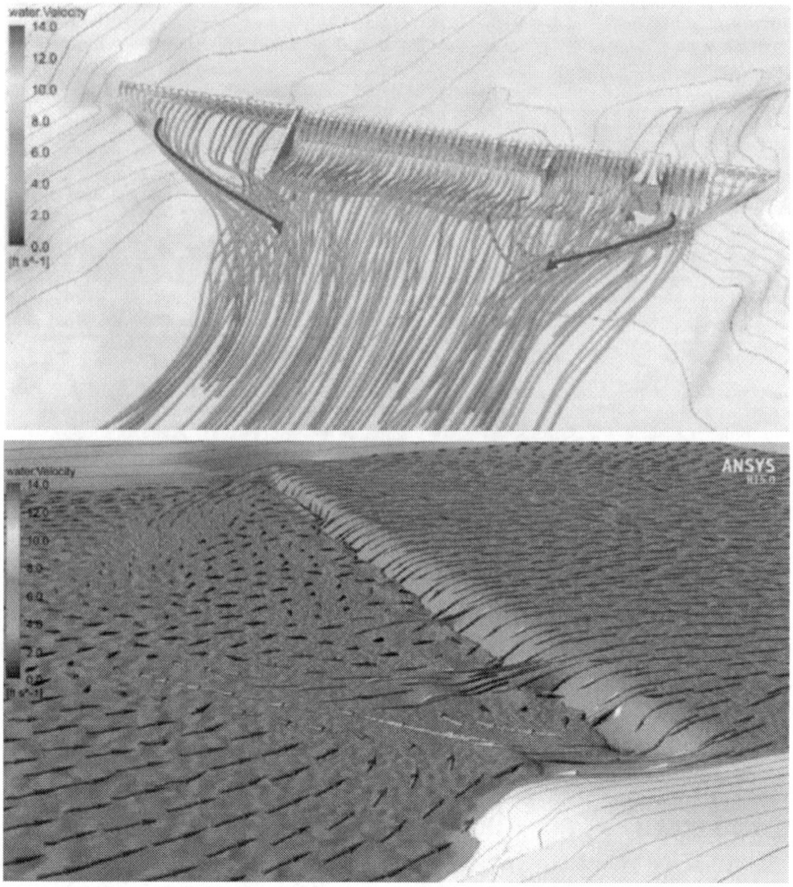

Figure 16. Streamlines at 4,000 cfs with gate closed.

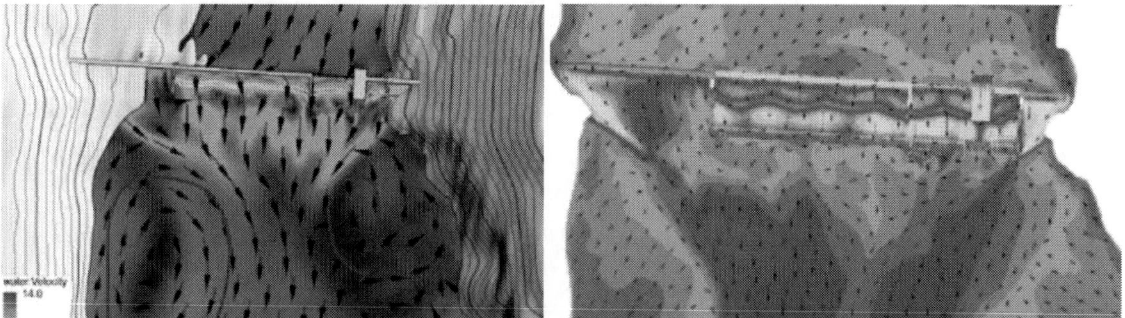

Figure 17. Velocities mid-depth (left) and near bed (right) at 4,000 cfs with gate closed.

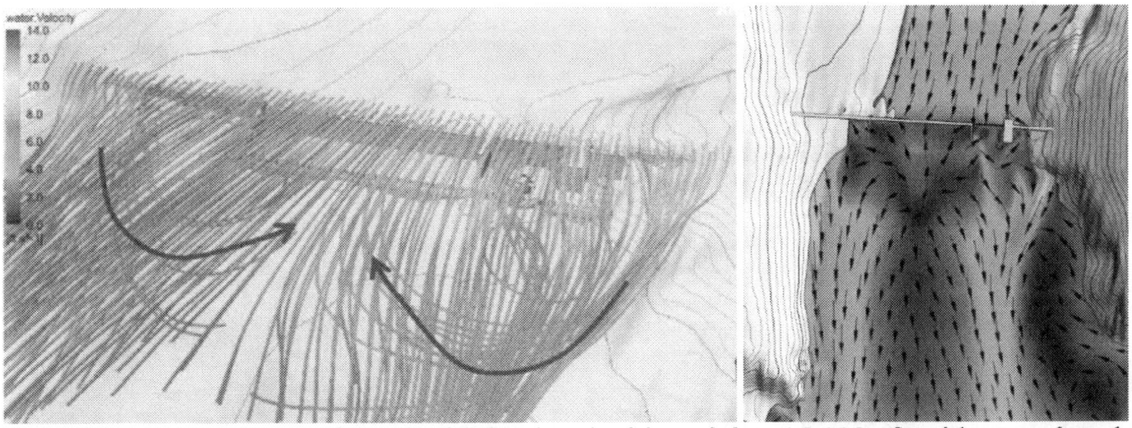

Figure 18. Streamlines (left) and mid-depth velocities (right), 15,000 cfs with gate closed.

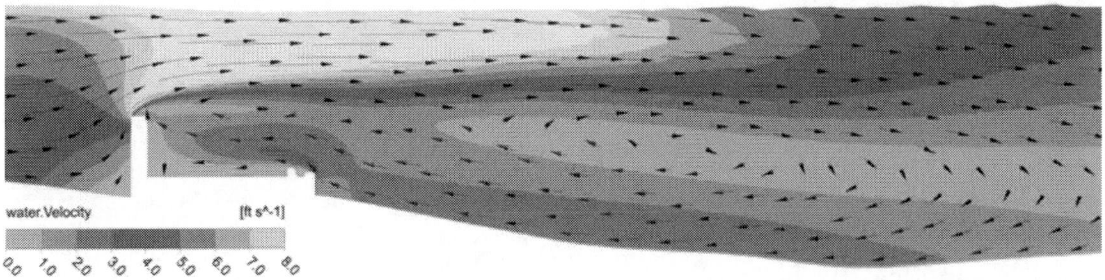

Figure 19. Profile showing upstream flow within stilling basin at 30,000 cfs

Figure 20 shows the near-bed velocities with the gate up for flows of 4,000 cfs, 7,000 cfs, 15,000 cfs and 30,000 cfs. By 7,000 cfs, high velocity flows (pink) sweep laterally across the toe of the stilling basin from the left wingwall and left back eddy. The flow is actually helical as it interfaces with vertical eddies at the toe and within the stilling basin. The high helical velocities against the stilling basin toe key have high potential to mobilize sediment and are located where a cavity was scoured under the stilling basin, causing the stilling basin to crack and settle. The cavity and scour hole were backfilled with grouted riprap, but peripheral undermining (Figures 7 to 9) continues. When the gate is closed, flows around 7,000 cfs may produce the most scour near the gate toe key.

By 15,000 cfs, high-velocity flow on the left has moved downstream near the deepest bathymetry (compare to Figures 13 and 15). Similar flows may have originally formed the deep scour hole; comparative bathymetry indicates that the hole's location migrates over time. By 30,000 cfs, the side eddies become displaced and virtually all surface flow is parallel. The largest near-bed velocities are concentrated on the bank revetments and toes; flows of this magnitude and larger may account for much of the bank damage.

Within the range of relatively frequent historical floods, the graphical results of the CFD modeling for existing conditions corresponded well with observed hydraulic behavior and helped explain the known scour patterns on the banks and within the channel. The next step was to modify the validated existing-conditions models to explore various design alternatives to modify the dam and reduce scour potential.

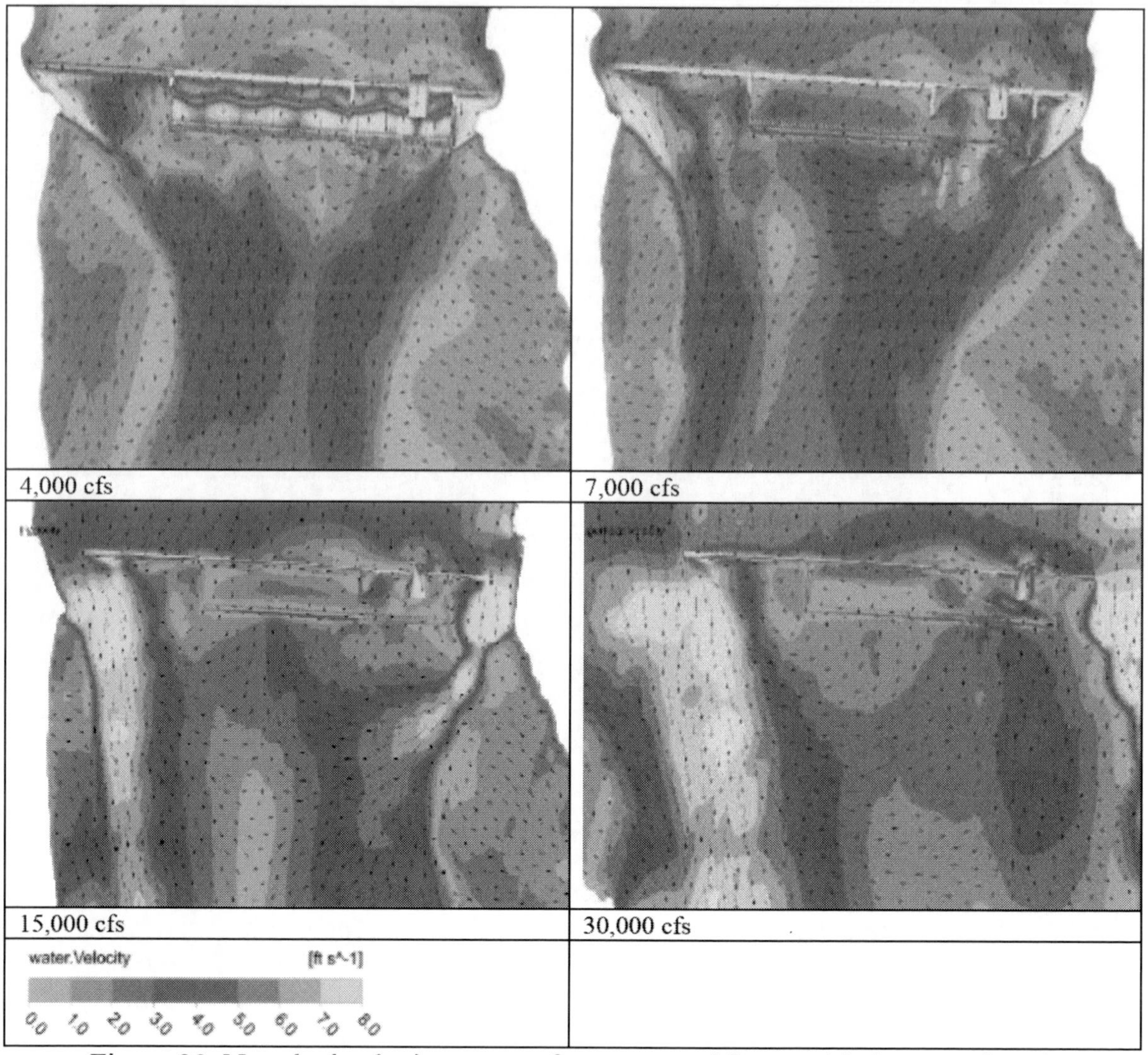

Figure 20. Near-bed velocity vectors for a range of flows with the gate closed.

LCD CFD RESULTS – DESIGN MODIFICATIONS

The final design configuration is illustrated on Figure 21. Primary modifications consisted of a straight row of cantilever piles, extending the stilling basin to integrate with the new piles, replacing the undermined side aprons with stilling basins, lowering and resizing the stilling basin in front of the gate, smoothing the right wingwall transition, and training flow with five divider walls (including the two gate walls) and two extension fins that extended the outermost training walls downstream 25 to 30 feet using sheet piles.

Divider Walls and Extension Fins

CFD modeling showed that the divider walls and downstream extension fins were particularly effective at reducing near-dam, near-bed velocities by intercepting lateral underflows. Figure 22 shows an example of pronounced helical flows within the stilling basin margins. In this early alternative, helical flows were drawn laterally by eddies plunging under the gate flow. Note that flow readily jumped submerged dividing walls. When jumping a wall, the flow tended to plunge toward the bottom, increasing near-bed velocities along the sheet-pile cutoff compared to no walls. Diverse CFD runs showed that the location, height and length of the divider walls and fins were all important, and performance varied with flow rate. When properly placed and sized, the dividing walls and extension fins helped turn the flow downstream, disrupting near-dam underflow.

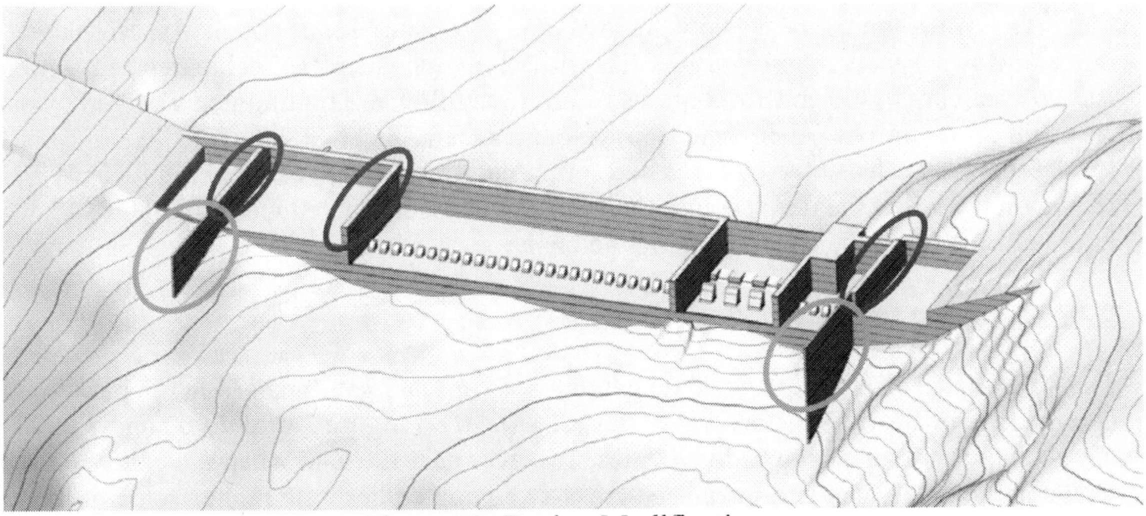

Figure 21. Design Modifications

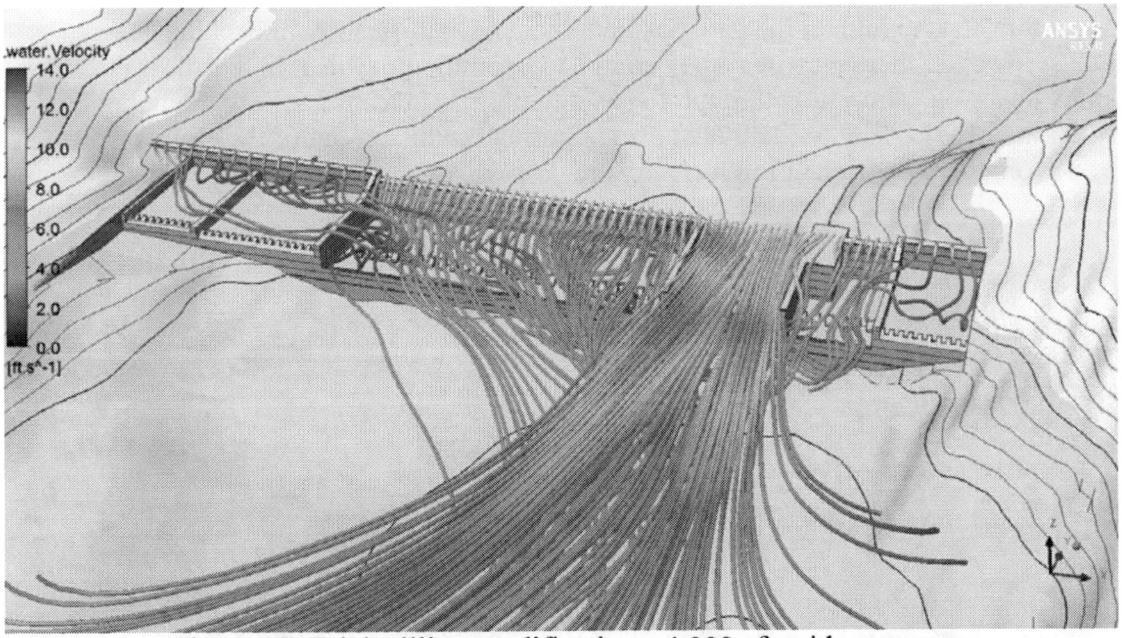

Figure 22. Trial stilling modifications, 4,000 cfs with gate open

Many extension fin configurations were evaluated. In general, an increase in the number of fins improved flow alignment. To manage costs, a final configuration of two fins was selected that minimized the number of fins while providing good overall protection against scour between the fins. Combined with extending the gate walls, this also eliminated the need for a symmetric gate entrance upstream. An isometric view of this configuration is shown above on Figure 21. The basin walls that help split flows are indicated in red and the fins are indicated in green. The right fin extends 25 feet downstream of the wall that separates the right extension-wall aprons, and the left fin extends 30 feet downstream from the left counterfort. The fins and their upstream divider walls were each raised 2 feet higher than the upstream dam crest to reduce the tendency for high-velocity flows to skip over the walls and plunge toward the channel bottom.

The effect of fins on the near-bed velocities (and associated scour potential) across a range of flows is shown on Figure 23. The fins reduce the near-bed velocities between the fins downstream of the endsill. Regions of improved flow are highlighted with red ovals. At 4,000 cfs overall near-bed, near-dam velocities are reduce. At 7,000 cfs the hot (pink) spot at the sheet piles at the gate has been eliminated. At 15,000 cfs the high velocity region on the left at the deep scour hole has been significantly reduced. At 30,000 cfs, high velocities near the dam and along the bank toes has been reduced.

Gate Operation

The CFD model was used to evaluate gate-closed (up) and gate-open (down) operations over a range of flows. Near-bed velocities are shown on Figure 24 for 4,000 and 7,000 cfs. At these flow rates, the bed velocities are lower near the dam when the gate is closed. Opening the gate accentuates near-bed cross-sweeping eddies with higher velocities along the central stilling basin sheet-pile cutoff, and could potentially induce scour.

At 15,000 cfs and higher, the gate position does not significantly impact the velocity field. However, there may be some benefits to opening the gate at higher flow rates to allow lower-elevation flow through the gate to help displace underflow, reducing cross-sweeping near-bed eddies. It should also help flush sediment out of the reservoir, which may promote deposition at low-velocity areas downstream of the dam.

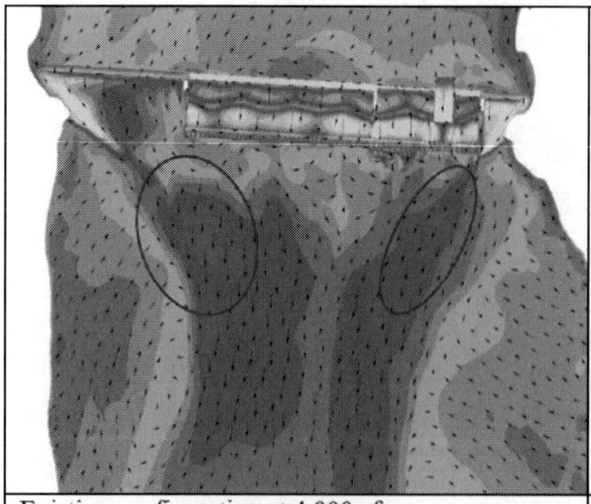
Existing configuration at 4,000 cfs

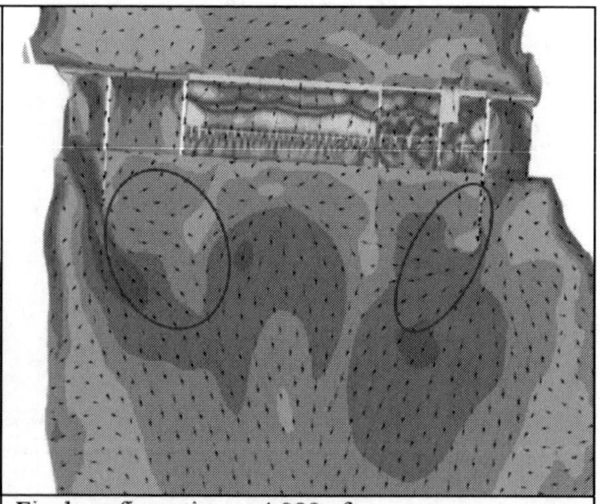
Final configuration at 4,000 cfs

Figure 23 (1 of 4). Near-bed velocities, existing and final configurations with gate closed.

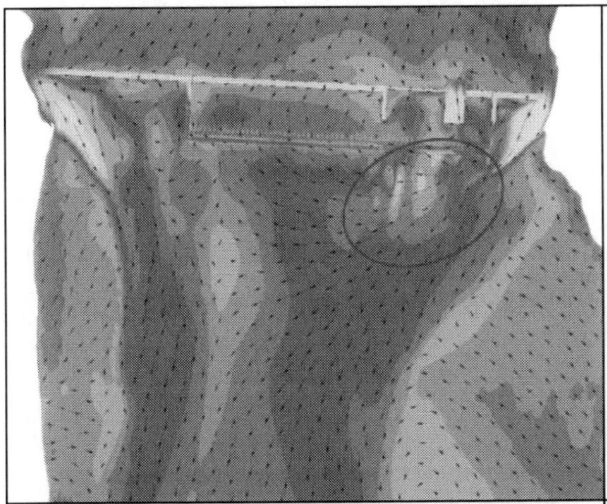

Existing configuration at 7,000 cfs

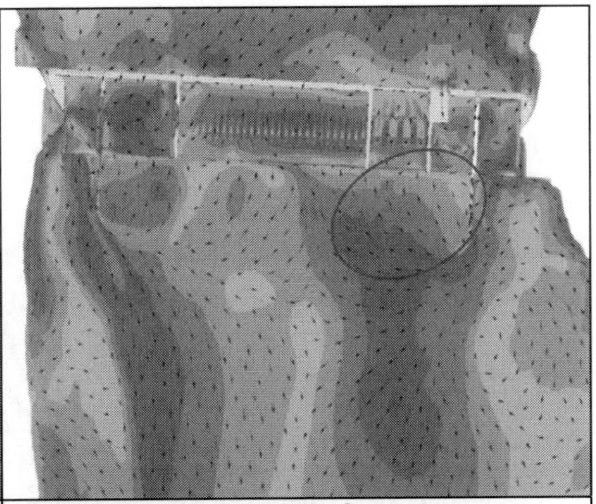

Final configuration at 7,000 cfs

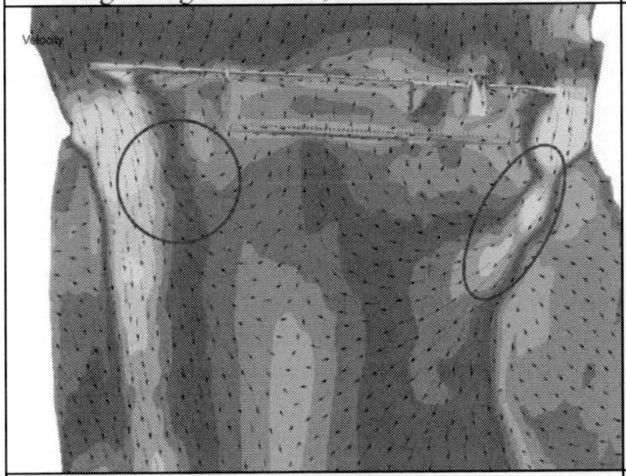

Existing configuration at 15,000 cfs

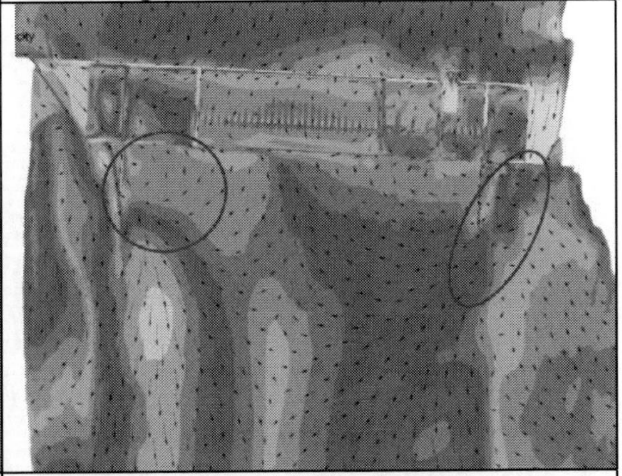

Final configuration at 15,000 cfs

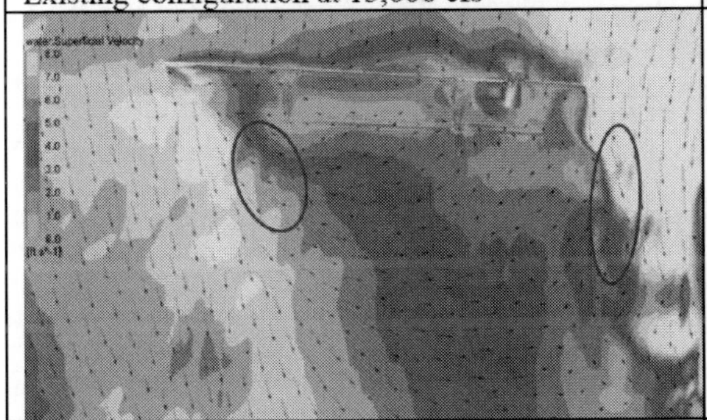

Existing configuration at 30,000 cfs

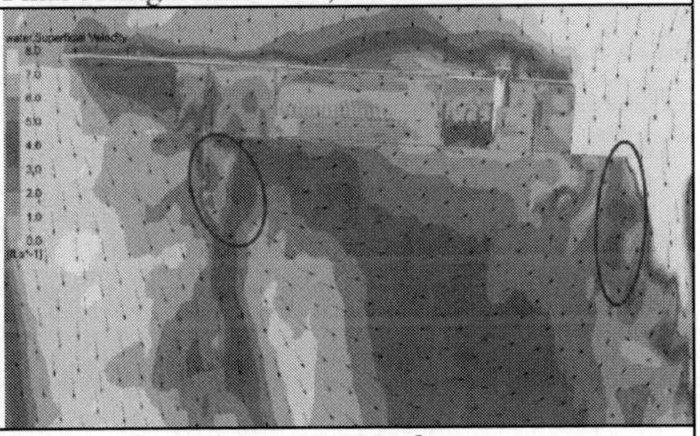

Final configuration at 30,000 cfs

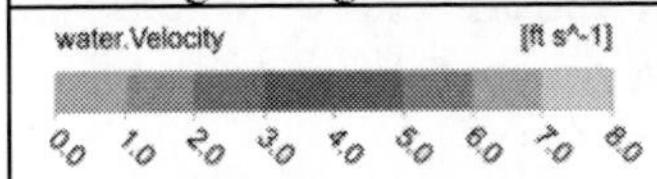

Figure 23 (continued). Near-bed velocities, existing and final configurations with gate closed.

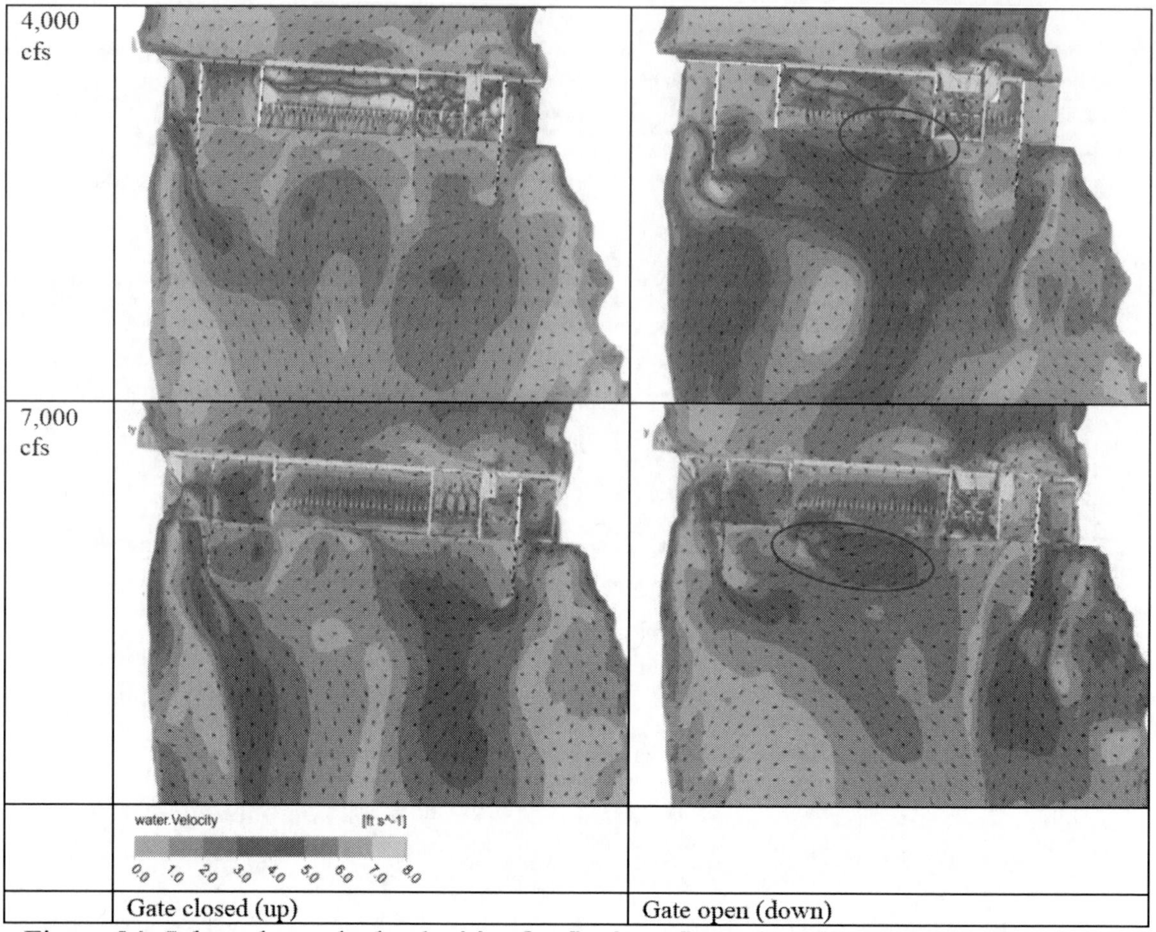

Figure 24. Selected near-bed velocities for final configuration with gate closed and open.

Design Scour Evaluation

A large number of different scour configurations may be possible downstream of LCD, depending on flow rates, durations and sediment loads during future floods, and how initial scour may change future flow patterns. CFD modeling demonstrated that near-bed velocities are sensitive to relatively small changes in both dam and channel geometry. Only a few scour configurations could practically be evaluated using CFD.

A key objective of selected scour comparisons was to determine if scour might be self-limiting by lowering near-bed velocities as channel depths increase, approaching a static state of equilibrium; or if local scour might promote high-velocity local eddies that could be progressive, or self-feeding, becoming more prone to scour as material is removed. This analyses was useful for identifying areas of greatest concern for installing or preserving revetments.

Independent of CFD, a design scour surface was postulated after analyzing bathymetric depths, patterns and changes over time, historical damage and flanking failures, subsurface sediment profiling and the existing shape of the channel and its banks. The first CFD scour scenario evaluated a total maximum scour down to the design scour surface at all locations. This is an unlikely scenario in which maximum scour is reached simultaneously at all locations, but provides a general sense of velocity reduction with channel depth.

The near-bed velocities at 30,000 cfs, which approximates the threshold velocity for greatest hydraulic attack on the existing revetments, are shown in the first row of Figure 25 for the existing surface and an approximation of the maximum design scour surface. The deeper design scour surface resulted in lower near-bed velocities. This is most evident downstream of the left wingwall and downstream of the right apron stilling basin. Some moderately high velocities remain outside the right fin and extending toward the middle of the channel from the end of the right fin; however, the extent and magnitude of these velocities have generally been reduced and kept at some distance from the cantilever sheet piles, limiting the potential for failure. Nevertheless, consideration should be made for protecting revetments where velocities are highest.

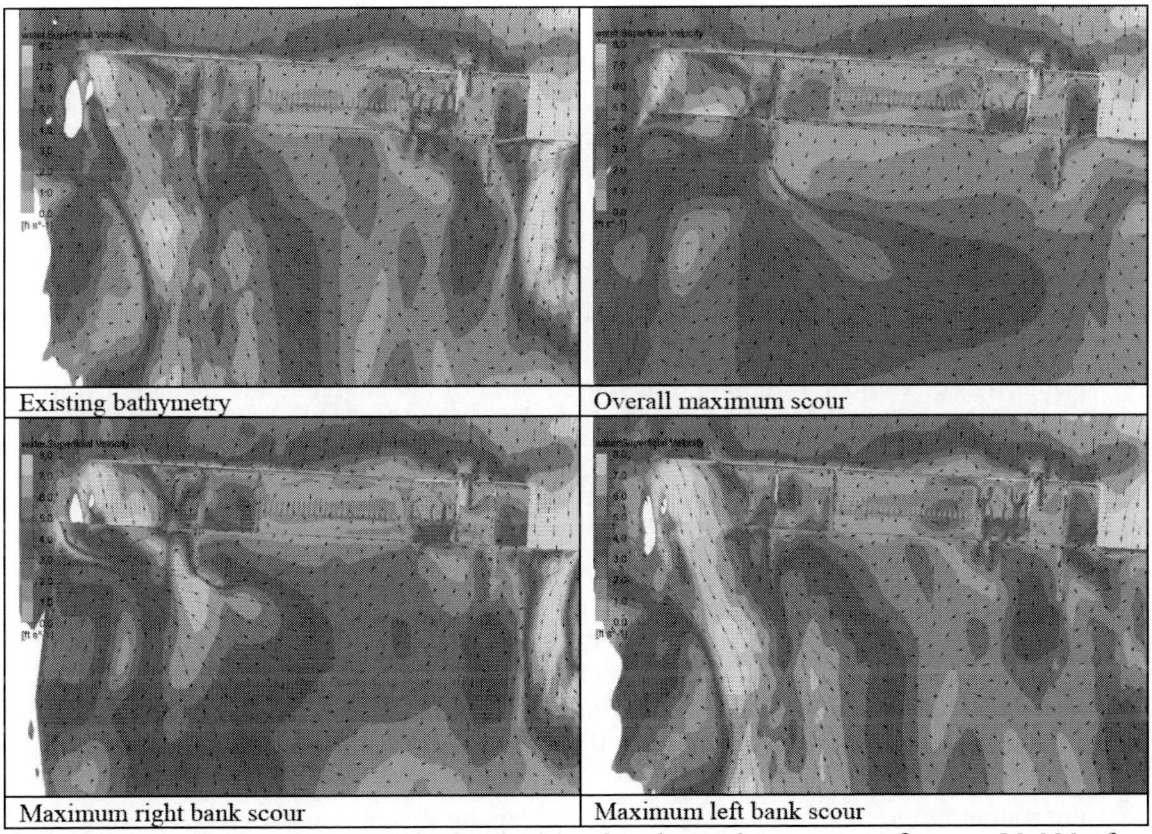

Figure 25. Comparative near-bed velocities for alternative scour surfaces at 30,000 cfs.

The second scour scenario entailed isolated scour where high velocities occur outside the right wingwall along the right bank. The existing and post-scour near-bed velocities along the right bank are shown in the top left and bottom left quadrants of Figure 25, respectively. The velocity on the left bank remains largely unchanged in this scenario. The high velocity region on the right bank decreases in size, but higher velocities develop near the dam inside and downstream of the right fin. This suggests that, should scour progress in this area, it may not be self-limiting within a reasonable depth, and revetment protection around the fin could be important for design longevity.

The third scour scenario entailed isolated scour where high velocities occur outside the left wingwall along the left bank. The existing and post-scour near-bed velocities along the left bank are shown in the top left and bottom right quadrants of Figure 25, respectively. The near-bed velocities on the left bank become low as scour progresses, suggesting a self-limiting condition. Also, scour on the left bank may slightly increase near-bed velocities on the right bank. These observations, along with previous observations, are conditioned on the extension fins remaining in place. Therefore, while less important on the left side, revetment protection around both fins appears warranted.

CONCLUSIONS

The following conclusions resulted from the CFD analysis of the Lane City Dam. Results were obtained for a variety of configurations and scenarios.

1. CFD modeling is a highly adaptive tool that offers a wide range of insights in diverse applications. In applying CFD to LCD, observed flow patterns were duplicated and high velocity hot spots were confirmed at locations with historical scour.
2. Scour damage at different locations was likely caused at different flow rates, ranging from relatively frequent flows around 7,000 cfs and higher. Lower flows through the open gate may help keep the deepest scour hole relatively free of deposition after a scour event.
3. At higher flow rates, flow from the banks becomes entrained below surface flow. This promotes high velocities near the riverbed and potentially near the dam. This is primarily due to flow sweeping in from the sides of the channel, and is not generally evaluated in stilling basin design for prismatic channels.
4. Various design options were evaluated and a final configuration was proposed which showed significant reduction in near-dam, near-bed velocities over a wide range of flow rates.
5. Primary hydraulic features of the final proposed design were raised divider walls and their extension walls, as well as modification of stilling basins, particularly downstream of the gate and the extension walls. The extension fins were especially useful at reducing lateral sweeping flows near the dam, but this relied on careful placement and integration with upstream divider walls that were raised higher than the dam crest to limit the potential for flows to drop over the wall and plunge toward the channel bottom.
6. For moderate floods of 4,000 and 7,000 cfs, raising the gate improved hydraulic conditions near the dam. At flows of 15,000 cfs and higher, some hydraulic benefit may result from lowering the gate to flush sediment out of the impoundment.
7. Many different scour configurations may be possible downstream of LCD, and therefore it is only practical to evaluate a few scour configurations using CFD. Based on preliminary CFD results, the following tentative conclusions were reached:

a. Broad scour across the entire channel may tend to be self-limiting by near-bed velocity reduction. Under these conditions, extensive failure of the sheet-pile wall was judged to be unlikely.
b. Similarly, if only the left side of the channel were to scour, generally outside the left extension fin, scour would likely become self-limiting, approaching scour equilibrium. However, this is only true if the sheet-pile fin remains in place. Therefore, revetment to protect the sheet-pile fin appears warranted.
c. If only the right side of the channel were to scour, generally outside of the right sheet-pile fin, near-bed, near-dam velocities may persist or shift, promoting progressive scour. Therefore, some protection of both the right extension fin and the right bank revetments near the dam appears to be warranted.

ACKNOWLEDGEMENTS

The authors would like to thank LCRA and its fantastic team that supported this paper and design of LCD and the Lane City Reservoir Project. About the authors:

Duane McClelland is a registered water resources engineer with CH2M in Seattle, Washington, who has provided leadership in design, planning and management for more than 25 years. For the past 18 years, he has specialized in water resources engineering, including hydraulic and hydrologic modeling and design, dam design and safety, feasibility and alternatives assessments, emergency preparedness, and multifaceted planning and decision support. Mr. McClelland has engineered small to large storage projects and hydraulic structures, coordinated teams of consultants in multidisciplinary water-resources planning, characterized project benefits and risks, and ensured the quality of complex hydraulic and hydrologic analyses. He offers strengths in group and process facilitation to help achieve solutions to multidimensional challenges.

Daniel Morse is a hydraulic engineer with CH2M in Corvallis, Oregon. He has 5 years' experience as a graduate research assistant in fluid mechanics at Oregon State University and 7 years as a hydraulic engineer for CH2M. He has primarily been involved utilizing CFD analysis for water, wastewater, and water resources projects. He has also developed CFD methods in specialized applications such as vortex identification in pump stations and sludge modeling in secondary clarifiers.

Nathan Gullo is a registered civil engineer at the Lower Colorado River Authority (LCRA) in Austin, Texas with 16 years of planning, design, and operations experience on dams, raw water, treated water, and wastewater facilities. He has served as the lead engineer managing the LCRA's safety of dams program for the past four years. Mr. Gullo has extensive experience as the owner's engineer for planning, design, operation, and management of water-resources systems and facilities for public utilities. This work has included independent hydrologic and hydraulic modeling as well as collaborative and review efforts with engineering consulting firms.

Jason Eichler is a registered civil engineer at the Lower Colorado River Authority (LCRA) in Austin, Texas with 19 years of experience performing preliminary engineering, design, and operations of water resources, water, and wastewater facilities.

He has served as the lead engineer for LCRA's Lane City Reservoir project for the past three years. As lead engineer, Mr. Eichler has been responsible for coordinating various engineering and operations teams during the design phase of the project.

REFERENCES

Ferziger, J.H., Peric, M., "Computational Methods for Fluid Dynamics", 2002, Springer-Verlag.

Hanna, L.J., "Mitigation of Stilling Basin Abrasion Damage," unpublished presentation at Dam Safety 2014, Association of State Dam Safety Officials, September 24, 2014, San Diego, California.

Menter, F. R., "Zonal Two Equation k-ω Turbulence Models for Aerodynamic Flows", AIAA 1993, Paper 93-2906.

Menter, F. R., "Two-Equation Eddy-Viscosity Turbulence Models for Engineering Applications", AIAA Journal, August 1994, 32 (8): 1598–1605

SEDIMENT TRANSPORT ANALYSIS IN SUPPORT OF PRADO BASIN SEDIMENT MANAGEMENT PROJECT

Craig P. Baxter, PE[1]
Nathan Scheevel, PE[2]
George W. Annandale, D.Ing, PE, D.WRE, F.ASCE[3]

ABSTRACT

Prado Dam is an earthen fill dam on the Santa Ana River in Riverside County near Corona, CA. The dam was constructed in 1941 by the US Army Corps of Engineers (USACE) as a flood control structure. Since that time, the dam has acted as a sediment trap. This has had the effect of significantly reducing the sediment load to the Lower Santa Ana River (LSAR). The storage capacity of Prado Basin has also been reduced, as anticipated during initial planning of the dam.

The Orange County Water District (OCWD) in conjunction with USACE is studying the feasibility of a complex sediment management project at Prado Basin. The project will be executed by removing sediment from within Prado Basin where the Santa Ana River flows enter the basin. The sediment would then be transported downstream of the dam at which point it will be re-entrained in water discharges from the dam. Potential benefits of the project include:

- Increased storage capacity in Prado Basin
- Improved aquatic habitat due to coarsening of the Santa Ana River bed
- Reducing incision in the Santa Ana River downstream of Prado Dam
- Potential increased infiltration rates in the OCWD groundwater recharge reach of the LSAR

One-dimensional sediment transport analyses in support of the project have been performed over the last several years. The effects of the project on sediment transport and deposition patterns has been estimated for approximately a 50-mile stretch of the Santa Ana River from the San Bernardino/Riverside County line to the Pacific Ocean. The simulation estimates the coarsening of the river bed upstream of Prado Basin, aggradation downstream of Prado Dam, and changes throughout the entire modeled reach of the river.

INTRODUCTION

The headwaters of the Santa Ana River are located in the San Bernardino Mountains in San Bernardino County, CA and the river flows approximately 75 miles to the southwest

[1] Senior Project Engineer, Golder Associates Inc., 44 Union Blvd, Suite 300, Lakewood, CO, 80228, Craig_Baxter@golder.com.
[2] Owner, Scheevel Engineering, LLC, P.O. Box 28745, Anaheim, CA, 92809, nathanscheevel@yahoo.com
[3] Program Leader, Golder Associates Inc., 44 Union Blvd, Suite 300, Lakewood, CO, 80228, George_Annandale@golder.com

to the Pacific Ocean. The river includes tributaries flowing from the San Gabriel Mountains to the north and the San Jacinto Mountains to the south.

Historically, the lower Santa Ana River watershed became a productive agricultural area supporting the citrus growing industry. During the 1920's and 1930's the area's population continued to grow as did flooding concerns. The construction of Prado Dam was authorized by the Flood Control Act of 1936. In 1938 a major flood resulted in 68,400 acres inundated and 19 deaths. This flood reinforced the previously identified need for Prado Dam. Figure 1 below shows the Santa Ana River watershed.

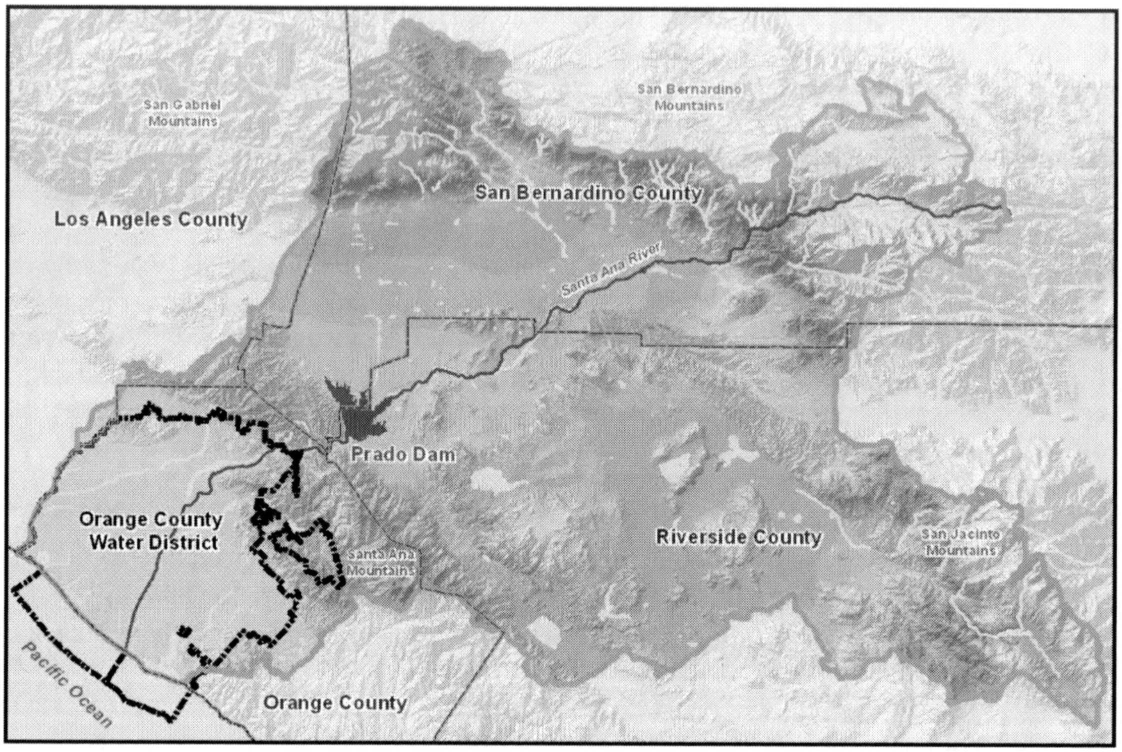

Figure 1. Santa Ana River Watershed Map

The US Army Corps of Engineers (USACE) constructed Prado Dam approximately two miles west of Corona, CA on the Santa Ana River. Construction on the earth fill embankment dam was completed in May 1941. Since construction, the natural sediment transport of the river has been disrupted by the dam which does not allow for sediment continuity to the downstream reach of the river.

Congress authorized the Santa Ana River Mainstem Project in 1986 and construction of the project began in 1989. The primary purpose of the project was to improve Prado Dam and the Santa Ana River Mainstem in general to provide adequate flood risk management features. Various project phases have been constructed between 1989 and present day. Several of the project phases designed and constructed in the recent past (10 years or so) have incorporated features to counteract matters arising from sediment transport issues. Focused efforts to analyze and plan for long-term sediment management solutions (specific to this report) began in the early to mid 2000's. Two separate, but connected, efforts by the USACE and OCWD to study and analyze sediment management solutions are ongoing. They include the Prado Basin Sediment Management Demonstration Project and the Prado Basin Feasibility Study.

Current Santa Ana River Issues

Upstream of Prado Dam. The discontinuity in sediment load has resulted in the accumulation of sediment and aggradation within the reservoir created by Prado Dam, Prado Basin. Since 1941, approximately 50,000 acre-feet of storage has been lost to sediment deposition according to a 2008 survey of Prado Basin, the last survey of the basin. Approximately 25,000 acre-feet (of the total 50,000 acre-feet) of deposition has occurred at elevations below the approved water conservation elevation, thereby reducing the region's local water supply potential. In addition to the loss in storage volume, the accumulation of fine sediments has diminished riverine habitat for the federally listed endangered Santa Ana Sucker fish, which is best suited for channels with steeper beds made up of gravel and cobble sized sediments.

Downstream of Prado Dam. The release of sediment starved water from the dam has created scour issues in the Lower Santa Ana River (LSAR) downstream of the dam resulting in degradation and incision of the channel immediately below the dam resulting in risks to civil infrastructure and loss of habitat. The Orange County Water District (OCWD) performs groundwater recharge operations within the river downstream of the Dam. Historically he LSAR bed has been composed primarily of sands which act as a natural filter medium, trapping silts and clays near the surface of the riverbed allowing subsequent flows to easily transport the low permeability silts and clays out of the recharge reach of the LSAR. A sand river bed provides a very efficient self-cleaning recharge facility. The natural supply of sediment (primarily sands) has been cut off from the LSAR, allowing a portion of the clogging sediments (silt and clay) to infiltrate deeper into the riverbed beyond the reach of flows that would have otherwise transported the silt and clay past the recharge reach. The reduction in storage capacity within Prado Basin caused by sedimentation has decreased the annual water conservation volume benefit and water conservation flow releases to the recharge reach which combined with the decreased infiltration capacity, exacerbates the lowered aquifer recharge flow. Beach sand replenishment is also reduced due to the lack of sand moving through the LSAR.

Studied Corrective Measures

OCWD in conjunction with USACE is studying the feasibility of several corrective measures to address the above issues. After conceptual analysis of each of the options, four options for sediment management within Prado Basin and in the Santa Ana River were analyzed further. All of these options include the removal of sediment from the upstream end of Prado Basin, transport to a location just downstream of the Prado Dam outlet works, and re-entrainment in the Santa Ana River within the dam outflow.

OCWD is planning a Sediment Management Demonstration Project (SMDP) which is a short-term sediment re-entrainment project. The goal of the demonstration project is to remove up to 500,000 cubic yards (yd^3) of sediment from Prado Basin and re-entrain it in the LSAR over a period of about 1 year. The re-entrainment period will be subject to storm water availability and releases from Prado Dam. Monitoring in the LSAR will take place during the SMDP to analyze how the LSAR responds to the influx of sediment.

The Feasibility Project, jointly managed by OCWD and USACE will be a larger scale project with the goal of re-entraining larger quantities of sediment (on average approximately 1,000,000 yd^3 annually) over longer durations (potentially up to 50 years).

Potential Sediment Management Benefits

The potential benefits of the Projects address all of the current issues laid out above. The storage capacity of Prado Basin will be maintained (or possibly increased) with the removal of sediment. Maintaining the storage capacity has the dual benefit of increasing the useful life of the Dam and increasing water conservation in the region. The removal of sediment will create a steeper channel bed which will be more conducive to gravel and cobble sediment sizes and restoring the Santa Ana Sucker habitat upstream of the dam.

Downstream of the dam, the sediment laden flow could help to stop and potentially reverse channel incision. Fine sediment sizes could flow into the recharge reach and help to sustain or restore the infiltration rates. Groundwater recharge in the LSAR relies heavily on the sand filter medium found naturally in the riverbed. The lack of replenishment sediment to the recharge reach of the LSAR will have severe long term impacts to the region's water supply. Renewed sediment transport to the LSAR (due to the sediment re-entrainment) will result in ocean and beach replenishment as well.

Benefits common to areas upstream and downstream of the dam include the preservation of civil infrastructure. Multiple projects have already been constructed, or are currently underway, to protect bridges, pipelines, freeways and river banks from excessive aggradation upstream of Prado Dam and excessive degradation downstream of Prado Dam. Sediment management would help preserve the constructed improvements and extend the life of assets that are currently at risk.

Objectives of Sediment Transport Analysis

One-dimensional sediment transport modeling using HEC-RAS was conducted in three phases in support of the SMDP and Feasibility Project. For the SMDP, the existing and proposed conditions in the LSAR were analyzed. For the Feasibility Project, the sediment transport was analyzed for the proposed condition in the LSAR as well as for the existing and proposed condition in the Santa Ana River upstream of Prado Basin.

The goal of the modeling was to study the river's response to the increase in sediment load in the LSAR as well as the removal of sediment in the Santa Ana River upstream of Prado Basin. The locations and magnitudes of significant sediment deposits, sediment sizes depositing in certain areas, slope changes, and changes to flooding risk due to deposited sediments were reported.

SEDIMENT TRANSPORT MODELING FOR PRADO SEDIMENT MANAGEMENT DEMONSTRATION PROJECT

Introduction

The purpose of the SMDP sediment transport modeling is to analyze the effect in the LSAR of an increased sediment load due to re-entrainment. Field implementation of the project would include extensive monitoring and data collection which would be used in future modeling efforts to provide refined recommendations to implement long-term sediment management for the system.

Base Model Setup

Geometry. Model geometry was obtained by USACE. The geometry includes all the bridges and drop structures present from Prado Dam to the Pacific Ocean. The geometry was broken into three reaches for the purpose of this analysis due to the presence of inflatable rubber dams used to divert flow from the river at two locations. Because of the dynamic nature of the rubber dams, a rating curve is required to accurately represent the sediment transport at the rubber dams which can only be input at the downstream end of a model geometry. Figure 2 shows the modeled reach and several bridge crossings.

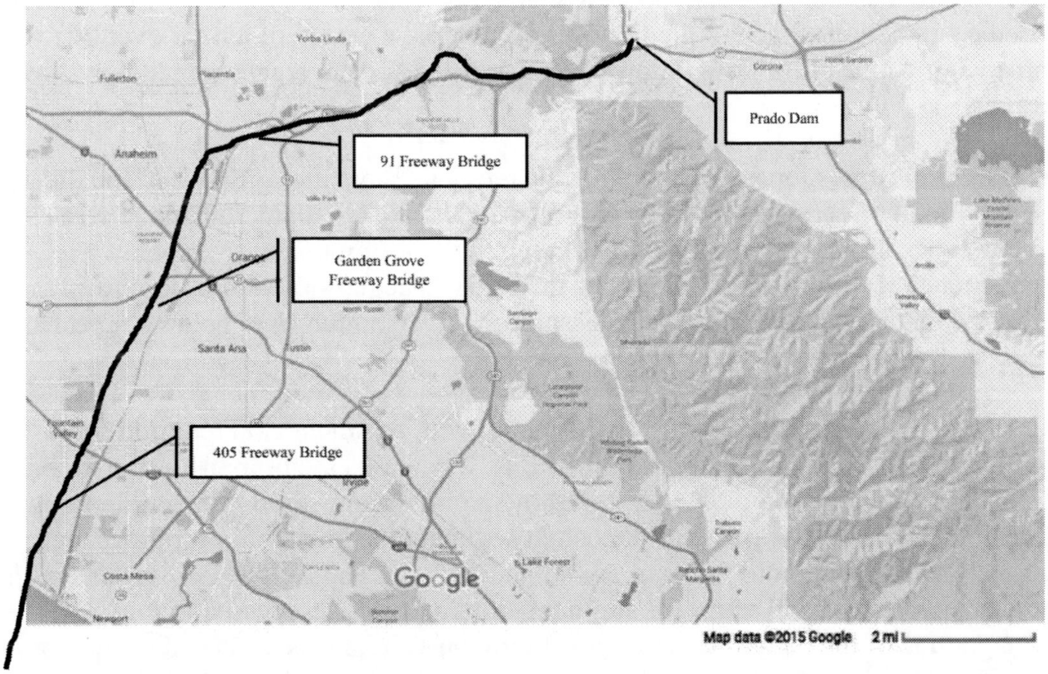

Figure 2. LSAR Reach Map

Hydrology. The flow from Prado Dam, controlled by USACE, is the only inflow hydrology used in the HEC-RAS model. Inflows from the catchment downstream of

Prado Dam were not included in the model at this point. Tetra Tech and HDR (2010) indicate that the watershed area contributing to the Santa Ana River downstream of the dam is two orders of magnitude smaller than the watershed area upstream of Prado Dam. Neglecting flows from the downstream part of the catchment is therefore deemed reasonable.

Historical daily flow data from Prado Dam is available from 1980 through the beginning of 2010. These flow records were analyzed to establish five pulse flow scenarios. The pulse flows will be used to model sediment re-entrainment downstream of the dam to augment sediment loads in the SAR. These sediments will be dredged from the Prado Basin and will be placed just downstream of the dam prior to re-entrainment. It was assumed for the purpose of the analysis that when the pulse flows are released from the reservoir the augmentation sediment will be introduced into the river for conveyance downstream. The pulse flow magnitudes that were selected were 500, 750, 1,250, 2,000 and 5,000 cfs. They will be released under controlled conditions over periods of several days until all the augmentation sediment has been re-entrained. In order to determine the fate of the augmented sediment after it has been re-entrained into the river by the pulse flows it was necessary to develop representative flow records in the river for use after the augmentation period simulation.

The historical daily flow data were therefore further analyzed to identify water years representing 75% (Wet), 25% (Dry), and 50% (Median) exceedance water years from October 1 to September 30. The flow records successively representing these conditions are the water years 2002/2003 (wet), 1987/88 (dry), and 1999/2000 (median). These flows were added after the pulse flows to create one year of historic daily flow data for use in the simulations. Each of the flow series covered a period of a little over one year, commencing with a pulse flow followed by the full year of average annual daily flow series.

Operating procedures for the inflatable dams previously mentioned leads to the diversion of all flow in the river when the flow from Prado Dam is less than 500 cfs. There are very few times throughout the year where there will be flow in the river below the recharge reach in all three flow scenarios modeled. The periods where flow from the dam is less than 500 cfs were modeled assuming 0.1 cfs in the river below the recharge reach.

Sediment. The sediment file provides the model with sediment data required for execution of the sediment transport simulations. It contains the properties of both bed material and incoming sediment loads. Sediment grain sizes, grain size distributions, sediment mass density, porosity of deposited sediment, bed material properties, and sediment loads flowing into the river are contained in the input files. For this study, the natural sediment load and sediment augmentation load were expressed in tons per day. The magnitude of the natural sediment load into the system was specified using a rating curve relating water flow and sediment load provided in the 1988 USACE Santa Ana River General Design Memorandum (USACE 1998). The input sediment rating curve is as follows:

$$Q_s = 0.003Q_w^{1.42} \tag{1}$$

Where Q_w is in cfs and Q_s is in tons/day. The augmentation load was specified as a percent concentration in the flowing water. The incoming sediment gradation was specified from a recently completed geotechnical investigation including borehole data in the deposited sediment of Prado Basin.

Several equations are available in HEC-RAS to calculate sediment transport, bed sorting, and fall velocity. Bed sorting calculations are required to simulate how the bed material composition may change over time. For example, if the bed material becomes coarser due to the preferential removal of fine material it may become armored. Once armored it protects the sub-surface sediments and limits the amount of sediment that can be removed from the riverbed for transportation. The fall velocity of the sediment is required to determine its mobility, i.e., how easy is it to settle the.

For this analysis the Engelund-Hansen sediment transport function, the Exner sorting equation, and the Rubey fall velocity equation were selected. The Engelund and Hansen (1967) total load equation was selected because of its simplicity and suitability to calculate sediment transport in sandy river conditions. The equation is a stream-power based relationship using commonly available parameters (e.g., grain size, flow velocity and bed shear stress). It has been identified as one of the most accurate total load sediment transport equations for a wide range of non-cohesive sand sizes (USACE 2010). Annandale (2007) showed that the Engelund and Hansen (1967) equation responds appropriately as a total load predictor to changes in total turbulence as opposed to only responding to changes in near-bed turbulence (bed load) or only to turbulence in the water column above the near-bed region (suspended load).

The Exner sorting equation is effective in predicting armoring of the bed material. For this reason, it was selected for the Santa Ana River where armoring has already been found to occur in some river reaches. The Rubey fall velocity method is suitable for silt, sand, and gravel sizes with specific gravities of around 2.65, similar to materials found in the Santa Ana River.

One of the limitations of the HEC-RAS sediment transport modeling software is that specification of only one sediment gradation per cross section is allowed. This means for example that differing gradations cannot be specified for the surface layer and the sub-surface layer at one particular cross-section. This is an important observation because the sediment properties on the surface and sub-surface often differ especially if the channel has armored.

Predictive model setup

The predictive assessment was used to evaluate the effects of augmenting sediment loads by introducing 500,000 yd^3 of sediment into the SAR.

Location of Sediment Introduction. Locations and methods to introduce the augmented sediment load are still under consideration. From a modeling point of view, it is assumed that the sediment will be introduced into the SAR at a location just downstream of Prado Dam.

Sediment Concentration. Based on engineering judgment, it was deemed reasonable to assume that a sediment concentration of 1% would be appropriate for the re-introduction of sediment. This is a conservative assumption, based on engineering judgment, of a sediment concentration that will not cause sediment to fall out of suspension immediately after re-entrainment and create excessive deposits in the river. As this assumption may affect acceptable levels of sediment concentrations in the river water it may be varied to optimize the project. During project implementation the concentration will be purposely varied to see how the river responds to re-entrainment operations.

Hydrology and Sediment Augmentation . The historical flows range between 100 and 5,000 cfs. The hydrology used for the predictive runs are the same years of data run for the three respective base case models with the addition of the pulse flows and corresponding durations added to the beginning of each run when sediment augmentation is assumed to take place. At this time, the pulse flow range that was used in the simulations varies: 500, 750, 1,250, 2,000, and 5000 cfs. It should be noted that OCWD does not anticipate dam operations will be modified to provide pulse flows as modeled here. Sediment re-entrainment will take place throughout normal dam operations as flows allow for a period of 72 hours followed by a period of 24 hours without re-entrainment to mimic the episodic nature of storm flows in the watershed and help meet water quality requirements in LSAR flows during re-entrainment. The pulse flows with re-entrainment were modeled to show the minimum amount of time re-entrainment can take place and the effects on the re-entrained sediment after a reasonable amount of time of normal flow conditions. The natural sediment load is still assumed to be flowing from the dam during re-entrainment flows and is added to augmented sediment loads which is specified from the previously given rating curve equation..

The duration of these flows was dictated by the time required to re-entrain 500,000 yd^3 of sediment at a 1% concentration by weight. The practical implementation of this re-entrainment assumption is likely overly aggressive but was used in the modeling effort to provide the most conservative results when considering flood risk impacts.

Results

The re-entrainment during the SMDP does not have a long term effect on the existing profile of the LSAR as shown in Figures 3 and 4 below. Figure 3 shows the base case results for the wet year analysis with little change to the profile as expected. Figure 4 shows the profile after re-entrainment of 500,000 yd^3 with a 1,250 cfs pulse flow followed by one year of inflow using the wet year series. Any deposition that occurs after re-entrainment using the pulse flows is scoured away with the following year of clear flow from the dam. Similar results are seen for the other pulse flow and hydrology scenarios.

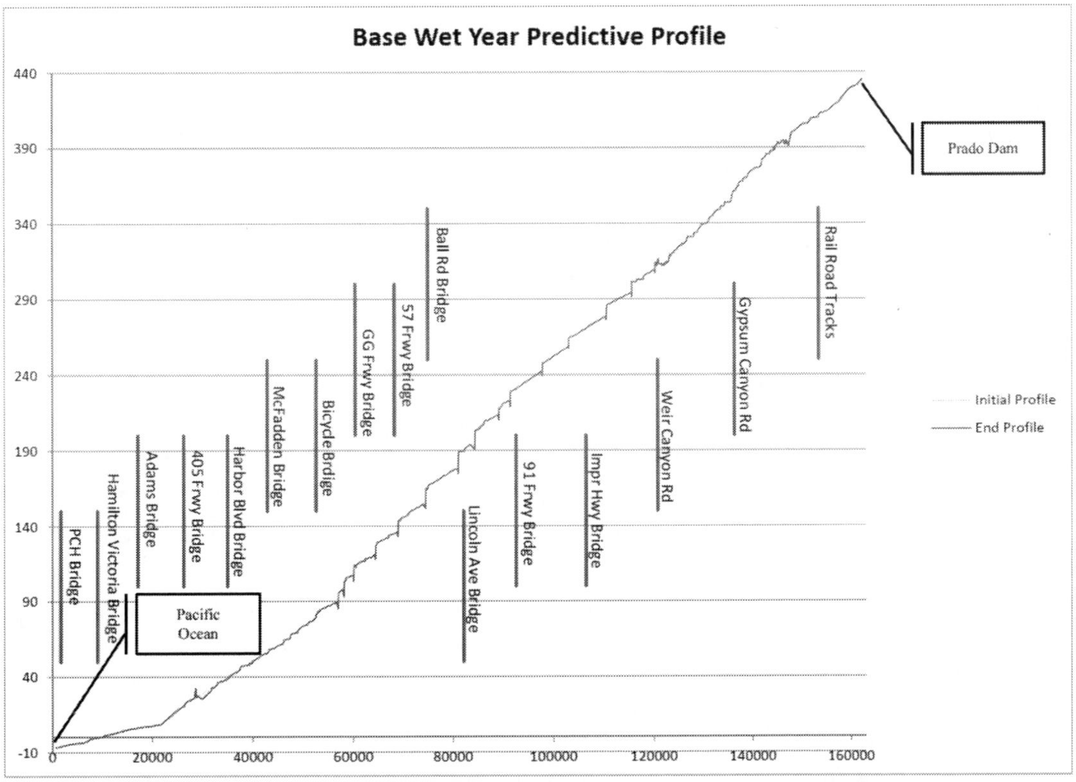

Figure 3. SMDP Wet Year Base Case Invert Change

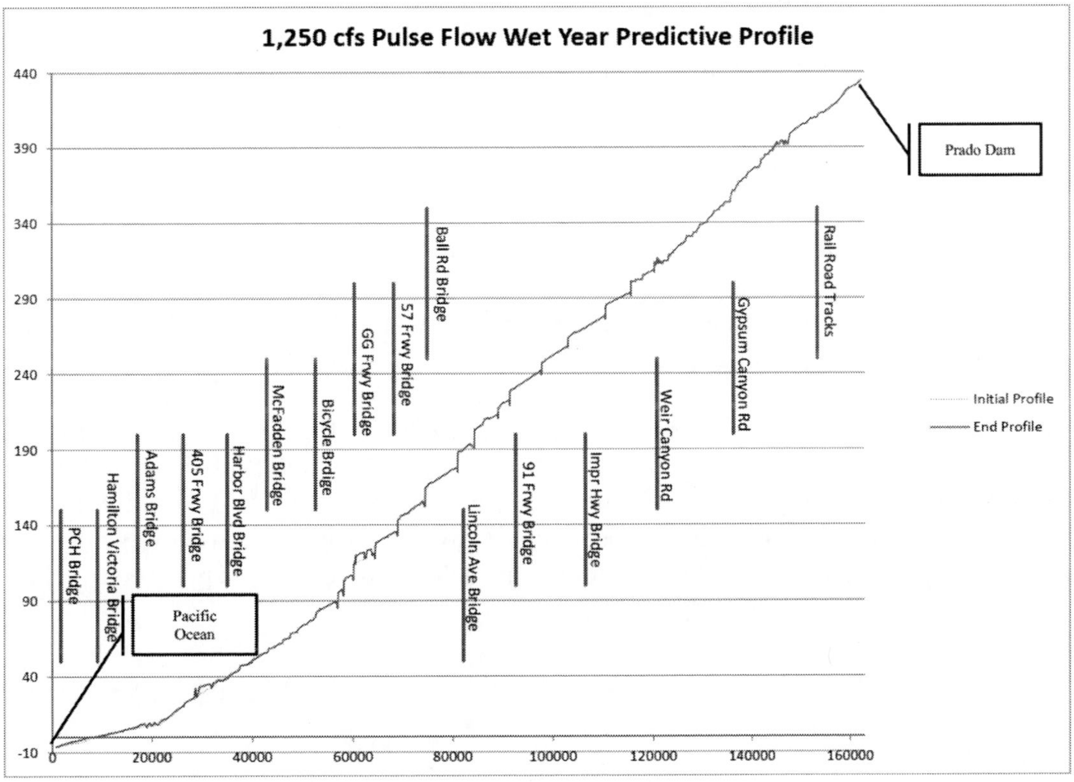

Figure 4. SMDP Wet Year Predictive Case Invert Change

SEDIMENT TRANSPORT MODELING IN LSAR FOR PRADO FEASIBILITY PROJECT

Introduction

The Feasibility Project expands on the demonstration project in that the re-entrainment occurs annually over a period of many years. The model geometry, sediment transport equations, sediment introduction location, and sediment augmentation concentration is the same as in the SMDP modeling described above.

Model Setup

Hydrology. The outflows from Prado Dam for the Feasibility Project vary from the assumptions in the SMDP. An artificial future flow series based on historic flow records and projected future upper watershed conditions was developed by USACE and was used for the inflow to the LSAR. The pulse flows were removed and it was assumed that re-entrainment will occur during normal flow conditions through the year. To date, a ten year range of daily flow values from the USACE flow series has been used which includes the highest flow in the series of approximately 21,000 cfs.

Sediment. The sediment gradation used for the inflow for the Feasibility Project varies from the SMDP project in that it is not based on borehole data. The modeling effort was designed to look at two sizes; fine and medium sand. Fine and medium sand gradtions were selected based on a general review and wide sampling of sediment size data sources in the upper river and watershed, which are most likely to enter Prado Basin in the future. An artificial gradation for the two sizes was developed and were used as the inputs for the model. A one percent sediment concentration was used for the Feasibility Project as in the SMDP.

Results

The re-entrainment during a 10-year Feasibility Project scenario has a much greater effect on the LSAR profile than the SMDP re-entrainment schedule. The consistent re-entrainment leads to much more deposition especially upstream of Weir Canyon Road as shown in Figure 5 below for the fine sand gradation.

The result shows that the profile changes much more in the first five years than in the subsequent five years for some portions of the river which indicates the slope of the LSAR has reached an equilibrium for the incoming sediment load. The medium sand model results show similar patterns with more deposition just downstream of the re-entrainment point due to the larger sediment sizes dropping out of suspension quickly.

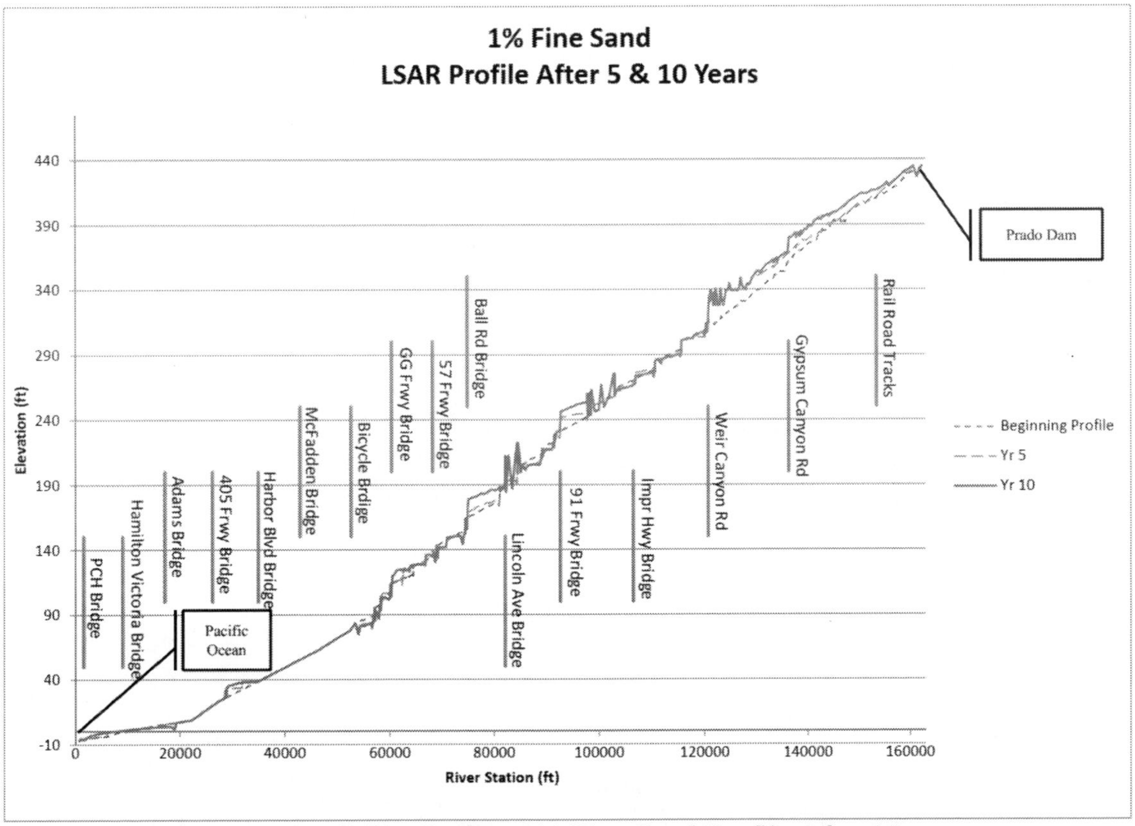

Figure 5. Feasibility Project 1% Fine Sand Profile After 10-years

SEDIMENT TRANSPORT MODELING UPSTREAM OF PRADO DAM FOR PRADO FEASIBILITY PROJECT

Introduction

In the Santa Ana River upstream of Prado Basin, the sediment will be excavated for re-entrainment into the LSAR. The excavation will create a sediment trap which will increase the slope of the river as it enters Prado Basin. The goal of the modeling is to analyze the effects on sediment deposition/scour and bed grain size of creating a sediment trap and increasing the slope of the channel into the trap. Both base and trap models were developed for comparison.

The extent of the sediment transport models are from the San Bernardino County/Riverside County line on the upstream end and the upstream end of Prado Basin approximately 2 miles below River Road bridge on the downstream end for approximately 20 miles. Base case and trap scenario models were constructed for three scenarios. Each model was run for a selected 10-year period.

Model Setup

Topography. Detailed 1-foot contour interval topography within Prado Basin extends far enough east to be used as the model geometry from the downstream end of the models to approximately the I-15 bridge crossing. Supplemental DEM data with 10-foot contour interval topography from the USGS was used for the remainder of the model geometry. The DEM data does not have enough resolution to accurately represent the low-flow main channel or leveed portions of the river where it is used and these features were manually entered to HEC-RAS. A main channel width was estimated from aerial imagery and a depth of 5 feet was assumed. Leveed portions of the river were observed in aerial imagery and were assumed to be 15 feet high.

For the trap scenarios, a transition channel and sediment trap was manually entered in place of the existing channel geometry in the lower portion of the model. Various sediment trap scenarios (areas, depths and shapes) are under consideration for the feasibility project. To transition flow in the river to the trap, a transition channel connecting the existing channel invert to the trap bottom is needed. The transition channel modeled has a slope of 0.33 percent, 80-foot bottom width, 4H:1V side slopes, and an 18.5-foot depth. The modeled reach and selected bridge crossings is shown on Figure 6.

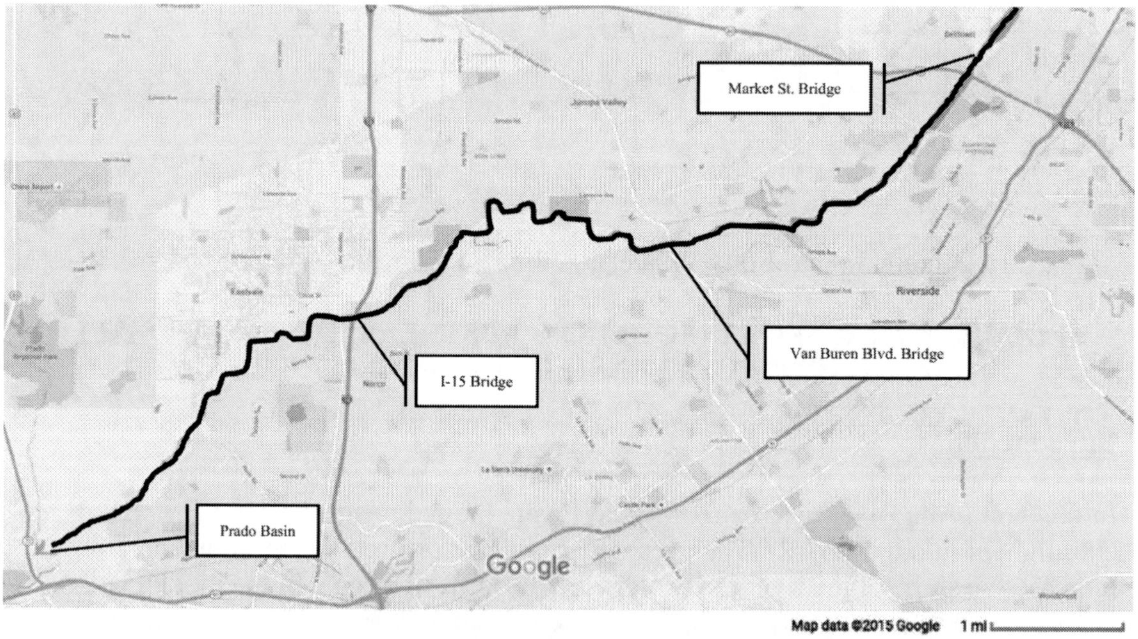

Figure 6. Upper Santa Ana River Modeled Reach

Hydrology. The U.S. Army Corps of Engineers has developed two future flow series as inflow to Prado Basin based on historical data modified for predicted watershed and precipitation changes. A ten year flow series was used for the purpose of this modeling effort.

Reservoir Water Level. Currently the Prado Basin is operated to have a water conservation level of 498 feet from September through February and 505 feet from March through August of any given year. It is under consideration to change the operating rules to have a water conservation level of 505 feet from October through August and 498 feet in September only. Environmental and permitting requirements called for analysis of the depositional patterns in the river under base case and sediment trap scenarios with and without the change to the proposed water conservation level.

Sediment. The base and trap models utilized limited data available for the bed material through the modeled reach. A particle size distribution at River Road Bridge (located within the upper zone of Prado Basin) was provided which was developed with sediment that had been dredged from the river. The particle size distribution for the dredged material is generally fine sand (D_{50}=0.4 mm). A paper looking at the habitat of the Santa Ana Sucker (Thompson et al., 2010) gives a general sediment size in the upper reach of the modeled river. A particle size distribution was created using this information for the upstream most reach and has a majority of gravel and cobbles. The remaining portions of the modeled reach were divided into four segments and particle size distributions were created based on an interpolation of the upper most reach and lower most reach particle size distributions.

Incoming Sediment Load. The USGS gage at Prado Park contains about 1,700 sediment concentration records over about 4 years coupled with water flow data. This data was used to develop a sediment load rating curve (sediment load in tons per day versus water flow in cubic-feet per second) for use as an input to the HEC-RAS model. The particle size distribution used for the incoming sediment load in the base and trap models was the same as was provided at River Road Bridge, composed of fine sand.

Results

In general, the transition channel and sediment trap increases the stream power within the transition channel leading to a coarsening of the river bed. The river bed degrades and coarsens upstream of the transition channel for some distance but generally stops downstream of the Hamner Avenue crossing. Common geomorphology knowledge dictates that the degradation should continue upstream. The rate at which it continues upstream is difficult to predict but it is possible that the degradation could eventually adversely affect bridge piers. It is recommended that the potential for upstream migration be investigated further prior to implementation of the trap and transition channel. If implemented, the SMDP would provide additional data and insight as to the extent and rate of bed degradation upstream.

The results show that the reservoir water level does not affect deposition or scour within the transition channel or trap significantly between the higher and lower reservoir water level models. Deposition occurs at the downstream end of the transition channel throughout all runs when higher reservoir water elevations are encountered. When water levels decrease, the deposition is transported deeper into the Basin as is the current condition.

The sediment trap was not fully incorporated into the HEC-RAS model due to the complexity of the hydraulics and sediment transport within the trap. The model assumes that, depending on the water surface in the reservoir, suspended sediment flows out of the downstream end of the reach and is unaccounted for after that. The model assumes (and is the intent of the project) that after the transition channel and trap are constructed, there would be annual sediment removal operations which result in adequate storage volume in the trap to not constrain the transport of sediment at the downstream end of the model reach. In practice, if the sediment which collects in the trap is allowed to remain without removal, the deposition will continue upstream to a point where it will eventually affect sediment transport within the transition channel. If this occurs, the transition channel may aggrade more on an annual basis due to the buildup of sediment in the trap blocking the flow of sediment from the transition channel.

The results after 10 years of modeling can be used to develop trends over longer periods of time if the transition channel and trap configuration were to be maintained. In general, assuming the bottom of the trap elevation remains constant, the transition channel will continue to degrade which will cause the river slope to flatten some. This will cause the river upstream of the transition channel to degrade as well and steepen to achieve a stable slope along the entire reach.

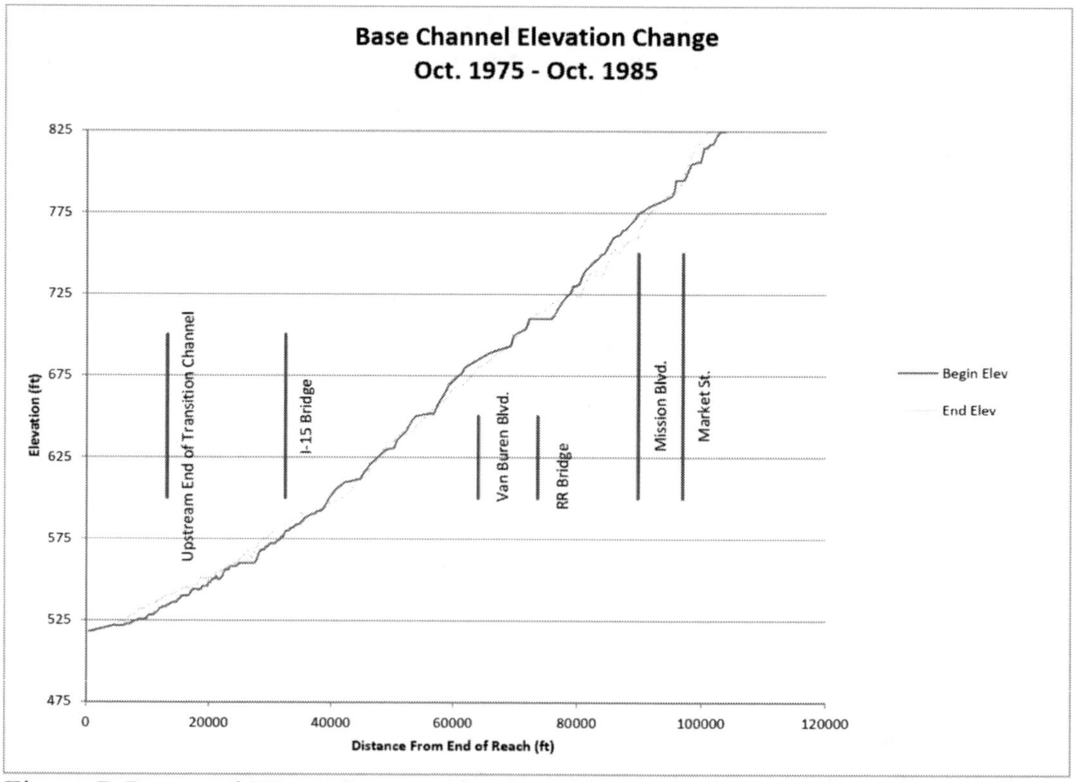

Figure 7. Proposed Water Conservation Level Base Profile Upstream of Prado Basin

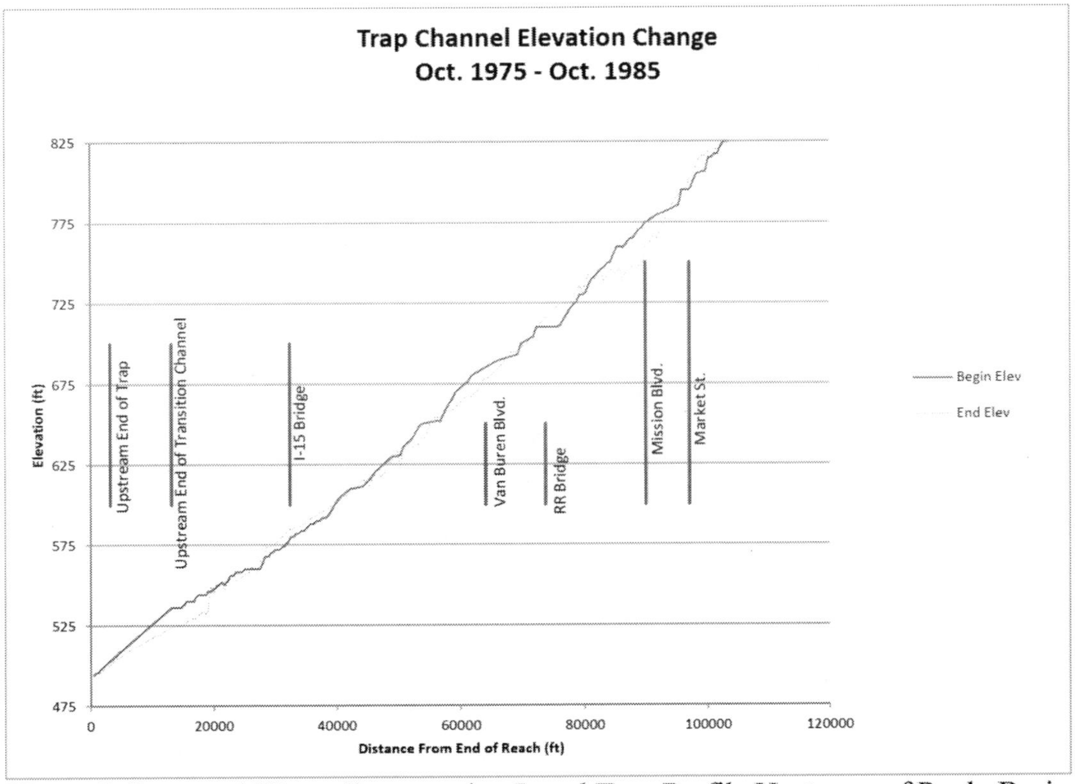

Figure 8. Proposed Water Conservation Level Trap Profile Upstream of Prado Basin

CONCLUSION

In general, the proposed SMDP and Feasibility Project will achieve the potential benefits outlined above. Upstream of the dam, an inherent benefit of sediment removal in Prado Basin will be increased storage and model results suggest that the river bed will coarsen due to the increased bed slope which will help to restore Santa Ana Sucker fish habitat. In the LSAR, model results to date suggest that aggradation will occur immediately downstream of Prado Dam reversing the channel incision and reduction of habit that has occurred since the construction of the dam. Fine sediments will also make their way into the OCWD recharge reach and to the ocean with the benefit of helping sustain infiltration rates in the recharge reach and replenishing beach sand.

Potential future work for the Feasibility Project includes longer term modeling both upstream and downstream of Prado Dam. Additional modeling in the LSAR could analyze different re-entrainment rates to study the changes in the river's response to varied sediment loads as well as continue to look at longer term re-entrainment to study whether the depositional patterns and size fraction of deposits changes over longer periods. Upstream of the dam, the migration of the channel degradation that formed from the steepened channel section in the trap could be studied more if the modeling period were extended to analyze the risk to civil infrastructure upstream of Prado Basin.

ACKNOWLEDGEMENT

The authors would like to recognize Greg Woodside, Executive Director of Planning and Natural Resources, at OCWD for his support and guidance throughout the project.

REFERENCES

Annandale, G.W. 2007. Identification of the True Nature of Commonly Used Sediment Transport Equations, 32nd IAHR World Congress, "Harmonizing the Demands of Art and Nature in Hydraulics", 1-6 July, Venice, Italy

City of Santa Ana. History of the Santa Ana River. Santa Ana River Vision Plan.

Engelund, Frank, and Eggert Hansen. 1967. A Monograph on Sediment Transport in Alluvial Streams. Monogr, Denmark Tech Univ., Hydraul Lab, 62 pp, 30 figures, 2 tables, 31 references.

Golder Associates Inc. (Golder) 2014. Appendix B – HEC-RAS Sediment Transport Modeling Santa Ana River. Project 1300314. Submitted July 14.

Scheevel Engineering. 2015. Prado Basin Feasibility Study: Reach 9 – Projected Sedimentation Trends (Final) *Technical Memorandum.* January 21, 2015.

Tetra Tech and HDR. January 2010. The Santa Ana Regional Interceptor Pipeline Relocation Project: Scour Study of Santa Ana River below Prado Dam – Final Report.

Thompson A.R., et al. 2010. Influence of Habitat Dynamics on the Distribution and Abundance of the Federally Threatened Santa Ana Sucker, *Catostomus Santaanae,* in the Santa Ana River. *Environmental Biology of Fishes.* DOI 10.1007/s10641-010-9604-2.

U.S. Army Corps of Engineers (USACE). 1988. Santa Ana River Design Memorandum No. 1: Phase II GDM on the Santa Ana River Mainstem Including Santiago Creek, Volume 3: Lower Santa Ana River (Prado Dam to Pacific Ocean).

U.S. Army Corps of Engineers (USACE). 2010. *HEC-RAS River Analysis System User's Manual.* Version 4.1. January.

U.S. Army Corps of Engineers (USACE). 2014. Engineering Design Appendix (Draft) Prado Basin Ecosystem Restoration Study. June 2014.

VERIFICATION OF SEDIMENT YIELD CALCULATIONS IN THE ARID SOUTHWEST

Brian Wahlin, Ph.D., P.E., D.WRE[1]
Jesse Piotrowski, P.E., CFM[2]
Chuck Davis, P.E., CFM[3]

ABSTRACT

The Powerline, Vineyard Road, and Rittenhouse (PVR) Flood Retarding Structures (FRS) were designed and constructed by the Soil Conservation Service (now the Natural Resources Conservation Service or NRCS). These dams currently provide significant flood control and erosion protection for downstream commercial, agricultural, and residential properties within Pinal County and Maricopa County covering a large portion of the Phoenix metropolitan area in Arizona. Due to significant concerns regarding aging infrastructure, land subsidence, earth fissuring, revised hazard classifications, and current dam safety standards, the ability of the PVR structures to provide the required level of protection is questionable. Based on these dam safety concerns, the Flood Control District of Maricopa County (FCDMC) has determined that rehabilitation or replacement of the dams is required.

These PVR structures were originally designed as Significant Hazard dams and have since been re-classified as High Hazard dams. The reservoirs were designed with a sediment storage pool to allow for future reservoir sedimentation behind the structures. The sediment yield calculations for the PVR structures were updated to verify the sediment storage pool size. Because of the uncertainty associated with sediment yield calculations, the FCDMC desired to have a secondary method of verification showing that the calculated sediment yield was reasonable. However, no historic surveyed sedimentation data are available for the PVR structures. Even if there were surveyed sediment data, these data may not have been useful because of the coincident subsidence that has been occurring in the area. Thus, published reservoir sedimentation data for similar sized reservoirs were used to determine if the calculated sediment yields were reasonable. The Reservoir Sedimentation (RESSED) Database—a product of the U.S. Federal Government's Advisory Committee on Water Information, Subcommittee on Sedimentation—was utilized for this comparative analysis.

INTRODUCTION

The Powerline, Vineyard Road, and Rittenhouse Flood Retarding Structures (FRS) were designed and constructed by the Soil Conservation Service (now the Natural Resources

[1] WEST Consultants, Inc., 8950 S. 52nd St, Ste 210, Tempe, AZ 85284, (480) 345-2155, bwahlin@westconsultants.com
[2] WEST Consultants, Inc., 8950 S. 52nd St, Ste 210, Tempe, AZ 85284, (480) 345-2155, jpiotrowski@westconsultants.com
[3] Beaver Creek Hydrology, 907 National Ave, Lexington, KY 40502, (480) 717-2441, chuck@beavercreekhydrology.com

Conservation Service or NRCS) and are located just upstream of (i.e., to the east of) the Central Arizona Project (CAP) canal in Pinal County, Arizona (see **Figure 1**). The dams were designed to protect portions of eastern Maricopa County from flood waters originating in the Superstition Mountains. The dams currently provide significant flood control and erosion protection for downstream commercial, agricultural, and residential properties within Pinal County and Maricopa County. The reservoirs were designed for the ½ Probable Maximum Flood (PMF) level of protection with a 100-year sediment storage pool to allow for future reservoir sedimentation behind the structures (Stantec, 2014).

Due to significant concerns regarding aging infrastructure, land subsidence, earth fissuring, and current dam safety standards, the ability of the PVR structures (i.e., Powerline, Vineyard Road, and Rittenhouse) to maintain the current level of protection is questionable. Based on these dam safety concerns, the Flood Control District of Maricopa County (FCDMC) has determined that rehabilitation or replacement of the dams is required. The PVR FRS Rehabilitation or Replacement Planning Project identified the selected alternative for each structure as:

- Replacing Powerline FRS with a channel that discharges into the Vineyard Road FRS;
- Raising and rehabilitating Rittenhouse FRS; and
- Raising and rehabilitating Vineyard Road FRS.

NEED FOR THE STUDY

Previous sediment yield analyses had been performed on the PVR structures to estimate the amount of sediment that could be expected to accumulate within the impoundment area (i.e., the size of the sediment storage pool). The sediment storage pool is considered "dead storage" and does not count towards the flood control capacity of the three dams. Thus, if the sediment storage pool is artificially too high, then the dams will have to be raised higher than necessary, resulting in additional (and unnecessary) costs to rehabilitate the dams.

Several studies have been done to produce curves showing the sediment yield as a function of drainage area. A summary of these studies is show in **Figure 2**, taken from Branson et al. (1981). Note that there is a wide range of sediment yield for a given drainage area even though all these studies were based on measured data. The variability in the measured sediment yield values is another indication of the uncertainty and complexity of predicting sediment yield.

Construction of the original dams was completed in the late 1960s. The sediment accumulation in the sediment storage pools over the last 45+ years has been significantly less than what was predicted by the sediment yield calculations. Thus, the FCDMC desired to revisit the previous sediment yield studies and, more importantly, verify that the sediment yield studies produce reasonable estimates of the 100-year sediment storage volume needed for the three dams.

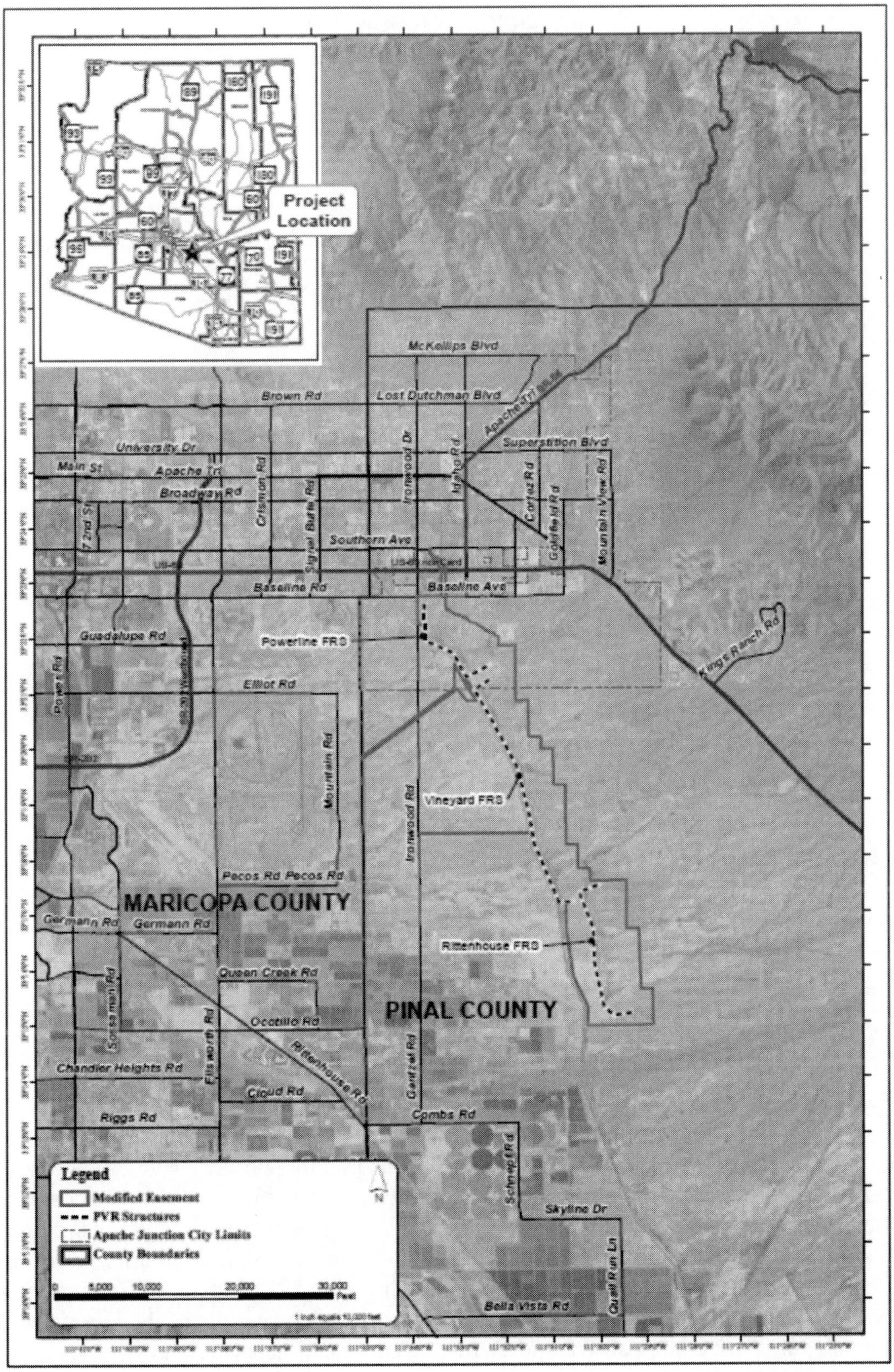

Figure 1. Project location map (Kimley-Horn and Associates, 2012)

In light of the uncertainty surrounding sediment yield calculations and the lack of observed sediment impounded behind the PVR structures, the FCDMC desired to have the sediment yield calculations verified in some manner.

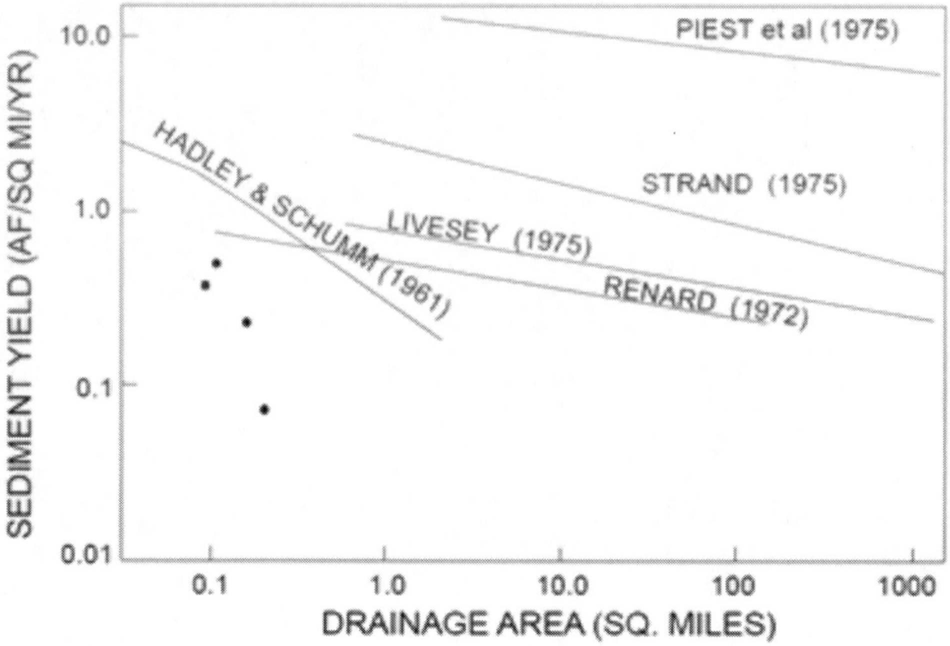

Figure 2. Sediment yield as a function of drainage area (Branson et al., 1981)

PREVIOUS SEDIMENT YIELD REPORTS

Original Designs

The original watershed work plans for the three PVR structures included sediment volume analyses in the planning and design; however, the methodology used to determine the required sediment volume is unknown. The dams were designed for a 50-year design life (which still meets the current NRCS standards). The original design sediment volumes are 175 acre-feet for Powerline FRS, 178 acre-feet for Vineyard Road FRS, and 175 acre-feet for Rittenhouse FRS as shown in **Table 1** and **Table 2** (multiplied by two because the original design was for a 50-year design life).

Arizona State Land Department Study

The Arizona State Land Department (ASLD) sponsored the Desert Drive Study to provide updated sediment yield computations for future conditions at the three dams (JE Fuller Hydrology and Geomorphology, 2008). The total sediment load was estimated in this study using Modified Universal Soil Loss Equation (MUSLE, Williams and Berndt (1977)) to estimate wash load yield in conjunction with a U.S. Bureau of Reclamation (USBR) correction factor to account for bed load yield. The results of the Desert Drive Study are summarized in **Table 1** and **Table 2**. These estimates are significantly higher than the original sediment storage design.

First FCDMC Sediment Yield Study

In 2010, the FCDMC completed future conditions sediment yield estimates for the three dams (Flood Control District of Maricopa County, 2010a, 2010b, 2010c). This FCDMC study utilized MUSLE for calculating the wash load yield and the Zeller-Fullerton equation (1983) for calculating bed load yield (using hydraulic results from several representative cross-sections in the watershed). The hydrology used for this study was the future conditions hydrology generated from the ASLD study (JE Fuller Hydrology and Geomorphology, 2008). The results of the first FCDMC study are also summarized in **Table 1** and **Table 2**. These values are lower than the ASLD study, but still significantly higher than the original design values for the dams.

Second FCDMC Sediment Yield Study

In 2014, the FCMDC updated the 2010 future conditions sediment yield study (Flood Control District of Maricopa County, 2014). The main difference in the second FCDMC study compared to the first is that the steep portions of the watershed in the Superstition Mountains were ignored in the MUSLE calculations. As with the first FCDMC study, the hydrology used for this study was the future conditions hydrology generated from the ASLD study (JE Fuller Hydrology and Geomorphology, 2008). The results for the second FCDMC study are summarized in **Table 1** and **Table 2**. These values are more in line with the original design estimates.

WEST Sediment Yield Study

In 2015, WEST took the FCDMC wash load yield values (Flood Control District of Maricopa County, 2014) and updated the bed material load calculations using the new future conditions hydrology (Stantec, 2015) to determine the total sediment yield for the Powerline, Vineyard Road, and Rittenhouse Structures (WEST Consultants, 2015). A summary of the results is shown in **Table 1** and **Table 2**. Due to updated bed material load yield calculations, the total sediment yields have increased for Powerline and decreased for both Vineyard Road and Rittenhouse FRS compared to the 2014 FCDMC study. As with the 2014 FCDMC study, WEST's results are similar to the original design values.

Table 1. 100-year net sediment yield (acre-feet) delivered to Powerline, Vineyard Road, and Rittenhouse FRS from previous studies (future conditions hydrology)

Structure	Drainage Area (mi^2)	Sediment Contributing Area (mi^2)	Original Design	JE Fuller (2008)	FCDMC (2010)	FCDMC (2014)	WEST (2015)
Powerline	46.2	34.0	350	2,754	810	298	406
Vineyard Road	53.7	53.7	356	3,816	1,442	761	603
Rittenhouse	47.1	47.1	350	3,886	714	461	456

Table 2. Annual sediment yield (in ac-ft/mi^2/yr) delivered to Powerline, Vineyard Road, and Rittenhouse FRS from previous studies (future conditions hydrology)

Structure	Drainage Area (mi^2)	Sediment Contributing Area (mi^2)	Original Design	JE Fuller (2008)	FCDMC (2010)	FCDMC (2014)	WEST (2015)
Powerline	46.2	34.0	0.10	0.81	0.24	0.09	0.12
Vineyard Road	53.7	53.7	0.07	0.72	0.27	0.14	0.11
Rittenhouse	47.1	47.1	0.07	0.83	0.15	0.10	0.10

VERIFICATION OF SEDIMENT YIELDS

The five different sediment yield analyses that have been conducted for the PVR structures yield significantly different results. Because of this high variability, the FCDMC desired to verify the sediment yield calculations with measured sediment yield data. However, there are no historic surveyed sedimentation data available for the Powerline, Vineyard, or Rittenhouse structures. Even if there were surveyed sediment data, these data would not be useful because of the coincident subsidence that has been occurring in the area. Thus, WEST reviewed published reservoir data for similar sized reservoirs and watersheds with similar geomorphology to determine if the calculated sediment yields are reasonable. WEST utilized the Reservoir Sedimentation Database (RESSED) for this comparative analysis.

RESSED is a public database published by the Advisory Committee on Water Information (ACWI), which is part of the United States Geological Survey (USGS). The database is maintained by ACWI's Subcommittee on Sedimentation (SOS). RESSED was developed in March 2009 and remains a work in progress (Advisory Committee on Water Information, 2014). RESSED's predecessor, RESIS-II, included information on over 1,800 surveyed reservoirs throughout the United States. RESIS-II included data spanning from 1755 to 1993 (Ackerman et al., 2009). RESIS-II data was imported into RESSED by the SOS. Survey data collected after 1993 and made available to the SOS has been added to RESSED.

At the time of WEST's study (2015), RESSED included climate, hydrologic, topographic, and sediment data for 1,835 reservoirs throughout the United States. Data availability is on a reservoir-by-reservoir basis; all RESSED data variables are not available for most of the reservoirs in the database.

The RESSED database (in XML format) was downloaded in October 2014. WEST only considering reservoirs and contributing watersheds that were similar to the three reservoirs in the current study. Reservoirs were initially filtered using the following criteria:

- No reservoir in the RESSED database with a contributing sediment drainage area smaller than Powerline FRS (34 mi^2) was considered;

- Because the primary goal of the RESSED analysis was to determine surveyed sedimentation rates, reservoirs without sediment survey data were not considered; and
- Only reservoirs in the southwestern portion of the United States were considered. The initial filtering included reservoirs in CA, NV, UT, AZ, CO, NM, OK, and TX.

This filtering resulted in 147 reservoirs (8% of database) being selected for analysis from the RESSED database. Using the survey information inside RESSED, the amount of storage loss (in acre-feet) between each successive reservoir survey was calculated per reservoir. Note that for most reservoirs, there were more than two sediment surveys. The amount of storage loss was computed between each successive survey as well as the total loss over the entire survey period of record. It was assumed that this reservoir storage loss was equal to the amount of sediment delivered to the reservoir.

The storage loss per year as a function of sediment contributing area for the 147 reservoirs considered from the RESSED database is shown in Figure **3**. As can be seen, there are many data points in vertical alignment, indicating that there was more than one reservoir survey for a particular reservoir. The vertical scatter represents the high degree of variability in storage loss that was calculated for different survey periods. Note that Figure **3** displays the calculated sediment yields for Powerline, Vineyard Road, and Rittenhouse from the previous and current studies. In general, the calculated values from all the previous studies fit within the scatter of data shown in the RESSED database with the FCDMC 2014 sediment yield estimates being on the low side. A similar graph for just the Arizona reservoirs in the RESSED database is shown in Figure **4**. Again, the various calculated values of sediment yield for Powerline, Vineyard Road, and Rittenhouse follow the general trend of the data, with the FCDMC 2014 sediment yield results being on the low side.

Sediment contributing drainage area is not the only contributing factor to sediment yield and looking at just this factor can be misleading. For example, some of the reservoirs that were included in the initial filtering were actually debris basins. Obviously, the material delivered to a debris basin will have much different characteristics than the sediment delivered to the Powerline, Vineyard Road, and Rittenhouse structures. In addition, the sediment yield to a dam will also depend on factors like basin slope, annual precipitation, vegetation amount, basin elevation and orientation, climate, etc. Over 95% of the reservoirs included in the initial filter differ from the PVR structures in many of these characteristics.

After initial filtering, WEST reviewed the list of 147 reservoirs and applied a second filter based on:

- structure type and design (i.e., detention basin or flood control reservoir);
- contributing watershed size;
- the existence of upstream reservoirs;
- climate;
- geographic location; and
- watershed topographic characteristics.

This analysis resulted in only two reservoirs with comparable watershed characteristics: Millet Swale Dam near Snowflake, AZ, and Upper Rio Hondo Watershed Site #1 near Capitan, NM. The watershed characteristics for these two dams along with Powerline, Vineyard Road, and Rittenhouse FRS are shown in Table **3**. Note that the watershed characteristics were obtained from the RESSED database, the Environmental Protection Agency's (EPA) MyWaters database, and the USGS StreamStats database.

The average measured sediment yield for Millet Swale Dam and Upper Rio Hondo Watershed Site #1 are significantly higher than the WEST (2015) computed values for the Powerline, Vineyard Road, and Rittenhouse of 0.12, 0.11, and 0.10 ac-ft/mi^2/yr, respectively. The measured sediment yield for Upper Rio Hondo Watershed Site #1 agrees well with the U.S. Army Corps of Engineers' (USACE) typical sediment yield values of 0.3 to 0.5 ac-ft/mi^2/yr (U.S. Army Corps of Engineers, 1974) while the measured sediment yield for Millet Swale Dam is higher.

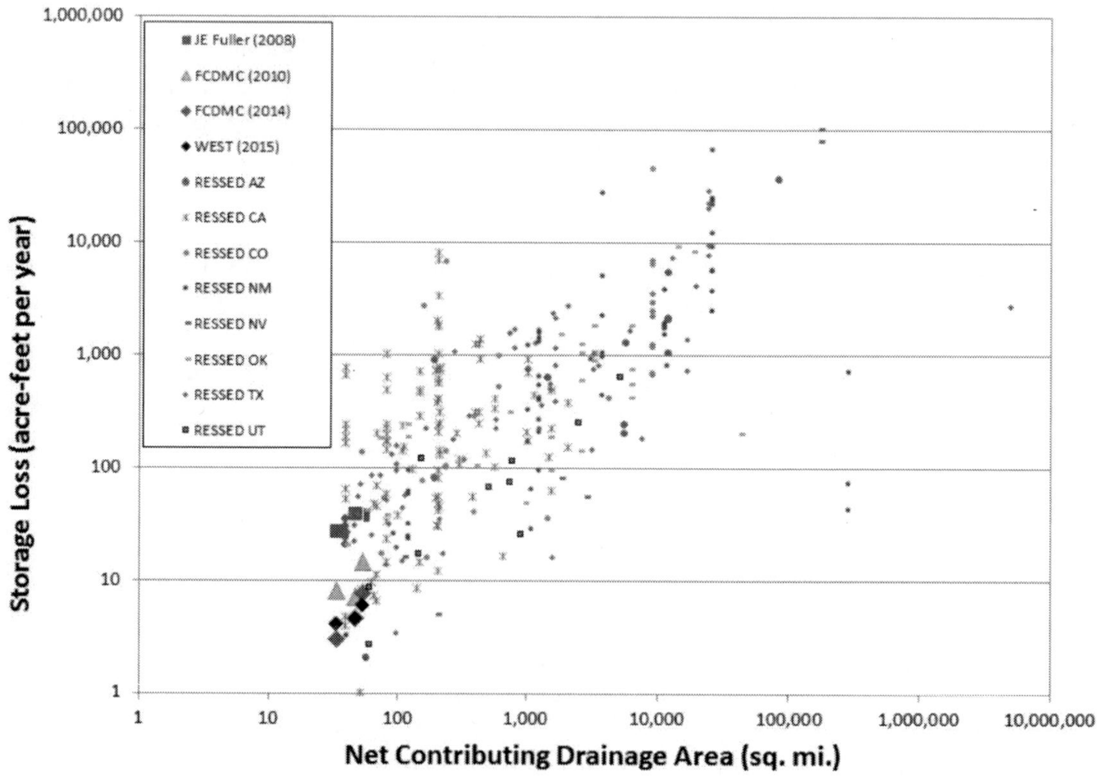

Figure 3. Southwestern reservoir storage loss as a function of sediment contributing area

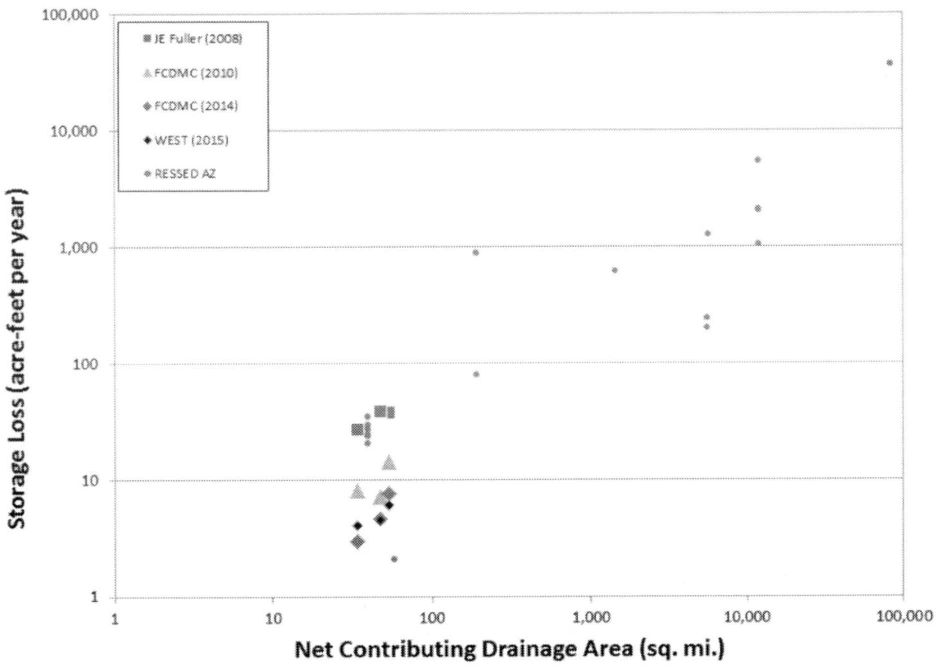

Figure 4. Arizona reservoir storage loss as a function of sediment contributing area

Table 3. Watershed characteristics of various reservoirs

Site	State	RESSED ID	Sediment Contributing Area (mi^2)	Annual Rainfall (in)	Annual Runoff (in)	Watershed Slope (%)	Average Measured Sediment Yield (ac-ft/mi^2/yr)
Millet Swale	AZ	61002	39.7	11.7	0.3	1.3	0.67
Upper Rio Hondo Site #1	NM	58005	121.6	19.9	0.75	3.7	0.41
Cave Creek	AZ		121.0	16.1		19.3	0.24
Embudo Arroyo	NM		20.7	15.0			0.07
North Diversion Channel	NM		101.0	15.0			0.21
Walnut Gulch	AZ		57.5	12.8		2.0	0.08
Powerline	AZ		34.0	12.0		4.2	0.12[4]
Vineyard Road	AZ		53.7	12.0		3.1	0.11[4]
Rittenhouse	AZ		47.1	12.0		4.2	0.10[4]

[4] These are the calculated values from WEST (2015) but are listed in the table for ease of comparison.

Other Measured Sediment Yield Data

The FCDMC Hydraulics Manual (Flood Control District of Maricopa County, 2013) lists 21 different sediment yield measurements for reservoirs in Arizona, New Mexico, and California. A vast majority of these reservoirs are stock tanks with very small drainage areas, making a comparison to the Powerline, Vineyard Road, and Rittenhouse structures problematic because sediment yield rates from smaller watersheds are typically higher than from larger watersheds. There were only three of the dams in Arizona listed that had a similar watershed area: Cave Creek Dam, Spook Hill FRS, and Saddleback FRS. The source documentation for Saddleback FRS was examined and it appears that the measured annual sediment yield value reported in the FCDMC's Hydraulics Manual (Flood Control District of Maricopa County, 2013) of 0.08 ac-ft/mi^2/yr is actually a calculated value and not a measured value. The source documentation for Spook Hill FRS could not be obtained; therefore it was not considered in this analysis. The source documentation for Cave Creek Dam (U.S. Army Corps of Engineers, 1974), with a drainage area of 121.0 square miles, indicated that the measured annual sediment yield was 0.24 ac-ft/mi^2/yr. This value is higher than the WEST (2015) computed values for all three PVR structures. The watershed characteristics for the Cave Creek Dam also appear in Table **3**.

The FCDMC's Hydraulics Manual (Flood Control District of Maricopa County, 2013) does list two sites in New Mexico with similar watershed areas: Embudo Arroyo and the North Diversion Channel. The measured annual sediment yield for these two sites was based on data from the Albuquerque Metropolitan Arroyo and Flood Control Authority's (AMAFCA) silt haul records (Mussetter, Lagasse, and Harvey, 1994). Embudo Arroyo has a drainage area of 20.7 square miles and a measured annual sediment yield of 0.07 ac-ft/mi^2/yr (Mussetter, Lagasse, and Harvey, 1994). The North Diversion Channel has a drainage area of 101.0 square miles and a measured annual sediment yield of 0.21 ac-ft/mi^2/yr (Mussetter, Lagasse, and Harvey, 1994). The WEST (2015) sediment yield calculations are lower than the North Diversion Channel but are slightly higher than Embudo Arroyo. Only limited watershed characteristics were available for these Embudo Arroyo and the North Diversion Channel (see Table **3**).

Multivariable Regression Analysis of RESSED Database

To truly account for the different watershed characteristics within the RESSED database when estimating sediment yield, a multivariable regression analysis should be done. Fortunately, Kane (2003) already completed such an analysis. Note that Kane defines a term called "specific yield" with units of ac-ft/year and specific degradation (SD) as the sediment yield divided by drainage area (with units of ac-ft/mi^2/yr). Kane's specific degradation is what has been referred to in this paper as annual sediment yield. Using multivariable regression, Kane (2003) developed five different equations for annual sediment yield (specific degradation) as a function of:

- Precipitation, R (mm);
- Drainage basin area, A (km^2);
- The product of precipitation and drainage basin area, AR;

- Watershed slope, S (%); and
- Slope, vegetation (V), and the product of precipitation and drainage basin area.

Note that vegetation was accounted for through a vegetation factor, V, which is dimensionless. The average value for V for shrubland is reported as 0.037 (Kane, 2003); this value was used as a representative value for this study of the PVR structures. Using Kane's equations, the annual sediment yield (i.e., SD) was calculated for Millet Swale Dam, Upper Rio Hondo Dam Site #1, Cave Creek Dam, Embudo Arroyo, North Diversion Channel, and the Powerline, Vineyard Road, and Rittenhouse structures. **Table 4** lists the results of the regression analysis along with the measured annual sediment yields (i.e., SD). Kane's regression equations under-predict the annual sediment yield for Millet Swale Dam and Upper Rio Hondo Watershed Site #1. The regression equations that do not consider watershed slope predict the sediment yield fairly well for Cave Creek Dam (but still under-predict). The regression equations predict the sediment yield within 5% to 25% (depending on the regression variables used) for the North Diversion Channel. Sediment yields are significantly over-predicted for Walnut Gulch and Embudo Arroyo. In terms of the Powerline, Vineyard Road, and Rittenhouse structures, the regression equations agree well with the WEST (2015) calculated sediment yield. The regression equation utilizing S, V, and AR performed the best with approximately a 10% difference. In general, the regression equations are slightly higher than the WEST (2015) calculated sediment yield.

Table 4. Summary of RESSED regression analysis

Site	SD from R (ac-ft/mi^2/yr)	SD from A (ac-ft/mi^2/yr)	SD from AR (ac-ft/mi^2/yr)	SD from S (ac-ft/mi^2/yr)	SD from S, V, and AR (ac-ft/mi^2/yr)	Measured SD (ac-ft/mi^2/yr)	WEST Computed SD (ac-ft/mi^2/yr)
Millet Swale	0.15	0.21	0.13	0.27	0.16	0.67	
Upper Rio Hondo Site #1	0.26	0.19	0.21	0.20	0.18	0.41	
Cave Creek	0.22	0.19	0.17	0.03	0.02	0.24	
Embudo Arroyo	0.20	0.23	0.18			0.07	
North Diversion Channel	0.20	0.20	0.16			0.21	
Walnut Gulch	0.17	0.21	0.14	0.25	0.15	0.08	
Powerline	0.16	0.22	0.14	0.18	0.11		0.12
Vineyard Road	0.16	0.21	0.13	0.21	0.12		0.11
Rittenhouse	0.16	0.21	0.13	0.18	0.11		0.10

CONCLUSIONS

Five separate sediment yield estimates were determined for the PVR structures, with a large amount of variability between the studies. Determining a reasonable sediment yield for the watersheds in the study was crucial in order to minimize the cost of rehabilitating the dams. Thus, it was desired to verify the sediment yield results. Ideally, the verification would have been completed with historic surveyed sedimentation data available for the Powerline, Vineyard, and Rittenhouse structures. However, this data does not exist.

Even with the lack of historic sediment data, verification of the sediment yield estimates could be accomplished by analyzing the RESSED database and comparing the computed sediment yield results to measured values in the RESSED database for reservoirs with similar watershed characteristics.

In completing this analysis, it was noted that all of the previous sediment yield studies for the PVR structures fall within the wide scatter of observed data from the RESSED database. In addition, two reservoirs from the RESSED database that had similar watershed characteristics to the Powerline, Vineyard Road, and Rittenhouse structures were identified: Millet Swale Dam and Upper Rio Hondo Watershed Site #1. Despite having similar watershed characteristics, both of these reservoirs had a significantly higher measured sediment yield than the computed values for the PVR structures.

Regression equations of the RESSED data were used to estimate the sediment yield as a function of area; rainfall; the product of area and rainfall; slope; and slope, vegetation, and the product of area and rainfall. These regression equations agree well with the calculated sediment yields for Powerline, Vineyard Road, and Rittenhouse FRS.
Based on the verification of the sediment yield estimates, the FCDMC has adopted the WEST (2015) sediment yield estimates for the PVR structures and is using these values to size the sediment storage pools in the rehabilitation of these structures.

ACKNOWLEDGEMENTS

The authors would like to acknowledge the staff at the FCDMC that reviewed this work, which includes Dr. Bing Zhao, PE; Mr. Raj Shrestha, PE, CFM; and Ms. Felicia Terry, PE, CFM. Their comments and insight greatly improved this work. The authors would also like to acknowledge the staff at Stantec (the prime firm on the PVR rehabilitation project), which includes Mr. Mike Gerlach, PE, and Mr. Bob Eichinger, PE, for their leadership on this interesting project.

REFERENCES

Ackerman, K. V., Mixon, D. M., Sundquist, E. T., Stallard, R. F., Schwarz, G. E., and Steward, D. W. (2009, March 10). *RESIS–II: An Updated Version of the Original Reservoir Sedimentation Survey Information System (RESIS) Database.* Retrieved November 17, 2014, from http://pubs.usgs.gov/ds/ds434/

Advisory Committee on Water Information. (2014, April 8). *The Reservoir Sedimentation Database (RESSED)*. Retrieved October 31, 2014, from http://water.usgs.gov/osw/ressed/index.html

Branson, F. A., Gifford, G. F., Renard, K. G., & Hadley, R. F. (1981). *Rangeland Hydrology*. Denver, CO: Society for Range Management, Range Science Series 1.

Flood Control District of Maricopa County. (2010a). *Sediment Yield Estimation for Powerline Flood Retarding Structure*. Phoenix, AZ: Flood Control District of Maricopa County.

Flood Control District of Maricopa County. (2010b). *Sediment Yield Estimation for Vineyard Road Flood Retarding Structure*. Phoenix, AZ: Flood Control District of Maricopa County.

Flood Control District of Maricopa County. (2010c). *Sediment Yield Estimation for Rittenhouse Flood Retarding Structure*. Phoenix, AZ: Flood Control District of Marciopa County.

Flood Control District of Maricopa County. (2013). *Drainage Design Manual for Maricopa County, Arizona: Hydraulics*. Phoenix, AZ: Flood Control District of Maricopa County.

Flood Control District of Maricopa County. (2014). *Sediment Yield Analysis for Powerline, Vineyard Road and Rittenhouse Flood Retarding Structures*. Phoenix, AZ: Flood Control District of Maricopa County.

JE Fuller Hydrology and Geomorphology. (2008). *Desert Drive Area Study Volume I: Existing Conditions Hydrology*. Phoenix, AZ: Arizona State Land Department.

Kane, B. (2003). *Specific Degradation as Function of Watershed Characteristics and Climatic Parameters*. Fort Collins, CO: Ph.D. Dissertation, Colorado State University.

Kimley-Horn and Associates. (2012). *Level II Hydrology and Hydraulics Technical Memorandum*. Phoenix, AZ: Flood Control District of Maricopa County.

Mussetter, R. A., Lagasse, P. F., and Harvey, M. D. (1994). *Erosion and Sediment Design Guide*. Albuquerque, NM: Albuquerque Metropolitan Arroyo and Flood Control Authority.

Stantec. (2014). *Powerline, Vineyard and Rittenhouse FRS - Background*. Phoenix, AZ: Flood Control District of Maricopa County.

Stantec. (2015). *Powerline, Vineyard Road, and Rittenhouse Flood Retarding Structures Hydrologic Analysis*. Phoenix, AZ: Flood Control District of Maricopa County, FCD 2013C004.

U.S. Army Corps of Engineers. (1974). *New River and Phoenix City Streams.* Los Angeles District: Arizona Design Memo No. 2, Hydrology, Part 1.

WEST Consultants, Inc. (2015). *PVR Sedimentation Pre-Design Analysis.* Phoenix, AZ: Flood Control District of Maricopa County, FCD 2013C004.

Williams, J.R., and Berndt, H.D. (1977). "Sediment Yield Prediction Based on Watershed Hydrology." *Transactions of the American Society of Agricultural Engineers*, 20(6), pp. 1100-1104.

Zeller, M.E., and Fullerton, W.T. (1983). "A Theoretically Derived Sediment Transport Equation for Sand-Bed Channels in Arid Regions." *Proceedings of the D.B. Simons Symposium on Erosion and Sedimentation*, R.M. Li and P.F. Lagasse, eds.

POST- REMOVAL RESERVOIR EROSION: ELWHA DAM AND GLINES CANYON DAM REMOVALS, WASHINGTON

Jennifer Bountry, P.E., M.S.[1]
Andrew Ritchie[2]
Timothy Randle, P.E., PhD [3]

ABSTRACT

Removal of Elwha and Glines Canyon Dams in northwest Washington State occurred between September 2011 and October 2014, utilizing the river's natural stream power to establish a new reservoir landscape. Of the estimated 21 million m^3 of reservoir sediment, 60 percent has been released downstream as of September 2015. Flood peaks were less than a 2-year flood during the three years of dam removal and between a 2 and 3-year flood during the year following dam removal. After the river eroded down to the pre-dam valley bottom, a new "quasi-equilibrium" floodplain has developed through lateral erosion. Modeling and recent erosion trends indicate 20 to 30% of sediment will remain over the long-term in the reservoir landscape. While the rate of sediment erosion was closely tied to the rate of dam removal during the first three years, the final phase of erosion is now dependent on the timing and magnitude of natural floods. In this paper, we present predictions for future reservoir sediment erosion based on erosion trends during the first four years of removal and current reservoir conditions. We predict that once a given flood magnitude has occurred the river will require a larger flood than has previously occurred to initiate the next increment of erosion. In absence of a large flood, we anticipate sediment recruitment in the former reservoirs to be largely restricted to reworking of depositional features within the newly developed floodplain. During dam removal, downstream turbidity remained at least an order of magnitude higher than background during the majority of the year. Post-removal turbidity should be of similar order of magnitude to background levels, and only elevated during flows greater than the mean annual flow.

INTRODUCTION

While dams continue to be an essential component of water resource management throughout the world, dam removal has become a viable consideration for aging or abandoned infrastructure in disrepair (USSD, 2015). Dam removal may be utilized as a tool to restore lost ecosystem processes when the lost operational benefits of a dam and reservoir are no longer needed, can be met through alternative means, or the costs to address safety and infrastructure repairs exceed the cost of decommissioning. As of 2014, 1,185 dams have been removed in the United States (American Rivers, 2015). Several other dams have been removed throughout the world including dams in Australia, Canada, France, Italy, Libya, Norway, Romania, and Sweden (ICOLD, 2012). The cost of dam removal has been documented to range from tens of thousands of dollars to

[1] Sedimentation and River Hydraulics Group, Bureau of Reclamation, Denver, CO, jbountry@usbr.gov
[2] National Park Service, Port Angeles, WA, andrew_ritchie@nps.gov
[3] Sedimentation and River Hydraulics Group, Bureau of Reclamation, Denver, CO, trandle@usbr.gov

hundreds of millions. Sediment management is among one of the greatest drivers in decision making and can represent a large portion of the dam removal cost. Many dam removal projects avoid release of sediment to minimize potential impacts downstream. While sediment release can have temporary impacts that must be tolerated or mitigated, the release of stored reservoir sediment can provide enormous benefit to jumpstarting the restoration of the ecosystem in the downstream sediment starved reaches. The removal of two dams on the Elwha River in northwest Washington State provides an opportunity to study a project where the river was allowed to erode the largest volume of reservoir sediment to date associated with a U.S. dam removal project.

The Elwha and Glines Canyon Dam removals occurred between September 2011 and October 2014 on the Elwha River (Figure 1). The Elwha River Ecosystem and Fisheries Restoration Act (Public Law 102-495) was authorized by the U.S. Congress in 1992. The intent of dam removal was to restore fish passage past the dam sites, opening up to 113 km of habitat in the upper watershed. Elwha Dam was constructed in 1913, 7.9 km upstream from the river mouth. This 32-m-high concrete gravity dam had 30 m of head and formed Lake Aldwell, which had a storage capacity of 10 million m^3 at the time of removal. Glines Canyon Dam was completed in 1927, 21 km upstream from the river mouth. This 64-m-high concrete arch dam had 59 m of head and formed Lake Mills, which had a storage capacity of 32 million m^3 at the time of removal (50 million m^3 when first built) (Bountry et al., 2011).

Topographic surveys and sediment surveys determined a sediment volume of 16.1 million m^3 was stored in the upstream Lake Mills and 4.9 million m^3 was stored in the downstream Lake Aldwell at the start of dam removal (Bountry et al., 2011). Drill borings, dive surveys, and sediment probes determined approximately half of the stored sediment was sand, gravel, and cobble and the other half was silt and clay (Gilbert and Link, 1995). Wood of all sizes and litterfall (pine needles, leaves, etc.) were also present in the sediment deposits. The Elwha Restoration Act required mitigation for released sediment including the construction and operation of water treatment plants, flood protection, and upgrades to wells which accounted for about 90 percent of the total project cost of $325 million. Based on a field drawdown experiment in 1994, numerical modeling, and a physical model study of dam removal rates, drawdown increments of 4.6 m were selected with 14-day hold periods in between. Three additional hold periods of 4 to 6 weeks were incorporated during times when fish were expected to migrate along the Elwha River.

The average annual flow of the Elwha River is 43 m^3/s, with peak floods occurring during late fall through winter, and a spring snowmelt runoff. Flood peaks range from a 2-year frequency of 400 m^3/s to a 100-year frequency of 1,240 m^3/s (Duda et al., 2011). The lowest flows typically occur in September, with a 10-year recurrence interval 7-day low discharge of 7 m^3/s (Duda et al., 2011). Elwha River floods are often flashy with rapid increases on an hourly time scale that can increase river stage several meters. During the four years since the start of dam removal, approximately 60% of the reservoir sediment has been eroded and released downstream. Now that both dams are gone, the last phase of the adaptive management program is predicting the fate of the remaining reservoir sediment. In this paper, we present predictions for future reservoir sediment erosion based on erosion trends during the first four years of removal and current reservoir conditions.

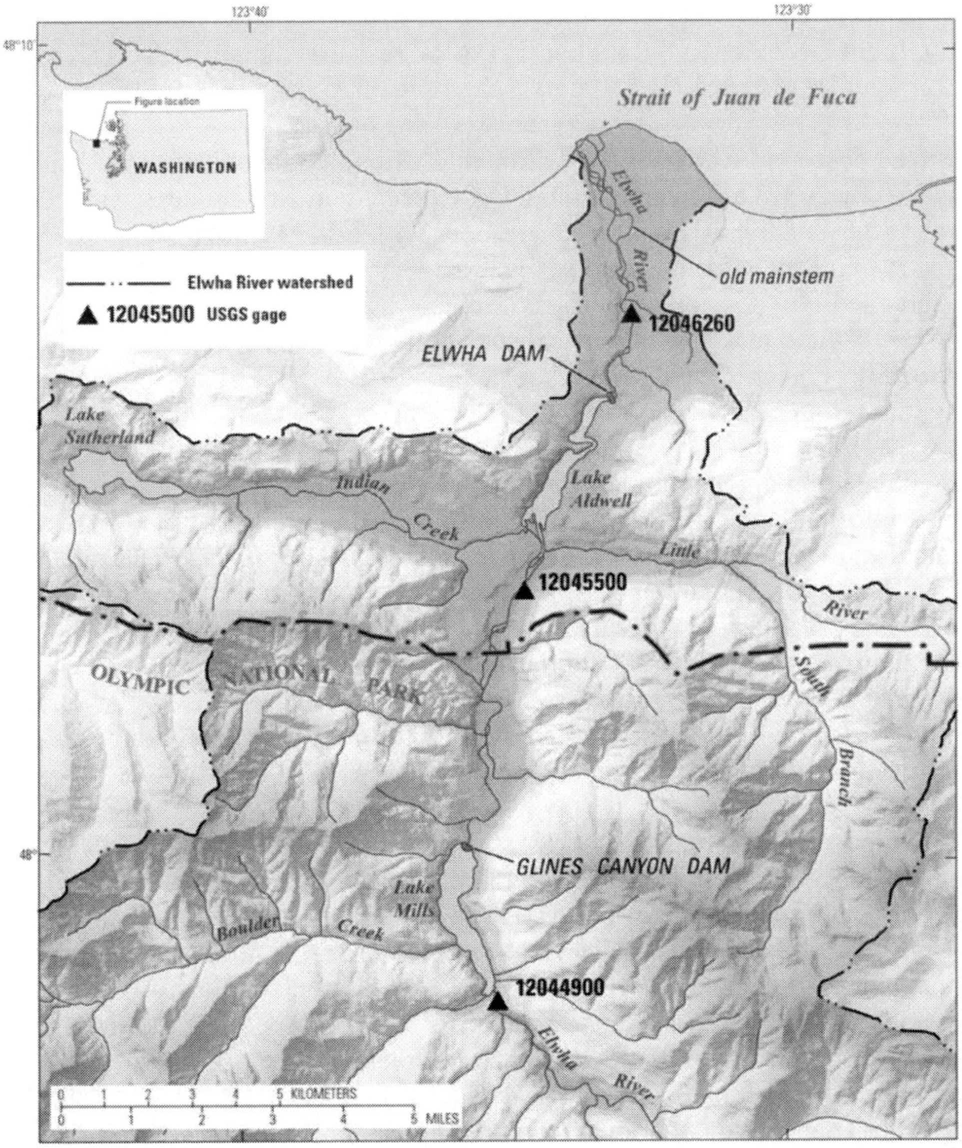

Figure 1. Project location map and USGS gage locations (Randle et al., 2015).

METHODS

Elwha River discharge is based on measurements at the U.S. Geological Survey gage 12045500, located between Glines Canyon Dam and Elwha Dam. Flood frequency analysis is provided by Duda et al. (2011). Flood frequency values at each reservoir were estimated by using the ratio of drainage areas at the reservoir and the gage location. Suspended sediment data are based on daily averages of total suspended sediment (TSS) measurements collected at the surface water diversion intake approximately 5.6 km upstream from the mouth.

Past reservoir erosion volumes were based on volumetric change between measured topographic surfaces derived from Structure-from-Motion photogrammetry (Westoby et al., 2012), Light Detection and Ranging (LiDAR), real-time kinematic (RTK) global navigation satellite systems (GNSS) data, sonar data, or a combination of data sources.

We predicted post- removal reservoir sediment erosion tied to a given flood frequency interval. We developed a conceptual model of expected erosion processes based on literature synthesizing physical reservoir response after dam removal, combined with river erosion trends documented during the three years of the project. The conceptual model accounted for cohesion properties of remaining reservoir sediment deposits based on August 2014 field mapping of vertical stratigraphy at twelve locations. We used a core sampler to collect lakebed mud samples for laboratory analysis of size gradation, plasticity index, and bulk density.

To make the prediction quantitative, we mapped areas vulnerable to erosion in a geographic information system (GIS). We limited erosion predictions to terraces greater than 2 m above the river at low flow, because most low-relief surfaces are similar to the pre-dam valley bottom. Potential erosion volumes were computed by multiplying the mapped areas by the thickness of the remaining reservoir sediment down to the estimated pre-dam valley bottom. A pre-dam contour map was available for Lake Mills. Pre-dam elevations had to be estimated for Lake Aldwell based on the pre-dam valley slope, valley bottom elevations identified in drill holes within the reservoir sediment, and exposed tree-stumps after river erosion (Randle et al., 2015).

We based vulnerability to future reservoir sediment erosion on trends of recent reservoir sediment erosion, river planform and alignment, pre-dam surface exposure, and modeled velocity vectors during a range of floods. River channel migration patterns and trends were estimated based on aerial imagery collected before and after flood events. A two-dimensional hydraulic model, SRH-2D, was utilized to predict velocity vectors during a range of flood peaks for the latest topographic surface representing conditions in late fall (low-flow) prior to the winter flood season. Additional erosion potential was identified at locations where the river had not yet completed incision through the reservoir sediment down to the pre-dam valley bottom and had not yet achieved the pre-dam slope.

We used average annual sediment load to normalize the reservoir sediment erosion volumes. Average annual sediment load is based on the total measured reservoir sediment volume in the upstream Lake Mills at the time of dam removal (2010) divided by the 84 years of storage. The watershed upstream of Lake Mills is federally protected wilderness with no infrastructure that would alter the sediment load. Because Lake Mills only trapped a portion of the fine sediment (silt and clay), which was half of the total volume, a trap efficiency curve was used to adjust the average annual incoming load. The fine sediment portion of the loads was increased by 20% to estimate the average portion passing through the reservoir and not included in the total 2010 volume (Magirl et al., 2015).

RESULTS

Reservoir Erosion during First Four Years

The majority of the 12.1 million m^3 of sediment exported from the two reservoirs occurred in response to base level lowering at the dam sites, even in the absence of floods. Annual discharge was typical during dam removal, but the 2-year flood was not exceeded until the first winter flood season after dam removal was completed. Annual sediment volumes exported from the reservoir areas into the downstream river have far exceeded background sediment loads, ranging from 25 to 30 times the average annual load in year 2, and then reducing to 5 to 10 times the average annual load in years 3 and 4. Specific erosion trends for each reservoir are presented below.

Lake Aldwell had an early drawdown in late spring 2011 and was then lowered as planned until there was no longer a lake in spring 2012 (Figure 2 and 3; Randle et al., 2015). The majority of incision occurred during the first year of removal (October 2011 to October 2012), but minor incision occurred in year 3 and again in year 4 up to 1 m. Incision lagged in the upstream portion of the reservoir, particularly in the fluvial reach extending 0.6 km upstream of the wide reservoir pool. The lag of the final incision phase is likely due to the river initially flowing over a pre-dam terrace that slowed the rate of erosion due to a large number of exposed tree stumps. Deposition was observed in year 1 where the river eroded delta sediment and redeposited the sediment in the receded lake during each hold period. The most extensive lateral erosion occurred during dam removal in the first year, and in the fourth year when a ~3-yr flood occurred resulting in additional removal of sediment terraces. With a stable base level and no large floods, lateral erosion continued in year 2 and 3 but at a much smaller magnitude and at fewer locations than the first year. In year 2, Lake Aldwell also received a large flux of sediment from the upstream Lake Mills which further increased the sediment supply, resulting in net deposition along the active river channel. The erosion of sediment terraces in year 4 largely occurred along the outside of meander bends where the river was pointed into a remaining terrace where the floodplain was relatively narrower than a wider upstream segment.

Lake Mills was lowered according to planned increments in the first year (September 2011 to October 2012), followed by one year of no removal (hold) in water year 2013 (Figure 4; Randle et al., 2015). Initially, the river incised and reworked coarse sediment deposits consisting mostly of sand and gravel. During this phase, the morphology was braided and shifted rapidly during flows greater than the average annual flow, which occurred just under half of the days in year 1. The lateral erosion resulted in valley wall to valley wall erosion across the exposed delta sediment. As sediment eroded, the river redeposited a new delta within the remaining lake resulting in net deposition in the downstream portion of the reservoir. Once the lake was lost in October 2012, dam removal was put on hold for nearly all of year 2 to allow time for the river to flush the released sediment downstream to the sea. Lateral erosion was most extensive when the river eroded through unconsolidated sand and gravel deposits containing numerous large wood pieces that locally accelerated lateral erosion. As the river incised into more consolidated lakebed mud layers, the river transitioned from a wide, braided channel with many actively eroding flow paths, to a much narrower meandering channel. Lateral erosion primarily occurred along the outside of meander bends. The transition from wide and braided to a more single thread morphology propagated from upstream to downstream, largely due to incision into the lakebed mud layer during the one-year hold period.

The river first incised into the lakebed mud layer in the upstream portion of the reservoir within the delta between late November and early December 2012 (year 2). The lakebed mud layer was on average 7.6 m thick above the predam valley bottom in the delta. During the year-long hold period, the river progressively flattened in slope between the dam and the downstream end of the delta, resulting in a maximum incision of 9 m. The headcut stalled when it intercepted a pre-dam terrace or coarser cobbles and boulders in the upstream delta or tributaries. Incision extended an additional 1.6 km upstream of the wide reservoir pool into a narrow bedrock canyon. The river incised into lakebed muds in the middle of the reservoir during late March to early April 2013 (year 2). The river did not incise into lakebed muds in the downstream portion of the reservoir until after dam removal was resumed in year 3, between October 2013 and January 2014. As a result the river transitioned to intermittent locations of lateral erosion along meander bends first in the upstream portion of the reservoir (see year 3 in Figure 4), and then in the downstream half of the reservoir in year 4. By the end of dam removal, the river had largely reached the pre-dam valley bottom within a couple meters. However, the last round of incision (up to 4.5 m) lagged in the upstream-most portion of the Lake Mills delta until larger floods occurred post-dam removal.

During year 1, minor increases in downstream suspended sediment occurred from fine sediments being released past the Elwha Dam site as the shrinking Lake Aldwell lost the ability to trap fine sediment (see bottom graph in Figure 2). After the Lake Aldwell delta prograded downstream to Elwha Dam during spring 2012, the first increment of coarse sediment was released past Elwha Dam. This event increased total suspended sediment during the spring snowmelt season. Downstream sediment levels then reduced as river discharge dropped below the average annual flow. Downstream sediment levels increased after the Lake Mills delta reached Glines Canyon Dam. During the one-year hold period at Lake Mills (year 2), the river continued to erode reservoir sediments and long periods of elevated suspended loads occurred until river discharge dropped below the average annual flow in fall 2013. A downstream water treatment plant required improvements during this hold period to manage the high levels of sediment, including sand, gravel, silt, clay and organic material that were diverted into the treatment facility. Downstream sediment levels again picked up in winter 2013-2014 when the remainder of Glines Canyon Dam was removed. However, sediment levels again reduced during the low flows of summer and fall 2014. During winter 2014-2015, downstream sediment levels increased again as the largest floods since the start of dam removal occurred, but then quickly dropped back down during the remainder of the water year. The spring snowmelt in 2015 was a record low, so downstream sediment levels remained low during the summer and fall of 2015.

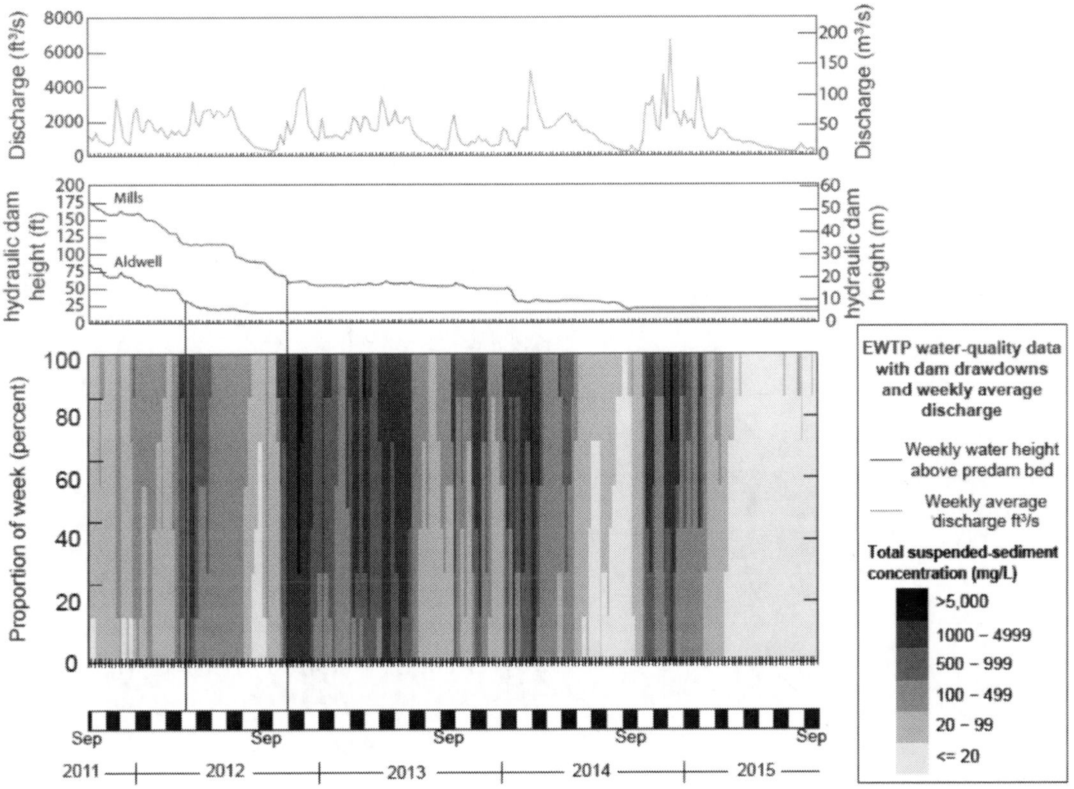

Figure 2. Temporal changes in weekly average river discharge (top), hydraulic height of reservoir stages above the river bed (middle), and daily total suspended sediment concentrations in the downstream river below both dam sites (bottom). Vertical lines in middle graph represent when the Lake Aldwell or Lake Mills coarse delta had prograded to the Elwha or Glines Canyon Dam sites.

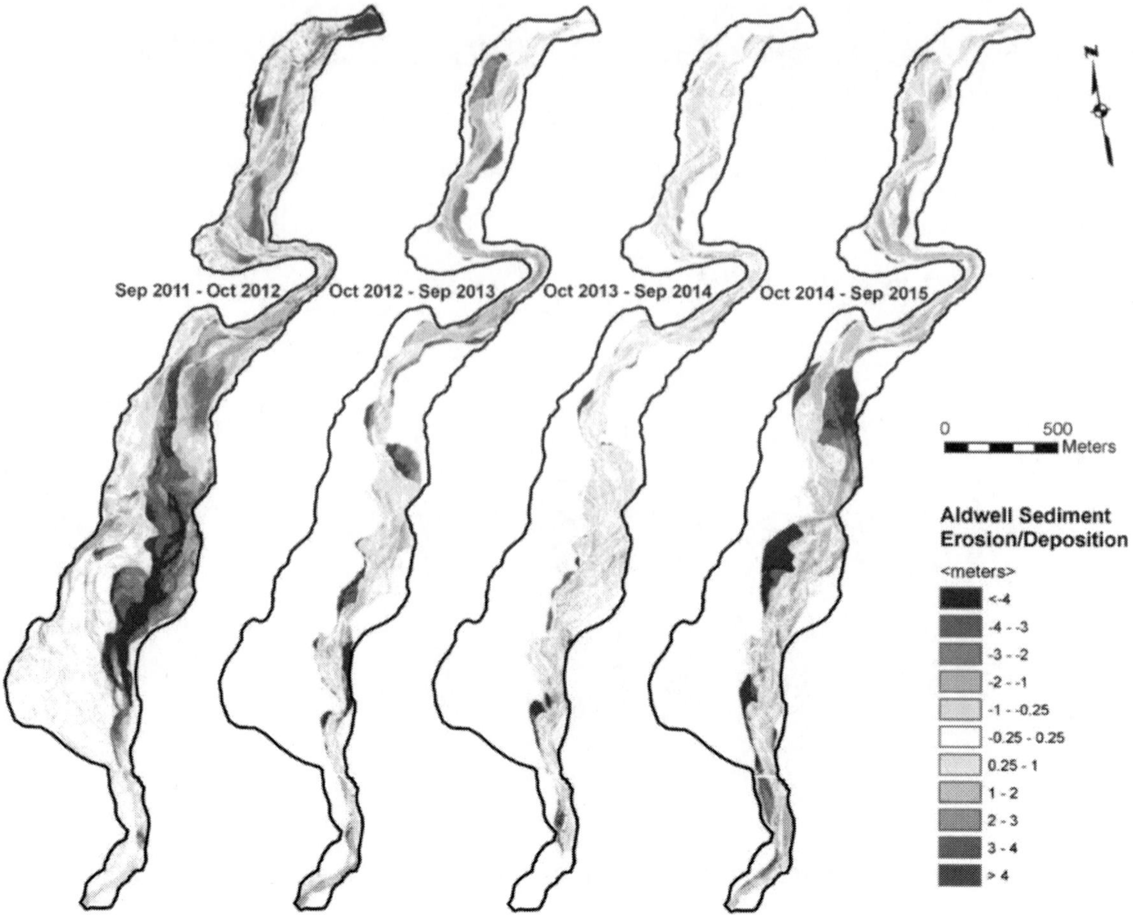

Figure 3. Annual erosion and deposition thickness in Lake Aldwell during the first year of dam removal and three years post-removal (flow is south to north).

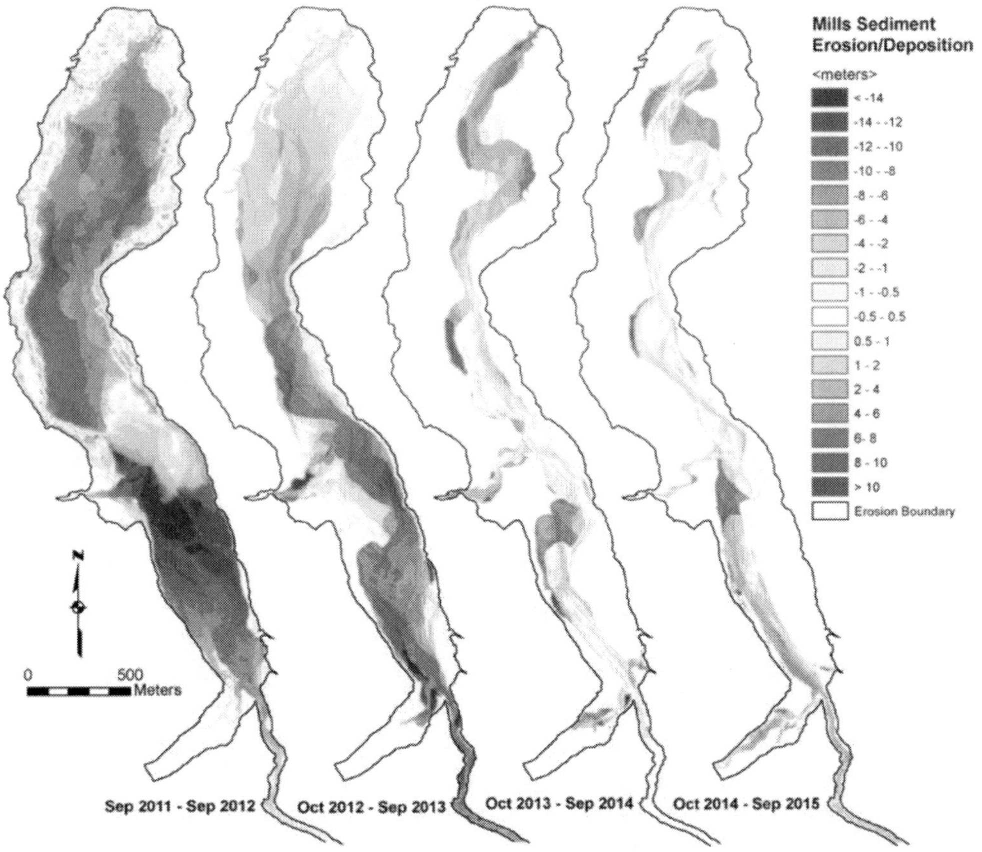

Figure 4. Annual erosion and deposition thickness in Lake Mills during the first three years of dam removal and one year post-removal (flow is south to north).

Future Reservoir Erosion Predictions

We mapped reservoir sediment erosion predictions for year 4 (water year 2015) and year 5 (water year 2016) in Lake Aldwell and Lake Mills (Figure 5 and Figure 6). By the end of year 1 for Lake Aldwell and the end of year 3 for Lake Mills, the river was predominantly eroding terraces composed of lakebed muds capped by a coarse sand and gravel layer (with some cobbles). Therefore, the cohesive properties of the reservoir sediment samples were used to estimate erosion which was then compared with the locations of actual erosion. The sediment terraces banks of lakebed mud were nearly vertical and the fine sediment in the field could be easily rolled in a ball without breaking apart. There were also numerous organic layers noted in the banks that may have contributed to the cohesion properties. Clay percentage within the lakebed mud ranged from 7 to 32, and laboratory testing did not find any plasticity characteristics that would indicate strong cohesion properties (Table 1 and Table 2). Although the lakebed muds did provide enough cohesion for meander bends to form, all locations sampled in fall 2014 experienced some amount of lateral erosion by the river during winter 2014-2015 (see previous section and Figures 3 and 4). This indicated the fine sediment properties may have slowed or limited the rate of lateral erosion where stream velocities were slow or parallel to the bank, but did not have enough cohesion to prevent erosion where velocities were high or pointed into the bank, especially along the outside of meander bends during floods.

Table 1. Sediment sample results from Lake Aldwell collected August 26, 2014.

X (m)	Y (m)	Z (m)	Site	Bulk Density (g/cm^3)	% Sand	% Silt	% Clay	D_{10} (mm)	D_{50} (mm)	D_{90} (mm)
295501	122492	57	1a	1.10	46.7	43.4	9.8	0.004	0.056	0.179
295552	122734	55	2a	1.18	54.2	34.8	11.0	0.004	0.075	0.268
295660	122947	54	3a	1.00	8.9	74.6	16.5	0.002	0.016	0.059
295658	122949	54	3b	1.01	28.6	60.9	10.5	0.004	0.037	0.106
295731	122993	53	4a	1.12	18.6	58.0	20.9	0.002	0.013	0.283
295731	122994	54	4b	1.09	14.7	68.9	16.4	0.002	0.017	0.079
295733	122997	54	4c	-	20.6	57.2	22.2	0.002	0.012	0.101
295973	123543	50	5a	1.06	7.3	75.2	17.6	0.002	0.014	0.053
295857	123641	49	6a	1.16	8.5	64.4	27.1	0.002	0.009	0.055
295857	123640	50	6b	0.96	23.5	61.2	15.3	0.002	0.027	0.094

Table 2. Sediment sample results from Lake Mills collected August 27, 2014.

X (m)	Y (m)	Z (m)	Site	Bulk Density (g/cm^3)	% Sand	% Silt	% Clay	D_{10} (mm)	D_{50} (mm)	D_{90} (mm)
293475	113180	152	1a	1.04	4.8	75.2	20.0	0.002	0.010	0.041
293475	113180	153	1b	1.07	12.6	69.3	18.1	0.002	0.013	0.072
293282	113329	153	2a	0.97	6.8	72.3	20.8	0.002	0.011	0.050
293174	113999	146	3a	0.87	15.0	67.9	17.1	0.002	0.015	0.079
293393	114793	136	4a	0.93	7.2	71.5	21.3	0.002	0.014	0.054
293393	114793	136	4b	1.00	0.1	67.9	32.1	0.001	0.006	0.017
293397	114789	141	4c	1.02	1.0	77.9	21.1	0.002	0.009	0.026
293339	114530	140	5a	1.18	74.3	18.6	7.1	0.006	0.152	0.377
293339	114530	140	5b	1.04	2.4	70.9	26.8	0.002	0.008	0.030

Erosion predictions for water year 2015 were mapped where the hydraulic model indicated relatively higher velocities at channel constrictions or along the outside of meander bends where the channel was expected to migrate into a remaining terrace, based on current position and recent erosion trends (Figure 5). The water year 2015 erosion predictions were intended to represent a 2- to 5-year flood. Predictions were compared to measured erosion during four high-flow periods: September 30 to November 14, 2014 with a flood peak of 224 m^3/s (~ half of 2-year flood); November 14, 2014 to January 16, 2015 with a flood peak of 447 m^3/s (~3-yr flood); January 16 to March 3, 2015 with a similar magnitude flood peak of 411 m^3/s (~ 2-year flood); and March 3 to July 2015 with a flood peak of 94 m^3/s (after July, flows became very low with no erosion). The general locations of erosion during 2014-2015 were correctly predicted, but the actual extent of erosion varied depending on where the river migrated. Despite the differences between the predicted and actual erosion, the total erosion volume predicted in Lake Mills was the same order of magnitude as measured, representing about 7 years of average annual sediment load. In Lake Aldwell, we underestimated the extent of meander migration along the downstream portion of the delta and in a backwater area upstream of a bedrock constriction. In both locations, the river flowed along a sharp curve and directly into the terrace bank, resulting in one to two channel widths of lateral erosion. Lateral erosion in water year 2015 confirmed that erosion is highest where river meanders migrate into reservoir sediment terraces, particularly in relatively narrow reaches that had not yet developed a quasi-equilibrium floodplain width. Additionally, the 3-year flood peak had the largest erosion area, and the subsequent 2-year peak had less erosion. The smaller peaks in early November 2014 and during the 2015 spring snowmelt had minimal erosion.

For water year 2016, about 3 years of average annual sediment load was predicted to erode from Lake Mills and about 1 year of average annual load was predicted to erode from Lake Aldwell during a 5 to 10-year flood peak (Figure 6). Actual discharge values are still under review, but on November 13, 2015 an approximately 5- to 10-year flood occurred, and then on November 17, 2015 a 20-year flood occurred. Aerial imagery collected the day after the second and larger flood peak shows actual river erosion and was compared to the water year 2016 predictions. The general locations of actual erosion are comparable to predicted erosion areas, except for two locations in the downstream half of Lake Mills where the river migration did not progress as much as expected. Erosion volumes had not yet been computed at the time of writing this paper, but will be available for comparison in the near future.

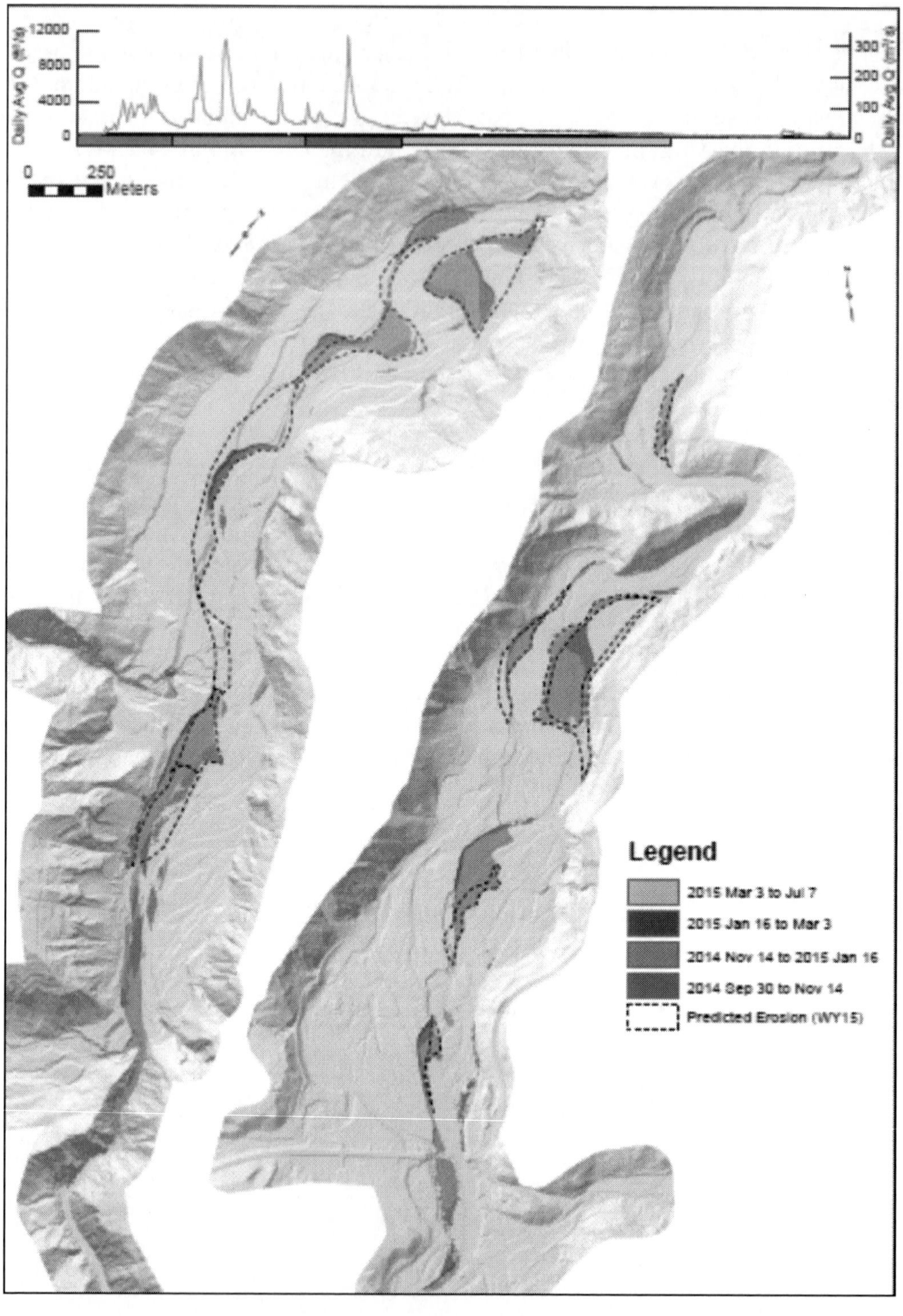

Figure 5. Predicted erosion areas for water year 2015 (black dash outline) versus actual sediment terrace erosion during four time periods in Lake Mills (left) and Lake Aldwell (right) in winter 2014-2015. Mean daily hydrograph for water year 2015 shown at top of figure.

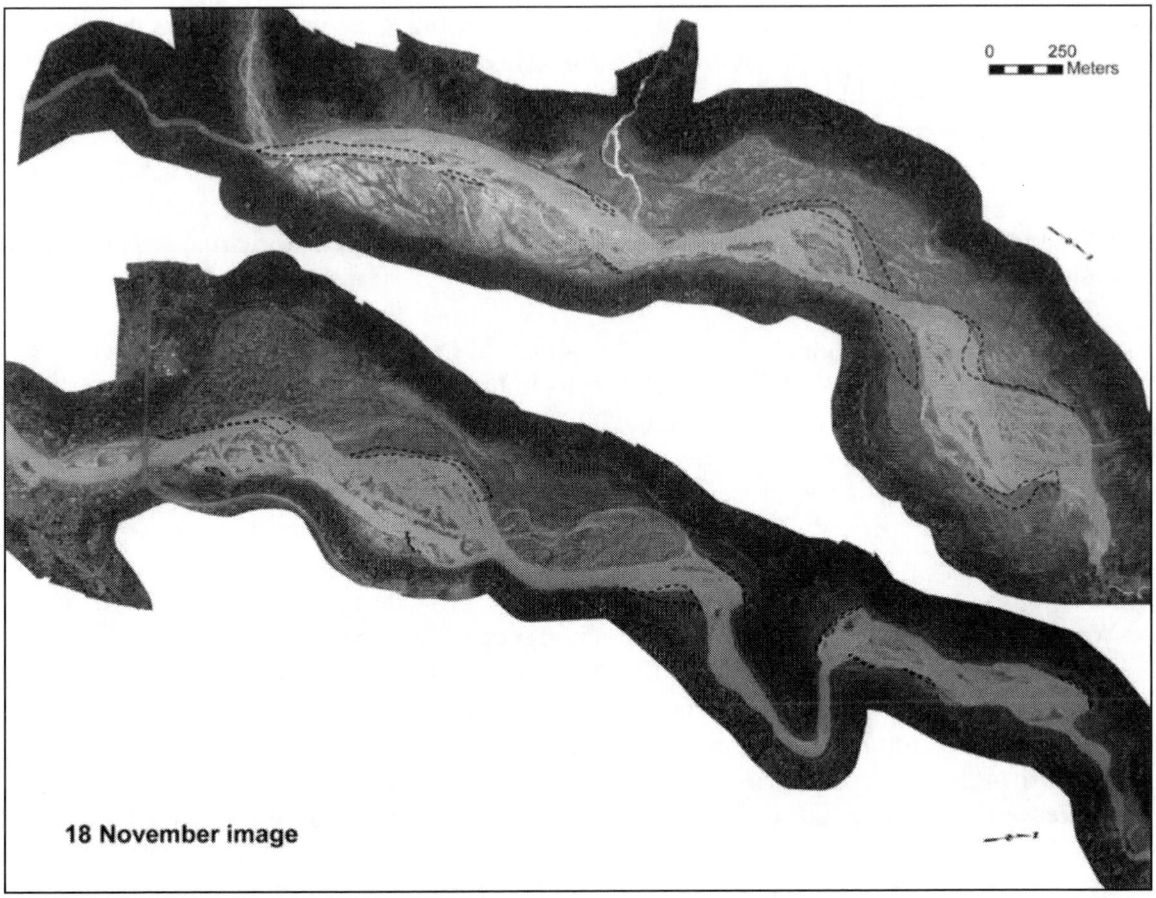

Figure 6. Predicted erosion areas for water year 2016 (black dash outline) versus actual erosion (light colored wetted channel) during a 20-year flood as visible on a November 18, 2015 aerial imagery background for Lake Mills (top) and Lake Aldwell (bottom).

DISCUSSION

Phased dam removal was designed to maximize sediment erosion and redistribution during dam removal. Prior to removal, numerical models predicted 40 to 60 percent of Lake Aldwell and Lake Mills sediment would be released past the dam sites into the downstream river during three years of dam removal and as a result of three large floods post removal. Original estimates in 1994 were based on a reservoir sediment volume of 13.8 million m^3 (Bountry et al., 2011). By 2010, the reservoir sediment volume had significantly increased to 21 million m^3 due to an additional two decades of incoming sediment including a major flood in 2007. Measurements indicate that during three years of removal and one-year post removal with a 3-year flood, the river eroded about 60% of reservoir sediment, the upper end of the original model predictions (Figure 7).

As part of the adaptive management program, the numerical model for Lake Mills was updated to reflect the final sedimentation and actual erosion through November 2013 (Randle et al., 2015b). The updated model predicted about 70 to 75% of the reservoir sediment volume would be eroded. The timing of model erosion predictions varied depending on the occurrence of large floods post-removal (four hydrology ranges were utilized). The model logic assumes that once a given flood magnitude occurs the river has largely done the erosion possible for that flood level. To initiate the next increment of erosion, a larger flood magnitude is necessary. Local exceptions to this logic may occur where a sediment terrace is particularly vulnerable because a river meander is directed into it, such that even a smaller magnitude flood is capable of eroding it. Therefore, the model logic was updated to incorporate migration of the river centerline until the river was coincident with the pre-dam channel alignment shown on historic maps. This update resulted in the increased percentage of erosion.

As an alternative prediction method, an exponential trendline was fit to the reduction in sediment volume at Lake Mills, which results in total sediment erosion up to 80 to 85%. However, it is unlikely reservoir sediment erosion will continue at the same rate because as the floodplain widens, the ability of the river to erode remaining sediment terraces is expected to diminish. Therefore, based on the combination of methods we predict 20 to 30% of the reservoir sediment will remain over the long-term in the former reservoir areas.

We also qualitatively evaluated erosion occurrence in the first alluvial reference reach upstream from the two reservoirs. This reference reach is known as Geyser Valley, and has approximately 1.5 to 2 times the slope of the pre-dam valley within Lake Mills and Lake Aldwell. Aerial image interpretation between flood peaks indicates that floods larger than the 2-year flood are required to initiate significant river migration within the reference reach. When large floods did occur, the river was observed to erode into forested islands and, in some cases, small segments of older terraces. However, the majority of erosion occurred within the floodplain providing further evidence that once a quasi-equilibrium floodplain develops, erosion will dramatically decrease in the former reservoirs.

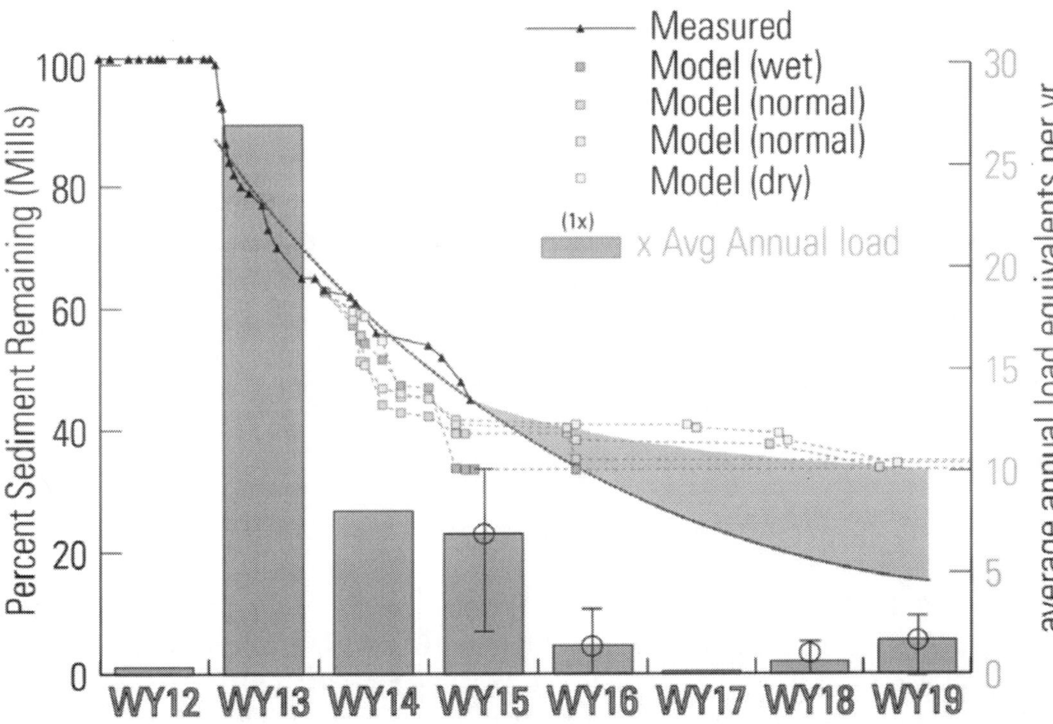

Figure 7. Measured reservoir erosion in Lake Mills (black triangles) compared with the most updated numerical model predictions (blue squares) and an exponential trendline (red line). Reservoir erosion shown in yellow bars in terms of average annual load equivalent for measured periods through water year 2014, and model predictions through water year 2019.

CONCLUSION

Through September 2015, we measured 12.1 million m^3 of reservoir sediment released past the dam sites or about 60% of the total stored sediment at the start of removal. Post-dam-removal erosion has been dominated by locations where river meanders are migrating into remaining sediment terrace banks. The sediment terraces are particularly vulnerable where they form a relatively narrow channel width. We predicted that a few large floods were needed post-dam removal to complete significant erosion within the former reservoirs and develop a quasi-equilibrium floodplain. We incorporated logic that once a given flood magnitude occurs, it would generally take a larger flood peak to progress the erosion and development of the floodplain. We made predictions for water year 2015 that were near the same order of magnitude as measurements for Lake Mills, and slightly under predicted volumes from Lake Aldwell. Areas where erosion was underpredicted were at locations of tight meander bends. For water year 2016, we predicted the equivalent of 4 years of average annual sediment load to erode during a 5- to 10-year flood peak. Provisional data indicate a 5- to 10-year flood did occur in November, followed by a 20-year flood. In general, predicted erosion areas matched actual erosion for these floods. Volumes of erosion are not yet available. In the absence of large floods, we expect reworking of floodplain depositional features to be more dominant than erosion of remaining reservoir sediment terraces. Now that three large floods greater than a 2-year flood have occurred post-removal, the portion of remaining reservoir sediment still vulnerable to erosion during future floods is decreasing with time.

ACKNOWLEDGEMENTS

The Elwha sediment trend analysis is a collaborative effort that utilizes data and analysis from many agencies and universities that builds upon the adaptive management program funded through the National Park Service, and research efforts funded through USGS, NOAA, Bureau of Reclamation, USFWS, Elwha Tribe, University of Washington, and Seattle University. Special thanks to Heidi Hugunin, Anna Geffre, Gary Smillie, Brian Cluer, Kurt Wille, Chris Magirl, Chris Curran, Jon Warrick, and George Pess for technical contributions.

REFERENCES

American Rivers, 2015, electronic information accessed at AmericanRivers.org, September 2015.

Bountry, J.A., Ferrari, R., Wille, K., and T.J. Randle, 2011, 2010 Survey Report for Lake Mills and Lake Aldwell on the Elwha River, Washington, SRH-2010-23, U.S. Department of the Interior, Bureau of Reclamation, Denver, CO, June 2011.

Duda, J.J., Warrick, J.A., and C.S. Magirl, 2011, Coastal Habitats of the Elwha River, Washington—Biological and Physical Patterns and Processes Prior to Dam Removal

Gilbert, J. and Link, R., 1995. "Alluvium Distribution in Lake Mills, Glines Canyon Project and Lake Aldwell, Elwha Project, Washington," Elwha Technical Series PN-95-4, U.S. Bureau of Reclamation, Pacific Northwest Region, Boise, ID, 60 pages.

International Committee on Large Dams (ICOLD), January 2012, ICOLD dam decommissioning guidelines.

Magirl, C.S., Hilldale, R.C., Curran, C.A., Duda, J.J., Straub, T.D., Domanski, M., and J.R. Foreman, 2014, Large-scale dam removal on the Elwha River, Washington, USA: Fluvial sediment load, Geomorphology, http://dx.doi.org/10.1016/j.geomorph.2014.12.032

Randle, T.J., Bountry, J.A., Ritchie, A.C., and Wille, K.B, 2015a, "Large-scale dam removal on the Elwha River, Washington, USA: erosion of reservoir sediment," Journal of Geomorphology Volume 246, 1 October 2015, Pages 709-728.

Randle, T.J., Bountry, J.A., and K. Wille, 2015b, Elwha reservoir sediment modeling in a GIS framework, 10th Federal Interagency Sedimentation Conference and 5th Hydrologic Modeling (SEDHYD), Reno, NV.

U.S. Department of the Interior, 1995a, Elwha River ecosystem restoration— Final environmental impact statement: Olympic National Park, National Park Service, 215 p.

U.S. Department of the Interior, 1995b, Alluvium distribution in Lake Mills, Glines Canyon Project and Lake Aldwell, Elwha Project, Washington: Bureau of Reclamation, Pacific Northwest Region, Elwha Technical Series PN-95-4, 60 p.

U.S. Department of the Interior, 1996, Sediment analysis and modeling of the river erosion alternative: Bureau of Reclamation, Elwha Technical Series PN-95-9, 136 p.

United States Society on Dams, 2015, Guidelines for Dam Decommissioning Projects, Dam Decommissioning Committee, http://www.ussdams.org/15Decommissioning.PDF.

Westoby, M.J., Brasington, J., Glasser, N.F., Hambrey, M.J., and J.M. Reynolds, 2012, Structure-from-Motion photogrammetry: A low-cost, effective tool for geoscience applications, Geomorphology, Volume 179, 15 December 2012, pages 300-314.

UNIT HYDROGRAPH PEAKING FACTORS AND THEIR APPLICATION TO PMF SIMULATIONS

Michael Bartles, P.E.[1]
Matthew Fleming, P.E.[2]

ABSTRACT

Within the U.S. Army Corps of Engineers (USACE), current guidance for dam design and operation requires the safe passage of an Inflow Design Flood (IDF) with adequate freeboard. The IDF is commonly comprised of the Probable Maximum Flood (PMF) along with assumed antecedent conditions for the moisture state in the watershed and the initial pool storage.

Unit hydrograph theory is the most commonly used hydrologic modeling approach that converts excess precipitation to a point runoff hydrograph. Unit hydrograph theory assumes a linear precipitation – runoff response without consideration of several complicated hydrologic processes that can occur. When using unit hydrograph theory, the ordinates of a runoff hydrograph from a two inch/hour excess precipitation event are assumed to be two times larger than a one inch/hour excess precipitation event. As such, unit hydrograph transform parameters are dependent upon the magnitude of the events that were used for calibration. It is essential to calibrate to extreme events when investigating events on the order of the PMF. However, in most cases, even the most extreme observed events are much less intense than the PMF.

Errors in timing and peak magnitude can arise when simulating the PMF due to the assumption of linear runoff response and the lack of observed events on the order of the PMF. In an attempt to use conservative runoff parameters, guidance has been in use within the USACE community for approximately 50 years requiring unit hydrograph peaking factors between 1.25 and 1.5 when simulating the PMF. Commonly, the applicability of a unit hydrograph peaking factor of 1.25 or 1.5 is not explicitly analyzed.

New tools are under development at the Hydrologic Engineering Center (HEC) that can better evaluate the non-linear runoff response associated with increasingly large rainfall amounts. These tools will allow for more accurate extreme event simulations and a more accurate determination of potential hydrologic hazards within dam design and operation.

INTRODUCTION

When a dam impounds water upstream of a populated area, a distinct hazard to that area from a possible failure of the dam is created. As such, extreme care is required in every

[1] Hydraulic Engineer, U.S. Army Corps of Engineers, Institute For Water Resources, Hydrologic Engineering Center, 609 Second Street, Davis, CA 95616; 530-302-3706; Michael.D.Bartles@usace.army.mil.

[2] Chief, Hydrology & Hydraulics Technology Division, U.S. Army Corps of Engineers, Institute For Water Resources, Hydrologic Engineering Center, 609 Second Street, Davis, CA 95616; 530-302-3701; Matthew.J.Fleming@usace.army.mil.

phase of the engineering design, construction, and operation of dams. Current U.S. Army Corps of Engineers (USACE) policy dictates that dam design and operation should allow for the safe passage of an Inflow Design Flood (IDF) with adequate freeboard (U.S. Army Corps of Engineers, 1991). An appropriate IDF is dependent upon the dam's uses, size, and relevant upstream and downstream factors. In most cases, this IDF is determined from the Probable Maximum Precipitation (PMP) that can occur over the watershed upstream of the dam. The Probable Maximum Flood (PMF) is then determined through an iterative process of centering and aligning the PMP within the contributing watershed. Finally, the IDF is determined through the use of varying antecedent conditions and/or ratios of the PMF inflow, depending upon site conditions. Instead of an IDF, most USACE dams were originally designed using a Spillway Design Flood (SDF) that made use of site-specific transposed and maximized historical storm events.

PMF and SDF hydrographs have almost exclusively been determined using simplified, lumped, empirical rainfall-runoff methodologies. In fact, throughout the world, dam safety modeling guidelines specify or recommend the use of simplified hydrologic routines (Institute of Hydrology, 1999 and The Institution of Engineers, Australia, 2001). One of these simplified routines is unit hydrograph theory, which is the most commonly employed method used to convert excess precipitation to a point runoff hydrograph. During the mid-1900s, when the majority of USACE dams were designed and built, unit hydrograph theory helped simplify the complicated rainfall-runoff process so computations could be performed by hand.

It is common knowledge that unit hydrograph theory transforms excess precipitation to runoff hydrographs without detailed consideration of complicated internal processes (Szilagyi, 2007 and Harrison, 1999). As such, the empirical equations and parameters that go into many commonly employed synthetic unit hydrograph methods tend to have limited physical significance. To "ground truth" these methods and parameters, parameters are adjusted through a model calibration process where model outputs are compared to observed data in order to achieve an adequate "fit". Normally, calibrated modeling parameters are dependent upon the magnitude of the event being investigated. For instance, according to Sherman, who originally proposed the unit hydrograph concept, the unit hydrograph of a watershed is "…the basin outflow resulting from one unit of direct runoff generated uniformly over the drainage area at a uniform rainfall rate during a specified period of rainfall duration" (Sherman, 1932). This implies that ordinates of any hydrograph resulting from a quantity of runoff-producing rainfall of unit duration would be equal to corresponding ordinates of a unit hydrograph for the same areal distribution of rainfall, multiplied by the ratio of rainfall excess values. However, due to differences in areal distributions of rainfall and hydraulic reactions between large and small precipitation events, the corresponding unit hydrographs have not been found to be equal, as implied by unit hydrograph theory (Minshall, 1960).

These realizations must be combined with the fact that most precipitation events used when calibrating hydrologic models are normally much less intense than the PMP for a given area. In an attempt to use conservative runoff parameters and counteract the inherent linearity of unit hydrograph theory, guidance has been followed within the

USACE community for approximately 50 years requiring the use of unit hydrograph peaking factors between 1.25 and 1.5 when simulating the PMF event (U.S. Army Corps of Engineers, 1959). In most PMF investigations, the applicability of these rules of thumb is not thoroughly analyzed. For instance, it is unknown whether a 25% peaking factor over or under-predicts the true unit hydrograph of a watershed in response to the PMP. Similarly, it is unknown whether a 50% peaking factor over or under-predicts the true unit hydrograph.

PREVIOUS INVESTIGATIONS

Minshall extracted unit hydrographs for multiple precipitation events with varying intensities for a small watershed in southern Illinois (Minshall, 1960). This observed data demonstrated a non-linear runoff response (i.e. faster runoff response and greater peak flow rate) as precipitation intensities increased.

The application of numerical models to better understand the relationship between precipitation and runoff has been investigated. England tested the applicability of a physically-based model when simulating events on the order of the PMF (England, Velleux, & Julien, 2007). England also used the same model to constrain flood frequency curves for the watershed above Pueblo Dam on the Arkansas River in Colorado (England, Klawon, Klinger, & Bauer, 2006). Abdullah used a physically-based modeling code to simulate extreme precipitation events on the order of the PMF within three different watersheds in Malaysia (Abdullah, 2013).

The non-linear runoff response for the contributing watershed to Prompton Dam in the Upper Delaware River watershed in Pennsylvania was also investigated using a physically-based rainfall runoff model (Bartles, 2014). This analysis predicted a large non-linear runoff response that varied in response to excess precipitation intensity. Simulation results using the physically-based rainfall runoff model were compared against previously-completed empirical modeling results. Large differences in runoff rates, volumes, and timing were predicted to occur during the PMF event.

MODELING PURPOSE

While all of the aforementioned analyses attempted to quantify the runoff response of specific watersheds due to extreme precipitation events, none of them answered these pertinent questions:

1.) How do PMF inflow hydrographs developed using unit hydrograph theory compare against results obtained using 2D runoff approaches?
2.) How do differences in PMF inflow hydrographs translate to differences in pool elevations when routed through a reservoir/dam?
3.) Are peaking factors between 1.25 and 1.5 appropriate for use when simulating the PMF event?

Due to its relative simplicity, flexibility, and accuracy, unit hydrograph theory will likely continue to be frequently used within dam safety studies to calculate runoff hydrographs.

Additionally, the increasing use of stochastic and period of record simulations to better inform hydrologic loading estimates will preclude the use of computationally-intensive runoff computations due to time and/or funding constraints. Therefore, it is pertinent to investigate and potentially adapt unit hydrograph transform methods and parameters to site-specific and event-specific conditions when simulating extreme events.

Recent advances in hydrologic and hydraulic modeling capabilities have begun to allow modelers to investigate and predict the non-linear runoff response associated with increasingly large rainfall amounts. The Hydrologic Engineering Center's River Analysis System (HEC-RAS) now has the ability to transform excess precipitation into runoff with a two-dimensional scheme that allows for non-linear routing effects (Hydrologic Engineering Center, 2015). This same two-dimensional routing scheme is also being implemented within the Hydrologic Modeling System (HEC-HMS) (Hydrologic Engineering Center, 2015).

Using these new modeling abilities, more appropriate site-specific unit hydrograph peaking factors were ascertained for Ball Mountain Dam, which is a high-risk flood control dam within the state of Vermont. Ball Mountain Dam was constructed by USACE in October 1961 as a single-purpose flood control project on the West River, approximately 29 miles upstream from the confluence of the West and Connecticut River, as shown in Figure 1. The total drainage area for the watershed above Ball Mountain Dam is approximately 172 square miles (sq. mi.). The West River watershed is conducive to rapid runoff due to mountainous terrain with few natural or artificial ponds. Another USACE project, Townshend Dam, is located nine miles downstream of Ball Mountain Dam. Together, these two flood control projects are operated to reduce flood stages on the West and Connecticut Rivers (New England Division, 1990).

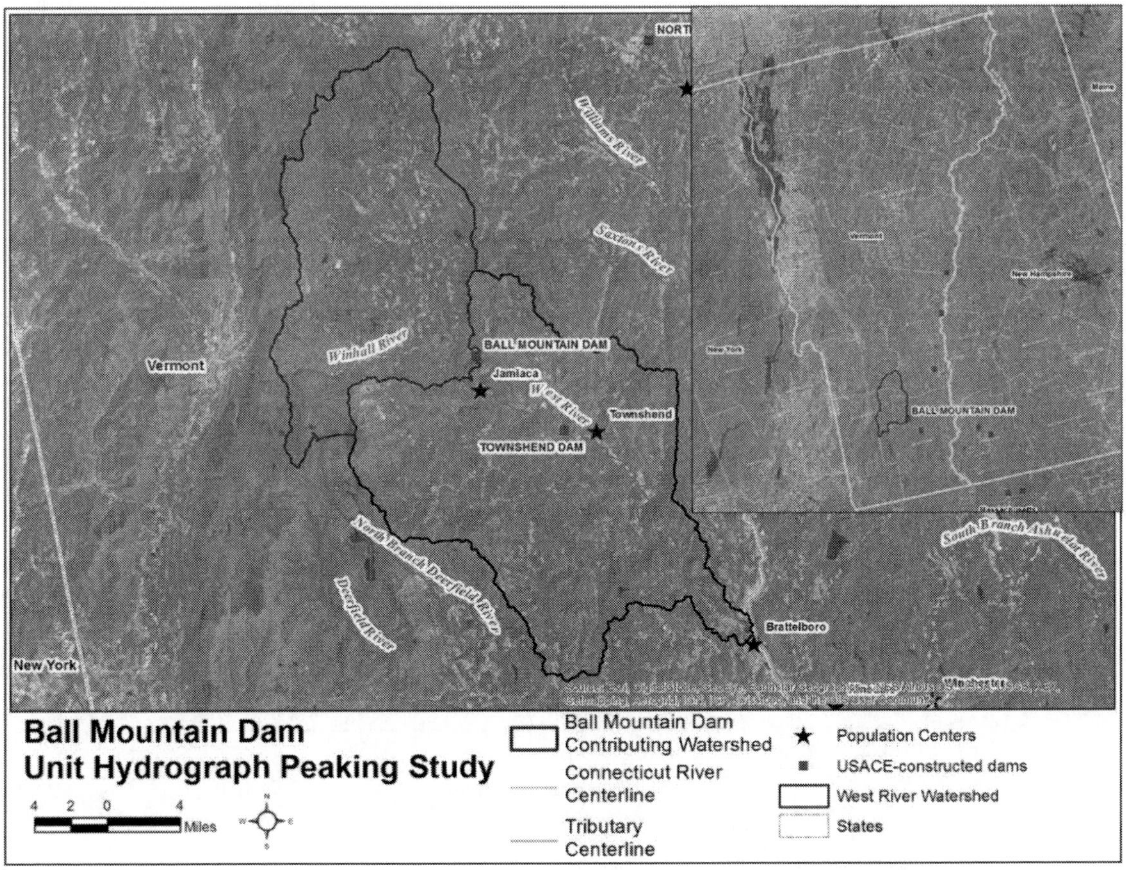

Figure 1. Location of the Study Area within Vermont

Ball Mountain Dam consists of a rolled earth fill embankment with rock slope protection approximately 915 feet in length, with a 265 feet maximum section height above the streambed. The dam crest has an elevation of 1051.6 feet NAVD88 which was designed to provide 30 feet of spillway surcharge plus five feet of freeboard. The top width of the embankment is 20 feet with side slopes varying from 1:1.75 to 1:2.5 (vertical:horizontal). The storage capacity for Ball Mountain Lake totals 54,450 acre-feet when filled to the spillway crest, equivalent to about 5.9 inches of runoff from the 172 sq. mi. drainage area. The resulting pool extends approximately 6.5 miles up the West River and covers a surface area of 810 acres. Additional information related to Ball Mountain Dam is shown in Table 1.

Table 1. Pertinent Features of Ball Mountain Dam

Reservoir	
Conduit Invert Elevation	805.1 feet NAVD88
Permanent Pool	830.1 feet NAVD88
Conservation Pool	870.1 feet NAVD88
Spillway Crest	1016.6 feet NAVD88
Embankment	
Type	Rolled earth and rock fill embankment
Length	915 feet
Top Width	20 feet
Top Elevation	1051.6 feet NAVD88
Maximum Height	265 feet
Spillway	
Location	Right (West) Abutment
Type	Uncontrolled, ogee weir and chute spillway
Length	235 feet
Crest Elevation	1016.6 NAVD88
Surcharge	30 feet
Capacity	150,000 cfs at 1,146.6 ft NAVD88
Outlet Works	
Type	Circular concrete-lined tunnel
Tunnel Inside Diam.	13.5 feet
Tunnel Length	864 feet
Gate Number	3
Gate Size	5'8" X 10'0"

Ball Mountain Dam has been classified as a Dam Safety Action Classification (DSAC) II structure. Within USACE, the DSAC rating system spans five classes (I – V). A DSAC II rating is applied to a structure when failure could begin during normal operations or be initiated as the consequence of an event.

In order to estimate the PMF inflow to Ball Mountain Dam, an HEC-HMS model was created for the contributing area to Ball Mountain Dam. Modeling processes that were used within the HEC-HMS model included Green Ampt infiltration, Modified Clark unit hydrograph transform, Recession baseflow, and Muskingum channel routing (Hydrologic Engineering Center, 2015). Prior to simulating the PMF event, the HEC-HMS model was calibrated and validated to several historic events. Following calibration/validation, the PMF event was simulated using a 72-hr duration, site-specific PMP (Omaha District, 2015).

When investigating events on the order of the PMP, it is essential to calibrate to extreme events, as was done within the HEC-HMS modeling investigation. However, in most cases, even the most extreme observed events are much less intense than the PMP. For instance, during the August 2011 event (remnants of Hurricane Irene), the greatest excess precipitation rate observed was approximately 0.5 inches/hour (distributed over the entire contributing drainage area). Conversely, the site-specific PMP has a peak excess precipitation intensity of approximately 4 inches/hour, as shown in Figure 2.

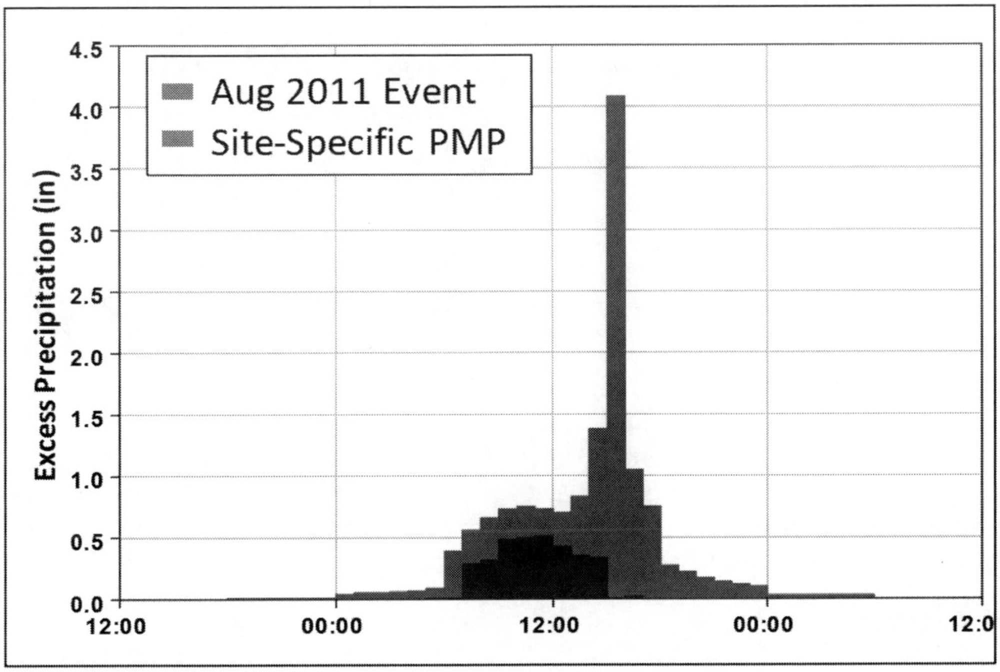

Figure 2. Excess Precipitation Comparison

MODELING DOMAIN

An HEC-RAS two-dimensional (2D) model was created for the West River watershed above Ball Mountain Dam using the same contributing drainage area as the HEC-HMS model. All boundary cell faces along the West River watershed divide were assumed to be no-flow boundaries while the downstream face was assigned a normal depth boundary condition to allow water to exit the modeling domain. Due to possible backwater influences from placing the normal depth boundary along the dam crest, the model domain was altered to allow the boundary condition to reside at a more favorable location slightly upstream from the dam crest. The normal depth boundary was assigned a friction slope of 0.01 ft/ft, which was representative of the stream channel slope in this area.

A nominal 500 ft x 500 ft computation point spacing was initially used within the HEC-RAS 2D modeling domain which resulted in the creation of approximately 18,500 cells. Within the contributing drainage area, elevations range from 800 ft near the dam site to approximately 4000 ft in the headwaters, as shown in Figure 3. The National Elevation Dataset 10m Digital Elevation Model that was used to delineate subbasins within the HEC-HMS modeling domain was used as the "background" elevation dataset within the HEC-RAS 2D model. This elevation dataset was used to assign cross-sectional profiles to each 2D cell face as well as an elevation-volume relationship for each 2D cell.

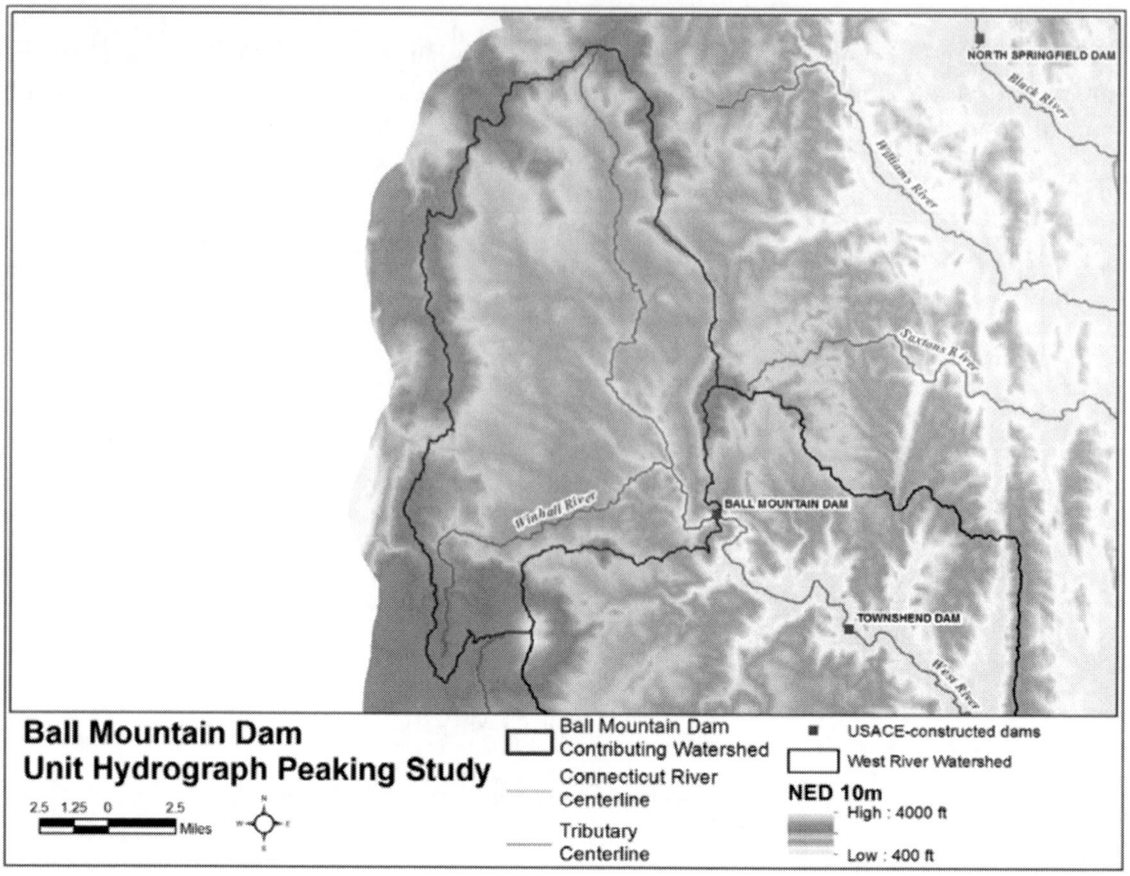

Figure 3. Elevations within the Study Area

Multiple simulations were performed to determine if the initial computation point spacing of 500 ft x 500 ft was appropriate. This spacing was found to adequately balance accuracy and simulation time. Also, multiple simulations were performed to determine the sensitivity of the results to a chosen computation time step. The initial time step of one-minute was reduced until runoff hydrographs were stable and no longer fluctuated with decreases in the time step. In addition to the initial one-minute computation time step, time steps of 30-seconds, 15-seconds, and 5-seconds were simulated. A final computation time step of 30-seconds was selected for further use.

Land uses within the modeling domain range from low intensity developed areas to emergent wetlands as derived from the NLCD 2006. These land uses were used to assign roughness values for overbank areas within the HEC-RAS 2D routing scheme. A polygon demarcating major watercourses was identified and used to delineate the stark flow roughness differences between overbank areas and stream channels. Initial roughness estimates were made using literature values and further adjusted during model calibration (Feldman, 2000). The initial estimates are shown in Table 2.

.

Table 2. NLCD 2006 Land Uses and Initial HEC-RAS 2D Roughness Estimates

NLCD 2006 ID	Type	Initial Roughness	NLCD 2006 ID	Type	Initial Roughness
11	Open Water	0.01	52	Shrub/Scrub	0.2
21	Developed, Open Space	0.1	71	Grassland/Herbaceous	0.4
22	Developed, Low Intensity	0.1	81	Pasture / Hay	0.25
23	Developed, Medium Intensity	0.1	82	Cultivated Crops	0.3
24	Developed, High Intensity	0.1	90	Woody Wetlands	0.25
31	Barren Land	0.1	95	Herbaceous Wetlands	0.25
41	Deciduous Forest	0.4	-	Stream Channels	0.03
42	Evergreen Forest	0.4			
43	Mixed Forest	0.4			

MODEL SIMULATIONS

Similar to the HEC-HMS model, the HEC-RAS 2D model was calibrated to a historical event to ensure that the HEC-RAS 2D model was able to recreate an observed runoff event with acceptable accuracy. Also, during model calibration, the initial estimates of several model parameters were altered to better reflect actual watershed conditions. Model parameters were adjusted to minimize the difference in computed and measured inflow to Ball Mountain Dam. In all cases, restart files were created and used to guarantee that no input excess precipitation was left within the modeling domain that should have run off.

August 2011 Event

In addition to changing model parameters, the August 2011 event (remnants of Hurricane Irene) was simulated to gauge the effects that various simplifications have upon the results. HEC-RAS does not currently employ infiltration computations that precede or accompany the runoff routing computations. Therefore, the precipitation that was input as a boundary condition to the HEC-RAS 2D model was in fact excess precipitation that had already been subjected to infiltration computations within HEC-HMS. Also, the input precipitation was evenly distributed throughout the model domain (HEC-RAS does not currently allow for the input of distributed precipitation within a single 2D area). To create an appropriate, spatially uniform, excess precipitation hyetograph for input to the HEC-RAS model, excess precipitation hyetographs from each of the five subbasins within the HEC-HMS model were extracted. Using an area-weighting scheme, an averaged, excess precipitation hyetograph was created for use within the HEC-RAS 2D model.

Overland roughness values were reduced to achieve an adequate fit to the observed data. The HEC-RAS 2D results are compared against the HEC-HMS results as well as the observed/computed inflow to Ball Mountain Dam in Figure 4. The results are further tabulated within Table 3. It should be noted that the HEC-RAS 2D results are solely comprised of excess precipitation; there was assumed to be no baseflow contribution to the inflow hydrograph to Ball Mountain Dam within the HEC-RAS 2D model. Also, infiltration (which was computed by the HEC-HMS model) was assumed to occur immediately following rainfall; once rainfall was converted to excess precipitation, it was not subjected to infiltration losses again.

The rising limb from the HEC-RAS model displayed a lag of approximately one hour as compared to the observed/computed inflow while the time of peak flow recession limb matched well. The computed peak inflow rate was within 2% while the computed inflow volume was within 18% of the computed/observed values, respectively.

Table 3. August 2011 Event HEC-RAS 2D Simulation Results

Simulation	Peak Flow (cfs)	Time of Peak	Volume (in)
Observed	28,225	28Aug2011 19:45	4.01
HEC-HMS	28,534	28Aug2011 19:45	3.57
HEC-RAS	28,823	28Aug2011 19:30	3.28

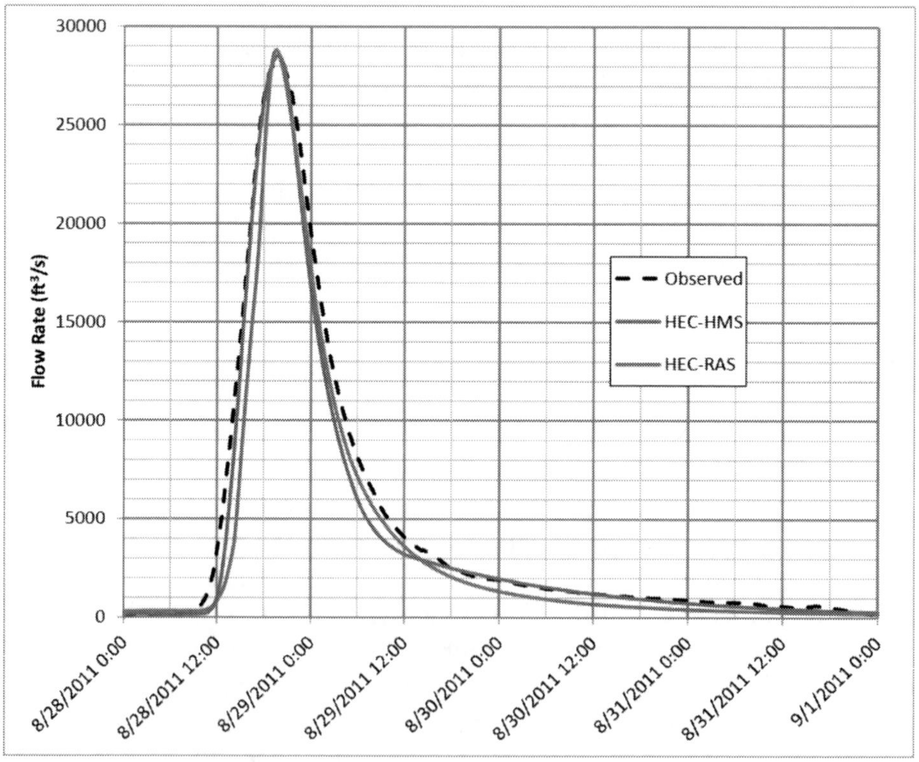

Figure 4. August 2011 Event HEC-RAS 2D Simulation Results

UNIT HYDROGRAPH EVENTS

Unit hydrograph simulations were executed within the HEC-RAS model to compare against the HEC-HMS results for varying durations. These simulations were performed due to two current limitations when simulating the August 2011 historical event within

HEC-RAS: 1) spatially uniform excess precipitation within a single 2D area is required and 2) no infiltration computations are performed within the HEC-RAS 2D model. While providing supplemental model calibration, these simulations also allowed for further investigation of the dynamic runoff response of the contributing watershed above Ball Mountain Dam.

One inch of excess precipitation with (no infiltration or baseflow) for durations of 0.25-, 0.5-, 1-, 2-, 3-, 6-, and 12-hrs was uniformly applied over the entire modeling domain in order to create varying duration unit hydrographs. The initial roughness estimates presented in Table 2 were reduced to better match the HEC-HMS derived unit hydrographs. The resulting unit hydrographs were collected along the "outlet" cell faces and are shown in Figure 5, Figure 6, Figure 7, and Figure 8 and tabulated within Table 4.

The 0.25-, 0.5-, 1-, 2-, and 3-hr unit hydrographs from the HEC-RAS 2D model display a "double-peak" shape which is indicative of at least two discrete flood waves arriving at Ball Mountain Dam. This effect can be explained by the two primary sub-watersheds which, together, comprise the total contributing watershed to Ball Mountain Dam: West River (approximately 110 mi^2) and Winhall River (approximately 60 mi^2). The first, smaller, peak is comprised of runoff originating within the Winhall River watershed while the second, larger, peak originates within the West River watershed. This effect is less pronounced for the 6-hr and 12-hr unit hydrographs due to the longer unit precipitation durations. However, the unit hydrographs derived from the HEC-HMS model do not display a double-peaked shape for any durations indicating that the flood waves from the West River and Winhall River subbasins are predicted to arrive at the same time.

Table 4. Unit Hydrograph Simulation Results

Simulation	Peak Flow (cfs)	Time to Peak (hrs)	Volume (in)
HEC-HMS 0.25hr UH	13,932	4.0	1.00
HEC-HMS 0.5hr UH	13,918	4.3	1.00
HEC-HMS 1hr UH	13,771	4.5	1.00
HEC-HMS 2hr UH	13,263	5.3	1.00
HEC-HMS 3hr UH	12,619	6.0	1.00
HEC-HMS 6hr UH	10,620	8.3	1.00
HEC-HMS 12hr UH	7,572	13.5	1.00
HEC-RAS 0.25hr UH	13,633	4.3	0.99
HEC-RAS 0.5hr UH	13,617	4.4	0.99
HEC-RAS 1hr UH	13,553	4.8	0.99
HEC-RAS 2hr UH	13,309	5.4	0.99
HEC-RAS 3hr UH	12,914	6.1	0.99
HEC-RAS 6hr UH	11,299	8.0	0.99
HEC-RAS 12hr UH	7,959	12.5	0.99

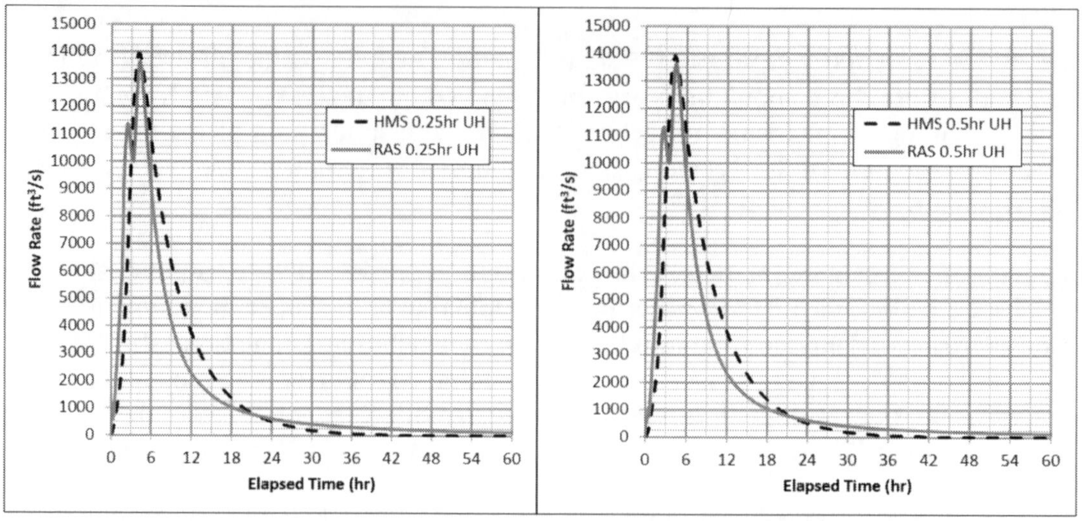

Figure 5. 0.25-hr and 0.5-hr Unit Hydrograph Simulation Results

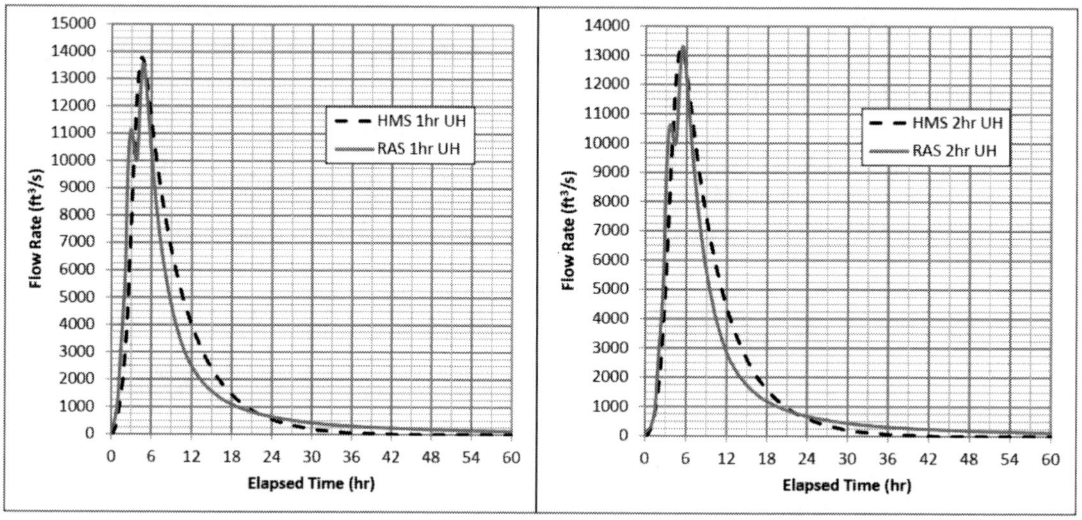

Figure 6. 1-hr and 2-hr Unit Hydrograph Simulation Results

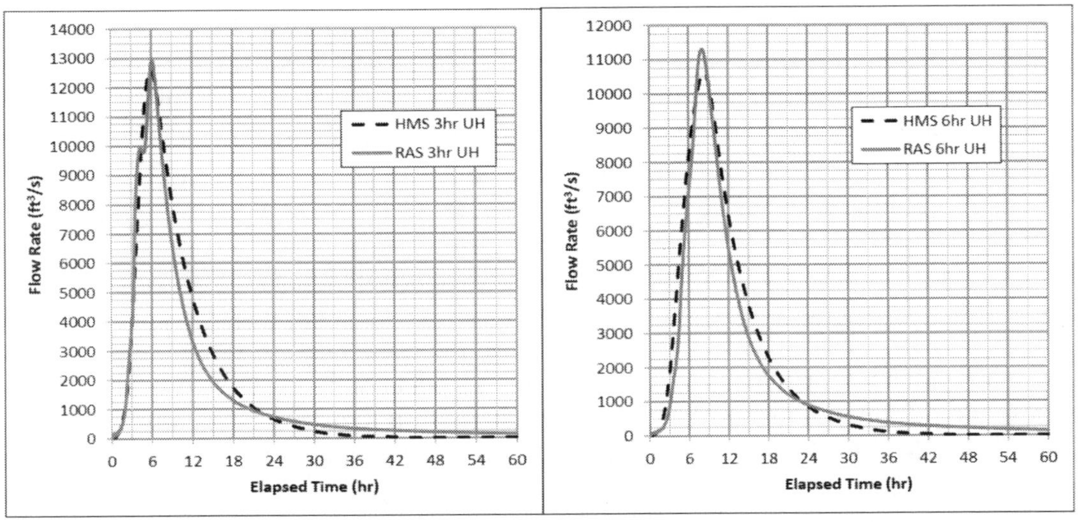

Figure 7. 3-hr and 6-hr Unit Hydrograph Simulation Results

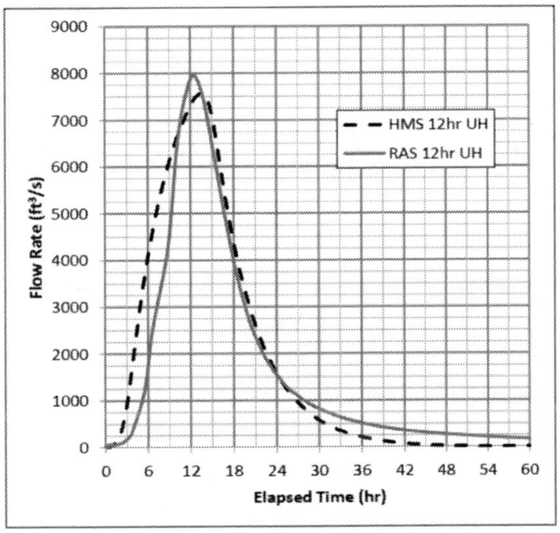

Figure 8. 12-hr Unit Hydrograph Simulation Results

SITE-SPECIFIC PMP ROUTING

The overland and channel roughness parameters from the Aug 2011 event and unit hydrograph simulations were averaged to produce a final set of overland and channel roughness parameters for using within the HEC-RAS 2D model. The HEC-RAS 2D model was then used to estimate the PMF inflow hydrograph. First, excess precipitation was tabulated for all of the subbasins within the Ball Mountain Dam HEC-HMS model when the site-specific PMP was applied. Using an area-weighting scheme, an averaged excess precipitation hyetograph was then created. This PMP excess precipitation hyetograph was then input as a precipitation boundary condition to the HEC-RAS 2D model.

The site-specific HEC-RAS PMF inflow hydrograph was then compared against the HEC-HMS routed results using unit hydrograph peaking factors of 1.0, 1.25, 1.5, 1.75, and 2.0, as shown in Figure 9. The Clark Unit Hydrograph transform and Muskingum routing parameters within the HEC-HMS model were decreased uniformly until the resulting 1-hr unit hydrograph peak inflow rate to Ball Mountain Dam was increased by a factor of 1.25, 1.5, 1.75, and 2.0, respectively. No baseflow was added to the HEC-HMS inflow hydrographs to better compare against the HEC-RAS 2D model.

The HEC-HMS routings over-predict flows along a significant portion of the rising limb of the inflow hydrograph while under-predicting flows at/near the peak when compared to the HEC-RAS PMF inflow hydrograph. Also, the HEC-HMS routings over-predict flows along a large portion of the receding limb. This is due to the use of a single unit hydrograph peaking factor for the entire duration within the HEC-HMS simulation whereas the HEC-RAS 2D model allowed for dynamic changes in runoff and flood wave routing throughout the entire simulation. Unit hydrograph peaking factors in excess of 2.0 resulted in unrealistic Clark Unit Hydrograph parameters and were therefore not simulated.

However, it is plainly evident that the HEC-RAS routing results in much greater runoff concentrations than any of the HEC-HMS simulations when using a single unit hydrograph peaking factor, as shown in Table 5. These differences are significant considering that the HEC-RAS routing used a spatially-uniform excess precipitation pattern while the HEC-HMS simulations made use of a spatially distributed storm that was aggregated and applied to each of the five subbasins.

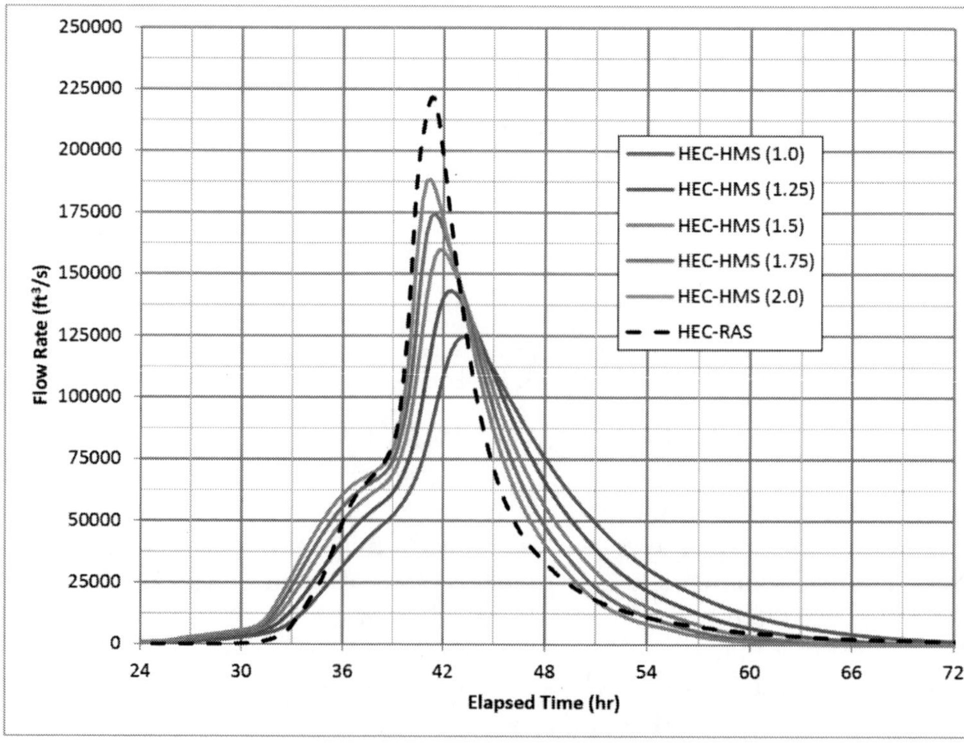

Figure 9. Site-Specific PMF Inflow Hydrographs

Table 5. Site-Specific PMF Inflow Results

Simulation	Peak Flow (cfs)	Time to Peak (hrs)	Volume (in)
HEC-HMS (1.0)	124,811	43.3	14.32
HEC-HMS (1.25)	143,107	42.5	14.32
HEC-HMS (1.5)	159,728	41.8	14.32
HEC-HMS (1.75)	174,656	41.5	14.32
HEC-HMS (2.0)	188,271	41.3	14.32
HEC-RAS	221,636	41.3	14.0

Note: HEC-HMS and HEC-RAS routing results do not include baseflow contributions

The four different PMF inflow hydrographs were then routed through a structure representing Ball Mountain Lake within a simplified HEC-HMS model. The antecedent pool elevation was set to be equivalent to the spillway crest. Outflows were simulated through the ungated spillway and the outflow conduit, which was operated according to the reservoir regulation manual. These routings showed that when a unit hydrograph peaking factor of 2.0 is applied throughout the entire simulation, not only were inflows overestimated along the rising limb of the hydrograph, but pool elevations were overestimated as well. However, peak pool elevations attained when using a unit hydrograph peaking factor of 2.0 are still less than the peak pool elevation caused by routing the HEC-RAS 2D PMF inflow hydrograph, as shown in Figure 10. The results are further detailed within Table 6.

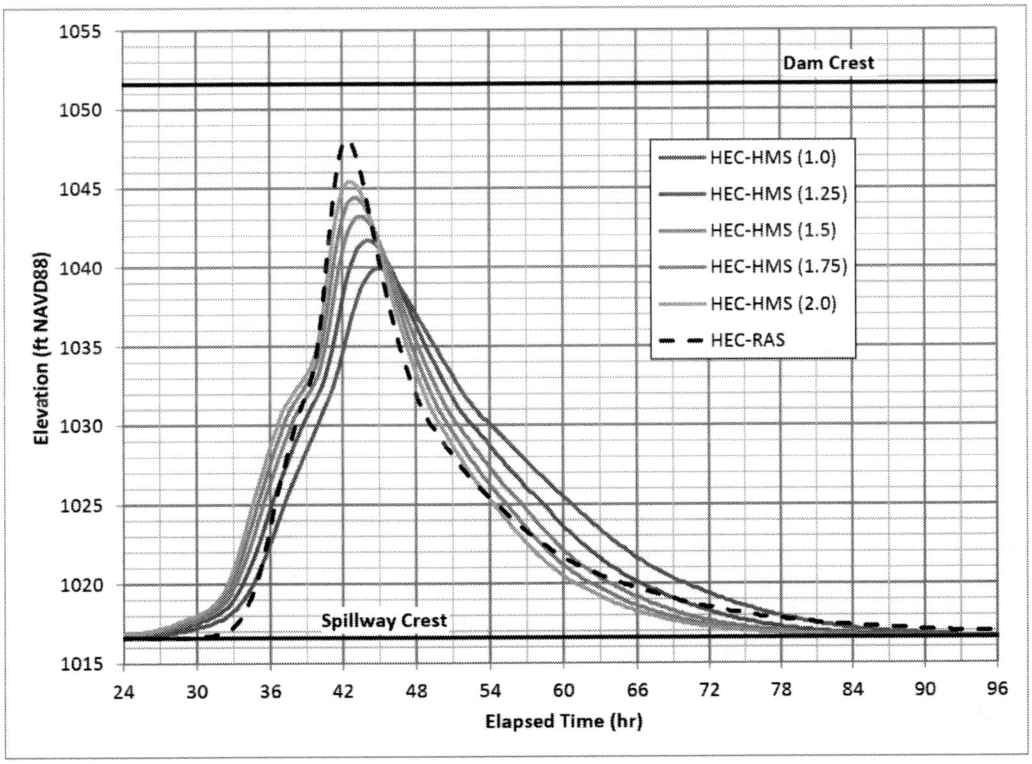

Figure 10. Site-Specific PMF Pool Elevations

Table 6. Site-Specific PMF Peak Pool Elevations

Simulation	Peak Pool Elevation (ft NAVD88)	Time to Peak (hrs)
HEC-HMS (1.0)	1039.9	44.8
HEC-HMS (1.25)	1041.7	44.0
HEC-HMS (1.5)	1043.2	43.3
HEC-HMS (1.75)	1044.4	43.0
HEC-HMS (2.0)	1045.4	42.5
HEC-RAS	1048.1	42.5

CONCLUSIONS

Using the results shown within Figure 9, it is possible that a larger peaking factor could be applied to the current study to better match the peak flow rate estimated by the HEC-RAS 2D model. However, this would result in unrealistic Clark Unit Hydrograph parameters while further overestimating flow rates on the rising and receding limb of the PMF inflow hydrograph, as compared to the HEC-RAS 2D model results. This would consequently negatively impact pool elevations attained when routing the PMF inflow hydrograph.

While the first two questions posed as part of this study were investigated, additional research is necessary to evaluate the continued use of unit hydrograph peaking factors between 1.25 and 1.5. Further research is underway to better examine this question as well as several other key aspects of the use of unit hydrograph theory during extreme event simulations. These research topics will investigate the use of variable unit hydrograph peaking factors using excess precipitation rates as well as more advanced two-dimensional rainfall runoff routing routines and techniques. As such, this research would benefit from additional site-specific analyses using 2D rainfall runoff modeling. Also, to truly evaluate the applicability of peaking factors between 1.25 and 1.5, unit hydrographs caused by an extreme rainfall event (on the order of the PMP) will be extracted from observed flow data and then compared against unit hydrographs due to "normal" (i.e. less extreme) events at the same gage location.

In the short term, HEC-HMS will be modified to allow for the use of excess precipitation varying unit hydrograph parameters. Specifically, Clark unit hydrograph parameters (time of concentration, T_c, and watershed storage coefficient, R) will be allowed to vary during each computational time step based upon user-input tables relating T_c and R to excess precipitation. These site-specific relationships will need to be developed using modeling techniques that are similar to this analysis (i.e. HEC-RAS 2D modeling). In the long term, HEC-HMS will allow for the use of the same 2D routing routine that is currently employed within HEC-RAS as an additional runoff transform method. These two modifications to HEC-HMS will allow for a scalable and dynamic runoff routing approach that will be beneficial to a range of applications, such as flood forecasting, dam and levee safety, and hydrologic uncertainty analyses.

ACKNOWLEDGEMENTS

Multiple USACE employees offered assistance throughout this research particularly Dana Moses, P.E. and Dr. Nawa Pradhan. The lead developers of the HEC-RAS and HEC-HMS software packages are Gary Brunner, P.E. and Dr. William Scharffenberg, respectively. Their contributions to this research were extremely valuable. Special modifications to the HEC-RAS software that contributed to the success of this research were made by Alexander Kennedy and Mark Jensen. Special thanks go to David Margo, P.E. and Dr. John England, P.E. for their continued support regarding this research.

REFERENCES

Abdullah, J. (2013). *Distributed Runoff Simulation of Extreme Monsoon Rainstorms in Malaysia Using TREX.* Fort Collins, Colorado: Colorado State University.

Bartles, M. (2014). *Hydrologic Analysis of Prompton Dam Using a Physically-Based Rainfall Runoff Model.* Villanova, PA: Villanova University.

England, J. F., Klawon, J. E., Klinger, R. E., & Bauer, T. R. (2006). *Flood Hazard Study Pueblo Dam, Colorado.* Denver, Colorado: U.S. Bureau of Relamation.

England, J. F., Velleux, M. L., & Julien, P. Y. (2007). Two-dimensional Simulations of Extreme Floods on a Large Watershed. *Journal of Hydrology*, 229 - 241.

Harrison, J. (1999). PMF and the Unit Hydrograph - Piquiing More Than Interest. (pp. 373-388). Lexington, KY: Association of State Dam Safety Officials.

Hydrologic Engineering Center. (2015). *HEC-RAS River Analysis System User's Manual, Version 5.0.* Davis, CA: U.S. Army Corps of Engineers.

Hydrologic Engineering Center. (2015). *Hydrologic Modeling System HEC-HMS Version 4.1, User's Manual.* Davis, CA: U.S. Army Corps of Engineers.

Hydrologic Engineering Center. (2015). *Probable Maximum Flood (PMF) Study for Ball Mountain Dam.* Davis, CA: U.S. Army Corps of Engineers.

Institue of Hydrology. (1999). *Flood Estimation Handbook (FEH) - Procedures for Flood Frequency Estimation.* Wallingford, Oxfordshire, United Kingdom: Institute of Hydrology.

Minshall, N. E. (1960). Predicting Storm Runoff on Small Experimental Watersheds. *ASCE Journal of the Hydraulics Division, HY8*, 17-37.

New England Division. (1990). *Ball Mountain Lake Spillway Design Flood Review Using HMR 51.* Concord, MA: U.S. Army Corps of Engineers.

Omaha District. (2015). *Ball Mountain Dam Site-Specific PMP Study.* Omaha, NE: U.S. Army Corps of Engineers.

Sherman, L. K. (1932). *Streamflow from Rainfall by the Unit Graph Method.* Eng. News Rec.

Szilagyi, J. (2007). Analysis of the Nonlinearity in the Hillslope Runoff Response to Precipitation Through Numerical Modeling. *Journal of Hydrology*, 391-401.

The Institution of Engineers, Australia. (1999). *Australian Rainfall and Runoff (AAR).*

U.S. Army Corps of Engineers. (1959). *EM 1110-2-1405 (Superseeded).* Washington, D.C.: USACE.

U.S. Army Corps of Engineers. (1991). *Inflow Design Floods for Dams and Reservoirs, Engineer Regulation 1110-8-2.* Washington, D.C.: USACE.

AREAL REDUCTION FACTORS — COLORADO FRONT RANGE AND ANALYSIS OF THE 2013 STORM

Doug Hultstrand[1]
Bill Kappel[2]

ABSTRACT

Information about extreme precipitation is of interest in hydrologic engineering applications such as dam design, river management, and rainfall-runoff-relations. These require knowledge on the spatial and temporal variability of precipitation over an area. In order to obtain areal average values for hydrologic modeling purposes, point rainfall amounts are often transformed to average rainfall amounts over a specified area. This is addressed using depth-area curves which require the use of areal reduction factors (ARFs).

The Colorado Department of Transportation (CDOT) Flood Hydrology Committee tasked Applied Weather Associates (AWA) to derive 24-hour ARFs for the Front Range of Colorado for area sizes of 1- to 1000-sqmi. In addition, basin specific ARFs for the September 2013 rainfall event were calculated for four basins (Boulder Creek, St. Vrain Creek, Big Thompson River, and Thompson River basin). This study was initiated due to areal limitations and potential issues associated with NOAA Atlas 2 ARF curves.

AWA analyzed storm events along the Front Range of the Rocky Mountains extending from northern New Mexico through southern Canada, including the September 2013 event. Each storm event utilized in the analysis represented meteorological and topographical characteristics that were similar to each other and to the September 2013 event. These storms were selected to derive storm specific ARFs which represented to the meteorological and topographical characteristics of the four basins. The individual storm ARFs were utilized to derive a site-specific set of 24-hour ARF values to be used in the hydrologic analysis of four basins along the northern Front Range of Colorado.

INTRODUCTION

Information about extreme precipitation is of interest for a variety of purposes, which include meteorological and hydrologic engineering applications such as dam design, river management, and rainfall-runoff-relations. These entail knowledge on the spatial and temporal variability of precipitation over an area. In order to obtain areal average values for an area, point rainfall amounts (in inches) are transformed to average rainfall amounts over a specified area (in square miles). These issues are addressed using depth-area curves which require the use of ARFs. The derivation of ARFs is an important topic that has been dealt with using several methodologies.

[1] Applied Weather Associates, PO BOX 175, Monument, CO 80132, 720-771-5840, dhultstrand@appliedweatherassociates.com.
[2] Applied Weather Associates, PO BOX 175, Monument, CO 80132, 719-488-4311, billkappel@appliedweatherassociates.com.

The National Ocean and Atmospheric Administration (NOAA) defines an ARF as the ratio between area-averaged rainfall to the maximum depth at the storm center (NOAA Atlas 2, 1973). The most common sources for generalized ARFs and depth-area curves in the United States are from the NOAA Atlas 2 (NOAA Atlas 2, 1973) (Figure 1), and the U.S. Weather Bureau's Technical Paper 29 (U.S. Weather Bureau, 1957-60). Examples of site specific ARFs and depth-area curves are referenced in the NOAA Technical Report 24 (Meyers and Zehr, 1980) for the semi-arid southwest, the NOAA Technical Memorandum Hydro- 40 (NOAA Hydro-40, 1980) for the semi-arid southwest, and the city of Las Vegas, Nevada (Gou, 2011).

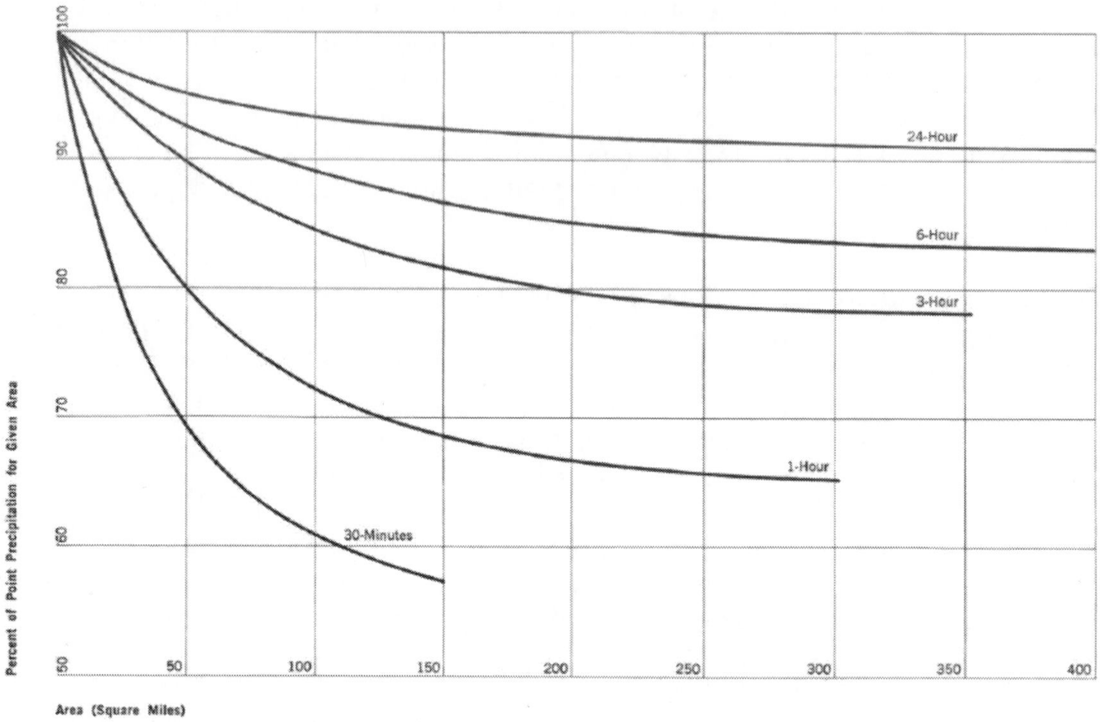

Figure 1. NOAA Atlas 2 Volume 3 ARF curves.

METHODS

There are two common methods for deriving ARFs: geographically fixed and storm centered. Geographically fixed ARFs originate from rainfall statistics, whereas storm centered ARF values are based on discrete rainfall events. Geographically fixed ARFs relate the precipitation depth at a point to a fixed area. The representative point is the mean of annual maximum point rainfall values at gauged points located within the network (U.S. Weather Bureau, 1957-60; NOAA Atlas 2, 1973; Osborn et al., 1980). This is a hypothetical point rather than a point for a particular location. The areas within the network are known beforehand and are both fixed in time and space (U.S. Weather Bureau, 1957-60; Osborn et al., 1980). With geographically fixed ARFs, the storm center does not correspond with the center of the location and does not need to fall within the area at all (Omolayo, 1993). Geographically fixed ARFs are based on different parts of different storms instead of the maximum point values located at the representative storm centers. A geographically fixed ARF is calculated as:

$$ARF_{Fixed} = \frac{\frac{1}{n}\sum_{j=1}^{n}\hat{R}_j}{\frac{1}{k}\sum_{i=1}^{k}\left(\frac{1}{n}\sum_{j=1}^{n}R_{ij}\right)} \tag{1}$$

where $\hat{R}_j$ is the annual maximum areal rainfall for year *j*, R_{ij} is the annual maximum point rainfall for year *j* at station *i*, *k* is the number of stations in the area, and *n* is the number of years.

The storm centered ARF does not have a fixed area in which rain falls but changes dynamically with each storm event (NOAA Atlas 2, 1973; Gou, 2011). Instead of the representative point being an average, the representative point is the center of the storm, defined as the point of maximum rainfall. Storm centered ARFs are calculated as the ratio of areal storm rainfall enclosed between isohyets equal to or greater than the isohyet value to the maximum point rainfall at the storm center. A storm centered ARF is calculated as:

$$ARF_{center} = \frac{\overline{R}_i}{R_{center}} \tag{2}$$

where $\overline{R}_i$ is the areal storm rainfall enclosed between isohyets equal to or greater than the isohyets, and R_{center} is the maximum point rainfall at the storm center.

AWA calculated ARFs using a storm centered depth-area approach based on gridded hourly rainfall data from the Storm Precipitation Analysis System (SPAS). SPAS has demonstrated reliability in producing highly accurate, high resolution rainfall analyses during hundreds of post-storm precipitation analyses (Tomlinson and Parzybok, 2004; Parzybok and Tomlinson, 2006). SPAS has evolved into a hydrometeorological tool that provides accurate precipitation data at a high spatial and temporal resolution for use in a variety of sensitive hydrologic applications. AWA and METSTAT, Inc. initially developed SPAS in 2002 for use in producing storm centered Depth-Area-Duration (DAD) values for Probable Maximum Precipitation (PMP) analyses. SPAS utilizes precipitation gauge data, "basemaps" and radar data (when available) to produce gridded precipitation at time intervals as short as 5-minutes, at spatial scales as fine as 1-km^2 and in a variety of customizable formats. To date, (December 2015) SPAS has analyzed over five-hundred storm centers across all types of terrain, among highly varied meteorological settings and with some events occurring over 100-years ago. For more detailed discussions on SPAS and DAD calculations refer to (Tomlinson et al., 2003-2013, Kappel et al., 2012-2014).

September 2013 Basin ARFs

The September 2013 can be classified as an upslope synoptic storm event associated with an area of low pressure to the east/southeast causing the air to flow into the Front Range (upslope) from the Midwest and Southern Plains. This air was forced to lift by both interaction with the terrain and the lift associated with the storm system. The storm event exhibited low to moderate intensity rainfall that occurred over long durations and contained periods of higher intensity rainfall. A detailed description of the meteorology associated with the storm can be found at http://coflood2013.colostate.edu/meteo.html. The Colorado September 8-17, 2013 rainfall event was analyzed using the SPAS (SPAS number 1302) for use in several PMP and hydrologic model calibration studies (Figure 2). The hourly gridded rainfall data, based on gauge adjusted radar data, were used to derive basin specific ARFs. Four basins (Table 1) located along the Colorado Front Range were used to derive the 24-hour basin specific ARFs. The SPAS DAD program was used to derive basin specific 24-hour depth-area values. The point maximum (1-mi^2) 24-hour rainfall (within each basin) was selected as the storm center. The maximum average basin 24-hour rainfall depth for standard area sizes (1-, 10-, 25-, 50-, 100-, 200-, 300-, 400-, and 500-mi^2) up to the basin total area were calculated. The point maximum and maximum areal averages depths were used to calculate the basin specific ARFs.

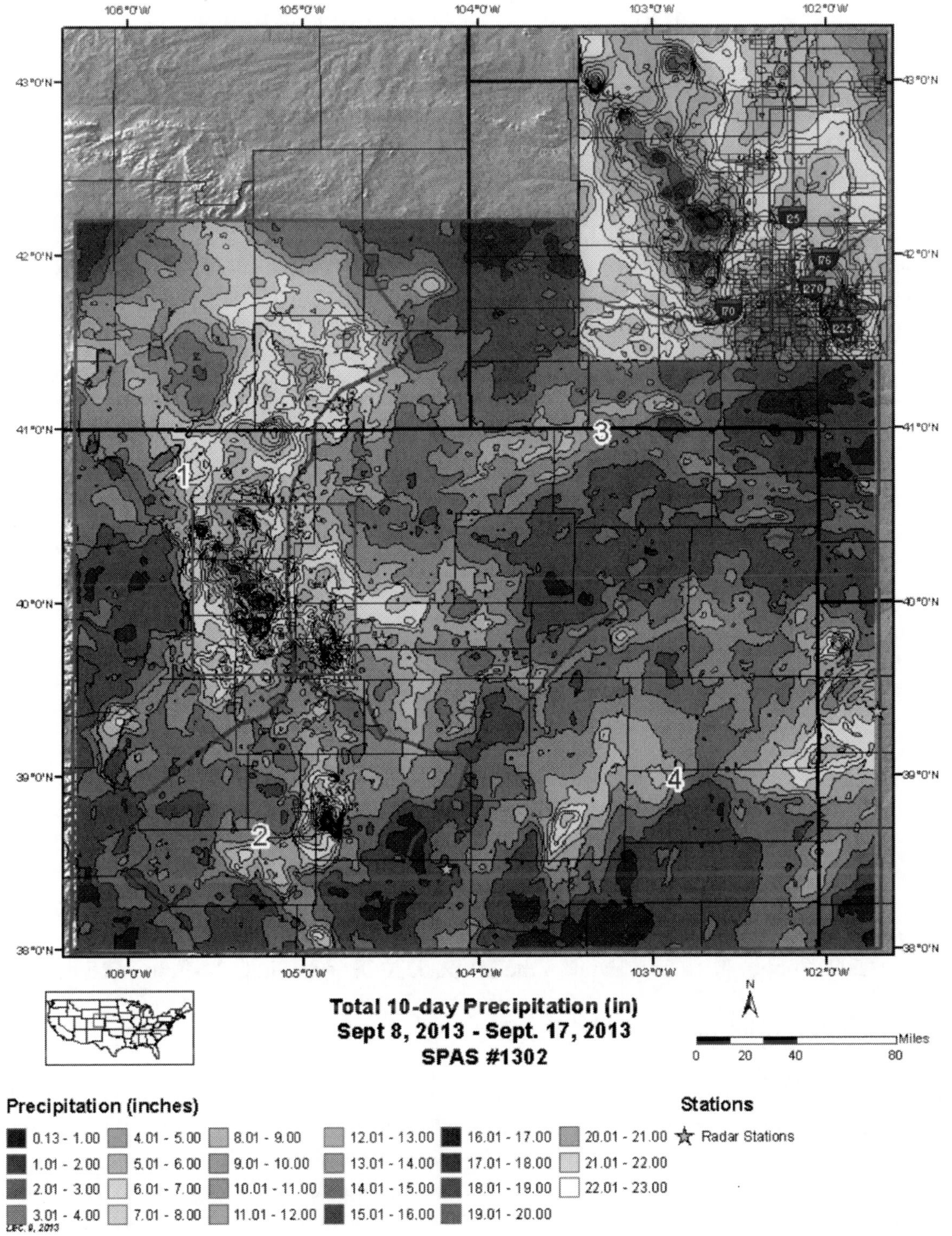

Figure 2. SPAS total rainfall for the Colorado September 8-17, 2013 storm event.

Table 1. Basin specific 24-hour ARFs for the September 2013 storm event.

Basin	Area (mi^2)	ARF
Boulder Creek	446	0.352
St. Vrain Creek	982	0.384
Big Thompson River	630	0.357
Thompson River	827	0.355

The four calculated basin specific 24-hour ARFs for the September 2013 event were compared to NOAA Atlas 2 24-hour ARF curve and to the HMR 55A Orographic C 24-hour ARF curve (Hansen et al., 1988) (Figure 3). Table 1 shows the basin specific 24-hour ARF values. As expected, the four September 2013 basin ARF values have a significantly larger reduction in rainfall than published NOAA Atlas 2 and HMR 55A ARFs.

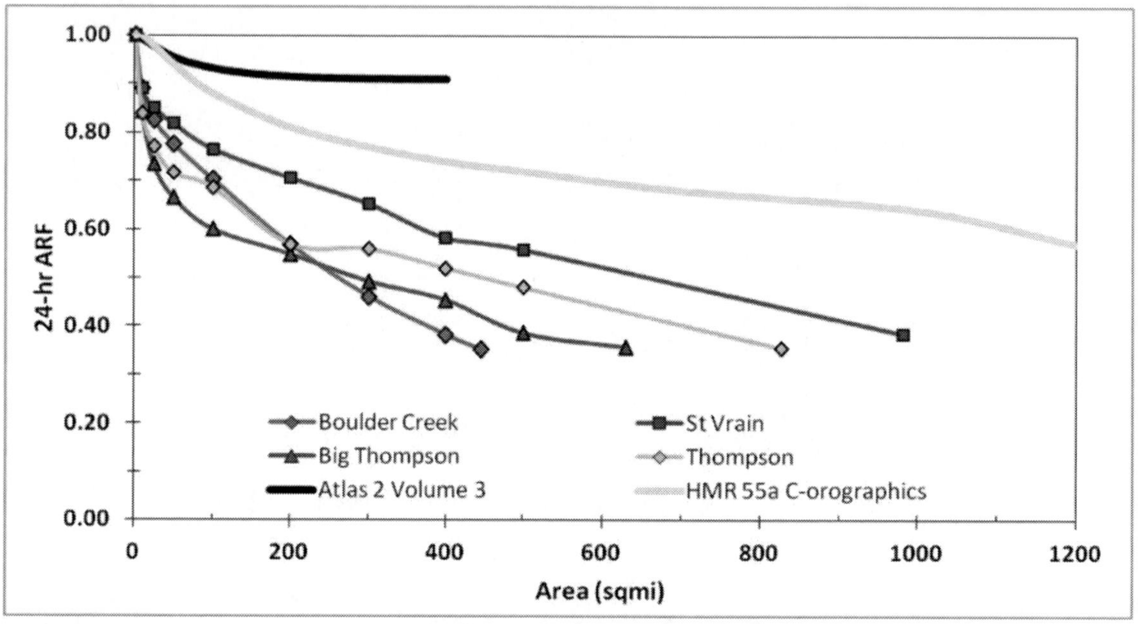

Figure 3. Basin specific 24-hour ARFs for the September 2013 event compared to NOAA Atlas 2 24-hour ARF curve and to the HMR 55A Orographic C 24-hour ARF curve.

Colorado Front Range ARFs

Initially, 28 SPAS storm center DAD zones were identified to have occurred over similar meteorological and topographic regions as the September 2013 storm event that occurred along the Colorado Front Range (Figure 4). The initial list was refined to nine storm centers that had storm characteristics representative of an upslope synoptic event similar to the four basins analyzed in this study. Each storm event utilized in this analysis represented meteorological and topographical characteristics that were similar to each other and similar to the September 2013 event. All storms were of the synoptic type (aka HMR 55A General Storm). Each were associated with an area of low pressure to the east/southeast causing the air to flow into the Front Range (upslope) from the Midwest and Southern Plains. This air was forced to lift by both interaction with the terrain and the lift associated with the storm system. All nine events used exhibited low to moderate intensity rainfall, which occurred over long durations, interspersed with periods of higher intensity rainfall. Storm events removed from the initial list were representative of shorter duration, higher intensity storms, i.e. local storms/thunderstorms or occurred in significantly different topographical settings. This allowed the ARF data derived during this analysis to represent the same storm type and meteorological setting as occurred during the September 2013 event. The final set of nine storm centers (Table 2 and Figure 5) were used to derive 24-hour storm center ARFs.

The point maximum (1-mi^2) 24-hour rainfall (within each SPAS DAD zone) was selected as the storm center. The maximum average 24-hour rainfall depth for standard area sizes (1-, 10-, 25-, 50-, 100-, 150-, 200-, 250-, 300-, 350-, 400-, 450-, 500-, 700-, and 1000-mi^2) were calculated. The point maximum and maximum areal averages depths were used to calculate each events specific ARFs. Based on the nine events, an average ARF for each area size was calculated. Several other ARF curves were created for comparison purposes: maximum, minimum, +1-sigma, 85% confidence, 90% confidence, and 95% confidence. Based on discussions with the CDOT flood review committee and Nolan Doesken (Colorado State Climatologist), the 85% confidence ARF ($ARF_{85\%}$) was selected as the best representation of ARFs along the Colorado Front Range. The 85% confidence limit ARF was selected based on several justifications. Similar use of the 85% percentile was employed in the HMRs in determining various Depth-Area and Depth-Duration relationships. Further, during the site-specific Probable Maximum Precipitation (PMP) study for Lewis River, WA, the 85% was used to determine which Depth-Duration relationship were appropriate for deriving PMP values at durations other than 24-hours. That study was accepted for use by Federal Energy Regulatory Commission (Tomlinson et al., 2011). In addition, the 85% ARF curve is similar to independent study in HMR 55A (see Figure 6 and Table 3 below). Finally, the 85% ARF curve adds a level of conservatism compared to using the average ARF which is typical in most ARF studies. The final equation used to represent Colorado Front Range 24-hour ARFs is:

$$ARF_{85\%} = 0.646 + 0.354 * \exp(-kA) \quad (3)$$

where $ARF_{85\%}$ is the 85% confidence ARF, k is a decay coefficient, and A is storm area in square miles. The average ARF curve and final 85% confidence ARF curve are shown in Figure 6. The NOAA Atlas 2 ARF curve and HMR 55A Orographic C curve are also shown for comparison (Figure 6 and Table 3).

Table 2. Final SPAS storm centered locations with similar meteorology and topography as the September 2013 storm event used to derive 24-hr ARFs.

ID	SPAS ID	Storm Location	Dates	Latitude	Longitude	Max Precipitation	HMR 55A CLASS	HMR 55A SUBUNIT
1	1211	Gibson Dam, MT	Jun. 6-8, 1964	48.3541	-113.3708	19.16	Orographic	Orographic "A"
2	1251	Lake Maloya, NM	May 17-21, 1955	37.0090	-104.3410	14.82	Orographic	Orographic "E"
3	1252	Waterton Red Rock, AB	June 14-21, 1975	49.0875	-114.0458	14.46	Orographic	Orographic "A"
4	1253	Big Elk Meadow, CO	May 3-8, 1969	40.2700	-105.4200	20.01	Orographic	Orographic "C"
5	1302	Northeast Colorado	Sep. 8-17, 2013	40.0150	-105.2650	20.41	Orographic	Orographic "C"
6	1320	Calgary, AB	Jun.19-22, 2013	50.6350	-114.8550	13.78	Orographic	Orographic "A"
7	1325	Savageton, WY	Sep. 27-Oct. 1, 1923	43.8458	-105.8042	17.56	Nonorographic	Min. Nonorographic "A"
8	1335	Warrick, MT	Jun. 5-10, 1906	48.0791	-109.7041	13.69	Orographic	Orographic "A"
9	1338	Spionkop Creek, AB	Jun. 4-7, 1995	49.1708	-114.1625	14.48	Orographic	Orographic "A"

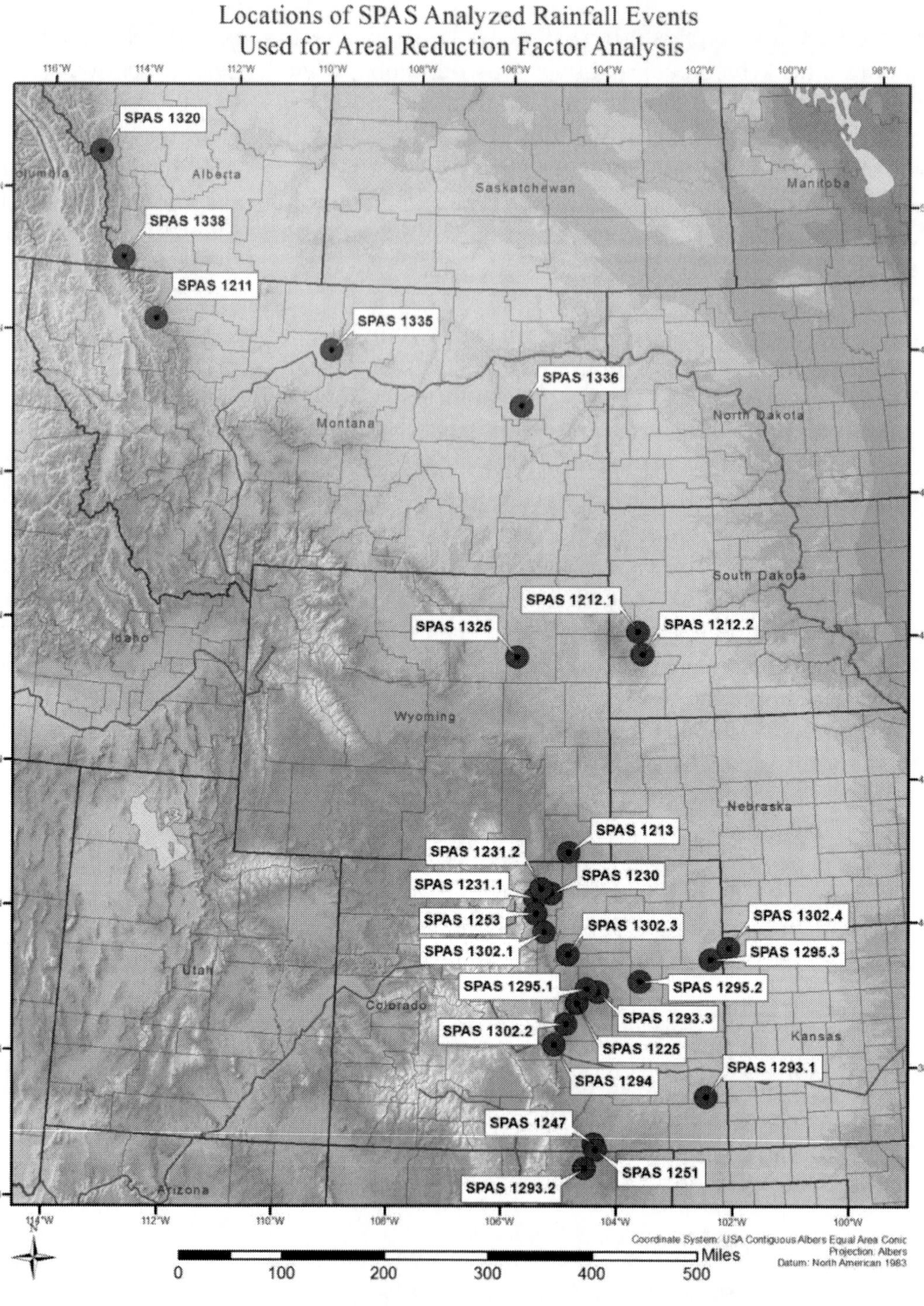

Figure 4. Initial 28 SPAS storm center locations with similar meteorology and topography as the September 2013 storm event.

Locations of SPAS Analyzed Rainfall Events
Used for Areal Reduction Factor Analysis

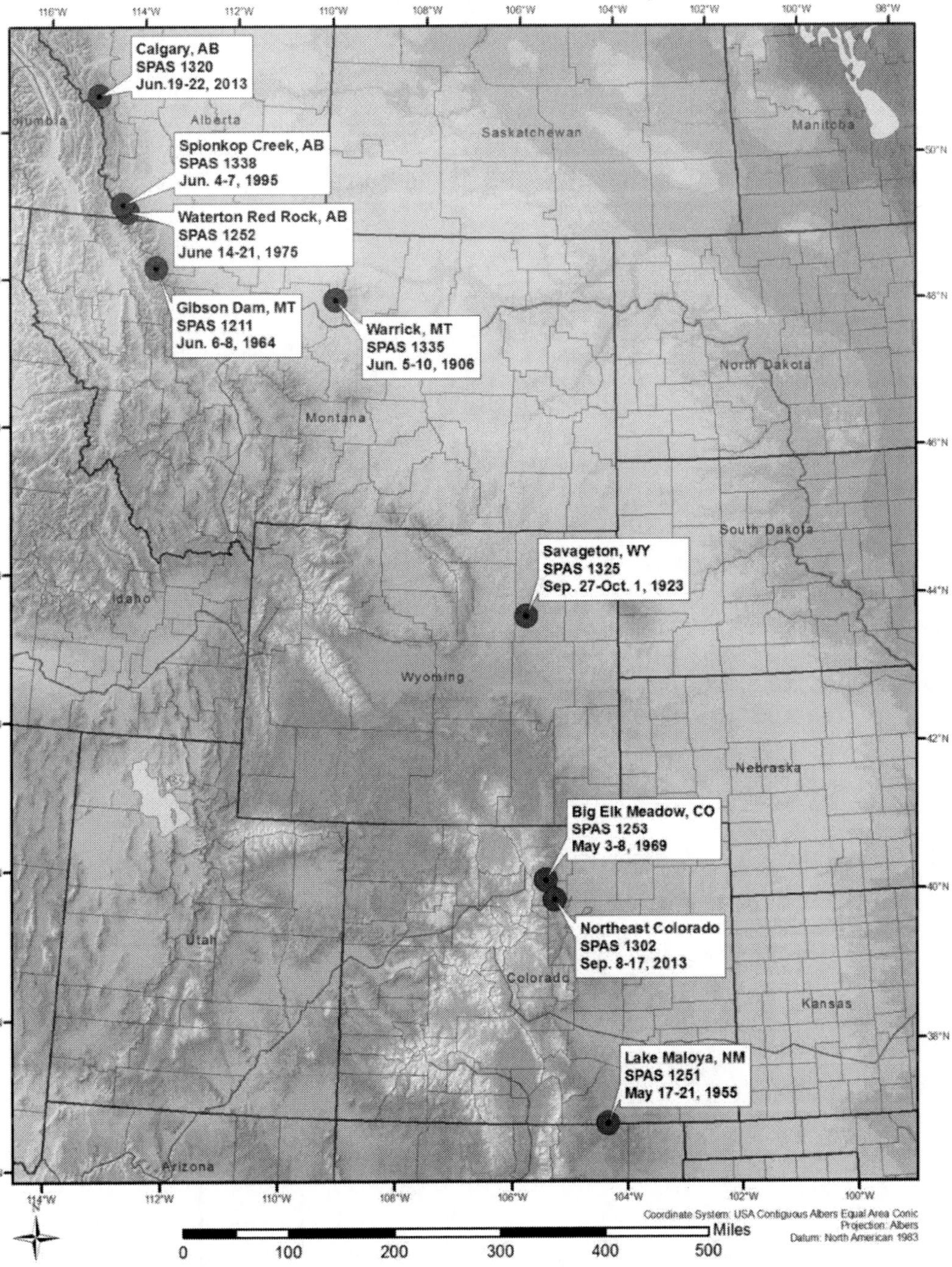

Figure 5. Final SPAS storm center locations used to derive 24-hr ARFs.

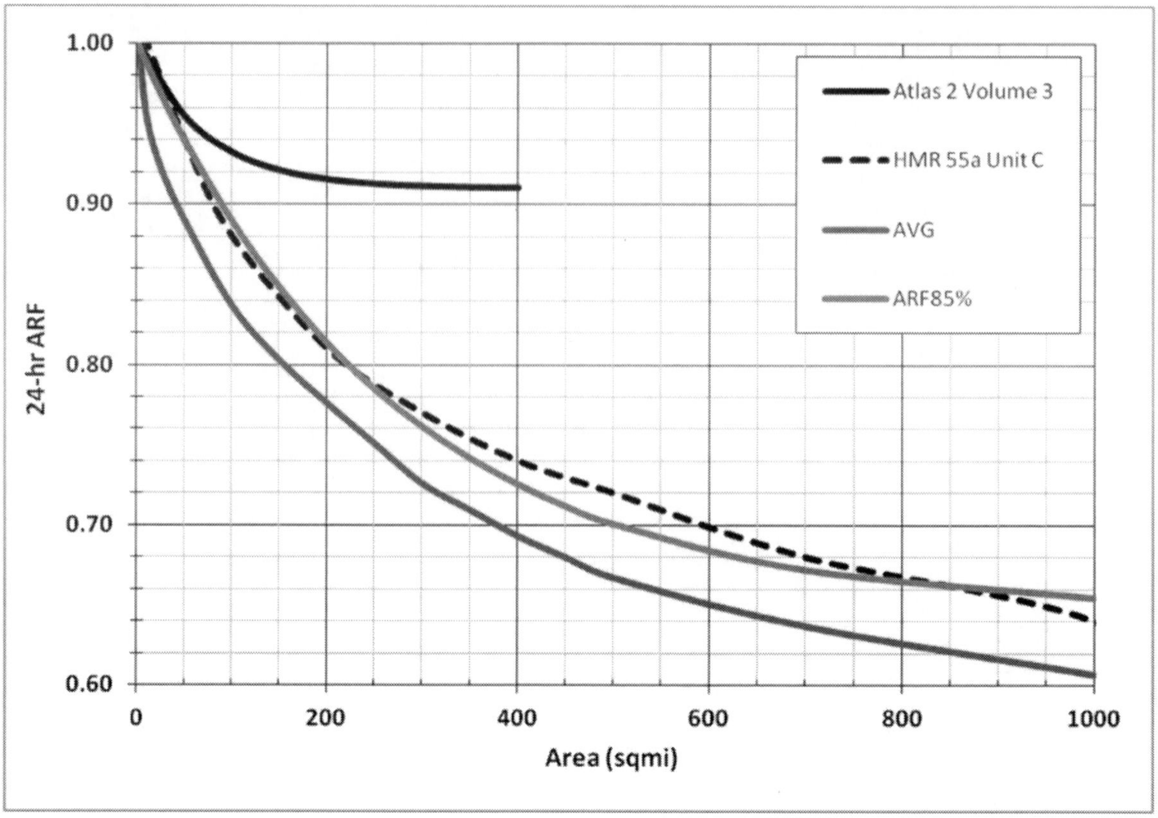

Figure 6. The average 24-hour ARF curve and final 85% confidence 24-hour ARF curve. The NOAA Atlas 2 24-hour ARF curve and HMR 55A Orographic C 24-hour ARF curve are shown for comparison.

Table 3. Comparison of 24-hour ARF values. AVG is the average ARF, ARF85% is the 85% confidence ARF, HMR 55A is HMR 55A Orographic C ARF, and Atlas 2 is NOAA Atlas 2 ARF.

*** General Storms 24-hr ARF				
Area (sqmi)	AVG	ARF85%	HMR 55a	Atlas 2
1	1.00	1.00	1.00	1.00
10	0.95	0.99	1.00	-
25	0.92	0.97	0.97	-
50	0.89	0.94	0.94	0.95
100	0.84	0.89	0.88	0.93
150	0.80	0.85	0.85	0.92
200	0.78	0.81	0.81	0.92
250	0.75	0.78	0.79	0.91
300	0.73	0.76	0.77	0.91
350	0.71	0.74	0.76	0.91
400	0.69	0.73	0.74	0.91
450	0.68	0.71	0.73	-
500	0.67	0.70	0.72	-
700	0.64	0.67	0.68	-
1000	0.61	0.65	0.64	-

RESULTS

The final derived $ARF_{85\%}$ values created significantly larger reductions in point rainfall as compared to NOAA Atlas 2. Because results of the Phase I CDOT September 2013 Flood Study are not being changed as part of this work, a smooth transition between NOAA Atlas 2 24-hour ARF and the derived 24-hour $ARF_{85\%}$ is needed for Phase II basins. The largest basin used in Phase I was 315-mi^2 and the smallest basin used in Phase II was 446-mi^2. In order to maintain consistency between Phase I results and Phase II results, a linear transition was applied between NOAA Atlas 2 315-mi^2 ARF value and $ARF_{85\%}$ 500-mi^2 (Figure 7 and Table 4). Based on the areal limitations of NOAA Atlas 2, the larger point precipitation reductions based on $ARF_{85\%}$, and maintaining consistency with Phase I study the linear transition between NOAA Atlas 2 315-mi^2 ARF value and $ARF_{85\%}$ 500-mi^2 was chosen for application of Phase II of the CDOT September 2013 Flood Study. In addition, application of this transition in the hydrologic modeling for the four basins investigated showed good agreement and acceptable results. The final 24-hour $ARF_{85\%}$ curve is compared to the four basin specific 24-hour ARF curves for the September 2013 event (Figure 8).

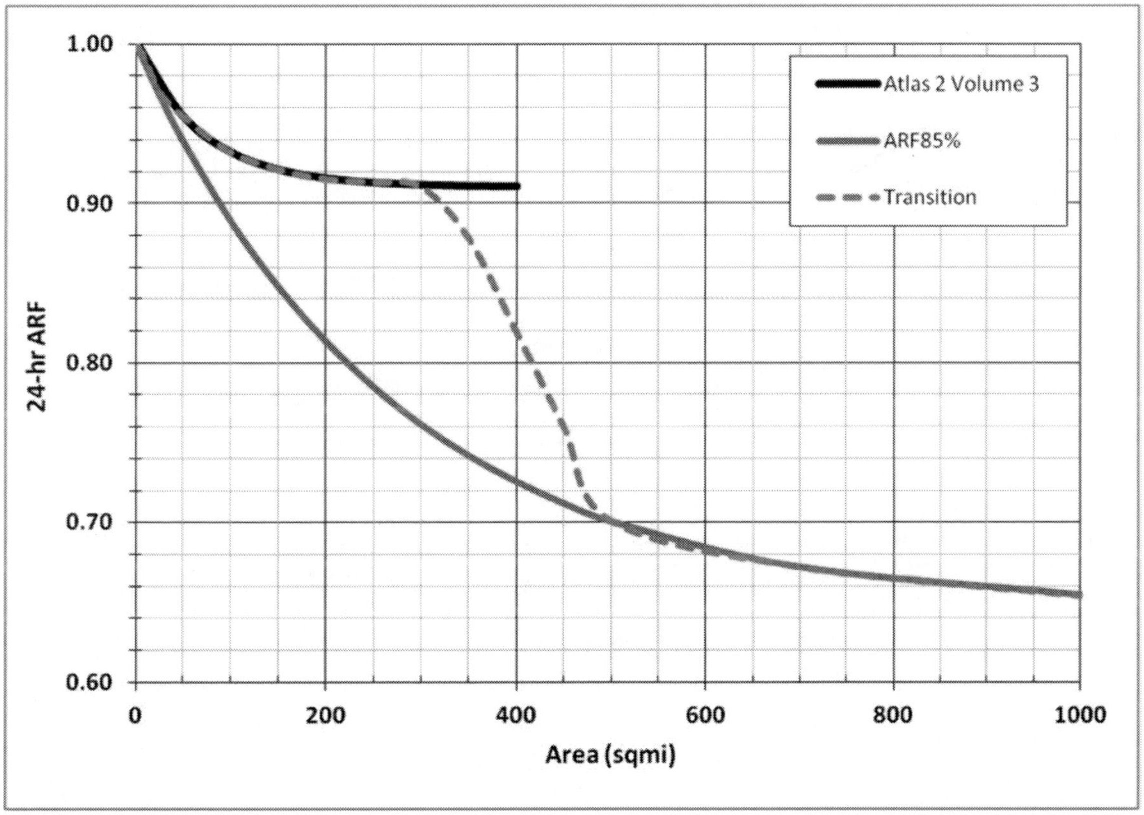

Figure 7. Final 24-hr ARF curve with transition between NOAA Atlas 2 and AWA $ARF_{85\%}$.

Table 4. Comparison of final 24-hour ARF values. ARF85% is the 85% confidence ARF. Transition is the transition between NOAA Atlas 2 and ARF85%, and Atlas 2 is NOAA Atlas 2 ARF.

*** General Storms 24-hr ARF			
Area (sqmi)	ARF85%	Transition	Atlas 2
1	1.00	1.00	1.00
10	0.99	0.99	-
25	0.97	0.97	-
50	0.94	0.95	0.95
100	0.89	0.93	0.93
150	0.85	0.92	0.92
200	0.81	0.92	0.92
250	0.78	0.91	0.91
300	0.76	0.91	0.91
350	0.74	0.88	0.91
400	0.73	0.82	0.91
450	0.71	0.76	-
500	0.70	0.70	-
700	0.67	0.67	-
1000	0.65	0.65	-

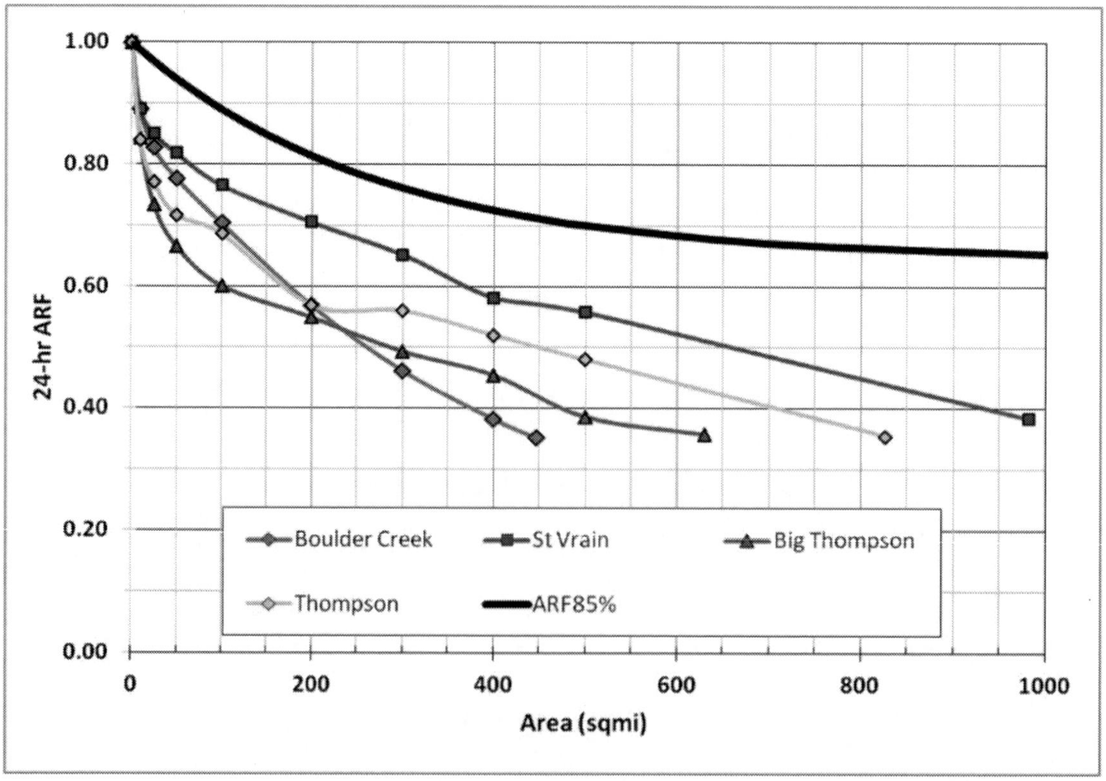

Figure 8. 24-hour ARF curve compared to basin specific ARFs for the September 2013 event

CONCLUSION

The final 24-hour $ARF_{85\%}$ values create significantly larger reductions of point rainfall at larger area sizes as compared to NOAA Atlas 2. These are based on actual storms that have occurred along the Front Range of the Rockies and of similar storm type as the September 2013 event. These updated ARF values produce more realistic and representative point to areal reductions for synoptic storm events along the Colorado Front Range. The 24-hour $ARF_{85\%}$ curve is only representative and applicable for large synoptic and orographic storm events similar to the September 2013 storm event in Colorado. Future hydrology and engineering flood studies should utilize a more site and duration specific ARF curve based on procedures applied in this study and storms specific to a given locations. This investigation has shown that the generalized ARF curves provided in NOAA Atlas 2 are not necessarily representative of spatial rainfall accumulations along the Colorado Front Range.

REFERENCES

Guo, James C.Y.,2011. Storm Centering Approach for Flood Predictions from Large Watersheds, second review, ASCE J of Hydrologic Engineering, May 2011.

Hansen, E.M., Fenn, D.D., Schreiner, L.C., Stodt, R.W., and J.F., Miller, 1988: Probable Maximum Precipitation Estimates, United States between the Continental Divide and the 103rd Meridian, *Hydrometeorological Report Number 55A (HMR 55A)*, National Weather Service, National Oceanic and Atmospheric Association, U.S. Dept of Commerce, Silver Spring, MD, 242 pp.

Kappel, W.D., Hultstrand, D.M., Tomlinson, E.M., and G.A., Muhlestein, August 2012: Site-Specific Probable Maximum Precipitation (PMP) Study for the Tarrant Regional Water District-Benbrook and Floodway Basins, Ft Worth, TX.

Kappel, W.D., Hultstrand, D.M., Tomlinson, E.M., Muhlestein, G.A., and T.W. Parzybok, March 2013: Site-Specific Probable Maximum Precipitation (PMP) Study for the Indian Point Energy Center, Peekskill, NY.

Kappel, W.D., Hultstrand, D.M., Muhlestein, G.A., Tomlinson, E.M., and D. McGlone, April 2014: Site-Specific Probable Maximum Precipitation (PMP) Study for the North Umpqua Basin, Oregon, prepared for PacifiCorp.

Kappel, W.D., Hultstrand, D.M., Muhlestein, G.A., Tomlinson, E.M., and D. McGlone, March 2014: Site-Specific Probable Maximum Precipitation (PMP) Study for the Duck River Basin, Alabama, prepared for Utilities Board of the City of Cullman.

Kappel, W.D., Hultstrand, D.M., Muhlestein, G.A., Steinhilber, K., and D. McGlone, July 2014: Site-Specific Probable Maximum Precipitation (PMP) Study for the College Lake Basin, Colorado, prepared for Colorado State University.

Myers, V.A. and R.M. Zehr, 1980. A Methodology for Point-to-Area Rainfall Frequency Ratios. NOAA Technical Report NWS No. 24, National Weather Service, Silver Spring, MD.

National Oceanic and Atmospheric Administration, 1973. National Weather Service, NOAA Atlas-2: Precipitation-Frequency Atlas of the Western United States, Silver Springs, MD, 1973.

NOAA Hydro-40 (1984). "Depth-Area Ratios in the Semi Arid Southwest United States," NWS NOAA TM Hydro-40, US Department of Commerce, Silver Spring, Maryland.

Omolayo, A.S., 1993. On the transposition of areal reduction factors for rainfall frequency estimation. *Journal of Hydrology*, (145), 191-205.

Osborn, H.B., L.J. Lane, and V.A. Myres, 1980. Rainfall/watershed relationships for southwestern thunderstorms, *ASAE 23*: 82-91.

Parzybok, T.W., and E.M. Tomlinson, 2006. A New System for Analyzing Precipitation from Storms, *Hydro Review*, Vol. XXV, No. 3, 58-65.

Tomlinson, E.M., and T.W. Parzybok, 2004: Storm Precipitation Analysis System (SPAS), proceedings of Association of Dam Safety Officials Annual Conference, Technical Session II, Phoenix, Arizona.

Tomlinson, E.M., Ross A. Williams, and Parzybok, T.W., September 2003: Site-Specific Probable Maximum Precipitation (PMP) Study for the Great Sacandaga Lake / Stewarts Bridge Drainage Basin, Prepared for Reliant Energy Corporation, Liverpool, New York.

Tomlinson, E.M., Kappel W.D., Parzybok, T.W., Hultstrand, D., Muhlestein, G., and P. Sutter, December 2008: Statewide Probable Maximum Precipitation (PMP) Study for the state of Nebraska, Prepared for Nebraska Dam Safety, Omaha, Nebraska.

Tomlinson, E.M., Kappel, W.D., Hultstrand, D.M., Muhlestein, G.A., and T. W. Parzybok, November 2011: Site-Specific Probable Maximum Precipitation (PMP) Study for the Lewis River basin, Washington State.

Tomlinson, E.M., Kappel, W.D., and Parzybok, T.W., March 2011: Site-Specific Probable Maximum Precipitation (PMP) Study for the Tarrant Regional Water District, Texas.

Tomlinson, E.M., Kappel, W.D., Hultstrand, D.M., Muhlestein, G.A., S. Lovisone, and Parzybok, T.W., March 2013: Statewide Probable Maximum Precipitation (PMP) Study for Ohio.

Tomlinson, E.M., Kappel, W.D., Hultstrand, D.M., Muhlestein, G.A., S. Lovisone, and T. W. Parzybok, July 2013: Statewide Probable Maximum Precipitation Study for Arizona.

U.S. Weather Bureau, 1957-60. Rainfall Intensity-Frequency Regime, Part 1—The Ohio Valley; Part 2—Southeastern United States. U.S. Weather Bureau Technical Paper No. 29, U.S. Department of Commerce, Washington, DC.

EVALUATING EFFORT LEVELS OF INFLOW DESIGN FLOOD STUDY METHODS FOR DAMS

S. Samuel Lin[1]
Jason E. Hedien[2]

ABSTRACT

The current state-of-the-practice for inflow design flood (IDF) selection can be implemented by executing three hydrologic safety analysis methods for a dam and its appurtenant structures. They are (1) prescriptive assignment of IDF based on dam hazard class; (2) incremental consequences approach (ICA); and, (3) risk informed decision making (RIDM) approach. The development of these methods has evolved over time as more comprehensive tools for assessing a dam's hydrologic adequacy have become available.

This paper briefly reviews each IDF method. The key differences between each method are illustrated to better understand various analysis effort levels. Incorporating present knowledge with the practical lessons learned, a case study for hydrologic risk mitigation or reduction is presented that highlights the differences between each method, the differing results produced, and the effort and complexity involved with implementing each of the three methods.

The level of analysis effort and complexity of each method for best practices are discussed in this paper; however, specific technical guidance on which method to use is not provided. It is hoped that this paper provides a general understanding regarding IDF selection approaches so the proper method for selecting a dam's reasonable IDF result can be identified according to the specific needs of the dam owner and regulator.

This paper solely represents the authors' viewpoints not necessary their entities' policy and/or practice.

INTRODUCTION

Watershed contributions of extreme precipitation and runoff to reservoirs can endanger the safety of the dams that retain them. The adequacy of hydrologic safety of dam structures is thus a primary concern to public safety. The development of numerical models and various analytical methods are used to evaluate spillway performance limitations under flood hazard loading conditions to estimate the dam's overall response effects. If the required safety standard is not met, then engineered and/or non-engineered mitigation measures should be implemented as remedial measures.

---

[1] Civil Engineer, Federal Energy Regulatory Commission, 888 First St. N. E., Washington, DC 20426, Shyangchin.Lin@FERC.Gov
[2] Principal Geotechnical Engineer, MWH Americas, Inc., 350 N. Orleans, Suite 1301, Chicago, IL 60654, jason.e.hedien@mwhglobal.com

The prevailing hydrologic safety standard for dam safety purposes is the inflow design flood (IDF). The basis of determining the IDF consists of conducting an inundation study to assess dam breach consequences and then selecting an appropriate IDF through using one of three methods: (1) prescriptive assignment of IDF based on dam hazard class; (2) incremental consequences approach (ICA); and, (3) risk informed decision making (RIDM) approach. FEMA (2013a) and Lin (2015) describe and demonstrate each of the three methods, present the differences resulting from application of these methods, and discuss observations from case studies of their application, respectively.

Depending on specific regulatory hydrologic safety requirements, the resulting IDF of any method, regardless of their possible different results, can be used for justifying the required hydrologic safety of the dam, managing hydrologic risks to downstream (and upstream) populations and/or properties, and developing mitigation measures as needed during the dam design or operation phase.

This paper briefly describes the above three state-of-the-practice IDF selection methods, including information on the technical capability and resource requirements for their application. The overview of the three IDF methods includes: (1) clarifying IDF deterministic vs. probabilistic methodologies; (2) comparing the merits of standard vs. in-depth refined (or detailed) methodologies; and (3) highlighting the positive aspects of IDF selection process thru dam breach assumptions and consequence assessment.

FUNDAMENTAL CONCEPTS AND PRACTICE PRINCIPLES

Three basic types of dam breach events include overtopping and non-overtopping resulting in dam failure, and sunny day dam failure. IDF studies focus on hydrologic overtopping and non-overtopping breach events. The selected IDF represents the flood hydrograph which enters the reservoir and is used to design and/or modify a specific dam and its appurtenant works for its hydrologic safety; particularly for sizing the spillway and outlet works, and for evaluating maximum storage, height of dam, and freeboard requirements.

Provision of adequate spillway capacity is required on the basis of three scenarios: safely passing the IDF without overtopping, overtopping causing failure of the dam without increasing downstream consequences by creating a greater flood event or additional risks to the public, or withstanding overtopping over the dam without initiating failure (not included here). For instance, safely passing the IDF without compromising the integrity of the dam structure requires an adequate spillway system be designed to release the required IDF without overtopping the dam and initiating a failure mode leading to breaching the dam and uncontrolled release of the impoundment.

For an IDF study, it is essential to perform and provide an assessment of the potential downstream adverse consequences due to dam failure caused by various hydrologic loading conditions. In assessing the consequences, the likelihood of loss of human life and property damage is evaluated using dam failure analysis resulting inundation scenarios and sound engineering judgment.

IDF SELECTION AND STUDY METHODOLOGIES

The Federal Emergency Management Agency (FEMA) published an updated guidance document entitled "Selecting and Accommodating Inflow Design Floods for Dams" in 2013 (FEMA, 2013a). This regulatory guidance document provides the basis evaluating spillway adequacy. The main objectives of the document are (a) to recommend appropriate implementation procedures for the three prevailing approaches mentioned above for selecting and accommodating the IDF for dams based on current and accepted practices, and (b) to provide a flexible framework within which both federal and state agencies can develop and update their specific guidelines according to their varied goals and resources. Important updates and distinctions regarding both selection and accommodation of the IDF are included in FEMA (2013a), such as eliminating dam size in dam hazard classification, treatment of existing "grandfathered" dams, eliminating use of percentage of PMF for assigning the IDF, recognition of uncertainty in analyses, etc.

Two major IDF selection methodologies are categorized on the basis of three types of hydrologic, flood and dam hazards: (1) deterministic, which aims to demonstrate that a dam can safely pass the IDF that is defined based on a regulatory standard (prescriptive method) or determine the IDF above which dam failure may occur without increasing downstream consequence (ICA method); and, (2) probabilistic, which aims to provide an estimate of the risk presented by those three types of hazards (RIDM approach). Figure 1 provides a comparison of each of two main methodologies.

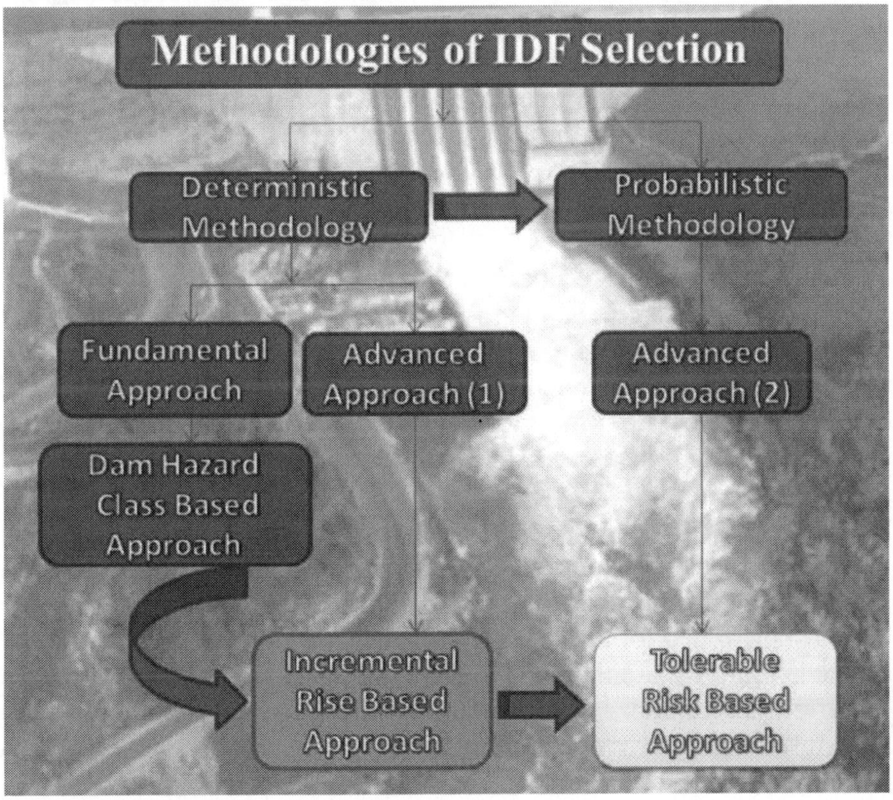

Figure 1. Two Methodologies of Inflow Design Flood Selection

An IDF determined by the prescriptive approach for a specific project may be a gross estimation of critical hazards to a dam, mainly because the approach does not consider the condition of the dam structure. The dam failure assumption is made only for a worst downstream consequence. If that is the case, the IDF value can then be refined by applying the other two approaches (ICA or RIDM) to determine whether a lesser flood could be adopted without increasing incremental hazards upstream and/or downstream. The prescriptive and both ICA and RIDM approach are considered as fundamental and advanced approaches, respectively.

Deterministic and probabilistic methodologies use the same basic modeling approach for performing a dam breach analysis that relates the various parameters likely to affect breach flood routing and consequence assessment. However, they differ in the figures used to represent these parameters. Thus, it is important to apply the concepts for each IDF method consistently for that particular IDF study method. For example, potential failure modes (PFMs) used as a basis for the ICA or RIDM methods should not be directly applied when using the prescriptive method. Also, the ICA method should not directly use the prescriptive method's conservative dam failure assumptions.

LEVELS OF EFFORT FOR IDF STUDIES

General Study Procedures

Procedures for IDF studies usually start from evaluating "hydrologic hazards". A range of flood events may be evaluated starting from relatively more frequent, smaller peak events to infrequent, larger events. By trial and error, dam breach flood magnitudes are evaluated in order of increasing flood inflow beginning with a certain flood magnitude event and proceeding to an upper magnitude flood, even to the PMF. The simulated inflow is loaded to the reservoir for reservoir flood routing and the dam breach outflow hydrograph as the input to available river flood routing software will then be used to compute water surface profiles and associated flooding conditions downstream.

Specifically, each of three IDF approaches follows a process for dam failure consequence assessment: (1) determine a critical dam failure scenario to predict a breach outflow hydrograph and route the hydrograph through downstream critical sections while considering simultaneous downstream critical tailwater conditions; (2) define the area flooded by the dam breach flood wave and evaluate inundation impacts (typically by flow depth/velocity in areas of interest); and, (3) estimate economic damages and potential life loss consequences.

For the purpose of a sensitivity analysis, the analysis of the dam failure with a range of breach configurations and subsequent routing of the breach flood wave along the river reach should be performed for the selected impact area. For more accurately reflecting the hydrologic/hydraulic conditions at the dam and flood routing, the model flow network should begin from a certain distance upstream of the dam and use reliable bathymetric data to accurately reflect the hydraulic conditions in the river/reservoir for flood routing.

Dam Breach Modeling and Consequence Assessment

Key Considerations for Available Dam Breach Software. FEMA P-946 (FEMA, 2013b) describes several software packages used for dam breach hydrograph generation and downstream hydraulic routing for the purposes of hazard classification, the development of Emergency Action Plans (EAPs), and assessment of potential risks. Included are recommendations for selecting modeling software based on a "tiered approach" considering the type of results needed or detail required, the "level of effort" that can expended, and the "potential" for loss of life and economic damages that can result from a dam failure. Cook et al. (2015) also provides guidance for selecting dam breach inundation software.

Selection of dam breach inundation software depends upon the following key determinants: (a) level of effort required proportional to the decision made from the end products (e.g., precision of hazard classification, IDF, or EAP); (b) available resources (e.g., capability, time, cost, etc.); and, (c) acceptability of uncertainty (acceptable accuracy of the model). The level of complexity chosen for dam breach modeling is dependent on the needs for consequence assessment. Increased complexity of dam breach modeling and consequence analysis can provide increased confidence on the study results. The level of analysis should correlate the desired sophistication and accuracy of the analyses with the scale and complexity of the dam and downstream area under investigation.

FEMA (2013b) uses a tiered approach to dam breach analysis for inundation mapping in EAP development as shown in Table 1. In general, the HEC-RAS (1-D model) is a popularly used hydraulic flood routing model for well channelized areas. However, the application of a 2-D model using "shallow water equations" is more reliable for the inundation estimation when flood flow enters a wide, flat area (i.e. non-channelized area) or floodplain with a small channel. More detailed modeling may be warranted for high-hazard potential / large-sized dams, dams with a large population in the affected area, or dams with significant downstream hydraulic complexities such as major diversion structures, split flows, or the potential for a series of dam failures such as a "domino effect". Three tier levels of analysis efforts are designated as basic (i.e. screening or simple) (Tier 1), intermediate (Tier 2), and advanced levels (Tier 3) as shown in Table 1.

Assessing Dam Failure Consequences. Comprehensive information is needed to define and estimate consequences for dam failures. Common factors used for consequence assessment includes flow depth and velocity, time of dam failure, and flood wave arrival. A discussion of approaches to consequence assessment is outside the scope of this paper, and the reader is referred to presentations included in the proceedings of the National Dam Safety Program Technical Seminar of February 2015 (NDSPT, 2015) and RCEM (2014).

Table 1. Tiered Approach for Dam Breach Inundation Study

Tier Level	Applicable to Dam Hazard Potential / Size	Breach Parameter and Hydrograph Estimations and Downstream Routing
Tier 1 – Basic level	First level screening for any hazard potential dams / small size dams	Empirical equations; Simplified models, HEC-HMS, or 1-D HEC-RAS (Steady-State) model, etc.
Tier 2 – Intermediate level	Significant-hazard potential / intermediate size dams or High-hazard dams with limited population at risk	Empirical equations; HEC-HMS or HEC-RAS (Steady or Unsteady modeling) 1-D (well channelized areas) or 2-D (wide flat areas) models, etc.
Tier 3 – Advanced Level	High-hazard potential / large size dams with sufficient population at risk	Empirical equations, physically models, a probabilistic approach using a Monte Carlo analysis; HEC-RAS Unsteady or 2-D models

IDF Study Levels of Effort

The tiered approach to dam breach analysis described previously is also applicable to levels of effort required for IDF studies. As the sophistication of the modeling increases, so does the level of effort, time, and cost necessary to conduct the analysis. Although the amount of detail and degree of complexity grow with each level of effort, the degree of uncertainty in the results decreases. A step-by-step process can be followed while recognizing the needed resources and/or capabilities. The levels of effort are as follows:

- Level of Effort 1: Modeling in a dam hazard classification framework (prescriptive method)
- Level of Effort 2: Modeling in an incremental consequence framework (ICA method)
- Level of Effort 3: Modeling in a probabilistic risk framework (RIDM)

Table 2 shows how the IDF study levels of effort compare with the tiered approach to dam breach analysis described earlier.

Table 2. IDF Study Levels of Effort

Levels of Effort (or Complexity)	IDF Study Method	Tiered Approach of Dam Breach Inundation Study
1	Prescriptive	Tier 1 – Basic Analysis
2	Incremental Consequence	Tier 2 or 3 – Intermediate or Advanced Analysis
3	Risk-informed Decision Making	Tier 3 – Advanced Analysis

Level of Effort 1: Use of Prescriptive Method. The purpose of the "Level 1" approach is to provide some meaningful information about downstream exposure to conservative hypothetical dam failure parameters without having to perform relatively complex modeling or calculations. This lowest level of effort is relatively simple to perform and relies on the use of inundation depth and/or flow velocity to determine the degree to which a particular asset or area is exposed to the effects of extreme events. It usually starts from the worst probable maximum flood (PMF) to classify dam hazard based on the estimated consequences for the IDF selection.

For some cases, the goal of a Level 1 study might be to capture the sensitivity of a particular asset or area to the effects of river level rise rather than predict a particular value of flooding depth or wave height, flow velocity, etc. In this manner, the Level 1 approach can be used as a screening tool to identify impacted areas or structures. The IDF determined using the prescriptive approach is typically more conservative (higher flood flows) than the ICA or RIDM approaches. If there is a lack of sufficient data, resources, expertise, or time to perform a more rigorous study, or there is no need of detailed inundation information, a Level 1 study may be sufficient. A Level 1 study may also be sufficient if the dam can safely pass the IDF determined using the prescriptive approach without uncontrolled release of the impoundment.

Level of Effort 2: Use of ICA Method. The IDF selection using the ICA is to determine the flood above which there is a negligible increase in downstream inundation depth, flow velocity, and/or consequences due to failure of the dam when compared to the same flood without dam failure. This process is continued until the flood of greatest magnitude that causes incremental consequences is identified.

IDF determination via the ICA method must go through two stages of analysis based on the existing condition and hypothetical upgraded condition of the dam under study. The incremental water level increase downstream is estimated using the difference between non-dam breach and dam breach consequences under the same flood scenario. Typically, the downstream water level for the non-dam breach scenario is determined from flood wave routing assuming the scenario of any dam overtopping without failure. The hydraulic performance of a hypothetical non-failure scenario for many cases could differ from that of an assumed upgraded dam condition under which the study flood can safely pass. As a result, the estimated incremental rise could be overestimated due to a greater flow discharge. Thus, an iterative process based on assumed upgraded dam conditions is needed to further refine the initial IDF and rationalize the ICA result. Namely, the ICA method must go through an iterative process while checking any new PFMs introduced by an upgraded spillway scheme (Lin, 2015). The resulting IDF may be further refined by taking into account the hypothetically added spillway capacity.

As a rule of thumb, the consequences of failure are considered acceptable when the incremental effects of failure on downstream structures are approximately two feet or less. However, the two-foot increment is not an absolute decision-making point. Usually, sensitivity analyses and engineering judgment based on the value of inundation depth * flow velocity are the means used in making final decisions.

Level of Effort 3: Use of RIDM Method. A Level 3 study characterizes downstream exposure in terms of probability and risk by estimating hydrologic, flood, and dam hazards. The RIDM approach uses a probability-driven process to combine a similar deterministic ICA process with risk estimation. Strengths of RIDM include: (a) better understanding of project risks; (b) provides dam owners and regulators the ability to assess the marginal value of increasing levels of flood protection; and, (c) balances capital investment in risk reduction across a number of different PFMs.

RIDM includes a site-specific evaluation of the probabilities of a full range of extreme hydrological events and performance of the dam during those events, and evaluates in more detail the social, economic, and environmental consequences of failure. Thus, RIDM is a tool for evaluating hydrologic events in a risk-based context and the level of effort is in proportion to information made available to assess dam safety issues and risk mitigation measures.

The RIDM process can include specifically assessing individual incremental life safety risk using probability of loss of life, or societal incremental life safety risk expressed as a probability distribution of potential life loss or expressed as an average annual life loss (ALL). As a reference, some agencies use two major tolerable risk indices to justify the IDF selection including average annual failure probability (AFP) of a dam (e.g., AFP $\leq 1.0 \times 10^{-4}$/year), and a resulting ALL (e.g., ALL $\leq 1.0 \times 10^{-4}$ lives/year).

However, probabilistic hazard analyses have faced a variety of challenges. For instance, the rainfall-runoff watershed model algorithms contain uncertainties due to incomplete understanding of the underlying hydrological processes. Clearly, it is very difficult, if not impossible, to capture the total uncertainty in the estimation of extreme floods. Thus, flood characterization should include both (a) a best estimate of the annual exceedance probability of floods of different magnitudes and (b) a description of the uncertainty in such results.

In addition, the structural integrity and properties of a dam structure are sources of cognitive epistemic uncertainty (i.e. knowledge uncertainty) and non-cognitive aleatory uncertainty (i.e. natural variability) that are often not addressed in risk analysis practice. Similar to the ICA method, PFMs are identified usually based on existing conditions of a dam and its environments, which, however, are dynamic. The potential failure modes analysis (PFMA) result doesn't account for the potential changes in the future on a multiple facets of impact, such as dam conditions, upstream/downstream developments, climate change, etc. Hence, RIDM approach needs periodical safety review like prescriptive and ICA methods, especially when any dam safety related incidents take place, a follow-up PFMA may be required.

<u>Technical Complexity and Analysis Effort</u>

The time and effort required to complete a Level 3 study is significantly greater than a Level 1 or 2 study, while the effort required to complete a Level 2 study is greater than a Level 1 study. Accordingly, the technical capabilities and associated costs required to complete a Level 3 study are higher than a Level 1 or 2 Study. But with an increase in time, effort, and cost comes a significant reduction in uncertainty and a narrower range of possible answers. The reduction in uncertainty in Level 3 studies is mostly attributed to use of the dam failure and breach hydraulic or computational fluid dynamics models.

Selection of an IDF study Level of Effort may start at Level 1. However, determination of whether Level 2 or 3 studies may be beneficial to determining a proper IDF for a given dam should not be strictly a function of dam hazards for some cases. To establish a level of hydrologic safety protection for any dam systems, irrespective of the dam hazard classification, proper consideration of the IDF in the full context of the economic and other benefits of the dam to the owner and public may be applicable, particularly for a low hazard dam.

CASE STUDY OF APPLICATION OF IDF METHODOLOGIES

Case Overview

Hedien (2013) provides a case history of two concrete core-wall earth embankments on the same river. The dams are referred to as Dam A (upstream dam) and Dam B (downstream dam), respectively. Dam A and Dam B are separated by approximately 12-river miles.

The hydrologic hazard to both dams is overtopping and breaching of the earth embankments. The IDF was evaluated using each of the three IDF methods described in this paper, with levels of effort and complexity increasing from Level 1 through Level 3 as described previously. The results are summarized in Table 3.

Table 3. Inflow Design Flood Study Results

Tandem Dams	**IDF by Prescriptive Method (kcfs)** (i.e. the cold season PMFs)	**IDF by ICA: Assumed Overtopping Failure Depths**		**RIDM Approach (kcfs)**
		10' at Peak Reservoir Level (kcfs)	2' at Peak Reservoir Level (kcfs)	
Dam A	223	129	70	53
Dam B	230	230	135	60

For this case history, the accepted tolerable risk levels utilized in the RIDM approach were: AFP $\leq 1.0 \times 10^{-4}$/year; ALL $\leq 1.0 \times 10^{-4}$ lives/year; and AFP and ALL assessed to be as low as reasonably possible (ALARP).

IDF Study Levels of Effort

Inundation mapping performed in support of the IDF study indicated that failure of either dam has the potential for loss of life and both dams are classified as high-hazard. Thus, the IDF determined by the prescriptive method (Level 1 effort) was equivalent to the cold season rain-on-snow PMF for both dams. Given that the existing spillway capacity for Dam A and Dam B are 39 kcfs and 45 kcfs, respectively, it was apparent that both dams have insufficient spillway capacities to pass the determined IDF by the prescriptive method by a significant margin. Results of the Level 1 studies indicated that significant modifications to both dams would be required to safely pass the resulting PMFs. In the case of Dam A, most of the existing dam structure would have needed to have been replaced with a large multiple bay gated spillway section. The need and desire to perform more sophisticated Level 2 studies was clear.

Level 2 studies using the ICA approach indicated that incremental consequences were minimized at a lower inflow flood level for Dam A. Thus, application of Level 2 studies showed a reduction in the IDF at Dam A (with no impact on the IDF for Dam B). Although the extent of the modifications to Dam A was reduced, the estimated costs for modifying both dams to pass the IDF determined from the Level 2 studies were still considered significantly high and there was interest in exploring the use of RIDM to see if the IDF for both dams could be reasonably lowered further.

The time, cost, and complexity to evaluate the dams using RIDM was large. Use of RIDM required development of a comprehensive PFMA, a risk assessment workshop to further develop the controlling PFMs and the event tree for the risk assessment, assignment of a conditional probability of occurrence or system response probability (SRP) for each PFM, and determination of the AFP and consequences (potential for property damage and life loss) under the range of flood inflow events considered significant to the performance of the dams. Investigation of the effectiveness of various risk reduction measures to accommodate the selected IDFs was also carried out as part of a risk reduction assessment (RRA) for the dams, with the goal of achieving the most reasonable combination of risk reduction cost effectiveness with respect to capital cost and on-going operation and maintenance costs while allowing the dams to safely pass the resulting IDF.

Selected IDFs for Two Dams

The RIDM approach assesses both the probability of the flood loading and the probability of the resulting adverse response to evaluate AFP of each dam. The estimated total risk is the summation of risk from the identified PFMs. As shown in Table 3, the IDFs determined by RIDM were further reduced for both dams due to the ability to introduce factors such as evacuation effectiveness, population distribution, and a more rigorous dam breaching analysis into the study. Following subsequent reviews of the risk assessment results, the additional spillway capacity to be added to each dam was increased somewhat so that the computed warm season PMF (58,000 cfs) could be safely passed during the summer months, when the population that could be impacted by a failure of one of the dams is the greatest. The spillway capacity at Dam B was increased somewhat further to accommodate increased flood flows from Dam A in the event Dam A fails (mitigate a domino failure of Dam B). The final IDFs of 60 kcfs and 65 kcfs were adopted for Dam A and B, respectively in order to increase more margin of hydrologic safety.

Remedial Actions

Mitigation measures utilized for both dams to meet the RIDM determined IDF (increasing the spillway capacity at each dam) were significantly less costly to implement. The spillway capacity expansion for Dam A consists of lowering the crest of an existing gravity overflow section at the dam and installing a gated ogee spillway in five new spillway bays. The spillway capacity expansion for Dam B consists of adding an ungated emergency overflow spillway atop the crest of the embankment section of the dam. Construction of the spillway capacity addition at Dam B was completed in 2013. Construction of the spillway capacity addition at Dam A is scheduled to be completed in 2016.

A Key Finding

Furthermore, a key finding of the RIDM approach was the importance of a well-developed and coordinated emergency action plan (EAP) so that the population at risk can be evacuated ahead of overtopping and failure of the dams. Evacuation effectiveness was determined to be an important parameter when estimating life loss consequences in the risk assessment, and demonstration of the ability to evacuate a significant portion of the population at risk (PAR) is imperative to risk mitigation for high hazard dams like these. The owner has developed an actively exercised and communicated EAP so that evacuations can be well coordinated during flood events well before the IDF is reached.

SUMMARY AND CONCLUSION

1. The IDF study characterizes from hydrologic hazard to flood and dam hazards. The state-of-art and practice of an IDF study is a multidisciplinary endeavor of hydrology and hydraulics. An adequate case-by-case IDF selection is considered as an essential, primary dam safety process for public safety. The IDF, selected either deterministically or probabilistically, is required for hydrologic safety of dams. This paper describes the selection of an adequate IDF for spillway design through three methods along with their analysis levels of effort: (a) prescriptive method; (b) incremental consequences approach (ICA); and, (c) risk informed decision making (RIDM) approach. Each method is technically defensible with credible explanations and interpretations; however, the methods progressively increase in required effort and complexity in order to provide refined IDF estimates for hydrologic safety evaluations and detailed results and information to guide risk mitigation decisions.

2. Levels of effort increase progressively from the prescriptive method (Level 1), to the ICA method (Level 2), to the RIDM approach (Level 3). The prescriptive assignment method minimizes effort and complexity. However, the resulting IDF may be not be optimal with respect to minimizing incremental consequences or providing information to evaluate uncertainties of dam failure modes due to hydrologic loadings. The ICA or RIDM approaches can be applied for refining IDF results for some cases. The ICA method more explicitly accounts for the incremental consequences from a dam failure under hydrologic loadings. The more complex RIDM method evaluates a full range of risks and provides further refinement to the IDF achieved by using prescriptive or ICA method.

3. To assure or bring the dam into regulatory compliance, Level 1 study may be appropriate when limited information is available or there are limited resources or time

available for performing a more detailed study. Level 1 study is often performed first since it may be that an existing dam may be able to safely pass the IDF without modification, or mitigation measures can be implemented with little or no significant cost to the owner. Level 2 and Level 3 studies are appropriate when (a) more detailed information on the impacts of the IDF to upstream and downstream consequences is desired, (b) uncertainties in operation, dam behavior, or other factors exist that may significantly impact the resulting IDF, or (c) there is the potential to identify a smaller value for the IDF to minimize possible remedial measures.

ACKNOWLEDGEMENT

The authors want to extend their appreciation to those who assisted either directly or indirectly by providing their inputs/comments in this study and paper.

REFERENCES

Cook, Shane, etc. (2015). "Selecting Dam Breach Inundation Software", The Journal of Dam Safety, Vol. 13, Issue 1

Lin, S. Samuel (2015). "Justifying Inflow Design Flood for Dam Hydrologic Safety: Holistic Comparison of Methods", Federal Interagency SEDHYD Conference, Reno, Nevada, April

FEMA NDSPTS #22 (2015). "Dam Breach Modeling and Consequence Assessment"

FEMA Document P-94 (2013a). "Selecting and Accommodating Inflow Design Floods for Dams", August 2013

FEMA Document P-946 (2013b). "Federal guidelines for Inundation Mapping of Flood Risks Associated with Dam Incidents and Failures", July 2013

Hedien, J. E. (2013). "RIDM Approach for IDF Selection and Accommodation for Dams: A Practical Application Case Study", NRC Probabilistic Flood Hazard Assessment (PFHA) Workshop, January 2013

2015 USACE INSTRUMENTATION AND MONITORING PROGRAM INTERNAL PEER REVIEW RESULTS

Georgette Hlepas PE, PhD[1]
Travis Tutka, PE[2]

ABSTRACT

The United States Army Corps of Engineers (USACE) is evaluating their instrumentation monitoring data and processes to further explore concerns noted in a Dam Safety Program Independent External Peer Review that was conducted in 2013. Thirty-eight dams were chosen as an initial sample of various dam conditions, heights, hazard potential, and geographical locations. The main goal of this internal peer review is to aid USACE Districts in improving program effectiveness with particular focus on adequate data collection, interpretations, and reporting.

District evaluators reviewed the instrumentation and monitoring program management, instrumentation folios, data acquisition and processing methods, data evaluation and reporting documents, and identified data anomalies or potentially concerning trends for various projects at each District. Design assumptions, installation details, past performance issues, maintenance records, and field condition data were provided in historical documentation to aid in fully understand instrumentation results. Funding for instrumentation monitoring is a recurring limitation, but the reviews have identified many low cost best practices. Leveraging technology requires some initial expense, but it is believed that the gains in efficiency will return that investment. Most importantly, the added timeliness and confidence in monitoring data are priceless. The review team has identified good examples of annual instrumentation reports and data acquisition, storage, and processing methods. At the conclusion of the reviews, lessons learned and common issues were documented and presented herein.

INTRODUCTION

A review of instrumentation monitoring programs for 38 dams within the USACE has identified several weaknesses and some solutions. USACE is continuing the review effort on additional dams to further understand ways to improve USACE stewardship of over 700 dams. This document details the review process, findings, and recommendations for the path forward.

EXTERNAL PEER REVIEW BACKGROUND

USACE underwent a Dam Safety Program Independent External Peer Review (IEPR) in 2013. One of the review comments addressed instrumentation and monitoring. The comment is as follows:

[1] US Army Corps of Engineers, Dam Safety Modification and Mandatory Center of Expertise, 312-846-5457, georgette.hlepas@usace.army.mil.
[2] US Army Corps of Engineers, Risk Management Center, 314-288-7035, Travis.C.Tutka@usace.army.mil.

> "The data acquisition phase of a project instrumentation program appears adequate but there are concerns that the data are not always reviewed and evaluated in a timely manner by experienced engineers. Lack of timely data reviews can lead to unacceptable levels of risk. Interviews uncovered a lack of reliable exception reporting, where reading changes beyond predefined limits are not flagged for priority reporting and response. Failure to adequately collect, interpret, and report findings in a timely manner is an urgent shortcoming."

The USACE identified three actions in response to this comment. The first was an update to the Engineering Manual 1110-2-1908 Instrumentation and Monitoring of Embankment Dams and Levees (expected to be published in 2016). The revised manual includes expanded information regarding the importance of data evaluation in a timely manner, monitoring frequency and thresholds, as well as the importance of adequate data collection, interpretation, and reporting by trained staff. The second was the development and implementation of an instrumentation and training course that covered appropriate evaluation techniques and timeliness of data collection, review, and reporting. That course has been added to USACE PROSPECT training offerings in 2016 and is planned to be offered annually. The third USACE action was to perform a more in-depth internal peer review of a sampling of dams at multiple Districts from across the USACE. The main goal of all these actions is to improve dam monitoring program effectiveness with particular focus on adequate data collection, review, interpretation, and reporting. This paper focuses on the internal peer review efforts.

INTERNAL PEER REVIEW PROCESS

In 2015, an Internal Peer Review Team (IPRT) was assembled that was comprised of members of the USACE Dam and Levee Performance Monitoring Committee. The IPRT selected USACE Districts and projects for review and developed the review process in coordination with senior leadership.

USACE has a Headquarters office and 8 Regional Division Offices known as MSCs. Figure 1 indicates the geographic locations of each Division. One District in each MSC was reviewed with the following exceptions: the Mississippi Valley Division review was postponed until 2016 due to funding limitation; the Pacific Ocean Division (POD) and Transatlantic Division (TAD) were not included in the reviews based on their limited number of dams. POD has one dam whose instrumentation has been extensively evaluated with a Dam Safety Modification Study and TAD has no dams.

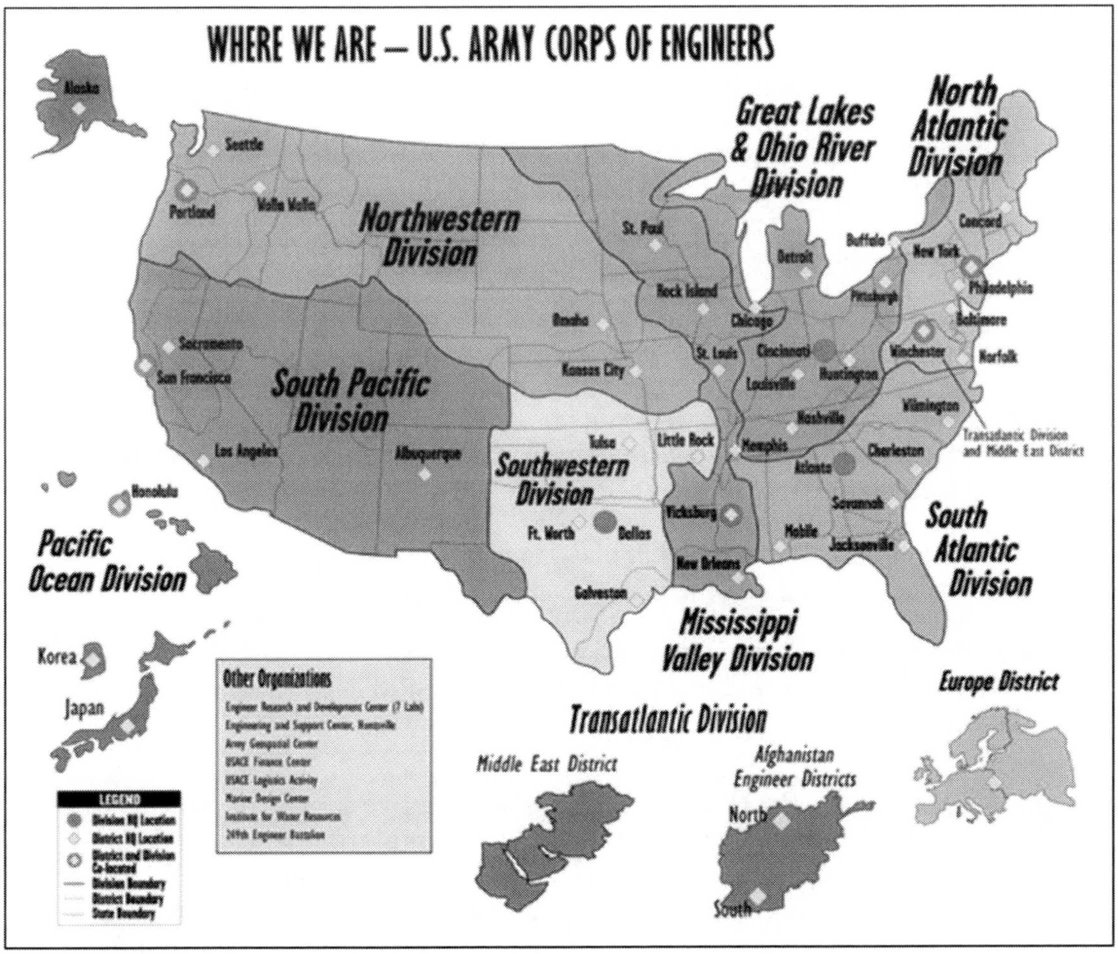

Figure 1. Geographic locations of USACE Divisions and Districts

USACE utilizes Dam Safety Program Management Tools (DSPMT) to track many dam safety activities at each individual dam, including instrumentation monitoring. Scores are assigned for various dam safety activities. For instrumentation and monitoring, the score is based on the Dam Safety Program Manager's answers to the following questions:

- Are all instrumentation readings current as scheduled AND has all instrumentation data been reviewed and interpreted by a qualified individual/Engineer?
- All scheduled instrumentation readings and interpretations have been documented in an annual instrumentation report and furnished to their Major Subordinate Command (MSC)?
- Instrumentation readings and interpretations since the last Periodic Inspection (PI) or Periodic Assessment (PA) are given a thorough assessment in the PI or PA report that is submitted to their MSC?

- Have procedures been implemented for evaluating data accuracy and addressing unusual readings?
- Are all essential instrumentation functional, and none needs replacement?
- Adequacy of instrumentation has been evaluated as part of the PI or PA AND is considered adequate for monitoring critical failure modes?
- Instrumentation threshold values or ranges for irregular readings are established?
- Have surveillance procedures for more frequent readings and inspections been established for higher pools and/or unusual loading events?
- Do you have adequate gages to support water management operational decisions including precipitation, stage and discharge information for developing basin responses to multiple inflow scenarios?
- Do you have adequate gages in place to monitor consequences within the watershed and support emergency evacuation planning?

A score is automatically generated based on the response provided. Districts that had the lowest instrumentation and monitoring score in DSPMT per MSC were selected for the initial peer review.

For efficiency purposes, the IPRT selected a representative sample of projects from each District that included a range of high to low risk projects. Where similar risk dams were identified, projects with greatest dam heights were selected. Dam Safety Program Managers (DSPMs) at each District were provided the selected list of projects to verify that the projects were a representative sample. Substitutions were permitted where DSPMs identified alternative projects that were more representative of their instrumentation program. The initial selections included 10 projects per District, with the exception of the first District that was reviewed. The first District had a limited number of projects with the majority being navigation projects with identical instrumentation and monitoring programs. After the initial review, the IPRT narrowed the number of projects per District to 7. The reduced number permitted a more thorough review of each selected project without compromising the intent to review the instrumentation and monitoring program as a whole. Appendix A includes a list of Districts and projects reviewed.

Districts were each requested to provide existing project documents to the review team . Documents requested included: Surveillance and Monitoring Plans, Instrumentation Design Documentation Reports, Instrumentation Folios (inventories, drawings, and installation details), last two Periodic Inspections or Periodic Assessment, Issue Evaluation Studies and Dam Safety Modification Reports, Instrumentation Annual and Interim Reports, Instrumentation Inventory, and data displays/plots.

The documentation was reviewed prior to arrival on site. The team used a checklist (provided to the District prior to arrival) to facilitate the review process and provide consistency in the reviews. The checklist used is included as Appendix B. The checklist facilitated the evaluation of the following areas:

A. Instrumentation and Monitoring Program Management
B. Instrumentation Folio
C. Data Acquisition and Processing/Plotting
D. Data Evaluations and Reporting
E. Reporting Elements for Various Types of Monitoring

The review team visited each District Office for four days. In-briefings were held on the first day describing the review process and detailing the review schedule for the week. The review team met with District staff with instrumentation responsibilities on each dam to discuss questions that arose during documentation review and to clarify some of the details of the instrumentation program. Engineering technicians and DSPMs were included in the discussions on an as-needed basis and the review was "open" to individuals interested in attending.

After each individual project discussion, the peer review team, independent of the District local staff, discussed their findings and came to a consensus on recommendations and comments for each individual project. After review of all individual projects, the peer review team, independent of the District local staff came to a consensus on general review comments and recommendations applicable to the District Instrumentation and Monitoring Program as a whole.

Recommendations for individual projects as well as the District Instrumentation and Monitoring Program were assigned priority codes to provide a general sense of priority. The same recommendation was sometimes made for both the Instrumentation and Monitoring Program as a whole as well as on multiple dams, often with varying priority codes. This was intentional as the priority code for a given recommendation took into consideration the risks at the project and the significance of the recommendation to performance monitoring and the Instrumentation and Monitoring Program as a whole. Table1.1 includes a description of the Priority Codes (PCs)

Table 1. 1 Priority Codes and Descriptions

Priority Code	Descriptions
1	Major comment/recommendation that needs to be addressed immediately
2	Significant comment/recommendation that needs to be addressed as soon as practicable.
3	Minor comment/recommendation to be considered for implementation

Prior to review team departure, out-briefings were held with District personnel to discuss the review findings. In addition, a comprehensive document summarizing the review intent, process, and conclusions and recommendations for the District Instrumentation and Monitoring Program as a whole, as well as for each individual project was provided. The review notes taken on the checklists for each project, the summary of documentation reviewed, and list of personnel that met with the review team for each project were included in the documentation.

A follow-up with each reviewed District will be held one year following the review to evaluate progress on recommended actions and identify best practices that may help other Districts. This follow-up will not be a re-review. Districts will be annually requested to provide a status towards addressing findings. The intent is to capture innovative and effective monitoring solutions and aid in determining the effectiveness of the review process.

IDENTIFIED GOOD PRACTICES

The following good practices were noted during the peer review process:

- Lead engineers and technicians assigned for individual projects,
- Methods to ensure instrumentation and monitoring programs were being monitored and reviewed at the appropriate frequencies,
- Use of databases to store and process data,
- Use of Geospatial Information Systems (GIS),
- Good communication practices between engineers and operations personnel, and
- Incorporation of informative annotations on plots.

Lead engineers and technicians assigned to individual projects: In most Districts, a lead engineer is assigned to specific projects and is responsible for their instrumentation and monitoring. In a couple of Districts, a lead engineer is assigned to a region (or basin) and is responsible for all the projects within that area including the instrumentation and monitoring. Having the lead engineer responsible for projects within the same region/basin facilitates an understanding of the projects interdependency (if any), the geology of the area, and efficiency in site visits and visual monitoring. One District included a back-up engineer for each project (or region of projects). This is considered good practice in that it facilitated having a knowledgeable person available should the lead engineer become unavailable. This is especially important during an emergency event. Furthermore, most Districts had at least one technician responsible for ensuring data is entered, validated, and graphed at the prescribed time intervals. This support permits the lead engineer to focus more on the evaluation aspects of the instrumentation and monitoring program. It is considered good practice for more than one personnel to be trained as to the responsibilities of the technician for contingency purposes.

Methods to ensure instrumentation and monitoring programs were being monitored and reviewed at the appropriate frequencies: Ensuring data is collected, validated, and evaluated in a timely fashion can be a challenge, but is vital to an instrumentation and monitoring program. Several Districts have implemented good practices to aid in ensuring these challenges are being surmounted. One District implemented a master spreadsheet which documented all their dams' data collection frequencies, when data are collected, any anomalies identified, and the resultant action. Another District implemented the practice of including instrumentation and monitoring program efforts on the lead engineer's and engineering technician's personnel performance evaluation. A third District implemented a database which documented the date of data collection, any anomalies identified by the engineering technician, as well as a "comments field" for the lead engineer to provide feedback as to any action performed. Another District implemented a routine in which the Dam Safety Officer (DSO) signed the annual instrumentation reports to signify that they have been reviewed. The DSO is the senior engineer within the District responsible to coordinate dam safety issues and activities such as emergency action plans, dam safety training, and management of project documentation.

Use of databases to store and process data: Historical standard practice was to use spreadsheet software to store project instrumentation data. However, as the number of instrumented projects, as well as the number of instruments and amount of data collected increases, spreadsheets become inadequate. Good standard practices for projects with multiple instruments and a relatively large amount of data is the use of a database to store and process data. Most Districts reviewed have implemented (or were in the process of implementing) a data management tool to store, process, and plot data such as WinIDP by URS or Microsoft Access. These types of database tools permit storage of large data sets as well as the ability to relate data, perform queries and calculations, and easily plot information. The use of database tools permits flexibility in data storage, user management (limited editing ability and limited user access), multiple simultaneous users, and overall reduction in human error and increase in plotting efficiency.

Use of a Geospatial Information System (GIS): Several projects, typically those with higher hazard classifications, also implemented the use of GIS for identifying locations of their instrumentation along with other pertinent project data such as borehole locations, specific project features, and visual distress locations. This is considered good practice for any project as it aids in visualizing locations of instruments, distress locations, and subsurface stratigraphy as related to the structure and each other to better understand dam performance. Additional attribute data, such as details of the instrumentation, most recent readings, and hyperlinks to associated reports are easily added to the system to further aid in evaluation. Bentley's gINT software is another tool that works relatively well with GIS applications and can be used to geospatially locate subsurface data and aid in the overall presentation and understanding of the project details and instrument responses.

Good communication practices between engineers and operations personnel: A couple of Districts had demonstrated good communication practices between engineers and operations personnel. This includes regular communication of data collection, data concerns requiring additional readings and visual observation of the area around the instrument, and results of any evaluations. This aids operations personnel understanding the importance of the data being collected, the project performance status, and signs of distress to remain vigilant in observation. Some projects' staff periodically visit their projects and obtained readings along with the data collectors and visually inspected distress locations. This activity facilitates firsthand knowledge of the instrument location, function, aids in identifying potential reasons for anomalies, and aids in understanding overall project performance over time.

Incorporation of informative annotations on plots: Several projects incorporated annotations on plots to facilitate a better understanding of instrument response. For example, an indication of "bottom of hole" on piezometer plots is used to aid in identifying when a piezometer is reading dry. "Automation" or "calibration" indicates when an instrument is modified from a manual to an automated device or serviced to better identify and explain anomalous data.

INTERNAL PEER REVIEW SUMMARY OF FINDINGS

There are several key aspects of the Instrumentation and Monitoring Program that were found to be deficient at several Districts. These aspects are considered important in instrumentation and monitoring programs and for understanding project performance and risks. In some cases, the deficiencies are specifically in violation of requirements within the USACE Safety of Dams - Policy and Procedures Engineering Regulation (ER 1110-2-1156). In addition, the internal peer review validated the findings of the 2013 IEPR comment that "Failure to adequately collect, interpret, and report findings in a timely manner is an urgent shortcoming." The deficiencies include the lack of:

- Project Specific Surveillance and Monitoring Plans,
- Instrumentation Inventory and Folios,
- Instrument evaluations,
- Dam monitoring program evaluations,
- Appropriately established reading frequencies and thresholds,
- Timely data review and evaluations
- Appropriate training plans, and
- Identification and response to potentially concerning data.

Project Specific Surveillance and Monitoring Plan: Very few projects have developed a comprehensive surveillance and monitoring plan (or instrumentation and monitoring plan) as required by ER 1110-2-1156. Most have some of the pertinent information on hand, however the information is not readily accessible and portions are documented in various locations such as in Periodic Inspection (PI) reports, Emergency Action Plans (EAPs), or in historic project documentation (most of which are outdated). Surveillance and Monitoring Plans should include both visual and instrumentation monitoring requirements and be periodically updated throughout the life of the project. The rational for the project requirements, based on project performance monitoring needs and potential failure modes, should be clearly identified in the plan. ER 1110-2-1156 outlines some of the key requirements of a Monitoring Plan. The USACE Instrumentation and Monitoring of Embankment Dams and Levees Engineering Manual (EM 1110-2-1908) also includes recommendations. Some key items to be included in the plan are as follows:

a. Selected locations for instrumentation or visual monitoring as related to potential failure modes and performance monitoring,
b. Frequency requirements of instrument data collection and visual inspection,
c. Data processing, data evaluation, and reporting processes,
d. Thresholds and action levels (pool events, instrument readings, etc.),
e. QA procedures and methods to document erroneous/anomalous data,
f. Document locations (network, Projectwise, Instrumentation Database, etc.),
g. Roles & responsibilities (including defining roles between the Engineering and Operations staff),
h. Yearly budget considerations for program and instrument maintenance, and
i. Visual monitoring checklist.

The last listed item, visual monitoring checklist, is an item found in reviewed project documentation for many projects. However, the visual monitoring checklists, in almost all cases, are not project specific and are not utilized as part of the monitoring program. In these cases, it was recommended that the checklists be updated to include project specific features and distress locations and be incorporated as part of the surveillance and monitoring efforts. A record of locations of distress with both qualitative and quantitative data from previous visual observations may aid observers in quickly identifying if distress locations are worsening or stabilized. Checklists should be reviewed and updated as distress locations are identified.

<u>Instrumentation Inventory and Folios:</u> Most projects have a partial inventory of instrumentation with much of the basic information such as instrument surface location and type (i.e. piezometer, inclinometer, weir, etc.). However, there is a lack of comprehensive inventories that include other pertinent information such as coordinates, datum, instrument status (i.e. active, abandoned, stand-by), reason for abandonment, maintenance history, reading frequency, etc. In addition, most projects did not have an Instrumentation Folio with scaled plan and profile views indicating the following:

a. Dam and all appurtenant features and structures with appropriate construction details
b. Location of instruments, sensor, and ADAS (as applicable in x,y,z)
c. Locations of distress locations
d. In situ materials and stratigraphy/geology (profile view)
e. Appropriate labels and legends
f. Coordinate system and stationing

This information is critical to successful execution of instrument monitoring. The harder the information is to obtain, the less effective the monitoring program will be.

<u>Instrument Evaluations</u>: In general, project engineers and/or technicians provide instrument status updates and data validation. The focus of instrumentation reviews is primarily on identifying which instruments are functional and which data are anomalous. At several dams, instrument evaluation did not consider 1) the instrumentation response with respect to geologic and stratigraphic context, 2) potential failure mode or performance parameters being monitored, 3) the effects of environmental influences, 4)

both short term and long term trends, or 5) indicate whether the instrument responses are indicative of expected project performance or an indication of potential distress.

Recommendations to provide comprehensive instrumentation evaluations in consideration of potential failure modes were made to most Districts. Plotting instrument responses on cross-sections to visually identify instruments that are monitoring embankment, foundation, or rock conditions, and their associated responses, was also recommended. Further, grouping instruments with respect to potential failure modes was recommended to aid recognition of developing concerns.

Environmental influences, such as pool and tailwater fluctuations, precipitation events, and construction activities, are not routinely plotted nor evaluated. Consideration of these influences on instrument response is key in understanding the subsurface and surface behavior of a project.

In many cases, data analyses did not consider both short term and long term behavior. Short term plots, during a high water event, for example, may indicate that an instrument has a particular response during specific events. Without review and evaluation of long-term trends, it is difficult, if not impossible, to identify responses that are changing over time. Evaluation of long-term trends may aid in identifying the effectiveness of an instrument and may indicate that maintenance is required.

Program Evaluations: Most project evaluation summaries reviewed did not include a conclusion on whether or not the project is performing as expected, nor if there are any concerning trends. They also did not indicate whether the installed instrumentation is appropriate, if additional instruments are needed, or if instruments should be abandoned, repaired, or replaced. In many cases, instruments are being monitored, but an understanding of the purpose of the instrument is unknown. In several cases, the usefulness of the instrument was outlived and yet monitoring continues.

In a few instances, recommendations for instrumentation and monitoring program modifications had been made, however long periods of time passed without action or response to the recommendations. In these cases, it was recommended that items be re-reviewed to ensure that they are being pursued with the appropriate level of urgency and a timeline should be developed for actions. If no action is taken, the reasoning should be documented. Once these items are addressed, it was recommended that actions taken be documented and included in annual and interim reports as appropriate.

Appropriately established readings frequencies and thresholds: Instrument monitoring frequencies should be established based on the frequency required to understand any developing behavior that is intended to be captured, the expected rate of response, and the project operation and project hazard classification (DSAC). Many Districts reviewed have set an identical monitoring schedule for all projects without consideration of these factors. EM 1110-2-1908 offers recommendations for establishing appropriate monitoring frequencies for various common instruments. All instrument monitoring frequencies should be reviewed as part of the Annual Instrumentation Report findings to ensure that the frequency of data collection is appropriate for each project and instrument.

Thresholds should be established in consideration of the accuracy of the instrument, historical response, and design criteria. Most projects have not established thresholds for each instrument. Some projects have established a pool elevation which triggers an increased reading frequency. However, the instrument reading or rate of change in a set of readings for each individual instrument that would trigger a response is not established. Thresholds should be reviewed on an annual basis to ensure appropriateness. Multiple levels of thresholds are also encouraged. For example, establishing one threshold value (or rate of change) that results in increase in monitoring frequency and a second value that results in an emergency action.

Timely data review and evaluations: Most District programs have standard operating procedures for data review. However, many of these Districts did not have a method to verify that data was reviewed and validated in a timely fashion. In a few cases, one individual was responsible for validating data of a large number of projects. It was difficult to determine if the data validation and collection was being performed in a timely fashion as there was no tracking mechanism. For these Districts, it was recommended that a data tracking mechanism be employed indicating 1) the dates of required data collection, data validation, and data evaluation, 2) signature and name of individual having completed each task, 3) time stamp of task completion.

Another major deficiency identified in many of the Districts reviewed was the lack of completion and submission of Annual Dam Safety Instrumentation Program Review and Project Performance Review reports. Annual instrumentation reports are required to be completed and furnished to the MSC per USACE regulations. More frequent evaluation and reporting should be implemented as appropriate. It was recommended to these Districts that the reporting process, responsibilities, and time requirements be documented and tracked to ensure evaluations are completed in a timely fashion.

Appropriate Training Plans: During the review process, it was uncovered that most Districts did not have a training plan in place for personnel involved with the instrumentation and monitoring program. In some cases, project lead engineers have insufficient experience in instrumentation data collection and have minimal experience with the project. Many engineers did not visit the site on a minimum of a yearly basis and did not have first-hand knowledge of the current project conditions and general locations of the instruments.

Most Districts have a Dam Safety Training event that is held on a 5-year frequency for each project. However this training does not typically include project specific information on the instrumentation at the site or the monitoring requirements. It was recommended that this training be updated to include project specific information such as instrument installation information, purpose of the instruments, possible sources of reading error, appropriate data collection procedure, data collection and reporting frequency, and threshold and responses.

It was also recommended that the project engineer perform yearly site inspections and participate in instrumentation readings. It was also recommended that a formal training plan be developed and documented for new and existing engineers and technicians.

Identification and response to potentially concerning data: Several projects reviewed had data that indicated a potential dam safety concern. In some cases, the data had gone unnoticed prior to the review. In other cases, the potentially concerning data was identified in the documentation; however no course of action was identified. There was not enough information or time available during the review to fully evaluate the data. The review team noted the potentially concerning data and included recommendations for actions to better understand the dam performance.

CONCLUSIONS AND PATH FORWARD

The instrumentation and monitoring programs at thirty-eight dams were peer reviewed. The review was conducted in response to an external review that found the USACE Dam Safety Instrumentation Programs lacked "timely review of data by experienced engineers" and "failure to collect, interpret, and report findings in a timely manner is an urgent shortcoming." This peer review sought to further understand the underlying issues to aid Districts in improving monitoring program effectiveness with a particular focus on adequate data collection, interpretation, and reporting.

The internal peer review validated the statements by the external review team based on the results of those Districts reviewed. However, it is noted that the Districts that were reviewed were those that had the lowest scores in DSPMT for instrumentation. It is uncertain if the findings for these Districts will be common throughout the USACE.

Recommendations were made to the Districts on areas of improvement. Recommendations were assigned priority codes to aid Districts in prioritizing improvements. Labor, funding, and IT security constraints have hampered District use of instrumentation and monitoring their dams and pose a challenge to implementing improvements. However, those challenges must be managed to assure good stewardship of the dams in the USACE portfolio.

Peer reviews are scheduled to continue in 2016. A different set of Districts have been selected for review, however this time the focus is on Districts that have demonstrated higher scores in DSPMT for instrumentation. The main goal is to identify if the deficiencies identified in the first 6 lower scoring Districts are the same in the higher scoring Districts. In addition, it is expected that the higher scoring Districts likely have good practices in place that should be recognized and implemented more broadly across USACE.

The internal peer review of District instrumentation and monitoring programs is not considered to be a long-term continuous activity and might not be conducted at all Districts. There currently is not a plan to repeat reviews at completed Districts. However, Districts that have been reviewed will be contacted one year post-review to track progress on recommended actions.

The peer review is considered an effort to "jump start" and supply an initial example to Districts as to how to review their projects. EM 1110-2-1908 discusses long-term evaluation of instrumentation and monitoring programs and this review is considered part of that effort. It is intended that Districts continue this type of review process and potentially employ the internal peer review checklist as a guide for their own quality assurance. Having an instrumentation lead in each District that ensures that the instrumentation program meets certain requirements and reviews instrumentation reports for each project is recommended. It is also recommended that each district perform this type of review as part of the PA process as part of the instrumentation program evaluation. Potentially, DSPMs or project engineers from districts that have been through the internal peer review process can aid in training other Districts in their divisions.

Additionally, it is recommended that the ER 1110-2-1156 be updated to provide more detail of MSC annual report oversight. Proposed draft language is as follows:

> "Upon receipt of annual instrumentation reports at MSC, reports shall be reviewed for any monitoring program concerns. MSC and the District DSPM will engage others as needed to investigate those concerns and respond with level of urgency warranted by available data and knowledge of potential failure modes."

Future instrumentation and monitoring PROSPECT training courses will incorporate the evaluation peer review process and findings reinforced in the course objectives. The internal peer review checklist (Appendix B) has been made available to the dam safety community. An example template for a surveillance and monitoring plan as well as example annual instrumentation reports have also been made available.

ACKNOWLEDGEMENTS

The authors would like to thank the additional USACE internal peer reviewers, Andrejs Dimbirs, Randy Welch, David Blackmore, and Philip Smith, for their time and effort in performing the peer reviews.

REFERENCES

USACE, ER 1110-2-1156, Safety of Dams - Policy and Procedures, 2014

USACE, EM 1110-2-1908, Instrumentation and Monitoring of Embankment Dams and Levees, 2016 (pending publication)

APPENDIX A: DISTRICTS AND PROJECTS REVIEWED

Division	District	Projects Reviewed
South Atlantic	Jacksonville	Herbert Hoover Dike
		Portuguese Dam
		Structure 80
North Atlantic	New England	Ball Mountain Dam
		Union Village Dam
		Mansfield Hollow Dam
		North Hartland Dam
		Littleville Dam
		Colebrook River Dam
		Knightville Dam
Southwest	Tulsa	Pine Creek Lake
		Keystone Lake
		Canton Lake
		Denison Dam
		Wister Lake
		Broken Bow Lake
		RS Kerr
South Pacific	Albuquerque	Abiquiu Dam
		Santa Rosa Dam
		Trinidad
		Two Rivers Dam (Diamond A & Rocky)
		John Martin Dam & Reservoir
		Cochiti Lake
		Conchas Dam
Lakes and Rivers	Louisville	Patoka Lake Dam
		Rough River Lake Dam
		Cave Run Lake Dam
		Brookville Lake Dam
		Taylorsville Lake Dam
		Lock and Dam 52
		Ceasar Creek Dam
Northwest	Walla Walla	Lower Granite Dam Levee System (Lewiston Levees)
		Mill Creek Diversion Dam
		Lower Monumental Lock and Dam
		Mill Creek Storage Dam
		Lucky Peak Lake
		Dworshak Dam and Reservoir
		McNary Levee System

APPENDIX B: 2015 INSTRUMENTATION AND MONITORING PROGRAM REVIEW CHECKLIST

A. Instrumentation and Monitoring Program Management

- Does the District have an appropriate project level Surveillance and Monitoring Plan (SMP) (a programmatic document that governs the performance monitoring, data management and the data quality assurance procedures).

- Is the relationship between the monitored parameters and PFMs made clear?

- Are appropriate locations for instrument installation (x,y,z) identified.

- Are project details such as initial fillings, historical low and high pool levels identified?

- Are any pertinent modifications to the dam and changes in operations highlighted?

- Is appropriate instrumentation identified for monitoring performance and PFMs?

- Is frequency of monitoring for each performance parameter and PFM appropriately identified for normal operating conditions, high water events, and unusual events?

- Is there a visual observation document (checklist/form) to assist in documenting and inspecting the dam?

- Are appropriate thresholds identified for each instrument?

- Are appropriate actions to be taken once a threshold is met or exceeded identified (e.g. verification of readings, increased readings frequency, upward reporting)?

- Are appropriate roles and responsibilities of personnel identified for:
  - Data collection
  - Instrument system maintenance
  - Data entry
  - Data Processing
  - Data Interpretation and Analysis
  - Date Reporting
  - Instrumentation Program Management

- Are the training requirements for the personnel identified?

- Are policies and procedures in place to assure Quality Control of Data Collection, Storage, and Access (Data Management Procedures)

- Is the SMP evaluated and modified in relation to the potential failure modes, as part of the PI, PA, IES, IRRMP, and DSMS processes.

B. Instrumentation Folio (Inventory of Instrumentation and ADAS)

- Is each active instrument related to performance monitoring or a documented potential failure mode as per SMP?

- Is there a list of all instruments and ADAS at the site with details such as:

— Installation dates?
— Model and serial number?
— Coordinates and elevation of the instrument surface location?
— Coordinates and elevation of sensing point?
— Identification of status (abandoned, active, broken, etc.)?
— Is the reason for any inactive status instrument identified/documented?

- Do plans and maps identify:

— Locations of instruments (x,y)
— Location of ADAS
— Appropriate labels and legends?
— Coordinate system and stationing?
— Dam and all appurtenant features and structures?
— Drill holes and Instrument ID's cross-references?

- Do cross-sections show:

— Location of instrument and sensor (x,y,z)
— Appropriate labels and legends?
— Reference construction detail drawings
— In situ materials and stratigraphy/geology?
— Ground surface elevations?

- Is there a maintenance schedule that shows:

— Maintenance and calibration requirements for each instrument?
— Is the schedule being adhered to?
— Is the maintenance status of the instrument documented?

- Are field borehole logs and /or as-built drawings from drilling used to illustrate instrument sensor location relative to geologic stratigraphy and embankment zoning?

C. **Data Acquisition and Processing/Plotting**

- Do individual instrument data collection and visual monitoring frequencies

— follow the SMP for normal operations
— follow the SMP for unusual pool levels/events

- Are visual observations appropriately documented (e.g. photos, maps, flags)?

- Are data collection/visual monitoring personnel appropriately trained?

- Is there a data quality control program in-place? (e.g. are recent readings available to the data collector, are repeat readings taken)

- Are there procedures in place for recognizing anomalous data or data that meets or exceeds thresholds during collection?
  - Are procedures regularly followed
  - Are procedures appropriate to assure that action is taken in a timely fashion?

- Is collected data processed in a timely fashion?

- Are personnel performing processing appropriately trained?

- Are the equations and associated constants necessary to reduce raw data for each instrument appropriately documented?

- Is the datum used for data processing different from the construction survey datum? If so, is it accounted for in the calculations?

- Are anomalous or erroneous data appropriately identified and documented? If masked,
  - has reason for masking been included
  - has personnel who masked the data been identified
  - has the masked data been reviewed by engineer/geologist

- Are there notations in the database concerning the quality of instrumentation data, history of associated maintenance, calibration date and method, and the skill level and experience of the person who made the instrument reading or observation?

- Are survey-data precision and accuracy along with the datum and coordinate system used identified?

- Are environmental influences appropriately included on plots including
  - Pool and tail water
  - Precipitation
  - Construction activities
  - Ambient/In-situ temperatures

- Are types of graphs used appropriate (e.g. time/history, absolute deformation over time, relative deformation)? *Refer to atta chment for summary of review recommendations for various types of monitoring.*

- Are performance thresholds included on plots or otherwise documented such that they are obvious during data reviews?

- Are pertinent features such as crest and spillway elevation included on appropriate plots?

- Is format and scale used for plotting appropriate for each instrument/instrument grouping?

- Is instrumentation appropriately grouped for each PFM or parameter monitored?

- For critical elements, are scale drawings (typically with a 1:1 aspect ratio) used to show selected results from measured performance while also depicting the dam's geometry, materials, and foundation geology.

D. Data Evaluation and Reporting

- Are evaluations performed in a timely matter?

- Are personnel performing evaluations appropriately trained?

- Are evaluations performed in relation to PFM or other performance monitoring?

- Does existing evaluation include a d etailed comprehensive narrative assessment that places all da ta in a m eaningful structural and geologic contex t that address es both the behavior of the dam and the monitoring system.

- Are discussions and displays organized around behavior topics such as leakage, deformation and piezometric pressure?

- Are drawings, tables, and graphical displays that illustrate the observations and findings included?

- Are graphical displays used for all data and are they self explanatory?

- Is unsatisfactory performance, especially behavior related to potential failure modes, highlighted?

- Are reports of the evaluation distributed to the appropriate reviewers/chain of command in a timely manner and as per the SMP?

- Have there been any thresholds met and/or exceeded?

— Are these events documented?
— Have appropriate actions taken place as per SMP?
— Has action occurred in a timely fashion?

- Does reporting result in an assessment of effectiveness and performance of individual instruments and the instrument arrays?

- Are aspects of measurements and observations that can provide an advanced warning of failure mode initiation highlighted?

- Are conclusions concerning the future usefulness of all monitoring elements based upon instrument and dam performance in view of the potential failure modes reported?

- Are the overall results of visual observations reported for timeless, adequacy of coverage and intermittent phenomenon such as wet spots?

- Are reports available on how unexpected observations have been addressed?

- Is an Annual Instrumentation Report Submitted and for background does it reference a more detailed and comprehensive report?

- Is data quality and adequacy feedback regularly provided to data collectors and DSO.

- Are data evaluation results regularly provided to data collectors and inspectors?

- Is the meaning of the data evaluation results clearly communicated to data collectors and
Inspectors following evaluation?

E. Summary of Review and Reporting Elements for Various Types of Monitoring

E1. Surface Water Quantity and Quality

- Are all flow quantities that are, or potentially are, a function of reservoir surface elevation included in the data?
— Are other influencing factors, such as tail water elevation, precipitation, snow melt and groundwater measured?
— Are increasing flow responses highlighted?
— Are drainage features that appear to lose capacity or convey less expected flow over time highlighted?

- Are there any measurements or evidence of material transport by seepage flow?

- If erosion of water soluble salts is a concern,
— does collected flow water quality differs from reservoir quality?
— is reservoir water quality aggressive toward any foundation materials?

E2. Deformation and Displacement

- Are indications clear on whether deformation is over time, decreasing, steady, accelerating or unstable?

- Are cyclic deformations predictable as functions of season or reservoir operation?

- Are differential deformations that could lead to cracking highlighted?

- Are the locations for expected maximum and minimum deformations or displacements identified for individual features such as the dam, dike, spillway wall, etc?

E3. Uplift, Piezometric Pressure, and Ground Water Level

- Are pressures or water levels that are greater than expected highlighted?

- Can the distribution of pressure be delineated enough to know if it conforms to the expected pattern?

- Are trends of increasing or decreasing pressure responses to reservoir surface level highlighted?

- Is the effectiveness, and any change in effectiveness that has occurred over time, of seepage cutoff and drainage features indicated?

- Are seepage exit gradients in erodible materials where pressures are measured near seepage exits estimated?

- Is uplift on planes of suspected weakness such as lift lines, joints, shale seams and foundation contact estimated?

E4. Seismic Monitoring

- Has it been determined that the dam is in a seismic zone and if so, what level?

- Are there records of measured ground motion accelerations or displacements?

E5. Non-routine Performance Investigations

- If parameters were measured using non-routine monitoring,

— Are the method of measuring, the accuracy standards used, and reason for method documented?
— Has the method implemented been documented?
— Is the interpretation of the data and evaluation appropriate?

THE HYDROGEOLOGICAL MODEL OF BOONE DAM

Scottie Barrentine, P.E.[1]
Andrew Dodson, P.E.[2]
Lindsay McMichen, P.G.[3]

ABSTRACT

In response to the development of an apparent seepage and piping issue at Boone Dam, over 46 boreholes were advanced and over 100 new vibrating wire sensors were installed to monitor performance of the dam, the foundation, and the surrounding geology. The exploration and instrumentation aided in confirming the existence of a complex seepage, piping, and internal erosion issue. This investigation provided a wealth of information which was used to develop a hydrogeological model of the site conditions including the various strata, their characteristics and properties, and how they relate to the performance of the structure. This model serves as an indispensable tool in the development of a comprehensive remedial program for the dam. This paper will provide an overview of the development of the hydrogeological model for the Boone Dam site with an emphasis on the use of instrumentation and monitoring data to evaluate the performance of the structure.

INTRODUCTION

Boone Dam is located on the South Fork Holston River near Johnson City, Tennessee. The dam was built by the Tennessee Valley Authority (TVA) from 1950 to 1952. The western half of the dam is a concrete structure, with spillway gates and a three-unit powerhouse. An earthen embankment forms the eastern half of the dam. The project was constructed to provide flood control, hydroelectric power generation, water supply, water quality, recreation and aquatic ecology. The project retains the Boone Reservoir, which under normal operations has an annual pool variation of approximately 27 feet, which provides for seasonal flood storage. The total drainage area for the reservoir is 1,840 square miles, including 666 square miles of uncontrolled drainage area located below the Wilbur and South Holston projects.

Boone Dam is founded entirely on limestone formations. Prior to construction, numerous karst-related features, including springs, depressions, and sinkholes were documented in vicinity of the site. Original geologic investigations further confirmed the presence of an extensive karst network in the dam foundation and reservoir rim.

As originally constructed, the embankment section was approximately 750 feet long at crest elevation (El.) 1400, and a crest width of 20 feet. The original configuration included a homogeneous rolled earth fill embankment with an upstream slope of 3 feet horizontal to every 1 foot vertical (3H:1V) and downstream slopes of 2H:1V to a berm at

[1] Tennessee Valley Authority (TVA), 1101 Market Street, Chattanooga, TN 37402, 423-751-6781, slbarrentine@tva.gov
[2] TVA, 1101 Market Street, Chattanooga, TN 37402, 423-751-7723, abdodson@tva.gov
[3] TVA, 1101 Market Street, Chattanooga, TN 37402, 423-751-8078, lcooper@tva.gov

El. 1365 and 2.5H:1V below the berm. The embankment is in line with the concrete dam for a distance of 60 feet, then curves downstream to its terminus in the right abutment.

Modifications to the embankment to allow the dam to safely pass the Probable Maximum Flood (PMF) were completed in 1984. The modification included raising the embankment crest with an impervious core and rockfill slopes 8.5 feet to El. 1408.5, with a modified crest width of 16 feet. Figure 1 shows a typical embankment section with the PMF modifications.

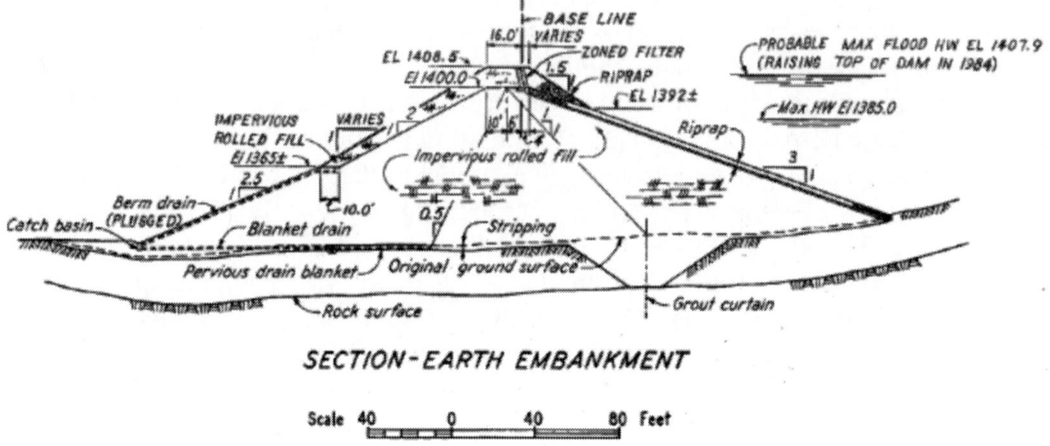

Figure 1. Typical Embankment Section.

The embankment is equipped with a pervious drain blanket which consists of fine filter material. The blanket covers approximately the middle third of the downstream half of the foundation and is completely surrounded by engineered fill. The blanket drains into concrete pipes which convey flow to catch basins at the control building plaza and at the powerhouse access road.

The embankment is also equipped with a cutoff trench to rock under the upstream portion. The overburden was excavated to the top of rock, which, due to the up-ended bedding of the underlying limestone, was a series of dramatic rock pinnacles, cavities, and overhangs. This rock was cleaned off as thoroughly as possible by hand and clay fill was tamped into and around the varied rock surfaces that comprise the bottom of the cutoff trench excavation. No concrete was used. Judging by the construction photos, this process was completed in a series of phases, by which a portion of the trench would be excavated, cleaned and backfilled before moving on to an adjacent section of the trench. The sides of the trench were designed with a slope of 1.5H:1V, but it is apparent from construction photos and as-built drawings slopes much steeper were excavated and left open for a period of months. An as-built section showing the deviation of the actual construction of the cutoff trench from the design is shown in Figure 2. A grout curtain which extends approximately 75 feet below the rock surface was installed in the center of the cutoff trench. A construction photograph depicting conditions encountered during preparation of the cutoff trench is provided as Figure 3.

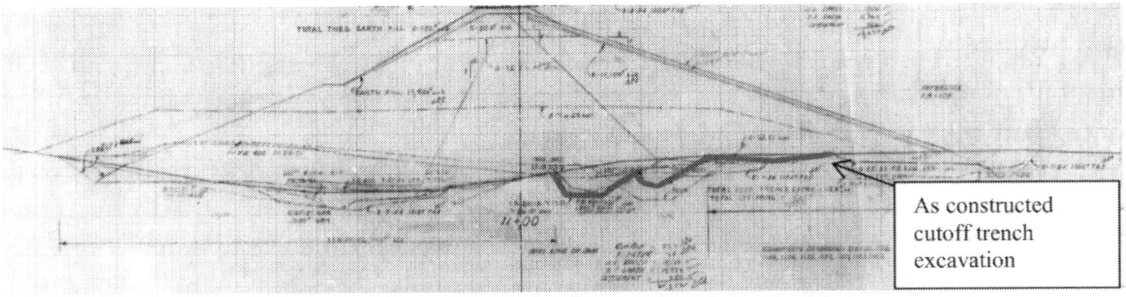

Figure 2. As-built Section at Station 11+00.

Figure 3. Construction Photograph of Cutoff Trench.

Signs of seepage became evident on the embankment almost immediately following initial reservoir impoundment. Numerous wet areas and depressions have been noted throughout the history of the project. On October 20, 2014 a large sinkhole appeared near the downstream toe of the earthen section, followed by turbid seepage in the tailrace area six days later. These events triggered emergency response actions by TVA, which included immediate lowering of the headwater level, 24-hour surveillance and monitoring of the project, construction of temporary filter berm, a thorough review of historical records, and implementation of an extensive subsurface exploration and instrumentation program.

ELEMENTS OF THE GEOLOGICAL MODEL

Geologic Setting

Boone Dam is located in the Valley and Ridge physiographic province of the Southern Appalachians. This physiographic province is characterized by long, alternating, linear-trending valleys and ridges. Although faults and folds are very common throughout the region, topography is primarily controlled by the relative resistance of the Paleozoic age sedimentary rocks that form the region's bedrock. More siliceous rocks such as sandstone and chert tend to hold up ridges while rocks more prone to mechanical (shale) or chemical weathering (limestone) tend to form the valley floors. The Valley and Ridge province of east Tennessee is represented by rocks of the entire Paleozoic era, except Permian age. Geologic formations at the Boone Dam range from Middle Cambrian to Middle Ordovician age representing portions of the Conasauga and Knox Groups.

The majority of the foundation for the concrete structures, the entire embankment foundation, and the right reservoir rim are comprised of the Conococheague formation. The Conococheague is the basal member of the Knox Group and consists predominantly of a light to medium gray, fine grained mottled and banded limestone with a few thin beds of chert and dolomite.

The dam is located on the northwest limb of a broad syncline that trends northeast-southwest. Faulting is associated with folding and many of the major faults are thrusts running parallel to fold axis. The largest faults are found to the northwest and southeast of the site, however, minor fault splays and associated folding are found underneath the site especially along and within a few hundred feet of the right rim grout curtain.

Generally, the strike of bedding at the site is N 25°E to N 30°E with moderate dips of about 45° to the southeast. The rock under the dam site and vicinity is extensively jointed. Most jointing developed during folding and faulting, and is therefore related to the direction of folding and tectonic stresses. Three types of joints have been identified at the site: strike, dip and oblique. The complex geologic structure (beddings, joints, folds, and faults) at the site has resulted in an interconnected network of discontinuities that provided preferential pathways for karst development within the limestone foundation. Additionally, karst development along the moderately inclined bedding planes has resulted in a highly irregular top of rock surface consisting of pinnacles of rock separated by 20- to 30-foot deep crevices.

Description of Geologic Elements

For the purpose of discussion in the context of the hydrogeological model, the four geological elements at Boone Dam are defined in ascending order below:

- **Competent Bedrock** is the deeper limestone rock which is less permeable, containing fewer cavities and fissures.
- **Epikarst** is the upper highly weathered and pinnacled portion of the limestone bedrock together with the soft foundation soils present between rock pinnacles.
- **Foundation Soils** consist of a highly heterogeneous mixture of residual and alluvial materials. Residual materials are predominantly clay with some chert gravel while alluvial materials are comprised of gravel, pebbles, and cobbles in a clay matrix.
- **Embankment material** consists of well compacted, high to moderate plasticity clay with occasional lenses of sand and gravel.

A geologic profile of the embankment and the geological elements is provided as Figure 4.

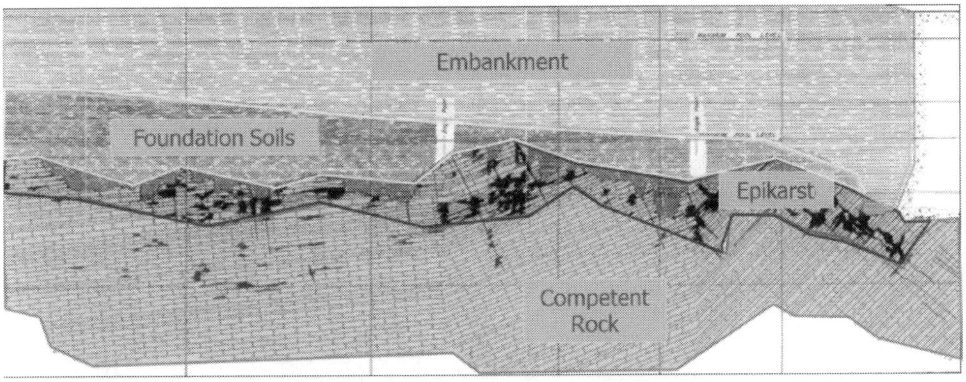

Figure 4. Geologic Profile of Embankment.

PREEXSITING DRAINAGE NETWORK

Based on the efforts of the historic data review, the exploratory drilling, and instrumentation program it is clear that an extensive karstic drainage network existed at the dam site before the construction of the dam. This is particularly evident in the preconstruction site topography, the existence of many springs and sinkholes in the area, and the consolidation grouting that was required in preparation for the construction of the embankment.

Preconstruction Site Topography

The preconstruction site topography is typical of a surface that is being shaped not only by the weathering of wind and water from the atmosphere, but also the subsurface transportation of material by water moving through a carbonate aquifer drainage system. Numerous closed depressions are evident on the preconstruction topographic contour maps, as well as current topography that has been collected as part of the exploration, analysis, and remediation efforts using LIDAR techniques. Near the parking lot in the right abutment, multiple drainage paths that would otherwise continue down-gradient are observed to suddenly disappear in the vicinity of a depression. These are examples of disappearing wet weather conveyances and streams. Figure 5 illustrates surface expressions of the extensive preexisting drainage network at the dam site.

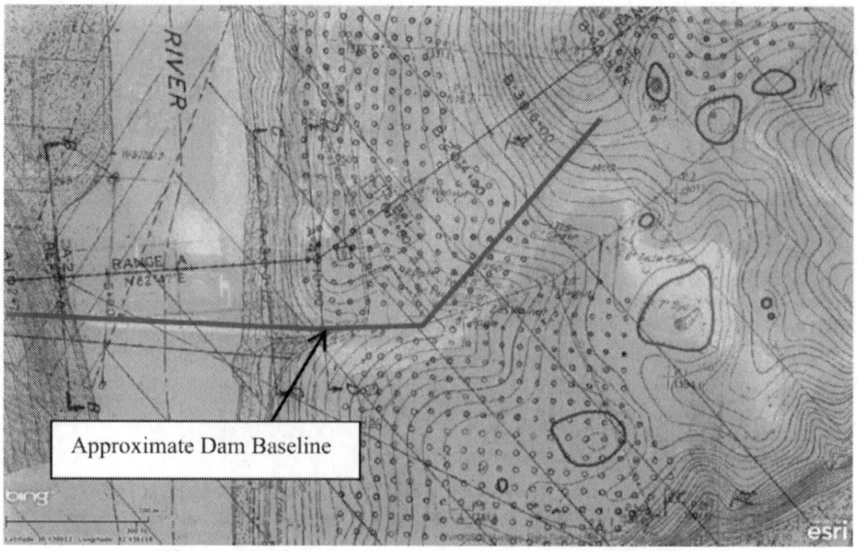

Figure 5. Preconstruction Site Topography.

Preconstruction Springs and Sinkholes

The closed depressions, or sinkholes, capture surface water runoff where it is percolated through overburden material to the karstic aquifer in areas of recharge. The water then flows through the karstic aquifer drainage system to a point of discharge where the water daylights as a spring. A striking example of such a spring is evident on the preconstruction site topographic maps at the right bank of the river just under the first few blocks of the concrete portion of the dam near the concrete-embankment interface. The preconstruction topography indicates the presence of a pronounced draw at this location. Figure 6 shows a photograph of the area before the dam was constructed. An impression of the depth and extent of this feature can be gained by noticing that only the tops of the trees can be observed at this location, and in the way the former inhabitants constructed their road around it. The severe weathering and erosion in this area suggest that this location is the discharge point for a source of water much greater than the relatively minor estimated area of 30 acres of surface area that drain to this location.

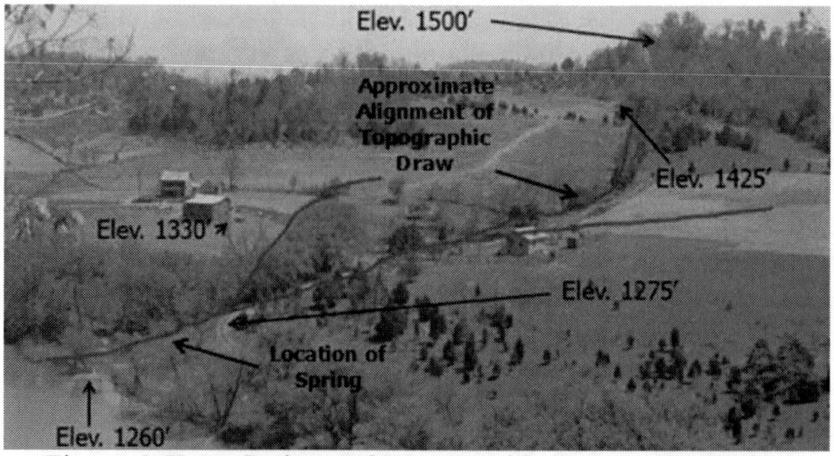

Figure 6. Karst Spring and Topographic Draw at Dam Site.

Construction Consolidation Grouting

Another indicator of the extent to which the existing karstic drainage system at the site was developed is the consolidation grouting program conducted during construction the embankment dam. During excavation of the cutoff trench, voids were encountered in the overburden near the top of rock surface. Consolidation grouting was performed to ensure adequate support of the embankment. Grout holes were drilled on 50-foot centers across the entire footprint of the embankment and extended approximately 10 feet into rock. A clay-cement grout consisting of one part cement to twelve parts clay (1:12) by volume was injected at a maximum pressure of 10 pounds per square inch (psi) to treat the voids. Figure 7 provides a contour map of the consolidation grout takes (by bags of cement). Note that in some locations grout takes reached nearly 800 bags of cement, or approximately 10,000 cubic feet of grout.

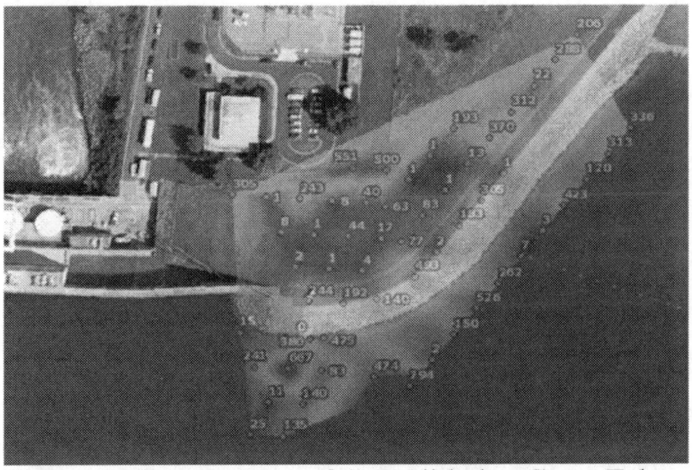

Figure 7. Contour Map of Consolidation Grout Takes.

RIGHT RIM INFLUENCE

The right rim is the area east of the dam and includes the area around the right abutment, the right reservoir rim, and the high ground to the east of the dam. A grout curtain was installed through the right rim during the original construction to prevent reservoir rim seepage through the existing geologic strata and around the dam site. Extensive weathering, folding, faulting, and carbonate dissolution exists in the right rim. Evidence of this was obtained during the drilling and grouting operations. The right rim is thus extremely porous with many open channels and cavities that are interconnected to create a free-flow carbonate aquifer. This aquifer is generally referred to in the hydrogeological model as the right rim. The extents of the right rim aquifer have not been fully defined at this point, but the influence of the right rim on the performance of the dam has been made clear through evaluation of piezometric data, review of historic documentation, recent exploration, and inspection observations. The right rim provides a constant source of head under the embankment. Recharge to the right rim causes surges in the epikarst and foundation of the dam which have continued to develop the existing karstic drainage network. Figure 8 provides a geologic profile, looking upstream, of the right rim just beyond the right embankment abutment, note the depth and extent of weathering.

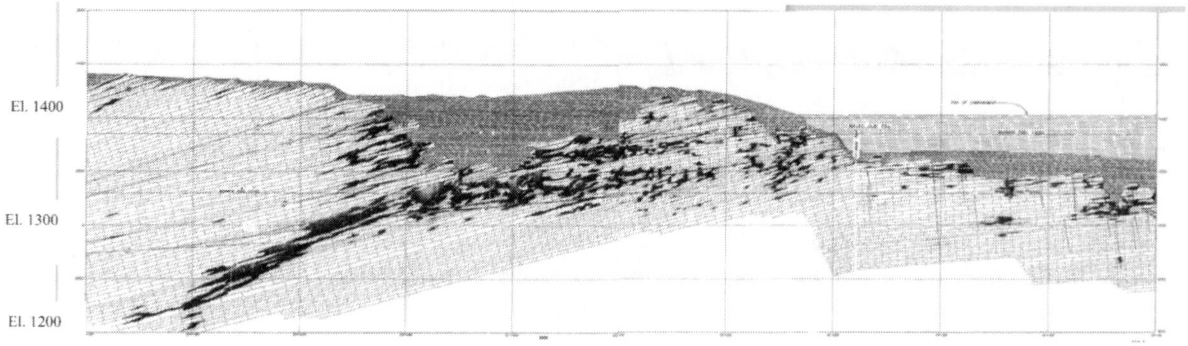

Figure 8: Geologic Profile of Right Rim.

Piezometric Data

Prior to the exploration effort initiated in response to the 2014 sinkhole, very limited piezometric data was available for the project. Two piezometers along a single section near the concrete dam interface were the sole source of piezometric data until 2013. Six additional standpipe piezometers were installed in 2013. All of the existing instrumentation was monitored manually on a monthly frequency. A comprehensive instrumentation program was implemented following the discovery of the 2014 sinkhole. Nested installations of standpipe and vibrating wire piezometers were installed to monitor piezometric pressures in the embankment, foundation soils, epikarst, and competent rock. All project instrumentation was automated and integrated into an Automated Data Acquisition System (ADAS) to collect readings at 15 minute intervals. The automated, near real-time piezometric data has been an invaluable tool for evaluating the performance of the dam and development of the hydrogeological model.

In regards to the role of the right rim in the model, the piezometric data has demonstrated heads significantly higher than headwater, responsiveness to high infiltration events, and connection of the right rim aquifer to the foundation of the dam.

A number of the piezometers in the right rim exhibit pore pressures that are significantly above current headwater elevation. In general, piezometric heads observed in the right rim range from El. 1370 to 1380, or a maximum of almost 30 feet above the current average reservoir elevation of 1352.

Following significant infiltration events, notable increases were observed in piezometers located in the right rim and at the downstream toe of the dam. Responses in the right rim piezometers are generally rapid and sharp. Responses in the toe piezometers exhibit a muted, more delayed response, although still of significant magnitude. Figure 9 illustrates a seven-foot rise of the piezometric surface at the toe of the dam following several days of rainfall. It should be noted that during dry periods, antecedent rainfall events are required to provide adequate infiltration to initiate the piezometric response.

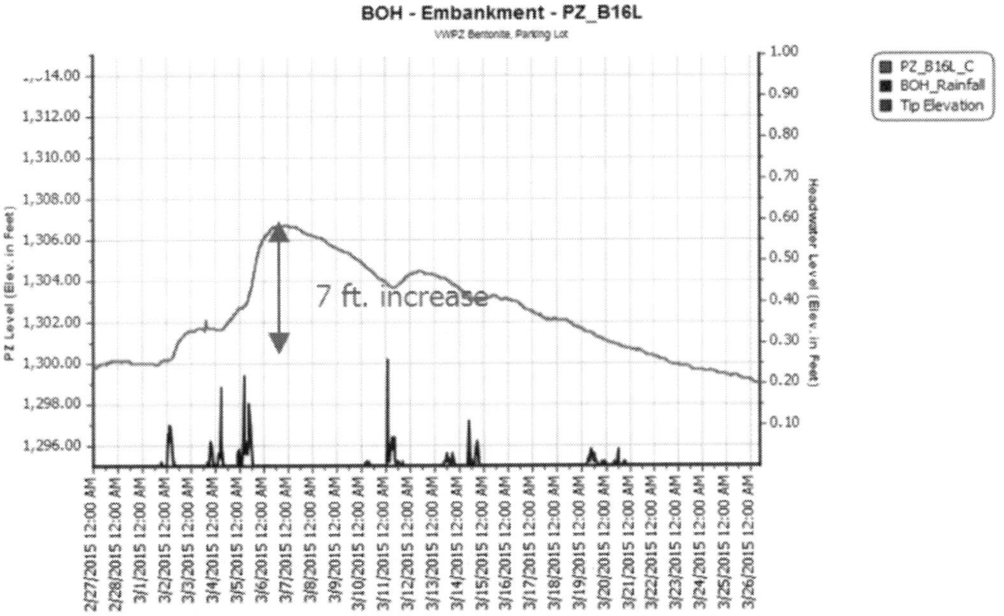

Figure 9. Piezometric Response of Toe Piezometer to Significant Infiltration Event

Additional evidence of the right rim contribution to the seepage flows at Boone Dam is illustrated by a notable offset of piezometer data when plotted against headwater elevations. As observed in Figure 10, piezometer B-23, located under the crest of the dam, fluctuates at lower elevations during dry periods and higher elevations following high infiltration events.

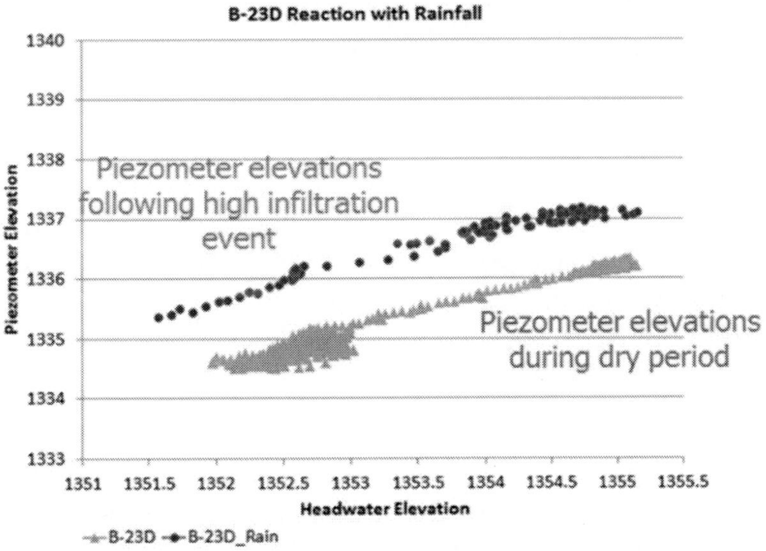

Figure 10. Piezometric Response of Crest Piezometer to High Infiltration Event

Spring Blanket Drain

Detailed review of historical documentation revealed the presence of a drain blanket beneath the left downstream portion of the embankment, in the location of the preconstruction spring. The drain was constructed to deal with excessive spring flows encountered during construction. Figure 11 provides a plan view of the spring blanket drain. Piezometers located in the vicinity of the spring blanket drain generally follow tailwater, but also exhibit response to high infiltration events.

Figure 11. Plan View of the Spring Blanket Drain.

Confined Aquifer Behavior

During the drilling operations from the crest at boreholes B-28, B-30, and mid-slope at boring B-32, an unexpected reaction of the water levels in the boreholes was noted. The water in the drill string rose approximately 20 feet in the drill casing in a matter of minutes after penetrating the epikarst layer. Piezometric response in the epikarst was confirmed after the evaluation of the surrounding piezometers.

Seeps on Upstream Toe

On March 5, 2015, dam safety inspections discovered multiple muddy seeps along the upstream toe of the dam. As depicted in Figure 12, the location of these seeps is significantly below the normal pool elevation. Close monitoring of these seeps since discovery has provided evidence that suggests the seeps only flow following high infiltration events. As the seeps are above current reservoir elevations, they are fed by the higher heads from the right rim.

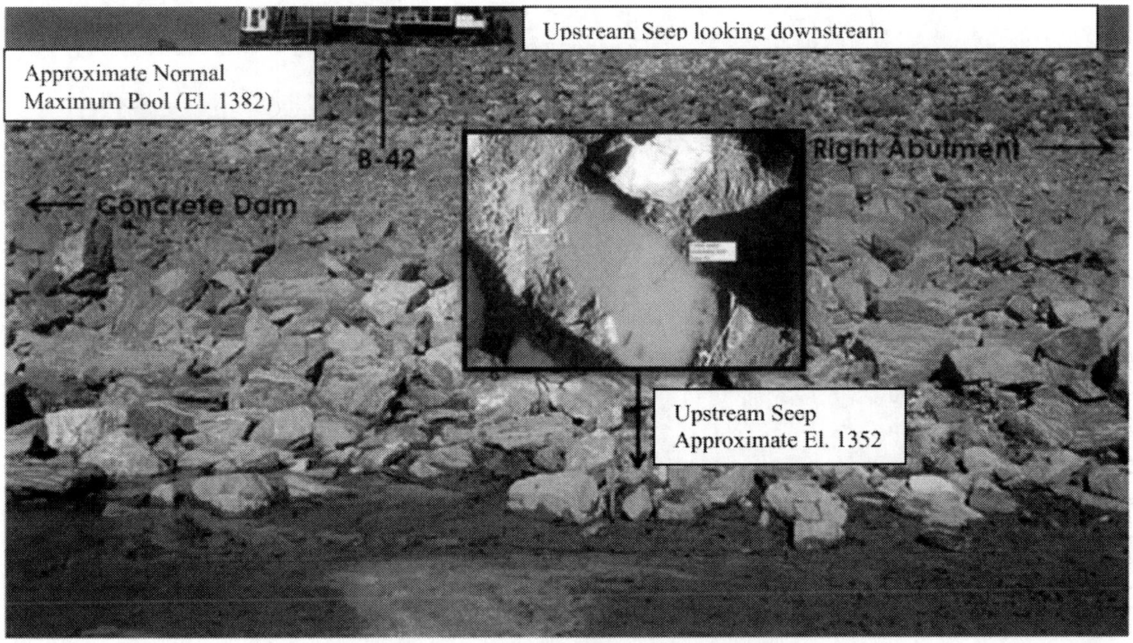

Figure 12. View of Upstream Seep.

RESERVOIR INFLUENCE

In addition to the contribution of right rim, the reservoir also plays an important role in the hydrogeological model of Boone Dam. Over the life of the project, increased gradients from the reservoir have exploited weaknesses in the cutoff trench and tied headwater to the drainage network. Piezometric data, dye tests, and current subsurface exploration, all provide supporting evidence of the reservoir contribution.

Piezometric Data

Once the resolution of the piezometric data was enhanced by the installation of new instrumentation and increased reading frequency facilitated by instrumentation automation, close correlation to headwater became apparent in many of the piezometers. Specifically, a number of piezometers located downstream of the grout curtain, on the crest and downstream embankment, respond to minor fluctuations of headwater with very short lag times. The largest and quickest responses were observed in piezometers which monitor pore pressure at the top of rock under the dam. This behavior suggests deterioration has occurred within the cut off trench resulting in increased permeability.

Another benefit provided by the automated instrumentation is the collection of temperature data. Groundwater temperatures measured in the right rim and the eastern portion of the embankment are consistent with expected ground temperatures for the region. Water temperatures measured in the reservoir, especially during winter months, are notably cooler. As shown in Figure 13, temperature data plotted as a contour map appears to indicate the intermixing of the right rim and reservoir seepage flows and highlight a potential zone of increased reservoir seepage near the curved portion of the embankment.

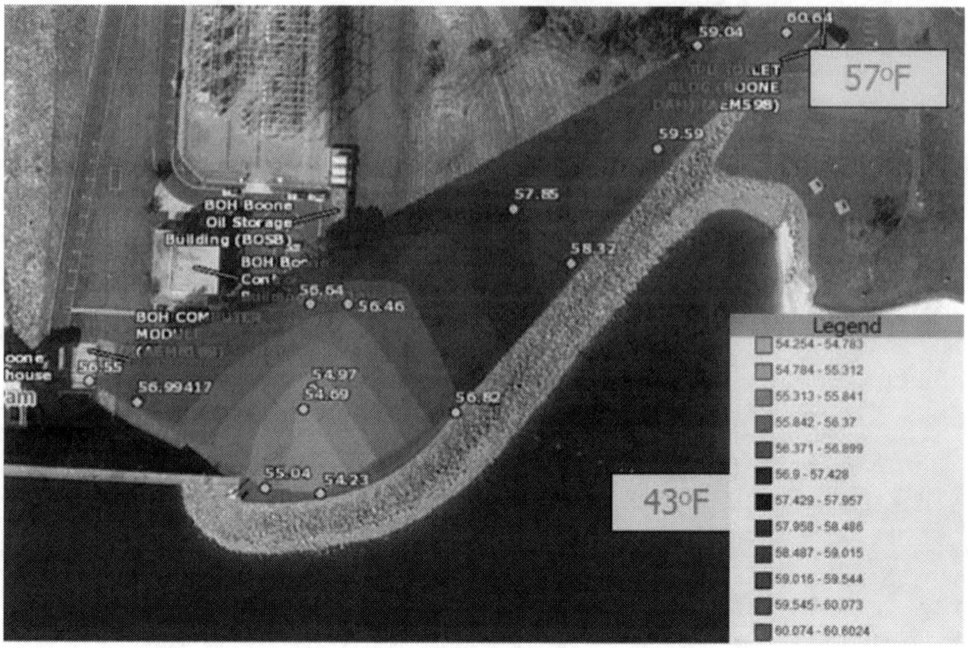

Figure 13. Piezometric Temperature Contour Map.

Dye Tests

Multiple dye trace tests have been conducted to evaluate the seepage flows at the site. Rhodamine, Fluorescein, and PTSA dyes were injected at select locations and monitoring was performed using carbon packs, fluorometers, and visual observation. One such test was performed from upstream of the cutoff trench and grout curtain. In boring B-44, located about 150 feet upstream of the dam on the right reservoir rim, approximately one-half gallon of fluorescein dye was injected and flushed with 38,000 gallons of water. Within 48 hours, dye concentrations were detected in two fluorometers downstream of the cutoff trench in borings located on the crest of the dam, B-42 and B-28. Dye was also detected after about 70 hours in the reservoir. The rapid movement of dye across the cutoff trench provides additional evidence of reservoir seepage through the embankment's compromised cutoff.

Current Subsurface Exploration

Further evidence of the deterioration of the cutoff was discovered inadvertently during installation of a piezometer on the crest of the dam. While grouting to backfill the annular space for the instrument, grout was observed in the upstream reservoir. Although several potential paths for the grout migration were considered, such as through a defect in the embankment material or though the upper zone of riprap, the most likely path was judged to be through the foundation cutoff. Figure 14 provides a schematic of the potential grout paths from the boring to the reservoir. Field observations suggest that the level of grout in the borehole never rose far enough to travel the alternate pathways. To verify this connection to the reservoir, a dye test was performed in an adjacent borehole. Dye was injected at the crest of the dam and was detected in the reservoir fluorometer in less than 24 hours.

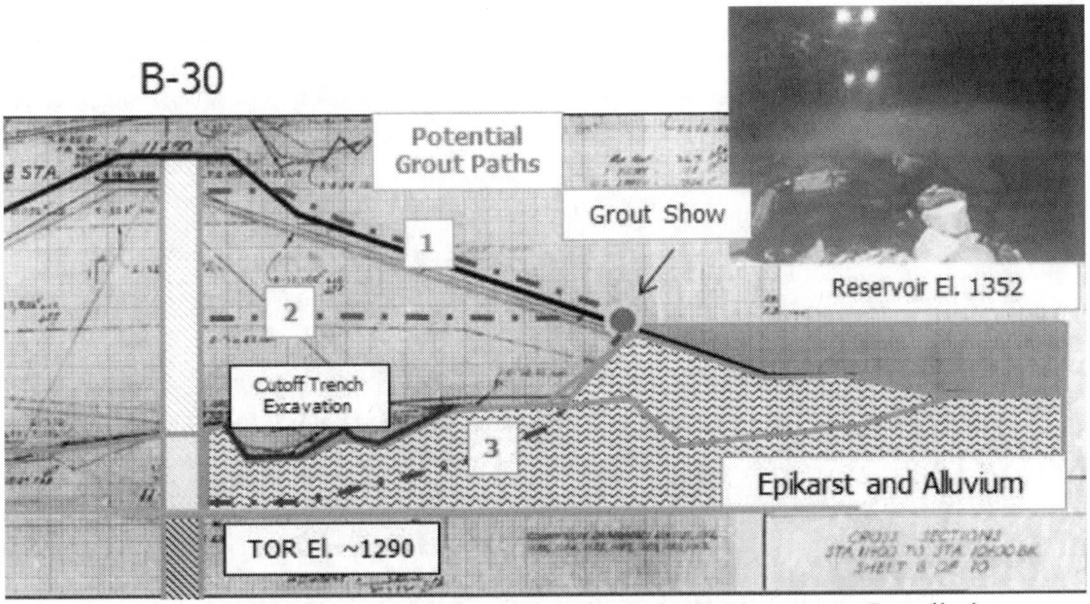

Figure 14. Potential Grout Paths to Reservoir from Piezometer Installation.

CONCLUSION

Seepage flows at Boone Dam are complicated not only by the challenging geologic conditions present at the site, but also by the contribution from multiple seepage sources. The seepage is controlled by the source with the greatest head, which varies depending on fluctuations of both the reservoir and the right rim aquifer. Therefore, the potential exists for large fluctuations in the system dependent on water surface elevation in both the right rim and the reservoir.

The available information indicates that an extensive drainage network in the epikarst has existed at the Boone Dam site since before the dam was constructed. An aquifer system in the area that is now the right rim, provides a constant source of head in the epikarst under the embankment. Precipitation on the right rim causes recharge of the aquifer system and significant fluctuations in the water table elevation in the epikarst under the embankment. This has continued the development of the drainage networks. The construction of the dam and subsequent filling of the reservoir increased the head in existing drainage networks. This resulted in an increase in gradients, which has exploited weaknesses in the cutoff trench and tied headwater to the drainage networks. Defects in the cutoff trench have developed and could result in higher seepage rates, concentrated flows, and accelerated soil erosion into the epikarst. Verification of the size and location of the pathways that comprise the network is difficult. Figure 15 illustrates the hydrogeological model of Boone Dam.

Figure 15. Hydrogeological Model of Boone Dam

In applying the knowledge gained from the evaluation of the conditions at TVA's Boone Dam, a list of key tools utilized to gain the knowledge and extent of conditions have been developed. These key tools were essential in the analysis of Boone Dam and development of the hydrogeological model. The extent of conditions should be utilized to identify other dams in which internal erosion could be underdevelopment.

Key Tools

The use of modern Geographic Information Systems (GIS) allows for analyzing multiple decades of inspection data and construction records. Construction records can be referenced into the database by drawing coordinates and overlaid with current instrumentation data, boring data, and inspection data. This tool is powerful for providing engineers with an efficient way to obtain and understand behavior of a dam.

The automated network of nested open standpipe and grouted vibrating wire piezometers is essential in understanding the behavior of a dam with a complex foundation system. The monitoring system at Boone Dam provides headwater, tailwater, piezometer readings, weather data, and displacement data every 15 minutes for the underlying bedrock, epikarst, foundation soils, and embankment fill. Without an advanced monitoring system, the impacts of the high infiltration precipitation events would not have been recognized. This system allowed for analysis of responses to drilling, grouting, and changes in headwater so that an understanding of potential flow paths were developed.

Where known karst features are present, dye tracing is a valuable tool for confirmation of hypothesized flow paths. Dye tracing is an essential tool in karst conditions due to the complexity and irregularity of the flow paths.

Construction records and photographs are extremely valuable and owners should never compromise these records, if available. The core trench photographs and preconstruction records where key indicators of the existing karst conditions at the site and the potential weaknesses in the construction practice.

Extent of Conditions

While advantageous for slope stability, earthen dams with cutoff trenches upstream of the axis of the dam can be subjected to high gradients if the foundation rock has a high drainage capacity. Even though constructed of well compacted clay, core trenches can be susceptible to internal erosion when proper drainage and filters are not utilized.

In areas where higher elevations are present in the abutments or in karst regions, surface runoff and groundwater seepage should be considered in modeling of seepage flows in dams.

RELEARNING HOW TO LOOK AT PIEZOMETRIC DATA FOR SEEPAGE EVALUATION

Richard S. Olsen[1] PhD PE
Isaac J. Stephens[2] PE

ABSTRACT

Examining field piezometric data from earth dams provides the direct link to understand seepage flow and potential seepage related problems. However, there are several issues that must be overcome to achieve optimum evaluation. The first issue is that field measured piezometric data are typically plotted in ways that can't illustrate seepage problems. A new method for plotting piezometer data versus reservoir level is introduced that can be used as a tool when evaluating piezometric data for situations ranging from foundation layer obstructions to formation of seepage piping.

INTRODUCTION

Pore pressure, seepage and their resulting effects are the number one issues influencing dam safety in terms of geotechnical engineering. Examining field piezometric data from earth dams provides the direct link to understand seepage flow and potential seepage related problems. Seepage problems can be high pore uplift pressures, soil piping (and pipe channel) generation, decreased strength for granular soils due to high pore pressure, etc.

There are several piezometric issues that must be overcome to achieve optimum evaluation. The first issue is that field measured piezometric data is typically plotted in ways that can't illustrate seepage problems. Another issue is that computer based seepage modeling today is too often the first (and also the only) step for evaluating observed piezometric data or to predict seepage behavior.

There are many requirements to lessen the potential for seepage issues at a dam, such as; to have (and follow) a technical dam safety program, require regular field inspections (with qualified inspectors) together with field personnel trained to recognize abnormal behavior, monitoring field installed instruments (such as piezometers, inclinometers, settlement markers, pressure cells, etc), evaluate field measured data, and finally to have an emergency action plan in case of problems. This paper will focus on issues related to evaluating piezometers, generally installed from the crest of an earthen dam and into a foundation sand layer.

---

[1] Acting Principle Geotechnical Engineer, Headquarters, US Army Corps of Engineers (USACE), 441 G Street NW, Washington DC, Richard.S.Olsen@usace.army.mil
[2]Research Geotechnical Engineer, USACE Research and Development Center (ERDC) Geotechnical and Structures Laboratory (GSL), 3909 Halls Ferry Rd, Vicksburg MS, Isacc.J.Stephens@usace.army.mil
Permission to publish was granted by Director, Geotechnical & Structures Laboratory

Unexpected seepage observed at the soil ground surface is the most direct indicator that seepage (and associated seepage velocities) may have generated underground cavities and/or piping holes. Regardless of the cause of seepage, it is a response to a problem that can escalate from a warning sign to a dangerous dam stability issue very quickly. Geotechnical engineers don't directly measure seepage flow in situ nor can we directly measure the generation of a water filled cavity (or seepage piping). Our best approach is to measure pore pressures inside critical soil layers and study the data using a wide range of evaluation tools.

BASICS OF SEEPAGE

Civil engineering departments at universities have taught seepage evaluation theory for over a hundred years. Teaching seepage concepts in the early 1900s started as over-generalized closed form equations (based only on Darcy's theory), then in the 1950s in terms of how-to make flow nets, and finally in the last 20+ years toward computer based seepage modeling. Civil engineering students today are taught basic seepage theory, then how to hand draw flow nets, and finally how to run computer seepage modeling software. Theory and hand drawn flow nets seem to be forgotten by the time the student becomes a mature practicing engineer.

Geotechnical engineers working with water control structures need to understand seepage concepts beyond what is taught at the graduate level. Many additional seepage concepts are important to understand, such as early/late pinching, leakage influence, flow/static differences, permeability independences, etc. This paper describes methods for basic data evaluation and plotting of piezometer data.

Seepage is defined as water flowing through soil generating a differential pore pressure condition. For this discussion, the driving pore pressure is the reservoir water level (head) and the lower pore pressure is potentially the tail water level. The tail water can also be a perched water level inside an earth dam or foundation inside the downstream embankment of an earth dam, or a ground water table in the natural soil abutment of a dam.

This paper describes a new procedure for pore pressure evaluation. Field measured piezometric data is today typically plotted in ways that can't illustrate seepage problems. The plotted slope (and intersection) of reservoir level versus piezometric level data provides important insight to infer seepage related performance.

We, as a profession, have no basic procedures for visualization of piezometric data. We understand that high pore pressures in the foundation along a downstream section of an embankment are bad, but assume that seepage computer modeling will provide the answers. We generally have little knowledge about how to set up acceptable piezometric trends – we therefore have difficulty selecting piezometric trends that would indicate a red flag condition.

SEEPAGE EVALUATION

Seepage (or water flow through soil) is simply caused by a water pore pressure drop between two points in the ground (called pore pressure differential) as shown in Figure 1. The pore pressure differential is caused by high reservoir pore pressure (also called the head level) on the left to the low water pressure on the right side (termed tailwater level). Therefore, differential head (i.e. decrease pore pressure along a seepage length) causes water flow. The soil path can be idealized as a simple pipe as shown representing a uniform thickness soil layer with an unchanging shape from left to right. Pore pressure at any point along the soil path will decrease as illustrated with the four measurement points: each measurement point has an equivalent water level in a water standpipe. Equivalent water level is also illustrated in the figure with a water well standpipe. The pore pressure can be defined either as a gauge pore pressure or as an elevation pore pressure. Gauge pore pressure is simply the pore pressure measured at a point. Elevation pore pressure is an important property in seepage evaluation and is the elevation level of the equivalent standpipe water level. In all cases the head, tail, and piezometers are reported as the elevation of the water level, and in the case of the piezometer as the elevation water level in a represented standpipe with the sensor tip inside a critical sand layer.

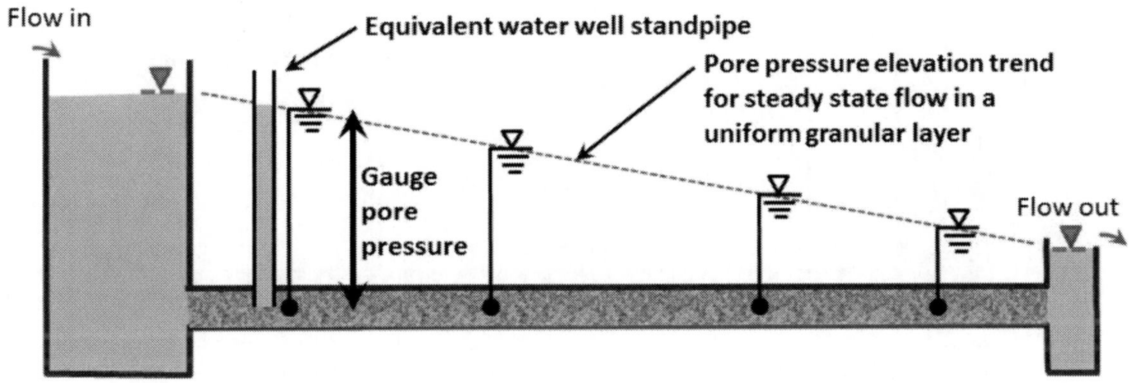

Figure 1. Pore pressure for an equivalent sand layer

Seepage velocity (or velocity of water in soil) is defined by Darcy's law using two parameters; soil permeability (which is related to soil type) and pore pressure differential. The only difference between sand and gravel in Figure 2 is water velocity. While the water flow velocity is much greater for gravel, compared to sand, the trend of elevation pore pressure from head to tail is identical. Therefore, sand/gravel type does not change the shape of the elevation pore pressure trend – This idealized soil pipe is soil type independent as long as the soil type remains constant. Also, pipe dimensions (vertical and width) are also independent as long as the vertical shape is constant from the head water level to tailwater (or equivalent) level.

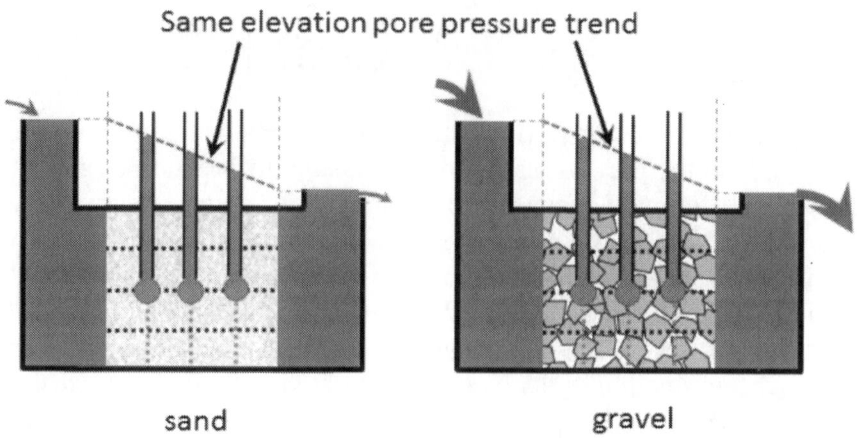

Figure 2. Comparison of pore pressure trend for sand and gravel layers

If the water flow path is blocked then seepage stops and there is no decrease in pore pressure due to seepage flow. No flow is because the pore pressure remains constant and equal to the head pressure. As shown in Figure 3, if there is no change in water pore pressure between two lateral points (i.e. no decrease in pore pressure) then there is no water flow – Seepage flow can only occur when there is differential pore pressure.

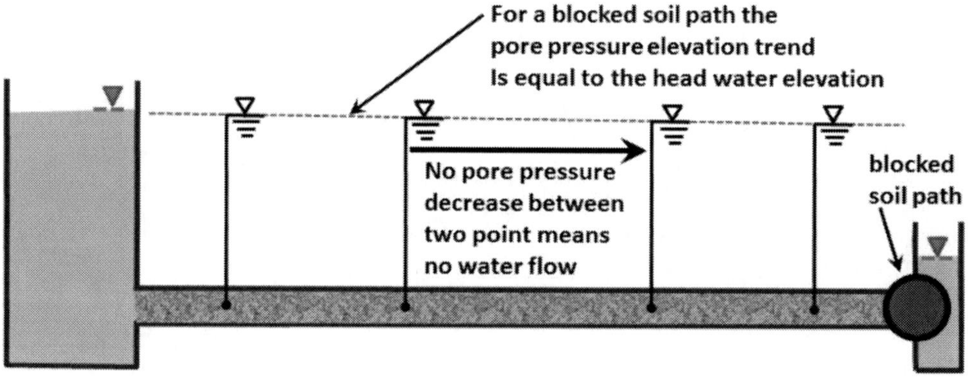

Figure 3. Pore pressure response for a blocked sand layer.

SEEPAGE NETS

Seepage flow nets have been a fundamental part of geotechnical engineering for at least 50 years. There are several basic behaviors for all seepage flow nets, no flow boundaries, water ingress face, water exit face, flow lines (contours), equal potential elevation pore pressure contours and finally material characteristics. Clay layers can easily be considered boundaries when the primary seepage soil layer is clean sand – generally because this assumption does not change the answer.

Seepage in geotechnical engineering is classically described using flow nets as shown in Figure 4. Differential pore pressure is defined as ΔH. Water flow is along seepage lines and equal potential lines define contours of equal elevation pore pressure. In this example there are 8.3 equal potential elevation contours and 4 flow lines defining this flow net (i.e. 4 flow zones). A pore pressure loss equal to ΔH/8.3 occurs as seepage along a flow lines pass each equal potential line. Figure 5 illustrates the elevation pore pressure level (shown as purple) for each equal potential contour as the water flows from the entry location to the exit – in this case it's difficult to see any curvature of elevation pore pressure. Figure 6 includes a cutoff wall into the foundation – the curvature of the plotted elevation potential line (in red) along the seepage path is more obvious. Also plotted, for illustration is the elevation potential line for the interface between the embankment and foundation (in yellow).

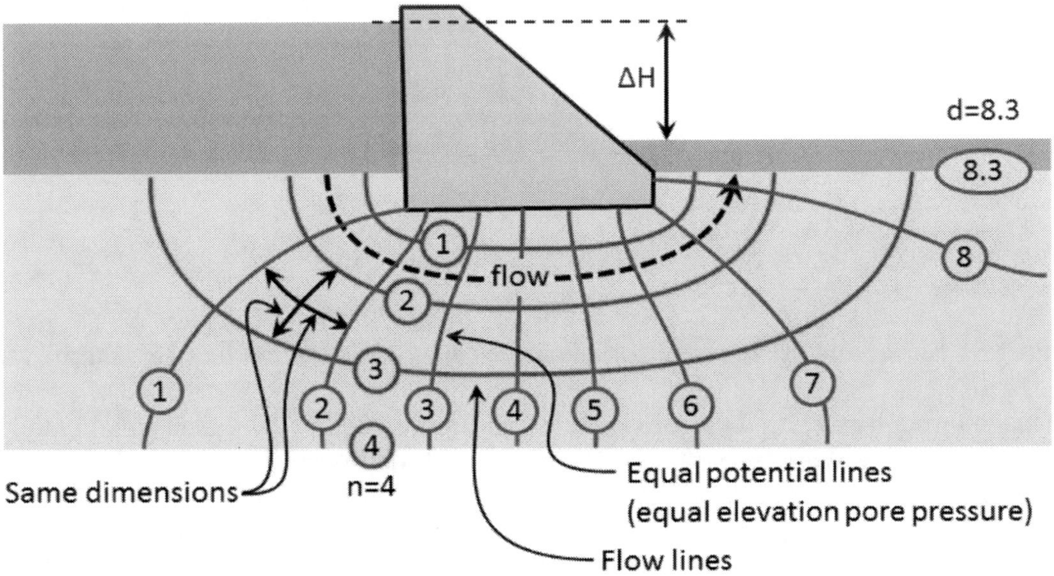

Figure 4. Simple flow net construction

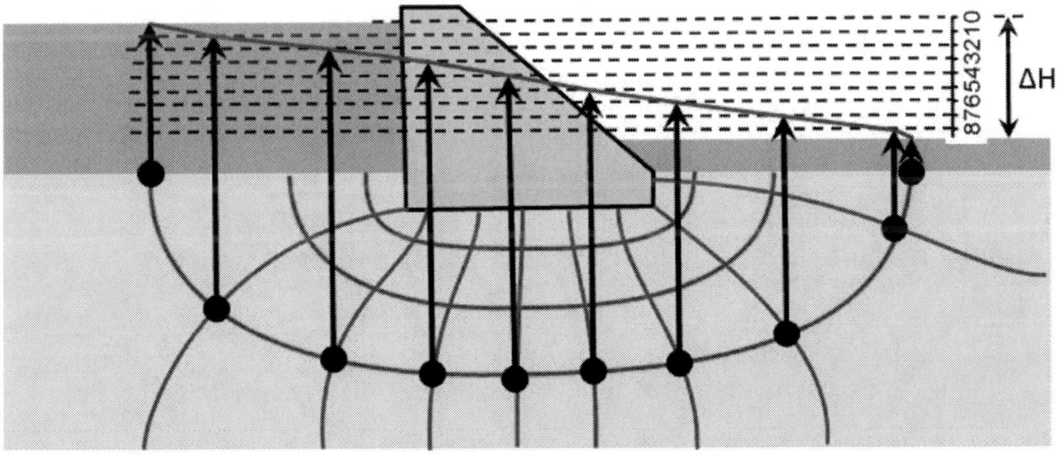

Figure 5. Determining elevation pore pressure along a flow line.

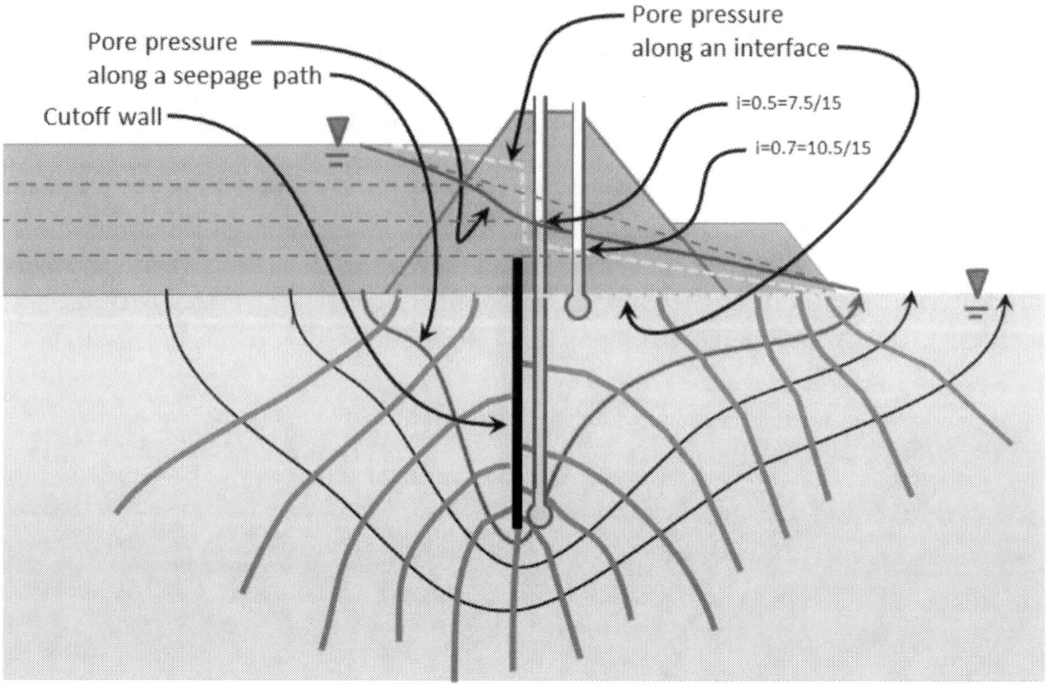

Figure 6. Elevation pore pressure for a complex flow net situation having a cutoff wall

Seepage through an earth dam core is shown in Figure 7 with flow lines, equal potential lines, and a free water table surface. The piezometer on the left has an equal elevation potential level equal to 19% loss of ΔH. The equal potential line for this piezometer intersects the free water table surface at an elevation pore pressure equal to 19% loss of the reservoir ΔH.

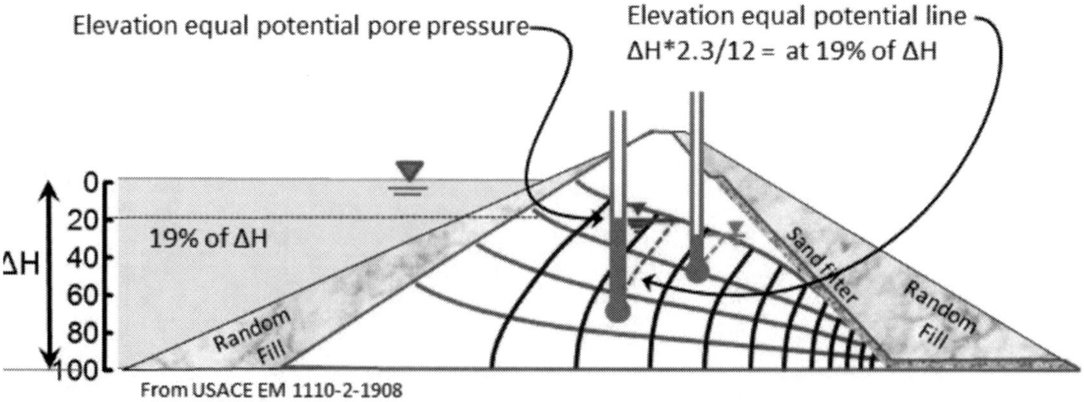

Figure 7. Elevation pore pressure for piezometers in an earth dam clay core.

An equivalent soil pipe can also be represented using flow nets as shown in Figure 8. The slope of elevation pore pressure level versus location along the pipe is constant for any flow line and at any equal potential line.

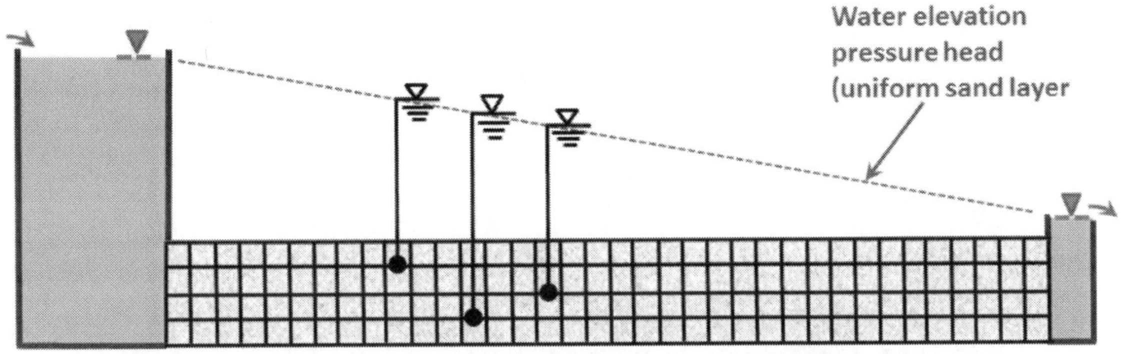

Figure 8. Flow nets representing an equivalent uniform sand layer

EARLY TO LATE PINCHING

Pinching is just as it sounds, there is some form of constriction of a uniform sand layer causing a change in pore pressure and velocity distribution along the seepage path. Pinching of the uniform sand layer in Figure 9 causes a change in the elevation pore pressure (red line) distribution. Specifically, pinching will cause higher pore pressures before the pinch and lower pore pressures after the pinch. The trend of elevation pore pressure with length for the initial zone (points H to P) is equal to the trend for the final zone (points B and T) because the equivalent pipe conditions are equal. In this case the "pinch" is at the middle of the equivalent pipe. Within the area of the pinch there is an increased trend of pore pressure loss with incremental length ($\Delta P/\Delta L$), between point P and B. The pinch is also causing a higher velocity flow rate (at point S) compared to initial and final zones (points H to P and points B to T). It should be noted that the volume flow (i.e. cm^3/sec) will decrease due to pinching compared to a uniform non-pinched condition. Figure 10 illustrates the individual calculations and elevation pore pressure values for a pinch at the midpoint. Note that the highest $\Delta P/\Delta L$ trend is at the middle of the pinch having the smallest equivalent pipe size and largest seepage velocity.

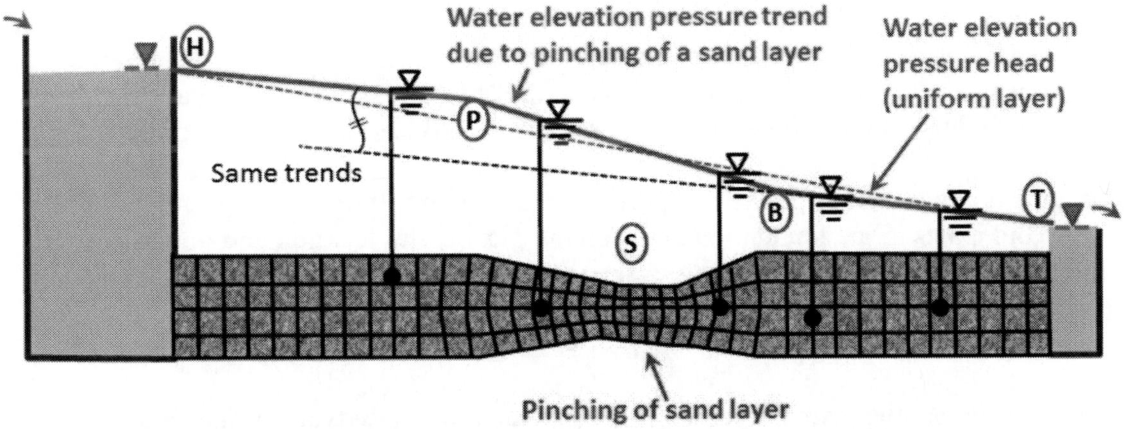

Figure 9. Defining a pinch condition in a uniform sand layer

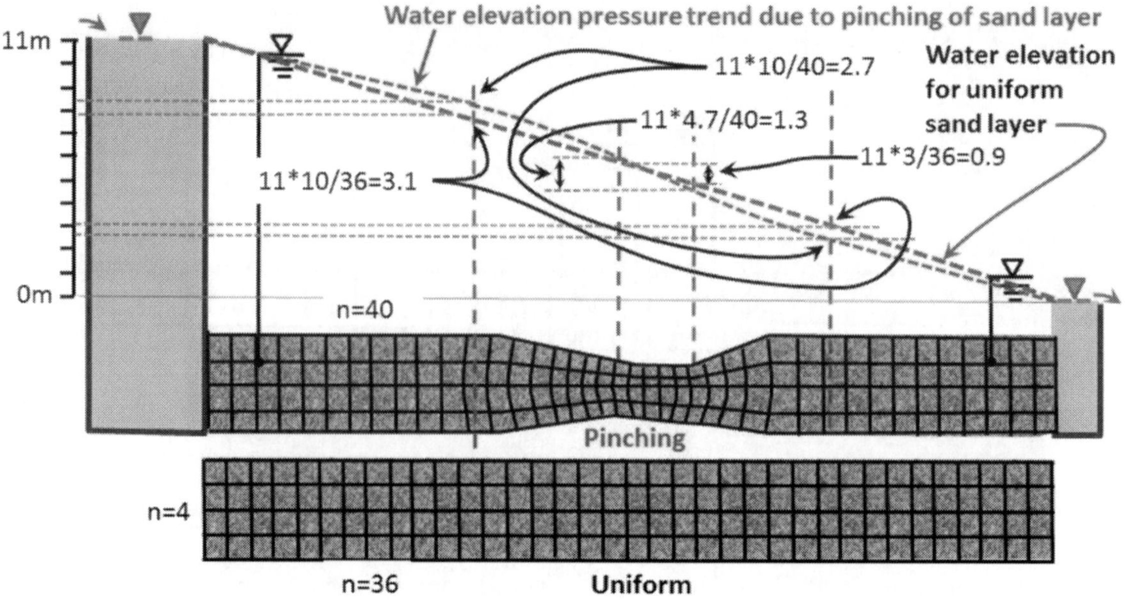

Figure 10. Details for pinching in a uniform sand layer.

RESERVOIR LEVEL/PIEZOMETER DATA PLOTTING METHOD

The main portion of this paper deals with methods for evaluating piezometric data from earth dams, generally obtained from piezometers installed from the crest. Most earth dams have crest based piezometers (into the foundation), maybe a few piezometers on benches at mid slope on the downstream side, and only a few dams have piezometers at the downstream slope toe. In many cases the only piezometer is located at the crest and there is very little information in publications about how to evaluate this data. Generally, the most basic evaluation approach is comparing the measured pore pressure to the reservoir level. When the pore pressure measurements are relatively close to the reservoir level (and there is no definition for "relatively") there are several potential problems:

1) A void exists in the foundation between the reservoir and piezometer location, or
2) Blockage of the foundation layer between the crest and downstream toe, which if unblocked due to high differential pore pressure, could lead to seepage piping.

On the other hand, a low measured pore pressure could infer:

1) A void exists in the foundation between the piezometer location and downstream toe,
2) There is transient or unstable blockage in the foundation layer between the reservoir and crest, that if unblocked could lead to high differential pore pressure, and could lead to seepage piping.

The issue is that there are no procedures for evaluating these types of piezometer measurements, and modeling point measured pore pressures with computer seepage software has no value because there are so many unknowns.

Evaluating measured pore pressures from foundation piezometers installed from the crest is a very complex issue. The key is to initially assume a uniform foundation sand layer and then add simplistic constraints such as; late/early "pinching," late/early "grabbing," initial/critical generated seepage piping, and finally seepage cutoff wall construction. This paper will provide tools to evaluate how measured pore pressure can be influenced by layer constriction, which can occur due to seepage piping, and what to expect after seepage wall construction.

There are two commonly utilized methods for plotting piezometric data;
1) time history of measured piezometric data plots, and
2) reservoir level versus measured piezometric plots.
This paper will focus on the latter method. In previous sections of this paper the discussion focused on how piezometric pore pressure varies from the dam upstream toe to the downstream toe. In many cases with earth dams there is only a single piezometer installed at the crest. For the purposes of this paper the assumption will generally be:
1) there is only one installed piezometer, and
2) that the water level at the toe is constant.
There will be some discussion about more than one piezometer location (in the upstream to downstream direction) and about how the tailwater level can vary.

Figure 11 shows the visualization progression from field realistic to the final representative format of plotting that will be used for most of the remainder of this paper. Note that the sand layer is bounded by impermeable upper and lower boundaries.

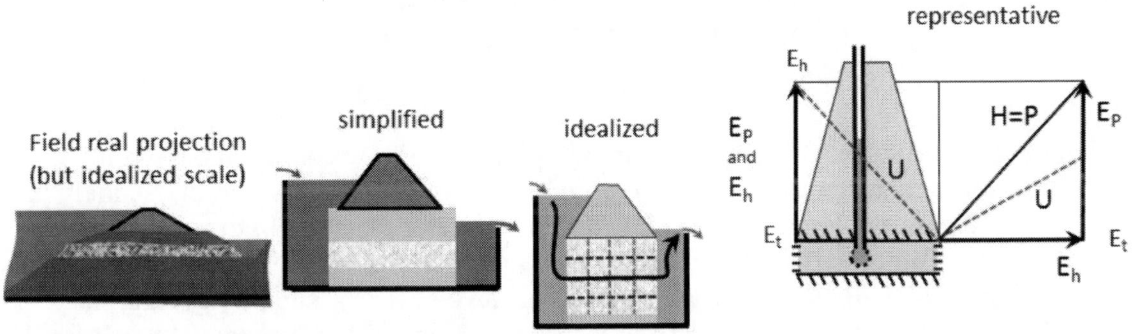

Figure 11. Visualization for changing from a field realistic to a representative section

Figure 12 illustrates the conventions that will be used in this paper. Specifically the different elevation values (E_H, E_T, E_S, and E_P) are standardized for all plots. Other parameters such as Seepage length (L_S), piezometer distance (or item of interest) offset from the tail end of the idealized sand seepage (D_s), and word description (i.e. Pore Pressure Elevation Trend) are defined. An over generalized piezometer is used throughout this paper to represent an elevation piezometer measurement (E_P) for a piezometer sensor elevation of E_P. Most plots will be over generalized representations of seepage flow through a sand layer. In most cases there will be at least two plotted lines (for an idealized uniform sand layer and a situational anticipated condition) from the left (head) to right (tail) sand layer boundary. A blue dashed line will always represent an equivalent uniform idealized sand layer, without pinching or other issues. An over generalized earth dam is shown only for illustration. It would be unstable at this shape. The sand layer is idealistically shown having vertical face surfaces together with a "grill" face: this grill (having a permeability of sand) allows the sand layer ends to have vertical faces.

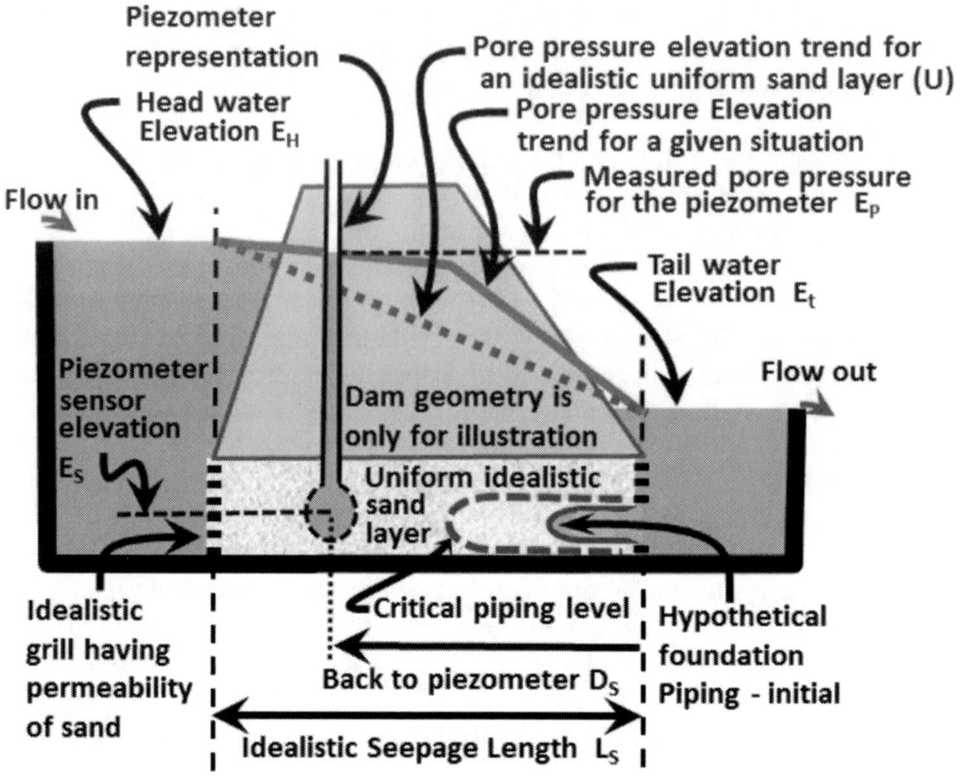

Figure 12. Defining parameters for the seepage evaluation technique.

This section introduces the concept of trending to a zero flow condition based on reservoir level versus piezometer level. Figure 13 shows two types of plots, namely the L-P plot (i.e. seepage length versus elevation piezometer plots) on both sides and the R-P plot (i.e. reservoir (head) elevation versus elevation piezometer plot) in the middle. This figure shows two L-P plots of different reservoir levels. The assumption is that the foundation sand is uniform. The plot in the middle is in terms of reservoir (head) level (on the X axis) versus measured elevation piezometer level (on the Y axis).

All R-P plots will have a line representing Head equal to Piezometer with annotation arrow added showing the following text "R=P (no soil or no flow)". The "no soil or no flow" statement infers that if the measured piezometer is equal to the reservoir then, a) either there is no soil between the reservoir and piezometer (i.e. no soil statement) or b) there is no water flow because the reservoir level is equal to the tail (downstream) level (or a perched water table) consequently there is no driving pore pressure differential to cause water flow. Both of these concepts will be fully described in this paper. Perched water tables (in or under the dam) and/or influence of water table from adjacent natural soil abutments can over shallow the influence of the tail (downstream) water level.

The middle R-P plot is constructed using both L-P plots (red for high water and green for low water). The high head water example on the left will be used to illustrate the construction of one data point for the trend to zero flow condition. The example starts with the reservoir level (point A) which is projected to the R-P line shown by point B. The procedure continues with a vertical line drawn downward from point B. The piezometer level (point C) is then projected laterally from the L-P plot to intersection (point D) using the vertical line from point B. The point D can also represent a typical field measured piezometer data point. This procedure is repeated for the low head water condition on the right (using green lines) to establish point E. A line (with an arrow) can now be generated through points D and E to intersect the R=P line at point G, which represents a condition of "no flow." having zero differential pore pressure. At point G the head water level is equal to the measured piezometer level representing no differential pore pressure and consequently no potential for seepage flow. The line defined by item F is generally termed the R-P trend line slope (and typically signified with a nearby "U" symbol) to a no flow condition point. This is for a uniform sand layer and this trend line will always be plotted with a blue dashed line. The piezometer for this example is in the middle of the section which should be approximately similar to installed piezometers from crests of earth dams.

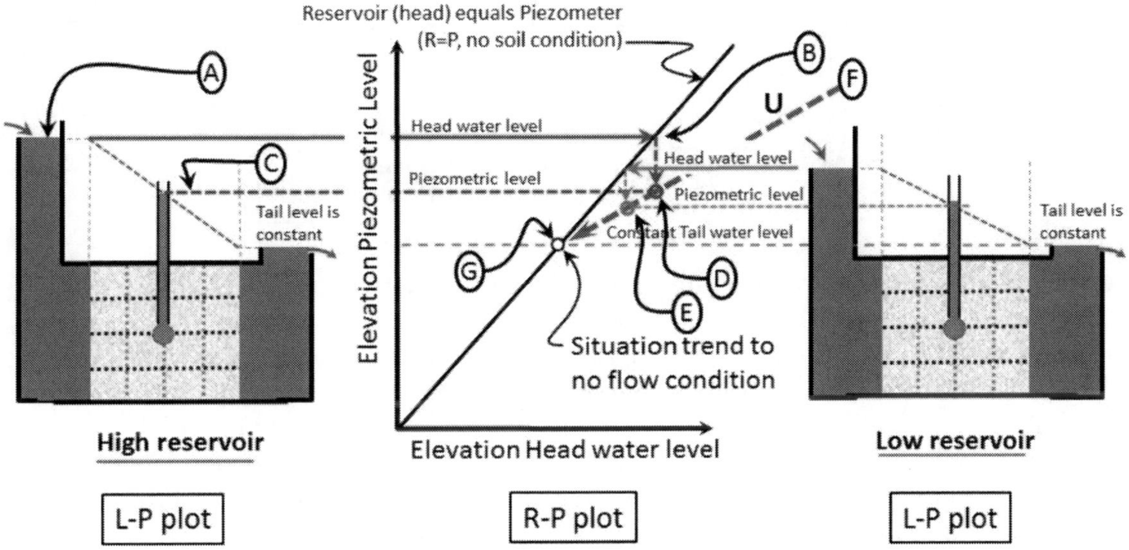

Figure 13. Construction of the R-P trend line to a no flow condition.

Figures 14 and 15 (derived from Figure 13) defines the standardized plotting format for combined L-P and R-P plots in this paper. The two plots are, 1) piezometric level versus location from the downstream toe to upstream toe on the left, and 2) reservoir (head) level versus measured elevation piezometric level on the right side.

Figure 14 illustrates piezometer data trend for a uniform sand layer condition. The L-P plot on the left side is in terms of the relative piezometer location (as D_S to L_S defined as R_S) on the X axis versus measured pore pressure (E_P) on the Y axis for a given reservoir (head) level (E_H). This figure is to explain how to represent the uniform "U" line on the standard L-P and R-P combined plots. For this (and many other) L-P plots the piezometer into the foundation sand is assumed to be in the middle of the dam and therefore R_S is equal to R_{SM}=0.5. The assumption for this uniform sand layer is that half the differential pore pressure between E_P and E_T has been dissipated (i.e. ΔH/2) and the piezometer measurement is equal to E_{PMU}. This behavior for the L-P and R-P plots are shown using the standardized blue dashed lines. If the piezometer is not in the middle of a section then the trend slope for a uniform sand layer has the designation of a_{PU}.

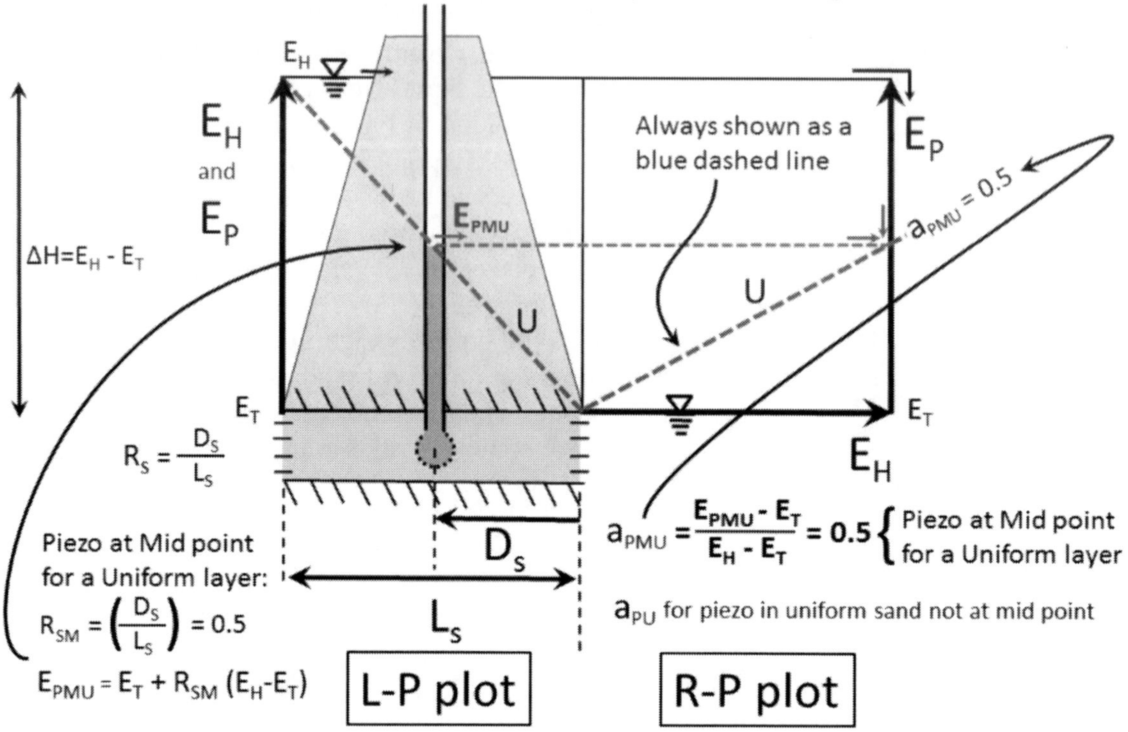

Figure 14. Description of a uniform sand condition in terms of L-P and R-P plots

Figure 15 differs from Figure 14 by showing other potential situations which are different than for a uniform sand layer. An example constrain situation (i.e. early grab) is shown (in red) together with the resulting plot on the L-P plot and trend line on the R-P plot. The trend line slope is defined as a_P. Also shown are example piezometer data points and the corresponding trend line (item T) to a zero flow point condition which infers that the tail level is not influencing the behavior.

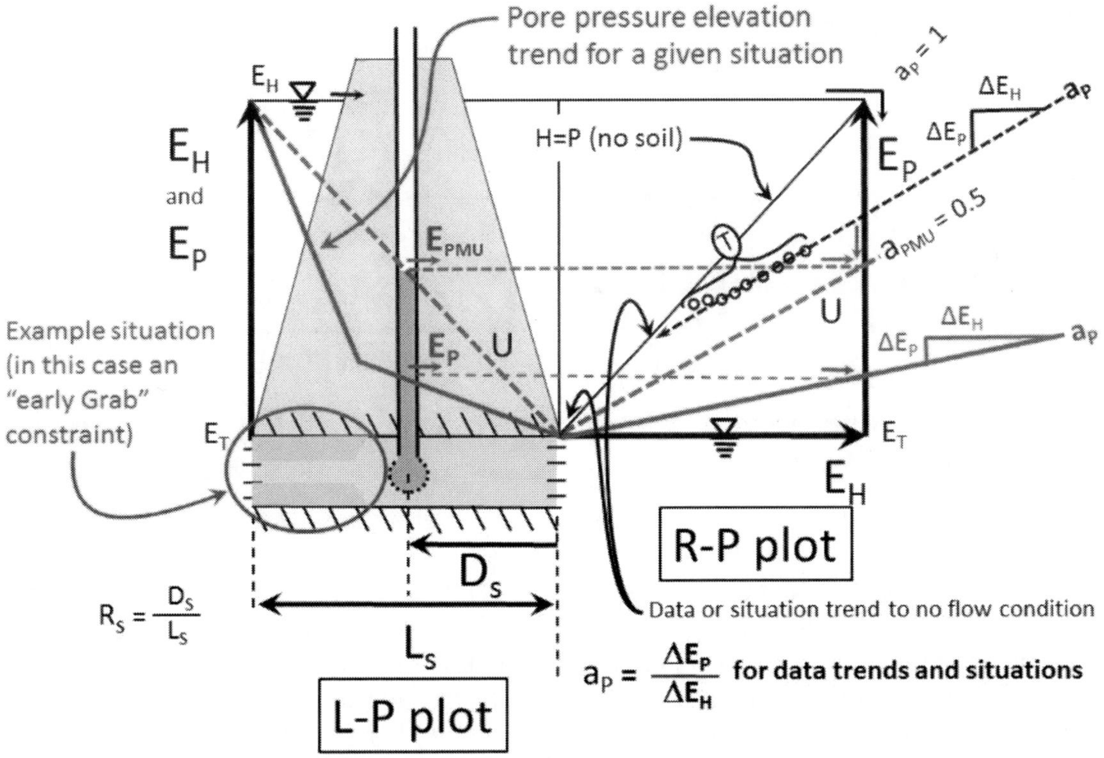

Figure 15. Description of other types of situations using the L-P and R-P plots

The first of a series of field example situations is shown in Figure 16 which simply represents two piezometers at different lateral locations, not at the standardized center location. R_S is defined as the ratio of D_S versus L_S. Both example piezometers have names representing the percent location from the right side (i.e. Piezo 65 and Piezo 21) with corresponding R_S values of $(R_S)_{65}=0.65$ and $(R_S)_{21}=0.21$. The a_{PU} trend line slopes on the R-P plot equal the corresponding R_S lines as shown. So, the a_{PU} trend slope angle equals the relative location of the piezometer in terms of the seepage length R_S – this can be important when evaluating piezometer data.

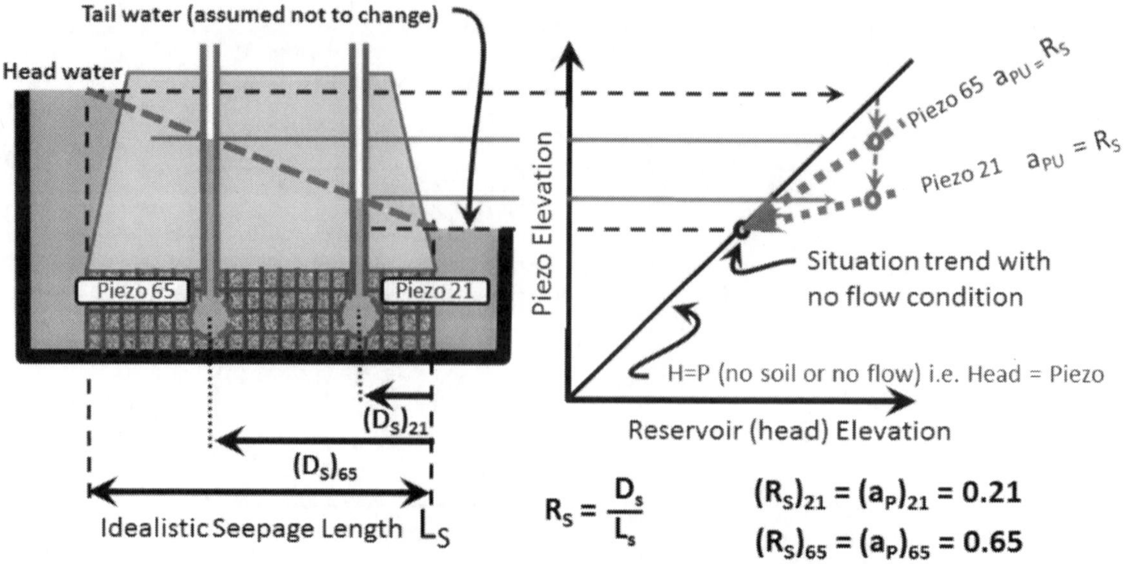

Figure 16. Two piezometers and resulting R-P trends lines.

The next example introduces how pinching of a uniform sand layer influences the trend line for R-P plots. Figure 17 shows three examples of late pinching at different locations and different degrees of constriction, approximating 20, 50, and 80 percent constriction. Late pinching is defined simply as pinching occurring on the downstream side of the foundation sand layer. Note in the L-P plot (left side) how the elevation pore pressure versus lateral distance line slope is the same before and after the constriction, this is because the sand type and path on both sides are identical. The resulting trends in the R-P plot are shown for the three examples plus for 0% and 100% constriction. The level of constriction for a late pinch increases the trend slope a_P level.

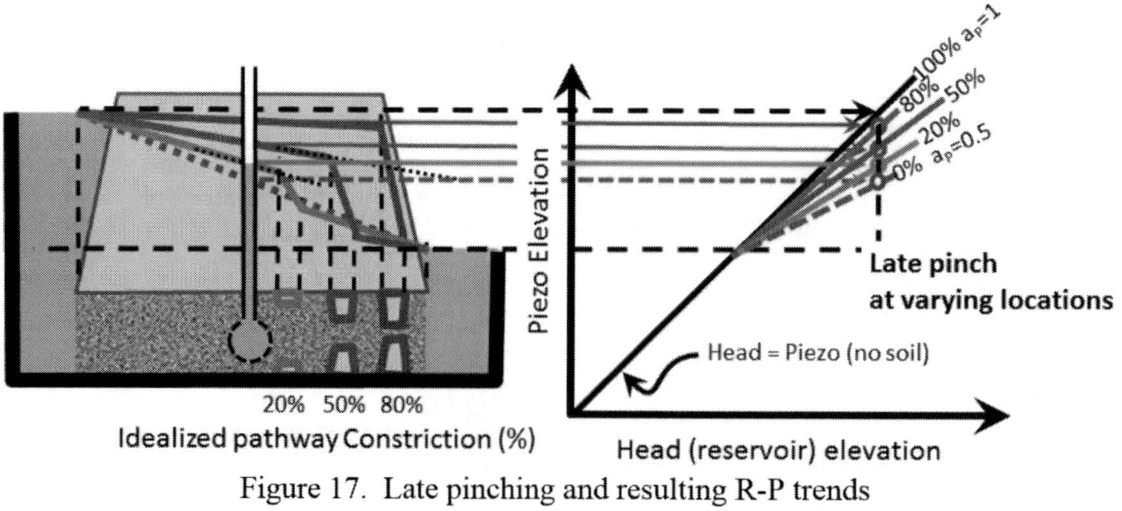

Figure 17. Late pinching and resulting R-P trends

Figure 18 shows early pinching all occurring at the same location and that pinching is upstream from the piezometer location. Note how increasing constriction for an early pinch situation decreases the R-P trend slope a_P which is opposite of the late pinching trend. This observation is because of the piezometer location in reference to the pinch location.

Figure 19 is for late pinch, similar to Figure 17) except that all pinching is occurring at the same location. The resulting R-P trends are also similar.

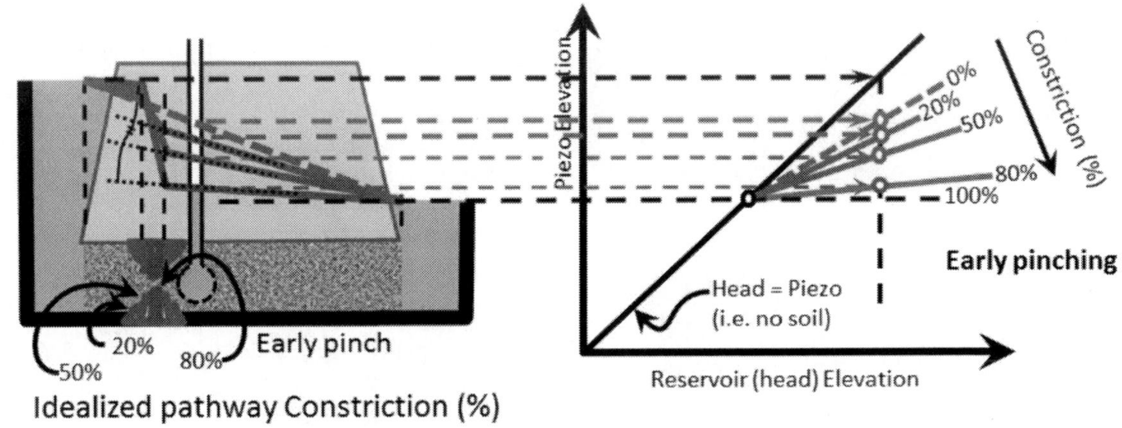

Figure 18. Early pinching and resulting R-P trends.

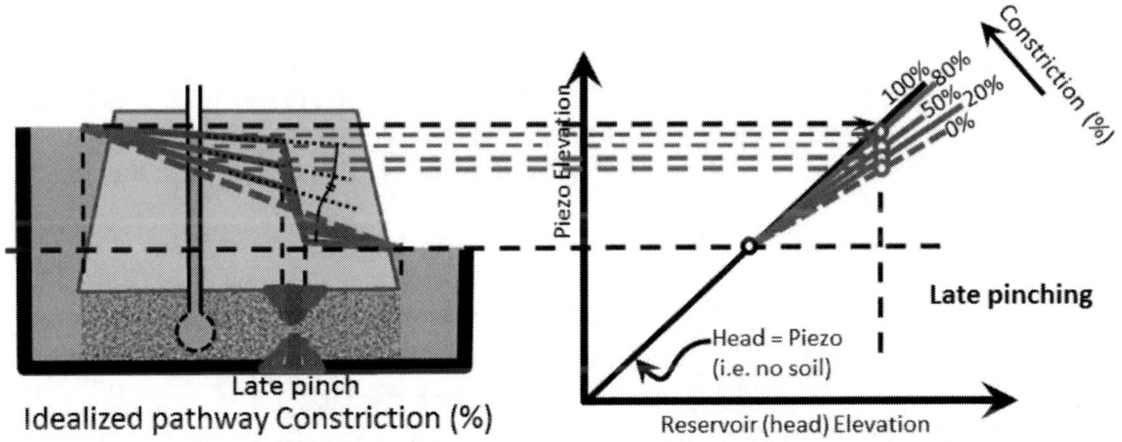

Figure 19. Late pinching but at the same location, with resulting R-P trends.

Figure 20 introduces a new concept, similar to pinching, called a grab as illustrated graphically in the figure. In this case grab constricts the layer for the entire length as shown. In this case the grab starts from the toe side (downstream side) and progresses toward the reservoir. The resulting R-P slope a_P is much more complicated than for the pinch situation. As the grab progresses (from item A, to item B and finally to item C) the R-P slope a_P initially increases then decreases. Actually, the a_P for a no grab situation (uniform sand) is $a_{PMU}=0$ (see R-P plot) then as the grab length increases (item A to B) the resulting a_P level increases. As the grab length goes past the location of the piezometer (item C) the result is that a_P slope will decrease. When the grab length extends the total length (i.e. back to a uniform sand layer situation) the resulting a_P is again equal to a_{PMU}. The R-P behavior is complex but can be explained by examining the pore pressure plotted shapes in the L-P plot.

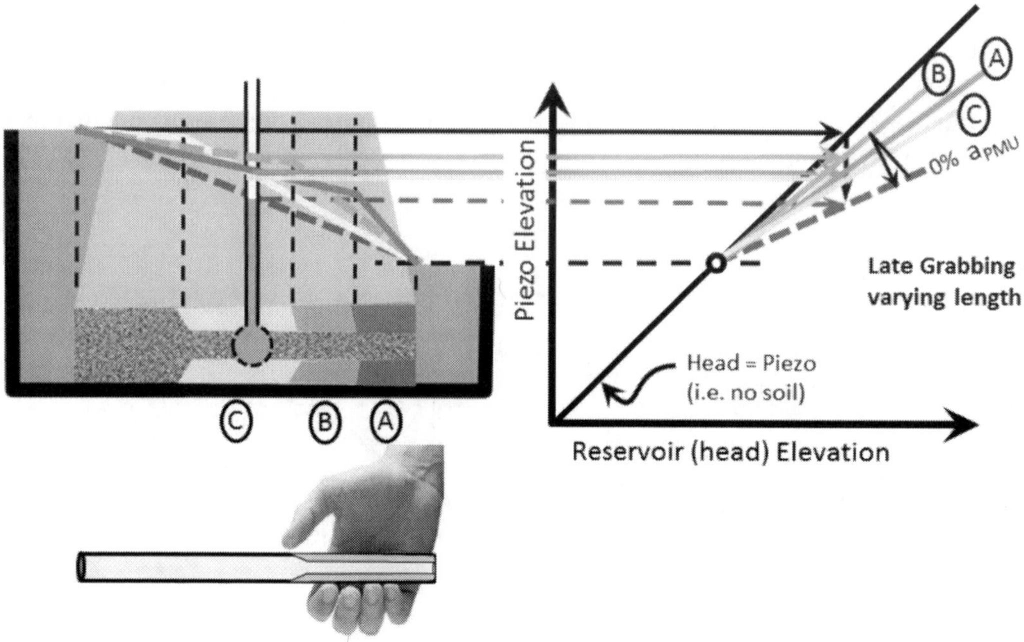

Figure 20. Defining Grab and resulting complex R-P trends.

Figure 21 illustrate two cutoff wall constructions, an early and late wall with the piezometer near the center of the sand layer. The resulting R-P trends slopes are shown at the right. Figure 22 is similar to Figure 21 but includes an early pinch. An early pinch and early cutoff wall have similar behavior which is to decrease the R-P trend slope a_G level. Figure 23 is again similar to the previous plots but for this case is for a late pinch.

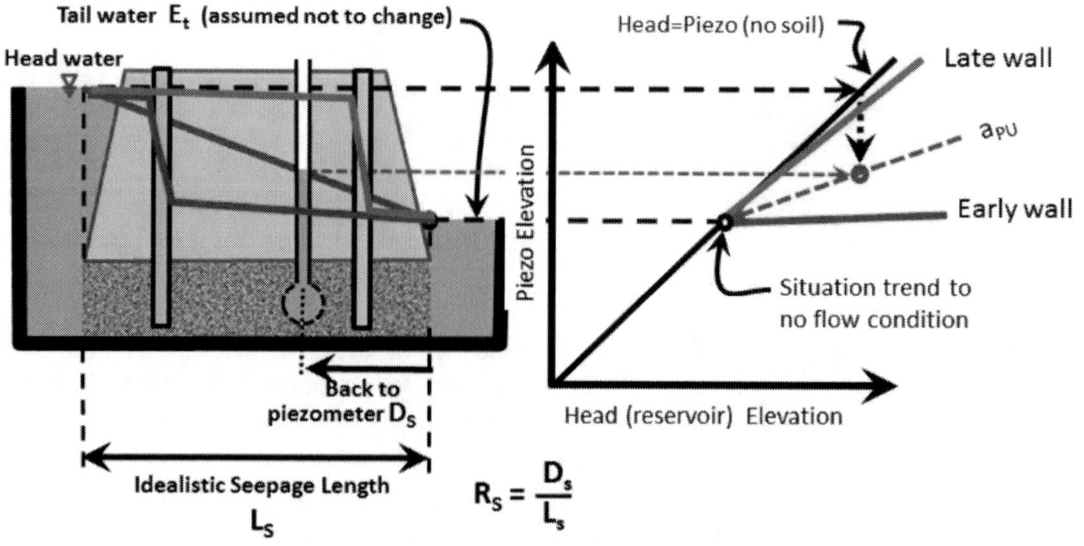

Figure 21. Cutoff wall construction and resulting R-P trends.

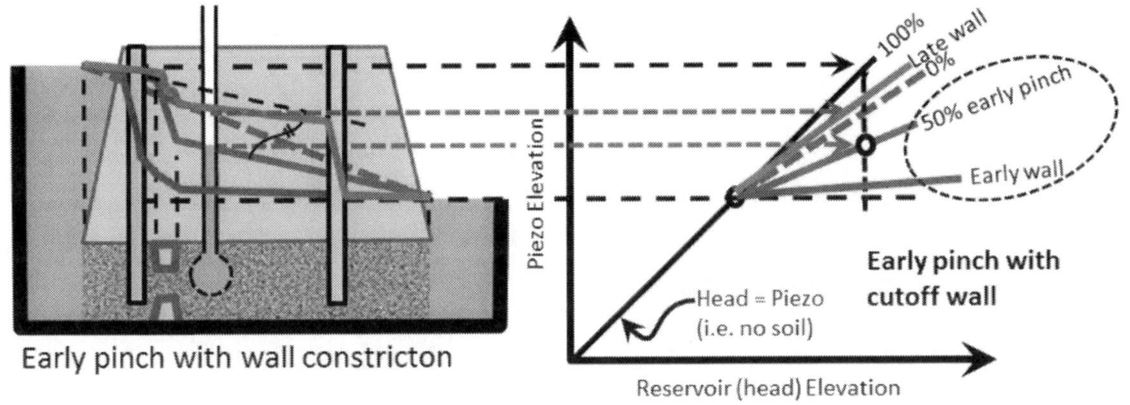

Figure 22. Cutoff wall construction with an early pinch and resulting R-P trend.

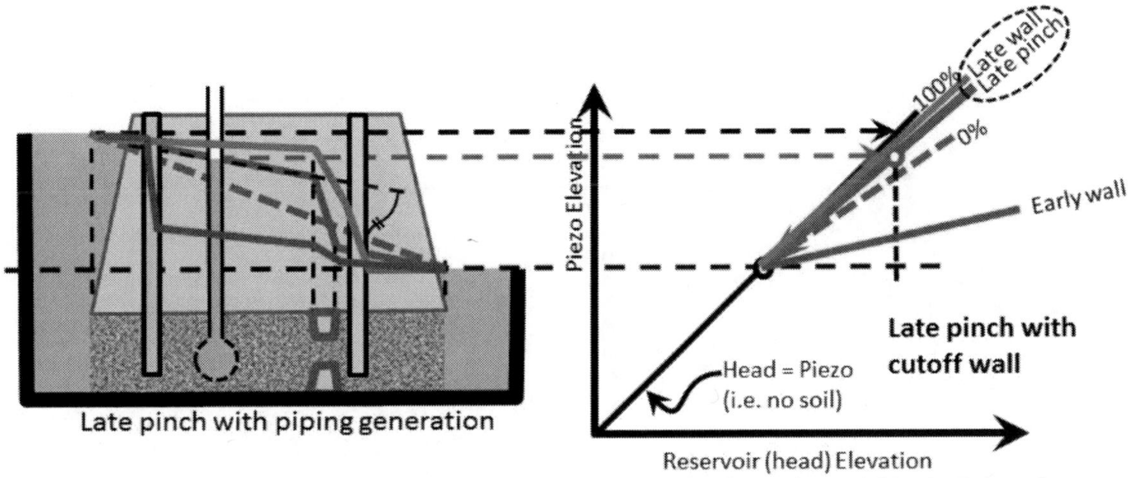

Figure 23. Cutoff wall construction with a late pinch, and resulting R-P trends.

Figure 24 introduces the concept of a seepage pipe progression from initial formation to a critical situation. Increasing seepage pipe formation (from initial to critical) results in the R-P slope trend a_P to decrease initially from a_{PMU} to lower values as shown. Seepage piping formation will always decrease a_P. When a_P is zero there is a seepage pipe extending from the upstream tail area to the piezometer sensor. Figure 25 is similar to Figure 24 except that the piezometer location is near the upstream reservoir. The R-P trend is similar but the shape is different. Figure 26 includes an early pinch together with increased seepage piping. The R-P trend slope in this case starts at the pinch trend slope (shown as 50%) and decreases with increased seepage piping criticalness. Finally, Figure 27 shows a late pinch with increased seepage piping. It is difficult to differentiate initial piping and late pinching for this condition. However, as seepage piping progresses past the late pinch the R-P trend slope a_P drops dramatically, as shown. Figure 28 shows a late grab together with seepage piping formation. The R-P trend slope a_P for this grab situation is very high. The decrease of a_P is dramatic as the seepage piping length approaches the grab length, as shown.

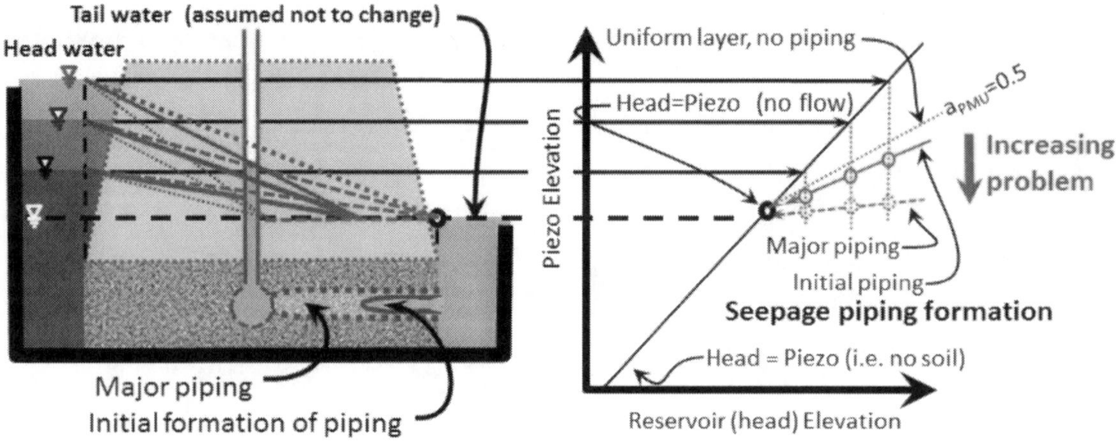

Figure 24. Seepage piping and resulting R-P trends.

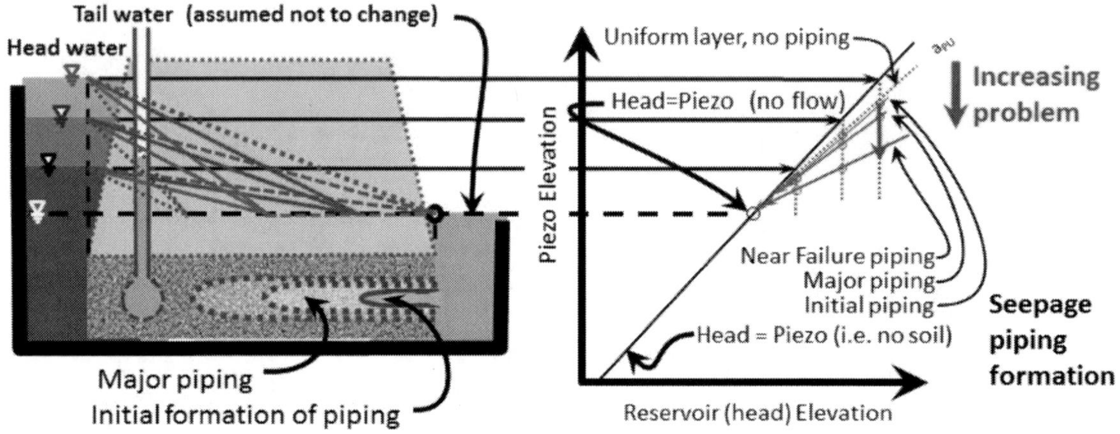

Figure 25. Seepage trends with piezometer located near reservoir side.

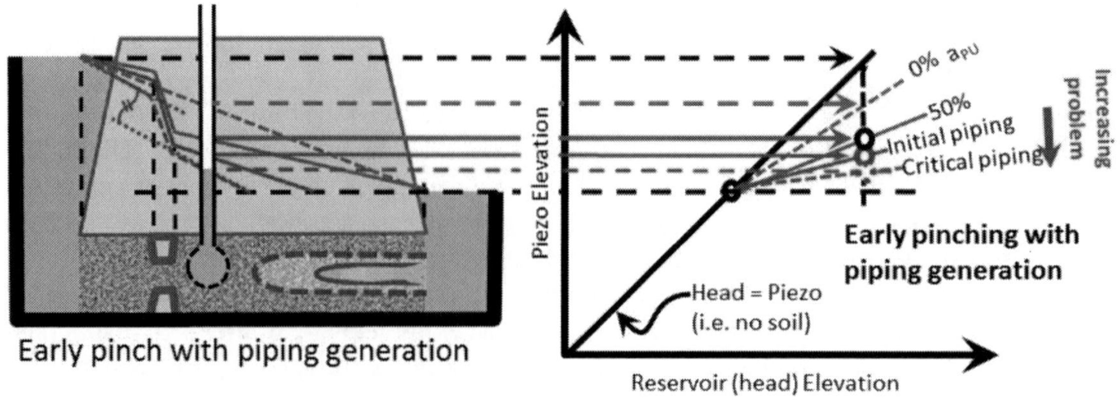

Figure 26. Seepage piping together with early pinching, and resulting R-P trends.

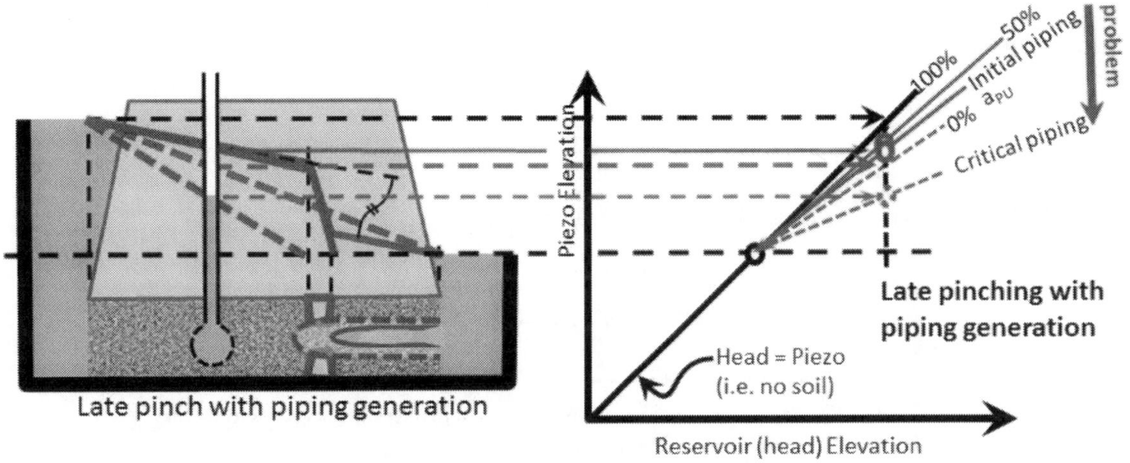

Figure 27. Seepage piping and late pinch, and resulting R-P trends.

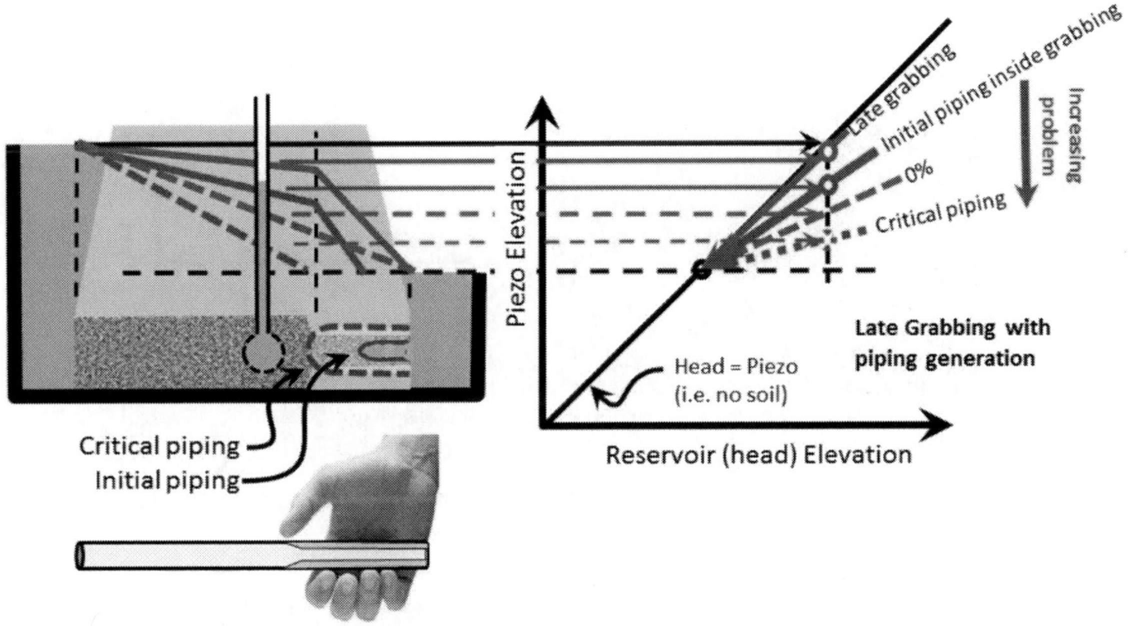

Figure 28. Defining late grab and seepage piping and resulting R-P trends.

Figure 29 is the accumulation of all the above situations. This plot is only for a piezometer at the midpoint of the foundation sand layer. The resulting plot, termed the L-AP plot, shows how a_P changes for given situations (i.e. pinching or grab) from the reservoir side to the toe side. An example is the best means for explaining this plot. The example is for a late pinch with approximately 80% constriction, shown by item A. The resulting measured piezometer (at the midpoint along the sand layer) is shown by item B. Item C on the R-P reflects item B level, resulting in $a_P = 0.7$ (item D). This a_P is placed on the L-AP plot on the Y axis at item E. Item F is the intersection of item E (i.e. item D) and the location of the late pinch (item A). As should be expected, the L-AP plot for a grab is more complex than for a pinch. In all cases, increased seepage piping will decrease the a_P as shown by the red arrow.

Evaluating reservoir piezometer data is always complex, especially when there is only a single piezometer located at the crest. Figure 29 as well as Figures 16 to 28 are only tools to aid in evaluating piezometer data.

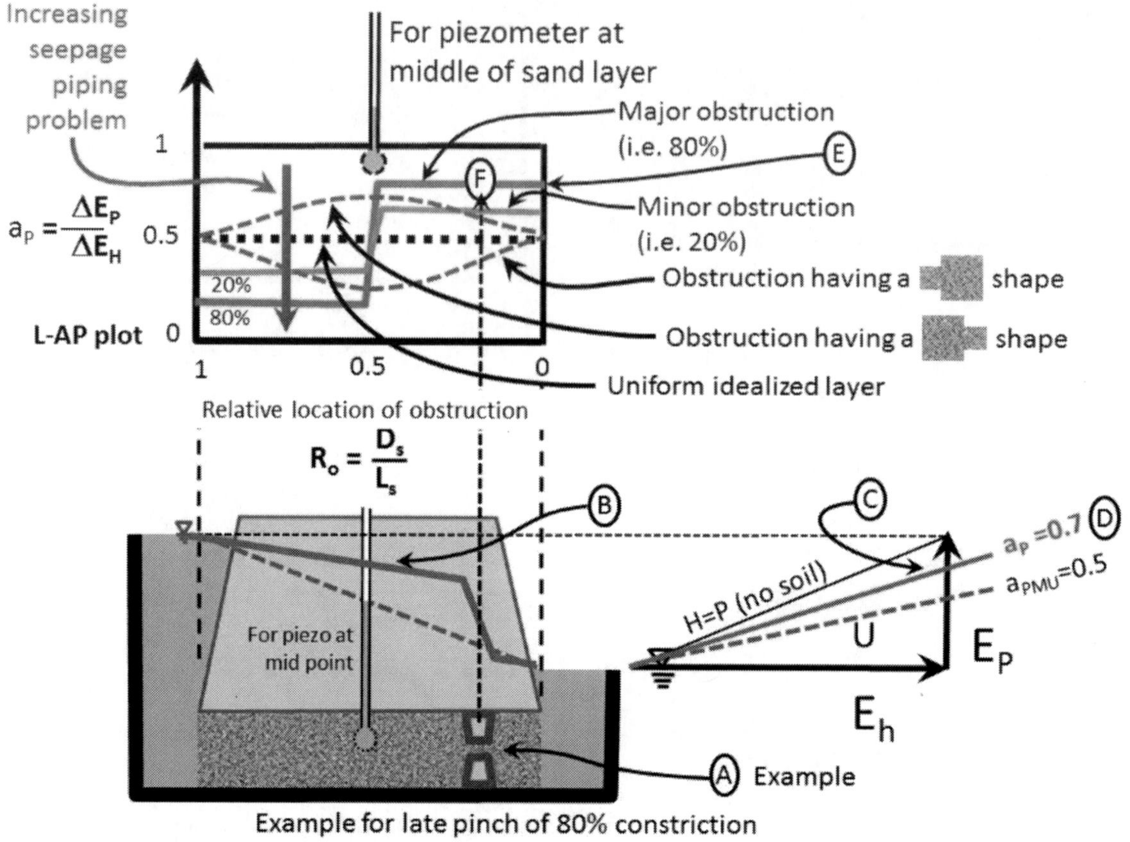

Figure 29. Defining the L-AP plot as a tool for evaluating piezometric data.

The next several plots (Figures 30 to 34) show R-P trending to a zero flow condition (at the H=P line) using data from USACE sources and others. Further evaluation for these examples was not performed because more information is required such as; tail level, location of piezometers, foundation type, geology, etc. In most cases the trend to a zero flow condition has an elevation much higher than the tail level, inferring a pinched water table in the dam or seepage from the adjacent dam abutment. These examples are only to illustrate how existing data can be used to construct the R-P plot using L-P data. Figure 35 is shown to illustrate that adding the H=P line shows that the dramatic observed changes (items A) have a slope equal to H=P. It is unknown if it's important but a good observation.

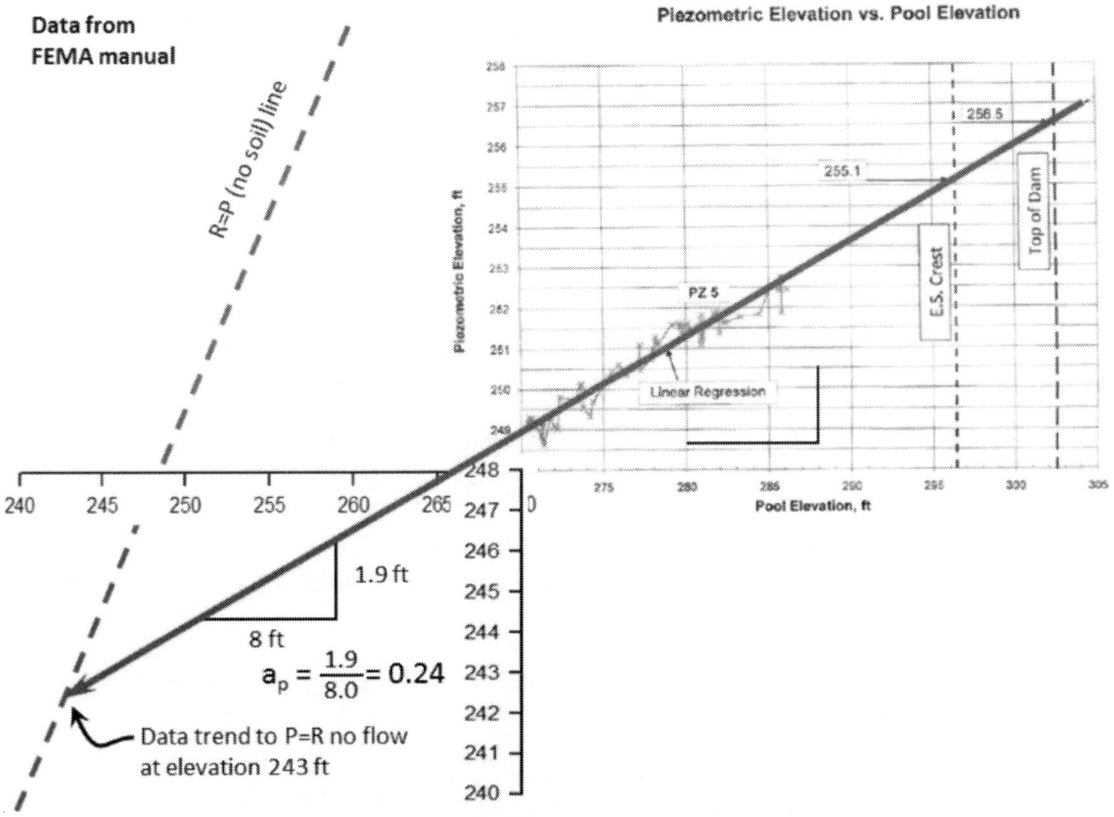

Figure 30. Data example from FEMA P-1032 (figure 8-8)
to determine the zero flow condition.

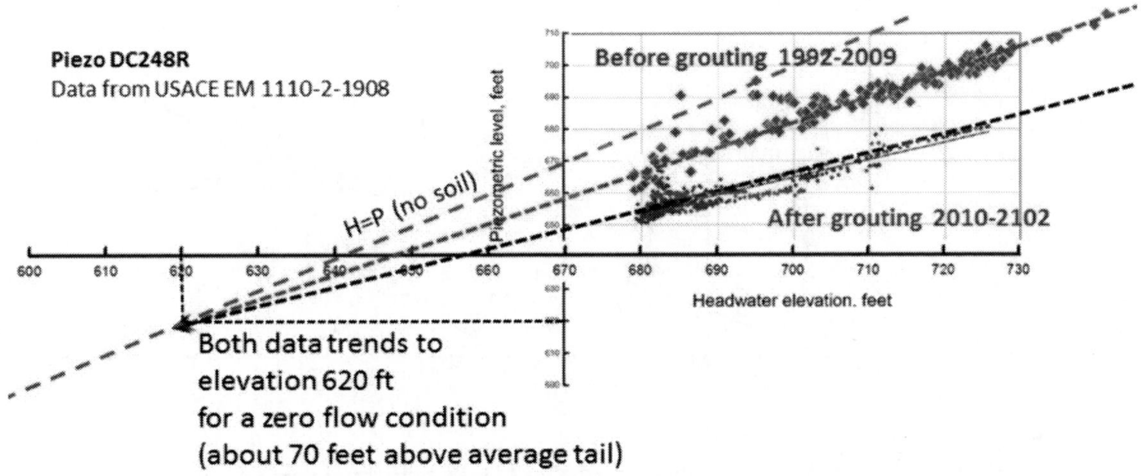

Figure 31. DC248R data example from USACE manual
to determine the zero flow condition.

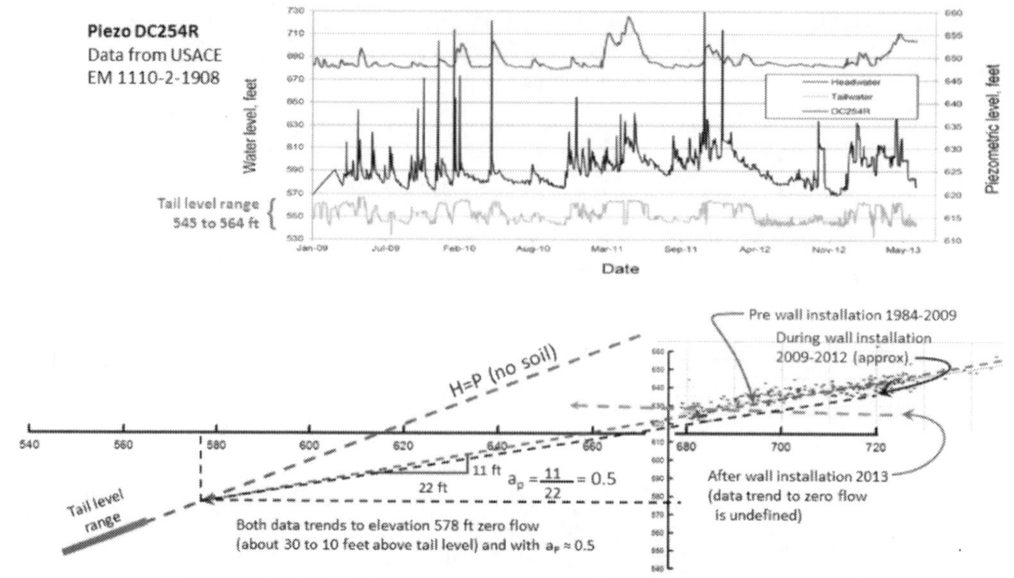

Figure 32. DC254R data example from USACE manual to determine the zero flow condition.

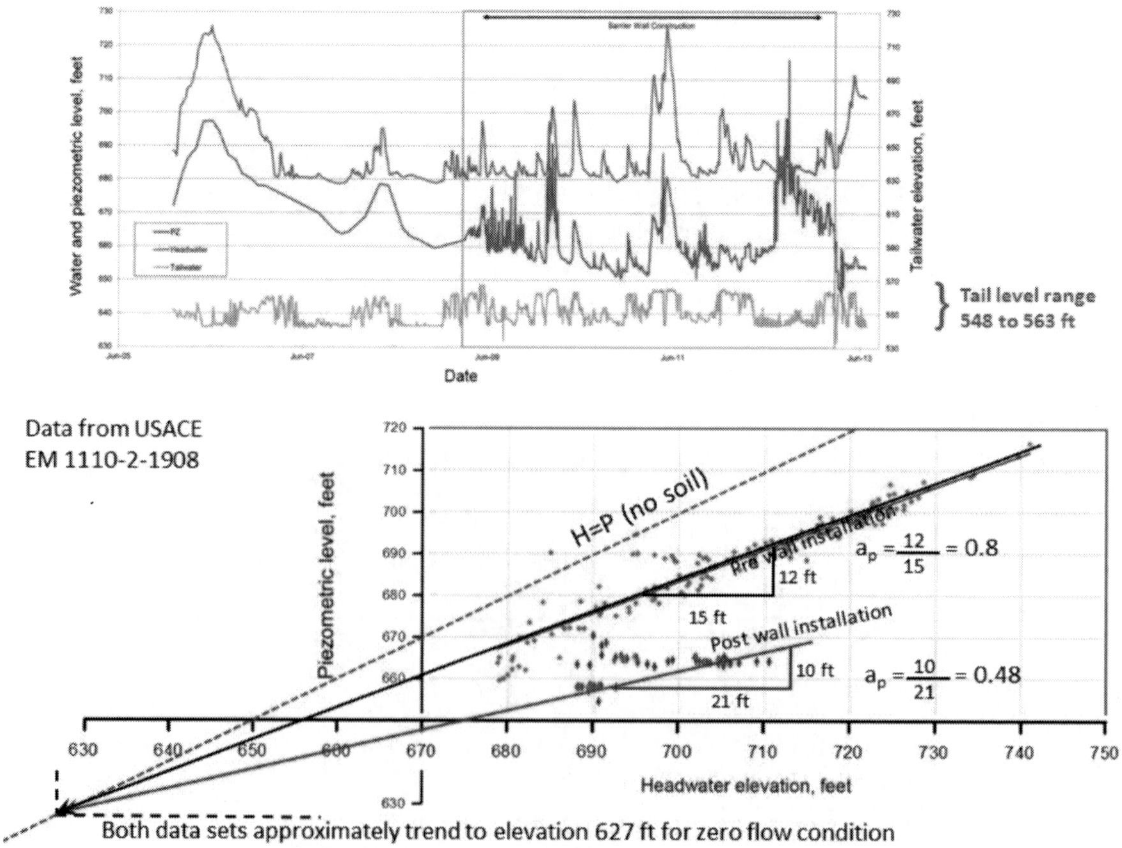

Figure 33. Data example from USACE manual to determine the zero flow condition.

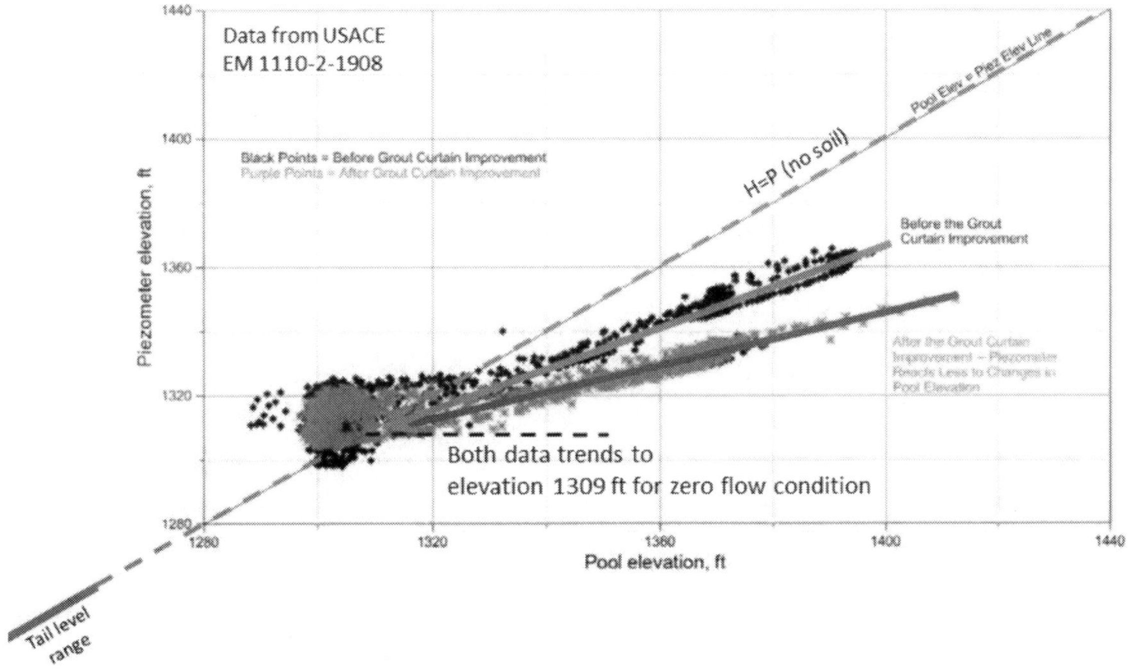

Figure 34. Data example from USACE manual to determine the zero flow condition.

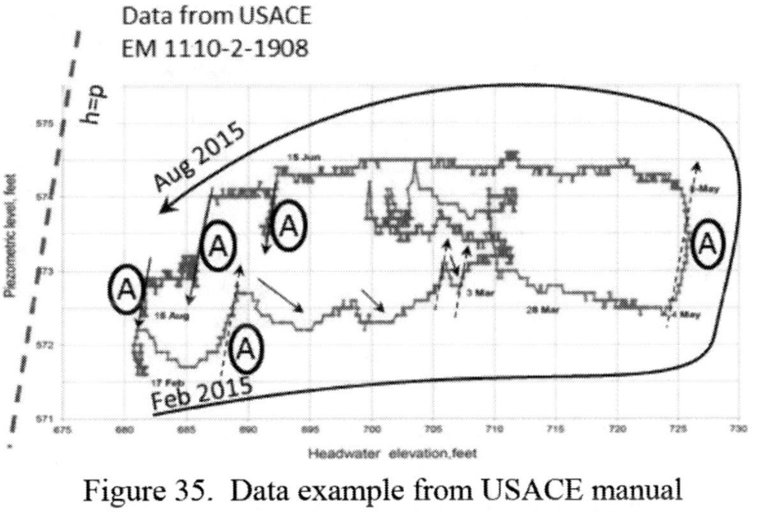

Figure 35. Data example from USACE manual
to show the potential usefulness of the H=P line.

MAKING PIEZOMETER VERSUS RESERVIOR PLOTS

Making plots of piezometer measurements versus reservoir level must be more than just computer software plotting the data and copying the plot into a report. Previous sections described how to critically examine the measured pore pressure data. Figure 36 specifically illustrates the minimum information required. Always include the piezometer information at the top left of the plot to allow the reader to quickly understand critical information: This block of text is required to allow a plot to be standalone. Use different plotting symbols for different situations, such as maybe pre construction of a cutoff wall, before (and after seepage), etc. Add the R=P line and the trend slope line pointing to the zero flow on the P=R line. For many situations a separate plot may be required to show the two additional lines (i.e. the R=P line and the trend slope line pointing to the zero flow).

Another suggestion is to plot time histories of piezometer data at the same scale as for the reservoir (head) and tail levels. Having the same scale is important when comparing how piezometer data is changing in reference to reservoir and tail levels. If required, the reservoir and tail levels can be plotted separately above and below the piezometer but all three having the same scale.

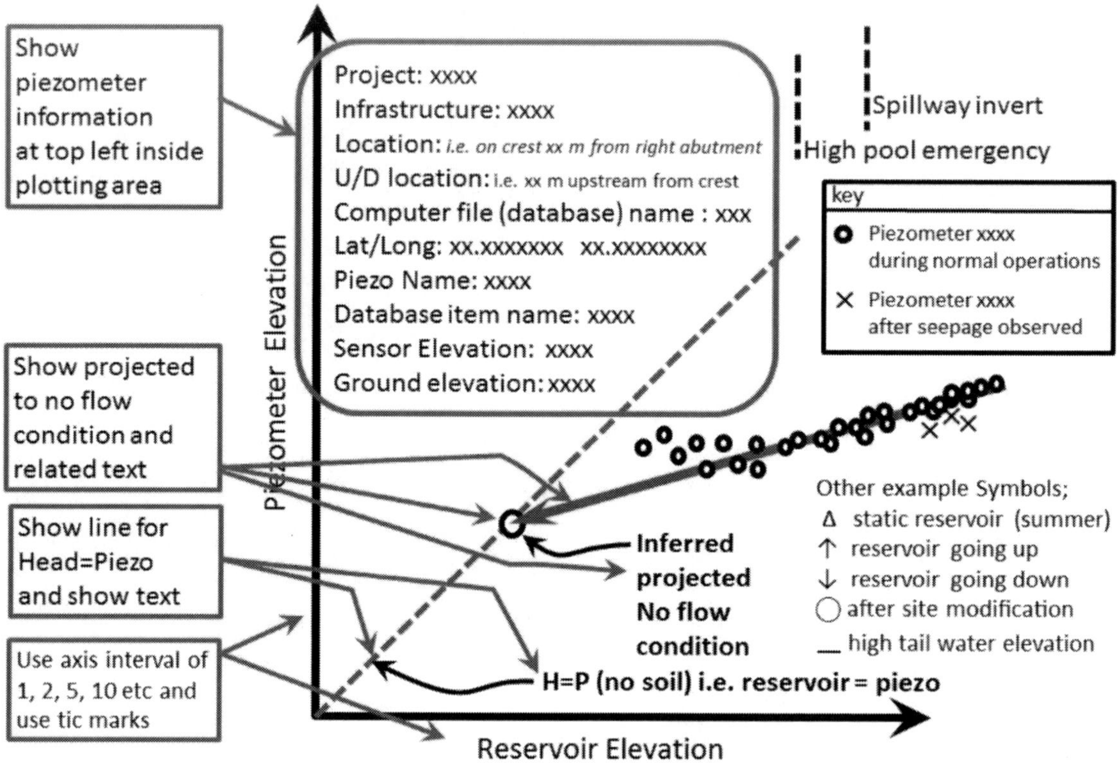

Figure 36. Required information for a R-P plot.

An important part of piezometer data evaluation is to provide clear site maps of piezometer locations and other site information. Figure 37 combines a modified site map together with the time histories of several piezometers. Figure 38 shows in concept how to include a photo of a dam together with a simplified cross section showing the piezometer locations.

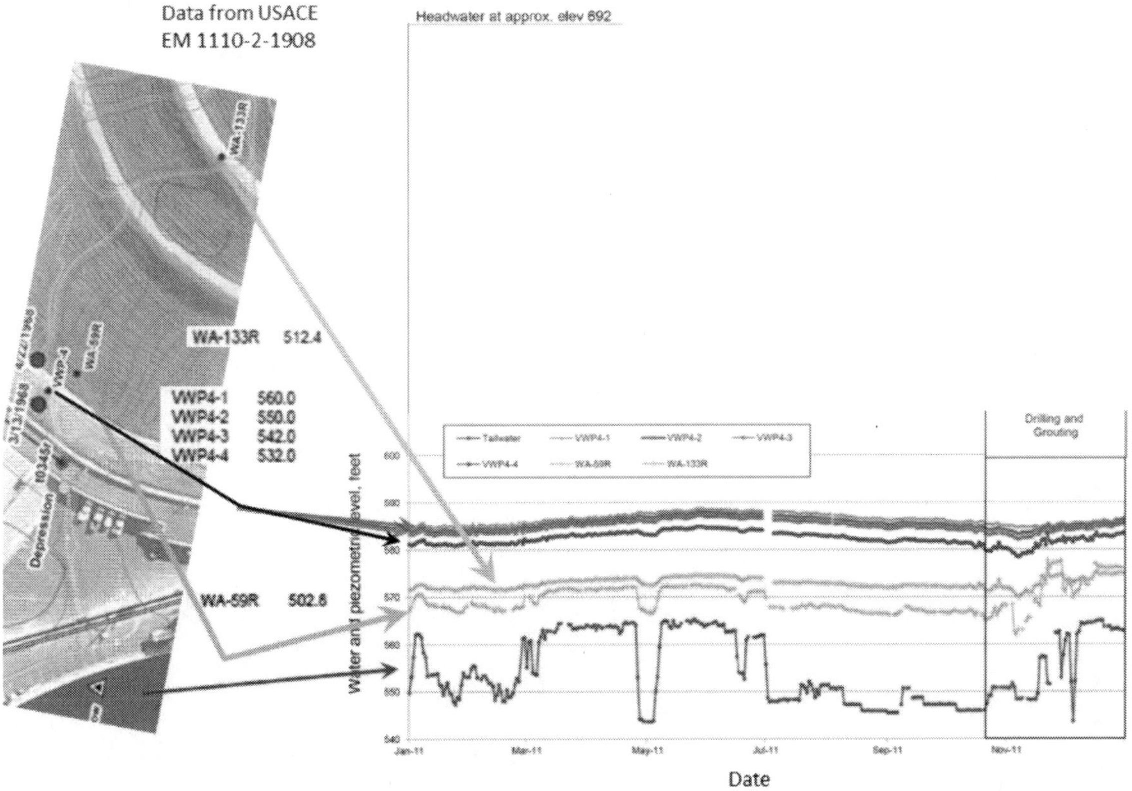

Figure 37. USACE information to show a site map and time histories of piezometer data

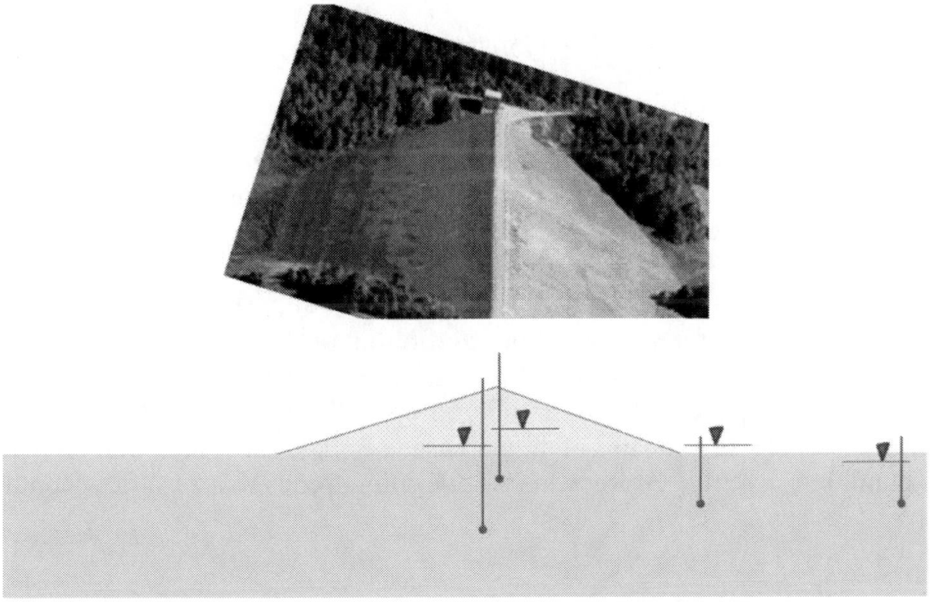

Figure 38. Illustration showing a photo of a dam and cross section with piezometer locations

When performing computer seepage modeling it's important to graphically show how permeability changes with different soil layers. Most commercial software provide an option to include a table showing soil layers and corresponding input values. Presenting critical data in tables does not allow the engineer to easily visualize how the soil layers differ. Figure 39 shows two depth/elevation plots next to a software generated embankment plot. For this example, the two plots are for permeability and D_{10}. For both plots there is a difference between measured values and engineering assigned values.

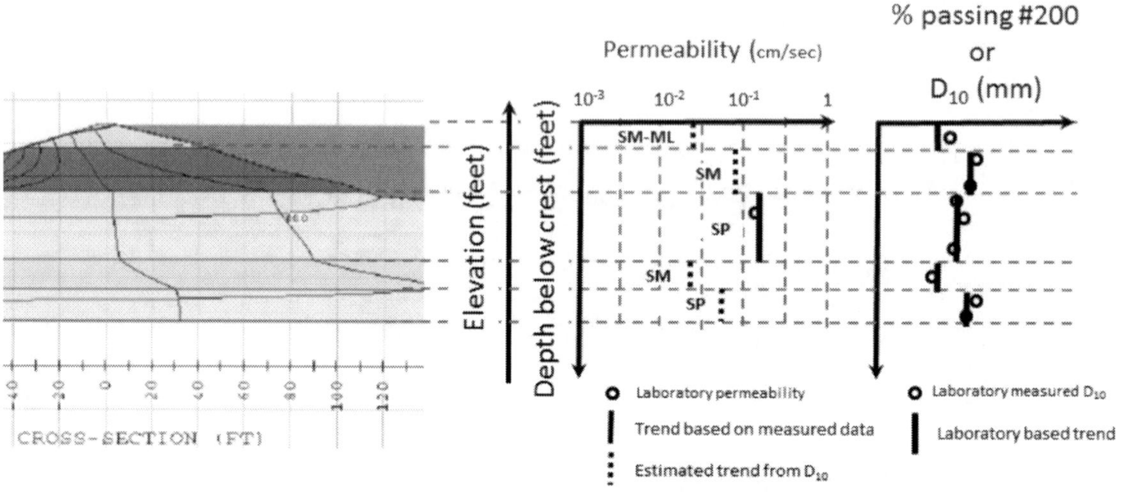

Figure 39. Plots that should be added to computer seepage modeling output

CONCLUSIONS

Piezometric data can be one of the best indicators of seepage in earthen embankments, but the evaluation for earth dams can often be difficult because of the many interrelated factors. Factors such as the character of permeable layers, micro geology in critical foundation materials, the location and character of seepage obstructions all have a role in piezometric response to seepage. A fundamental understanding of seepage forces and flow are essential to correctly interpret the significance of piezometric changes over time. This paper provides several new tools for interpreting piezometer data in terms of reservoir levels. A standard way of plotting piezometric data is proposed and trends that generally indicate seepage problems are identified. When established pore pressure trends drop it's likely an indicator of an increased seepage flow and/or an increased seepage condition severity. A procedure for defining trends to a zero flow condition is described.

ACKNOWLEDGEMENTS

This paper does not reflect recommendations, endorsement, or policy of the U.S. Army Corps of Engineers. The paper was initially submitted for conference review in 2015 October. Full size images are posted at USSD2016.geostaff.net

REFERENCES

USACE, Engineering Manual EM-1110-2-1908, "Instrumentation and monitoring of Embankments Dams and Levees. Draft 2015 Sept 04.

FEMA, "Evaluation and Monitoring of Seepage and Internal Erosion" - Interagency Committee on Dam Safety (ICODS), FEMA P-1032 /May 2015

EVALUATION AND MONITORING RESPONSE TO UPSTREAM SLOPE FAILURE AT AN EMBANKMENT DAM

Jay N. Stateler, P.E.[1]
Jeff Wormer, P.E.[2]

ABSTRACT

On September 23, 2014, a slide was detected in embankment material blanketing pervious right abutment rock at an embankment dam. The response to the slide included slope stability evaluation work, site exploration work, development of new Potential Failure Modes (PFMs) for the dam in light of the slide, and conducting risk analysis work. This work allowed the situation and dam safety risks to be better understood. The response to the slide also included enhanced and expanded monitoring for possible additional slope movements, and intensive efforts to gather data regarding abutment water pressures, toe drain flows, and visual observations, to allow for timely detection of any anomalous performance that could indicate initiation/development of a PFM. Changed water pressure conditions in the right abutment were detected, presumably due to disruptions to the blanketing materials in the slide area. Dam safety risks under normal operating conditions were determined to exceed Bureau of Reclamation (Reclamation) criteria. Consequently dam modification work is planned for the right abutment area of the dam.

INTRODUCTION AND BACKGROUND INFORMATION

The dam is located in the United States and is a rolled, zoned embankment dam whose original construction was completed in 1961. [Note: The name and specific location of the dam is not provided in light of security considerations.] The dam has a structural height of 162 feet and its crest is at elevation 5527 feet. The dam has a reinforced concrete overflow spillway structure at its right abutment, and the dam's outlet works is located directly beneath the spillway, sharing the same stilling basin. Figure 1 shows a plan view of the dam's configuration as it was originally constructed. Figure 2 shows a profile view along the centerline of the dam m.

When the initial Potential Failure Modes Analysis (PFMA) meeting was held regarding the dam in April 1995, one of the meeting participants commented at the start of the meeting that the dam is a dam only a mother could love. To say it has had a checkered history is putting it mildly.

Reservoir filling began in early 1962. In November 1962, a sinkhole developed on the downstream face of the dam near the left groin that was approximately 10 feet in diameter and 7 feet deep. The cause of the sinkhole was speculated to be the result of piping of alluvial foundation materials and/or shell materials of the dam embankment

[1] Civil Engineer, Bureau of Reclamation, Denver, Colorado, jstateler@usbr.gov
[2] Geotechnical Engineer, Bureau of Reclamation, Denver, Colorado, jwormer@usbr.gov

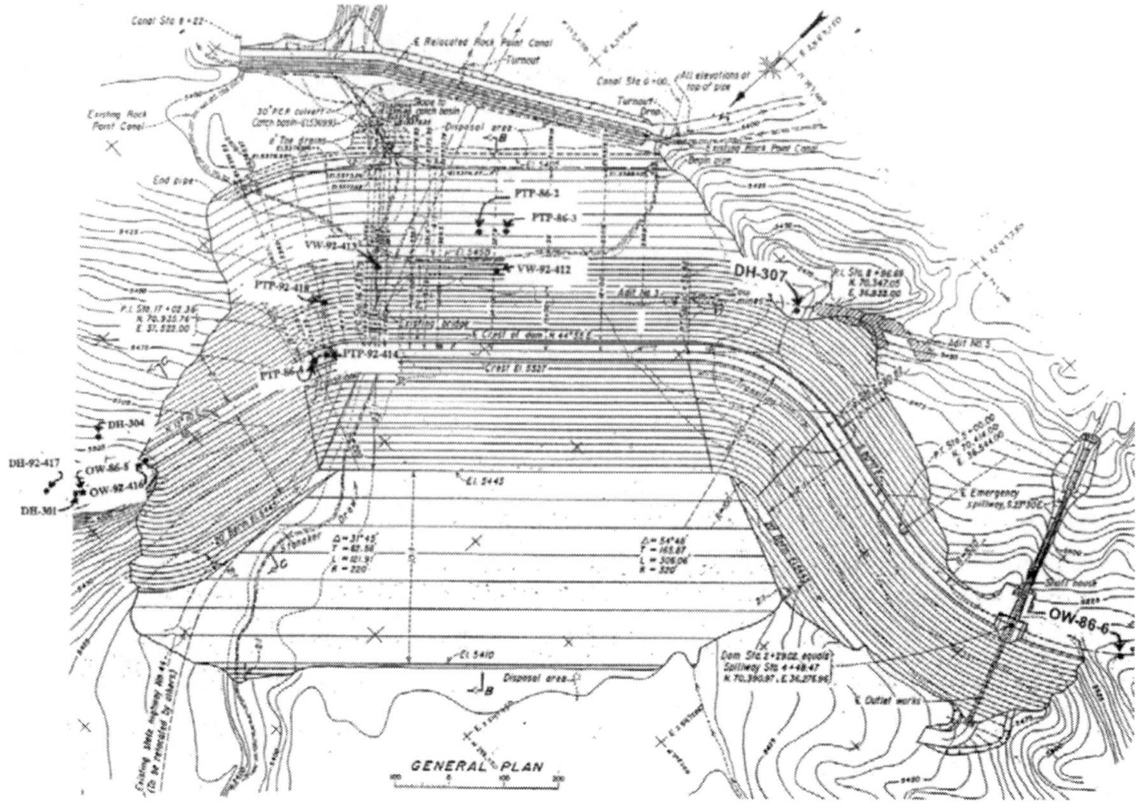

Figure 1. Plan View of the Dam as Originally Constructed.

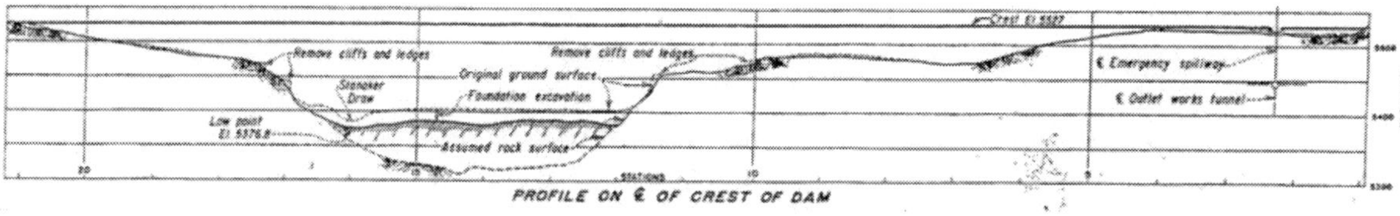

Figure 2. Profile View of the Dam Along the Centerline of the Dam Crest.

(zone 3) resulting from seepage through the joints in the bedrock of the left abutment. The continued removal of the finer fraction of the zone 3 material resulted in the stoping of the zone 3 material and formation of the sinkhole. Additional sinkholes developed in the same general area in June and October 1963, and again in August 1965. Increasing seepage was occurring at the left toe drain throughout this time period, and the seepage was transporting large quantities of sand that were deposited in the catch basin at the toe drain outfall. On May 11, 1963, seepage was first reported to be observed exiting from bedrock near the left groin on the dam. In light of all these observations, exploratory drilling was embarked upon at the damsite. Dye testing was also performed in an effort to better understand the seepage flow paths. As a result of the development of the sinkholes, and the results of the exploratory drilling and dye testing, it was decided to commence a grouting program in the left abutment. The grouting work was performed between December 15, 1965 and April 20, 1966, and it was successful in greatly reducing the seepage flows through the left abutment. No other sinkholes or other evidence of seepage problems at the left abutment, or in the adjacent embankment, have been observed since completion of the grouting program in 1966.

Modifications to the dam to reduce the risk of liquefaction-induced stability failure were completed in 1994. The modifications consisted of excavating and removing low-density alluvial foundation soils to a depth of about 30 feet at the downstream toe of the dam, using dynamic compaction to densify alluvial foundation soils below the bottom of the excavation, backfilling the excavation with compacted soil, and then constructing a stability berm to about 50 feet above the original ground surface in the downstream area (top elevation of 5450 feet).

As the reservoir was being drawn down in preparation for the 1994 seismic stability dam modification work, a slope failure occurred at the upstream left abutment where relatively impervious zone 1 embankment material had been placed to blanket (and therefore control seepage flow through) the upstream left abutment area. The zone 1 material was placed at a slope of 2H:1V at the time of original dam construction, which is steeper than the 2.5H:1V upstream slope of the dam. Figure 3 shows a typical detail regarding placement of the impervious zone 1 material to blanket abutment areas. As can be seen looking at Figures 1 and 2, blanketing was done regarding both dam abutments.

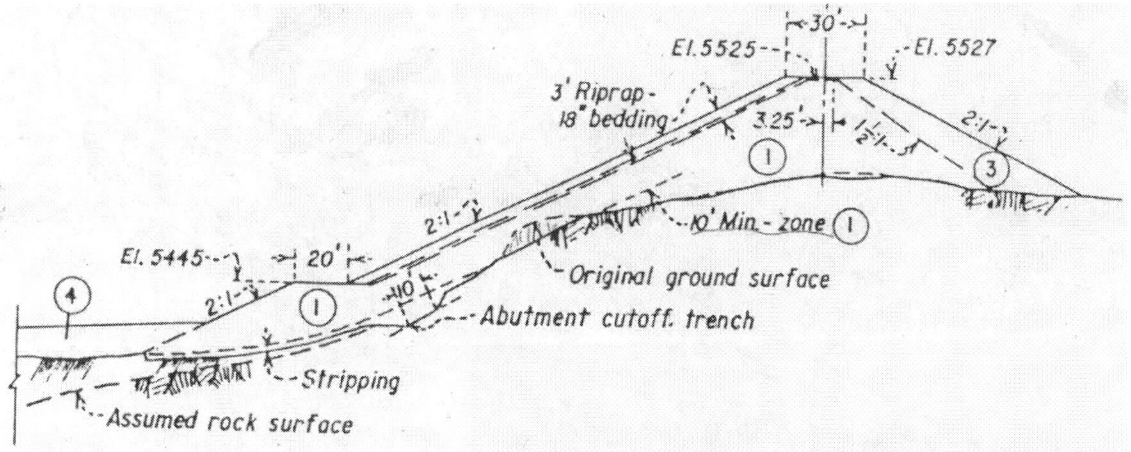

Figure 3. Section View of the General Embankment Blanketing Details at the Abutments.

Evidence of the slope failure in the zone 1 material at the left abutment was detected April 7, 1994. After investigation, it was concluded that the slope failure was the result of a steep embankment slope in combination with rapid drawdown of the reservoir. Exploratory drilling was performed in the area, and inclinometers were installed in an effort to identify the slide plane. Although suspicious materials were encountered (zones of swelling montmorillonite or bentonite clays), the failure plane could not be definitively defined. It was decided to purposely draw down the reservoir at rates of up to 1 foot per day to induce further slide movement such that the inclinometers could allow identification of the slide plane. Extensive visual and instrument observations were performed during this time. When the water level had dropped 17 feet to elevation 5463, movement was detected in the inclinometers, as well as in surface measurement points that had been installed. As a result of this movement, the failure plane was identified and the design of the remedial fix commenced.

The repair at the left abutment consisted of: (1) removing and stockpiling the existing overlying riprap, (2) removing embankment material to below the depth of the slide plane, (3) placing a 10-foot-width of zone 1 impervious earthfill material on the excavated slope, (4) placing a large volume of sands, gravels, and cobbles on top of this earthfill material to create a weighted berm with a 3.5H:1V slope that would be stable under rapid drawdown loading conditions, and (5) covering the berm with a 3-foot-thick layer of riprap. The repaired slope has performed satisfactorily since its construction, with no additional cracks, slumps or other indications of movement being detected in this area.

2014 SLOPE FAILURE

On September 23, 2014, a fisherman noticed scarps and evidence of a slope failure at the upstream right abutment area of the dam, where zone 1 impervious blanket material covers right abutment rock, as shown in Figure 4. (Figure 4 is similar to Figure 3, except no berm is present at the toe of the slope in Figure 3.) This slope failure was generally

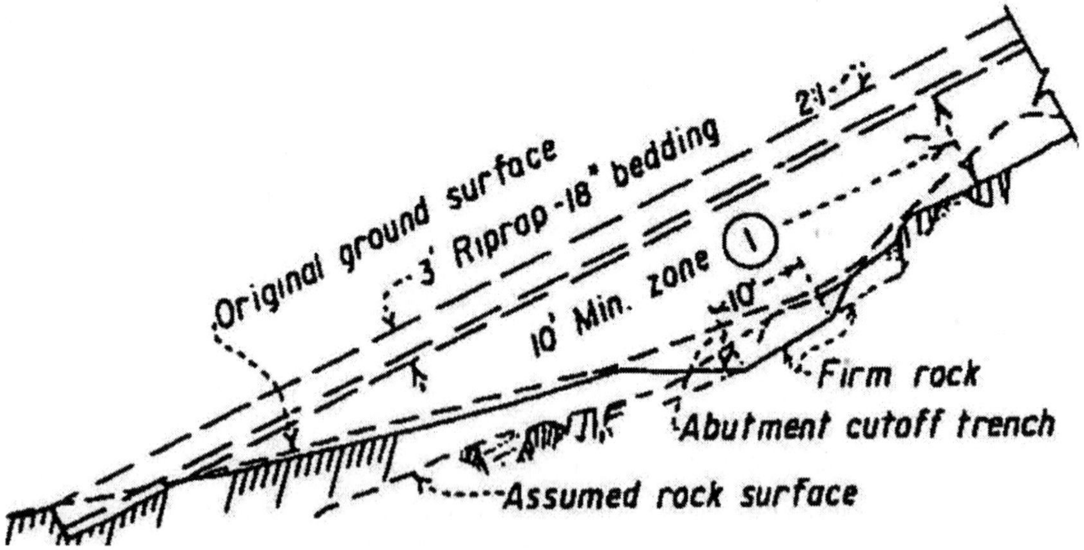

Figure 4. Section View of Embankment Blanketing at in the Area of the 2014 Slide.

similar in nature and size to the 1994 slope failure that occurred at the left abutment of the dam as the movement occurred in the zone 1 material following a significant but not atypical drawdown of the reservoir. Figure 5 shows the right abutment slide area, and Figure 6 shows an aerial view of the dam and damsite, with the location of the slide area highlighted.

Figure 5. View of Scarps (highlighted by arrows) and Slide Area – Right Abutment Area.

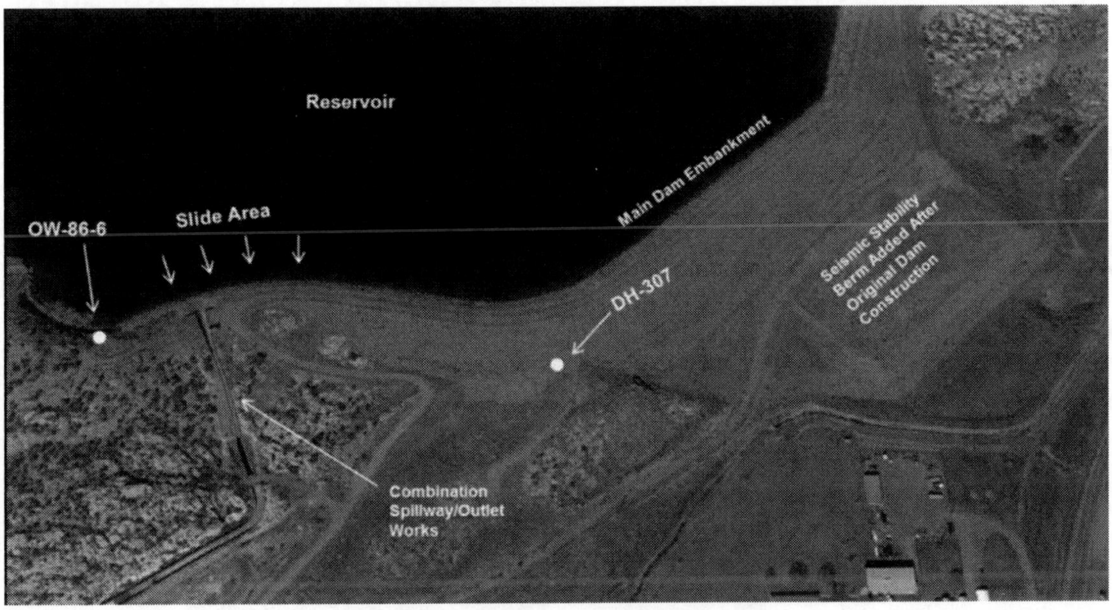

Figure 6. Aerial View of Slide Area and Damsite in General.

As is evident in Figure 5, at the right portion of the slide area, there are two scarps. This suggests that movement first occurred at a lower elevation, which then apparently triggered movement at a higher elevation.

Figure 7 shows reservoir level data and rainfall data for 33 months prior to the detection of the slide (January 2012 to September 2014), as well as water level data for observation well OW-86-6A that is located at the right end of the dam (see Figure 1, OW-86-6). Figure 8 shows an expanded view of this same information for the time period from January 2014 to September 2014. Two water pressure monitoring instruments are located at OW-86-6, with OW-86-6A being the lower instrument and OW-86-6B being the upper instrument. Looking at Figures 7 and 8, it is noteworthy that OW-86-6A typically closely tracked reservoir levels, but for the two monthly readings immediately prior to when the slide was detected, readings for this instrument stayed anomalously high.

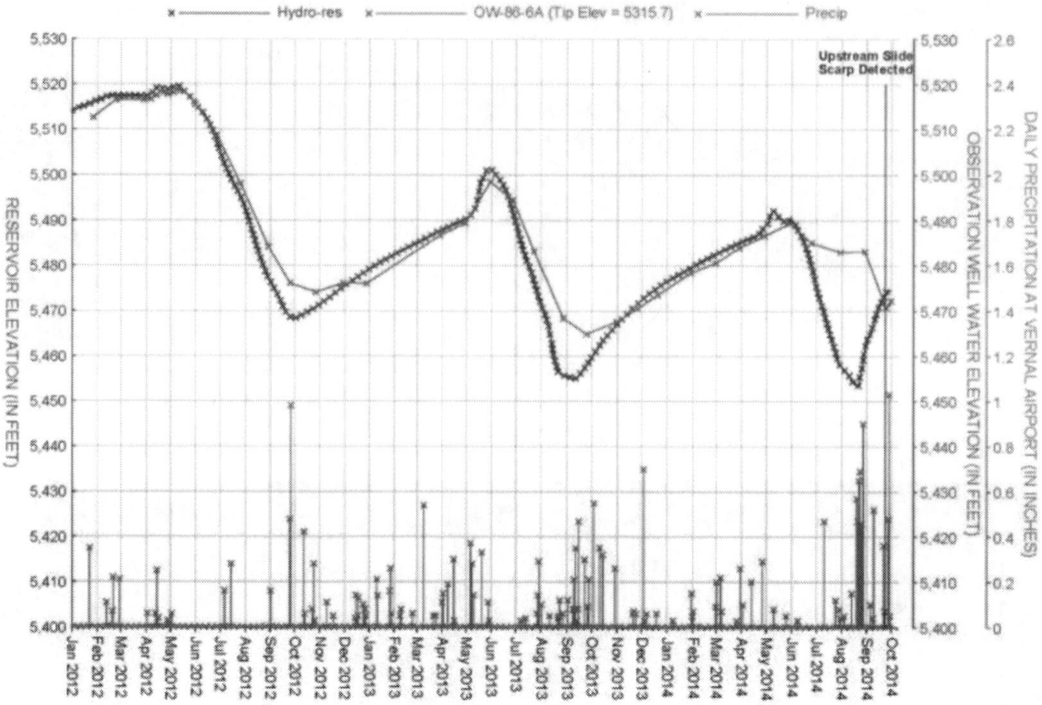

Figure 7. Rainfall Data, Reservoir Level Data, and Water Pressure Data from Observation Well OW-86-6A from 2012 until the 2014 Slope Failure

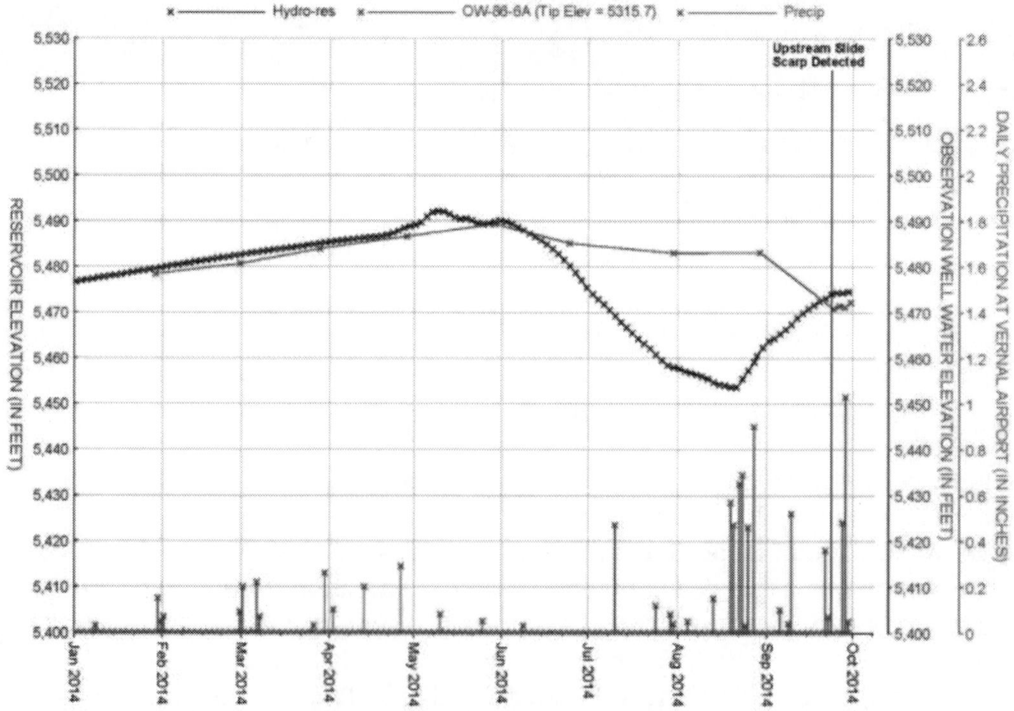

Figure 8. Rainfall Data, Reservoir Level Data, and Water Pressure Data from Observation Well OW-86-6A for 2014 Prior to the 2014 Slope Failure

PRE-SLIDE DAM SAFETY MONITORING PROGRAM

Prior to the 2014 slope failure at the right abutment, the active dam safety monitoring program for the dam consisted of ongoing visual inspections, toe drain seepage flow monitoring at the 2 toe drain outfall locations, water pressure monitoring at 20 instruments (whose locations are shown on Figure 1), embankment settlement and deflection monitoring at 30 embankment measurement points located on the main dam embankment, and embankment compression and foundation settlement monitoring at one internal vertical movement (IVM) installation at the main dam embankment. None of these instruments are located in the area of the damsite where the 2014 slide occurred, and only three of the water pressure monitoring instruments, OW-86-6A, OW-86-6B, and OW-DH307, are located somewhat proximate to the area of the slide. These three instruments all have their influence zones in foundation rock (as opposed to embankment material). Instruments OW-86-6B and OW-DH307 provided only "dry" readings in the months prior to detection of the upstream slide, indicating water levels at these instrument locations were below the tip elevations of these instruments. Actual water levels were only detected at instrument OW-86-6A, as shown on Figures 7 and 8.

GEOLOGY

The geologic formations at the right abutment of the dam consist of: (1) an upper sandstone formation, (2) a shale formation, and (3) a lower sandstone formation. The upper sandstone formation consists of a shaley to clean sandstone with limestone, calcareous shale, siltstone, and coal beds. It has high secondary permeability because some units within this formation are intensely fractured with closely to moderately spaced joints. Most of the seepage problems that developed upon first reservoir filling at the left abutment of the dam were associated with units of the upper sandstone formation.

The shale formation consists of a dark gray siliceous shale with interbedded montmorillonitic (bentonite) clay beds. It is soft (clay beds) to hard (shale) poorly indurated and fissile, laminated and moderately to intensely fractured with variably spaced joints that are tight to partly open. The 2014 right abutment slide area is approximately at the elevation of the contact between the upper sandstone formation and the underlying shale formation, with the scarps, especially the higher scarp, being located within the highly permeable upper sandstone formation. Beneath the shale formation is the lower sandstone formation which is composed of beds of moderately cemented, porous and friable sandstone and pebble conglomerate. The influence zone of observation well OW-86-6A is located in this unit.

Several abandoned coal mine adits existed on the downstream right abutment of the dam at the time of original construction. Some of these adits were reportedly backfilled with sand, gravel and cobbles during dam construction.

RESPONSE TO THE 2014 SLIDE

Immediately following detection of the slide on September 23, 2014, filling of the reservoir, that had been ongoing, was halted, and the outlet works intake structure was checked for possible blockage. A more comprehensive dive inspection of the intake structure was performed on October 6, 2014, that indicated that there was no evident blockage or damage to the structure, and no upstream, toe bulging as a result of the slide. Sets of lath stakes/markers were placed across various points along the scarps in the disrupted slope to monitor for possible further slide movements. Some of these installations were subsequently damaged and lost during the winter of 2014-2015, and all but one of them became submerged by reservoir water when reservoir filling occurred in 2015, but based on the available data for these installations, no appreciable movements have been recorded at them to date. Also, an aerial survey was performed to provide elevation contour information for the disrupted area of embankment material at the right abutment.

Stability analysis work was performed that concluded that the blanketed area of the right abutment has a minimum upstream steady-state safety factor of 1.2 using the best estimate for long-term strengths and assuming long-term operations (normal water surface elevation 5520.5). The stability analysis indicated that for rapid drawdown from the normal water surface elevation of 5520.5 feet to elevation 5445.6 (the outlet works intake elevation), using the best estimates of material properties, the right abutment impervious blanket has a safety factor of about 0.8.

A one-day team risk analysis was performed on October 9, 2014 to assess the situation. The following four new Potential Failure Modes (PFMs) for the dam were identified by the team due to the occurrence of the right abutment slide:

- PFM 1 - Internal erosion by seepage flow of zone 1 (core) embankment material at the contact between the zone 1 and the right abutment bedrock at the main dam embankment due to seepage entering and travelling through a scarp at the right abutment, and then traveling laterally through the abutment to reach the zone 1/abutment contact area at the main dam embankment where the internal erosion would initiate and develop.
- PFM 2 - Expansion or progression of the right abutment slide results in a lowering of the dam crest elevation and allows for overtopping of the dam embankment.
- PFM 3 - Internal erosion by seepage flow at the right abutment due to flow entering and travelling through the scarp, and then initiating internal erosion that would develop through the right abutment.
- PFM 4 - Dam overtopping by a storm with the outlet works unavailable for release because it is blocked by a subsequent slope failure at the right abutment.

The one-day team risk analysis found no significant risks associated with PFM 2, 3, and 4, but noteworthy risks associated with PFM 1. Assuming the reservoir would be allowed to fill to elevation 5499 (25 feet above the reservoir level at the time of the

October 9, 2014 risk analysis), to allow the 2015 water irrigation needs to be satisfied for the project's beneficiaries, the dam failure risks associated with PFM 1 were estimated to be slightly above Reclamation's risk criteria. In light of this situation, the following decisions were made during a meeting on October 10, 2014:

- Exploratory drilling would be conducted at the damsite to gather more information about the foundation materials of the right abutment and to allow installation of new water pressure monitoring instruments in key locations: (1) related to the slide area, and (2) related to PFM 1.
- New, more robust slope monitoring installations would be provided in the slide area.
- Reservoir filling would be allowed to resume at a rate of 0.2 foot/day. When the reservoir level reached 5495 feet, which would put water on portions of the scarps, the reservoir level would be held steady for 30 days to look for evidence of unsatisfactory dam performance. If no such evidence was apparent, filling would be allowed to continue to elevation 5499.
- Instrumented and visual monitoring of the dam and damsite would be conducted three times per week.

Twenty-four new slope movement monitoring installations were installed in the slide area, with the majority of these installations being completed in early November 2014. These monuments were installed in groups of three, as shown schematically in Figure 9. Checks for distance changes between the reference monuments at the dam crest and the upper and lower monitoring locations on the slope in the slide area allow monitoring for new slope movements. There has been no evidence of such movement, based on the data obtained from these installations to date.

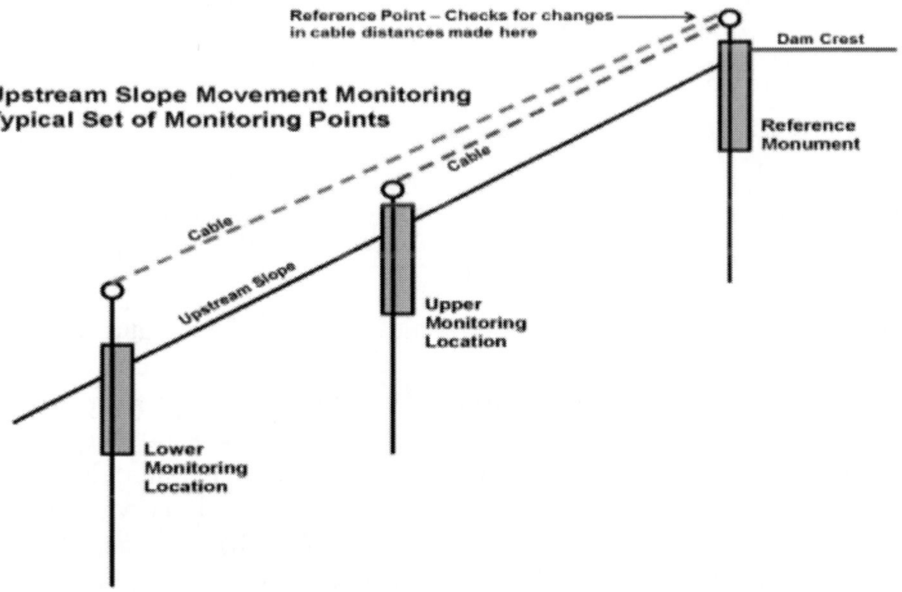

Figure 9. Upstream Slope Movement Monitoring Installations – Right Abutment Slide Area

Reclamation personnel carried out exploratory drilling at the damsite in the fall and winter of 2014-2015 to gain greater understanding of the foundation materials in the right abutment area and the area where PFM 1 might initiate, and to allow installation of new water pressure monitoring instruments in these areas. Figure 10 shows the plan view locations of these new open-standpipe piezometers, which were designated PT-14-1 through PT-14-7. PT-14-2, -6, and -7 have their influence zones in bedrock, PT-14-1, and -5 have influence zones at the bedrock/embankment contact, and PT-14-3 and -4 have their influence zones in embankment material above the bedrock contact. Figure 11 shows the available data collected to date from these instruments, along with data for the other water pressure monitoring instruments in the right abutment area. Since PT-14-1 through PT-14-7 were not installed until after the right abutment slide occurred, they cannot provide information about changed water pressures in the abutment and dam embankment resulting from the right abutment slide.

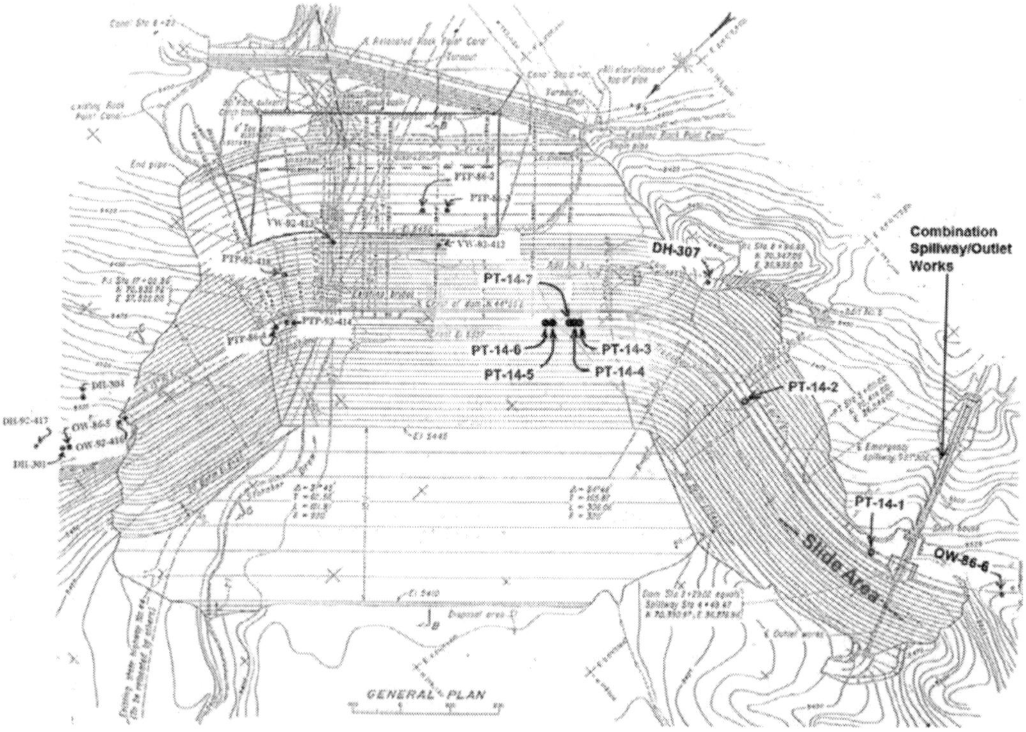

Figure 10. Water Pressure Monitoring Instruments, Including the New Piezometers Installed in 2014-2015 (PT-14-1 through PT-14-7)

Pump out testing of PT-14-5 and PT-14-6 in mid-May 2015 (see Figure 11) showed these instruments to be located in relatively pervious areas, since they recovered to pre-test water pressures within just a few days. Since the influence zone of each of these instruments is located either partially or totally in bedrock, this is not surprising as the secondary permeability of bedrock sandstone is envisioned to be high. Pump out testing of PT-14-4 showed this instrument to be located in a relatively impervious area, since its recovery towards pre-test water pressures was slow. Since the influence zone of this instrument is located in core material of the dam embankment, this is in line with expectations.

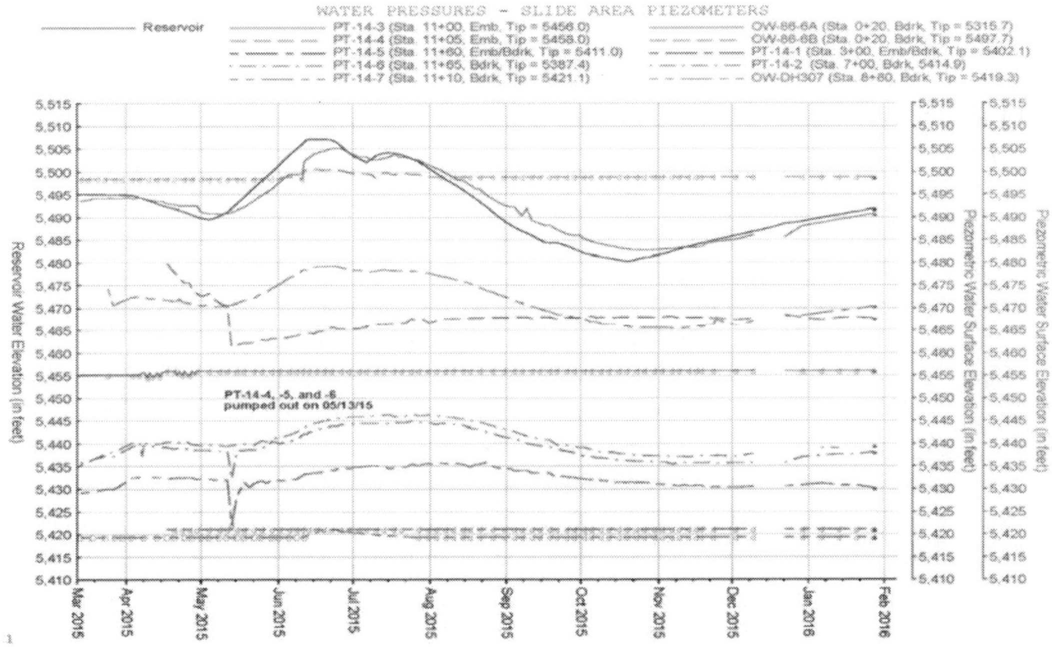

Figure 11. Data from Water Pressure Monitoring Instruments at or near the Right Abutment

Foundation water pressures beneath the dam crest decrease as one moves from the right end of the dam (OW-86-6A and -6B) to PT-14-1, and then to PT-14-2, and then to the cluster of new instruments in the area of the main dam embankment. That said, the pressure drop from PT-14-2 to PT-14-6 is very small (only about 1 foot) over a distance of roughly 450 feet, and the pressures of these two instruments closely track one another, as can be seen on Figure 11. This indicates a very high degree of hydraulic connection between these locations in the right abutment. While not desirable, this is not unexpected, given the high secondary permeability in the right abutment sandstone formations, and the strike and dip of the foundation formations at the damsite (the strike is subparallel to the dam centerline at the main body of the dam, and the formations dip steeply in the downstream direction).

As can be seen on Figure 11, when the 30-day hold period at reservoir elevation 5495 was concluded (with no indications of unsatisfactory performance), lower amounts of water available for filling the reservoir, in combination with reservoir releases that were occurring, resulted in falling reservoir levels. However, significant rainfall occurred from mid-May to mid-June in 2015 that allowed filling of the reservoir to resume. A decision was made to allow the reservoir to continue to fill to a level above the planned limit of 5499 feet, so long as there were no indications of unsatisfactory performance. This took place. This was done so that the reservoir drawdown that would occur in the summer and fall of 2015 would not need to extend to very low reservoir levels (occurring at rapid drawdown rates) that could lead to new slide movements.

The intensive monitoring of the dam revealed no unusual or anomalous observations from the visual monitoring efforts. No new seepage flows were observed. No change from historical performance has been seen regarding the toe drain flows at the dam. However, unexpected data from observation well OW-DH307 in mid-June 2015 brought a halt to the "extended" reservoir filling, as discussed below.

Figure 12 shows the data collected since the slide was detected from just the water pressure monitoring instruments that were in place prior to discovery of the slide (OW-86-6A, OW-86-6B, and OW-DH307). OW-DH307 had always been "dry," even under full reservoir conditions (reservoir at elevation 5520.5), but following the occurrence of the slide it had anomalous periods of "non-dry" readings in late October and early November, 2014, and then again from June 12, 2015 to the present. On Figure 12, the "green line" shows the OW-DH307 data to an exaggerated scale so that the anomalous data are easier to view and assess. Since OW-DH307 is located near the downstream toeof the dam, the foundation water pressures recorded by it, though anomalous, are much lower than the foundation water pressures seen beneath the dam crest (recorded by the new instruments PT-14-2, -5, and -6). The influence zone for OW-DH307 is in the upper sandstone formation, above the shale formation.

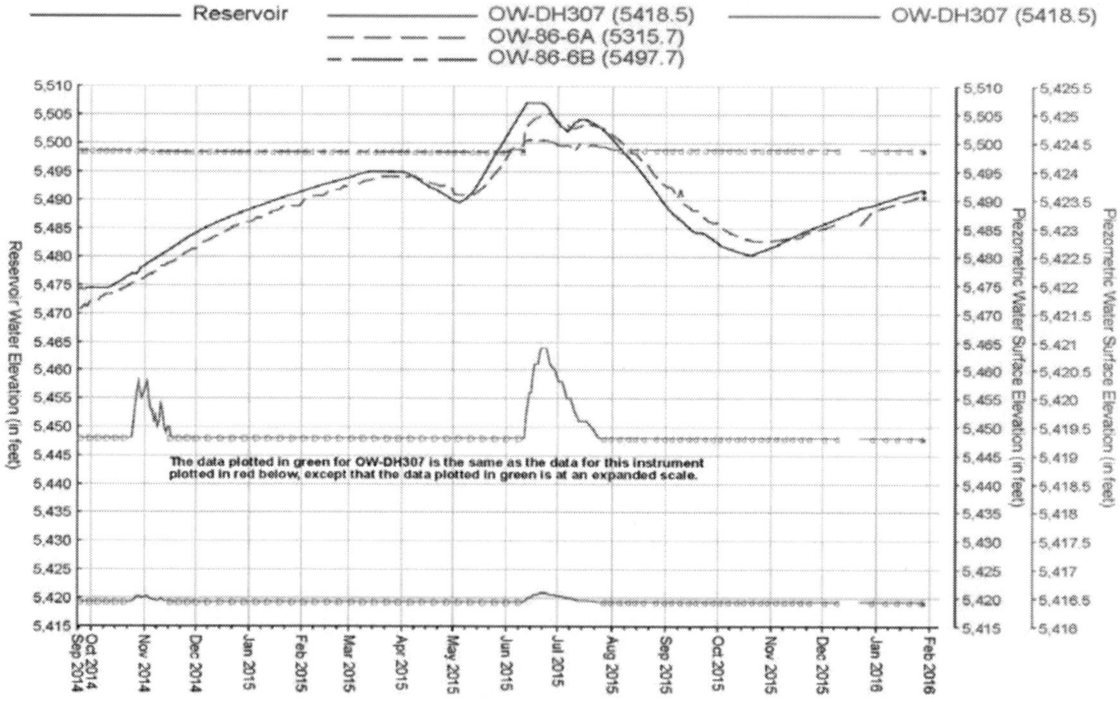

Figure 12. Data from Water Pressure Monitoring Instruments at the Right Abutment that were Present When the 2014 Abutment Slide Occurred

OW-DH307's water level suddenly rose about 1 foot on October 27, 2014, gradually though somewhat erratically declined over the next 19 days, and then went back to being continuously "dry" starting on November 16, 2014. This period of anomalous data has no correlation with the rising reservoir levels at the time. It looks like the instrument somehow suddenly got charged with water, and then the excess pressure gradually dissipated. The top of the instrument's influence zone is only 10 feet below the ground surface, so it is not inconceivable that surface water somehow reached the influence zone and impacted the instrument's data. However, an 8-foot-thick bentonite "plug" was placed in the drill hole above the top of the influence zone to prevent such an occurrence, and no such anomalous data have ever been seen before October 2014 at this instrument. No firm explanation for the 19-day period of "non-dry" readings in October and November of 2014 at this instrument was developed at the time of its occurrence. Daily monitoring of this instrument was performed during this time period of "non-dry" readings.

On June 12, 2015, a second period of anomalous "non-dry" readings began at OW-DH307, which also triggered daily monitoring of this instrument. Unlike the first period of anomalous "non-dry" readings, this second period of anomalous data is clearly related to the reservoir level. Filling of the reservoir was halted on June 12, 2015 in response to OW-DH307 once again showing water. When significant lowering of the reservoir level began to occur on June 26, 2015, water levels in OW-DH307 began to drop, confirming that reservoir water levels were driving the water levels seen at OW-DH307. To reiterate, prior to October 2014, OW-DH307 had always been "dry," even under full reservoir conditions. In June 2015, with the reservoir level 13.5 feet below the full reservoir elevation of 5520.5 feet, water was seen in OW-DH307. This situation strongly suggests that the disruption of the upstream slope at the right abutment area caused by the 2014 slide opened up seepage pathways through the blanketing embankment material placed upstream of the abutment rock, and resulted in increased water pressures in the right abutment (at least in the vicinity of OW-DH307).

EVALUATION

The high water pressures in the right abutment at the time of the slide could have resulted in a decrease in the effective stress and a subsequent loss in strength in the right abutment blanketing embankment material. As indicated on Figures 7 and 8, OW-86-6A showed high water levels at the time the slide occurred that had not declined in tandem with the decline in the reservoir level. This disconnect was historically unprecedented. Why this occurred in 2014 is not known for sure, but dam operating personnel did indicate that roadwork had been done on the right abutment road to the dam crest shown on Figure 6 in the summer 2014. This work did involve the use of dynamic compaction equipment. It is not clear that this has any association with the changed performance of OW-86-6A, but it represents information that was uncovered during investigation of the occurrence of the 2014 slide. Beside considerations of abutment water pressures, the clay that comprises the right abutment blanket may have lost strength over time due to softening. The decreased strength in the right abutment blanket material due to high water pressures in the abutment sandstone rock, and strength softening of the clay, likely resulted in destabilization of the right abutment blanket slope and sliding at the approximate contact of the upper sandstone formation and the underlying shale formation.

Looking at Figure 8, it is hard to believe that the slide actually occurred in late September 2014. One month earlier, the reservoir was being drawn down to a very low level at the same time that heavy rainfall was occurring (which would have added some weight to the slide mass driving force). By late September, the reservoir level was rising, and had risen roughly 20 feet above its low level for the year. Rainfall that was occurring in late September was much less than was recorded 30 days earlier. Nonetheless, personnel that operate a park associated with the dam's reservoir, and regularly view the upstream face of the dam, indicated that the slide was not apparent until September 23, 2014.

In retrospect, the "non-dry" condition seen at OW-DH307 in October and November 2014 may have been a "true indication" of the abutment water pressure at this instrument at that time. It is possible that these historically unprecedented water levels at this instrument may have resulted in the blow-out of material that was backfilled in an abandoned coal mine adit in the downstream right abutment, thus relieving the abutment pressures in the area and resulting in water levels in OW-DH307 dropping and returning to a "dry" condition. Another possible explanation for the drop in water pressures at OW-DH307 is that the water passing through a damaged area of the blanket moistened the embankment material, causing it to swell and self-heal.

A week-long team risk analysis was commenced on April 13, 2015 to follow up on, and explore in greater detail, the risk analysis work previously conducted on October 9, 2015. The same four new PFMs were again evaluated, and no new PFMs were brought forward during this effort. The risk analysis was charged with evaluating the situation where the reservoir was at its normal, full operating level of elevation 5520.5 without repair of the slide, and found the risks associated with PFM 1 to be 1-2 orders of magnitude above Reclamation risk criteria. Consequently, it is planned that dam modification work will be carried out in 2017 to remediate the disrupted areas of the right abutment embankment so as to achieve a tolerable risk situation that allows normal operation of the dam to re-commence thereafter.

CONCLUSIONS

The response to the 2014 right abutment slide proceeded in a logical manner that allowed: (1) assessment of the situation, (2) a determination of the dam safety risks posed by the situation, and (3) reservoir operations and water deliveries to occur in 2015, to the extent consistent with dam safety. Slope stability evaluation work, site exploration work, development of PFMs, and conducting risk analysis work allowed the situation and dam safety risks to be better understood. Enhanced and expanded monitoring of slope movements, abutment water pressures, toe drain flows, and visual observations allowed for timely detection of any anomalous performance that could indicate initiation/development of PFM 1 or any other PFM of concern.

One cautionary note is that the dearth of water pressure monitoring instruments in the right abutment area of the dam at the time of the slide posed difficulties with respect to assessing pre-slide versus post-slide abutment water pressures. Everything pretty much came down to one instrument, OW-DH307, which historically had always been dry. (OW-86-6A, which is located beyond the right end of the dam, historically had reflected reservoir levels, so it provided no insight regarding pre-slide versus post-slide abutment water pressures.) As a consequence of this lack of water pressure monitoring instruments in the right abutment area, the impact of the slide on the possible development of PFM 1 could not be readily evaluated from the available information, until OW-DH307 eventually showed non-dry readings. These non-dry readings did indicate increased abutment water pressures, apparently due to the scarps of the slide. However, the provision of a modest network of water pressure monitoring instruments in the right abutment at the time the dam was constructed would have greatly benefited evaluation efforts with respect to the 2014 right abutment slide.

ON CALERO POND: EVALUATION OF DOWNSTREAM WETLAND AND POTENTIAL DAM SEEPAGE

Matthew Powers, PE[1]
Bal Ganjoo, PE[2]
Mark Freitas, PE, GE[3]

ABSTRACT

Calero Main Dam is an 840-ft long, rolled earthfill structure with an upstream concrete face slab. The maximum height of the Main Dam is about 90 feet. The project site was dramatically altered when the dam was originally constructed in the 1930's and later in the early 1980's when the reservoir outlet works were modified. Seepage has occurred at the Calero Main Dam since shortly after construction and the reservoir filling in 1937. Historically, seepage has occurred at the Calero Main Dam on the downstream right abutment, the downstream right face and groin, and the downstream left face of the dam when the reservoir level is high. Approximately 450 feet downstream of the toe of the dam is a small pond which is currently part of a delineated wetland.

It had been postulated that seepage from Calero reservoir contributes to or is linked to the presence of the pond (i.e., seepage from Calero Reservoir is a contributing factor to the formation and presence of the pond). As part of planned seismic improvements at Calero Main Dam, an engineering assessment was performed to evaluate the potential influence of reservoir seepage at Calero Main Dam on the existence of the pond. Results of this assessment will be used to inform design of any necessary seepage collection systems associated with the embankment retrofit and aid in the understanding of longer-term impacts to surrounding habitat.

The assessment of the downstream pond utilized historic records and photography relevant to Calero Dam, topographic survey, subsurface exploration and installation of ground and surface water monitoring instruments, and evaluation of available piezometric data. The paper discusses the assessment methodology and presents a concluding opinion on the contributing sources to the formation and presence of the downstream pond area.

INTRODUCTION

As part of planned seismic improvements at Calero Main Dam, The Santa Clara Valley Water District (District) is addressing a postulation that seepage from Calero reservoir contributes to or is linked to the presence of a pond situated about 450 feet downstream of the toe of the Main Dam. The pond area is part of a delineated wetland below Calero Dam and is adjacent to the Calero valve yard, and the low-level outlet release to Calero

[1] GEI Consultants, Inc., 180 Grand Avenue, Ste. 1410, Oakland, CA 94612, mpowers@geiconsultants.com
[2] Santa Clara Valley Water District, 5750 Almaden Expressway, San Jose CA 95118, bganjoo@valleywater.org.
[3] GEI Consultants, Inc., 180 Grand Avenue, Ste. 1410, Oakland, CA 94612, mfreitas@geiconsultants.com

Creek. The pond area, Calero Main Dam, and other relevant site features are shown on Figure 1. The dam is located southeast of San Jose in Santa Clara County, California.

Seepage has occurred at the Calero Main Dam since shortly after construction and the reservoir filling in 1937. Historically, seepage has occurred at the Calero Main Dam on the downstream right abutment, the downstream right face and groin, and the downstream

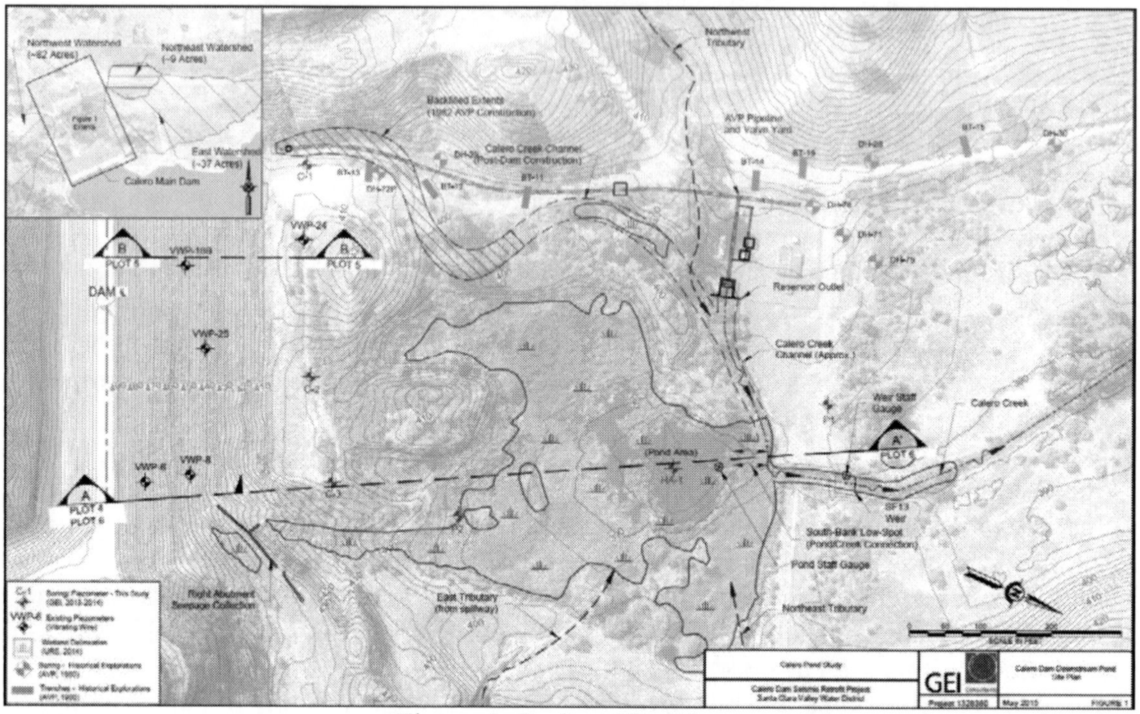

Figure 1. Project Site Map.

left face of the dam when the reservoir level is high. A grout curtain was installed in 1991 to reduce seepage through the right abutment foundation; however, the grouting program was only partially successful. A previous evaluation performed by the District (Nelson, 2007) provides an in-depth review and analysis of historic seepage patterns and previous mitigation measures at the Calero Main Dam. There are currently reservoir operation restrictions due to inadequate seismic stability of the Calero Main Dam and, as such, reservoir levels have not been at a sufficiently high level for seepage to be observed.

The objective of this engineering assessment is to evaluate the potential influence of reservoir seepage at Calero Main Dam on the existing pond downstream from the toe of Calero Dam (i.e. has seepage from Calero Reservoir been a contributing factor to the formation and presence of the pond?), provide a concluding opinion on the contributing sources to the formation and presence of the downstream pond area. Results of this assessment will be used to inform design of any necessary seepage collection systems associated with the embankment retrofit and aid in the understanding of longer-term impacts to surrounding habitat.

EXISTING DOCUMENTS AND DATA REVIEWED

Available documents from the District's files were reviewed for this engineering assessment as listed in the references. This information included construction history and as-built documentation, geologic and geotechnical conditions encountered during construction activities, wetland delineations downstream of the Main Dam, photos taken during construction, aerial photographs of the dam site, and records of the water level within Calero Reservoir and Calero Creek. Instrumentation data provided by the District was also compiled for this study and used in project analyses. Study data included historic records of Calero Reservoir level and flows within Calero Creek, pond water levels, rainfall, and piezometer readings from existing Main Dam piezometers and piezometers installed in the vicinity of the pond area. Evaluation of available information and data is discussed in the following sections.

FIELD WORK PERFORMED FOR STUDY

Field work performed for this study included site reconnaissance, subsurface exploration, installation and monitoring of new standpipe piezometers, and topographic survey of the pond area features. In addition, an updated survey and delineation of wetland areas was completed previously as part of the overall seismic retrofit project (URS, 2014).

Standpipe piezometers P-1 and P-2 were installed in the vicinity of the downstream pond to allow monitoring of groundwater levels within the alluvial soils overlying bedrock. A topographic survey of the downstream pond area and vicinity was completed while the pond was dry and accessible by the survey crew. Survey data was used to refine topographic contours in the vicinity of the pond, correlate piezometer and staff gauge measurements to elevation datum, and to aid in project evaluations.

BACKGROUND AND SITE HISTORY

Background

Calero Main Dam is an 840-ft long, rolled earthfill structure with an upstream concrete face slab. The maximum height of the Main Dam is about 90 feet. The crest of the dam is at El. 492.9[4]. The crest is approximately 20 feet wide and is gravel paved. The most recent survey of the dam crest indicates a minimum elevation of 492.2 feet. Calero Main Dam was designed and constructed as a zoned earthfill dam with an impervious zone in the upstream half and a pervious zone in the downstream half. Zoning within the dams was achieved by using relatively finer-grained materials in impervious zone and coarser-grained materials in the pervious zone; however, the actual distinction of the grain size distribution between the two zones within the embankment is only slight. The upstream and downstream slopes are approximately 2.5H:1V. The upstream slope is covered with a 4-inch-thick reinforced concrete slab for wave and erosion protection.

[4] Unless noted, elevations listed in this report correspond to the North American Vertical Datum of 1988 (NAVD88).

The spillway is located in a natural saddle about 900 feet west of the Main Dam, with a weir crest length of 84.4 feet. The outlet works consist of a concrete box with a trash rack structure 481 feet long, 36-inch diameter concrete jacketed steel outlet pipe under the Main Dam. Flow is controlled by a hydraulically operated slide gate valve located in the intake structure.

Calero Main Dam was constructed on Calero Creek starting in January 1935 and completed by the end of 1935. From 1936 to 1962, Calero Reservoir stored water every year, with the exception of 1948 and 1961, when it was completely empty the whole year. Historic operations captured winter runoff and released stored flows through the spring to achieve an empty reservoir by mid-summer for most years. In 1962, a concerted effort was made to maintain year round minimal storage of approximately 5,000 acre-feet in the reservoir to sustain the local fishery (Titus, 2003). Current seismic operating restrictions require the reservoir level to be maintained below elevation 467.4, approximately 19 feet below the spillway lip.

Review of the available historic documentation indicates that the project site was dramatically altered when the dam and reservoir were originally constructed and later in the early 1980's when a new reservoir outlet works was built as part of the modifications associated with construction and connection to the Almaden Valley Pipeline. Borrow material for the construction of the Main Dam embankment appears to have been excavated from two primary sources including alluvial material from the valley floor and diabase rock removed from the knoll on the right abutment of the dam.

Historic Review of Calero Creek and Pond Area

Prior to dam construction, in the vicinity of the present day embankment footprint, the Calero Creek channel was located along the base of the right abutment knoll. Downstream of the dam location, the channel meandered slightly within the valley as it flowed towards McKean Rd. The historic course of Calero Creek is shown on historic New Almaden Quadrangle maps (USGS, 1919) and photographs taken at the site prior to dam construction (Tolman, 1934).

Figure 2, adapted from Tolman, 1934, shows a panorama of the Calero Creek valley downstream of the future Main Dam site. The panorama shows the former alignment of Calero Creek, the right abutment hillslope, and the adjacent drainage tributaries in the vicinity of the Main Dam site and right abutment. The following description of the Calero Creek Valley prior to dam construction is offered in Titus, 2003:

"Topography in adjacent, unaltered areas and historic photographs suggest Calero Creek was a meandering channel with riparian borders prior to [borrow] excavation and dam construction. The pond and wetland basin probably did not exist, or was in a very different shape and form before excavation in 1935. There are hints on historic photographs that a small wetland basin of emergent marsh existed along Calero Creek near this location prior to dam construction."

Figure 2. Calero Creek Valley, Downstream of Main Dam Site (Titus, 2003).

Following completion of dam construction, the District constructed a stream flow monitoring station in 1938 along Calero Creek below the Main Dam, approximately 200 feet upstream of McKean Road. In 1976, the monitoring station (SF13) was relocated upstream to its current location below the pond and historic borrow pit area. Downstream of the monitoring station weir, Calero Creek flows away from the dam site and beneath a bridge along McKean Rd. A photograph of the monitoring station and weir at SF13 is shown on Figure 3.

Figure 3. Calero Creek Monitoring Station and Weir at SF13.

Main Dam Outlet Modifications due to Construction of the Almaden Valley Pipeline

The original Calero Dam outlet works, constructed in 1935, consisted of a concrete box inlet structure and a 36-inch-diameter concrete-encased steel pipe under the Main Dam which discharged to a stilling basin at Calero Creek, located at the downstream left toe of the dam. In 1982 the outlet works were modified to allow water to be pumped into the reservoir from the State Water Project South Bay Aqueduct. The modifications included extending the outlet pipe to the District's raw water transmission system approximately 700 feet downstream from the toe of the dam. The design also provided a new stream release facility approximately 650 feet downstream of the original outlet, with a new turnout and stilling basin outlet to Calero Creek. Approximately 240 linear feet of the historic Calero Creek channel below the old outlet works was backfilled during these modifications. Below the new reservoir outlet facility, Calero Creek now flows through a dredged channel before joining with the former Calero Creek channel downstream of the pond area.

Historic Review of Downstream Pond

Review of the available historic documents and photography indicates that an area of low topography was created downstream of the main dam resulting from borrow excavations for Main Dam construction. Figure 4 shows annotated limits of the borrow excavation area downstream of Calero Main Dam.

Figure 4. Downstream Borrow Area Limits (Titus, 2003).

Another historic site photograph, shown in Figure 5, shows water present in the area of borrow excavation and low topography dating back to January 1936, shortly after construction of the embankment was completed and prior to the reservoir filling with water and construction of the original Calero Creek weir.

Figure 5. Ponded Water within Downstream Borrow Area (Marlaive, 1936).

Review of historic aerial photography dating back to 1939 indicates that an area of ponded water downstream of the Main Dam is a consistent feature through the current life of the dam. The surface area of the pond varies somewhat over time, but is present regardless of reservoir level maintained behind the Main Dam. As documented by Titus, 2003, the pond was present even when Calero Reservoir did not have surface water, including the dry seasons from 1935 through 1962 and most of 1948 and 1961.

DOWNSTREAM POND AREA SURFACE AND SUBSURFACE CONDITIONS

Surface Conditions

Downstream of the Main Dam is a delineated wetland approximately 5 acres in area, including the area of ponded water. The wetland is bordered by tule and rush marsh with scattered stands of willow swamp (URS, 2014). As discussed above, the entire area surrounding the pond was a designated borrow area for the original Main Dam construction, and the local grade is lower than the surrounding areas. Within the dam site area, three intermittent tributary sources flow into the pond area and Calero Creek downstream of the reservoir outlet. The identified tributaries include the northwest tributary and historic Calero Creek channel, the northeast tributary, and the east tributary. These tributary sources and channels are shown on Figure 1.

The northwest tributary contains seasonal flows from a small watershed to the west of the pond area. These flows join with the Calero Creek channel just downstream of the existing reservoir outlet. The northeast tributary originates in a small meadow east of the pond which drains into the pond area. The east tributary transports reservoir overflows from the spillway during occasional spill events. No spills have occurred since 2006, in part due to recent seismic stability operational restrictions. The eastern drainage channel below the bottom of the concrete lined spillway and plunge pool has been greatly incised and locally down cut from past reservoir spill events.

On the northwest side of the pond, there is a natural or constructed berm between Calero Creek and the pond. During site reconnaissance, a low spot along this berm was observed near the 90 degree bend in Calero Creek and confirmed by project survey work. From review of historic photographs, it appears likely that the berm may have been breached or overtopped and eroded in the past by pond overflows from reservoir releases into the spillway. Figure 6 is a historic photograph taken at the dam site in 1956 which shows spillway flow entering the pond area and pond flow overtopping a small berm into Calero Creek.

Figure 6. Pond Area During Spill Event of 1956 (Titus, 2003).

It is likely that historic pond overflows in this area have partially eroded the berm creating a hydraulic connection between the creek and the pond area. Topographic contours on Figure 1 illustrate the observed low spot along the Calero Creek berm.

Subsurface Conditions

Subsurface conditions downstream of the dam in the vicinity of the pond were compiled from review of available historic documentation, including records of historic explorations and observations during construction as well as from conditions encountered during recent site explorations. In general, the area around the pond downstream of the dam consists of soil material that was largely disturbed and left behind during construction borrow excavations. Soil thickness above the underlying bedrock ranges from approximately 6 to 20 feet thick, generally thickening towards the middle of the valley, away from the adjacent hillslopes. The soil consists of fill, alluvial, and colluvial material; generally clayey sand and sandy clay with varying amounts of gravel. These soils are directly underlain by siltstone and shale bedrock of the Franciscan Formation. Compared to the soil material above, the bedrock is relatively impermeable. Groundwater, where encountered in recent explorations (GEI, 2013) was generally found between 5 to 10 feet below the existing ground surface, within the alluvial and colluvial soils above the bedrock horizon.

ENGINEERING EVALUATION AND HYDROGEOLOGIC ANALYSIS

During this study, several possibl e sources of water that coul d contribute to the presence of the downstream pond were considered and evaluated. These sources include:

Precipitation and Tributary Inflow: Precipitation within the surrounding watersheds and resulting inflow from tributary channels, including spillway discharges,
Outlet Releases: Releases from Calero Reservoir into Calero Creek,
Main Dam Seepage: Main Dam under seepage and seepage through the right abutment,
Groundwater: Local groundwater levels in the vicinity of the pond, or
A combination of several or all of the above sources.

A discussion of each of these possible sources is provided below, including a summary of the evaluations and assessments performed:

Precipitation and Tributary Inflows

As discussed previously three tributary sources contribute periodic surface flow to the pond, including the northwest, northeast, and east tributaries. The watersheds associated with each of the three tributary channels are relatively small[5]; however, the eastern tributary is associated with the reservoir spillway and periodically conveys significant reservoir overflows into the pond.

In years of significant precipitation, spillway releases from the reservoir result in flows into the eastern tributary drainage and contribute large pulses of water into the pond over a period of several days. Figure 7 shows the Calero Reservoir water surface elevation relative to the spillway crest[6] and indicates the historic spill events which have occurred at Calero Dam.

This data indicate that a total of 15 spill events have historically occurred at Calero Dam, with the last spill event recorded in 1996. Over the 79-year life of the dam (1936-2014), the average spill frequency occurs once every 5.3 years.

Data from monitoring of Calero Creek levels show that spillway releases and significant rainfall events are linked to flooding in the downstream pond. Figure 8 shows the water surface elevation of Calero Creek downstream of the Main Dam, monitored at SF13.

[5] Estimates of Watershed Areas: northwest tributary: ~82 acres, northeast: ~9 acres, and east: ~37 acres, excluding the reservoir.
[6] The spillway crest was raised by approximately 1.4 feet in 1962.

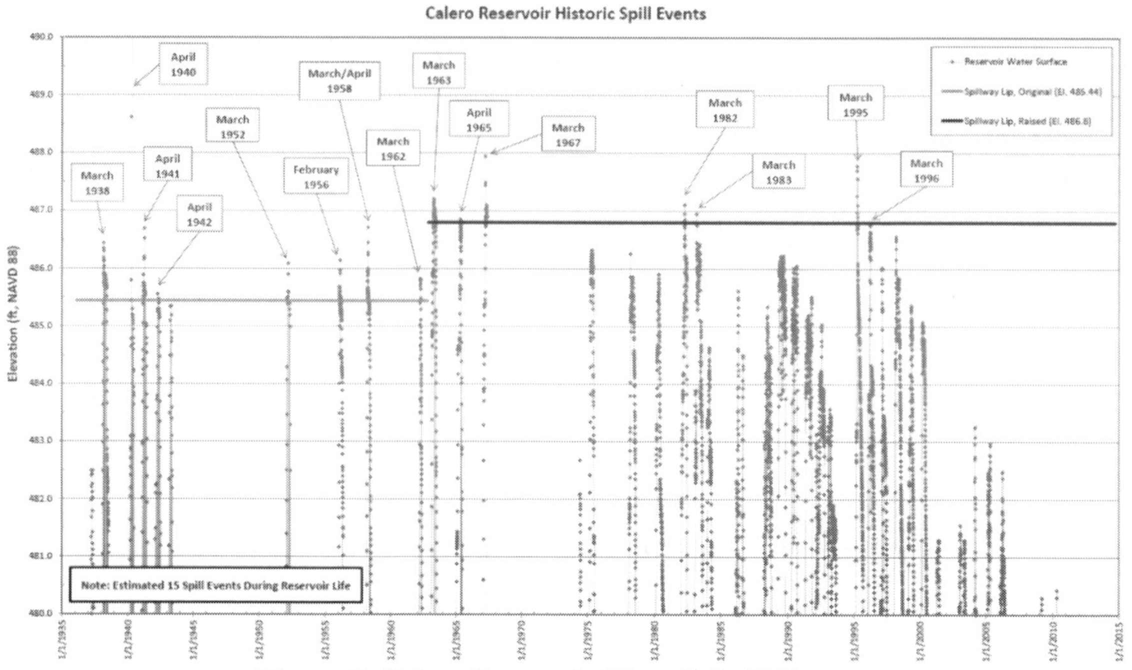

Figure 7. Calero Reservoir Historic Spill Events.

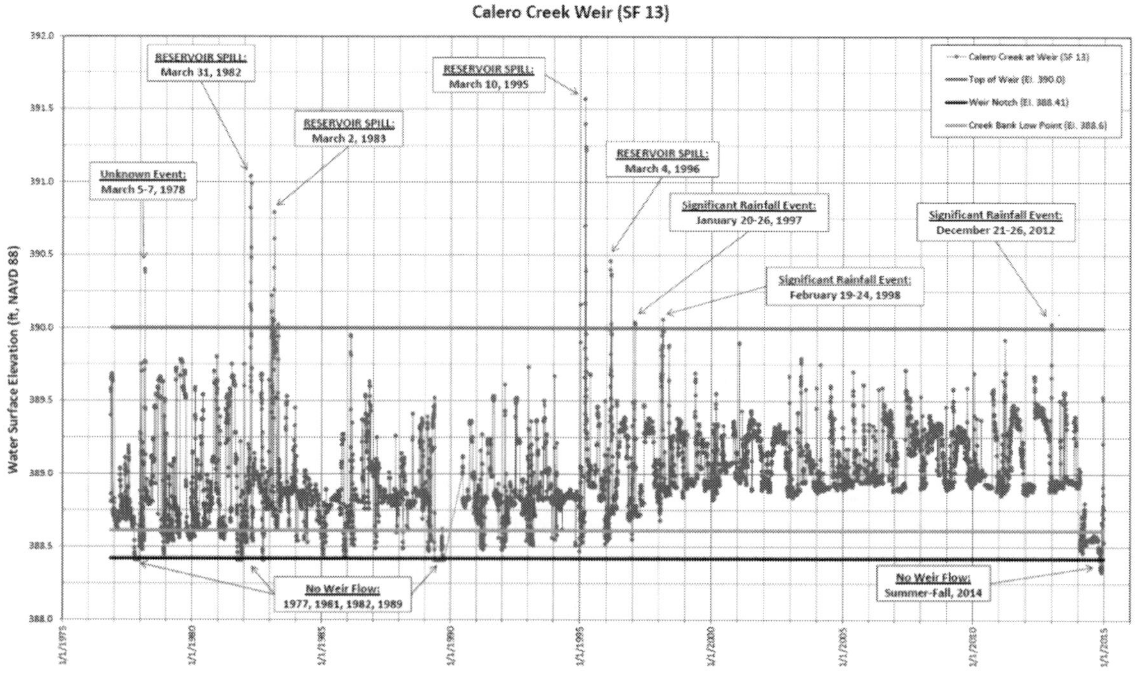

Figure 8. Calero Creek Level

Data from SF13 shows that since weir relocation in 1976 there have been eight instances when the structure was overtopped by high flows. Of these eight events, four are associated with recorded spill events (1982, 1983, 1995, and 1996), and three are attributed to significant rainfall events recorded at the Calero Dam rainfall gauge

(RF128), totaling approximately 3" or more (1997, 1998, and 2012). The remaining single event (March, 1978) is not associated with a recorded spill event and pre-dates available rainfall data. It is likely that the March 1978 weir overtopping was caused by a rainfall event, given the early-spring timeframe.

Tributary flow into the pond area was observed during a recent site visit on December 17, 2014, following a period of significant rainfall at the site. Flows within the northwest tributary channel were visually estimated at 50-60 gpm, or approximately 0.1 cfs. The flow rate in Calero Creek at that time, recorded at SF13, was approximately 20 cfs. Thus, flow from the northwest tributary constituted approximately 0.5% of the total flow rate in Calero Creek at that time, with the remainder attributed to outlet releases from the reservoir. Tributary flows in the northeast and east tributaries were judged to be significantly less than that of the northwest tributary, given the smaller size of the associated watersheds and observations of channel dimensions and evidence of flow during a subsequent site visit (12/31/14). The reservoir level was maintained below the Seismic Stability Evaluation (SSE) operating restriction, well below the spillway lip (i.e., no spillway flow).

Therefore, based on the available information and site observations, it appears that precipitation and associated tributary flow periodically contributes minor amounts of flow to the pond during significant winter storm events. In years of significant precipitation, spillway releases from the reservoir contribute large but temporary pulses of water to the pond area, causing flooded conditions within the pond.

Outlet Releases

As shown above in Figure 8, data from SF13 indicates that, in general, outlet releases from Calero Reservoir into Calero Creek have historically maintained a water surface below the top of the v-notch weir (El. 390.0') and above the surveyed low-spot along the Calero Creek south bank (El. 388.6'). Additionally, the data shows that since 1997, the water surface in Calero Creek has not dropped below the south bank low-point until this past year (2014). Calero Creek level data confirms a fairly consistent hydraulic connection between the water surface within Calero Creek and the pond. This connection is also apparent in many of the historic aerial photographs of the pond area. Figure 8 shows that there are only four recorded instances (1977, 1981, 1982, and 2014) where the water surface in Calero Creek dropped below the invert of the weir, indicating no flow over weir.

Due to the extended local drought conditions, the District closed the outlet to Calero Creek on April 7, 2014 in order to retain water within Calero Reservoir. Following the outlet closure, the surface water connection between the creek and the pond was lost as the creek level dropped below the south bank low-spot. The pond water surface continued to drop and eventually dried up completely by June 24th, 2014.

The outlet remained closed until December 17, 2014, when the District opened the outlet valve in order to maintain the reservoir level below SSE operating restrictions. Prior to opening of the outlet, the pond was effectively empty, except for a small pool of water in the low spot of the pond, below the base of the pond staff gauge. Water levels within Calero Creek were at, or just below the invert elevation of the weir, and below the south bank low-point (i.e. no flow into the pond or over the weir). The outlet was opened at approximately 7AM and within three hours the pond was nearly full. At that time a flow rate of approximately 20 cfs was recorded at SF13, corresponding to a water surface of El. 389.6, well above the Calero Creek south bank low-point. A photograph of Calero Creek from the December 17, 2014 site visit is shown in Figure 9.

Figure 9. Calero Creek and Low-Spot Following Reopening of Outlet, 12/17/14.

Figure 10 shows the water su rface of both Ca lero Creek (at SF13) an d the pond (at the staff gauge) between January 1, 2014 and January 1, 2015.

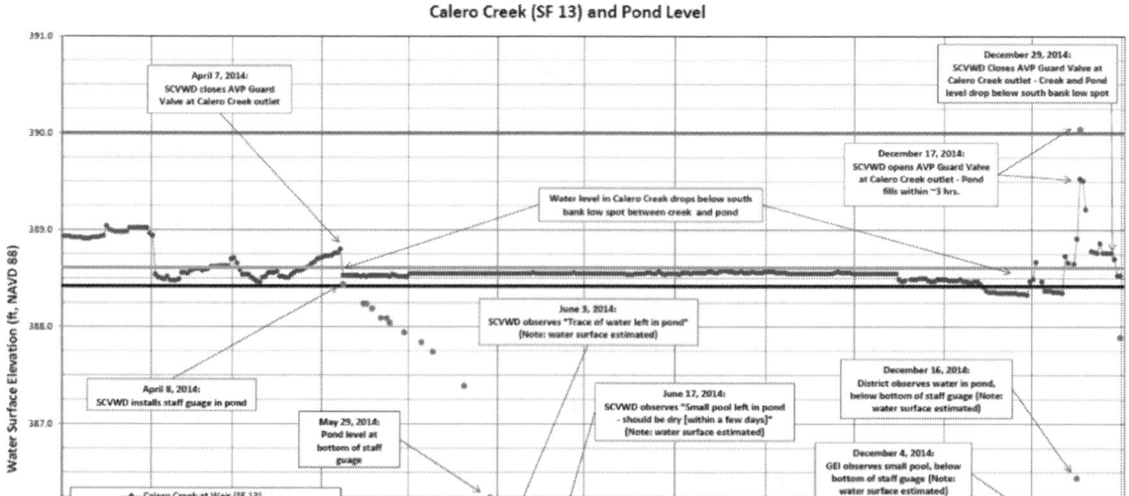

Figure 10. 2014 Calero Creek and Pond levels.

The plot shows that prior to the outlet closure, the water surface within Calero Creek was above the bank low-point. Once the outlet was shut, Calero Creek immediately dropped below the channel low point and remained below the low-point until the outlet was reopened. During that period, the pond level steadily dropped and eventually dried up completely within approximately 2 months of the outlet closure. Prior to reopening the outlet, a minor amount of water was observed to have accumulated in the lower areas of the pond, as a result of precipitation from several prior winter rainfall events and groundwater accumulation. When the outlet was reopened, the water surface within Calero Creek immediately rose above the south bank low-point, and water was observed flowing into the pond area, in addition to flowing downstream over the weir. Within three hours, the pond was nearly full.

Therefore, given recent site observations and associated data, it is apparent that outlet flows and the water surface within Calero Creek have a significant influence on the water surface within the pond. Historically, outlet flows maintain a fairly consistent hydraulic connection between the Calero Creek and the adjacent pond.

Main Dam Seepage

Evaluation of dam seepage considers both right abutment seepage and under-seepage. It should be noted that right abutment seepage refers to surface expressions of seepage which have been historically observed at Calero Main Dam. Under-seepage refers to the potential for seepage beneath the dam occurring through material within the dam foundation.

Right Abutment Seepage: Surface expressions of seepage have been observed at the Calero Main Dam since shortly after construction and the reservoir filling in 1937. Historically, seepage expressions have been observed at the Calero Main Dam on the downstream right abutment, the downstream right face and groin (right face), and the downstream left face (left toe) of the dam when the reservoir level is high. This seepage has been attributed to reservoir water migrating through the fractured right abutment knoll which exits on the downstream face of the right abutment and downstream face of embankment (Nelson, 2007). Previous studies (URS, 2014; Titus, 2003) have indicated a connection between seepage water originating through the right abutment and the delineated wetland areas below the right abutment and dam embankment. It appears that this seepage reaches the pond area by means of surface and groundwater flow.

Although there is an apparent connection between right abutment seepage and the downstream wetland areas, the contribution of this source to the presence of the downstream pond is judged to be insignificant. Seepage expressions are only apparent when the reservoir water level is above a threshold elevation of approximately EL. 470'. When observed, historic estimates of dam seepage are minimal and temporal in nature, reported to range from 0 to 10 gallons per minute (gpm) with periodic spikes in flow rates of up to 20-60 gpm (SCVWD, 2013). Furthermore, it is suspected that these recorded spikes may have been biased by precipitation or surface runoff at the time of measurement, and thus overestimate the actual magnitude of seepage flow at that time.

Review of historic aerial photographs indicates that the pond is a consistent feature throughout the life of the dam and reservoir, regardless of the reservoir level maintained behind the Main Dam and observations of seepage along the right abutment or embankment. Prior to improvements in the early 1980's which connected Calero Reservoir to the larger District water supply system, the Calero Reservoir level fluctuated seasonally and would often drain completely. This occurred every dry season from 1935 to 1962, and the entire years of 1948 and 1961 (Titus, 2003). More recently, since 2011, the reservoir level has been maintained below the seepage threshold as a result of SSE operating restrictions, and no seepage has been observed at the dam. Despite the lack of observed seepage expressions both recently and historically, the pond has remained in existence downstream of the Main Dam, with the exception of the 2014 summer season, when the outlet vault was closed.

Therefore, based on historic seepage flow data and aerial photographs of the pond area, it appears that right abutment seepage does not provide a significant contribution to the presence of the downstream pond.

<u>Main Dam Under-Seepage:</u>
Calero Main Dam was designed and constructed as a zoned earthfill dam with an impervious zone in the upstream half and a cut-off trench extending into the bedrock foundation. Both design features appear to be performing as intended as foundation under seepage has never been observed or identified as a potential deficiency at the Main dam. To confirm that significant under seepage is not occurring beneath Calero Main dam and contributing to the presence of the downstream pond, data from piezometers VWP-6, VWP-8, VWP-18B, VWP-24, and VWP-25 was evaluated relative to historic reservoir levels. These piezometers are installed within the foundation alluvium, beneath the embankment fill and downstream of the foundation cutoff trench; the same soil unit in which the downstream pond area is situated. Significant underseepage and a connection between the reservoir level and downstream alluvium would be indicated by fluctuations in piezometric levels corresponding to changes in reservoir elevation. The data from each of these piezometers spans from December 2011, shortly after the instruments were installed and readings stabilized, through September, 2014.

Figure 12 shows piezometric levels within the foundation alluvium along two separate cross-sections including the corresponding reservoir elevation. The locations of the vibrating-wire piezometers and reference cross-sections are shown on Figure 1.

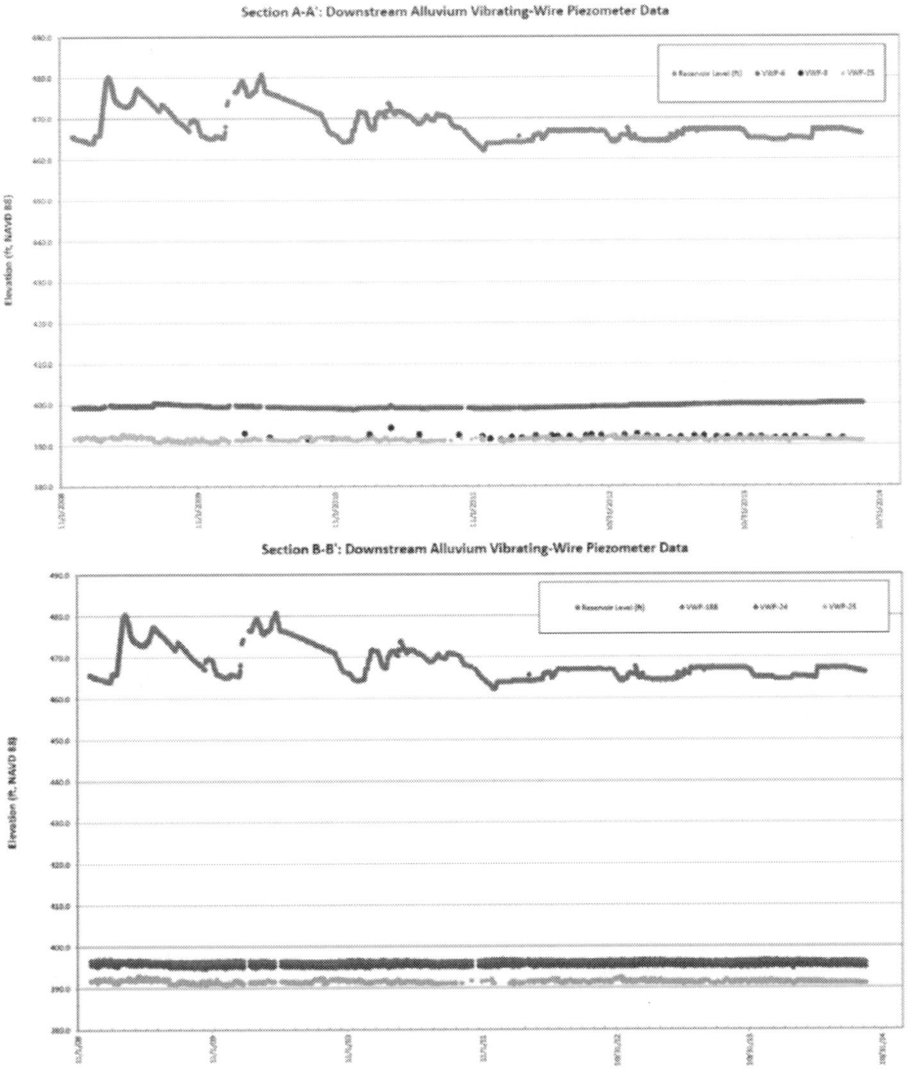

Figure 12. Downstream Alluvium Groundwater Levels; Sections A-A' and B-B'.

Figure 12 indicates that piezometric levels within the foundation alluvium, downstream of the cutoff trench, do not fluctuate with changes in the reservoir level. In general, the piezometric surface within the downstream foundation alluvium remains relatively constant, ranging between El. 390-400' depending on the particular instrument.

Therefore, based on historic observations and available piezometer data, it appears that significant under-seepage at Calero Main dam is not occurring and thus is not a contributing source of water to the downstream pond. Future efforts to control or reducing right abutment seepage will likely not dewater the pond to any substantial extent.

Groundwater

Groundwater downstream of the dam and in the vicinity of the pond was evaluated through review of available subsurface geologic information and recent hydrologic data in the vicinity of the pond. The evaluation focused on the alluvial material above bedrock; the soil unit in which the downstream pond area is situated. Geologic information included logs from recent and historic explorations performed downstream of the Main Dam. Hydrologic data included existing (VWP-6 and VWP-8) and recently installed piezometers (P-1 and P-2) which monitor groundwater levels within the downstream alluvium, water levels monitored at Calero Creek (SF13), and the water level within the pond (measured at the staff gauge).

To evaluate groundwater conditions observed in the vicinity of the downstream pond, a subsurface profile was constructed to illustrate interpreted geologic conditions and hydrologic data recorded during recent monitoring. The alignment of the evaluation subsurface profile is shown in plan view on Figure 1 (Section A-A'). The profile extends from the Main dam embankment to approximately 100' downstream of the weir structure within Calero Creek (SF13), transecting the pond area and low spot along the Calero Creek south bank. The evaluation subsurface profile is shown on Figure 13, including both a geologic profile and hydrologic profiles constructed from recent monitoring.

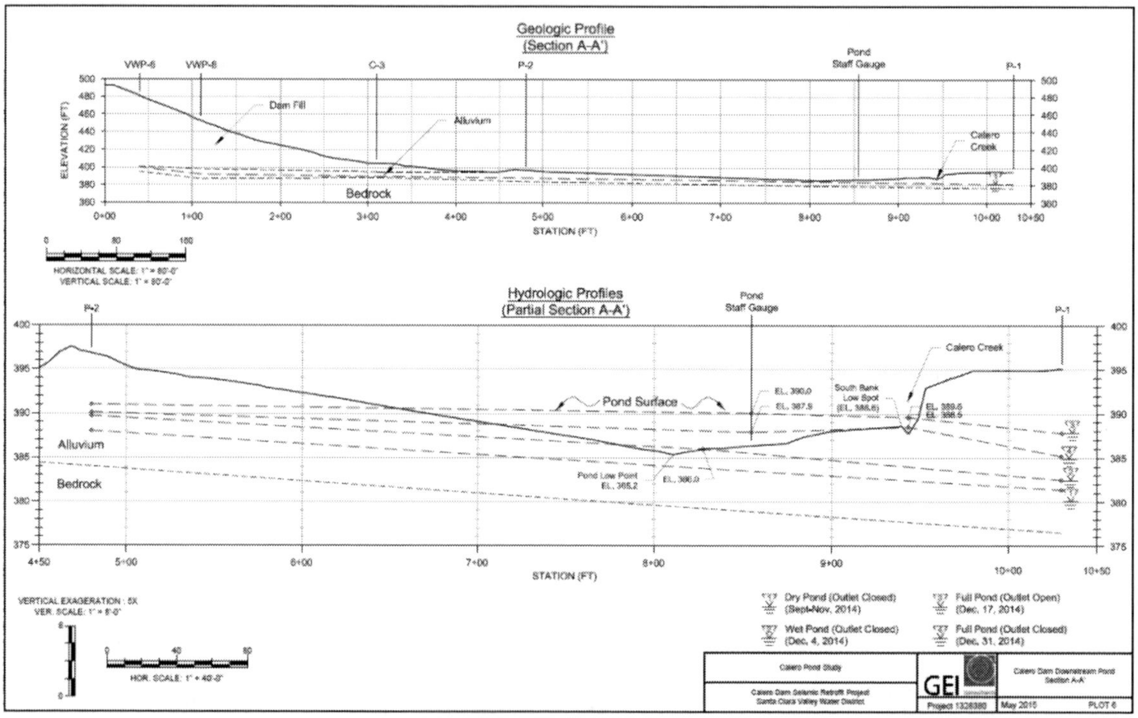

Figure 13. Calero Pond Geologic and Hydrologic Profiles.

As shown in the geologic profile on Figure 13 and discussed in detail above, the pond area is formed by a depression created during borrow excavations for dam construction. Subsurface geologic conditions downstream of Calero Main Dam in the vicinity of the pond predominantly consist of alluvial and colluvial soils; generally clayey sand and sandy clay with varying amounts of gravel, that were disturbed during construction borrow excavations. Soil thickness varies somewhat but in general is relatively shallow (less than ~20 feet). This soil layer is directly underlain by relatively impermeable siltstone and shale bedrock material of the Franciscan Formation. Groundwater is encountered within the alluvial and colluvial soils, above bedrock.

The recent monitoring of groundwater and pond levels indicates four distinct hydrologic conditions which were observed during the Fall-Winter season of 2014. Figure 13 shows these four hydrologic profiles constructed from piezometer readings and observations of water levels within the pond and Calero Creek. The four hydrologic conditions are summarized below:

1. Dry Pond (Outlet Closed): This condition represents the fall dry period from September through November, 2014, when the reservoir outlet was closed, Calero Creek had no flow, and the pond was completely dry. During this period, the groundwater surface was observed below the bottom of the pond area depression. The condition, shown as Hydrologic Profile 1, was constructed from measurements and observations recorded on October 24, 2014; the approximate lowest groundwater level recorded.
2. Wet Pond (Outlet Closed): This condition represents the early winter period from the end of November through December 16, 2014. The reservoir outlet remained closed, and Calero Creek had little to no flow, but several significant rainfall events had occurred at the site in late November and early December. Cumulative rainfall at the site through December 16, 2014 totaled approximately 11.5 inches, as recorded at the site (RF128). During this period, the groundwater surface rose above the bottom of the depression (i.e. ponded water), but below the bottom of the pond staff gauge and the low spot along the Calero Creek south bank (i.e. no connection to Calero Creek). The condition, shown as Hydrologic Profile 2, was constructed from measurements recorded on December 4, 2014; when ponded water was observed.
3. Full Pond (Outlet Open): This condition represents the winter season with the reservoir outlet to Calero Creek open. Due to significant winter rainfall in the area, the District released water in order to maintain the reservoir below the current SSE restricted level. With the reservoir outlet open, the water level in Calero Creek rose above the south bank low spot, with an approximate flow rate of 20 cfs. Water was observed flowing from Calero Creek into the pond which was near full. This condition, shown as Hydrologic Profile 3, was constructed from measurements and observations recorded on December 17, 2014; approximately three hours after the reservoir outlet was opened. This condition represents the highest groundwater levels recorded.
4. Full Pond (Outlet Closed): This condition represents the winter season with the reservoir outlet closed. With the reservoir stabilized below the current SSE restricted level, the District closed the outlet for water storage purposes. Calero Creek was observed flowing over the weir at a level just below the south bank low spot elevation. The pond was relatively full. This condition, shown as Hydrologic Profile 4, was constructed from measurements recorded on December 31, 2014.

During dry periods with no flow within Calero Creek or tributary channels to the pond area, the groundwater level can drop below the bottom of the pond depression, causing it to dry up completely (Hydrologic Profile 1). With the outlet remaining closed, local precipitation and tributary inflow began to recharge the shallow groundwater surface, forming a small pond after sufficient groundwater accumulated to raise the water level above the bottom of the depression (Hydrologic Profile 2). When the reservoir outlet to Calero Creek was open and significant flows were released, the creek level rose above the south bank low spot almost instantaneously, creating a hydrologic connection between Calero Creek and the pond. Thus, water levels in the pond were observed to be largely controlled by reservoir outlet releases (Hydrologic Profile 3). When the reservoir outlet was again closed, the level of Calero Creek dropped below the south bank low spot and the surface hydraulic connection to the pond was cut off (Hydraulic Profile 4). Pond levels remained near full shortly after outlet was closed. As drier conditions persisted and groundwater levels lowered, the pond level was also observed to drop.

Based on the observed hydrologic conditions it appears that the pond area is influenced by local groundwater levels. The water surface within the pond reflects the local groundwater table when the outlet remains closed for a sustained period of time (i.e. several months). Seasonal tributary inflows accumulate in the pond area and effectively raise the groundwater surface. Reservoir outlet releases to Calero Creek; however, generally control the water surface in the creek and pond area and have a significant and lasting impact on local groundwater levels. Although not observed during the period of this study, it is assumed that spillway releases into the pond area would have a similar effect on local groundwater levels as outlet releases.

Conclusions - Opinion on Contributing Sources to the Downstream Pond Area

Based on the results of our evaluation, it is apparent that water in the pond is maintained by near-perennial discharges to Calero Creek, seasonal discharges from three tributary sources, and shallow groundwater conditions which are intersected by the pond area low topography created during historic borrow activities. Thus, it is our opinion that water in the pond is maintained by a combination of three of the possible sources considered including Precipitation and Tributary Inflow, Outlet Releases, and Groundwater. Each of these sources varies in its overall contribution of water to the pond; however, it appears that reservoir outlet releases into Calero Creek provide the most significant contribution to water levels within the pond. We also judge that Main Dam under seepage and right abutment seepage do ***not*** provide a significant contribution to the presence of the downstream pond.

Based on these findings, the collection and conveyance of internal embankment and right abutment seepage, including seepage control design features associated with the planned seismic improvements of the Calero Main Dam, should not have a significant impact on the future presence of the downstream pond area. Should there be a desire to preserve the pond area and associated wetland habitat as part of the retrofitted embankment, the overall design and future site configuration should account for the identified sources which contribute to maintaining seasonal water levels within the pond.

REFERENCES

Geosyntec Consultants 2012, Supporting Techni cal Information Document (STID) for Calero Main and Auxiliary Dams, Santa Clara Valley Water District.

GEI Consultants, Inc. (GEI) 2013, Draft Geotechnical Data Mem orandum, Calero Dam, Calero and Guadalupe Dams Seismic Retrofits Project.

Marliave, C. 1936, Geol ogical Report on Calero Dam Situated on Calero Creek in Santa Clara County, State of California Departm ent of Public W orks, Division of Water Resources.

Nelson, J. L. 2007, Calero Main Dam (No. 72-3) Forensic Geologic Report, Santa C lara Valley Water District.

Santa Clara Valley W ater District (SCVWD) 1982, Map a nd Construction P lan for Almaden Valley Pipeline, Unit II, Phase II, from Almaden Valley Pipeline, Unit II, Phase I to Calero Reservoir Outlet Works, 135 sheets.

Santa Clara Valley W ater District (SCV WD) 2013, Surveillance R eport, July 2012 through July 2013, Calero Dam Facilities No. 72-3, Santa Clara County, prepared for the California Division of Safety of Dams.

Tolman, C.F. 1934. Geological report on the Calero, Alm aden, and Guadalupe dam sites: unpublished consultant re port on-file at th e Santa Clara Valley W ater District, San Jose, CA.

URS Corporation, 26 O ctober 2012, Seism ic Stability Evaluation of C alero Main Da m and Calero Auxiliary Dam, Final Report SSE1B-B.

URS Corporation, 12 June 2014, Technical Me morandum Re: Calero Da m Geotechnical Exploration Plan Area Aquatic Resources, Submitted to GEI Consultants, Inc.

United States Geological Survey, 1919. New Almaden Quadrangle Map, San Jose CA.

Titus, D.M. 2003, Calero Da m Seepage Geotechnical Exploration Project, Delineation of the Waters of the United States, Santa Clara Valley Water District.

MAINTAINING PERFORMANCE — REHABILITATION OF CRITICAL DRAINS AT CRAB ORCHARD DAM

David Snyder, PE[1]
Robert Saber, PE[2]
David Hibbs[3]

ABSTRACT

Crab Orchard Dam is an earthfill embankment dam constructed in 1939 at the Crab Orchard National Wildlife Refuge in Marion, Illinois. The dam embankment is 57 feet high and approximately 3,800 feet long. Modifications were initiated in 1988 to increase freeboard and spillway capacity, flatten the downstream slope and install a drainage system that included a chimney drain, a horizontal drainage blanket, a collector drain at the toe, five inspection wells, eleven relief wells, and a weir for measuring combined seepage flow from both the relief wells and toe drain. Twelve piezometers were also installed to monitor performance of the embankment and drain system.

This paper describes the ongoing assessment of the dam, relief well system and toe drain system, which started in 2010. Aspects of the assessment include seepage and slope stability modeling; uplift/blowout analysis; toe drain inspection; evaluation of piezometer performance and establishing threshold levels; and relief well inspection, pump testing, rehabilitation and monitoring. Unique aspects of the project include a relatively long duration of piezometer monitoring, occasional development of artesian pressure in some piezometers in the vicinity of relief wells and a relatively high influence of tailwater levels to uplift/blowout factors of safety. The performance of the relief well system and relative success of the rehabilitation program will be summarized. The methods used at Crab Orchard Dam can be used in evaluations of other similar dams so that risk can be adequately defined and proper allocation of funding can be applied where it is most needed.

INTRODUCTION

Project Location and History

Crab Orchard Dam is an earthfill embankment dam located within the Crab Orchard National Wildlife Refuge in Williamson County in southern Illinois, about 10 miles east of the town of Carbondale. A project location map is shown in Figure 1. The dam is classified as a large, high hazard dam with a population at risk of 10,000 people. The dam embankment is 57 feet high and approximately 3,800 feet long. Crab Orchard Dam

[1] Senior Geotechnical Engineer, Gannett Fleming, Inc., P.O. Box 67100, Harrisburg, PA 17106, dmsnyder@gfnet.com
[2] Chief Geotechnical Engineer, Gannett Fleming, Inc., P.O. Box 67100, Harrisburg, PA 17106, rsaber@gfnet.com
[3] Project Manager, United States Fish and Wildlife Service, 5275 Leesburg Pike, Falls Church, VA 22041, dhibbs@fws.gov

was constructed in 1939 as part of the federal Work Projects Administration program to provide work for unemployed Americans during the Great Depression. When World War II began, an ordinance facility was built on the land surrounding the reservoir, and a water treatment plant was constructed near the reservoir to serve the ordinance plant. In 1949, U.S. Fish & Wildlife Service (USFWS) was authorized to obtain the dam, reservoir, and surrounding land from the War Department to provide habitat for fish and wildlife and recreation for local residents.

Modifications to construct a new outlet works were completed in 1980, and extensive dam and spillway modifications were initiated in 1988 and completed in 1991. These modifications included increasing freeboard by raising the embankment, increasing spillway capacity by constructing a fuse plug auxiliary spillway, flattening the downstream slope, and installing a new drainage system. Components of the new drainage system included a chimney drain, a horizontal drainage blanket, a collector drain at the toe, five inspection wells, eleven relief wells, a weir for measuring combined seepage flow from both the relief wells and toe drain, and seven piezometers for monitoring performance of the system. The design of dam modifications was completed by the Bureau of Reclamation Technical Service Center.

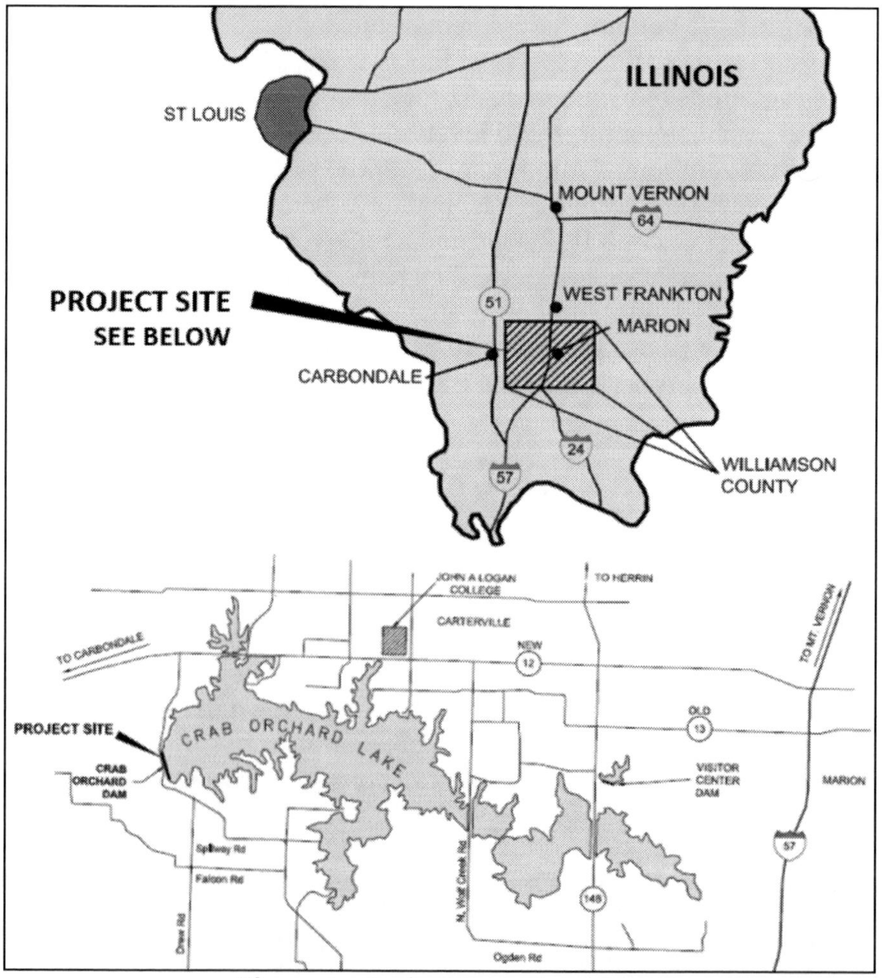

Figure 1. Project Location Map

Geology

Crab Orchard Dam is founded on approximately 30 feet of glacial till consisting of lean to fat clays. The glacial till contains numerous deposits of silty sands that are the product of the meandering nature of Crab Orchard Creek. Below the glacial till is a layer of silty sand approximately 30 feet thick. Below the silty sand layer is the sandstone bedrock. The bedrock that underlies the surficial soils beneath the embankment is about 60 to 80 feet deep and consists of Pennsylvanian period shale with local interbedded siltstone and sandstone. The potential for piping in the foundation could be associated with the saturated and soft conditions observed in the old stream channel downstream of the embankment. This condition is attributed to the upward flow of seepage from the artesian conditions in the silty sand layer.

Original Design and Construction

There are no calculations or analyses available for the original 1939 design or construction. There is no information regarding foundation excavation and treatment for the original dam. There is only meager information on the design and construction of the structures.

Crab Orchard Dam was first designed within the Division of Land Utilization of the Resettlement Administration, a short-lived Franklin Roosevelt era public works agency responsible for many varied types of construction projects. The design took advantage of a relatively thick impervious blanket of clayey soil that extends far upstream from the dam site. The prevalence of clayey soils in the vicinity provided ample material from which to construct the embankment. The embankment was originally a homogeneous dam, effectively working as a wide impervious core.

The embankment is 57 feet high and has an average crest width of 20 feet. The dam crest is approximately 3,800 feet long at elevation 420.5 (NGVD29). The upstream and downstream embankment slopes are 2.5:1 (H:V). According to limited sampling obtained for a previous stability analysis, boring logs describe the original embankment samples as "clay (CL-CH), about 80% medium to high plasticity fines, and 20% fine sand, stiff to very stiff". An 8-foot-wide by 6-foot-deep rock toe drain located along the downstream toe was grouted during modifications in the 1990's.

Release facilities at the dam include a service spillway at El. 405.0 feet at the left abutment, a fuse plug spillway in the right abutment at elevation 415.5, and a low-level outlet works at the right end of the service spillway. The service spillway consists of a concrete ogee crest and an unlined downstream channel. The emergency spillway consists of a 924-foot-wide earthen channel with three earthen fuse plug sections separated by concrete splitter walls. The fuse plug sections have varying crest and pilot channel elevations with each section having a zoned embankment with sand and gravel, a central clay core and 2H:1V side slopes. The original outlet works, located near the right abutment of the dam, were abandoned by concrete plugging in 1980. The current outlet works were constructed at that time and consist of a concrete structure at the right end of the service spillway, with discharge controlled by three sluice gates. Discharge from the outlet works passes into the service spillway discharge channel.

Modifications

The Bureau of Reclamation, Engineering and Research Center, Denver, Colorado performed an examination of Crab Orchard Dam and issued a Safety Evaluation of Existing Dams (SEED) Report in 1980. The SEED report identified problems with the existing structure and listed several recommendations considered to be directly related to the safety of the dam and appurtenances. The major deficiency as noted was the structure's inability to contain or safely pass the inflow design flood. Additionally, the failed drain system located at the toe of the dam and the lack of a filtered drain were also identified within the report. Subsequent to the SEED report, further investigations were performed and designs and specifications prepared for Crab Orchard Dam Modification.

Between 1988 and 1991extensive Safety of Dams Modifications were completed. These modifications were designed to provide adequate freeboard, collect seepage and reduce uplift pressures in the foundation, provide means to pass the inflow design flood, and prevent erosion of the upstream dam embankment face. The principal components of the work performed included the following:

1. Raising the existing dam embankment crest approximately 5 feet to provide adequate freeboard.
2. Grouting of the original rock toe drain along the downstream toe.
3. Constructing internal embankment and foundation drainage systems, including eleven relief wells, to collect seepage and reduce uplift pressures.
4. Installation of seven new Porous Tube Piezometers (PTP's).
5. Excavating an auxiliary spillway channel and constructing a fuse plug embankment to pass the inflow design flood.
6. Sealing off the abandoned outlet works and removing the intake tower to eliminate the possibility of uncontrolled releases.

The fill materials used in the embankment raise are composed of a lean clay and clayey silt with up to about 10 percent fine sand. The borrow source for the dam raise was the material excavated to construct the auxiliary spillway, which consisted mostly of silty clay. The blanket/trench drain aggregate filter sand used was a blend of three different materials consisting of concrete sand, extra fine mortar sand, and crushed limestone. The three materials were blended using a modified asphalt aggregate blending plant. The percentage of each source material used was adjusted to obtain the desired final gradation.

SCOPE OF ONGOING SEEPAGE MONITORING AND DRAIN ASSESSMENTS

Gannett Fleming has assisted U.S. Fish & Wildlife Service with inspections, piezometer and drain monitoring and assessments at Crab Orchard Dam from 2006 through recent relief well and toe drain evaluation and rehabilitation projects completed in 2015. The work has included development of standard operating procedures for piezometer data collection, review and analysis of piezometer data, development of piezometer threshold limits, and recommendations for automated data collection. The team has developed a proactive approach to seepage monitoring to ensure that any deterioration of seepage conditions can be detected in a timely manner.

In 2010, an Emergency Action Plan (EAP) Event Level 1 was activated due to a flood event at Crab Orchard reservoir. The reservoir pool level peaked at 2.3 feet above the crest of the service spillway during this event. Some piezometer readings taken by the refuge staff during and after the event exceeded the threshold values, and artesian flow was observed at the top of the riser pipe in some of the piezometers. Based on these observations, it was suspected that portions the relief well system were not functioning as designed. Therefore a series of additional investigations and evaluations of the drainage system were initiated. An evaluation of the performance of the relief wells individually or as a system had not been completed since the time of their installation in 1989.

Gannett Fleming performed a preliminary field investigation of the relief wells, piezometers and collector drain in August, 2011 to evaluate their performance and functionality. This investigation included video camera inspection of the toe drain pipe.

Seepage and slope stability analyses were performed to further assess the design and performance of the drainage system in 2014. Available reports and analyses were reviewed, and relevant data and conclusions were summarized. The performance of the existing relief wells and drain system was evaluated using the results of the seepage analysis and historical piezometer and drain flow data. Slope stability of the dam was evaluated based on seepage conditions.

A video camera inspection of the toe drain pipe system was performed again in 2014 to evaluate whether the previously-observed structural pipe damage had progressed since the 2011 inspection. The relief wells were rehabilitated in August 2015, and a section of toe drain pipe was replaced in October 2015.

SEEPAGE AND DRAIN SYSTEM ASSESSMENTS

2010 Instrumentation Review

The storm event that caused the May 3, 2010 EAP Event Level 1 resulted in a peak reservoir elevation of 407.3 feet, or 2.3 feet above the crest of the service spillway. Piezometer readings were taken by the refuge staff, and five piezometers indicated abnormal readings compared to threshold values, which had been developed based on historic readings and engineering judgment. No seepage measurements or seepage observations were recorded during this event.

Piezometers PTP-3, PTP-5 and PTP-6A are located in the coarse-grained foundation layer at the downstream toe of the embankment, as shown in Figure 2. When the piezometer caps were removed from PTP-3 and PTP-5 at 7:00 am on May 3, 2010, the water levels were observed by the refuge staff to be at or near the top of the piezometer pipes but not flowing. Later during the day, water was observed flowing from PTP-6A. It was reported that Piezometers PTP-5 and PTP-6A had experienced similar artesian flow in the past during similar elevated reservoir levels. Later on May 3 and 4, water began to flow from piezometers PTP-3 and PTP-5, and this was the first event where piezometer PTP-3 experienced artesian flow.

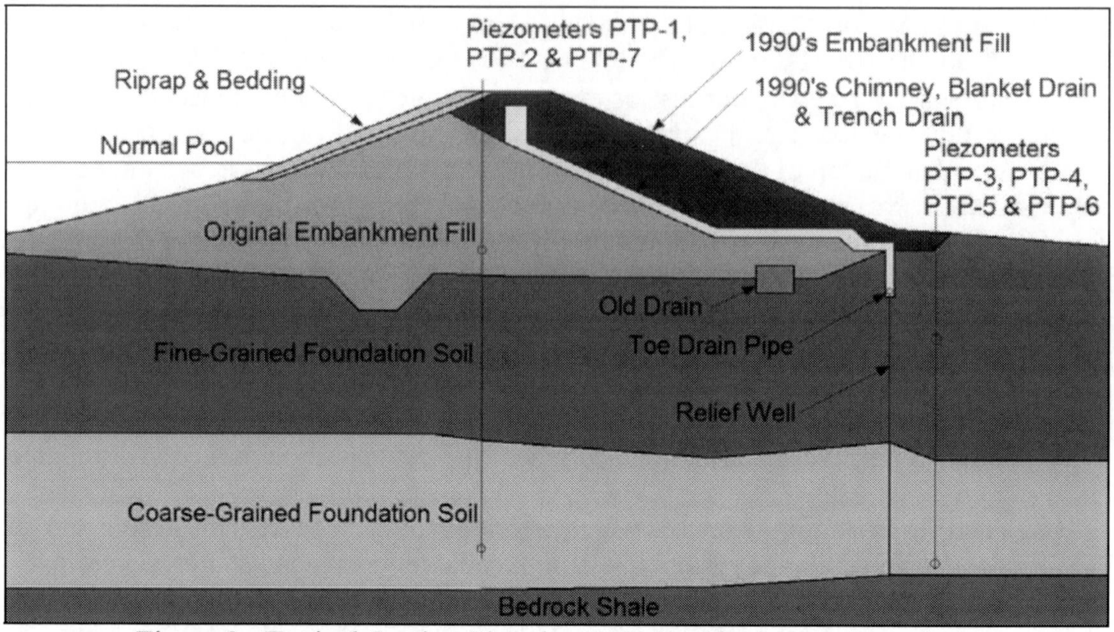

Figure 2. Typical Section Showing Relative Piezometer Locations
Note: Piezometers are located at various stations along the dam.

Piezometer PTP-4B is located in the fine-grained foundation layer at the downstream toe of the embankment. The phreatic elevation of Piezometer PTP-4B during the May 3, 2010 event was 388.0 feet, which was 3 feet higher than the upper threshold for reservoir levels up to elevation 407.0 feet. The phreatic elevation remained above the threshold level for almost three months until it resumed normal historical levels in August, 2010. High readings in this piezometer indicate that the relief well system near this location may not have been functioning as designed.

On May 3, 2010, the phreatic elevation of PTP-7B, which is located in the original embankment fill as shown in Figure 2, was approximately 7 feet lower than normal, and 5.5 feet below the lower threshold elevation. The area at the toe of the dam near the piezometer was continuously monitored. Readings returned to normal later that day, when the reservoir level receded to elevation 404.7 feet.

As a result of the piezometer readings outside of threshold levels, testing of the relief wells was recommended. A recommendation was made in the 2010 SEED Inspection Report to test the relief wells in accordance with *EM-1110-2-1914, Design, Construction, and Maintenance of Relief Wells* (U.S. Army Corps of Engineers, 1992), including pumping tests to determine the yield of each relief well for comparison to yields obtained after construction. Other assessments of the relief wells were recommended.

2011 Drain System Field Assessments

Relief Wells. Gannett Fleming performed a field investigation of site relief wells in August, 2011. The field investigation included pump tests in two relief wells, soundings and video camera inspection of all relief wells, and water chemistry testing.

Results of pumping tests performed immediately after construction of the relief wells were compared to the pumping tests performed in 2011, and the observed total seepage rates indicated that the relief well system had deteriorated significantly since installation.

The relief well video inspections indicated the interior of the relief wells were generally in good condition, but were showing signs of deterioration. Figure 3 is an example photograph from the video inspection showing typical conditions. Zones of incrustation were observed along the inside of the well screen in some relief wells. The thickness of sludge in the bottom of the relief well manholes ranged from 0.1 to 2.2 feet.

Laboratory testing of water samples indicated heavy bacterial activity and extremely heavy iron oxide present within the relief well casing. The testing of the aquifer sample indicated low bacteria activity, heavy clay particulate matter, low iron oxide, the presence of sulfate reducing bacteria, and high amounts of total organic carbon. These factors all contribute to a reduction in well efficiency.

Figure 3. Example of Typical Relief Well Condition from 2011 Investigation

Compatibility analyses between the relief well filter pack and the coarse-grained foundation soil indicate that the filter pack is slightly coarser than recommended based on U.S. Army Corps of Engineers' (USACE) filter design criteria (EM 1110-2-2300), but should limit movement of coarse-grained material into the filter.

The relative effectiveness of the relief well system immediately after installation could not be evaluated due to a lack of useful piezometer data from that time period. A review of available piezometer data from 1989 to 1990 indicated that the relief wells may have never been as effective at reducing pore pressures in the coarse-grained foundation soil layer as intended by the design. If the relief wells performed as designed, piezometers would not experience artesian head under any reservoir or tailwater conditions. If the relief wells were fully effective, then the excess head in the coarse-grained foundation soil layer would be negligible, which would eliminate upward seepage flow through the fine-grained confining layer.

Toe Drain Pipe. Gannett Fleming performed a field investigation of the toe drain system in August, 2011. The field investigation included water chemistry testing, video camera inspection of the pipe system, cleaning and flushing of the pipe system and evaluation of sediment deposits at the outfall weir.

The toe drain pipe video inspection in 2011 indicated that the pipe was generally in good condition with the exception of a section of pipe that was partially collapsed and deformed between relief well RW-5 and Inspection Well #3, which are shown in Figure 5. An example of this deformation is shown in Figure 4. Various sags and minor cuts were identified in this reach. Very few areas of hard scale or deposits were observed in the pipe. Tree roots were identified within the drain pipe outfall between Inspection Well #3 and Inspection Well #5, and the tree root masses obstructed flow through approximately half of the length of this pipe. The inlet and outlet pipes connecting the relief wells ranged from 5 to 100 percent full of sludge. The weir box contained 0.8 feet of sludge behind the v-notch weir plate.

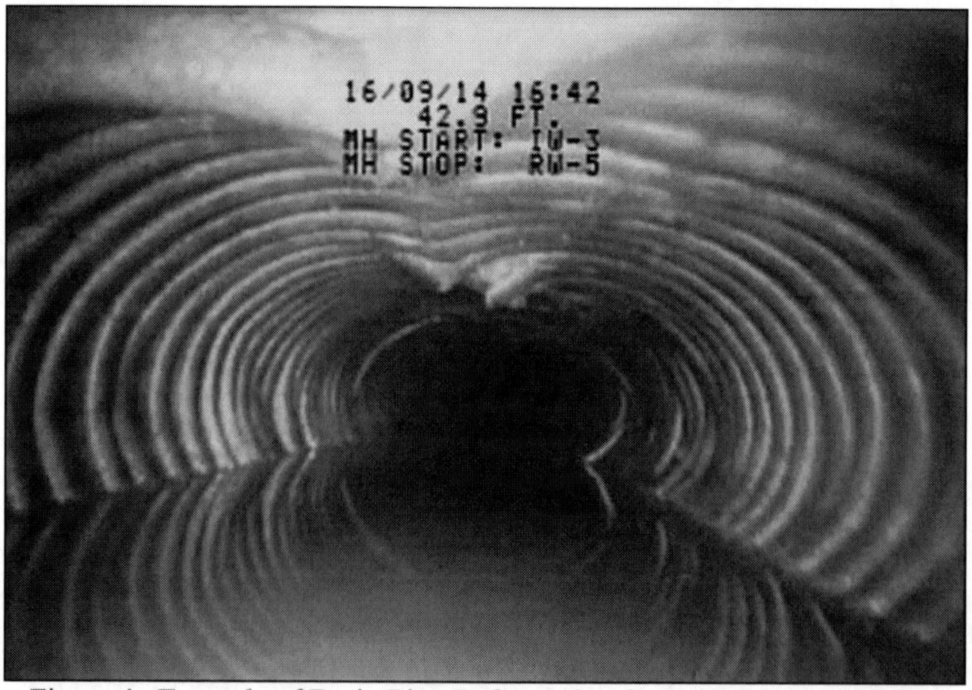

Figure 4. Example of Drain Pipe Deformation from the 2011 Inspection

On August 13, 2011, prior to cleaning the toe drain system, a typical discharge of about 4 gpm was measured at the outfall below the weir box. On August 16, 2011, upon completion of the cleaning of the drain system, the discharge at the outfall below the weir box was 7.5 gpm. No meaningful drain system discharge measurements were available between the original construction of the wells in 1990 and 2011 for comparison. However, the total flow from the relief wells at the end of the original yield tests conducted in 1989 was reported to be 56 gpm. The discharge after the 2011 toe drain pipe cleaning was therefore only 13 percent of the original discharge.

Compatibility analyses indicated that the drain aggregate is slightly coarser than the recommended filter material for the embankment soils based on USACE criteria. In situ embankment fill gradations were not available, so borrow area gradations were used. The drain pipe perforations are three times larger than the maximum allowed by USACE criteria, which could result in finer aggregate particles migrating into the pipe. The 2011 inspection indicated an accumulation of sludge and possibly some sediment deposits in the pipes. However, the toe drain system appeared to be performing well to control seepage through the embankment and foundation under observed conditions.

2014 Seepage Analyses

Development of Cross Sections. Cross section were developed at the locations of existing piezometers and areas of interest, including Station 24+00 (at the previously abandoned low-level outlet works), Station 29+00, Station 33+00 (at the existing drainage outlet at Manhole #5), Station 36+50 and Station 40+50. Section locations are shown in Figure 5, and a typical subsurface profile is represented in Figure 6.

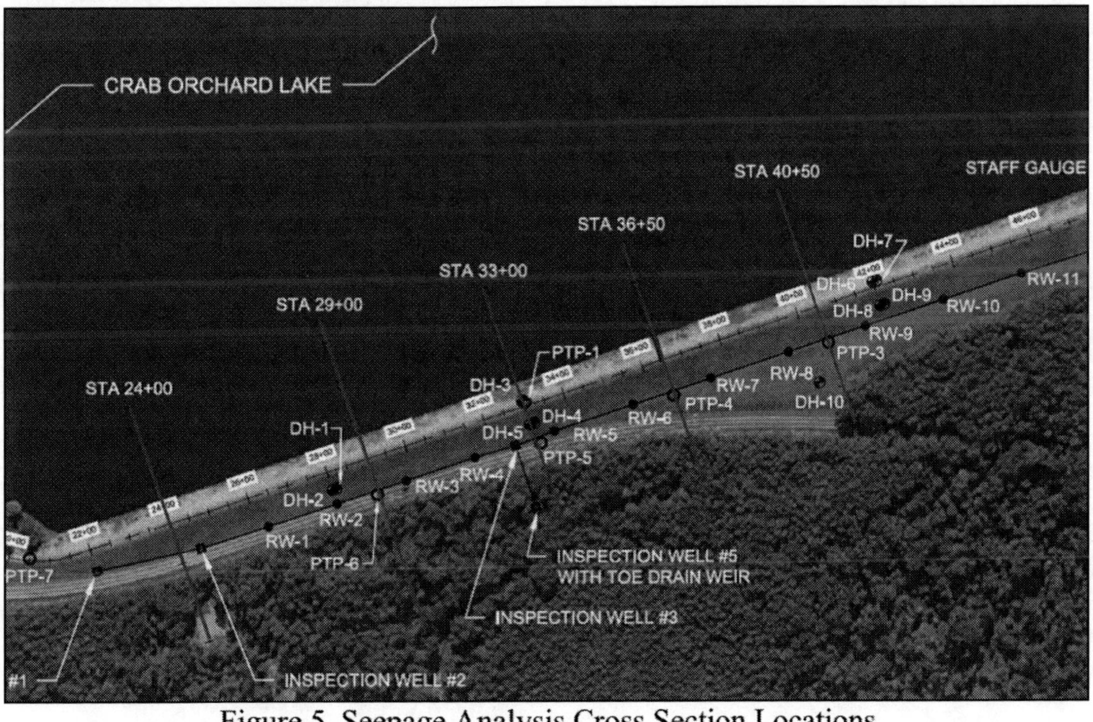

Figure 5. Seepage Analysis Cross Section Locations

Historic boring logs and design drawings were used to determine the embankment and drain geometry and subsurface soil strata depths at the five cross sections. The scope of work did not include new borings or site exploration. The lake bottom ground surface upstream of the dam was estimated based on design drawings and lake chart data.

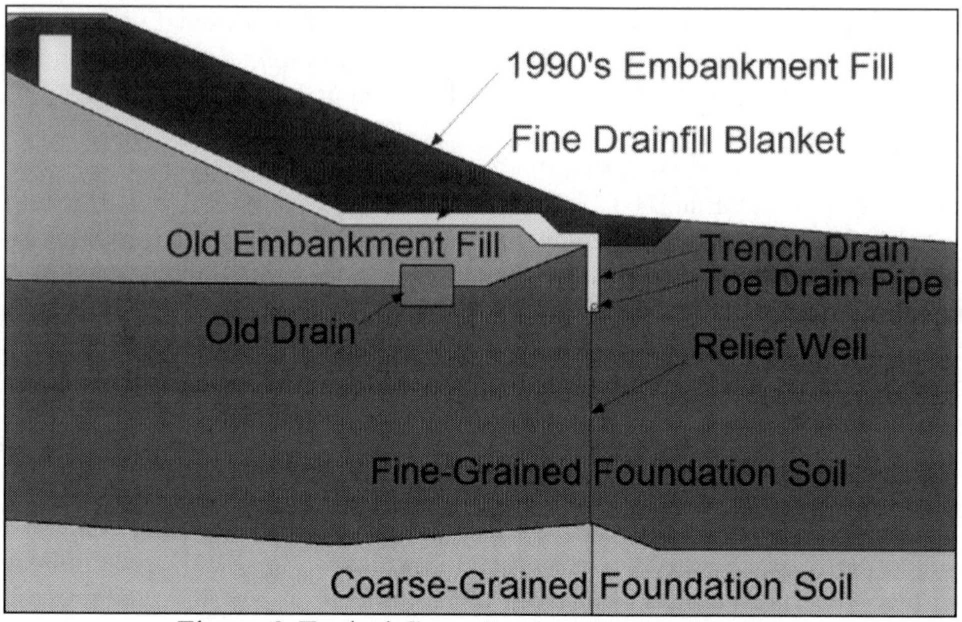

Figure 6. Typical Cross Section (Station 33+00)

Modeling Approach. Historic and recent field permeability testing data, including falling head testing in piezometers in 2011, were evaluated and summarized to determine layer permeabilities. Relief well pump test data and grain-size correlations were also used to estimate the permeability of the coarse-grained foundation layer. Design values are summarized in Table 1. Limitations of the historic documents included lack of grain-size and permeability data of embankment fill and fine-grained foundation soil.

Table 1. Design Permeability Values Used in the Seepage Analysis

Stratum	Horizontal Permeability (cm/s)
New Embankment (1990's)	1.0E-06
Original Embankment (1940's)	1.0E-06
Foundation Soils - Fine-Grained	1.0E-06
Foundation Soils - Coarse-Grained	4.8E-03
Shale Bedrock	6.7E-05

Seepage analyses were performed at each of the cross sections based on four cases, each with unique reservoir and tailwater levels. Cases 1 and 4 modeled normal reservoir and design high water levels respectively. Cases 2 and 3 modeled observed reservoir and tailwater levels during recent documented high water events on April 27, 2011 and May 3, 2011 respectively. Since tailwater is not monitored at the site, tailwater elevations for the seepage modeling were estimated based on hydraulic modeling and photographs taken during inspections.

Seepage modeling was conducted using the SEEP/W software that is part of the GeoStudio® 2007 software suite, version 7.18, by GEO-SLOPE International, Ltd. of Calgary, Alberta, Canada. Steady state seepage conditions were assumed for all analyses. The toe drain pipe and relief wells were incorporated into the seepage model boundary conditions to model the subsurface flow regime.

Each of the four reservoir pool and tailwater elevation cases was analyzed both with and without the relief wells. This was achieved by performing the analysis with the relief well boundary conditions applied, and again with the relief well boundary conditions removed. Typical results from the seepage modeling are shown in Figures 7 and 8. The blue dashed line is the phreatic surface result from the model, and the labeled contours are total head contours.

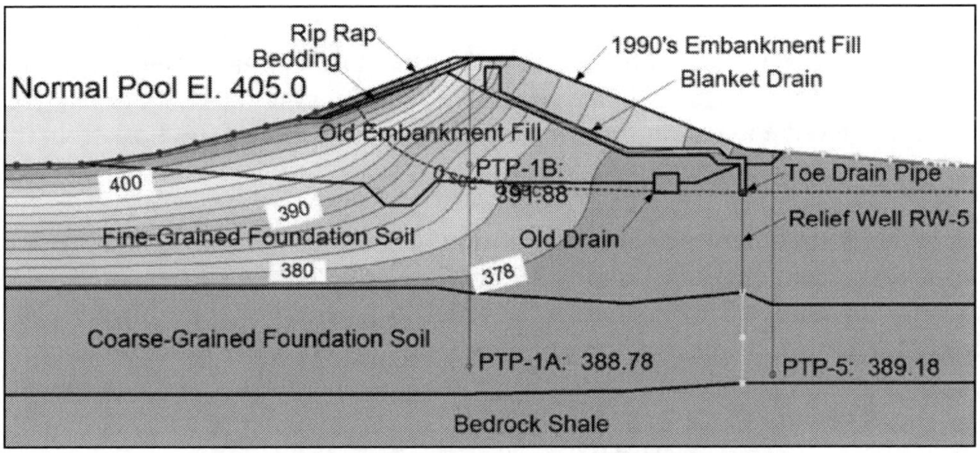

Figure 7. Typical Seepage Model Results with Relief Well Boundary Conditions

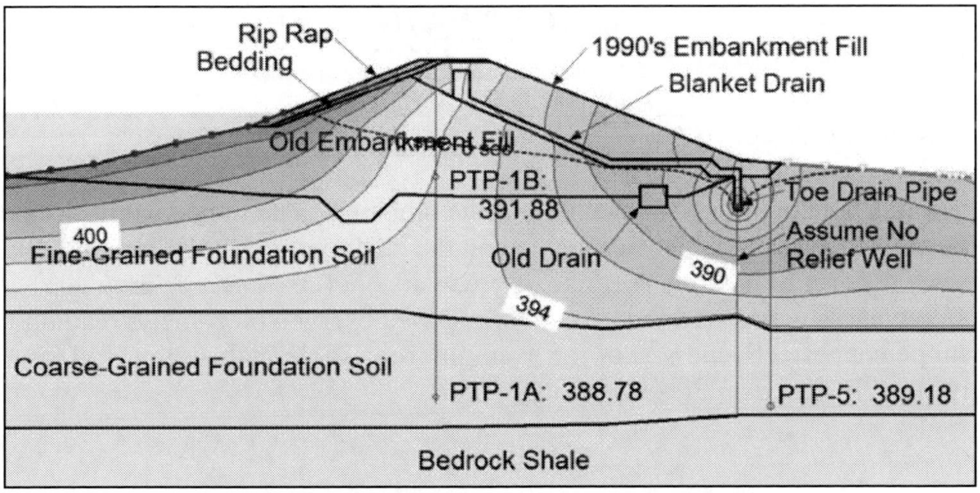

Figure 8. Typical Seepage Model Results without Relief Well Boundary Conditions

The seepage model is two-dimensional, and therefore it models the relief well as existing infinitely into the page and not as discrete relief wells spaced 200 feet apart. The head loss in the well and mounding of water levels between the wells is not accounted for in the model. The results of the model with relief wells would more accurately reflect seepage conditions in close proximity to the wells. However, the mounding between relief wells is relatively small based on piezometer data.

Seepage Model Results. The results of the seepage analyses for cases with relief wells were fairly consistent for the different cases in that they showed significant head losses within the embankment and fine-grained foundation soils. Head losses in the coarse-grained foundation soil were considerably smaller; generally less than 1 foot of head loss in more than 800 feet of length (upstream to downstream).

The models provide insight as to how seepage conditions change when the relief wells effects are included or ignored. Seepage flow rates from the models were summarized for each section and hydraulic case. According to the model, if the relief wells are effective at passively collecting seepage and conveying it vertically to the toe drain pipe, then the total seepage flow rate into the toe drain system for steady state normal pool conditions is relatively high. If the relief wells are not functioning, then seepage flow rates at the ground surface and stream channel downstream of the dam are relatively high, and total flow into the toe drain system is much less since there is no flow from the relief wells.

By comparing actual piezometer readings from the site to anticipated piezometer readings based on 100 percent effective relief wells or completely ineffective (0 percent effective) relief wells, the relative performance of the relief well system was evaluated. For example, at Station 33+00 Case 1, the predicted and observed phreatic elevations at Piezometer PTP-1A (located in the coarse-grained foundation soil near the centerline of the dam) are shown in Table 2.

Table 2. Example Comparison of Seepage Model Results & Observed Piezometer Levels

Station 33+00, Case 1 (Normal Pool)	**Piezometer PTP-1A Phreatic Level**	
	Predicted Elev. (Feet)	**Observed Elev. (Feet)**
Relief Wells 100% Effective	377.22	-
Actual Observed	-	388.78
Relief Wells 0% Effective	394.28	-

Observed piezometer phreatic elevations were compared to predicted values from the seepage model for each piezometer located on the analyzed cross sections and for each of the cases where piezometer measurements are available (i.e., Cases 1 through 3). Relief well effectiveness was quantified by converting the observed piezometer reading into an apparent percent effectiveness. For the example from Table 2, the relief well effectiveness, E, based on Piezometer PTP-1A is,

$$E = \frac{(394.28 - 388.78)}{(394.28 - 377.22)} * 100 = 32.2\% \quad (1)$$

The distance of each piezometer to the nearest relief well was considered when evaluating the results of the relief well percent effectiveness. Based on the seepage model results and observed piezometer readings, the relief wells are typically between 20 and 50 percent effective at reducing pore pressures in the coarse-grained foundation soil at the centerline and toe of the dam.

Uplift/Blowout Evaluation Results

The terms heave, uplift, blowout and piping are often used synonymously to describe the potentially adverse effects of seepage exiting at the toe of an embankment (Pabst, et al, 2013). However, they are not synonymous, and they describe different mechanisms related to seepage at the toes of dams. "Heave" is a condition in cohesionless soils involving vertical seepage forces acting on the soil grains (Pabst, et al, 2013). "Uplift and blowout" is evaluated when a relatively low-permeability layer is present under a dam, and pressures can become large in an underlying pervious stratum (Pabst, et al, 2013), as is the case at Crab Orchard Dam. When the uplift pressure in the underlying pervious stratum exceeds the downward pressure of the relatively low permeability confining layer, uplift pressures on the bottom of the confining layer may cause rupture of the layer, which is referred to as blowout. If blowout occurs, seepage can flow through the ruptured soil.

It may be pertinent to evaluate "heave" at Crab Orchard Dam if alluvial material in the original stream channel (or old stream meanders) are composed of cohesionless sandy materials and are present near the ground surface at the site. However, no information on the characteristics or properties of the alluvial material could be located, nor any evidence that the alluvium, if present, was left in place prior to stream channel backfilling for dam construction and raising. Therefore heave was not evaluated.

A factor of safety greater than 1.5 for uplift of a fine-grained confining layer is recommended in *EM 1110-2-1901, Seepage Analysis and Control for Dams* (U.S. Army Corps of Engineers, 1993). The 1988 design of the relief wells was based on a target minimum factor of safety of 1.5 against what the designers called "heave" in the original stream channel. The factor of safety was based on the escape or exit gradient calculated through the fine grained foundation soil and an assumed critical gradient of 0.8. The critical gradient is the gradient at which flotation of particles occurs. The critical gradient for sand is approximately 1.0 based on *EM 1110-2-1901* (USACE, 1993), and Turnbull and Mansur (1961) indicate that sand boils have been observed at Mississippi River levees when gradients were 0.5 to 0.8. The calculated factors of safety in the 1988 relief well design were 1.6 for high pool conditions (El. 415) and 1.7 for normal pool conditions (El. 405).

Uplift/blowout safety factors were calculated in 2014 at each cross section for two cases:

- Observed pore pressures in the coarse-grained foundation stratum (based on observed piezometer data)
- Pore pressures in the coarse-grained foundation stratum from the seepage model with relief wells assumed to be 0 percent effective (relief wells neglected)

Factors of safety were not calculated for the cases where the relief wells were modeled as 100 percent effective, since the relief wells would completely relieve any excess pressure on the bottom of the fine-grained layer, eliminating any force to cause instability.

The piezometric head in the coarse-grained layer is nearly uniform across the site. Since the uplift factor of safety is primarily dependent on the differential pressure acting on the confining layer and the thickness of the confining layer, the critical section with respect to uplift/blowout is Station 29+00 where the thickness of the confining layer is estimated to be as low as 30 feet because of the presence of the stream channel located relatively close to the dam. At the stream channel, differential pressure is influenced by tailwater elevation, so the variability in factor of safety in Cases 1 through 4 include the effect of tailwater. Calculated uplift factors of safety for Station 29+00 are presented in Table 3.

Table 3. Factors of Safety against Uplift/Blowout for Station 29+00

Case		Factor of Safety Uplift/Blowout
Case 1	Without relief well	1.4
	Observed Piezometer level	1.6
Case 2	Without relief well	2.0
	Observed Piezometer level	2.9
Case 3	Without relief well	2.9
	Observed Piezometer level	> 4
Case 4	Without relief well	3.6

Recommended minimum factor of safety = 1.5
There are no observed piezometer levels for Case 4 because this flood case has not occurred.

As shown in Table 3, the minimum calculated safety factor for uplift/blowout based on **observed** piezometer levels for Case 1 was 1.6 at the original stream channel. The Case 1 seepage analysis performed **without relief wells** indicated a safety factor of 1.4. Therefore, further deterioration of the relief wells will likely result in a factor of safety for uplift/blowout below the recommended minimum value of 1.5. Increasing the effectiveness of the relief wells would reduce piezometric pressure in the coarse-grained material and improve the safety factor.

The calculations for evaluating uplift/blowout are conservative in that they neglect the confining layer shear strength, which resists the rupture failure mechanism (Pabst, et al, 2013). However, the calculations also do not account for defects such as rodent burrows, desiccation cracks, improperly abandoned borings, old stream meanders, and other similar features. Evidence of uplift and blowout might include moderate to severe ground surface movement, sand boils or concentrated vertical seepage.

Piezometer Thresholds. Piezometer threshold values were developed based on the results of the uplift/blowout factor of safety calculations. Since the factor of safety is based on the differential head that is acting on the confining layer, the piezometric level to provide a target factor of safety is a function of the tailwater level. To provide a factor of safety greater than or equal to 1.5 against uplift/blowout at the original stream channel, the piezometric level in the coarse-grained foundation soil layer must be less than 18.1 feet above the tailwater level observed at the time of the piezometer readings.

The stability of the downstream embankment slope at each of the five cross sections was also evaluated using the results of the seepage analyses for the four design cases. Relief wells were neglected in these seepage analyses to conservatively estimate the phreatic surface and pore pressures throughout the model for use in the slope stability model. Calculated minimum factors of safety of the existing downstream slope under steady state seepage conditions varied from 1.6 to 1.9 for Cases 1 through 4. The minimum recommended factor of safety from *EM 1110-2-1902, Slope Stability* (U.S. Army Corps of Engineers, 2003) varies from 1.4 to 1.5 for this case.

DRAIN SYSTEM REHABILITATION

As previously discussed, the toe drain pipe system was cleaned during the inspection and assessment in 2011. Based on the subsequent assessment results and corresponding recommendations, relief well and drain system rehabilitation was conducted in 2015.

Relief Well Rehabilitation

Relief well testing, rehabilitation and inspection work was performed in April and May, 2015. The purpose of the relief well rehabilitation was to further evaluate existing conditions and functionality of the system, complete rehabilitation of all eleven relief wells using mechanical and chemical re-development methods, and re-evaluate the post-rehabilitation condition and functionality of the system. The pre- and post-rehabilitation evaluations of relief well condition and functionality were completed using well capacity pumping tests and video inspections.

The work included:

- Pre-rehabilitation specific capacity tests (all relief wells)
- Scrub brush cleaning (all)
- Air-lift pumping/surging developments (select relief wells)
- Surge block plunger developments (all, see Figure 9)
- Chemical development treatments with phosphoric acid and dispersion polymer (all)
- Post-rehabilitation specific capacity tests (all)
- Video inspections (all)
- Disinfection treatment with sodium hypochlorite (all)

The relief well specific capacity calculated from pumping tests decreased from 96 percent shortly after the relief wells were constructed in 1989 to 33 percent in 2015. The pumping rate to achieve the same drawdown elevation had also decreased for all of the relief wells.

Figure 9. Relief Well Surge Block Plunger Development

Specific capacity tests indicated that the 2015 mechanical and chemical rehabilitation efforts increased 10 of the 11 relief well capacities by 50 to 96 percent. There was no increase in the post-rehabilitation specific capacity value calculated for RW-9. The pumping rate to achieve the same drawdown also increased for all wells except RW-9.

Toe drain system outlet flow rate increased to 13 gpm as a result of the relief well mechanical and chemical rehabilitation. Phreatic elevation in some piezometers decreased by as much as 3 feet, but some piezometer levels did not show measurable change.

Toe Drain Pipe Rehabilitation

The section of toe drain pipe described as partially collapsed and deformed between relief well RW-5 and Inspection Well #3 in the 2011 video camera inspection report was replaced in October, 2015.

Gannett Fleming designed and subcontracted the replacement of 40 feet of toe drain pipe with solid-wall HDPE with the same nominal diameter. Slotted perforations were used to increase the open area per foot of pipe, and slot width was sized to meet current criteria, since the existing circular perforations are oversized. The drain aggregate below the pipe invert was left in place, and drain aggregate around the pipe was replaced with an aggregate of similar gradation. Dewatering of the excavation was achieved by pumping from nearby relief wells while blocking and rerouting toe drain pipe flows between RW-5 and Inspection Well #3.

Post-Rehabilitation Evaluation

Following the 2015 relief well rehabilitation, piezometer data were utilized to re-evaluate uplift and blowout factors of safety. Based on piezometer data collected at approximately normal pool, the relief well rehabilitation improved relief well effectiveness from between 20 to 50 percent pre-rehabilitation to between 40 and 60 percent post-rehabilitation. Four of the five piezometers monitoring the underlying coarse-grained foundation soils had post-rehabilitation piezometric levels from 1.7 to 3.0 feet lower than pre-rehabilitation levels. The fifth piezometer (PTP-4A) was slightly higher after rehabilitation. The minimum factor of safety for uplift and blowout for the critical section (Sta. 29+00), based on observed piezometer readings, increased from 1.6 to 1.9. As previously discussed, the recommended factor of safety for uplift and blowout is 1.5, and the calculated factor of safety for the seepage model condition of no relief wells was 1.4.

The piezometer threshold limits developed prior to the 2015 relief well rehabilitation, which were based on the historical range of piezometer readings and the calculated piezometric levels that provide the minimum recommended factor of safety against uplift/blowout of 1.5, are still appropriate and were not revised. However, additional guidance levels for the monitoring system were recommended to evaluate the performance of the relief well system and the need for further inspection or maintenance. Guidance levels were established for the coarse-grained foundation soil piezometers and the drain system outfall discharge based on the new data.

Guidance values were developed for the coarse-grained foundation soil piezometers based on the difference between pre-rehabilitation and post-rehabilitation levels at normal pool. The guidance values were determined based on an increase in phreatic elevation of 25 percent between the measured post-rehabilitation and pre-rehabilitation phreatic elevations at normal pool. The guidance values are only for normal pool conditions and should not be compared against readings taken when the reservoir level is 0.2 feet or more above or below normal pool.

Guidance values for the drain outfall were developed based on the variation in drain flow measured at Inspection Well #5. The pre-rehabilitation discharge was 2 gpm and the post-rehabilitation discharge was 13 gpm. The guidance value was determined based on a reduction in drain discharge of 25 percent between the measured post-rehabilitation and pre-rehabilitation outflow at normal pool.

In *EM 1110-2-1914* (USACE, 1992) the USACE recommends that if pumping tests indicate a reduction in specific capacity of more than 20 percent compared to that measured at installation, then the wells should be evaluated to determine what remedial measures should be employed. Based on *Water Well Maintenance and Rehabilitation* (The Pennsylvania State University, College of Agricultural Sciences, 2015), rehabilitation should occur when a decrease of 25 percent or more in well yield is observed. This guidance also indicates that "Delaying rehabilitation procedures can significantly increase costs and in some cases make rehabilitation impossible."

RECOMMENDED MONITORING AND EVALUATION

It was recommended that piezometer levels and toe drain flow rates at Crab Orchard Dam be monitored monthly. However, the frequency of monitoring increases to daily readings immediately before and for one week or more after events resulting in unusually high reservoir levels. It was recommended that reservoir pool and tailwater levels be recorded when monitoring piezometers or flow rates.

It was recommended that the systems of threshold and guidance values for piezometer levels and drain flow rates be used to assess the monitoring data. Relief wells should be inspected when piezometer guidance values are exceeded or when toe drain flow rate decreases below the guidance value. The 2015 relief well and toe drain work concluded with a recommendation to develop long-term relief well maintenance procedures and schedules based on responses to cleaning, and the relief wells should be inspected and tested at least every six years in accordance with procedures and recommendations in *EM 1110-2-1914* (USACE, 1992).

Tailwater levels have a significant influence on the piezometric levels and factors of safety against uplift/blowout, but tailwater elevations at the site have not been consistently monitored. Installation of two staff gages in the vicinity of Inspection Well #5 and the toe drain outlet was recommended to allow tailwater monitoring whenever piezometers and seepage data are collected. It was also recommended that the area along the drain system outfall pipe be maintained to prevent growth of trees and shrubs to eliminate root infiltration and facilitate observations.

A follow-up video camera inspection of the toe drain pipe system is scheduled for October, 2016 to verify the performance of the 2015 toe drain repair.

CONCLUSION

Crab Orchard Dam is a high hazard dam and a valuable resource for the Crab Orchard Wildlife Refuge. Modifications were initiated in 1988 to increase freeboard and spillway capacity, flatten the downstream slope and install a drainage system including relief wells and a new toe drain. Significant and credible potential failure modes related to seepage have been identified, but no significant seepage problems have been observed. A proactive approach to monitoring of site piezometers and drain system performance has been implemented to detect seepage-related problems as early as practical. The elevated and artesian piezometer levels observed during the 2010 flood event raised the suspicion that the relief well system was not functioning as designed. This initiated a series of investigations to evaluate the performance of the dam and drainages system, particularly the relief wells.

Field assessments, seepage modeling and uplift calculations indicated that the relief well system was not performing as originally designed, and the effectiveness of the system was deteriorating. An ongoing relief well and toe drain rehabilitation program was initiated in 2014 and included testing, rehabilitation and inspection of relief wells, as well as replacement of a section of partially collapsed and deformed toe drain pipe.

The rehabilitation of the relief wells confirmed that the effectiveness of the relief wells had deteriorated based on specific capacity of the well, measured piezometric levels and measured drain system outlet flow. The specific capacity of ten of the eleven relief wells increased by 50 to 96 percent, and the toe drain system outlet flow rate increased from 2 gpm measured just prior to the rehabilitation to 13 gpm after the rehabilitation. Water levels have been reduced in some of the piezometers by as much as 3 feet.

Piezometric levels and flow measurements will be measured periodically and evaluated to determine if effectiveness of the relief wells is deteriorating and if maintenance or rehabilitation is necessary. A long-term maintenance schedule and procedures for the relief wells will be determined based on the piezometer and drain flow levels observed post rehabilitation.

The monitoring of the piezometers at the site was essential for determining that drainage system was not performing as designed and was deteriorating. Piezometers were also utilized to assess the safety of the dam against heave. The piezometers and drain outflow measurements were beneficial in evaluating the effect of the relief well rehabilitation and will be utilized to evaluate the need for future maintenance and rehabilitation and to maintain the performance of Crab Orchard Dam.

REFERENCES

Pabst, M., et al, *Heave, Uplift and Piping at the Toe of Embankment Dams*, ASDSO Journal of Dam Safety, Vol. 11, Issue 2, 2013.

The Pennsylvania State University, College of Agricultural Sciences, Cooperative Extension, Water Facts #32, Water Well Maintenance and Rehabilitation, 2015.

Turnbull, W. J., and Mansur, C.I., "Investigation of Underseepage - Mississippi River Levees." Transactions, ASCE, Volume 26, Part I, pp. 1429-1539, 1961.

United States Army Corps of Engineers, EM 1110-2-1901, Seepage Analysis and Control for Dams, September, 1986.

United States Army Corps of Engineers, EM 1110-2-1902, Slope Stability, October, 2003.

United States Army Corps of Engineers, EM 1110-2-1914, Design, Construction, and Maintenance of Relief Wells, May, 1992.

United States Army Corps of Engineers, EM 1110-2-2300, General Design and Construction Considerations for Earth and Rock-Fill Dams, July, 2004.

United States Department of Homeland Security, Federal Emergency Management Agency, Filters for Embankment Dams, Best Practices for Design and Construction, October, 2011.

United States Department of Homeland Security, Federal Emergency Management Agency, Technical Manual: Plastic Pipe Used in Embankment Dams, November, 2007.

FOUNDATION PROTECTION DURING POST TENSIONED ANCHOR CONSTRUCTION

William B. Stiller, PE[1]
Y. Erwin Supranata, PE, PhD[2]
Stephen J. Mockler, PE[3]

ABSTRACT

Hawks Nest Dam, owned and operated by Brookfield Renewable Energy Group, is located on the New River in West Virginia. Hawks Nest Dam is a concrete gravity dam approximately 836 feet long with 14 Stoney-type spillway gates. The dam structure is comprised of 31 blocks, constructed independently, and interconnected by vertical shear keys. The 8th Part 12 Safety Inspection Report recommended evaluation of sliding stability along the sandstone bedding planes and shale partings in areas where downstream scour and undercutting had been historically observed. This stability assessment determined the structure did not meet FERC required safety factors for sliding, and due to the rapid tailwater rise during flooding, was controlled by the normal loading condition. A total of 49 multi-strand tendon anchors were designed and installed for improving the stability of the structure. Since the "normal" loading condition presents the biggest deficit with regards to required sliding stability factor of safety, the impact of drilling and grouting of the structure to the stability was of concern during the construction of the post-tensioned anchors. Close monitoring of foundation pressures during drilling and grouting was incorporated along with an instrumentation and monitoring plan developed specifically for construction. Prior to construction, additional observation wells were installed to increase coverage for monitoring change in foundation uplift pressure during anchor installation. Threshold values for existing and new instrumentation were also developed to assist monitoring change in foundation conditions during construction. The concern for foundation pressurization also impacted the consolidation grouting process. In cases where the contractor needed to pressurize consolidation grout to obtain a grout tight hole, a detailed grouting plan with pressure limitations was required. The implementation of instrumentation and monitoring plan, coupled with a detailed consolidation grouting plan, enabled the contractor to obtain grout tight holes without jeopardizing the stability of the structure.

INTRODUCTION

The original stability analysis for Hawks Nest Dam was performed in 1930. As part of the first independent inspection in 1969, the structural stability of the dam was reanalyzed with respect to overturning and sliding. Although the stability of the dam was found to be

[1] HDR Engineering of the Carolinas,440 S. Church Street, Suite 900, Charlotte, NC 28202, brad.stiller@hdrinc.com.
[2] HDR Engineering of the Carolinas,440 S. Church Street, Suite 900, Charlotte, NC 28202, (704) 338-6834, erwin.supranata@hdrinc.com.
[3] Brookfield Renewable Energy Group, 75 State Street, Suite 2701, Boston, MA 02109, (857) 313-7679, Stephen.Mockler@brookfieldrenewable.com.

adequate, installation of relief wells was recommended in order to increase the factor of safety against overturning. The relief wells were installed in 1972.

In 1986, the stability of the dam was reanalyzed as part of the fourth independent inspection report. Three analyses (normal, unusual, and extreme) were performed for Blocks 26 and 13 of the dam. The analyses were based upon an interface friction angle (ϕ) = 45° and cohesion (c) = 500 psi.

The FERC indicated that cohesion of 500 psi was too high without supporting data. In response, D'Appolonia, (2001) developed a c=0 set of analyses for Blocks 26 and 13 of the Hawks Nest Dam. As a result of the 2007 Eighth Part 12 Safety Inspection further stability evaluations were required to include an evaluation of rock erosion at the spillway toe, and stability with respect to sliding along a shale seam in the foundation rock.

The resulting stability analysis submitted showed the need for structural modification of monoliths 5 to 31 to increase the stability factor of safety to within the FERC required criteria. A plan for the installation of 49 post tensioned anchors, with 28 to 50 strands each, and the resultant change in the stability calculations was submitted to the FERC in a letter dated March 25, 2013 and was subsequently approved in a letter by FERC dated July 26, 2013.

As the tailwater elevation increases rapidly with the increase of flow, the governing stability condition was deemed the normal operating condition for Blocks 5 to 8, with the intermediate flood controlling the stability of the remainder. See Figure 1 for a typical cross section of Blocks 5 to 8. The installation of post tensioned anchors involves high pressure drilling, pressurized grouting and pressurized water testing; therefore, additional precautions were required for the construction phase of the project. These additional precautions included the installation of additional relief wells and the development of a Construction Dam Safety Surveillance and Monitoring Program (CDSSMP).

The construction of the required structural modifications, consisting of post tensioned anchors, was completed in 2014.

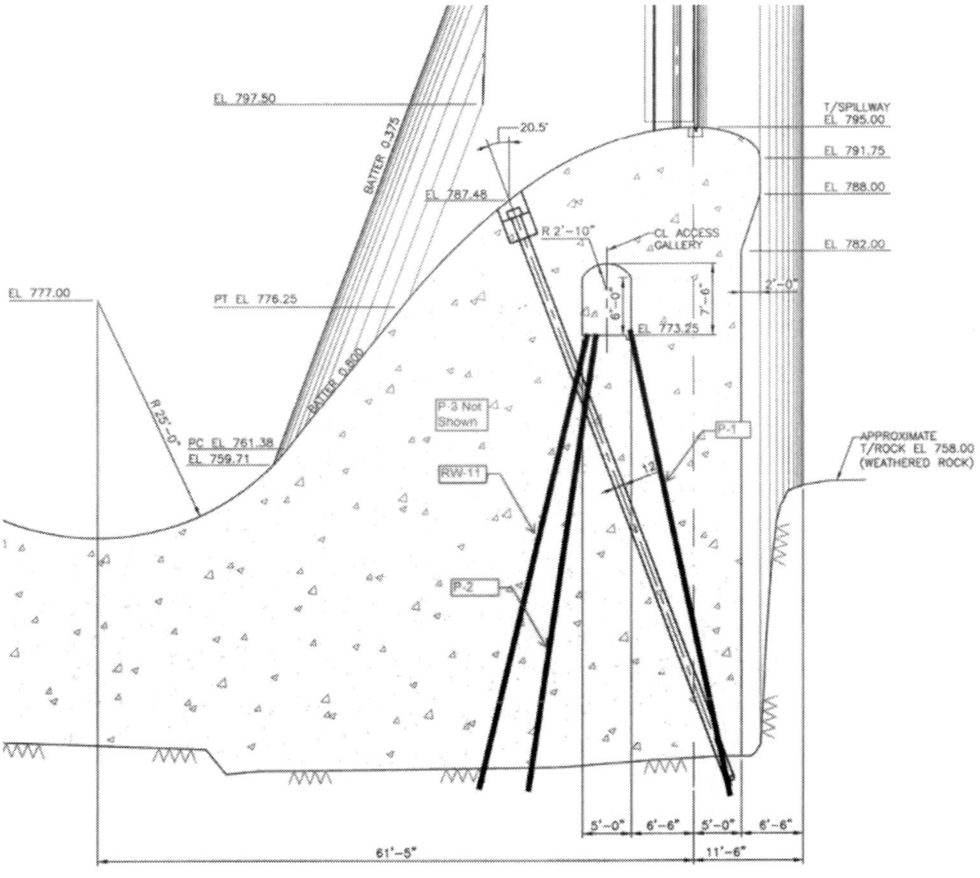

Figure 1. Cross Section of Block 7

CONSTRUCTION INSTRUMENTATION

Potential Failure Mode Analysis (PFMA) in conjunction with the anchor construction was conducted in July 2012. The PFMA identified six potential failure modes (PFMs) related to the anchor construction. One of the PFMs involves the potential for injecting grout into the foundation drains through fractures in the rock during construction. The grout could seal off the foundation drain and cause an increase in uplift pressure. The increased uplift pressure may lead to the sliding failure, breach, and uncontrolled release of the reservoir. The risk reduction measures recommended for this PFM include frequent instrumentation monitoring and installation of additional observation wells prior to anchor construction.

To implement the risk reduction measures suggested by the PFMA, twelve new observation wells were installed prior to the construction. The observation wells have dual functions: to relieve uplift pressure and to enable monitoring the change in foundation and joint conditions. New observation wells (OW) along with existing relief wells (RW) and piezometers (P) were used as monitoring instruments in the CDSSMP. The instrument locations for Blocks 5 to 16 are shown on Figure 2.

The CDSSMP was developed to establish instrument monitoring frequency, threshold levels, and evaluation procedure to asses the dam safety during the construction. The CDSSMP focused on the close monitoring of the foundation pressures during the various stages of the construction process. The use of existing relief wells, piezometers, and new observation wells enable a comprehensive monitoring of the spillway monolith blocks.

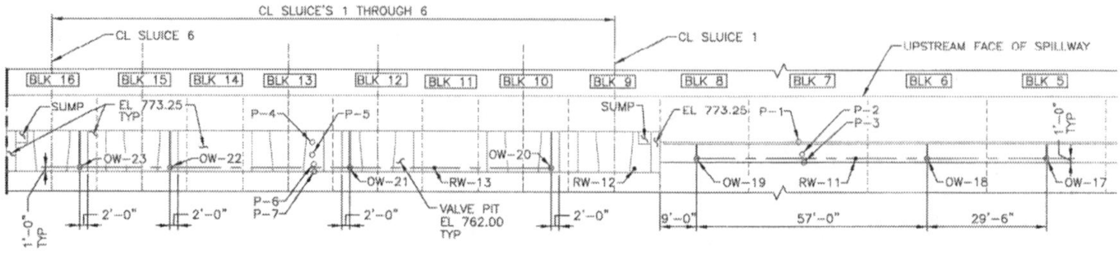

Figure 2. Construction Instrumentation for Blocks 5 to 16

The new observation wells were required to be monitored on a daily basis for a minimum of a week, to establish baseline readings, prior to the start of drilling associated with the installation of the post tensioned anchors. High and low threshold values for the new observation wells were calculated as average plus and minus two standard deviations of the measured water levels, respectively. The threshold values were updated periodically to incorporate new data. The existing instrumentation readings were evaluated based on the pre-construction high and low threshold values as well as their historical maximum and minimum readings.

The Specification required that when water testing or consolidation grouting of boreholes within a specific monolith of the spillway was being performed, the instrumentation of the block and the adjacent blocks were to be closely monitored and documented. If, during the consolidation grouting and water testing process, readings were observed to be outside of the established high or low threshold values, operations would cease and the Owner notified.

CONSTRUCTION LIMITATIONS

The installation of post tensioned anchors typically requires the use of percussion air hammers for the drilling of the anchor holes, pressurized water to test for water tightness of the finished drill hole, and pressurized grout to seal holes not meeting water tightness criteria. The pressures associated with drilling were assumed to be relieved by the cutting return line. The water test implemented met the minimum criteria of Post-Tensioning Institute (PTI). And as higher pressure tended to increase the flow rate out of the hole, the Contractor agreed with limiting the water test pressure to 5 psi at the top of the drill hole.

The pressure associated with consolidation grouting, while typically monitored during construction of post tensioned anchors, was restricted for the post tensioned anchor construction at Hawks Nest. The Specification restricted consolidation grouting by not allowing the simultaneous consolidation grouting of adjacent boreholes whether in the same or adjacent monoliths. The Specification also limited the pressurization of consolidation grout to 5 psi above the top of the drill hole unless permitted in writing by the Owner.

As construction commenced, the effectiveness of consolidation grouting at 5 psi was questioned by the Contractor. The Contractor encountered multiple locations where the consolidation grouting and redrilling process had gone through multiple iterations without a passing water test. As such the water testing and consolidation grouting process was revisited. Based on discussions with the contractor and review of the previous water testing and consolidation grouting records, the primary source of water loss was speculated to be in the rock mass below the foundation contact.

The water testing and consolidation grouting process was revised to first water test the rock below the foundation. Before allowing an increase in grouting pressure, an effort was made to develop a grout mixture that was more effective at sealing the hole. As part of this effort, the use of admixtures such as an anti-washout agent, super plasticizer and retarder were permitted to increase the performance of the grout. Additionally, the specified Type I/II cement was replaced with a finer Type III cement. The results of the water test provided guidance as to the consolidation grout mixture to be used. Measured loss through the drill hole less than 10 gallons in 10 minutes was approached with a grout mixture incorporating a water to cement ratio (w/c) of 1.0 for the first 30% of the theoretical hole volume and increasing cement content to a w/c of 0.67 for the remainder. Water tests resulting in over 100 gallons were approached with a w/c 0.5 mixture.

The first consolidation grout was placed with a packer set just above the foundation contact point. Consolidation grout was placed via tremie tube from the spillway. Consolidation grout was metered with operations halting when a consistent grout return was observed, or when 200% of the theoretical volume of the drill hole had been placed. If a consistent return was observed prior to reaching the volume limit, grouting operations were switched from tremie to pressure grouting with a limit of 5 psi surcharge pressure. Consolidation grout was continued to be placed under pressure until the grout take halted or the volume limit had been achieved.

Prior to drilling out the consolidation grout, the borehole above the packer (spillway concrete) was water tested. The hole was then redrilled to the design depth and the entire hole water tested (assuming the spillway concrete passed the water test). If the hole still did not pass the water test, the consolidation grouting and water testing process was repeated, with minor variations to address the rare event of the spillway concrete not being water tight. The biggest variance for the second and additional rounds of consolidation grout placement was an increase in the allowed top of hole pressure. An increase from 5 psi to 10 psi was allowed. The basis for allowing this increase was that the lower pressure consolidation grout would have filled the large cracks and fissures in the concrete and rock. This would reduce the distance at which foundation pressures would be impacted by pressurized grout.

MONITORING IMPACT

The pressure restrictions with grouting and water testing resulted in only a handful of situations that triggered a stop work situation. When staff monitoring the gallery instrumentation notified the Contractor of readings outside of the established limits, the ongoing work was halted. The individual blocks with instrumentation outside of the established limits were evaluated for stability using the readings measured in the gallery along with the actual headwater and tailwater elevations. The sensitivity of the calculated stability was evaluated for change in the instrumentation readings. A new, temporary instrumentation limit was then established for the specific site conditions being encountered. This temporary modification to the instrument threshold value provided the Contractor with enough allowances to complete the installation of the anchor.

An example of this program in action is associated with the drilling of anchor 7B. The Contractor had initially drilled a 5.75" diameter pilot hole with the directional drill and had just finished reaming the hole from 5.75" diameter to the 12" final diameter. The gallery attendant during the drilling did not note any readings outside threshold values. During the morning instrumentation readings, instruments P-1, P-2 and RW-11 all exhibited readings in excess of the threshold values.

RW-11: Slow drip increased to 21 seconds per quart
P-1: Depth to water level decreased from 12.4 feet to 2.5 feet
P-2: Depth to water level decreased from 10.36 feet to 4.1 feet

See Figures 1 and 3 for the location of the subject instrumentation in relation to Anchor 7B. See Figure 4 for a graph of instrument P-1 readings.

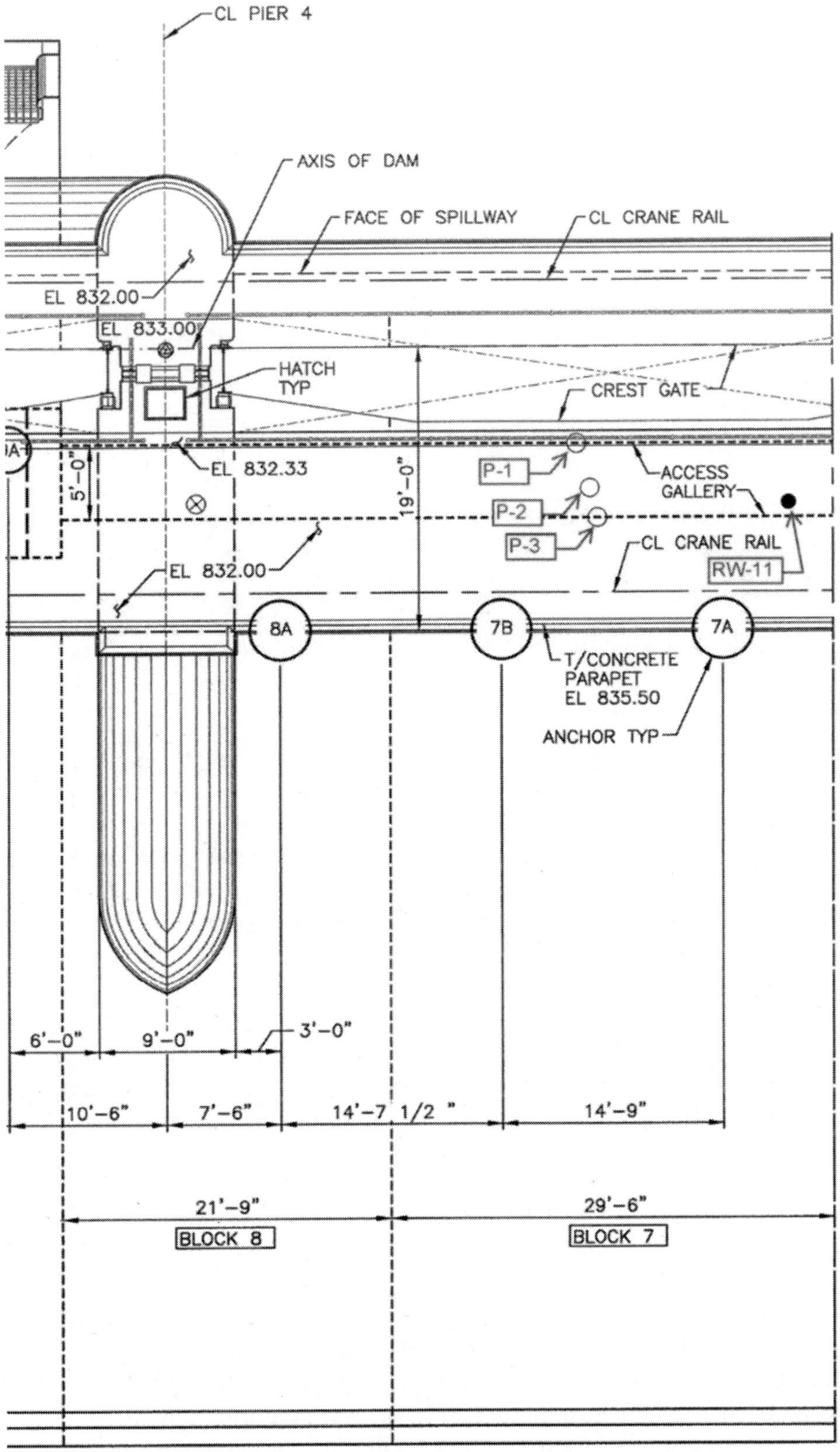

Figure 3. Plan View of Block 7 with Instrumentation

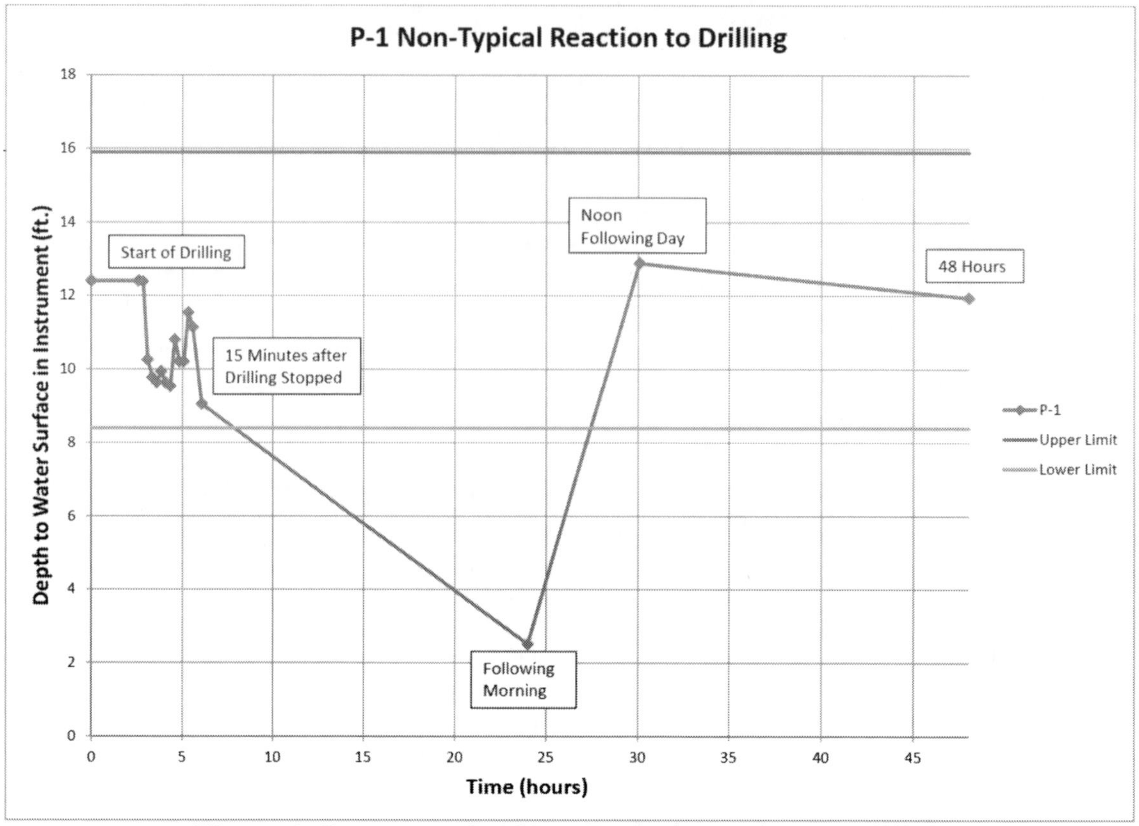

Figure 4. Instrument P-1 Non-Typical Reaction to Drilling

Upon notification of these changes, the Contractor was instructed to halt work in this and the adjacent blocks until additional evaluation could be performed. HDR performed stability analysis for Block 7 utilizing the measured values from instruments P-1, P-2 and P-3 and confirmed the overall change in foundation uplift pressure resulted in a slight reduction in the sliding factor of safety that would not impact the stability of the structure. Based on the stability evaluation result, the high threshold value was updated using the latest readings and the work was allowed to continue. The subsequent construction activity proceeded with consideration that continued pressure increase over the updated high action value would cause the halting of the work. While the subsequent construction activity in Block 7 did result in variation of the instrumentation readings, the updated high threshold value was not exceeded.

CONCLUSION

The typical post tensioned anchor installation relies upon high pressure drilling and pressurized grouting that can increase uplift pressures at the base of the structure during construction. The typical post tensioned anchor installation, however, is normally required to improve stability for unusual loading conditions, including extreme flood and/or earthquake. The anchor installation at Hawks Nest required a higher level of monitoring and construction restrictions to protect the structure being anchored to maintain stability during normal operating condition.

By expanding upon the existing gallery instrumentation to provide monitoring of all key monoliths, the ability to monitor changes in the foundation conditions was established. By establishing limits to the gallery instrumentation and closely monitoring instruments during construction, a higher level of protection was provided to the structure. In the few situations where the established instrumentation limits were encroached upon, stability evaluations based on actual instrument readings were able to provide temporary construction allowances that facilitated the successful installation of the anchor.

In addition to the increased monitoring effort, restrictions were placed on the pressurization of consolidation grout to further protect the foundation. These limits did require modification during the construction process to provide the Contractor with a better means to attain the required water tight hole while maintaining the foundation protection. These modifications resulted in the Contractor being able to attain a water tight hole, typically with less than two rounds of consolidation grouting and redrilling.

By collaboration with the Owner and Contractor, prior to the onset of construction, reasonable and attainable instrumentation and construction practices can be established to effectively protect the stability of a structure during the installation of post tensioned anchors.

ACKNOWLEDGEMENTS

The authors would like to thank Brookfield Renewable Energy Group for their permission to share the challenges and success of the recent post tensioned anchor installation project at the Hawks Nest Spillway.

BUCHANAN DAM — KEEPING CURRENT FOR EIGHT DECADES

Victor M. Vasquez, P.E.[1]
Doug Witkowski, P.E.[2]
M. Leslie Boyd, P.E.[3]
Monica Masters[4]
Layne Bukhair, P.E.[5]

ABSTRACT

The Lower Colorado River Authority (LCRA) began taming the Colorado River in Texas in the mid-1930s. Its goals were to control flooding and use the river as a productive resource for water and hydropower. The chain of six dams that make up the Highland Lakes successfully accomplished these goals.

Buchanan Dam, the first Highland Lakes dam, is the most complex structure in LCRA's dam portfolio. At 145 feet tall and stretching more than two miles, Buchanan Dam is considered the longest multiple-arch dam in the nation. The dam consists of two multiple-arch sections, three gated spillways, five non-overflow sections, and one uncontrolled spillway. Completed in 1938, its construction was one of the first successful dam projects to help revitalize the Texas economy. As Buchanan Dam has endured many droughts and floods, LCRA has kept a watchful eye on the dam to maintain its investment and protect the public. Today, Lake Buchanan serves as a primary water supply for one of the fastest growing regions of the US.

As a result of the dam failures reported nationally in the '70s and '80s, LCRA embarked on a journey to protect its aging dam infrastructure and address changing dam safety regulations. Following an early form of portfolio risk management, LCRA established a modern dam safety program and implemented a series of engineering and construction projects to take Buchanan Dam into the 21st century. One hundred-ten post-tensioned anchors were added at non-overflow sections to address higher flood loads and potential instability and twenty-eight hoists were added to gated spillways to enable faster, safer responses to severe weather. Currently, structural repairs are addressing concrete spalling on the multiple-arch sections. Simultaneously, dewatering systems consisting of stoplogs and traveling cranes are being added to the spillways to prevent loss of the reservoir in the event of gate failure. Once complete, this innovative dewatering system will facilitate structural strengthening and repair of thirty-seven tainter gates over the next 10 years.

This paper summarizes LCRA's dam safety efforts and presents Buchanan Dam as a model example of how to address evolving dam safety regulations while preserving valuable benefits of water, hydroelectric power, flood control, and recreation.

[1] Senior Project Manager, Freese and Nichols, Inc., Austin, TX, vmv@freese.com
[2] Principal Engineer, Lower Colorado River Authority, Austin, Texas, doug.witkowski@lcra.org,
[3] Senior Engineer, M.ASCE, Freese and Nichols, Inc., Austin, TX, mlb@freese.com
[4] Senior Project Manager, Lower Colorado River Authority, Austin, Texas, monica.masters@lcra.org
[5] Design Engineer, P.E., Freese and Nichols, Inc., Austin, Texas, llb@freese.com

INTRODUCTION

The Lower Colorado River Authority (LCRA) owns and operates six dams on the Colorado River in Central Texas. The dams from upstream to downstream are Buchanan, Inks, Wirtz, Starcke, Mansfield, and Tom Miller as shown in Figure 1. The dams create the Highland Lakes; Buchanan, Inks, LBJ, Marble Falls, Travis, and Austin respectively. These dams were constructed between the 1930's and 1950's and have provided a reliable water supply, flood control, and a renewable source of power production for almost eighty years. The dams supply water to support a population now exceeding one million people as well as businesses, power plants, and agriculture. The growth and changes in the area being served by the dams for almost 80 years are truly remarkable. This growth also comes with the massive responsibility of proper operation and maintenance to preserve public safety and the well-being of the vast populace served by the dams.

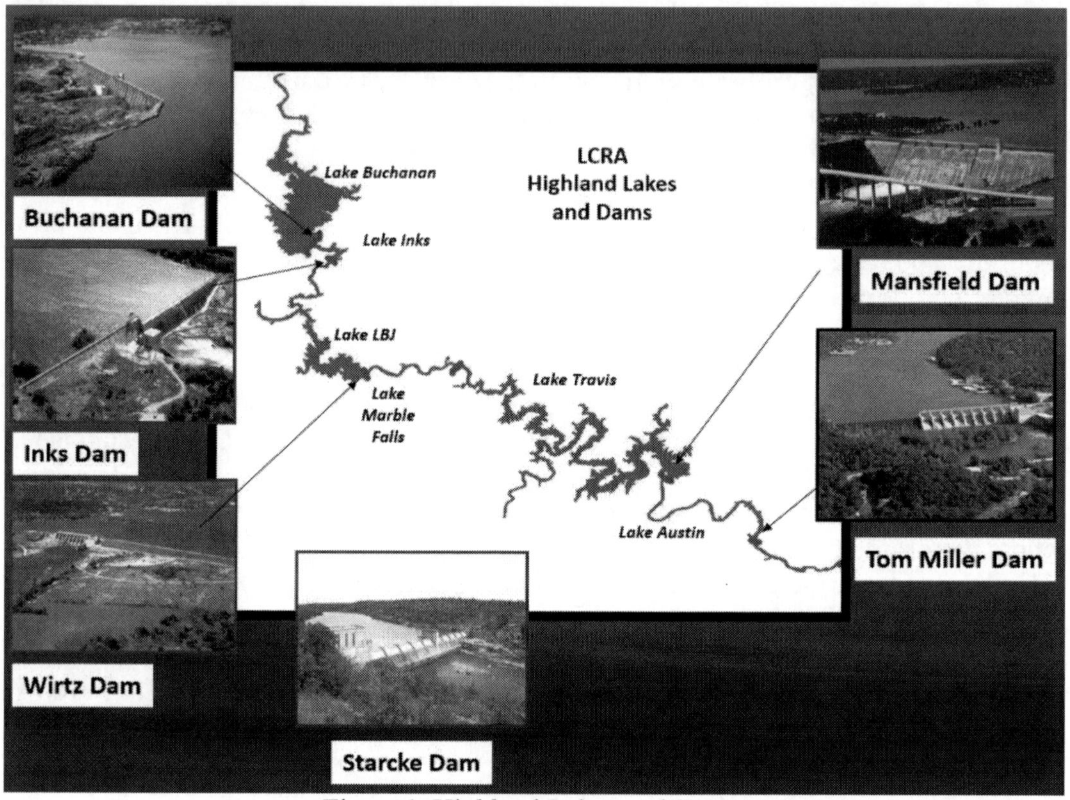

Figure 1. Highland Lakes and Dams

LCRA has been a model dam owner throughout its history. During the first four decades, inspections were regularly performed by LCRA personnel with occasional technical support of state and federal agencies as needed. External dam safety oversight by the State of Texas began in the late 1960's when it implemented an inspection program of existing dams with State inspectors visiting the Highland Lakes Dams in the early 1970's. The results of the inspections repeatedly indicated that the dams were well maintained and showed no signs of distress.

Following numerous dam failures from across the country, the U.S. Army Corps of Engineers (USACE) developed the National Dam Inspection Act, which was authorized in 1972. Federal guidelines for dam safety were enacted and implemented in the late 70's. Soon thereafter, Texas began a statewide safety evaluation of existing dams for the USACE. This dam safety evaluation was different from past inspection programs because it involved deciding which existing dams might experience a bigger flood than the standard design project flood. The State of Texas and USACE identified several dams in the state as unsafe under the newly determined probable maximum flood (PMF). These programs resulted in significant changes in dam safety standards and the first comprehensive Texas Dam Safety rules in 1986.

As a result of the dam safety evaluation, the State of Texas and USACE listed two LCRA dams, Wirtz and Buchanan, as high risk dams. In response, LCRA implemented a Dam Modernization Program soon thereafter to further evaluate all Highland Lakes Dams starting with detailed inspections, PMF re-evaluations, and stability analyses. The findings indicated that, except for Mansfield Dam, five of the Highland Lakes Dams did not meet new dam safety standards. These findings led to a risk-based prioritization of more advanced engineering studies, preliminary design, final design, and construction projects. Over a period of 15 years culminating in 2005, LCRA led dam safety improvements and operational upgrades worth approximately 50 million dollars in construction to modernize and bring its dams into compliance with state dam safety regulations.

During the modernization program, LCRA established a dam safety program with internal policies and procedures for operations and maintenance based on state and federal guidelines. The program was modeled after the U.S. Bureau of Reclamation (USBR) approach for dam safety through comprehensive facility reviews (CFRs). Since 2001, this program has identified new operational and safety improvements for safe upkeep of the dams, water preservation, and compliance with state dam safety regulations. LCRA has used this new information to implement new studies and embark on new construction projects for safe upkeep of its dams.

History of Buchanan Dam

Buchanan Dam, the first and most complex structure in LCRA's portfolio of dams, has been through a unique journey and provides the best example of the challenges that LCRA has faced to keep current with dam safety regulations, address maintenance issues, enhance operations, and respond to extreme weather including floods and droughts. The dam is located in both Burnet and Llano counties, approximately 12 miles west of the City of Burnet, Texas. The dam provides hydroelectric power and also creates a large water supply reservoir for the region. Lake Buchanan, which stores 875,566 acre-ft at normal pool level, is also a key recreational area for the surrounding community.

Construction of Buchanan Dam, initially named Hamilton Dam, began in November of 1931. The construction was initiated by a Texas subsidiary of the Chicago-based Insull Utility Company and performed by Fegles Construction Company under the supervision of Fargo Engineering Company. The construction was suspended in 1932 as a result of the Insull Utility Company going bankrupt. In 1934, the LCRA was formed and resumed construction with initial technical assistance by USBR. The design and construction effort was ultimately led by LCRA with project completion accomplished in 1937 (see Figure 2 to Figure 5). The dam was named after U.S. Rep. J.P. Buchanan, who helped secure federal funds to complete the project.

When completed, Buchanan Dam was the largest multiple-arch dam in the world. The dam is categorized as a "Large-High-Hazard" dam and must be able to safely pass the PMF. The dam is over 2 miles long and consists of multiple concrete arches; a powerhouse, gated spillways, an uncontrolled spillway, and an embankment, as shown in Figure 6 and Figure 7. The multiple concrete arches are divided into two distinct sections. 35-ft arch and 70-ft arch sections. The spillway closest to the powerhouse (southernmost spillway) is referred to as the 7-gate spillway with each gate measuring 40 feet wide and 25.5 feet high. The second spillway (to the north) is referred to as the 14-gate spillway with each gate each measuring 33-feet wide and 15.5-feet high. The third spillway (northernmost and adjacent to the emergency spillway concrete overflow section) is referred to as the 16-gate spillway with gate and pier dimensions identical to those of the 14-gate spillway. As noted above, all these sections were built over a prolonged period of time under two different construction efforts, thus making the dam even more unique.

Figure 2. Buchanan Dam 70-ft arch section during construction

Figure 3. Buchanan Dam impoundment, 35-ft arch section soon after construction

Figure 4. 7-Gate Spillway during its first flood release in 1938

Figure 5. 14-Gate Spillway during its first flood release in 1938

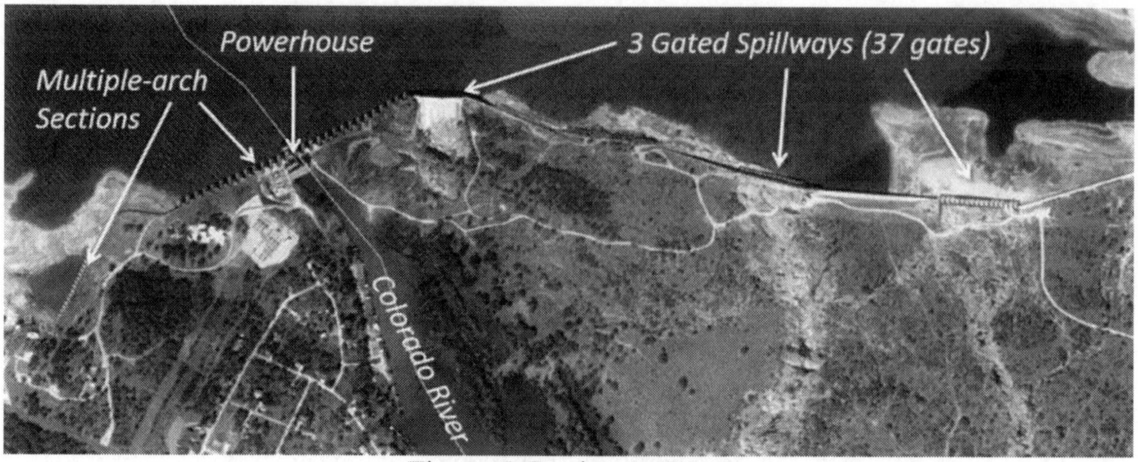

Figure 6. Buchanan Dam

Figure 7. Buchanan Dam

DAM MODERNIZATION PROGRAM

The Dam Modernization Program was LCRA's response to changing dam safety regulation at the state and federal levels and findings from the National Inspection Act. LCRA's goal was to address concerns with hydraulic adequacy and lower than desirable stability safety factors. LCRA successfully performed initial evaluations of the identified concerns, and after inspections and analysis for various flood conditions, the results disproved some problems and confirmed others as follows:

- The spillways and non-overflow sections failed to meet safety factors for stability starting at the 500-year flood.
- The ability to timely operate the 37 tainter gates with four traveling hoists was identified as critical for dam safety.

Based on their findings, LCRA developed a plan of action for Buchanan Dam which included additional engineering analysis, design of structural modifications, preparation of formal emergency action plan, improvement of weather forecasting and warning procedures, and public education about flood management.

Preliminary Engineering Analysis and Design

Preliminary engineering studies associated the low safety factors to new flood levels, increased hydrostatic and hydrodynamic loads, low concrete unit weight, and absence of drains for uplift reduction. Beginning in the mid 1990's, LCRA retained Freese and Nichols, Inc. (FNI) to assist with engineering analysis and design for the concrete gravity spillways and non-overflow sections.

As part of engineering design, additional studies and analysis were conducted to better define the stability issues at Buchanan Dam. State-of-the-art computer modeling for design storms were used to refine flood events up to the PMF and associated system demands. Physical hydraulic modeling, as shown in Figure 8, were also extensively used to more accurately evaluate hydraulic performance and scour potential downstream at the spillways. The results were a lower PMF level and refinement of hydrodynamic forces on the structures. Other key aspects of the design effort included geotechnical investigations and installation of monitoring instrumentation to develop site-specific parameters for the dam and its foundation. In combination with advanced structural and stability analysis, the engineering evaluations helped refine the necessary level of upgrades and stabilization improvements. The result was an optimized approach to dam safety improvements that eliminated stabilization of the spillways and reduced the stabilizing forces required at the non-overflow sections.

Seven stabilization alternatives were proposed for the non-overflow gravity sections that were deemed unstable. Environmental permitting constraints, construction costs, and life-cycle cost were considered in the evaluation. Based on all the analysis, post-tensioned anchors were recommended for stabilizing the gravity non-overflow sections.

Figure 8. Physical Models of 7-Gate Spillway

During the PMF evaluation, LCRA also recognized that its ability to safely pass the PMF was limited by the number of traveling hoists at each spillway: one hoist at each of the 14-gate and 16-gate spillways and two hoists at the 7-gate spillway. The traveling hoists required LCRA personnel to be on the spillway during a storm and fully raise one gate before proceeding to the next gate. This process was repeated as needed to manage the flood water and resulted in slower than desired gate operations. To improve gate operations, two additional hoists and a backup generator were added to the system in 1999, resulting in two traveling hoists at each spillway as shown in Figure 9.

Figure 9. 16-Gate Spillway with two traveling hoists

Final Design and Construction

The final design phase for the gravity non-overflow sections consisted of three major tasks: a comprehensive geotechnical field investigation, a final stability analysis, and a final design including development of plans and specifications. The geotechnical investigation included over twenty additional boreholes at key locations, additional dam safety instrumentation, and specialized laboratory testing. Refined material properties and uplift pressures were developed for use in final stability analysis and development of required anchor loads. A test anchor program was performed as part of the design to evaluate the construction techniques for the project and assist the specification preparation. The program demonstrated that the most important aspects of the anchoring system included the quality of the materials, the cleanliness of the tendons in the anchoring zone, an extensive quality control and inspection program, and the uniform seating of the anchor head wedges.

A total of ninety-six 5-strand anchors and forty-two 21-strand anchors with double corrosion protection using corrugated sheaths were installed at the non-overflow sections (See Figure 10 and Figure 11). Construction was completed for the post-tensioned anchoring project in January 2001.

Upon completion of the Buchanan Dam Anchoring Project, the Modernization Program continued for a few more years until Tom Miller Dam was modernized in 2004. In the ensuing six years, LCRA focused on its dam safety program with rigorous inspections using the USBR inspection model. The next chapter for Buchanan Dam came by way of revised Texas Dam Safety rules, findings from CFRs, and Mother Nature in the form of more than 17 inches of rain in 5 hours.

Figure 10. Anchor steel tendon installation

Figure 11. Anchor corrosion protection sheath installation

DAM SAFETY PROGRAM

LCRA implemented a dam safety program in 2000 that followed USBR's comprehensive facility review guidelines. USBR, under agreement with LCRA, completed the first CFR at Buchanan Dam in 2002. The CFR revisited some operation and maintenance recommendations from the Dam Modernization Program regarding the structural adequacy of the tainter gates, which led to subsequent climbing inspections and structural analysis of the gates at the three spillways. The CFR also provided recommendations for a phased stress analysis of the multiple arch sections using finite element modeling to support detailed visual inspections performed by LCRA.

A second CFR was performed at Buchanan Dam in 2007, and the risk assessment reinforced the following:

- The need to evaluate the multiple arch sections with an emphasis on concrete section loss due to spalling and the buttress head capacity.
- The need to enhance spillway operations and strengthen the gates including means to dewater each spillway bay for gate repairs.
- The need to update hydraulic loads on the dam in light of 2006 State Design Storm Guidelines.

LCRA embarked on new studies beginning in 2007 as a result of the CFR findings and recommendations to further enhance Buchanan Dam. The most significant efforts were related to the multiple-arch internal stresses, tainter gate structural capacity, gated spillway operational limitations, and dewatering system needs.

Corbel Peer Review and Physical Structural Modeling

Structural analyses following the 2007 CFR reported concerns with the adequacy of the arch sections and buttress heads for normal pool and PMF (probable maximum flood) loading. The concerns were related to localized tension and shear between the arches and light reinforcement in the corbels that could be subject to brittle failure. Repairs to address the concerns were estimated at $30 million. The potential public risk, the significance of the proposed construction cost, and the fact that the corbels had not demonstrated any significant structural problems in over 70 years justified further consideration and research. LCRA contracted with FNI to further evaluate the reported concerns and conduct additional structural analysis.

The evaluation began with a peer review study to assess the existing information against the definition of a corbel based on applicable structural codes. Referred to as corbels in the CFR, the buttress heads are the structural elements of the buttresses on which the barrel arches rest (See Figure 12 and Figure 13). The study concluded that the code requirements were very restrictive and limited in application to accurately evaluate the corbels at Buchanan Dam. Two-dimensional finite-element models using various software packages were developed next to evaluate stresses for the corbels. The model resultes were deemed inconclusive. FNI concluded that further evaluation of the corbels was warranted to more accurately assess the corbel capacity given the inadequacy of standard analysis techniques, the inconclusive nature of several other non-standard analysis techniques, and the successful performance history of the dam.

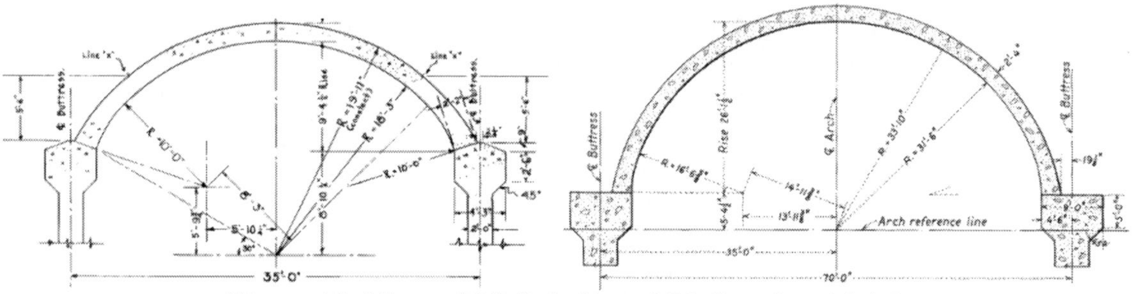

Figure 12. View of 35-ft (left) and 70-ft arches (right)

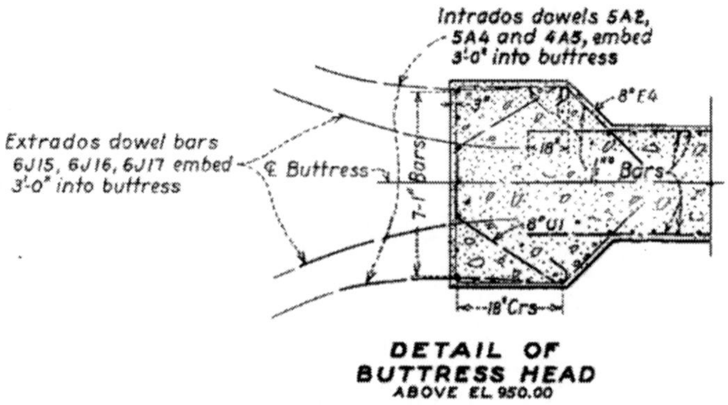

Figure 13. Corbel

A corbel evaluation program was developed to provide a cost-effective method of determining the pertinent failure mechanisms, establish if there are reasons for concern regarding the behavior of the existing structure, and determine if extensive repair can be avoided. The program consisted of above and underwater visual examinations, material testing, physical corbel model testing, and additional structural analyses.

A visual examination of the corbels was performed to assess if significant cracking had occurred that could be related to tensile or shear stresses. The purpose of the examination was to find any obvious signs of cracking or deterioration along the upstream face of the corbels between the arches. The visual observations consisted of first-hand observations, review of photographs of the upstream faces of the corbels above the water level, and an underwater examination using a remotely operated vehicle with a video camera. There were no obvious signs of cracking that could suggest tensile stresses or structural deficiencies along the upstream face between the arches.

Samples of concrete were tested to establish baseline strengths for an "as-constructed" condition. FNI performed sampling from multiple areas given the lapse in the original construction. The reinforcing steel strength was not explicitly stated in the original design documentation. Therefore, the steel strength was based on historical usage, research papers, calculated stresses listed on the original drawings, and tests of reinforcement samples taken from the from old construction debris. The results indicated that current day lower grade rebar could be used to simulate the in-place reinforcement.

Physical models were built to replicate a lightly reinforced corbel section with a configuration representative of the dam. The physical model study was performed in Ferguson Laboratory at the University of Texas at Austin and consisted of constructing two half-size specimens of a representative corbel section (See Figure 14). The specimens were tested to evaluate the possible interaction of the tension in the corbel face and the corbel's shear resistance. The models were instrumented with strain gages and load cells prior to testing. Horizontal and vertical loads were applied to the model to simulate the combination of moment and shear forces representative of the actual dam arch reactions.

Figure 14 - Test setup and crack propagation after testing to failure.

The results of the physical tests were then used to develop analytical models replicating the geometry, loads, and material properties of the corbels. The models were then utilized to evaluate the performance of the corbels at various locations. Findings from the analytical models indicated that the corbels provide adequate strength to resist the anticipated loads if considered plain concrete elements. Furthermore, static analyses of the corbels as well as the tests showed that the capacity lost when the concrete cracks is more than compensated for by the steel reinforcement. In addition, the locations of the cracks formed were not expected to reduce the shear capacity. The original design was ultimately determined adequate by the testing and analysis, so no corbel repairs or modifications were recommended. It was recommended that observations of the upstream face of the corbels be a part of the periodic dam inspections.

Tainter Gate Inspection and Structural Analysis

As follow up to the 2002 CFR recommendations, LCRA contracted with an engineering consultant in 2006 to perform hands-on inspection and detailed structural analysis of the gates at Buchanan Dam. The gate analysis was performed following USACE guidelines. The tainter gate members were found in good condition, but coatings systems were failing. The analysis indicated that several steel members were overstressed under operational and extreme loads when trunnion friction was included. The gate structural analysis concluded that, although the gates have operated successfully over the life of the project, they should be strengthened and/or replaced as soon as practical in order to attain an acceptable factor of safety and prevent loss of significant storage in the event of gate failure. The 2007 CFR confirmed that the risk of gate failure justified actions by LCRA. Actions included the repairs of the radial gates and/or installation of emergency stoplogs to prevent uncontrolled release of the reservoir due to gate failure.

Figure 15. Small Gates at 14 and 16-gate spillways

Figure 16. Large Gates at 7-gate spillway

LCRA contracted with FNI to peer review the structural gate analysis and findings. FNI concluded that the structural analysis were sound and concurred with the findings. FNI recommended that final design be initially focused on one gate to evaluate repair procedures before committing to a program involving the remaining 36 gates. The development of a dewatering system was also recommended to facilitate the repairs over multiple years and prevent a sudden reservoir loss in the event of gate failure.

Gated Spillway Operational Improvements

The Dam Modernization Program recognized the importance of an effective gate operating plan and provided one additional hoist at each spillway to manage flood releases including the PMF as defined by state regulations from the late 1990s. The 2006 gate structural analysis recommended that additional studies be conducted to evaluate the need for more hoists to prevent gate overtopping during extreme events. Also in 2006, Texas issued a new set of Design Storm Guidelines that adjusted the probable maximum precipitation (PMP) temporal distributions and required that multiple durations be reviewed in search of a critical PMP. The peak rainfall intensities were also less under the new guidelines and the temporal distribution was front-end loaded to better represent very severe events that can occur in Central Texas.

LCRA retained FNI in 2007 to perform a hydrologic and hydraulic (H&H) evaluation analysis in light of the new state requirement. The analysis also included an assessment of the potential impacts of severe PMP events occurring on or close to Lake Buchanan, which might cause a rapid rise in the reservoir and result in gate overtopping prior to gates being opened using traveling hoists. The analysis showed that operations with normal pool levels were not able to keep the gates from being overtopped under most of the local PMP scenarios reviewed. Therefore, overtopping of these gates prior to opening would likely put unacceptable structural loads on the gates. FNI recommended the enhancement of the gate operating system to allow prompt gate operations and safely pass the local PMP events. The improvements included additional hoists and gate extensions.

While the H&H study was underway and the 2007 CFR report was just submitted, a severe rainfall event occurred in the immediate vicinity of Starcke Dam, which is relatively close to Buchanan Dam. The event produced over 19 inches of rainfall in the overnight hours and, when crews were called out, roads were already closed and the access road to Starcke Dam was washed away. This event reinforced the need for a quick response to a severe localized rainfall event at any of the LCRA dams and the benefit of remote operational capabilities of the spillway gates at Buchanan Dam. Therefore, LCRA began the Buchanan Dam Spillway Optimization to strengthen the gates and enhance flood operations.

SPILWAY OPTIMIZATION

The severe "rain bomb" that occurred on the Starcke Dam watershed reinforced the need for the most significant operational enhancement of Buchanan Dam. LCRA tasked FNI to improve the spillways through hoist addition and gate strengthening. A gate dewatering system was requested to facilitate gate strengthening and long term maintenance. LCRA was to perform all fabrication and construction for the project with its staff and forces. The design and construction of these projects began in 2007 and have been performed simultaneously. The overall uniqueness of this project lies in how LCRA, FNI, and contractors became partners in the design and construction of this project. Though not considered a design-build project in the traditional sense, this project has had a similar dynamic to design-build partnerships. By working together, LCRA and its team have dealt with shifting demands, including weather conditions and the availability of materials and labor, to successfully implement a multi-year and multi-discipline project.

Pilot Gate Examination and Repair

In light of the potential cost of rehabilitating 37 gates, LCRA implemented a pilot study starting in October 2008 that consisted of rigorous examination of one large gate as part of the repairs. The pilot study goal was to provide lessons learned on repairs for implementation on the rest of the gates.

Gate 2 of the 7-gate spillway was selected for the pilot program (see Figure 17 to Figure 20). Per the analyses, the gate improvements consisted of the following:

- Replacing existing members with like-sized members of higher strength steel using bolts in lieu of rivets,
- Adding plates where member replacement was not feasible or practical,
- Cleaning the entire gate to bare metal and applying new protective coatings
- Refurbishing the existing lifting lugs and rollers,
- Weld repairs to heavily corroded portions of the skin plate,
- Adding a 1-foot tall gate extension (splash and debris guard) to the top of the gate.
- Replacing side and bottom seals.

A needle-beam cofferdam was used as the dewatering system for the spillway bay. This dewatering system incorporated components from several different LCRA installations in order to develop a fast and economical cofferdam system. This allowed LCRA to make progress on gate repairs while simultaneously developing a permanent dewatering system for all spillways.

Trunnion pins were removed from Gate 2 to assess the pin and bushing surfaces for distress or damage. Contact surfaces were in excellent condition suggesting that frictional resistance was within assumed values in the analysis. Test results of steel coupons from existing gate members indicated the presence of low levels of lead, the steel was considered weldable by current day standards, and the yield strength was in line with the analysis. Special containment systems were required to remove existing coatings and apply the new coatings.

Figure 17. Needle beam dewatering system in place

Figure 18. Scaffolding and containment for gate repairs

Figure 19. Trunnion pin removal

Figure 20. Trunnion Pin.

The Gate 2 gate repairs were completed in November 2009. At this time, drought conditions had developed in the Colorado River Basin and limited the work that could be accomplished at the 7-gate spillway. The low lake level allowed access to the upstream side of the 14-gate and 16-gate spillways, thus facilitating work in these spillways. Therefore, LCRA focused efforts on the hoist addition and the permanent dewatering system at the 14-gate and 16-gate spillways. All gate repairs were put on hold awaiting the installation of the permanent dewatering systems.

Hoist Addition

FNI performed a detailed study to assess the hydraulic performance of the spillways at Buchanan Dam during extreme local rain events in line with new State H&H guidelines. As part of the study, FNI evaluated the response capability of LCRA crews during an extreme rain event to reach the spillways and operate the hoist. The study also evaluated the hoist condition, speed to open gates, and traveling time to the next gate. The study indicated that the existing hoist capabilities were inadequate to handle some of the predicted extreme inflows similar to the one experienced by Starcke Dam. The study indicated that additional hoists would be needed on two of the three spillways to enhance operational flexibility in response to extreme events. Thus, LCRA embarked on adding 26 new dedicated hoists to Buchanan Dam beginning in 2010. The 14-gate spillway required 12 new hoists and the 16-gate spillway required 14 new hoists. The LCRA also decided to provide remote operation of the new hoists in the event dam personnel could not reach the site.

LCRA performed an extensive evaluation on how to implement the hoist addition project and concluded that the best way to expedite the project was to have FNI prepare technical specifications for the new hoists and for LCRA to implement all purchasing and installation. LCRA awarded the contract for hoist fabrication in September 2010. While the hoists were fabricated, LCRA proceeded to sandblast and completely recoat the hoist bridges at both small spillways to alleviate the need for painting the bridge after new hoists were installed.

Figure 21. Hoists on bridge

Figure 22. Installation of new hoist

LCRA crews performed the installation of all 26 hoists including the electrical improvements to the power supply systems for the hoists while FNI provided engineering support during construction. LCRA engineers also designed the control systems for remote operations. The hoist addition was completed in August 2012 providing Buchanan Dam with significantly improved spillway operations to manage severe storms. Unlike the old system, all small gates now can be operated remotely from the LCRA River Operations Center in Austin or from several locations in and around Buchanan Dam. This will ensure that if personnel access is hampered or lost, LCRA can operate Buchanan Dam in a safe manner.

Figure 23. New hoists on 16-gate spillway

Dewatering System

It was not uncommon in the era of Buchanan Dam's construction for a dam to be without a tainter gate dewatering system. However, not having a dewatering system makes it difficult to perform gate maintenance or repairs regularly and safely. The absence of such a system can also discourage cycling gates when lake levels are critical and a fear exists of opening a gate and not being able to close it. The Buchanan Dam CFRs not only determined that the tainter gates needed repairs and structural strengthening to meet current dam safety criteria, but also emphasized the need for a gate dewatering system to facilitate the repairs.

In response to the need for a dewatering system, LCRA tasked FNI with developing dewatering alternatives. LCRA, fully aware of the Folsom Gate failure in 1995, wanted to prevent a similar occurrence at Buchanan Dam. The focal point of the alternative study was to develop a system that could deploy in an emergency situation. The study evaluated needle beam systems, floating bulkhead systems, and conventional stoplog systems. However, none of these had the capability of deployment in an emergency. Thus, FNI developed a system of stoplog segments pinned together at hinges and with wheels to address the expected high friction forces during installation in front of a failed gate. Stoplogs would be stored individually at each bay and transported with a traveling monorail hoist for assembly at the bay to be dewatered.

The dewatering system was designed for installations in non-flowing and flowing conditions. Computational fluid dynamic modeling was used to evaluate its performance under flowing water conditions. Due to uncertainty in the CFD results, a physical model study was built at the Utah Water Research Laboratory (UWRL) to evaluate the effects of different operating conditions including gate failure. The model study evaluated if the system could in fact be deployed in a flowing water condition, if additional loads would be imparted to the lifting mechanism, and if the necessary pier modifications would cause any hydraulic issues on spillway capacity.

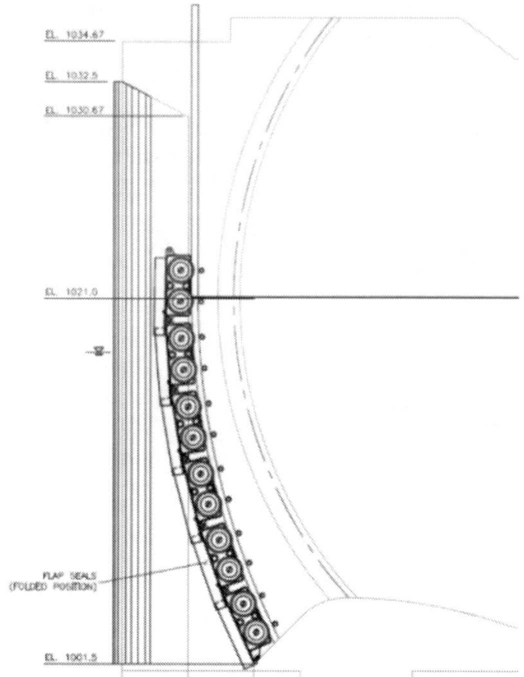

Figure 24. Stacking for small gate stoplogs

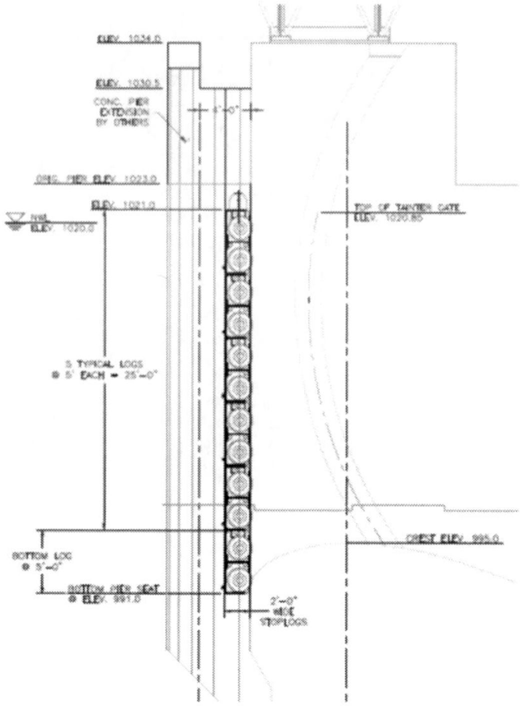

Figure 25. Stacking for large gate stoplogs

Figure 26. Stoplogs in Testing apparatus

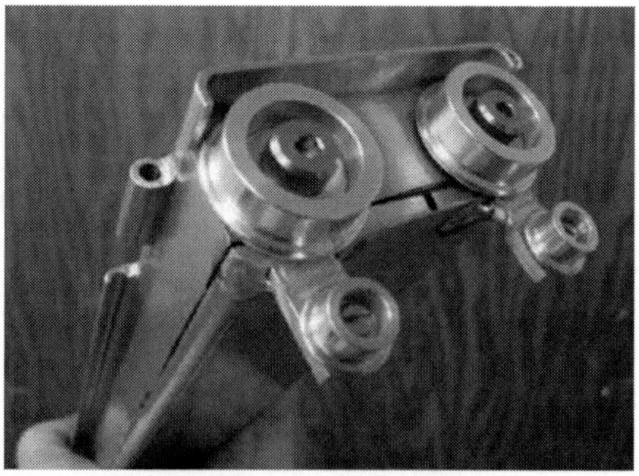

Figure 27. Model stoplog

The model study showed that the stoplogs did not experience forces in excess of their static weight for all the spillways. The stoplogs operated smoothly and did not experience vibrations or other problematic conditions when being deployed. The stoplogs seated correctly for each reservoir level tested and the stoplog guide rails had little effect on the flow and produced only minor spillway discharge reduction. The 7-gate spillway stoplogs resulted in minor flow oscillations during certain operating conditions, but the oscillations were short and did not result in a force increase.

The dewatering system final design began in early 2010 following the successful verification of the stoplog concept. FNI led the design and LCRA led the construction, which included fabrication at their railcar maintenance facility and performing all installation of the handling superstructure.

Stoplogs

The dewatering system for the 14-gate and 16-gate spillways, completed in 2014 (see Figure 28 and Figure 29), utilizes stoplogs that descend along a curved steel guide track bolted on the face of each pier. The stoplogs are comprised of Hollow Structural Sections (HSS) tubes. Each stoplog includes lifting shackles and a dogging system for storage and stacking. The individual stoplogs are stored on the spillway, one per bay. There is a special bottom log but the other logs are interchangeable. In order to dewater a gate, six stoplogs must be stacked together as a unit. Each stoplog is designed to support the entire weight of the system beneath it and each stoplog acts as a lifting beam. The logs are transported by a traveling monorail hoist.

Figure 28. Small gate stoplogs

Figure 29. Small Gate Stoplog

The dewatering system for the 7-gate spillway, currently under construction, utilizes stoplogs that descend in slots at each pier. The stoplogs are comprised of built-up sections of structural steel using wide flanges and plates. Each stoplog includes lifting devices, counterweights, and a dogging system for storage and stacking. The individual stoplogs are stored on the spillway, one per bay. There is a special bottom log but the other logs are interchangeable. In order to dewater a gate, six stoplogs must be stacked together as a unit. Each stoplog is designed to support the entire weight of the system beneath it and each stoplog acts as a lifting beam. The logs are transported by a traveling gantry hoist.

Figure 30. Large Gate Stoplog

Pier Modifications

The noses of the spillway piers required modifications to provide the stoplogs a mechanism for travel and an adequate sealing surface for the side seals while maintaining an efficient hydraulic shape (see Figure 31). For the 14-gate and 16-gate spillways, FNI developed a blunt nose pier with small elliptical transitions which required concrete be removed from the existing nose to provide an adequate keyway for a mechanical lock. In addition, bolted steel guide tracks were developed to provide a traveling mechanism for the stoplogs' decent and support. For the 7-gate spillway, slots were cut into the piers to function as guides and to secure the stoplogs in place. The tops of the piers were also raised to accommodate handling and storage of the deployment system.

Figure 31. 14-gate spillway guide rail (left) and 7-gate spillway slot (right)

Deployment System

The deployment system at the 14-gate and 16-gate spillway consists of a steel superstructure constructed on top of the existing piers. The piers each have a HSS frame that extends from the upstream to downstream portion of the pier. A large monorail beam (S-shape) spans the 37 feet from pier to pier, and runs the entire length of the spillway. A two-hook under running 25-ton monorail hoist is required for handling of the stoplogs. The deployment system travels carrying one stoplog at a time. The process of stacking logs together proceeds until all six logs are pinned in one bay. The heavy lift for the deployment system will occur over the piers when all six logs are on the guide rails. Conducting the heavy lift over the piers reduced the monorail beam size by avoiding heavy loads at center-span of the monorail. The small gate deployment system is shown in Figure 32 to Figure 34.

Figure 32. 14 and 16-Gate Spillways Deployment system frame

Figure 33. Traveling Hoist

Figure 34. 14-Gate spillway

The deployment system at the 7-gate spillway, currently under construction, consists of two steel carrier beams bearing on concrete extensions of the existing piers. The beams run the full length of the spillway as shown in Figure 35 and Figure 36. A large twin hook hoist on a gantry frame is required for handling the stoplogs. The deployment system travels on the beams carrying one stoplog at a time. The process of stacking logs together proceeds until all six logs are pinned in one bay. The heavy lift for the deployment system will occur over the piers when all six logs are in the slots, thus reducing the beam size by avoiding heavy loads at center-span.

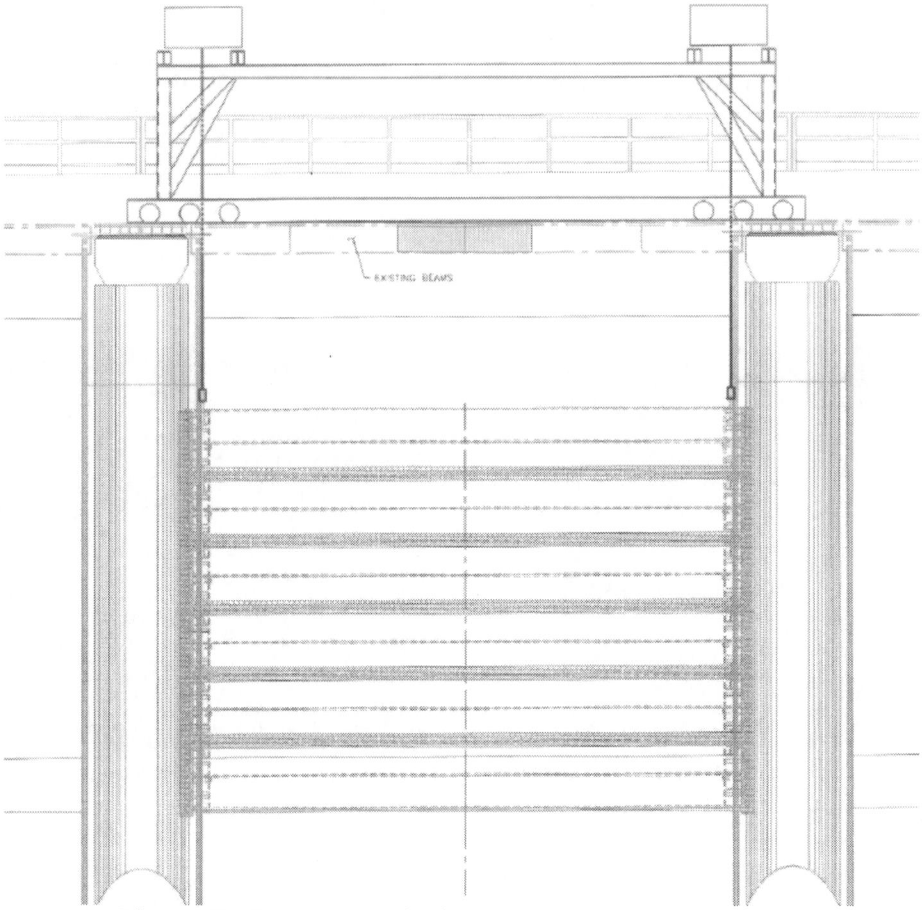

Figure 35. Large gate deployment system arrangement

Figure 36. Carrier beams (left) and traveling gantry crane (right)

CONCLUSION

LCRA has been steward of the six Highland Lakes dams for over 80 years. They are actively engaged in maintenance of and improvements to their entire dam system. They are steadfast in their goals but maintain flexibility when needs arise. Buchanan Dam is one of the best examples of responding to changing needs and safety standards in a responsible manner.

LCRA's efforts have been noted by industry and recognized with awards such as the Association of State Dam Safety Officials (ASDSO) Regional Award of Merit and the Structural Project of the Year awarded by the American Public Works Association, Texas Chapter.

REFERENCES

1. Freese and Nichols, Inc., Dam Safety Evaluation Project, Task Order B, Reconnaissance Investigation Interim Report, August 1992.

2. Freese and Nichols, Inc., Buchanan Dam Stabilization Final Design Report, Task Order P, 1997.

3. Bureau of Reclamation, Comprehensive Facility Review for Buchanan Dam, March 2002.

4. HDR, Radial Gate Inspection and Analysis, July 2006.

5. Freese and Nichols, Inc., Buchanan Dam Radial Gate Inspection and Analysis, May 2007.

6. URS Corporation, Buchanan Dam Comprehensive Facility Review, June 2007.

7. Freese and Nichols, Inc., Lake Buchanan Gate Operations Review, June 2007.

8. URS Corporation, Buchanan Dam Structural Analysis 35-Foot and 70-Foot Multiple Arch Sections, September 2008.

9. Freese and Nichols, Inc., Buchanan Dam Optimization Reports, Volume I and II, November 2009.

10. Utah Water Research Laboratory, Physical Hydraulic Model Study of the Buchanan Dam Spillway Dewatering System, February 2010.

11. Freese and Nichols, Inc., Design Basis Memorandum Buchanan Dam Dewatering System, June 2010.

12. Freese and Nichols, Inc., Buchanan Dam Corbel Evaluation Report, September 2010.

13. Freese and Nichols, Inc., Completion Memorandum, Buchanan Dam Dewatering System, 14-16 gate Spillways, February 2015.

RISK MANAGEMENT FOR DAM CONSTRUCTION PROJECTS — AN OWNER'S PERSPECTIVE

Susan S. Hou, P.E.[1]
Tedman Lee, P.E.[2]
Alan Johanson, P.E.[3]
Daniel L. Wade, P.E., G.E.[4]

ABSTRACT

San Francisco Public Utilities Commission (SFPUC), serving 2.6 million customers in the San Francisco Bay Area, owns and operates 10 dams and 3 hydropower plants within the Hetch Hetchy Regional Water System. Calaveras Dam, an existing 90-year old 220-ft tall hydraulic fill dam, is one of the most important facilities within the system as it impounds water to form the largest local Bay Area reservoir. In 2001, the SFPUC, in cooperation with the California Department of Water Resources Division of Safety of Dams (DSOD), lowered the reservoir to about 40 percent capacity due to seismic safety concerns with the existing dam. In 2002, planning began for a project to replace the existing dam, and construction began for the Calaveras Dam Replacement Project (CDRP) in 2011.

The dam is sited in a complex geologic setting that is seismically active. Furthermore, the construction site is located in a highly sensitive environmental setting with numerous threatened and endangered species. Therefore, a risk register was developed and used during the planning, design, and environmental review stages of the project and was very effective at identification and mitigation of major risks. However, a more formalized risk management program was not implemented until the construction phase, and as a result, a number of risks that were recognized during the pre-construction phases of the project were not fully characterized and quantified until the construction phase. New risks were introduced and risks that were thought to be relatively minor during the pre-construction phases ended up being larger than anticipated. By highlighting some of the major project risks, this paper discusses the owner's perspective, including lessons learned and recommended best practices in the implementation of an effective risk management program for the construction of a dam.

INTRODUCTION

Considerable discussion has been given to the topic of "risk" within the dams industry in recent years. In particular, most of the "risk" discussion centers around risk-based

---

[1] Senior Project Manager, San Francisco Public Utilities Commission, 525 Golden Gate Ave., 9th Floor, San Francisco, CA 94102, USA, (415) 551-4666, shou@sfwater.org
[2] Pipeline Section Manager, Engineering Management Bureau, San Francisco Public Utilities Commission, 525 Golden Gate Ave., 12th Floor, San Francisco, CA 94102, USA, (415) 551-4863, tlee@sfwater.org
[3] Manager, Construction Management Bureau, San Francisco Public Utilities Commission, 525 Golden Gate Ave., 6th Floor, San Francisco, CA 94102, USA, (415) 554-1506, ajohanson@sfwater.org
[4] Director, Water System Improvement Program, San Francisco Public Utilities Commission, 525 Golden Gate Ave., 6th Floor, San Francisco, CA 94102, USA, (415) 554-1853, dwade@sfwater.org

approaches for analyzing and managing dam safety. However, there appears to be much less discussion about managing project risk related to the construction of dams. Given that dams are arguably one of the most complex structures in the world to plan, design, permit and build, the authors seek to share their experiences on the Calaveras Dam Replacement Project (CDRP) to encourage additional discussion and knowledge sharing on this very important and salient topic.

HETCH HETCHY REGIONAL WATER SYSTEM

The Hetch Hetchy Regional Water System, owned and operated by the San Francisco Public Utilities Commission (SFPUC), is an engineering marvel that has a colorful history. Portions of the system within the Bay Area date to the late 1800's and early 1900's when Spring Valley Water Company (Spring Valley), a private company later acquired by San Francisco, started developing water resources within the greater San Francisco Bay Area, including the construction of Calaveras Dam and Reservoir among other significant assets that remain part of the system today. At the time the dam was completed in 1925, it was the largest earth and rockfill dam in the world.

By 1925, San Francisco was constructing a much larger reservoir in the Sierra Nevada mountain range. After the devastation of the 1906 'Great' Earthquake, San Francisco gained public support to supplement local Bay Area supplies with imported water from the Sierra Nevada mountain range. In 1913, Congress passed the Raker Act, ending a controversial political battle involving John Muir, founder of the Sierra Club, and paved the way for construction of Hetch Hetchy reservoir within Yosemite National Park. San Francisco completed Hetch Hetchy Reservoir in the early 1930's and acquired the assets of Spring Valley, including Calaveras Dam and Reservoir around the same time.

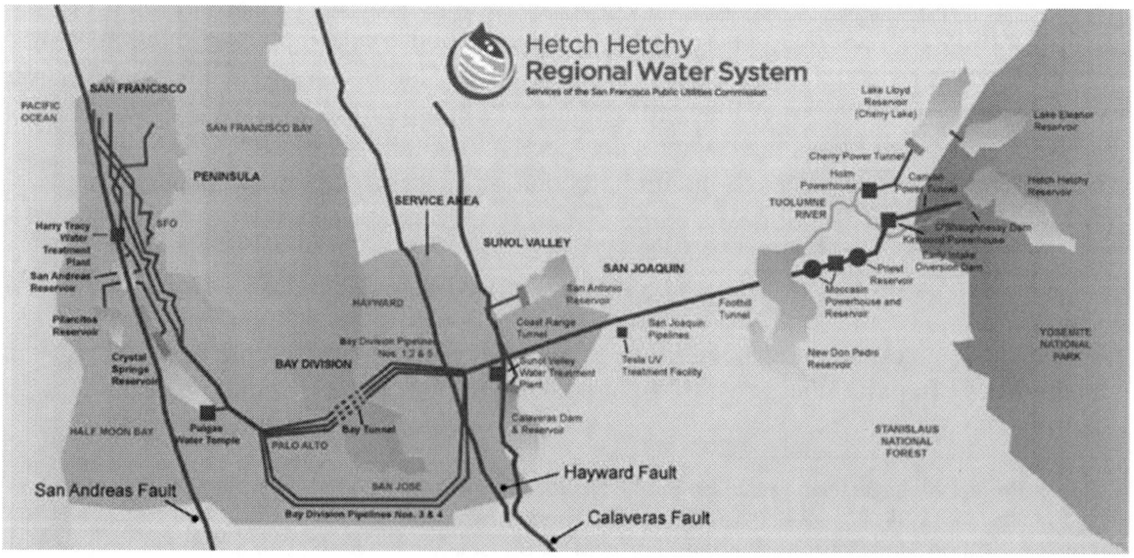

Figure 1. Hetch Hetchy Regional Water System

Today, approximately 85 percent of San Francisco's water travels by gravity from Hetch Hetchy Reservoir to San Francisco, a total of 167 miles, crossing three major active earthquake faults during the journey before reaching taps in San Francisco and the surrounding Bay Area (see Figure 1). Nevertheless, as the largest local Bay Area reservoir, Calaveras Reservoir supplies 66 percent of the local yield to the system, making it a vital asset to SFPUC's overall water supply portfolio, especially in an extended drought scenario or in the event Hetch Hetchy supplies become disrupted or otherwise unavailable in an emergency, such as an earthquake or water quality event due to wild fire.

WATER SYSTEM IMPROVEMENT PROGRAM (WSIP)

In 2002, recognizing the need for major upgrades to the aging water system, the SFPUC, together with our wholesale partners represented by the Bay Area Water Supply and Conservation Agency (BAWSCA) initiated the Water System Improvement Program (WSIP) to meet level of service (LOS) goals for seismic and delivery reliability, water quality and water supply. The WSIP is a $4.8 billion multi-year program including 83 projects over seven counties from the Central Valley to San Francisco. The WSIP includes a broad range of improvements to the system, such as upgrades to and addition of new water treatment, transmission (pipelines, tunnels, pump stations), and storage (dams, reservoirs, tanks) facilities. It is the largest capital program ever undertaken by San Francisco, and one of the largest water capital improvement programs in the United States.

CALAVERAS DAM REPLACEMENT PROJECT (CDRP)

Calaveras Dam, an existing 90-year old 220-ft tall hydraulic fill dam, is one of the most important facilities within the system as it impounds water to form the largest local San Francisco Bay Area reservoir. In 2001, the SFPUC, in cooperation with the California Department of Water Resources Division of Safety of Dams (DSOD), lowered the reservoir to about 40 percent capacity due to seismic safety concerns with the existing dam. In 2002, under the overall umbrella of the WSIP, planning began for a project to replace the existing dam, and construction began for the Calaveras Dam Replacement Project (CDRP) in 2011.

The CDRP includes the following major elements (See Figure 2):

- A new 220-foot tall earth and rockfill dam to replace the existing dam
- A new spillway and stilling basin
- A new intake tower and shaft, where the drain line and three adits from the existing facility will be connected to the new shaft and fitted with new valves and SCADA controls
- A new 78-inch outlet conduit to connect to the existing downstream portal of the 72-inch outlet tunnel and extend the outlet works downstream beneath the new dam
- New fish screens added to the existing adits of the intake tower
- New access roads, electrical buildings and other appurtenant facilities

As with many dam sites, the construction site is located in a highly sensitive environmental setting with numerous threatened and endangered species. Also as with many dams, the foundation is sited in a very complex geologic setting with a nearby active fault, and has a number of inactive faults and shear zones running through the project site. An additional unusual challenge for the CDRP is that the Franciscan Mélange Complex geologic unit at the site contains Naturally Occurring Asbestos (NOA), which affects many aspects of the planning, design and construction of the project.

The CDRP is the largest and one of the most challenging projects in the WSIP. Needless to say, it also carries the largest risks in the program. This is demonstrated by the fact that the original project budget prior to construction was approximately $420M, and the original construction schedule was four years. However, due to major risks that have been realized during construction, the project is now forecasted to be over $700M and scheduled to be completed by the end of 2018 (approximately seven years). Although some of these risks were unavoidable, this paper explores some of the lessons learned by the WSIP and the CDRP project team in managing risks for this highly complex and challenging project.

Figure 2. Aerial Photo of the CDDP Construction Site

WSIP RISK MANAGEMENT PROGRAM

Prior to the WSIP, the SFPUC was typically spending less than $50M per year on repairs and replacement of existing facilities. However, SFPUC management recognized that the organization needed to radically change its ways of doing business in order to successfully implement a $4.8 billion program. Therefore, procedures were written for all aspects of the program to ensure consistency among project teams. This took quite some time, especially since a number of procedures proved to be inadequate at first, and needed to be revised after a trial run.

One of the procedures that took considerable time to develop and fully implement was the WSIP Risk Management Program. In the early stages of the WSIP, the program was somewhat informal and was left to the discretion of each individual project team to implement as they saw fit. Project teams were required to have a risk register, but this took many forms among different project teams and lacked consistency throughout the program. Not surprisingly, many project teams took a "check the box" attitude towards risk management – in other words, once a risk register was created during the planning or design phase of a project, some project teams tended to put the risk register "on the shelf" and effectively forget about the risks until the project was bid for construction. SFPUC management realized that a formal program was needed for the construction phase of the program, especially for some of the larger and more complex projects. Therefore, in 2009, the SFPUC launched a formal WSIP Risk Management Program. Although the WSIP only requires teams to implement the formal program during the construction phase of projects, the SFPUC has recently decided to require future project teams to formally manage risk during the planning and design phases of projects to gain the benefits of early and decisive mitigation prior to construction.

The objective of the WSIP Risk Management Program is to provide a systematic way for different project teams within the WSIP to manage construction risk in a consistent manner. Risks may have positive (opportunity) or negative (threat) impacts. Because the SFPUC's goal for the WSIP Risk Management Program is to track impacts with respect to project budget and schedule, only potential negative impacts are tracked formally. Every identified risk is assigned an "owner" who has the responsibility of ensuring that mitigation measure(s) are developed and implemented to avoid, eliminate or minimize schedule and budget impacts to the project, and ultimately the program. The risk management program is, at its core, designed to be proactive. Project teams are required to understand and proactively manage project risks, and risk registers and mitigation plans are re-visited and updated monthly.

Each project forms a Risk Management Team at the beginning of construction, including representatives from each of the major functions, typically consisting of the Project Manager, Construction Manager, Designer, Scheduler, Cost Estimator, Quality Assurance Inspector, Environmental Compliance Manager, Safety Personnel and the Construction Contractor. The first task for this team of experts is to identify project risks within the different risk categories and include them in a risk register. Risk categories typically include a broad spectrum of risks, including contractual compliance issues such as timely and accurate documentation of field changes, technical challenges such as differing site conditions, environmental compliance challenges such as nesting birds, quality issues such as concrete curing temperature control, schedule challenges such as weather delays, and safety challenges such as inspector vehicles mixed with construction vehicles on haul roads. It is important to engage the key project team members that perform different functions on the project in this effort at the beginning of the construction, preferably during mobilization to help ensure risks of all areas will be identified, characterized and quantified early in the construction phase.

The risk register is a powerful tool that allows the project team to proactively manage risks identified for the project. Once risks are identified and clearly defined, the Project team will collaboratively assess each of the risks by estimating their probability of occurrence, and associated potential cost and schedule impacts. Cost and schedule impacts are developed as three point estimates which include "minimum" exposure, "most likely" exposure and "maximum" exposure if the risk were to occur. Once the risk assessment is done, plans to avoid or mitigate the risk are established for the Project team to implement and proactively manage and reduce the risk exposure. In cases where a risk cannot be mitigated or avoided, a contingency plan is developed.

To understand the total project's risk exposure, a Monte Carlo simulation is performed. The results provide total project risk exposure at different confidence levels, which allows the organization to allocate appropriate contingency for the project given an acceptable confidence level established for the program. Under the WSIP, the goal is to assign enough contingency to cover risk exposure at the 80 percent confidence level. However, the risk register is a living document, and the CDRP Project team has been continuously managing the project by reviewing, reassessing and updating the risk register on a monthly basis. This means that when risks are added or probabilities of occurrence are adjusted, the level of contingency needs to be re-assessed. This has proven to be a powerful communication tool to help the WSIP secure additional contingency funding for the CDRP as several large risks have been realized over the course of the project.

Due to the size and complexity of CDRP and to get an early start in managing risk, an informal risk management program was implemented by the project team prior to the construction phase. During the design phase, a number of risks were identified, however not all were fully characterized, due in large part to the fact that the Construction Manager and Construction Contractor were not yet available to participate in the risk assessment process. Although independent constructability reviews were performed, one of the WSIP's "lessons learned" is that there is no substitute for early Construction Manager and Construction Contractor involvement in the risk assessment process. Although this can prove to be challenging for public contracting where traditional design-bid-build methodology is often employed, the SFPUC has embraced alternate delivery methods on many recent projects to better utilize the contracting community in the early project development to gain their input to mitigate risks.

For those risks that were fully characterized, mitigation measures were put into place by including design elements to address them. Although considerable amount of money was spent for the changes in design elements, the benefits to avoid and mitigate the risks far outweighed the cost of dealing with the consequences of the impacts if they were not mitigated. Other risks that were identified and thought to be relatively minor during the design phase, but not fully characterized, were often found to be much more substantial when the risk was realized during the construction phase.

EXAMPLES OF RISK MANAGEMENT ON CDRP

The CDRP Project team has found that the risk register is a powerful tool for managing the project's budget and schedule. However, even with a formal and fully developed risk management process, some of the project risks could not be fully avoided or mitigated. The remainder of this paper will highlight some examples of risks experienced on the CDRP to demonstrate how a formal and proactive risk management process has benefited the project, but to also demonstrate where there have been lessons learned. This will lead to concluding remarks on recommended best practices in the implementation of an effective risk management program for the construction of a dam.

Risk Example No. 1: Differing Site Condition for Disposal Site 3 Dike Foundation

The dike at the toe of Disposal Site 3 (DS-3) was designed to span an existing draw on the west side of the reservoir that established a location for the disposal of dam excavation materials. Prior to construction of the dike, unsuitable materials found under the base of the dike would need to be removed. During design, from the limited geotechnical information available, the estimated depth of this material was assumed to be five feet below grade. To gain additional geotechnical information at this sensitive location would require obtaining environmental permits. The combined durations required to obtain permits, perform core boring and analyze the data may have pushed the start of critical construction work on the overall schedule beyond a summer shutdown window and had the potential to add a year to construction duration. Weighing this potential risk to the construction schedule and the additional potential change order costs, a decision was made to have the contractor perform the borings to confirm the extent and depth of the unsuitable material. A plan was formulated to control the costs of changed conditions at this site during construction through the use of unit price quantities.

Based on the subsurface drilling performed by the contractor following construction notice-to-proceed (NTP), the results over much of the length of this 1,500 foot long dike roughly conformed to expectations; however, a 150 foot long segment revealed unsuitable sediment to depths approaching 40 feet—compared to an assumed depth of five feet. Further complicating these findings were the following factors:

- Construction of the dike and preparation of the disposal site were on the critical path. Any time delays would necessarily extend the construction project.
- The entire schedule was driven by planned shutdowns. Thus, any schedule delays had to be fully recovered prior to the shutdown or risk extending the schedule by an additional year.
- Initial disposal of "clean" materials not containing NOA would be key to later project work that would involve disposal of NOA containing material – which had to be placed above the normal waterline. Thus, finding an alternate location for

current spoils would cause major problems down the road when excavation of NOA-containing materials was to begin.

- Rock for the construction of the dike was planned to be obtained from the on-site rock borrow source. However, full development of the planned quarry was years away and only a limited amount of material was readily available. Thus any increase in quantity would have a major impact on how the work was accomplished.
- Additional excavation beyond the assumed five feet would encroach on the reservoir, potentially requiring costly in-water work or options such as deep soil mixing to improve the soft foundation soils beneath the planned dike footprint.
- Although unit prices were available in the contract, the complexity of the work and the required schedule sequence limited their application. Further complicating matters, the reservoir levels were being held low to facilitate this specific construction effort. With the rainy season approaching, substantial delays had the potential to extend the project for a full year. The owner, contractor, CM and designers were able to review multiple avenues to reduce schedule and in the end were able to limit the project delay to approximately four months. The total cost of the delay, including both hard and soft costs, was about $15 million, amounting to almost 50 percent of the original contingency– all in the first several months of construction.

In this example, a risk was identified early during the design phase; however a full understanding of the consequences of the risk, to cost and especially to schedule, was not well understood prior to the risk being triggered during construction. It is the opinion of the authors that had a more formal risk process been in place during the design phase, the various experts required to be members of the Risk Management Team may have had the opportunity to fully recognize the extent of the impacts of this risk, which may have led to different decisions during design about taking the necessary time to obtain the permits and collect the additional geotechnical data prior to construction.

Risk Example No 2: Differing Site Condition in Left Abutment

During the project design phase, difficult and complex site geology was identified as a risk to the project. More than 11,000 linear feet of geotechnical cores were obtained during the design phase of the project. Preliminary evaluations of the left abutment revealed existing, undisturbed slopes that were steeper than 1 horizontal to 1 vertical (1H:1V), including near vertical cliffs approaching 20 feet. These evaluations provided confidence that the design slope (1.3H:1V) would be stable and supported the detailed stability assessment completed by the design team.

During the project construction phase, the excavation of the Left Abutment slope above the future spillway revealed geologic features that brought excavation slope stability into question (See Figure 3). Supplemental geotechnical investigations in previously inaccessible areas of the left abutment identified large ancient landslide complexes (Slides A and B) that could affect the stability of the left abutment. Ultimately, the final

Figure 1 Photo Revealing Geologic Features in the Left Abutment

slope needed to be redesigned and flattened to approximately 2H:1V to ensure long-term stability of this slope above the future dam and spillway.

This change in final slope triggered additional changes – specifically, additional excavation volume, double handling of material because of constrained site conditions, re-sequencing of construction activities and construction of a strand tieback reinforced wall to provide the Contractor a staging area for the construction of the upper spillway structure.

The 2H:1V revised final slope resulted in 1.6 million cubic yard (cy) of additional excavation that needed to be hauled and permanently disposed. In addition, approximately 1.4 million cy of left abutment material that was intended to be placed directly into the downstream shell of the new dam now had to be hauled to a temporary storage site, then double-handled for later placement into the new dam.

The time needed for additional excavation, hauling and double-handling of material required the project team to re-sequence construction activities to not impact planned shutdown windows.

Construction of a "strand tieback reinforced wall" in the upstream shell was needed in order to provide a staging area to the contractor for the construction of the upper spillway structure and foundation grouting activities on the left abutment upon the removal of Slide B in the upstream shell.

During the construction of the strand tieback reinforced wall, it was subsequently discovered that (a) the existing rock mass in the southern end of the strand tieback reinforced wall was more fractured than anticipated and (b) the contact of Slide B was deeper than what was originally shown and anticipated. These conditions resulted in additional work that was not included in the original change order.

Even though the risk of slope instability due to differing site conditions in the left abutment was identified in a general sense during the design phase of the project, the ancient landslides that were discovered during construction were not anticipated, and had considerably more extreme consequences than any differing site condition mitigation scenario had previously considered. This example demonstrates that even well intended risk mitigation planning is not always a "silver bullet" and some risks that are not known prior to construction do sometimes occur. In this case, although the project team demonstrated tremendous teamwork to develop a solution within a short timeframe during construction, this issue nevertheless resulted in significant delay and added cost to the project.

Risk Example No. 3: Access Constraints Caused by Bald Eagle

For the preferred option for transportation of the clay core borrow material found at the southern end of the reservoir to the location of the new dam, the contractor is required to pioneer a haul road on the west side of the reservoir. Pairs of nesting bald eagles have been observed at Calaveras Reservoir for the past 10 to 12 years. One historic nest was located approximately 30 feet from the projected location of the West Haul Road (see Figure 4).

Figure 2 Bald Eagle Nest

The CDRP Final Environmental Impact Report (FEIR) required that a 660 foot diameter construction buffer be established around a bald eagle nest once an eagle pair is observed displaying reproductive behavior. The required monitoring period is between January 1 and August 31 or until the eaglet(s) fledge(s). With the closest historical nest 30 feet from the proposed West Haul Road, the FEIR identified the need for an eagle nest "take" permit. Concurrent with the beginning of construction, an eagle nest take permit was obtained from the US Fish and Wildlife Service (USFWS). This permit allowed for the placement of cones to exclude habitation on two eagle nests previously occupied.

Currently the eagle pair has been returning to a nest located more than 660 feet from the West Haul Road. Monitoring continues during the nesting season as required by the USFWS permit.

To avoid delay to the embankment work which is slated to begin in the fall of 2016, the project plans to actively monitor the nesting eagles in the vicinity of the West haul Road during the spring and summer of 2016 to determine if there are potential behaviors that would impact the project.

Evaluation of this risk continues during the construction phase. This impact could either force the project to use the more costly barge option or as a worst case a delay to the critical path that would result in a one year delay to the project.

This is an example of the importance of an early and proactive approach to risk mitigation during not only the planning and design phases, but also during the construction phase. Careful planning in the pre-construction phases of the project allowed for mitigation plans and options to be provided in the design. However, during construction, risks are not static and it is essential that risks and their associated mitigation measures are continually assessed and modified as needed.

Risk Example No. 4: Controlling Emissions of Naturally Occurring Asbestos (NOA)

In addition to the environmental challenges caused by the special status species present on the project site, the presence of natural occurring asbestos (NOA) in the Franciscan Mélange Complex geologic formation within portions of the site creates another major challenge with respect to health and safety risk.

The CDRP site contains two primary geologic formations, namely highly fractured sandstones (Temblor Sandstone) and hard intact blue schists, graywackes, and greenstones embedded within a matrix of siltstone, shale and serpentinite (Franciscan Mélange Complex). In order to build the new dam, the Franciscan Mélange Complex needs to be excavated and removed for the preparation of the dam and spillway foundation, and the excavated material such as blue schist and greenstones will be processed for the building of the upstream shell of the dam embankment.

NOA, specifically chrysotile and amphiboles, has been documented in serpentinite, blueschists, and greenstone within the Franciscan Mélange Complex rocks. These NOA-containing materials, when disturbed, may become airborne with dust generated during construction operations. In general, NOA is a long thin fiber and when inhaled for a prolonged duration, can lead to a range of adverse health-effects including asbestosis which is a chronic, degenerative lung disease; lung cancer; and mesothelioma which is a rare form of cancer that develops in the lining of the lung. Since asbestosis does not typically result from environmental asbestos exposure (e.g., Nicholson 1986, HEI-AR 1991), the potential for asbestos-related cancer is the primary concern for protection of workers and the public at the CDRP.

Figure 5. Controlling of NOA Emission Using Water Hoses

Protecting public health is SFPUC's top priority, including workers at the project site, residents in the general vicinity, and visitors and park workers in the regional park near the site . During the design phase, considerable investigative work was performed to delineate the presence of NOA on the project site to ensure a good understanding was acquired prior to construction. Top industry experts were hired to guide the development of plans for both the environmental review process as well as incorporation of NOA-related requirements into the plans and specifications for construction.

In addition, a comprehensive baseline air quality monitoring program was implemented to document the existing baseline concentrations, types, sizes and dimensions of the asbestos fibers in the atmosphere in the vicinity of the site before construction started. Together with the geologic unit information and anticipated construction activities, an extensive dust monitoring program called a Comprehensive Air Monitoring Program (CAMP) and dust control requirements were developed during design phase. The CAMP was developed based on the regulations and procedures established for assessing risk from exposure to toxic substances by the federal, state and regional governmental entities, including United States Environmental Protection Agency (USEPA), California Air Resources Board (CARB), and Bay Area Air Quality Management District (BAAQMD). The CAMP was designed to assure that no airborne asbestos would leave the site at concentrations sufficient to cause unacceptable exposure to people who may visit, recreate, work, or live in the areas surrounding the CDRP site. As a commitment to the public, CDRP has defined an acceptable level of risk that is above and beyond the guidelines provided by the regulatory agencies. Protocol requires that if airborne NOA is detected at a level above the acceptable level of risk, immediate corrective actions would need to take place to bring the levels within the acceptable level of risk.

With the above measures in place, the project team together with the contractor have been following the CAMP and the dust control plan to closely monitor and manage the levels of asbestos generated from the site (see Figure 5). Given the level of uncertainty and anticipated level of effort required to avoid the risks of exceeding the established threshold, a substantial amount of budget was set aside in the bid documents as an allowance to provide resources for out-of-scope mitigation measures that may be necessary to address the risk.

While the project team has been able to control the NOA emissions on the site to date, the experience gained from the project over the past four years has prompted the team to continuously evaluate and test new mitigation measures to control NOA in anticipation of the upcoming activities to mine hard rock from within the Franciscan Complex in Borrow Area B for the upstream shell of the dam. This example highlights that risk mitigation plans are living documents that need to be continuously evaluated, monitored and improved over the life of the project.

Risk Example No. 5: Tight Shutdown Window for Outlet Works

The planned 2016 shutdown window for the outlet works is tight. Activities include installation of several sections of 78-inch diameter pipeline, installation of temporary 48-inch diameter pipeline, dam foundation treatment, installation of a grout curtain, grouting below the existing outlet works, test fills for dam embankment, installation of filter blanket and toe drain system, downstream filters, upstream rock shell, core filter, refurbishment and re-installation of a 72-inch fixed cone valve at the new stilling basin location, and electrical work at various buildings and vaults. All work must be complete on time so that the outlet works can go back into service prior to the 2016-17 winter season.

Figure 3 Installation of the New 78-inch Outlet Pipeline

Of these activities, the project team identified the connection of the new 78-inch outlet pipeline (see Figure 6) to the existing pipeline and the refurbishment of the fixed cone valve as having the most uncertainty, and thus, risk. Should there be problems with either activity; the resulting delay may push past the end date of the shutdown.

In an effort of mitigate risk, the SFPUC was able to secure a short shutdown window in advance of the planned shutdown to perform investigative work related to these two activities. Issues discovered, if any, could be acted upon prior to the start of the 2016 shutdown.

For the outlet works pipeline tie-in connection of the new outlet pipe to the existing pipe, the tie-in point was excavated and surveyed to verify the tie-in point coordinates. The survey showed that there was a one-foot horizontal offset between the tie-in point shown on the contract drawings and the existing pipe in the field. To accommodate this unforeseen offset, the project team directed the contractor/pipe manufacturer to lengthen a short cut-to-fit section of 78-inch diameter steel pipe.

For the existing fixed cone valve (FCV) that is specified to be refurbished and re-installed on the low level outlet, the refurbishing subcontractor was invited to inspect the FCV prior to taking possession of the FCV during the planned shutdown. The FCV was operated in the dry and the subcontractor was able to take various measurements of the valve seat from the inside of the FCV. The inspection also revealed a pair of bent screw jacks. The subcontractor felt confident that the screw jacks could be repaired, but as a backup plan, they intend to purchase a new pair.

This proactive approach was not expensive and it helped avoid risks that could have had a huge impact and delay to the project if they were not discovered until the contractor was in the middle of the critical 2016 shutdown.

CONCLUSION

The informal risk register developed and used during the pre-construction phase of the CDRP successfully mitigated some of the project risks that would have otherwise been realized during construction. However, due to the lack of resources supporting the preliminary effort, a number of risks that were recognized during the pre-construction phase were not fully characterized and quantified until the construction phase started. With the more formalized and comprehensive risk management strategy implemented during construction, a number of these risks were better characterized and quantified. This allowed the project team to better understand the risks so that mitigation measures could be developed to avoid or mitigate the risks either partially or fully.

A variety of risk mitigation strategies were employed on the CDRP. As demonstrated by several examples presented in this paper, some risks were well-understood and fully mitigated, enabling the team to be well prepared to avoid the risks from triggering, such as risk example no 5, “tight shutdown window”. In other cases, risks were identified but not fully understood; therefore, the preliminary mitigation measures proved to be inadequate when risks were triggered, such as risk example no. 1 “differing site conditions for the disposal site 3 dike foundation". To address this type of risk, the entire team needs to be encouraged to be collaborative, innovative and expeditiously adapt to the reality of the situation to come up with solutions to address the issue and minimize the impact as much as possible. Finally, some risks, such as risk example no. 4 "controlling NOA emissions" and example no. 3 "access constraints caused by bald eagle", require that mitigation measures be continuously reviewed, evaluated and adjusted during the course of the project, given what has been learned to date. If these risks are realized, contingency plans must be put in place to help reduce impacts to the minimum practicable.

The owner of the project can have a huge influence on the success of the risk management program by bringing the necessary leadership to "set the tone" for proactive risk management from project inception in the early planning stages, design, construction, startup and commissioning. This includes bringing the entire team of experts together to identify risks in all areas in the early stages of the project. This is not only good policy for successful budget and schedule controls, but is also a tremendous communication tool that can be utilized to help stakeholders better understand the project and the commensurate risks, which can help to obtain early buy-in and ownership of the risks. Whether a risk is identified during an early phase of the project or in construction, it is very important to provide transparency to stakeholders so that realistic expectations may be established.

Examples of risks presented in this paper demonstrate that a proactive project team can help avoid some risks from occurring, and can significantly reduce the impact of other risks that are realized. They also highlight the importance of continuously and rigorously managing the risks throughout the course of the project so that risks can be mitigated to the extent possible, or until they expire and are no longer threats to the project. While it is critical to mitigate risk, it is also important to ensure that the mitigation cost will not outweigh the cost of dealing with the consequence should the risk become realized.

CLOVER AND JUNCTION DAMS: RISK-BASED SEISMIC EVALUATION OF TWO SLAB AND BUTTRESS DAMS

Scott Jones, PE, PhD[1]
Steve O'Brien, PE[2]
David Hughes, PE[3]

ABSTRACT

Clover and Junction Dams are central spillway slab-and-buttress dams located on the East Kiewa River between the townships of Mt. Beauty and Falls Creek in Northeast Victoria, Australia. The dams impound Clover Reservoir and Lake Guy, respectively, and are used exclusively for power generation. Previous safety reviews and structural assessments of the dams concluded that neither dam meets modern dam design standards and acceptance criteria for a High 'C' Consequence Category, the hazard classification for these two dams in accordance with ANCOLD 2012 [*1.*]. These conclusions were based largely on the results of response spectrum analyses of the two dams and the confirmed presence of Alkali-Aggregate Reaction (AAR). Based on these conclusions, remedial works, including infilling the slab-and-buttress dams with mass concrete, were recommended to improve the safety of the dams and reduce the risk of dam failure.

In 2013, as part of a tender design for the proposed remedial works, URS reviewed the background information on the dam and concluded that infilling the dams with mass concrete was not the preferred remediation measure for these dams. Visual observations and a review of the data from AAR testing indicated that the concrete deterioration was not severe enough to warrant rehabilitation measures. Additionally, preliminary analyses of the dams performed during the tender phase using simplified finite element models indicated that additional linear and nonlinear time-history analyses would likely support constructing significantly less expensive rehabilitation measures for these dams. As a result, a risk-based seismic evaluation of the slab-and-buttress dams was undertaken to better understand the potential failure modes and consequences of failure of these two dams.

This paper discusses an assessment of the AAR observed at the dams, the progressive analysis approach recommended to support the risk assessments, preliminary analysis results from response spectrum analyses, the preliminary risk assessment outcomes, and recommendations for additional analyses based on those outcomes.

INTRODUCTION

Clover and Junction Dams are part of the Kiewa Hydroelectric Scheme in the Australian Alps in Northeastern Victoria, Australia. Both dams are located on the East Kiewa River, with Clover Dam located approximately 6 miles (10 km) south of Mt. Beauty and

[1] Senior Civil/Structural Engineer, AECOM, scott.jones@aecom.com
[2] Principal Engineer, AECOM, steven.obrien@aecom.com
[3] Principal Engineer, AECOM, david.k.hughes@aecom.com

Junction Dam located approximately 2.5 miles (4 km) south of Clover Dam. The primary use for both dams is power generation. Clover Dam was constructed between 1948 and 1953 and impounds Lake Clover, which provides storage for the West Kiewa Hydroelectric Power Station. Junction Dam was constructed between 1939 and 1944 and impounds Lake Guy, which provides storage for the Clover Hydroelectric Power Station.

Clover Dam comprises a central reinforced concrete buttress spillway, flanked on both sides by spillway and non-overflow mass concrete gravity abutment sections, with a maximum height of 82 feet (25 m). A view from downstream and an upstream profile of the dam are shown on Figure 1. The buttress section is 150 feet (45.7 m) in length, with five buttresses, equally spaced at 25 feet (7.6 m) on center, between the gravity sections forming six bays. The reinforced concrete buttresses, with a thickness of 2 feet, 6 inches (0.76 m), support upstream and downstream reinforced concrete facing slabs and a central walkway strut, all of which provide lateral stability to the buttresses. The upstream slabs vary in thickness from 2 feet, 6 inches (0.76 m) at the crest to 4 feet, 9 inches (1.45 m) at the toe, and are keyed into the upstream edges of the buttresses. The

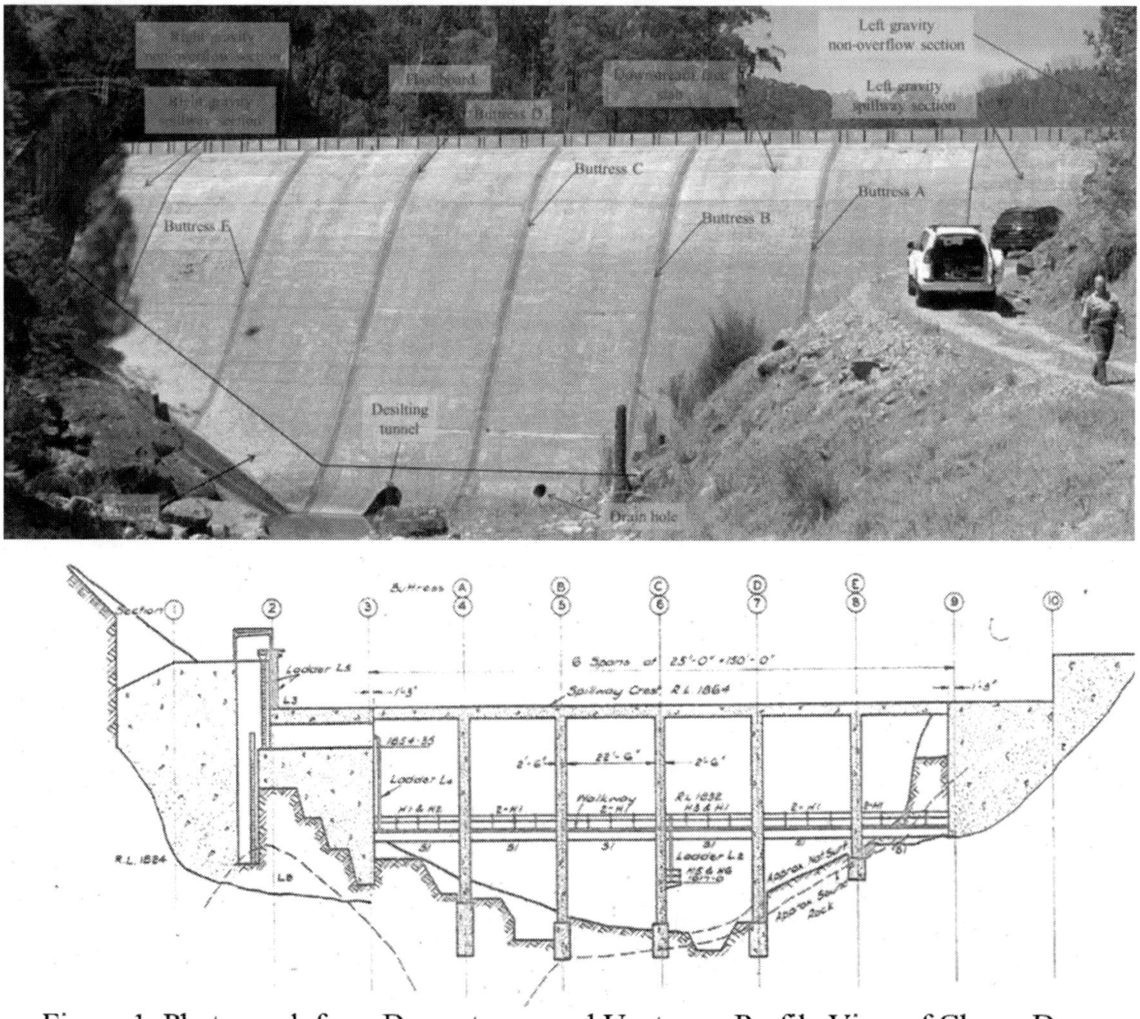

Figure 1. Photograph from Downstream and Upstream Profile View of Clover Dam

struts are simply supported in sockets formed in the buttresses. The gravity spillway sections are 25 feet (7.6 m) long. The overall crest length of the dam is approximately 250 feet (76.2 m).

Junction Dam is a reinforced concrete buttress dam with a maximum structural height of approximately 141 feet (43 m) above the lowest foundation level and a crest length of 429 feet (130.8 m). The dam consists of three sections: a 140-foot- (36.6-m-) long, ogee-crested spillway with flashboards near the left abutment, and non-overflow sections on the right and left sides of the spillway. A view from downstream and an upstream profile of the dam are shown on Figure 2. The dam comprises 19 buttresses, equally spaced at 20 feet (6.1 m) on centers forming 18 bays. The variable thickness reinforced concrete buttresses support upstream reinforced concrete facing slabs and several rows of reinforced concrete struts, all of which provide lateral stability to the buttresses. The upstream slabs vary in thickness from 1 foot, 2 inches (0.36 m) at the crest to 3 feet, 11 inches (1.19 m) at the toe, and are keyed into the upstream edges of the buttresses and supported on corbels. The struts continuously reinforced into and through the buttresses.

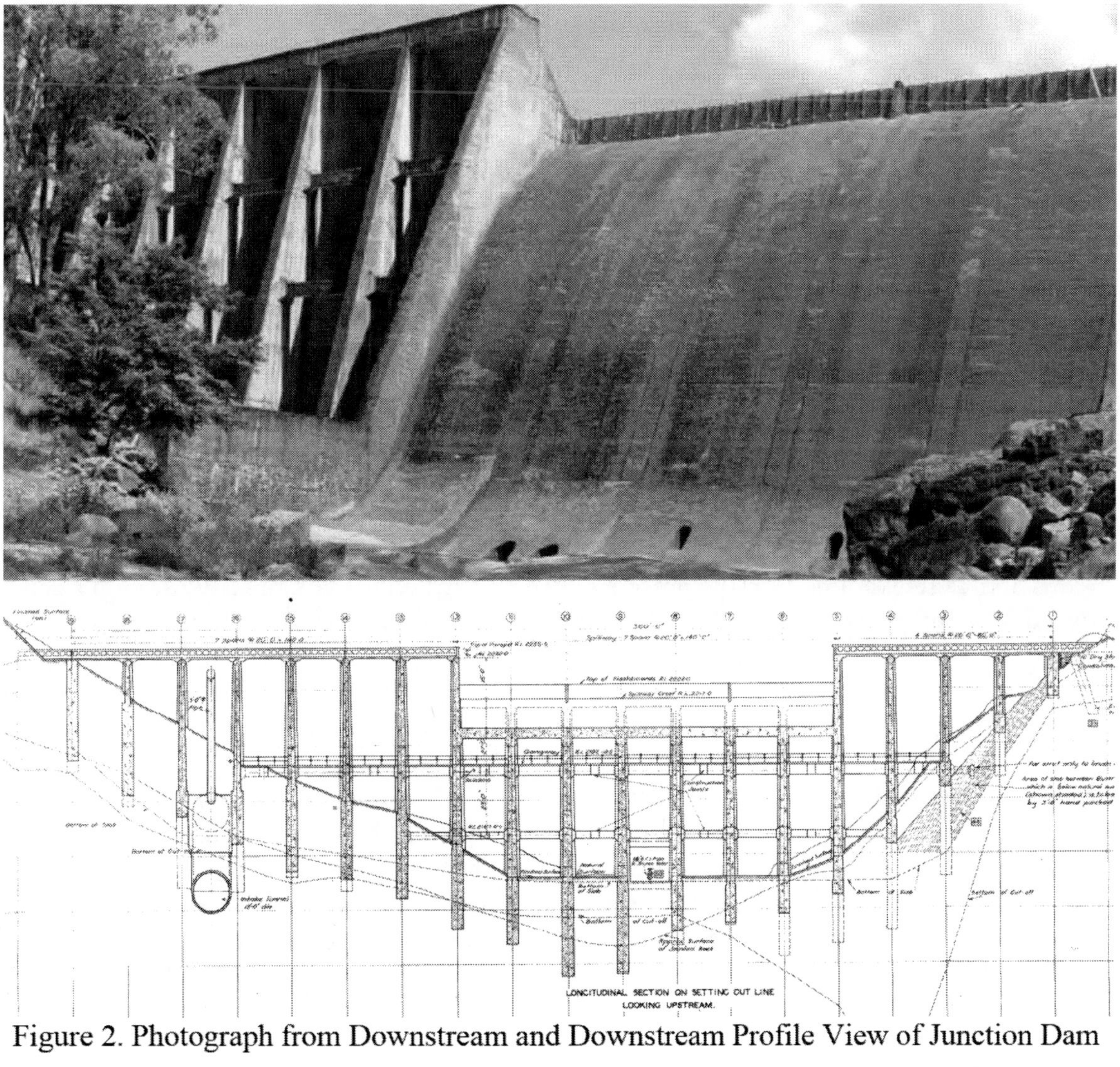

Figure 2. Photograph from Downstream and Downstream Profile View of Junction Dam

In 2013, URS teamed with Fulton Hogan to prepare a bid for a design-build project to infill the buttresses on both Clover and Junction Dams. The solicitation for the design-build project was based on comprehensive studies on the dams to evaluate alkali-aggregate reaction (AAR) in the concrete, seismicity in the region, and seismic stability of the dams. A review of the information provided with the solicitation revealed that the risks for these dams may have been overstated for several reasons, including the following:

1) The impacts of AAR on the structural integrity of the dams may have been overestimated.
2) The seismic loading (response spectra) appeared to overestimate the peak ground and spectral accelerations for northeastern Victoria.
3) The structural analyses were limited to response spectrum analyses and the analysis of Junction Dam appeared to overestimate the demands on the slabs.

Based on this understanding, the team agreed that infill of the buttresses may not be warranted and a bid was submitted to undertake additional investigations in support of an updated dam safety review and risk assessment of the dams. The work includes running an updated Probabilistic Seismic Hazard Assessment (PSHA) for the two dam sites, performing finite element analysis of the dams to evaluate the ability of the dams to withstand the updated seismic loading, additional concrete coring and AAR investigations to confirm the previous investigations and improve understanding of the AAR growth mechanisms, and perform a risk assessment of the dam to evaluate compliance with Australian dam safety guidelines. The following sections describe the AAR data review and planned investigations; a comparison with seismic loading recommended in the current PSHA against that used in previous studies; and a summary of structural analyses performed to date and path forward for additional structural evaluations.

ALKALI-AGGREGATE REACTION (AAR)

Investigations into AAR at both Clover and Junction dams were conducted by ARRB Transport Research in 1997 [*2.*] and further investigations conducted in 1999 [*3.*]. As part of the investigations a series of concrete cores were taken from both Junction and Clover dams and were examined for the presence of AAR, tested for compressive strength and tested for splitting tensile strength. Results from compressive strength tests yielded a range of compressive strength from 5200 psi (36 MPa) to 7100 psi (49 MPa). Results from tensile strength tests showed splitting tensile strengths in the range of 7 to 10 percent of the compressive strengths, which is typical for concrete of this strength [*4.*].

The monitoring results for the crest survey pins for Clover and Junction Dams show that horizontal movements are generally very minor, typically less than 10mm over 30 years. The measured vertical movements are as much as to 30 mm (in an upwards direction) over a 30 year period with the rate of movement being fairly consistent over the monitoring period. This equates to strain rates in the order of 20 to 30 $\mu\varepsilon$/year. Strain rates of this magnitude are typically assessed as representing mild AAR expansion.

The investigations [2.] [3.] confirmed the presence of AAR in both structures and suggested that the amount of soluble alkali available in the concrete appeared adequate for AAR to continue. Based on this information, additional observations of cracks in the dam were made during recent field investigations. The key observations are as follows:

1) Crack patterns observed at Clover Dam, primarily in the vertical faces of the gravity sections but also in the slabs and buttresses, appear to be consistent with AAR growth, with growth more advanced at the upstream and downstream faces than the centers of the gravity sections and buttresses. Figure 3 shows a photo of cracks and a map of the cracks in the left abutment gravity section.
2) Crack patterns in the downstream ends of the buttresses for the non-overflow sections at Junction Dam show evidence of typical AAR crack patterns. Figure 4a shows a photograph of this cracking. There is limited evidence of AAR elsewhere in Junction Dam. There is some cracking in the buttress and movement relative to the dam crest bridge at the right end of the spillway that could be attributed to AAR or thermal loading.

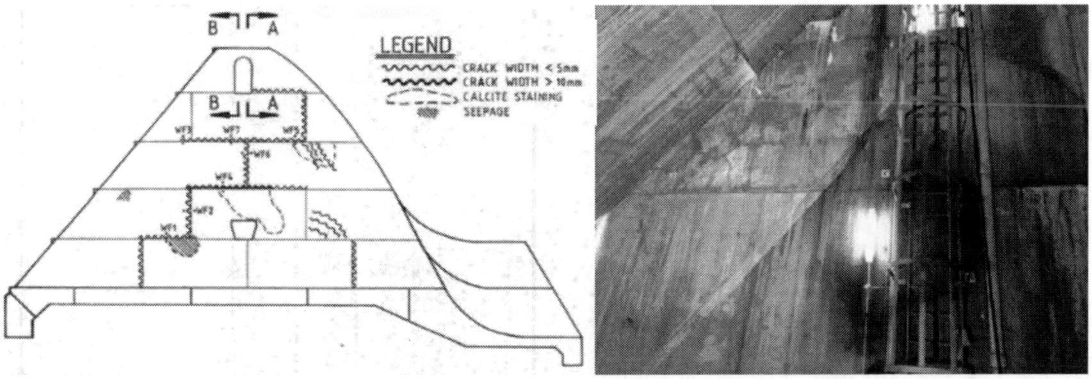

(a) (b)

Figure 3. Crack Map and Photograph of Left Abutment Gravity Section from Walkway Strut.

(a) (b)

Figure 4. (a) Photograph of Downstream Edge of Non-overflow Buttresses and (b) Photograph of Offset between Buttress 10 and Dam Crest Bridge.

Based on these observations, the following recommendations have been made and are in process of being implemented:

1) Extract concrete cores from the faces and center of the gravity sections and buttresses at Clover Dam to determine extents of AAR and evaluate if continued AAR growth is expected.
2) Install instrumentation on Buttress 10 at the right end of the spillway at Junction Dam to determine if the observed offset is attributable to AAR or thermal loading.

In the case of both dams, continued monitoring of survey monument movement and crack growth over time is recommended.

SEISMIC LOADING FOR EVALUATION

The solicitation for the design-build of concrete infill for the buttress dams was based on a 2007 stability evaluation of Clover Dam [*5.*], a 2008 stability evaluation of Junction Dam [*6.*], and a 2011 revised stability evaluation of both dams [*7.*]. The conclusion that infill of the buttress dams with concrete was largely driven by a significant increase in seismic loading from the original evaluations and the revised evaluation. Figure 5 shows the response spectra used in the original evaluation and the revised stability evaluation, which was the basis for the recommendation to infill the buttress dams with mass concrete.

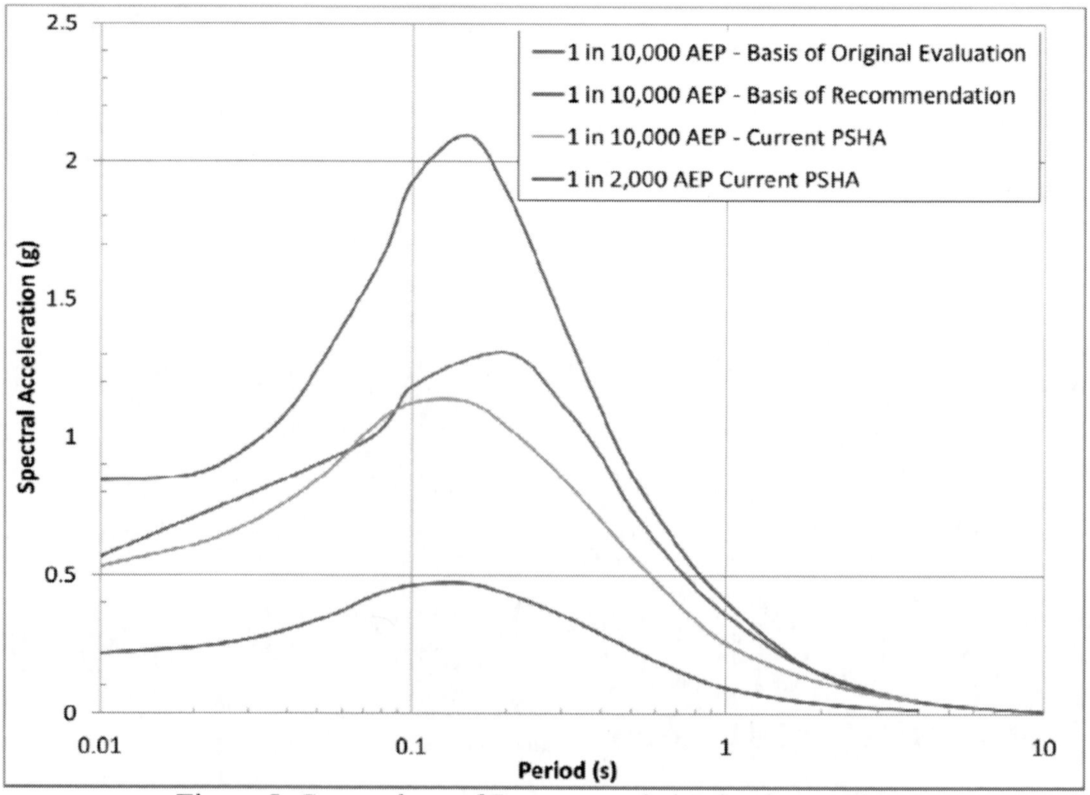

Figure 5. Comparison of Recommended Seismic Loading.

An independent PSHA was performed as part of this study, and the response spectrum for the 1 in 10,000 Annual Exceedance Probability (AEP) earthquake was found to be comparable to the response spectrum used in the original evaluation, as shown on Figure 5 (Current PSHA). The 1 in 10,000 AEP earthquake was selected as the Maximum Design Earthquake (MDE) for the original evaluations based on a hazard classification of High B in accordance with ANCOLD guidelines [*8.*]. A recent re-evaluation of hazard classification, which is not further discussed in this paper, indicated that a hazard classification of High C would be more appropriate for these dams. In this case, a 1 in 2,000 AEP MDE can be used if the dam satisfies all other risk assessment criteria of ANCOLD. The response spectrum for the 1 in 2,000 AEP earthquake is also shown on Figure 5. Based on these findings, structural stability evaluations were undertaken for the 1 in 2,000, 1 in 5,000, and 1 in 10,000 AEP earthquakes to support a risk assessment of Clover and Junction Dams.

STRUCTURAL EVALUATION OF DAMS

A structural evaluation employing a progressive analysis approach has been undertaken to evaluate the existing dams for the updated seismic loading in support of the risk assessment of Clover and Junction Dams. The approach involves the following steps:

1) Perform response spectrum analysis (RSA) of the dams for the 1 in 10,000 AEP earthquake. If the results of the RSA indicate potential overstressing of key elements of the dam, perform RSA for the 1 in 5,000 AEP and 1 in 2,000 AEP as necessary to develop a risk profile based on RSA.
2) If the RSA indicates potential overstressing, perform linear time history analysis (THA) of the dams for the events where overstressing was indicated.
3) If the linear THA indicated potential for failure based on demand-to-capacity ratio evaluations, perform a nonlinear THA to assess the response of the damaged structure and potential for failure of key elements.

As indicated above, the evaluation starts with simple linear analysis and progresses to more complex analyses only if the results of the simpler analysis methods indicate potential for damage to key structural elements. The analyses have been undertaken using the finite element software ANSYS [*9.*].

The finite element model for each dam included a significant portion of the foundation in addition to dam. The foundation in the model extends approximately one dam height in all directions in accordance with USACE guidelines [*10.*]. The upstream and downstream slabs were simulated using 4-noded linear elastic shell elements. The buttresses, struts, foundation and backfill were modelled using 8-noded solid brick elements. The reservoir interaction during seismic loading was modeled using direction mass elements with masses assigned based on Zangar's method [*11.*]. For the RSA, the slab-to-buttress connections and contraction joints were modeled as pinned using line contact elements. The dam to foundation contact was modeled as fixed due to the trenching of the buttresses into the foundation rock. The finite element model for

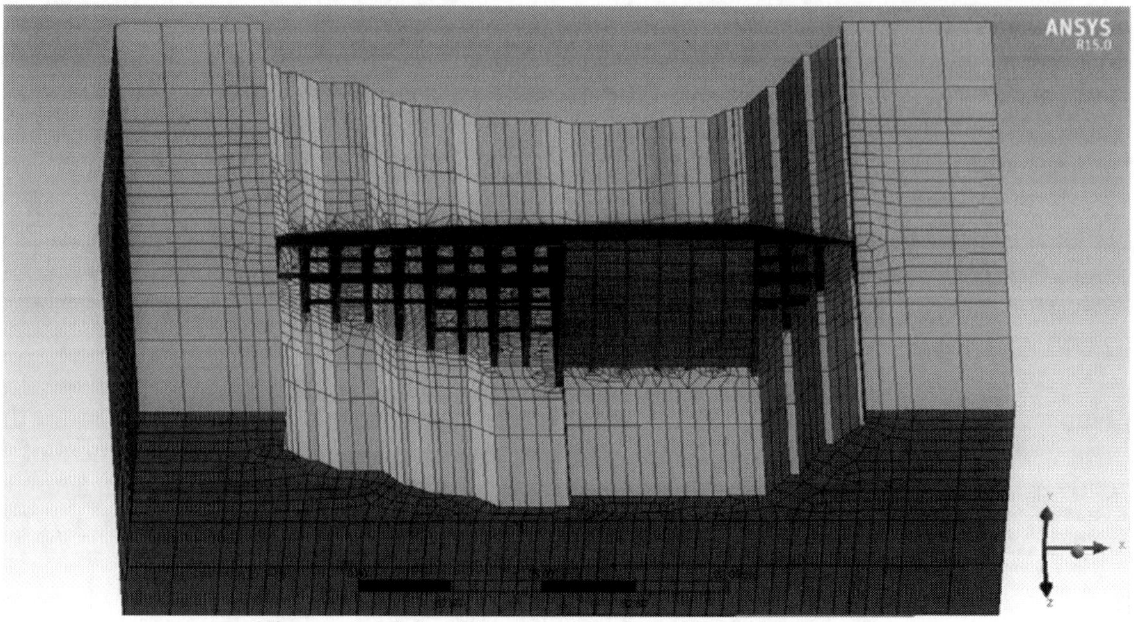

Figure 6. Finite Element Model of Junction Dam.

Junction Dam is shown in Figure 6. The model for Clover Dam was developed similar to that for Junction Dam.

The RSA for the 1 in 10,000 AEP for Clover and Junction Dams were run using the Complete Quadratic Combination Method. Results from the analysis indicated the following:

- The slabs at Clover and Junction Dams have adequate capacity to withstand the 1 in 10,000 AEP earthquake, as shown in Figures 7 and 8, respectively.
- The buttresses at Clover Dam have adequate capacity to withstand the 1 in 10,000 AEP earthquake, as shown in Figure 9.
- The buttresses at Junction Dam may not have adequate capacity to withstand the 1 in 10,000 AEP earthquake, as shown in Figure 10a. The buttresses at Junction Dam are adequately designed to withstand the 1 in 2,000 AEP earthquake, as shown in Figure 10b.
- Preliminary evaluation of results for the shear keys and walkway struts at Clover Dam for the 1 in 10,000 AEP earthquake indicates adequate capacity to resist the 1 in 10,000 AEP earthquake.
- Preliminary evaluation of the results for the struts at Junction Dam for the 1 in 10,000 AEP earthquake indicate they may not have adequate capacity to withstand the 1 in 10,000 AEP earthquake.

Based on these findings, no further analysis appeared to be required to characterize the response of Clover Dam to the 1 in 10,000 AEP earthquake. A linear THA will be undertaken to better understand the response of Junction Dam to the 1 in 10,000 AEP earthquake. In both cases, the results of the RSA support the conclusion that infill of the buttresses is not warranted.

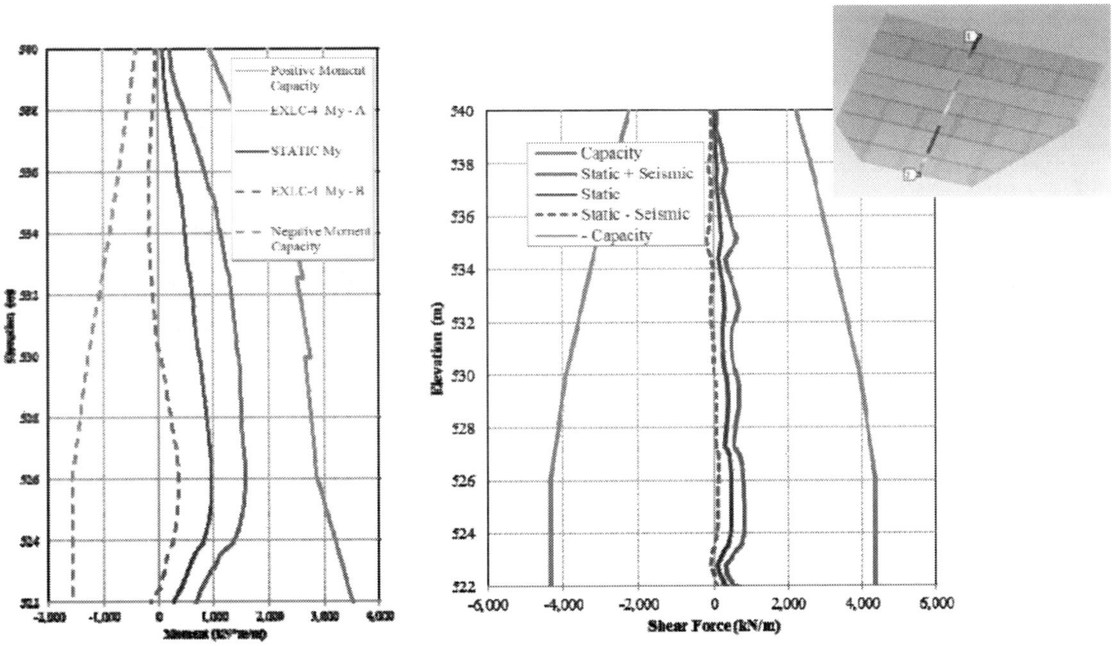

Figure 7. Moment and Shear Demand Versus Capacity for the Upstream Slab at the Maximum Section of Clover Dam during the 1 in 10,000 AEP Earthquake.

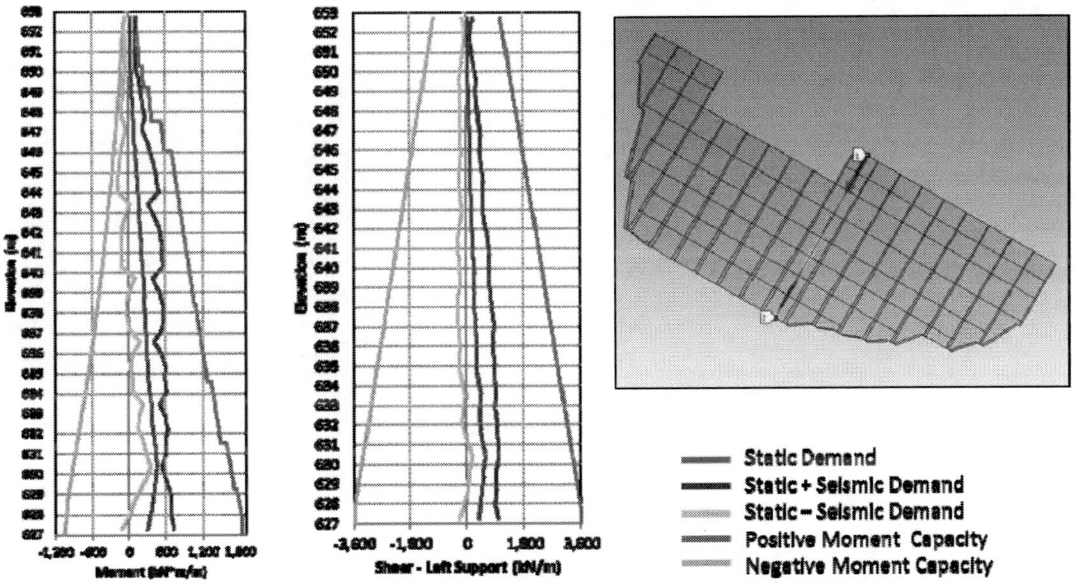

Figure 8. Moment and Shear Demand Versus Capacity for the Upstream Slab at the Maximum Section of Junction Dam.

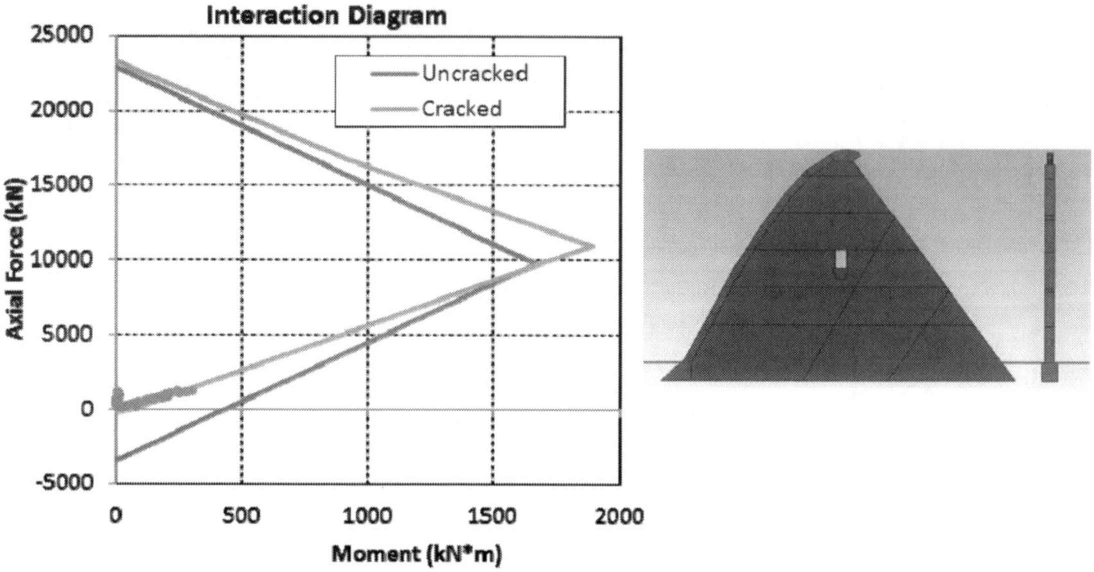

Figure 9. Moment Interaction Diagram for the Maximum Height Buttress at Clover Dam.

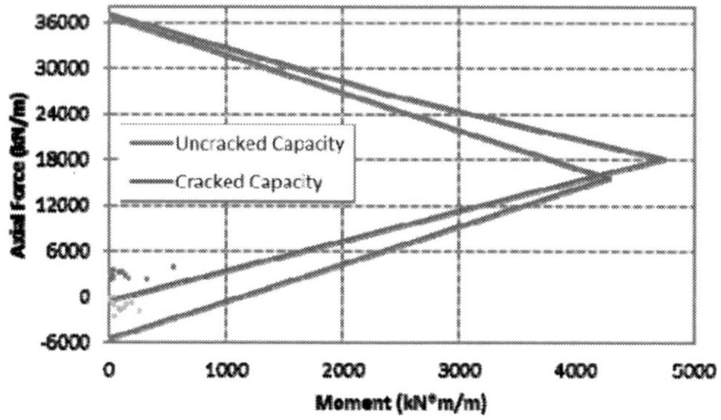

(a)

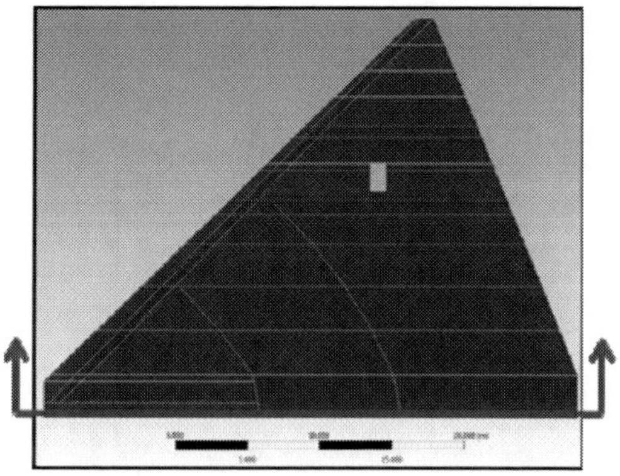

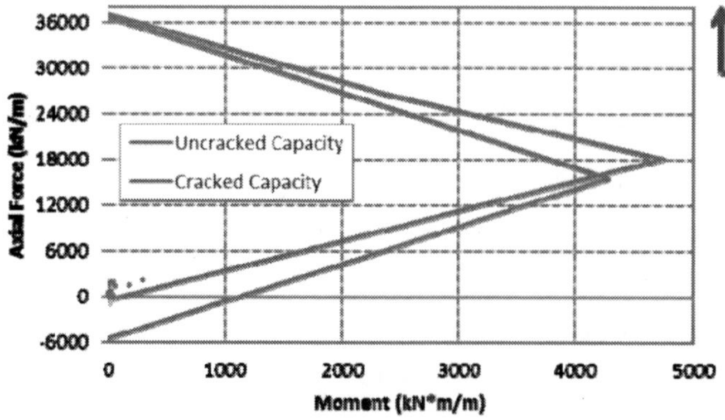

(b)

Figure 10. Moment Interaction Diagrams for the Maximum Height Buttress at Junction Dam during the (a) 1 in 10,000 and (b) 1 in 2,000 AEP Earthquakes.

Furthermore, the RSA for both dams indicates adequate performance for the 1 in 2,000 AEP earthquake. Based on these findings, a risk assessment may indicate that no modifications are warranted for either dam. The linear THA for the 1 in 10,000 AEP earthquake at Junction Dam is expected to provide the information required to complete the risk assessment and determine if additional analysis or repairs to Junction Dam are warranted.

CONCLUSIONS

The studies on Clover and Junction Dams described in this paper were undertaken to support the preliminary conclusions that infilling of the buttress dams with concrete was not warranted. The following are the key findings of the studies to date:

- The concrete investigations and monitoring of survey monuments to date indicate behavior representative of mild alkali-aggregate reaction (AAR). Additional concrete coring and instrumentation are in progress and will be compared with previous studies to better understand the nature of AAR. Continued monitoring is planned for both dams.
- The seismic loading used to support the conclusions that the dams be infilled with mass concrete appears to have been overly conservative. An updated seismic hazard assessment indicated reductions in load between 30 and 50 percent for the 1 in 10,000 AEP earthquake. The 1 in 2,000 AEP earthquake, which is an approximately 60 percent reduction from the updated 1 in 10,000 AEP earthquake, is warranted for these dams based on an updated hazard classification.
- Structural evaluations of the dams using the updated seismic loading indicate that infill of the dams with concrete is not warranted for Clover and Junction Dams. In fact, the response spectrum analyses indicate no modifications are likely to be required for Clover Dam to satisfy the 1 in 10,000 AEP earthquake and time history analysis is warranted to determine if modifications would be required to allow Junction Dam to safely pass the 1 in 10,000 AEP earthquake.

These findings will be factored into a risk assessment of Clover and Junction Dams to support conclusions regarding the need for modifications to the dams.

REFERENCES

[1.] ANCOLD, Guidelines on Consequence Categories for Dams, October 2012.

[2.] ARRB Transport Research, Investigation of Concrete Cores from Junction and Clover Dams for the Identification of AAR in the Concrete, Vermont South, Victoria, Australia, October 1997.

[3.] ARRB Transport Research, Further Examination of Concrete Cores from Junction and Clover Dams for the Characterisation of AAR-affected Concrete, Vermont South, Victoria, Australia, October 1999.

[4.] Raphael, Jerome M., Tensile Strength of Concrete, Title 81-17, ACI Journal, March-April 1984.

[5.] GHD, Report for Clover Dam Safety Review: Stage 2 – Detailed Structural Analysis, Melbourne, Victoria, Australia, March 2007.

[6.] GHD, Junction Dam Safety Review Final Report, Melbourne, Victoria, Australia, December 2008.

[7.] GHD, Final Report for Clover and Junction Dams Further Seismic Assessment, Melbourne, Victoria, Australia, January 2011.

[8.] ANCOLD, Guidelines on Risk Assessment, October 2003.

[9.] ANSYS, ANSYS Mechanical Version 15.0, 2013.

[10.] U.S. Army Corps of Engineers, Time-history Dynamic Analysis of Concrete Hydraulic Structures, EM 1110-2-6051, December 2003.

[11.] U.S. Department of the Interior, Bureau of Reclamation, Hydrodynamic Pressures on Dams due to Horizontal Earthquake Effects, Denver, Colorado, 2006.

NONLINEAR SEISMIC EVALUATION AND ACCEPTANCE CRITERIA FOR ARCH DAMS

Yusof Ghanaat[1]

ABSTRACT

Seismic performance of concrete arch dams has traditionally been assessed by stress checks using a linear elastic analysis. The design of new dams and evaluation of existing dams built in highly active seismic regions often require advanced nonlinear analyses. There are no established criteria for the nonlinear evaluation of arch dams. The paper attempts to provide a methodology for performing a nonlinear analysis and acceptance criteria for assessing the results.

Arch dams are constructed as individual cantilever blocks separated by vertical contraction joints with or without shear keys. The contraction joints are grouted to create a monolithic structure to resist operating loads. Water loads induce compressive arch stresses that keep the contraction joints closed ensuring a monolithic dam behavior. Severe earthquake ground motions, however, can overcome the hydrostatic compressive stresses and generate net tensile forces across the joint causing them to open and close repeatedly. In addition to the joint opening, the paper also discusses other major nonlinear mechanisms such as cracking at lift joints and foundation contacts as well as the stability of thrust blocks and kinematically moveable rock wedges as coupled systems in combination with the dam. Further, the paper considers the effects of dam-water and dam-foundation interactions and elaborates on the application of a compliant seismic input to excite the dam supported by a foundation model with mass, damping, and non-reflecting boundaries.

The paper presents an example dam to demonstrate the application of the nonlinear seismic evaluation, the acceptance criteria, and the compliant seismic input. The nonlinear analysis of the example dam uses advanced contact surfaces with failure criteria to predict potential cracking and sliding along lift joints and foundation contacts beneath the thrust block and the dam.

INTRODUCTION

Arch dams are constructed as a serious of cantilever monoliths separated by vertical contraction joints. The contraction joints provide a mechanism to prevent the formation of tensile cracking that could have developed if the dam was constructed monolithically. This way, during construction, each monolith is free to undergo a volumetric shrinkage and temperature drop without restraint from adjacent monoliths. Later, the contraction joints are grouted under high pressure to bind the monoliths together so the dam will behave as a monolithic structure. Under operating loads, hydrostatic pressures further increase compressive stresses across the contraction joints, enabling a truly monolithic behavior concerning static loads. A strong earthquake ground shaking, however, can

[1] Quest Structures, Inc., 3 Altarinda Road, Suite 203, Orinda, CA 94563 yghanaat@QuestStructures.com

overcome the static compressive stresses and generate net tensile forces across the joint causing them to open and close repeatedly. The paper describes this and other major nonlinear mechanisms that are likely to occur in an arch dam and provides a nonlinear evaluation methodology and acceptance criteria to deal with them.

CURRENT ACCEPTANCE CRITERIA IN PRACTICE

Seismic performance of arch dams has traditionally been performed using simple stress checks from the linear elastic analysis combined with engineering judgment. A recently developed damage threshold approach (Ghanaat 2002 and USACE 2003) provided a systematic interpretation and evaluation of the results of linear elastic analysis considering stress demand-capacity ratios (DCR), cumulative inelastic duration (CID), and spatial extent of overstressed areas (OSA). The threshold approach uses a series of linear time-history analyses to determine if the damage is low enough to accept the linear analysis results, or a nonlinear analysis should be conducted to quantify the damage. The threshold approach accomplishes this by comparing estimated CIDs and OSAs for DCRs of 1 to 2 with the established threshold values (Figure 1). The DCR is the ratio of the estimated stress (arch, cantilever, or principal) to the static tensile strength of the concrete. The CID represents the total duration of stress excursions above a certain DCR in the range of 1 to 2. Note that the CID is a measure of energy and accounts for the magnitude, duration, and the number of high-stress excursions, and not just the number of stress cycles used in the traditional stress checks. The OSA is the ratio of the surface area of an overstressed region above a certain DCR to the dam face surface area indicating overstressing (i.e., upstream or downstream).

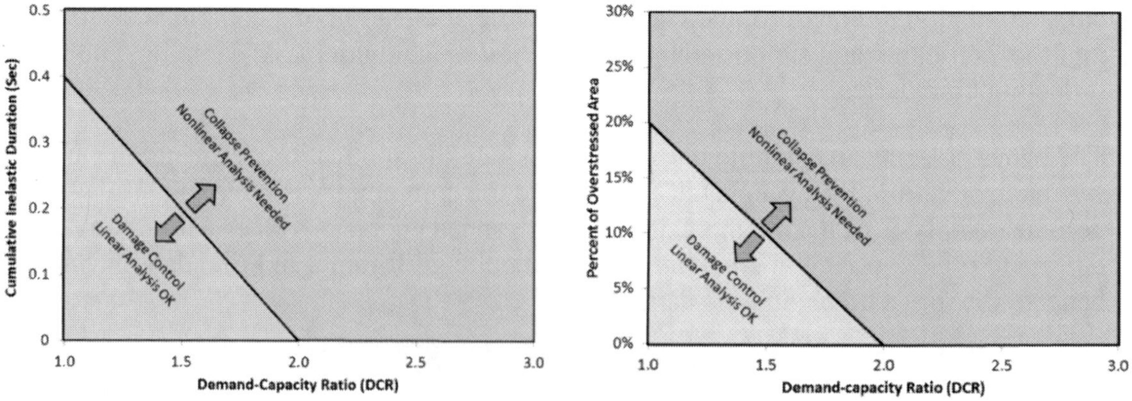

Figure 1. Damage thresholds for CID and OSA.

According to the damage threshold criteria, stress DCRs of less than 1 indicate an acceptable linear-elastic or nearly linear-elastic response with minor or no joint openings and cracking. Stress DCRs of 1 to 2 indicate acceptable performance if OSAs and CIDs remain below the threshold expressed by:

$$\text{OSA} < 0.2\,(2 - \text{DCR}) \quad (1)$$

$$\text{CID} < 0.4\,(2 - \text{DCR}) \quad (2)$$

In such a case, the dam is likely to experience some opening at the contraction joints and cracking at lift joints and the foundation contact, but these nonlinear responses are not severe enough requiring a nonlinear analysis.

Finally, the damage is considered severe and should be evaluated by nonlinear analysis if stress DCRs are greater than two or stress DCRs are less than two, but CIDs exceed the threshold.

ARCH DAMS NONLINEAR MECHANISMS

When linear time history analysis results do not meet the damage threshold criteria, a nonlinear time history analysis may be required to evaluate the stability of the dam during the ground shaking and to quantify the damage to assess the post-earthquake stability under static loads. Although there are many sources of nonlinearities in an arch dam, the aim of a nonlinear analysis should be to capture the dominating nonlinear mechanisms that significantly impact the dynamic response of the dam. The construction of arch dams as separate monoliths combined with rock discontinuities present in the foundation and abutments of the dam make the occurrence of several major types of structural nonlinearities more likely in the response to earthquake ground motions. These include openings at the contraction joints, cracking at the lift joints and foundation contact, sliding at the thrust blocks and foundation contact, or along the planar discontinuities and rock wedges.

Contraction Joint Opening

The nonlinear mechanism that occurs most commonly in arch dams arises from opening and movement of the contraction joints. Under the static condition, the hydrostatic pressures of the reservoir water force close the contraction joints enabling the dam to resist static forces as a single continuous unit. During the dynamic response to a severe earthquake, however, bending and upstream motion of the arch tends to open the contraction joints, as net tensile forces develop across the joints that are known to possess very little or no tensile resistance. It is evident that a linear analysis that neglects the possibility of the joint opening could produce misleading results. Specifically, tensile stresses indicated by a linear analysis in the arch direction cannot be transmitted across the monolith joints.

One of the earliest studies of the contraction joint opening phenomenon was a shaking table test of a segmented arch ring model that simulated a horizontal slice of a prototype dam (Niwa and Clough, 1982). The test successfully reproduced the joint opening that expected to occur in the earthquake response of arch dams, but it also indicated intensified compressive stresses at the joint faces due to a reduction in joint contact area.

The joint opening was also observed on two occasions at the 365-ft-high Pacoima Arch Dam located in Southern California. First, during the 1971 M6.6 San Fernando Earthquake and then a second time during the 1994 M6.8 Northridge Earthquake (Ghanaat 2004 and County of Los Angeles 1994). Later a nonlinear analysis of the dam using recorded ground motions at the dam site predicted contraction joint opening and lift joint cracking with a reasonable prediction of the dam crest acceleration (Mojtahedi and Fenves, 1995).

Cracking at Lift Joints and Foundation Contact

The second type of nonlinearity that may result from an intense earthquake is horizontal cracking of the cantilever monoliths. Such cracking is most likely when the dam is deflecting upstream so that the monolith joints open. At such a time, the arch action is eliminated, and cantilever bending provides the structural resistance. Considering that the cantilever blocks are built by placing mass concrete in lifts, the ability of cantilever blocks to resist such bending will depend on the strength of the lift joints. The strength of the lift joints depends not only on the concrete strength but also to a large extent on methods of construction. A proper preparation and cleanup before placing fresh concrete upon the life joint surfaces could result in high-quality lift joints whose tensile strength approaches the parent concrete strength. Questionable cleanup and construction quality control could lead to unbonded or significantly lower-strength lift joints. Weak lift joints could also be the result of deterioration due to seepage and freeze-thaw cycles. Consequently, it is reasonable to assume that the horizontal cracking more likely would occur along the lift joints. The typical linear dynamic response analysis does not predict the cracking of the concrete at the lift joints, and no estimate is made of the movements that may occur in the post-earthquake condition.

The cracking may also occur at or beneath the foundation contact. Such cracks result from high bending stresses along the footprint of the dam and also beneath the thrust blocks that provide abutment support. The cracking may occur at the concrete-rock interface, or most likely in the fractured/jointed rock below having little or no tensile resistance. Again, the typical linear dynamic analysis does not evaluate such cracking and the effects they might have on the stability of the dam. Instead, high tensile stresses from the linear analysis are commonly interpreted to be partly fictitious and assumed to be relieved by slight opening of the joints and fissures present in the rock.

Thrust Block and Rock Wedge Stability

Abutment thrust blocks and kinem atically moveable rock wedges in the foundation and abutments of an arch dam are two f eatures of potential stability c oncern. Stability analysis of these features is comm only evaluated separately from the dam by applying dam thrusts as additional external loads.

A thrust block supported by the foundation rock with an inadequate shear strength could be susceptible to sliding along the f oundation contact under severe earthquake ground motions. Any m ovements of an abutm ent thrust block could significantly affect stress distributions with the dam, t hus affecting the m agnitude and orientation of dam thrusts exerted on the thrust block. This dynam ic interaction, which m ay also involve opening and closing of the joint between the dam and th rust block, is nonlinea r in nature and is best handled using a coupled nonlinear analysis that incl ude both the dam and thrust block in the same model.

The potential sliding of rock wedges within the foundation and abutments that are in contact with the dam is, even more, critical to dam safety. A factor rarely considered in the design of older dams. It is essential to make a proper assessment of the geologic structure of the dam foundation and abutments identify potential failure planes and major rock wedges that are kinematically capable of movements. The rock wedge sliding may occur along one potential shear plane or the line of the intersection of two of such planes formed beneath the dam by combinations of major faults, shears and/or joint sets. Investigation of rock wedges could be quite involved going beyond the sliding mode of failure along the concrete-rock interface that commonly considered. Depending on the criticality, size, location, and geometry of the rock wedge, the stability analysis may warrant a coupled nonlinear analysis with the dam. In a coupled analysis, the wedge is included in the dam model to allow updating of the thrust forces as the wedge moves and the dam deflects and experiences joint opening/cracking and movements during the earthquake ground shaking.

NONLINEAR MODELING

The major nonlinear mechanisms discussed above are all limited to the known discontinuities and cracking along weak surfaces in the dam and foundation rock. Nonlinear analysis procedures are now available that can adequately deal with such nonlinearities while the rest of the structural system is assumed to behave linearly. Initial nonlinear calculations have been made using specialized computer programs (ADAP-88 1989, and QDAP 1999) to account for the contraction joint opening (Quest 1991; Fenves and Mojtahedi 1993; Mojtahedi and Fenves 1995; Bureau and Ghanaat 2000; Quest 2011). In recent years, more advanced calculations have been made using commercial computer programs with capabilities to account for the contraction joint opening, effects of shear keys, and prediction of cracking followed by sliding along the joint and cracked surfaces (e.g., Noble and Nuss 2004; Quest 2014). The prediction of cracking, however, has proven to be more difficult and sensitive to the failure criteria assumptions.

The overall modeling of the dam-water-foundation system for nonlinear analysis of joint and cracking discontinuities is identical to that developed for linear analysis, except for the addition of nonlinear contraction joints, lift joints, dam-foundation contact, and other potential sliding surfaces.

Dam Model

For modeling of the contraction joint opening and lift joint cracking, the body of the dam is discretized using linear-elastic three-dimensional (3D) solid elements with concrete material properties. The mesh lines are oriented in the horizontal and vertical directions to facilitate modeling of the contraction joints and lift joints. The mesh fineness could be identical to that commonly used in a linear-elastic analysis, but five or more elements through the dam thickness are necessary to capture the partial joint opening and cracking. Regarding the contraction joints, it is preferable to model all joints to avoid excessive opening at any one joint, which might alter potential failure mechanisms. However, if computational efforts dictate less number of joints, consider including, at least, three to five joints at critical locations. For example, the joints may be placed at the dam center, the quarter span points, and near the upper abutment regions where large openings might occur. One to three lift joints may be sufficient to capture horizontal cracking that may develop at locations of high tensile cantilever stresses. Modeling cracking at the foundation contact is essential to deal with high tensile bending stresses in this region that cannot exist due to the inability of the fractured/jointed rock below to resist any significant tension.

Figure 2 shows an overall view of the finite-element model of an example arch dam. The example dam consists of a 315-foot-high double curvature arch with thirteen 50-foot-wide cantilever monoliths with keyed contraction joints, a thrust block, and a joint partially separating the arch from the foundation. The 3D model includes the arch dam with the thrust block and spillway section on the left abutment, the foundation rock, and the impounded water. Figure 3 depicts the thrust block and arch dam model with cantilever monoliths numbered from 1 to 13 and contraction joints from A to M. The concrete mesh lines are shown in Figure 2 but removed from Figure 3 for clarity. The model includes all thirteen contraction joints with an upper lift joint and the foundation contacts beneath the dam and the thrust block using different nonlinear contact surfaces described below. Figure 3 shows the contact surfaces as black rectangular grids.

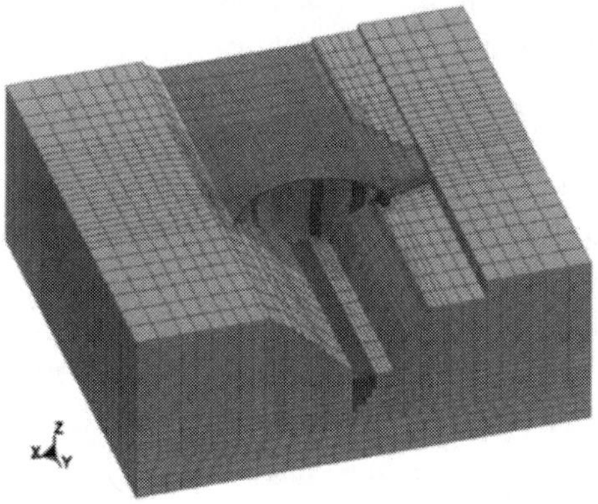

Figure 2. Overall dam-water-foundation Model

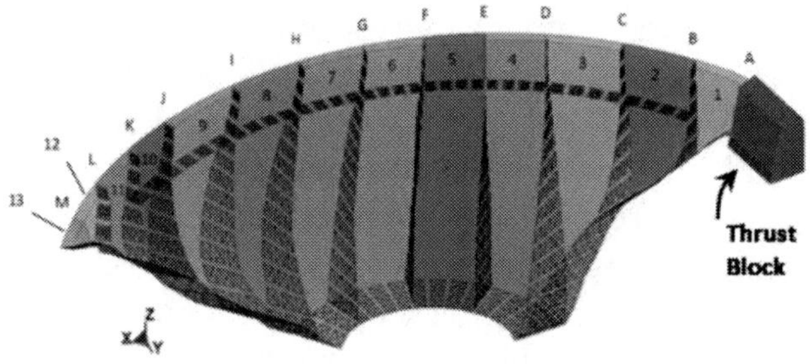

Figure 3. Dam and Thrust Block Model with Nonlinear Contact Surfaces.

Sliding Gap-Friction Contacts. A suitable contact surface for modeling the contraction joints is the gap friction type that resists compression and shear forces but not tension and cohesive shear. The gap friction contact allows the joint to open when net tensile forces develop across the joint and to slide when shear forces exceed the frictional shear resistance. Shear keys, when present, limit the upstream and downstream movements of the joints. The resistance from the shear keys may be necessary for the application of dead loads to individual monoliths and during strong earthquake ground shaking. Modeling of the shear keys is discussed below under subheading "Shear keys."

Sliding Contact with Failure Criteria. The lift joints and concrete-rock contact surfaces initially possess some tensile and shear resistance that keep them intact under static loads, but could experience cracking under high stresses of strong earthquake ground motions. Such cracking may form sliding surfaces if the cracking continues and propagates through the width of the lift joints or the concrete-rock contact. A nonlinear modeling of such surfaces, therefore, necessitates a type of contact element that recognizes the initial tensile and shear strength of the surfaces, has tension and shear failure capabilities, and can model the frictional sliding that follows. A tiebreak contact available in the computer program NL-DYNA is such a contact surface that was used to model the lift joint and foundation contacts beneath the dam and thrust block in the example dam (Figure 3). The LS-DYNA tiebreak contact allows modeling of interfaces that transmit compressive, tensile, and shear forces until failure and then fails in tension or shear when tensile and/or shear strength are exceeded. Once the failure occurs tensile resistance and cohesion vanish, and the tiebreak contacts behave similarly to the traditional compression-only contacts where resistance is limited to frictional forces. The tiebreak failure criterion has normal and shear components and is expressed by:

$$\left(\frac{\sigma_n}{f_t'}\right)^2 + \left(\frac{\sigma_s}{c}\right)^2 \geq 1 \tag{3}$$

Where σ_n and σ_s are the normal and shear stress demands at the interfaces, and f'_t and c are the tensile and cohesive strengths. According to this criterion, in the absence of shear, tensile cracking occurs when tensile stresses reach the tensile strength of the interface; and in the absence of tensile stresses, the interface breaks in shear when shear stresses reach the cohesive strength of the interface. When both tensile and shear stresses are present, the tensile cracking occurs at tensile stresses less than the tensile strength, and shear breaks at shear stresses lower than the shear strength.

Shear Keys. Ideally, it is possible to incorporate the shear keys in the finite-element model of the dam. However, capturing the shear key geometry is cumbersome and could drastically increase the computer time and resources. One simple approach to model the shear keys is to add discrete nonlinear springs across the contraction joints. Figure 4 describes in detail how such discrete elements work in conjunction with the gap-friction contact surfaces (Noble and Nuss 2004). The nonlinear springs are added in a cross pattern across the contraction joints from top to bottom. They are oriented and modeled as compression members to provide shear resistance in the radial direction (upstream-downstream), compression resistance normal to the joint, but no tension. When the contraction joint is in tension, the springs provide no resistance and allow the joint to open freely. However, when the contraction joint undergoes shear deformation, one spring experiences compression and provides radial shear resistance (same as the keys) while the other becomes ineffective in tension. In compression, the springs are given a stiffness consistent with the shear stiffness of the keys. This approach was judged appropriate for the example dam (Figure 4), which included 12-inch-deep shear keys angled steeply at 63 degrees.

The discrete springs work well for the rectangular keys with openings less than the depth of the key, but they are not a true representation of beveled shear keys. The geometry of beveled keys makes the opening and connectivity across the contraction joint complicated. Any amount of joint opening allows the monoliths to move freely in the upstream-downstream direction until they come in contact with the keys. The discrete springs cannot capture this behavior. A beveled key, however, can be modeled using a special contraction joint model developed at Lawrence Livermore National Laboratory (Noble and Solberg 2004) to simulate the effects of sinusoidal joint surfaces (shear keys).

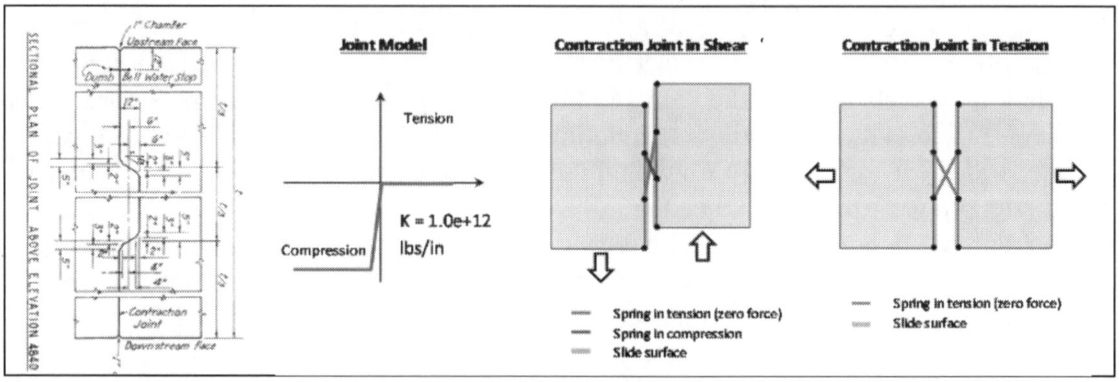

Figure 4. Sketches demonstrating contraction joint model with shear keys (adapted from Noble and Solberg 2004).

Foundation Model

The geometry and finite-element mesh of the foundation rock for the nonlinear analysis are identical to that of a standard linear analysis. Except that it is prudent to include the inertia and damping effects of the foundation to account for wave propagation and influence of the canyon topography. The foundation model with mass should include transmitting boundaries to allow for radiation damping. In the example dam model, the transmitting boundaries were applied at the bottom and all four sides of the foundation model to transmit seismic waves propagating away from the dam in the three orthogonal x, y, and z directions. The use of transmitting boundaries, however, complicated the application of seismic input, since acceleration time histories could not be used directly to excite the model. The transmitting or nonreflecting boundaries are excited using traction time histories, as described below.

Impounded Water Model

For the nonlinear analysis, it is desirable to model the effects of hydrodynamic pressures of the impounded water more accurately and also to account for the radiation damping. In the example dam model (Figure 2); this was accomplished by modeling the impounded water using compressible fluid elements with water properties and including a transmitting boundary at the upstream extent of the fluid domain. The fluid modeling of the impounded water allows for pressure wave propagation and dynamic interaction between the dam and water and accounts for the geometry and flexibility of the dam and the reservoir bottom. The example fluid model included zero friction contact surfaces between the water and the dam to allow water to slide freely against the dam.

NONLINEAR ANALYSIS

Seismic Input

Seismic input for the nonlinear evaluation of dams consists of a set of five or more response-spectrum compatible acceleration time histories (USACE 2003). It is desirable to use linear scaling or match the acceleration time histories loosely to the design response spectra to preserve characteristics of the recorded time histories as much as possible. The use of non-reflecting boundary conditions, however, does not allow for the direct application of the acceleration time histories as the seismic input. Instead, the acceleration time histories should be converted to equivalent forces or traction time histories before they are applied to the sides and bottom of the foundation model.

The required input traction stress histories are obtained from the acceleration time histories in two steps. First a one-dimensional wave propagation (deconvolution analysis) is performed to obtain in-depth ("within") velocity time histories at elevations of the foundation boundary nodes. Then the within velocities are converted to shear (σ_s) and normal (σ_n) stress histories using Equations (4) and (5), respectively.

$$\sigma_s = 2(\rho\, c_s)\, v_s \tag{4}$$

$$\sigma_n = 2(\rho\, c_n)\, v_n \tag{5}$$

Where ρ is the mass density of the rock, c_s and c_n are the shear and compression wave velocities in the rock, and v_s and v_n are the horizontal and vertical within velocities at each foundation boundary elevations.

For the example dam model, the above process was verified by subjecting the foundation model without the dam and canyon to stress histories applied on the sides and the bottom of the model. The resulting acceleration time histories on the top of the model were then compared with the input acceleration time histories developed for the project.The verification analysis was performed separately for each component of the ground motion. Figures 5 to 7 illustrate the verification process and the results for one of the components of the ground motion. Similar analyses were repeated for the foundation model with the canyon but not the dam. The results (not shown) indicated that the magnitude, phasing, and waveform of the ground motion vary across the canyon, factors that were captured in the seismic analysis of the example dam.

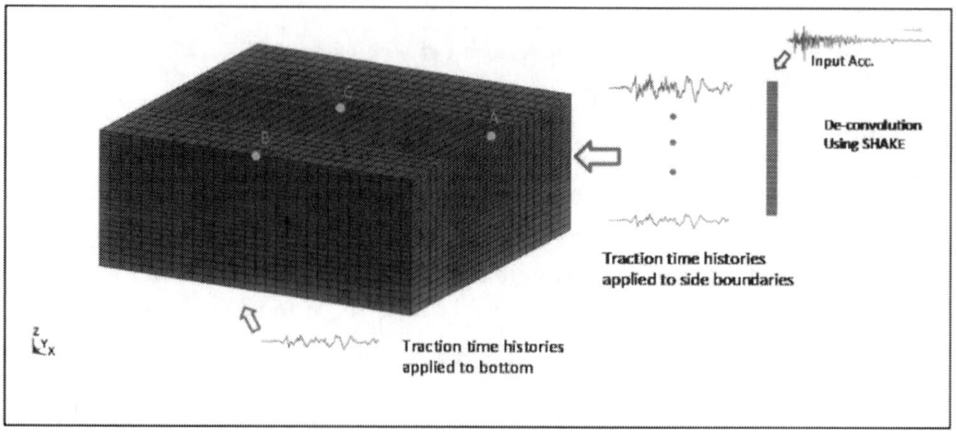

Figure 5. Foundation model without dam and canyon to verify input traction time histories.

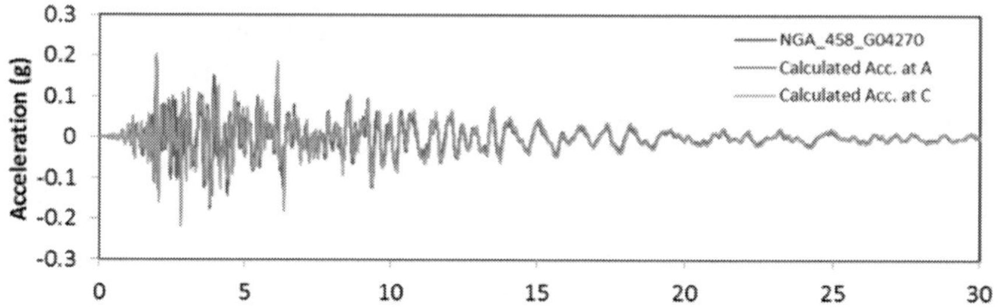

Figure 6. Comparison of input and computed acceleration time histories at Points A and B.

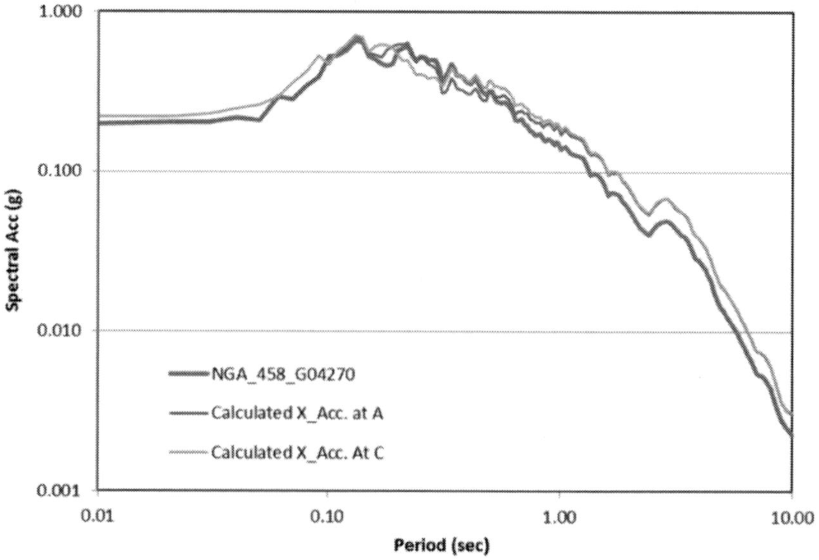

Figure 7. Comparison of response spectra of input and computed acceleration time histories at Points A and B.

Nonlinear Response Evaluation

The nonlinear response evaluation starts with the application of static loads as the initial condition followed by the solution of coupled equations of motion for seismic loading. The example dam was evaluated using the computer program LS-DYNA. The static loads were applied as quasi-static loads using five-second ramps, followed with the application of seismic loads for the duration of ground shaking. A 5% elastic damping was assumed for the concrete. Additional energy loss due to the joint opening/cracking and sliding was provided through the nonlinear mechanisms. The nonlinear analysis was carried out by explicit step-by-step integration of the coupled equations of motion for five sets of ground motions using conservative baseline material parameters. Additional sensitivity analyses were performed for one of the ground motion time histories to investigate the influence of the tensile and shear strengths of the dam-foundation contact on the results. Samples of results including the joint opening, cracking, and sliding are discussed below.

Joint Opening. Figure 8 shows that the contraction joints repeatedly open and close during the earthquake excitation. However, the maximum joint opening is very small in magnitude and duration. Each joint opening lasts only a few milliseconds and lacks the energy needed to inflict damage and instability. The maximum joint opening for all load cases was 0.14 inches at the crest.

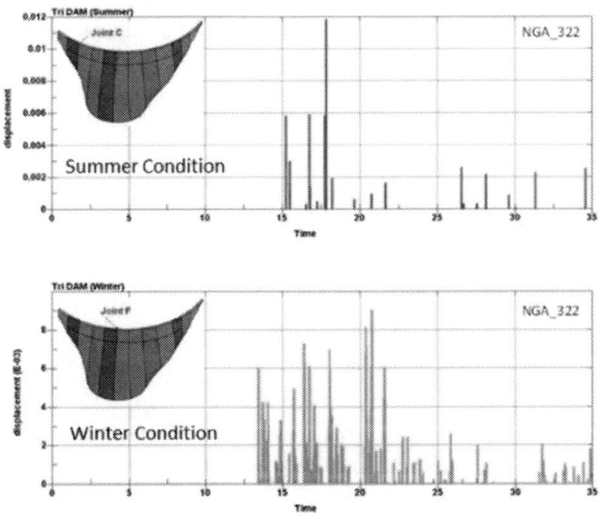

Figure 8. Seismic joint openings for summer and winter conditions.

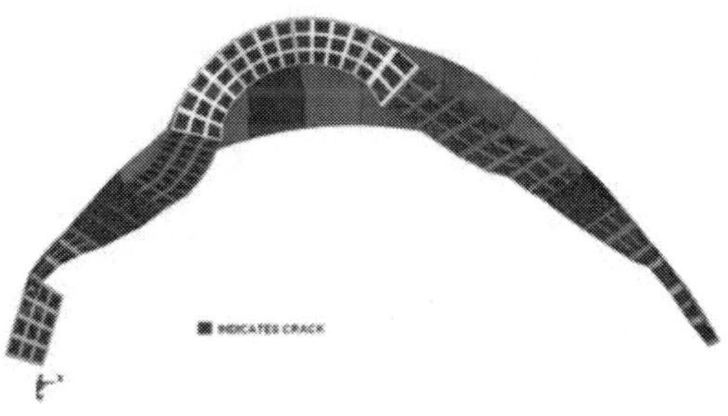

Figure 9. Fully cracked concrete-rock contact.

Cracking. The results show minor and localized lift joint cracking for all seismic load cases. The liftjoint cracks occur at the intersection of contraction joints in several locations on the upstream and downstream faces of the dam. They are shallow and spotty and do not propagate through the dam thickness or laterally along the axis of the dam. The thrust block suffered 8% cracking at the base under one of the ground motions and 100% under the other four. All ground motions indicated nearly 100% cracking at the dam-foundation contact. The results also suggest that in summer, seismic cracks develop along the entire arch and thrust block contacts with the rock (Figure 9) but in winter only beneath the entire arch and not the thrust block. The widespread cracking at the foundation contact was the result of assumed low tensile and shear strengths and the presence of a construction joint with zero tension resistance at the base of the dam.

<u>Sliding</u>. The nonlinear analyses of the example dam indicated cracking along the entire foundation contact followed by sliding of the thrust block and dam monoliths. However, the magnitudes of sliding displacements even for conservatively assumed strength parameters were small with no consequence on the stability of the dam. The maximum sliding displacement were 0.75 inches for the thrust block (Figure 10), 1.13 inches at the river channel (Figure 11), and 1.26 inches at the top of the right abutment (Figure 12).

The thrust block sliding occurred for four of the five baseline summer load cases and none of the three sensitivity load cases. The sliding for each ground motion starts at a different time, but the permanent sliding displacement at the end of the ground shaking is about the same for all four load cases (Figure 10). The results also suggest that the thrust block slid when the dam had fully cracked and slid on the left abutment, thus exerting significantly larger forces to the thrust black than would have been predicted by a linear analysis. Nevertheless, the sliding of the thrust block or the dam is considered unlikely since the dam is securely keyed into the abutments and the shear strength at the rock contact is expected to be much higher than the values used in the analysis.

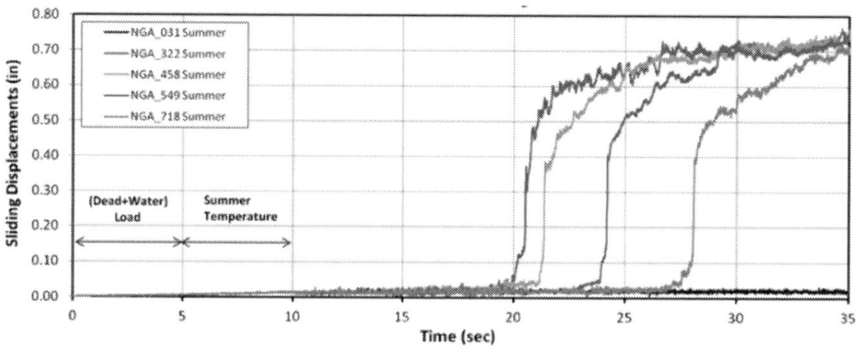

Figure 10. Sliding displacements of Thrust Block.

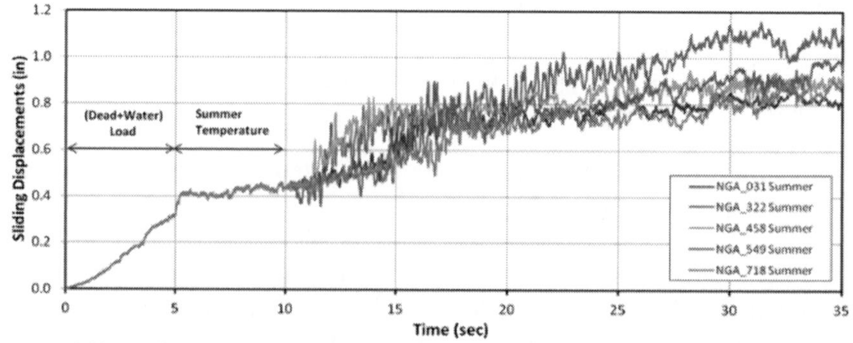

Figure 11. Sliding displacements of the dam on a construction joint at the bottom.

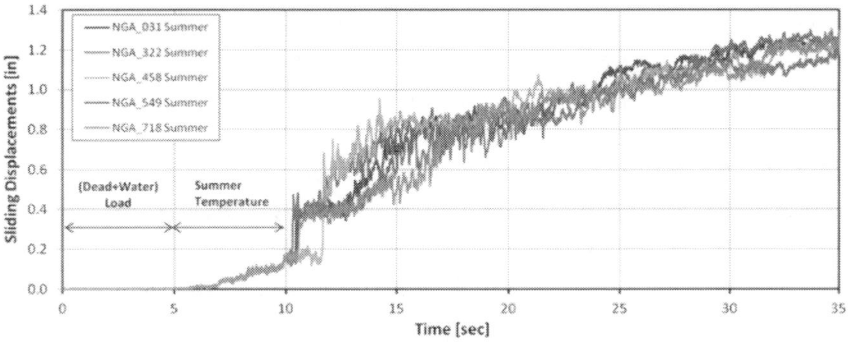

Figure 12. Sliding Displacements of the top cantilever on right abutment.

Compressive Stresses. The stress results were investigated to determine maximum compressive stresses for comparison with the compressive strength of the concrete. A contour plot of the principal compressive stresses (Figure 13) clearly shows that high compressive stresses occur at the contraction joints and near the intersection with the cracked lift joint (green colors). The Solid Element 73,925 identified on the contour plot shows blue paint near the crest, indicative of the peak compressive stress of -409,910 psf (2,846 psi). A stress plot of this element shows that only compressive stresses develop at the joint and that a single peak stress of -409,910 psf (2,846 psi) is significantly less than the compressive strength of the concrete (5,000 psi). Overall, the magnitudes of high compressive stresses vary in the range of 2,000 to 2,846 psi. They are well within the compressive strength of the concrete; thus, meet the acceptance criteria.

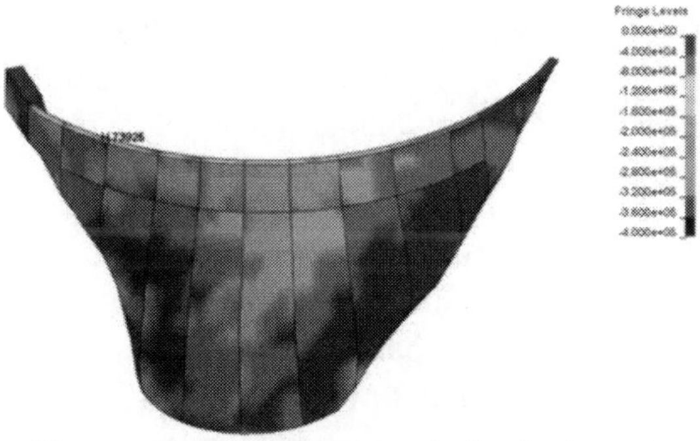

Figure 13. Compressive principal stress contours(psf)

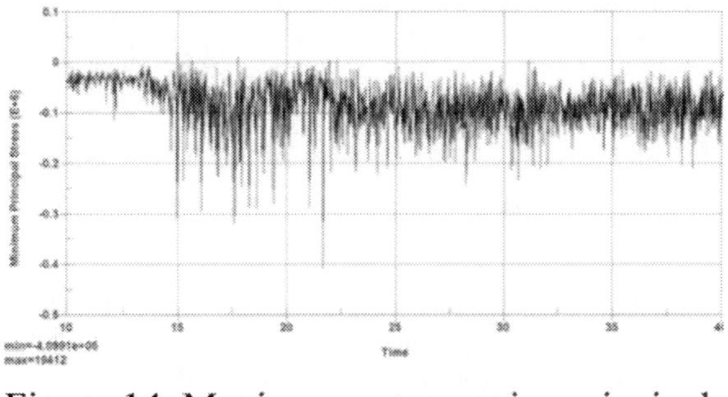

Figure 14. Maximum compressive principal stress history (psf)

Tensile Stresses within Monoliths. Nonlinear joint opening and tensile cracking relieve tensile stresses at the contraction joints, dam-rock interface, and lift joints that have been modeled for cracking. However, tensile stresses may still remain within the cantilever monoliths away from the contraction joints and cracked sections. Tensile stresses within the cantilever monoliths were checked and found to be small and within the tensile strength of the concrete.

Performance of Shear keys. For the example dam, the shear keys were expected to perform satisfactorily due to relatively low ground motion and negligible contraction joint openings. Nevertheless, the shear strength of the shear keys was computed using a fracture mechanics approach (Kaneko et al. 1993a and 1993b) and compared with the shear demands from the nonlinear analysis. Figure 15 illustrates the shear key interaction diagram, which is a plot of the Kaneko et al. (1993a and 1993b) equations for the concrete compressive strength of 4,000 and 5,000 psi. Note that the shear strength reaches a peak value of about half the concrete compressive strength when the normal stress approaches half the concrete compressive strength. Normal and shear stress histories of a concrete element attached to the top shear key were retrieved and displayed as scattered plot in Figure 15 to assess the integrity of the shear keys. The comparison clearly demonstrates that the shear demands of the shear keys are significantly lower the shear capacity of the shear keys, thus meet the criteria.

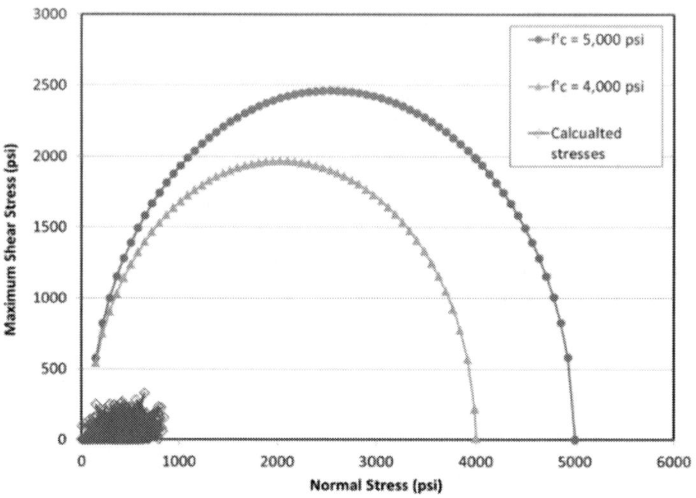

Figure 15. Comparison of shear demands with shear capacities of shear keys.

ACCEPTANCE CRITERIA

There are no established criteria for assessment of the results of nonlinear analysis of arch dams. The universally accepted prescriptive rules for the seismic performance of dams are that the dam should remain stable without the sudden and uncontrolled release of water during and after a major earthquake event. This implies that the dam can experience some level of damage, but should maintain its water retention capability. The prescriptive rules, however, neither characterize the type of damage nor quantify the level of damage. As such, the following is an attempt to develop a set of criteria for assessment of the nonlinear analysis of major types of structural nonlinearities expected to occur in an arch dam in response to earthquake ground motions.

Joint Opening

For the keyed contraction joints, the amount of joint opening at the crest level should be limited to the depth of the shear keys. A full opening at the crest level can still keep the monoliths interlocked since the joint opening lessens and diminishes at lower elevations. This requirement is to limit movements of the monoliths to ensure they remain interlocked and stable.

In the absence of shear keys, seismic performance depends on whether or not fully opened joints and unstable blocks develop. A fully opened joint condition can potentially form unstable blocks bounded by opened joints on each side and a cracked lift joint at the bottom. In a partially opened joint condition, the monoliths are likely to remain stable through wedging action or partial contact with one another.

Tensile Stresses

The joint opening diminishes high tensile arch stresses near the contraction joints and lowers their magnitudes to the level of tensile strength of the joints, or nearly zero. However, high tensile arch stresses may remain within the monoliths away from the joints.

Tensile arch stresses within the monoliths away from the joints should be less than the tensile strength of the parent concrete to avoid vertical cracks. Tensile cantilever stresses should be limited to the tensile strength of the lift joints. Otherwise, contact surfaces should be introduced at the lift joints to allow for lift joint cracking until the stress criterion is met.

Compressive Stresses

Joint opening and cracking tend to increase compressive stresses due to subsequent load redistributions. Compressive stresses should, therefore, be checked against the compressive strength of the concrete to avoid concrete crushing and ensure that the capacity of the concrete to carry compression is not exhausted.

Shear Keys

The integrity of shear keys should be evaluated by comparing their shear demands with shear capacities. This may be accomplished using the fracture mechanics approach proposed by Kaneko et al. (1993a and 1993b) or other procedures.

Nonrecoverable Movements

The magnitude of non-recoverable movements, if any, should be small to ensure the post-earthquake stability. The acceptable nonrecoverable movements should be determined considering the post-earthquake uplift pressures, drainage operability, and reduction of shear strength parameters. The post-earthquake uplift pressures elevate to the full water head due to cracking and potential partial or full drainage blockage if the permanent movements are large enough to cover the drain holes. The shear strength parameters of the sliding surfaces could drop to residual values if the movements are large enough to affect surface roughness.

CONCLUSIONS

The paper provides a summary of the current seismic performance evaluation of arch dams involving a damage threshold approach in combination with the linear-elastic dynamic analysis. The damage threshold approach identifies cases for which the linear analysis would be acceptable as well as those requiring a nonlinear analysis. The paper then discusses major nonlinear mechanisms that should be recognized and included in a nonlinear analysis. These include the opening and closing of the contraction joints, cracking at lift joints and foundation contacts, followed by unrecoverable movements. Further, the paper provides general modeling guidelines for each of these nonlinear mechanisms. An example dam was provided to demonstrate the modeling approach and produce results for discussion. The paper concludes with proposing acceptance criteria for evaluation of the nonlinear analysis results, concerning the amount of contraction joint opening, the formation of unstable blocks, tensile cracking and compression crushing of the concrete, and unrecoverable movements that could affect the post-earthquake stability.

REFERENCES

ADAP-88 (1989): Fenves G.L., Mojtahedi, S., and Reimer, R.B., "A Computer Program for Nonlinear Earthquake Analysis of Concrete Arch Dams," the University of California at Berkeley, Earthquake Engineering Res earch Center, Report No. UCB/EERC-89/12, November 1989.

Bureau, G. and Ghanaat, Y. (2000). "Seism ic Evaluation of a Historic Arched G ravity Dam," Proceedings of the 17th annual conference of the A ssociation of State Dam Safety Officials, Providence, RI, Sept. 27-29, 2000.

Fenves, G.L. and Mojtahedi, S. (1993). "E arthquake response of an arch dam with contraction joint opening," Dam Engineering, 4, 2, 63-89. 22

Ghanaat, Y. (2002). "S eismic Performance and Damage Criteria for Concrete Dam s," Proceedings of the 3 rd US-Japan Workshop on Advanc ed Research on Earthquake Engineering for Dams, San Diego, California, June 22-23, 2002.

Kaneko, Y., J.J. Connor, T.C. Triantafillou, and C.K. Leung, "F racture Mechanics Approach for Failure of Concrete S hear Key, I: Theory, "*J. of Engineering Mechanics, ASCE,*" 119(4), 681-700, 1993a.

Kaneko, Y., J.J. Connor, T.C. Triantafillou, and C.K. Leung, "F racture Mechanics Approach for Failure of Concrete Shear Key, II: Veri fication, "*J. of Engineering Mechanics, ASCE,*" 119(4), 701-719, 1993b.

Mojtahedi, S., and Fenves, G.L. (1995), "Effect of Contraction Joint Opening on Pacoima Dam in the 1994 Northridge Earthquake," Earthquake Engineering Research Center, University of California, Berkeley, September 1995.

Niwa, A. and Clough, R.W. (1982). "Nonlinear Seismic Response of Arch Dams," J. of Earthquake Engineering and Structural Dynamics, March 1982.

Noble, C.R., and Nuss, L. (2004), "Nonlinear Seismic Analysis of Morrow Point Dam," 13th World Conference on Earthquake Engineering, Vancouver, Canada, August 204.

Noble, C.R. and Solberg J. (2004), "Nonlinear Seismic Analysis of Morrow Point Dam," Report No. UCRL-TR-202545, Lawrence Livermore National Laboratory, Livermore, CA, March 2004.

QDAP (1999), "An enhanced version of ADAP-88 with joint opening and cracking at lift joints and foundation contact," Quest Structures, Orinda, California.

Quest (1991), "Nonlinear Earthquake Analysis of Portugues Dam," a study conducted by Quest Structures for the US Army Corps of Engineers Jacksonville District, Florida for initial Portugues Arch Dam design, June 1991.

Quest (2011), "Nonlinear Analysis of Salinas Arch Dam Considering Contraction Joint Opening and Lift Joint Cracking," a study conducted by Quest Structures for the City of San Luis Obispo, California, 2011.

USACE (2003). Engineer Manual EM 1110-2-6051, "Time History Dynamic Analysis of Concrete Hydraulic Structures," US Army Corps of Engineers, Washington, DC, December 2003.

Quest (2014), "Linear and Nonlinear Seismic Evaluations of Donnells Arch Dam," a study conducted for Tri-Dam Project Water Utility Company, May 2014.

DEVELOPING TIME HISTORIES WITH ACCEPTABLE RECORD PARAMETERS FOR DILLON DAM

Dina Bourliea Hunt, P.E.[1]
Justin Beutel, P.E.[2]
Christine Weber, P.E.[3]

ABSTRACT

Dillon Dam project is located in Summit County, Colorado. Dillon Dam is an earthfill structure constructed on the Blue River in 1962, rising 231 feet above the riverbed. At over 257,000 acre-feet, it contains the largest reservoir in the Denver Water system, diverting water into the South Platte River Basin via the Harold D. Roberts Tunnel.

After completing a seismic hazard analysis and calculating a design response spectrum, time histories were developed for use in finite element modeling of Dillon Dam. Several record parameters (i.e. Arias intensity, peak ground velocity, cumulative absolute velocity, and significant duration) were estimated from published articles. Time histories were then preselected from the NGA-West2 Pacific Earthquake Engineering Research Center database using magnitude, distance, and pre-calculated record parameters as part of the search criteria. The overall goal of spectral matching is to achieve a fit as close as possible to the design response spectrum, while not deviating from the Fourier amplitude spectrum over the spectral frequency range of interest, nor changing the non-stationary phasing of the spectrally matched time history. Performing spectral matching using a conventional approach led to unacceptable record parameters, therefore an alternative method of reducing peak ground acceleration scale factors was used to perform the spectral matching. This alternative method resulted in a spectrally matched time history with record parameters closer to the range of design values.

This paper presents a summary of the published references used to estimate record parameters, a discussion of the process used to pre-select the time history records, and the trade-offs between achieving a good fit to the design response spectrum and corresponding record parameters.

---

[1] MWH Global, 1340 Poydras Street, Suite 1420, New Orleans, Louisiana 70112, 510-329-9672, dina.b.hunt@mwhglobal.com
[2] Pacific Gas and Electric Company, 245 Market Street, MC N11D, San Francisco, California 94105, 415-973-5615, jv1w@pge.com
[3] MWH Global, 1560 Broadway 18th floor Denver, Colorado 80202, 303-291-2249 christine.t.weber@mwhglobal.com

BACKGROUND

Located in Summit County, Colorado, Dillon Dam is an earthfill structure constructed on the Blue River in 1962, rising 231 feet above the riverbed. At over 257,000 acre-feet, it contains the largest reservoir in the Denver Water system, diverting water into the South Platte River Basin via the Harold D. Roberts Tunnel. A finite element dynamic deformation analysis was to be performed for the dam, thus requiring a design seismic event to be defined and the time histories to be developed for use in the deformation model.

Probabilistic and deterministic seismic hazard analyses were completed to compute a design response spectrum. Given the results of these analyses, it was decided that a deterministic response spectrum would be the basis for development of time histories. The controlling source for Dillon Dam was identified as a M7.0 event occurring on the Gore Range Frontal Fault, located approximately 5.2 km west of the dam. The fault trace for the Gore Range Frontal Fault (USBR, 2014) is shown in red on Figure 1.

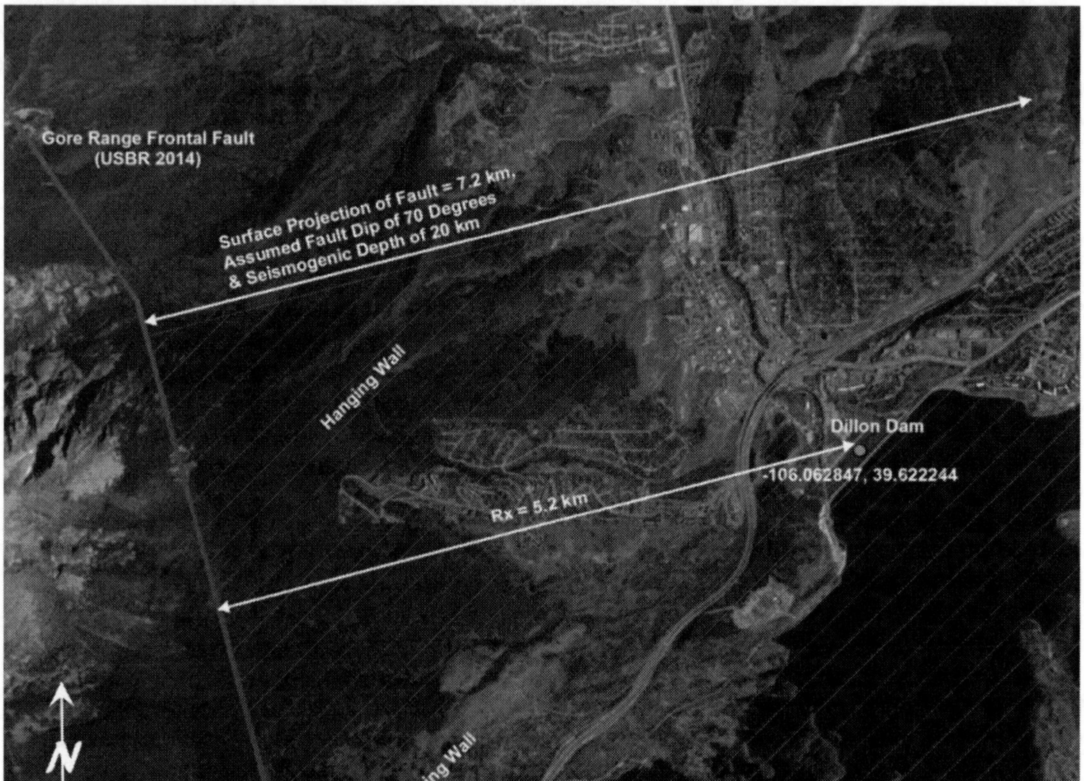

Source: Esri, DigitalGlobe, GeoEye, i-cubed, Earthstar Geographics, CNES/Airbus DS, USDA, USGS, AEX, Getmapping, Aerogrid, IGN, IGP, swisstopo, and the GIS User Community

Figure 1. Plan View of Dillon Dam and Gore Range Frontal Fault

Deterministic Response Spectrum

The development of a deterministic response spectrum included selecting ground motion prediction equations (GMPEs) and associated weighting.

Ground Motion Prediction Equations. The Pacific Earthquake Engineering Research (PEER) Next Generation Attenuation (NGA) West2 relationships were used to estimate ground motions at Dillon Dam. The NGA West2 relationships draw upon a larger dataset and are therefore an improvement on than the previous NGA West relationships. The five relationships are as follows:

- Abrahamson, Silva, and Kamai (2014) – ASK14
- Boore, Stewart, Sehan, and Atkinson (2014) – BSSA14
- Campbell and Bozorgnia (2014) – CB14
- Chiou and Young (2014) – CY14
- Idriss (2014) – I14

The main inputs for the GMPEs are magnitude, distance, and shear wave velocity. The distance parameters required as input for the deterministic analysis are also shown on Figure 1 (plan view) and Figure 2 (profile view).

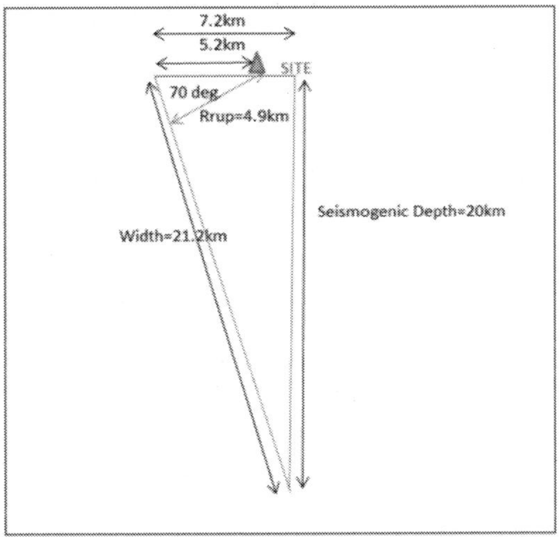

Figure 2. Cross Section sketch of Gore Range Frontal Fault and Dillon Dam (not to scale)

Since Dillon Dam is located on the hanging wall of the Gore Range Frontal Fault (Figure 1 and Figure 2), only four of the five relationships are considered applicable for the site: ASK14, BSSA14, CB14, and CY14. The I14 model was not used because it does not consider the effects of the hanging wall in the calculation of ground motions (Idriss, personal communication, September 30, 2014). The BSSA14 GMPE does not specifically include a hanging wall term, but the effect of the hanging wall is implicitly captured by the use of the Joyner-Boore distance term (R_{JB}), which is the closest distance of the site to the surface projection of the rupture.

Additional discussion regarding the applicability of the BSSA14 model to hanging wall sites can be found in the PEER Report "Hanging-Wall Scaling using Finite-Fault Simulations" by Donahue and Abrahamson (2013). Plots comparing all NGA West2 GMPEs for a range of magnitudes were included in the Donahue and Abrahamson (2013) report and are provided in Figure 3.

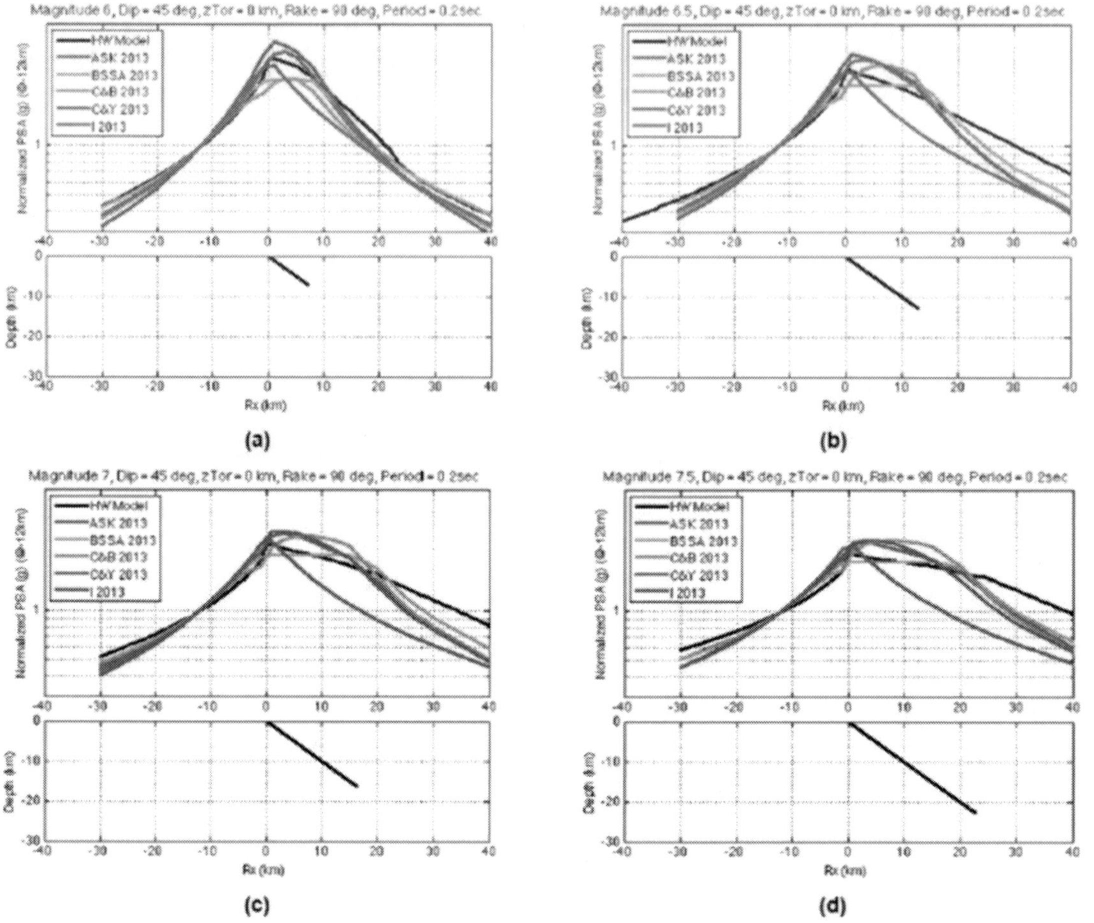

Figure 3. Comparisons of the hanging-wall results for the NGA West2 Models, from Donahue and Abrahamson (2013)

Shear Wave Velocity. The NGA West2 relationships require the input of V_{S30}, the average shear-wave velocity (m/s) over a subsurface depth of 30 meters. This parameter can either be measured or inferred, with a decrease in uncertainty if measured. The shear wave velocity at Dillon Dam was measured in-situ by Stokoe et al. (2008), using spectral analysis of surface waves (SASW). The SASW measurements were performed for finite element modeling purposes and not necessarily to obtain data for up to 30 meters of the foundation. The following equation was used to calculate the V_{S30} starting at an elevation of 8,720 feet.

$$V_{S30} = \frac{30}{\Sigma\left(\frac{d}{V_S}\right)} \qquad \text{(Equation 1)}$$

where, d = thickness of layer and V_S = shear wave velocity

Only those profiles that extended to depths greater than elevation 8,720 feet were used. The shear wave velocity was estimated by extending the last available layer in the profile to depths of 30 meters (98 feet), adding some conservatism into the calculation since shear wave velocity normally increases with depth. Based on the results of the field tests, a V_{s30} of 751 m/s was used to characterize the rock.

Deterministic Results. The deterministic inputs are summarized in Table 1. The resulting design response spectrum was calculated by taking the equally weighted average of the four GMPEs. The design (target) response spectrum, along with the response spectra calculated from the four GMPEs, are shown in Figure 4.

Table 1. Deterministic Inputs

Deterministic Input Parameter	**Definition**	**Value**
M_W	Moment magnitude	7
R_{RUP} (km)	Closest distance to coseismic rupture	4.9
R_{JB} (km)	Closest distance to surface projection of coseismic rupture	0
R_X (km)	Horizontal distance from top of rupture measured perpendicular to fault strike	5.2
R_{y0} (km)	The horizontal distance off the end of the rupture measured parallel to strike	0
V_{S30} (m/s)	The average shear-wave velocity over a subsurface depth of 30 m	751
Fault Mechanism	-	Normal fault
Dip (deg)	Average dip of rupture plane	70
Z_{TOR} (km)	Depth to top of coseismic rupture	0
Z_{HYP} (km)	Hypocentral depth from the earthquake	10.4
$Z_{1.0}$ (km)*	Depth to Vs=1 km/sec	0.048
$Z_{2.5}$ (km)*	Depth to Vs=2.5 km/sec	0.615
W (km)	Fault rupture width (km)	21
V_{s30} Flag	-	Measured
Region	-	Global

*$Z_{1.0}$ and $Z_{2.5}$ estimated from V_{S30}

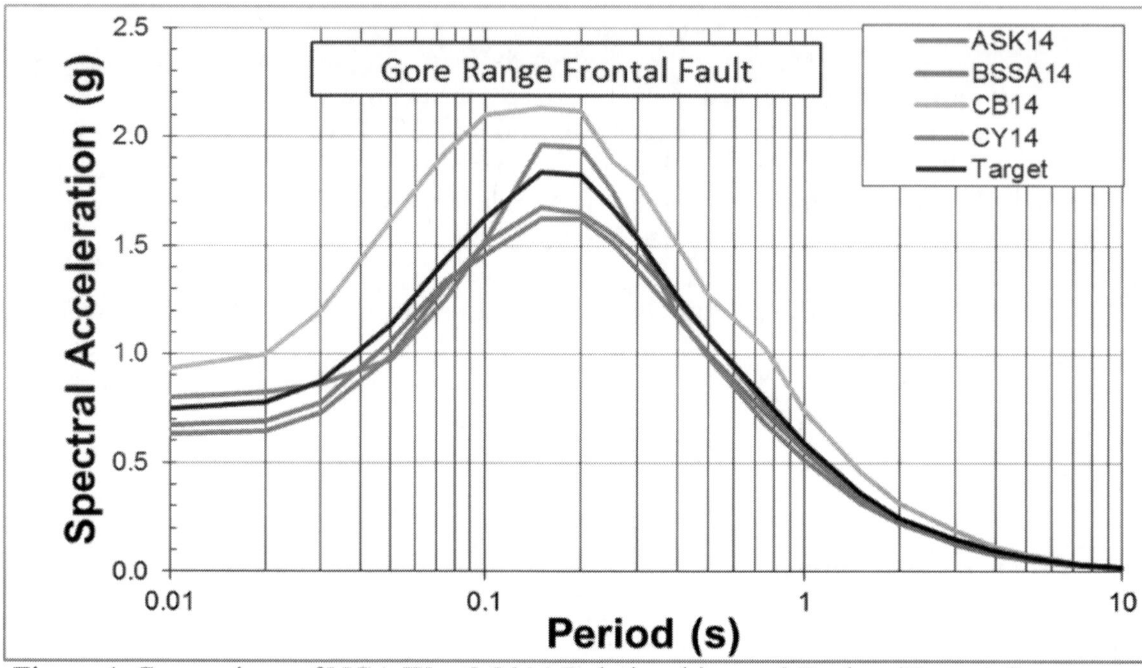

Figure 4. Comparison of NGA West2 2014 Relationships and Design (Target) Spectrum

SELECTION CRITERIA FOR TIME HISTORIES

The search criteria were defined based on the design earthquake event, as defined in Table 1. Time histories were selected from the PEER strong motion database. A search of the NGA West2 database was performed on October 16, 2014 using the initial search criteria discussed below. Earthquakes downloaded from the PEER strong motion database are referred to as seed motions.

Initial Search Criteria

1. **Magnitude** – Range considered: M6.75-7.25
2. **Distance (R_{RUP})** – Range considered: 0-15 km
3. **Significant Duration** (this criterion was used to reduce the number of potential seed motions) – Range considered: 11-22 seconds (5-95% Arias intensity).

The 5-95% significant duration for the design earthquake was estimated using a formula developed by Dobry et al. (1978). Using Equation 2, the estimated 5-95% significant duration (5-95% Arias intensity) for a magnitude 7.0 event is approximately 15.6 seconds.

$$D = 10^{(0.432M - 1.83)}$$

where, D = significant duration, in seconds
and M = magnitude

(Equation 2)

The 5-95% significant duration was also estimated using a method developed by Kempton and Steward (2006) that is based on the site-specific V_{s30}. Kempton and Steward (2006) developed their model to estimate the significant duration and uncertainty, calculations for the median and ±1 standard deviation (16th and 84th percentile) are also presented.

The 5-95% significant duration calculated using these two models are summarized in the Table 2. 2. Note that these values were used to narrow down the potential seed motions.

Table 2. Significant Duration (5-95%)

Reference	**Significant Duration (s)**		
	16th Percentile	**Median**	**84th Percentile**
DIN78 (5-95%)	-	15.6	-
KS06 (5-95%)	9.2	14.2	22.1

References: **DIN**: Dobry, Idriss, and Ng (1978); **KS06**: Kempton and Steward (2006).

4. Scale Factor – Less than 4

Reviewed Parameters

The four initial search criteria narrowed down the number of potential seed motions to 27 motions. From those 27 motions, record parameters (the Arias intensity, 5-75% significant duration, peak ground velocity (PGV), and cumulative absolute velocity) were reviewed to verify the values fell within the 16th to 84th percentiles. These record parameters were reviewed prior to spectral matching. The methods used to estimate the record parameters are summarized below.

1. Arias Intensity

The range of Arias intensity for the design event was estimated using three different equations: 1) Campbell & Bozorgnia (2012), 2) Travasarou et al. (2003), and 3) Abrahamson (personal communication, 2014). The sources and results are presented in Table 3. There is some difficulty in estimating the Arias intensity, as can be seen in the difference between the 16th and 84th percentile see Table 3. Campbell and Bozorgnia (2012) point out that Arias intensity exhibits significantly greater amplitude scaling and aleatory uncertainty as compared to the cumulative absolute velocity.

Table 3. Arias Intensity

Reference	Arias Intensity (m/s)		
	16th Percentile	Median	84th Percentile
C&B12	0.8	2.1	5.7
TBA03	0.5 1.6		4.8
A14	0.8	1.7	4.0

2. Peak Ground Velocity

The peak ground velocity for the design event was estimated using the NGA West2 spreadsheets for the ASK14, BSSA14, CB14, and CY14 models. The calculated average of the median PGV was approximately 36.2 cm/s and the calculated 84th percentile of the PGV was approximately 66.1 cm/s. The range of PGV values is provided in Table 4.

Table 4. Peak Ground Velocity

Reference	Peak Ground Velocity (cm/s)		
	16th Percentile	Median	84th Percentile
NGA West2	19.9	36.2 66.1	

3. Cumulative Absolute Velocity

The relationship developed by Campbell and Bozorgnia (2010) was used to calculate the Cumulative Absolute Velocity (CAV) for the design event. The range of CAV values is presented in Table 5.

Table 5. Cumulative Absolute Velocity

Reference	Cumulative Absolute Velocity (g-s)		
	16th Percentile	Median	84th Percentile
CB2010	0.7	1.2 1.9	

4. Significant Duration (5-75%)

The Brookhaven Model (Silva et al. 1996) was used to estimate the 5-75% significant duration for each of the time histories. The rupture distance and magnitude for the Gore Range Frontal Fault was used as input to the Brookhaven Model. The estimated range of 5-75% significant duration is provided in the Table 6.

Table 6. Significant Duration (5-75%)

Event	Horizontal Duration (5-75%), seconds		
	16th percentile	Median	84th percentile
Gore Range Frontal Fault M7.0, Distance 4.9	3.4	6.0	10.6

Iterations and Comparisons

The record parameters discussed above are summarized in Table 7. A range is given when more than one reference was used.

Table 7. Summary of Record Parameters

Record Parameter	PGA (g)	PGV cm/s	Arias Intensity (m/s)	CAV (g-sec)	Significant Duration (s)	
					5%-95%	5%-75%
16th Percentile	0.23	19.9	0.5-0.8	0.7	9.2	3.4
Median	0.42	36.2	1.6-2.1	1.2	14.2-15.6	6.0
84th Percentile	0.75	66.1	4.0-5.7	1.9	22.1	10.6

After the record parameters for the potential seed motions were verified to fall within the range of values summarized in Table 7, the response spectra for both horizontal motions were visually compared to the target spectra, and candidates were selected for spectral matching. During this process, only one station was chosen for each earthquake record. For example, several stations recorded the Loma Prieta event and were included in the acceptable motions, but after one station resulted in a good spectral match, the ground motions from other stations were not reviewed. The purpose of this was to provide variability in the developed time histories. Overall, a total of ten motions were selected for spectral matching.

SPECTRAL MATCHING AND INITIAL SCALING FACTORS

Time histories were developed using spectral matching techniques. Spectral matching adjusts the time series in the time domain by adding wavelets to the initial time series. A formal optimization procedure for this type of time domain spectral matching was first proposed by Kaul (1978) and was extended to simultaneously match spectra at multiple damping values by Lilhanand and Tseng (1987, 1988). This time domain procedure has good convergence properties and, in most cases, preserves the non-stationary character of the reference time history which is a key goal in the spectral matching process. The program RSPM v99 (Abrahamson, 1993) was used to perform the spectral matching calculations.

Of the ten time histories selected for spectral matching, the three time histories that maintained a good fit to the design response spectrum and similar displacement to the original input motion (no extraneous wavelets added) after spectral matching were selected as the final time histories. The three motions chosen: Darfield, New Zealand M7.0 (PEER ID 6927); Loma Prieta, California M6.93 (PEER ID 768); and the Chuetsu-oki, Japan M6.8 (PEER ID 4866).

Normally, a time history is scaled to the target peak ground acceleration (PGA) prior to spectral matching. However, when the selected time histories were scaled to the peak ground acceleration, the results after spectral matching resulted in Arias intensity, cumulative absolute velocity, and peak ground velocity that were generally in excess of the 84th percentile values calculated for the design event. To reduce these parameters, the time histories were scaled to a factor of the target PGA. By modifying the initial scaling factor prior to spectral matching, a decrease in the Arias intensity and CAV is normally observed (personal communication Richard Armstrong, October 17, 2014).

The scale factors and resulting PGA are summarized in Table 8.. Each of these factors was evaluated and the resulting motion that maintained an acceptable significant duration was selected as the final spectrally matched motion.

Table 8. Scale Factor for Peak Ground Acceleration

Factor	**Starting PGA of SEED Record**
0.7	0.53
0.8	0.60
0.9	0.68
1.0*	0.75*

*Note: *Corresponds to target PGA*

As shown in Figure 5, the CAV and Arias intensity show a decrease when reducing the initial motion PGA prior to spectral matching; however, the significant duration also decreases. The black line in Figure 5 is plotted at the lowest acceptable significant duration (5-95%) of 13 seconds.

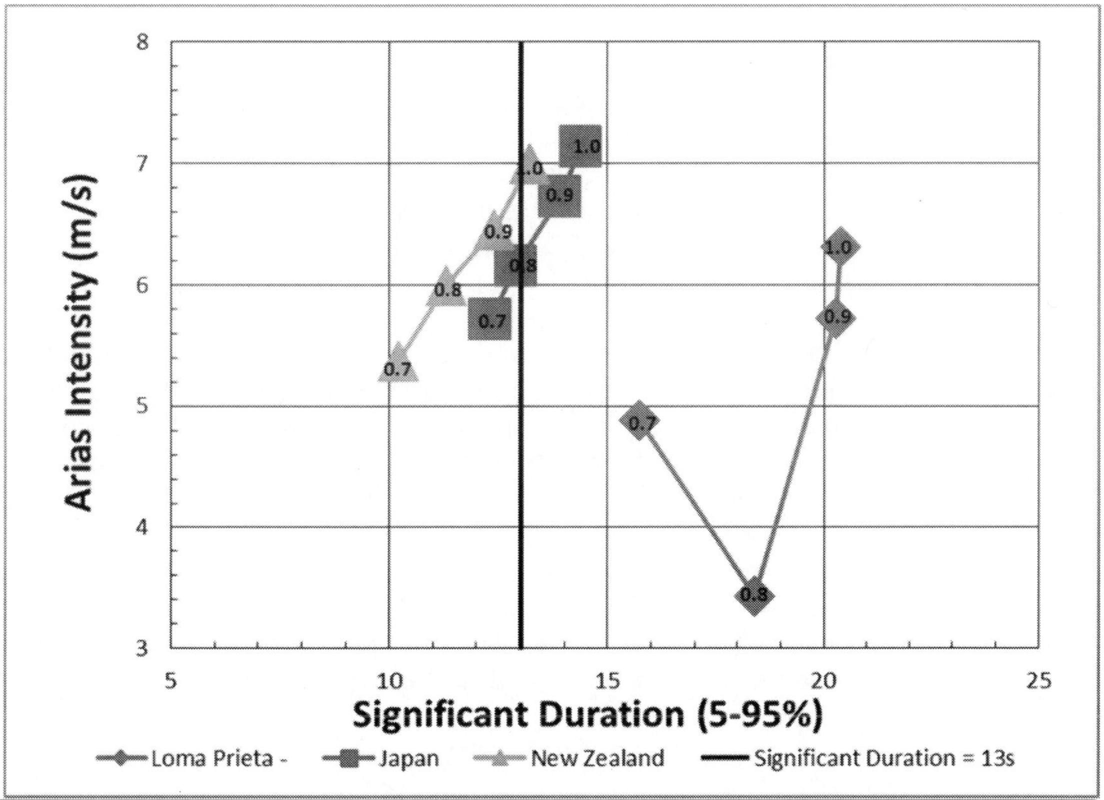

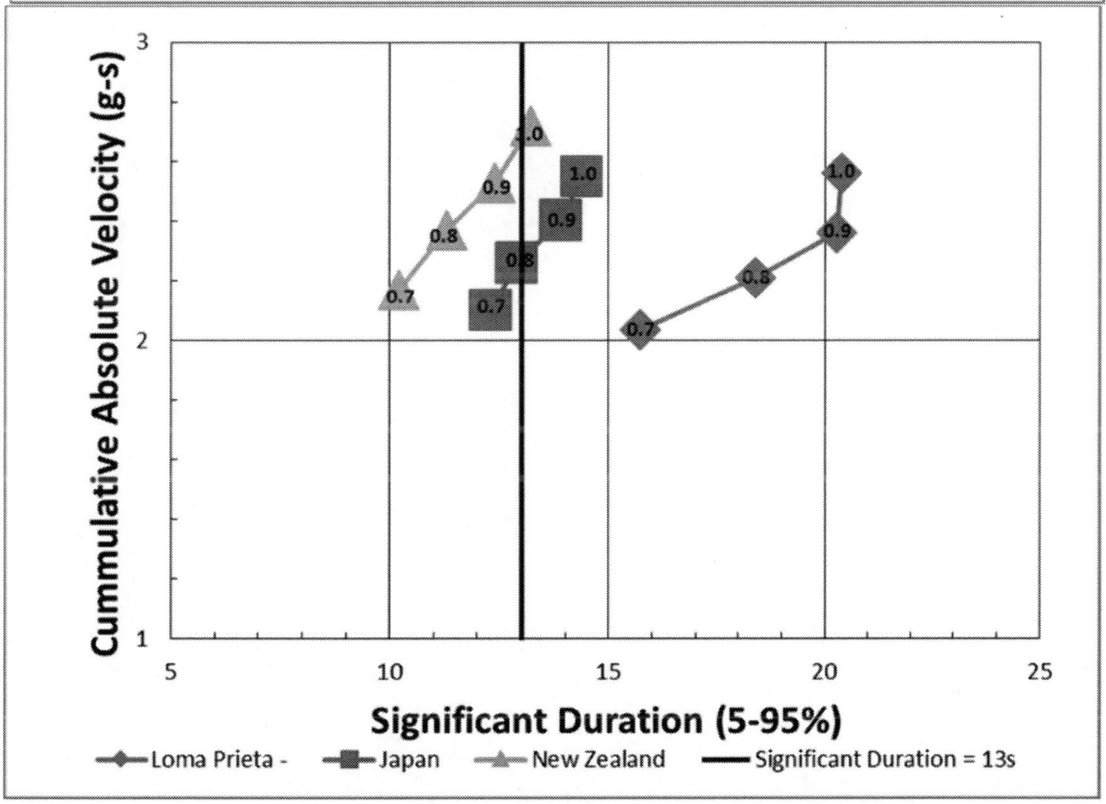

Figure 5. Record parameters after spectral matching for 4 scale factors after scaling to PGA (0.7, 0.8, 0.9, and 1.0)

Final Selections and Plots

The final selected motions had a magnitude range from 6.8 to 7.0, distance range (R_{RUP}) 7.11-14.34 km, significant duration (5-95%) range of 12.9 to 13.9 seconds, and initial scale factor of 1.7 to 2.4 (see Table 9). The other record parameters (Arias intensity, peak ground velocity, and cumulative absolute velocity) were also calculated before and after spectral matching for comparison purposes and summarized in Table 9. Modifications to the initial scale factor of the SEED record were successfully implemented on the three final records.

The final selected scale factors including a comparison to the default "scale to the PGA" are illustrated for each motion in Figure 6 through Figure 8. The top plot is the match to the target design response spectrum and the bottom plot is the normalized displacement time history. As shown in Figure 6, the match to the "0.7 x PGA" (green line) matches closely with the "scaled to the PGA" (red line), for both the response spectrum plot and normalized displacement time history plot. For the Darfield event, presented in Figure 7, the adjustment to the PGA prior to spectral matching had very little effect on the end spectrum and normalized displacement time history. Of the three final time histories, the Chuetsu-oki event (Figure 8) probably had the worst fit to the displacement time history after changing the initial scale factor, but overall the fit is still acceptable.

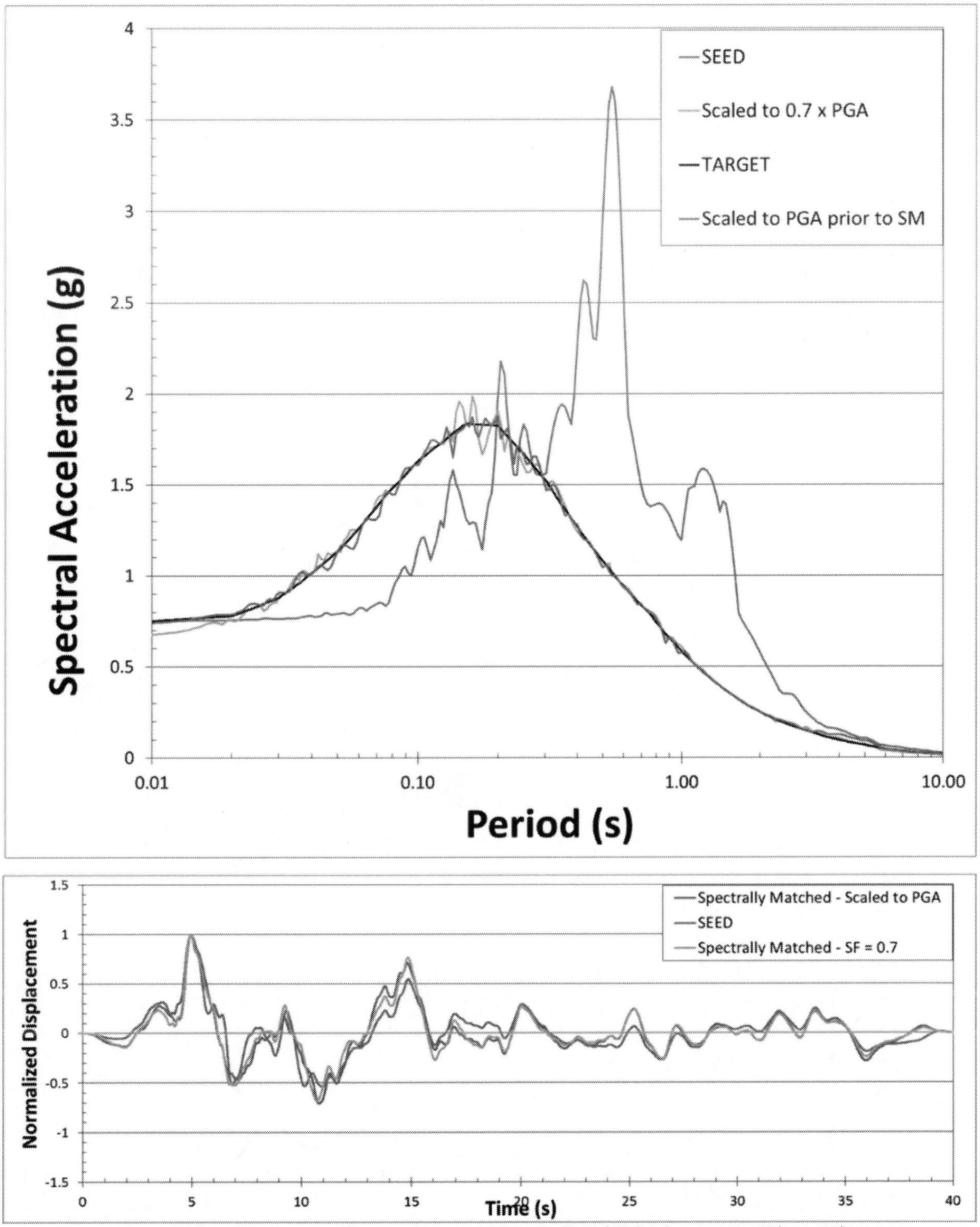

Figure 6. Loma Prieta Response Spectrum and Normalized Displacement Time History – Comparison of seed and spectrally matched motions with different initial scaling factors

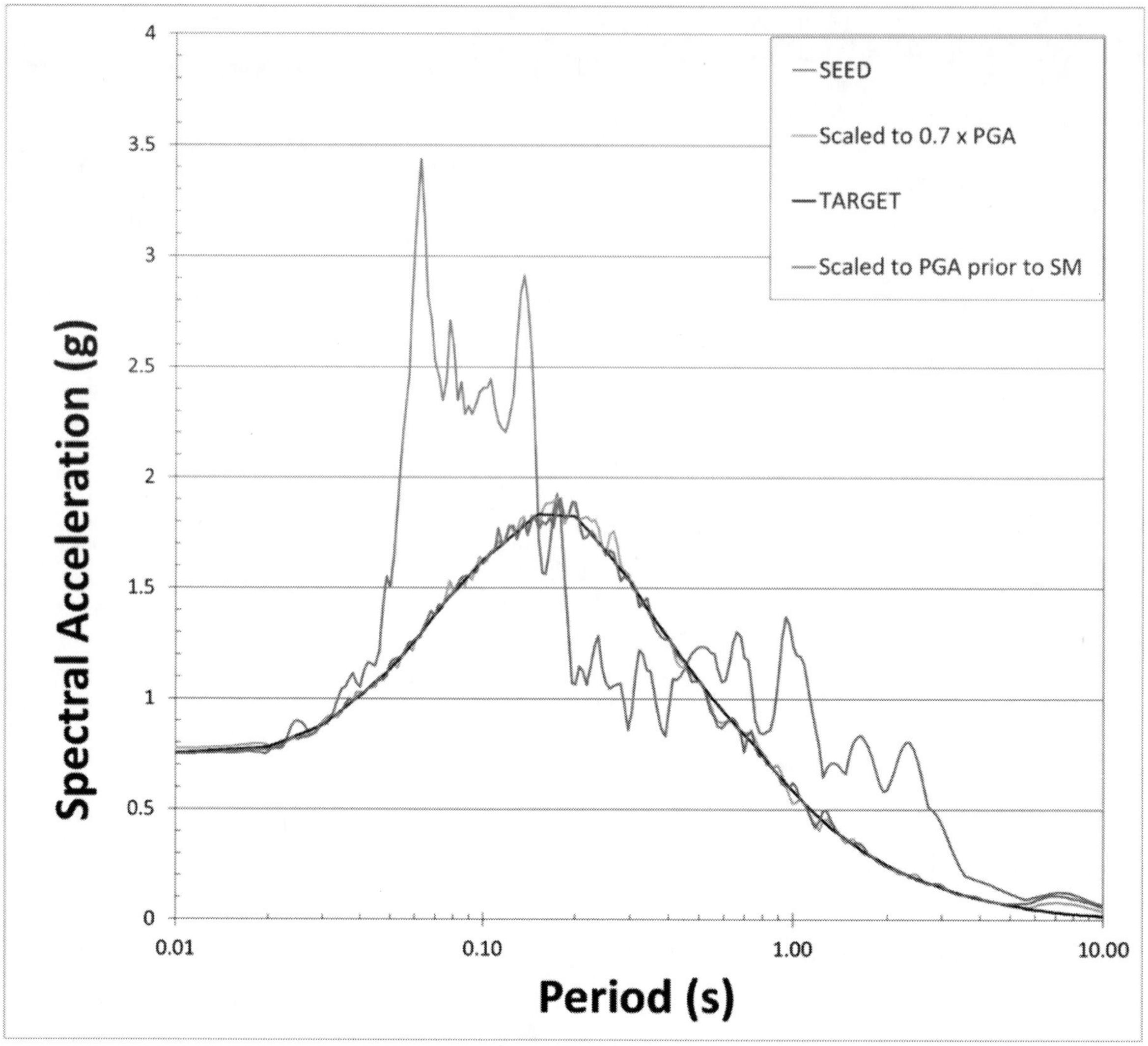

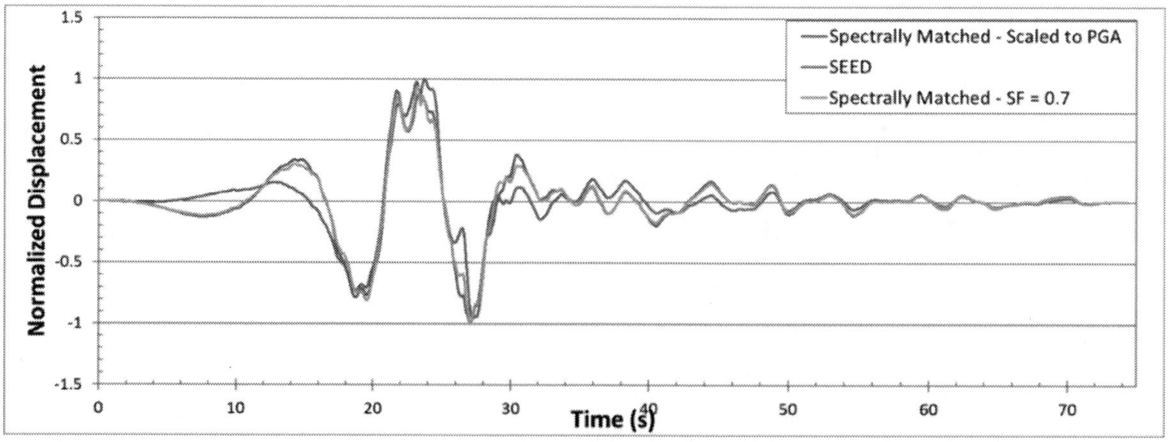

Figure 7. Darfield, New Zealand Response Spectrum and Normalized Displacement Time History – Comparison of seed and spectrally matched motions with different initial scaling factors

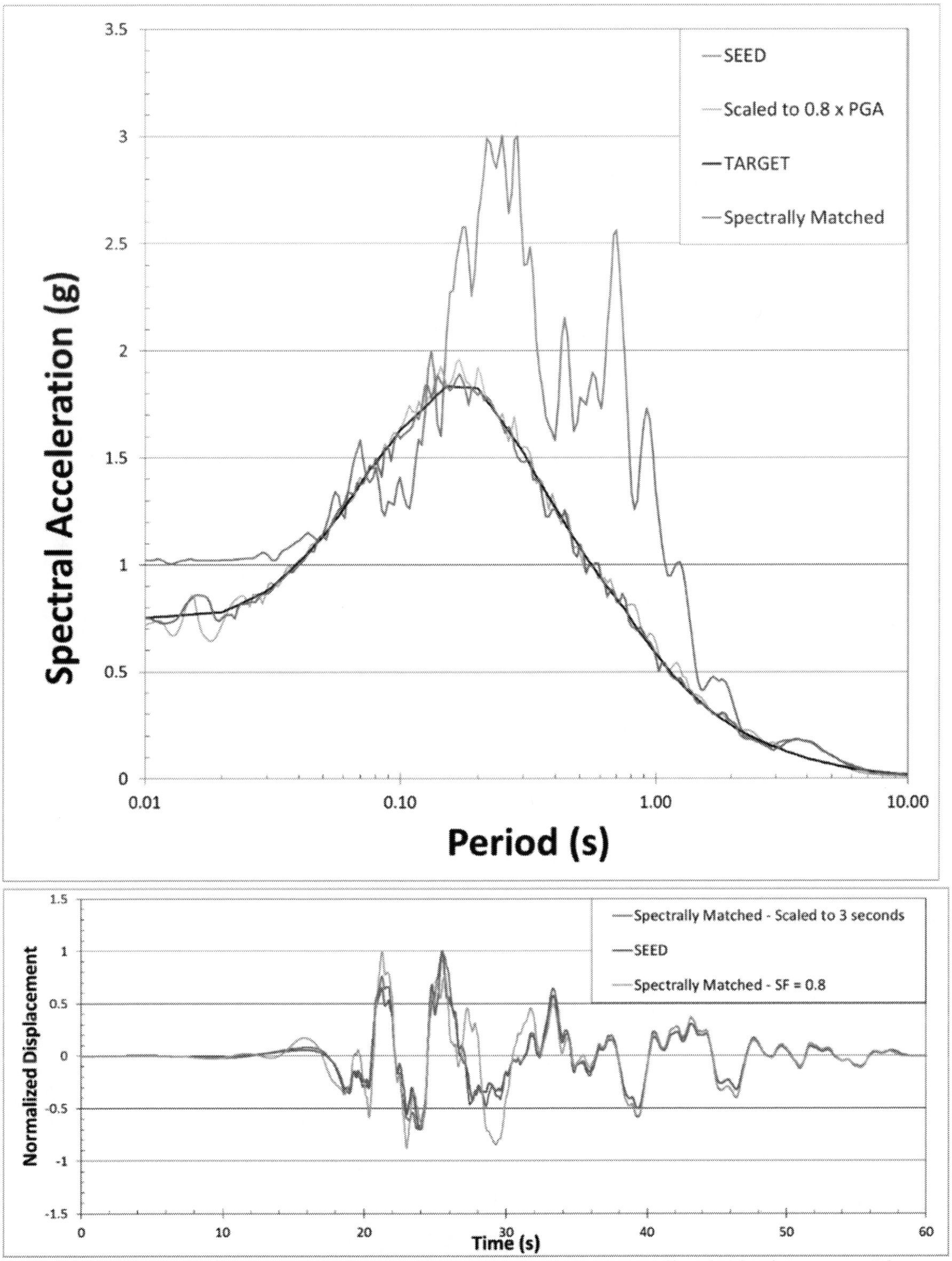

Figure 8. Chuetsu-oki, Japan Response Spectrum and Normalized Displacement Time History – Comparison of seed and spectrally matched motions with different initial scaling factors

In summary, there are tradeoffs between achieving a good spectral fit and displacement time history while maintaining acceptable record parameters. However, by choosing a different initial scale factor and not the standard practice of "scale to the PGA" prior to spectral matching, the engineer can control the resulting record parameters, such as Arias intensity and CAV. As more time history records become available (e.g. PEER NGA West2 database), the input record parameters should be considered when developing time histories for both scaled and spectrally matched motions.

Table 9. Record Parameters before and after Spectrally Matching

Event Title	Event ID	R_{RUP}	V_{S30} (m/s)	Station		PGA (g)	PGV(cm/s)	Arias Intensity (m/s)	CAV (g-sec)	Predominant Period (sec)	Predominant Freq. (Hz)	Significant Duration (s)		No. Cycles
												5%-95%	5%-75%	
Darfield, New Zealand (M7.0)	6927	7.11	263	LINC N67W	Seed	0.39	56.6	2.7	1.6	2.48	0.40	10.8	7.8	83
					Spectrally Matched (SF=2.1)[1]	0.75	67.6	7.0	2.7	0.73	1.4	13.2	9.4	41
Loma Prieta, Californi a (M6.93)	768	14.34	222	Gilroy Array #4 - 090	Seed	0.22	38.5	1.0	0.9	0.53	1.90	14.8	5.4	30
					Spectrally Matched (SF=2.4)[1]	0.68	72.2	4.9	2.0	0.32	3.08	15.7	7.0	33
Chuetsu-oki, Japan (M6.8)	4866	11.75	338	Kawanishi Izumozaki -NS	Seed	0.36	38.6	1.8	1.3	0.90	1.12	13.9	5.8	28
					Spectrally Matched (SF=1.7)[1]	0.72	52.7	6.2	2.3	0.33	3.07	12.9	5.7	32

Source: NGA West2 PEER Database

Note: SF = Scale Factor applied prior to spectral matching.

REFERENCES

Abrahamson, N.A., W.J. Silva, and R. Kamai (2014) *Summary of the ASK14 Ground-Motion Relation for Active Crustal Regions*. Earthquake Spectra. Volume 30, Issue3. August.

Boore, P.M., J.P. Stewart, E. Seyhan, and G.M. Atkinson (2014). *NGA-West2 Equations for Predicting PGA, PGV, and 5%-Damped PSA for Shallow Crustal Earthquakes*. Earthquake Spectra. Volume 30, Issue 3. August.

Campbell, K. and Bozorgnia, Y. (2010). A Ground Motion Prediction Equation for the Horizontal Component of Cumulative Absolute Velocity (CAV) Based on the PEER-NGA Strong Motion Database, Earthquake Spectra, Volume 26, No. 3, pages 635–650, August 2010

Campbell, K. and Bozorgnia, Y. (2012). A comparison of Ground Motion Prediction Equations for Arias Intensity and Cum ulative Absolute Velocity Developed Using a Consistent Database and Functional Form, Earthquake Spectra, Vol. 28 No. 3, 931-941.

Campbell, K. W., and Y. Bozorgnia, (2014). *NGA-West2 Ground Motion Model for the Average Horizontal Components of PGA, PGV, and 5%-Damped Linear Acceleration Response Spectra*. Earthquake Spectra. Volume 30, Issue 3. August

Chiou, B. S., and R.R. Youngs, (2014). *Update of the Chiou and Youngs NGA Model for the Average Horizontal Component of Peak Ground Motion and Response Spectra*. Earthquake Spectra. Volume 30, Issue 3. August.

Derouin, S.A., Klinger, R.E. (2014). Revise d Fault Characterization for Green Mountain Dam Seismic Hazard Analysis. T echnical Memorandum No. 85-8333000-2015-05-A, Seismotectonics and Geophysics Group, Bureau of Reclamation, Denver, Colorado, 57pp

Dobry, R., Idriss, I. M., & Ng, E. (1978) . Duration characteristics of horizontal components of strong-motion earthquake record s. Bulletin of the Seismological S ociety of America, 1487-1520.

Donahue, J. L., & Abraham son, N. A. (2013). Pacific Earthquake Engineering Research Center. Retrieved from PEER: http://peer.berkeley.edu/publications/peer_reports/reports_2013/webPEER-2013-14-Donahue.pdf

Idriss I. M., 2014. An NGA E mpirical Model for Estimating the Horizontal Spectral Values Generated By Shallow Crustal Eart hquakes. Earthquake Spectra. Volum e 30, Issue 3. August.

Kaul, M. K. (1978). Spectrum -consistent time-history generation, Journal. Engineering Mechanical. Div. 104, 781–788.

Kempton, J. J., & Stewart, J. R. (2006). Pr ediction Equations for Significant Duration of Earthquake Ground Motions Considering Site and Near Source Effects. Earthquake Spectra, 985-1013.

Lilhanand, K., & Tseng, W. S. (1987). Generation of synthetic time histories compatiable with multiple-damping response spectra. SMiRT-9, Lausanne, K2/10.

Lilhanand, K., & Tseng, W . S. (1988). Deve lopment and application of realistic earthquake time histories com patible with m ultiple damping response spectra. Ninth World Conference Earthquake Engineering Vol. II , (pp. 819-824). Tokyo, Japan

Pacific Earthquake E ngineering Research (PEER). (2014). Pacific Earthquake Engineering Research Center Database. Retrieved October 16, 2014

RSPMATCH (v.99, 1999)

Silva, W., Abrahamson, N., Toro, G., & Costantino, C. (1996). Description and Validation of the Stochastic Ground Motion Mo del. El Cerrito: Pacific E ngineering and Analysis.

Stokoe, K.H., Yuan, J., Lin, Y-C. and Jung, M.J. (2008). Field Data Summary: SASW Field Testing at Dillon Da m, Colorado. Geotechnical Report No. GR08-19. Geotechnical Engineering Center, University of Texas at Austin. Prepared for GEI Consultants, December 2008.

Travasarou, T., Bray, J. and Abraham son, N., 2003. Empirical Attenuation Relationship for, Earthquake Engineering & Structural Dynamics 32, 1133–1155.

SEISMIC DEFORMATION ANALYSES OF EMBANKMENT DAMS: A REVIEWER'S CHECKLIST

Ross W. Boulanger[1]
Michael H. Beaty[2]

ABSTRACT

Developing confidence in non-linear seismic deformation analyses of embankment dams, and the insights they may offer, is dependent on thorough internal or external reviews of the analysis procedures and results. Difficult geologic characterizations, selection and calibration of constitutive models, determination of material parameters, general limitations of numerical modeling, and development of input ground motions that appropriately reflect the seismic hazard – each of these aspects individually and collectively are important factors affecting the quality of the analysis results. Current analysis practices cover a wide range in approaches, which can make it difficult for dam owners, regulatory agencies, technical reviewers, and engineering supervisors to assess the appropriateness of any particular analysis. The nature of this task does not lend itself to standardization, but improved documentation practices are an important step toward facilitating thorough reviews and improving standards of practice. The purpose of this paper is to provide a checklist of questions that a reviewer of an NDA study may reasonably ask as part of any internal or external review process.

INTRODUCTION

Non-linear deformation analyses (NDAs) to evaluate seismic performance of embankment dams have become increasingly common in engineering practice (e.g., Beaty and Perlea 2011). NDA studies, while routinely performed for large or high-risk embankment dams, have also proven valuable for a wide range of other embankment dam and seismic engineering projects. NDAs can provide insights that simpler analysis methods cannot, such as the patterns of deformations, influence of progressive failure mechanisms, interactions between embankments and embedded structures, dependence on ground motion characteristics, or relative effectiveness of remediation options. The insights afforded by NDAs should, in principle, lead to increased confidence in decisions regarding risk assessments or economical design of remedial measures.

The increased use of NDAs has, however, been accompanied by concerns regarding the repeatability of results. There are examples where different engineers have analyzed the same embankment dam, with apparently similar representations of the site characterization and seismic loading conditions, and produced results that have ranged from being in reasonable agreement to being so dissimilar they would lead to different decisions on the nature of any remedial actions. Evaluating these differences is challenging due to the complexity of the analyses as well as the somewhat unpredictable

[1] Professor, Department of Civil & Environmental Engineering, University of California, Davis, CA 95616, rwboulanger@ucdavis.edu
[2] Principal Engineer, Beaty Engineering LLC, Beaverton, OR, 97007, mhb@beatyeng.com

paths that each analysis may follow. The ability to resolve or understand the reasons for these differences in NDA results is important for establishing confidence in these types of analyses (e.g., Boulanger et al. 2015).

Reviewers and regulators can have difficulties evaluating the suitability of a particular analysis because of wide variations in the available analysis methods, the procedural details of calibrating the constitutive models and executing the analysis, and the level of documentation provided in engineering reports. Increasing the confidence that owners, regulators, and engineers can have in the standard practice and performance of NDA studies will require improved guidance (or protocols) for use of these analysis methods and higher standards for documentation of the methods and results. These improvements will facilitate the necessary internal or external review process and can lead to improved insights and benefits for all involved.

Other factors affecting the quality and value of an NDA study include the organization of the team (necessary range of skills; effective communications; review process) and the reasonableness of the budget and schedule. The effort spent in performing a high quality NDA study can add value over the long-term for dam owners by, for example, reducing undue levels of conservatism and avoiding schedule delays that could develop if the NDA results are later questioned by external reviewers or regulatory agencies. Improved guidance on NDA documentation requirements or expectations will accordingly help facilitate better definition of budgets and schedules for NDA studies.

The purpose of this paper is to provide a checklist of questions that a reviewer of an NDA study should consider asking as part of any internal or external review process. The questions posed herein are not exhaustive, but rather represent the authors' perspectives on many of the details that can significantly affect NDA results. Not all of the aspects presented below will be relevant or significant to all projects, but most should be considered or addressed in engineering analyses and reports representing the highest levels of practice.

The questions are grouped into twelve categories, spanning from numerical details to the judgments drawn from the NDA results. The technical issues behind each question are not discussed, as the scope of issues covered in an NDA study must draw from a large body of technical literature. Similarly, opinions on the best technical approaches are not provided because of length restrictions for providing context on each issue. The focus is instead on suggesting what should be addressed in reports summarizing NDA studies, with the aim that the first step toward advancing the standard of practice is improving the transparency with which actual practices are documented. It is the authors' hope that the checklist provided herein will help both consumers and producers of NDA studies for critical facilities.

REVIEW QUESTIONS

Questions that should be considered when reviewing an NDA study are presented in twelve categories. The questions are intended to encompass a range of issues that can significantly affect NDA results, but they cannot be viewed as exhaustive because any given project or analysis procedure can raise other questions of equal importance. Instead, it is hoped that these questions illustrate the type and nature of questions that should be addressed in engineering studies and reports at the highest levels of practice.

The engineering effort involved in addressing these questions is not large for a well-executed analysis study given the increased value it provides. In most cases, the answers involve brief statements documenting the basis of the work that was done and transparent presentation of intermediate results which were likely already available. Much of the information can be easily obtained if the need for transparent documentation is anticipated at the beginning of the analysis process. The inclusion of such details, however, can determine if the consumers of the analyses can independently evaluate the results or exercise their own judgment regarding the insights they provide.

The process of addressing these questions, like the process of an NDA study, is likely to be iterative. NDA studies frequently benefit from a cycle of updates or refinements to the input parameters based on the insights gained as the study progresses. For example, the results of the initial analyses may indicate that certain parameters have a greater influence on the computed response than originally anticipated, such that additional effort to refine the estimates of those parameters is warranted. In fact, addressing the following questions as the NDA study progresses can help identify when such a cycle of refinement may or may not be warranted. Thus, the order of the following questions should not be viewed as implying a linear process for either the NDA study or its review.

Aspect 1: Seismic Failure Modes and Important Behaviors

Have the potential failure modes associated with seismic loading been summarized or discussed?

- Performance objectives for life safety and operations should be summarized to provide context for the NDA study.
- For NDA studies in support of risk analyses, the failure modes addressed in the analysis should be identified.
- The potential for liquefaction, cyclic softening, or strain softening in all materials should be discussed.
- All loading conditions concurrent with seismic loading, including reservoir levels and phreatic surfaces, should be summarized.
- The existence of conditions that might accentuate embankment cracking should be discussed; e.g., steep abutments, variations in foundation stiffness, or presence of embedded structures.
- The potential effects of embankment deformations on embedded or appurtenant structures should be discussed.
- The dependence of potential failure modes on embankment deformations should be summarized; e.g., overtopping, seepage erosion through cracks, piping in the dam and/or foundation, seepage erosion into or around damaged or deformed

embedded structures, increased pore pressures following damage to drainage elements, and possible post-earthquake operating restrictions.

Have the engineering behaviors important to those failure modes been adequately identified? Are the limitations in the NDA model's ability to simulate the key engineering behaviors acknowledged?

- The choice of a simpler versus more complex analysis model should be justified based on the degree to which the key behaviors can be modeled and the degree to which a more refined analysis may or may not affect final decisions regarding risk mitigation efforts or remedial works.
- Limitations in the numerical model's ability to simulate certain behaviors should be acknowledged and how they may affect the performance assessment should be discussed. For example, current two-dimensional models cannot address most cracking mechanisms and so crack depths must be estimated by other means.

Aspect 2: Validation Record for the Numerical Modeling Procedure

Have the constitutive models and numerical modeling procedures been fully described and validated?

- The record of validation of the numerical modeling approach against case histories or physical models of embankments, levees, or similar structures should be briefly summarized or referenced. Validation results depend on the numerical procedures, constitutive models, and the approach used to determine (or calibrate) all input parameters.
- Any record of comparisons between the adopted numerical modeling approach and other analysis approaches should be noted. For example, the first time a user applies a new constitutive model or numerical approach, the results should be compared to those obtained using more established procedures.
- Details of how the numerical modeling approach was applied in the referenced validation or comparison studies should be summarized, if available. The extent to which the same details were followed in the application NDA should be discussed.

Aspect 3: Site Characterization Basis for Material Parameters

Has the basis for all material parameters been described and related to the site characterization?

- The input parameters for all components of the numerical model should be listed in tables, including all geotechnical materials, structural elements, and any interface elements.
- The basis for all soil and rock constitutive model parameters should be discussed with reference to the site characterization data. A geologically based description

of the site stratigraphy should be provided. The site characterization report(s) should be included as part of the NDA review process.

- Either the site characterization report or NDA report should provide geotechnical cross-sections that include plots of the in-situ test data (i.e., fence diagrams). The sections should include a geologically based interpretation of the stratigraphy with identification of subzones having common properties.
- Key summaries of site data should be provided to support selection of material subzones and their respective material properties; e.g., profiles of shear wave velocity and penetration resistances binned in various ways to evaluate potential spatial trends and distributions.
- The basis for selecting representative material properties for each subzone should be documented. For example, presentation of cumulative distributions for penetration resistances in each model subzone and the selection of a representative percentile for analysis purposes.

Have the key material properties and their uncertainties been identified and adequately addressed?

- Key parameters affecting the response, and the relative magnitude of these effects, should be identified through sensitivity studies.
- The adequacy of the site characterization for support of the analysis models should be discussed. For example, the potential for additional explorations to affect final decisions should be addressed based on the geologic setting, available site characterization data, the likelihood of additional explorations to reduce uncertainty in the input parameters, and the sensitivity of the analysis results to reasonable variations in the estimated material properties.

Aspect 4: Calibration and Evaluation of the Constitutive Model

Are the constitutive models able to approximate the important soil behaviors for both the static and dynamic phases of the analysis? Have the strengths and limitations of the constitutive model been assessed and documented?

- Comparisons of constitutive model element behavior to laboratory test results and/or published relationships should be provided.
- Comparisons should include relevant element behaviors such as:
  - Monotonic drained and undrained shearing.
  - Shear modulus reduction and damping curves.
  - Cyclic resistance ratio versus number of uniform loading cycles for several overburden stresses and initial static shear stress ratios spanning the ranges of interest.
  - Cyclic undrained stress-strain responses for representative conditions, specifically including cases with and without initial static shear stresses.
- Constitutive model behaviors should be documented in the report, a manual, or some other accessible document (e.g., Beaty and Byrne 2011, Boulanger and

Ziotopoulou 2015). The strengths and limitations of the constitutive model as applied to the current analysis should be identified.

Aspect 5: Numerical Modeling Procedures

Have the modeling approaches for the static and dynamic phases of the analyses been documented and are the approaches reasonable?

- The documentation of the modeling procedures should include a description, assessment, and/or justification of the following items:
    - The boundary conditions for the lateral edges and base of the model for both static and dynamic loading phases.
    - The inclusion or exclusion of transient seepage effects during dynamic loading.
    - The inclusion or exclusion of large deformation effects during dynamic loading.
    - The viscous damping parameters (e.g., Rayleigh damping type, amount, frequency, and basis for frequency).
    - The appropriateness of the element size with respect to the geometry, materials, and loading (e.g., the ability of the element mesh to transmit waves in the frequency range of concern or to adequately represent gradients of stress, strain and pore pressure).
    - The porosity of all materials and the bulk modulus of the pore fluid. These should be consistent with those used in the single element validation/calibration studies.
    - Any special analysis features that were used (e.g., multi-stepping in FLAC, automatic or manual remeshing, modification of ill-behaved surface elements) or significant non-physical adjustments made to improve solution time.
- Static modeling procedures are often simplified but must be sufficient to create a reasonable initial stress state for the dynamic analysis. The dam construction and reservoir filling sequence should be included as required by the unique nature of the project and modeling procedure.
- The method of applying the input ground motions should be consistent with the modeling objectives and the selected time series. If a compliant base is used for both the x and y directions, the efforts used to control base bending should be identified.
- The effects of large geometry changes should be discussed, particularly if the analyses were performed using a small-strain formulation.
- The importance of the drainage assumptions often depends on hydraulic conductivity contrasts and layering. The inclusion of transient seepage during the dynamic analysis requires extra attention to the site characterization.
- The assumptions regarding strain-rate effects and the potential for softening in clay-like soils should be described and justified.

- When the estimated strain magnitude is used to control the element response (e.g., for strain softening materials), the effect of element size on the estimated strain magnitudes should be assessed.
- The estimated residual strength for any liquefied zone should be directly considered in the analysis, such as through a post-shaking stability analysis. The criteria used to identify elements that have liquefied during the analysis should be identified (e.g., peak excess pore pressure ratio and/or shear strain). The residual strength in sand-like zones should be capped to the drained strength in each element. Any special steps needed to perform a post-shaking stability analysis after completion of the dynamic phase of the analysis should be described.
- If ground improvement or other remedial measures are included in the analysis, the modeling of these features should be described in detail and any limitations to the analysis identified (e.g., approximation of 3D effects, assumptions regarding densification and/or drainage, modeling of potential bending effects or strain incompatibilities).
- If discrete structural elements are present, the method of structural modeling and any resultant approximations should be discussed. For structural or ground improvement elements that are not continuous in the out-of-plane direction, assumptions regarding the estimate of drag loads on those elements should be presented.
- The analysis should be performed in reasonable agreement with the validation and calibration studies, and this should be apparent from the supplied documentation. Any discrepancies should be identified and assessed.

Aspect 6: Input Ground Motions

Are the selected ground motions consistent with both the seismic hazard evaluation and the numerical model?

- The primary earthquake scenarios contributing to the seismic hazard should be identified and addressed by the selected time series. Significant parameters and characteristics of the selected earthquake recordings should be described, such as source earthquake and magnitude, type of fault rupture, distance to fault rupture, site conditions, peak acceleration, peak velocity, Arias intensity, and duration.
- In most cases, the V_{S30} of the target response spectrum should be similar to the corresponding V_{S30} of the rock underlying the dam. Any discrepancy between the response spectrum and site condition should be addressed in the development of the input ground motions.
- The number of input time series evaluated in the analysis should be sufficient to allow a reasonable estimate of the average deformation response. This is often considered to require three or more spectrally matched motions, or seven or more motions scaled to the frequency range of interest, for each earthquake scenario. If three spectrally matched motions produce significantly different estimates of behavior, use of additional records is often needed.
- If input motions have been spectrally matched, the method of spectral modification should be referenced and a comparison plot of the target and

individual response spectra should be provided. Plots of acceleration and velocity histories before and after matching should also be provided, particularly for ground motions that include directivity effects.

- The method of applying the input ground motions to the model should be consistent with the modeling objectives and the selected time series. If the input motions have been deconvolved to the base of the model, the method for deconvolution should be described and the spectra compared to the spectra for the outcrop motions. If the input motions are applied to a fixed base without deconvolution, the potential impact of the additional conservatism should be assessed. The inclusion or exclusion of vertical input motions should be identified.
- The input ground motions should be baseline corrected. Alternatively, if there is a modest baseline drift, displacement results can be reported as displacements relative to the base.

Aspect 7: Initial Static Stress Conditions

Are the estimates of the initial static stress state reasonable?

- The static stress state (e.g., σ'_{yy}, σ'_{xx}, τ_{xy}), pore pressure (u), coefficient of lateral earth pressure K_o (where $K_o = \sigma'_{xx}/\sigma'_{yy}$), and static shear stress ratio α (where $\alpha = \tau_{xy}/\sigma'_{yy}$) should be documented with contour plots.
- Equipotential plots for hydraulic head should also be provided as an aid in evaluating the steady state seepage solution. The pore pressure head distribution should be compared to available piezometric measurements if available.
- The initial stress state, as documented on the contour plots, should be consistent with the anticipated variations across material zones and beneath slopes (e.g., no unreasonable arching of stress between zones, no areas of excessively high K_o, no unreasonable artifacts of modeling procedure, no random fluctuations). Simplified approaches for static analysis have been known to introduce local stress concentrations that are not realistic.
- If the initial pore pressure conditions were imposed assuming a ground water table and hydrostatic conditions, the potential impacts of this approximation on the dynamic response should be discussed.
- Contour plots of total unit weight should be provided to ensure the proper mass is included both above and below the phreatic surface. If elements above the phreatic surface fully desaturate during a steady-state seepage analysis, some analysis programs will compute unit weights for these elements that are less than typical moist unit weights.

Aspect 8: Dynamic Response

Is the estimated dynamic response reasonable, and has it been presented in sufficient detail to allow for this assessment?

- The following information should be provided and assessed for at least one input motion:
  - Acceleration time series for a few key points (e.g., crest, downstream face, upstream face, toes, top of rock, base of model).
  - Acceleration response spectra for the above points.
  - Displacement time series for the above points.
  - Spectral amplification ratios for key points, such as the crest relative to the bedrock outcrop motion. Alternatively, Fourier amplification spectra can be used to evaluate predominant response frequencies.
  - Deformed shapes with contours of shear strains, excess pore water pressure ratios, and displacements. Deformed shapes with displacement vectors can also be valuable.
  - For cases with undrained strength loss (e.g., residual strength or remolded strength), contour plots of the final strength distribution.
  - Displacement, moment, and shear envelopes for embedded beam-type structures.
  - Stress-paths and stress-strain response for representative elements.
- Response summaries should be provided for each set of modeling conditions and suite of ground motions (e.g., peak crest acceleration and displacements, ground surface displacement profile, peak response of embedded structural elements).
- The full suite of ground motions should be evaluated using best-estimate input parameters.
- Significant differences in the character of the estimated response over the range of input motions and modeling conditions should be identified and assessed.
- The magnitude of base bending should be described if a compliant base is used for both the x and y directions.
- If remedial measures such as ground remediation are included in the analysis, the response of these measures as well as their impact to the dam response should be clearly described.
- Comparisons of the estimated responses to field observations from past earthquakes and, where possible, to the results from other applicable analysis techniques (e.g., two-dimensional equivalent linear response analyses) should be provided.

Aspect 9: Post-Shaking Deformations

Is the post-dynamic stability analysis reasonable, and has it been sufficiently documented to permit this assessment?

- The following information should be provided and assessed for the same input ground motion(s), nodal points, and elements detailed in Aspect 8:

    - Deformed shapes with contours of shear strains, excess pore water pressure ratios, and displacements.
    - Stress-path and stress-strain response for representative elements.
    - Displacement time series continued for the same key points examined during shaking.
- Pockets of non-liquefied zones that may stabilize the slope should be identified. These pockets may indicate a potential brittleness in the response (i.e., small changes in the predicted response could induce large increases in displacements), or they may result from limitations in the constitutive model for addressing the effects of initial static shear stress. The potential impact of the zones, and the reliability of the estimates, should be assessed.
- Zones showing dilative tendencies with negative pore pressures should be identified and their influence on the post-earthquake stability should be discussed and assessed.
- Analyses that were unable to reach static equilibrium due to excessive grid distortion should be noted.

Aspect 10: Parametric Analyses

Have sufficient parametric analyses been performed to identify the behaviors and parameters of most importance to the computed responses?

- The parameters evaluated in the sensitivity studies should be those expected to have a stronger effect on computed responses; the selection of those parameters should be guided by, in part, by an understanding of the initial analysis results.
- Direct comparisons of results from the parametric analyses and those from the best-estimate input parameters should be provided. An assessment of the relative importance of each parameter should be made.
- The set of parametric analyses should include some or all of the following items depending on the characteristics of the dam and the initial response estimates:
    - Sensitivity to properties of the liquefiable materials: e.g., CRR values, shear moduli (velocity profiles), peak strengths, residual strengths.
    - Sensitivity to properties of the non-liquefiable materials: e.g., shear moduli (velocity profiles), drained and undrained shear strengths, post-cyclic strengths.
    - Sensitivity to procedural issues such as the liquefaction "triggering" criteria used to decide if residual shear strengths are imposed.
    - Sensitivity to modest changes in any remediation elements (e.g., depth and/or length of soil mix grids).
    - Sensitivity to any other input parameters having the potential of significantly affecting the dam response.
    - The polarity (or sign) of the input ground motion, and the effect of the vertical component of the input motion.
- A simple fragility evaluation should generally be performed. The relationship between dam displacement and the amplitude of the input motion can be assessed using best estimate properties. The objective is to identify if a modest increase in

loading above the selected design level produces a significant increase in dam response.

Aspect 11: Uncertainties and Limitations

Have the uncertainties and limitations in the numerical model been identified and addressed?

- Mechanisms such as cracking, particularly in the direction transverse to the dam axis, are outside the capabilities of most numerical models. The report should recognize and account for this limitation. A discussion should be provided on locations and materials susceptible to cracking.
- An assessment of apparent bias towards an overly optimistic or pessimistic selection of the best-estimate inputs at each step of the process should be provided.
- The potential significance of 3-dimensional effects should be discussed.
- The impact of simplifications or uncertainties in the site characterization should be discussed.
- As possible, the analysis results should be compared to relevant responses from case histories (e.g., observed deformations, peak crest accelerations).
- The range of estimate responses considering the full suite of ground motions should be assessed and best-estimate response values provided.
- Undefined uncertainties in the model estimates should be considered in the discussions of dam performance and potential risk. It has been common practice for high quality analysis programs to consider best estimates of displacement to be between 0.5 and 2.0 times the best estimate of the analysis. A factor larger than 2.0 may need to be considered if the analysis shows the displacements are unduly sensitive to one or more input parameters.

Aspect 12: Reasonableness of Conclusions

Has the case been made, in plain language, for the conclusions and recommendations drawn from the results of deformation analyses?

- The case for the overall conclusions or recommendations should be concisely summarized with reference to the essential aspects of the geology, material characteristics, dam configuration, seismic hazard, estimated response, and key sources of uncertainty.
- The case should include discussion of whether additional information (e.g., exploration or analysis) could reasonably be expected to change the outcomes. The parameters found to have the greatest impact on the estimated performance of the dam should be identified. An evaluation of reasonable variations in these values on the estimated response should be included.

CONCLUSION

Nonlinear deformation analyses (NDAs) are a valuable tool for evaluating soil and soil-structure systems affected by liquefaction, cyclic softening, and/or large seismic loads. A high-quality NDA study can provide more realistic assessments of deformation mechanisms and potential failure modes than are possible with simplified equivalent-static or equivalent linear analysis methods.

A suggested checklist of questions was provided for participants and reviewers of NDA studies. The questions are not exhaustive but represent the authors' perspectives on many of the details that can significantly affect NDA results for embankment dams. These details should be addressed or considered in engineering analyses and reports representing the highest levels of practice. An additional objective was to foster discussion on items that should be addressed in engineering reports summarizing NDA studies. It is hoped that advancing the standard of practice for documentation will help advance the standard of practice for NDA studies and ultimately benefit both consumers and producers of NDA studies.

ACKNOWLEDGMENTS

The opinions expressed in this paper draw from the practices of numerous colleagues whom the authors have been fortunate to work or interact with over many years. The authors also appreciate the valuable comments and suggestions provided by Dr. Rambod Hadidi and Mr. Andrew Dinsick during their review of the draft paper.

REFERENCES

Beaty, M. H., and Byrne, P. M. (2011). UBCSAND constitutive model: version 904aR. Documentation report: UBCSAND constitutive model on Itasca UDM Web Site, February.

Beaty, M.H. and Perlea, V.G. (2011). Several Observations on Advanced Analyses with Liquefiable Materials. In Proc., 31st USSD Annual Meeting and Conference, United States Society on Dams, San Diego, April 11 – 15, 2011. (pp. 1369-1397).

Boulanger, R. W., Montgomery, J., and Ziotopoulou, K. (2015). "Nonlinear deformation analyses of liquefaction effects on embankment dams." Perspectives on Earthquake Geotechnical Engineering, A. Ansal and M. Sakr, eds., Geotechnical, Geological and Earthquake Engineering 37, 247-283, Springer, DOI 10.1007/978-3-319-10786-8_10.

Boulanger, R. W., and Ziotopoulou, K. (2015). "PM4Sand (Version 3): A sand plasticity model for earthquake engineering applications." Report No. UCD/CGM-15/01, Center for Geotechnical Modeling, Department of Civil and Environmental Engineering, University of California, Davis, CA, 112 pp.

LESSONS FROM EARTHQUAKE LAKE IN PREPARING FOR SEISMICALLY-INDUCED LANDSLIDES

Roger L. Kay, P.E.[1]

ABSTRACT

A magnitude 7.3 earthquake, the strongest in recorded Montana state history, struck the Madison River valley just west of Yellowstone National Park on August 17, 1959, at 11:38 p.m. The earthquake triggered a 40-million ton landslide of soil and rock, resulting in 28 fatalities and completely blocking the Madison River to heights ranging from 200 to 400 feet and ¾-mile in width from upstream to downstream. Hebgen Dam, which impounds approximately 325,000 acre-feet of water and is located approximately 6 miles upstream of the landslide, was overtopped by multiple seiche waves following the earthquake and was extensively damaged due to extreme subsidence in the vicinity of the dam. At the request of the Governor of Montana, the U.S. Army Corps of Engineers was directed to assist in stabilizing the landslide so that Hebgen Dam could be drained for inspection and repair. Over the next 2 months, the crest of the landslide was lowered and a channel armored down the face of the landslide to reduce the risk of catastrophic failure of the landslide mass. The slide mass with its engineered spillway remain in place today, with a permanent lake behind it named "Earthquake Lake."

Much of the western United States is susceptible to seismic events. A large seismic event in this area could result in landslides blocking river valleys as well as structural damage to multiple dams. Dam owners in this area should take into account the very real possibility that emergency drawdowns of seismically damaged dams may not be rapid due to downstream obstructions. Lessons learned from the Earthquake Lake response may offer emergency planners and dam owners guidance in how to avoid catastrophic failures of future landslides blocking river valleys.

SETTING

The Madison River originates in northwest Wyoming, within the boundaries of Yellowstone National Park. The Madison River flows northwest and enters Hebgen Lake, which is formed by Hebgen Dam, constructed at the entrance to Madison Canyon. Hebgen Dam is an 88-foot high earth-fill structure constructed by Montana Power Company in 1914 impounding approximately 386,000 acre-feet of water. Hebgen Dam is utilized to regulate flow into the Madison and Missouri hydropower dams operated by NorthWestern Energy (http://www.northwesternenergy.com). From Hebgen Dam, the river flows in a southwesterly direction through the Madison Canyon into the Missouri Flats region, where it eventually joins the Gallatin and Jefferson Rivers near Three Forks, Montana forming the Missouri River. All of the area along the Madison River from Yellowstone National Park to the mouth of Madison Canyon resides within Gallatin National Forest.

---

[1] Hydraulic Engineer, Missouri River Water Management, U.S. Army Corps of Engineers, 1616 Capitol Avenue, Omaha, NE 68102, roger.l.kay@usace.army.mil

The Madison Canyon is a deep narrow gorge cut through the Madison mountain range and is composed of pre-Cambrian rocks, primarily dolomite, gneiss, schist, amphibolites and quartzite. The Madison Range fault extends from Big Springs, Idaho northward about 70 miles to near Ennis, Montana along the western edge of the Madison Range; this fault has been active as recently as a few hundred years ago, with a scarp 10-30 feet high extending from near Ennis to several miles south of the Madison Canyon mouth, with the downthrust on the valley side of the fault. Numerous faults extend into the blocks on either side of this fault, including the Hebgen and Red Canyon faults, both of which are near the epicenter of the 1959 earthquake (Alden, 1953; Hadley, 1964; Ross and Nelson, 1964; Witkind, Hadley and Nelson, 1964) as shown in **Figure 1**.

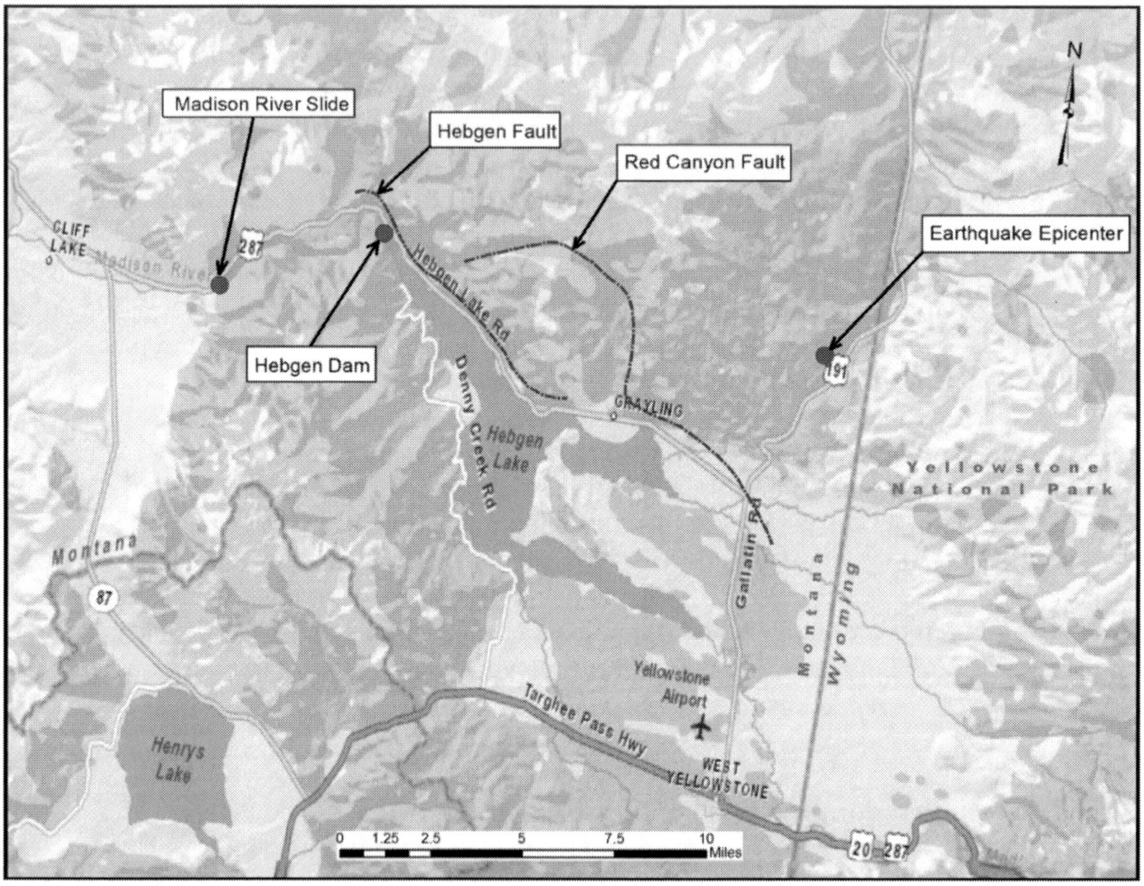

Figure 1. Vicinity Map.

EVENT

A magnitude 7.3 earthquake, the strongest in recorded Montana state history, struck the Madison River valley just west of Yellowstone National Park on August 17, 1959, at 11:38 p.m. The earthquake triggered a 40-million ton landslide of soil and rock from the south wall of the Madison Canyon, resulting in 28 fatalities and completely blocking the Madison River near the mouth of the Madison Canyon to heights ranging from 200 to 400 feet above the valley floor and ¾-mile in width from upstream to downstream. Hebgen Dam, which impounds more than 300,000 acre-feet of water and is located approximately 6 miles upstream of the landslide, was overtopped by multiple seiche waves for more than 12 hours following the earthquake and was extensively damaged due to extreme subsidence (up to 22 feet) in the vicinity of the dam. A number of other surficial landslides occurred throughout the region. Subsidence associated with the earthquake was widespread, occurring over an area approximately 15 miles north-south and nearly twice as long east-west; approximately 200 square miles had subsidence in excess of one foot. Numerous aftershocks were observed, with at least two per day for more than two months following the initial quake, with the strongest aftershocks in the days following the quake measuring more than magnitude 6 (USACE, 1960a; Hadley, 1964).

RESPONSE

The large slide at the mouth of Madison Canyon, coupled with the extreme subsidence along the northern shore of Hebgen Lake, destroyed large portions of Highway 287, the only means of egress from Madison Canyon, trapping more than 200 people (along with their automobiles, trailers and other camping equipment) between the slide and Hebgen Dam. Initial response from the U.S. Forest Service was to organize rescue teams, as well as to bulldoze emergency trails around the breaks in Highway 287 along Hebgen Lake; by the end of the first day, 167 people had been rescued from the canyon. The Montana Civil Defense Director contacted the Garrison District (USACE) office via phone on the morning of August 18, with the District dispatching two observers via small plane to fly over the affected area; however, due to damage to communications and roads and the initial emphasis on rescue efforts, only fragmentary communication occurred between the state and USACE officials. On August 20, MG Barney, Missouri River Division Engineer, and staff from Omaha flew to the site, and subsequently met with Montana Civil Defense and representatives of other agencies the following day to discuss possible emergency measures at the site of the slide. Following the meeting, MG Barney engaged Woodward, Clyde, Sherard, and Associates to make an appraisal of the situation and the stability of both Hebgen Dam and the slide. Montana's governor, J. Hugo Aronson officially requested assistance from the Corps through the Secretary of Defense on August 22 to investigate and remove the serious threat to life and property, which was beyond the capabilities of local and state governments. This initiated the emergency response work by USACE, which can be divided into three phases, described briefly below (USACE, 1960a).

Phase 1, Spillway Construction, August 22 through September 10

Following receipt of Governor Aronson's request, MG Barney directed the Garrison District Engineer, LTC Hogrefe, to proceed with the investigation and necessary emergency measures. A project office was established in West Yellowstone, Montana that same day by LTC Hogrefe. The initial report from Woodward, Clyde, Sherard, and Associates was received that same day; they concluded that Hebgen Dam, with approximately 330,000 acre-feet of water be lowered to reduce the risk of failure, due to what they deemed the unsafe condition of the dam. They also advised that the slide mass had a favorable distribution of materials and that construction of a spillway through the lowest part of the mass be commenced, utilizing bulldozers (see **Figure 2** for initial construction of this spillway). The consultants also recommended an immediate survey of the slide mass and test borings be made at multiple locations to investigate the possibility of piping under the slide mass. The following day, surveys of the slide mass began, and construction of access roads to the top of the slide began with two bulldozers. A preliminary spillway alignment was laid out on August 23, with a level run completed by August 25, tying all benchmarks between the canyon mouth and Hebgen Dam to a common datum. Drawdown of Hebgen Dam to an elevation below spillway crest also commenced on August 23, with discharges initially increased from 1,000 cfs to 2,200 cfs, followed by a subsequent increase to 2,850 cfs on August 26.

Figure 2. Looking Downstream at Slide and Initial Spillway Construction, September 3, 1959; Slide Scar on Left (USACE, 1960a).

After several days, and a more careful inspection of Hebgen Dam, it was determined that Hebgen Dam was not likely to fail, although it was still desired to drain Hebgen Lake as much as possible to facilitate a thorough inspection of all damage. It was also desired to have the drawdown complete prior to winter, as formation of frazil and anchor ice was a well-known problem along the Madison River in cold weather. Preliminary computations based on available topographic information and an assumed inflow of 2,850 cfs showed that a water surface elevation of 6450 ft would be reached about September 4; the elevation of 6450 ft was subsequently selected as the spillway crest elevation through the slide mass. LTC Hogrefe ordered a number of technical studies to check the preliminary results of Woodward, Clyde, Sherard and Associates, as well as to obtain more detailed information than was possible for the preliminary investigation. These technical studies were later published by USACE (1960b).

The Montana Chapter of Associated General Contractors (AGC) was consulted on August 23 to determine availability of construction equipment. Construction contract negotiations were begun on August 25, and by August 27, 23 bulldozers were on-site moving rock for spillway construction. At the peak of construction activity in mid-September, construction equipment included 27 bulldozers, 5 shovels, 29 end-dump trucks, 3 end-loaders, 1 patrol grader, 2 truck cranes, 1 dragline, 1 water truck, 5 light plants, 1 tractor-mounted drill, and one compressor, drill and blasting outfit. A total of 190 individuals were required to operate and maintain this equipment over 2 daily shifts of 11½ hours each; the ½-hour between shifts was reserved for equipment maintenance and repair.

The initial spillway construction consisted of a spillway channel 250 feet wide, with a crest elevation at 6450 ft extending longitudinally approximately 900 feet, followed by a slope at 5% for 600 feet, and then transitioning to a 10% slope for a distance of 1,900 feet to the toe of the slide. The 5 and 10% slopes were to be armored with the best material that could be salvaged from the slide mass, with a depth of seven to ten feet of quartzite. In order to allow more time for spillway construction, releases from Hebgen Dam were gradually reduced from 2,850 cfs to 2,100 cfs on September 4, followed by a further decrease to 1,300 cfs. By September 6, the water surface upstream of the slide had reached elevation 6539, so releases from Hebgen Dam were further reduced to between 700 and 1,000 cfs, matching inflow to Hebgen Dam. Even with the reduced flows, the lower 1,200 feet of the proposed spillway construction could not be completed prior to commencement of spillway flow. Based on observations, it was determined that the lower 1,200 feet consisted of very large rock and was the least likely to be unstable; therefore, the lower 1,200 feet of the spillway channel were left in its original condition at a slope of approximately 14%. Water reached the spillway crest on the morning of September 10, although a continuous sheet of water over the spillway crest and down the channel was not achieved until later in the afternoon. **Figure 3** shows a comparison of profiles through the slide mass along the spillway alignment at various times of construction and after construction.

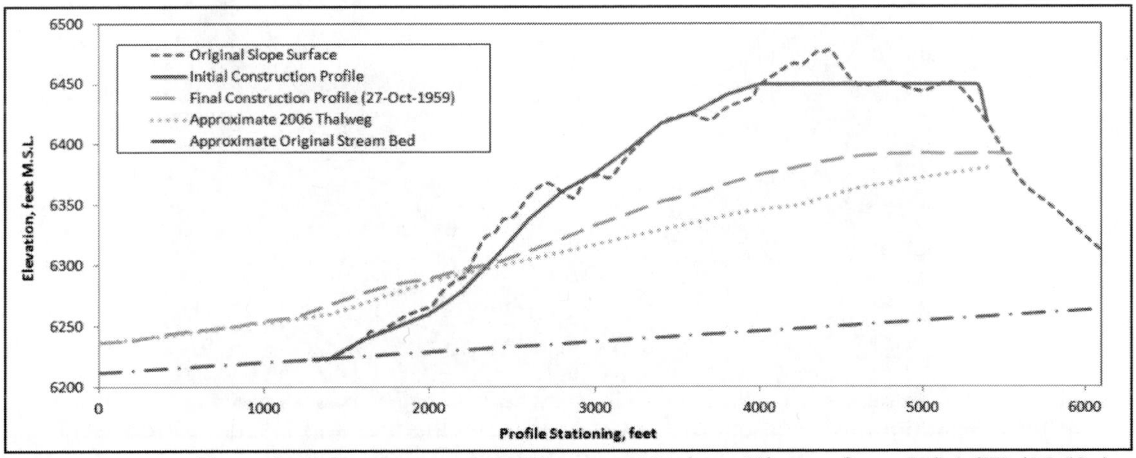

Figure 3. Profiles of Slide Mass and Spillway at Various Times, from USACE (1960a) and Chase and McCarthy (2012).

Due to the unusual character of the work, and prior to initiation of spillway flow, a board of consultants was appointed to review the work accomplished and to advise on any further investigations or additional remediation. The board was comprised of: Dr. Arthur Casagrande, Dr. L.G. Straub, Mr. E.B. Burwell, and Mr. L.C. Steele. Although the board was unable to formally meet until late October, individual consultants were able to visit the site at various times to advise on plans and procedures (USACE, 1960a).

Phase 2, Maintenance of Initial Spillway, September 10 to September 25

Following commencement of spillway flow, the spillway channel was to be monitored closely with flows from Hebgen Dam gradually increased until stability of the spillway channel could be ascertained. However, the finer slide materials were rapidly eroded, particularly in the steeper downstream reach, quickly resulting in formation of a deep gully that began advancing upstream (see Figure **4**). In an effort to arrest this erosion, large quantities of quartzite and dolomite were taken from the ridge along the northern half of the slide mass and placed at critical locations along the spillway channel. The largest rocks available were used for this effort; many individual rocks were larger than could be hauled by truckload, so these larger rocks were pushed downslope by a bulldozer (or multiple bulldozers). The largest rocks had to be drilled and blasted with dynamite before they could be moved. Some individual rocks placed in the channel were estimated to be as large as 100 tons apiece; in order to move these large rocks, special steel rock sleds were designed and constructed on-site, with up to 3 tractors required to pull the loaded rock sled. These large rocks were used to form several rock drop structures, which helped to reduce the effective slope between drop structures and slow the progression of headcut (see **Figure 5**).

Figure 4. Lateral Movement of Lower Spillway Channel and Headcut Progression, September 16, 1959 (USACE, 1960a).

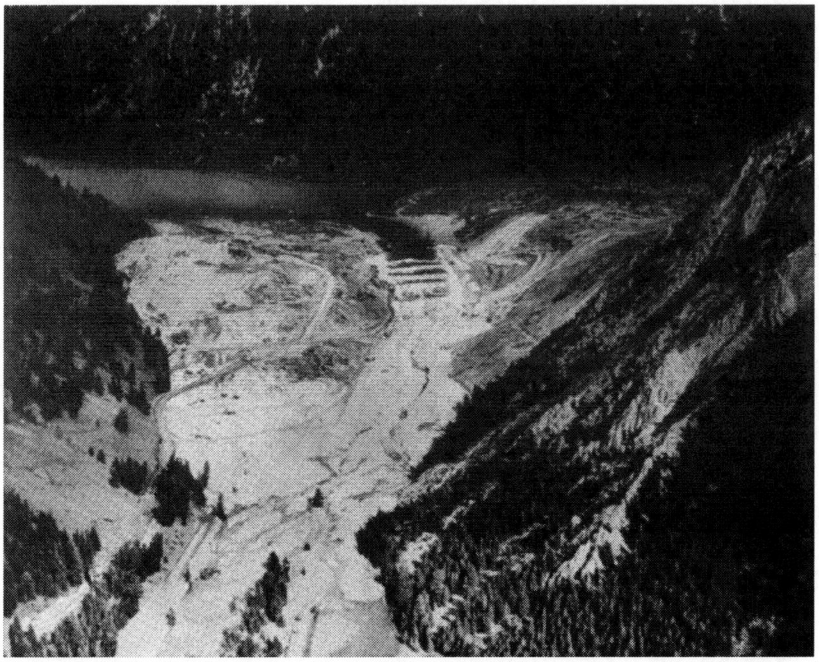

Figure 5. Rock Sills in Place to Arrest Headcut Progression, September 22, 1959 (USACE, 1960a).

It soon became apparent that the quantity of large quartzite and dolomite was not unlimited; rather, much of the material below the surface was shattered and ground into fine particles. In addition, many large rocks broke apart when they were moved, rendering these rocks useless for slowing the erosion in the spillway channel. Some consideration was given to obtaining suitable rock from suitable quarry sites in the nearby mountains, but this idea was rejected due to the expense and time involved. Two members of the board of consultants, Dr. Casagrande and Mr. Steele, visited the site on September 18. Both felt it would be necessary to lower the spillway crest, and thereby reduce the channel slope, to reduce the erosion taking place in the spillway channel to a level that would require minimal maintenance. Dr. Casagrande suggested that the crest be breached by a pilot channel, and allowing water to erode the remainder of the channel; he estimated that erosion would stop when the channel slope reached 4%, or a lowering of the crest by 50 feet. Mr. Steele agreed that it would be safe to allow excavation of the channel via erosion, but thought it prudent to test a small pilot channel first, so that complete control could be retained at all times. The other two members of the board of consultants visited the site on September 24 and 25 and both agreed that the spillway crest should be lowered. Dr. Straub recommended a 50 ft wide pilot channel be excavated in the center of the spillway, with enough slope to induce erosion, as well as a cofferdam at the upstream end of the channel to prevent flow over the sides from carrying material back into the channel as fast as it was excavated.

Due to the inundation of Highway 287 upstream of the slide, as well as the extensive damage to the highway along Hebgen Lake, travel between the project office in West Yellowstone and the slide area required travel over two mountain passes, as well as use of an unimproved road. With the rapid approach of winter weather and the concern with roads between project office and work site, the project office was moved to Ennis, Montana, approximately 43 miles downstream of the slide, on September 20 and 21. Although this increased one-way travel by about 10 miles between project office and work site, all travel was now on paved roads (USACE, 1960a).

Phase 3, Lowering Spillway Crest, September 25 through October 29

Computations showed that lowering the spillway crest by 50 feet would greatly reduce the volume of water behind the slide mass, thereby significantly reducing the downstream flood risk in the event of rapid failure of the slide. Bulldozers and rippers were used beginning on September 25 in an attempt to loosen the material on the spillway crest, to see if the flowing water would carry the material away. Unfortunately, only very fine material was transported by the flowing water, so a modification to the plan proposed by Dr. Straub was employed. A 50-foot wide pilot channel with slope of about 1.5% was excavated beginning at a point about 300 feet downstream of the original break in slope from the spillway crest upstream to the spillway channel entrance about 1,300 feet upstream. Material was stockpiled at various points along the pilot channel such that an emergency cofferdam could be constructed in the event that the downstream erosion progressed too rapidly.

Dr. Straub again visited the project site between September 30 and October 2; he suggested that flows from Hebgen Lake be increased to 2,600 cfs to further assist in the scouring of the spillway channel. The increase in flow resulted in rapid erosion of the slide material, to the point where part of the spillway structure was bypassed about halfway down the slope, with rapid erosion of fine slide material leading to lateral movement of the channel 200 feet to the north and vertical downcut of more than 40 feet. Significant effort was put into stopping this erosion and diverting the flow back into the original spillway channel, with control of the flow achieved on October 6, thereby preventing a disastrous discharge of the lake.

The lowering of the spillway crest then proceeded according to plan. A dragline worked from a berm alongside the pilot channel just above the water surface, working from upstream to downstream. The excavated material was hauled away by truck, and once the dragline reached the downstream end of each cut, the dragline then returned to the spillway channel entrance and repeated the cycle of excavating material from upstream to downstream. Experience showed that a slope of 1.5% was too low to effectively erode material, while a slope of 3% allowed degradation of the channel to proceed as rapidly as mechanical lowering of the spillway crest. Some of the larger rocks that had been previously placed were too large to be removed by dragline and had to be drilled and blasted before being removed. As the spillway crest was lowered, very large blocks of canyon wall that survived the slide intact were encountered, and this material could only be removed through considerable time and effort. Due to the high velocity of flow in the channel, it was impossible to accurately measure the spillway channel thalweg. For this reason, work on lowering the spillway crest continued until the lake behind the slide mass reached elevation 6400, at which time the thalweg at the entrance to the spillway channel was estimated to be at elevation 6392 (see **Figure 6**).

Figure 6. Spillway Channel Following Crest Lowering, Note Incision of Spillway Channel and Downstream Fan Formation, October 22, 1959 (USACE, 1960a).

The hydraulic removal of material from the spillway channel resulted in large amounts of sediment transport past the original toe of the slide mass. Although the finer materials remained in suspension, much of the larger material deposited in and along the channel downstream from the slide, resulting in a large fan. Development of this fan damaged the access road to the work site and required maintenance. By the time the emergency work ceased, the channel at the lower end of the slide and through the fan had reached a slope of about 3%, which appeared to be relatively stable with a discharge of 3,000 cfs.

Throughout this course of action, periodic reports were provided to the members of the board of consultants, to apprise them of progress made towards lowering the spillway crest. The only formal meeting of the board of consultants (minus Dr. Casagrande) occurred on October 29 at the Ennis project office. After a site visit and review of all data collected, the consultants concluded that the serious flood threat had been removed and the mission of the Corps of Engineers had been accomplished. The final report of the board of consultants is included in USACE (1960b) (USACE, 1960a).

Post-Construction

Following completion of construction on the slide, releases from Hebgen Lake were increased to 3,800 cfs until November 13, at which time releases were decreased to about 1,000 cfs due to flooding on the Missouri River near Townsend, Montana, due to ice effects caused by cold weather. Releases were increased and decreased in accordance with weather conditions until Hebgen Lake was drawn down about 28 feet, with only about 24,700 acre-feet of storage on January 5, 1960 (USACE, 1960a).

High flows on the Madison River in 1970 led to severe erosion of the spillway channel and Highway 287 road embankment downstream from the slide area. Subsequent investigations by USACE concluded that significant erosion through and downstream of the spillway channel could be expected for flows in excess of 3,500 cfs. This led the owners of Hebgen Dam to restrict releases to not exceed this 3,500 cfs threshold, taking into account local tributary flows. However, when the 3,500 cfs threshold was exceeded by flooding in 1993, 1996 and 1997, no road damage was reported. Average channel slopes of approximately 3% were noted through the slide mass by 2006 (see **Figure 3**), although localized slopes of 4.6% were noted (Chase and McCarthy, 2012). The slide mass with its engineered spillway remain in place today, with a permanent lake behind it named "Earthquake Lake."

LESSONS LEARNED

Large areas of the United States are subject to high seismic risk. According to the National Inventory of Dams (USACE, 2015), approximately 1,339 dams are located in areas with a 2% probability of exceeding 0.6g PGA in the next 50 years as determined by Petersen, et al (2014) (see **Figure 7**). Of these 1,339 dams, approximately 927 are located in western states, with approximately 553 rated as high hazard. Many of these dams are located in areas with high topographic relief, so it stands to reason that there is some likelihood of future seismically-induced landslides occurring downstream of a dam that may be damaged in the same seismic event. Of course, the likelihood of damage to any individual dam will vary significantly, depending on dam construction and seismic load, as will the likelihood of a seismically-induced landslide, but it would be prudent for dam owners and emergency planners to be prepared for the possibility of a landslide concurrent with seismically-induced damage to a dam structure. The USGS Landslide Overview Map (Radbuch-Hall, et al, 1982) could be one means by which dam owners could assess the potential for landslides downstream of their specific dam.

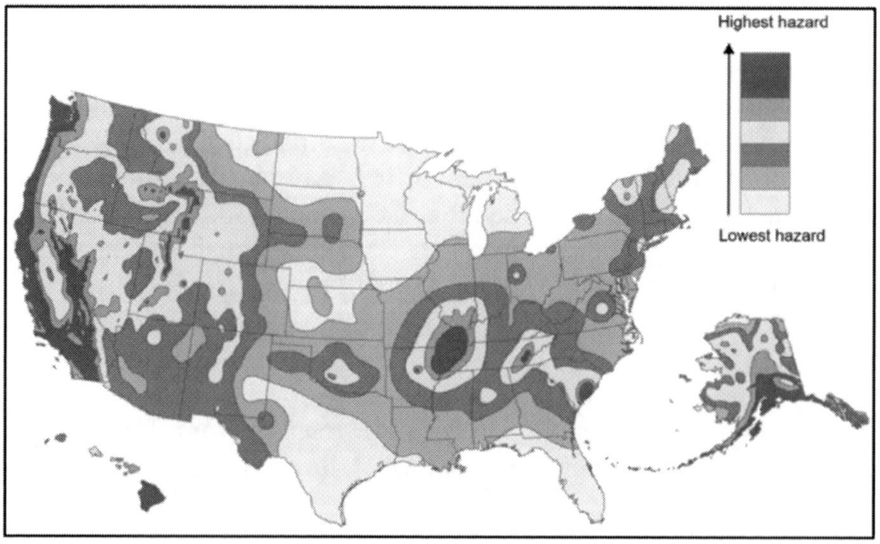

Figure 7. Simplified Seismic Hazard Map, PGA with 2% Exceedance in 50 Years.

Costa and Schuster (1987) note that geologically active areas typically contain abundant material, such as sheared or fractured bedrock, and experience triggering events, such as earthquakes, necessary to initiate landslides. Evans, et al (2011) suggest that a significant percentage of rockslide dams are emplaced due to seismic events and that roughly one-third to one-half of all landslide dams fail. The longevity of natural landslide dams is heavily influenced by the rate of inflow into the lake behind the natural dam, and once filled, by the resistance to erosion of the natural dam's materials (Schuster and Evans, 2011). Costa and Schuster (1987), among others, report that the vast majority of landslide dams that fail do so due to overtopping, and that of those that fail, approximately half fail within 10 days and two-thirds fail within 1 month of emplacement. Failure of naturally formed dams have resulted in some of the largest floods on Earth and can have profound geomorphical effects on the downstream landscape (O'Connor and Beebee, 2009). Sudden catastrophic failure of a landslide dam has the potential to cause large loss of life; Lee and Dai (2011) indicate at least 5,500 deaths due to failure of landslide dams in China since 1900, and as many as 100,000 lives lost in the failure of the Dadu River landslide dam in Sichuan, China in 1786.

Initial Assessment of Remedial Measures and Hydrologic Design

Seismically-induced damage to a large man-made dam upstream of a natural landslide dam may add to the urgency to mitigate the natural landslide dam, as was the case with Hebgen Dam upstream of the Madison Canyon slide. The danger of sudden partial or complete failure of the man-made dam poses significant risk to the downstream natural landslide dam and the population below both dams. A decision must be quickly reached as to how rapidly to drawdown the reservoir behind the damaged dam if the safety of the damaged dam cannot be quickly ascertained, so as to reduce the risk of failure. However, this decision must be balanced against how quickly water behind the downstream landslide will rise, so as to provide adequate time to construct engineering works to safely pass inflow over or around the landslide. It is imperative that assessments of both the damaged dam and landslide be conducted as soon as possible by highly qualified personnel, as was done following the 1959 earthquake (this assessment was fortuitously augmented by the availability of USGS personnel who were performing geologic mapping of the area at the time of the earthquake, USACE (1960b)). Multiple engineering and scientific disciplines should be represented in this assessment to ensure the best possible decisions are made in regards to the recommended path forward.

The decision to construct a spillway over the Madison Canyon slide mass was made before any detailed hydraulic computations could be made on channel size or stability or the time to fill the area behind the landslide. Although multiple initial assessments came to the same conclusion regarding constructing a spillway over the slide mass, there does not appear to have been any serious consideration as to the constructability of such works other than "... the elevation of [the crest] should be lowered as much as practicable within the time available..." (USACE, 1960b). It is critical that hydrologic data, such as inflow to the dam and between the dam and landslide, as well as the area-capacity curve of the area upstream of the landslide be determined as quickly as possible. Although detailed topographic mapping was available for determining an area-capacity curve behind the Madison Canyon slide, it was not available for several days following the slide. The availability of such elevation data online (e.g., USGS, 2015) means area-capacity relationships can be estimated readily for such emergencies, rather than after construction starts. Field collection of hydrologic data, such as rate of rise of the pool behind the landslide, should begin as soon as possible. Multiple computations of spillway channel capacity and velocities can be rapidly performed with modern analytic tools and should be utilized heavily to better inform the decision to construct a spillway over the landslide mass, so as to minimize the likelihood of an unstable spillway channel forming, as occurred with the Madison Canyon slide.

Availability of Equipment Following Seismic Event

In its post-flood report, USACE (1960a) noted that there normally would not have been as much equipment and construction personnel idle and available for the spillway construction if there had not been an ongoing strike at the mines in Butte, Montana. The region in which the 1959 earthquake occurred was sparsely populated; thus there were few conflicting demands on this idle equipment. It is likely that following a large seismic event, there will be multiple competing demands for available construction equipment, which may limit the ability to mitigate a landslide in a timely manner. This demand may be compounded if there are multiple landslides involved; Costa and Schuster (1987), Evans, et al (2011), and Cui, et al (2011), among others, cite several examples of a single seismic event triggering dozens, even hundreds of landslides blocking river valleys. Cui, et al (2011) describe a risk assessment matrix for prioritizing landslide mitigation based on the slide height, storage capacity behind the landslide, and the composition of materials forming the landslide. Any pre-planning for response to a seismic event with widespread damage should include development of similar risk matrix that allows all hazards to be quickly evaluated to assist in readily identifying those areas in which initial construction activities should be focused beyond rescue missions.

Deployment of Proper Equipment

Pace of construction was at times hampered by inadequate identification of landslide materials to be encountered, leading to procurement of larger shovels and draglines later than they were needed, leading to delays in lowering of the spillway crest (USACE, 1960a). Although timely detailed investigations of a landslide may not be possible prior to initiation of construction, an investigation of materials encountered in other landslides such as the Madison Canyon slide may make it possible to infer what materials may be encountered below the surface, allowing for more timely procurement of properly sized equipment. It is noted that during the construction of the spillway at the Madison Canyon slide, that rocks sleds were designed and constructed on-site to facilitate movement of large rocks with the equipment available (see **Figure 8**). It would be prudent to engage construction contractors at regular intervals during such construction in order to elicit ideas that may facilitate construction activities or overcome construction obstacles.

Figure 8. Tractor Pulling Rock Sled Designed and Constructed On-Site for Moving Large Rock (USACE, 1960a).

Maintenance of Access Roads

Access to the Madison Canyon slide proved to be problematic, particularly towards the end of construction activities. The degradation of the spillway channel and subsequent transport of slide materials led to the formation of a sizeable debris fan at the toe of the slide, which led to repeated washing out of the only access road (USACE, 1960a). Since narrow canyons are most susceptible to formation of sizeable landslide blockages, and narrow canyons usually only have one road leading up and down the canyon, access is likely to be an issue for future landslides. Emergency response to a future landslide needs to take access into account and provide protective measures (if possible) to prevent access roads from being washed out. The post-flood report by USACE (1960a) suggested that construction of an access road from Hebgen Dam upstream of the slide may have led to avoiding access problems. However, construction of access from the upstream side of the slide may be difficult and time-consuming depending on site topography and availability of equipment.

Coordination

Response to the Madison Canyon slide involved a high level of cooperation between multiple agencies, including U.S. Army Corps of Engineers, U.S. Forest Service, U.S. Geological Survey, Montana Highway Department, local Sheriff, local Civil Defense, as well as the Red Cross and Montana Power Company (owner of Hebgen Dam); a number of other entities were involved in the response as well, but not to the degree as the aforementioned. The level of cooperation included coordination of activities, acquisition of rights-of-way, communication, establishing an emergency warning system, collection of data, and liaison with the media. The post-flood report prepared by USACE (1960a) does not provide detail as to which, if any, agency took the lead in all of these activities, but it is recommended that one agency provide the coordination of all efforts and establish liaison with all other agencies to ensure a high level of coordination between all activities. It is also recommended that roles and responsibilities be clearly defined and agreed upon by all agencies at the start of the emergency response, although these roles and responsibilities should be periodically reviewed to ensure the best possible response, especially as new entities become involved.

CONCLUSION

The strongest earthquake in recorded Montana history occurred on August 17, 1959, resulting in a landslide that caused 28 fatalities and the complete blockage of the Madison River, as well as extensive damage to the Hebgen Dam upstream of the landslide. Initial response focused on rescue of more than 200 individuals trapped between the landslide and Hebgen Dam. Initial inspections of the damage to Hebgen Dam raised concerns that it may fail, which would potentially fail the downstream landslide dam, resulting in a catastrophic flood in the Madison River valley downstream. A prompt multi-agency response ultimately resulted in successfully stabilizing the flow of the Madison River over the crest of the landslide, as well as the drawdown of Hebgen Lake to allow a thorough inspection of seismic-induced damage. The response to this emergency provides a number of lessons learned that can be applied by emergency responders and dam owners in the event of future seismic events that involve heavy damage to a dam and a valley-blocking landslide downstream of the dam. Application of these lessons learned may result in avoiding a catastrophic failure of the landslide and/or dam.

REFERENCES

Alden, W.C., 1953, Physiography and Glacial Geology of Western Montana and Adjacent Areas, U.S. Geological Survey Professional Paper 231, 192 pp.

Chase, K.J. and McCarthy, P.M., 2012, Lateral and Vertical Channel Movement and Potential for Bed-Material Movement on the Madison River Downstream from Earthquake Lake, Montana, U.S. Geological Survey Scientific Investigations Report 2012-5024, 39 pp.

Costa, J.E. and Schuster, R.L., 1987, The Formation and Failure of Natural Dams, U.S. Geological Survey Open-File Report 87-392, 39 pp.

Cui, P., Han, Y., Chao, D., and Chen, X., 2011, Formation and Treatment of Landslide Dams Emplaced During the 2008 Wenchuan Earthquake, Sichuan, China, in S. G. Evans, R.G. Hermanns, A. Strom, G. Scarascia-Mugnozza (eds.), *Natural and Artificial Rockslide Dams, Lecture Notes in Earth Sciences*. Springer, Berlin, pp. 295-321.

Evans, S.G. Delaney, K.B., Hermanns, R.L., Strom, A., and Scarascia-Mugnozza, G., 2011, The Formation and Behavior of Natural and Artificial Rockslide Dams; Implications for Engineering Performance and Hazard Management, in S. G. Evans, R.G. Hermanns, A. Strom, G. Scarascia-Mugnozza (eds.), *Natural and Artificial Rockslide Dams, Lecture Notes in Earth Sciences*. Springer, Berlin, pp. 1-76.

Hadley, J.B., 1964, Landslides and Related Phenomena Accompanying the Hebgen Lake Earthquake of August 17, 1959: U.S. Geological Survey Professional Paper 435-K, 32 pp.

Lee. C.F. and Dai. F.C., 2011, The 1786 Dadu River Landslide Dam, Sichuan, China, in S. G. Evans, R.G. Hermanns, A. Strom, G. Scarascia-Mugnozza (eds.), *Natural and Artificial Rockslide Dams, Lecture Notes in Earth Sciences*. Springer, Berlin, pp. 369-388.

O'Connor, J.E. and Beebee, R.A., 2009, Floods from Natural Rock-Material Dams, in D.M. Burr, P.A. Carling and V.R. Baker (eds.), *Megafloods on Earth and Mars*. Cambridge University Press, Cambridge, United Kingdom, pp. 128-171.

Petersen, M.D., Moschetti, M.P., Powers, P.M., Mueller, C.S., Haller, K.M., Frankel, A.D., Zeng, Y., Rezaeian, S., Harmsen, S.C., Boyd, O.S., Field, E.H., Chen, R., Luco, N., Wheeler, R.L., Williams, R.A., Olsen, A.H., and Rukstales, K.S., 2014, Seismic-hazard maps for the conterminous United States, 2014: U.S. Geological Survey Scientific Investigations Map 3325, 6 sheets, scale 1: 7,000,000, http://dx.doi.org/10.3133/sim3325

Radbruch-Hall, D.H., Colton, R.B., Davies, W.E., Lucchitta, I., Skipp, B.A., and Varnes, D.J., 1982, Landslide Overview Map of the Conterminous United States, U.S. Geological Survey Professional Paper 1183, U.S. Geological Survey, Washington.

Ross, C.P. and Nelson, W.H., 1964, Regional Seismicity and Brief History of Montana Earthquakes: U.S. Geological Survey Professional Paper 435-E, 6 pp.

Schuster, R.L. and Evans, S.G., 2011, Engineering Measures for the Hazard Reduction of Landslide Dams, in S. G. Evans, R.G. Hermanns, A. Strom, G. Scarascia-Mugnozza (eds.), *Natural and Artificial Rockslide Dams, Lecture Notes in Earth Sciences*. Springer, Berlin, pp. 77-100.

U.S. Army Corps of Engineers, 1960a, Madison River, Montana, Report on Flood Emergency, Madison River Slide, Vol. I – Main Report, Omaha, Nebraska.

U.S. Army Corps of Engineers, 1960b, Madison River, Montana, Report on Flood Emergency, Madison River Slide, Vol. II – Appendixes, Omaha, Nebraska.

U.S. Army Corps of Engineers, 2015, National Inventory of Dams, http://nid.usace.army.mil/.

U.S. Geological Survey, 2015, The National Map, http://viewer.nationalmap.gov/.

Witkind, I.J., Hadley, J.B., and Nelson, W.H., 1964, Pre-Tertiary Stratigraphy and Structure of the Hebgen Lake Area: U.S. Geological Survey Professional Paper 435-R, 9 pp.

NONDESTRUCTIVE TESTING OF POST-TENSIONED ANCHORAGES AT J.T. MYERS AND MARKLAND DAMS

Robert F. Lindyberg, PhD, PE [1]
Afshin Karshenas, PhD [2]
Jeff Lillycrop [3]
Charles E. Wiggins [4]
Larry L. Dalton II, PE [5]
Matthew Watts [6]
Brandon Compton, PE [7]

ABSTRACT

The U.S. Army Corps of Engineers (USACE) has a large inventory of dams that use trunnion anchor rods to secure the Tainter gate's trunnion beam to the dam. The traditional method of measuring rod tension (i.e., lift-off testing) has been a difficult, costly, and risk-laden undertaking. To address these concerns, FDH Velocitel has developed the innovative nondestructive dispersive wave test method for in-situ measuring of a trunnion rod post-tension force. This test method relies primarily on the relationship between the fundamental frequency of the vibration of the cantilever portion of the rod and its diameter, cantilever length, post-tensioned force, and elastic properties. Initial use of this method required the gathering of lift-off data for a limited number of trunnion rods to develop an empirical model that establishes these parametric relationships. However, due to the safety concerns with lift-off testing, it is necessary to develop and validate fundamental, mechanics-based theoretical models from which these parametric relationships can be developed. To eliminate the need for performing the lift-off tests completely, finite element analysis (FEA) is used to build 3D models of the post-tension trunnion rods and extract the modal frequencies for constructing the empirical model.

In close collaboration with the USACE Engineer Research and Development Center (ERDC) in Vicksburg, MS, a series of eight (8) dams were identified for the initial demonstration and validation of this test method. This paper presents the testing and initial results from the first two dams in this series tested in the Louisville District: John T. Myers Locks and Dam, and Markland Locks and Dam.

INTRODUCTION

Trunnion anchor rods support the trunnion girders of Tainter gates, which are considered the most suitable type of gate for controlled spillways of dams (EM 1110-2-2702). Figure 1 shows the Tainter gates of Markland and John T. Myers dams, respectively.

[1] Vice President, FDH Velocitel, Robert.Lindyberg@fdhvelocitel.com
[2] Senior Research Scientist, FDH Velocitel, Afshin.Karshenas@fdhvelocitel.com
[3] Technical Director, CHL, U.S. Army Engineer Research & Dev. Center, Jeff.Lillycrop@usace.army.mil
[4] Associate Technical Director, U.S. Army Engineer Research & Dev. Center, Charles.E.Wiggins@usace.army.mil
[5] Technical Manager, Inland Navigation Design Center, USACE, Larry.L.Dalton@usace.army.mil
[6] Structural Engineer, USACE, Matthew.A.Watts@usace.army.mil
[7] Research Engineer, FDH Velocitel, Brandon.Compton@fdhvelocitel.com

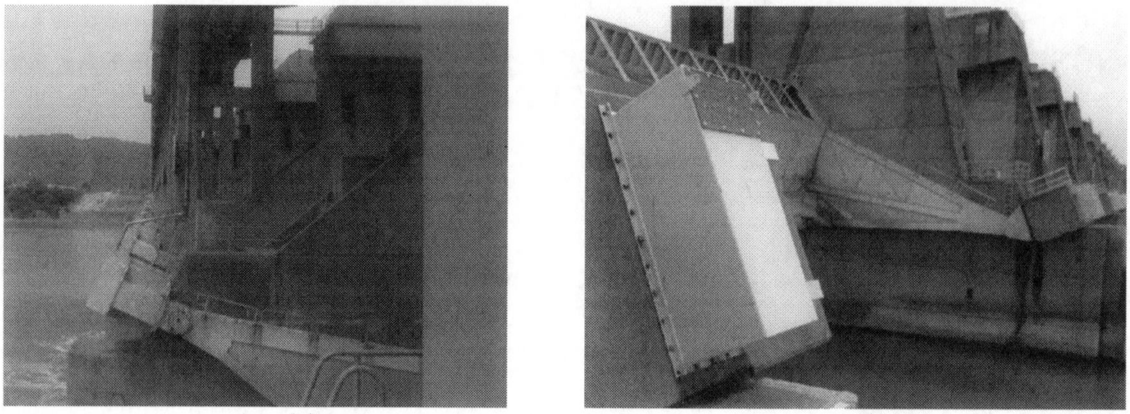

(a) (b)

Figure 1. Tainter Gates of (a) Markland Dam and (b) John T. Myers Dam

Trunnion rods play an important role in providing the structural integrity of Tainter gates. Failure of the trunnion rods can lead to catastrophic failure of the Tainter gate, which may cause significant human and financial losses. A review of the trunnion rod inspections in 14 dams through the mid-1990s by the U.S. Army Corps of Engineers Institute for Water Resources (USACE IWR), Risk Management Center, identified 156 failed rods (2% of the rods inspected) (Schaff 2009). Therefore, development of an effective inspection method of the trunnion rods would be crucial to prevent the catastrophic failure of Tainter gates and associated losses.

Currently, the traditional lift-off test is used to measure the post-tension force in the trunnion anchor rod. This test method is time consuming, expensive, and poses a significant human safety concern. Even a modest over-tensioning of the anchor rod in this test may result in abrupt rod failure, with the broken rod flying out of the anchorage at a high rate of speed and force. The result of an abrupt failure is shown in Figure 2, where the failed rod penetrated a 0.24 in. thick steel cover box.

Figure 2. Trunnion Rod Failure

This paper presents a summary of an innovative nondestructive dispersive wave (DW) technique developed by FDH Velocitel to evaluate the post-tension loads in the trunnion anchor rods. The DW test method is faster and safer than the lift-off test to measure the post-tension loads in the dam's trunnion rods. This nondestructive technique uses the dynamic responses of the anchor rods to impact instruments for predicting the post-tension loads in the rods. The evaluation of the post-tension forces in the trunnion rods of two USACE dams (i.e., John T. Myers and Markland dams) by the DW method are presented in the following sections.

J.T. Myers Locks and Dam is located on the Ohio River about 3.5 miles downstream from Uniontown, Kentucky, and is operated and maintained by the U.S. Army Corps of Engineers, Louisville District. The dam was opened in December 1969 and has 10 Tainter gates of 32'x110' (http://www.lrl.usace.army.mil/). Markland Dam is also operated by the USACE, Louisville District, and crosses the Ohio River, connecting Gallatin County, Kentucky, and Switzerland County, Indiana (https://en.wikipedia.org). The construction of Markland Dam, consisting of 12 42'x42' Tainter gates, started in April 1959 and was completed in June 1964 (http://www.lrl.usace.army.mil/).

Trunnion Rod Anchorage Mechanisms

The two common anchorage mechanisms used in the USACE dams' trunnion rods are as follows:

- Threaded insert wedge with a threaded grip nut, Figure 3(a)
- Gripping wedge directly embedded in conical holes within the trunnion rod anchor plate

In J.T. Myers Dam, the post-tension rods' forces are transferred to the embedded steel plate in the concrete trunnion girder by the threaded insert wedges with threaded grip nuts, Figure 3 (b). Markland Dam utilizes the gripping wedge directly embedded in the trunnion rod anchor, as shown in Figure 4.

(a) (b)

Figure 3. (a) Threaded Insert Wedge with a Threaded Grip Nut (b) J.T. Myers Gripping System and Steel Plate

(a) (b)

Figure 4. (a) Gripping Wedge (b) Gripping Wedge Directly Embedded in the Trunnion Rod Anchor Plate of Markland Dam

DIPERSIVE WAVE TEST PROCEDURE AND DATA COLLECTION

In the DW test method, the dynamic responses of the end cantilever part of the rods and the physical measurements of the rods (i.e., cantilever length and diameter) are used to develop an empirical model relating dispersive wave frequencies to the rod tensions. Lift-off tests performed on the representative number of rods provide the post-tension loads necessary to calibrate the empirical model. This empirical model can be used to estimate the post-tension forces in the remainder of the rods within the dam that did not undergo a lift-off test. To measure the dynamic response of the cantilever end, the trunnion anchor rod is instrumented with two accelerometers (one at the free end of the rod and one on the rod next to the anchorage mechanism), as shown in Figure 5.

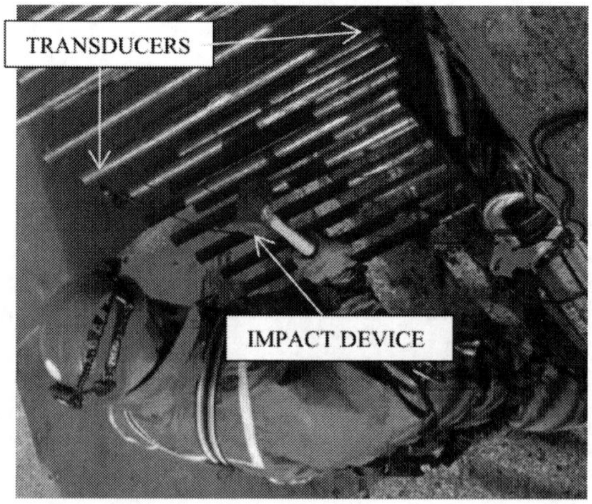

Figure 5. DW Test of a Trunnion Rod at J. T. Myers Dam

Subsequently, the middle of the cantilever section of the rod is struck with the impact devices. In the following sections, the procedures to measure the lift load and dynamic responses of the trunnion rods are presented.

Modal Frequencies Results

Frequency spectra of the DW test results are calculated from the time-domain acceleration data captured by the accelerometers attached to the rods. A typical time domain and a frequency domain spectrum of a rod dynamic response are shown in Figure 6 and Figure 7, respectively. The frequency spectrum of the rod is calculated by custom FDH Velocitel software, applying the Fast Fourier Transform (FFT) to the time domain response of the trunnion rod.

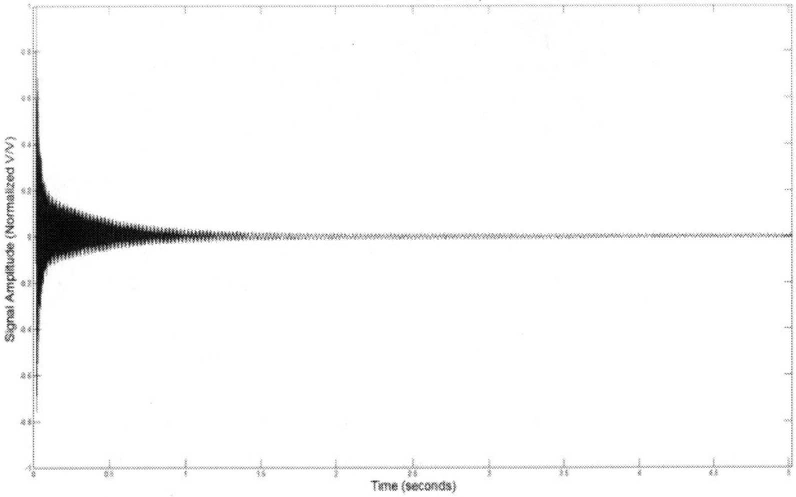

Figure 6. Time History of Acceleration for a Typical Rod

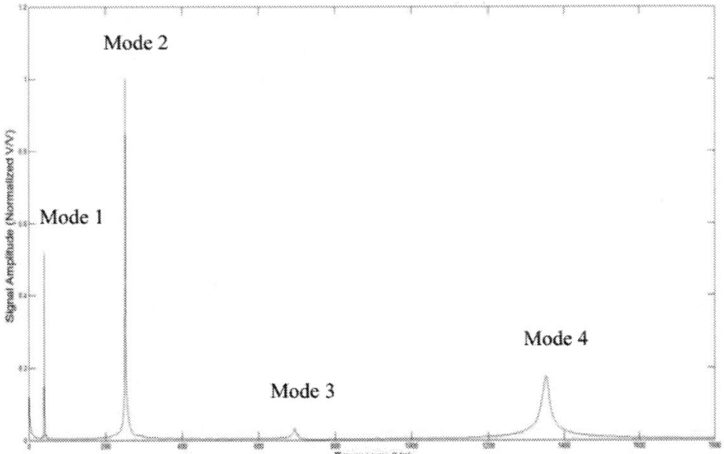

Figure 7. Frequency Domain for a Typical Rod DW Test

The modal frequencies determined from the field tests are used to build an empirical model relating the trunnion rod modal frequencies to length, diameter, and post-tension load. In the following section, the procedure for the calculation of the lift-off loads is presented.

Lift-Off Test Procedure

The custom made FDH Velocitel lift-off test setup is demonstrated in Figure 8.

Figure 8. Custom FDH Velocitel Lift-Off Test Setup

The lift-off test assembly, Figure 8, consists of (1) the chair, (2) LVDT assembly for measuring the relative displacement between the rod and the chair, (3) hydraulic jack, (4) steel plate between the load cell and jack to prevent possible load concentration on the load cell, (5) load cell, and (6) gripping wedges and locking nut.

Figure 9 shows a typical load versus displacement curve for the J.T. Myers Dam lift-off tests. The curve consists of four distinct regions: the initial nonlinear segment followed by a steep linear region and the third region, which is nonlinear, followed by a linear region with a lower slope compared to the second region.

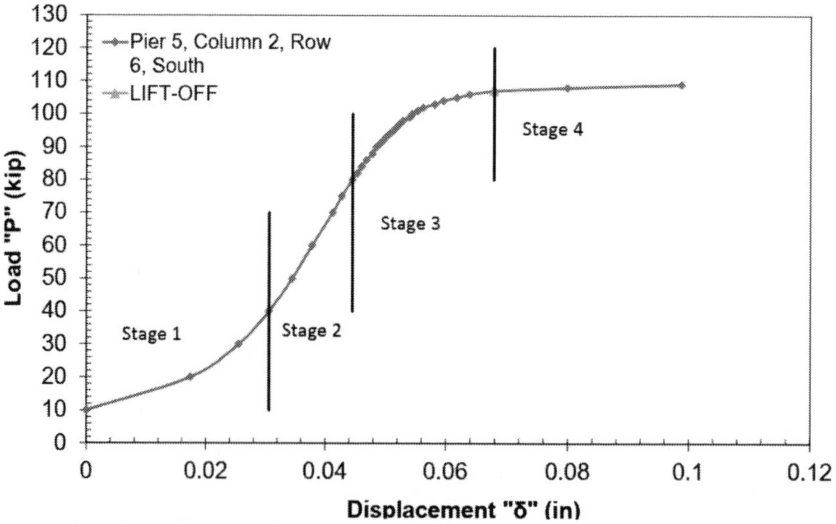

Figure 9. Typical Lift-Off Load Versus Displacement Curve for J.T. Myers Dam In-situ Test

To specify the accurate lift-off load and the location of the lift off on the displacement vs. load curve, a simulation of the lift-off test was performed on the prototype beam setup at the FDH Velocitel research laboratory for the J.T. Myers Dam gripping mechanism, i.e., threaded insert wedge with a threaded grip nut. The load-displacement curve for the prototype lift-off test is presented in Figure 10.

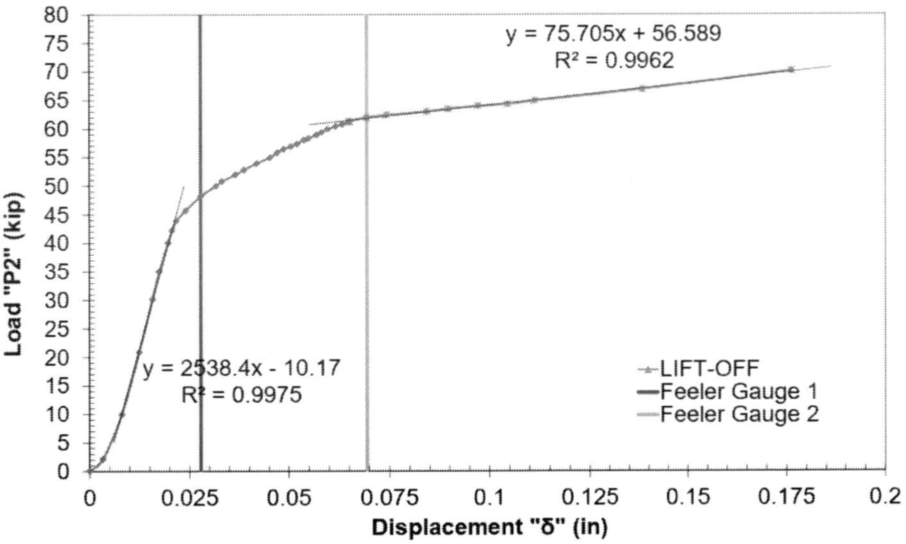

Figure 10. Load-Displacement Curve for Prototype Lift-Off Test

In this figure, the relationship between load and deflection, similar to the field lift-off tests, consists of the four distinct regions. Mechanically, within the first region, the gripping wedge is locking in the nut at the end of the lift-off setup. In the second stage, the external portion of the rod is extending as the applied load increases, while remaining well below the internal tension of the trunnion rod. The third stage is a transition stage where load is removed from the gripping mechanism and transferred to the bearing plate on the dam or the prototype beam, as it is in this case, through the chair. As a result, the mechanical connection between the rod, insert, and grip nut is gradually disengaged, and the contact area between the grip nut and the steel plate is decreased. Within the fourth region, the entire length of the rod is extending as the applied load increases beyond the internal tension of the trunnion rod.

The vertical feeler gauge 1 line in Figure 10 indicates the point at which a 0.015" feeler gauge could be placed between the grip nut and the steel plate at the top outer point of the grip nut circumference. The vertical feeler gauge 2 line indicates the point that the 0.015" feeler gauge could be placed between the grip nut and the bearing plate from any direction around the rod. At this point, the internal load has equalized with the external load and displaced the grip nut from the bearing plate 0.015". As such, the magnitude of the tensile load contained within the trunnion rod can be determined as the load just prior to the feeler gauge 2 line, which, according to Figure 10, is also the location where the transition from the third region and the fourth region appears. By applying this knowledge to the field data gathered at J.T. Myers Dam, the magnitude of the internal tension was determined by identifying the location on the load-displacement curve where the transition from the third region to the fourth region occurs.

The Markland Dam anchorage mechanism consists of a gripping wedge directly embedded in the trunnion rod anchor plate, which is different from the gripping mechanism on J.T. Myers Dam. In most cases, as presented in Figure 11, the wedge would remain stuck inside the conical plate hole after the load had surpassed the internal tension within the trunnion rod due to a bond between the surface of the wedge and the surface of the conical hole in the plate.

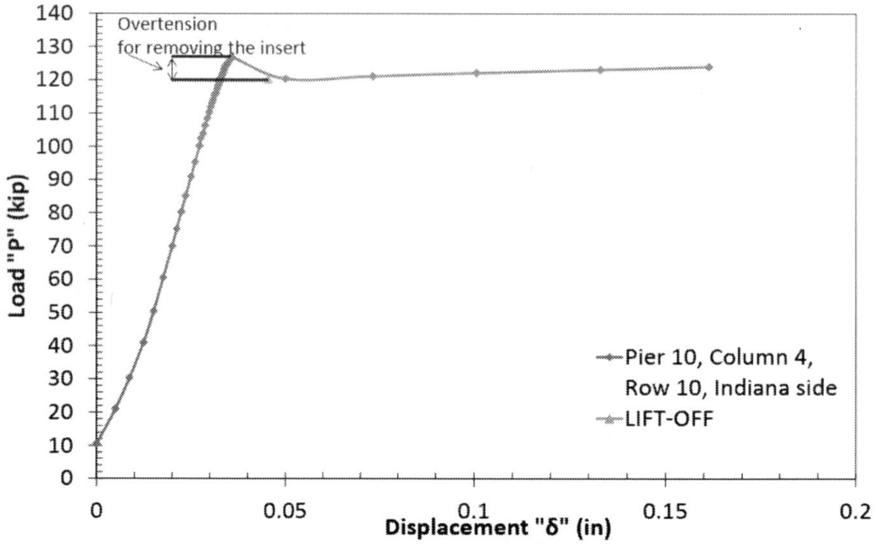

Figure 11. Typical Markland Dam Lift-Off Load-Displacement Curve

In these cases, the applied load is increased to break this bond, causing the displacement to increase and the load to decrease significantly from the previous point as the load carried by the bond is redistributed into the trunnion rod. The lift-off load is calculated utilizing the principle of conservation of work (FDH Engineering 2013)

EMPIRICAL MODELS FOR MARKLAND AND J.T. MYERS DAMS

The data for creating the empirical model for Markland Dam is presented in Table 1. A schematic view of Markland Dam's trunnion rod arrangement is shown in Figure 12.

Table 1. Data for Empirical Model of Markland Dam

Rod	Pier	Column	Row	Side	Lift-off (kip)	D (in)	L (in)	DW F_1 (Hz)	DW F_2 (Hz)
IP12C1R3	12	1	3	Indiana	116.81	1.2575	30.623	36.72	229.509672
IP12C1R4	12	1	4	Indiana	115	1.264	30.5094	36.42	228.864496
IP12C2R3	12	2	3	Indiana	116.1	1.254	30.7795	36.23	226.068497
IP12C2R4	12	2	4	Indiana	116.1	1.256	30.5991	37.19	229.756037
IP12C2R5	12	2	5	Indiana	118.31	1.2525	30.4113	37.48	233.014425
KP12C1R4	12	1	4	Kentucky	122.2	1.258	29.3308	39.96	249.00039
KP12C2R6	12	2	6	Kentucky	118.9	1.258	29.2965	39.86	248.622894

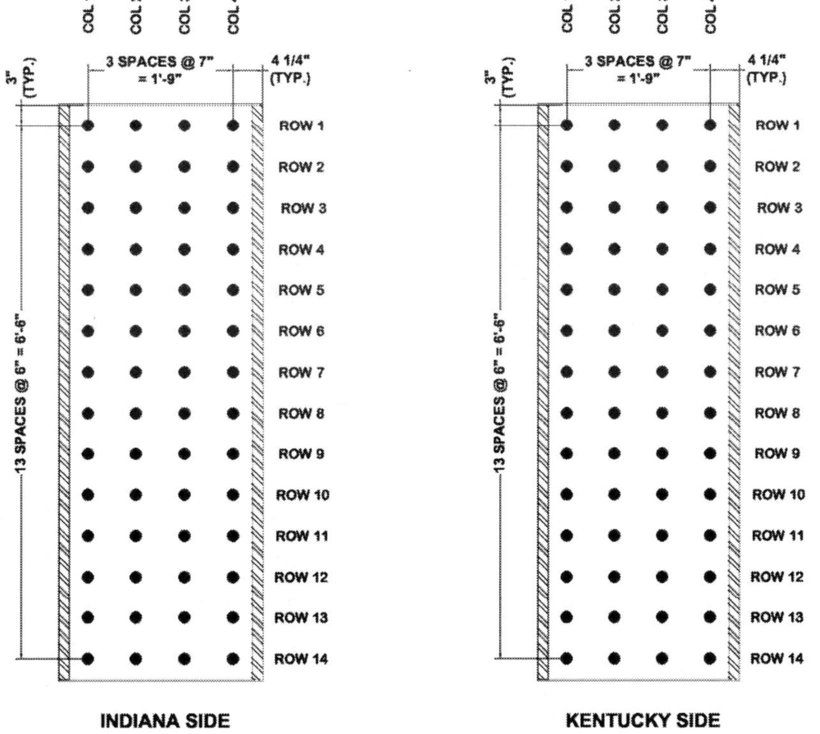

Figure 12. Markland Dam Trunnion Rod Arrangement on Each Pier

Multiple regression analysis is used to find the best equation for predicting the post-tension forces (i.e., modal frequencies in the trunnion rod). Several combinations of the variables, which are cantilever length, diameter, and post-tension force (lift-off load), were examined to find the best equation that adequately fits the data. The best model was selected based on having the highest value of R^2 and the most significant terms (i.e., low p-value for all independent variables). The model presented in equation (1) is selected as the best function for predicting the post tension for Markland Dam.

$$T = \frac{-0.679 \times F_1 \times L^2}{D} + \frac{0.109 \times F_2 \times L^2}{D} + 6.826 \times F_1^2 - 0.175 \times F_2^2 + 5.404 \qquad (1)$$

Where:
F_1 = the first vibration frequency
F_2 = the second vibration frequency
T = post-tension force in the trunnion rod
D = Diameter of the rod
L = Length of the cantilever part of the rod

The J.T. Myers Dam data are presented in Table 2. The trunnion rod arrangement of J.T. Myers Dam for the external piers (i.e., pier 1&11) and the internal piers (pier 2 through

10) are illustrated in Figure 13. As shown in Table 2, the external piers have a larger rod's diameter than the internal piers.

Table 2. Data for Empirical Model of J.T. Myers Dam

Rod	Pier	Column	Row	Side	Lift-off (kip)	D (in)	L (in)	DW F_1 (Hz)	DW F_2 (Hz)
P6C4R12N	6	4	12	North	111.1	1.140625	27.14063	41.23847	255.4973
P6C2R12S	6	2	12	South	109.2	1.140625	27.15625	41.43715	256.3477
P6C4R10N	6	4	10	North	111.1	1.140625	27.1875	40.72189	252.231
P11C3R11N	11	3	11	North	120.5	1.266	27.203	46.134	285.093
P11C3R9N	11	3	9	North	123	1.266	27.204	45.77637	282.9393
P6C4R8N	6	4	8	North	109.5	1.140625	27.23438	41.10336	255.9741
P6C2R9S	6	2	9	South	106.8	1.140625	27.3125	40.53911	251.1978
P6C1R10S	6	1	10	South	107	1.140625	27.32813	41.09542	254.5039
P11C4R12N	11	4	12	North	123.5	1.266	27.344	45.71279	282.7803
P6C2R10S	6	2	10	South	104.6	1.140625	27.40625	40.08611	248.5911
P11C3R8N	11	3	8	North	120	1.266	27.422	44.92601	278.6875
P10C1R10S	10	1	10	South	101	1.140625	27.4375	39.89538	247.9315
P3C2R4S	3	2	4	South	109.1	1.125	27.4375	41.008	253.7171
P3C2R7S	3	2	7	South	106.9	1.125	27.4375	40.53116	252.0084
P3C1R2S	3	1	2	South	109.7	1.125	27.45313	40.253	250.411
P11C2R4S	11	2	4	South	115.5	1.281	27.454	45.1088	278.9418
P10C3R9N	10	3	9	North	99	1.140625	27.54688	40.14969	249.6958
P3C2R12S	3	2	12	South	107	1.125	27.5625	40.14969	249.5925
P3C3R10N	3	3	10	North	104.5	1.140625	27.5625	40.76163	251.8257
P3C3R8N	3	3	8	North	103.3	1.140625	27.5625	40.69805	250.4031
P3C4R11N	3	4	11	North	100.6	1.140625	27.57813	40.35632	249.8865
P10C1R11S	10	1	11	South	104.5	1.140625	27.59375	40.1179	248.0507
P3C2R10S	3	2	10	South	105.2	1.125	27.59375	40.53116	251.5157
P3C2R8S	3	2	8	South	106	1.140625	27.625	40.53116	251.8574
P11C2R10S	11	2	10	South	116	1.281	27.641	44.48096	275.1748
P10C3R4N	10	3	4	North	109.5	1.140625	27.71875	40.43579	249.2507
P11C2R12S	11	2	12	South	116	1.281	27.734	44.34586	275.2463
P10C1R6S	10	1	6	South	104.5	1.140625	27.73438	39.48212	245.5393
P3C4R1N	3	4	1	North	105.2	1.140625	27.76563	39.76822	247.3275
P10C3R11N	10	3	11	North	107.5	1.140625	27.8125	39.8318	246.5487
P10C1R5S	10	1	5	South	107.5	1.140625	27.9375	39.80796	246.0003
P10C1R4S	10	1	4	South	107	1.140625	28	39.55364	243.9817

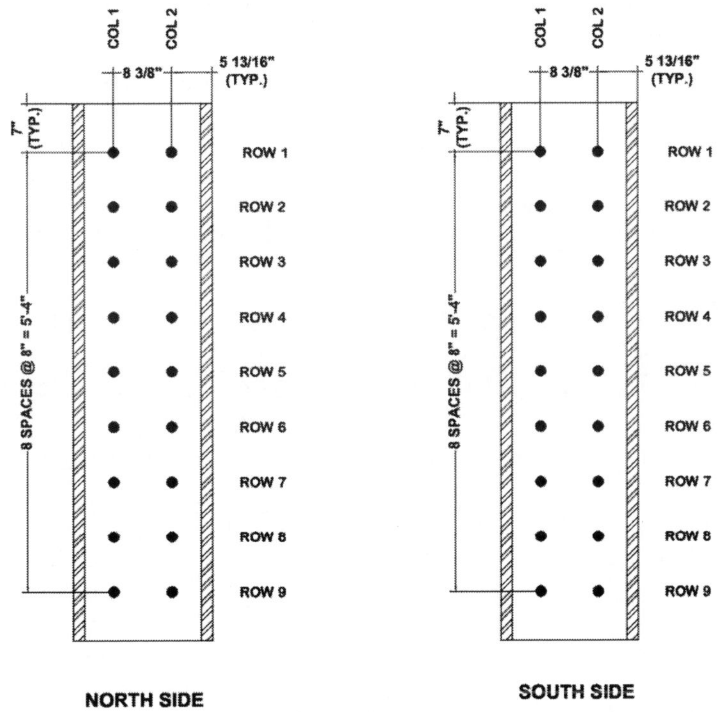

(a)

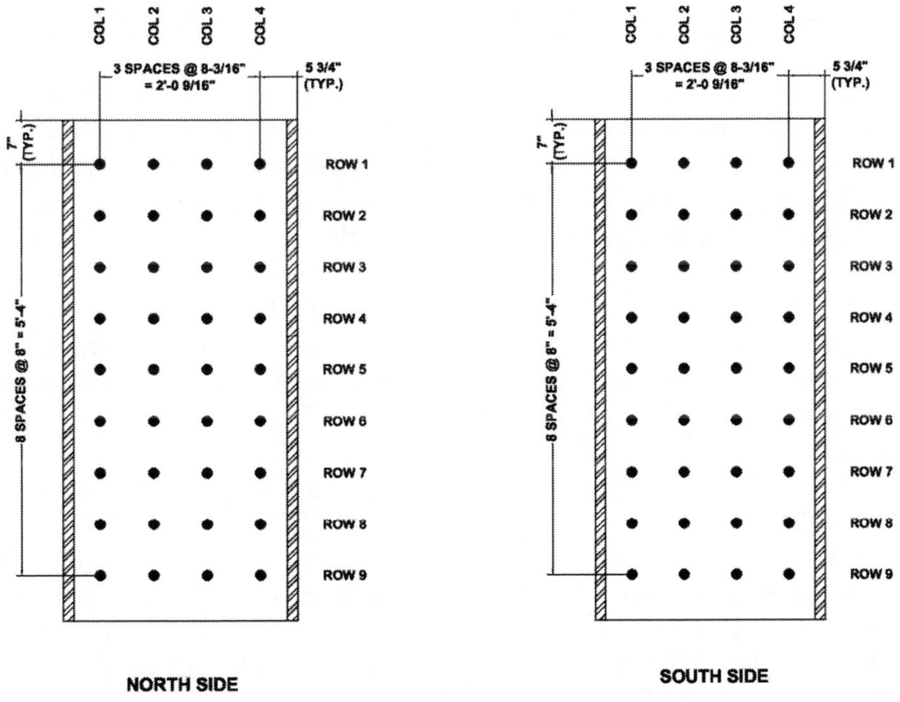

(b)

Figure 13. Trunnion Rod Arrangement of J.T. Myers Dam: (a) External Piers and (b) Internal Piers

The best fitted equation with R^2 equal to 0.817, which is shown in equation (2), is selected for predicting the post-tension force in the trunnion rods of J.T. Myers Dam.

$$T = \frac{0.00624 \times F_1 \times L^2}{D} - 0.1865 \times \frac{L^2}{D} + 62.73693203 \qquad (2)$$

Where:
F_1 = the first vibration frequency
T = post-tension force in the trunnion rod, i.e., lift-off load (kip)
D = Diameter of the rod
L = Length of the cantilever part of the rod

Regression parameters for the Markland and J.T. Myers dams are presented in Table 3.

Table 3. Empirical Model Regression Parameters

Dam	Terms	R^2	Standard Error	P-value
Markland	C	0.996180543	0.259403965	0.862146144
	F_1^2			0.018484184
	F_2^2			0.018593218
	$F_1 x L^2/D$			0.01827351
	$F_2 x L^2/D$			0.017926134
J.T. Myers	C	0.817171	2.773282356	0.180908916
	$F_1 x L^2/D$			0.001915604
	L^2/D			3.79851E-12

The difference between the empirical models of the Markland and J.T. Myers dams can be due to the different trunnion rod anchorage mechanisms used in each dam. However, more tests on each type of anchorage mechanism would be necessary to verify the model validity across multiple dams. Therefore, the presented models are valid for the range of data used for the development of the empirical models.

In Table 1 and Table 2, only the partial sets of lift-off and DW data for Markland and J.T. Myers dams' trunnion rod tests are presented. At the time of this paper, more field data were still being collected and analyzed, and would be used in the construction and modification of the final empirical models.

FINITE ELEMENT ANALYSIS

To build the DW empirical model, the post-tension forces in a representative number of rods in a dam should be known. Therefore, it is necessary to perform the lift-off tests on a limited number of rods in each dam. To eliminate the need for performing the lift-off tests completely, a finite element (FE) analysis is used to build a three-dimensional (3D) model of the post-tension trunnion rods, and extract the modal frequencies of the post-tension rods from the FE model for constructing the empirical model. The general purpose finite element program ABAQUS is used for construction of the FEA model of the trunnion rods. The preliminary results of the FE model for the trunnion rods within the J.T. Myers and Markland dams are presented in this section. Both FE models utilized field measurements to determine the individual rod diameters, exterior rod lengths (cantilever lengths), and insert protrusions from the grip nuts or trunnion girder plates. Furthermore, the dimensions of the anchorage mechanism parts (insert, grip nut, and wedge) were determined using digital calipers on similar parts either recovered from the field or purchased from the manufacturer. Finally, original drawings of the trunnion rod design for each dam provided information on the rod spacing, bearing plate thickness, and the internal length of each trunnion rod.

Markland FE Model

The Markland FE model uses the gripping wedge directly embedded in the anchor plate mechanism to be modeled in 3D, as presented in Figure 14. Based on the original shop drawings, a W-10 wedge part is used in Markland Dam, which the manufacturer (Howlett) still produces. Once the part was purchased, the necessary measurements of the wedge were obtained and implemented in the model by sketching the cross-section of the wedge, revolving the cross section and cutting out the opening and grooves in the wedge as presented in Figure 15. Once the geometry was constructed, trunnion rod tension (lift-off load), boundary conditions, and contact interactions were applied according to the lift-off results. Finally, an eigenvalue calculation was performed by the FE software to determine the natural frequencies of the trunnion rods modeled. The preliminary results of the Markland FE model fundamental frequencies compared to the Markland DW fundamental frequency results are presented in Figure 16 with respect to lift-off load.

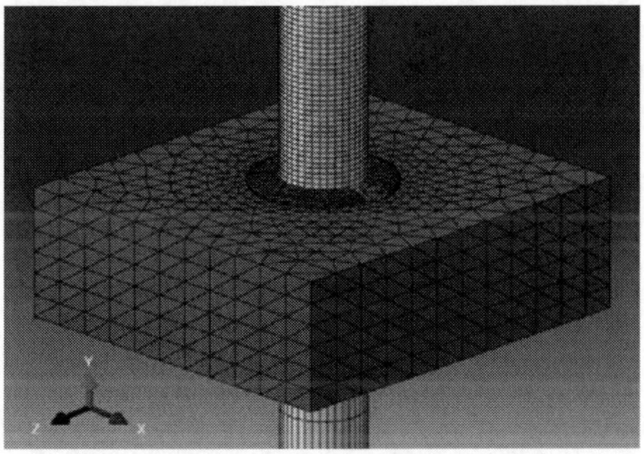

Figure 14. Wedge/Plate Anchorage Mechanism Modeled in FE Software

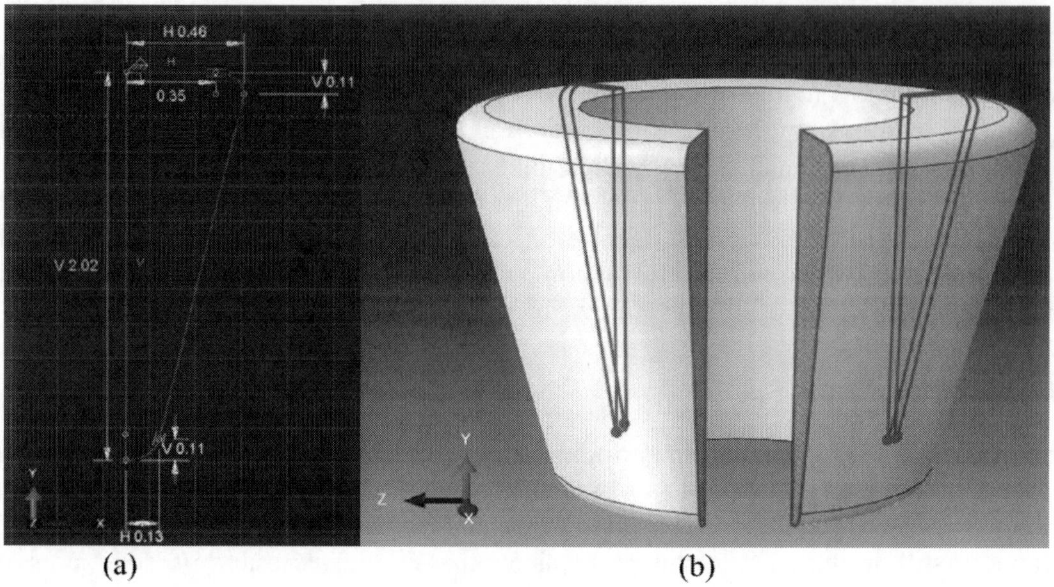

Figure 15. (a) Wedge Cross Section Prior to Revolving It into a 3D Shape and (b) the Final 3D Wedge after Cutting Out the Opening and Grooves of the Part

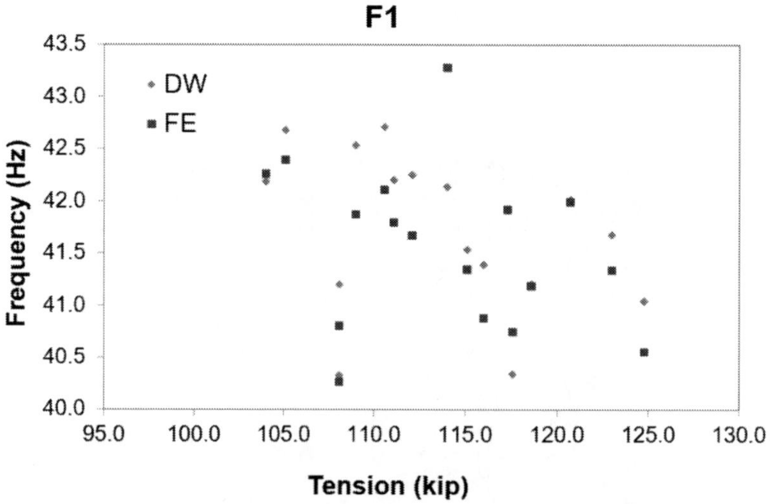

Figure 16. Markland Dam FEA and Dispersive Wave Frequencies Versus Post-Tension Forces (Lift-Off)

J.T. Myers FE Model

The J.T. Myers FE model required a threaded insert wedge with a threaded grip nut anchorage mechanism to be modeled in 3D, as presented in Figure 15. This particular anchorage mechanism is no longer being manufactured, so ordering a part for FE modeling purposes was not an option. Fortunately, an insert and grip nut from J.T. Myers Dam were recovered during a site visit and were used to construct the model. Following the recovery of the parts, the necessary measurements of the insert and grip nut threads were obtained and implemented in the model by sketching the cross-section of the threads, revolving the cross-section of the threads with a measured pitch, and cutting out the opening and grooves in the insert and grip nut, as presented in Figure 16, Figure 17, and Figure 18. Once the geometry was constructed within the FE model, the trunnion rod tension (lift-off load), boundary conditions, and contact interactions were applied to the J.T. Myers' model according to the lift-off results. Finally, an eigenvalue calculation was performed by the FE software to determine the natural frequencies of the trunnion rods modeled. The preliminary results of the J.T. Myers' FE model fundamental frequencies compared to the J.T. Myers' DW fundamental frequency results are presented in Figure 21 with respect to trunnion rod tension.

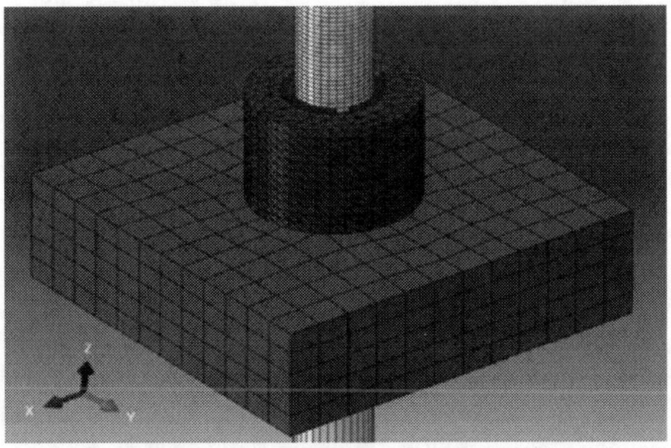

Figure 17. Insert/Grip Nut Anchorage Mechanism Modeled in FE Software

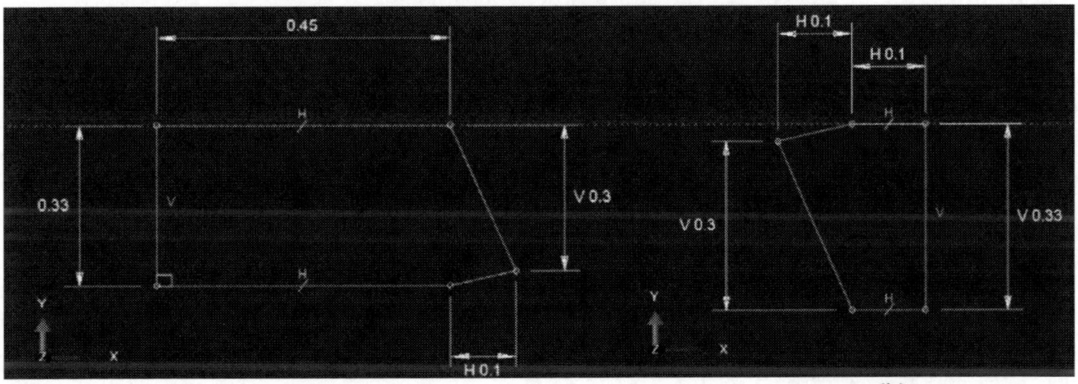

(a) (b)

Figure 18. (a) Grip Nut Thread and (b) Insert Thread Prior to Revolving the Sketch with a Pitch to Create the 3D Parts

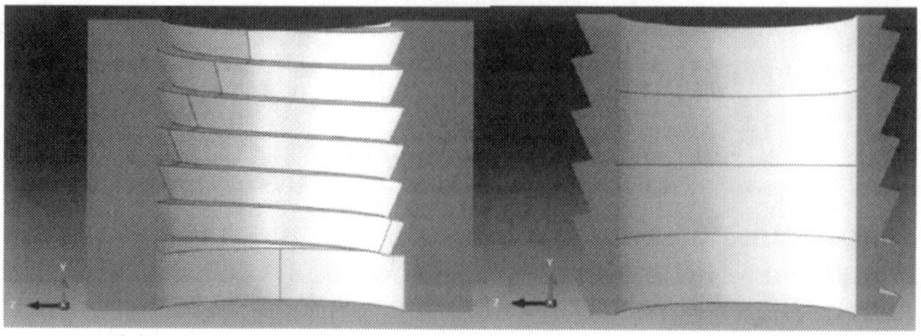

(a) (b)

Figure 19. Cross-Sections of the (a) Grip Nut and (b) Insert after Revolving the Sketched Threads

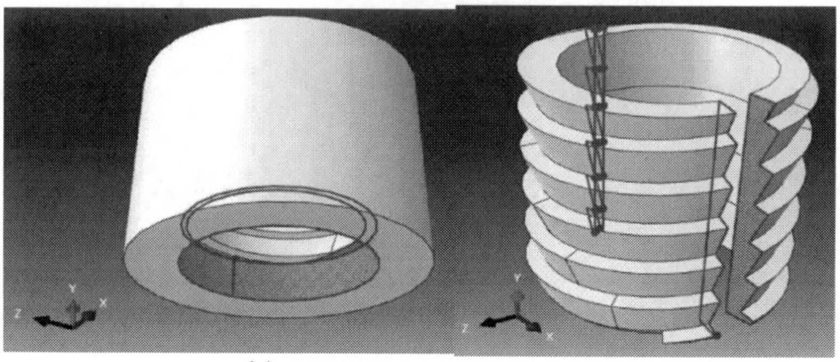

(a) (b)

Figure 20. 3D Figures of the (a) Grip Nut and (b) Insert with the Opening and Groove Cut Out

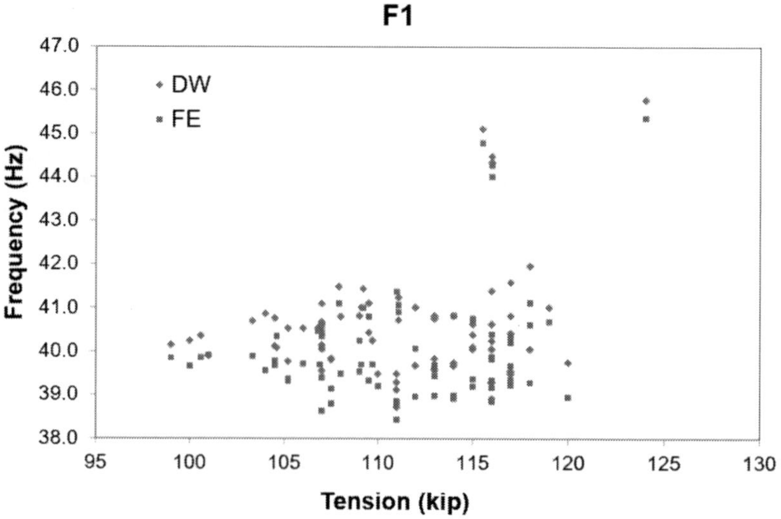

Figure 21. J.T. Myers Dam FEA and Dispersive Wave Frequencies Versus Post-Tension Forces (Lift-Off)

The maximum difference between the predicted frequencies by FEA and the DW frequencies for Markland Dam is 2.7% and for J.T. Myers is 6.85%. Also, the load versus frequency plots show that the FE models are consistently less stiff than the DW data, indicating either the modulus of elasticity should be increased or the interaction properties should be stiffened to provide a closer fit.

CONCLUSION

The preliminary results of the trunnion rods' post-tension forces investigation of Markland and J.T. Myers dams show that the trunnion rods' post-tension forces can be accurately predicted using the empirical models developed by combined in-situ DW (dispersive wave) frequencies and the lift-off tests of the rods. These models would be valid in the ranges of the data used for the development of the empirical models.

In addition, the trunnion rods of Markland and J.T. Myers dams are modeled using the general purpose finite element software ABAQUS. The initial FEA models of Markland and J.T. Myers dams' trunnion rods confirm the technical feasibility of the combined DW-FEA methods for predicting the trunnion rods' post-tension forces. It is expected that with additional data from testing more dams, the lift-off tests can be abandoned.

Future dams to be tested in this program, including West Point Dam (USACE SAM), Cannelton and Newburgh (USACE LRL), Hannibal Dam (USACE LRP), and Greenup and Willow Island (USACE LRH), will provide the appropriate ranges of data to build accurate models (i.e., empirical and FEA) for predicting the post-tension forces in the trunnion rods.

REFERENCES

FDH Engineering (2013), Greenup Tainter Gate anchor Analysis, Final Submittal, Black & Veatch Project No. 42475, prepared for U.S. Army COE - Huntington District.

Schaff, D. (2009). Tainter Gate Trunnion Anchorage: Database of Inspection, Testing, and Incidents. IWR, Risk Management Center.

U.S. Army Corps of Engineers (2000), Design of Spillway Tainter Gates, Engineering Manual, EM 1110-2-2702, Department of the Army, Washington, DC.

http://www.lrl.usace.army.mil/Missions/CivilWorks/Navigation/LocksandDams/JohnTMyersLocksandDam.aspx

https://en.wikipedia.org/wiki/Markland_Locks_and_Dam

USACE PORTLAND DISTRICT SPILLWAY GATE REHABILITATION PROGRAM

Matthew Craig, PE[1]
William Fortuny, PE[3]
Matt Hess, PE[5]

Laurie Ebner, PE[2]
David Hamernik, PE[4]
Mark Sawka, PE[6]

ABSTRACT

The US Army Corps of Engineers' (USACE) Portland District (District) has 90 spillway radial Tainter gates. There are 42 gates at 11 Willamette Valley Basin projects, 5 gates at 2 Rogue Basin projects and 43 gates at 2 Columbia River projects. Since 2001 various studies, inspections and incidents have prompted the District to become concerned about the structural integrity and mechanical and electrical reliability of these gates. In 2008, load induced buckling observed on three of the four Tainter gate end frames at Foster Dam led to emergency repairs. In December 2009, additional observations of buckled Tainter gate strut arms at Dexter Dam required repairs and created urgency with respect to documenting risk to the downstream population at all District projects. In 2010, these events led to the implementation of a comprehensive assessment of the Tainter gates in the Willamette Valley and Rogue Basin projects. The assessment included a gates specific potential failure modes analysis (PFMA), structural, mechanical and electrical assessments, identification of interim risk reduction measures (IRRMs), and analysis of impacts of the IRRMs. Based upon findings, the District developed a tool for prioritizing projects for gate repair. Since identifying the issues with Tainter gates, Portland District has implemented IRRMs, taken a systems-based approach to repairs, and implemented a capitalization program to improve gate reliability. Repairs are complete or underway at 36 of the 42 Tainter gates in the Willamette Valley, a study is underway to address reliability of regulating outlets, and future studies are planned for Columbia River spillway gates.

INTRODUCTION

The US Army Corps of Engineers' Portland District has 90 spillway Tainter gates. There are 42 gates at the Willamette Valley projects, 5 gates at the Rogue Valley projects, and 43 gates at the Columbia River projects.

[1] US Army Corps of Engineers, 333 SW 1st Avenue, Portland, OR 97204, 503-808-4846, matthew.craig@usace.army.mil
[2] US Army Corps of Engineers, 333 SW 1st Avenue, Portland, OR 97204, 503-808-4880, laurie.l.ebner@usace.army.mil
[3] US Army Corps of Engineers, 333 SW 1st Avenue, Portland, OR 97204, 503-808-4794, william.b.fortuny@usace.army.mil
[4] US Army Corps of Engineers, 333 SW 1st Avenue, Portland, OR 97204, 503-808-4943, david.r.hamernik@usace.army.mil
[5] US Army Corps of Engineers, 333 SW 1st Avenue, Portland, OR 97204, 503-808-4955, matthew.k.hess@usace.army.mil
[6] US Army Corps of Engineers, 333 SW 1st Avenue, Portland, OR 97204, 503-808-4900, mark.j.sawka@usace.army.mil

Table 1 provides an inventory of the District's Tainter Gates. Since 2001 there have been various studies, inspections, and incidents that have prompted the District to become concerned about the structural integrity and mechanical and electrical reliability of these gates. There are several factors that led to this conclusion.

- The consequences, both life safety and economic, of an unreliable spillway Tainter gate system are significant.
    - Life safety consequences are greatest at the Willamette and Rogue River projects where a full pool release through a spillway bay exceeds downstream channel capacity. These projects have large volume reservoirs and little or no other means to evacuate the reservoir.
    - Economic consequences dominate on the Columbia River projects, due to lost hydropower and navigation. Life safety consequences are considered.

- Design criteria have been updated as knowledge expanded due to operational experience. The majority of the District's gates are deficient when evaluated under the current design guidance. New guidance includes effects of loads such as trunnion friction and single sided hoisting. Structural analysis indicates the gates are overstressed when operated, even under normal pool levels.

    - Maintenance was deferred in previous years leading to frozen trunnion pins, sheared keeper plate bolts and increased friction.
    - Fabrication deficiencies have been identified during inspections. The primary fabrication deficiency is poor weld quality and cracked welds.
    - Operational use changes have occurred. These operational changes include spilling with small gate openings for improvements to water quality (temperature control), spilling for fish passage, and spilling for power load balancing resulting from wind power generation being added to the grid.
    - Inspections have identified numerous deficiencies including: localized buckled members, wire rope failures, seized trunnion pins, cracks in steel in gate members, and deformed members.

Table 1. Portland District Tainter Gates and Status.

Project	Total Number of Gates	Placed In Service	Design Guidance Used (Assumed)
Willamette Valley Basin Projects			
North Santiam Subbasin			
Detroit	6	1950	1945
Big Cliff	3	1951	1945
South Santiam Subbasin			
Green Peter	2	1962	1950
Foster	4	1964	1950
McKenzie River Subbasin			
Cougar	2	1959	1950
Blue River	2	1965	1950
Middle Fork Willamette Subbasin			
Hills Creek	3	1957	1950
Lookout Point	5	1951	1945
Dexter	7	1953	1950
Fall Creek	2	1963	1950
Coast Fork Willamette and Long Tom Subbasins			
Fern Ridge	7	1940	Unknown
Rogue Valley Projects			
Applegate	2	1988	1966
Lost Creek	3	1976	1966
Columbia River Projects			
The Dalles	23	1952	1950
John Day	20	1966	1950

Tainter Gate Assessment

In 2010 the current reliability and safety of the District Spillway Gate Systems was assessed due to the following events:

2008 – Buckled Tainter gate strut arms at three of the four Tainter gates at Foster Dam, which led to emergency Tainter gate repairs.

2009 – Wire rope failure at two of three gates at Big Cliff, which led to emergency wire rope replacement.

2009 – Buckled Tainter gate strut arms at Dexter Dam where emergency Tainter gate repairs were implemented.

The assessment was done using risk-based tools and a risk-informed decision making process. The goal of this analysis was to inform the District of the risk associated with the operability and dependability of spillway Tainter gates. By establishing the relative risk at each of these projects, a prioritization of risk mitigation and IRRMs can be established. IRRMs are a temporary approach to reduce Dam Safety risks while long-term solutions are being pursued. In establishing IRRMs, the prevention of loss of life is the first and foremost objective, followed by prevention of catastrophic economic or environmental losses. A systems-based repair and capitalization program was implemented to improve gate reliability. Repairs are complete or underway at 30 of the 42 Tainter gates in the Willamette Valley.

PROJECT OPERATIONS AND INTERIM RISK REDUCTION MEASURES

Most of the projects in the Willamette Basin requiring IRRMs due to Tainter gates are flood storage projects (exception two re-regulation dams, Big Cliff and Dexter) and are regulated seasonally in accordance with a Water Control Diagram, including the Flood Control Rule Curve, which is not normally exceeded except to regulate floods.

The Conservation Season occurs from February through October, and is divided into the Conservation Storing Season and the Conservation Release Season. The general operation of the Willamette Valley storage projects is to fill the reservoir during the Conservation Storing Season (February through May). During this period, the filling rate of the reservoirs coincides with decreasing flood potential, as determined by analysis of historic floods at this time of year. Minimum reservoir releases are maintained throughout the Conservation Storing Season.

During the Conservation Release Season (May through October) water is released to maintain a minimum outflow from the projects, and to meet minimum flow requirements at downstream control points on the main stem Willamette River. Near the end of the Conservation Release Season (usually after Labor Day) releases at the reservoirs are increased to draft to the minimum flood control elevations (generally by 1 December).

The Major Flood Season in the Willamette Basin is November through January, although some flooding has occurred in February. During this period, the general reservoir operation is to pass inflow or a minimum outflow as long as flows at downstream control points are below flood regulation goals (bank full levels). In a flood control operation, reservoir outflows are reduced and water is stored in the reservoirs until the peak of the flood has passed and downstream flows have receded below the flood regulation goal levels. Outflows from the reservoirs are then increased to evacuate water stored during the flood control operation.

To reduce risk associated with failure of the Tainter gates, gate operating restrictions were implemented at nine projects in order to reduce risk on an interim basis until the gates could be rehabilitated. Operation of the gates with the pool near the top of the gate can mobilize large friction forces within the trunnion, resulting in additional bending in the struts. If these forces are large enough, the gate structure could catastrophically fail resulting in a sudden release of water from behind the gate. These sudden releases would be significantly larger than typical, for example, the downstream river could be flowing at 1000 cfs then suddenly jump to 30,000 cfs due to a failed gate.

Analysis showed that operating the gates at these nine projects with the pool above the upper girder on the gate could overstress the struts. The gate operating restrictions significantly reduce the likelihood that the Tainter gates would be operated with the pool above the upper girder. It is possible to operate at a higher pool elevation than the elevation of the upper girder by raising the gates as the pool rises (the gates would need to track with the pool). Gate operating restrictions at a project remain in place until rehabilitation work on the Tainter gates at the project are complete. The use of gate tracking forces the project to pre-maturely release flows that might not have been released until the downstream flow receded.

GATE REPAIR PROGRAM

Although structural concerns initiated the spillway gate rehabilitation program effort, Portland District realized through the PFMA process that a systems-based approach was necessary for spillway gate rehabilitation. The District focused on four factors to ensure reliability and safety:

- Structural strengthening
- Improving mechanical reliability
- Improving electrical reliability
- Implementing a regular operational testing, maintenance, and inspection program

Structural Strengthening

The structural problem encountered was spillway Tainter gates required strengthening due to unaccounted loads (typically trunnion friction) and new design criteria. The objective of the program was to strengthen the gates to meet the current design criteria. Typically, the overstressed members were the struts, ribs and horizontal girders. The criteria used to strengthen the gates came from USACE EM 1110-2-2702 and USACE ETL 1110-2-584.

Finite element models (FEM) were built for each gate including separate models for quality control purposes. Load combinations were run from the criteria to determine the worst case loading and to ultimately determine which members were overstressed. At the same time, steel samples were taken from gates in order to verify yield and ultimate strengths as well as chemical composition for weldability determination.
The most significant issue involved in the strengthening program was to address the large machinery load (Q) that occurs in a few of the load combinations as seen in Table 2. The maximum applied machinery load is generally two to five times larger than the required lifting load. This load occurs when the gate is jammed within its bay in a non-moving single-sided hoisting event (LC 5) or when the gate is completely open bearing against its gate stops (LC 6) while the hoist is still operating. To overcome this issue a load limiting device (LLD) was incorporated into the design which limited the hoisting load to what was required to operate the gate under its worst case loading (LC 3, single-sided hoisting, gate moving) in order to achieve the flood control mission.

Welding and fabrication issues from the 1950's and '60's needed to be addressed. Typically, complete joint penetration welds were not back-gouged, which left a line of indications at the joint's mid-depth and effectively made them partial penetration welds. Therefore, a fit-for-service weld analysis was performed to ascertain if the joint had adequate capacity. Intersecting welds that cause restraint within the joint and that can lead to fracture were drilled out to eliminate a brittle failure mode (Figure 1).

Table 2. Load Factors for Tainter Gates.

Load Cases			Loads/Load Factors							
			D	G	Hs	Hd	Q	EV	IM	EQ
Limit State	Description	Case	γD	γG	γHs	γHd	γQ	γEV	γIM	γEQ
Strength I	Gate Closed	1	1.2	1.6	1.4Hs2	0	1.2Q1	0	0	0
Strength I	Gate Closed	1	1.2	1.6	1.4Hs2	1.2(1)	1.2Q2	0	1.6(1)	0
Strength II	Usual Operation	2.a	1.2	1.6	1.4Hs3	0	1.4QFs2+ 1.0QFt2	0	0	0
Strength II	Usual Operation	2.b	1.2	1.6	1.4Hs2	1.2(1)	1.4QFs2+ 1.0QFt2	0	1.6(1)	0
Strength II	Unusual Operation	3	1.2	1.6	1.4Hs2	0	1.4QFs2+ 1.0QFt2	0	0	0
Extreme I	Gate Closed	4.a	1.2	1.6	1.4Hs3	0	1.2Q2	0	0	0
Extreme I	Gate Closed	4.b	1.2	1.6	1.4Hs1	0	0	0	0	1.0EQ
Extreme I	Gate Jammed	5	1.2	1.6	1.4Hs2	0	1.2Q3	0	0	0
Extreme I	Gate Opened	6	γD2(2)	γG2(2)	0	0	1.2Q3(1)	1.3W(1)	0	1.0EQ(1)

Notes: (1) Select one at a time
(2) Use max or min values, whichever produces maximum effects

Figure 1. End Frame at Casting Intersecting Welds, Fall Creek Dam, Lowell, Oregon.

Various repair options were considered for each individual project that took into account schedule, cost, constructability and the risk of unknown field conditions. Generally, the strut arms, ribs and horizontal girders required strengthening. The ribs and girders were reinforced by either additional bracing or increasing the section. Repair options considered for strengthening of the strut arms included: cover plates, complete end frame replacement (struts and strut bracing) and additional bracing. The addition of bracing (to decrease the effective length of the strut arms) was typically not a viable option. The strut arms see significant axial and weak axis bending forces and relatively small strong axis bending forces. Therefore, even if the strong axis of the strut arms could be effectively braced the additional capacity gained would not do much to bring down the stress level in the section. Strong axis bracing of the strut arms is not practical since bracing would be required to cross the width of the spillway to the gate's adjacent end frame or attach to the horizontal girders. The location of the existing weak axis bracing from the original design is typically at its optimal location. Therefore, the section's axial capacity is 'capped off' and any additional bracing that would decrease the strut's effective length would not increase section capacity.

There are numerous benefits to being able to repair a gate on seal by reinforcing the strut arms with cover plates. This saves time and money and lessens the impact on other disciplines. Cover plates have been implemented at Willamette Valley projects such as Fall Creek and Lookout Point Dams outside of Lowell, Oregon, and Foster Dam outside Sweet Home, Oregon. The strengthening of the Fall Creek Dam spillway gates may be seen in Figure 2. However, there are issues that need to be evaluated if this option is to be implemented. For instance, cover plate termination before the casting was required for the Fall Creek project. It is not desirable to weld to the castings due to potential cracking concerns and melting of the trunnion bushing from welding heat. An analysis was performed in order to determine if the built-up section immediately upstream of the steel casting (not reinforced with a cover plate) had adequate strength (Figure 1 and Figure 2). Fatigue is an issue that needs to be addressed for cover plates. Welded cover plate terminations have an undesirable fatigue category associated with them, which can ultimately lead to fracture of the base metal member.

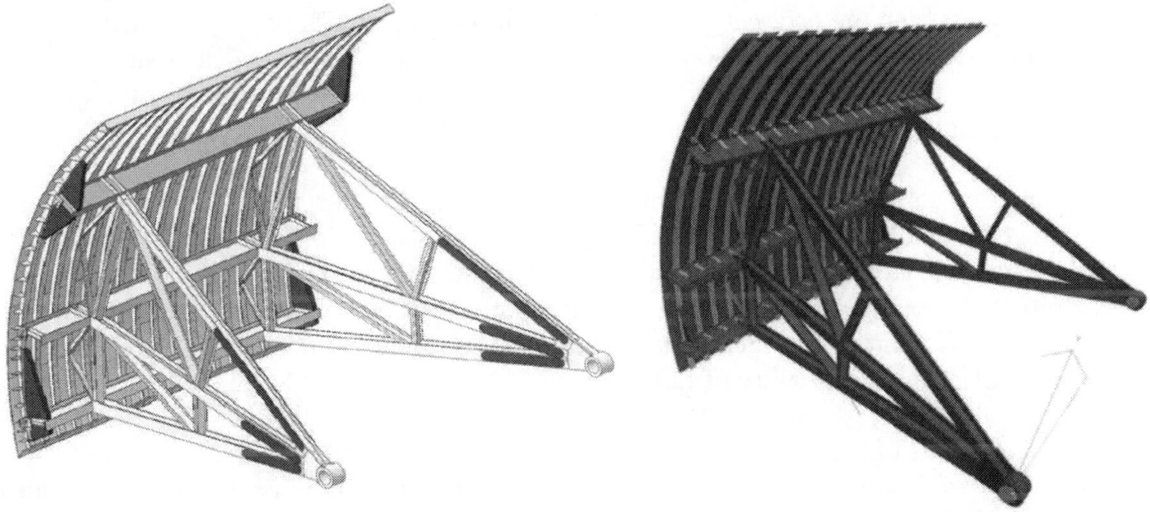

Figure 2. Strengthening of Fall Creek Dam Spillway Gates. Drawing Isometric Left and FEM Isometric Right.

Complete end frame replacement including the trunnion hub, is another strengthening option which was implemented at Dexter Dam Spillway gates outside of Lowell, Oregon (Figure 3), Big Cliff Dam outside of Detroit, Oregon, and Green Peter and Foster Dams outside of Sweet Home, Oregon. This option is obviously more expensive than cover plates and also requires additional work and schedule consideration since the gate will be taken off seal. However, this option minimizes the risk of unexpected field conditions and results in a new end frame that utilizes modern steel and welding techniques, which increases long term reliability.

Figure 3. New End Frame at Dexter Dam Spillway Gates, Lowell, Oregon.

Electrical Modifications for Reliability

The electrical assessment was conducted to establish a probability of failure for the electrical components associated with the operation of spillway Tainter gates at Portland District projects. For the purposes of the electrical assessment, the probability of failure is defined as the likelihood the electrical equipment will not perform as intended when operated, which will prevent the spillway gate from opening. It does not take into account the reasons behind the necessity for operation, the consequences of inoperability, or the timeframe required to troubleshoot and repair the failed component. The goal is to obtain a probability of failure below 1.0E-04 for each gate individually and also for the gate system at each project.

At each project, the electrical equipment was subdivided into three primary categories: (1) power sources; (2) power distribution; and (3) control equipment. A probability of failure for each category of equipment was derived by developing a fault tree based on the project's electrical configuration and distribution system, as well as its local control equipment. The three individual values were then summed to provide an overall project spillway electrical probability of failure.

From the results of the District reliability study it was observed that the dominant probability of failure lies with the local control equipment, which is also the easiest and least expensive to repair or replace. The second most likely to cause a failure of the gate to operate is the loss of distribution equipment due to the lack of redundancy in dam power feeders at most of the District's sites. The least likely to cause a failure to operate is the loss of all power sources which includes utility power, loss of medium voltage switchgear in the powerhouse, and emergency backup generators.

In order to increase the reliability of the District spillway gate electrical systems the following modifications have been made as part of the gate rehab projects:

Controls: Replace entire control systems for all spillway gates. Convert all local spillway gate control systems to deadman style pendant controls, provide multiple pendants for each site that can be used on all new gate systems across District sites. This pendant style system provides redundant pushbutton controls and thus increases reliability. Also, by converting to the deadman style the controls were simplified and the number of components used in the control systems was reduced by eliminating incremental limit switches and timing relays, which reduces the overall probability of failure.

Power Distribution: Replace existing dam feeder and install/replace redundant dam feeder. It is also important to ensure each dam feeder is physically separated from one another, i.e. each feeder either takes different route to dam or is in separate raceway. For example, two feeders fed from the same distribution panel with a single bus are not considered to be redundant. The goal is to ensure that a single fault in a piece of distribution equipment or physical damage to a single raceway will not cause a loss of both feeders.

Power Supply: Replace switchgear and/or associated feeder breakers for main and redundant dam feeders. Replacing or rehabilitating the emergency backup generators is also considered on each of the rehab projects. In addition to these changes, adding a portable generator supply receptacle at the spillway decks, which allows the ability to utilize trailer mounted generators for emergency spillway operation, has also been done, thus adding a fourth available source of power.

One other recommendation to come out of the electrical reliability assessment, which has not been addressed due to operational constraints at a majority of the District's sites, is increasing the frequency of operations. An increase in operations would improve the reliability of the gate systems, but there are operational constraints that do not enable the gates to operate except when the reservoir levels are down below spillway crest, which is typically only during flood season.

In addition to increasing the reliability of the spillway gate systems, the modifications that have been made greatly increase the operability and maintenance of the systems. The reduced hardware and simplified control systems have decreased the time it takes for operators to be trained on these systems. Additionally, troubleshooting these systems is simplified as a result of the reduced amount of hardware.

Mechanical Modifications for Reliability

The mechanical objective of each gate rehab project has been ensuring the hoist and trunnion bearing systems are in a condition that will result in safe and reliable gate operation for the next 50 years of service. To obtain this objective each system has been evaluated for compliance with current design criteria and the potential for continued reliable operation. The scope required to obtain this objective at each site has been full replacement of the mechanical hoisting and trunnion bearing systems.

Replacement of some hoist systems has been necessitated by gate weight increases from the structural retrofit. However, even locations with negligible weight increases have required replacement to ensure compliance with more conservative, updated mechanical criteria. In addition, most original hoist systems have met or exceeded their intended service life, are starting to show signs of degraded reliability, and are not expected to provide continued reliable service for a 50 year lifecycle. Full system replacements have dramatically increased the reliability of District spillway gate systems. In addition, other improvements, such as designing to more conservative criteria, improved material selections, and design simplifications have also resulted in reliability improvements.

The main design criteria changes that have occurred since the original District gate systems were installed have been the assumptions of how overload hoisting forces are distributed between sides of a gate. Original systems were designed assuming that a maximum overload condition would result in a 50/50 load split between sides of the gate. Gate operating experiences have shown this assumption is not valid. When gates are jammed between piers they tend to flex and skew resulting in differential hoisting loads between sides of the gate. Current USACE criteria requires designing for overloads with a 70/30 load split between sides of a gate. Portland District designers have gone a step further and have been designing District hoist systems for a 100/0 load split to match the structural design criteria. Designing to this more conservative overload assumption has been made feasible with the use of load limiting devices, which are discussed below. Other design changes have helped to improve assembly, alignment, and installation of the new hoist systems. This includes C-face mounting of drive system components and skid mounting hoist assemblies with electrical controls. Both of these changes have helped facilitate shop assembly/alignment and have minimized the onsite installation time where the gate is out of service, Figure 4.

Figure 4. Shop Assembly and Testing of Skid Mounted Hoists.

The largest change in material selections have been on trunnion bearing systems. Portland District has utilized fabric reinforced composite self-lubricated trunnion bearing materials to improve the reliability of the trunnion bearing lubrication. Spillway Tainter gate trunnions across the US have had issues with seized or "frozen" trunnion bearings. This often results from insufficient grease being pumped across the full running surface of the bearing. Self-lubricated materials have the primary lubricant (solid lubricants) incorporated into the parent bearing material. The lubrication is always present on the bearing running surface and does not rely on grease pumps, grease lines, maintenance funding, and other external factors for successful performance. PTFE seals have been utilized with the self-lubricated materials to exclude water and contaminants and ensure low coefficients of friction, Figure 5.

Figure 5. Self-Lubricated Trunnion Bearing and Thrust Washer.

Load Limiting Devices

Changes to more conservative structural and mechanical design criteria have driven a need to optimize how the hoist system is controlled in overload cases. The purpose of a spillway gate is to control flow. When an overload condition is experienced a gate is still capable of performing the function of controlling flow until it becomes jammed between piers. When a gate jams there is no value added in subjecting the gate to additional load. The District has been able to optimize hoist and gate sizing by utilizing load limiting devices (LLD) to prevent unnecessary load after a gate becomes jammed.

To determine appropriate load limits for a hoist system Portland District has investigated each gate being rehabbed with a 3D FEM of the gate. Deflections caused by the gate skewing under single-sided hoisting are investigated in the model and reaction forces are observed where the gate would be in contact with the pier faces. Next, a coefficient of friction is applied to the pier reactions resulting in a drag force that is used to quantify the theoretical maximum single-sided hoisting load required to operate the gate. An uncertainty factor is then applied to the theoretical hoist load to account for efficiencies of the mechanical system as well considering the accuracy range of the LLD.

Portland District has performed this analysis on many gate systems and found that most require a motor overload limit between 130%-140% of the motor full load torque (FLT) to prevent unnecessary hoist loads. Motors used for hoisting application are typically selected as a NEMA design B or D. Per NEMA MG-1 design D motors are required to have a max torque no less than 275% of the full load torque with actual values commonly between 300%-350% of FLT. Design B motors are required to have a max torque no less than 205%-225% of the FLT (depending on hp rating and RPM for sizes common to spillway gates) with actual values commonly between 250% - 275% of FLT. These common hoist motor designs have a tremendous amount of overload potential above what adds value to the goal of controlling flow. Portland District has started utilizing LLD's to lower the max torque the system can deliver to keep designs feasible and economical.

Portland District has investigated various LLD's for spillway Tainter gate hoist systems and has authored a White Paper on this subject. Key features of these devices include:

- Reliable operation (devices that don't operate reliably are unacceptable)
- Overload protection for 100% of the operating time (no protection at inrush is unacceptable)
- No disengagement of the load (devices are load holding and never disconnect the motor or brake from the load which could potentially drop the gate)
- Fast reaction time (time between sensing overload and system shut down are minimal)

After evaluating LLD's against a typical spillway gate application there were three devices that had clear advantages over others. These top LLD's are:

- Custom wound motors – A motor with a custom winding design that meets the normal operating and overload characteristics specific to the application. The motor is designed with a reduced overload value that prevents overload of the system.
- Torque Transducers – A device that measures drive train torque and triggers a relay to shut the hoist system down when allowable values are exceeded.
- Torque Switch – A shaft coupling with a ball detent system calibrated to move a coupling plate axially when a critical torque value is reached (without disengagement of power-train load). The axial movement of the plate can be used to activate a switch or proximity sensor to shut the hoist system down when allowable values are exceeded.

Portland District has successfully installed custom wound motors and is evaluating them for long term use. In addition, the District is also in the process of installing and evaluating torque transducers for successful performance.

OPERATION AND MAINTENANCE CHANGES

USACE Northwestern Division introduced a policy (Corps of Engineers, Northwestern Division, Operational Testing and Inspections, Periodic Inspections, and Evaluations of CENWP Spillway Gates) in 2012. This policy requires regular trunnion bushing lubrication, full cycle open/close tests, full mechanical/electrical component inspections, and hands-on structural inspections. This has been fully implemented across the District to ensure gate reliability and safe operation. Results of the testing and inspection are included in project Periodic Inspection reports.

PRIORITIZATION OF REHABILITATION WORK

In order to develop a long-term plan for repairing the District's spillway gates, a process to prioritize gate repair at the project level was developed. A matrix was developed to consider the various criteria that would be used to prioritize repairs and provided a tool to rank the projects based on these criteria. Each criterion had a metric defined for it and a score was given for each project. It was decided that the various criteria should not all have the same weight in terms of determining the priority of repair; sensitivity analyses were performed to assess the impact of varying the weight of each of the factors. The criteria for prioritization included:

Engineering Reliability: these criteria were based on the results of structural, mechanical and electrical reliability assessments. Rehabilitation of projects with lower reliability systems were prioritized over higher reliability projects.

Economic Impacts: these criteria were based on the economic impacts of the gate operating restrictions to the project missions (flood damage reduction, hydropower and recreation). Rehabilitation of projects with high economic impacts were prioritized over those with lower impacts.

Operations Impacts: these criteria were based on the operations manpower impacts of existing mechanical and electrical reliability deficiencies and from the gate operating restrictions. Rehabilitation of projects with larger operation impacts were prioritized over those with lower impacts.

Environmental Impacts: these criteria were based on the impacts of the gate operating restrictions on the ability to refill the reservoirs in the spring, fish and wildlife habitat, and water quality. Rehabilitation of projects with larger environmental impacts were prioritized over those with lower impacts.

CONSTRUCTION WINDOWS

The flood protection projects do not have bulkheads or the provisions to install bulkheads so identifying construction windows is a balancing act between of all of the project functions. It's assumed the gates must be hydraulically isolated for eight weeks, per gate, during the structural repairs. Outside of these windows the gate must be able to be on seal and hydrostatically hold full pool. Other work activities involving tainter gate repairs (i.e. mechanical/electrical) can be done with the gate hydrostatically loaded as long as they do not affect or have an impact on load paths to the gate during structural gate modifications. It is necessary to maintain sufficient outlet capacity for the flows associated with the particular time period (smallest requirements during the summer and largest requirements during the winter). Ideally all spillway gates should be available during flood season and one or two gates available during the rest of the year. Short duration outages can most likely be scheduled when the pool is below the spillway crest but will need to be coordinated with Reservoir Regulation and Water Quality Section (EC-HR) on a case by case basis and will depend on weather conditions and forecast.

The factors that play into identifying the construction window are: Flood Damage Reduction, Water Supply, Recreation, Power Generation and Environmental Concerns. The analysis was done using HEC-ResSim, the analyses used the 73 year Period of Record daily flows to obtain statistical information on pool elevations, releases, and impacts to BiOp flows. The entire Willamette Basin is modeled so construction at multiple projects was included in the analysis.

For each project a specific construction window (specific weeks and year) has been identified. During structural repairs the project releases are managed aggressively during a flood event and when evacuating stored water after the event. A goal of regulation will be to get the pool back down below spillway crest to minimize delays to the Contractor since some repairs cannot be conducted while the gate is hydraulically loaded.

Given the construction window, a Special Operations Request must be submitted to and approved by EC-HR prior to construction window start. Final coordination with EC-HR and WVP must occur immediately prior to beginning construction activities and throughout the work in or near the lake.

CONSTRUCTION WITHOUT BULKHEADS

Of the Willamette Basin Projects, only Big Cliff, Foster, and Dexter have bulkheads to allow for dewatering of the gates during construction. For the other projects, preventative maintenance typically occurs during the flood season when reservoir elevation is below the spillway crest. During major construction to structurally strengthen the gates, reservoir elevation could rise above the spillway crest and load the gate. Two approaches were taken to facilitate construction. For Hills Creek, Blue River, and Lookout Point projects, strengthening could be accomplished by increasing section modulus with welded plate. Construction could be accomplished while the reservoir was below crest. See Figure 6. For the Green Peter project, full end frame replacement was required. In this case bracing was fabricated and installed to support the radial skin plate. This allowed partial loading of the gates while the end frames were removed and replaced. See Figure 7.

Figure 6. Cover Plate Strengthening

Figure 7. End Frame Removal w/Braced Skin Plate

REHABILIATATION PROGRAM

In 2009, Portland District initiated a capitalization program to rehabilitate all spillway gates in the Willamette Valley. Table 3 provides the progress to date of the District's rehabilitation program. Progress on the spillway gate work has occurred at a faster rate than many in the District initially expected.

Funding sources for the work was from USACE Operating and Maintenance (O&M) program and the Federal Columbia River Power System Large Capitalization Program. Processes to obtain O&M funding for the spillway gates include the prioritization of District needs associated with the spillway gates and other critical infrastructure needs. The prioritization process is informed based on the risk posed by the infrastructure deficiency for which the funds are needed. Throughout the year, the District identifies and prioritizes infrastructure needs to guide the District's annual budget request. Based on the risk posed by unreliable spillway gates, funding to rehabilitate the spillway gates is high on the list of priorities. In addition, the District engages in a regional project prioritization process that includes prioritization of critical infrastructure needs among the five Northwestern Division District's within USACE. Portland District's spillway gate deficiencies have also been high on the regional list, enabling funding to be obtained for the rehabilitation program.

Table 3 Portland District Spillway Gate Rehabilitation Program Status

Project		Strengthening	Mechanical and Electrical Reliability
WILLAMETTE VALLEY BASIN			
Detroit	6 Gates	Design underway ; 2017 budget request	In Design; 2017 budget request
Big Cliff	3 Gates	Repairs completed in 2013	Repairs completed in 2013
Green Peter	2 Gates	Construction underway, repairs to be complete in 2017	Construction underway, repairs to be complete in 2017
Foster	4 Gates	Repairs completed in 2011	Repairs completed in 2011
Cougar	2 Gates	Design underway; Construction Contract will be awarded in 2016	Design underway; Construction Contract will be awarded in 2016
Blue River	2 Gates	Construction underway, repairs to be complete in 2017	Construction underway, repairs to be complete in 2017
Hills Creek	3 Gates	Construction underway, complete in 2017	Construction underway, complete in 2017
Lookout Point	5 Gates	Construction underway, complete in 2016	Construction underway, complete in 2016
Dexter	7 Gates	Repairs complete in 2014	Repairs Complete
Fall Creek	2 Gates	Repairs complete	Repairs Complete
Fern Ridge	6 Gates	Analysis Indicates Strengthening is not required	Mechanical and Electrical Repairs are Planned
ROGUE VALLEY			
Project		**Strengthening**	**Mechanical and Electrical Reliability**
Applegate	2 Gates	Analysis Indicates Strengthening is not required	Not required
Lost Creek	3 Gates	Analysis Indicates Strengthening is not required	Mechanical and Electrical Repairs are being planned
COLUMBIA RIVER			
Project		**Strengthening**	**Mechanical and Electrical Reliability**
The Dalles	23 Gates	In planning phase	In planning phase
John Day	20 Gates	In planning phase	In planning phase

FUTURE WORK

Columbia River Spillway Gates

Once the Willamette and Rogue projects are complete, the effort will be expanded to include the Columbia River projects. Although potential life safety consequences at these projects is unlikely since the instantaneous change in discharge would not be significant, the potential economic consequences are greatest at the Columbia River projects, where hydropower and navigation capabilities could be impacted in the event of Tainter gate failure.

Regulating Outlets Reliability

There are 34 reservoir regulating outlets (ROs) in the Willamette and Rogue Basin Projects. The regulating outlets, just as the spillway gates, are integral to the flood damage risk reduction mission. Recent work on ROs within the District revealed a high likelihood of systemic reliability issues. Inability for the gates to function as designed could lead to dam safety and, consequently, life safety issues for downstream stakeholders. The regulating outlets are operated under hydraulic pressure. Any oil leaks could have significant environmental impacts. Additionally, the ROs may be used for downstream temperature control as an interim measure as part of the fish passage improvement plan. Loss of use would severely impact fish listed under the Endangered Species Act in the Willamette and Rogue River Basins. Efforts to assess the condition of each RO and provide a comprehensive evaluation and report on their condition have just begun.

Trunnion Anchors

The trunnion is often anchored to the concrete through steel anchor rods. USACE is currently researching and developing means to test the integrity of these anchors.

REFERENCES

ETL 1110-2-584, Design of Hydraulic Steel Structures, April 30, 2014, USACE

EM 1110-2-2702, Design of Spillway Tainter Gates, January 1, 2000, USACE

EM 1110-2-62610, Mechanical and Electrical Design for Lock and Dam Operating Equipment, June 30, 2013, USACE

NEMA MG1, 2011, National Electric Manufacturers Association

LESSONS LEARNED FROM THE SPILLWAY GATE FAILURE AT FOLSOM DAM

Jerzy Salamon, PE, PhD[1]

ABSTRACT

In 1995, failure of spillway Gate No.3 at Folsom Dam during normal operation resulted in an uncontrolled release of the reservoir. Although no significant downstream damage occurred, reservoir storage was partially lost due to the gate failure.

In response to the failure of the gate, the Bureau of Reclamation (Reclamation) formed a multi-disciplinary, multi-agency Forensic Team to investigate and determine the causes of the failure. The team attributed the gate failure to the collapse of the arm frame, with excessive friction at the gate trunnion due to corrosion at the surface of the trunnion pin being important factors. Comparable loading conditions were not accounted for in design practices during the 1950's.

The failure of the spillway gate at Folsom Dam initiated a national dam safety program to inspect and evaluate the structural integrity of radial gates within state and federal jurisdiction. The outcome from the program had significant impact on the change to the design practice of radial gate structures.

The case study and lessons learned from the failure of the gate at Folsom Dam are presented in this paper. In addition, a summary of the forensic team's investigation is provided in the paper along with the author's additional remarks made based on the advanced structural analysis results of the original gate structure.

INTRODUCTION

Project Information

Folsom Dam is located on the American River, approximately 20 miles northeast of the City of Sacramento, California. The dam was designed and constructed by the United States Army Corps of Engineers (USACE) between 1948 and 1956 and was transferred to Reclamation for operation and maintenance in May 1956. The Main Dam consists of a straight concrete gravity structure across the river channel flanked by long earthfill wing dams.

A new auxiliary spillway, equipped with six submerged gates, is currently under construction as a part of the Joint Federal Project. After completion in 2017, the new spillway will increase the discharge capacity at Folsom Dam.

[1] Technical Specialist for the Waterways & Concrete Dams Group I at the Bureau of Reclamation, P.O. Box 25007, Denver, Colorado, 80225, USA, tel. 303-455-3219, e-mail: jsalamon@usbr.gov

Spillway Gates

The Main Dam concrete section has an overflow spillway equipped with eight crest type radial gates (also called Tainter gates) numbered 1 through 8 from right to left looking downstream: five service gates and three emergency gates. The spillway gates are operated by electric chain hoists. A 140-ton gantry crane at the crest of the dam is used for installation of stoplogs at the spillway bays.

The radial gate leaf consists of a skinplate reinforced by vertical tee-type stiffeners. The leaf is attached to four horizontal girders. The girders are supported by four strut gate arms. The gate arms are angled in plan view and are braced in-plane between bottom struts and out-of-plane by the transverse bracing between both gate arms. Tested material properties of the original Gate No. 3 members were found to meet the composition requirements for both the ASTM Designation A36 and A7 specifications with the yield limit of 33,000 lb/in^2, except for the second strut from the top in the right gate arm where the yield strength was 4 percent below the minimum required strength for A7 steel. The steel used in the fabrication of the gate structure appears to be adequate in original Gate No. 3.

Spillway gate trunnion assemblies consist of casted bronze bushings, forged carbon steel pins, and casting blocks. The gate load transmits through casting blocks to bearing blocks and then to the 6-inch by 16-inch and 60-foot-long steel trunnion anchors. A bronze bearing plate allows relative movement between the gate trunnion assembly and the concrete embedded bearing block. The trunnion pin (forged steel SAE 1045, BH 108 min.) is hollow, with a 32-inch-outer-diameter, 24-inch-inner-diameter and greased 1-inch-thick bronze trunnion bushings.

SPILLWAY GATE INCIDENT

General

On Monday, July 17, 1995, Gate No. 3 was scheduled to be opened 3 feet to maintain 6,000 ft^3/s flow in the river during a powerplant shutdown. The reservoir water level corresponded to 43.5 feet of water head at the spillway gates. Since May 1958, Gate No. 3 had not been opened more than 2.0 feet with a maximum head of no more than 40 feet.

The operation procedure involved automatically stopping the gate at 6 inches and again at an opening of 1-foot. The auto-stop function was overridden (normal procedure) with no stop being made at a 2-foot opening [5].

Reporting Gate No.3 Failure

Operation of Gate No. 3 on July 17 began at 7:58 in the morning. As the gate opening approached 2.4 feet, the gate operator noticed an "unusual vibration" and stopped the gate hoist motor. The operator stated that at this time, "the gate swung open slowly, like a door hinged on the left side and water was pouring around both sides of the gate leaf."

The breach of the gate was restrained by the right hoist chain (Figure 1). The time from the operator's initial awareness of the vibration to observing gate displacement and uncontrolled flow of water was estimated to be no more than 5 seconds [5].

Figure 1. Failure of Gate No. 3 at Folsom Dam, July 1995.

Implications of the Gate Failure

The failed gate released a peak flow up to 40,000 ft^3/s; well below the rated downstream safe channel capacity. The flow was moderated mostly by operations on Nimbus Dam. The gate failure did not result in any loss of life or significant downstream damage. Downstream flooding due to the gate failure was limited to the park areas within the American River flood channel. Uncontrolled release through the spillway resulted in a 36-foot drawdown and loss of approximately 40 percent of the reservoir before the breach was closed with stoplogs [5].

Forensic Investigations

Following the failure of Gate No.3, Reclamation formed a multi-disciplinary, multi-agency Forensic Team to investigate and determine the causes of the failure. More than 30 different types of tests, examinations, and analyses were performed to assist the Forensic Team in determining the cause of the gate failure.

According to the forensic investigation report [5], collapse of the gate was initiated by a failure of the bolted connection at the upper end of the diagonal strut brace nearest to the trunnion in the right arm frame (Figure 2). Once the initial diagonal brace failed, the internal forces were transferred to the adjacent brace connections, which subsequently failed. Immediately following the brace connection failure, the lower most strut (strut No. 4) buckled downward leading to buckling of the remaining arm struts and resulting in the loss of structural integrity of the right arm (Figure 2).

Causes of the Gate Failure. The Forensic Team identified two main causes of the gate failure:

- Insufficient stiffness and strength in critical structural gate arm members – The forensic report [5] indicated that the brace bolted connections in the structural diagonal bracing members were the critical weak links in the entire gate structure and the failure of the gate initiated at the Joint 1 bolted connection (Figure 2).

- Increased trunnion friction due to corrosion of the steel trunnion pins - According to the failure investigation for trunnion fixture tests, rust occurred mainly on the loaded side of the pin surface [6]. With corrosion of the trunnion pin the friction coefficient may have been as high as 0.3 at the time of gate failure.

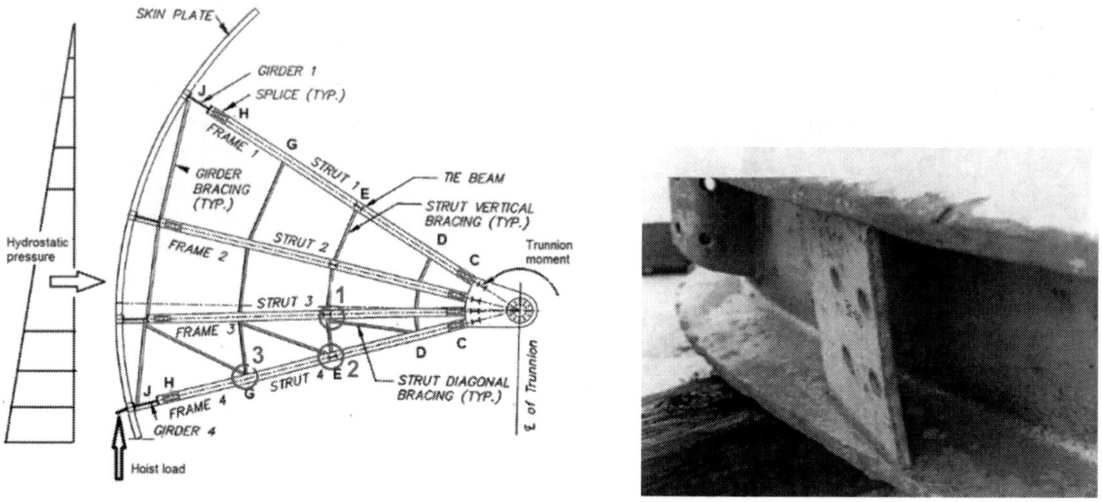

Figure 2. Right gate arm with indicated diagonal brace connection failure (left) and the gusset plate at the Joint 1 (right).

Other Contributing Factors to Gate Failure. Factors that may have contributed to the gate failure included:

- Corrosion at Gate Arm Members and Bolts - Considerable rusting of the gate arm members and their joints, especially at locations of debris collection, was identified during climb inspections of the structures. Although corrosion occurred in the joints of arm braces, it was considered in the forensic investigations to have a minor effect on the failure of Gate No.3.

- Corrosion of the hoist chain or the chain to gate connection - Although corrosion of the chain links occurred, it was considered in the forensic investigations to have a minor effect on the failure of Gate No.3

Non-Contributing Factors. Several other factors were investigated by the Forensic Investigators but were ruled unlikely to contribute to the gate failure. Some of these factors included: low-stress high-cycle fatigue, poor quality of welding, use of standard strength bolts, and potential gate vibrations.

Corrective Actions after Folsom Gate Incident

After failure, Gate No. 3 was replaced with a similar structure, with the exception of larger wide-flange sections selected for the arm struts, and an additional bracing system provided in the gate arms between the top two struts. The bronze trunnion bushings were replaced with self-lubricating bushings to maintain the trunnion coefficient of friction below 0.1. Subsequently, the other seven radial gates at Folsom Dam were modified to updated design standards.

Screening level evaluation of 78 radial gates in Reclamation's inventory identified fifteen gate structures that did not pass the screening criteria and were subject to further structural investigations.

EVALUATING STRUCTURAL INTEGRITY OF FOLSOM GATES

Forensic investigations and the following research by Reclamation and others [7][8] was associated mostly with the operational and mechanical causes of Gate No.3 failure. Although, a structural evaluation of the gate structure using Finite Element (FE) analysis was performed to determine critical trunnion friction moments at which the gate failure initiated, some additional observations could be formulated from the 20-year time perspective.

In this section, the author offers complementary considerations that provide additional insights on the failure of Gate No.3.

Discussing Gate Failure Sequence

Forensic Investigations. The explanation of Gate No.3 failure from the Forensic Report was based on the FE analysis results performed by McClellan AFB 1995 and the strength laboratory tests performed on the diagonal brace connection at Joint 1 (Figure 2) [4]. For the average 80,000 lbf capacity of the four ¾-inch standard bolt connection obtained in the test, the McClellan FE simulation of the gate failure determined that the corresponding friction coefficient at the gate trunnion was 0.25. The subsequent study by Reclamation concluded that the bolted connection at Joint 1 could have failed with the friction coefficient of 0.22 to 0.28 [5].

Evaluating Gate Failure. A new non-linear three-dimensional FE simulation of the gate structure was performed for the purpose of the research presented in this paper. The analysis confirmed that the strength capacity of the bolted connection at Joint 1 was reached for the trunnion friction coefficient of 0.28. The analysis showed that the Interaction Ratio (IR), as defined in [10], was about 1.0 at the most stressed Strut #3 trunnion end for the corresponding trunnion friction coefficient of 0.28. The results of the analysis indicated that the gate failure could have initiated through the failure of the diagonal brace connection as well as through the buckling of Strut #3 at the trunnion end. The estimated bending strength of Strut #3 would be even lower when the splice at the trunnion end (Section C-C in Figure 2) is considered in the analysis. For a typical splice at Strut #3 and Strut #4 (Figure 3), two ¾-inch-thick splice plates were bolted to a 1¼-inch-thick flanges using 1-inch-diameter bolts in standard holes. The strength of the cross-section at the splice is lower in bending when compared with the actual cross-section of 12Wx133 Struts #3 and #4 (Figure 3). In addition, full engagement of the splice plates in transferring forces requires some deformations at the connection, leading to increase in strut eccentricity and consequently higher second-order bending moment in the member. These all will result in IR above 1.0 and consequently it leads to the conclusion that the failure of Strut #3 at the splice most probably occurred even before the strength capacity of the diagonal brace connection in Joint 1 was reached.

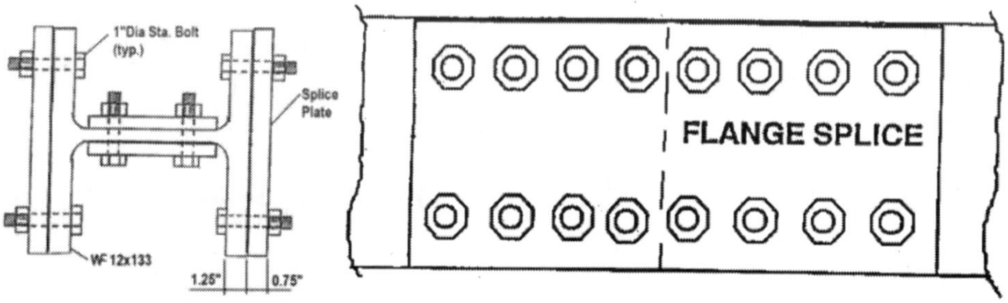

Figure 3. Typical strut splice.

Photographs in Figures 4 and 5 taken during the post-failure Gate No.3 examination [4] appear to support the conclusion from the FE analysis that the gate failure could have initiated at Strut #3. The arrows in Figure 4 show permanent deformations of the flanges at the Strut #3 splice. The inside flange of the trunnion assembly (left photo) is clearly bent in the opposite direction to the deformation of the adjacent Strut #3 flange (right photo). This would indicate the local buckling of the inside flange at the top at the splice.

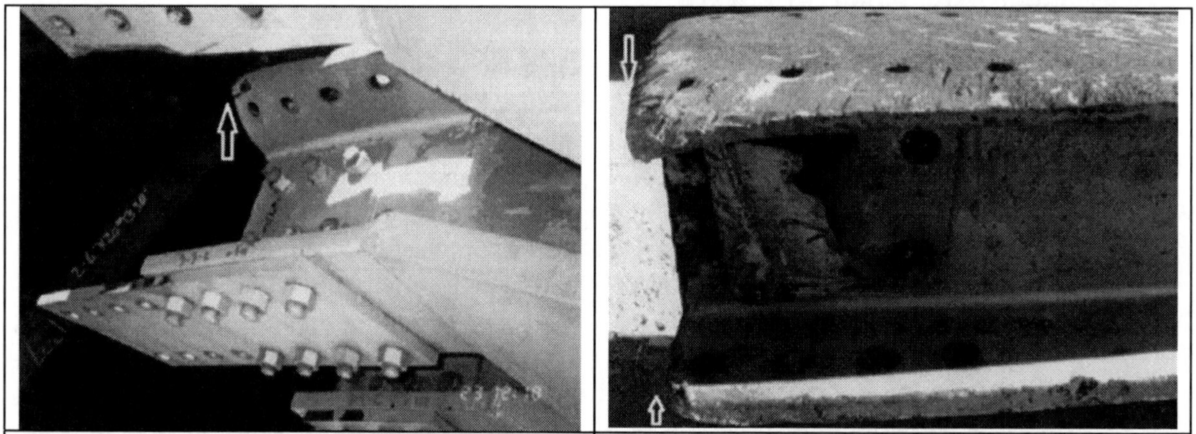

Trunnion, outside view, Strut #3, Joint 3-C	Right Strut #3 (Top), at Joint 3-C

Figure 4. Failure of the splice connection (Joint 3-C) at Strut #3 (Reclamation 1996-1).

Broken splice plate in Strut #4 (Figure 5) and sheared bolts for the remaining splice plates indicates that the splice connections was the weak cross-section in Struts #3 and Strut #4.

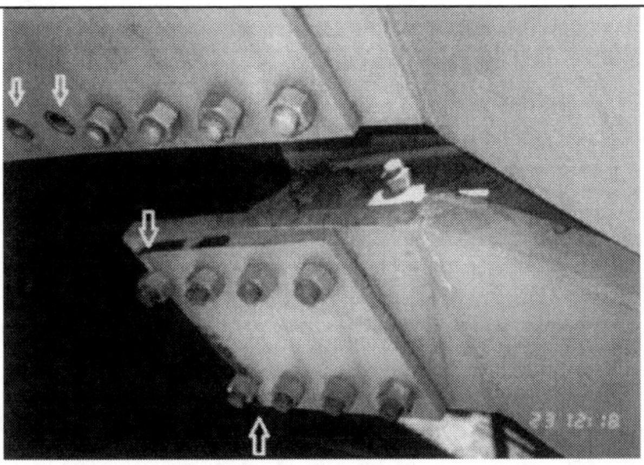

Trunnion, Strut #4 Splice at Joint 4-C	Right Strut #4 (Bottom), Splice at Joint 4-C

Figure 5. Failure of the splice connection (Joint 4-C) at Strut #4 (Reclamation 1996-1).

Trunnion Friction Moment

Forensic Investigations. Inspections of Gate No.3 trunnions indicated corrosion at the bearing surface of the carbon steel pins but no corrosion on the non-bearing part. Approximately 33 percent and 20 percent of the cylindrical surface area of the bushing was covered with rust at the right and left trunnion pins, respectively. Laboratory examination and testing of pin samples indicated that there was no evidence of galvanic corrosion, however, the hydroxide oxidation (rust) of the steel surface in the presence of water and air was observed. The bronze bushings had some scratches and wear patterns expected for a gate that had been in operation for almost 40 years [6].

Corrosion of carbon steel trunnion pins resulted in increased friction between pins and bronze bushings. The test data showed the friction coefficient of 0.15 determined for the new gate was in a range between 0.22 - 0.28 at the time of the gate failure. A reduced frequency of lubrication and lack of weather protection (at both ends of the trunnion pin, where gaps between the trunnion hub and the bearing housing allowed rainwater, spray, and water vapor to enter) increased the rate of corrosion over the years [5].

Evaluating Gate Trunnion Friction. The tribological examination of the trunnion bearing [6] reported that rust, dirt, and water was found in the space between the pin and the bushing. The rust, as much as 0.006 inches thick, occupied the clearance.

Evaluating combinations of various fabrication dimensions and tolerances of the trunnion assembly components (Table 1) and deformations of the 1-inch-thick bronze bushing (reduction of the bushing ID) when installed in a 7-inch-thick trunnion hub, clearance of 0.003-0.009 inch could be expected at the time when the bearing was initially assembled. The additional compression at the bearing sliding surface caused by the increased volume of the rusted material could result in higher static friction in the trunnion bearing or even bearing seizure. This could lead overtime to even an higher trunnion friction moment than what was estimated in the forensic investigations.

Table 1. Fabrication dimensions and tolerances of gate trunnion bearing components.

Diameter	Fabrication Dimensions	Fabrication Tolerances
Trunnion Pin O.D.	31.989 / 31.991 inch	Clearance +0.009 /+0.014 inch
Trunnion Bushing I.D.	32.000 / 32.003 inch	
Trunnion Bushing O.D.	34.017 / 34.018 inch	Press-fitting -0.015/- 0.018 inch
Hub I.D.	34.000 / 34.002 inch	

Structural Design Criteria

Design Criteria Overview. The design of the Folsom Dam gates and several other radial gates in 1950's was based on the USACE Engineering Manual (EM) published in 1950 [2]. The manual presents design requirements for radial gates, describes a graphical method for determining the arm strut loads, recommends typical details for the gate components, and refers to USACE EM [1] for the working stress design criteria.

In the USACE's engineering practice, friction at the gate trunnion was not accounted for in the structural design of the gate arm frames until the new design criteria for "Design of Spillway Tainter Gates" (EM 1110-202702, Aug 1966) were updated, which was ten years after completion of the Folsom Dam. The new criteria recommended the coefficient of friction of 0.3 for the normal gate operation conditions.

In original design calculations for gates at Folsom Dam [3], only the axial load from the hydrostatic and hoist chain load was considered in the design of arm struts. The arm struts were modeled as a pin supported columns, carrying axial loads only. Since the bending moment in the arm struts could not be easily computed in 1950's, the strut members were given arbitrary sections. It appears that the bracing arrangement was determined based on the experience and judgment of the designer.

Additional Observations on the Gate Design at Folsom Dam. It appears that the arrangement of the bracing system, size of the brace member and their connections were arbitrarily selected for the gate structure at Folsom Dam since the original design computations [2] do not contain any calculations supporting the design of these brace members. In addition, the extremely small angles of approximately 14 degrees between the cross braces and the arm struts at Joint 1 (Figure 2) resulted in a high tension in this member.

Hollow trunnion pins usually result in larger ID when compared with solid pins (32-inches outer diameter, trunnion pins were installed at the Folsom Dam spillway gates). For the same type of the bushing, a larger pin diameter may lead to higher frictional moments at the gate trunnion and consequently larger bending moment applied to the gate arms.

LESSONS LEARNED

Failure of Gate No.3 at Folsom Dam initiated a dam safety evaluation program of radial spillway gates within state and federal jurisdiction. Investigations included inspections of radial gates, determination of proper loadings, definition of the acceptance criteria, and structural evaluations of the gates. Several of gates required retrofitting, but the lessons learned from these evaluations and from the Gate No. 3 failure resulted in better understanding of the radial gate behavior, resulted in the more precise criteria used in current design practice [9] [10].

Some observations made during the study of the Gate No.3 failure combined with other author's experience related to the design of new radial gates and the structural analysis of existing radial gates are summarized below.

Trunnion Pin Material

Several existing radial gates equipped with carbon steel pins are still in use. In some such gates, vibrations during periodic gate operations and indications of increased friction at the gate trunnions are reported. A typical sign of potential corrosion at the trunnion bearing is the rust "leaking" from the bearing assembly. Figure 6 shows a typical indication of potential corrosion in the trunnion bearing for a gate constructed in 1960's.

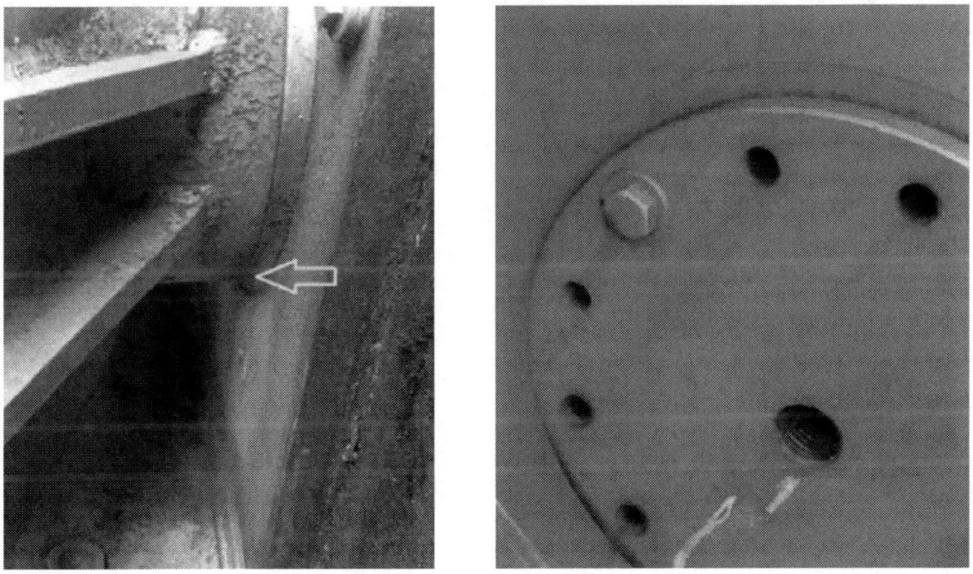

Figure 6. Sign of corrosion at the gate trunnion bearing equipped with carbon steel pin.

In the current practice of new radial gate design, stainless steel trunnion pins with self-lubricating bushings are commonly specified. The first eliminates potential for pin corrosion and the second reduces risk of the increased friction when grease type bushings are not properly designed. For the trunnion bearings with carbon steel pins, the bearing assembly must be sealed against the intrusion of water and dirt to maintain proper gate performance.

Fracture Critical Gate Components

The weight of gate arms was only 18.5 percent of the total cumulative weight of the service spillway gates at Folsom Dam. If the cross sectional area of all arm members were increased on average by 15 percent, this would result in a total weight increase of the gate of only 2.7 percent, which is less than approximately one percent increase of the total cost of the gate fabrication. However, if only selected fracture critical arm components were oversized, the increase in the cost would be even smaller. Extra capacity provided to the arms would significantly increase the safety of the entire gate structure and would reduce probability of the failure.

It is highly recommended that all fracture critical members (defined here as the members where failure leads to a breach of the gate structure) are identified for all radial gates and the arm frames of new radial gates are designed with an extra capacity to account for corrosion allowances and unexpected loads during gate's extended life of service.

Structural Analysis

Loads. In general, normal gate operations are considered the governing loading conditions in the design of radial gate structure. In such case, bending moments in the gate arms generated by the self-weight, eccentricity of the gate struts, as well as the bending moment resulting from friction at the trunnion bearing, are the critical loads to be combined with the reservoir pressure. Second-order effects are essential in the analysis of the arm frames at each level of the design for accurate determination of the forces in the struts and bracing members.

Force Redistribution in Arm Members. The highest bending moment in all arm struts of Gate No.3 occurred at the trunnion end where the strut splices were provided (Section C-C in Figure 3). In addition, the axial force in Strut #3 increased by 20 percent and in Strut #4 lowered by 25 percent as the trunnion friction moment corresponding to the coefficient of friction of 0.28 was applied. The results of the analysis showed that the force redistribution in the arm frame could be significant and cannot be neglected in the structural analysis of radial gates.

Strut Splices. Splices of the arm struts require special attention from the gate structural engineers. For the bearing type connection of W-shapes reinforced by splice plates bolted to the strut sections, local buckling of the strut flanges needs to be considered in the stability analysis of arm frame.

Stability Analysis of Gate Arms. Structural evaluation of radial gates requires the stability analysis to be performed for the whole structure, each of its components, and for the local buckling of the members.

Fatigue of Gate Components. Opening and closing cycles of the radial gates may lead to low-cyclic high-stress fatigue of the gate members or their connections over extended operation life of the structure. In general, the arm members and their connections, subject to typical cyclic loading during gate operations or vibrations, will not create fatigue conditions if the developed stress range is within the limit of the static allowable stress. For stresses exceeding the design limits, a fatigue analysis needs to be considered [9].

CONCLUSION

Failure of spillway gate at Folsom Dam initiated a comprehensive investigation program that resulted in better understanding of the behavior of radial gates, which lead to more precise criteria for the structural design of the gates. The most current design criteria and the state-of-practice analysis for radial gate structures developed by the USACE and Reclamation (USACE 2014 and Reclamation/USACE 2015) integrated several lessons learned from the failure of Gate No. 3 and the documents reflect the most current developments in the structural engineering.

REFERENCES

[1] Stress and Criteria for Structural Design – Working Stresses for Hydraulic Structures, Engineering Manual, Part CXXI, Chap.1, US Army Corps of Engineers, August 1947

[2] Crest and Head Gates - Tainter Gates, Engineering Manual, Part CXXVII, Chap.2, US Army Corps of Engineers, November 1950

[3] Supplemental Design Analysis Tainter Gates, Folsom Reservoir Project, US Army Corps of Engineers, January 1952

[4] Special Examination Review of Operation and Maintenance Program, Spillway Tainter 3, Folsom Dam, Bureau of Reclamation, April 1996

[5] Forensic Report on Spillway Gate 3 Failure, Folsom Dam, Bureau of Reclamation, November 18, 1996

[6] Folsom Dam Spillway Gate 3 Failure Investigation Trunnion Fixture Tests, Bureau of Reclamation, July 1997

[7] Spillway Tainter Gate Failure at Folsom Dam, by Todd R., Waterpower Conference, California 1999

[8] Recent Findings for Folsom Dam Tainter-Gate Failure, by Ishii N., Anami K., Knisely C., Academic Publications for Hydraulic Gates, July 2005

[9] Design of Hydraulic Steel Structures, US Army Corps of Engineers, Engineering Manual, ETL 1110-2-584, June 30, 2014

[10] Failure of Radial (Tainter) Gates under Normal Operational Conditions, Best Practice Guidelines, Section VII-1, US Army Corps of Engineers / Bureau of Reclamation, July 2015

A FINITE ELEMENT APPROACH TO PREDICT THE STABILITY OF GRAVITY DAMS LAYING ON ROUGH ROCK FOUNDATIONS

Sylvain Renaud, Jr. Eng., M.Sc.A.[1]
Najib Bouaanani, Eng., Ph.D.[2]
Benjamin Miquel, Eng., Ph.D.[3]

ABSTRACT

Owners of high-capacity concrete dams are required to regularly assess and monitor the structural behaviour of these structures under the effect of various loads, especially their sliding stability. Therefore, an appropriate understanding of contact at dam-rock interface is crucial as it might have a direct impact on dam owners' decisions relevant to rehabilitation strategies. A literature review is conducted and mainly shows that very limited published research detailing the numerical modeling of large scale irregular concrete-rock contact is available. Previous experimental work demonstrates the effect of roughness on shear strength of small scale unbounded concrete-rock interfaces. This effect can be associated with a non-null apparent cohesion at the interface. This paper proposes an efficient finite element procedure to better characterize the effects of shear strength against sliding of irregular dam-rock interfaces. The proposed technique implements special finite elements developed to account for the effects of cohesion along the dam-rock interface, as well as for tensile strength effects where dam-rock interface is bounded. These special elements are combined to conventional contact elements that can model pure sliding and rocking. The development and implementation of the proposed procedure are described in detail, and then applied to assess the stability of an existing concrete gravity dam with irregular dam-rock interface. The effects of key parameters, i.e. friction angle, tensile strength and cohesion are highlighted. Recommendations about the importance of including such parameters in concrete dam stability analyses are provided.

INTRODUCTION

Stability assessment of aging concrete dams is crucial to ensure their safe structural behavior. Dam-rock interface is generally considered as a potential weakness plane that may trigger the instability of a concrete dam through cracking and sliding under static or dynamic loads (Lo and Grass 1994, USACE 1995, FERC 2002, CDA 2006, USBR 2009). The outcome of a dam stability analysis might directly impact decisions relevant to its rehabilitation, generally associated with high costs and significant operational issues. However, field observations show a relatively good structural behavior of dam structures, which may reflect the conservatism of current stability analysis methods (Nuss

[1] Dept. of Civil, Geological and Mining Eng., Polytechnique Montréal, QC, Canada. Email: sylvain.renaud @polymtl.ca
[2] Dept. of Civil, Geological and Mining Eng., Polytechnique Montréal, QC, Canada. Email: najib.bouaanani@polymtl.ca
[3] Division of Expertise in Dams, Hydro-Québec, Montreal, QC, Canada. Email: miquel.benjamin@hydro.qc.ca

et al. 2012). The Mohr-Coulomb criterion is recommended by most guidelines to assess the sliding stability of concrete gravity dams over joints using the gravity method (USACE 1995, FERC 2002, CDA 2006, USBR 2009). According to this approach, the Sliding Safety Factor *SSF* is determined assuming a linear distribution of normal stresses at dam-rock interface, which is simplified as a unique plane corresponding to a possible failure mode. Such a unique plane is difficult to determine for irregular dam-rock interfaces, which are quite common in actual projects. Therefore, the validity of the obtained results using the gravity method might be affected in such cases. Furthermore, uncertainties remain about the appropriate input parameters for Mohr-Coulomb failure criterion, namely cohesion c, tensile strength f_t and friction angle ϕ. Most standards assume the conservative case of an unbounded dam-rock interface with no cohesion and no tensile strength (FERC 2000, ICOLD 2004, CDA 2006). However, shear tests conducted on dam-rock core samples drilled from dam sites indicate that cohesion and tensile strength can reach 0.5 to 2 MPa (EPRI 1992, Lo et al. 1990, Lo and Grass 1994). A friction angle of $\phi = 55°$ is generally considered in dam sliding stability analyses, although experimental studies reported friction angles close to 35° at unbounded dam-rock interfaces (EPRI 1992, ICOLD 2004). Moreover, previous experimental work showed that undulations and asperities of an unbounded concrete-rock interface, i.e. with null tensile strength $f_t = 0$, can significantly affect its shear behavior by increasing shear strength (Seidel and Haberfield 2002, Johansson 2009, Gutiérrez 2013). Fig. 1 illustrates a typical model representing the variation of measured peak shear strength as a function of applied normal stress during shear testing of an unbounded joint (Patton 1966, Ladanyi and Archambault 1970). This nonlinear failure curve can be linked to the classical Mohr-Coulomb criterion by drawing straight lines tangent to the curve at various applied normal stresses, as illustrated in Fig. 1. These lines provide an estimate of the apparent cohesion $\tilde{c}$ and apparent friction angle $\tilde{\phi}$ under a given normal stress (Bouchard 1991). These parameters are referred to as *apparent* to be differentiated from the *true* cohesion and friction at concrete-rock interface.

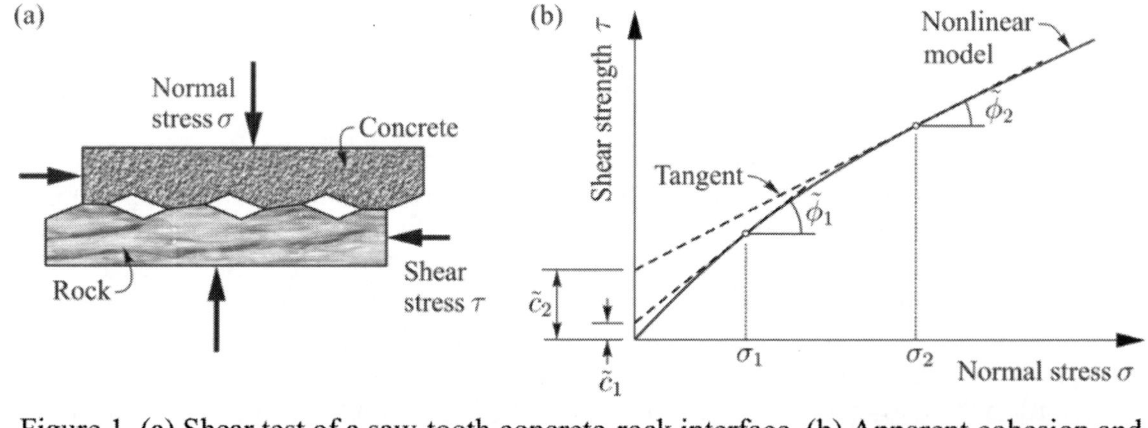

Figure 1. (a) Shear test of a saw-tooth concrete-rock interface, (b) Apparent cohesion and friction angle based on nonlinear model of strength curve of concrete-rock interface.

In this context, it is recommended to conduct parametric analyses to evaluate the effects of variations in cohesion (or apparent cohesion), tensile strength and friction on the stability of dams with irregular dam-rock interfaces. In this paper, we propose a practical technique that can be used efficiently by practicing dam engineers to carry out such parametric analyses. For illustration purposes, the developed methodology is applied to an actual gravity dam monolith laying on a stepped dam-rock foundation subjected to an earthquake. A discussion of the effects of dam-rock geometric irregularity on dam sliding stability is presented, as well as the influence of the mechanical properties at dam-rock interface.

PROPOSED METHOD

Basic Assumptions

A dam-reservoir-foundation system as illustrated in Fig. 2 is considered. Nonlinearities are localized at concrete-rock interface to investigate the stability of the dam against sliding along this interface. Although some coupling between cohesion and tensile strength at dam-rock interfaces has been observed experimentally (Lo et al. 1990, Lo and Grass 1994), the modeling approach developed here assumes that these two parameters are independent to give the analyst more freedom in specifying most appropriate values for the dam being studied, eg. in case of available experimental data. However such flexibility calls for caution from the analyst who bears the responsibility of appropriately selecting analysis parameters and assessing the conservatism or unconservatism of the obtained results with respect to safety evaluation of each dam studied.

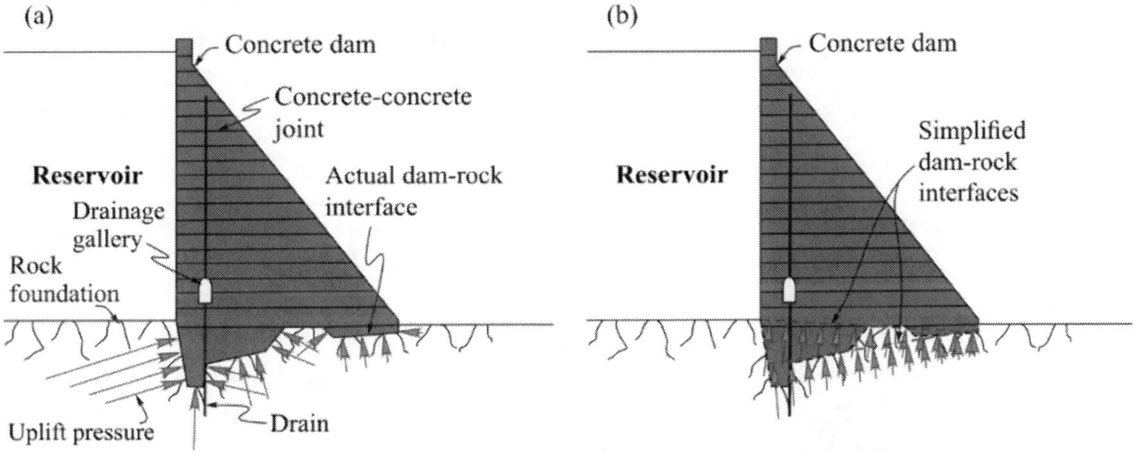

Figure 2. Gravity dam subjected to earthquake loads: (a) dam-rock interface with irregular geometry; (b) dam-rock interface with simplified geometry.

Treatment of Nonlinearities and Failure at Dam-Rock Interface

Sliding and uplift can be observed at irregular dam-rock interfaces. The sliding at the dam-rock interface can be divided into two phases. The first sliding phase is controlled by basic frictional contact elements. The second sliding phase is triggered when cohesive strength at the interface is exceeded. Denoting ϕ the friction angle at the dam-rock interface, and τ and σ the shear and normal stresses, respectively, the ratio ζ_0 is introduced

$$\zeta_0 = \frac{\tau}{\sigma\ \tan(\phi)} \tag{1}$$

First phase sliding at dam-rock interface is then governed by the Mohr-Coulomb rupture criterion, implying that sliding occurs as soon as $|\zeta_0| < 1$. The dam-rock joint is at zero state of shear stress at this stage. Eq. (1) does not take account of the influence of cohesion c on the behavior of the dam-rock interface. For this purpose, we introduce the parameters

$$\zeta_c = \frac{\tau}{\sigma\ \tan(\phi) + c} \tag{2}$$

and

$$\zeta = \begin{cases} \zeta_c \text{ as long as } |\zeta_c| < 1 \\ \zeta_0 \text{ otherwise} \end{cases} \tag{3}$$

After the first sliding phase controlled by Eq. (1), shear stresses at the dam-rock interface will increase gradually due to cohesion, as illustrated in Fig. 3. This process continues until shear fracture of cohesive links at the joint occurs when $|\zeta| \geq 1$, which triggers a second sliding phase associated with post-fracture of the joint. Eq. (3) implies that if $|\zeta_c|$ becomes larger than unity at an instant t_1 corresponding to fracture of cohesive links, then $\zeta = \zeta_0$ during the rest of dam motion for $t > t_1$, i.e. post-fracture. Therefore, the degradation of the area over which cohesion is effective could be tracked during sliding. Similarly to sliding, the uplift at the dam-rock interface can be divided into two phases. The first uplift phase can be controlled by basic frictional contact elements, while the second is triggered when tensile strength at the interface is exceeded. Denoting η the normal gap between the dam and the rock, and assuming that compression stresses are positive, the basic contact between the dam and the rock foundation can be expressed as

$$\eta\,\sigma = 0; \qquad \sigma \geq 0; \qquad \eta \geq 0; \tag{4}$$

Therefore full contact can be expressed as $\eta = 0$ and $\sigma > 0$ while first phase uplift at dam-rock interface occurs as soon as $\eta > 0$ and $\sigma = 0$. The dam-rock joint is at zero state of tension stress at this stage. Eq. (4) does not take account of the influence of tensile strength f_t on the behavior of the dam-rock interface. When tensile strength is considered, there is full contact between the dam and the rock foundation as long as $\eta = 0$ and $\geq -f_t$, while second phase uplift occurs as soon as $\eta > 0$ and $\sigma < -f_t$, which corresponds to tension fracture of the dam-rock joint, as illustrated in Fig. 4.

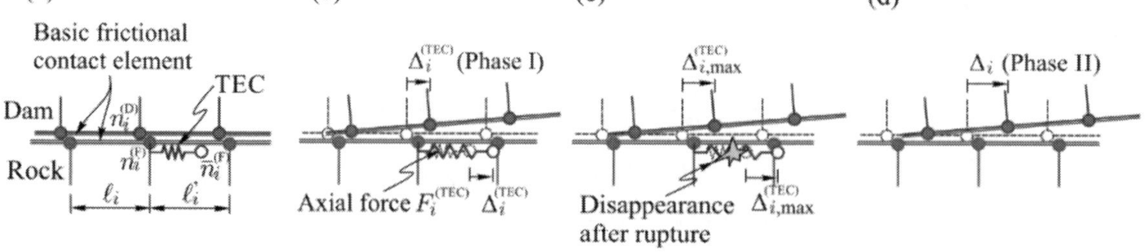

Figure 3. Non-linear sliding response at dam-rock interface: (a) bounded condition; (b) first sliding phase; (c) disappearance of the TEC; and (d) second sliding phase.

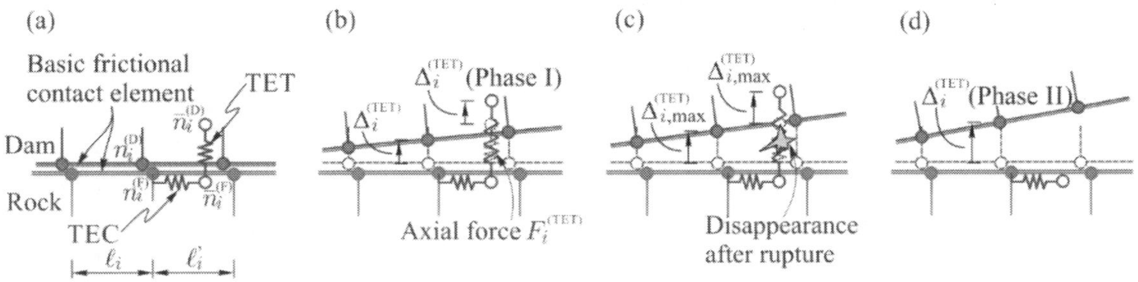

Figure 4. Non-linear uplift response at dam-rock interface: (a) bounded conditions; (b) first uplift phase; (c) disappearance of the TET; and (d) second uplift phase.

Once $\eta > 0$ and $\sigma < -f_t$, say at an instan t t_1 during seismic excitation, the subsequent contact conditions at the dam-rock joint at ins tants $t > t_1$ are governed by Eq. (4), i.e. tension post-fracture phase. As for cohesion, the degradation of the tensile strength at the dam-rock interface can also be tracked during dam motion.

Modeling of Dam-Rock Interface

Sliding and uplift at dam-rock interface without cohesion and tensile strength will be modeled using basic frictional contact elements. The gravity dam is considered as the contactor block which can slide or rock over the target block, represented by the rock foundation. Contactor and target interfaces are created to simulate contact conditions. Three behaviors can be observed: (i) the dam and rock foundation are bounded, (ii) the gap at the unbounded dam-rock interface is open, and (iii) the gap at the unbounded dam-rock interface is closed. In the latter case, the contactor interface slides over the target interface and a compression force is generated. As shown in Fig. 5, two series of N_I coincident nodes are created on each interface: (i) nodes n_i^{D}, $i = 1 \ldots N_I$, at the dam contactor interface, and nodes n_i^{F}, $i = 1 \ldots N_I$, at the rock foundation target interface.

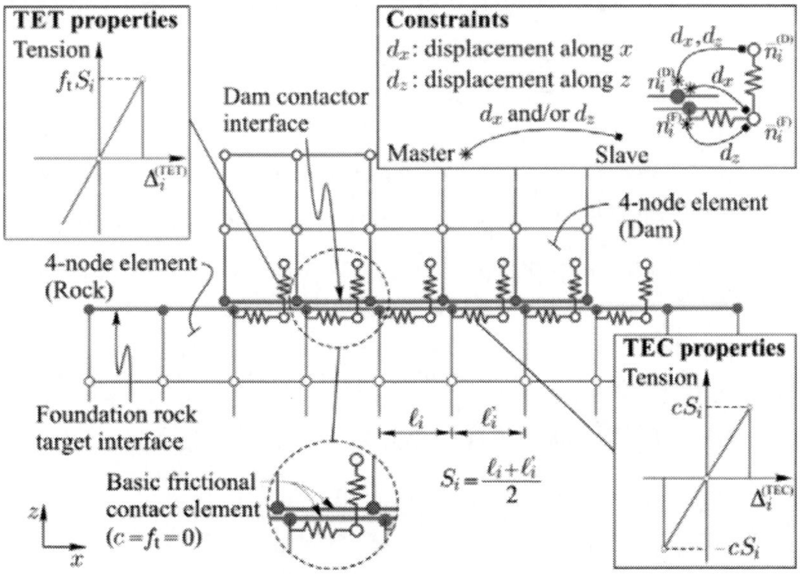

Figure 5. Properties of the proposed finite elements, i.e., TEC and TET, to model cohesion and tensile strength at dam-rock interface.

To model cohesion and tensile strength at dam-rock, two types of interface truss elements are introduced next: (i) a Truss Element for Cohesion modeling, denoted hereafter as TEC, and (ii) a Truss Element for Tensile strength modeling, denoted hereafter as TET. For this purpose, two series of nodes $\bar{n}_i^{\mathrm{D}}$ and $\bar{n}_i^{\mathrm{F}}$, $i = 1 \ldots N_I$, are created. TECs connect nodes n_i^{F} to nodes $\bar{n}_i^{\mathrm{F}}$, while TETs connect $\bar{n}_i^{\mathrm{F}}$ to nodes $\bar{n}_i^{\mathrm{D}}$, as illustrated in Fig. 5. Nodes $\bar{n}_i^{\mathrm{F}}$ are positioned in a manner that TECs remain parallel to dam-rock interface during sliding, while nodes $\bar{n}_i^{\mathrm{D}}$ are positioned so that TETs remain perpendicular to dam-rock interface. These geometrical conditions are imposed through the following constraints as shown in Fig. 5: (i) each node $\bar{n}_i^{\mathrm{D}}$ is constrained to have the same displacements as node n_i^{D}, (ii) each node $\bar{n}_i^{\mathrm{F}}$ is constrained to have the same displacement parallel to the dam-rock interface as node n_i^{D} and perpendicularly to the dam-rock interface as node n_i^{F}. Truss elements were adopted as they constitute basic elements available on most finite element software. TECs and TETs are implemented using bilinear materials including a rupture option which triggers the disappearance of the elements once maximum elastic strength is reached. $\Delta_i^{(\mathrm{TEC})}$ and $\Delta_i^{(\mathrm{TET})}$ denote the relative displacements of node n_i^{D} with respect to node n_i^{F}, respectively, along the parallel and perpendicular directions of the dam-rock interface, as illustrated in Fig. 3. The frictional strength at the dam-rock interface is given by $\sigma \tan(\phi)$ as when cohesion is neglected. Therefore, as soon as $|\zeta_0| \geq 1$ is satisfied, the first sliding phase occurs and each node n_i^{D} moves from node n_i^{F} by a distance of $\Delta_i^{(\mathrm{TEC})}$ as shown in Fig. 3 (b). Each TEC then sustains a shear force $F_i^{(\mathrm{TEC})}$ at the dam-rock interface, given by

$$F_i^{(\mathrm{TEC})} = \frac{E_i^{(\mathrm{TEC})} A_i^{(\mathrm{TEC})}}{L_i^{(\mathrm{TEC})}} \Delta_i^{(\mathrm{TEC})} = [\tau - \sigma \tan(\phi)]\, S_i \tag{5}$$

where $E_i^{(\mathrm{TEC})}$, $L_i^{(\mathrm{TEC})}$ and $A_i^{(\mathrm{TEC})}$ denote the TEC modulus of elasticity, length and cross-section, respectively, and S_i is the tributary area associated with each node i considering a unit width of the dam monolith. Using Eq. (5), the following equation can be obtained

$$\frac{E_i^{(\mathrm{TEC})} A_i^{(\mathrm{TEC})}}{L_i^{(\mathrm{TEC})} S_i} \Delta_{i,\max}^{(\mathrm{TEC})} = c \tag{6}$$

where $\Delta_{i,\max}^{(\mathrm{TEC})}$ denotes the maximum displacement that can be sustained by a TEC. We note that the inclusion of $\Delta_{i,\max}^{(\mathrm{TEC})}$ in the formulation of TECs implies that the modeled cohesion corresponds to the gap between the peak strength and the residual strength at the dam-rock interface. As soon as $\eta > 0$ and $\sigma = 0$, the first uplift phase occurs and each node n_i^{D} moves from node n_i^{F} by a distance of $\Delta_i^{(\mathrm{TET})}$ as shown in Fig. 4 (b). Therefore, each TET, with modulus of elasticity $E_i^{(\mathrm{TET})}$, length $L_i^{(\mathrm{TET})}$ and cross-section $A_i^{(\mathrm{TET})}$, sustains a tensile force $F_i^{(\mathrm{TET})}$ at the dam-rock interface, given by

$$F_i^{(\mathrm{TET})} = \frac{E_i^{(\mathrm{TET})} A_i^{(\mathrm{TET})}}{L_i^{(\mathrm{TET})}} \Delta_i^{(\mathrm{TET})} = -\sigma S_i \text{ if } \sigma < 0 \text{ (tension)}; \qquad F_i^{(\mathrm{TET})} = 0 \text{ otherwise} \tag{7}$$

Using Eq. (7), the following expression can be obtained

$$\frac{E_i^{(\mathrm{TET})} A_i^{(\mathrm{TET})}}{L_i^{(\mathrm{TET})} s_i} \Delta_{i,\max}^{(\mathrm{TET})} = f_\mathrm{t} \tag{8}$$

in which $\Delta_{i,\max}^{(\mathrm{TET})}$ denotes the maximum displacement that can be sustained by a TET.

Verification Example of the Proposed Modeling Procedure

In this section, the proposed modeling procedure using TECs and TETs is verified against analytical predictions based on the Mohr-Coulomb failure criterion. For this purpose, we study the sliding of two solid blocks separated by a contact interface as illustrated in Fig. 6. For illustration purposes, four contact interfaces with different mechanical properties are considered. Table 1 contains the friction angle ϕ, cohesion c and tensile strength f_t of the studied contact interfaces denoted as CI1 to CI4. The modulus of elasticity of the upper and lower blocks are 21.375 GPa and 200 GPa, respectively, and their Poisson's ratios are 0.164. ADINA (2015) is used to model the basic frictional contact between the blocs, while TECs and TETs are implemented to account for cohesion and tensile strength. For simplicity, TECs and TETs are assumed to have the same modulus of elasticity, respectively, i.e. $E_i^{(\mathrm{TEC})} = E^{(\mathrm{TEC})}$, and $E_i^{(\mathrm{TET})} = E^{(\mathrm{TET})}$. All TECs and TETs are also assumed to have the same length, i.e. 0.25 m in this case. The maximum displacements of all the TECs are selected to be the same. The same applies to the TETs. Furthermore, $\Delta_{\max}^{(\mathrm{TEC})}$ (respectively $\Delta_{\max}^{(\mathrm{TET})}$) is selected as small as possible so that displacements before sliding (resp. before uplift) remain negligible. In the example studied, $\Delta_{\max}^{(\mathrm{TEC})} = 0.2$ mm and $\Delta_{\max}^{(\mathrm{TET})} = 0.05$ mm are used. Inter-nodal distances are generally variable. Therefore, the TEC and TEC cross-sections have to be adapted according to Eqs. (6) and (8). Finally, TECs and TETs are implemented in ADINA (2015) using bilinear materials including a "death upon rupture" option which triggers the disappearance of the element once maximum elastic strength is reached. The upper block is subjected to a uniform normal load pressure σ and a uniformly distributed shear load τ, while the base of the lower block is fixed, as indicated in Fig. 6. Six load cases, denoted by LC1 to LC6, and shown in Table 2, are considered.

Table 1. Mechanical properties of the simplified contact interfaces studied.

Contact interface	Friction angle ϕ (°)	Cohesion c (kPa)	Tensile strength f_t (kPa)
CI1	45	100	0
CI2	45	0	50
CI3	45	100	50
CI4	0	100	50

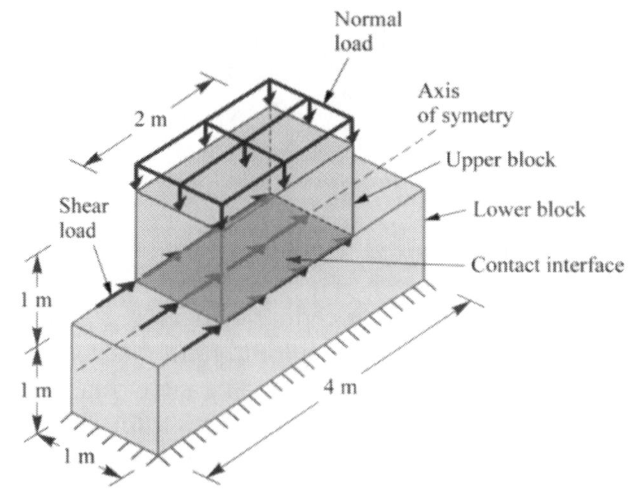

Figure 6. Verification example of two sliding blocks separated by a contact interface with cohesion and tensile strength.

Table 2. Considered load cases.

Load case	Normal load σ (kPa)	Shear load τ (kPa)
LC1	0	Linearly varying from 0 to 400
LC2	50 (*Compression*)	Linearly varying from 0 to 400
LC3	100 (*Compression*)	Linearly varying from 0 to 400
LC4	Linearly varying from 200 (*Compression*) to -200 (*Tension*)	0
LC5	Linearly varying from 200 (*Compression*) to -200 (*Tension*)	50
LC6	Linearly varying from 200 (*Compression*) to -200 (*Tension*)	100

For load cases LC1 to LC3, i.e. constant normal load σ and varying shear load τ, the shear strength τ_f corresponding to the onset of sliding of the upper block, i.e. failure, is the peak shear strength given by the Mohr-Coulomb criterion

$$\tau_f = \sigma \tan(\phi) + c \tag{9}$$

Similarly, for load cases LC4 to LC6, i.e. varying normal load σ and constant shear load τ, the normal stress σ_f corresponding to the onset of sliding of the upper block is obtained as

$$\sigma_f = \max \left(-f_t; \frac{\tau - c}{\tan(\phi)} \right) \tag{10}$$

The values of σ_f and τ_f obtained using the TEC and TET modeling procedure are compared to those based on Mohr-Coulomb failure envelope in Fig. 7. A very satisfactory agreement between both numerical and analytical solutions is clearly observed. This excellent agreement of the proposed model with the predictions of Mohr-

Coulomb failure criterion was confirmed by similar verification tests conducted considering various other load cases and contact interface parameters.

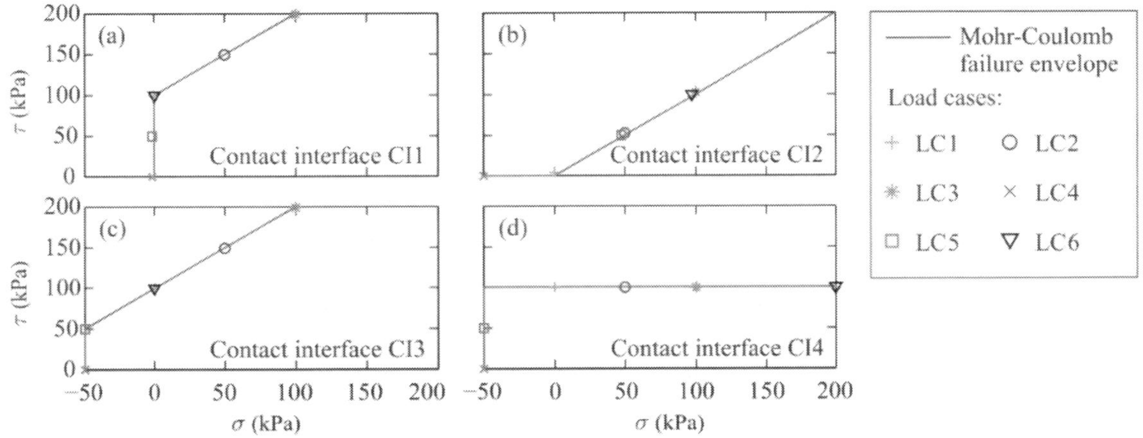

Figure 7. Comparison of numerical results obtained using the proposed modeling procedure to analytical Mohr-Coulomb failure envelope: (a) contact interface CI1; (b) contact interface CI2; (c) contact interface CI3; (d) contact interface CI4.

APPLICATION TO THE STABILITY OF AN EXISTING CONCRETE DAM

In this section, the methodology presented above is applied to investigate the effects of geometric irregularity and shear strength at dam-rock interface on the seismic response of an actual concrete gravity dam in Quebec. The studied dam monolith has a height of 27 m and a width varying from 6.1 m at the crest to 18 m at the base as shown in Fig. 8 (a). Two simplified dam sections are also considered, as illustrated in Figs. 8 (b) and (c), to assess the effects of stepped dam-rock interface. For brevity of notation, the dam with the actual stepped geometry of the dam-rock interface is designated by $D^{(S)}$, and the dams with simplified dam-rock interfaces modeled as an inclined and a horizontal plane are designated by $D^{(I)}$ and $D^{(H)}$, respectively. We note that the simplified inclined dam-rock interface is required when assessing dam seismic stability using the gravity method. The principal sliding directions for dams $D^{(S)}$, $D^{(I)}$ and $D^{(H)}$ are horizontal, inclined, and horizontal, respectively, as shown in Fig. 8. A modulus of elasticity $E_{s1} = 21.375$ GPa is considered for all the dam structure, except the area around the drainage gallery (Fig. 8) where a modulus of elasticity $E_{s2} = 22.775$ GPa is used to account for the presence of steel reinforcement. A Poisson's ratio $\nu = 0.164$, and a density $\rho_s = 2295$ kg/m^3 are also adopted for the dam concrete. The previously mentioned concrete mechanical properties were obtained experimentally for the actual dam studied. A mass density $\rho_r = 1000$ kg/m^3 is considered for water and the dam rock foundation is assumed rigid.

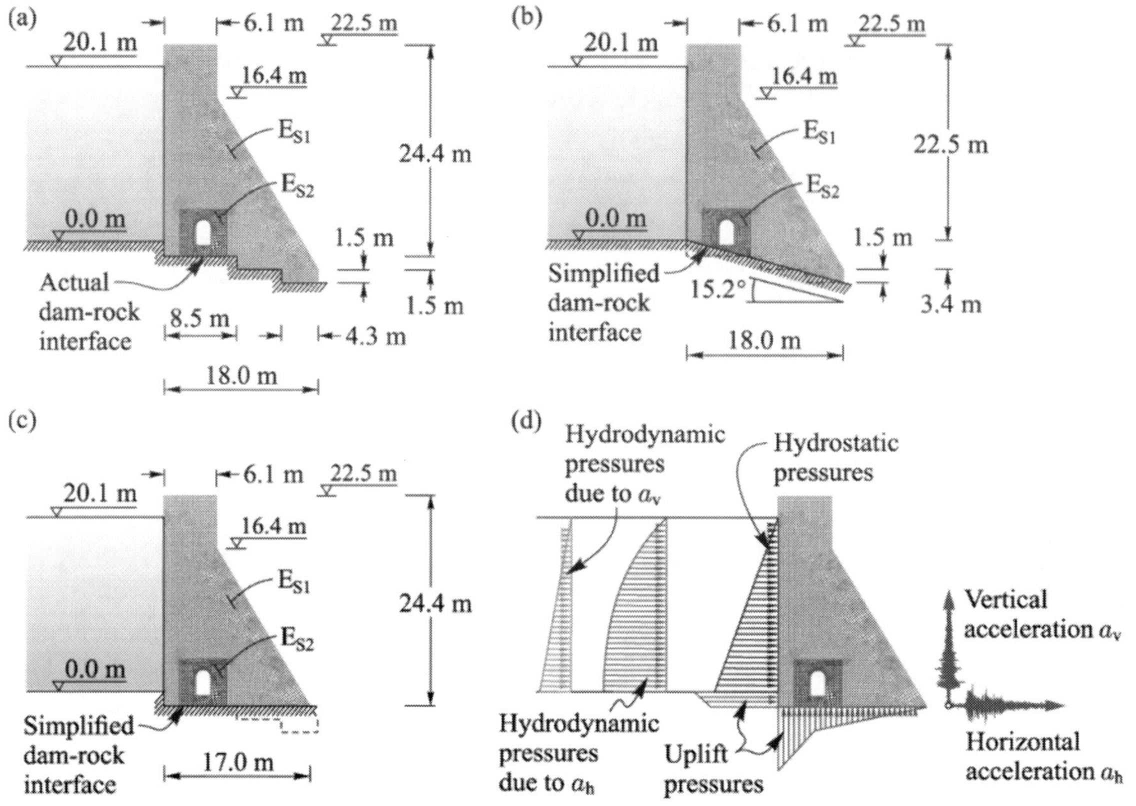

Figure 8. Studied gravity dams: (a) Dam $D^{(S)}$ with stepped dam-rock interface; (b) Dam $D^{(I)}$ with simplified inclined dam-rock interface; (c) $D^{(H)}$ with simplified horizontal dam-rock interface; and (d) applied loads considered.

Finite element models of the dams studied are built using ADINA (2015). The dam and rock foundation are modeled using 4-nodes plane strain finite elements. Basic frictional contact elements programmed in ADINA (2015) are used to model first sliding and uplift phases at dam-rock interface. The TEC and TET properties are the same as in the verification example. The horizontal and vertical components of the Imperial Valley ground motion (1940) at station El Centro are considered in this work. Prior to earthquake loads, the dams are subjected to static gravity loads applied gradually through a ramp. A Rayleigh damping equivalent to a viscous damping $\xi = 5\%$ is adopted for the concrete dam (USBR 2009). Water is assumed incompressible and hydrodynamic loads are implemented using Westergaard added-masses (Westergaard 1933). Uplift pressures along the irregular dam-rock interface are determined according to USACE (1995). A drain is located at 6 m from the upstream dam face and a drain efficiency of 66.7 % is considered. As recommended by many guidelines (USACE 1995, FERC 2002, CDA 2006), uplift-pressures are assumed constant during earthquake shaking, independently of possible cracking. The foundation is assumed massless and infinitely rigid.

RESULTS AND DISCUSSION

The dam response indicators studied in the following section are: (i) the horizontal residual displacement d_r of the gravity dam determined at its heel, and (ii) the following sliding safety factor SSF proposed to account for the geometric irregularity of dam-rock interface

$$SSF = \frac{[(W - E_{\mathrm{V}})\cos(\alpha) - (H_{\mathrm{S}} + E_{\mathrm{H}} + H_{\mathrm{D}})\sin(\alpha) - U_{\mathrm{n}}]\tan\phi + c\,A_{\mathrm{c}} + f_{\mathrm{t}}\,A_{\mathrm{t}}}{[H_{\mathrm{S}} + E_{\mathrm{H}} + H_{\mathrm{D}}]\cos(\alpha) + [W - E_{\mathrm{V}}]\sin(\alpha) + U_{\mathrm{t}}} \tag{11}$$

where W is the weight of the dam, α is the slope between the horizontal axis and the principal sliding direction, U_{n} and U_{t} are the projections of the resultant force of uplift pressures normal and tangent to the principal sliding direction, respectively, H_{S} and H_{D} are the resultant forces of hydrostatic and hydrodynamic pressures, respectively, E_{H} and E_{V} are the horizontal and vertical earthquake inertia forces, and A_{c} et A_{t} are the areas where cohesion and the tensile strength are still active against sliding movement, updated at each time step, respectively. According to Mohr-Coulomb criterion, no sliding occurs as long as $SSF > 1$. First, a friction angle $\phi = 55°$ is considered, while cohesion c and tensile strength f_{t} are selected to be close to 0 as recommended by most guidelines (FERC 2000, ICOLD 2004, CDA 2006), i.e. in this case $c = 0.01$ MPa and $f_{\mathrm{t}} = 0.01$ MPa. The location of a point at the dam-rock interface can be defined by a distance L_{p} corresponding to the length along the dam-rock interface between the heel of the dam and the point of interest.

Fig. 9 illustrate σ_{n} and σ_{t}, the normal and tangential stresses to the principal sliding direction, respectively, determined along the dam-rock interface for dams $\mathrm{D}^{(\mathrm{S})}$, $\mathrm{D}^{(\mathrm{I})}$ and $\mathrm{D}^{(\mathrm{H})}$, subjected to Imperial Valley (1940) ground motion. Normal compressive stresses σ_n are defined as positive. Fig. 9 reveals the high sensitivity of stress distributions to the geometry of dam-rock interface. It is indeed observed that stress distributions along the simplified dam-rock interfaces of dams $\mathrm{D}^{(\mathrm{I})}$ and $\mathrm{D}^{(\mathrm{H})}$ are clearly different from those at the stepped dam-rock interface of dam $\mathrm{D}^{(\mathrm{S})}$.

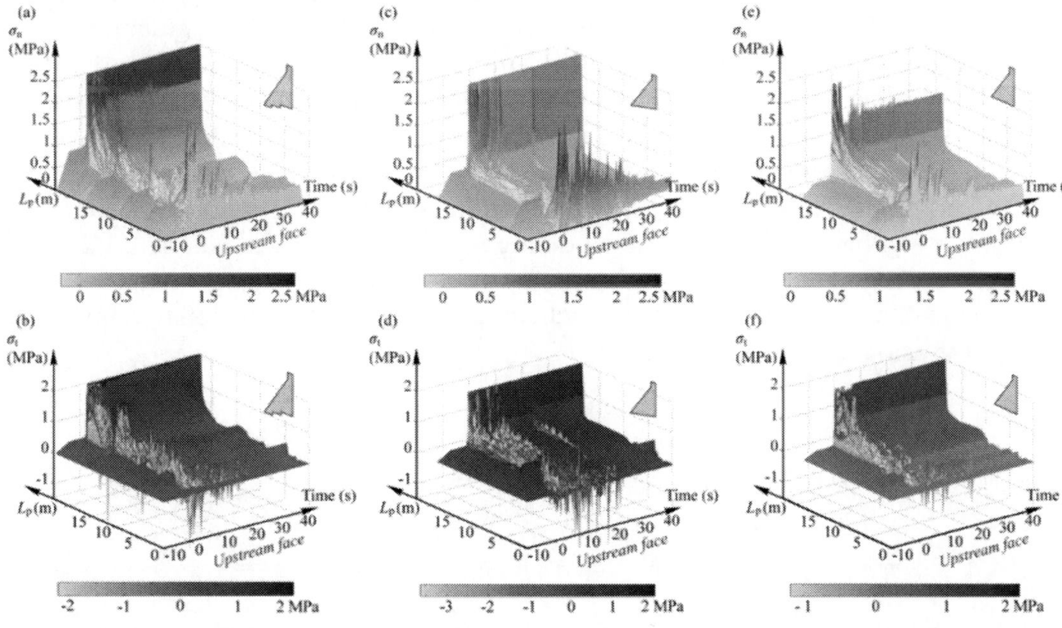

Figure 9. Seismic response at dam-rock interface: (a) σ_{n} and (b) σ_{t} for dam $\mathrm{D}^{(\mathrm{S})}$; (c) σ_{n} and (d) σ_{t} for dam $\mathrm{D}^{(\mathrm{I})}$; and (e) σ_{n} and (f) σ_{t} for dam $\mathrm{D}^{(\mathrm{H})}$.

Then a parametric study is conducted to assess the effects of different values of friction angle ϕ, cohesion c, and tensile strength f_t on the stability and sliding of dams $D^{(S)}$, $D^{(I)}$ and $D^{(H)}$ subjected to Imperial Valley (1940) earthquake. For this purpose, 18 models with various friction, cohesion and tensile strength parameters are created for each dam geometry, i.e. a total of 54 models. For the sake of brevity, only the results of 12 models are presented in this paper. The other results can be found in Renaud et al. (2016).
The parameters corresponding to the 12 models are given in Table 3. The values of these parameters are selected judiciously to highlight their relative importance on the seismic stability of the studied gravity dams. The orders of magnitudes are based on experimental data of dam-rock mechanical properties reported in the literature (EPRI 1992, Lo et al. 1990, Lo and Grass 1994). However, the conducted parametric analyses assume no coupling between cohesion and tensile strength to evaluate the effects of their variations independently. Table 3 presents the obtained horizontal residual displacement d_r corresponding to each case analyzed. As can be seen, for a given set of parameters ϕ, c, and f_t, maximum differences between the residual displacements of dams $D^{(S)}$, $D^{(I)}$ and $D^{(H)}$ are observed when cohesion is very low, i.e. $c = 0.01$ MPa. In such case, the residual displacement d_r at the heel of dam $D^{(I)}$ is significantly larger than those corresponding to dams $D^{(S)}$ and $D^{(H)}$. Therefore, the dam-rock interface of dam $D^{(I)}$ may be substantially weaker in terms of strength than that of dams $D^{(S)}$ and $D^{(H)}$, for the same set of values ϕ, c, and f_t. As cohesion increases from 0.01 MPa to 0.5 MPa or 1 MPa, the residual displacements of the three dams decrease considerably to reach practically the same value, i.e $d_r = 0.1$ mm. By comparing the responses of small-cohesion models M1, M2 and M3 or M10, M11 and M12, we note that tensile strength can be effective in reducing residual displacements even when dam-rock interface does not contain vertical segments. However, it is important to note that, for illustration purposes, the ratio c/f_t of cohesion to tensile strength adopted for models M2, M3, M11 and M12 is chosen voluntarily low to emphasize the effects on seismic stability. The important influence of friction on residual displacements can also be clearly seen by comparing the results of models M10, M11 and M12 to those of models M1, M2 and M3. The residual displacements of models M10, M11 and M12 of dam $D^{(I)}$ could not be obtained as the low friction angle of $\phi = 35°$ is sufficient to trigger dam instability even under static gravity loads. These observations, although based on a limited number of values of friction angle ϕ, cohesion c, and tensile strength f_t, are sufficient to clearly show the trends of variation effects of these parameters on the stability of gravity dams $D^{(S)}$, $D^{(I)}$ and $D^{(H)}$. For a given dam project, the modeling procedure proposed can be used effectively to perform extensive simulations by systematically varying strength parameters at dam-rock interface over practical ranges to assess their relative influence considering various ground motions.

Table 3. Friction angle ϕ, cohesion c, and tensile strength f_t corresponding to each model used in the parametric study of dams $D^{(S)}$, $D^{(I)}$ and $D^{(H)}$ subjected to Imperial Valley (1940) earthquake and obtained final horizontal residual displacement d_r (mm).

	Dam-rock interface parameters			Residual displacement d_r (mm)		
Model	ϕ (°)	c (MPa)	f_t (MPa)	$D^{(S)}$	$D^{(I)}$	$D^{(H)}$
M1	55°	0.01	0.01	8.8	25.4	3.9
M2	55°	0.01	0.50	6.4	22.4	3.0
M3	55°	0.01	1.00	0.5	5.8	0.1
M4	55°	0.50	0.01	0.5	0.1	0.1
M7	55°	1.00	0.01	0.1	0.1	0.1
M9	55°	1.00	1.00	0	0.1	0
M10	35°	0.01	0.01	84.0	—*	143.8
M11	35°	0.01	0.50	54.1	—*	77.9
M12	35°	0.01	1.00	9.1	—*	8.6
M13	35°	0.50	0.01	0.1	0.1	0.1
M16	35°	1.00	0.01	0.1	0.1	0.1
M18	35°	1.00	1.00	0	0.1	0

* Dam unstable under static gravity loads.

For illustration purposes, Fig. 10 presents the sliding stability factors and horizontal residual displacements obtained considering models M1, M2, M3, M10, M11, and M12 of dam $D^{(S)}$. The results in Table 3 and Fig. 10 show that, as expected, the dam is more stable as higher friction angle, cohesion and tensile strength are included. The SSF corresponding to $\phi = 55°$ and $\phi = 35°$ presented in Figs. 10 (a) and (c), respectively, confirm the predominant role of friction angle in the stability of the dam. When cohesion is negligible, Figs. 10 (b) and (d) show that the residual displacement d_r varies from 0.4 mm (M3) to 77.3 mm (M10) for dam $D^{(S)}$. Fig. 10 also demonstrates that a higher tensile strength f_t leads to a larger Sliding Safety Factor SSF due to the resulting action against destabilizing lateral forces at vertical dam-rock sections of the interface. This can also be interpreted from Eq. (11) which contains the term $f_t\, A_t$. The degradation of dam-rock interface during seismic loading also involves energy dissipation. Fig. 11 shows the values of the Sliding Safety Factor of an upper dam joint SSF_{UJ} (at a concrete-concrete interface) located at a height of 14.15 m from the bottom of the reservoir. SSF_{UJ} is computed based on linear stresses from models M9 and M10 of dam $D^{(S)}$, the strongest and weakest dam-rock interfaces of this monolith, respectively. For illustration purposes, two cohesion values, i.e. $c_{UJ} = 0.466$ MPa and $c_{UJ} = 0.233$ MPa, are used at this upper joint where uplift pressures are also considered. The results show that SSF_{UJ} for model M9 reaches lower values than the SSF_{UJ} of model M10. This indicates that the stronger is the dam-rock interface, the more seismic energy is transmitted to the dam, the larger are the stresses in an upper dam joint, and the lower is the sliding safety factors SSF_{UJ} at such an upper joint. Therefore, it is important to carefully model dam-rock interfaces for the assessment of the seismic stability along a concrete-concrete joint located at a certain height above dam base. For example, underestimating the tensile strength of a dam-rock interface, as usually done in practice when no test results are available, may lead to non-conservative results related to the seismic stability of an upper part of the dam monolith.

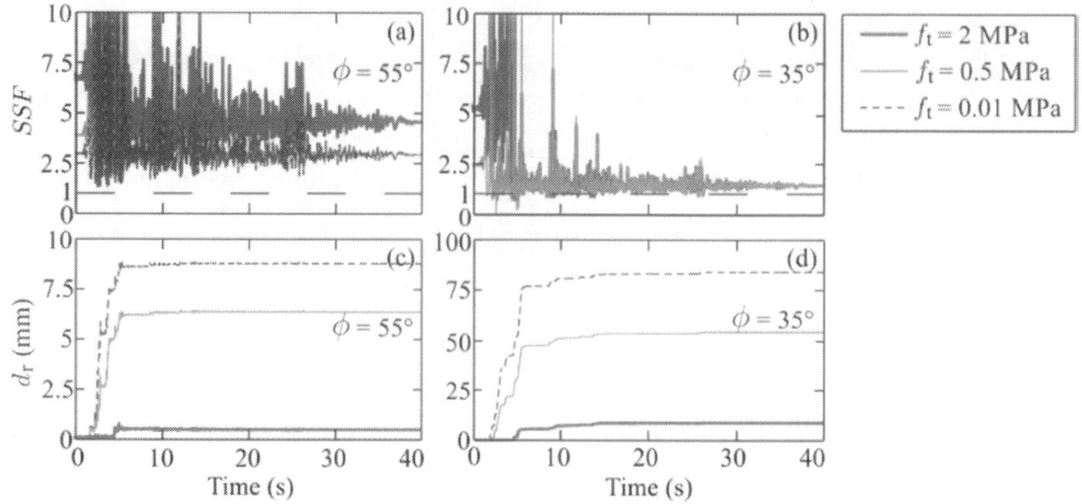

Figure 10. Effects of tensile strength f_t on the sliding safety factor SSF and residual displacement d_r of dam $D^{(S)}$ subjected to Imperial Valley (1940) ground motion, with $c = 0.01$ MPa: (a) and (c) with $\phi = 55°$; and (b) and (d) with $\phi = 35°$.

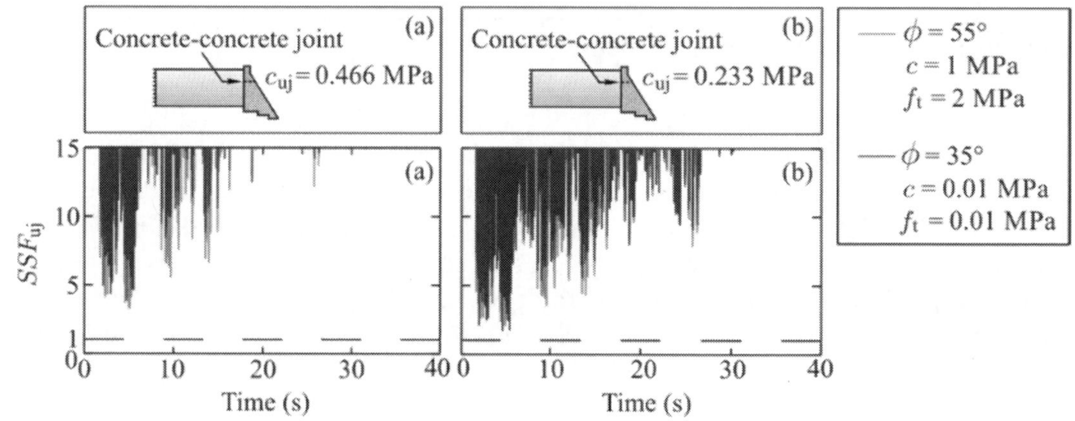

Figure 11. Effects of friction angle ϕ, cohesion c, and tensile strength f_t at the dam-rock interface of dam $D^{(S)}$ on the sliding safety factor SSF_{UJ} at a dam concrete-concrete joint characterized by: (a) a cohesion $c_{UJ} = 0.466$ MPa; and (b) $c_{UJ} = 0.233$ MPa.

CONCLUSION

This paper proposed an efficient finite element method to better characterize the effects of the shear strength against sliding of irregular dam-rock interfaces, while accounting for the effects of friction, cohesion, and tensile strength. To account for such parameters in a practical manner, simplified truss elements were developed and combined to conventional contact elements modeling pure frictional sliding and rocking. The implementation of the proposed methodology was described in detail and applied to an actual gravity dam with a stepped dam-rock foundation subjected to a ground motion. The results emphasize the important role of cohesion, friction angle and tensile strength in dam seismic stability. Tensile strength should be included in seismic stability assessment, namely when dam-rock interface contains vertical segments. We showed that a simplified dam-rock geometry should be used with caution as is may lead to inaccurate results. This work emphasizes the importance of appropriately modeling dam-rock interface and selecting relevant strength parameters, a process that calls for caution from the analyst who bears the responsibility of assessing the conservatism or unconservatism of the obtained results with respect to safety evaluation of each dam studied. Research work is under progress to experimentally validate the numerical models presented.

ACKNOWLEDGEMENTS

The authors would like to acknowledge the financial support of the Natural Sciences and Engineering Research Council of Canada (NSERC), the Québec Funds for Research on Nature and Technology (FRQNT), and Hydro-Québec.

REFERENCES

ADINA. (2016). *Theory and Modeling Guide, Report ARD 06-7*, ADINA R&D Inc.

Bouchard, S. (1991). *Stabilité des ouvrages miniers* (in French). Québec: Éditions O. Germain.

CDA (Canadian Dam Association). (2006). *Dam safety guidelines*, Edmonton, Alberta, Canada.

EPRI (Electric Power Research Institute). (1992). "Uplift pressures, shear strengths and tensile strengths for stability analysis of concrete gravity dams." *EPRI TR-100345*, Palo Alto, CA.

FERC (Federal Energy Regulatory Commission). (2002). *Engineering guidelines for the evaluation of hydropower projects - Draft chapter III: Gravity Dams.*

Guttiérez, M. (2013). "Shear resistance for concrete dams." Ph.D. Thesis, Norwegian University of Science and Technology, Trondhiem.

ICOLD (International Commission on Large Dams). (2004). *Sliding Safety of Existing Gravity Dams - Final Report.*

Johansson, F. (2009). "Shear strength of unfilled and rough rock joints in sliding stability analyses of concrete dams." Ph.D. Thesis, Royal Institute of Technology of Stockolm.

Ladanyi, B., and Archambault, G. (1970). "Simulation of shear behavior of a jointed rock mass. " In *11th US Symposium on Rock Mechanics*, American Rock Mechanics Association.

Lo, K. Y., Lukajic, B., Wang, S., Ogawa, T., and Tsui, K. K. (1990). "Evaluation of strengh parameters of concrete-rock interface for dam safety assessment." *Canadian Dam Safety Conference*, Toronto, 71-94.

Lo, K. Y., and Grass, J. D. (1994). "Recent experience with safety assesment of concrete dams on rock foundation." *Canadian Dam Safety Conference*, Winnipeg, Manitoba, 231-250.

Nuss, L. K., Matsumoto, N., and Hansen, K. D. (2012). "Shaken, but not stirred - Earthquake performance of concrete dams." *32nd Annual USSD Conference*, New Orleans, Louisiana, United States of America.

Patton, F. D. (1966). "Multiple modes of shear failure in rock." In *1st ISRM Congress*, International Society for Rock Mechanics.

PEER (Pacific Earthquake Engineering Research Center) - Strong Motion Database, *http://peer.berkeley.edu/smcat/*, Date accessed: 15 September 2012.

Renaud, S., Bouaanani, N., and Miquel, B. (2016). "Critical Appraisal of Common Simplified Assumptions in Seismic Stability Analyses of Gravity Dams. " *Journal of Performance of Constructed Facilities*, 10.1061/(ASCE)CF.1943-5509.0000843.

Seidel, J. and Haberfield, C. (2002). "Laboratory testing of concrete-rock joints in constant normal stiffness direct shear. " *ASTM Geo. Test. J.*, 25(4), 391–404.

USACE (United States Army Corps of Engineers). (1995). "Gravity Dam Design." *Engineering monograph EM1110-2-2200.*

USBR (United States Bureau of Reclamation). (2009). *Risk Analysis for Concrete Gravity Dams. Chapter 12.*

Westergaard, H. M. (1933). "Water pressure on dams during earthquakes." *Transactions ASCE*, 98, 418-472.

APPLICATIONS OF FLOW-3D FOR STABILITY ANALYSES OF CONCRETE SPILLWAYS AT FERC PROJECTS

Elise N. Dombeck, P.E.[1]

ABSTRACT

The engineering guidelines published by the FERC recognize that stability analyses of concrete gravity spillways are complicated by the hydrodynamic forces resulting from changes in flow speed and direction. The depth of flow, pressure distribution along the spillway profile, and tailwater conditions are not easily solved without rigorous analysis or physical model studies. Procedures are given in Chapter 3 of the FERC Engineering Guidelines for a simplified method of determining hydrodynamic forces; however, they are not applicable for cases in which the tailwater is not swept away from the spillway toe. The most common stability analysis method for such cases includes the following assumptions for the determination of hydrodynamic forces: ignore the weight of the nappe, and use 60% of the expected tailwater height. These assumptions can lead to an under-estimation of the factor of safety against sliding, which could result in costly remediation of projects with significant stabilizing nappe and tailwater forces. This paper focuses on the use of CFD modelling in estimating the hydrodynamic forces on concrete gravity spillways and the model's application in evaluating a structure's stability. A two-dimensional model representing a unit width of an ogee spillway at two FERC-licensed projects was developed within FLOW-3D for two loading conditions: Probable Maximum Flood (PMF), and an intermediate flood where tailwater is not swept away from the toe. The resulting hydrodynamic forces were incorporated into a two-dimensional gravity stability analysis, and the results showed that the sliding factor of safety for both projects under both loading conditions was greater than that estimated using the conventional stability analysis method.

INTRODUCTION

The catastrophic consequences of dam failure have been demonstrated throughout history with events such as the sliding failure of the concrete gravity Austin Dam in Pennsylvania, which resulted in the loss of 78 lives, and the sliding failure of the concrete gravity Bouzey Dam in Belfort, France, which resulted in the loss of more than 100 lives (USBR, 1998). These case histories emphasize the importance of the continual evaluation of the safety and performance of our existing infrastructure. With the implementation of the Part 12D inspection and independent consultant review program, the Federal Energy Regulatory Commission (FERC) requires complete review of all pertinent engineering analyses for each jurisdictional project every five years. This provides the opportunity for independent consultants and FERC staff to reevaluate the stability analyses for each project against the most recent engineering guidelines and current state of dam safety

---

[1] Civil Engineer, Division of Dam Safety and Inspections, Federal Energy Regulatory Commission, Address: 3700 Crestwood Parkway, Duluth, GA 30096. Email: elise.dombeck@ferc.gov. Phone: 678-245-3008.

practice, ensuring compliance with current factors of safety against sliding and overturning failures.

The most common and simple approach to performing stability analyses on spillway structures begins with the traditional gravity method, which calculates the steady-state forces acting on the dam, usually completed for a unit width. The hydrostatic loading and uplift forces resulting from reservoir and tailwater loading are relatively easy to quantify; however the hydrodynamic forces resulting from water changing speed and direction as it flows over a spillway can become complicated. The depth of flow, pressure distribution along the spillway profile, and tailwater conditions are not easily solved without rigorous analysis or physical model studies. Procedures are given in Chapter 3 of the FERC Engineering Guidelines for a simplified method of determining hydrodynamic forces resulting from nappe pressures (FERC, 2002); however, they are not applicable for cases in which the hydraulic jump is submerged and tailwater is not swept out away from the spillway toe. The common approach to quantifying the hydrodynamic forces on a spillway for such cases, and hereto forth referred to as the conventional method, includes the following assumptions: ignore the weight of the nappe, and use 60% of the expected tailwater height in determining the tailwater resisting forces. In the case of projects with significant stabilizing tailwater forces, these assumptions can lead to an under-estimation of the factor of safety against sliding, which could result in costly and unnecessary remediation to achieve compliance with the current criteria for dam stability. However in other cases, such as those projects where uplift pressures exist along the ogee crest, low bucket pressures exist at the toe, and little tailwater is present, the assumption of 60% tailwater and no hydrodynamic force calculations could produce unconservative results, or results in which the factor of safety against sliding is over-estimated. With the advancements in computing power and the increasing ease of use of finite element modeling and computational fluid dynamics (CFD), numerical modelling programs are replacing physical models and complicated calculations associated with the hydrodynamic loading conditions as the latest state of the engineering practice.

Several papers have already been published focusing on the accuracy of CFD analyses to model ogee spillway flow and pressure characteristics, comparing numerical model results to physical model results. The conclusions presented in (Johnson, 2006), for example, establish that CFD analyses can be sufficiently accurate, economical substitutes for physical models to obtain data necessary to perform stability analyses of gravity structures, particularly under high tailwater conditions. This paper, then, will focus on evaluating the benefits of using a CFD model as an alternative for the computation of hydrodynamic forces to be utilized in a two-dimensional stability analysis. A CFD model was developed for the ogee spillway at two FERC-licensed projects, and the results of the model were used to complete a stability analysis and calculate the factor of safety against sliding for each project under two loading conditions: the Probable Maximum Flood (PMF), and an intermediate flood in which tailwater is not swept away from the toe. The sliding factors of safety were compared to the factors of safety calculated using the currently-accepted practice of ignoring hydrodynamics forces acting on a submerged spillway and using 60% of the resisting tailwater force.

BACKGROUND

Gravity Method for Stability Analysis

The traditional gravity method for stability analysis of concrete gravity structures is merely a static analysis of the forces applied to a two-dimensional cross-section of the dam. Under normal loading conditions with minimal flow over the crest of the gravity spillway, these forces include: hydrostatic loading from the reservoir, hydrostatic loading from silt pressure, hydrostatic loading from tailwater, the weight of the dam itself, and uplift along the foundation contact. Figure 1 presents a free-body diagram of these forces on a typical gravity spillway under normal loading conditions.

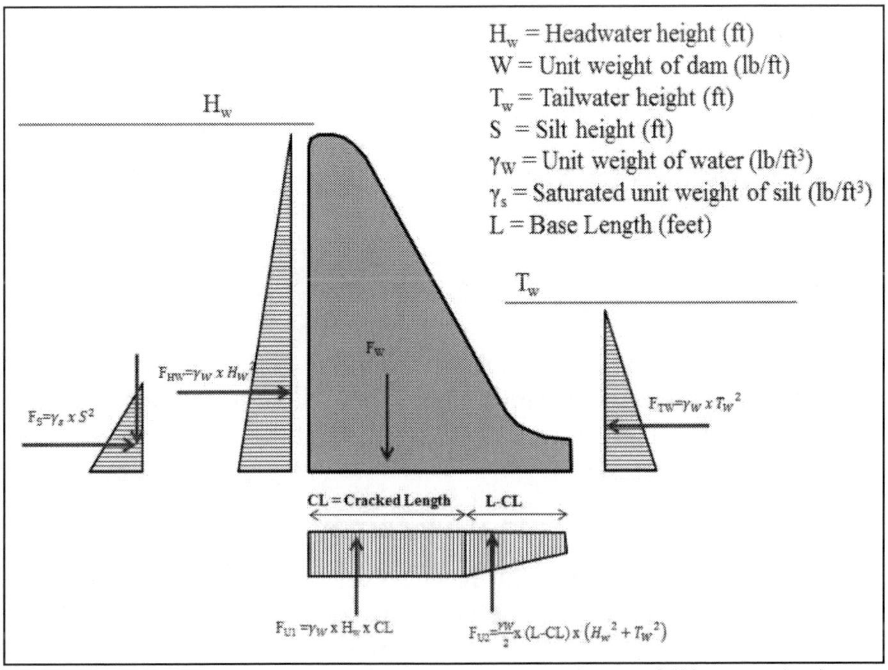

Figure 1. Typical Forces Applied to Gravity Stability Analysis

The uplift applied along the foundation contact is assumed to vary linearly from headwater pressure at the heel to full tailwater pressure at the toe for an uncracked base length. The gravity method of analysis assumes that total length of the base is in compression for an uncracked base length, and the pressure along the base from the resultant forces is triangularly distributed. The resultant force therefore acts at the one-third point on the uncracked base length. To determine whether or not the base cracks under a particular loading condition, the resultant forces are calculated and the horizontal distance of the resultant from the toe is multiplied by 3 to determine the length of the base in compression. If this length is not the same as the base length of the dam, then the base is cracked and a new crack length is assumed. The cracked-base analysis is completed in an iterative fashion, assuming full headwater pressure along the cracked length (shown schematically in Figure 1), until the calculated length of the base in compression matches the remaining uncracked base length.

The FERC Engineering Guidelines require that "High" and "Significant" hazard potential dams meet or exceed the minimum required factor of safety against sliding, given below, for a series of loading conditions. This table is shown with two differing assumptions with regard to the value of "cohesion" used in the stability formula: (1) cohesion is assumed to exist (with some basis for the assumed value); or (2) cohesion is neglected.

Loading Condition	**Factor of Safety (FS)**	
	Cohesion	**No Cohesion**
Usual (Normal)	3.0	1.5 (Worst Static Case)
Unusual (PMF)	2.0	1.3
Post Earthquake	1.3	1.3

As Chapter 3 of the FERC Engineering Guidelines points out, one of the main sources of uncertainty in the stability analysis of gravity dams is the assumed value of the cohesion parameter for the interface between the foundation and the structure as it is difficult to quantify through testing. For this reason, this paper will compare results to the no-cohesion minimum factors of safety. With the crack length and steady-state forces determined, the factor of safety against sliding can be computed using the following equation:

$$FS = \frac{(\Sigma F_{VER} - \Sigma F_{HOR} \times \tan \alpha)}{(\Sigma F_{HOR} + \Sigma F_{VER} \times \tan \alpha)} \times \tan \phi$$

Where F_{VER} are the forces acting on the structure in the vertical direction in pounds, F_{HOR} are the forces acting on the structure in the horizontal direction in pounds, φ is the friction angle of the foundation in degrees, and $\propto$ is the slope of the dam foundation in degrees.

Hydrodynamic Forces

For loading conditions where significant flow occurs over the crest of the spillway, hydrodynamics forces should be considered in the performance of a stability analysis in addition to the forces discussed above. Ogee weirs are often chosen for the crest shape of gravity overflow spillways due to their high discharge efficiency at a single given design head. At the design head, the ogee shape results in near atmospheric pressure over the crest section. The pressures and discharge over an ogee-crested dam, however, are dependent on a variety of factors such as crest and spillway geometry, upstream flow depth, and downstream tailwater depth that make determination of the flow and pressure distribution along the spillway profile challenging. For tailwater that is swept out away from the spillway toe, Appendix 3A of Chapter 3 of the FERC Engineering Guidelines has established a simplified method for calculating a nappe profile and pressure distribution at incremental points along the spillway to determine a resultant force (FERC, 2002). In cases where the tailwater is greater than the calculated conjugate depth of the hydraulic jump, the current state of the practice relies on assuming 60% of the tailwater height for calculation of resisting forces and neglecting the weight of the nappe. This 60% reduction appears to come from the formulas for calculating the flow geometry along a straight drop spillway; the flow from a straight drop spillway results in a jet that plunges into a straight-bottom channel. The jet causes a hydraulic jump to occur downstream, and results in an upstream water depth against the drop spillway equal to about 60% that of the conjugate depth of the hydraulic jump (Chow, 1959). This paper utilizes FLOW-3D, a commercially-available CFD program, to estimate the pressure distribution along the spillway to calculate the resultant hydrodynamic force to be utilized in a stability analysis.

METHOD

The FLOW-3D computer model, developed by Flow Science Inc., uses the finite volume method to solve the complete Navier-Stokes equations for three-dimensional flow models. The model applies a Cartesian grid, or mesh, to the model domain and solves for cell-averaged flow parameters at discrete time steps. Because the model uses a volume-of-fluid (VOF) technique, each cell is considered to have one of the following six conditions: completely solid, partly solid and partly fluid, completely fluid, partly fluid and partly empty, partly solid and partly empty, and empty. User defined geometry files are applied to each of the mesh cells to determine the fractional volume blocked by the input geometry.

To develop a FLOW-3D model, three primary forms of inputs are required: geometric inputs, boundary conditions, and mesh inputs. Geometric files can either be imported in the form of a stereolithographic (STL) file, or primitive shapes can be created and grouped within the model interface. For this study, Autodesk's AutoCAD software was used to create solid shapes extruded from two-dimensional cross-sections. The cross-section information for each project was obtained from the stability study of record contained within the Supporting Technical Information (STI) document required by FERC to be compiled for each project subject to the Part 12D inspection program. In order to minimize computation time and modeling effort, two-dimensional models representing a unit width of the spillways were developed instead of a full three-dimensional simulation. The two projects selected were chosen based on their relatively long spillway lengths, lending themselves well to the assumption of symmetrical flow across a cross-sectional slice.

The models were oriented the same for both projects; the x-axis is perpendicular to the spillway crest and corresponds to the upstream to downstream direction, the y-axis runs parallel to the spillway crest, and the z-axis represents elevation. Since the mesh is not dependent on the geometry shape, meshing in FLOW-3D merely requires the definition of a start and end coordinate, and either cell size or number of cells per x-, y-, and z-direction. Mesh cells were defined as cubes, with the size of the cells the same in each of the three dimensions. Three meshes were defined to allow for smaller mesh sizes to be used around the spillway geometry, and coarser meshes to be located upstream and downstream of the spillway where changes in the flow surface are not as abrupt. Each adjacent mesh area had cell size ratios of 2:1 such that the change in cell size in any direction is not more than twice that of the neighboring mesh. The upstream mesh boundary was located 50 feet upstream of the heel of the spillway to allow for weir drawdown effects. The finer mesh was located in between the two coarser mesh blocks and was defined such that the spillway geometry was completely contained within the finer mesh. In order to ensure that enough room was left downstream of the toe for a full hydraulic jump to develop, the downstream mesh boundary was located 400 feet downstream of the spillway toe. The floor of the mesh was set to the toe elevation, while the ceiling of the model mesh was located 10 feet above the PMF reservoir elevation.

The upstream and downstream boundary conditions were set as specified fluid elevations corresponding to the headwater and tailwater elevations estimated for the specified loading condition as presented in the STI document, respectively. Because the purpose of this analysis was to evaluate the difference in the results of the two separate stability analysis methods, no additional analysis was performed to verify the headwater and tailwater rating curves. The right and left boundaries in the y-direction were set to symmetrical based on the assumption that the modeled spillways are long enough to have negligible three-dimensional effects at the location of maximum height from their abutments. The lower z-boundary was set as a wall, and the upper boundary was assumed to be at atmospheric pressure with no fluid. Because nested mesh blocks were used, the finer mesh located in the middle of the model utilized a boundary condition of symmetrical in the upstream and downstream x-direction and the same boundary conditions previously specified in the y- and z-directions.

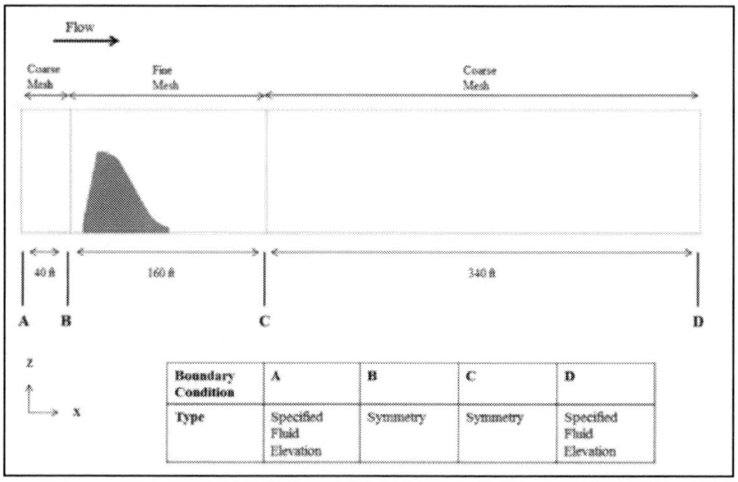

Figure 2. Boundary Conditions and Mesh Block Locations in the X-Z direction

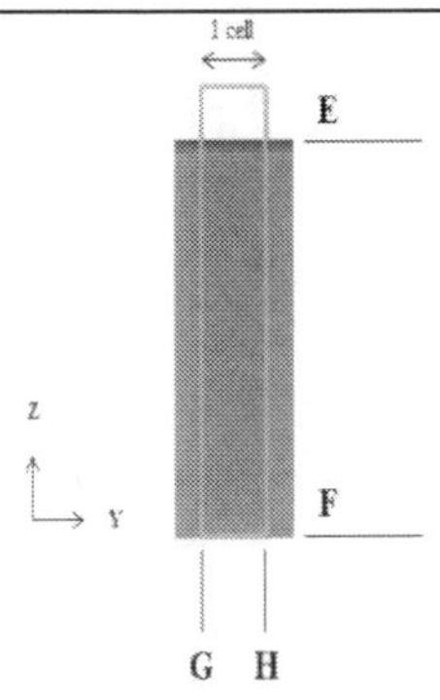

Boundary Condition	E	F	G	H
Type	Specified Fluid Fraction	Wall	Symmetry	Symmetry

Figure 3. Boundary Conditions and Mesh Block Locations in the Y-Z direction

Several simulations were developed for each project in order to minimize computation time. The first simulation utilized a coarse mesh of 4-foot cubes upstream and downstream of the spillway, and 2-foot cubes around the spillway geometry. The subsequent simulations were restart simulations, utilizing the results from the last time step of the previous simulation as the initial condition and refining the mesh around the spillway structure until the minimum mesh of 0.5-foot cubes was achieved. The model simulation time was set such that the model reached a steady-flow state before the end of each simulation. The water-surface profile, pressure distribution, and total fluid force acting on the spillway component for the last time step of the final simulation were taken as the steady-state results to be utilized in the stability analysis.

The FLOW-3D model produces a pressure head and water surface elevation at discrete points along the spillway face. A neutral file was developed to extract the pressure and water surface elevation at specific points along the spillway face located at uniform increments of approximately 0.5 foot, equivalent to the mesh size. A spreadsheet was then developed to sum the total force on the spillway due to the fluid forces acting at each discrete point, and a resulting total fluid force in the x- and z-direction was obtained as well as a location that the resultant acts. The resulting total fluid force represents the hydrostatic loading of the reservoir, the hydrodynamic forces due to the discharge over the spillway, and the hydrostatic loading of the tailwater. A separate spreadsheet was developed to perform the gravity stability analysis as outlined in the previous section and compare the hydrodynamic loading results to those results calculated following the conventional stability analysis assuming 60% tailwater and ignoring the weight of the nappe. Additional static loads such as silt loading and the weight of the structure concrete were obtained from the stability analysis of record. No cohesion was applied to the dam foundation interface, and the stability analysis assessed only sliding along the concrete-rock interface.

RESULTS

The water surface profiles and pressure distributions resulting from the numerical analysis using the FLOW-3D software program, as shown in Figures 4 and 5, for the PMF loading condition at both projects confirm that the high tailwater predicted in the downstream channel will submerge the spillways, resulting in tailwater resisting forces higher than that calculated assuming 60% of the predicted tailwater elevation.

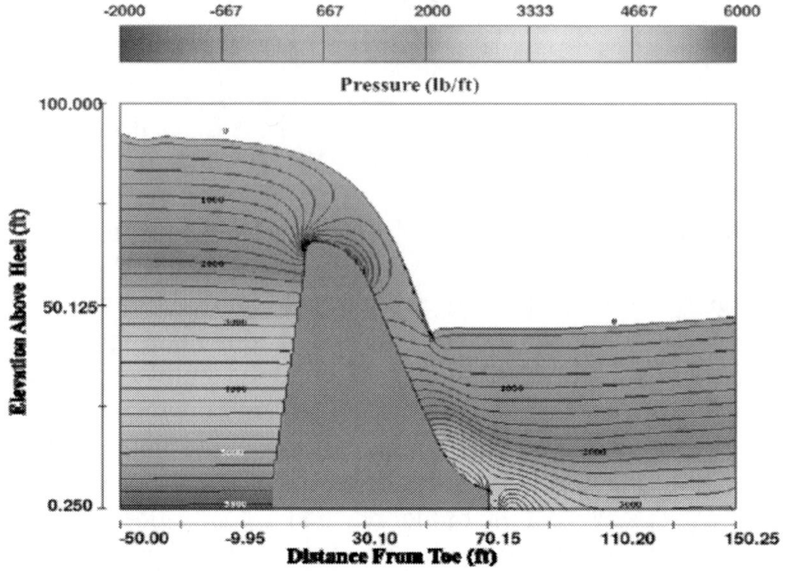

Figure 4. Pressure Distribution and Water Surface Elevation Results from PMF Loading Project A.

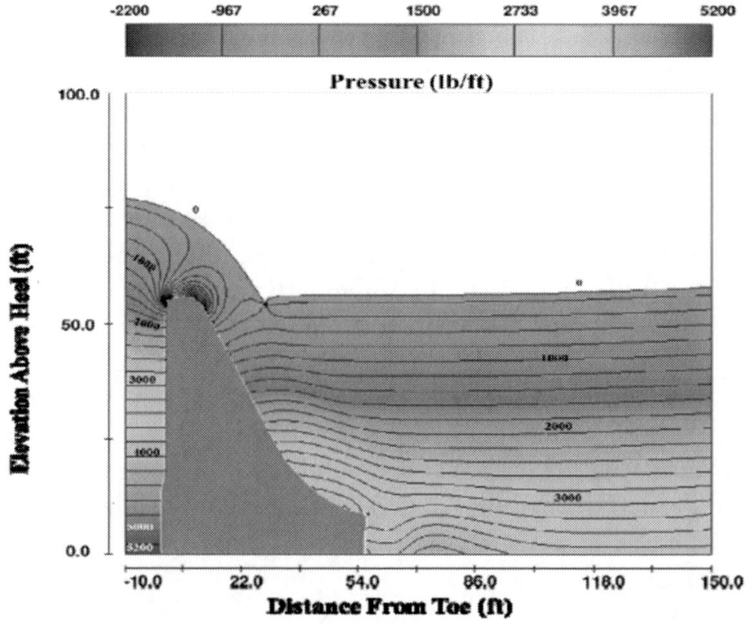

Figure 5. Pressure Distribution and Water Surface Elevation Results from PMF Loading at Project B.

The pressure distribution along the spillway profile computed within FLOW-3D agrees with the general profile that is expected for ogee spillways experiencing discharges greater than the design flow. Figures 6 and 7 below illustrate that the pressure is nearly equal to hydrostatic head upstream of the dam and decreases to below atmospheric pressure along the crest of the ogee. The pressure then increases steadily from below the crest along the spillway profile until the pressures reach a maximum within the spillway bucket. Because significant negative pressures are reached along the spillway crests, it can be concluded that the example spillways were designed for loading events much less than the PMF inflow. This illustrates the importance of continuing to reevaluate projects as the engineering state of practice and engineering guidelines change. For older dams that were designed for inflow design floods less than the PMF, the PMF loading condition may result in significant upward forces along the crest and significant downward forces in the spillway bucket. If tailwater forces are not present, the significant uplift at the crest coupled with the large downward bucket forces could result in a significant overturning moment about the spillway toe.

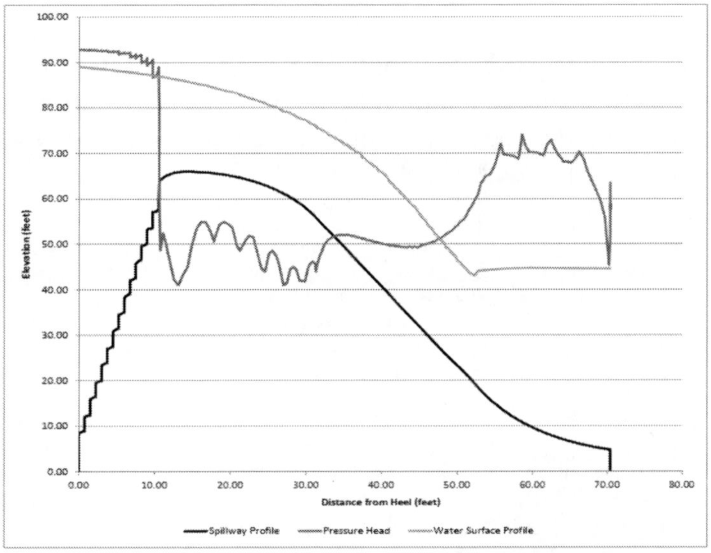

Figure 6. Comparison of Pressure Distribution and Water Surface for PMF Loading at Project A.

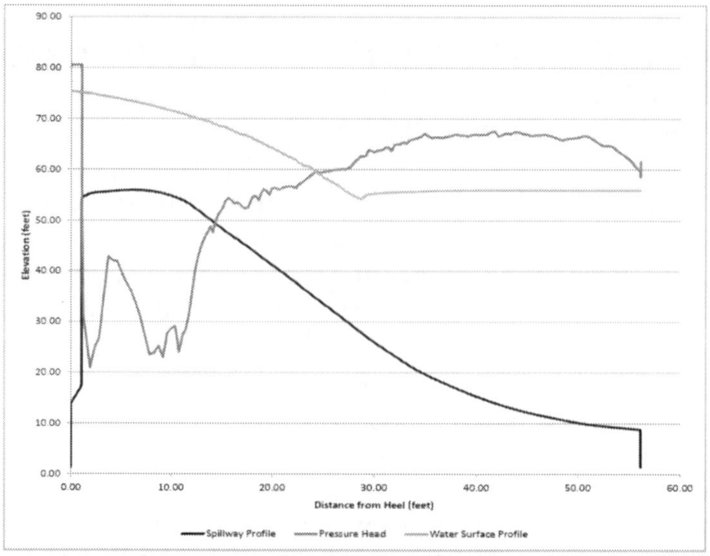

Figure 7. Comparison of Pressure Distribution and Water Surface Profile for PMF Loading at Project B.

For the intermediate loading cases, the numerical analysis results from FLOW-3D show that the tailwater conditions force a hydraulic jump to occur within the spillway bucket of both projects. Figures 8 and 9 show the water surface profile and pressure distribution for the intermediate loading case for Projects A and B.

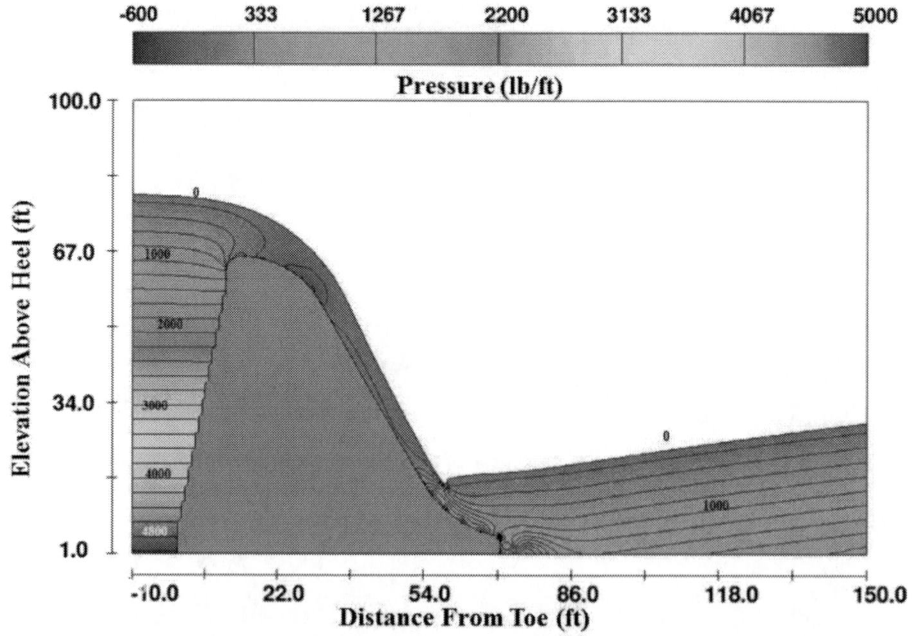

Figure 8. Pressure Distribution and Water Surface Elevation Results from Intermediate Loading at Project A.

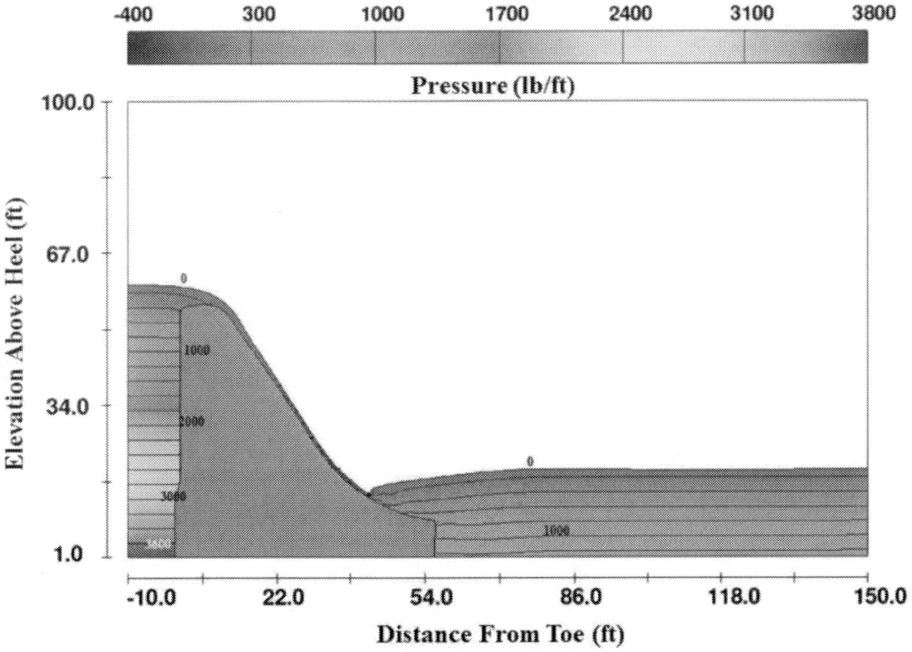

Figure 9. Pressure Distribution and Water Surface Elevation Results from Intermediate Loading at Project B.

The pressure distribution along the spillway profile for both projects, as shown in Figures 10 and 11, show that the pressure is near atmospheric along the crest of the ogee, increasing slightly along the length of the spillway, and exceeding hydrostatic head in the spillway bucket where the flow rapidly changes direction. A significant negative pressure lift at the crest was not observed, signaling that the flow rates for the intermediate loading conditions do not exceed the designed capacity.

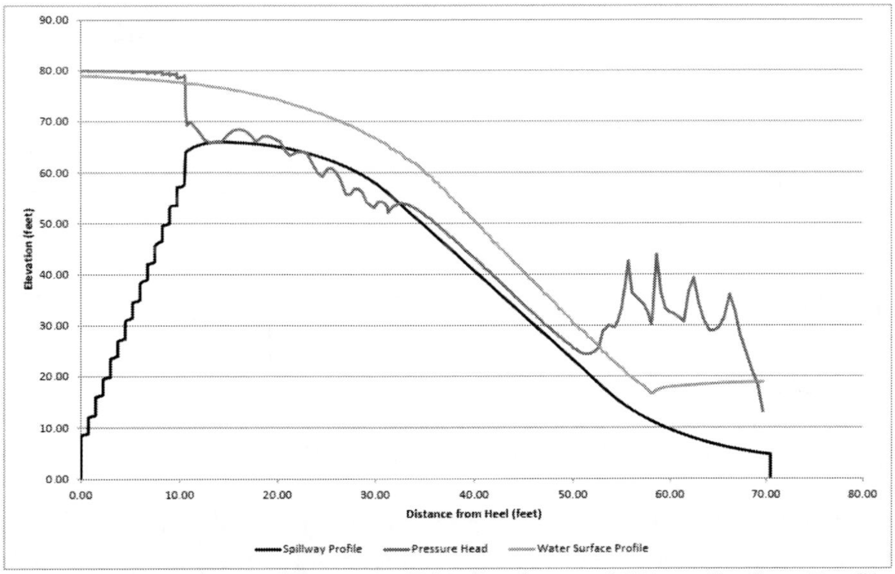

Figure 10. Comparison of Pressure Distribution and Water Surface Profile for Intermediate Loading at Project A.

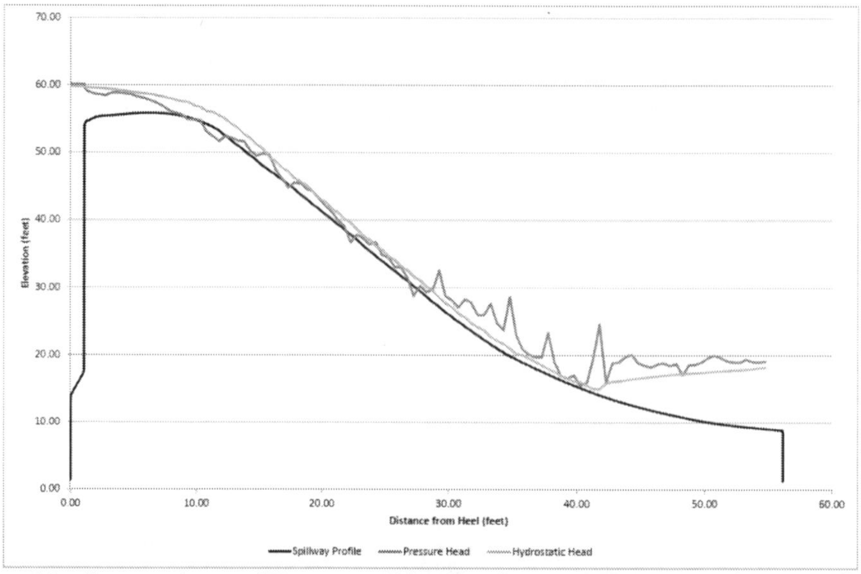

Figure 11. Comparison of Pressure Distribution and Water Surface Profile for Intermediate Loading at Project B.

The pressure distributions predicted by the FLOW-3D model along the spillway profiles were used to compute the total fluid forces exerted on Water Surface Profile projects’ spillways, and the resulting forces were compared to those used in a conventional analysis. The results are summarized in Table 1. For both projects, the total fluid force acting in the x-direction, which includes the hydrostatic force of the reservoir, the hydrostatic force of the tailwater, and the hydrodynamic forces, was decreased over the conventional approach by using FLOW-3D. Similarly, the total fluid force in the y-direction was increased for both projects, resulting in a more stable structure.

Table 1. Comparison of Total Fluid Forces from Numerical Model to Conventional Analysis.

		Project A			**Project B**		
		Numerical (kips/ft)	**Conventional (kips/ft)**	**Δ%**	**Numerical (kips/ft)**	**Conventional (kips/ft)**	**Δ%**
PMF	**Total Fluid Force (x)**	145.2	205	-29	67.3	139.6	-52
	Total Fluid Force (y)	-101.7	-77.4	+31	-89.3	-41.2	+117
Inter.	**Total Fluid Force (x)**	160.0	181.7	-12	93.9	109.3	-14
	Total Fluid Force (y)	-56.4	-44.0	+28	-15.1	-2.1	+620

The sensitivity of the resulting total fluid forces to changes in the size of the mesh was evaluated by comparing the results of the model simulation using 2-foot cubes with the results of the model simulation using 0.5-foot cubes. The results, given in Table 2, show that minimal change was noted between the model results using the two different mesh sizes. The 0.5-foot mesh, thus, does not require further refinement.

Table 2. Comparison of Total Fluid Forces for Different Mesh Sizes.

		Total Fluid Force (X)			Total Fluid Force (Y)		
		2-foot Mesh	0.5-foot Mesh	Δ%	2-foot Mesh	0.5-foot Mesh	Δ%
A	PMF Loading	144.1	145.2	1	-107. 1	-102.0	5
	Intermediate	157.9	160.0	1	-63.3	-56.4	12
B	PMF Loading	71.7	67.8	-6	-81.2	-88.9	-9
	Intermediate	96.6	93.9	-3	-9.9	-15.1	-35

The results of these hydrodynamic forces were then input into a two-dimensional gravity stability analysis to determine the safety factor against sliding for both projects along the concrete-rock interface. All stability parameters, such as the friction angle along the concrete-rock interface and the unit weight of concrete, were kept the same between the two stability analyses, varying only the total fluid force calculation to evaluate the stability benefits of the numerical FLOW-3D model over the conventional model. Tables 3 through 6 below present the results of both stability analyses performed for both projects under the PMF loading condition. For the PMF loading condition, the FERC recommended factor of safety against sliding assuming no cohesion at the base is 1.3. The forces described in the tables below correspond with those presented in Figure 1.

Table 3. Conventional Stability Analysis Results for Project A assuming 60% Tailwater Elevation and Ignoring Hydrodynamic Forces for PMF Loading.

Force	HORIZONTAL FORCES		VERTICAL FORCES		MOMENT ABOUT TOE
	Force	$\overline{Y}$	Force	$\overline{X}$	
Spillway Weight, F_W			-394	27.2	10741
Silt Force, F_S	6	7.3	-2	1.0	43
Uplift Along Cracked Base, F_{U1}			407	35.2	-14343
Uplift along Uncracked Base, F_{U2}			0	70.4	0
Headwater Force (1), F_{HW}	246	26.9	-42	4.5	6811
Tailwater Force (2), F_{TW}	-41	12.0	-35	60.2	1615
RESULTANT FORCE	211	0	-66	73.3	
Crack Length:				70.4	
Sliding Safety Factor:				**0.45**	

Table 4. Total Fluid Forces and Conventional Stability Analysis Results for Project A utilizing the Hydrodynamics Forces Calculated by FLOW-3D for PMF Loading.

Force	HORIZONTAL FORCES		VERTICAL FORCES		MOMEN T ABOUT TOE
	Force	$\overline{Y}$	Force	$\overline{X}$	
Total Fluid Forces, $F_F = F_{HW} + F_{TW} + F_{HYDN}$	146.0	34.3	-101	42.3	9597
RESULTANT FORCE:	152	0	-101	66.9	
Crack Length:				10.6	
Sliding Safety Factor:				**0.95**	

Table 5. Conventional Stability Analysis Results for Project B assuming 60% Tailwater Elevation and Ignoring Hydrodynamic Forces for PMF Loading.

Force	HORIZONTAL FORCES		VERTICAL FORCES		MOMENT ABOUT TOE
	Force	$\overline{Y}$	Force	$\overline{X}$	
Spillway Weight, F_W			-257	19.7	5060
Silt Force, F_S	11	9.2			96
Uplift Along Cracked Base, F_{U1}			283	28.0	-7935
Uplift along Uncracked Base, F_{U2}			0		0
Headwater Force (1), F_{HW}	187	23.2			4335
Tailwater Force (2), F_{TW}	-47	12.9	-41	60.2	1876
RESULTANT FORCE	150	0	-16	225.1	
Crack Length:				56.0	
Sliding Safety Factor:				**0.15**	

Table 6. Conventional Stability Analysis Results for Project B utilizing the Hydrodynamics Forces Calculated by FLOW-3D for PMF Loading.

Force	HORIZONTAL FORCES		VERTICAL FORCES		MOMENT ABOUT TOE
	Force	$\overline{Y}$	Force	$\overline{X}$	
Total Fluid Forces, $F_F = F_{HW} + F_{TW} + F_{HYDN}$	67	30.7	-89.3	44.8	6057
RESULTANT FORCE:	78	0	-70.3	51.8	
Crack Length:				41.2	
Sliding Safety Factor:				**1.30**	

The conventional stability analysis results for both Projects A and B show that the structures are computationally unstable under PMF loading with a fully cracked base of the structure. The numerical model stability results using the hydrodynamic forces calculated by FLOW-3D, however, show a significantly higher factor of safety against sliding than the conventional analysis. Based on the results of the numerical model stability analyses, Project A would still require rehabilitation to be in compliance with the FERC Engineering Guidelines, but the additional forces necessary to meet the minimum factor of safety against sliding would be less if the design were performed using the results of the numerical model. The use of a numerical model for the determination of fluid forces at Project B results in Project B meeting FERC stability criteria, and no further rehabilitation would be necessary.

While the two examples presented above suggest that a reduction in tailwater of 60% is overly conservative by under-estimating the factor of safety against sliding, the comparison between the numerical model stability results and a stability analysis result using 100% tailwater as shown in Table 7 illustrates that a submerged hydraulic jump does not justify the use of 100% tailwater in completing a stability analysis. For the PMF loading condition at both projects, the sliding safety factor is over-estimated as compared to the numerical model results using full tailwater depths to calculate the resisting forces at the spillway toe. This result highlights the importance of performing a detailed spillway profile analysis to justify the use of tailwater forces greater than that proposed by the conventional approach.

Table 7. Comparison of Stability Analysis Results using Numerical Model Approach, Conventional Approach, and Full Tailwater Approach for PMF Loading.

		Project A			Project B		
		Num. (kips/ft)	Conv. (kips/ft)	Full TW (kips/ft)	Num. (kips/ft)	Conv. (kips/ft)	Full TW (kips/ft)
PMF	Resultant Force (x)	152	211	138	78	150	67
	Resultant Force (y)	-101	-66	-182	-70	-15	-116
	Sliding Safety Factor	1.0	0.5	1.9	1.3	0.2	2.5

The stability analysis results for the intermediate loading condition at both projects, as shown in Table 8, demonstrate that the resultant forces calculated based on the conventional approach are in closer agreement with the numerical model, however the conventional approach still overestimates the total fluid forces in the x-direction and underestimates the total fluid forces in the y-direction. Results of the numerical model for both projects show a close agreement between the total fluid forces and resultant forces for both the numerical approach and the full tailwater approach.

Table 8. Comparison of Stability Analysis Results using Numerical Model Approach, Conventional Approach, and Full Tailwater Approach for Intermediate Loading.

		Project A			Project B		
		Num. (kips/ft)	Conv. (kips/ft)	Full TW (kips/ft)	Num. (kips/ft)	Conv. (kips/ft)	Full TW (kips/ft)
Inter.	Total Fluid Force (x)	160	182	161	94	109	101
	Total Fluid Force (y)	-56	-44	-61	-15	-5	-14
	Resultant Force (x)	166	187	167	104	120	111
	Resultant Force (y)	-207	-195	-212	-131	-120	-129
	Sliding Safety Factor	1.8	1.5	1.8	1.8	1.4	1.7

CONCLUSION

This paper presented a comparison of the conventional method for calculating the stability of a submerged gravity spillway structure versus the use of a numerical analysis of structural loading for two FERC-licensed projects. The results illustrate that the use of a CFD model for loading assumptions of structures with high tailwater conditions under the critical loading condition may result in a higher factor of safety than results using a conventional stability analysis approach. With these results, substantial cost savings may be realized for projects that require rehabilitation to meet FERC minimum factor of safety criteria against sliding. While some cases may result in no action necessary (like Project B), results presented above also illustrate that the use of CFD may not entirely remove the need for rehabilitation (like Project A). A two-dimensional analysis can be developed economically and quickly using commercially available CFD programs and may be a good entry-level analysis tool to evaluate the stability of projects with significant stabilizing forces due to tailwater conditions. The use of CFD models should be thoroughly evaluated on a case-by-case basis by professional engineers that are well-versed in CFD software programs and hydraulic theory.

REFERENCES

Chow, Ven Te (1959). *Open-Channel Hydraulics*. McGraw-Hill, New York, NY.

Federal Energy Regulatory Commission, FERC (2002). "Engineering Guidelines for the Evaluation of Hydropower Projects". Washington, D.C.

Johnson, Michael C and Savage, Bruce M (2006). "Physical and Numerical Comparison of Flow over Ogee Spillway in the Presence of Tailwater". *Journal of Hydraulic Engineering* 132 (12), 1354-1357.

U.S. Bureau of Reclamation, USBR (1998). "Concrete Dams: Case Histories of Failures and Nonfailures with Back Calculations". Report No. DSO 98-05, Denver.

THE STABILITY OF MASONRY WALLS AND DAMS: A REVIEW OF THE HISTORICAL DESIGN OF CONCRETE DAMS

Kevin Griebenow, PE[1]

ABSTRACT

Masonry walls are one of the oldest and easiest ways to build structures. We have learned over time that to improve the likelihood that a masonry wall will not fall down, all of the wall is to be in compression. Likewise, we have been building dams to store or alter river flows since at least 3000 BC, most like masonry walls. Built in 1500 BC, the Marib dam in Yemen is a superb example of a "wall" dam reaching a height of 20 meters.

A dam can be described as a water retaining surface with the remainder of the structure supporting this surface. Initially, the designers and builders of dams depended on its overwhelming mass to support and resist the hydrostatic load, i.e. a gravity dam. The determination of the necessary mass for a gravity dam was initially based on past experience, followed by geometric and then stability analysis. The stability of a gravity dam is compared to a masonry wall in terms of the role of thrust, the middle third, and preventing tension.

The first structural dams (arch, multi-arch and slab and buttress) were constructed based on past experience. Driven in part to be more economical, structural dams required a more detailed understanding and analysis for their forms to withstand the hydrostatic loading. For arch dams, this structural analysis started with the "cylinder formula", followed by the constant angle and trial-load method. A buttress dam shares characteristics of gravity and arch structures; it relies primarily on gravity for stability but requires less material to achieve this "mass". Examples of these dam structures, some constructed in the nineteenth century, will show the evolution of the design and construction developed by early civil and structural engineers. This history will show a small part of the evolution of "state of the art" along with that this history of progression should not be considered irrelevant but necessary to understand and improve our profession.

INTRODUCTION

I have been a part of the dam safety community for almost 25 years learning about the middle third or kern of the base early on. I knew that the resultant should fall in the middle third of the base of the dam. Knowing this and understanding the ramifications if it falls outside of it are two different things. This is my journey of learning the history of keeping the line of thrust in the middle third and the implications of this in design of gravity dams and the dichotomy with arch and buttress dams.

---

[1] Federal Energy Regulatory Commission, Chicago, IL Kevin.Griebenow@ferc.gov. Paper and opinions are those of the author and should not be inferred to be an official FERC position

We have been building masonry walls for thousands of years. We probably have been building dams to divert and store water for almost as long. The first dam for which we have reliable evidence was built 5,000 years ago in Jordan. With a height of 66 feet (20 meters) and parts of the control structures still standing, the Marib dam in Yemen was built about 3,500 years ago. With these masonry dams constructed like a wall, the understanding of counteracting the force of water by the dam's weight, thrust and keeping the line of thrust in the middle third would not occur until the 1850's. How did this occur and what was the understanding in dams resisting the forces of water?

Figure 1. Remnants of the Marib Dam, Yemen.

The early builders tried to avoid tension in building with masonry. A masonry wall, limited by the compressive strength of bricks and stone[2], could be constructed to a height of 7,000 feet before the material at the base would be crushed. However, the Great Pyramid and the highest spires are slightly higher than 500 feet. Yet, many walls in buildings have fallen down due to a lack of stability, not compression, with the stress at the base a small fraction of the material's compressive strength.

A gravity dam's design is to "accumulate as much material as possible." Structural dams (arch and slab/buttress dams) rely on their shape to support the impervious layer. The

[2] Assumed density of 120 lbs per cubic foot with a compressive strength of 6,000 psi

refinement of these alternative structural solutions was partially driven by economics and increased understanding of the structural forces and the ability to solve them.

GRAVITY DAMS

Walls to Triangular Cross Section

The lack of understanding of how a gravity dam worked is evident in many designs through the mid 1800's. Gravity dams varied in cross section from being a straight up and down wall to a triangle with the slope on the downstream or UPSTREAM face. Such a dam designed and built in 1837 with a sloping upstream face was unstable upon filling (Figure 2). Buttresses were installed on the downstream wall to stabilize and save the dam.

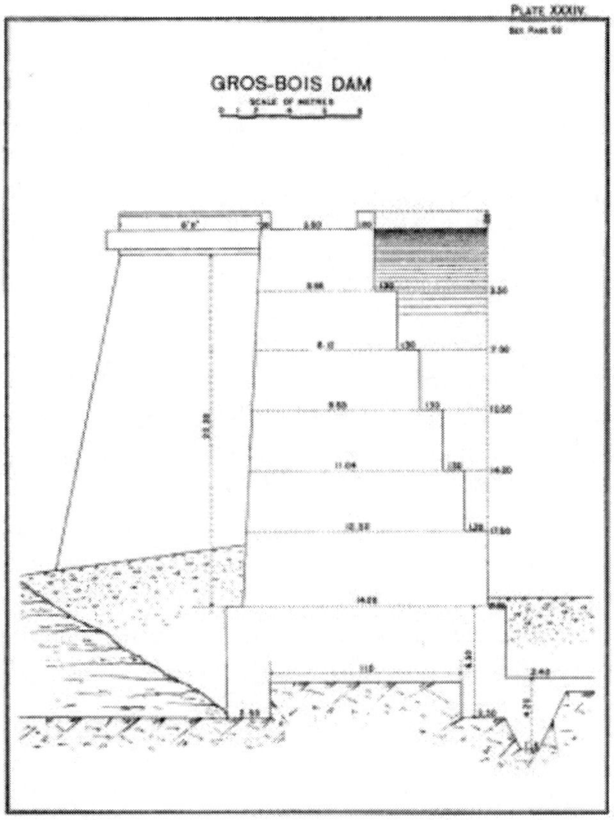

Figure 2. Gros-Bois 1 Reservoir is on the right. Buttresses on the left were added as it was unstable upon filling (Wegmann, Plate XXXIV).

A French engineer, Augustin Torterne de Sazilly, showed in 1850 that the most efficient cross section for a gravity dam is a triangle with the slope on the downstream face. He developed the "profile of equal resistance" of the dam's weight verses the hydrostatic force (Figure 3). These profiles were generated with an empty reservoir for the upstream profile and with the reservoir at the maximum level for the downstream profile. This cross section and method was put to the test with the construction of a 183-foot high dam in the Furan River gorge. Construction of this dam with 52,000 cubic yards of masonry was completed in 1866. Not fully trusting gravity to do all of the work, it was also slightly arched across the narrow gorge. Shortly after its construction, it appeared in design books as the first masonry dam "built according to correct scientific principles." It is still in use today for flood mitigation and water supply.

Building the Ultimate Dam

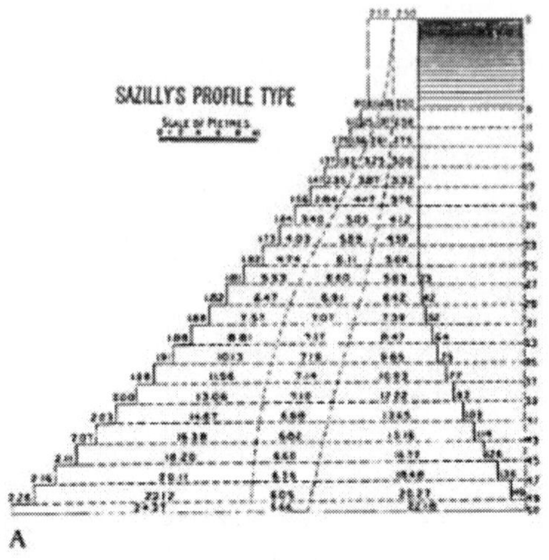

2.8 *(A) DeSazilly's "Profile of Equal Resistance" for masonry gravity dams. (B) Edward Wegmann's "Practical Profile No. 3." (C) The cross section of Croton Dam, showing the "Croton Profile."* (Wegmann, The Design and Construction of Dams, *1911)*

Figure 3. DeSazilly's "Profile of Equal Resistance"

The concept of the middle third followed this work in the 1870s. This concept was developed by the Scottish professor William J. M. Rankine[3]. He determined it was essential to avoid tension in a masonry dam. To reach this state of having no tension stresses, he noted this would be met if the resultant was inside the middle third for all of the horizontal joints. If the resultant continues to move ending outside of the face of the wall, it will tip up and fall down or "overturn". These concepts for designing a dam were gathered in *The Design and Construction of Dams*, by Edward Wegmann[4] in 1888. Using this concept to design the new Croton Dam in New York, the "Croton Profile" became the standard for gravity dams following its construction.

[3] Better known for his thermodynamics and lateral earth pressure work

[4] Donald Jackson calls this book "the world's most prestigious book on dam engineering. p. 238

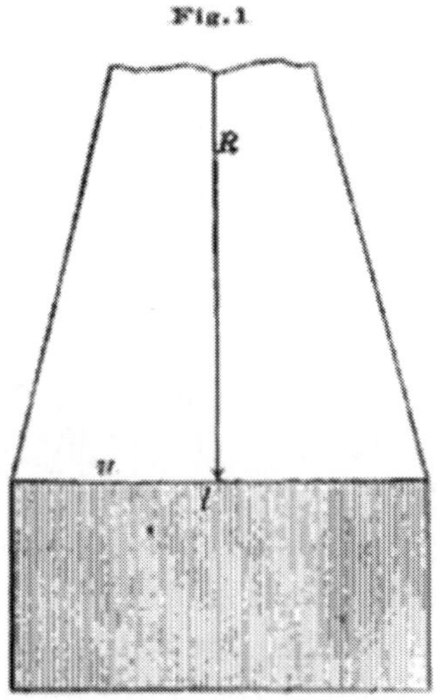

Figure 4-1. Foundation under equal compression pressure with the resultant in the middle

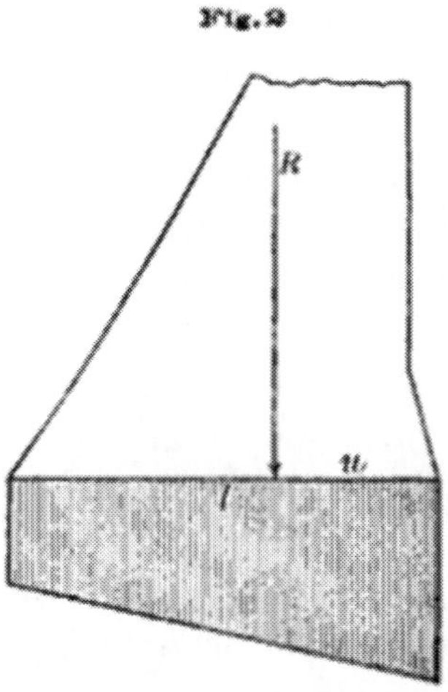

Figure 4-2. Foundation pressure with the resultant inside the middle third.

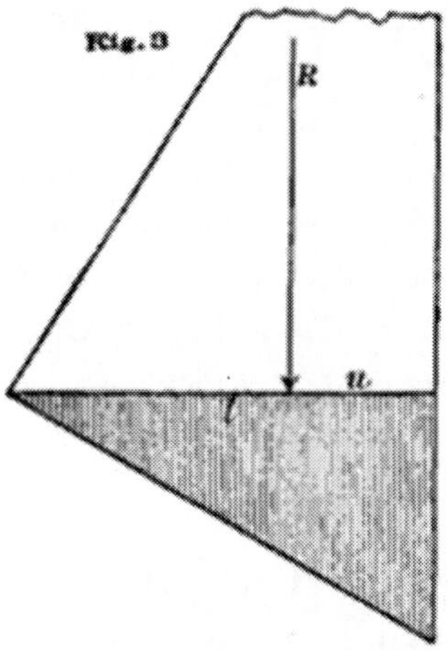

Figure 4-3. Foundation pressure with the resultant at the edge of the middle third.

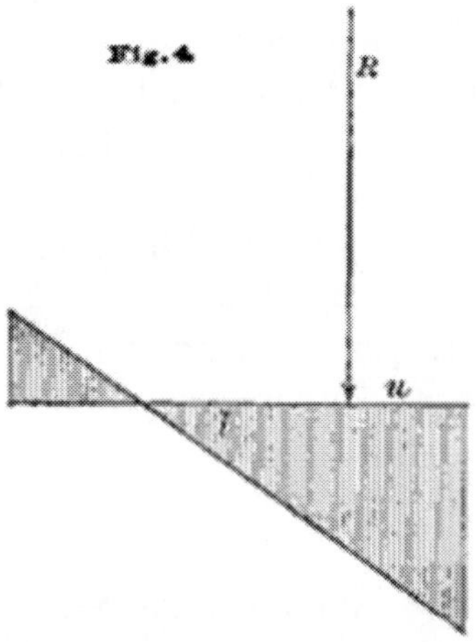

Figure 4-4. Foundation pressure with the resultant outside of the middle third. A small portion of the foundation is in tension.

Figure 4 Progression of foundation reaction as the resultant moves from the middle towards the edge of the wall.

Uplift Forces

For the next milestone dam completed in 1875, Belgian engineers were concerned about a "novel" force called uplift pressure created by the flow of water seeping through the foundation. The size and mass of the Gileppe dam was enormous due to the assumption of the uplift force being equal to the whole dam being submerged. With this overly conservative assumption, the base of the dam was 132 percent of the height of the dam. Shortly after this dam was completed, uplift was included in the design for the 80-foot high Alfeld Dam in Germany. With construction starting in 1883, the uplift was assumed to vary between the headwater and would decrease to the tailwater level. The construction manager, George Deacon thought the uplift should be assumed to be full reservoir for the upstream third and tailwater pressure for the downstream two-thirds (Figure). His forward thinking also improved the situation when he had foundation drains installed.

Attention to this novel force we now take for granted was thought to be necessary for only "very seamy foundations, but in most cases this force will be exerted on only a very small area of the base of the dam." The thinking was that the dam must be in direct contact with most of the foundation and uplift would not be present there. This attention to uplift changes in the 8th edition of Wegmann's book published in 1926, but this force is "impossible to estimate accurately". The importance of this issue is acknowledged with the failure of the Bouzey Dam in France in 1895, the Austin Dam in Texas in 1900 and the Austin Dam in Pennsylvania in 1911, being built on a pervious foundation.

In *A History of Dams*, Nicolas Schnitter writes that it takes more than a generation to understand and accept the full force of uplift. I had been lead to believe that it gained acceptance following the failure of the St. Francis Dam on March 12, 1928. Some believed uplift played a role in its failure, which would fit this time frame. However, this understanding and acceptance is not so clear even in a 1944 book on designing dams. In *Engineering for Dams*, it states that "there is perhaps no other factor in the design of masonry dams and which so little is known and about which there are so many differences of opinion." The equation in this book for total uplift force starts with the now typical assumption of the uplift pressure linearly decreasing from headwater to tailwater. The magnitude of the uplift is reduced by the effective pressure[5], "c", by 33 to 66 percent (Eq1). It is also includes a term "ζ" similar to drain efficiency called the hydrostatic pressure intensity factor.

$$\text{Total Uplift } W_u = c w_2 [h_2 + 1/2 \zeta (h_1 - h_2)] A \qquad (1)$$

The thought behind this reduction term "c" was that uplift would only act on the base of the dam not in direct contact with the rock foundation. Terzaghi found the value of c to be at least 0.95 for concrete, or the uplift pressure is not acting on 5 percent of the base. Conservative assumptions for calculating uplift are recommended by Hinds et. al. are to use a "c" value of 1.0. This 1944 discussion includes the statement "a considerable amount of shearing and tensile strength adds materially to the stability of the dam and may be sufficient to counterbalance deficient uplift assumptions."[6]

---

[5] c = percentage of joint or base of a dam block subjected to hydrostatic uplift pressure, w_2 = unit weight of water, ζ = proportion of the net head remaining to be dissipated downstream of the grout curtain, h_1 = headwater depth, h_2 = tailwater depth, A = area of the base. Engineering for Dams.
[6] Volume 2 P. 268

This factor greatly impacting the design appears to have reached a consensus with the publishing of papers in 1949 and 1951. These papers concluded that uplift pressure affects the entire base of the structure, only the magnitude could be reduced through grouting and drainage.

Foundation Drainage

With the assumption of uplift forces making the gravity dam lighter and less efficient in resisting the driving water force, the idea to relieve or reduce this force was attempted at the 40 meter high Vyrnwy Dam (Figure 5). The construction of this 1882 dam included foundation drains. The uplift assumed was full reservoir pressure to the line of the drains and then tailwater pressure over the remainder of the base. The Bureau of Reclamation does consider uplift and foundation drainage in the design for the Arrowrock dam. The world's first large concrete dam at its time of construction from 1911 to 1915, the dam's foundation included a grout curtain and drainage holes. From the 1949 paper on uplift mentioned above, this design feature to reduce uplift was not standard.

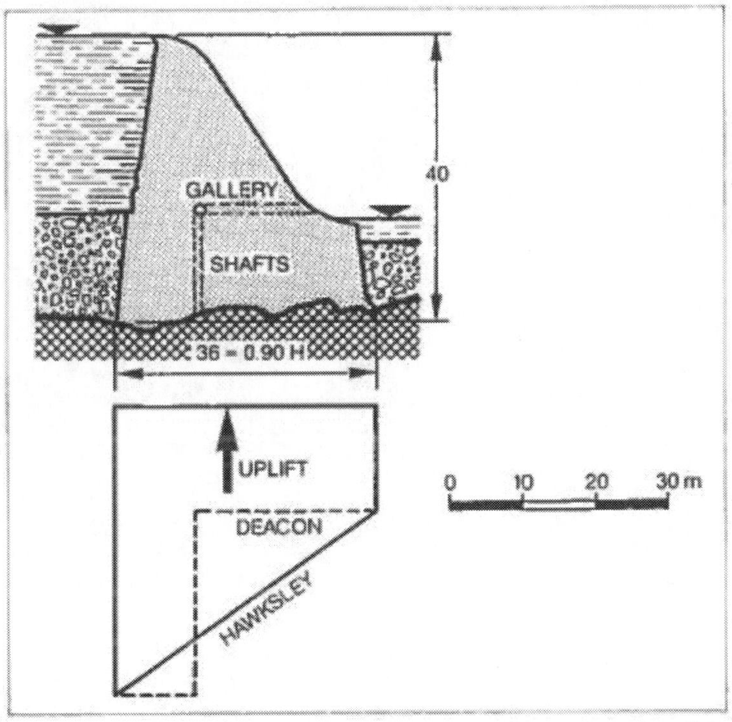

Figure 5. Cross section of Vyrnmy dam (1882) and assumed uplift pressures.

BUTTRESS DAMS

Used by Roman engineers, a buttress dam uses less mass compared to a gravity dam. With a vertical upstream face, up to 40% of the mass can be removed without affecting stability. We now understand it also has less uplift due to a smaller footprint. With an incomplete or evolving understanding of uplift, the introduction of these dams was not related to this advantage. Additional reduction in necessary mass is realized if the upstream face is sloped with a 45 degree face. The British built Mir Alam Dam, which is a multiple arch dam in India in 1804. This multi-arch 12 meter high dam was constructed with 21 semicircular vertical arches having variable spans with the maximum being 51 meters. Amazingly, 100 years go before another one is constructed.

In the United States, John S. Eastwood becomes the expert on the design and construction and promoter of buttress dams. In 1908, he designed and oversaw the construction of Hume Lake dam taking 114 working days to construct. The dam is 20 meters high and had 12 circular arches. His design had 10 % of the mass of an equivalent gravity dam. He constructed 12 of his multiple arch designs culminating in 1918 with a 41 meter high Hodges dam near San Diego (Figure 6).

Figure 6 . Lake Hodges Dam (photo credit Phil Konstantin)

Near the east coast of the United States, flat-slab buttress dams were constructed starting in 1903. Developed and patented by Nils F. Ambursen, this 3 meter high dam was constructed in 18 days near Watertown, New York. More than 200 of this design are completed in the next 17 years. The speed and minimal use of materials in this design is very similar to the multiarch design.

What curtails the construction of multiarch buttress dams is fellow engineers depicting them as being too light, ironically after the failure of the St. Francis dam, a gravity dam. This effort to shut down Mr. Eastwood was also helped by the failure of the Gleno Dam in 1923 even though it was due to "faulty foundation" and the "unbelievable poor quality of masonry"[7]. Eastwood continued to rally support for his multiarch dams on the basis that they are a much better value than a gravity dam up to his death in 1924 (Jackson, 1995).

ARCH DAMS

Arch dams were constructed by the Romans, Mongolians and Spanish. The first dam designed using stresses was completed in 1854 near Aix-en-Provence, France. The designer, François Zola, assumed the 43 meter high dam was made up of a stack of arches, all acting separately. The stress in each arch was calculated by the 'cylinder formula' where the stress is equal to the hydrostatic pressure and radius of the arch divided by the arch's thickness (1).

$$Q = (P \times R)/T \qquad (2)$$

The analysis of this dam first appears in a French text in 1872. Edward Wegmann[8] in the 1888 first edition of "The Design and Construction of Dams" gives the dam the spotlight it deserved. At about the same time as the Zola dam was being built, a very slender dam was built near Parramatta, Australia. This 16 meter high concrete dam was 4.6 meters thick at the base.

Figure 7. Zola Dam, completed in 1854 near Aix-en-Provence, France

[7] p. 232 Building the Ultimate Dam.

[8] Of interest concerning the water retaining surface - p. 38 Design and Construction of dams: "Concrete is considered by many engineers to be too pervious a material to be placed in a dam. "

This method of analysis is greatly improved upon by the Bureau of Reclamation for the Pathfinder and Buffalo Bill dams in Wyoming. The loads and deflections of the vertical vault of arches were divided with a cantilever in the middle of the dam. This sharing of the load between the arches and the cantilever reduced the loads on the lower arches but the upper arches carried additional loads to support the cantilever.

Even with the economies of time and materials inherent with arch and buttress dams, the gravity dam remains the design of choice for concrete dams in the 20th century. However, all of the concrete dams, gravity, arch, and buttress, built in the United States in this time period is fewer than the number of embankment dams according to the USSD.

CONCLUSION

We have been building all kinds of dams: gravity, buttress, arch and embankment, for several millennia. Prior to 1850, these dams were constructed throughout the world based on trial and error and past experience. The typical material for gravity, buttress and arch dams was masonry. Over time, masons had learned to build a masonry wall without cracks, perhaps without knowing that tension was the reason for the cracks.

Preventing tension in a dam's masonry wall was clearly stated by Rankine in 1872 as one of three[9] design rules. The book *Design and Construction of Masonry Dams,* by Wegmann in 1888, starts with "Chapter 2, Distribution of pressure in a Wall of Masonry". It describes that if the eccentric resultant falls outside of the middle third, "then part of the base will have to sustain tension." However, the implication of uplift pressure, first raised in 1875, was not fully accepted by the dam engineering community until 1950.

Dam designs based on analysis take off with the determination of the ideal profile for a [masonry] gravity dam by de Sazilly in 1850. The analysis of an arch dam was initiated just after this in 1854. Buttress dams take off in the United States in 1903 due to two people, John Eastwood and Nils Ambursen. However, buttress dams fall out of favor due to their appearance of being too light.

EPILOGUE

Professor Henry Petroski[10] develops a thought experiment on designing based on past successes. The Titanic ship was heralded as an "unsinkable" success even before the ship was launched. Professor Petroski poses the question what if it had not hit the iceberg and reached New York safely? The success of the design would have been "proven". The more times it crossed the Atlantic the more confident the ship's captain, owner, and potential passengers would have become in its extraordinary seaworthiness. Other

---

[9] First rule: predetermined maximum compressive stresses should not be exceeded in any direction at any point; second rule: no tensile stresses should occur, the 'middle third rule' being observed to deal with this requirement; and third: a dam should be heavy enough to prevent sliding.

[10] Professor of Civil Engineering and History, Duke University

steamship companies would have emulated Titanic's success and design. But they would have made design improvements: technical, economic, or commercial advantages. Larger, faster and more opulent ocean liners would have been built. With these "improvements" the ships would be more vulnerable to an accident and with the potential for a large tragic event (e.g. cutting back on the number of life boats). His conclusion is that "successful design and change comes not from emulating success and trying to better it, but from learning from and anticipating failure, whether actually experienced or hypothetically imagined."

The Titanic thought experiment is reinforced by the "old dictum that more is learnt from one unsuccessful structure than from dozens of successful ones."[11] This statement is tied to the failure of the Bouzey dam near Epinal France in 1895. The failure of this dam reinforced the need for a solid foundation and to not allow tensile stresses to develop (or keeping the resultant in the middle third). Additionally, the serious investigation of uplift pressures was undertaken after this failure. Maurice Lévy assumes cracking that would allow water to penetrate and cause a vertical force that would combine with the horizontal driving force potentially overturning the dam. Still, 'Lévy's uplift rule' developed by the French following this investigation in 1897 was not mandatory.

Professor Petroski also theorizes about building on successful designs culminate in a failure due to the designers not understanding the underlying assumptions. He found that pushing the design envelope to the point of failure for bridges occurs about every 30 years, or about a generation of engineers. Is that what was occurring with the designing of dams starting circa 1850s? It takes several dam failures caused by uplift before it was accepted as a required consideration in design.

Finding that it was about 70 years[12] from the time uplift pressure was considered in design to the time it was universally accepted was startling and hard to believe given our current understanding. This time period of about two generations of engineers to conclude that 100 percent of the uplift force needs to be included does align with Professor Petroski's theory mentioned above. This finding raises the question of what design assumptions are we ignoring or not treating correctly based on successful dams.

In the same light, we should review the progression of dams based not on the successful ones, but on the lessons learned from the ones that were not successful. George Santayana wrote the famous phrase "those who cannot remember the past are condemned to repeat it" when discussing progress. The collection *Dams of the United States, A Pictorial Display of Landmark Dams*, is a great celebration of these successful dams. However, a book that brings the unsuccessful dam designs together and the lessons learned would be of greater value to the dam design community.

[11] P. 204-205 A History of Dams, by Norman Smith

[12] 1880-1950

REFERENCES

Billington, David P., Jackson, Donald C., Melosi, Martin V. *The History of Large Federal Dams: Planning, Design, and Construction in the Era of Big Dams.* U.S. Department of the Interior, Bureau of Reclamation.

Creager, William P., Justin, Joel D., and Hinds, Julian. *Engineering for Dams.* J. Wiley and Sons, New York, 1945.

Gordon, J. E. *Structures: Or Why Things Don't Fall Down.* New York: Penguin Books, 1978

Harza, L.F., "The Significance of Pore Pressure in Hydraulic Structures," ASCE Transactions 114 (1949), pp. 193-298

Jackson, Donald C. *Building the Ultimate Dam: John S. Eastwood and the Control of Water in the West.* Lawrence: University Press of Kansas, 1995.

Keener, Kenneth, B., "Uplift Pressures in Concrete Dams," ASCE Transactions 116 (1951), pp. 1218-1264.

Konstantin, Phil by Philkon – Figure 6 Own work. Licensed under CC BY-SA 3.0 via Commons - https://commons.wikimedia.org/wiki/File:LakeHodgesDamByPhilKonstantin.jpg#/media/File:LakeHodgesDamByPhilKonstantin.jpg

Petroski, Henry. *Success through Failure : the paradox of design.* Princeton University Press, 2006

Petroski, Henry. *To forgive design : understanding failure.* Belkap Press of Harvard University Press, 2011

Santayana, George, *The Life of Reason*, 1905

Schnitter, Nicholas J. *A History of Dams: The Useful Pyramids.* Rotterdam, Netherlands: A. A. Balkema, 1994

Smith, Norman, *A History of Dams*, First Edition, Citadel Press, Inc., 1972

USSD, Boyer, Douglas, Editor in Chief, Wiltshire, Richard, Co-editor, Tarbox, Glenn, Technical Editor. *Dams of the United States. A Pictorial Display of Landmark Dams.* 2013

USSD, Kollgaard, Eric B. and Chadwick, Wallace L., editors. *Development of Dam Engineering in the United States.* Pergamon Press, 1988

Wegmann, Edward, Jr. *The Design and Construction of Dams.* 1st, 5th and 8th Editions. New York: John Wiley and Sons Inc., 1888, 1907 and 1927.

A NEW AND ADVANCED DAM PERFORMANCE AND RESPONSE INTEGRATION PROGRAM

Ziyad Duron, PhD[1]
John Yen, M.S., P.E.[2]

ABSTRACT

Modern practice requires engineers to predict behavior of aging infrastructure under conditions that are unlikely to occur during their lifetimes. This requirement is particularly notable in dam safety where hydro facilities provide energy, recreation, and environmental preservation in areas of the US and Canada under strict regulatory constraints. The ability to predict how a dam could perform under hazard loading conditions typically hinges on an analyst's ability to drive a numerical model to failure. If the unknowns associated with understanding what constitutes failure are considered large, the unknowns associated with the construction, validation, and understanding of how a complex numerical model actually fails may be larger. This paper describes a new strategy for addressing these unknowns through the use of performance based techniques developed for concrete dams. These techniques involve the use of existing surveillance and monitoring programs, new diagnostic performance based testing techniques, and the development of performance indicators – that include fragility curves – for monitoring, tracking, and predicting a dam's performance. Case studies from a variety of concrete dams are described, and samples of performance indicators and fragility curve estimates are also presented. Recommendations for changing the way engineers think and learn about how a dam performs are provided.

INTRODUCTION

Modern advances in computer technology during the 1990s led to the development of sophisticated software packages for solving previously intractable civil engineering problems. These advances enabled engineers to produce visual representations (that included animations of deformation, stress, and strain distributions) which contributed to new insights into how large structures might handle extreme or hazard loading conditions. In the area of dam safety, these analysis packages took on an even greater role as engineers developed predicted behaviors that contributed to decision making processes, in areas of regulatory compliance and prioritization of resource allocation by dam owners.

The March 2011 Tōhoku earthquake and subsequent tsunami off the Pacific coast of Japan highlighted the vulnerability of an asset thought to be well protected and understood in terms of its ability to withstand hazard loading conditions. By contrast, recent experience with the failure of a supporting monolith in the spillway of a large

[1] Jude and Eileen Laspa Professor of Engineering, Harvey Mudd College, Claremont, CA, 91711, Ziyad_Duron@hmc.edu
[2] Dam Safety Consultant & Chair, Dam Safety Advisory Board, Southern California Edison, Rosemead, CA, 91770, John.Yen@sce.com

hydroelectric facility illustrates the vulnerability of an asset in the absence of a hazard loading condition. One asset fails under hazard loading, and the other fails under seemingly normal operating conditions – and the question is asked – how can we be sure that we have the information and knowledge regarding how these assets perform to avoid or even mitigate such failures in the future?

The application of modern computational modeling and analysis tools to any system, particularly those whose original design life has been exceeded, presents a number of interesting challenges. For example, the development of credible analytical and numerical models must reflect current conditions inside and around the system, and should also provide evidence of confirmed behavior. Further, records of how the system has performed over its lifetime should be factored, somehow, into the development of any model – whether analytical or numerical – to ensure that predicted behavior can be traced to documented performance in the field. These challenges are particularly difficult for dam safety engineers who must use their models to predict how dams might fail under conditions that may actually never occur during their lifetimes. To do this, engineers are often required to base their predictions on results obtained from computational investigations in which models fail in a manner intended to represent what the actual dam would experience under the presumed, anticipated hazard loading conditions. How this is done, and what the subsequent evaluations mean within the context of existing operating conditions can vary widely. At least where dam safety is concerned, the trend has been toward an increasing reliance on the use of complex numerical models and sophisticated analysis techniques for developing the information upon which our knowledge is based.

This paper describes a new approach for improving our understanding of how dams actually function and perform. A performance based approach is presented for monitoring and tracking dam performance that builds on existing surveillance and monitoring programs, and new diagnostic field test procedures are described that form the basis of developing predictive capabilities of dam performance. The premise of this approach is that improved knowledge of how dams perform will translate directly to improved techniques – analytical and numerical – for predicting how dams will fail under hazard loading conditions. Such an approach will be successful if decision making and resource allocation are significantly and positively affected under normal and emergency operating conditions across all levels of the dam owner's organization.

WHAT DO WE KNOW ABOUT HOW DAMS PERFORM?

Figure 1 provides a context for thinking about what we actually know about how dams perform. Shown is a response path for a generic concrete dam as a function of intensity measure meant to characterize a hazard loading condition. The response begins at low intensity level and follows a linear (elastic) path that reaches a transition region as the hazard's intensity measure continues to increase. As the hazard continues to grow, the dam's performance exits the transition region and follows a trajectory that will ultimately fail the dam. But how do we know this? And, what exactly does it mean to fail the dam?

We can start by suggesting that under normal, everyday conditions, we would expect a dam to be operating near the lower end of its linear path, and it is here where we have the highest confidence in our knowledge of how the dam performs. As we move up the linear path, our confidence may not be so high. For instance, traditional forced vibration testing of a dam can produce steady state indications of the dam's fundamental (i.e. eigen) properties acquired under excitation levels that usually exceed normal (i.e. ambient) operating levels. Most dam owners, however, are not confident in their ability to use a dam's fundamental properties gained through traditional field monitoring techniques in evaluating how the dam might fail. In fact, as computational techniques have improved, the apparent need for validation against field test data has significantly declined. Eigen properties are no longer viewed as useful to a numerical investigation that is aimed at predicting failure since failure is assumed to take place well outside the dam's linear elastic range. But is this really the case? Let's continue up the linear path.

Dams, like many high hazard systems, are operated in ways that preserve safety margins, mitigate the need for repeated and extensive repairs, and that provide a comfortable margin against failure. Under these conditions, safe operations can be demonstrated but they do not encompass the dam's full operating capabilities, nor do they guarantee that a failure will not occur. What we require is improved knowledge of a dam's full range of performance so that effective emergency action plans (EAPs) can be developed, and uncontrolled rapid release of water (URRW) avoided. As a first step toward this (practical) goal, the dam owner should attempt to identify the Transition Region highlighted in Figure 1. Questions regarding how this transition region is located remain unanswered, and procedures for approaching the region by extending the performance along the linear path into the transition region have not been fully developed.

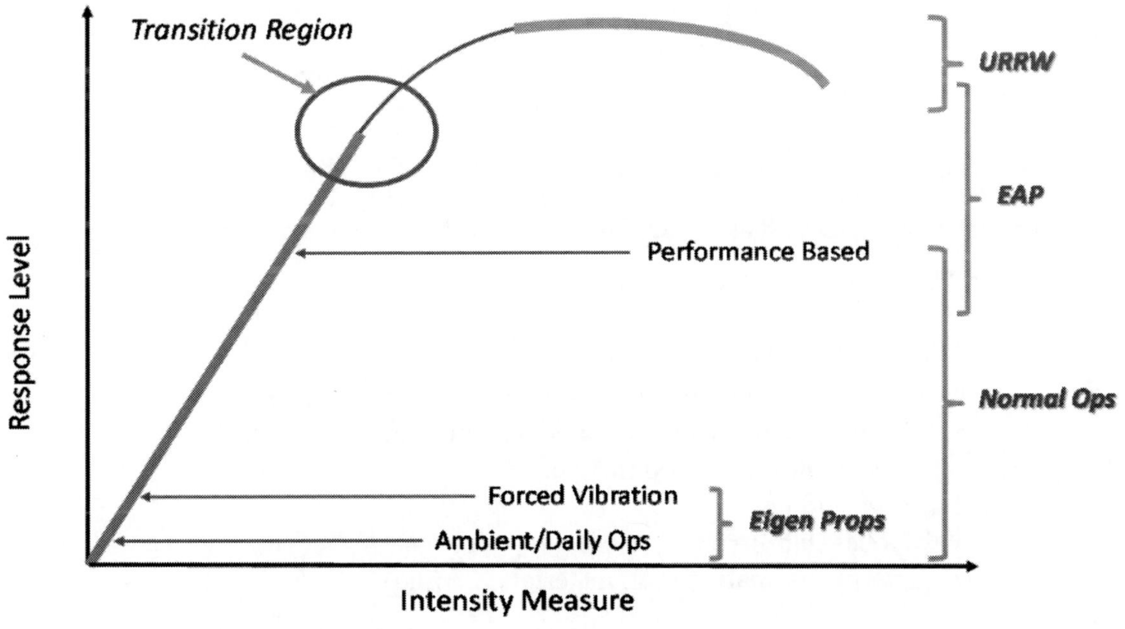

Figure 1. A Dam's presumed full range of performance.

Another context for asking what is known about a dam's performance is shown in Figure 2. Shown is a typical (seismic, 10000 year Return Period) hazard spectrum that ranges from levels at 0.1Hz to the PGA level at 100Hz. The figure highlights three (frequency) decades or regions for which a dam's performance must be considered. For instance, a dam's behavior within the low frequency range is typically associated with long period response that can include considerations of base sliding, near rigid body response behavior in individual monoliths, and it is here where overall stability of the dam is usually considered. In the mid-frequency range, other behavior and patterns in a dam might occur that include relative movements in the dam and between monoliths, behavior that leads to cracked sections and elevated stress and strain over segments of the dam, and in areas of the dam where components such as arches and buttresses are joined. A dam's behavior in the high frequency range is less well understood, and in fact, apart from the connection to the PGA, a typical numerical analysis of a large concrete dam does not address a dam's behavior outside the mid frequency range.

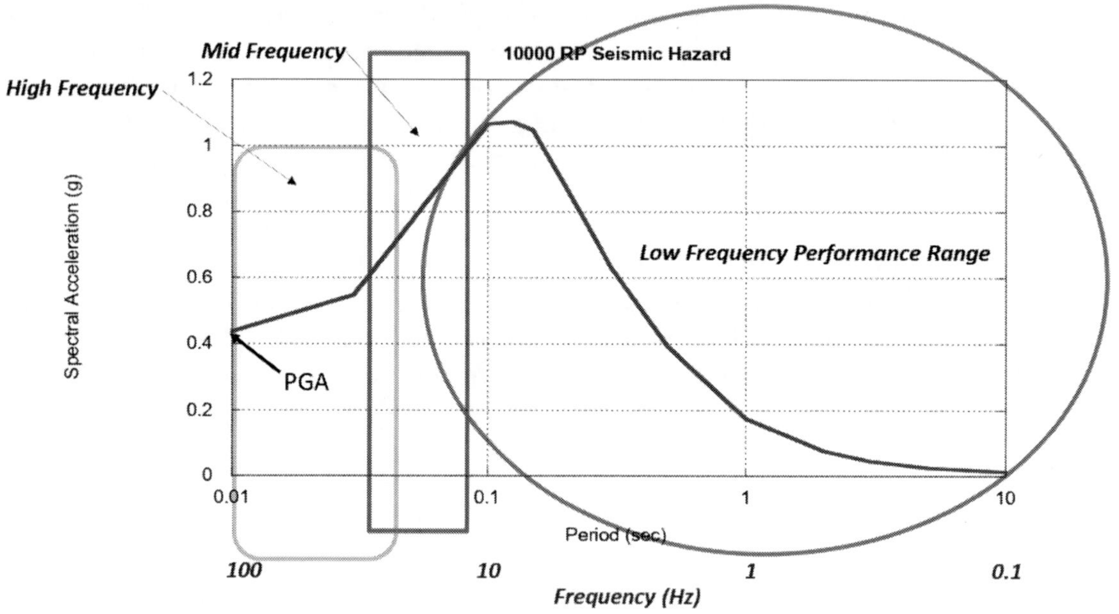

Figure 2. Performance regions for a dam against a hazard spectrum.

A grouping of behavior and associated mechanisms by frequency range might be organized as follows.

Low Frequency (.1Hz – 15Hz)

- Near Rigid Body motions (small gradient behavior)
- Separation inside/along boundary interfaces
- Cracking (long period induced)
- Base Sliding
- Global stability
- Energy dissipation (e.g. material damping)

Middle Frequency (15Hz – 30Hz)

- Relative monolith motions (larger gradient behavior)
- More Cracking (shorter period induced)
- Stiction (stick – slip behavior)
- Cyclical opening/closing (cracks, interfaces)
- Energy dissipation (e.g. component absorption)

High Frequency (30Hz-100Hz)

- Connection motions (high gradient behavior)
- More Cracking (mechanisms differ from above)
- Energy dissipation (combination of above mechanisms)

It is interesting that while analysts and engineers agree that scenario spectra may be used to define the dam's entire hazard spectrum since no single seismic event is likely to produce the hazard itself, different numerical models that target the dam's behavior in each of the frequency ranges are not usually developed or even considered. For example, the mechanisms that assist energy dissipation in the dam during excitation below 15Hz are usually different than those present above 30Hz. But, what evidence exists to support this? Consider the spillway at Wanapum Dam, which is shown in Figure 3.

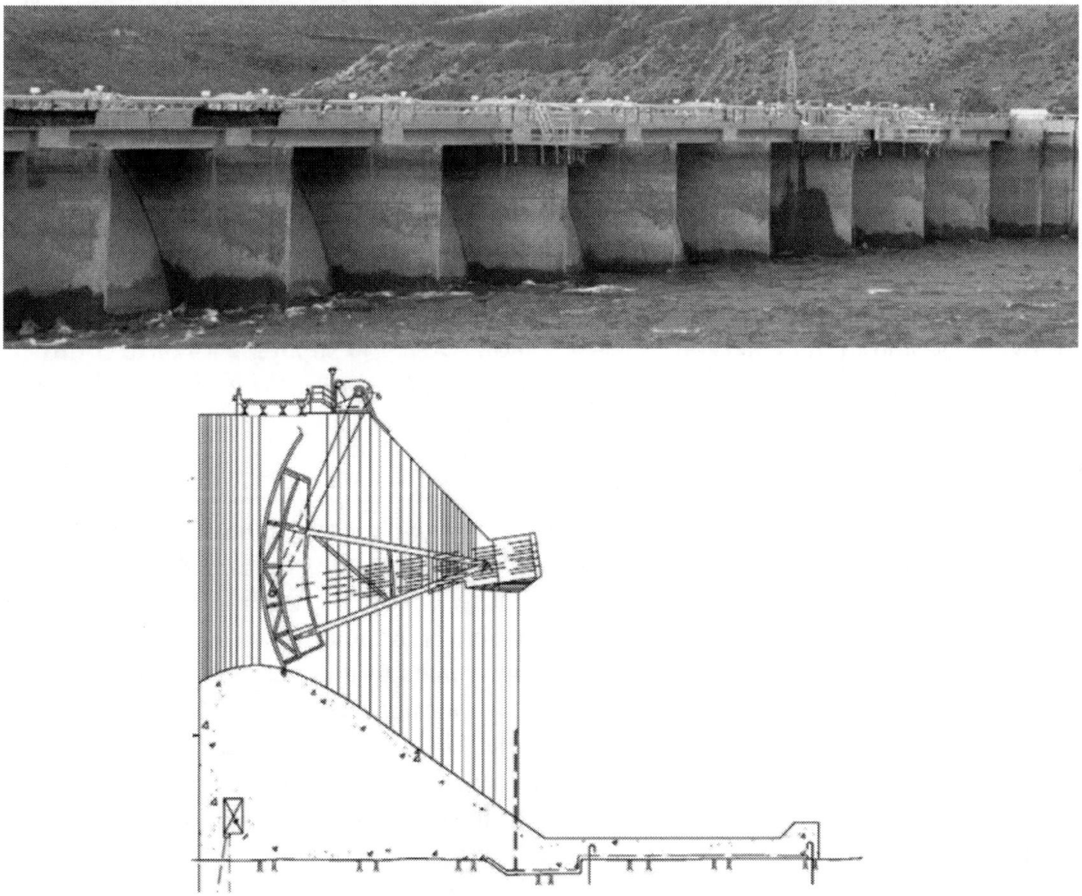

Figure 3. Wanapum Dam spillway and drawing that indicates a typical pier-monolith configuration.

A field investigation conducted on the spillway produced the spectral responses shown in Figure 4.

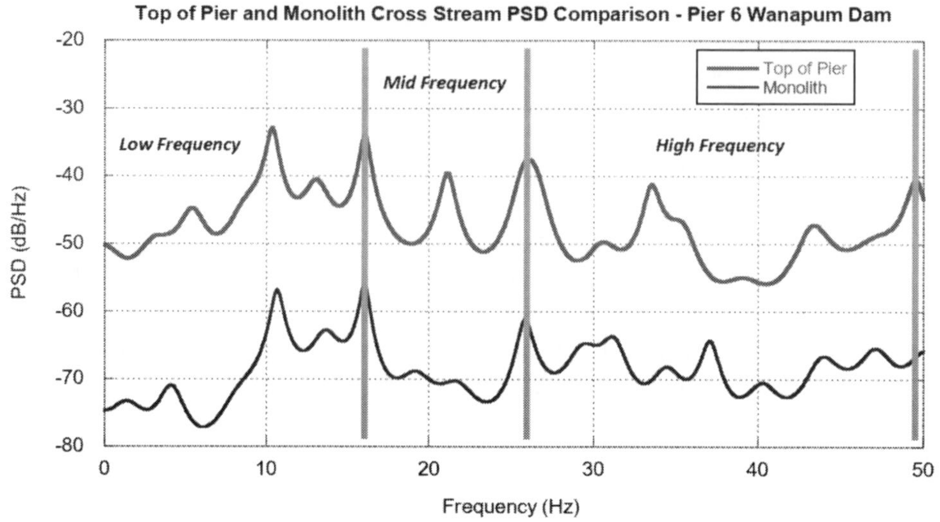

Figure 4. Spectral response on a spillway pier and support monolith.

The responses shown in terms of Power Spectral Density (PSD) indicate how the root mean square (RMS) response levels vary as a function of frequency and can be used to infer the dynamic interactions between the pier and its support monolith. For example, in the low frequency range, the spectral pier and monolith responses look very similar and this suggests that the pier acts as a source of excitation that drives steady state response in the monolith. In the mid frequency range, there is a noticeable spectral peak in the pier response near 22Hz but when the monolith response is examined, there is an absence of a corresponding peak at 22Hz. In this case, the pier acts as a tuned vibration absorber that significantly reduces the monolith response. So, in one case the pier drives the monolith movement, and in the other it absorbs it. In one case, the support monolith is moving along with the pier, and in the other it is acting as a motionless (i.e. rigid) base for a vibrating pier. How is energy dissipation affected?

The dynamic interactions between the pier and its monolith dictate how energy is dissipated and suggest different mechanisms engage depending upon whether the pier drives the monolith into steady state motion, or whether the pier is moving on top of a near motionless monolithic support. In one case, energy dissipation through the use of modal or equivalent damping may suffice, whereas in the other case, the modeling of the interface between the pier and its monolith may need to include mechanisms that allow cracking and sliding inside and across the pier/monolith interface. Regardless of what mechanisms are actually included in the model, this level of characterization suggests that different energy dissipation mechanisms engage depending upon the excitation characteristic (e.g. frequency) that dominates during loading. Where the larger question of how a dam fails is concerned, does this "subtle distinction" matter?

ARE WE MISSING SOMETHING WHEN WE PREDICT HOW A DAM FAILS?

A modern evaluation of the risks and vulnerabilities of a dam or of any high hazard asset typically focuses on how that dam or asset might fail. The emphasis on understanding failure is warranted when one considers the significant consequences of a large dam failure, both economically and – more importantly – in terms of potential injury and loss of life. Significant resources are allocated in terms of computational tools, funding, and expertise to arrive at well understood descriptions of how a dam might fail, and in some cases, where the probability for failure is high, decisions are made for remediation or retrofit. These decisions are driven by scenarios that are characterized as extreme, far outside normal operating conditions which is probably why connections between normal operating conditions and conditions during extreme events are not typically considered. Is something missing when these connections are not made? In the remaining portion of this paper, the authors will try to show why a more thorough understanding of a dam's normal behavior can be relevant to understanding how the dam will behave under extreme loading conditions.

PERFORMANCE BASED EVALUATIONS – THE MISSING LINK

The gap between a dam's normal operating conditions and what is predicted under extreme events is normally considered quite large. The presumption is that normal behavior cannot be extrapolated to behavior under extreme conditions, and it is this presumption which leads to the missing link and the following claim.

Effective integration of performance into surveillance and monitoring programs can produce real-time indicators of actual behavior under normal and safe operating conditions that can be extended into the early stages of emergency conditions.

Two case studies are presented that illustrate how performance can be used to develop a fuller understanding of how critical aspects of a dam perform under normal conditions, and in a second example, how fragility can be used to evaluate conditions that represent unacceptable states or conditions that actually far short of catastrophic failure.

Case Study 1 – Foundation Discharge/Charge (Uplift) under a Large Concrete Dam

Uplift pressure in the foundation under a dam plays a large role in evaluating whether the dam can withstand flood, seismic, or any loading event that mobilizes a significant portion of the reservoir water. For each of these load cases, uplift is typically described as a linearly varying distribution as shown in Figure 5.

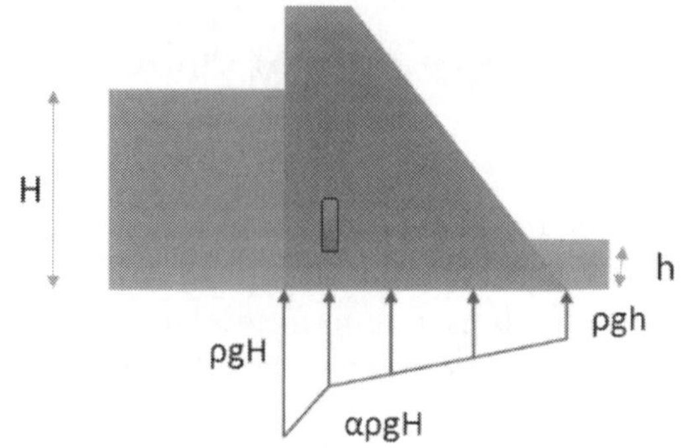

Figure 5. Typical uplift distribution under a concrete dam.

The distribution varies linearly from full uplift at the heel to a reduced pressure associated with the drainage system's effectiveness (α) to a level associated with tailwater at the dam's toe. Stability analyses usually vary the magnitude of the uplift at the heel in search of a configuration that destabilizes the dam, which is then incorporated into an evaluation of whether the dam is able to handle uplift loads that are expected to develop under the hazard events considered. But what do we really know about how uplift develops under a dam during extreme events and is there anything during normal conditions that can aid in our assessment?

A large concrete gravity dam is instrumented with vibrating wire piezometers that record pressure readings at 15mins intervals. The reservoir behind the dam is cycled annually to meet recreation and water conveyance needs which results in a water elevation change of 140ft each year. Drawdown and filling of the reservoir provides an opportunity to monitor the foundation discharge and charge behavior by examining the hysteretic behavior shown in Figure 6.

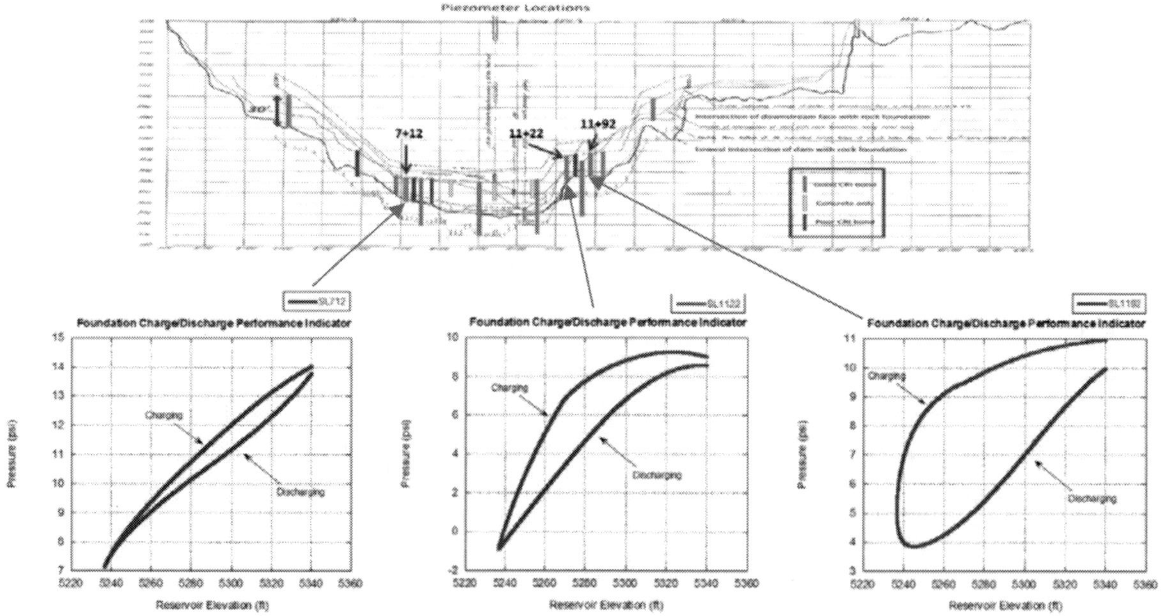

Figure 6. Foundation discharge and charge behavior under a large concrete dam.

Pressure as recorded by the piezometer instrumentation system is examined as a function of reservoir elevation change to produce the hysteretic response. The response at station SL712 (far left in Figure 6) suggests that whether during drawdown or filling, the rate of decrease or increase in uplift under the dam in this region can be considered as nearly equal. At station SL1192 (far right in Figure 6) the hysteretic loop widens as a result of a time delay between the change in reservoir elevation and the corresponding change in pressure. Furthermore, as the reservoir elevation approaches and stays near its lowest level, the pressure nearly doubles and then increases at a smaller rate as the reservoir is filled. The hysteretic loop indicated in between these two stations at SL1122 shows how the time delay may develop as behavior is examined moving from right to left under the dam.

The characteristics associated with how the foundation pressure discharges and charges as reservoir elevation is changed are fundamental to understanding how uplift is generated and distributed under the dam. These results are Performance Indicators that can be used to monitor and track behavior during each reservoir cycle to produce a foundation pressure profile under the dam that can be relied on when anomalous behavior or impending hazard conditions develop.

Consider the case of an impending flood condition. If the foundation pressure profile is known to include time delays in certain regions under the dam, these profiles can be extended into the early stages of the flood condition and included in considerations of field operations and emergency action plans. Further, if a dam exhibits significant time delays between foundation pressure response and reservoir elevation change, how does this affect modeling for stability evaluations? Is it possible that the traditional inclusion of uplift using linear distributions under the dam can lead to unnecessarily conservative findings?

Case Study 2 – Vulnerability Evaluation using Fragility Estimates of Dam Performance

Risk assessments of dams are aimed at identifying potential failure modes (PFMs) that can lead to damage and an uncontrolled rapid release of water (URRW) which are included in considering how to manage (i.e. reduce or eliminate) associated risks. Key outcomes of the risk assessment include detailed descriptions of the sequence of events that contain clear definitions of failure and favorable and/or adverse factors for each PFM. From the dam owner's perspective, it may be difficult to imagine how these scenarios can develop based on what they see in terms of normal, day to day operations, and a tool that assesses the probabilities of the intermediate events that make up the PFM scenario could be quite helpful. For example, consider the PFM described below.

> During a seismic event, significant ground motions occur at or near the dam site to excite large dynamic response behavior in the dam. The seismic excitations have large magnitudes across a sufficiently wide frequency spectrum to transfer large dynamic loads into the dam. These dynamic loads contribute to regions of elevated tensile or compressive stress states that lead to significant cracking or concrete failure. Failure occurs as individual monoliths are released from their underlying foundation rock and

begin to slide, closing off substantial numbers of relief or drain holes allowing uplift to build and destabilizing clusters of monoliths leading to total dam failure and uncontrolled rapid release of water (URRW).

As written above, the PFM takes place only after a series of events have occurred and the last event before the reservoir is released is associated with the closing off of the relief holes in the dam. Using a performance based approach, the following investigation could shed light onto the probability of compromising the drainage system during a seismic event.

PERFORMANCE BASED TESTING AND FRAGILITY ESTIMATION

A new field testing technique for large concrete dams has been developed that can lead to fragility estimates for the dam and its subsystems. Details of the technique are included in Goldkamp et al. (2014) and an overview of the analysis techniques employed in the development of the fragility estimates are found in companion papers Menda et al. (2016), Zhao et al. (2016), and Spenser et al. (2016). The Performance Based Testing (PBT) technique employs the use of a cold-gas-thruster (CGT) excitation source to deliver a short duration, large amplitude, impulse load to the dam that induces large amplitude, broad frequency transient behavior. The CGT mounted to the crest of a large concrete dam during a recent test is shown in Figure 7, and transient (acceleration) responses acquired as a result of the CGT blast are shown in Figure 8.

Figure 7. Cold Gas Thruster (CGT) mounted on the crest of large concrete gravity dam during blast.

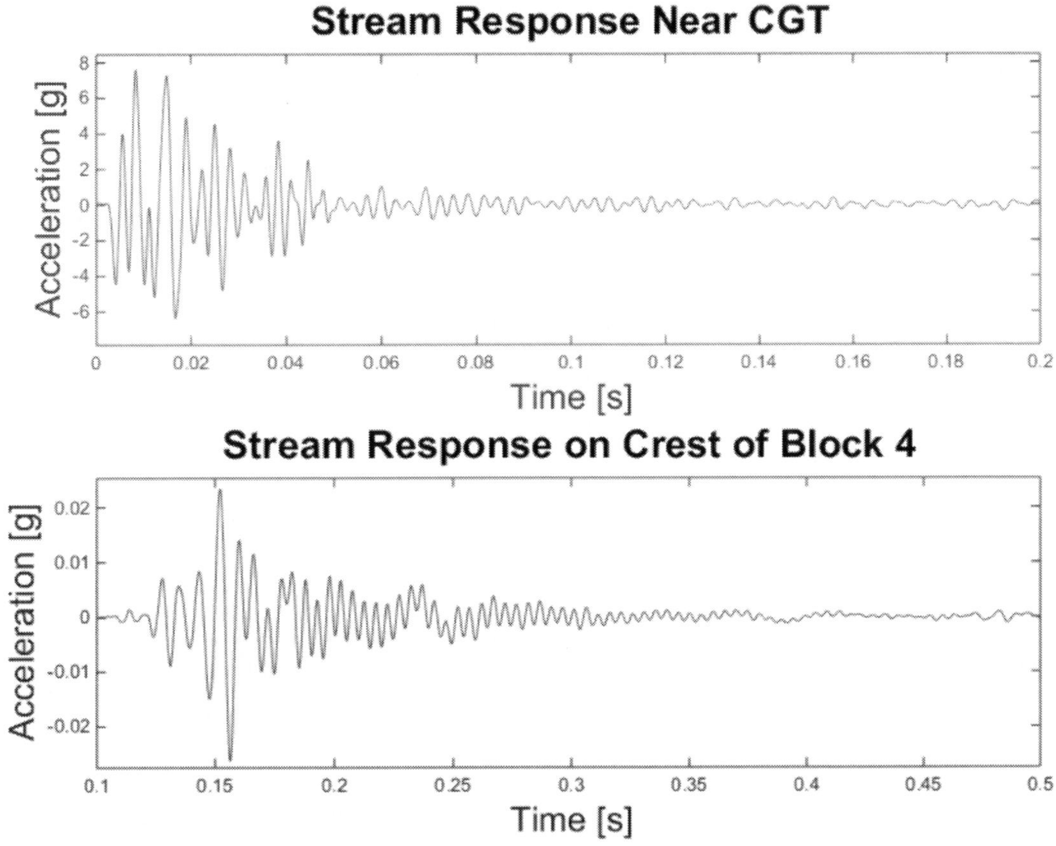

Figure 8. Measured transient responses on the crest of large concrete dam induced by a CGT blast.

The underlying concept of PBT relies on the delivery of an ideal load that produces an ideal response that is then convolved into predicted behavior. A graphical representation of the mathematical technique that produces predicted responses of the dam under seismic loading is illustrated in Figure 9.

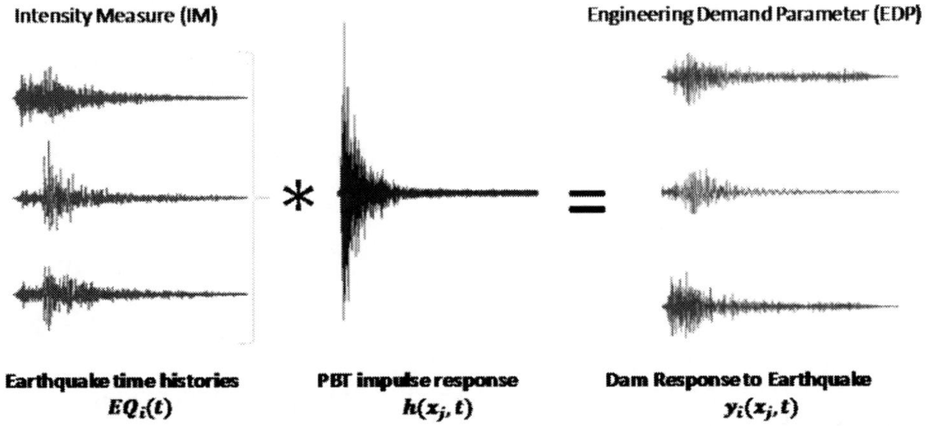

Figure 9. Predicted seismic dam response using PBT responses.

The technique illustrated in Figure 9 can lead to the development of fragility estimates that predict probabilities of exceeding damage criteria for each event in the PFM scenario. For the case study here, the damage criterion of interest is associated with base sliding that closes off the relief holes in the dam since it is the last event in the described sequence of the PFM. Therefore, assuming that the relief holes are 3inch diameter holes, fragility estimates that predict probabilities of incremental base sliding up to 3 inches are obtained as follows.

A performance based definition of fragility is given as

$$Fragility = P(Damage|EDP) \times P(EDP|IM)$$

where $P(Damage|EDP)$ is the conditional probability that damage will occur for a given Engineering Demand Parameter ($EDP, or\ response\ parameter$), and $P(EDP|IM)$ is the conditional probability that the EDP will occur for a given Intensity Measure (IM) that describes the seismic event. Without loss of generality and for this particular case study, let's assume that the dam's relief holes are capable of remaining functional during base sliding of no more than 1.5inches.

Using the techniques illustrated in Figure 9, the response of the dam subject to seismic loading can be estimated for a large number of hazards (in this example 296 records were evaluated) to produce the scatter (blue dots) of maximum displacements shown in Figure 10. Highlighted in Figure 10 is the maximum displacement criterion of 1.5inches from which probabilities of exceedance can be obtained, as indicated in the bin containing a total of 80 occurrences of maximum displacements for the seismic events described by the Intensity Measure (IM)[3]. At this IM, there are a total of 33 predicted maximum displacements in the dam that exceed the damage criterion of 1.5inches which corresponds to a 41% probability that the criterion will be exceeded for a seismic event of equal IM. Using the same counting approach for each of the bins indicated, the fragility estimate shown in Figure 11 is obtained.

[3] For this example, seismic events are characterized by their Pseudo Energy defined as $\int_T v \cdot v \, dt$, where v is the velocity of the seismic event of duration T.

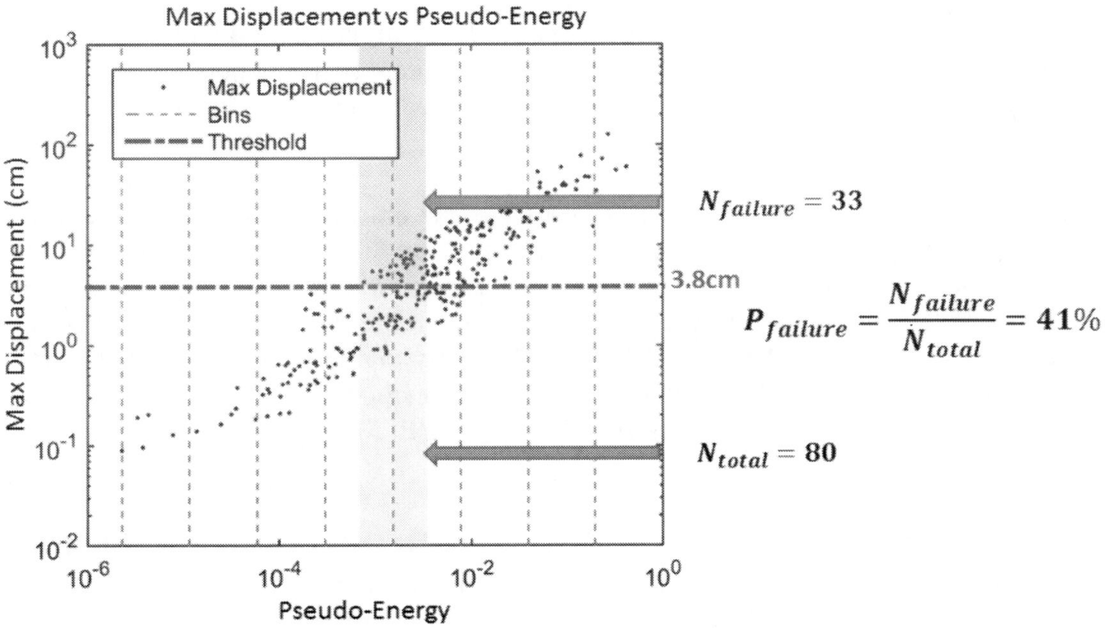

Figure 10. Predicted maximum displacements associated with hundreds of potential seismic loading events for the dam.

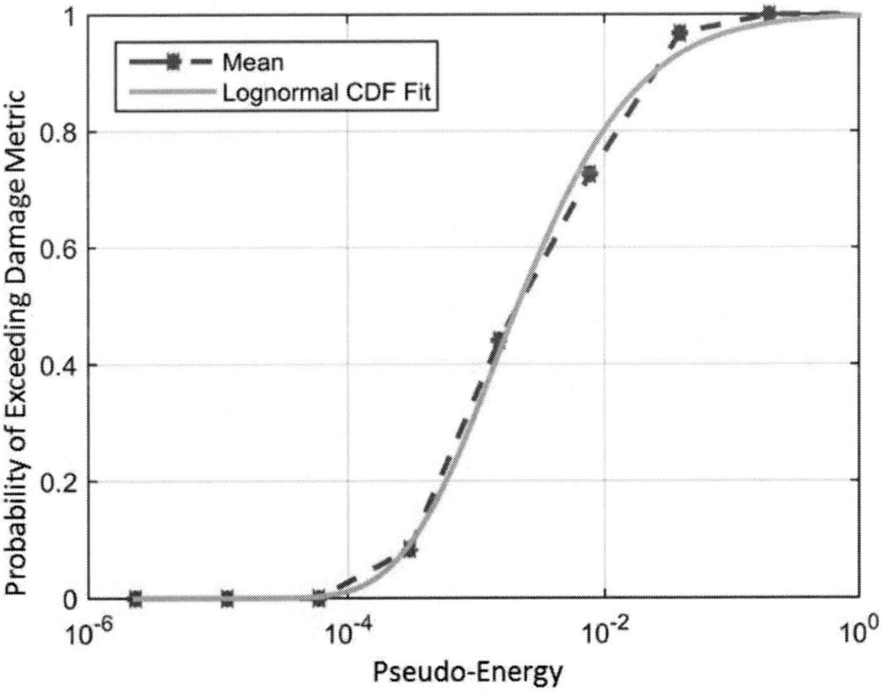

Figure 11. Estimated fragility curve that predicts the probability of exceeding a predetermined damage criterion.

Armed with this information, the dam owner is better positioned to consider the individual probabilities of the events described in the PFM scenario. For instance, the dam owner may not be willing to accept probabilities larger than 10% of exceeding the damage criterion (i.e. maximum displacement in the dam of 1.5inches) for this dam. Based on the predicted fragility estimate, seismic events corresponding to this probability of exceedance can be identified as well as their return period which will lead directly to higher confidence in risk estimates for the individual event and for the sequence leading to the PFM. The portion of the fragility estimate that draws attention is associated with seismic events characterized by an IM level of 10^{-4} and the subsequent events that raise the probably of exceedance for this damage criterion above 10%, a level that for many dam owners would represent the upper bound of tolerable response. Relating the IM for each event to its Return Period (RP) will lead directly to improved risk estimates for the events that make up the PFM scenario and this information will enhance the understanding of the actual risk associated with each PFM, and ultimately of the dam and its subsystems[4].

OVERVIEW OF SCE'S ADVANCED DAM PERFORMANCE AND RESPONSE INTEGRATION PROGRAM

The Dam and Public Safety Group at The Southern California Edison Company (SCE) is developing a new program to enhance its existing surveillance and monitoring capabilities. The ultimate objective of this initiative is to improve SCE's ability to gather and evaluate relevant information regarding a dam's performance for decision making and resource allocations during both normal and emergency conditions. The major components of the program are shown in Figure 12.

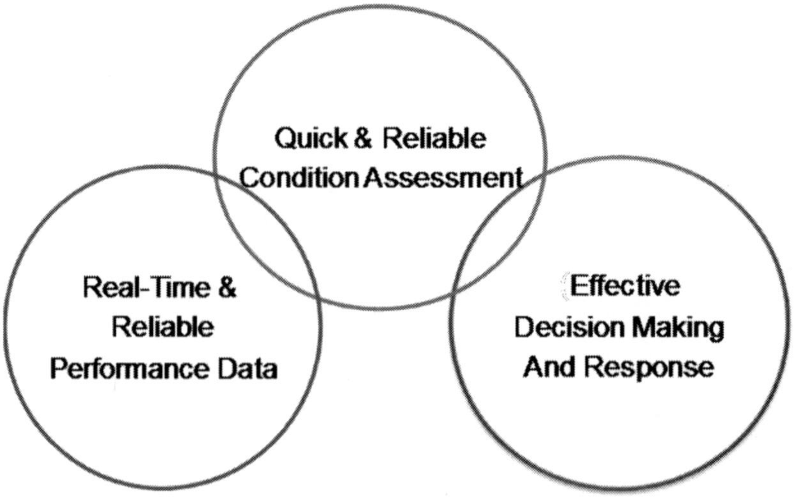

Figure 12. Components of SCE's Advanced Dam Performance and Response Integration Program (ADPRIP).

[4] Additional work is required to relate IM to RP that will be reported in future publications.

Existing instrumentation and monitoring programs will be enhanced to achieve the following technical objectives.

- To acquire real-time performance data for *monitoring* a dam's fundamental behavior characteristics and known performance indicators.
- To develop Performance Indicators (PIs) for *tracking* relative changes in trends, hysteretic behavior, magnitudes, and delays.
- To develop diagnostic procedures for *predicting* a dam's response to hazard loading conditions.

The program relies on the acquisition of performance data acquired over short intervals to provide (near) real-time indications of response behavior. The short interval acquisition rates are important for understanding the behavior across wider frequency bands and for tracking changes in the dam's performance. SCE will conduct a review of its existing instrumentation and monitoring systems to ensure alignment with behavior characteristics and mechanisms that are relevant to PFM scenarios. Contrary to traditional instrumentation and monitoring programs which rely on threshold and alarms for signaling anomalous response (*reactive monitoring*), the ADPRIP is designed to produce key Performance Indicators for a dam that improve the dam owner's confidence in how the dam performs and that provide early warning indications (*proactive monitoring*) of changes that could lead to unacceptable states or configurations in the dam and its support systems. SCE believes Performance Indicators provide a more effective means for describing how a dam functions, and can be used to translate data into information that enhances confidence in decision making. Samples of what Performance Indicators are able to track are shown in Figure 13 for one of SCE's large concrete dams.

Figure 3. Sample Performance Indicators (PIs) for a large concrete dam.

Performance Indicators (PIs) can be identified to support almost any decision making process by producing indicators of actual behavior under normal and safe operating conditions that can be extended into the early stage of emergency conditions. In this manner, PIs can be effective for tracking changes in the dam and for making important decisions during and after the emergency. Instrumentation systems and their readings are combined with modern digital signal processing (DSP) techniques to assure data quality, accuracy, and that produce reliable PIs. The Case Studies presented above included PIs for monitoring foundation discharge/charge (uplift) and for predicting probability of exceedance for individual events in a PFM scenario. Additional PIs are presented in Anderson et al. (2015).

SUMMARY

Traditional surveillance and monitoring programs are not fully integrated into a dam owner's decision making processes. Many programs in service today were originally installed for monitoring dams during initial filling, and were certainly not intended to provide the type of information and capabilities being demanded by today's sophisticated assessments and evaluations. Recent calls for aligning monitoring programs with a dam's known PFMs fall a bit short of the actual objectives for improving a dam owner's ability to make proactive, effective decisions when needed. The reason for this potential shortfall, in the authors' opinion, is based on the need to develop advanced monitoring and diagnostic techniques that highlight key behavioral characteristics in the dam and that produce Performance Indicators that can track changes in these characteristics. The premise for this is that the PFMs scenario rely on a sequence of events, many of which engage key characteristic behavior in the dam which, when monitored and tracked, should provide early warning of impending conditions that could lead to adverse conditions. SCE plans to develop strategies and frameworks for building on its existing monitoring programs by conducting a thorough review of its instrumentation systems including acquisition and data assurance operations for the purpose of integrating performance requirements onto these systems and for identifying needs not currently addressed. In addition to this review, SCE will develop a series of guidelines for developing Performance Indicators (PIs) to aid in their calculation and interpretation. Finally, SCE plans to work with a select group of dam owners to identify effective decision making strategies that incorporate and leverage performance based evaluations.

Performance based monitoring and a fully integrated decision making process that considers performance during normal operating conditions as indications of behavior during the early stages of emergency conditions will lead to higher confidence in the authorization and allocation of technical as well as financial resources. Over time, this will lead to improved safety of high hazard assets.

REFERENCES

Anderson, S., K. Menda, M. Zhao, and Z. Duron. (2016). "Selecting an Intensity Measure for Fragility Curves Derived from Performance-Based Testing." Denver, Colorado, 36th Annual USSD Conference.

Anderson, S., P. Orme, Z. Duron, R. Nash, and K. Marshall. (2015). "Evaluating Relative Condition of Repeated Structural Elements with Performance-Based Testing." Louisville, Kentucky, 35th Annual USSD Conference.

Goldkamp, M., T. Savage, J. Edelman, P. Orme, Z. Duron, and M. Knarr (2014). "Performance-Based Testing of A Large Concrete Dam," San Francisco, California, 34th Annual USSD Conference.

Menda, K., M. Zhao, S. Anderson, and Z. Duron. (2016). "A Performance-Based Approach to Developing Fragility Curves for Structures Subject to Seismic Loading Conditions," Denver, Colorado, 36th Annual USSD Conference.

Zhao, M., S. Anderson, K. Menda, and Z. Duron. (2016), "Selecting a Performance Measured for Fragility Curves Derived from Performance-Based Testing." Denver, Colorado, 36th Annual USSD Conference.

PRACTICAL ANALYSIS AND ASSESSMENT OF AAR-AFFECTED DAMS AND HYDROELECTRIC PLANTS

Dan D. Curtis, M.A.Sc., P.Eng.[1]
Lingmin (Frank) Feng, Ph.D., P.Eng.[2]
Gurinderbir S. Sooch, M.A.Sc., P.Eng.[3]
Jiqin (Tom) Zheng, M.A.Sc., P.Eng.[4]
John Fletcher, C.E.T[5]

ABSTRACT

Hatch has been involved in many hydro projects where alkali aggregate reaction (AAR) has affected dams and power plants. AAR causes the concrete to expand and this leads to issues including reduction in spillway and intake gate clearances, concrete cracking, high compressive and shear stresses in concrete substructures, stressing of turbine embedded parts and turbine/generator alignment problems. During the course of these investigations, a practical finite element program (GROW3D) has been developed to analyze the effects of concrete expansion on these structures. The program is based on relatively simple engineering principles yet it has proven to be a reliable tool in the prediction of the structural behavior of dams and power plants. The objective of this paper is to present selected results from recent analyses performed on the Mactaquac Dam and power plant which is located in New Brunswick, Canada. The stress-dependant concrete growth law used in GROW3D has recently been modified to better match the observed behavior of the Mactaquac intake structure response to slot cutting. The new concrete growth law was subsequently used in a global finite element analysis of the Mactaquac powerhouse and the results of the analysis provided an excellent match to measurements of structural and mechanical AAR-induced deformations. Finally, the powerful effect of differential concrete expansion in structural elements is examined using simple analytical methods.

INTRODUCTION

The Mactaquac Generating Station is located in New Brunswick, Canada, approximately 20 km upstream of the provincial capital of Fredericton on the St. John River. The station is the largest hydroelectric generating facility in the Maritime Provinces and was constructed in stages between 1964 and 1980.

[1] Senior Project Manager, HATCH Ltd, 4342 Queen St, Suite 500, Niagara Falls ON Canada L2E 7J7 +1 905 357 6998 dcurtis@hatch.ca .
[2] Senior Civil/Structural Engineer, Stress Analysis Specialist, HATCH Ltd, 4342 Queen St, Suite 500, Niagara Falls ON Canada L2E 7J7 +1 905 374 0701 Ext 5296 lfeng@hatch.ca .
[3] Civil/Structural Engineer, HATCH Ltd, 4342 Queen St, Suite 500, Niagara Falls ON Canada L2E 7J7 +1 905 374 0701 ext 5252, gsooch@hatch.ca .
[4] Senior Civil Engineer, HATCH Ltd, 4342 Queen St, Suite 500, Niagara Falls ON Canada L2E 7J7 +1 905 374 0701 Ext 5415 jzheng@hatch.ca .
[5] AAR Project Manager, NB Power, 451 Route 105 Keswick Ridge, NB Canada E6L 1B2 +1 506 462 3864 jfletcher@nbpower.com

The concrete of the structures are affected by alkali-aggregate reaction (AAR) in the specific form of alkali-silica reaction (ASR). This reaction causes the concrete to expand if unrestrained, or to develop large compressive forces, if restrained. Since 1985, various innovative remedial measures have been undertaken to mitigate the effects of concrete expansion on the structures and equipment. These have included various slots within the structures cut with a diamond wire saw, anchoring of structures, periodic grouting and modifications to existing gates and electro-mechanical equipment. The concrete structures are monitored with an extensive instrumentation system. The concrete material properties are carefully monitored using a long-term testing program and periodic specialty testing. The design, analysis and implementation of remedial measures in the water retaining structures and powerhouse have been presented in other references, for example, Charlwood et al (1992), Thompson et al (1994), Curtis (1995) and Codrington et al (2007).

Figure 1 shows the main 5-bay spillway, 6-unit intake and powerhouse.

Figure 1. Spillway Gates 6 to 10 are Left to Right; Intakes 1 to 6 from Right to Left

Figure 2 shows a section through the intake and powerhouse.

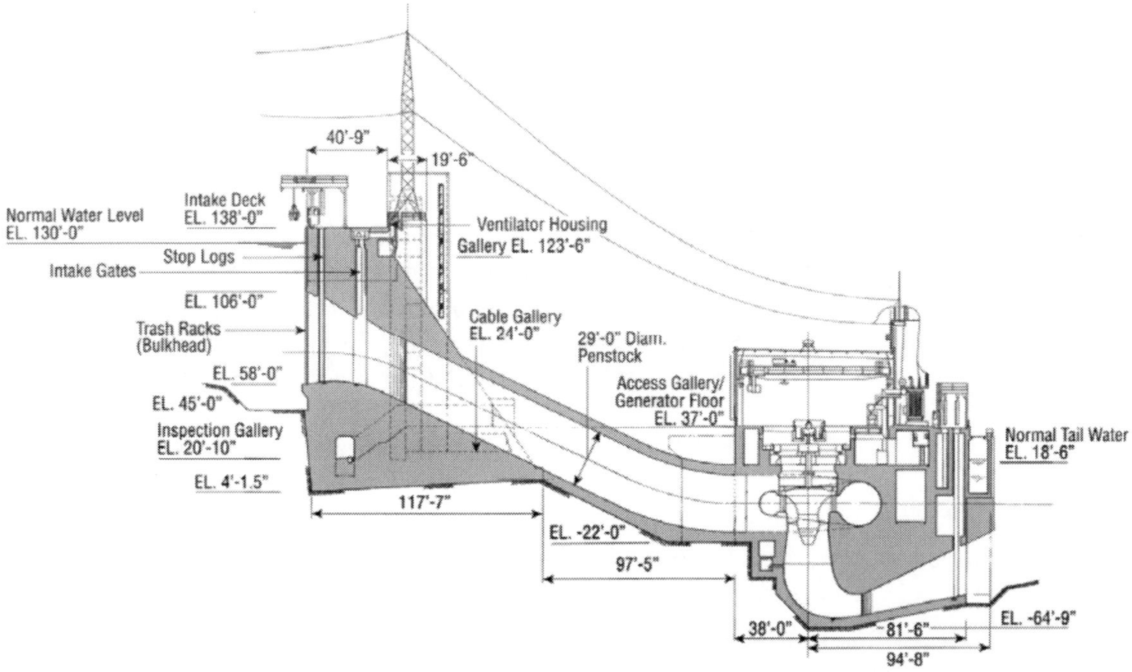

Figure 2. Intake Section and Powerhouse

In 1985, Spillway Gate 10 was found to be jammed closed as a result of AAR-induced concrete expansion of the intake and spillway east end pier. In 1988, the first of several diamond wire slot cuts was installed between intake units as a means of controlling deformations. The initial finite element modeling predicted adverse intake water passage pier concrete cracking, and large post-tensioned anchors were installed to stabilize the structure if this cracking did occur. Slots were cut in the intake in 1988, 1989 and 1992. The slots were first cut with a 10-mm diamond wire saw. As concrete growth continues, periodic re-cutting of slots is required and most current cuts are made with a 15-mm diamond wire saw. The intake slots are cut periodically in order to prevent the ASR deformation from accumulating across the entire intake thus leading to large tilting of an intake block and/or the spillway east end pier (EEP) (as observed in 1985).

The powerhouse also suffers from AAR and the resulting expansion is causing turbine/generator alignment issues, development of large stresses in the turbine embedded part including the stay vanes, and large concrete stresses in the draft tube piers. In 1995, transverse slot cuts were installed in the unit block concrete between each unit of the powerhouse. In 1996, a program of cutting penstocks and installing couplings was initiated by working on one penstock per year. The cutting of the penstocks and installation of couplings allowed the upstream portion of the powerhouse to tilt upstream thereby reducing alignment issues, reducing stay vane stress buildup and reducing concrete stresses in the draft tube piers. The initial response to these remedial measures is described in Curtis (2001).

The short- and long-term deformation response of the intake and powerhouse is monitored with a large number of instruments including numerous extensometers, joint meters and inverted pendulums. The intake has inverted pendulums located on either side of the slots, a plumbline in the spillway EEP and invar rod extensometers running longitudinally in the upper galley (at el 123'-6" in Figure 2). The powerhouse has a similar set of instruments; however a very detailed monitoring program has also been established to monitor the turbine/generator movements. Further details can be found in Thompson et al (1995).

It must be recognized that the rates of concrete expansion at Mactaquac are in the range of 2 to 3 times larger than any other typical dam/power station affected by AAR. The maximum measured concrete growth rate is in the range of 130 to 150 micro-strain/year. As a result, the intake is almost 9 inches higher than originally constructed and the height of the intake gates had to be increased to accommodate this new dimension.

INCORPORATION OF CONCRETE GROWTH MATERIAL MODEL IN ANSYS

This section describes the important features of the GROW3D subroutines which Hatch has incorporated in ANSYS. The following is described:

- creep effects in AAR-affected concrete
- stress dependent concrete growth law
- temperature dependent concrete growth rate.

Creep Effects

The instantaneous strain occurring during the application of the stress is calculated by:

$$\varepsilon(t_o) = \frac{\sigma(t_o)}{E(t_o)} \tag{1}$$

Where

$\sigma(t_o)$ = the concrete stress at age t_o
$E(t_o)$ = the concrete modulus of elasticity at age t_o
t_o = the age of concrete at the time of loading.

Under sustained stress, concrete strain increases with time due to creep. The total strain – instantaneous plus creep – at time t is

$$\varepsilon(t) = \frac{\sigma(t_o)}{E(t)}[1 + \varphi(t, t_o)] \tag{2}$$

Where

$\varphi(t,t_o)$ is the creep coefficient, which is a function of the age at loading, t_o, and the age t for which the strain is calculated. The creep coefficient φ represents the ratio of creep to the instantaneous strain.

$E(t)$ is the concrete modulus at time t.

Therefore, the creep strain under a constant stress is obtained by:

$$\varepsilon_c(t,t_o) = \frac{\sigma(t_o)}{E(t)}\varphi(t,t_o) \tag{3}$$

For normal concrete, from Appendix A, Equation A.3.1 of Ghali and Favre (1994), the coefficient for creep at time t for age at loading t_o is given by:

$$\varphi(t,t_o) = \frac{(t-t_o)^{0.6}}{10+(t-t_o)^{0.6}}\varphi_\mu \tag{4}$$

Where

φ_μ = the ultimate creep coefficient after a very long time (10,000 days) for age at loading, t_o

For $t - t_o$ = 5 years = 1,826.25 days

$$\varphi(t,t_o) = \frac{1826.25^{0.6}}{10+1826.25^{0.6}} = 0.90 \tag{5}$$

Therefore, 90% of the ultimate creep occurs in the first 5 years after the loading is applied.

φ_μ is a function of the characteristic compressive strength of concrete f_{ck}, the ambient relative humidity and average thickness of the member or its volume-to-surface ratio. Ghali and Favre (1994) provide graphs for $\varphi(t,t_o)$ with various f_{ck}, relative humidity and nominal size of the member. For dam structures, usually f_{ck} is low, the relative humidity is high and volume-to-surface ratio is high. Therefore, Figure A.11 from Ghali and Favre (1994) with f_{ck} = 3,000 psi, RH=80% and nominal size h_o=40 inches is used for a mass concrete structure, i.e., the largest available size is used. From Figure A.11 of Ghali and Favre (1994), the φ_μ for various t_o is listed in Table 3.1-1 of Ghali and Favre (Table 1 below). When φ_μ versus log (t_o) is plotted, the three points are almost on a straight line. Therefore, φ_μ can be calculated from the following formula:

$$\varphi_\mu(t_o) = -0.4559 * \log(t_o) + 1.222 \tag{6}$$

Where

t_o = the age of concrete at loading in unit of year.

Table 1. Ultimate Creep Coefficient for Dam Structure

t_0 (day)	t_0 (year)	log10(t_0) (year)	φ_u
360	0.98562628	-0.006287724	1.227
700	1.91649555	0.282507815	1.089
1400	3.8329911	0.583537811	0.958

The stresses in an AAR affected concrete structure change with time as a result of continued concrete expansion. As indicated above, the creep strain is proportional to the applied stress. This linear relationship allows superposition of the creep strains due to stress changes. Therefore, when the magnitude of the applied stress changes with time, the total creep strain of concrete due to the applied stress is obtained by:

$$\varepsilon_c(t) = \frac{\sigma(t_o)}{E(t)}\varphi(t,t_o) + \int_{\sigma(t_o)}^{\sigma(t)} \frac{\varphi(t,\tau)}{E(t)}\,d\sigma(\tau) \quad (7)$$

Where

t_o and t = the ages of concrete when the initial stress is applied and the strain is considered
τ = an intermediate age between t_o and t
$\sigma(t_o)$ = the initial stress applied at age t_o
$d\sigma(\tau)$ = the stress change applied at age τ
E(t) = the concrete modulus of elasticity at time t
$\varphi(t,t_o)$ = the creep coefficient at time t for loading at age t_o
$\varphi(t,\tau)$ = the creep coefficient at time t for loading at age τ.

In the AAR analysis using GROW3D, the time step is taken as one year. In order to calculate the creep strain, the six stress components at each gauss point of each element are saved as state variables in ANSYS. However, there is a limit on the number of time steps for which stresses can be stored due to limitations on computer resources. As indicated previously, over 90% of the ultimate creep strain occurs in the first five years after the stress is applied. Therefore, for practical purposes, only the stress components for the last five years preceding the current time step are saved for the calculation of creep strain.

AAR affected concrete has larger creep coefficient than normal concrete. To account for this, a factor CAAR is introduced which is the ratio of creep coefficient of AAR concrete over normal concrete. Thus, for AAR concrete, the creep coefficient is given by:

$$\varphi(t,t_o) = CAAR * \frac{(t-t_o)^{0.6}}{10+(t-t_o)^{0.6}} * \varphi_u \quad (8)$$

By comparing the analysis results of AAR projects using the current nonlinear time dependent creep model with the previous linear time independent creep model, it is found that equivalent results are obtained with CAAR varying from 1.0 to 2.0.

<u>Stress-Dependent Concrete Growth Law in ANSYS</u>

The stress dependent concrete growth law is input to ANSYS using the user programmable features within ANSYS. It is noted that GROW3D was developed using measured displacements and concrete stresses from the Mactaquac Generating Station in New Brunswick, Canada. It was found that variation of concrete expansion strain rates with the log of stress provided a very good correlation with field measurements. In the field of rock mechanics dealing with swelling rocks, a similar relationship was found (as referenced in Curtis 1995).

The stress dependent growth law characterizing AAR-induced growth of concrete is given as follows:

$$\dot{\varepsilon}_{gi}(t) = \dot{\varepsilon}_{g0}(t) \qquad \text{for} \quad \bar{\sigma}_i < \sigma_0$$

$$\dot{\varepsilon}_{gi}(t) = \dot{\varepsilon}_{g0}(t) - K\left[\log\left(\frac{\bar{\sigma}_i}{\sigma_0}\right)\right] \qquad \text{for} \quad \sigma_0 < \bar{\sigma}_i < \sigma_{ltm}$$

$$\dot{\varepsilon}_{gi}(t) = 0 \qquad \text{for} \quad \bar{\sigma}_i >= \sigma_{ltm} \qquad (9)$$

Where

$\dot{\varepsilon}_{gi}(t)$ = concrete growth strains in principal directions at time (t)

$\dot{\varepsilon}_{g0}(t)$ = unrestrained concrete growth rate at low stress at time (t)

K = slope of the line defining the concrete growth rate versus the log of stress

σ_i = the three principal stresses (i = 1 to 3)

σ_0 = a low compressive stress cut-off, whereby at lower compressive stresses or any tensile stress, the concrete growth rate is set at the unrestrained growth rate, and at larger compressive stresses, the concrete growth rate is reduced according to the above logarithmic growth law

σ_{ltm} = the cut-off stress level beyond which no concrete expansion is assumed to occur.

At each time interval, the concrete growth strains are computed based on the previous stress state. The concrete growth strain increments are resolved in the direction of the principal stresses. These strain increments are the initial strain loads (internal loads) used in the finite element analysis of a given time step.

Temperature-Dependent Growth Rate

It is noted that GROW3D has been updated to include a temperature dependant concrete growth law and it has been used on other projects but currently not on Mactaquac. The temperature dependant growth law is important when calibrating to stress measurements taken at different times of the year.

UPDATED INTAKE SLOT CUTTING ANALYSIS

The slot cutting program at Mactaquac was recently updated in an effort to better optimize the timing of slot cuts in the intake. A 3D finite element model was developed and analyzed using the latest version of GROW3D running in ANSYS. It is noted that during the course of the updated model calibration, the concrete growth law was changed to allow a larger concrete growth rate in the direction of tensile stress. The updated concrete growth law and finite element model is shown in Figure 3.

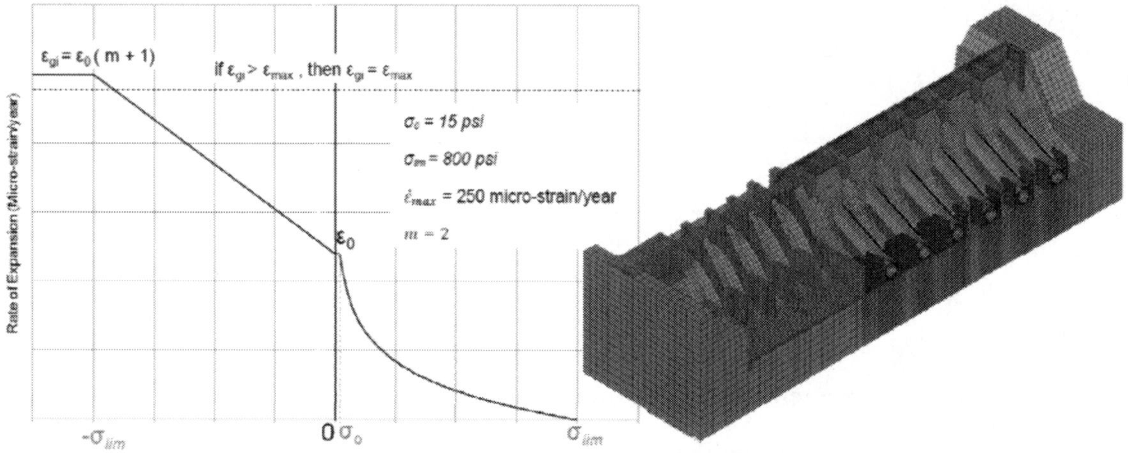

Figure 3. a) Update Concrete Growth Law b) Intake and Spillway FE Model

Figure 4 shows the location of the slot cuts in the water retaining structures and summary information on timing and re-cutting of the slots. The intake slots are cut down to roughly the same elevation at the floor of the intake water passage (el 60).

	NBW	WEP	EEP/6	5/6	4/5	3/4	2/3	1/2	0/1
Year of first cut	1993	1993	2000	1988	1989	1992	1989	1992	2015
No. of Re-cuts	8	5	2	7	5	4	6	6	0
Total sum of saw wire diameters used (mm)	146	88	45	121	99	83	111	103	30

Figure 4. Slot Cut Locations in the Water Retaining Structures Including Cut Summaries

The geometry of the intake and spillway at Mactaquac is such that concrete expansion has a compound effect on the spillway EEP as shown in Figure 5. In the early stages of the conceptual design of remedial measures, a large strut was proposed to replace the spillway deck with a massive reinforced concrete beam. This would eliminate the need for slot cuts in the intake. However, the size of the strut was limited due to the required flood handling capability of the spillway. In addition, the strut would not stop the push from the spillway ogee into the intake Unit 6 water passage. Therefore, after considerable analysis and review, it was decided that slots should be cut in the intake and have the upper portion of the intake blocks behave independently. The design of the original slot cuts at Mactaquac in the late 1980s was undertaken using an elastic equivalent temperature model to model concrete expansion. The enhanced creep behavior of AAR-affected concrete was simulated using an effective modulus approach. The model was subsequently refined to include the simulation of concrete cracking and this model predicted significant cracking of the intake water passage piers as a result of slot cutting. The intake slot cut design included the provision of post-tensioned anchors to stabilize the structure against sliding in the event that adverse cracks developed in the water passage piers.

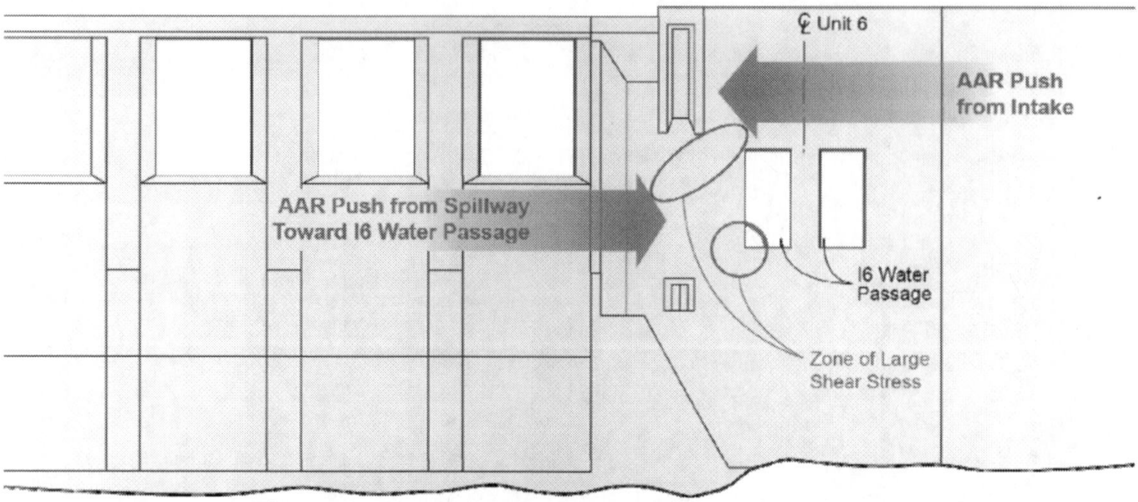

Figure 5. Interaction Between Intake and Spillway

The updated slot cut scheme (year 2015 to 2030) for Mactaquac Intake Structures has been developed to reduce the following:

- Total number of required slots
- Movement of the intake water passage piers
- Development of the tensile and shear stresses in the intake passage piers.

The initial slot cutting schedule was developed to maintain the slot open and this was done prior to the development of GROW3D. In the current update, some sensitivity analyses were undertaken using GROW3D and it was found that the ideal time to re-cut a slot was when the open width of the slot had reduced to about 2 to 5 mm at el 80 at a middle section in the plane of the intake gate slots. It should be noted that el 80 corresponds to the elevation of the ogee crest of the Main Spillway and it was also considered necessary to keep the EEP/6 slot open at this elevation. The upper portion of the slot is allowed to close for a short period of time without having a significant effect on the water passage pier deflections and stresses.

As mentioned earlier, a balanced slot cut plan is required to manage the deformations and stresses in the intake water passage piers. To illustrate this further, the slot cuts EEP/6 and 0/1 are discussed in more detail.

The pre-slot (1988) and post-slot (2010) computed longitudinal displacements of the intake are plotted in Figure 6. From Figure 6, the deflection of the spillway EEP towards the spillway has been reduced by the slot cutting program. As shown in Figure 4, the EEP/6 slot was not cut until the year 2000. The concrete growth in the main spillway rollway caused longitudinal movement of spillway concrete at and below el 80 which was directed towards the Intake 6 water passage with its floor at el 60. This movement towards Intake 6 resulted in a shear stress of about 400 psi at the base of the intake piers (as shown in Figure 7). An inspection of the Intake 6 water passage showed some initial cracking had developed as a result of these movements. Therefore a slot cut was recommended at the EEP/6 and it was installed in the year 2000. Figure 7 shows Unit 6 shear stresses were reduced considerably as a result of the EEP/6 slot cut, i.e., to around

200 psi by the year 2005. Therefore the slot achieved the desired result. It should be noted that the shear strength of AAR-affected structural elements is typically higher than normal mass concrete because the buildup of compressive stresses (restrained expansion) overcomes the tensile stresses associated with large shear stresses. This effect is discussed in more detail, and with an example, in Curtis et al (2005).

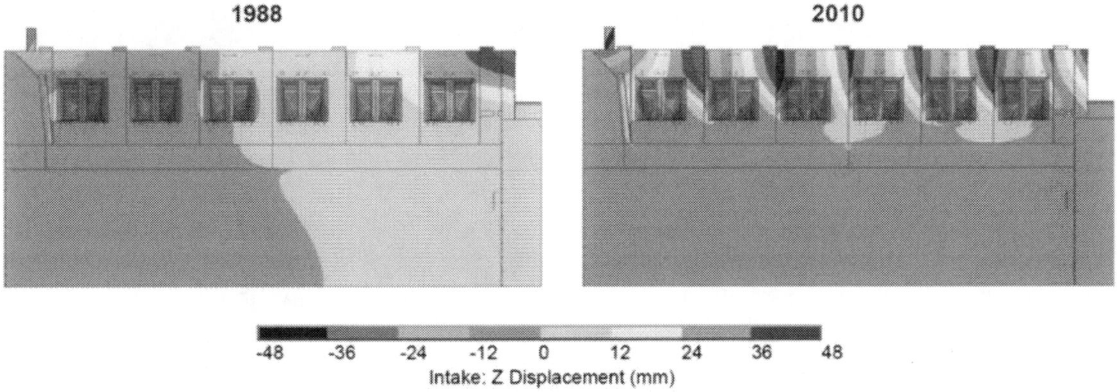

Figure 6. Computed Longitudinal Displacements of the Intake – Looking Downstream

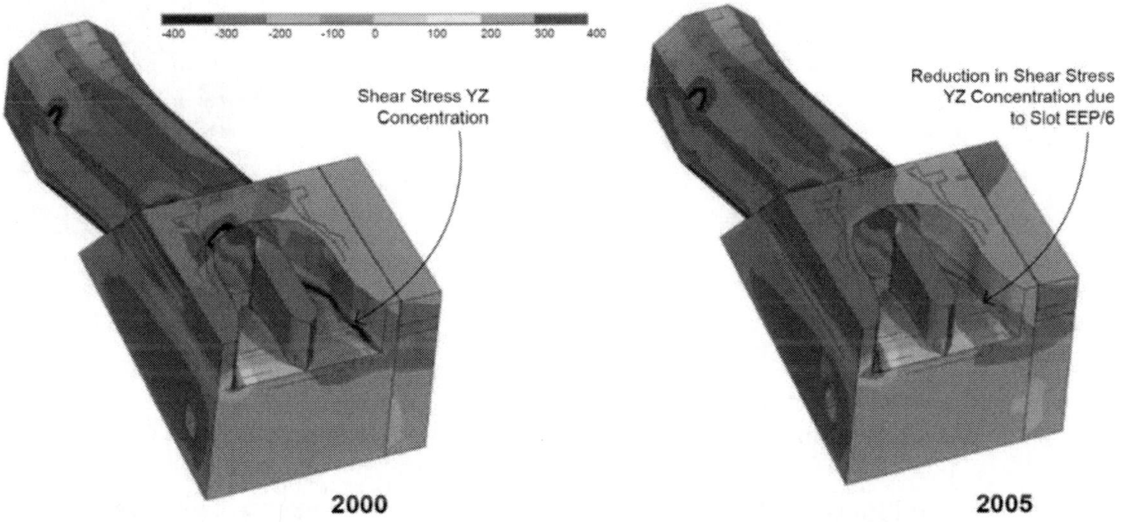

Figure 7. Intake 6 – Computed Shear Stresses in the Water Passage

In order to illustrate the response to slot cutting, the computed and measured response at Intake 1/2 is provided. Figure 8 presents the computed slot width change for Slot 1/2 as a function of time, Figure 8a shows the computed open width and Figure 8b shows both the elongation of Intake 2 block and the slot closure at Slot 1/2. These measurements are made using invar rod extensometers in the upper gallery of the intake. The upper gallery is shown in Figure 2. From Figure 8b, the measurements and computed response is in excellent agreement.

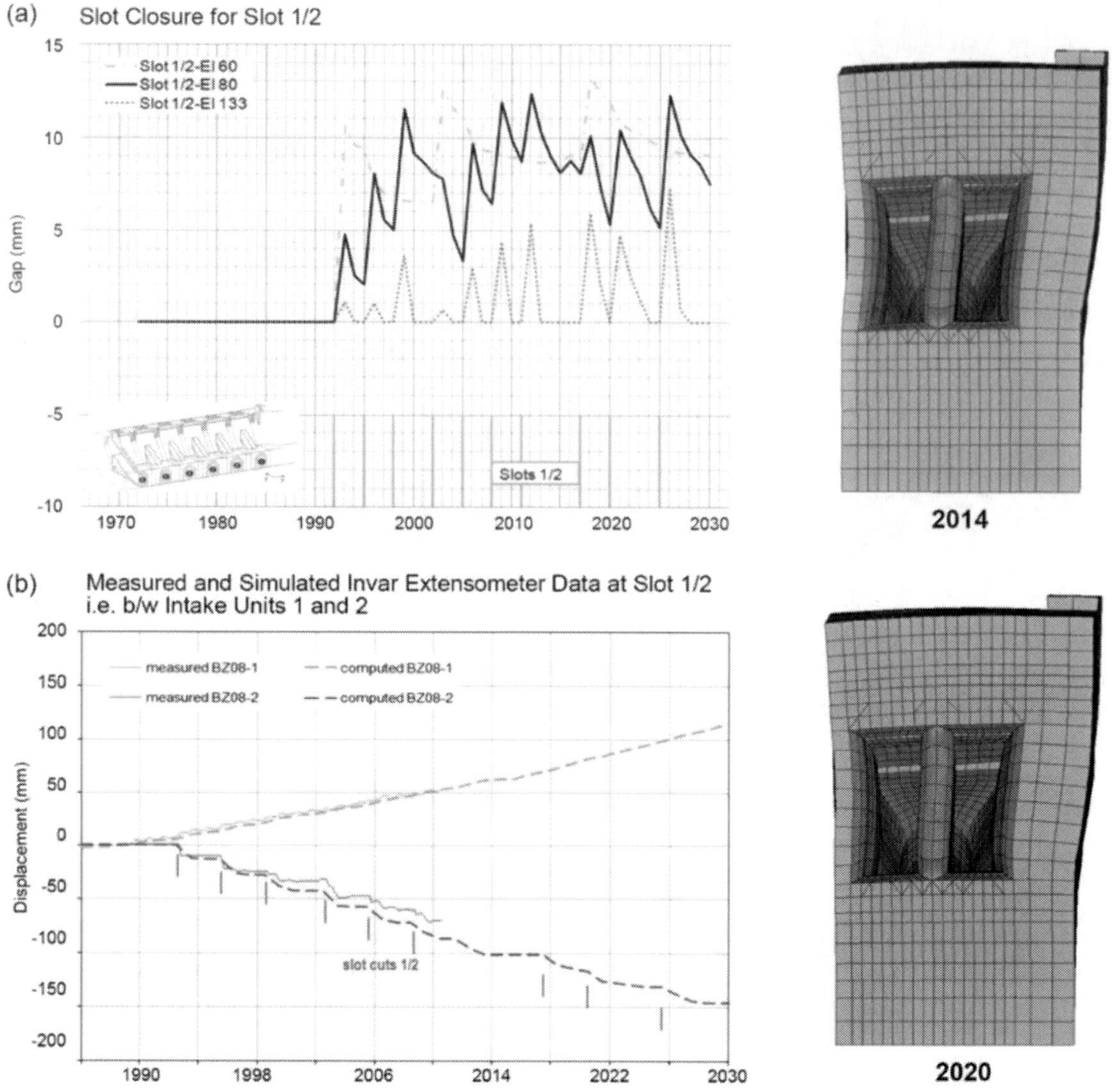

Figure 8. a) Computed Slot Width Versus Time
b) Measured and Computed Intake Block and Slot Closure at Slot 1/2
(Right) Deformed Shape of Intake 1 before and after 2015 Slot Cut

As indicated in Figure 4, the Intake 0/1 slot was initially cut in 2015. This was done to relieve the relatively large shear stresses that had developed in the Intake water passage piers. Figure 8 shows the deformed shape of the Unit 1 Intake block before and after the 2015 slot cut. The 0/1 slot cut caused an immediate shear stress reversal from -400 psi to 100 psi in the Unit 1 Intake piers as shown in Figure 9.

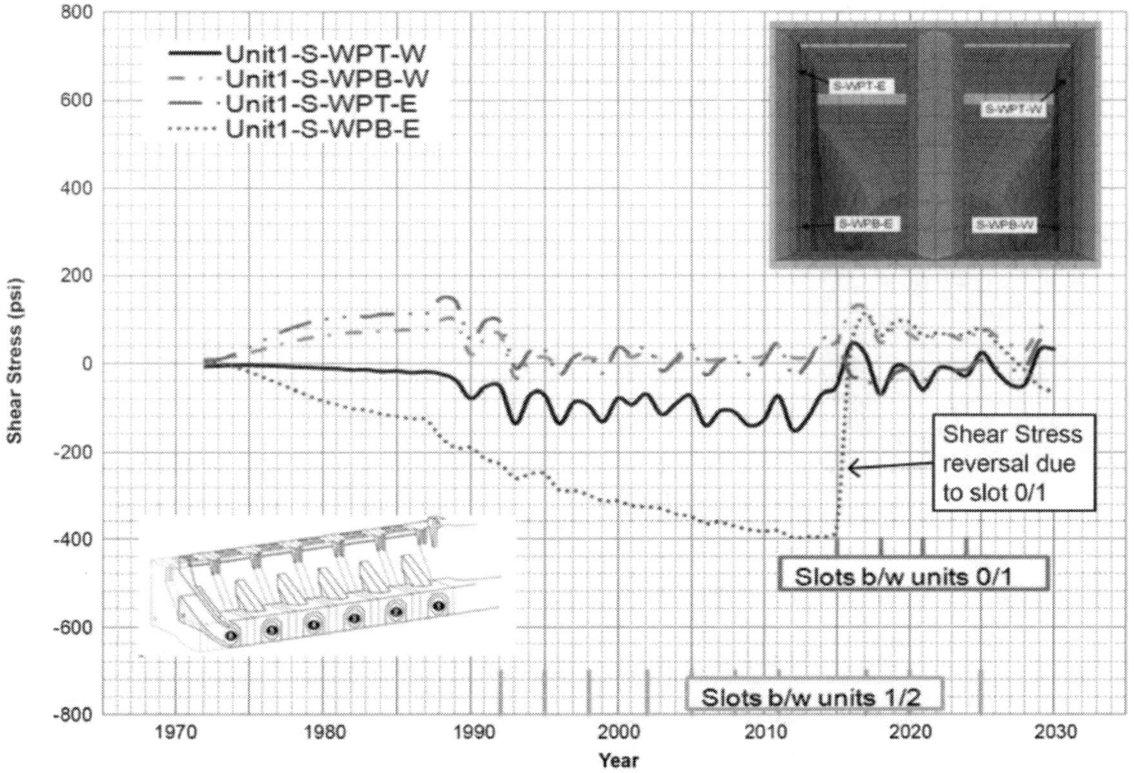

Figure 9. Intake 1 Water Passage Shear Stresses Versus Time

AAR EFFECTS IN THE POWERHOUSE

The first sign of AAR-induced deformations at Mactaquac was the opening of the Line J joint at the Turbine Floor and Line F under the generator floor in the powerhouse. The locations of the Line J joint and Line F are shown in Figure 10. Figure 10 also shows some key instrumentation which will be used to illustrate the effects of differential vertical concrete expansion on behavior of the tailrace substructure. From Figure 10, invar rod Extensometer BZ-27 measures an elongation rate of 4.5 mm/yr under the generator floor and about 90% of this is directed in the downstream direction (based on other instrumentation not shown). A large portion of this downstream displacement is due to an AAR-induced "curl" of the tailrace piers.

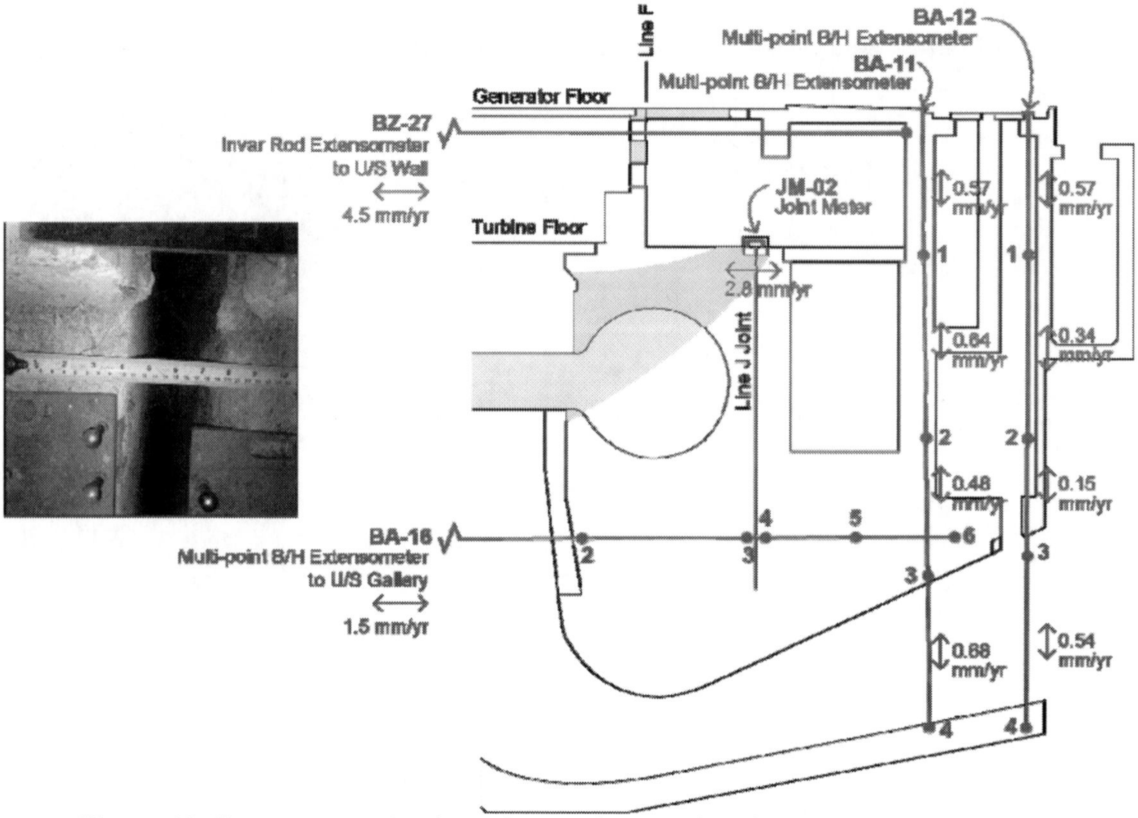

Figure 10. Downstream Portion of Powerhouse Showing Key Instrumentation

For this review, instrumentation at Unit 2 was selected for data processing to illustrate the "curl" behavior mechanism discussed above. Figure 10 shows the selected instruments at Unit 2. From Figure 10, the multi-point Borehole Extensometers BA-11 and BA-12 measure the vertical displacement increments between anchor points in the tailrace substructure. Although the el -29 gallery is not shown in Figure 10, Borehole Extensometer BA-16 measures the elongation from the el -29 gallery (see Figure 2 to locate el -29 gallery) to just beyond BA-11. The displacement at the el -29 gallery is essentially zero due to foundation rock restraint hence BA-16 measurements at Anchor 6 represent downstream translation.

The differential expansion in each incremental height at similar elevations can be used to compute downstream deflections assuming Anchor Point 4 is the fixed end of a cantilevered column. The rotation and displacement of a cantilevered column subjected to differential expansion through its depth (d) is given by

$$\theta_B = \frac{\Delta s \times L}{d} \tag{10}$$

$$\Delta_B = \frac{\Delta\varepsilon \times L^2}{2d} \tag{11}$$

Where

L and d are the length of cantilever beam and depth, respectively
$\Delta\varepsilon$ is the differential vertical expansion.

The accumulated horizontal downstream displacement using the extensometer data over the four vertical segments as summarized in Figure 10 is

$$\Delta_i = \Delta_{i-1} + \theta_{T(i-1)} \times L_i + (\theta_i \times L_i)/2 \tag{12}$$

Where

i is the current segment starting at the base (Anchor 4 in Figure 10)
Δ_i is the displacement at the top of the current segment
$\theta_{T(i-1)}$ is the accumulated rotation at the top of the last segment.

Table 2. BA-11, BA-12 and BA-16 Extensometers – D/S Deflection Caused by Differential Vertical Strain and the el -25 Translation

			BA-11			BA-12			D/S Dsp due to differential Vert Growth in Pier					Add BA-16 Translation	
Location	Total Length	Increment Length	Rate	Increment Rate	Strain Rate	Rate	Increment Rate	Strain Rate	Differential Strain	Theta	Total Theta	D/S Disp	Elev	Incremental D/S Displacement	Total D/S Dsp
	(m)	(m)	(mm/yr)	(mm/yr)	(με/yr)	(mm/yr)	(mm/yr)	(με/yr)	(με/yr)	μ-Radians	μ-Radians	(mm)	(ft)	(mm/yr)	(mm/yr)
C-1	5.95	5.95	0.57	0.57	95.8	0.57	0.57	95.8	0.0	0.000	171.111	2.64	37.0	1.02	4.04
C-2	14.30	8.35	1.21	0.64	76.6	0.91	0.34	40.7	35.9	66.667	171.111	1.63	17.5	1.15	3.02
C-3	19.75	5.45	1.69	0.48	88.1	1.06	0.15	27.5	60.6	73.333	104.444	0.48	-9.9	0.37	1.87
C-4	26.60	6.85	2.37	0.68	99.3	1.60	0.54	78.8	20.4	31.111	31.111	0.11	-27.8	BA-16 = 1.5	1.5
	Vertical Rate		5.84	Vertical Rate		4.14				0.0	0.0	0.00	-50.3		0

From Table 2, the downstream displacement rate at the tailrace deck level is 4.04 mm/yr with 1.5 mm/yr due to translation at el -25 and the remainder (2.5 mm/yr) is due to differential vertical concrete growth of the tailrace substructure. The computed 4.04 mm/yr is quite close to that measured by BZ-27 (4.5 mm/yr) shown in Figure 10. The rate of Line J joint opening is less than that measured by BA-27 because the unit block concrete around the scroll case is expanding due to AAR but the dry generator floor is not. It is of interest to note that the stresses in the tailrace piers caused by differential vertical expansion would be relatively small, and this phenomenon would be difficult to stop without causing issues elsewhere in the substructure.

A global finite element model of the powerhouse has been developed for all six units as shown in Figure 11. The model has been calibrated to a massive amount of structural and turbine/generator unit parameters data totaling more than 300+ calibration measurement items. It is of interest to note that the modified concrete growth law shown in Figure 3 was developed to improve the Intake updated slot cut model calibration and when it was applied to the powerhouse a remarkable improvement in the calibration was achieved.

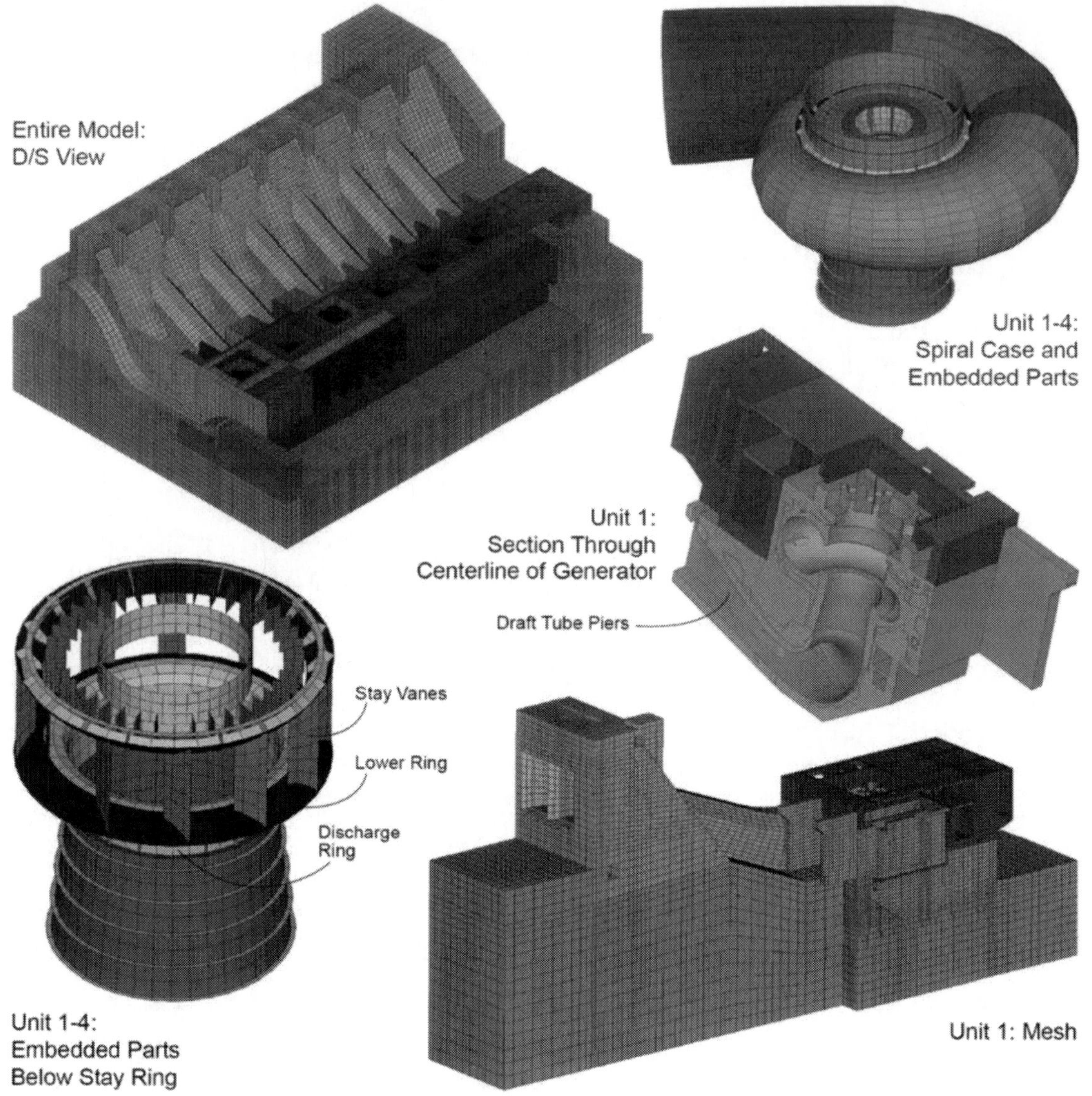

Figure 11. Global Finite Element Model and Selected Components

Concrete growth in the powerhouse was causing the embedded turbine components to deform in an oval shape and this was affecting clearances in the cross-flow direction where the discharge ring diameter was reducing at a steady rate. This was causing issues with blade tip clearances, and extensive grinding and re-working of the units was required. Transverse slots were cut between units as shown in Figure 12. The computed response to the transverse slots is shown in Figure 13 and the improvements in the critical cross-flow direction have been confirmed by measurements. As noted previously, the penstocks were cut and couplings installed from 1996 to 2001 at one unit per year. The penstock cuts improved the downstream "racking" of the turbine stay ring and reduced the stresses in the draft tube piers. However, due to space limitations the penstock cuts are not discussed further herein.

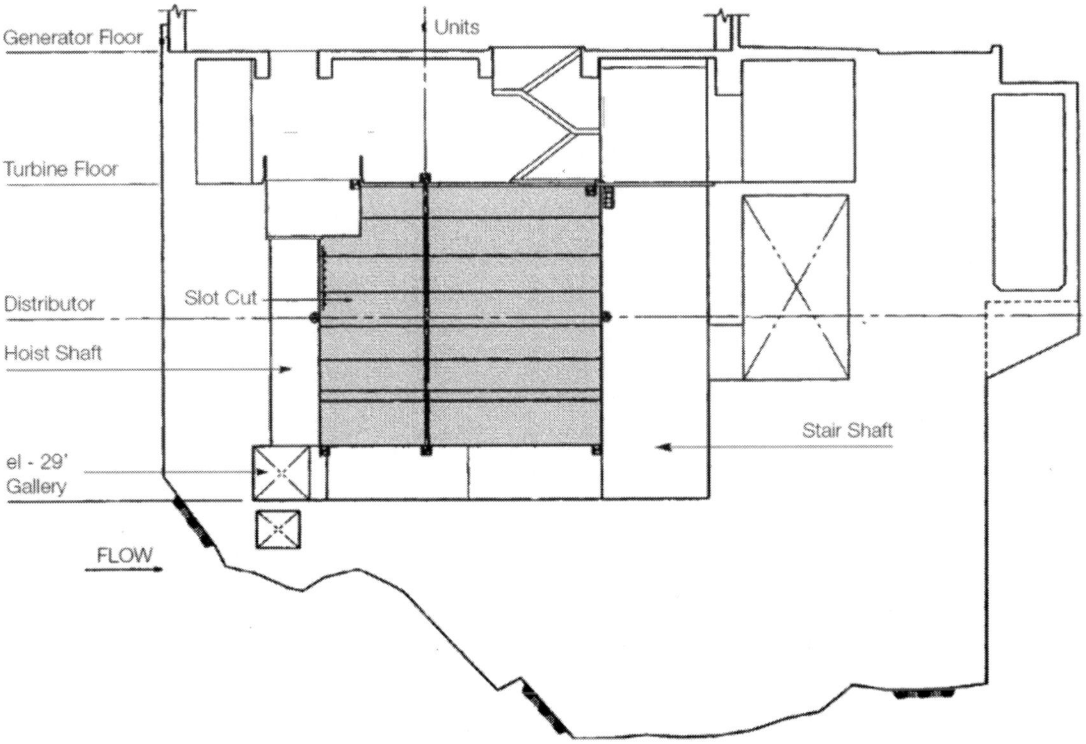

Figure 12. Extent of Transverse Slots Between Units

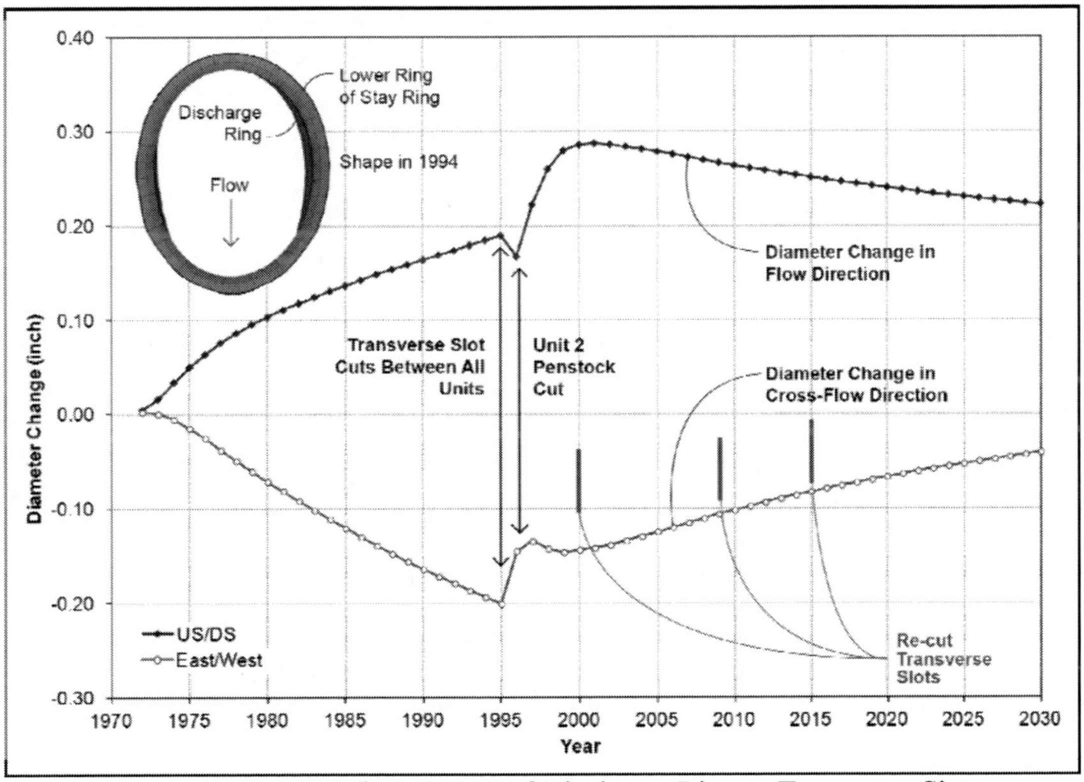

Figure 13. Computed Response of Discharge Ring to Transverse Slots

The stresses in selected turbine stay vanes were measured in 1994 and the computed stresses from GROW3D compared well to these measurements. The location of the stay vanes is shown in Figure 11. The computed vertical stresses in 1994 and 2016 are contoured in Figure 14. The stay vane material is modeled with a non-linear stress-strain law and some yielding of the stay vanes has occurred. However, the plastic strains are well below the ultimate strain for this material. It is noted that one of the unit parameter measurements is measurement of the height change of the leading and trailing edge of the stay vanes. The model calibrated very well to these measurements therefore the computed stresses are considered to be quite realistic. There is concern for potential cracking of the stay vanes given the high stresses and their thick cross-section. It is noted that the stay vanes are stressed in a net tensile stress due to the upward pull of the concrete on the upper ring of the stay ring.

Finally, the draft tube piers are becoming highly stressed due to a downstream dipping thrust caused by restrained AAR expansion and the shear stress induced by the downstream translation of the concrete above the draft tube roof (as measured by BA-16, see Figure 10). GROW3D predicts a maximum compressive stress in the draft tube piers of about 1,800 psi. It also predicts only a small increase in stress going from the present to 2030 due to stress-dependent and concrete creep effects. The stresses in the draft tube piers were measured in 1989 and the maximum principal stress was 1,200 psi compression dipping in the downstream direction. The intermediate principal stress was 700 psi compression. GROW3D matched the maximum principal stress very well but slightly underestimated the intermediate principal stress. The stress of the draft piers was checked using the procedure given by Balakrishnan and Murray (1988) and the piers were found to be acceptable to the year 2030 and beyond assuming concrete compressive strength loss of up to 25%. It is noted that concrete strength is closely monitored.

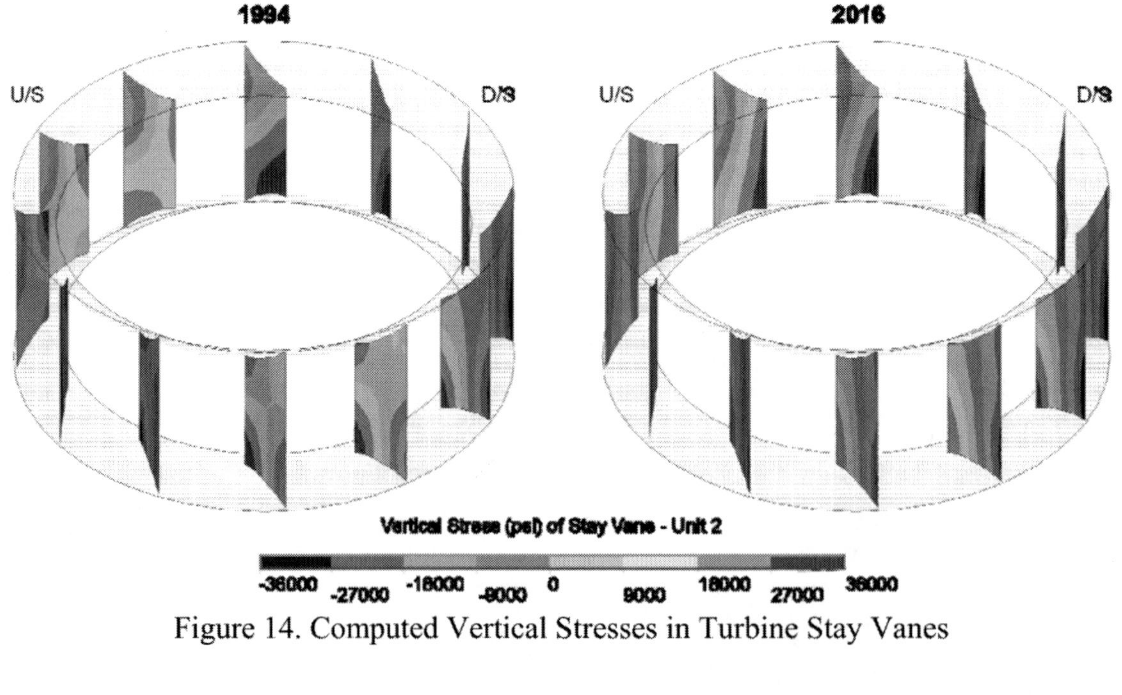

Figure 14. Computed Vertical Stresses in Turbine Stay Vanes

CONCLUSIONS

The following conclusions are drawn from this work:

- The concrete expansion rates at Mactaquac are very large at about three times the rate observed at many other AAR-affected dams and powerhouses.
- Slot cuts installed in the Intake and Powerhouse are used to improve clearances for spillway gates and turbine embedded parts ovaling and unit alignment issues. The ovaling effect causes significant stressing of the embedded parts and also the inner and outer headcovers of the turbine.
- The geometry of the intake and spillway is such that alternatives to slot cutting were not available.
- The geometry of the powerhouse was such that slot cuts could be installed relatively easily between existing stairwells and hoist shafts and the benefit was immediate and ongoing.
- Small amounts of differential vertical concrete expansion can cause significant horizontal deflection of the concrete piers without causing large stresses in the piers themselves.
- The concrete growth law was modified to simulate increased concrete growth rate in the direction of tensile stresses. This modification greatly improved the GROW3D model calibration for the Intake and the Powerhouse. It is noted that there is no technical literature to support this as laboratory testing of this effect would be difficult, however the observations at Mactaquac certainly support its use. This modified growth law may be project specific, as we have not used it on any other AAR-affected projects.
- Steel stresses in the stay vanes and concrete stresses in the draft tube piers are quite large and close monitoring of these structural elements is required.
- GROW3D has proven to be a practical tool for the analysis of AAR-affected dams and powerhouses. More sophisticated scientific models exist with considerable input requirements and it is doubtful that this data would ever be available. Also, many of these scientific models do not properly account for concrete creep; hence they likely overpredict concrete stresses.
- The results of these analyses are currently being used to help update the estimate of the remaining service life of these structures.

ACKNOWLEDGEMENTS

The authors would like to thank NB Power for their permission to publish this paper. We would also like to gratefully acknowledge the diligent work and input by NB Power engineers, technicians and plant staff. In addition, we acknowledge the helpful input provided by the ongoing Independent Review Board and the past contributions of Robin Charlwood who directed the project for over 20 years from about 1985.

REFERENCES

Balakrishnan, S. and Murray, D.W., 1988. "Strength of Reinforced Concrete Panels", Canadian Journal of Civil Engineering, Vol. 15, pp 900–911.

Charlwood, R.G., Solymar, Z.V. and Curtis, D.D., 1992. "A Review of Alkali Aggregate Reactions in Hydroelectric Plants and Dams." Proceedings for the International Conference of Alkali-Aggregate Reactions in Hydroelectric Plants and Dams, CEA and CANCOLD, Fredericton, pp 1–29.

Codrington, J.B., May, T.H. and Curtis, D.D., 2007. "Spillway Gate Rehabilitation and Intake Bulkhead Design for Mactaquac GS." CDA 2007 Annual Conference, St John's, NL, Canada.

Curtis, D.D., 2000. "Analysis of the Structure Response to Recent Slot Cutting at Mactaquac Generating Station", 11th International Conference on Alkali-Aggregate Reaction, Quebec City, QC, Canada, pp 1283–1291.

Curtis, D.D., 1995. "Modeling of AAR Affected Using the GROW3D FEA Program." Proceedings for the Second International Conference on Alkali-Aggregate Reaction in Hydroelectric Plants and Dams, USCOLD, Chattanooga, Tennessee, pp 457–478.

Curtis, D.D., Davis, J.D., Rahman, S. and Powell, R.C., 2005. "Updated Assessment of Concrete Growth Effects on a TVA Dam". The 25th USSD Annual Meeting and Conference was held in Salt Lake City, Utah.

Ghali, A. and Favre, R., 1994. Concrete Structures Stress and Deformations, 2nd Edition.

Thompson, G.A., Steele, R.R. and Coulson, D.M., 1995. "Management of Concrete Growth at Mactaquac Generating Station", Proceedings for the Second International Conference on Alkali-Aggregate Reaction in Hydroelectric Plants and Dams, USCOLD, Chattanooga, Tennessee, pp147–159.

Thompson, G.A., Charlwood, R.G., Steele, R.R. and Curtis, D.D., 1994. "Mactaquac Generating Station Intake and Spillway Remedial Measures." Proceedings for the Eighteenth International Congress on Large Dams, Durban, South Africa, Vol. 1, Q-68, R-248, pp 342–368.

USING THE TRIAL LOAD METHOD TO OPTIMIZE THE FEASIBILITY DESIGN OF WATANA DAM

Aled Hughes, PE[1] Glenn Tarbox, PE[2]
Brian Sadden, PE[3] Bryan Carey, PE[4]

ABSTRACT

The proposed Susitna-Watana Hydroelectric Project will be a hydroelectric project on the Susitna River, Alaska providing approximately 2800 GWh annually. Watana Dam will be a 705 ft. high curved gravity RCC dam with an estimated concrete volume of 5.67 million cubic yards (Mcy). Development of the selected configuration involved numerous iterations with each revised layout being analyzed to ensure a robust engineering design.

In the era before the development of computers, the structural analysis of arch dams was carried out using the Trial Load Method (TLM), first applied by the U.S. Bureau of Reclamation (USBR) for the design of Hoover Dam in the 1930s. The manual calculation was a laborious and time consuming process typically taking engineers months to complete. In the 1970s software was written by the USBR, using the 3-dimensional calculation method. The software, Arch Dam Stress Analysis System (ADSAS), successfully reduced the analysis time from months to minutes. Developments in computing power have further reduced the analysis time.

The Finite Element Method (FEM) has latterly become the industry standard method for the analysis of large concrete structures, including curved gravity dams, providing detailed information on the stresses within, and anticipated deflections, of the structure. While a powerful tool, a FEM model does not readily accommodate geometric adjustments; changes are a time consuming process which prolong the early stages of feasibility design where the focus is the verification of, and optimization of, the dam configuration. Automated routines for FEM models have been presented in several papers from the 1990s onwards but use is limited.

This paper provides a recapitulation of the TLM and describes how it was successfully used on the Susitna-Watana Hydroelectric Project to accelerate the iterative optimization of the dam configuration at an early stage, thereby allowing the FE methodology to be focused on optimized configurations. The results of the TLM was compared with the FEM output to demonstrate the accuracy of the method and also highlights its usefulness for the analysis of thicker section RCC curved gravity dams.

---

[1] Lead Civil Engineer, MWH Americas, Inc., 2353 130th Ave NE, Bellevue, WA 98005. Aled.R.Hughes@mwhglobal.com
[2] Global Dams Practice Leader, MWH Americas, Inc., 2353 130th Ave NE, Bellevue, WA 98005. Glenn.S.Tarbox@mwhglobal.com
[3] Project Manager, MWH Americas, Inc., 2353 130th Ave NE, Bellevue, WA 98005. Brian.E.Sadden@mwhglobal.com
[4] Engineering Manager, Alaska Energy Authority, 813 West Northern Lights Blvd., Anchorage, AK 99503. bcarey@aidea.org

TRIAL LOAD METHOD

History of Trial Load

Concrete dams in the early 20th century were gravity structures and primarily constructed using masonry materials. Early practitioners of concrete dam design began constructing gravity dams with horizontal curvature to provide an additional margin of safety. The cylinder formula was used for some of these early designs. However, it had been realized for some time that the cylinder formula did not produce correct results because it made no allowance for rib shortening, temperature changes, or the effect of cantilever action. Several investigators and practitioners wrote reports and professional papers projecting more rigorous treatments of arch design, beyond cylinder theory.

Professor William Cain (1847-1930) wrote a series of reports in which he presented formulas describing the properties and behavior of voussoir arches. His formulas became known as the "Cain Formulas." The formulas were used by early practitioners in the design of several arched gravity dams in the early 20th century including Lake Spaulding in 1919.

Lars R. Jorgensen (1876-1938) introduced his concept of a constant-angle/variable-radius arch dam with the design of Salmon Creek Dam in 1913. In 1921 F. A. Noetzli introduced the idea of trial and error by using a tentative division of water load between the arch and cantilever elements.

With the planned construction of many arch dams of great height, it became imperative to develop a complete theory of design. The cooperative efforts of the Engineering Foundation created by the American Societies of Civil, Mining and Metallurgical, Mechanical, and Electrical Engineers, and others had brought about the necessary funds, talents and other resources to establish a Committee on Arch Dam Investigation. Financing of an experimental dam included a large contribution of funds from the Southern California Edison Company along with other contributions from power companies, banks, municipalities, construction firms and engineers, totaling $115,500. An important part of the Committee's achievements was the construction of a test dam in 1926 on Stevenson Creek in California. It was 60 feet in height, 7.5 feet thick at the base, reducing to 2 feet at mid height up to the crest. The crest length was 140 feet and the upstream face was vertical. The dam was defined by a constant radius of 100 feet over its entire height. The dam was intentionally conceived to be ultra-thin to yield large deflections necessary for comparison with physical models and analytical solutions.

The experimental results of the Stevenson Creek Test Dam demonstrated conclusively the inadequacy of the cylinder formula and laid the foundation for the development of more sophisticated methods of analysis, culminating in the introduction and general acceptance of the TLM.

The TLM of analysis was developed at the USBR between 1926 and 1935 based on results from the Stevenson Creek Test Dam and was used to design Boulder Dam, 1936.

An article by C. H. Howell and A. C. Jaquith in 1929 was one of the first thorough treatments of the TLM with complete radial adjustments. The USBR provided a thorough treatment of the method in its Boulder Canyon Final Reports in 1938.

Methodology

The TLM is based on the concept that an arch dam is composed of a system of horizontal arches and vertical cantilevers, that each system carries a proportion of the total load, and that the deflections of the arches and cantilevers are equal at all points of intersection between the two systems. The dam is considered a continuous structure supported by a rigid foundation.

The requirements for a correct solution of the stress problem are defined by the Kirchhoff uniqueness theorem in the theory of elasticity. The requirements are as follows:

1) The dam is built of a homogeneous, isotropic and uniformly elastic material.

2) The framework of arch and cantilever elements selected to represent the dam body must be in equilibrium under the forces and stresses which act upon it.

3) Each of the deflected arch and cantilever elements must fit with the adjacent element at each point of intersection.

4) The stresses or displacements at the boundaries of the dam body must conform to the stresses or displacements imposed.

Under these conditions Kirchhoff proved that it is impossible for more than one stress system to exist and the system satisfying these conditions is unique. The requirement expressed in (2) is commonly called condition of equilibrium; in (3) condition of continuity; and in (4) the boundary conditions.

The method consists of proportioning the water load between the system of arches and cantilevers such that the deflections at all points of intersection of the two systems are equal following a number of successive trial loading sequences starting first with loads applied to the dam radially, then tangentially (normal to the radial). Horizontal deflections parallel to the extrados radii and normal thereto are the radial and tangential deflections, respectively. Thirdly twist load or horizontal moments about a vertical axis are applied to the arches and cantilevers to rotate them into angular agreement. Usually two rounds of loading sequences, radial,

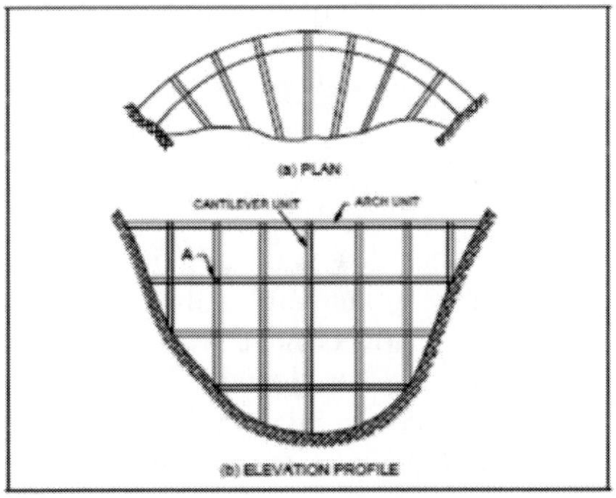

Figure 1. Typical Elements Used in the Trial Load Method

tangential and twist and then ending with a third radial adjustment will result in the dam structure being in sufficient deflection agreement and accuracy to then compute stresses based on the proportioned loads. A detailed explanation of the TLM is contained in Design of Arch Dams, USBR, 1977.

ADSAS

ADSAS (Arch Dam Stress Analysis System) was a special project established within Reclamation in the 1960's to computerize the hand calculations required to conduct a complete Trial Load analysis of a concrete dam. A small group of engineers and computer specialists within the Concrete Dams Section were assembled to pursue the goal. The project took approximately 8 years to complete.

The time required to make a complete three dimensional stress analysis of a dam by hand required two engineers working approximately six months or one person-year. Following completion of the mission to computerize Trial Load analysis, the required time was a matter of minutes using a mainframe CDC 6600/6400 computer. New capabilities were added in subsequent years including three-centered geometry, pads, and dynamic analysis of earthquake loads. ADSAS was subsequently converted to a desktop computer application and today a complete 3-dimensional linear-elastic stress analysis can be completed in seconds.

Limitations

ADSAS has limitations and the introduction of the Finite Element Method (FEM) of structural analysis in the late 1960's provided for the inclusion of geometry and parameters beyond those that could be considered in ADSAS. Capabilities not available in the ADSAS software but present in FEM include more realistic stress distribution in the dam, more flexibility in adapting to irregular geometries, incorporation of openings in the dam, foundation mass and mechanical properties, inclusion of contraction joints and other discontinuities such as cracks, non-linear material properties, Alkali Silica Reactivity (ASR), thermal analysis, reservoir and foundation structure interaction, stage construction, and implicit and explicit equation solvers.

SUSITNA-WATANA HYDROELECTRIC PROJECT

History

The Susitna-Watana Hydroelectric Project has been studied by the State of Alaska for many years, principally in the early 1980s, through the Alaska Power Authority (APA). Extensive engineering, environmental and economic studies were performed which resulted in a draft Application being submitted to FERC in 1983 for a license to construct a major power project. In March 1986 the Project was put on hold by APA primarily because of the low cost and ample supply of natural gas in the Cook Inlet basin.

In 2010, the Alaska Legislature authorized a goal that the State should obtain 50 percent of its electric generation from renewable and alternative energy sources by 2025; hydropower is considered a renewable resource in Alaska. The Susitna-Watana project was deemed to be the appropriate development to achieve the target.

Watana Dam Type Selection Evolution

The 1983 draft license application proposed an earth core rockfill dam (ECRD) for Watana. At that time, the only types of dam structures that had been constructed of a similar height were concrete thin arch dams, concrete gravity dams, and earth core rockfill dams. Progression in dam engineering design and construction since 1983 has resulted in concrete faced rockfill dams (CFRD) and roller compacted concrete (RCC) dams being candidates for the Susitna-Watana development.

The Watana Dam proposed in the 1980s was based on a two stage development. Stage 1 would be a reservoir with a normal maximum water level (NMOL) of El. 2000 ft. which would increase to El. 2185 ft. upon completion of Stage 2. The Stage 2 development was referred to as the "High" dam; the Stage 1 height was adopted for the studies described in this paper.

The current design development for the Watana dam commenced with a linear configuration at the centerline identified in the 1980s studies. Results of geotechnical investigations performed at that time, including drilling, sampling, laboratory testing, seismic lines and geologic mapping were available, so there was substantial geotechnical information available to the current designers.

Given the progression in dam engineering experience noted above, three alternative dam types were considered in the feasibility assessment of the site: ECRD, CFRD, and RCC. The findings of the assessment of the various projected layouts concluded that a safe and economic arrangement was found to be based on a RCC dam with a surface powerhouse at the toe of the dam. The initial RCC dam configuration, Layout 1, from which the design development commenced included a straight axis with a vertical upstream face and a downstream face sloping at 0.85H:1V. The volume of RCC required in this layout was approximately 5.11 Mcy.

Following submission of the preliminary layout to the regulators, development of the layout commenced with the aim of reducing the concrete volume, while maintaining stability and safety, thereby reducing cost and minimizing the construction schedule. The project site is remote and would necessitate a significant investment in infrastructure to enable transportation of materials, plant, and construction workers to and from the site. A reduction in imported materials could noticeably reduce the capital cost and any savings in the time of construction would allow generation to commence earlier to initiate revenue flows.

Given the topographic and geologic conditions at the selected site, and experience at other projects, introduction of a curve in the dam plan was investigated to assess the possibility of steepening the downstream face, beyond that for a gravity structure. The inclusion of a curve in the dam axis would enable transfer of limited loading to the abutments. Unlike a gravity dam where the structure could be analyzed as a 2-D structure, the assessment of load transfer to the abutments required a 3-D structural analysis.

In order that any development option selected would not preclude a future generation from raising the dam to the "High" dam height, the initial analysis of the first curved layout (Layout 2) was performed for a NMOL of El. 2185 ft. The results of the analysis were used to identify a suitable configuration that was subsequently analyzed with the selected NMOL of El. 2000 ft.

Application of ADSAS

ADSAS was used in advance of FE analysis to quickly perform multiple analyses of different geometries for the dam. Fourteen different dam geometries were analyzed including single centered and three centered configurations and some with a single centered curve and straight buttresses. The straight dam (Layout 1) that had been proposed as the initial option was one of the fourteen geometries analyzed.

The results of the ADSAS analysis provided a level of confidence that the selected dam cross section would be capable of withstanding the expected seismic loads.

Material Properties and Loads

Material properties used in the ADSAS analysis were:

- Sustained modulus of elasticity of concrete: 4,000,000 psi
- Poisson's ratio for concrete: 0.25
- Unit weight of concrete: 150 pcf
- Compressive strength of concrete: 4,500 psi
- Deformation modulus of the foundation rock: 4,000,000 psi
- Poisson's ratio for rock: 0.33

At this conceptual stage of analysis, dead loads and static water loads were included. Thermal, sediment, ice and seism ic loads were not considered, but the examination of the output of the analysis was made with the understanding that the concrete stresses could be somewhat higher than generated.

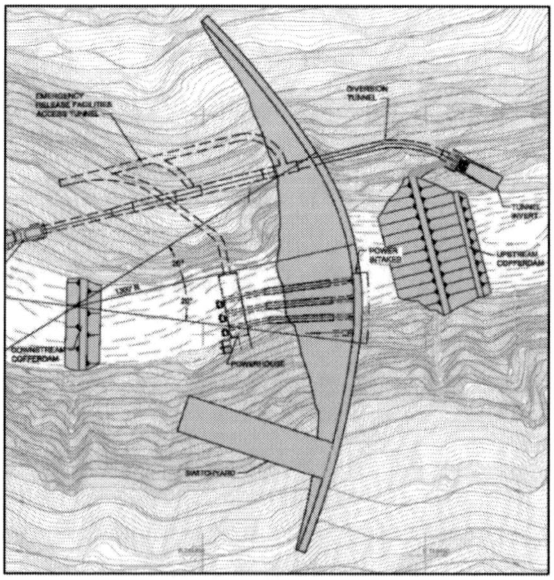

Figure 2. Layout 2 Configuration

Each run of ADSAS provided upstream and downstream arch, cantilever and principal stresses and a volume of concrete. The most promising arrangement was chosen for a more detailed analysis, using FE analysis for both static and dynamic loading.

The first conceptual layouts developed were based on a single curvature, of 1,800-ft. radius, chosen from experience and with regard to the projected rock contours at the dam foundation. The radius was selected with respect to a final crest elevation of El. 2025 ft. Horizontal angles of intersection between the downstream face of the dam and the contour line of competent rock of 30° or less was used as the guideline for establishing the orientation and abutment conditions.

The arrangement of cantilever and arch elements was determined to create an almost symmetrical layout. The locations of the cantilever and arch elements were determined by firstly locating the cantilever element positioned at the tallest point of the dam (crown cantilever). The spacing of the arch elements was selected to create a uniform distribution between the lowest point of the excavation and the dam crest proposed to the regulator (crest El. 2025 ft.). The lowest arch element was located at El. 1500 ft., about 140 feet above the lowest foundation level. The thickness of the dam below this elevation means that there would be minimal transfer of the stresses from the cantilever to the arch elements. The distance between the lowest arch element and El. 2025 ft. was divided into five sections resulting in the arch elements being spaced at 105 feet. A further two arch elements were included in the model, one at the crest of the projected ultimate development, the other located equidistant between the current projected dam crest elevations and the ultimate crest. At the abutments of the arch elements, cantilever elements were positioned.

The first ADSAS analysis indicated unacceptable values of tensile stresses in the horizontal (arch) elements at the downstream face. To address the unacceptable tensile stresses, the radius of curvature was reduced to 1,300 ft. with a transition to a 4,000-ft. radius at each flank. The three-centered arrangement was used in all subsequent runs of ADSAS, with a NMOL of El. 2185 ft. As noted 14 alternatives were analyzed, each option was analyzed for the potential ultimate development height of the dam, and with downstream face slopes ranging from 0.3H:1V to 0.85H:1V. Ninety-eight permutations were considered in total, and the dam configurations analyzed are summarized in Table 1.

Table 1. - Analyzed Plan Configurations for Layout 2 (high dam)

DAM	Radii (feet)		Center arc angle (degrees)	Crest length (feet)
	Middle	Outer arms		
A	1800	Straight	91.3	4543
B	2000	Straight	79.8	4335
C	2200		-	4350
D	2400		-	4251
E	4000		-	3979
F	1300	4000	30	4241
G	1500	4000	30	4171
H	1700	4000	30	4150
I	1800	4000	30	4133
J	1300	4000	40	4426
K	1500	4000	40	4391
L	1700	4000	40	4287
M	1800	4000	40	4252
N	Straight (999,999 ft. radius)		-	3805

*Dam N was Layout 1

After the analysis of the "high" dams, configurations F and J were identified as preferred alternatives, with a downstream face at 0.5H:1V. A steeper downstream face resulting in a thinner vertical cross section would result in a significant transfer of stresses from the vertical to the horizontal plane resulting in the dam no longer performing solely as a gravity structure. These two favored alternatives, along with Dam N (Layout 1) were then assessed for the dam configuration being studied at that time (NMOL El. 2000 ft.) for downstream slopes ranging from 0.4H:1V to 0.6H:1V.

Both dams F and J were satisfactory alternatives as analyzed. The preliminary stress analysis did not take into account seismic or thermal loads which could affect the stress values negatively. However, the effect of these additional loads would be less pronounced for Dam J, and it was therefore selected as the configuration for further analysis.

The selection of alternative J for "Layout 2" consisted of a central portion with a curved axis of 1,300 ft. radius which changes to a 4,000 ft. radius at a tangent point on the abutments. The downstream face is a uniform slope of 0.5H:1V and the upstream face is vertical. The central portion is defined by a phi angle of 40°. The dam crest was at El. 2075 ft. since the NMOL had been increased by 50 ft. to El. 2050 ft. The estimated RCC volume for the Layout 2 dam was 5.325 Mcy.

The static vertical stresses at the upstream and downstream face of the Crown Cantilever and horizontal stresses along the extrados and intrados of the arches intersecting the Crown Cantilever as calculated by ADSAS for Layout 2 are shown below in Figure 3.

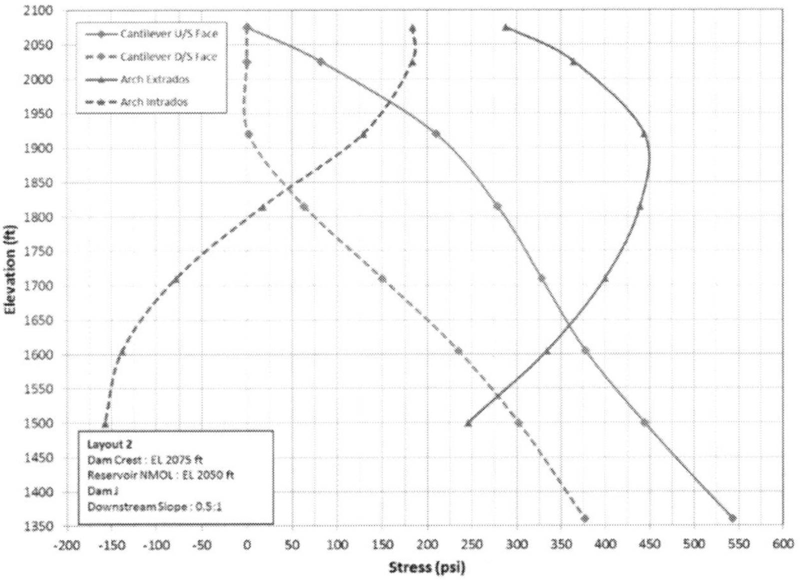

Figure 3. Layout 2 stresses at the Crown Cantilever (Static)

Finite Element Static Analysis Comparison

Finite Element Analyses using normal static loading were performed using ANSYS and were compared with the results of the ADSAS analysis. Some deviations in the stress distribution between the methodologies were observed that were attributed to the way dead load is considered. The TLM assumes that the dam is built as separate vertical cantilever blocks resulting in the dead load being transferred vertically to the cantilevers with no "arch" action. Elastic FE analysis considers the dam weight as a single block and arch stresses are developed within the dam body due to the dam weight. The two-step odd-even cantilever method, described in FERC Engineering Guidelines, Chapter 11 – Arch Dams, was used to analyze the dam for dead loading conditions. The dead load stresses determined using the odd-even cantilever method are in agreement with the ADSAS analysis.

The preliminary FE analysis of Layout 2 verified the static results of the ADSAS analysis while recording slightly higher than anticipated stresses during the dynamic loading condition. Figure and Figure 4 show the comparison in static stresses calculated using the ADSAS and FE programs and show that the FE cantilever stresses, using the odd-even cantilever method are in close agreement with the ADSAS results. The ADSAS program estimated a higher maximum arch stress at the extrados compared with the both FE approaches.

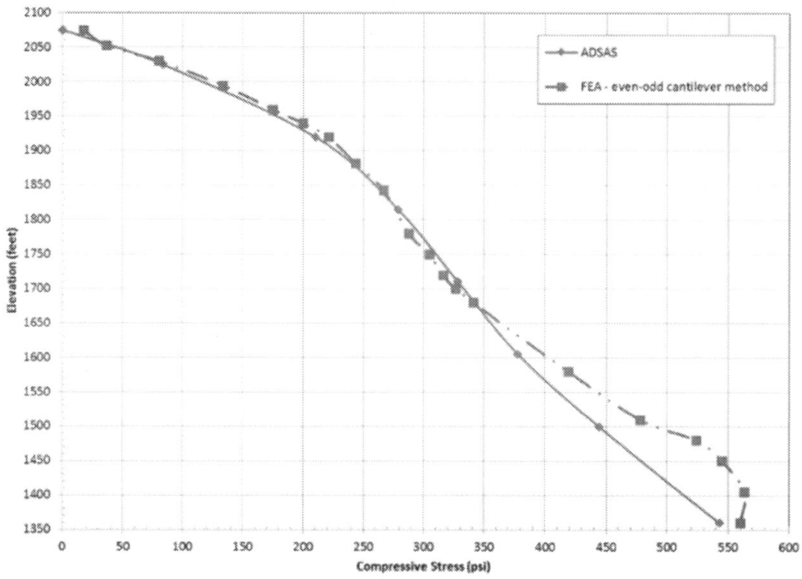

Figure 4. Layout 2 Upstream Face Crown Cantilever Stresses (Static) Comparison

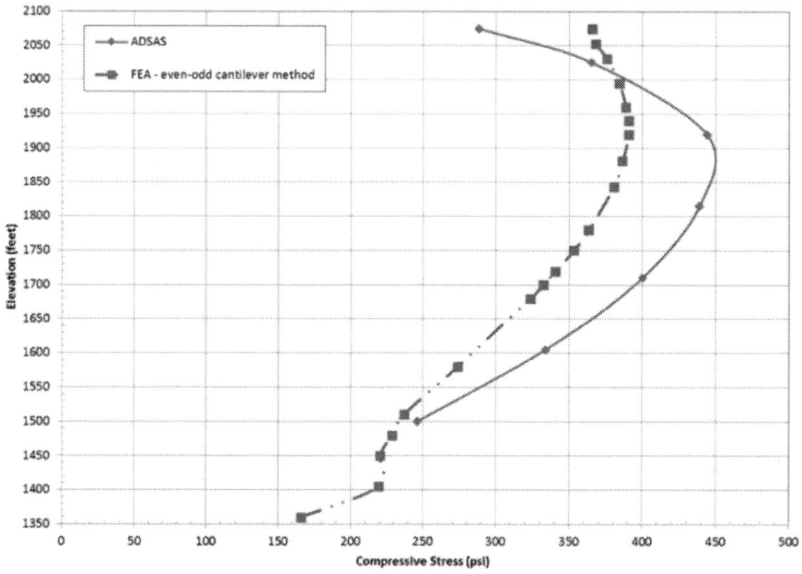

Figure 4. Layout 2 Extrados Arch Stresses (Static) Comparison at the Crown Cantilever

Following the presentation of the preliminary results of Layout 2 to the regulator and to a Board of Consultants convened for the project, it was decided to analyze a configuration which increased the gravity contribution to the stability of the structure.

Layout 3 - Single Curvature Gravity Dam

A single curvature gravity dam was developed to increase the dam's reliance on gravity for stability. A radius of 3,500 ft. was adopted so that the single curvature structure was positioned symmetrically in the valley and so that it performed predominantly as a gravity structure, with limited benefit realized by load transfer to the abutments. The maximum cross section of the dam was analyzed in 2-D using the Computer Analysis of Dams (CADAM) software to determine the section properties necessary to satisfy FERC stability criteria. CADAM demonstrated that a vertical upstream face and a downstream face sloping at 0.85H:1V was appropriate to satisfy the criteria and was then analyzed using ADSAS and FE analysis. The estimated RCC volume for Layout 3 was 6.38 Mcy.

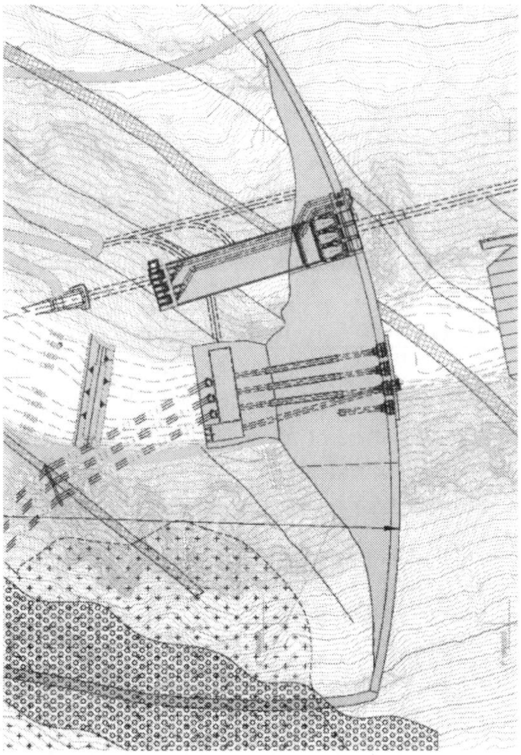

Figure 5. Layout 3 Configuration

The ADSAS analysis showed that the vertical cantilever stresses are compressive at the upstream and downstream faces for the entire height of the dam. Figure 6 shows the static vertical stresses at the upstream and downstream faces of the Crown Cantilever and horizontal stresses along the extrados and intrados of the arches intersecting the Crown Cantilever as calculated by ADSAS. The maximum vertical compressive stress is less than 37 percent of the allowable compressive stress.

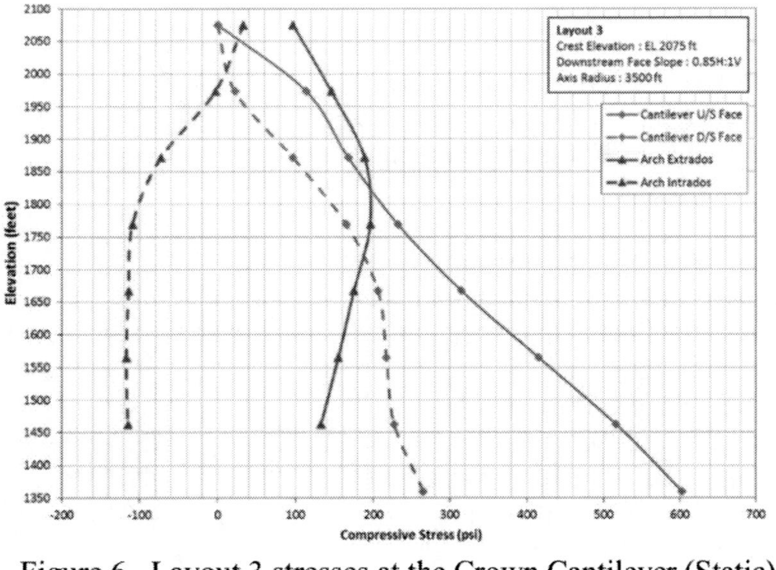

Figure 6. Layout 3 stresses at the Crown Cantilever (Static)

The analysis also showed that there is some transfer of load horizontally to the abutments. This demonstrates that the curved axis reduces the loads carried by cantilevers, and the horizontal "wedge action" resulting from the curvature of the dam increases the sliding stability of the dam.

One of the important attributes of Layout 3 was the 20% increase in the RCC volume, compared to Layout 2. This would have resulted in an unacceptable one year increase in construction time. The design team therefore revisited the geometry with the goal of reducing the concrete volume while not compromising the stability of the dam.

Layout 4 - Single Curvature Gravity Dam with Straight Axis Abutment Thrust Blocks

While satisfying all of the stability criteria, Layout 3 was acknowledged to be a very conservative configuration. It was recognized that if the total RCC volume could be reduced to approximately 5.5 Mcy or less, a full season of construction could be eliminated and substantive reductions in project cost achieved.

Using the original comparison of volumes for the high dam calculated by ADSAS, a compromise among Dams A, D, and E appeared promising, so a layout was prepared for a single radius for the central part of the dam of 2,600 ft., together with straight gravity sections on either end as shown in Figure 7. The dam axis changes to a straight line at a tangent point and continues to the abutments.

The curved section has a downstream face slope of 0.7H:1V while the straight gravity has a 0.85H:1V slope on the downstream face. All portions have a sloping upstream face (0.1H:1V) below El. 1770 ft. transitioning to a vertical face above. The change in downstream face slope reflects the reliance on gravity at the abutments. The estimated RCC volume was 5.47 Mcy, a 16% reduction compared with Layout 3.

Similar to Layout 3 the ADSAS analysis showed that the cantilever stresses are compressive at the upstream and downstream faces for the entire height of the dam. The cantilever stresses within the dam are more evenly distributed than Layout 3.

The static vertical stresses at the upstream and downstream face of the Crown Cantilever and horizontal stresses along the extrados and intrados of the arches intersecting the Crown Cantilever as calculated by ADSAS are shown in Figure 8. The revised configuration results in slightly greater transfer of loading to the arches compared with Layout 3. The stresses within the arches at the extrados are compressive for the full height of the

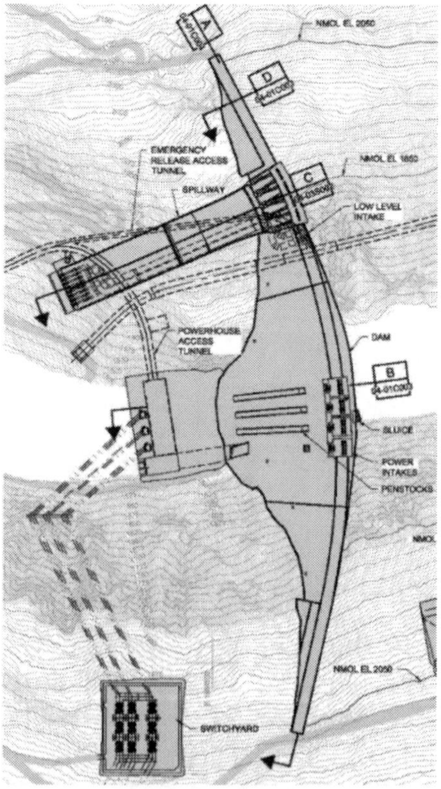

Figure 7. Layout 4 Configuration

crown cantilever while the along the intrados the stresses transition from compressive to tensile about 160 ft. below the crest. The analysis shows that the upper section of the dam, where the arch effect is most noticeable, has a larger compressive component than Layout 3. The stresses are low, relative to the allowable stresses but demonstrate that the dam shape is more efficient in accommodating the applied loads. The analysis confirmed that the dam cross-section was stable under static loading conditions.

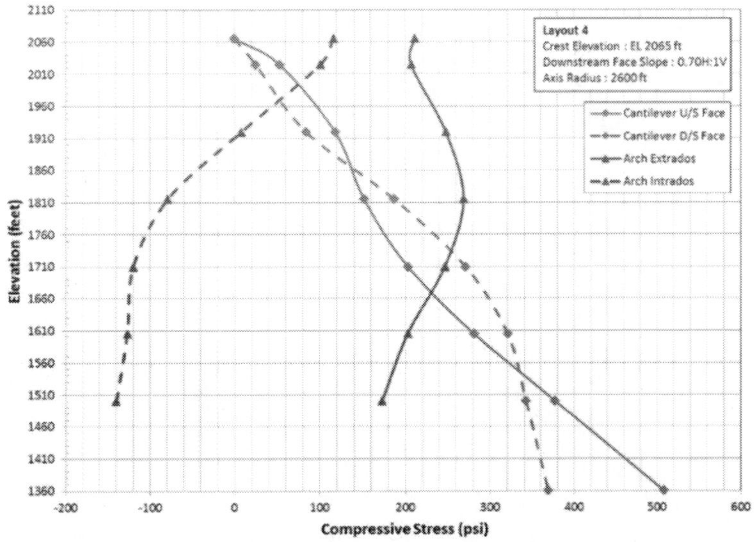

Figure 8. Layout 4 stresses at the Crown Cantilever (Static)

<u>Layout 4 (modified) - Single Curvature Gravity Dam with Straight Axis Thrust Blocks</u>

For final feasibility analysis, the model of the Layout 4 configuration was modified to increase the width of the dam crest to 45 ft., and instead of a straight downstream face, the downstream face of the curved section of the dam was modified to a single radius curve. The revised geometry reduced the estimated RCC volume by 5% to 5.21 Mcy, which is only 2% greater than the volume estimated in Layout 1. The dam geometry shown in plan and cross-section for the modified Layout 4 is shown in Figure 9 and Figure 10, respectively.

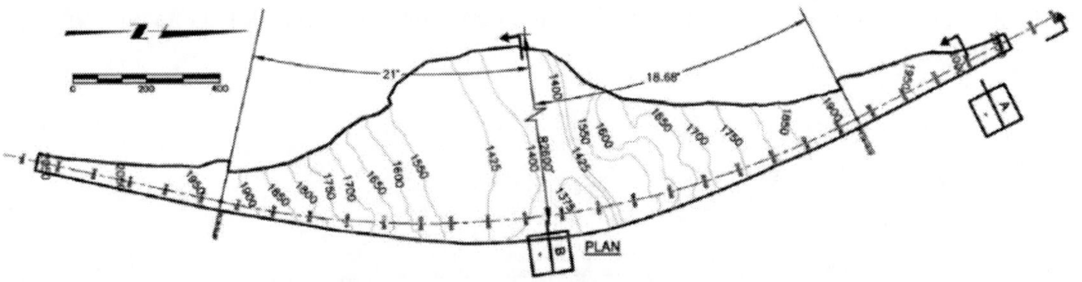

Figure 9. Layout 4 (modified) plan layout

Layout 4 (modified) was the subject of a robust dynamic analysis using seismic criteria derived from the Site Specific Seismic Hazard Analysis which had been performed for the project site. After the shape and cross section were updated, the final FEM analysis of Layout 4 (modified) was performed, including fluid structure interaction, and introducing mass (and damping) into the dam foundations using LSDYNA. The introduction of these aspects had the expected result of lowering tensile from those predicted by the dynamic ANSYS analysis.

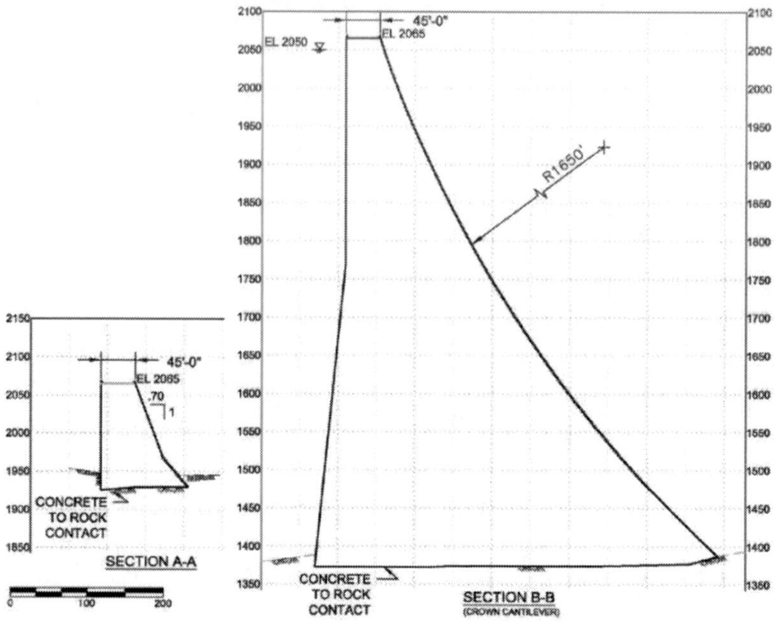

Figure 10. Layout 4 (modified) Crown Cantilever cross section

This final analysis confirmed the feasibility of the layout for the Watana dam, although some optimizing remains to be performed during detailed design. Among the aspects to be addressed during the next stage of design are the sequence of construction over each construction season and the effect on the thermal performance of the dam, the ease of construction of the geometry and other relevant matters.

The models created for the analysis and design facilitated the creation of a 3D CAD rendering which is shown in Figure 11. The excavated area upstream of the dam is the proposed quarry location which will be submerged when the reservoir is at NMOL.

Figure 11. Artist Rendering of Layout 4 (modified)

CONCLUSION

Although the ADSAS software was developed in an earlier era of computer technology its use on the Susitna Watana feasibility study demonstrated that it remains a powerful tool for quickly performing initial assessments of curved concrete dam configurations; particularly useful where RCC curved gravity dams of thicker section are being considered.

Use of the trial load method, enabled the initial analysis of 98 alternative configurations of the curved gravity RCC Watana Dam within a two week period. The ability to quickly analyze numerous alternatives enabled a viable dam configuration to be selected early in the design process.

In addition, the information derived from this initial analysis, and from subsequent trial load analyses of proposed configuration iterations was used to confidently refine the geometry of each alternative design layout before applying the FE Method. Much of the investigation of the geometry was therefore completed economically before a candidate configuration was tested using the broader array of capabilities and power (but with higher cost and longer time) of the Finite Element Method of analysis. Inefficient use of time and money on iterative but complex FEM was averted.

ACKNOWLEDGEMENTS

The authors are grateful to Alaska Energy Authority for granting approval for the preparation of this paper.

REFERENCES

USACE User's Guide: Arch Dam Stress Analysis System (ADSAS), August 1997

USBR, Design of Arch Dams, 1977

Alaska Energy Authority Susitna-Watana Hydroelectric Project, Pre-Application Document, 2011

Alaska Energy Authority Susitna-Watana Hydroelectric Project, Engineering Feasibility Report, 2015

FERC, Engineering Guidelines for Evaluation of Hydropower Projects, Chapter 11 – Arch Dams, October 1999.

F. A. Noetzli, discussion of William Cain's paper, "The Circular Arch Under Normal Loads," Transactions, American Society of Civil Engineers, vol. LXXXV (1922), p. 233.

Lars R. Jorgensen, "The Constant·Angle Arch Dam," Transactions, American Society of Civil Engineers, vol. 78 (1915), p. 1322.

F. A. Noetzli, "Gravity and Arch Action in Curved Dams," Transactions, American Society of Civil Engineers, vol. LXXXIV (1921), p. 1.

Eric B. Kollgaard and Wallace L. Chadwick, "Development of Dam Engineering in the United States," 16th ICOLD Congress, San Francisco, (1988).

C. H. Howell and A. C. Jaquith, "Analysis of Arch Dams by Trial-Load Method," Transactions, American Society of Civil Engineers, vol. 93. No.1712 (1929).

"Trial· Load Method of Analyzing Arch Dams," Bulletin no. 1, vol. V, Technical Investigations. Boulder Canyon Final Reports, Bureau of Reclamation, (1938).

S.Timoshenko and J.N. Goodier, "Theory of Elasticity," McGraw-Hill Book Company, Inc., (1951).

LINEAR AND NONLINEAR SEISMIC ANALYSIS FOR THE POST TENSIONING SYSTEM DESIGN OF THE WANAPUM DAM SPILLWAY MONOLITHS

Juan E. Quiroz, Ph.D., P.E. [1]
Bashar S. Sudah, P.E. [2]
Bryan J. Brumley, P.E. [3]
Richard E. Barrie, P.E. [4]
David P. Thompson, P.E. [5]

ABSTRACT

This paper presents analyses and evaluation methodologies used to assess the seismic stability of the Wanapum Dam spillway structure, located on the Columbia River south of Vantage, Washington. Early in 2014, a horizontal crack was discovered in one of the 120 foot tall spillway structure monoliths that extended the entire 65-foot monolith width. Investigations determined that the crack propagated below the pier reinforcement to the downstream side of the monolith. Repairs using post-tensioned anchors were completed recently. This paper presents the evaluation of the seismic performance of the repaired monolith under the Maximum Design Earthquake (MDE). To assess the state of the structure prior to the application of earthquake loads static stress analyses and transient thermal analyses were performed to identify potential failure mechanisms due to the cyclical thermal loading and the development of hydrostatic pressure within the crack. The seismic analysis was then performed for the MDE with a magnitude of 7.1 and a peak ground acceleration (PGA) of 0.42g.

Seismic analysis of the Wanapum cracked spillway monolith, indicates the importance of performing non-linear analyses capable of capturing dominant mechanisms that can be used effectively to assess the stability of the structure and to evaluate potential failure modes.

INTRODUCTION

Wanapum Dam is owned and operated by Public Utility District No. 2 of Grant County, Washington (the District). Wanapum Dam is located in south-central Washington State on the main stem of the Columbia River that separates Grant County from Kittitas County. The dam is situated at river mile 415, about midway between Seattle and Spokane. Wanapum Dam is part of the overall Priest Rapids Hydroelectric Project that includes Priest Rapids Dam, located about 15 miles downstream of Wanapum Dam.

---

[1] MWH Global, 350 N. Orleans St., Suite 1301, Chicago, IL 60654, Juan.E.Quiroz@mwhglobal.com
[2] Brookfield Renewable Energy Group, Los Angeles, CA 90017, Bashar.Sudah@brookfieldrenewable.com
[3] MWH Global, 2353 130th Ave NE, Suite 200, Bellevue, WA 98005, Bryan.J.Brumley@mwhglobal.com
[4] MWH Global, 350 N. Orleans St., Suite 1301, Chicago, IL 60654, Richard.E.Barrie@mwhglobal.com
[5] MWH Global, 2353 130th Ave NE, Suite 200, Bellevue, WA 98005, David.P.Thompson@mwhglobal.com

During a routine survey of the spillway structure monoliths, large displacement was observed in Monolith No.4 that significantly exceeded historic values. Follow up evaluations conducted by a dive team revealed that a crack had formed on the upstream face of Monolith No.4 at approximate El. 485 feet. As a result, the reservoir was lowered to reduce the load acting on the monolith and the uplift pressure within the crack. Available construction records indicated the location of the observed crack coincides with the construction joint elevation corresponding to the termination point of the vertical pier reinforcement. Concrete core explorations were carried out to evaluate the condition of the monolith and evaluate the presence of the observed opening at El. 485 and the extent of the crack into the ogee.

Remedial designs for the monolith were performed in accordance with the requirements of the Federal Energy Regulatory Commission (FERC) *Engineering Guidelines for Evaluation of Hydropower Projects*, including evaluations of normal, unusual and extreme loading conditions, for which all acceptance criteria have been met. The repairs using post-tensioned anchors were completed in 2015.

The analyses and evaluation methodologies used to assess seismic stability of the Wanapum Dam spillway structure are presented in this paper. These included static analyses, transient thermal analyses, and the evaluation of the seismic performance of the repaired monolith under the Maximum Design Earthquake (MDE). The results of the thermal model indicate that the monolith experiences annual cyclic thermal stresses that vary from upstream face compressive stresses during the warm weather season to tensile stresses during the cold weather season. The stresses estimated from this analysis were combined with the other loads to determine design stresses to remediate the structure.

Based on structural analyses considering static plus thermal loads, an anchoring scheme using post- tension rock and concrete anchors was developed to satisfy stability criteria. The seismic performance was then evaluated using dynamic, finite element, computer modeling techniques. The seismic analysis was then performed for the MDE with a magnitude of 7.1 and a peak ground acceleration (PGA) of 0.42g. The seismic evaluation considered three spectrally matched, deterministic seismic events; two crustal events and one interface (subduction zone) event. In addition, the design considered a fourth time history that was developed by scaling to the crustal response spectrum. Based on the total structure downstream movement as a result of initial seismic modeling, additional anchors were required to limit stream direction deformations and the analysis was repeated. Post-earthquake stability requirements in accordance with FERC guidelines were also satisfied with the anchors installed.

WANAPUM SPILLWAY

The project is owned and operated by the District, who is the original developer of the Project. The dam was completed in 1963 and consists of left and right earth embankments, left and right bank fish ladders (concrete gravity sections), concrete gravity spillway with 12 radial gates, future units intake section (concrete gravity section) and a 10-unit reinforced concrete powerhouse with a reinforced concrete erection bay. Overall, the dam is 8,637 feet long and has a maximum height of 186.5 feet.

The Wanapum spillway is a reinforced concrete structure with 13 monoliths. A concrete stilling basin slab extends beyond the spillway to provide energy dissipation and hydraulic control. Each radial gate is 50 feet wide by 67 feet high. Each of the monoliths (2 to 12) are 65 ft. wide, with a center pier with one-half of a gate bay on each side, and is separated from their adjacent monoliths by unkeyed and ungrouted contraction joints. Monoliths Nos. 1 and 13 comprise the right and left ends of the spillway, respectively, and contain only one-half of a gate bay. A typical monolith section is presented on Figure 1.

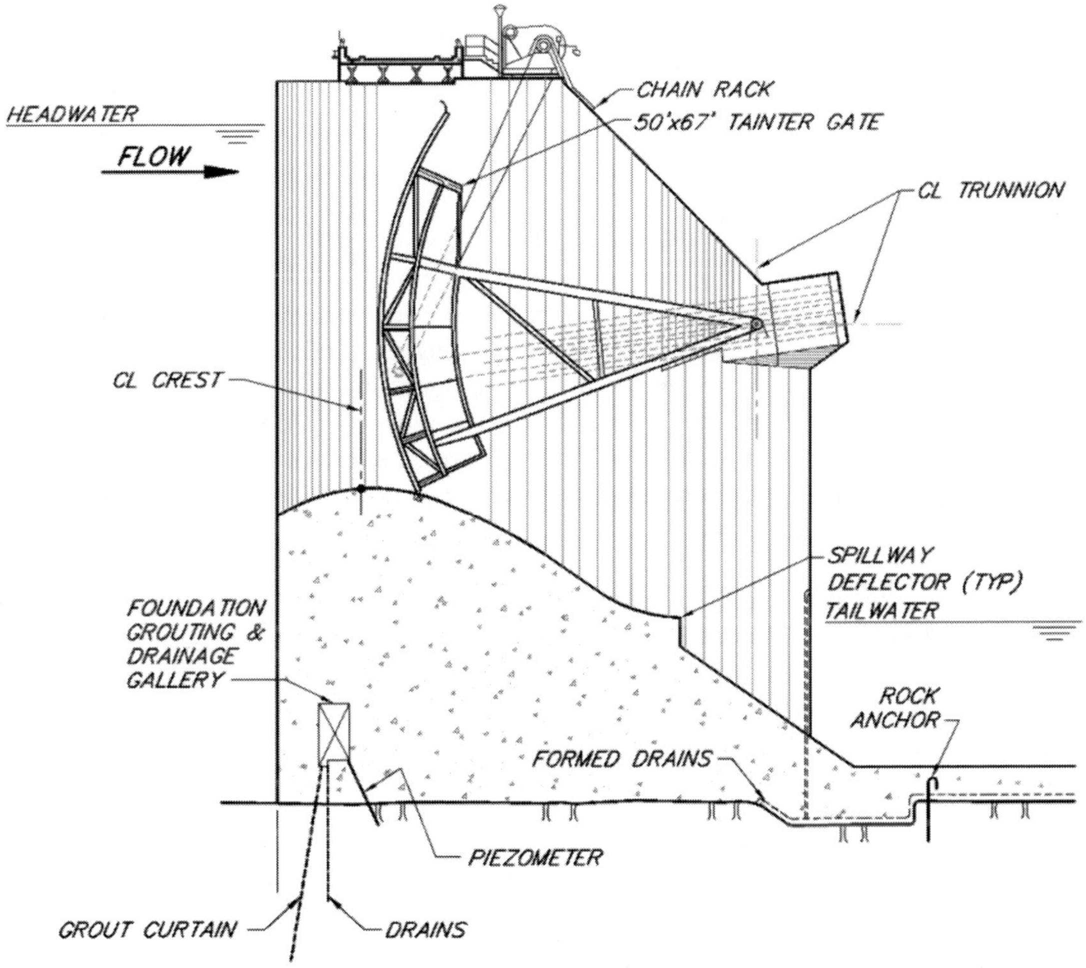

Figure 1. Typical Monolith – Cross-sectional View

Material Properties

The material properties used for rigid body and finite element analyses are listed on Table 1. No field investigations or verifications were performed as part of these analyses, except for rock parameters.

Table 1. Material Properties

Spillway Concrete	Unit weight	150 pcf
	Poisson's Ratio, n	0.2 [3]
	Static Young's Modulus, E_c	57,000 $(f'_c)^{1/2}$
	Dynamic Young's Modulus, E_d	1.0 E_c
Rock Foundation	Poisson's ratio of rock mass (massive basalt),	0.22
	Static Young's Modulus of rock mass, E_c	3,000,000 psi
	Dynamic Young's Modulus of rock mass, E_d	1.0 E_c
Rock/Concrete Interfaces	Friction Angle	56.4 deg
	Cohesion (when considered)	25.6 psi
Reinforcement	Existing rebar yield stress, f_y	40,000 psi

APPROACH FOR ANALYSIS

The spillway remediation and post-tensioned anchor design was evaluated using three-dimensional finite element analysis (FEA). Load cases included dead load, silt load, thermal load, uplift pressure within existing crack and foundation interface and external water load acting on the structure. Seismic analyses incorporated sliding planes along cracks, lift joints and along foundation plane. Figure 2 shows the crack profile within Monolith No. 4. Post-earthquake case was performed considering damage that may have occurred during the earthquake.

The key requirements for stability are that force (sliding) and moment (overturning) equilibrium shall be maintained with adequate factors of safety along the crack surface and at the foundation contact. The limits of concrete strength, bearing capacity of foundation and concrete/foundation interface strength are evaluated with appropriate factors of safety. Other considerations such as the effects of thermal stress, loss of tensile strength at lift joints and changes in uplift pressure were considered in evaluating potential failure mechanisms.

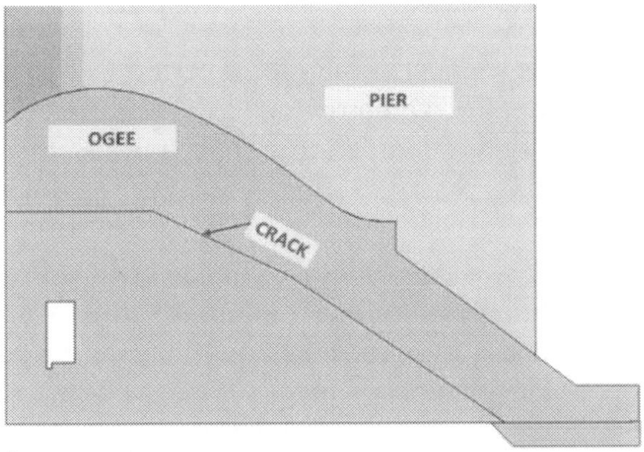

Figure 2. Monolith No. 4 Crack Location

The FEA model was used to evaluate seasonal thermal behavior and its effects on lift joints and internal water loads. The results of the thermal model indicate that the monolith experiences annual cyclic thermal stresses, which vary from upstream face compressive stresses during the warm weather season to tensile stresses during the cold weather. It has been concluded from thermal FEA that during the winter, the concrete between the crack and the ogee surface, at the contraction joint, tends to separate locally as a result of the temperature differential between the ogee surface and interior crack. Therefore, winter temperature effects were considered for the remedial design.

For Normal Operating Conditions:

First, material properties were calibrated with respect to seasonal deflection measurements. Then, a static analysis was performed for each load case including winter temperatures. The anchors were adjusted to develop a zone of compression from contraction joint to contraction joint at El. 485, upstream of existing inclined crack. With the anchor layout, the sliding stability requirements along existing crack were confirmed. These sliding safety factors were calculated by using finite element analysis and reducing friction angle incrementally until sliding was initiated. Concrete stresses and cross-stream deformations were verified to be within acceptable limits for all load cases.

For MDE and Post-Earthquake Conditions:

Dynamic analysis were performed considering full headwater pressure in the open crack upstream of the compression zone as determined from the normal headwater operating condition in winter. The post-earthquake analysis considered damage to the structure and was performed with winter temperature loads.

LOAD CASES

Water and Uplift Pressures

Hydrostatic and silt pressures were included along with the resulting gate loads. For seismic load case, the uplift condition along the foundation is to remain unchanged during the earthquake. For the post-earthquake load case, loss of foundation drain efficiency was considered.

Thermal Loads

The transient thermal analysis was performed for a period of five years to investigate the temperature distribution within the structure due to the seasonal temperature and water temperature variations. The starting temperature for the transient analysis in both the concrete and rock was set to the average annual temperature at the site.

Minimum winter temperature was determined to be the critical thermal case for Monolith 4. The temperature distributions within the structure resulting from the seasonal variation at the site were applied to a structural model as body loads to be combined with dead, hydrostatic, uplift and post-tension anchor loads.

As a result of winter thermal analysis, the upstream end of the horizontal El. 485 joint that was not in compression was pressurized with headwater pressure. A trapezoidal uplift distribution was assumed from headwater at the tip of the thermal opening to tailwater at the downstream side of the ogee.

Seismic Motions

The design considered three spectrally matched, deterministic seismic events; two crustal events and one interface (subduction zone) event. In addition, the design considered a fourth time history that was developed by scaling to the crustal response spectrum. The earthquakes chosen for the analysis are the Chi-Chi, Cape Mendocino, the Cascadia interface, and the Iwate motions. Each earthquake consists of three components of motion (stream, cross-stream and vertical). All three components are applied to the structure concurrently for analysis. The seismic analysis was then performed for the MDE with a magnitude of 7.1 and a peak ground acceleration (PGA) of 0.42g.

Based on the total structure downstream movement as a result of initial seismic modeling, additional anchors were required to be added to limit stream direction deformations of the structure and the model was re-run.

ACCEPTANCE CRITERIA

Since Wanapum Dam is classified as a high hazard potential dam, the acceptance criteria was based on FERC guidelines (FERC, 2002). The spillway's behavior during the earthquake motion and stability under post-earthquake static loading considering damage from the earthquake was evaluated. The seismic behavior of the monolith with installed post-tension anchors was evaluated to verify that no additional failure modes were present. The response of the repaired monolith during the earthquake motion should approach monolithic behavior with the installed anchors.

During the earthquake, seismic induced stresses could exceed the strength of materials. If the dynamic analysis indicates that tensile cracking, shear displacements, or rocking are likely to occur, post-earthquake static allowable capacities, such as friction angle are downgraded accordingly. If dynamic analysis indicates that shear failure will occur, then residual shear strengths are used. No cohesion was assumed at the concrete/rock interface and along any cracked concrete lift joint.

FEA MODEL

Mesh

The model incorporates 68,800 nodal points and 64,300 elements. Figure 3 presents an overall view of the mesh. The concrete portion (17,700 elements) includes the pier, ogee and stilling basin slab. There are 4,700 elements that make up the contact surfaces and anchors. The monolith is supported on a foundation block (rock) consisting of 41,800 rock elements. The foundation extends away from the spillway about one and a half times the height of the structure. The foundation rock is included to provide appropriate support stiffness to the structure. A detailed evaluation of the foundation itself is not included in this study.

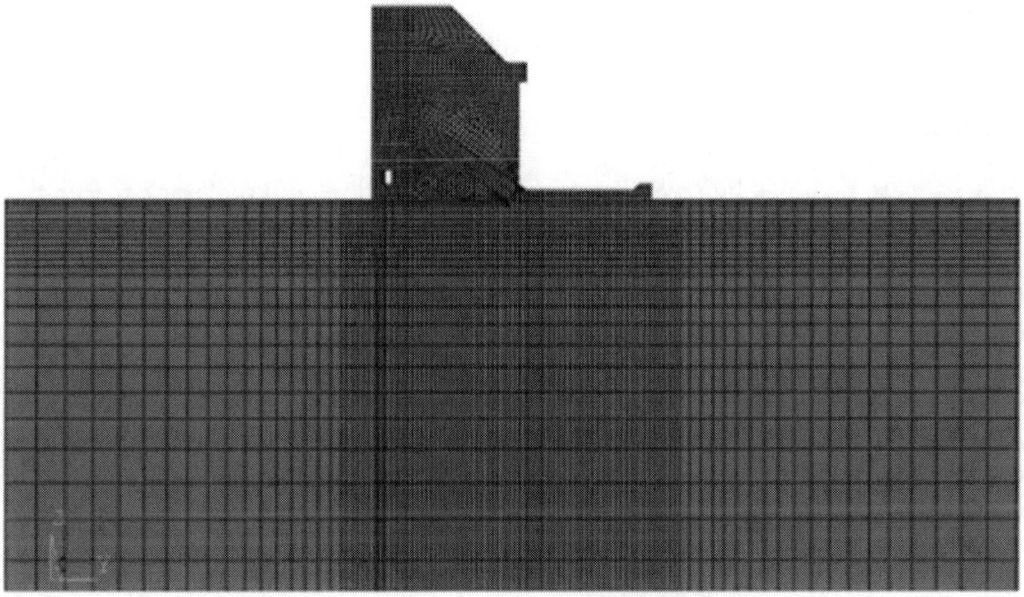

Figure 3. Finite Element Model - Concrete

The spillway and foundation is modeled with eight node brick elements. The material modeling is linear elastic. Nonlinearity in the model is provided by discrete joint or interfaces. Main gate features are modeled as presented in Figure 4.

Figure 4. Finite Element Model – Gate Features

Interfaces

Interfaces were included between the ogee and foundation, and the crack within the model. The properties of these interfaces are presented in Table 1. The crack orientation in the model incorporates the latest field data and it is modeled to converge with the horizontal interface at the foundation. A concrete/foundation interface is modeled at El. 450 starting at the upstream face of the ogee and extending through the toe of spillway and continuing under the stilling basin slab as shown in Figure 5. The interfaces are modeled such that the ogee above the crack has the capability to slide with respect to the concrete below the crack and that the concrete structure has the capability to slide along the foundation.

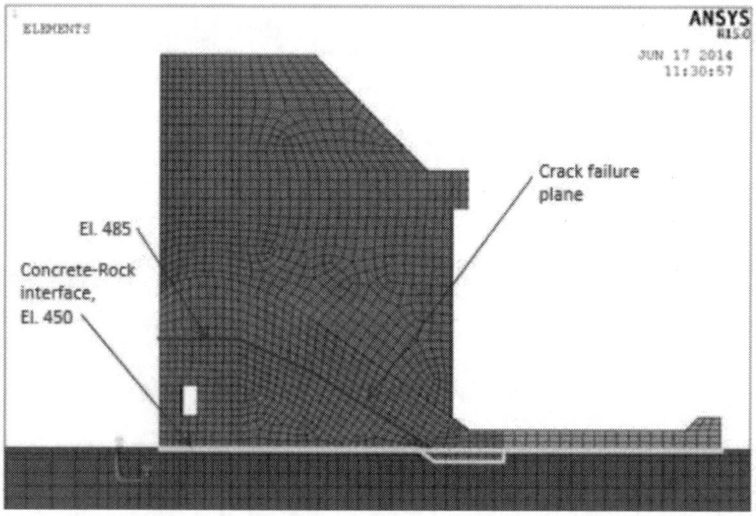

Figure 5. Finite Element Model – Interfaces

Boundary conditions

The base of the foundation block is restrained against any movement and the sides of the foundation block are placed on "rollers" which prevent movement perpendicular to the planes. The ogee sides in the model are located at the contraction joints and are not restrained by adjacent monoliths.

Post-tensioned Anchors

The post-tensioned anchors were modeled using tension-only truss elements. The cross sectional area and elastic material properties for each of the anchors and tendons were included in the model. The coefficient of thermal expansion was also included with a temperature load to initiate the tension in the truss elements that corresponds to 60% of the ultimate tensile strength.

Thermal Analysis & Calibration

The seasonal temperature distributions from the transient thermal analysis for the average annual ambient and water temperatures were used to verify that the model behavior is within an acceptable range, as compared to the deflection measurements from the available survey data for monuments located on top of the piers, and to validate whether the material properties used in the structural design analyses were adequate.

FEA vs Rigid Body Analyses

To validate the FEA model, safety factors against sliding for static loads obtained from the RBA were compared to those from the FEA model. Sliding safety factors for FEA approach were computed by reducing the joint friction incrementally until sliding was initiated. The comparison of sliding safety factor obtained from RBA and FEA revealed that the results were rather close to each other and, therefore, validated the finite element approach for sliding stability. Additional base pressure comparisons were performed and stability analysis for final design were performed using FEA because of the ability to include thermal effects as it provides a better representation of the pressure distribution at the base and lift joints.

ANCHORS/REPAIR WORK

Post-tension anchors for Monolith No. 4 were defined through an iterative design process. The model with the anchors are presented in Figure 6. Anchors were used to provide adequate safety factors per FERC guidelines (FERC, 2002) for concrete stresses including sliding evaluation. Three 61-strand tendon anchors at the upstream end of Pier No. 4, eight 3-inch diameter rock bar anchors at the upstream side of the ogee, four 2 ¼-inch diameter concrete anchors above the gallery, eight 3-inch diameter concrete anchors upstream of the gate near the contraction joints and fourteen 3-inch diameter concrete anchors installed perpendicular to the face of the downstream ogee were required to satisfy overturning and sliding stability requirements for operating load conditions. To satisfy sliding stability requirements and to limit downstream movement during the design seismic events, twenty (20) 3-inch diameter downstream ogee anchors were provided. The addition of these anchors limit total downstream movement to less than 1.5 inches at the base of the structure and to approximately 0.5 inch of differential movement from above and below the crack. Post-earthquake stability requirements in accordance with FERC guidelines have been satisfied with the anchors installed.

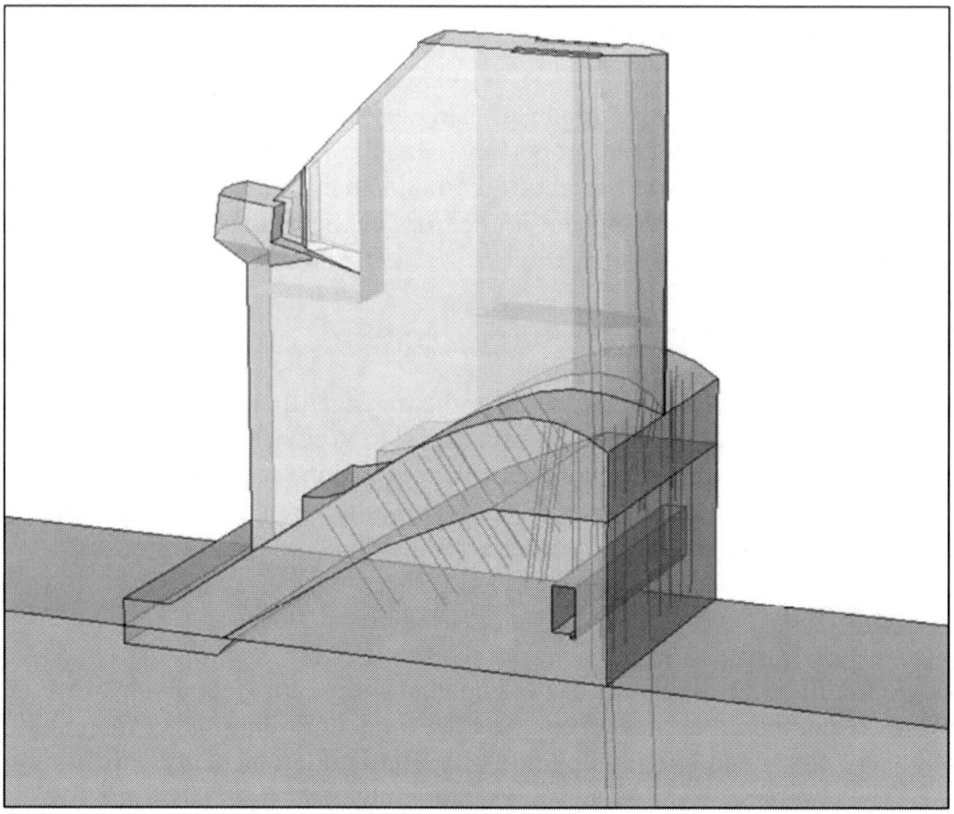

Figure 6. Finite Element Model – Post-Tension Anchors

SEISMIC ANALYSIS RESULTS

Seismic Displacements

The seismic displacements from the four earthquakes are summarized in Table 2 below. Cross-stream values are instantaneous pier displacements. Stream direction values are residual displacements due to sliding. The maximum displacement in the cross stream direction occurs for Earthquake 4 (0.95 inches), and the maximum sliding distance occurs for Earthquake 1 (1.91 inches). The structure's response to Earthquake 3 is very minimal and considered insignificant.

Table 2. Earthquake Displacements

Direction	Earthquake 1 (Chi Chi)	Earthquake 2 (Cape Mendocino)	Earthquake 3 (Interface)	Earthquake 4 (Scaled Iwate)
	Maximum Displacement (in)	Maximum Displacement (in)	Maximum Displacement (in)	Maximum Displacement (in)
Cross-Stream	0.93	0.94	0.124	0.95
Stream	1.91	1.37	0.03	1.63

Nonlinear seismic analyses indicate sliding along the crack, and along the base of the structure in the downstream direction. The stream direction deformations represent sliding deformation. Figure 7 depicts the deformed shape at the end of the earthquake.

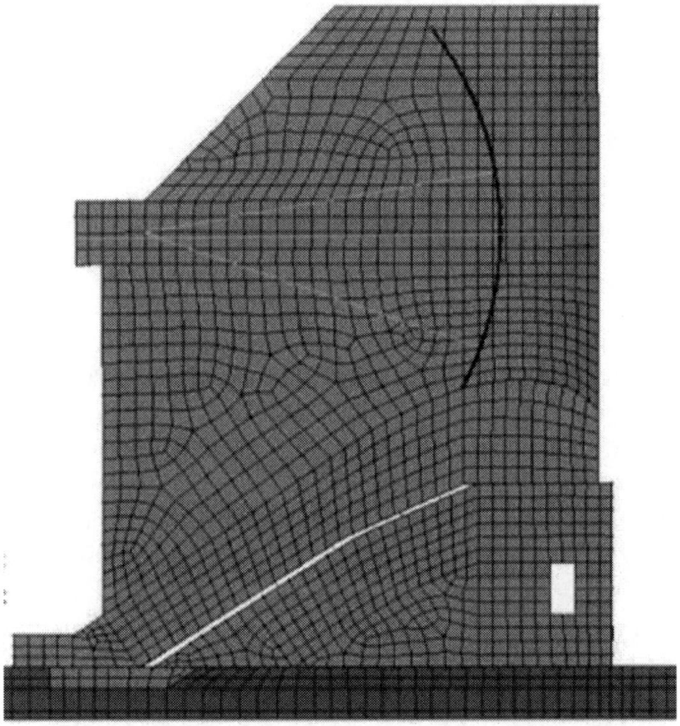

Figure 7. Deformed Shape – End of Earthquakes

Sliding in the seismic analysis occurs at the crack interface in addition to the structure base (concrete-rock interface). Figure 8 shows the relative downstream displacement at the upstream end of the crack at El. 485, and the upstream end of the concrete-rock interface at El. 450 for the Chi Chi Earthquake.

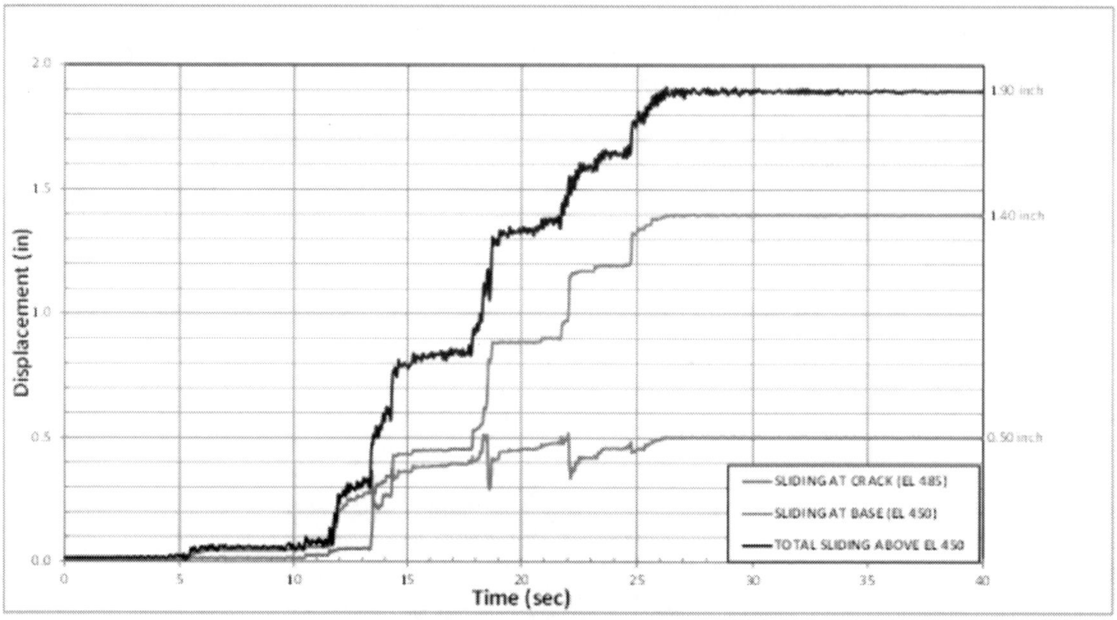

Chi Chi Earthquake

Figure 8. Relative Displacements at Crack and Concrete-Rock Interface

Anchor Forces

Maximum anchors forces for the two crustal earthquakes are presented in Table 3. The anchor forces for the interface event are much smaller. Anchors on the pier and on the ogee generally remain below their initial jacking force (0.8 x fpu) during the earthquake. The upstream anchors have seismic loads which occur for a short duration during the earthquakes. Therefore, the anchors on pier are considered to be sufficient to maintain the integrity of the ogee for the earthquakes evaluated. Similarly, the downstream ogee anchors also remain below the initial jacking force and are also considered to be sufficient to resist the sliding forces

Table 3. Maximum Anchor Loads during Earthquake

Anchor Group	0.8 fpu (kips)	EQ-1 (Chi Chi)	EQ-2 (Cape Mendocino)	EQ-4 (Scaled Iwate)
Pier Tendons	2,853	2,175	2,174	2,162
US Ogee Anchors (3")	776	630	616	595
US Ogee Anchors (2 ¼ ")	491	418	408	387
DS Ogee Anchors (3")	776	736	776	812

Post-Earthquake Stability

During the earthquake sliding occurs at the foundation due to the dynamic motion. As a result, the friction angle at the rock-concrete interface was reduced to 45 degrees for post-earthquake analysis. Downstream sliding of the monolith above the crack increased the tension in the downstream ogee anchors during the seismic event. The downstream displacement has also relaxed the upstream ogee anchors. The forces in the anchors were adjusted accordingly for the post-earthquake analysis.

Downstream earthquake displacement of about 0.5 inches presented in Figure 8 resulted separation along the inclined portion of the crack. Friction is only resisted on the upstream and downstream horizontal crack planes. A shear plot at the crack interface is presented in Figure 9.

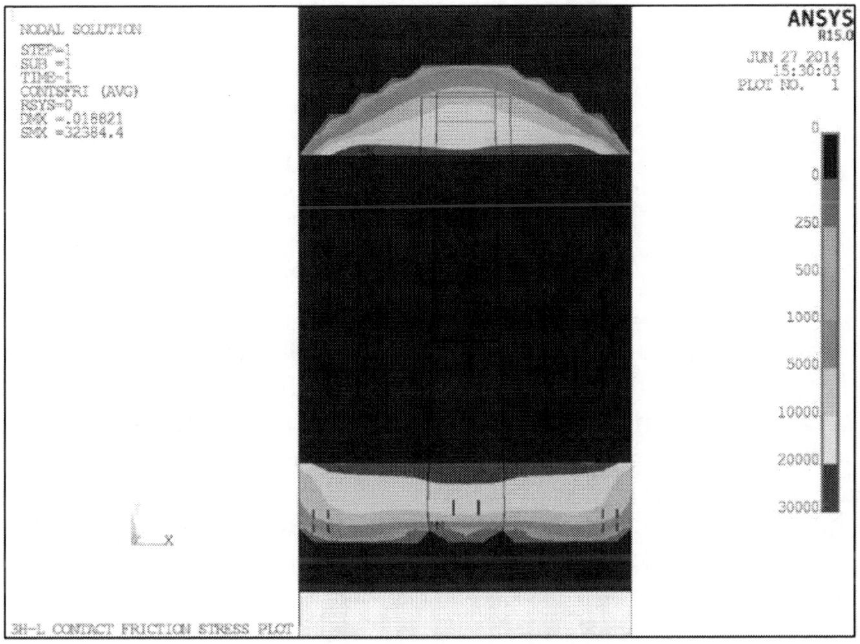

Figure 9. Post-Earthquake Friction Stress at Crack (psf)

Sliding stability requirements along existing crack were evaluated using FEA by reducing the friction angle incrementally until sliding was initiated. Safety factors are computed by comparing the design friction angle with the friction angle prior to sliding. The post-earthquake sliding evaluation is presented on Figure 10.

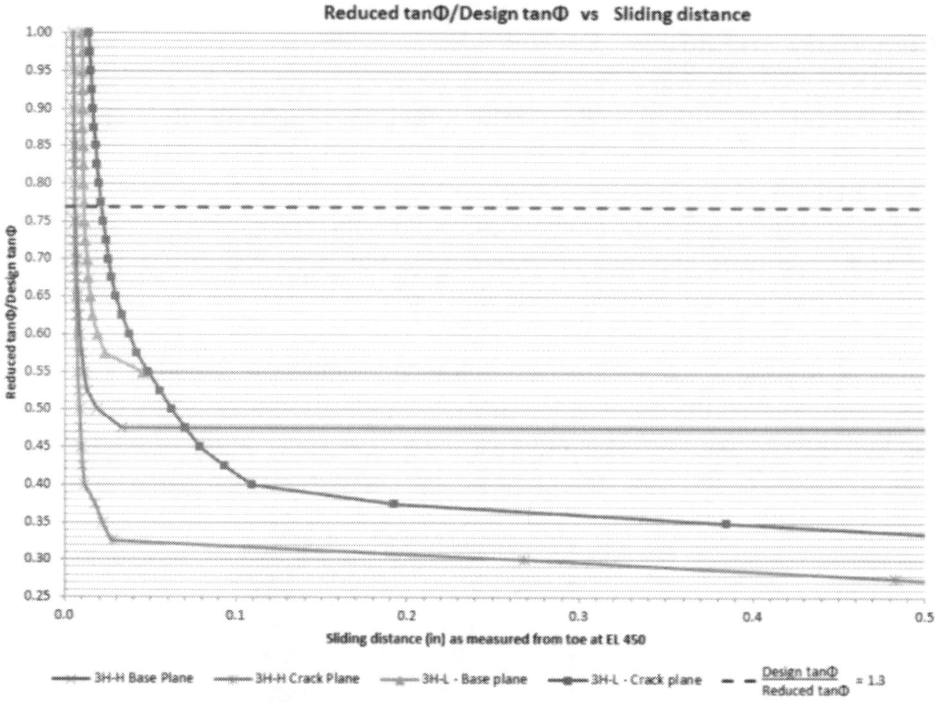

Figure 10. Post-EQ Sliding Evaluation

CONCLUSION

Several analyses were conducted to assess stability of the spillway monolith. The results indicated the importance of performing non-linear analysis to understand the global structural behavior including the interfaces at the base and within the ogee, and allowed the evaluation of dominant mechanisms. The downstream sliding at both interfaces was instrumental for the determination of the post-tensioning system not only for the static cases, but also for control of seismic deformations and for the post-earthquake stability. The stability of the structure can be assessed more rigorously as well as the potential failure modes.

REFERENCES

ACI 207.2R-07, Report on Thermal and Volume Change Effects on Cracking of Mass Concrete, September 2007.

ACI 318-11, Building Code Requirements for Structural Concrete, 2011.

ANSYS Mechanical, A General Purpose Finite Element Program, Version 15.0, 2012.

Bruce, The Stabilization of Concrete Dams by Post-Tensioned Rock Anchors: The State of American Practice, Ground Anchorages and Anchored Structures, Proceedings of the International Conference, Institution of Civil Engineers, 1997.

Douglas and Arthur, A Guide to the Use of Rock Reinforcement in Underground Excavations. Report 101, 1983.

FERC, Engineering Guidelines for Evaluation of Hydropower Projects, Chapter 3, Gravity Dams, October 2002.

Mills-Bria, B., Nuss, L., O'Connell, D., & Harris, D., State-of-Practice for the Nonlinear Analysis of Concrete Dams at the Bureau of Reclamation. Denver, CO. Bureau of Reclamation, 2006.

PTI, Acceptance Standards for Post-Tensioning Systems, 1998.

PTI, Recommendations for Prestressed Rock and Soil Anchors, Fourth Edition, 2004.

USACE, Rock Foundations, Engineering Manual 1110-1-2908, 1994.

USACE, Time History Dynamic Analysis of Concrete Hydraulic Structures, Engineering Manual 1110-2-6051, 2003.

USACE, Earthquake Design and Evaluation of Concrete Hydraulic Structures, Engineering Manual 1110-2-6053, 2007.

Zicko, Bruce and Kline, The Stabilization of Gilboa Dam, New York, Using High Capacity Rock Anchors: Addressing Service Performance Issues, International Conference on Ground Anchorages and Anchored Structures in Service, Institution of Civil Engineers, 2007.

STATIC AND DYNAMIC MASS CONCRETE MATERIAL PROPERTIES OF A CONCRETE GRAVITY DAM

Soheil Razavi Darbar, Ph.D., P.Eng.[1]
David Queen, P.Eng.[2]
Chris Hatton, P.Eng.[3]
Timothy Dolen, P.E.[4]
Katie Bartojay, P.E.[5]

ABSTRACT

A concrete testing program of large-diameter specimens was conducted to obtain material properties of existing mass concrete of the Ruskin Dam in British Columbia, for use in a 3D Finite Element Model (FEM) seismic analysis. Static compression, static tension and dynamic tension tests were performed on parent concrete, and specimens containing lift lines. The static compression and tension test results are compared with the historical United States Bureau of Reclamation (USBR) mass concrete properties database of concrete dams (ACIS) from 1905 to 1993. The material properties of mass concrete for the Ruskin Dam (completed in 1930) are consistent with mass concrete of dams constructed in the 1920s to early 1930s. From the static tests, the average tensile strengths of the parent concrete and the lift line specimens are 3.6% and 2.7% of the average tested concrete compressive strength, respectively. The dynamic tensile tests consisted of cyclic ramped amplitude and constant amplitude testing to failure for both parent concrete and lift line specimens. The cyclic ramped loading resulted in a 30% increase in the concrete tensile capacity. The failure strength in constant amplitude cyclic loading tests resulted in a 20% to 40% decrease in the concrete tensile capacity.

The intent of the testing program was to guide the selection of mass concrete material properties for the numerical analysis of an aging concrete dam, considering the concrete test results for Ruskin Dam, the historical mass concrete test results and design codes.

INTRODUCTION

The Ruskin Dam, completed in 1930, is a concrete gravity dam located in Mission, British Columbia, Canada (Figure 1). The dam has a crest length of 120 m and height of 58 m at its deepest point. It consists of six monoliths and one buttress wall and is topped

[1] Structural Team Lead, BC Hydro, 6911 Southpoint Drive, Burnaby, BC V3N 4X8, Canada, 604-528-7988, soheil.razavidarbar@bchydro.com
[2] Specialist Engineer, BC Hydro, 6911 Southpoint Drive, Burnaby, BC V3N 4X8, Canada, 604-528-2918, david.queen@bchydro.com
[3] Structural Engineer, BC Hydro, 6911 Southpoint Drive, Burnaby, BC V3N 4X8, Canada, 604-528-2019, Chris.Hatton@bchydro.com
[4] Dolen and Associates LLC, P.O. Box 447 Loveland, Colorado, 50539, tpdolen@yahoo.com, 970-292-8583
[5] Bureau of Reclamation, P.O. Box 25007, 86-68180, Denver, Colorado 80225-0007, 303-445-2374, kbartojay@usbr.gov

by a gated ogee-shaped spillway crest and bridges between the spillway piers. There are non-overflow sections on each side of the spillway.

Due to the downstream population at risk the dam is classified as an extreme consequence structure according to the Canadian Dam Safety Guidelines. The dam site has a high seismic hazard and a major seismic upgrade is underway. The first phase of the upgrade replaces the piers, spillway gates and spillway bridges.

Figure 1. Ruskin Dam and Powerhouse

A second phase of the facility upgrade is to investigate the overal seismic stability of the dam. To complete this work, a three-dimensional Finite Element Model (FEM) consisting of the dam, foundation and reservoir will be used to quantify the seismic demands and to investigate remedial measures, if required.

Previous concrete testing did not provide reliable concrete properties such as adequate representation of maximum aggregate size for the upper dam body, thus it was deemed appropriate to conduct large-diameter concrete specimen testing to obtain concrete properties in conformance with ASTM C39. A comprehensive concrete-testing program on large size concrete samples was developed of which the testing and results are presented in this paper. The ultimate value of the testing program will be evident in the analysis and design of potential remediation of the dam body.

BACKGROUND

Seismic Assessment of the Ruskin Dam

Previous seismic assessments of the Ruskin Dam identified stability concerns and proposed rehabilitation work to satisfy the high seismic demands caused by the Maximum Design Earthquake at the site. The current, more detailed, engineering work is intended to reduce uncertainty that may have been present in previous studies. To accomplish this, improved three-dimensional FEM modelling techniques, updated seismic hazard input, and concrete material properties based on testing will be incorporated.

<u>Previous Concrete Testing</u>

Original construction records indicate the concrete mix design included maximum aggregate of 152 mm with a compressive strength of 14 MPa at 28 days. Actual test results from the construction period are not available.

Concrete tests were performed in 1996 on samples taken from the dam (Li, 1996). Two - 178 mm ø and 14 - 153 mm ø horizontal cores were taken from inside the drainage gallery. These cores provided 22 specimens that were tested for compressive strength, shear strength, and modulus of elasticity, among other properties as shown in Table 1.

Table 1. 1996 Dam Body Concrete Test Results

Test	# of Specimens	Average Value
Compressive Strength	10	27.8 MPa
Cohesion	7	4.64 MPa
Static Modulus of Elasticity	5	27 GPa

Other concrete testing was conducted on samples recovered from cores drilled in the right abutment wall for the installation of post-tensioning anchors to be used during the current construction project. The available cores were 70 mm OD from the pilot hole and 263 mm OD, with the pilot hole removed, for the anchor hole. The handling and testing did not follow exacting procedures, but testing was conducted due to the availability of the material. Graphical results of the previous concrete testing is shown in Figure 6.

The test results may provide some indication of the concrete properties but are not considered appropriate for mass concrete with 152 mm maximum aggregate. For concrete with 152 mm maximum aggregate core diameters of 457 mm ø are recommended to obtain representative test results.

TESTING

<u>Sample Collection</u>

Phase 1 Seismic Upgrade construction activity at Ruskin Dam provided an opportunity to retrieve large concrete samples during demolition of the existing piers. Pier EP2 (Existing Pier #2), shown in Figure 2, was selected as the concrete was expected to be representative of the dam body. Original construction documentation did not indicate different mix designs and the aggregate gradation was similar to that seen in the mass concrete in the dam (Figure 3). A total of nine concrete blocks were selected, shown in Figure 2 outlined in yellow that included some blocks with lift joints (shown in blue).

Figure 2. EP2 of Ruskin Dam – Blocks Selected for Coring

Figure 3. Concrete Aggregate – Pier EP1 Aggregate

Coring of the blocks was done off-site to collect samples for static compression, static tension and dynamic tension tests, as shown in Table 2.

Table 2. Samples Tested

	Static Compression 457 mm ø	Static Tension 457 mm ø	Dynamic Tension 305 mm ø
With lift line	0	8	6
Parent (no lift line)	8	8	6

The cores were shipped in custom-made crates, padded with foam, wrapped in wetted burlap then wrapped in plastic to keep the moisture as high as possible during storage on site and shipping.

Testing Facility

The concrete samples were tested at the United States Bureau of Reclamation's (USBR's) Concrete, Geotechnical and Structural Laboratory in Denver, Colorado. Samples were prepared for testing and stored in a 185 m^2 100% relative humidity and 23°C fog room prior to testing. The test equipment used (Figure 4) included a 25 MN Baldwin hydraulic compression/tension machine (one of only two such large capacity machines in the United States) for compression tests, a 2.7 MN Instron hydraulic compression/tension machine for tension tests, and a 222 kN USBR custom designed compression/tension machine for dynamic tests.

Figure 4. USBR Concrete Specimen Testing Equipment

Static Compression Tests

The compression tests were performed on eight 457 mm ø specimens using ASTM C39 Standard Test Method for Compressive Strength of Cylindrical Concrete Specimens and ASTM C469 Static Modulus of Elasticity and Poisson's Ratio of Concrete in Compression (ASTM, 2009). The specimen ends were capped with gypsum cement and two sets of diametrically opposed axial (longitudinal) and lateral (transverse) strain gages were bonded to the surface to measure deformation. Maximum applied load at failure, compressive strength, Young's Modulus, Poisson's Ratio and axial strain at failure were reported.

Static Direct Tension Tests

Sixteen 457 mm ø specimens (eight parent and eight lift line) were tested using a method developed at the Bureau of Reclamation, USBR 4914 Direct Tensile Strength, Static Modulus of Elasticity, and Poisson's Ratio of Cylindrical Concrete Specimens in Tension (Concrete Manual, 1992). The top and bottom of each concrete specimen were lightly sandblasted to remove laitance, and end-loading plates were adhered to the two cylinder ends using a two-part structural epoxy.

Once aligned in the machine, with lift lines centered within the specimen, the specimens were instrumented with two sets of diametrically opposed axial (longitudinal) bonded strain gages to measure deformation, used to determine the axial strain at failure. Due to strain data becoming unreliable above 90%, the reported value coincides with the strain at 90% ultimate tensile stress. Maximum applied load at failure, direct tensile strength, calculated failure strain (ε) and the fracture location were reported.

Dynamic Tension Tests

The dynamic (cyclic) tension test procedure was developed, collaboratively between the Bureau of Reclamation and BC Hydro, to replicate the response of the concrete specimens under dynamic loading. The testing apparatus and procedure allowed the specimen to be subjected to cycles of rapid tension and compression loading to predetermined amplitude. The dynamic test procedure consisted of a ramped amplitude test and a constant cycle amplitude test, see Figure 5.

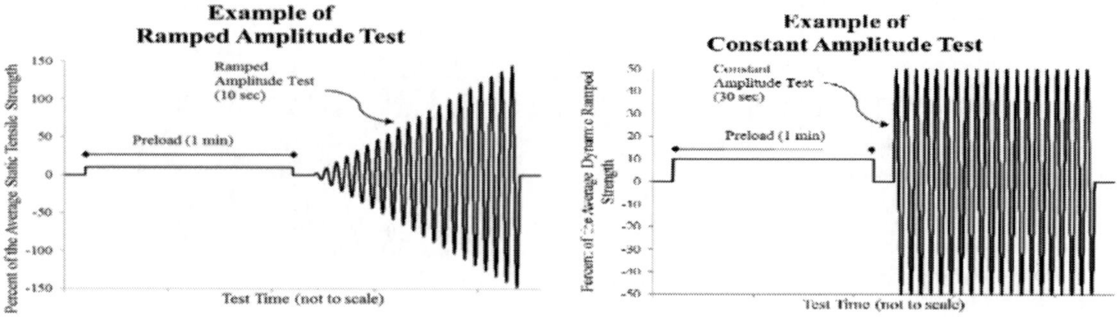

Figure 5. Schematic Dynamic Testing Signal Examples

Each test began with a one-minute preload of about 2 MPa. The ramped amplitude test increased the load magnitude with each compression/tension cycle for 10 seconds or until failure. The rate of the loading was set to reach 150% of the average tested static tensile strength of the concrete in 10 seconds. The constant amplitude test applied a cyclic tension/compression load equal to 50% of the average dynamic concrete strength from the ramped tests. The test procedure began with a constant load at 50% of average ramped dynamic tensile strength for a period of 30 seconds with added increases of 10% for additional 30 second intervals until failure. The loading frequency for dynamic test was 8 Hz to reflect the natural frequency of the Ruskin Dam, as determined via ambient vibration testing (Kemp, 1996).

The USBR Vibration Laboratory's custom designed and fabricated dynamic testing equipment includes a servo-hydraulic test assembly, supplied by a hydraulically powered actuator, atop a 2,224-kN seismic mass isolation system. Load is applied by a double-acting hydraulic actuator having a force rating of 222-kN and a total stroke of ± 28 mm. A linear variable differential transformer (LVDT) and load cell were used to measure displacement controlled loading.

Twelve 305 mm ø specimens (six parent and six lift line) were prepared (similar to USBR 4914) and conditioning, and data acquisition were the same for all of the tests. Time (relative to the start of data acquisition), displacement, load and strain were recorded at an acquisition rate of 1000 samples per second using an antialiasing filter set to 420 Hz.

TEST RESULTS

Static Compression Results

The compression tests conducted on the parent concrete are summarized in Table 3.

Table 3. Static Compression Test Results of Parent concrete

Test	Comp Strength (MPa)	Elastic Modulus (GPa)	Poisson's Ratio	Fracture Strain (με)
C-1	36.0	43.9	0.23	1019
C-2	37.0	41.3	0.23	1063
C-3	35.7	41.4	0.25	1125
C-4	36.1	34.4	0.25	1240
C-5	31.5	34.5	0.25	1143
C-6	41.5	41.3	0.25	1432
C-7	37.6	32.4	0.23	1596
C-8	40.9	40.3	0.24	1321
Average	**37.0**	**38.7**	**0.24**	**1242**

Comparison of the large-diameter specimen results with the results from previous tests is illustrated in Figure 6.

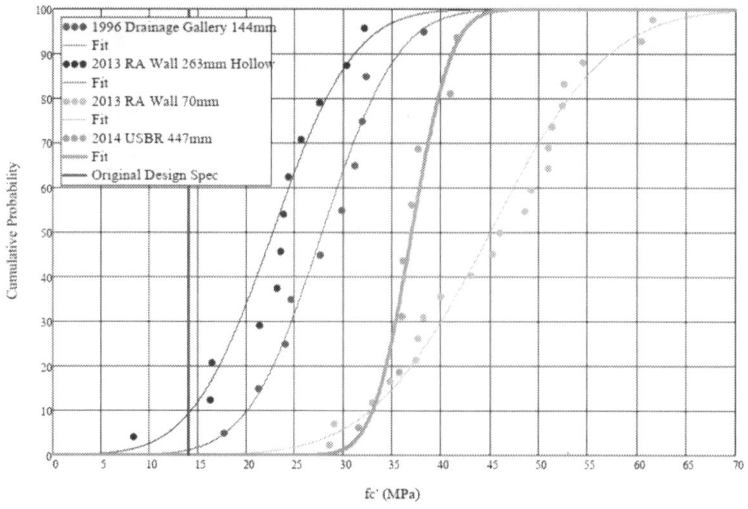

Figure 6. Frequency Distribution of Concrete Compressive Strength Test Results for Ruskin Dam

The average compressive strength of the concrete sampled was 37.0 MPa with a corresponding elastic modulus of 38.7 GPa. This indicated a higher compressive strength than that obtained from the 152 mm cores (Li, 1996) and significantly higher than the original design strength.

Static Tension Results

The tensile resistance of the parent and lift line specimens are summarized in Table 4.

Table 4. Static Tension Test Results

Parent Concrete		
Test	Tensile Strength (MPa)	90% Axial Failure Strain (με)
T-1	1.60	73
T-2	1.35	33
T-3	1.25	54
T-4	1.20	41
T-5	1.10	34
T-6	1.20	65
T-7	1.60	61
T-8	1.40	49
Average	**1.35**	**51**

Lift Line Specimens		
Test	Tensile Strength (MPa)	90% Axial Failure Strain (με)
T-9	1.00	20
T-10	0.75	36
T-11	1.00	55
T-12	1.00	57
T-13	1.05	43
T-14	0.90	62
T-15	1.30	33
T-16	1.10	35
Average	**1.00**	**43**

The lift line specimens had an average tensile strength 26% lower than the parent concrete. The average tensile strength of the parent concrete is about 3.6% of the average compressive strength of the parent concrete.

Dynamic Tension Results

The dynamic tension test results are summarized in Table 5 and Table 6. With the limited number of specimens, the intent of the tests was to provide insight into the behavior of concrete under dynamic loading.

Table 5. Ramped Dynamic Test

Test No.	Time to Failure (Sec)	Cycles	Maximum Tensile Stress (MPa)	
D-1	6.1	48.8	1.10	Parent
D-2A	NA (2 tests needed)		1.54	Parent
D-2B	8.3	66.4	1.67	Parent
D-3	8.9	71.2	1.70	Parent
D-12	7.3	58.4	1.57	Parent
Average	**8.2**	**65.6**	**1.65**	Parent
D-4	1.3	10.4	0.63	Lift Line
D-5	8.6	68.8	1.59	Lift Line
D-6	6.4	51.2	1.13	Lift Line
Average	**7.5**	**60**	**1.36**	Lift Line

The ramped tests indicate that the parent concrete has an ultimate dynamic tensile strength of approximately 125% of the static tensile strength; this may be due to loading rate effects. The lift line specimens show much more variability under the same loading, with all specimens failing at the observed lift lines.

Table 6. Constant Dynamic Test

Test No.	Time to Failure (Sec)	Cycles	Maximum Tensile Stress (MPa)	
D-7	1.2	10.4	0.75	Parent
D-8	2.9	68.8	0.75	Parent
D-9	4.2	51.2	0.74	Parent
Average	**2.8**	**60**	**0.75**	Parent
D-10	50.0	400	0.81	Lift Line
D-11	44.2	353.6	0.81	Lift Line
Average	**47.1**	**376.8**	**0.81**	Lift Line

It was observed that under constant dynamic loading the constant amplitude dynamic tests were observed that, mass concrete can fail as a result of low cycle fatigue at a significantly lower stress range. This is an important observation as a Maximum Design Earthquake can induce a large number of tensile cycles and thus a lower tensile resistance is possible.

HISTORICAL PROPERTIES OF MASS CONCRETE

The Ruskin Dam core tests were compared to historical properties expected for mass concrete dams contained in the ACIS mass concrete database (Dolen 2005, 2011), which is populated with data primarily from USBR historical tests. Since the properties of mass concrete dams evolved throughout the twentieth century, the estimates reflect the state-of-practice at the time of construction for different age subgroups. Initially, dams were constructed using relatively rudimentary techniques, equipment and materials. As dams became larger and more complex, the construction techniques also evolved. Therefore, the average properties from the Ruskin Dam tests can be compared to material properties of dams with comparable construction methodology and quality.

Ruskin Dam, constructed in 1929 to 1930, falls in the transition time frame of more deliberate mass concrete mixture proportioning methods of the 1920s and introduction of better equipment for mixing, placing and consolidating mass concrete in the mid-1930s. Before approximately 1920, the proportioning methods for mass concrete were less defined, and the water content of mixtures was not consistently controlled. Mixtures from this era had generally lower strength and more variability. After about 1920, Abrams "water cement law" was applied to concrete mixtures and better strength and quality were obtained (Abrams, 1918). The development of modern equipment and methods for manufacturing and placement of mass concrete took a great leap forward beginning with the Hoover Dam in the mid-1930s, resulting in higher quality mixing and placement of mass concrete.

Of particular significance to Ruskin Dam is the introduction of the mass concrete internal vibrator in the U.S. in the mid-1930s. Before internal vibration, the slump of mass concrete was higher in order to increase workability. Higher water contents were necessary to achieve a slump of 75 mm to 125 mm so that the concrete could effectively be placed and consolidated. Mixtures of this consistency had a higher tendency of bleeding and settlement of aggregate, resulting in non-uniform strength in individual placement lifts. Also, the surface water content of the top of placement lifts had more surface "laitance," which is mortar with weak strength that inhibits bonding between subsequently placed lifts of mass concrete. In the mid to late 1930s, pressurized, wet sandblasting became more common and both the strength of lift lines and percent of bonded lift lines improved. Ruskin Dam was constructed just before changes in the state-of-the-art of improved lift surface preparation.

Average Strength Properties of Mass Concrete

Material properties adapted from the USBR ACIS database (including the Ruskin Data) are shown in Table 7 and Table 8. The weighted average, long-term compressive strength and elastic properties of mass concrete from 1905 to 1993 are shown in Table 7, and the average direct tensile properties of mass concretes from 1908 to 1993 are shown in Table 8. . Reported tensile data is shown for both "parent" concrete with no lift lines and for bonded lift line specimens. Typically, lift line failures were about 20% lower than those that failed away from the lift line or in parent concrete. On average, the direct tensile strength of parent concrete is about 4.8% of the average compressive strength, whereas the direct tensile strength of lift lines is about 3.8% of the compressive strength.

Compressive Strength. The average compressive strength for mass concrete cores from Ruskin Dam is about 37 MPa. Based on historical data shown in Table 7 and Figure 7, this would be considered above average. The modulus of elasticity is about 38.7 GPa and is also higher than the historic average. The modulus of elasticity is influenced primarily by the strength of the concrete, the modulus of the aggregate and the volume of aggregate.

Table 7. Average compressive strength and elastic properties of mass concrete strength of parent concrete and bonded lift lines[6], adapted from USBR/ACIS[7]

Era	Average Compressive Strength (MPa)	Modulus of Elasticity (GPa)	Poisson's ratio	Number of tests	Density (Kg/ m^3)
1905-1930	26	22.0	0.21	264	2380
1920-1930	34	26.3	0.21	88	2443 1/
Ruskin Dam	37	38.7	0.24	8	2570
1934-1993	36	36.0	0.19	309	
All concrete	31	29.0	0.19	573	2424

[6] Average from 1920 to 1941; considered more representative of none, air-entrained mass concrete mixture proportioning methods similar to Ruskin Dam.
[7] Prepared by Dolen and Associates using a sampling of USBR data.

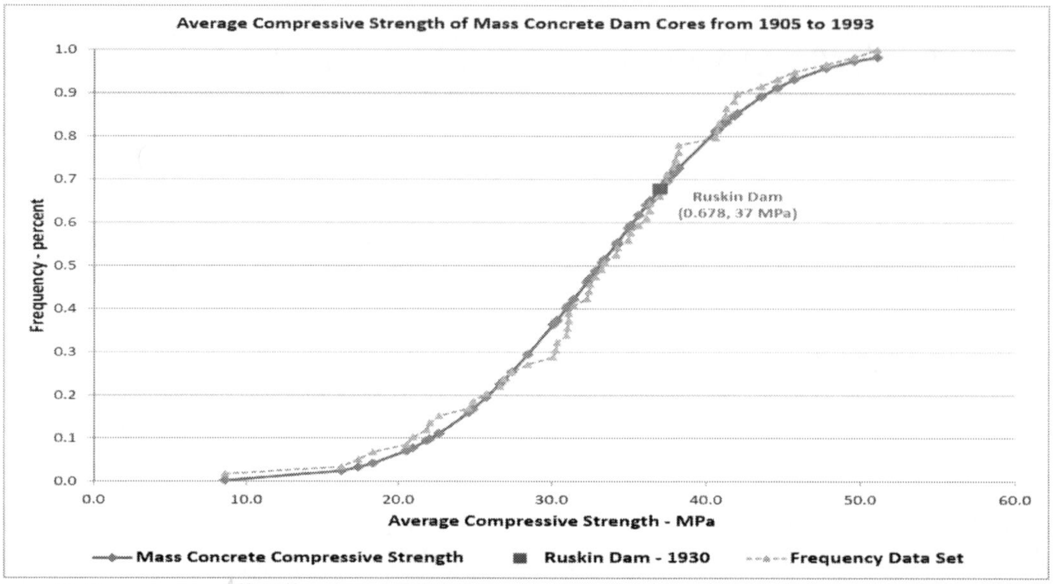

Figure 7. Frequency distribution of USBR mass concrete core compressive strength tests from 1905 to 1993 and average of 447-mm diameter mass concrete cores from Ruskin Dam[7]

Direct Tensile Properties of Parent Concrete and Bonded Lift Lines. The average direct tensile strength of parent concrete for Ruskin Dam is about 1.35 MPa and ranges from 1.1 MPa to 1.6 MPa. The average direct tensile strength for a sampling of USBR dams from 1905 to 1930 is about 1.5 MPa, as shown in Table 8 and and Figure 8. The direct tensile strength of Ruskin mass concrete is about 10% to 20% less than average mass concrete of similar ages and compressive strength. Many parent concrete direct tension breaks were on the underside of coarse aggregates, which is fairly common for mass concrete. This may be influenced by accumulation of bleed water and entrapped air on the underside of the larger aggregates, common for a mix with a high W/C ratio and high slump.

Table 8. Average direct tensile strength of parent concrete and bonded lift lines, adapted from USBR/ACIS Database[7]

	Average tensile Strength (MPa)	Standard deviation (MPa)	80% exceed (MPa)	Number of tests	Tensile strength – percent of average comp strength %	Tensile strength – percent of parent tens strength %
Parent Concrete						
All parent concrete	1.50	0.53	1.07	239	4.8	
1905-1930	1.47	0.53	1.00	74	5.7	
1920-1930	1.67	0.53	1.25	45	4.9	
Ruskin Parent	1.35				3.6	
1934-1964	1.53	0.54		53	4.4	
1965-1993	1.51	0.53		112	3.9	
Bonded Lift Lines						
All bonded lift lines	1.19	0.54	0.73	162	3.8	79
Bonded lift lines (LL failure)	1.05	0.59		72	3.4	70
All 1905-1930 [50] 1/	0.77	0.49	0.46	36	3.0	52
All 1920-1930	1.36	0.49	0.95	19	4.0	81
Ruskin Lift Lines	1.00				2.7	74
1934-1964 2/ [85]	1.13	0.49		16	3.2	74
1965-1993 3/ [74]	1.34	0.57		27	3.4	88

1/ Lift surface cleaning primarily by brooming and washing. [Estimated percent bonded lift lines for each subgroup].
2/ Lift surface cleaning primarily by green cutting and sandblasting.
3/ Lift surface cleaning primarily by sandblasting or high-pressure, water blasting.

The average direct tensile strength of Ruskin Dam lift lines compared to historic trends is shown in Table 8 and Figure 8. The average direct tensile strength of Ruskin Dam lift lines is about 1.0 MPa, about 14% less than the historic average for all tests. Not all of the historical tests of lift lines have failed at the lift line. The Ruskin Dam tests failed primarily at the lift line and the average strength is comparable to historic tests that had similar failure modes (at lift lines) – 1.05 MPa. The average mass concrete lift line strengths for a sampling of USBR dams from the 1905 to 1930 and the 1934 to 1993 subgroupings are approximately 0.77 MPa, and 1.26 MPa, respectively. The Ruskin Dam lift line strength falls between the two groups, although it is still less than what would be expected based on the relatively high compressive strength.

The average lift line strength for the Ruskin samples is approximately 74% of the parent concrete direct tensile strength, and the average lift line strength for historic mass concrete dams is approximately 79% of its parent concrete strength. This indicates the lift line strength performance is fairly typical compared to historic mass concrete. The Ruskin lift line strength is approximately 2.7% of the parent concrete compressive strength, while the historical average for all dam concretes in the database is about 3.8% of the compressive strength. However, the Ruskin Dam lift line performance is nearly the same as for other mass concrete from 1905 to 1930 subgroup, which averaged about 3% of average compressive strength. The fact that all Ruskin lift line specimens failed at lift line locations indicates the top of lift bleeding and/or lift surface preparation influenced and reduced direct tensile strength of the lift line samples.

The direct tensile strength of parent concrete and lift lines from Ruskin Dam is lower than typical historical mass concrete. This is expected since the dam was constructed before the advent of internal vibration of mass concrete and use of sandblasting for lift surface cleaning.

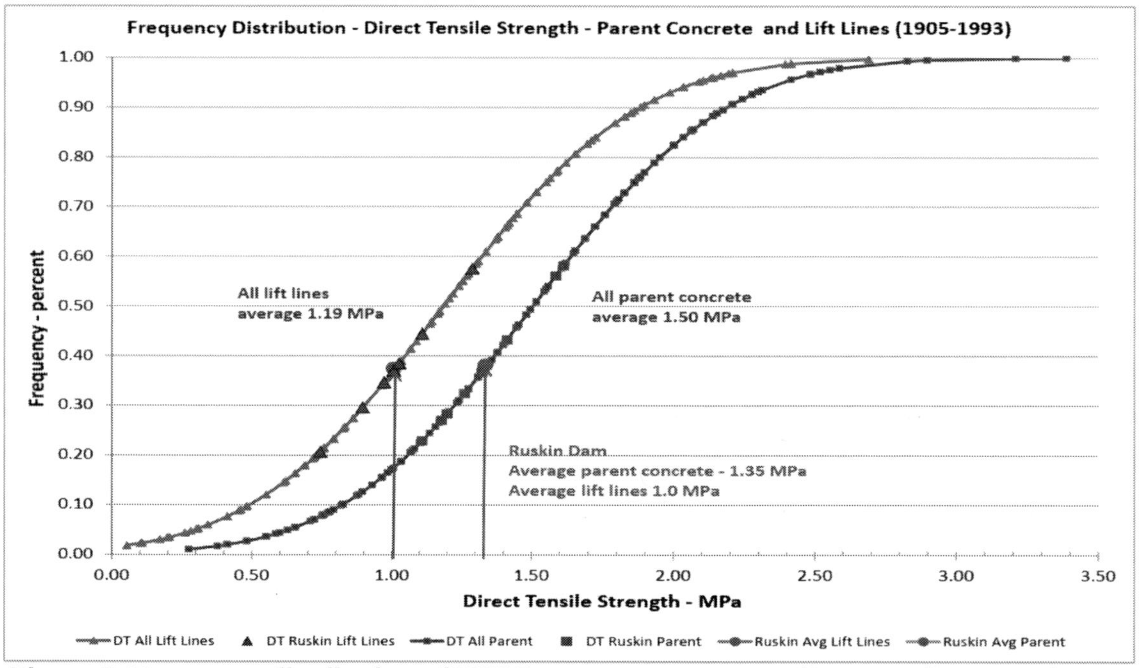

Figure 8. Frequency distribution of USBR mass concrete core direct tensile strength tests of parent concrete and lift lines from 1905 to 1993 and 447-mm diameter mass concrete core tests from Ruskin Dam[7]

Historical Dynamic Properties of Mass Concrete

There is limited documented, historical information on dynamic properties of large-diameter mass concrete specimens in direct tension. Saucier and Carpenter reported the dynamic direct tensile strength of 200- by 400-mm cylindrical cores from mass concrete with 75 mm NMSA [Saucier and Carpenter, 1977]. These tests were used as a basis of comparison to the Ruskin Dam testing. Jackmauh and Munoz reported on dynamic tensile strength of 300- by 900-mm mass concrete cast cylinders with 200 mm NMSA [Jackmauh and Munoz, 1987]. Both test programs include static compressive strength, and Saucier and Carpenter report static direct tensile strength and cyclic dynamic direct tensile strength at frequencies ranging from 1 to 10 Hz. Saucier and Carpenter concluded that there was little difference in rate of loading for the various test conditions with one exception; saturated cores showed about a 33% increase in direct tensile strength from the static to the dynamic loading. The Ruskin tests were all conducted with saturated cores and under this premise, some increase under dynamic load rates can be expected. The tests by Jackmauh and Munoz report dynamic loading to failure (single stroke) at about 5 Hz. The values of the historical dynamic tests are summarized in Table 9.

Table 9. Average dynamic and static direct tensile strength of parent concrete cores and cylinders, USBR 1977 and 1987[7]

Saucier & Carpenter – 200 x 400-mm drilled cores tested about 90 to 180 days (75 mm NMSA)								
Frequency	Average static compressive strength	Average dynamic direct tensile strength	No. of tests	Average static direct tensile strength	No. of tests	Direct tensile strength percent of compressive strength		Notes
Hz	MPa	MPa		MPa		Dynamic	Static	
5–10	41	2.10	10	2.0	5	5.1	5.1	
1–5	21	1.66	10	1.64	10	7.9	7.8	Not previously tested
1 Hz	21	1.75	8	1.64	10	8.5	7.8	After 20-25 tension cycles
1 Hz	21	1.85	7					
5 Hz	21	1.72	7					
10 Hz	21	1.84	5					
Saucier & Carpenter – 200 x 400-mm cores drilled cores tested about 1 year – 75 mm NMSA								Change
Tested @ 1 year wet	19.4	1.8	5	1.3	5	8.9	6.9	+133%
Tested @ 1 year dry	23.5	1.7	5	1.6	5	7.3	6.9	+106%
Jackmauh & Munoz – 300 x 900-mm cast cylinders tested after 2 to 4 years – 200 mm NMSA								
5 Hz	29.3	2.1	25			7.3		
5 Hz	32.3	2.3	23			7.0		

The average dynamic tensile strength for Ruskin parent concrete is about 1.65 MPa and for lift line specimens, 1.36 MPa. These values are 4.5% and 3.7% of the average static compressive strength, respectively. The average Ruskin tensile strength is considerably less than the results of the reference data. However, the reference concretes were manufactured from good quality aggregates under modern laboratory conditions. Thus, the lower dynamic tensile strengths are not unexpected as the static tensile strengths were also low compared to reference compressive strength.

When compared to the average static direct tensile strength, the average dynamic tensile strength for parent concrete and lift lines is about 130% of the average static strength. This compares very favorably to the results of Saucier and Carpenter for the saturated moisture conditioning. (It should be noted that, overall, the static dry test in the Saucier and Carpenter test specimens tested higher than the saturated specimens, but they increased only 6% under dynamic load conditions compared to the 33% increase in the saturated specimens.)

The dynamic tests results are plotted over the static test frequency distribution charts for parent concrete strength and lift line strength, as shown in Figure 9 and Figure 10, respectively. For parent concrete, the frequency distributions for 1903–1930 and 1905–1993 plot essentially the same. The three Ruskin Dam dynamic tests and average dynamic tensile strength of parent concrete plot near the two highest static tests. The two lift line tests plot in the top 25% of comparative (1905–1930) static tests and middle third of all (1905–1993) static tests.

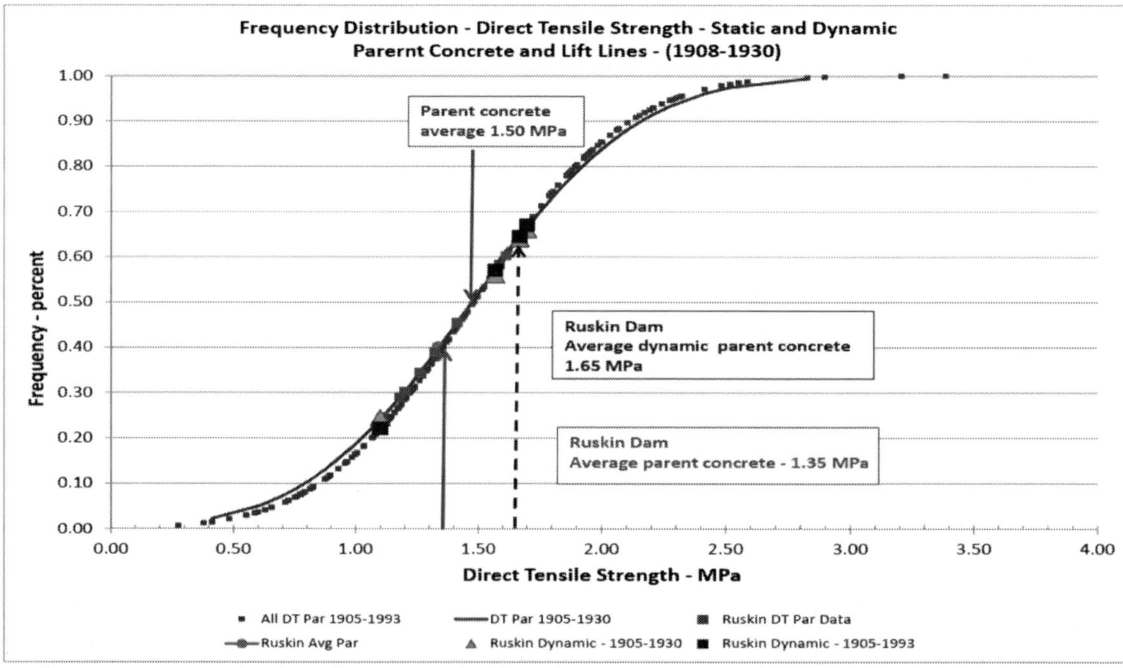

Figure 9. Dynamic Tensile Strength of 297 mm ø, Parent Concrete Cores from Ruskin Dam, Plotted over the Static Frequency Distribution Charts for 1905-1993 USBR Tests of Mass Concrete Cores with Various Diameters[7]

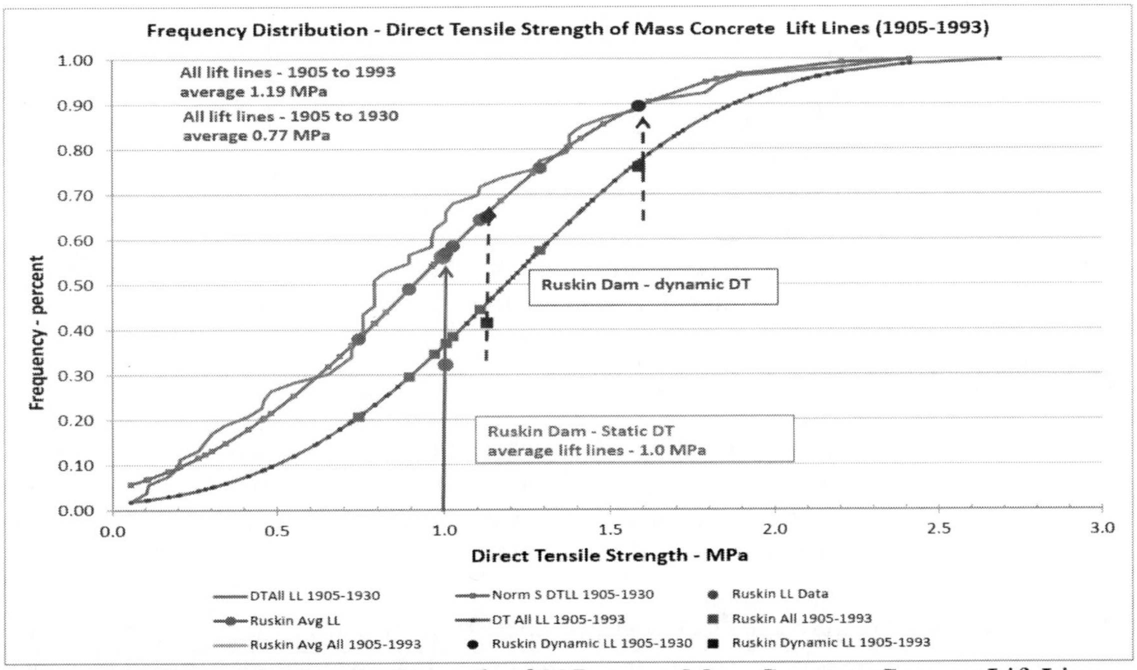

Figure 10. Dynamic Tensile Strength of 297 mm ø, Mass Concrete Cores at Lift Lines from Ruskin Dam, Plotted over the Static Frequency Distribution Charts for 1905-1993 USBR Tests of Mass Concrete Cores with Various Diameters[7]

DISCUSSION

Static Compression

The compressive strength of the Ruskin Dam mass concrete has increased from the original design strength of 14 MPa over the last 80 years. Using the International Federation for Structural Concrete Model Code (CEB-FIP 2010) prediction equation, the resulting present day compressive strength of 32 MPa is obtained. The average tested compressive strength is compared to the CEB-FIP predicted strength in Table 10, illustrating the values are within 15%.

Table 10. Compressive Strength Comparison

	Compressive Strength (MPa)
Tested Cores	37
CEB-FIP	32

Comparison of concrete stiffness values from CEB-FIP and other applicable codes to the tested values is summarized in Table 11. The calculated stiffness values are within 20% of the actual tested values, indicating some consistency between code equations and test results.

Table 11. Compressive Elastic Modulus Comparison

	Elastic Modulus (GPa)
Tested Cores	39
CEB-FIP (aged)	32
ACI	34
CSA	34

Actual concrete mix strength may have been higher than the design strength of 14 MPa that CEB-FIP formulation would not anticipate in the calculations.

Static Tension

The static tension strength of concrete is typically related to the compressive strength by formulas included in standards and codes. Calculated values are compared to the average test strength in Table 12. The calculated values vary widely and are highly dependent on the compressive strength used.

Table 12. Parent concrete tensile strength comparison between codified and tested values

	Tensile Strength Based on Design Strength (MPa)	Tensile Strength Based on Tested Strength (MPa)
Tested Specimens	1.35	
USBR Standard of Practice	0.9	1.8
CEB-FIP 2010	1.7	3.3
USACE (EM1110-2200)	2.5	4.9

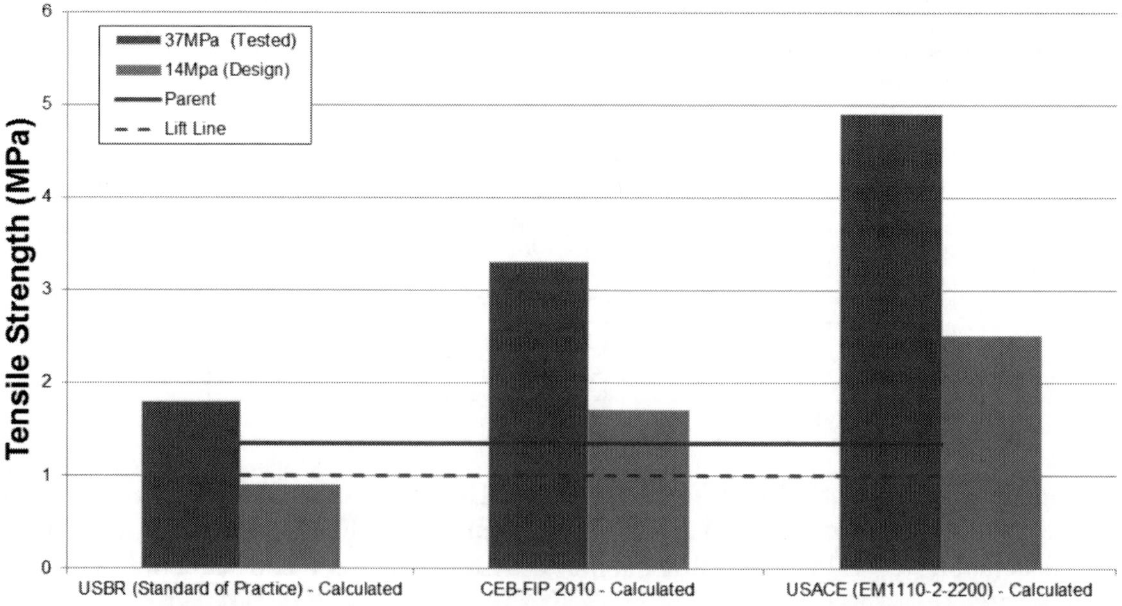

Figure 11. Calculated Concrete Tensile Strength Compared to Ruskin Dam Test Results

Dynamic Tension

The comparison of the dynamic tests to the corresponding average static tension tests shown in Table 5 illustrates an increase of 22% and 36% of parent concrete and lift line samples respectively. The comparison in Table 13 shows that the average relationship between the tested dynamic and static sample is in agreement with the accepted convention of industry standards and codes for the recommended increase in tensile strength of concrete under dynamic loading.

Table 13. Dynamic ramped tensile strength increase

	Ramped Dynamic Tensile Strength Increase (relative to static strength)
Ramped Amplitude Tested Specimens	22% (Parent) – 36% (Lift Line)
Standard of Practice	30%
CEB-FIP 2010	33%
USACE (EM1110-2200)	33%

In contrast, the constant amplitude tests resulted in lower stress failures of 44% less and 59% less than the average static tensile strength of the samples. While the number of samples tested was limited, the results provide a general indication that the mass concrete tested potentially has a lower tensile capacity for lower amplitude constant stresses over a longer duration.

CONCLUSION

The tensile properties of mass concrete are a key parameter in determining the predicted seismic response of concrete gravity dams in high seismic zones. Estimating these properties based on code equations can significantly over predict the tensile capacity of older mass concrete such as the samples obtained from Ruskin dam. This could have led to under prediction of the dam's response during strong shaking and possibly could have impacted the selection of retrofit strategies.

Furthermore these test demonstrated that while the tensile strength increases during rapid ramped loading, it was observed that under constant amplitude cyclic loading a reduction in the tensile strength was observed. For seismic events with significant strong shaking duration, this is an important observation which will guide the selection of material parameters for the simulation of the dams nonlinear response to Maximum Design Earthquake events.

The authors hope that the dynamic tensile test results obtained as part of this study add to the very limited data set of large aggregate mass concrete and that other owners are encouraged to carry out similar test on their aging concrete dams in order to establish a better data base such that code estimates of tensile strength can be improved in future. More research in this area is still needed to further the understanding of the tensile behaviour of large aggregate mass concrete under cyclic tensile loading.

ACKNOWLEDGEMENTS

The development of the testing project discussed in this paper was initiated by Andreas Felber Ph.D., P.Eng., who recognized the need and value of this work to the project.

The authors of this paper appreciate BC Hydro Dam Safety for the opportunity of working on this interesting project.

REFERENCES

Abrams, Duff, A., Design of Concrete Mixtures, Structural Materials Research Laboratory, Lewis Institute, Chicago, Illinois, 1918.

ACI. Building Code Requirements for Structural Concrete (ACI 318-11). (2011).

ASTM International, (ASTM C39, ASTM C469) West Conshohocken, PA, 2009.

Boulder Canyon Project Final Reports, U.S. Dept. of the Interior, Bureau of Reclamation, Denver, CO, 1940.

Canadian Standards Association (CSA). CAN/CSA-A23.3-04 (R2010). (2010).

CEB-FIP. fib Model Code for Concrete Structures 2010. (2013).

Dolen, T.P., Materials properties model for aging concrete, Bureau of Reclamation Dam Safety Program Report No. DSO-05-05, Denver, CO, 2005.

Dolen, T.P., Selecting Strength Input Parameters for Structural Analysis of Aging Concrete Dams, USSD Annual Meeting, San Diego, CA, 2011.

Jackmauh, Francis J., and Munoz, Richard L., Rapid Load Testing of Mass Concrete (Final Draft), U.S. Dept. of the Interior, Bureau of Reclamation, Denver, CO, August, 1987.

Kemp, Bradley Gary. Ambient Vibration Assessment of Ruskin Dam Dynamic Properties. Thesis. University of British Columbia, 1996.

Li, Jim. Testing Concrete Cores From Ruskin Dam. Powertech Labs Inc. July 1996.

Saucier, K. L. and Carpenter, L., Dynamic Properties of Mass Concrete, ASTM Symposium on Dynamic Geotechnical Testing, Denver CO, June 28, 1977, *ASTM Special Technical Publication 654*, ASTM, Philadelphia, PA, pp.163–178.

U.S. Dept. of the Interior, Bureau of Reclamation, Denver, CO, Concrete Manual, Part 2, Ninth Edition. A Water Resources Technical Publication, 1992.
United States Bureau of Reclamation. State-of-Practice for the Nonlinear

Analysis of Concrete Dams 2013. (2013).

U.S. Army Corps of Engineers (USACE). EM1110-2200 Gravity Dam Design. (1995).

USING RISK ASSESSMENT TO INFORM LEVEE SYSTEM EVALUATION DECISIONS FOR THE NATIONAL FLOOD INSURANCE PROGRAM

David A. Margo, M.S., P.E.[1]
Lea G. Adams, M.S., P.E.[2]
Scott E. Shewbridge, PhD, P.E., CEM[3]
Richard Varuso, PhD, P.E.[4]

ABSTRACT

The U.S. Army Corps of Engineers (USACE) is currently updating its policy for the evaluation of levee systems in support of the National Flood Insurance Program (NFIP). This update is an outcome of a joint effort between USACE and the Federal Emergency Management Agency (FEMA) to align their processes so that information developed by USACE for the Inspection of Completed Works (ICW) program can also be used to evaluate the NFIP accreditation requirements. Risk assessments provide a suitable framework to achieve this alignment. Risk assessments provide a more comprehensive understanding of the flood hazard and associated risks by considering a full range of flood events, performance of the levee under these various loadings, and the potential consequences. Outcomes of a risk assessment can be used to prioritize levee safety program activities, inform accreditation decisions, identify flood risk mitigation actions, and support risk communication. This paper will summarize key components of the policy update along with the results of a parametric analysis that was conducted to better understand how flood hazards are influenced by the combination of flood loading and levee performance.

BACKGROUND

National Flood Insurance Program

The National Flood Insurance Program (NFIP) was authorized by the Congress of the United States in 1968 to mitigate flood losses. The NFIP is administered by the Federal Emergency Management Agency (FEMA) within the Department of Homeland Security (DHS). Communities may voluntarily participate in the program by adopting and enforcing floodplain management regulations that reduce future flood losses. In exchange, the Federal Government provides federally backed flood insurance as a financial protection against flood losses. In 2015, more than 20,000 communities participated in the NFIP (FEMA, 2015). The NFIP also identifies and maps flood hazards to provide the information needed to implement floodplain management

---

[1] Lead Civil Engineer, Corps of Engineers Risk Management Center, 1000 Liberty Avenue, Pittsburgh, PA 15222, david.a.margo@usace.army.mil, 412-667-6629.
[2] Chief, Water Resources Division, Corps of Engineers Hydrologic Engineering Center, 609 Second Street, Davis, CA 95616, lea.g.adams@usace.army.ml, 530-756-1104.
[3] Senior Geotechnical Engineer, Corps of Engineers Risk Management Center, 12596 West Bayaud Ave., Suite 400, Lakewood, CO 80228, scott.e.shewbridge@usace.army.mil, 720-201-9299.
[4] Senior Program Manager, Corps of Engineers Risk Management Center, P. O. Box 60267, New Orleans, LA 70160, richard.j.varuso@usace.army.mil, 504-862-2984.

regulations and establish flood insurance rates. The standard for the NFIP is the base flood which is defined as a flood that has a 1% annual chance of exceedance (aka the 100 year flood).

Levee System Evaluation and Accreditation

A levee system is comprised of one or more levee components (e.g. earthen embankment, concrete floodwall, closure structure, pump station, etc) that collectively reduce flood hazard and risk over a range of flood magnitudes within a defined floodplain referred to as the leveed area. According to the National Committee on Levee Safety (NCLS, 2009), the United States is estimated to have more than 100,000 miles of levee systems. These levees were built to manage flood risk for millions of people, trillions of dollars in property value, and thousands of critical infrastructure facilities. The United States Army Corps of Engineers (USACE) maintains a federal interest in nearly 15,000 miles of levees representing more than 2,500 levee systems.

The NFIP considers the reduction in flood hazard provided by the levee system when mapping leveed areas. Requirements for the evaluation of levee systems for the NFIP are provided by the code of federal regulations, 44 CFR 65.10 (Mapping of Areas Protected by Levee Systems, 1986). Levee system evaluation provides a process for levee owners to prepare and document whether or not the levee system provides adequate reduction of the flood hazard using data and evidence that is accurate and supported by analyses in accordance with sound engineering practices. A levee system may be accredited by FEMA when sufficient documentation is provided by the levee owner to demonstrate that the levee system meets minimum design, operation, and maintenance standards such that the levee system provides adequate reduction of the flood hazard. Accreditation means that FEMA will recognize the reduction in the flood hazard provided by the levee system on flood hazard maps. A leveed area associated with an accredited levee system is generally mapped as a moderate flood hazard and therefore the leveed area is not subject to mandatory floodplain management regulations or flood insurance. Leveed areas associated with non-accredited levee systems are mapped based on specific characteristics of the levee system (FEMA, 2013). For a non-accredited levee system, the portions of the leveed area that could be flooded during an occurrence of the base flood as a result of either breach or overtopping of the levee are mapped as a special flood hazard. These special flood hazard areas are subject to mandatory floodplain management regulations and flood insurance.

TRANSITION FROM DETERMINISTIC TO PROBABILISTIC

Estimating and mapping the flood hazard for the NFIP in a floodplain without any levees is relatively straightforward. Analyze the hydrologic conditions; make an estimate of the peak discharges and stages associated with the base flood; and map areas flooded by the base flood as a special flood hazard. Estimating and mapping the flood hazard in a leveed area is inherently more complicated due to the interdependencies between flood magnitude and levee performance. The leveed area could be flooded due to a breach prior to levee overtopping or as a result of overtopping. A breach could also occur under a range of flood loadings that are less than, greater than, or equal to the base flood.

Deterministic Evaluation of Levee Systems

When the first interim levee evaluation policy was issued by the Federal Insurance Administration (now FEMA) in 1981, it was recognized that the flood hazard in a leveed area was a function of both the uncertainty in flood elevations and the uncertainty in the performance of the levee itself (NRC, 1982). However, common engineering practices at the time did not readily support adoption of a probabilistic or risk informed framework to explicitly consider these uncertainties. As a result, the requirements for evaluation of a levee system defined by the code of federal regulations include a suite of deterministic analyses (Mapping of Areas Protected by Levee Systems, 1986). The standard is based on common engineering practices for levee design and construction that existed at the time of publication. The objective of the deterministic standard is to provide assurance that a leveed area has less than a 1% annual chance of experiencing flooding. Levee height, erosion, stability, settlement, and other applicable design criteria are evaluated against engineering design and construction standards. If a levee system meets all of the individual standards under a base flood loading event, then it is assumed that the flood hazard is adequately reduced to something less than a 1% annual chance. A levee that meets the standards can be recommended for accreditation and the leveed area mapped accordingly. Conversely, a levee system that does not meet all of the criteria when subjected to a base flood loading event is assumed to pose a flood hazard to the leveed area that is greater than a 1% annual chance. A levee that does not meet the standard cannot be recommended for accreditation and the leveed area would be mapped as a special flood hazard. While this approach provides a reasonable solution to address a complex problem, it does not explicitly assess the flood hazard in the leveed area. Currently available risk assessment methodologies can support a more direct estimate of the flood hazard in leveed areas.

Motivation for Change

The Flood Protection Structure Accreditation Task Force was formed by USACE and FEMA in response to Section 100226 of Public Law 112-141. A primary charge to the Task Force was to align the information and data collected for the USACE Inspection of Completed Works (ICW) program so it is sufficient to satisfy levee evaluation requirements and inform levee accreditation decisions. Recommendations made by the Task Force in November 2013 included using risk assessments conducted by USACE to assess levee performance for NFIP purposes. (USACE/FEMA, 2013) Congress adopted the recommendation in Section 3014 of the Water Resources Reform and Development Act of 2014, Public Law 113-121, by requiring that at least one program activity in the USACE ICW program provide adequate information to reach a levee accreditation decision. A framework to implement the recommendations of the Task Force was formally adopted by USACE and FEMA in a November 2014 Memorandum of Understanding (USACE/FEMA, 2014). This framework includes using USACE risk assessments for NFIP purposes.

USACE has been using various forms of risk and uncertainty analysis to evaluate levees since the early 1990s. Initially, risk analysis was implemented to address challenges aimed at deterministic levee freeboard standards. Freeboard achieved different levels of flood hazard reduction at different locations due to site specific geographic and hydrologic uncertainties. USACE believed that application of risk analysis held great promise in addressing the freeboard issue and similar issues related to uncertainties in levee design analysis. A risk analysis approach was adopted by USACE for flood damage reduction studies in 1995 with the publication of Engineer Regulation 1105-2-101. A National Academy of Sciences Committee was formed in response to the Water Resources Development Act of 1996 to study the use of risk analysis by USACE. The Committee was charged with evaluating the impact of risk analysis on project formulation, economic justification, and engineering standards. The Committee concluded that risk analysis methods were a significant step forward and that there should be no turning back from this approach (NRC 2000). The use of risk analysis by USACE for levee evaluation expanded significantly and rapidly more recently with the formation of the Risk Management Center and the Levee Safety Program. The use of risk analysis to evaluate levee flood hazards and risks is now standard practice within USACE. A logical next step is to expand the use of risk analysis for NFIP purposes.
Today, risk assessments are the common practice used by USACE to evaluate and manage flood hazards and risks for both the dam and levee safety programs. Risk assessments leverage the analytical methods of traditional engineering practice, probabilistic methods, consideration of consequences, and the expertise of engineers, review boards, and decision makers. Risk assessments support a more explicit estimate of the flood hazard to better inform accreditation decisions. The identification of potential failure modes and potential breach locations also provides information that supports FEMA's mapping process for non-accredited levees.

Probabilistic Evaluation of Levee Systems

The use of risk assessment to inform levee evaluation decisions requires the adoption of new evaluation processes, guidelines, and metrics. Each time USACE conducts a risk assessment, a levee evaluation will be performed to inform accreditation and NFIP mapping decisions. The code of federal regulations, 44 CFR 65.10 (Mapping of Areas Protected by Levee Systems, 1986), provides the authority for FEMA to accept a USACE risk assessment evaluation as the basis for an accreditation decision because USACE is a federal agency with responsibility for levee design.

The framework for using risk assessments and establishing new guidelines for evaluation decisions includes three important elements. First, levee evaluation decisions will consider the flood hazard (i.e. probability of flooding) and not the flood risk (i.e. probability of flooding and consequences). This provides consistency with the primary mission of the NFIP which is to identify and map flood hazards. Second, levee evaluation decisions will explicitly consider the combined effect of flooding and levee performance. This aligns with the approach used in risk assessment and supports a more credible estimate of the flood hazard by directly evaluating the flood hazard in a leveed area. Third, the evaluation process will consider all possible flood magnitudes instead of only evaluating performance under a base flood loading. This also supports a more credible estimate of the flood hazard by considering the multiple pathways that could lead to flooding in the leveed area.

The primary evaluation metric for levee evaluations based on a risk assessment will be the annual flood hazard for the leveed area. The flood hazard will be assessed by estimating the annual probability of levee breach or overtopping. A levee system can receive a positive evaluation and be recommended to FEMA for accreditation when the annual probability of breach or overtopping for the levee system is less than 1% with a reasonable degree of assurance. Achieving reasonable assurance using risk assessment will require estimation and consideration of the uncertainty in the flood hazard estimate. Reasonable assurance can be evaluated by estimating the probability or confidence that the annual flood hazard is less than 1%.

The deterministic guidelines in the code of federal regulations currently achieve reasonable assurance via minimum factors of safety, freeboard, and other applicable design criteria (Mapping of Areas Protected by Levee Systems, 1986). The levee evaluation guidelines and most engineering guidelines in general are structured to provide a relatively high degree of assurance. For example, current USACE criteria for levee height requires a minimum of 90% assurance that the base flood will not overtop the levee. Because of the high degree of assurance required, some levee systems may be rejected for accreditation even though the annual flood hazard may be less than 1%. This is done to ensure that few poor quality levees with potential flood hazard greater than 1% can be accepted for accreditation. Engineers often use the term "conservative" to characterize this type of approach. However, such an approach might not be "conservative" from other perspectives. If the standard is too stringent, then the rejection of relatively good levees may result in misallocation of resources across the national portfolio of levees. A standard that is too stringent could also result in disparities between leveed and non-leveed floodplains. The assurance guidelines for annual flood hazard must consider some of these tradeoffs between engineering conservatism and other impacts.

Existing deterministic guidelines and past levee evaluation decisions could provide another potential benchmark for an assurance guideline. There are several challenges with this approach. First, existing information that would support a direct comparison between existing deterministic guidelines and proposed probabilistic guidelines is limited. An investment in collecting and evaluating additional data would be needed over time to fill this gap. Second, it is unknown whether the existing guidelines provide an appropriate balance between engineering conservatism and other impacts. Comparison with past decisions requires some confidence that the decision was correct in the first place. Does three feet of freeboard provide the right amount of flood risk reduction to support credible floodplain management and flood insurance investment decisions at both a local and national scale? This is an open question that requires further evaluation.

The use of risk assessment and adoption of a new flood hazard assurance metric for levee evaluation provides an opportunity to explore these issues in more depth. Over time, increased knowledge and experience will support continued improvement in the credibility and consistency of levee evaluation decisions. Risk assessments provide the type of information needed to explicitly evaluate these issues and the impacts of the various tradeoffs. As a first step, USACE has performed a parametric analysis to provide some initial insights on the flood hazard assurance metric.

PARAMETRIC ANALYSIS

The flood hazard for a levee system can be estimated by a convolution of the discharge frequency function, the stage discharge function, and the system response function. The discharge frequency function defines the annual chance exceedance of peak annual discharge. The stage discharge function defines peak stage as a function of peak discharge. The discharge frequency and stage discharge functions can also be defined with uncertainty. The system response function defines the probability of levee breach or overtopping as a function of peak stage. The uncertainties associated with levee performance are usually reflected by the system response function itself. The computation framework is illustrated in Figure 1. This computation framework is the same as that used by the USACE Flood Damage Analysis (HEC-FDA) software package. The computation procedures are described in more detail in the HEC-FDA user manual (USACE, 2014). For efficiency purposes, a customized spreadsheet that implements the HEC-FDA computation framework was developed and used for this parametric study.

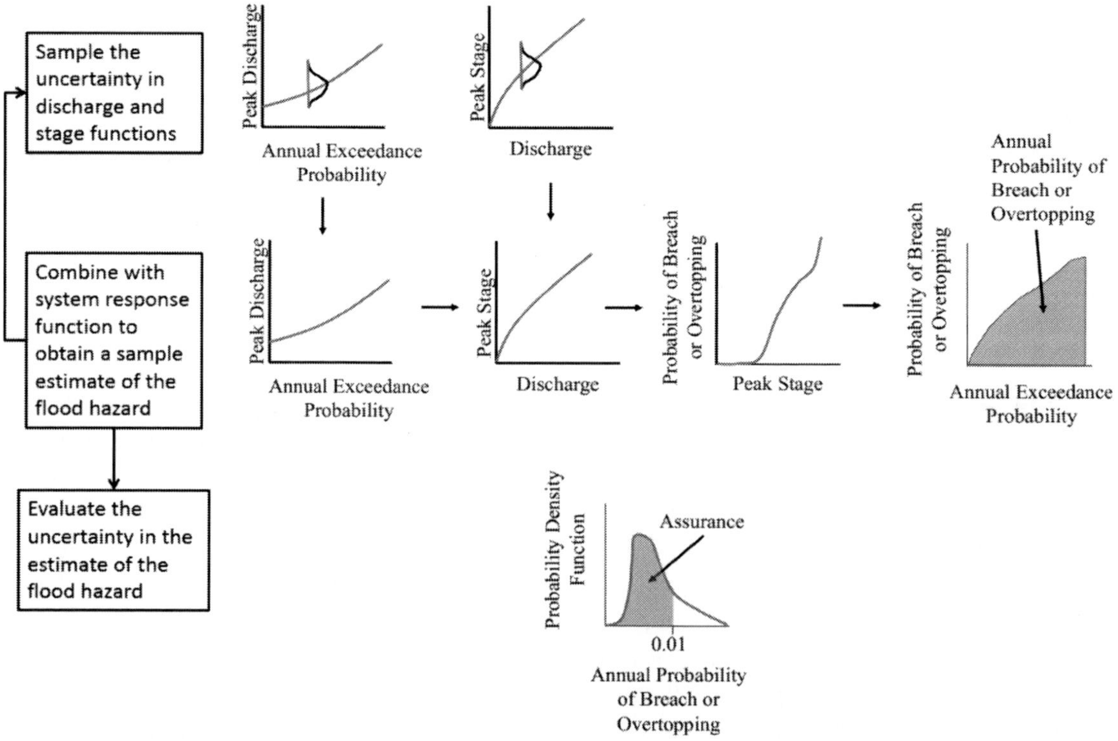

Figure 1. Estimating Flood Hazard for a Levee System

A single estimate of the annual probability of breach or overtopping without uncertainty can be calculated by combining the computed discharge frequency, stage discharge, and system response functions to obtain a probability of breach or overtopping frequency function. This function can be integrated to obtain a single estimate of the annual probability of breach or overtopping. Uncertainty can be added to the analysis using monte-carlo simulation methods. The discharge frequency and stage frequency functions can be sampled many times and combined with the system response function to obtain many estimates of the annual probability of breach or overtopping. From these estimates a distribution reflecting the uncertainty in the annual probability of breach or overtopping can be obtained and the assurance can be estimated as the probability that the annual probability of breach or overtopping is less than 1%.

Discharge Frequency Function

For the parametric analysis, the Log Pearson Type III distribution defined in Bulletin 17B (USGS, 1982) is used to define the discharge frequency function. The distribution and its uncertainty are defined by a mean, standard deviation, skew, and equivalent record length. Because the mean is a scaling parameter that does not change the shape of the discharge frequency function, variations in the mean do not need to be considered in this parametric analysis. The standard deviation influences the slope and the skew influences the concavity of the discharge frequency function. By normalizing the discharge frequency function, a representative range of shapes can be defined by a fixed value for the standard deviation and varying the skew. Discharge is normalized by the 2 year and 10,000 year discharges as shown in Equation 1. The 2 year flood defines the lower bound of the parametric analysis which is assumed to be equal to a bankfull discharge and the toe of the levee. The 10,000 year flood defines the upper bound of the numerical calculations. The equivalent record length was varied in the study because it influences the uncertainty in the annual probability of breach or overtopping estimate. A shorter record length results in greater uncertainty. Three discharge frequency shapes and three equivalent record lengths are used to define nine discharge frequency functions used for this parametric analysis. The parameters are listed in listed in Table 1. The normalized discharge frequency function shapes are shown in Figure 2.

$$Q' = \frac{Q - Q_2}{Q_{10,000} - Q_2} \tag{1}$$

Table 1. Discharge Frequency Function Parameters

Parameter	Values
Mean	4
Standard Deviation	0.3
Skew	-0.5, 0, 0.5
Equivalent Record Length	30, 60, 100

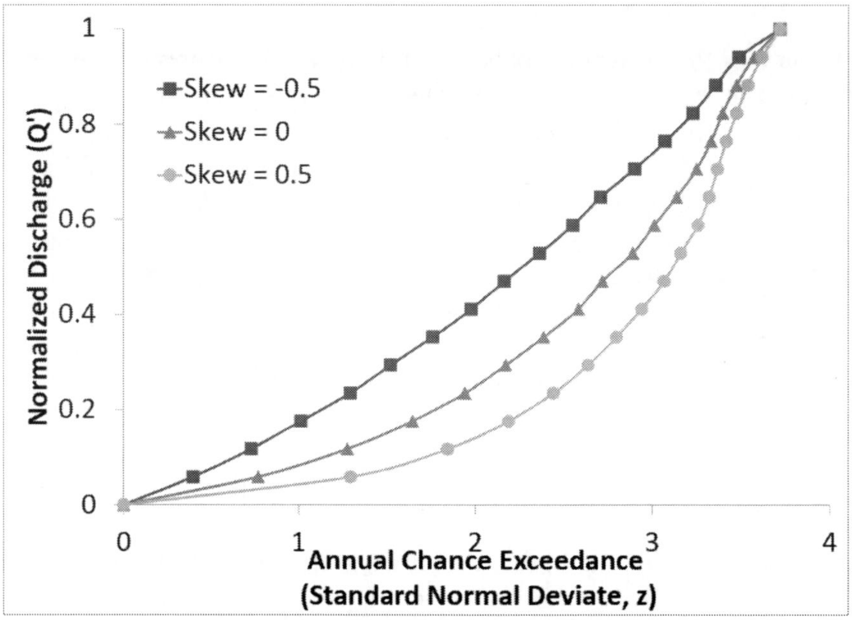

Figure 2. Normalized Discharge Frequency Functions

Stage Discharge Function

A trapezoidal prismatic channel is assumed to define the stage discharge function. The channel shape illustrated in Figure 3 represents a typical channel and floodplain configuration and can be defined by the channel bottom width (b), channel depth (d), channel side slope (s_c), floodplain width (w), floodplain side slope (s_f), channel roughness (not shown), and stream slope (not shown). Variations in any of these parameters can influence the shape of the stage discharge function. If the discharge and stage are normalized, these variations become small enough to support selection of three normalized stage discharge function shapes corresponding to each of the three discharge frequency function shapes. The discharge is normalized using equation 1 and the stage is normalized by the 2 year and 10,000 year stages using equation 2. The normalized stage discharge functions are shown in Figure 4. The uncertainty in stage at a given discharge is defined by a normal distribution with mean equal to the computed stage. The standard deviation at the 1% stage was assumed equal to values of 0.25, 0.5, and 0.75 feet to represent scenarios with low, moderate, and high uncertainty.

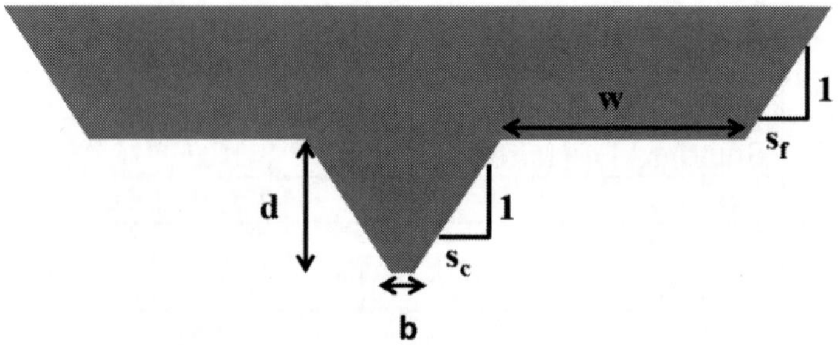

Figure 3. Prismatic Channel Dimensions

$$S' = \frac{S - S_2}{S_{10,000} - S_2} \quad (2)$$

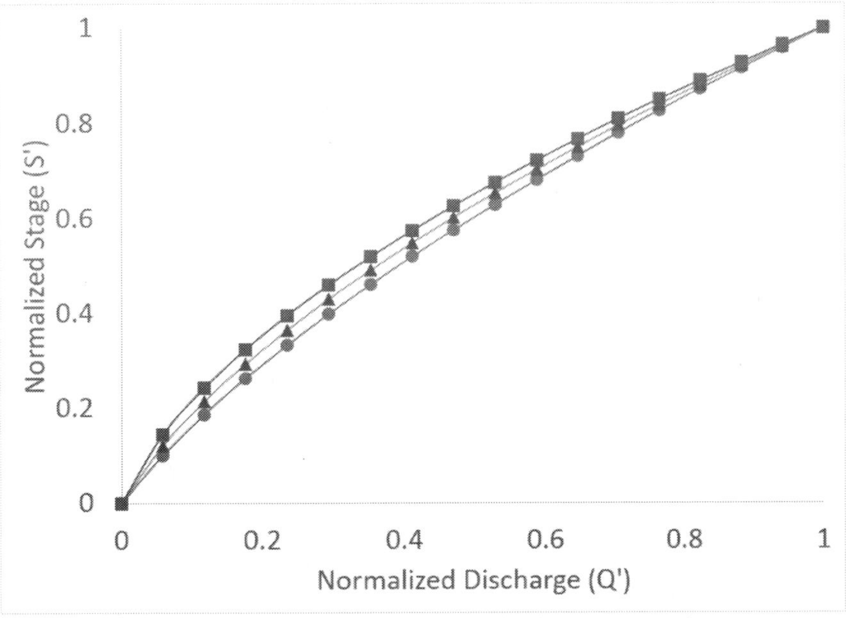

Figure 4. Normalized Stage Discharge Functions

System Response Function

A truncated normal distribution is used to define the system response function. The mean of the distribution describes the best estimate of the stage at which the levee will breach and the standard deviation describes the uncertainty about the best estimate. The mean is computed so that the system response function has a zero probability of breach at the toe of the levee and a user defined probability of breach at the computed 1% annual chance exceedance stage. The standard deviation is estimated from the mean by assuming a coefficient of variation. The system response is assumed equal to 1 for all stages above the top of levee to capture the overtopping events. Nine values for the probability of breach at the 1% stage and three values for the coefficient of variation are used to generate a total of 27 system response functions. Table 2 lists the parameters used for the system response functions. A few example system response functions are illustrated in Figure 5.

Table 2. System Response Function Parameters

Parameter	Values
Probability of Breach at Computed 1% Stage	0.001, 0.002, 0.005, 0.01, 0.02, 0.05, 0.1, 0.2, 0.5
Coefficient of Variation (CV)	0.2, 0.4, 0.6

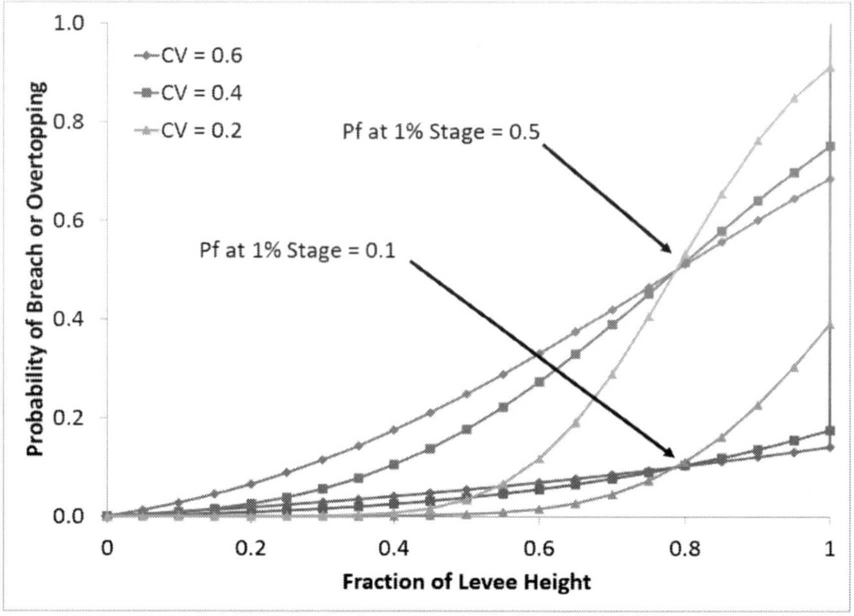

Figure 5. Example System Response Functions

Levee Height

A total of 243 unique levee configurations were obtained from the 3 discharge frequency functions, 3 stage discharge functions, and 27 system response functions. Each of these combinations was evaluated for 7 different levee heights to obtain a grand total of 1701 unique scenarios. The levee heights selected for the parametric analysis were based on overtopping assurance levels of 50%, 60%, 70%, 80%, 90%, 95%, and 99%. Overtopping assurance is estimated using the same computation framework in Figure 1 and assuming the probability of levee breach prior to overtopping is zero. A higher overtopping assurance is analogous to having a greater amount of freeboard. Figure 6 illustrates the relationship between overtopping assurance and freeboard for three combinations of discharge frequency and stage discharge.

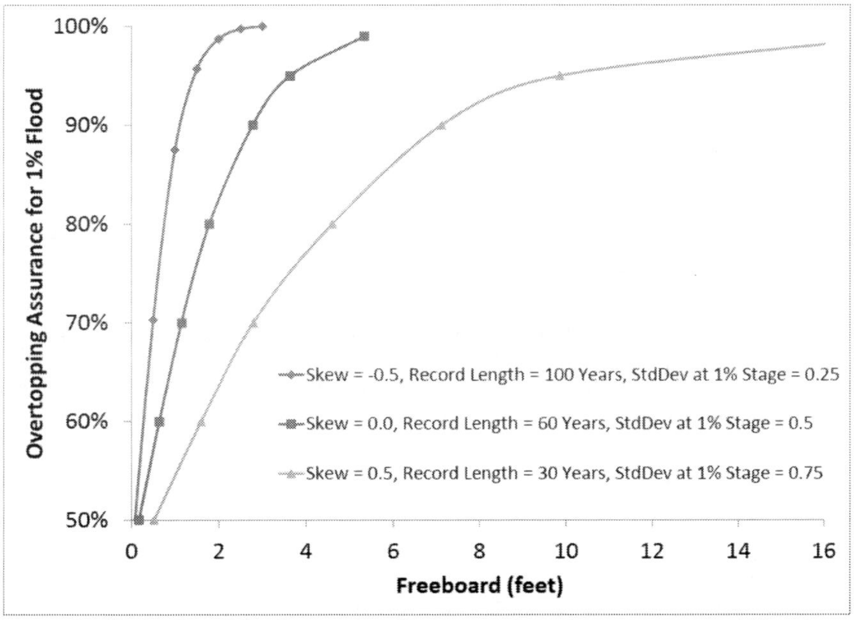

Figure 6. Relationship Between Overtopping Assurance and Freeboard

Results and Observations

Figure 7 illustrates a representative summary of the results for 324 of the 1701 scenarios. The plot shows the relationship between the probability of breach at the 1% stage and the assurance that the flood hazard is less than 1% for levees with overtopping assurance values of 50%, 70%, 90%, and 95%. Variations due to the uncertainty in the system response function defined by the coefficient of variation are shown by the different line colors. Variations due to the uncertainty in the discharge frequency and stage discharge functions are shown by the different symbols. The results suggest that assurance begins to decrease significantly as the probability of breach at the 1% stage increases above a value of about 0.01. These results could inform selection of a threshold value for reasonable assurance based on the combined effect of levee height and levee performance. For example, assurance levels greater than 70% appear to be reasonable and achievable given an overtopping assurance greater than 80% and a probability of breach at the 1% stage less than 0.01.

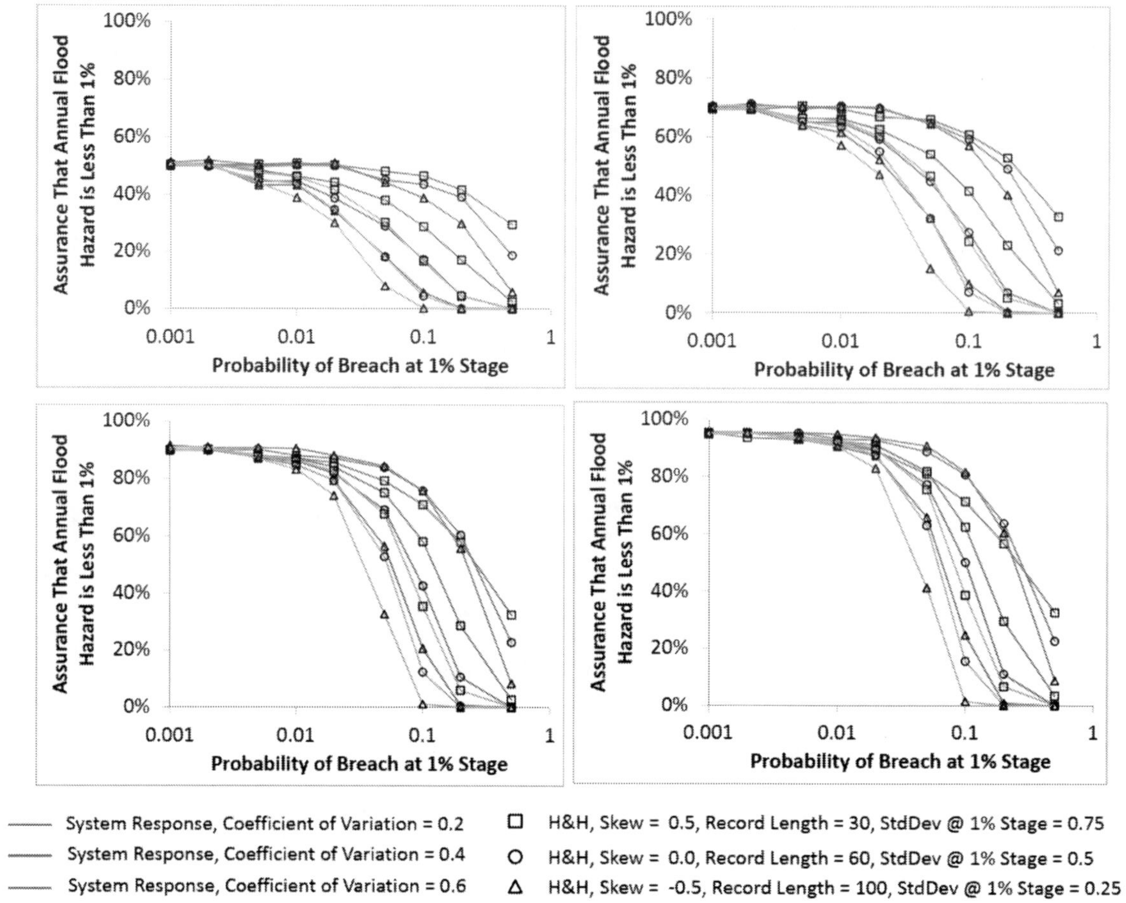

Figure 7. Assurance for Various Scenarios

Figure 8 shows the relationship between the best estimate of the annual flood hazard computed without uncertainty and the assurance that the annual flood hazard is less than 1% computed with uncertainty. The plot contains data points for all 1701 scenarios. These results could inform selection of a threshold value for reasonable assurance based on the relationship between the computed flood hazard and the assurance for the flood hazard. For example, an assurance threshold of 90% that might result in a computed annual flood hazard as infrequent as 0.002 (500 year) might be considered to be too stringent. Alternatively, an assurance threshold of 75% that might result in a computed annual flood hazard of about 0.005 (200 year) might be more reasonable. The results also provide an opportunity to develop a threshold based on annual flood hazard for screening level analysis or when uncertainty information is not readily available.

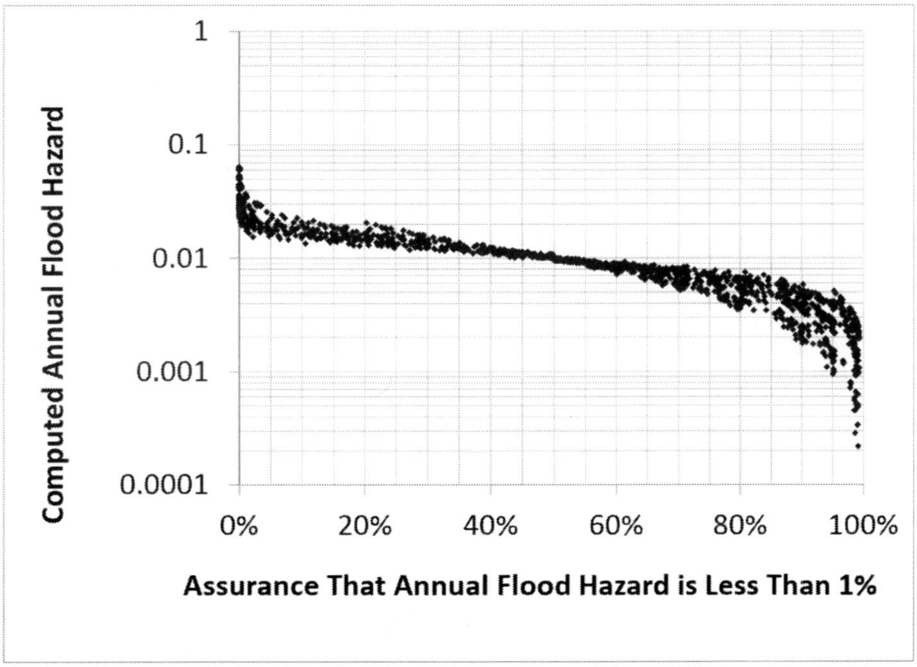

Figure 8. Relationship Between Computed Annual Flood Hazard and Assurance

Discharge frequency and stage discharge functions were obtained for four levee systems in the USACE portfolio to check the reasonableness of the parametric analysis. The system response function with a coefficient of variation of 0.4 was selected for this comparison. The HEC-FDA software package was used for the analysis of the USACE levee systems and the results were compared against results obtained from the parametric analysis. Figure 9 shows a comparison between the USACE and parametric levee systems. The results indicate reasonable consistency between the parametric analysis and the four levees evaluated. Despite the small sample size, the results suggest that the parametric analysis might be representative of levee systems in the USACE portfolio.

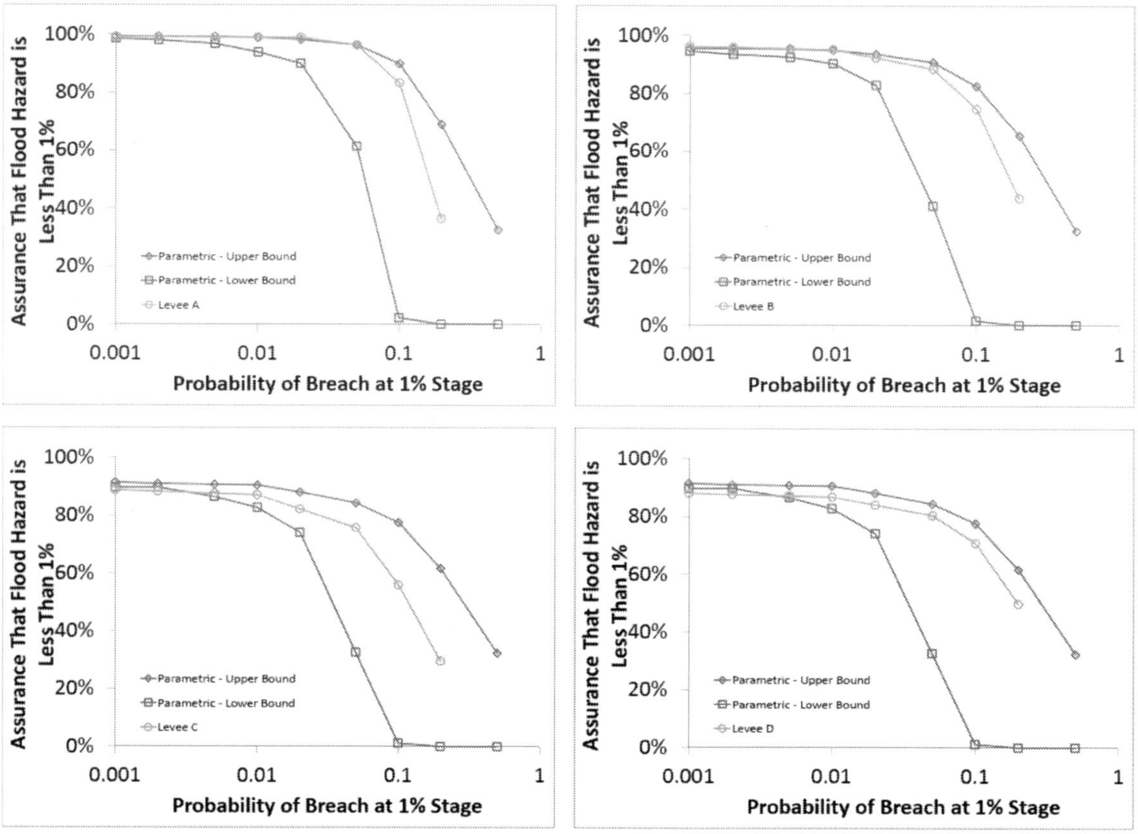

Figure 9. Comparison of Parametric Levees with Levees in USACE Portfolio

PATH FORWARD

The USACE levee safety program has transitioned to using risk assessment and risk management principles to guide credible and consistent decision making. The next evolution of this transition is to use risk assessment information to inform levee evaluation and accreditation decisions. This change will be reflected in an update to USACE policy for the evaluation of levee systems in support of the NFIP. The updated policy supports improved alignment of USACE and FEMA processes and will also provide a more credible and consistent characterization of the flood hazard to support improved floodplain management, floodplain mapping, and flood insurance decisions. An additional benefit is that the information from a risk assessment aligns with FEMA's process for mapping non-accredited levees. Initial selection of a threshold for reasonable assurance will be informed by the results of the parametric study and the expert judgment of engineers and decision makers. As experience is gained in using this new approach, the guidelines will be refined.

The use of risk assessments for NFIP decisions provides several opportunities for the future. The use of probabilistic and risk informed methods supports many of the recommendations made recently by the National Research Council (NRC, 2013). These recommendations include using a modern risk informed approach that includes consideration of flood frequency analysis, levee performance, systems analysis, inundation assessment, and consequence analysis. With a modern risk informed approach, a gradation of flood hazards that vary from property to property could be estimated and applied to establish flood insurance rates. Results of risk analysis can also be used to inform flood risk communication strategies. Shared solutions at the local, state, and federal level can be more readily identified and implemented with the use of risk assessments.

REFERENCES

National Research Council (1982). A Levee Policy for the National Flood Insurance Program. Retrieved 1 October 2015 from http://www.nap.edu/catalog/19600/a-levee-policy-for-the-national-flood-insurance-program.

United States Geologic Survey (1982). Guidelines for Determining Flood Flow Frequency. Retrieved 1 October 2015 from http://water.usgs.gov/osw/bulletin17b/dl_flow.pdf.

Mapping of Areas Protected by Levee Systems, 44 CFR 65.10 (1986).

National Research Council (2000). Risk Analysis and Uncertainty in Flood Damage Reduction Studies. Retrieved 1 October 2015 from http://www.nap.edu/catalog/9971/risk-analysis-and-uncertainty-in-flood-damage-reduction-studies.

National Committee on Levee Safety (2009). Recommendations for a National Levee Safety Program. Retrieved 1 October 2015 from http://www.leveesafety.org/rec_reporttocongress.cfm.

United States Army Corps of Engineers (2010). Engineer Circular 1110-2-6067, USACE Process for the National Flood Insurance Program (NFIP) Levee System Evaluation. Retrieved 1 October 2015 from http://planning.usace.army.mil/toolbox/library/ECs/EC1110-2-6067_31Aug2010.pdf.

Federal Emergency Management Agency (2013). Analysis and Mapping Procedures for Non-Accredited Levee Systems. Retrieved 1 October 2015 from http://www.fema.gov/media-library-data/20130726-1922-25045-4455/20130703_approachdocument_508.pdf.

Federal Emergency Management Agency (2013). Analysis and Mapping Procedures for Non-Accredited Levee Systems, New Approach. Retrieved 1 October 2015 from https://www.fema.gov/media-library/assets/documents/33587.

National Research Council (2013). Levees and the National Flood Insurance Program: Improving Policies and Practices. Retrieved 1 October 2015 from http://www.nap.edu/catalog/18309/levees-and-the-national-flood-insurance-program-improving-policies-and.

United States Army Corps of Engineers and Federal Emergency Management Agency (2013). Flood Protection Structure Accreditation Task Force: Final Report. Retrieved 1 October 2015 from http://www.usace.army.mil/Portals/2/docs/civilworks/levee/FINAL_FPSATaskForceReport.pdf.

United States Army Corps of Engineers and Federal Emergency Management Agency (2014). Memorandum of Understanding for Alignment of Levee Activities, Information, and Activities. Retrieved 1 October 2015 from http://www.usace.army.mil/Portals/2/docs/civilworks/levee/FEMA_USACE_MOU_Signed_20141113.pdf.

United States Army Corps of Engineers (2014). HEC-FDA, Flood Damage Reduction Analysis, User's Manual. Retrieved 1 October 2015 from http://www.hec.usace.army.mil/software/hec-fda/documentation/CPD-72_V1.4.pdf.

Federal Emergency Management Agency (2015). The National Flood Insurance Program Community Status Book. Retrieved 1 October 2015 from http://www.fema.gov/national-flood-insurance-program/national-flood-insurance-program-community-status-book.

USING TOLERABLE RISK GUIDELINES IN THE CALIFORNIA DELTA

Lawrence H. Roth[1], P.E., G.E.
Jessica Ludy[2], CFM

ABSTRACT

Tolerable risk guidelines (TRG) recognize that society is willing to live with some risk in order to secure certain benefits. TRG supports policy formulation, informs risk management decisions, enables evaluation of trade-offs, and promotes clear and transparent risk communications. The TRG process is useful in allocating scarce resources, achieving equity, and promoting efficiency. This paper describes the TRG process being used to inform levee improvement investment priorities in the California Delta, the confluence of the Sacramento and San Joaquin Rivers, and keystone of the state's water system. That 25 million Californians and four million acres of farmland depend on the Delta for some or all of their water is a testament to its importance to the state's economy. Much of the Delta is below sea level and is protected by 1,100 miles of levees that have water against them year-round. People, water supply, and infrastructure are dependent on levees for flood risk reduction. Delta levees, however, are threatened by earthquakes; by seepage through aged, poorly engineered embankments; and by overtopping made worse by subsidence and sea level rise. Catastrophic levee failure would be devastating to people, water supply, infrastructure, and California's economy. Flood risk in the Delta cannot be eliminated, and design standards specifying levels of flood protection address only the hazard while ignoring consequences. TRG recognizes that the Delta is complex, the future is uncertain, and that absolute protection is not possible. Application of TRG by decision-makers may enable reducing risk in the Delta to tolerable levels through prudent levee investments.

INTRODUCTION

At the heart of California, the Sacramento and the San Joaquin Rivers flow into the California Delta, an estuary comprising 60 major islands formed by 1,100 miles of levees in an area of about 1,300 square miles, much of which is below sea level (see Figure 1). Levees in the Delta are critical to protecting people, property, natural resources, and infrastructure systems of statewide importance. Twenty-five million Californians and four million acres of irrigated farmland receive some or all of their water from the Delta. Levee failure would cause devastating flooding, threaten lives, damage property and infrastructure, disrupt Delta communities, and cause harm to water supplies and the Delta ecosystem.

[1] Principal Engineer, Arcadis U.S., 101 Creekside Ridge Court, Roseville, CA 95678, larry.roth@arcadis.com.
[2] Water Resources Specialist, Arcadis U.S., 100 Montgomery Street, San Francisco, CA 94104, Jessica.ludy@arcadis.com.

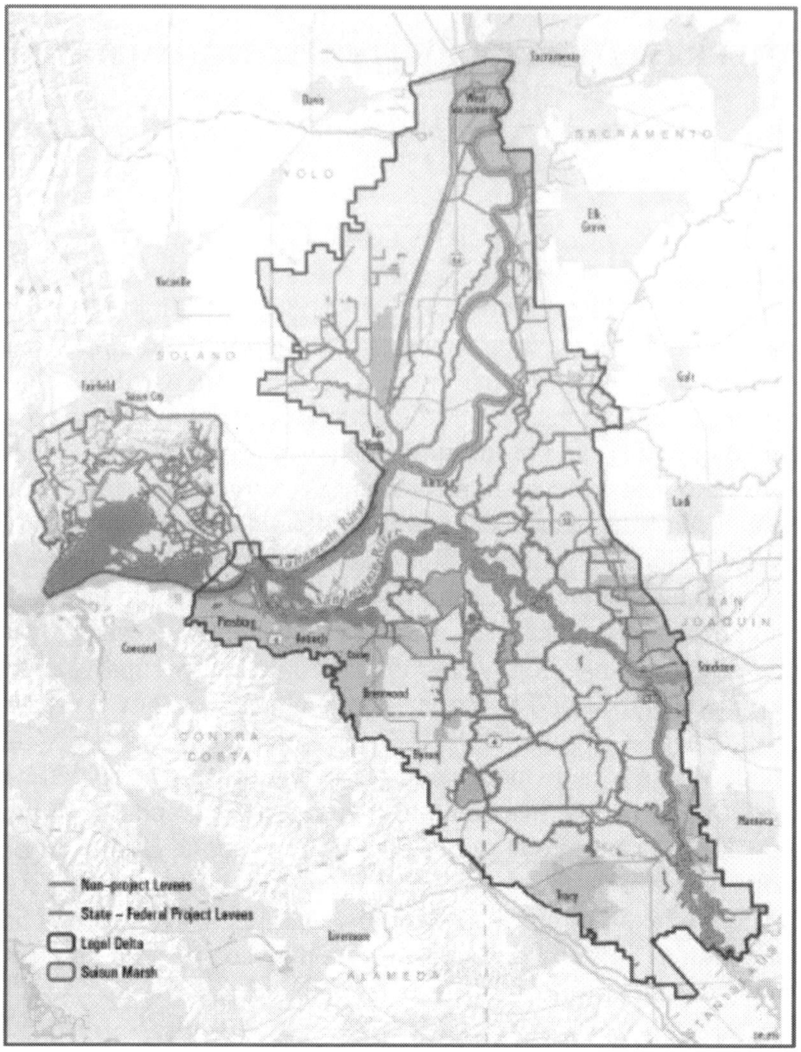

Figure 1. The California Delta (Council 2013)

Since 1900, there have been 158 levee failures that have flooded Delta islands (DWR 2009). While levee maintenance and improvements over the past 30 years have helped strengthen levees, the State of California has not developed a long-term strategy to guide future investments of the limited funding available to improve levees. Unfortunately, without improvement, the probability of levee failure will only increase with deteriorating levee conditions, sea level rise, and the uncertainty associated with climate change.

Spurred by the devastation caused by Hurricane Katrina on the Gulf Coast, the California legislature passed a legislative package in 2007 to increase awareness of the fragility of the state's levee system and to allocate additional funds for emergency repairs. Two years later, the legislature passed the Delta Reform Act of 2009 mandating that the Delta Stewardship Council determine *strategic levee investments* to reduce risks to *life, property, and State interests* in the Delta (California Water Code sections 85305(a) and 85306).

The state's primary interests in the Delta are embodied in legislation defining the *coequal goals* of water supply reliability and ecosystem restoration, two goals often viewed in conflict with one another. The Delta Reform Act also requires that the coequal goals *must be achieved in a manner that protects and enhances the Delta as an evolving place.* This means recognizing the unique values of the Delta's communities, its agricultural importance, its cultural and historical heritage, its recreational opportunities, and its scenic nature (California Water Code 29702).

Funding for levee repairs and improvements is scarce and an investment strategy to reduce flood risks in the Delta must be cost-effective. Since eliminating risk is impossible, the sensible goal should be to reduce flood risk to *as low as reasonably practicable* (more on this later). In contrast to the nation's longstanding focus on the "one-percent approach" (44 Code of Federal Regulations 65.10), the use of Tolerable Risk Guidelines (TRG), which has been applied to risk management in other industries for many years and to flood risks in many other countries, offers a framework in which to allocate limited resources while also considering the question, "*how safe is safe enough?*"

THE CALIFORNIA DELTA

The California Delta is the largest estuary on the west coast of the Americas, generally bound by Sacramento and West Sacramento to the north, by Tracy to the south, and by Antioch and San Pablo Bay to the west. The Delta is characterized primarily by farmland and over a thousand miles of winding channels and sloughs whose levees were first built in the 1850s. In normal years, the Delta receives extensive runoff of rain and snowmelt from the Sacramento and San Joaquin River watersheds, which cover just over one-third of California.

Before European settlers arrived in California in the 1500s, the Delta was a vast estuary and wetland complex with over 540 square miles of freshwater wetlands and 380 square miles of salt marsh that provided rich and productive habitat for more than 500 species of fish and wildlife. The pre-1800s Delta, often referred to as the American Serengeti, flooded frequently creating a vast, seasonal inland sea. In recent years, however, the ecosystem of the Delta has been in dramatic decline and several species face extinction if steps are not taken soon. Today, less than five percent of its former salt marsh and wetland habitat remains.

During and following the Gold Rush, its rich potential for agriculture was exploited and today's Delta is largely man-made, characterized by an extensive levee system built to reclaim land for farming, improve navigation, and stave off flooding. A century ago the land surface was near sea level. Years of levee building and farming have caused consolidation, compaction, and oxidation of the peat and organic soils, and levees built on them vary in quality from adequate to fragile. The islands continue to subside and most land is now as much as 20 feet below sea level. Notably, the levees function as dams with water against them year-round.

Not unlike other parts of the country, levees that were built for one purpose now serve many others; at their onset, the Delta's levees enabled reclaiming land to farm, but they now provide flood risk reduction for:

1. Nearly 500,000 people, including urban areas of Stockton, Sacramento, and West Sacramento.
2. Water supply reliability and protection from salt-water intrusion for users in the Delta, for users that divert from the Delta, and for extensive exports to the San Joaquin Valley and Southern California.
3. Critical infrastructure systems (water supply facilities, highways, railroads, natural gas pipelines, power lines, etc.) that crisscross the Delta and that are important to the entire state.
4. Important habitat and managed wetlands, and for channel-margin habitat critical to fish.
5. Delta as a place, including its agricultural-based economy, its cultural and historic heritage, and its recreational opportunities.

Each of these is of interest and importance to the State of California: threats to lives and property; disruption of water supply; harm to the ecosystem; and damage to Delta as a place. The state, however, has limited resources. How should the state prioritize its investments to complement investments made by local reclamation agencies to improve Delta levees to reduce risk to these state interests?

LEVEES AND RISK

The Delta flood risk reduction system is critically dependent on levees, which are susceptible to failure from ground shaking, overtopping, and seepage. Levee stability is made worse by continuing subsidence and sea level rise. Catastrophic levee failure would be devastating to people, property, water supply, infrastructure, and the environment. William Hammond Hall, California's first state engineer, is credited with saying, "there are two types of levees—those that *have been* overtopped by floodwaters, and those that *will* be overtopped by floodwaters" (emphasis added). If there are consequences from flooding, it is inescapable that the risk is real and that it must be managed.

Eliminating flood risk, however, is impossible (NAS 2013). The National Academy of Sciences (NAS) defines *Residual Risk* as the risk that remains after considering the mitigating effects of structural, non-structural, and other risk reduction measures (NAS 2013); it is the risk that remains even after all reasonable and practicable measures have been implemented. Unfortunately, many people, including well-meaning decision-makers, perceive that levees eliminate risk. The NAS report reminds us that residual risk is always present, because no levee is fail-safe.

What is risk? *Risk* is generally defined as the potential for loss or injury. To assess risk in the case of levees, we need to know:

1. What is the hazard from river stage or earthquake, and how likely is it to occur?

2. How will the levee perform in the face of the hazard?
3. Who and what are in harm's way?
4. How much harm will be caused if the levee fails to perform adequately?

Put simply: *Risk* = *Probability* of an event x *Consequences* of that event.

We are all familiar with risk, and we make decisions to manage risk every day – for example: should we wear a seatbelt (or not) when driving?; what car insurance deductible should we choose?; and when, based on our own technological prowess, should we purchase an extended product warranty such as Apple Care? Yet, when it comes to flood management, communities in the United States have historically focused almost exclusively on the one-percent-chance flood (44 Code of Federal Regulations 65.10), a level-of-protection (LOP) approach.

In the U.S., the one-percent-chance LOP approach overwhelmingly drives most development as well as design and investments in levees. Some communities, however, have recently adopted higher freeboard requirements than those specified in 44 Code of Federal Regulations 65.10; for example:

1. Urban areas in California are now required to provide protection from the 0.5 percent annual chance flood, or 200-year LOP. (California Government Code Section 65007(n)).
2. New York City's Office of Long Term Sustainability (2013) post-Hurricane Sandy report, "Stronger, More Resilient New York," recommends protection of critical infrastructure against water levels with a 0.2 percent annual chance of occurrence, or 500-year LOP.
3. The Red River Basin Commission (2011) in the U.S. and Canada recommends 1/750-year LOP (0.13 percent) for critical infrastructure, and 1/500-year LOP (0.2 percent) for urban metropolitan areas.

While these design standards are greater than the standard set in the National Flood Insurance Program (NFIP), all are tied to a water surface elevation associated with a flood hazard; they remain focused on the probability, or hazard part of the risk equation while ignoring the consequences part.

The NAS (2013) defines LOP as the flood recurrence interval that a specific structure is designed to withstand. Current practice regarding LOP in the Delta is based on standards or guidelines generally promulgated by federal or state agencies. Figure 2 illustrates current levee guidance as described in the *Delta Plan* (Council 2013).

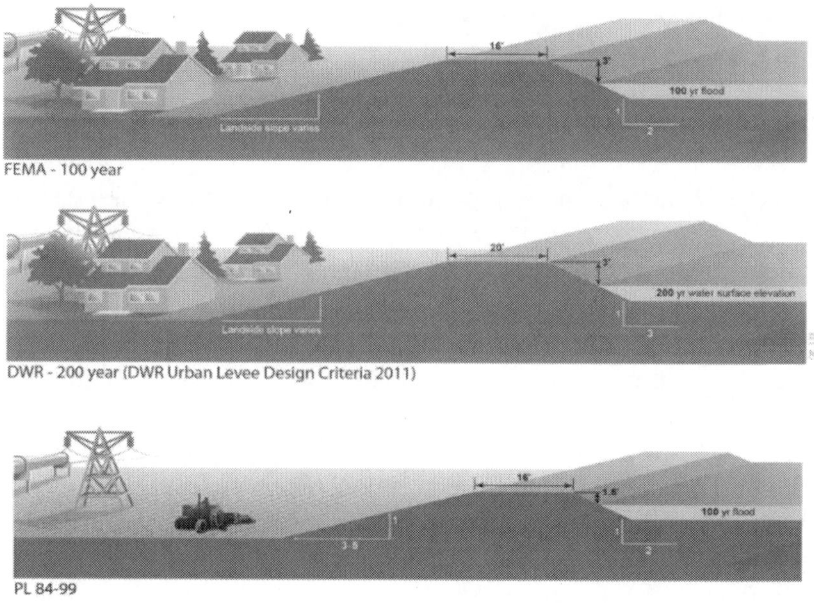

Figure 2. Current Levee Guidance in the Delta (Council, 2013)

Public Law 84-99 (PL 84-99) guidelines represent a minimum requirement established by the U.S. Army Corps of Engineers (USACE) for levees in its Rehabilitation and Inspection Program (USACE 2009). Levees (and Delta islands) that meet this standard may be eligible for federal funding for emergency assistance, and for levee rehabilitation after flooding. The PL 84-99 standard for levee geometry defines a minimum levee height and a factor of safety for slope stability, but is not generally associated with a specific LOP. In 1987, the USACE developed a Delta-specific standard based on the Delta's soil and peat foundation conditions (Council 2013).

The Federal Emergency Management Administration (FEMA) 100-year criterion is intended to reduce the frequency of flooding of low-lying lands by providing one-percent Annual Exceedance Probability (AEP) LOP. This criterion, often referred to as the insurance standard, provides that levees must meet flooding that is the basis of FEMA's flood insurance rate maps under the NFIP (44 Code of Federal Regulations 65.10). Levees must meet prescribed requirements for freeboard, slope stability, seepage, erosion, and settlement. In communities where levees meet these requirements, new housing or commercial developments on the protected side of the levee are not required to meet federal flood-proofing standards, and owners can obtain federally guaranteed mortgages without purchasing flood insurance.

The California Department of Water Resources (DWR) Urban Levee Design Criteria establishes criteria for levee height and geometric design, and includes requirements for freeboard, slope stability, seepage, erosion, settlement, and seismic stability (DWR 2012). The criteria are intended to provide protection against a flood that has a 0.5

percent chance of being equaled or exceeded in any given year (200-year LOP). Flood-prone urban or urbanizing areas with 10,000 or more residents must meet this standard by 2025 (California Government Code 65865.5 (a) (3)).

Depending on the circumstances, each of these standards or guidelines has been considered for use in the Delta as LOP to be attained or sustained. The PL 84-99 criteria, however, are disaster rehabilitation guidelines based primarily on levee geometry with little or no consideration of performance of either the levee or its foundation. The FEMA 100-year criterion is an insurance standard that includes requirements regarding levee performance. The DWR Urban Levee Design criterion is a levee design standard that must be used for urban areas in California.

Earlier, we posed the question: how should the state prioritize its investments in Delta levee improvements to reduce risk to state interests? Unfortunately, applying LOP methodology does not provide for prioritizing investments in ways that are open, transparent, and quantifiable as discussed below.

The one-percent approach (and the use of LOP in general) poses many challenges to decision-makers because LOP focuses on the hazard and ignores the consequences of flooding. This may unintentionally communicate that flood risk can be eliminated. In addition, LOP standards and guidelines are not flood safety standards – they are not developed after meaningful discussions of "*how safe is safe enough*?" Specifically in the Delta, should levees protecting large populations (e.g., the Pocket Area of Sacramento) be designed to the same LOP as levees protecting agricultural land?

In addition, LOP standards and guidelines are generally not performance-based, and they do not account for the possibility of floods larger than those prescribed by their water surface metric. LOP standards and guidelines generally fail to account for critical issues such as: sea level rise and increased water surface elevations; changing precipitation patterns due to climate change; increased consequences from population growth and development in flood-prone areas; and increased costs for rebuilding infrastructure that was designed for today's flood protection systems.

Of critical importance, LOP standards and guidelines fail to address residual flood risk. The most insidious residual risk is a disaster of epic proportions that is completely unexpected, sometimes termed a *Black Swan* (Taleb 2007). Black Swans 1) are outliers, outside our normal expectations; 2) carry extreme impacts; and 3) lead us to concoct explanations for their occurrence after the fact, trying to make them explainable or predictable.

LOP standards and guidelines typically exclude the possibility of a Black Swan. To put this in the context of flood risk: if we had truly expected hurricanes like Katrina and Sandy, would we have been better prepared? Application of tolerable risk guidelines (TRG) can help us consider the possibility of a Black Swan as we evaluate residual risk. Specifically for the Delta, how should we account for a rare, but very large earthquake on a nearby fault? And, how should we account for the possibility of massive flooding caused by a rare, but plausible atmospheric river precipitation event?

TRGs offer an informed and appropriate basis on which to formulate flood risk management policy and to prioritize and allocate scarce resources for levee improvements. TRGs can inform discussions on "*how safe is safe enough*?" and create a sound basis for better understanding of risk, improved communications of risk and risk-related issues, transparency in risk decision-making, and overall enhanced management of risk.

TOLERABLE RISK

Since risk cannot be eliminated, *tolerable risk* can be viewed as a range between every-day, broadly acceptable risks like driving a car, to intolerable risks, such as mass casualties or societal disruption from a major natural or man-made disaster (USACE 2011). Figure 3 illustrates that there is a range of risks that may be tolerable between those broadly acceptable and those intolerable – i.e., risks that are not negligible, yet are risks that people are willing to accept in order to secure certain benefits (HSE 2001 and USACE 2010).

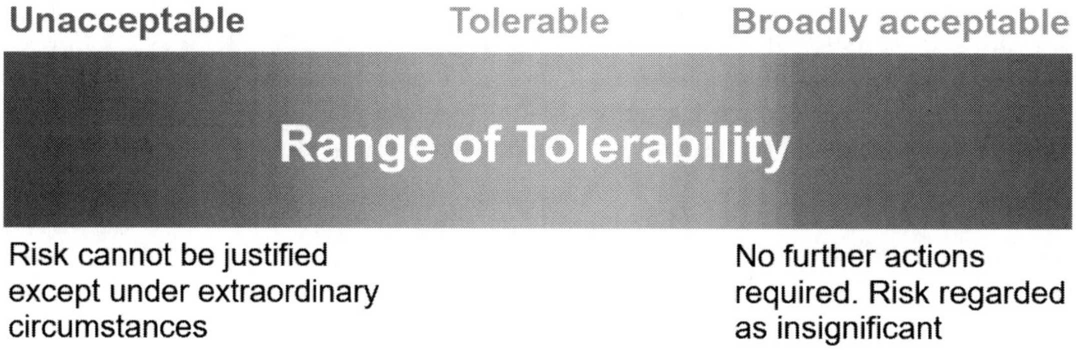

Figure 3. Range of Tolerable Risks

The basic principles of tolerable risk are that life safety is paramount, that risk cannot be ignored, and that absolute safety cannot be guaranteed. Our goal should be to reduce risk to *as low as reasonably practicable,* or ALARP[3]. To meet this goal, TRG enables us to consider:

1. *Individual Risk:* The probability of harm to individuals and the things they value.
2. *Societal Risk:* The probability of adverse consequences that impact society as a whole. Society is increasingly averse to risk as the scale of consequences increase.
3. *Efficiency:* The need for society to use available resources to achieve the greatest benefit.

[3] ALARP is defined as what can be reasonably achieved without spending an inordinate amount of time, money, and resources to reduce risk relative to benefits.

4. *Equity:* The right of individuals to be protected, and the right that the interests of all are treated with fairness.

Where the LOP approach focuses on the hazard, TRG enables consideration of risk and the consequences of flooding in the Delta on life, property, water supply, the ecosystem, and Delta as a place. In contrast to *flood control* or *flood protection*, applying TRG is a paradigm shift in flood risk management where decisions are informed by a clear understanding of both the probabilities and the consequences of flooding (Sayers et al. 2013). Where funding is limited, tolerable risk guidelines (TRG) also provide a way to rationally allocate scarce resources, to manage flood risk using alternatives such as non-structural options, and to determine what actions should be taken first.

Application of TRG to levees is consistent with the best available science in risk management and is widely used in other countries; for example, the Dutch flood investment priorities are risk-informed (Eigenraam et al 2014). Recent national policy in the Netherlands establishes a tolerable risk limit for loss of life (in contrast to LOP-based guidelines) that limits the risk to which both individuals and society as a whole can be exposed (Delta Programme 2015). In the Netherlands, the goal is that no individual should have more than a 1 in 100,000 annual chance of dying from a flood.

Following Hurricane Katrina, the National Committee on Levee Safety (NCLS) recommended that life safety must be considered paramount, and that tolerable risk guidelines be used to inform decisions on levee safety (NLCS 2009). The USACE and the U.S. Bureau of Reclamation (USBR) apply tolerable risk principles to management of dam safety and to aid in determining how to spend maintenance and improvement funds (USACE 2011 and USBR/USACE 2015). In addition, the Water Resources Reform and Development Act of 2014 requires that levees managed in the federal system use the levee safety action classification to stipulate actions to reduce risk. This action classification system, used by the USACE for its levees, is guided by principles of tolerable risk. Specifically, levee systems that pose unacceptable risks to life safety are classified as needing urgent action to reduce risks to tolerable levels.

Application of the concept of tolerable risk to decision-making on flood risk reduction offers many advantages. As adopted from the USACE (2011), tolerable risk:

1. Enables the decision-making process to focus on the most serious risks and to prioritize actions accordingly.
2. Enables taking actions that are commensurate with risks, which improves cost-effectiveness.
3. Enables rational consideration of both structural and non-structural options to reduce risk as well as consideration of benefits that cannot be readily monetized.
4. Promotes consistency in that similar approaches can be taken in similar circumstances to achieve similar ends.
5. Promotes equity in that everyone's interests and safety can be treated with fairness.

6. Promotes efficiency by using available resources to achieve the greatest benefit.
7. Provides transparency in the decision-making process.
8. Enables clear communication of the risks and risk-based decisions to those most affected.
9. Provides clarity on who is responsible for identifying, managing, and communicating risk.

USING TOLERABLE RISK GUIDELINES IN THE DELTA

Using TRG helps characterize the risk to state interests in the Delta and to evaluate and compare trade-offs amongst investment options in a manner generally compatible with that proposed by the USACE (2011). Areas with greater risk tend to be areas with greater consequences, and using TRG will encourage taking actions to reduce risks to tolerable levels. The USACE's use of TRG in its levee safety program is focused on life safety. Similarly, applying TRG in the Delta enables attention to life safety while also considering risk to other state interests. In stark contrast to application of LOP thinking, using TRG enables decision-makers and stakeholders to compare actions to reduce risk of:

1. Life loss and property damage.
2. Water supply disruption.
3. Harm to the ecosystem and habitat.
4. Damage to Delta as a place.

For example, if life safety is given the highest priority, then reducing risks to the islands with the highest Expected Annual Fatalities (EAF) to as low as reasonably practicable (i.e., to a tolerable level of risk to life safety) is a logical choice on which to focus investments. On the other hand, if life safety and reduced property damage are both high priorities, the TRG process can be used to identify those investments that provide balanced benefits to both life safety *and* property protection.

The process of applying TRG to support decision-making involves four basic steps. The first step is assessing, mapping, and characterizing flood risks. Based on the risk equation, we need to know: what the hazards are; how likely the hazards may occur; and who and what is threatened (people, assets, infrastructure, etc.). We also need to know how the structure will perform in the face of those hazards. In many cases the focus may be on life loss and property damage. The risk of flooding to life safety is measured in EAF, and risk of property damage is measured in expected annual damages, or EAD. EAF and EAD are important metrics in the TRG process.

In the Delta, however, this step also includes assessing, mapping, and characterizing flood risk to state interests as required by legislation. To calculate risk of water supply disruption, of harm to the ecosystem, and of damage to Delta as a place requires identification of metrics similar to those used in EAD and EAF. For example, we can assess risk of harm to the ecosystem using a metric that considers changes to habitat caused by flooding, and to habitat lost or gained by the process of improving levees. The fact that we can include a gain in habitat (say, by constructing a new setback levee to provide for improved channel-margin habitat) illustrates the flexibility that TRG offers to address unique objectives that may be defined by stakeholders and those affected by risk reduction efforts.

Preparation of risk maps can illustrate islands in the Delta with the highest risks, including those islands, if any, that pose unacceptable risks to life safety. An example risk map is shown in Figure 4, which is an EAF risk map for the Netherlands. Maps can also help distinguish between individual risks and societal risks. Specifically, areas where there is higher relative risk (for example, areas where mass casualties or widespread damages could be high) can be a proxy for societal risk. In contrast, areas with lower risk to life safety or flood damage are likely more important for individual risk than societal risk. Using TRG, risk maps can be prepared for the Delta showing risk to life and property, to water supply disruption, to ecosystem harm, and to damage to Delta as a place.

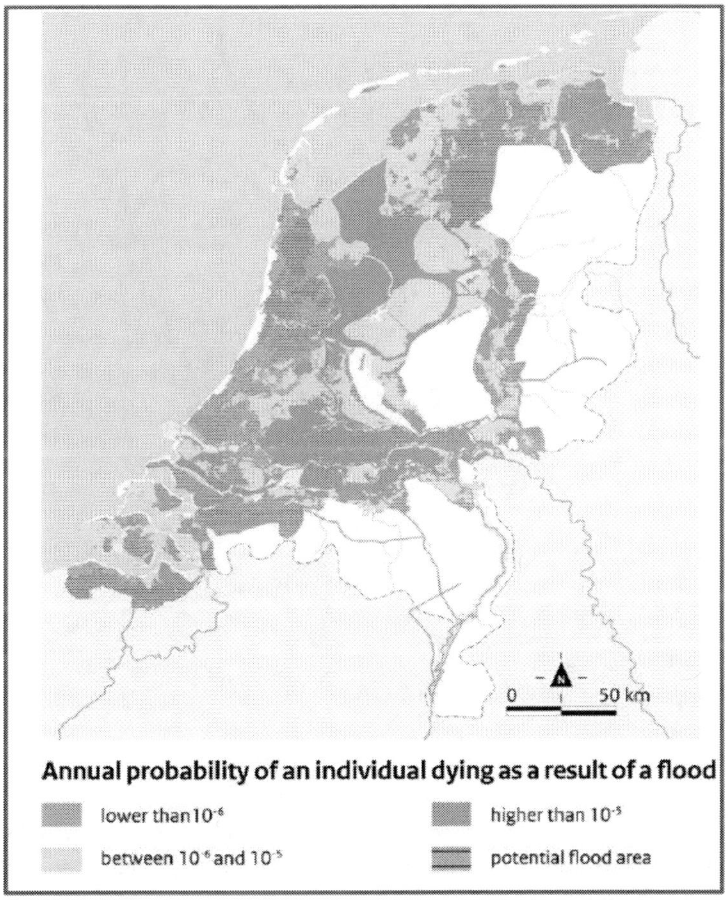

Figure 4. Annual Probability of an Individual in the Netherlands Dying as a Result of a Flood (Delta Programme 2015)

Another way to graphically view risks is on F-N plots as shown in Figure 5. F-N plots are a graphical representation of societal risk that relate the annual probability of failure to fatalities or dollars lost. Figure 5 illustrates several examples of societal risk including plane crashes, car accidents, and mine pit slopes (an occupational safety hazard). The dashed red, dashed blue, and solid green sloping lines are indicators of tolerable risk limits that may be applied through judgment. The blue dashed line represents the tolerable limit for fatalities used by USACE and USBR for dams.

Figure 5 shows where the pre-Katrina New Orleans Hurricane Protection System plots, well above and to the right of the blue dashed line indicating actions were justified to reduce risk before Katrina struck. The plot also illustrates that pre-Katrina risks in New Orleans could have been reduced by building stronger and taller levees (move down on the plot), or by non-structural means (move to the left on the plot), for example by improving evacuation. Moving left reduces risk to people by taking them out of harm's way. Moving down reduces risk to people and property by reducing the probability of a damaging event. Both actions to reduce risk are being implemented post-Katrina.

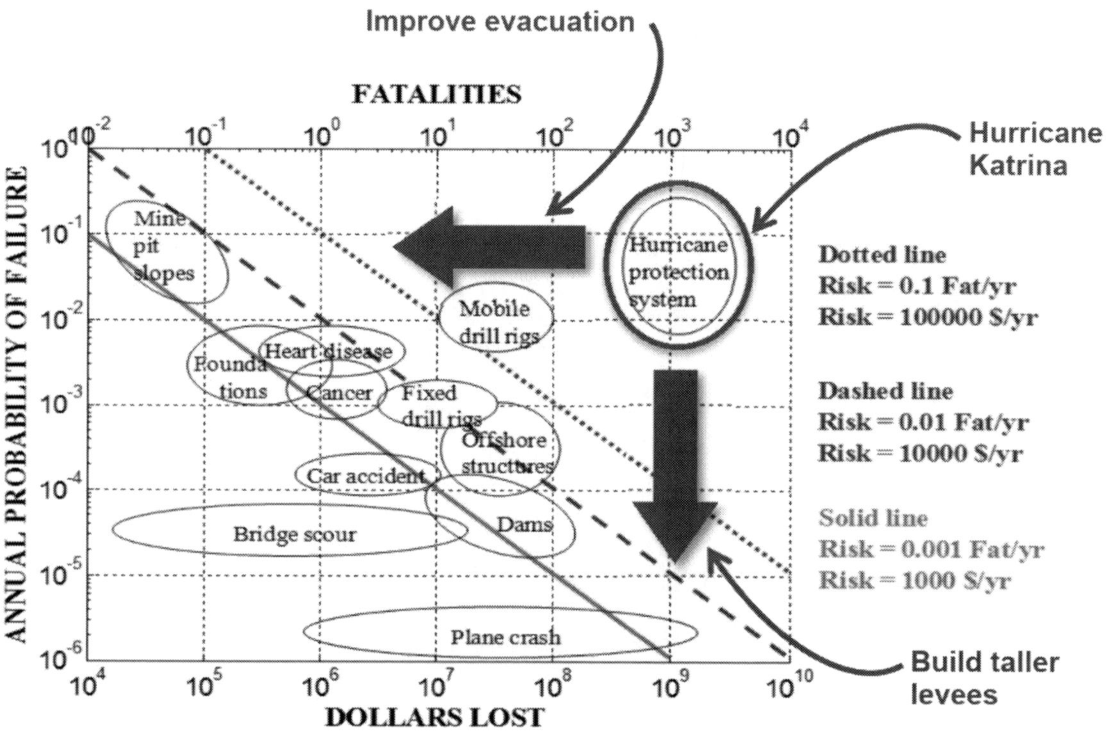

Figure 5. F-N Plot (Briaud 2015)

F-N plots also illustrate a key aspect of TRG – decision-makers and stakeholders must decide together on the limit of tolerable risk, the amount of risk they are willing to accept. Little guidance is provided in the literature for the limit of tolerable risk, except in the case of fatalities where guidance provided by UASCE and USBR for dams is useful. In the Delta, decision-makers and stakeholders will need to determine what the tolerable risks are for life loss and property damage, for water supply reliability, for harm to the ecosystem, and for damage to Delta as a place to guide selection and prioritization of improvements to reduce risk.

After risks have been calculated and limits of tolerable risk have been identified, the second step in the TRG process involves identifying options, both structural and non-structural, that are available to reduce risks to state interests. Working together, decision-makers and stakeholders can identify and evaluate measures that include, but are not limited to, building and improving levees and floodwalls including setback levees; flood-proofing structures; raising base elevations; adopting new land use policies; acquiring land; relocating assets; improving emergency preparedness and evacuation procedures; and encouraging improved insurance coverage.

In the third step, each risk-reduction option is evaluated first for its ability to reduce risks to tolerable levels, and next for its cost-effectiveness. This can only be done using TRG. LOP methodology does not involve calculating risk, and unless we know the risk, we cannot judge the effectiveness of measures that might be available to reduce risk. The evaluation of risk-reduction options by decision-makers and stakeholders enables comparison of trade-offs, such as the impacts to water supply reliability, the Delta's ecosystem, and to Delta as a place. Comparing trade-offs across various investment scenarios helps inform a comprehensive risk reduction strategy.

The fourth step is implementation of a program for continuous review and evaluation of flood risks. Using the TRG process enables us to consider future conditions, including sea level rise, climate change impacts on hydrology, continuing island subsidence, and future development patterns. This offers an advantage over LOP methodology, which focuses on a levee with geometry that accommodates the one-percent water surface elevation at one point in time, regardless of whether farmland or a hospital is behind it. If new data suggest the one-percent flood is much greater than what the models show, infrastructure built today may more quickly become inadequate. In assessing future risks, however, we can calculate the impact that various water surface elevations and land use decisions will have on risk. If future conditions pose unacceptable risks to life safety or other state interests, then policy and investment decisions can be made today that reduce the risk in the future.

CONCLUSION

Executive Order 13960 (Federal Register 2015) specifies the Federal Flood Risk Management Standard and intends to improve the "resilience of communities and federal assets against the impacts of flooding." Implementation regulations for federal agencies are forthcoming, and the order states that, given the risks posed by climate change, compliance in designing for protection against the one-percent water level may be achieved largely by increasing freeboard. However, it also offers that the base flood used for design could be informed by using:

> "...a climate informed, science approach that uses the best-available, actionable hydrologic and hydraulic data and methods that integrate current and future

changes in flooding based on climate science. This approach will also include an emphasis on whether the action is a critical action as one of the factors to be considered when conducting the analysis."

The TRG approach meets the intent of the Executive Order by considering the criticality of the consequences, facilitating a better understanding of the risks associated with climate change, and presenting actionable alternatives along with information on cost-effectiveness and on risk-reduction benefits to inform decisions. On the other hand, using the LOP approach, recreates many of the same challenges we face today and will be less adaptable to future changing climate and development conditions. TRG offers the opportunity to spend limited investments wisely and to build resilient communities.

Catastrophic levee failures in the Delta would be devastating to people, water supply, infrastructure, the ecosystem, and California's economy. Flood risk in the Delta cannot be eliminated, and design standards specifying levels of flood protection address only the hazard while ignoring consequences. The TRG process prioritizes life safety while recognizing that the Delta is complex, the future is uncertain, and absolute protection is not possible. Application of the TRG process enables assessing risks to state interests in the California Delta and prioritizing investment options across its 1,100 miles of levees to reduce those risks in an open, transparent, and cost-effective manner that promotes efficient and equitable use of resources.

REFERENCES

Briaud, Jean-Louis, Texas A&M University, personal communication, 2015.

California Department of Water Resources (DWR). 2009. Delta Risk Management Strategy. February.

California Department of Water Resources (DWR). 2012. FloodSAFE, Urban Levee Design Criteria. Accessed March 3, 2015. http://www.water.ca.gov/floodsafe/leveedesign/ULDC_May2012.pdf.

Delta Programme. 2015. Working on the Delta: Decisions to Keep the Netherlands Safe and Livable. Accessed January 15, 2015. http://www.deltacommissaris.nl/english/Images/Delta%20Programme%202015%20English_tcm310-358177.pdf.

Delta Stewardship Council (Council). 2013. *The Delta Plan.* Available from: http://deltacouncil.ca.gov/sites/default/files/documents/files/DeltaPlan_2013_CHAPTERS_COMBINED.pdf.

Eijgenraam, C., J. Kind, C. Bak, R. Brekelmans, D.k den Hertog, M. Duits, K. Roos, P. Vermeer, and W. Kuijken. 2014. *Economically Efficient Standards to Protect the Netherlands Against Flooding. Interfaces*, Vol. 44, No. 1, January–February, pp. 7–21.

Federal Register. 2015. Executive Order 13690 of January 30, 2015. *Establishing a Federal Flood Risk Management Standard and a Process for Further Soliciting and Considering Stakeholder Input.* Volume 80, No 23. Accessible: http://www.gpo.gov/fdsys/pkg/FR-2015-02-04/pdf/2015-02379.pdf

Health and Safety Executive (HSE). 2001. *Reducing Risks, Protecting People. HSE's Decision-making Process*, Risk Assessment Policy Unit, HSE books, Her Majesty's Stationery Office, London, England, 2001. ISBN 0 7176 2151 0 (www.hse.gov.uk/risk/theory/r2p2.htm).

New York City Office of Long Term Sustainability. 2013. *Plan NYC: Stronger more Resilient New York.* Accessed January 1, 2015. Available from: http://www.nyc.gov/html/sirr/html/report/report.shtml.

National Committee on Levee Safety (NLCS). 2009. *Recommendations for a National Levee Safety Program*. Washington, DC.

National Research Council of National Academy of Sciences (NAS). 2013. *Levees and the National Flood Insurance Program: Improving Policies and Practices.* National Academies Press, 500 Fifth Street, NW, Washington, DC 20001.

Red River Basin Commission. 2011. *Long Term Flood Solutions for the Red River Basin.* Accessed January 21, 2015 from http://www.redriverbasincommission.org/Comprehensive_Report_12-15-11_FINAL.pdf

Sayers, P., Y. L.i, G. Galloway, E. Penning-Rowsell, F. Shen, K. Wen, Y. Chen, and T. LeQuesne. 2013. *Flood Risk Management: A Strategic Approach.* Paris, UNESCO.

Taleb, Nassin Nicholas, *The Black Swan*, Random House, New York, New York, 2007.

U.S. Army Corps of Engineers (USACE). 2009. Public Law 84-99, *Rehabilitation Assistance for Non-Federal Flood Control Projects*, Department of the Army, Corps of Engineers, Omaha District, Emergency Management Branch.

U.S. Army Corps of Engineers (USACE). 2010. *Proceedings of the Workshop Exploration of Tolerable Risk Guidelines for the USACE Levee Safety Program*, Institute for Water Resources Report 10-R-9.Washington, DC: USACE.

U.S. Army Corps of Engineers (USACE) 2011. Engineering Regulation (ER) 1110-2-1156. *Engineering and Design. Safety of Dams-Policies and Procedures.*

U.S. Department of the Interior, Bureau of Reclamation and U.S. Army Corps of Engineers (USBR/USACE). 2015. "Best Practices in Dam and Levee Safety Risk Analysis." Version 4. July.

GEOTECHNICAL EVALUATION OF THE LEVEES IN THE CENTRAL VALLEY, CALIFORNIA

Richard Millet, PE, GE[1]
Steve Mahnke, PE, GE [2]
Mary Perlea, PE[3]
Ray Costa, PE, GE[4]

ABSTRACT

From 2006 through 2015, the California Department of Water Resources (DWR) initiated an unprecedented evaluation program for levees located within the California Central Valley in the Sacramento San Joaquin River Basin. The program successfully brought individuals and local, state, and federal agencies together as a team with a common goal of reducing flood risk. In-depth investigations and analyses were performed for 415 miles of levees protecting population centers of greater than 10,000 people under the Urban Levee Evaluations (ULE) program. To allow multiple analysis teams to simultaneously implement consistent evaluations, a comprehensive program technical guidance document was developed based on DWR's Urban Levee Design Criteria document and the U.S. Army Corps of Engineers criteria for levee design and construction. Evaluations included freeboard and erosion analyses, seepage and waterside and landside stability evaluations, and analyses of liquefaction potential and anticipated seismic performance. The multi-year ULE program was implemented in a 5-step process: 1) historical data collections, 2) initial field investigation, 3) preliminary analysis, 4) supplemental field investigation, and 5) final analyses and report. Over 400 individual levee reaches (0.5 to 1.5 miles long) were assessed. For reaches of urban levees that did not meet criteria, deficiencies wee identified, conceptual remediation alternatives developed, and screening level cost estimates prepared. This paper summarizes the results of these evaluations and the resulting proposed remedial alternatives and their cost.

INTRODUCTION

Central Valley Flood Control System

California's Central valley stretched from Shasta County in the north to Kern County in the south covering an area approximately 450 miles long and 40 to 60 miles wide. The valley is drained by California's largest river systems: the Sacramento River flowing from the north, and the San Joaquin River flowing from the south. The two rivers converge with the Mokelumne, Calaveras and Consumes rivers in the Sacramento-San Joaquin Delta before flowing into the Pacific Ocean (Figure 1).

[1] AECOM, 2870 Gateway Oaks Drive, Suite 150, Sacramento, CA 95833, richard.millet@aecom.com
[2] California Department of Water Resources, 3464 El Camino Avenue, Sacramento, CA 95821, gmahnke@water.ca.gov
[3] United States Army Corps of Engineers, 1325 J Street, 10th Floor, Sacramento, CA 95814, mpperlea@gmail.com
[4] Geotechnical Consultant, Granite Bay, CA 95827, rcosta@gmail.com

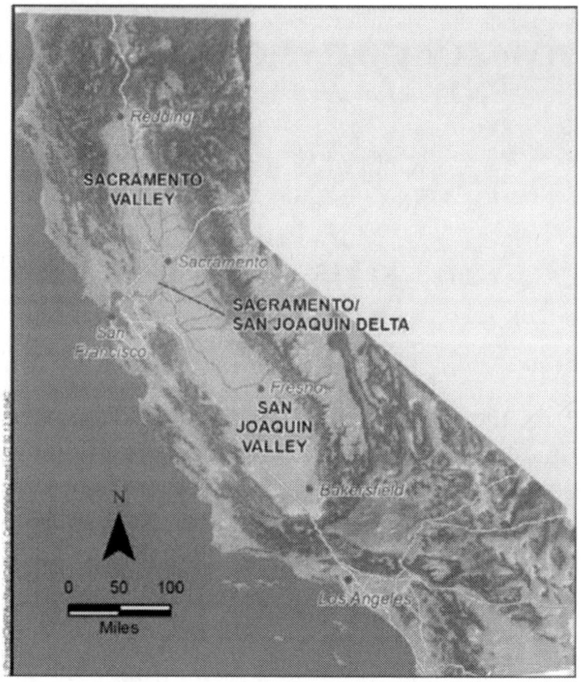

Figure 1. California Central Valley

The Central Valley is one of the most productive agricultural regions in the world, comprising approximately one-sixth of the nation's irrigated land and producing approximately one-fourth of the nation's foo supply. The Central Valley is also home to a rapidly expanding urban population of the nation's most populous state. Since 1980, the population of the Central Valley has nearly doubled from 2 million to 3.8 million people, and the Census Bureau projects that this number will increase to 6 million people by 2020.

The Central Valley Flood Control System protects people, property and infrastructure from flooding on the Sacrament and San Joaquin rivers. Over 1 million people and 1.9 million acres of cultivated land with an annual production value of over $2 billion are protected behind project levees. Approximately 200,000 structures with an estimated value of $64 billion rely on this system, as well as the last remnants of extensive riparian forests and wetlands that once flourish along California's rivers prior to European settlement.

Today, the Central Valley Flood Control System includes multiple reservoirs with flood detention space, approximately 1,600 miles of levees and a series of overflow weirs and bypass channels. These facilities were originally constructed by or incorporated into a federally designated flood control project. The State of California is the non-federal sponsor of the system and accepted the assurances of operating and maintaining these federal projects under the authority of the Flood Control Act of 1944. In accordance with State law, most of the operation and maintenance responsibilities have been delegated to local reclamation districts. These 1,600 miles of state/federal project levees are referred to as Project Levees. In addition, Project Levees are augmented by approximately 520 miles of appurtenant non-project levees (sometimes referred to as private or local levees) that directly impact basins protected by Project Levees.

Historically, the Central Valley was often an immense island sea subject to seasonal, valley-wide flooding. During winter storms and spring snowmelt, the Sacramento River overflowed its banks into large, low-lying basins that slowly drained back into the rivers or Delta sloughs. The San Joaquin River waters spread into broad floodplains and wetlands adjacent to the river. Flooding of the basins and floodplains provided water storage that attenuated peak flows to downstream areas, encouraged percolation into groundwater and supported diverse riparian and extensive tule/emergent wetland communities. Ove the course of a century of reclamation work, the rivers were gradually contained behind levees that kept flows from flooding rural communities and farmland. The resulting construction of the modern Central Valley Flood Control System enabled the rapid expansion of the agricultural economy that dominated the landscape today and enabled urbanization of rural communities.

CHALLENGES

Over the years, major storms and flooding have taken many California lives, caused significant property losses and resulted in extensive damage to public infrastructure. Moreover, a combination of factors described below has put public safety and the State's financial stability at risk for an even greater calamity in the future.

Land Use: Escalating development in floodplains increases the potential for flood damage to homes, businesses and communities. The original flood protection system was successfully designed to protect pioneer farms. Today this system also protects thousands of homes, businesses and other structures.

Inadequate Design and Construction: Early levee design and construction (which started over 150 years ago) are inadequate by modem standards (Figure 2). Construction was often accomplished with dredges and horse-drawn equipment with little consideration for material properties or placement procedures. Many of these non-engineered levees were simply grandfathered into the flood control system without adequate investigation or analyses of their condition.

Figure 2. Levee Construction in the 1920s

Deferred Maintenance: California's levee system has been further weakened by deferred maintenance. Sedimentation of channels and erosion damage to levees are the results of maintenance that has been deferred due to financial, physical, environmental and regulatory obstacles.

Erosion: The historic decision to scour out Sacramento River channels of mining debris by setting levees close to the channel was successful; much of the sediment has now been flushed from the system. However, an unintended consequence is that high velocity flows continue to scour not just sediment, but channel banks and levees. As a result, many levees have been critically damaged and many more will continue to erode.

State and Federal Financial Roles: Federal funding for effective flood prevention and management programs has been reduced. Between 2006 and 2014, California spent over $600 million on flood remediation projects that are normally funded by the federal government.

Liability: Court decisions have resulted in greater State flood damage liability. The November 2003 *Paterno vs. State of California* decision found that when a public entity operates a flood control system built by someone else, it accepts liability as though the State had planned and built the system. The *Paterno* ruling held the State responsible for defects in a Yuba County levee foundation that existed when the levee was constructed by local agricultural interests in the 1930s.

Not surprisingly, flood risk for the Central Valley ranks as among the highest for any developed country. And within the U.S., the greater Sacramento area is subject to the lowest level of protection for major cities, including pre-Katrina New Orleans (Figure 3).

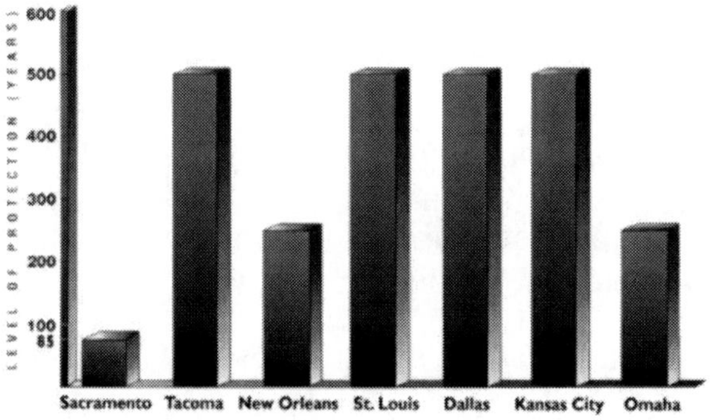

Figure 3. Flood Protection for Major U.S. Cities

FLOOD SAFE

To meet these challenges, the California DWR launched FloodSAFE California in 2006, a multi-faceted program to improve public safety through integrated flood management. FloodSAFE improves integrated flood management in California through a system-wide approach while reducing flood risk at regional and local levels. FloodSAFE is funded by almost $5 billion in bond measures.

SYSTEM ANALYSIS

A major deliverable of FloodSAFE is the Central Valley Flood Protection Plan (CVFPP). This document described a sustainable, integrated flood management plan that reflects a system-wide approach for protecting areas of the Central Valley currently receiving flood protection via existing facilities in the Central Valley Flood Control System. DWR prepared an initial CVFPP for adoption by the Central Valley Flood Protection Board in 2012, and plans to update the plan every five years thereafter.

This system-wide approach includes the hydrologic, hydraulic, economic, ecosystem and geotechnical elements (Figure 4) of flood management in the Central Valley. Key to this approach is an understanding of the reliability of the levee system. The DWR Levee Evaluation Program (LEP) provides this important input. State legislation has defined specific outcomes for the CVFPP; notable among these laws is the mandate that urban communities (populations greater than 10,000 people) achieve 200-year level of flood protection by 2025. Begun In late 2006, the LEP is now complete.

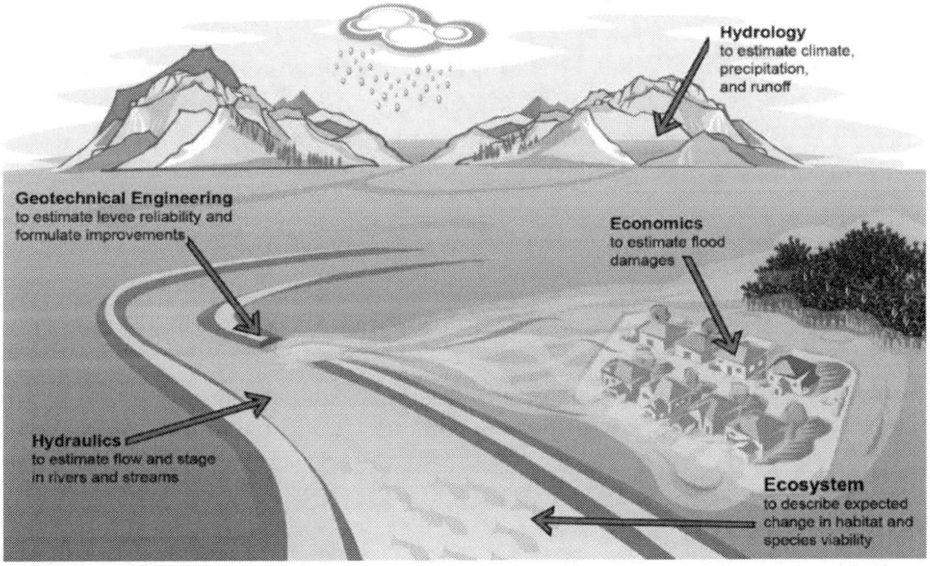

Figure 4. System Evaluation

The purpose of the LEP was to evaluate State/federal project levees and appurtenant nonproject levees to determine if they meet defined geotechnical criteria or, as appropriate, identify deficiencies and develop and cost remedial measure(s) to meet those criteria. Goals of the program were to:

- Support the CVFPP and Central Valley Floodplain Evaluation and Delineation (floodplain mapping) projects, federal and local flood management projects, local FEMA certification efforts and the legislative mandate of urban, 200-year flood protection by 2025.
- Provide geotechnical data, analyses results and conceptual alternatives and planning level costs for remediating levees.
- Improve geotechnical information exchange methods among State, local and federal flood management agencies.
- Identify levees that require immediate, critical repairs.

LEVEE EVALUATION PROGRAM

While there are a number of technical research and policy issues encompassed by DWR's Levee Evaluations Program (LEP), this paper focuses on the geotechnical exploration , testing, analysis, conceptual remedial design and resulting planning cost estimates from the assessment of 415 miles of urban levees (levees protecting 10,000 people or more) in the major population centers of the Central Valley—the Urban Levee Evaluation Project (ULE).

URBAN LEVEE EVALUATION

From 2006 through 2015, DWR's Division of Flood Management with the assistance of URS Corporation (now AECOM), working closely with the United States Army Corps of Engineers (USACE) and local stakeholders, conducted an innovative and unprecedented evaluation program to assess the existing conditions of urban levees in California's Central Valley.

The Urban Levee Evaluations (ULE) Project identified levees with deficiencies, which led to developing a comprehensive approach to reduce flood risk in all urban areas of California's Central Valley. ULE addressed 327 miles of Urban Project levees divided into 18 study areas and 88 miles of Urban Non-Project levees divided into 15 study areas. As noted, these urban levees protect populations of 10,000 or more.

ULE EVALUATIONS

Figure 5 shows the approximately 425 miles of project and appurtenant non-project levees that comprise the group of urban levee systems in the Central Valley. The overall strategy for ULE is impacted by two legislative and executive actions. California Senate Bill 5, signed in 2007, mandates that communities with populations of 10,000 people or more must be protected by levees that provide a minimum of 200-year protection and the

Governor's 2006 Emergency Order S-18-06 fast-tracked ULE with the goal of quickly identifying geotechnical deficiencies for repair.

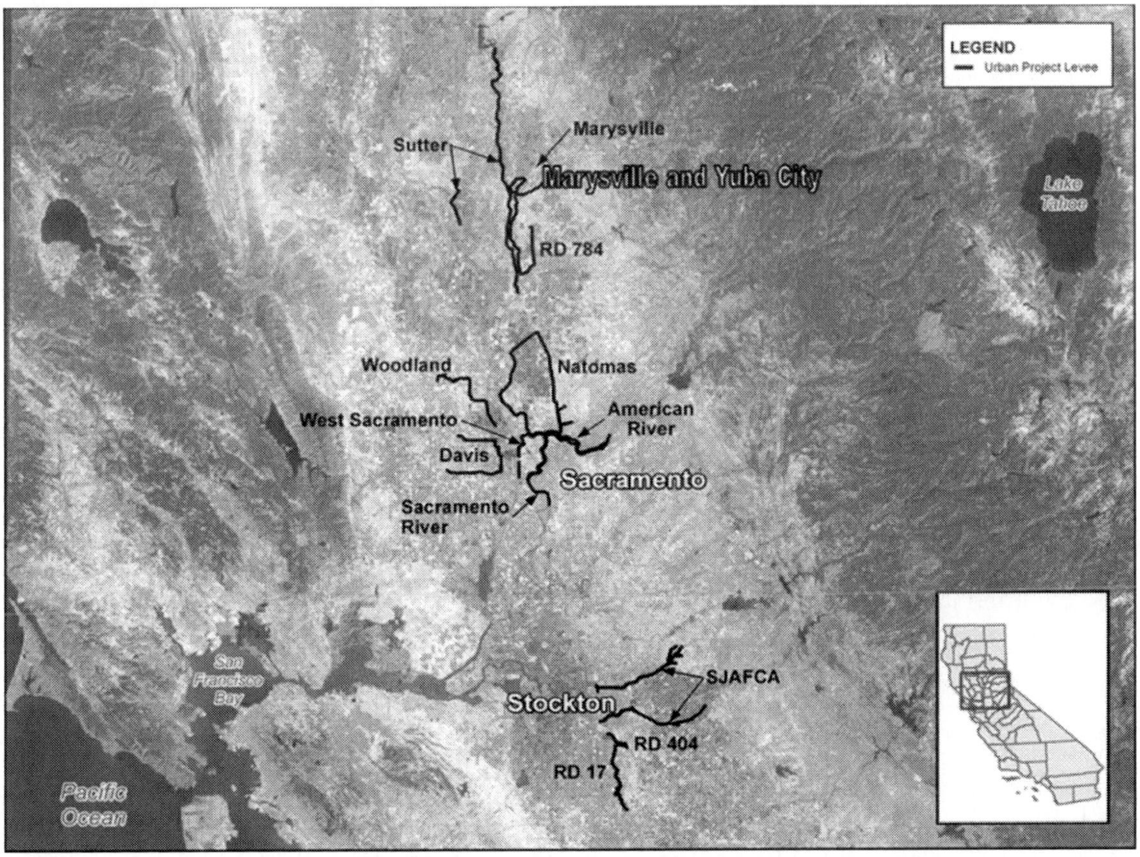

Figure 5. Levee Task Area Map

In order to implement the ULE project in a consistent process, AECOM, with input from DWR, its independent consultants and board of consultants, and USACE Sacramento District, developed a comprehensive "Guidance Document" (URS, 2015). This document outlines and defines a step-by-step comprehensive process for implementation of levee evaluations and remediation, including how to define site characterizations and material properties and then perform seepage and stability analyses.

In addition, because little but subjective assessment tools were available to assess levee erosion, a three-tier Erosion Susceptibility Process (ESP) was developed in the Guidance Document. This process was developed with the assistance of a special technical panel headed by Professor Jean-Louis Briaud of Texas A&M University. This ESP allows for consistent and repeatable assessment of the risk of an erosion break during a 200-year flood event.

Finally, in conjunction with the requirement of the ULDC, an assessment of seismic vulnerability of the 415 miles of urban levee was completed using a project-specific liquefaction and seismic deformation process, again documented in the Guidance Document.

ULE CRITERIA

Evaluation and assessment of urban levees was based on the State of California's *Urban Levee Design Criteria* (ULDC) (DWR 2012). This document was developed following the USACE levee design criteria adjusted for the Central Valley special issues and conditions. ULDC established specific criteria to address the following issues under applicable design water surface elevations:

- Inadequate freeboard, resulting in overtopping and erosion.
- Underseepage due to the presence of pervious foundation soils beneath the levee.
- Through seepage due to the presence of pervious soils in the levee embankment.
- Instability on a levee's waterside or landside slopes due to the presence of soft or weak soils and/or high seepage forces.
- Seismically-induced deformations and/or liquefaction.
- Significant erosion of a levee's waterside slope and undercutting of the waterside toe, leading to instability and a breach of the levee embankment.
- Undetected or man-made defects (e.g. penetrations or animal burrows, construction problems) generating uncontrolled, concentrated seepage, internal erosion and eventual levee breach.

WATER SURFACE ELEVATIONS

Water surface elevations (WSEs) define important loading conditions for the ULE assessments; however, due to ongoing changes in hydrologic models, input data and climate change, WSE will always be a moving target. As a result levee performance was evaluated at a range of WSE, resulting in a rating curve of levee performance that will support planners and designers over a wide range of future changes in hydraulic and hydrolic conditions. The three WSE used to evaluate urban levees are described below.

55/57 Profile: When California took over federal levees, the State made assurances to operate and maintain levees at the design WSE. This WSE is referred to as the 55/57 profile, a shorthand term to describe the 1957 water surface profile for the Sacramento River Flood Control System and the 1955 profile for the San Joaquin Flood Control System. Both of these water surface profiles are commonly accepted as the liability limit for the State, following the legal principle that State assurances produced an expectation that levees would perform at the 55/57 WSE.

200-year WSE: The 200-year WSE is the legally mandated standard for urban levees. Urban areas must achieve protection from a 200-year flood event by 2025. Geotechnical performance is evaluated at this WSE, which is based on the best available hydrologic and hydraulic information.

Top-of-Levee: A brittle failure of a levee can cause extensive loss of life in an urban setting. Urban levee evaluations were conducted from the standpoint that levee failure should only occur due to overtopping. Under this approach, sudden, high-head failures would be avoided, allowing a high percentage of an affected population to get out of harm's way. Urban levees are therefore evaluated at the WSE corresponding to the levee crest or 200-year WSE plus 3 feet, whichever is lower.

ULE IMPLEMENTATION

As noted, Urban levees were broken into 33 areas and evaluated using the 5-step process outlined in Figure 6.

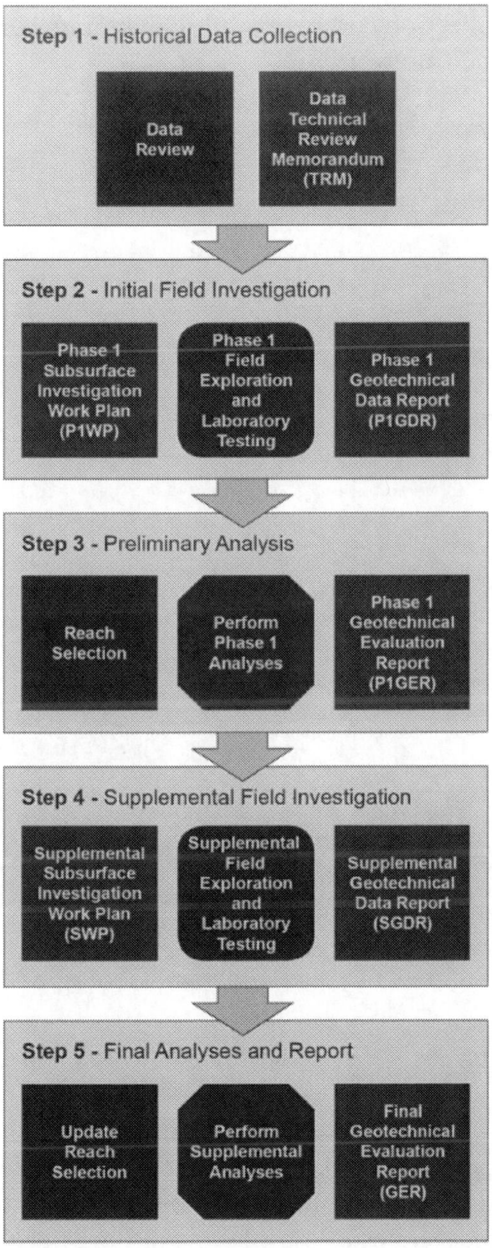

Figure 6. 5-Step Process

Historical Data Collection

In each of the urban levee study areas, historical data focusing on initial construction, construction upgrades, and remediation and past flood performance, was collected from multiple sources. Sources included newspapers and historical reports from USACE, DWR, and stakeholders. In addition, more that 100 personal interviews were conducted as a follow-up to the data review. Over ten thousand records and reports were scanned and registered in GIS for electronic recall. This data was also prepared for inclusion into the California Levee Database and the USACE National Levee Database. In addition to historical data, geomorphic studies were implemented resulting in the preparation of 785 square miles of surficial geomorphic maps (see Figure 7 for example of a typical map). The results of the data collection and geomorphic mapping were used to plan the initial field investigations for each study area.

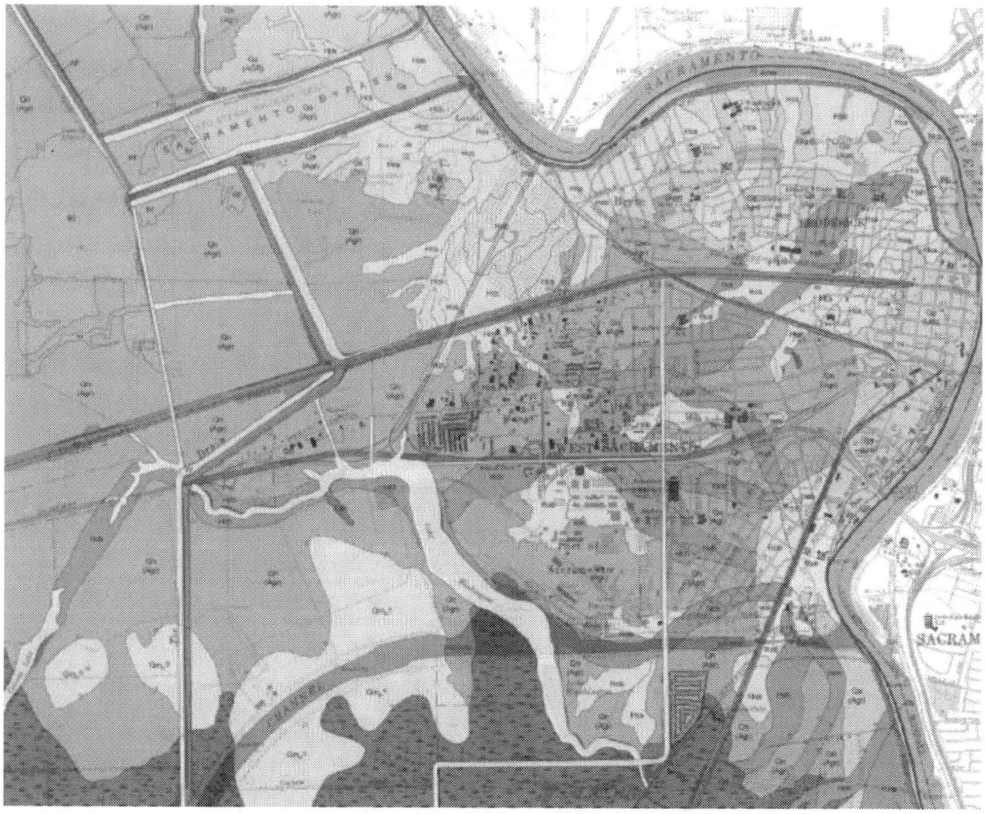

Figure 7. Typical Surficial Geologic Map

Initial Field Investigation

Based upon the results of Step 1, a site-specific phase 1 field investigation was planned for each levee study areas and in total included: 510 linear miles of low-altitude light detection ranging (LiDAR) topographical mapping; 95 linear miles of riverine multi-beam bathymetry; 455 linear miles of a triple line airborne geophysical electromagnetic surveys along the levee alignment; 1,000 cone penetration tests; and 550 sample soil test borings along levee crests.

LiDAR

LiDAR data was collected via low-altitude helicopter and provided significant advantages over other methods of topographic data collection, including:

- Ability to focus the survey along a levee alignment
- Higher data accuracy than with traditional fixed-wing LiDAR surveys
- Video and aerial photographic imagery collected simultaneously with topographic data
- Participation of project personnel on helicopter flights to direct surveys and ensure appropriate data was collected
- Collection of a large amount of topographic data in a limited period of time
- Smooth transition to GIS-based digital terrain models (DTM) with minimal ground truthing requirements due to the high density of survey points (90 per square meter)
- No right-of-entry issues

Bathymetry

As shown in Figure 8, LiDAR data was supplemented as available with bathymetric data to produce a complete model of a levee's prism and adjacent water channel. Bathymetric data was collected using multi-beam sonar technology, which facilitated collection of shallow water data and production of three-dimensional models.

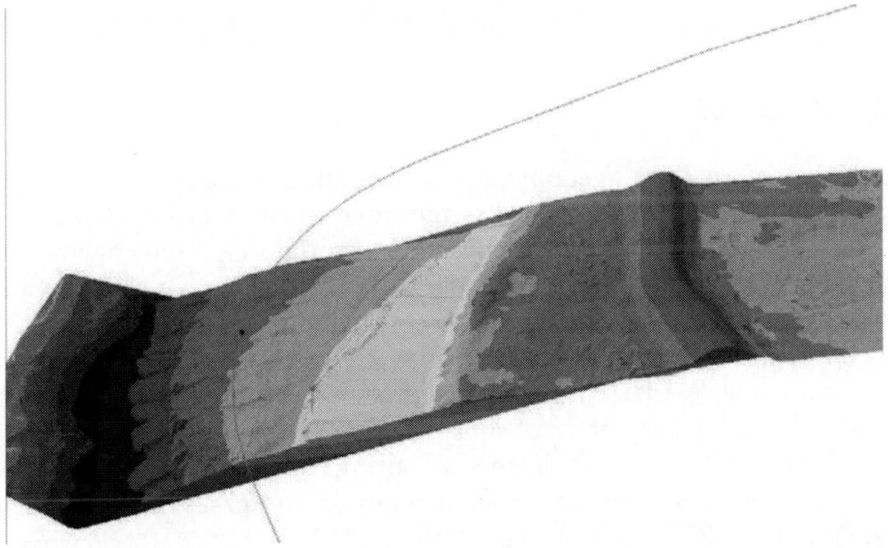

Figure 8. Topographic Data Collection from LIDAR and Bathymetric Surveys

Fugro Airborne conducted a helicopter electromagnetic (HEM) survey for the 455 miles of urban areas in the fall of 2007 to measure the vertical and horizontal variation of resistivity in the levee foundation materials.

Helicopter Electromagnetic Survey

The HEM survey was conducted in three parallel passes over the levees using fixed ground stations to position and document survey lines.

Resistivity values of soil are affected by porosity, water content, mineral content, and grain size distribution. Typically, sands have high resistivity and clays have low resistivity. The contrasts shown on different resistivity profiles (Figure 9) were evaluated to identify continuous, discontinuous, or anomalous zones of lower resistivity (generally clays) or higher resistivity (generally sands and gravels).

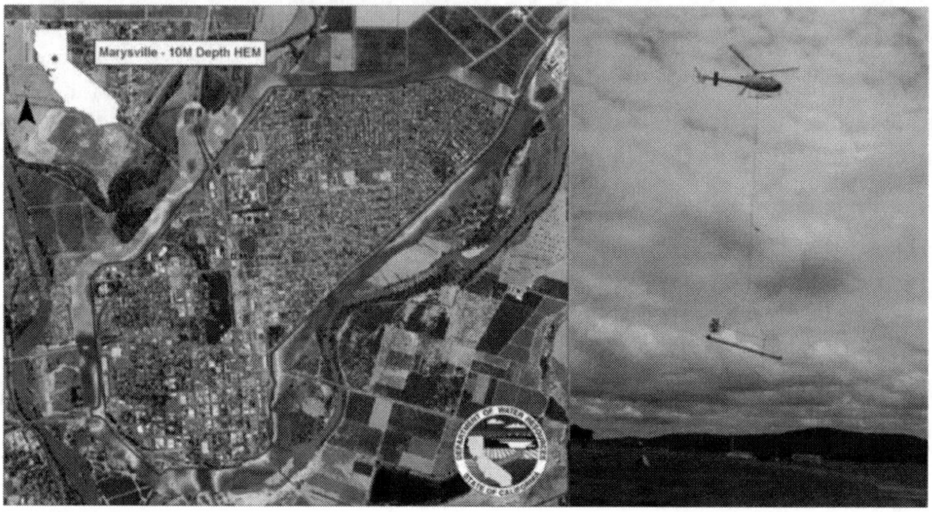

Figure 9. HEM Survey and HEM Survey Equipment

Intrusive Investigation

The results of geomorphic and geophysical investigations were evaluated to identify areas of potential anomalies that needed further investigation through intrusive field investigations. Intrusive investigations consisted of soil borings and cone penetrometer test (CPT) soundings. Soil boring samples were obtained using punch-core (mud-rotary) and sonic drilling methods. Sonic and punch-core sampling provided a means of obtaining continuous core samples from borings. In general, mud-rotary borings were separated about 1 mile apart. CPT soundings or sonic borings were spaced at about 1,000 foot intervals, providing a signature of subsurface conditions between mud-rotary borings. CPT exploration was correlated with borings approximately every mile. Selected soil samples from borings were then tested in a laboratory for a suite of geotechnical properties, ranging from index to strength tests.

Subsurface data collected from intrusive and non-intrusive field investigations provided sufficient information to develop an understanding of subsurface conditions in each urban study area and allow preliminary analyses to be performed.

Step 3: Preliminary Analyses

Based upon the data collected in Step 1 and Step 2, each urban study area levee was divided into multiple reaches based upon geometry and geotechnical and hydrologic conditions of the levee. For each levee reach, representative geotechnical cross sections with initial stratigraph material properties was prepared. Preliminary seepage and stability analyses were then made following the detailed protocol outlined in the Guidance Document. Over 400 geotechnical cross sections were analyzed. The results of these preliminary analyses were compared to past performance data collected and inconsistency and data gaps were identified. These inconsistencies and data gaps were used to plan Step 4, Supplemental Field Investigation.

Step 4: Supplemental Field Investigation

As noted, plans for supplemental investigations for each study are were based on closing data gaps identified in Steps 1, 2, and 3 to allow final screening level assessment and evaluation of urban levees.

The supplemental investigations included 590 cone penetrations test and 920 soil test borings. Whereas the preliminary investigation borings and cones were conducted along the levee crests, the majority of these supplemental investigations were performed along the landside levee toe. Levee waterside explorations were excluded due to access and environmental permit restrictions. Again, soil samples were obtained from the borings and over 14,000 laboratory tests were made to characterize the embankment and foundation conditions of the levees.

During this supplemental investigations step, an unprecedented special soil testing program was designed and implemented under the direction of Dr. Deme Koutsoftas. This testing program addressed the soil strength and permeability characteristics for in situ condition at very low effective stresses. This condition occurs in levee embankments and near surface foundation fine-grain soils which were typically control the stability and seepage conditions of the levees. Based upon the special testing program results, presumptive strength and hydraulic conductivity values for multiple Central Valley soil types were established for use in analyses of levees.

Step 5: Final Analysis

In the final phase of the ULE Project, using the combined information obtained in Steps 1 through 4, the 33 urban study areas wee segmented into 440 reaches and/or subreaches. For each reach or subreach a representative geotechnical cross section was established including stratigraph and material properties.

Based upon the cross sections and the designated water loading, each levee reach was assessed as to whether it met or did not meet ULDC criteria for under seepage, through seepage, landside stability, waterside stability, freeboard, and erosion risk. Over 4,000 computerized SeepW and SlopeW analyses were completed.

In addition to the static assessment, a seismic vulnerability evaluation was performed using a 200-year return period seismic event. This assessment was driven by the ULDC criteria based upon either frequently loaded levees[5] or intermittently loaded levees[6]. The results of the evaluation were used to classify intermittently-loaded levees as having high, medium, or low vulnerability with respect to post-seismic flood protection ability.

Seismic vulnerability assessment classified 290 miles of levee as having low seismic vulnerability, and 88 miles of levee as having high seismic vulnerability. No seismic remediations were selected as seismic remediation is not required for intermittently-loaded levees.

Based upon the results of these final analyses, an assessment was made for each reach or subreach as to whether the levee met or did not meet ULE criteria for each of the criteria.

For each reach or subreach that did not meet ULE criteria, typically two conceptual remedial alternatives were evaluated and costed. Based upon total cost, practicality, and environmental considerations, a preferred alternative was selected. Figure 10 presents a summary of the total miles that did not meet criteria by each deficiency. As noted, the primary deficiencies for the 200-year WSE were under seepage and/or seepage-driven landside stability.

Typical conceptual static remediation alternatives consist of: installing cutoff walls along the centerline of the levees to address seepage and stability deficiencies; placing waterside rock slope protection for erosion and waterside slope deficiencies; and localized embankment raises for freeboard repair. The graph below depicts the total length of each type of repair in the study area. In total, 281 miles of urban levee requires some remediation (Figure 11).

The cost estimating for ULE used a project-specific screening-level Class 4 cost estimate process based upon 2013 dollars. Class 4 estimates are not design-level cost estimates. However, they do include both construction cost and owners' "soft" costs, such as permitting, legal, environmental mitigation, and contingency. The total remediation cost is approximately $3 billion. Figure 12 presents a breakdown of the conceptual station remediation cost by repair type for all reaches in the urban study area that do not meet ULE criteria.

[5] A frequently-loaded levee is defined in the Urban Levee Design Criteria (DWR, 2012) as "a levee that experiences a water surface elevation of one foot or higher above the elevation of the landside levee toe at least once a day for more than 36 days per year on average."

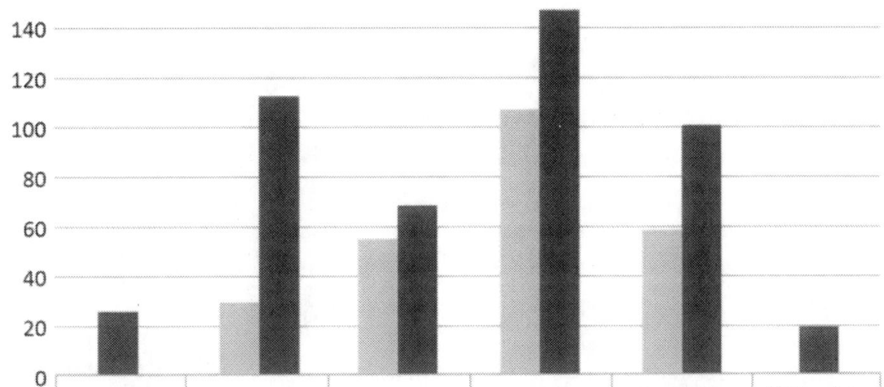

	Erosion Risk*	Freeboard*	Through Seepage	Under-Seepage	Landside Slope Stability	Waterside Slope Stability
1955/57 Design WSE	25.5	30	55.2	107.5	59.8	17.5
200-Year WSE	25.5	112.7	68.3	147.4	100.7	19.6

** Erosion and freeboard are not typically deficient across an entire reach. Where deficiencies were noted, conceptual remedial alternatives were evaluated.*

Figure 10. Total Miles of Levee that Do Not Meet Static ULE Criteria

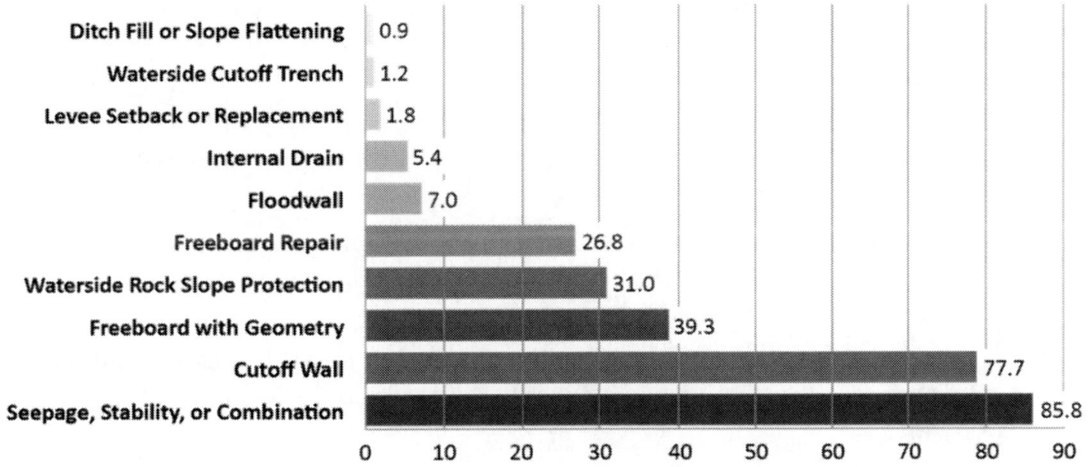

Figure 11. Conceptual Static Remedial Alternatives

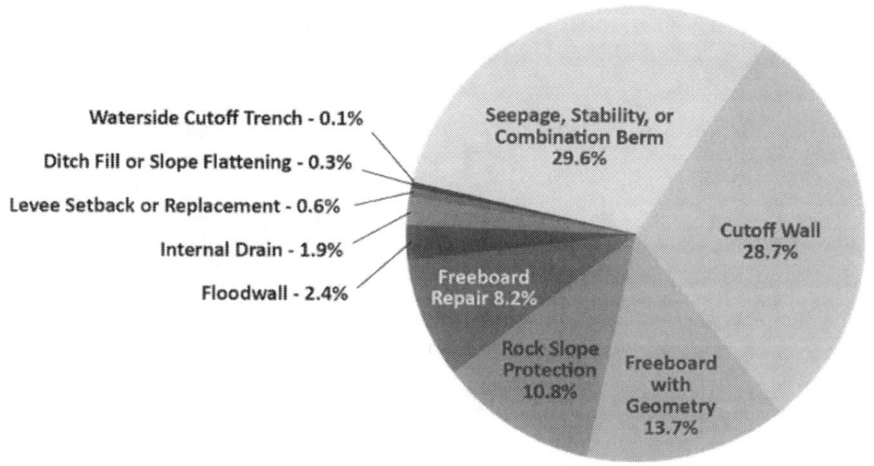

Figure 12. Total Static Remediation Costs ($3,252M)

FUTURE VALUE TO THE CENTRAL VALLEY FLOOD PROTECTION PROGRAM

Upon completion of the program in April 2015, ULE has provided:

- An electronic compilation of over 10,000 documents identifying locations of past performance incidents along the urban levee systems; all information georeferenced in a web-based accessible system. This has provided a comprehensive legacy of retrievable electronic data documenting history and analyses that will be used for many years to come.
- Systematic characterization of over 415 miles of urban levees as meeting or not meeting ULDC performance criteria for freeboard, erosion, seismic risk stability, seepage.
- Fifty-two comprehensive data, assessment and conceptual remedial design reports for the 33 urban study areas which will allow local flood agencies and stakeholders the ability to prioritize, plan and budget future repairs.
- A comprehensive Guidance Document for the evaluation of levees that will provide a proven path for the effective and efficient assessment of levees in California and nationwide.
- Enhance public perception of engineering through a public website (available at http://dwr-lep.com) constructed to provide access to the complete results of ULE.
- Established the current condition of urban levees and identified needed repairs to reduce flood risk and protection for more than a million people and for more than $64 billion in property and infrastructure.
- By identifying the locations with possible deficiencies in the system and recognizing the limited available funding, the ULE allows establishment of a priority bases for DWR, local flood agencies and stakeholders to make decisions to cost-effectively maximize flood risk reduction in their communities.

REFERENCES

California Department of Water Resources (DWR). 2012. *Urban Levee Design Criteria*. May.

URS. 2015. *Guidance Document for Geotechnical Analyses*. April.

PHASED INVESTIGATION APPROACH FOR A PARTIAL SLOPE FAILURE ON THE RIO GRANDE

Isaac J. Stephens, PE[1]
Lucas A. Walshire, PE[2]
Joseph B. Dunbar, PhD, RPG[3]

ABSTRACT

A partial slope failure occurred on a section of levee along the Rio Grande following a significant drop in water level in the spring of 2014. A geotechnical investigation was performed by the U.S. Army Engineer Research and Development Center (ERDC) to determine the causes for the failure and provide remediation alternatives. In order to provide the best possible understanding of subsurface conditions that contributed to the failure, the site investigation was performed using a phased approach. Historic borings, local geology, and groundwater records were reviewed to guide the placement of strings of cone penetrometer test (CPT) soundings. These soundings were then used to develop both stratigraphic and strength cross-sections. This information was used to target borings to collect soil samples of critical layers for laboratory testing and to install monitoring instrumentation. While the concept of using data gained from each phase to inform the scope of subsequent phases is not new, the authors believe it is an underutilized approach to site investigation that can yield great benefits. The application of this approach in the context of this particular investigation is discussed and examples of how this approach improved the results of the overall investigation plan are given.

INTRODUCTION

Geotechnical engineers are often asked to provide solutions to complex and variable problems, often with limited data. When failures occur, a good understanding of the subsurface conditions and the factors that contributed to the failure allow for effective remediation solutions to be designed. When those factors are poorly understood, it can lead to overly conservative and non optimized design at best, and at worst, to an under conservative and unsafe design. The investigation of a partial levee slope failure is used here to demonstrate how the kind of understanding needed for a good design can be obtained.

In the spring of 2014, transverse cracks were discovered running along the crest and toe of a levee along the Rio Grande near Brownsville, Texas. By July 2014, nearly a foot of slippage at the levee crest and two prominent cracks at the levee toe were observed. The cracks extended approximately 600 feet in length and appeared to be deep seated. The cracks were located less than 150 feet downstream of an international bridge spanning the

---

[1] US Army Corps of Engineers, Engineer Research and Development Center (ERDC), Vicksburg, MS, isaac.j.stephens@usace.army.mil
[2] US Army Corps of Engineers, ERDC, Vicksburg, MS, lucas.a.walshire@usace.army.mil
[3] US Army Corps of Engineers, ERDC, Vicksburg, MS, joseph.b.dunbar@usace.army.mil
Permission to publish was granted by Director, Geotechnical & Structures Laboratory

Rio Grande and immediately adjacent to the U.S. Port-of-Entry (POE). Figure 1 shows the extent of the cracks in red and their relationship to the facilities described. Initially, the impact on the levee stability as well as to the infrastructure mentioned was unknown but of serious concern. The levee owner requested that the Engineer Research and Development Center (ERDC) assess any dangers posed by the cracking as well as determine the cause of the cracking and recommend remediation options.

Figure 1. Project location map, levee cracking shown in red.

Initial information was sparse, but provided two possible contributing factors to the failure. The first was that the river level had recently dropped significantly, and the second was that a levee improvement project had recently been completed, raising the crest of the levee by approximately 3 feet. However, these two factors alone did not completely explain the failure. The cracking was relatively localized and many more miles of levee had been raised and experienced the same draw-down event. The purpose of ERDC's investigation was to determine if and how these factors contributed to the cracking and partial slope failure and whether any other factors played a role. By understanding what made this particular reach different from the miles of levee that did not fail, an effective and appropriate remediation could be designed.

SITE HISTORY

Communities within the Lower Rio Grande Valley (LRGV) are protected today by what was originally a patchwork of levees built by city and county governments during the late-1800s and early-1900s. Between 1900 and 1939, the Rio Grande River over-flowed 23 times within the LRGV, with hurricanes hitting the area in 1910, 1913, and 1933 (Stubbs et al. 2003). During the depression years, the Texas border counties within the

LRGV were unable to maintain the piecemeal network of private, local, and county levees because of the poor economy. Repeated flooding on the Rio Grande led the local governments within the LRGV to petition the federal government to take over the existing levee system and provide comprehensive flood control protection.

After the levees came under federal control, the entire LRGV flood control system was rebuilt during the 1930s and 40s (USIBWC 1992). These early levees were built according to local construction practice in the LRGV, using soils obtained from the nearby floodplain, and generally correspond to a standard levee section as defined by the U.S. Army Corps of Engineers (USACE) in USACE (2000): less than 25 ft in height, with side slopes flatter than or equal to 1V:2H. Sections of the levee system have subsequently been refurbished since then due to urban land use and environmental changes and to meet the new project flood requirements. The most recent modification involved raising the height of the levee an additional 3 ft and regrading the levee slopes over a 12 mile stretch to meet the new project flood requirements. Construction of the levee raise was completed in October 2013.

SITE INVESTIGATION

A phased approach to geotechnical site investigation is when efforts to understand the subsurface environment are taken in a logical stepwise manner. Information gained in each step is used to focus subsequent data gathering efforts. This approach results in improved quality and relevance of the data gathered.

A phased approach does not mean that a plan is made up as you go, rather, a framework or structure is planned with a sufficient number of tests, borings or effort kept in reserve to examine more closely information deemed important by information gleaned from the more general study. In this way, plans for subsequent phases are left sufficiently flexible to allow information gained in previous stages to guide subsequent work. Although described here as distinct phases, in practice many tasks and efforts have some amount of overlap or iteration between tasks.

Initial Site Visit and Desk Study

Often a desk study is performed before a site visit, but for this investigation, as is sometimes the case, the desk study and initial site visit were carried out nearly in parallel. After a brief review of the limited information available and of aerial images of the area, a site visit was conducted that included gathering more information for the desk study. The site visit included a walking survey followed by considering potential contributing factors and failure modes, which was then followed by a more in-depth inspection of the area. Important observations included that the extent of the ground movement was relatively localized and that it did not seem to involve the nearby bridge piers or adjacent POE facilities, that the ground displacements were consistent with a block type failure, and that where the floodplain met the river there was an almost vertical slope of fine sand to sandy silt material.

The site visit also uncovered several other possible contributors that had not been mentioned previously. One was a fire hydrant that was located behind the levee near the center of where the cracking was. The hydrant was old and partially buried and it was not known if it was currently pressurized or not. If it was pressurized and leaking, the leaked water could be creating excess pore pressures in the slope or foundation leading to the slide. Another source of water and potential excess pore pressures was Lake Brown which had a water surface level 10-15 feet higher than the river. Although not directly behind the section of levee that had developed cracks, porous soil layers could have been transmitting lake-level pore pressures beneath the levee. The site visit also revealed that one of the bridge piers had been retrofitted with some sort of support frame (Figure 2). While this wasn't a contributing factor to the cracking, its presence served as a red flag warning of possible foundation issues.

A comprehensive review of the design and construction documents for the levee raise was made to understand the design assumptions and construction activities in the study reach.

Figure 2. Retrofitted pier support structure indicates probable foundation issue.

Nearly one full day of the site visit was spent collecting and review historic maps, charts, and photography to characterize the evolution of the river and subsequent land use changes within the study area through time. Important to this effort were visits to local county and municipal engineering and utility departments for plans and utility maps as well as to the local historical society for early photographs of the river front, and the U.S. National Park Service for early historic map data.

While this type of information is not always available at all sites, efforts to uncover such information can be time well spent. In this investigation, these efforts were some of the most productive when the amount of data collected is compared to the cost of collecting it. Some of the key results included information about how the city regulated Lake Brown water levels and determining that although the fire hydrant was pressurized, it had been recently tested for leaks and none had been found.

Cross-channel borings for the nearby bridge were located from several previous geotechnical investigations which revealed a wealth of information about the foundation soils in the area. The previous investigations revealed that the purpose for the pier modification was because the pier had been founded on a soft clay layer and had begun to settle. The support frame had been built on piles extending past the weak layer and into stiffer clay. It was possible that the same soft layer could be responsible for the levee cracking.

Geology/River History

An important part of the desk study was to determine historic land use changes and activities that have occurred in the study reach. A primary goal was to identify significant land use changes and activities that may have contributed to the levee instability within the study reach.

Geographically, both Lake Brown and the nearly right-angle Rio Grande course have been prominent and stable features within the study area for the past 170 years. The stability of the channel alignment at this location is noteworthy considering the numerous abandoned oxbows and courses within the floodplain of the Rio Grande (Brown et al. 1980; Bureau of Economic Geology 1976).

Review of historic maps showing the previous courses of the Rio Grande indicates that the area on the waterside of the levee between the toe and river channel contains younger alluvial fill as compared to that on the landside of the levee.

The river bank extents in 1930 were mapped using historic aerial photos and overlaid over present day imagery. Yellow lines in Figure 3 show the location of the 1930 river channel relative to present-day. The presence of the younger alluvial fill on the waterside of the levee indicates that there may be weak zones present in the area providing support to the levee.

Figure 3. Current river channel compared to 1930 channel (yellow) shows recent alluvial fill in study area.

Subsurface Investigation

For any investigation requiring more than a few borings, it can be advantageous to initially perform CPT soundings for stratigraphic evaluation and strength profiling and then use well-placed borings for soil sample retrieval based on the CPT stratigraphic evaluation. Used this way, CPT soundings were used identify critical, problem, or weak soil layers which were then flagged for soil sampling using borings. The boring program was designed to retrieve targeted soil samples and place instruments into the ground (i.e., water-level measurement devices and inclinometers). Understanding the inherent strengths and weaknesses of CPTs and borings allows the Engineer to maximize their effectiveness by allowing the strength of each method to complement the other.

One of the main strengths of CPT, in addition to continuous soil information with depth and in-situ undrained strengths, is its remarkable speed and cost when compared to traditional boring. In this investigation, 30 CPT soundings were performed for a cost comparable to 3 or 4 borings. This added level of resolution allows for increased confidence that "minor geological details" (see Terzaghi 1929) have not been overlooked.

Proper layout of CPT soundings is critical for stratigraphic evaluation, which requires an understanding of the costs of CPT soundings and soil sample borings and, preferably, a general understanding of the site geology. Random placement of CPT soundings makes each CPT sounding an individual data point rather than part of a group. Also, a grid pattern of CPT soundings has limited usefulness unless computerized 3D visualizations are anticipated. CPT soundings should be arranged in strings or lines to draw accurate cross sections. The location and arrangement of cross sections should be based on the requirements of the project and site geology. The stratigraphic interpretation of the CPT-based cross sections can be used to properly locate borings for soil sample retrieval.

For this investigation, the CPT soundings were performed in 2 phases; one immediately following the other. In the first phase, four lines or strings of soundings were made perpendicular to the levee with 3 three soundings each (Figure 4). Three strings were made through the failure zone, one through the center and one on each extent of the cracking. The purpose of these soundings was to determine if weak or permeable layers existed beneath the area that had cracked, and if so, what was the general extent of such layers. One line was made downstream of the failure zone between Lake Brown and the river to determine if there were any permeable layers that could be transmitting lake water pressures under the levee. Additionally, one sounding was made upstream of the bridge, approximately 500 feet from the failure zone to get a sense of the variability of foundation conditions between failed and non-failed sections of levee.

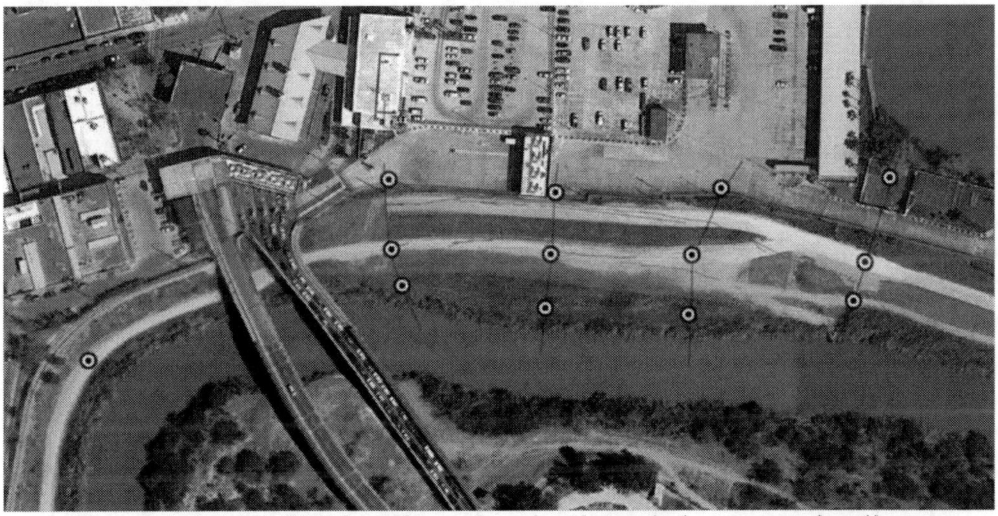

Figure 4. Initial CPT locations showing intended cross section lines.

The results of these soundings were evaluated on-site as each string of soundings was completed. The first phase indicated that the area was underlain by silts and clays and that the presence of a hydraulically charged permeable layer was unlikely. It was also discovered that limestone cobbles, likely placed as rip-rap or shore protection were present at shallow depths along the river bank. This prevented the CPTs from penetrating more than 6-10 feet. Based on this information, the locations of the second set of soundings were determined. These test locations were used to “fill-in” gaps from the first set of soundings and better define specific cross section strings. Because of the cobbles encountered by the first set of CPTs, no further soundings were attempted along the river bank, with the understanding that this information would be gathered during the drilling phase.

For the second phase of testing, additional soundings were added to the cross section running perpendicular to the center of the failure zone as well as soundings to define a cross section along the toe of the levee. Figure 5 shows how these additional soundings were laid out to produce high resolution cross sections.

Figure 5. Secondary CPT locations (in red) were placed to better define key cross sections.

During the CPT tests, whenever sandy layers were encountered, quick dissipation tests lasting less than 7 minutes were performed to determine hydrostatic water levels. These tests showed groundwater levels and gave an indication of continuity or connectivity of sand layers encountered.

Upon completion of the CPT field tests, cross sections were created to show soil stratigraphy and strengths. Figure 6 shows the cross section of predicted soil classification running perpendicular to the levee in the center of the area of interest. This cross section and others showed silts and clays inter-layered with the only sand layer 50 to 60 feet deep. This sand layer was found to be too deep and too dense for excess pore pressures leading to low effective stress to be a likely contributor to the levee instability. However, when the same cross section was plotted in terms of undrained strength (S_u) normalized with effective confining pressure (p') (Figure 7), extremely low S_u/p' values were identified ranging from 0.2 to 0.4 that start near the levee toe and extends toward the river. This weak clay layer is at approximately the same elevation as the soft layer that was found to be responsible for the settlement of the nearby bridge pier.

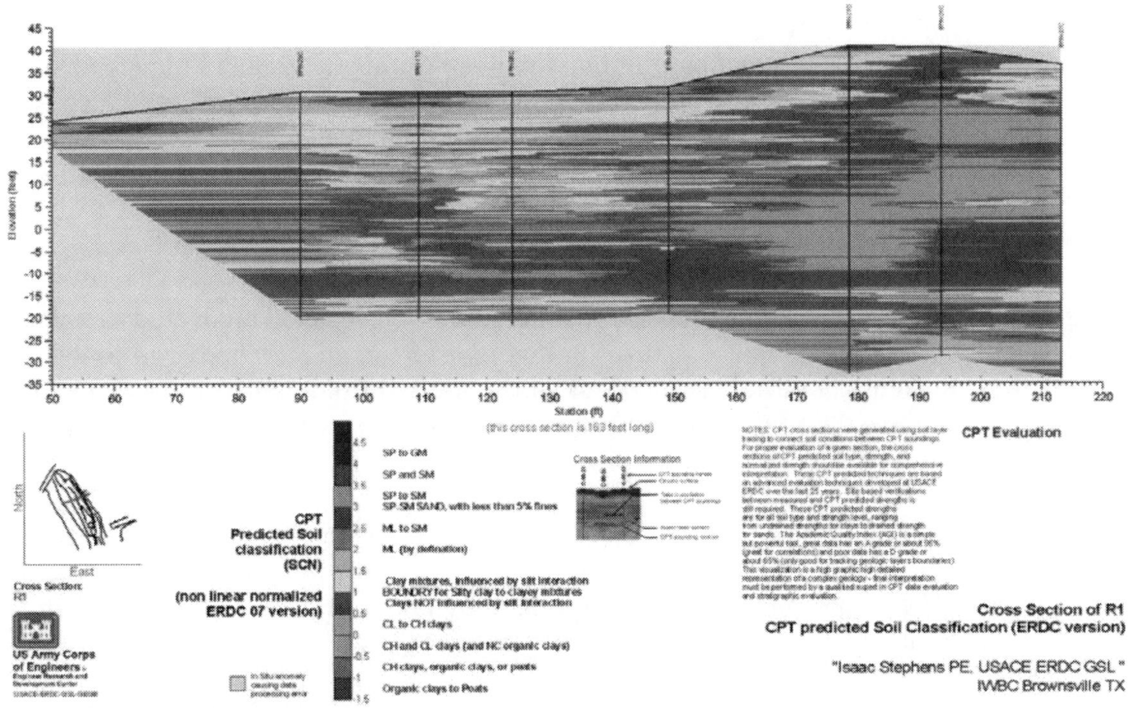

Figure 6. CPT predicted soil classification cross section at right angle to levee

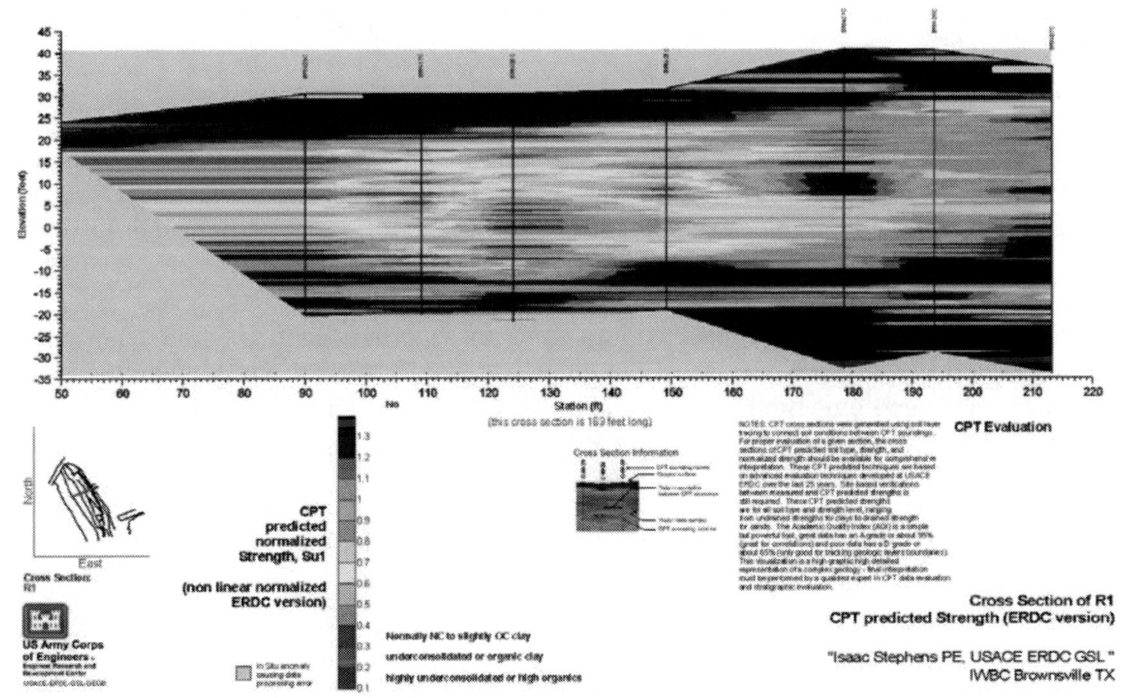

Figure 7. Normalized strength profile shows weak layer at 15 ft elevation.

Boring and Soils Testing

The next phase of the investigation program involved drilling to collect samples for soil testing and to install monitoring wells and inclinometer casing. Borings were placed at the levee crest and toe and near the river bank along the same line as the CPT cross section was developed. Additional borings were also placed upstream and downstream of this line to obtain stratigraphic and soils data that couldn't be obtained by CPT due to the presence of limestone cobbles at shallow depths.

Engineering properties were measured in the laboratory on selected recovered samples and included soil texture, grain-size distribution, moisture content, Atterberg Limits, and shear strength. Natural moisture contents of soils collected from the river bank and levee toe were at or above their liquid limits at the depths corresponding to the weak layer identified in the CPT investigation.

Monitoring wells and lake and river water level data provided information to confirm the impact of Lake Brown's water elevation on the stability of the levee system. The lack of connected highly permeable lenses was confirmed by flatter delayed well responses to river and lake fluctuations.

Inclinometers were installed as part of the boring program because even though the long term measurements from these devices would not be available in time for this site investigation, the minor incremental cost of adding theses devices while borings were being drilled, was justified based on their usefulness to monitor the long term stability of the site and determine the effectiveness of any remediation programs.

Development of Geologic Model

The data gathered in each phase of the investigation, which included regional geology, river level and course history, historic borings, CPT and boring data, and soil testing information, were all combined to provide a clear picture of the conditions that lead to the partial slope failure and levee cracking.

This information was first synthesized into hand drawn cross sections by an experienced geologist familiar with each phase of the investigation (Figure 8). These were then put into a site wide geologic model.

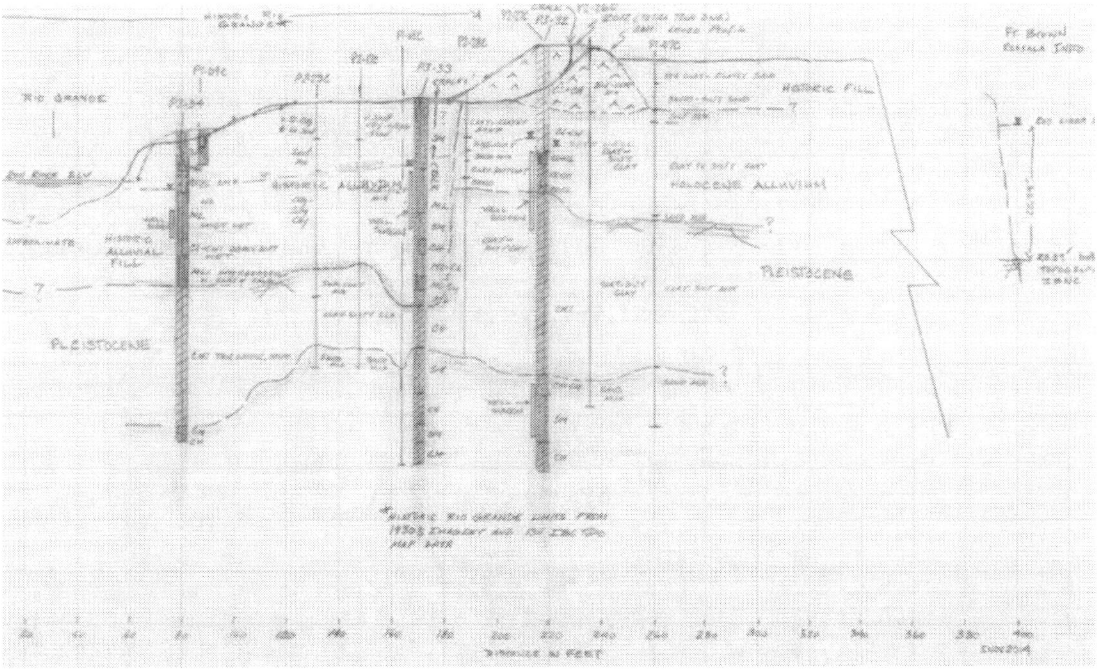

Figure 8. Hand drawn cross section incorporating data gathered from all phases of investigation.

Because of the way the investigation had been structured, this information provided a high resolution, high confidence picture of the underlying levee conditions.

The comprehensive geologic model was used to develop slope stability models that were used to evaluate potential remediation alternatives. This effort is beyond the scope of this paper, but is described in detail by Walshire, et al. (2015).

CONCLUSIONS

A comprehensive site investigation was performed to determine the cause and develop remediation alternatives to a partial slope failure on a levee on the Rio Grande. A site visit and search for historic information lead to the identification of possible contributing factors and resulted in valuable information about the site history and previous geotechnical investigations. CPTs were performed and cross sections were developed showing soil stratigraphy and strength profiles. Soil borings were performed to collect soil samples and install wells and inclinometer casings for continued site monitoring.

The results of the field and laboratory testing indicated the presence of soft alluvial sediments in the foundation material located between the river and levee. The soft material was found to have natural water contents above the liquid limit indicating it had very low shear strengths and this was further corroborated by the results of laboratory undrained unconsolidated (UU) tests. CPT testing in this area correlated the cones tip and side resistance to a minimum undrained strength ratio (Su/p') of near 0.2 and corresponded to a soft zone. This weak layer, combined with the added weight of fill from the levee raise and a rapid drawdown event were responsible for the levee instability.

The phased site investigation approach resulted in improved quality and relevance of the data gathered. The comprehensive geologic model developed through this investigation allows remediation design to be performed with a high degree of confidence of the conditions that lead to the partial slope failure. This improved understanding allows for a more optimized design and avoids the risk of an under conservative or unsafe design.

ACKNOWLEDGEMENTS

The authors would like to thank the U.S. International Boundary and Water Commission (IBWC) and particularly Jose Nunez for recognizing the value of this investigative approach. We are also extremely appreciative for the ongoing support of Dr. William Grogan and Dr. Maureen Corcoran of the Geotechnical and Structures Laboratory.

REFERENCES

Brown, L. F., Brewton, J. L., Evans, T. J., McGowen, J. H., White, W. A., Groat, C. G., and Fisher, W. L., 1980. "Environmental Geologic Atlas of the Texas Coastal Zone – Brownsville-Harlingen Area," Bureau of Economic Geology, University of Texas, Austin

Bureau of Economic Geology, 1976. "Geologic Atlas of Texas, McAllen-Brownsville Sheet," scale 1:250,000, University of Texas, Austin, TX

Glossop, Rudolph. "The rise of geotechnology and its influence on engineering practice." *Geotechnique* 18.2 (1968): 107-150.

Look, Burt G. *Handbook of geotechnical investigation and design tables*. CRC Press, 2014.

Olsen, Richard S., and John V. Farr. "Site characterization using the cone penetrometer test." *Use of in Situ Tests in Geotechnical Engineering*. ASCE, 1986.

Olsen, Richard S. "Rapid Reconnaissance of Major Landslides." *Geo-Strata—Geo Institute of ASCE* 5.4 (2004): 16-18.

Simons, Noel Edward, Bruce Keith Menzies, and Marcus C. Matthews. *A short course in geotechnical site investigation*. Thomas Telford, 2002.

Stevens, William R., and Bill E. Zehrbach. "Inclinometer data analysis for remediated landslides." *GEOTECHNICAL SPECIAL PUBLICATION* (2000): 126-137.

Stubbs, M. J., Rister, M. E, Lacewell, R. D., Ellis, J. R., Sturdivant, A. W., Robinson, J. R. C., and Fernandez, L., 2003. "Evolution of Irrigation Districts and Operating

Institutions: Texas, Lower Rio Grande Valley,"TR-228, Texas Water Resources Institute, Texas A&M University, College Station, TX.

Terzaghi, Karl. "Effect of minor geologic details on the safety of dams." American Society of Civil Engineers, 1929.

U.S. Army Corps of Engineers, 2000. "Design and Construction of Levees," EM 1110-2-1913, Washington, D.C.

U.S. International Boundary and Water Commission, 1992. "Status of Conveying Capacity of the Lower Rio Grande Flood Control Project," International Boundary and Water Commission, United States Section, El Paso, TX

Walshire, Lucas A., et al. *Geotechnical Evaluation of the Brownsville Levee Cracking and Partial Slope Failure*. ERDC Draft Technical Report Engineer Research and Development Center Vicksburg, MS, 2015.

REDUCING UNCERTAINTIES WHEN ASSESSING THE STABILITY OF ROCKFILL DAMS: CHARACTERIZING ANALYSIS PARAMETERS

Daniel B. Rogers[1]
Kyle R. Blakley[2]
Jeffrey S. Dingrando[3]
Jeff W. Munsey[4]

ABSTRACT

In 2012, the Tennessee Valley Authority (TVA) began the process of re-assessing the stability of its portfolio of hydroelectric and flood control dams, aided by select consultants. As part of the assessment, four large rockfill embankments were evaluated. These embankments, ranging between 175 and 318 feet in height, were constructed at three distinctly different points in history with respect to standards of practice for design and construction. The first dam was constructed from 1941 to 1942, the intermediate pair of dams from 1941 to 1950, and the final considered dam from 1966 to 1970. The rockfill material was quarried from local bedrock at each site. The assessment of each dam included research of historic documents, development and execution of a field exploration and laboratory testing program, development of material parameters, and stability analysis.

This paper reviews the rockfill and foundation characterization process by presenting the geologic conditions, historic design and construction documentation, construction processes and uncertainties associated with these considerations. Additionally, the means and methods of field exploration, field testing and laboratory testing are discussed along with how each mitigates or reduces potential uncertainty associated with selecting analysis parameters and conducting analyses. Finally, the historical data is combined with the results of exploration and testing to derive material parameters for seepage and stability analyses. Of particular interest are the evolving design and construction practices used as the engineering community gained experience with rockfill dams.

INTRODUCTION

Background

In 2012, the Tennessee Valley Authority (TVA) began re-assessing their fleet of river system dams. Nottely, Watauga, South Holston, and Tims Ford Dams are large, rockfill embankment dams which were evaluated as part of the program. The rockfill shells of the embankments consisted of locally quarried bedrock particles ranging in size from gravel

[1] Stantec Consulting Services Inc., 11687 Lebanon Rd. Cincinnati, OH 45241, 513-842-8200, Daniel.Rogers@stantec.com
[2] Stantec Consulting Services Inc., 11687 Lebanon Rd. Cincinnati, OH 45241, 513-842-8200, Kyle.Blakley@stantec.com
[3] Stantec Consulting Services Inc., 1409 N. Forbes Rd. Lexington, KY 40511, 859-422-3000, Jeff.Dingrando@stantec.com
[4] Tennessee Valley Authority Safety, River Management & Environment, 400 W. Summit Hill Dr. Knoxville, TN 37902, 865-632-4777, jwmunsey@tva.gov

to six-foot diameter or larger boulders. Assessment and characterization of the rockfill materials presented various challenges due to the limited sampling and testing methods available for materials of this size and composition. Through careful consideration of historic data, construction photos, limited field and laboratory test results, and published data, the rockfill materials were characterized in a manner appropriate for analysis of eight seepage conditions and approximately 50 load cases for slope stability evaluation.

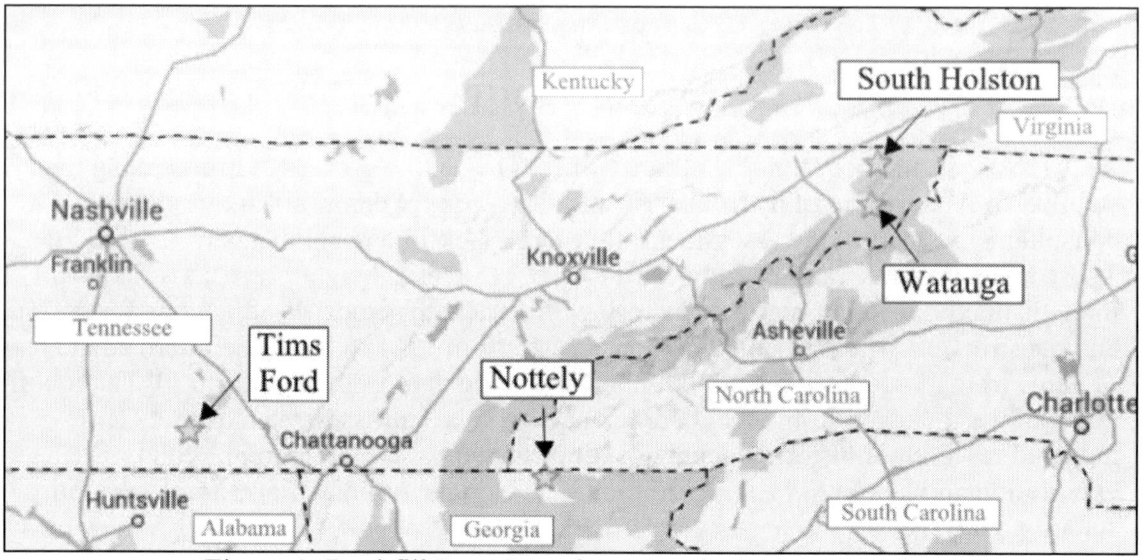

Figure 1. Rockfill Dam Location Map (basemap: Google)

Rockfill Dams

Nottely Dam. Nottely Dam is located in Union County, Georgia, approximately 11 air miles southwest of the town of Murphy, North Carolina and 2.3 river miles upstream from the Georgia-North Carolina State line on the Nottely River. Figure 2 shows the typical cross section geometry of Nottely Dam. The dam is a 2,600-foot long rockfill embankment with a rolled earth fill core. Construction was completed in 1942 with a crest Elevation of 1,794.0 feet. The main embankment was later raised to Elevation 1807.5 feet in 1986 to address an updated Probable Maximum Flood.

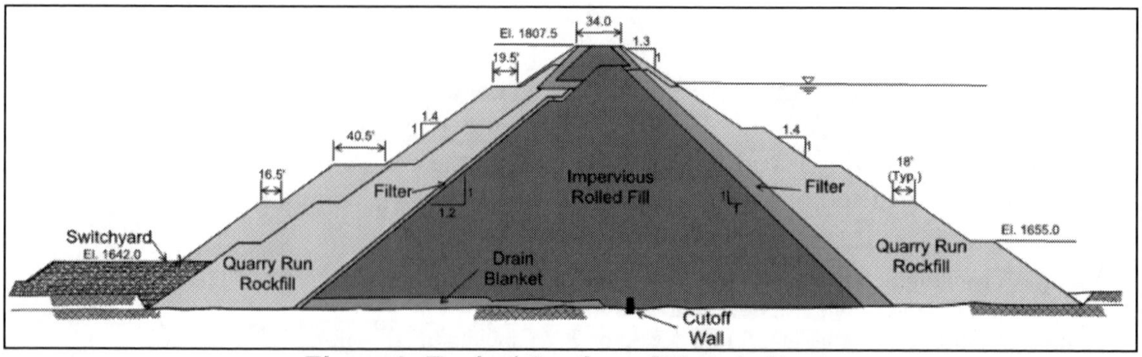

Figure 2. Typical Section of Nottely Dam

Bedrock at the Nottely Dam site belongs to the Pre-Cambrian Carolina Gneiss formation. Petrographic studies showed that various schists and quartzites, along with altered arkose, and pseudodiorite rocks occur in the formation at the site. The upstream and downstream slope faces were observed to be covered with large (three feet in diameter) rockfill. The quartzite was noted to be more durable than the schist. Fifty percent of the stone placed was required to be greater than 800 pounds, as indicated by historical documentation. Two on-site quarries were used to generate the stone for the rockfill. The quarries were within 0.2 miles of the dam, one upstream and one downstream. The quarried rockfill consisted of between 50 and 80 percent quartzite and 20 and 50 percent schist.

Two drilling methods were used to mine material for rockfill. Several holes advanced by well drills were loaded with Nitramon powder to produce material. Additionally, several pneumatic percussion drilled holes were loaded and blasted with dynamite. Processing of blasted material was very limited prior to placement (TVA 2011).

<u>Watauga Dam.</u> Watauga Dam is located about 5 miles downstream from Butler, Tennessee and 10 miles upstream from Elizabethton, Tennessee on the Watauga River. Figure 3 shows the typical cross section geometry of Watauga Dam. The dam is approximately 900 feet long, 300 feet high and 1,200 feet wide. It consists of a rockfill embankment with an impervious rolled earth core that was constructed between 1942 and 1948.

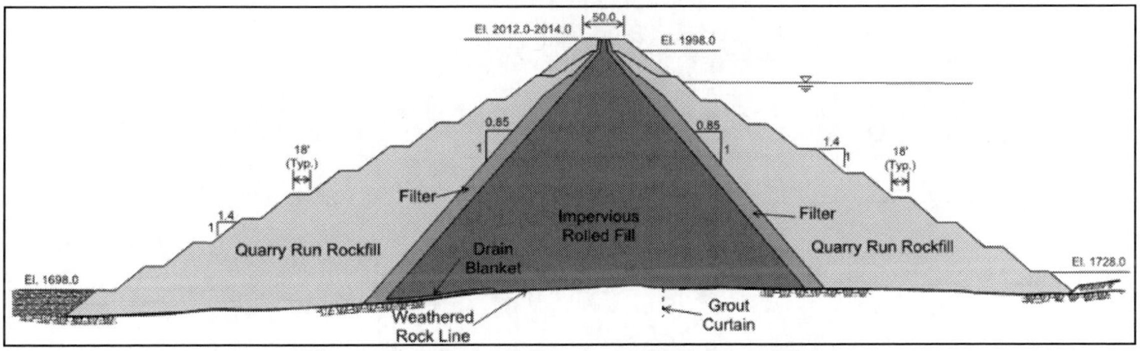

Figure 3. Typical Section of Watauga Dam

The dam is founded directly on bedrock of the Snowbird or Lower Unicoi formations from the Cambrian geologic period, with the exception of the upstream toe. These formations are comprised of metamorphic quartzite, arkose, and occasional beds of shale. The shale has reportedly undergone metamorphism making it harder than what is typically associated with shale, but not hard enough to be considered slate. The upstream toe of the dam is underlain by the Shady Formation, which is described as dolomite and limestone. The Iron Mountain Fault passes under the dam and thrusts the Unicoi Formation over the Shady Dolomite. The bedrock at the dam is characterized by bedding planes dipping steeply in the downstream direction. It is reported that the brittle quartzite broke readily under the stresses applied during folding and faulting, resulting in a mass of interlocking blocks bounded by joint and bedding surfaces (Leland and McMinn, 1943).

The rockfill was obtained from Cardens Bluff on the left bank of the Watauga River, downstream from the dam site. Drill bits from the wagon drills were destroyed due to the hardness and abrasiveness of the rock. Drill bit manufacturers and metallurgists were consulted for modifications, but even those did not improve production. Wagon drilling operations were replaced by coyote blasting to produce the rockfill. Coyote blasting uses large charges fired in small adits or tunnels at the floor level in the face of the quarry. Three large blasts provided the amount of rockfill needed to complete the dam (TVA 1950).

South Holston Dam. South Holston Dam is located approximately 7.5 miles southeast of Bristol, Tennessee and 110 miles northeast of Knoxville, Tennessee. Construction began in 1942 and was completed in 1951, with a five-year suspension of construction during World War II. **Error! Reference source not found.** shows the typical cross section geometry of South Holston Dam. The main dam is a rockfill embankment with a rolled earthen core. The dam is approximately 285 feet tall with a crest length of about 1,600 feet.

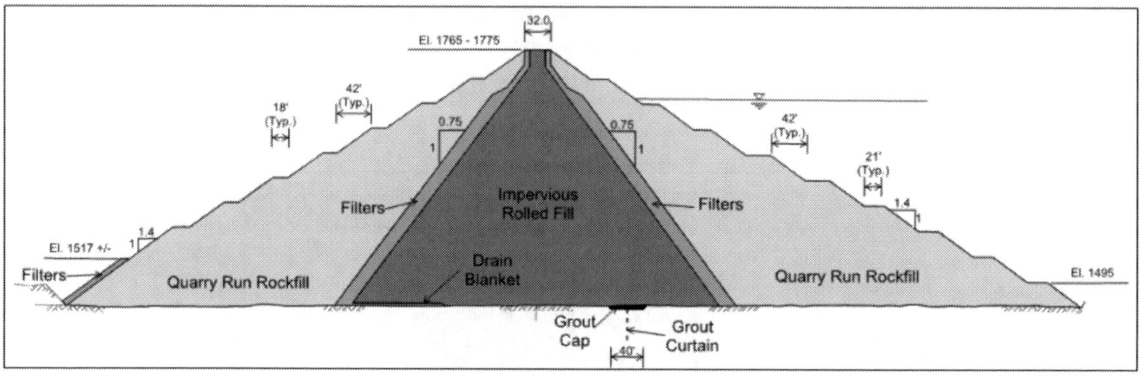

Figure 4. Typical Section of South Holston Dam

The dam is founded on the Middle Ordovician-aged Tellico formation. Local thickness of the layer is unknown due to extensive deformation. The Tellico formation consists of lenses and beds of sandstones, sandy shales, and conglomerates. The content of the shale and sandy shale in the formation ranges from 6 to 71 percent with a highly irregular thickness, ranging from 1 inch to greater than 60 feet. The sandstone is a hard, bluish gray rock containing about 25 percent calcium carbonate cementing matrix and 60 percent water worn quartz sand. The conglomerate occurs in beds and lenses and consists of pebbles and cobbles of limestone, quartzite, sandstone, and chert well cemented in a calcareous sandstone matrix. The rockfill used for the shells consisted mainly of calcareous sandstone with minor amounts of shale. A point of concern leading into field work at the site was the size and condition of the rockfill used as the dam shell. In areas where the shale was exposed, it appeared to be severely weathered and, in some cases, degraded to a fine-grained, soil-like material that had sloughed down onto the mid-slope benches. The sandstone was observed in sizes as small as sand particles up to as large as six feet in diameter (Rice and Allred, 1952).

The initial quarry used in 1942 was about 3,000 feet downstream of the dam. Approximately 166,400 cubic yards of usable rock was mined before work was suspended. When work was resumed after WWII, the downstream quarry was used for filter materials. Rockfill was then obtained from the Tellico sandstone and conglomerate on a steep hill immediately upstream of the dam. Four coyote blasts were completed between 1947 and 1950 to produce rockfill for the shells. One of the blasts used 1.33 million pounds of nitramon, the largest blast fired for construction or quarry purposes at the time. The blast was recorded on seismographs as far away as St. Louis, Missouri and Ottawa, Canada, and was used by multiple agencies to study properties of the earth's crust. As a whole, the rockfill was slightly weathered. Athens shale, which weathered quickly upon exposure, was discovered in the quarry underlying the Tellico and could not be used for construction. From 1948 to 1951, 580,000 tons of rockfill was removed from the quarry (TVA 1958).

Tims Ford Dam. Tims Ford Dam is located at river mile 133.3 on the Elk River, about ten miles downstream from Winchester, Tennessee. Tims Ford Dam is comprised of a primary embankment, a spillway, and a secondary embankment. **Error! Reference source not found.** shows the typical cross section geometry of the primary embankment at Tims Ford Dam. The primary embankment is an approximately 1,421-foot long, 175-foot high rock fill embankment with an upstream inclined impervious rolled earth core constructed between 1966 and 1970.

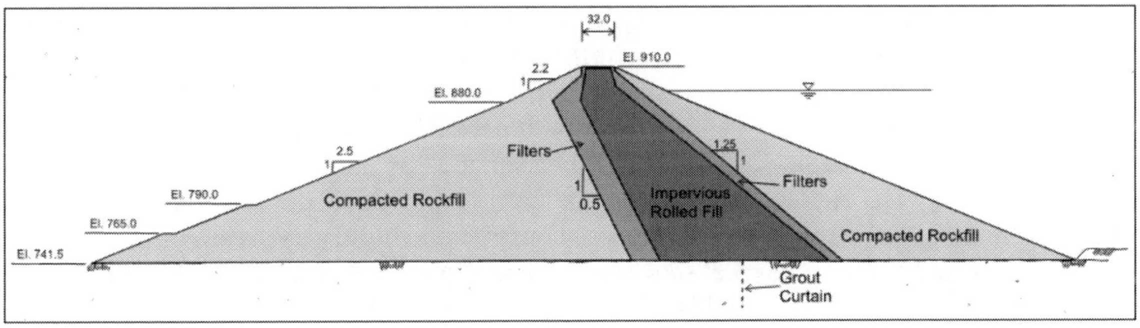

Figure 5. Typical Section of Tims Ford Dam

The dam was constructed upon limestone bedrock that was cleaned of overburden prior to fill placement. Along the axis of the dam, the foundation is composed primarily of two types of limestone, both of which belong to the Cathey-Leipers Formation. Below elevation 850 feet, the limestone is fine grained, slightly argillaceous to very shaly. There is very little weathering along vertical joints; however, in the top 15 feet of rock, a few weathered bedding planes appear between shale-limestone contacts. The rockfill surface conditions on the slopes of the dam indicate the size of the rock placed along the slope is widely variable in size, ranging from less than one-inch to approximately three feet in diameter, with occasional larger diameters. Upon inspection in 2013, the rockfill surface appeared to be weathered due to freeze/thaw effects.

Approximately 1,670,000 cubic yards of rockfill were required to complete the dam. A quarry developed into a hill at the upstream entrance to the spillway approach channel was mined for approximately 822,000 cubic yards. Rock excavation in the quarry was generally done in lifts ranging in depth from 14 to 25 feet to maintain benches accessible to drilling and other excavation equipment. The balance of the rockfill shell material was obtained from excavation for other features of the dam (spillway, powerhouse, penstock). The quarry operations produced durable limestone rockfill, mostly from the Cathey-Leipers formation. Other formations which provided rockfill included Fort Payne, Brassfield, and Fernvale Limestones (TVA 1977).

HISTORY

Evolving Design and Construction Processes

Nottely Dam. Construction of the project began in July 1941 and was completed in December 1943. It was completed quickly due to wartime needs. Foundation areas for the rockfill under the right and left sections of the main dam were prepared differently. Upstream of the baseline, the foundation areas were excavated to firm rock before fill placement. Downstream of the baseline, the area was stripped only of organic matter. The rockfill was separated from the earth core with a 2-layer filter on the upstream side, and a single layer filter on the downstream side. In general, rockfill placement followed the placing of the earth fill, with the earth fill constantly higher than the rockfill. The rockfill was dumped in place by dump trucks and leveled by bulldozers. Initially, rockfill was dumped in 10-foot lifts, the surface of each was scarified before placing the next. Above elevation 1,685 (the second bench), the fill was placed in 30-foot lifts (the total height between benches). Continuous sluicing of the rock as it was dumped from the trucks washed out the small amount of fines and facilitated the bearing of rock on rock. The bulk of the rockfill was moved over a six month period, with 680,000 cubic yards placed. An additional 12,268 cubic yards of rockfill was placed by hand to dress the slopes to design grades by December 1942 (additional ten months). A reinforced concrete cutoff wall (without underlying grout curtain) was constructed slightly upstream of the baseline. It was excavated two to three feet into sound impervious rock and was six feet tall and three feet wide overall (TVA 1948).

Watauga Dam. The dam was constructed from 1942 to 1948. Operations were suspended mid-construction due to World War II efforts taking manpower and budgetary priority. The total rockfill required for construction was 1,771,900 cubic yards. The rock was handled in the quarry area by diesel powered shovels and hauled using rear-dump trucks. The rockfill was constructed concurrently with the earth fill in order to provide support. Fines were washed out by sluicing of the rockfill material after being placed, which also aided in compaction and settlement. Before starting a new rockfill lift, the underlying material was scarified to permit escape of water and fines in the material. The size of the rockfill placed along the slope appears to consist of a widely variable size ranging from less than an inch to greater than 6 feet in diameter. Rock placement was continuously inspected (TVA 1950). A grout curtain was drilled on approximately the upstream quarter point of the width of the impervious fill. A two-foot thick by eight-foot wide grout cap was placed in a slight trench along the grout line (TVA 1951).

South Holston Dam. Construction operations began in February 1942 and were completed in November 1950. In October 1942, the project was placed on hold due to World War II. The shells of the rolled-earth core dam consist of 3,383,700 cubic yards of quarry run rockfill placed between 1942 and 1950. The rock was placed in 30-foot lifts. Some modifications to the design were made on a portion of the upstream slope because of poor quality rock (it contained a high percentage of shale and decomposed sandstone from a section of the upstream quarry). Because of the large amount of rock required, quarrying and placement operations took more than two years to complete. Compaction was accomplished by passage of equipment and sluicing to facilitate compaction. Before starting a new rock lift, the underlying surface was scarified to assure a bond and to permit permeation of water and fines. Large scale direct shear tests of internal friction angles were conducted on the various quarry rock encountered. Tests on the hard and sound sandstone showed friction angles of slightly over 45 degrees, while the more decomposed sandstone and shale proved closer to 30 degrees (TVA 1958). The cut-off grout curtain was drilled in two phases. A shallow grout curtain was constructed using low pressure grouting to consolidate the upper zone of highly weathered rock. The treatment of deeper foundation areas under higher pressures was completed after the low-pressure grouting (TVA 1958).

Tims Ford Dam. The dam was constructed between 1966 and 1970, and had a more consistent and deliberate construction schedule than the earlier rockfill dams. The placement of the rockfill was divided into two stages. The first stage began in August of 1967, near the downstream toe of the dam, for the purpose of conducting a compaction test. Two small vibratory compactors were used initially to compact the rock, but were considered inadequate. A large vibratory compactor was substituted in late August for compaction of the test sections. Across the site, rockfill was placed in direct contact with cleaned and prepared bedrock. Rockfill/bedrock interface friction tests were conducted shortly after beginning placement to verify that as placed material was consistent with design assumptions. Results indicated interface friction angles between 26 and 33 degrees, generally increasing with rockfill density. Rockfill adjacent to and beneath the downstream filter required two-foot lifts according to construction specifications. The remaining portion of the downstream rockfill area was built in three-foot lifts. Compaction for the rockfill material was provided by the larger 15-ton vibratory roller from the compaction testing. Placement of the rockfill specified that the three-foot lifts were to receive a minimum of two passes of the roller and achieve a minimum density of 120 pounds per cubic foot. Of the four sites evaluated, this is the only one that used compaction of the rockfill beyond sluicing techniques. The total amount of rockfill placed was 1,669,810 cubic yards. A grout curtain was located by projecting a 0.8 vertical to one horizontal slope from the baseline on the crest to its intersection with the rock surface. Some holes were angled up to 30 degrees in an attempt to intercept vertical or near-vertical joints (TVA 1970).

Several design and construction techniques evolved between the earliest subject dam (Nottely) and the most recent of the subject dams (Tims Ford). Early designs at Nottely used single layer filters for portions of the dam, while the later dams utilized two-layer filters. Design slopes were able to be better optimized by the addition of field scale shear testing on internal and interface friction angles. Construction practices evolved from simple sluicing of dumped quarry rock to achieve compaction to later requiring thinner lifts and vibratory compaction equipment. The design of the foundation cutoff walls evolved from the early dams to the later dams. Additionally, documentation of the completed work improved as time progressed. Understanding of this evolution assisted in characterizing the rockfill during analyses.

CHARACTERIZATION

Parameters Required

Several stability conditions were evaluated. The load cases included static drained and undrained conditions (selected pool levels), seismic (pseudostatic) undrained conditions, and consolidated-undrained rapid drawdown conditions. To run the seismic load cases, the pseudostatic coefficient was derived from a site-specific ground response analysis. To complete the seepage and slope stability analyses, several parameters of the rockfill were required:

- unit weight (with specific gravity, natural water content, void ratio),
- saturated permeability (including anisotropy ratio),
- unsaturated permeability (with saturated/residual volumetric water content),
- shear wave velocity,
- drained and undrained shear strength parameters for static conditions, and
- undrained shear strength parameters for seismic conditions.

There were multiple challenges in characterizing these parameters for the rockfill materials.

Issues Encountered

Drilling and Sampling. Because of the size of the rockfill at the four dams, drilling and sampling through the material was initially a concern. Data was needed, not only to obtain the parameters of rockfill listed above, but also to characterize the clay core materials below the rockfill shells. Conventional augering or casing advancement methods were not feasible for one or two reasons at each dam:

1. The rockfill consisted of metamorphic rock that was too hard or abrasive to penetrate efficiently (Watauga and Nottely).
2. Traditional casing advancement methods would have required sizable (four to six-inch) tooling in order to create a penetration large enough for the planned instrumentation (All four subject dams).

Another drilling concern for penetration through the rockfill was the integrity of the underlying filter layers. It was not desirable to clog the filters with drill slurry, nor create a vertical seepage path through the filters that separated the rockfill shells from the clay cores. These penetrations could create locally elevated phreatic levels or increase potential for piping of the core material.

Environmental Concerns. Environmental concerns were considered as they related to traditional casing advancement techniques; there was an inability to circulate coring water to the surface due to the high permeability of the rockfill. The possibility of coring water flowing through the rockfill and into the pool, creating a turbid discharge, was not acceptable from an environmental control standpoint.

Selected Testing, Drilling, and Sampling Method(s). To combat the issues mentioned above, the upper portion of each of the borings completed through the rockfill was advanced using sonic drilling techniques. The sonic technique was advanced until the clay core or bedrock was encountered. Sonic drilling (with water as the lubricating medium, when required) was also used during continuous sampling of the filter material. The amount of water needed to advance the sonic drill was much less than that needed to advance traditional drill casing; therefore the environmental concern was not as great. After reaching the target stratum, 4-inch diameter steel casing was seated a minimum of 6 inches into the underlying strata in preparation for more traditional sampling. Methods used within the casing included mud-rotary drilling and rock coring. The casing seal was monitored (based on fluid volume), and the seating depth was increased up to 30 feet when required to maintain an adequate seal.

Sampling in the rockfill was accomplished in targeted zones using the sonic drilling equipment. This method returned a continuous sample; however, the samples were disturbed and suitable only for index testing and creating reconstituted specimens. In some instances, shallow, hand-dug test pits were utilized to verify construction geometry of the rockfill and filters. The pits also allowed samples of the filter material to be obtained from below the rockfill.

In addition to drilling the rockfill material, geophysical testing was conducted at both South Holston Dam and Tims Ford Dam. The geophysical testing selected was in situ Multi-channel Analysis of Surface Wave (MASW) testing. The results of the MASW testing were interpreted and provided greater confidence in the uniformity of the placed rockfill along with quantitative data for shear wave velocity.

Laboratory Testing. Testing on rockfill samples was limited to SDI testing to assist in gauging the durability of the rockfill particles. Equipment to test the full gradation samples of rockfill was not available; therefore such samples were not obtained. "Scalped" samples, or samples with an artificially limited particle size, were not considered to be representative of the rockfill, because traditional testing methodology requires the maximum particle size to be one sixth of the size of the testing apparatus (ASTM D4767). The subject dams had maximum particle sizes of approximately two to six feet. With no feasible method to determine a representative gradation of the rockfill, and the very large particle size, a suitable scalped sample was not able to be obtained.

Available Information

Rather than solely using traditional laboratory testing to determine the necessary parameters for analyses, several other sources were utilized. TVA kept extensive records of historic design and construction reports, photographs, and field and laboratory testing results. Each of the four dams had differing historic data available, and was therefore characterized uniquely. It was noted that the quantity and quality of the records available improved as time progressed. The historic data was combined with limited field and laboratory testing and published rockfill data in order to assign the necessary parameters.

Unit Weight. Direct measurements of the rockfill unit weight were not feasible due to the large range of encountered particle sizes (sand-size to 6-foot diameter boulders). Phase relationships were used to estimate the necessary unit weight parameters. Specific gravity was generally determined from laboratory testing on the filter materials. This was appropriate because the aggregate was generally mined from the same sources as the rockfill. For characterization at Nottely, the moisture content was taken as the average from SDI samples. Historic data was used in several cases which indicated unit weights assumed in previous analyses or from field testing during construction. For example, the moist unit weight used for the Tims Ford Dam analysis was based on averages from 79 density tests conducted as part of the field construction quality control program (TVA 1977). Published data, such as that provided by Leps (1970) was also used to estimate void ratios in some cases. After deciding on several of these parameters, the dry unit weight, moist unit weight, and saturated unit weight of the materials could be determined using the assumptions along with phase relation calculations.

Saturated Permeability. The saturated permeability parameters were determined in various ways for the rockfill at the four dam sites. In some cases, the vertical permeability was assumed to be equivalent to the results of constant head permeability tests conducted on the filter materials. At South Holston, one piezometer was installed in the rockfill layer in order to monitor the impact of tailwater backup into the rockfill zone. Because the piezometer was there, it allowed for a slug test to be conducted. Slug test information was limited however, because the water flowed out of the instrument faster than the level in the standpipe could be read. This provided a quantifiable lower boundary for permeability. In the case of Tims Ford (TVA 1967), six percolation tests were conducted during the construction phase in the rockfill material after it had been placed. One of the six indicated that the vertical permeability was slightly lower than the others. The lower result was attributed to excessive crushing of the shale material within the rockfill. The remaining five tests were limited to the capacity of the water meter being used, which was 30 gallons per minute. USBR guidance was also used in order to classify the anisotropy ratio (assumed to be 1.0 for rockfill materials because of the very coarse-grained nature). In each case, the permeability of the rockfill was estimated; however was not a limiting parameter in the seepage analyses.

Shear Wave Velocity. In order to derive the pseudostatic coefficient for the seismic stability analyses, the shear modulus and shear wave velocity were necessary for the site specific ground response analysis. The parameters were estimated either by in situ testing at the time of exploration or from a correlation provided by Seed and Idriss (1970). MASW testing was conducted on the rockfill at South Holston and Tims Ford from the rockfill surface of the benches on the downstream slope of the dams. The test results indicated that with the selected energy source, the shear wave velocity could be measured to approximately 50 feet below the surface at South Holston, and 145 feet deep at Tims Ford. The compacted rockfill at Tims Ford allowed energy to travel deeper into the mass than the dumped rockfill at South Holston. At Watauga and Nottely, MASW testing was not conducted; therefore the shear modulus (G_{max}) was estimated using the equation suggested by Seed and Idriss (1970):

$$G_{max} = 1{,}000K_{2,max}(\sigma'_m)^{0.5} \tag{1}$$

The correlation factor, $K_{2,max}$, was estimated based on the void ratio of the rockfill layer as discussed above. The mean effective stress (σ'_m) of the rockfill varies depending on the cross section. The shear wave velocity for the rockfill zone was calculated for each cross section using the equation given below:

$$V_s = SQRT(G_{max}/\text{unit density}) \tag{2}$$

Drained and Undrained Shear Strength. Because of the very large particles (cobbles and boulders) in the rockfill, direct measurements for the strength of rockfill were not feasible. Instead, the drained strengths were estimated using the published data presented by Leps (1970).

Leps (1970) compiled the database of rockfill test results in Figure 6, and showed the effects of normal pressure on drained friction angle. The data are plotted on a semi-log graph of secant friction angle versus normal pressure. The data shows that at lower normal pressure, the decrease in friction angle is greater with increasing normal pressure than at higher pressures. The three trend lines in Figure 6 were defined by Leps (1970) for the range of typical rockfill; the trend lines can be represented by the following equations:

- Low density, poorly graded rockfill, with weak particles (unconfined compressive strength between 500 and 2,500 psi)

$$\phi'_{sec} = 41.8° - 6.8°\log(\sigma'_n/P_a) \tag{3}$$

- Average rockfill, with average strength particles (unconfined compressive strength between 2,500 and 10,000 psi)

$$\phi'_{sec} = 47.1° - 6.8°\log(\sigma'_n/P_a) \tag{4}$$

- High density, well graded rockfill with strong particles (unconfined compressive strength between 10,000 and 30,000 psi)

$$\phi'_{sec} = 52.1° - 6.8°\log(\sigma'_n/P_a) \tag{5}$$

Here, ϕ'_{sec} is the secant, effective stress friction angle (slope of a tangent line to the Mohr's circle of stress at failure, with zero cohesion intercept), σ'_n is the effective normal pressure, and P_a is one atmosphere of pressure in the units of σ'_n (P_a = 14.69 psi = 2,116 psf = 101.3 kPa).

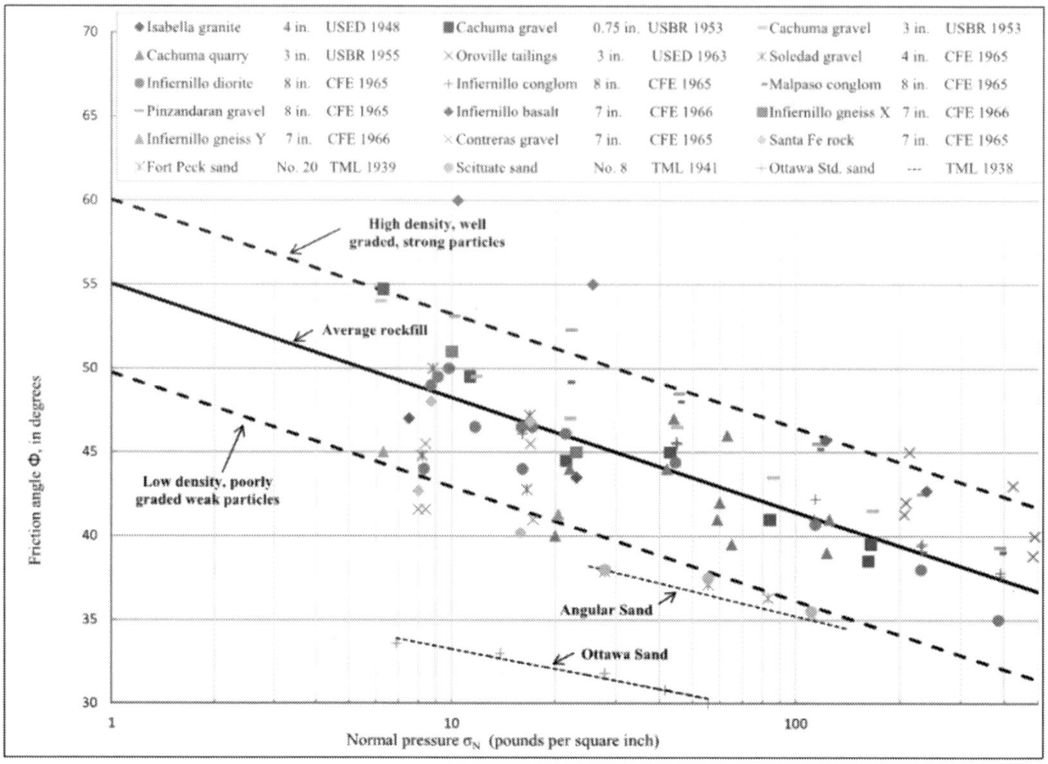

Figure 6. Shearing Strength of Rockfill from Large Triaxial Tests (after Leps, 1970)

After considering the historic construction methods, geological, geotechnical, geophysical, and laboratory information, an appropriate curve was selected for each facility. Laboratory testing that was considered included SDI testing on rockfill and/or parent material, and unconfined compression strength (UC) testing on parent material.

- South Holston: Average rockfill was chosen for the downstream shell to account for the inclusion of shale in the rockfill. Other historic and testing information indicated strong particles. The low density rockfill curve was selected for the upstream shell, representing a weaker rockfill due to inclusion of higher percentages of weathered shale and weakly cemented sandstone particles.

- Nottely: Average rockfill was chosen. Generally, there was greater than 92 percent retention after four cycles of SDI testing. The UC strength tests results varied widely (1,590 psi to 39,280 psi).

- Tims Ford Dam: Average rockfill was assigned. The rockfill is well-graded, was constructed with good compaction, and UC strength of the parent material was high (10,290 psi to 18,810 psi). Field scale shear testing was conducted during initial construction on this site for the rockfill/bedrock interface. However, these results were not indicative of the internal friction angle of the rockfill itself.

- Watauga Dam: The upper curve for high density, well-graded rockfill with strong particles was utilized. SDI testing generally indicated greater than 90 percent retention after four cycles. The UC test results were between 10,240 psi to 40,380 psi.

The equations by Leps account for the reduction in strength due to increasing overburden pressure, and predict a range of secant friction angles. The values obtained were comparable to the typical strengths of rockfill as published by Marachi (1972).

The rockfill layers at each location were determined to be coarse grained, free-draining materials that would not develop excess pore pressures or an undrained loading condition. For the static and pseudostatic undrained analyses, strength parameters were assumed to be equal to the drained strength parameters.

ANALYSES

The amount of site data collected over the course of the dam evaluations allows for more risk and fragility based analyses to be completed in the future. An analysis option could have been to use only historical data, assumed properties, and very limited testing to justify that the dams met minimum requirements for a standards based approach. However, because significant efforts went into historic review, current testing, and more realistic assumptions, a framework was developed for doing better risk-based analyses in the future.

Software Used

The seepage and slope stability analyses were conducted using GeoStudio 2007. SEEP/W, a finite element code tailored for modeling two-dimensional groundwater seepage problems in soil and rock, was used for seepage analysis. Slope stability was evaluated using conventional, limit equilibrium methods as implemented in the GeoStudio SLOPE/W 2007 software. Available from GEO-SLOPE International, Ltd., of Calgary, Alberta, Canada (www.geo-slope.com), SLOPE/W is a special-purpose computer code designed to compute the stability of earth slopes using two-dimensional, limit equilibrium, method of slices analyses. With SLOPE/W, the distribution of pore water pressures within the earth mass may be mapped directly from the results of a SEEP/W analysis.

While the SEEP/W analysis required permeability, anisotropy, and unsaturated seepage parameters for each soil type in a cross section, the completion of the analyses indicated that the actual values did not have a large impact on the seepage model's prediction of pore water pressures. Because of the large difference in permeability between the core material of the dams and the rockfill shells, the predicted pressures did not change significantly with changes to the rockfill seepage parameters.

The SLOPE/W analysis, however, relies heavily on the shear strength properties of the rockfill zones for determining the factor of safety against slope instability. Because the rockfill zones make up such significant portions of the dams, the strength of those zones has a significant impact on the failure surface and factors of safety determined. To utilize the functions as described by Leps (970), the equations require input as shear strength versus normal stress (shear/normal) functions for the rockfill. The shear/normal function is input as a table of points rather than a conventional Mohr-Coulomb envelope. The shear strength was calculated via the equation by using a range of normal stresses that coincides with the normal stresses expected for the cross section. This was a simple exercise; the calculation was completed in a spreadsheet, and then copied into the SLOPE/W program material properties.

The shear/normal function allows the analyses to be more representative of in-situ conditions at both low and high normal stresses. If a traditional Mohr-Coulomb model were used, the friction angle may be estimated as approximately 40 to 45 degrees. By using the shear/normal function, friction angles as high as 52 degrees for the upper bound Leps model can be modeled for the rockfill near the surface. This eliminates some of the shallow failures that are incorrectly predicted by the software model when using a single, lower friction angle. It is also more representative of deeper failure surfaces by applying a reduced friction angle at higher normal stresses, where the overburden pressure can cause crushing or smoothing of the contact points between rockfill particles.

Rapid Drawdown

Although not typically considered for rockfill because of how well drained the material is, the rapid drawdown analyses were run due to the magnitude of the difference in projected headwater conditions at the subject dams. The shear/normal function is able to be applied in the static and pseudostatic slope stability analyses. However, SLOPE/W does not allow the shear/normal function to be used for rapid drawdown analyses. Material strengths must therefore be classified using Mohr-Coulomb strength parameters using an appropriate cohesion and friction angle derived from the shear/normal function. Mohr-Coulomb strength parameters were estimated in three steps.

1. The median value of normal stresses was determined from the base of slices on the failure surface of the drained-strength, normal pool condition. Only slices through the rockfill were evaluated to find the median value.
2. A value was determined at which 2/3 of the normal stresses in the range were lower. This number was used for consistency with the United States Army Corps of Engineers' method for assigning shear strength parameters.
3. The values determined in Steps 1 and 2 were then used to create a trendline within the shear/normal function range. This developed an apparent cohesion and friction angle for the rockfill.

It was assumed that the rapid drawdown failure surface would look similar enough to the normal pool, drained strength failure surface that the normal stress range would be appropriate for the estimation.

CONCLUSION

When seepage and slope stability analyses are required for a rockfill dam which has limited existing data available; drilling, sampling, and characterization of the rockfill can be challenging. By carefully examining historic construction, testing, and inspection records, initial assumptions can begin to be made to facilitate analysis. Following the historic documentation review, plans can be made in an attempt to gather information that is lacking. Drilling and sampling techniques should be examined to limit the negative environmental and dam safety impacts which could develop from the process. These impacts can include drilling fluid releases into the headwater or tailwater pools or creating penetrations through the filter zones. After careful consideration of historic data, construction photos, geophysical testing, limited field and laboratory test results, and published data, it is possible to characterize the rockfill materials in a manner appropriate for seepage and slope stability evaluation. Additionally, in the case of the subject dams discussed, the types and amounts of data that was reviewed and collected creates a solid framework for doing risk-based analyses in the future. By using good engineering judgment based on available data, it is possible to more accurately characterize a rockfill facility with a more robust analysis using strength functions for rockfill materials rather than typical assumptions for Mohr-Coulomb parameters. Table 1 summarizes the differences between a typical Mohr-Coulomb strength representation, and a non-linear strength representation.

Table 1. Rockfill Strength Characterization Summary

Strength Parameter Data	Strength Parameter Characterization	
	Traditional Mohr-Coulomb	Non-Linear Function (Leps)
Type of φ’	Single friction angle	Friction angle varies as a function of overburden pressure
Value of φ’ (degrees)	35 – 45	35 – 52
Value of Cohesion (psf)	0	0

REFERENCES

Grant, Leland F. and McMinn, Paul M., (1943). “The Geology of the Watauga Project.” Tennessee Valley Authority, May 1943.

Leps, T. M., (1970). “Review of Shearing Strength of Rockfill.” *J. of the soil mechanics and Foundations Division*, ASCE. Vol. 96, No. SM4, July, pp. 1159-1170.

Marachi, N. D., Chan, C.K., and Seed, H.B. (1972). “Evaluation of Properties of Rockfill Materials.” *J. of the Soil mechanics and Foundations Division*, ASCE, Vol. 98, No. SM1, January, pp. 95-114.

Rice, Elgin R. and Allred, Paul W., (1952). "Construction Report No. VIII – South Holston Project – Foundation Treatment." Tennessee Valley Authority, March 11, 1952.

Seed, H. B., and Idriss, I. M. (1970). "Soil Moduli and Damping Factors for Dynamic Response Analyses." *Report EERC 70-10,* Earthquake Engineering Research Center, University of California, Berkeley.

TVA, (1948). "The Hiwassee Valley Projects, Technical Report No. 5, Volume 2: The Apalachia, Ocoee No. 3, Nottely, and Chatuge Projects." Tennessee Valley Authority, 1948.

TVA, (1950). "Watauga Project Construction Report No. XIV, Main Embankment." Tennessee Valley Authority, October 13, 1950.

TVA, (1951). "Watauga Project Construction Report No. 19-300-VII-12 #2." Tennessee Valley Authority, January 15, 1951.

TVA, (1954). "The Tims Ford Project on the Elk River – Report No. 54-4 – A Feasibility Report." Tennessee Valley Authority, June 1954.

TVA, (1958). "The Upper Holston Projects, Technical Report No. 14 – Watauga, South Holston, Boone and Fort Patrick Henry." Tennessee Valley Authority, 1958.

TVA, (1967). "Tims Ford Project Information for Consultants – Report No. 54-134-3." Tennessee Valley Authority, November 20, 1967.

TVA, (1970). "Tims Ford Project Information for Consultants – Report No. 54-134-6." Tennessee Valley Authority, June 17, 1970.

TVA, (1977). "Tims Ford Project Final Construction Report." Tennessee Valley Authority, March 1977.

TVA, (2011). "Nottely Project Supporting Technical Information (STI)." Tennessee Valley Authority, September 2011.

AUTHOR INDEX

Aberle, Jochen....1531
Adams, Lea G....749
Alicescu, Vlad....1463
Allerton, William....1035
Amos, Peter....829
Anderson, Spenser....1483, 1559, 1655
Annandale, George W....221
Arnold, Justin....1639
Arrocha, Jaime....1341
Auld, Bernie....1191
Ball, Andre....1365
Barrentine, Scottie....335
Barrentine, Scottie L....1379
Barrie, Richard E....711
Bartles, Michael....269
Bartojay, Katie....727
Baxter, Craig P....221
Beaty, Michael H....535
Beduhn, Robert....1
Bennett, Tony....1173
Benson, Brian....829
Bentler, David J....849, 939
Beriswill, Jeffrey A....1267
Beutel, Justin....515
Bishop, Matt....1497
Bisnett, Rachael V....1133
Blakley, Kyle R....813
Bliss, Mark....1283
Booth, Derek....1001
Bouaanani, Najib....611
Boulanger, Ross W....535
Bountry, Jennifer A....251
Boyd, M. Leslie....445
Boyer, Bradley T....165
Boyer, D....1017, 1035
Brandon, Thomas L....849, 939
Brumley, Bryan J....711
Bukhair, Layne....445
Bulger, Jonathon R....1395
Byrd, Aaron R....123
Calcagno, Frank....1161
Campbell, Dave....11
Cannon, Robert....1365
Carey, Bryan....695
Castillo-Rodríguez, Jesica Tamara....1053
Charlwood, Robin G....1119
Chowdhury, Khaled....1297
Christensen, Dan....107
Coats, J. Tyler....165
Compton, Brandon....563
Cooper, Paul J....1379
Costa, Ray....781
Craig, Matthew....581
Crookston, B. M....79, 165
Crowder, James R....165
Cui, Yantao....1001
Curtis, Dan D....675
Dalton II, Larry L....563
Daly, Steven F....1507
Darbar, Soheil Razavi....727
Davidheiser, Steven....1251
Davis, Chuck....237
De Puy, Maximiliano....1341
De Rubertis, Kim....1017, 1145
Deere, Don....1419
Depin, Nickolas J....1267
Dingrando, Jeffrey S....29, 813
Dirdal, Melinda....11
Dodson, Andrew....335
Dolen, Timothy....727
Dombeck, Elise....627
Downer, Charles W....123
Dunbar, Joseph B....799
Duren, Angela....1507
Duron, Ziyad....657, 1483, 1559, 1655
Dworak, Frank....139
Ebner, Laurie....581
Eichinger, Robert A....149
Eichler, Jason A....199
Engemoen, William....1069
Escuder-Bueno, Ignacio....1053
Esterhuizen, Jacob....1419
Feng, Lingmim....675
Fernandez-Delgado, Gabriel....1419
Fiedler, William R....1069, 1097
Fleming, Matthew....269
Fletcher, John....675
Foos, William F....1161
Forrest, Michael....969
Fortuny, William....581
Franceschi, Pastora....1341
Franz, Raymond J....1395
Freitas, Mark....395
Gagnon, J. Hawkins....1365
Ganjoo, Bal....395
Gentzler, Seth....1001
Gerlach, Michael....149
Ghanaat, Yusof....497
Gilbert, Daniel A....29, 1379
Giovando, Jeremy....1521
Goetz, Chris....1341
Goodell, Christopher....107
Gowda, Santhosh G....1603
Grant, Brian....179
Graulau-Santiago, Jaime A....123
Griebenow, Kevin....643
Groeneveld, Joe....1639
Gullo, Nathan M....199
Haberman, Ruth....97
Hamernik, David....581

AUTHOR INDEX

Hamilton, David 29
Hannes, Danielle Tripp 1201
Hansen, Kenneth D. 1447
Harder, Leslie F. 883
Harris, David W. 955
Hasan, Husein 1217
Hatton, Chris 727
Heath, Benjamin 1191
Hedien, Jason E. 303
Hess, Matt 581
Hibbs, David 415
Hiller, Priska Helene 1531
Hlepas, Georgette 315
Hou, Susan 969
Hou, Susan S. 469
Hughes, Aled 695
Hughes, David 485
Hultstrand, Doug 287
Hunt, Dina Bourliea 515
Irving, Derek 1341
Jaeger, Robert A. 865, 909
Jamieson, Stephen L. 1201
Johanson, Alan 469
Jones, Adam N. 1119
Jones, Scott 485
Kappel, Bill 287
Kara, Redha 1463
Karshenas, Afshin 563
Kay, Roger L. 547
Kazyak, Brian J. 1395
Keel, K. 79
Kees, S. 79
Keizer, Richard A. 909
Kim, Wonnie 1133, 1541
Kirschbaum, Daniel 1639
Kjosavik, Fredrikke 1531
Kleiner, David 45
Kline Jr., Robert A. 1251
Kovacich, Joseph R. 1395
Krohmer, Brian 1217
Kynett, Michael 1297
Laboon, John H. 1097
Lan, Frank 179
Laugier, Frédéric 61
Lee, Tedman 469, 969
Lia, Leif 1531
Lillycrop, Jeff 563
Lim, Yeo Howe 1591
Lin, S. Samuel 303
Lindyberg, Robert F. 563
López-Soto, Jamie 927
Lothes, Amanda 1251
Ludy, Jessica 765
Macleod, Bruce 1267
Mahnke, Steve 781
Maki, Ian P. 865
Malvick, Erik J. 985
Margo, David A. 749
Marshall, Kevin 1119
Martin, Kristen M. 985
Martinez, Gabriel 1541
Masters, Monica 445
May, Michael R. 1621
McClelland, Duane 199
McMichen, Lindsay 335
Meier, Dan 1341
Mejia, Lelio 1341
Menda, Kunal 1483, 1559, 1655
Meyer, Don B. 1621
Millet, Richard 781, 1297
Miquel, Benjamin 611
Miranda, Roberto 1341
Mirghasemi, Ali Asghar 1591
Mishalanie, David 1119
Mockler, Stephen J. 435
Moen, Keith 1639
Montalvo-Bartolomei, Axel 927
Montgomery, Donald J. 1133
Morales-Torres, Adrián 1053
Morrison, Michael A. 1575
Morse, Daniel 199
Munsey, Jeff W. 813
Muto, Matthew 1083
Neff, Keil 1191
Ng, Carman 969
Nolan, Cecily M. 1235
Nottis, Aaron 1217
Nuss, Larry K. 1447
O'Brien, Steve 485
Oestreich, Cynthia 107
Olsen, Richard S. 351
Owens, Brian 1497
Pate, Kim 1365, 1639
Paxson, Greg 11
Perlea, Mary 781, 1297
Pinchard, Thomas 61
Piotrowski, Jesse 237
Powers, Matthew 395
Pradhan, Nawa Raj 123
Queen, David 727
Quiroz, Juan E. 711
Randle, Timothy J. 251
Rauch, Alan F. 1379
Renaud, Sylvain 611
Rice, John D. 909
Riker, Richard E. 1419
Ritchie, Andy 251
Roadifer, John 969, 1001
Robbins, Bryant A. 927
Robblee, G. 79

AUTHOR INDEX

Rogers, Daniel B. 813
Rogers, Gary 1365
Rosculet, Dan 1463
Roth, Lawrence H. 765
Rutherford, James H. 1639
Saber, Robert 415
Sadden, Brian 695
Salamon, Jerzy 599
Salter, Paul 1341
Savage, Bruce 97
Sawka, Mark 581
Scheevel, Nathan 221
Schenk-McFarland, Tara 1283
Schug, David 1341
Schulte, Bernard 1575
Schweiger, Paul G. 1161
Scott, Gregg A. 1083
Serota, Tali 1173
Shewbridge, Scott E. 749
Sheydayi, Peter 1001
Shoghli, Bahareh 1591
Siddappa, Gopi 1603
Sills, George L. 883
Simpson, Mark 1267
Skahill, Brian E. 123
Snider, Frederic 1365
Snyder, David 415
Snyder, Kevin 1119
Sooch, Gurinderbir S. 675
Stanard, Tina 97
Stateler, Jay N. 379, 1017
Stephens, Isaac J. 351, 799, 927
Stiller, William B. 435
Storino, Louis 1395
Sudah, Bashar S. 711
Supranata, Y. Erwin 435
Sutley, Garrett 955
Tarbox, Glenn 695
Tenney, Ray D. 1621
Thompson, David P. 711
Thompson, Mark J. 849, 939
Toose, James 1341
Tournier, Jean Pierre 1463
Triplett, Dennis 1217
Tullis, Blake 97
Tutka, Travis 315
Urban, Robert 1341
Vandenberge, Daniel R. 849
Vargas, Boris Caro 1217
Varuso, Richard 749
Vasquez, Victor 97
Vasquez, Victor M. 445
Vermeulen, Julien 61
Von Gersdorff, Nicolas 1083
Waddell, Robert 179
Wade, Daniel 969
Wade, Daniel L. 469
Wahlin, Brian 237
Walberg, Francke 1297
Walshire, Lucas A. 799
Wan, Cheuk Y. 1235
Warms, Mikell 107
Watts, Matthew 563
Weber, Christine 515
Welshans, April L. 29
Werner, John 1639
Weston, David M. 123
Whitten, Zachary 149
Wiesner, S. 1321
Wiggins, Charles E. 563
Williams, Luke 1497
Witkowski, Doug 445
Wohletz, A. 1321
Wormer, Jeff 379
Yen, John 657, 1083
Zambrano, Carlos 1541
Zhao, Mo 1483, 1559, 1655
Zheng, Jiqin 675
Zusi, Michael 179

United States Society on Dams (USSD)
1616 Seventeenth Street #483
Denver, CO 80202
USA

ISBN 978-1-5108-2942-8

9 781510 829428

CURRAN ASSOCIATES INC.
proceedings.com

36th Annual USSD Conference 2016

Celebrating the Value of Dams and Levees – Yesterday, Today and Tomorrow

Denver, Colorado, USA
11 - 15 April 2016

Volume 2 of 2

36th Annual USSD Conference 2016

Celebrating the Value of Dams and Levees – Yesterday, Today and Tomorrow

Denver, Colorado, USA
11 - 15 April 2016

Volume 2 of 2

ISBN: 978-1-5108-2942-8

Printed from e-media with permission by:

Curran Associates, Inc.
57 Morehouse Lane
Red Hook, NY 12571

Some format issues inherent in the e-media version may also appear in this print version.

Printed by Curran Associates, Inc. (2016)

For permission requests, please contact United States Society on Dams (USSD)
at the address below.

United States Society on Dams (USSD)
1616 Seventeenth Street #483
Denver, CO 80202
USA

Phone: 303-628-5430
Fax: 303-628-5431

gene@ussdams.org

Additional copies of this publication are available from:

Curran Associates, Inc.
57 Morehouse Lane
Red Hook, NY 12571 USA
Phone: 845-758-0400
Fax: 845-758-2633
Email: curran@proceedings.com
Web: www.proceedings.com

United States Society on Dams

USSD

Celebrating the Value of Dams and Levees — Yesterday, Today and Tomorrow

36th Annual USSD Conference
Denver, Colorado, April 11-15, 2016

Gross Dam is a concrete gravity arch dam in Boulder County, owned by Denver Water. A planned modification will raise the dam 131 feet with a downstream roller compacted concrete buttress, to a final height of 471 feet. The new dam and reservoir will provide an additional 77,000 acre-feet of municipal water storage. Denver Water expects to start final design engineering in late 2016, with construction set to begin in 2018.

U.S. Society on Dams

Vision

A world class organization dedicated to advancing the role of dam and levee systems and building the community of practice.

Mission

USSD, as the United States member of the International Commission on Large Dams, is dedicated to:

ADVOCATE: Champion the role of dam and levee systems in society.

EDUCATE: Be the premier source for technical information about dam and levee systems.

COLLABORATE: Build networks and relationships to strengthen the community of practice.

CULTIVATE: Nurture the growth of the community of practice.

FOREWORD

Included in these Proceedings are papers presented during the **36th USSD Annual Meeting and Conference**, held April 11-15, 2016, in Denver, Colorado. A separate book includes abstracts for each paper, as well as abstracts for presentations not included in the Proceedings.

The theme of the 36th USSD Conference was ***Celebrating the Value of Dams and Levees — Yesterday, Today and Tomorrow.*** The Conference technical program was organized by several USSD Committees under the leadership of **Rodney Eisenbraun**, RJH Consultants, Inc. Dams and levees have provided and will continue to provide significant benefits to our society. Many of our existing dams and levees provide greater benefits today than they did when they were first constructed and they will be needed to provide even greater benefits to address the challenges of the future. USSD has been and continues to be a source for balance and reason — neither ignoring the negative impacts nor their benefits. USSD and its members have worked and continue to work to identify and advance real solutions that minimize impacts and maximize the benefits of dams and levees. The papers included in these Proceedings address the Conference Theme, and other critical technical and management topics for the dam professional.

Technical Sessions during the Conference focused on best practices and new technologies related to concrete dams, embankment dams and levees; dam safety; public safety and security; construction and rehabilitation; hydraulics; foundations, monitoring; and seismic and environmental issues.

The papers in the Proceedings were selected from abstracts submitted in response to a Call for Papers, and include both oral and poster presentations. Authors are specialists with broad experience from government agencies, utilities, academia, water districts, consulting firms and private industry.

The Conference Organizing Committee extends thanks and appreciation to the **Bureau of Reclamation**, Host of the 36th Annual Meeting and Conference.

Special thanks are also extended to the Committee Members who selected the abstracts and reviewed the technical papers, and to the authors who prepared the papers included in the Proceedings.

CONTENTS

Conference Theme

Is This Community Worth Saving? 1
Robert Beduhn, HDR

Unintended Consequences of Dam Removal 11
Melinda Dirdal, **Greg Paxson** and **Dave Campbell**, Schnabel Engineering

The Kentucky River: Managing 19th Century Infrastructure to Meet 21st Century Needs 29
Daniel A. Gilbert, **Jeffrey S. Dingrando** and **April L. Welshans**,fMGS Stantec Consulting Services, Inc.; and **David Hamilton**, Kentucky River Authority

Historical Perspective — Dams, Hydro, and Multi-Purpose Projects 45
David Kleiner, MWH

Hydraulics

Use of Labyrinth Piano Key Weir (PKW) Spillway to Improve Dam Safety for a Large Set of Dams 61
Frédéric Laugier, **Julien Vermeulen** and **Thomas Pinchard**, EDF-CIH

Ensuring Water Supply Reliability with Innovative PK Weir Spillway Design 79
G. Robblee, **S. Kees** and **B.M. Crookston**, Schnabel Engineering; and **K. Keel**, Town of Hillsborough

Importance of Site Considerations for Labyrinth Spillway Hydraulic Design — Upper Brushy Creek Dam 7 Modernization 97
Tina Stanard and **Victor Vasquez**, Freese and Nichols, Inc.; **Ruth Haberman**, Upper Brushy Creek Water Control and Improvement District; **Blake Tullis**, Utah State University; and **Bruce Savage**, Idaho State University

Automated Incremental Hazard and Sensitivity Analysis of a Hypothetical Dam Breach Model 107
Mikell Warms, **Dan Christensen** and **Christopher Goodell**, WEST Consultants, Inc.; and **Cynthia Oestreich**, Aista Utilities

GSSA Modeling for Analysis of Flood Design Features at the Picayune Strand Restoration Project 123
Charles W. Downer, **Jaime A. Graulau-Santiago**, **Brian E. Skahill**, **David M. Weston**, **Nawa Raj Pradhan** and **Aaron R. Byrd**, Corps of Engineers

Using the Variable Infiltration Capacity Model in a Semi-Stochastic Format for Hydrologic Hazard Estimates for Dam Safety Risk Analysis 139
Frank Dworak, Bureau of Reclamation

Dynamic Desert Floods at a Dry Dam 149
Robert A. Eichinger, **Zachary Whitten** and **Michael Gerlach**, Stantec Consulting Services, Inc.

Cultivating Ingenuity and Safety In Alabama: The Taming of Lake Ogletree Reservoir 165
James R. Crowder, **Brian M. Crookston**, **Bradley T. Boyer** and **J. Tyler Coats**, Schnabel Engineering

Replacing Ruby Dam Outlet Uses Computational Fluid Dynamics to Model Energy Dissipation 179
Frank Lan, **Robert Waddell** and **Michael Zusi**, AECOM; and **Brian Grant**, Montana DNRC

Reducing Scour with Flow Splitters, Extension Fins and Computational Fluid Dynamics 199
Duane McClelland and **Daniel Morse**, CH2M; and **Nathan M. Gullo** and **Jason A. Eichler**, LCRA

Sediment Transport Analysis in Support of Prado Basin Sediment Management Project 221
Craig P. Baxter, Golder Associates Inc.; **Nathan Scheevel**, Scheevel Engineering, LLC; and **George W. Annandale**, Golder Associates Inc.

Verification of Sediment Yield Calculations in the Arid Southwest 237
Brian Wahlin and **Jesse Piotrowski**, WEST Consultants, Inc.; and **Chuck Davis**, Beaver Creek Hydrology

Post-Removal Reservoir Erosion: Elwha Dam and Glines Canyon Dam Removals, Washington 251
Jennifer A. Bountry, Bureau of Reclamation; **Andy Ritchie**, National Park Service; and **Timothy J. Randle**, Bureau of Reclamation

Unit Hydrograph Peaking Factors and their Application to PMF Simulations 269
Michael Bartles and **Matthew Fleming**, Corps of Engineers

Areal Reduction Factors — Colorado Front Range and Analysis of the 2013 Storm 287
Doug Hultstrand and **Bill Kappel**, Applied Weather Associates, LLC

Evaluating Effort Levels of Inflow Design Flood Study Methods for Dams 303
S. Samuel Lin, Federal Energy Regulatory Commission; and **Jason E. Hedien**, MWH Americas, Inc.

Monitoring

2015 USACE Instrumentation and Monitoring Program: Internal Peer Review Results 315
Georgette Hlepas and **Travis Tutka**, Corps of Engineers

The Hydrogeological Model of Boone Dam 335
Scottie Barrentine, **Andrew Dodson** and **Lindsay McMichen**, Tennessee Valley Authority

Relearning How to Look at Piezometric Data for Seepage Evaluation 351
Richard S. Olsen and **Isaac J. Stephens**, Corps of Engineers

Evaluation and Monitoring Response to Upstream Slope Failure at an Embankment Dam 379
Jay N. Stateler and **Jeff Wormer**, Bureau of Reclamation

On Calero Pond: Evaluation of Downstream Wetland and Potential Dam Seepage 395
Matthew Powers, GEI Consultants Inc.; **Bal Ganjoo**, Santa Clara Valley Water District; and **Mark Freitas**, GEI Consultants Inc.

Maintaining Performance — Rehabilitation of Critical Drains at Crab Orchard Dam 415
David Snyder and **Robert Saber**, Gannett Fleming, Inc.; and **David Hibbs**, U.S. Fish and Wildlife Service

Foundation Protection during Post Tensioned Anchor Construction 435
William B. Stiller and **Y. Erwin Supranata**, HDR; and **Stephen J. Mockler**, Brookfield Renewable Energy Group

Environment

Buchanan Dam — Keeping Current for Eight Decades 445
Victor M. Vasquez, Freese and Nichols, Inc.; **Doug Witkowski,** Lower Colorado River Authority; **M. Leslie Boyd**, Freese and Nichols, Inc.; **Monica Masters**, Lower Colorado River Authority; and **Layne Bukhair**, Freese and Nichols, Inc.

Risk Management for Dam Construction Projects — An Owner's Perspective 469
Susan S. Hou, **Tedman Lee**, **Alan Johanson**, and **Daniel L. Wade**, San Francisco Public Utilities Commission

Seismic

Clover and Junction Dams: Risk-Based Seismic Evaluation of Two Slab and Buttress Dams 485
Scott Jones, **Steve O'Brien** and **David Hughes**, AECOM

Nonlinear Seismic Evaluation and Acceptance Criteria for Arch Dams 497
Yusof Ghanaat, Quest Structures, Inc.

Developing Time Histories with Acceptable Record Parameters for Dillon Dam 515
Dina Bourliea Hunt, MWH Global; **Justin Beutel**, Pacific Gas and Electric Company; and **Christine Weber**, MWH Global

Seismic Deformation Analyses of Embankment Dams: A Reviewer's Checklist 535
Ross W. Boulanger, University of California, Davis; and **Michael H. Beaty**, Beaty Engineering LLC

Lessons from Earthquake Lake in Preparing for Seismically Induced Landslides 547
Roger L. Kay, Corps of Engineers

Concrete

Non-Destructive Testing of Post-Tensioned Anchorages at JT Myers and Markland Dams 563
Robert F. Lindyberg and **Afshin Karshenas**, FDH Velocitel; **Jeff Lillycrop**, **Charles E. Wiggins**, **Larry L. Dalton II** and **Matthew Watts**, Corps of Engineers; and **Brandon Compton,** FDH Velocitel

USACE Portland District Spillway Gate Rehabilitation Program 581
Matthew Craig, **Laurie Ebner**, **William Fortuny**, **David Hamernik**, **Matt Hess** and **Mark Sawka**, Corps of Engineers

Lessons Learned from the Spillway Gate Failure at Folsom Dam 599
Jerzy Salamon, Bureau of Reclamation

A Finite Element Approach to Predict the Stability of Gravity Dams Laying on Rough Rock Interfaces 611
Sylvain Renaud and **Najib Bouaanani,** Polytechnique Montréal; and **Benjamin Miquel**, Hydro-Québec

Applications of FLOW-3D for Stability Analyses of Concrete Spillways at FERC Projects 627
Elise Dombeck, Federal Energy Regulatory Commission

The Stability of Masonry Walls and Dams: A Review of the Historical Design of Concrete Dams............643
Kevin Griebenow, Federal Energy Regulatory Commission

A New and Advanced Dam Performance and Response Integration Program............657
Ziyad Duron, Harvey Mudd College; and **John Yen**, Southern California Edison

Practical Analysis and Assessment of AAR-Affected Dams and Hydroelectric Plants............675
Dan D. Curtis, **Lingmim (Frank) Feng**, **Gurinderbir S. Sooch** and **Jiqin (Tom) Zheng**, Hatch, Ltd; and **John Fletcher,** NB Power

Using the Trial Load Method to Optimize Feasibility Design of Watana Dam............695
Aled Hughes, **Glenn Tarbox** and **Brian Sadden**, MWH Americas, Inc.; and **Bryan Carey**, Alaska Energy Authority

Linear and Nonlinear Seismic Analysis for the Post Tensioning System Design of the Wanapum Dam Spillway Monoliths............711
Bashar S. Sudah, **Juan E. Quiroz**, **Bryan J. Brumley**, **David P. Thompson** and **Richard E. Barrie**, MWH

Static and Dynamic Mass Concrete Material Properties of a Concrete Gravity Dam............727
Soheil Razavi Darbar, **David Queen** and **Chris Hatton**, BC Hydro; **Timothy Dolen**, Dolen and Associates LLC; and **Katie Bartojay**, Bureau of Reclamation

Levees

Using Risk Assessment to Inform Levee System Evaluation Decisions for the National Flood Insurance Program............749
David A. Margo, **Lea G. Adams**, **Scott E. Shewbridge** and **Richard Varuso**, Corps of Engineers

Using Tolerable Risk Guidelines in the California Delta............765
Lawrence H. Roth and **Jessica Ludy**, Arcadis U.S.

Geotechnical Evaluation of the Levees in the Central Valley, California............781
Richard Millet, AECOM; **Steve Mahnke**, California Department of Water Resources; **Mary Perlea**, Corps of Engineers; and **Ray Costa**, Geotechnical Consultant

Phased Investigation Approach for a Partial Slope Failure on the Rio Grande............799
Isaac J. Stephens, **Lucas A. Walshire** and **Joseph B. Dunbar**, Corps of Engineers

Embankment Dams

Reducing Uncertainties when Assessing the Stability of Rockfill Dams: Characterizing Analysis Parameters 813
Daniel B. Rogers, **Kyle R. Blakley** and **Jeffrey S. Dingrando**, Stantec Consulting Services Inc.; and **Jeff W. Munsey**, Tennessee Valley Authority

Don't Judge a Rockfill Dam by its Name Alone 829
Brian Benson and **Peter Amos**, Damwatch Engineering Ltd

Shear Strength of Compacted Beaumont Clay for Consolidated-Undrained Conditions 849
Mark J. Thompson and **David J. Bentler**, CH2M; **Thomas L. Brandon**, Virginia Polytechnic Institute and State University; and **Daniel R. VandenBerge**, Tennessee Technological University

Estimating the Peak Friction Angle of Sandy Soils In Situ with State-Based Overburden Normalized CPT Tip Resistance 865
Robert A. Jaeger, GEI Consultants, Inc.; and **Ian P. Maki**, California Department of Water Resources

Decades of Foundation Deterioration Threatening Public Safety 883
George L. Sills, George Sills Geotechnical Engineering Consultant, LLC; and **Leslie F. Harder**, HDR

An Empirical Model for Estimating Internal Erosion Critical Gradients for Inclined Exit Face Conditions 909
Richard A. Keizer, GEI Consultants, Inc., **John D. Rice**, Utah State University; and **Robert A. Jaeger**, GEI Consultants, Inc.

Laboratory Measurements of Critical Gradients of Cohesionless Soils 927
Bryant A. Robbins, **Axel Montalvo-Bartolomei**, **Jamie López-Soto** and **Isaac J. Stephens**, Corps of Engineers

Reliability Slope Stability Analysis for Rapid Drawdown 939
Mark J. Thompson and **David J. Bentler**, CH2M; and **Thomas L. Brandon**, Virginia Polytechnic Institute and State University

Embankment Dam Analysis Parameters Estimated from 100 Years of Laboratory Tests 955
David W. Harris, David Harris Engineering and Assistance, LLC/Allegiance Consulting, LLC; and **Garrett Sutley**, Bureau of Reclamation

Calaveras Dam Replacement Project — Providing a Seismically-Reliable Water Source 969
John Roadifer and **Michael Forrest**, AECOM; and **Daniel Wade**, **Susan Hou**, **Tedman Lee** and **Carman Ng**, San Francisco Public Utilities Commission

Dam Decommissioning

Non-Linear Modeling of Sediment Retention Structures for Dam Removal Project.......985
Kristen M. Martin and **Erik J. Malvick**, California Department of Water Resources

Recent Progress Towards Removal of Matilija Dam 1001
Seth Gentzler and **John Roadifer**, AECOM; **Derek Booth** and **Yantao Cui**, Stillwater Sciences; and **Peter Sheydayi**, Ventura County Water Protection District

Dam Safety

Risk Analysis Work for Dams — "Big Picture" Considerations..................................1017
Jay N. Stateler, Bureau of Reclamation; **Kim de Rubertis**, Consulting Engineer; and **Doug Boyer**, Federal Energy Regulatory Commission

Implementing a Risk-Informed Approach to Dam Safety in a Regulatory Environment..1035
Douglas D. Boyer and **William Allerton**, Federal Energy Regulatory Commission

Use of Quantitative Risk Results to Inform Dam Safety Governance: Practical Cases in Europe..1053
Ignacio Escuder-Bueno, Universitat Politècnica de València; **Adrián Morales-Torres**, iPresas; and **Jesica Tamara Castillo-Rodríguez**, Universitat Politècnica de València

The Role of Risk Assessment in the Bureau of Reclamation's Dam Safety Program..1069
William R. Fiedler and **William Engemoen**, Bureau of Reclamation

Semi-Quantitative Portfolio Risk Assessment for Prioritizing Dam Safety Activities..1083
John Yen, Matthew Muto and **Nicolas Von Gersdorff,** Southern California Edison Company; and **Gregg A. Scott**, Scott Consulting, LLC

Bureau of Reclamation Risk-Informed Approach to Selecting the Inflow Design Flood..1097
William R. Fiedler, Bureau of Reclamation; and **John H. LaBoon**, Consulting Civil Engineer

The Wanapum Spillway Monolith Cracking Incident Dam Safety Management Implications..1119
Robin G. Charlwood, Robin Charlwood & Associates; **Adam N. Jones** and **Kevin Snyder**, HDR; **Kevin Marshall** and **David Mishalanie**, Grant County Public Utility District No. 2

Panama Canal Borinquen Dams and Third Set of Locks Operations and Maintenance Strategy 1133
Rachael V. Bisnett, **Wonnie Kim** and **Donald J. Montgomery**, MWH

Safety in Numbers? 1145
Kim de Rubertis, Consulting Engineer

Public Safety and Security at Dams Are Not Mutually Exclusive 1161
William F. Foos, **Frank Calcagno** and **Paul G. Schweiger**, Gannett Fleming, Inc.

Emergency Action Planning — Canadian Practice and Lessons Learned at Ontario Power Generation 1173
Tali Serota and **Tony Bennett**, Ontario Power Generation

The Criticality of Engagement Before, During and After a Dam Safety Emergency 1191
Bernie Auld, **Benjamin Heath** and **Keil Neff**, Tennessee Valley Authority

Construction

Stagecoach Spillway Replacement and Performance at Electra Lake 1201
Danielle Tripp Hannes and **Stephen L. Jamieson**, W. W. Wheeler and Associates, Inc.

High Capacity Anchors on Existing Concrete Dams — Application to Two TVA Dams 1217
Husein Hasan and **Aaron Nottis**, Tennessee Valley Authority; **Brian Krohmer**, Barnard Construction Company; and **Dennis Triplett** and **Boris Caro Vargas**, Nicholson Construction Company

Lessons Learned on Post-Tensioned Trunnion Anchorage — Folsom Auxiliary Spillway 1235
Cecily M. Nolan and **Cheuk Y. Wan**, Corps of Engineers

Uncovering and Addressing a History of Significant Incidents After an Annual Dam Safety Inspection 1251
Steven Davidheiser and **Robert A. Kline, Jr.**, Gannett Fleming, Inc.; and **Amanda Lothes,** Newport News Waterworks

Emergency Repairs to Prevent a Piping Failure in the Lake Manatee Dam 1267
Jeffrey A. Beriswill and **Nickolas J. Depin**, Amec Foster Wheeler Environment & Infrastructure, Inc.; and **Mark Simpson** and **Bruce MacLeod,** Manatee County Utilities Department

Jet Grouting for Echo Dam Seismic Remediation 1283
Tara Schenk-McFarland and **Mark Bliss**, Bureau of Reclamation

Improvements to Seepage Cutoff Wall Specifications Considering Lessons Learned from Construction 1297
Michael Kynett, Corps of Engineers; **Khaled Chowdhury** and **Richard Millet**, AECOM; **Mary Perlea**, Corps of Engineers; and **Francke Walberg**, AECOM

Foundations

Levee Assessment Tool: Using Limited Data to Inspect/Evaluate Levees for Various Failure Modes Against Established Criteria 1321
A. Wohletz and **S. Wiesner**, Kleinfelder

Fault Investigations During Borinquen Dam 1E Foundation Panama Canal Expansion 1341
David Schug, **Paul Salter**, **Dan Meier**, **Robert Urban**, **Chris Goetz**, **Lelio Mejia** and **James Toose**, AECOM; and **Derek Irving**, **Pastora Franceschi**, **Maximiliano De Puy**, **Jaime Arrocha** and **Roberto Miranda**, Autoridad del Canal de Panamá

3D Modeling at Boundary Dam, WA: Visualizing a Geologically Complex Site 1365
Robert Cannon and **Frederic Snider**, Schnabel Engineering; **Kim Pate** and **Andre Ball**, Seattle City Light; and **Gary Rogers** and **J. Hawkins Gagnon**, Schnabel Engineering

Underseepage and Heave Mitigation at Douglas Saddle Dam No. 1 1379
Daniel A. Gilbert, **Alan F. Rauch** and **Paul J. Cooper**, Stantec; and **Scottie L. Barrentine**, Tennessee Valley Authority

A Deep Seepage Cutoff for the Thornton Composite Reservoir 1395
Joseph R. Kovacich, MWH; **Louis Storino**, Metropolitan Water Reclamation District of Greater Chicago; **Raymond J. Franz** and **Jonathon R. Bulger**, Hayward Baker Inc.; and **Brian J. Kazyak**, MWH

Designing a Hybrid Dam on a Geologic Profile with Highly Variable Engineering Properties 1419
Richard E. Riker and **Jacob Esterhuizen**, CH2M; **Don Deere**, Deere and Ault Consultants, Inc.; and **Gabriel Fernandez-Delgado**, University of Illinois

Lunch Presentation and Poster Session

Things You Wanted to Know About Individual Consulting, but Were Afraid to Ask 1447
Larry K. Nuss, Nuss Engineering; and **Kenneth D. Hansen**, Consulting Engineer

Construction of La Romaine Complex in Northern Quebec, Canada: Behaviour of Asphalt Core Rockfill Dams 1463
Vlad Alicescu, **Jean Pierre Tournier**, **Redha Kara** and **Dan Rosculet**, Hydro-Québec

Selecting an Intensity Measure for Fragility Curves Derived from Performance-Based Testing 1483
Spenser Anderson, **Kunal Menda**, **Mo Zhao** and **Ziyad Duron**, Harvey Mudd College

Retrofit and Closure of CCR Impoundments to Meet New Regulations 1497
Matt Bishop, **Brian Owens** and **Luke Williams**, Amec Foster Wheeler

Antecedent Snow Covered Area Analysis for Inflow Design Floods 1507
Steven F. Daly and **Angela Duren**, Corps of Engineers

Evaluation of New Storage with Optimal Hydropower: Weiser-River Basin, Idaho 1521
Jeremy Giovando, Corps of Engineers

Field Tests of Placed Riprap as Erosion Protection Against Accidental Overtopping and Leakage 1531
Priska Helene Hiller, **Fredrikke Kjosavik**, **Leif Lia** and **Jochen Aberle**, Norwegian University of Science and Technology

Design-Build Construction of the Panama Canal Borinquen Dams 1W, 2W, and 2E 1541
Gabriel Martinez, **Wonnie Kim** and **Carlos Zambrano**, MWH

A Performance-Based Approach to Developing Fragility Curves for Structures Subject to Seismic Loading Conditions 1559
Kunal Menda, **Mo Zhao**, **Spenser Anderson** and **Ziyad Duron**, Harvey Mudd College

Implementing Lessons Learned from Wheeler Lock Collapse on New Lock Construction Projects 1575
Michael A. Morrison, Tennesee Valley Authority; and **Bernard Schulte**, Corps of Engineers

Assessing and Evaluating the Operation of Pressure Cells in Embankment Dams: A Case Study of Five Iranian Embankment Dams 1591
Bahareh Shoghli, University of North Dakota; **Ali Asghar Mirghasemi**, University of Tehran; and **Yeo Howe Lim**, University of North Dakota

Stability of Slopes in Embankment Dams 1603
Gopi Siddappa and **Santhosh G. Gowda**, P.E.S. College of Engineering

Ritschard Dam Monitoring: Awareness Leads to Learning which Leads to Action 1621
Ray D. Tenney and **Don B. Meyer**, Colorado River Water Conservation District; and **Michael R. May**, AECOM

Boundary Dam Spillway No. 2 Modifications 1639
John Werner and **James H. Rutherford**, Hatch Associates Consultants Inc.; **Kim Pate** and **Daniel Kirschbaum**, Seattle City Light; **Joe Groeneveld**, Hatch Associates Consultants Inc.; **Keith Moen,** HDR; and **Justin Arnold**, Alden Laboratories Inc.

Selecting A Damage Measure for Fragility Curves Derived from Performance-Based Testing.....1655
Mo Zhao, **Spenser Anderson**, **Kunal Menda** and **Ziyad Duron**, Harvey Mudd College

DON'T JUDGE A ROCKFILL DAM BY ITS NAME ALONE

Brian Benson, CPEng[1]
Peter Amos, CPEng[2]

ABSTRACT

Faced, compacted rockfill dams are generally viewed as inherently robust because their design is based on high strength, free draining embankment characteristics. This paper presents evidence that not all faced, compacted rockfill dams can be expected to behave with the robustness generally attributed them. Questions regarding the embankment characteristics of an asphalt faced rockfill dam (AFRD) arose during a detailed dam safety review.

The dam was designed and built in the late 1960's to early 1970's during the heyday of faced, compacted rockfill dams' development and popularity. Design and construction records for the dam were relatively good. Until recently, the dam has been viewed as an idealized, compacted dirty rockfill dam. Targeted geotechnical site investigation followed a dam safety review to address uncertainties about embankment characteristics. The findings indicated the embankment materials are more characteristic of what would be found in an earthfill embankment. This key finding significantly changed the dam's predicted performance and failure modes. The increased likelihood that leakage through the facing could saturate the embankment resulted in the dam being vulnerable to leakage induced saturation instability. Finer materials at the dam toe behind the rockfill facing stone increased the dam's vulnerability to concentrated leakage toe unravelling.

This case reminds dams' professionals that actual conditions may be different than those implied by design and type, and assessment based on investigation and first principles is essential.

INTRODUCTION

This paper examines the characteristics and expected performance of an existing embankment dam. It challenges a commonly held convention in embankment dams engineering and reminds dam engineers to base engineering judgments on evidence and 1st principle thinking.

The advent of compacted rockfill embankments replacing dumped and sluiced rockfills in the 1960's was a milestone for modern development of embankment dams. Good quality, compacted rockfill are expected to result in a robust embankment. Such rockfills have high strength, modulus and permeability, and are free draining with high resistance

[1] Principal Engineer, Damwatch Engineering Ltd, Level 8 80 The Terrace Wellington 6140 New Zealand, +64 4 381 1300, brian.benson@damwatch.co.nz
[2] Managing Director, Damwatch Engineering Ltd, Level 8 80 The Terrace Wellington 6140 New Zealand, +64 4 381 1300, peter.amos@damwatch.co.nz

to large through leakage. Many such rockfill dams have been successfully built and operated since.

Nonetheless, the robustness associated with 'rockfill dams' depends on their material characteristics, not their name. This paper presents the case of a dam that was widely considered to be a typical compacted rockfill, until recently. This dam is currently in the remediation process and its details shall remain confidential.

Case study is presented in this paper. This dam's development history is reviewed. Examination by dam safety review, site investigation and engineering assessment is presented. The dam has been re-characterized with resulting significant changes to its inherent conditions and expected performance. Its embankment is more like a poorly compacted earthfill than a rockfill.

As a result of re-characterization to an earthfill embankment, the dam has been identified as being deficient for two potential failure modes: 1) Earthquake induced cracking leakage causing toe unravelling; and 2) earthquake induced cracking leakage causing embankment saturation and instability.

DAM DEVELOPMENT

Damsite

The damsite is situated in mountainous terrain across an incised river valley and two contiguous adjoining saddles. In concert the dam is approximately 1 km in length.

The damsite is in an area of moderate seismicity. Several moderate and numerous small earthquakes have been recorded in the last 150 years.

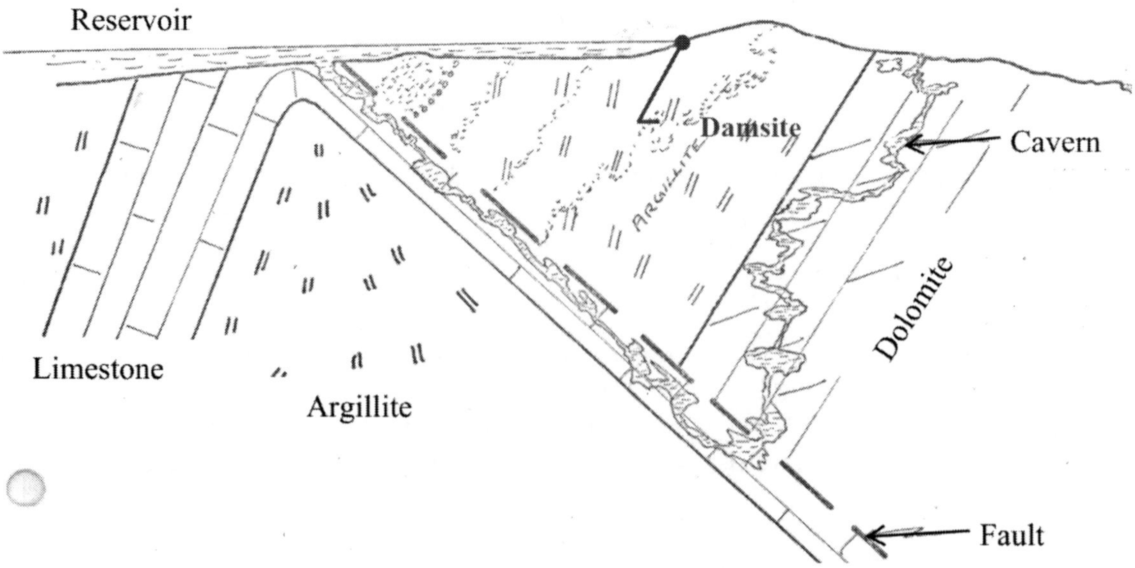

Figure 1. Original Geologic Section through Damsite

The foundation and abutments along the dam alignment are predominantly composed of steeply dipping weathered mudstone (argillite) with minor interbedded sandstone. Bedding is generally near orthogonal to the dam alignment. Joints are mostly tight and closed. The original geologic cross section through the dam and adjacent reservoir is presented in **Figure 1**.

Dam Design

Embankment design followed rockfill dam design practices of the late 1960's to early 1970's era. The design section is presented below in Figure 2.

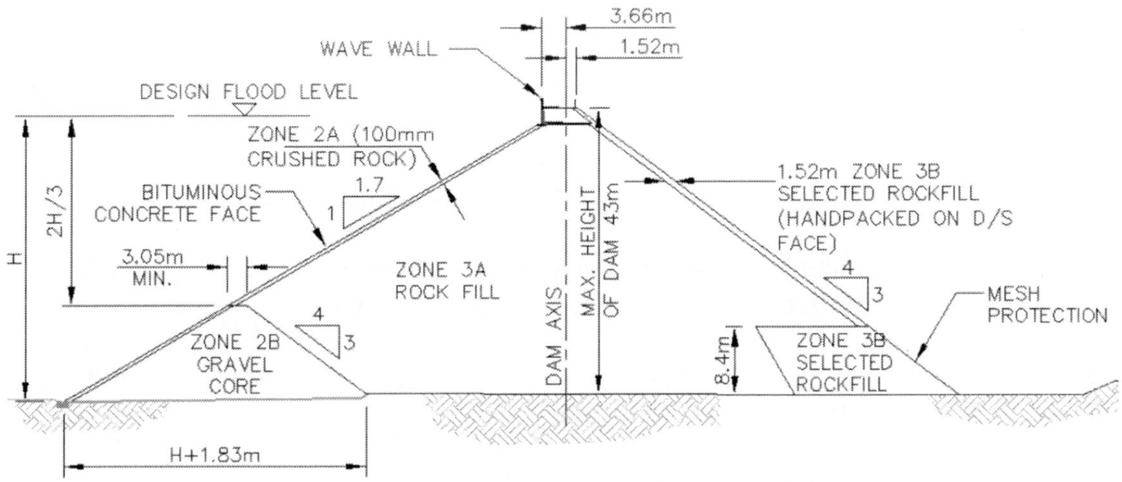

Figure 2. Design Cross Section

The main components are outlined below.

- The downstream slope has a 4H:3V slope.
- The upstream slope facing was subject to cost comparison between reinforced concrete and asphalt. Asphalt facing was selected based on cost savings. The upstream slope was set to 1.7H:1V as the practical steepness limit for adapting road paving equipment to place the asphalt facing.
- Below the upstream asphalt facing a crushed dolomite zone acts like a paving base course to support the asphalt pavement. Its nominal thickness was limited to 200 mm to minimize costs.
- A concrete plinth forms the parametric joint between the upstream face and foundation along the upstream toe.
- At the start of construction a gravel core, Zone 2B, was added to dam design. It was intended to lengthen the seepage path around the plinth, thereby lessening hydraulic gradients and internal erosion potential. Material for Zone 2B was borrowed near the damsite and consists of quartzite sandy gravel with silt.
- Overtopping protection during construction on the downstream slope in the river section consisted of terraced steel gabions filled with stone.

- A parapet wall along the upstream crest provides wave and flood protection, while also reducing the embankment crest width.
- By design a downstream toe and slope facing of select rockfill was intended for erosion protection.
- Earthquake design was accounted for by using a 0.15g pseudo-static seismic load, as was typical of the time. The rockfill embankment was considered inherently safe against this load.

Construction

The hunt for a nearby good quality rockfill source proved very challenging. The best material near the damsite was a highly variable, weathered mudstone. There was debate about the suitability of this material. Construction had started when an engineering geologist from the Dam Design Division visited the damsite and inspected the prospective mudstone quarry. The engineering geologist stated the weathered mudstone was weak and not suitable as the rockfill source for the body of the dam. He recommended the dam be redesigned as an earthfill dam with a 2H:1V downstream slope if the weathered mudstone were to be used for the main embankment zone.

To settle the debate about whether the proposed weathered mudstone was suitable for use as rockfill a compaction trial was conducted. Variables in the trial included lift thickness and the number of passes using a 10 ton roller compactor. It was decided to use the mudstone. Four passes with a 10 ton roller on a 0.9m thick lift was specified with the intention to limit particle crushing to the upper 150mm.

The upstream plinth and downstream toe areas beneath the embankment were stripped clean of extremely and highly weathered rock. Several interbedded mudstone units were exposed. Tightly spaced, closed joints generally orthogonal to the dam axis were exposed along the plinth. Cut-off grouting was performed, but grout takes were minor, indicating the foundation was relatively impermeable.

Mudstone quarrying and placement were highly variable. Minor to major material degradation was experienced due to extraction, handling and placement and compaction. In the first year of dam construction earthworks continued through the wet winter period. This resulted in removing and wasting large volumes of saturated, poorly drained compacted mudstone materials.

The dam was built before the first hydraulic excavators were developed. This meant placing materials at both the upstream and downstream edges of the embankment were largely done by hand. On the upstream edge, the weathered mudstone fill was not rolled on the slope. Crushed quartzite base course nominally 200 mm thick was then dumped, hand raked and rolled up and down the slope. On the downstream slope dozers pushed the more durable, larger mudstone cobbles and boulders to the edge where they were hand pitched to surface the downstream face.

The asphalt facing was constructed after the completion of the embankment. Asphalt was placed in adjoining vertical panels using standard road paving equipment cabled to a support structure on the crest. Three asphalt layers were placed over the lower part of the embankment, while two layers were placed over the top portion.

1st Filling – End of Construction

Embankment settlements following dam completion during first filling were greater than expected. As the new reservoir's level neared its normal operating level for the first time the upstream asphalt facing cracked and leakage ensued. The leakage exacerbated the embankment settlements. A plot of crest settlement versus time following construction is shown in **Figure 3**. Plots of crest settlement vs. time for other embankment dams composed of compacted, moderately strong rockfill are also shown on Figure 3 for comparison.

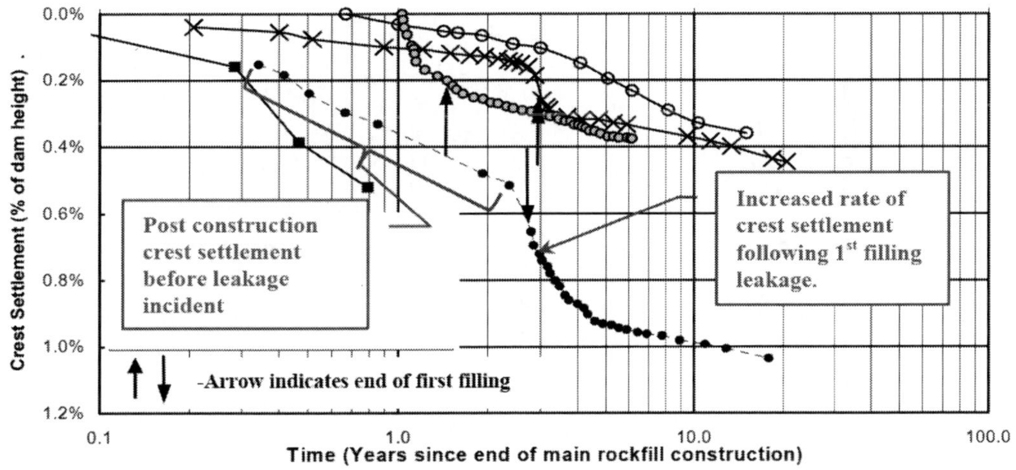

Ref: (Hunter G., 2003)

Figure 3. Crest Settlement vs. Time of Well Compacted Rockfill of Medium to High Strength CFRDs

The cracking was subsequently stemmed by divers fixing rubber sheets and barge placement of a silty sandy gravel blanket along the upstream slope high enough to cover the cracks. The alluvium used for the gravel core Zone 2B was also used for the silty sandy gravel blanket. It was placed from a barge using a clamshell bucket lowered to just above the asphalt facing's surface. It is in very loose condition. Leakage reduced to an acceptable level and no further remedial measures have been needed since.

DAM SAFETY REVIEW

A Dam Safety Review (DSR) of this dam was performed in 2011-12. This reviewed consisted of reviewing all available design and construction documentation and a site inspection, then assessing the dam against current dams' standards, guidelines, and practices. Engineering analyses and studies as required to meet the objective were also performed in conjunction with the DSR.

The seismicity for the damsite was updated as part of the DSR. Seismic hazard relevant to this dam is dominated by a fault with its surface expression about 3 ½ km from the dam. This reverse thrust fault dips steeply with its projection about 8 km directly beneath the dam (illustrated in **Figure 1**). Earthquake specialists concluded this fault could be active within the dam safety context timeframe; i.e. 10,000 years.

There are seismograph records of small earthquakes in the vicinity of the damsite, but none for rupture along the dominant fault. The seismicity study produced probabilistic mean acceleration spectra in **Figure 4** below. The spectra include 5% damping and normal component, forward directivity effects. Topographic effects were considered insignificant.

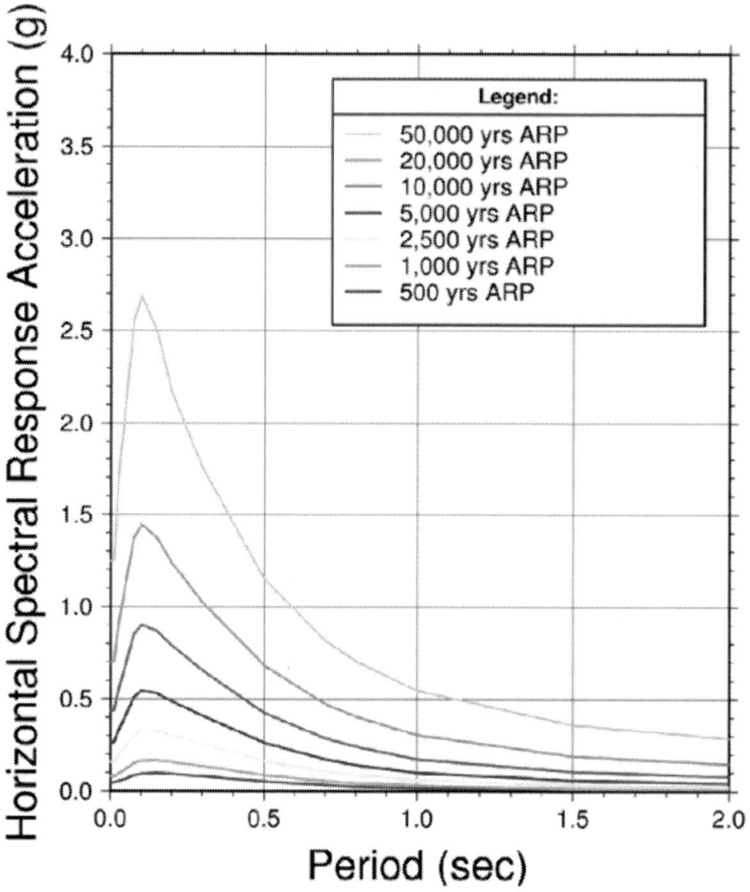

Figure 4. Acceleration Response Spectra for dam

A walkover site inspection was done in conjunction with the DSR. In several locations along the slope the upper facing of mudstone cobbles and boulders were removed to reveal compacted silty sandy gravel below. A photograph of a subsequent test pit showing this condition is presented in Figure 6 of the Geotechnical Investigation Section of this paper.

Key findings from the DSR are listed below.

1. Dam design was done according to the standards of the time.

2. The choice of upstream membrane material was based on cost comparison. The asphalt and concrete facing were considered technically equivalent.

3. Material availability and wet construction conditions forced the use of mudstone as rockfill. The original design team struggled with characterizing the mudstone material as a rockfill versus an earthfill. In the end it was considered appropriate for rockfill.

4. Zoning considered material compatibility according to the standards of the time. However, Zone 2A is not compatible with Zone 3A according to current dam engineering practice.

5. There were no seismic induced deformation considerations.

6. The Stress-strain behavior of the asphalt facing, and its interaction with other zones of the dam, was not well understood or appreciated during design. There was a lack of appreciation for potential asphalt cracking at the Zones 2B-3A interface due to settlement. The asphalt facing is vulnerable to earthquake induced cracking.

7. Design of the 'select rockfill' Zone 3B along the downstream slope and toe did not eventuate in construction.

8. The silty sandy gravel blanket seal overlying the first filling asphalt facing cracks is very loose and subject to liquefaction at modest earthquake ground motions.

9. The embankment toe is susceptible to potential unravelling if large, concentrated leakage flows were to occur from earthquake induced cracking of the upstream asphalt facing.

10. Geotechnical investigation was recommended to assess the embankment characteristics and its performance. Following geotechnical investigation, potential failure modes assessment was recommended.

GEOTECHNICAL INVESTIGATION

Geotechnical investigation of the dam was undertaken following the DSR recommendation. The investigation consisted of:

- Six 150mm core diameter sonic borings drilled vertically from the crest a nominal depth into the foundation to assess embankment conditions;
- One 30m sonic boring drilled into the foundation adjacent to the toe for shear wave velocity determination;
- Eleven test pits excavated into the toe to characterize downstream toe conditions;
- Two hand dug test pits along the slope to determine downstream slope conditions;
- Downhole shear-wave velocity profiles of four embankment borings and the one 30m foundation boring;
- Four infiltration tests to assess permeability of the Zone 3A embankment materials;
- Large shear box lab tests to assess embankment materials shear strength; and
- Lab gradation tests to provide gradings of the finer embankment materials.

Overall the embankment materials encountered by borings were finer than previously understood. Embankment materials encountered were highly heterogeneous, reflective of variability of the mudstone quarry materials. Layering due to placement and compaction was semi-regular. A core photograph clearly depicting a construction lift joint is presented in Figure 5.

Three types of mudstone embankment materials were clearly observed in the sonic drill core. One has a highly degraded silty sandy gravel texture that was largely unaffected by sonic coring. A second has a sandy gravel with silt and cobbles texture that was minimally degraded by sonic coring (larger particles were degraded). A third was predominantly cobbles and boulders that exhibited significant sonic coring degradation as evidenced by fresh fines (readily visible in Figure 5).

Downstream toe conditions varied considerably. In many places silty sandy gravel soils were present beneath a stone facing. At some locations coarse mudstone cobbles extended to about 1 m thickness. In the river section of the dam toe conditions consisted of gabions filled with dolomitic stone have a $d_{50} \sim 80$ mm. A photograph of toe conditions representative of silty sandy gravel beneath facing stone is shown in Figure 6.

EMBANKMENT RE-CHARACTERIZATION

Design and as-built drawings (see **Figure 2**) depict an embankment composed of four main zones:

1. Zone 2A – Crushed aggregate beneath the asphalt concrete facing
2. Zone 2B – Gravel core
3. Zone 3A – Bulk of embankment composed of dirty rockfill
4. Zone 3B – Select rockfill along downstream slope and toe.

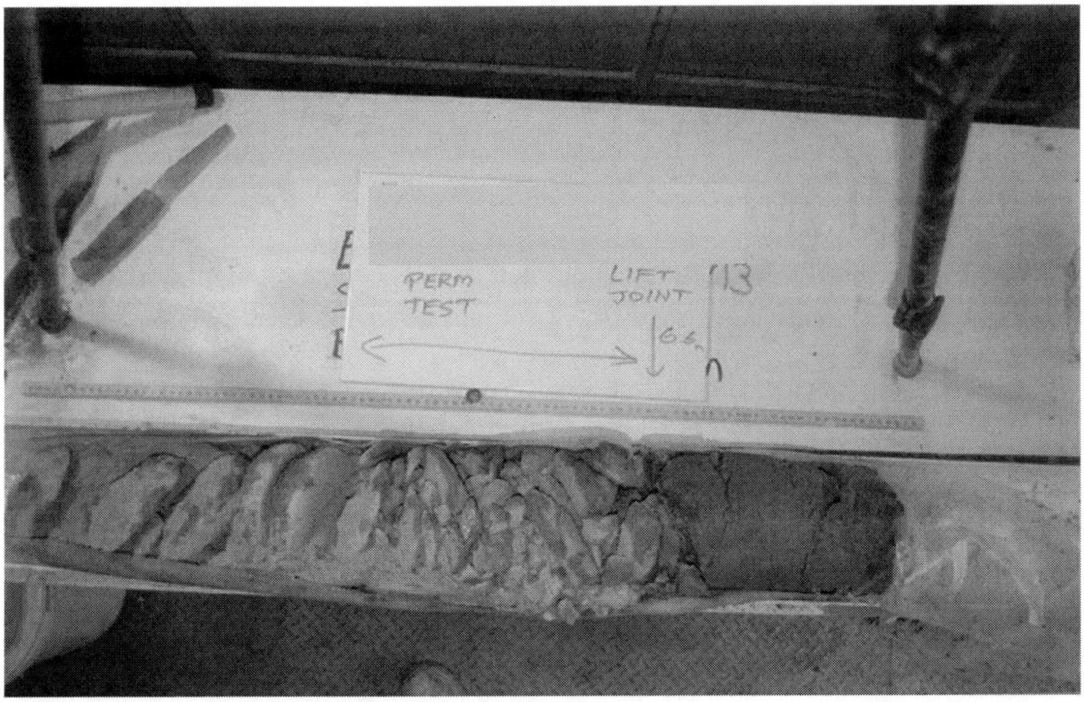

Figure 5. Core Photograph Depicting Embankment Construction Lift Joint

Figure 6. Silty Sandy Gravel Downstream Toe Conditions beneath Stone Facing

Embankment conditions have been re-assessed based on detailed review of available construction records, interviews with design and inspection engineers involved with construction, and geotechnical investigation. Three of the four zones above have been re-characterized as being significantly different than previously understood. Characterization of these four zones is presented in the remainder of this section.

Zone 2A

Zone 2A lies beneath the asphalt facing along the upstream face. By design it was composed of crushed sandy gravel with small cobbles up to 100 mm.

Use of crushed materials was minimized because of it relatively high cost; reportedly six times the unit cost of the alluvial gravels used in Zone 2B. Lack of hydraulic excavators at the time of construction resulted in considerable handwork. The construction report outlined a complex, step-by-step procedure resulting in the sequence of materials placement, depicted on Figure 7.

Zone 2B

Zone 2B is sandy gravel with silt material shown on as-builts abutting the asphalt facing in the lower part of the embankment. Its purpose is to extend the potential seepage path around the plinth, thereby reducing the gradient and potential erosion. The alluvial gravels used for Zone 2B were abundant and near the damsite. Zone 2B was reportedly overfilled, but otherwise is understood to be as reported in the drawings.

Zone 3A

Overall Zone 3A is highly heterogeneous and variable. However, one general consistency is highly degraded mudstone at the top of successive construction lifts. The lower part of each lift is more variable, depending on the materials' durability. Spatially, Zone 3A materials are highly heterogeneous due to the highly variable durability of mudstone coming from the quarry.

Zone 3A was interpreted to be composed of five distinct materials based on visual classification, gradation testing and construction methods. All five materials are predominantly composed of weathered mudstone of variable hardness and weathering with variable amounts of compaction and degradation.

1. Crushed Material – Silty sandy gravel and silty gravelly sand with occasional small cobbles. It is generally 'dense' with a blocky structure. This material was interpreted to be the crushed upper part of argillite fill lifts. Construction reporting stated compaction was controlled to limit crushing to the upper 150 mm of material.
2. Semi-crushed Material - The semi-crushed material is sandy gravel with silt and occasional cobbles. It is generally 'loose' to 'medium dense'. The semi-crushed material is generally poor quality argillite that experienced less compaction

destruction than the crushed material. Much of it occurs below the crushed material, typically in layers less than 1 m thick.

3. Dirty Rockfill - The dirty rockfill is a coarse material with a rockfill fabric. Gradationally it is gravel and cobbles with silt, sand and occasional boulders. The spaces between coarse particles have some voids along with the finer materials. The dirty rockfill is generally higher quality argillite and experienced less compaction destruction than the semi-crushed material.
4. Semi-compacted Zone 3A – To construct Zone 2A along the upstream slope as thin as practical a small width of mudstone fill was placed alongside normally constructed, Zone 3A fill lifts. Its location is illustrated in Figure 7. It is interpreted to be 'semi-compacted' based on the reported construction placement methodology. It is interpreted being sandy gravel with silt and cobbles.
5. Uncompacted Zone 3A – Along the downstream slope the compactors could not safely compact the outer edge, resulting in an uncompacted zone estimated to be ½m to 1m wide. Its location is depicted in Figure 8. This zone was inspected during the 2013 Site Investigation. It is similar to dirty rockfill material as described above.

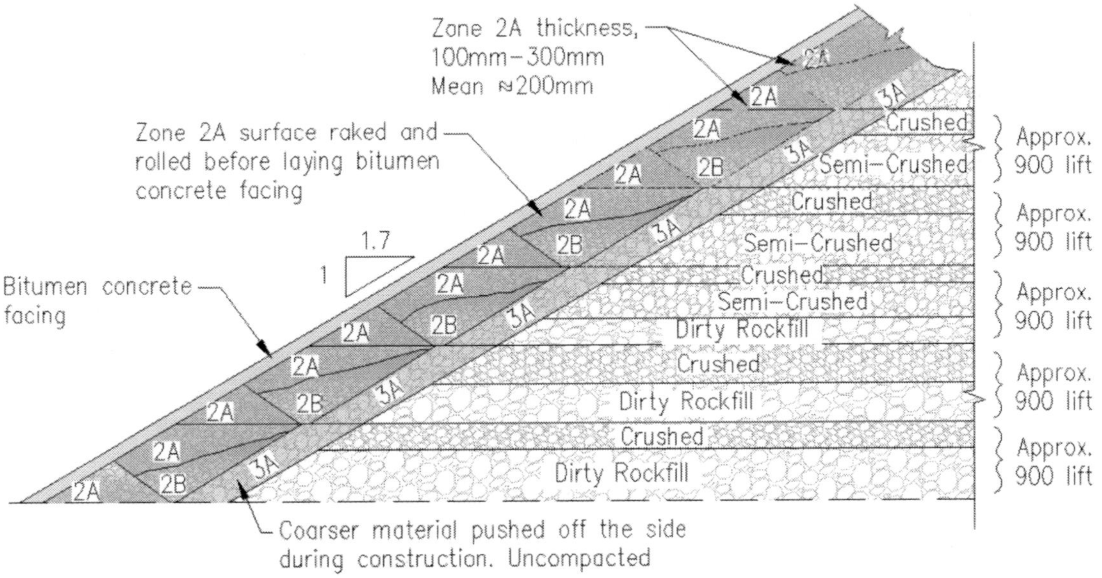

Figure 7. Upstream Slope Embankment Characterization

The bulk of the embankment is composed of crushed, semi-crushed and dirty rockfill materials in Zone 3A. The fraction of each of these three materials was estimated by summing the sonic core lengths through Zone 3A. The results were:

- Crushed Material = 52%
- Semi-crushed Material = 33%
- Dirty Rockfill = 15%.

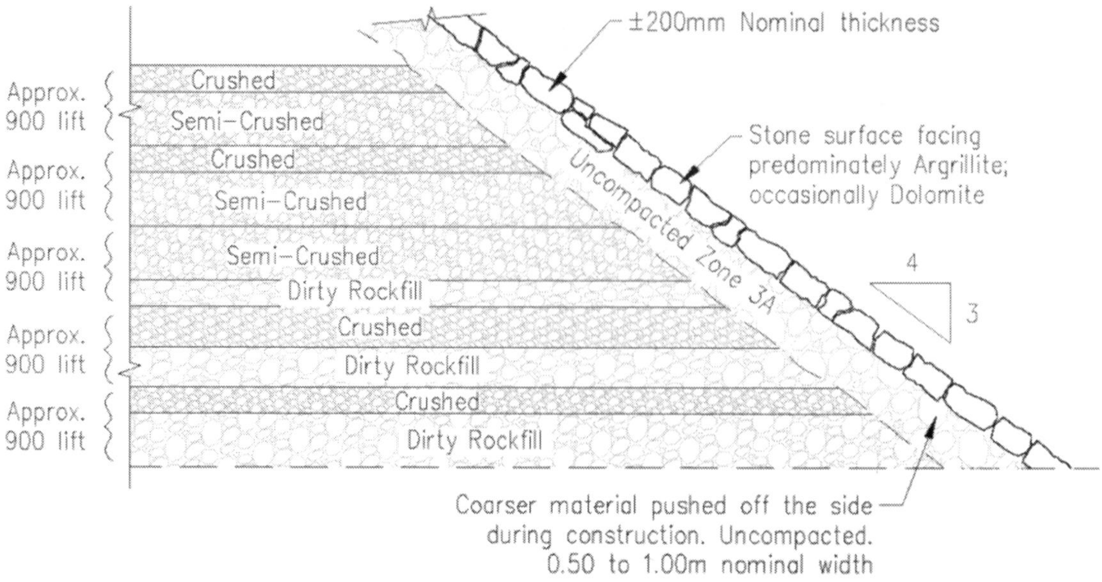

Figure 8. Downstream Slope Embankment Characterization

Zone 3B

Construction records state the select rockfill zone, Zone 3B, was not realized in practice. Hand dug test pits and a statement by a former construction engineer indicate the downstream slope is dressed with a thin veneer of stone facing. Between the compacted Zone 3A materials and the stone facing an uncompacted Zone 3A zone exists.

The visible stones covering the downstream slope are generally only one stone thick above the toe area. Construction documents report hand pitching of these stones. This stone facing has a nominal thickness of about 200 mm. In the toe area away from the gabions the stone thickness typically increases to 400 mm to 600 mm. The downstream stone facing and uncompacted Zone 3A locations are illustrated in Figure 8.

Gradings were performed on a number of samples of crushed and semi-crushed materials from Zone 3A materials penetrated by the sonic borings. These gradings are presented in **Figure 9**. Observations of recovered core indicated little sonic particle degradation of the crushed and semi-crushed materials. Significant particle degradation was observed within the dirty rockfill portions of the recovered core. This degradation is readily visible in Figure 5.

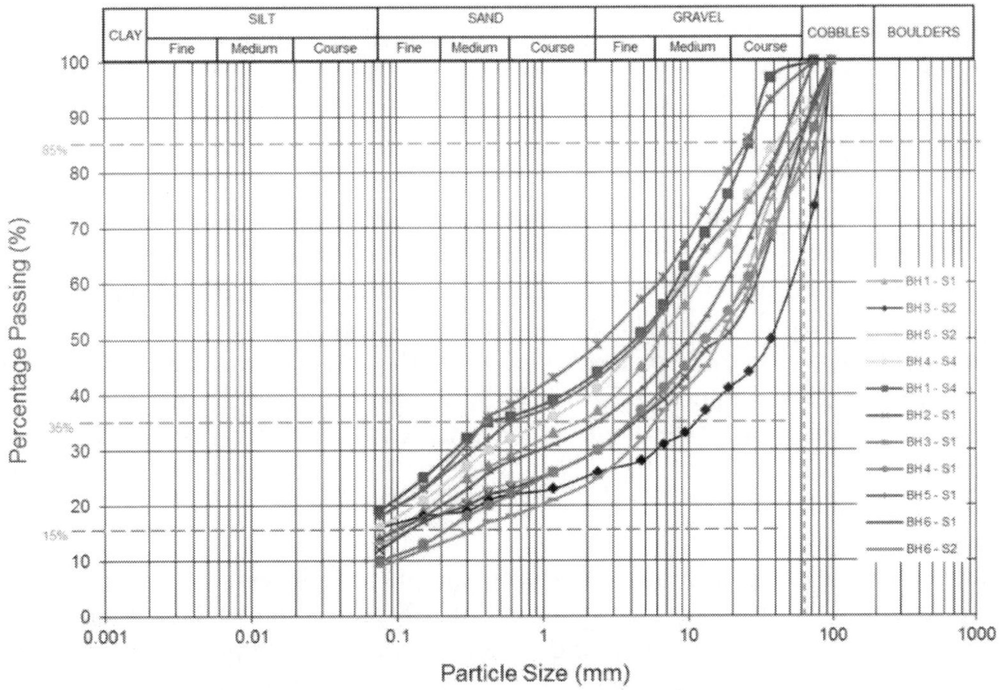

Figure 9. Gradings of Zone 3A Crushed and Semi-crushed Materials

Embankment Re-Characterization Summary

Figure 10 presents the re-characterized cross section of the maximum dam section. The main characteristics are outlined below.

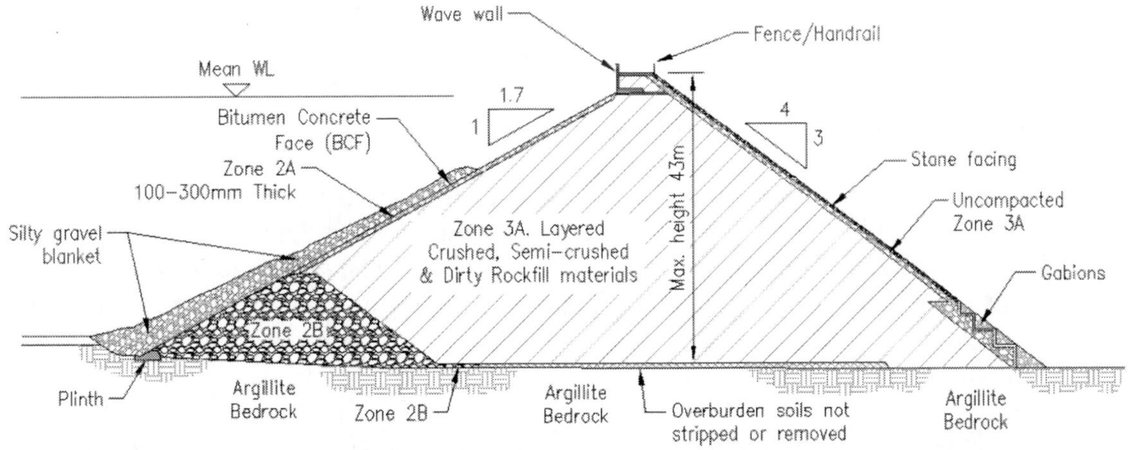

Figure 10. Re-Characterized Cross Section of Maximum Dam Section

1. The body of the embankment (Zone 3A) is more characteristic of an earthfill rather than a rockfill. This characterization is based on material composition, soil fabric and

embankment settlement behavior.

Rockfill fabric is defined as a coarse matrix of stone within which the stones maintain point to point contact, and there are unfilled voids between the stones. Leps (Leps, 1973) reported rockfill containing more than 30% finer than 25mm particles should be treated as an earthfill.

Soil fabric is dominated by the matrix of particles smaller than 25mm. Where soil contains large gravel, cobbles or boulders, most particle contacts are between particles less than 25 mm and these particles dominate the soil's behavior. An estimated 85% of compacted Zone 3A consists primarily of the earthfill type materials of crushed and semi-crushed soils. Review of these materials' gradings shown in **Figure 9** indicates 57% to 87% passing 25mm, with one exception having 43% passing 25mm. This is much finer than Leps' 30% finer than 25mm threshold between rockfill and earthfill. Thus, these materials are judged to be earthfill rather than rockfill.

Settlement behavior exhibited by the embankment following construction is also consistent with earthfill rather than rockfill. Figure 3 shows the settlement records of the embankment compared to other rockfill dams in the same region. Over the first ~3 years prior to first filling and leakage, crest settlement was on the order of 0.55% of the dam height. This amount of settlement is more consistent with compacted earthfill dams.

2. Zone 2A is composed of crushed sandy gravel with cobbles to 100mm, alluvial sandy gravel with silt (Zone 2B material), and uncompacted Zone 3A materials. Details of Zone 2A is shown in **Figure 7**. All these materials were assessed to be potentially internally unstable. The outboard surface of Zone 2A was compacted with rollers on the slope before constructing the asphalt facing.

3. The Zone 3B Select Rockfill along the downstream slope and toe were not realized in construction. Mudstone from the quarry was highly variable and was not successfully distinguished and sorted to form Zone 3B.

   Graders and dozers spread the quarried mudstone to the downstream limits. Roller compacters could not effectively compact the edge of the fill, leaving a region of uncompacted Zone 3A material parallel to the slope. The stone facing along the downstream slope was hand pitched. The uncompacted Zone 3A has an estimated $d_{50} \sim 75$mm. For unravelling considerations the stone facing is considered ineffective.

   The downstream toe in the deepest section is surfaced with overtopping protection gabions filled with uniform stone having a $d_{50} = 80$mm. Outside the deepest section, the downstream toe has variable conditions with d_{50} ranging between 75mm and 190mm.

POTENTIAL FAILURE MODES AND EXPECTED PERFORMANCE

The correct characterization of the dam has a major bearing on the potential failure modes of the dam. The potential failure modes (PFMs) of concern at this dam pertain to earthquake induced cracking of the asphalt facing and crest joints:

1. Earthquake induced cracking leakage causing toe unravelling
2. Earthquake induced cracking leakage resulting in embankment saturation causing downstream slope instability.

Earthquake Induced Cracking Leakage Causing Toe Unravelling

Assessing potential toe unravelling resulting from earthquake induced cracking and leakage was done by estimating leakage and comparing to expected unravelling resistance. Earthquake loads were partitioned according to Annual Return Period (ARP). For each ARP, crest deformations were estimated by Bray & Travasarou's empirical method (Bray & Travasarou, 2007). Normalized crest settlements were calculated to determine Damage Class, and then correlated to estimated crest region maximum crack widths according to Pells and Fell's case study empirical correlations (Pells & Fell, 2002). Given estimated crack widths for a range of earthquake ground motions, corresponding leakage discharges were calculated by two dimensional finite element modelling and analysis using SEEP/W®. The unit width discharge results were used to estimate discharge concentrations based on abutment and foundation geometry.

Downstream toe unravelling was assessed using empirical discharge vs. unravelling correlations from Norwegian model dam testing research. The empirical relationship by Bartsch and Nilsson (Bartsch & Nilsson, 2006) correlates mean particle size to discharge for a given slope angle. Toe conditions at the maximum section of the dam consist of gabions filled with stone, as illustrated in Figure, having a d_{50} = 80mm. Conditions behind the gabions are hypothesized to be loose, uncompacted Zone 3A mudstone with a d_{50} up to 150mm. A plot of estimated concentrated leakage and unravelling resistance over the range of earthquake loads represented by Annual Return Period (ARP) is presented in Figure 10.

Earthquake Induced Cracking Leakage Resulting in Embankment Saturation Causing Downstream Slope Instability

A fundamental design premise for faced rockfill dams is the embankment body remains drained and unsaturated. In order to saturate this embankment from earthquake induced cracking and leakage, cracking needs to be adequately pervasive, and the permeability of Zone 3A needs to be no more than that of Zone 2A. Zone 2A's materials consist of crushed sandy gravel. The construction period gradings suggest it is likely suffusive. The crushed and semi-crushed materials dominate the permeability of Zone 3A. Their permeability was estimated based on insitu infiltration tests. Without Zone 2A suffusing the permeability of Zone 2A was estimated to be one order of magnitude greater than Zone 3A. If Zone 2A were to suffuse its permeability would increase on the order of ½ to 1 order of magnitude. Thus, it is likely saturation can occur should adequate cracking of the asphalt facing occur from earthquake shaking and resulting embankment deformations.

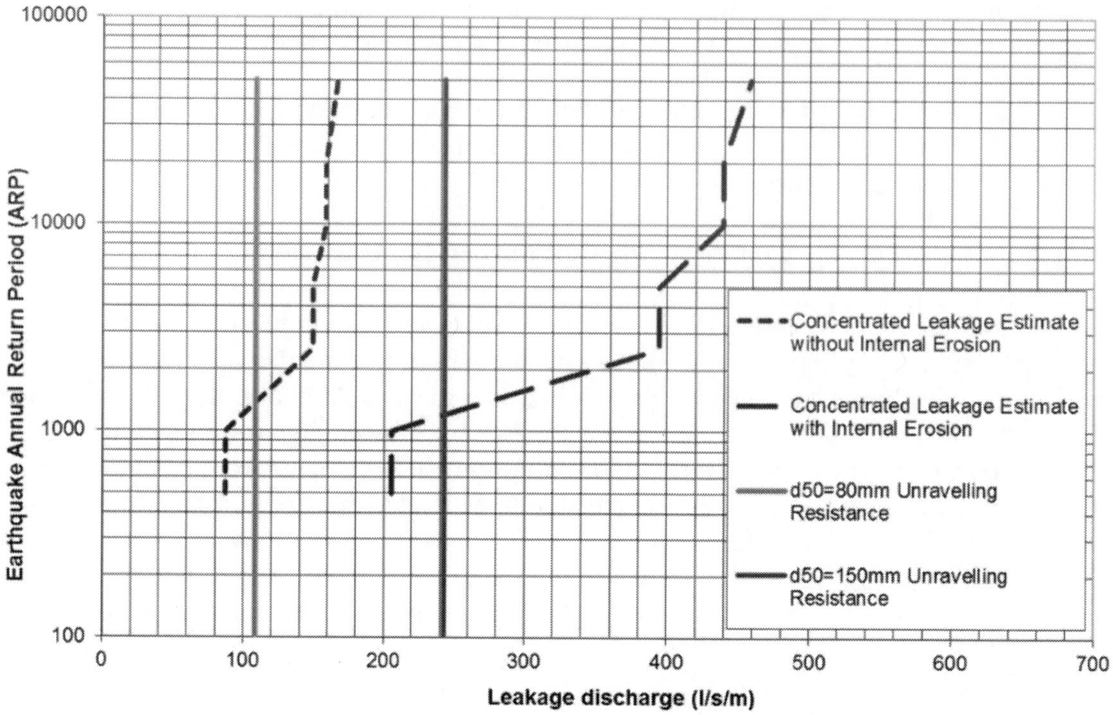

Figure 10. Estimated Concentrated Leakage & Unravelling Resistance vs. Earthquake ARP

Post-earthquake downstream slope instability was assessed for three potential post-earthquake embankment saturation conditions: 1) unsaturated; 2) low level saturation (judged to be similar to the 1st filling leakage condition); and 3) high level saturation. Various cracking scenarios were modelled and analyzed using SEEP/W®. Isolated cracks through the upper and middle parts of the upstream asphalt facing the embankment are predicted at 2500 ARP earthquake ground motions. SEEP/W® indicated high level saturation of the embankment for the 2500 ARP cracking model. Cracking increases at ground motions above 2500 ARP. Thus, high level embankment saturation is predicted for earthquake ground motions of 2500 ARP and greater.

Post-earthquake slope stability was analyzed using the Morgenstern and Price limit equilibrium method, and SLOPE/W® software. The results for these three saturation cases for the maximum dam section are presented in **Table 1**. The SLOPE/W® model illustration for the high level saturation condition is presented in Figure 11.

Table 1. Post-Earthquake Slope Stability for various Embankment Saturation Conditions

Embankment Saturation Level	Factor of Safety
Unsaturated	1.22
Low	1.00
High	0.59

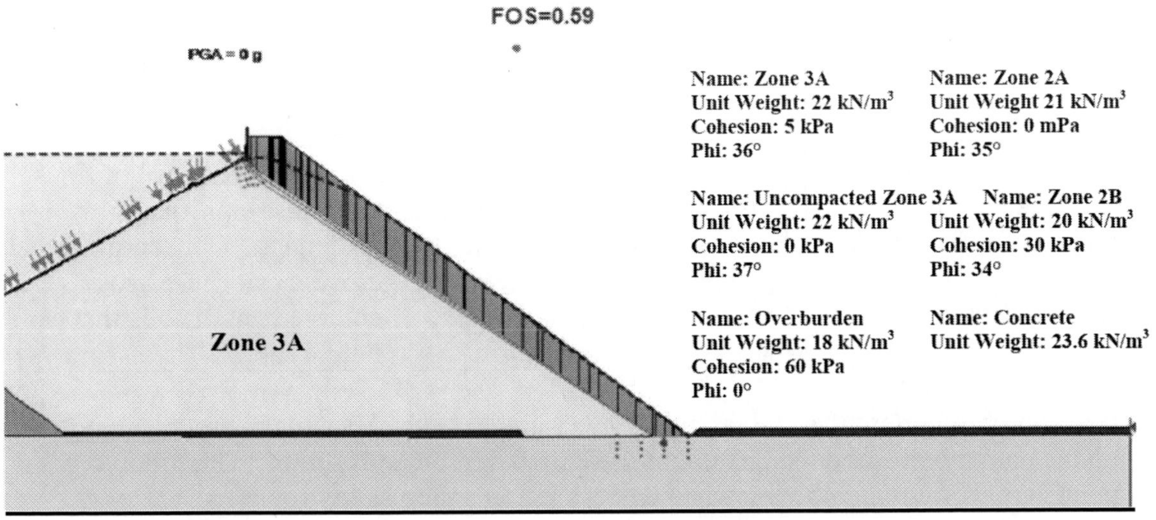

Figure 11. Post-Earthquake Static Stability, High Level Embankment Saturation

CONCLUSIONS

This project began as a dam safety review with particular focus on seismic safety, with a seismic hazard dominated by a nearby active fault. The construction records were good and perhaps overly trusted in the past. Fortunately, we elected to verify the embankment materials and this caused a major change in the assessment of the dam, in particular its seismic resistance.

If seismic assessment was based on the assumed well-drained, faced rockfill dam type, the seismic resilience of this dam would be seriously over-estimated. Expected performance of a typical faced rockfill embankment dam and faced earthfill embankment dam is summarized in **Table 2**.

Table 2. Rockfill and Earthfill Faced Dams Performance for Cracking & Leakage

Potential Failure Mode	General Rockfill Embankment Performance	General Earthfill Embankment Performance
Upstream face membrane cracking and leakage causing unravelling.	Typically able to discharge high throughflow without unravelling.	Prone to unravelling at significantly lower throughflow.
Upstream face membrane cracking and leakage causing saturation induced instability.	Rockfill inherently well drained, alleviating saturation induced instability.	Prone to saturation depending on materials and relative permeability conditions within the embankment. Full embankment saturation reduces downstream stability by roughly one-half.
Upstream face membrane cracking and leakage causing internal erosion leading to settlement or gross enlargement.	Fines within the rockfill fabric may washout, but settlement not expected to be large. Collapse settlement upon wetting can occur. Gross enlargement mechanism will not form.	If an unfiltered exist exists, significant removal of fines can cause significant settlement. Concentrated leak erosion can cause gross enlargement in soils that can hold a roof.

It is vitally important that the safety reviewer understands the actual materials that make up the dam, not the assumed nature of the dam based on convention. The following points are worth remembering when embarking on a dam safety review:

1. Actual conditions of embankment dams may differ from that reported by as-builts, construction reports and convention.

2. There is a human tendency to accept convention, sometimes with little to no direct supporting evidence.

3. The term rockfill has the connotation of a free draining, inherently stable embankment. It has been, and continues to be used, as a conventional term.

4. Understanding the characteristics of the embankment earth materials is essential to understanding their expected behavior.

5. A change in the characteristics of an embankment dam can have significant change in its expected performance. In the case of this dam, its expected post-earthquake performance is deficient because two potential failure modes became much more likely as a result of the changed understanding of the embankment's characteristics.

6. As a dam investigator, use evidence including observations, investigations and performance indicators to assess conditions based on 1st principles. Construction photos and discussions with people actually involved in the construction are

invaluable. This is good forensic practice.

7. Although estimating earthquake induced deformations have improved markedly in recent times, there is scant case history evidence and guidance in the literature for predicting cracking in embankment dams. We encourage further study in this area.

ACKNOWLEDGEMENTS

The authors wish to thank the dam owner for their permission and support for this paper.

REFERENCES

Bartsch, M., & Nilsson, A. (2006). *Function Analysis and Strengthening by Adding a d/s Berm.* Vattenfall Power Consultant AB.

Bray, J., & Travasarou, T. (2007). Simplified Procedure for Estimating Earthquake-Induced Deviatoric Slope Displacements. *ASCE Journal of Geotechnical and Environmental Engineering*, 381-392.

Hunter G., G. J. (2003). *The Performance of Concrete Face Rockfill Dams.* Sydney: UNICIV Report R-413.

Leps, T. M. (1973). Flow Through Rockfill. In *Embankment Engineering - Casangrande Volume* (pp. 87-107). New York: Wiley and Sons.

Pells, S., & Fell, R. (2002). *Damage and Cracking of Embankment Dams by Earthquakes and the Implications for Internal Erosion and Piping.* Sydney: UNICIV.

SHEAR STRENGTH OF COMPACTED BEAUMONT CLAY FOR CONSOLIDATED-UNDRAINED CONDITIONS

Mark J. Thompson[1]
David J. Bentler[2]
Thomas L. Brandon[3]
Daniel R. VandenBerge[4]

ABSTRACT

Soil shear strength is a principal input to slope stability evaluations of new and existing earth dams. The selection of drained and undrained shear strength parameters for cohesive soils is usually made on the basis of results from consolidated-undrained triaxial compression tests with pore pressure measurement. This paper discusses consolidated-undrained triaxial testing and strength characteristics of compacted clays. Fundamental concepts for shearing behavior of compacted clay are recounted in this paper using the consolidated-undrained test data for compacted Beaumont clay, which were collected to support the design of an off-channel reservoir in the Texas Gulf Coast. Key topics of the discussion include: (1) specimen preparation and triaxial testing of compacted clay; (2) effects of failure criterion selection on shear strength characterization; (3) specimen volume change that occurs during saturation and consolidation; (4) effects of compaction conditions and compaction method on drained and undrained shear strength, and (5) influence of laboratory compaction versus field compaction on soil behavior.

INTRODUCTION

Compacted clays are encountered in dams, highway embankments, and other engineered fills. Notwithstanding the extensive use of compacted clay in construction, technical literature addressing the testing and behavior of compacted clay is disproportionately sparse. Shear strength characterization of compacted clay is most often restricted to undrained strength of clay in its compacted state, determined by unconsolidated-undrained (UU) triaxial compression tests, which result in a ϕ-c characterization of the partially saturated soil. Representative effective-stress shear strength parameters for various compacted soil types are sometimes encountered in literature, since drained shear strength parameters are recognized as being largely unaffected by compaction conditions. However, the undrained strength of saturated compacted clay is central to slope stability analyses for rapid drawdown and seismic loading conditions of dams and levees, yet is scarcely mentioned.

[1] CH2M, 9191 South Jamaica Street, Englewood, CO 80112; PH (720) 286-0481; Email: mark.thompson@ch2m.com
[2] CH2M, 9191 South Jamaica Street, Englewood, CO 80112; PH (720) 286-4896; Email: david.bentler@ch2m.com
[3] Virginia Polytechnic Institute and State University, 22 Patton Hall, Blacksburg, VA 24061; PH (540) 231-4454; Email: tbrandon@vt.edu
[4] Tennessee Technological University, Prescott Hall, Box 5015, Cookeville, TN 38505; PH (931) 372-3257; Email: dvandenberge@tntech.edu

Consolidated-undrained (CU) triaxial compression testing methods are commonly used to determine drained and undrained shear strength parameters. This test is particularly suited to the estimation of shear strength parameters for slope stability analyses required for most earth dams. Effective procedures have been developed for CU tests on saturated clays, but tests on compacted clays involve additional factors for which procedures are not so well established (VandenBerge et al. 2014). Furthermore, the changes in state that occur during saturation and consolidation phases of CU tests are often overlooked. The over-emphasis on compaction conditions, as opposed to conditions prior to shearing, mislead engineers regarding the actual factors affecting undrained strength of compacted clay for consolidated-undrained conditions. In addition, little attention has been given to the differences in behavior resulting from field compaction of clay, as opposed to laboratory compaction.

This paper presents consolidated-undrained shear strength of compacted Beaumont clay. Testing considered a wide range of compaction conditions, and included strength evaluation for both laboratory and field-compacted clays. In this presentation, governing principles regarding shear strength characterization of compacted clay are demonstrated. These principles are directly applicable to selection of shear strength parameters for engineering analyses and should also be considered when developing technical specifications for earthwork projects.

BACKGROUND

This section provides a background for understanding the context in which shear strength of Beaumont clay was investigated. The Lane City Reservoir project is described, as well as compaction characteristics of Beaumont clay.

Lane City Reservoir, Texas

The Lower Colorado River Authority (LCRA) is expanding water supply in the Texas Gulf Coast by constructing Lane City Reservoir – the first significant new water supply reservoir developed in the lower Colorado River basin in decades. Lane City Reservoir is an off-channel reservoir with storage capacity of 40,000 acre-feet, formed by a continuous embankment dam approximately 40 feet high and over 5 miles in length.

The off-channel reservoir is located in Wharton County, Texas. The site is underlain by Beaumont Formation, which is locally a sequence of clay and sand deposited by ancestral Colorado River streams in a deltaic environment. Clay deposits tend to be highly plastic (plasticity index up to 60), heavily overconsolidated, and expansive. The native clay comprises the 5.7 million cubic yards of borrow material for constructing the embankment dam. This paper's contents originate from a geotechnical investigation of the engineering behavior of Beaumont clay for the purpose of designing the reservoir embankment. Specifically, rapid drawdown and steady-state seepage slope stability design criteria govern the reservoir embankment's water side and land side slope configurations, respectively. Therefore, shear strength of compacted Beaumont clay for consolidated-undrained conditions is especially relevant for the project.

Compaction Characteristics of Beaumont Clay

Standard and modified Proctor compaction tests were conducted on bulk samples throughout different stages of the project. These tests identify the maximum dry unit weight and optimum moisture content for the compactive efforts applied to the soil samples. These compaction characteristics are shown in Figure 1. Comparison of compaction characteristics for both compactive efforts is presented in Table 1.

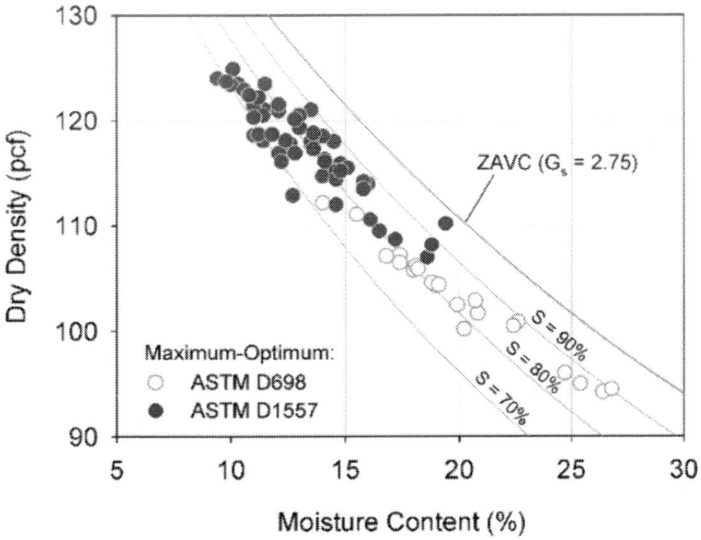

Figure 1. Standard and Modified Proctor Compaction Test Results for Beaumont Clay

Table 1. Comparison of Compaction Characteristics Based on Two Compactive Efforts

Compaction Characteristic Based on Modified Effort (ASTM D1557)	Equivalent Compaction Characteristic Based on Standard Effort (ASTM D698)
Optimum moisture content	4 to 8 percent higher than for modified effort
95 percent relative compaction	105 to 109 percent relative compaction
90 percent relative compaction	100 to 104 percent relative compaction

Following preliminary engineering of the reservoir, the anticipated compaction criteria for the project was changed to reference maximum dry density based on modified compactive effort instead of standard compactive effort. Therefore, unless otherwise stated, the term relative compaction in this paper refers to the ratio of sample dry density to the maximum dry density from a Proctor compaction test using modified compactive effort. In addition, the discussed behaviors are applicable to well compacted clays with acknowledgement that other factors may govern the behavior of clay compacted to lower standards (e.g., 95 percent relative compaction based on standard compactive effort).

CONSOLIDATED-UNDRAINED TESTING OF COMPACTED CLAY

This section describes the shear strength testing program for the Lane City Reservoir project, preparation techniques for laboratory-compacted clay specimens, and consolidated-undrained triaxial testing methods. The effect of failure criterion on drained and undrained shear strength parameter determination is also presented.

Shear Strength Testing Program

The project geotechnical investigation included extensive shear strength testing of Beaumont clay to identify the strength parameters most applicable to various slope stability analyses. Unconsolidated-undrained and consolidated-undrained triaxial compression tests were conducted on intact and laboratory-compacted clay specimens. In addition, direct shear and torsional ring shear tests were conducted on remolded clay to determine fully softened and residual shear strength parameters, respectively. This paper addresses consolidated-undrained triaxial testing and compacted clay shear strength.

Accompanying the geotechnical investigation, a test embankment program was executed for the project, wherein natural clay borrow materials were excavated, processed, and placed as embankment fill using a range of earthwork methods. Identifying material processing, placement, and compaction methods needed to achieve earthwork specifications were central goals for the test embankment program. In addition, a portion of the test embankment program was completed using consistent placement and compaction methods to produce fill for post-construction sampling and shear strength testing of field-compacted clay using consolidated-undrained triaxial testing methods.

In total, 57 CU tests were conducted laboratory-compacted clay specimens, and 34 CU tests were conducted on field-compacted clay specimens. The relative compaction and relative moisture content of CU test specimens are shown in Figure 2. Testing of laboratory-compacted clay for establishing the basis of design strength parameters considered a wide range of relative compaction and moisture content conditions. Compaction conditions typically fell between the line of optimums and the zero air-voids curve. Field-compacted clay samples at a range of relative compaction and moisture content conditions were also recovered. Field-compacted clay was generally observed to be at a higher compacted degree of saturation than the laboratory-compacted specimens.

Specimen Preparation

Compacted clay specimen preparation began with the processing of bulk samples until the material passed through a No. 10 sieve. A mechanical food processor was used to expedite the processing of large quantities of soil. Sample material was then moisture conditioned to specified moisture contents and the material allowed to mellow for a period of at least 16 hours before specimen compaction.

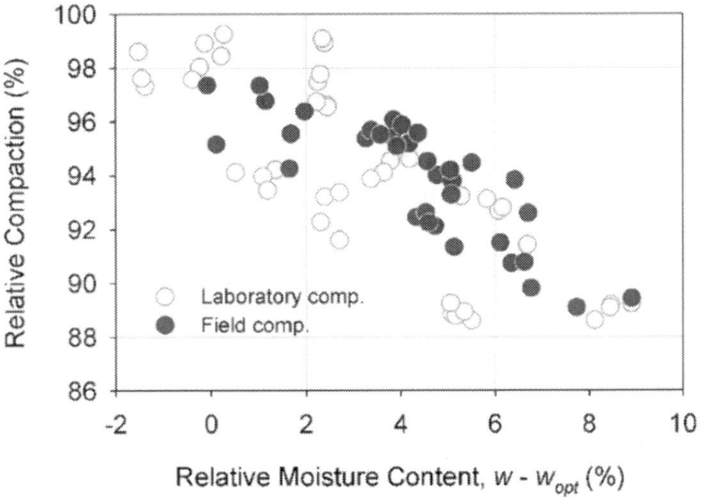

Figure 2. Relative Compaction and Moisture Content Deviation from Optimum for Compacted Clay Specimens Used in Consolidated-Undrained Triaxial Testing

Three different compaction techniques were used to prepare compacted clay specimens for CU testing: hand-tamping, Proctor compaction; and Harvard miniature compaction. These methods are shown in Figure 3.

Figure 3. Triaxial Specimen Compaction by Hand-Tamping, Proctor Compaction, and Harvard Miniature Compaction Device

Initially, a *hand-tamping* method was used. Equal quantities of processed soil were compressed into eight lifts of equal thickness by rod-tamping in a 2-inch mold. This procedure provided specimens at low dry density and high moisture content compaction conditions that were mostly free from visible voids. Specimens prepared using this technique to relative compaction values greater than 85 percent and moisture content drier than about 6 percentage points above the optimum moisture content were rejected because of an abundance of visible voids.

A limited number of specimens were prepared using *Proctor compaction*, wherein processed soil was compacted into a 1/30 ft^3 Proctor mold with a Proctor hammer. Six equal quantities of soil were compacted in the mold with the individual lift thicknesses equal to one-sixth of the mold height. After compacting the Proctor sample to specified moisture and density conditions, the sample was extruded from the mold and trimmed down to a 2-inch-diameter specimen. Trimming of the hard clay was difficult, but this method produced compacted clay with no visible voids. Relative compaction of the trimmed specimens was generally 2 to 4 percentage points above the average relative compaction of the Proctor samples.

Laboratory compaction of CU specimens was typically accomplished by a *Harvard miniature compaction* apparatus, which produced specimens having diameter of 1.4 inches. Specimens were compacted in 10 equal lifts, each receiving the same number of applications of the spring-actuated tamper. Trial-and-error was used to determine the number of tamps required to produce samples at the specified moisture contents and dry unit weights. Tolerances for dry unit weight and moisture content of the specimens were ±1 pcf and ±1 percentage point, respectively. If the unit weight determined immediately after compaction in the Harvard miniature apparatus did not fall within the tolerated limits, a new specimen was compacted.

The Harvard miniature compaction apparatus is usually characterized as providing *kneading* compaction, wherein the tamper is able to penetrate the soil lift and shear the clay. However, during Harvard miniature compaction to relative compaction values typically ranging from about 88 to 100 percent at moisture contents just slightly higher than the plastic limit, significant kneading action was not observed. The tamping end of the compaction device did not substantially penetrate and shear the soil within the mold.

Testing Procedures

Consolidated-undrained tests were conducted following the requirements of ASTM D4767 with added procedures following VandenBerge et al. (2014) and VandenBerge et al. (2015).

Saturation. During back-pressure saturation, test specimens were initially subjected to an effective stress equal to 70 percent of the final isotropic consolidation stress. When specimen swell exceeded 2 percent, the effective stress was increased to as much as the final effective consolidation stress. The primary means to test for a saturated condition was measurement of the pore water pressure parameter B. A minimum B-value of 0.97 was set as the criterion for accepting a specimen as effectively saturated. When this minimum B-value could not be achieved, a pore-pressure stability check was conducted by closing the specimen drainage lines and monitoring the pore water pressure. Pore-pressure stability checks lasted for 30 minutes. If pore water pressure changed less than 2 percent during the check, then the specimen was accepted as effectively saturated.

Consolidation. The vast majority of test samples swelled during back-pressure saturation with the effective stress equal to the assigned effective consolidation stress (usually less than or equal to 3 ksf). Accordingly, distinct consolidation phases of the test were not required. Rather, the back-pressure saturation phase continued until the specimen was considered to be saturated and the volumetric strain stabilized. Where consolidation to the assigned effective consolidation stress was required, the consolidation was performed by increasing the effective stress in a single step.

Shearing. The axial strain rate for specimen shearing was determined using the procedure given in ASTM D4767 using the time for consolidation from the specimen consolidation phase and an estimate of the failure strain. In the absence of consolidation curves from which to estimate an appropriate strain rate, a conservative rate was selected to ensure a reasonable measurement of pore pressures. The axial strain rates for specimen shearing were generally equal to 0.15%/hr. Some specimens were sheared as fast as about 0.25%/hr, while some other specimens were sheared as slow as 0.07%/hr. These axial strain rates are slower than usually employed for conducting consolidated-undrained triaxial compression tests. In the absence of certainty about the ability to measure well-distributed sample pore pressures, conservative strain rates are recommended.

Effective Stress Measurement. For triaxial test specimens tested at effective consolidation stresses less than or equal to 500 psf, a high resolution differential pressure transducer was used to measure effective stresses during shearing. These low consolidation stresses were used because they were representative of stresses observed in slope stability analysis. The transducer measurement of pore-water and cell pressures permitted a more accurate determination of the effective stress for samples with small consolidation stresses.

Failure Criteria and Effects on Shear Strength Characterization

Failure criteria for consolidated-undrained triaxial tests must always be considered when interpreting and reporting test results. Common failure criteria include: (1) maximum effective principal stress ratio (maximum obliquity), (2) zero excess pore-water pressure (pore pressure parameter, A = 0), (3) maximum deviator stress, or (4) deviator stress at limiting values of axial strain (e.g., 5 percent, 15 percent). Shearing data for two specimens of a CU test series are provided in Figure 4. Failure points based on the aforementioned failure criteria are shown to demonstrate their effect on shear strength characterization.

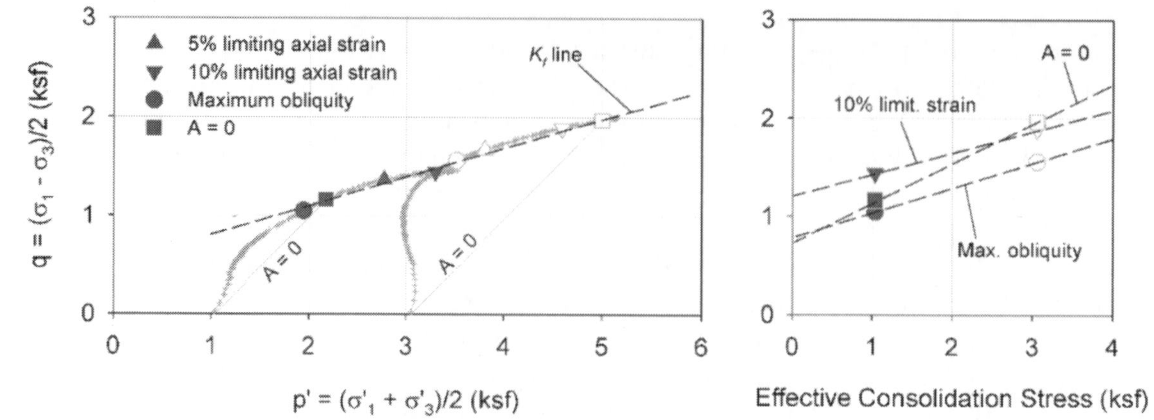

Figure 4. Effect of Failure Criteria on Drained and Undrained Strength Parameters

The failure points in effective stress-path space for the two specimens follow the same K_f line, demonstrating that effective shear strength parameters are not significantly influenced by selection of a failure criterion. Conversely, the characterization of undrained strength for compacted clays is highly affected by the selected failure criterion. The undrained strength, s_u, associated with the maximum effective principal stress ratio is generally the most conservative, because maximum obliquity occurs at relatively low axial strains and deviator stresses continue to increase with increasing axial strain. Defining failure by zero excess pore-water pressure typically allows for high undrained strength but prevents undrained strength from relying on negative pore pressures, which are more common at lower consolidation stresses. For the example in Figure 4, A = 0 gives highest s_u at σ'_{3c} = 3 ksf (lower tendency for dilation) but relatively low s_u at σ'_{3c} = 1 ksf (higher tendency for dilation). Limiting axial strain is a common failure criterion that provides intermediate undrained strength. Limiting axial strain is often used as the failure criterion with an intent to limit deformation, however consideration should be given whether this may permit negative pore pressures to contribute significantly to the strength.

Failure criterion for the strength characterization presented herein was selected as being maximum deviator stress with limiting axial strain of 10 percent. For most tests, the limiting strain value of 10 percent governed the failure condition, rather than maximum deviator stress.

SHEAR STRENGTH OF COMPACTED BEAUMONT CLAY

This section reviews shear strength characterization of laboratory and field-compacted Beaumont clay. The dataset indicates the effects of compaction conditions and compaction method on drained and undrained shear strength of compacted clay.

Specimen Volume Change During Saturation and Consolidation

Lovell and Johnson (1981) and VandenBerge et al. (2014) discussed the volume change that occurs during saturation and consolidation phases of a CU test on compacted clay.

Compacted clay specimens prepared at low relative compaction values or subject to high effective stresses tend to consolidate after saturation and application of isotropic consolidation stress. Conversely, specimens prepared at high relative compaction values tend to swell during saturation. The volumetric strain is also strongly affected by the compacted degree of saturation and effective consolidation stress. Figure 5 shows the volumetric strain for various relative compaction and as-compacted degree of saturation conditions. Data are only shown for samples subjected to an effective consolidation stress equal to about 1 ksf. Despite the variation, the highest swell potential (negative volumetric strain) is observed for laboratory-compacted specimens at higher values of relative compaction and at as-compacted degrees of saturation less than about 0.9.

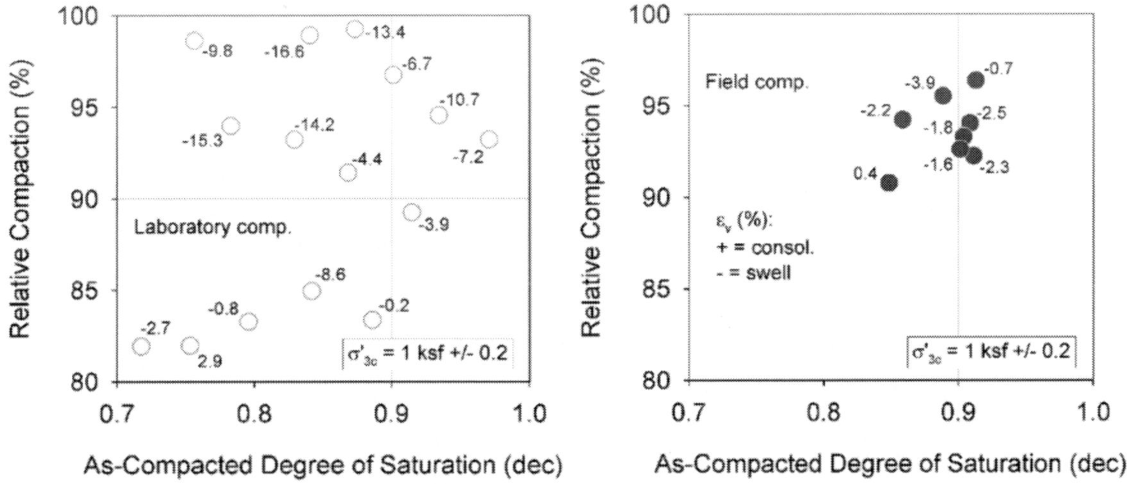

Figure 5. Volumetric Strain during Saturation and Consolidation for Laboratory-Compacted Specimens (left) and Field-Compacted Specimens (right)

Field-compacted clay from the test embankment program saturated at effective consolidation stresses of 1 ksf and higher did not swell more than 4 percent (see Figure 5). The difference in volume change response to saturation for laboratory and field-compacted clay suggests an effect of compaction method. Seed et al. (1960) showed that increasing shear strain imposed during compaction creates a more dispersed soil structure, which is generally less susceptible to swelling upon introduction of water. Dispersed structure is also created by compaction at relatively high moisture contents. Compared to the kneading compaction induced by field compaction by means of a Caterpillar 825 tamping foot roller, the observed laboratory compaction using the Harvard miniature compaction apparatus induced comparatively little shear strains. The tamping end of the Harvard miniature compaction device did not penetrate lifts of laboratory-compacted specimens. Rather, compaction with the Harvard miniature compaction apparatus at the specified moisture-density conditions might instead be described as tamping or pseudo-static compaction. The more flocculated soil structure of laboratory-compacted clay makes these specimens more susceptible to volume changes during saturation. This difference in volume change response generally results in lower final moisture content in the saturated state for field-compacted clay and, ultimately, results in higher undrained strength observed for field-compacted clay than laboratory-compacted clay.

Figure 6 shows volumetric strain as a function of relative compaction and relative moisture content. The plots show that increasing relative compaction and decreasing moisture content lead to increased swell of laboratory-compacted clay. In addition, increasing effective consolidation stress reduces the amount of net swell in laboratory-compacted clay specimens. Volumetric swell of field-compacted clay, however, did not exceed about 5 percent, which was significantly less than laboratory-compacted clay at similar relative compaction and moisture content conditions.

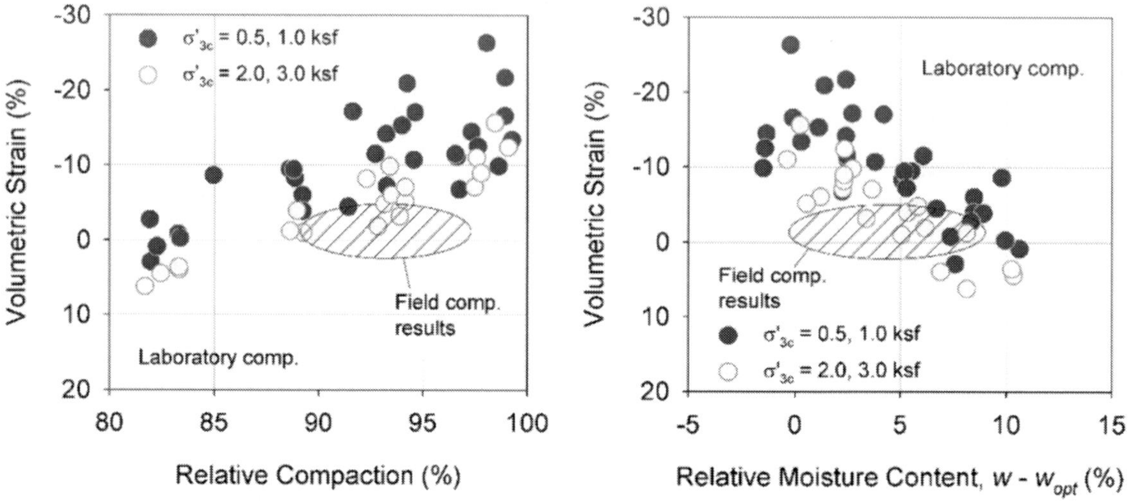

Figure 6. Effect of Relative Compaction, Relative Moisture Content, and Effective Consolidation Stress on Volumetric Strain during Saturation

The swelling of laboratory-compacted CU specimens during back-pressure saturation resulted in void ratios during shearing being significantly different than the as-compacted void ratios. Figure 7 shows a comparison of these void ratios for laboratory-compacted clay samples. The plot indirectly indicates the tendency for well-compacted clays (low as-compacted void ratios) to swell during saturation depending on effective consolidation stress and other factors such as moisture content and initial degree of saturation. The differing swell response of laboratory-compacted clay resulted in a poor correlation of as-compacted void ratio and the void ratio at shearing. Practically, Figure 7, which is based only on laboratory-compacted specimens, suggests that working to achieve very high degrees of compaction is unwarranted, since the swell potential of the compacted clay will result in significantly higher void ratio during shearing. Conversely, the magnitudes of swell for field-compacted clay specimens were lower, such that as-compacted void ratio and void ratio during shearing both generally ranged from roughly 0.5 to 0.65 with some level of correlation along the *line of unity*.

Factors Affecting Drained and Undrained Shear Strength

Seed et al. (1960) described how differences in compaction condition, method of compaction, and clay structure cause large differences in stress-strain characteristics, pore-water pressures, and effective stresses in compacted clays, but only small changes in maximum effective principal stress ratio or strength parameters expressed in terms of effective stresses. The CU test data for Beaumont clay from the Lane City Reservoir investigation illustrate this guiding principle.

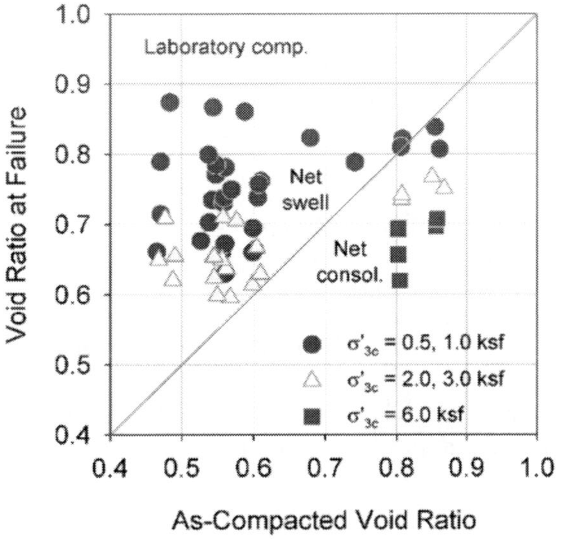

Figure 7. Comparison of As-Compacted Void Ratio and Void Ratio after Consolidation for Laboratory-Compacted Specimens

Effective principal stress ratios at failure are plotted against effective minor principal stress at failure in Figure 8. No difference is observed between results for laboratory- and field-compacted clay. Furthermore, the highly varied compaction states and test stress conditions of the specimens do not result in considerable scatter. Instead, a singular relationship is identified, whereby the effective principal stress ratio at failure follows the five-parameter, double rectangular hyperbola relationship shown in Figure 8. Variability about this relationship is most likely attributed to variation in the soils tested throughout the program, for which liquid limit ranged from 44 to 71 and plasticity index ranged from 31 to 56.

The stress-path points at failure are shown in Figure 9. The figure differentiates laboratory-compacted samples from field-compacted samples, but the differences in compaction method and compaction states are not shown to affect the failure envelope. Failure points for field-compacted specimens fall within the scatter of failure points from laboratory-compacted clay. Also shown in Figure 9 is a nonlinear K_f line, which was derived from the regression function presented in Figure 8. The dramatic rise in $(\sigma'_1/\sigma'_3)_f$ at low σ'_{3f} gives rise to a cohesion intercept or requires the use of a curved failure envelope. The gradual decrease in $(\sigma'_1/\sigma'_3)_f$ with increasing σ'_{3f} leads to nonlinearity of the K_f line. A power-function shear strength envelope converted to p-q space is shown in Figure 9 along with two sets of drained shear strength parameters, ϕ' and c', which could be applicable to the two respective stress ranges.

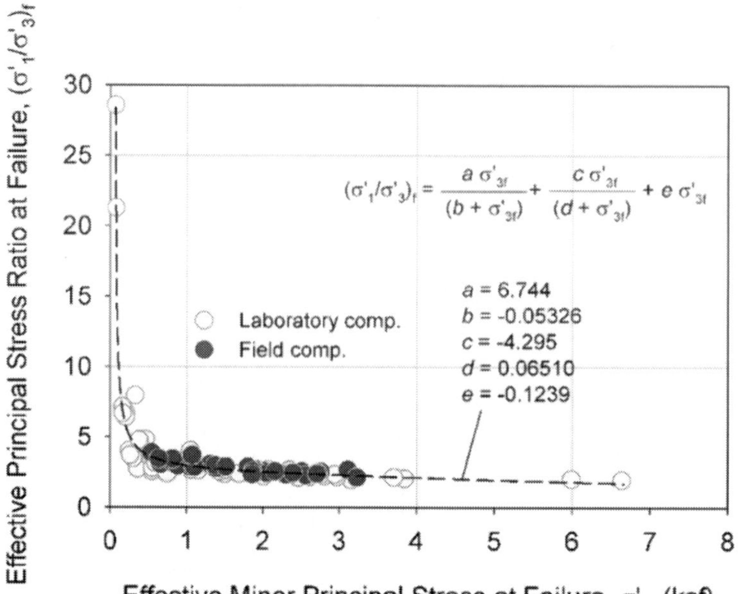

Figure 8. Relationship between Effective Principal Stress Ratio at Failure and Effective Minor Principal Stress at Failure for All Specimens

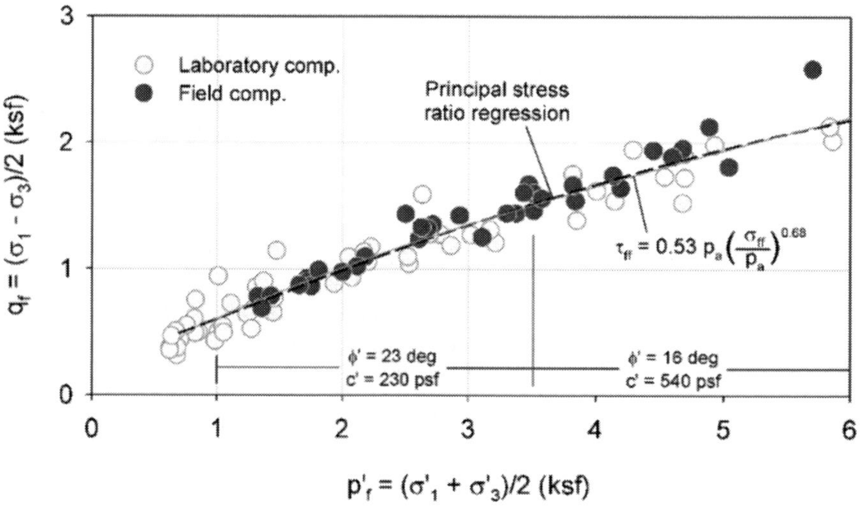

Figure 9. Drained Shear Strength Parameters Based on Stress-Path Points at Failure

The undrained shear strength of natural clays has long been recognized to depend on soil structure and void ratio, which are directly related to stress history. Overconsolidated clays encountered at low void ratios, for example, tend to exhibit higher undrained strength than normally consolidated clays at similar effective stresses due to increased tendency for dilation. Similarly, the undrained shear strength of compacted clay depends primarily on the void ratio during shearing and, to an extent, the soil structure resulting from certain compaction conditions and compaction methods.

For a saturated clay specimen (S = 100%), the moisture content at failure is related to the void ratio during shear through the phase relationship in Equation (1).

$$\omega = \frac{e}{G_S} \tag{1}$$

The relationship between undrained strength and moisture content at failure is shown in Figure 10. The data presented for each effective consolidation stress include results for both laboratory and field-compacted specimens. For a given effective consolidation stress, decreasing moisture content (decreasing void ratio) produces higher undrained strength. Furthermore, the observed relationships are not affected by compaction method.

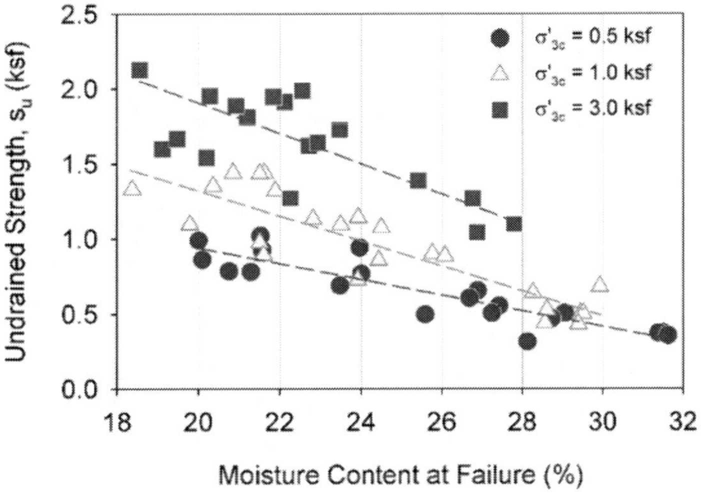

Figure 10. Effect of Moisture Content at Failure on Undrained Shear Strength

Undrained shear strength is more conventionally plotted against effective consolidation stress in Figure 11. Data for laboratory and field-compacted specimens are again presented. As previously demonstrated, lower moisture content at failure produces higher undrained strength compared to higher moisture content at failure. This condition prior to shearing definitively governs the undrained strength. The compaction conditions affect the undrained strength through the effect the compaction conditions have on the moisture content at failure. As seen in Figure 6, volumetric strain (during saturation and consolidation) increases as relative compaction increases and decreases as the initial saturation increases. The connection between the initial conditions, effective consolidation stress, and final water content is complex and cannot be expressed in a simple relationship. These interactions are also dependent on the shrink-swell characteristics of the soils, making the trends even more difficult to identify. From the project dataset, no simple, direct relationships between undrained strength, relative compaction (as compacted), or compacted moisture content were identified. This lack of correlation is not surprising when the lack of correlation between as-compacted void ratio and void ratio after consolidation (see Figure 7) is considered.

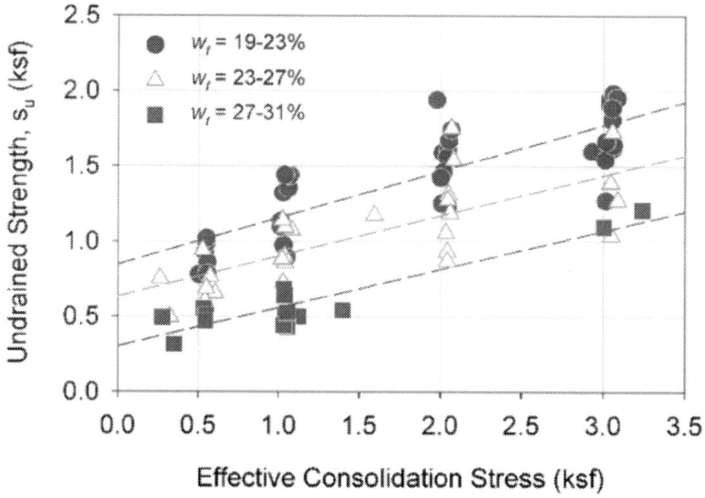

Figure 11. Effect of Moisture Content at Failure on Undrained Shear Strength

Effect of Compaction Method on Behavior of Compacted Clay

Comparison of shear behavior presented in Figure 12 reiterates the principal factor affecting undrained shear strength of compacted clay, which is the moisture content at failure. Specimens A and B were compacted at the same moisture content to the same approximate dry density but by laboratory and field compaction methods, respectively. The laboratory-compacted specimen swelled about 10.7 percent compared to only 2.5 percent for the field-compacted specimen. The field-compacted specimen was therefore sheared at a lower moisture content with greater tendency for dilation. Accordingly, the undrained strength for Specimen B was nearly twice that for Specimen A.

Specimens C and D compacted using laboratory and field compaction methods, respectively, exhibited different compaction conditions. However, the differing swell response at the same effective consolidation stress resulted in the moisture content at failure to be the nearly equivalent for the two specimens. The stress-strain and pore-water pressure responses were observed to be similar. Thus, while some effect of compaction method on undrained strength might be expected, the moisture content (or void ratio) during shearing is the predominant factor affecting undrained strength.

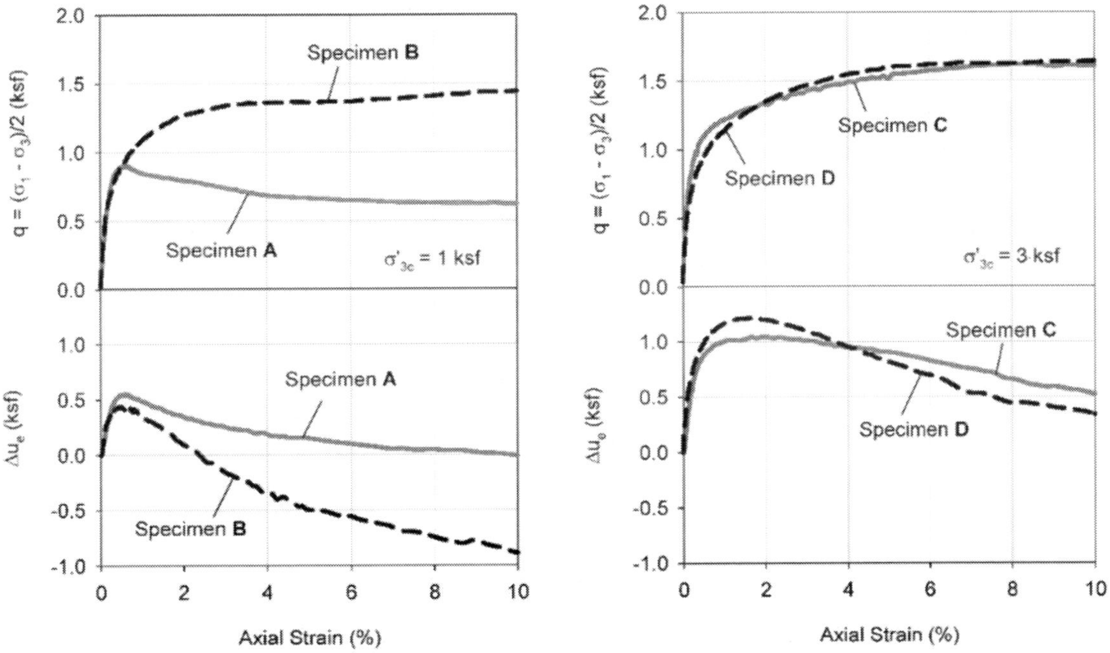

Specimen Detail	Specimen A	B	C	D
Compaction method:	Laboratory	Field	Laboratory	Field
Compacted dry density:	111.6 pcf	110.3 pcf	111.1 pcf	107.6 pcf
Compacted moisture content:	18.3%	18.4%	15.0%	19.9%
Compacted void ratio:	0.54	0.56	0.54	0.59
Consolidation stress, σ'_{3c}:	1.037 ksf	1.072 ksf	3.057 ksf	3.061 ksf
Volumetric strain:	-10.7%	-2.5%	-5.1%	-2.1%
Void ratio at failure:	0.70	0.59	0.62	0.63
Moisture content at failure:	25.8%	21.6%	22.7%	22.9%

Figure 12. Comparison of Shear Behavior of Laboratory and Field-Compacted Clay at Various Compaction and Consolidated Conditions

CONCLUSIONS

The shear strength characteristics of compacted clays are important in the engineering of a variety of structures, particularly dams and levees. Gaps in understanding, or misconceptions, about the actual factors affecting the strength of these materials for consolidated-undrained conditions can lead to either unsafe or inefficient designs. Thorough investigation and application of accepted soil mechanics principles, however, provide engineers with the ability to make reasonable decisions in selecting shear strength parameters for engineering analyses. This paper presented an investigation of the shear strength of Beaumont clay for consolidated-undrained conditions. The findings support a number of generalizations regarding the strength of well compacted clays:

- Drained shear strength parameters are not significantly affected by compaction conditions or compaction method.

- Compacted clays, particularly highly plastic clays, are susceptible to swelling during the saturation phase of CU tests. The swell increases with increasing relative compaction and decreasing degree of saturation. The volume change also depends on effective consolidation stress and overall shrink-swell potential of the clay.
- Undrained strength is primarily related to the void ratio (or moisture content) during shearing. This condition depends on compaction conditions and the volume change that occurs during saturation and consolidation of the specimen. For a given effective consolidation stress, compacted clay at lower shearing moisture content exhibits higher undrained strength due to increased tendency for dilation.
- The observed volume change and undrained strength behaviors discussed in the paper suggest that clay compacted to lower relative compaction at higher moisture content sometimes exhibit higher undrained strength than clay compacted to higher relative compaction at lower moisture content.
- Development of technical specifications for earthwork projects should consider the comprehensive behaviors of compacted clay so that assessments about the likely project performance are not misconceived.

REFERENCES

Lovell, C. W. and Johnson, J. M. (1981). "Shearing behavior of compacted clay after saturation." *Laboratory Shear Strength of Soil, ASTM STP 740*, R. N. Yong and F. C. Townsend, Eds., American Society for Testing and Materials, pp. 277-293.

Seed, H. B., Mitchell, J. K., and Chan, C. K. (1960). "The strength of compacted cohesive soils." *Proceedings of the Research Conference on Strength of Cohesive Soils*, ASCE, Boulder, CO, pp. 877-964.

VandenBerge, D. R., Brandon, T. L., and Duncan, J. M. (2014). "Triaxial tests on compacted clays for consolidated-undrained conditions." *Geotechnical Testing Journal*, Vol. 37, No. 4, ASTM, pp. 705-716.

VandenBerge, D. R., Duncan, J. M., and Brandon, T.L. (2015). *Consolidated-Undrained Strength of Compacted Clay, CGPR #83*. Center for Geotechnical Practice and Research, Virginia Tech, Blacksburg, 128 pp.

ESTIMATING THE PEAK FRICTION ANGLE OF SANDY SOILS IN SITU WITH STATE-BASED OVERBURDEN NORMALIZED CPT TIP RESISTANCE

Robert A. Jaeger, PE, PhD[1]
Ian P. Maki, PE [2]

ABSTRACT

Relationships for estimating the peak friction angle of sandy soils from cone penetration test (CPT) tip resistance (q_t) often involve correcting q_t to a reference stress of one atmosphere with an overburden normalization (or correction) factor. Existing relationships between q_t and peak friction angle have typically been derived from a database of field tests and accompanying laboratory tests on undisturbed soil samples. Published in situ datasets of q_t and peak friction angle are not well-constrained due to limitations with respect to sand characteristics and stress range. Development of relationships between q_t and peak friction angle based on these datasets thus often requires extrapolation. This paper presents a relationship between peak friction angle and state-based overburden-normalized q_t. A state-based overburden normalization approach is adopted because it provides both a rational basis for interpreting the peak friction angle in sandy soils from q_t, and a critical state-based method for extrapolating these relationships to conditions outside the range currently covered by available peak friction angle data. Finally, a design relationship between state-based overburden-normalized q_t and peak friction angle is developed based on published results from in situ and laboratory tests on undisturbed sand samples, calibration chamber cone penetration tests, laboratory tests on calibration chamber sands, and a numerical model of cone penetration resistance.

BACKGROUND

Relationships between cone penetration test (CPT) data and soil properties have been developed over the years through a combination of empirical and theoretical studies (e.g., Wroth 1984, Been et al. 1986, Kulhawy and Mayne 1990, Yu 2004, Mayne 2006, Robertson 2009). These studies have shown that the CPT tip resistance (q_t), sleeve friction (f_s), and pore pressure measurements (e.g. u_2) are affected to various degrees by a range of in situ conditions and soil properties. Thus the accuracy of relationships for predicting soil properties depends on the degree to which the various factors that influence q_t, f_s, and u_2 can be isolated and reliably accounted for.

Previous studies have shown the q_t value in sand primarily depends on the sand's strength (typically characterized in terms of peak friction angle, ϕ'_{peak}), stiffness, and dilatancy. Sand strength, stiffness, and dilatancy have been shown to depend on density, stresses, fundamental soil characteristics (i.e., mineralogy, grain size distribution, angularity), and environmental factors (i.e., aging, and cementation). Been and Jefferies (1985) and

[1] Senior Engineer, GEI Consultants, Rancho Cordova, California, CA 95670, rjaeger@geiconsultants.com.
[2] Design Engineer, Department of Water Resources Division of Safety of Dams, California, CA 95818, Ian.Maki@water.ca.gov.

Bolton (1986) both demonstrated the combined influence of confining stress and void ratio on strength and dilatancy can be accounted for with a single parameter, albeit different expressions were used in each publication. The single parameter used by Been and Jefferies (1985) was the initial state parameter ξ_0, which is the difference between the initial void ratio (e_0) and the void ratio on the critical state line (e_{cs}) at the same initial mean effective stress, p'_0 (Figure 1). Therefore it is logical that a relationship exists between q_t, peak friction angle, and ξ_0.

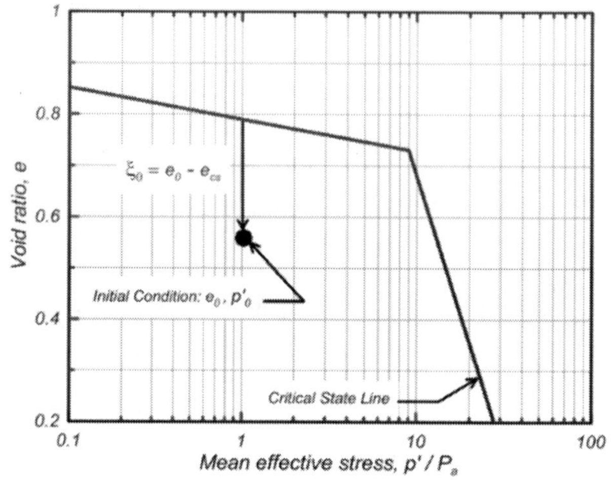

Figure 1. Example bilinear critical state line of sand and definition of initial state parameter

This paper presents a relationship between q_t, stress, and ϕ'_{peak} with a state parameter-based (or state-based) overburden normalization of cone penetration resistance. First, background on a framework for state-normalization of cone penetration resistance is presented. Second, two commonly used semi-empirical relationships between ϕ'_{peak}, stress, and density are then reviewed. Third, these relationships are then used to develop a state-based relationship between q_t, stress, and ϕ'_{peak} based on (1) a database of experimental in situ CPT data with laboratory testing on undisturbed sand samples (Mayne 2006), (2) a database of calibration chamber cone penetration and laboratory test data for four well-studied clean sands (Golder Associates 2011), and (3) a numerical model of cone penetration resistance based on cavity expansion analysis (Salgado et al. 1997a,b). Finally, the developed relationship between q_t, stress, and ϕ'_{peak} is compared to other commonly used relationships in geotechnical practice.

FRAMEWORK

Relationships between Cone Tip Resistance, Soil Density, and Stress

As discussed by Maki et al. (2014a), cone penetration resistance can be estimated as a function of σ'_{v0}, total vertical overburden (σ_{v0}), and ξ_0:

$$\frac{q_t - \sigma_{v0}}{P_a} = ae^{b\xi_0}\left(\frac{\sigma'_{v0}}{P_a}\right)^{m_\xi} \quad (1)$$

where (q_t - σ_{v0}) is the net tip resistance ($q_{t,net}$), parameter m_ξ provides a correlation between $q_{t,net}$ and σ'_{v0}, parameter b provides a correlation between $q_{t,net}$ and ξ_0, parameter a is a fitting parameter, and P_a = atmospheric pressure (1 atm or 101.3 kPa). The parameters a, b, and m_ξ depend on various soil characteristics (e.g., grain minerology, grain size, grain angularity, soil density) and environmental factors (e.g., aging, cementation).

The variation of q_t with σ'_{v0} in Eq. 1 can be isolated from the variation with density using an alternative representation of Eq. 1 in the form of a tip resistance that is projected (or normalized) to a standard reference stress while all other factors are held constant. A commonly used reference stress of one atmosphere was adopted for the current study. The current study also adopts the normalizations developed by Boulanger (2003), Boulanger and Idriss (2004), and Idriss and Boulanger (2008) to either ξ_0 or D_{R0}. The stress normalization approach in terms of ξ_0 can be expressed as:

$$Q_{1\xi} = \frac{q_t - \sigma_{v0}}{P_a} C_{N\xi} = ae^{b\xi_0} \quad (2)$$

$$C_{N\xi} = \left(\frac{P_a}{\sigma'_{v0}}\right)^{m_\xi} \quad (3)$$

where $Q_{1\xi}$ is the net tip resistance (in atmospheres) that a soil at a given ξ_0 would have at a reference stress of $\sigma'_{v0} = P_a$, and $C_{N\xi}$ is the mapping function that accounts for the effects of stress on $q_{t,net}$ for the same soil at the same ξ_0. The choice of reference stress thus is conceptually arbitrary, but $\sigma'_{v0} = P_a$ is typically in the stress range of practical interest and is commonly adopted for interpretation of in situ test results.

Maki et al. (2014a) demonstrated $Q_{1\xi}$ was strongly correlated to ξ_0 for each of four clean sands tested in calibration chambers. Based on the results presented by Maki et al. (2014a), the parameters a and b can be reasonably assumed to be constant for all stresses provided that all other environmental factors are held constant. The relationships between $Q_{1\xi}$ and ξ_0 for different sands would, however, depend on the specific soil characteristics and environmental factors which manifest themselves as different a and b values.

The adopted state-based stress normalization (Boulanger 2003) uses stress normalizations in terms of D_{R0} or e_0 as an intermediate step. In terms of D_{R0} or e_0, the stress normalization of $q_{t,net}$ can be expressed as:

$$Q_{1e} = \frac{q_t - \sigma_{v0}}{P_a} C_{Ne} \quad (4)$$

$$C_{Ne} = \left(\frac{P_a}{\sigma'_{v0}}\right)^{m_e} \quad (5)$$

where Q_{1e} is the net tip resistance (in atmospheres) that a soil at a given e_0 or D_{R0} would have at a reference stress of $\sigma'_{v0} = P_a$, C_{Ne} is the mapping function that accounts for the effects of stress on $q_{t,net}$ for the same soil at the same e_0 or D_{R0}, and m_e is a fitting exponent. In much of the literature on foundation and liquefaction engineering (i.e. Liao and Whitman 1986, Youd et al. 2001, Idriss and Boulanger 2008), Q_{1e} is commonly written as either q_{t1} or q_{c1N} and C_{Ne} is commonly written as C_N; the addition of the subscript e is intended to identify the underlying assumption of normalization at the respective e_0 or D_{R0}.

Boulanger (2003) developed the following regression for the fitting exponent m_e based on numerical analyses of cone penetration testing by Salgado et al. (1997b):

$$m_e = 0.784 - 0.521 D_{R0} \tag{6}$$

Boulanger (2003), and later Idriss and Boulanger (2008), developed the following regression between D_{R0} and Q_{1e} (or q_{c1N} as they are practically equal in sands):

$$D_{R0} = 0.465 \left(\frac{Q_{1e}}{C_{dq}}\right)^{0.264} - 1.063 \tag{7}$$

where C_{dq} is a parameter ranging from 0.64 to 1.55 based on numerical analyses by Salgado et al. (1997b) using properties of clean sands tested in calibration chamber CPTs (Boulanger 2003). Maki et al. (2014a) reviewed calibration chamber CPT data in four clean sands (described later in this paper) and found $C_{dq} = 1.3$ provided the best fit to the data.

Boulanger and Idriss (2004) developed a regression of the mapping function $C_{N\xi}$ based on the Bolton (1986) stress-strength-dilatancy relationship. The Boulanger and Idriss (2004) $C_{N\xi}$ factor can be written as:

$$C_{N\xi} = C_\xi C_{Ne} \tag{8}$$

where,

$$C_\xi = \left(\frac{D_{R0} - \Delta D_{R,CS} + 1.063}{D_{R0} + 1.063}\right)^{3.788} \tag{9}$$

$$\Delta D_{R,cs} = \frac{1}{Q - ln\left(100 \frac{1+2K_O}{3} \frac{\sigma'_{v0}}{P_a}\right)} - \frac{1}{Q - ln\left(100 \frac{1+2K_O}{3}\right)} \tag{10}$$

where Q is an empirical factor found to be about 10 for quartz and feldspar sands (Bolton 1986) and K_0 is the ratio of horizontal to vertical initial effective stresses. As described by Boulanger (2003), the parameter $\Delta D_{R,cs}$ defines the difference in relative density on the critical state line between the in situ stress and the reference stress (one atmosphere).

Examples of relationships between $q_{t,net}$, σ'_{v0}, and either ξ_0 or D_{R0} using the form in Eqs. 2-10 as proposed by Boulanger (2003) and Boulanger and Idriss (2004) are shown in Figures 2a and 2b. The relationships in Figure 2a show $q_{t,net}$ versus σ'_{v0} for the same sand at six different initial values of ξ_0 (-0.2, -0.15, -0.1, -0.05, 0.0, and 0.05) based on $C_{dq} = 1.3$, $Q = 10$, $R = 1$, $K_0 = 0.45$, $e_{max} = 0.89$, and $e_{min} = 0.60$. These $q_{t,net}$ versus σ'_{v0} graphs are slightly curved in a log-log scale. These results also illustrate how the stress exponent m_ξ will decrease with increasing density for sands (i.e., more negative ξ_0). The corresponding dependence of $C_{N\xi}$ on density is shown in Figure 3 for ξ_0 values corresponding to D_{R0} of 35%, 55%, and 75% at $\sigma'_{v0} = 1$ atm.

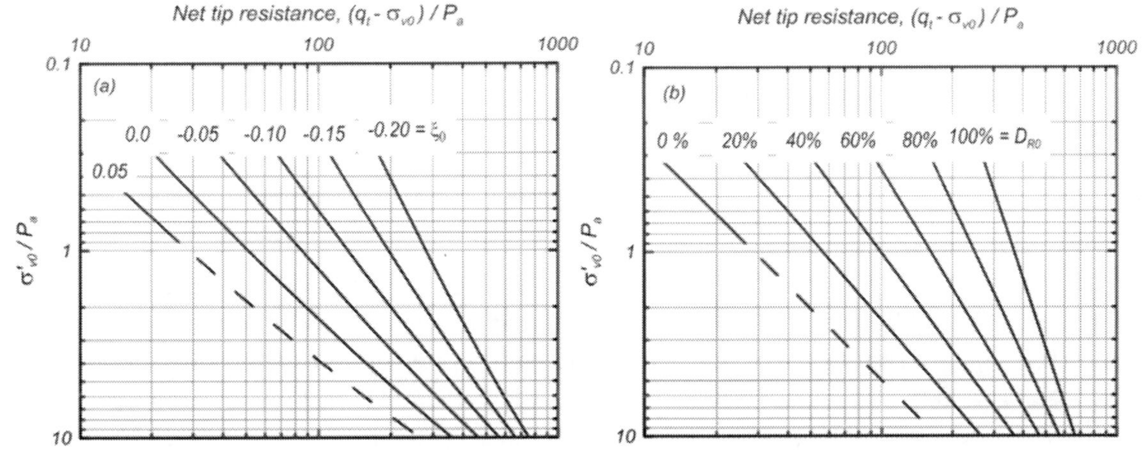

Figure 2. Relationship between net cone penetration resistance, initial vertical effective stress, and (a) initial state parameter, and (b) initial relative density

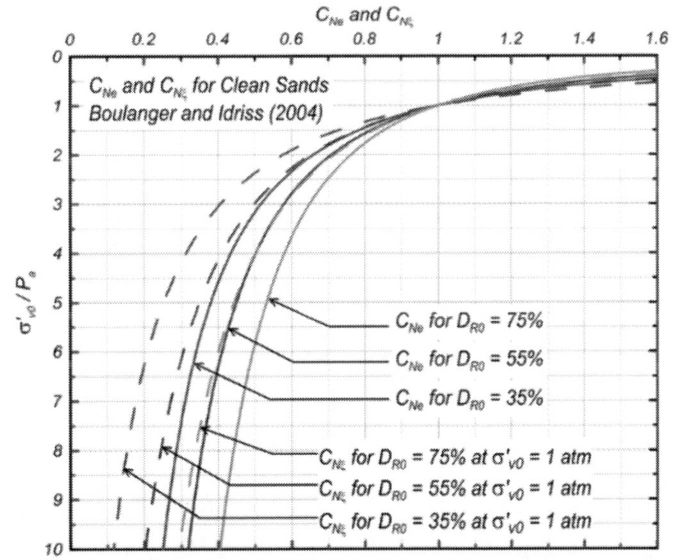

Figure 3. C_{Ne} and $C_{N\xi}$ relationships as recommended by Boulanger and Idriss (2004)

The relationships in Figure 2b show $q_{t,net}$ versus σ'_{v0} for a sand at six different D_{R0} of 0%, 20%, 40%, 60%, 80%, and 100% based on $C_{dq} = 1.3$. The slopes of these $q_{t,net}$ versus σ'_{v0} graphs, in a log-log scale, correspond to the stress exponent m_e. Similar to the trends for constant ξ_0, the stress exponent m_e decreases with increasing D_{R0} for sands (i.e., steeper lines in Figure 2b). The corresponding dependence of C_{Ne} on D_{R0} is shown in Figure 3 for D_{R0} of 35%, 55%, and 75%.

Relationships between Peak Friction Angle, Density, and Stress in Sands

Several frameworks have been proposed for estimating the drained shear strength of coarse grained soils (e.g., Been and Jefferies 1985, Bolton 1986, Collins et al. 1992). Generally these relationships compute the peak friction angle (ϕ'_{peak}) as a function of the critical state friction angle (ϕ'_{cs}), e_0, p'_0, and material-specific parameters. As previously discussed, p'_0 and e_0 can also be described in a single term, ξ_0. In terms of ϕ'_{cs} and ξ_0, Collins et al. (1992) suggested the following fit to the data presented by Been and Jefferies (1985):

$$\phi'_{peak} = \phi'_{cs} + A\left(e^{-\xi_0} - 1\right) \geq \phi'_{cs} \tag{11}$$

where A is a soil type parameter in the range of 0.6 to 0.95 (in radians). Figure 4a illustrates the relationship between peak friction angle and σ'_{v0} as a function of ξ_0 for $K_0 = 0.5$, $A = 0.8$, and $\phi'_{cs} = 32$ degrees, which are representative of the sands evaluated by Collins et al. (1992).

Similar to Been and Jefferies (1985) and Collins et al. (1992), Bolton (1986) identified the peak friction angle of sands is dependent on both void ratio and stress. The relationship presented by Bolton (1986) for triaxial compression is given in terms of critical state friction angle, mean effective stress at failure, and initial void ratio (or relative density):

$$\phi'_{peak} = \phi'_{cs} + 3\left[D_{R0}\left(Q - \ln(p'_f)\right) - R\,\right] \geq \phi'_{cs} \tag{12}$$

where R is typically taken as 1.0, Q is generally assumed equal to 10 for quartz and feldspar sands, and p'_f is the mean effective stress at failure in kPa (herein assumed to be twice the initial vertical effective stress, per Kulhawy and Mayne 1990). Bolton (1987) later recommended limiting the use of Eq. 12 to p'_f greater than 150 kPa when data at lower p'_f values is unavailable. Based on the recommendation by Bolton (1987), Eq. 12 can be rewritten as:

$$\phi'_{peak} = \phi'_{cs} + 3\left[D_{R0}\left(Q - \ln\left(\max(p'_f, 150\text{ kPa})\right)\right) - R\,\right] \geq \phi'_{cs} \tag{13}$$

Figure 4b illustrates the Bolton (1986, 1987) relationship (Eq. 13) between ϕ'_{peak}, σ'_{v0}, and D_{R0} for $Q = 10$, $R = 1$, and $\phi'_{cs} = 32$ degrees. Note, that for a constant D_{R0}, ϕ'_{peak} is generally not constant and changes with stress.

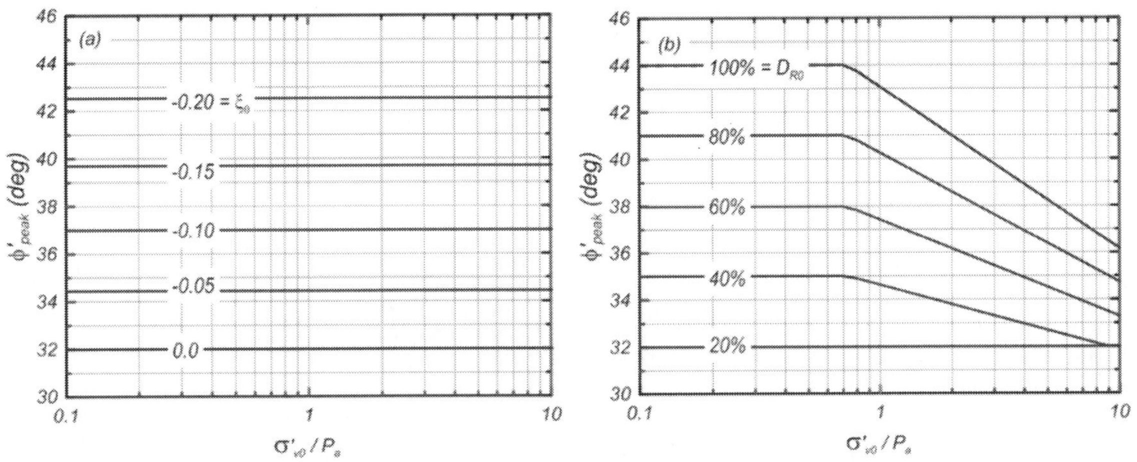

Figure 4. Relationships between ϕ'_{peak} and σ'_{v0} with the relationships by (a) Collins et al. (1992) after Been and Jefferies (1985), and (b) Bolton (1986, 1987)

DATA SOURCES

The relationship between cone penetration resistance and peak friction angle is examined with data from two sources: (1) a database of experimental in situ CPT data and laboratory tests on undisturbed sand samples, and (2) calibration chamber CPT data. The experimental in situ CPT data and laboratory tests on undisturbed sand samples were taken from a database developed by Mayne (2006). The calibration chamber CPT data was taken from Jefferies and Been (2006) and reanalyzed by Maki (2012) and Maki et al. (2014a).

In Situ Tests

The database of experimental in situ CPT data and laboratory tests on undisturbed sand samples compiled by Mayne (2006) was used for the present study. Mayne's database included tests on 15 sands from China, Japan, Canada, Norway, and Italy (Table 1). As shown in Table 1, 10 sands were natural deposits, 3 sands were hydraulic fills, and 2 were tailings. Fourteen of the 15 sands were obtained with frozen sampling techniques; the remaining sample was obtained with a Mazier tube. 11 of the 15 sands are clean sands (less than 5% fines) and the other four sands have fines contents between 8 and 15%. The particle shapes varied from subrounded and subangular to angular (Mayne 2006).

The sands in the Mayne (2006) dataset typically had vertical effective stresses near one atmosphere. Fourteen of the 15 in situ tests had σ'_{v0} values between 0.42 atm and 1.8 atm, with the remaining test at $\sigma'_{v0} = 5.16$ atm for Mildred Lake sand (10% fines). Thus, the Mayne (2006) dataset for clean sands is relatively constrained for vertical effective stresses between 0.5 to 2.0 atmospheres, but is not well-constrained for vertical effective stresses greater than about two atmospheres.

The peak friction angles of the Massey, Kidd, LL Dam, Mildred Lake, and J-pit sands were revised from the values listed by Mayne (2006). The revised values were selected based on the authors' review of the available laboratory data for the sands and the reported friction angles in the original references (Wride and Robertson 1997a,b,c). The peak friction angles for the Massey, Kidd, and Mildred Lake sands decreased by 0.1°, 0.2°, and 0.6°, respectively. The peak friction angles of the LL Dam and J-pit sands were increased by 0.3° and 2.0°, respectively.

Table 1. Characteristics of Undisturbed Sand Samples (adapted from Mayne 2006)

Sand (Location)	Deposit Type	Sampling Method	Fines Content	D_{R0}	σ'_{v0} (atm)	q_t (atm)	ϕ'
Clean Sands (11 of 15 Sands)							
Kowloon (China)	Hydraulic Fill	Mazier Tube	1%	46.5%	1.80	100.0	38.1°
Yodo (Japan)	Natural Alluvial	Frozen	1.9%	60.2%	1.02	177.7	42.4°
Yodo (Japan)	Natural Alluvial	Frozen	0.3%	51.9%	1.23	112.7	38.4°
Yodo (Japan)	Natural Alluvial	Frozen	2.1%	37%	1.43	126.4	39.1°
Natori (Japan)	Natural Alluvial	Frozen	0.2%	77.2%	0.87	193.4	40.9°
Tone (Japan)	Natural Alluvial	Frozen	3.8%	69.1%	0.84	134.5	41.7°
Edo (Japan)	Natural Alluvial	Frozen	0.4%	44.3%	0.51	79.0	39.7°
Massey (Canada)	Natural Alluvial	Frozen	<5%	32.5%	1.2	58.5	36.6° [a]
Kidd (Canada)	Natural Alluvial	Frozen	<5%	29.8%	1.6	86.4	37.1° [a]
Holmen (Norway)	Natural Alluvial	Tube/frozen	2%	30.5%	1.1	29.1	33.2°
Gioia Tauro (Italy)	Natural Coarse Sand	Frozen	0.7%	42%	0.42	113.0	41.5°
Sand with ≥ 5% Fines (4 of 15 Sands)							
LL Dam (Canada)	Tailings	Frozen	8%	40.3%	1.0	39.4	39.4° [a]
Mildred Lake (Canada)	Hydraulic Fill	Frozen	10%	43.6%	5.16	167.6	39.0° [a]
Highmont Dam (Canada)	Tailings	Frozen	10%	37.4%	1.38	51.6	41.5°
J-pit (Canada)	Hydraulic Fill	Frozen	15%	42.7%	0.55	15.1	34.7° [a]

[a]Value reported by Mayne (2006) revised based on original reference.

Calibration Chamber Tests

Calibration chamber cone penetration test data for Ticino, Ottawa, Monterey, and Hokksund sands (data from Baldi et al. 1982; Been et al. 1986; Harman 1976; Parkin et al. 1980; Baldi et al. 1986; Tringale 1983; Huntsman 1985), as compiled by Jefferies and Been (2006), were obtained from Golder Associates (2011). The calibration chamber tests were performed under initial vertical effective stresses of 40 to 720 kPa and with K_0 of 0.3 to 1.4. The diameters of the calibration chambers ranged from 0.76 m to 1.5 m, with chamber-to-cone-diameter ratios of 21 to 48.

Relevant soil properties of the evaluated calibration chamber sands are provided in Table 2 below. The four sands are similar in that they are clean, uniformly graded sands with median grain sizes ranging from 0.37 to 0.53 mm, coefficients of uniformity ranging from 1.5 to 2.0, minimum void ratios of 0.49 to 0.60, and maximum void ratios of 0.79 to 0.91. Critical state parameters, including critical state friction angle and the slope and intercept of the critical state line, are also listed in Table 2.

Table 2. Relevant Properties of Calibration Chamber Test Sands

Property	Ticino sand	Ottawa sand	Hokksund sand	Monterey sand
Median grain size, D_{50}[a]: mm	0.53	0.53	0.39	0.37
Maximum void ratio, e_{max}[a]	0.89	0.79	0.91	0.82
Minimum void ratio, e_{min}[a]	0.6	0.49	0.55	0.54
Critical state friction angle, ϕ'_{cs}[a]	30.9°	28.4°	32°	32°
Slope of CSL in e – $\log_{10}(p')$ space for semi-log idealization, λ_{10}[a]	0.056	0.028	0.054	0.029
Reference void ratio on CSL, defined at p' = 1 kPa, Γ_1[a]	0.986	0.754	0.934	0.878
Bolton (1986) Parameter Q for R=1[b]	10.5	10.5	11.0	10.4
Angularity[c]	subangular to angular	rounded to well rounded	subangular to angular	subrounded

[a]Data from Jefferies and Been (2006)
[b]Interpreted from data compiled by Jefferies and Been (2006)
[c]Data from Kulhawy and Mayne (1990)

The Bolton (1986) parameter Q (assuming R=1) was estimated for each of the four calibration chamber sands from isotropically-consolidated, drained triaxial compression tests, as tabulated by Jefferies and Been (2006). Sufficient laboratory data on the calibration chamber sands was available to preclude the limit recommended by Bolton (1987). The Q values for each sand were estimated by fitting Eq. 12 to the data for each sand (Table 2). The selected Q values are slightly higher than the Q = 10 recommended by Bolton (1986) for quartz and feldspathic sands, but the resulting peak friction angles are within the range of data presented by Bolton (1986).

Calibration chamber corrections (CCCs) are applied to calibration chamber test q_t values to convert them to free-field conditions (e.g., Houlsby and Yu 1990, Schnaid and Houlsby 1991, Parkin and Lunne 1982, Been et al. 1986, Salgado et al. 1998). The corrections are necessary for evaluating the free-field q_t values due to effects of the artificial boundary conditions imposed by the calibration chamber.

CCCs by Been et al. (1986) and Salgado et al. (1998) are evaluated in this paper. The CCCs for Been et al.'s (1986) method were taken as the values listed in the database obtained from Golder Associates (in reference to Jefferies and Been 2006). The CCCs by Salgado et al. (1998) were published as plots for given D_{R0} values and boundary conditions over a range of stresses and diameter ratios. These plots had maximum and minimum D_{R0} values of 30% and 90% respectively; any D_{R0} values outside of these limits involved extrapolation.

The calibration chamber cone penetration test data were additionally normalized to a reference K_0 using the lateral earth pressure correction factor (C_{NH}) proposed by Salgado et al. (1997b). Following the recommendations of Salgado et al. (1997b) and Maki et al. (2014a), the C_{NH} factor was calculated as

$$C_{NH} = \sqrt{\frac{K_{0,ref}}{K_0}} \tag{14}$$

where $K_{0,ref}$ is the reference K_0 value. A C_{NH} factor based on $K_{0,ref}$ of 0.5 was applied to all tests in this paper.

RESULTS

Relationship between $Q_{1\xi}$ and Initial State Parameter

$Q_{1\xi}$ values were calculated for the in situ and calibration chamber cone penetration tests using Eqs. 2-10. The relationship between $Q_{1\xi}$ and ξ_0 for the in situ CPTs was compared to the $Q_{1\xi}$ and ξ_0 relationships developed by Maki et al. (2014a) for clean sands in calibration chamber tests with the Salgado et al. (1998) CCCs (Figure 5a) and Been et al. (1986) CCCs (Figure 5b). Due to limitations in the available in situ sand data, only 6 of the 15 in situ CPT data points are shown.

Five of the in situ CPT data points are in good agreement with the calibration chamber data from Maki et al. (2014a). The sixth point is the J-pit sand, which is a hydraulically filled tailings material with the highest fines content in the dataset (15% fines). This sand plotted significantly below the clean sand calibration chamber cone penetration test data for both CCCs. The J-pit sand point is believed to plot lower than the others because of higher compressibility due to its higher fines content relative to the other sands.

This comparison emphasizes the $Q_{1\xi}$-ξ_0 relationships developed from calibration chamber testing are generally supported by the in situ sand data, albeit limited to a narrow range of ξ_0. Further, the in situ $Q_{1\xi}$ versus ξ_0 data appear to support the observation of Maki et al. (2014a) that there is a unique relationship between the two parameters with the form shown in Eq. 2. Lastly, this comparison illustrates the potential differences in the $Q_{1\xi}$-ξ_0 relationship between different sands, which can be attributed to differences in the mineralogy, gradation, and particle characteristics of the sands.

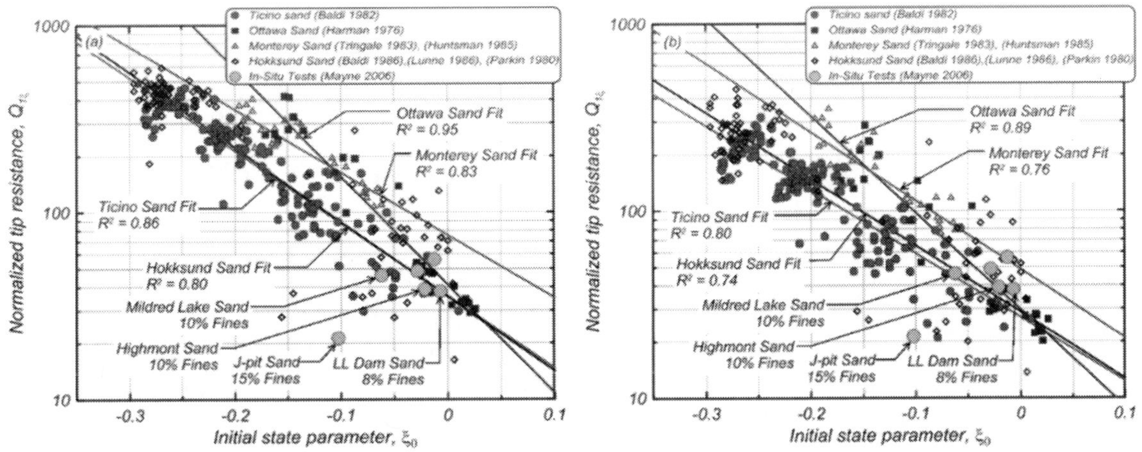

Figure 5. Relationships between $Q_{1\xi}$ and ξ_0 for in situ data and calibration chamber data with (a) Salgado et al. (1998) CCCs and (b) Been et al. (1986) CCCs

State-Based Relationships with Peak Friction Angle

The relationship between $Q_{1\xi}$ and ϕ'_{peak} was plotted for the in situ data from Mayne (2006) and the calibration chamber data presented by Maki et al. (2014a) with CCCs by Salgado et al. (1998) (Figure 6a) and CCCs by Been et al. (1986) (Figure 6b). The ϕ'_{peak} values for the calibration chamber data were estimated from the Bolton (1986) equation (Eq. 12) with each sand's ϕ'_{cs}, Q, and R values. For comparison, two lines representing the combination of the Idriss and Boulanger (2008) approach for estimating q_t and the Bolton (1986) approach for estimating ϕ'_{peak} were also added using $C_{dq} = 1.3$, $Q = 10$, and $R = 1$ for ϕ'_{cs} of 28° and 35°.

The calibration chamber data with the Been et al. (1986) CCCs plot to the left of the data with the Salgado et al. (1998) CCCs. The trend is a result of the Salgado et al. (1998) CCCs predicting a higher $Q_{1\xi}$ than Been et al. (1986) for a given ξ_0 or ϕ'_{peak}. The 11 in situ clean sand data points plot above or along the upper bounds of the calibration chamber data with the Salgado et al. (1998) and Been et al. (1986) CCCs, respectively. Differences between the in situ and calibration chamber data can be attributed to differences in the mineralogy, gradation, particle characteristics of the sands, and CCCs.

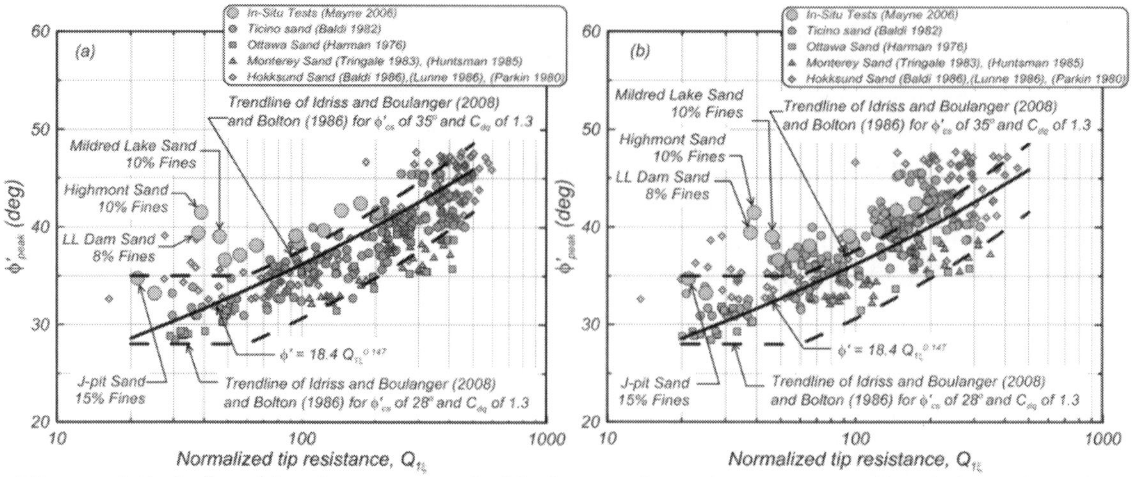

Figure 6. Relationships between peak friction angle and state-normalized tip resistance for in situ data and calibration chamber data with (a) Salgado et al. (1998) CCCs and (b) Been et al. (1986) CCCs

Interestingly the in situ clean sands and calibration chamber clean sands are approximately parallel for both sets of CCCs. This is expected based on the relationship presented by Been and Jefferies (1985) between ξ_0 and ϕ'_{peak}, which were shifted for different sands based on differences in their respective ϕ'_{cs} values. Although the ϕ'_{cs} values are generally not known for the in situ sands, consideration that the ξ_0 values of the Canadian sands were close to zero suggests the ϕ'_{peak} values may closely resemble each sand's ϕ'_{cs} values, which would suggest ϕ'_{cs} values around 33° to 35°.

The four in situ sands with 8-15% fines plot above the in situ and calibration chamber clean sands for both CCCs. These four sands were generally deposited under different conditions than the in situ and calibration chamber clean sand data: either from tailings basins (Mildred Lake and J-pit) or hydraulically placed fills (LL Dam and Highmont Dam). Additionally, the LL Dam and Highmont Dam sands have angular particles and contain appreciable amounts of mica, feldspar, and illite (Wride and Robertson 1997c), which would lead to increased friction angles and reduced $Q_{1\xi}$ values as compared to the other less compressible sands evaluated in this study (i.e., Jaeger 2012, Jaeger et al. 2014, Mayne 2014).

A representative fit was developed based on the in situ data, calibration chamber data with both CCCs, and a semi-empirical trend combining the Idriss and Boulanger (2008) approach for estimating q_t with $C_{dq} = 1.3$ and the Bolton (1986) approach for estimating ϕ'_{peak} with Q = 10, R = 1, and $\phi'_{cs} = 33°$. The selected fit can be approximated with the following regression:

$$\phi'_{peak} = 18.4 Q_{1\xi}^{0.147} \tag{15}$$

The above regression falls below the in situ sand data points, but provides a balanced fit to the entire calibration chamber dataset considering both the Salgado et al. (1998) and the Been et al. (1986) CCCs.

The relationship between ϕ'_{peak} and $Q_{1\xi}$ (or ξ_0) shown in Figure 6 includes several sands with varying grain sizes, shapes, and mineralogy, leading to a range of critical state friction angles (28° to 32° for the calibration chamber data, unknown for the in situ data). The proposed trendline would be expected to translate vertically in Figure 6 with changing critical state friction angles based on the relationships presented by Been and Jefferies (1985), Collins et al. (1992), and Bolton (1986) (Eqs. 11 and 12).

The calibration chamber data and proposed relationship are replotted in Figure 7 as (ϕ'_{peak}-ϕ'_{cs}) versus $Q_{1\xi}$. As noted above, the selected $Q_{1\xi}$-ϕ'_{peak} relationship proposed in this paper was derived based on a ϕ'_{cs} of 33° and can be rewritten as:

$$\phi'_{peak} = 18.4Q_{1\xi}^{0.147} + (\phi'_{cs} - 33°) \tag{16}$$

Comparison of this regression to the calibration chamber data in terms of (ϕ'_{peak}-ϕ'_{cs}) shows the selected relationship falls in the middle of the data for the Salgado et al. (1998) CCCs, but plots within about the lower-third of the data with the Been et al. (1986) CCCs. The trend of lower $Q_{1\xi}$ values at the same ξ_0 (or equivalently ϕ'_{peak}-ϕ'_{cs} per Eqs. 2 and 11) for the Been et al. (1986) CCCs compared to the Salgado et al. (1998) CCCs was also observed in Figures 5 and 6. This indicates that the proposed relationship in Eq. 15 can be rewritten as in Eq. 16 for sands with critical state friction angles different from 33°.

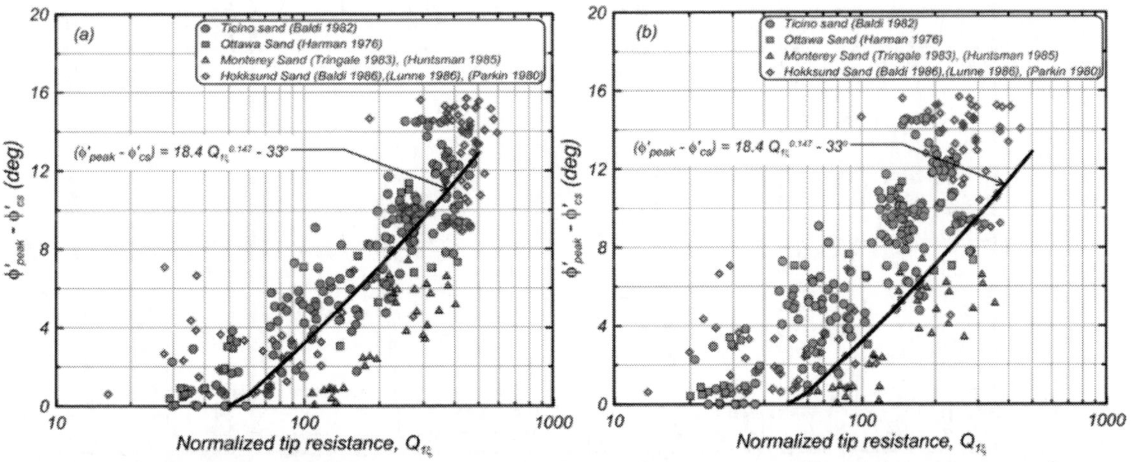

Figure 7. Relationships between (ϕ'_{peak}-ϕ'_{cs}) and state-normalized tip resistance for calibration chamber data with (a) Salgado et al. (1998) CCCs and (b) Been et al. (1986) CCCs

Variation of Peak Friction Angle with Stress and Relative Density

The variation of the proposed state-based method to with σ'_{v0} overburden stress and D_{R0} (or Q_{1e}) was examined for three hypothetical CPT soundings (Figure 8a) with constant Q_{1e} values of 60, 100, and 200 (Figures 8b,c,d). Based on Eq. 7, these hypothetical CPT soundings would pertain to D_{R0} values of about 20%, 40%, and 70%, respectively. Peak friction angles estimated with the proposed method are compared to results from two commonly used relationships for CPT data (Kulhawy and Mayne 1990; Robertson 2010) and the Bolton (1986) relationship (Eqs. 13). Peak friction angles calculated with the Robertson (2010) relationship were performed assuming the sleeve friction measurement (f_s) was 0.2% of the cone tip resistance (i.e., $f_s = 0.02q_t$), corresponding to the clean sand region of the Robertson (2009) classification chart. The Bolton (1986,1987) approach assumed $Q = 10$, $R = 1$, and $K_0 = 0.5$. A ϕ'_{cs} value of 33° is assumed for the proposed method, as well as for the Robertson (2010), and the Bolton (1986,1987) methods.

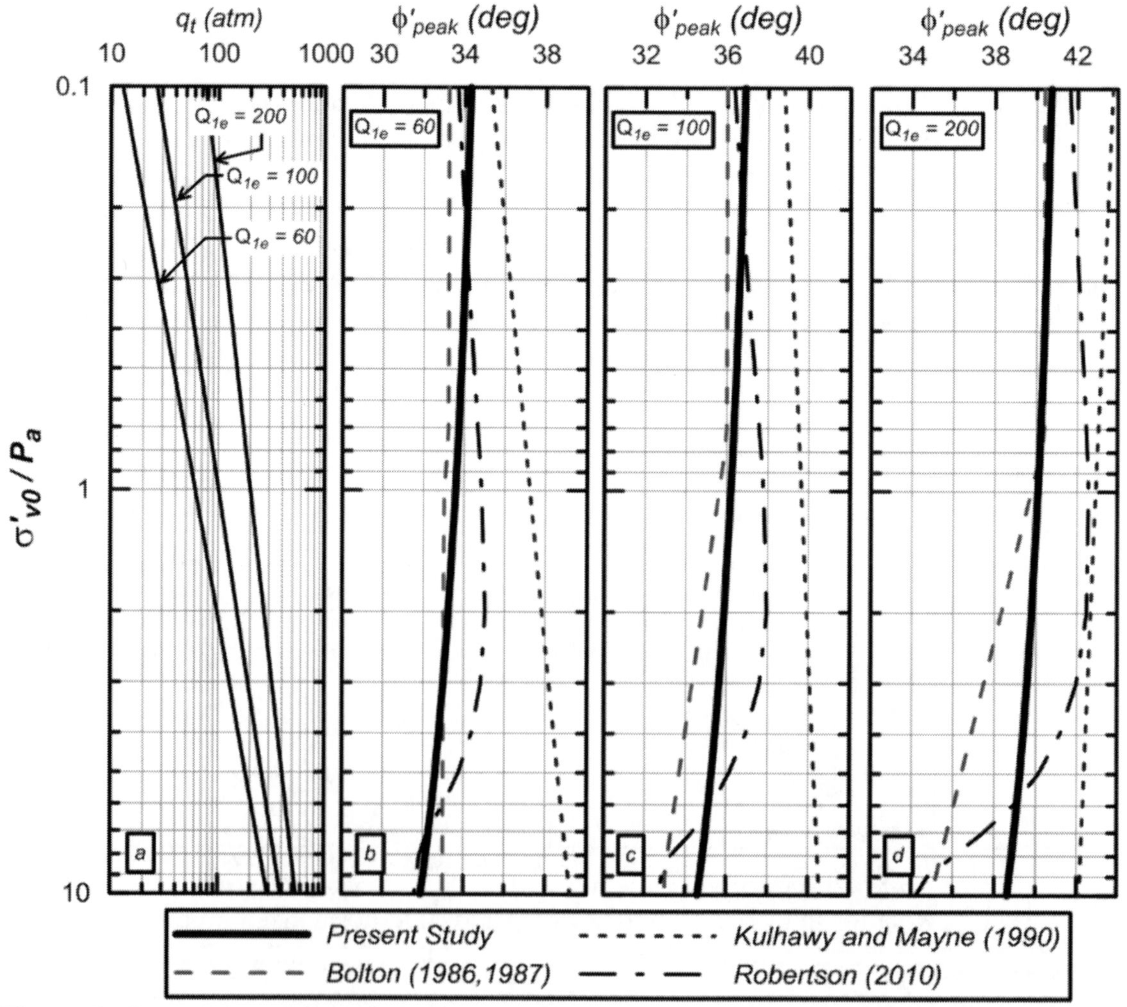

Figure 8. Comparison of relationships between ϕ'_{peak} and σ'_{v0} for three hypothetical CPT soundings (shown in subplot a) with (b) $Q_{1e} = 60$, (c) $Q_{1e} = 100$, and (d) $Q_{1e} = 200$

As shown in Figures 8b-d, the proposed method predicts peak friction angles that decrease with increasing σ'_{v0}, consistent with the observations and relationships by Been and Jefferies (1985) and Bolton (1986,1987). For each of the three profiles, ϕ'_{peak} was found to decrease by up to 2.5° for the loose CPT (Q_{1e} = 60) and up to 2° for the dense CPT (Q_{1e} = 200). The reduction in ϕ'_{peak} with increasing σ'_{v0} predicted by the proposed method occurred because (1) ξ_0 increases with stress for a fixed e_0 or D_{R0}, and (2) ϕ'_{peak} decreases as ξ_0 increases.

The Bolton (1986,1987) relationship exhibited trends similar to the proposed method for all three Q_{1e} profiles: for a given D_{R0}, ϕ'_{peak} decreased as σ'_{v0} increased. The ϕ'_{peak} values are approximately equal between the Bolton (1986,1987) relationship and the proposed method at σ'_{v0} less than 10 atm for Q_{1e} = 60 and at σ'_{v0} below about one atmosphere for Q_{1e} = 100 and 200. For σ'_{v0} above one atmosphere, the Bolton (1986,1987) relationship predicted ϕ'_{peak} values about 3.5° lower than the proposed relationship.

The Kulhawy and Mayne (1990) relationship predicted ϕ'_{peak} values about 1° to 7° higher than the proposed relationship. Additionally, the Kulhawy and Mayne (1990) correlation exhibited a reverse trend (i.e., ϕ'_{peak} increased as σ'_{v0} increased) for Q_{1e} = 60 and Q_{1e} = 100, but exhibited a more typical trend (i.e., ϕ'_{peak} decreased as σ'_{v0} increased) for Q_{1e} = 200. The reverse trend exhibited is caused by differences in the stress normalization exponents between the Kulhawy and Mayne (1990) approach (assumes m_e = 0.5) and the three hypothetical CPTs (m_e values specified per Eqs. 6 and 7).

The Robertson (2010) relationship predicted ϕ'_{peak} values within 2° to 3° of the proposed method for all three hypothetical CPTs. Interestingly, the Robertson (2010) relationship exhibited parabolic trends of ϕ'_{peak} versus σ'_{v0} , where ϕ'_{peak} increased up to σ'_{v0} of 1-2 atm and decreased for higher σ'_{v0}. The parabolic trend exhibited by the Robertson (2010) relationship can be attributed to differences in the stress normalization exponents between the Robertson (2009, 2010) approach (m varies with σ'_{v0} and soil behavior type index I_c) and the three hypothetical CPTs (m_e values specified per Eqs. 6 and 7). However despite the difference in trends, the overall values between Robertson (2010) and the proposed approach were largely in agreement.

CONCLUSIONS

A relationship between peak friction angle and state-based overburden-normalized q_t was developed based on in situ test results, laboratory test results, calibration chamber cone penetration test results, and a numerical model of cone penetration resistance. A state-based overburden normalization approach was adopted because it provides both a rational basis for interpreting the peak friction angle in sandy soils from q_t, and a critical state-based method for extrapolating these relationships to conditions outside the range currently covered by available peak friction angle data. The results indicated state-based overburden-normalized CPT tip resistance and peak friction angle are strongly correlated, as expected based on critical state soil mechanics and published trends from laboratory testing of sands. The proposed relationship also compared favorably with well-accepted empirical relationships between CPT tip resistance and peak friction angle.

ACKNOWLEDGEMENTS

The authors are grateful for the data made available by Mayne (2006) and the calibration chamber data provided by Jefferies and Been (2006) through Golder Associates. The authors would also like to thank Professor Jason DeJong, Dr. Joseph Weber, and Messrs. Michael Paster, Stephen Verigin, and Justin Dominguez for their valuable comments and insights, which improved the paper. The opinions expressed in this paper are solely those of the authors, and do not necessarily represent the views of GEI Consultants, Inc. or the California Department of Water Resources, Division of Safety of Dams.

REFERENCES

Baldi, G., Bellotti, R., Ghionna, V., Jamiolkowski, M. and Pasqualini, E. (1982). "Design parameters for sand from CPT." In *Proc. ESOPT 2,* Amsterdam.

Baldi, G., Bellotti, R., Ghionna, N., Jamiolkowski, M. and Pasqualini, E. (1986). "Interpretation of CPTs and CPTU's, 2nd Part." In *Proc 4th International Geotechnical Seminar,* Nanyang Technological Institute, Singapore, 143–156.

Been, K., and Jefferies, M. G. (1985). "A state parameter for sands." *Géotechnique,* 35(2): 99–112.

Been, K., Crooks, J. H. A., Becker, D. E., and Jefferies, M. G. (1986). The cone penetration test in sands: part I, state parameter interpretation. *Géotechnique, 36*(2): 239-249.

Bolton, M. D. (1986). "The strength and dilatancy of sands." *Géotechnique*, 36(1): 65-78.

Bolton, M. D. (1987). "Discussion on The strength and dilatancy of sands." *Géotechnique*, 37(2): 219-226.

Boulanger, R. W. (2003). "High overburden stress effects in liquefaction analyses." *J. Geotechnical and Geoenvironmental Eng.,* ASCE 129(12), 1071–082.

Boulanger, R. W. and Idriss, I. M. (2004). "State normalization of penetration resistance and the effect of overburden stress on liquefaction resistance." *Proc 11th SDEE and 3rd ICEGE*, Berkeley, CA.: 484-491.

Collins, I. F., Pender, M. J., and Yan, W. (1992). "Cavity expansion in sands under drained loading conditions." *Int. J. Numer. Anal. Meth. Geomech.*, 16: 3–23.

Golder Associates (2011). CPT database, http://www.golder.com/cn/en/modules.php?name=Publication&sp_id=200&page_id=212 October 2011

Harman, D.E. (1976). "A statistical study of static cone bearing capacity, vertical effective stress, and relative density of dry and saturated fine sands in a large triaxial testing chamber." *M.Sc. Thesis*, University of Florida.

Houlsby, G. T. and Yu, H. S. (1990). "Finite element analysis of the cone pressuremeter test." In Pressuremeters, London: Thomas Telford, pp. 221-229.

Huntsman, S.R. (1985). "Determination of in situ lateral pressure of cohesionless soils by static cone penetrometer." *Ph.D. Thesis*, University of California at Berkeley.

Idriss, I. M. and Boulanger, R. W. (2008). "Soil liquefaction during earthquakes." Monograph MNO-12, Earthquake Engineering Research Institute, Oakland, CA, 261 pp.

Jaeger, R. A. (2012). "Numerical and Experimental Study on Cone Penetration in Sands and Intermediate Soils." *Ph.D. Thesis*, University of California at Davis.

Jaeger, R. A., DeJong, J. T., Boulanger, R. W., and Maki, I. P. (2014). "Effects of state parameter, fines content, and overburden stress on CPT resistance in silty sands." In *Proc. 3rd International Symposium on Cone Penetration Testing*, CPT'14.

Jefferies, M. G. and Been, K. (2006). "Soil Liquefaction – A Critical State Approach", Taylor and Francis Group.

Kulhawy, F. H. and Mayne, P. W. (1990). "Manual on estimating soil properties for foundation design." Report EL-6800. Electric Power Research Institute, Palo Alto, CA.

Liao, S. C. and Whitman, R. V. (1986). "Overburden correction factors for SPT in sand." *J. Geotech. Engrg.*, 112(3): 373–377.

Maki, I. P. (2012). "State normalization of cone penetration resistance." *M.S. Thesis,* University of California, Davis.

Maki, I. P., Boulanger, R. W., DeJong, J. T., and Jaeger, R. A. (2014a). "State-Based Overburden Normalization of Cone Penetration Resistance in Clean Sand." *J. Geotech. Geoenviron. Eng.*, 140(2).

Maki, I. P., Boulanger, R. W., DeJong, J. T., and Jaeger, R. A. (2014b). "Overburden normalization of CPT data in sands to clays." In *Proceedings 3rd International Symposium on Cone Penetration Testing*, CPT'14.

Mayne, P. W. (2006). The Second James K. Mitchell Lecture Undisturbed sand strength from seismic cone tests. *Geomechanics and Geoengineering: An International Journal, 1*(4), 239-257.

Mayne, P. W. (2014). "Interpretation of geotechnical parameters from seismic piezocone tests." In *Proc. 3rd Int. Symposium on Cone Penetration Testing*, CPT'14.

Parkin, A., Holden, J., Aamot, K., Last, N., and Lunne, T. (1980). "Laboratory investigations of CPTs in sand." *NGI Report S2108–9,* 9 October 1980.

Parkin, A. and Lunne, T. (1982). "Boundary effects in the laboratory calibration of a cone penetrometer in sand." *Proc. 2nd European Symposium on Penetration Testing,* Amsterdam, Vol. 2, 761–768

Robertson, P. K. (2009). "Interpretation of cone penetration tests – A unified approach." *Can. Geotech. J.*, 46(11), 1337–1355.

Robertson, P. K. (2010). "Estimating in-situ state parameter and friction angle in sandy soils from CPT." In *Proc. 2nd Int. Symposium on Cone Penetration Testing*, CPT'10.

Salgado, R., Mitchell, J. K., and Jamiolkowski, M. (1997a). "Cavity expansion and penetration resistance in sand." *J. Geotech. Geoenviron. Eng.*, 123(4), 344–354.

Salgado, R., Boulanger, R. W., and Mitchell, J. K. (1997b). "Lateral stress effects on CPT liquefaction resistance correlations." *J. Geotech. Geoenviron. Eng.*, 123(8), 726–735.

Salgado, R., Mitchell, J. K., and Jamiolkowski, M. (1998). "Calibration chamber size effects on penetration resistance in sand." *J. Geotech. Geoenviron. Eng.*, 124(9), 878–888.

Schnaid, F. and Houlsby, G. T. (1991). "An assessment of chamber size effects in the calibration of in-situ tests in sand." *Géotechnique,* 41(3): 437-445.

Tringale, P.T. (1983). "Soil identification in situ using an acoustic cone penetrometer." *Ph.D. Thesis*, University of California at Berkeley.

Wride, C. E. and Robertson, P. K. (1997a). "CANLEX: Phases I and III data review report (Mildred Lake and J-Pit Sites)."

Wride, C. E. and Robertson, P. K. (1997b). "CANLEX: Phase II data review report (Massey and Kidd Sites, Fraser River Delta)."

Wride, C. E. and Robertson, P. K. (1997c). "CANLEX: Phase IV data review report (LL Dam and Highmont Dam Sites, Highland Valley Copper Mine)."

Yu, H.S. (2004). "In situ soil testing: from mechanics to interpretation." *J.K. Mitchell Lecture,* In *Proc. ISC-2 on Geotechnical and Geophysical Site Characterization.* Millpress, Porto, 3-38.

DECADES OF FOUNDATION DETERIORATION THREATENING PUBLIC SAFETY

George L. Sills, PE[1]
Leslie F. Harder, Jr., PhD, PE, GE[2]

ABSTRACT

Sand boils which pipe and move materials during a flood represent an extremely serious condition. During the last 50 years, very little has changed in the methods we use to fight or evaluate progressive piping associated with sand boils during high-water events.

The mechanics of this problem appear to be related to the removal and degradation of the soil around the pipe. The seepage gradient in the soil increases as the piping progresses from the protected side to the unprotected side of the levee and the flow to the pipe then increases. The rate of erosion and soil transport also then increase unless the sand boil is ringed and the gradient thereby reduced, or until the flood stage recedes. During subsequent high-water events, additional boils may occur at these same locations and cause the pipes to lengthen and/or widen. This degradation may lead to earlier and more rapid development of the sand boil during future floods.

Very little has changed over the last 50 years in modeling this damage because we have no way to measure or mathematically model the foundation damage caused by sand boils. Most designers ignore the damage when computing a design solution and designers today commonly model the foundation as though it is completely intact and undamaged by past piping. Research has been performed by J. H. Schmertmann, L. Nagy, J. Rice, and others concerning the dangers of ignoring these pipes while designing seepage remediation measures. Two case studies of existing levee piping problems that were remediated with standard methods are discussed. At one of these locations, the levee failed during a high-water event despite being remediated shortly before the event. A safer design approach that will improve public safety is also proposed and discussed.

INTRODUCTION

General

There are several potential modes of failure that a levee may be subject to during a flood event. Other than overtopping, probably the most notable is foundation piping leading to foundation and slope instability. The first evidence of this mode of failure is often a small sand boil landward of the levee. Sand boils are typically "*controlled*" by temporarily ringing the boil with sand bags in an effort to reduce the seepage gradient, thereby reducing the flow and seepage velocity. Following the event, people often think they have successfully prevented a failure with the sand bag ring and that "*all is well.*"

[1] Manager of George Sills Geotechnical Engineering Consultant, LLC, Vicksburg, MS
[2] Senior Professional Associate, HDR Engineering Inc., Folsom, CA

Experience shows, however, that while a failure may have been averted for this event, the foundation has actually been damaged and a failure may occur at the same location during a subsequent flood at a lower river stage. The occurrence of a sand boil during a flood indicates a very serious condition and that the foundation may have been damaged enough that levee instability may result in the next major flood event.

Sand Boil Development

Sand boils may start as an area of seepage and soft ground landward of the levee, or be identified as an area of small pin boils. As the river stage increases, these pin boils may coalesce and form a sand boil, which starts to carry fines. Fines are removed with the flowing water and may not be readily observable in the immediate area around the sand boil. Figures 1 and 2 depict a sand boil piping initially fine material and then forming a cone of coarser sandy soil around the boil as piping progressed.

Figure 1. Photograph of initial formation of a sand boil.

Ringing the boil with sand bags is the standard course of action to lower the seepage gradient and stop the loss of soil particles from the foundation (see Figure 3).

Figure 2. Photograph of piped sandy material forming a ring or cone around boil.

Figure 3. Photograph of sand boil ringed with sand bags.

Underseepage piping problems are commonly associated with the presence of a semi-pervious top stratum or blanket of soil overlying a more-pervious aquifer. A top stratum is a surface or near-surface deposit of soil with a lower hydraulic conductivity than the underlying aquifer that tends to confine seepage in the underlying soil layers and result in elevated seepage pressures landward of the levee. What is not well known is why sand boils start where they do. One theory is that there was a defect in the top stratum that caused the seepage to localize. The defect may have been a thin spot in the top stratum or desiccation cracks in the surface blanket or holes formed by roots, animal burrows, or manmade activities. The surface defect may only need to have been about 2 or 3 feet deep to provide a preferred path for seepage through the top stratum and to then form a sand boil. A defect that extends only 20 to 40 percent of the thickness of the top strata can cause a concentration of seepage at the defect and initiate the loss of fines.

Another theory is that the increasing water pressure under the top stratum flexes the top stratum upward and causes cracks which concentrate the flow (Ozkan, 2003). Some witnesses report seeing bulges in the top stratum surface prior to development of the sand boil. Shown in Figure 4 are photographs of top stratum bulging in the bottom of a drainage ditch near the landside toe of a levee in the Sacramento-San Joaquin Delta, together with an accompanying sand boil. This theory works well with the idea that the more often flooding occurs, and the more often that bulging/flexing of the top stratum is repeated, the more likely a sand boil will form. Repeated flexing may finally cause cracks and preferred seepage paths through the top strata. Likewise, repeated hydraulic stressing along an animal burrow or a decayed root hole may eventually open up the anomaly and allow for the flow of water.

Ozkan (2003) also reported that:

"Sherard et al. (1963) investigated the mechanics of piping in earth and earth-rock dams. As water flows, the pressure head is dissipated in overcoming viscous drag forces, which resist the flow through the small pores. The seeping water also generates erosive forces and tends to drag the soil particles with it as it travels through the pervious layer. If the seepage erosive forces are greater than the erosion resisting forces, the soil particles are washed away and piping starts. If the soil has some cohesion, a small tunnel or pipe can form at the downstream exit face of a seepage path. Once piping starts, the flow in the pipe increases due to the decreased resistance to flow, piping accelerates, and the small tunnel or pipe lengthens. Van Zyl and Harr (1981) stated that the analysis of piping erosion was almost impossible due to control by discontinuities."

A further complication is that if the aquifer is not saturated at the start of the flood, there will be a time delay to saturate the aquifer and build up pressure (Ozkan, 2003). This can be an important factor in determining why a levee may not fail during one flood but may fail during a later flood.

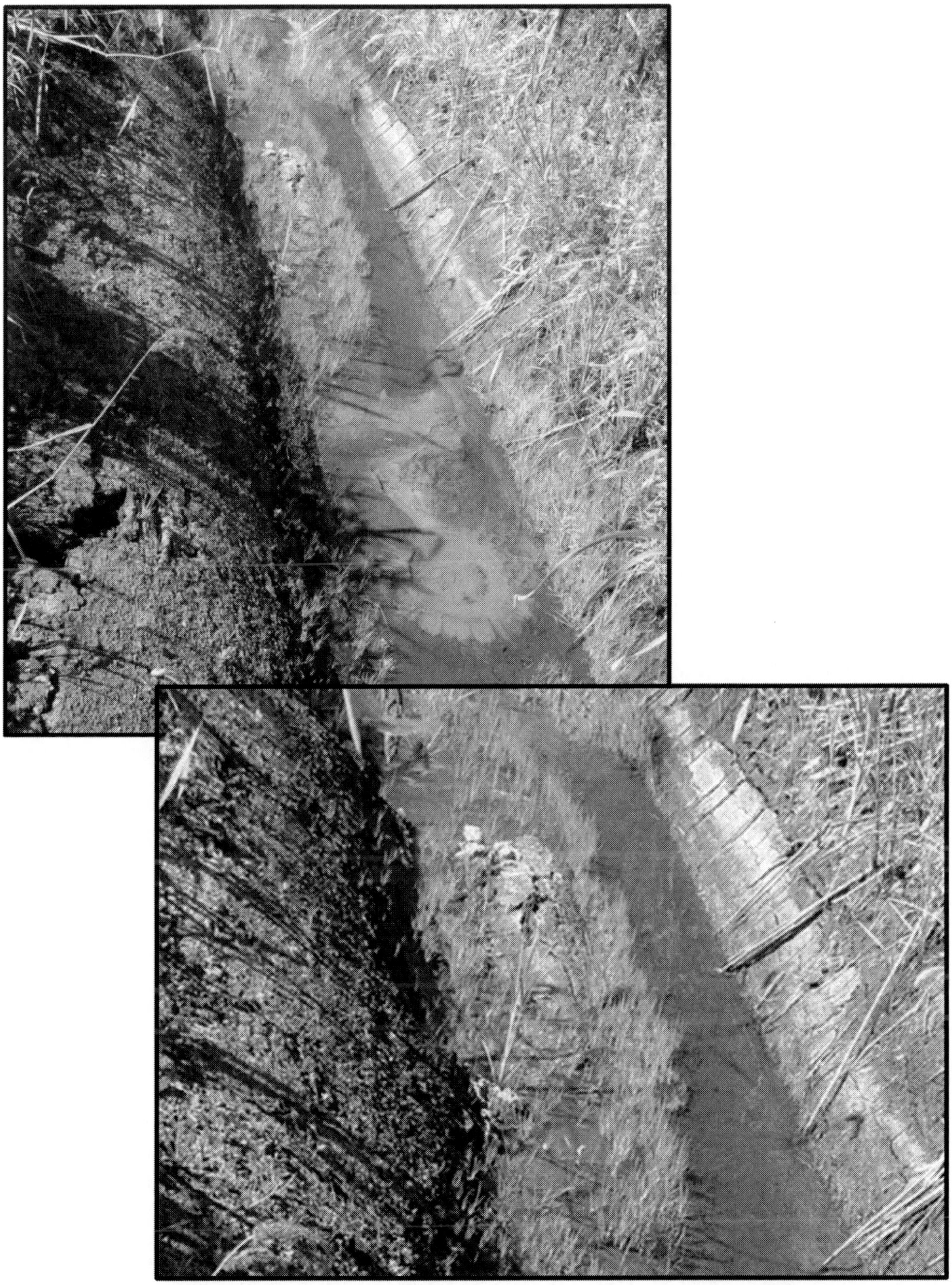

Figure 4. Photographs of bottom bulging and sand boil at base of drainage ditch near the landside toe of a levee in the Sacramento-San Joaquin Delta in California.

CASE HISTORIES

Buck Chute, MS

For more than 40 years, sand boils and piping have been occurring at Buck Chute, an agricultural area along the Mississippi River waterway in Central Mississippi approximately 20 miles northwest of Vicksburg, MS (see Figure 5). The geology of the area has been studied in detail and Llopis and Smith (2003) described the local geology as:

> *"Buck Chute lies within the Yazoo Basin of the Mississippi Alluvial Valley. The Yazoo Basin consists mainly of Holocene meander belt and backswamp environments that average about 100 ft in thickness (Saucier, 1994). The levee and the problem area of seepage is located within the meander belt between two abandoned channels on what is most likely former point bar deposits and younger overbank deposits. Sand, silt and clay would be expected to occur in certain predictable sedimentary sequences as a result of fluvial processes in the meandering river. "*

Buck Chute was a channel that connected Eagle Lake Pass with Eagle Lake and was probably formed as a crevasse splay into Eagle Lake during a period of flooding. Historical records suggest that this channel was active until at least 1934.

Ever since 1973, and possibly before, Buck Chute has been used by the Vicksburg District to educate young engineers as to what a sand boil is and to allow the engineers to view sand boils in various stages of development. Figure 6 shows the locations of sand boils over time in the Buck Chute area.

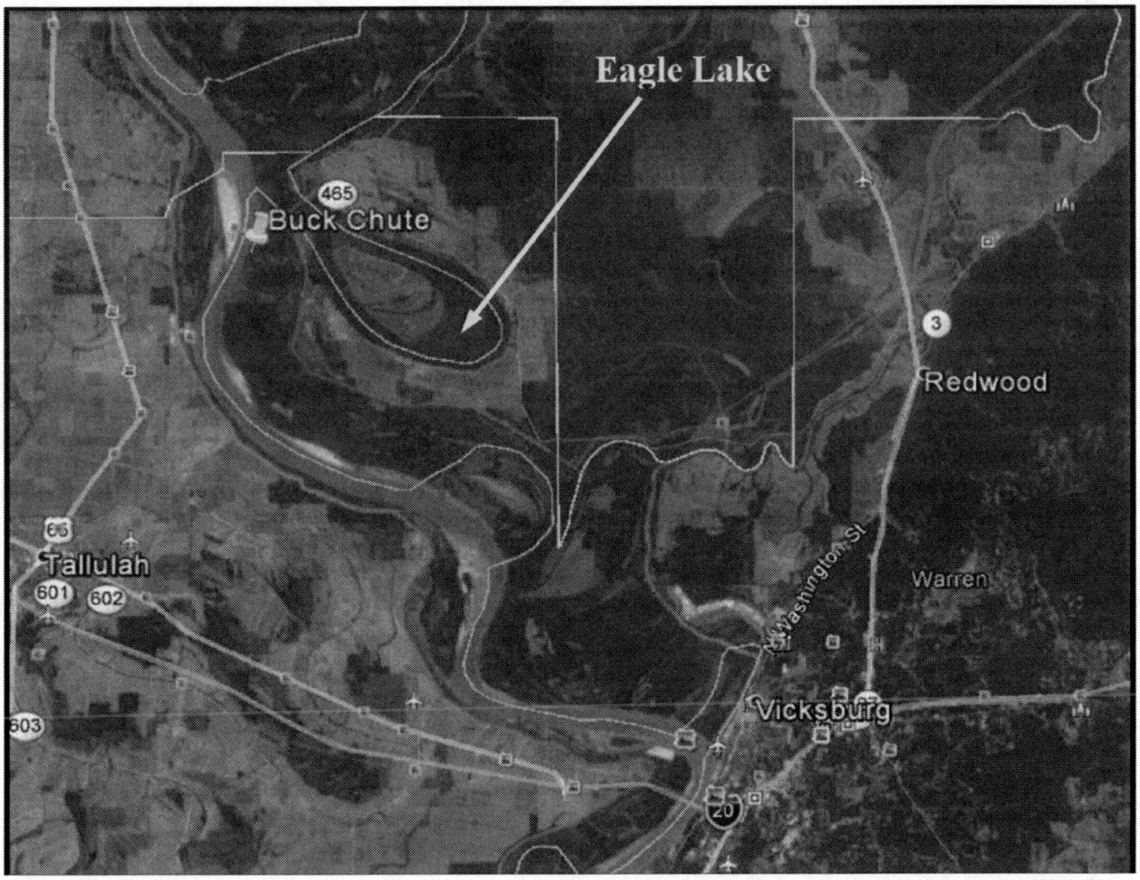

Figure 5. Location map for Buck Chute, Mississippi (adapted from Google Earth).

Before the high water in 1997, it was standard procedure to simply observe the sand boils at Buck Chute and not ring them or take any remedial actions as they were flowing. However, as reported in the U. S. Army Corps of Engineers (USACE) Flood Situation Report dated March 11, 1997, several sand boils were active during the 1997 flood event. This document reported boils moving sand at Buck Chute with the river at Elevation 43.5 feet on the Vicksburg gauge. That year the decision was later made to ring these boils for the very first time in history. Figures 7 and 8 show some of the containment rings around sand boils during the 1997 flood event. The day after these boils were contained, the sandbagged areas filled with sand and the heights of the sandbagged containments had to be raised. This indicated that substantial amounts of foundation material had been removed due to piping.

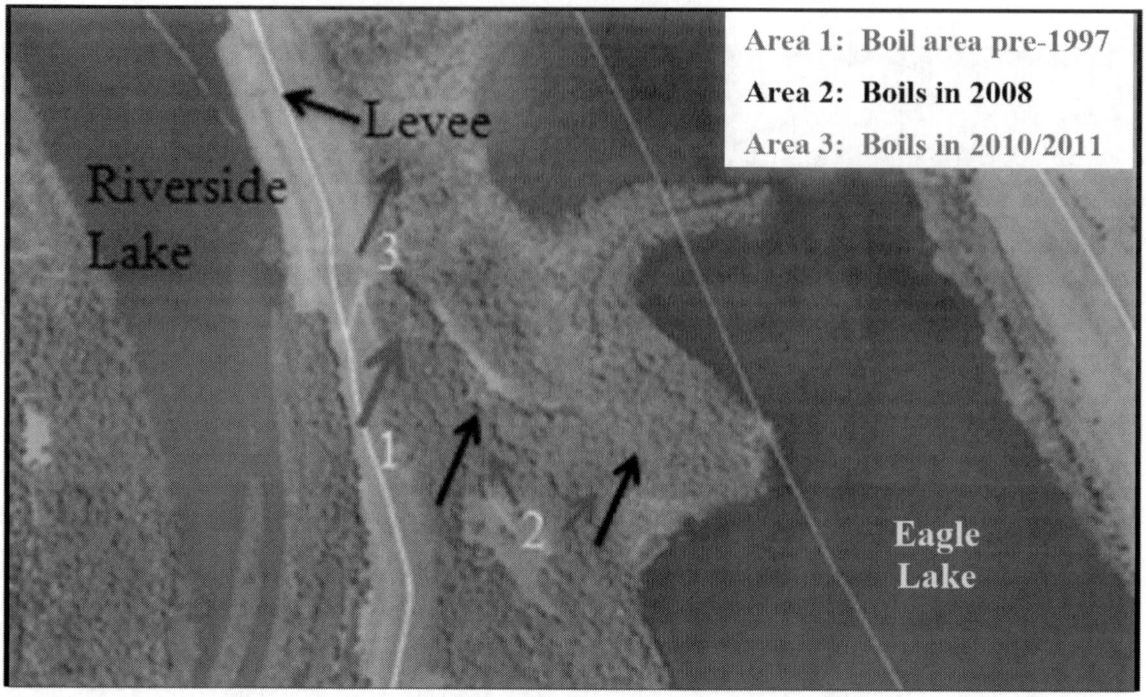

Figure 6. Aerial photograph of Buck Chute seepage/boil areas
(from Google Earth 12/30/2010).

Figure 7. Photograph of containment of sand boils at Buck Chute
during 1997 high water.

Figure 8. Photograph of sand boil containments at Buck Chute during 1997 high water.

After the 1997 high-water event, the Vicksburg District installed 6 relief wells near the location of the 1997 sand boils. Even after these wells were installed, the area continued to experience sand boils and problems. The Towns TechNote (Llopis and Smith, 2003) prepared by USACE Engineer Research and Development Center (ERDC) reported that the Vicksburg District was concerned that "*future flooding episodes will cause additional deterioration to levee integrity possibly leading to levee failure*." They also reported on the "*observed unusual subsidence immediately around the wells.*" Because of the subsidence occurring around the wells, the District had the wells cleaned with air jets in November 1999. During this operation, some of the wells were found to have almost 4 feet of sediment within the wells (these sediments would have been foundation material that had piped through the well filter and into the well pipe).

On June 4, 2002, a group of employees from the Vicksburg District made a site visit to Buck Chute and documented their findings in a Memorandum for Record (2002). They reported that they had measured flows from the wells on May 29th (river stage at Elevation 44.5 feet at Vicksburg gauge) and on June 4th (river stage at Elevation 45.3 feet at Vicksburg gauge) and that the flows on June 4th were "significantly higher" than the well flows measured six days earlier. They also reported that "*numerous boils were discovered approximately 200 yards southeast of the relief wells and ending approximately 200 yards beyond the point of origin following the bank of Eagle Lake. Boils were readily identified by the presence of iron stained effluent. Many boils had a characteristic ring of fine sand deposits. The majority of the boils ranged in size between one and two inches in diameter*" (Note that Eagle Lake is an old meander lake located immediately landward of the relief well area (see Figures 5 and 6).

After these problems and concerns were documented, the Vicksburg District installed 3 additional relief wells in 2007. However, in the following year during the high-water event of 2008, numerous sand boils were discovered and required sandbagging of the boils to reduce the material being piped from the foundation. Figures 9 through 12 present photographs depicting the conditions during the 2008 high-water event.

Figure 9. Photograph of the seepage area within the relief well field at Buck Chute during the 2008 high-water event (4/28/2008).

Figure 9 shows the heavy seepage around and between the wells during the 2008 flood event. During the 2008 event, the river crested at a stage of 51.0 feet on the Vicksburg gauge. The river went above flood stage on March 28th and returned below flood stage on May 10th. Several new boils occurred within the relief well fields during this time and the photograph in Figure 10 presents a view of one of these new boils. Even with these wells flowing and relieving pressure, new boils were occurring within the well field and landside of the wells. These new boils often were ringed with sand bags in an effort to reduce the piping of materials (see Figures 11 and 12).

Figure 10: Photograph of sand boil within relief well field at Buck Chute (4/28/2008).

Figure 11. Photograph of new boil landward of relief wells at Buck Chute during 2008 high-water event (04-29-08).

Figure 12. Photograph of USACE employees inspecting boils located landward of relief wells at Buck Chute during 2008 high-water event (04-29-08).

During the 2010 high-water season, the river crested three separate times at Buck Chute:

- Elevation 41.0 feet on February 11th
- Elevation 40.86 feet on April 10th
- Elevation 42.77 feet on May 23rd

During the 2010 high-water season, sand boils were found and sandbagged for the first time along the levee just north of the previous Buck Chute sand boil areas (see Figure 6 for locations of the various sand boil areas). Heavy seepage was occurring around the relief wells in 2010, and the wells were flowing. As may be noted from the data shown above, the river during this high-water season never reached flood stage (Elevation 43 feet) or even previously sustained flood elevations. Nevertheless, new sand boils were occurring in 2010 in an area that had never before experienced them during even higher river stages (see Figures 13 and 14).

Figure 13. Photograph of new seepage/boil area at Buck Chute in 2010.

Figure 14. Photograph of new seepage/ringed sand boil at Buck Chute in 2010.

This new area of seepage and boils that developed in 2010 created new concerns and the Vicksburg District planned to provide additional remediation to this problem area. The District had planned to install 30 additional relief wells at the toe of a new 220-foot-wide drained seepage berm in the area of the new boils. However, before this work could be completed, the record 2011 Flood (Elevation 57.1 feet) occurred and the adjacent landward Eagle Lake had to be flooded and its elevation raised to lower the differential head within the boil area to save the levee from failure (see location of Eagle Lake relative to the Buck Chute seepage area in Figures 5 and 6). After the 2011 Flood, the District was able to construct its planned remediation. However, after the drained seepage berm and relief wells were installed during the fall and winter of 2011, the river stage reached Elevation 36.73 feet. As shown in the photographs presented in Figures 15 through 18, only some of the wells within the new well field were actually discharging water. As can be seen in these figures, some of the relief wells were flowing large amounts while adjacent wells were not flowing at all. In addition, the internal drain in the drained seepage berm was also not flowing water. What was causing these inconsistencies in the relief well flows? An additional question that should be asked is why some of the relief wells were discharging so much water with the river stage well below the flood stage elevation of 43 feet.

In summarizing the Buck Chute area, numerous flood events over time have resulted in major sand boils and piping. For much of this time, little to no remediation was implemented and a badly deteriorated foundation resulted. Even after remedial measures were implemented after several decades of seepage distress, the remedial measures do not appear to be effective. This is because of the apparent inability of the newly installed relief wells to adequately relieve the excess underseepage condition in the damaged foundation.

In theory, the foundation conditions, seepage entry condition, and other design conditions within the relief well field should be fairly uniform. Therefore, the flow from the wells should also be relatively uniform. However, as depicted by the photographs, the flows were anything but uniform, and the wells were flowing more water than a designer would calculate for the relatively low river stages. This indicates a significantly damaged aquifer. The only reasonable explanation is that some of these wells are located very close to pre-existing pipes and are being fed water at a much closer entry point whereas other wells are located much farther away from the pipes and the damage to the foundation.

Figure 15. New relief well near middle of relief well field flowing the highest amount with others nearby not flowing (Winter 2011 high-water stage @ Elevation 36.73 feet).

Figure 16: New relief well adjacent to relief well shown in previous figure barely flowing (Winter 2011 high-water stage @ Elevation 36.73 feet).

Figure 17. Two of the many new relief wells located at the landward toe of the new drained seepage berm that never flowed during the Winter 2011 high-water event.

Figure 18: Two new relief wells located at the landward toe of the new drained seepage berm: the first one flowing a significant amount, the second one barely flowing. The internal drain within the drained seepage berm was also not flowing (Winter 2011 high-water stage @ Elevation 36.73 feet).

Kaskaskia Island Levee, IL

During the Mississippi River Flood of 1973, a levee protecting Kaskaskia Island near Kaskaskia, Illinois, experienced major seepage and sand boil problems. This levee system was then improved with the addition of a 100-foot-wide seepage berm and other improvements which included a levee raise in the early 1990s. After these improvements, and during the Mississippi River Flood of 1993, this newly improved levee protecting Kaskaskia Island failed at a river stage before the levee was overtopped. A prior evacuation order was instrumental in preventing loss of life on the island. It was eventually determined that the 1993 failure occurred in the general location where sand boils had occurred during the 1973 flood. During the subsequent 1993 flood, sand boils again occurred just beyond the 100-foot berm (see Figure 19) and resulted in the eventual failure of the levee.

Figure 19. Photograph of flood-fight personnel ringing the sand boil along the Kaskaskia River Levee just prior to failure during the 1993 Flood (from USACE).

While investigating the performance of levees during the 1993 Flood, Ozkan (2003) reported:

"Mansur, et al. (2000) reviewed studies carried out since the 1940's on underseepage, piping, and sand boil formation in the Mississippi River Valley. The Mississippi River floods of 1993 produced seepage under some levees which resulted in dramatic levee

failures in the Kaskaskia Island Levee District in Illinois (Mansur, et al., 2000). A sand boil and subsurface piping caused the Kaskaskia Island levee to fail, flooding the entire levee district."

It is likely that the sand boils and piping that occurred in 1973, and probably during prior events, resulted in damage to the aquifer at this location, and that without having a reasonable method to account for such damage; this may have resulted in a hidden foundation defect being left under the improved levee and seepage berm. Later, during the 1993 flood, the reactivation of the sand boils resulted in rapid piping and loss of the levee.

DEFINING THE PROBLEM

Where is the Pipe?

As of the date of this paper, there are no reliable methods to determine how many pipes may exist in a foundation, how far they have backward eroded toward the water source, what are the diameters of the pipe or pipes, or any other pertinent information about the pipes that would be necessary to properly model the damaged foundation. Therefore, most designers just model an *undamaged* foundation. This lack of understanding and unconservative modeling can and will lead to future levee failures.

Ignoring Past Piping and Foundation Damage

The case histories previously summarized show that sand boils over time can damage the foundation beneath a levee and result in making the levee much more vulnerable to potential piping failures in subsequent flood events. Schmertmann (2000), Nagy (2012), and Keizer and Rice (2015) have all published research that demonstrates how critical it is to understand what is occurring within a foundation in the areas of these seepage pipes. Schmertmann (2000) reported:

"*Piping by its nature happens progressively and the engineer needs to try to understand the point-by-point progression of the safety factor.*"

Unfortunately, current designers typically model the foundation as "*undamaged*" and determine a gradient for that condition without any consideration of past piping damage (see Figure 20). The authors of this paper propose that we name the calculated gradient in such models an "***apparent gradient***" and define it as the gradient calculated for an "*undamaged foundation.*" If an area has only received minor seepage or "pin boils" during past floods, then this calculated gradient may be an appropriate representation of the true gradient for that area. However, if the area had been subjected to major sand boils that have removed foundation material and had sustained any backward erosion within the foundation, then this calculated gradient would **NOT** represent the actual gradient within the foundation.

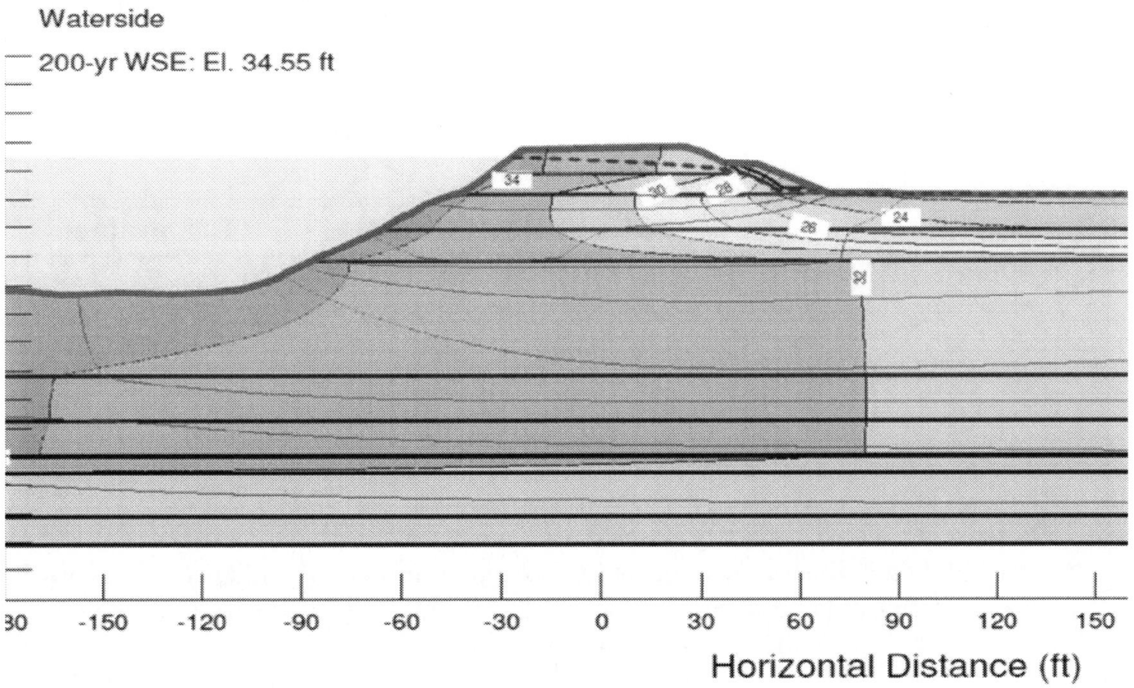

Figure 20. Typical example calculation of seepage using computer program SEEP/W for an undamaged foundation.

Schmertmann (2000) reported that locally near an internal erosion pipe, two-dimensional (2D) seepage has flow lines converging to the pipe and pipe head in a three-dimensional (3D) pattern. This causes a large increase in gradients and a consequent large decrease in effective stresses around the pipe head vs. the pre-pipe conditions. The increased horizontal and new vertical gradients loosen the soil at the pipe head sufficiently that the particles at the pipe head move downstream. The loss of particles at the pipe head causes the pipe to advance waterwards, producing a new flow and gradient concentration at the pipe head, etc., in an approximately continuous process.

Numerous other research studies support this with Keizer and Rice (2015) stating:

"With the flow concentration the behavior becomes a complex three-dimensional mechanism of fluidization and transverse piping coming into play."

So an illustration of a flow net under a levee with a pre-existing pipe and chimney above an isotropic 25-foot-thick aquifer would show concentrated seepage flows and gradients into the pipe and look more like Figure 21 below (it should be noted that this is a very simple representation using a 2D model for a 3D problem).

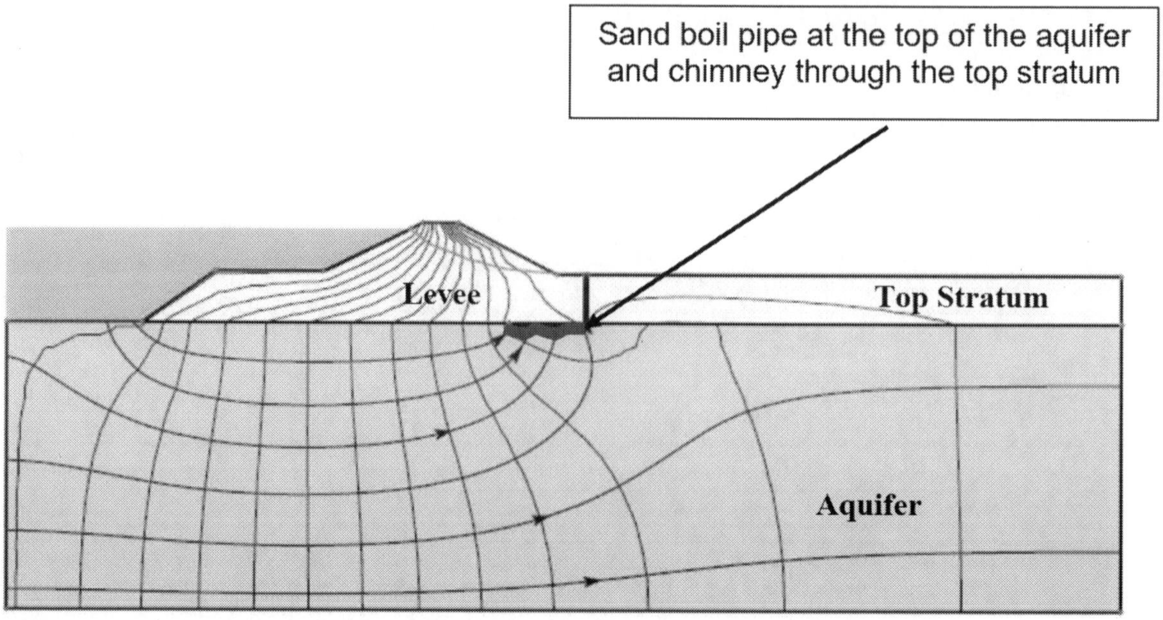

Figure 21. Flow net illustrating concentrated seepage flows and gradients into an erosion pipe and chimney above a 25-foot-thick isotropic aquifer (for illustration use only - illustration furnished by Dr. John Rice)

When designers look at the flow net in Figure 21, it is should be obvious that without cutting off these internal erosion pipes, a proposed repair becomes very complicated and may not be effective. Also, without knowing the number, size, or riverward extent of the pipe or pipes adds significantly more complexity to the problem. Schmertmann (2000) stated:

"Internal erosion can be a major threat to the safety of embankment dams, yet the mechanisms involved are not well understood and no analytical techniques are available which are comparable to those used to calculate a factor of safety against shear failure. The problem sometimes receives relatively little attention, yet it may be the greatest hazard to the safety of many embankment dams."

This hazard probably holds truer for levees than dams because we as engineers have not had sufficient resources to prevent or remediate as much severe piping in levees as we have in dams. Therefore, there are miles of levees that may become critical given more flood events.

How to Determine if this is an Issue?

Obviously, the issue of past sand boils and internal piping is not an issue for brand-new first-time systems and should only be considered for existing systems which have sustained significant piping damage during past high-water events. In order to address this in a meaningful way, designers must be given reliable data from the levee owner/maintainer as to how the system has performed during past floods. Performance data during any high-water event are critical. Photographs and locations of any performance issues must be documented, together with the river stage at which they occurred. This type of documentation will allow designers to determine if the seepage resulted in a major sand boil or boils moving material or if it was just heavy seepage or just pin boils. Currently there are no criteria that would instruct a designer as to when a sand boil was becoming a considerable threat to public safety. Hopefully soon, the USACE will be able to develop guidance on this subject.

In the interim, the authors suggest that designers consider the following approach: when investigating a levee system for safety issues, determine at what river stage sand boils are becoming active. Then, run a seepage model using this river stage and an undamaged foundation model. This will provide the designer an "***apparent gradient***" for that section. When this gradient is calculated at a value of 0.3 or less, immediate remediation measures should be undertaken. This is because seepage distress is occurring at a very a low apparent gradient and is indicative that the foundation has sustained significant piping damage in the past. Recommended remediation measures should be as discussed below.

HOW TO REMEDIATE THESE PROBLEMS

Nagy (2012) stated:

"At present, knowledge concerning the process of sand boil formation and the internal and external factors leading on to soil failure is limited. We cannot define parameters of the criteria of sand boil formation, we cannot unambiguously determine in advance the location and the water level at which disaster is to be expected and ***we cannot recommend any other form of prevention but increasing the length of travel of the flow or closing the tube of the boil*** *when flood levels keep increasing."*

With this in mind, and assuming the presence of pre-existing pipes, it should be obvious that the installation of relief wells alone into a damaged aquifer will unlikely be adequate to address future underseepage problems. This is because the design of relief wells does not consider the preferred seepage paths along pre-existing pipes or damaged aquifers. Using a berm without considering previous piping damage is also very risky because the designer does not know the extent of any pre-existing pipes and if the piping is sever. Thus, a piping failure could still be possible as in the case of the Kaskaskia Island Levee. To ensure public safety, the pre-existing pipes must be cut off in a manner to reduce the potential for any future piping of material and to force the seepage to flow through the undamaged foundation again. This prevents the use of these pipes as a preferred seepage path in future flood events. The recommended remediation approach for severely damaged levee foundations is as follows:

1. Construct a fully penetrating cutoff wall through the aquifer and any past piping damage (assuming that an underlying aquaclude exists beneath the aquifer that is within reasonable depth to extend the bottom of the cutoff wall into). This is shown conceptually in Figure 22.

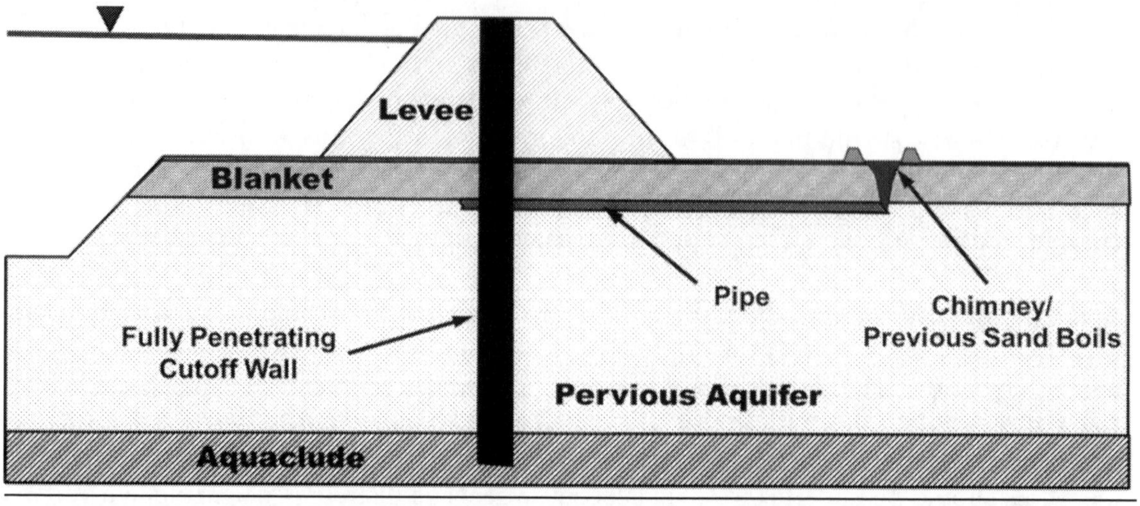

Figure 22. Recommended remediation approach using a fully-penetrating cutoff wall through levee to address both past piping damage and original underseepage problems

<u>or</u>

2. Construct a partially penetrating, or hanging, cutoff wall (being sure to be at least 15 to 25 feet below the bottom of the top stratum where pipes may be located) to mitigate the effects of previous piping damage to the foundation. A partially penetrating, or hanging, cutoff wall, however, is unlikely to remediate the original seepage deficiencies which lead to the creation of the sand boils and piping in the first place. Accordingly, supplemental underseepage remediation measures such as properly designed relief wells or a seepage berm should be constructed to further reduce exit gradients to acceptable levels (see Figure 23).

   This approach, however, must be used with caution when the top stratum or blanket may contain continuous lenses of pervious material, such as those created by crevasse splay deposits, that are separated from the underlying aquifer. In such instances, excess seepage pressures equivalent to almost full river stage might result in such a blanket sublayer immediately waterward of the shallow cutoff wall. This would then pose a potential stability problem. Therefore, the potential for such conditions must be considered and, if present, either extra measures such as thickening the landside seepage berm or an additional partially penetrating wall through the levee crown may be necessary if such an approach is to be used.

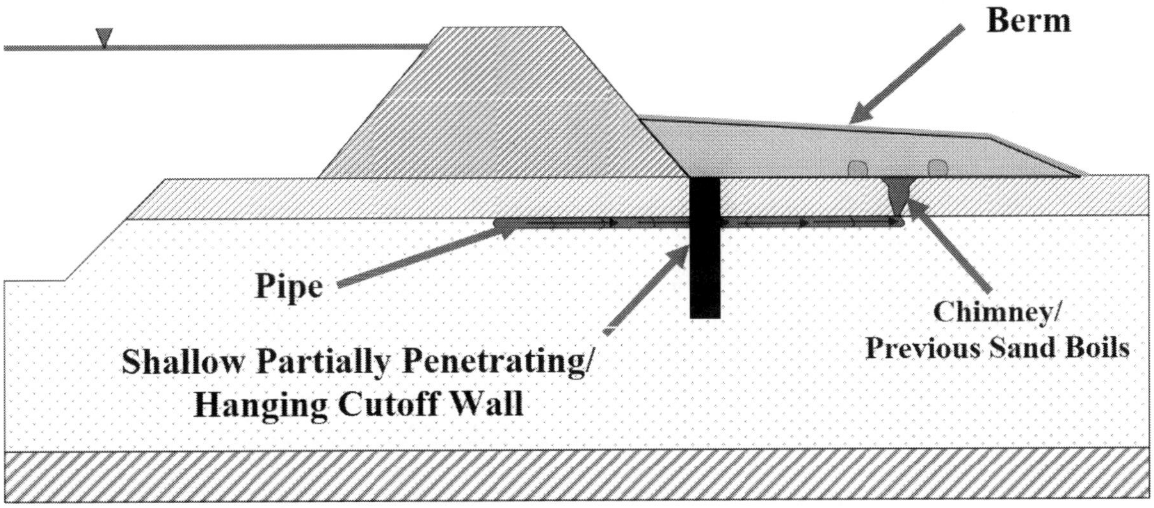

Figure 23. Recommended remediation approach using a partially-penetrating cutoff wall and a seepage berm for levee foundations damaged by piping during previous flood events.

CONCLUSION

Levees along most of the waterways within the United States have seen numerous flood events since being constructed. Each of these events takes a toll on the system's integrity. In some cases, repeated production of sand boils and piping have severely damaged the levee foundation making it more susceptible to failure in future flood events. Evaluating a flood event may seem like a relatively simple task at first glance; however, in a detailed study, the overwhelming complexity of the problem becomes apparent.

One of the variables to consider is the flood stage. If the river rises to a given elevation, an engineer is tempted to compare flood problems with this event to another flood event having a similar flood stage. This comparison can be misleading. How long did it take the river to reach this flood stage? How long was the river at this stage? Was the area very wet, or was there a dry spell before the flood event? Looking at these questions, one realizes that the seepage conditions in the levees and their foundations are related to not only stage, but also other factors such as duration and what were the saturation and seepage conditions prior to the flood. These factors and the history of deterioration play a major role in comparing flood events.

Schmertmann (2000) reported that everyone who has studied the piping problem has realized its complexity and difficulty. He stated that the study of piping involves the interaction of the disciplines of soil mechanics, fluid mechanics, and sediment transport. While working at the Vicksburg District Corps of Engineers in the 1990s, the first author started to notice and document many areas where sand boils had been a problem at certain river stages, and which had become worse and had initiated boiling at lower stages in subsequent events. Buck Chute, Mississippi, was one of the main areas where this phenomenon was occurring. It was becoming quite apparent that piping was occurring during each flood event, and that the soil being removed would leave behind significant voids and pipes that would be subjected to seepage forces in subsequent floods. Also, the soil arching around these pipes would prevent healing to the pre-flood state. Therefore, the soil's porosity and hydraulic conductivity were increasing along these pipes together with gradients in subsequent events.

As discussed by Schmertmann (2000), where a number of laboratory tests were conducted on piping through sands, time has no healing effect on piping. Once a sand boil and a pipe have started during a flood, then a similar flood experienced later would continue the pipe further towards the source of seepage. Depending on the duration and magnitude of the floods, the total formation of the pipe through the foundation could take many flood events, which could take many years to occur. However, this would lead one to assume that in certain circumstances a pipe can remain more or less open below the ground surface long after the flood event that created it. This poses a great threat to levees because of the seriousness of the problem (or development of the pipe), which can go undetected until failure occurs in subsequent high-water events.

The problem is simple in concept, but actually very complex. Initially a levee is subjected to a head high enough to start a sand boil. The sand boil pipes some of the material from the foundation/embankment. The first time this situation occurs, there are a number of different methods to calculate the response. Now, the real difficult part of the problem begins. The next flood/high-water event on the levee then causes the sand boil to reactivate. More material is removed this time, and the seepage length to the source has been reduced. How can this be measured or mathematically determined? As of this date, no accurate means is available. Also, how many more flood events can the system take before the pipe is extended to the source and a failure occurs? This question cannot be answered today. However, we now know that given enough large flood/high-water events, the pipe will continue to extend until failure of the levee or dam occurs.

Two bottom-line points:

1. To ensure public safety, designers must account for the presence of any pre-existing pipes and related foundation damage.
2. Because of very limited research about this subject, more research is badly needed.

In the interim, the authors recommend that designers responsible for the design of remediation measures for levee systems have a very good understanding of the performance of the system during past floods. If the system has sustained severe piping in the past, the designers should design a partially penetrating cutoff wall to a minimum depth below where the pipes may exist in order to mitigate the past piping damage, or a fully penetrating wall if the depth is not too excessive. If a partially penetrating, or hanging, cutoff wall is designed to mitigate the effects of previous piping erosion, the designer should then address the original seepage deficiency and calculate the exit gradient for this section assuming an undamaged aquifer. Then a standard form of remediation (relief wells or seepage berm) could be designed to address the original seepage deficiencies of the levee.

REFERENCES

Keizer, R and J. Rice (2015), "*Evaluating The Mechanics Of Piping Erosion Initiation Through Laboratory Modeling*", Proceedings of the Annual Conference of the United States Society on Dams (USSD), Louisville, KY April 2015.

Llopis, J. L. and Smith, E.W. (2003), "*Geophysical Investigation of Chronic Seepage, Buck Chute, Eagle Lake, Mississippi*", USACE ERDC Towns TechNote.

Mansur C.I., Postol G., and Salley J.R. (2000), "*Performance of Relief Well Systems along Mississippi River Levees,*" Journal of Geotechnical and Geoenvironmental Engineering, American Society of Civil Engineers, 126(8), pp. 727-738.

Nagy, L. (2012), "*Characterization of Sand Boils with Grading Entropy„*" Proc. Nat. Sci, Matica Srpska Novi Sad, № 122, 73—88, 2012

Ozkan, Senda (2003), "*Analytical Study on Flood Seepage under River Levees.*" Dissertation, Louisiana State University.

Schmertmann, J.H. (2000), "*The non-filter factor of safety against piping through sands,*" ASCE Geotechnical Special Publication No. 111, Judgment and Innovation, edited by F. Silva and E. Kavazanjian, ASCE, Reston, VA.

USACE, Vicksburg District (1997), Flood Situation Report March 11, 1997, On Record at Vicksburg District Corps of Engineers.

USACE, Vicksburg District (2002), Memorandum for Record June 5, 2002, On Record at Vicksburg District Corps of Engineers.

AN EMPIRICAL MODEL FOR ESTIMATING INTERNAL EROSION CRITICAL GRADIENTS FOR INCLINED EXIT FACE CONDITIONS

Richard A. Keizer, Jr.[1]
John D. Rice[2]
Robert A. Jaeger[3]

ABSTRACT

Internal erosion represents a complex combination of several mechanisms related to the failure or near failure of dams and levees. Of the known mechanisms, heave is the most common used in design when calculating critical gradients. Current geotechnical engineering practice uses a theoretical model for predicting heave based on the ratio of the soil buoyant unit weight and the unit weight of water. There does exist a theoretical model which accounts for exit face inclination which is derived from the heave model, but it does not consider actual trends in soil behavior at various inclinations. It would therefore prove useful to have an empirical model for comparison against and validation of theoretical models.

This paper presents an empirical model in contrast to the theoretical equations used to predict the critical gradient to cause heave which are currently used in engineering practice. It considers actual trends observed during laboratory testing to develop a prediction of critical gradients based on exit face inclination. The observed trends are consistent between different sandy soil types, and agree with the theoretical prediction when considering a horizontal exit face. This method considers a soil's specific gravity, void ratio, and angle of repose to calculate a critical gradient. Using this model with existing methods, engineers can make better informed decisions for criteria, design, and analysis when calculating critical gradients.

INTRODUCTION

Internal erosion is the process where soil particles within an embankment or its foundation are carried downstream by seepage flow. Several different internal erosion mechanisms have been identified, including backward erosion, concentrated leak erosion, contact erosion, suffusion, and heave (ICOLD 2013).

- Backward erosion (or piping) occurs when soil particles detach from an exit face and a pipe is formed through or underneath a structure transporting particles downstream.
- Concentrated leak erosion (or CLE) occurs along the sides of an open crack or in the interconnecting voids in a continuous permeable zone, where the shear stress induced by flows exceeds the critical value.

---

[1] Staff Geotechnical Engineer, GEI Consultants, Inc., Rancho Cordova, CA, rkeizer@geiconsultants.com

[2] Associate Professor, Utah State University, UMC 4110, Logan, UT, john.rice@usu.edu

[3] Senior Engineer, GEI Consultants, Inc., Rancho Cordova, CA, rjaeger@geiconsultants.com

- Contact erosion develops at the horizontal boundary of a fine soil overlying a coarse soil where the fine soil is washed into the coarse soil by horizontal flow.
- Suffusion is the migration of the finer soil particles out of a soil matrix eventually decreasing the strength of the soil and increasing the likelihood of failure due to higher seepage velocities and higher gradients caused by preferential flow paths.
- Heave is the phenomenon where a mass of soil is lifted causing near failure or failure of an earth structure (ILH 2013).

Internal erosion is usually assessed based on hydraulic gradients. Critical gradients are defined as the gradients required to initiate the internal erosion mechanism. Despite the known differences among internal erosion mechanisms, calculating critical conditions for individual mechanisms is difficult due to unknown subsurface conditions which could largely influence results. Defects or other unique conditions may allow mechanisms like CLE and backward erosions to initiate at conditions less than critical. Due to the unknown nature of actual subsurface conditions, heave calculations are commonly used for evaluating the potential of internal erosion.

Methods to evaluate internal erosion have been developed based on laboratory testing. Line of creep methods were developed in the early 1900s (Bligh 1910, Lane 1934) to calculate critical hydraulic heads required to progress a pipe underneath an impermeable structure. Flume tests were performed in the 1980s and 1990s where methods were developed to calculate critical conditions to initiate backward erosion (Sellmeijer 1988, Sellmeijer and Koenders 1991, Koenders and Sellmeijer 1992, Schmertmann 2000). Although the testing performed has given us a better understanding of internal erosion initiation and continuation, the tests performed cover a limited number of cases and methods are constantly being refined with additional testing (Sellmeijer et al, 2012).

The findings presented in this paper are based on laboratory testing conducted at Utah State University. Tests were performed on soil samples with horizontal and inclined exit faces to evaluate the mechanisms associated with the initiation of backward erosion. Among the evaluation of the testing results, behaviors associated with the heave mechanism were identified and a model for predicting gradients to initiate these behaviors are proposed in this paper.

CURRENT METHODS

There are three theoretical methods of calculating critical gradients for heave which are typically used in current engineering practice. The first method, and the method of which the other methods are based is Terzaghi's derivation (Terzaghi 1922, Terzaghi and Peck 1948) of the heave mechanism. This method compares the average vertical hydraulic gradient to the critical gradient, i_{cr}, needed to initiate heave. Terzaghi derived this expression based on vertical seepage through uniform material with a horizontal exit face. Terzaghi defined the critical gradient through the following equation,

$$i_{cr} = \frac{\gamma_b}{\gamma_w} \tag{1}$$

where γ_b is the buoyant unit weight, and γ_w is the unit weight of water.

The second method was developed by the United States Army Corps of Engineers (USACE) using a simplified method to predict a factor of safety against heave using an effective seepage length based on the blanket thickness upstream and downstream (USACE, 1993). The factor of safety against uplift can be calculated by,

$$F = \frac{Z_{bd}}{h}\frac{\gamma'}{\gamma_w}\frac{L_1+L_2+L_3}{L_3} \tag{2}$$

The effective upstream length, L_1 (m), is defined as

$$L_1 = \sqrt{\frac{k_f}{k_{bu}} Z_{bu} d} \tag{3}$$

where k_f is the horizontal permeability of the pervious foundation (m/s), k_{bu} is the vertical permeability of the upstream blanket (m/s), Z_{bu} is the thickness of the upstream blanket (m), and d is the thickness of the pervious aquifer (m).

The effective downstream blanket length, L_3 (m), is defined as

$$L_3 = \sqrt{\frac{k_f}{k_{bd}} Z_{bd} d} \tag{4}$$

where k_{bd} is the vertical permeability of the downstream blanket (m/s), and Z_{bd} is the thickness of the downstream blanket (m).

The length of the impervious core or levee base is L_2 (m), and the head above the downstream toe is h (m). The head at the toe of the structure can be determined by,

$$h_0 = \frac{hL_3}{L_1+L_2+L_3} \tag{5}$$

and considering the downstream blanket, the critical head can be found using Terzaghi's heave method by,

$$h_c = Z_{bd}\frac{\gamma_b}{\gamma_w} \tag{6}$$

A critical gradient can be calculated from the factor of safety determined with Equation (2), and the actual gradient observed in the analysis model. This simplified model is also assuming flow through the blanket is vertical, flow through the pervious underlying material is horizontal, the earth structure itself is impervious, and the exit face is horizontal.

These methods can give an estimation of a critical gradient or critical head to initiate heave by under-seepage, but neither method considers non-vertical seepage, specifically the "leaker" case where the earth structure is susceptible to through-seepage.

The third method was published in 2013 by O'Leary (O'Leary et al, 2013) which outlined methods previously developed to calculate backward erosion (often referred to as piping) critical gradients for non-horizontal exit face conditions. The paper correctly points out errors in previously developed methods of critical gradient calculation when considering an inclined exit face, and suggests an equation developed by Kovács (1981) to calculate critical gradients for given values of internal friction angle, slope angle, seepage direction, and unit weight. This critical gradient can be calculated by the following equation,

$$i_{cr} = \left(\frac{\gamma_b}{\gamma_w}\right)\left(\frac{\tan(\varphi')\cos(\beta)-\sin(\beta)}{\cos(\beta-\alpha)+\tan(\varphi')\sin(\beta-\alpha)}\right) \quad (7)$$

where γ_b is the buoyant unit weight, γ_w is the unit weight of water, φ' is the drained internal friction angle, β is the slope angle from the horizontal, and α is the seepage angle (horizontal seepage is 0, downward seepage is positive, and upward seepage is negative).

In the paper, O'Leary indicated this equation calculates a critical gradient for backward erosion; however, the original derivation as shown in Equation 1 is actually being modeled given a non-horizontal exit face, and non-vertical seepage. This equation may be better used modeling the heave mechanism from where the initial derivation originated. The model represents a macro-mechanism of internal erosion rather than a micro-mechanism like backward erosion particle detachment. Individual particle detachment depends on the size, angularity, gradation, relative density, and cohesion of the soil (Schmertmann 2000, ICOLD 2013, ILH 2013) which is not considered in Equation 7.

Table 1 compares the three methods of critical gradient calculation to each other and outlines the characteristics of the individual methods.

Table 1. Comparison of Gradient Calculation Methods

Calculation Method	Calculates	Most Significant Variable	Limitations
Terzaghi's	Critical Gradient	Soil unit weight	Theoretical model. Does not consider exit face/seepage conditions
USACE's Simplified	Factor of safety	Blanket thickness	Theoretical model. Does not consider exit face/seepage conditions
O'Leary's	Critical Gradient	Slope angle	Theoretical model

Though these methods are useful for calculating possible critical gradients for the heave mechanism, they are all theoretical, and do not exhibit similar trends in the laboratory testing performed in this study for inclined exit faces, or calculate results similar to studies previously performed by Sellmeijer or Schmertmann.

OBSERVED LABORATORY TRENDS

A testing apparatus and procedure was developed by Fleshman (2012) to observe critical gradients with vertical flow through sandy soils using a 2-inch diameter sampler cell. The primary objectives of the study were to better understand the backward erosion "piping" mechanism through sandy soils, and to identify and evaluate stages which initiated failure. Using the same methods, Keizer (2014) tested sandy soils at inclined exit face conditions using the 2-inch diameter sampler, and a larger 4-inch diameter sampler. Detailed procedures and results of the tests are presented in previous papers (Fleshman and Rice 2013, Swainston-Fleshman and Rice 2014a, 2014b, Keizer and Rice, 2015). Properties of the soils tested can be seen in Tables 2 and 3. More data and properties related to the soils tested can be found in Keizer (2014).

Table 2. Properties and characteristics of the soils tested.

Soil Type	Sphericity	Roundness	Specific Gravity	Average Grain Size (D_{50} mm)	Coefficient of Uniformity	Min & Max Void Ratio$_{\psi}$
Graded Ottawa Sand	0.9	0.7	2.65	0.373	1.76	**0.48-0.72**
20-30 Ottawa Sand	0.9	0.7	2.66	0.574	1.51	**0.49-0.71**
Garnet Sand	0.3	0.1 – 0.3	4.05	0.155	1.25	**0.96-1.54**
Graded Angular Sand	0.5-0.7	0.3	2.63	0.305	2.18	**0.63-1.15**
Graded Zirconium Beads	1.0	1.0	3.84	0.373	1.57	*0.46-0.60*
Uniform Zirconium Beads	1.0	1.0	3.81	0.312	1.33	*0.52-0.67*
Uniform Glass Beads	1.0	1.0	2.49	0.503	1.20	*0.52-0.65*

Ψ– Min & Max Void Ratio measured by **ASTM** and *Modified* Relative Density tests

Table 3. Angle of Repose, failure angle, and internal friction angle of tested soils.

Soil Type	Loose Angle of Repose*	Failure Angle£	Friction Angle†
Graded Ottawa Sand	35.0°	40.0°	38°
20-30 Ottawa Sand	35.3°	40.0°	38°
Garnet Sand	39.0°	42.0°	44°
Graded Angular Sand	37.6°	41.3°	44°
Graded Zirconium Beads	24.0°	29.0°	35°
Uniform Zirconium Beads	25.3°	30.3°	33°
Uniform Glass Beads	21.6°	28.5°	28°

* – Loose Angle of Repose for dry soil using dry raining technique.
£ – Failure Angle determined in lab by slowly inclining soil sampler until failure without seepage.
† – Internal Friction Angle measured using direct shear testing

Keizer (2014) and Keizer and Rice (2015) identified four stages of failure in soil samples with horizontal and inclined exit face conditions. The four stages and corresponding gradients are: 1) first visible movement (i_m), 2) boil formation (i_b), 3) heave (i_h), and 4) total failure (i_f) as shown in Figures 1 through 4. These stages are described in detail below.

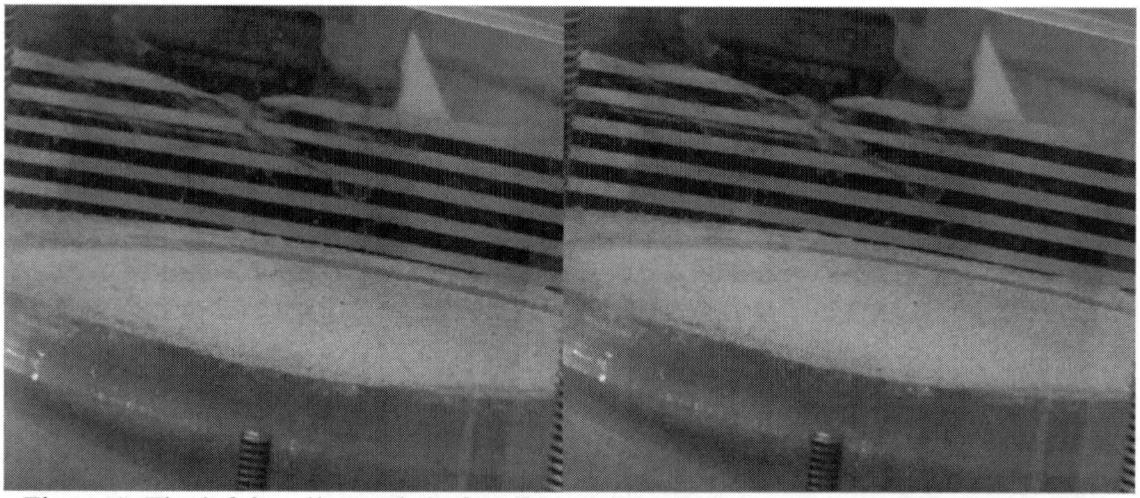

Figure 1. The left is soil sample before first visible movement (i_m), on the right is after.

Figure 2. Sand boil (i_b) in garnet sand soil sample.

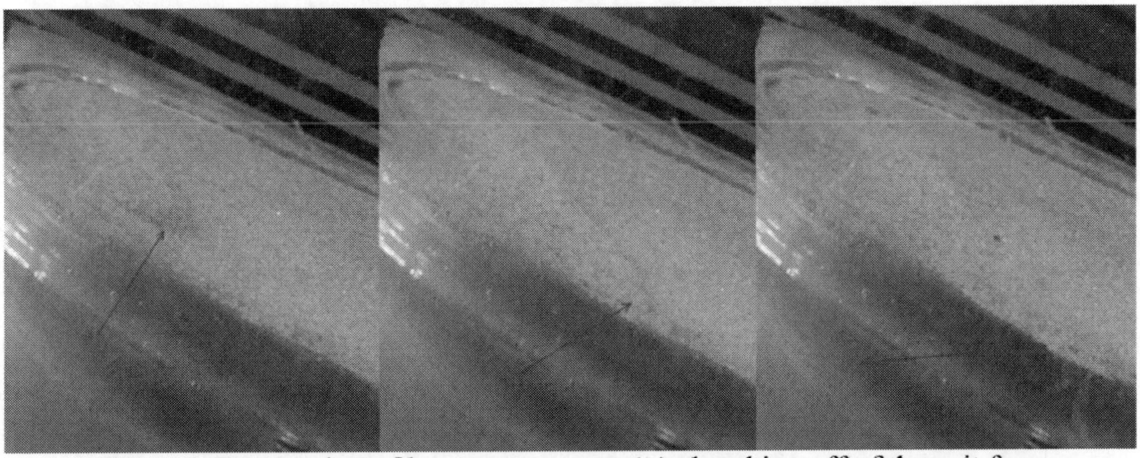

Figure 3. Progression of heave movement (i_h) sloughing off of the exit face.

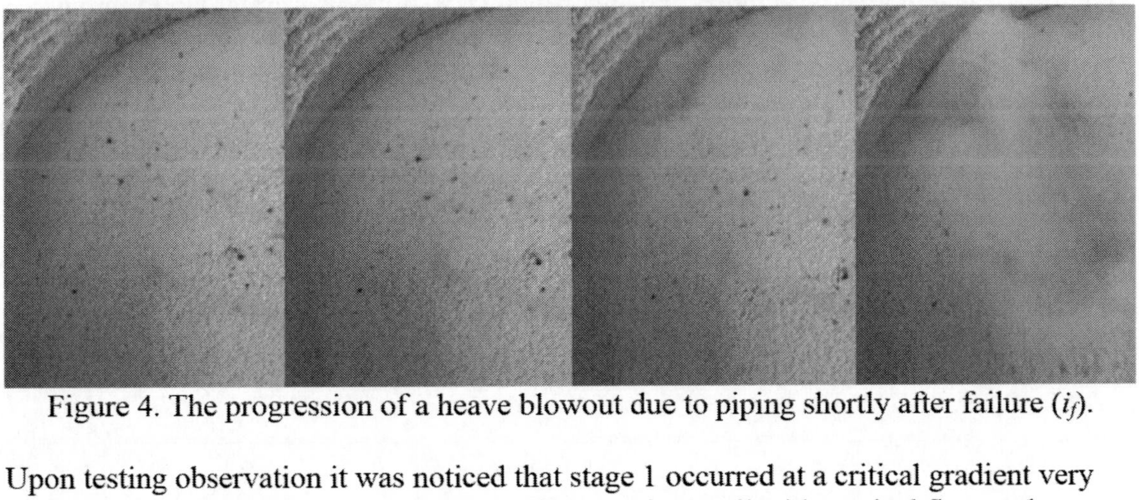

Figure 4. The progression of a heave blowout due to piping shortly after failure (i_f).

Upon testing observation it was noticed that stage 1 occurred at a critical gradient very similar to what Equation 1 would predict for that given soil with vertical flow and horizontal exit face conditions. This behavior was also confirmed in the previous research (Fleshman 2012, Fleshman and Rice 2013, Swainston-Fleshman and Rice 2014a, 2014b).

As the exit face inclination was increased, the gradient required to cause stage 1 decreased in every soil type with a similar trend. In the following figures, the gradients are normalized to Terzaghi's critical gradient as shown in Equation 8,

$$i_{crNorm} = \frac{i_m}{\left(\frac{\gamma_b}{\gamma_w}\right)} \tag{8}$$

where i_m is the observed gradient at stage 1. The normalized critical gradients are plotted against the percentage of the loose angle of repose (PAOR) for the corresponding soil type, allowing soils with different angle of repose values to be compared to each other.

$$PAOR = \frac{Exit\ face\ slope\ angle}{Loose\ angle\ of\ repose} \times 100 \tag{9}$$

Six inclination angles were used in the tests: 0 PAOR – Horizontal exit face (no inclination), 25 PAOR – 25% of loose angle of repose, 50 PAOR – 50% of loose angle of repose, 75 PAOR – 75% of loose angle of repose, 97 PAOR – 97% of loose angle of repose, Failure angle – determined in lab by slowly inclining soil sampler until failure without seepage. The Failure Angle varied between 108 and 132 PAOR and was influenced by angularity, grain size, gradation, and specific gravity of the soil tested.

Figure 5 shows i_m normalized to Terzaghi's critical gradients (i_m/i_c) for all tests run plotted against the PAOR at stage 1. The data exhibits a 2nd order polynomial trend with an R^2 value of 0.88. According to the trend, a decrease in critical gradient would not be expected at small inclination angles (up to around 20 PAOR). After 20 PAOR the critical gradient decreases rapidly, approaching zero for the failure angle (between 108 and 132 PAOR). The plot also shows maximum and minimum trends indicated by the faint trendlines. The soils which typically exhibited values above the main trendline were the angular or graded soils. Lower values were generally observed with the rounded beads.

The data was evaluated for bias between the 2-inch and 4-inch samples. The data from Figure 5 was replotted with the 2-inch and 4-inch tests plotted separately in Figure 6.

There is a slightly higher trend for the smaller (2-inch) diameter sampler which could be attributed to increased edge effects. The trends for the smaller and larger samplers give R^2 values of 0.90 and 0.88, respectively. The difference between the two sampler trends is marginal, both starting very close to what Terzaghi's prediction would indicate for a horizontal exit face, and both ending at a zero critical gradient near 118-119 PAOR.

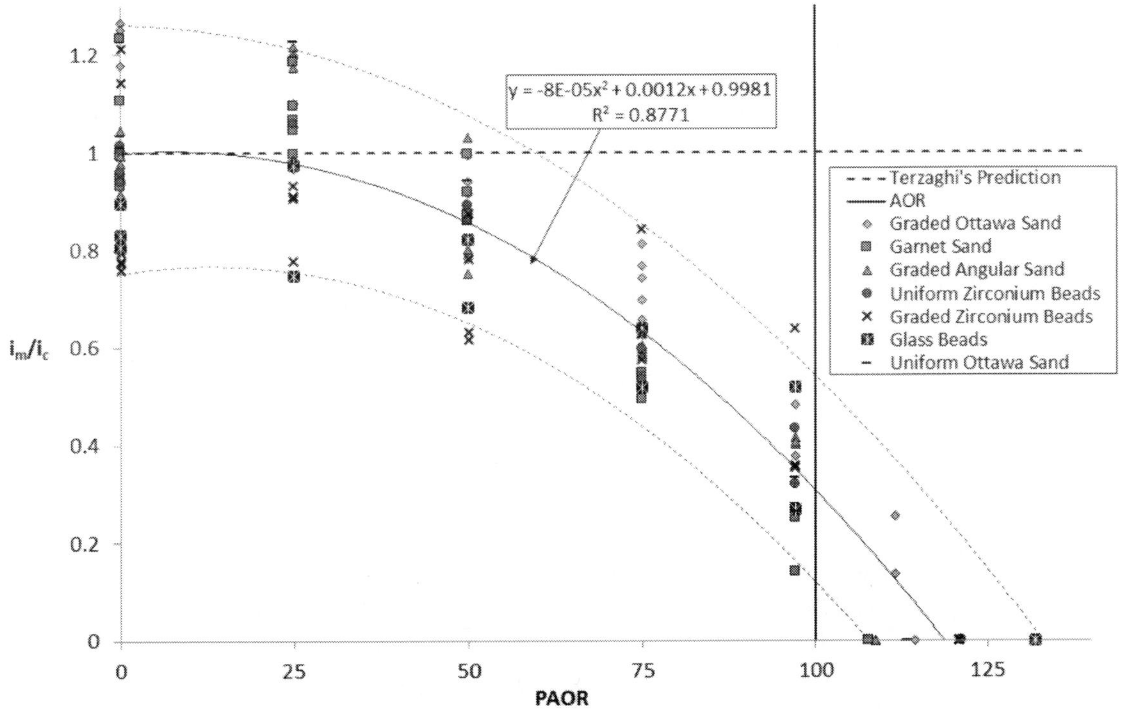

Figure 5. i_m/i_c versus PAOR for all laboratory tests.

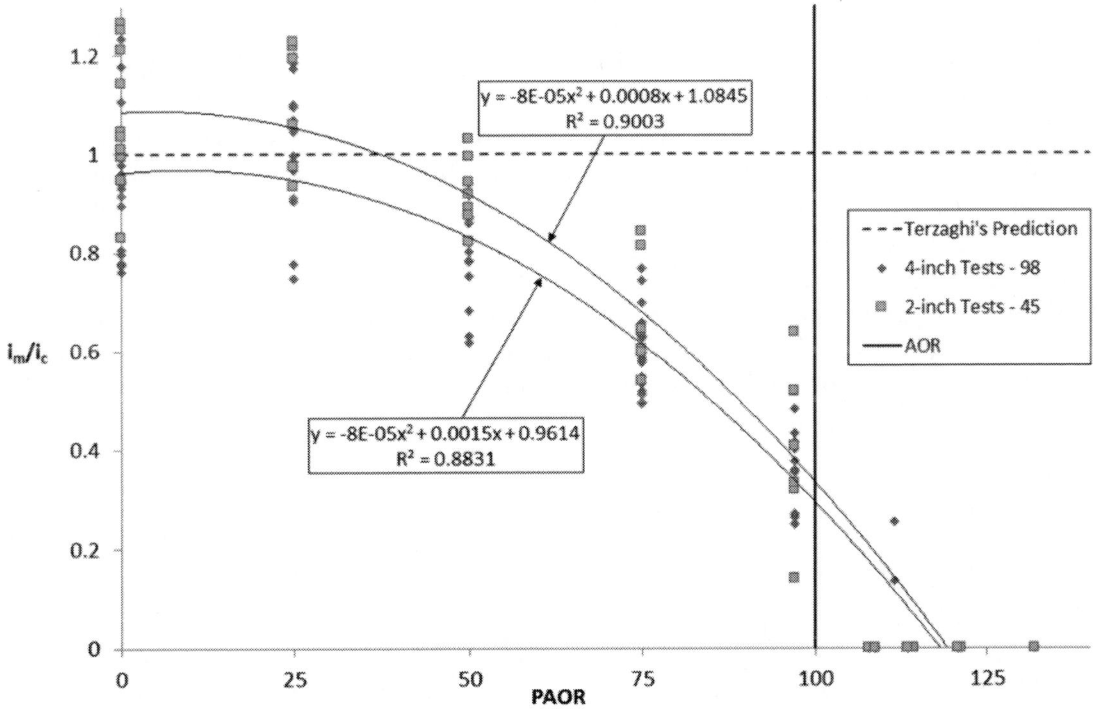

Figure 6. The effect of different size samplers on i_m/i_c versus PAOR.

PROPOSED MODEL

The proposed model assumes the soil critical gradient will exhibit a downward second order polynomial trend when plotting the PAOR versus gradient. The critical gradient is calculated using the following equation,

$$i_{cr} = a * PAOR^2 + b * PAOR + c \quad \text{(for PAOR} \geq 8.8) \tag{10a}$$

$$a = -1.8 \times 10^{-6}(\gamma) + 1.16 \times 10^{-4} \tag{10b}$$

$$b = 1.9748 \times 10^{-5}(\gamma) - 0.0016 \tag{10c}$$

$$c = \frac{\gamma_b}{\gamma_w} \tag{10d}$$

where γ is the unit weight of the soil (saturated), c is Terzaghi's critical gradient (Equation 1), and PAOR is the slope angle in percent of the loose angle of repose.

Figure 7 plots i_m/i_c with the proposed model. The proposed model is based on a couple of conservative simplifications. The first simplification is at a horizontal exit face assuming that the highest possible critical gradient matches Terzaghi's model in Equation 1. The second simplification is that the critical gradient is assumed to be 0 at the loose angle of repose. Both of these simplifications are conservative due to possible hardened crust at the surface likely exhibiting higher resistance to heave by way of arching, higher unit weight, and a likely higher failure angle than the loose angle of repose. Both cases where the simplifications are made, laboratory tests showed higher critical gradients, which can be attributed to the soil arching and higher density than the loose soil state. Using these simplifications, the trend agrees precisely with Terzaghi's prediction when assuming a horizontal exit face.

As shown in Figure 7, none of the data points fall below the proposed model at 75 PAOR or greater. This depicts the conservative nature by using the loose angle of repose rather than the internal friction angle of the soil, especially at the steeper inclination angles.

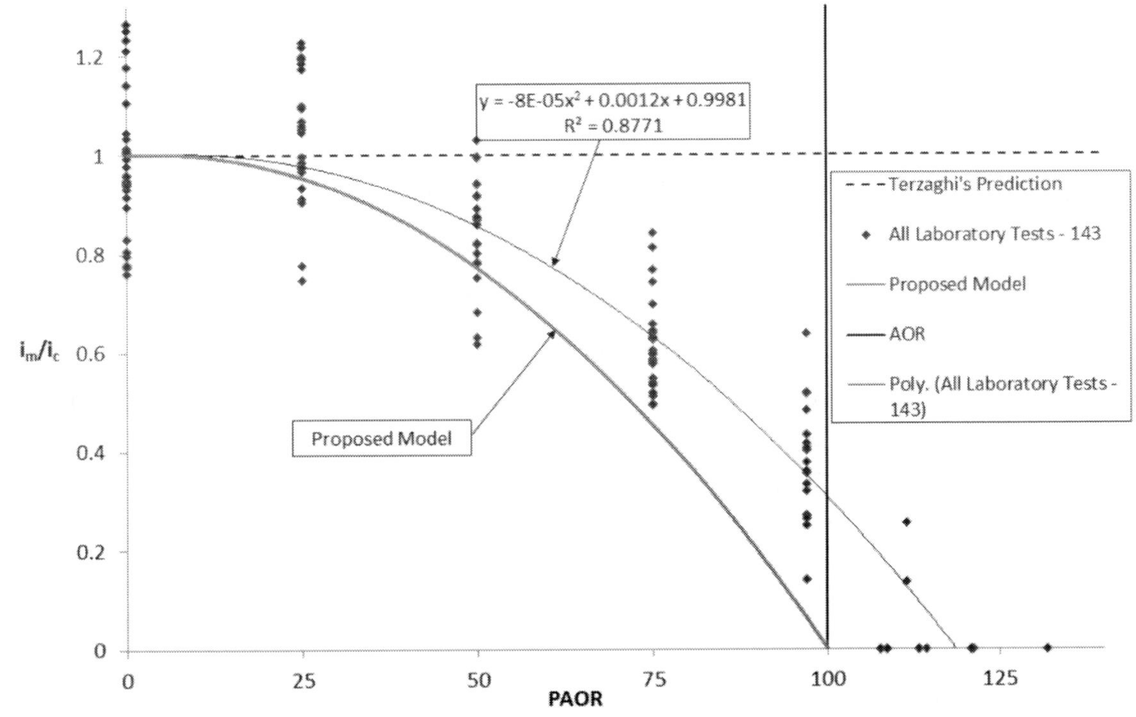

Figure 7. Proposed model plotted with laboratory trends

Due to the proposed model's 2nd order polynomial trend, there is a small increase in critical gradient at very small values of PAOR which do not represent the lab trends or any other methods. The proposed trend is only representative of the lab trends with PAOR ≥ 8.8. For values of PAOR < 8.8, it is suggested that Equation 1 or Equation 2 should be used to calculate critical gradients for heave. Table 4 shows minimum exit face inclination (ß) values to be used for soils with given AOR's.

Table 4. Table of minimum slope values (ß) used in proposed model.

Soil AOR	Minimum ß (AOR x 0.088)
24°	2.1°
26°	2.3°
28°	2.5°
30°	2.6°
32°	2.8°
34°	3.0°
36°	3.2°
38°	3.3°
40°	3.5°
42°	3.7°
44°	3.9°

Figure 8 is the same plot as Figure 7, except showing the difference between the different diameter samplers.

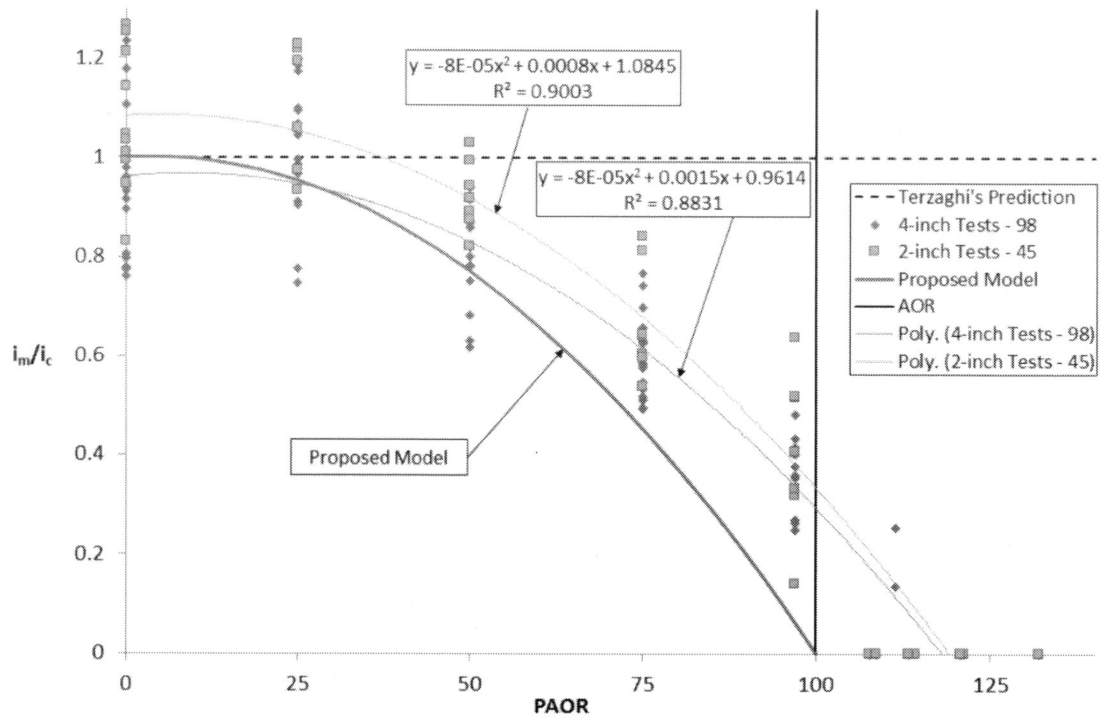

Figure 8. Proposed model plotted with different sampler trends.

As shown in Figure 8, most of the data points falling below the proposed model line are from the larger diameter tests. It is important to note that stages 1 through 3 are all visually observed during testing, and correlated with the gradients recorded. Stage 1 was especially difficult to identify with the 4-inch diameter sampler due to the larger surface area to observe. This could be a source for error in determining the stages prematurely. The stronger fit for the 2-inch diameter sampler despite having the smaller data pool is likely due to the ease in determining when stage 1 actually occurred during testing.

Assumptions

The first, and overlying assumption when calculating a critical gradient using most methods for internal erosion mechanisms, is that no subsurface abnormality or defect exists (voids, trees with roots, culverts, or anything which could lead to a concentrated leak). These methods assume a uniform homogeneous soil. A second assumption is that the soil is saturated at the exit face and throughout the soil matrix where heave is likely to occur. This will likely be the case if seepage is occurring through the exit face. The third assumption is that the model assumes flow perpendicular to the exit face. This could be observed as a limitation, because all of the laboratory data is based on flow perpendicular to the exit face; however, when assuming perpendicular flow with any given gradient, conditions with the maximum detachment seepage force and the minimum normal resistance exist. Figures 9a and 9b show the seepage force with corresponding normal forces.

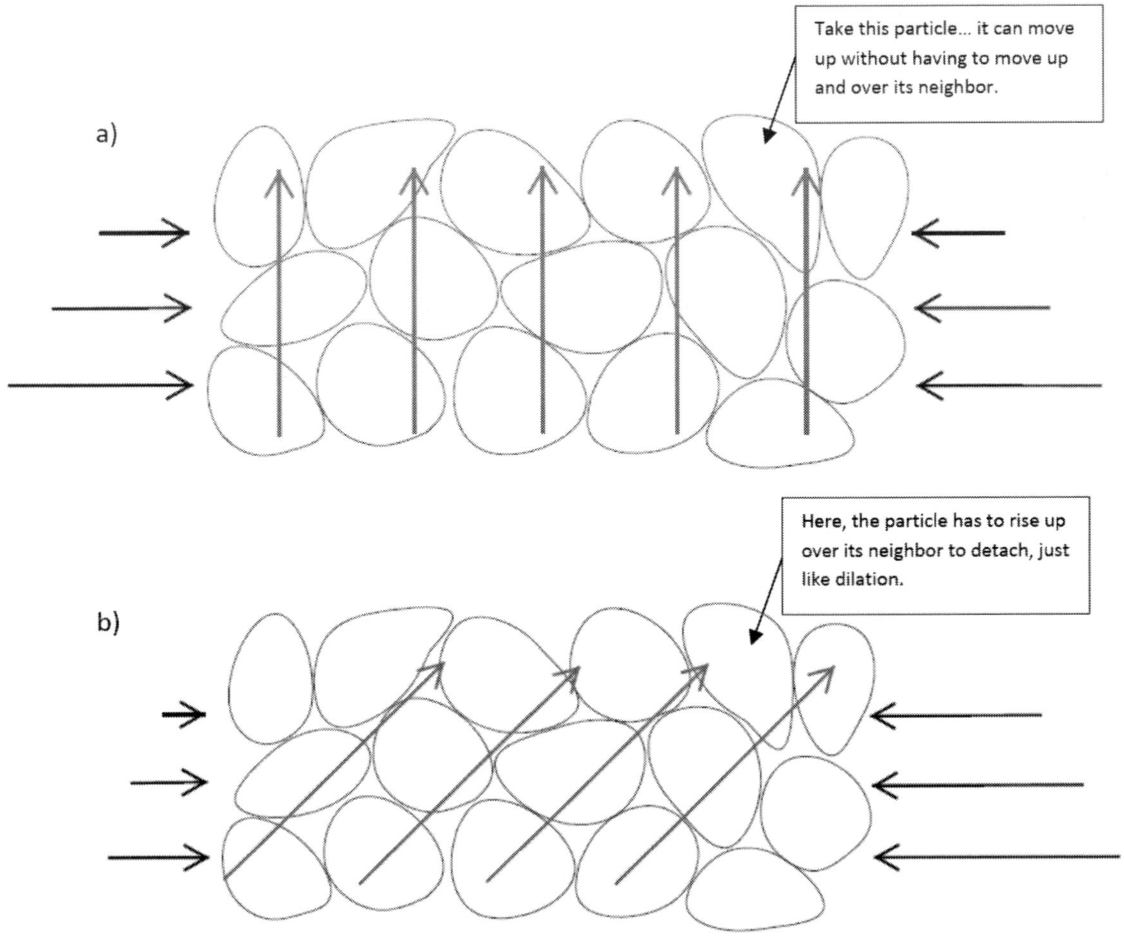

Figure 9. a) Diagram showing perpendicular flow.
b) Diagram showing inclined flow.

Particle detachment is easiest in the perpendicular flow diagram (Figure 9a) because particles have fewer kinematic constraints. The maximum resistive forces exist when flow becomes parallel to the exit face, thereby increasing the likelihood of arching/interlocking between soil grains (Figure 9b). Also, the more parallel the seepage, the smaller the force component in the perpendicular (detachment direction) direction becomes.

Limitations

The laboratory testing used to develop this model were performed on cohesionless sandy soils, and for that reason the trends observed do not necessarily apply to cohesive soils. Another limitation is that only one seepage angle was tested with respect to each slope angle (perpendicular to the exit face). However, the perpendicular flow condition may better represent a critical condition with respect to particle detachment, as described in the previous section and shown in Figures 9a and 9b.

Application to Field Conditions

If this proposed method of gradient calculation is to be applied to two-dimensional models, some finite distance of which a gradient is calculated should be set forth. O'Leary's method does not specify over what distance to calculate the design gradient when using a finite element program, and results can vary significantly where calculating a gradient over small distances, sometimes resulting in unrealistically large values. The gradients calculated in these laboratory tests were all calculated over a 5-inch soil sample, however, calculating a gradient over 5-inches in a finite element model could perhaps result in an unrealistically large result. Duncan et al. (2011) suggests that an average value of hydraulic gradient over a length of 1-foot is preferred. For the application of this proposed model, when calculating an design gradient it is recommended that an average hydraulic gradient over a minimum distance of 1-foot be used.

COMPARISON TO OTHER METHODS

Figure 10 plots O'Leary's method calculated for various angles of seepage flow. O'Leary's method is plotted with five different scenarios corresponding to five different seepage angles. The plot assumes an internal friction angle of 38 degrees, a unit weight of $2\gamma_w$, and an AOR of 35 degrees.

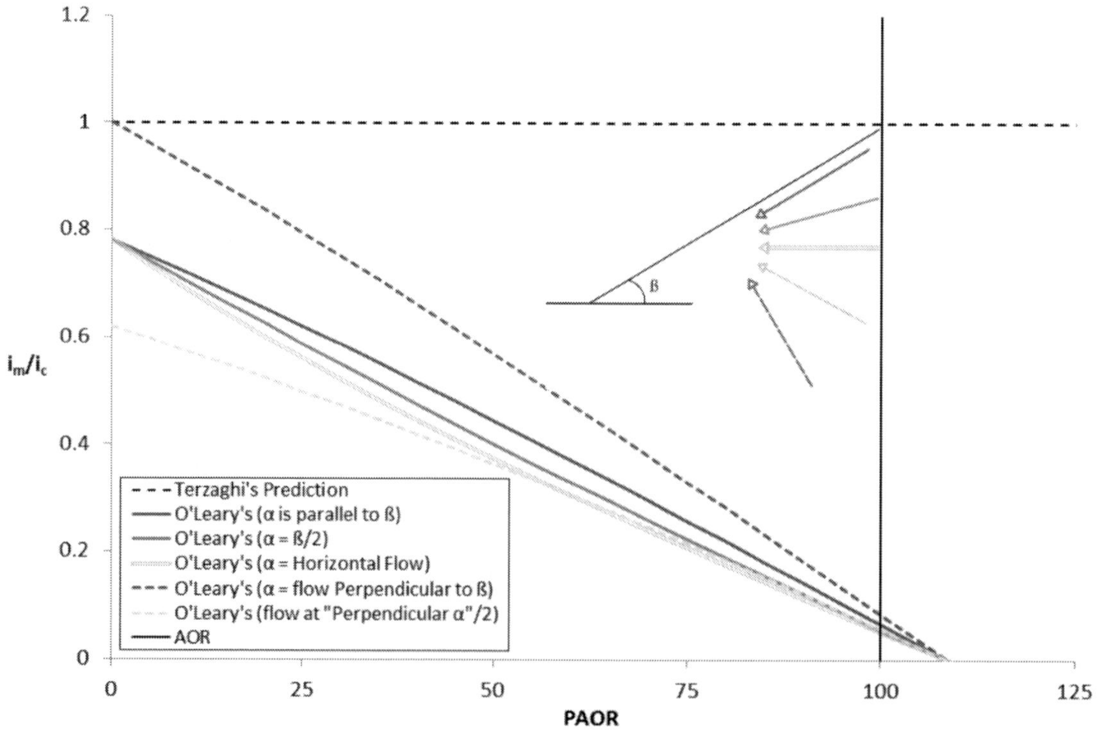

Figure 10. O'Leary's method at various angles of seepage flow.

For situations where the flow angle α is parallel to ß, half of ß, and horizontal, the trend only marginally deviates from each other, with a maximum difference at approximately 50 PAOR. However, once the flow angle α becomes negative (flows upward), there is no similarity in the trends. This method also shows that for a horizontal exit face, heave will occur at smaller critical gradients when the flow angle is not vertical, which means a smaller force component in the vertical direction is required for particle detachment. This trend seems counter-intuitive when considering a free-body diagram, or Figures 9a and 9b. The trend seen for perpendicular flow seems much more intuitive, starting at Terzaghi's prediction for a horizontal exit face and decreasing almost linearly to a zero gradient at the internal friction angle.

Figure 11 plots O'Leary's and Terzaghi's predictions versus all laboratory tests in this research. With regards to O'Leary's method, this is assuming perpendicular flow to the exit face, an internal friction angle of 38 degrees, a unit weight of $2\gamma_W$, and an AOR of 35 degrees. Figure 12 adds the proposed model to the plot in Figure 11.

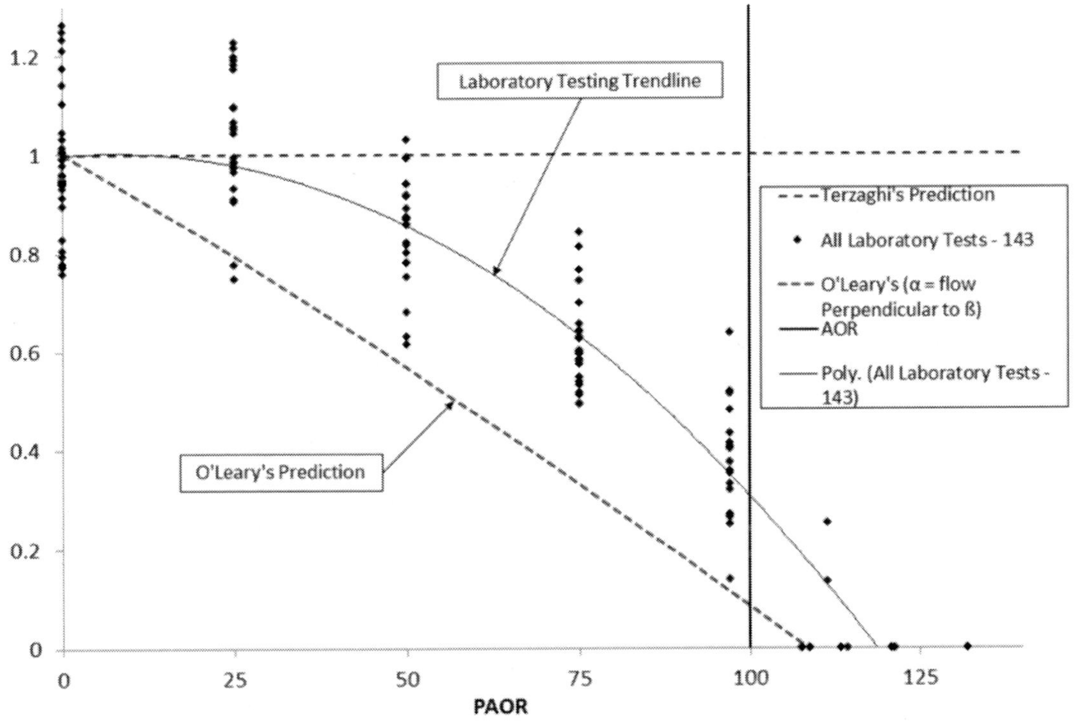

Figure 11. O'Leary's and Terzaghi's predictions versus all laboratory data.

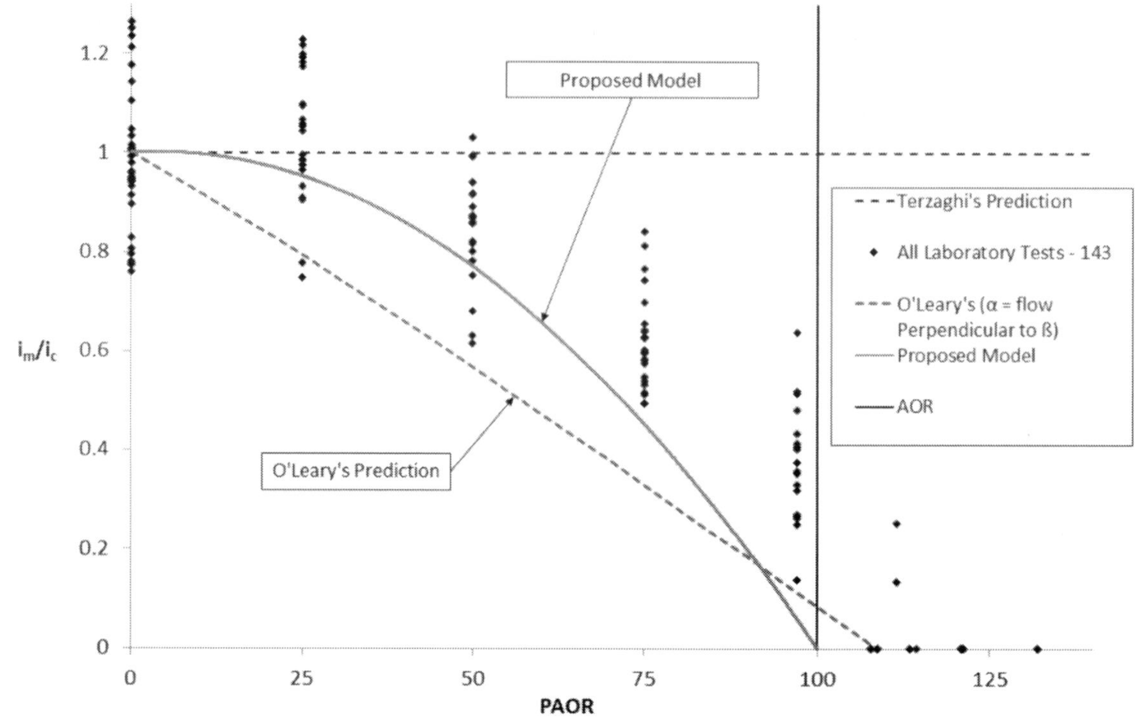

Figure 12. Proposed model compared to Terzaghi's and O'Leary's methods.

As previously explained, Terzaghi's method does not account for exit face inclination when calculating a critical gradient for heave, as shown by the constant line. O'Leary's method shows an almost constant decrease in critical gradient from Terzaghi's prediction at 0 PAOR, to the internal friction angle in terms of PAOR (approximately 109 PAOR). It can be seen between Figures 11 and 12 that neither method accurately follows the trends observed in this research.

CONCLUSIONS

This paper presents an empirical model which can be used to estimate critical gradients for the heave mechanism of internal erosion where inclined exit face conditions exist with sandy soils. It is based on laboratory studies performed at in the Earth Structures Laboratory at Utah State University. Current methods used in engineering practice today are summarized, and the results are compared to the proposed model. Existing theoretical methods are overly conservative compared to the laboratory data obtained in this study. This empirical model better identifies critical conditions and can be used to reduce undue conservatisms in the existing analytical approaches. Engineering judgement should be used when calculating critical gradients for the heave mechanism, and if used, this proposed model should be compared to methods currently used in geotechnical engineering practice.

ACKNOWLEDGEMENTS

This material is based upon work supported by the National Science Foundation (NSF) under Grant CMMI 1131518. Any opinions, findings, and conclusions or recommendations expressed in this material are those of the authors and not necessarily the views of NSF.

REFERENCES

Bligh, W. G. (1910). "Dams, barrages, and weirs on porous foundations." *Engr. News* 64(26), 708-710.

Duncan, J. M., O'Neil, B., Brandon, T., and VandenBerge, D. R. *Evaluation of Potential for Erosion in Levees and Levee Foundations.* CGPR #64, Virginia Tech Center for Geotechnical Practice and Research, 2011.

Fleshman, Mandie, 2012, "Laboratory Modeling of Critical Hydraulic Conditions for the Initiation of Piping", Master's Thesis, Utah State University, 2013.

Fleshman, M. and Rice, J., 2013, "Constant Gradient Piping Test Apparatus for Evaluation of Critical Hydraulic Conditions for the Initiation of Piping", *ASTM Geotechnical Testing Journal*, 36(6), pp 1-14.

ICOLD (2013). *Internal Erosion of Existing Dams, Levees and Dikes, and their Foundations – Volume 1: Internal Erosion Processes and Engineering Assessment.* Bulletin 164 (2013), International Commission on Large Dams (ICOLD), Paris, France.

ILH (2013). *The International Levee Handbook.* CIRIA (2013), London, UK.

Keizer, Richard Allen Jr., 2014, "Laboratory Study on the Effects of Exit Face Inclination of Critical Gradients", Master's Thesis, Utah State University, 2015.

Keizer, R. and Rice, J. (2015). *Evaluating the Mechanics of Piping Erosion Initiation through Laboratory Modeling*. Proceedings of USSD Annual Conference, April 13-17, 2015. Louisville, KY.

Koenders, M. A., and Sellmeijer, J. B. (1992). "Mathematical model for piping." *J. Geotech. Engr.*, ASCE, 118(6), 943-946.

Kovács, G., (1981), *Seepage Hydraulics*, Elsevier Scientific Amsterdam.

O'Leary, T., Vroman, N., Williams, C., Pabst, M., (2013). "Critical Horizontal Seepage Gradients." *ASDSO Journal of Dam Safety*, 11(3), pp 33-39.

Schmertmann, J. H. (2000). "The no-filter factor of safety against piping through sands." Geotechnical Special Publication No. 111, *Judgment and innovation*, F. Silva and E. Kavazanjian, eds., ASCE, Reston, Va., 65-133.

Sellmeijer, J. B. (1988). "On the mechanism of piping under impervious structures." PhD thesis, Technical University, Delft, The Netherlands. TR diss. 1670.

Sellmeijer, J. B., and Koenders, M. A. (1991). A mathematical model for piping. *Applied Mathematical Modelling,* Vol. 115, 646-661. Butterworth-Heinemann.

Sellmeijer, J. B., Lopez De La Cruz, J., Van Beek, V. M., Knoeff, J. G. (2012). *Fine tuning of the backward erosion piping model through small scale, medium scale and IJKdijk experiments.* European Journal of Environmental and Civil Engineering.

Swainston-Fleshman, M., and Rice, J. (2014a). *Evaluating the Mechanics of Piping Erosion Initiation through Laboratory Modeling*. Proceedings of USSD Annual Conference, April 7-11, 2014. San Francisco, CA.

Swainston-Fleshman, M., and Rice, J. (2014b). *Laboratory Modeling of the Mechanisms of Piping Erosion Initiation*. ASCE Journal of Geotechnical and Geoenvironmental Engineering, 140-6.

Terzaghi, K., 1922, "Der Grundbrunch an Stauwerken und Seine Verhutung [The Failure of Dams and Its Prevention]," *Die Wasserkraft*, Vol. 17, pp. 445-449.

Terzaghi, K., and Peck R. B. (1948). *Theoretical Soil Mechanics in Engineering Practice*. Wiley, New York.

USACE (1993). *Seepage analysis and control for dams*, EM 1110-2-1901, US Army Corps of Engineers, Washington DC, USA.

LABORATORY MEASUREMENTS OF CRITICAL GRADIENTS OF COHESIONLESS SOILS

Bryant A. Robbins, PE[1]
Axel Montalvo-Bartolomei, PE[2]
Jamie López-Soto[3]
Isaac J. Stephens, PE[4]

ABSTRACT

Internal erosion refers to mechanisms by which soil particles are eroded from beneath or within embankments due to the forces of moving water. The initiation of internal erosion can be readily prevented through the use of properly designed filters and drainage features in embankments; however, many existing embankments do not include these features. For these structures, the potential for internal erosion to occur is frequently assessed through risk analyses. As part of these assessments, the expected gradients at the site are compared to critical values of hydraulic gradients as determined by laboratory testing to evaluate the probability of internal erosion. Due to the limited critical gradient data available in the literature and the importance of these data on risk assessments, the U.S. Army Corps of Engineers has undertaken a multi-year laboratory testing program to confirm critical gradient values in the literature, evaluate the uncertainties surrounding critical gradient measurements, and extend the ranges of soil types for which critical gradients have been measured. Initial results were obtained from testing uniform soils ranging in size from fine sand to fine gravel. Results indicate that relative density has a significant influence on the critical gradient for all of the soils tested in this study.

INTRODUCTION

Internal erosion refers to any mechanism by which soil particles are eroded from beneath or within embankments due to the forces of moving water. Internal erosion mechanisms can be subdivided into four distinct processes: (1) contact erosion, (2) concentrated leak erosion, (3) suffusion (internal instability), and (4) backward erosion piping (BEP). Only BEP will be discussed as part of this research. For discussion on the other three mechanisms, the authors suggest reviewing Bonelli et al. (2013).

BEP refers to a process by which soil particles are gradually eroded from the foundation of a structure as shown in Figure 1. The erosion process initiates at a downstream, unfiltered exit and progresses gradually beneath a structure, forming a "pipe". If a pipe is allowed to progress entirely beneath a structure connecting the upstream reservoir directly to the unfiltered exit, it can potentially lead to failure of the structure. BEP accounts for approximately one third of all internal erosion related failures (Richards and

[1] U.S. Army Corps of Engineers, Engineer Research and Development Center, Vicksburg, MS, Bryant.a.robbins@usace.army.mil
[2] U.S. Army Corps of Engineers, Engineer Research and Development Center, Vicksburg, MS
[3] U.S. Army Corps of Engineers, Engineer Research and Development Center, Vicksburg, MS
[4] U.S. Army Corps of Engineers, Engineer Research and Development Center, Vicksburg, MS

Reddy, 2007). Therefore, the potential for failure due to BEP is regularly assessed as part of risk assessments of dams and levees.

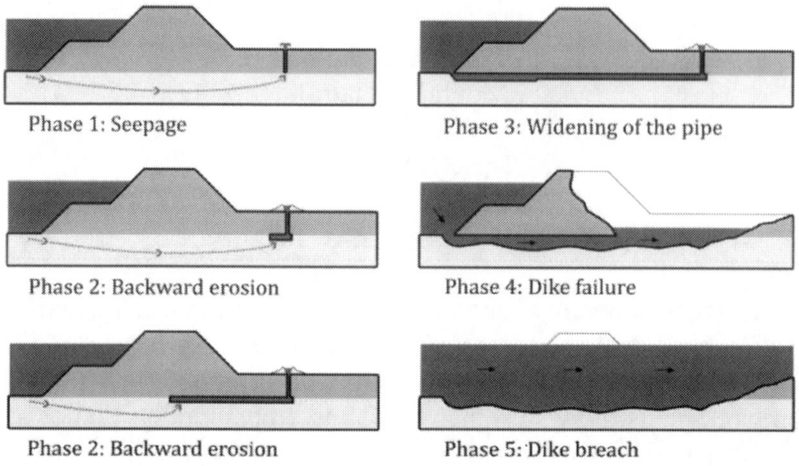

Figure 1. Illustration of the process of backward erosion piping (BEP). (van Beek et al., 2011)

In order to assess the probability of BEP, the potential for BEP progression beneath the structure must be determined. The continual progression of erosion is dependent on sufficient pressure gradients existing along the entire erosion path to erode the materials present. A "point by point" method of assessing the potential for BEP progression was developed by Schmertmann (2000). This method compares the anticipated horizontal pressure gradients in the field to critical values of gradient as determined by laboratory testing, and is commonly used during risk assessments in the United States. Unfortunately, the range of materials for which laboratory critical gradient measurements exist is limited. To expand upon available information in the literature, the U.S. Army Corps of Engineers has undertaken a multi-year laboratory testing effort to confirm critical gradient values in the literature, evaluate the uncertainty surrounding critical gradient measurements, and extend the ranges of soil types for which critical gradients have been measured. The first phase of this laboratory testing effort is presented herein, in which horizontal critical gradients of uniform, cohesionless materials were measured. Small-scale testing was conducted on fine-to-coarse sand, while large-scale testing was conducted on fine gravel. The following sections present an overview of the laboratory equipment, soil properties of the tested materials, the testing procedure, and the test results.

EQUIPMENT

The laboratory equipment constructed for this study was designed based on the small-scale flume constructed by van Beek et al. (2011) and the large-scale flume constructed at University of Florida (Townsend and Shiau, 1986 and Schmertmann, 2000). The large-scale flume was constructed to a width fifty percent larger than the flume used by Townsend and Shiau in order to accommodate the diameter of the coarser particles tested in this study. An overview of the laboratory flumes, overall equipment arrangement, and instrumentation is provided separately for both the small-scale and large-scale flumes below.

Small-Scale Experimental Setup

The small-scale flume (Figure 2) consisted of a rectangular box constructed from 2.54-cm-thick acrylic. The internal flume dimensions are approximately 82-cm long, 30-cm wide, and 10-cm tall. One of the side walls of the flume had 20 ports for installing pressure transducers. Honeywell® 26PC pressure transducers were connected to 14 of the pressure transducer ports to continuously monitor the hydraulic gradient within the sample. The upstream end of the flume was connected to a constant head, overflow tank that was supplied water from a faucet for lower permeability sands. For coarse sands, a 1,920-L water reservoir and 1.5-HP pump were required to obtain the necessary flow rates to induce erosion. A flow meter was connected to the water supply to measure the inflow. Pressure and flow data were continuously logged with a data acquisition system at a frequency of 1 Hz. The downstream end of the flume was attached to an overflow tail-water tank to obtain a constant head, downstream boundary condition on the sample.

The small-flume was attached to a frame that rotated such that the flume could be positioned vertically for sample preparation and rotated into a horizontal position for testing. This feature ensured that an integral contact was obtained between the sand and the acrylic boundary. The downstream end of the sample was supported by a perforated terminal wall wrapped in filter fabric. The terminal wall was 6-cm tall, such that a 4-cm tall slope formed at the angle of repose on the end of the sample when rotated to the horizontal position. The downstream terminal wall was supported by 6 springs attached to the end cap. The downstream terminal wall, end cap, and overflow tank were installed while the sample was still oriented vertically, and the terminal wall was firmly held in place by two latches. For further details on the small-flume apparatus, refer to Robbins et al. (2016a).

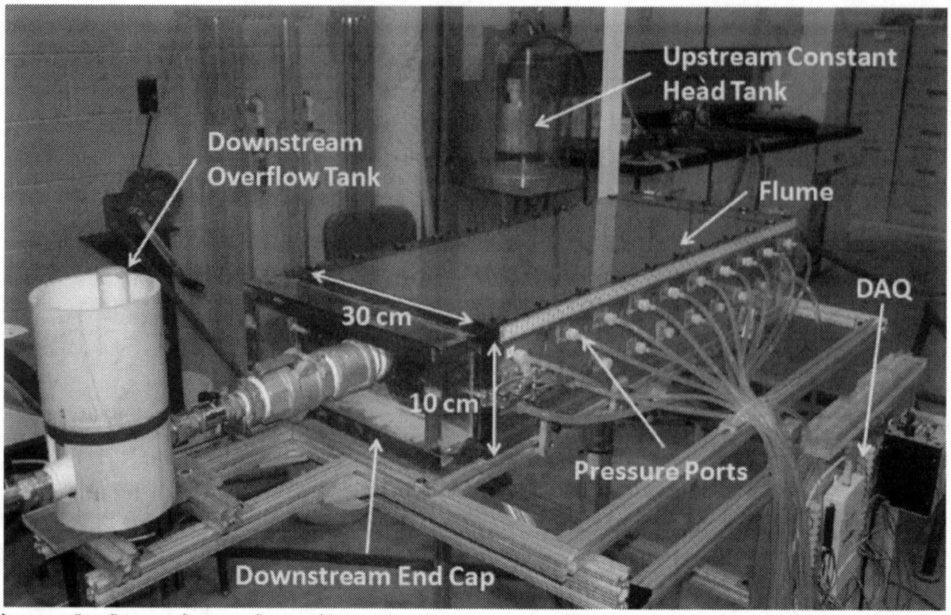

Figure 2. Overview of small-scale, laboratory flume apparatus and equipment.

Large-Scale Experimental Setup

The large-scale flume was similar in design to the small-scale flume in that a sloped exit face was used with a downstream boundary condition established by an overflow tank. The flume walls were constructed of marine-grade plywood supported by a timber frame. The inside dimensions of the flume were 2.4 m in length, 0.45 m in width, and 0.30 m in depth. An overview of the flume apparatus is shown in Figure 3. One side wall of the flume contained 14 ports for pressure measurements. Each port was connected to both a pressure transducer and a manometer such that the pressure transducer readings could always be verified by the manually read manometers.

Due to the size of the flume, rotating from a vertical to horizontal position would have been quite difficult. Therefore, a pressure bladder was installed in the bottom of the flume. By using the pressure bladder, integral contact between the sample and the acrylic top could be ensured. The acrylic top consisted of 1.9-cm-thick acrylic and was fixed into place on top of the flume with a series of clamps as shown in Figure 4.

Water was supplied to the upstream end of the flume by tandem 1.5-HP pumps connected to two reservoirs. The main reservoir was 1,920 L in volume, and the auxiliary reservoir was 960 L in volume. The tandem pumps were connected to a single inflow line containing a flow meter such that the flow rate could be continuously recorded throughout the test along with the pressure measurements. Further details regarding the large-flume apparatus are provided in Robbins et al. (2016b).

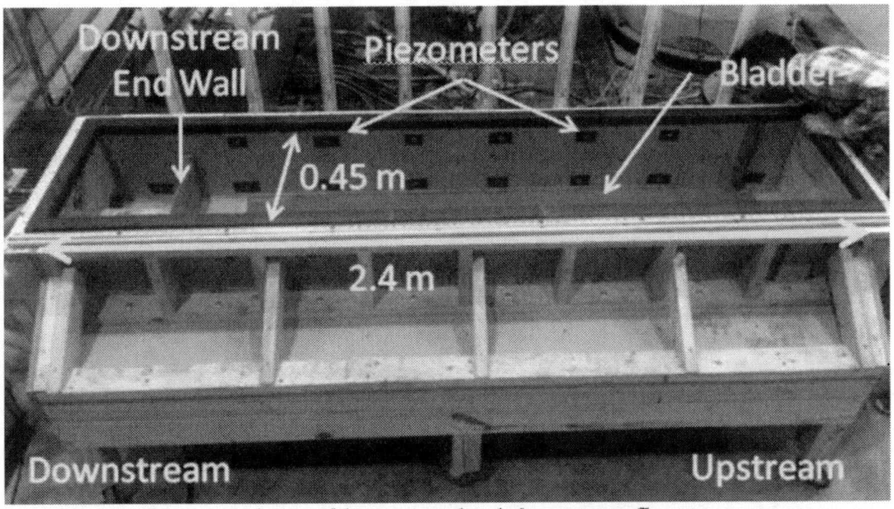

Figure 3. Overview of large-scale, laboratory flume apparatus.

Figure 4. Overview of large-flume system during testing.

MATERIALS TESTED

Eight uniform sands were tested at varying densities in the small flume. The small-scale flume was used for testing the sand because a significant number of tests were desired, and the small-flume testing could be conducted in a rapid and continuous manner. The large flume was used for performing BEP tests on a fine gravel material. The median grain size and coefficient of uniformity for each soil used in this testing program are provided in Table 1.

Table 1. Median grain size and coefficient of uniformity for materials tested.

Soil Name	D_{50} (mm)	C_u
40/70 Sand	0.304	1.42
30/50 Sand	0.429	1.42
20/40 Sand	0.609	1.38
16/30 Sand	0.855	1.41
12/20 Sand	1.127	1.32
8/16 Sand	1.570	1.33
8/12 Sand	2.287	1.27
6/9 Sand	2.517	1.25
Pea Gravel	7.228	1.67

TEST PROCEDURES

While the small-flume and large-flume tests were similar in theory, the different designs required slightly differing sample preparation and testing procedures. The following sections describe the procedures followed for the small- and large-flume testing, respectively.

Small-Flume Procedures

The small-flume was rotated to the vertical position, and the downstream end of the flume was removed such that soil could be manually placed in the flume. The flume was filled approximately halfway with water. Prior to placing any sand in the flume, all submerged flume components (ports, transducer lines, and upstream reservoir) were checked to ensure no air was present. If air was discovered, measures were taken to remove the air. Ensuring saturation of the system prior to sand placement was the only way to ensure air would not interfere with the test.

Once all air was removed from the flume, sand could begin to be placed manually. The sand was carefully pluviated into the flume from a hand scoop or hopper. For loose samples, the sand was continuously placed, avoiding vibrations, until the entire sample volume was filled. For dense samples, the sand was pluviated into the flume in 10-cm layers with each layer being vibrated to the desired density prior to placing the next layer.

After the sand was placed, the terminal wall and end cap were placed on the flume. Water was manually added through the downstream outflow valve until all air was removed from the fittings and flume. At this point, the downstream overflow tank was attached to the flume and saturated. All required hoses were attached, and the flume was rotated to the horizontal position for testing.

Prior to initiating flow, all pore pressure transducers were zeroed to the downstream tail-water level to ensure a consistent datum for pressure measurements. An initial gradient was then applied across the sample (set to a value of 0.05) and measures were taken to ensure all instrumentation (pressure, flow, and video) were responding correctly. The gradient was gradually increased in intervals until particle movement was observed. Once particle movement was observed, the gradient was increased in increments of 0.01 until active and continuous BEP erosion was observed. The gradient across the sample, as determined by a least squares, linear regression of the pressure transducers, was recorded as the critical average gradient for the sample. Figure 5 illustrates the steps involved in the small flume procedure from sample preparation through BEP.

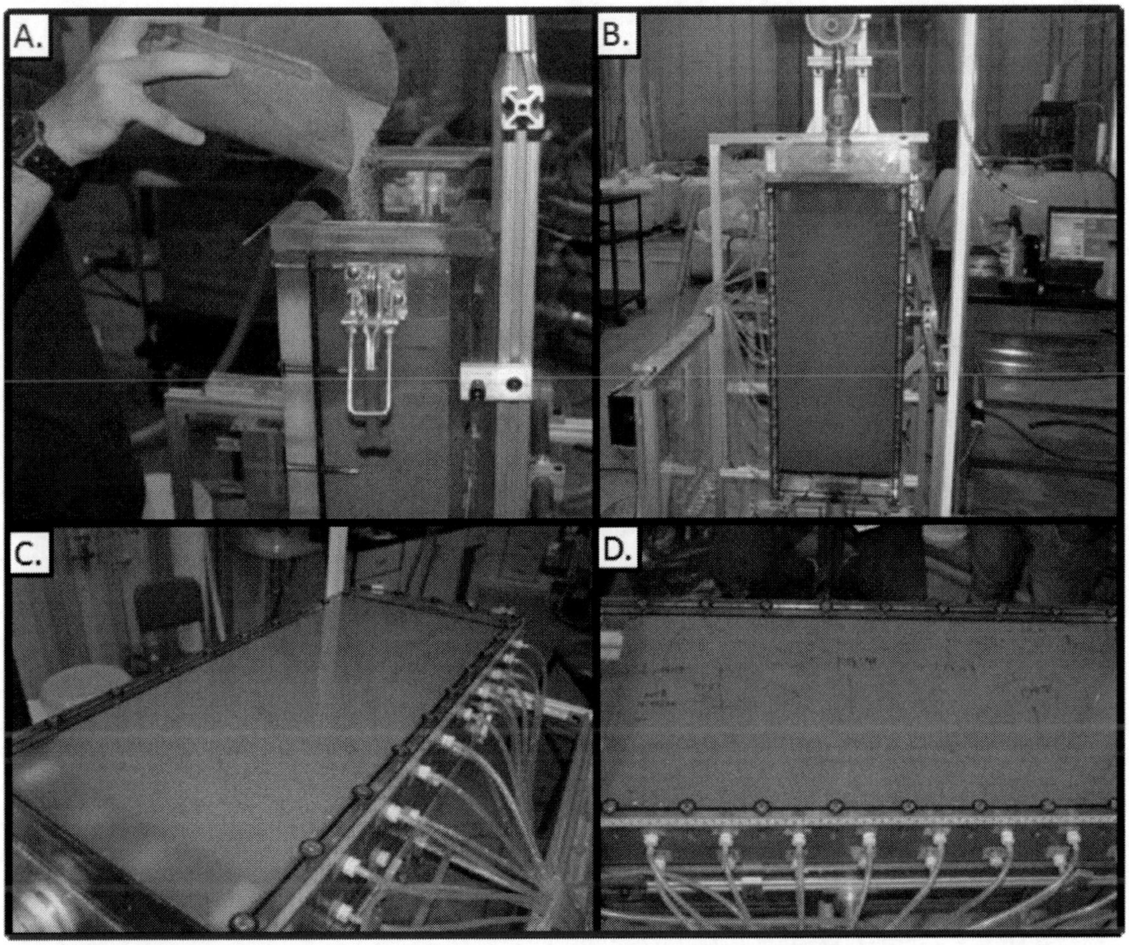

Figure 5. Photographs of small-flume testing procedures: (A) Sample placement, (B) Device assembly and saturation, (C) Test initialization, and (D) Final BEP erosion path.

Large-Flume Procedures

The acrylic top of the large flume was removed completely such that full access to the interior of the flume was possible. The bladder was positioned on the bottom of the flume and covered with a vinyl protective cover to ensure that the bladder was not punctured during testing. The gravel was then placed in lifts in the flume in a dry state. While it was necessary to water pluviate the sand to ensure no air became entrapped in the sample, the gravel was coarse enough that it was easy to saturate the material after placement. For loose samples, the gravel was continuously placed and spread in even lifts until the entire sample had been placed. For dense samples, each lift (following the third lift) was rodded with a steel rod and vibrated with a plate to densify the gravel. The first three lifts were not rodded to avoid puncturing the bladder. After each lift placement, the exit slope was manually formed and shaped with a screed. Photographs illustrating typical sample construction and final sample configuration are provided in Figures 6A and 6B, respectively.

Once the sample was placed, the acrylic top was placed on the flume and clamped down until a 5-mm gap remained between the gravel and the acrylic. This gap was left alone until the flume was saturated to ensure air did not become entrapped between the gravel and the acrylic. The flume was then slowly filled with water until all air had been displaced. A small amount of vacuum was applied to all pressure transducer lines to saturate the system. Once all air was removed from the flume and transducer lines, the acrylic top was slowly and evenly clamped down until it came into contact with the gravel. The acrylic top had two bleed valves downstream to allow air to escape while the top was being clamped into place. The bladder was then pressurized to a seating load of 34.5 kPa. Once the sample came into equilibrium under the seating load, as indicated by no further volume change in the bladder, the bladder pressure was reduced to 19 kPa and held constant during testing. 19 kPa was selected as the testing pressure for the bladder as it resulted in a contact pressure on the acrylic of approximately 14 kPa.

Prior to initiating testing, the pressure transducers were zeroed to the tail-water elevation, and all transducer lines were again checked for any air bubbles. If no air was present, the pump was then turned on, and the inflow valve was slowly opened until flow was observed in the tail-water tank. The flow was adjusted until an initial gradient of 0.05 was obtained. At this initial value, all instrumentation (pressure transducers, flow meter, and video) was examined to ensure proper operation. The gradient was then gradually increased in intervals until particle movement was observed. Once particle movement was observed, the gradient was increased slowly in increments of 0.01 until BEP occurred. The value of the average gradient across the sample at the initiation of BEP, as determined by a least squares linear regression of the pressure transducer measurements, was recorded as the critical average gradient for the test. Photographs of the entire gravel flume testing process, from sample preparation to piping, are provided in Figure 6.

Figure 6. Photographs of large-flume testing procedures: (A) Sample placement, (B) Final sample configuration, (C) Test initialization, and (D) Final BEP erosion path.

TEST RESULTS

The eight sands were only tested in the small flume. For each sand, at least two tests were conducted in a dense state and two tests were conducted in a loose state. This resulted in a total of 32 small-flume tests. A linear trend was observed between critical average gradient and relative density for all sands tested. From this trend, the expected value of the critical gradient for 0% relative density and 100% relative density could be determined. The results are shown in Figure 7. From the test results, the influence of sand density and grain size on critical gradient is observed. Density appears to be a more significant factor in predicting critical gradient than grain size.

Similar small scale testing was conducted by Van Beek et al. (2011) on a sand with a d_{50} of 0.132 mm (Baskarp sand). From a best fit linear trend of the results, the expected values of critical gradients at 0% and 100% relative density were calculated to be 0.33 and 0.51. While quite similar for dense sand, the Baskarp sand had higher critical gradients in a loose state than those in this study. The higher value may be attributed to the coefficient of uniformity of the Baskarp sand being higher (C_u=1.6) than the sands tested in this study.

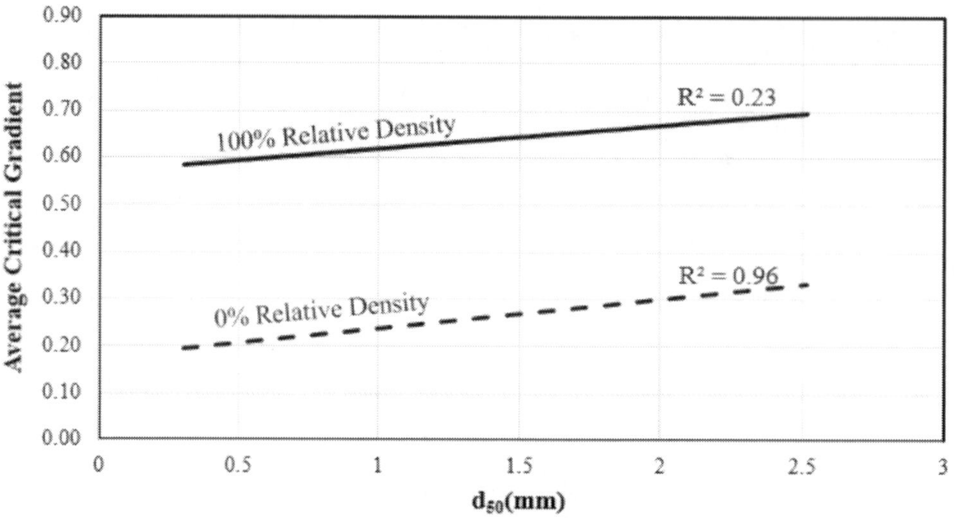

Figure 7. Small-flume test results for eight, uniform sands tested. Results illustrate the influence of grain size and density on critical gradient.

A total of sixteen tests were conducted on the gravel in the large flume with the primary variable being porosity (density). The resulting trend of average critical gradient as a function of porosity is shown in Figure 8. Again, the sample density is found to be significant, with density alone being able to account for a fifty percent increase in critical gradient.

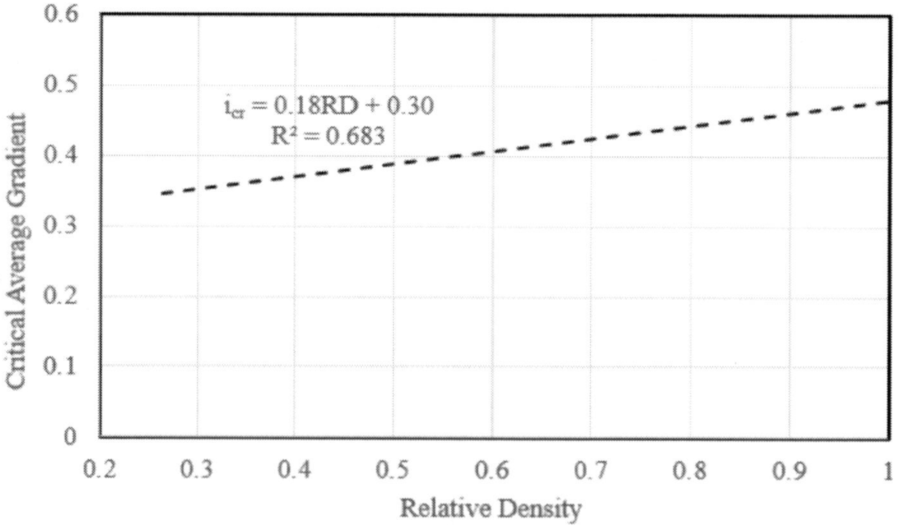

Figure 8. Mean trend of results of sixteen large-flume BEP tests on pea gravel.

CONCLUSIONS

Results of an experimental study are presented in which small-scale and large-scale laboratory tests were conducted to determine the average critical gradients for BEP of various soils. Small-scale testing was conducted on eight uniform sands. Results indicate that the critical, average gradient is influenced by both soil density and grain size, with density being more influential. The average critical gradient varied from 0.20 for loose, fine sand to 0.70 for dense, coarse sand. Measured values of critical gradient for fine gravel were found to vary from 0.36 in a loose state to 0.48 in a dense state in the large scale experiments. It must be recognized that the values for small- and large-scale flume testing must be adjusted for scale effects prior to applying the results of this study to the field.

ACKNOWLEDGEMENTS

The authors gratefully acknowledge the financial support of the U.S. Army Corps of Engineers for the work reported in this paper. Additionally, the authors would like to say thank you to Dan Leavell for his substantial help in developing the test devices and conducting many of the tests.

REFERENCES

Bonelli, S. (2013). *Erosion in Geomechanics Applied to Dams and Levees*. (S. Bonelli & F. Nicot, Eds.). Hoboken, NJ, USA: John Wiley & Sons, Inc. doi:10.1002/9781118577165

Richards, K. S., and Reddy, K. R. (2007). Critical appraisal of piping phenomena in earth dams. *Bulletin of Engineering Geology and the Environment*, *66*(4), 381–402. doi:10.1007/s10064-007-0095-0

Robbins, B.A., Montalvo-Bartolomei A.M., López-Soto, J.F., and Stephens, I.J. (2016). Laboratory Measurements of Horizontal Critical Gradients in Uniform Sands. Vicksburg, MS. U.S. Army Engineer Research and Development Center. Manuscript in Preparation.

Robbins, B.A., Stephens, I.J., Leavell, D.A., López-Soto, J.F., Montalvo-Bartolomei, A.M. (2016). Laboratory Piping Tests on Fine Gravel. Vicksburg, MS. U.S. Army Engineer Research and Development Center. Manuscript in Preparation.

Schmertmann, J. H. (2000). The No-Filter Factor of Safety Against Piping Through Sands. In F. Silva & E. J. Kavazanijian (Eds.), *Judgement and Innovation* (pp. 65–133). American Society of Civil Engineers.

Townsend, F., and Shiau, J.-M. (1986). *Analytical and Experimental Evaluation of Piping and Filter Design for Sands*. Gainseville, Florida.

Van Beek, V. M., Knoeff, H., and Sellmeijer, H. (2011). Observations on the process of backward erosion piping in small- , medium- and full- scale experiments. *European Journal of Environmental and Civil Engineering*, *15*(8), 1115–1137. doi:10.3166/EJECE.15.1115-1137.

RELIABILITY SLOPE STABILITY ANALYSIS FOR RAPID DRAWDOWN

Mark J. Thompson[1]
David J. Bentler[2]
Thomas L. Brandon[3]

ABSTRACT

Reliability slope stability analysis is increasingly being used to evaluate the combined effects of analysis uncertainties and to guide risk-informed decision making on dam projects. This paper describes recent experiences with applying reliability methods to rapid drawdown slope stability analysis in the design of earth dams. A brief background discussion on reliability slope stability analysis is included. The paper also addresses factors that affect reliability calculated with the first-order second-moment method, as well as methods for establishing target reliability. The methods are demonstrated with an example analysis performed during design of an off-channel reservoir. The adoption of reliability methods into the design of earth dams as a supplement to conventional deterministic analyses, as presented in the paper, has successfully supported the confirmation of design criterion selection, prioritization of geotechnical investigation objectives, and communication of project risks to dam owners. These benefits are achieved with only modest increase in effort during project engineering.

INTRODUCTION

As a significant number of geotechnical design procedures have begun to incorporate elements of parameter uncertainty, probability, and risk-informed decision making; the design of earth slopes is still primarily completed using a deterministic, working stress framework. In this framework, slope stability analyses use limit equilibrium or stress reduction methods that incorporate parameters somewhat conservatively selected. Computed factors of safety are then compared against deterministic slope stability design criteria established from a history of observing the performance of slopes designed and constructed to different standards. Managing the risk of unsatisfactory performance is intrinsic to this process. However, notwithstanding the general success of this traditional approach, dam engineers encounter situations where uncertainties and risks are desired to be more explicitly addressed. Even though reliability techniques do not remove uncertainty and do not alleviate the need for engineering judgment (Baecher and Christian, 2003), reliability methods are rational means to quantify and incorporate uncertainty into slope analysis and design.

[1] CH2M, 9191 South Jamaica Street, Englewood, CO 80112; PH (720) 286-0481; Email: mark.thompson@ch2m.com
[2] CH2M, 9191 South Jamaica Street, Englewood, CO 80112; PH (720) 286-4896; Email: david.bentler@ch2m.com
[3] Virginia Polytechnic Institute and State University, 22 Patton Hall, Blacksburg, VA 24061; PH (540) 231-4454; Email: tbrandon@vt.edu

Reliability slope stability analysis can take different forms with a corresponding range of effort needed for successful application. The most effective applications of reliability methods are those in which the relative probabilities of failure for design alternatives are evaluated, or those in which the methods are used to highlight the effects of analysis uncertainties. In general, attempts to determine the absolute probability of failure are much less successful (Christian et al., 1994). Other appropriate applications of reliability slope stability analysis include:

- Identifying analysis parameters most significantly contributing to risk of failure;
- Prioritizing financial resources for improving or further studying the most sensitive parameters;
- Performing cost-benefit analysis of design or rehabilitation alternatives;
- Conducting feasibility studies or preliminary design when limited geotechnical data are available for confidently selecting design parameters;
- Evaluating the appropriateness of deterministic design criteria for uncommon problems; and
- Communicating project risks to owners.

This paper expounds on lessons learned from recent experiences with applying reliability methods to slope stability analysis in the design and evaluation of earth dams. Many of the lessons presented in this paper are oriented around rapid drawdown slope stability, yet the methods and benefits are also applicable to other slope stability loading conditions. Since a comprehensive review of reliability slope stability analysis methods is much broader in scope than this paper, the specific objectives of this paper are simply to highlight the reliability methods most likely to be applied by engineers and to draw attention to certain analysis mechanics that could affect the computation or interpretation of reliability. With only a modest increase in effort during project engineering, reliability slope stability analysis can add significant value to a project as a supplement to conventional analyses.

RELIABILITY METHODS FOR SLOPE STABILITY ANALYSIS

Reliability analysis procedures for slope stability vary in assumptions, limitations, and complexity. Most procedures fall into one of two general categories: approximate methods or Monte Carlo simulation. Christian et al. (1994), Tang et al. (1999), Duncan (2000), El-Ramly et al. (2002), Baecher and Christian (2003), Wolff et al. (2004), and Sleep and Duncan (2014) have discussed in detail the application of these reliability methods to geotechnical problems. Readers are referred to this sampling of references for underlying theory and examples of reliability analysis in practice.

The two approximate methods of reliability analysis which are, in the authors' opinion, most easily and most commonly applied to slope stability analysis include the first-order second-moment (FOSM) method and the Hasofer-Lind method. These methods typically require analysis inputs in the form of expected values, standard deviations, and correlation coefficients. The methods produce reasonable and consistent results for (nearly) linear functions and random variables with relatively small coefficients of variation (Wolff et al., 2004). A reliability analysis using an approximate method produces an estimate of reliability as its primary result. The results also indicate which of the random variables contribute most significantly to probability of failure. In this regard, first-order second-moment reliability methods have frequently been used by the authors to conduct sensitivity analyses in a probabilistic framework, often making the case for supplemental geotechnical investigation of specific analysis parameters.

Monte Carlo simulation is a simple computational procedure used to conduct a reliability analysis. A Monte Carlo analysis incorporates the same parameters as approximate methods, except that the entire distribution of a random variable may be assumed. The ability to characterize random variables as having any distribution makes Monte Carlo methods more attractive when random variables are not normally distributed or when coefficients of variation are large. Monte Carlo trials are conducted using input parameters based on normalized random variables. The performance function – the factor of safety for slope stability analysis – is then characterized as a random variable with certain distribution based on results from the trials. The probability of failure is roughly estimated as the percentage of trials returning a factor of safety less than unity.

Monte Carlo simulation is the most commonly encountered probabilistic technique in commercial slope stability software programs, oftentimes making Monte Carlo simulation a logical choice for conducting a reliability analysis. Still, some software programs do not accommodate sufficient user-control over the evaluation. Therefore, familiarity with multiple reliability methods is beneficial. Moreover, using more than one method is encouraged for situations where a reliability analysis is informing a risk-based dam design or dam safety decision.

IMPLEMENTATION OF THE FOSM METHOD

This section presents some considerations for conducting reliability slope stability analyses using the first-order second-moment method. The potential effects of analysis implementation methods on the analysis results are demonstrated, highlighting the importance of analysis details in trying to produce reasonable estimates of reliability.

The three-stage rapid drawdown analysis procedure by Duncan et al. (1990) is recommended and used herein for conducting reliability slope stability analysis for rapid drawdown, despite some inherent conservatism in the approach. Multi-stage undrained analyses are preferred over effective stress analyses since shear-induced pore pressures and pore pressures due to rapid changes in loading are not easily predicted. The three-stage procedure is also preferred over the multi-stage undrained finite element analysis approach (VandenBerge and Brandon, 2014). The finite element method does not lend itself to easily conducting many analyses, nor does the strength reduction approach allow for isolating a pre-defined slip surface for which to evaluate either factor of safety or reliability.

Hypothetical Embankment Dam

A simple, hypothetical embankment dam is shown in Figure 1. The example depicts a 100-foot-high compacted clay embankment constructed on a clay foundation at Elevation 0 feet. Interior and exterior slopes are 3:1. The interior slope includes a 70-foot-wide bench at Elevation 40 feet. A 10-foot-wide chimney drain and 5-foot-thick blanket drain control seepage through the dam. Normal pool is at Elevation 80 feet, and the reservoir is drawn down to varying levels (down to Elevation 20 feet in Figure 1). Material parameters for the model are provided in Table 1. For complete drawdown from Elevation 80 feet to Elevation 0 feet, the most likely value of the factor of safety, F_{MLV} – the computed factor of safety with each parameter set equal to its mean value – equals about 1.31, which corresponds to the *deep* slip surface in Figure 2 extending into the foundation.

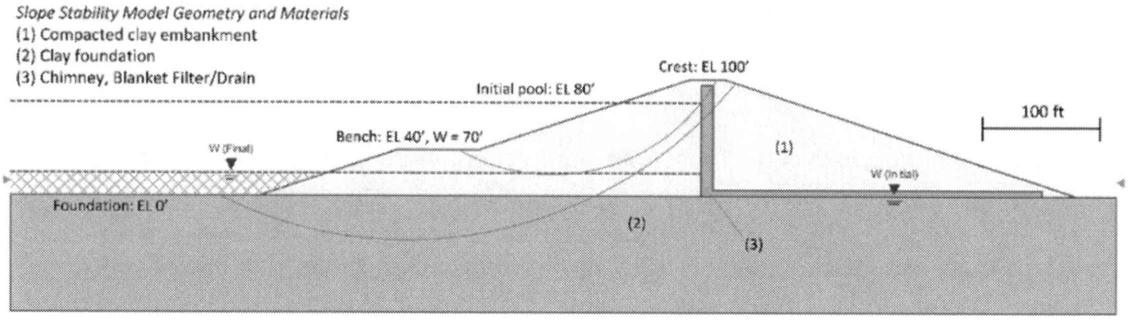

Figure 1. Hypothetical Embankment Dam for FOSM Example

Table 1. Material Parameters for Hypothetical Embankment Dam

Material	Unit Weight (pcf)	φ' (deg) MLV	φ' (deg) σ	c' (psf) MLV	c' (psf) σ	ϕ_{CU} (deg) MLV	ϕ_{CU} (deg) σ	c_{CU} (psf) MLV	c_{CU} (psf) σ
Clay embankment	125	28	2	300	100	12	1	1000	330
Clay foundation	125	28	2	0	0	12	1	700	200
Chimney, blanket	125	36	0	0	0	NA	NA	NA	NA

Note: the calculation of slope and intercept parameters for the failure shear stress versus failure plane consolidation stress envelope from ϕ_{CU} and c_{CU} is described by Duncan et al. (1990).

Estimating the Variance of Factor of Safety

The first-order second-moment method of reliability analysis amounts to an effort to describe the factor of safety, *F*, as a random variable with a known or assumed probability distribution defined by its mean (most likely value) and variance, σ_F^2. The variance of *F* is proportional to the partial derivatives of *F* with respect to each of uncertain variable *x*. The following sections describe considerations for making accurate estimates of the partial derivatives.

Fixed Slip Surfaces for Analysis. A first consideration for estimating the partial derivatives of F relates to the specification of the slip surface for which to calculate the factor of safety. A later section discusses slip surface selection. Once a potential slip surface has been identified, the slip surface must be fully specified for calculation of factor of safety associated with varied parameters x. Where the slip surface is not fully specified and held constant, the search algorithms to find the global minimum factor of safety is likely to find a slightly (or significantly) different slip surface. This difference in slip surface, which represents a unique performance function for the slope, could be accompanied by a difference in factor of safety attributed to the different performance function, i.e. slip surface, and not values of the analysis parameters. Accordingly, the difference in F for two different values of an analysis parameter to estimate a partial derivative by central difference approximation could be inaccurate. The likely effect of not holding constant the slip surface over the entire reliability analysis is overestimation of the partial derivatives of F, which leads to significant underestimation of slope reliability. Grossly erroneous reliability estimates can be produced if the variation in x to estimate a partial derivative results in a radically different slip surface than for F_{MLV}.

Central Difference Approximation of Partial Derivatives. Estimates of the partial derivatives of F are usually made by a central difference approximation since simple analytical solutions for real slope stability problems rarely exist. Application of the first-order second-moment method by Duncan (2000) assumes that the central difference approximation is evaluated with x varying by plus and minus one standard deviation of x. This approach simplifies the calculation of σ_F. However, a more precise estimate of the partial derivative can be made by varying x by some smaller percentage of σ_x. Slope stability reliability was estimated for the *shallow* slip surface in Figure 2 by evaluating the central difference approximation with x varying by σ_x, $0.5\sigma_x$, $0.2\sigma_x$, and $0.1\sigma_x$. Results are summarized in Table 2. For the presented example, varying x by up to σ_x does not provide a significantly different estimate of reliability than approximating the partial derivative over a smaller range about the mean values of x. Much more uncertainty in the estimated probability of failure results from the assumed statistical distribution of F than how the variance of F is determined.

Table 2. Effect of Variable Change on Evaluation of Partial Derivatives

$\pm\Delta x$	F_{MLV}	σ_F	P_f (Normal)	P_f (Lognormal)
1.0 σ_x	1.59099	0.26144	0.011895	0.002862
0.5 σ_x	1.59099	0.26147	0.011902	0.002865
0.2 σ_x	1.59099	0.25790	0.010967	0.002534
0.1 σ_x	1.59099	0.25783	0.010948	0.002527

Note: rapid drawdown from Elevation 80 feet to Elevation 0 feet

Convergence Criteria and Factor of Safety Precision. As with any slope stability analysis, factor of safety precision and convergence criteria for a reliability analysis should be reviewed. Central difference approximations of the partial derivatives of F are proportional to the difference between two calculated factors of safety, such that using too few decimal places in the factor of safety can result in either overestimation or underestimation of the partial derivative. The number of decimal places needed for accurate approximation of the partial derivative also depends on how the uncertain variables x are varied, since using smaller Δx results in smaller difference in F from F_{MLV}.

Table 3 shows the effect using different factor of safety precisions on the estimate of failure probability. In these calculations, uncertain variables were varied by plus and minus $0.1\sigma_x$. For the example, failure probabilities are all similar for each of the factor of safety precisions used. However, other experiences have shown more significant differences in results. Using more decimal places is, therefore, recommended over fewer decimal places, particularly since this practice involves no extra effort.

Table 3. Effect of Factor of Safety Precision on Slope Reliability

Decimal Places	F_{MLV}	σ_F	P_f (Normal)	P_f (Lognormal)
2	1.59	0.25495	0.010329	0.002321
3	1.591	0.25710	0.010760	0.002462
4	1.5910	0.25794	0.010975	0.002536
5	1.59099	0.25783	0.010948	0.002527

Note: rapid drawdown from Elevation 80 feet to Elevation 0 feet

Slip Surface Selection

When evaluating the stability of a slope, there are unlimited potential slip surfaces with corresponding factors of safety. In a deterministic analysis, the critical slip surface is that with the minimum factor of safety. However, this critical slip surface does not always represent the slip surfacing having highest probability of failure.

Slope stability analyses were conducted for the *shallow* and *deep* slip surfaces shown in Figure 1. Conditional probability of failure and F_{MLV} for different drawdown magnitudes are provided in Figure 2. For drawdown from Elevation 80 feet to about the bench elevation, F_{MLV} for both slip surfaces are about the same – the shallow slip surface being slightly lower. For drawdown to pool elevation below the bench, the deep slip surface gives a lower factor of safety.

Reliability analyses tend to show that conditional probability of failure is higher for the shallow slip surface than for the deep slip surface, even though F_{MLV} is higher than for the deep slip surface. The variance of F (and therefore the probability of failure) is higher for the shallow slip surface, because as the number of uncorrelated variables (different materials) is reduced (in this case, by not considering the foundation clay), σ_F becomes more similar to σ_x. With fewer variables considered in the analysis, the probability of having the low values of the variable being the operative input parameter is higher than the probability of having the low values of two or more variables being operative. Therefore, a number of different slip surfaces representing different performance functions should be evaluated in a reliability analysis. As the presented example demonstrates, simply using the slip surface giving lowest factor of safety does not always produce the highest probability of failure.

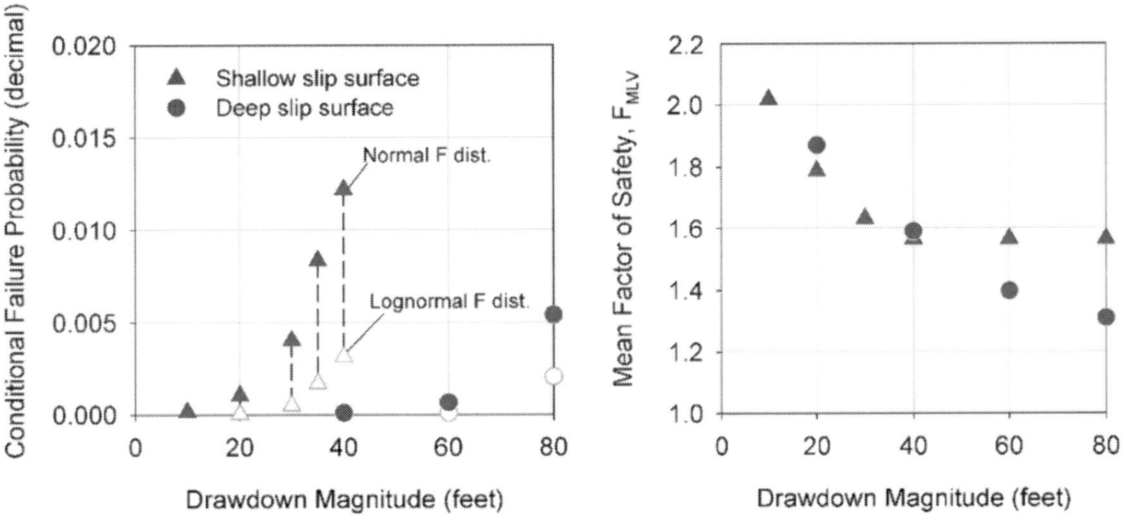

Figure 2. Evaluation of Failure Probability and Factor of Safety for Two Slip Surfaces

The authors note that rapid drawdown slope stability analyses of small to moderate height dams on soil foundations often show critical slip surfaces that extend into the foundation. However, few *deep* rapid drawdown failures have been documented. A slope stability reliability perspective supports the suggestion that shallow slip surfaces contained within the embankment dam are often more critical than deep slip surfaces, regardless of the computed factors of safety for the reasons described above. This observation is further supported by experiences with landslides, which are often attributed to weakness or geologic factors associated with a single soil unit, not the operative properties for all materials contained in the slope.

TARGET RELIABILITY

Reliability analyses enable assessment of slope stability, inclusive of analysis uncertainties, and support owner/engineer decisions, wherein tolerable risks are usually balanced against investment of often-limited resources for improving performance and mitigating hazards. Computed reliability from such an analysis is a relative estimate of the expected slope response under the considered loading conditions. Slopes with a high reliability index are not susceptible to failure or unsatisfactory performance, whereas slopes with a low reliability index are more prone to perform poorly and present rehabilitation problems. A very low reliability index is indicative of a potential hazard. Two approaches for interpreting computed reliability index or establishing target reliability for design include: (1) reference values, and (2) Bayesian reliability updating method. These approaches are described in the following sections.

Reference Values

The simplest approach for interpreting slope stability reliability is to compare computed results against reference reliability indices. Expected performance levels with corresponding reliability index and failure probability based on USACE guidance is provided in Table 4. Christian et al. (1994) and Loehr et al. (2005) have both suggested that typical target failure probability for slope design purposes is 0.001. Target failure probability could range up to about 0.01 for slopes of lesser importance or with lesser consequence of failure. Santamarina (1992) summarized acceptable P_f based on a survey of engineers involved in slope stability analysis (see Table 5).

Table 4. Reliability Index Corresponding to Expected Performance (after USACE, 1997)

Expected Performance Level	Reliability Index, β	Failure Probability, P_f
High	5	0.0000003
Good	4	0.00003
Above average	3	0.001
Below average	2.5	0.006
Poor	2	0.023
Unsatisfactory	1.5	0.07
Hazardous	1.0	0.16

Table 5. Acceptable Probability of Failure of Slopes (after Santamarina, 1992)

Condition	Acceptable P_f
Temporary structures: no potential life loss, low repair cost	0.1
Minimal consequence of failure: high cost to reduce P_f (e.g., open pit mine)	0.1 – 0.2
Minimal consequence of failure: repairs can be done when time permits, repair cost less than cost of reducing P_f	0.01
Existing large cut on interstate highway	0.01 – 0.02
Large cut on interstate highway to be constructed	< 0.01
Lives may be lost when slope fails	0.001
Acceptable for all slopes	0.0001
Unnecessarily low	< 0.00001

Bayesian Reliability Updating

Factor of safety design criteria for rapid drawdown loading varies with the estimated frequency of rapid drawdown loading. Reservoirs operated with frequent drawdown occurrences – pumped storage reservoirs, for example – generally demand a higher minimum factor of safety than reservoirs for which rapid drawdown is a rare event. A reliability updating approach for estimating a frequency-of-loading-based target reliability can be used to assess the appropriateness of selected deterministic design criterion for rapid drawdown applied to a specific reservoir and operating plan.

This target reliability approach uses the premise that the reliability estimate for a particular slope design should exceed the reliability associated with a target performance for the dam. Target performance is the number of discrete loading events expected in the design life with acknowledgement that the desire to prevent failure implicitly requires that each event has a success outcome. For this scenario, the number of events, N, equals the number of successes, α. The number of failures, β, in N events is zero. Success and failure are defined by $F>1$ and $F<1$, respectively. Target slope reliability is then estimated based on the number of successes and failures given by this target performance.

Reliability, R, is itself a random variable whose characteristics are prescribed by some probability distribution having R_{min} equal to 0 and R_{max} equal to 1. The reliability distribution is taken as a beta variate. When the number of successes exceeds the number of failures, the distribution is skewed left, where reliabilities favor higher probabilities. The expected value (i.e., mean) of R is estimated by Equation 5.

$$\bar{R} = E[beta(R\_\alpha, \beta)] = \frac{\alpha+1}{\alpha+\beta+2} \qquad (5)$$

For $R_{min} = 0$, $R_{max} = 1$, and $\beta = 0$, the expected value R can be computed for any number of events that all result in consecutive successes (i.e., $\alpha = N$). Figure 4 shows this relationship. Confidence limits can also be extended to the Bayesian estimate of reliability following Harr (1987). For different values of $N = \alpha$, the expected value of R and the upper limit reliability, R_U, for different confidence limits can be calculated (see Figure 3). A target reliability can, therefore, be calculated for a given number of anticipated loading events and required confidence.

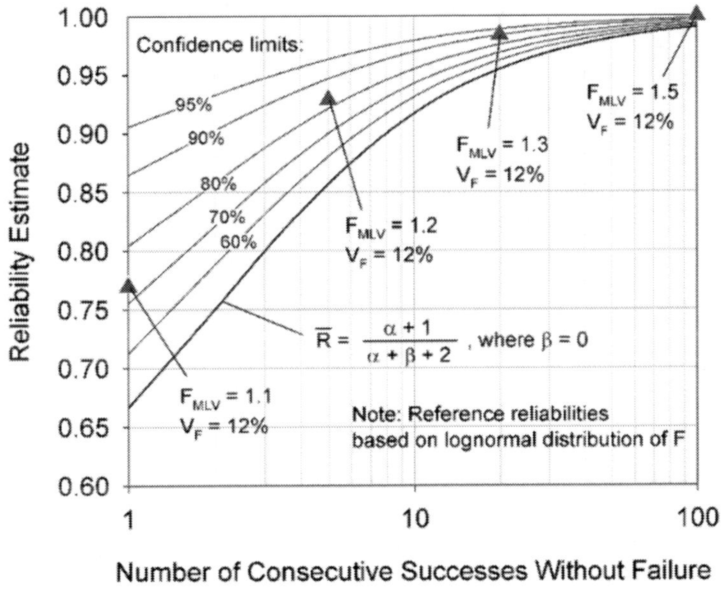

Figure 3. Target Reliability with Reference Reliability Estimates Corresponding to Deterministic Design Criteria

The Bayesian approach to estimating reliability is similar to the logic behind suggested design criteria for deterministic slope stability analysis and design. The reasonableness of the Bayesian probability approach to estimate target reliability considering frequency of loading is supported by considering conventional rapid drawdown slope stability design criteria used for different types of dams. These design criteria include minimum factors of safety, as follows:

- $F > 1.1$, for dams experiencing minimum change in pool or drawn down only one or two times during the design life.
- $F > 1.2$ or 1.3, for dams subject to more regular rapid drawdown loading.
- $F > 1.4$ or 1.5, for pumped storage schemes wherein dams are subject to daily, weekly, or monthly drawdown loading.

The reliability calculated from the deterministic design criteria as *most likely values* of the factor of safety are also shown in Figure 3. The reliability calculation assumes: (1) coefficient of variation of F, V_F, is equal to 12 percent (a reasonable value), and (2) lognormal distribution of F. The approximate number of consecutive successes and corresponding reliability values are shown in Figure 3 relative to the expected Bayesian reliability and confidence limits associated with consecutive successes. As the frequency of loading increases, the required reliability also increases to account for the increased *exposure* to the loading. The reference reliability values (triangle symbols) representing conventional design criteria follow a similar frequency-based reliability trend as the proposed approach.

Limitations of this comparison include:

- The number of consecutive successes (without failure) applied to each of the qualitative loading frequency scenarios is approximate and subjective.
- The calculated reliabilities are based on a coefficient of variation equal to 12 percent. Slopes and foundations having higher or lower variation would give a different calculated reliability.
- The calculated reliabilities assume a lognormal distribution of F. In most instances, higher probability of failure and lower reliability would be obtained assuming a normal distribution of F.
- The calculated reliabilities assume that the deterministic factor of safety design criteria are most likely values of F. Implementation of the design criteria usually involves some conservatism in selecting shear strength for conducting the rapid drawdown loading. Therefore, the reliabilities shown in Figure 3 are absolute minimum reliabilities that should be achieved.

While strict application of the Bayesian approach for developing a rigid, reliability-informed design criterion is not proposed, the comparison does support the approach as a rational tool for evaluating probabilities of failure calculated using reliability methods.

EXAMPLE: OFF-CHANNEL RESERVOIR

Reliability slope stability analysis was used in the engineering of Lane City Reservoir (LCR), which is an off-channel reservoir in the Texas Gulf Coast. The 40,000 acre-foot reservoir is formed by a continuous embankment dam approximately 40 feet high and over 5 miles in length. Given this length of embankment dam, modest changes to slopes resulted in substantial changes to project cost estimates. With rapid drawdown loading controlling the interior slope configuration, engineering effort was invested to more fully understand the applicability of deterministic design criterion and the comparative risk of failure associated with alternative interior slope configurations.

Parameter Sensitivity Analysis

During preliminary engineering of the project, prior to acquiring significant geotechnical data with which to conduct design analyses, reliability slope stability analyses were performed to corroborate deterministic analyses using assumed material parameters and to identify the analysis parameters having most significant effect on slope reliability. Table 6 contains the material parameters and assigned values for a sensitivity analysis, which was conducted in a reliability framework using the first-order second-moment method. The results in Table 6 are for a 41-foot-high, homogeneous (clay) embankment with 3.5:1 interior slope. F_{MLV} =1.39 corresponds to the circular slip surface giving minimum factor of safety, which was fixed for subsequent analyses.

Table 6. Preliminary First-Order Second-Moment Calculation with Central Difference Approximation Assuming $\Delta x_i = \sigma_{xi}$

Material Parameter	μ_x	σ_x	$(\Delta F/2)^2$ ($\times 10^{-8}$)
Clay fill (interior slope)			
Saturated unit weight, γ_{sat} (pcf)	135	4	1,400
Drained strength parameters, ϕ'-c' (deg, psf)	23-240	3-80	95,000
Undrained strength parameters, ψ-d (deg, psf)	19-355	3-110	550,000
Random fill (exterior slope)			
Unsaturated unit weight, γ_t (pcf)	135	4	33
Drained strength parameters, ϕ'-c' (deg, psf)	28-0	3-0	110
Clay foundation			
Saturated unit weight, γ_{sat} (pcf)	125	4	17,000
Drained strength parameters, ϕ'-c' (deg, psf)	28-0	3-0	300,000
Undrained strength parameters, ψ-d (deg, psf)	17-600	3-200	2,000,0000
Chimney, blanket drain			
Unsaturated unit weight, γ_t (pcf)	125	4	4
Drained strength parameters, ϕ'-c' (deg, psf)	36	3	220 E-08
	F_{MLV} = 1.39	$\sigma_F = [\Sigma(\Delta F/2)^2]^{0.5}$ = 0.1707	

The tabulated values in Table 6 indicate which analysis parameters contribute most significantly to the probability of failure. The example shows quantitatively that uncertainty in shear strength parameters for clay fill and clay foundation materials most notably affect σ_F. In particular, undrained strength parameters are the critical parameters for the reliability of the embankment. This reliability analysis justified detailed investigation of the strength characteristics of the foundation and of compacted Beaumont clay.

A reliability analysis to assess the sensitivity of analysis parameters requires evaluation of multiple slip surfaces, not just the slip surface giving minimum factor of safety with mean values of the parameters. Non-circular slip surfaces were also considered. In addition, the authors observed that changes in mean values of the analysis parameters resulted in slight to substantial changes to the critical slip surface. With these different performance functions, the effect of parameter uncertainty may vary and could affect interpretation of the critical parameters.

Shear Strength Characterization

Shear strength parameters for the dam foundation were established from consolidated-undrained triaxial compression tests conducted on undisturbed samples trimmed to 2-inch diameters. Strength data from the nearby South Texas Project Main Cooling Reservoir (MCR) were also considered, since the MCR was built in a similar geologic setting as the LCR. Stress path failure points based on failure criterion of maximum deviator stress with 10 percent limiting axial strain are shown in Figure 4a. The graphical N-sigma rule (see Sleep and Duncan, 2014) was applied to the data, where $\pm 2\sigma$ bound the data ($n = 73$) and the MLV parameters divide the range in half. Drained shear strength parameters directly define the $K_c = K_f$ envelope for the three-stage rapid drawdown analysis procedure. Undrained shear strength parameters ϕ_{CU} and c_{CU} are presented in Figure 4a, however, these parameters were converted to ψ and d, which define the $K_c = 1$ envelope.

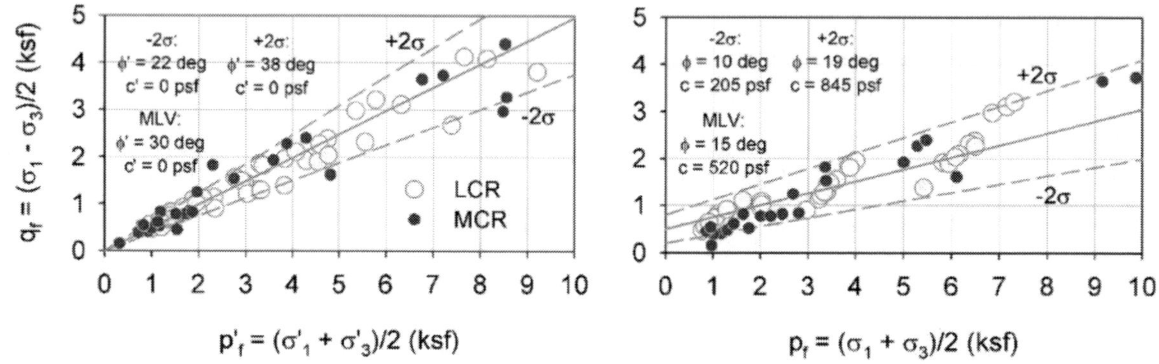

(a)

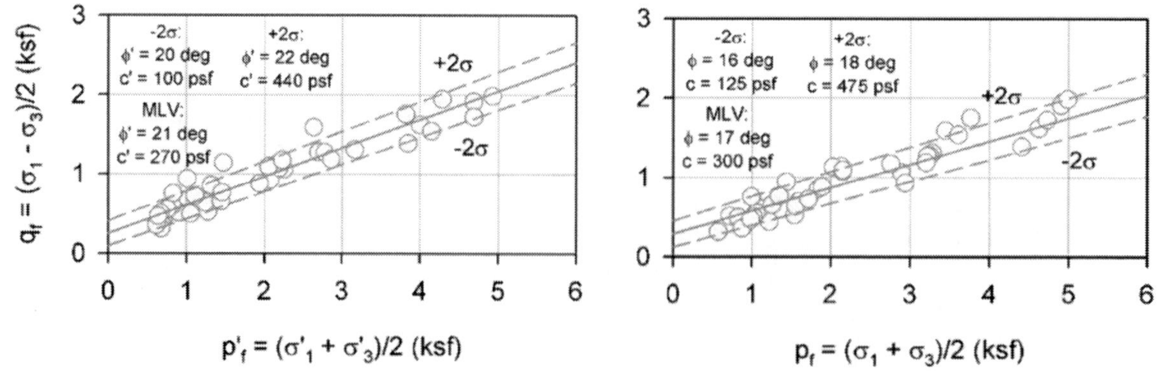

(b)

Figure 4. Drained and Undrained Shear Strength Parameters for: (a) Foundation Clay and (b) Compacted Clay Embankment

Compacted clay embankment strength parameters were developed similarly, except that consolidated-undrained triaxial tests were conducted on laboratory-compacted clay specimens prepared at various compacted moisture-density conditions using a Harvard miniature compaction apparatus. *Most likely values* of shear strength parameters with $\pm 2\sigma$ bounds are shown in Figure 4b. Additional details of the testing and characterization of shear strength of compacted Beaumont clay are provided in Thompson et al. (2016).

Reliability Slope Stability Results

Simulation of future reservoir operation using historical hydrological records indicated that rapid drawdown would be a common occurrence throughout the life of the project. Frequent partial drawdown and regular full drawdown events are expected. Considering this anticipated reservoir operation, a target reliability of 0.993 for the interior slope was selected. This reliability was used to inform engineering judgment and as a benchmark value for evaluating the results of reliability slope stability analyses using first-order second-moment methods. Strict adherence to this "required" reliability was not proposed for the following reasons:

- Uncertainty in the distribution of factor of safety (for example, normal or lognormal).
- The three-stage rapid drawdown analysis procedure is implicitly conservative with respect to using drained or undrained shear strength parameters – those which produce the lowest factor of safety.
- The rapid drawdown slope stability analysis is based on pore-water pressures generated by water tables, whereas actual pore-water pressures in the embankment and foundation are lower because of head loss that occurs with seepage through the clay materials. Higher consolidation stresses associated with lower pore-water pressures would give factors of safety higher than those computed.
- Saturation of the embankment and foundation clay is anticipated to take many years, such that the number of drawdown events qualifying as true rapid drawdown events – drawdown loadings that meet the assumptions of the analysis – is less than the assumed exposure.

A slope stability model for the Lane City Reservoir embankment is shown in Figure 5. The final interior configuration included a steepened 2H:1V section near the crest, 3H:1V slopes, and a small bench approximately 28 feet below the crest. The benched slope configuration was selected to accommodate an interior access road. Deterministic rapid drawdown slope stability analysis using *design* shear strength parameters based on the U.S. Army Corps of Engineers' two-thirds rule was conducted for the proposed interior slope subject to complete drawdown from maximum normal pool to the slope toe. The deterministic factor of safety was equal to 1.31, and the critical non-circular slip surface extended into the foundation (see Figure 5).

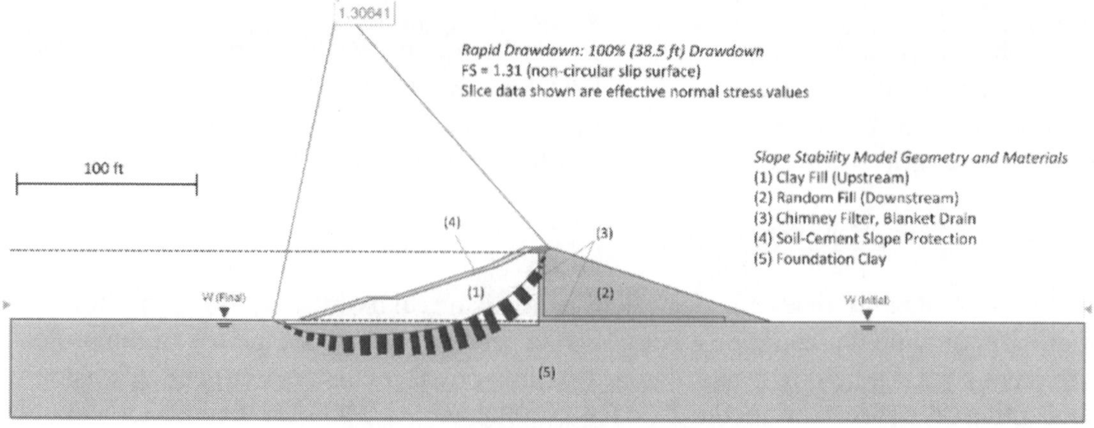

Figure 5. Deterministic Factor of Safety and Critical Slip Surface for Off-Channel Reservoir Embankment Dam

Reliability analyses were conducted for the proposed interior slope using the first-order second-moment method, wherein reliability index and failure probability were evaluated for specified loading conditions defined as a percent drawdown from normal maximum pool. Figure 6 shows the fragility curve resulting from the first-order second-moment analysis. The 3:1 interior slope with bench subject to the anticipated frequency of rapid drawdown loading was considered to be an acceptable slope configuration, because the highest calculated probability of failure for full drawdown (that is, 0.0036 for normal distribution of *F*) is less than the maximum allowable value of 0.007, as suggested by the target reliability of 0.993. Considering a lognormal distribution of *F*, the highest conditional probability of failure is less than 0.001, which corresponds to above average performance and acceptable risk when lives may be lost upon slope failure (see Tables 4 and 5). Therefore, satisfying the deterministic design criterion for rapid drawdown loading of factor of safety exceeding 1.3 was accepted as being appropriate for this project.

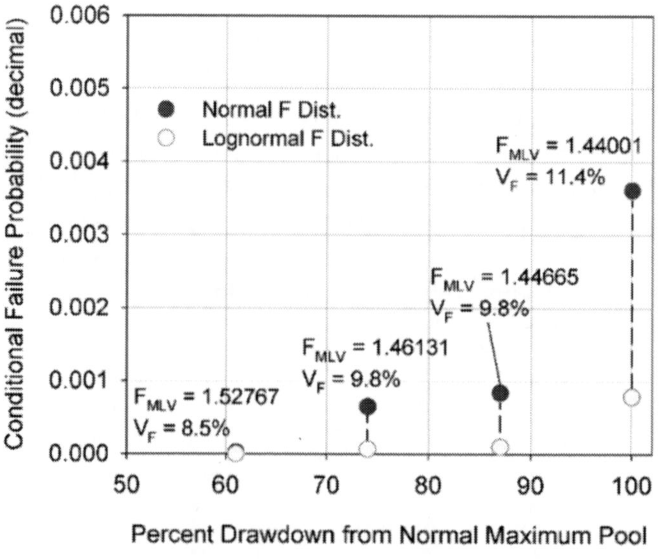

Figure 6. Fragility Curve for 3:1 Interior Slope with Bench

CONCLUSIONS

Deterministic slope stability analysis is conventionally used to design embankment dams. The relative consistency associated with this established practice offers a number of benefits to engineers, technical reviewers, and project owners. Nevertheless, reliability analysis is promoted as a rational and effective means to consider the effects of analysis uncertainties and better interpret risks of failure or unsatisfactory performance. The benefits of adopting reliability methods into stability evaluations of new and existing dams are particularly valuable for large projects, because: (1) including compounding conservatisms in a deterministic analysis creates significant (and possibly unnecessary) project cost increases; (2) an engineer's perception of a conservative design may be misinformed regarding the actual factors affecting risk of failure and could therefore give a misleading sense of stability; and (3) ignoring or discounting the effects of analysis uncertainties without proper evaluation may generate liabilities for engineers and project owners alike. As demonstrated in this paper, simple reliability methods can be applied to slope stability evaluation and design to address these deficiencies with only modest increase in effort during project engineering.

REFERENCES

Baecher, G., and Christian, J. (2003). *Reliability and Statistics in Geotechnical Engineering.* John Wiley & Sons Ltd. Hoboken, NJ.

Christian, J., Ladd, C., and Baecher, G. (1994). "Reliability applied to slope stability analysis." *Journal of Geotechnical and Geoenvironmental Engineering*, ASCE, Vol. 120, No. 12, p. 2180-2207.

Duncan, M., Wright, S., and Wong, K. (1990). "Slope stability during rapid drawdown." H. Bolton Seed Memorial Symposium, Vol. 2, p. 253-272.

Duncan, M. (2000). "Factors of safety and reliability in geotechnical engineering." *Journal of Geotechnical and Geoenvironmental Engineering*, ASCE, Vol. 126, No. 4, p. 307-316

El-Ramly, H., Morgenstern, N., and Cruden, D. (2002). "Probabilistic slope stability analysis for practice." *Canadian Geotechnical Journal*, NRC Press, Vol. 39, p. 665-683.

Harr, M. (1987). *Reliability-Based Design in Civil Engineering*, McGraw-Hill, New York.

Loehr, J. E., Finley, C. A., and Huaco, D. (2005). *Procedures for Design of Earth Slopes Using LRFD.* Report No. OR 06-010, University of Missouri-Columbia and Missouri Department of Transportation, p. 80.

Santamarina, J., Altschaeffl, A., and Chameau, J. (1992). "Reliability of slopes: incorporating qualitative information." *Transportation Research Record: Journal of the Transportation Research Board*, No. 1343, National Academies, Washington, D.C., 1-5.

Sleep, M., and Duncan, J. M. (2014). *Manual for Geotechnical Engineering Reliability Calculations, CGPR #76, 2nd Ed.* Center for Geotechnical Practice and Research, Virginia Tech, Blacksburg, 75 pp.

Tang, W., Stark, T., and Angulo, M. (1999). "Reliability in back analysis of slope failures." *Journal of Soil Mechanics and Foundations*, Japanese Society of Soil Mechanics and Foundation Engineering, October.

Thompson, M., Bentler, D., Brandon, T., and VandenBerge, D. (2016). "Shear strength of compacted Beaumont clay for consolidated-undrained conditions." *Proceedings, 2016 USSD Conference*, April.

USACE (1997). *Introduction to Probability and Reliability Methods for Use in Geotechnical Engineering*, U.S. Army Corps of Engineers, ETL 1110-2-547, September.

USACE (2006). *Reliability Analysis and Risk Assessment for Seepage and Slope Stability Failure Modes for Embankment Dams*, U.S. Army Corps of Engineers, ETL 1110-2-561, January.

VandenBerge, D., and Brandon, T. (2014). "Multi-stage rapid drawdown analysis using the finite element method." *Proceedings, 2014 ASDSO Conference*, September, CD-ROM.

Wolff, T., Hassan, A., Khan, R., Ur-Rasul, I., and Miller, M. (2004). *Geotechnical Reliability of Dam and Levee Embankments*, report prepared for U.S. Army Corps of Engineers, ERDC/GSL CR-04-1, September.

EMBANKMENT DAM ANALYSIS PARAMETERS ESTIMATED FROM 100 YEARS OF LABORATORY TESTS

David W. Harris, PhD, PE[1]
Garrett Sutley[2]

ABSTRACT

The analysis of embankment dam performance is important in support of design of the dam and in the evaluation of current and future conditions. Laboratory tests have been used to isolate strength and volumetric parameters for dams. Two of the most common tests used since the early 1900's are the conventional triaxial test and the 1-D consolidation test. Triaxial test results are shown from over 250 case histories as stress-strain curves and Mohr-Coulomb parameters. The data ranges primarily from the 1910 to 1990 period. In addition, the use of mean pressure versus deviator stresses are shown for clay, silt, sandy silt, and sandy clay cases which are the most common in the data library. Consolidation data is also shown for the same materials. The data is presented as a whole for the reader to use as a basis for analyses. Recommendations are made on ranges of data properties, and for the choice of typical properties for initial/planning studies. There is not a strong correlation of properties with time but correlations are shown with soil type. The data presented in this paper gives a baseline of comparison for structures. The measurement of actual properties is recommended for critical structures,

INTRODUCTION

The U.S. Bureau of Reclamation (USBR) maintains 475 dams; about two thirds of this inventory is embankment dams. Founded in 1902 to "Reclaim the Arid West," USBR designed and developed water supply facilities to store and deliver water to homestead farmers, towns and cities. Figure 1 summarizes the pattern of construction through the 20th century (Kramer and McDaniels, 1994, Harris, 2015). This data set highlights construction data for a large percentage of the inventory, but is not exhaustive. Obvious from the figure is that the largest collection of data is from 1940-1979.

Also available for a large percentage of the dams are laboratory data with tests for various parameters. Of some interest are results as a function of the year tested. Of greater interest is data by soil type which shows parameters used with various analysis methods. This paper will present data organized by selected soil type and in formats common in analysis methods.

[1] David Harris Engineering and Assistance, LLC /Allegiance Consulting, LLC, 2490 Antelope Circle, Elizabeth, CO 80107, davidWharris@comcast.net, 303-748-9163
[2] Garrett Sutley, U.S. Bureau of Reclamation, PO Box 25007, Denver, CO 80225, gsutley@usbr.gov, 303-445-3047

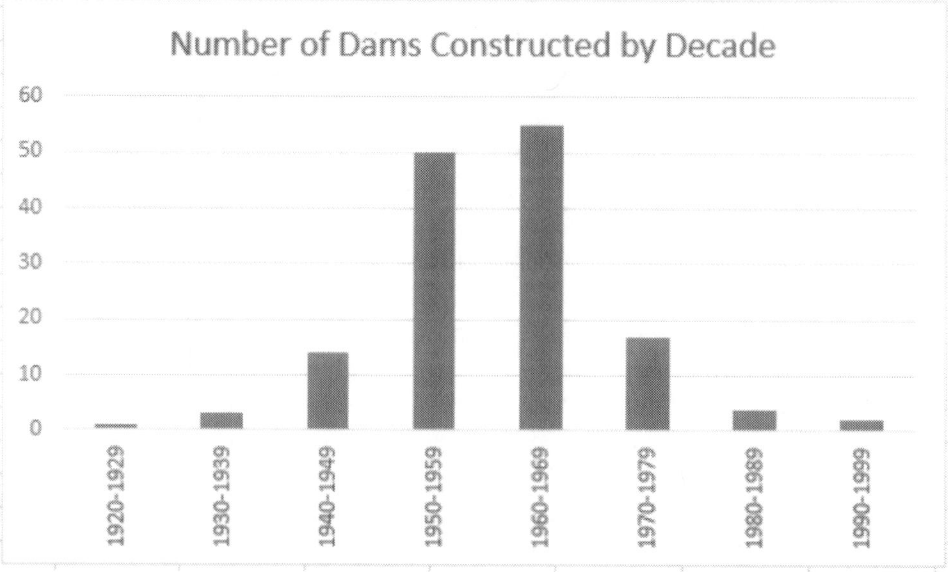

Figure 1. USBR selected constructed facilities

CHARACTERISTICS OF EMBANKMENT MATERIALS

The type of soil used as earthfill varies by site and design. Some dams designed by USBR are zoned with a central clay impervious zone, while others are generally homogeneous. Figure 2 shows the data library used for presentation in this paper with soil types. The largest data sets are clay, sandy clay, sandy silt, and silt. These will be the focus of this paper.

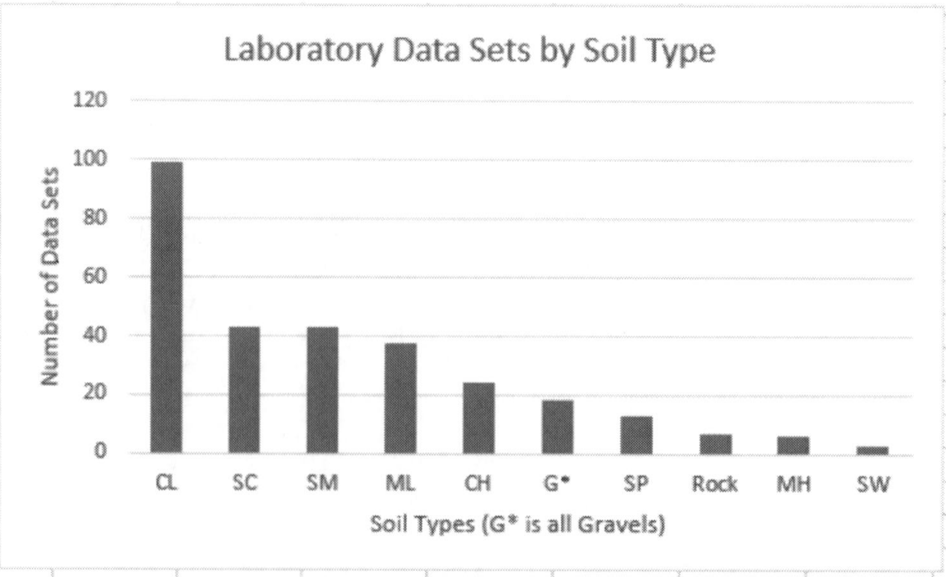

Figure 2. Data library distribution of soil types (ASTM D2487-11)

Figure 3 shows moisture content ranges for the various soil types in the data library. A few values fell above 40% and are not shown for a more accurate visualization of the majority of these physical property tests. The clay materials have the widest range of moisture contents between 5-40%. The sandy clay, silt and sandy silt materials which are similar generally range between 5-30%. The moisture contents shown are generally from the maximum density. A few may be at -2% maximum density which is the USBR preferred placement density.

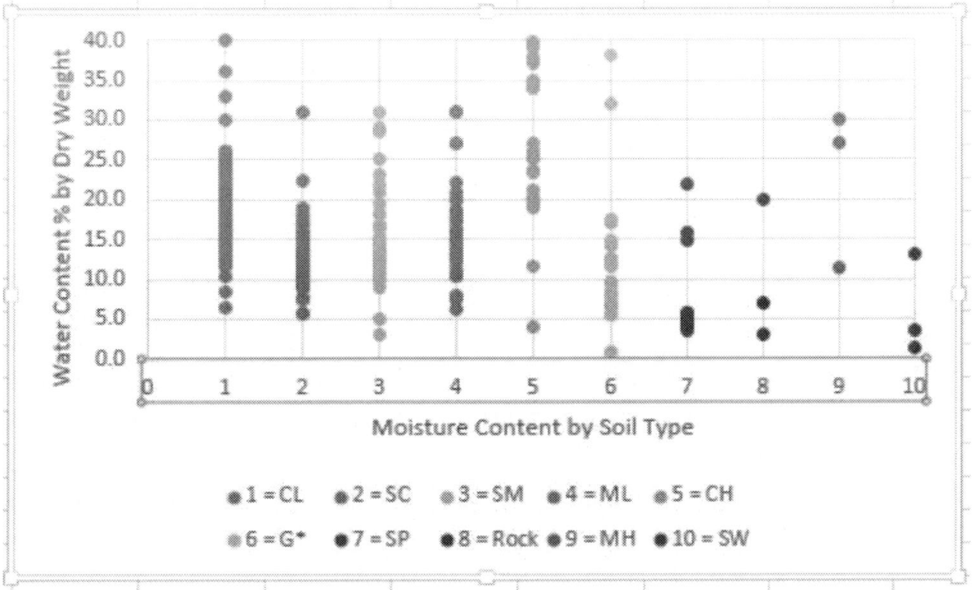

Figure 3. Data library - moisture distribution of soil types

For the plastic materials, Figure 4 shows the plasticity index for the materials.

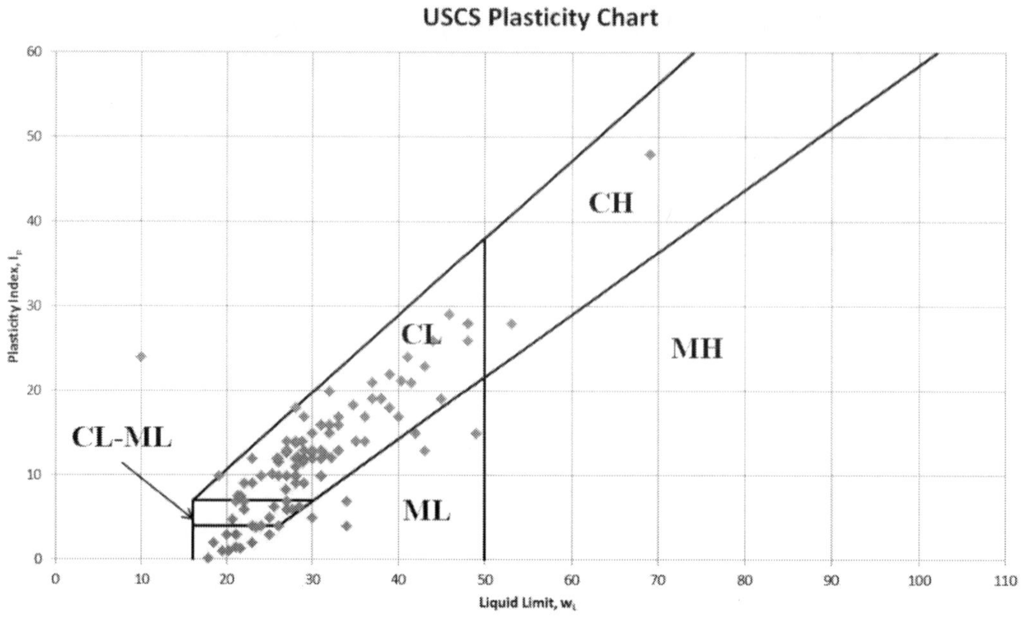

Figure 4. Data library – Plasticity of Soils

ANALYSIS PARAMETERS DERIVED FROM LABORATORY TESTS

Different parameters used in numerical modeling are needed and readily found from laboratory tests. This section discusses some selected models.

Mohr Coulomb and Drucker Prager

Perhaps the most common and well known strength model for geotechnical analysis uses Mohr-Coulomb parameters: c- cohesion, and ϕ - Angle of Internal Friction. These parameters are usually found using conventional triaxial testing either drained or undrained, or can be found using direct shear methods. Effective stress or total stress surfaces can be constructed. Values in this paper show effective stress values resulting from triaxial tests with pore pressure measurements and corrections. Definitions of terms are shown in Figure 5. Effective stress circles generally plot nearer the origin owing to the correction or shift from the pore pressure term. Conversely, total stress circles generally plot at greater x-axis values. Generally, effective stress ϕ values are higher than total stress values, and generally total stress c values are greater than effective stress. The term von Mises is used here as the maximum shear value measured in a test, creating a horizontal line graphically. The Mohr Coulomb plots end at the von Mises value. The use of effective and total stress is discussed in detail by the USSD Committee on Materials for Embankment Dams (USSD, 2007) and is not included here. Also discussed in the same reference is the use of an envelope for total stress analysis, using minimum values of c and ϕ as shown in Figure 5 by the points labeled "Min". The bilinear fit creates the recommended surface for the total stress cases.

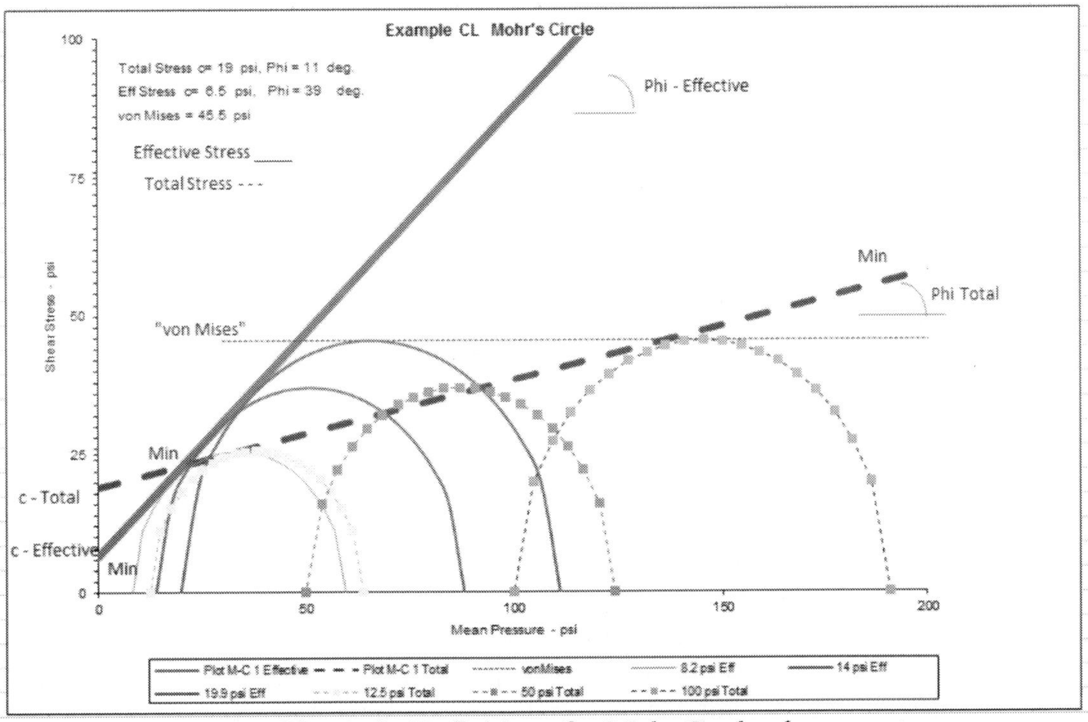

Figure 5. Definitions for Mohr Coulomb parameters

Figure 6a shows effective stress values for approximately 50 different site tests with clay materials, while Figure 6b shows effective stress values for silts for approximately 30 different site tests. A clear difference is readily seen in that the clay values have a wider spread while the silt values show very little variation. Upper and lower bound Mohr Coulomb lines are shown as a heavy dashed line which include 80% of the cases.

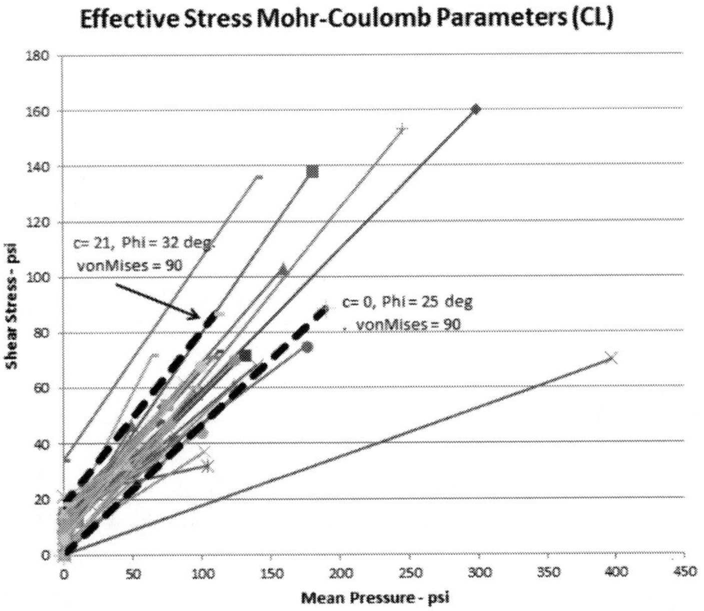

Figure 6a. Effective Stress Mohr Coulomb plots for CL materials

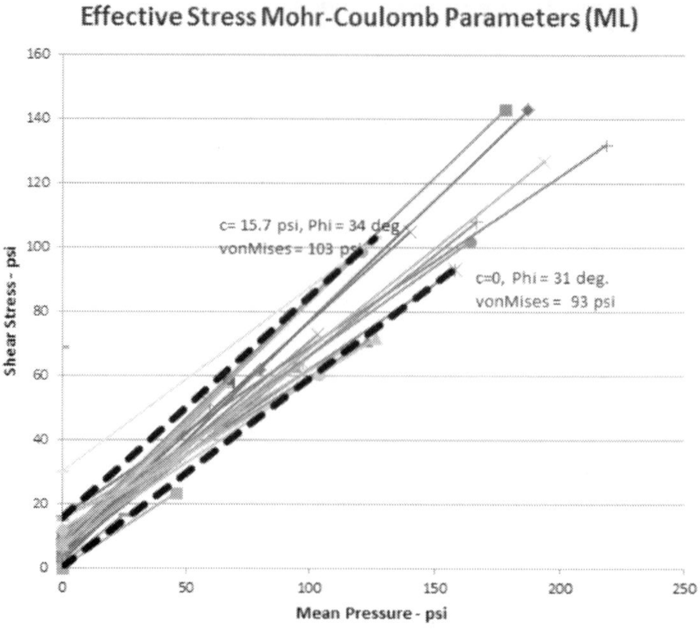

Figure 6b. Silt Mohr Coulomb Effective Stress

Table 1 extends the concept shown in Figure 6 showing effective and total stress values with the range of values and the boundary lines that contain 80% or more of the values for the four major soil types. For clay materials the data range is larger than the bounding region to enclose 80% of the values. This is also true of the silt materials. Conversely, the SC materials produce a small range of values making the full range and 80% bounds nearly the same.

Table 1. Full range of Mohr Coulomb data, and Range of data to contain 80% of the data

	Total Stress							
	Data Minimum		Data Maximum		Lower Bound 80%		Upper Bound 80%	
USCS	c	Φ	c	Φ	c	Φ	c	Φ
CL	0	6	34	36	0	15	20	25
SC	0	30	19	37	0	30	19	34
SM	0	14	32	39	0	25	20	35
ML	0	8	25	36	0	24	20	34
	Effective Stress							
	Data Minimum		Data Maximum		Lower Bound 80%		Upper Bound 80%	
	c	Φ	c	Φ	c	Φ	c	Φ
CL	0	10	34	36	0	25	18	32
SC	0	30	19	37	0	30	20	35
SM	0	25	22	39	0	32	15	38
ML	0	27	30	30	0	30	16	34

It is noted, as shown in Figure 7, that the clay values generally are site dependent with no clear correlation to other properties. Figure 7 shows water content at maximun density versus strength as an example with a R^2 test value of 0.048. However, the data is denser for smaller values of maximum shear strength. As noted later water content does effect the shear strength when comparing results within a given site (set of laboratory tests).

Drucker Prager modeling uses a straight line fit to testing data. Mohr-Coulomb parameters can be converted to the Drucker Prager parameters using equations as suggested by Desai (Desai and Siriwardane, 1984):

$$\alpha = \frac{2 \sin\phi}{1.73(3-\sin\phi)} \tag{1}$$

$$\acute{k} = \frac{6c \cos\phi}{1.73(3- \sin\phi)} \tag{2}$$

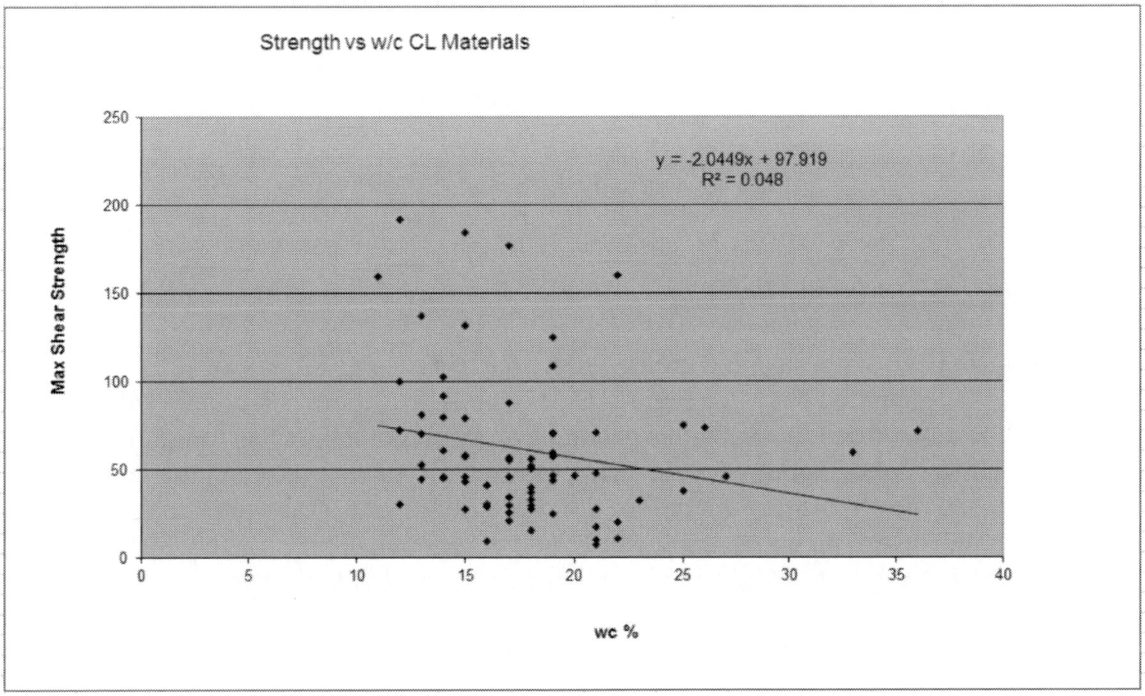

Figure 7. Clay values – Water Content versus Maximum Shear Strength (vonMises values)

Conventional Triaxial Tests

Mohr Coulomb values as shown above can be determined using different confining pressures and measured maximum loads. Data may also be collected continuously during a stress strain test. The maximum value can be based on stress, typically maximum stress, or strain criteria. Typical values are shown in Figure 8 for SM materials. Most modeling techniques do not use stress-strain values directly but convert to a method that recognizes the effect of confining on strength. An example of this approach is shown in the next section as Mean Stress – Deviator modeling.

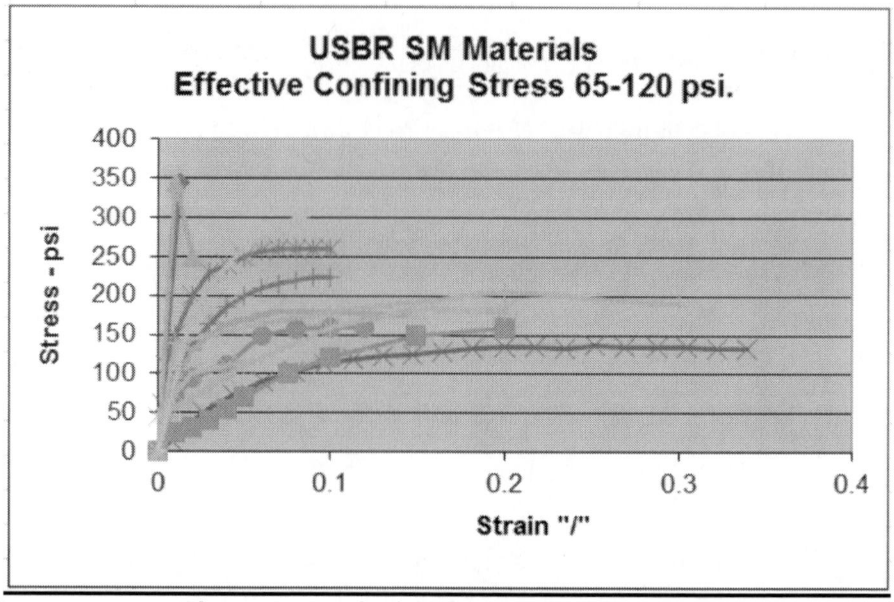

Figure 8. Typical SM Stress-Strain values

Mean Stress – Deviator Modeling

Models have been suggested that create failure surfaces using 3-D mean stress and a deviator, as highlighted below in Figure 9. Shown in this figure is laboratory data in the heavier line and a trend line fit with a second order polynomial and a straight line. The polynomial fit is quite good with a R^2 value of 0.9975. As will be seen later the linear fit is better for higher pressures. As before the range of all measured values in the data base will be bounded with an upper and lower bound containing approximately 80% of the data range. For this data fit a fitting equation will be used of the form:

$$\text{Deviator Stress} = A\,x^2 + B\,x + C \qquad (3)$$

Where x = mean stress

Of note is the effect of the coefficients; C adjusts the y axis intercept, B adjusts the slope of a straight line and generally adjusts the rise of the polynomial, while A adjusts the end point of a polynomial, increasing A raises the endpoint. The polynomial is convenient as it can be easily used with excel or other programs. Note that a second order polynomial has an inflection point after which values decrease. The coefficients published here are limited to 200 psi. mean stress.

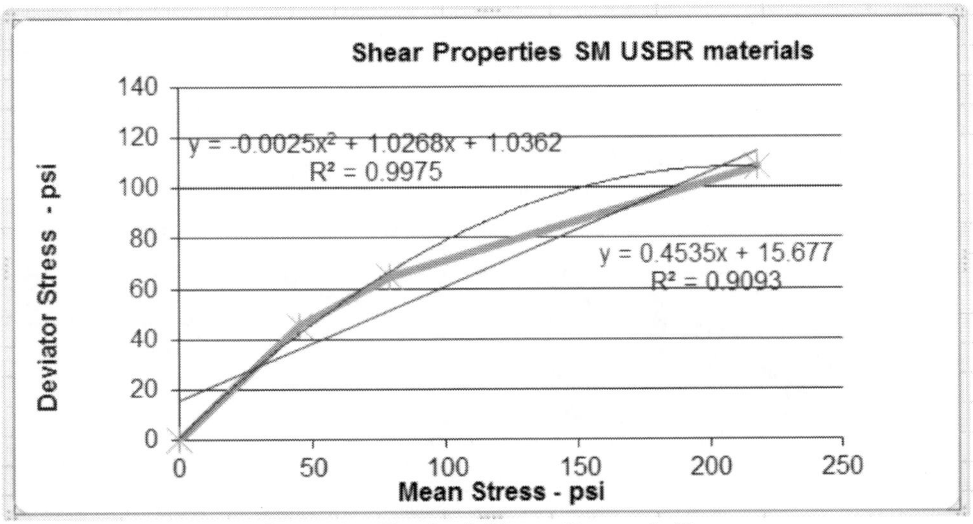

Figure 9. Typical Shear Strength Data

Figure 10 shows data for the SM materials with solid lines and symbols, a polynominal fit for lower and upper bounds of the data below 200 psi with large symbols and a dashed line, and a linear fit (A=0) for the lower and upper bounds of the data above 200 psi and up to 450psi with dashed lines. The linear equation fits the higher values more closely. This densely populated graph demonstrates the range of data that occurs when many tests and sites are presented together. This paper suggests upper and lower bounds for materials as a check for typical values to use in scoping analyses. It is clear that with a wide possible range of values final evaluation of critical structures should use site specific test values.

Table 2 summarizes the four USCS types with the polynominal (lower values) and linear (higher values) fitting equation coefficients. The data range for the higher stress data is smaller oweing mainly to the fact that less values are available. Without additional data linearity is not suggested as a material property. As before the clay values show the greatest range with the SC values showing the lowest range.

Figure 7 showed a poor correlation of water content %, w/c, with maximum shear strength when comparing all values, sites, from the data base. Figure 11 shows the effects of varying the confining pressures and water content when testing soil materials from a single site. As can be seen in the figure values are asymototic to a nearly single value for a given moisture content. Also easily seen from the figure is the effect of water content with increased moisture causing lower values.

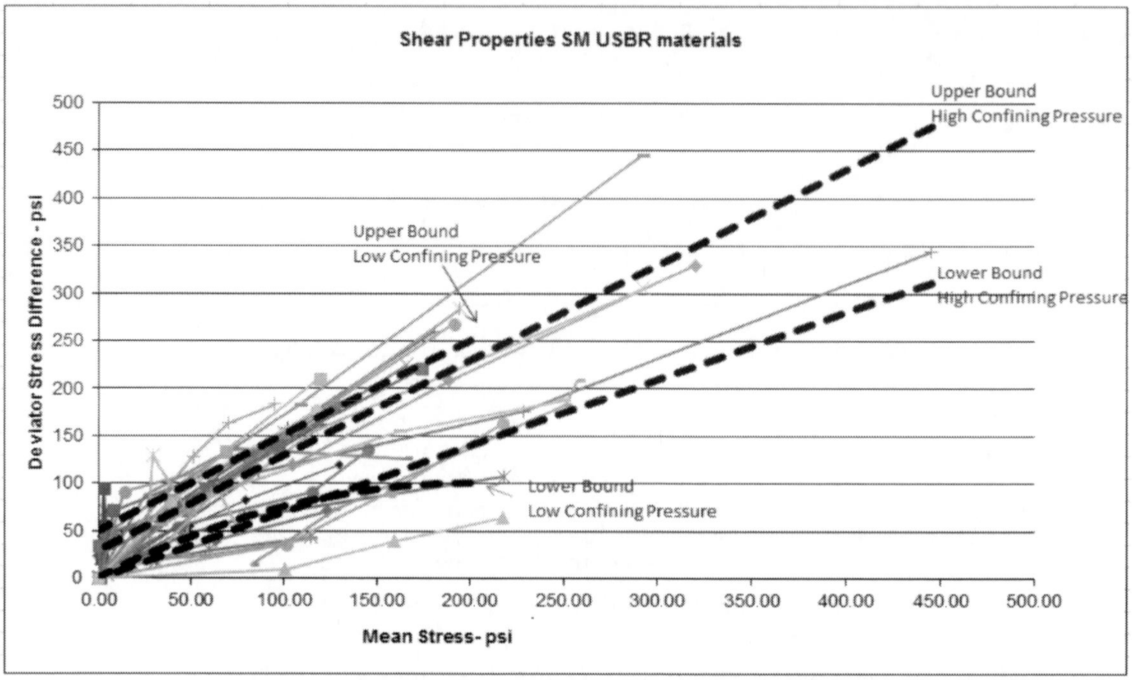

Figure 10. Shear Strength Data for SM Materials
(Laboratory with light solid lines, Ranges in heavy dashed lines)

Table 2. Shear Strength Fitting Equation Boundaries for Selected Soil Types

	Polynominal Fit Shear = A*x*x+B*x+C ; x = Mean Stress											
	Lower Bound x<200psi			Upper Bound x<200 psi			Lower Bound x> 200 psi			Upper Bound x> 200 psi		
USCS	A	B	C	A	B	C	A	B	C	A	B	C
CL	-0.002	0.1	0	-0.0001	1	20	0	0.6	0	0	1	20
SC	-0.002	0.07	0	-0.0001	1	20	0	0.5	0	0	1	20
SM	-0.002	1	0	-0.002	1.5	20	0	0.7	0	0	1.3	20
ML	-0.0002	0.6	0	-0.00005	1.3	20	0	0.7	0	0	0.85	20

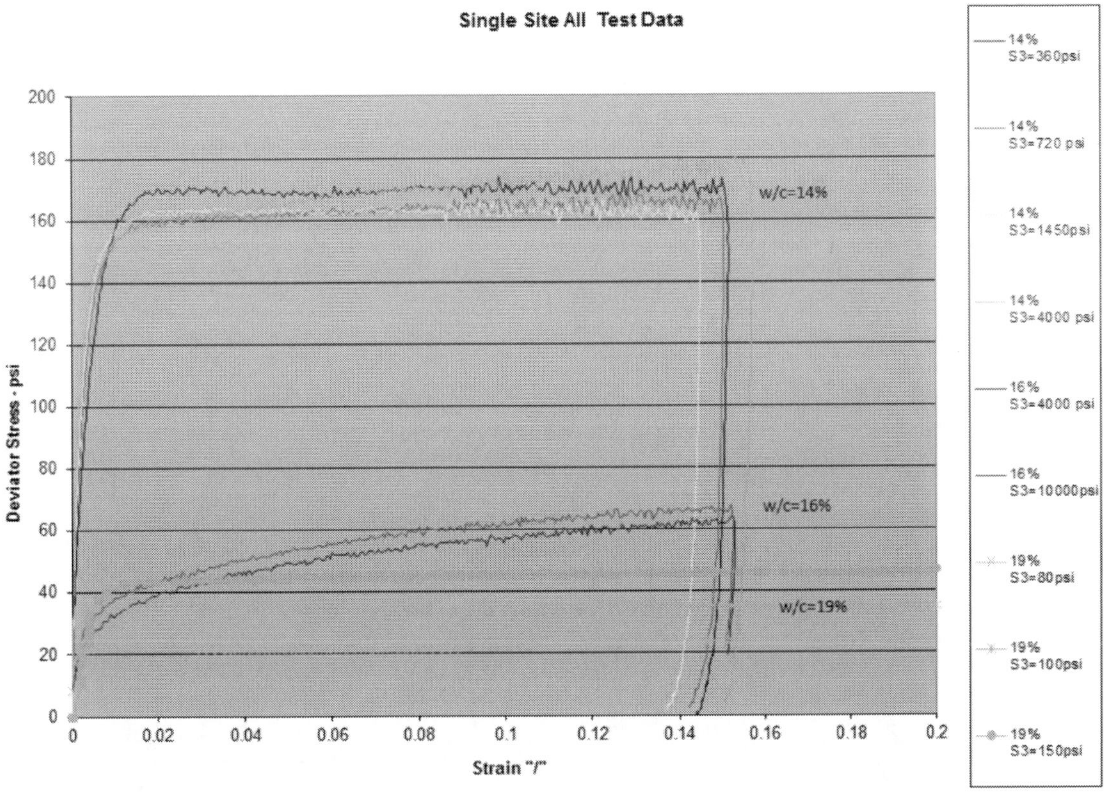

Figure 11. Water content,w/c %, and Confining Pressure Effect on Shear Strength

<u>Volumetric Modeling</u>

Modulus and volumetric effects from laboratory testing can be related using an Equation of State. Typical data can be generated by measuring the volume changes during the pressureization of the confining test confining pressure, with a zero lateral strain test in a conventional triaxial cell, or from a consolidation test. Typical data is shown in Figure 12. In Figure 12 it can be seen that in most cases the volume changes fairly easily with lower pressures, and then stiffens with a somewhat linear and steeper slope. The first part of the curve represents the compression of the void space up to the break point. The steeper curve represents voids closed with the compression curve caused by compression of water in the void space and the soil structure. This general pattern will again be tabulated with the four soil types, Table 3. In this case the fit will use a 3 point, two lines fit.

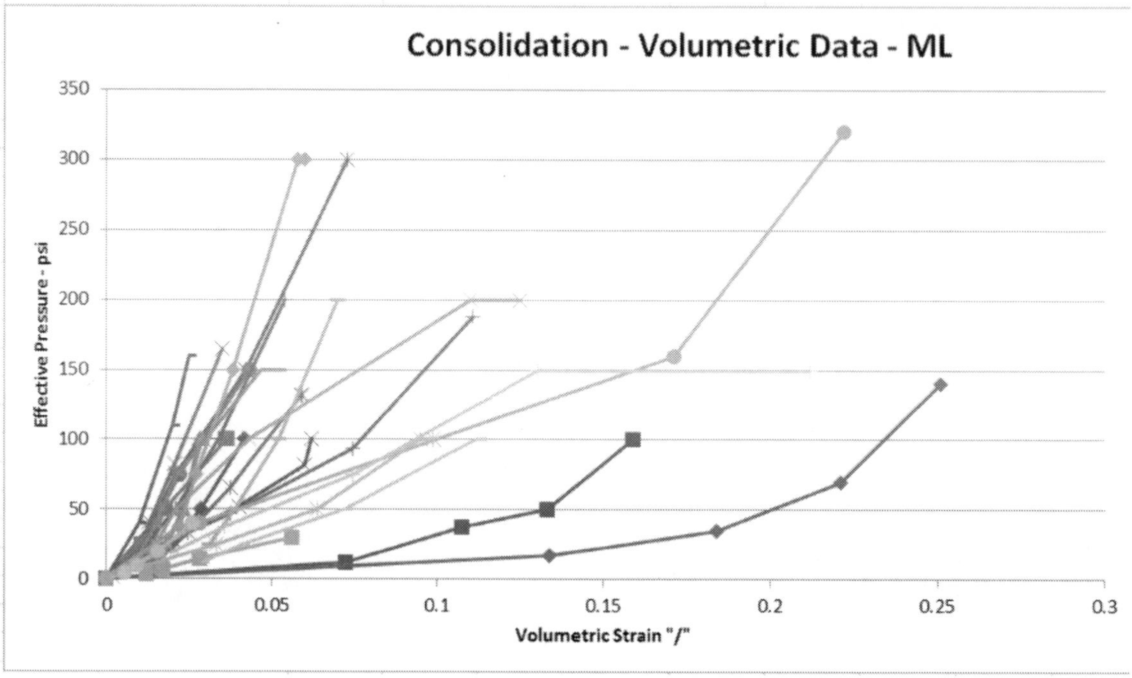

Figure 12. Volumetric data – ML materials

Table 3. Range of Volumetric Values

USCS	CL				SC				SM				ML			
	Lower		Upper		Lower		Upper		Lower		Upper		Lower		Upper	
	Vol Strn	Mean Strs	Vol Strn	Mean Strs	Vol Strn	Mean Strs	Vol Strn	Mean Strs	Vol Strn	Mean Strs	Vol Strn	Mean Strs	Vol Strn	Mean Strs	Vol Strn	Mean Strs
Origin	0	0	0	0	0	0	0	0	0	0	0	0	0	0	0	0
Break Point	0.15	100	0.04	150	0.05	100	0.03	100	0.07	100	0.02	75	0.07	50	0.05	150
Upper Point	0.19	400	0.08	400	0.09	300	0.05	200	0.1	300	0.04	300	0.17	200	0.07	300

Other Models

Many models exist other than those suggested here. Any discussions in this paper are not recommendations or endorsements of any particular model or analytical computer codes. This presentation focused on some of the more common and simpler models. Many of the other models use the same laboratory tests as suggested here as a basis. For example, many of the early Cap Models (Desai and Siriwardane, 1984) use a failure shear surface similar to this description, and the cap is fitted to volumetric data as presented here.

ESTIMATING PROPERTIES FOR SCOPING STUDIES

The data library values can be used for scoping studies, critical structures should use site specific properties. With limit equilibrium studies or studies using Mohr Coulomb parameters a lower bound can be picked using Table 1, or a range can be used for parametric or risk studies. If the water content is known or easily estimated, Figure 7 can be used to estimate the maximum shear strength. Although Figure 7 shows a wide range, there are values that cluster for most water content values and a conservative value can be chosen or a range chosen for parametric studies. The maximum shear strength can be used to bound stresses that are calculated.

For models using a failure surface such as the Mean Stress – Deviatoric Stress model the fitting equation, equation 3, with Table 2 can be used. Volumetric models can choose from the ranges suggested and Table 3 can be used.

CONCLUSION

Dams have been constructed with soils placed under strict construction standards. Laboratory tests to help design field procedures and test existing conditions have been extensive. The majority of test data available from the last century is derived from conventional triaxial tests and from consolidation tests.

Materials vary because of many parameters, e.g. type of soil, origin, moisture content, placed density, confining effects. Ranges of values were presented for four different soil types. Clays showed the greatest range in measured properties with sandy silt materials showing the lowest range.

Various analytical input parameters can be calculated from the standard laboratory tests. Values for Mohr Coulomb parameters are shown. For models that use a failure surface and/or Equation of State a fitting equation is suggested and coefficients given to provide the surfaces. A range of volumetric values is shown.

Ranges of values can be used as a check of ongoing analyses to see how the input values compare to hundreds of tests made over the last 100 years. Upper and lower bounds can be used to produce parametric studies and can be used to form various levels of risk. For initial or scoping analyses, with no measured data, reasonable values from previous work can be used.

Critical structures should be studied with measured, site specific, properties.

ACKNOWLEDGEMENTS

The data library shown in figures and tables was drawn from the archives of the U.S. Bureau of Reclamation. It is greatly appreciated that this data was allowed for use in the figures and to provide a discussion of how data can range throughout typical jobs. Technical review and comments were provided by reviewers organizing the USSD conference.

REFERENCES

American Society of Testing Materials (ASTM), 2011, Standard Practice for Classification of Soils for Engineering Purposes (Unified Soil Classification System), ASTM D-2487-11, West Conshohocken, PA.

Desai, C.S. and H.J. Siriwardane, 1984, Constitutive Laws for Engineering Materials with Emphasis on Geologic Materials, Prentice-Hall, Englewood Cliffs, NJ.

Harris, David W., 2015, Data Summary of Reclamation Embankment Dams Laboratory Data, Unpublished (For Official Use Only).

Kramer, R.W. and T. N. McDaniels, 1994, "Maximum Sections and Earthwork Control Statistics of Earthfill Dams built by the Bureau of Reclamation," National Technical Information Service, Springfield, VA.

USSD, 2007, Committee on Materials for Embankment Dams, Strength of Materials for Embankment Dams, 1616 17th St, Denver, CO.

CALAVERAS DAM REPLACEMENT PROJECT — PROVIDING A SEISMICALLY RELIABLE WATER SOURCE

John Roadifer[1]
Michael Forrest[2]
Daniel Wade[3]
Susan Hou[4]
Tedman Lee[5]
Carman Ng[6]

ABSTRACT

Calaveras Dam is a major component of the San Francisco Public Utilities Commission (SFPUC) Regional Water System. Since 2001, in response to seismic stability concerns from the California Division of Safety of Dams about this 90-year-old hydraulic fill dam, the SFPUC has lowered water levels in Calaveras Reservoir to about 39 percent of its 96,850-acre-foot capacity. The dam is located 0.3 miles from the Calaveras Fault, which is capable of a generating a magnitude 7¼ maximum credible earthquake (MCE) with a peak ground acceleration of 1.1 g. The dam site, which is located in the active and geologically complex Coast Range, contains multiple secondary faults (classified as conditionally active or inactive), Tertiary sedimentary bedrock, a diverse suite of Franciscan assemblage rocks, and numerous active, dormant, and inactive landslides.

To restore reservoir capacity, significant planning and design of a new replacement dam, spillway and outlet works were undertaken starting in 2003. Besides improving seismic reliability and restoring reservoir capacity, the project incorporates accommodation for continued reservoir operation during construction and potential future enlargement.

The project, during design and current ongoing construction, has faced and overcome many geologic and seismic challenges. A central focus of the project was to design the dam and appurtenances to withstand the large MCE ground motions, which are to remain functional after the design earthquake. Seismic stability of 400-foot-high spillway cuts in variably weathered and fractured rock conditions was also a major consideration. During construction, several small bedrock faults that were discovered in the foundation required the design of measures for the outlet works to accommodate potential sympathetic movement. When completed in 2018, this project will provide a reliable source of high quality water to SFPUC's 2.6 million residential, commercial, and industrial customers.

---

[1] Sr. Project Manager, AECOM, 1333 Broadway, Suite 800, Oakland, CA 94612, (510) 874-1732, john.roadifer@aecom.com
[2] Vice President, AECOM, 1333 Broadway, Suite 800, Oakland, CA 94612, (510) 874-3012, michael.forrest@aecom.com
[3] Director, Water System Improvement Program, San Francisco Public Utilities Commission, 525 Golden Gate Ave., San Francisco, CA 94102, (415) 554-1853, dwade@sfwater.org
[4] Project Manager, San Francisco Public Utilities Commission, 525 Golden Gate Ave., San Francisco, CA 94102, (415) 551-4666, shou@sfwater.org
[5] Project Engineer, San Francisco Public Utilities Commission, 525 Golden Gate Ave., San Francisco, CA 94102, (415) 551-4866, tclee@sfwater.org
[6] Assistant Project Engineer, San Francisco Public Utilities Commission, 525 Golden Gate Ave., San Francisco, CA 94102, (415) 551-4867, cng@sfwater.org

INTRODUCTION

Calaveras Dam and Reservoir are located in the geologically active eastern San Francisco Bay region that includes well-defined active faults such as the Calaveras, Hayward, Greenville, and Concord Faults (Figure 1). The reservoir inundates Calaveras Valley, which is located near the intersection of the northern and central Calaveras faults and can be described as a "releasing bend"; a topographic and structural depression containing several north-striking active fault strands arranged in a left-stepping *en echelon* pattern.

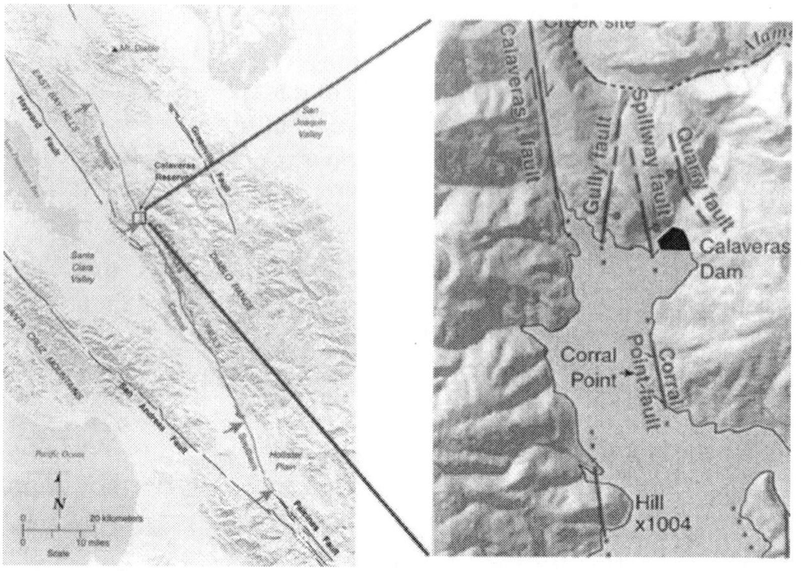

Figure 1. Calaveras Fault and Secondary Faults near Calaveras Dam.

The 210-foot-high existing dam was constructed starting in 1913 using hydraulic filling methods. In 1918, when the dam was nearing completion, a very large slope failure in the dam embankment occurred in the upstream direction. Between 1923 and 1925, the dam was reconstructed on top of the failed dam material. The reconstructed embankment consisted of weak rockfill shells and a rolled clay fill above a zone of hydraulic fill. In the 1970s, the upstream and downstream shells of the dam were buttressed with rockfill following a stability reevaluation. Studies in late 1990s found that large deformations of the dam during seismic shaking were likely. In 2001, in response to seismic stability concerns by California Division of Safety of Dams (DSOD), the SFPUC lowered water levels in Calaveras Reservoir to about 39 percent of its 96,850-acre-foot capacity.

The purpose of the project is to replace the existing dam with a new dam that meets current seismic safety standards. Upon completion, the DSOD restriction would be lifted and the reservoir capacity would be fully restored. The project objectives are summarized below:

- Re-establish water delivery reliability.
- Restore water supply and reservoir capacity to pre-2001 level of 96,850 acre-feet.
- Improve seismic reliability such that the dam and appurtenant works remain functional after the design earthquake.

- Construct replacement dam that could accommodate potential future enlargement.
- Continue reservoir operation during construction.
- Maintain high water quality.

SEISMIC SOURCE AND GROUND MOTION PARAMETERS

The most significant seismic source to the project in terms of strong ground shaking is the central Calaveras Fault ($\geq$ 9 mm/yr slip rate) located about 1,500 feet west of the dam site (Figure 1). The Calaveras Fault exhibits prominent geomorphic expression along its entire active length of 80 miles, and has generated small to moderate earthquakes during the past two hundred years of recorded history.

Secondary faults in the vicinity of Calaveras Dam include the Quarry, Spillway, and Gully faults (Figure 1). Activity of the secondary faulting was characterized based on 1) geomorphic evidence, 2) fault trenching investigations, and 3) comparison with the characteristics and morphology of known active faults in the vicinity and then classified in accordance with the California Division of Safety of Dams (DSOD) guidelines (Fraser, 2001). Based on these three lines of evidence, the Quarry Fault and Spillway Fault were determined to not have experienced surface fault rupture within the past 35,000 years and therefore inactive. Although no evidence was observed to indicate otherwise, the Gully Fault is considered to be conditionally active because trenching investigations were not carried out for this fault.

Consistent with the DSOD established guidelines for selecting the level of ground motions to be used in the seismic design of dams (Fraser and Howard, 2002), the 84th percentile ground motions were used for design. The calculated 84th percentile peak horizontal acceleration adjusted for directivity due to the near-field distance of the dam to the Calaveras Fault is 1.1 g for a MCE moment magnitude **M**7¼ on the central Calaveras Fault. The directivity adjustment was made using the modifications of Somerville et al. (1997) and Abrahamson (2000). Acceleration time histories for seismic analyses were developed for the recommended design response spectrum using time histories selected from a database of past earthquake records and then modified to match the recommended design response spectrum.

EMBANKMENT

During conceptual engineering several alternatives for remediation of the existing dam and several alternative dam types for a replacement dam downstream of the existing dam were screened and evaluated. Based on those analyses, it was determined to design a replacement earth- and rockfill dam. The replacement dam is founded directly on bedrock. The bedrock foundation at the left abutment will be Temblor Sandstone and Franciscan Complex at the valley bottom and right abutment. The upper 40 feet of the left abutment of the dam is a concrete gravity wall.

Replacement Dam Section

The replacement dam section (Figure 2) is based on materials available at the site and on results of stability and seismic deformation analyses discussed below. The crest width of the 220-foot-high dam will be 80 feet and is controlled by the core width required for the potential future enlargement up to 386,000 acre-feet (NMWS elevation 890 feet). The dam section consists of a rockfill upstream shell (Zone 5) and an earthfill downstream shell (Zone 4) supporting a central clay core (Zone 1). A 15-foot-thick chimney filter (Zone 2) and 10-foot-thick drain (Zone 3) separate the core from the downstream shell. The core is separated from the upstream shell by an 8-foot-thick upstream filter (Zone 2A) and an 8-foot-thick coarse filter (Zone 5A). A blanket drain system underlies the downstream shell. The blanket drain is a filter/drain/filter (Zone 2/3/2) on the abutments and a filter/drain/rockfill/filter (Zone 2/3/5/2) in the valley bottom. The upstream slope of the dam is protected against erosion by a layer of riprap (Zone 6).

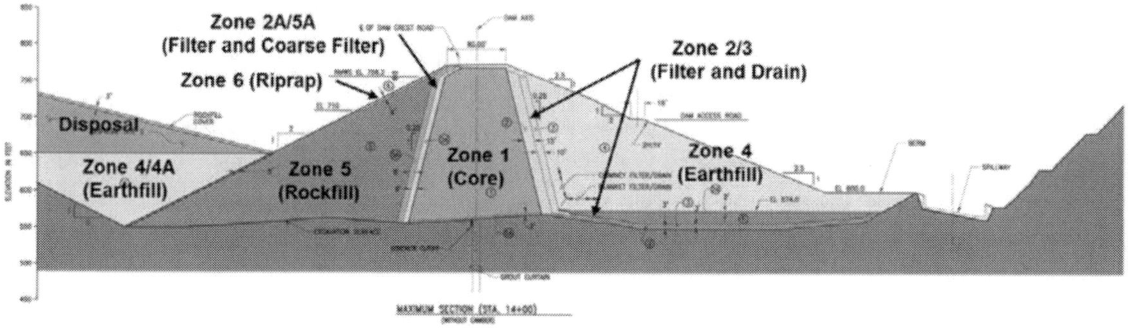

Figure 2. Section of Replacement Calaveras Dam

Core, earthfill, rockfill, coarse filter, and riprap materials are being sourced from on-site excavations and borrow areas. The filter, upstream filter, and drain materials are being imported from commercial sources.

Stability and Seismic Deformation Analysis

The safety of the dam embankment must be maintained by preventing extensive cracking, crest settlement that would impair freeboard, and excessive deformation in critical zones such as filters and drains. Stability and seismic deformation analyses were performed to confirm that settlement and deformation would not impair the safety of the dam.

The space between the existing dam and the replacement dam will be filled with earthfill and disposal material (Figure 2). However, for the purposes of design, the assumption was made that the replacement dam would not rely on the existing dam or the fill between the dams. Engineering properties for the embankment materials were based on the results of laboratory testing of samples obtained during the field investigations. The engineering properties for the Temblor Sandstone, a calcareous feldspathic marine sandstone with thin shale and conglomerate interbeds, and the Franciscan Complex bedrock, a block-in-matrix rock, were based on the foundation conditions observed in the field investigations and the results of laboratory tests as discussed in Roadifer and

Forrest, (2012). The foundation excavated thus far confirms the assumptions made for the strength of the dam foundation.

Reservoir conditions that were analyzed for stability included end of construction, long-term steady state seepage, probable maximum flood, rapid drawdown, and static stability immediately following the design earthquake. The minimum required factors of safety established for the design were met for all conditions and sections of the dam analyzed.

Seismic deformation analyses were performed using two approaches; a Newmark-type analysis method using QUAD4M (Hudson, et al., 1994) to develop average horizontal accelerations for each slide mass considered, and a nonlinear finite element method using the computer program FLAC (Itasca, 2005).

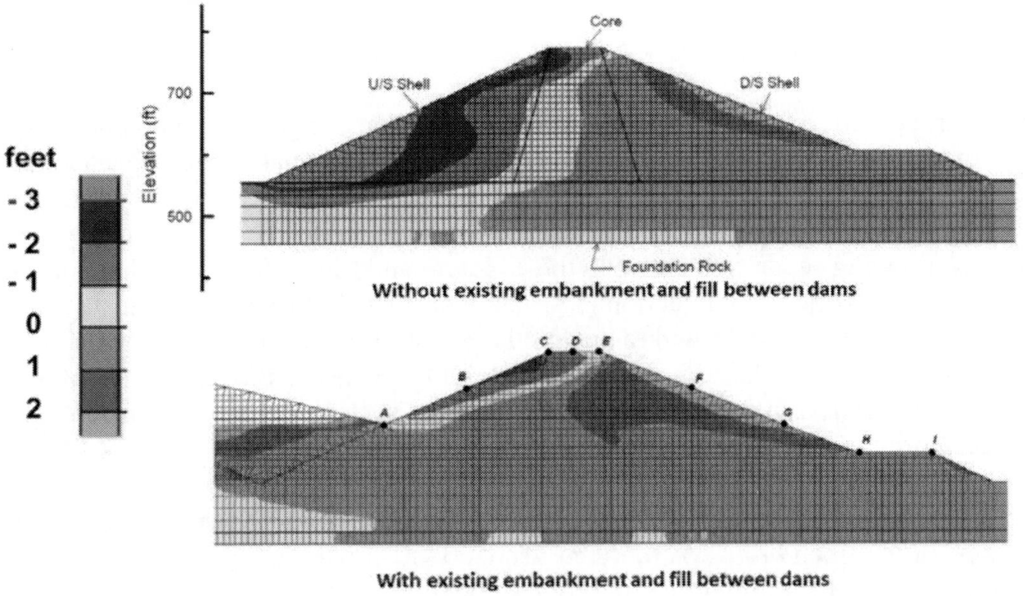

Figure 3. Maximum Section - Horizontal Displacement Results of FLAC Analysis

Seismic deformation analyses indicated horizontal displacements of up to about 3.0 feet in the near-surface upstream slope and up to 2.0 feet in the downstream slope for the design earthquake (Figure 3). The analyses also indicated settlement at the crest of less than 4 feet, which is substantially less than the nearly 16 feet of freeboard for the dam. Although the replacement dam design does not rely on the existing dam or the fill between the two dams, the deformation of the replacement dam will be influenced by them. For this case, the estimated horizontal deformation of the upstream slope and settlement of the crest are reduced to 1 to 2 feet and 2 to 3 feet, respectively (Figure 3).

Left Abutment Gravity Wall

The upper 35 to 40 feet of the left abutment of the replacement dam is founded against a gravity wall that is also the right wall of the spillway adjacent to the crest of the dam (Figure 4). Design of the gravity wall required consideration of the interaction of the rigid concrete structure with the deformable embankment during shaking under the MCE. Of particular concern was designing the gravity wall to prevent rotation about its toe that might allow an opening to develop at the concrete/foundation contact where piping and erosion of the core into the foundation below the gravity wall could be initiated.

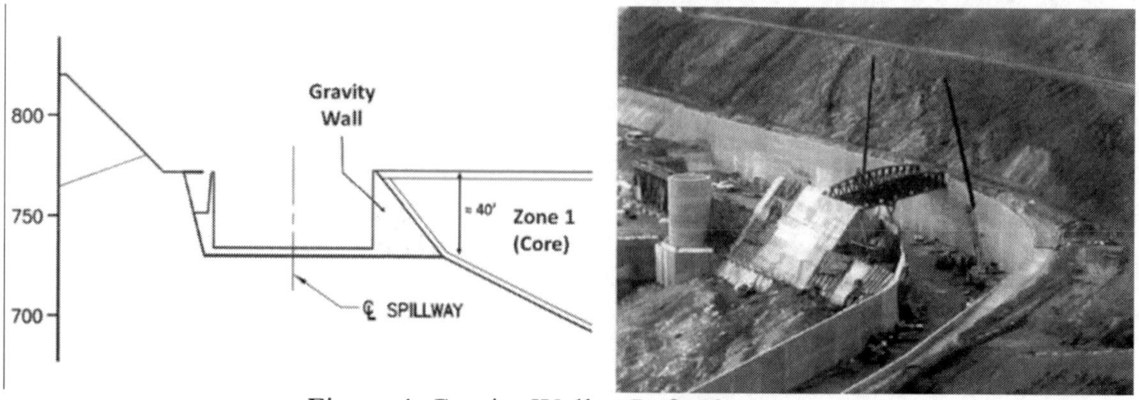

Figure 4. Gravity Wall at Left Abutment.

Initial limit equilibrium stability analyses indicated that the gravity wall was stable for all cases examined including sliding and overturning for earthquake loads. Pseudostatic analysis for the design earthquake found that the resultant of forces was within the base of the gravity wall, but outside of its kern, suggesting that the wall could lean towards the spillway allowing separation between the concrete and foundation at the heel of the wall. A dynamic soil structure interaction (SSI) analysis was performed to estimate the magnitude of the separation and to design the restraint needed to address the separation. The computer code FLAC Version 5.0 (Itasca, 2005) was used for the SSI analysis on the maximum section of the wall. Without a restraining system, the maximum separation of the heel from the rock foundation was determined to be between 0.2 and 0.4 feet.

For the purpose of designing a restraining system, the maximum separation at the heel was limited to 1 inch or less. A moment of 3,500 kip-ft was calculated using the results of the SSI analyses, to restrain the separation to the maximum allowable value. Based on this moment, a restraining system was designed consisting of five rows of 20-foot-deep epoxy coated grade 75 thread bars anchors spaced at 5 feet on center in each direction (Figure 5). Each anchor was grouted into 6-inch-diameter holes using cement grout with 28-day grout strength of 4,000 psi. Design of the anchors considered yielding of the steel bars, failure of the rock to grout bond, failure of the steel to grout bond, and failure of the rock mass. The two rows closest to the toe (located closest to the spillway) are No. 14 bars and the three rows closest to the heel are No. 18 bars.

Confirmation SSI analysis for the designed restraint system indicate instantaneous separation and permanent separation at the heel of less than 0.7 inches and 0.3 inches, respectively. Because the model indicates a potential residual separation between the base of the wall and the foundation, a waterstop system using a 12-inch waterstop installed along the perimeter of the base of the gravity wall in trench sawn into the foundation sandstone was incorporated into the design to prevent potential particle movement through this opening (Figure 5).

Figure 5. Gravity Wall Anchors and 12-inch Waterstop.

SPILLWAY CUT SLOPES

The excavation for the spillway includes two tall excavation slopes (see Figure 6); a 420-foot-high slope having an overall slope of 1.3H:1V in Observation Hill just to the west of the crest structure and upper chute, and a 450-foot-high slope having an overall slope of 1H:1V in Hill 1000 to the north of the stilling basin.

Figure 6. Locations of Observation Hill and Hill 1000.

The cut in Observation Hill is entirely in the Temblor Sandstone. The upper approximately 150 feet of the Hill 1000 cut is in Temblor Sandstone with the remaining cut in the Franciscan Formation. Stability and seismic deformation analyses were performed for both slopes. Engineering properties for the rock materials in the slopes were based on the foundation conditions observed in the field investigations and the results of laboratory tests. Characterization of the Temblor Sandstone in both slopes and for the Franciscan complex in the stilling basin slope were performed using standard rock mass engineering procedures.

Stability analyses during final design included consideration of both large rotational failures through the highly fractured rock mass and kinematic analyses of large wedges that could potentially form. The minimum factor of safety (FS=1.5) were determined to be met for global stability as shown for the critical failure surfaces (see Figure 7). Seismic deformation analyses were performed using a Newmark-type analysis method. In the analyses, QUAD4M was used to develop average horizontal accelerations for each slide mass considered within the slopes including the most critical slide masses shown on Figure 7. The results of the analyses indicated acceptable horizontal displacements of 1.3 feet in Observation Hill and 0.6 feet in Hill 1000 for the design earthquake.

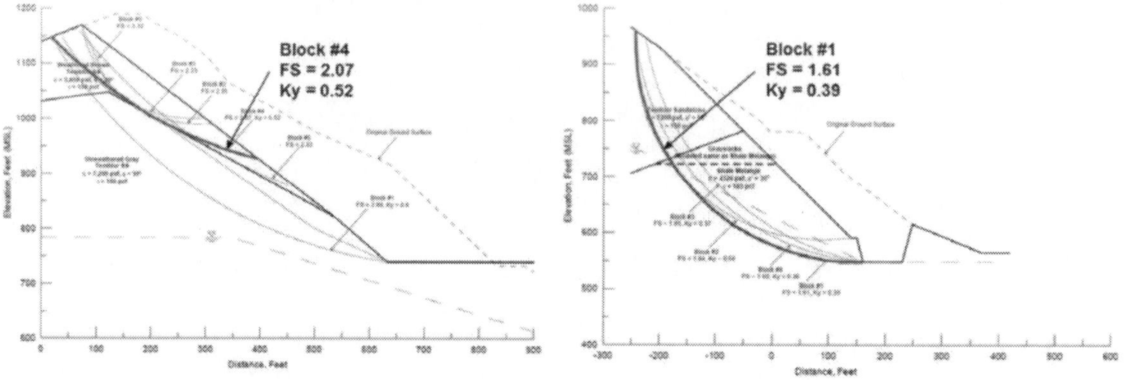

Figure 7. Psuedo-Static Analyses Results for Observation Hill and Hill 1000.

The design included the need for protection of portions of the spillway cut slopes with shotcrete and rock dowels to avoid unraveling and failure of smaller rock blocks, particularly in the weathered and fractured portions of the Temblor Sandstone and Franciscan Complex.

Observation Hill - Conditions Uncovered During Construction

During 2012, as the Observation Hill cut was being initiated, a potential landslide was identified by the Contractor's geologist. The potential landslide was investigated by drilling twenty-six borings and associated optical and acoustic televiewer data and one large-diameter auger boring. The results of the investigation confirmed the presence of the potential landslide (designated Slide A) and a second, likely younger, large landslide complex on the lower east flank of Observation Hill (designated Slide B) (Forrest, et al, 2014). Slide B is a highly weathered, fractured mass of coherent to partially articulated bedrock that may have moved downslope as a result of the steep valley sidewall and varied mass wasting processes along the Calaveras Creek valley margin. The distribution of the landslide complexes indicated that some of the Slide A material would remain in any likely revision of the left abutment permanent cut slope, whereas all of the Slide B material would be able to be removed during construction.

In order to meet static stability criteria, reduce seismic deformations, and remove much of the Area A landslide mass from the northern half of the Observation Hill excavation, a design with an overall slope of 2H:1V (including benches) was modeled. Stability analyses showed that the flattened slope meets static stability criteria with a minimum factor of safety of 1.52. Seismic deformations were calculated to be in the range of 2 to 4 feet, which are considered to be acceptable for the left abutment slope. Figure 6 shows the as-constructed flattened slope of Observation Hill with the Slide A remnant.

Hill 1000 - Conditions Uncovered During Construction

The exposed face of the 450-foot-high stilling basin cut slope during excavation as of July 2015 2014 is shown on Figure 8. The excavation has proceeded as designed with 10-foot-wide benches separated by 50-foot-high 0.75H:1V slopes for an overall slope of 1H:1V down the Bench G. The slope between Bench G and Bench H is 0.5H:1V. A temporary cut having a slope of 0.25H:1V extends nearly 50 feet down to the base of the stilling basin. As shown on Figure 8, the upper portion of the slope is Temblor sandstone and the lower portion of the slope is Franciscan Formation. The Franciscan Formation portion of the slope consists of a mélange shale matrix containing extensive quartz veins and blocks of hard greenstone, blueschist, and greywacke. The Franciscan Formation was generally protected with bolts and shotcrete except for a large greenstone block and large blueschist block that were left untreated (Figure 8). The dimensions of the blocks within the Franciscan Formation range from inches to hundreds of feet. The foliation in the Franciscan Formation at this location is generally oriented favorably into the slope generally dipping steeply (greater than 80 degrees) to the NE and striking NW. Locally, the foliation varies where it wraps around blocks.

Figure 8. Stilling Basin Cut Slope.

As the Franciscan rock was exposed in the excavation, the condition of some portions of the slopes at Bench G appeared to be potentially weaker than had been observed in the borings drilled for final design and the large greenstone block appeared to be pinching out. The shear strength of the weaker portions of the rock mass were re-evaluated based on the observations of the excavated slope using the procedures of Marinos and Hoek (2000). Sensitivity analyses considering geological strength index (GSI) values ranging from 30 to 35 were used in the re-evaluation as compared to a GSI of 45 used in the original analyses. Based on the analyses, which included planned stabilization (No. 8, Gr

75 bars, 24 feet long, 12 feet on-center), it was determined that the factor of safety for slip involving the two lowest benches were slightly below the minimum required factor of safety of 1.5 for a low-bound GSI value of 30. Increasing the size of the rock dowels from No. 8 to No. 11 raised the factor of safety to greater than the minimum required. Slips greater 2 benches in height were judged to not be likely given the three-dimensional effects of the large blueschist and greenstone blocks in the slope.

As the excavation reached Bench H, portions of the shale matrix appeared to become more characteristic of mélange (Mel_{13}) in the right abutment of the dam foundation. It was decided that the stability of the lower slope should be re-evaluated using the block-in-matrix (BIM) approach that was used for the dam foundation (Roadifer and Forrest, 2012). Estimates of block proportions in the slope, assuming a minimum block size of nine inches, generally ranged between 25% and 40% with a value of 30% being selected for analyses. The shear strength used to re-evaluate the stability of the two lower benches is shown as Mel_{SB} in Figure 9.

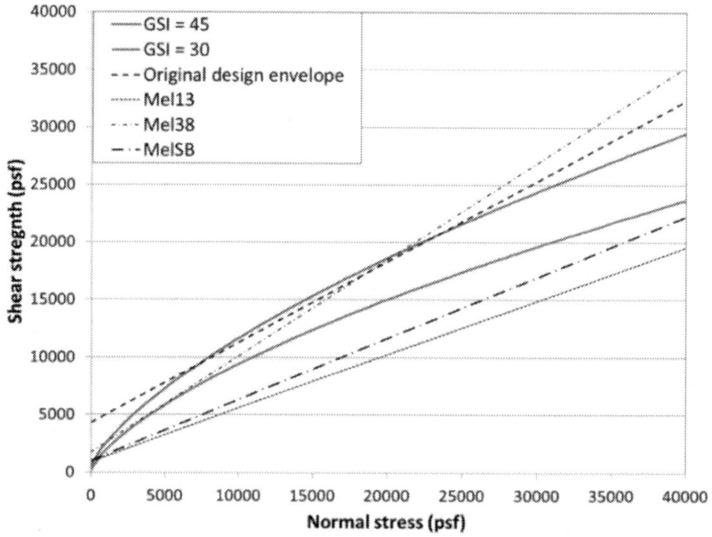

Figure 9. Shear strength for mélange in lower stilling basin cut.

Based on the re-evaluation, it was determined that two additional rows of 36-foot-long No. 11 dowels should be installed on a 12' x 12' pattern. In addition, the length of No. 11 rock dowels in the 0.25H:1V cut were increased from 24 feet to 36 feet and the spacing reduced from a 12' x 12' pattern to a 8' x 8' pattern. The potential seismic deformation of the overall cut was judged to not be impacted by these stabilization measures for local stability at the base of the excavation.

OUTLET PIPE MODIFICATIONS FOR SYMPATHETIC MOVEMENT OF BEDROCK FAULTS IN DAM FOUNDATION

During construction, several bedrock faults were identified along the outlet pipe alignment in the left abutment of the dam foundation as shown on Figure 10. The bedrock faults, which are located between the inactive Quarry Fault and Spillway Fault (Figure 1) were conservatively considered to be capable of sympathetic movement given the proximity to the active Calaveras Fault and the potential consequences if the 78-inch-diameter ½-inch-thick outlet pipe were to be ruptured during a large earthquake on the Calaveras Fault. The potential sympathetic displacement of the faults was estimated to be about 6 inches using the relationship of displacement to magnitude and distance from the primary fault developed by Peterson et al (2011). Potential axial and vertical displacements at each fault location along the pipe alignment were then estimated based on the orientation and sense of movement of each fault. These estimated potential displacements were considered in the development of the mitigation measures discussed below.

Figure 10. Minor Bedrock Faults in Left Abutment of Dam Foundation.

Because the project is under construction and critical path activities including excavation and outlet pipe fabrication were going to be impacted, the design of modifications needed to be developed prior to completion of excavation before the geologic conditions were able to be fully understood. Following completion of excavation, the design of some modifications were revised as the understanding of the geology became clearer. The goal of the mitigation measures developed for the outlet pipe was not to prevent damage to the outlet pipe and its reinforced concrete encasement, but to minimize damage due to sympathetic movement of the faults and make repairs of damage to the pipe feasible. The mitigation measures included modifications to the foundation for the pipe encasement, the pipe encasement, the 78-inch-diameter pipe, and the embankment fill over the pipe encasement as described below.

<u>Quarry Fault Zone</u>

The QFZ crosses the outlet pipe very near the downstream toe of the dam at the left abutment. Being downstream of Zone 1 (core) of the dam, there is no reservoir induced water pressure acting on the outlet pipe, which will be pressurized during normal operations. Therefore, rupturing of the pipeline could result in high pressure water jetting from the pipe and eroding the toe of the dam.

Mitigation measures under the pipe encasement included overexcavation of the QFZ and backfilling with a jointed concrete block having the purpose of allowing differential movement of the concrete across the 10-foot width of the QFZ (Figure 11). The two joints in the concrete backfill are a 3/4-inch-thickness of bentonitic material made up of four Volclay Panels®. Deformation of the pipe encasement vertically or laterally is allowed through the use of a 12-inch-thick layer of crushable cellular concrete (compressive strength of 50 to 100 psi) that starts at the location of the upstream Volclay Panel® joint and extends downstream to the fixed cone valve vault. Differential movement where the pipe enters the fixed cone valve vault is provided through a Romac Style 400 coupling that allows the pipe to deform axially in compression or extension.

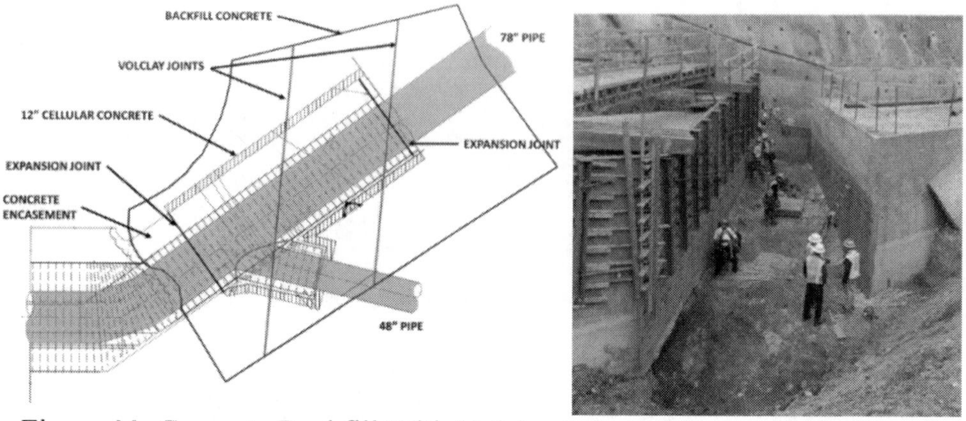

Figure 11. Concrete Backfill With Volclay Joints at Quarry Fault Zone.

In the event that deformation of the pipe is sufficient to cause a tear in the pipe wall, a riprap zone has been included on the outside of the encasement to that will collect and direct leakage into the spillway stilling basin during operation of the pipe. Inspection of the pipe and repairs, if needed, following dewatering of the outlet system, could be made from the inside the pipe or from outside the pipe by excavating the small amount of fill covering this portion of the pipe.

L5 and L6 Bedrock Faults

Like the QFZ, the outlet pipe crossing of the L-5 and L-6 faults will be downstream of the core of the dam and will therefore not be subject to reservoir induced water pressure. Rupturing of the pressurized pipeline could result in high pressure water jetting from the pipe into the blanket drain under the downstream shell of the dam.

Mitigation measures under the pipe encasement were limited to treating the narrow faults by over excavating 3 times the width and backfilling with concrete. Mitigation measures to the pipe included changing slip joints near the faults to single V butt joints, shrouded slip joints further from the faults, and cutting off construction access manholes and 2-inch nozzle to allow the pipe to deform axially within the encasement. Deformation of the pipe (wrinkling) due to vertical or lateral movement is allowed. In the event that deformation of the pipe is sufficient to cause a tear in the pipe wall, two layers of a TegraSeal TripleSeal waterproofing membrane are included on the outside of the encasement to reduce leakage from the pipe into the blanket drain system (Figure 12). The waterproofing membrane is a 5 layered system (polypropylene scrim/bentonite clay/HDPE membrane/bentonite clay/polypropylene scrim) providing a very low permeability barrier that can accommodate the potential deformation without rupturing. Two layers are used to make the system more robust by allowing slippage between the layers to occur during deformation. Inspection of the pipe and repairs, if needed, would be made from inside the pipe following dewatering of the pipe.

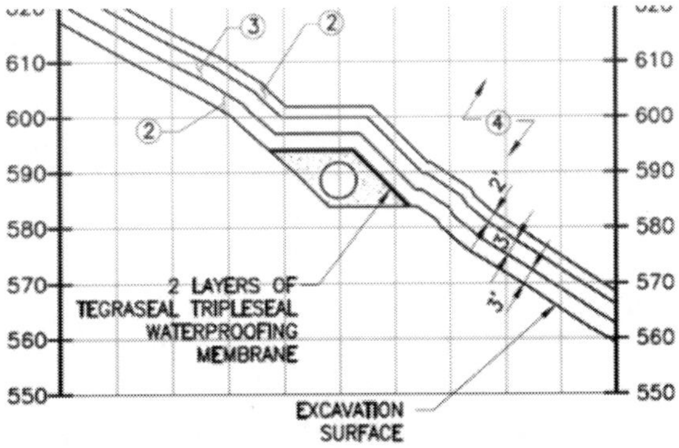

Figure 12. Embankment Fill Over Concrete Encasement at L5/L6.

L2, L3 and L4 Faults

The outlet pipe crosses the L-2 and L-3 faults upstream of the core of the dam and the L-4 fault within the core of the dam and will have reservoir induced water pressure acting on the outside of the pipe. Rupturing of the pipe at the L-2 and L-3 would not likely have a major impact on normal operations. Rupturing of the pipe at the L-4 fault in the core during normal operation could expose the core to an open conduit that would allow piping of the core. Leakage under reservoir pressure into the pipe could occur following dewatering of the ruptured pipe for inspection. The volume and force of the leakage through ruptures could inhibit efforts to repair the pipe.

Mitigation measures in the encasement foundation at the L-4 fault crossing included an overexcavation and jointed concrete backfill having the purpose of shifting the location of the L-4 fault from inside of the core (Zone 1) foundation to 20 feet upstream of the core foundation (Figure 13a). The joint in the concrete backfill is a 15/16-inch-thick bentonitic material made up of five Volclay Panels®. Following completion of excavation for the concrete backfill, it was found that the L-4 fault did not extend into the Franciscan Formation and cross through core foundation at depth as had been previously though. Therefore, the concrete backfill design was simplified as shown on Figure 13b.

Mitigation measures to the pipe design are similar to those described for the L-5 and L-6 faults except that the pipe was thickened from ½-inch to ¾-inch in the vicinity of the L-2 fault and at the Volclay Panel® joint in the L-4 concrete backfill to increase the resistance of the pipe against buckling failure due to reservoir induced pressure that could occur if the pipe were to be deformed vertically or laterally.

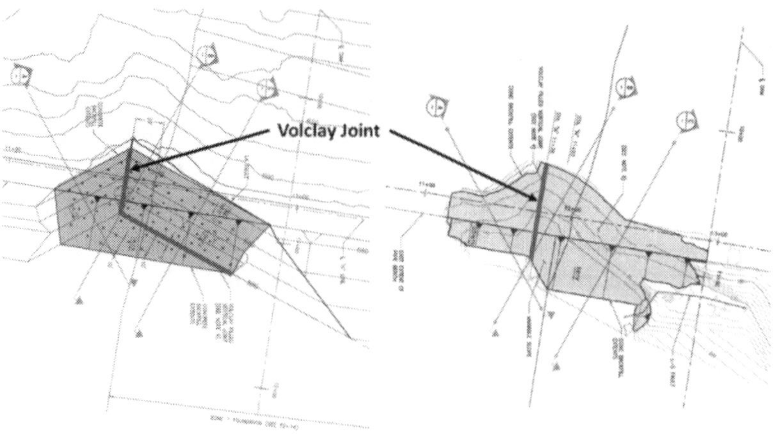

Figure 13. L3/L4 Backfill Concrete – a. Pre-excavation and b. Post-excavation.

In the event that deformation of the pipe is sufficient to cause a tear in the pipe wall, a filter protected impervious layer of Zone 1A core material is included on the outside of the encasement at the L-4 fault to reduce leakage into the pipe (Figure 14). The concrete encasement at the L-2 fault is treated similar to the L-5 and L-6 faults. Inspection of the pipe and repairs, would be made from inside the pipe following dewatering of the pipe.

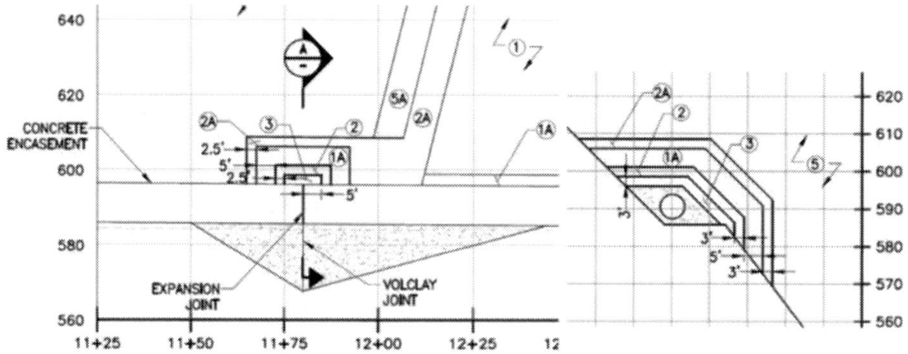

Figure 14. Embankment Fill Over Concrete Encasement at L3/L4.

CONCLUSIONS

This paper presents some of the challenges the highly seismic environment at Calaveras Dam has presented both during final design and construction. Modification to the outlet pipe during construction based on the complex geologic conditions as they were uncovered was particularly challenging. However, the efforts made by the Contractor, the CM Team, SFPUC, and the designer have resulted in a design that will provide a reliable source of high quality water to SFPUC's 2.6 million residential, commercial, and industrial customers in the Bay Area. The project is scheduled to be completed in 2018.

ACKNOWLEDGEMENTS

The contribution from Erik Newman (AECOM) on geotechnical topics is gratefully acknowledged. In addition, the authors of this paper acknowledge the substantial efforts and contributions made to the formulation, approval, and implementation of the engineering solutions presented in this paper by the staffs of the Designer (AECOM and SFPUC), the San Francisco Planning Department, the Construction Manager (Black and Veatch), the Contractor (Dragados, Flatiron, and Sukut Joint Venture), the Calaveras Technical Advisory Panel, and the California Division of Safety of Dams.

REFERENCES

Abrahamson, N., (2000). Effects of rupture directivity on probabilistic seismic hazard analysis, *Sixth International Conference on Seismic Zonation*, vol. 1, 151–156.

Fraser, W.A., (2001). Fault activity guidelines of the California Division of Safety of Dams. *California Division of Mines and Geology Bulletin* 210:319–325

Fraser, W.A., and Howard, J.K., (2002). Guidelines for use of the consequence–hazard matrix and selection of ground motion parameters. Unpublished memorandum. California Division of Safety of Dams. 9 pp.

Forrest, M., Roadifer, J., Wade, D., Hou, S., and Tang, G. (2014). Update on the Calaveras Dam Replacement Project, Alameda County, California, Proceedings of the 34th USSD Conference, San Francisco, California, April 2014.

Hudson, M., Idriss, I.M., and Beikae, M., (1994). QUAD4M, A computer program to evaluate the seismic response of soil structures using finite element procedures and incorporating a compliant base. Center for Geotechnical Modeling, Department of Civil and Environmental Engineering, University of California, Davis, California. May.

Itasca (Itasca Consulting Group, Inc.), (2005). *FLAC, Fast Lagrangian Analysis of Continua, Version 4.0, User's Guide.* Minneapolis, Minnesota.

Marinos, P, and Hoek, E. (2000). *GSI A Geologically Friendly Tool for Rock Mass Strength Estimation*, Proceedings GeoEng2000, Melbourne, Technomic Publishers.

Petersen, M., Dawson, T., Chen, R., Cao, T., Wills, C., Schwartz, D., and Frankel, A. (2011). Fault displacement hazard for strike-slip faults, Bulletin of the Seismological Society of America, v. 101, p. 805-825.

Roadifer, J.W. and Forrest, M.P., (2012). "Characterization and treatment of mélange and sandstone foundation at Calaveras Dam." *Proceedings of GeoCongress 2012: State of the Art and Practice in Geotechnical Engineering*, Oakland, California.

Somerville, P.G., Smith, N.F., Graves, R.W., and Abrahamson, N.A., (1997). Modification of empirical strong ground motion attenuation relations to include the amplitude and duration effects of rupture directivity. *Seismological Research Letters* 68:199–222.

NON-LINEAR MODELING OF SEDIMENT RETENTION STRUCTURES FOR DAM REMOVAL PROJECT

Kristen M. Martin, PE[1]
Erik J. Malvick, PhD, PE, GE[2]

ABSTRACT

The California Division of Safety of Dams (DSOD) regulates over 1,200 dams. Its responsibilities include independent analyses of dam construction projects for safety, including dam removals. DSOD recently approved the removal of the 106-foot-tall San Clemente Dam, a concrete arch dam retaining a reservoir that had nearly filled with sediment. DSOD determined the dam was hydrologically and seismically deficient, requiring remediation to make the dam safe. The owner ultimately chose to remove the dam. A key component of the removal project was containment of the sediments for public and environmental safety during periods of flood or high seismic loading. This was achieved by stockpiling the sediments in one reservoir arm and containing them between two earthfill structures. The project also included diversion and environmental restoration of the river around the sediment stockpile and the dam removal. The owner implemented a design-build contract due to the project scale. The project site is in a high seismic zone, so liquefaction is a concern for the long-term sediment containment. As a result, we analyzed the seismic stability and deformation mechanism of the containment structures using the software FLAC. To accommodate the design-build process and complete a thorough evaluation, we developed a quickly adjustable FLAC template file from the initial design concept. The template file allowed us to alter inputs to change slope geometry, material extents, and material properties as the design evolved. The template gave us the ability to evaluate design changes in as few as two days. Sensitivity analyses considered uncertain boundary conditions, material properties, and construction variability for the two structures. We modeled 17 design configurations using up to 10 ground motion time histories. Ultimately, our evaluations provided confidence that the two structures would be stable. The embankment template file will be used by DSOD for future projects because it allows our analyses to proceed efficiently and consistently.

INTRODUCTION

The California Division of Safety of Dams' (DSOD) mission is to protect people against loss of life and property from dam failure. DSOD regulates the design, construction, and maintenance of over 1,200 dams in California. This includes oversight of dam enlargements, repairs, alterations, and removals. DSOD performs independent analyses to understand dam performance and prevent dam failures. These analyses include structural, hydrologic, and geotechnical evaluations.

[1] Design Engineer, California Division of Safety of Dams. 2200 X Street, Suite 200, Sacramento, CA 95818. (916) 227-2170. Kristen.Martin@water.ca.gov.
[2] Design Technical Specialist, California Division of Safety of Dams. 2200 X Street, Suite 200, Sacramento, CA 95818. (916) 227-6742. Erik.Malvick@water.ca.gov.

The San Clemente Dam Removal is the largest dam removal in California history and was completed in 2015. In addition to removing the dam, the project included rerouting a river and stabilizing liquefiable reservoir sediment. DSOD's primary role was to ensure that the dam did not fail during removal. We also reviewed other aspects of the project and provided comments to preclude an obvious threat to the downstream population after the dam was removed. Our primary concern was ensuring the stockpiled sediment and the remaining liquefiable reservoir sediments underlying the stockpile do not flow downstream in a large seismic or flood event after dam removal.

DSOD's role was complex because the dam removal project was implemented as a design-build contract, and the final design of all components was not established before construction began. Most projects DSOD reviews and approves follow a design-bid-build process where a design is finalized and approved by DSOD prior to going out to bid for construction. For this project, however, the design for each project component was submitted for our review independently after the contract had been awarded. This led to a tight schedule requiring our quick evaluation of the designs as they were submitted.

This paper presents an overview of DSOD's design review of the San Clemente Dam Removal project with an emphasis on the analyses we performed for the two sediment retention structures. The design concept and material characteristics are summarized. The decision to develop a template model using the software FLAC (Itasca 2011) is also described. We recognized early in the design process that the tight design to construction schedule would limit our ability to sufficiently complete our review as the design changed. The template approach allowed us to invest a large amount of work up front while the project was in a conceptual stage to provide flexibility and efficiency to change the finite element model and input properties as the design evolved. The ability to adjust the design ultimately allowed us to complete analyses of various designs in as few as two days.

PROJECT HISTORY

San Clemente Dam was a thin arch concrete dam constructed in 1921 immediately downstream of the confluence of the Carmel River and San Clemente Creek near Carmel Valley, California (Figure 1). The reservoir had 2 arms formed by the inundation of the Carmel River and San Clemente Creek and controlled a 126 mi^2 watershed. The dam was owned by the California American Water Company (CalAm).

The dam was 106 ft tall, 300 ft long, 7 ft wide at the crest, and 17 ft wide at the base. The upper 25 ft of the dam was reinforced. The spillway structure was located at the center of the dam and had the capacity to pass 27,500 cfs when the reservoir level was at the dam crest (Figure 2a). When constructed, the reservoir had the capacity to store 1,425 acre-feet of water, which was used to supply water for the Monterey peninsula.

During its lifetime, the reservoir accumulated an estimated 2.2 million yd^3 (1,355 acre-feet) of sediment that occupied 95% of the reservoir, leaving 70 acre-feet of storage remaining (State Coastal Conservancy 2011). This sediment accumulation limited the

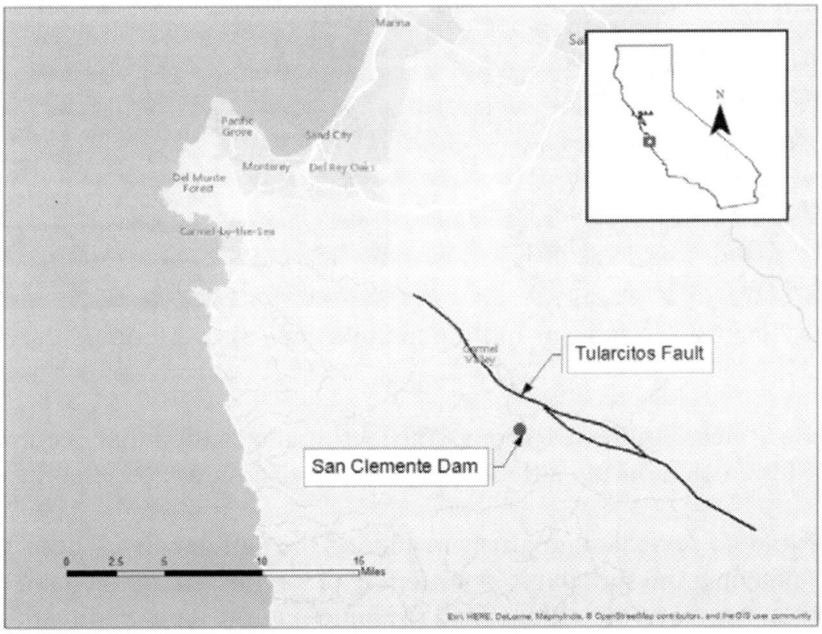

Figure 1. Project Location.

hydrologic surcharge capacity of the reservoir and rendered the low-level outlet works inoperable. Additionally, the hydraulic capacity of the spillway was limited compared to the runoff from the watershed, leading to spilling and overtopping of the dam (Figure 2b).

<u>Dam Deficiencies</u>

In 1931, the dam was reportedly overtopped for 10 days by a maximum of 3 feet, which resulted in considerable erosion of loose and weathered rock. More extensive erosion would be expected if the dam overtopped during a design storm, which would have compromised the stability of the dam.

Because of the dam's size, the reservoir's volume, and the population downstream, DSOD required the dam to be evaluated for a Probable Maximum Flood (PMF). In 1981, DSOD evaluated the hydrologic capacity of the dam using the PMF based on Hydrometeorological Report 36 (U.S. Weather Bureau 1969). This evaluation found that the PMF would overtop the dam by 14 ft.

In 1989, DSOD requested that the owner reevaluate the seismic stability of the dam, incorporating recent seismic activity in the area. In 1992, the owner completed a seismic and hydrologic stability reevaluation of the dam, which confirmed the stability of the dam was deficient (Woodward-Clyde 1992). At that time, the design earthquake (the maximum credible earthquake) was a Magnitude 6.7 event on the nearby Tularcitos Fault that would shake the dam with a peak ground acceleration (PGA) of 0.7-g. This earthquake would overstress the dam and cause severe structural damage to the dam and

Figure 2. (a) San Clemente Dam spilling in 2008 and (b) overtopping during a storm.

an uncontrolled release of the reservoir. Considering the erosion that occurred in 1931, the potential for erosion during a flood event would also reduce the stability of the dam.

The owner explored several alternatives to address the deficiencies. These alternatives included (1) notching the spillway, (2) stiffening or strengthening the dam, (3) removing sediment to recover reservoir storage, and (4) removing the dam. Sediment removal would have economic and environmental impacts because of the large sediment volume and limited site access. In addition, improving fish passage was an environmental goal. The owner ultimately opted to remove the dam and stabilize the sediments in-place to abate the deficiencies and restore fish passage up the Carmel River for the threatened central coast steelhead trout. The design-build contractor selected for the project was Granite Construction (Watsonville, CA), together with Kleinfelder (Oakland, CA).

DAM REMOVAL PROJECT DESCRIPTION

The 2.2 million yd^3 of reservoir-deposited sediments are generally loose, unconsolidated, saturated sands and silts with limited clays and organic matter. Given the high seismicity in the area, these sediments are liquefiable. The primary challenge of the removal project was stabilizing these sediments in the event of an earthquake.

The Carmel River is much larger than San Clemente Creek. To minimize the volume of material to be moved, sediments from the smaller San Clemente Creek arm of the reservoir were excavated and stockpiled on top of the sediments in the Carmel River arm, forming a sediment stockpile (Figure 3). This stockpile was stabilized at its upstream and downstream extents by compacted earthfill structures within the former Carmel River channel. The upstream and downstream compacted earthfill structures are known as the Diversion Dike (DD) and Stabilized Sediment Slope (SSS), respectively. Since the stockpile occupies this reach of the former Carmel River channel, the river was rerouted around the stockpile into San Clemente Creek.

A 450-ft-long Reroute Channel (RRC) was excavated through a granitic ridge separating the two streams at the upstream extent of the historical reservoir. The DD now diverts the

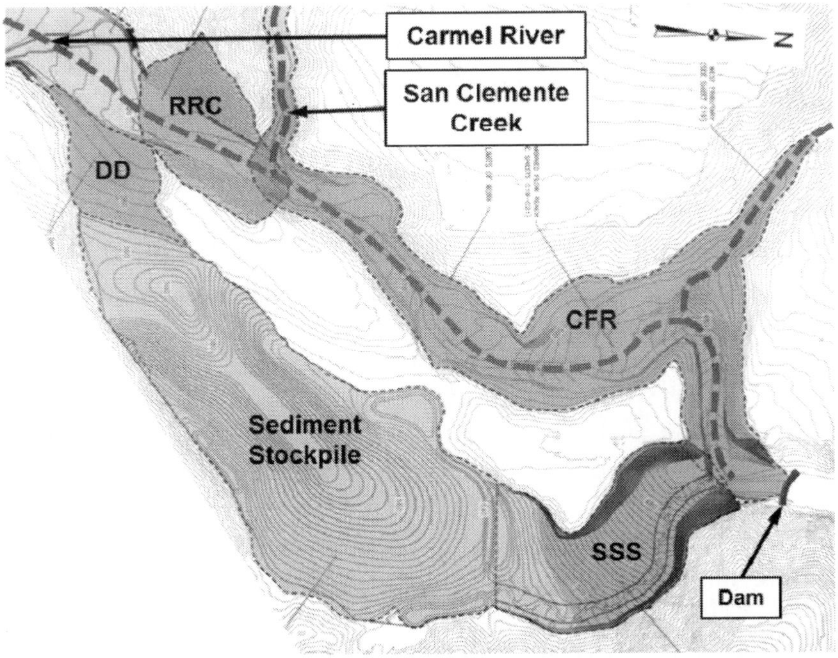

Figure 3. Plan view of Project Components.

Carmel River away from the stockpile and through the RRC into San Clemente Creek. The DD and SSS were constructed using earthfill sourced from the RRC excavation.

Rerouting the Carmel River relocated the confluence of the waterways from immediately upstream of the dam to approximately 3,000 feet upstream of the dam. The reach of San Clemente Creek between the RRC and the former dam site now carries combined flows from the Carmel River and San Clemente Creek; this reach is identified as the Combined Flow Reach (CFR). The peak flow in the Carmel River is approximately 10 times that in San Clemente Creek, so the project included erosion protection for the CFR.

The dam was removed once sediment was stockpiled and the Carmel River was rerouted.

DSOD reviewed the RRC, DD, SSS, and dam demolition. Since the reservoir was nearly full of sediment, the project has a minimal effect on flood control. Sedimentation will likely resume to pre-dam conditions in the Carmel River downstream of the dam. Removing the dam eliminated the stability deficiencies and improved fish passage.

Environmental Impact

In California, the potential environmental impacts of a project must be studied and declared before a project can be completed. The California Environmental Quality Act (CEQA) requires that public agencies identify environmental impacts and avoid or mitigate potentially significant impacts. DSOD is a Division within the California Department of Water Resources (DWR). DWR was the primary public agency involved in the project and acted as the CEQA lead agency to develop an Environmental Impact

Report to identify and mitigate the project's environmental impacts. The Environmental Impact Report was also written to function as an Environmental Impact Statement to satisfy the National Environmental Policy Act requirements for this project.

DD and SSS Design

The DD and SSS function primarily to stabilize and contain the sediment stockpile in the Carmel River arm of the former reservoir.

The DD (Figure 4) spans the former Carmel River channel at the upstream end of the stockpile and also serves to divert the Carmel River into the RRC. At the DD, the in-situ sediment and underlying alluvium have a total thickness of 55 ft. The foundation of the DD was constructed by excavating through the sediments and alluvium to bedrock and replacing them with compacted earthfill up to the surface of the sediments. The base of the foundation at bedrock is approximately 120-ft square with excavated slopes in the sediment of 3H:1V. The abutments of the DD are the rock slopes along the river channel. The 40-ft-high embankment portion of the DD was constructed on top of the compacted foundation using similar earthfill. The upstream and downstream slopes of the DD embankment are 3H:1V and 1.5H:1V, respectively. Sediment from the San Clemente Creek arm of the reservoir was stockpiled against the downstream slope of the DD embankment.

The SSS (Figure 5) spans the former Carmel River channel immediately upstream of the dam and former confluence with San Clemente Creek. The rerouted Carmel River now flows parallel to the downstream toe of the SSS. The SSS is a compacted earthfill buttress founded against the in-situ sediments at a 3H:1V slope. The SSS has a foundation and abutments consisting of bedrock. The crest of the SSS was constructed to match the grade of the in-situ sediments and alluvium it is stabilizing, which are 75-ft thick at its location. The sediment stockpile was placed and compacted upstream of the SSS to a maximum 5% grade to limit static loading on the SSS.

Within the SSS, a zone of compacted sediment was placed against the in-situ sediments to provide a firm surface for the structural fill. A blanket drain was placed at the base of the SSS to drain water from the stockpile. The downstream slope of the SSS is 6H:1V.

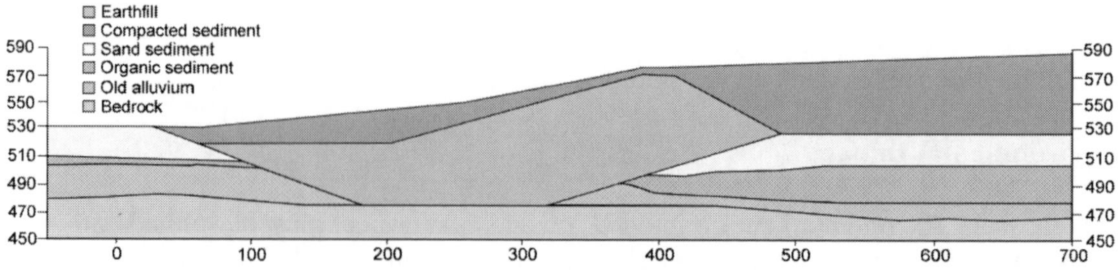

Figure 4. Section view of Diversion Dike.

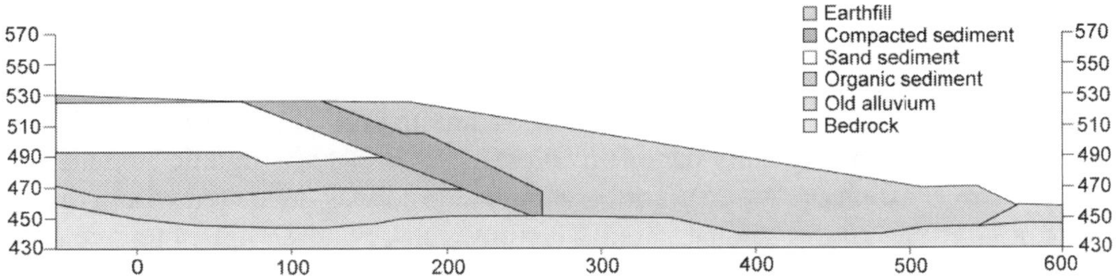

Figure 5. Section view of Stabilized Sediment Slope.

INDEPENDENT EVALUATION

The DD and the SSS must be seismically stable to prevent the liquefiable sediments from being released and flowing downstream during an earthquake. As a result, we reviewed the seismic stability to verify the structures would not fail and potentially block the river or exacerbate flooding. The reservoir sediments are loosely deposited silts and sands with some organics that are identified as potentially liquefiable and susceptible to strength loss using methods described in Malvick et al. (2014). FLAC was used to model the dynamic response and deformations of the liquefiable soils.

In order to complete our analyses, we estimated the seismic load, characterized the soils, reviewed the design concept, and developed our modeling approach.

Seismicity

The controlling seismic event for the San Clemente Dam site is a Magnitude 6.7 earthquake on the Tularcitos Fault, 2 km NE of the site. The ground motion parameters are the mean of the 50th percentile estimates from all five 2008 NGA West attenuation relationships; the 50th percentile was chosen based on the reduced hazard of the completed project. Ten ground motion time-histories were developed using a target PGA of 0.49g, an associated target response spectrum, and an estimated Arias Intensity of 2.2 m/s (Travasarou & Abrahamson 2003, Watson-Lamprey & Abrahamson, 2006). The time histories were developed using spectral matching for some motions and amplitude scaling for the remaining motions with the intent to meet the target parameters on average.

Material Characterization

We characterized the material properties for the sediments and alluvium that would be left in place, the recompacted sediments for the stockpile, and the engineered fill.

Sediment and Alluvium. Characterization of the in-place alluvium and sediment focused on estimating their resistance to liquefaction and magnitude of potential dynamic strength loss using methods we developed at DSOD that are consistent with Malvick et al. (2014). Multiple geotechnical investigations have been conducted over the years, including one by the design-build engineer in 2013. The sediment and alluvium were characterized as three separate soil units identified as Old Alluvium, Organic Sediment, and Sandy Sediment from bottom to top.

The Old Alluvium is actually a geologically younger alluvial unit consisting of cobbles, gravels, and sand that was identified by the design-build engineer as "Old" because it predated the dam and reservoir. The Organic Sediment unit consists of various sands, silts, and clays that are visually darker in color, have up to 15% organic content by mass, and have low plasticity ($PI < 15$). This organic material was deposited within the reservoir after the dam was constructed. The uppermost Sand Sediment unit consists mostly of poorly graded sand and was largely free of organics.

Soil properties for the sediment and alluvium are shown in Table 1. The liquefaction resistance and subsequent residual strength of the reservoir-deposited sediments was evaluated from SPT blow-counts corrected using recommendations in Idriss and Boulanger (2008), $(N_1)_{60\text{-}cs}$. SPT testing was not conducted in the Old Alluvium in 2013 and considered unreliable from earlier investigations. Cores from this unit recovered in 2013 show the interstitial sand to be similar to overlying sediment sands. Representative values for dry unit weight, γ_d, were determined from relatively undisturbed samples obtained during exploration. Additional properties, such as $k_{2,max}$ (Seed et al 1986) and residual shear strength ratio ($S_r/\sigma'_{v,0}$), came from estimates based on field observations and measured blow-counts.

<u>Engineered Fill</u>. There were 3 types of engineered fill for this project (Table 2). Rock coring was performed within the RRC alignment to characterize the earthfill materials for the DD and SSS. Design parameters reported in Table 2 are based on laboratory testing, correlations, and engineering experience. This fill is ripped rock that classifies as well-graded gravel with silt (GW-GM).

A shortage of earthfill material became apparent during construction, which required an additional borrow source from the same ridge, closer to the dam. The material properties were similar.

The SSS also includes a zone of compacted sediment that was constructed from reworked Sand Sediment unit material. Stockpile material was sourced from all of the sediment units in San Clemente Creek and was compacted on top of the existing sediments in the

Table 1. Sediment and Alluvium Soil Properties

Zone	PI	$(N_1)_{60\text{-}cs}$ (bpf)	k_{2max}	γ_d (pcf)	Fines content (%)	$S_r/\sigma_{v,0}'$
Sand sediment	NP	4-6	23	70	≤ 10	0.05
Organic sediment	15	12	20	50	30 – 80	0.10
Old Alluvium	NP	10	80	95	≤ 10	0.09

Table 2. Properties of engineered fill used for DSOD analyses. Note the earthfill represents the range of strengths found over all borrow sources.

Zone	γ_d (pcf)	ϕ' (°)	c' (psf)	ϕ (°)	c (psf)
Earthfill	120	37-40	0	--	--
Compacted sediment (SSS)	105	35	0	30	0
Stockpile	105	32	300	20	500

Carmel River. This compacted stockpile material is not structural fill, but it contributes to the loads on the SSS and DD.

Modeling Approach

We initiated non-linear numerical modeling during the conceptual design phase after the site investigation was complete. Since liquefaction is expected to occur in the in-place saturated reservoir sediments when subjected to seismic loads, finite element modeling using FLAC was conducted to evaluate the deformations, the liquefaction response, and the failure mechanism of each structure.

We anticipated the need for a quick and comprehensive evaluation process with sensitivity analyses and several design iterations. Developing our FLAC models to be easily modified would allow us to adjust the model within the time constraints and expedite our review once design and construction began. We invested significant time up front to develop a template file using the FLAC scripting language, FISH. This template allowed specified input parameters for all geometric and material properties to be adjusted immediately as the design was changed, reducing the time needed for mesh adjustments.

The designs of both the DD and SSS were changed several times during construction due to analyses results, subsequent comments during design, and material changes and volume shortfalls during construction.

Schematics of the DD and SSS models are shown in Figures 6 and 7, respectively. The non-linear deformation modeling sequence began with building a geometric mesh, assigning soil properties, initiating a phreatic surface, and establishing static equilibrium. Adjustable geometric inputs included slopes (S_i), widths (W_i), and thickness and heights of in-situ sediments and compacted fill (h_i). Provisions were made to easily add additional zones that were not included in the original design by outlining the zonal dimensions. We also ensured that a sufficient number of elements were present in each soil layer to accurately model liquefaction.

Liquefiable materials were modeled using the PM4 Sand constitutive model (Ziotopoulou and Boulanger, 2013) while a Mohr-Coulomb model was used for the remaining

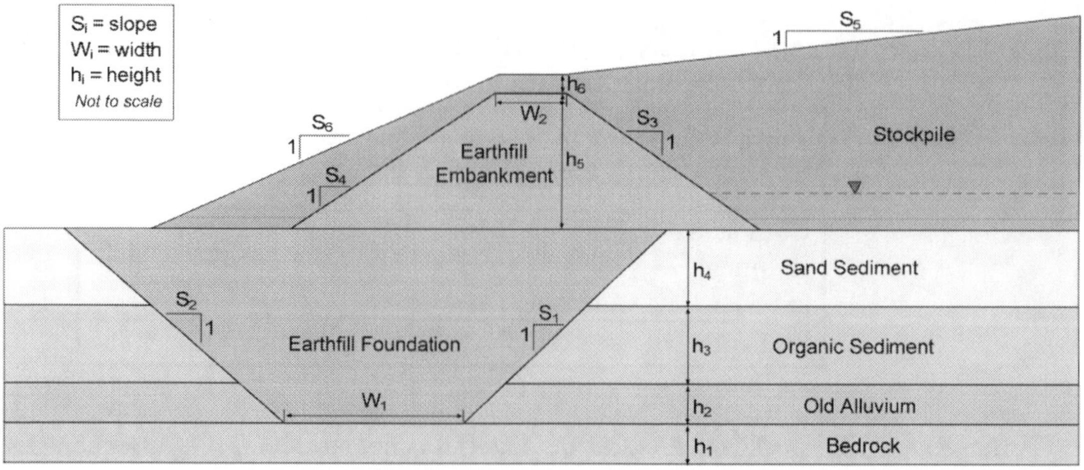

Figure 6. Model of Maximum Section of Diversion Dike.

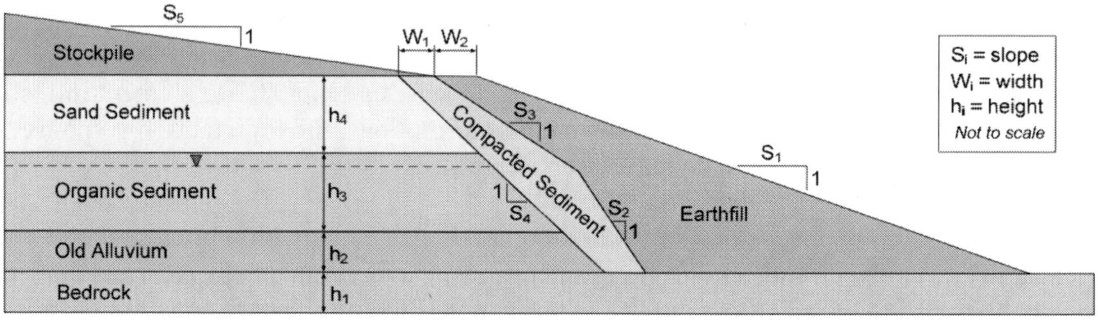

Figure 7. Model of Maximum Section of Stabilized Sediment Slope.

materials. For template flexibility, adjustable material inputs include total and effective soil strengths, density, shear moduli, and saturation. For liquefiable soils inputs include SPT blow count and residual strength.

Once the inputs were set, we initialized the stress state in the model by applying gravity and piezometric boundary conditions. We based piezometric conditions on contract requirements for stability.

After we established hydrostatic equilibrium, we converted the liquefiable soils to the PM4 Sand constitutive model using the parameters in Table 1. We calibrated PM4 Sand for each soil type using a single element FLAC mesh to simulate cyclic simple-shear behavior. The constitutive model was calibrated to reach 3% double-amplitude shear strain in 15 cycles of loading based on the liquefaction triggering correlation in Idriss and Boulanger (2008) for $(N_1)_{60\text{-}cs}$. The Organic Sediment unit, however, was calibrated using 30 cycles to mimic the more liquefaction-resistant behavior observed in laboratory cyclic simple-shear testing conducted during the 2013 site investigation.

Dynamic Loading. The models were then ready for dynamic loading. Boundary conditions included a compliant base and free-field nodes on the left and right of the model to capture the expected site response. The simulations were conducted under undrained conditions to facilitate a practical modeling time.

The DD and SSS models were each subjected to the 10 time-histories and their inverses for a total of 20 evaluations each (40 total). The maximum excess pore pressure ratio, r_u, was tracked for each element (e.g. Figures 8c and 8f). At the end of dynamic loading, elements with $r_u > 70\%$ were considered to be liquefied and converted to the Mohr-Coulomb constitutive model with undrained strengths assigned using the residual strength ratios, $S_r/\sigma_{v,0}'$ (Table 1). The models were then run through a post-seismic static period under drained conditions to capture additional deformation or flow failure.

Design Changes and Sensitivity Analysis. During construction, the excavated material from the RRC was finer than assumed. This fill was initially characterized as a dirty rockfill at exploration, but graded as a sandy, silty gravel (GW-GM) at the time of construction. We adjusted our estimate of this material's strength within the models and reevaluated the stability of the DD and SSS. In addition, several factors contributed to a fill volume shortfall. The design-build engineer proposed multiple solutions to make up for the shortfall, which we modeled quickly to assess for acceptability. The solutions included redesigns of both fill material properties and geometric configuration of the zones. The flexible geometry inputs in our template file and our previous evaluations allowed us to quickly make confident conclusions about the changing configurations. In one case, we validated the final design configuration in less than two days of FLAC modeling.

Sensitivity analyses helped establish our confidence as we often varied material properties, design assumptions, loads, and construction quality to observe their influence on the failure mechanism. We also ran the models with varying boundary extents to ensure that our mesh could adequately capture potential lateral spreading.

Finally, we evaluated multiple scenarios to estimate the influence poor compaction near the interfaces between the fill and in-situ sediments could have on each structure. We accomplished this by reducing the widths (i.e. W_1 in Figures 6 and 7) of the compacted fill zones at the DD and SSS by 30 ft and 20 ft, respectively.

RESULTS

The template file we used to script various model configurations allowed us to complete evaluations of each variation of the SSS and DD designs quickly. This enabled us to respond to the dam owner with minimal impact on the construction schedule. We ultimately evaluated 17 configurations; the final configuration for each model is shown in Figures 8a and 9a.

Our main concern for safety was that the DD and SSS do not experience a liquefaction-induced flow failure that could block the river channel or result in an uncontrolled release of stockpiled sediments. Because there is inherent uncertainty with any model, we did not expect the results to be a precise indicator of dynamic performance. Rather, we qualitatively determined that the broader failure mechanism indicated by shear strain contours does not indicate a flow failure (Figures 8b and 9b).

Our dynamic analyses showed that the DD and SSS experience horizontal deformations less than 4 ft at the toe of slopes and vertical settlements less than 1 foot at the head of slopes. We concluded that neither structure demonstrated flow failure based on the absence of deformations during the post-earthquake static analyses. Figure 10 shows example time history results for the DD and SSS models shown in Figures 8 and 9, including deformations at the head and toe of the structures. The maximum shear strains indicated that a lateral spread mechanism developed, but the limited deformations indicated that the DD and SSS adequately restrain movement.

As a result, we expect the DD and SSS to retain the liquefiable sediments in the design earthquake. The actual deformations should be of the same order of magnitude due to the benefits the three-dimensional site topography will provide at each structure, such as the curvature of the canyon that will allow the rock abutments to limit the magnitude of deformations that can occur.

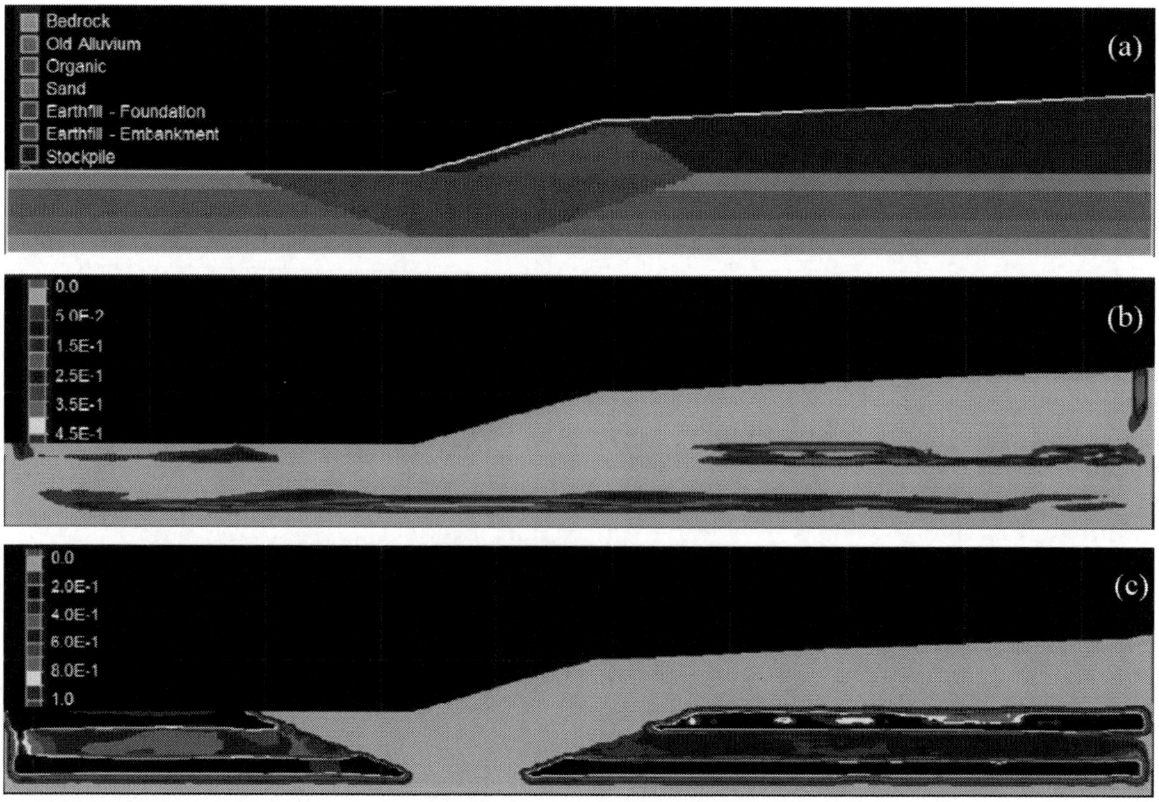

Figure 8. FLAC model showing (a) DD mesh configuration, (b) DD maximum shear strain distribution, and (c) DD maximum pore pressure ratio during dynamic loading

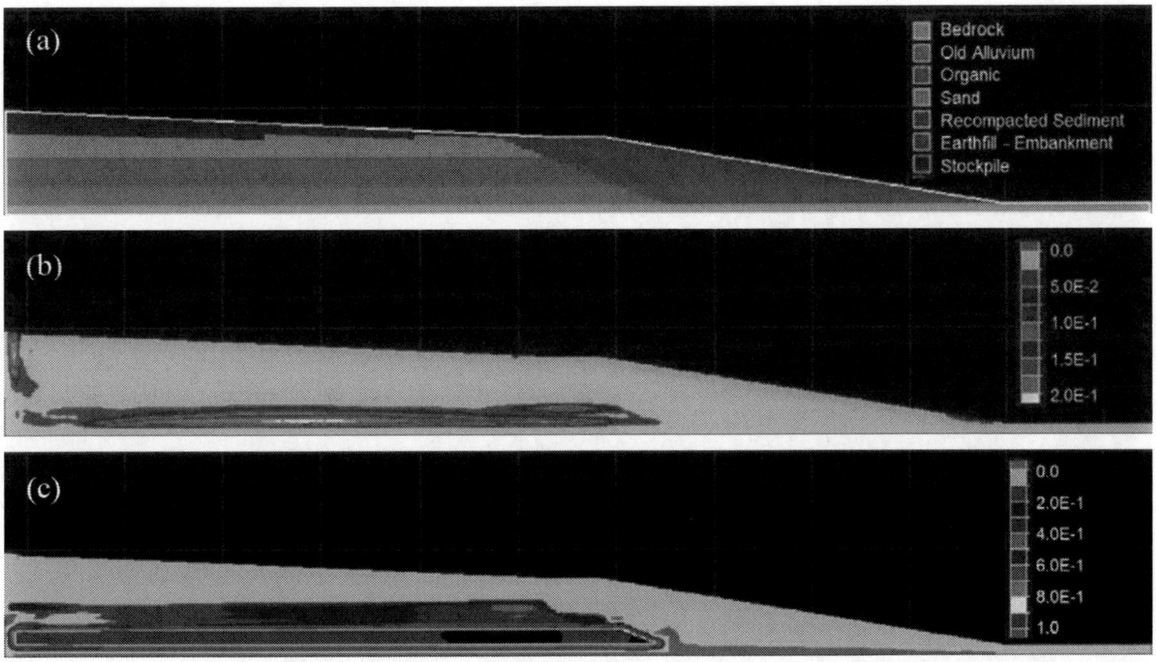

Figure 9. FLAC model showing (a) SSS mesh configuration, (b) SSS maximum shear strain distribution, and (c) SSS maximum pore pressure ratio during dynamic loading

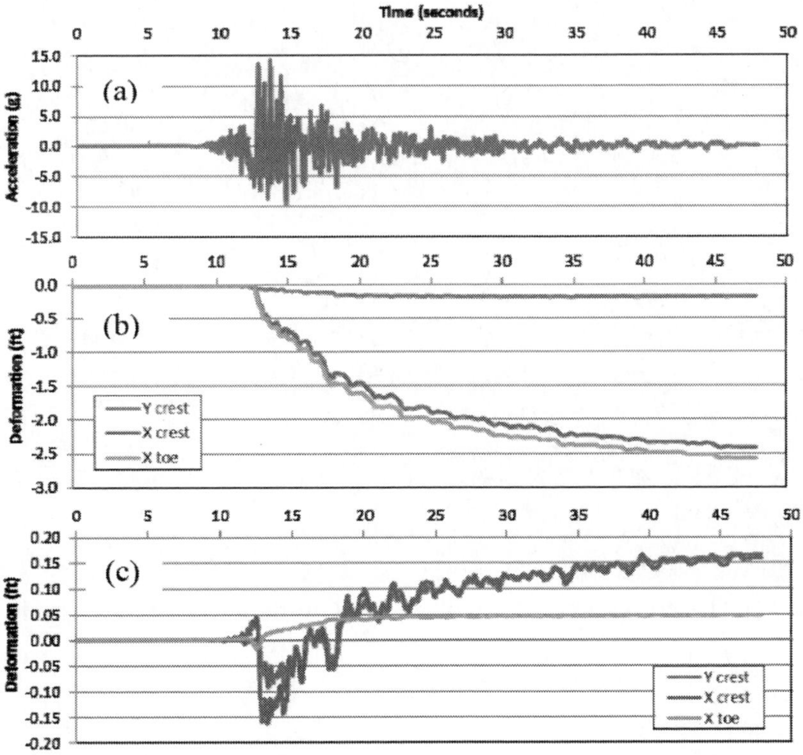

Figure 10. Records of (a) one ground motion time history and associated horizontal, x, and vertical, y, deformations from the FLAC models of (b) the DD and (c) the SSS.

CONCLUSION

The San Clemente Dam removal was a complex project that mitigated known hydrologic and seismic deficiencies of the dam (Figure 11). DSOD reviewed multiple aspects of the project as part of the dam removal. Containment of reservoir sediments was the most critical feature of the project, which necessitated an independent seismic evaluation of the Diversion Dike and Stabilized Sediment Slope using the software FLAC.

The challenges of accommodating the design-build process and a compressed project schedule drove us to develop a template file that allowed for rapid adjustment of the numerical mesh and the material properties to provide a confident and thorough evaluation as the design evolved.

This flexible modeling approach was successful for this project. This approach could be adapted in the future for other design-build projects if a significant effort is made prior to the design phase. However, such an approach may be limited if a design concept is uncertain. This approach could also be used broadly for more typical finite element evaluations to promote consistency within an organization and its engineering analyses.

Figure 11. The completed removal of San Clemente Dam looking downstream.

ACKNOWLEDGEMENTS

This work was supported by the California Department of Water Resources Division of Safety of Dams. We would like to thank California American Water Company, the owner of San Clemente Dam. The opinions expressed in this paper are solely those of the authors, and do not necessarily represent the views of the California Department of Water Resources, Division of Safety of Dams.

REFERENCES

Idriss, I., and Boulanger, R. (2008). *Soil Liquefaction during Earthquakes, Monograph MNO-12*, EERI, Oakland, CA, 261 pp.

Itasca (2011). *FLAC – Fast Lagrangian Analysis of Continua, Version 7.0*, Itasca Consulting Group, Inc., Minneapolis, Minnesota.

Malvick, E.J., Armstrong, R.J., and Martin, K. M. (2014). Estimating the Dynamic Shear Strength of Soils for Dam Seismic Safety Analysis. *ASDSO Journal of Dam Safety,* 12(3), 9-20.

Seed, H. B., Wong, R. T., Idriss, I. M., & Tokimatsu, K. (1986). Moduli and damping factors for dynamic analyses of cohesionless soils. *Journal of Geotechnical Engineering,* 112(11), 1016–1032.

State Coastal Conservancy (2011). San Clemente Dam Removal Project: Implementation phase. Staff Recommendation. Project No. 07-004-03 (http://scc.ca.gov/webmaster/ftp/pdf/sccbb/2015/1506/20150625Board03D_San_Clemente_Dam.pdf).

U.S. Weather Bureau, (1969). *Interim Report, Probable Maximum Precipitation in California, Hydrometeorological Report Number 36.* U.S. Department of Commerce, Washington, DC, 202 pp.

Woodward-Clyde Consultants (1992). *Seismic and Flood Stability Evaluation, San Clemente Dam.*

Ziotopoulou, K., and Boulanger, R. W. (2013). Calibration and implementation of a sand plasticity plane-strain model for earthquake engineering applications. *Journal of Soil Dynamics and Earthquake Engineering*, 53, 268-280.

RECENT PROGRESS TOWARDS REMOVAL OF MATILIJA DAM

Seth Gentzler[1]
John Roadifer[2]
Derek Booth[3]
Yantao Cui[4]
Peter Sheydayi[5]

ABSTRACT

The Matilija Dam and reservoir are located approximately 16 miles upstream of the Ventura River mouth, near Ventura, California. Since its construction in 1947, the 168-foot-high, arched concrete Matilija Dam has blocked the transport of an estimated 8 million cubic yards (mcy) of sediment from naturally moving downstream to the Pacific Ocean. This has resulted in losses to almost all reservoir storage, downstream sand- and gravel-sized materials necessary to promoting downstream wildlife habitat, and sediment needed to maintain beaches at Surfer's Point. The dam also prevents southern California steelhead (*Oncorhynchus mykiss*) from reaching upper Matilija Creek.

In the early 2000's, Ventura County Watershed Protection District and the US Army Corps of Engineers evaluated several alternatives for dam removal, publishing a Final EIS/EIR in 2004. Their preferred alternative included substantial mechanical handling of sediment and infrastructure: slurrying an estimated 2 mcy of fine sediment from reservoir deposits just upstream of the dam to a downstream disposal location, excavation of a channel through the upper reservoir deposits, removal of the dam, and downstream improvements to address reintroduction of natural full annual sediment loads.

In 2014, VCWPD contracted with AECOM and Stillwater Sciences to reevaluate the dam removal concepts previously considered and to develop new concepts. The present paper discusses the concepts considered and evaluation of those concepts for key biologic, sediment transport, water quality, construction cost, and schedule objectives. These concepts emphasize the rapid erosion and transport of impoundment sediment by natural high flows without direct handling of the deposit.

---

[1] Vice President, AECOM, 1333 Broadway, Suite 800, Oakland, CA 94612, (510) 874-3018, seth.gentzler@aecom.com.
[2] Sr. Project Manager, AECOM, 1333 Broadway, Suite 800, Oakland, CA 94612, (510) 874-1732, john.roadifer@aecom.com
[3] Senior Geologist/Geomorphologist, Stillwater Sciences, PO Box 904, Santa Barbara, CA 93102, (206) 914-5031, dbooth@stillwatersci.com
[4] Senior Hydraulic Engineer, Stillwater Sciences, 2855 Telegraph Avenue, Suite 400, Berkeley, CA 94705, (510) 848-8098 x120, yantao@stillwatersci.com
[5] Project Manager, Ventura County Water Protection District, Engineering Services Division, 800 S. Victoria Avenue, L#1670, Ventura, CA 93009, (805) 654-2016 , Peter.Sheydayi@ventura.org

INTRODUCTION

Since its construction in 1947, the 168-foot-high, arched concrete Matilija Dam has blocked the transport of nearly 8 million cubic yards (mcy) of fine and coarse sediment, preventing it from naturally moving downstream to the Pacific Ocean (see Figure 1). This has resulted in loss of the reservoir's original function of water storage for agricultural needs, and it has also resulted in loss of downstream sand and gravel-sized materials necessary to promote habitat for a variety of wildlife species, loss of sediment needed to maintain beaches at Surfer's Point, and increased erosion of the Ventura River streambed. The dam, with its non-functioning fish ladder, also prevents southern California steelhead (*Oncorhynchus mykiss*) from reaching upper Matilija Creek, which, prior to dam construction, was the most productive spawning and rearing habitat for this species in the Ventura River system. Without dam removal, an estimated total of 9 mcy of sediment will be trapped behind the dam before the entire natural annual sediment load of Matilija Creek begins to be carried over the dam, likely within the next few decades. While such a scenario would eventually begin to address sediment deprivation of the downstream reaches, leaving the dam in place would not address fish passage beyond the dam and impacts to upstream habitat.

In the early 2000's, Ventura County Watershed Protection District (VCWPD) and the US Army Corps of Engineers (USACE) evaluated several alternatives for dam removal and published a Final Environmental Impact Statement/Environmental Impact Report (EIS/EIR, USACE 2004). They arrived at a preferred alternative (Alternative 4b) that involved slurrying an estimated 2 mcy of fine sediment from the Reservoir sub-area just upstream of the dam to a downstream disposal location, removing the dam in one season, excavating a channel through the remaining coarse sediment, and protecting the lower seven feet of the channel banks with soil cement, while allowing 10-year and greater discharges to remove the accumulated sediments above the seven-foot level. At some future date, the soil cement would be removed, allowing the remaining accumulated sediment to be flushed through the river system.

Subsequently, in 2009 and 2010, the Matilija Dam Fine Sediment Study Group (FSSG) was convened to address concerns over cost and constructability of the downstream disposal options for the fine sediment. In particular, they evaluated temporary upstream disposal of the fine sediment as an alternative solution.
In 2014, VCWPD contracted with AECOM and Stillwater Sciences (the Consultant Team) to evaluate a range of concepts including those documented in previous documents, concepts developed by the FSSG, and new concepts. A shortlist of six dam removal concepts was identified and screened based on selected key criteria (URS and Stillwater Sciences, 2014a) to arrive at the three preferred concepts discussed below. All concepts analyzed during this phase of work assume that the incoming coarse sediment load will move through the former reservoir area and address concerns about downstream beach erosion. Following the evaluation, up to two alternatives are anticipated to move forward into preliminary design at a future date.

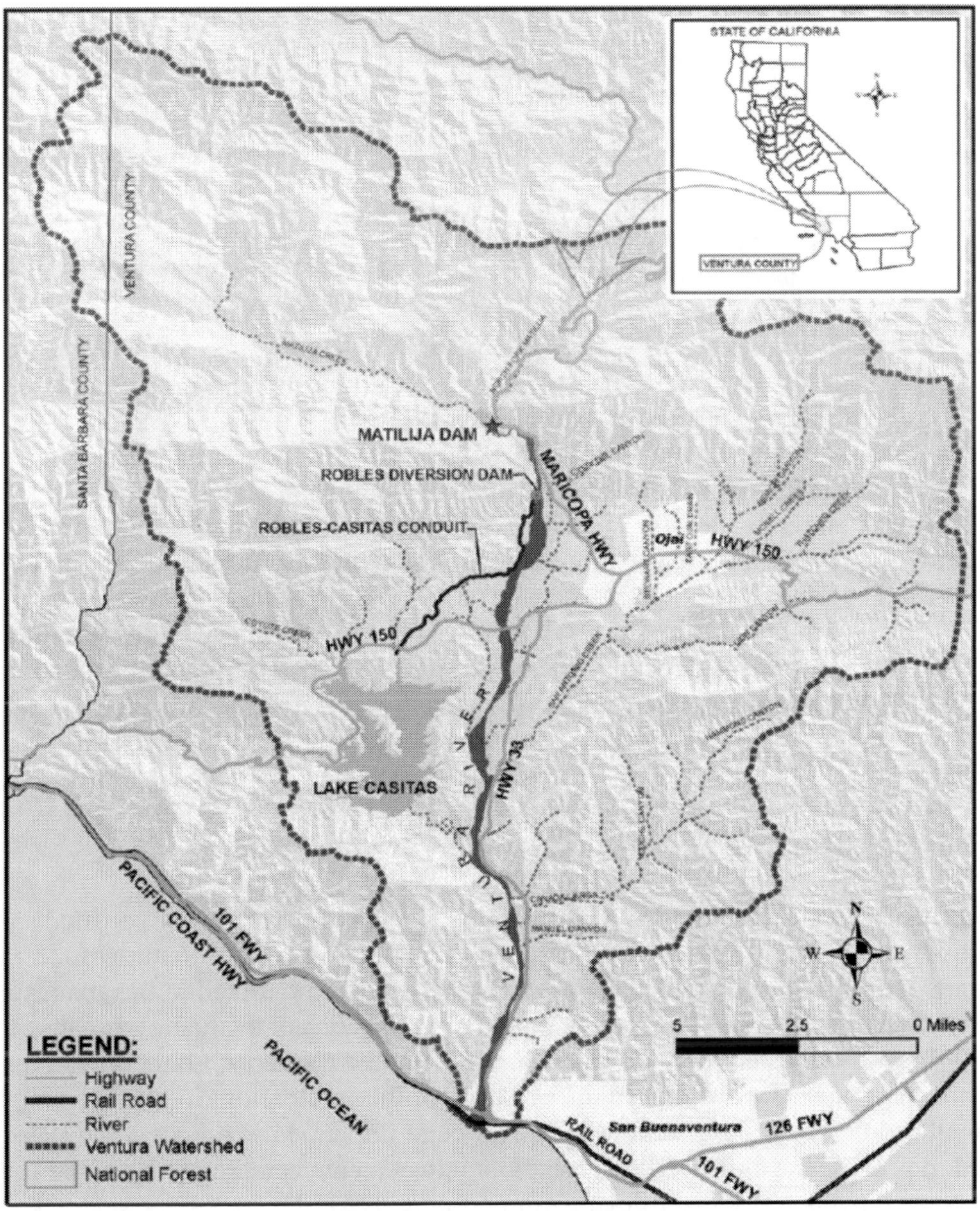

Figure 1. Project Location Map.

PROJECT OBJECTIVES & EVALUATION CRITERIA

Project Objectives

In the January 2013, project design objectives were refined by the project stakeholders and documented as shown below:

1. Biological Objectives:
    a. restore effective fish passage within one year after the dam is removed;

b. minimize to the degree feasible ecological impacts from project implementation;
c. restore aquatic/riparian/upland habitats in the reservoir area;
d. encourage beach nourishment, as feasible.

2. Cost-Effectiveness Objectives:
   a. reduce cost/maximize cost-effectiveness;
   b. develop a feasible removal plan within the context of the federally authorized project.

3. Sediment Transport and Water Quality Objectives:
   a. maximize mobilization of fine material during high flow events;
   b. reduce mobilization during low flows;
   c. minimize artificial or permanent stabilization material (e.g., riprap or soil cement) in the project area and environmentally sensitive areas;
   d. minimize project-related turbidity increases and nutrient inputs to Casitas reservoir, connected to the Ventura River downstream of Matilija Dam by Robles Diversion Dam and a canal;
   e. maintain existing level of safe supply of water, provided by Casitas reservoir and local groundwater wells, to customers of all water districts.

These objectives are the basis of formulation of the current phase of technical work presented here.

Project Objectives

Based on subsequent discussions with the project stakeholders, the above-listed objectives were used to develop a refined list of evaluation criteria to allow for comparison of the short-listed concepts developed and summarized in this manuscript. Certain objectives listed above are not included in the current evaluation because they are not significantly different between the three short-listed concepts. Other objectives were determined to be outside of the technical nature of this evaluation. In total, the objectives for steelhead passage and health, water quality and implementation costs and risks were judged to provide the clearest discrimination between concepts, resulting in the following pertinent and manageable list of evaluation criteria (Table 1).

The approach to evaluating the dam removal concepts involves quantification of the various evaluation criteria listed above, using relevant tools and empirical data available from analogous dam removals elsewhere. After the criteria are populated, the dam removal concepts can be ranked under each criterion individually and then cumulatively. The goal for the evaluation is to utilize the criteria and cumulative rankings to select those concepts that best meet the project objectives and anticipated funding constraints.

Table 1. Evaluation Criteria

	Evaluation Category	Evaluation Criteria Description
A.	**Biological Resources**	
	Steelhead Passage	Time to achieve steelhead passage through dam/reservoir area
	Steelhead Health	Scale of severity of impacts for suspended sediment to steelhead
	Ecological Health	Acres of vegetation communities impacted by actions at dam/reservoir
	Ecological Health	Acres of future vegetation communities within reservoir area
B.	**Technical & Implementation**	
	Cost	Rough Order of Magnitude Construction Cost
	Risk Management	Inherent risk and/or flexibility to manage risk and react to unanticipated conditions
C.	**Sediment Transport & Water Quality**	
	Water Supply	Dry cycle percent reduction in Casitas reservoir water storage levels

CONCEPT DESCRIPTIONS

In coordination with the project stakeholders, three concepts were chosen from six initial dam removal options screened by the Consultant Team (URS and Stillwater Sciences, 2014a). The three short-listed concepts are primarily focused on methods for managing the fine sediment and organic material accumulated in the Reservoir sub-area (see Figure 2). All three concepts involve natural transport of most or all of the coarse sediment from the reservoir to the downstream reach.

The following descriptions and evaluation of the transport of fine sediment out of the impoundment area make reference to two discrete phases of erosion and transport. From an initial, unchanneled condition (Figure 3a), "Phase I" erosion occurs while the fine sediments in the Reservoir and Delta sub-areas remain accessible to the flow (i.e., the flow is in direct contact with the fine sediment deposits) (Figure 3b); "Phase II" erosion occurs once fine sediment is no longer directly accessible to the flow (Figure 3c). During Phase I erosion, the high-energy flow powered by the large discharge and steep gradient will cut through the fine sediment deposits quickly. Net transport rates are limited only by the discharge of water, and the maximum anticipated total suspended sediment (TSS) concentration, estimated on the basis of empirical data, reach as high as 10^6 mg/l. While it is not anticipated that fine sediment cohesiveness would be a limiting factor for Phase I erosion given the extremely high forces associated with the flushing flow, future fine sediment characterization could help confirm this assumption.

Phase II erosion will occur at rates more typical of fluvial erosion and transport, with a somewhat indeterminate but declining rate of additional sediment input from local mass failures of the bank, and sheet wash and rilling of the exposed upper surface of the fine sediment deposits adjacent to the incised channel.

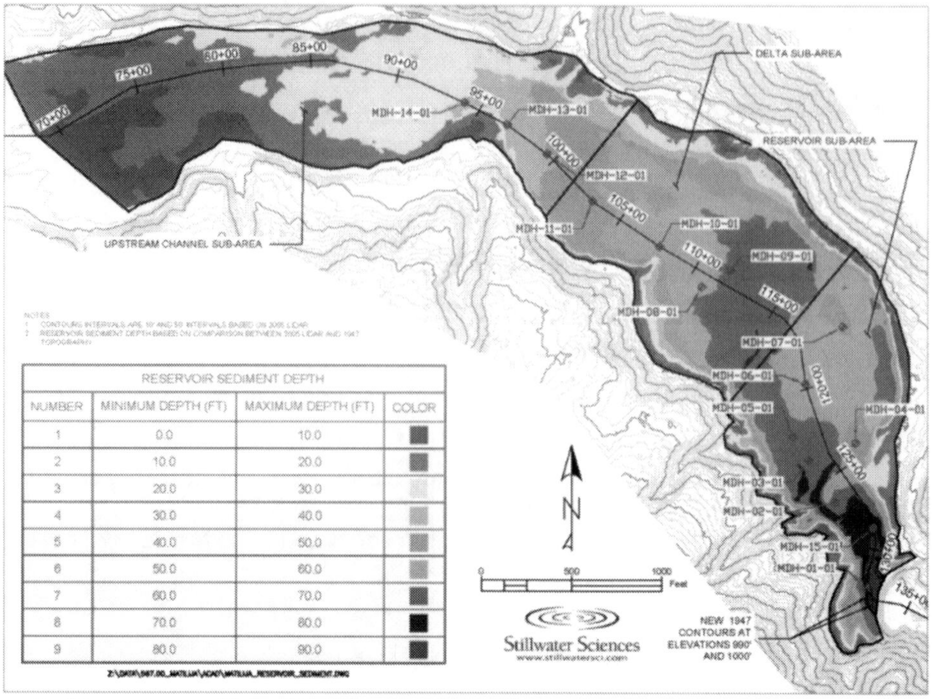

Figure 2. Plan view of Matilija Reservoir sediment deposits and borehole locations

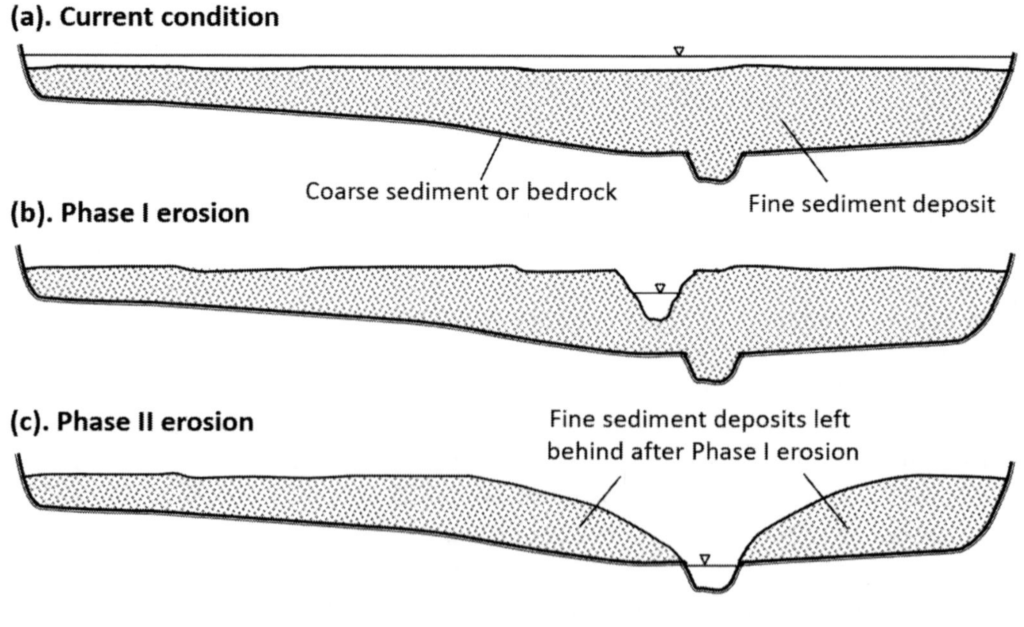

Figure 3. Sketches showing the two phases of fine sediment erosion. (a) Current, pre-incision condition; (b) Phase I erosion when fine sediment is directly accessible to the flow, presenting a virtually unlimited supply of sediment with transport limited only by the capacity and rate of discharge; and (c) Phase II erosion when fine sediment is no longer directly accessible to the flow.

The following sub-sections provide a detailed description of the three short-listed concepts being evaluated in this manuscript.

Dam Removal Concept – 1: Containment Berm with High Flow Bypass

The primary objective of this dam removal concept would be to use one high flow event, having a minimum average daily flow of approximately 1,700 cubic feet per second (cfs) (equivalent to about a 4-year peak discharge on Matilija Creek; Stillwater Sciences 2014b), to quickly erode and transport as much fine sediment and organic material as possible out of the reservoir and through the downstream reaches to the ocean. As described above, channelization and sediment transport during the high flow event would occur in two phases: Phase I sediment transport would occur as the channel cuts through the sediment resulting in the highest sediment loads during the high flow event, and Phase II sediment transport would occur as some sediment from the banks of the down cut channel and adjacent upland areas continue to erode. Channelization through the Reservoir sub-area would be allowed to form naturally through transport of coarse sediment to the downstream reaches over time. Fish passage during the channelization process might need to be monitored, and if necessary, construction equipment might be required to maintain effective passage, although recent history on other dam-removal projects suggests that this outcome is very unlikely.

This dam removal concept is intended to minimize the duration and associated high turbidity impacts to downstream water diversions and ecology by concentrating the majority of fine sediment erosion and transport into a single storm event. Due to the uncontrolled nature of the sediment erosion and the large amounts of organic debris that may be present within/above the accumulated sediment, future consideration (i.e., additional studies or development of contingency plans) of high organic debris loads and associated flood control risk (during the fine sediment erosion process) would be necessary to address the potential for significant downstream impacts, likely limited to the relatively narrow reaches upstream of Robles Diversion Dam, especially at locations that are naturally confined by bedrock outcrops or valley walls.

In order to set the project up to "wait" for a high flow event, the dam would be removed, while flows smaller than the minimum high flow event would bypass the lower reservoir and dam site. The following major features would need to be implemented and are shown conceptually in Figure 4:

- Bypass Tunnel: Prior to the high flow event, smaller flows would be routed around the reservoir to prevent them from eroding accumulated sediment that could cause repeated high turbidity impacts to the downstream reach. A flow bypass would be constructed to divert flows up to the peak flow (3,000 cfs) for the minimum high flow event, plus some additional flow capacity as a safety factor. The flow bypass would be accomplished by constructing a bypass tunnel to connect into North Fork (NF) Matilija Creek, as shown in Figure 3.

- Upstream Cofferdam: The cofferdam would be constructed across the reservoir just upstream of the current pool. This location is purposely placed as far downstream as possible to minimize both the length of the bypass tunnel and the volume of localized storm water flow that would pass over the fine sediment prior to occurrence of the high flow event. The cofferdam would divert Matilija Creek flows into the upstream tunnel portal (bypass entrance). The quality of the bypassed water would be similar to flows already seen at Robles Diversion and other downstream diversions.

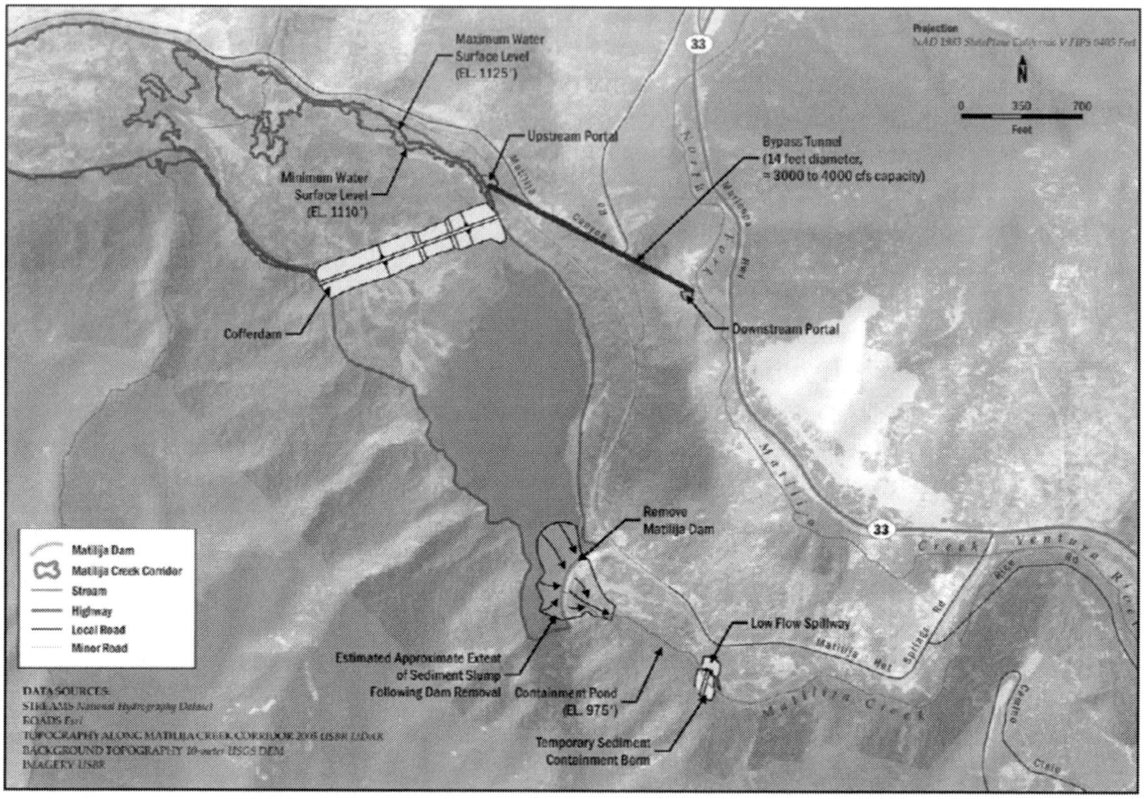

Figure 4. Dam Removal Concept 1 – Containment berm with High Flow Bypass.

- Temporary Sediment Containment Berm: The berm would be constructed approximately 950 feet downstream of the dam and 2,200 feet upstream of the confluence of Matilija Creek and NF Matilija Creek. Materials for construction of the containment berm would be obtained from the alluvial deposits in the Matilija Reservoir area. The purpose of the containment berm is to allow the dam to be removed during the dry season while preventing fine sediment in the reservoir from moving too far downstream, thereby limiting initial downstream impacts.

- Single-Season Dam Removal: The dam would be removed during the same season as construction of the bypass tunnel, cofferdam, and containment berm. It is desirable to remove the dam with as little fine sediment handling as possible, i.e., without dredging or excavating sediment upstream of the dam to access the dam for demolition, in order to minimize costs. The method of removal of the portion of the dam that is below the sediment that would require the least fine sediment handling is blasting. Removal of the portion of the dam above the fine sediment level could be done using either a hydraulic hoe ram or blasting methods.

Dam Removal Concept – 2A: Uncontrolled Orifices/2B: Optional Gates

The objective of this dam removal concept (DRC-2A) would be to erode and transport as much fine sediment as possible from the reservoir, while minimizing costs and time associated with large bypass/containment structure construction and sediment removal. The fine sediment mobilization would be achieved by allowing flow through two uncontrolled orifices, whose opening would be coordinated to coincide with a sufficiently high flow event (as summarized for DRC-1). The orifices are sized to pass the minimum high flow event in an unpressurized condition and are located to maximize mobilization of the Reservoir sub-area sediments from the upstream face of the dam. However, if the unpressurized capacity of the orifice is exceeded, it could result in reduced reservoir sediment erosion and transport for the duration of the capacity exceedance. The diameter and number of tunnels could be increased to increase the magnitude of the high flow event (something greater than a 4-year event) that could pass through the reservoir without choking the tunnels and forming a temporary reservoir.

Channelization through the Reservoir sub-area would be allowed to form naturally through transport of coarse sediment to the downstream reaches over time. Fish passage during the channelization process may need to be monitored, and if necessary, construction equipment may need to be mobilized to adaptively manage and maintain effective passage.

If, following the high flow event, a survey of the reservoir indicates a significant potential risk for further Phase I sediment transport during a future storm, DRC-2B would add gates at the upstream end of the orifices to allow for a second flushing during a future high flow event. Installation of the gates after the first high flow event, if determined to be needed, would allow flexibility to "wait" for another similar event to transport the remainder of the fine sediments, while allowing for cleaner downstream water diversions during the waiting period (as the reservoir fills up and spills). In this document, DRC-2B is evaluated as a separate concept to document the range of impacts associated with concepts that initially use uncontrolled orifices.

Similar to DRC-1, future consideration (i.e., additional studies and/or development of contingency plans) of high debris loads and associated flood control risk in the narrow reaches upstream of Robles Diversion Dam would be necessary during the fine sediment erosion process. In addition, a minor risk for plugging the orifices by debris during the fine sediment erosion process exists for DRC-2.

The following major features would be implemented, and are shown conceptually in Figure 5:

- Installation and Operation of Uncontrolled Orifices: Two twelve-foot diameter tunnels would be bored through the base of the dam from the downstream face up to a safe and stable distance (conceptually estimated to 10 feet) from the upstream face. The two tunnels sloping downstream at a 5 percent grade would conceptually have a combined maximum unpressurized capacity of about 3,300 cfs. The tunnels would be mined through the dam using drilling and concrete breaking, wire sawing methods, or possibly micro-blasting techniques. The remaining 10 feet of concrete that would connect through to the upstream side of the dam would be set up for rapid mobilization of a controlled blasting operation. A series of blast holes would be drilled to within 1 to 1½ feet of the upstream face of the dam. The open holes would be plugged and secured until a high transport storm is approaching.

- Installation of Optional Gates (DRC-2B): In the event that substantial fine sediment remained after the first high flow event, an assessment would be needed to determine if a portion could be permanently stabilized in place (naturally or by mechanical means), or whether additional creek flows or larger future high flow events would mobilize the remaining material causing some level of unacceptable downstream impact. Potentially, the upstream 3 to 5 feet of each orifice could be over-excavated (using the same drilling method described above) to a diameter of 14 feet, steel sluice gate thimbles could be installed in the over-excavated tunnels, and gates could be installed on upstream side of dam.

- Dam Removal: Under this concept, the dam would be removed in a single phase in the dry season, following the determination that sufficient sediment was removed from behind the dam. Dam demolition would be by blasting similar to what was described for DRC-1. The concrete from the demolition would be broken up and hauled to a recycling plant.

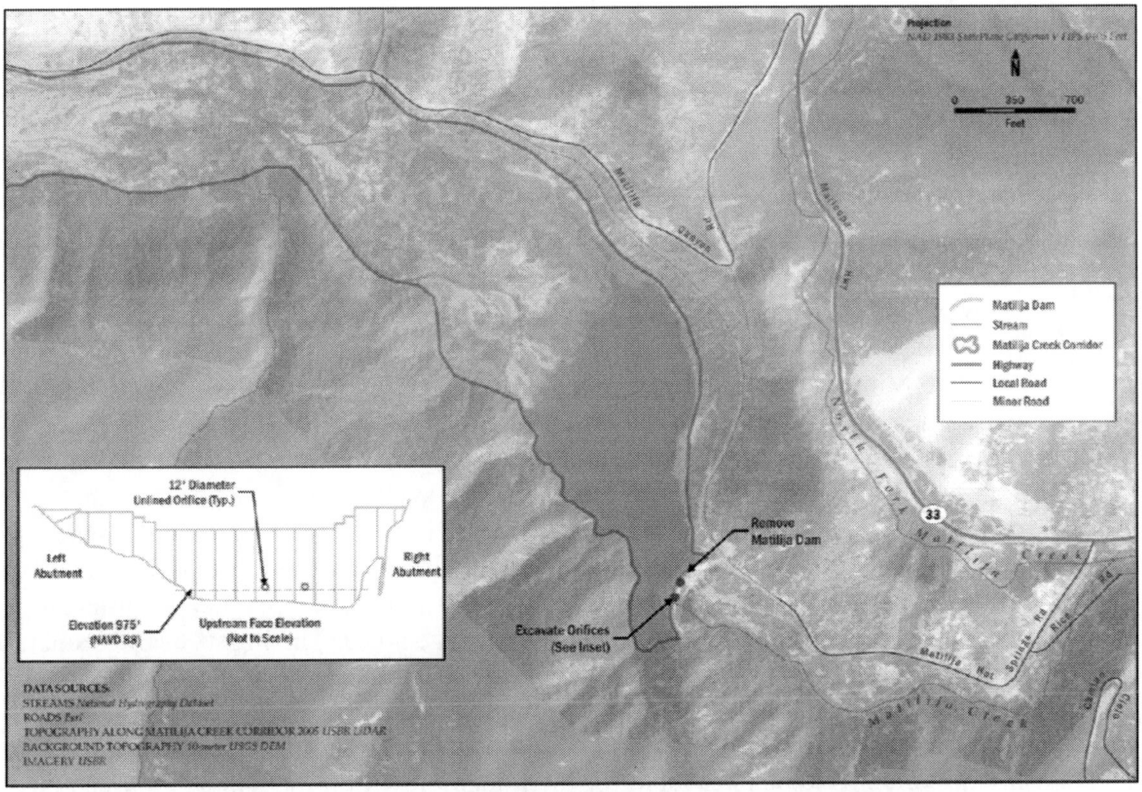

Figure 5. Dam Removal Concept 2 – Uncontrolled Orifices.

<u>Dam Removal Concept – 3: Temporary Upstream Storage of Fine Sediment</u>

The objective of this dam removal concept is to provide a concept that has some similarity to the EIS/R Alternative 4b but reduces the volume of excavated sediment and eliminates the need for a dredge-and-slurry system. These objectives are achieved by handling and temporary stabilization of a portion of the accumulated sediment within the reservoir and allowing limited release of a portion of the accumulated sediment. This could be accomplished by excavating a wide enough channel down to the pre-dam level from the dam to a point within the upstream reservoir that mobilization of fine sediment from the channel slopes would be no greater than during Phase II sediment transport for DRC-1 and DRC-2. The limited erosion of coarse sediment during Phase I would also be modestly further reduced due to the excavation of a pilot channel through the sediment to the upstream end of the reservoir. Excavated sediment would be temporarily stored upstream of the dam at elevations that would allow for inundation and potential transport during ten-year or greater storm events. The dam would be demolished and removed during the final season of channel excavation, with the goal of providing limited fish passage through the site immediately after dam removal.

The portion of the channel invert rising from the pre-dam surface to near the existing sediment level would be temporarily stabilized with erosion protection. The intent of the temporary stabilization would be to prevent down cutting of the channel and mobilization of fine sediment within the Delta sub-area of the reservoir during low flows until the minimum high flow event occurs, at which time a post-flush channel would develop through the remaining upstream sediment similar in size and location to DRC-1 and DRC-2.

The following major features would be implemented and are shown conceptually in Figure 6:

- Channel Excavation: The channel excavation can be broken into three sections; a downstream section excavated to the pre-dam surface, a transition section, and an upstream pilot channel section. The downstream section, which would extend from the dam upstream approximately 2,350 feet, would be excavated along the pre-dam Matilija Creek thalweg. The channel bottom would be 60 feet wide, which is similar to anticipated width of the post-flush channel following Phase I transport of sediment for DRC-1 and DRC-2. The side slopes in the excavated sediment would be 3H:1V, consistent with the stable side slopes of incised reservoir deposits at other recent dam removal projects. The estimated in-situ volume of sediment to be excavated from the downstream section is 1,230,000 cy. The alignment of the transition section and upstream pilot channel sections are located to the west side of the temporary storage areas. The invert of the transition section would rise at a 5 percent grade for approximately 1,100 feet to the invert elevation of the pilot channel that is capable of containing a 4-year storm event. The pilot channel would extend to near the upstream end of the accumulated. The in-situ volumes to be excavated for the transition within the Delta sub-area and the pilot channel are 460,000 cy and 120,000 cy, respectively.

- Temporary Storages Areas: The temporary storage areas would be constructed with base elevations that would be within the 10-year flood levels to allow for some degree of downstream sediment transport during storms exceeding a 10-year storm event. While long-term erosion of the temporary storage areas would be a continued source of sediment, the incremental increase in suspended sediment loads from the storage areas during flooding would likely be small relative to the natural sediment load of Matilija Creek. The total capacity of the temporary storage areas is approximately 1,570,000 cy.

- Dam Removal: Following excavation of the channel, the dam would be removed in its entirety. The dam would likely be demolished using blasting methods similar to what was described for DRC-1, with the concrete being broken up and hauled to a recycling plant.

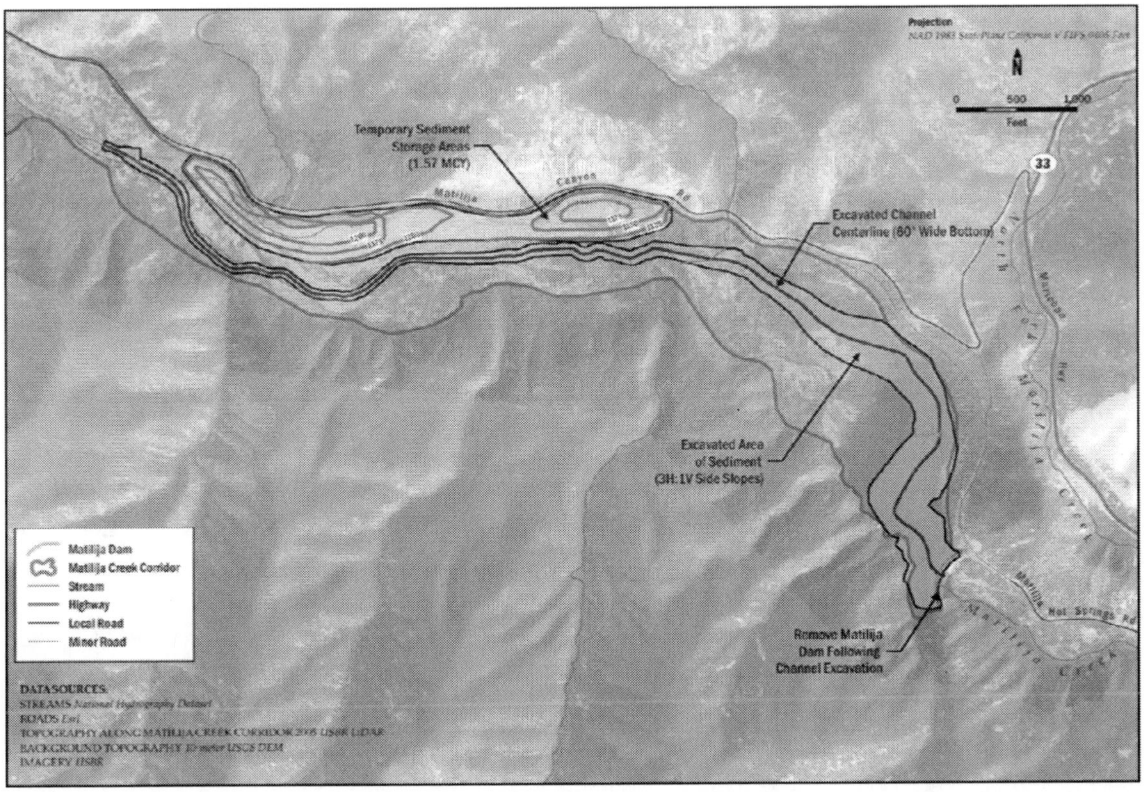

Figure 6. Dam Removal Concept 3 – Temporary Upstream Storage of Fine Sediment.

RESULTS AND CONCLUSION

Table 2 below summarizes the results of the various analyses completed to quantify the evaluation criteria for each dam removal concept. Weighting factors were not used for this evaluation since all criteria were deemed similar in importance. DRC-2 was separated into two separate concepts for evaluation: DRC-2A (without orifice gates) and DRC-2B (with optional gates).

Based on the evaluation completed, DRC-2A (without gate) ranks first overall, primarily due to ranking lowest in terms of cost, risk, and impact to existing vegetation. It also does not rank last in any evaluation category. The gates could be added as a contingency measure (to adaptively manage fine sediment), in the event that the fine sediment mobilization and associated impacts vary significantly from that anticipated.

Table 2. Evaluation Summary

DRC	Steelhead Passage (yrs)	Steelhead Health (severity scale; Newcombe and Jensen 1996)	Impact to Vegetation (acres)	Post-Project Vegetation[1]	ROMCC (Millions)	Risk (rank)	Water Supply (drop in lake level)	Total Eval. Score
1	3 to 6 (3)	13 (2)	25 (2)	50 (1)	$40.4 (3)	4	4-6% (2)	17 (4)
2A	2 to 5 (2)	13 (3)	20 (1)	50 (1)	$18.5 (1)	3	4-6% (2)	13 (1)
2B	3 to 9 (4)	13 or more (4)	20 (1)	50 (2)	$20.4 (2)	1	4-6% (2)	16 (3)
3	2 to 5 (1)	Minimal (1)	69 (3)	40 (3)	$49.7 (4)	2	Minimal (1)	15 (2)

Relative rank shown in parentheses (1 is best)
[1] Wetland and riparian vegetation

DRC-3 ranks second, with its high impact to existing vegetation (large footprint), high cost and lower long-term vegetation habitat benefit, pulling it below DRC-2A in overall score.

DRC-2B ranks third, primarily due to the harm that subsequent gate closure(s) could do to steelhead health and passage, and the potential for the longest implementation duration.

DRC-1 ranks fourth given the selected criteria, primarily due to the fact that it only ranks first in one of the criteria, generally failing to significantly differentiate itself from the other concepts, while carrying the greatest risk and relatively high cost.

Based on the evaluation results, DRC-2A and DRC-3 will move forward into the preliminary design phase, pending subsequent project funding.

ACKNOWLEDGEMENTS

The work summarized in this manuscript would not have been possible without the support and input from many individuals, their organizations, and several advisory committees. We would like to express our gratitude to those listed below:

- Sam Jenniches, California State Coastal Conservancy
- Brian Cluer, National Marine Fisheries Service, National Oceanic and Atmospheric Administration
- Blair Greimman, United States Bureau of Reclamation
- Paul Jenkin, Surfrider Foundation
- Steve Wickstrom, Casitas Municipal Water District

- Neil Cole, Casitas Municipal Water District
- Technical Advisory Committee, Matilija Dam Ecosystem Restoration Project
- Design Oversight Group, Matilija Dam Ecosystem Restoration Project

REFERENCES

Newcombe, C.P., Jensen, J.O.T. 1996. Channel suspended sediment and fisheries: A synthesis for quantitative assessment of risk and impact. North American Journal of Fisheries Management, 16 (4), pp. 693-727.

Stillwater Sciences, 2014a. Task 2.2, Watershed and Impoundment Area Sediment Characterization, Matilija Dam Removal, Sediment Transport, and Robles Diversion Mitigation Study, prepared for the Management Team, Matilija Dam Ecosystem Restoration Project, June 3.

Stillwater Sciences, 2014b. Task 2.1, Hydrologic Assessment, Matilija Dam Removal, Sediment Transport, and Robles Diversion Mitigation Study, prepared for the Management Team, Matilija Dam Ecosystem Restoration Project, June 3.

URS and Stillwater Sciences, 2014a. Draft Initial Options Screening Report, prepared for the Ventura County Watershed Protection District, September 9.

URS and Stillwater Sciences, 2014b. Hydrologic Assessment for Water Supply, Matilija Dam Removal, Sediment Transport, and Robles Diversion Mitigation Project, prepare for the Ventura County Watershed Protection District, September 10.

U.S. Corps of Engineers (USACE), 2004. Final Environmental Impact Statement/Environmental Impact Report (EIS/EIR), September.

RISK ANALYSIS WORK FOR DAMS — "BIG PICTURE" CONSIDERATIONS

Jay N. Stateler, P.E.[1]
Kim de Rubertis, P.E.[2]
Doug Boyer, P.E.[3]

ABSTRACT

Risk analyses performed relative to dam safety evaluations clearly benefit the understanding of the potential failure modes (PFMs), and allow assessment of PFMs that are difficult or impossible to evaluate using deterministic techniques. Risk analyses help to highlight the "drivers" of risk for a dam, key information associated with those "drivers," and opportunities to reduce risk that may be low-cost and straightforward to implement. Risk analyses for an inventory of dams can be very beneficial for prioritizing dam safety activities relative to those dams, if the risk analyses are carried out with consistency in both personnel and processes used. Most people who have been involved with risk analysis work have few illusions about the absolute "accuracy" of the numbers produced, but consistency in number development can produce useful relative information among dams. Our ability to identify the PFMs used as the foundation for risk analysis work is necessarily limited by our experience and ability to anticipate things that could go wrong (including human factors, equipment issues, etc.). Therefore, effective dam surveillance and monitoring programs should be in place to detect anomalous performance associated with unidentified PFMs, as well as with "unknown unknowns" regarding the dam and dam site. Risk analysis work needs to be approached with full understanding and knowledge of the limitations of the work, the limitations and uncertainties of the input information that is available, the guidelines the results risk analysis will be compared to (and why those are appropriate guidelines), and the method that will be used to determine when the risks are considered tolerable. Applying experience and judgment is crucial regarding the beneficial use of risk analysis results.

INTRODUCTION

Risk analysis work, in various forms, has been carried out with respect to dams for a number of years, and it has been intensively employed in some organizations for well over 10 years. While risk analysis efforts for dams are constantly evolving, stepping back and looking at the "big picture" with respect to this activity can be useful. This paper attempts to do this by collecting and sharing input and thoughts gathered from individuals with differing perspectives on this activity, who have been "in the trenches" with respect to risk analysis work.

If there was an infinite amount of money, and if desired dam modification work could be accomplished overnight, then the need for utilizing risk analysis methods at dams would not be as compelling as it currently is. We would simply design and modify dams to

[1] Civil Engineer, Bureau of Reclamation, Denver, Colorado, jstateler@usbr.gov
[2] Consulting Engineer, Cashmere, Washington, kimderubertis@gmail.com
[3] Geotechnical Engineer, Federal Energy Regulatory Commission, Portland, OR, douglas.boyer@ferc.gov

handle: (1) flood loadings associated with the Probable Maximum Flood (PMF), (2) seismic loads associated with Maximum Credible Earthquakes (MCEs), (3) the most extreme and severe reservoir operations scenarios that can be imagined, and (4) all other conceivable loading conditions. Then we would have in place a process for continually and immediately modifying all dams so that they are always in line with the "current state-of-the-practice" and the latest design standards. Obviously this is not the world we live in. Using risk analysis methods allows us to be more efficient and effective in dealing with and addressing the dam safety risks we actually face.

The fundamental and foundational activity for risk analysis work is to rigorously develop PFMs for the dam under study. A commonly referenced method for doing this is the Federal Energy Regulatory Commission (FERC) Potential Failure Modes Analysis (PFMA) approach, although other similar approaches are also employed to achieve a comparable result. The PFMA helps people think critically about a dam's vulnerabilities, how it could fail, and what factors could contribute more or less to the failure process. The following elements are important with respect to having a PFMA that is rigorous and thorough in developing PFMs:

- Gather all available, relevant information about the design, construction, and performance of the dam, and all relevant inspection, analysis, and evaluation information. If apparently key information cannot be readily found, make efforts to hunt down and find this information, to the extent reasonably possible.
- Have several PFMA participants thoroughly review the gathered information.
- Have diversity of input to the PFMA discussions, including all relevant technical disciplines (geologist, seismologist, flood loadings specialist, etc.) as well as field personnel who are familiar with the dam's performance and operations.
- Have an open, investigative attitude in PFMA discussions, with participants unafraid to share information, even if they do not have an understanding of its possible significance.
- Thoroughly document the insights and ideas put forth during the discussions for future reference.

The results of a PFMA can be used to:

- Identify investigations/studies that are needed to answer questions/unknowns exposed during the PFMA effort that are important with respect to a PFMA evaluation.
- Establish the timing and frequency of future dam safety inspections.
- Establish an effective and focused surveillance and monitoring program.
- Achieve a better understanding of warning times associated with the various PFMs, which can be used to develop an improved Emergency Action Plan (EAP).
- Develop plans for intervention efforts that can be used to prevent possible dam failure.
- Identify training needs for operations staff, management staff, and others.

With the completion of a conscientious effort to thoroughly and exhaustively develop the PFMs for a dam, risk analysis methods can then be used to delve deeper into each PFM:

- Identify all the steps necessary to get from PFM initiation to ultimate dam failure.
- Identify the levels of earthquake loads, flood loads, reservoir levels, etc. that should be considered in the risk analysis effort, and develop probability of occurrence information for them.
- Develop an event tree that incorporates the information developed regarding the first two bullets above, such as shown in Figure 1 below.
- Develop probability of occurrence information for all the event tree nodes, which can be discrete values, or can be expressed as distribution functions incorporating a range of possible values.
- Develop consequence estimates associated with each event tree branch that results in dam failure. The consequence evaluation typically focuses on potential loss of life, but can relate to other important considerations as well, such as property damage and environmental impacts.
- Develop an annualized probability of dam failure value for each PFM and for the dam as a whole, including confidence limits.
- Develop annualized consequence probability values for each PFM and for the dam as a whole, including confidence limits. Again, loss of life is typically included as one of the consequence categories evaluated, but other categories can be evaluated as well.

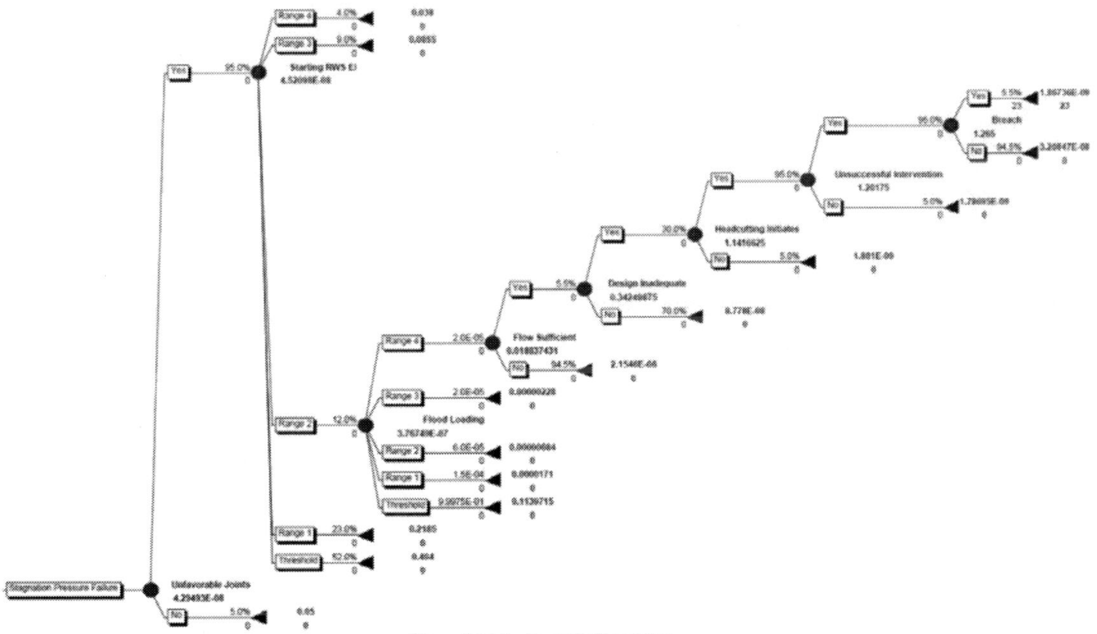

Figure VI-1-3 – Example Event Tree

Figure 1. Example Event Tree. From Reference [1], Chapter VI-1 – Stagnation Pressure Failure of Spillway Chutes.

Presentation of risk analysis results can be in the form of discrete "best estimate" values on a plot of annualized failure probability versus consequences (such as shown in Figure 2 below). The values are typically presented with confidence limits.

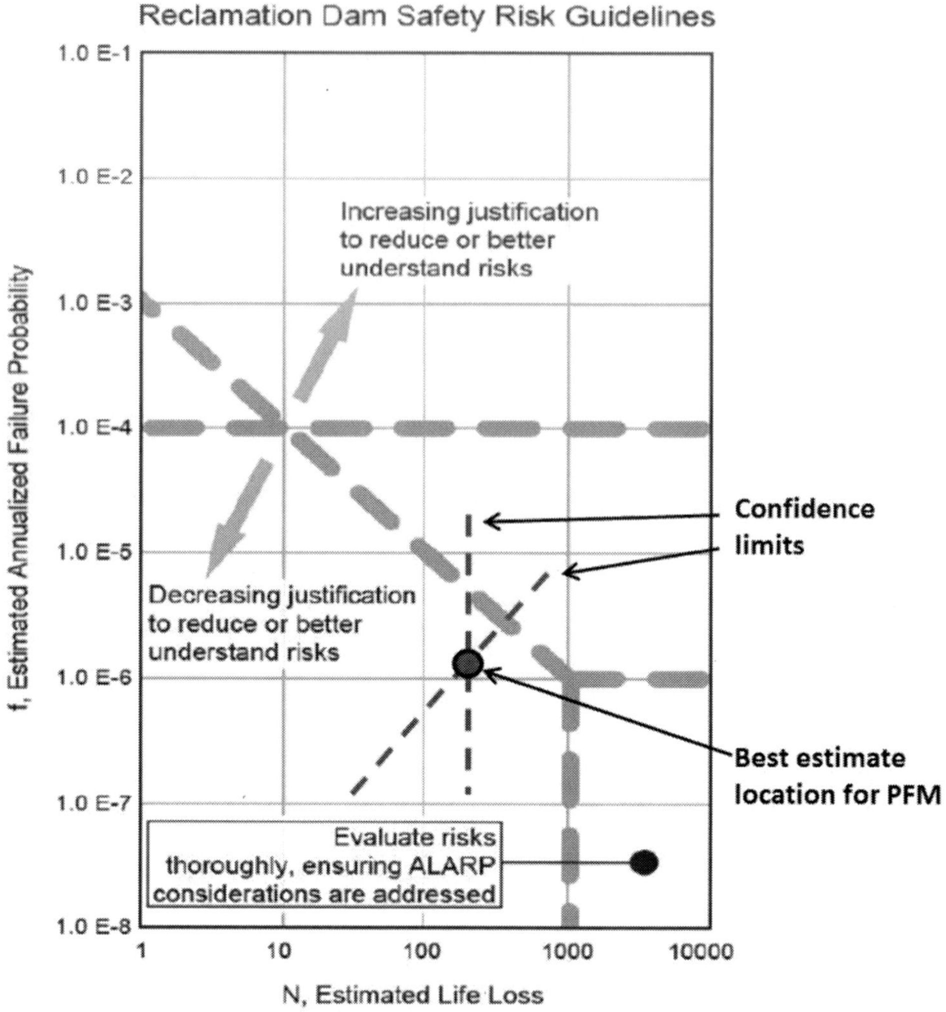

Figure 7. Reclamation's Dam Safety Public Protection Guidelines [9]

Figure 2. Example Form for Presenting Risk Analysis Results, from Reference [1], Chapter IX-1 – Risk Assessment/Management, with example PFM added.

Risk analysis results can also be presented in more qualitative ways, such as in a table where the probability of PFM occurrence is one dimension, consequences is the other dimension, and each PFM is placed in an appropriate box in the table that spans the proper range of values for each dimension. Figure 3 provides an example of this presentation form.

Detailed discussion of risk analysis methods is available and is not the topic of this paper. This paper will consider some general, "big picture" questions with respect to risk analysis work.

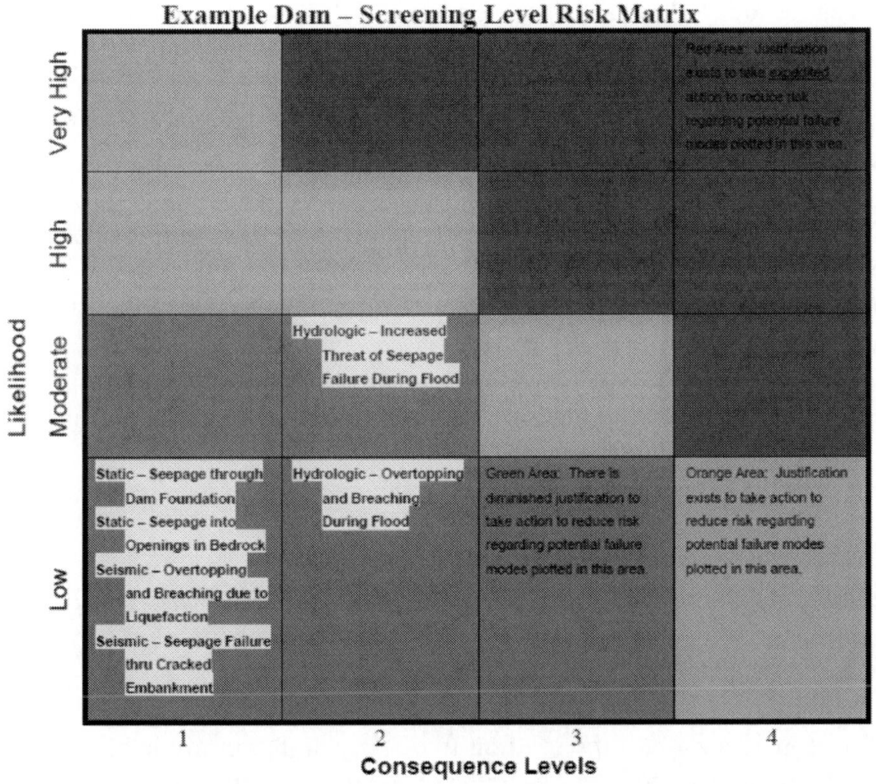

Figure 3. Example Qualitative Presentation of Risk Analysis Results

STRENGTHS AND BENEFITS OF RISK ANALYSIS WORK

What is risk analysis work effective in doing?

- Comparing the dam safety risks of a portfolio of dams for the purpose of prioritizing which dams and issues to address first (when the work is done in a consistent and equitable way). This allows for efficiency in reducing dam safety risks, and effectiveness in using limited funding.

- Enhancing the understanding of a PFM, by identifying all the steps necessary to get from initiation to uncontrolled release of reservoir water. Often this can bring to light some things about a PFM that previously had not been thought of, and can identify some things that could go wrong, which were not previously recognized. The understanding gained regarding a PFM is typically much greater than what is achieved by performing a deterministic analysis.

- Analyzing PFMs that are difficult or impossible to evaluate using deterministic techniques (e.g., internal erosion, operational PFMs, etc.). Allowing an effective evaluation of PFMs where no 'standards' or deterministic guidelines exist.

- Understanding what PFMs really drive the dam safety risks at a dam, so that efforts can be focused toward addressing those PFMs. In other words, helping to distinguish between critical, important, and minor dam safety issues.

- Bringing to light opportunities to take actions, which may be relatively simple, that can reduce the risks of dam failure from a PFM.

- Identifying key information that greatly impacts the assessment of risk for a PFM, which promotes close study and evaluation of the key information, which in turn improves the understanding of the potential for development of the PFM.

- Aiding in the development of an effective dam safety monitoring program to address the PFMs, since the goals of the monitoring work are better understood.

- Helping dam operating personnel better understand the dam safety issues regarding their dam(s).

- Providing transparency with respect to the evaluation of PFMs. Factors that were considered, and were central to judgments made, are documented and available for review. Uncertainty is presented and quantified.

- Including consideration of the loading frequency and magnitude of consequences, in addition to the performance/response of the dam, in the overall evaluation of a PFM.

- Helping to "build the case" regarding the need for action with respect to a PFM, or alternatively, that action is not needed. The risk numbers and their confidence limits can be an important part of the argument for action, or that there is no need for action. Engineering judgment also needs to be part of the decision process, and other factors may impact the decision process as well.

- Helping to define what "As Low As Reasonably Practicable" (ALARP) realistically means for a PFM. Opportunities to reduce risk, and their impacts on risk reduction, can be effectively assessed using event tree information produced by risk analysis efforts and engineering judgment. Conclusions regarding ALARP can be effectively and rigorously documented and justified.

- Allowing the comparison of risks associated with a portfolio of dams that includes low-hazard dams. Considerations regarding consequences can focus on property damage and economic losses, where the potential for loss of life appears low or negligible, or when property damage/economic losses need consideration along with (and in conjunction with) considerations of potential loss of life.

- Allowing the comparison of risks associated with dams to other infrastructure risks (such as bridges) and other societal risks to aid macro-level funding allocation decisions.

LIMITATIONS REGARDING RISK ANALYSIS WORK

The Numbers

When people are asked "What is risk analysis not effective in doing?" the discussion often quickly moves to "the numbers." Most people that have been involved with risk analysis work have few illusions about the accuracy of the numbers produced. Sentiments as extreme as "The numbers are "*#*$%*#&!" are sometimes expressed, but more commonly the response is at least a bit more nuanced. The range of sentiments that might be heard includes:

- Risk analysis may not be effective at determining accurate values for failure probability or risk. Risk estimates are best used for relative comparisons. For use in decision-making, rigorous development of risk numbers along with engineering judgment is needed to effectively "build the case" for decisions to be made.

- Analysis of a PFM can break it into its constituent steps effectively, but then estimating the likelihood of some of the steps can seem like pulling numbers out of thin air if limited information and supporting evaluation work are available. The numbers sometimes seem very subjective.

- Risk analysis work is not effective in producing much useful information or insight for a particular dam if the risk numbers are based off historical information for dams in the same basic category, particularly when the database for the historical information is limited. The risk analysis evaluation should be site-specific to produce meaningful information.

- Experienced, reasonable people can produce very different estimates of risk. Having consistency in personnel, and most particularly having the same individuals review all risk analysis work done by an organization, can allow meaningful relative comparisons to be made between dams for the purpose of work prioritization, and can allow valuable information to be developed and used for decision-making.

- Often there are some participants in risk analysis work whose experience is such that they are not capable of providing useful/beneficial risk estimates that benefit the overall risk estimate values developed by the risk team.

- We put in the numbers, look at the result, and if we don't like it we may go back and change some numbers to get to where we think the risks should be.

- Bias can enter the equation, including anchoring bias, where the first estimated values heard by the team can influence subsequent estimates made. Bias toward a specific end result can also exist. The potential for bias of all types is well-known to experienced facilitators (leaders) of risk analysis work, who generally do everything in their power to find ways to reduce bias impacts on risk analysis results.

- For seepage-related PFMs for embankment dams, which history tells us is a very important category regarding dam failures, defensibility of calculated failure probability values appears to present an issue because a lack of consensus exists with respect to developing an estimated value for the "Initiation Node" of the standard seepage event tree. The U.S. Army Corps of Engineers (USACE) approach to this critical node and the Bureau of Reclamation (BOR) approach to this node, as presented in Chapter IV-4 of their joint "Best Practices in Dam and Levee Safety Risk Analysis" guidance [1], employ completely different methods. This would appear to be indicative of a problem area. Differing USACE and BOR approaches exist with respect to other areas of this guidance as well. (Perhaps this is why the guidance is called "Best Practices" as opposed to "Best Practice.") Divergent approaches used by USACE and BOR make it difficult for the broader dam safety community to know how best to approach these matters.

Winston Churchill famously said that "Democracy is the worst form of government, except for all the others." Similar thoughts come to mind when thinking about using numbers to express the results of risk analysis efforts. Providing numbers, including confidence limits, provides much more information and "value" than attempting to capture the results with words. Even when using a more qualitative approach, such as shown on Figure 3, quantitative anchoring is required so that the appropriate box is selected.

Definition of Potential Failure Modes (PFMs) for a Dam

The PFMA is the foundation upon which risk analysis work is built. If the definition of the PFMs is poorly done, the risk analysis effort clearly will be compromised, no matter how diligently the risk analysis work is carried out. Conversely, many would note that a PFMA that is well done may be able to provide the bulk of the benefits associated with a full-blown risk analysis effort.

Stepping back to a larger point, the definition of PFMs for a dam is generally limited by the knowledge gained from past experience. We cannot completely identify all the ways that loss of reservoir control could occur because we just cannot imagine some events. PFMs that involve a chain of events can be difficult to identify and bring forward. This is not a limitation of risk analysis work itself, but rather a limitation regarding us humans as we perform PFMAs and risk analyses. The recent serious dam incident at Wanapum Dam occurred after three PFMA sessions and numerous dam safety inspections failed to discover the problem. PFMA work obviously has limited effectiveness with respect to identifying "unknown unknowns."

Consideration of Human Factors

This topic relates to the previous topic discussed. Sometimes the development of PFMs for a dam, and the follow-on risk analysis effort, focus intently on the physical characteristics of the dam and dam site, and may not give appropriate consideration to the possibility that things may not go as planned. Gates, back-up power generators, etc. have not been regularly exercised as required, such that when the crucial moment arrives, equipment does not operate as expected. Routine maintenance has been deferred, operating instructions are not correct, old equipment fails, etc. are all possibilities that could lead to serious problems. In the heat of the moment during a flood, after an earthquake, or in any other unusual and potentially stressful situation, dam operating personnel may inadvertently make an operational mistake that initiates a chain of events that could lead to dam failure. The notion that PFMA and risk analysis work can identify all such possibilities represents an unconservative assumption.

Information developed by the Association of State Dam Safety Officials (ASDSO) indicates that between January 1, 2005 and June 30, 2013 (8.5 years), there were 173 dam failures in the United States. During this same time period, there were 587 incidents that "without intervention, would likely have resulted in dam failure." This demonstrates the significant role that intervention can play with respect to preventing dam failures, and the other side of the coin that should be fully considered with respect to human factors as they relate to risk evaluation work.

The "unsuccessful detection and intervention" node of the standard seepage event tree (and some other event trees as well) really involves a number of sub-steps that should be considered:

- Detection – the problem is observed by someone in a time frame that can allow for actions to prevent dam failure
- Detection – the situation is recognized as a problem and action is attempted to alert someone that action needs to be taken
- Detection – the attempt to alert someone in a timely fashion is successful
- Response – responding personnel understand the situation
- Response – responding personnel can mobilize people, equipment, and materials in a timely factor to attempt to address the situation
- Response – the situation can be brought under control (i.e. dam failure can be averted) given what the situation amounts to, and the people, equipment, and materials that can be mobilized to address it

It is noteworthy that the "Best Practices in Dam and Levee Safety Risk Analysis" guidance [1] includes the following statement:

> "It is USACE practice to prepare risk estimates for both with and without intervention to address the need for transparency in the risk estimate and to possibly evaluate the effectiveness of interim risk-reduction measures."

This practice provides important focus on the "unsuccessful detection and intervention" node, which clearly is beneficial regarding fully assessing the prospects for being successful with intervention efforts to prevent dam failure.

Judgment

The limitations of the results of risk analysis work are often not well-understood by whose making decisions based on those results. It is common to see a "standard" confidence (uncertainty) interval of plus or minus one order of magnitude when risk results are presented, such that it would appear that perhaps the estimation of the confidence interval is perfunctory. Sensitivity studies can be carried out to get a good sense regarding the confidence that can be placed in the estimates. Those using the results of risk analysis work need to know about limitations and deficiencies regarding key information that could impact the results, assumptions that were made and the likelihood that those assumptions are valid, other information that would affect how the risk analysis work should be viewed, and the limitations of the work.

Risk analyses are not a "black box" where the inputs are developed and submitted, magic occurs, and out comes the answer. Like anything else, the results of a risk analysis can be used incorrectly and even abused. It is important that the results are carefully considered, including careful thought about what they really mean. Risk analyses do not tell us what to do. Interpretation of the risk analysis results is critical for developing the proper next course of action.

Other Limitations

Additional thoughts regarding "What is risk analysis not effective in doing?" include:

- Establishing consistent criteria for design once it has been decided that dam modification work is necessary, based on the results of risk analysis work.

- Making the case for some routine activities. In the big picture, routine monitoring and maintenance can seem to have only minor benefit. For example, the risk reduction benefits of mowing grass at the toe of a dam, to be able to effectively observe this area, are tough to quantify. But most people would view this as a necessary activity, and it would represent a level of negligence if this was not routinely performed.

COST-EFFECTIVE RISK ANALYSIS ACTIVITIES

What aspects of risk analysis work produce great benefits without consuming much time and effort?

- Identification of PFMs and their descriptive development produce great benefits without much time and effort. Understanding the PFM and each of the steps in its progression is crucial for a meaningful dam safety evaluation of a PFM.

- Quickly going through an event tree to assess vulnerabilities and aspects that are providing protection against PFM development. A detailed understanding can be developed quite rapidly that can validate or refute existing engineering judgment about vulnerabilities. Interim risk reduction measures can be quickly identified.

- Qualitative or semi-quantitative risk analyses can quickly identify those issues that are more likely to contribute to a better understanding of the problem and can focus future investigations/evaluations/analyses on developing/obtaining the needed key information.

- When done correctly, screening level risk analyses can provide a valuable first-cut basis for relative ranking and prioritization of dams in a portfolio regarding future dam safety work activities.

RISK ANALYSIS ACTIVITIES WITH LOWER COST-EFFECTIVENESS

What aspects of risk analysis work produce relatively limited benefits, particularly considering the time and effort associated with the activity?

- Time consuming analyses are occasionally performed in preparation for a risk analysis. Sometimes these analyses are associated with PFMs that turn out to be insignificant contributors to the total risk. Often it is not easy to predict ahead of time the outcome of the analyses or risk estimates, but a screening-level risk analysis can help reduce the potential for spending a lot of time and money on issues that are not critical. However, it should be noted that spending time and money on non-critical issues is not necessarily a wasted effort. The case for why a potential issue is not critical needs to be made.

- Writing risk analysis reports can be very time consuming. Streamlining the report process in some manner could reduce costs without losing the results and conclusions that come out of the meetings.

- In some instances, some people believe that they know what the most critical issues are before a risk analysis is conducted and persuading them of different results can require substantial time and effort. This is not always of limited value as some bad decisions of great significance can be avoided. But at times, it borders on a significant waste of time to get some personalities to overcome their preconceived notions and accept the results of a team risk analysis.

- There would seem to be no reason to pursue a detailed risk analysis when low-cost, readily-implementable risk reduction measures are available.

- Big efforts to develop highly refined loss of life estimates can be a high-cost, low-benefit endeavor. Spending significant time and money in this area shows we are trying to do our best relative to this very important topic, but may at times produce little or no incremental benefit and may lead to delusions that we really have a handle on what will happen. There are many variables associated with how an actual dam failure situation will play out, so making enormous efforts to get the "right answer" for the best estimate may not substantially impact the final decision and can be an ineffective use of resources.

- Beating the evaluation of a PFM to death when the information available for the evaluation is limited trends to a low-benefit, high-cost exercise as the activity drags on.

- Drilling down into any one thing, whether it be loadings, system response, or consequences, without a full understanding of the potential benefits is ill-advised. A few key factors often substantially influence the risk results. Those items should be identified, and sufficient effort should be made to quantify and understand those parameters, with due consideration of uncertainty. Spending significant time and money on factors that have little impact on the risk analysis outcome is obviously not the best use of resources.

USACE guidance regarding risk analysis work includes the following discussion:

> "The methodology contained in the best practices manual and supporting toolboxes provides a suite of scalable assessment approaches that provide information to promote critical thinking and guide a risk assessor's judgment. These methods can be tailored and used with varying degrees of effort (time and cost) to provide the appropriate level of accuracy and rigor required to make credible risk informed decisions. It is important to understand that every decision does not require a high level of rigor, detail, and accuracy in the risk estimate in order to be credible."

This guidance seems like good concluding discussion regarding the topic of this section.

SYNTHESIS OF CONSIDERATIONS

Cost

Grumbling can occasionally be heard about (1) all the evaluation and analysis work that may be required before embarking on a risk analysis, (2) the length of time and number of people involved in a risk analysis effort, and (3) all the time and effort associated with preparing a post-meeting comprehensive report about the activity, that captures the numbers as well as the rationale, logic, differing opinions, etc. regarding the numbers. It is definitely true that the work can be expensive and time-consuming. However, the magnitude of the questions being addressed and the amounts of money that may be saved through use of the process typically dwarf the cost concerns. It generally is money very well spent.

A thorough, well-done PFMA can set the table for an efficient and effective follow-on risk-analysis effort. Time invested in planning and thinking about an upcoming risk analysis can potentially yield substantial cost savings by eliminating some expensive, low-value activities. Preliminary screening-level, perhaps qualitative, risk analysis work can also help focus the effort, eliminate low-cost activities, and perhaps identify low-cost risk reduction measures that could defer or eliminate the need for a more involved evaluation. Planning for the development of the needed consequence information, with costs and benefits in mind regarding the exercise, can pare expenses while still providing information appropriate for the risk analysis.

The Numbers

Most would agree that the validity of the absolute "best estimate" values developed in risk analysis work is open to debate, to put it mildly. Perhaps the numbers are within an order of magnitude of the "true" values, but who knows and how is one to know? It is not scientifically testable. On the other hand, numbers are much more useful than words. Numbers clearly present specific information and meaning, while describing levels of risk in words may be subject to varying interpretations by different readers. In the end, the best that can be hoped for is to endeavor to develop the most representative and "accurate" numbers possible, eliminating bias as best as possible, and founding the estimates on site-specific information and evaluations to the extent possible. The process should be transparent so that the basis for the estimates, and their confidence limits, are clear to all. Regardless, the decision-making process should not focus on the numbers generated from the risk analysis, but rather the information gained from the risk analysis process that generates the building blocks that "build the case" for the conclusions and potential recommendations.

To the extent the profession can develop consistent approaches and methods for risk analysis work, the processes used and results produced will have more value. Where this cannot be achieved, comparing results developed by varying methods for the purpose of making dam safety decisions on prioritization of work would not be appropriate. Consistency in process and personnel involved can allow an organization to do useful risk analysis work that allows effective prioritization of dam safety work, based on the relative numbers, even if the accuracy of the absolute values of the individual numbers is doubtful. Currently the Bureau of Reclamation and U.S. Army Corps of Engineers have "joint operations" with respect to risk analysis work, but have taken different approaches to a number of key matters, which creates difficulties for the larger dam safety community.

Risk analysis work often generates a large quantity of numerical estimates, which can be overwhelming. It would be suspected that trying to develop so many numbers would, over the course of one or more days, numb the participants in the activity to the point that the quality of the work would deteriorate with time. While it is reasonable to wonder about this, the experience of the authors suggests that actually this is not a major problem. The activity is often very interesting to participate in. What sometimes seems to pose a bigger problem is the review of the avalanche of numbers. Fairly obvious errors can sometimes slip through the review process, and with all the numbers before them, reviewers sometimes just look at the final result and see if they can accept it. Focusing on the final result for reasonableness is a good thing. Does the result make sense? In a macro sense, what are the key factors driving the risk numbers? Do the principal drivers seem reasonable, both in what they are and how they impact the results? Assessments of this sort are fundamental to the effective use of risk analysis for dams. However, these assessments rely on the validity of the numbers, so great care is needed in the development and review of the numbers to ensure that the fundamental assessments and judgments made from the developed information are based on error-free event trees having the "best possible" estimates.

Limits of Experience Can Limit Effectiveness

Dam safety evaluation work took a tremendous step forward with the development and use of the PFMA approach. A well-executed PFMA for a dam intensively looks at all the information available for the dam in question, the dam's performance history, and the performance history of dams similar in nature to the dam in question to develop the dam's credible dam safety threats (the PFMs that are based on the dam's vulnerabilities). When done well, the PFMA can develop an extremely valuable compilation of PFMs for a dam, and include PFMs that have equipment or human factor components, and potentially include PFMs that no one has ever raised or considered before. The PFMs are the foundation upon which the risk analysis work is based. The potential problem area here is that experience is an important teacher regarding the development of PFMs for a dam. It is harder to develop PFMs regarding situations that have not been experienced. This is particularly true regarding the category "Things Just Don't Go as Planned." Procedures have errors, people make mistakes, equipment does not work as planned, design computations are incorrect, etc. This latter factor arose as a key issue regarding the Wanapum Dam spillway situation (which was not a PFM that had been previously identified for the dam, as noted previously).

The PFMA process cannot be perfunctory. Pre-meeting preparation and reviews of information need to be thorough. Participants need to approach the activity with a fresh perspective and an open mind. Everything should be on the table and open for discussion. Participants need to feel comfortable bringing up things they do not understand, but think the PFMA team should be aware of. But even with the best of PFMA meetings, there are those pesky "unknown unknowns" about a dam and dam site that a PFMA may not have any reason to suspect being present. Therefore, it needs to be recognized that even the best PFMA, and therefore also the best follow-on risk analysis work, may not be considering some important PFMs which may present significant dam safety risk. Consequently, conscientiously prepared and executed surveillance and monitoring plans, that are designed to pick up PFMs that have not been recognized for a dam, are essential. What prudent monitoring should be performed for general health monitoring purposes, and to pick up anomalous performance due to "unknown unknowns"? What aspects of the dam and dam site that we potentially do not know about could cause real problems? Effective monitoring is needed regarding these considerations.

Tension Between Consistency in Process and Fresh Thinking

Consistency in process and personnel can allow an organization to do useful risk analysis work that allows effective prioritization of dam safety work. With respect to developing an estimated value for the "Initiation Node" of the standard seepage event tree, the Bureau of Reclamation approach involves reference to the tables shown in Figure 4 below. These tables are based on historical rates of initiation of internal erosion at Reclamation embankment dams, and the values shown were set 2-4 times higher than actual historical experience, to account for situations where initiation may have occurred, but the process never "progressed" far enough for the evidence of initiation to be detected. Extensive guidance is provided regarding selecting an appropriate "best estimate" values regarding the ranges presented in these tables. Figure 5 provides a small sample of this qualitative guidance, to get a sense of this guidance information.

Probability of Initiation of Internal Erosion: Reclamation Approach

Table 26-17. Proposed Best Estimate Values of Annual Probabilities of Initiation of Internal Erosion by Category

Type of Internal Erosion	Range of Initiation Probability
Embankment only	$3x10^{-4}$ to $1x10^{-3}$
Foundation only	$2x10^{-3}$ to $1x10^{-2}$
Embankment into foundation	$2x10^{-4}$ to $1x10^{-3}$
Into/along conduit	$4x10^{-4}$ to $1x10^{-3}$
Into drain*	$5x10^{-4}$ to $2x10^{-3}$

*Note: Into drain values were adjusted downward given the limited number of instances where vulnerable drains are part of the inventory

Table 26-18. Proposed Best Estimate Values of Annual Probabilities of Initiation of Internal Erosion by Mechanism

Type of Internal Erosion	Range of Initiation Probability
Backward Erosion Piping	$5x10^{-4}$ to $2.5x10^{-3}$
Internal Migration	$1x10^{-3}$ to $5x10^{-3}$
Scour	$1x10^{-3}$ to $5x10^{-3}$
Suffusion/Suffosion	$6x10^{-4}$ to $2.5x10^{-3}$

Figure 4. From Reference [1], Tables IV-4-11 and -12 (table numbers recently changed, but the content is the same). Proposed Best Estimate Values of Annual Probabilities of Initiation of Internal Erosion – Reclamation Approach – From "Best Practices in Dam and Levee Safety Risk Analysis," Chapter IV-4. Internal Erosion Risks for Embankments and Foundations.

FACTORS INFLUENCING THE LIKELIHOOD OF INITIATION OF INTERNAL EROSION THROUGH AN EMBANKMENT DAM				DATE: JULY 2012
Factor	**Influence on Likelihood / Relative to Reclamation Historical Base Rates (see notes)**			**Comments**
	Less Likely	**Neutral**	**More Likely**	
Seepage				The presence or absence of seepage may not be known with certainty. Episodic seepage could be an indicator that an internal erosion pathway is repeatedly opening and closing. Evidence of material transport in seepage flow would indicate near certainty that erosion is occurring.
Presence of seepage	No seepage (low probability for a concentrated leak)	Insignificant seepage; or seepage possible but unseen	Seepage significant	
Seepage fluctuations	Long-term steady rate of seepage unrelated to reservoir level	Seepage fluctuates with reservoir, but at a predictable rate	Seepage is increasing over time at the same reservoir level; or seepage is episodic or surging	
Soil Erodibility (adapted from Sherard, 1953)	Well-graded material with clay binder, (10<PI<15), well or poorly compacted Much less likely if plastic clay, PI > 15, well or poorly compacted	Well-graded, low plasticity material, (6<PI<10), well or poorly compacted	Well-graded, cohesionless material, (PI<6), well or poorly compacted Uniform, fine cohesionless sand, (PI<6), well or poorly compacted	Sherard's guidance is intended to represent the relative erodibility of a range of soil types with different PI and different compaction efforts. **Use this guidance in conjunction with base rate statistics such as *87% of Reclamation incidents were in soils with PI<7*.** Dispersive soils are not included here; dispersive soils can be much more erodible and rates of initiation should be adjusted to reflect dispersion potential.
Cracks – located on the crest, or in test pits that expose the upper part of the impervious zone.	No cracking observed when large areas, or all, of the top of the core is exposed	No cracking observed on the crest or in limited test pits exposing the core	Transverse cracks on the surface of the core and/or, extensive, open longitudinal cracking. Much more likely for transverse cracks that extend across the core, and extend below reservoir water level being considered	At Reclamation it has not been standard practice to excavate test pits at the crest of the dam to expose the impermeable zone. Therefore, the potential for cracking generally relies on observations of cracks at or near the crest, and on other factors (in this table) that could increase or decrease the potential for cracking.
Sinkholes or depressions	No observations of sinkholes or depressions, including on upstream slope areas that are normally submerged.	Minor depressions on the upstream or downstream slopes that developed slowly and do not change over time.	Observations of sinkholes or depressions on the crest, upstream slope, or downstream slope that appear suddenly or change with time.	Sinkholes and depressions are important observations but are not always associated with an internal erosion potential failure mode. Localized settlement of limited loose zones in the embankment could result in sinkholes or depressions. Wave action on riprap can also create localized depressions. If sinkholes or depressions are observed, look for nearby conduits, toe drains, coarse graded materials or other anomalies that could allow for material transport. Sinkholes and depressions could result from seepage through internally unstable soils that allow material

Figure 5. Sample Excerpt – Table 1 - Factors Influencing the Likelihood of Initiation of Internal Erosion Through an Embankment Dam – Bureau of Reclamation – For use with "Best Practices in Dam and Levee Safety Risk Analysis," Chapter IV-4. Internal Erosion Risks for Embankments and Foundations.

The Reclamation approach to this "Initiation Node" certainly has the virtue of narrowing the discussion and promoting consistency at this event tree node (as well as almost always multiplying in a nice low number into every event tree, that lowers the apparent risk computed). The danger is that it can constrict the range of values that are truly in play, anchoring the discussion to very low numbers right at the start. In contrast, a USACE paper titled "Anatomy of a Risk Assessment" presented at the 2015 Association of State Dam Safety Officials (ASDSO) Conference employed much higher initiation values (falling between 0.1 to 0.9) for the PFM depicted in Figure 6, indicating a much wider range of "best estimate" values were in play for this study.

Uniqueness is the hallmark of dams. This uniqueness makes the task of developing "accurate" risk numbers for dams especially challenging. Laudable efforts to promote consistency in risk analysis work can have the unintended consequence of constraining the activity and potentially diminishing the "fresh perspective" and "fresh thinking" that can be so important to the work and beneficial to the value of the effort.

Review Process

The importance of a thorough, consistent review process cannot be emphasized enough. After all the planning and investment in resources to perform a risk analysis, reviews for compliance with process, risk methodology, critical assumptions, and consistency with

Failure Mode Sketch and Event tree nodes

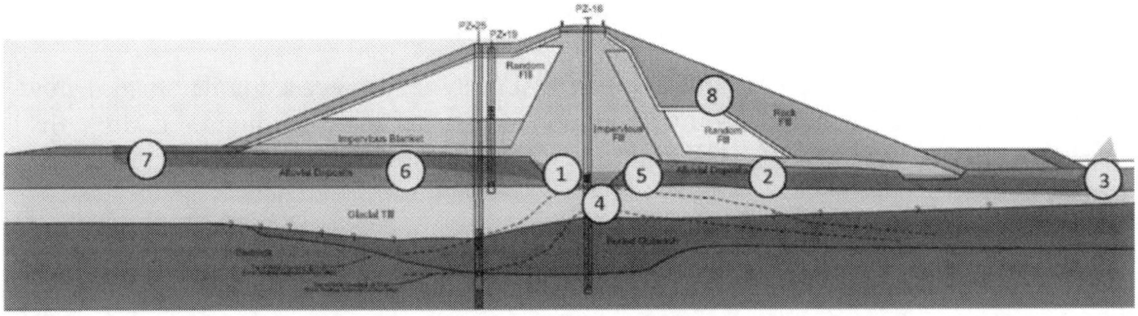

1. Construction or Geologic Defect at end of Cutoff Trench – Neutral
2. Continuous Alluvial Pathway US and DS of Cutoff – virtually certain
3. Unfiltered Seepage Exit at Toe Pond – virtually certain
4. Initiation of Erosion of Alluvial Soils (Pool dependent)
5. Alluvial Soils Fail to Act as Flow Limiter (unlikely)
6. Ability to Support Roof (Neutral)
7. Ability to Emerge or Bypass Upstream Impervious Stripping Fill (virtually certain)
8. Breach of Dam (virtually impossible to likely – Pool dependent)

Figure 6. Example USACE Presentation Regarding a Seepage-Related PFM

the results compared with other similar dams is vitally important. Concurrent or participatory reviews throughout the risk analysis process are encouraged.

CONCLUDING THOUGHTS

Dam designers from the beginning of time have considered the risks involved and have adapted their designs to accommodate them. How to do that was the result of centuries of empirical work. That work resulted in the successful design and construction of most of today's existing dams.

Failing to recognize the limits of knowledge in the context of dam safety is a very significant concern. Numbers produced in risk analysis work have the potential to blind the profession with a false sense of confidence. Effective dam surveillance and monitoring programs need to be in place not only to address the identified PFMs and general performance and "health" monitoring needs, but also to detect anomalous performance associated with unidentified PFMs and with "unknown unknowns" regarding the dam and dam site.

No one should consider doing a risk analysis just for the sake of doing a risk analysis. There needs to be a reason. There must be an upcoming decision for which input and insight is sought, and gaining incremental knowledge is desired. The activity needs to be approached with full understanding and knowledge of the limitations of the work, the limitations of the input information that is available, what guidelines the results risk analysis will be compared to (and why those are appropriate guidelines), and the method that will be used to determine when the risks are considered tolerable. Assessing whether risks are tolerable needs to consider the risks in relation to an organization's guideline values and the "As Low As Reasonably Practicable (ALARP)" standard, and the level of uncertainty in the evaluation, but also the financial risks, the risks to the integrity of the organization, and potential liability and legal factors.

Risk analyses for dams can provide a valuable tool to gain increased understanding and insight, if they are embarked upon with a good understanding regarding their strengths and capabilities, as well as their limitations. They can provide valuable input to dam safety decision-making.

REFERENCES

[1] Bureau of Reclamation and U.S. Army Corps of Engineers, "Best Practices in Dam Safety and Levee Safety Risk Analysis," most recent update dated June 20, 2015.

IMPLEMENTING A RISK-INFORMED APPROACH TO DAM SAFETY IN A REGULATORY ENVIRONMENT

Douglas Boyer, PE, CEG[1]
William Allerton, PE[2]

ABSTRACT

The Federal Energy Regulatory Commission (FERC) is currently developing risk-informed decision making (RIDM) concepts to integrate risk into our dam safety program. This concept was initiated in 2009 and was a specifically identified objective in the 2010 FERC Strategic Plan. This strategy continues to be included in FERC's most recent (2014) strategic plan. In part, the strategy reads:

> *"Risk-informed decision making (RIDM) provides the capability to assess non-traditional failure modes, . . . focus inspections and surveillance on the specific potential failure modes and monitoring programs at projects, and guide remediation projects to provide an overall reduced level of risk to the public. FERC will continue to develop the necessary risk assessment guidelines, procedures and policies, and to train Commission staff, dam owners, and consultants in risk assessment procedures, methodologies and tools."*

A number of other federal agencies have developed and are actively using risk-informed approaches in their dam safety programs, including the Bureau of Reclamation and US Army Corps of Engineers. For the most part, these dam owning agencies use in-house resources to perform risk analyses, risk assessments, and risk management. FERC's role is as a regulator and not an owner; therefore, the approach and execution of risk, among other things, will be different than the other federal agencies.

This paper discusses how FERC is addressing these and other challenges with regard to implementing RIDM.

INTRODUCTION

Risk, if used properly, can be a very powerful tool to aid in understanding in more detail the critical aspects of a dam's design, construction, analysis, and performance and the potential consequences of a dam failure. The approach to risk analyses can improve the prioritization of dam safety actions to focus limited resources on what's important, to direct risk reduction evaluations to those measures that are more effective, and to better

[1] Chief, Risk-Informed Decision Making Branch, Federal Energy Regulatory Commission, Division of Dam Safety and Inspections, 850 SW Broadway, Suite 550, Portland, OR 97205, (503) 552-2709, Douglas.Boyer@ferc.gov

[2] Director, Federal Energy Regulatory Commission, Division of Dam Safety and Inspections, 888 First Street, NE, Washington, DC 20426, (202) 502-6025, William.Allerton@ferc.gov

inform routine dam safety activities such as dam safety inspections, surveillance and monitoring, emergency action plans, and others.

Some challenges to integrating RIDM into the FERC dam safety program include:

1. Educating internal dam safety staff and upper management in risk concepts and risk management practices, including the value and limitations of risk.
2. Developing risk policies and guidelines as regulatory documents, including guidelines for the tolerability of risk.
3. Developing training curriculum for the many hundreds of dam owners, consultants, and internal staff.
4. Developing a realistic implementation plan that takes into account funding and resource constraints.

BACKGROUND

Since 2009, FERC has been pursuing and evaluating the development of risk principles and concepts for dam safety (Regan and Boyer, 2009). Early work by FERC included participating by invitation with the USACE in the development of their dam safety policy and procedures document (USACE, 2014), attendance at key interagency dam safety risk briefings, and in the development of a first-level risk screening tool referred to as screening level portfolio risk analysis (SLPRA). More recently, FERC participated with other federal agencies in the development of the Federal Guidelines for Dam Safety Risk Management (FEMA, 2015), recently published by FEMA. FERC recognizes the value that risk approaches and tools can provide in conducting dam safety activities and in making better informed dam safety decisions. As one of the leaders in the development of potential failure modes analyses (PFMA), FERC sees risk as a logical extension of the PFMA process.

Risk methodologies and experiences have matured to the point where, as a dam safety regulator, FERC is able to apply these concepts in a regulatory environment.

The RIDM guidelines are available on the FERC website at http://www.ferc.gov/industries/hydropower/safety/initiatives/risk-informed-decision-making.asp

PRINCIPLES

Several fundamental principles have guided FERC in the development of RIDM guidelines. Some of these include:

1. To provide goal-oriented policy and guidance rather than narrowly defined and prescriptive policies. Minimum regulatory expectations must be defined where possible.

2. To provide the guidance within FERC's existing mission and defined statutes (18 CFR 12).

3. To provide the guidance, where possible and when appropriate, within the framework of previously established federal and industry dam safety risk guidance. This includes the Federal Guidelines for Dam Safety Risk Management (FEMA, 2015) and other references by the Bureau of Reclamation (BOR, 2011) and US Army Corps of Engineers (USACE, 2014).

4. To understand that there are limited resources that have the requisite dam safety risk experience that can contribute initially to these efforts. These include limited resources internally to FERC and also externally within licensees' staff and available industry. Given this initial constraint, we must understand that development of this plan will not happen overnight and that training of personnel, staff, and industry will be required, all of which will take time.

5. To be truly effective, risk must become an integral part of the dam safety program and not simply something that is bolted on to the existing structure of the program. Risk not only informs decision making regarding dam safety modifications, but also provides valuable insights to focus routine dam safety activities, including:

    a. Dam safety inspections, including the frequency and timing of inspections
    b. Surveillance and monitoring
    c. Emergency action plans (EAPs)
    d. Design review activities
    e. Others

RISK-INFORMED DECISION MAKING IMPLEMENTATION PLAN

The FERC's RIDM implementation plan includes the following key elements:

1. Development of RIDM guidelines (risk guidelines)
2. Development of risk methodologies
3. Identification and performance of pilot risk studies
4. Development and execution of risk training classes/workshops
5. Communication of the RIDM plan

Each of these elements is discussed in the sections below.

RIDM Guidelines

At the drafting of this paper, the RIDM guidelines have been drafted and are in the review process. The information presented in this paper is based on the draft version of the RIDM guidelines and is subject to change based on review comments and other revisions. The RIDM guidelines include four chapters:

Chapter 1 – Introduction to Risk-Informed Decision Making
Chapter 2 – Risk Analysis
Chapter 3 – Risk Assessment
Chapter 4 – Risk Management

This structure of the guidelines mirrors the key elements included in the Federal Guidelines for Dam Safety Risk Management (FEMA, 2015) as shown in Figure 1.

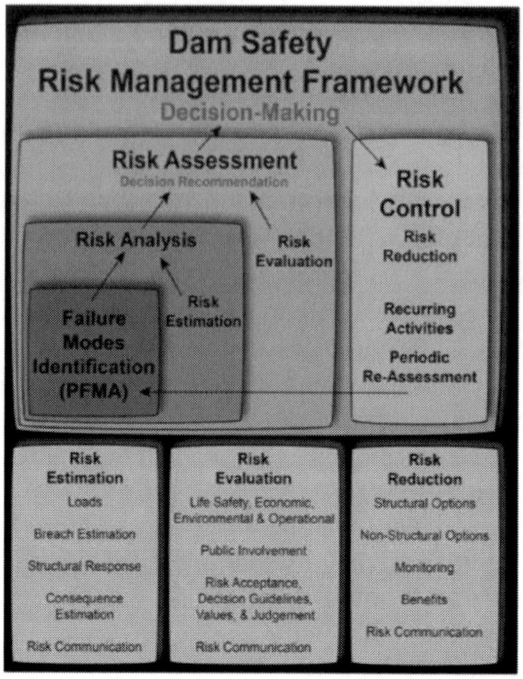

Figure 1. Dam Safety Risk Management Framework (FEMA, 2015)

Chapter 1 – Introduction to RIDM. The first chapter provides a brief discussion on the benefits of risk in aiding decision making, defines the guidance principles, provides definitions of key risk terms, and provides an overall glossary of terms. Risk terminology can be quite challenging and even within the United States some agencies/organizations/individuals use risk terms differently. This chapter establishes the terminology that FERC will use. Key terms and definitions include:

Risk – A measure of the probability and severity of an adverse effect to life, health, property, or the environment.

Risk Analysis - Risk analysis is the use of available information to estimate the risk to individuals or populations, property or the environment, from hazards. Risk analyses generally contain the following steps: scope definition, hazard identification, and risk estimation. The risk analysis process involves the scientific characterization of what is known and what is uncertain about the present and future performance of the dam system under examination.

Risk Assessment –The process of making a decision recommendation on whether existing risks are tolerable and present risk measures are adequate, and if not, whether alternative risk reduction measures are justified or will be implemented. Risk assessment incorporates the risk analysis and risk evaluation phases.

Risk Evaluation - The process of examining and judging the significance of risk. The risk evaluation stage is the point at which values (societal, regulatory, legal, and owners) and value judgments enter the decision process, explicitly or implicitly, by including consideration of the importance of the estimated risks and the associated social, environmental, economic, and other consequences, in order to identify and evaluate a range of alternatives for managing the risks.

Risk Management – The systematic application of management policies, procedures and practices to the tasks of identifying, analyzing, assessing, communicating, mitigating, and monitoring risk.

Chapter 2 – Risk Analysis. Chapter 2 provides guidance on what risks are to be provided and what general framework and methodologies are to be used to develop risk estimates. The chapter defines the various types of risks, including incremental risk, non-breach risk, and residual risk. The chapter also defines four levels of risk analysis, including:

Level 1 - Screening Level Risk Analyses
Level 2 - Periodic Risk Analyses
Level 3 - Semi-Quantitative Risk Analyses (SQRA)
Level 4 - Quantitative Risk Analyses (QRA)

Each level of risk analysis provides a different set of tools and methods that are proportionate in terms of the level of effort required, details considered, and confidence in their outcomes. These levels of risk analyses provide a suite of scalable approaches that provide information to promote critical thinking and guide a risk analyst's judgment. The risk analysis methods applied to each level are scalable and can be applied with varying degrees of effort (i.e., time, resources, and cost) to provide the appropriate level of accuracy, rigor, and confidence required to make credible risk-informed decisions. It is important to understand that every decision does not necessarily require a high level of rigor, detail, and accuracy in the risk estimate in order to support a credible decision. These risk analysis levels vary in purpose and therefore in the data required, detail and robustness of analysis, and in the uncertainty and confidence in the results. However, in all cases the level of detail should only be what is needed to support the decision(s) that will be informed by the risk analysis. The analysis should be as simple as it can be, but not simpler.

These levels of risk analysis are accomplished using either quantitative or qualitative approaches.

Additional topics in the chapter include:

- The roles and responsibilities and qualifications of key individuals performing risk analyses are defined. Key individuals include the risk facilitator, subject matter experts, software operator, and note taker/report writer.

- Risk analysis execution, including the planning, preparation, and communication that's required for a risk analysis; conducting a risk analysis, risk software, and risk methodology.

- The importance of uncertainty, confidence, and sensitivity in developing risk estimates.

- The definition and requirements of ALARP (as low as reasonably practicable). ALARP is that principle that states that risks, lower than the limit of tolerability, are tolerable only if risk reduction is impracticable or if the next increment of risk reduction is not cost effective compared to the improvement gained (ICOLD, 2005). This concept is shown in Figure 2.

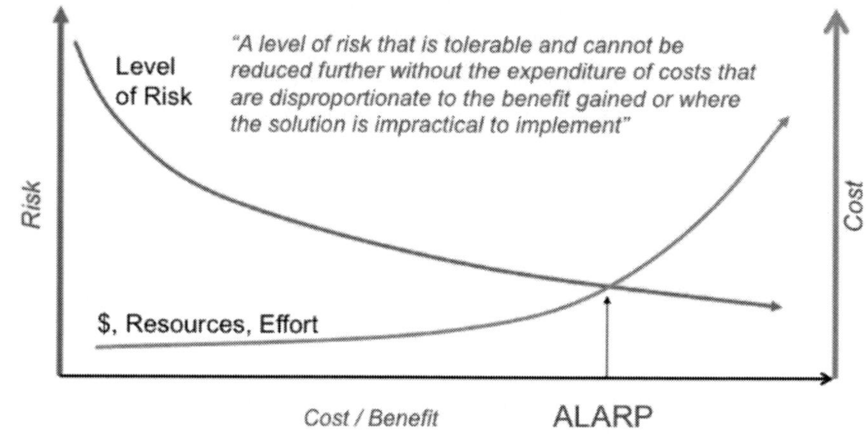

Figure 2. ALARP, visually displayed.

There is no mathematical formula or 'seven-step process' available to determine ALARP. Instead, there are a number of factors that must be evaluated and considered to judge ALARP. To make a judgment on whether risks are ALARP, the following factors must be considered (adapted from NSW, 2010):

1. The cost-effectiveness of the risk reduction measures.
2. The level of risk in relation to the established risk guidelines.
3. The disproportion between the sacrifice (i.e., money, time, trouble and effort) in implementing the risk reduction measures and the subsequent risk reduction achieved.
4. Any relevant recognized good practice.
5. Societal concerns as revealed by consultation with the community and other stakeholders.

Finally, the chapter includes a discussion of the information that must be included in a risk analysis report and a general risk analysis report outline is provided. The chapter emphasizes the importance of good documentation. Understanding the basis of the risk estimates is as important as or more important than the risk numbers. The dam safety case is a logical and objective set of arguments used to advocate a position that either additional safety-related action is justified or that no additional safety-related action is justified (FEMA, 2015). The dam safety case should cite the most compelling information that supports the risk estimates and the overall findings and also discuss the uncertainties that were identified in the risk analysis.

Chapter 3 – Risk Assessment. This chapter focuses on a risk-informed approach to make decisions. This method has the advantage of providing a more consistent basis for decision making. Also, since it is risk-informed, rather than risk-based, it allows for other important factors to be considered in the decision, beyond a sole reliance on numerical risk estimates.

In order to understand the assessment of risks, a brief background on tolerable risk is provided in the chapter. Tolerable risk is (by the International Committee on Large Dams (2005) adapted from HSE (2001) and Bowles (2007) defined as:

1. Risk society is willing to live with so as to secure certain benefits
2. Risk society does not regard as negligible or something it might ignore
3. Risk that society is confident that are being properly managed by the owner
4. Risk the owner keeps under review
5. Reduces still further as practicable

The concept of tolerable risk is illustrated in Figure 3.

Figure 3. Generalized and Project Specific Tolerability of Risk Framework (Adapted and Modified from HSE, 2001 and USACE, 2014)

Four types of risk measures are defined:

1. Life safety risk – which includes incremental and non-breach risk within the context of tolerable risk guidelines.

2. Annual probability of failure (APF).

3. Economic considerations – which includes incremental and non-breach risk.

4. Environmental and other non-monetary consequences - which includes incremental and non-breach consequences.

Three types of incremental life safety risk guidelines are also defined.

1. Individual incremental life safety risk using probability of life loss for the identifiable person or group by location that is most at risk of loss of life due to dam breach.

2. Societal incremental life safety risk, which is expressed in two different ways:

    a. Probability distribution of potential life loss
    b. Average Annual Life Loss (AALL)

Interim individual and societal risk guidelines (AALL) are shown on Figures 4 and 5, respectively.

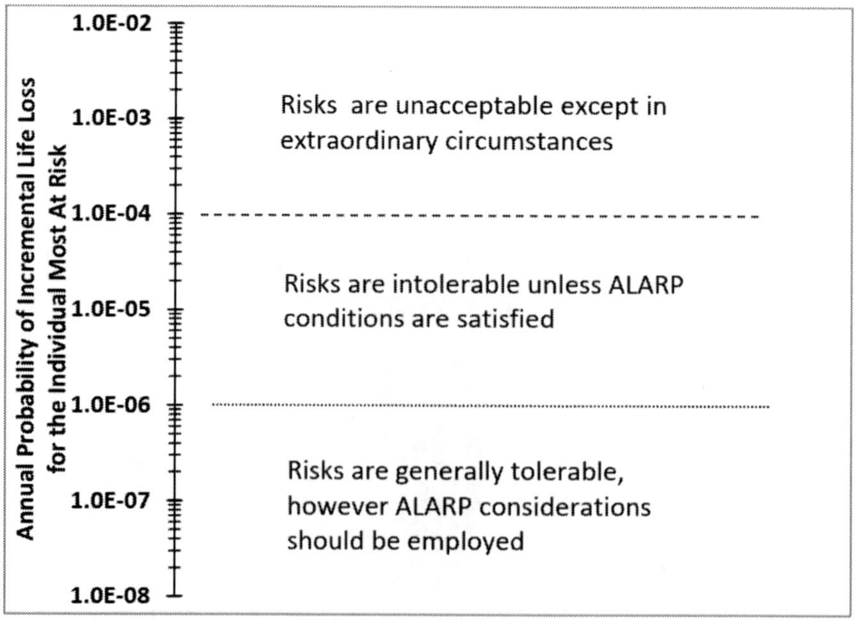

Figure 4. Individual Life Safety Guidelines for Incremental Risk

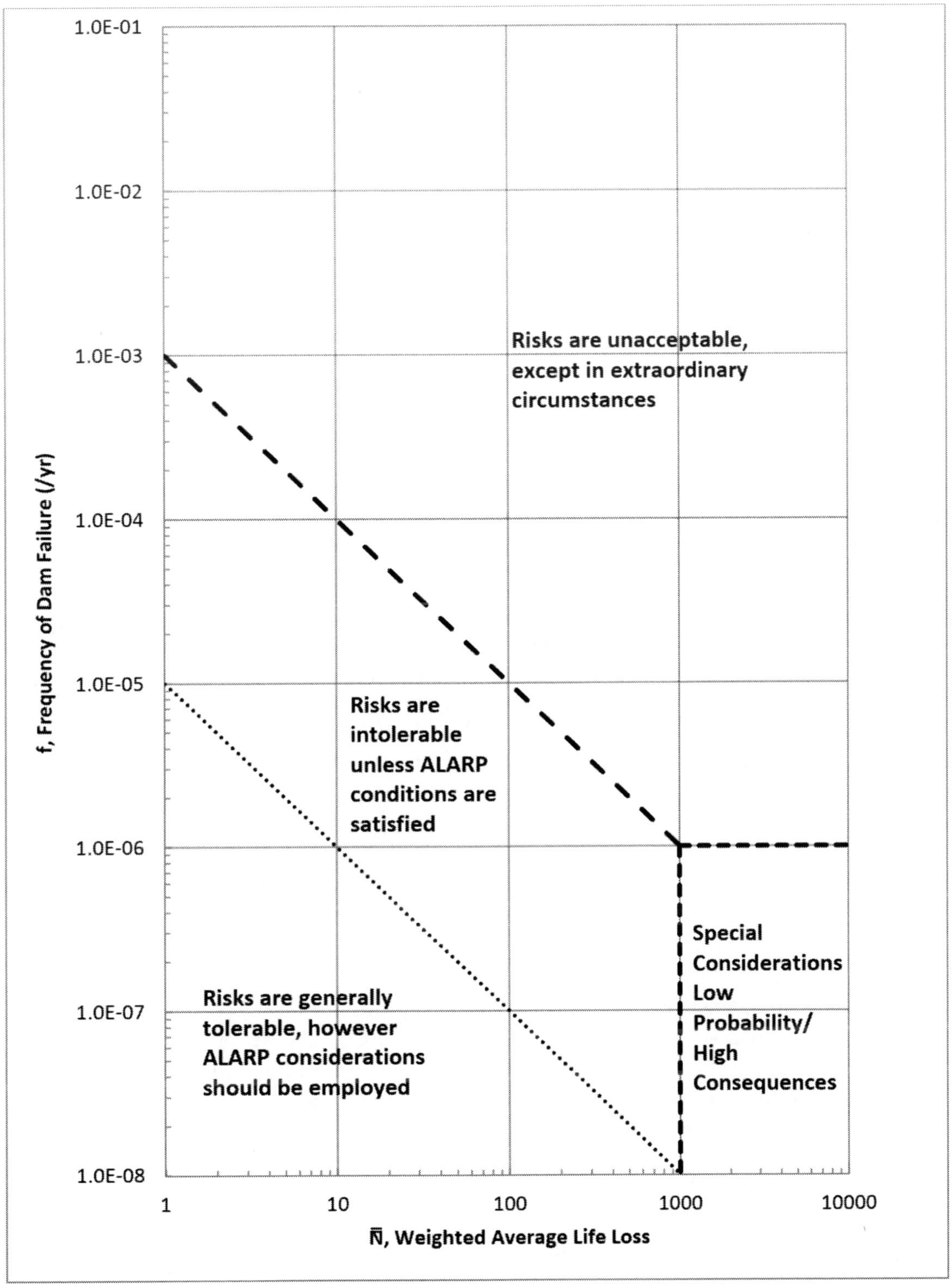

Figure 5. f-N Chart for Displaying Average Annual Life Loss for Incremental Risk

The incremental life safety risk is to be evaluated against all three life safety guidelines.

As can be seen from Figures 4 and 5, ALARP will play a large role in assessing the tolerability of risk. Another way of describing or thinking about tolerable risk is that, after hearing all the facts and information related to an issue or issues on a dam or project, an organization decides that further action is not reasonably practicable. When a judgment is made that risks are as low as reasonably practicable, this is often determined by comparing the effectiveness of reducing risk further (evaluated by considering the cost to further reduce risk and the amount of risk reduction achieved, and then comparing it to other risk reduction actions implemented). If the costs to achieve an additional level of risk reduction are disproportionate, the current risk may be as low as reasonably practicable.

The evidence based on the ALARP factors defined in Chapter 2 will be used to evaluate ALARP. As such, it is incumbent on the dam owner to make sure the documentation and evidence to support the ALARP factors is clearly presented in the risk report and the case is clearly made in the report as to whether ALARP considerations for each potential failure mode are met or not. The use of these factors will be used to inform and not to prescribe the outcomes of an ALARP evaluation.

The results of the risk analysis and the assessment of the tolerability of those risks are key ingredients in the RIDM process. It has been widely recognized that procedures and data available for dam safety risk analysis, while mostly quantitative, do not provide precise numerical results. Therefore, relying solely on the numeric estimates in comparison to hard line criteria (sometimes referred to as "risk-based" evaluation) would not be appropriate (USBR, 2011). The assessment of risks and the corresponding decisions are generally more complex than can be portrayed using only the numerical results of a risk analysis. FERC and the other federal dam safety agencies using risk-informed approaches have chosen to use a more "risk-informed" approach where additional information is included to support the assessment and case for proposed actions (or non-action). The intent in the assessment process is to use the entirety of the information available to build and support the case to take a particular action (or to take no action).

It cannot be emphasized enough – FERC's RIDM guidelines are not intended to be used as rigid decision-making criteria to declare a facility "safe" or "unsafe" based solely on a risk estimate. Since the numbers are only approximate measures of risk, and since the risk guidelines themselves are not rigid, additional reasoning is essential to justify the risk estimates and the recommended actions. The case is intended to present rationale in a formal and methodical manner to persuade decision-makers to take responsible action (or to justify no action).

The case is a logical set of arguments used to advocate either the position that additional safety-related action is justified, or that no additional safety-related action is justified at any given (current) time (USBR, 2011). The arguments string together key evidence regarding the three basic risk components (i.e., load probability, response probability, and consequences) so as to convince decision-makers that the dam's existing condition and ability to withstand future loading, the risk estimates, and the recommended actions are all coherent. Since uncertainty is inherent in each claim, the arguments should also address whether confidence is high enough for the conclusions to stand on the basis of existing evidence. The case should be carefully crafted so that all descriptions and terms are easy to understand by the prime audience, all arguments are cogent and coherently developed, all references are easily accessible, and all conclusions are fully supported and follow logically from the arguments (USBR, 2011).

When making a decision regarding future actions, FERC will consider the risk estimates, the issues most influencing the risks, the sensitivity of the risks to particular inputs, the cost of additional actions, and the potential for reducing uncertainty.

Chapter 4 – Risk Management. Risk management encompasses activities related to making risk-informed decisions, prioritizing evaluations of risk, prioritizing risk reduction activities, and making program decisions associated with managing a portfolio of facilities (FEMA, 2015). Risk management is greatly facilitated and enhanced by having the knowledge base supplied by the risk analyses and risk assessment inputs for the dams within the portfolio. Such knowledge allows a logical and consistent basis for substantiating and prioritizing risk reduction activities and/or making program decisions associated with managing a portfolio of facilities. Risk management, because it uses the findings from a risk assessment process, includes considering the environmental, social, cultural, ethical, political, and legal factors (FEMA, 2015). The primary goal of risk management is to implement actions to accept, further monitor or evaluate, control, or reduce risk, while considering the cost and benefits of any actions taken (FEMA, 2015).

The chapter includes discussions on portfolio risk management, risk reduction measures, and routine dam safety risk management.

A Dam Safety Risk Classification (DSRC) system is established that provides consistent and systematic guidelines for appropriate actions to address the dam safety issues and deficiencies of FERC-regulated dams, as shown in Table 1. FERC-regulated high and significant hazard potential dams will be assigned a DSRC informed by the probability of failure and the incremental risk.

The DSRC table presents different levels and urgencies of actions that are commensurate with the different classes of the safety status of FERC-regulated dams. These actions range from immediate recognition of a situation with very high urgency requiring extraordinary and immediate action (DSRC I) to situations with no urgency other than to continue normal operations and dam safety activities (DSRC V).

An important element of risk management is the identification of effective risk reduction actions. These actions can take many forms, from monitoring to engineering studies to immediate corrective actions. These actions can be temporary or permanent. Actions may need to be staged to be both efficient and effective. Interim measures may be required to provide more immediate short-term risk reduction while other risk reduction measures are being investigated and evaluated.

Interim risk reduction measures (IRRM) are dam safety risk reduction measures that are intended to act as temporary or interim measures until such time as more permanent remediation measures can be effectively implemented. An interim risk reduction measures plan (IRRMP) must be developed for DSRC I, II and III dams. The DSRC table (Table 1) provides the actions and characteristics for each DSRC, including preparation of an IRRMP, considerations for preparation of the plan, and example interim measures.

The results of the risk analyses and the knowledge gained from performing the risk analysis on individual potential failure modes should be deliberately incorporated into the routine dam safety activities, including dam safety inspections, EAPs, dam safety surveillance and monitoring plans/reports, etc. Significant investment is required in understanding potential failure modes and the vulnerabilities from each potential failure mode during a risk analysis. That information should be used to focus and inform dam safety inspections, assist in the understanding of consequences and the EAP and information conveyed to emergency management agencies (EMAs), and provide a better assessment of the existing surveillance and monitoring activities and the potential for revisions to location, frequency of reading, instrumentation type, etc. for each potential failure mode.

Risk communication is a fundamental part of the risk framework and is integrated into the risk analysis, assessment and management steps and ensures that the decision makers, other stakeholders and affected parties understand and appreciate the process of risk and in doing so can be fully engaged in and responsible for risk management. It must begin early and continue throughout the portfolio risk management process, including the dissemination of information of any adverse impacts of the risk reduction actions and how those impacts can be mitigated, and is essential to risk-informed management. Public education is included under the umbrella of risk communication.

Risk communication is a foundational element in a successful dam safety program. Licensees should have the philosophy that they will ensure communication regarding potential inundation hazard and consequences, and that solutions are open, transparent and understandable to the public.

Communicating risk to the public is the responsibility of the licensee. An open, interactive and ongoing dialogue is critical. Communicating risk is as important as assessing and managing risk. Today's risk communication goes beyond just communicating technical information; it includes recognition of important cultural values and ideas that affect decisions. Research has shown that communicating recommended actions to the public is an effective way to change behavior. In emergency situations, communicating the immediate hazard is important and, in most cases, local authorities will be communicating about the imminent danger.

Table 1. FERC Dam Safety Risk Classification (DSRC) Table

Urgency of Action (DSRC)	Description	Characteristics	Potential Actions
I – VERY HIGH	An active potential failure mode is in process or the likelihood of a failure is judged to be extremely high, such that immediate actions are necessary to reduce risk.	**CRITICALLY NEAR FAILURE**: There is direct evidence that failure is in progress, and the dam is almost certain to fail during normal operations if action is not taken quickly. OR **EXTREMELY HIGH RISK**: Combination of life or economic consequences and likelihood of failure is very high with high confidence.	• Take immediate action to avoid failure • Communicate findings to potentially affected parties • Implement IRRMs, including operational restrictions • Ensure that the emergency action plan is current and functionally tested for initiating event • Conduct heightened monitoring and evaluation • Expedite investigations and actions to support long-term risk reduction • Initiate intensive management and situation reports
II – HIGH	Potential failure mode(s) are judged to present very serious risks, either due to a very high probability of failure or due to very high life loss, that justify an urgency in actions to reduce risk.	**RISK IS HIGH WITH HIGH CONFIDENCE, OR IT IS VERY HIGH WITH LOW TO VERY LOW CONFIDENCE**: Failure could begin during normal operations or be initiated as a consequence of an event. The likelihood of failure from one of these occurrences, prior to taking some action, is too high to delay action.	• Communicate findings to potentially affected parties • Implement IRRMs, including operational restrictions, as warranted • Ensure that the emergency action plan is current and functionally tested for initiating event • Conduct heightened monitoring and evaluation • Expedite investigations and actions to support long-term risk reduction • Expedite confirmation of classification
III – MODERATE	Potential failure mode(s) appear to be dam safety deficiencies that pose a significant risk of failure and actions are needed to better define risks or to reduce risks.	**MODERATE TO HIGH RISK**: Confidence in the risk estimates is generally at least moderate, but can include facilities with low confidence if there is a reasonable chance that risk estimates will be confirmed or potentially increase with further study.	• Implement IRRMs, including operational restrictions, as warranted • Ensure that the emergency action plan is current and functionally tested • Conduct heightened monitoring and evaluation • Prioritize investigations and actions to support long-term risk reduction • Prioritize confirmation of classification as appropriate
IV – LOW	Potential failure mode(s) appear to indicate a potential concern, but do not indicate a pressing need for action.	**LOW RISK**: The risks are low to moderate with at least moderate confidence, or the risks are low with low confidence, and there is a potential for the risks to increase with further study.	• Ensure that routine risk management measures are in place • Determine whether action can wait until after the next Part 12D Before the next Part 12D Report, take appropriate interim measures and schedule other actions as appropriate • Give normal priority to investigations to validate classification, but do not plan for risk reduction measures at this time
V – NO	Potential failure mode(s) do not appear to present significant risks and there are no apparent dam safety deficiencies.	**VERY LOW RISK**: The risks are low to very low and are unlikely to change with additional investigations or studies.	• Continue routine dam safety risk management activities and normal operations and maintenance

Communication is important in all aspects of dam safety within a licensee's organization, with emergency management officials, with FERC, with the public, and with other local, state, and federal agencies. However, communication about the work associated with risk is particularly important because of the fears, sentiments, perceptions, and emotions surrounding the word "risk" and the use of risk analysis in engineering. Thus, it is important to understand and have a good plan for communicating risk, including:

1. What information is available at a given dam related to potential failure modes and how the information is considered in a risk analysis
2. How risk will be considered by an organization
3. What the results of the risk analyses are
4. What decisions were reached and what risk remains

Risk Methodology

FERC is responsible for the development, dissemination, and interpretation of methodology guidance for use in conducting dam safety risk analyses. The Best Practices for Dam and Levee Risk Analyses (BOR/USACE, 2015) has been developed jointly between the USACE and the BOR for the purpose of summarizing the overall philosophy, methods, and approach to risk analysis for dam safety. Unless otherwise directed by FERC, the risk analyses must use the methodologies included in the Best Practices manual to determine the loads, the conditional probability of failure associated with each potential failure mode, and the consequences associated with each potential failure mode. The BOR/USACE Best Practices manual is generally maintained and updated on an as-needed basis. The current version of the Best Practices manual may be obtained from BOR at http://www.usbr.gov/ssle/damsafety/risk/methodology.html or from the USACE.

Risk methodologies for Level 2 Periodic Risk Analyses are currently under development by FERC.

Pilot Studies

Pilot risk studies have been initiated to test the methodologies, policies, and procedures provided in the draft risk guidance documents. The purposes of the risk pilot studies are to:

1. Provide a clearer understanding of the outstanding dam safety issues at each dam and provide a more confident path to resolution of those issues through RIDM.
2. Evaluate the RIDM policy and processes, including identifying potential shortcomings in the interim guidelines.
3. Evaluate the risk methodologies, including looking for holes or gaps in methodology/guidance, identifying potential challenges or short comings in existing methodologies, and identifying potential tools that would be beneficial for future risk analyses.
4. Identify future training needs for risk implementation.

5. Obtain input from licensees and consultants on the overall processes and methodologies.

It is the expectation that the pilot studies will be performed by teams that are familiar with dam safety risk analysis and are not intended to be training opportunities for those not familiar with dam safety risk concepts and practices.

Training

Training of internal staff and external personnel in risk policies, procedures, and methodologies will be a herculean task. Although risk topics have been a subject of many technical presentations and workshops at professional conferences, very few individuals have actually participated or even observed a quantitative risk analysis. Training will required the development of basic, intermediate, and advanced risk analysis course material and how the results of risk analyses and risk assessments will be used in dam safety decision making.

Table 2 includes some early thoughts on the types of curriculum that may be needed for both internal FERC staff and external personnel.

Table 2. FERC Risk Training Curriculum

Risk Training Curriculum
Base Courses
Overview of Risk Analyses
Best Practices in Dam Safety Risk Analyses
Level 2 Risk Analyses
Level 3 Risk Analyses
Facilitation
Loadings and Consequences
Hydrologic Loading
Seismic Loading
Consequences
Potential Failure Modes and Risks
Internal Erosion Mechanics
Internal Erosion Risks
Overtopping/Overwash/Erosion of Soil and Rock
Seismic Analysis of Concrete Structures and Gates
Seismic Analysis of Embankments
Operational Risks
Risk Analysis
Potential Failure Modes and Event Tree Construction
Risk Analysis Software Tools
Portrayal of Risks to Support Decisions

The content for some of the courses listed in Table 2 has already been developed by FERC and other federal agencies. For example, the content for the Best Practices in Dam Safety Risk Analysis has been developed by BOR and USACE and the content for the

Internal Erosion Mechanics has been developed by USACE. The development of the course materials for the additional training classes will take time and effort.

In addition, FERC must develop training modules associated with the implementation and execution of the RIDM guidelines.

Communication

As is always the case, communication is a key element of any successful plan. FERC is committed to providing ongoing development of RIDM information. This information will be available through a variety of communication channels, including:

- FERC-Division of Dam Safety and Inspections website
- Status updates at professional society meetings and workshops
- Annual letter to licensees
- e-mail announcements
- Other letters to licensees

CONCLUSIONS

The Federal Energy Regulatory Commission (FERC) is currently developing RIDM concepts to integrate risk into our dam safety program. Risk methodologies and experiences have matured to the point where, as a dam safety regulator, FERC is able to apply these concepts in a regulatory environment. Many challenges exist to integrating RIDM into the FERC dam safety program, including development of risk guidelines; training of risk policies, procedures, and methodologies; and communication. Risk pilot studies will be used to help evaluate the RIDM policies and processes, including identifying potential shortcomings in the interim guidelines and to obtain input from licensees and consultants on the overall RIDM processes and methodologies.

REFERENCES

Bureau of Reclamation, "Dam Safety Pubic Protection Guidelines", Dam Safety Office, Denver, Colorado, August 2011. Available at: http://www.usbr.gov/ssle/damsafety/documents/PPG201108.pdf

Bureau of Reclamation and U.S. Army Corps of Engineers, "Best Practices in Dam and Levee Safety Risk Analysis", Denver, Colorado, July 2015. Available at: http://www.usbr.gov/ssle/damsafety/Risk/methodology.html

Bowles, D.S., "Tolerable Risk for Dams: How Safe is Safe Enough?" US Society on Dams Annual Conference, Philadelphia, Pennsylvania, March 2007.

Federal Emergency Management Agency, "Federal Guidelines for Dam Safety Risk Management", FEMA P-1025, Washington, DC, January 2015. Available at:

http://www.fema.gov/media-library-data/1423661058965-58dfcecc8d8d18b7e9b2a79ce1e83c96/FEMAP-1025.pdf

Health and Safety Executive, "Reducing Risks, Protecting People," Her Majesty's Stationery Office, London, UK, 2001. Available at: www.hse.gov.uk/risk/theory/r2p2.pdf

International Commission on Large Dams, "Risk Assessment in Dam Safety Management. A Reconnaissance of Benefits, Methods and Current Applications," Bulletin 130, 2005.

New South Wales Government Dams Safety Committee 2006, "Risk Management Policy Framework for Dam Safety", New South Wales, Australia, 22 August 2006.

Regan, Patrick J. and Douglas D. Boyer, Risk Informed Decision Making in a Regulatory Context, Association of State Dam Safety Officials Annual Meeting, 2009.

U.S. Army Corps of Engineers, "Safety of Dams – Policy and Procedures", ER 1110-2-1156, Washington, DC, March 2014. Available at: http://www.publications.usace.army.mil/Portals/76/Publications/EngineerRegulations/ER_1110-2-1156.pdf

Victoria Department of Sustainability and Environment, "Guidance Note on Dam Safety Decision Principles," Melbourne, Australia, May 2012.

USE OF QUANTITATIVE RISK RESULTS TO INFORM DAM SAFETY GOVERNANCE: PRACTICAL CASES IN EUROPE

Ignacio Escuder-Bueno, PhD[1]
Adrián Morales-Torres, MEng[2]
Jesica Tamara Castillo-Rodríguez MEng[3]

ABSTRACT

Large dams are critical infrastructures whose failure could produce important societal and economic consequences. This failure has a probability, although it is usually very low. Therefore, these infrastructures have an associated risk that must be analyzed and managed. Risk analysis is a suitable methodology to inform dam safety governance, since it integrates available information about hazards, system vulnerability and consequences. In particular, results of quantitative risk models can be used to prioritize dam safety investments.

In this sense, iPresas software is a useful tool to compute quantitative risk results. It integrates all aspects of risk analysis into the general framework of event trees. In addition, within this software, quantitative risk results can be used to recommend prioritization sequences of potential dam safety investments. These prioritization sequences are obtained combining investments' costs and risk reduction indicators, based on risk reduction principles of equity and efficiency.

In recent years, iPresas software has been used to compute risk and to obtain prioritization sequences in different systems of dams from different public and private owners in Europe. In this paper, the main characteristics, challenges and uncertainties of the risk analysis process for several cases are described. These cases illustrate the usefulness of quantitative risk results to prioritize investments in dam safety programs.

INTRODUCTION

The world has evolved significantly in recent decades in the way of understanding dam safety and flood risks related to dam response. Large dams are critical infrastructures whose failure could produce important societal and economic consequences. These infrastructures have an associated risk that must be analyzed and managed. Modern societies depend on the correct and efficient behavior and operation of dams. Therefore, as society demands higher levels of safety and reliability of dams, the integration of design, construction and operation of dams within a risk management framework is required to mitigate threats in an efficient and effective way.

---

[1] Tenured Professor, Universitat Politècnica de València (UPV). Research Institute of Water and Environmental Engineering (IIAMA), Camino de Vera s/n. 46022 Valencia. Spain; +34 96 387 98 93; iescuder@hma.upv.es
[2] CTO Projects and Development, iPresas (Spin-off UPV), Avenida del Puerto, 180 1-B. 46023 Valencia. Spain. adrian.morales@ipresas.com
[3] Researcher, Universitat Politècnica de València (UPV). Research Institute of Water and Environmental Engineering (IIAMA), Camino de Vera s/n. 46022 Valencia. Spain. jecasrod@upv.es

The concept of integrated risk management has acquired importance in recent years and current dam safety management strategies require incorporating aspects such as sustainability, resilience and public participation.

In the United States, following the journey initiated by the United States Bureau of Reclamation (USBR) in the nineties, the United States Army Corps of Engineers (USACE) and the Federal Emergency Regulatory Commission (FERC) have implemented risk-based dam safety management processes.

At the European level, the application of risk analysis techniques has emerged as a paradigm shift, enhancing dam safety and flood risk assessment and management. The European Flood Directive, published in 2007 (European Parliament, 2007), and the European Directive on Critical Infrastructure Protection, published in 2008 (European Union, 2008), are examples of such change. Both directives remark the need for identifying, characterizing and analyzing risk for establishing strategies for flood risk and critical infrastructure management.

The state-of-the-practice worldwide recognizes the benefits of risk analysis as a suitable methodology for supporting decision-making for critical infrastructure management, encouraging other improvements in operation and maintenance, surveillance or emergency preparedness.

In Spain, national regulation includes since 2008 the need for considering risk management as a key driver for establishing dam safety strategies, following the example of other countries. Therefore, in recent years there has emerged a new perspective for dam safety management which incorporates risk analysis, which aims at combining both the traditional (essentially deterministic) and the risk-informed dam safety approach, focusing on all risk components associated with a potential dam failure or uncontrolled release.

In addition, tools have been developed for supporting risk calculation and analysis. As an example, iPresas software (which allows risk calculation by means of event trees and influence diagrams, first developed at the Polytechnic University of Valencia [UPV] and now being developed and upgraded by iPresas, a UPV spin-off company) is a useful tool to compute quantitative risk results. It integrates all aspects of risk analysis into the general framework of event trees. In addition, quantitative risk results can be used to recommend prioritization sequences of potential dam safety investments. These prioritization sequences are obtained combining costs of measures and risk reduction indicators, based on principles of equity and efficiency.

In this paper, an overview of the dam safety management framework in Spain is provided since it has inspired the recent application of risk analysis techniques to several pilot cases in Europe. Lessons learned from these pilot cases support the usefulness of quantitative risk results to inform dam safety management and governance.

RISK-INFORMED DAM SAFETY GOVERNANCE IN SPAIN

The Context

Spain is geographically located in the southwest of Europe, with an area of 505,182 km^2 and a population of roughly 47 million inhabitants. Spain´s total water resources are estimated as 112 km^3 per year (CICCP, 2006), resulting in about 2,700 m^3/person/year (As a comparison, the average of European Union countries is 3,200 m^3/person/year).

These water resources are highly irregular in time and have an uneven geographic distribution. Spain has a long tradition of dam construction. According to ICOLD (International Commission on Large Dams), Spain ranks ninth in the world in terms of the number of large dams[4], with a registered number of large dams of 1,082 (Castillo-Rodríguez et al., 2015). Including smaller dams, the total number of dams in Spain rises to 1,225, with most of them built in the last 60 years.

Dams have played an important role in Spain regarding flood protection, hydroelectric production and water resources management for decades. Dams have been built across the whole country (Figure 1) in both upstream and downstream areas of river catchment basins, aiming to cope with highly irregular water inputs, frequent drought events, and floods that usually occur rapidly (i.e., flash flooding).

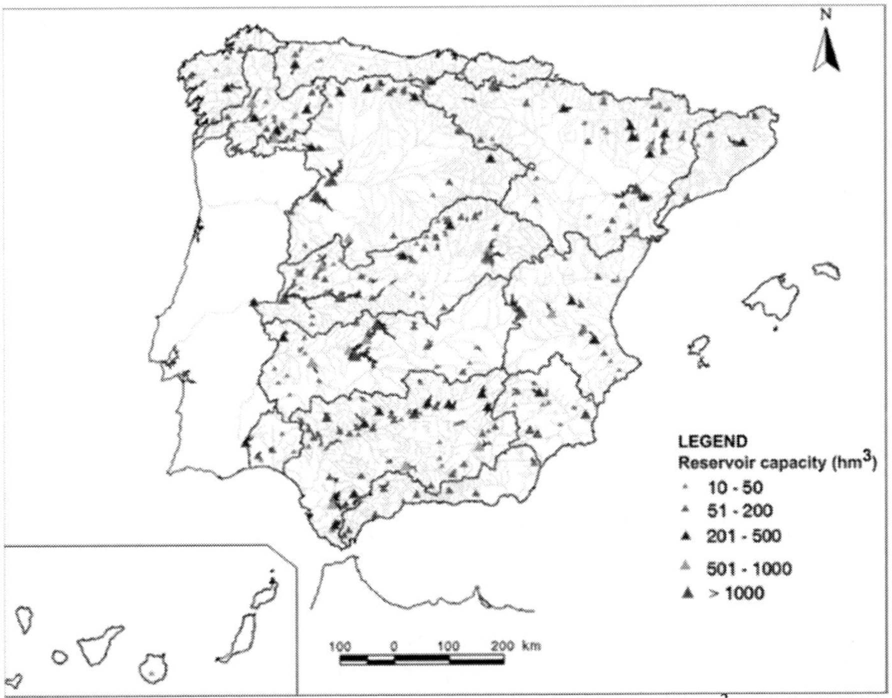

Figure 1. Dams in Spain and storage capacity in hm^3 (Note: 1 hm^3 = 1,000,000 m^3)
Source: MAGRAMA, 2015.

[4] The ICOLD Register of Dams includes dams above 15 m high or ranging from 5 to 15 m high with a storage capacity above 3,000,000 m^3.

Given the aforementioned facts, there exist a large number of dam owners in Spain, including public and private holders. Around one-third of the total number of Spanish large dams are owned and operated by the Spanish Ministry of Food, Agriculture and Environment (MAGRAMA), through the surrogated authorities given to the River Basin Authorities (RBAs). River Basin Authorities are public entities with autonomy of action and their own legal personality. With more than 1,400 dams (including also small dams and ponds), the functions and responsibilities of dam owners regarding dam safety management have evolved as a result of a changing dam safety regulation framework.

Overarching national and European legislation influencing, one way or another, dam safety management includes the Water Act of 1985 (and subsequent upgrades), the National Hydrological Plan Act of 2011 (and subsequent upgrades) and the Water Framework Directive (WFD) of 2000 and its transposition to Spanish legislation.

More related to dam safety issues are the more recent European Directives on Floods and Critical Infrastructure, as well as their transposition to the Spanish legislation. Directive 2007/60/CE acknowledges and explicitly requires that risk analysis must be used as a tool for flood risk reduction, through the development of flood risk management plans at river basin level and conducting flood risk analyses including dam failure.

More particularly, the history of dam safety regulation in Spain has been influenced by two dam failure events: the failure of Ribadelago Dam (Zamora, Spain), in 1959; and the failure of Tous Dam (Valencia, Spain), in 1982. Both failures were landmark events in the national dam safety regulation framework.

The current dam safety regulation context in Spain includes two standards: the "Instruction for the Design, Construction and Operation of Large Dams" (Instrucción para el Proyecto, Construcción y Explotación de Grandes Presas), published in 1967; and the "Technical Regulation for Dam and Reservoir Safety" (Reglamento Técnico sobre Seguridad de Presas y Embalses), published in 1996. The approval of the 1996 Technical Regulation did not revoke the 1967 Instruction. This fact has been one of the problems more intensely discussed by the Spanish dam safety community and both standards remain applicable.

The Technical Regulation was approved one year after the adoption of the Basic Directive of Civil Protection Planning against Flood Risk (1995). Up to this point, dam safety management had focused almost exclusively on the infrastructure (system response); however, the 1995 Civil Protection Directive incorporates the consideration of both flood risk components (probability and consequences). Therefore, the 1995 Civil Protection Directive represented a milestone in the dam safety regulatory framework, requiring the implementation of Dam Emergency Action Plans (DEAPs) for those that have previously been classified as Category A or B based on its potential risk (considering the possibility of adverse potential consequences in case of failure or uncontrolled dam release).

The last update of the Spanish Public Water Regulation (16 January 2008) added a new chapter which deals with dam safety, including the need for considering risk management as a key aspect for dam safety, whose main objective is to unify all dam safety criteria to be applied to all dams. Consequently, three Technical Standards were published in 2011, but not approved so far, that will replace both the 1967 Instruction and the 1996 Technical Regulation.

In addition to the aforementioned regulatory documents, a set of Technical Guides for Dam Safety has been published by the Spanish National Committee on Large Dams (SPANCOLD) of the International Commission on Large Dams (ICOLD). These guidelines include the state-of-the-art in several dam safety aspects and provide recommendations for accomplishing current standards. Published in 2012, the Technical Guide on Dam Safety, Operation of Dams and Reservoirs, entitled "Risk Analysis applied to Dam Safety Management" (SPANCOLD, 2012), is a recent milestone in the Spanish context and describes the general process for implementing risk analysis in order to inform decision-making and prioritization of risk reduction measures for dam safety management. This guideline can be considered a significant contribution towards integrated risk-informed dam safety management.

In view of the foregoing, there is a need for analyzing and managing risks as a result of a new context, influenced by societal, economic and regulatory aspects such as:

- A new dam safety regulatory context at the national level, influenced by European legislation on flood risk and critical infrastructure management;
- A societal demand of higher levels of safe ty and better justification of the use of private and public funds;
- The need for prio ritizing corrective act ions for risk reduction to optim ize resources;
- The need f or optimizing water re sources as w ell as the n eed to inc rease dam regulation capacity;
- The difficulty of building new dam s, mainly due to social and environmental reasons, that will predictably extend dam operational life; and
- Dam aging and the need for allocating resources for dam m aintenance and rehabilitation.

The Challenge

In this new context, dam owners and operators face the challenge of managing and reducing risks, taking into account multiple principles such as equity, efficiency, effectiveness, sustainability and transparency.

Therefore, existing and new knowledge, tools, methods and practices have been (and should be) developed aiming to support dam owners and operators in decision-making. On this point, science and governance should meet to solve this challenge.

The application of risk analysis techniques for supporting dam safety governance represents a good example of how dam safety actions can be prioritized based on quantitative risk results. The current state-of-the-art on risk analysis and existing knowledge has demonstrated through recent pilot cases that the challenge of managing and reducing risks has turned into an opportunity for better dam safety governance.

The Opportunity

The application of risk analysis techniques to dam safety management has involved a paradigm shift, but dam safety risk analysis is not new. Following the failure of Teton Dam in 1976, the USBR implemented a dam safety program in 1978, in accordance with national dam safety legislation.

This program was in line with their mission "to protect, to the extent practicable, people from risks posed by dams." It represented the first step in a journey towards risk-informed dam safety management. Risk-based approaches for dam safety analysis were presented by the Australian National Committee on Large Dams (ANCOLD) in 1994, through the Dam Safety Management Guidelines, which were later revised in 2003.

Risk-based approaches were also introduced by the United States Society on Dams (USSD) in 2003, in a White Paper on Dam Safety Risk Assessment (Bowles, 2003). In addition, the International Commission on Large Dams (ICOLD) published its Bulletin 130 in 2005, including concepts such as the As-Low-As-Reasonably-Practicable (ALARP) criterion and the Cost-Per-Statistical-Life-Saved indicator (CPSLS).

Risk-informed dam safety management requires outcomes from risk analysis and risk assessment for supporting decisions. Risk analysis allows the dam owner, as well as the public, to understand the system and to document all information of the system.

Among existing methodologies for dam safety risk analysis (Harrald et al., 2006), several examples of complete and quantitative risk analysis can be found. In the Spanish context, the framework for risk-informed dam safety management published by the Spanish National Committee on Large Dams (SPANCOLD, 2012) includes the analysis of all potential combinations of hazard events, system response and potential consequences that may result from dam failure or mission disruption through the application of risk modeling and the use of event tree analysis for risk calculation. Related work on risk assessment methods where event tree analysis has also been applied can be found in the literature (McGill et al., 2007; and Serrano-Lombillo et al., 2011).

The publication of the SPANCOLD Technical Guide emerged from the collaborative effort of several institutions and experts, incorporating results from research projects that had been developed in recent years. These projects resulted in, among other outcomes, tools for risk calculation, analysis and assessment, and prioritization of risk reduction measures (e.g., the software tool iPresas Calc, which allows risk calculation by means of event trees and influence diagrams).

Efforts from a wide sector of the Spanish dam safety community are now allocated to enhance and promote risk-informed dam safety management through research and development, practical applications, training and communication. The aim is to achieve improved dam safety governance, based on efficiency, transparency, and sustainability principles.

Several pilot cases have been carried out in recent years. In addition, other risk analyses have also been conducted in Europe, inspired by experiences in Spain.

PILOT CASES OF RISK-INFORMED DAM SAFETY GOVERNANCE IN EUROPE

Summary

A set of pilot cases is summarized in this section of the paper. These cases have included dam risk analyses at small and large scales (e.g., at the portfolio level). For all cases, a risk analysis was conducted for analyzing the current situation of the dam and different risk reduction measures. Risk analyses followed the process proposed by the SPANCOLD Technical Guide on Dam Safety (SPANCOLD, 2012),

Quantitative risk results, obtained from risk models developed using the software tools iPresas Calc and iPresas Manager[5], were used for evaluating the existing risk (by comparison with international tolerability recommendations).

Risk outcomes for different scenarios (e.g. before and after implementing each measure) and implementation and maintenance costs of analyzed risk reduction measures are used to estimate risk indicators. The iPresas Manager software tool estimates the values of different indicators for identifying different sequences of measures. Among them, the EWACSLS (Equity Weighted Adjusted Cost per Statistical Life Saved) indicator is used for obtaining the optimal sequence of measures (SPANCOLD, 2012). This indicator is a modified version of the ACSLS indicator, obtained by dividing the latter by an equity factor that considers the dam failure probability before and after implementing the analyzed measure, compared to the maximum individual risk of 10^{-4}. This equity factor is incorporated to give more importance to those measures that reduce the dam failure probability to values lower than 10^{-4}. The EWACSLS is therefore a mixed criterion of efficiency and equity.

Table 1 shows a summary of the main characteristics of the pilot cases conducted in recent years.

[5] For further information on these tools, the reader is referred to www.ipresas.com. A simplified version, iPresas HidSimp, is available for downloading.

Table 1. Summary of pilot cases.

Dam owner	Ownership	Dams	Dam location	Dam type	Dam purpose	Year of conducted risk analysis
Iberdrola Generación S.A.	Private	1	Spain	Buttress dam	Water supply and hydropower	2010
Catalan Water Agency	Public	1	Spain	Concrete gravity dam	Water supply and irrigation	2010
Duero River Authority	Public	27	Spain	Portfolio	Portfolio of dams	2010-2013
Elforsk	Private	1	Sweden	Embankment dam	Hydropower	2011-2012
Korporata Elektroe-nergjetike Shqiptare (KESH)	Public	1 5	Spain Albania	Concrete gravity dam Embankment dams	Hydropower and irrigation	2014-2015
Canal Isabel II Gestión	Private	1	Spain	Concrete gravity dam	Water supply	2014-2015
Government of Extremadura	Public	2	Spain	Concrete gravity dam Embankment dam	Irrigation	2014-2015

Examples

Three of the pilot cases shown in Table 1 are briefly described in this section.

Example 1: Concrete Gravity Dam in Spain. This dam, which is 31.5 m high as measured from the dam's foundation and includes a 30-m long uncontrolled spillway, is owned by the Government of Extremadura (a wide region located in western Spain).

The set of risk reduction measures, planned by the Government of Extremadura, included the following actions: (i) implementation of the dam emergency action plan; (ii) improved impermeability system; (iii) upgrading of monitoring systems; and (iv) lowering the spillway crest level.

Results from the risk analysis and evaluation show that the current dam failure probability is higher than the maximum allowable values suggested by tolerability recommendations. Based on the EWACSLS indicator (SPANCOLD, 2012), the first three risk reduction measures within the optimal sequence are: (1) lowering the spillway crest level by 1.65 m; (2) upgrading of monitoring systems; and (3) implementation of the dam emergency action plan. These three measures would be highly justified in terms of efficiency and equity. Based on conclusions of this analysis, the first two measures have already been implemented and the third alternative is pending approval by national authorities.

Results from the uncertainty analysis (considering the impact on the risk results of variations in the hydrology data, system response and available warning times on consequence estimations) did not show significant changes to the existing risk or to the recommended optimal sequence of actions. Figure 2 depicts results from the uncertainty analysis in an fN graph. The solid lines represent the sequence of fN pairs from the current situation to the implementation of the last risk reduction measure. Two cases are represented: the best estimate case and the 5% percentile case.

This analysis represented the first step of an ongoing process towards improved dam safety governance by the Government of Extremadura. First actions have been undertaken and future safety programmes will be based on this risk-informed approach.

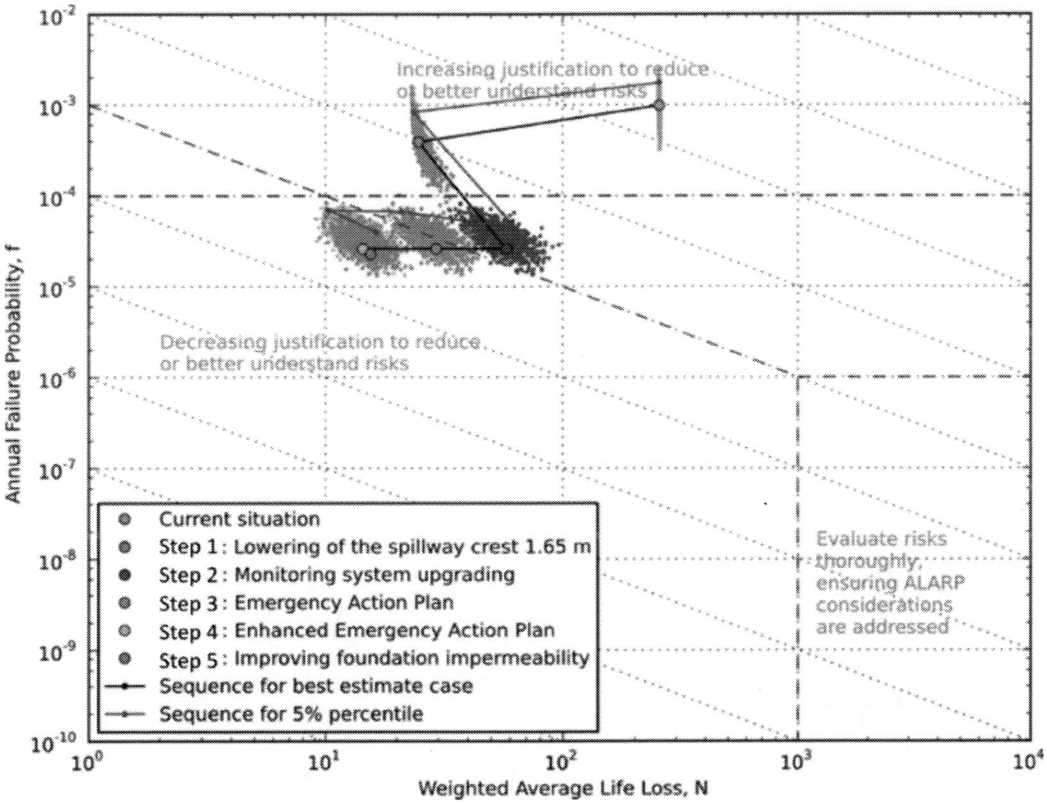

Figure 2. Results from the risk analysis of Example 1: current situation and recommended sequence of risk reduction measures. (Adapted from Setrakian et al., 2015).

Example 2: Portfolio of Dams in Spain. The work conducted by the Duero River Authority in recent years represents one of the most complete and documented international examples for analyzing a portfolio of dams (Ardiles et al., 2011). A risk analysis of 27 large dams (located in the northwest region of Spain) was performed. After analyzing risk for all of the dams within the portfolio, a set of more than 80 different risk reduction measures was considered to establish the optimal strategy for dam safety management. This set of measures includes specific measures for each dam, although some measures presented similar characteristics, such as the implementation of dam emergency action plans. However, the impact on portfolio risk reduction of each measure differs due to the broad spectrum of obtained risk outcomes (i.e. dam failure probabilities range from 10^{-3} to values lower than 10^{-10}). Based on the EWACSLS indicator, first actions were mainly related to those measures associated to dams with failure probabilities above 10^{-4}. Impact on societal risk at the portfolio level was analyzed, and the prioritization sequence showed that after implementing the first five actions, societal risk would shift from 0.3 lives/year to 0.06 lives/year (with a total cost of 3 M€) and economic risk would be reduced from 1.7 M€/year to 0.25 M€/year.

Figure 3 shows portfolio societal risk (in lives/year) in the vertical axis and total implementation cost (in M€) for each step within the prioritization sequence.

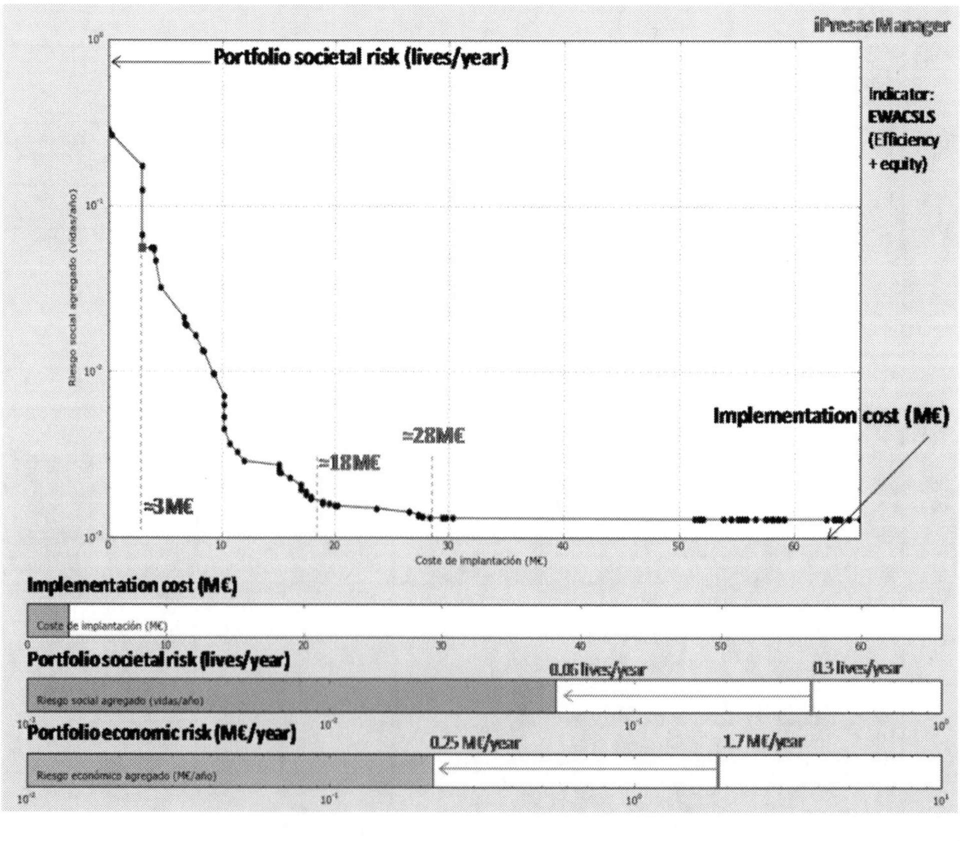

Figure 3. Results from risk analysis of Example 3: each point represents an action (step) taken to reduce the portfolio societal risk (lives/year) within the prioritization sequence (Castillo-Rodríguez et al., 2015).

Each dot of the graph represents societal risk and total implementation cost of risk reduction measures for the given step. The horizontal bars at the bottom of Fig. 3 show implementation cost, societal and economic risk for the situation after implementing the first 5 measures (red dot). It is noted that after implementing the first 30 measures within the sequence (with a total implementation cost of 28M€), a residual risk level would be achieved.

Example 3: System of Dams in Albania. Risk analyses conducted in Albania during the period 2014-2015 included the following dams on the Drin River (the longest river in Albania with a length of 285 km and a catchment area of 13,650 km^2): Fierze Dam (162 m high); Koman Dam (120 m high); and the 'Vau i Dejes' system of dams (three dams ranging from 21 to 60 m high). All dams, with a total reservoir capacity of 3.8 km^3, These are managed by KESH (Korporata Elektroenergjitike Shqiptare) and the main purpose is hydroelectric production which yields an annual average production rate of 3,700 GWh.

Risk analyses were conducted to inform decision-making on dam monitoring, surveillance and emergency management (Moralo García et al., 2015). In addition, specific training activities were developed for enhancing knowledge transfer on dam safety and risk management issues.

The quantitative risk model performed in iPresas Calc was used to evaluate the effect of different risk reduction measures. The obtained prioritization sequence based on the EWACSLS indicator is shown in Figure 4.

Step	Measure
1	Emergency Action Plan
2	High maintenance program of gates
3	New Spillway 5 in Fierze
4	New Spillway 5 in Koman
5	Restoration of Spillway 3 in Fierze
6	New limit water level in Koman
7	Rehabilitation of bottom outlet in Qyrsaq

Figure 4. Obtained sequence of risk mitigation measures (Escuder-Bueno et al., 2016).

Figure 5 shows the itinerary followed by each dam in the fN graph when the obtained sequence is followed.

Results showed that the key actions to reduce risk in the Drin River cannot only rely on building new structural measures. The development of a dam safety culture which would include the implementation of an Emergency Action Plan, revision of Rules on Discharges from Spillways, a better control and maintenance of the gates, better compilation and analysis of monitoring data and clearer flood routing rules will have a very significant impact in risk control.

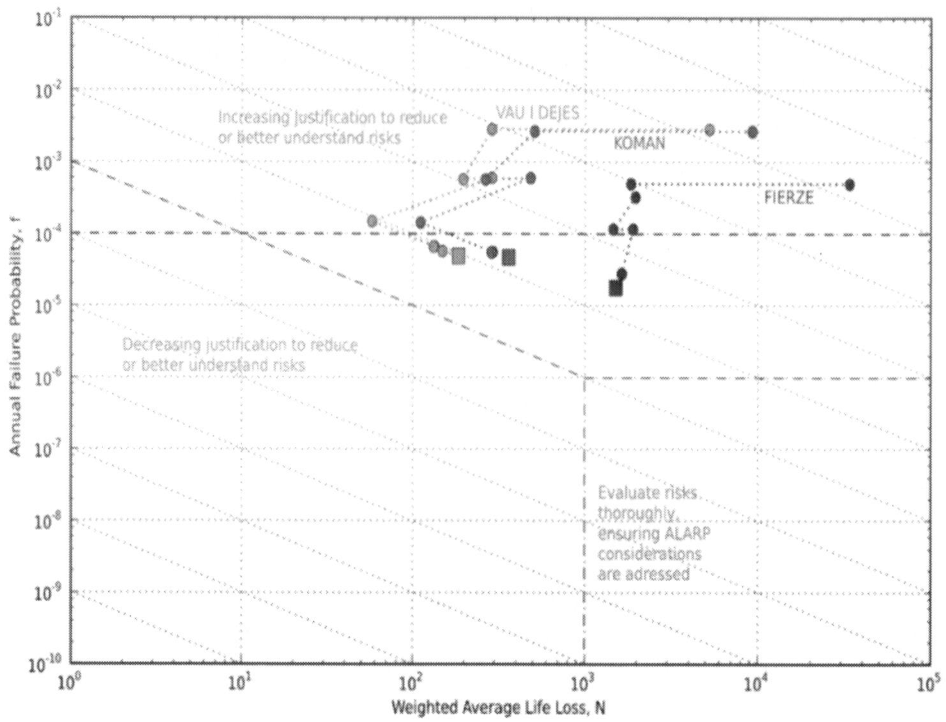

Figure 5. Variation of risk in a fN graph following the proposed implementation sequence (Escuder-Bueno et al., 2016).

Lessons Learned

Lessons learned from recent pilot cases in Europe have revealed the benefits of incorporating risk analysis techniques to dam safety management. These benefits have an impact on, but are not restricted to, the following areas:

- Knowledge of the system: improved understanding of dam systems, deeper knowledge of the dam, identification of knowledge gaps, uncertainty analysis, etc.;
- Risk communication: improved risk communication to the public and increased risk awareness. Decisions are justified, defensible and transparent;
- Emergency management: improved procedures and flood risk management plans; and,
- Operation and maintenance: improved information for defining activities for dam operation and maintenance, prioritization of investments including efficiency and equity principles and better knowledge of business risks, enhancing business sustainability in both the short and long term.

All of the aforementioned benefits improved dam safety governance from a risk-informed perspective.

THE VALUE OF INTEGRATING DIFFERENT SOURCES OF FLOOD HAZARD: FROM DAM SAFETY TO FLOOD RISK MANAGEMENT

The benefits of risk analysis and the added value of integrating information through risk analysis techniques go beyond the dam safety management discipline.

As proposed by Castillo-Rodriguez et al., (2014), risk analys is techniques allow information from m ultiple hazards to be integrated into flood risk analysis and management. Dam failure events and uncontroll ed releases represent one of the possible sources of flood hazard. Similarly to the overall process proposed for dam safety risk analysis by the SPANCOLD Technical Guide, and applied in recent pilot cases, a generic risk model architecture is proposed for flood risk analysis from multiple hazards.

This integrative approach was applied to a case study in Spain (including flooding from the failure of a concrete gravity dam located upstream of the analyzed urban area). Risk outcomes, represented on an FN graph, are shown in Figure 4.

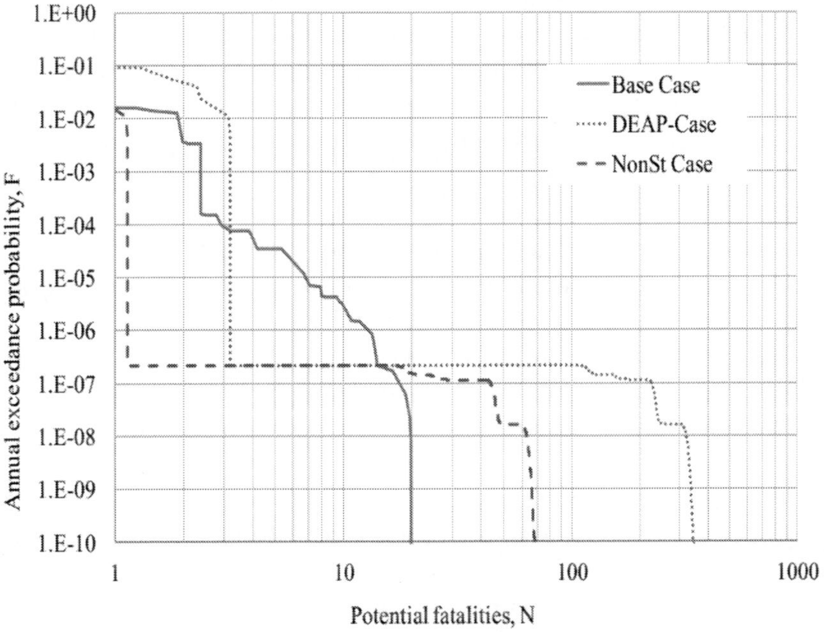

Figure 4: F-N graph for a case example of integrated flood risk analysis including dam failure. Three scenarios are shown: Base Case; DEAP; and non-structural measures. Source: (Castillo-Rodriguez et al., 2014)

Results show the cumulative annual exceedance probabilities (on the vertical axis) for each level of potential consequences by integrating three sources of hazard: pluvial flooding; river flooding; and flooding from dam failure.

Three scenarios are depicted in Figure 4: the Base Case (current situation); the scenario after implementing the DEAP; and a third case after implementing non-structural measures of improved emergency management and public risk awareness. A significant impact on flood risk reduction was found for the DEAP case and the scenario with non-structural measures (i.e., "high probability" flood events are reduced compared to the Base Case).

This is an example of a recent flood risk analysis that was performed by integrating dam response, river and pluvial flooding. Ongoing research efforts should focus on the analysis of more complex systems, where flood risk depends on system response of multiple infrastructures such as dam-levee systems. On this point, quantitative risk results from integrated dam-levee risk analysis may support decision-making for prioritizing risk reduction actions and for establishing combined operational procedures.

CONCLUSIONS

The application of risk analysis techniques provides a logical framework for better understanding of the dam system.

The added value of integrating all information regarding the dam system is inherent to any dam risk analysis, even in the case of purely qualitative analyses. Information from inspection and monitoring, operation and maintenance, operational rules, emergency action plans, etc. is reviewed, discussed, updated and, in a quantitative stage, incorporated into the risk model representing the system. Dam owners and operators benefit from this process, gaining a better knowledge of the system.

The dam safety management context in Europe is evolving as a result of new regulation and best practices. Recent experiences have shown how risk analysis has arisen as a tool to guide dam owners towards an integrated dam safety management and governance.

In future years, dam owners will face new challenges driven by a changing society, the environment and regulatory context. Risk analysis provides dam owners with information that is essential to establish strategies for dam safety management.

In this context of highly dynamic and complex scenarios, it is worth remarking that an increased capacity for effective risk governance will only be possible with the extensive and willing participation of all concerned actors, including international collaboration as a way of improving, validating and maximizing the benefits of enhanced risk governance.

Successful pilot cases can be found at the national and international scale for which the benefits of applying risk analysis techniques have been proven, supporting the prioritization of risk reduction measures and improving dam knowledge, operation and maintenance.

Three case examples have been summarized in this paper. In all cases, dam owners have adopted the first actions defined by obtained prioritization sequences of risk reduction measures. Results from risk analyses supported decision-making on short-term actions for dam safety management (e.g. for the first case, lowering the spillway crest level and upgrading of monitoring systems have already been implemented, reducing dam failure probability from 10^{-3} to values in line with international tolerability recommendations; for the second case, decommissioning of one of the dams in under consideration based on obtained risk outcomes).

Apart from illustrating the usefulness of quantitative risk results to prioritize investments in dam safety programs, recent pilot cases, summarized in this paper, represent the first steps towards improved, robust and efficient dam safety governance. The application of novel tools for dam risk analysis has proved to be of high interest for dam owners. Results have been used in prioritizing risk reduction actions and managing a portfolio of dams.

ACKNOWLEDGEMENTS

This research was conducted within the framework of the INICIA project, funded by the Spanish Ministry of Economy and Competitiveness (BIA2013-48157-C2-1-R, 'Methodology for investment assessment on water cycle infrastructures based on risk and energy efficiency indicators', 2014-2016).

REFERENCES

Ardiles, L., Sanz, D., Moreno, P., Jenaro, E., Fleitz, J. and Escuder, I.: RISK ASSESSMENT AND MANAGEMENT FOR 26 DAMS OPERATED BY THE DUERO RIVER AUTHORITY (SPAIN), in 6TH INTERNATIONAL CONFERENCE ON DAM ENGINEERING, edited by C.Pina, E.Portela, and J.Gomes, Lisbon., 2011.

Bowles, D. S.: SUMMARY OF USSD EMERGING ISSUES WHITE PAPER ON DAM SAFETY RISK ASSESSMENT: WHAT IS IT? WHO'S USING IT AND WHY? WHERE SHOULD WE BE GOING WITH IT?, 2003.

Castillo-Rodriguez, J. T., Escuder-Bueno, I., Altarejos-García, L. and Serrano-Lombillo, A.: The value of integrating information from multiple hazards for flood risk analysis and management, Nat. Hazards Earth Syst. Sci., 14, 379–400, doi:10.5194/nhess-14-379-2014, 2014.

Castillo-Rodríguez, J. T., Morales-Torres, A. and Escuder-Bueno, I.: A RISK-INFORMED JOURNEY TOWARDS IMPROVED DAM SAFETY GOVERNANCE IN SPAIN, in Second International Dam World Conference, edited by C. Pina, E. Portela, L. Caldeira, A. Batista, I. Dias, and R. Santos, pp. 61–70, ISBN 978–972–49–2274–4, LABORATÓRIO NACIONAL DE ENGENHARIA CIVIL (LNEC), Lisbon., 2015.

Escuder-Bueno, I., Jovani, A., Moralo-García, J. and Alonso Muñoz, J. M.: Risk Governance Implementation for a Cascade System on Drini River, Albania, in 84th ICOLD Annual Meeting, International Commission on Large Dams (ICOLD) [Accepted for publication]., 2016.

European Parliament: DIRECTIVE 2007/60/EC OF THE EUROPEAN PARLIAMENT

AND OF THE COUNCIL of 23 October 2007 on the assessment and management of flood risks, , (L 228), 27–34, 2007.

European Union: COUNCIL DIRECTIVE 2008/114/EC of 8 December 2008 on the identification and designation of European critical infrastructures and the assessment of the need to improve their protection, , 75–82, 2008.

Harrald, J. R., Renda-Tanali, I., Shaw, G. L., Rubin, C. B. and Yeletaysi, S.: Review of risk based prioritization/decision making methodologies for dams, Washington., 2006.

McGill, W. L., Ayyub, B. M. and Kaminskiy, M.: Risk Analysis for Critical Asset Protection, Risk Anal., 27(5), doi:10.1111/j.1539-6924.2007.00955.x, 2007.

Moralo García, J., Alonso Muñoz, J. M., Baztán Moreno, F. J., Escuder-Bueno, I., Morales-Torres, A. and Castillo-Rodríguez, J. T.: FORTALECIMIENTO INSTITUCIONAL DEL PROGRAMA DE SEGURIDAD DE PRESAS, APLICACIÓN DE LA METODOLOGÍA DE ANÁLISIS DE RIESGOS Y REDACCIÓN DE PLANES DE EMERGENCIA DE LAS PRESAS DE FIERZE, KOMANI Y VAU I DEJES DE KESH (ALBANIA), in X JORNADAS ESPAÑOLAS DE PRESAS, SPANCOLD (Comité Nacional Español de Grandes Presas), Seville (Spain)., 2015.

Serrano-Lombillo, A., Escuder-Bueno, I., Membrillera-Ortuño, M. G. De and Altarejos-García, L.: Methodology for the Calculation of Annualized Incremental Risks in Systems of Dams, Risk Anal., 31(6), 1000–1015, doi:10.1111/j.1539-6924.2010.01547.x, 2011.

Setrakian, M., Escuder-Bueno, I., Morales-Torres, A. and Simarro Rey, D.: Safety investments in Horcajo dam (Spain): A process informed by the application of SPANCOLD guidelines on risk analysis, in XXV ICOLD Congress, Stavanger, Norway, Q99, International Commission on Large Dams (ICOLD)., 2015.

SPANCOLD: TECHNICAL GUIDE ON RISK ANALYSIS APPLIED TO MANAGEMENT OF DAM SAFETY, edited by Spanish National Commitee on Large Dams, Spanish Professional Association of Civil Engineers, Madrid., 2012.

Spanish Professional Assoc. of Civ. Eng. (CICCP), 2006. Dams in Spain, ISBN-84-380-0318-4, 2006.

THE ROLE OF RISK ASSESSMENT IN THE BUREAU OF RECLAMATION'S DAM SAFETY PROGRAM

William R. Fiedler, PE[1]
William Engemoen, PE[2]

ABSTRACT

The Bureau of Reclamation (Reclamation) relies on information from risk analyses and risk assessments to guide all aspects of its dam safety program. Results of a potential failure modes (PFMs) analysis and the risk analysis estimates are used to focus the monitoring program that is established at a dam. Risk analysis estimates along with the supporting dam safety case are used to assess risk at each dam and make decisions on whether additional actions are required to better understand or reduce dam safety risk. Risk analyses are conducted as part of the ongoing Comprehensive Reviews (CR) conducted for all high and significant hazard dams on an 8-year cycle or as part of an Issue Evaluation (IE), where compelling dam safety issues are evaluated in depth. As part of the risk assessment process, dams are assigned a Dam Safety Priority Rating (DSPR). The DSPR for each dam along with the risk estimates and the dam safety case are used to prioritize dam safety actions across the entire portfolio of Reclamation dams. When performing corrective action studies, the risk reduction achieved by each alternative is a key parameter used to select the preferred alternative. Finally, during risk analyses and detailed evaluation of specific PFMs, limitations and deficiencies of current analysis and methodologies are often identified. This has led to focused research to improve current methods. This paper will detail how risk information is used throughout Reclamation's Dam Safety Program.

INTRODUCTION

Reclamation has been using risk-informed decision making for over 20 years. While early efforts in risk activities were limited to selected dams, current practice is to utilize risk information for evaluating all dams and in making dam safety decisions. Some level of risk analysis has been performed and reviewed multiple times for all significant and high hazard dams within Reclamation's inventory. This is done as part of the CR process for all significant and high hazard Reclamation dams on an 8-year cycle. The CR process relies on a multi-disciplinary team to review all key dam safety records and perform an inspection of the dam, evaluate recent performance of the dam (based on instrumentation data and regular visual inspections), review and update normal operation, hydrologic and seismic loadings as necessary, review and revise life loss consequence estimates and update the PFMs and risk estimates for the dam. While the risk analysis results provide a critical input into a decision on whether additional dam safety actions will be pursued, other considerations such as the uncertainties related to the risk estimates and the

---

[1] Lead Civil Engineer, Geotechnical Services Division, Bureau of Reclamation, PO Box 25007, Code 86-68300, Denver CO 80225-0007, wfiedler@usbr.gov
[2] Lead Civil Engineer, Geotechnical Services Division, Bureau of Reclamation, PO Box 25007, Code 86-68300, Denver CO 80225-0007, wengemoen@usbr.gov

confidence in the risk estimates are also key considerations in the dam safety decision process.

For dam safety issues that require more study and evaluation, an IE is performed. The IE will often involve some form of geologic/geotechnical investigations, material testing, updating of seismic and/or hydrologic hazard studies, higher level engineering analyses and studies and a team risk analysis. Decisions are influenced by the risk estimates, which will typically reflect less uncertainty than what exists at the CR level, but the uncertainty and confidence in the risk estimates are again key parameters in making dam safety decisions.

If the conclusion of an IE is that risks are well defined with reasonable confidence and the actions provide increasing justification to take action to reduce risk, corrective actions will typically be pursued. During this process, a suite of alternatives will usually be considered. Decisions need to be made on which alternatives will be carried to a higher level of study and ultimately which alternative is selected as the preferred alternative. One of the parameters considered in evaluating which alternatives to select is the risk reduction achieved with each alternative. Once a dam safety modification is completed, a team will review the completed work and conduct a risk reduction verification exercise where the revised risk for the completed modifications is estimated.

There are a number of ways that risk assessments are incorporated into dam safety decisions that are made in Reclamation's Dam Safety Program. The following are typical decisions that are informed by risk information:

1. Decisions on the monitoring program for all high and significant hazard dams.
2. Decisions for an individual dam on whether dam safety actions are warranted (can be done at either CR or IE level).
3. Decisions on which dam safety action to pursue (can be done at either CR or IE level).
4. Decisions on prioritization of dam safety actions (the initial dam safety studies conducted as part of an IE, followed by corrective action studies to identify potential dam safety actions for known dam safety deficiencies and dam safety modifications to reduce dam safety risks are all prioritized).
5. Decisions on which dam safety corrective action alternative to select.
6. Decisions on whether adequate risk reduction was achieved at the completion of modifications.

In addition to the above, the risk analysis/risk assessment process has led to recognition that better tools are needed to analyze dams and improve risk analysis methodology. Examples of some of these improvements will be described at the end of this paper.

MONITORING PROGRAMS FOR SIGNIFICANT AND HIGH HAZARD DAMS

The long term monitoring program for dams should be efficient and effective. If these objectives are not met, money will be spent unnecessarily and critical components of the dam's performance may be overlooked. One way to accomplish this objective is to focus the instrumentation and monitoring program on the PFMs that represent significant risk at a given dam. This allows those responsible for the monitoring to focus resources on collecting the key information, while not being distracted by tasks that produce little relevant information relative to PFMs. The following decision tree has been used in evaluating instrumentation and establishing the monitoring programs for Reclamation dams:

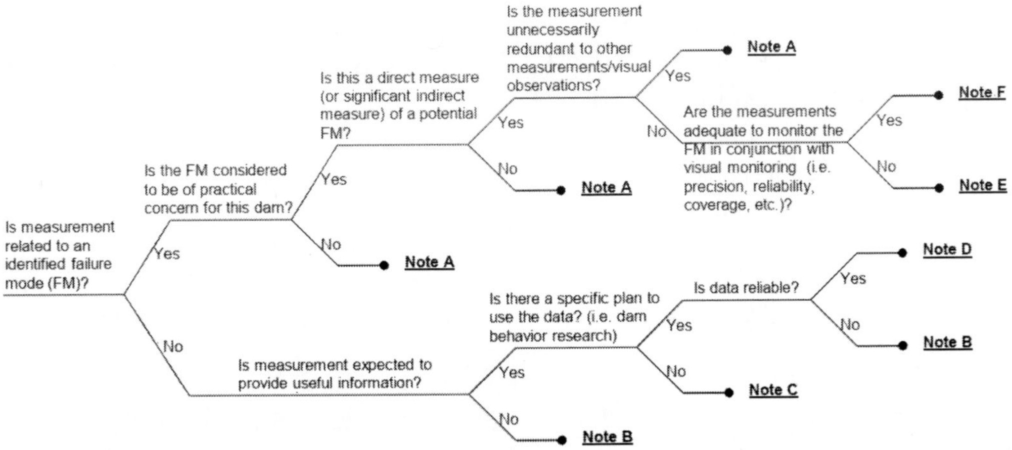

Note A – Read on the minimum frequency necessary to confirm proper functioning of the instrument
Note B – Decommission and/or abandon the instrument
Note C – Stop measuring, and either place the instrument on standby or decommission/abandon the instrument
Note D – Read at the frequency justified by the planned use of the data consistent with the resources dedicated to the planned research
Note E – Identify additional visual monitoring and/or measurement consistent with the need for analyzing or reducing risk
Note F – Read on a frequency consistent with use of the measurement in analyzing or reducing risk

Figure 1. Instrumentation Review Decision Tree

DAM SAFETY EVALUATION OF AN EXISTING DAM

Whether at the CR level or as part of an IE, at the conclusion of a dam safety study, Reclamation makes dam safety decisions for the dam under consideration. If the risks are low and the confidence in the risk estimates is high, the decision to not take any dam safety actions (other than the routine monitoring, inspections and future Comprehensive Reviews) will likely be made. If the risks indicate justification to take action to reduce risk, dam safety recommendations that detail specific dam safety actions will be made. One of the key inputs into the dam safety decisions at the end of a dam safety study is the risk estimates related to the key PFMs for a given dam. Risk estimates for individual PFMs, as well as, the total risk are plotted on Reclamation's f-N chart shown on Figure 2. Two measures of risk are portrayed. The Annualized Failure Probability is the product of the probability of the loading (normal operations, flood, or earthquake) and the probability of dam failure given the loading. The Annualized Life Loss is the product of the Annualized Failure Probability and the life loss estimated to result from dam failure. The f-N chart portrays threshold values for increasing justification to take action to reduce risk related to the two measures of risk, which are a threshold value of 1E-4 for the Annualized Failure Probability, and a threshold value of 1E-3 for the Annualized Life Loss.

Figure 3 shows an example f-N plot that indicates a dam with risks above threshold guidelines, indicating increasing justification to take actions to reduce or better define risks. Figure 4 shows an example f-N plot that indicates a dam with risks below threshold guidelines, indicating decreasing justification to take actions to reduce or better define risks. Uncertainty about the total risk estimate is depicted by the dashed lines through the total risk marker in Figures 3 and 4.

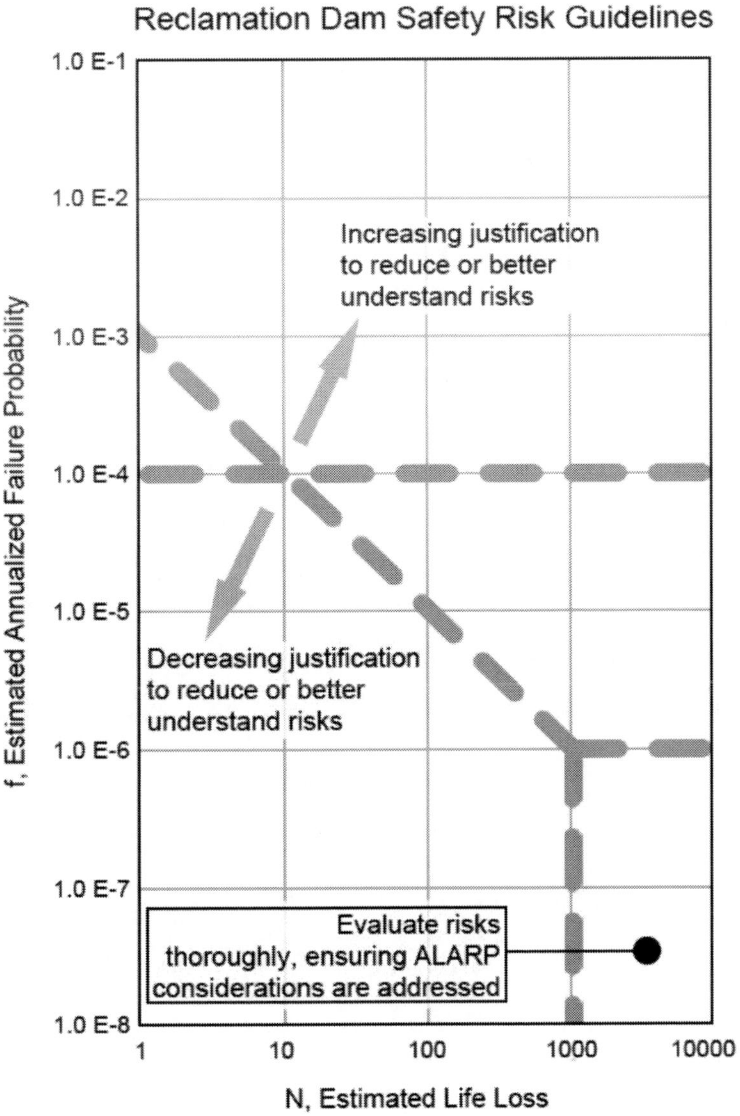

Figure 2. Reclamation's f-N Chart

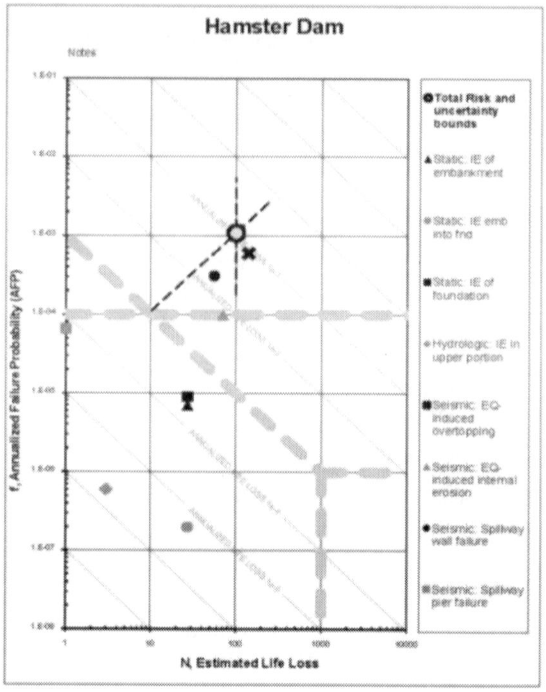

Figure 3. Risks for Dam with Increasing Justification to Take Action

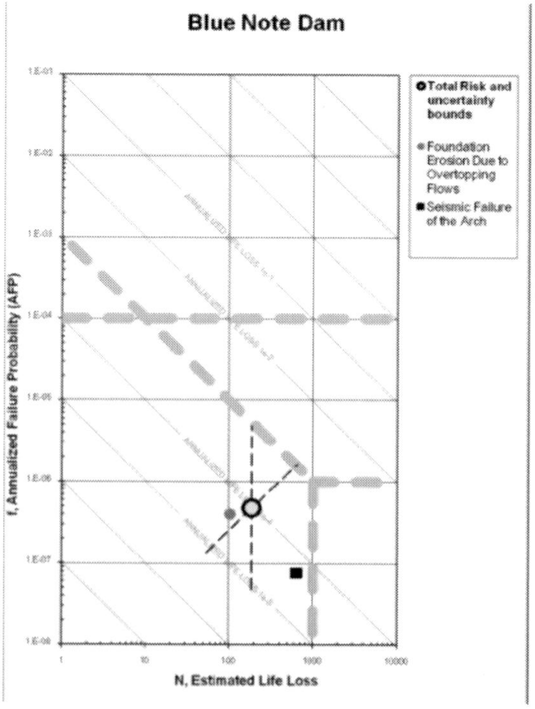

Figure 4. Risks for Dam with Decreasing Justification

DETERMINING APPROPRIATE DAM SAFETY ACTIONS

The ultimate objective of dam safety actions is to reduce risk to the downstream public. In situations where the confidence in the risk estimates is low, potential actions (if any) should be focused on increasing the confidence. For situations where there is high confidence in high risk estimates, actions should be focused on reducing those risks. Figure 5 illustrates the general process of how potential actions can lead to gaining confidence in the risk estimates.

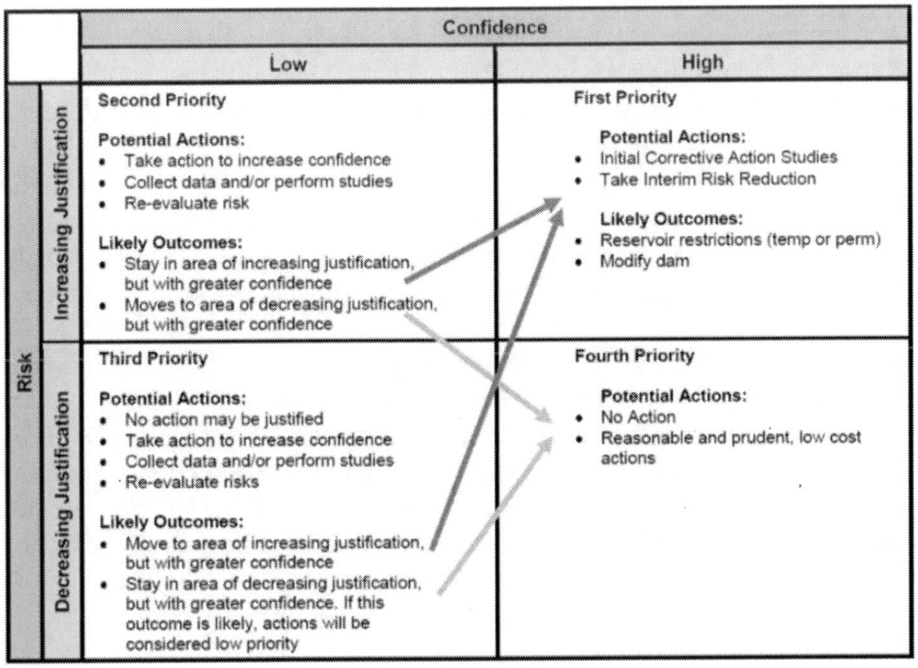

Figure 5. Risk, Confidence, and Actions

Once the risks have been estimated and the level of risk has been considered, technical teams recommend appropriate actions, given the portrayal of risks. There are six potential actions that can be supported with risk results:

1. The estimated risk is in the area of decreasing justification to reduce or better understand the risk, and confidence is high so that no further actions or studies are deemed necessary at the present time.
2. The estimated risk is in the area of decreasing justification to reduce or better understand the risk, but the confidence is low and there is the potential that additional information could increase the confidence and estimated risk to the extent that risk reduction actions may be justified.
3. The estimated risk is in the area of increasing justification to reduce or better understand the risk, but the confidence is low and there is the potential that additional information could increase confidence. Ideally, this would lead to better defined risk estimates that may increase, decrease or remain the same.

4. The estimated risk is in the area of decreasing justification to reduce or better understand the risk and confidence is high, but relatively low cost, reasonable and prudent actions are recommended nonetheless.
5. The estimated risk is in an area of increasing justification to reduce or better understand the risk and confidence is high so that no further studies are necessary before moving to a Corrective Action Study and possibly implementing interim risk reduction measures.
6. The estimated risk is in an area of very low probability and very high consequences such that to satisfy "As Low as Reasonably Practicable" (ALARP) principles thorough study of the risks and possible risk mitigation alternatives is appropriate.

Each of these potential actions requires that the case be established with respect to two main issues: First, the technical team must persuade decision-makers that risks are such that actions are justified (or not). Second, a case must be made for the confidence and uncertainty in the risk estimates and whether additional exploration, investigation, or analysis has the potential to change the perceived risk such that it falls in a different category. It is the rationale and structure of the case that helps convince decision makers whether the risk numbers generated and the actions recommended are reasonable.

The level of risk analysis, including data, technical studies, and related engineering analyses, needs to be factored into building the case. For example, some risk analyses are completed by individuals rather than teams, and may rely on screening-level data, or preliminary analyses. There may be a staged approach where recommendations are first made, for example, in improvements to monitoring, collection of additional data, or performance of additional analyses to reduce uncertainty or improve confidence in risk estimates.

An important part of building the case is the use of sensitivity analyses. By varying key parameters in the engineering analyses, teams can gain insights into which properties or assumptions are most important to the analysis. This in turn builds confidence in analyses results, or helps bracket the uncertainty by defining the range of expected performance based on a range of parameter assumptions. As with engineering analyses, sensitivity also plays a key role in risk analysis. Providing a range of probability estimates for key events, and determining how important the varying values are, helps define the confidence in the risk estimates. Sensitivity analyses can also provide insight in the best path to reducing uncertainty and improve confidence in the risk estimates, whether it should be to focus on the loadings, the structural response or the consequences. For evaluating the structural response, a number of approaches may be viable. These may include additional geologic investigations, additional material testing or application of refined analysis methods. A sensitivity analysis can provide direction on the most effective way to reduce uncertainty.

PRIORITIZATION OF DAM SAFETY ACTIONS

Reclamation has established a system for categorizing all the high and significant hazard dams in its inventory. This rating system provides a first cut when prioritizing dam safety actions, which may include further investigations/studies/analyses, corrective action studies and dam modifications. The rating system involves assigning a DSPR category to each dam. The rating system relies on risk estimates for the dam but other considerations apply as well. The following describes considerations for each of the five DSPR categories.

DSPR 1 – Immediate Priority

This category is reserved for cases where extremely high annualized life loss or annualized failure probability is combined with high confidence. The assignment of a DSPR 1 category would be appropriate for facilities where an active failure mode is in progress or when the likelihood of failure is judged to be extremely high. Immediate interim risk reduction measures followed by long term risk reduction measures would be appropriate.

DSPR 2 – Urgent Priority

This category is reserved for cases where the annualized life loss or annualized failure probability is judged to be very high with high confidence or is suspected of being very high to extremely high, but with low to moderate confidence. In the former case, implementing interim and long term risk reduction measures would be appropriate. In the latter case, timely confirmation of the risk would be appropriate, and implementing interim risk reduction measures may be appropriate. The primary difference between a DSPR 1 and a DSPR 2 category is that DSPR 1 will typically indicate an "emergency" situation that calls for immediate action due to the possibility of an impending failure. DSPR 2 facilities have very high risks or likelihoods of failure, but are not in "imminent" danger.

DSPR 3 – Moderate to High Priority

This category is reserved for annualized life loss risks or failure probabilities estimated to be moderate to high (near, and generally above, the guideline values), with moderate to high confidence. For cases with high confidence, long term risk reduction action would be appropriate. For cases with moderate confidence, confirmation of the risks would be appropriate as soon as they can be scheduled within the other priorities. Interim actions that are reasonable and prudent would be appropriate. DSPR 3 categorizations indicate that the facility has potential dam safety deficiencies with significant risks or probabilities of failure to justify actions to better define or reduce the risk.

DSPR 4 – Low to Moderate Priority

This category includes those cases where the annualized life loss risks and failure probabilities are estimated to be low (typically, but not always, below the guideline values), but with low confidence such that collection of additional information has the realistic potential to move the estimates into the area of increasing justification to reduce risk. In addition, this category also includes facilities where estimated risks and failure probabilities are moderate to high, but with low confidence such that additional information may lower the estimated risks. DSPR 4 facilities may have "potential" concerns, but the issues typically will not reflect a pressing need for action. Additional studies to better define the risks should be scheduled as time and budget permit. It is anticipated that in most cases these can wait until after the next comprehensive review of the dam.

DSPR 5 – Low Priority

This category includes those dams for which the annualized life loss and failure probability are estimated to be low to very low with high confidence such that the estimates are unlikely to change with additional investigation or study. Normal dam safety risk management activities (including monitoring), operations, and maintenance would continue.

DSPR System Usage

Since Reclamation has finite financial resources available to address dam safety issues, it is critical to not only identify future actions but also to identify the priority or the time frame associated with these actions. The priority for initiating actions to address risks depends in part on available resources and on the risks throughout Reclamation's dam inventory. The intent is to make the greatest reduction in risk throughout the inventory of Reclamation dams within the resource limitations of the program, while at the same time assuring that no dam presents an unreasonably high risk in the short term.

The DSPR system (and associated subcategories) forms the initial basis for prioritization after which priorities may be adjusted for other reasons. Ultimately, the case must be made as to which DSPR category represents each dam as part of the risk analysis and risk assessment activities. Within each DSPR category and subcategory, everything else being equal, the actual numerical values may be used to set priorities. However, risk is not the sole piece of information used to set priorities, as other information and unique opportunities can affect the prioritization queue. Other factors may include (but are not limited to) the confidence in the risk estimates, the number of PFMs driving the risk, the type of loading condition(s) driving the risk, and the costs of additional actions to reduce or better define the risks.

SELECTION OF PREFERRED DAM SAFETY CORRECTIVE ACTION

For a dam where corrective actions have been justified, multiple structural and non-structural alternatives are typically generated during a corrective actions study. Factors that influence the selection of a preferred alternative are many, and include such considerations as the cost of the modification, the lost project benefits and environmental consequences of a corrective action (both during construction and after completion), construction risks, and the amount of risk reduction achieved. Risk reduction involves estimating the risk given the corrective actions, and comparing the reduced risk to the baseline (existing condition) risk. Cost and degree of risk reduction are generally the most important factors in selecting a preferred alternative. There are various metrics that can be used to compare costs and degree of risk reduction, such that various alternatives can be compared relatively. At Reclamation, a general rule of thumb is that a corrective action should reduce risks to about an order of magnitude below Public Protection Guidelines (Reclamation, 2011) threshold values. This often means that the technically preferred alternative is one that reduces risk (both Annualized Failure Probability and Annualized Life Loss) to roughly an order or more below guidelines and has a lower cost than other alternatives that also reduce risk to those levels. It is, however, important to recognize that other key factors, such as the confidence in both the risk estimates and the ability of the corrective actions to effectively and efficiently provide a safe facility, are very important in the selection of the preferred action.

RISK REDUCTION VERIFICATION

After corrective actions have been selected and designed, risk analysis continues to play a role on a Reclamation dam safety project. Typically, additional risk analyses are performed at the end of construction and at the end of reservoir re-filling.

Risk reduction is evaluated at the end of construction for a couple of reasons. For one, designs may have changed during construction due to changed site conditions or perhaps due to value engineering options proposed by the contractor. In addition, construction activities may have provided new data and a resulting change in interpretation of key site characteristics. This may be particularly applicable to a corrective action that exposes portions of the dam or foundation and thus provides direct observation of conditions that would otherwise be postulated based on more limited exploration data. This risk reduction effort after construction should consider all new information (if any) learned during construction to provide an updated estimate of the risks at the facility.

Another important milestone for verification of risk reduction is after the completion of reservoir re-filling. The re-filling of the reservoir is a test of the modified dam's performance, and provides an opportunity to observe any unexpected behavior. This is particularly important for corrective actions designed to mitigate internal erosion, as internal erosion is a particular threat at normal high pool levels and during first filling.

IDENTIFICATION OF NEEDED IMPROVEMENTS IN ANALYSIS METHODS AND RISK ANALYSIS METHODOLOGIES

The risk analysis process leads risk analysis teams to think critically about: assumptions made in engineering analyses; the limitations of numerical analyses; uncertainties in loadings, structural response and consequence estimates; and whether or not the detailed mechanisms that will lead to dam failure are well understood. As a result of risk analyses, the need for better approaches and methods is often identified. Reclamation has responded to a number of these issues, including those listed below.

1. One of the critical events in the evaluation of internal erosion PFMs is the initiation of erosion. Previously, judgement of the team was the basis for estimating probabilities for this event. Reclamation reviewed the details from documented internal erosion incidents within our inventory of dams and developed base frequency rates for the initiation of erosion. This information has become very valuable to risk analysis teams as a starting point for estimating the probability of this key event.

2. From 1999 to 2014, Reclamation relied on DSO-99-06 (Reclamation, 1999) to guide the life loss estimates as a result of dam failure. One of the limitations of this methodology is that high severity was not well defined and the use of this category for selecting fatality rates was discouraged for embankment dams. In 2014, Reclamation updated its methodology with the new Reclamation Consequence Estimating Methodology (RCEM; Reclamation, 2014). The new approach does not limit the selection of fatality rates to prescribed categories, but relies on a graphical approach where flood severity is considered on a continuum. This has resulted in higher fatality rates being applied to some embankment dams, where warning time is limited.

3. Due to the identification of increased potential for large seismic loadings at many Reclamation dams, risks for many seismic PFMs have increased. A number of dams have gated spillway crest structures that are severely overstressed during large earthquakes. Previous finite element analyses with linear material models would often indicate very large overstresses of the reinforced concrete. Risk analysis teams often had a sense that the results of the analysis were overstated but the teams did not have a good basis for adjusting their risk estimates. Over the last several years, non-linear reinforcement concrete material models have been incorporated into the seismic analyses of spillway crest structures. The non-linear models account for cracking of concrete and the changed response of the structure as it is damaged and typically indicate greater capacities than those predicted by linear models. These refined models provide a better basis for estimating the risk of seismic failure.

4. Over the last several years, there have been a few dams where reservoir landslides have been connected to a potential failure mode for the dam. If a large volume landslide were to fail rapidly, there is the potential for these landslides to create a large wave that could overtop the dam and lead to downstream consequences (either with or without the breach of the dam). Risk analysis teams have struggled with identifying the potential for landslide movement under earthquake loading and in predicting the volume and rate of the landslide failure. A research program is underway to develop better tools for monitoring and analyzing reservoir landslides.

CONCLUSIONS

The use of risk analysis and risk assessment in Reclamation's Dam Safety Program has expanded considerably over the last twenty years. Baseline risk estimates are now available for all high and significant hazard dams in our inventory, and these estimates are reviewed and updated regularly. Risk information is considered in all dam safety decisions that are made. Risk assessments are used to help make dam safety decisions on individual dams as well as to manage the entire portfolio of dams.

REFERENCES

Bureau of Reclamation, *A Procedure for Estimating Loss of Life Caused by Dam Failure*," Dam Safety Office Report DSO-99-06, Denver, Colorado, September 1999.

Bureau of Reclamation, *Interim Dam Safety Public Protection Guidelines*, Dam Safety Office, August 2011.

Bureau of Reclamation, *RCEM – Reclamation Consequence Estimating Methodology – Interim Guidelines for Estimating Life Loss for Dam Safety Risk Analysis*, Denver, Colorado, February 2014.

SEMI-QUANTITATIVE PORTFOLIO RISK ASSESSMENT FOR PRIORITIZING DAM SAFETY ACTIVITIES

John Yen[1]
Matthew Muto[2]
Nicolas Von Gersdorff[3]
Gregg A. Scott[4]

ABSTRACT

Southern California Edison Company owns and operates more than 20 high hazard dams for hydropower production. The Potential Failure Modes Analyses (PFMA) conducted under Federal Energy Regulatory Commission guidelines (FERC) were extremely useful in better understanding dam safety issues associated with the facilities, but stopped short of providing an understanding of the risk profile. In order to better understand the corporate risks associated with the portfolio of dams, a semi-quantitative risk assessment process was initiated. This process included setting up a risk matrix portraying both likelihood of failure and consequences of failure. Categories of failure likelihood were anchored to historical dam failure rates, and probabilistic hazard analyses were used to define failure likelihood for large floods or earthquakes. Similarly, consequence categories were defined for various levels of consequences, with life safety being paramount. Each identified potential failure mode (PFM) for each dam was then classified according to the appropriate likelihood and consequence categories and plotted on the risk matrix. A confidence rating was also assigned to each PFM rating. With this information, the highest risks and largest uncertainties could easily be identified, and dam safety activities could be appropriately prioritized. The risk matrix is treated as a living document, and as new information becomes available, the risks are updated and activities reprioritized.

INTRODUCTION

Southern California Edison Company (Edison) has invested considerable resources in performing PFMA for all of its dams, in accordance with FERC requirements. This process went a long way toward better understanding of the dams and their potential vulnerabilities, but stopped short of providing a good portrayal of the portfolio risk profile under which the Company is operating with regard to dam safety.

Risk, by definition, includes both likelihood of failure and consequences of failure. As the consequences become larger, society in general becomes more averse to them, and the likelihood of the consequences must become smaller. It also makes sense from a Company liability perspective that events with high consequences should have a lower likelihood of occurring.

---

[1] Southern California Edison Company, Rosemead, CA, John.Yen@sce.com
[2] Southern California Edison Company, Rosemead, CA, Matthew.Muto@sce.com
[3] Southern California Edison Company, Rosemead, CA, Nicolas.Von@sce.com
[4] Scott Consulting, LLC, Lakewood, CO, Scott.Consulting@comcast.net

The portfolio risk assessment for Edison's high hazard dams incorporated and built upon the body of work compiled during the PFMA's. A semi-quantitative approach was taken, whereby likelihood and consequences categories were assigned to the identified PFMs, with the categories roughly anchored to historical failure rates and set up to roughly correlate with dam safety risk guidelines established by other organizations throughout the world. Probabilistic loadings were also considered. A risk matrix was used to portray increasing failure likelihood levels on the vertical axis and increasing consequence levels on the horizontal axis. Those PFMs plotting toward the upper right hand corner of the matrix have an estimated high likelihood of developing and high consequences, and thus high risk. Similarly, those plotting toward the lower left hand corner of the matrix represent low risk. By classifying all PFMs, and plotting them on the risk matrix the full risk profile of the inventory was portrayed. The confidence in the assigned categories was also noted, which helped in determining and prioritizing follow-on dam safety activities.

BACKGROUND AND PURPOSE

During the PFMA process, all of the available background material on each of the dams was collected and thoroughly reviewed by a core team of dam safety specialists and operations personnel. Based on this review, potential vulnerabilities and associated PFMs were identified and thoroughly described. Factors making each PFM more or less likely to develop were captured and key factors were noted. Potential consequences associated with each failure mode were also evaluated and described. The process resulted in classifying each identified PFM into one of four categories, considering both likelihood and consequences of failure simultaneously. However, this categorization scheme is too coarse to provide a good picture of the risk profile associated with Edison's inventory of dams. Therefore, the portfolio risk assessment program has been undertaken to more fully portray the risks associated with Edison's dams while taking advantage of and building upon the considerable investment already made by Edison in completing the PFMA program.

The purpose of the portfolio risk assessment is to: (1) obtain an initial risk profile for the inventory of dams, (2) facilitate an evaluation of risk reduction needs and opportunities, (3) provide a basis for prioritizing dam safety activities, (4) establish a "launch point" for risk management activities, and (5) provide input to and justifications for the budgets and General Rate Case (GRC). This will enable the Company to better understand the risks associated with its dams and appurtenant structures, thus enhancing the management of the dams for effective power generation, protection of public safety, and meeting environmental and regulatory requirements. A priority list established by the risk assessment program forms a key input for future engineering studies, safety evaluations, operations and maintenance (O&M) improvements, Emergency Action Plan (EAP) improvements, and retrofit projects.

SEMI-QUANTITATIVE RISK ASSESSMENT METHODOLOGY

Risk Guidelines

In establishing the procedure for categorizing PFMs according to risk, it is instructive to examine the risk guidelines for dams established by other organizations. Five organizations have established such guidelines: (1) the Australian National Committee on Large Dams (ANCOLD, 2003), (2) the Australian State of New South Wales Dam Safety Committee (NSW DSC, 2006), (3) the Canadian Dam Association (CDA, 2007), (4) the United States Army Corps of Engineers (USACE, 2014) and (5) the United States Bureau of Reclamation (USBR, 2011). These guidelines are all similar. The USBR risk guidelines are summarized in figure 1, also known as an "fN" diagram. The likelihood of failure is portrayed on the vertical axis, and the incremental consequences of failure are portrayed on the horizontal axis.

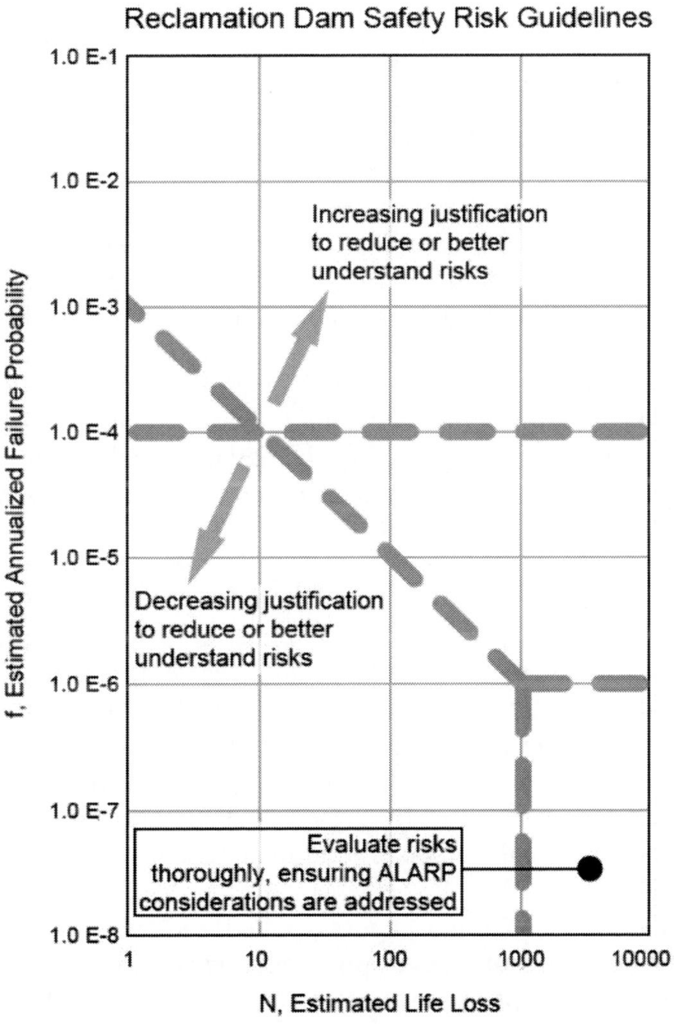

Figure 1. Summary of USBR Risk Guidelines

In general, society expects that as the consequences (e.g. number of fatalities) increase, the likelihood of those consequences should decrease. From a Company liability perspective, it also makes sense that the likelihood of severe consequences should be small in comparison to the likelihood of minor consequences. Thus, the risk guidelines require decreasing likelihood of failure as the consequences increase, represented by a diagonal line with negative slope on the fN diagram in figure 1.

In addition to societal risk guidelines, all four organizations suggest that a level of safety is needed even if the consequences are not large. For the USBR guidelines, this is represented by an Annualized Failure Probability less than 1 in 10,000, as shown by the dashed horizontal line in figure 1. From a Company liability perspective, failure of a dam could result in negative publicity and loss of public confidence, regardless of the consequences. This guideline helps ensure that the chance of any dam failure is low and that the risk to any individual from dam failure is a small part of the total risk they are exposed to.

Failure Likelihood

For the purposes of the portfolio risk assessment, failure was defined as an uncontrolled rapid release of reservoir water. The likelihood of failure is a function of both the likelihood of the loading condition that could lead to the failure, and the likelihood of failure given the loading condition. Potential failure modes have been identified for loading conditions associated with normal operations, floods, and earthquakes. For normal operating conditions, the loading is frequent. However, for flood or earthquake PFMs, the likelihood of the loading could be small. Therefore, the likelihood of the loading needs to be considered in the risk evaluation.

Seismic Hazard. Edison has performed seismic hazard analyses for all of its sites. This resulted in seismic hazard curves for each dam, including equal hazard response spectra (at 5% damping) representing earthquakes with annual exceedance probabilities from 1/500 to 1/10,000. The equal hazard response spectra can be used to estimate an approximate probability for loadings that have been analyzed or whose response can be approximated, taking into account the natural period of the dam structure where appropriate. It should be noted that the zero period spectral acceleration from the response spectrum (approximated by the value where the curves intersect the left-hand vertical axis) represents the peak horizontal ground acceleration for a given annual exceedance probability. These curves provide a sense of the likelihood of experiencing different levels of seismic loading at each of the dam sites. Although there was considerable discussion regarding the Maximum Credible Earthquake (MCE), mostly because analyses had been performed for these hypothetical events, there was no expectation that these represented upper bound loadings.

Hydrologic Hazard. Similarly, a set of approximate flood frequency curves was developed. It should be noted that developing these hazard curves requires experience in dealing with the probability of large remote flood events. Regional regression analyses are seldom robust enough to serve as adequate hydrologic hazard assessments. The use of Paleo-flood information or stochastic event flood modeling is more appropriate. For the portfolio risk assessment, the Probable Maximum Flood (PMF) studies were regarded as a notional representation of conditions approaching the upper limit of possible inflows.

Anchoring to Historical Failure Rates. It is important to calibrate (to the extent possible) the likelihood and consequence categories and descriptors to be meaningful within the context of risk guidelines, traditional analyses, failure mode assessment, and historical dam performance. Examination of historical dam failure rates indicate that dams fail at a rate of roughly 1 in 10,000 per dam year of operation, depending on the failure mode and age of the structure (Von Thun (1985), Hatem (1985), MK Engineers (1988), Foster et al (1998), Douglas et al (1998)). This forms the basis for evaluating failure likelihood for a given PFM under normal operating conditions. For example, if the key factors (those factors that carried the most weight) affecting the PFM are weighted toward adverse (more likely), the annual failure probability is probably greater than 10^{-4}. If weighted toward favorable (less likely), then the annual failure probability is probably less than 10^{-4}. Similarly, the estimated annual exceedance probability of a flood or earthquake that is likely to cause failure indicates the approximate likelihood of seismic or hydrologic failure.

Given this background information, the following likelihood categories and descriptors were used for approximate evaluations relative to existing risk guidelines.

Failure Likelihood Categories.

- **Very High** – There is direct evidence or substantial indirect evidence to suggest it has occurred and/or is likely to occur. Or, a flood or an earthquake with an annual exceedance probability more frequent than 1/1,000 would likely cause the PFM.

- **High** – The fundamental condition or defect is known to exist, indirect evidence suggests it is plausible, and key evidence is weighted more heavily toward likely than unlikely. Or, a flood or an earthquake with an annual exceedance probability between 1/1,000 and 1/10,000 would likely cause the PFM.

- **Moderate** – The fundamental condition or defect is known to exist, indirect evidence suggests it is plausible, and key evidence is weighted more heavily toward unlikely than likely. Or, a flood or an earthquake with an annual exceedance probability between 1/10,000 and 1/100,000 would likely cause the PFM.

- **Low** – The possibility cannot be ruled out, but there is no compelling evidence to suggest it has occurred or that a condition or flaw exists that could lead to its development. Or, a flood or an earthquake with an annual exceedance probability more remote than 1/100,000 would likely cause the PFM.

- **Remote** – Several events must occur concurrently or in series to trigger failure. Most, if not all of the events are very unlikely; potential failure is negligible or non-credible.

In a general sense, each category represents an "order of magnitude" range in failure probability. The dividing line between those PFM with approximately greater than a 1/10,000 chance and those with less than a 1/10,000 chance of occurrence (on an annual basis) is between the descriptors for "Moderate" and "High".

Failure Consequences

Dam failure consequences can take many forms, including loss of project benefits, destruction of downstream property, loss of life, environmental damage, and socio-economic impacts. For the purposes of the portfolio risk assessment, the decision was made to focus on the potential for loss of life, with the idea that the broader socio-economic, environmental, and property damages would be generally commensurate.

An evaluation of dam failure case histories indicates that the number of fatalities is primarily dependent on: (1) the population at risk (PAR) within the dam break inundation boundary, (2) the severity of the flooding, and (3) the amount of warning time the PAR has to evacuate the area (Graham, 1999). Other critical considerations include the degree to which the PAR understands the severity of the damaging flows which are about to arrive, and the availability and clarity of possible evacuation routes. The PAR can be broadly categorized by transient activity and the size of the towns and development within the inundation zone. The severity of flooding is a function of the potential destruction to structures, infrastructure and large natural objects (such as trees) within the flood plain. The warning time is a function of when the warnings are issued and the time it takes for the flood wave to reach the PAR. For example, with flood monitoring and forecasting, it may be possible to warn and evacuate the downstream population prior to overtopping failure. On the other hand, a sudden failure under normal operating conditions may catch the downstream population by surprise.

Considering all of these aspects of consequence evaluation, the following broad consequence categories were defined.

Consequence Categories.

- **Level 0** – No significant impacts to the downstream population other than temporary minor flooding of adjacent roads or land. Or downstream discharge results from an intended release at a designated emergency release location, to help ensure discharge occurs from this designated location rather than causing failure of larger structures.

- **Level 1** – Downstream discharge results in limited property and/or environmental damage. Although life-threatening releases occur, direct loss of life is unlikely due to severity or location of the flooding, or effective detection and evacuation.

- **Level 2** – Downstream discharge results in moderate property and/or environmental damage. Some direct loss of life is likely (1-10), related primarily to difficulties in warning and evacuating recreationists/ travelers and small population centers.

- **Level 3** – Downstream discharge results in significant property and/or environmental damage. Large direct loss of life is likely (10-100), related primarily to difficulties in warning and evacuating recreationists/ travelers and smaller population centers, or difficulties evacuating large population centers with significant warning time.

- **Level 4** – Downstream discharge results in extensive property and/or environmental damage. Extreme direct loss of life can be expected (>100) due to limited warning for large population centers and/or limited evacuation routes.

In a general sense, each category represents an order of magnitude range in consequences, with loss of life likely in Categories 2 through 4. Inundation studies and maps formed the basis for the consequence categorization.

<u>Risk Matrix</u>

A "risk matrix" was established to portray the risk associated with the identified dam safety PFMs, with likelihood of failure on the vertical axis (using cell divisions corresponding to the likelihood categories previously defined) and consequences on the horizontal axis (using cell divisions corresponding to the consequences categories previously described) similar to the "fN" diagram (USBR, 2010). Figure 2 illustrates the risk matrix.

A likelihood category and consequence category was assigned to each identified PFM, and then they were plotted in the appropriate cell of the matrix. The decision was made not to include Remote likelihood or Level 0 consequences in the matrix since there were many PFMs brought up during the PFMA sessions and summarily ruled out either because the chance of failure or consequences were thought to be negligible. It was anticipated that many (but not all) of these would fall into the Remote Likelihood and/or Level 0 Consequences cells, and that the matrix would become cluttered with these PFMs. Eliminating them from the matrix allowed the team to focus on the more important PFMs. However, all PFMs were discussed in the workshop and when ruled out, the reasons for their exclusion were documented.

For the purposes of this risk assessment report, the risk matrix was used to delineate approximate risk guidelines. Risk increases going diagonally from the yellow cells (figure 2) toward the upper right hand corner of the matrix; hence there is increasing justification to take action to address the risks, and the urgency of addressing PFMs increases. Similarly, risks decrease moving diagonally from the yellow cells toward the lower left hand corner of the matrix, and the need and urgency to address them also decreases. However, risks plotting in the green-shaded cells should be kept under review and properly managed. This requires continued monitoring and evaluation. The yellow cells represent the approximate boundary of established guidelines, and consideration should be given to addressing PFMs that plot in the yellow or red cells.

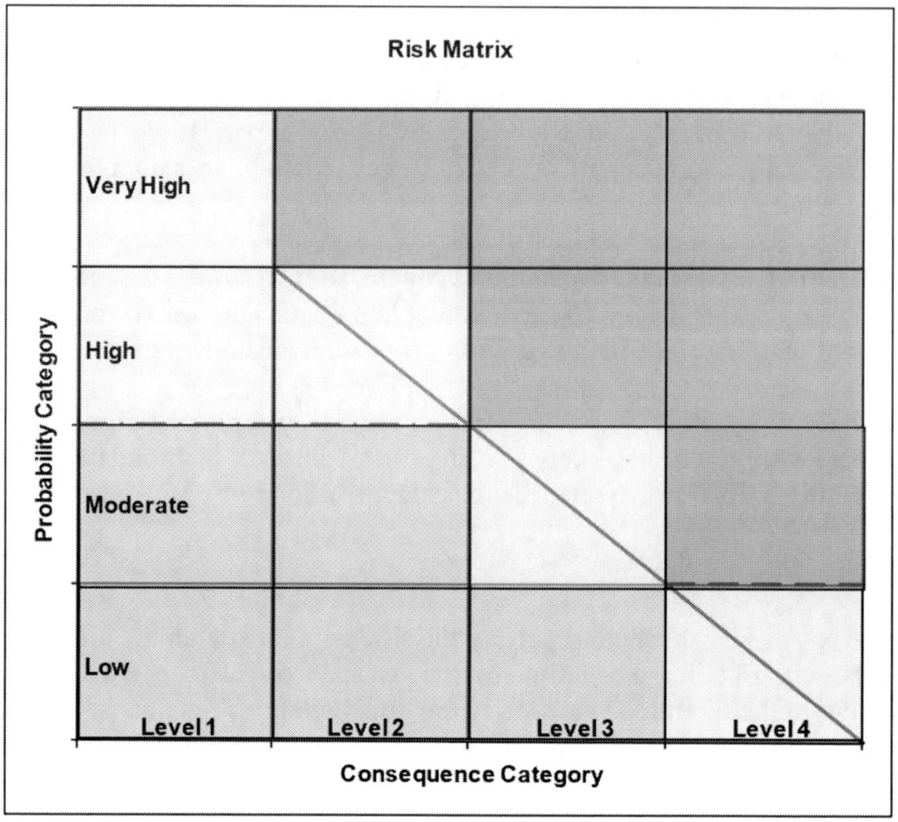

Figure 2. Risk Matrix

RISK ASSESSMENT PROCESS

General Procedures

A worksheet was set up for each dam prior to the initial workshop, listing each PFM identified and described in the PFMA reports. Columns were set up in the worksheets for Likelihood and Consequence categories, the confidence in each, and the rationale for each. The following confidence categorization scheme was adopted. (The category names were selected to avoid confusion with the likelihood category names.)

Confidence Categories.

- **Good** – Confidence in the rating is high; it is unlikely that additional information would result in a change to the rating.

- **Poor** – Confidence in the rating is low; additional information could very well result in a change to the rating.

- **Medium** – In between Good and Poor; it is highly uncertain whether additional information would result in a change to the rating.

Confidence in the risk category ratings was considered to be an essential part of the risk assessment. A PFM with a Poor confidence rating, particularly if high risks were indicated, would probably require additional investigations or analyses before taking action. On the other hand, with a Good confidence rating, it may be appropriate to go directly to risk reduction action.

A "dry run" was held to determine the optimal setup for the initial workshop. It was decided that three computers/projectors would be used. The first would have background information on the dams (photographs, statistics, seismic hazard curves, hydrologic information, flood routing results, and inundation information), the second would have the PFMA reports and supporting reference information (including drawings and results of analyses that had been performed), and the third would have the worksheets, to be filled out by the recorder during the workshop.

During the workshop, as each dam was addressed, the following steps were taken:

- The basic statistics for the dam were reviewed (type of dam, height of dam, reservoir volume, etc.) and photographs of key features of the dam were shown.

- The flood information and flood routings were reviewed. Of particular interest were the frequency of the flood of record, the zero freeboard spillway capacity flood frequency, the projected frequency of the PMF, and the peak inflow/ projected reservoir level for the 1/10,000 annual exceedance probability flood.

- The seismic hazard curves were reviewed. Of particular note were the ground motions associated with the 1/10,000 earthquake, the approximate annual exceedance probability of the MCE ground motions, and the ground motions used in any seismic analyses along with their approximate exceedance probabilities.

- The breach inundation studies were reviewed (both "sunny day" and flood scenarios) including probable impacts to downstream dams, powerhouses, roads and bridges, recreation areas, and permanent structures.

Then, the following steps were taken for each PFM listed in the PFMA report. Occasionally a PFM was added that was not identified in the PFMA report when one of the team members made a case for including it, and the same process was used for these as well.

- The PFM and its full description were reviewed. It was important to understand the failure mode from initiation, through step-by-step development, to breach of the dam so that all participants had a common understanding of what was being estimated. It was also important to understand what the breach (or uncontrolled release of the reservoir) entailed in terms of size and rapidity, as this had a direct bearing on the consequences and applicability of available inundation studies.

- The factors making each PFM more likely and less likely were reviewed by the team, including analysis results and associated load probabilities where applicable, and the key factors were identified.

- The team was then queried as to whether the evidence was weighted more toward "unlikely" or "likely", or what exceedance probability event would likely cause failure, considering the discussion that had just occurred. Each team member made their likelihood estimates individually at first. A candidate likelihood category was elicited from team members, along with the reasoning behind their estimates. This prompted discussion among team members. After the discussion died down, the facilitator summarized what had been said, proposing a "consensus" likelihood category and the reasoning why it made sense, and then asking if there were any objections. If objections were raised, additional discussion ensued, and the process was repeated.

- When no objections were raised, the recorder captured the information. Then the team was queried as to the level of confidence they had in the assigned category, prompting additional discussion and finally a consensus confidence category and rationale, which were also recorded on the worksheet.

- A similar elicitation process was repeated to arrive at a consensus consequence category for each PFM. It was especially important during this process to note differences between the likely breach flows associated with a PFM, and what had been assumed in the breach inundation study. In many cases, the breach outflow associated with a PFM would be considerably less than assumed in the inundation study.

- Once a PFM had been discussed and classified according to likelihood and consequences, it was listed on a "sticky note" and placed on a large blank risk matrix posted on the wall. A different color was used for each dam.

PORTFOLIO RISK SUMMARY AND RISK-INFORMED PRIORITIES

The fundamental objective of this study was to help Edison determine the need for, the order of, and the urgency in accomplishing risk verification or risk reduction actions for their dams. This risk management question involves consideration of not only the likelihood of occurrence and magnitude of consequences, but also the confidence in the estimates, the cost of remediations or other actions, the relative risk reduction achieved and other factors related to FERC jurisdictional dams. The success of the portfolio risk assessment was due in a large part to the high quality PFMA reports available for the Edison dams.

In the risk assessment process, the existing forecasting, monitoring and Emergency Action Plan (EAP) procedures are taken into account in defining the risks. Therefore, unless otherwise recommended based on the results of the risk assessment, it is extremely important to continue with these activities, and if any are dropped, the risks should be reassessed. Periodic visual inspections, flood forecasting, monitoring, and post-earthquake inspections continue to be appropriate risk management activities in all cases.

The results of the portfolio risk assessment for one of Edison's hydro divisions are shown graphically in figure 3. For the eleven dams and dozens of PFMs that were identified and evaluated, 13 PFMs were identified that are potentially above or border-line to the approximate risk guidelines. All of these were classified with a Poor or Medium confidence level. Thus, the results indicate an effective dam safety program is in place at Edison. Nevertheless, there are some areas where the Company will be taking action to verify and/or reduce the risks.

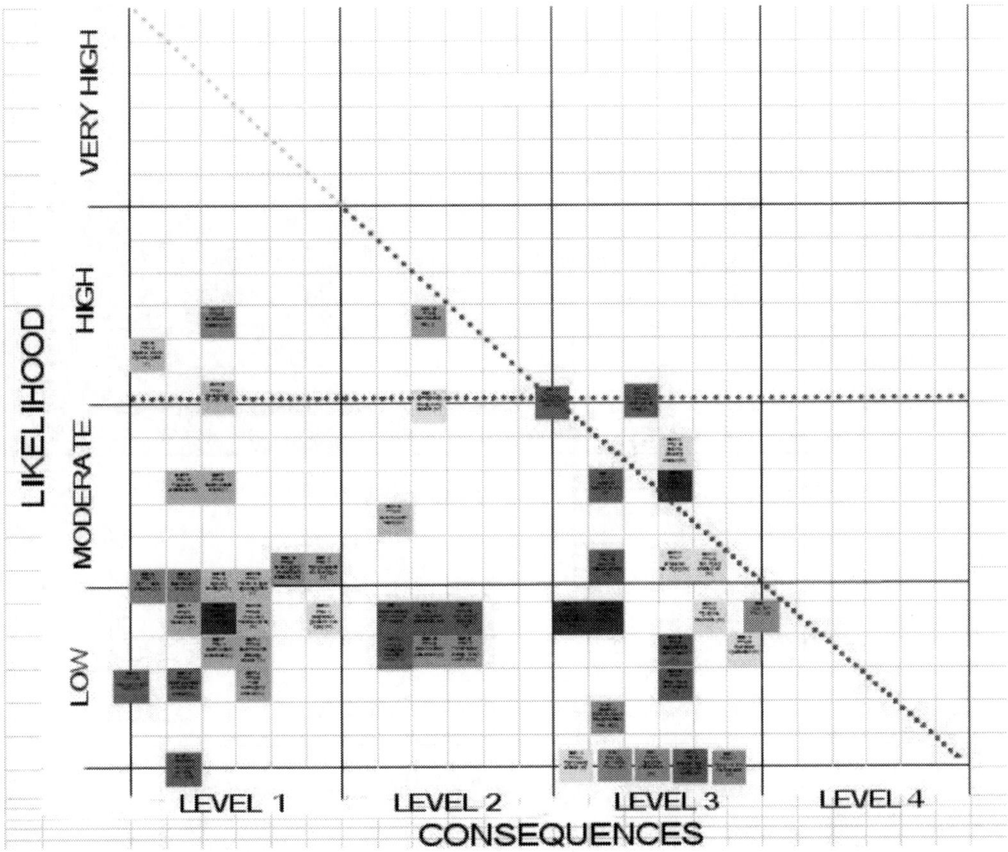

Figure 3. Risk matrix for one of Edison's hydro divisions. Each color represents a different dam and each square represents a different PFM. Level 0 consequences and Remote Likelihood are not shown

In order to help prioritize and define the follow-on actions, five risk-informed priority groups were identified as follows (and summarized in figure 4 relative to the risk matrix). Increasing likelihood and/or consequences categories would indicate higher risk priorities within each group.

Priority 1 Actions

These actions would be associated with reducing high risks associated with those PFMs that exceed both societal risk and failure probability guidelines, with a confidence rating of Good. These fall in the High likelihood or higher at consequence Level 3 or 4. These would be PFMs for which risk reduction actions would likely be appropriate. The urgency in taking action increases as the risks plot closer to the upper right-hand portion of the risk matrix.

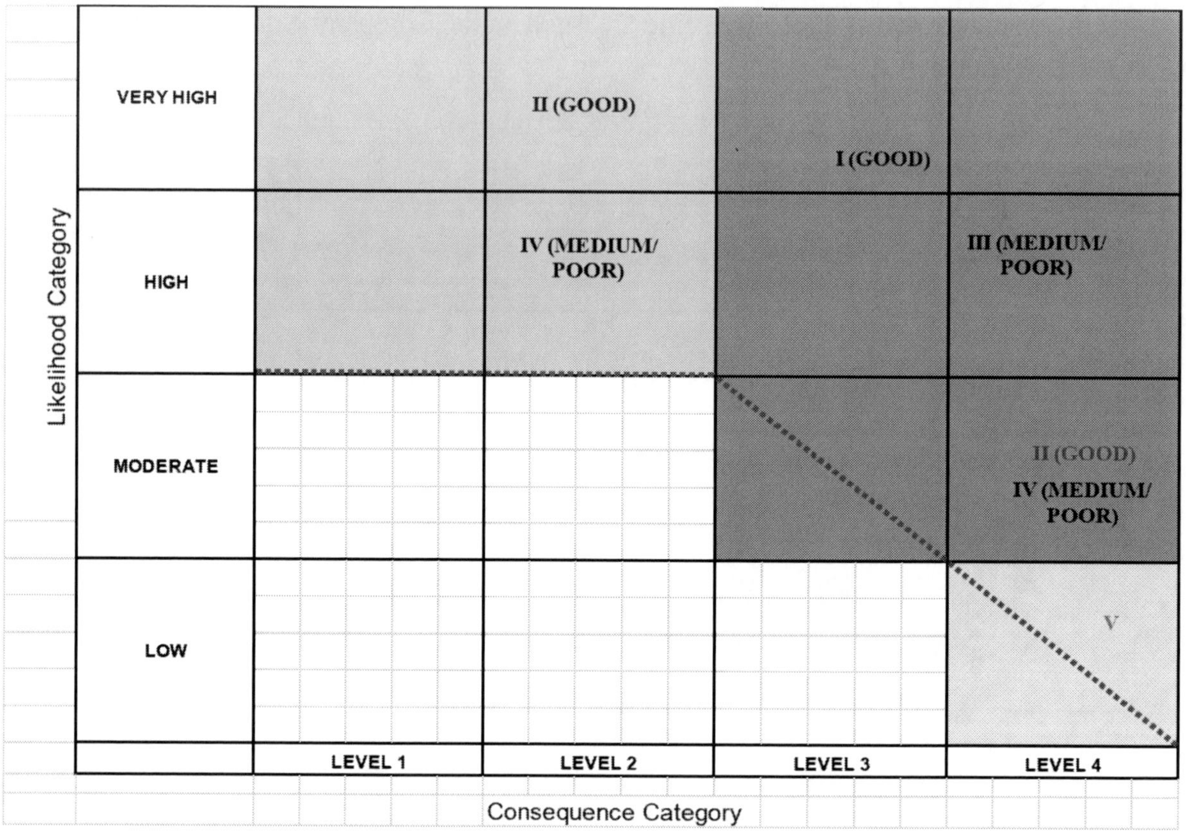

Figure 4. Priority Action Categories

Priority 2 Actions

These actions would be associated with reducing high risks associated with those PFMs that exceed either societal risk or failure probability guidelines with a confidence rating of Good. These would be PFMs that fall in the High likelihood category or higher at consequence Levels 1 or 2, and failure modes that fall in the Moderate likelihood category at consequence Levels 3 or 4.

Priority 3 Actions

These actions would be associated with verifying high risks associated with those PFMs that exceed both societal risk and failure probability guidelines, with a confidence rating of Medium or Poor. These fall in the High likelihood or higher at consequence Level 3 or 4. The urgency in verifying risks increases as the risks plot closer to the upper right-hand portion of the risk matrix.

Priority 4 Actions

These actions would be associated with verifying high risks associated with those PFMs that exceed either the societal risk or probability of failure guideline with a confidence rating of Medium or Poor. These would be PFMs that fall in the High likelihood category or higher at consequence Levels 1 or 2, and failure modes that fall in the Moderate likelihood category at consequence Levels 3 or 4.

Priority 5 Actions

These actions would be those associated with critically examining those PFMs with Low likelihood ratings but very high Level 4 consequences, at any confidence rating. The likelihood of these PFMs is already low, and driving them lower or verifying that they are lower could be difficult. Yet, the consequences of failure are so high that these PFMs should be examined and cost effective means of better defining and reducing risk should be considered.

The actual order in which these studies and actions are undertaken is based on several factors, including: (1) the estimated risk from the portfolio risk assessment, (2) the ease (effort and cost) with which they can be accomplished (i.e. accomplish the "low hanging fruit" first), (3) building on the momentum of work already in progress to address FERC requirements, and (4) taking care of all dam safety issues at once for dams that are scheduled for retrofit, where possible.

ONGOING ACTIVITIES AND UPDATES

Treating the risk matrices as "living documents" helps demonstrate how the dam safety risk is changing over time and the value dam safety activities are providing to the Company. The risk matrix and portfolio risk understanding is updated as dams are modified or new information is obtained through testing, analysis, or observed performance, and through a program of regular risk assessment workshops for individual dams.

CONCLUSIONS

A portfolio risk assessment building upon the results of the PFMA's for Edison's dams has been an effective means of better understanding the risk profile under which the Company's hydropower facilities were operating. Starting with a thorough understanding of the PFMs was key to the success of the risk assessment efforts. The results of the risk assessment helped to identify potentially high-risk failure modes and helped to prioritize follow-on activities to better understand and reduce the risks. Updates and revisions are essential as new information is available.

REFERENCES

ANCOLD (Australian National Committee on Large Dams), "Guidelines on Risk Assessment," ANCOLD, Sydney, New South Wales, Australia, 2003.

Bureau of Reclamation/U.S. Army Corps of Engineers, "Best Practices in Dam Safety Risk Analysis," Technical Service Center, Denver, Colorado, Version 4.0, July 2015.

Bureau of Reclamation, "Public Protection Guidelines for Dam Safety," Dam Safety Office, Denver, Colorado, August 2011.

Canadian Dam Association, "Dam Safety Guidelines," Canadian Dam Association, Edmonton, Alberta, Canada, 2007.

Douglas, K.J., M. Spannagle, and R. Fell, "Analysis of Concrete and Masonry Incidents," UNICIV Report No. R-373, The University of New South Wales, Sydney, Australia, 1998.

Foster, M.A., R. Fell, and M. Spannagle, "Analysis of Embankment Dam Incidents," UNICIV Report No. R-374, The University of New South Wales, Sydney, Australia, 1998.

Graham, W.J., "A Procedure for Estimating Loss of Life Caused by Dam Failure," Report DSO-99-06, Bureau of Reclamation, Denver, Colorado, September 1999.

Hatem, G.A., "Development of a Data Base on Dam Failures in the U.S. – Preliminary Results," Thesis, Stanford University, 1985.

Morrison-Knudsen Engineers, Inc., "Verde River Risk Assessment Study, Phase I Report," Arizona, 1988.

New South Wales Government Dams Safety Committee 2006, Risk Management Policy Framework for Dam Safety. New South Wales, Australia, 22 August 2006.

U.S. Army Corps of Engineers, "Safety of Dams – Policy and Procedures," Engineering Regulation ER-1110-2-1156, USACE, Washington, D.C., March 2014.

Von Thun, J.L., "Application of Statistical Data from Dam Failures and Accidents to Risk-Based Decision Analysis on Existing Dams," Bureau of Reclamation, Denver, Colorado, 1985.

BUREAU OF RECLAMATION RISK-INFORMED APPROACH TO SELECTING THE INFLOW DESIGN FLOOD

William R. Fiedler, PE[1]
John H. LaBoon, PE[2]

ABSTRACT

The Bureau of Reclamation (Reclamation) relies on information from risk analyses to guide all dam safety decisions. The selection of the Inflow Design Flood (IDF) is a key decision that is necessary when designing a new dam or when modifying an existing dam due to hydrologic risks. Reclamation developed a design standard in November 2013 that outlines the methodology for selecting an IDF. Design Standards No. 14, Appurtenant Structures for Dams (Spillways and Outlet Works), Chapter 2, Hydrologic Considerations (Reclamation, 2013) outlines this risk-informed approach. Specifically, Chapter 2 provides methods for sizing dams, spillways, and outlet works based on the selection of an IDF, along with methods for evaluating dam freeboard (above the maximum design reservoir water surface). This paper outlines the approach identified in the Reclamation's design standard and concludes with an example application of the methodology.

INTRODUCTION

Overtopping flow is a component event of many or even most potential failure modes resulting from floods. Dams and levees have been overtopped by a few inches to more than a foot without breaching, while other structures have failed quickly. Overtopping is a failure mode of concern since Costa (1985) reported that of all dam failures as of 1985, 34 percent were caused by overtopping, 30 percent due to foundation defects, 28 percent from piping and seepage, and 8 percent from other modes of failure. Costa (1985) also reports that for earth/embankment dams only, 35 percent have failed due to overtopping, 38 percent from piping and seepage, 21 percent from foundation defects; and 6 percent from other failure modes. One of the most infamous and devastating dam failures was the failure of South Fork Dam in 1889. The embankment dam failed during a large flood event and caused the death of over 2200 people downstream (Frank, 1988). While the number of concrete dams that have failed due to flood overtopping is less than that for embankment dams, there are a number of documented case histories where this has occurred. Examples include Sweetwater Dam in California, which failed in 1895 (Hansen, 2007), and the Secondary Dam of Sella Zerbino in Italy, which failed in 1925 (www.molare.net).

---

[1] Lead Civil Engineer, Geotechnical Services Division, Bureau of Reclamation, PO Box 25007, Code 86-68300, Denver CO 80225-0007, wfiedler@usbr.gov
[2] Civil Engineer, 11301 W. Berry Ave, Littleton CO 80127, 303-979-6683, john.h.laboon@gmail.com

HYDROLOGIC HAZARDS AND DESIGN FLOODS

Historically, dam design and analysis methods have focused on selecting a level of protection based on the evaluation of flood loadings. Traditionally, the protection level has been the Probable Maximum Flood (PMF). Currently, Reclamation uses quantitative risk analysis to assess the safety of dams, recommend safety improvements, and prioritize expenditures. Quantitative risk analysis, from a hydrologic perspective, requires an evaluation of a full range of hydrologic loading conditions and possible dam failure mechanisms tied to consequences of a failure. This risk approach may result in the conclusion that acceptable dam safety risks can be achieved with an IDF smaller than the PMF. However, for both existing and new dams and their appurtenant structures, there may be compelling non-risk factors (such as minimal cost increases and maintaining public trust) to design for the PMF even if risk considerations support designing for a flood event less than the PMF.

The flood loading input to a dam safety risk analysis prepared by Reclamation is a Hydrologic Hazard Curve (HHC) that is developed from a Hydrologic Hazard Analysis (HHA). HHCs are peak flow and volume probability relationships. These HHCs are presented as graphs and tables of peak flow and volume (for specified durations) versus annual exceedance probability (AEP). The range of AEPs that is displayed on these graphs is intended to be sufficient to support the decision making needs of the organization. Presently, suites of frequency flood hydrographs ranging from 0.01 AEP (1/100-year return period) to approximately 0.00001 AEP (1/100,000-year return period) are developed to help size existing modifications to dams and their appurtenant structures, along with sizing new dams and their appurtenant structures. Frequency flood hydrographs for AEPs less than (more remote than) this range are developed as needed for specific potential failure modes (PFMs) or for particular hydrologic risk evaluations. The maximum frequency flood peak and/or volume should not exceed the current critical PMF peak and/or volume (for definition of critical PMF, refer to Flow Chart 1A or Flow Chart 1B).

Inflow Design Floods

The IDF is defined as the maximum flood hydrograph or a range of flood hydrographs for a given return period, used in the design of a dam and its appurtenant structures, particularly for sizing the dam, spillway and outlet works. Ranges of IDF hydrographs are considered in order to encompass load uncertainties, and AEP ranges, along with variations in initial reservoir levels are used in flood routings to determine maximum reservoir water surface (RWS) elevations. Features are designed to safely accommodate floods up to and including the IDF. The IDF is selected to achieve acceptable levels of hydrologic risk at a dam, which is typically an iterative process. The IDF will be equal to or smaller than the current critical PMF.

The concept of an IDF is more straightforward when a single flood hydrograph is provided for each return period flood. In this case, the largest of the hydrographs that can be passed is the IDF, and the IDF can be defined by the return period associated with that flood.

If a sophisticated flood study is conducted and hundreds or thousands of hydrographs are generated for each return period flood (through a Monte Carlo simulation), the IDF may become more difficult to define. Judgment may be required to select a representative return period that defines the IDF. Approaches might include selecting or estimating the return period for which all generated hydrographs can be safely passed, or selecting a return period for which a set percentage (say 80 or 90 percent) of the generated hydrographs can be safely passed. Once the IDF is selected, additional freeboard (robustness study) is evaluated and selected to establish the top of dam elevation or the top of parapet wall elevation.

Flood Variables, Load Ranges, and Initial Reservoir Levels

In order to evaluate specific hydrologic-related PFMs, flood frequency and flood routing information is needed on variables including flood peaks, volumes, durations, and elevations (e.g., maximum RWS). An example maximum RWS frequency curve (with uncertainty) is shown in Figure 1. These relationships integrate variations in frequency flood hydrograph ranges, load uncertainties, and initial reservoir levels and reflect the result of flood routings.

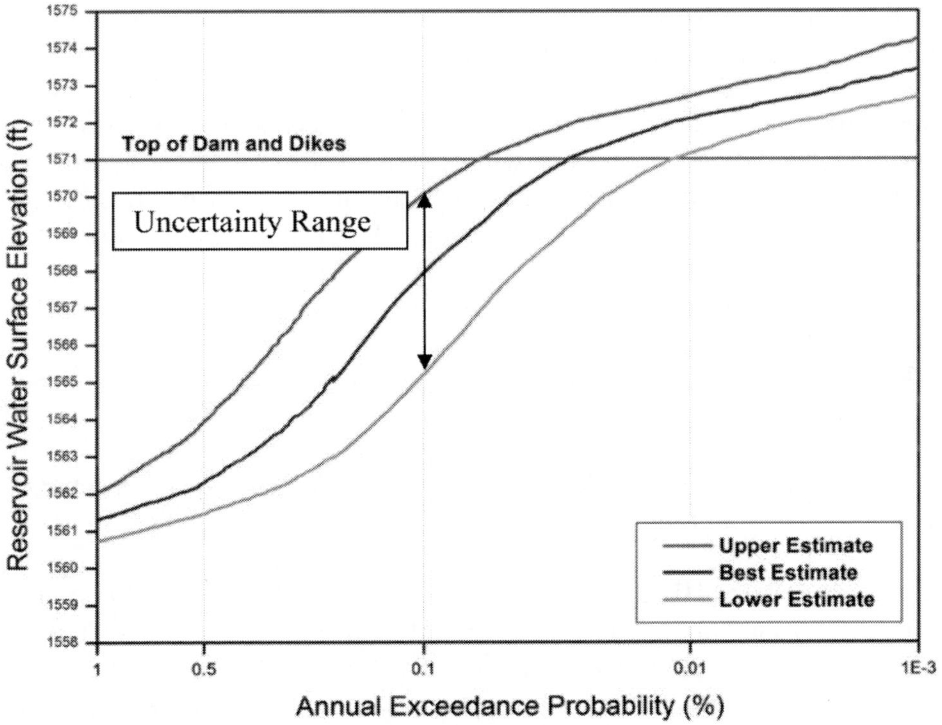

Figure 1. Example: Maximum reservoir water surface elevation frequency curve with uncertainty.

Identifying Hydrologic Data Needs

Hydrologic hazard data needs are determined on a case-by-case basis. Key hydrologic hazard data requirements include: streamflow data, historical flood and paleoflood data, and precipitation and extreme storm data. These data types are described in various sources, including Reclamation's *Guidelines for Evaluating Hydrologic Hazards* (Reclamation, 2006) and Chapter II-2, "Probabilistic Hydrologic Hazard Analysis," of Reclamation's and the U.S. Army Corps of Engineer's *Dam Safety Risk Analysis Best Practices Training Manual* (Reclamation, 2015).

Developing HHCs for risk analysis involves using the length of record and type of data to determine the extrapolation limits for hydrologic hazards. The length of record is often 50 to 100 years at a particular site (i.e., at-site). At-site data are defined as measured or obtained data within the watershed upstream of the dam or dam-site of interest. Data sets are significantly expanded by using regional information and using space-for-time substitution concepts (regional data from similar sites and similar flood events can be used to expand the data set of hydrologic events and increase the confidence in extrapolations to extreme event). Regional data (or regional analysis) consists of pooling streamflow, paleoflood, and precipitation data from many sites around the location of interest to substantially increase the information on extreme floods that is used to estimate HHCs. Paleoflood data are estimates of extreme floods, or limits on floods that usually have occurred over the past hundreds to thousands of years determined from geologic evidence along rivers or streams. Details on data sources, data types, and extrapolation ranges for hydrologic hazards are listed in Reclamation's *Guidelines for Evaluating Hydrologic Hazards* (Reclamation, 2006).

Hydrologic hazard data needs vary and depend on the available information, hydrologic hazard method, level of study, and decision(s) being made. The flood hydrologist works with the designer of record and design team to determine appropriate hydrologic hazard data needs commensurate with the HHA method and design level of study.

No single HHA approach is capable of providing the needed characterization of extreme floods over the full range of AEPs required for quantitative risk analysis. Results from several methods and sources of data should be combined to yield a HHC. In ideal situations, Reclamation uses multiple methods to estimate HHCs due to the significant extrapolation of the flood frequency relationships and the uncertainties involved in the analysis. For more information about the multiple methods used by Reclamation, refer to the *Guidelines for Evaluating Hydrologic Hazards* (Reclamation, 2006).

Figure 2 is a generalized HHC showing the relationship between the PMF and frequency floods. The HHA and resulting HHCs are used to develop flood hydrographs, considering both the peak inflow and the flood volumes for a given return period flood. The frequency flood hydrographs are then routed through the reservoir and information regarding the maximum RWS and spillway discharges during the flood routing is used in subsequent risk analyses.

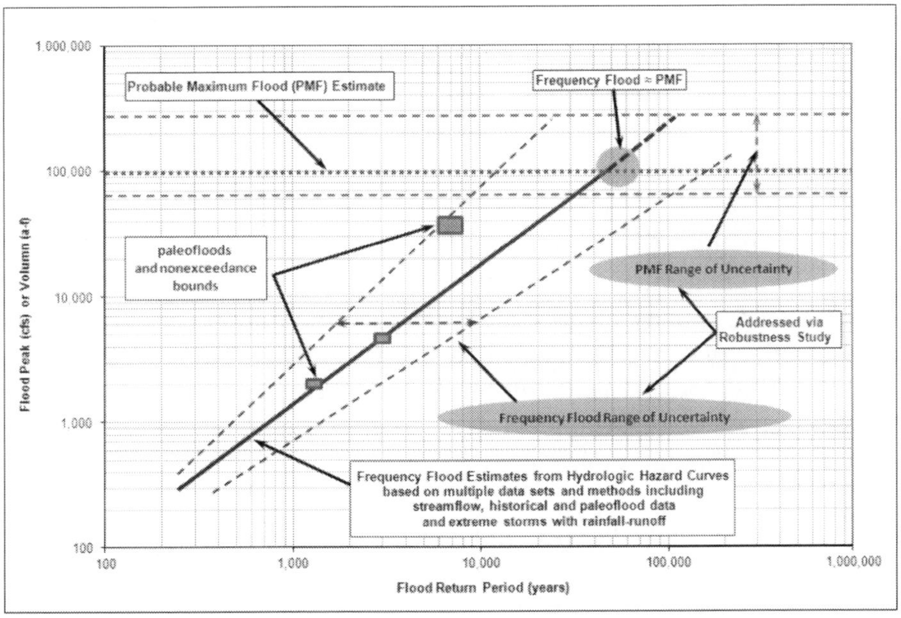

Figure 2. Generalized HHC schematic (relationship between the PMF and frequency floods).

RECLAMATION'S CURRENT PROCESS FOR SELECTING THE IDF

For both modifying existing and constructing new dams, along with appurtenant structures such as spillways and/or outlet works, the following generalized selection process is used:

- Clearly identify credible PFMs for all loading conditions (static, seismic, and hydrologic).

- Estimate baseline total risks for all credible PFMs involving all loading conditions (includes loads, responses, and consequences).

- Establish initial design loading conditions:
    - For modifications to an existing dam and appurtenant structures, the IDF is equal to or less than the current critical PMF, and the design earthquake event is typically equal to or less than the 50,000-year return period.
    - For designing a new dam and appurtenant structures, target design flood and earthquake may have similar ranges as noted for modifications to an existing dam and appurtenant structures. For new dams, there may be great cost efficiency in mitigating more remote risks than there would be for modifying existing dams. Initial flood and earthquake loadings are selected to maintain total risks to a level of decreasing justification to take action (i.e., less than Reclamation's risk guidelines).

- Confirm or revise initial flood and earthquake design levels. If the initial frequency flood level results in total risk plotting in an area of the f-N chart indicating decreasing justification to take action to reduce risks, the frequency flood level may be acceptable. Note: whether or not a flood level is judged as acceptable will be unique to each condition/situation and will be recommended by the designer of record and concurred with by Reclamation management (decision-makers). Also, risks below Reclamation guidelines do not ensure that a chosen frequency flood level is acceptable. Other risk and non-risk factors such as uncertainty, confidence, cost, physical constraints, etc., will have a bearing on identifying acceptable risks. If the total risk is unacceptably high, a more remote frequency flood event should be assumed, and total risks should be re-estimated until the total risks are in an area of the f-N chart indicating decreasing justification to take action to reduce risks. Total risk could range from just below, to more than one order of magnitude below Reclamation guidelines (Reclamation, 2011).

- Select freeboard above the IDF-induced maximum RWS elevation based on robustness study.

The following flow charts summarize the IDF selection process for an existing dam. A similar approach is used for selecting an IDF for a new dam and details are provided in the design standard (Reclamation, 2013). Small pictures of the flow charts appear beside the subsequent text to provide the reader with a visual image of the portion of the flow chart under consideration in the text (highlighted by red dashed lines).

Flow Chart 1A – Significant and High Hazard Existing Dams:

Process for Evaluating Baseline Risks (Hydrologic Loading)

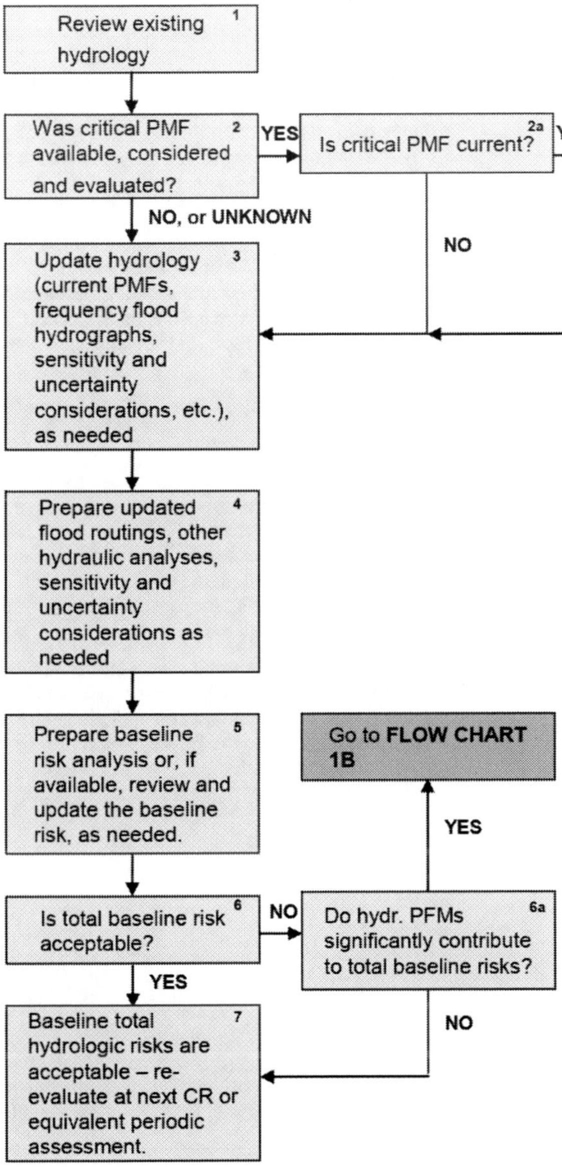

NOTES:

Acceptable total risks – typically, total risks are below Reclamation guidelines; however, total risks below guidelines does not ensure that they are always acceptable. Other risk and non-risk factors (uncertainty, confidence, ALARP, cost, physical constraints, etc.) will have a bearing on identifying acceptable total risks.

Critical PMF – based on flood routings, typically the controlling PMF that will induce the highest maximum RWS above other PMF induced maximum RWSs. In some cases, other factors may dictate selection of a PMF as the critical event that are not based on the highest maximum RWS.

PFM – potential failure mode.

PMF – Probable Maximum Flood.

Robustness considerations – used to address uncertainties by determining freeboard above the IDF-induced maximum RWS which involves evaluating site-specific "what-if" scenarios such as hydrologic uncertainties, mis-operation of spillway gates, debris blockage, wind-generated waves, etc.

Total risks – the sum of the annualized failure probability (AFP) estimates for all credible PFMs and the sum of the annualized life loss (ALL) estimates for all credible PFMs, which are considered separately.

ALARP – As Low as Reasonably Practicable

Flow Chart 1B – Significant and High Hazard Existing Dams:

Process for Selecting the Inflow Design Flood (IDF)

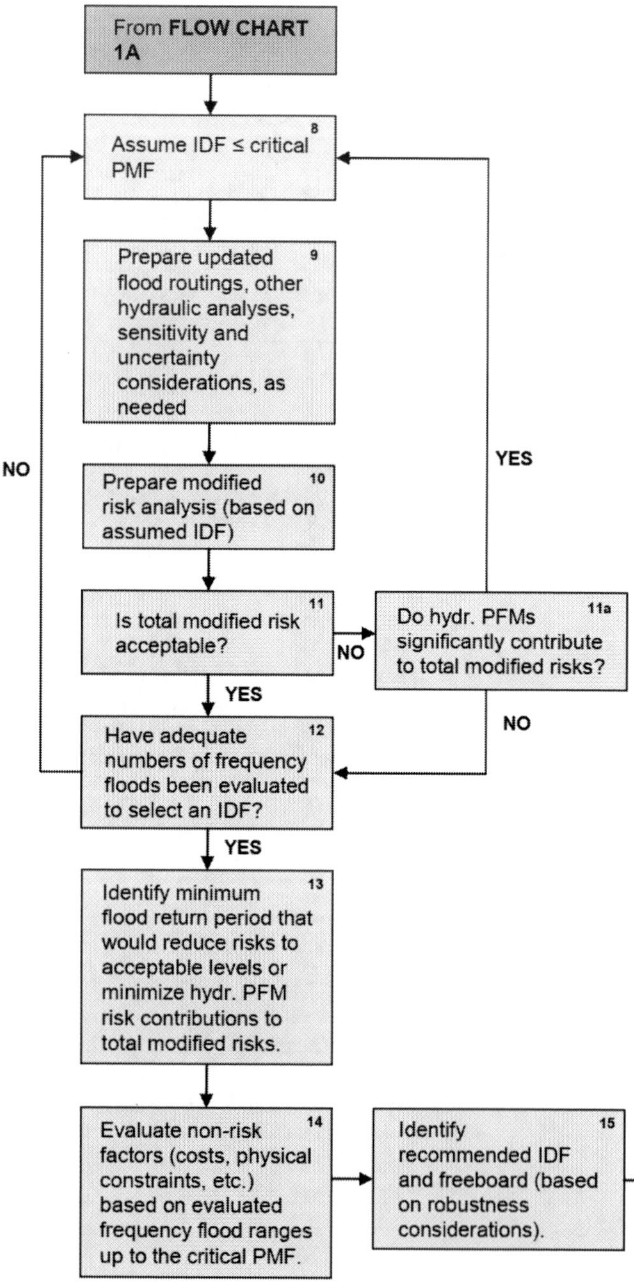

NOTES:

Acceptable total risks – typically, total risks are below Reclamation guidelines; however, total risks below guidelines does not ensure that they are always acceptable. Other risk and non-risk factors (uncertainty, confidence, ALARP, cost, physical constraints, etc.) will have a bearing on identifying acceptable total risks.

Critical PMF – based on flood routings, typically the controlling PMF that will induce the highest maximum RWS above other PMF induced maximum RWSs. In some cases, other factors may dictate selection of a PMF as the critical event that are not based on the highest maximum RWS.

PFM – potential failure mode.

PMF – Probable Maximum Flood.

Robustness considerations – used to address uncertainties by determining freeboard above the IDF-induced maximum RWS which involves evaluating site-specific "what-if" scenarios such as hydrologic uncertainties, mis-operation of spillway gates, debris blockage, wind-generated waves, etc.

Total risks – the sum of the annualized failure probability (AFP) estimates for all credible PFMs and the sum of the annualized life loss (ALL) estimates for all credible PFMs, which are considered separately.

Existing Dams

The following text discusses Flow Chart 1A (Significant and High Hazard Existing Dams: Process for Evaluating Baseline Risks) and 1B (Significant and High Hazard Existing Dams: Process for Selecting the IDF).

1. Review and Evaluate Existing Hydrology/Hydraulics. This discussion addresses **Flow Chart 1A, boxes 1, 2, 2a, 2b, 2c, and 3**.

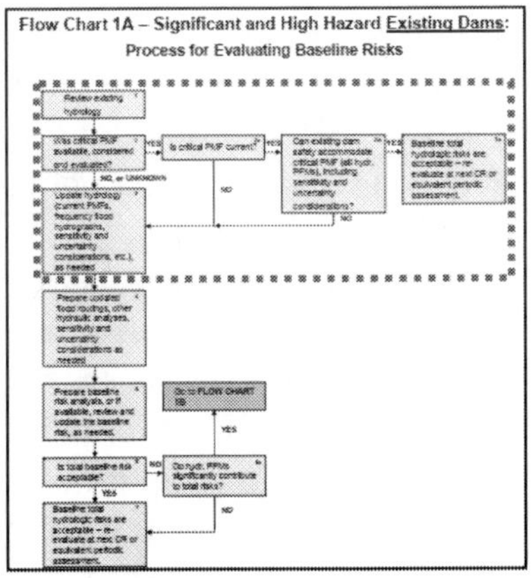

An initial screening/evaluation of existing hydrology is made **(box 1)**, which will help determine if the current critical PMF is available and was considered and evaluated for the existing dam **(boxes 2 and 2a)**.

If the current critical PMF is not available or is unknown, updated hydrology is needed **(box 3)**. The type and amount of updated hydrology will be site specific.

If the current critical PMF is available, a determination is made as to whether the dam can safely accommodate the current critical PMF, including analyses (typically flood routings and water surface profiles) needed to evaluate all hydrologic PFMs, such as dam overtopping, and PFMs related to spillways. It should be pointed out that a range of starting RWSs with the maximum starting RWS either at the top of active conservation or top of joint use storage, whichever is higher, should be used for the hydraulic analyses. The minimum starting RWS would typically reflect historical reservoir operations in terms of how low the reservoir might reasonably be during the time of year for the flood events.

If the existing dam can safely accommodate the current critical PMF, including sensitivity and uncertainty considerations **(box 2c)**, baseline total hydrologic risks are acceptable, and further evaluation is not warranted.

2. Update Hydrology/Hydraulics. This discussion addresses **Flow Chart 1A, boxes 3 and 4**.

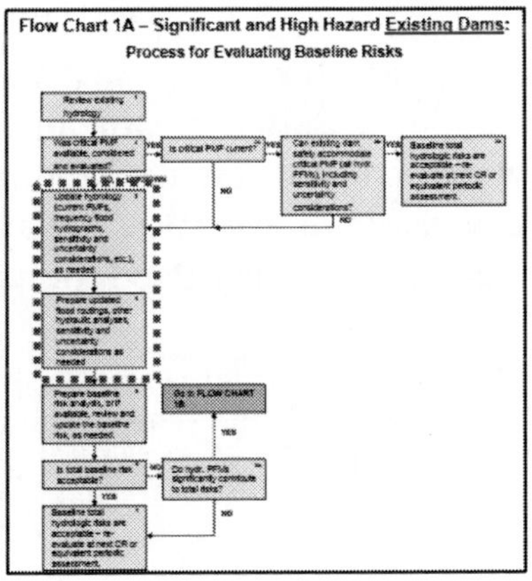

If it is determined that the current critical PMF has not been developed or routed through the existing dam or the existing dam cannot safely accommodate the current critical PMF, then some level of updated hydrology will likely be needed **(box 3)**. Once the updated hydrology is available, flood routings would be performed to determine the response of the dam to a range of flood loadings up to and including the current critical PMF **(box 4)**. It should be noted that sensitivity and uncertainty considerations come into play.

3. Prepare Total Baseline Risks. This discussion addresses **Flow Chart 1A, boxes 5, 6, and 6a**.

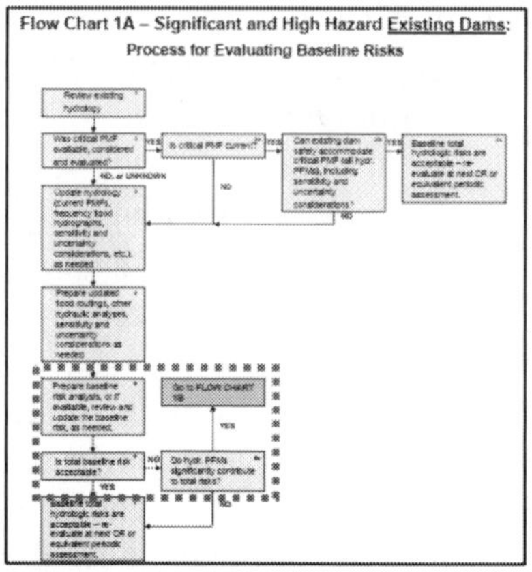

All credible PFMs for all loading conditions (static, seismic, and hydrologic) are identified and evaluated for the existing dam. A quantitative risk analysis using current risk methodology is prepared or, if it already exists, is reviewed and updated, if needed **(box 5)**.

If the total risks (the total annualized failure probability or the total annualized life loss) are not acceptable **(box 6)**, further evaluation of the hydrologic PFM risk contributions to the total risks is made **(box 6a)**. The term "not acceptable" typically means the total risk is in an area on the f-N chart indicating increasing justification to take action to reduce risks. If the total risks are acceptable **(box 6)**, further evaluation of risks associated with static and seismic PFMs are not warranted.

4. Actions if Total Baseline Risks are Unacceptable. This discussion addresses **Flow Chart 1A, boxes 6, 6a, and 7**.

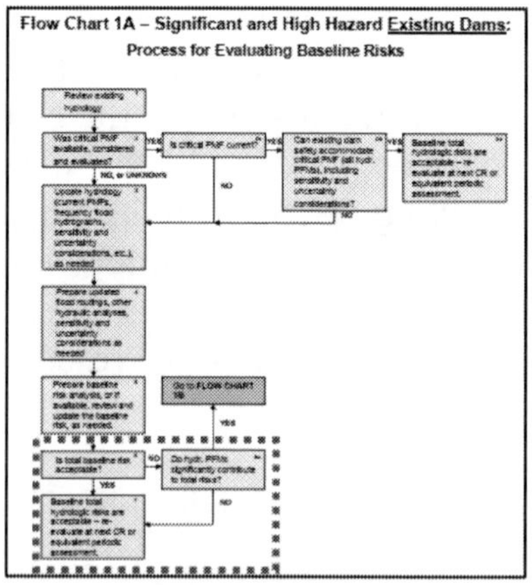

If it is determined that the hydrologic PFM risks do not significantly contribute to the total baseline risks **(box 6a)**, the process continues to the conclusion that no further evaluation of the hydrologic risks are needed.

If the hydrologic PFM risks are determined to significantly contribute to the total baseline risks **(box 6a)**, further evaluation of nonstructural and structural modifications that could reduce risks is made, which leads to the selection of an IDF (Go to **Flow Chart 1B**). This effort generally involves evaluating a suite of possible modifications, including but not limited to: reservoir restriction, breaching the dam, raising the dam, increasing the discharge capacity of existing appurtenant structures (spillways and/or outlet works), overtopping protection, and/or constructing new appurtenant structures (spillways and/or outlet works). The evaluation of these possible modifications would require appropriate analyses and designs, including flood routings up to and including the current critical PMF.

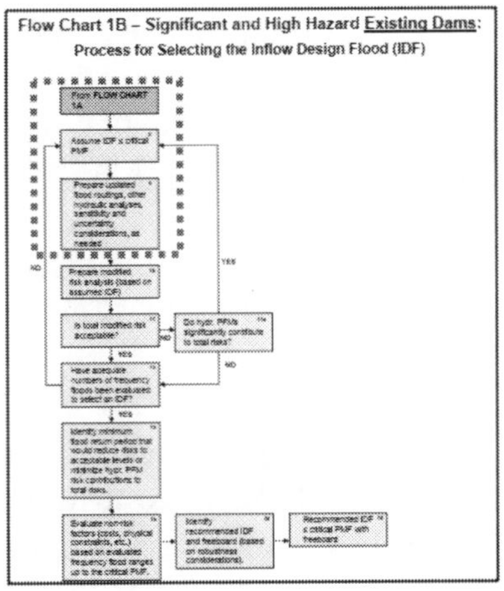

5. Actions if Hydrologic PFMs Significantly Contribute to Total Baseline Risks. This discussion addresses **Flow Chart 1B, boxes 8 and 9**.

As a starting point, the current critical PMF (determined from flood routings) or the maximum (most remote) frequency flood available is the assumed IDF **(box 8)**. This will be the maximum hydrologic loading condition and will set upper limits of modifications. As an alternative, in lieu of initially equating the IDF to the current critical PMF, the initial assumed IDF can be estimated by selecting a frequency flood equal to, or more remote than, the result of dividing the mean annualized life loss by the mean incremental life loss for the hydrologic PFMs.

Updated flood routings and other hydraulic analyses would be performed to determine the response of the modified dam to the assumed IDF **(box 9)**. The flood routings and other hydraulic analyses are used to select the type, size, and location of the modified dam and/or appurtenant structures (spillways and/or outlet works). It should be noted that sensitivity and uncertainty considerations are addressed prior to selecting an IDF and are integrated into the baseline risk analysis. Also, the flood routings would use a range of starting RWSs.

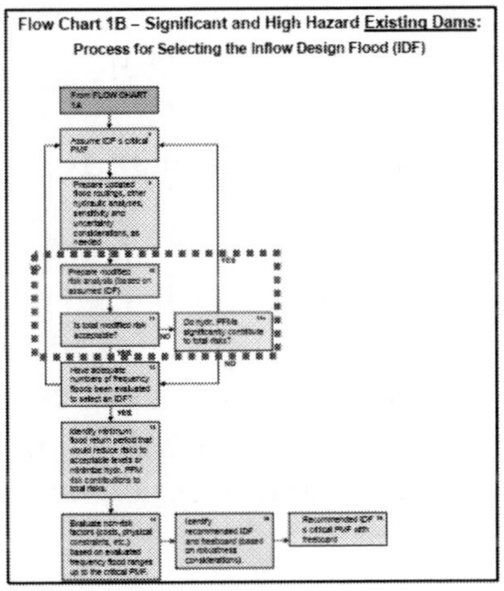

6. Prepare Total Modified Risks. This discussion addresses **Flow Chart 1B, boxes 10, 11, and 11a**.

All credible PFMs for all loading conditions are identified and evaluated for the modified dam and/or appurtenant structures. A modified quantitative risk analysis (based on the assumed IDF) using current risk methodology is prepared (**box 10**).

If the total modified risks (the total annualized failure probability or the total annualized life loss) are not acceptable **(box 11)**, further evaluation of the hydrologic PFM risk contributions to the total risks is made **(box 11a)**. However, it should be noted that risks below Reclamation guidelines are not necessarily acceptable. Other risk and non-risk factors such as uncertainty, confidence, cost, physical constraints, etc., will have a bearing on identifying acceptable risks.

If the total modified risks are acceptable **(box 11)**, a determination is made as to whether adequate numbers of frequency floods have been evaluated to select an IDF **(box 12),** which is discussed in more detail in the next step. There may be situations where the total modified risk is in an area of the f-N chart indicating decreasing justification to reduce risk, but a decision is made that further risk reduction is appropriate (for example, a modest increase in modification cost would significantly reduce total risks).

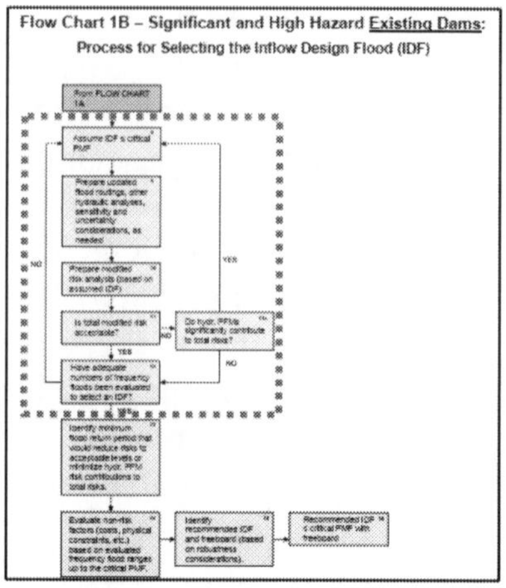

7. Actions Related to Total Modified Risks; Hydrologic PFMs Risk Contribution to Total Modified Risks; and Whether Sufficient Information to Select an IDF Exists. This discussion addresses **Flow Chart 1B boxes 11a and 12,** and recycling through **boxes 8 through 12**.

If it is determined that the hydrologic PFM risks do not significantly contribute to the total modified risks **(box 11a)**, a decision is made as to whether adequate representation of frequency floods have been evaluated to select an IDF **(box 12)**. Multiple assumed IDFs equal to a range of frequency floods will be used to estimate associated modified risks. This effort is needed to identify total risk trends, which are used to determine acceptable risks. If there is not enough information to identify acceptable risks, part of Flow Chart 1B (**boxes 8 through 12**) is repeated until sufficient information to select an IDF has been developed. Part of Flow Chart 1B will involve reassessing and redesigning the nonstructural and structural features. If there is sufficient information to select an IDF (**box 12**), then the minimum flood return period that would reduce risks to acceptable levels is identified **(box 13)**, which is discussed in more detail in the next step.

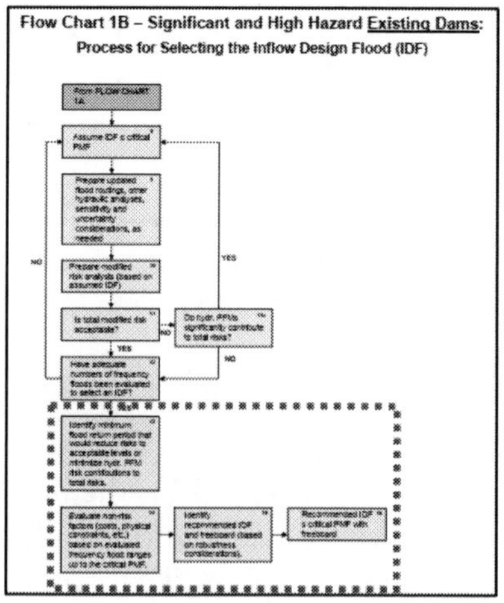

8. Actions if There is Sufficient Information to Select an IDF. This discussion addresses **Flow Chart 1B, boxes 13 through 16.**

From the previous effort (**box 12**), the minimum flood return period that would reduce total modified risks to acceptable levels is identified **(box 13)**. It should be highlighted that evaluation of the static and seismic risk contributions will be needed to identify the minimum flood return period that would reduce total modified risks to acceptable levels.

Prior to recommending an IDF, further evaluation takes place to assess the non-risk factors such as cost, physical constraints associated with the dam, policies, etc. **(box 14)**. These factors could change (reduce or increase) the identified minimum flood return period that would reduce risks to acceptable levels or minimize hydrologic PFM risk contributions to total risks.

Using the frequency flood level identified as the recommended IDF (**box 14**), a robustness study will be conducted to evaluate the uncertainties associated with the flood, operations, future events, etc. **(box 15)**. The existing dam with modifications and robustness considerations (which is used to determine freeboard requirements) should safely accommodate a frequency flood equal to or less than the current critical PMF. This frequency flood is the recommended IDF **(box 16)**.

Address Uncertainties with Freeboard (Robustness) Study (Existing and New Dams)

Once an IDF has been selected, uncertainties are incorporated into the design process, which is referred to as a "robustness study." These uncertainties may be related to the method of estimating the IDF return period and the size and shape of the IDF hydrograph, reservoir and dam operations, gated spillway operations, reduction of spillway discharge capacity due to debris, and other mechanisms and future events associated with upstream and downstream developments. These uncertainties represent variables that cannot be reasonably defined with current methodologies. Studies are underway at Reclamation to develop better tools to predict these variables and have a basis for numerically estimating their likelihood in the future. To account for these uncertainties, plausible "what-if" scenarios are evaluated. These scenarios could create an elevated RWS above the IDF-induced maximum RWS and might include:

1. **Misoperation of gated spillways.** This what-if scenario relates to human error associated with how gates are opened and closed. Examples might include gate operations that are not in accordance with operating documents such as the Standing Operating Procedures (SOP). To account for this what-if scenario, it is typically assumed that gates are either not opened enough or there is a delayed operation which can be approximated by reducing the discharge capacity.

- **Mechanical/electrical malfunction of gated spillways.** This what-if scenario relates to potential malfunction of gates that is not associated with human error. Even with regular maintenance of equipment and appropriate attendance by an operator, mechanical and/or electrical failure could occur. To account for this what-if scenario, it is typically assumed that one or more gates remain closed or not opened enough, which can be approximated by reducing the discharge capacity.

- **Change in hydrology (methodology, watershed changes, and climate changes).** This what-if scenario relates to potential future changes in the hydrologic data and methods used to develop flood loadings. To account for this what-if scenario, the size (inflow peak and/or volume) of the IDF may be increased and routed through the reservoir and dam. Percentages should be verified in consultation with a hydrologist, and other percentages could be considered.

- **Debris blockage of hydraulic structures (spillways, outlet works, etc.).** This what-if scenario relates to potential blockage of the spillway and/or outlet works by debris during a flood event, which can result in reduced discharge capacity. General guidelines would initially assume a 10- to 50-percent reduction in total discharge capacity, with possible adjustments based on spillway geometries and potential debris loads.

- **Failure of upstream dams.** This what-if scenario relates to evaluating the potential failure of upstream dams during a flood event. This is typically limited to upstream Reclamation dam failure impacts on the dam being evaluated. However, on a case-by-case basis, upstream non-Reclamation dam failure impacts on the dam being evaluated may be considered.

- **Change in upstream and/or downstream consequences.** This what-if scenario relates to potential future increases in upstream and/or downstream population at risk and subsequent increase in economic damage and potential life loss.

- **Flood events that differ from the original basis for the IDF (i.e., series of storms versus one storm).** This what-if scenario relates to potential future changes in the hydrologic data and methods used to develop flood loadings. To account for this what-if scenario, different types of flood hydrographs (such as scaled historical events, thunderstorm events, series of storms, etc.) with different shapes (such as front-end loaded, with/without antecedent event, etc.) as those used for the IDF could be evaluated.

- **Wind-generated waves (runup and setup).** This what-if scenario relates to assessing the potential wave sizes that could be generated by winds occurring over the normal, maximum, and intermediate water surface elevations.

Results of the robustness study will be considered when recommending freeboard requirements for either an existing or new dam, along with maximum discharge potential for hydraulic structures such as spillways. In some cases (such as changes in hydrology), the maximum discharge potential could exceed the maximum design discharge, which could lead to potential hydraulic and/or structural issues that should be further evaluated as part of the design. The robustness study will indicate estimated maximum RWS elevations for a number of different scenarios. This result should be used to guide, but not dictate, the recommended freeboard. Other considerations that will influence the final freeboard requirements associated with the IDF-induced maximum RWS for the modification of an existing dam (for considerations regarding a new dam, see the design standard (Reclamation, 2013)) include:

- **For an existing embankment dam being modified,** a minimum freeboard (above the IDF-induced maximum RWS) should be recommended based on the robustness study results with consideration given to the cost for various levels of protection, sensitivity to various and critical uncertainties, and other relevant factors.

- **For an existing concrete dam being modified,** a minimum freeboard (above the IDF-induced maximum RWS) should be recommended based on the robustness study results. If overtopping failure of a concrete dam is judged to be remote, minimal or no freeboard may be appropriate (however, if the reservoir is normally kept near the top of dam, additional freeboard may be considered to prevent waves from lapping over the dam on a regular basis).

EXAMPLE DAM F (EXISTING DAM, IDF LESS THAN CURRENT CRITICAL PMF)

The following example demonstrates how the IDF selection process can be applied to an existing dam.

Background

Dam F and reservoir are approximately 20 miles northeast and upstream of a medium sized city. Construction of the dam was completed in 1976. The reservoir provides a total storage capacity of 1,084,780 acre-feet at the original design maximum RWS elevation 475.4.

- **The main dam and two dikes**, including:
  - **A concrete dam** having a structural height of 340 feet, a crest width of 30 feet, a crest length of 1,400 feet, and crest elevation of 480.5 feet
  - **Two embankment dikes**, with heights of 60 and 105 feet, crest widths of 30 feet, crest lengths of 740 and 2,060 feet, and crest elevations of 480.5 feet
  - **A gated overflow spillway** with crest elevation of 418.0 feet is located in the center of the concrete dam and is controlled by eight 42.0- by 50.0-foot service radial gates. The rated discharge capacity of the existing spillway is 468,000 ft^3/s at RWS elevation 475.4.
  - **The outlet works** consists of eight 5- by 9-foot-gated sluice outlet conduits located through the overflow section of the concrete dam. The outlet works can pass 28,600 ft^3/s at reservoir elevation 427.0.

The current critical PMF has a peak inflow of 906,000 ft^3/s and a 120-hour volume of 3.2 million acre-feet. Also, frequency flood hydrographs have been prepared. It has been determined that a frequency flood with a return period of 500,000 years had a similar size (peak and volume) to the current critical PMF. It was also determined that the total baseline risks were unacceptably high, and there was increasing justification to reduce risks. Flood routings indicated that floods with a return period of about 25,000 years, about 25 percent of the current critical PMF, would overtop the dam and dikes, which in turn could lead to failure and uncontrolled release of the reservoir into the downstream urban area. The flood-induced overtopping PFM was by far the largest contributor to the total risk. Therefore, the focus of a subsequent Corrective Action Study was on actions that reduce the risk associated with hydrologic PFM risk contributors.

Inflow Design Flood Selection

Estimate of Baseline Risks

As previously noted, baseline risks for Dam F were unacceptably high, and there was increasing justification to reduce risks. The total estimated baseline risks included total annualized failure probability (AFP) of 6.0E-4 and total annualized life loss (ALL) of 1.5E-2, with the overtopping PFM risks making up 91 percent of the total AFP and 89 percent of the total ALL. Key factors in estimating baseline hydrologic risks are summarized below:

- To evaluate PFMs, appropriate data were needed. In the case of hydrologic PFMs, a HHA was prepared, which developed frequency hydrographs ranging from 25,000-year to 500,000-year return periods. In addition to the frequency floods, scaled historic floods and off-season PMFs were prepared.

- Flood routings for a range of starting RWSs were then performed to identify operational responses in terms of maximum RWSs, maximum discharges, and freeboard or overtopping conditions. Two starting RWSs that bound the minimum and maximum limits of flood operations were used. These RWSs included the maximum flood control reservoir elevation of 388.4 feet (which provided 670,000 acre-feet of flood control storage) and the maximum normal RWS (top of joint use storage) of 466.0 feet. Given that there are flood damage reduction requirements associated with reservoir operations, fairly complex flood routings were required.

- Downstream consequences were estimated; then, event trees were used to estimate baseline risks for each PFM.

Estimate of Modified Risks

To achieve sufficient risk reduction (from baseline conditions), potential modifications focused on increasing the discharge capacity, increasing the flood surcharge, or a combination. Various nonstructural and structural alternatives were identified and evaluated through the following process:

- Each alternative was sized by assuming a range of frequency floods as the IDF. Specifically, a range of frequency floods with return periods ranging between 25,000 and 500,000 years was used to hydraulically size the alternatives for various flood operations. Alternatives considered included dam/dike breach and removal, dam/dike raise, constructing a new auxiliary spillway on the left abutment of the concrete dam (either fuseplug or gated auxiliary spillway), and nonstructural alternatives, such as reservoir storage restrictions.

- Flood routings for the same range of starting RWSs as previously noted were then re-done to identify operational responses in terms of maximum RWSs, maximum discharges, and freeboard or overtopping conditions.

- Downstream consequences were re-evaluated; then, event trees were used to estimate modified risks for each PFM.

- A comparison of total baseline and total modified risks was made to assess the level of risk reduction. A key finding was that substantial risk reduction could be achieved with an IDF less than the current critical PMF. Table 1 and figure 1 summarize the comparison of baseline and modified total risks. Of note is the range of IDF return periods (>140,000 years to 500,000 years) that would result in total risks being in an area of reduced justification to take action.

- Other non-risk factors influencing the evaluation and selection of alternatives were modification costs, constructability considerations, and potential changes to operations after modifications. Of these non-risk factors, cost was a key factor for the selection of the IDF (see table 1). It was determined that significant cost increases resulted if an IDF with a return period greater than 250,000 years was selected. Additionally, constructability issues came into play (i.e., additional construction seasons would be required for implementing modifications that could accommodate IDFs with return periods greater than 250,000 years).

Table 1. Summary of dam F alternatives risks, costs, and modifications

Flood return period (years)	Total risks			Alternative costs	Comment
	AFP	Life lost	ALL		
25,000	6.0E-04	25	1.50E-02	$0	Baseline (existing)
50,000	1.0E-04	25	2.50E-03	$10,000,000	Small dam/dike crest modifications.
100,000	5.0E-05	25	1.25E-03	$17,000,000	Modest dam/dike crest modifications.
250,000	2.3E-05	25	5.80E-04	$30,000,000	Modest dam/dike crest modifications, modify existing spillway, new modest sized auxiliary spillway
500,000	8.0E-06	25	2.00E-04	$75,000,000	Significant dam/dike crest modifications, modify existing spillway, new large sized auxiliary spillway

Based on the previous discussion, the 250,000-year flood event, about 65 percent of the current critical PMF, was selected as the IDF, and it will be safely accommodated once the following modifications are made at Dam F:

- **New auxiliary spillway.** – Construct a new 6-bay top-seal radial gated auxiliary spillway on the left abutment of the concrete dam. The discharge capacity is 312,000 ft^3/s at RWS El. 477.5 feet.

- **Dam and embankment dikes.** – A water-retaining parapet wall would be located on the dam/dike crests to accommodate (impound) the higher maximum RWS.

Figure 3 shows the risk reduction for Dam F modifications.

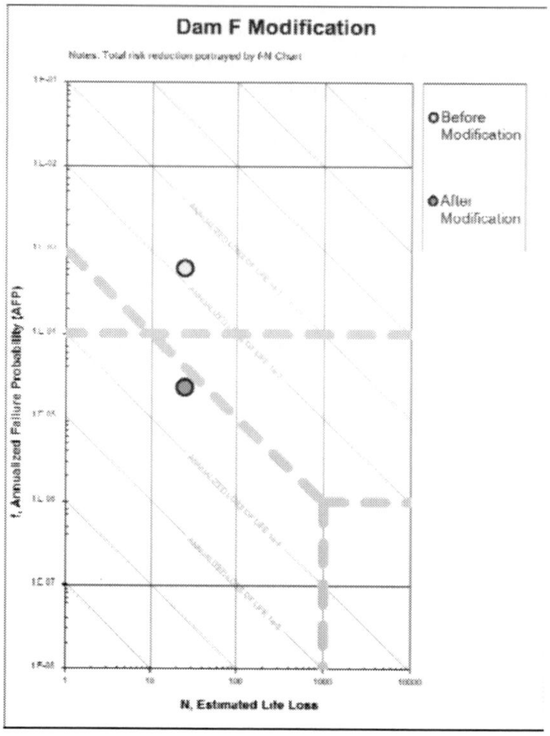

Figure 3. f-N chart portraying risk reduction for Dam F modifications.

With these modifications in place, the total baseline risks (ALL of 1.50E-2 and AFP of 6.0E-4) would be reduced to the total modified system risks (ALL of 5.8E-4 and AFP of 2.3E-5). Both the total modified system AFP and total modified system ALL are below Reclamation's guidelines and in an area of the f-N chart indicating decreasing justification to take action.

Robustness (Freeboard) Study

To address uncertainties associated with the method of estimating the IDF, reservoir and dam operations, and future events that could affect risk estimates, plausible "what-if" scenarios were evaluated and used to establish freeboard requirements between the IDF-induced maximum RWS and the dam/dike crests. These what-if scenarios represent plausible events that could occur, but events for which probabilities cannot be easily estimated. For Dam F, the what-if scenarios evaluated included:

- Scaled historic floods and off-season PMFs
- Inoperable existing spillway gates and/or new auxiliary spillway gates
- Misoperating in terms of delayed releases
- Future 10 percent increased hydrologic loading

- Wind-generated waves

Figure 4 shows the results of the robustness study for Dam F modifications.

Based on flood routing results, 3 feet of freeboard would accommodate most of these what-if scenarios (without overtopping of the dam and/or dikes). Of note (as can be seen from figure 4), 3 feet of freeboard would not accommodate all what-if scenarios - in particular, 110 percent of the IDF, the IDF plus two inoperable spillway gates (one existing and one new gate), and the IDF in combination with misoperation by limiting releases to 160,000 ft^3/s until RWS elevation 475.5 is reached. It was decided that most of the uncertainties represented by these what-if scenarios would be addressed by 3 feet of freeboard, and three what-if scenarios that resulted in more than 3 feet above the IDF-induced maximum RWS were remote enough to not pursue raising the dikes and the concrete dam. The other what-if scenario that exceeded 3 feet above the IDF-induced maximum RWS was the maximum wind-generated waves of 3.2 feet. It was concluded that additional freeboard was unnecessary given that the waves would be intermittent and damage (if any) would be minimal. The parapet walls on the dam and dikes were designed to provide 3 feet of freeboard and the costs are reflected in Table 1.

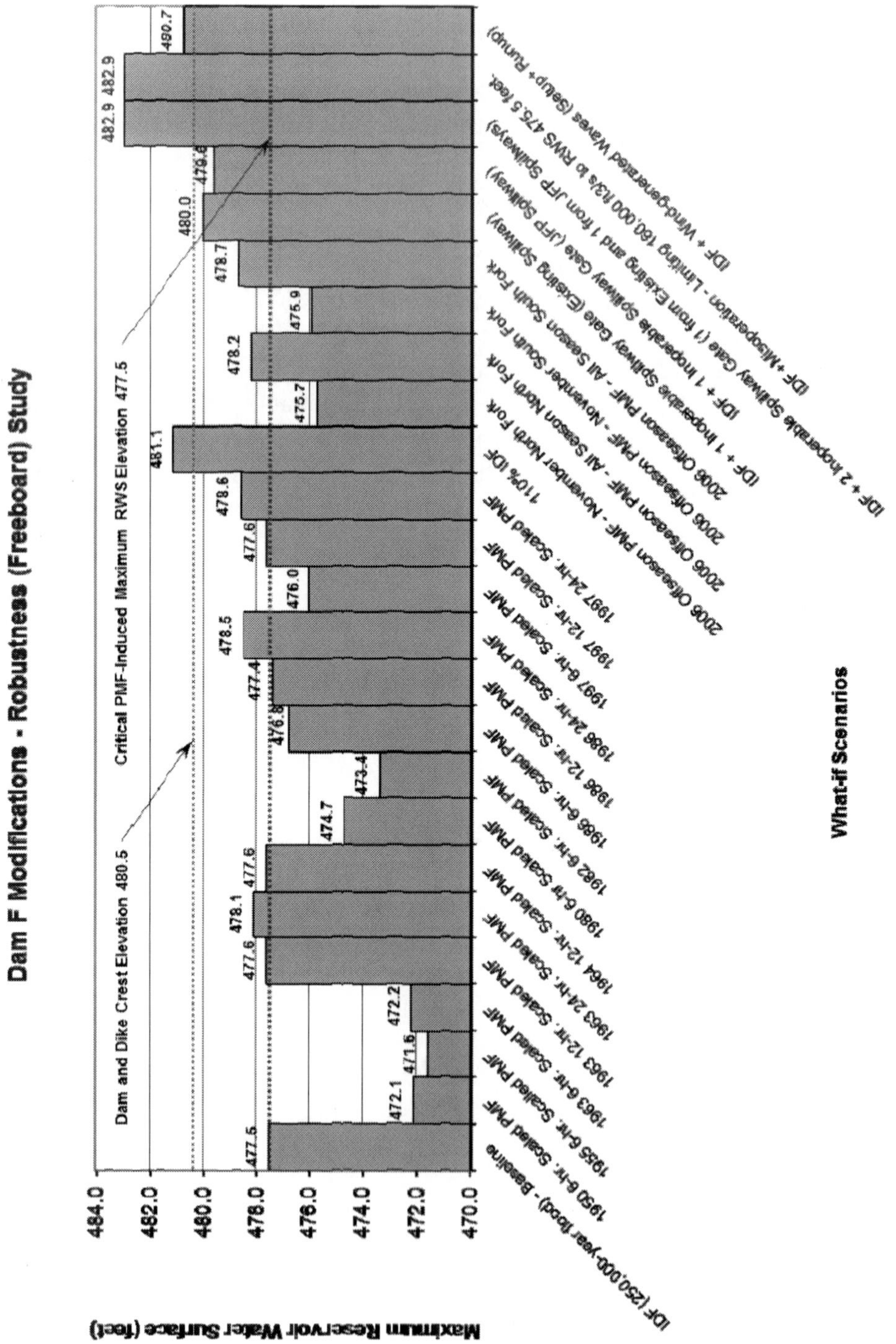

Figure 4. Robustness study results for Dam F modifications.

Findings

For Dam F modifications, the IDF has been equated to a frequency flood with a return period of 250,000 years. Additionally, 3 feet of freeboard above the maximum RWS associated with the IDF will be required to address uncertainties associated with the hydrology, modification, operations, and future events.

CONCLUSIONS

Through the use of risk-informed decision making, Reclamation has been better able to effectively manage the dam safety program. It allows for more efficient use of the limited dam safety funds and helps identify the critical dam safety issues within our inventory. The inflow design flood selection process that is outlined in Design Standard No. 14, Chapter 2, incorporates risk methodology into the design process and provides a basis for optimizing the selection of the IDF.

REFERENCES

Bureau of Reclamation, *Guidelines for Evaluating Hydrologic Hazards*, Technical Service Center, Denver, Colorado, June 2006.

Bureau of Reclamation, *Interim Dam Safety Public Protection Guidelines*, Dam Safety Office, August 2011.

Bureau of Reclamation, Design Standards No. 14, Appurtenant Structures for Dams (Spillways and Outlet Works), Chapter 2, Hydrologic Considerations, Technical Service Center, Denver, Colorado, 2013

Bureau of Reclamation, U.S. Army Corps of Engineers, *Best Practices in Dam and Levee Safety Risk Analysis Training Manual*, Version 4.0, July 2015.

Costa, J. E. , ''Floods from dam failures.'' U.S. Geological Survey, Open-File Rep. No. 85-560, Denver, 54, 1985.

Frank, W., "The Cause of the Johnstown Flood," Civil Engineering Magazine, May 1988.

Hansen, B., Ste mming the Flow: The Sweetwater Dam , *Civil Engineering Magazine*, June 2007.

www.molare.net.

THE WANAPUM SPILLWAY MONOLITH CRACKING INCIDENT DAM SAFETY MANAGEMENT IMPLICATIONS

Robin G. Charlwood, Ph.D., P.E.[1]
Adam N. Jones, P.E.[2]
Kevin Snyder, P.E.[3]
Kevin Marshall, P.E.[4]
David Mishalanie, P.E.[5]

ABSTRACT

Opening of a crack with vertical displacement of up to 2 inches along a submerged lift joint within Spillway Monolith No. 4 was observed at Wanapum Dam in February 2014, after 50 years of service. This resulted in the need to immediately lower the reservoir by 27 feet to stabilize the spillway structure and maintain dam safety. After completion of a Forensic Investigation/Root Cause Analysis (FI/RCA) and development of a remedial measures design, stabilization measures have been completed and the project was progressively returned to full service. The total impact of this incident was substantial when the costs of investigation, design and construction, environmental and cultural impact mitigation, loss of generation, water supply and irrigation mitigation, loss of recreational use and the need for full-time security patrols in the 38-mile long reservoir while the reservoir level was reduced are considered.

The project is owned and operated by a large, experienced and conscientious organization, Grant County Public Utility District No. 2, with substantial engineering and operational resources. The project is licensed by FERC and is subject to their dam safety guidelines and requirements, including Part 12D inspection requirements. These requirements were expanded by FERC in 2002 with the introduction of the Dam Safety Performance Monitoring Program (DSPMP) which, in addition to 5-year inspections by an Independent Consultant, included a Potential Failure Modes Analysis (PFMA) and the development of a formal Dam Safety Surveillance and Monitoring Plan (DSSMP). All of these measures have been fully applied at Wanapum Dam.

This paper provides a brief description of the incident, summarizes the findings of the Forensic Investigation/Root Cause Analysis of the observed cracking and reviews relevant aspects of the design process and the subsequent application of the FERC DSPMP to the project. Lessons are identified regarding dam safety management practices that could help prevent similar incidents at this and other projects.

---

[1] Principal, Robin Charlwood & Associates, PLLC, 1842 Foliage St., Freeland, WA 98249, and Past Chairman, ICOLD Committee on Concrete Dams; Tel: 425-478-1642; Email: robincharlwood@gmail.com
[2] Senior Engineer, HDR Engineering, Inc., 970 Baxter Boulevard Suite 301, Portland, ME 04103; Tel: 207-239-3820 ; Email: Adam.Jones@HDRInc.com
[3] Vice President | Assistant Department Manager Hydropower Services, HDR Engineering, Inc., 2379 Gateway Oaks, Suite 200 | Sacramento, CA 95833; Tel:: 916.679.8762; Email: kevin.snyder@hdrinc.com
[4]Hydro Director at Grant County Public Utility District No.2, Moses Lake, WA; Tel: (509) 793-1536 ;Email: Kmarsha@gcpud.org
[5] Dam Safety Supervisor at Grant County Public Utility District No.2, Moses Lake, WA; Tel: (509) 754-6622 ;Email: Dmishal@gcpud.org

INTRODUCTION

Wanapum Dam is located on the Columbia River in central Washington, and includes an 832-foot-long concrete gravity spillway containing 12 radial gates, each 50 feet wide and 67 feet high. At the time the dam was constructed in the early 1960s, the radial gates were the largest in the world. Wanapum Dam is a part of the Priest Rapids Project and began operations in 1963. Figure 1 shows a plan of the dam which includes seven major structures. Figure 2 shows a view of the spillway from downstream. Figure 3 shows a cross section of the spillway showing the location of the crack in Monolith 4.

In late 2013, GCPUD personnel observed an increase in leakage at the bottom of Gates 3 and 4, which was initially attributed to seal damage or debris. On February 24, 2014, a hydro mechanic observed misalignment of the upstream and downstream curbs and handrails located on the roadway on top of the spillway. Alignment surveys determined that the top of Pier 4 had moved downstream approximately 1 1/2 inches (0.04 meters). Divers inspected the upstream face of the spillway and observed a 65-foot (19.8-meter)-long by 2-inch (0.05-meter)-high open crack along a horizontal lift joint, spanning the full length of Monolith 4 at elevation (El) 485 feet (147.8 meters), as seen in Figure 5. Cracking and spalling of the contact surfaces with the adjacent monoliths were observed on the upstream face of the spillway above El 485 feet (147.8 meters) and on the downstream face of the ogee. Inspection of Monolith 4 below El 485 feet (147.8 meters) and the adjacent monoliths found no comparable differential displacements within or between the other monoliths.

Observation of the displacement resulted in a series of actions, including:

- Activation of the Wanapum Dam Emergency Action Plan;
- Formation of an Independent Board of Consultants to review and assess ongoing investigations, studies, analyses and spillway repair alternatives from a dam safety perspective;
- Drawdown of the reservoir (Lake Wanapum) by approximately 27 feet (8.4 meters);
- Numerous environmental and cultural mitigation activities;
- A Forensic Investigation and Root Cause Analysis (FI/RCA);
- Design and construction of spillway stabilization measures; and
- Modifications to the Wanapum Spillway surveillance and monitoring program.

While the incident was not considered a dam failure, in that there was not an uncontrolled release of the reservoir, it was recognized as a significant dam safety incident with significant environmental, cultural and economic impacts. GCPUD took appropriate, timely action to modify dam operations and surveillance procedures to maintain project safety, identify the root cause, and remediate the project so that the reservoir could be restored to its normal operating level in early 2015.

Two companion papers reviewed details of the Forensic Investigation/Root Cause Analysis of the observed cracking (Adam Jones et al, 2015) and the response and remediation (Dave Mishalanie, et al 2015).

This paper focuses on lessons learned regarding dam safety management practices.

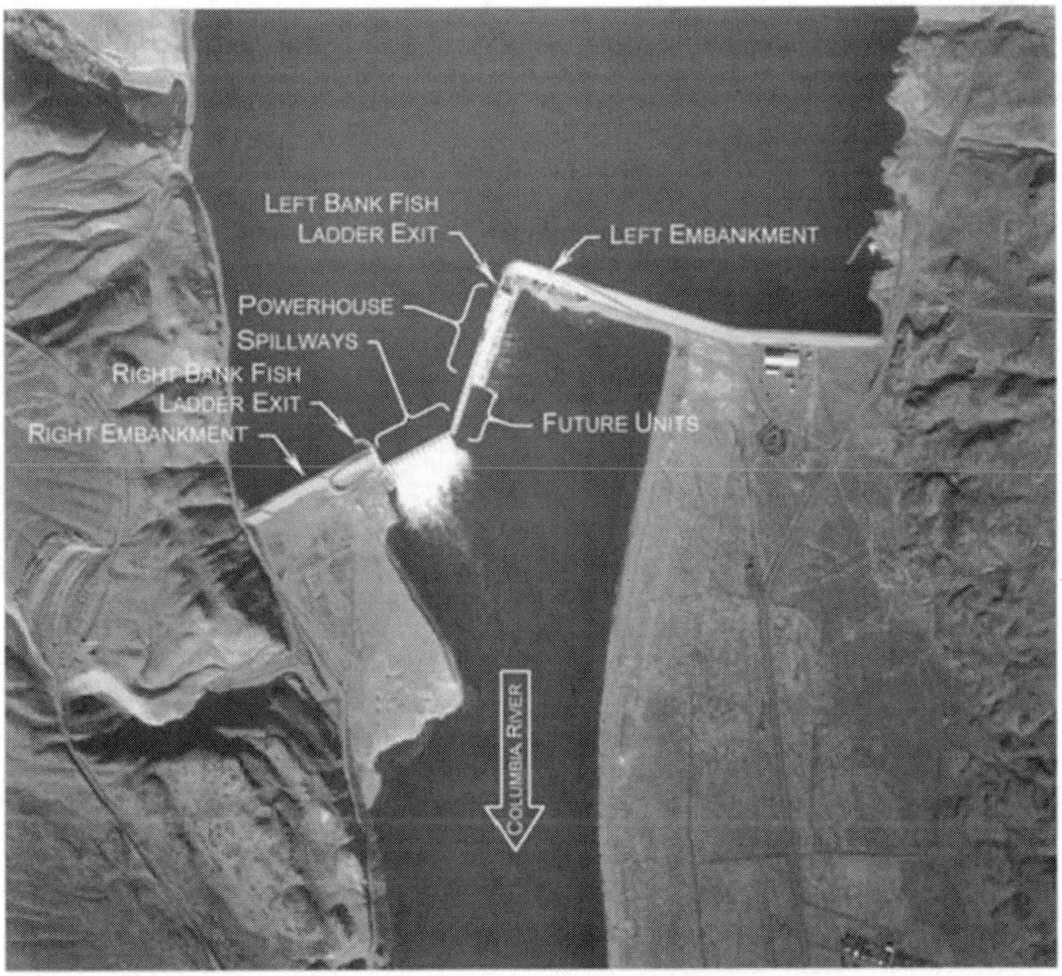

Figure 1. Aerial View of Wanapum Project showing Spillway Location

Figure 2. Downstream View of Wanapum Spillway

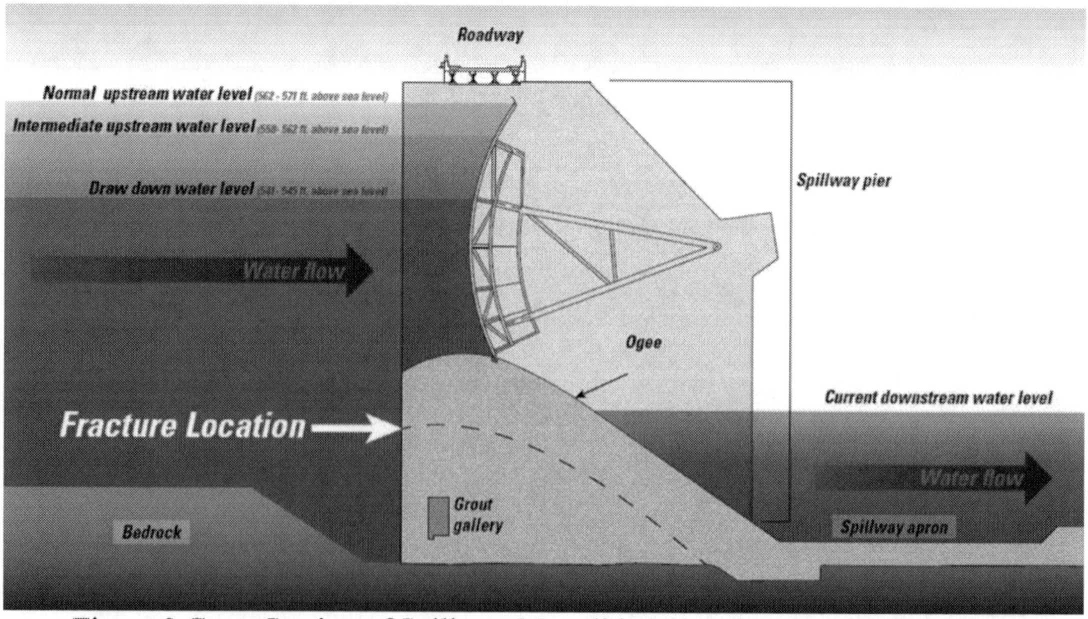

Figure 3 Cross Section of Spillway Monolith 4 Showing Location of the Crack

RESULTS OF THE FORENSIC INVESTIGATION/ROOT CAUSE ANALYSIS

The cracking initiated along a lift joint in the mass concrete of the ogee at elevation 485 ft., immediately below the termination of reinforcement in the spillway pier. The lift joint was found to be in a state of low tensile stress at the upstream face under normal loading and then this stress condition was subjected to additional local annual cyclic thermal stresses which significantly increased the tensile stress at the face each winter. No significant, recent extreme load events had occurred, although the spillway had been exposed to a moderate earthquake, surcharge loading, and spillway dewatering, all of which had temporarily increased the stresses on the spillway, but occurred many years before the movement was observed. Although analyses indicated that the normal tensile stresses at the upstream face of the spillway were relatively low, the section was shown theoretically to be unstable if cracking did occur and uplift occurred in the lift joint.

The FI/RCA concluded that the crack initiation was the result of tensile stress under normal loading, exacerbated by thermally-induced seasonal tensile stresses and tensile strength loss as a result of exposure to repeated the relatively high-amplitude, cyclic thermal stresses. The crack then propagated as a result of hydrostatic uplift intrusion.

At the time the dam was constructed in the early 1960s, the radial gates were the largest in the world. The design calculations for the original layout of the trunnion anchors were fully checked and the stability and stress results met the designer's no-tension criteria at the base. The lift joint that cracked was also shown to be in compression under normal loading. It was also shown that the stability of the entire monolith was dependent on an effective drainage system limiting the uplift pressures on the foundation contact. A grout curtain and drainage system was therefore installed at the contact. No drainage was provided at the lift joints above the foundation presumably on the basis that the lift joint was in compression and uplift penetration was not expected.

The inclination of the trunnion anchors was adjusted to raise the anchor heads above the tailwater at a late stage of the design. A "check calculation" was performed to assess the stresses on the critical lift joint with the modified trunnion anchor location which was now above this lift joint and no longer provided a stabilizing compressive force component at that elevation. These "check calculations" showed that the critical lift joint would be in compression. In addition, similar calculations were used to check that the vertical reinforcement could be terminated above this lift joint.

Unfortunately, one key page of these "check calculations" contained numerical errors and was not marked as having been checked. Construction then proceeded on the basis of the new trunnion anchor and pier reinforcing layout.

Corrected rigid body stability calculations by the FI/RCA team showed that the lift joint was in a state of low tensile stress, approximately 40 psi, at the upstream face under normal loading. Furthermore, if cracking was assumed at this lift joint and uplift pressures extended into the crack, the monolith was predicted to be unstable.

FERC DAM SAFETY PERFORMANCE MONITORING PROGRAM

Failure Modes Analysis (PFMA) has been a center piece of FERC's Dam Safety Performance Monitoring Program (DSPMP) since 2002. A PFMA was performed for Wanapum Dam in 2004.

The PFMA Workshop took place over a period of four days. The project was considered in seven parts for the PFMA. These included the: Powerhouse; Future Unit Intakes; Spillway and Gates; Left Gravity Structure/Fishway, the Right Gravity Structure/Fishway and the Erection Bay; Right Embankment; Left Embankment and the River Closure Section.

Some perspectives on the 2004 Wanapum PFMA:

- 48 PFMs and 9 Other Considerations were identified in all 7 structures. This was considered by the participants to be very thorough, if not excessive!
- PFMs were identified for each structure within a process of considering major loading conditions one by one: Flood, Earthquake and Normal Operating conditions and recording the discussion in an established tabular format.
- 10 PFMs were identified for the Spillway. These included no Category I PFMs, 3 Category II PFMs, 3 Category III PFMs and 4 Category IV PFMs. These included issues such as undermining, failure to open gates, trunnion anchor failures, loss of drain effectiveness, flood capacity, etc.

  <u>However, the monolith lift joint cracking PFM that occurred in February 2014 was not identified</u>!

  The PFMA "Major Findings and Understandings" noted that: "Stability of all of the major concrete structures has been recently re-evaluated using current analysis methods and all structures were found to meet current FERC Guidelines criteria in their "as designed" condition."

  It was also noted that: "The MCE analyses of the Spillway monoliths show potential cracking at the foundation contact near the heel. It was suggested to assess potential for rupture of the grout curtain and resulting effects on uplift and stability."

Observations on the PFMA process and Wanapum PFMA include:

- The Core Team was a broadly experienced group of dam engineers lead by one of the most experienced Facilitators in the FERC program with three senior dam safety engineers from other utilities as observers. Several operating staff attended throughout the Workshop sessions.

- What other mechanisms and/or PFMs could have been considered? With the benefit of hindsight again, perhaps:
  - Stability calculation errors?

- Design deficiencies, eg. no internal drains?
- Thermal effects?
- Cyclic/fatigue?
- Others…?

Recognizing that this PFMA missed what transpired to be the critical PFM, how could the PFMA process be improved generally?

- The key point is obviously to find ways to avoid missing important PFMs!
- Initially, this may be more important than being really precise about the description and analysis of the PFMA mechanism.
- Some recent PFMAs have started the Workshop stage with a free thinking "Brainstorming" session which is then reviewed and only selected PFMs developed in detail.
- With encouragement by the Facilitator, "Brainstorming" may free up the group to keep "thinking of the things they haven't thought of" without concerns of running out of time or exhausting everyone by going through too many PFMs in detail.
- Once the important PFMs have been identified then fully describing the sequence of development is definitely important to provide a rational basis for analysis.
- More training and documentation of case histories would help. This should and could be done in such a way that it doesn't stop free thinking based on engineering fundamentals.
- Perhaps special attention should be applied to features that are precedent setting and extending the state of practice?
- Reviewing or reanalyzing every element of every structure for all foreseeable loading conditions as part of the PFMA process is not practical. Diligent review of the existing drawings and analyses, and comparison with incident case histories of similar structures may provide clues regarding areas that should be assessed in more detail. When done appropriately, this would provide the appropriate focus to allow significant issues to be identified.

DAM SAFETY SURVEILLANCE AND MONITORING

The District monitored movement of concrete monoliths and embankments relative to established baselines using alignment monuments. The baselines were established by monuments on the right and left banks.

Spillway monolith movement data was collected quarterly up to 2009. In the absence of an identified stability PFM for the spillway monoliths, it was suggested that the reading frequency could be reduced to annually, at the same time as the embankment settlement survey, or post-earthquake, GCPUD collected spillway offset data bi-annually after 2009 to capture seasonal variations.

An Engineering Technician performs data collection and prepares time-history graphs. The Engineering Technician works closely with the Dam Safety Engineer on checking and plotting of data. Data is reduced, graphed, and evaluated by the Engineering

Technician (ET) and the Dam Safety Engineer usually within a week. All ETs were cross-trained to read the various surveillance instruments at Wanapum Dam..

However, it was not the surveillance and monitoring program that initiated the discovery of the offsets and cracking in February 2014. A District mechanic noticed an offset in the curb and that the conduit rail on the spillway crest at Monolith 4 appeared bent or damaged as shown in Figure 4.

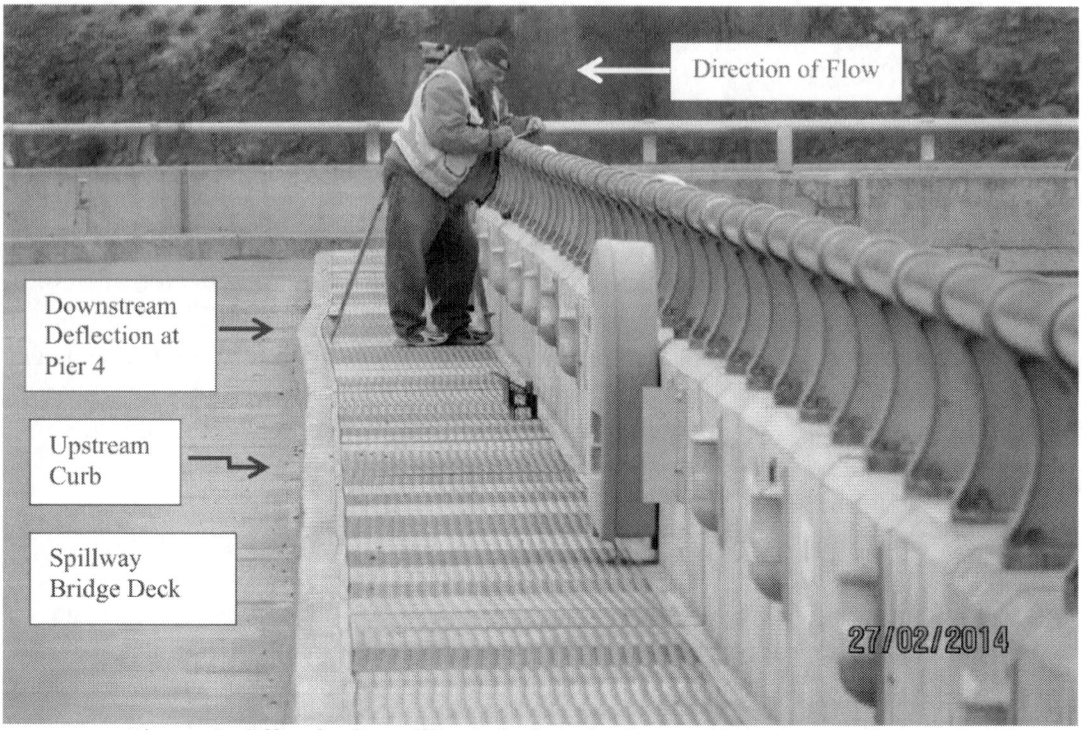

Figure 4. Offset in the spillway deck curb observed by operations personnel.

This lead to the investigations by the District engineering team and discovery of the underwater cracking.

Offset data collected since 1991, when the baseline survey was initiated, was plotted and included in the DSSMR reports as shown in Figure 5 which also shows reservoir head for correlation purposes.

In Figure 5 the offset data for Pier 4 is shown and an "Illustrative Pier Threshold Value" of 0.06 ft, added for discussion purposes herein. The Threshold value shown is simply the maximum value of offset that occurred in the period 1991 to 2004 when the DSPMP program was introduced and the PFMA performed.

Comments on the monitoring:

- No PFM had been identified that directly required monitoring of the spillway monolith offsets. Consequently there was continuing debate on the need for this monitoring. However, this was done as a matter of good practice or what is sometimes called "Basic health monitoring". The value of this type of monitoring is clearly evident.

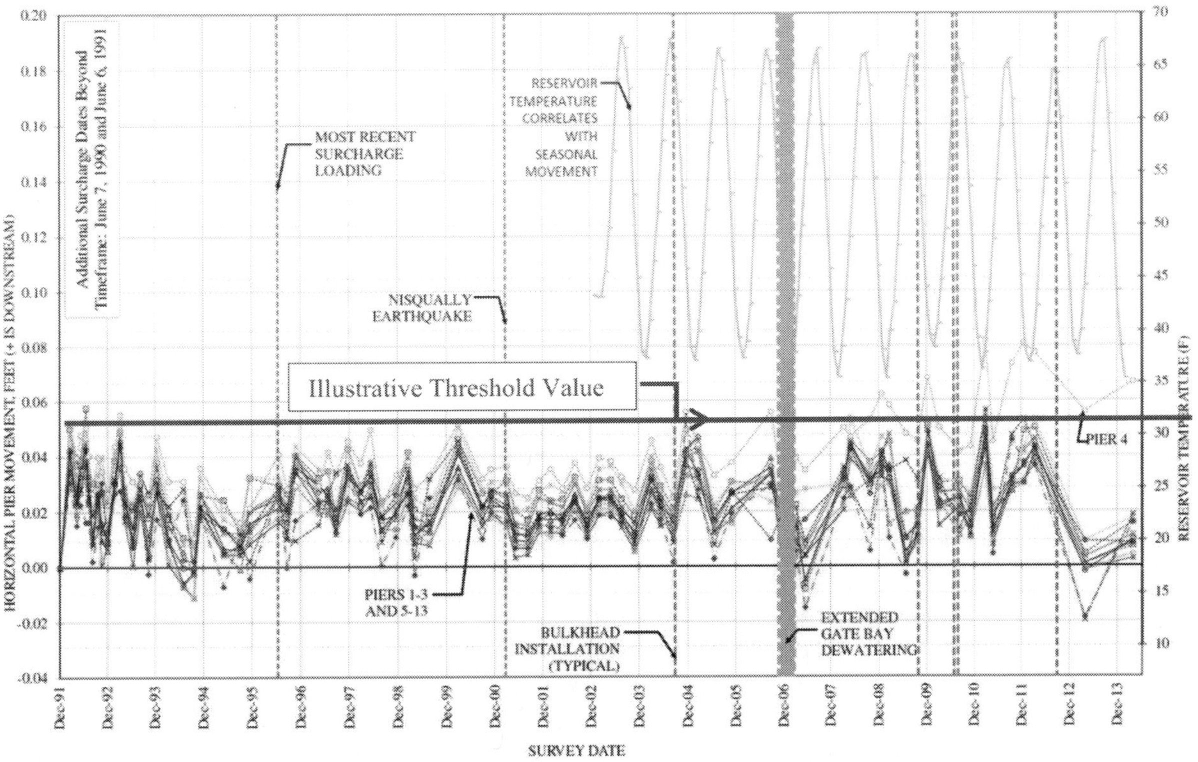

Figure 5. Monitoring Data

- The absence of an identified PFM may have diluted the apparent value of a Threshold Value which in fact may have assisted earlier identification of anomalous behavior.
- With the benefit of hindsight, there does appear to be something different happening with the Pier 4 data relative to data for the other piers from 2006 to 2011 although the data was within the range of historical data.
- Review of the offset data relative to a threshold level of 0.092 ft may not have identified a significant emerging trend in monitoring data until February 2011.
- The FERC DSPMP comments on applications of thresholds and action levels all sound very reasonable in principle. Would a more elaborate threshold have been helpful? Maybe not. The data varied substantially between and within each year. Setting a cyclic threshold band would be more complicated and may or may not not be a practical option.
- GCPUD had a team of an ET and Engineer who read and evaluated the surveillance and monitoring data which is typical good practice.
- The most recent Part 12 Inspection prior to the 2014 incident took place in 2009. It was not until 2011 that the data anomaly was clearly apparent.

- An underwater inspection program may have identified the issue at some stage. Finding a crack and then recognizing its significance would have been a tough ask for a diver, until after the gross movement occurred
- Several of the ETs and operators attended the PFMA in 2004 and would have had familiarity with the identified PFMs but this would not have sensitized them to the incident mechanism which was not identified or discussed as a normal load case scenario.

REVIEW OF SUPPORTING TECHNICAL INFORMATION (STID)

In 2004 in preparation for the PFMA and Part 12 D Inspection, GCPUD prepared a comprehensive Supporting Technical Information document which provided a summary description of the project, surveillance and monitoring systems and other required topics including a summary of the stability analyses.

FERC Chapter 14 DSPMP section 7 requires the Part 12D Independent Consultant to assess the contents of the "Supporting Technical Information" document compiled by the licensee. FERC states that:

> *The STI document should include information needed to understand and confirm the underlying assumptions and the conclusions of the analyses of record supporting the assessment of the safety of the Project. In each section, where appropriate, the Independent Consultant shall make a clear statement that they have reviewed the pertinent analyses and evaluations along with the underlying assumptions and that they have concluded that the assumptions and methods of analysis or evaluation were appropriate for the structure, were applied correctly and are appropriate given current guidelines and the state of dam safety practice.*

Section 8 reviews the required contents of the STID and states:

> *Because every dam and hydroelectric project is unique, it is not possible to list here all the various items that are required to adequately detail stability or stress assessments of the project water retaining structures. It will be the responsibility of the Licensee to include all information necessary for the reader to understand the assumptions, methods of analysis, and load cases assessed for each project structure. Stability and stress analyses for each structure shall be summarized graphically for ease of understanding. The following types of information should be provided:*
> *8.1 General*
> *8.1.1 Listing of credible load cases analyzed (including water levels for dam, canal and flume analyses or pressure for penstock and flowline analyses)*
> *8.1.2 Statement of the method of analysis used and the computer program adopted.*
> *8.1.3 Properties of materials based on site specific tests or assumptions (state which). Include representative test data and summary sheets.*
> *8.2 For each gravity structure and load case:*
> *8.2.1 Graphic free body diagram (cross-section) of each structure showing:*
> *• the assumed self-weight of the cross section*

- *all applicable loads including, as appropriate:*
- *assumed uplift pressure distribution*
- *silt loads*
- *headwater and tailwater loads*
- *point loads*
- *ice load*

8.2.2 Key elevations
8.2.3 Key lateral dimensions
8.2.4 Piezometer and drain locations
8.2.5 Foundation shear strength parameters
8.2.6 Minimum cohesion to meet stability criteria
8.2.7 Negative crest pressures
8.2.8 Concrete unconfined and splitting tensile strength test results
8.2.9 AAR potential or evidence
8.2.10 Failure planes investigated, etc.

From the viewpoint of achieving a comprehensive set of PFMs, the contribution of the items on the FERC list of issues to be addressed will vary depending on the circumstances. In the case of the Wanapum incident, with hindsight, clear identification of the failure planes within the monolith was critical and is clearly something that should be considered in future safety reviews.

During the course of the post-incident Forensic Analysis/Root Cause Analysis some of the original design stability calculations were recovered by the designers from storage and examined. These 1959 and 1960 calculations addressed overall spillway monolith stability and also included checks on stability at critical lift joints within the ogee section. The availability of these calculations in the designer's record management system was helpful and is a credit to the designers.

The original design calculations were not referred to in the STID, Owners seldom have these, but a comprehensive summary of the various re-analyses performed as part of the subsequent Part 12D Inspections was provided. No mention was made of any of the Part 12 D Inspections specifically addressing stability at the lift joints above the foundation in the ogee section. Comments focused on the foundation contact. The 2004 PFMA and Part 12D Inspection noted the reliance on the grout curtain and drainage at the contact and recommended a check on the drain efficiency requirements in a post-earthquake condition. The concern was whether the loss of function of the grout curtain and the drains could result in increased uplift pressures for the post-earthquake condition and reduced stability safety factors. The 2009 IC performed the recommended studies and found that the spillway structure met stability criteria without the grout curtain/drain system.

In the follow-up to the Wanapum incident, the following general questions have been asked by various observers about the contents of a STID and the approaches to reviews of stability calculations:

- If they are available, should the original design calculations be reviewed and copies appended to the STID?
- In reviewing previous stability analyses, is there a need for spot checks at critical locations and if so should these be based on PFMs or based on engineering judgment or some other basis?
- In cases where drainage is required at the foundation contact for stability, should this trigger attention to the possible need for drains at lift joints?
- If tension is allowed at lift joints under static loading should the possible need for drainage be considered in case cracking occurs?
- What is the significance of cyclic thermal stresses on lift joints with regard to local tensile stresses?
- Should the potential effects of repeated thermal cycling on lift joints be considered as a concrete fatigue mechanism?
- To what extent should available reinforcing drawings be examined and the locations of termination of reinforcing examined?

The following responses are offered by the authors (again, with the benefit of hindsight):

- If available, then it is suggested that the original design calculations should be reviewed and copies appended to or referenced in the STID.
- In particular, if major changes were made in the final stages of the design and layout that could affect stability were identified then it would of course be beneficial to check that the implications of these were appreciated and fully considered.
- Normally, when reviewing previous stability analyses, attention to specifics is guided by experience. However, in cases where the design was precedent setting, such as the large gates at Wanapum, then experience may be insufficient and there could be a need for spot checks at critical locations. From a safety systems analysis viewpoint, ideally, these checks should be focused on issues identified as PFMs. This really reinforces the key role of a comprehensive and rigorous PFMA although this will be challenging in precedent setting designs.
- In cases where drainage is required at the foundation contact for stability, then this should certainly trigger attention at lift joints or potential failure planes within the dam body.
- If tension is allowed at lift joints under static loading then the possible need for drainage should be considered in case cracking occurs as part of a program of parametric or "what if" calculations to address uncertainties in the analyses.
- Cyclic thermal stresses on lift joints will normally be very localized near the upstream face of a gravity structure. The potential effects of these with regard to local tensile stresses will depend on the normal state of stress and the sensitivity of the section to uplift pressure penetration in the event that a crack develops. These are not explicitly considered in normal design and safety assessment practice but deserve more attention.
- There is only very limited data available on the potential effects of repeated thermal cycling on lift joints as a concrete fatigue mechanism. It has the potential

to exacerbate the effects of thermal cycle stresses in critical situations. This mechanism deserves further research.

- In situations where structural stability relies on reinforcement, such as relatively thin section spillway piers, the locations of termination of reinforcing are critical.

RISK INFORMED DECISION MAKING (RIDM)

The introduction of PFMAs to FERC's dam safety program allowed qualitative assessment of failure modes that were not amenable to numerical analysis or criteria and allowed surveillance and monitoring programs and emergency action plans to be focused on specific failure modes. After ten years of experience with PFMA FERC is proposing that the practice of risk analysis in dam safety now has the potential to be a major aid to decision-making in the interest of public safety.

- Would RIDM thinking help prevent a Wanapum type of situation? The PFMA process is central to the RIDM process and certainly focusses attention on the key issue, identifying critical PFMs.
- Would addressing uncertainty more explicitly help? The key issue at Wanapum appears to be that the key PFM was missed. Consideration of aleatory (randomness) uncertainty in loads or material properties would not have helped.
- The issue is one of epistemic uncertainty, the process of acquiring knowledge or data and the process of using it.
- Would improving awareness of historic failures help? Yes, if they were similar. This would be improving knowledge and broaden the basis for identification of PFMs.
- Similarly, improved knowledge of the behavior of concrete structures, particularly long term and ageing behaviors, would help identify underlying or initiating conditions or events.
- The PFMA process provides a practical and readily applicable process for identifying critical PFMs. Focusing on seeking completeness in the set of credible PFMs appears to offer the most immediate opportunity for avoiding situations similar to the Wanapum incident.

CONCLUSIONS & RECOMMENDATIONS

PFMA Process:

- PFMAs: The PFMA approach is certainly helpful BUT is dependent on completeness of set of PFMs.
- The challenge is to "think of the things that haven't been thought of".
- Give emphasis to the "Brainstorming" stage of PFMA without limitation of number of PFMs to try to get all possible PFMs identified.
- Detailed definition of sequence of development of the PFMs is important for subsequent analysis and monitoring but identifying all of the critical PFMs is the essential first step.

Part 12 Inspections and Reviews of STID:

- Critical review of identified PFMs and checking for any new PFMs should be emphasized.
- Access to design calculations would be beneficial – include on DVDs attached to STID when available.
- In cases of precedent setting designs, spot checks on key calculations would probably be helpful.
- Give special attention to lift joints within the mass concrete in cases where drainage is required at foundation contacts and consider the need for internal drainage in such cases.
- Give special attention to the location and manner of reinforcing termination when stability is controlled by reinforcing.

DSSMP:

- Data collection is good but informed and thorough data evaluation is essential.
- Clear plots formatted to a scale so that abnormalities are apparent are essential.
- Define and monitor to appropriate thresholds whenever possible.
- There is still no substitute for informed visual observations. Wanapum demonstrated that!

ACKNOWLEDGEMENTS

The authors appreciate the willingness of GCPUD to openly and objectively discuss a difficult incident and hope that the lessons learned will help improve dam safety generally.

REFERENCES

Adam Jones, Kevin Snyder, Robin Charlwood, Kevin Marshall, and Dale Campbell, "Wanapum Spillway Crack Development Root Cause Analysis", ASDSO, Annual Meeting, New Orleans, September 2015

Dave Mishalanie, Greg Rollins, Rebecca Simpson, Ian Hunter and David Thompson, "The Wanapum Dam Spillway Incident: Response, Remediation and Lessons Learned", ASDSO, Annual Meeting, New Orleans, September 2015

PANAMA CANAL BORINQUEN DAMS AND THIRD SET OF LOCKS OPERATIONS AND MAINTENANCE STRATEGY

Rachael V. Bisnett, P.E.[1]
Wonnie Kim, P.E.[2]
Donald J. Montgomery, P.E.[3]

ABSTRACT

The new Third Set of Locks for the Panama Canal has been designed to optimize reliability, availability, maintainability, and safety (RAMS) of operation. As part of the Third Set of Locks, the Borinquen Dams will allow navigation access through the Pacific Approach Channel to the new Pacific Post-Panamax Locks (Pacific Locks). Together with the Pacific Locks structures, the Borinquen Dams form the waterway of the Panama Canal and retain Gatun Lake.

Due to the significant consequences caused by interruptions to operations, the Borinquen Dams and the Third Set of Locks have been designed to a high level of reliability. To achieve this goal, operational requirements have been integrated with risk-based maintenance management techniques including fault tree analysis (FTA), failure mode effects and criticality analysis (FMECA), and application of reliability-centered maintenance (RCM) principles. Because the civil works components comprise the major water-retaining structures of the lock and canal system, it is a challenge to devise a comprehensive maintenance program – considering both planned and unplanned actions – that achieves the RAMS objectives without disrupting vessel transit operations or compromising personnel and vessel safety. This paper presents a unique approach to developing a comprehensive maintenance management program employing RAMS principles for the civil works components of the Third Set of Locks. The approach can serve as a template for other civil and structural systems where optimizing life-cycle performance is paramount.

INTRODUCTION

The Borinquen Dams and Third Set of Locks for the Panama Canal have been designed to optimize reliability, availability, maintainability, and safety (RAMS) of operation. To achieve this goal, operational requirements have been integrated with risk-based maintenance management techniques. These techniques include fault tree analysis (FTA), failure mode effects and criticality analysis (FMECA), and application of reliability-centered maintenance (RCM) principles. This integrated approach to performing required surveillance and maintenance practices while maximizing

[1] MWH, 350 N. Orleans St., Suite 1301, Chicago, IL 60654, 312-831-3188, Rachael.V.Bisnett@mwhglobal.com
[2] MWH, 1560 Broadway, Suite 1800, Denver, CO 80202, 303-291-2246, Wonnie.Kim@mwhglobal.com
[3] MWH, 1560 Broadway, Suite 1800, Denver, CO 80202, 303-291-2169, Donald.J.Montgomery@mwhglobal.com

availability of the civil systems for lockages enables optimal management of the asset throughout its life cycle.

While these techniques are typically applied to critical electrical, mechanical, and process control systems, the same principles also apply to the civil works components – embankment dams, lock structures, water saving basins, and civil site features.

PROJECT OVERVIEW

The Panama Canal Third Set of Locks Project will add a third lane to the existing Panama Canal locks to allow Post-Panamax size ships to traverse the Canal, greatly expanding shipping through the isthmus. The Third Set of Locks Project consists of a new lock complex at both the Atlantic and Pacific entrances to the Canal, which will allow vessels to move between Lake Gatun and sea level, an elevation difference of about 27 m.

The Borinquen Dams, comprised of four embankment dams, form the Gatun Lake approach channel to the Pacific set of locks and separate the waters of Gatun Lake from Miraflores Lake. The Borinquen Dams have a total combined length of almost 4 km, of which Dams 1W, 2W, and 2E are approximately 2.9 km. Dams 1W, 2W, and 2E, are being constructed under a design-build contract awarded to the construction joint-venture Grupo Unidos Por El Canal (GUPC) and the design joint-venture Consultores Internacionales LLC (CICP), of which MWH is the lead designer. Dam 1E is being constructed under a separate contract. The term "Borinquen Dams" from this point forward shall refer to Dams 1W, 2W, and 2E.

The new Pacific Lock Complex is located adjacent to the existing Miraflores Locks. Upstream of the new Pacific Locks is the new 218-m wide Pacific Approach Channel (PAC) formed by the Borinquen Dams (see Figure 1), which will retain Gatun Lake, the main waterway and water source for the operation of the Panama Canal. The Borinquen Dams will allow vessels to enter the Pacific Locks from Gatun Lake at approx. El. +27 m, which is nearly 10 m above the Miraflores Lake entrance to the adjacent Miraflores Locks.

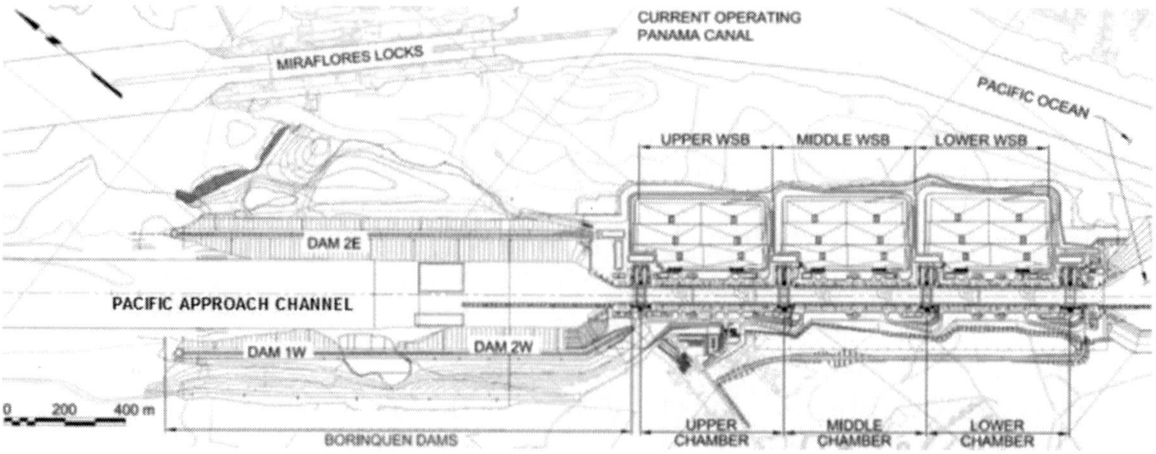

Figure 1. Third Set of Locks Project – Pacific Site Plan

The Borinquen Dams have been designed and constructed to achieve an extremely high degree of reliability since significant damage or failure of the dams could lead to the release of Gatun Lake and interruption to the Canal operation. The performance criteria and conceptual design of the dams were developed by the Owner, Autoridad del Canal de Panama (ACP), and reflect the utmost importance of reliability over the 100-year design life of the Project. Throughout the design-build process, several modifications were made to the conceptual design in order to arrive at a final design that accommodates construction methods, construction sequencing, material availability and site conditions. Key features of the Borinquen Dams are summarized in Table 1 and Figure 2.

Table 1. Summary of Key Dam Features – Borinquen Dams 1W, 2W, and 2E

Dam Type	Zoned basalt rockfill dam with central clay core
Crest Elevation (without camber)	El. 32 m PLD
Top of Core Elevation (without camber)	El. 29 m PLD
Crest Width	30 m
Max. Dam Height (above foundation)	35 m
Dam Length	0.6 km (1W), 0.9 km (2W), 1.4 km (2E)
Dam Fill Volume	7 million m^3 (approx.)
Dam/Core Freeboard (above NMOP)	4.87 m / 1.87 m
Inboard Slopes	2.87H:1V
Outboard Slopes	2.5H:1V (1W & 2W), 2.5H:1V to 3H:1V (2E)

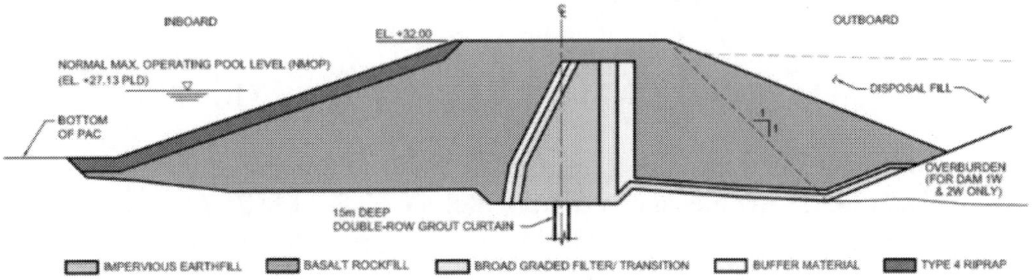

Figure 2. Typical Dam Section

Also retaining Gatun Lake and the Canal, are the locks complexes. Together, both the Borinquen Dams and the locks complexes must be operated reliably in order to prevent release of Gatun Lake and interruptions to Canal operations. Both the Atlantic and the Pacific locks complexes contain three lock chambers, which are 55 m wide and 427 m long, separated by lock heads (LH) with rolling gates. Adjacent to each lock chamber is a water saving basin (WSB) designed to save and reuse approximately 60% of the water used in a lockage cycle. The existing Panama Canal locks feature two identical lanes which allow one lane to be taken out of operation to accommodate maintenance or repairs. The Third Set of Locks however, consists of a single lane.

Although the Third Set of Locks includes redundancies in the filling and emptying systems (e.g. culverts and valves) and the gates, reliability, availability, and ease of maintenance of the civil components is critical to successful operation.

AVAILABILITY REQUIREMENTS

Each lock complex is required to be operational at least 99.6% of the time each month. Therefore, the development of maintenance philosophy is driven not only by the characteristics of the system, but also by the availability requirements. System control software has the greatest effect on availability, largely because unplanned software outages are likely to occur, and therefore need to be more carefully factored into availability computations.

The design of civil works utilizing international design standards and appropriate factors of safety, coupled with dedicated construction quality control, is a primary reason that unplanned outages due to civil components were considered to have a negligible influence on availability computations. Subjectively, unplanned outages due to civil works could be described as very unlikely to virtually impossible with corresponding probabilities of occurrence of 0.01 to 0.001 (USACE and USBR, 2012). Natural occurrences such as significant seismic events are categorized as force majeure events, so outages due to such events are similarly not incorporated into the availability computations.

On the other hand, civil works components contribute substantially to the scheduled outage portion of availability computations. As discussed in more detail within this paper, scheduled outages for civil works arise from planned inspections of normally inaccessible (e.g. submerged) components and owing to the reliability-centered maintenance (RCM) philosophy that any maintenance designed to prevent an outage is condition based.

MAINTENANCE MANAGEMENT TECHNIQUES

Maintenance management techniques utilized in the development of the Borinquen Dams and Third Set of Locks maintenance program include FTAs, FMECAs, and application of RCM principles. A system breakdown structure (SBS) was developed to organize each locks complex into systems, subsystems, and components for application of the maintenance management practices. The Third Set of Locks Project is comprised of eleven (11) major systems and eighteen (18) other systems. Figure 3 presents an example SBS for the Borinquen Dams.

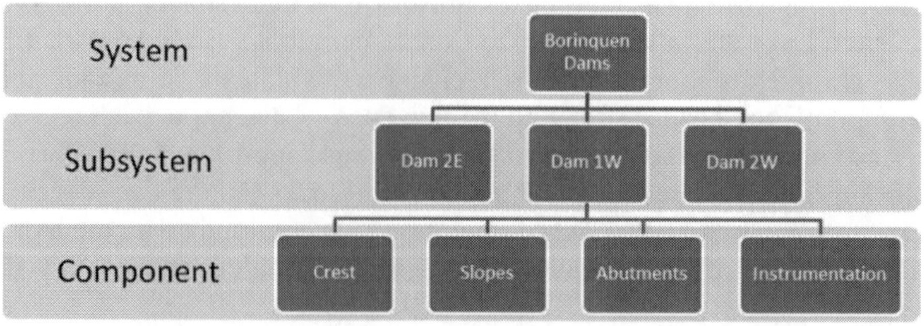

Figure 3. Example SBS for the Borinquen Dams

Fault Tree Analyses

The FTA is a "top-down" evaluation of a system or sub-system that begins with failure or malfunction of the system and ends with identification of root causes contributing to the system failure. Logic gates are used to define the path through the fault tree. Frequently, probability of failure of the system or subsystem will be computed using a fault tree that identifies the probability of occurrence and propagation of a root cause; the sum of probabilities in a FTA describes the probability of failure of the system or sub-system.

During the design process, fault trees are particularly beneficial to identify critical components of a system that may require additional analyses, redundancies, or special monitoring during operation. A FTA was performed for each subsystem determined by the SBS. A conceptual example of the Borinquen Dams subsystem fault tree is provided in Figure 4.

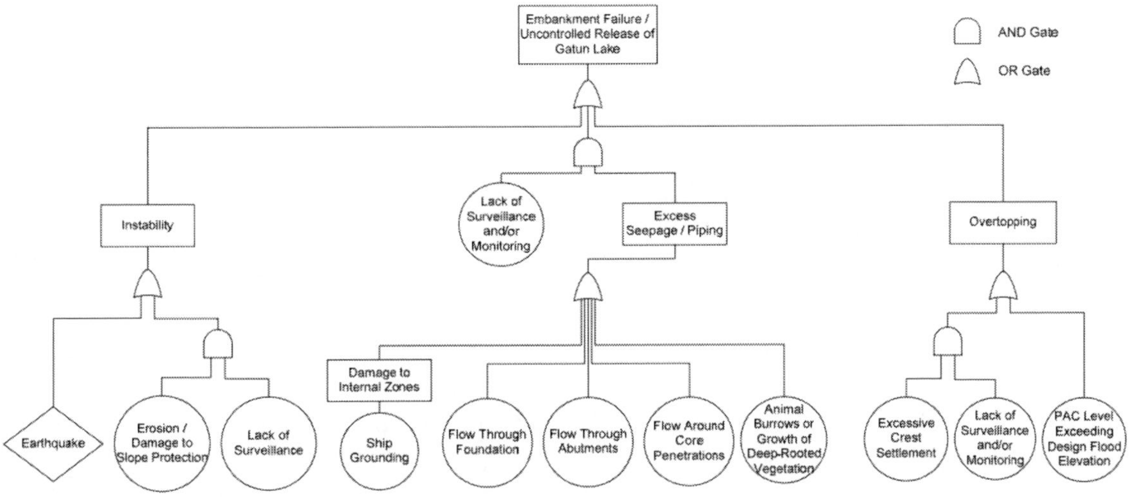

Figure 4. Borinquen Dams Conceptual Fault Tree Analysis

The top event in this fault tree describes the subsystem failure of "embankment failure or uncontrolled release of Gatun Lake". The top event was determined by the basic function of the subsystem; the function of each dam is to retain the PAC (and therefore Gatun Lake).

The "or" gate below the top event indicates that the top event could occur as a result of the occurrence of more than one series of events. The "or" gate leads to the three failure mechanisms. Below the failure mechanisms are the root causes that could each potentially initiate failure. Lack of surveillance and/or monitoring is coupled to either the root causes or higher event using an "and" gate. The condition-based maintenance program is designed to monitor the system for evidence of the potential root causes and perform early intervention to prevent a worsening that could lead to failure.

Another example fault tree for the lock chambers subsystem is provided in Figure 5. The top event in this fault tree describes the subsystem failure of "lock unavailable". The function of the lock chambers is to maintain a certain hydraulic geometry such that vessels can transit the canal. The "or" gate leads to "geometry change" which is an intermediate event that describes the failure progression. Geometry change can be caused by lack of inspection coupled ("and" gate) with one of the following root causes: accumulation of sediment in the lock chambers; erosion or scour of the chamber floors or walls; differential movement of the lock chamber monoliths; or elevated water in the backfill. The condition-based maintenance program is similarly utilized for the lock chambers.

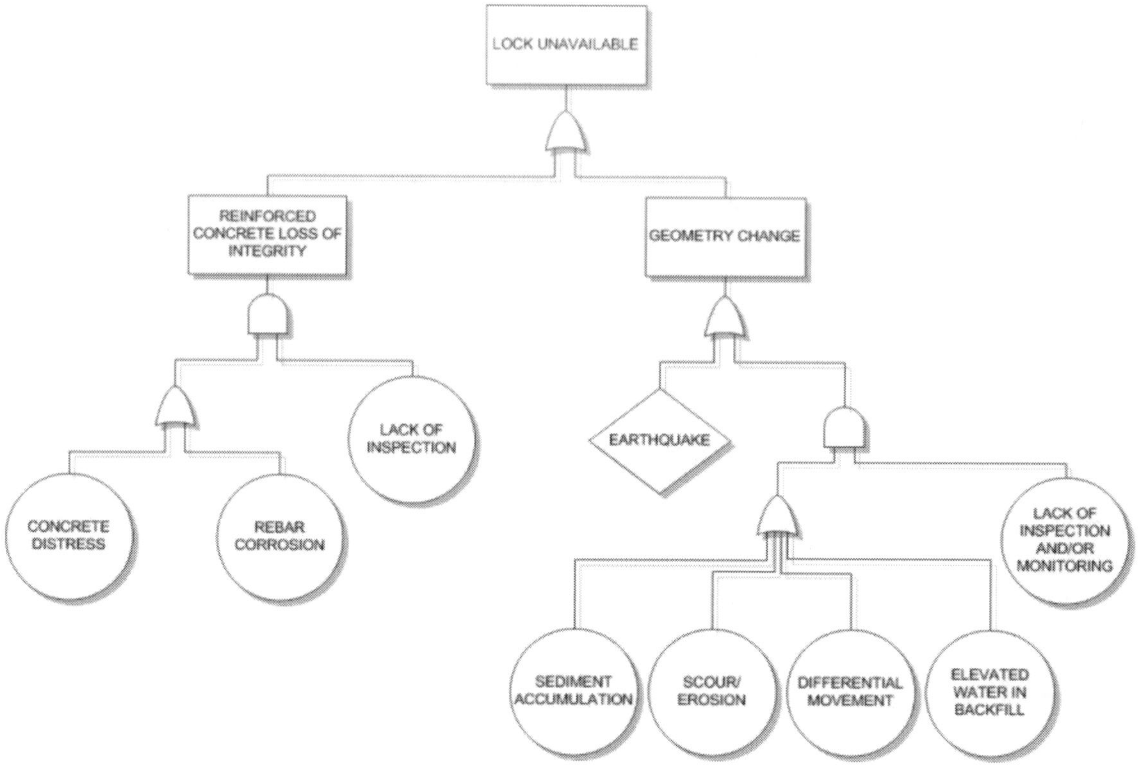

Figure 5. Lock Chambers Conceptual Fault Tree Analysis

Failure Modes, Effects, and Criticality Analyses

The FMECA describes the effects, or indicators, and criticality of the root causes, or failure modes, identified in the FTA. In the context of civil works, the indicators of a failure mode provide guidance for operations and maintenance (O&M) staff to identify the occurrence of a failure mode which is either in progress or has already occurred. Criticality can be defined in many ways, which may vary by project and stakeholder concerns. For the civil works of the Third Set of Locks Project, criticality was ranked on a scale of I through IV in accordance with US Department of Defense MIL-STD-882D (2000) guidance which considers injury or death of employees or the public; economic loss; and environmental consequences. A Category I event has catastrophic consequences, while a Category IV event is negligible.

The assignment of criticality varied for the system or subsystem. For the Borinquen Dams, the criticality ranking for each failure mode was assigned on the basis of estimated economic loss for repairs, as it is appreciated that should any root cause be permitted to fully develop into a dam failure or uncontrolled release of Gatun Lake, the results would be catastrophic.

Taking the flow through the foundation root cause as an example from Figure 4, the indicators of such a condition include piezometer response indicative of rising phreatic surface and depressions, sinkholes, or heaving identified in the slopes. Mitigating measures that would make development of the root cause into a failure mode unlikely include the minimum frequency of instrument monitoring which would allow early detection, foundation treatments performed during construction (e.g. grout curtain and dental concrete) and downstream filter. The criticality for the root cause was assigned a ranking of II.

Reliability-Centered Maintenance

The objective of RCM is to optimize the maintenance of an asset. RCM strategies include run to failure maintenance; fixed time maintenance; and condition based maintenance. The effects/indicators and criticality of the failure modes identified from the FMECA are used to inform the selection of RCM strategy for each asset.

Run to failure maintenance is applicable to those components where early detection of a failure or deterioration is not possible; or failure of the component is not critical to overall system availability. Fixed time maintenance is applicable to those components that have known durations of effectiveness such that regular intervals of maintenance efforts can be scheduled to prevent deterioration or failure of the component. Condition based maintenance relies on inspections to determine the condition of components and requires maintenance efforts to be scheduled as necessary based on the observed conditions.

For the Borinquen Dams and other civil works of the Third Set of Locks, condition based maintenance is most applicable. In the case of the foundation seepage/piping example from Figure 4, instrumentation monitoring and surveillance inspections will be utilized to assess the condition. Based on the trends and magnitude of piezometer readings and observations during inspections, the necessity and type of corrective actions can be evaluated.

MAINTENANCE MASTER MANUAL

Maintenance requirements for Third Set of Locks systems and components are grouped into preventive or corrective maintenance programs (manuals) based on the results of the FTA, FMECA, and RCM determination.

Preventive maintenance for the Project's civil components includes inspections, periodic maintenance, predictive maintenance (through monitoring instrumentation), and documentation of the condition of system components. Corrective maintenance typically incorporates necessary maintenance identified through the inspections and monitoring. Both preventive and corrective maintenance are opportunities for intervention to prevent a root cause from propagating up the fault tree toward a failure mode.

Preventive Maintenance - Inspections

Fixed-time inspections are part of the preventive maintenance program and consist of a program of routine and comprehensive inspections.

Routine inspections are performed daily by operations staff as a part of their normal duties. Routine inspections are considered the first-line of defense against potential development of failure modes. The success of the routine inspection program relies on the experience of operations personnel identifying something that looks different or out of place from what they are accustomed to seeing on a daily basis. If something out of the ordinary is noted, the staff member is required to document the observation for follow-up.

Comprehensive inspections are more detailed inspections performed by knowledgeable and trained engineers at standard time intervals. During a comprehensive inspection for normally visible (i.e. not submerged) features, the inspector uses a checklist to walk through a standard series of observations for each feature. The checklists are constructed so that any positive ("yes") response requires additional comments, description, and follow-up. Additionally, care is taken to particularly observe any previous repairs or positive responses for a feature.

Comprehensive inspections for normally submerged features of the dams and locks require specialized tools to avoid unwatering which is nearly impossible for the dams and extremely undesirable for the locks. Side-scan sonar is ideally suited since it is a widely available technology that provides visibility of underwater features and can be deployed with only minor planned interruption to lock operations.

Preventive Maintenance – Monitoring

Preventive maintenance also includes necessary instrumentation monitoring to evaluate performance of the systems. Instrumentation for the dams includes piezometers, survey monuments, and accelerometers located at critical sections of the dams. Instrumentation used to monitor the locks includes piezometers for backfill water levels adjacent to the lock walls, water level measurement wells for chamber water levels, and survey monuments for the concrete structures. Certain instruments are connected to an automatic data acquisition system that will record readings at set intervals, while many instruments will require manual readings.

The instruments each have action levels and threshold values established to aid in the interpretation of readings following guidance from Chapter 14 of the US Federal Energy

Regulatory Commission's (FERC) Engineering Guidelines for the Evaluation of Hydropower Projects (2005).

A reading that reaches an action level indicates a significant departure from the normal range of readings, and prompts an action. A reading that reaches a threshold value is indicative of the design threshold for the structures; and readings exceeding threshold values have the potential to impact the safety and stability of the structures. The data is evaluated to look for individual exceedance values as well as trends indicative of declining performance.

Corrective Maintenance

Corrective maintenance tasks are contingent on the findings from preventive maintenance inspections and instrumentation monitoring.

For example, corrective maintenance to address erosion or damage to slope protection on the dams (see Figure 4) includes repairing or replacing riprap; corrective maintenance to address accumulated sediment in the lock chambers (see Figure 5) would be to perform dredging within the lock chambers to remove the accumulated sediment.

As stated above, corrective maintenance is triggered if an action level or a threshold value is reached. In most cases, the corrective maintenance is defined as performing a comprehensive inspection of the feature(s) in question and increasing the instrument reading frequency for verification of the readings.

MAINTENANCE MASTER SCHEDULE

The maintenance master schedule compiles the planned maintenance tasks for each system into a schedule that can be rolled-up to form an overall maintenance schedule for the lock complex. For additional functionality, the schedule could be resource loaded with preventive and corrective maintenance crew and equipment to ensure adequate resource and budget allocations.

Inspection cycles were developed in consideration of the standard of care and quality control employed during construction. Therefore, comprehensive inspections are typically performed annually for easy to inspect or access systems. A longer time between inspections, once every five or more years, is planned for systems that are more difficult to access and have monitoring installed to evaluate the structure performance in the interim.

Monitoring cycles were developed based on previous project experience and in general accordance with the recommendations presented in Chapter 9 of the FERC Engineering Guidelines (1991). The monitoring frequency considers the type of instrument, criticality of measurement, and length of project operations (since certain readings, such as settlements, should decrease over time as the structures adjust to the imposed loads).

The schedule also captures the possibility of additional inspections or monitoring that would occur in the event of significant or unusual events (e.g. large earthquake).

PARALLELS TO OTHER TECHNIQUES

The US Army Corps of Engineers (USACE) and US Bureau of Reclamation (USBR) have transitioned to a risk-informed decision-making (RIDM) approach for dam safety. The Best Practices in Dam and Levee Safety Risk Analysis (2012) outlines the agencies' approaches to risk analysis and risk assessment.

The basis of the approach is a potential failure mode analysis (PFMA). The PFMA exercise is a facilitated team exercise used to identify and evaluate potential failure modes (PFMs) for a structure or project. After identification, the PFMs are described from initiation through progression to the point of failure and resulting impacts are summarized. This approach can be outlined in an event tree.

An event tree is a "bottom-up" evaluation of a system. An event tree begins with the failure of a critical component of the system, and traces the propagation path that is required to cause system failure. Probabilities can also be assigned to event trees, with the likelihood of occurrence of each step contributing to the probability of system failure due to failure of a specific system component.

The event tree traces a route similar to that which could be followed from a root cause up to the top event in a fault tree. The USACE and USBR (2012) Best Practices in Dam and Levee Safety Risk Analysis provide a number of event tree examples and guidance for the use of event trees in evaluating risk to a structure or system.

Event trees are particularly useful for evaluating an existing system where knowledge of the system behavior and performance over time can inform the selection and evaluation of credible PFMs. Factors that make the occurrence of an event along the tree more or less likely can be captured in the assigned probabilities.

Potential consequences from full propagation of a failure mode are accounted for in the calculation of project risk. Finally, the calculated risk is compared to the tolerable risk to inform decision making on repair and upgrade decisions.

The overall RIDM approach closely parallels the FMECA and RCM principles. The fundamental goals of the techniques are: to prevent an undesirable consequence through the use of engineering judgment and timely intervention; and to provide appropriate system information to allow educated decision making.

REFERENCES

FERC (1991). Engineering Guidelines for the Evaluation of Hydropower Projects, Chapter 9 – Instrumentation and Monitoring.

FERC (2005). Engineering Guidelines for the Evaluation of Hydropower Projects, Chapter 14 – Dam Safety Performance Monitoring Program.

USACE and USBR (2012). Best Practices in Dam and Levee Safety Risk Analysis.

US Department of Defense (2000). Standard Practice for System Safety, MIL-STD-882D.

SAFETY IN NUMBERS?

Kim de Rubertis, PE, LG, D.GE, F.ASCE[1]

ABSTRACT

Using numbers to evaluate dam safety is ingrained in our current engineering practice. This paper describes why numbers are necessary but not sufficient in evaluating safety and in conveying risk to the public we serve. Therefore, engineering judgment is advocated to test the reasonableness of numbers and to address safety issues not amenable to rational calculation such as the likelihood of human error leading to a loss of reservoir control.

Engineering judgment is an acquired skill, born of experience. It is a way of thinking anchored in common sense. It may or may not be informed by numbers. It benefits from visual observation and is likely to use words instead of numbers to express probabilities. It recognizes that predicting human behavior is not possible, and it seeks to avoid human error by training and practice. Engineering judgment provides a platform from which the reasonableness of safety estimates may be assessed.

HISTORY

How did we get to our current reliance on numbers? And is there really safety in numbers?

"I have great respect for the past. If you don't know where you've come from, you don't know where you're going." Maya Angelou

Figure 1. Proserpina Dam

Figure 2. Kavir Dam

For more than 5,000 years, our ancestors constructed dams to store and divert water for irrigation and domestic use. Dams were conceived without numbers as we know them. It was the Age of Empiricism when folks built, learned from what did and didn't work, and then modified their designs until stable structures were achieved. Ancient dams survive and some still are in service. Constructed by Romans more than 1,000 years ago,

[1] Consulting Engineer, P.O. Box 506, Cashmere, WA 98815, kimderubertis@gmail.com

Proserpina Dam in Spain is a notable example. Of a similar age, Kavir Dam, one of the first known arch dams, was built by Mongols in an earthquake prone region of Iran and still survives. The ancients may not have been called engineers, but they acquired the critical faculty of judgment through experience. A respect for shape and mass fueled their ability to advance the profession prior to the arrival of numbers.

The dawn of using numbers , as we know them, arrived in the 19th century. In 1853 De Sazilly proposed a "profile of equal resistance" for a concrete gravity dam, marking the birth of the "middle third rule."

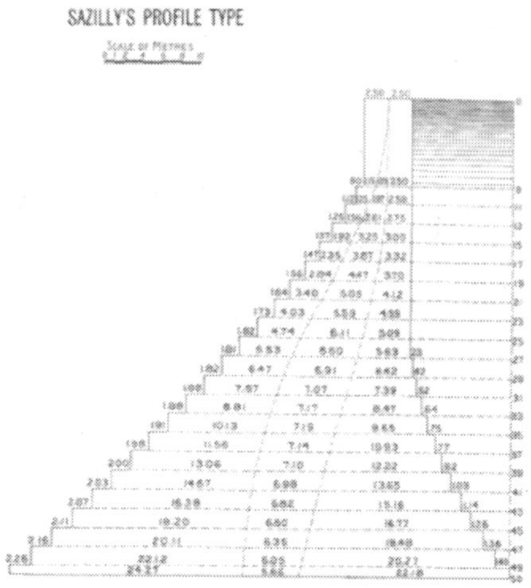

Figure 3. Sazilly's Profile

Figure 4. Furens Dam

Delocre modified De Sazilly's profile to save a bit of masonry and then designed and constructed the 50-meter-high Furens Dam in France in 1858. It still is in service today.[i]

Zola Dam was the first arch dam for which the stresses were estimated as if the dam was part of a cylinder. It was completed in France in 1854.

Darcy introduced the theory of flow through porous media in 1856. Embankment dam designs began to evolve with a better understanding of how the designs would have to respond to the prospect of seepage. Lower San Leandro Dam in California was started in 1875 as a zoned embankment with deep cutoffs. It was raised in 1890 to a height of 150 ft and still serves as a reserve water supply. Rockfill dams gained popularity throughout the 20th century. Knowledge about embankment behavior was advanced by Terzaghi (seepage), Casagrande (filters), Fellenius (slope stability), and many others. Many embankment dams were constructed, building upon increasing understanding of how their materials behave, adding to the scope of engineering judgment.

Gravity dam designs continued to improve. Uplift's effect on behavior was recognized and designs began to address the needs for positive cutoff and drainage. In 1914-5, both Elephant Butte and Arrowrock dams were designed with both foundation and internal dam body drains.

In the 1960s several countries introduced ideas for rapid construction of concrete dams. This led to the advent of roller compacted concrete (RCC) construction which the USACE championed with the construction of Willow Creek Dam in Oregon in the early 1980s. Improvements in design and construction practice since then have promoted RCC as a favored material for concrete dams worldwide.

Arch dam designs gained attention in the late 1800s and then evolved throughout the 20th century. Timoshenko's theory of plates and shells and Boussinesq's theory of elastic half-space fed design improvements. The crown cantilever method of estimating stresses employed a single cantilever with bounding arches, and this method was used to evaluate Bear Valley Dam in California. Variations on the crown cantilever method were applied to many designs, notably Noetzli's designs for USBR's Buffalo Bill and Pathfinder dams. Proud of his achievement, Noetzli applied for and was granted a patent for his design.

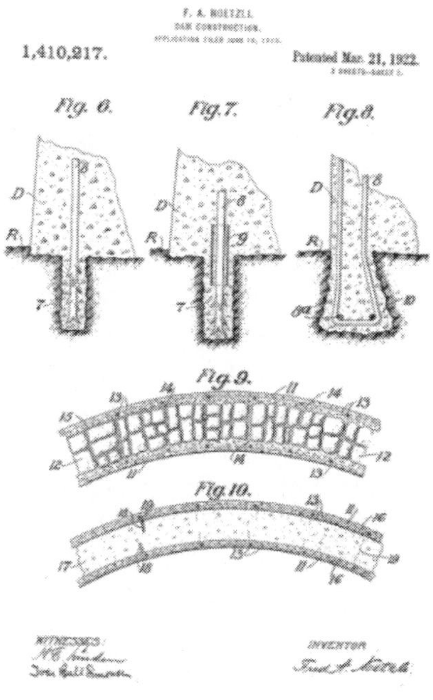

Figure 5. Noetzli's Patent Application

Figure 6. Mareges Dam

Design innovation continued with advances such as Jorgensen's configuration reducing the radius toward the base, creating a "constant angle" design. Coyne improved on prior designs with Mareges Dam in France and introduced the concept of the double curvature arch dam.

Working at USBR, Howell and Jaiquith formalized the computation for both arches and cantilevers that ultimately became the trial load method.[ii]

Now we have reliable analytical tools capable of estimating responses to the loads that a dam (or appurtenant works) must safely resist -- provided that we understand the distribution of the properties governing behavior and the mechanics of response. Engineering judgment underpins our understanding.

As improved understanding of how dams respond to loads evolved, so did standards for evaluating dam safety. One of the earliest works to find level ground among all agencies with jurisdiction was the 1979 publication of "Federal Guidelines for Dam Safety."[iii] Subsequently USACE, USBR, FPC (now FERC), and many states produced guidance documents outlining the acceptable margins of safety for responses to loads. Standards-based acceptance criteria were born, and they are still in use as the underlying rationale for determining if a dam is safe enough in its environment. How these criteria are documented varies with the agency having jurisdiction; however, acceptance criteria (required factors of safety) are generally consistent among the agencies. Numbers underpin the analyses employed to compare estimated behavior with acceptance criteria.

In the past few years, agencies with jurisdiction have enhanced traditional standards-based criteria by requiring dam safety decisions to consider risk. To date, USACE and USBR have established risk assessment guidelines. It is important to note that standards-based criteria still lie at the heart of risk-informed decision making. The following excerpt from USACE's ER 11102-1156[iv] illustrates the role of traditional acceptance criteria:

"The standards-based or essential guidelines approach is included in the risk-informed approach to the dam safety program and dam safety program decisions will now be risk-informed."

UNCERTAINTY

Using numbers demands an appreciation of the uncertainties numbers may present. There are the epistemic uncertainties that involve the limits of knowledge. Uncertainties in the physical conditions of a dam and its foundation are aleatory. Both types of uncertainties are subject to estimates of their correctness and applicability as functions of probability. Probability is purely subjective -- best characterized as one's degree of belief, requiring the exercise of engineering judgment.

Aleatory Uncertainty

How soil and rock respond to load presents a difficult challenge in evaluating safety. Recent work by Christian and Baecher[v] summarized the uncertainties with soils as both aleatory (spatial and temporal variation) and epistemic (parameter error and model bias). Similar uncertainties are applicable to rock. Both soil and rock behave differently in the natural (undisturbed) state and in the fill or excavated (disturbed) states. Depending on the property of interest, either soil or rock may exhibit continuous or discrete variable behavior. Evaluating the safety of an existing dam requires a look-back at the original design properties, the dam's service life during which the original design properties may have changed, and perhaps the need for sampling and testing to inform a decision regarding likely response to a load not previously experienced.

Often the most vexing problem in evaluating safety of an existing dam is the limited amount of data for the properties affecting actual behavior. Data about one dam are not transportable to another dam. Having data without knowing the actual mechanics controlling behavior can lead to constructing erroneous models. This is especially tricky if zones of weakness control behavior.

Epistemic Uncertainty

There are several ways in which the limits of knowledge affect our ability to evaluate dam safety. Our predictions about extraordinary events rely on our knowledge of the past. In that regard, we are prisoners of history. The White Queen scolds Alice for not understanding to eat jam only on Thursday. To which Alice replies: "It's not Thursday. I can't remember things before they happen." To which the White Queen retorts: "It's a poor sort of memory that only works backward."[vi]

Figure 7. Alice and the White Queen

Flood and earthquake predictions look backward to see ahead, but they do not fully recognize the randomness of nature. The engineer making judgments about a dam's safety relies upon estimates of flood and earthquake loadings provided by experts specializing in hydrology and seismicity. The analyses involved in deriving the loading estimates involve a multiplicity of variables that affect the outcomes. There often is stunning disagreement among experts about what the loading estimates should be. Nature regularly provides surprises that exceed prior estimates. More often, centuries pass without loads approaching predictions.

Recognizing the limits to what can be known with certainty requires engineering judgment as the key to making good decisions about dam safety.

We imagine that our memory works both ways, but the reality is we can only use the past to guide our predictions of the future.

Predicting the magnitude and duration of extraordinary events, both floods and earthquakes, rest on past history and their probability of occurrence in the future.

Floods

H. C. Riggs's 1961 paper entitled "Frequency of Natural Events"[vii] was one of the first flood studies to develop the relationship between magnitude, design life, and probability of exceedance by using a cumulative frequency curve.

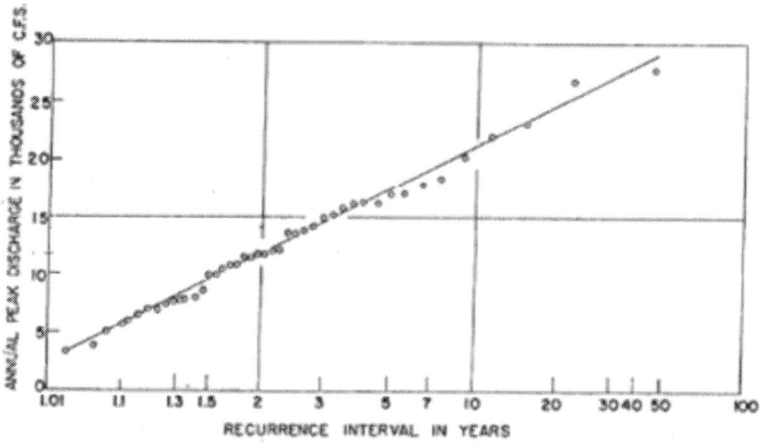

FIG. 3.—CUMULATIVE FREQUENCY CURVE

Figure 8. Cumulative Frequency

The current state of practice estimates maximum precipitation and flooding that employs a multitude of continuous and discrete variables. The estimates of precipitation and flow from historical records are examined both deterministically (precipitation) and stochastically (flood).

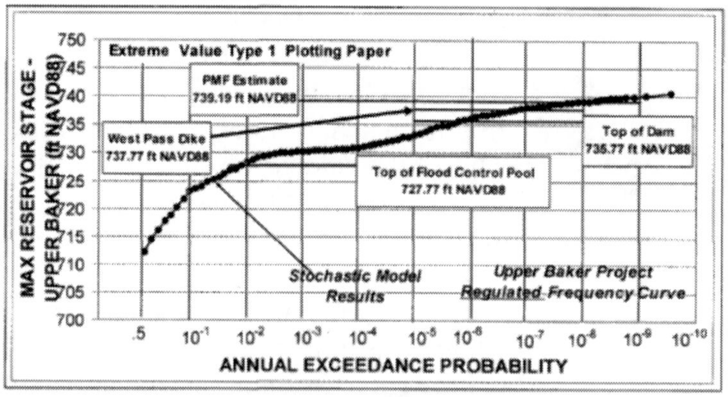

Figure 9. Annual Exceedance

Flood recurrence intervals are estimated from cumulative frequency. A dam's safety is then measured as its ability to safely pass a selected Inflow Design Flood, the magnitude and duration of which is based on an evaluation of failure consequences. Floods approaching or exceeding estimated maximums occur (Pearl River 1979, Colorado 2014, South Carolina, 2015). Current estimates may not capture changing climate. They may not consider loss of land cover that decreases runoff time, resulting in floods with higher peaks. Salas *et al* address the difficulty of making reasonable estimates in their review "Uncertainty in the PMP and PMF":

". . . in recent decades, there has been a growing concern (expressed in literature) regarding the uncertainties involved in estimating such extreme events. . ."[viii]

Earthquakes

Dams are designed and maintained to resist earthquakes. Earthquake study probably began after the November 1, 1755, Lisbon Earthquake, a disastrous event with an estimated magnitude approaching 9.0. Through the years, many attempts have been made to predict magnitude and frequency of damaging earthquakes, especially in earthquake-prone areas such as California and China.

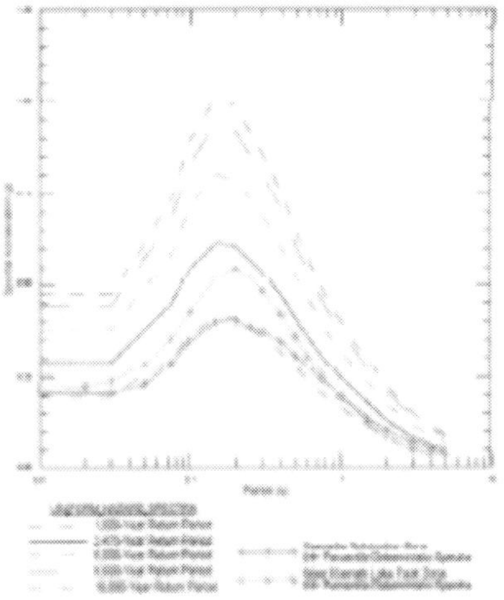

Figure 10. Conditional Mean Spectra

Current practice is based on earthquake history and the physical attributes of faults capable of generating ground motions. These are blended in an analysis that combines both deterministic and probabilistic estimates of the extent and duration of earthquake-induced strong shaking from each earthquake source. Conditional mean spectra (CMS) are developed for various estimated recurrence intervals depending on hazard to the public.

Akin to the uncertainties of the PMP and PMF are those of earthquakes. Clarence Allen observed:

". . . some of scientists are now challenging this simple concept (elastic rebound), *arguing from laboratory experiments and theoretical models as well as from some seismological and geological observations, that the friction on a fault at seismogenic depths (that is, the depth at which rupture initiates is so complicated and so unpredictable, that something akin to* ***chaotic*** *behavior may be more typical than* ***systematic*** *behavior – not only in the time intervals between large earthquakes, but also in their magnitudes, rupture lengths, and locations along a specific fault system. And this, if true, is certainly not good news for those of us involved in hazard assessment, where the time and size of the last large earthquake has often been an important parameter in our prognostications for the near future."*[ix]

Allen made those remarks 20 years ago, concluding with a challenge to the strongly held belief that tectonic processes do not change with time. From what we know today, he was prescient.

What Numbers Can Do

Numbers provide the means by which we seek to understand uncertainty, the principal question of which is: "how will these materials behave given the expected loading conditions?"

Numbers estimating properties drive the input. Analyses employ input numbers to produce the output, an estimate of behavior. Measurements, tests, and references providing the basic physical properties such as unit weight, shear strength, permeability, and deformability can be input to generate analyses.

Fundamental differences exist among the properties of concrete, steel, soil, and rock. The properties of concrete and steel do not vary spatially; however, the properties of soil and rock may vary widely over short distances. How those variations are distributed is a key question for analysis.

Analysis requires a model of the dam and its foundation. For concrete dams, the focus is on the properties of the foundation that promote or threaten stability. An important example is rock wedges capable of movement that may deprive a concrete dam of support. Modeling an embankment focuses not only on the properties of the embankment but also on the foundation. Uncertainty in the physical and spatial distributions of the properties may produce model error. That uncertainty is captured in the required factors of safety for reliability analyses.

Reliability Analyses. The standards-based approach to predicting behavior utilizes both data and estimates of the distribution of the data. The probability of the spatial distribution of properties is based on data and visual observations.

Which properties can we estimate with confidence? The easy ones are unit weights, geometry, and reservoir level.

Which loads can we estimate with confidence? The easy ones are the weight of the dam and the reservoir water. The difficult ones are the loads applied by floods and earthquakes.

Which responses to load can we estimate with confidence? The responses of concrete and steel are not difficult. Their properties are well understood as well as their likely response to loads. Difficulties arise when we connect a dam with the ground and estimate the response of soil and rock.

We estimate the number(s) for each property of interest based on test results and experience with similar materials. If there are no clear differences in the spatial distribution of strengths, a single estimated strength often is applied throughout. Simplification introduces the prospect of error. Analysis becomes difficult if spatial differences are evident because the probabilities of those differences must be taken into account.

Foundation surfaces routinely are simplified as flat and sloping planar surfaces, ignoring important foundation assets that contribute to sliding resistance. Tensile capacity along concrete-rock contacts and roughness are routinely ignored to produce a non-linear model relying on friction alone to resist sliding. "Proving" tensile capacity in the foundation of an existing dam is difficult, expensive, and doubtful of an acceptable resolution because of the work required to obtain a statistically valid sample. Alternatively, the exercise of engineering judgment would suggest that properly placed concrete will strongly adhere to a properly prepared rock surface. An examination of the properties of the foundation and concrete, foundation preparation, and the care taken during construction will support judgment.

Risk Analyses. Steven Vick, in "Degrees of Belief, Subjective Probability and Engineering Judgment,"[x] presents the case for risk analysis as the alternative to reliability analysis.

Risk analysis differs from reliability analysis in several respects, the first being that while it may make use of results from one or more geomechanical models, it does not necessarily rely on these models directly. Risk analysis provides the ability to address multiple failure modes – not just one – whether geomechanical models are available to represent them or not. It goes beyond where most models stop, to the full progression of failure processes and the factors affecting this progression – be they technical, human, or otherwise."

An analysis identifies an initiator, and then follows the effect of that initiator through any progression of events that could lead to a loss of reservoir control -- the failure mode. At each step in the progression, the probability of continued progression is evaluated against its consequences. The risk is then defined as the product of the probability of failure and its consequences.

Developing reliable geomechanical models is difficult.

Leonard Mladinow observes: "*We also use our imagination to take shortcuts to fill gaps in patterns of nonvisual data. As with visual input, we draw conclusions and make judgments based on uncertain and incomplete information, and we conclude, when we are done analyzing the patterns, that our 'picture' is clear and accurate. But is it?"*[xi]

Baecher and Christian highlighted some of the difficulties in their "Geotechnical Reliability – Ten Unresolved Problems." Two of their findings about internal erosion illustrate difficulties, one in the exercise of judgment and another in the limit of knowledge.

- *"Typical event trees for internal erosion include physical processes that are intuitively reasonable but unobservable, and add a somewhat theological aspect to predictions."*
- *"The major problem is that we lack a physical model for predicting internal erosion from first principles. The issue is not one of applying probabilistic methods to a problem whose physics is well understood. It is a problem for which the basic physics of failure has not been described adequately."*

Rothman and Sudarshan summarized debates about fundamental physics in words that speak to our state of knowledge: *"The innocuous interchange of the words 'theory' and 'model' evidently feeds a confusion of models with the real world. The Earth is treated as a billiard ball until that idealization does not work, then more structure is brought in.*

Can this process result in the World Equation, the Equation of the Universe? No, it results in equations of the latest model.'"[xii]

Risk analysis has gained some acceptance in evaluating dam safety. It is best accomplished by experienced practitioners in the process of expert elicitation. Results from risk analyses performed by inexperienced personnel are not likely to produce reliable outcomes suitable for decision-making.

What Numbers Can't Do

A key element in any dam safety monitoring program is visual surveillance. The likelihood that visual surveillance, augmented by instrumented monitoring, would fail to detect a progression in a potential failure mode is unlikely.

"Monitoring of every dam is mandatory, because dams change with age and may develop defects. There is no substitute for systematic intelligent surveillance. But monitoring and surveillance are not synonymous with instrumentation." Ralph Peck

Experience clearly demonstrates that human error initiates most, if not all, dam incidents and failures.[xiii] Numbers are not reliable predictors of human behavior. Barring an earthquake or flood, experience demonstrates that a dam safety incident or loss of reservoir control seldom, if ever, occurs without some physical manifestation of the potential failure progression.

A good example is the Big Bay failure in Georgia.[xiv] The authors noted: *"From the time of first filling of the dam, completed in 1993, until the failure in 2004, the dam exhibited several distress indicators and other warning signs."* Evaluation of the Big Bay failure cited 11 separate human factors contributory to the failure concluding: *"If these combined deficiencies in human factors – which interacted with physical factors in complex ways . . . had not been present, it is very unlikely that the dam would have failed."*

Careful review of instrumented measurements is another key to evaluation. Another good example is the recent incident at Wanapum Dam in Washington, where a trend of increasing movement in a monolith presaged the crack that led to the discovery of errors in design and workmanship. Unexpected behavior went unnoticed for several years. Design errors went unnoticed for more than 50 years.

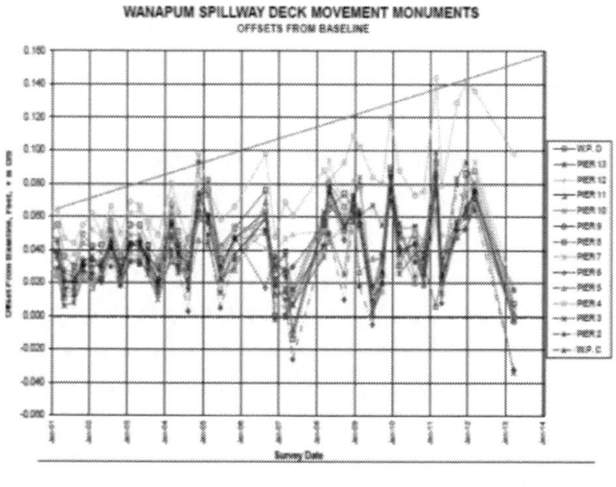

Figure 11. Wanapum Deck Movements

Three contemporary and widely publicized failures are poster children for human error – Swift No. 2, Taum Sauk, and Silver Lake. All three were regulated projects with long histories of inspections by owners, regulators, and consultants. All three project exhibited evidence of problems long before they failed.

Pat Regan once likened the game of dam safety to a three-legged stool and noted that all three legs (regulator, owner, and consultant) have to function properly to guard safety. Given contemporary experience, one might ask: "Who's minding the store?" Although each project had accepted reliability analyses, none of the failures would have been prevented by risk analysis. Understanding human error and avoiding it is an elusive goal not reached by numbers alone. Engineering judgment is required.

James Reason points to the success of the airline industry in creating an organizational culture of safety.[xv] Creating a culture of safety for every dam requires leadership from every owner. Until more attention and appropriate funding are applied to the issue of reducing human error in dam safety, incidents and failures are likely to continue.

Analytical Error

Analyses support decision making, and understanding the capabilities and limitations of analyses is essential in making appropriate decisions. There is a tendency among dam safety professionals to ascribe ultimate value to analyses, not recognizing the uncertainties that accompany the analyses.

Hynes and Vanmarcke performed field testing of the reliability of embankment performance prediction that demonstrated the wide spread of expert opinion using the same parameters for analysis.[xvi] Experience has shown that the likelihood of any analysis being incorrect exceeds the likelihood that it is correct. A recent example demonstrates the lack of agreement produced by differences in analytical approaches.

Faced with prediction of higher ground motions during the design earthquake, an owner commissioned a surge tank stability analysis. The owner's engineer analyzed the expected behavior of the surge tank during an earthquake, concluding that it would fail in base shear. The analysis was submitted to the regulator who rejected it, citing that the mode of failure was moment instability. The owner sought a third opinion. Another consultant was hired and returned a verdict that the surge tank would fail in buckling. One thing became clear: the surge tank is likely to fail in the design earthquake, never mind the mode of failure. Each analyst brings a degree of belief, a personal exercise of engineering judgment based on an evaluation of the facts in evidence. Overconfidence is a weakness not easily overcome.[xvii]

Building consensus through expert elicitation is favored to reduce the likelihood of analytical error and offer an opportunity to find the best estimate of how a dam will behave under load. The limits of knowledge will always constrain outcomes. Heuristics and biases unrelentingly affect analyses. Analytical results must be viewed with caution. Most dam incidents and failures occur independent of the accuracy of numerical analyses.

Ralph Peck offered an opinion regarding analyses that is worthy of consideration. *Our concentration on investigating the properties of the materials of which dams are made, and on the technical analysis of the anticipated behavior, should be matched by attention to the nontechnical and human factors that are no less a part of this branch of engineering."*

Risk and the Public

Recent advances in utilizing risk as a guide to decision-making measure the importance of a decision based on the probability of occurrence and the number of fatalities anticipated – the two essential numbers in the fN diagram. Tolerable risk is measured by its position on the fN diagram.

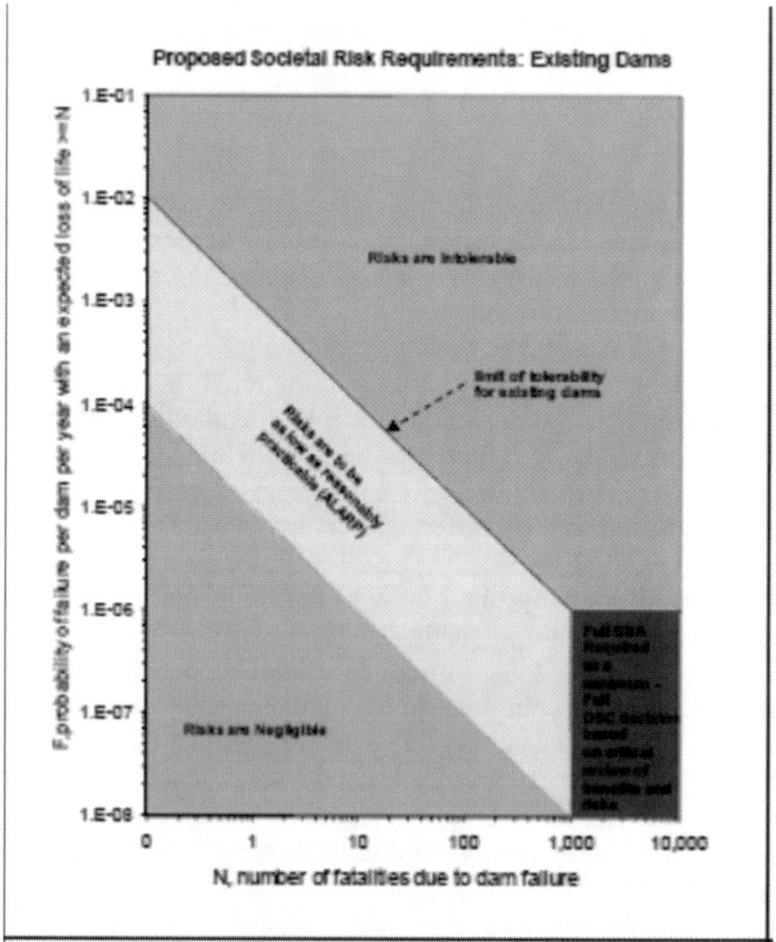

Figure 6 - New South Wales Dam Safety Committee Societal Risk Plot

Figure 12. Typical fN Diagram[xviii]

There are two troubling aspects of this approach to dam safety. The first is how risk is conveyed to the folks who are likely to die if a dam fails. A good summary is presented by Herkert in "Ethical Risk Assessment: Valuing Public Perceptions."[xix]
"The conventional view of risk communication, held by many engineers, is that risk communication need consist merely of "educating" the public to endorse expert judgment concerning which risks are acceptable and which are not. Under this model of risk communication, the experts have a corner on the truth of the matter; the only problem is to see that the public is properly informed of the experts' views."

Paul Slovic in "Perception of Risk" wrote:

". . . there is wisdom as well as error in public attitudes and perceptions. Lay people sometimes lack certain information about hazards. However, their basic conceptualization of risk is much richer than that of the experts and reflects legitimate concerns that are typically omitted from expert risk assessments. As a result, risk communication and risk management efforts are destined to fail unless they are structured as a two-way process."[xx]

In general, we have an uninformed public, and we have not conveyed risk to the point where the public is prepared to act in the event of an emergency. Many people who are PAR (people at risk) in the analyses don't know that they live downstream from a dam capable of causing immense damage if it fails. Have the PAR participated in establishing the limits of "tolerable risk?"

Lewisville Lake Dam presents a contemporary example of public reaction to the risks to 431,000 PAR from dam failure.

"Dallas doesn't sit along a coast line, but it does face a catastrophic flooding threat from aging Lewisville Dam, about 30 miles north of the city. When Lake Lewisville is full, the dam holds back 2.5 billion tons of water. If the dam were to fail, the loss of life and consequences for North Texas' economy would be a catastrophe of biblical proportions. Walls of water would tear through parts of Lewisville, Carrollton, Farmers Branch, Irving all the way to downtown Dallas, which could be under 50 feet of water.

It's time to square with people. The region deserves a more robust discussion of the dam's condition. Repairing this regional lifeline could cost hundreds of millions of dollars, and North Texas needs to be ready to respond financially. Residents need the chance to understand what's at stake."[xxi]

Videos of the streets in Tokyo during the aftershocks of the Tohoku earthquake showed pedestrians and traffic calmly observing traffic controls while the buildings all around them were swaying. Would our population respond as calmly or would it panic? The steps to reduce the risk to the public during a serious dam safety emergency beg resolution.

As difficult as it might be, we have an obligation to inform the public we serve about the risk posed by the works we defend. For the most part, we haven't done that.

CONCLUSION

The key to knowing if there is safety in numbers relies on having a clear understanding of what numbers can and cannot do.

Numbers are necessary, but not sufficient, in our duty to hold public safety paramount. Numbers alone, without the benefit of engineering judgment, will not serve that duty.

"While all these new things enable us to do our work smarter, faster, better, and with improved understanding of the materials with which we work, there is one vital, but not so new component of successful geotechnical engineering that is now more important than ever before: that, of course, is sound ***engineering judgment****. Despite all the new tools, data and information sources, computational aids, and guidance documents, soils and rock still come in many forms. Their properties may change with time and environment, the boundaries are usually uncertain, no two projects are ever the same, and surprises often lie beneath the surface."*[xxii]

Is there safety in numbers alone? No.

ACKNOWLEDGEMENTS

The author is grateful to Kathleen Clarkson P.E. (FERC PRO), Ralph Grismala P.E. (ICF), and Pat McCarty P.E. (Tacoma Power) for their thoughtful reviews and helpful comments on this paper.

---

[i] Smith, Norman (1972) . *A History of Dams*, Citadel Press, Inc., Secaucus, NJ.

[ii] Chen, Shen-Hong (2015), *Hydraulic Structures*, Springer-Verlag, Berlin

[iii] Code of Federal Regulations (Revised 1977), Title 18, Conservation of Power and Water Resources.

[iv] US Army Corps of Engineers (2014), *Engineering and Design, Safety of Dams – Policies and Procedures*, Regulation No. 1110-2-1156.

[v] Christian, John T. and Baecher, Gregory (2014). *Geotechnical Reliability – Ten Unresolved Problems*, Geo-Strata, American Society of Civil Engineers, Reston, VA.

[vi] Carroll, Lewis (1871). *Through the Looking-Glass*, Macmillan, London, England.

[vii] Riggs, H.C. (1961). *Frequency of Natural Events* , ASCE Journal of the Hydraulics Division, Vol. 87.

[viii] Salas, Jose D. et al (2014). *Uncertainty of the PMP and PMF*, Taylor & Francis, New York.

[ix] Allen, Clarence R. (August 1995) in *Earthquake Spectra, Volume 11, No. 3*, Division of Geological and Planetary Sciences, California Institute of Technology, Pasadena, CA.

[x] Vick, Steven G. (2002), *Degrees of Belief, Subjective Probability and Engineering Judgment*, American Society of Civil Engineers, Reston, VA.

[xi] Mladinow, Leonard (2008). *The Drunkard's Walk,* Random House, New York, NY.

[xii] Rothman, Tony and Sudarshan, George (1998). *Doubt and Certainty*, Perseus Books, Reading, MA.

[xiii] de Rubertis, K. and van Donkelaar, C. (November 2014). *Look Both Ways – Dam Safety and Human Error*, USSD Newsletter.

[xiv] Myers, D., Ferguson, K., Alvi, I., and Baker, M. (2015). *Big Bay Dam*, The Journal of Dam Safety, Volume 13, Issue 2.

[xv] Reason, J., *Managing the Risks of Organizational Accidents (1997).* Ashgate Publishing Limited, Surrey, England.

[xvi] Hynes, N. and Vanmarcke, E., *Reliability of Embankment Performance Prediction (1976).* Proceedings of the ASCE Engineering Mechanics Division Specialty Conference, Waterloo, Ontario, Canada, University of Waterloo Press.

[xvii] Fischoff, B., Lichtenstein, S., Slovic, P., Derby, S., and Keeney, R. (1981). *Acceptable Risk.* Cambridge University Press, New York, NY.

xviii Federal Guidelines for Dam Safety Risk Management

[xix] Herkert, J., (1994). *Ethical Risk Assessment: Valuing Public Perceptions.* IEEE Technology and Society Magazine, Vol. 13.

[xx] Slovik, P. (1987), *Perception of Risk.* Science 236.

[xxi] Editorial. (January 21, 2016), Dallas Morning News, Dallas, TX.

[xxii] Mitchell, J. (2015). *What's New in Geo?. In* Geo-Strata, American Society of Civil Engineers, Reston, VA.

PUBLIC SAFETY AND SECURITY AT DAMS ARE NOT MUTUALLY EXCLUSIVE

William F. Foos, CPP, PSP[1]
Frank Calcagno, R.G.[2]
Paul G. Schweiger, P.E.[3]

ABSTRACT

Dams have become a popular destination point for an increased number of fisherman, boaters, waders, thrill-seekers, and other visitors. This increase in public visitation creates an ever-greater potential for vandalism and crime, property damage, injury, and even accidental death that can translate into costly litigation for dam owners.

Concepts of public safety and site security need not be mutually exclusive. If anything, a joint approach to evaluating and addressing both of these concerns can maximize the return on a dam owner's investment. Solutions to address security at a dam site can also be used to address public safety, while common solutions used in addressing public safety can enhance the overall security at the site. The steps and information needed to assess the risks associated with public safety and site security are also similar. Solutions to address both issues often emphasize the need to provide public warnings, limiting public access to features at the dam that create dangerous conditions and/or are critical to the mission of the dam, and effectively monitoring unauthorized activity around the dam.

This paper presents concepts from the authors' experience, on an approach in conducting dam site assessments that address both site security and public safety concerns at dams. We will present an approach for conducting a common site assessment that can be used to simultaneously evaluate security and public safety at dams. Additionally, we will present examples of mitigating strategies that could be used to minimize the risks associated with injury, accidental death, property damage, and related litigation.

BACKGROUND

The traditional managerial structure of organizations is to have one department dedicated to safety and another department dedicated to security. In many instances, large organizations require a separation of responsibilities, and when applying this structure of management at a project site such as a dam, the two functions quickly become very disjointed. Without a well thought-out approach to site assessments, an organization's Safety Manager and Security Manager will likely be competing for the same piece of the operational funding pie. In many cases, public safety at dams assumes a secondary role

[1] Senior Director of Security and Safety Services, Gannett Fleming, Inc. Camp Hill, PA 17011, wfoos@gfnet.com
[2] Senior Security Advisor/Engineering Geologist, Gannett Fleming, Inc. Fairfax, VA 22030, fcalcagno@gfnet.com
[3] Vice President, Manager of Dams and Hydraulics Section, Gannett Fleming, Inc. Camp Hill, PA 17011, pschweiger@gfnet.com

and escapes the responsibility of the organization's Safety Manager, who may be more focused on worker and industrial safety. As a result, public safety is left in the hands of the project management or operational staff. In these cases, project management staff must "borrow" from their shrinking operational budget to allocate funds for public safety. This typical managerial structure used by many dam owners is very inefficient and a recipe for disaster in that operational demands and O&M costs can easily overshadow the needs for an effective public safety program.

As such, there is no unified response to this problem from the start of the design phase through construction and ultimately into operation. Funding for separate projects will be spread among different supervisors; there may be multiple funding requests for similar structures that were conceptualized and designed separately. In the extreme, a death could result that was easily preventable with proper planning.

"$1.8 Million Settlement reached in Deaths of Two Men due to court ruled negligent, careless, and intentional release of water without proper advanced warning." (August 10, 2012 NYInjuryLawBlog.com)

The lack of standards on Public Safety for Dams in the United States is not an excuse for addressing clear and present dangers created by the operation of dams and hydropower facilities. The focus over the past 14 years since the tragic events of 9/11 has shown an incredible focus on security at high hazard dams, but the fact of the matter is that a dam owner is more likely to face an accidental injury or death to the public than experience a tragic event caused maliciously by an outside terrorist organization. The intent of this paper is to bring the dam owner's focus back to what is more likely to occur (and addressing this higher probability event), while at the same time planning for the worst-case event that could potentially occur (either through complementary security features or through contingency planning). A balanced approach needs to be struck between the needs for public safety at dams and the need to protect the project assets through traditional security measures. These two requirements are not mutually exclusive, and with proper planning and coordination, they can be far more mutually supportive than is typically envisioned.

PUBLIC SAFETY AND SECURITY

Today, the Federal Energy Regulatory Commission (FERC) manages the adjudication of hydropower licenses for over 1300 utility organizations throughout the United States. Nearly every license contains recreational requirements placed on dam owners to provide adequate recreational use of the dam and hydropower facilities, including the reservoir and the surrounding project lands. These requirements may often funnel public access to the immediate areas near the dam and around potentially dangerous zones. The most common recreational activities include swimming, fishing, and boating activities. From a strict security perspective, these activities can be interpreted as both a pro and a con to the project security.

Pros - public access provides an opportunity for the public to monitor unusual activity and report it to the proper authorities. As well, the presence of "witnesses" to a potential crime, theft, or act of terrorism may serve as an active deterrent. In many cases, project personnel are occupied with other duties, either off site or near the site, and can become distracted from the activities of the public in and around the dam. The advantage of allowing public access is that the dam owner can utilize individuals who use the benefits of the project repeatedly and who have become familiar with others using the facilities. They are quick to identify an individual visiting for the first time, and they are eager to protect the facility that has become important to them.

Cons – a false sense of security could be created with an over-familiarity with public presence, where an adversary may blend within a crowd of innocent recreationalists. Large events could create a crowd mentality where no one reports and/or does anything to stop the activity of a malicious individual. It also becomes increasingly difficult for dam site security personnel to monitor unusual activity in an area where the presence of the public is common.

Regardless of your views on public access, the public will often develop a sense of entitlement to enter the site, even if it is fenced off or is part of a restricted zone. Water activities are very attractive to people, regardless of the hazards they may present or the measures put in place by the facility owners to protect those people willing to enter. The approach we suggest is that the dam owner embrace the fact that the public will be present and tackle public safety and security as an integrated process. A balanced approach is possible.

Security Perspective

The fundamental purpose of security is to protect the critical assets at a site. In security, there is a concept called "Protection in Depth" or "Concentric Circle of Defense." Protection in Depth means that to accomplish a goal, an adversary must be required to avoid or defeat a number of protective devices in sequence. For example, an adversary might need to defeat one sensor and penetrate two separate barriers before gaining entry to a process control room or a filing cabinet containing sensitive data. The actions and amount of time required to penetrate each of these layers may not necessarily be equal, and the effectiveness of each layer may be quite different, but each will require a separate and distinct act by the adversary moving along the path. The effect produced on the adversary is one of uncertainty due to the differing layers of protection reducing their chances of successfully navigating the various security features.[4] The ability to provide Protection in Depth includes identifying zones of concern within different perimeters. An exclusion zone is an area within a perimeter demarcation, such as a fence line, that is defined to limit access; an exclusion zone can further be defined inside with restricted zones which are identified areas to be controlled with limited access to only a few permissible cleared individuals. The definition presented here provides identifying

[4] ASIS International, *Protection of Assets*, Alexandria, Virginia, USA; Section 2.1.1

exclusion zones and then restricted zones where required within the exclusion zone.

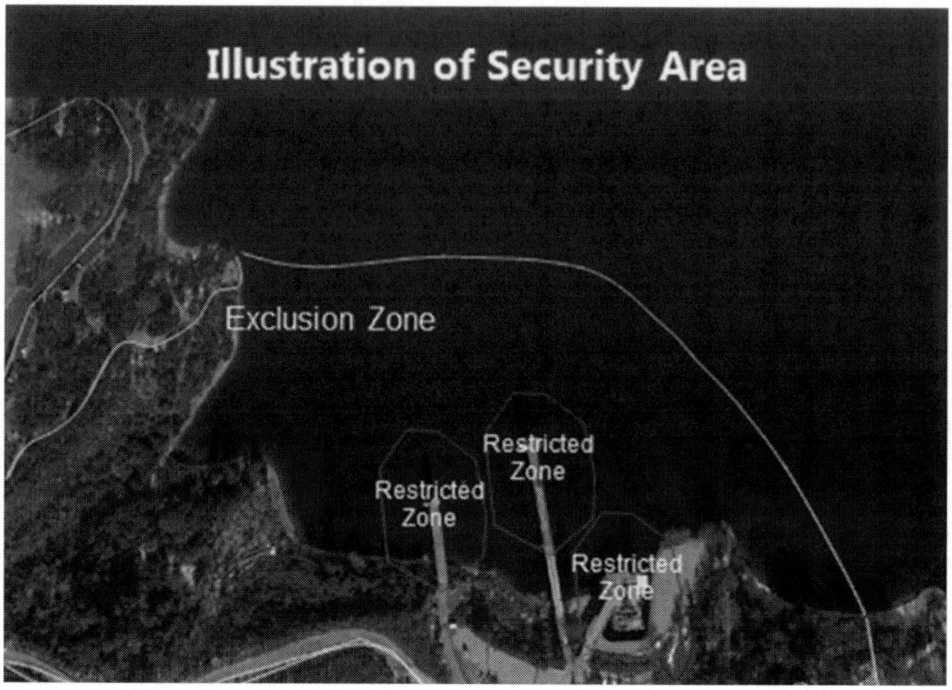

Public Safety Perspective

The fundamental purpose of public safety is to protect the public, including trespassers, from dangerous situations and hazardous areas. With water retention structures such as dams and hydropower plants, hazardous zones are categorized as either a Danger Zone or a Warning Zone. The ***Danger Zone*** is the area of the dam where the hazard is likely to be realized if an individual is exposed to the hazard. In most cases, the Danger Zone is located within a Warning Zone. The ***Warning Zone*** is considered that area that is or can be influenced by the Danger Zone.[5]

As an illustration of this concept, a spillway structure is within a Warning Zone whether the gates are opened or closed. The Danger Zone is the area downstream that is likely to become hazardous if the spillway is opened and water is released. The concept of zoning can additionally be applied upstream of the spillway.

[5] Canadian Dam Association, Public Safety Around Dams Workshop 2015.

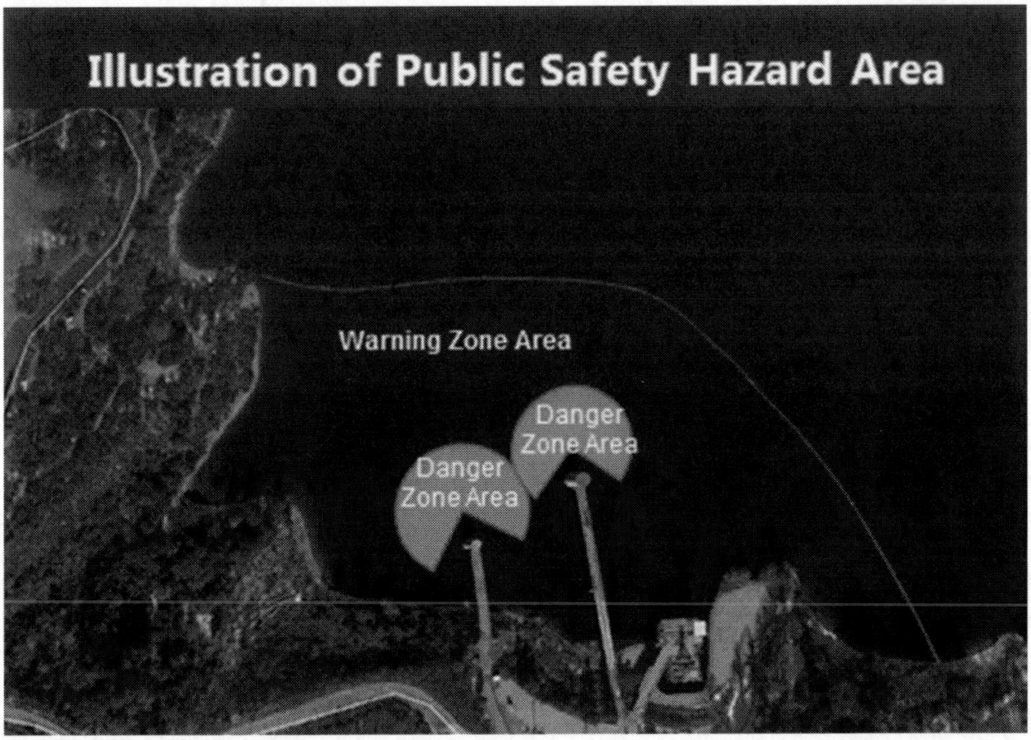

Mutually Supportive. Using the two principles previously mentioned and blending them together, we can begin to see that various zones can be described lending further clarification for Warning Zones to also be described as exclusion zones from a security perspective. This area of exclusion within the defined perimeter of the Warning Zone can provide a buffer area where individuals that venture into this exclusion area do so either at their own risk and/or with a demonstrated intent to cause harm. The Danger Zone from a public safety perspective could be further defined from a security prospective as a restricted zone. Someone entering this zone provides clear demonstration of an intent to initiate malicious activity or to enter an area that may place them in physical harm. The hazard they are exposing themselves to in many cases also provides them access to vulnerable features of the dam creating a security risk. (Such as the spillway, intake structure, outlet works, and dam crest.)

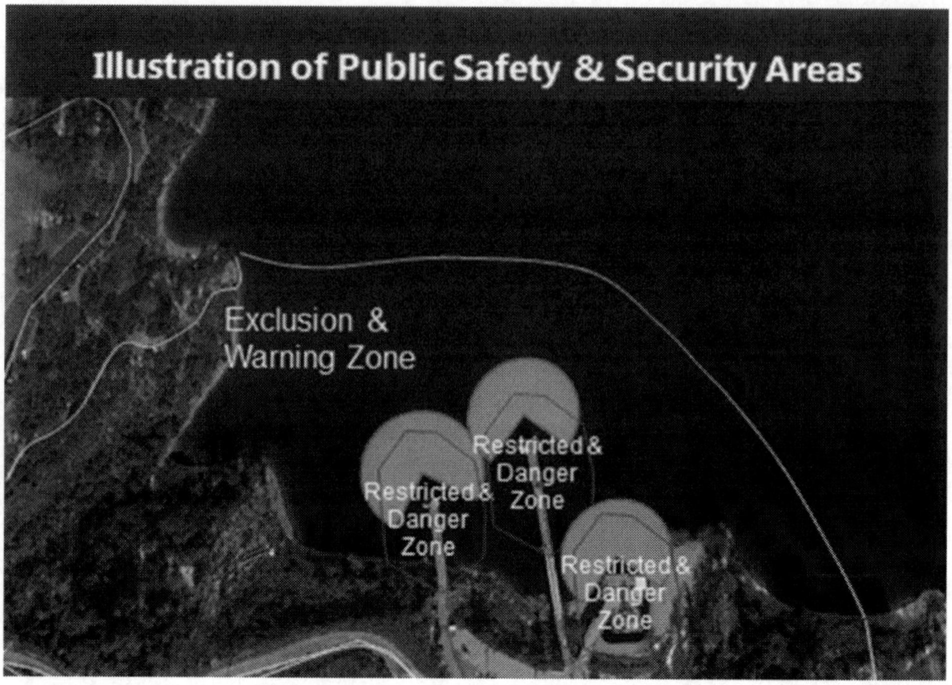

Unified Approach

The following steps provide a structural analytical approach in applying a unified site assessment of a dam site for public safety and security purposes.

Step 1 – Identify and cross reference the critical assets of the project site and hazardous areas.

Table 1 provides an example of how hazardous area information might be captured and defined for critical features and components of a dam.

Table 1. Example of Defining Hazardous Zones

Critical Features of a Dam/Hydropower Plant	Hazardous Area[6]
Spillway	Upstream 40 meters / downstream 100 meters of the Spillway
Hydropower Intake	Upstream 40 meters
Hydropower Outlet	Downstream 30 meters
Switchyard	20 meters around the switchyard (Restricted Area)

[6] Hydrologic analysis in many cases will need to be conducted to determine the area impacted by the feature of the dam.

Step 2 – Further categorize and identify the hazardous areas as Danger and Warning Zones.

Example – Spillway: Danger Zone is 10 meters upstream and 50 meters downstream

Warning Zone is 40 meters upstream and 100 meters downstream

Step 3 - Compare distance requirements provided by the hazardous zone classification with security detection/assessment and response time requirements.

Make adjustments to the zone classification distances if it is necessary to increase the distance of the Warning Zone area to provide additional response time for emergency or law enforcement responders.

Step 4 - Assess current safety and security features in use (if any) and adjust / modify or enhanced these measures to address the newly defined distance hazard zone classification.

Example – Spillway:

- Boom currently in place needs to be moved further upstream to restrict access to the Danger Zone.
- Warning signs adjusted to provide early warning before the public enters a zone of impact (or entrapment) from the release of water from the spillway.
- Cameras (if present) need to be replaced and/or adjusted to provide full viewing coverage of the identified zones, with some detection capability integrated with video image to alert the dam operator or security personnel when the public approaches these hazardous zones.

MITIGATION APPROACH

The idea of mutually beneficial protective designs may seem like a new concept to the dam and hydropower industry, but the concept has been used for some time. Since 1971, Crime prevention through environmental design (CPTED) [7] is an approach to try and influence the behavior of the public and direct their activities to certain approved areas. It is intended to deter bad behavior and provide an environment to encourage conforming behavior from the public.

[7] The term CPTED was first used by C. Ray Jeffery in his book Crime Prevention through Environmental Design (Thousand Oaks, CA: Sage Publications, Inc., 1971). Later Timothy Crowe used it in his book Crime Prevention Through Environmental Design (Boston: Butterworth-Heinemann, 1991). The CPTED definition given here is used by the National Crime Prevention Institute and was enhanced by Randall Atlas in his book 21st Century Security and CPTED (Boca Raton, FL: Taylor & Francis Publisher, 2008).

Crime Prevention Through Environmental Design (CPTED)

CPTED is a set of management tools targeting the following[8]:

- **Places.** Physical environments (such as office buildings, parking garages, parks and public spaces, multifamily apartment buildings, warehouses, schools, houses of worship and shopping centers) designed to produce behavioral effects that reduce the opportunity for certain types of crime and the fear of those crimes.
- **Behavior**. Some locations tend to create, promote, or allow criminal activity or unruly behavior, while other environments elicit compliant and law-abiding conduct.
- **Design and use of space.** Redesigning a space or using it more effectively can encourage desirable behavior and discourage crime and related undesirable conduct.

CPTED is congruent with the security mission of deterring, detecting, and delaying potential offenders, but also in effecting public behavior for the benefit of personal safety. Many of the measures employed through CPTED are commonly employed at dam and hydropower project sites, and include:

- **Mechanical measures**. Also referred to as target hardening. Mechanical measures include physical security hardware and technology (such as locks, security screens on windows, fencing and gating, key control systems, intrusion detection, video surveillance, and barriers).
- **Human and organizational measures.** These include Community Watch, security officer patrols and posts, police officer patrols, and any person or group serving as a capable guardian with the ability to observe, report, and intervene.
- **Natural measures.** Natural CPTED measures include having a well-defined entrance and arranging work sites, staging areas, and assets for unobstructed lines of sight. Natural measures provide specific guidance for the use of space; examples include architectural landscaping, ditches, berms, bollards, planters, moats, and visibility-enhancing actions such as lighting and shrub and tree trimming. Even when supplied by mechanical equipment (lamps), lighting is classified as a natural surveillance component.[9]

The measures are applied with several concepts in mind:

- **Natural access control.** The idea is to employ both real and symbolic barriers—including doors, fences, and shrubbery—to define and limit access to a building or other space. For example, to deter theft/buglers from entering lower-story windows, one could plant dense, thorny bushes near the windows or install window locking devices or an alarm system.
- **Natural surveillance**. Increasing visibility by recreational public and casual observers increases the detection of trespassers or misconduct at a facility. For instance, if a high wooden fence blocks the view of a loading dock, the lack of

---

[8] ASIS Protection of Assets / Table of Contents / Physical Security / Chapter 3 - Crime Prevention Through Environmental Design / 3.1 - CPTED Theory

[9] Ibid. 3.1.1CPTED Fundamentals

visibility may actually invite thieves. Conversely, the use of chain-link fencing that allows an unobstructed view of the area by workers or passers-by may discourage thieves. Windows, door viewers, mirrors, and other design features that improve visibility fall under natural surveillance.

- **Natural territorial reinforcement.** This is the process of establishing a sense of ownership, responsibility, and accountability in property owners, managers, or occupants to increase vigilance in identifying trespassers. For example, the use of small edging shrubbery along sidewalks marks the territory of individually-owned property and discourages trespassers from cutting through. Also, people pay more attention to and defend a particular space if they are provided psychological ownership of it. Territorial reinforcement measures, which may be physical or symbolic, tell people they are in a defined space. Color, texture, surface variations, signage, and way-finding systems are all part of territoriality and boundary setting. Thus, it is possible, through real barriers (fences, booms, buoys, and walls) and symbolic markers (warning signage, low hedges, and low wood picket fences) to encourage tenants or employees to defend the property from individuals with undesirable intentions. Such reinforcement is termed natural because it results from normal, routine use of the environment.

In addition to the preceding classic principles, the following concepts are also considered through a CPTED approach:

- **Management and maintenance.** For spaces to look well-cared for and crime-free, they must be maintained. The "broken windows" theory (Wilson & Kelling, 1982) suggests that leaving broken windows or other decay markers (e.g., graffiti, trash, or abandoned furniture) unattended or unrepaired can lead to the impression of abandonment and increase crime opportunity, as no capable guardian is perceived or observed by the criminal. A parked car left too long with one broken window invites the breakage of additional windows. Maintenance of a building, including lighting, paint, signage, fencing, walkways, and the rapid repair of any broken items, is critical for demonstrating that someone cares about the building and is responsible for the upkeep, and by implication, for its security as well.

- **Legitimate activity support.** Some places are difficult to protect by the nature of their location or other geographic features. In such instances, legitimate activity support is essential. A "hot" fishing spot within a hazardous waterway below a dam might be eliminated if park rangers (clearly known by the public to be responsible for keeping the area clear) place a substation in the area, or if maintenance staff persistently patrol (inspect) within view of the area of concern, providing legitimate activity support.[10]

The concepts of CPTED are really simple approaches to produce public behaviors that the dam owner would like to see occur, but also allow employees and security staff to quickly identify nonconforming behavior demonstrated by the public.

[10] Ibid 3.1.1 CPTED Fundamentals

Other Considerations

According to the Canadian Dam Association (CDA website), *"...in Canada, more people have died in accidents around dam sites than from structural failure of dams. This is a matter of considerable concern to dam owners, who are seeking guidance on how they can meet their responsibilities to protect public safety."* Also stated in the CDA website (Public Safety), *"It is worth noting that large or high dams command a respect from members of the public that may be greater than that which exists for low structures. However, low head and small diversion dams can be equally or more hazardous, because many of the hazards associated with these structures are not readily apparent."* This is similar to the experience in the United States. A statistic uncovered by research presented by Dr. Bruce Tschantz, professor emeritus of the University of Tennessee, is that during the past 30 years there have been more than four times as many fatalities from drowning at dams than there have been deaths resulting from dam failures (191 reported drownings at dams versus 40 deaths from dam failures)[11]. According to the Association of Dam Safety Officials website (Public Safety around Dams), *"This year alone [2015], at least 25 people have drowned and at least 10 seriously injured in low-head dam incidents across the U.S."* The victims include expert swimmers, rescue workers and other professionals equipped with state-of-the-art life-saving and rescue gear who were aware of the hazards but still succumbed to the lethal hydraulic forces. Low-head dams are often overlooked, especially with respect to public safety, and must be addressed to properly protect the public.

Although the mitigating strategies for improving site security and public safety presented herein are primarily focused on managing access to the site through effective exclusion and monitoring features and procedures, there are cases when structural modifications to the facility can completely eliminate the hazard. This is particularly true for the hazardous conditions at low-head dams and spillway outlets where the hazardous hydraulic roller that can develop can be eliminated by modifications to the downstream face of the dam. Reshaping the downstream face of dams has been one of the most commonly used modification alternatives to address the hazardous hydraulic roller. Details for this modification approach are described in another paper by the authors[12].

Effective signage is critical at dams to provide sufficient warning to the public. For further information on this topic, the reader is referred to the FERC report, "Safety Signage at Hydropower Projects, October 2001" (FERC website), and the CDA "Guidelines for Public Safety around Dams (2011)." In addition to guidance on signage, booms, and buoys, the CDA publication provides a managed system approach to public safety, a discussion of the public safety hazards associated with dams, a step-by-step risk assessment methodology and templates, and a description of typical risk treatment or control measures.

---

[11] "Hidden Dangers and Public Safety at Low Head Dams," B.A. Tschantz, & K.R. Wright, Journal of Dam Safety, ASDSO, vol. 9, Issue 1, 2011.
[12] "Saving Lives while Improving Fish Passage at Killer Dams", Paul G. Schweiger, Journal of Dam Safety, ASDSO, Vol. 9, Issue 2, 2011.

Enforcement goes a long way in ensuring that the public obeys signage and restricted zones. When word gets out to the local community that noticeable fines are being issued for trespassing, fewer people will be willing to assume that financial risk. If speeding tickets were not issued, would fewer people obey the posted speeds?

Many dam owners believe it is not cost-effective to expend large security expenditures for protection against a very low-probability terrorist attack. Considering the current threat conditions historically seen within the United States, this belief may possibly be justified. However, the dam owner is cautioned to use proper risk assessment methods to gauge the potential consequences against a level of acceptable risk, both to that of the dam owner and the community where the dam resides, to accurately determine what level of security is needed. However, the basics of security can be implemented and constructed relatively easily at any dam to address the real threat of public safety that confronts all dam owners. This is especially important when considering the needs for liability management. Security procedures and installations for higher-level threats, up to and including the highest threat level reasonably considered, can be planned on an "as needed basis" where phased hardware security additions, more stringent procedures and protocols, temporary closures and barriers, temporary armed guard contracts, etc. can be added to the basic security features already used onsite to control the public. These actions should be well planned, easily initiated, and well documented in a dam facility site-specific security and public plan.

When addressing public safety, most dam owners immediately consider the obvious: spillway or generation (high flow) releases, falling hazards, and switchyard dangers. However, a prudent dam owner should also consider many other issues, such as:

- Fluctuating water levels (both on the reservoir and downstream of the dam),
- Strong currents and undertoes,
- Submerged hydraulic jumps or "hydraulic rollers,"
- Strainers and excessive seepage paths that can trap swimmers,
- Open spillways which may not be visible from above the dam,
- Submerged hazards,
- Steep river banks with no exit,
- Improper siting and design of boat launches and portages,
- Winter ice (thin or dangerous reservoir ice and falling ice near structure overhangs),
- Rock-falls along appurtenant structures and access roads,
- Bridges, cables, and transmission lines (boat clearance and/or electrocution),
- Slippery surfaces at or near vertical drops and waterways (notably in high-mist areas),
- Improperly installed boat barriers (sags) that can entrap boats and swimmers due to high flow currents,
- Engineered (and signed) safe havens for the protection of employees and the public in the event of an active shooter or act of domestic violence occurring onsite or in an established recreational facility.

This paper provides a unified approach for two programs that share obvious similarities (namely those of public safety and facility security). It should be noted, however, that a third program should not be neglected; traditional dam safety concerns can also benefit through an integrated design of security and public safety features. Consider how security cameras can not only monitor for the presence of persons of malicious intent and trespassers entering hazardous areas, but that those same cameras can also be utilized for dam safety monitoring and verification of potential or impending dam failure, especially if tied into the monitoring program of identified potential failure modes. Roving patrols can combine efforts in monitoring for security, public safety, and dam safety inspections with a small amount of cross training. Obviously, one profession cannot become an expert in another field, but basic indicators of dam safety distress (unusual seepage, cracks or settlement) can be taught to security patrols. Similarly, dam safety inspectors and operators can be mindful of suspicious activity and public intrusions into hazardous areas. Consider combing frequent operational maintenance checks to include both public safety and security checks, as well as dam safety. In this way, one series of hardware installations, or scheduled patrols, can monitor for three separate programs, thus reducing onsite operational costs.

SUMMARY

The approach to managing these two (or possibly three) programs should be considered as complementing programs that provide mutual support from a public safety, security, and perhaps dam safety, perspective. A common assessment approach, assessing public safety needs, dam safety needs, and security needs will provide for a simultaneous optimum design, resulting in one design package rather than multiple ones. Consider mitigation strategies that work to minimize or eliminate risks associated with injury, loss of life, and property damage as well as protecting against a low-probability adversary.

There are many challenges faced by dam owners today regarding public access, public safety, and facility security needs. But the advantage of approaching public safety and security as a unified process is that these programs can complement each other, and that design and construction costs can be combined, and ideally be done only once. The project will be safer for the public and the final design will result in a system that allows dam operators and personnel to more easily and efficiently monitor the activities occurring at and near the site. The result is that the system enables the operations staff to better assess and respond to developing hazardous situations either caused by public negligence, thrill-seekers, criminals and vandals, or by a deliberate malicious attack.

EMERGENCY ACTION PLANNING — CANADIAN PRACTICE AND LESSONS LEARNED AT ONTARIO POWER GENERATION

Tali Serota, MDEM[1]
Tony Bennett, P.Eng[2]

ABSTRACT

Unlike many other jurisdictions, Canada does not have a federal agency providing regulatory oversight of dam safety; rather each province and territory is responsible to manage dams within their jurisdiction. In fact only four of the ten Canadian provinces and three territories have enacted dam safety regulations. The absence of regulations in Canada does, however, offer a unique role for the Canadian Dam Association (CDA) and their system of Guidelines and Technical Bulletins to provide industry leadership in developing a system of good practice which the provinces and territories have promoted.

Specifically, in the area of emergency management for dam safety, the CDA Guidelines are the only guidance available to dam owners. While three of the provinces do require dam owners to prepare emergency preparedness and response plans (EPRPs), in general, EPRPs are voluntarily prepared as part of a dam owner's initiatives to demonstrate due diligence in managing risks.

Ontario Power Generation (OPG) has integrated an all hazards emergency management approach as part of its Dam Safety Program; capturing key elements within a managed system to ensure that a state of readiness is maintained, both within the organization as well as with stakeholders and community first responders. OPG's Dam Safety Program offers a good example of the implementation of the CDA Guidelines by a dam owner.

This paper presents the Canadian landscape in regards to emergency management for dam safety and examines the program which OPG has implemented, including initiatives with stakeholder communities to improve their response. Lessons learned from developing EPRPs, conducting exercises and training activities are discussed in the context of the OPG's continuous improvement initiatives.

INTRODUCTION

In its most basic form, an effective risk management practice identifies hazards, eliminates them where practical, or monitors the hazards and develops strategies to mitigate the impacts of the residual risks. Figure 1 (adapted from the Report of the Intergovernmental Panel on Climate Change) serves to illustrate the various strategies available to manage the residual risk, amongst which is the implementation of an emergency management system. Within this paper emergency management will refer specifically to the Preparedness, Response and Recovery phases, as well as encompass

[1] Ontario Power Generation, Toronto, ON, Coordinator, Dam & Public Safety tali.serota@opg.com
[2] Ontario Power Generation, Toronto, ON, Director, Dam & Public Safety tony.bennett@opg.com

other risk mitigation strategies available to dam owners shown in Figure 1 as they relate to emergency management.

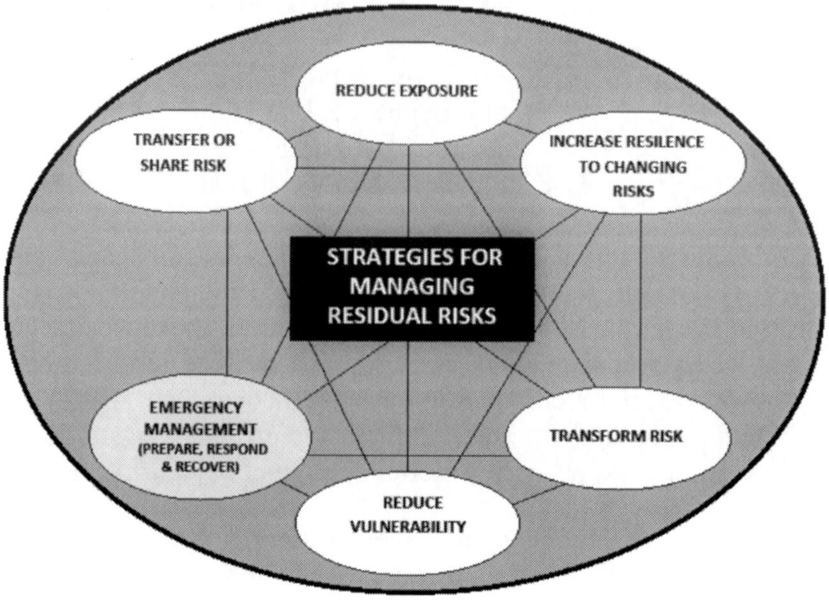

Figure 1. Strategies for Managing Residual Risk

Canadian Landscape

In Canada, similar to other jurisdictions, it is a dam owner's responsibility to prepare EPRPs to protect their assets and to form the basis from which downstream communities and other dam owners/stakeholders can prepare their plans, aimed at the protection of the public, property and the environment.

Only four Canadian Provinces; Alberta, Quebec, British Columbia and Ontario, have enacted regulations addressing dam safety, three of which require owners to have EPRPs (i.e. Ontario does not require EPRPs). Having a regulation does not imply, however, that all dam owners are placing the appropriate emphasis on maintaining their EPRPs and state of readiness. Recent dam safety regulatory compliance audits conducted in Alberta (AG Report, 2015), Quebec (AG Report, 2015) and British Columbia (Hatch, 2015) indicate that full compliance is not being achieved. Similarly, ASDSO has reported that not all dams in the United States under State Dam Safety Program jurisdiction, that require an EPRP have them in place; with 75% compliance reported during their annual review in 2015 (ASDSO, 2015).

Even in the absence of a regulatory requirement, dam owners have developed EPRPs as part of their own due diligence initiatives to manage risk. This is common practice amongst a number of the large dam owners in Canada, which includes Provincial and Federal agencies, as well as organizations with a mature safety culture. However, a survey of dam owners in 2015 by the Canadian Dam Association (CDA) Emergency Management Working Group indicated that there remain a significant number of dams in Canada without EPRPs (CDA, 2015).

Developing an EPRP is an important step towards mitigating the potential impacts of a dam failure for both the owner and downstream communities. However, maintaining a state of readiness requires that responders are familiar with the plan through training and exercises and that there is a commitment to regularly update the EPRP. While it is common in Canada for dam owners to have an EPRP the CDA survey indicated less than 30% of dam owners were conducting annual exercises of their EPRP's and only 30% answered that they met regularly with the First Responders in the downstream communities to discuss the plans (CDA, 2015); putting into question the full effectiveness of those plans.

The International Joint Commission (IJC) reported in its 1998 review of dam safety practices for structures located on boundary waters between Canada and the United States that there is an over-reliance on the due diligence efforts of dam owners and that there is a need for further engagement of government in regards to the *"...establishment and regular testing of emergency actions plans which take account of eventualities and include detailed notification procedures..."*(IJC, 1998). A 2006 IJC status report further noted that Canadian regulators were lagging the practice specified by the US FERC in regards to development of EPRPs for dams (IJC, 20016).

The fact that most large dams in Canada are owned by Provincial hydropower organizations, coupled with few dam failures in Canada which have resulted in loss of life, offers a partial explanation for the lack of regulations governing dam safety. The historical record in Canada is one of five dam failures resulting in a total of eleven fatalities (Bennett, Serota 2013). The failure of the Mount Polley Tailings Dam in August 2014 and other incidents involving dam safety, though not resulting in loss of life, have served to refocus both Canadian regulators and the public on improving oversight, and in all likelihood resulted in the Alberta and Quebec Auditor Generals' including dam safety within the scope of their 2015 reports on Provincial activities. Quebec's Nicolet Commission on the 1996 Saguenay Flood, during which a number of dams failed and extensive property damaged ensued, was the last major driver in Canada for dam safety regulatory review; resulting Quebec (2002) and BC (2000) enacting regulations.

Figure 2. Recent Dam Failures in Canada resulting in Regulatory Reviews
Left, Testalinden Dam Failure, Oliver BC, June 13, 2010 – loss of 5 homes and valuable cropland; and Right, Mount Polley Tailings Dam Failure, Likely BC, August 4, 2014, environmental impacts from tailings release (Photo credits CP)

CANADIAN DAM ASSOCIATION — EMERGENCY MANAGEMENT

Since being issued in 1995, the CDA Dam Safety Guidelines have served as defacto standards for many Canadian jurisdictions, and have provided guidance for dam owners pursuing good practice. Sections of the Guidelines have undergone revisions (1999, 2007 and 2013) in order to keep pace with industry developments; however the section dealing with Emergency Preparedness (Section 4.0) has remained similar in content when compared to the original 1995 version, prompting the CDA in 2015 to initiate a Working Group to review the need for an update.

Reviews of emergency management guidelines for other hazardous industries, as well as those prepared by international jurisdictions dealing with dam safety, suggest that the CDA Guidelines may be lagging good practice in certain areas of emergency management and are in need of updating (Bennett and Serota, 2013). Table 1 provides a high level comparison of the CDA Guidelines with other guidance available in North America, illustrating some of the deficiencies that need to be addressed.

Table 1. Comparison of North American Guidelines

EPRP Guidance (requirements for Preparedness, Response and Recovery)	Guidelines for Dams			General Industry	
	CDA[1] Guidelines	FERC[2] Guidelines	FEMA[3]	Public Safety Canada[4]	CSA Z1600[5]
General Contents	✓	✓	✓	✓	✓
EPRP Format	X	✓	✓	✓	✓
Activation Guidelines	✓	✓	✓	N/A	N/A
Inundation Mapping	X	✓	✓	N/A	N/A
Recovery Planning	X	X	X	✓	✓
Review Period	✓	✓	✓	✓	✓
Drills/ Exercises	✓	✓	✓	✓	✓

Ref. 1. Canadian Dam Association - Dam Safety Guidelines (2007)

Ref. 2. FERC Chapter VI, Emergency Action Plan Guidelines (2007)

Ref. 3. FEMA 64, Federal Guidelines for Dam Safety: EAPs for Dam Owners (2013)

Ref. 4. Public Safety Canada Emergency Management Planning Guide (2011)

Ref. 5. CSA Emergency Management and Business Continuity Programs (2008)

\The CDA Dam Safety Committee, as part of its stewardship role of the Dam Safety Guidelines, established an Emergency Management Working Group to update the technical guidance. The Working Group, as part of its initial task, held a workshop in February 2015 to consider the needs for emergency planning and management related to both conventional water-retaining dams and mine tailings dams (CDA, 2015). Representatives at the workshop included dam owners, operators, regulatory bodies, provincial and federal government agencies, consultants and emergency management specialists.

An outcome of the workshop was a listing of products that the CDA should consider developing to serve the needs of its members, including Canadian regulators. The products included:

- Technical Bulletin on Emergency Management
- Development of a standard format for a dam breach inundation maps
- Exercise planning templates
- EPRP templates for owners of small dams (where loss of life is not anticipated)
- Benchmarking and self assessment tools for owners
- Educational materials to aid stakeholder communities in completing HIRA's addressing dam failure
- Development of case examples of EPRPs

The new CDA Technical Bulletin on Emergency Management will align with Canadian industry standards and guidance as well as internationally recognized documents. An important aspect of the new Bulletin will be an explicit recognition of the role that resiliency plays in managing residual dam safety risks. As an example, outlining processes for community engagement can be a cornerstone to identifying longer term goals such as removing flow restrictions or improving evacuation routes, that work in the interests of both the community and the dam owner. This addresses risk mitigation strategies beyond just emergency management such as; reducing vulnerability, reducing exposure and sharing risk, as illustrated in Figure 1.

OPG EMERGENCY MANAGEMENT PROGRAM

Ontario Power Generation (OPG) established a formal Dam Safety Program (DSP) in 1986 to manage the safety of their 238 dams and 65 hydropower facilities. The Program encompasses traditional dam safety work, as well as emergency management and public safety around dams with the goal of ensuring that OPG's dams are designed, constructed, operated and maintained in compliance with all regulations, and in the absence of a regulation, to meet the guidelines of the Canadian Dam Association or other recognized international practice.

As part of the DSP management system, OPG has developed detailed governance in the form of internal standards, guidelines and procedures. The governance includes requirements to prepare EPRPs for all dams classified as Extreme, Very High or High (according to CDA terminology) or where having an EPRP in place would result in mitigating property or environmental impacts associated with dam failure. Associated with these plans is a comprehensive maintenance, training and exercise program. As a result OPG has 24 EPRP's in place as they are generally river based plans encompassing several dams in the watershed in one plan. These EPRPs involve OPG interfacing with over 200 external organizations, including; First Nations, municipalities and government agencies, as official copy holders of the plans with dams that cross international and provincial boundaries and with the potential to impact infrastructure critical to community safety. They require a system of exercises and training for OPG employees to ensure a state of readiness is maintained to respond to an event.

OPG Emergency Management Standards

OPG has developed two internal standards which describe the overarching emergency management system covering its hydropower and thermal facilities, as well as a specific standard addressing dam safety EPRPs.

The Standard for Hydro Thermal Operations – Emergency Management and Communications Structure (DS-STD-11) is based on the all hazards approach, referencing the hazards listed by the Ontario Office of the Fire Marshall and Emergency Management. It serves to operationalize the Incident Command System (ICS) as described in Canadian Standards Association Z1600 (2014) for the Hydro Thermal Operations Business Unit at OPG. The ICS provides clearly defined Roles and

Responsibilities for all of the response structures within OPG through all the phases of emergency management (mitigation, preparedness, response and recovery).

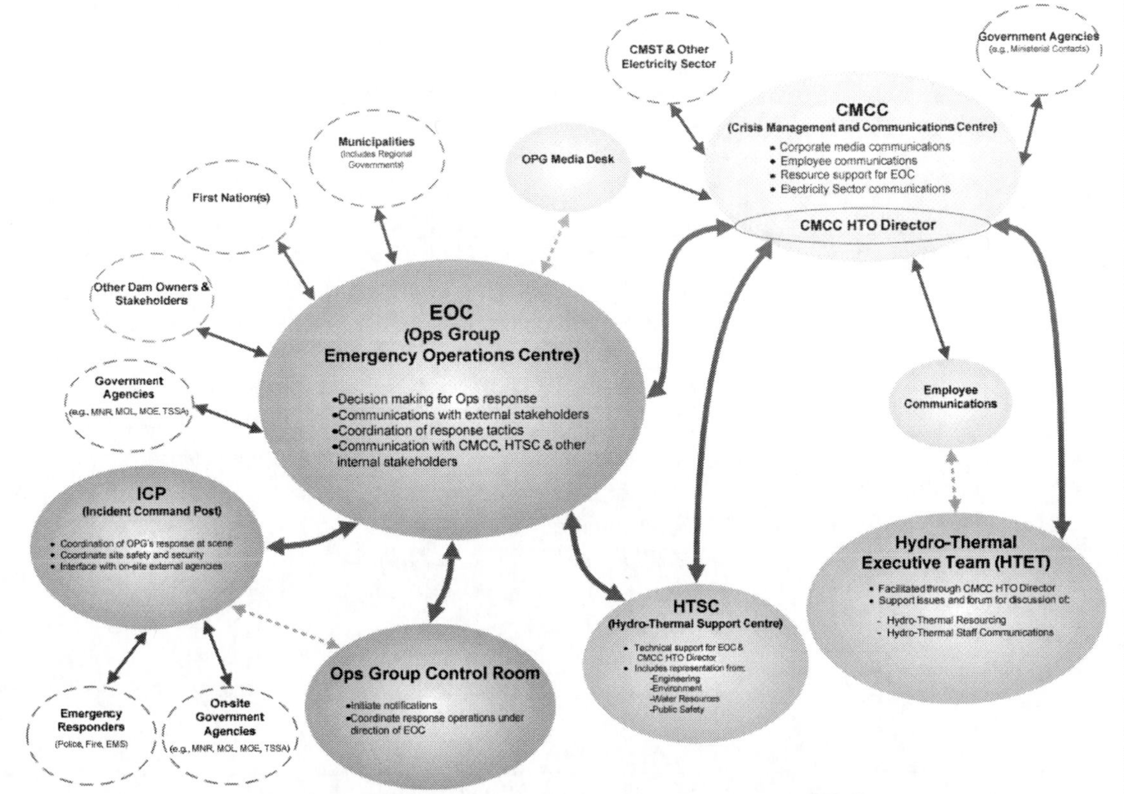

Figure 3. Emergency Response Centres at OPG

Figure 3 illustrates the interoperability between OPG's emergency response centres and with key external stakeholders who would be engaged in responding to an emergency at

one of OPG's hydropower or thermal facilities, including a dam safety event. Each of these response centres has predefined objectives for managing the event that align with their primary paths of communication. The objectives of OPG's emergency response centres can be broadly characterized as tactical, support or strategic. Table 2 provides a summary of the objectives and primary communication paths for each of the response centres.

Table 2. OPG Emergency Response Centres			
Response Centre	**Primary Objectives**		**Primary Communication Paths**
Incident Command Post(s) (ICP)	Tactical	▪ Establish command and control at incident scene and manage on-site response ▪ Interface with 1st Responders on-site ▪ Maintain site safety	▪ On-site with OPG personnel and 1st Responders ▪ Situational updates with the EOC & Control Room ▪ Receives direction on response from EOC
Operating Control Room	Tactical	▪ Event verification ▪ Maintain operating control and water management ▪ Initiate notifications to internal and external stakeholders	▪ Initiates notifications until EOC established ▪ Direction from EOC ▪ Situational updates with EOC, ICP and other operators in the watershed
Emergency Operations Centre (EOC)	Strategic Tactical Support	▪ Directs/ Integrates response and recovery operations for OPG's ICPs, Control Rooms and with external EOCs and stakeholders ▪ Maintains situational updates with external stakeholders and the ICP's as well as with internal response centers	Situational updates with: ▪ External EOCs and Stakeholders ▪ ICP(s) & Control Room ▪ HTTSC ▪ HTET ▪ CMCC ▪ Media Desk
Media Desk	Tactical	▪ Prepare and deliver media releases and briefings	▪ CMCC and EOC ▪ External Media/Public
Hydro Thermal Technical Support Centre (HTTSC)	Support	▪ Provides technical support (e.g. engineering, environment, water management) to the EOC as well as the CMCC	▪ EOC ▪ CMCC
Hydro Thermal Leadership Team (HTET)	Support	▪ Support the EOC in the response and recovery operations when additional resources are required	▪ Receives situational updates from CMCC Hydro Thermal Director
Corporate Crisis & Communication Centre (CMCC)	Strategic	▪ Integrates response and recover operations for events affecting multiple EOC's ▪ Executive Briefings ▪ Business Continuity	▪ Senior levels of Government (Provincial and Federal) ▪ OPG Board and Senior Executives

<u>OPG's Application of the ICS Emergency Response Structure</u>

The application of the Incident Command System (ICS) is common amongst all emergency management organizations with the generalized structure being defined in Canadian Standards Association CSA Z1600 (2014). However, each organization needs to customize the ICS in alignment with the nature of their operations and needs. OPG has developed detailed profiles of roles and responsibilities based on the ICS which are defined in the internal standard DS-STD-11 and a series of job aids. Implementation of the system ensures that responders are focused on their specific tasks, while relying on a formal communication structure to maintain situational awareness through briefings and access to communications systems. Figure 4 illustrates the response structure specific to OPG's EOC, along with a summary of the primary functions for each of the positions.

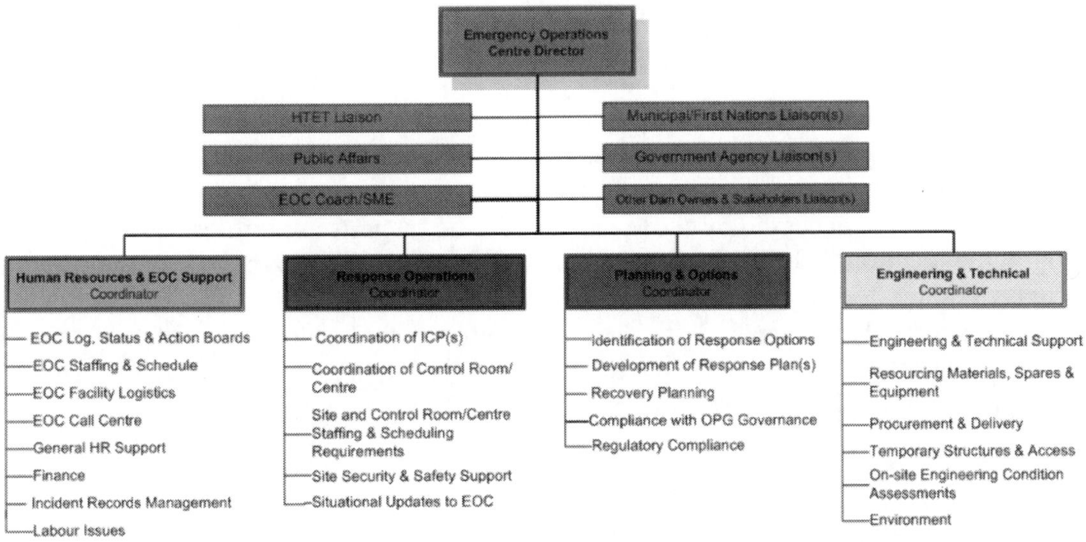

Figure 4. OPG EOC Emergency Response Structure

The ICS system is premised on the principle that it is flexible and scalable based on the specific response needs of the incident. In practice, the positions illustrated in Figure 6 can be fulfilled by staff acting in multiple roles, or there may be a Coordinator in each of the positions being supported by teams of individuals. The needs will vary both with the nature of the incident, as well as the stage of the response. In the illustration shown, the EOC Director is responsible for assigning the Coordinator positions to individuals, as well as making the decision on whether the circumstances warrant additional appointments to the Liaison positions. Each Coordinator is then tasked with deciding what resources they require to fulfill their role, and in turn reach out to assemble their support team. The main consideration is that all of the responsibilities for each of the Coordinator positions have been addressed.

With a pre-defined response structure, it is clear who is responsible for each element of emergency response and recovery, including the communication paths. The benefits of a properly implemented ICS structure are; clarity on roles and responsibilities, a focused effort on specific response goals, consistent messaging, avoidance of duplication of effort, and depth in resourcing. Consistency in the implementation of the ICS framework at OPG has the added benefit of creating depth in the response staff (i.e. staff from other regional areas can be brought in to support the response during a sustained event).

Another unique aspect of how OPG has chosen to implement ICS is the creation of a Coach position in each of the response centres. The Coach position is typically assigned to staff who are well versed in EPRPs and emergency management as their duties focus on the functionality of the response centre and the roles of the individual Coordinators. The Coach positions are also accountable to ensure that situational updates are held, both routinely as well as at key decision points, to ensure integration of effort between the response centres and external stakeholders. This position has shown to be invaluable, especially at the early stages of a response, as there is oftentimes a tendency for staff engaged in the response to drift back to their day-to-day accountabilities and communication paths which may run counter to effective emergency communications.

In order to facilitate communications and manage actions arising from the event, OPG has in parallel implemented WebEOC communications software. The main benefit of the system is that it maintains the event information in real time, ensuring that accurate information can be accessed by those engaged in the response regardless of their geographic location. While this commercial software is quite flexible, it is important to recognize that it does require customization to align with the ICS framework and response centres that are specific to the organization. OPG customized reference pages, termed 'Boards', that have been developed for each of the response centres covering; i) Status, ii) Actions; iii) Significant Events, and iv) Position Logs.

OPG's Implementation of Recovery Planning

Despite Recovery being recognized as one of the so called 'pillars of emergency management', along with Mitigation, Preparedness and Response, it is often overlooked in the discussion of emergency management for dams.

In addressing stages of emergency management, OPG has highlighted two stages of Recovery; Interim Recovery and Final Recovery. Interim Recovery plans need to be developed prior to OPG's stand down of an EOC, ensuring; site safety and security is maintained, surveillance is established to verity that conditions remain stable, communication paths with external stakeholders are established to support their on-going response, and data collection is initiated to create a record of the event and manage liabilities which may arise. Final Recovery plans focus on the return to normal operations, and may include additional considerations focused on the assessment of damages to OPGs operations or equipment, as well as to third party property, arising from the incident. Final Recovery could also involve establishing terms of reference for an investigation into the circumstances of the incident in order to develop lessons learned.

As an added note, depending on the magnitude of the impacts of the incident there may be a strong link between OPG's Final Recovery Plans and elements of OPG's Business Continuity plans when establishing what is often termed the 'new norm'.

OPG's Standard for Dam Safety EPRPs

OPG's emergency management program addressing dam failure was developed in the 1990's as an exercise in due diligence. It was not until the Quebec Dam Safety Act was passed in 2000 that regulations requiring EPRPs was established and even then it only covered a relatively small number of OPG dams on the Ottawa and Abitibi River watersheds which were located in the Province of Quebec. Currently, dam safety requirements established by the Province of Ontario (2011) and Parks Canada (2011) having jurisdiction over the remainder of OPG's dams, do not require EPRPs to be in place.

In the absence of any regulatory requirement, and predating the development of the original version of the CDA Dam Safety Guidelines in 1995, OPG established an internal Standard for Dam Safety EPRPs (DS-STD-08). This document establishes such elements as:

- Contents of an EPRP
- Distribution, maintenance and document management requirements
- EPRP Activation Guidelines
- Templates for Notification Charts
- Defines the expected actions of key external stakeholders
- Guidance on conducting routine meetings with external stakeholders
- Training requirements for OPG staff involved in response to a dam emergency
- Requirements for Call Tests, Tabletops and Full Scale Exercises

The Standard ensures consistency in the development and management of OPG's 26 Dam Safety EPRPs.

Training and Exercises

OPG has established a variety of mechanisms to ensure that a state of readiness is maintained by the organization to respond to a dam safety incident. These include formal courses designed to train staff in the emergency response process and their specific roles, development of job aides for each of the positions potentially engaged in a response, creating opportunities to participate in exercises as players, controllers, observers, coaches, as well as to play a role on the actual exercise design team. This is all intended to ensure staff is competent in their response roles and that there is an adequate level of redundancy, sufficient to support the response to a sustained event.

OPG's Standards establish the training objectives and requirements for each of the positions and include such aspects as the timeframe for staff new to their role to be trained in EPRPs and for the periodic requalification. The Standards also establish goals for staff to routinely participate in exercises.

OPG's comprehensive exercise program includes Call Tests, Tabletop Exercises and Full Scale Exercises with the following providing a summary of the requirements related to Dam Safety EPRPs:

- Call Tests, including a brief scenario, are to be conducted annually for all Dam Safety EPRPs and must include all internal and external contacts
- Tabletop Exercises are to be conducted:
  - as a minimum one annually per Regional Operating Center (there are 5 Centers)
  - once every three years for all EPRPs where there is a potential for loss of life
  - prior to each Full Scale Exercise as a form of pre-test
- Objectives are established for each Tabletop Exercise, with typical examples including:
  - Familiarize internal and external EPRP plan holders with OPG's EPRP
  - Verify the priority for notifications made in OPG's EPRP is consistent with stakeholders expectations
  - Ensure clarity around roles and responsibilities amongst all stakeholders
  - Document any gaps identified in the process or obstacles which may arise in implementing OPG's EPRP or the responders Plans
  - Fully implement three-way communication amongst the participants
  - Familiarize stakeholders with any updates/revisions to OPG's Dam Safety EPRP
- Full Scale Exercises are to:
  - be carried out annually within the Hydro Thermal Operations group and engage all five Regional Operations Centres in order to facilitate the lessons learned process
  - have a detailed terms of reference prepared in advance
  - engage external EPRP copyholders in the exercise
  - have specific objectives established for each of the emergency response centres and include evaluation criteria against which the success of achieving the objective can be assessed
  - have an After Action Report prepared, documenting the lessons learned in the form of Findings and Opportunities for Improvement

Full Scale Exercises involve significant planning, training and preparation in order for the exercise to be successful. OPG has been conducting a minimum of one Full Scale Exercise annually since 1998. These exercises have proved invaluable in exposing community vulnerabilities, such as constrictions in evacuation routes or encroachments on flood plains, which may create the opportunity for these issues to be addressed through infrastructure investments and future planning.

Figure 7 illustrates the approach adopted by OPG for conducting a Dam Safety Full Scale Exercise. Although the diagram is specifically focused on Full Scale Exercises, other types of exercises would be considered and executed in the "exercise execution" stage for verification (call test) or training (tabletop) purposes.

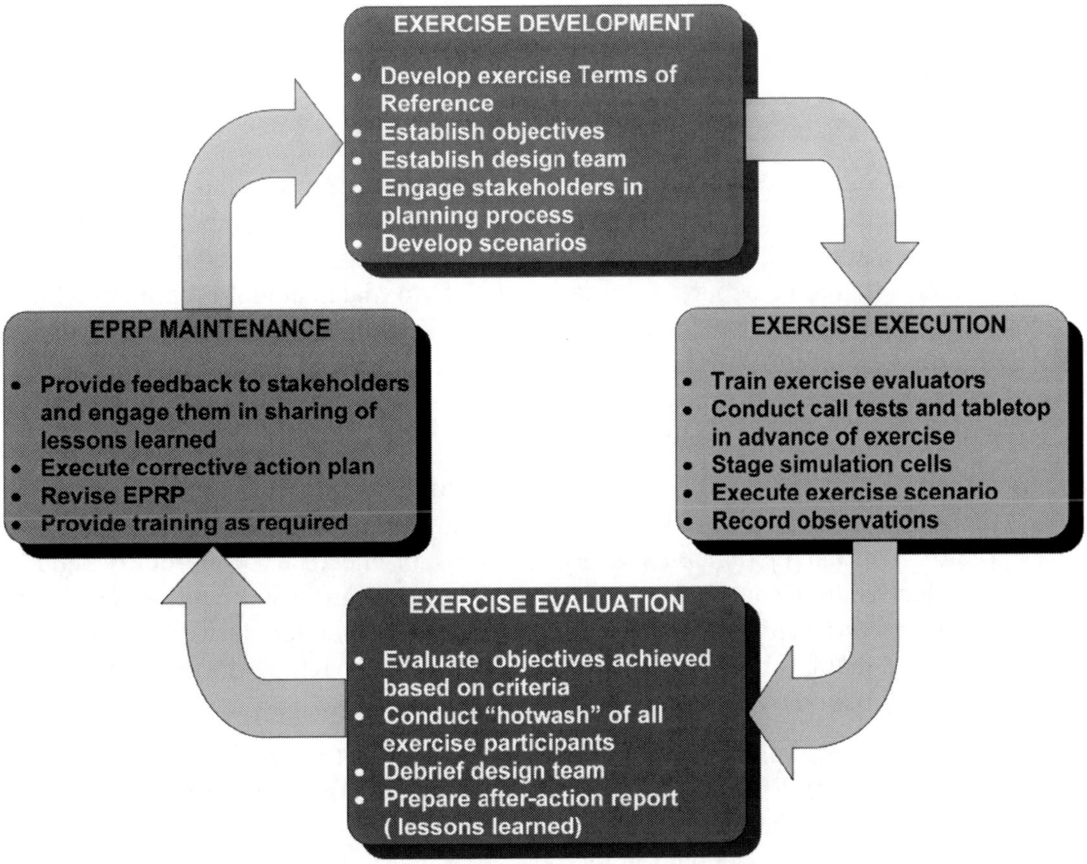

Figure 7. OPG Full Scale Exercise Process

SUMMARY OF LESSONS LEARNED FROM EXERCISES BY OPG

While each exercise has specific lessons learned based on the scenario, some of the general observations which may be useful to others engaged in the development of exercises have been;

i. **Build Up to Your Exercise** – exercises are intended to test the plan, not the people. In order for exercises to be successful, staff need to be prepared to participate otherwise staff may be discouraged and fail to appreciate the value of conducting the exercise. It is important therefore to ensure that staff training is current for their role in emergency response; following which a Tabletop Exercise should be conducted to build confidence and understanding of the roles. The development of a 'Players Handbook' has also proven valuable in establishing expectations and setting the context for participation in the exercise.

ii. **Resource the Design Team** – actual dam safety incidents rarely follow a predictable path or only engage a single discipline in the response. A hydrologic event that leads to a dam overtopping would involve water management, an assessment of the civil structures and the flow control equipment, and will in all likelihood result in downstream environmental impacts. It follows therefore that in order to design a realistic exercise scenario that fully engages the participants it is necessary to resource the design team with a broad spectrum of expertise from across the organization.

iii. **Create Realism** – in order to maximize the learning gain and actually test the EPRP the exercise scenario needs to simulate the actual response and events. Realism can be achieved through such means as; basing injects on previous incidents that have occurred, initiating injects to match actual timelines (which may require them to be delivered in the days leading up to the full play), and utilizing 'actors' to simulate media or the public. An example would be to use hydrologic data from a historical storm and issue the rainfall/ runoff pattern inject in the days leading up to the actual full play of the exercise.

iv. **Engage Your "B Team"** – one of the objectives of any Full Scale Exercise should be to develop organizational resiliency, which implies engaging the 'B Team'. This can be done in a variety of ways, but needs to be explicitly stated as an objective to ensure that it takes place. The 'B Team' can be introduced as Observers or Evaluators, or a more experienced responder can be teamed with an employee that has received the training to provide coaching during the exercise as they walk their way through the learnings.

v. **Make Your Stakeholders Equal Partners** – given that the purpose of an EPRP is to ensure First Responders are prepared to take action based on events at the dam it is critical that they participate to the extent that they are able in the exercise. This means engaging them in the preparation of the Terms of Reference and encouraging their participation throughout.

While it may be difficult for some stakeholders to participate in the exercise design process, they should be encouraged input on elements of their plans which they would like to be included in the exercise. The Design Team may also need to be creative in terms of how the stakeholders participate with options ranging from; i) full participation, ii) conducting their own Tabletop Exercise during the Full Scale Exercise, iii) staffing a Simulation Cell, or iv) providing some pre-coaching to the Design Team to realistically create a Simulation Cell in their absence if they are not able to participate in the exercise.

vi. **Simulate the Public and the Media** – one of the biggest challenges during an actual event is to set expectations of the public and the media as to what can be accomplished in managing water levels and flows and mitigating the impacts. Simulating the public and media provides the opportunity to test an owner's protocols and decision making processes to ensure they are transparent to the public and balance the various interests of both upstream and downstream communities. It can also be a good means to exercise security procedures as oftentimes the public will attempt to access the site during actual events.

Social Media injects are also becoming an essential component of an exercise scenario considering the widespread public use during natural disasters. Traditional media releases may involve multiple levels of messaging approval, whereas social media requires a more immediate and less formal response. Providing mock social media injects provides communications staff with a modern, realistic situation that they should be prepared to respond to.

vii. **Develop Evaluation Criteria Matching Your Objectives** – given the level of effort involved in designing and executing a Full Scale Exercise it follows that there needs to be clear objectives established for each of the response centres and criteria that provides transparency in assessing whether the objectives are achieved or if a corrective action plan needs to be developed to address a gap. Training the Exercise Evaluators and Observers on the evaluation criteria such that it is carried out consistently and objectively will allow assessments to be made not only on individual exercises, but to also determine if stepped improvements are being achieved from exercise to exercise.

Despite the maturity of OPG's Dam Safety and Emergency Management Programs, in each case improvements to the processes, communication protocols and technical resources have arisen from the execution of Full Scale and Tabletop exercises; for this reason it remains a cornerstone to the continuous review and revision to OPG's Dam Safety EPRPs.

CONCLUSION

Being a responsible dam owner implies much more that meeting the minimum requirements of a regulation or an industry issued guideline. It requires commitment to continuous improvement and managing risks to as low as is reasonably achievable. Even in the absence of a legislative requirement many dam owners have committed to developing EPRPs to guide their response and inform community first responders of the hazards that may occur. This is an important step in meeting the due diligence test; however maintaining a state of readiness means ensuring that both internal staff and the community first responders are trained in the response procedures and are able to work in an integrated manner to mitigate the impacts of a dam failure. This can be a bigger challenge than the effort involved in documenting the plan.

In order to make real stepped improvements in how dam safety risks are managed, dam owners and regulators need to embrace the longer term goal of building community resiliency. To achieve this goal they will need to engage communities in understanding the risks posed by dams; with stakeholder meetings, tabletop and full scale exercises offering valuable opportunities improve the awareness. Lesson learned reports arising from the exercises should consider the community vulnerabilities and seek to close the gaps through such mechanisms as infrastructure upgrades and better land use planning which take into consideration dam failure risks. In Canada, the Canadian Dam Association's Emergency Management Working Group is currently addressing these concepts in the development of a Technical Bulletin, along with other tools intended to assist dam owners and regulators in managing the risks.

REFERENCES

Association of State Dam Safety Officials, Inc., 2015, "Annual Report on State Dam Safety Programs".

Bennett, T., Serota, T., "Examining Dam Safety Guidelines For Emergency Management In The Context Of Industry Standards And Practice", Canadian Dam Association Montreal Conference Proceedings, 2013

Canadian Dam Association – Dam Safety Guidelines, Section 4 (2007)

Canadian Standards Association, Emergency Management and Business Continuity Programs (Z1600, 2014)

Federal Emergency Management Agency 2013 – Federal Guidelines for Dam Safety: Emergency Action Planning for Dam Owners

Federal Energy Regulatory Commission (FERC) 2007 – Engineering Guidelines for the Evaluation of Hydropower Projects, Chapter 6: Emergency Action Plans

International Joint Commission (IJC), 1998. "Unsafe Dams?", Ottawa, Canada

International Joint Commission (IJC), 2006. "An Update of the IJC's 1998 Report, Unsafe Dams? – Seven Years Later – What has Changed?",Ottawa, Canada

Managing the Risks of Extreme Events and Disasters to Advance Climate Change Adaptation, Report of the Intergovernmental Panel on Climate Change (IPCC), Cambridge, 2012

Ontario Ministry of Natural Resources and Forestry (MNRF) 2007 – Lakes and Rivers Improvement Act

Ontario Power Generation – Standard for Dam Safety Emergency Preparedness and Response Plans (DS-STD-08, R8)

Ontario Power Generation – HTO Emergency Management and Communication Structure (DS-STD-11, R4)

Parks Canada, Directive for Dam Safety Program of Parks Canada Dams and Water-Retaining Structures, 2009

Public Safety Canada 2011 – Emergency Management Planning Guide

Report of the Auditor General of Alberta, 2015, Edmonton, Canada.

Report of the Auditor General of Quebec to the National Assembly, 2015, Canada.

Report on Synthesis and Analysis of Submissions to the Chief Inspector's Orders, Hatch, 2015, Canada.

Serota, T., Bennett, T., "Emergency Action Planning In Action: Practices and Lessons Learned at OPG", Canadian Dam Association Banff Conference Proceedings, 2014.

Summary Report: Workshop on Emergency Management for Dams – Ottawa, February 24-25, 2015, Canadian Dam Association, 2015.

THE CRITICALITY OF ENGAGEMENT BEFORE, DURING AND AFTER A DAM SAFETY EMERGENCY

Bernie Auld, PE[1]
Benjamin Heath, PE[2]
Keil Neff, PhD., PE[3]

ABSTRACT

With 75,000 structures in the United States National Inventory of Dams reaching or exceeding design life, the likelihood of failure and incidents impacting populations and economies increases. While dams provide significant benefit compared to risk of failure, risks should be evaluated and preparedness efforts planned. With a critical review by the Tennessee Valley Authority (TVA) Dam Safety program, emergency preparedness and response was evaluated through surveys, interviews, and industry guidelines for dam safety. This validated TVA's engagement in response efforts with municipalities and provided opportunities for improving communication, technologies, and procedures between and within agencies.

TVA's Interim Risk Reduction Measures (IRRMs) drives the ability to reduce risk to the public and provides continuous monitoring. As part of the IRRMs, TVA has constructed a Dam Failure Warning System at Pickwick Landing to alert residents downstream, should a seismic event occur. This system includes a seismic recorder, time domain reflectometry (TDR), and live composite video that permits 24-hour viewing of potential seepage or breach conditions based on its heat signature.

Information related to emergency communication and response was incorporated into Hydrologic Engineering Center - Flood Impact Analysis (HEC-FIA) consequence models to evaluate emergency preparedness efforts to reduce loss-of-life from dam failure. The review verified the criticality of engagement with emergency managers for improvement of preparedness and response efforts in emergencies while aiding to protect humanity, infrastructure, and agency reputation. This paper reflects TVA's actions to develop and maintain engagement efforts and provide accurate warnings in the event of dam failure.

INTRODUCTION

With the signing of the original TVA Act by President Roosevelt in 1933, Congress charged the Tennessee Valley Authority (TVA) with ensuring the economic, environmental, social, and physical wellbeing of the people in the areas that it serves. Protecting the wellbeing of people continues to be a top priority for TVA. The concept

[1] Dam Safety Program Manager, Tennessee Valley Authority - TVA, 1101 Market St., Chattanooga, TN, 37402, bauld@tva.gov
[2] Emergency Preparedness Engineer, Tennessee Valley Authority - TVA, 400 W. Summit Hill Dr., Knoxville, TN, 37902, btheath@tva.gov
[3] Hydrology Engineer, Tennessee Valley Authority - TVA, 400 W. Summit Hill Dr., Knoxville, TN, 37902, kjneff@tva.gov

of protecting the community recently became apparent when TVA discovered there were potential issues with seismic stability of a dam located on what is considered one of the most active earthquake faults in the country.

From the beginning of its formation, TVA established a unique problem-solving approach to fulfilling its mission: integrated resource management. For each issue TVA faces, whether power production, navigation, flood control, malaria prevention, reforestation, or erosion control, it is studied in its broadest context. TVA weighs each issue in relation to the whole picture and has held fast to its strategy of integrated solutions, even as issues change over time. This is exactly the approach TVA undertook when it determined a dam failure warning system was necessary to protect the downstream community should deformation of the dam occur from an earthquake. It is very important throughout the process that the community is aware of this possibility.

For this reason, TVA focused efforts for addressing recently identified dam safety issues in a very rural area where access after a seismic event would be difficult. An appropriate action plan was developed immediately after the seismic risk was identified. TVA assembled a team consisting of emergency responders, local leaders, and the National Weather Service (NWS) to explain the risk and comprehend community needs should a dam failure occur. Understanding the threats and consequences prior to an emergency event aids in the mission to respond correctly and efficiently. Having the necessary information, for example, contact numbers, inundation areas, and the knowledge of how each first responder will be apprised of the situation is critical to this mission. For this reason, engagement with public servants can dramatically reduce the loss-of-life potential in the event of a dam failure and starts with the initial perception of the first responder interaction process.

HISTORY OF SOUTH EMBANKMENT DAM

Located at Tennessee River Mile (RM) 206.7 in Hardin County, Tennessee and 100 miles directly east of Memphis and the New Madrid Fault, Pickwick Landing is one (1) of nine (9) multipurpose main river projects owned and operated by TVA. Construction began in 1935, with current operating objectives that include flood control, augmentation of flows for navigation and water supply, hydropower production, water quality, recreation, and aquatic ecology. The annual pool elevation variation is approximately six (6) feet, with an additional four(4) feet during flood events. The Pickwick Landing Reservoir (see Figure 1) is a critical juncture for navigation for both the Tennessee River and Tombigbee Waterway.

The South Embankment Dam (SED) is 4,380-feet long, with a maximum height of 65-feet, at a crest elevation of EL 436.5. Since the soil within the flood plains was about 55-feet in depth, it was impractical to consider excavation and placement of concrete structures to connect to the abutments. Therefore, hydraulic fill construction methods were employed to construct the SED. This method was also selected because 1) dam-building activities could occur throughout the year minimizing construction period, and

2) soil was available from on-site excavation of soil material from the lock, reducing cost and effort.

Figure 1. Pickwick Landing Dam and Reservoir

A Hydraulic Fill Dam (refer to Figure 2) is one in which the dredged material may be pumped along the rim within two (2) outer edge ridges where the material settles. The courser material settles first at the edges, while the fine particles migrate to the middle to form a watertight central core. This method does not require compaction of the material. Embankment dams are no longer built by this method because the cost of compacted rolled earth has decreased with the development of larger, more economical earth moving equipment. When hydraulic fill dams are located within certain seismic zones, they may deform due to earthquake liquefaction during seismic activity, which is a resultant of the un-compacted soils within the dam. Seismic improvements are often required of hydraulic fill dams to meet current dam safety standards.

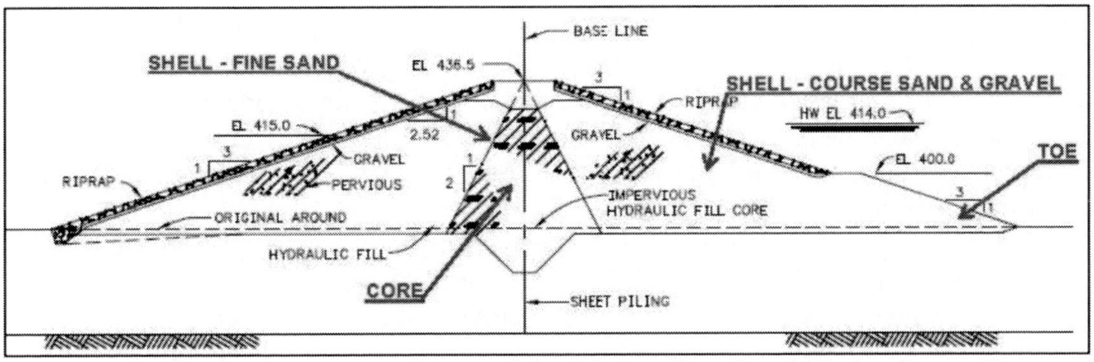

Figure 2. Typical Cross-Section at Pickwick Landing South Embankment Dam

Although it is impossible to anticipate when or what kind of seismic event might occur, history of the New Madrid Fault provides a glimpse of what TVA would need to be prepared for, should history repeat itself. According to the Missouri Department of Natural Resources, the New Madrid Seismic Zone is the most active seismic area in the U.S. east of the Rocky Mountains and covers portions of five (5) states (Fuller, 1912).

Between December 16, 1811 and February 7, 1812, three (3) to five (5) major earthquakes (believed to have been magnitude 7.0 or larger) occurred in the New Madrid Seismic Zone. Due to factors such as the harder, colder, drier, and less-fractured nature of the rocks in the earth's crust in the central United States, earthquakes in this region shake and damage an area approximately 20 times greater than earthquakes in most other active seismic areas (Missouri Department of Natural Resources, 2015). To put this in perspective, history maintains that the Mississippi River actually ran backwards for several hours causing an upheaval 15 miles south of New Madrid creating Reelfoot Lake (Fuller, 1912).

Understanding the New Madrid Seismic Zone and having a historical reference as to the magnitude of previous earthquakes in the area, allows TVA to evaluate what the impact of similar seismic events could have on the dam today. TVA determined at the design earthquake (3,000-yr period, magnitude 7.8 at 160 km, PGA=0.21 g) that the SED strength will be lost in the areas such as the hydraulic fill core, portions of the upstream and downstream hydraulic fill shells, upstream fill, and silty sand foundation layer. If these portions of the SED were to fail, then the impacts would likely result in the loss-of-life at the dam, as well as additional loss-of-life and property damage downstream. At the same time, disruption to river navigation and other secondary consequences could impede economic development of the entire Tennessee Valley. Even if the instability to the slope did not cause a breach to occur, it would probably still be necessary to lower the reservoir causing an economic impact to the area, which would also be considered unacceptable to TVA.

At the time of construction, little was known about building dams for earthquake safety. A seismic stability assessment concluded that the SED is susceptible to deformation. Based on these analyses, the embankment does not satisfy current TVA criteria for Design and Evaluation of New and Existing Dams for post-earthquake stability including deformation of less than two (2) foot in both upstream and downstream directions (note: Deformation analysis does not take into account cracks developing in the dam, which could also lead to a breach).

While conditions at the dam have not changed, there is a remote chance that Pickwick Landing could experience a large seismic event occurring along the New Madrid Fault that could damage the SED. It is noted that such a seismic event would be considered a major earthquake capable of widespread damage. TVA has been working to achieve the following goals concurrently:

- Protection of the Public Using Consequences Assessment
- Interim Risk Reduction Measures (IRRM)

    - Communication, Education, and Awareness
    - Emergency Action Plan (EAP) Supplement
    - Enhance Surveillance and Monitoring Systems
- Feasibility Study Program to address Seismic Modernization
- National Environmental Protection Act (NEPA) Permitting
- Design and Construction of Mitigation Measures

<u>Protection of the Public Using Consequences Assessment</u>

A breach of the SED would result in a flood similar to recent and re-occurring floods experienced by Savannah, Tennessee (RM 190 and 16.7-miles downstream of the SED) residents, which is consistent with a 10-year frequency event (0.1% annual chance). However, the flood wave would arrive much faster with a more rapid water surface increase than previously experienced. Although the flood wave attenuates approximately 25-miles downstream of the SED due to the significant alluvial overbank areas, populations immediately downstream are at significant risk. To address the flood wave condition, TVA has constructed and maintains a Dam Failure Warning System to alert residents and protect the community in areas downstream of the SED. Beginning in the Fall of 2014, TVA initiated an early winter drawdown of the Pickwick Landing Reservoir due to seismic stability concerns. The announcement of this information was carefully planned and the affected first responders were personally notified through face-to-face educational meetings. Due to the success of these meetings, TVA received complete support from each of these stakeholders. The education process for these first responders was a high priority and as new information develops, it is distributed to these stakeholders. These meetings confirm TVA's dedication to minimize the impact to this community while attaining the best solution for seismic stability.

A consequences assessment was conducted for a greater comprehension of the spatial and timing implications to the population at risk (PAR) and enhance planning of the IRRMs. For this analysis, Hydrologic Engineering Center River Analysis System (HEC-RAS) models were developed and incorporated into the Flood Impact Analysis (HEC-FIA) models to estimate and assess downstream consequences. For the hypothetical breach of the embankment subsequent to a significant seismic event, a sunny-day breach was modeled utilizing assumptions postulated from the stability analysis and understanding of the failure mode.

Multiple scenarios were simulated in the HEC-RAS model to understand the sensitivities of pool elevation, time of day, seasonality, breach geometry and breach formation time. Hydraulic results, structure, agricultural and HAZUS (geographic information system-based natural hazard developed by the Federal Emergency Management Agency) population data was incorporated into the HEC-FIA model to estimate downstream loss-of-life and economic consequences. Depth-damage functions for structures and contents of residential, commercial, industrial, and agricultural properties also were incorporated into the model. To approximate loss-of-life from the PAR, warning diffusion, warning issuance, and mobilization curves were developed to represent current and IRRM conditions and incorporated each into the model. HEC-FIA model results were used to evaluate the effectiveness of a surveillance and monitoring system and Supplemental EAP, and communicate the risk with internal and external stakeholders.

Interim Risk Reduction Measures (IRRM)

While TVA continues additional field investigations and analyses for seismic stability modernization of the SED, it was determined that the reservoir could be returned to normal summer operation levels, upon implementation of IRRMs. The IRRMs continuously monitor the dam and include an enhanced notification process for the region impacted by this failure mode. TVA is working with local emergency management agencies (EMA) to educate the PAR, along with media and others to raise awareness of the warning system, educate the Pickwick community and provide dam safety status information.

The first step in notifying the potentially affected communities was to determine the extents of the inundated regions. Once this information was identified by TVA's Hydrological Staff, the next step was to determine who and what would be impacted. TVA's Geographic Information System (GIS) staff provided a list of structures located within this area. Due to the interaction with the first responders, TVA acquired local knowledge from the county road superintendent as well as information from the tax assessor. These resources were able to reduce the number of affected structures by approximately 20%, based on mislabeling of structures or knowledge of storm water infrastructure improvements. This resulted in a significant decrease in the PAR and aided in reducing the overall loss-of-life potential.

With a confirmed list of affected addresses, TVA undertook several actions which included: notifying the individuals within the inundation regions, providing a means for these households to be warned, and providing dam safety communication, education, and awareness to the community.

Communication, Education, and Awareness. A Community Outreach Program was initiated in multiple steps, beginning with an Evacuation Guidance and Maps brochure created by TVA and in conjunction with local EMA's and the NWS. This brochure describes the condition of the SED, provides a map of the inundation regions, describes the early warning system and components, and includes supporting information such as contact information and emergency preparedness. TVA mailed this brochure with a free weather radio voucher and a return envelope to each address. The vouchers that were returned included appropriate shipping addresses to where radios were distributed.

While TVA works to modernize the SED to meet current standards, it has initiated additional activities including newsletters, brochures, public meetings, posters, exhibits at local venues, a web site (https://www.tva.gov/Energy/Our-Power-System/Hydroelectric/Pickwick-Reservoir), and an email address where TVA employees can respond directly to citizens' questions (pickwickalert@tva.gov). These informal presentations as well as question/answer sessions have provided an open dialogue between the community and TVA that has prevented unwarranted false information and misrepresentations.

Emergency Action Plan (EAP). TVA annually distributes EAPs to first responders, which includes contact information, dam failure inundation maps, and a host of other pertinent information. In addition to the EAP, TVA provided local EMAs with a Supplemental EAP which included mapping that provides scenario specific inundation zones to identify populations, critical facilities, and people needing special assistance to enable residences, businesses, schools, and other facilities downstream of the dam to quickly evacuate. The distribution of this information aids in the relationship process as a TVA employee annually meets with each stake holder and reviews the EAP to address any questions or concerns that are elevated. These informal meetings establish and maintain relationships that will be crucial if a dam safety emergency were to occur.

Additionally, the Department of Homeland Security (DHS) developed the National Incident Management System (NIMS) in March 2004, which aids in the ability to engage with first responders due to the common language that is used within the industry. The adoption of this language into every aspect of the emergency services industry, including dam failures, is critical to ensure effectiveness and efficiency. TVA implemented NIMS in 2005, which provides TVA's employees the ability to confidently communicate during a time of crisis both internal to TVA and external to all response organizations.

Enhanced Surveillance and Monitoring Systems. As TVA determined the most effective approach to this project, public safety was the highest priority. Therefore, it was determined that the following criteria would be integrated into the project's system.

The Dam Failure Warning System included:

- ❑ Robust, Reliable, and Redundant Methods and Technologies
- ❑ Remote Monitoring Units (RMU)
- ❑ Automated systems triggered by strong motions (Seismic Recorder)
- ❑ Instruments to detect earthquake-induced damage the length of the dam (TDR)
- ❑ Live composite video feeds
- ❑ The Warning System will permit monitoring by operators in the TVA River Forecast Center 24/7
- ❑ United States Army Corps of Engineers (USACOE) Lock operations verification 24/7
- ❑ NWS will send an alert in the event of deformation or breach condition
- ❑ Tone Alert System used by NWS
- ❑ Provide Weather Radios to homes within inundation regions
- ❑ Radio backup

To ensure all data and components were installed to properly operate within a secure network, TVA contracted with Panduit to provide the systems integration of all instrumentation and equipment required by TVA for this project. Working together, TVA and Panduit ensured that the proper steps were taken on the forefront, allowing the project to move forward with all the necessary specialists involved to allow the Dam Failure Warning System to be installed and operating within two (2) months, which was ten (10) months earlier than initially projected. This early completion was a positive accomplishment for TVA in that they were able to protect the public with a warning system sooner than anticipated.

Panduit RMU Enclosures. The enclosures incorporated:

- internal thermal and moisture monitoring for data systems
- shock sensors to notify of attempted vandalism of the enclosure
- analog to digital data conversion with data collection/transmission
- harsh environment reliable data switches
- radio backup (for all instrumentation excluding composite video)
- allowable access for non-electrical trained personnel
- badge access control
- data backup via ups

within a heavy gauge stainless steel structure.

Seismic Recorder. To facilitate rapid notification following strong ground shaking, an automated system to retrieve event triggered seismic waveform data from a digital data logger at remote sites was installed.

Time Domain Reflectometry. TDR monitors landslide and embankment stability. The basic principle of TDR is similar to that of radar. An electrical pulse is sent down the coaxial cable. When the pulse encounters a break or deformation in the cable, it is reflected. The reflection shows as a "spike" on the characteristic cable signature. The relative magnitude and rate of displacement and the location of the zone of deformation can be determined immediately and accurately (www.kanegeotech.com).

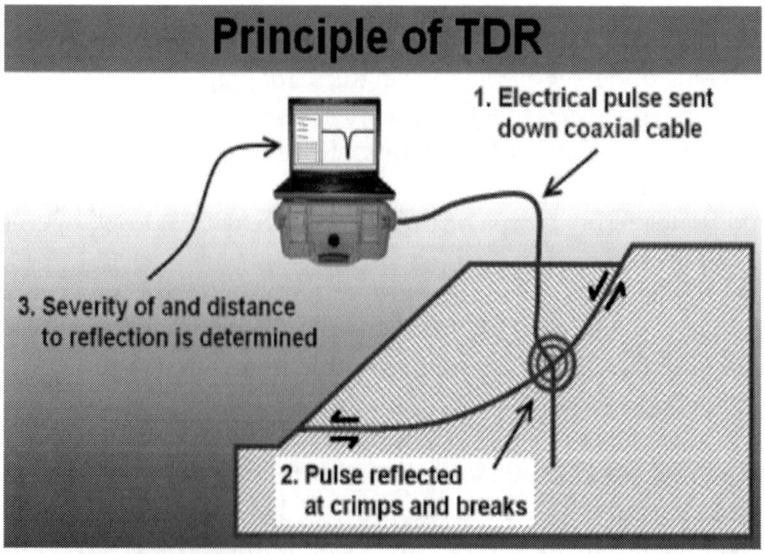

Figure 3. Typical TDR placement and operation per Kane GeoTech, Inc.

TVA's designed placement of the TDR cables was horizontally along the upstream crest, downstream crest, and downstream toe of the embankment (approximately three (3) miles total length). These locations permit the monitoring of deformation along the entire length of the SED (approximately one (1) mile)

Thermal Imaging Cameras. Images from infrared radiation or heat emitted from objects are formed within thermal imaging cameras. Since all objects have a heat signature, thermal imaging cameras can operate at all times, and even in complete darkness. Because thermal imaging cameras work by "seeing" heat rather than reflected light, they can convert the temperature of objects into shades of gray. On a cold day a person may be distinguished as a lighter object than the surrounding area because they are warmer and on a hot day the person may be distinguished as darker since they are cooler than their surroundings. The same is true for water. A composite video provides a high quality light reflecting visual with a thermal imaging overlay, which permits viewing of a potential seepage or breach condition within the embankment whether night or day based on its heat signature (see Figure 4).

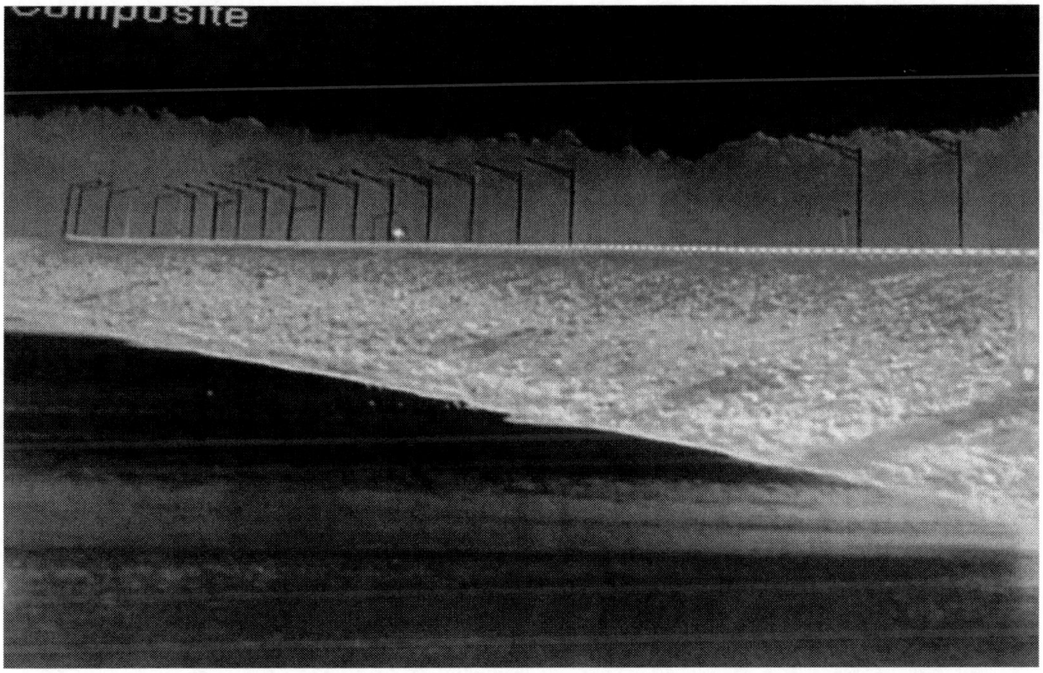

Figure 4. Composite Video - Upstream Reservoir at South Embankment Dam

Operators in the TVA River Forecast Center located in Knoxville, Tennessee (327-miles away from the SED) monitors the Dam Failure Warning System. This monitoring, along with operations verification via the USACOE Lockmaster (via satellite phone) occurs 24-hours a day, 7-days a week. The system provides specific actions for evaluating alerts from monitoring equipment to prevent false notifications. Should a verified deformation or breach condition occur, the NWS would be immediately notified and send an alert through their Tone Alert System. TVA provided Weather Radios to every resident that returned the radio voucher, approximately 160 of the 250 structures. To assure redundancy, the warning system also is equipped with radio backup for all instrumentation excluding composite video.

STATUS

While dam stability analysis and designs are complex technical efforts, TVA is currently evaluating alternatives for the embankment modernization program. Given the multiple purposes of Pickwick Landing and the long-term benefits it provides, TVA will require a thorough analysis for any modifications to assure protection of the public and that the benefits are not compromised by the modifications to the embankment. TVA also takes great effort to comply with the NEPA and the Endangered Species Act (ESA), which requires significant ongoing regulations to assure that all conditions are met.

CONCLUSION

Since the installation and implementation of the Enhanced Surveillance and Monitoring Systems, TVA has been able to verify that the system works as it was intended. TVA's IRRMs will continue to play a significant role moving forward with its various projects. The fact that this type of system had never been implemented prior to this project and that TVA was able to implement the system in two (2) months rather than the initial one (1) year estimate, is yet another example of TVA's dedication to the communities it serves. It goes without saying that TVA hopes that it will never have to utilize the Dam Failure Warning System at Pickwick Landing, but should a seismic event occur, TVA has made every effort to prepare the surrounding community, first responders, and the TVA staff on how to effectively respond in an emergency situation.

REFERENCES

Fuller, Myron L. 1912. *Bulletin 494 - The New Madrid Earthquake*. U.S. Geologic Survey, Washington D.C.

Missouri Department of Natural Resources. *Facts about the New Madrid Seismic Zone*. http://dnr.mo.gov/geology/geosrv/geores/techbulletin1.htm, accessed October 2015.

STAGECOACH SPILLWAY REPLACEMENT AND PERFORMANCE AT ELECTRA LAKE

Danielle Tripp Hannes, P.E.[1]
Stephen L. Jamieson, P.E.[2]

ABSTRACT

Stagecoach Spillway is the service spillway for Electra Lake, the primary water storage reservoir for the Tacoma Hydroelectric Project, located near Durango, Colorado. The Tacoma Hydroelectric Project is owned and operated by Public Service Company of Colorado (PSCo) doing business as Xcel Energy. The original Stagecoach Saddle Dam was constructed in the early 1900s with a wooden plank spillway constructed over the dam in the 1960s. The Stagecoach dam and spillway were replaced in 2013 with a two-cycle, reinforced concrete, labyrinth weir spillway with a homogeneous earth embankment dam on both sides of the spillway. The dam and spillway were substantially complete in 2013, however first fill did not occur until 2015 because of a major leak in the Cascade Flowline that fills Electra Lake. This paper will discuss the some of the dam removal, construction challenges, and initial monitoring and performance of the dam and spillway.

The design and on-site resident engineering was performed by the W. W. Wheeler and Associates, Inc. (Wheeler). Key subconsultants on Wheeler's engineering team included Kumar and Associates, Inc. (geotechnical engineering) and SM&RC Structural Engineers, Inc. (structural engineering). ASI Constructors, Inc. (ASI) was selected by PSCo as the construction contractor.

INTRODUCTION

The Tacoma Hydroelectric Project is located approximately 20 miles north of Durango, CO as shown on Figure No. 1. Water for operation of the facilities originates from three drainage basins: Cascade Creek, Little Cascade Creek, and Elbert Creek. The main water storage for the project is Electra Lake, which is formed by impounding water behind Terminal Dam on the main channel of Elbert Creek. Terminal Dam is a large, high hazard-potential, rockfill dam with a structural height of about 78 feet and a crest length of 1,270 feet. Terminal Dam is located on the south end of Electra Lake. Aspaas Dam and Stagecoach Spillway are appurtenant structures to Electra Lake and these structures are located at the north end of the lake. Electra Lake impounds about 22,364 acre-feet of water at the normal high water line elevation of 8377.0 feet.

[1] Professional Engineer, W. W. Wheeler and Associates, Inc., 3700 S Inca Street, Englewood, CO 80110, danielle.hannes@wwwheeler.com
[2] Principal Engineer, W. W. Wheeler and Associates, Inc., 3700 S Inca Street, Englewood, CO 80110, steve.jamieson@wwwheeler.com

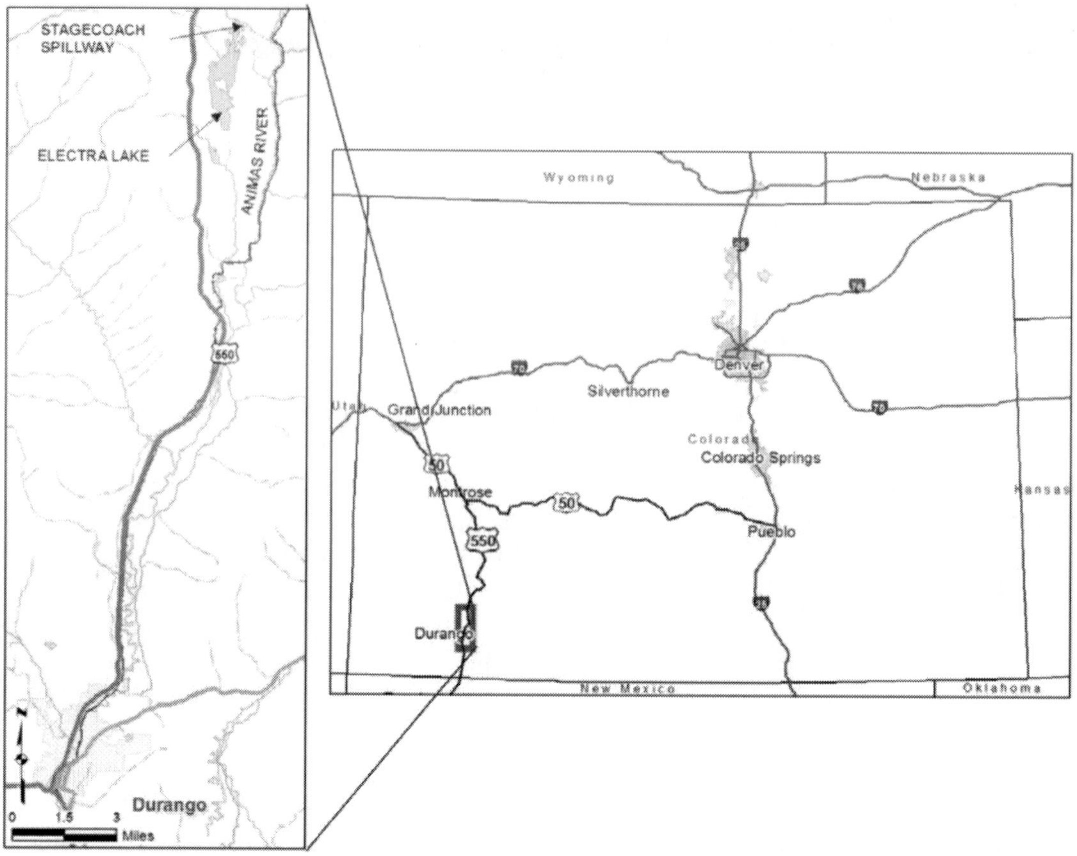

Figure 1. Location Map

Water from Cascade Creek is diverted and conveyed through the Cascade Creek diversion facilities into Little Cascade Creek, where it is diverted into Electra Lake through the Aspaas Dam diversion channel. Aspaas Dam is a small, low hazard-potential, diversion dam located on Little Cascade Creek at the north end of Electra Lake. The dam was rebuilt as a zoned earthfill structure in the early 1980s. The small pool upstream of the dam impounds about 24 acre-feet at the normal operating level, Elevation 8377.0, which is equivalent to the normal high water line in Electra Lake.

Stagecoach Dam is a large, high hazard-potential, structure located on the north end of Electra Lake as shown on Figure No. 2. The dam is considered to be a high hazard-potential because, if it failed with storage at the normal high water line, approximately 12,000 acre-feet of water would be released into Little Cascade Creek and the Animas River that would endanger lives in several residences and ranches along the Animas River.

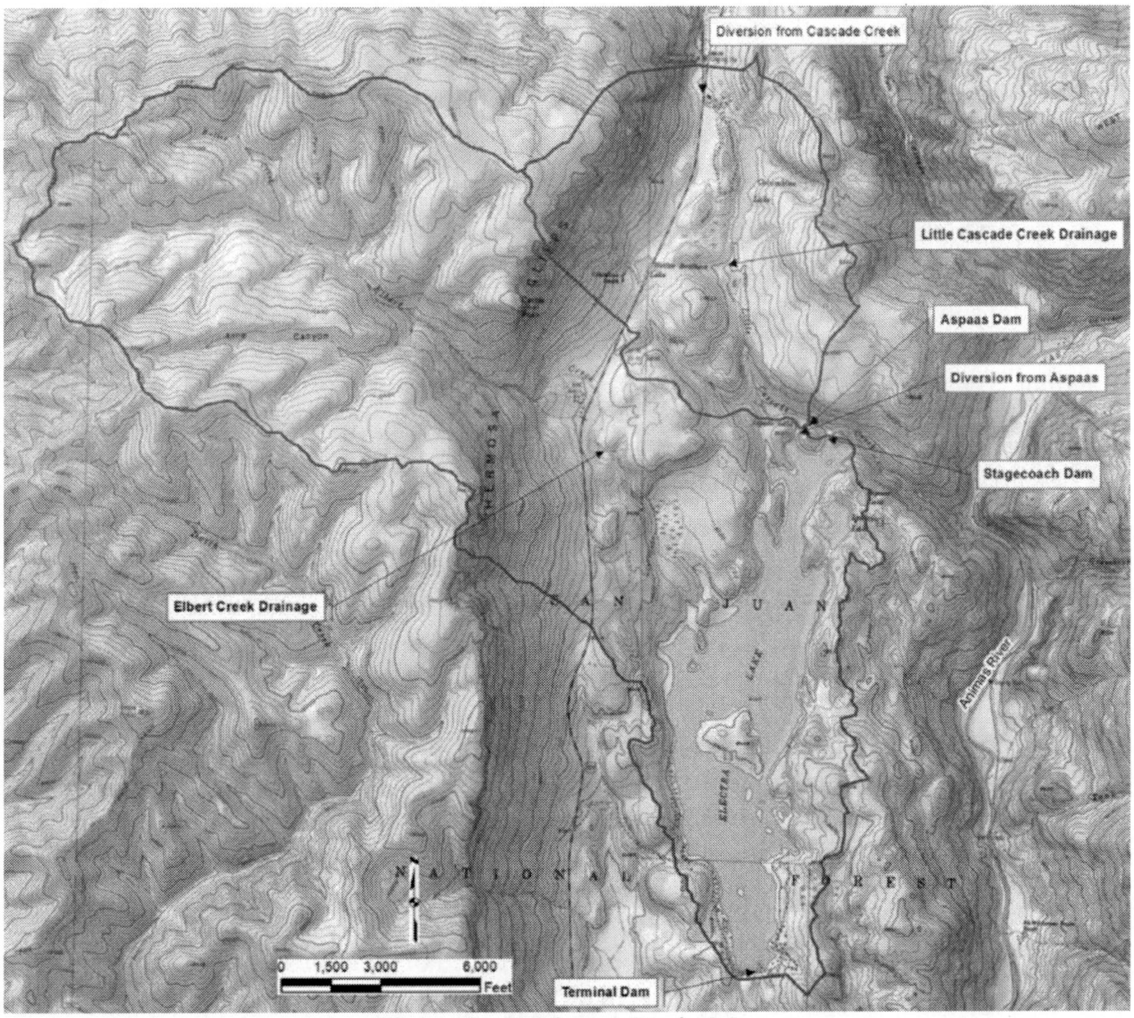

Figure 2. Drainage Basin Map

The original Stagecoach Dam was a timber crib structure, with a wooden upstream seepage cutoff constructed in a low saddle of Electra Lake. In the 1960s, the spillway for Electra Lake was moved from the left abutment area of Terminal Dam to Stagecoach Dam, where a wooden plank and timber spillway was installed over the existing dam crest. Dam safety evaluations conducted at Stagecoach Dam in 2009 and 2010 resulted in the conclusion that the dam and spillway are in a state of severe deterioration and that the Stagecoach Dam was very near the end of its operational life.

The Stagecoach Spillway Replacement project involved demolition of the existing Stagecoach Dam and Spillway and replacing the spillway with a two-cycle, reinforced concrete, labyrinth weir with embankment backfill on both sides of the new spillway. The crest of the labyrinth weir was constructed to Elevation 8377.0, which is consistent with the old spillway crest and the existing normal high water line in Electra Lake. The width of the new labyrinth weir spillway is 70 feet, with an effective labyrinth weir length of 140 feet. The height of the labyrinth weir walls is 12 feet above a reinforced concrete base slab. The spillway base slab is founded on and anchored to bedrock with eleven post-tensioned rock anchors. The spillway backfill embankment was constructed as a homogeneous embankment dam with a minimum crest width of about 15 feet and a minimum elevation of 8382.8 at the top of the embankment crest base course. The upstream slope of the spillway embankment backfill was constructed at a slope of about 2.5H: 1V (horizontal: vertical) and is protected with riprap. The downstream slope was constructed at a slope of about 2H: 1V. The spillway embankment includes a gravel filter drain along the downstream toe and toe drain system that is designed to collect seepage through the embankment backfill as well as seepage through the natural abutment soils.

Electra Lake is typically maintained at its normal high water line of Elevation 8377.0 during the summer for as long as water supplies permit. When inflows are low, normally in the winter, storage in Electra Lake is drawn down to maintain generation at the Tacoma Powerhouse. The reservoir is refilled in the spring from snowmelt runoff, with a majority of the water coming from Cascade Creek. In April 2014, Xcel began filling Electra Lake following the construction of Stagecoach Dam using the Cascade Flowline. In early June 2014 a major leak was discovered in the Cascade Flowline and the flowline was not operational throughout 2014. As a result of a wet spring and summer season and temporary repairs to the Cascade Flowline, Electra Lake was able to fill to approximately 2 feet below the crest of the labyrinth weir in 2015.

NEW SPILLWAY

The objective of the Stagecoach Spillway Project was to eliminate a four foot reservoir restriction at Electra Lake imposed by the Federal Energy Regulatory Commission (FERC) and to minimize the number and significance of future potential failure modes associated with the replacement structure. The labyrinth spillway dimensions are shown on Figure 3.

Spillway discharges were calculated using both the Lux equation (Lux and Hinchcliff, 1985) and the Tullis equation (Tullis et al., 1995). Both of these equations are commonly used to calculate labyrinth spillway discharges. Discharges calculated with both equations were within 6-percent of one another when more than 2 feet of water was calculated to be flowing over the weir. The Lux method was used to establish the spillway discharge capacity curve because it is slightly more conservative at higher discharges. The Lux equation is considered to be reasonable for the Stagecoach Spillway because the key characteristics of the spillway are within the modeling bounds of the flume studies used to develop the equation. The aspect ratio (W/P) for the Stagecoach Spillway is 2.9 and Lux recommends an aspect ratio greater than 2.0. The head to weir height ratio (H/P) is 0.34 and both Lux and Tullis recommend a maximum H/P ratio of 0.9.

Conservatism was in included in the hydraulic design of the spillway because the spillway design flood is based on a Probable Maximum Flood (PMF) developed from Hydrometeorological Report No. 49 (HMR-49). The Probable Maximum Precipitation (PMP) values estimated with HMR-49 are generally considered to be conservatively high when compared with site-specific meteorology studies. The resulting spillway discharge during the PMF event was 4,000 cubic feet per second (cfs). Additional conservatism was included in the design because the residual freeboard above the maximum water surface associated with routing the PMF through the Stagecoach Spillway is 2.2 feet, which exceeds the Colorado Dam Safety Branch guidance of one foot.

The labyrinth weir was rotated clockwise by about 15 degrees to take advantage of a bedrock saddle that was identified during the subsurface explorations (Kumar, 2012). This slight rotation minimized concrete and excavation volumes without sacrificing spillway hydraulic efficiencies because of the large approach channel depths relative to the water depths over the labyrinth weir crest.

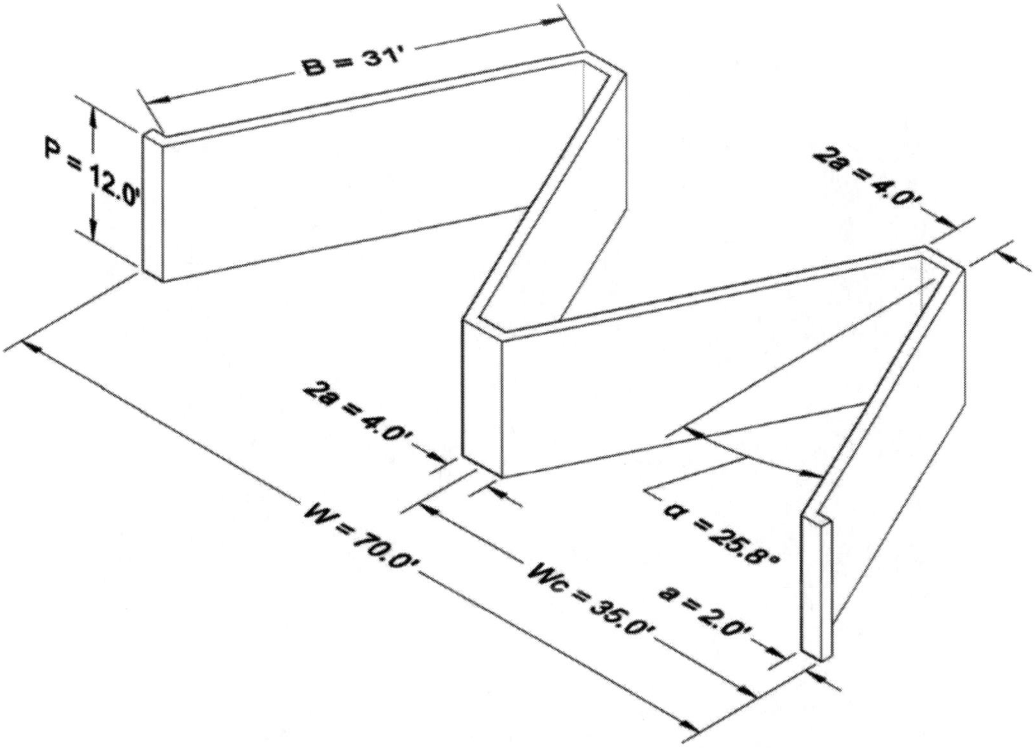

Figure 3. Stagecoach Spillway Labyrinth Spillway Dimensions

CONSTRUCTION CHALLENGES

Site Access

One of the more significant construction challenges associated with this project involved improving a 2.1-mile-long access road to the Stagecoach Spillway site from Highway 550. See Figure No. 4 for the access road alignment. Prior to construction, PSCo obtained easements from private land owners and a Special Use Permit (SUP) for the road improvements on U.S. Forest Service lands from the San Juan National Forest. Although the access road had not been used for decades, the route was generally aligned along a route used to transport construction equipment and materials to the site in the 1980s when Aspaas Dam was reconstructed. The 1980s road alignment was slightly modified in 2013 to provide a construction access road for the new Stagecoach Spillway construction. The access road improvements also included constructing a road fill in the right abutment of Aspaas Dam to keep construction traffic away from the toe of the dam; constructing an entrance to the road off of Highway 550 that minimized the road grades from the highway; and construction of a temporary channel crossing over the diversion channel between Aspaas Reservoir and Electra Lake. As part of the U.S. Forest Service SUP, timber road mats were installed in wetland areas.

Due to construction ending in mid-November 2013, the timber road maps were not removed until the summer of 2014. The road reclamation was completed in the spring of 2014. As of 2015, the access road has attained the required revegetation success. Figure Nos. 5 and 6 show the placement of the timber road mats during construction and the access road condition in October 2015, respectively.

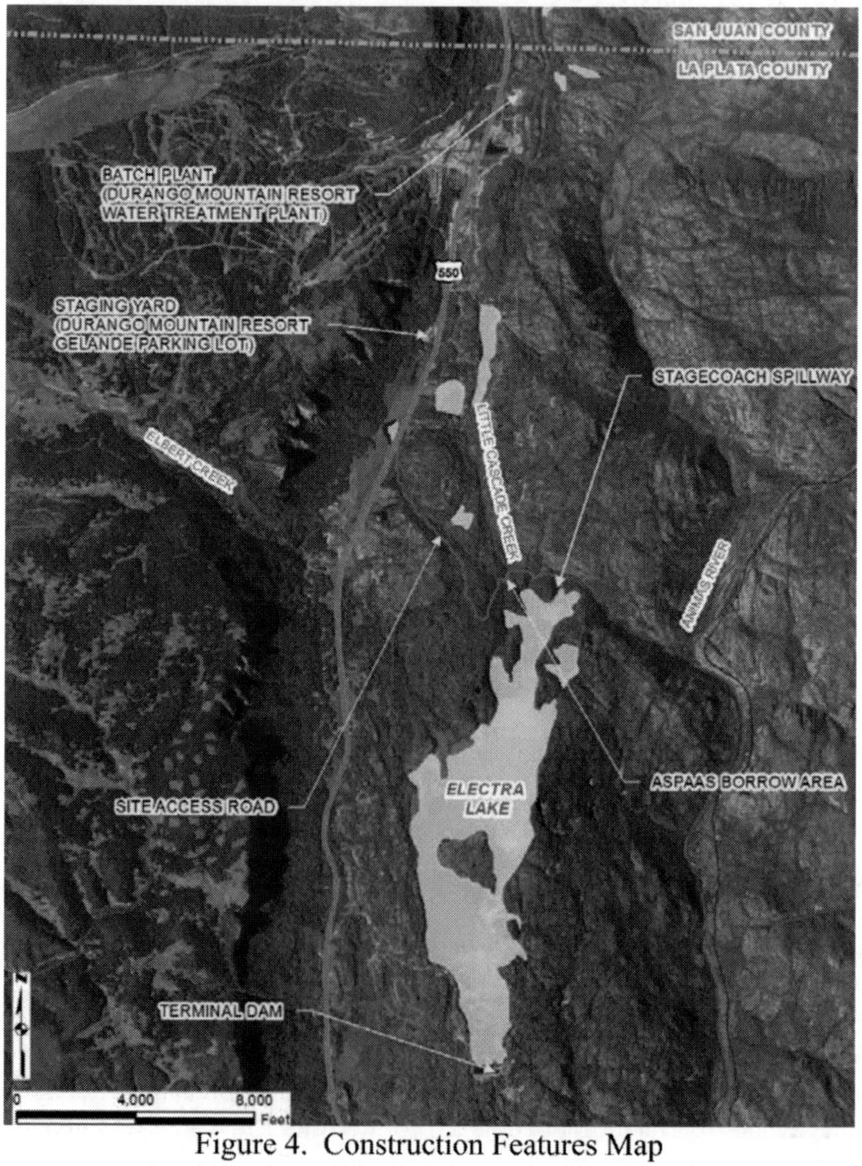

Figure 4. Construction Features Map

Figure 5. View of timber road mat placement. Photo taken 6/28/13.

Figure 6. View of the wetland area where the timber mats were placed in October 2015.

Some of the most significant challenges associated with this project were scheduling challenges. The planned start date for the project was June 1, 2013, but the U.S. Forest Service, Special Use Permit for the project was not provided until June 26, which caused an initial delay of about one month. This delay shortened the already limited construction season at an elevation of nearly 8400 feet. ASI worked six to seven days a week to recover time in the schedule and to overcome other schedule delays.

Unexpected Conditions

Demolition of the timber spillway and dam began on July 19, 2015. The spillway planking, spillway sidewalls, and support timbers removed from the original dam structure were hauled to a staging area on the access road and were eventually disposed of off-site. During the demolition of the old timber spillway and dam, a one- to two-foot-deep void was observed under the surface planking, likely due to timber structure decomposition and settlement. The wooden support structure for the original timber crib dam appeared to be almost completely deteriorated and very little wood was observed in the dam below the spillway planking. The small amount of timber left in the dam was mainly wooden fragments. The upstream cut-off consisted of vertically-placed wooden planks with asphalt paper and bentonite placed against the downstream face of the planks. Black organic soils were observed to be intermixed with approximately 3-inch-diameter to 1.5-foot diameter riprap inside the timber crib structure. Although there was no documentation of the purpose of the black organic soils, these soils may have been lake sediments that were excavated and placed in the dam to provide additional seepage control. Some of these features of the demolished dam are shown on Figure No. 7.

The black organic soils and riprap from the original timber crib dam were temporarily stockpiled separately upstream and left of the new spillway as shown on Figure No. 8. The black organic soil from the old timber crib dam was tested by Kumar and Associates, Inc. and was not considered to be suitable for embankment fill materials because of the amount of organic materials in the soil. The organic soils were also tested for nutrients and it was determined that there was not enough nutrients in the black soil for it to be used as a suitable topsoil for site reclamation. The majority of the black organic soils were hauled to the Aspaas Borrow Area to rebuild and contour the slopes in the borrow area.

Figure 7. View of upstream face of the original Stagecoach Dam

Figure 8. View of temporary stockpile of black organic soils from Stagecoach Dam

Available borrow areas

As part of the initial excavation of soil for the Stagecoach Spillway, soils were hauled to the Aspaas Borrow Area, where the soils were screened over a 3-inch grizzly screen to produce a stockpile of soil that met 3-inch minus requirements for the spillway embankment backfill. The amount of fill soil available at the Aspaas Borrow Area was limited and ASI used a significant amount of the fill soil from the Aspaas Borrow Area to construct the access road and the roadway access at Highway 550, so the amount of adequate on-site soil became a concern during construction. ASI was able to produce enough fill soils to complete the spillway embankment by screening soils from the upstream right abutment area and soils excavated from the spillway discharge channel.

Concrete Challenges

Due to the haul distance for commercial concrete, ASI set-up a temporary batch plant for mixing backfill concrete and structural concrete near the Durango Mountain Resort Water Treatment Plant as shown on Figure No. 4. Due to the specified requirement to place the main spillway slab concrete at temperatures less than 65 degrees Fahrenheit (F), ASI used a water chiller for the concrete water during placement of the spillway slab. ASI also misted the fine and coarse aggregate stockpiles at the batch plant to keep the structural concrete spillway slab placement temperatures below 65 degrees Fahrenheit. The low concrete placement temperature was specified to minimize cracking in the monolithic spillway slab. During colder weather near the end of the project, ASI used a heater for the water to help maintain structural concrete temperatures above 50 degrees Fahrenheit for the spillway walls.

Following the removal of the old Stagecoach Dam and Spillway and after the surficial bedrock was removed, the bedrock surface was observed to be very hard and irregular, which significantly increased the cohesion that was assumed at the bedrock concrete interface during the spillway design. ASI used a hoe ram mounted on the large excavator to remove bedrock that was above the elevation of the bottom of the spillway slab. Consistent with the design plans, the top of the bedrock surface sloped from left to right across the spillway foundation. The bedrock under the spillway slab was hand cleaned with shovels and then cleaned with air and water pressure to remove loose soil and rock from the bedrock surface.

Due to the irregularity of the bedrock, with no observed rock overhangs, dental concrete was not used in the foundation. As a result, only backfill concrete was placed under the spillway slab and the backfill concrete was vibrated in staged lifts in the deeper holes and pockets in the foundation bedrock. This approach was determined to be preferable over the specified approach of placing a small amount of dental concrete and creating numerous cold joints in the backfill concrete. The slush grout placing operation is shown on Figure No. 9.

Figure 9. View of slush grout application immediately prior to backfill concrete placement

PROJECT PERFORMANCE

The new spillway and dam were substantially complete in 2013. However, the initial reservoir fill did not occur until 2015 due to a leak in the Cascade Flowline that fills the reservoir. As a result of a wet spring and summer season and temporary repairs to the Cascade Flowline that were made in 2014, Electra Lake was able to fill to within two feet of the crest of the labyrinth weir in 2015.

Per the initial fill schedule, PSCo has been diligently observing the spillway and dam and reading the instrumentation per the initial fill plan that was reviewed and approved by the Colorado Dam Safety Branch and FERC. PSCo, on average, made a site visit to Stagecoach Dam twice a week during the summer of 2015. Access to the Stagecoach Dam is now available only by boat or hiking along the reclaimed access road. The instrumentation at the spillway and dam include four piezometers, two toe drain outfalls, and two embankment settlement points.

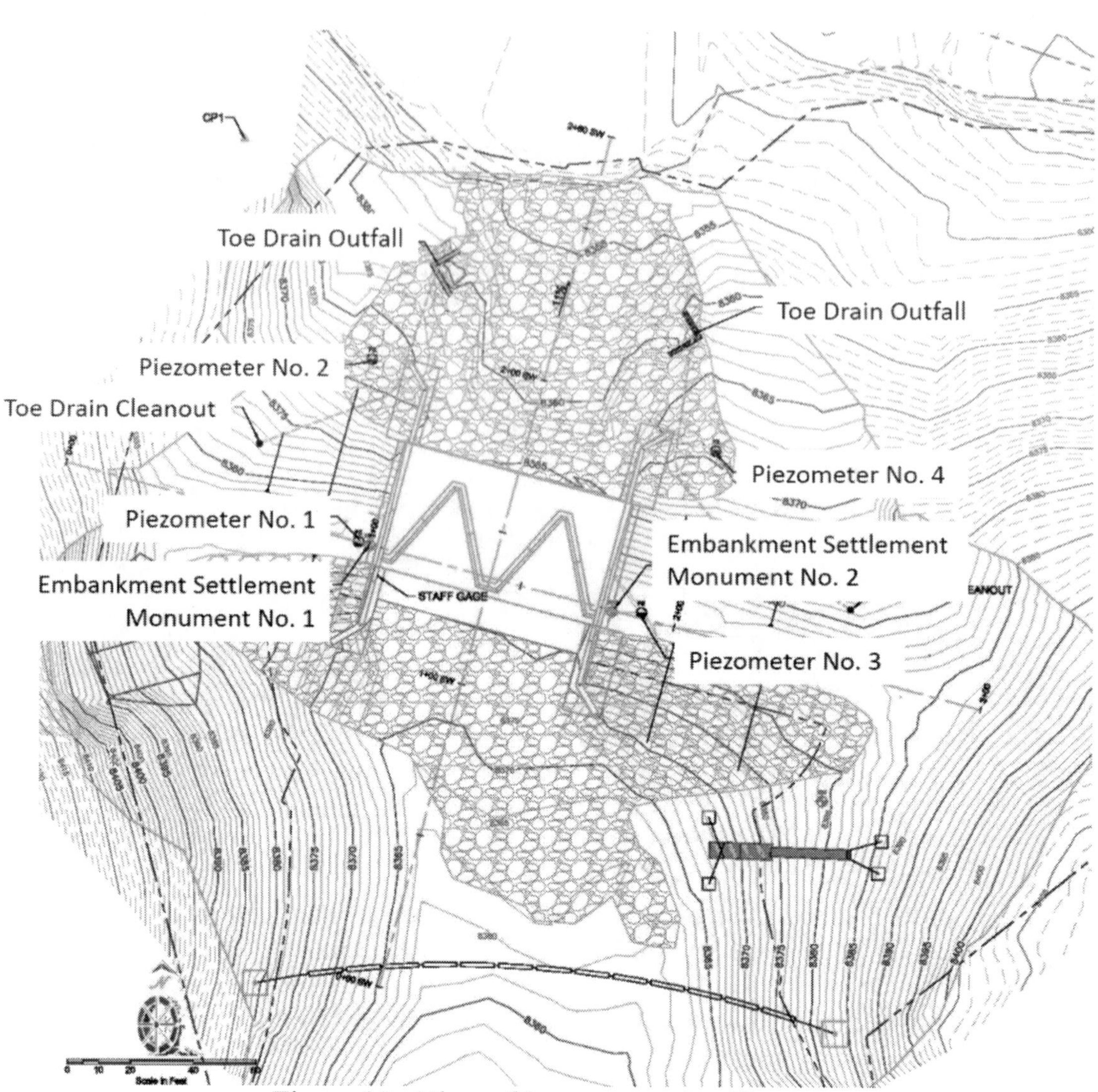

Figure 10. View of Instrumentation Plan

An inspection of the dam and spillway was performed by Wheeler on October 23, 2015 to observe the condition and performance of the structure following the initial reservoir filling period. The inspection was performed when Electra Lake was at Elevation 8375.2 feet, which is approximately 1.8 feet below the normal high water elevation of Electra Lake. Several, minor shrinkage hairline cracks were observed on the labyrinth weir walls, spillway training walls, and concrete slab during the inspection, but most of these were sealing by calcification. No dam safety issues were observed during the inspection. The Stagecoach embankment appeared to be functioning as intended. The spillway has not yet operated.

Four piezometers were installed in the embankment. One piezometer is installed on the crest and one piezometer is installed at the toe of the dam (downstream of the toe drain system) on both the left and the right sides of the embankment. The locations of the piezometers are shown on Figure No. 10. The piezometers on the dam crest showed a strong correlation with the reservoir elevation throughout 2015. There was approximately a two-month lag in the crest piezometers to reach a steady state, or equilibrium condition, in response to the reservoir elevation. Piezometer No. 1, the crest piezometer located on the left side of the embankment, generally displayed water surfaces approximately 1.8 feet to 2.6 feet higher than Piezometer No. 3, the crest piezometer located on the right side of the embankment. The bedrock was at an approximately constant elevation of 8360 feet through the entire dam section (upstream to downstream) along the left spillway wall. The bedrock along the right spillway wall is approximately 5 feet lower than the bedrock along the left spillway wall. The lower bedrock on the right side of the dam is the reason for the lower phreatic surface observed in Piezometer No. 3 located on the right embankment crest based on a revised two-dimensional seepage model developed by Kumar for the as-built bedrock and embankment conditions. The 2015 piezometer data is shown on Figure No. 11.

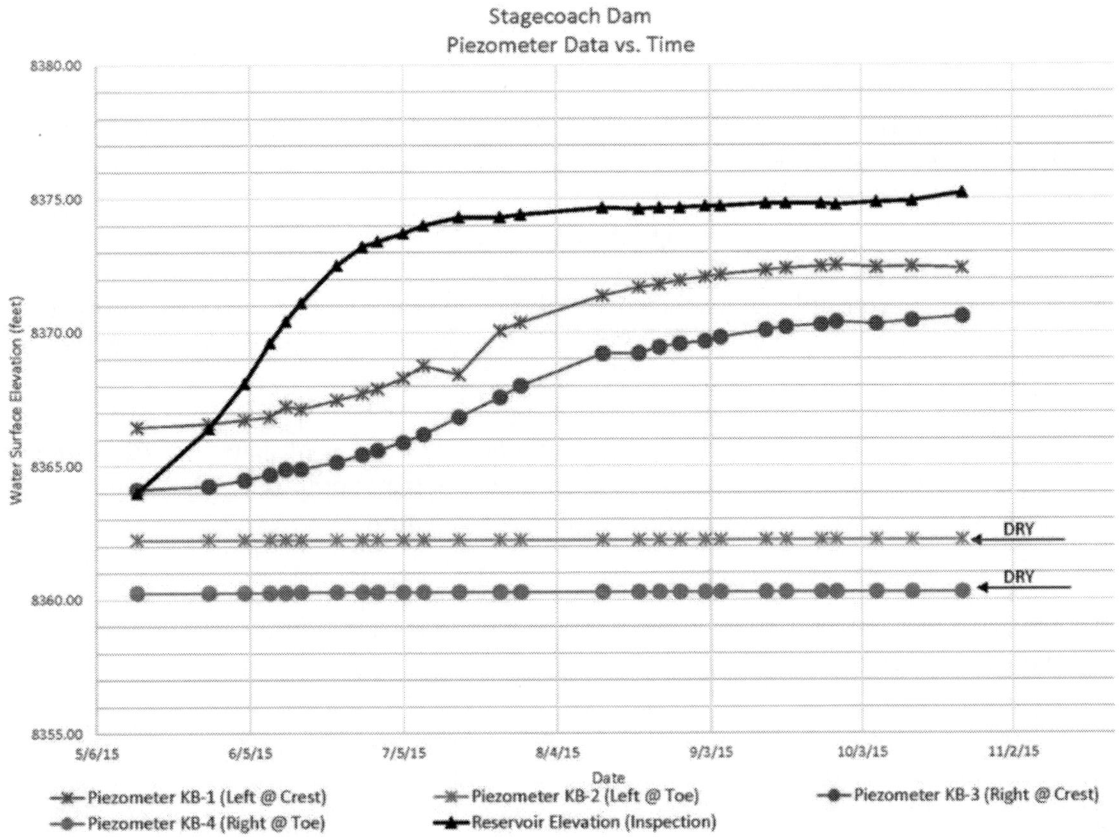

Figure 11. Piezometer Data during Initial Fill

The piezometers at the toe of the dam have remained dry throughout 2015. This is expected as these piezometers are located downstream of the toe drain system.

Two toe drain outfalls were installed at Stagecoach Dam located right and left of the spillway as shown on Figure No. 10. These toe drains were installed to measure seepage discharge through the left and right sides of the embankment. Throughout 2015 no flows were measured at the toe drain outfalls. Since flows were not observed in 2015 at the drains, it is expected that the seepage water is leaching into the bedrock prior to reaching the toe drain system. This is not a dam safety concern.

Two embankment settlement monuments (ESM) are located on the dam crest near the maximum embankment sections on the left and right side of the spillway. The locations of the embankment settlement monuments are shown on Figure No. 10. The monuments were surveyed in November 2013 and November 2015. The maximum settlement was 0.07-foot over a two-year span observed on the right side of the dam (where the bedrock is lower, resulting in a higher embankment section). Kumar's geotechnical report indicated an expected total settlement of the embankment to be less than 0.25-foot (Kumar, 2012) and the maximum settlement is well within this range. The maximum settlement rate is expected to occur directly following initial construction.

Xcel and their engineering staff will continue the monitor the performance of Stagecoach Dam and spillway in the upcoming years.

CONCLUSIONS

1. The Stagecoach Spillway Replacement eliminated the two PFMs associated with the existing structure and removed the FERC-required four (4) foot reservoir restriction at Electra Lake.

2. Weekly progress meetings between the owner, engineer, contractor, FERC, Colorado Dam Safety Branch, and other key stakeholders helped complete a challenging project within a very limited, high altitude construction season.

3. Based on observations made during the October 23, 2015 inspection and review of the 2015 instrumentation data, the dam appears to be functioning as intended. No dam safety issues were observed following the initial filling of the dam.

Figure 12. View of looking downstream and right at the completed Stagecoach Dam and Spillway.

REFERENCES

Colorado Division of Water Resources, (CO DWR, 2007), *Rules and Regulations for Dam Safety and Dam Construction*, Denver, CO, January 2007.

Falvey, Henry T. (Falvey, 2003), *Hydraulic Design of Labyrinth Weirs*, 2003.

Federal Energy Regulatory Commission (FERC, 2010) *Engineering Guidelines for the Evaluation of Hydropower Projects*, 2010.

Kumar and Associates, Inc. (Kumar, 2012), *Geotechnical Engineering Design Report, Stagecoach Spillway*, October 2012.

Lux III, F.L., and Hinchcliff, D. (Lux and Hinchcliff, 1985) *Design and Construction of Labyrinth Spillways*, Proc. 15th International Congress on Large Dams, Vol. 4, Q. 59, R. 15, 1985.

Paxton, G., Savage, B. (Paxton and Savage, 2006) *Labyrinth Spillways: Comparison of Two Popular U.S.A. Design Methods and Consideration of Non-Standard Approach Conditions and Geometries*, 2006.

Paxton, G. Campbell, D., Monroe, J. (Paxton et al., 2011) *Evolving Design Approaches and Considerations for Labyrinth Spillways*, USSD Conference, April 2011.

Trautner Geotech, LLC. (Trautner, 2010) *Field Study Test Results for Electra Lake Stagecoach Dam Spillway Evaluation Project*, January 6, 2010.

Tullis, J.P., Amanian, N., Waldron, D. (Tullis et al., 1995) *Design of Labyrinth Spillways*, Journal of Hydraulic Engineering, Vol. 121, No. 3, March 1995.

U.S. Department of Commerce, National Oceanic and Atmospheric Administration; U.S. Department of the Army, Corps of Engineers, (NOAA, 1977) *Hydrometeorological Report No. 49*, Silver Spring, MD, September 1977.

W. W. Wheeler and Associates, Inc. (Wheeler, 2009-1), *Supporting Technical Information (STI) for Tacoma Hydroelectric Project*, October 2009.

W. W. Wheeler and Associates, Inc. (Wheeler, 2009-2), *Stagecoach Dam Evaluation*, May 29, 2009.

W. W. Wheeler and Associates, Inc. (Wheeler, 2010), *Stagecoach Dam Site Investigations*, January 26, 2010.

W. W. Wheeler and Associates, Inc. (Wheeler, 2012-1), *Stagecoach Spillway Flood Hydrology Report*, May 2012.

W. W. Wheeler and Associates, Inc. (Wheeler, 2012-2), *Stagecoach Spillway Replacement, 50 Percent Design Summary Report,* August 2012.

W. W. Wheeler and Associates, Inc. (Wheeler, 2012-3), *Stagecoach Spillway Replacement, Construction Potential Failure Modes Analysis Report*, November 2012.

W. W. Wheeler and Associates, Inc. (Wheeler, 2013), *Stagecoach Spillway Replacement, Design Summary Report,* April 2013.

HIGH CAPACITY ANCHORS ON EXISTING CONCRETE DAMS — APPLICATION TO TWO TVA DAMS

Husein Hasan[1]
Aaron Nottis, P.E.[2]
Brian Krohmer[3]
Dennis Triplett[4]
Boris Caro Vargas[5]

ABSTRACT

High capacity anchors have recently been installed on a number of existing concrete dams across the country to respond to changing needs, conditions and use of the dams. Recently, two of the Tennessee Valley Authority's (TVA) Dams were strengthened using high capacity anchors. As part of TVA's systematic assessment of all of its dams, revised Probable Maximum Flood (PMF) levels and changes in the seismic demand necessitated the installation of fifty (50) high capacity anchors to one Dam (Dam #1) and twenty (20) to another Dam (Dam #2), with design loads up to 1,900 kips. Anchors were comprised of 54-strand 0.6 inch diameter anchors with total lengths varying from 100 feet to a maximum of 235 feet. Based on results from a 3D finite element analysis (FEA), the Non-Overflow and Spillway sections required anchors to meet the project design criteria. Customized working platforms which were designed and assembled in place, attached to the existing spillway to support the drilling and the related efforts. Anchor installations for the Spillway Monoliths were performed with a 250 foot crane situated on a barge located in the tailrace. A state-of-the-art structural monitoring program was installed to record real-time movements of the anchored monoliths before and during tensioning of the anchors. Three Elasto-Magnetic sensors were factory installed on separate strands for each anchor to permit long-term monitoring. An accelerated project schedule required the use of four drill rigs simultaneously. Despite the intense activity on the project and long working hours, the construction team performed more than 200,000 hours of work without a recordable injury and the project was successfully completed ahead of schedule.

The intent of this paper is to present an overview of the challenges related to the design, installation, and testing of permanent high capacity anchors in existing concrete dam structures. It will also outline details of the logistics required, in order to safely install anchors from an existing spillway. Emphasis will be placed on managing safety for delicate operations under an accelerated project schedule and with multiple pieces of equipment working simultaneously in a confined area.

---

[1] Senior Program Manager, Tennessee Valley Authority, hahasan@tva.gov, 865.632.4194
[2] Senior Program Manager, Tennessee Valley Authority, afnottis@tva.gov, 423.751.3438
[3] Project Manager, Barnard Construction Company, Inc., brian.krohmer@barnard-inc.com, 406.586.1995
[4] Operations Manager, Nicholson Construction Company, dennis.triplett@nicholsonconstruction.com, 412.677.2245
[5] Director of Business Development, Nicholson Construction Company, boris.carovargas@nicholsonconstruction.com, 303.926.0411 ext. 6505

INTRODUCTION

As part of a systematic assessment of all of its dams, the Tennessee Valley Authority (TVA) recently determined Probable Maximum Flood (PMF) levels and changes in the seismic demand necessitated the installation of fifty (50) high capacity anchors to one Dam (Dam #1) and twenty (20) to another Dam (Dam #2), with design loads up to 1,900 kips. Anchors were comprised of 54-strand 0.6 inch diameter anchors with total lengths varying from 100 feet to 235 feet. The newly calculated PMF elevation showed that both dams would be overtopped by the predicted design flood. Furthermore, seismic requirements for both dams have changed significantly since the original closure of the dams by 1943. To assess the impact of the changes in the PMF and seismic loading conditions and to determine any required modifications, a detailed, 3D non-linear finite element analysis was performed for each dam. A 3D model for each dam was considered a critical component in determining the extent of modifications required for both dams. Results from the 3D non-linear finite element analysis indicated the right Non-Overflow and Spillway Section of the dam required anchoring to meet the project design criteria. Post tensioned anchors and new reinforced concrete flood walls were required to ensure that the structural integrity of the dam(s) would not be compromised in the event of an earthquake or an extreme flood event and provide adequate protection for downstream facilities. This paper focuses on the construction-related activities associated with the modifications of both dams.

DAM #1 — INSTALLATION OF HIGH CAPACITY ANCHORS IN THREE SEPARATE AREAS OF THE DAM

The successful installation of high capacity anchors combines both geotechnical and structural expertise. The entire process is comprised of coring and drilling, grouting, installing the anchor components and finally stressing. Installing an anchor in an existing concrete structure presents several challenges and risks based on the age of the concrete and the difficulties in accessing precise as-built information. Dam #1 is a 75-year old lightly reinforced concrete structure of a low density of steel. Anchors were installed in three separate areas of the dam: the right non-overflow section, the spillway and the south training wall as shown in Figure 1.

Figure 1. Overview of Dam #1

<u>Installation of Anchors on the Non-Overflow section</u>

Thirty-two (32) vertical anchors were installed in the right non-overflow section from the top of the dam. Anchors were comprised of 54 tendon strands with lengths between 162 and 247 feet with 32.4 feet of bond length in limestone underlying the dam structure. Anchor lengths were designed such that the top of the bond zone between adjacent anchors was staggered a minimum of 5 feet to avoid concentrated stress regions associated with overlapping cones. Before proceeding to the actual full drilling depth, an 18-inch hole was cored to install the trumpet and sub-bearing plate. Next, a 15-inch pilot hole was drilled to a depth of approximately 18 feet to provide the most accurate tool alignment. The verticality of each hole was surveyed twice during the drilling process with a Down Hole Survey (DHS) gyro tool. Surveys were performed at depths near the interface between concrete and rock and final hole depth upon termination of drilling. A drilling tolerance of 0.75% was specified in all directions to avoid interference with the existing drainage gallery of the dam, adjacent anchors, existing grout curtain or other dam features.

After drilling was complete, each hole was tested for water head loss. Consolidation grouting was necessary in most of the anchor holes prior to approval by the engineer. Seventy five percent (24 out of 32) of the holes failed the initial water test and were required to be grouted and re-drilled. In total, more than 32,000 gallons of consolidation grout was placed. Following re-drilling, the hole was subjected to a second water head loss test and once passed, the external corrugated sheathing was installed and the pre-assembled anchor (including anchor head and three Dyna Force sensors on each anchor) was lowered into the hole. The interior and the exterior of the corrugated sheathing was grouted simultaneously using a 0.44W/C ratio grout. Once the grout had reached the minimum specified strength of 4,000 psi and the bearing plate was in place, the anchor was tested in accordance with Post-Tensioning Institute (PTI) Recommendations. All rigging and lifting operations were performed using a 275-Ton crane with a 174 feet boom and 253 feet long luffing jib; the heaviest anchor was more than 12,000 pounds. Additional details describing the lifting operations are described in the section "Dam #1 – Access, Platforms and lifting Operations." The anchors were tested to a maximum of 80 percent of the guaranteed ultimate tensile strength (GUTS) equating to 2,524 kips and locked off at 70 percent of GUTS or 2,215 kips using a 1,500 ton jack. Anchor stressing was sequenced to prevent differential or unbalanced stresses potentially created during stressing of the four anchors within each concrete block. A typical stressing operation setup is shown in Figure 2. Anchor sequencing also considered differential movement between the eight concrete blocks.

Figure 2. Overview of the stressing platform and hydraulic jack

In addition to multiple quality control checks during the process and the instrumentation and monitoring in place (as described in detail in the section "Dam #1 – Instrumentation and Monitoring"), it is important to emphasize state of the art industry practices were implemented by TVA for Safety and Environmental protection; for example, all the rod handling operations were performed with a Manipulator rather than manually; all drill spoils were flushed into a cyclone system and then directly collected into dumpsters as shown in Figure 3. In addition, all hydraulic hoses were wrapped and all equipment atop the dam had containments to avoid unexpected leaks into the reservoir.

Figure 3. Left: Drill rod handling with Manipulator/Right: Spoil collection system

Installation of Anchors on the Spillway

Installation of the eighteen (18) anchors on the spillway section brought an additional set of challenges. All anchors were drilled to a depth of 140 feet with 32.4 feet of bond length in the limestone. The anchor heads were situated below the level of the normal reservoir water level which increased the probability of encountering excess flow conditions, or artesian flow, with an open drilled hole. The 54-strand anchors at the spillway were shorter than those designed for the right non-overflow section. The spillway anchors were installed from a customized platform at an angle of 56.3° from the horizontal (as described in detail in the section "Dam #1 – Access, Platforms and Lifting Operations"). The installation, testing and verification process was similar to the anchors of the right non-overflow. More than 94% (17 out of 18) of the holes had to be grouted and re-drilled as they failed the initial water test.

The main difference between the holes drilled at the spillway and right non-overflow was the specified tolerance on the drilling. A 2% tolerance was specified for the axis perpendicular to the dam and 0.75% tolerance was specified in all other directions. Additionally, since the head of the spillway anchors was below the upstream normal reservoir level of the dam, artesian conditions were encountered and valves were mounted on the diverter assembly, as shown in Figure 4. The anchors were tested at 80 percent GUTS or 2,524 kips and locked off at 70 percent GUTS or 2,215 kips. Similar to the anchors on the non-overflow section, an alternate stressing sequence was implemented to minimize the differential loads and movements between the 10 concrete blocks (two anchors on the nine inner blocks and one anchor on each end-block). Specific safety procedures were implemented for evacuation on the working platform. Finally, due to the lack of access to the anchor heads after the demobilization of the work platform, the read out boxes for the Dyna Force sensors (two sensors were installed on two anchors) had cable extensions routed to the drainage gallery.

Figure 4. Valve mounted on the Diverter Assembly

<u>Installation of Anchors on the South Training Wall</u>

The third set of anchors installed were passive anchors using 1 ¾" Grade 150 bars. Twenty-seven (27) anchors were installed in total, dispersed on two levels, with depths ranging from 103 feet to 139 feet with 20 feet of bond length in rock. The design required holes to be drilled on an angle of 38° from horizontal. Due to the poor quality of the soil at the interface with the rock, all holes were cased and drilled using a duplex system. Two rows of anchors were situated vertically 10 feet apart and a special procedure was implemented to locate the working platform at the correct height to support the drilling, grouting, installing and stressing operations (the details of the working platform and the rigging process using both a barge-mounted crane and a crane on land are discussed in the section "Dam #1 – Access, Platforms and Lifting operations"). The rod mat equipped drill mast was suspended by the crane as the platform could not support heavy loads. Due to the passive nature of the anchors, they were tested to 60 percent of GUTS or 240 kips and locked off at only 1 percent of GUTS or 4 kips using a specific coupling and stressing protocol as shown in Figure 5.

Figure 5. Stressing of the passive anchor bars

DAM #2 — INSTALLATION OF HIGH CAPACITY ANCHORS ON THE NON-OVERFLOW SECTION

Dam #1 and Dam #2 were designed and constructed in a similar time period. However, the 3D analysis concluded Dam #2 required twenty (20) anchors on the right non-overflow section to satisfy project design criteria. The 54-strand vertical anchors were installed from the top of the dam. Anchors had designed lengths varying between 107 and 217 feet with 26.1 feet of bond length in limestone/dolomite underlying the dam structure. Similar to Dam #1, Dam #2 anchor lengths were designed such that the top of the bond zone between adjacent anchors was staggered a minimum of 5 feet to avoid concentrated stress regions associated with overlapping cones. The anchor work required additional set-up time compared to a similar installation on Dam #1 due to the inability of having a crane situated on the side of the dam or on a barge in the tailrace. Installation was performed using an uncoiler and a 30,000 pound Extreme 3034 Forklift to set the tendons in the hole, as shown in Figure 6.

Safety and environmental protection measures were required and similar to those implemented on Dam#1. Dam #2 also required a substantial amount of consolidation grouting before the hole passed the water head test. In total, more than 8,000 gallons of consolidation grouting were necessary for anchor holes to pass the specified water head loss test criteria. Similar to Dam #1, anchors were tested at 80 percent GUTS (2,524 kips) and locked off at 70 percent GUTS (2,215 kips).

Figure 6. Left: Installation of the tendons using an uncoiler - Right: stressing jack setup using the Extreme Forklift

DAM #1 — ACCESS, PLATFORMS AND LIFTING OPERATIONS

As mentioned above, the use of cranes as the primary means of accessing and supporting the anchor installation work was necessary due to the limited working areas and tight access restrictions.

Crane Operations on the Non-Overflow section

The non-overflow work was supported by a 275-ton crane with a luffing jib as shown in Figure 7. The crane was sized considering maximum anchor lengths, weights, and reach required to lift the anchors into place. Once the crane was sized, an earthwork stability analysis was performed to assist in engineering the crane pad and ensuring the toe of the dam was a suitable and stable foundation to support the loads.

Figure 7. Crane with luffing jib supporting non-overflow anchor work

Platform and Crane Operations on the Spillway Section

Many variables and options were considered in determining the best access approach for the eighteen (18) spillway anchors. Ultimately, a full-length working platform supported by a barge-mounted 200-ton crane situated within the tailrace was the most feasible and flexible option to pursue. The work platform was carefully designed taking into consideration factors such as drill and support equipment loads and dimensions, existing concrete conditions, and crane lifting capacities of each platform element. In addition, since the platform was anchored to the face of the spillway, the platform needed to be designed and erected to allow for rapid removal in the event the spillway gates needed to be opened in a spill situation. For these reasons, the platform was designed as a structural steel frame and hardwood crane mat decking as shown in Figure 8.

Figure 8. Platform steel frame erection

In conjunction with the work platform, the crane and barge layout was designed to handle all anticipated picks including platform erection and teardown, anchor installation, and picking the drill and other support equipment onto and off the platform. The barge consisted of sectional floats pinned together in a square-designed configuration with additional design calculations made for walking the crane onto and off of the barge along with mooring point load calculations. As shown in Figure 9 and explained below, an additional 200-ton crane was positioned on the left shore above the training wall to assist in both the spillway and training wall operations. Working in close proximity to each other above the water required careful coordination and robust safety and environmental controls.

Figure 9. Spillway and training wall support cranes

Platform and Crane Operations on the South Training Wall

As explained above and shown below in Figure 10, the two 200-ton cranes were used to support the training wall anchor work. In addition, a smaller working platform was secured to the face of the south training wall for personnel access. To ensure the loads imposed on the training wall by the crane would not cause structural damage, a structural analysis was performed. Existing steel trestles used during original construction of the dam were embedded within the dam, resulting in conflict with several of the anchor locations. Another investigation was performed, resulting in some successful re-design and modifications to the anchor layout and work platform configurations.

Figure 10. Training wall hanging mast drill

DAM #1 — INSTRUMENTATION AND MONITORING

Due to the sensitivity of the drilling and stressing operations on anchors of this size and magnitude into 75 year-old structures and the potential impact they can have if not performed correctly, it was important to have a series of real time monitoring instruments to control the tolerance of the drilling operations, the load of the anchors and the impact on the existing structure (3D displacement).

Instrumentation and Control for the Drilling Operations

The tolerance for deviation of the drilling hole was specified to 0.75% in every direction for every type of anchor, with the only exception of the spillway anchors of Dam #1 in the direction perpendicular to the dam (2%). To meet this strict requirement several controls were implemented:

- For inclined holes, a tool aligner was implemented for the initial setup of the rig and aligned with the theoretical angle with an accuracy of 0.1°
- Used DHS gyro tool to measure the as-built deviation of the hole at 2 distinct elevations:
    - at the interface between the existing concrete and the rock
    - once the hole was drilled at full depth

While the rod aligner only gave an indication at the start of the hole, it was very useful. A small angular error at the start of the hole could affect the entire hole and would not be an "as built measurement" as the DHS gyro, which can only measure the deviation after a minimum of 10 feet of drilling once the MEMs sensors are in the hole. A typical output of the downhole measurement for a vertical hole on the non-overflow section is shown in Figure 11.

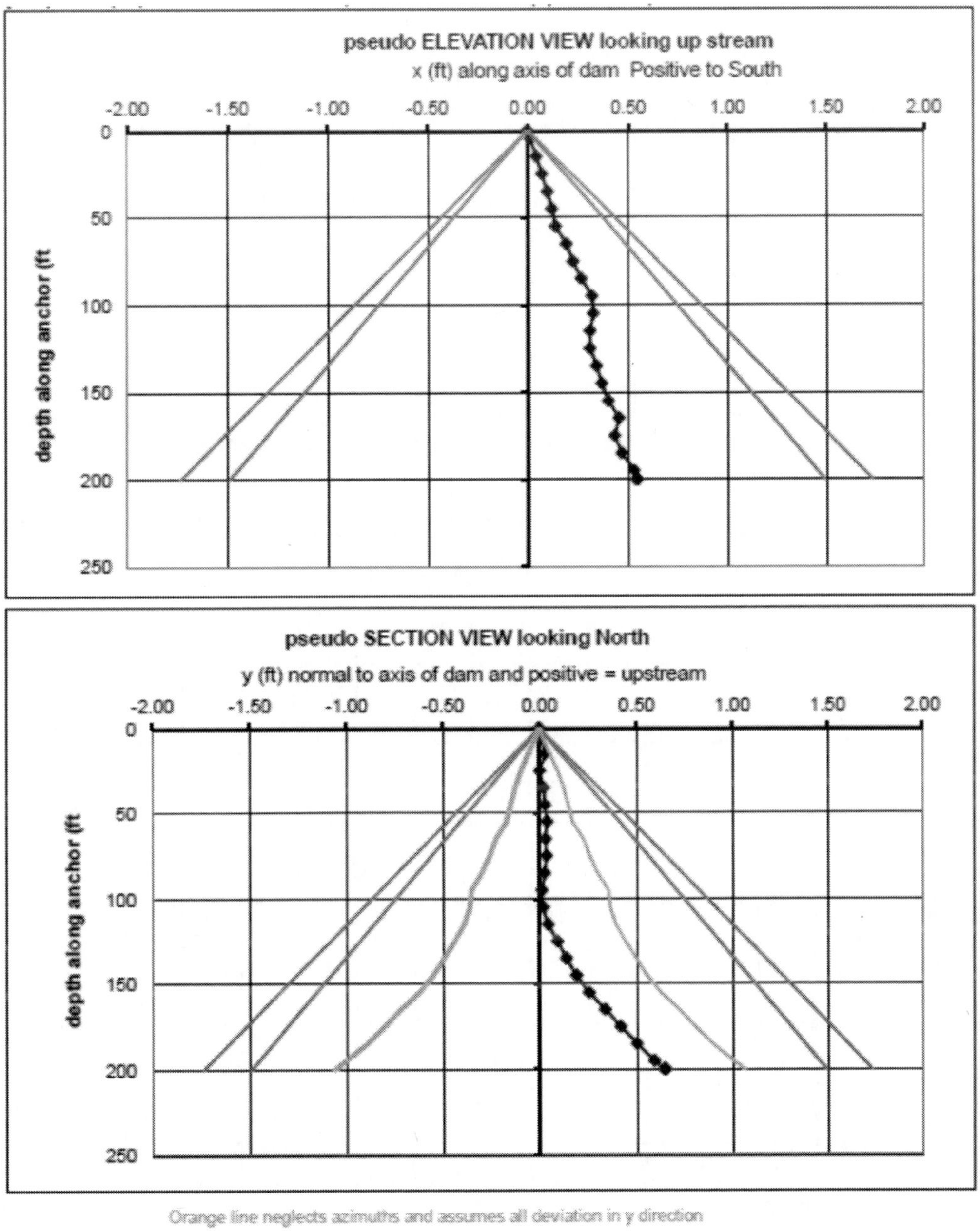

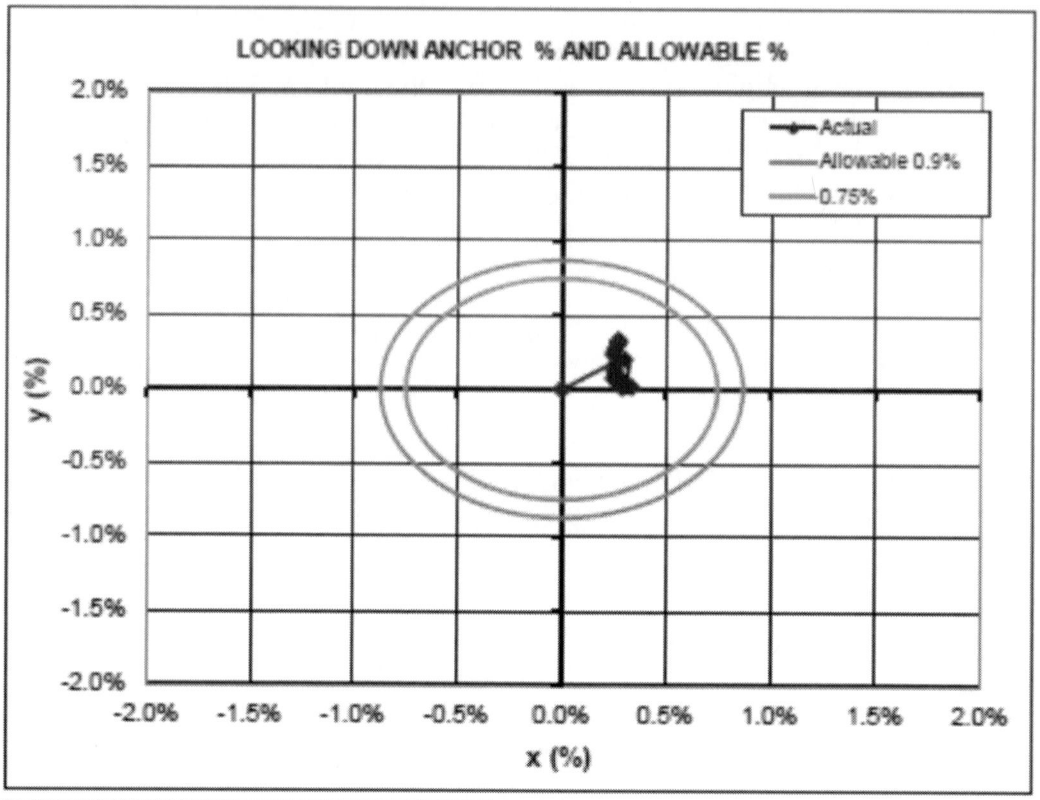

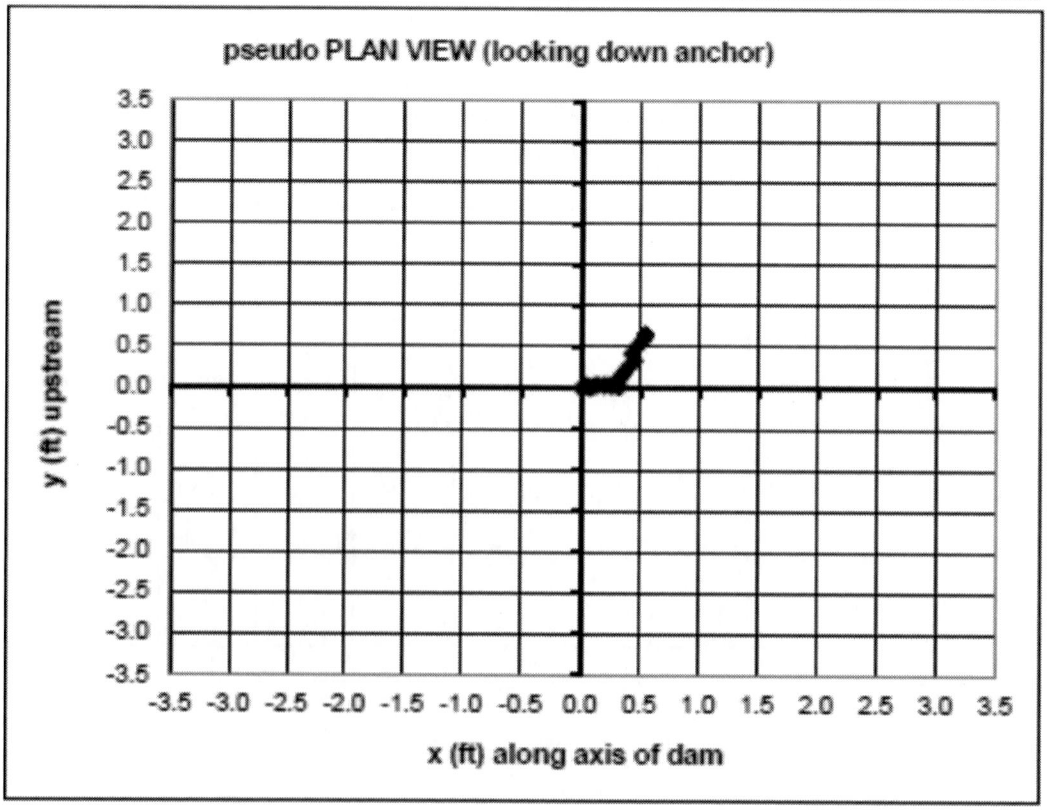

Figure 11. Borehole deviation measured on anchor of the non-overflow of Dam #1

Instrumentation and Monitoring of the Anchor Loads

The lock-off process followed the PTI Recommendations. The design load was 60 GUTS or equivocally 1,898 kips, the lock off load was 70 percent GUTS or 2,215 kips and the maximum test load was 80 percent GUTS or 2,524 kips. Every anchor was tested by increments measuring the tendon elongations and the jack pressure, as well as the load on each of the three Dyna Force sensors installed in the free length about 5 feet apart on three different strands. The Dyna Force sensors are commonly used on stay cables to retrieve the load on single strands through measurements of the permeability of the steel subjected to a magnetic field. Creep tests were also performed using load cells based on the PTI Recommendations. All the anchor elongations were within the theoretical range and all the anchors successfully passed the creep test. An example of the deflection curve is shown in Figure 12. Dyna Force sensor readings performed approximately one year after initial lock-off showed that all the loads were within 5% of the lock-off load (average 2% deviation). However, some readings obtained by the sensors during initial testing were not completely consistent with the load cells, which seemed to provide more accurate readings while the Dyna Force sensors gave a good indication of the "trend" of the load changes. Also, the cabled installation required by the Dyna Force sensors are susceptible to damage during handling and anchor installation and may not be ideal considering long-term performance reliability.

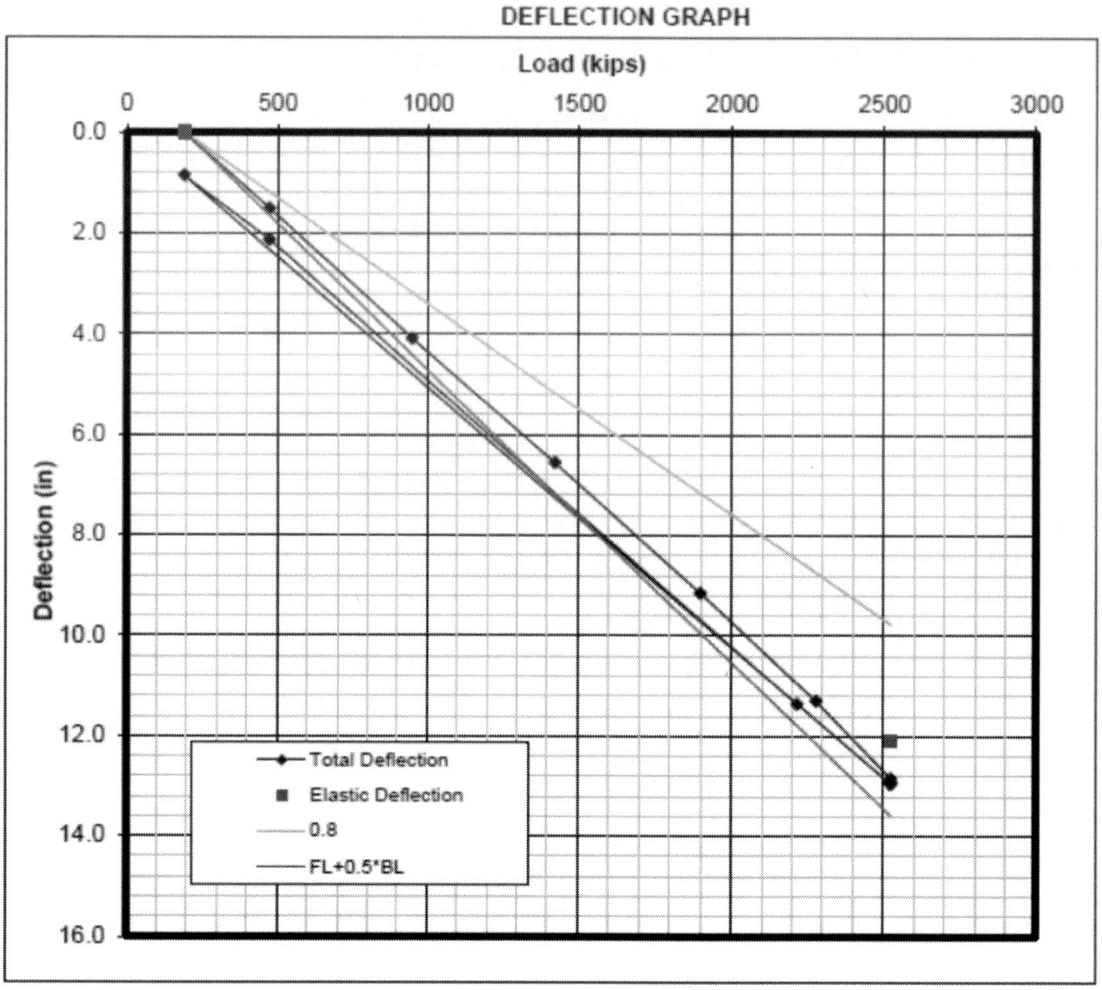

Figure 12. Results of an anchor proof test performed on an anchor of Dam #1

Real Time Movement Monitoring

Real-time structural monitoring was installed and implemented to capture 3D absolute movements of selected monoliths of Dam #1 and Dam #2 during testing of all anchors. Relative displacement and/or movement of concrete monoliths due to the stressing of the anchors was the primary focus of the monitoring. Two (2) Automatic Motorized Total Stations (AMTS) continually recorded x,y and z movements of Dam #1 on the right non-overflow and Spillway while one (1) AMTS continually measured similar movements on Dam #2 on the right non-overflow section as shown in Figure 13 below. Accuracy of the results were between 1 and 3mm in all directions for the differential movements of the blocks at a distance of 400 feet measuring a total of 58 prisms.

Figure 13. Real Time 3D movement monitoring of the non-overflow section of Dam #1 using an AMTS

No significant movements were recorded on Dam #1 during the work and the stressing of the anchors, which occurred between July 15th and August 5th, 2014. The Time Evolution of the 3D movements of anchor #1 of concrete block 9 is shown in Figure 14. In this Figure, x represents the movement parallel to the dam axis, z is perpendicular to the dam crest and parallel to the anchors. The graph shows the maximum displacement registered is less than 0.1 inch, thus, no significant movements were registered in the direction of the anchor.

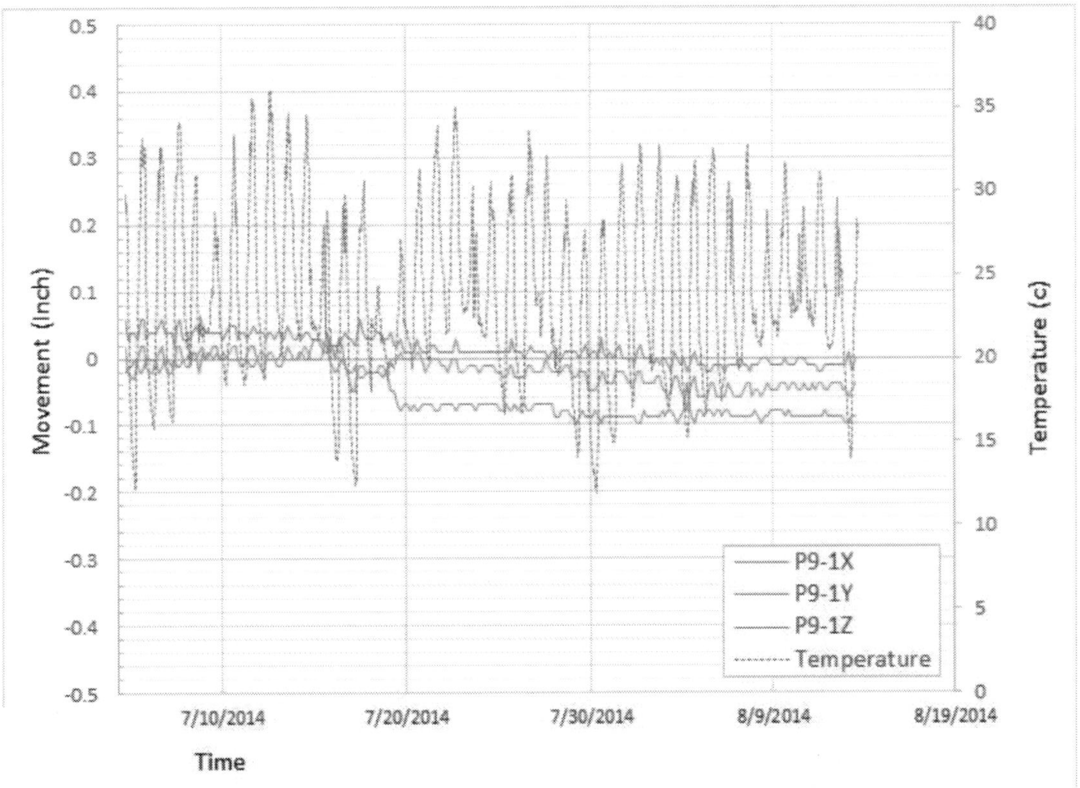

Figure 14. 3D movements of a prism on concrete block #9 of the right non-overflow section at Dam #1 before, during and after the stressing operations.

CONCLUSION

In addition to the technical success, the work on both dams was also an operational success, as it was completed on schedule and without a recordable lost time injury in more than 200,000 work hours. This achievement was made possible only through a close collaboration between the Owner, the General Contractor and the Specialty subcontractor during all the phases of the project. In particular, all parties were prompt to address unexpected conditions, such as the necessity to inject considerable amounts of consolidation grouting in the limestone and at the interface between the concrete and the rock.

LESSONS LEARNED ON POST-TENSIONED TRUNNION ANCHORAGE–FOLSOM AUXILIARY SPILLWAY

Cecily M. Nolan, PE[1]
Cheuk Y. Wan, PE[2]

ABSTRACT

Post-tensioned trunnion anchorage is the state of the practice to structurally support tainter valves and tainter gates under hydrostatic and seismic loads. It is therefore arguably the most important structural feature of the tainter valves and gates. However, there is no single document that provides design and construction guidance on trunnion anchorage. There are conflicts and, in some cases, lapses in documents commonly used for guidance: US Army Corps of Engineers' Engineering Manual, AASHTO design guidance and construction specification, and PTI's specification. In the Folsom Dam Auxiliary Spillway Control Structure project, new trunnion anchorage was recently installed for the new tainter valves inside the new control structure. This paper captures: lapses in guidance and guide specifications the designers faced in this project; design challenges faced by the designers such as details to enable proper grouting and allow post-grout inspection, and tracking losses and elongations; construction difficulties tackled by the contractor, including vertical grouting with standpipes, and vacuum grouting as a repair; quality assurance during construction, including on-site monitoring; methods to provide long term durability of the anchors; and recommendations for future designs, plans, and specs specifications in an attempt to minimize construction challenges thereby improving grouting and anchor reliability.

INTRODUCTION

The Folsom Dam Joint Federal Project Auxiliary Spillway (JFP) includes six submerged tainter gates (also known as tainter valves), approximately 23 ft wide by 40 ft high with a 39.5 ft radius. See Figure 1. At maximum pool, the head at the bottom of the gate is approximately 114 ft. Each gate includes a dedicated, single span steel trunnion girder that bears on the piers and is anchored by post-tensioned, fully threaded, fully grouted 3 in diameter rods. See Figure 2.

The trunnion anchorage design was based on geometric constraints, accessibility goals, and the Corps design criteria. Due to the geometric constraints and accessibility goals, the anchors are a total of 88 ft 9.5 in long. The anchor length is well beyond the 39.5 ft gate radius and the dead ends terminate deep within the structure, accessible by a gallery throughout the life of the project. With this length, couplers were required at mid-length, driving the duct size to 6 in to allow for the AASHTO LRFD Bridge Construction Specifications, (LRFDCONS), and PTI "Recommendations for Prestressed Rock and Soil

[1] US Army Corps of Engineers, Sacramento District, Senior Structural Engineer, Address: 1325 J Street, Sacramento, CA 95814 , Email: cecily.m.nolan@usace.army.mil Phone: (916)557-7472
[2] US Army Corps of Engineers, Sacramento District, Senior Structural Engineer, Address: 1325 J Street, Sacramento, CA 95814, Email: cheuk.y.wan@usace.army.mil Phone: (916) 557-7149

Anchors" (PTI) recommended ½ in annular space at the coupler and 1.5 in of annular space elsewhere. At the live ends, geometric constraints within the trunnion girder resulted in a smaller, 4 in diameter duct. See Figure 3 for the duct reducer and 4 in ducts before encasement in concrete.

Figure 1. Aerial View of Auxiliary Spillway Control Structure

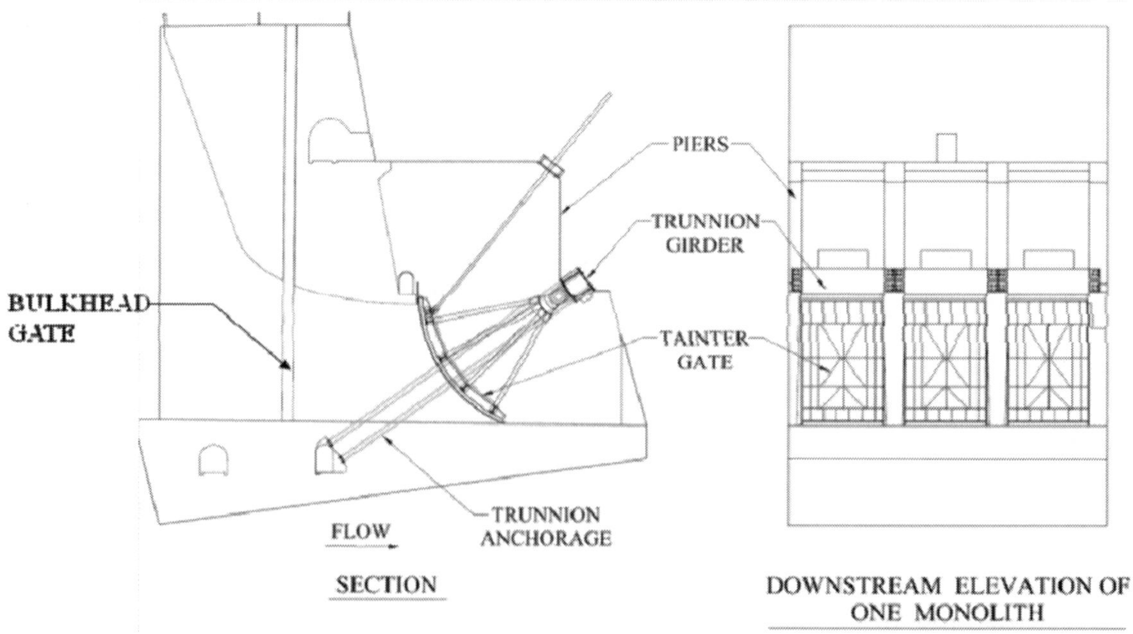

Figure 2. Tainter Gate, Trunnion Girder, and Trunnion Anchorage

Figure 3. HDPE duct reducer at Live End

The double corrosion protection system includes HDPE duct and grout, with additional protection of a zinc-based primer added to the surface of the bars and nuts during construction. During the design process, wax, grease, and grout were each considered as a corrosion inhibitor for an uncoated anchor bar. While wax and grease offered the benefit of facilitating a potential future bar replacement, the long term reliability was not considered to be on par with grout, so in keeping with the requirements in ETL 1110-2-584 "Design of Hydraulic Steel Structures" (ETL), grout was selected. Grout requirements were presented in the specifications and allowed for the Contractor to select the material.

This paper includes discussions on post-tensioning but focuses on grouting details and procedures. Discussions include specifications, construction management, observations and lessons learned in the field, and recommendations for future projects. Specifications include the Corps guide specification (which is the standard for all Corps trunnion anchor projects), the project specifications (which are a tailored version of the guide specifications), and other industry standards which were referenced in the project specifications or are included in this paper for discussion.

Each discussion point is organized as follows: Concept, Specifications, Procedure Employed, and Recommendations or Lessons Learned.

POST-TENSIONING

Overview

The Government designed the anchorage system and minimum lock-off load but left final detailing of the system to the Contractor. The anchor design was driven by requirements established in Appendix D of ETL: *"Anchorage groups shall be sized such that the long-term effective anchorage force provides a minimum compressive bearing stress between the girder and the pier of 200 psi while the system is subject to the maximum trunnion reaction due to service loads."* The Government specified bar diameter, tensile strength of material, embedded bearing plates, and provided basic details at the anchor ends. During construction, the Contractor refined these details and provided computations, shop drawings, and the Stressing Plan.

The design includes eight 3 in diameter 150 ksi anchors per gate trunnion. The project specifications listed the materials specification of ASTM A722 for the rods but increased the minimum yield strength from 120 ksi to 125 ksi.

Elongation Measurements

Concept: During stressing or post-tensioning, the bar elongates linearly with load. Continuous tracking of elongation and comparison to jacking stresses is beneficial in alerting the operator of excessive elongation that would indicate anchor yielding.

Specifications: Two key requirements for elongation measurements from LRFDCONS were included:

- 10.10.1.2: *"The jacking system shall provide an independent means by which the tendon elongation can be measured."*
- 10.10.1.4: *"When a discrepancy between gauge pressure and elongation of more than five percent in tendons over 50.0 ft long occurs, the entire operation shall be carefully checked and the source of error determined and corrected before proceeding further."*

Procedure Employed: The elongation measurement was performed using a gauge that was mounted on the jack frame and attached to the live end of the anchor. See Figure 4. This type of mounting is inherently simpler than an externally mounted gauge but does not provide absolute anchor elongation data. The gauge measures the change is distance between the distribution plate and the bar, which includes bar elongation due to tension and compression of the bearing surface. Bar elongation will be the majority of the reading, but ignoring the compression component will introduce error. Accuracy could be improved with pre- and post-tension surveys of the anchor live end.

Figure 4. Stressing Jack and Elongation gauge with Digital Readout

The means for satisfying compliance with LRFDCONS on 5% agreement between gauge pressure and elongation was to compare the actual elongation to theoretical, which is the industry standard. Only 53 of 96 anchors (55%) were in agreement within 5%, while the remainder was as high as 12.5%. Upon discovery of the disagreement, the work was paused to reevaluate elongations and procedures LRFDCONS. The manufacturer contended that the disagreement was not a concern and proceeded with work.

Recommendations and Lessons Learned: For JFP, the large diameter anchors were cold rolled during manufacture, which resulted in a less defined yield point. While lab test results showed the lower bound of the yielding-type behavior occurs well above the peak jacking force, lack of agreement between elongation and gauge pressure was a concern during stressing due to the possibility that excessive elongation at the early stages of yielding would be missed and, in a worst case scenario, anchor would fracture during tensioning. It is recommended, therefore, that the LRFDCONS requirement for independent measurement of elongation be enforced, especially for cold rolled rods. Accuracy of gauges and surveys of exposed ends of anchors before and after tensioning could also be helpful.

Jacking Force and Load Confirmation

Concept: The minimum lock-off for the anchors was required to be 103.5 ksi after instantaneous losses. The Contractor was responsible for determining the jacking force, estimating prestressing losses, and confirming the lock off load in the field in a variety of ways.

Guidance for confirmation of force for the trunnion anchors is complicated by the fact that the system shares similarities with ground anchors and post-tensioned concrete members. The function of trunnion anchors is similar to ground anchors, but the construction detailing is similar to a bridge pier or girder. Since trunnion anchors are not relying on grout bond, the Performance and Proof test requirements for ground anchors are inapplicable. Since post-tensioned concrete members are highly controlled structures with few uncertainties, LRFDCONS does not require verification testing. Trunnion anchors are critical, direct tension structures with perhaps less redundancy than typical post-tensioned concrete members, so verification tests are prudent.

Specifications: The project specifications required a Stressing Plan and stated that stressing must be performed in accordance with LRFDCONS, which outlined limits on the prestress allowed and requirements such as a Stressing Plan and Stressing Records. The jacking force could not exceed 90% of yield strength or 72% of ultimate strength, in accordance with LRFDCONS, but the jacking force was required to be determined by the manufacturer based on expected instantaneous losses to meet the lock-off load, calculated in accordance with AASHTO LRFD Bridge Design Specifications, 4th Edition (AASHTO).

The Corps guide specifications include verification tests, so that up to 10% of the rods would be tested to a "pull test design force" 7 to 14 days after tensioning. For JFP, the project specification included lift off tests for a minimum of 24 of the 96 rods. This was for assurance that the jacking load was adequate to achieve the desired lock-off load (while confirming the losses). In addition, 24 anchors have load cells on the dead end, for use during construction and future operations.

Procedure Employed: The manufacturer design calculations generally followed AASHTO, while placing the jacking force near the AASHTO limit and assuming anchor set (at the bearing plate on each of the bar) and coupler slippage as the instantaneous losses, while omitting friction losses and elastic shortening losses. Friction was deemed negligible due to the straightness of the rods and the oversized ducts along the majority of the length. Elastic shortening was omitted because concrete compression was negligible. Distribution plates and remaining details were designed for the ultimate bar capacity. Load cells and lift off tests (73 of 96 anchors, ultimately extended from the minimum of 24) were utilized to prove that the desired prestress were achieved for all anchors.

Recommendations and lessons learned: It is recommended that lift off tests be explicitly required for trunnion anchorage in order to confirm the predicted instantaneous losses and to verify lock-off load early in the sequence. Rely on LRFDCONS test requirements (Performance and Proof Tests) for ground anchors, if applicable.

Load cells provided instantaneous information about the jacking operation and were valuable in verifying prestress losses after jacking. Load cells were found to be useful (within the margin of error), but not as accurate as the results from the jack.

Equipment Accuracy

Concept: Accuracy of the stressing operation is important for control of a potentially dangerous activity. Overstressing a bar could lead to a bar fracture which would release a large amount of energy with personnel close at hand. Close monitoring is paramount.

High accuracy of all gauges should lead to compatibility between force and elongation readings, but the jack should be more reliable since it is a force dependent operation. Variations in geometry beneath each anchor (e.g., varying pier thickness) can affect the independence of the elongation measurement. The elongation of the bar can give a false sense of confidence, and the LRFDCONS 5% agreement between gauges becomes more difficult to achieve.

Specifications: Project specifications required accuracies of 0.01 in for the elongation measuring device (more stringent than 0.0625 in required by LRFDCONS) and 100 psi for the jack gauge.

Procedure Employed: The equipment utilized provided the specified accuracy, or better. The jack had a 1000 kips capacity while the maximum applied force was 740 kips. The ram size was approximately 100 in^2, which meant that the 100 psi accuracy requirement for the jacking gauge equated to a force of approximately 10 kips, with an equivalent bar stress of 1.5 ksi. The elongation gauge could be read to the 0.01 in, which equates to a tensioning force of 1.9 kips and an equivalent bar stress of 0.3 ksi. The elongation gauge was more refined than the jack gauge, which led to the initial perceived accuracy of the elongation measurement when it merely had high precision.

Recommendations and lessons learned: Accuracy requirements for elongation and jacking gauges should be tailored for the bar size (especially small diameter) to ensure that the level of error is acceptable. Consider the equivalent bar stress for each gauge. Since the jack is the primary means for assessing the bar stress and is utilized beyond the tensioning effort for lift off tests, the jack gauge should be more accurate. Comparing the equivalent bar stresses per Corps requirements of a 3 in bar to a 1 in bar, the need for gauge compatibility and improving jacking gauge accuracy is evident. The equivalent bar stresses are: 0.3 ksi for the elongation gauge accuracy, 1.5 ksi (for 3 in bar), and 3.5 ksi (for a 1 in bar due to smaller jack). Similarly, designers should ensure compatibility between gauge accuracy and precision of specified prestress.

It is also recommended to require digital gauges, complete with a laptop in the field to create a full stress versus elongation record, for stressing and lift off tests. This automation may be more costly but would simplify record-keeping, decrease errors and perhaps even precluding lift off tests by clearly showing decrease in effective prestress at lock off, using a time history.

GROUTING

Overview

During design, PTI was consulted for design of the corrosion protection system. Since the gates are a permanent structure in a non-aggressive environment with serious consequences of failure, the JFP anchor system merited a Class I Corrosion Protection. As a result, a double corrosion protection system, provided by HDPE ducts and grout, was selected.

Vertical grouting was accomplished through direct injections from within the inspection gallery. Injection ports were installed at the dead end and, as a contingency, at the mid length of the anchors, through the pier face. A thixotropic grout was selected, with an expected operating pressure of 50-75 psi (to overcome the 50 ft of head differential between dead and live ends) and a maximum pressure of 150 psi.

Pre-Grouting System Check: Air Pressure Tests

Concept: In keeping with the design intent of double corrosion protection, the duct system and all fittings and connections needed to be watertight so that both the primary (duct) and secondary (grout) lines of defense were reliable along the full anchor length. During construction, grout leakage due to non-tight ducts was the main concern since it leakage can possibly lead to an incompletely grouted duct. Grout leaking out of duct A can also find its way into as-yet adjacent un-grouted duct B, unknowingly blocking duct B so that its subsequent grout injection would be unsuccessful and the anchor corrosion protection could be compromised. Long term, a non-tight duct allows water seepage into the system during the life of the project, which could lead to anchor corrosion.

Air pressure tests are required by LRFDCONS to ascertain the tightness of the duct system to a limited degree. The test demonstrates that a peak pressure can be achieved within the system, indicating the overall tightness of the system. The test identifies leakage in the injection ports, vents, and any compromised fittings and connections.

Specifications: LRFDCONS requires both air and watertight ducts and duct fittings and the air pressure test. The air pressure test typically occurs on an empty duct, before the anchors and hardware are installed. If 50% of the 50 psi air pressure is maintained after 1 minute, the test is deemed a success and the system acceptable. If not, leaking connections are required to be repaired. LRFDCONS requires general procedures for the test and potential repairs but, as a general guideline, lacks explicit requirements. Since the project specifications refer to LRFDCONS, the air pressure test was required.

Procedure Employed: For this project, the air pressure tests occurred after the anchors and hardware. The anchors complicated the test since the bearing plates and hardware introduced additional air leakage sources that required caulking before the test. Preliminary tests revealed that the integrity of the injection port connections were the weakest component of the system in terms of sustained pressure, so the test procedure made provisions for the test pressure peaking at 30 psi while maintaining the 50% acceptance criteria. All ducts were determined to be tight.

Immediately before the grouting activity, each grout injection port was air flushed to confirm utility before commencing grout on a particular duct. Of the 96 total dead end injection ports and 96 intermediate ports, one intermediate port was found to be plugged immediately prior to commencement of grouting on that duct. Efforts to unplug were unsuccessful, due to the bends in the grout tube. Ultimately, no intermediate ports were used, but a plugged port could have compromised a grouting operation.

Recommendations: Based on the above, it is recommended that future contract specifications explicitly state the PTI corrosion protection category desired for the design. The duct work, including fittings and connections, should be designed and detailed for water tight construction. The LRFDCONS air pressure test should be explicitly required as well as requirements for a detailed procedure and detailed results of tests, including operability of ports. Procedure would describe locations of personnel during the procedure to assure that audible leaks would be observed, requirements to remedy any vulnerabilities identified during the test, and clear acceptance criteria.

Standpipes

Concept: One key topic in vertical grouting, and arguably the most important factor in trunnion anchor grouting, is the reservoir or standpipe at the live end. The live ends are particularly vulnerable to corrosion due to their proximity to oxygen and moisture. Undetected live end corrosion could, in a worst case scenario, fail an anchor or gate catastrophically since an anchor crack or decreased anchor cross section could be exploited and fail very quickly. A failed anchor would allow gate detachment (partially or fully) and allow some extent of uncontrolled releases and potential loss of life.

In vertical grouting, bleed water from within the grout rises to the top, and grout subsides from the top. To preclude voids at the live end due to either of these trends, a reservoir or standpipe is necessary to store excess grout and allow room for bleed water to rise.

Specifications: LRFDCONS states, *"In lieu of a positive shut-off, vertical or near vertical tendon ducts for grouting shall terminate in reservoirs at the upper-most point. The reservoir shall have sufficient capacity to store excess grout bleed water. Visible grout level shall be maintained in the reservoirs. The reservoirs shall be maintained until the grout has set."* PTI and Caltrans Prestressing Manual (Caltrans) offer similar guidance, additionally forbidding the level of the grout to drop below the highest point of the upper anchorage device but requiring a mitigating addition of grout to the standpipe. The Corps guide specification was silent on standpipes and subsidence. No codes offer on estimating the volume of standpipes or reservoirs or factors in determining the adequacy of a system.

Procedure Employed: Grout fluidity tests (ASTM C 939) were performed at the dead and live ends to confirm a) proper mixing and b) that the proper amount of "bad grout" was expelled before detaching the grout pump from the dead end injection port.

The standpipe design changed as grouting progressed and system experience was gained. The first day of grouting, during which only two ducts were grouted, small, flexible standpipes were provided. These could contain a very small volume of reserve grout that could compensate for subsidence at the live end.

Once it was clear that the grout was subsiding, the procedure was adjusted by changing the type of standpipe and employing means to preclude the grout from dropping below the base of the standpipe (i.e., the top of grout remained in sight). As Figure 5 shows, 2 in diameter PVC standpipes were installed with a slightly greater volume than the flexible tubes, and a worker was continually available to top off the standpipe as needed with expelled grout. This method increased the burden on the live end worker. In addition to receiving expelled grout every 20 minutes from a new duct, he monitored and maintained adjacent ducts, even on adjacent piers.

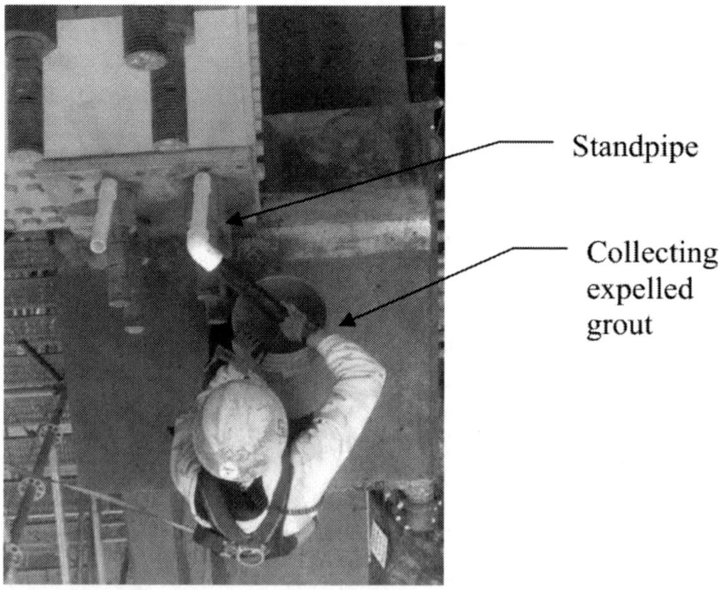

Figure 5. Grout Expelling and PVC Standpipes

The third iteration of the procedure improved the arduous process and eliminated the topping off procedure. Standpipes were increased to approximately 4 ft long (Figure 6), such that a larger volume of excess grout would be stored to overcome subsidence within the system. The geometry of the pipes included a few angles to snake around adjacent live end anchor rods. Thus, the worker could not maintain visual monitoring of the current grout level, so he used a tie wire as a sort of dip stick. In addition, the procedure was modified to maintain the grout injection capability for 10 minutes after completion so that additional injection of grout would overcome the initial subsidence. This injection would restore the grout level to the top of the standpipe, and the worker would continue to monitor the standpipes between capturing expelled grout for newly grouted ducts.

Figure 6. Lengthened PVC Standpipes

Recommendation: The Contractor should be required to consider and discuss in the Grout Procedure: the type of grout, the pressure and speed of the grouting process, and the resulting frothiness of the grout once expelled based on duct constrictions or other factors. This could affect the grout speed, volume of expelled grout, size of standpipes, or other procedures.

Despite PTI and Caltrans both permitting topping off standpipes, it is recommended that topping off be forbidden in favor of properly sized standpipes. Topping off relies on close attention of workers and management of expelled grout, which are more likely to allow errors than an adequately stocked standpipe. If a standpipe is topped off after grout has already subsided below the vent, the constriction of the vent and the air pressure beneath the vent preclude grout from dropping through the vent, leaving a void.

The specifications had no requirements on oversight or training where grout was expelled, and the project engineer was traveling from live end to dead end to perform fluidity tests with expelled grout. It is recommended that minimum qualifications or training be required for workers who collect the expelled grout and make judgments based on visual observations at the live end.

A more costly recommendation is to require a full scale mock up, as outlined in PTI Specification for Grouting of Post-Tensioned Structures (PTI-Spec), of the grout process to demonstrate the Contractor's means and methods and then work out any errors that may be detected. The mock up could also require a dissection of the live end, as a check that there are no voids in the grout. The decision to require this should be influenced by considerations such as the uniqueness of the system or vulnerability due to constrictions.

Post-Grouting Investigations and Repairs

Concept: While Specifications, qualified personnel, and construction oversight are the primary drivers for successful grouting, it is prudent to verify the quality of the effort shortly after grouting is completed and identify whether repairs are necessary. A standard method stated in codes for confirming success of the grouting operation is drilling through air vents to search for voids that may exist within the duct. Voids, which can admit moisture and promote anchor corrosion, can form in the systems that were not properly pressurized during grouting, with insufficient "bad grout" expelled at the live end, with excessive subsidence at live the end, or where topping off failed.

Specifications: LRFDCONS requires *"all anchorages and high-point vents"* to be *"drilled and probed 48 hours after grouting, until the Engineer is assured that no bleed water or subsidence voids exist,"* while the Corps guide specification was silent. Research of related industry standards shows that PTI-Spec requires post-grout inspection no less than 24 hours after grouting, while Caltrans requires investigations between 24 and 72 hours after grouting completion and filling of voids with secondary grout. LRFDCONS also requires *"vacuum grouting or other approved means"* for repairs if flaws are identified.

Procedure Employed: Probing occurred 2-8 days after grouting. It included removal of the air vent plug and insertion of a tie wire into the angled vent (Figure 7). The consequence of this delayed probing was an increase in grout compressive strength, which made the code-recommended drilling more difficult. Probing revealed that, in a small number of the ducts, grout did not completely fill the vent through the distribution plate. After the Government fully investigated the standpipes of these ducts, additional ducts were identified as questionable, based on observations such as Government observations that grout subsided greatly in a particular duct, or grout remaining in the standpipe (Figure 8) had noticeable air bubbles (hinting that perhaps voids would exist in the highly restrained area between the air vent and the anchor itself), etc. As a result, 9 ducts were then drilled to complete the investigation. Due to the detailing of the vent and bar, drilling creates a tiny window into the grout, and over-drilling could endanger the bar. Thus the drilling was accomplished by controlled drilling so that the bit extended just beyond the bottom of the vent hole, followed by peering into the hole with a flashlight and probing with tie wire, to identify whether a void existed.

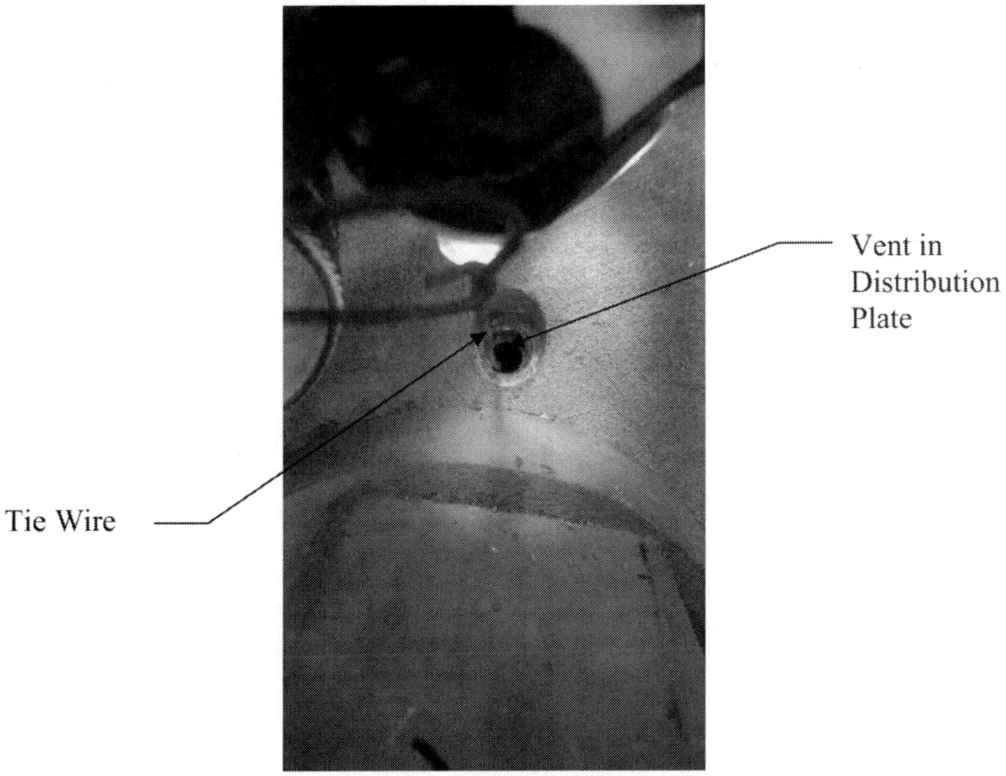

Figure 7. Probing into Vent

Figure 8. Cross Section of Grout in Standpipe

After the post-grouting investigations, it was concluded that a few ducts included voids. While no evidence of extensive voids existed due to visibility constraints imposed by the air vent detail, it was deemed prudent to perform vacuum grouting on the identified ducts. Vacuum grouting was preferred over tremie injection since it is the most foolproof way to ensure grout replaces air within the system. In such a constrained area, considering the small diameter of the vent hole and the inability to confirm that a tube truly reached the bottom of the void, vacuum grouting seemed the best choice. An alternative idea of injecting a more fluid plastic media was suggested, but grout was preferred to maintain homogeneity. It is worth noting that the results of the vacuum grouting confirmed that none of the air voids extended beyond the base of the drilled hole or the small void discovered during probing.

Recommendation: The Engineer should be intimately familiar with the grouting operation in order to make an informed decision as to whether drilling and probing is necessary.

The Contractor's written procedure should be required to explicitly state drilling and probing as part of the plan, to preclude potential delays due to disagreement as to the necessity of the effort. It should also detail what repair methods will be employed The Engineer should know preferences for maintaining grout (thereby requiring vacuum grouting regardless of suspected size of voids) or for allowing a fluid plastic material to be tremied in, or other.

Since the difficulty of drilling and probing was attributed to the vent design, consider such investigations during design or during the shop drawing development. An improved detail could have included greater annular space at the live end (or other investigation location) and allowed a deeper drill depth and/or a broader view into the duct.

CONCLUSION

The JFP trunnion anchor post-tensioning and grouting processes were considered a success, based on visual observation and verifications such as lift off tests, load cells, drilling and probing, and vacuum grouting. Even with success, the challenges presented opportunities for lessons to be learned in the process, which resulted in recommendations for future, similar projects. Since trunnion anchors are fairly unique features, applicable guidance can only be found in various non-unified industry standards. The Corps guide specification is a great baseline specification that could benefit from additional requirements for future projects. Most of the recommendations would result in additional requirements in specifications and would likely increase project cost, but with the criticality of intact trunnion anchors and the likelihood of stress corrosion failures of these important features, it would be money well spent.

ACKNOWLEDGEMENTS

The authors would like to thank David Neff (HDR, Inc.), Bradley Call (USACE, Sacramento), Sam Planck (HDR, Inc.), Travis Adams (USACE, Portland) and Philip Sauser (USACE, Philadelphia) for providing invaluable advice during the prestressing and grouting operations and/or for providing input to this paper. The authors would also like to acknowledge the excellent support provided by the JFP Construction Quality Assurance team (USACE, Sacramento) for their dedication to the project.

REFERENCES

AASHTO (American Association of State Highway and Transportation Officials). (2010). *AASHTO LRFD Bridge Construction Specifications*, Washington, DC.

AASHTO (American Association of State Highway and Transportation Officials). (2007). *AASTHO LRFD Bridge Design Specifications*, Washington, DC.

ASTM (American Society for Testing and Materials). (2012). "Standard Specification for Uncoated High-Strength Steel Bars for Prestressing Concrete." *ASTM A722*, West Conshohocken, PA.

ASTM (American Society for Testing and Materials). (2010). "Standard Test Method for Flow of Grout for Preplaced-Aggregate Concrete (Flow Cone Method)." *ASTM C939*, West Conshohocken, PA.

Caltrans (State of California Department of Transportation Engineering Services). (2005). *Prestressing Manual*, Sacramento, CA.

PTI (Post-Tensioning Institute). (2004). *Recommendations of Prestressed Rock and Soil Anchors*, Phoenix, AZ.

PTI (Post-Tensioning Institute). (2012). "Specification for Grouting of Post-Tensioned Structures." *PTI M55.1-12*, Phoenix, AZ.

U.S. Army Corps of Engineers. (2014). "Design of Hydraulic Steel Structures." *ETL 1110-2-584*, Washington, DC.

UNCOVERING AND ADDRESSING A HISTORY OF SIGNIFICANT INCIDENTS AFTER AN ANNUAL DAM SAFETY INSPECTION

Steven Davidheiser, PE[1]
Robert A. Kline, Jr, PE[2]
Amanda Lothes, EIT[3]

ABSTRACT

Diascund Dam is an important component of the water supply system for Newport News, Virginia. During the 2011 annual dam safety inspection, one 25-feet by 20-feet principal spillway slab was observed to be uplifted by a few inches along its downstream edge. Subsequent review of archive drawings revealed several deficiencies in the original design of the principal spillway which allowed uplift pressure to displace the slab in question. Review of the archived files revealed a history of similar incidents throughout the service life of the structure that began with two spillway slabs uplifting upon initial reservoir filling in 1963. After each incident, the implemented mitigation measures failed to identify and address the root cause of the problem, resulting in repeat in-kind failures. Following the 2011 safety inspection, the principal spillway was redesigned and reconstructed with several new features to correct the deficiency and address the potential failure mode.

This case study illustrates the value of conducting regular dam safety inspections to identify and mitigate dam safety problems before they become serious. Potential failure modes associated with reinforced concrete spillways founded on erodible foundations are presented and discussed along with solutions to rehabilitate this type of spillway. The value of maintaining archive records related to a structure's historical performance and repairs, as is the Newport News Waterworks Department's practice for all of their facilities, is also emphasized as good practice so that dam owners and their engineer(s) can troubleshoot incidents.

INTRODUCTION

Diascund Dam is a critically important component of the water supply system for Newport News, Virginia as it impounds Diascund Reservoir which is one of the most economical sources for delivery of raw water to a treatment facility, and the most economical source capable of supplying the full demand volume within the system. The dam is comprised of a 5,700-foot-long, 35-foot-high zoned earthfill embankment constructed across three branches of Diascund Creek (See Figure 1). The embankment is bisected by a 150-foot-wide concrete-lined principal spillway constructed on a natural hill between the two western branches of the creek. The 300-foot-wide grass-lined

[1] Gannett Fleming, Inc, 207 Senate Avenue, Camp Hill, PA 17011, 717-763-7211, sdavidheiser@gfnet.com.
[2] Gannett Fleming, Inc, 207 Senate Avenue, Camp Hill, PA 17011, 717-763-7211, rkline@gfnet.com
[3] Newport News Waterworks, 437 Waterworks Way, Newport News, VA 23608, 757-234-6776, alothes@nnva.gov

auxiliary spillway located at the embankment's right (looking downstream) abutment was widened to 600 feet in 2000, and its crest was modified to include a fuse-plug-type dike.

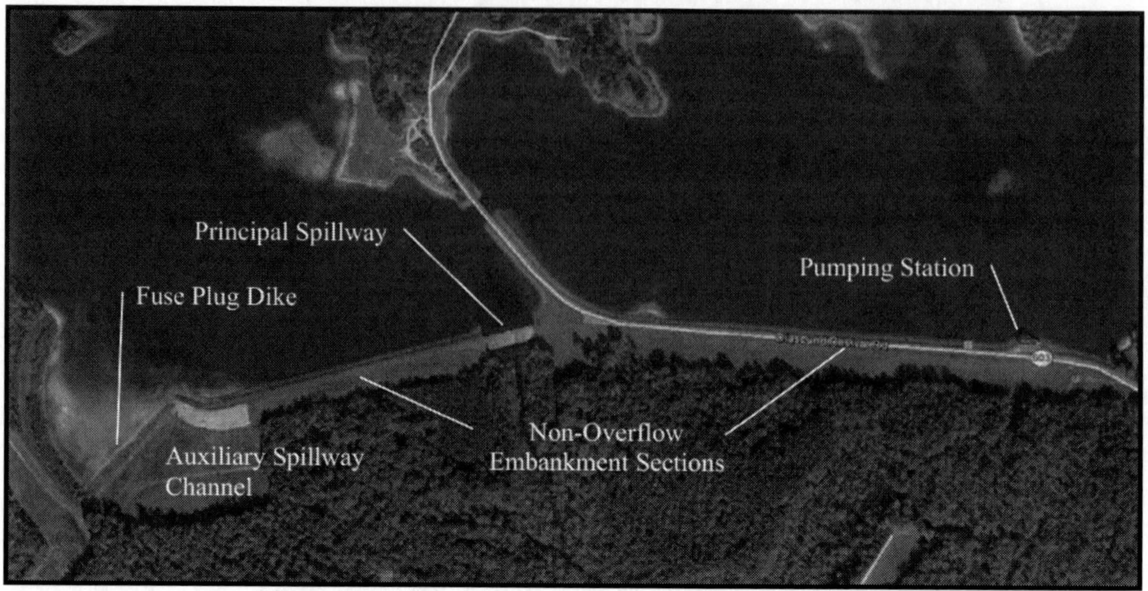

Figure 1. Diascund Dam (Source: Google Earth)

The downstream face of the embankment adjacent to the auxiliary spillway is armored with soil cement to protect it from auxiliary spillway discharges. A pumping station located along the embankment near the left abutment transmits raw water to either of two terminal reservoirs located at water treatment plants. The low-level outlet works consists of a 10-foot-wide by 8-foot-high conduit controlled by twin 4-foot-wide by 6-foot-high sluice gates located within the pumping station. This low-level outlet conduit served as the diversion conduit during dam construction. Original construction of the dam was completed in 1963, and the diversion gates were closed in late 1962, beginning impoundment. Water first flowed over the spillway crest on March 1, 1963. Construction of the Diascund Reservoir expanded the capacity of Newport News water supply system from 20 million gallons per day (mgd) to more than 30 mgd with the ability to increase capacity to approximately 50 mgd with little additional cost if demand continued to rise.

The principal spillway consists of steel-reinforced concrete slabs lining the forebay, crest, chute, and stilling basin and is bound on both sides by 15-foot-high concrete training walls. Underdrain pipes daylight through the vertical faces of chute blocks at the toe of spillway chute. Normal pool is at Elev. 24.84 feet and is controlled by 1-foot-tall wooden flashboards installed atop the rectangular concrete crest sill weir (crest Elev. 23.84). The Diascund Reservoir has a drainage area of 43.6 square miles, maximum storage capacity of 3,900 million gallons, and a surface area of 4.1 square miles at normal pool.

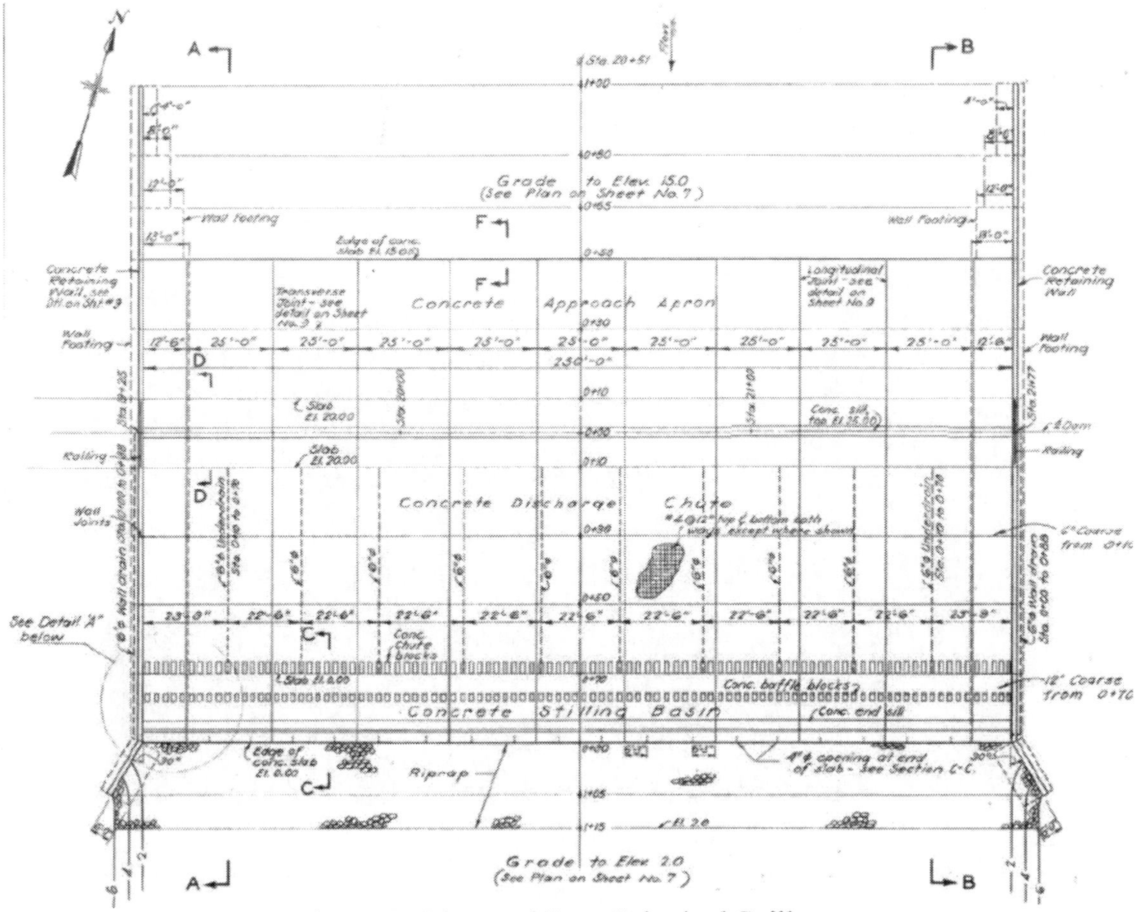

Figure 2. Diascund Dam Principal Spillway

Diascund Dam is classified as a "High Hazard" dam by the Virginia Department of Conservation and Recreation, Dam Safety (DCR Dam Safety) since failure of the dam would result in "probable loss of life and significant economic damage" due to the presence of private residences, businesses, and a major state highway, Route 60, located just downstream of the dam. DCR Dam Safety requires that high hazard dams be inspected every year, and inspections must be conducted by and the associated report signed and sealed by a licensed professional engineer every two years, commonly referred to as an engineer's inspection. Inspection reports must be completed and signed by an owner's representative in years that an engineer's inspection is not conducted, commonly referred to as an owner's inspection.

HISTORY

During the 2011 annual dam safety inspection, one of the slabs near the top of the chute was observed to be uplifted a few inches along its downstream edge as shown in Figure 3. To assess the severity and determine possible causes of this condition, archive documents including as-built drawings from original construction as well as records of repair and upgrade projects were requested from Waterworks and subsequently reviewed. This cursory review of archive documents revealed that the principal spillway had been constructed with several key design deficiencies including lack of waterstops in slab joints, lack of continuous steel reinforcement or dowels at slab joints, and an inadequately designed underdrain system. During the design phase of the spillway reconstruction project, additional historical documents including records of correspondence with consulting engineers were reviewed in detail. Through these detailed records, the following historical chain of events was discovered.

Figure 3. Uplifted Slab Noted in 2011 Inspection
(Note whitewater along lifted trailing slab edge, indicated by red arrow)

Original Construction

The underdrain system originally designed and constructed consisted of a sand drainage layer six-inches-thick between the chute slabs and prepared earthen subgrade. Six-inch-diameter porous concrete pipes were set in trenches 39-inches-deep with two-foot-bottom-widths spaced every 22.5 feet on-center across the spillway. The underdrain pipes discharged through the vertical downstream faces of the stilling basin chute blocks. The spillway chute and stilling basin were lined with twelve-inch-thick and 24-inch-thick steel-reinforced concrete slabs, respectively. Three-quarter-inch-wide longitudinal and transverse contraction joints were spaced every 25 feet and 20 feet, respectively, and sealed with mastic caulking.

During original filling of the reservoir, two slabs at the toe of chute near the right training wall uplifted. The location of these slabs is shown on Figure 4. The original design firm was consulted and concluded that failure of the caulking lead to leakage in the transverse joints on the downstream face of the spillway, which caused the two slabs to uplift. The original design firm further concluded there was "no leakage through the dam." Remedial repairs included grouting underneath the uplifted slabs; placing a concrete fillet to provide smooth transition between uplifted and surrounding slabs; and, presumably, re-caulking of the slab joints. The reservoir remained in service and, according to the records, continued to serve its function without further incident for eighteen years.

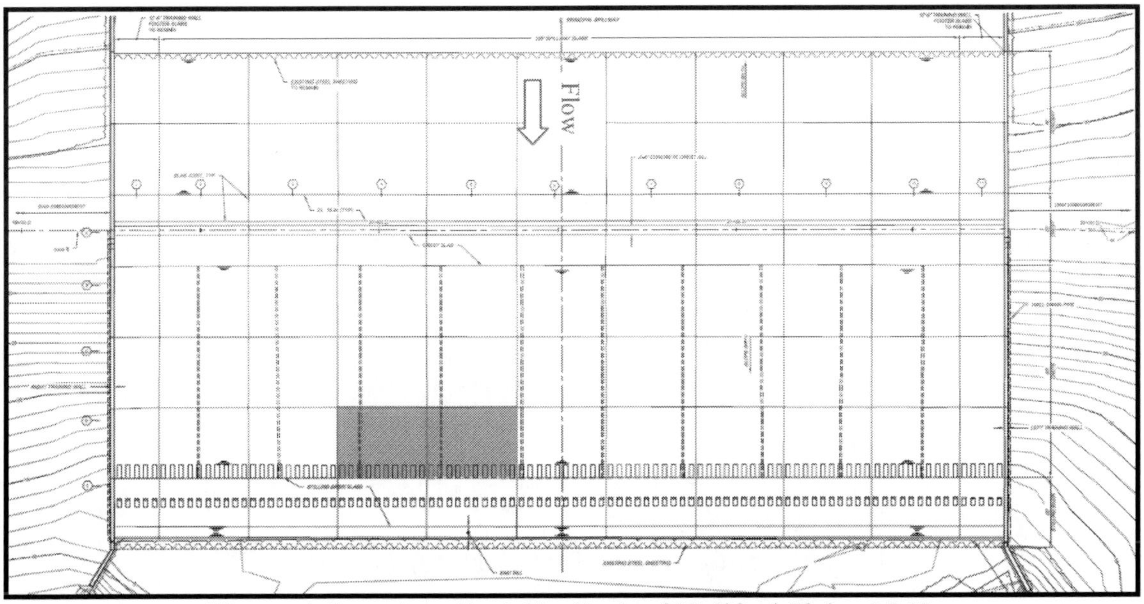

Figure 4. Location (Red Shading) of Uplifted Slabs, 1963

Mid-1980's

In October 1979, the original design firm was contacted due to the observation of water "bubbling up" through some of the chute and crest sill slab joints. At this time, sand deposits were also noted near the bottom of the chute however no repairs are known to have been made. In June 1981, the original design firm was again consulted when a slab was observed to be uplifted approximately 7 inches. The location of this slab is indicated with hatching on Figure 2. The design firm noted that water was leaking from the downstream edge of the uplifted slab, and also that a second pre-existing leak in the spillway chute had resulted in 10 to 20 cubic yards of material loss from beneath the spillway. The temperature of the water leaking at the two locations was noted to differ, and it was recommended that temperature readings be taken to determine which, if either, leak matched the water in the reservoir. The conclusion at the time was that the dam was not in immediate danger of failure. In January 1982, a field investigation concluded that five of the ten underdrains were partially or fully blocked and remediation planning began.

The initial plan for remediation of the spillway involved grouting beneath the spillway slabs, replacement of the caulk sealing where it had deteriorated within the slab joints, and replacement of the clogged underdrains. Only those slabs directly overlying the blocked underdrains were to be replaced to facilitate access to the underdrains. However in April 1983, while the design for the repairs was being developed, another slab uplifted over an underdrain that was believed to be functioning properly. Waterworks followed the recommendation of the design firm and expanded the project to include replacement of all of the underdrains and overlying spillway slabs. Measurements taken in May 1983 indicated that a total of seven slabs had moved by that time, as shown on Figure 5. Construction of the planned repairs began in November 1983.

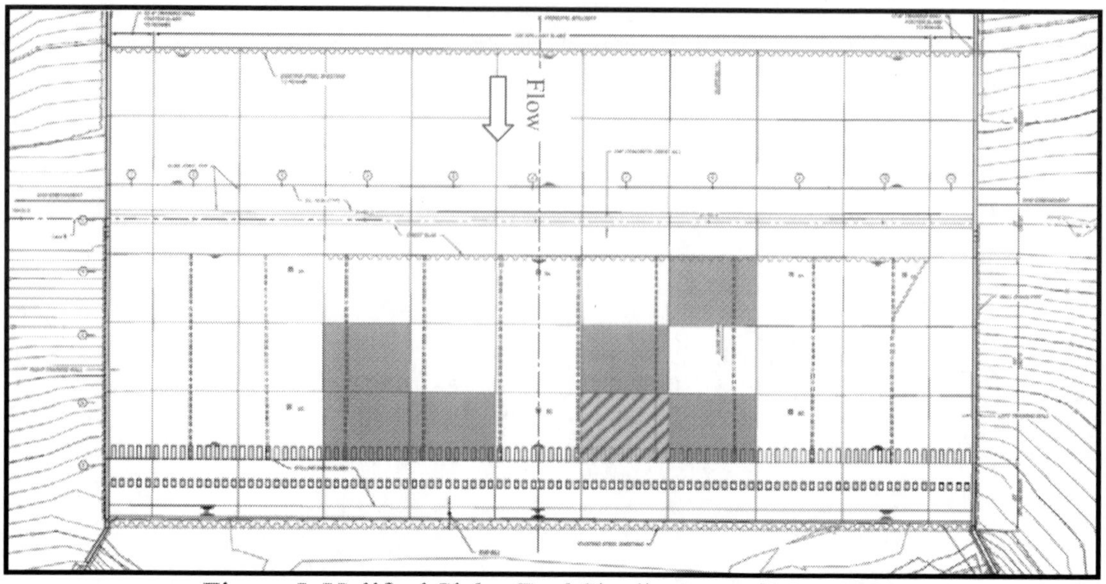

Figure 5. Uplifted Slabs (Red Shading) as of May 1983

The spillway repairs performed in 1983 represented an in-kind replacement of the spillway underdrain system. The only discernible differences between original construction and the 1984 repair drawings were requirements for grouting beneath the crest sill footer and installation of seven observation wells within the spillway chute. As shown on Figure 6, it appears that the original design details were re-used for the 1984 repair contract.

It is worth noting that during the discussions related to the conditions and planned remediation measures at the spillway, Waterworks offered to provide additional historic records related to past performance issues at the dam for review, if deemed useful. It is unknown whether or not the original design firm reviewed the historical files, but the conclusion was drawn that the uplifting of the slabs was due to failure of the joint sealant material. The original design firm offered assurance that current sealant products were superior to those available in the 1960's and would "provide many years of service."

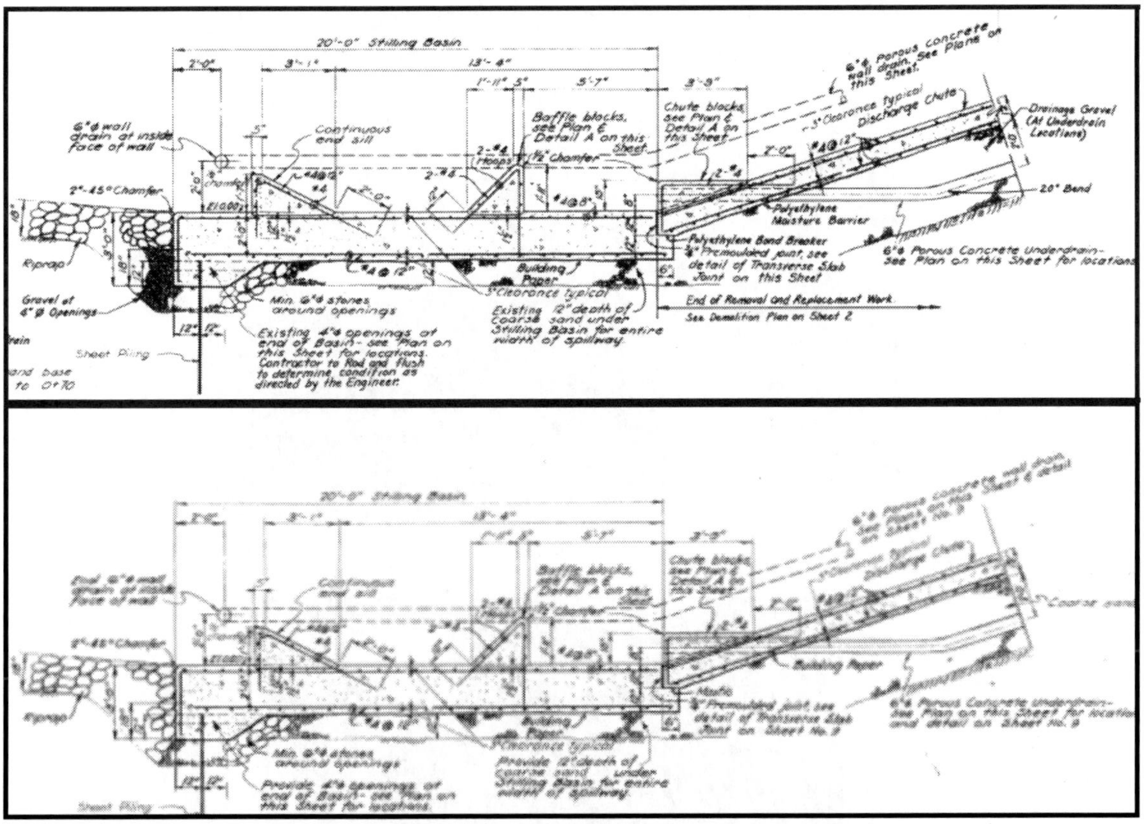

Figure 6. Comparison of Design Details Used on 1960 and 1984 Drawings

During the 1984 repairs, three to four-foot-wide "channels" were discovered beneath some of the slab joints extending from the crest sill to the toe of the chute, as shown in Figure 7. Due to these voids beneath the crest sill, a sheet pile cutoff was added immediately downstream of the crest sill to prevent continued loss of material should the upstream joint sealant fail in the future by forcing seepage down into the relatively impervious natural soil below the spillway. It does not appear that consideration was given to the possibility of seepage migrating laterally, creating a high gradient at the ends of the cutoff. Since no channels were discovered beneath the two right sill slabs, the cutoff was not installed the entire width of the spillway as shown on Figure 5. Following installation of the sheet pile cutoff, a seep was observed approximately ten feet below the crest sill along the left training wall footer during repair work as shown in Figure 8. An additional observation well was installed in the area, and the left-most underdrain pipe was modified to wye with a second leg extending to the corner of the slab as shown on Figure 9. Repairs were completed in May 1984, and the spillway performed without incident for fifteen years.

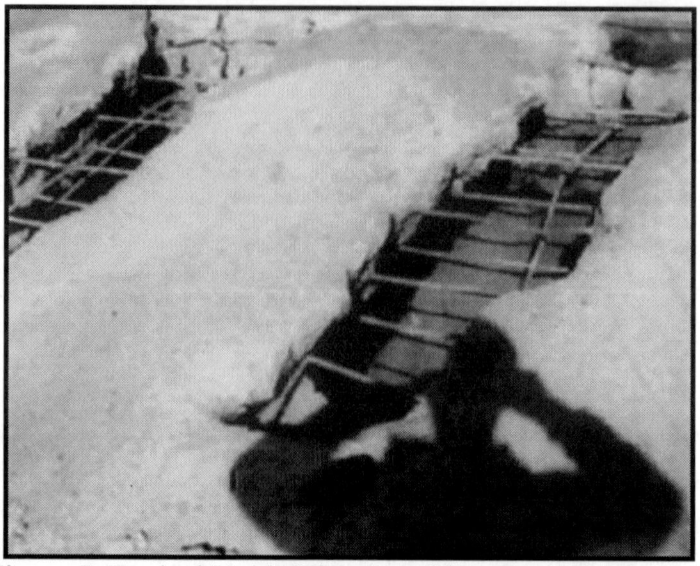

Figure 7. Typical Voids Discovered During 1984 Repairs

Figure 8. Seep Encountered during 1984 Repairs

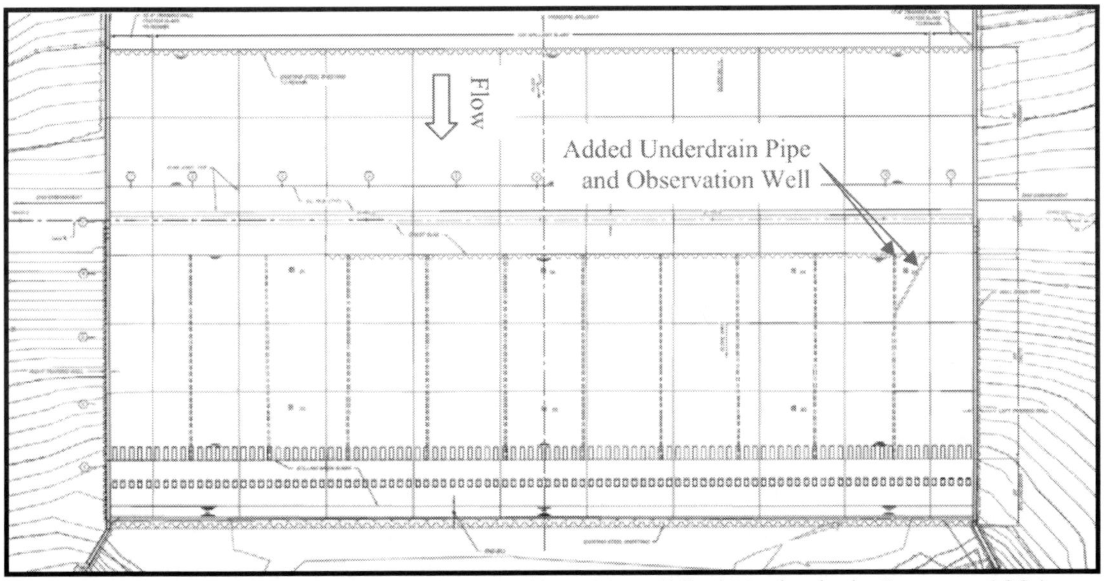

Figure 9. Additional Observation Well and Modified Underdrain Layout, 1984

Hurricane Floyd

On September 16, 1999, Hurricane Floyd made landfall in northern North Carolina and Virginia, producing significant rainfall and causing significant flooding throughout the region. Approximately seven inches (approximately 50-year 24-hour storm) of rain fell over the Diascund Dam watershed, causing the principal spillway crest to be overtopped by approximately four feet of water. Following Hurricane Floyd, one spillway slab at the toe of slope near the left training wall was uplifted as shown on Figure 10 and the joint sealant had been washed out of most of the slab joints. Repair details including replacement of the uplifted slab, repair of the underlying underdrain pipe, and re-sealing the slab joints were developed under an emergency contract and repairs began in early 2000.

Figure 10. Uplifted Slab following Hurricane Floyd, 1999

During the 2000 repairs, voids were discovered underlying adjacent slabs as well as the crest sill and left training wall footers. Repair details were modified, two additional spillway slabs were replaced, and the voids beneath the crest sill and training wall footer slabs were filled, presumably with grout. Remedial measures incorporated in response to the damage caused by Hurricane Floyd included installation of a two-layer filter drainage system beneath the replaced slabs comprised of four-inch-thick sand and eight-inch-thick gravel layers. The single six-inch-diameter porous concrete underdrain pipe in this area was replaced with three six-inch-diameter perforated high-density polyethylene (HDPE) pipes. Design details for these repairs were reviewed by the United States Bureau of Reclamation prior to implementation.

2011 Inspection

In early November 2011, Gannett Fleming performed the bi-annual "engineer's" inspection of Diascund Dam. During this inspection, one slab near the right side of the principal spillway chute was noted to be uplifted several inches along its downstream edge as shown in Figure 2. Review of historical inspection reports and photographs indicate that this condition had been present, but remained undetected, for at least eighteen years meaning that this slab had uplifted prior to Hurricane Floyd in 1999. Observation and identification of this uplifted slab in 2011 prompted Gannett Fleming to request archive design drawings from Waterworks to assess the potential severity and underlying causes of the uplifted slab. Following cursory review of the 1984 As-Built drawings, Gannett Fleming recommended to Waterworks that repairs be implemented to correct the noted design deficiencies. In response, Waterworks retrieved additional archive records for the dam which, upon review, revealed several past incidents that had occurred at the spillway and confirmed the need to implement corrective action.

POTENTIAL FAILURE MODES

Dams and spillways constructed on erodible foundations, as is the case with Diascund Dam, can provide long-term performance if designed and constructed properly, however particular care must be given to the design details to address several key potential failure modes that are unique to this type of spillway. The spillway structure / foundation interface can create a preferential seepage path that, if not properly controlled, can lead to internal erosion and/or piping failure of the foundation. Inadequate seepage cutoff at the upstream end of the structure, an inadequately sized underdrain system beneath the structure and lack of mechanical connections between spillway slabs can each contribute to a situation in which high uplift pressures can develop beneath spillway slabs possibly resulting in structural failure. Differential displacement of adjacent spillway slabs creates an open flow path and can lead to rapid erosion of the foundation material. These failure modes can occur very suddenly during an unusual loading condition, as was the case with the uplifting slab failure during Hurricane Floyd, however, they more commonly occur very slowly and remain undetected over a long period of time as was evidenced by the significant voids encountered during the 1980's repairs. The presence of the rigid concrete lining can often hide evidence of a progressing failure.

These potential failure modes have not been adequately addressed by the design features for many concrete spillways constructed on erodible foundations. It is important for dam inspectors to understand historical design and construction practices and their potential weaknesses, as well as the site-specific potential failure modes for a given dam.

COMPREHENSIVE SOLUTION

Design Phase

The scope of the 2015 repairs to the principal spillway includes grouting beneath the crest sill and training wall footer slabs to fill any existing voids; demolition and reconstruction of the downstream chute and stilling basin slabs; removal of the existing underdrain pipes and sand "filter" or drainage layer and installation of a new more robust underdrain system; removal of the observation wells located within the spillway chute, and installation of new overlay slabs atop the existing crest sill and training wall footer slabs. All of the concrete slab joints were constructed with steel dowels connecting adjacent slabs to preclude individual slab movement/displacement as well as polyvinyl chloride (PVC) waterstops to preclude surface waters from entering and surcharging the underdrain system as shown on Figure 11. The overlay slabs are doweled into the existing training wall footer slabs, and the perimeter of the spillway is sealed for water tightness with retrofit PVC waterstops installed on the vertical faces of the training walls and crest sill.

To reduce the potential for high hydraulic gradients around the ends of the sheet pile cutoff, four-inch-diameter perforations were installed in the sheeting below the bottom of the crest sill slab to convey seepage into the underdrain system uniformly across the entire width of the principal spillway. A compatible sand filter layer was installed over these perforations to prevent migration of soils.

The new underdrain system consists of a six-inch-thick layer of concrete sand to provide a filter preventing migration of underlying embankment or native soils and a twelve-inch-thick layer of coarse drain fill (gravel) to collect and safely convey the seepage downstream. Eighteen eight-inch-diameter perforated HDPE underdrain pipes are located beneath the spillway chute (two per 'column' of slabs). Each underdrain pipe has two discharge points, a high-level outlet and a low-level outlet, through the face of the chute slabs.

The stilling basin was also designed to be reconstructed with doweled and waterstopped joints as well as a more robust underdrain system consisting of a twelve-inch-thick sand filter layer and a twelve-inch-thick gravel drainage layer. Two eight-inch-diameter underdrain pipes are installed through the downstream cutoff wall of each stilling basin slab to allow free drainage of seepage beneath the stilling basin slabs. Note that this is the first time the stilling basin has been modified since the dam was built.

Figure 11. Typical New Slab Joint
(Note PVC Waterstop and Dowels between adjacent slabs)

Figure 12. Underdrain Installation

Construction Phase

The principal spillway reconstruction requires exposing the underlying foundation soil, making the dam highly vulnerable to failure during any large spillway flows. Due to its importance to water supply for Newport News, the reservoir could not be taken completely out of service and drained to facilitate construction. Rather, construction was completed behind a partially-full pool. To reduce the chances of flooding, the pool was drawn down and held approximately five feet below normal pool elevation, and Waterworks personnel monitored the reservoir level and weather forecast very closely. Regular adjustments to the low-level outlet gates and pumping (water supply) operations were made to prevent the reservoir level from rising. Additional protection against flooding was maintained by constructing the repairs in two phases, such that a portion of the spillway was available to safely pass significant flood flows without impacting the active vulnerable work zone. This phased approach to construction, combined with the lowered reservoir pool, provided sufficient storage and discharge capacity to safely pass a 100-year flood during construction.

During removal of the upper-most chute slab adjacent to the left training wall footer, a persistent seep was discovered just downstream of the crest sill. The location of this seep was similar to the seep encountered during the 1984 repairs. The left-most underdrain pipe was also revealed to extend behind the sheet pile cutoff, presumably in an attempt to collect and convey seepage thereby reducing the hydraulic gradient around the end of the sheet pile.

CONCLUSION

In general, concrete spillways on erodible foundations warrant very careful consideration during design and construction considering their inherent greater susceptibility to failure. Diascund Dam has had a long history of defects occurring in its principal spillway due to inadequate design features for a structure on an erodible foundation. The original design deficiencies leading to these defects were not identified and were repeated in subsequent repairs throughout its fifty-year service life. The intent of this case history is to inform dam owners and engineers regarding the vulnerabilities of concrete spillways on erodible foundations as well as illustrate the great importance of routine safety inspections, document record keeping, and forensic analysis of observed defects. With regard to the latter, it is imperative for dam owners and engineers to understand potential failure modes and to carefully consider and correct the design deficiencies causing dam defects, not just address the symptoms. In the case of Diascund Dam, under slightly worse conditions, these defects might have resulted in catastrophic failure of the principal spillway and uncontrolled release of the reservoir. Prior dam safety inspections also failed to identify conditions indicative of severe distress in the principal spillway. Owners and engineers performing these inspections should strive to understand a dam's original design and any subsequent modifications during its service life in order to identify existing or future unusual conditions. For Diascund Dam, the precipitous condition of its principal spillway was readily identified and remedied through inspection and remedial design by qualified personnel.

ACKNOWLEDGEMENTS

Thanks to the Waterworks staff who contributed to the success of the 2015 Diascund Dam Principal Spillway Reconstructed project, including Donald Rice, whose relentless review of archive documents proved invaluable throughout the investigation and design phases of this project, Kevin Phillips, Adrianna Dimperio, Amanda Lothes, George Baker, and Scott Dewhirst.

REFERENCES

City of Newport News Department of Public Utilities (June 1960). *Diascund Dam*. As-Built Drawings, Contract No. 1.

City of Newport News Department of Public Utilities (June 15, 1984). *Water System Improvements, Diascund Dam Spillway Renovation*. Record Drawings.

City of Newport News, Waterworks (January 18, 2000). *Hurricane Floyd Repairs, Diascund Dam, Slab Replacement – Joint Reconstruction, Flashboard Replacement.* Contract Drawings.

Crowder, Charlie C. (Director, City of Newport News, Department of Public Utilities, Newport News, VA). Letters to: Dudley B. Lewis and Robert C. Dolecki (Malcolm Pirnie, Incorporated, Newport News, VA). 1984.

J. C. Biggins (City Manager, City of Newport News, Newport News, VA). Letter to: James M. Pharr. April 4, 1963.

Lewis, Dudley B. (Malcolm Pirnie, Incorporated, Newport News, VA). Letters to: Crowder, Charlie C. (Director, City of Newport News, Department of Public Utilities, Newport News, VA). 1984.

Malcolm Pirnie Engineers, *Diascund Spillway Investigation,* April 1983.

EMERGENCY REPAIRS TO PREVENT A PIPING FAILURE IN THE LAKE MANATEE DAM

Jeffrey A. Beriswill, P.E.[1]
Nickolas J. Depin, P.E.[2]
Mark R. Simpson[3]
Bruce McLeod[4]

ABSTRACT

During the course of an in-depth supplemental inspection in 2013 of the Lake Manatee Dam near Bradenton, Florida serious concerns were raised concerning the structural integrity of the dam (both the service spillway and the adjacent zoned-earth embankment). In particular, evidence strongly suggested the presence of an active internal erosion and piping failure mechanism in the embankment and immediately adjacent to and underneath portions of the service spillway. Indications of this potential failure mechanism were manifest in visual and physical observations since the dam went into service in 1967 and were documented during the first official inspection in 1978. This evidence became more apparent in subsequent inspections by several other engineering groups over time.

In early 2014, a Phase I emergency repair was implemented based on the findings of the supplemental inspection. This project consisted of constructing a soil-cement slurry cut-off wall varying from 95 to 105 feet deep using a combination of the Trench Re-mixing Deep (TRD) device and Jet Grout (JG) columns. Approximately 3,000 lineal feet of TRD wall was installed through the upstream side of the embankment's clayey core. The Jet Grout wall was installed underneath the service spillway located within the central portion of the dam and as a transition between the TRD wall and spillway. Construction of the seepage cut-off wall took place from April 24, 2014 through September 11, 2014.

The Lake Manatee Dam is first described. The findings of the supplemental inspection including the visual and physical observation data integration technique is then presented, followed by a discussion of the construction of the emergency repairs.

---

[1] Amec Foster Wheeler Environment & Infrastructure, Inc., 1101 Channelside Drive, Suite 200, Tampa, FL 33602, Jeff.beriswill@amecfw.com
[2] Amec Foster Wheeler Environment & Infrastructure, Inc., 1101 Channelside Drive, Suite 200, Tampa, FL 33602, nickolas.depin@amecfw.com
[3] Manatee County Utilities Department, 4410 66th St. E., Bradenton, FL 33210, mark.simpson@mymanatee.org
[4] Manatee County Utilities Department, 17915 Waterline Road, Bradenton, FL 34202, bvmwater@verizon.net

INTRODUCTION

Amec Foster Wheeler Environment & Infrastructure, Inc. (Amec Foster Wheeler), was retained by Carollo Engineers, Inc. (Carollo) to complete a supplemental inspection and analysis of the Lake Manatee Dam located in Manatee County near Bradenton, Florida. This evaluation was intended to address certain concerns that had been previously identified in the Amec Foster Wheeler 2013 Annual Inspection Report and by others. A particular focus was placed on: the historical soil loss in the embankment, behind the spillway approach and training walls and underneath the stilling basin; slow moving water in the toe drains; and the condition of the service spillway approach channel. This work began in October of 2013.

During the course of this work and as a result of the analysis of historical data, serious concerns developed about the structural integrity of the dam. These concerns centered on both the service spillway and the adjacent zoned-earth embankment. In particular, there was strong evidence in both the historical record and in the recent field inspections to suggest that an active internal erosion and piping failure mechanism existed in the embankment and immediately adjacent to and underneath the service spillway. Based upon the historical record, the conditions that created such a failure mechanism appeared to have been activated immediately following construction and appeared to have progressed and become more severe over the years. As a result, emergency repairs were recommended and implemented in 2014.

PROJECT DESCRIPTION

The Lake Manatee Dam (Dam) is located near Bradenton in Manatee County, Florida and was constructed between 1965 and 1967. It is an in-stream reservoir on the Manatee River approximately 20 miles upstream of its discharge point into Tampa Bay. Its primary function is a freshwater supply for users in Manatee and Sarasota Counties. A photograph of the dam is provided on Figure 1.

Figure 1. Project Location Map.

The Dam consists of an approximately 4,700-foot-long zoned-earth embankment with its axis oriented about 35 degrees east of true north. Based upon NGVD 29 vertical datum, the present elevation of the crest of the dam ranges from approximately 51 to 55 feet. An aerial view of the dam and spillway structures is presented on Figure 2. The principal spillway (called the service spillway) is located near the southern end of the dam and just south of the original Manatee River channel. The normal operating pool is approximately El. 38 feet.

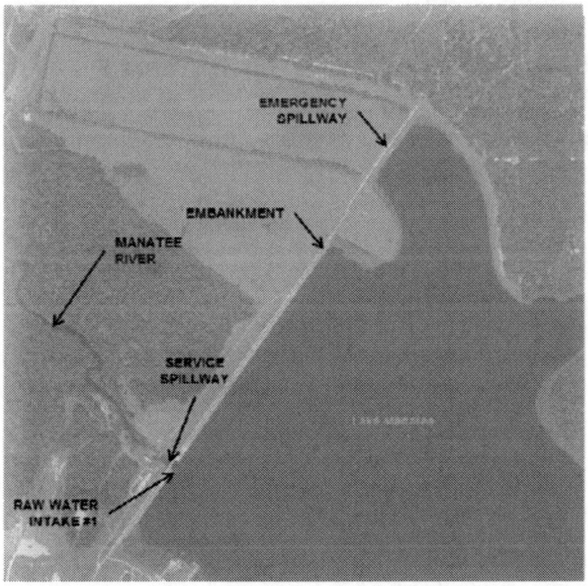

Figure 2. Dam Features

Reservoir Embankment

Figure 3 presents a typical cross section for the embankment from the original as-built drawings. The upstream slope was designed at 4 horizontal to 1 vertical (4H:1V) from the upstream toe up to a horizontal bench at El. 25 feet, NVGD 29. Above this bench; the upstream slope was designed at 3H:1V. The downstream slope was originally designed at 2.5H:1V in the upper portion and extended down to a horizontal bench at El. 30 feet, beyond which the embankment slope was designed at 3H:1V. In 2003, the embankment crest was raised by approximately 3 feet. During this raising, the centerline of the crest roadway was shifted a few feet upstream and soil-cement was placed over the upper portion of the newly created portion of the upstream slope. The downstream slope was flattened to 3H:1V

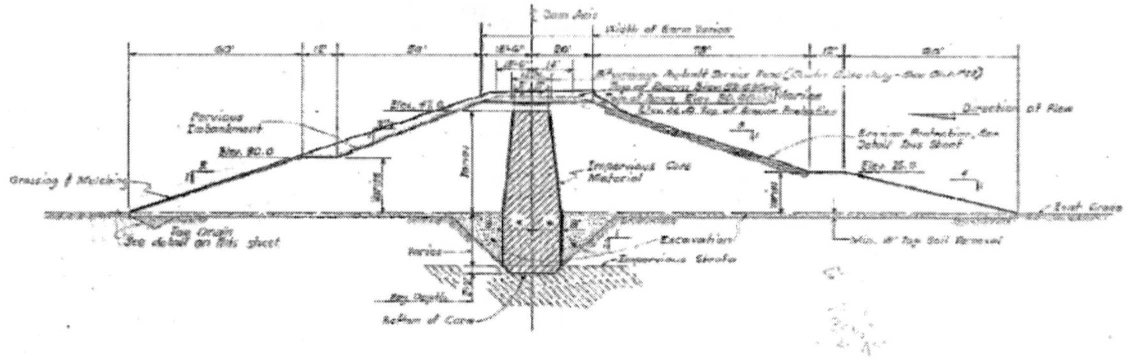

Figure 3. Typical Embankment Cross Section

A minimum 12-inch thick soil cement cover was provided on the upstream slope as a wave erosion protection measure. This cover extends from the lower horizontal bench at El. 25 feet up to the embankment crest. The original crest was designed to be approximately 27.5 feet wide with a 12-foot wide asphaltic concrete roadway along its centerline.

Service Spillway

The service spillway structure consists of three 15-foot radius tainter gates, each one spanning a 31.5-foot bay. Refer to Figures 4 and 5 for a plan view and cross section of the service spillway, respectively. The bottom of each tainter gate is set at El. 26.8 feet, NVGD 29, on top of an ogee section representing the lowest elevation to which the reservoir can be effectively lowered as there is no low-level outlet.

The spillway has upstream concrete approach walls and a concrete approach apron. It also has downstream concrete training walls and a stilling basin with an end sill for energy dissipation. The stilling basin floor slab is ground-founded and constructed of individual reinforced concrete sections. The first set of these sections immediately adjacent to the ogee section is 5 feet thick.

The next two sets of floor slab sections are 3.5 feet thick. The concrete approach walls and the downstream training walls have an inverted T-shape and together with the ogee section monoliths for the spillway are all founded on driven H-piles. The concrete stilling basin discharges over the end sill and into what was originally a bare-earth cut-soil channel connecting to the Manatee River. Subsequently, a concrete apron was placed to limit the amount of surface erosion.

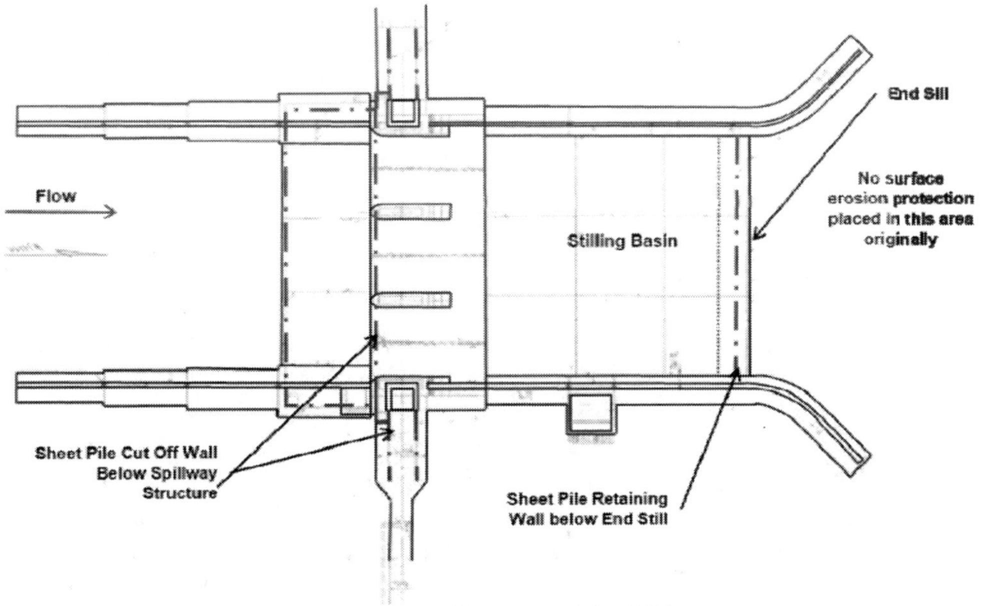

Figure 4. Service Spillway Original Plan

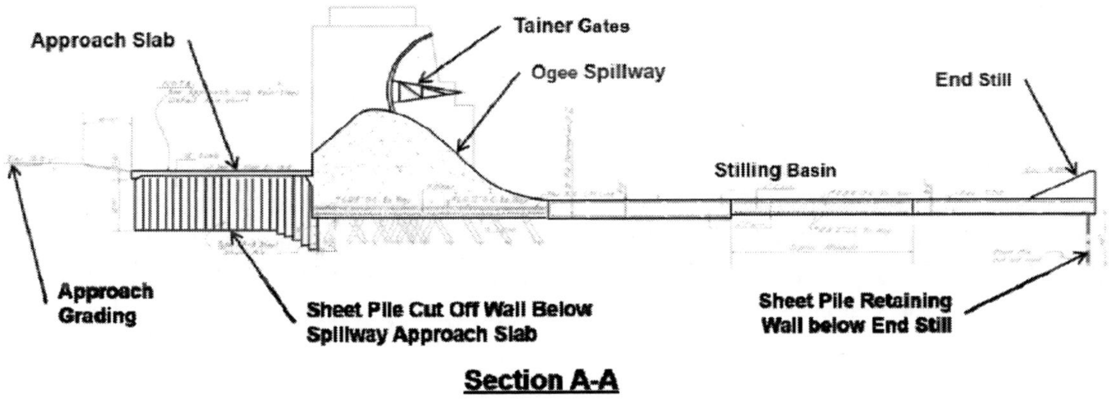

Figure 5. Service Spillway Original Cross Section

Emergency Spillway

The emergency spillway was constructed beyond the north end of the embankment in the mid-1980's. The excavated material from the emergency spillway channel was placed just south of the channel and decreased the effective length of the embankment from 4,700 feet to about 3,700 feet by filling in behind the embankment.

GEOLOGIC AND HYDROGEOLOGIC SETTING

The surficial soils at the project site consist of fine quartz sand with variable amounts of silts and clays. It generally extends as much as 20 feet below ground surface (bgs) at the project site.
These sandy soils form the water-bearing unit of the surficial aquifer system (SAS). It is a water table aquifer with unconfined conditions and a depth to water generally within 5 feet bgs. Surface water bodies such as the Lake Manatee Reservoir and local topography affect the potentiometric surface in the SAS. Hence, localized groundwater flow directions in the SAS in the vicinity of the Lake Manatee Dam are expected to be dominated by the presence of the reservoir and the embankment and the hydrogeology of the natural abutments on the extreme northern and southern ends of the embankment. Since the topography rises towards the east and away from the dam, the general direction for water flow in this aquifer is most likely from the east to the west.

The surficial sands are underlain by formations within the Hawthorn Group. This geologic unit is generally comprised of unconformably interlayered fossiliferous limestone and dolostone, quartz and phosphatic sand, silt, and clay. The Hawthorn Group possesses multiple high permeability layers interspersed between low permeability layers, which form the intermediate aquifer system (IAS). The IAS generally extends from approximately 50 to 450 feet bgs in the vicinity of the Lake Manatee Dam.

In general, the IAS contains three major permeable zones separated by less permeable zones. Two of these zones are present at the Lake Manatee Dam. These two zones are confined, and the low permeability confining unit between the surficial aquifer and uppermost permeable zone is located at an elevation between 10 to 20 feet along the embankment alignment.

Regional and site monitoring wells and piezometers indicate that piezometric water elevations (PWE's) in the IAS beneath the Lake Manatee Dam are affected not only by reservoir levels but also by regional ground water conditions. These PWE's can be artesian and in some cases may actually be higher than the reservoir elevation itself. Regional artesian pressures in the IAS could create vertical seepage gradients in the embankment and foundation of the dam in excess of those expected by the designers of the original seepage control system. Hence, these pressures may have accelerated the development and propagation of an internal erosion and piping failure mechanism within the dam.

FOUNDATION SEEPAGE CONTROL

Reservoir Embankment

The embankment is composed of a pervious, sandy outer shell with a central clayey core. This clayey core is 10 feet wide at the top and increases to 16 feet wide near the bottom, where it was keyed into a relatively impermeable stratum with a key-in depth of 2.5 feet.

This confining layer lies between the SAS and the IAS. It is part of the Hawthorn Group as described above.

A toe drain was installed to lower water pressures in the downstream shell. The original outfall for the toe drains was constructed without backflow prevention.
Hence, during high flow rate releases where the tailwater impounds above the invert of these drains, both the northern and southern toe drains are charged with tail water, thus increasing the rate at which the soils near the spillway and in the downstream toe area would saturate. When this toe drain was subsequently abandoned in place in 2003, the outfalls on both the north and south sides of the embankment were left unplugged and a portion of the toe drain is still open behind the face of each of these walls.

Service Spillway

The seepage cutoff system for the service spillway is also shown on Figure 4. It consists of a series of driven sheet piles that form a box around the edges of the approach slab and along the front edge of the ogee spillway sections. These sheet piles pass underneath the base slabs of the two approach wall monoliths closest to the Ogee section and are connected to another sheet pile cutoff wall that extends both to the north and south of the spillway for a distance of approximately 30 feet. A second set of cutoff sheet piles also extend to the north and south parallel to but downstream of the first set of sheet piles and on the opposite side of the dam crest. Two-foot thick reinforced concrete approach slabs with turned-down edges were poured over these two sheet pile walls for access to the spillway bridge over the gates. These two sheet pile cutoff walls to the north and south were designed to extend into the embankment's clay core.

A final sheet pile cutoff wall is located under the end sill of the stilling basin and runs north to south between the two base slabs of the spillway downstream training walls. It is important to note that there were no seepage cutoff features placed underneath the two downstream training walls.

The original pressure relief systems for the service spillway stilling basin and downstream training walls are shown on Figure 6. They consist of two underdrain systems beneath the stilling basin floor slab and a backfill drain system behind the two training walls. The first stilling basin underdrain system is directly connected by riser pipes through the training wall base slabs to the backfill drain system. The backfill drain system runs along the back of the training walls just above the base slab and drains upstream towards the ogee section. It is evacuated through a riser pipe and outfalls into the stilling basin with an invert at El. 11 feet, NVGD 29. The second stilling basin underdrain system runs underneath the end sill and drains directly into the downstream channel through 45-degree riser pipes running up through the floor slab and a portion of the end sill.

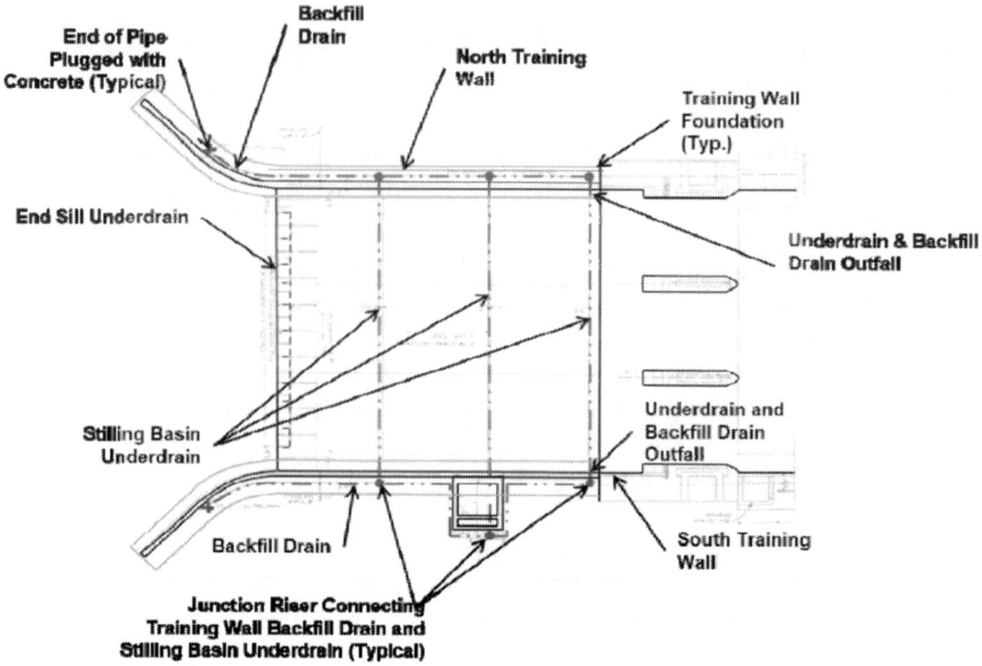

Figure 6. Service Spillway Pressure Relief System

Neither the approach walls nor the downstream training walls were designed with underseepage protection to prevent flow underneath their base slabs from upstream to downstream. Given that both the approach walls and the downstream training walls are pile supported, their base slabs can provide a roof for the development of a pipe if the soils underneath them are removed. The approach wall base slabs are partially protected from such a flow vector only by the sheet piling below the Ogee section of the spillway that is tied to the sheet pile cutoffs extending 30 feet north and south of the spillway. The downstream training wall base slabs have no such protection. Hence, material from underneath these training walls could be carried unchecked downstream along their entire axes if a void system were present because of the initiation of an internal erosion and piping feature. Materials underneath the approach wall base slabs could be carried downstream only after either going around or underneath the sheetpile cutoff walls.

HISTORICAL PERFORMANCE

Historical Aerial Photographs of Service Spillway

Six aerial photographs of the service spillway dating from 1973 to 2013 (see Figure 6) were closely reviewed. The first photograph in 1973 was taken five years before the first inspection by the Corps of Engineers, and the 2003 photograph was taken prior to dredging the river channel. Based upon the aerial photographs, the following observations can be made:

1) Significant erosion occurred on both the nor thern and southern discharge channel banks within 6 years of the dam going in to service and this erosion progressed

until the 1994 photograph where bank erosion protection had been added to the northern bank.

2) Sediments were accumulating in the river channel before 1973 and through 2003 when a dredging operation removed them.
3) After the dredging of these sediments in 2003, they have begun to accumulate again in the river channel as shown in the 2013 photograph.

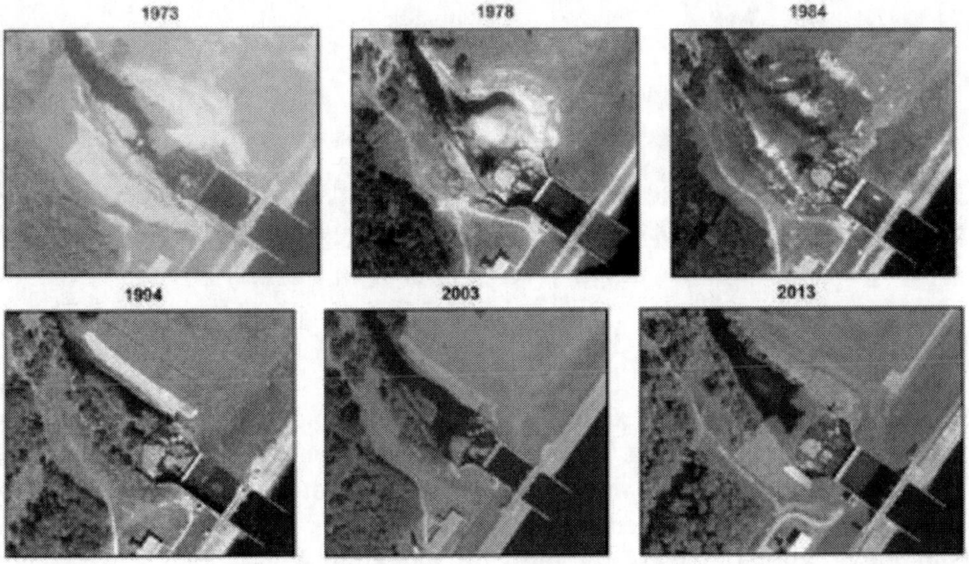

Figure 6. Historical Photographs of Service Spillway Discharge Channel

While not definitive, these aerial photographs are consistent with an active internal erosion and piping failure mechanism where a build-up of sediments in the discharge channel comes from eroded embankment and foundation materials that have been transported to the channel through pipes under the stilling basin, training wall base slabs and downstream apron.

Previous Dam Inspections

Since 1978 (eleven years after the dam was placed into service) inspections have been performed by engineers from eight different groups including the federal government and private consulting engineering firms. Specific observations that can be indicative of an internal erosion and piping failure mechanism were obtained based on a detailed review of the inspection reports. Each location where similar observations were made was given a number ranging from 1 to 18, as shown on Figure 7. The observations generally included voids, depressions, undermining/movement/deflections of structures, subsidence, loss of backfill material, low-strength material in soil borings, vortices in lake, and artesian pressure in piezometers.

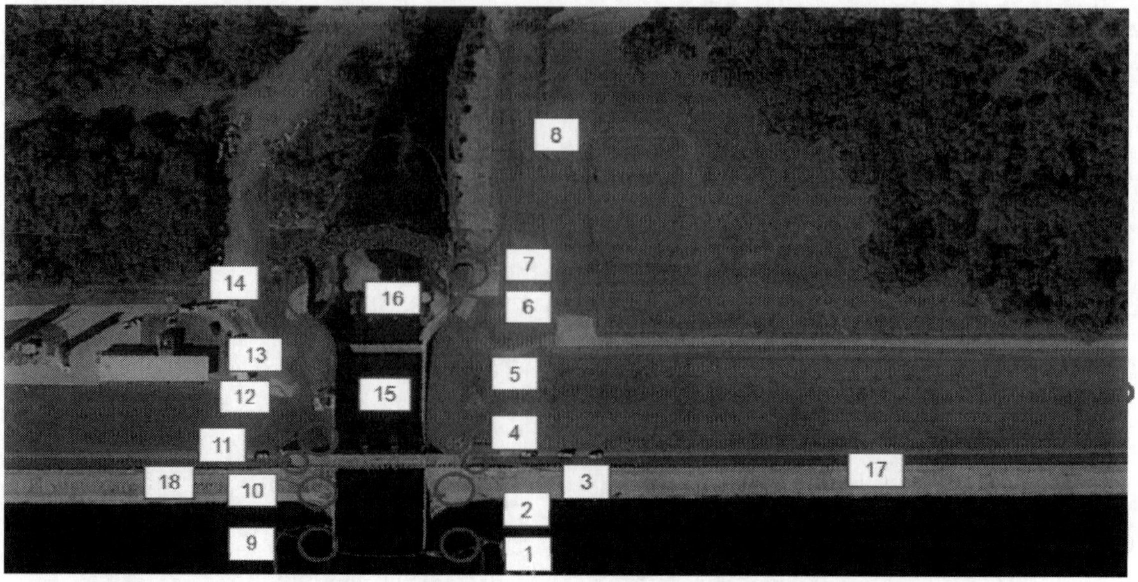

Figure 7. Locations of Observed Evidence of Internal Erosion/Piping

INTERPRETED FAILURE MECHANISM

The observations just summarized are consistent with the development and progression of an internal erosion and piping failure mechanism. The available data suggest that shortly after construction a pipe developed underneath both downstream training walls at their western edge and progressed back upstream towards the reservoir. The contribution factors to the initiation of the erosion probably including:

1) a hydraulically connected backfill drain and stilling basin underdrain system with inadequate backflow prevention;
2) a toe drain system with no backflow prevention;
3) an unlined discharge channel immediately adjacent to the end sill of the stilling basin;
4) no seepage cutoff under the downstream training walls;
5) a deep borrow pit excavated upstream of the dam near the south shore of the lake and within 1,500 feet of the service spillway that cut through the designed confining layer for the embankment;
6) regional artesian pressure condition in the aquifer immediately below this confining layer; and,
7) reportedly poor soil compaction processes, a pipe developed underneath both downstream training walls at their western edge and worked its way back upstream towards the reservoir.

CONCLUSIONS REGARDING THE STATE OF THE DAM AND RECOMMENDED ACTIONS

During the course of the Amec Foster Wheeler 2013 Supplemental Inspection and as a result of an analysis of historical data, serious concerns were raised about the structural integrity of the dam (both the service spillway and the adjacent zoned-earth embankment). In particular, evidence strongly suggested the presence of an active internal erosion and piping failure mechanism in the embankment and immediately adjacent to and underneath portions of the service spillway. Based upon the historical record and as-built drawings, the conditions that created such a failure mechanism appeared to have been activated immediately after construction and to have progressed and become more severe over the years.

In early 2014, Amec Foster Wheeler and Carollo met with representatives of Manatee County to review the historical data and preliminary findings of the 2013 Supplemental Inspection. During this meeting, Amec Foster Wheeler made the following short-term recommendations:

1. Consider lowering the reservoir;
2. Reestablish the seepage control function of the dam core (jet grouting, pressure injection, sheet piling, or similar); and,
3. Work with specialty contractors to collect necessary information and develop cost estimates.

These short-term recommendations were implemented through the Phase I emergency repairs completed in 2014, which involved the design and installation of a deep (95 feet to 105 feet) seepage cutoff wall under the service spillway and through the northern and southern embankments and extending to near the northern and southern abutments.

PHASE I EMERGENCY REPAIRS

The Phase I emergency repairs consisted of constructing a soil-cement cutoff wall varying from 95 to 105 feet bgs using a combination of the Trench Re-mixing Deep (TRD) device and Jet Grout (JG) columns. The various construction elements are discussed in the following sections.

General Site Work

The following site work operations associated with the cutoff wall construction were completed: 1) constructing a temporary work platform along the crest of the embankment to accommodate the TRD; 2) saw cutting the dam crest roadway in preparation for the TRD surge trench; 3) constructing a haul road across the riprap in the downstream river channel; and, 4) constructing temporary work platforms for jet grouting operations behind the two approach walls.

TRD Work Platform

A temporary 40-foot wide work platform was constructed for the TRD device by placing compacted earthfill on the downstream slope of the existing embankment. The embankment slope was steepened to between 2H:1V to 2.5H:1V to allow for the widened crest. The TRD cutoff wall was constructed 5 feet from the upstream (inside) edge of the crest road (refer to Figure 8).

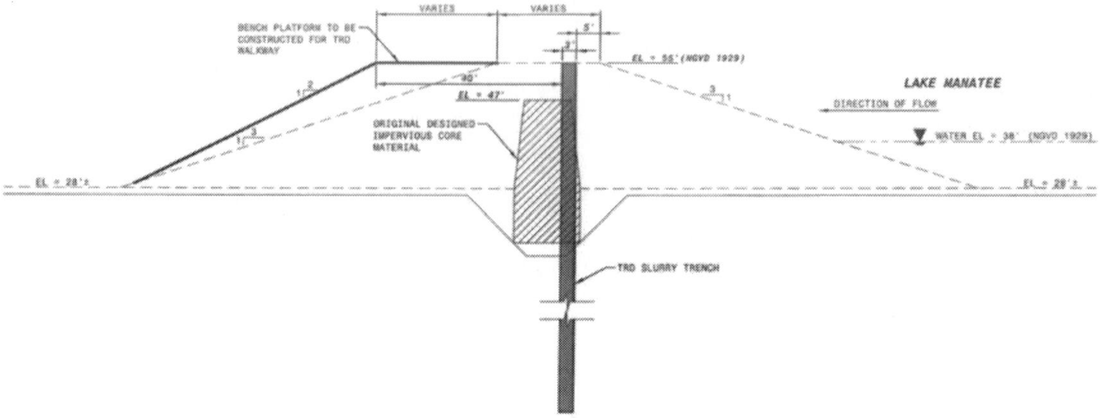

Figure 8. TRD Work Platform Typical Section

Saw Cutting of Dam Crest Roadway

The TRD device requires the construction of a surge trench to manage the soil/cement spoils created during construction of the seepage cutoff wall. During construction if voids are encountered during the installation process, this surge trench also functions as a source for soil/cement slurry to rapidly fill the voids. The trench was 3 feet wide and was excavated in 15-foot to 20-foot lengths as the TRD advanced along the alignment.

Haul Road Across Existing Rip Rap

In order to permit the movement of heavy equipment from southern embankment to the northern embankment without using the bridge over the service spillway, a limerock haul road was designed over the existing riprap in the river channel just downstream of the spillway end sill.

Jet Grout Work Platforms

Temporary work platforms for jet grouting were constructed behind the upstream approach training walls. Due to the footprint required to install the JG columns, small retaining walls were necessary along the downstream edges of these two temporary work platforms. These walls were constructed with steel H-piles 10 feet to 15 feet in length and wood lagging.

Jet Grout over Water Preparation

Divers cored through the approach slab along the proposed alignment of the JG wall. After the coring operation, a template was attached to the spillway approach walls, and standpipes for raising the returns from the JG operation were placed in the core holes, grouted in place with a Portland cement grout, and tack-welded to the template.

TRD Construction

The TRD wall was installed through the upstream side of the dam's clayey core to within approximately 100 feet of the service spillway. A 21-inch thick, 8-foot wide cutter was advanced to design depth and subsequently cut a continuous trench using "chainsaw" action. During cutting, soil from the trench was mixed with an injected cement/slag/water slurry to create a soil-cement mixture that hardened upon curing.

The trench was originally cut using a single pass of the T RD device. However, a hard limestone layer was encountered in the vicinity of the service spillway and along the wall alignment south of the spillway. Where the hard limestone layer was encountered, the wall was pre-drilled and the cutter made three passes through the trench. Three types of slurry mixes were used depending on the cutting method and trench soil type.

During the TRD wall construction, three voids were encountered within an approximately 15-foot length of wall. In each instance, the soil-cement mix in the surge trench in front of the TRD cutting head (approximately 3 feet wide, 5 feet deep, and 15 feet long) experienced a rapid drop in surface elevation. At two of the locations, the entire trench was emptied into the void encountered, while the soil-cement mix level dropped by 4 feet at the third location. In each of these instances, the TRD cutting head was held in position with the chain rotating and additional cement/bentonite slurry was pumped through the cutting head until the surge trench was again filled with soil/cement/bentonite mix. HBI estimated that the total amount of slurry required to fill each of these voids was on the order of 6 to 8 cubic yards.

Jet Grout Wall Installation

The JG wall was installed underneath the service spillway, as a transition wall between the TRD and spillway, and between the TRD wall and offset sections along that wall. The JG wall was constructed as a series of overlapping columns of stabilized soil. The columns were created by advancing a drill bit to the design depth of the wall and injecting a water/cement/bentonite mixture into the adjacent soils at 6,000 psi of pressure as the drill bit is rotated and withdrawn. The wall was typically designed with 4-foot diameter columns of 3-foot centers. Beneath the service spillway training walls, 8-foot diameter columns on 7-foot centers were used. The column diameter was controlled by injection pressure, bit rotation speed, and drill rod withdrawal rate.

The first 36 JG columns were constructed using the single cut method, where the drill bit is withdrawn from the hole while injecting slurry once.

Verification coring at randomly selected interstice locations between selected adjacent columns encountered unmixed soil and brought into question the viability of using the single cut method. As a result, the single cut method was abandoned and the remaining columns were installed with the double cutting method by mixing all or a portion of the columns twice.

JG operations within the interior of the service spillway were completed in approximately 25-feet of water, with the bottom of the JG wall about 60 feet below the spillway approach slab. Prior to initiating jet grouting in the spillway, the wall alignment was shifted 27 feet further upstream to avoid piles within the ogee and approach slab foundations. The final JG wall alignment is shown on Figure 9.

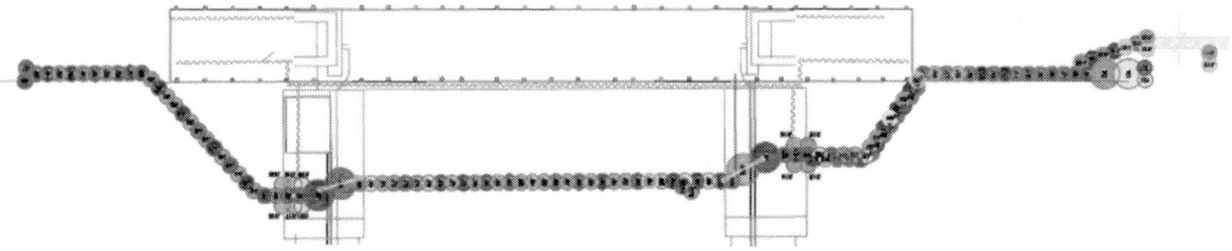

Figure 9. Final Jet Grout Wall Alignment

During the installation of these columns, the concrete approach slab started to heave due to previously undiscovered voids under the approach apron and under the action of the excess pressures created by the grouting operation. Several techniques were employed to reduce the amount of heave. However, a portion of the tongue-and-groove joint between the approach slab and northern approach wall base slab broke. The overwater JG operation was halted, and the slab was evaluated. Relief joints were cut through the approach slab at three locations to avoid the buildup of pressure beneath the slab, and the remainder of the JG columns overwater were subsequently installed.

During the installation of the JG columns, a loss of circulation of the treated soil mixture was observed a total of 23 times in 15 columns, or 12% of all of the JG columns installed on the site. The majority of this loss of circulation (87%) occurred on the north side of the spillway in 13 columns.

Quality Assurance and Quality Control

QC/QA testing by the HBI and Amec Foster Wheeler confirmed that the TRD wall was homogeneously mixed from the top to the bottom. Verification cores indicated that the JG columns were sufficiently mixed and overlapped. Representative samples taken from both operations confirmed that the seepage cutoff wall met the project specifications with regard to physical properties, including a minimum unconfined compressive strength of 50 psi and maximum permeability of 1×10^{-6} cm/sec after 28 days of curing.

CONCLUSIONS

The following conclusions may be drawn from this project:

- Piping and internal erosion within a dam is often a slow, steady progression up to the point of failure.
- Lack of continuity and familiarity in dam inspections can allow serious deficiencies to go unnoticed.
- The emergency repairs were completed on an accelerated schedule through a team approach between Manatee County, the Engineer, the Contractor, and Regulators.
- A well-qualified contractor is a great asset in responding to unanticipated conditions during construction.

REFERENCES

Amec Foster Wheeler Inspection (August 2012), “Lake Manatee Dam 2012 Annual Inspection Report”, Prepared for Carollo Engineers, Inc., Amec Foster Wheeler Project Number 300472.

Amec Foster Wheeler (July 2013), “Lake Manatee Dam 2013 Annual Inspection Report”, Prepared for Carollo Engineers, Inc., Amec Foster Wheeler Project Number 300472X1.

Amec Foster Wheeler (January 2015), “Lake Manatee Dam Supplemental Inspection Report”, Prepared for Carollo Engineers, Inc., Amec Foster Wheeler Project Number 300472X2.

Amec Foster Wheeler (May 2015), “Lake Manatee Dam Emergency Geotechnical Subsurface Investigation Data Report”, Prepared for Carollo Engineers, Inc., Amec Foster Wheeler Project Number 300472X2.

Amec Foster Wheeler (June 2015), “Lake Manatee Dam Emergency Geotechnical Subsurface Investigation Data Phase I Construction QC Report”, Prepared for Carollo Engineers, Inc., Amec Foster Wheeler Project Number 300472X2.

Andersen,Glen R., Simpson, Mark R. and McLeod, Bruce (2015), “The Importance of Integrating Historical Data in Preventing a Failure of the Lake Manatee Dam”, Journal of Dam Safety, Association of Dam Safety Officials, Volume 13, Issue 1.

Petuch, Edward J. and Roberts, Charles (2007), The Geology of the Everglades and Adjacent Areas, CRC Press, 212 pp.

JET GROUTING FOR ECHO DAM SEISMIC REMEDIATION

Tara Schenk-McFarland, P.E.[1]
Mark Bliss, P.E.[2]

ABSTRACT

Echo Dam, located near Coalville, Utah, is a facility that provides storage for irrigation and municipal use. Original construction of the dam was completed in 1931. Because of seismic concerns, the spillway crest structure and upper chute section, founded on old landslide and colluvial material, were recently replaced. Prior to replacement of these structures, jet grouting of a block of the loose and potentially liquefiable soils, directly beneath the spillway crest structure was selected as the preferred alternative to address the potential for settlement and cracking of the spillway crest structure. The replacement or alteration of the in-place soil reduces the likelihood of seismic-induced excess pore pressure development and subsequent settlements. Core drilling and sample recovery of the production jet grouted columns was completed to verify compliance with the specifications. It was recognized that due to the larger rock sizes within the deposits that full homogeneity of the jet grouted zone was unlikely to be achievable. However, the purpose of this jet grouted section was to strengthen enough of the soil so that widespread liquefaction does not occur. During the jet grouting production many instances of ground fracturing occurred. Fracturing was evident based on grout/air bubbles exiting the surface near some of the grout holes during jet grouting, air bubbles appearing in the spoil pits, and lifting of the ground surface in the vicinity of the monitor during the jet grouting process while the system was pressurized. In one instance, a section of the upstream spillway slab, which was left in place, was cracked due to pressurized grout escaping beneath the slab. The overall results of the jet grouting were satisfactory despite difficulties encountered. The cause of the fracturing is hypothesized and a recommendation is included to reduce the likelihood of this occurring on future jobs.

BACKGROUND

Echo Dam, located near Coalville, Utah, was constructed in 1931 and provides storage for irrigation and municipal use. The embankment is a zoned earth and rockfill structure with a structural height of 158 feet. A 40- to 50-foot deep cutoff trench, excavated to the top of the underlying bedrock, is located along the upstream toe of the dam. An approximate 15-17-foot-high concrete cutoff wall was constructed in the foundation cutoff trench, keyed into bedrock. A shear key, excavated to bedrock, was recently completed under the downstream toe to remove potentially liquefiable materials from the downstream embankment foundation. The shear key was backfilled with dense soils and an overlying berm was constructed. The modifications to the dam also included a rockfill berm on the upstream side of the dam. A new toe drain system was constructed beneath

[1] Civil Engineer; Technical Service Center; Bureau of Reclamation; Denver Federal Center, Bldg 67, PO Box 25007 (86-68313), Denver, CO 80225; tschenkmcfarland@usbr.gov
[2] Civil Engineer; Technical Service Center; Bureau of Reclamation; Denver Federal Center, Bldg 67, PO Box 25007 (86-68313), Denver, CO 80225; rbliss@usbr.gov

the overlying downstream berm to collect seepage from chimney drain and blanket filter/drainage zones included within the downstream berm. Potentially liquefiable materials were identified within the spillway foundation, and it was determined that the spillway walls and piers could structurally fail under strong seismic loading. Seismic modification to the spillway structure included strengthening the potentially liquefiable soils under the crest structure, demolition of the existing crest structure, and constructing a new spillway foundation drainage system, new spillway floor, walls, and piers, rehabilitating the radial gates, and placing a concrete overlay on a deteriorated section of the chute walls downstream from the crest structure.

Jet grouting of a block of the potentially liquefiable soils directly beneath the spillway crest structure was selected as the preferred alternative to address the potential for settlement and cracking of the spillway crest structure in response to seismic loading. This process is designed to strengthen these soils by both mixing the soils with cement to form soilcrete and in some cases, replacing the soil with cement grout. In either case, the replacement or alteration of the in-place soil will greatly reduce the likelihood of seismic-induced excess pore pressure development leading to liquefaction and subsequent settlement. A test section was included in the specifications contract so that the contractor could test the proposed jet grouting methodology (i.e. jetting parameters). Reclamation required core drilling the *in situ* jet grouted columns to evaluate mixing, uniformity, and strength. Core drilling was also intended to help define the actual jet grout column geometry formed in the ground. The production jet grouting work was predicated on the approval, by Reclamation, of the test section. In addition to the coring of the jet grouted columns in the test section, core drilling and sample recovery of the production jet grouted columns was completed to verify compliance with the specifications.

The soils beneath the crest structure, in order of shallowest to deepest include Quaternary older landslide (Qols), Quaternary colluvium (Qc), and Echo Canyon Conglomerate (Kec). Geologic investigations of the dam site included core drilling, auger drilling, Standard Penetration testing, grain size analyses and Atterberg limits testing. The Qols makes up the majority of the foundation to be jet grouted and consists of a broad mixture of lean clay (CL), clayey sand (SC), clayey gravel (GC), silty clayey sand (SC-SM), silty clayey gravel (GC-GM), with various amounts of silt, sand, gravel, cobbles, and boulders. Observations of Qols particle sizes were made based on the surface deposits of Qols on the left abutment of the spillway, and in excavation of the left side of the spillway during this modification. A thin zone of Qc underlays the Qols and overlies Kec. The original construction of the spillway cutoff wall included a stoped excavation, using mining methods, of a 2 x 5-foot tunnel along the top of Kec, through the Qc, backfilled with concrete. This served as a seepage cutoff in the saturated, permeable overburden. Limited records of this construction indicated the Qc consists of coarse sands, and gravels, cobbles, and some boulders.

PRODUCTION JET GROUT COLUMNS

The production jet grout column plan view layout is shown on figure 1, and the profile view is shown on figure 2. For column identification the rows of columns were labeled A through I starting with row A on the south (upstream) side.

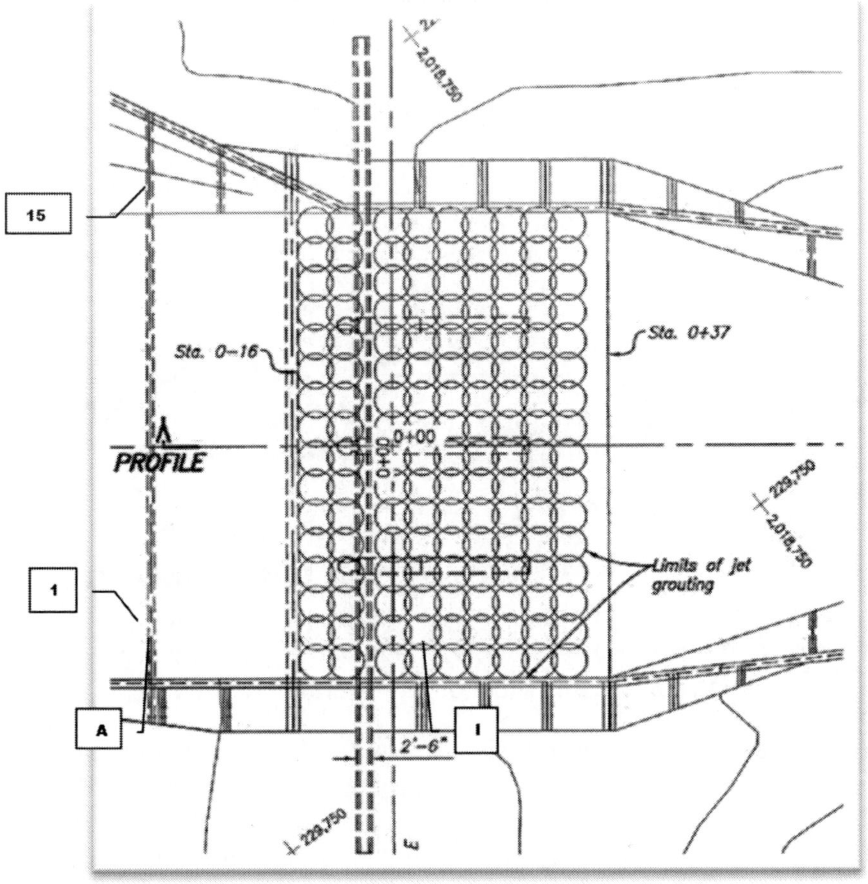

Figure 1. Plan view of jet grouting layout showing the row identifications starting with A on the upstream side through I on the downstream side and 1 on the right side through 15 on the left side.

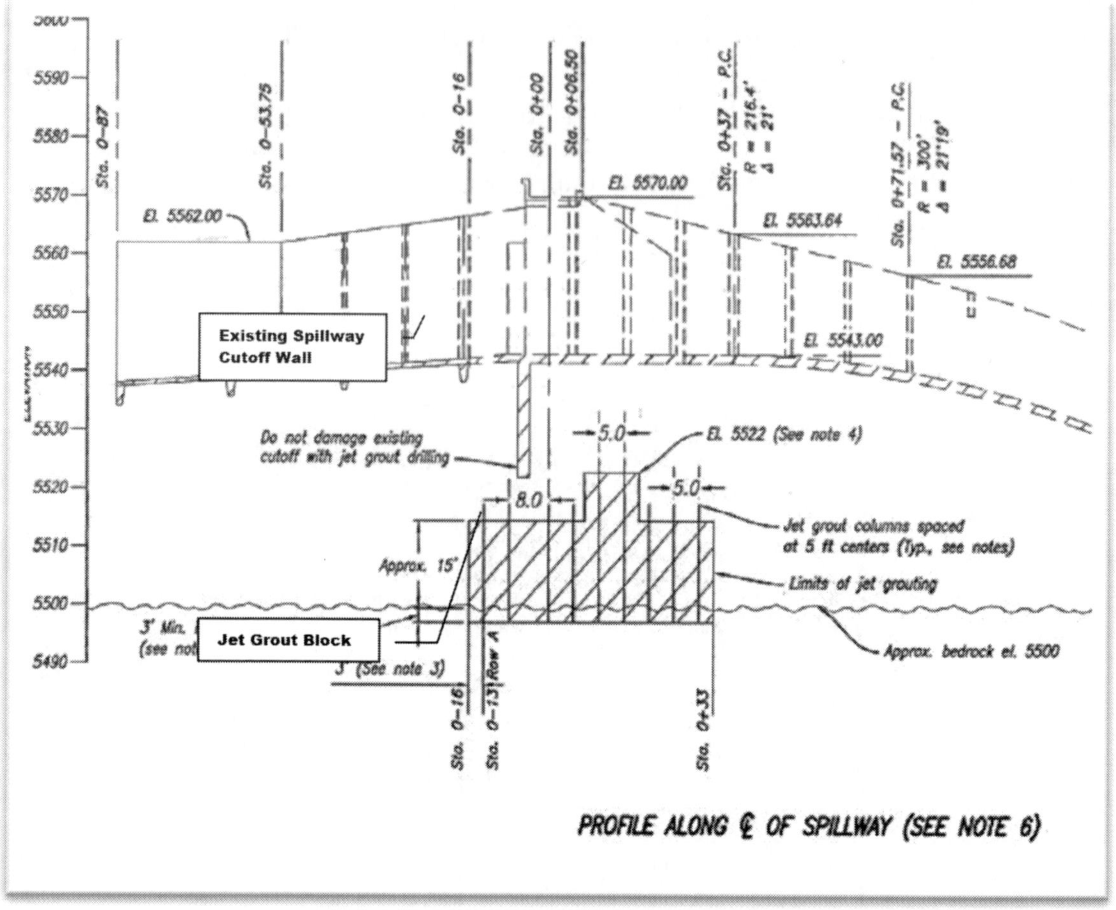

Figure 2. Profile view of jet grouting columns

Each column within each row was numbered 1 through 16 starting from east to west (i.e. A1, A2, A3, etc. and B1, B2, etc). The jet grout treatment targeted primarily the Qols soils below the groundwater table (saturated) that have low SPT blow counts. These are the soils that are most likely to develop high excess pore pressures and settle after strong ground shaking. The unsaturated soils within the Qols are less of a concern but the zone of jet grout treatment was extended well above the saturated zone to account for potential fluctuations in the groundwater levels.

Jet grout design parameters

Since the purpose of the jet grouting was to reduce the likelihood of excess pore pressure development and subsequent settlement during strong ground shaking, the standard for successful treatment was subjective (i.e. subjective judgment would be required to estimate the effectiveness of treatment by reviewing core recovery from the 15 verification drill holes). For the jet grouting to be considered successful, enough cement grout must be injected into the foundation soils to replace and/or mix with the *in situ* soils. With enough mixing and/or replacement, settlement and subsequent excess pore pressure will be minimized. In addition, it was felt that treatment would not have to be homogeneous throughout the entire block. Isolated zones of untreated soil that are discontinuous may still develop excess pore pressures and undergo some settlement, but the overall treated foundation block will provide support with the stronger, treated zones likely to arch over these isolated pockets. In addition, any untreated pockets would be of limited thickness and resulting settlements would be localized and small in magnitude compared with the thickness of the entire foundation. Another significant factor considered in the evaluation of the treatment effectiveness is that much of the foundation soil is non-liquefiable, as it is not saturated. Filter and drainage zones have been provided adjacent to the spillway walls and in the spillway foundation that will serve as a secondary line of defense against the progression of any seepage erosion that might initiate as a result of cracking and/or separation of the fill from the spillway crest structure walls caused by some settlement.

Based on the considerations above, the jet grouting specification paragraphs were written to require that treated soils have a minimum 28-day unconfined compressive strength (UCS) of 100 psi. This target value was believed to result in enough cement being injected into the soil matrix that adequate strength increase, and strain reduction during ground shaking, would be achieved. Final jet grouted columns were to be a minimum of 6-foot-diameter (on 5-foot centers) to create 1-foot overlap on the adjacent column, using a water to cement ratio by weight of 1.15:1. Other parameters such as grout and air injection pressure were not specified as these are typically determined by the jet grout contractor. It was known from original construction and final design explorations that the Qc layer exists for a few feet just above bedrock. Due to the potentially larger rock sizes in this zone, it was recognized that successful jet grouting of this zone would be difficult. In coarse materials such as this, larger rock sizes can cause a shadowing affect, resulting in inconsistent treatment coverage. However, homogeneous treatment of this zone was unnecessary for the following reasons:

1. The zone is limited in thickness and due to the coarse, broadly graded nature, likely to be dense enough that settlements would be minimal. In addition, in the geologic past, the Qc has had a large over burden of soil that would have resulted in some consolidation (settlement) of any low density zones. This would tend to reduce the magnitude of potential future settlement from strong ground shaking. If settlement were to occur, it is unlikely to have a significant impact to the overlying spillway structure due to the small overall thickness of the zone.

2. Jet grouting of this zone will result in significant volumes of cement being injected. It is anticipated that any cementation of the zone due to jet grouting will just add to the stability of the zone during strong ground shaking. Therefore, the jet grouting specification paragraphs were written to preclude the necessity of full and/or homogeneous treatment in this zone. To be more specific, the Qc zone was included in the jet grout treatment but there was no standard of successful treatment required in the specifications, such as homogeneous grout inclusion and/or minimum UCS.

3. The zone is coarse and known to have a high horizontal permeability. This may reduce its susceptibility to excess pore pressure development. This zone is now

laterally and hydraulically connected to the new spillway foundation drainage system at its downstream contact with the new spillway chute structure. Rapid dissipation of excess pore pressure may be possible through the drainage system.

The specification requirements also recognized the presence of zones within the Qols that contain soils with plasticity. Soils with even low plasticity are more difficult to treat uniformly with jet grouting. Therefore, the specifications recognized that some unmixed zones of soil were anticipated and the contractor would not necessarily be penalized for the presence of small unmixed zones. However, the decision that additional jet grouting would, or would not, be required was left up to evaluation and judgment by Reclamation engineers based on the results of the core drilling.

Required submittals from the jet grout Contractor included data related to the production grouting. The information requested by Reclamation included:

1. Identification of the drilling and jet grouting equipment.

2. Instantaneous recording of the jet grouting parameters was also requested including the depth, density of injected grout, cumulative injected volume of grout, grout injection pressure, grout injection flow rate, air injection pressure, air injection flow rate, and the lift rate and rotation speed of the jet grout monitor. These data were to be plotted with depth.

3. Total drilling time for each hole, total grout-injection time for each hole, total volume of grout injected in each hole, total weight of cement injected in each hole, top and bottom elevations of each constructed column.

4. Other pertinent information such as documentation of grout escape, ground heave or other unusual behavior.

Jet Grout Methods

Production jet grouting was accomplished with a double fluid system. The work space was limited, as shown in figure 3.

The starting grout mix conformed to the specification of a water:cement ratio of 1.15 (specific gravity = 1.47). The jet grout holes were pre-drilled into bedrock using duplex drilling which utilizes an internal drill string (downhole hammer or other bit) as well as an external casing that is advanced simultaneously. The use of air to remove the drill cuttings was originally proposed and rejected by Reclamation. Instead, the Contractor was allowed to use water as the drill fluid at a maximum pressure of 50 psi.

Figure 3. Jet grout treatment work area

According to the jet grouting method description submittal, grout injection pressures were a minimum of 4,000 pounds per square inch (psi) and ranged from 4,000 psi to 6,000 psi. The air injection pressures while grouting were to be a minimum of 50 psi with a range of 50 to 125 psi. Production jet grouting records submitted at the completion of grouting the test section were incomplete which limited Reclamation's ability to evaluate some of the issues that arose during the jet grouting production.

Treatment Results

Reclamation planned to drill 15 core holes at varying locations within the production jet grout area to verify the *in situ* treatment conditions. The recovered core from these holes was the primary quality assurance to judge the success of the jet grouting. The condition of the core was to be evaluated on uniformity, quality of soil/cement mixing (visual inspection), and unconfined compressive strength.

It is important to note that the limited working space within the jet grout treatment area, and the relatively short duration of the jet grouting work, made it infeasible to complete core drilling before the jet grouting was done. Therefore, assessments of the effectiveness of jet grouting had to be made during the verification core drilling process as core was being retrieved. This was critical in case the contractor wanted to make any changes to the jet grouting parameters in the middle of production grouting. The original plan called for 15 core holes. Since one of the proposed core holes was just outside of the jet grout zone (designed to determine if grouting extended beyond the treatment boundaries), it was eliminated at the end of production grouting to save time and money. Eliminating this hole also allowed the spillway contractor an earlier start for construction and placement of the new spillway concrete. It was decided that the data obtained in this hole was unnecessary for treatment verification.

Each verification core hole has a distinct label. The core hole label included the designation of the column (or columns) that the core hole intersected. As an example, DHPS13-D15/D16 designates the core hole drilled between columns D15 and D16 with the target location being at the intersection. Due to the close spacing of the jet grout columns, the actual distance from the center of any given column was not provided. Core holes with a single column designation were drilled near the center of the column. The following table lists each of the core holes, depth to bedrock, and the estimated percent core recovery within each column.

In column 3 of Table 1, the estimated percent of core recovery that contained grouted soil is based on subjective judgment of the engineer using drill core logs, results of laboratory testing of core for unconfined compressive strength, and photographs of the recovered core, taken for each core length.

Table 1: Qualitative Interpretation Results of Core Treated with Cement

Drill Hole Designation	**Depth to Bedrock (ft)**	**Estimated Percent Core Recovery in Jet Grout Zone *1/***
DHPS13-D5	43	75
DHPS13-D15/D16	41	99
DHPS13-E1/D1/E2/D2	Plugged off	70
DHPS13-I5/I6	42	93
DHPS13-D10/E10	Data missing	-
DHPS13-F7	42	73
DHPS13-E14	41 95	
DHPS13-G3	42	53
DHPS13-G11/G12	43 *2/*	83
DHPS13-G15/H15/G16/H16	41.5	<10
DHPS13-C3/D3	43	95
DHPS13-A14/B14	44	73
DHPS13-A3	43(?) Blocked off	86
DHPS13-A6	42.5	99

1/ Jet grout zone is defined as the design height of the jet grout column above bedrock. For rows A,B,C,D,G,H, and I the height is 15 feet. For rows E and F the height is approximately 22 feet.

2/ Actual depth to Kec (Echo Canyon Conglomerate bedrock) is uncertain due to loss of cobbles from the core barrel shoe and hole blockage that prevented further advancement.

Virtually all but two columns received a minimum of 70 percent treatment. Core hole DHPS13-G15/H15/G16/H16 had by far the least amount of cement in the recovered core (<10 percent). However, this core hole was located at the intersection of four columns. No significant treatment was anticipated in this interstitial zone in the design. Overall, the basis for a determination of successful treatment were 1) qualitative judgment of the extent of soil treatment and mixing using visual examination and core photographs and 2) quantitative evaluation of the select core specimens that were laboratory tested for unconfined compressive strength (UCS). The specifications requirement for the *insitu* jet grouted soil was that it have a 28-day UCS of 100 psi or greater. In addition, UCS testing of wet samples was performed during jet grouting. Wet samples are obtained by filling sample molds with the surfacing jet grout spoil flowing out of the hole during the jet grouting process. During jet grouting, the spoil is directed, via a small trench, into a pit where it can be removed after the initial set, to be hauled away for disposal. Although the results of the UCS testing on the wet samples is not related to a specification requirement, it does provide a general sense of mix quality and a range of strength that qualitatively can be related to the overall consistency of soilcrete being created in the subsurface columns. In almost all cases, the wet sample strengths will be higher than the actual cored sample due to the more homogeneous mixing of the eroded soil during flow up the annulus space and into the waste pit. This eliminates the discontinuities that may exist in the *in situ* soilcrete formed in the jet grouted column beneath the surface. These discontinuities and other features such as cobbles or high plasticity soils that may not get mixed well with the soilcrete mass, will typically lead to lower UCS values in the cored soilcrete samples compared to the wet samples.

Only two 28-day UCS tests did not meet the specification requirements, out of a total of 39 tests. These two tests indicated 28-day UCS values of 12 and 49 psi. All other tests exceeded the specification requirements with a range of 28-day UCS test results from 12 to 1835 psi; a mean value of 290 psi; and a median value of 193 psi. The high value of 1835 psi is an anomaly based on the data, as the large majority (90 percent) had 28-day UCS values less than 614 psi and 80 percent less than 458 psi. These test results indicate that the jet grouting achieved the required strength goals.

In addition to the core recovery evaluation and UCS testing results, the Contractor pumped almost 300,000 gallons of grout into the foundation according to the final jet grouting records. Using a grout mix of 1.15:1 (W/C ratio), this equates to approximately 1,075 tons of cement injected into the foundation zone to be treated. In jet grouting at past Reclamation projects, it has been estimated that the volume of spoil is approximately 60 percent of the volume to be treated. In this case, the Contractor estimated up to 50 cubic yards of waste per day would likely be generated. Since the spoil contains cement, and not all of the injected cement remains in the ground, some unknown quantity of cement is lost with the spoil. However, even with the loss of the cement contained within the spoil, the subsurface foundation treatment zone was mixed with hundreds of tons of cement within a relatively small volume of foundation soil.

ISSUES ENCOUNTERED

The primary concern that developed during the jet grouting process at Echo Dam was grout escaping beyond the effective column diameter due to ground fracturing. Based on visual evidence at the ground surface, this did not occur for all columns, but the problem was very widespread. At numerous times during the jetting process, air and grout could be seen bubbling from the surface locally around the monitor and/or lifting the ground surface in the same localized vicinity. The air and grout bubbling from the surface was generally confined to within a 6-8 foot radius from the monitor. In some cases, air and grout bubbles could be seen exiting from the surface of the jet grout spoil in the waste pit. It is believed that one of the key factors causing this issue, (if not the primary cause), was the lack of adequate annulus space around the monitor to allow for free-flow of the viscous jet grout/soil mix (spoil) to the surface. If the flow of this spoil is stopped, or the annulus becomes plugged, even for a short period, the bottom pressure in the hole can increase significantly. The contractor had elected to pre-bore the jet grout holes by conventional drilling methods using a rock bit and water circulation. This was primarily due to concerns for drilling through the coarse Qc zone with the jet grout monitor alone. Reclamation approved the contractor's proposal to pre-bore the holes. Each hole was filled with either a weak cement-bentonite slurry or biopolymer fluid to keep it open until the jet grouting of the hole was initiated. However, the diameter of the drill rods and the jet grout monitor and drill string were approximately the same. To assure an annulus space, cement grout was jetted at lower pressure during the down drilling of the jet grout monitor through the predrilled hole in order to enlarge the diameter and remove these cuttings. However, it is believed that this process did not always develop a significantly large enough annulus space beyond the predrilled diameter of the hole to assure free flow of spoil to the surface. A narrow annulus can create higher than expected frictional drag on the spoil as it is lifted out of the hole which can increase pressure at the point of jetting, and it can make it easier for a blockage to develop, increasing the subsurface pressure. When a blockage of the annulus space occurs, the pressure in the subsurface can increase immediately. In some cases, the spoil is so viscous, or dense, that the hole can become pressurized when the column of spoils becomes too heavy to be easily lifted out of the hole. This excessive pressurization of the subsurface during the jetting process can lead to fracturing. In most cases, the fractures are filled with grout or soilcrete due to these fluids forcing the fracture or following an air fracture. However, the issue of ground fracture beyond the jet grout column is undesirable and problematic, especially near an embankment and/or structure. In addition, ground fracture can occur almost instantaneously if the jet grout column becomes pressurized for the above reasons, and it may escape the notice of the operator and/or inspector.

During the jet grouting at Echo Dam, many instances of localized ground fracturing occurred as evidenced by the observations cited above. In one instance, during the grouting of column A2, grout was seen surfacing from the upstream, right spillway wall (left in place upstream of the new spillway crest structure) and at the contact joint of the upstream approach apron contact with the wall, indicating ground fracture had occurred (Figure 4).

At the same time, a crack opened up in the temporary protective soil cover on top of the upstream spillway concrete apron, running parallel to the crest structure. Cracking of this soil cover was an indication that the approach apron floor slab had likely heaved due to uplift pressure during the jet grouting of this column which was located very near to the slab. This was proven to be the case after the soil cover was removed. The crack in the approach apron floor slab can be seen in figures 5 and 6. Due to concerns for additional damage, jet grout column A2 was discontinued and grouting of the adjacent column, A1, was deleted from the work.

After completion of the jet grouting, thorough inspections of the new spillway crest structure foundation overlying the jet grouting treatment zone, and the upstream approach apron floor slab, were performed separately. The soil cover on the upstream approach apron was cleaned off to allow inspection of the slab. In preparation for construction of the new spillway crest structure and under drain system, the top 6.5 to 7 feet of soil was removed downstream of spillway station 0-16). The latter soil excavation was directly above the jet grout columns installed beneath the surface and just overlying the foundation zone where most of the ground surface disturbance appeared. This foundation surface was carefully inspected for any signs of ground fracture and/or grout stringers caused by grout infilling of a fracture. No visual evidence of ground fracturing or cracking was noted from the inspection.

Figure 4. View of jet grout that escaped along the inside right existing spillway wall and floor slab contact upstream of spillway station 0-16.

Figure 5. View looking upstream at crack in the upstream spillway approach apron and original concrete key footing (spillway station 0-16)

Figure 6. View of the upstream spillway approach apron showing longitudinal/ diagonal crack extending across the apron (view looking towards the right side of the spillway). Note, soil cover has been removed to expose the slab.

Inspection of the existing upstream approach apron cracking was performed by representatives of Reclamation's Technical Service Center (TSC) on August 21, 2013 and September 9, 2013. It was determined that grouting pressures had caused cracking of the slab and had caused the downstream end of the approach apron slab to be locally raised approximately one-half inch above the top of the shallow cutoff key and that the shallow cutoff key may have been lifted up to 1 inch in a localized zone (length of approximately 5 feet). The TSC representatives devised a repair plan to the damaged sections which consisted of the following:

1. Approximately 33 feet of saw cutting to remove damaged sections

2. Remove 23 cubic yards of concrete (including slab and foundation key in zone of uplift)

3. Install 3-inch PVC Retrofit Waterstop

4. Place 23 cubic yards of cast-in-place concrete

5. Epoxy grout 2 feet of crack to a depth of 2 feet

After completion of the repairs to the damaged concrete, the construction of the new spillway foundation filter and drainage system was completed and concrete placement continued without delays to the schedule.

LESSONS LEARNED AND CONCLUSIONS

Based on the problems encountered with the jet grouting, it is not likely that jet grouting would be recommended for this application on similar projects. It is believed that insufficient annulus space around the jet grout monitor was likely the major contributor to the problem by hindering the free flow of jet grout spoil cuttings to the surface.

In hindsight, it would have been prudent to ensure that the annulus space, created by the pre-boring operation, and the initial travel of the jet grout monitor to the base of the hole, resulted in a larger, more open annulus space. This issue could have been intercepted at the submittal phase in the contract and additional responsibility placed onto the contractor to ensure that sufficient annulus space was created for each jet grout column.

Other than the problems cited in the earlier sections, it appears from the foundation surface inspection that any fracturing that did occur was likely confined to the local area of the spillway crest structure. No signs of grout leakage, other than described above, were seen. The upper 6.5 to 7 feet of the foundation surface was removed after the jet grouting was completed and exposed for visual inspection. The visual inspection of the surface indicated an undamaged foundation with no obvious signs of fracture, other than that which cracked the section of the spillway apron and lifted the concrete key. A robust filter and drainage system was placed over the foundation surface under the new spillway footings and floor slab. Only full operation of the reservoir for an extended period will determine if an increase in seepage is seen in the spillway foundation. Prior to the spillway modification, the spillway underdrain system did not produce flow. Since jet grouting was completed, the reservoir has partially filled, placing approximately 3-4 feet of water onto the apron for a few weeks. During this time the new spillway drains remained dry. If the new underdrain system does produce flow as long-term operation of the full reservoir resumes in the next few years, ground fracture from the jet grouting may be a contributing factor. However, the spillway drainage system is designed to collect and filter any seepage and transport it to a proper discharge point. Therefore, if some seepage flow does occur, it is not anticipated to be an issue. Overall, the jet grout treatment, even with the issues described in this paper, is believed to have successfully reduced the potential for seismic-induced liquefaction and/or settlements of the spillway foundation. This will help to prevent cracking of the embankment and spillway structure due to differential settlement

REFERENCES

Bureau of Reclamation. Echo Dam – Corrective Action Alternative Study. Weber River Project, UT. Upper Colorado Region. Technical Service Center, Denver, CO. March 2010.

Bureau of Reclamation. Geological Data for Modification Decision Analysis for Echo Dam. Weber River Project, Utah. Upper Colorado Region, Provo, UT. Report No. G-551. Volume 1. February 2012.

Bureau of Reclamation. Geological Data for Modification Decision Analysis for Echo Dam. Weber River Project, Utah. Upper Colorado Region, Provo, UT. Report No. G-551. Volume 2. February 2012.

IMPROVEMENTS TO SEEPAGE CUTOFF WALL SPECIFICATIONS CONSIDERING LESSONS LEARNED FROM CONSTRUCTION

Michael Kynett, PE[1]
Khaled Chowdhury, PE, GE[2]
Richard Millet, PE, GE[3]
Mary Perlea, PE[4]
Francke Walberg, PE[5]

ABSTRACT

Cutoff walls are the most common seepage mitigation measure for levee improvement projects within the State of California's Plan of Flood Control and the Federal levees associated with the Sacramento River Project. Cutoff walls are widely accepted due to the reduced real estate and environmental impact, compared to other remedial measures such as berms, setback levees, etc. Within California's Central Valley, the State and Federal levees protect highly urbanized areas, valuable and productive agricultural lands, and contain some of the State's last remaining critical riparian habitat for many threatened and endangered species. Typical seepage cutoff walls utilized include soil-bentonite (SB) and slag cement-cement-bentonite (SCCB) constructed with the slurry support excavated trench method or soil-cement bentonite (SCB) constructed with an in-situ mixing method. Contract values for levee improvement projects in California's Central Valley have ranged from several millions of dollars to hundreds of millions of dollars and have extended up to tens of miles. Cutoff wall specifications that were originally developed for environmental or dam projects require modifications to address the long levee conditions. Also, the advancement of construction technologies requires adjustments in specification. Recently completed construction of several cutoff wall levee improvement projects has shown reoccurring limitations and inconsistencies among the commonly used technical specifications. Existing technical specifications generate potential issues associated with contract management as well as product quality and performance. These limitations were evaluated by practicing engineers involved in different stages of cutoff wall design and construction. This paper describes the specification areas that typically generate design and construction issues and provide recommended improvements to specifications.

---

[1] Levee Safety Program Manager, United States Army Corps of Engineers, Sacramento District, 1325 J Street, 11th Floor, Sacramento, CA 95814-2922, michael.n.kynett@usace.army.mil
[2] Project Manager, AECOM, 2870 Gateway Oaks Suite 150, Sacramento CA 95833, khaled.chowdhury@aecom.com
[3] Vice President and Levee Program Manager, AECOM, 2870 Gateway Oaks Suite 150, Sacramento CA 95833, richard.millet@aecom.com
[4] Ex-Levee Safety Program Manager, United States Army Corps of Engineers, Sacramento District, Levee Safety Section, Sacramento, perlea@msn.com
[5] Senior Technical Advisor, 8300 College Boulevard, Suite 200 Overland Park, KS 66210, francke.walberg@aecom.com

INTRODUCTION

California has over 13,000 miles of levees that protect residential and agricultural lands. Flood control systems on the Sacramento and San Joaquin Rivers, and their tributaries, in the Central Valley region of Northern California, consist of a large system of reservoirs, bypasses, and levees developed as a result of State and Federal legislation in 1911 and 1917, respectively. This system is referred to as the State Plan of Flood Control (SPFC), by the State of California, which consists of approximately 1,600 miles of State-federal "Project" levees. More than 700 miles of additional Delta levees are classified as "non-project" (DWR 2010) (Figure 1). Originally construction of these thousands of miles of levees began as early as the 1850s, by local land owners or entities as the need arose. Many of these early levees were eventually incorporated into the State-Federal flood protection project. The vast majority of levee construction occurred prior to the development of modern design and construction techniques, using clamshell dredges, suction dredges, and/or Fresno scrapers and even horse-drawn equipment (Figures 2 and 3). The majority of these levees were initially constructed to protect land for agricultural development; however, a significant portion of these areas have become or will soon be urbanized. This urbanization has historically occurred in locations right up to the levee footprint. This expedient form of historical levee construction, when combined with adverse riverine geomorphology, urban development, and productive agricultural regions, has led to a levee systems with severe seepage performance issues in a real-estate-constrained setting.

The United States Army Corps of Engineers (USACE), the California Department of Water Resources (DWR), and various other state and local agencies have been implementing improvement projects to structurally improve levee systems and reduce flood risk, since major flood events occurred in 1986 and 1997. These flood events exposed the vulnerability of the Sacramento and San Joaquin River levee systems to underseepage distress. Given the geological conditions, severe landside right of way constraints, and waterside hydrologic and environmental constraints, seepage cutoff walls are advantageous levee improvement measures. The first seepage cutoff walls installed along these systems were designed and constructed by USACE Sacramento District in the early 1990s. Since then, seepage cutoff walls have become widely accepted as a structural levee improvement measure, as they can be installed within an existing project's real estate right of ways, which are typically limited in heavily urbanized or high-value crop agricultural areas.

The common seepage cutoff walls are soil-bentonite (SB), soil-cement-bentonite (SCB), and cement-bentonite (CB). The most commonly used construction methods are slurry supported open trench and in-situ mixing. While the engineering design and construction practices have been in continuous development and evolution since the early 1990s, current specifications have exhibited limitations.

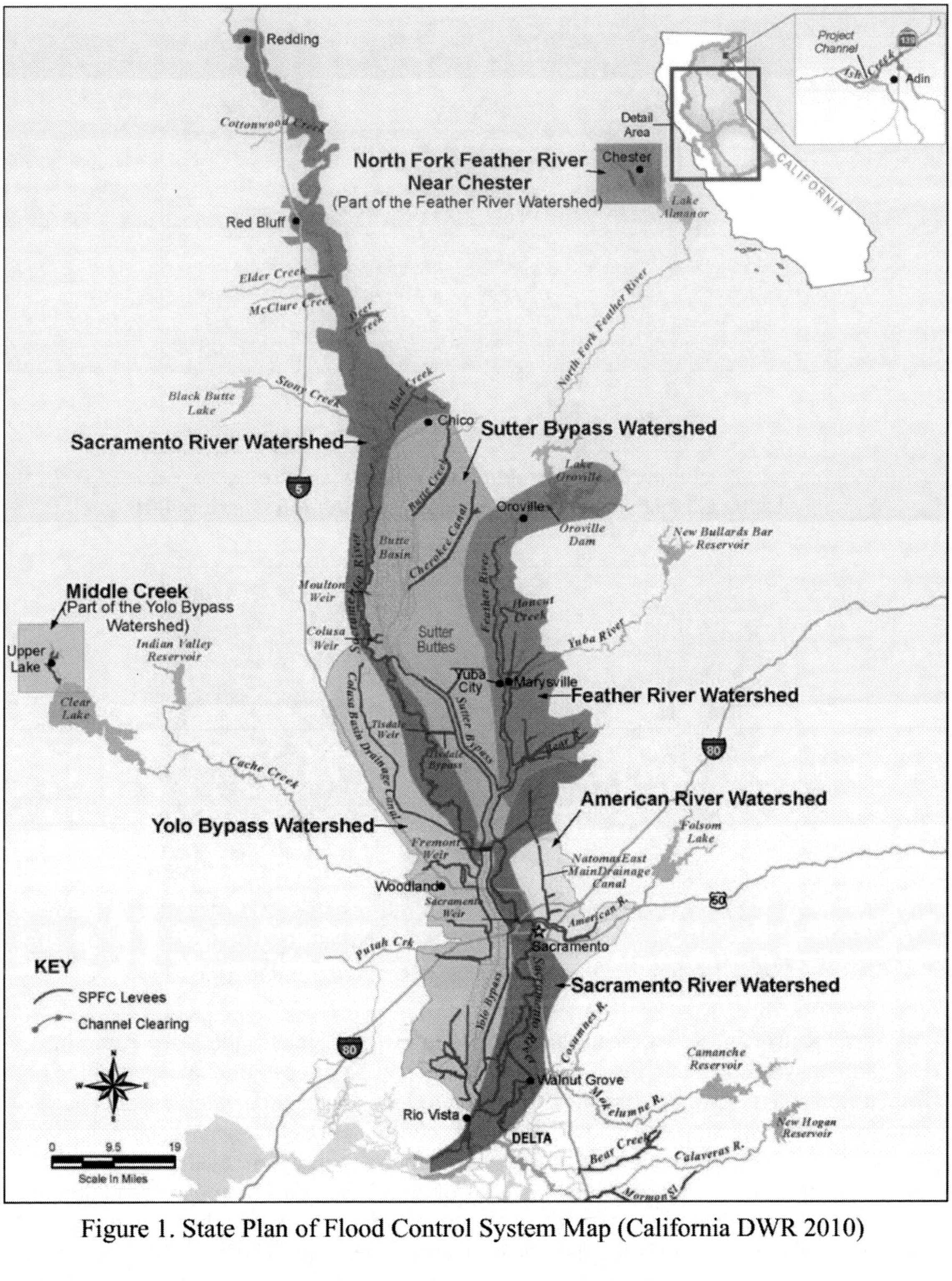

Figure 1. State Plan of Flood Control System Map (California DWR 2010)

Figure 2. Construction of Yolo Bypass Levee
Using Clamshell Dredge (Courtesy Reclamation District 900)

Figure 3. Construction of the Natomas Levee
Using Hydraulic Dredge (Courtesy Reclamation District 1000)

Hurricane Katrina in New Orleans and the Linda Levee Break on the Yuba River in California have shown that the consequences of levee failures could be catastrophic both from a financial and loss of life standpoint. Therefore, issues associated with seepage cutoff wall specifications may pose significant consequences to agencies, contractors, and the public protected by the levees. These specification issues were evaluated by the authors who have been involved in all stages of cutoff wall design and construction. This paper describes critical issues regarding design, construction oversight, construction methods, and considerations for specifications improvements. The paper's topics include: the applicability of the different types of cutoff wall materials and construction methods, quality control testing requirements, backfill gradation requirements, mixing procedures, and settlement criteria for backfill.

While this paper presents issues and recommendations for improved specifications related to design and construction of seepage cutoff walls for California levees, the discussion regarding these issues is still ongoing. These issues may be applicable to the seepage cutoff wall practice extending beyond the Central Valley. In August 2014, a workshop on seepage cutoff wall design issues was held at USACE Sacramento District. A total of 43 participants representing USACE, DWR, Deep Foundations Institute (DFI), and several consulting firms contributed in the discussion. A white paper summarizing the workshop discussions is currently under development. In June 2015, a Seepage Control Work Group meeting (a technical committee within DFI) was held during the Deep Mixing Conference in San Francisco, in which cutoff wall design and construction issues were further discussed with a broad group of 33 designers and contractors. The author's would like to also thank Dr. Stephen Jefferis and Gilbert Tallard for providing written comments to several of the topics presented in the DFI paper (Kynett et. al. 2015). This paper presents the opinions of the authors on various aspects of seepage cutoff wall specifications.

DISCUSSION TOPICS

The issues for seepage cutoff wall specifications are divided into four sub-groups for this paper:

- General area topics that are applicable to all types of walls
- The in-situ deep mixing method (DMM)
- SB slurry walls constructed by the conventional trench excavation method
- CB self-hardening slurry walls constructed by the conventional trench excavation method

The issues identified are first listed and then discussed in detail below.

General Issues

- Discrepancies in cutoff wall acceptance conditions
- Testing requirements of contractor's mix design
- Acceptance of previous project mix designs
- Percent bentonite reporting in mix design
- Bentonite slurry hydration and mixing
- Dry addition of constituents to surface backfill
- Qualifications of the contractor's crew and work plan
- Confining pressure for hydraulic conductivity testing
- Hydraulic Conductivity Testing
- Wet bulk sampling of backfill

- Post construction in-situ testing
- Limitations of backfill sampling for hydraulic conductivity testing

DMM

- Applicability of SB cutoff wall using DMM technique
- Upper limit of fines in slurry backfill
- Hydraulic fracture
- Explorations to confirm key-in depth and obtain samples for mix design

SB Slurry Walls

- Settlement criteria for backfill
- Trench slurry loss and levee degrade
- Trench bentonite slurry mixture

CB Self-Hardening Slurry Walls

- Use of slag cement and mix design for CB walls

SCB Slurry Walls

- Account for untreated zone in SCB wall

DISCUSSIONS AND RECOMMENDED APPROACH

The USACE, the California DWR, the Sacramento Area Flood Control Agency (SAFCA), the West Sacramento Area Flood Control Agency (WSAFCA), the Sutter Buttes Area Flood Control Agency (SBFCA), and several other local agencies have been leading the design and construction of seepage cutoff walls in an effort to improve levee performance in California. The agencies have designed, procured construction contractors, and overseen the construction of these cutoff walls through their own expertise and by enlisting the expertise of a host of local and national design and construction management consultants. The design specifications have evolved over time from the first cutoff walls constructed on the flood control system in the early 1990's by the USACE on the Sacramento River. The current specifications and construction QC/QA represent a continual evolution of the design and construction practice through several decades of design and construction experience. Resolution of the issues presented below represent the author's recommendation in regards to the next steps in that evolution based on lessons learned and recent technology.

General Issues

Discrepancies in Cutoff Wall Acceptance Conditions. In cutoff wall construction, the majority of the end products are not visually or physically tested. Also the effectiveness of the cutoff wall in levees to mitigate seepage potential cannot be tested directly as the wall being constructed. The effectiveness of the cutoff wall can be tested in indirect way at an uncertain future time during floods. Therefore, the procedures used to construct the wall are used as important indicators of the eventual quality of the wall and it is important to develop acceptance criteria that rely on both the procedure and products.

Cutoff wall acceptance criteria are currently split into general and acceptance criteria synopsis paragraphs. The general sub section lists final acceptance as based on several quality assurance (QA) and quality control (QC) testing requirements, or the entirety of the specification section, while the acceptance criteria synopsis sub-section lists final acceptance that is based on wall depth, wall thickness, continuity, and bulk sample permeability; other items may be included based on construction method. This approach presents the requirement of rigid adherence to specific quantifiable testing parameters, both those that measure method metrics and represents end product performance, and qualitative abstract properties (i.e., homogeneity) without upfront consequences for not meeting those metrics, nor the triggering of actions for mitigating those consequences.

These requirements have led to confusion for the contractor (on adherence) to specifications and for agencies (on enforcement) of specifications. Resolution of cutoff wall acceptance disagreements often requires substantial engineering judgment, and occasionally requires variation when implementing specifications. During recent Central Valley projects, contractors have formally or informally interpreted the importance of certain testing requirements over others in the specifications. The specification for cutoff wall construction contains several requirements such as soil backfill gradation and unit weight of the temporary trench slurry and backfill that significantly affect the quality of the final product and that should be used for wall acceptance, often these are viewed by the contractor as intermediate requirements not necessarily related to product acceptance. Construction practices dictate the necessity for increased clarity in specification requirements for items such as acceptance. While acknowledging that the specifications should not overly restrict the contractor from constructing the product, the specifications should be enforceable and not allow for acceptance of a product that does not meet all applicable requirements.

It is the authors' opinion that cutoff wall acceptance criteria should begin at the proposal phase, in prequalifying contractors, and carry on through construction observation, testing, and oversight, leading to acceptance based on the entirety of the work product, not simply test results from a soil sample whose representation of the final product includes several simplifications and assumptions. QA/QC programs should be viewed as the means to assure that, as the wall is being constructed, the materials, methods and procedures used will lead to an acceptable end product. Acceptance should be based on proving a particular construction method, and the contractor will achieve the desired end result in the conditions specific to the project. This can be achieved by proving that a particular construction method would achieve the design's intent. Specifications should require compliance with critical values for items needed for the wall's end performance (Millet et.al. 1990) such as wall permeability, continuity, strength, unit weight, thickness, and depth, while providing operating ranges for intermediate control items such as trench slurry properties.

Specifications should provide requirement thresholds and describe the consequences of not meeting these requirements. The intent of these proposed specification modifications is to clarify the design intent and importance of each paragraph, the results of failing to meet specification requirements, and removing uncertainties and variability while working toward more consistent, uniform, and predictable administration and enforcement of the specifications leading to a desired outcome. A contingency plan should be developed by the contractor as part of the construction plan. This contingency plan should be tailored to the cutoff wall construction method, specification requirements, site specific conditions, and contractor's available means. It will help the responsible agency instruct the contractor to implement a contingency plan, and will save time by avoiding unnecessary contractual complexities and delays. This provides an upfront and agreed upon plan for addressing potentially unacceptable products through other means, such as virtual coring, in-situ water testing, etc.

Testing Requirements of Contractor's Mix Design. A key issue for mix designs of cutoff walls which utilize site soils, either for surface mixed backfill or in-situ soil mixing, is the ratio of the components (cement/slag/bentonite/water) and, in the case of in-situ soil-mix methods, the amount of the components which is to be added per unit volume of in-situ soil. As the ratio of components to in-situ soil is expected to be somewhat variable over the course of the project, most current specifications require a range of mix ratios to be evaluated, by the contractor and prior to commencing work, due to the changes in soil naturally and tested during the mix design process. Current specifications do not require a complete suite of testing to characterize the soils Atterberg limits, gradation, and moisture content). It is well established that the plasticity of fines (materials passing through a #200 sieve) of the native materials excavated and used as backfill (for slurry supported trench walls) or in-situ mixing (for soil-mixed walls) has a large effect on the cutoff wall's final hydraulic conductivity.

Contractors and designers need to have a common understanding as to how much the native materials can be expected to affect the proposed quantity of bentonite in the mix design, based on the subsurface conditions and the cutoff wall installation method. The stratification, quantity, distribution, cementation, moisture content, plasticity, and/or fines content contribute to how effective native materials can be expected to reduce the hydraulic conductivity of the installed cutoff wall without the addition of sufficient quantities of bentonite. For instance, where the deep mix method is being used to install a cutoff wall through a relatively thick aquifer with little fines content (less than 12 percent) and the in-situ materials would not be expected to have a significant impact on the final hydraulic conductivity of the cutoff wall.

Water chemistry can also have adverse effects on the hydration of bentonite and cement. Current specifications require a maximum hardness, which is a combined measure of calcium and magnesium. While the cations affecting bentonite are magnesium, calcium, sodium, and potassium. For cement the important factor with regards to water chemistry is the salinity, which is typically not addressed in the specifications.

To understand the applicability of a mix design for a specific geologic formation, specifications should require characterization of the range of site materials and mix water. Soil types that may be encountered over the cutoff wall depth range and alignment must be evaluated, and the design mix component ratio prepared based on reasonable conservative conditions anticipated. In advance of construction, designers need to identify the likely site sources of water to be used in the mix and perform sufficient testing to identify and disclose to contractors the water chemistry. Reasonable requirements regarding site water should then be included in the specifications.
The designer should then prepare some assortment of trial mix designs in order to include their interpretation of reasonable bounds that the contractor's mix design should fall within based on the site specific soil conditions, groundwater, and source water.

Acceptance of Previous Project Mix Designs. Following issuance of notice to proceed, contractors often propose the use of a mix design and testing from a previous project with no or limited assessment as to their compatibility with site geotechnical and stratigraphic conditions,, in lieu of a new site/project specific testing and mix design. Frequently there is limited or no QA testing data to adequately verify the contractor QC results or to compare the previous project and current project conditions. This leads to uncertainty, and is dependent upon how well the previously-tested subsurface conditions, materials, and water match the current project site. However, site conditions with compatible existing proven designs have the advantage of field testing as opposed to bench scale laboratory testing, and the extensive quantity of production testing results.

In general, use of a mix design from a previous project should be carefully considered prior to acceptance or rejection. As the specifications typically call for a site specific mix design, the Agency/Owner and contractor should have a common and agreed upon understanding of mix design process during procurement, as the mix design process can be costly and time consuming. Project schedules and contract documents should consider the potential for mix designs to take several months. Bid documents should state that, if the contractor plans to use an existing mix design from a previous project, the contractor must state the intent to do so during the proposal phase, such that validity of such request could be verified in advance as part of the evaluation of contractor bids. It is the author's intent that the criteria for the contractor's use of existing mix designs would be stated in the project specific evaluation criteria and accounted for in the unit costs. The contractor should be required to submit a technical report prepared by a certified geotechnical engineer (California) or a Professional Engineer practicing in geotechnical engineering area with justification for use of mix design from a previous project. The rationale should explain why the mix design from previous project would provide satisfactory results for the current project including the most critical portions of the in-situ soil formation. This would also address time constraints imposed upon the construction management team when such a request is made following notice to proceed or during the submittal stage of the project. Contractors are likely to continue proposing use of previous mix designs, so the specifications and should include provisions to adequately address this topic.

Percent Bentonite Reporting in Mix Design. It is common to evaluate the effectiveness of bentonite use in a cutoff wall in terms of percent bentonite by dry weight of backfill (D'Applolonia 1980; Millet et. al, 1992). However, in absence of any guideline or requirement about reporting mix design, contractors usually report the mix design in a bentonite to water ratio. Further, many mix designs are inconsistent in their method of reporting mix quantities. For instance, a bentonite/water ratio might be given but a percent of total weight of cement or admixture might be reported. A ratio of 6 to 11 percent bentonite to water may give the illusion that bentonite content is high; however, these ratios may calculate less than 1 percent bentonite in the dry weight of backfill. Information about the gradation and plasticity of the mix design are also sometimes absent.

To eliminate misconceptions associated with reporting of dry constituents and component slurries, especially bentonite content, specifications should standardize methods for reporting trial mix designs, and include a comprehensive list of items to be reported and the reporting method for the trial mix designs. It is recommended that the specifications require trial mix designs to report the quantity of mix components (water/bentonite/cement/soil/admixture) in terms of percent of total weight, and soil-moisture state prior and after mix

Bentonite Slurry Hydration and Mixing. Current specifications require bentonite to be mixed and adequately hydrated with water to form the bentonite slurry mixture, which must meet certain material properties testing requirements. The current standard of practice is to mix bentonite with water using a high-shear mixer, which introduces the mixture into a pond or tank where a trash pump agitates the bentonite and water. Contractors have typically claimed "adequate" hydration in durations less than the required specification minimum of 8 hours, some specifications have required hydration times of up to 24 hours. Long duration hydration times are not always been enforced and can be difficult to monitor if enforced. Additionally, the process of introducing the bentonite into the water through a high-shear mixer is not always automated or measured but may be performed manually by an operator. The percent of bentonite by weight of water is a key component in a mix design for achieving the specified slurry viscosity and density, which are measures of workability.

The specifications need to include provisions for measured batching, if not automated batching, in order to achieve the required bentonite water mixture and specified slurry properties. While automated batching does not guarantee correct batching, measured (a system of calibrated scales and/or flow meters) should be required by the specifications and enforced by the Agency/Owner. For automatic batching it is important that the mixer scales and/or flow meters be calibrated prior to use. The mixture must then be continuously agitated to help develop the required/specified slurry properties.

For cement bentonite and slag cement, cement bentonite mixes, it should be remembered that bentonite reacts with cement and after a few days there will be no identifiable bentonite. The bentonite appears to act as nucleating sites for the deposition of cement hydrates and to achieve a low permeability it is essential that the bentonite is finely dispersed, i.e. sufficiently hydrated, swelled and dispersed to fine sizes. At constant bentonite addition rates, a very short hydration time will lead to a higher permeability mix. Longer hydration makes better use of the bentonite. While not the entire issue, at least in part, hydration affects that quantity of bentonite the contractor is required to use in order to achieve the same results and meet the expectations of the mix design.

It should be noted that some contractors add an additional (over the target mix) bentonite amount to lessens the hydration time and still achieve the specific slurry viscosity and eventual performance of the wall. As long as the bentonite slurry meets all identified specification requirements (i.e., viscosity, density, filtrate loss), the issues of "adequate" hydration should be considered as met. It is recognized that in this case the contractor may be trading increased bentonite usage versus construction time.

The Agency/Owner needs to evaluate whether and where they are willing to accept risk. By allowing the contractor to select their own hydration time, perhaps as part of a submittal in which the contractor provides the hydration time based on their specific equipment, some economy may be achieved in the process (the contractor chooses how to balance hydration time with bentonite quantity). However, the construction oversight, contractual disputes, and testing to prove whether adequate hydration was achieved may be onerous. This is, in part, a case where specifications have not evolved with construction practice and advancements in technology, leading to a disconnect between designers and contractors. The specifications need to be updated to account for changes in technology. As an alternative, designers could consult with industry experts to provide an updated hydration time based on the cutoff wall material, construction method, and variety of current mixing equipment, and compare that time with the conventional 8-hour hydration time-based method. Some have recommended that the minimum hydration time be reduced to between 2 and 4 hours when using a high shear mixer

Dry Addition of Constituents to Surface Backfill. The existing specifications allow the contractor to add dry constituents (bentonite, slag, or cement) to the backfill at the location of the surface mixing of the backfill, typically adjacent to the trench just prior to placement. This causes backfill to become non homogenous and lacks precise control of constituent percentages as required by the specifications. The specifications should not allow mixing of dry constituents directly into the backfill prior to placement in an open trench. If such mixing is required, a pug mill process should be specified.

Qualifications of the Contractor's Crew and Work Plan. The quality of a cutoff wall construction depends heavily on the quality of knowledge, experience, and workmanship of the contractor's crew. The specifications require use of experienced crew in cutoff wall installation. However, there is typically no requirement for additional crews based on multiple rigs or night shifts. For example, a contractor may win a project showing a single crew qualification for one shift. In reality, the contractor may work in multiple shifts and crews; however, the workmanship of these additional crews may not compare to single shift crew.

The specification should require the contractor to submit key staff qualifications on a shift basis if multiple shifts are likely for the project. For example, if the contractor chooses to use two shifts and two crews, then they should submit qualifications for each crew. Also, the specifications should require the contractor to provide a detailed, step-by-step procedure that covers all aspects of construction, including a shift/crew schedule to clarify the need for additional crew.

Confining Pressure for Hydraulic Conductivity Testing. It has been observed during several past projects that the laboratory testing of cast, wet grab samples, for hydraulic conductivity following ASTM D5084 can generate unreliable or contradictory results between projects or between quality assurance and quality control laboratories. The confining pressures specified for the permeability testing may cause lower or higher permeability to be calculated in the laboratory than what would be reasonably represented in the field. It becomes difficult to then interpret the meaning of the reported hydraulic conductivities.

Current local California practice is based off of criteria developed by Mr. Stephen J. Trautwein as part if the American River Project for the Sacramento District, U.S. Army Corps of Engineers. The effective confining pressure is limited to 10 psi. A "B" value of 0.9 or greater is specified. A gradient of 20 is used as the initial gradient due to the anticipated permeability of samples in accordance with ASTM D 5084. It is desired that the permeate be site water but should be clean water. The wet grab samples should be cast in 3x6 cylinders, allowed to cure at the project site for a period of 4 days in a moisture controlled environment. Some projects have required the samples to be stored submerged in water baths. The sample should be removed from the cylinder using three radial cuts into the plastic cylinder with limited to no reduction in sample size by the laboratory. The tested portion from the cylinder should come from the lower portion of the sample due to air entrapment and layering of the top portion of the sample. The maximum cell pressure should be 100 psi. A multi stage consolidation loading/seating pressure as per the table below should be used (Trautwein 2000).

Hydraulic Conductivity Testing. It should be noted that many different consolidation loading could be used for conducting hydraulic conductive laboratory tests on field QA/QC samples. We recommend the following (Table 1).

Table 1. Recommended Multi Stage Consolidation Loading for ASTM D5084 Testing

Stage	Cell Pressure (psi)	Back Pressure (psi)	Effective Confining Pressure (psi)
1	5	3 2	
2	10	8	2
3	20	15	5
4	30	20	10
5	40	30	10
6	60	50	10

The protocol has been used on many cutoff wall projects and thus will allow the design/engineering team to make reasonable comparative decisions.

The aquifer and blanket contact location can be considered as the critical location for confining pressure for hydraulic conductivity testing. The current requirement of the confining pressure can be considered appropriate for deep seepage cutoff walls. However, when conditions for shallower walls (less than about 30 feet) and conditions for a thin blanket (less than about 10 feet) are considered, confining pressures for hydraulic conductivity testing should be evaluated and specified, if needed.

Wet Bulk Sampling of Backfill. It is common for specifications to require wet bulk sampling of backfill (for slurry supported open trench) or mixed wall (DMM), depending on construction method, for testing. The wet bulk sample can provide a measure of installed product performance, verifying the overall success of the construction method. The specifications typically require a wet bulk sample obtained at three depth intervals, based on either a linear footage of cutoff wall constructed or for every rig shift. Sampling methods range from weighted bailers attached to a line at the ground surface or capped boxes attached to mixing equipment. Depending on the cutoff wall construction and sampling method, wet bulk sampling can either be relatively efficient with limited disruption to the construction process or very inefficient and time consuming.

DMM Sampler

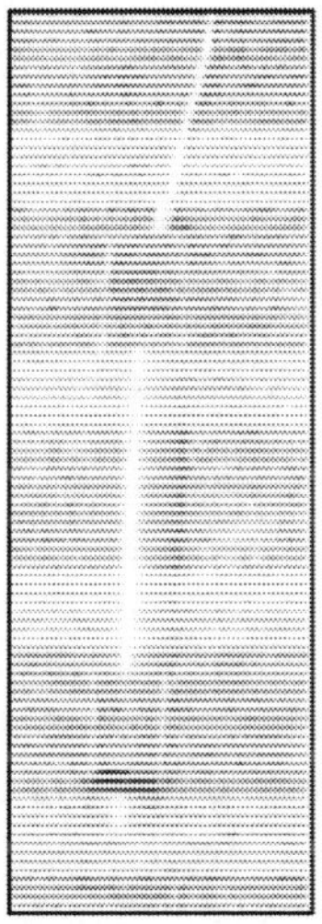

Open Trench Sampler

Figure 4. Comparison of Two Different Wet Grab Sampling Apparatus

The frequency of wet bulk sampling needs to be balanced with project specific considerations, such as variability in subsurface conditions, and disruption to the construction process. Designers should consider the subsurface conditions and anticipated production rate to determine a reasonable frequency of sampling, while contractors should continue to develop samplers that obtain representative samples which can be used frequently enough to meet project-specific needs. Specifying sampling per linear foot of wall constructed should be avoided, as it tends to burden the contractor with too-frequent sampling protocols, and produces sampling intervals unevenly across batches of material. Sampling per placed or batched volume is preferred.

Post-Construction In-Situ Testing. Even though post-construction in-situ testing is a common protocol in dam engineering, recent levee projects have not included post-construction testing. The difficulty in obtaining a representative test specimen for laboratory testing from completed cutoff walls was used as a rationale to eliminate the need for in-situ testing. However, some of the most disputed situations during recent Central Valley projects were resolved through in-situ testing. It should be also noted that the construction cost of the majority of levee projects involving seepage cutoff walls in Central Valley area comparable with mid to large size dam projects.

Given the limitations of sampling and laboratory testing, specifications should specify post-construction in-situ testing. This will encourage the contractor to use appropriate mix design and construction methods that are site-specific. Some of the more recent successful testing methods include in-situ hydraulic conductivity tests, cone penetration test (CPT) tests with pore water dissipation testing, and for CB and SCB cutoff walls physical coring and optical tele-viewer or video inspection (digital cores).

Limitations of Backfill Sampling for Hydraulic Conductivity Testing. The typical backfill gradation requirements are generally:

Sieve Size or Number	Percent Passing by Dry Weight
2-inch	100
No. 4 (4.76 mm/0.187-inch)	40 to 100
No. 40 (0.42 mm/0.0165-inch)	25 to 90
No. 200 (0.074 mm/0.0029-inch)	20 to 60

An observed issue with the current state of practice is identified below.

The backfill sample collection procedure for hydraulic conductivity testing is generally performed in accordance with ASTM D5084 and D2434 (withdrawn by ASTM in 2015), with some exceptions including "The retrieved sample shall be passed through a ¾ -inch sieve prior to cylinder fabrication." The removal of particles greater than ¾-inch is specified in ASTM D2434; this test procedure is limited to materials with less than 10 percent passing the No. 200 sieve. The ASTM D5084 test procedure is intended for materials that have a hydraulic conductivity less than 1×10^{-4} cm/sec, and specifies that the sample diameter shall be at least 6 times greater than the largest particle size. Slurry wall mix design testing should be performed on the coarsest portion of the aquifer; however, with the inclusion of bentonite (and cement), the resulting hydraulic conductivity is usually 1×10^{-6} cm/sec or less. Therefore, the exception clause of removing particles greater than ¾-inch is not consistent with the objectives of the slurry wall mix design. However, by referencing both ASTM test procedures, the Contractor is allowed to modify the gradation of the in-situ materials instead of using a larger size sample specimen. This modification of the in-situ gradation to a finer-particle biased gradation not only allows the Contractor to test unrepresentative conditions, it may also allow the

Contractor to use lower bentonite (and cement) content to meet required hydraulic conductivity.

DMM walls are usually used to mitigate seepage potentials from deep coarse-grained aquifer conditions (consisting of poorly-graded sand, gravel, and cobble). These deeper aquifers may be predominantly coarse-grained with very little or no fines. The screening of ¾-inch particles out of mix design testing sample may result in a completely different sample than in-situ conditions.

The specification gradation requirement also allows up to 40 percent of materials greater than ¾-inch, whereas the gradation for this coarser portion is not tested according to the procedures in ASTM D5084.

The testing on remolded 3-inch diameter by 6-inch length specimens is performed on portions containing particles smaller than ¾-inch. The discrepancy between the gradation requirement and sampling allows use of a gap-graded materials in the coarser than ¾-inch particles. It is generally assumed that the hydraulic conductivity of a composite material is controlled by the finer portions of the composite, when a minimum of 20 percent fines is present. However, if the portion of the gravels greater than ¾-inch is high and a gap graded gradation exist, this assumption may require some verification. As the current specifications do not restrict use of gap graded particles coarser than ¾-inch, and the resulting hydraulic conductivity is not tested, it cannot be verified whether there is a potentially un-conservative condition in the aquifer portions of the cutoff wall.

The composite should be well-graded to ensure fill-in of void spaces by finer particles. Existing sampling and testing procedures should be modified to allow testing of the actual backfill, including the coarser portions of the backfill. This may require use of larger test specimens and different equipment to account for coarser particles in the backfill.

The effect of a coarser portion in backfill on actual hydraulic conductivity should be evaluated based on the different proportions of coarser materials. The backfill gradation requirement should be modified, if needed. The in-situ soil should be used as much as possible as a backfill material. However, without proper testing procedures, there may be disconnect between hydraulic conductivity of the completed wall and hydraulic conductivity of the tests representing only the finer portions of the gradation.

Deep Mixing Method (DMM)

Applicability of SB Cutoff Wall Using DMM Technique. An SB cutoff wall using the conventional construction method differs from an SB cutoff wall using an in-situ mixing method in two major ways. In the conventional method, a filter cake is formed on the sides of the wall, which, depending on the filtrate loss properties, provides a thin layer of low hydraulic conductivity materials. The filter cake is believed to contribute positively to the overall horizontal hydraulic conductivity of the completed wall by reducing the overall hydraulic conductivity of the wall. There is also a requirement of backfill gradation, which includes a minimum of 20 percent fines content. However, in the case of an SB wall using an in-situ mixing technique, there may be little to no filter cake and the hydraulic conductivity relies on injected bentonite slurry mixed with in-situ soil. In such cases, the loss of slurry into the soil formation is an issue in absence of a filter cake. Coarse zones with little fines would seem to be particularly problematic where flow could continue until the bentonite is able to set or gel. These could be isolated zones in the column and could form voids where materials from above are slow to fill in the voids. Thus settlement has potential to continue over time and may not be related to time rate of settlement. Also, with commonly available DMM auger-type equipment, there is minimal vertical mixing in the completed columns. A long term erosion or removal of bentonite is also likely in a coarse-grain layer with low fines content.

Considering the uncertainty and difficulty associated with SB walls using a DMM technique in a deep aquifer with coarse-grained materials, a DMM wall should be limited to use only in in-situ geologic formations that have considerable fines content. A minimum fines content of 20 percent of the in-situ soil can be considered as a criterion for this application, as it will prevent wash out of bentonite in the surrounding coarse-grained formations. This minimum of 20 percent fines also matches with the fines content requirement for backfill when using a conventional trench method. In general, deep seepage cutoff walls using a DMM technique should be allowed only with addition of cementing agents such as Portland or slag cement to avoid the construction- and performance-related concerns.

Upper Limit of Fines in Slurry Backfill. The upper limit of fines in SB and SCB walls is currently 40 to 60 percent. Backfill with 60 percent fines may exhibit an increased amount of long-term settlement. The use of high fines content soil in slurry backfill may be acceptable for environmental applications where long-term settlement may not be a critical issue. However, it may be problematic for levees, where long-term settlement is an issue.

The upper limit of fines content in slurry backfill should be based on the in-situ conditions, wall depth, and effect of bentonite in potential long term settlement. As an alternative, if high fines content in excess of 40 percent is used and long term settlement is anticipated, the thickness of the cutoff wall starter platform should be increased from a minimum of 3 feet to as much as 6 feet and a camber should be added to accommodate potential long term settlement. An option could be to monitor long term settlement and take correction action, if needed.

Hydraulic Fracture. Fluid pressures (grout, water, air) used for deep soil mixing can greatly exceed pressures required to cause ground fracture. Usually the pressure is greatly reduced through the energy used to excavate or mix the soil at the injection point. However, since the deep soil mixing method injects a volume of fluid into the ground, sufficient spoil must be returned to the ground surface to avoid heave or the buildup of pressure that can result in ground fracture. Thus maintaining an unobstructed return is essential. Provided there is no obstruction the bottom pressure should be near the pressure caused by the weight of the column of mixed soil or fluid above. Injection of air has been found to be beneficial. The air is believed to aid in breaking up stiff, dense materials, and provides added lubrication. With air in the mixed soil column the density and weight of the column is reduced and the bottom pressure should be reduced. However, if too much air is injected, or if injected at too high a pressure, ground fracture could result.

Air injection is usually performed under request for information (RFI) from the contractor and the current specifications do not adequately address the issue of hydraulic fracture, the means to assure no obstructions above the injection point, and limitations on allowable fluid pressures and volume of air used.

Specifications for DMM should clearly address the issue of ground fracture since it can be detrimental to levee integrity and the foundation blanket. Monitoring of fluid pressures should be required including direct measurement at the point of injection. Use of air should be judiciously controlled including control of pressure and volume injected. The goal should be the minimum pressure and volume necessary to provide the specified end product. Use of air should not begin in the embankment and fine-grained blanket, or at a depth where potential for damage of foundation and blanket is high. If hydraulic fracture were to occur a site specific forensic engineering investigation would be needed by the agency's engineer, and a recommendation for remedial repairs would be developed. Typically, this has required excavation and reconstruction of the levee embankment. In some project areas, this can prove to be environmentally or cost prohibitive.

The specifications should also require the contractor to survey the levee and nearby structures and monitor during the air injection. This will motivate the contractor to use air pressure in a controlled manner, and would be similar to the current state of practice for blasting and deep excavation.

Explorations to Confirm Key-In Depth and Obtain Samples for Mix Design. For construction methods that install cutoff walls through in-situ mixing of native materials with injected water-bentonite-cement combinations [DMM, cutter soil mixing (CSM), trench remixing and deep wall (TRD)], the method does not produce confirmation of key-in to the semi-impermeable layer. To account for variability in the key-in depth, explorations at a frequent interval can be obtained or the depth into the key-in layer can be increased. The appropriate exploration interval and key-in depth should be based on project-specific subsurface conditions. The contractor is also required to representative or critical samples of in-situ materials for the cutoff wall mix design. Typically, specifications require the contractor to perform explorations at a regular intervals determined by the designer, to be adequate for determining variations in the depth and characterization of the key-in layer, balanced against a reasonable key-in depth. However, the exploration equipment, drill method, and sampling type are selected by the contractor, and the designer has only limited control over these selections. This results in obtaining information that may not meet the quality needs of the designer. In the presence of gravel or loose sand, sample recovery by auger or rotary drilling techniques may be inadequate; other exploration methods and sampling maybe required. These other methods may be more costly or the difficulty of drilling and sampling may not be immediately discernible or adequately conveyed to the contractor.

The specifications should include more detailed requirements regarding the drilling, sampling, and logging standards to establish the cutoff wall key-in depth. This will motivate the contractor to select an appropriate drilling method for obtaining samples for classification and mix design. As an alternative, the explorations could be performed in part by the designer. Regardless, the contractor should be required to obtain representative or critical samples, using the more detail requirements outlined above, for preparing a mix design. The exploration interval required for each project in the past has been recommended based on the site specific subsurface conditions relating to degree of variability in the materials expected to be encountered. This is typically related to the continuity, consistency, composition, and depth of the aquiclude layer, as well as the study area geomorphology. The exploration interval can be expected to range between 100 and 200 feet.

SB Slurry Walls

Settlement Criteria for Backfill. The specification currently provides minimum requirements for a settlement monitoring period and the frequency of settlement plate monitoring. However, it does not define acceptable settlement limits. Also, there is no discussion about the course of action if settlement continues. If settlement continues, a gap between the top of the cutoff wall and the base of the cutoff wall starter/working platform may eventually develop.

The specifications should include acceptance criteria to terminate settlement monitoring. This should clearly state the designer's authority in providing approval for the termination of settlement monitoring and placement of the permanent cap. The criteria for settlement should include a baseline monitoring period, a maximum settlement magnitude, required repair if the maximum values are exceeded, and provisions for extending the monitoring period as determined by the designer. Test sections to establish settlement period may be used as a guide; however, as settlement depends on site conditions and construction procedures, test section derived values should not be used as the only criteria for settlement monitoring.

Trench Slurry Loss and Levee Degrade. During the construction of slurry support trenches through levee embankments, there have been several previous incidents of water-bentonite slurry loss during construction. This led to a standardized practice of mandating a partial degrade of the levee embankment for the slurry supported open trench construction method. This is in part due to uncertainties related to the exact mechanism that caused the slurry loss. Theories about this loss range from hydraulic fracturing to slope instability to existing cracks to animal burrow holes to tree roots.

Figure 5. Trench Slurry Escape during Construction of American River Cutoff Walls

This is essentially a designer consideration. The standard practice of a one-third levee degrade is recommended to continue avoiding unwanted slurry loss and the potential need to degrade and reconstruct a levee embankment at disadvantageous times of the construction season (i.e., nearing flood season) to perform a repair. However, project specific cases may alleviate the need for a levee degrade, such as a substantially wide embankment.

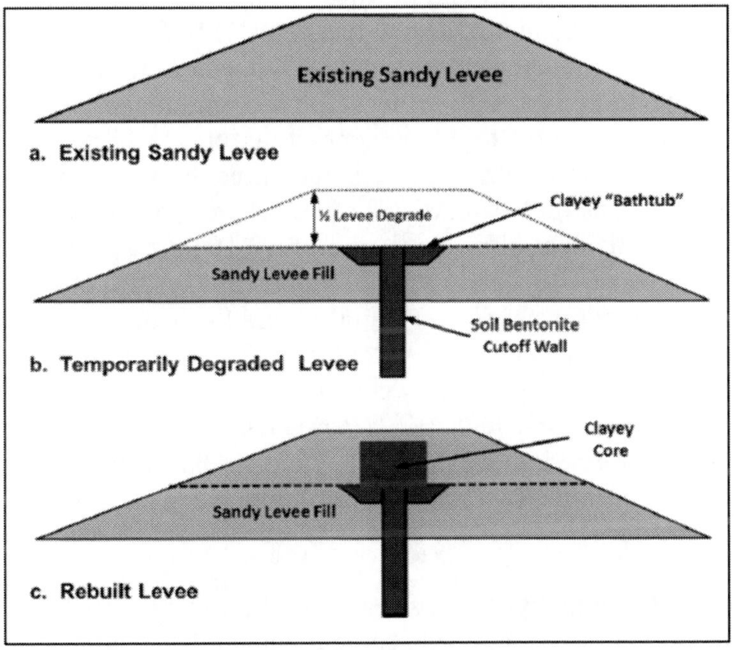

Figure 6. Typical Levee Degrade and Reconstruction (Harder 2013)

Trench Bentonite Slurry Mixture. Viscosity, filtrate loss, and density are the standard measures of quality of trench bentonite-water slurry. They directly or indirectly correlate to the ability of the slurry to stabilize the trench, form a filter cake, and be displaced by the SB backfill. In some specifications a maximum sand content is specified. Not specifying a maximum sand content requirement allows uncontrolled amounts of sand to enter the trench slurry during the excavation process. This affects the backfill gradation when the trench slurry is mixed with soil. This also allows for sand to settle out of the trench slurry and form lenses either along the backfill slope or in the keyway of the trench. This is evidenced by typically increasing sand content with increasing trench depth. The maximum sand content requirement may be difficult for the contractor to meet, requiring additional equipment, increasing the cost of the wall.

Sand content of the trench slurry effects backfill placement (density) and potential buildup on the face of the backfill slope and bottom of the excavated trench. The condition is not advantageous or desired. A maximum density and sand content should be specified and an operational de-sander should be available on-site in case density approaches the upper limits. It will help not to delay the progress and quality of the wall. Additionally, the two types of sounding devices should be used, one with a conventional ball weight and the other with a flat plate weight. An indication of sand accumulation can be shown by significant differences in trench depth measured between the two types of devices.

CB Self-Hardening Slurry Walls

Use of Slag Cement and Mix Design for CB Walls. For several recent levee improvement projects cement bentonite self-hardening slurry cutoff walls have been the preferred seepage improvement around utilities, bridges, or other restrictive site features. The regionally preferred typical mix design has come to include a repetitive combination of approximately 0.12 to 0.14 cementious material to water ratio, less than 1 percent Portland cement, 9 to 10 percent slag cement (ground, granulated blast furnace slag cement), 4 percent bentonite, 0.1 to 0.5 percent admixture, and 85 percent water (all percentages by total weight of slurry). This is in contrast to other cement-bentonite cutoff walls, such as the one constructed at A.V. Watkins Dam by the U.S. Bureau of Reclamation, which utilized a slurry consisting of a 0.18 portland (no slag cement was included in the mix) cement to water ratio, 14.5 percent Portland cement, 4.8 percent bentonite, and 80.6 percent water (Demars 2009), and the Tuttle Creek Dam cement bentonite slurry wall constructed by the U.S. Army Corps of Engineers, which utilized a 0.5 cement to water ratio, 16 percent slag cement, Portland cement, and bentonite, and 0.3 to 1.0 percent admixture (Bellew 2012). While the Tuttle Creek slurry wall was constructed for seismic remediation, the A.V. Watkins Dam cutoff wall was constructed through a relatively short but long dam to control underseepage and both serve as relevant comparisons for levee cement-bentonite seepage cutoff wall mix designs.

To improve hydraulic conductivity and strength of the completed cement-bentonite wall, Portland cement is often mixed with slag cement. Generally, increasing percentage of slag decreases hydraulic conductivity, but increases strength and brittleness. Set-up time for slag cement tends to be longer than for Portland cement, however contractors typically use admixtures (retarders/accelerators) to alter set-up time in the field. The consequences of long set-up times are the potential for increased accumulation of sand on the trench bottom, increased loss of slurry into open coarse sands and gravels, and associated settlement of slurry, displacements, breaking of bonds, and cracking. Specifications do not require that mix designs be optimized to achieve the required hydraulic conductivity while minimizing brittleness.

While the use of admixtures are typically required based on project specific conditions (pumping distance, excavation rate, trench depth, ambient temperature, excavated soils suspension in the slurry, etc.) in the current specifications do not adequately address their effects. It is understood by designers that the admixtures are overwhelmingly proprietary and that their exact chemistry and purpose may not be disclosed by contractors; however, the trial mix designs need to show that a credible range of admixture dosages results in acceptable slurry properties, or what dosage would be required to result in unacceptable properties.

Also, the focus on hydraulic conductivity requirement without a minimum strength requirement sometimes result in a mix design with excessive water and only a fraction of cement and bentonite content to meet the project criteria, which may have long term negative implications also.

The specifications should establish additional performance or reporting requirements of the mix design based on project specific conditions. This should include the set time of the mix, minimum strength prior to strain at failure.

The specifications need to require a monitoring program to assess slurry loss during installation, immediate settlement/consolidation and slurry loss following each rig shift, and the amount of bleed water. The intent of the monitoring is for the designer and contractor to verify the quality of the constructed product and also evaluate potential anomalies in the subsurface conditions that could affect the performance of the final product.

SCB Slurry Walls

Account for Untreated Zone in SCB Wall. An SCB wall using the conventional, slurry-supported open trench construction method with gravity slide backfill is typically used where increased wall strength is required. An SCB cutoff wall installed using this conventional method has been proven to contain a defective zone, also known as an "untreated" zone, in the bottom portion of the wall. This untreated zone is formed due to the set up of the SCB backfill material and the subsequent overtopping of trench bottom sediment and slough. This untreated zone may reduce the effectiveness of the cutoff wall and may also cause long term settlement.

The contractor should establish the thickness of the untreated zone and account for this zone considering the thickness and types of soil layers. The wall may need to be extended 3 to 5 feet depending on the depth of wall to account for the untreated zone. If the tie-in layer is thin and the untreated zone extends below this thin layer to a pervious layer, the wall should be extended further (i.e., by approximately 2 to 3 percent). The pay line item should not include quantities for the defective zone or a different rate should be established for this untreated zone.

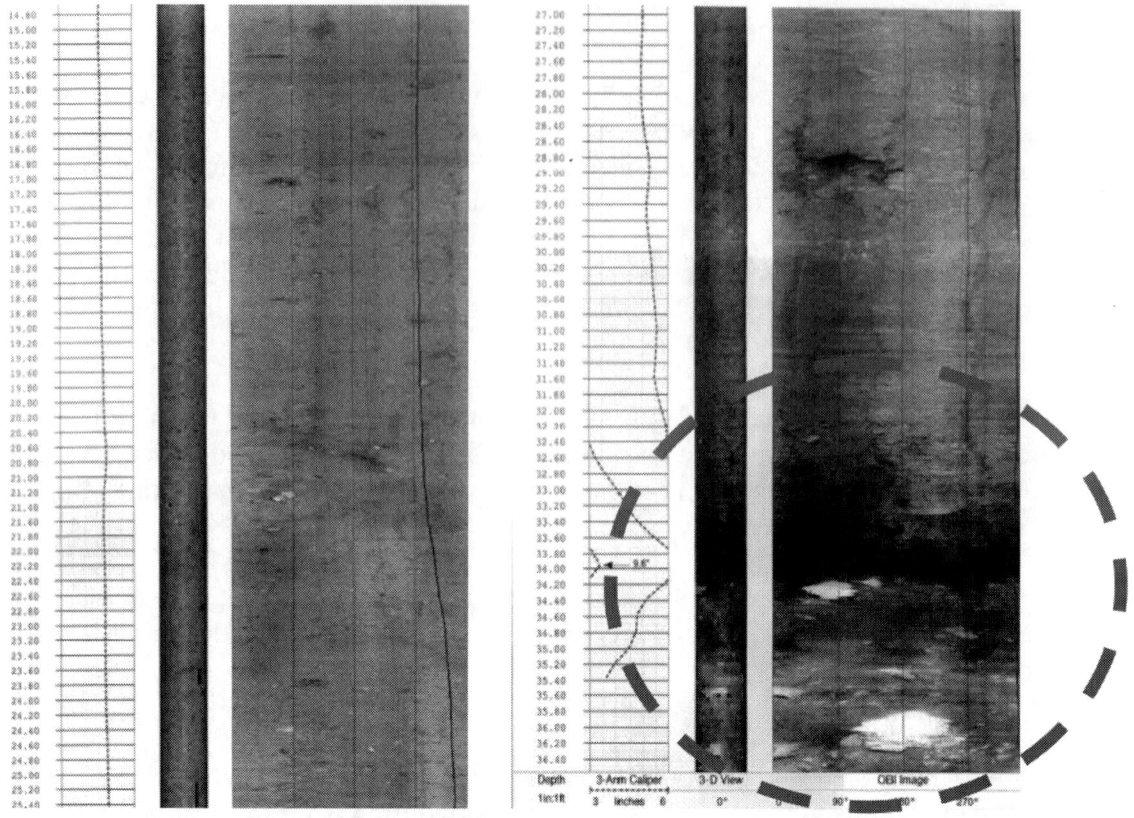

Figure 7. Borehole Optical Televiewer Log of Soil-Cement-Bentonite Cutoff Wall Showing Untreated Zone at Base Depth of Wall

CONCLUSION

Seepage cutoff walls are a significant levee rehabilitation measure used for reducing flood risk in the Central Valley of Northern California, but also represent an investment of hundreds of millions of dollars. While seepage cutoff wall design and construction practices in California have developed over time, recent projects have highlighted several significant issues. This paper discussed several critical issues, and recommended methods of addressing these issues regarding design, construction oversight, and construction of seepage cutoff walls. These issues show that designers need to develop relevant and consistently enforceable specifications, while contractors need to provide adequate proof of their methods to provide acceptable cutoff walls that will perform as intended during future flood events. The discussions in this paper may be beneficial to stakeholders in similar projects, such that the long-term performance of a seepage cutoff wall is not affected by ambiguity or the limitations of the specifications.

REFERENCES

Bellew, G., Koirala, A., Dillon, J., and Mathews, D. 2012 Tuttle Creek Dam Seismic Remediation with High Strength CB Slurry Walls. Grouting and Deep Mixing 2012: pp. 291-300.

D'Appolonia, D.J., 1980. Soil-Bentonite Slurry Trench Cutoffs. Journal of Geotech Eng. Div. ASCE. Vol. 106, No. GT4, April 399-417.

Demars, B., Pledger, C., Barrett, C., 2009. A.V. Watkins Dam Modification: Cement-Bentonite Slurry Cutoff Wall. USSD. 29th Annual USSD Conference. April.

Department of Water Resources (DWR). 2010. State Plan of Flood Control Descriptive Document for Central Valley Flood Management Planning Program.

Harder Jr., L., Weber, J., Williams, T., Chowdhury, K., 2013. Effect of Temporary Levee Degrade on the Slope Stability of Levees Following Construction of Soil Bentonite Cutoff Walls, A memo to USACE and DWR.

Jefferis, S. 2015. Email correspondence with attachments regarding Minutes of meeting of 25 August 2014 and Paper by Kynett, M., Chowdhury, K., Millet, R. Perlea, M. Walberg, W. (2015) Improvements to commonly used cutoff wall specifications..

Kynett, M., Chowdhury, K., Millet, R., Perlea, M., and Walberg, F.. Improvements to Commonly Used Cutoff Wall Specifications, 2015. Deep Mixing Conference, San Francisco, CA.

Millet, R.A., Perez, J.Y., and Davidson, R.R., 1992. USA Practice Slurry Wall Specifications 10 Years Later. ASTM STP 1129 (Slurry walls: design, construction, and quality control by David B. Paul, Richard R. Davidson, and Nicholas J. Cavalli, editors)

Trautwein, S., 2000. Evaluation of Laboratory Hydraulic Conductivity Testing Being Performed for the American River Levee Project, Contract 1, In Sacramento, CA.

LEVEE ASSESSMENT TOOL: USING LIMITED DATA TO EVALUATE LEVEES AGAINST ESTABLISHED CRITERIA

A. Wohletz[1]
S. Wiesner[2]

ABSTRACT

The California Department of Water Resources (DWR) recently conducted systematic geotechnical evaluations that were unprecedented in scope and extent, covering over 1,900 miles of levees spanning the length of the Sacramento and San Joaquin River watersheds. The Levee Evaluation Program (LEP) faced considerable challenges due to a lack of high-quality or site-specific information for approximately 1,500 miles of the levees evaluated in non-urban areas. To address these challenges, the Levee Assessment Tool (LAT) was created. This tool standardizes, integrates, and ranks readily available datasets (e.g., published soil survey and geologic data, geomorphology, past levee performance documentation, and visual levee inspections) assessed during the initial phase of the LEP (Phase 1). The outputs of the LAT are Weighted Hazard Indicator Scores (WHIS) for various levee failure modes (slope stability, underseepage, and through seepage). These outputs were used to indicate if individual levee segments met accepted and standardized levee criteria. LAT results were also compared with exploration and modeling data collected during the final phase of the evaluations (Phase 2). Our evaluation indicates that LAT results (Phase 1) are largely consistent with the results of the LEP Phase 2 exploration and modeling data. Consequently, we find that the LAT provides a valuable initial-approximation of levee geotechnical conditions that is beneficial to large, systematic assessments when site-specific data (soil borings, laboratory test data, and/or geophysical survey data) are unknown.

INTRODUCTION

Over 1,500 miles of non-urban levees in the Sacramento and San Joaquin River basins were recently evaluated under the California Department of Water Resources (DWR) Levee Evaluation Program (LEP). The spatial scope of these evaluations was unprecedented, as were the resulting data collection efforts. Data collection efforts were especially crucial for many of the levees evaluated under Non-Urban Levees Evaluation (NULE), where readily available data were often limited. To aid data collection efforts, historical consultants collected and digitized levee-related data for the DWR LEP. However, fundamental data regarding levee performance, construction, and composition were still limited for many NULE levees.

[1] Staff Geologist, Kleinfelder, Stockton, CA 95206, awohletz@kleinfelder.com
[2] Project Geotechnical Engineer, PE, GE, Kleinfelder, Stockton, CA 95206, swiesner@kleinfelder.com

NULE evaluations were conducted in two phases:

1. Phase 1: Geotechnical Assessment Report (GAR) (Kleinfelder, 2011; URS, 2011)
2. Phase 2: Geotechnical Overview Report (GOR) (Kleinfelder, 2015; URS, 2011)

Phase 1 assessments included desktop studies of readily available data accomplished with the Levee Assessment Tool, which standardizes, integrates, and ranks readily available data. Phase 2 included geotechnical drilling, laboratory analyses of soil samples, and seepage and slope stability modeling. In this study, we compare Phase 1 LAT results against Phase 2 modeling results. Our comparison is focused on levees in the San Joaquin River basin (Figure 1), as we are most familiar with these levees, having conducted many of the San Joaquin River basin Phase 1 and Phase 2 analyses. Our findings indicate the LAT is an effective tool for large, systematic levee evaluations where readily available levee performance, construction, and composition data may be limited.

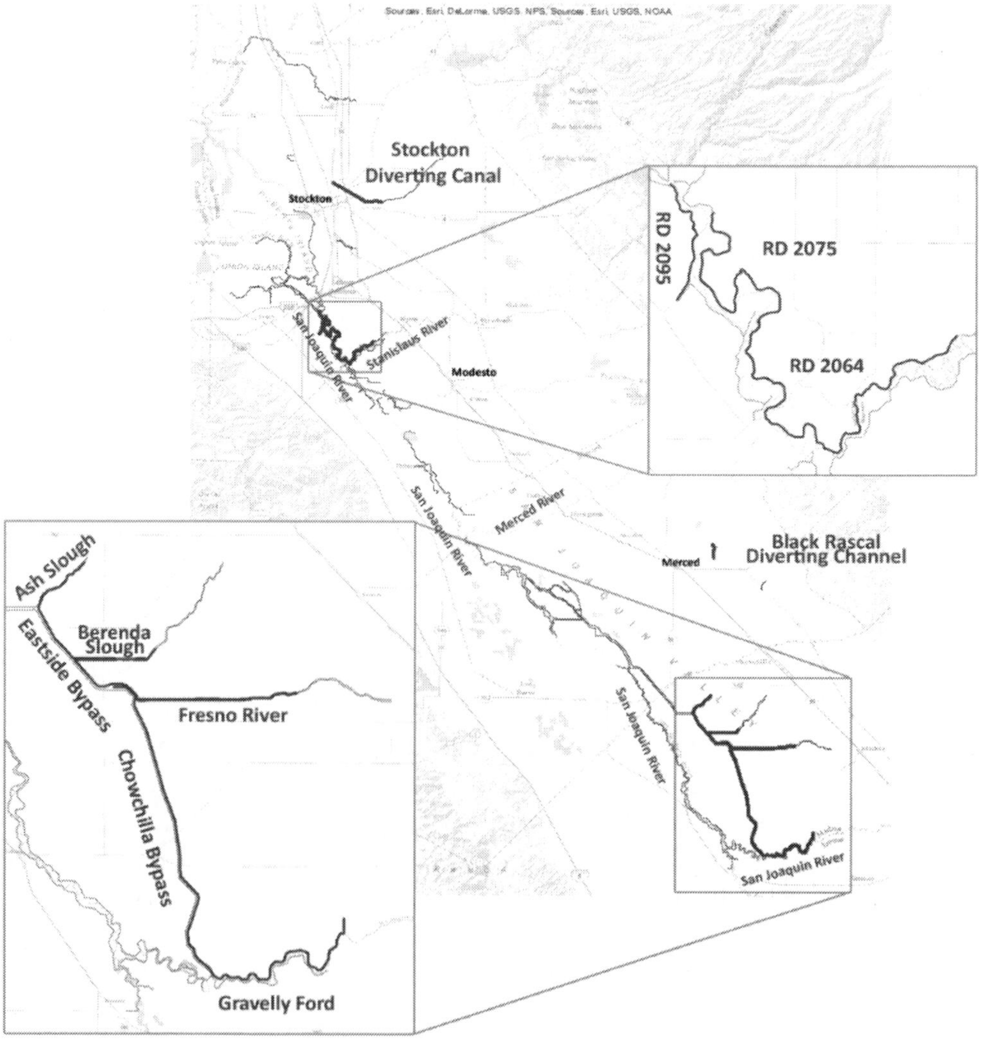

Figure 1. San Joaquin River basin levees in this comparison.

PHASE 1: LEVEE ASSESSMENT TOOL (LAT)

The LAT was developed to standardize, integrate, and rank available levee data. The LAT calculates a Weighted Hazards Indicators Score (WHIS; WHI scores for plural) for three failure modes: Underseepage, Through seepage, and Stability. WHI scores are compared with known performance data in order to categorize each failure mode.

LAT Data

Data entered into the LAT include:

1. Performance Data: Ratings for Underseepage, Landside slope stability, Through seepage, Erosion, Overtopping, and Breaches are entered based on historical performance documentation, DWR inspection logs, DWR flood damage inspection logs, site reconnaissance notes, maintenance personnel interviews, etc.
2. Hazard Indicators:
    a. Ratings for Through seepage and stability composition are entered based on soil condition from boring logs (if available), Natural Resources Conservation Service (NRCS) maps, site reconnaissance/inspection notes, etc.
    b. Geology: Ratings for Underseepage susceptibility are entered based on interpretation of geologic maps. Ratings for dispersive soils are entered based on NRCS Sodium Absorption Ratio (SAR) maps. Piping potential ratings are entered based on NRCS Piping Potential maps. Soft soil ratings are entered based on readily available surface data and historical maps.
    c. Other Indicators: Animal persistency ratings are entered based on DWR animal persistence rankings (if available), site reconnaissance notes, historical documents, and levee inspection data. Ditches and borrow pits noted near the levee toe are flagged based on site reconnaissance/inspection notes and topography data. Intact waterside blankets are flagged if clearly indicated by subsurface data. Performance issues caused by penetrations are flagged if indicated by available data, as are encroachments that may impact levee integrity. The number of penetrations below the Assessment Water Surface Elevation (AWSE) is entered based on levee inspection logs and site reconnaissance notes, and the Levee Maintenance Agency maintenance rating is entered based on the most recent Maintenance Deficiency Summary Report. Where Maintenance Deficiency Summary Reports are not available, engineering judgment is used.

   d. Geometry Data: Levee crest elevation, landside toe elevation, crest width, and AWSE[3] are entered based on topography data. From the entered data, the LAT calculates freeboard, levee height, levee base width, hydraulic head, and creep ratio.
   e. Anomalies: Anomalous conditions are flagged based on review of available data.
3. Mitigation and Past Breaches: Mitigated performance issues and past breaches are flagged based on review of performance data.

LAT Results

Failure modes are categorized based on the combination of WHI scores and past performance. A failure mode can be categorized as:

1. Hazard Level A: When water reaches the assessment water surface elevation (WSE), there is a low likelihood of either levee failure or the need to flood-fight to prevent levee failure.
2. Hazard Level B: When water reaches the assessment WSE, there is a moderate likelihood of either levee failure or the need to flood-fight to prevent levee failure.
3. Hazard Level C: When water reaches the assessment WSE, there is a high likelihood of either levee failure or the need to flood-fight to prevent levee failure.
4. Lacking Sufficient Data (Category LD): The segment is currently lacking sufficient data about past performance or hazard indicators to assign a hazard level, or there is poor correlation between past performance and hazard indicator scores.

The overall levee categorization is based on the worst individual failure categorization.

PHASE 2: SEEPAGE AND SLOPE STABILITY MODELING

Phase 2 evaluations were based on seepage and slope stability steady-state modeling supported by site-specific subsurface explorations and laboratory analyses of soil samples. Computer modeling was conducted with GeoStudio Seep/W and Slope/W software.

Modeling Data

Models were built with site-specific subsurface exploration data acquired as part of Phase 2. Subsurface data included borings drilled approximately every 5,000 feet and Cone

[3] For NULE evaluations, the AWSE was generally the Design WSE. Where the DWSE was unknown, it was assumed the levee was constructed to provide 3 feet of freeboard above the DWSE, so the AWSE was assumed to be 3 feet below crest elevation.

Penetration Tests (CPTs) advanced approximately every 1,000 feet. Laboratory tests of soil samples were conducted to support model parameter selection. Tests typically included soil classification, index, consolidation, and strength tests.

Model parameters included:

1. Saturated Hydraulic Conductivity: Estimated primarily from sieve and hydrometer data.
2. Anisotropy Ratio: Estimated from deposition or construction data.
3. Soil Unit Weight: Estimated from laboratory test data.
4. Internal Friction Angle: Estimated from relative density and Soil Unit Weight.
5. Cohesion: Estimated from program guidelines based on material type.

Sensitivity analyses were conducted and modeling results were compared with past performance (when available)Engineering judgment was then used to select final model parameters based on all available data.

Modeling Results

Modeling results were evaluated against underseepage, stability, and through seepage criteria. Levees were categorized as either meeting or not meeting criteria. For the purpose of the NULE evaluation, criteria included:

1. Underseepage:
   Blanket Condition: Average vertical exit gradients at the landside levee toe below 0.5 met criteria; gradients equaling or exceeding 0.5 did not meet criteria. Average vertical exit gradients below 0.8 at a point 150 feet landside of the toe met criteria; gradients equaling or exceeding 0.8 did not meet criteria. Between the landside toe and 150 feet beyond, gradient criteria were interpolated.
   Non-Blanket Condition: In non-blanket conditions, multiple factors were qualitatively assessed. Results were typically based on creep ratio (levee base width divided by hydraulic head) and foundation material. Creep ratio guidance was based on levee foundation. Creep ratios meeting criteria ranged from greater than 18 for very fine sand to silt foundations and greater than 9 for fine to medium gravel.
2. Stability: Factor of safety (FOS) equaling or exceeding 1.4 met criteria; FOS below 1.4 did not meet criteria.
3. Through Seepage: Multiple factors were qualitatively assessed. Results were based mostly on the height above the landside toe at which, and embankment material through which, the phreatic surface exited the landside slope. Generally speaking, a phreatic surface exiting the landside slope through potentially pipe-able material did not meet criteria.

Modeling results were evaluatedagainst past performance. Initial modeling results were generally consistent with past performance. In a few instances, modeling results did not indicate problems that past performance data clearly indicated (e.g., an exit gradient well below 0.5 where past performance indicated boils). In these cases, additional sensitivity analyses were conducted and parameters were adjusted accordingly.

SELECTING AND PREPARING DATA

Because Phase 1 LAT and Phase 2 modeling results are so different, direct comparisons are difficult. Stability comparisons are difficult, as the LAT specifically defines the stability failure mode as not caused by through seepage while Phase 2 stability modeling is heavily impacted by the AWSE and resulting phreatic surface through and pore water pressures within the levee embankment. Moreover, Phase 1 LAT and Phase 2 modeling analyses were conducted at different intervals and locations, and sometimes with different AWSEs.

Selecting Analyses for Comparisons

LAT-based assessments were conducted on a per-segment basis. Levee segments are generally consistent with maintenance agency levee units. Modeling based analyses, however, were conducted on a per-reach basis. For modeling analyses, segments were generally broken into multiple reaches based on stratigraphy, geometry, performance, etc. Therefore, LAT- and modeling-based results are not directly spatially comparable. For each LAT, segment-based assessment, there are generally multiple modeling, reach-based analyses.

LAT-based assessments were conducted at the most vulnerable, non-anomalous segment location indicated by limited data readily available at the time. Modeling analyses were conducted at locations that best represented the most vulnerable, non-anomalous reach conditions, based on much more extensive Phase 2 subsurface explorations and laboratory analyses. Therefore, depending on reach selection, LAT- and modeling-based analyses may differ considerably based on assessment intent and available data. Modeling analyses for reaches representing better-than-average conditions, for example, should indicate fewer adverse conditions than the segment-based LAT analysis.

LAT- and modeling-based analyses locations can be selected for comparison by selecting the modeling analysis location nearest the LAT analysis location, selecting the modeling analysis results that indicate the most adverse conditions, or averaging the per-reach modeling results into a single per-segment result. For this comparison, we selected both the nearest modeling analysis location and the location with the most adverse conditions. While the nearest modeling analysis location should usually represent the conditions assessed by the LAT, reaches were selected based on typical conditions at a relatively low resolution (typically greater than 0.5 to 1 mile); therefore, it is possible for a reach analysis location to represent conditions different than those for which the LAT location was selected. For this reason, modeling results indicating the most adverse conditions were also selected for comparison.

Special Conditions

In a few isolated cases, the nearest modeling analysis condition was not selected for comparison due to specific, anomalous conditions. For example, one modeling location was not selected despite the fact that it was nearest the LAT analysis location because model results differed from past performance. Despite numerous sensitivity analyses, the model results were unable to be reconciled with performance history, so analyses results were therefore largely based on documented past performance, and the discrepancy between model results and past performance was attributed to inadequate geotechnical data.

Additionally, the LAT categorization was altered for the Black Rascal Diverting Canal. In this case, the LAT categorization was an LD. This categorization was based on uncertainty in the WSE in a landside canal. With water in the canal, the assessment team and reviewers agreed that the categorization would be A, but there was no official documentation indicating the canal would be filled with water during high-water events, so the categorization was changed to LD. When modeled for Phase 2, however, it was assumed that the canal would be filled with water based on additional interviews with maintenance personnel. Therefore, the LAT categorization was updated to be based on the same canal WSE assumptions used in the Phase 2 model.

Correcting for High Segment-Density

The LAT was developed as a screening tool for large scale assessments where data may be limited. To compare LAT results with modeling results over a large-scale, regional perspective, various areas should generally be equally represented by segments. If, for example, one river is represented by four times as many segments as another river of equal length, the LAT's performance is disproportionately reliant on the overly-represented river. This disproportional segment representation could impact the way the LAT's performance is interpreted if higher segment-density is coincident with regional trends, creating a regional bias.

San Joaquin River Basin regions are typically equally represented by segments. However, due to an anomalous maintenance agency boundary, three tributaries (Ash Slough, Berenda Slough, and Fresno River) are split into twice as many segments as they otherwise would be. Moreover, both banks of Berneda Slough and Fresno River were analyzed, while in all other South NULE analysis areas, only one bank was analyzed. This high segment-density results in these 3 tributaries containing a total of 8 segments, while the rest of the South NULE study area contains just 10—nearly half the segments occur in a relatively small area. Each is largely a straight, man-made channel excavated through distal fan deposits, resulting in extremely complex stratigraphy and increased subsurface variability. To correct for the regional bias caused by the high segment-density in this area, Phase 1 and Phase 2 results for these tributaries were averaged such that each tributary is represented by one analysis.

ASWE Challenges

For the southern portion of the South NULE study area, Design WSE (DWSE) data were not available. Analyses were therefore conducted with an assumed WSE 3 feet below the top of the levee (btol). For the Phase 1 LAT assessments, the 3-foot btol AWSE was discretely calculated from levee crest elevation data at cross sections generated every 100 feet. LAT assessments were conducted where these cross section data were available, so the AWSE at the LAT assessment was usually exactly 3 feet btol. For the Phase 2 modeling analyses, however, the 3-foot btol was manually and conservatively "smoothed" based on engineering judgment to create a more realistic WSE. This usually resulted in a Phase 2 AWSE that was slightly higher than the Phase 1 AWSE; however, the "smoothing" process typically removed the highest Phase 1 AWSE areas that were caused by localized, anomalous variations in levee crest elevation.

For Project levees in the AWSE-impacted area, the difference between Phase 1 and Phase 2 AWSE is marginal. For non-Project levees, however, the difference is often substantial, as non-Project levees tend to be shorter and more variable. Because the AWSE impact is substantial, AWSE-impacted non-Project levees were removed from this comparison. This results in the removal of 2 non-Project Fresno River levees and 2 non-Project San Joaquin River levees.

COMPARING PHASE 1 AND 2 UNDERSEEPAGE RESULTS

Phase 1 LAT WHIS vs Phase 2 Model Average Vertical Exit Gradient

To compare LAT results with model results, LAT underseepage WHI scores were compared with average vertical exit gradients from Phase 2 models (Figure 2 below).

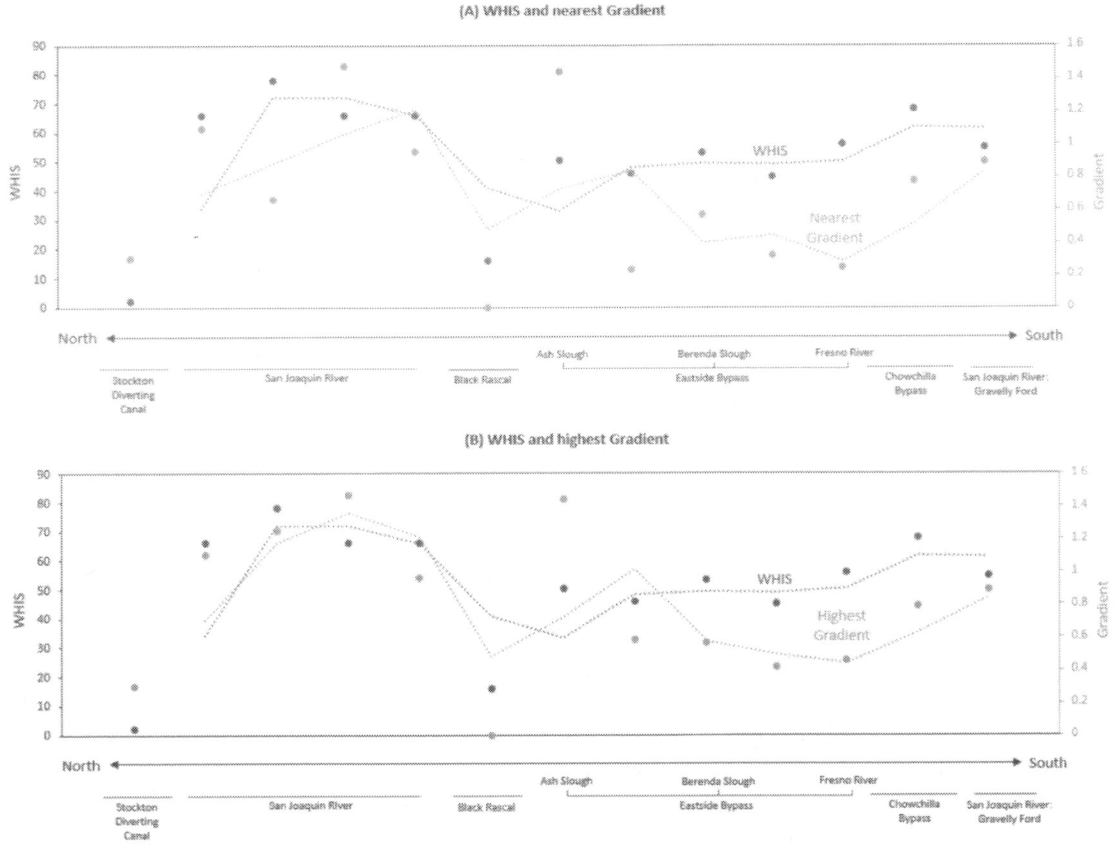

Figure 2. Underseepage (A) WHIS and nearest gradient (Y axis) per segment (X axis) and (B) WHIS and highest gradient (Y axis) per segment (X axis). Segments are arranged along the X axis from north (left) to south (right). Running average curves are added to better depict regional trends in WHIS and gradient.

Generally, the running average curves in Figure 2 (A) and (B) appear similar. WHIS and gradients are relatively low in the north along the Stockton Diverting Canal, high in the north along the San Joaquin River, lower toward the central portion of the graph along Black Rascal Diversion Canal, Ash Slough, Berenda Slough, Fresno River, and Eastside Bypass, and higher to the south along the Chowchilla Bypass and Gravelly Ford portion of the San Joaquin River. These results are generally consistent with levee performance. Numerous boils, breaches, and adverse through seepage conditions have been noted in the north along the San Joaquin River and in the south along the Chowchilla Bypass and Gravelly Ford portion of the San Joaquin River. Other portions of the study area have experienced comparatively few adverse conditions.

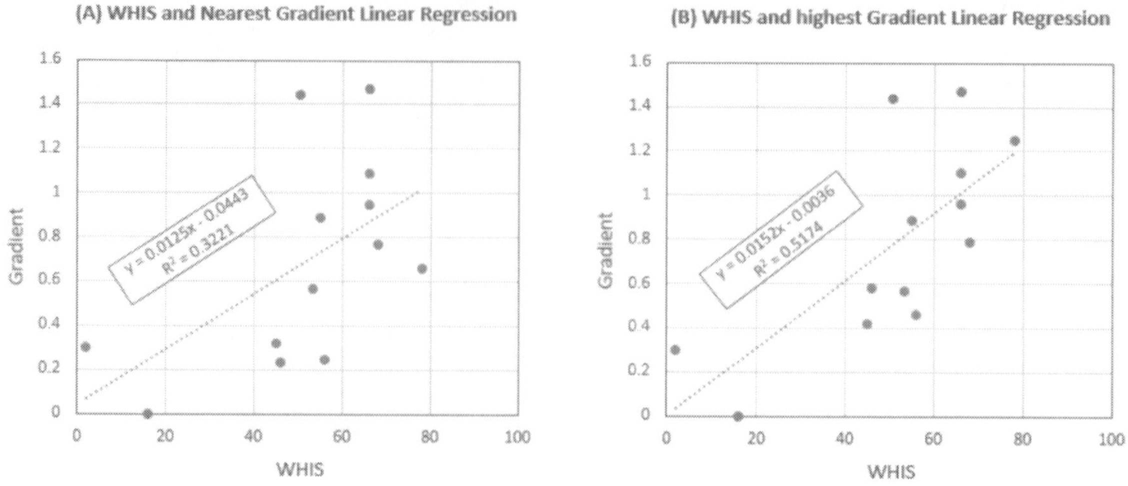

Figure 3. Linear regressions relating (A) nearest gradient with WHIS and (B) highest gradient with WHIS.

Figure 3 depicts linear regressions relating gradient with WHIS. While considerable variation is apparent, a correlation is clear. The R^2 for the nearest-gradient regression is about 0.32, which indicates that almost third of the variation in gradient can be mathematically calculated from WHIS, while the highest-gradient R^2 is 0.52, which indicates that just over half the variation in gradient can by mathematically calculated from WHIS. More importantly, no WHI scores below 45 are associated with gradients that did not meet criteria while no WHI scores above 56 are associated with gradients that met criteria. This suggests the LAT can function as an effective screening tool for large-scale assessments.

Phase 1 LAT Categorizations vs Phase 2 Underseepage Results

To better evaluate the effectiveness of the LAT as a screening tool, Phase 1 LAT underseepage categorizations were compared with evaluated Phase 2 modeling results. Phase 2 Modeling results were evaluated against a gradient criterion of 0.5. Generally, empirical data indicate boils are unlikely at gradients as low as 0.5, though this depends somewhat on blanket material (USACE, 2005). Therefore, Phase 2 modeling results that meet criteria (boils unlikely) are most comparable to Phase 1 LAT categorization A (low likelihood of levee failure or need for flood fighting). Phase 1 LAT categorizations B and C (moderate and high likelihood) are most comparable with Phase 2 modeling results that do not meet criteria. Phase 1 LAT categorization LD does not clearly compare with Phase 2 modeling results; however; there were no LD categorizations for this comparison. This comparison is depicted in Table 1 (below).

Table 1. Underseepage: Phase 1 LAT Categorizations and Phase 2 modeling results. Consistent results are shaded in blue.

	Area	Segment	Nearest Phase 2 Modeling Results		Most Adverse Condition Phase 2 Modeling Results	
			Phase 1 LAT Categorization	Phase 2 Modeling	Phase 1 LAT Categorization	Phase 2 Modeling
North	Diverting Canal	332 & 238	A	Meets Criteria	A	Meets Criteria
	RD 2095	258	C	Does not Meet Criteria	C	Does not Meet Criteria
	RD 2075	196	C	Does not Meet Criteria	C	Does not Meet Criteria
	RD 2064	195	C	Does not Meet Criteria	C	Does not Meet Criteria
	RD 2064	257	C	Does not Meet Criteria	C	Does not Meet Criteria
	Black Rascal	73	A	Meets Criteria	A	Meets Criteria
	Ash Slough	11 & 358	B	Does not Meet Criteria	B	Does not Meet Criteria
	Eastside Bypass	184	B	Meets Criteria	B	Does not Meet Criteria
	Berenda Slough	359 & 12 & 360 & 184	B	Does not Meet Criteria	B	Does not Meet Criteria
	Eastside Bypass	218	B	Does not Meet Criteria	B	Does not Meet Criteria
	Fresno River	27 & 28	B	Meets Criteria	B	Meets Criteria
	Chowchilla Bypass	261	C	Does not Meet Criteria	C	Does not Meet Criteria
South	Gravelly Ford	219	C	Does not Meet Criteria	C	Does not Meet Criteria

When compared with the nearest Phase 2 modeling results, 11 (85%) of the 13 segments for which Phase 1 LAT and Phase 2 modeling results can be compared are consistent. When compared with the most adverse conditions Phase 2 modeling results, 12 (93%) of the 13 results are consistent.

The regressions in Figure 3 and comparisons in Table 1 are based on averaged results, as described in the "Correcting for High Segment-Density" Section. Segments along Ash Slough, Berenda Slough, and Fresno River are otherwise over-represented in this analysis. This over representation is significant, because these levees are located along artificial channels excavated as straight lines through distal fan deposits. Subsurface variability therefore increases significantly while segment length decreases. Levees in this region are therefore much more difficult to assess, and over- or under-representation of this area impacts analysis results considerably. When results in this area are not averaged and are therefore over-represented, the R^2 for the nearest gradient decreases from 0.32 to 0.22, while the R^2 for the highest gradient decreases from 0.52 to 0.36. On the other hand, when results from this area are removed from this analysis, the nearest R^2 increases to 0.55 while the highest gradient R^2 increases to 0.75. This indicates that LAT and modeling results are far more comparable for analyses conducted in areas with less subsurface variability. The difference in Phase 1 LAT and Phase 2 modeling results in areas of high subsurface variability is likely caused by additional subsurface exploration data acquired in Phase 2. These data were not available in Phase 1 LAT assessments. These additional Phase 2 data are especially important in areas with increased subsurface variability.

COMPARING PHASE 1 AND 2 THROUGH SEEPAGE RESULTS

Phase 1 LAT WHIS vs Phase 2 Model Phreatic Surface Breakout Height

To compare Phase 1 LAT results with Phase 2 model results, Phase 1 LAT through seepage WHI scores were first compared against the height at which the phreatic surface exists the landside slope in Phase 2 models (Figure 4

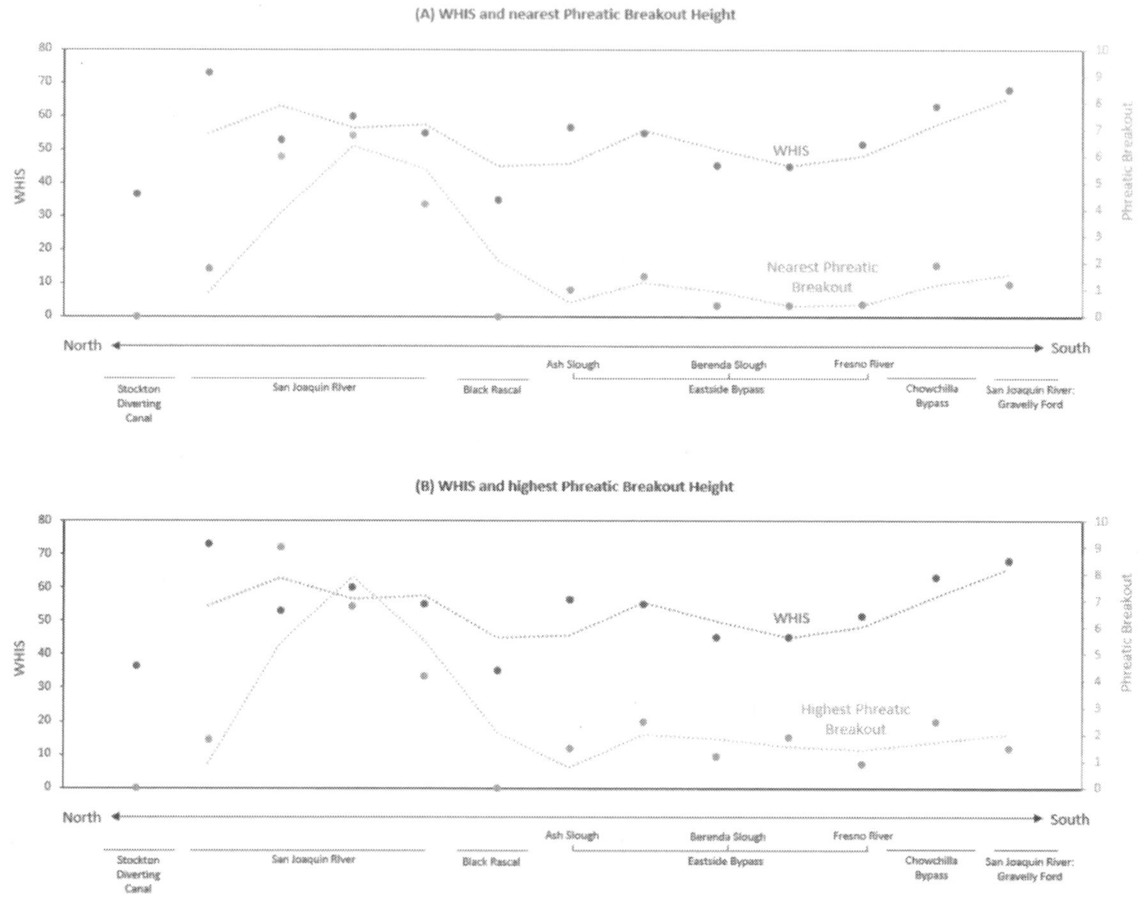

Figure 4. Through seepage (A) WHIS and nearest phreatic breakout (Y axis) per segment (X axis) and (B) WHIS and highest phreatic breakout (Y axis) per segment (X axis). Segments are arranged along the X axis from north (left) to south (right). Running average curves are added to better depict regional trends in WHIS and gradient.

Running average curves in Figure 4 (A) and (B) depict little apparent correlation between WHIS and either nearest or highest phreatic surface breakout height.

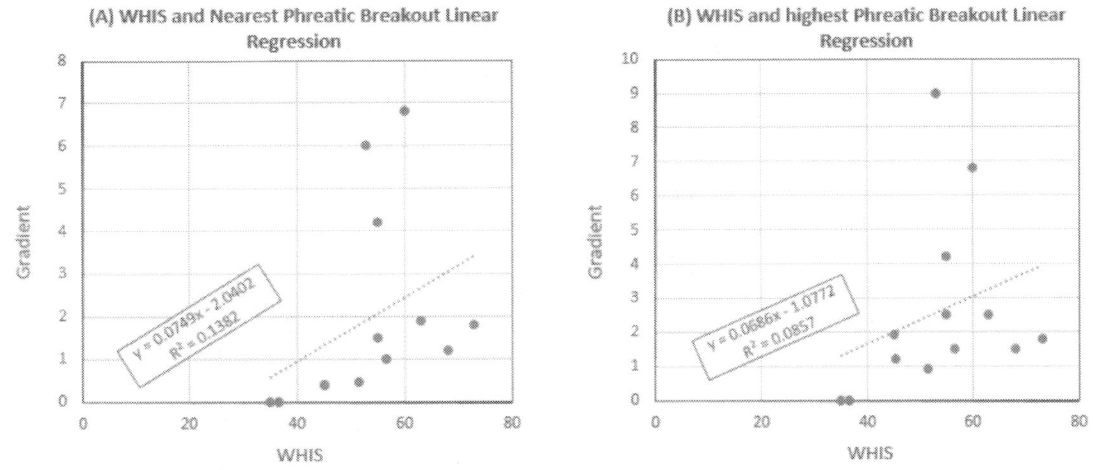

Figure 5. Linear regressions relating (A) phreatic breakout with WHIS and (B) highest phreatic breakout with WHIS.

Figure 5 depicts linear regressions relating phreatic-surface height with WHIS. The R^2 for the nearest-phreatic regression is about 0.14 and the R^2 for the highest-phreatic is about 0.09. These values indicate little to no correlation between WHIS and phreatic surface break out.

Phase 1 LAT Categorizations vs Phase 2 Through Seepage Results

To evaluate the effectiveness of the LAT as a screening tool, Phase 1 LAT through seepage categorizations were compared with evaluated Phase 2 modeling results. The Phase 2 phreatic surface breakout height was evaluated along with other factors, such as levee embankment composition and past performance. Therefore, phreatic surface breakout heights do not necessarily directly correlate with Phase 2 through seepage assessment results. The same phreatic surface breakout height may result in one levee meeting criteria and another levee not meeting criteria, depending on other factors.

To stay consistent with the underseepage comparisons, Phase 2 modeling results that meet criteria are considered most comparable to Phase 1 LAT categorization A (low likelihood of levee failure or need for flood fighting), while Phase 1 LAT categorizations B and C (moderate and high likelihood) are considered most comparable with Phase 2 modeling results that do not meet criteria. Phase 1 LAT categorization LD does not clearly compare with Phase 2 modeling results; however; there were no LD categorizations for this comparison. This comparison is depicted in Table 2 (below).

Table 2. Through Seepage: Phase 1 LAT Categorizations and Phase 2 modeling results. Consistent results are shaded in blue.

Area	Segment	Nearest Phase 2 Modeling Results		Most Adverse Condition Phase 2 Modeling Results	
		Phase 1 LAT Categorization	Phase 2 Modeling	Phase 1 LAT Categorization	Phase 2 Modeling
Diverting Canal	332 & 238	A	Meets Criteria	A	Meets Criteria
RD 2095	258	B	Does not Meet Criteria	B	Does not Meet Criteria
RD 2075	196	C	Does not Meet Criteria	C	Does not Meet Criteria
RD 2064	195	C	Does not Meet Criteria	C	Does not Meet Criteria
RD 2064	257	C	Does not Meet Criteria	C	Does not Meet Criteria
Black Rascal	73	A	Meets Criteria	A	Meets Criteria
Ash Slough	11 & 358	B	Meets Criteria	B	Does not Meet Criteria
Eastside Bypass	184	B	Meets Criteria	B	Does not Meet Criteria
Berenda Slough	359 & 12 & 360 & 184	B	Does not Meet Criteria	B	Does not Meet Criteria
Eastside Bypass	218	B	Meets Criteria	B	Does not Meet Criteria
Fresno River	27 & 28	C	Meets Criteria	C	Meets Criteria
Chowchilla Bypass	261	C	Meets Criteria	C	Does not Meet Criteria
Gravelly Ford	219	C	Does not Meet Criteria	C	Does not Meet Criteria

When compared with the nearest Phase 2 modeling results, 8 (62%) of the 13 segments for which Phase 1 LAT and Phase 2 modeling results can be compared are consistent. When compared with the most adverse conditions Phase 2 modeling results, 12 (93%) of the 13 results are consistent.

These results indicate that, despite the poor correlation between Phase 1 LAT WHI scores and Phase 2 phreatic surface break out height, the LAT can still be a useful screening tool for analysis of through seepage hazard. The difference in correlation strength between the WHIS- and categorization-based comparisons indicates that other factors considered during Phase 2 analyses have greater impact on Phase 2 results than phreatic surface breakout height.

The regressions in Figure 4 and comparisons in Table 2 are based on averaged results, as described in the "Correcting for High Segment-Density" Section. Segments along Ash Slough, Berenda Slough, and Fresno River are otherwise over-represented in this analysis. This over representation is potentially significant, because these levees are located along artificial channels excavated as straight lines through distal fan deposits. Subsurface variability therefore increases significantly while segment length decreases. Levees in this region are therefore much more difficult to assess, and over- or under-representation of this area impacts analysis results considerably.

When results in this area are not averaged and are therefore over-represented, the R^2 for the nearest phreatic surface breakout increases marginally from 0.14 to 0.2, while the R^2 for the highest phreatic surface breakout increases marginally from 0.09 to 0.13. When distal-fan area levees are removed from this comparison, the nearest phreatic surface breakout R^2 remains unchanged at 0.09 while the highest phreatic surface R^2 decreases to 0.06. These marginally increased or decreased R^2 values still indicate little to no correlation, but the marginal nature of the changes indicates that the over-representation of distal fan levees has little impact on the Through Seepage comparison. The lack of impact is consistent with known Project levee construction details. Project levees were generally constructed from Project borrow, typically located waterside of the levee in areas where Project explorations indicated relatively fine-grained materials. Accordingly, Project and Phase 2 explorations typically indicate less variability in Project levees than in Project levee foundations. Therefore, the bias from increased subsurface variability in distal-fan area levees has little impact in through seepage assessments.

COMPARING PHASE 1 THROUGH SEEPAGE AND PHASE 2 STABILITY RESULTS

Phase 1 LAT WHIS vs Phase 2 Model Factor of Safety

To further examine WHIS through seepage results, Phase 1 WHIS results were compared with Phase 2 calculated stability Factor of Safety (FOS). Phase 1 LAT assessment failure modes define stability independent of through seepage; Phase 1 stability does not include adverse conditions attributable to through seepage. Therefore, Phase 2 stability Factor of Safety (FOS) is comparable to Phase 1 through seepage assessment.

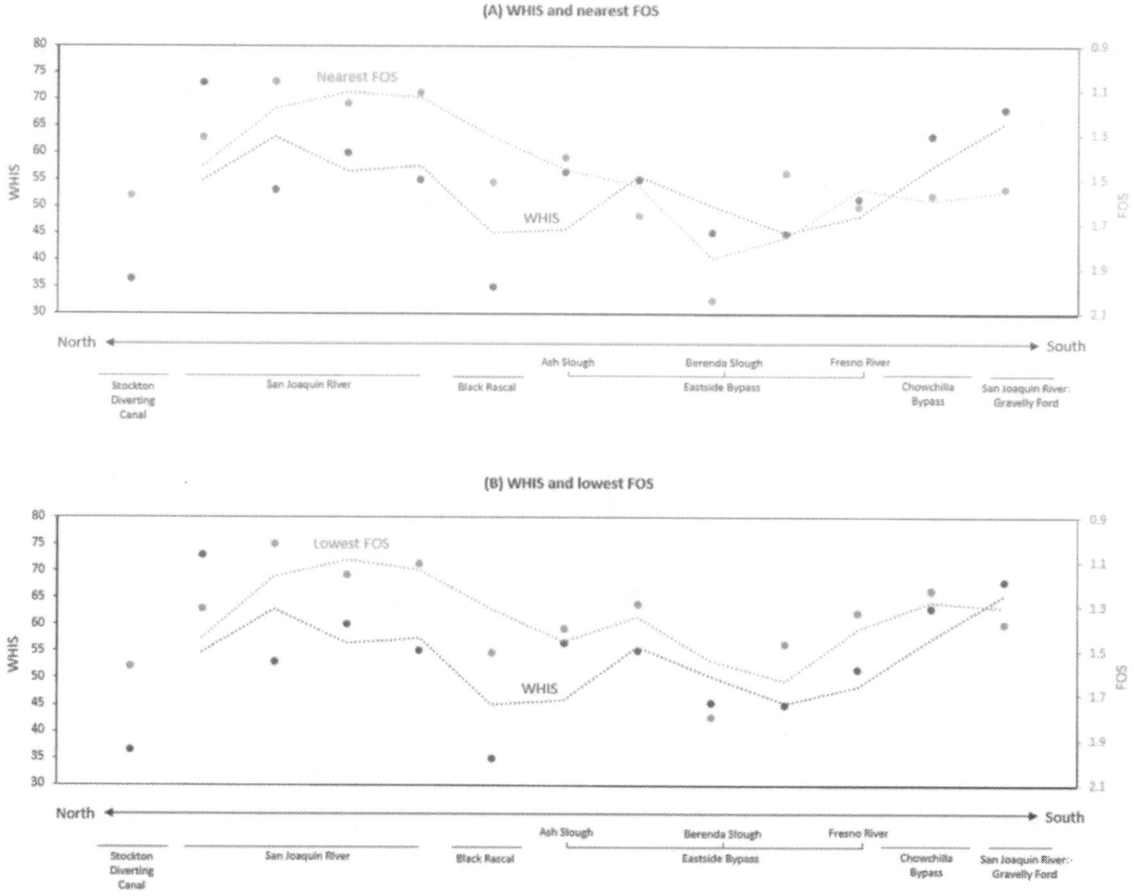

Figure 6. (A) WHIS and nearest FOS (Y axis) per segment (X axis) and (B) WHIS and lowest FOS (Y axis) per segment (X axis). Segments are arranged along the X axis from north (left) to south (right). Running average curves are added to better depict regional trends in WHIS and gradient.

Generally, the running average curves in Figure 6 (A) and (B) appear similar. WHIS (and FOS) are relatively low (high) in the north along the Stockton Diverting Canal, high (low) in the north along the San Joaquin River, lower (higher) toward the central portion of the graph along Black Rascal Diversion Canal, Ash Slough, Berenda Slough, Fresno River, and Eastside Bypass, and higher (lower) to the south along the Chowchilla Bypass and Gravelly Ford portion of the San Joaquin River. These results are generally consistent with levee performance. Numerous adverse stability conditions attributed to through seepage have been noted in the north along the San Joaquin River and in the south along the Chowchilla Bypass and Gravelly Ford portion of the San Joaquin River. Other portions of the study area have experienced comparatively few adverse conditions.

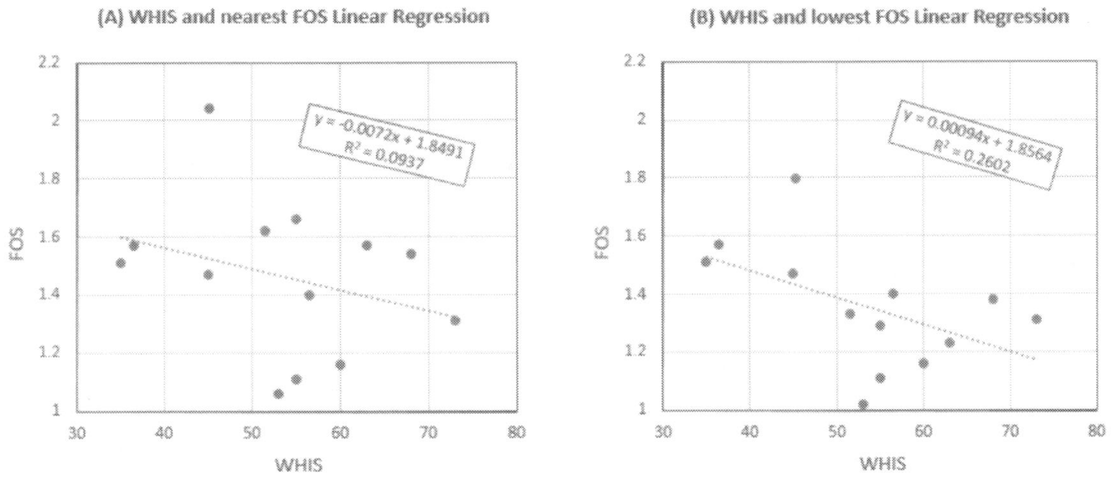

Figure 6. Linear regressions relating (A) nearest FOS with WHIS and (B) lowest FOS phreatic breakout with WHIS.

Figure 6 depicts linear regressions relating FOS with WHIS. The R^2 for the nearest-FOS regression is about 0.09 and the R^2 for the lowest-FOS is about 0.26. These values indicate little to no correlation between WHIS and the nearest FOS and a weak correlation between WHIS and the lowest FOS.

Phase 1 LAT Through Seepage Categorizations vs Phase 2 Stability Results

To better evaluate the effectiveness of the LAT as a screening tool, Phase 1 LAT through seepage categorizations were compared with evaluated Phase 2 stability modeling results. Phase 2 modeling results were evaluated against a FOS criterion of 1.4; FOSs exceeding 1.4 meet criteria while FOSs meeting or below 1.4 do not meet criteria. To stay consistent with underseepage and through seepage comparisons, Phase 2 modeling results that meet criteria (adverse conditions unlikely) are considered most comparable to Phase 1 LAT categorization A (low likelihood of levee failure or need for flood fighting) and Phase 1 LAT categorizations B and C (moderate and high likelihood) are considered most comparable with Phase 2 modeling results that do not meet criteria. Phase 1 LAT categorization LD does not clearly compare with Phase 2 modeling results; however; there were no LD categorizations for this comparison. This comparison is depicted in Table 3 (below).

Table 3. Phase 1 through seepage LAT Categorizations and Phase 2 stability modeling results. Consistent results are highlighted in blue.

	Area	Segment	Nearest Phase 2 Modeling Results		Most Adverse Condition Phase 2 Modeling Results	
			Phase 1 LAT Categorization	Phase 2 Modeling	Phase 1 LAT Categorization	Phase 2 Modeling
North ↑	Diverting Canal	332 & 238	A	Meets Criteria	A	Meets Criteria
	RD 2095	258	B	Does not Meet Criteria	B	Does not Meet Criteria
	RD 2075	196	C	Does not Meet Criteria	C	Does not Meet Criteria
	RD 2064	195	C	Does not Meet Criteria	C	Does not Meet Criteria
	RD 2064	257	C	Does not Meet Criteria	C	Does not Meet Criteria
	Black Rascal	73	A	Meets Criteria	A	Meets Criteria
	Ash Slough	11 & 358	B	Meets Criteria	B	Meets Criteria
	Eastside Bypass	184	B	Meets Criteria	B	Does not Meet Criteria
	Berenda Slough	359 & 12 & 360 & 184	B	Meets Criteria	B	Meets Criteria
	Eastside Bypass	218	B	Meets Criteria	B	Meets Criteria
	Fresno River	27 & 28	C	Does not Meet Criteria	C	Does not Meet Criteria
	Chowchilla Bypass	261	C	Meets Criteria	C	Does not Meet Criteria
South ↓	Gravelly Ford	219	C	Does not Meet Criteria	C	Does not Meet Criteria

When compared with the nearest Phase 2 modeling results, 8 (62%) of the 13 segments for which Phase 1 LAT and Phase 2 modeling results can be compared are consistent. When compared with the lowest FOS Phase 2 modeling results, 10 (77%) of the 13 results are consistent.

As with the Phase 1 WHIS vs Phase 2 phreatic surface breakout comparison, the regressions in Figure 6 and comparisons in Table 3 are based on averaged results, as described in the "Correcting for High Segment-Density" Section, to compensate for over-representation of distal-fan area levees.

When results in this area are not averaged and are therefore over-represented, the R^2 for the nearest FOS marginally increases from 0.09 to 0.13 while the lowest-FOS R^2 marginally increases from 0.26 to 0.29. When distal-fan area levees are removed from this comparison, the nearest-FOS R^2 decreases marginally to 0.04 and the lowest-FOS R^2 decreases marginally to 0.21. These marginally increased or decreased R^2 values still indicate little to no correlation between nearest FOS and WHIS, and a weak correlation between lowest FOS and WHIS, but the marginal nature of the changes indicates that the over-representation of distal fan levees has little impact on the Through Seepage comparison. Again, the lack of impact is consistent with known Project levee construction details. Project levees were generally constructed from Project borrow, typically located waterside of the levee in areas where Project explorations indicated relatively fine-grained materials. Accordingly, Project and Phase 2 explorations typically indicate less variability in Project levees than in Project levee foundations. Therefore, the bias from increased subsurface variability in distal-fan area levees has little impact in through seepage assessments.

CONCLUSIONS

This assessment compares LAT results with modeling results. The LAT is a new assessment tool developed for DWR NULE. LAT results have therefore not yet been systematically quantified and compared with modeling results.

Comparisons indicate that Underseepage LAT WHI scores correlate reasonably well with modeled average vertical gradients. Moreover, LAT underseepage categorizations correlate with overall modeling results. Levees that meet (don't meet) underseepage criteria correlate well with levees categorized hazard level A (B or C). Correlations are weaker along artificial channels cut in straight lines through distal fan deposits. This indicates the LAT performs better where subsurface stratigraphy is less variable.

Through seepage LAT WHI scores show little to no correlation with modeled phreatic surface exit heights but show weak correlation with modeled stability FOS. However, LAT through seepage categorizations correlate well with overall modeling results. Levees that meet (don't meet) through seepage or stability criteria correlate well with levees categorized hazard level A (B or C). Correlations do not weaken considerably along artificial channels cut in straight lines through distal fan deposits. This is expected, as exploration data and levee construction details indicate increased subsurface variability does not increase levee composition variability.

The correlations between LAT categorization and model results indicate the LAT is an effective screening tool. Overall, LAT categorization A correlate well with model results that meet criteria, and LAT categorizations B and C correlation well with model results that do not meet criteria. Moreover, these comparisons indicate the LAT can help predict some modeling metrics, as LAT underseepage WHI scores show moderate correlation with average vertical exit gradients.

REFERENCES

Kleinfelder, 2011. Geotechnical Assessment Report, DWR

Kleinfelder, 2015. Geotechnical Overview Report, DWR

URS, 2011. Geotechnical Assessment Report, DWR

URS, 2015. Geotechnical Overview Report, DWR

United States Army Corps of Engineers, 2005. Design Guidance for Levee Underseepage, Technical Letter 1110-2-569, Department of the Army

FAULT INVESTIGATIONS DURING BORINQUEN DAM 1E CONSTRUCTION PANAMA CANAL EXPANSION

David Schug[1], Paul Salter[2]
Dan Meier[3] Robert Urban[4]
Chris Goetz[5] Lelio Mejia[6]
James Toose[7] Derek Irving[8]
Pastora Franceschi[9] Maximiliano De Puy[10]
Jaime Arrocha[11] Roberto Miranda[12]

ABSTRACT

Borinquen Dam 1E is part of the new Pacific Access Channel (PAC), which will provide navigation access from the Gaillard Cut to the new Post-Panamax Pacific Locks. The 2.3 km long dam is a zoned rockfill dam 150 m wide and 30 m high at the crest, inclined at 3:1 (H:V). Excavations commenced February 2011. Final embankment construction was completed June 2015. Geologic mapping of the dam foundation and adjoining areas of the PAC was a collaborative effort between Autoridad del Canal Panama (ACP) and URS geologists. A key geologic objective was to confirm locations and activity of faults mapped at the dam during design. The design allowed for core widening at the anticipated location of the Pedro Miguel Fault at the south abutment and at a west branch of the Pedro Miguel Fault (believed at the time to be to be the main active trace of the fault) mapped about one third of the way north along the dam axis. As-built geologic mapping revealed a complex pattern of faulting associated with the Pedro Miguel Fault crossing the southeast half of the foundation, the PAC and the nearby Dam 1W foundation along a north-south trend. Trenching and age dating alluvium overlying faults crossing the Dam 1E foundation indicated the unbroken alluvium was latest Pleistocene to early Holocene age. Where significant faults cross Dam 1E, the core was widened to accommodate potential fault rupture. This paper provides an overview of fault investigations carried out during construction of Dam 1E.

[1] Principal Geologist, URS Corporation (merged with AECOM), San Diego, CA, david.schug@aecom.com
[2] Resident Geologist, AECOM, Borinquen Dam 1E Project Panama Canal Expansion Program, Orange, CA, paul.salter@aecom.com
[3] Assistant Resident Geologist, AECOM, Portland, OR, dan.meier@aecom.com
[4] Resident Geologist, AECOM. Santa Maria, CA, robert.urban@aecom.com
[5] Principal Geologist, AECOM, Santa Ana, CA, christopher.goetz@aecom.com
[6] Project Manager, AECOM, Oakland, CA, lelio.mejia@aecom.com
[7] Resident Engineer, AECOM, Borinquen Dam 1E Project Panama Canal Expansion Program, james.toose@aecom.com
[8] Supervisor, Geología, Autoridad del Canal de Panamá, Panama, DIrving@pancanal.com
[9] Supervisor, Geología, Autoridad del Canal de Panamá, Panama, PFranceschi @pancanal.co
[10] Manager, Geotechnical Section, Autoridad del Canal de Panamá, Panama, MDePuy@pancanal.com
[11] Geólogo, Autoridad del Canal de Panamá, Panama, JArrocha@pancanal.com
[12] Geólogo, Autoridad del Canal de Panamá, Panama, RMiranda@pancanal.com

INTRODUCTION

Borinquen Dam 1E is part of the works for the new Pacific Access Channel (PAC), a key element in the Panama Canal Expansion project, which also includes new Post-Panamax Pacific and Atlantic Locks and various other canal upgrades (Figure 1). Borinquen Dam 1E, along with several smaller dams, provides access from the Gaillard Cut to the new Pacific Locks (Figure 2). The dam extends approximately 2.3 km from near the northern gates of the 100-year-old Pedro Miguel Locks to the north side of Fabiana Hill. A cofferdam was installed along the southwest shoreline of Miraflores Lake to facilitate the construction of Dam 1E.

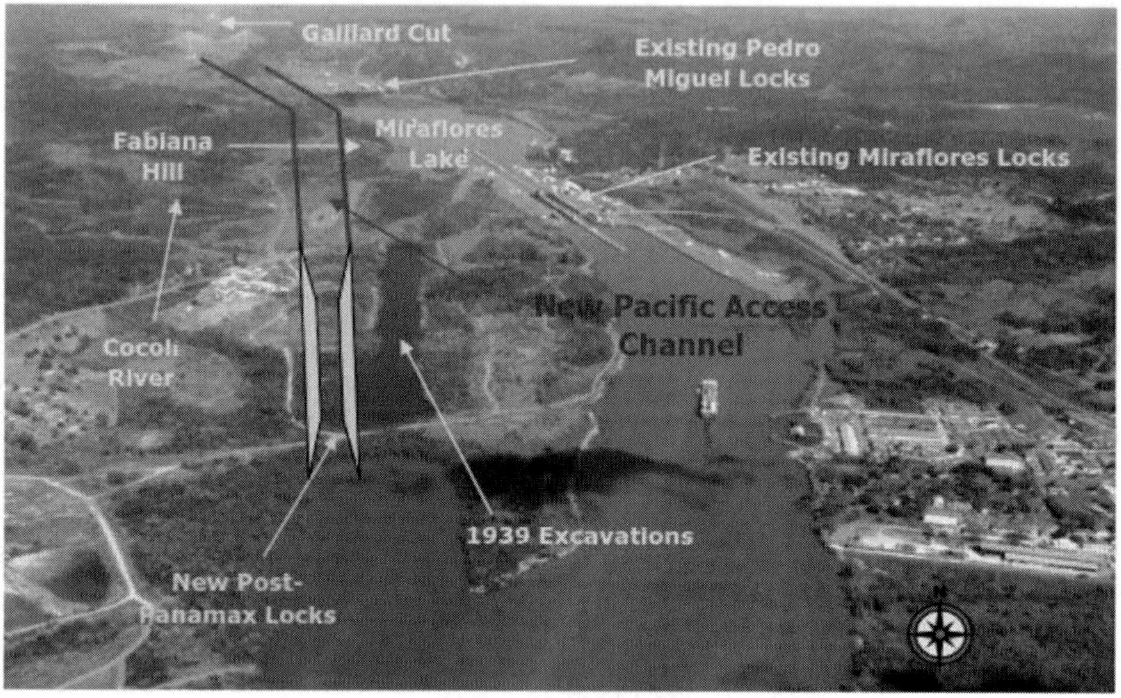

Figure 1. Overview of the new Pacific Access Channel

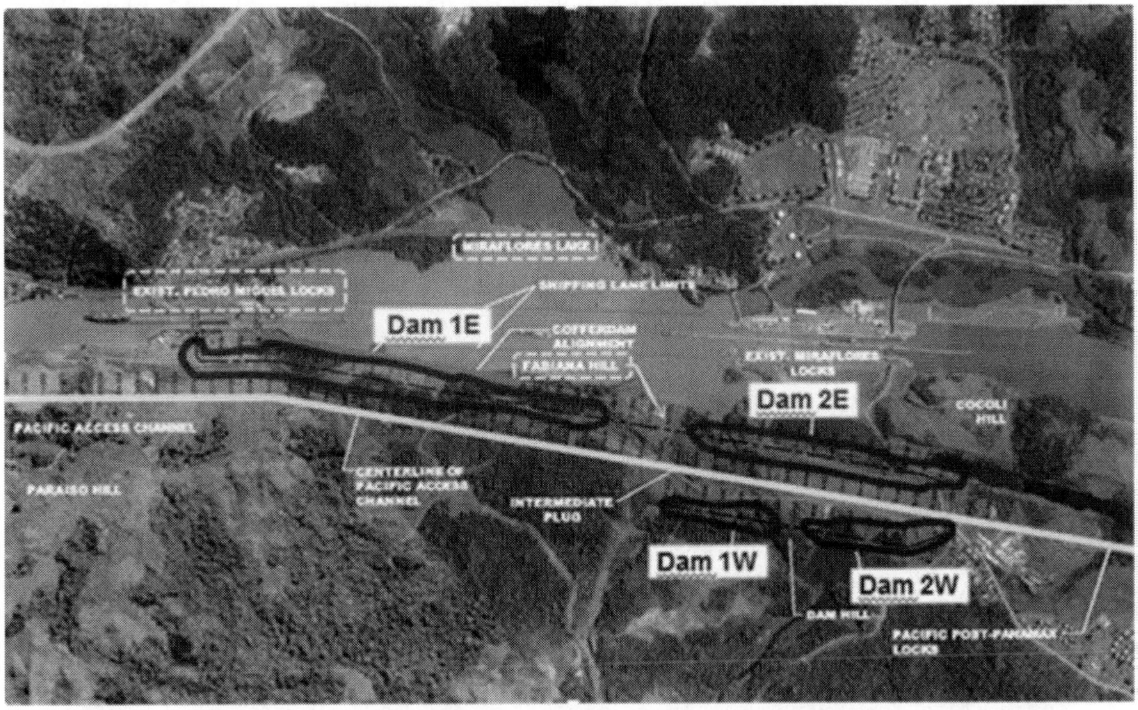

Figure 2. Aerial photo of the Borinquen Dams project area before construction.

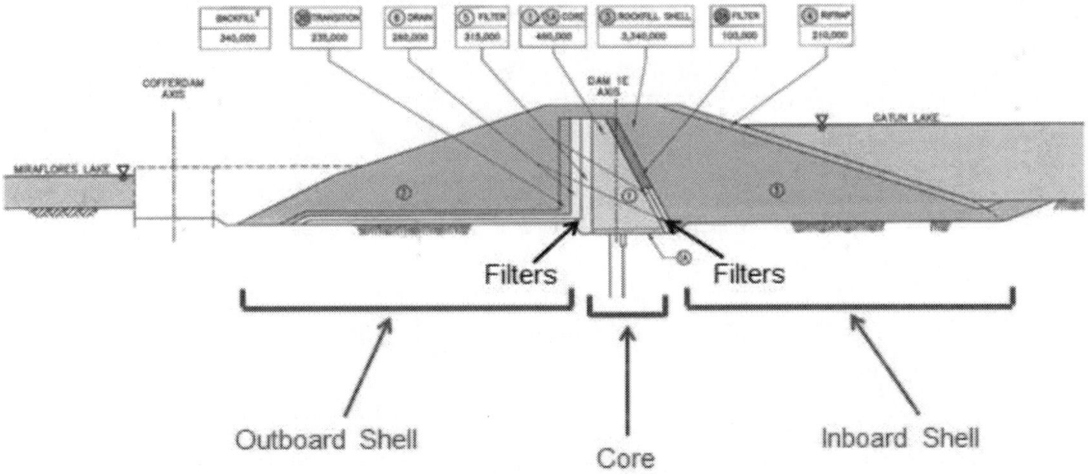

Figure 3. Dam 1E Cross Section

Dam 1E retains the PAC water level at about elevation 27.2 m, or 10.7 m above the minimum operating elevation of Miraflores Lake. The constructed width and height of Dam 1E varies, based on the depth of excavation required to achieve foundation objective, however the dam is generally 100 to 150 m wide at foundation grade and 25 to 30 m high. Figure 3 shows the zoned construction of Dam 1E, with rockfill embankments, chimney filters and a central clay core. Pre-construction groundwater levels at Dam 1E, which were as much as 18 m above the dam design foundation level (URS 2009), were lowered by construction dewatering.

There were numerous seismic design challenges, as described in Mejia and others (2011) including the potential for strike-slip fault rupture up to 3 m across the dam foundation. Based on fault trenching near Dam 1W, offsets of paleochannel margins were reported to range from 2.0 to 2.5 m for the most recent event on the Pedro Miguel Fault (Earth Consultants International, ECI 2007a,b). The seismic design criteria required that the thickness of the core, chimney filter and drains within 50 m of active fault locations be widened to accommodate the estimated fault displacement. The design allowed for widening at the anticipated location of the Pedro Miguel Fault at the south abutment and at a west branch of the Pedro Miguel Fault (believed at the time to be the main active trace of the fault) mapped about one third of the way north along the dam axis.

Foundation excavations commenced in February 2011, by initially excavating the inboard shell (adjacent to the PAC channel); then the dam core trench; and lastly, the outboard shell excavations. Foundation excavations in each of these areas required the removal of overburden fills, alluvium and weathered materials over broad areas. The core trench was excavated deeper than the adjacent inboard and outboard shells. Final embankment construction was completed June 2015 (Photo 1).

Photo 1. View along Dam 1E and the new PAC looking towards the Pedro Miguel Locks.

Dam foundation geologic mapping was undertaken jointly by Autoridad del Canal Panama (ACP) and URS geologists. Detailed investigations of faults near Borinquen Dam 1E, within and outside the PAC area were also conducted to corroborate locations of faults previously mapped and inferred during design.

TECTONIC SETTING

Panama is located between the Cocos, Nazca, Caribbean, and South American tectonic plates. Considering GPS data, Kellog and Vega (1995) suggested the existence of a rigid Panama-Costa Rica microplate that is moving northward relative to the stable Caribbean Plate. Seismicity studies by Adamek (1988) and other GPS studies by Trenkamp et al. (2002) suggest that that the Panama-Costa Rica microplate is low seismicity province that is presently undergoing minimal internal tectonic deformation. Conversely, Rockwell et al (2010) suggested that the tectonic deformation that began in the Miocene is ongoing and is distributed among faults throughout Panamá and western Colombia, resulting in internal deformation of the isthmus by folding and faulting. Rockwell et al. (2010) suggested that the Pedro Miguel fault, the Limón fault, and related faults comprise an active zone of faulting that extends from north-central Panamá southward for at least 40 km, and likely extending offshore into the Gulf of Panamá. The Pedro Miquel fault, which crosses the Panamá Canal between the Miraflores and Pedro Miguel Locks (Stewart et al. 1980), is the primary tectonic structure that has been the focus of a seismic hazard characterization for the canal expansion project.

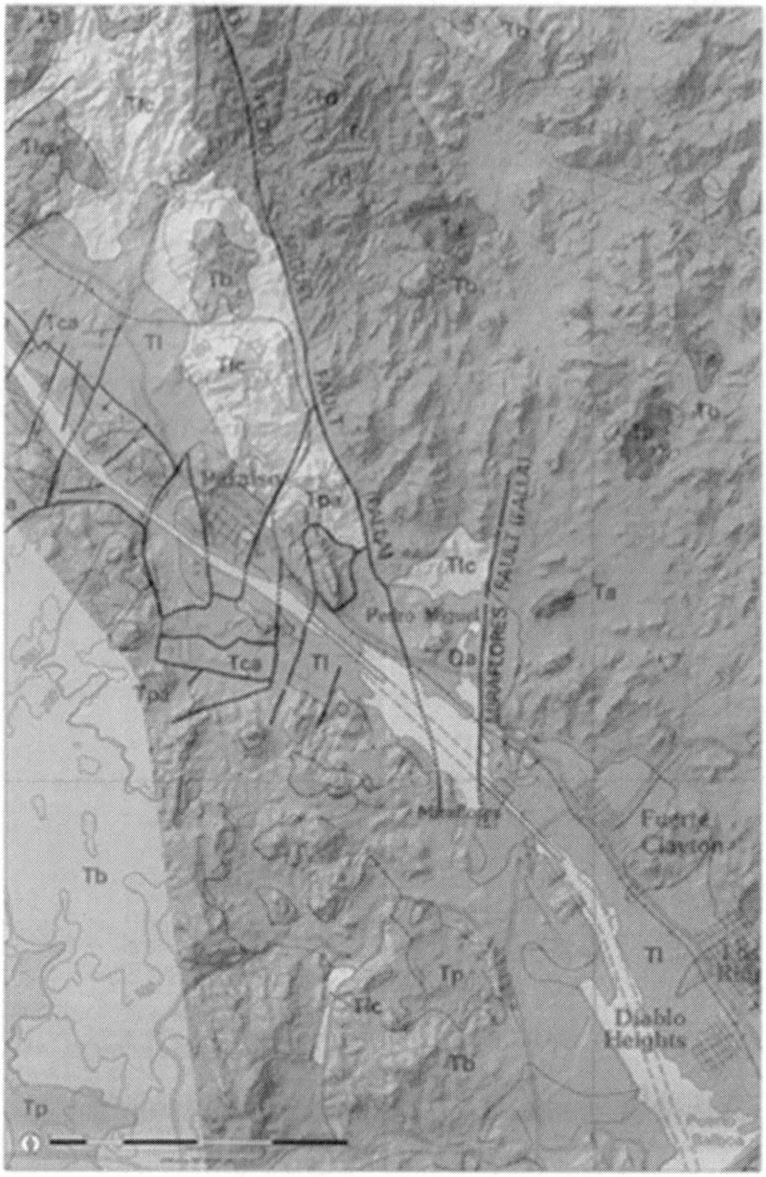

Figure 4. Geologic Map of the Canal Zone
(Stewart, et. al. 1980).

Figure 4 shows the location of the Pedro Miguel Fault as mapped by Stewart et al (1980). The Pedro Miguel fault was initially mapped in the Canal Zone as a bedrock fault that formed a structural boundary between several geologic units (Woodring, 1957, Stewart et al., 1980). North of the Canal, the fault was mapped as forming the contact between the Miocene Las Cascadas Formation and the Oligocene Panamá Formation. To the South of the Canal, the fault was mapped for only a short distance, forming the contact between Miocene basalt and Miocene La Boca Formation. As part of a seismic hazard characterization for the canal expansion project, the ACP retained Earth Consultant International (ECI) to conduct investigations to assess the precise locations and activity of the Pedro Miguel fault in the area of the proposed Borinquen dams (ECI, 2007a,b, 2010). Based on their investigations, ECI reported that the fault has a geomorphic signature indicative of a recurrent Holocene active, right lateral strike slip fault that includes linear valleys, ridges and escarpments, with several offset and deflected channels.

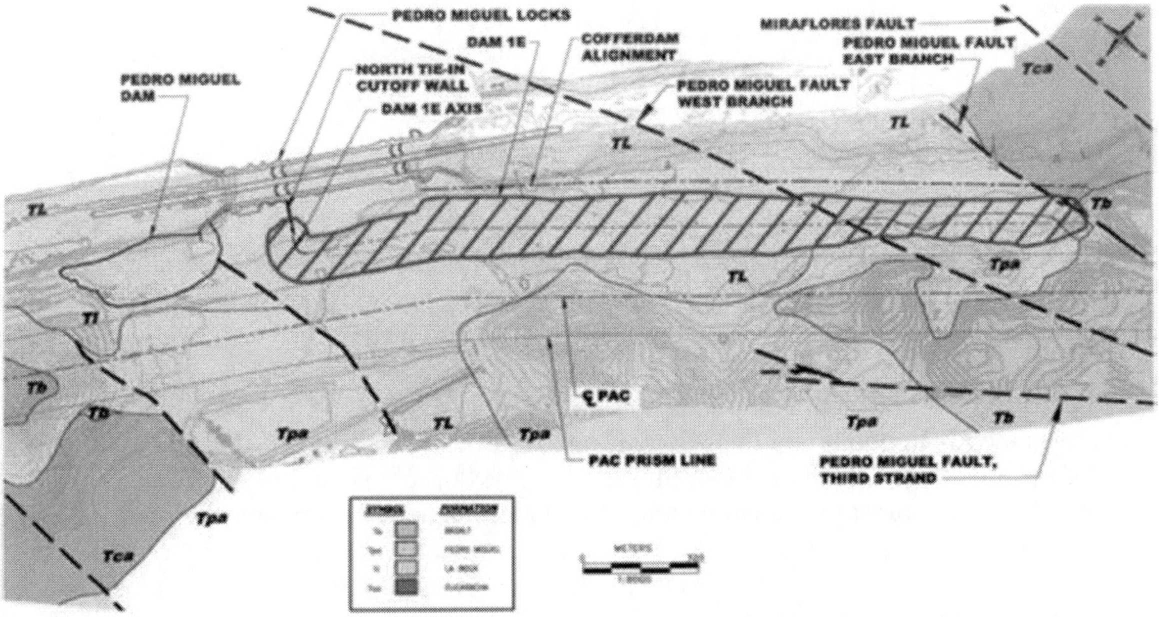

Figure 5. Bedrock Geologic Map of Dam 1E.

ECI mapped two strands of the Pedro Miguel fault in the area of the proposed dams. The more easterly strand, which coincides with the location mapped by Stewart and others (1980) is mapped along the northwest side of Fabiana Hill (the south abutment of Borinquen Dam 1E). This strand of the fault is informally referred to herein as the "Stewart Trace" .The more westerly fault strand of the Pedro Miguel fault is shown as diverging to the northwest from the Stewart Trace. ECI reported that this trace of the fault has produced 100 $\pm$ 20 meters of offset of two parallel, east-flowing Late Quaternary tributary drainages of the Coccoli River. ECI considered this previously unmapped strand of the fault to be the through-going "main strand" of the Pedro Miguel fault in the vicinity of the canal. As shown on Figure 5 the fault is mapped as intersecting Dam 1E about 600m northwest of Fabiana Hill.

In 2007, ECI performed fault trenching in the Rio Coccoli area, about a kilometer to the south of the proposed PAC, to characterize the activity of the PMF and to develop fault slip parameters for design of the proposed Borinquen dams. The trenching was done in the area of the two parallel offset drainages noted above. Based on the trenching investigations ECI concluded that the zone of Quaternary deformation is locally more than 100 m wide but the majority of the slip was accommodated by the "western strand" of the fault. ECI also concluded that the Pedro Miguel fault experienced at least two, and possibly up to three, late Holocene earthquakes in the past ~1,500 years and they surmised that the most recent event was the AD1621 earthquake that devastated the old capital of Panamá (Panamá Viejo). ECI (2007b) recommended that right lateral strike slip fault displacement of at least 3 m should be considered for design of the proposed dams. The 3 m displacement was related to the westerly trace "main strand" of the Pedro Miguel fault. The more easterly trace of the Pedro Miguel fault (the Stewart Trace) was judged as active but only having the potential to produce about one-third of the slip as on the westerly trace (about 1 m).

DAM SITE GEOLOGY

Pre-Construction (Design) Geologic Investigations

Pre-construction geological investigations carried out at Dam 1E by ACP, URS, and others included: reviewing historic data and aerial photographs, geologic mapping, test pits, borehole investigations (Pinilla and others, 2011); geophysics (Kauffman and others, 2008) and paleoseismic fault trenching, (ECI, 2007a,b; 2010).

The geologic assessment completed for the design of Dam 1E (URS 2008) identified the dam as mostly underlain by the Miocene age La Boca Formation (Woodring & Thompson 1949; Woodring 1964; Stewart et.al. 1980), consisting of stratified conglomerate, sandstone, siltstone and clay shale. Subaerial volcaniclastic rock agglomerate of the Miocene age Pedro Miguel Formation underlies the south end of the dam. Basalt within the dam footprint was mapped as small intrusive dikes and sills in the north and central portion of the dam, and as larger intrusive bodies and/or flows in the south abutment area (Fabiana Hill). The Pedro Miguel Fault and the fault deemed "Main Trace of the Pedro Miguel Fault" were mapped across the dam. Secondary faults were anticipated (URS 2008). Figure 5 shows the generalized bedrock geology and major faults anticipated during design.

Dam Foundation Geologic Mapping

The field methods used to map the foundation geologic data varied, depending on the dam foundation zone. The Contractor excavated defined parcels to an appropriate foundation objective grade as per the specifications and undertook final cleaning with a smooth-edged excavator bucket. This resulted in extensive exposures of the foundation bedrock geology. ACP survey crews marked a 5m x 5m survey grid with station and offset numbering on the cleaned foundation. ACP and URS geologists collaborated on identification of key geologic features and then used Trimble Yuma® GPS units to geologically map the cleaned inboard shell. The core trench was excavated into slightly weathered rock and was mapped by "hand" on mylar sheets. The outboard shell was similarly cleaned and mapped- parcel by parcel. A detailed geologic map of the dam foundation was compiled at a scale of 1:300. This level of geologic detail allowed documentation of displacement along faults mapped as crossing the dam, described further below.

Geologic Units

Interbedded La Boca Formation conglomerate, sandstone, siltstone, clay shale, ash flow, and lignite beds were mapped over the northern three-quarters of the dam foundation. La Boca Formation bedding in the dam foundation has an east-west trending strike and dips to the south at angles around 15° to 20° (Photo 2).

Photo 2. Dipping beds of La Boca Formation in the core trench.

Thinly to thickly bedded Pedro Miguel Formation consisting of stratified volcaniclastic sandstone and minor siltstone units were mapped at dam foundation in the south portion of the dam. The volcaniclastic sandstone units typically include fine to coarse sand-sized tuff and basalt fragments in a dark fine-grained volcaniclastic (tuffaceous) matrix.

Alluvium was removed during foundation excavation. Broad exposures of alluvium overlying the bedrock were mapped along the dam outboard margins (Photo 3). The alluvium typically consisted of clayey gravel channel deposits overlain by and interfingered with stratified fine-grained sandy silt units.

Samples of the alluvium were collected for age dating to help constrain fault activity, described below. Laboratory age dating analyses included optically stimulated luminescence (OSL) dating performed by the Luminescence Dating Laboratory at the University of Cincinnati. Radiocarbon (C-14) accelerator mass spectrometer (AMS) dating was performed by the University of Georgia, Center for Applied Isotope studies.

Photo 3. Fill/alluvium overlying La Boca Formation along the outboard cut slope.

FAULT INVESTIGATIONS

As the foundation excavation at Dam 1E progressed, ongoing geologic mapping and field observations indicated that faults considered for dam widening were overlain by potentially datable alluvium. Alluvium overlying faults was further investigated by trenching and sampling cut exposures. The objective of age dating alluvium was to demonstrate that faults at the dam have been inactive during the Holocene and therefore have a low likelihood to reactivate during the lifespan of the project.

Fault Investigations at Dam 1E

Numerous faults were mapped in the Dam 1E foundation. These range from minor, discontinuous, shears with little (a few cm) apparent displacement to faults that cross the entire width of the dam foundation, have thick (several meters wide) shear and gouge zones, and displacements of 10's of meters, or more. Dominant fault orientations are regionally north to northwest trending, with steep dips. The more significant faults had apparent right-lateral sense of displacement, with a few exceptions where apparent left-lateral displacements were mapped. Figure 6 shows the principal mapped faults that are continuous over significant distances in the dam foundation.

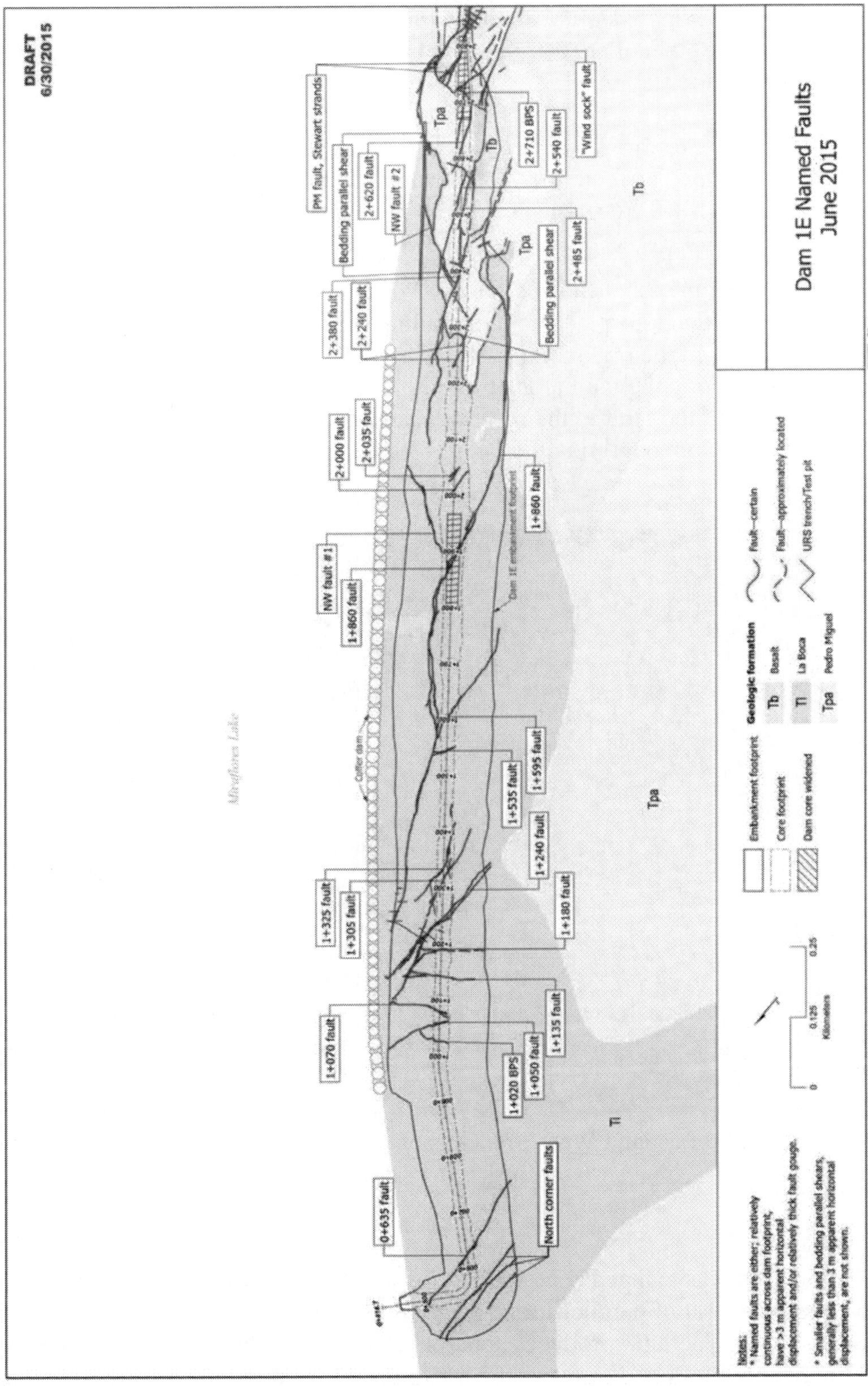
DRAFT
6/30/2015
Dam 1E Named Faults
June 2015
PM fault, Stewart strands
Bedding parallel shear
2+620 fault
NW fault #2
2+710 BPS
2+540 fault
"Wind sock" fault
2+380 fault
2+240 fault
Bedding parallel shear
2+485 fault
2+000 fault
2+035 fault
1+860 fault
NW fault #1
1+860 fault
1+535 fault
1+595 fault
1+325 fault
1+305 fault
1+240 fault
1+180 fault
1+070 fault
1+020 BPS
1+050 fault
1+135 fault
0+635 fault
North corner faults
Tpa
Tb
Tl
Geologic formation
Basalt
La Boca
Pedro Miguel
Embankment footprint
Core footprint
Dam core widened
Fault—certain
Fault—approximately located
URS trench/Test pit
Notes:
* Named faults are either: relatively continuous across dam footprint, have >3 m apparent horizontal displacement and/or relatively thick fault gouge.
* Smaller faults and bedding parallel shears, generally less than 3 m apparent horizontal displacement, are not shown.
URS

Figure 6

The principal faults observed in the Dam 1E foundation, in terms of continuity, thickness of gouge, and offset included the Pedro Miguel "Stewart Trace" Fault (at the south abutment), the 1+860 Fault;1+240 Fault and the 2+380 Fault (named for their intersection at the dam centerline station).

The Pedro Miguel Stewart Trace Fault crosses Dam 1E at the south abutment (Photo 4). The fault was mapped at about the same location as mapped by Stewart (Stewart, et al, 1980) at Fabiana Hill. Two strands of the fault were mapped designated the Stewart Trace Fault "East Branch" and the Stewart Trace Fault "West Branch" (URS, 2014a). The total displacement in early Miocene rocks across the faults is uncertain but there is evidence for significant displacement, including the juxtaposition of dissimilar bedrock types and changes in the orientation of bedding across the faults. Exposures of alluvium were sparse overlying the fault at Dam 1E; however foundation cuts at Dam 1W (along the continuation of the fault to the southwest across the PAC) produced extensive exposures of alluvium overlying the Pedro Miguel Fault, as described below.

Photo 4. East and West Branches of the Pedro Miguel "Stewart Trace" at the south end of Dam 1E.

The 1+860 fault was mapped continuously across the dam (Figure 7, Geologic Map of 1+860 Fault) as a set of parallel strands with varying thicknesses of sheared rock and gouge (Photo 5). The fault produces apparent right-lateral displacement of approximately 160m in distinct La Boca Formation beds (Figure 7). At the outboard toe of the dam the 1+860 Fault is overlain by several sequences of unbroken alluvium (Photo 6). Several trenches were excavated into alluvium overlying the 1+860 Fault (Figure 8). Alluvial samples were dated by AMS and OSL and place the age of the alluvium as latest Pleistocene to early Holocene (URS 2014b).

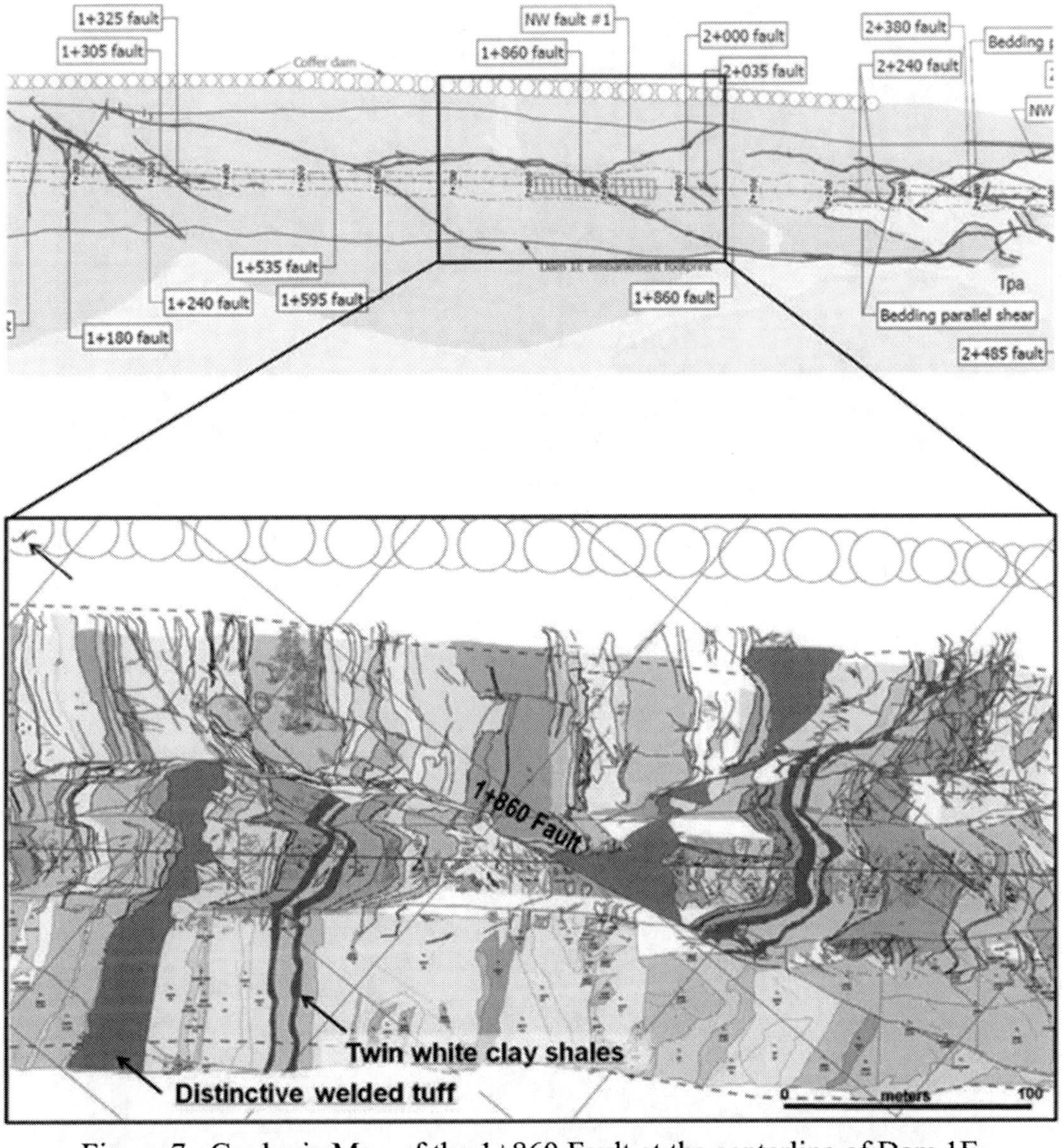

Figure 7. Geologic Map of the 1+860 Fault at the centerline of Dam 1E. Distinct marker beds in the La Boca Formation are offset 160m in a right lateral

Photo 5. View north along strike of 1+860 Fault in the outboard shell. Fault zone is about 2m+ wide at this location.

Photo 6: The 1+860 Fault was mapped below alluvial deposits in cuts bordering Dam 1E. Pedro Miguel Locks in background.

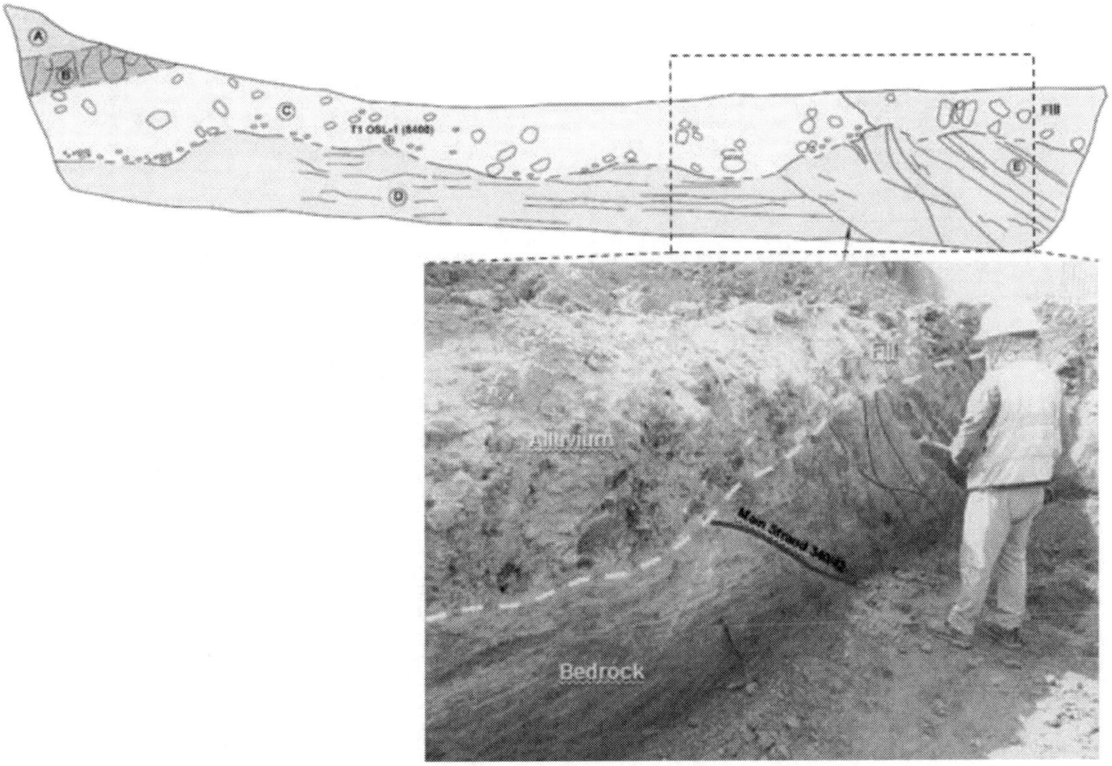

Figure 8. Geologic log of one of several trenches in alluvium across the 1+860 Fault. The unbroken alluvium was age dated as late Pleistocene to early Holocene.

The 1+240 fault was mapped continuously across the foundation. The strike is generally north-south, with moderate to steep dips to both the east and west. Approximately 300 m of right lateral displacement occurs across the 1+240 Fault in the PAC. At the outboard toe of the foundation excavation, strands of the 1+240 Fault are overlain by unbroken alluvium. Samples of the alluvium yielded late Pleistocene dates.

The 2+380 Fault was mapped over a length of several hundred meters in the core trench; the fault was mapped as terminating in the outboard shell. Near Fabiana Hill the 2+380 Fault either merges with a bedding plane shear or is offset by it. Several faults mapped at the south portion of the dam appear to branch or splay from more extensive faults mapped in the PAC; although none were through going across the foundation at Dam 1E.

The "Main Trace of the Pedro Miguel Fault" mapped by ECI (2007a,b) was not exposed in the foundation at Dam 1E. The anticipated location of the fault was based on trenches within the proposed footprint of Dam 1E (ECI 2010). Regular beds of La Boca Formation with distinct marker beds were mapped dipping at a low angle across the width of the dam (Photo 7 and 8). The foundation excavation was between 13 and 18 meters below the same area where paleoseismic trenches were excavated during design studies. The exposed geologic conditions indicated that a fault that would correlate with the Main Trace of the Pedro Miguel Fault does not exist at Dam 1E.

Photo 7. View of the core trench at Dam 1E with distinctive clay shale "red bed" exposed across the width of the dam. The "Main Trace of the Pedro Miguel Fault" had been mapped at this location during design studies.

Photo 8. View of initial excavation at Dam 1E with continuous beds of La Boca Formation across the cleared footprint in the same area where the "Main Trace of the Pedro Miguel Fault" had been mapped. The upper PAC cuts are in columnar

The PAC excavation West Bank cut slope produced extensive exposures of Miocene columnar basalt (Photo 8) at the mapped location of the Main Trace of the Pedro Miguel

Fault. The well-formed uniform cooling joint pattern in the basalt rock mass would be capable of revealing even a very minor fault. The absence of any significant fault in the columnar basalt further confirmed the Main Trace of the Pedro Miguel Fault did not exist (URS, 2012).

Fault Investigations at Dam 1W

At Dam 1W, the Pedro Miguel Stuart Trace Fault was well exposed in dental excavations of the foundation at the location mapped by Stewart et. al. (1980). The fault zone observed in the core trench was approximately 10 m wide where it juxtaposes basalt and Pedro Miguel Formation. The fault zone could be readily traced from the core trench to the adjacent back cut excavation slope (Photo 9). The 1:1 (H:V) cut slope provided an exposure of alluvium at least several meters thick overlying bedrock across several hundred meters of the back cut. The alluvium consisted of several meter thick basal gravel overlain by reddish-brown fine sand, silt and clay. Alluvial deposits were not displaced by the Stuart Trace Fault.. Figure 9 shows a geologic log of the alluvium/bedrock contact in the back cut at Dam 1W.

Photo 9. View of the Stewart Trace and Southeast Strand Faults in the excavation at Dam 1W.

Figure 9. Geologic log of Dam 1W back-cut exposure showing age dated basal alluvial gravels overlying the Stewart Trace of the Pedro Miguel Fault (fault shown in red). The top part of the slope is rock fill.

When the outer shell rockfill had been placed to within several meters below the alluvial contact elevation, the Locks contractor provided an excavator to help clear the slope and better expose the basal alluvial gravel contact. A series of progressive deepened notches and benched trenches were excavated to map the alluvial contact in detail. Samples of the alluvial gravel matrix and overlying fine-grained alluvium were collected for laboratory age dating including OSL and radiocarbon AMS. This detailed trenching and age dating performed at Dam 1W demonstrated that the "Stewart Trace Fault" does not displace basal alluvial gravels of early Holocene age (URS, 2012b).

A splay of the Pedro Miguel Fault was also exposed in the Dam 1W foundation, about 200 m south of the Stewart Trace Fault (Photo 9). This fault had been recognized in the PAC where it branches southeasterly from the Stewart Trace Fault, continuing into and across the south end of Dam 1W. This fault, named the "Southeast Strand Fault", was mapped into the dam back cut where it is overlain by the same alluvial gravel sequence. By benching and deepening the exposure with an excavator, it became apparent the "Southeast Strand Fault" had produced some minor apparent displacement (0.3 m) in the lower portion of the basal alluvium (Photo 10).

Photo 10. Minor fault offsets of basal alluvial gravel observed on the Southeast Strand fault at Dam 1W.

CONCLUSIONS

As-built geologic mapping of the Dam 1E foundation, as well as fault investigations performed by ACP and URS outside the PAC-4 boundaries (URS, 2012b, 2014a), revealed a complex pattern of faulting associated with the Pedro Miguel Fault Zone (PMFZ) crossing the southeast half of the Borinquen Dam 1E, the PAC and Dam 1W along a north-south trend (Figure 10). The PMFZ forms a system of sub-parallel faults which appear to be distributing tectonic strain across a zone that is at least several hundred meters wide in the vicinity of Dam 1E.

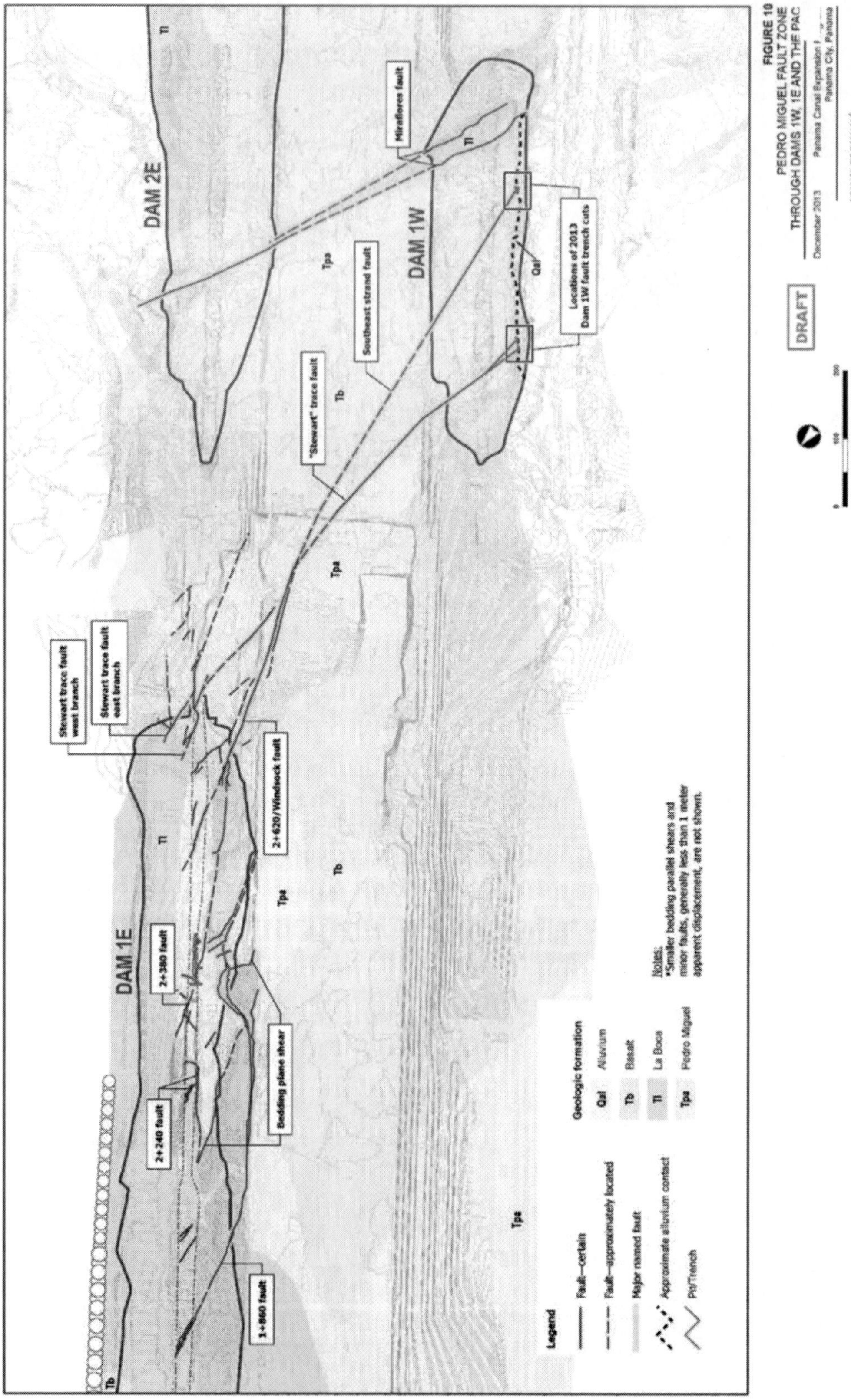
DAM 1E
DAM 2E
DAM 1W
Tpa
Tb
Tl
Qal
Miraflores fault
Southeast strand fault
"Stewart" trace fault
Locations of 2013 Dam 1W fault trench cuts
Stewart trace fault west branch
Stewart trace fault east branch
2+620/Windsock fault
2+380 fault
2+240 fault
Bedding plane shear
1+860 fault
Legend
Fault—certain
Fault—approximately located
Major massed fault
Approximate alluvium contact
Pit/Trench
Geologic formation
Qal Alluvium
Tb Basalt
Tl La Boca
Tpa Pedro Miguel
Notes:
*Smaller bedding parallel shears and minor faults, generally less than 1 meter apparent displacement, are not shown.
DRAFT
FIGURE 10
PEDRO MIGUEL FAULT ZONE
THROUGH DAMS 1W, 1E AND THE PAC
December 2013

Figure 10

South of Dam 1E, the PMFZ narrows slightly, extends across the PAC and crosses the footprint of Dam 1W as two strands referred to as the "Stewart Trace Fault" and the "Southeast Strand Fault" (ACP - URS, 2013). Trenching performed at Dam 1W (URS, 2012b) demonstrated that the "Stewart Trace Fault" does not displace basal alluvial gravels of early Holocene age. Trenching at Dam 1W also demonstrated that the "Southeast Strand Fault" has produced only minor apparent displacement (~0.3 m) of the same basal alluvial gravels of early Holocene age. Based on continuity of the fault zone between the two dams, it is reasonable that Holocene slip on the PMFZ is similar at Dam 1E to that observed at Dam 1W (URS, 2012b).

Considering actual fault locations encountered, and fault displacements measured in the Miocene bedrock, the core and chimney filters and drains of Dam 1E were widened at the south abutment to accommodate the Stewart Trace Fault (East and West branches). The dam core was also widened at the intersection with the 1+860 Fault, which was mapped along a 1,200m length of the dam. Activity on 1+860 Fault was unknown at the time the core trench was excavated, therefore the decision was made to widen the core at that time. Subsequent mapping efforts in the outboard shell provided an opportunity to age date unbroken alluvial deposits overlying the 1+860 Fault. As a result it was determined the 1+860 Fault was overlain by, and does not displace early Holocene dated alluvium at Dam 1E (URS, 2014b).

The 1+240 Fault was mapped extending across the Dam 1E footprint and extending into the PAC. Similar to the 1+860 Fault in style, the 1+240 Fault produced significant right lateral displacement in early Miocene La Boca beds. Trench logging and multiple age dates indicate the 1+240 Fault does not displace latest Pleistocene/early Holocene age alluvium. Therefore, the core of Dam 1E was not widened at the intersection with the 1+240 Fault.

The seismic design criteria required that the thickness of the core, chimney filter and drains be widened within 50 m of active fault locations. It was anticipated that widening would be required at the Pedro Miguel Fault (Stewart Trace Fault) and the Pedro Miguel Fault "Main Trace". With all the fault displacement information obtained during construction geologic mapping at Dam 1E, it was a prudent decision to allow for dam widening for potential displacement in areas of the PMFZ.

The PAC is a nearly 300 m wide and up to 70 m deep at the location of the Pedro Miguel Fault between Dam 1E and 1W. This provided an extraordinary opportunity to observe and map the major strands of the Pedro Miguel Fault at levels substantially below the original grade of the PAC. During excavation of the PAC, locations of the fault were observed, measured and surveyed (GPS) in temporary construction cuts that are now removed.

ACKNOWLEDGEMENTS

ACP and URS participated in geologic mapping, field review and map preparation during construction of Dam 1E, listed below. The authors appreciate the opportunity to have worked together at Boriquen Dam 1E.

ACP	URS
Derek Irving	Paul Salter
Pastora Franceschi	Daniel Meier
Jamie Arrocha	Robert Urban
Francisco Ponce	Christopher Goetz
Romy Coronado	Derek Rector
Roberto Campo	Adam Avakian
Mauricio Lacerda	Jason Moore
Roberto Miranda	David Schug
Eladio Almengor	Jason Castañeda
E.E. Portugal	

REFERENCES

Adamek, S., Frohlich,C., Pennington, W., 1988. Seismicity of the Caribbean-Nazca boundary; constraints on microplate tectonics of the Panamá region. J. Geophys. Res. 93, pp. 2053-2075.

Earth Consultants International, 2006, Geomorphic Evaluation of the Miraflores, Pedro Miguel, Azota and Caballo Faults, report to the Autoridad del Canal Panama.

Earth Consultants International, 2007a, Paleoseismic Trenching of the Pedro Miguel Fault in the Coccoli Located Immediately Southwest of the Panama Canal, report to the Autoridad del Canal Panama, February.

Earth Consultants International, 2007b, Quantitative Characterization of the Pedro Miguel Fault, Determination of Recency of Activity on the Miraflores Fault, and Detailed Mapping of the Faults through the Proposed Borinquen Dam Location, report to the Autoridad del Canal Panama, November.

Earth Consultants International, 2010, Additional Trenching of the Pedro Miguel and Miraflores Faults in Cocoli, immediately southwest of the Panama Canal, report to the Autoridad del Canal Panama, August.

Kirby M, Jones D, MacFadden B, 2008. Lower Miocene Stratigraphy along the Panama Canal and Its Bearing on the Central American Peninsula. PLoS ONE 3(7): e2791.

Kaufmann, R. D.; Irving, D. ; Yuhr, L. ; Casto, D. ; 2008, A geophysical investigation for the Panama Canal expansion; American Institute of Professional Geologists; Arvada, CO; USA; United States.

Mejia, L., and others, 2011. Design of the Dams of the Panama Canal Expansion, Proceedings of the United States Society on Dams 2011 Annual Meeting and Conference San Diego, California, April.

Neef, 1991, A clastic dike-sill assemblage in late Miocene (c. 6 Ma) strata, Annedale, Northern Wairarapa, New Zealand, 1991, New Zealand Journal of Geology & Geophysics, Vol. 34: 87—91.

Pinilla, R., Sampaco, K., and Irving, D. 2011. The Panama Canal's Third Set of Locks Project: Geologic Setting and Site Characterization. Geo-Frontiers 2011: pp. 2356-2365.

Rockwell, T., Gath, E., Gonzales, T., and others, 2010. Neotectonics and Paleoseismology of the Limon and Pedro Miguel Faults in Panama: Earthquake hazard to the Panama Canal, Bulletin of Seismological Society of America, v.100, no. 6, pp. 3097-3129

Stewart, R.H., Stewart, J.L., and Woodring, W.P., 1980. Geologic Map of the Panama Canal and Vicinity. Republic of Panama. U.S. Geological Survey Miscellaneous Investigations Series Map I-232, scale 1:100,000 1 sheet.

Trenkamp. R., Kellogg, J.N., Freymueller, J.T., Mora, H.P., 2002. Wide plate margin deformation, southern Central America and northwestern South America, CASA GPS observations J.S Am. Earth Sci. 15, pp. 157-171.

URS, 2008, Geologic Assessment of Design of New Borinquen Dams. Technical Memorandum prepared for Autoridad del Canal de Panama, July 2008.

URS, 2009, Task A.1.4, Foundation Materials – Dam 1E Geotechnical Interpretive Report (GIR). Report prepared for Autoridad del Canal de Panama, January 2009.

URS, 2012a, Geologic Memorandum Fault Investigations, Construction of Borinquen Dam 1E. URS Task Order # 21. URS Project No. 26818044.00021, April 2012 (Submittal No. C1106).

URS, 2012b, Geologic Assessment: Faults exposed at Dams 1E, 1W and the PAC. Task Order # 30 (Geologic Support to ACP). URS Project No. 26818044.00030, December 2012 (Submittal No. C1156).

URS, 2013, Geologic Review of Fabiana Hill for Dam 1E Grouting and Stability Assessment. Task Order # 34. URS Project No. 26818044.00021, June 2013 (Submittal No. C1201).

URS, 2014a, Assessment of Recency of Displacement of Pedro Miguel Fault Borinquen Dam 1E. ACP Task Order # 35. URS Project No. 26818044.00035, February 2014 (Submittal No. C1240).

URS, 2014b, Additional Fault Investigations, Including Station 2+310 to 2+715 and South Abutment, Borinquen Dam 1E, Expansion of the Panama Canal. ACP Task Order # 41. URS Project No. 26818044.00041, November 2014 (Submittal No. C1272).

URS, 2015a, Final Draft Report - Foundation Drilling and Grouting, Panama Canal Expansion Program, PAC-4 Borinquen Dam 1E, July 2015 (Submittal No. C1315).

URS, 2015b, Final Construction Report, Panama Canal Expansion Program, PAC-4 Borinquen Dam 1E, Rev. 1 May 2015 (Submittal No. C1271-003).

Wegner et. al., 2011, Magmatic history and evolution of the Central American Land Bridge in Panama since Cretaceous times. Geological Society of America Bulletin 2011; 123; 703-724.

Woodring WP, Thompson TF (1949) Tertiary formations of Panama Canal Zone and adjoining parts of Panama, Bulletin of the American Association of Petroleum Geologists 33, 223 – 247.

Woodring, W.P 1957, Geology and Paleontology of Canal Zone and Adjoining Parts of Panamá. U.S. Geol. Surv. Prof. Pap. 306-A.

3-D MODELING AT BOUNDARY DAM, WA: VISUALIZING A GEOLOGICALLY COMPLEX SITE

Robert Cannon, PG[1]
Frederic Snider, PG[2]
Kim Pate, PG,PE[3]
Andre Ball, PE[4]
Gary Rogers, PG[5]
J. Hawkins Gagnon, PG[6]

ABSTRACT

Boundary Dam is a 340-foot-high double curvature arch concrete dam in a steep, narrow canyon on the Pend Oreille River in northeastern Washington State. The dam is founded on the Metaline Limestone, which ranges from massive to severely brecciated and recrystallized. The site is criss-crossed by a number of fracture zones encountered in drill holes, exploratory adits, underground works, and mapped on the surface. The rocks also reveal several prominent joint sets. Original explorations at the site included over 150 borings, 700 packer tests, and exploratory adits to support the design of the dam foundation, the large underground powerhouse, the forebay, penstocks, and access and diversion tunnels. Recent work at the dam includes laser scanning of both aboveground and underground areas, and sonar bathymetric surveys of the reservoir and tailwater.

Rock bolting of surficial slide blocks and underground excavations was completed during construction and later. Continued concerns over the potential stability of exposed rock faces and the lack of an integrated understanding of the site geology prompted development of a CAD-based, 3-D geologic model of the dam site. The model integrates the recent laser scanning and bathymetric surveys with the original exploratory borings, adits, geologic mapping, water testing, and instrumentation. The intent of the model is to aid in the visualization and enhance our understanding of the geology and hydrogeology of the site, identify those Potential Failure Modes with the highest risk of occurrence, and to eventually upgrade the existing surveillance and monitoring program in light of the Potential Failure Modes.

INTRODUCTION

By the mid-1940's, it was already clear that Seattle City Light would have to develop additional electric capacity to support the growing population, and the utility was actively scouting potential hydroelectric dam sites throughout Washington State. The stretch of the Pend Oreille River where Boundary Dam was eventually built was identified as a possible dam site in 1945, but this area was a key strategic mining area for lead and zinc. A 1947 geological report noted concerns that damming this stretch of the Pend Oreille River would flood existing mines and prohibit future mine development. As a result, the

[1] Schnabel Engineering, 11-A Oak Branch Dr., Greensboro, NC 27407, rcannon@schnabel-eng.com
[2] Schnabel Engineering, 11-A Oak Branch Dr., Greensboro, NC 27407, fsnider@schnabel-eng.com
[3] Seattle City Light, 700 5th Ave, Suite 3200, Seattle, WA, 98124, kim.pate@seattle.gov
[4] Seattle City Light, 700 5th Ave, Suite 3200, Seattle, WA, 98124, andre.ball@seattle.gov
[5] Schnabel Engineering, 11-A Oak Branch Dr., Greensboro, NC 27407, grogers@schnabel-eng.com
[6] Schnabel Engineering, 11-A Oak Branch Dr., Greensboro, NC 27407, jgagnon@schnabel-eng.com

project was shelved until 1953, when the US Army Corps of Engineers re-evaluated the site using a 100-foot lower pool elevation and smaller reservoir. The Bureau of Mines had completed a regional boring and testing program in 1949 that demonstrated little connection between the river and the mines. The new reservoir elevation was considered unlikely to flood existing mines nor make future mining impossible, as the major mineralized zones lay several hundred feet below the river level. One hole drilled by a mine operator found artesian conditions at a depth of 2,000 feet, with ground water rising to the top of the hole and 100 feet above the existing river level.

With the lower pool elevation and reduced potential for mining impacts, several sites along this stretch of the Pend Oreille River were investigated by the USACE. The Corps selected the present Boundary Dam location as the preferred site based on the foundation geology, configuration of the narrow gorge, access, and available land for supporting construction activities. The site is 1 mile south of the Canadian Border, where the Pend Oreille Rivers runs northward into Canada, then turns west to join the Columbia River, as shown on Figure 1. On October 30, 1953, the City of Seattle filed an application for a preliminary permit with the Federal Power Commission (FPC). The permit was granted on August 30, 1954. The FPC issued a 50-year license on March 10, 1961. Construction of the dam started in 1964 with substantial completion in 1968. A recent photo of the dam is presented as Figure 2.

Figure 1. Location of the Boundary Dam Site (from Google Maps)

Figure 2. Boundary Dam as viewed from downstream

SITE GEOLOGY

The Boundary Project is located in the Northern Rocky Mountain geographic province at an average elevation of about 1,800 feet above sea level. The region is underlain by a thick sequence of Precambrian through Devonian metamorphic and sedimentary rocks. Quaternary continental glaciation left extensive glacial tills and thick unconsolidated fluvial and lake deposits throughout the region, including a number of well-defined alluvuial terraces and clay-rich lake bottom sediments. Since retreat of the glaciers in the late Pleistocene, the Pend Oreille River has excavated through the glacial deposits, exposing the bedrock along most of the river valley, and cutting deep canyons in the bedrock in several areas.

The dam project is founded on the Metaline Formation of late Cambrian age. The Metaline Formation is dominantly a fine-grained limestone, dolomitic limestone, and dolomite. The total thickness of the Metaline Formation exceeds 3,000 feet, and its color varies from shades of bluish-white, buff, cream, to grey to dark brown and black. Chert nodules are common in the upper 500 feet of the unit, and the lower section of the unit contains interbeds of limey shale. The regional strike near the dam site is roughly east-west, with a southerly dip of 40 to 50 degrees. Fracturing is common throughout the unit, reflecting a long history of tectonic deformation. The major joint sets are steeply dipping and generally trend either to the northeast or the northwest. The fracture zones range from narrow slickensided planes to wide brecciated zones. Other joints, many shallow dipping, occur throughout the unit with few apparent systematic trends. Brown iron staining is common along partially-healed and open fractures. Old solution collapse breccias are also common within the limestone and dolomite units, although they are typically re-cemented and solid.

Some solution cavities and minor caves and caverns are reported in the limestone, but are scarce in the project area. Despite the long deformational history, the Metaline at the site is generally sound rock which stands in near-vertical cliffs over 400 feet high. Extensive water pressure testing of the rocks at the site yielded only a few zones of high water takes associated with fracture zones, indicating that, in general, the rock mass is tight. Rock strength testing during site investigations indicate compressive strengths of the intact limestone range from 5,000 to over 16,000 psi. Figure 3 shows the exposed limestone rocks on the left abutment.

Figure 3. Metaline Limestone exposed on the left abutment of Boundary Dam

SITE INVESTIGATIONS

Geologic reports related to the geology and lead and zinc mining in the site area go back as far as the early 1940s. The USACE drilled a series of 26 borings in the general area in 1944 and 1945 as part of the original site selection investigations. Seattle City Light (SCL) completed additional drilling in 1961 through 1964 at the preferred site, including 106 vertical and angled coreholes. These 132 borings total over 28,800 feet of exploratory drilling, logging and testing. Water pressure testing was completed in nearly all the boreholes, providing data from over 700 packer tests. Of those 700 tests, only about 150 had water takes in excess of one Lugeon, and fewer than 65 tests exceeded 10

Lugeons, and those high-take tests were primarily shallow and along the known unhealed fracture zones.

In addition, a series of exploratory adits were excavated in each abutment and the area of the proposed underground powerhouse, just west of the dam. A number of exploratory borings, drains, and rock-anchor holes were drilled from these adits. Also, in-situ rock testing within the adits was completed using plate bearing tests, flat jack tests, and relaxation tests. Borings for dam instrumentation, including piezometers, extensometers and inverted pendulums, were logged and provide addition geologic data.

Surficial geologic mapping was also completed during the site investigations in the mid-1960s. A number of joints and fracture zones were mapped as faults exposed at the surface and within dam excavations and were mapped and characterized. Discontinuity surveys along both abutments were completed in 1994 by geologists rappelling down the canyon walls with the goal of identifying potentially unstable blocks. Also, SCL staff completed a program of laser scanning of the dam, abutments and underground works, and this data was used for computer-based discontinuity mapping in 2012.

SCOPE OF THE 3-D GEOLOGIC MODEL

Although the intact Metaline limestone and dolomitic rocks at the dam site are considered sound and competent, the number and varied orientations of the fracture zones and joints led to rock stability concerns in several areas of the project. Considerable care was given to orienting the access tunnels and machine hall excavations to be perpendicular to the main fractures. Even so, extensive rock bolting of surficial blocks and underground excavations was completed during construction and later to enhance the stability of the abutments and underground works. Several rounds of discontinuity mapping and kinematic analyses indicate that a few rock masses on the abutments may be potentially unstable. Uncertainties in the orientations, vertical extent and mechanical properties of the joints and fracture zones exposed at the surface, however, has precluded an accurate assessment of the size and stability of potentially removable blocks, a thorough kinematic analysis, or an adequate assessment of the risk posed by block failures.

Given these uncertainties, and continued concerns over the potential stability of exposed rock faces, coupled with a lack of an integrated understanding of the site geology prompted development of a CAD-based, 3-D geologic model of the dam site.

Specific goals of the 3-D geologic model include:

- Develop a computer-based 3-D topographic and bathymetric base model,
- Represent the surface traces of known fracture zones and joints sets,
- Extend the surface traces of the discontinuities to depth using subsurface data,
- Identify the lateral and vertical extent of mappable geologic units, and
- Identify potentially removable blocks that may pose safety or operational risks.

The model integrates the recent laser scanning and bathymetric surveys with the original exploratory data. The intent of the model is to enhance our understanding of the geology and hydrogeology of the site, address rock mass stability concerns, identify the geology-related Potential Failure Modes with the highest risk of occurrence, and to eventually upgrade and expand the existing surveillance and monitoring program in light of the Potential Failure Modes. Figure 4 illustrates the combined topographic and bathymetric data of the dam, abutments and downstream area, and a portion of the reservoir nearest the dam.

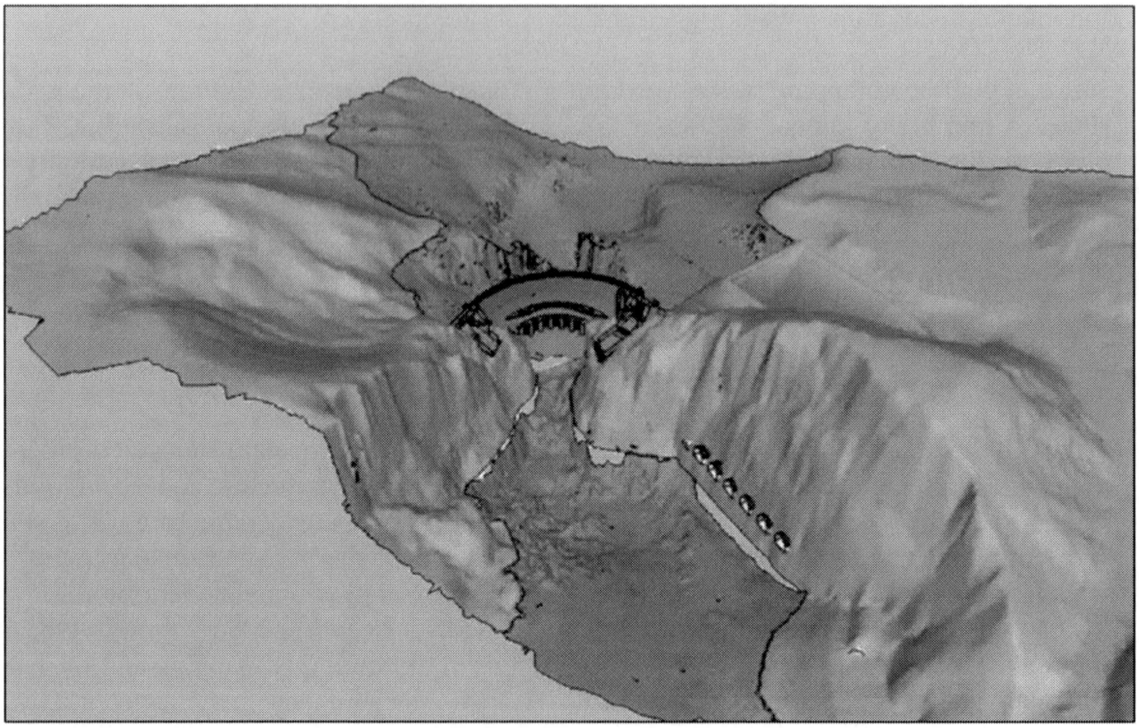

Figure 4. 3-D topographic and bathymetric model of the Boundary Dam area. Green areas are topography, blue areas are bathymetry. The underground powerhouse is in the rock mass where the six transformer tunnel portals are visible in left abutment area.

Geologic Mapping Data

The steep topography of the site canyon yields nearly continuous exposure of the Metaline Formation rocks along the dam foundation and on each of the abutments. Geologic mapping was completed during the original site investigations in the mid 1960s. Figure 5 presents the pre-construction geologic map of the dam site, highlighting what were identified as faults at the site. This information, coupled with refinement of the apparent fault trace locations using the recent point cloud data, aerial LiDAR data from the dam site and reservoir, and high-resolution photographs obtained using an unmanned aerial vehicle (UAV) has been incorporated into the 3-D model to delineate the surficial traces of these features.

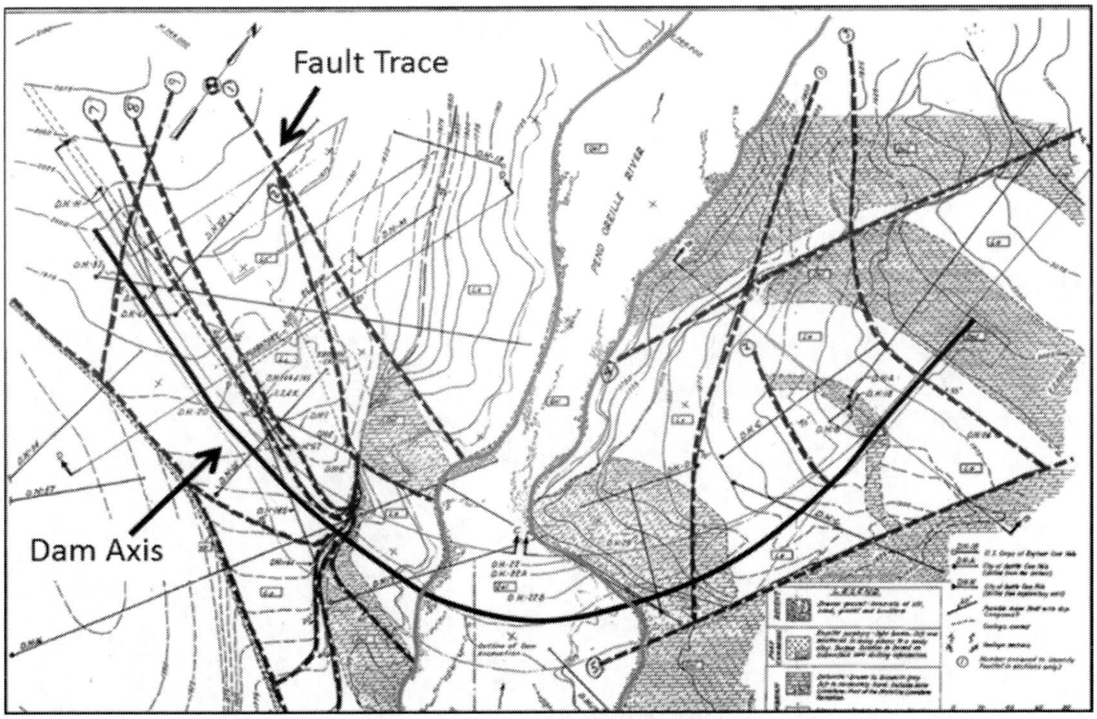

Figure 5. Pre-construction geologic map of the dam site. The approximate dam axis is shown in black, and mapped faults are shown as blue dashed lines. The river valley is outlined in light blue. Contour interval is 25 feet. Total relief is about 350 feet.

Borings and Stratigraphy

Between the early site selection investigations in 1944 and 1945 by the USACE and the 1961 through 1964 site investigations by SCL, a total of 132 exploratory borings were completed at the site. Although many of the borings were vertical, a large number of the borings were angled holes designed to intersect the mapped faults and joint sets. The total length of the borings was in excess of 28,800 feet, or about 5.5 miles. All borings were logged for lithology and structure, percent recovery, and drilling information such as loss of drilling water. Most of the borings were logged at a scale of 1 in=5 ft or less, such that the data set consists of several hundred pages of handwritten and typed logs.

For inclusion into the 3-D model, the paper boring logs were compiled and catalogued. The lithologies identified in the logs were grouped into a series of categories (such as massive limestone, brecciated and cemented limestone, limestone with chert nodules, dolomite, etc.) and the depth intervals of each lithology tabulated on Excel™ spreadsheets for conversion to intervals in the 3-D model space. Judgement had to be exercised during the categorization process, as the original geologists logging the cores varied in their descriptions of the lithologies. As such, conversion of the paper logs to the electronic spreadsheets had to be completed by our experienced geologists.

Other data tabulated from the boring logs included percent recovery, rock quality designation (RQD), color, and texture. Zones of drilling water loss or gain were also tabulated.

Figure 6 shows a small portion of the 3-D model boring data. Different colors were assigned to the various lithologies. Other symbol types, colors, and sizes were used to display the other borehole information. The various data types and categories were assigned to layers within the model, so that the investigator can display only one set of data, or select any combination to display.

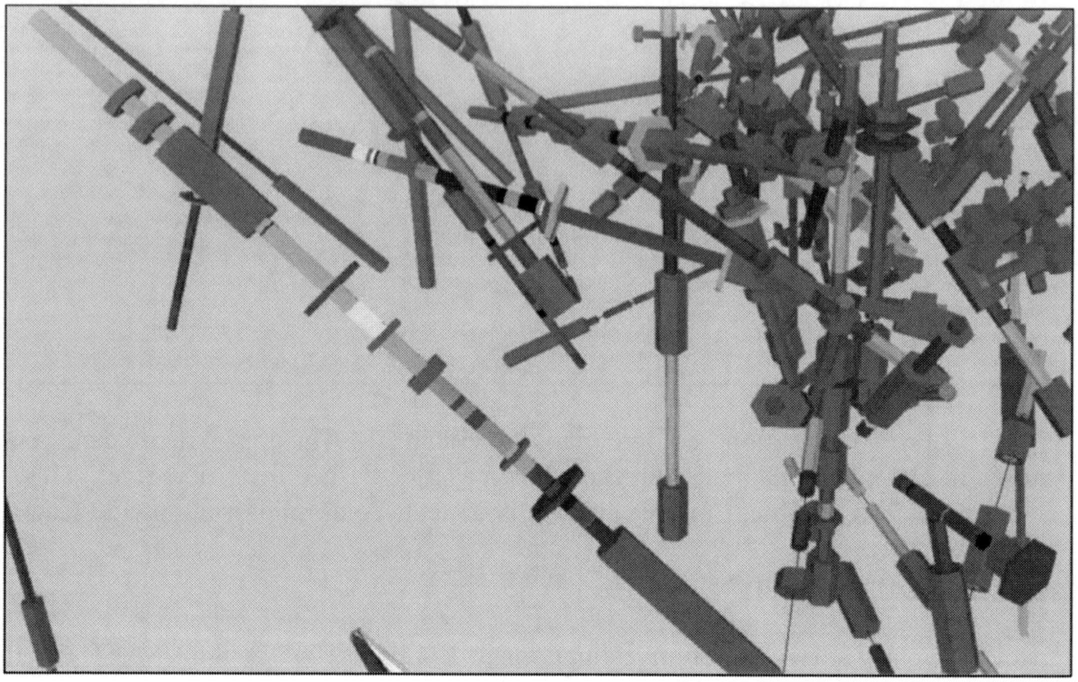

Figure 6. Visualization of borehole log data. Borings are shown in their actual position and orientation. Although difficult to interpret on this 2-D rendition, rotating in 3-D on the computer screen and turning the various data layers on and off allows interpretation and correlations.

Discontinuity Data from Site Borings

Incorporating discontinuity data from the boring logs into the 3-D model proved equally challenging. The vast number of individual joints and fractures noted in the logs made tabulation of this data infeasible and of limited usefulness. However, intervals longer than a foot or so along the cores logged as highly to severely fractured were tabulated on the spreadsheets to aid in the extrapolation of the surface fractures to depth. Unfortunately, the cores were not oriented, so that the true fracture orientations are not recoverable. Fortunately, some of the later instrumentation borings and recent drain and anchor holes were video logged, such that true fracture orientations can be extracted in certain areas. Figure 7 illustrates one example of the discontinuity visualization options.

Figure 7. Visualization of discontinuity mapping. View is underground beneath the left abutment, looking towards the dam. The six transformer tunnels and the powerhouse access tunnel are visible on left side of image. Each boring is shown as a black line. The gaps in the boring lines represent intervals originally logged as highly to severely fractured. This visualization can be rotated in 3-D to aid in identifying the continuity of surface fracture zones at depth.

Water Test Data

Nearly all of the borings completed at the site had water pressure testing done for the majority of the boring lengths. In a few cases, the entire hole length was tested at once, although in most cases packers were used to seal discrete intervals for testing. A total of 709 water pressure tests were completed during the two investigative phases. The test results were reported on the boring logs as gauge pressure at the surface and volume of water injected over time. In order to compensate for test interval depths, variations in pressures used, and make the data set comparable across the site, the water pressure test data was tabulated on electronic spreadsheets and takes were converted to Lugeons.

Despite the pervasive joints and fracture zones mapped at the site, fewer than 20% of the packer test intervals had water takes of 1 Lugeon or more. The highest take was 97 Lugeons, in a very shallow test interval in a known open fracture zone. The distribution of all the packer tests with takes of one Lugeon or more is presented on Figure 8. The packer test data was incorporated into the 3-D model, as shown on Figure 9.

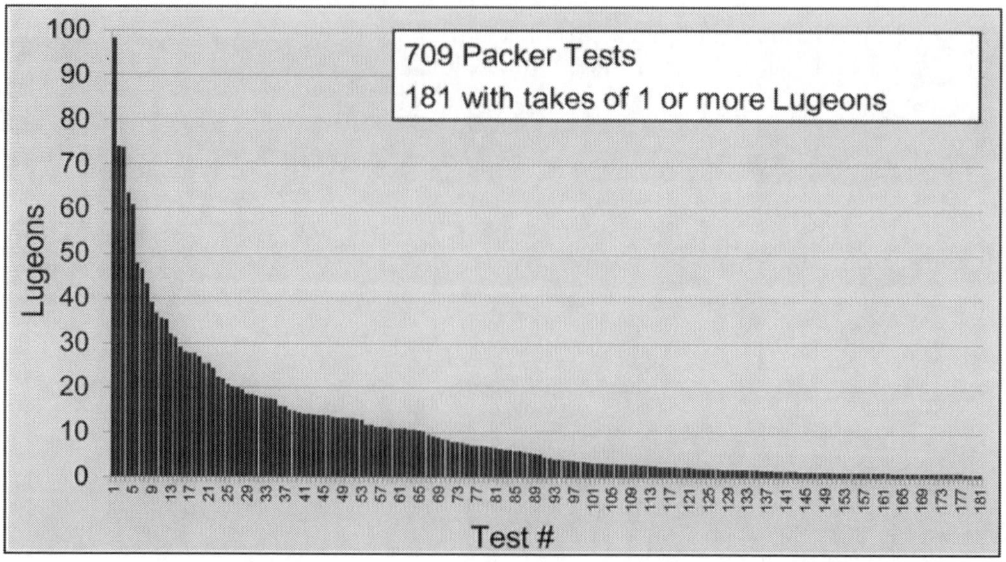

Figure 8. Distribution of the 181 packer tests with takes of 1 Lugeon or greater.

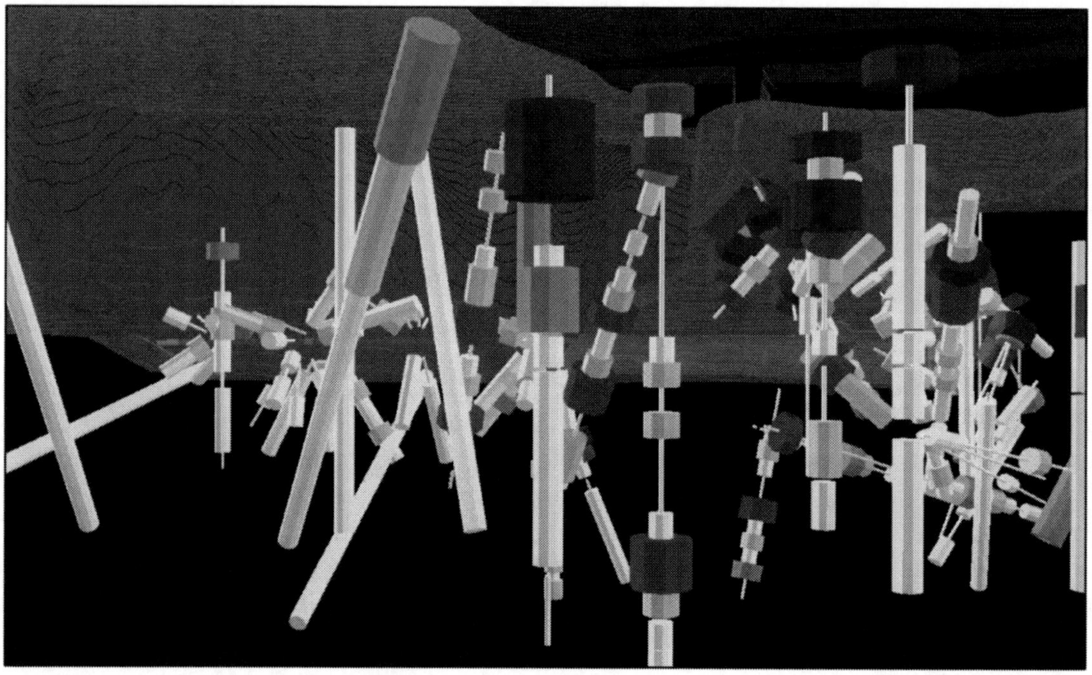

Figure 9. Visualization of the packer test data. Each set of stacked colored cylinders represents one borehole. The diameter of each cylinder is proportional to the water take in Lugeons. The length of each cylinder represents the interval tested in the borehole. Although difficult to interpret on this 2-D rendition, rotating in 3-D on the computer screen allows alignment of the data with lithology and structure data.

The water test information was used to help extrapolate the surface traces of the known fracture zones to depth and to identify open versus healed fracture zones. This information also supports the ultimate goal of identifying potentially removable blocks and the mechanical characteristics of their bounding surfaces.

INTEGRATING THE DATA SETS

Once the sitewide geologic mapping, borehole logs and water test data was integrated into the 3-D geologic model, the model was used to address the specific goals of the analysis outlined in the introduction of this paper. The surface traces of the mapped faults were extended to depth using the borehole logs and the maps, photographs and point clouds of the underground works. The cross-cutting relationships of the major fracture zones were established to allow termination or offsets of the zones to better characterize subsurface conditions. An example of this mapping is presented as Figure 10.

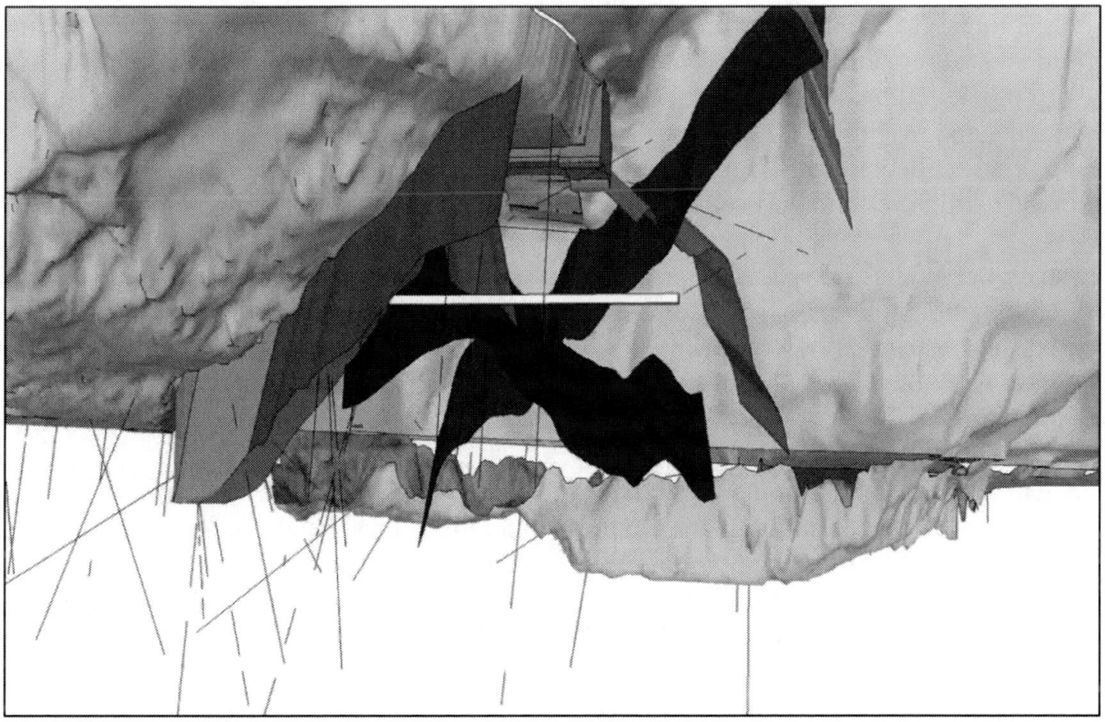

Figure 10. Example of fracture zone extension in the subsurface based on subsurface mapping and testing information.

The boring logs and water test data were also used to characterize the fracture zones, from completely healed to partially or fully open. Only the open or partially open fractures are deemed critical to the identification of removable block and stability analyses.

The value of the computer-based 3-D model became obvious at this point in the analysis, as the entire model could be rotated in space to help visualize alignments, and individual data layers could be turned on and off to either supplement the visualization or hide information that was not relevant or obscured the images. This mapping exercise was both interactive and collaborative, with both the team geologists and the CAD/GIS specialists working together to develop the final model.

DETAILED DISCONTINUITY MAPPING

Identifying potentially removable blocks that could pose dam safety or operational risks is based in part on the mapping of the 3-D configurations of the facture zones at the site. However, there are also a number of generally random joints in the rock mass, the orientations and extents of which are not fully represented in the site-wide 3-D geologic model. Therefore, a complete assessment of the potential for removable blocks near the dam includes more detailed assessment of discontinuities along both abutments.

The importance of discontinuity mapping on the dam abutments has been recognized for a long time. Several previous investigations completed at the dam include a 1994 study with geologists rappelling down the abutment slopes mapping, characterizing and taking attitude measurements of discontinuities along vertical transects from the river shore to the top of the abutments. A remote characterization of rock discontinuities on the abutments and machine hall access tunnel was also completed in 2012, using the survey point clouds collected by SCL. A simplified kinematic analyses was completed based on assumed geometries and generalized mechanical characteristics of the bounding surfaces.

These earlier studies served as a starting point for our analysis, which was based on the SCL survey point clouds, high-resolution photos of the abutments using a UAV, and limited field investigations as part of the 2015 FERC Part 12 inspection process. Original construction photos were also of great help in defining discontinuities in areas now occupied by the dam.

The geo-referenced SCL point cloud data was loaded into MapTek I-Site Studio™ software to map discontinuity planes expressed on the abutment surfaces and within the access tunnel. Integrating the point cloud data with the photography and our own field work allowed the definition of discontinuities bounding potentially removable blocks. Kinematic studies and more detailed stability analyses are planned.

An example of the discontinuity mapping is presented as Figure 11. Discontinuities are mapped either as discrete polylines traced on the surface, or as the polygonal outlines of exposed joint faces or slip planes. These polylines and polygons are then directly imported into the 3-D model, and attitude data are exported for plotting on stereronets and kinematic analyses. Wherever possible, the subsurface data from borings and adits is used to determine the extent of these planes in the subsurface and establish potential intersections that are not visible on the surface. These intersections then define the size and geometry of potentially removable blocks in the abutments and the site excavations. Water test data and the boring logs helped in determining the degree of cementation and characteristics of gouge or other fillings.

Figure 11. Example of point cloud data being used to define discontinuity planes on the exposed abutment wall.

CONCLUSION

The development of a computer-based 3-D geologic model using over 5.5 miles of boring logs, hundreds of packer tests, various scales and ages of geologic surface mapping, data from numerous adits and excavations, construction photos and recent laser point clouds is a non-trivial exercise. Nor is it just a matter of assigning a team of data-entry clerks to grind their way through the piles of information. To the contrary, each step of the model development required thoughtful collaboration between the expert team of geologists and CAD/GIS specialists. The team needed to identify and extract the required information, establish the formats for tabulating and QA'ing the data, define consistent scales and datums for the various datasets, and ensure that the data sets could be tracked, imported into the CAD model, checked, updated and corrected when necessary. The careful assigning of separate layers to the various datasets and subsets proved key to the visualization and analyses capabilities of the model. Finally, familiarity with the site and all the data sets was also critical, as the expert team could often view a modeled dataset and immediately assess if the visualization made sense.

Once the challenges were overcome, however, the 3-D model provided SCL and its consultants an integrated visualization of the otherwise scattered and unmanageable paper reports, boring logs, and test data for the project. The Boundary Dam 3-D Geologic Model met all of its initial goals, and has proven to be an even more valuable resource to the project than originally anticipated.

ACKNOWLEDGEMENTS

The authors from both Schnabel Engineering and Seattle City Light would like to acknowledge the valuable support provided by their respective teams of geologists, engineers, surveyors, and CAD and IT specialists in developing the 3-D geologic model. This paper only touches on the numerous labor-intensive data collection, evaluation and data management tasks that were required to complete the model. The excellent work completed by our support teams was key to meeting our immediate goals and to provide a well-documented base model for future expansion and analyses.

REFERENCES

Bechtel Leedshill, January 1968; City of Seattle, Board of Public Works, Department of Lighting: Boundary Project Design Report, Pend Oreille River, Washington, 253 pages.

Hong West & Associates, September 1994; Engineering Geology Investigation, Abutment Erosion Control System, Boundary Dam, Pend Oreille County, WA. 114 pages.

Klohn Crippen Berger, March 2012; Boundary Dam Project, Remote Characterization of Rock Discontinuities and Kinematic Analysis at the Boundary Dam. 74 pages.

UNDERSEEPAGE AND HEAVE MITIGATION AT DOUGLAS SADDLE DAM NO. 1

Daniel A. Gilbert, PE[1]
Alan F. Rauch, PhD, PE[2]
Paul J. Cooper, PE[3]
Scottie L. Barrentine, PE[4]

ABSTRACT

Douglas Saddle Dam No. 1 is a 100-foot tall, homogeneous clay embankment structure in eastern Tennessee. The largest of ten saddle dams in the Douglas Hydroelectric Project, Saddle Dam No. 1 was built in the 1940's by the Tennessee Valley Authority (TVA). The embankment dam is underlain by steeply dipping, thinly laminated, shale bedrock with strike running roughly perpendicular to the axis of the dam. Recently installed instruments in the shale measured artesian conditions at the toe of the dam during normal pool levels. To better understand the effects of underseepage in the foundation shales, a seepage model was developed and compared to the available piezometer readings. Historical flowrate data were used to model the efficiency of the dam's internal drainage features. The seepage model was used to extrapolate recorded pressures to predict piezometric levels that will develop during flood pools, when lake levels will be much higher than the pool of record. A gravity relief well system was designed to mitigate high groundwater pressures at the dam toe, which have the potential to cause soil heave and instability during flood pool events. The relief well system was installed with a unique, inclined arrangement to intercept the upended shale bedding and maximize effectiveness. Project challenges included a lack of piezometric data for elevated headwater pools, development of a representative seepage model, accelerated schedules that did not permit field well tests, and installation of the wells in the shale foundation.

INTRODUCTION

Saddle Dam No. 1 is located southeast of Douglas Dam in Sevier County, Tennessee. The Douglas project was constructed by TVA between 1942 and 1943. Saddle Dam No. 1 (Figure 1) is a compacted earth embankment with internal drainage features, and is the largest of the ten saddle dams associated with the Douglas reservoir.

Normal operating pool levels for Douglas Lake vary between elevations 954 feet in the winter and 994 feet in the summer. The pool-of-record at Douglas Dam is 1,002.4 feet, recorded in May 1984. The design probable maximum flood (PMF) level is 1,021.3 feet.

[1] Principal, Stantec, 1409 North Forbes Rd., Lexington, Kentucky 40511, Daniel.Gilbert@Stantec.com
[2] Sr. Principal, Stantec, 1409 North Forbes Rd., Lexington, Kentucky 40511, Alan. Rauch@Stantec.com
[3] Geotechnical Engineer, Stantec, 10509 Timberwood Circle Suite 100, Louisville, Kentucky 40223, Paul.Cooper@Stantec.com
[4] Civil Engineer-Geotechnics, Tennessee Valley Authority, Dam Safety Engineering, 1101 Market Street, Chattanooga, Tennessee 37401, SLBarrentine@TVA.gov

Figure 1. TVA's Douglas Saddle Dam No. 1

Cross Section and Geology

The maximum cross section of Saddle Dam No. 1 is presented in Figure 2. A sand and gravel drainage blanket, with a network of drain outlet pipes, was built under the downstream slope. The embankment is 1,920 feet long with a maximum height of about 100 feet. A paved public road crosses the dam crest at elevation 1,017 feet. A reinforced concrete floodwall, constructed in1987, raises the dam crest to elevation 1022.5 feet.

The Douglas project is situated on the French Broad River in the Valley and Ridge physiographic province of eastern Tennessee. The reservoir is located in a broad syncline with Knox Dolomite on both limbs and shale formations in the center (Laurence 1943). Saddle Dam No. 1 is founded predominately on the Athens Shale formation, with the Lenoir Limestone under a short length near the northern abutment. The foundation shales are thinly laminated and steeply inclined, with a dip that varies from about 45° to 90° across the dam site. Significantly, the bedding planes are nearly vertical under the maximum dam section and strike transverse to the dam axis.

A single-line grout curtain was installed during original construction. Vertical grout holes were typically spaced 25 feet apart and, to better intercept the shale bedding, inclined

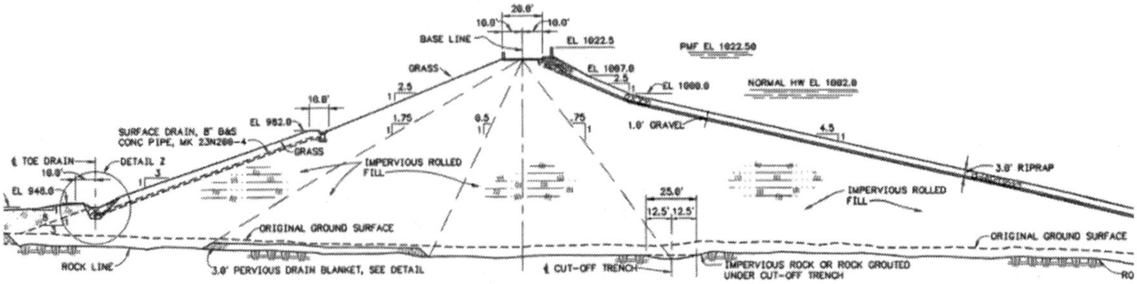

Figure 2. Douglas Saddle Dam No. 1 Cross Section

grout holes were drilled 100 feet apart. Several large grout takes were recorded. The foundation soils were removed under the dam footprint, and a 25-foot wide cutoff trench was excavated to sound rock or grouted rock. Artesian conditions were observed in some boreholes along Saddle Dam No. 1 during original construction, prior to reservoir filling.

Updated Stability Assessment

In 2012, TVA initiated work to perform new geotechnical explorations, associated laboratory testing, and updated stability evaluations for the Douglas project. Six borings, ten CPT holes, and six open standpipe piezometers were completed in 2013 at Saddle Dam No. 1. Artesian pressures were subsequently measured in the instruments located at the toe of the dam, even with the reservoir at the lower winter pool elevation. The measurements were consistent with historical data from an observation well downstream of the saddle dam.

These observations triggered further assessments and exploration work in 2014. Five additional borings were completed, and eight fully-grouted vibrating wire piezometers were installed. The exploration program was structured to ascertain piezometric conditions within the foundation near the toe of the dam, within the embankment, and near the blanket drain.

With the upended shale bedding and compacted clay embankment, most of the seepage at Saddle Dam No. 1 passes through the bedrock foundation. An improved seepage model was developed, with an attempt to match data from the newly installed piezometers. Projections were made of the piezometric conditions that might develop under the dam when Douglas Lake reaches probable maximum flood (PMF) levels.

High groundwater pressures in the shale could cause heaving of the ground surface at the toe of the dam during a PMF event. To mitigate this potential risk, TVA elected to construct relief wells and a stability berm at the toe of the dam. The relief wells, which drain by gravity, are inclined to intercept the near-vertical bedding of the shale. These modifications were completed in early 2015.

DRAINAGE BLANKET AND PIPES

A pervious drainage blanket and drain pipes were constructed under the downstream slope during original construction. The efficiency of this drainage system has a significant impact on groundwater pressures that develop at the toe of the dam. The drainage blanket was included in the seepage analyses, where a constant-head boundary condition was used to model flows from the outlet pipes. The appropriate head to assign in the drainage blanket, which varies for different pool elevations, was characterized using historical data on the flowrates from the system.

Layout of Drainage Blanket

The drainage blanket consists of 1.5 feet of gravel placed on exposed rock and then covered with 1.5 feet of sand. At the maximum dam section, the drainage blanket is approximately 125 feet wide. The drainage blanket follows the top of rock along the length of the dam, dipping to approximately 920 feet elevation at the lowest point under the tallest dam section and extending up to elevation 990 feet on both abutments.

Seepage collected in the drainage blanket is discharged via 12 lateral pipes, spaced at 20 to 100 feet and at differing elevations. The laterals are eight-inch diameter concrete pipes, laid with open joints that extend about 20 feet into the drainage blanket. The laterals flow to a 12-inch concrete pipe in a toe drain along the length of Saddle Dam No. 1. Video inspections in 2005 and 2011 revealed the accumulation of sediment in certain drain pipes, gaps and offsets at several pipe joints, and no flow emanating from some of the lateral pipes.

Water from the toe drain pipe is discharged to concrete-lined channels downstream of the dam, where flow rates are measured in two weirs (Weir 2 on the right side and Weir 3 on the left side). The maximum dam section (Figure 2) is located about midway between these two weirs. There are no lateral drain pipes over a length of approximately 200 feet in the central area of the dam, where the drainage blanket is about 20 feet below the lowest outlet. Above Weir 2, the toe drain pipes enter a catch basin at an invert elevation of 942 feet; the system will not discharge flows if the hydraulic head in the drainage blanket is below this level.

Historical Data on Drain Flowrates

Flow rates measured over 50 years at Weir 2 and Weir 3 are plotted in Figure 3. In addition to seepage from the blanket drain, the weirs capture flows from the toe drain, various near-surface drainage pipes, and catch basins on the mid-slope. This data provides an indication of seepage through the dam and drainage blanket, but significant interpretation is required to use this information effectively.

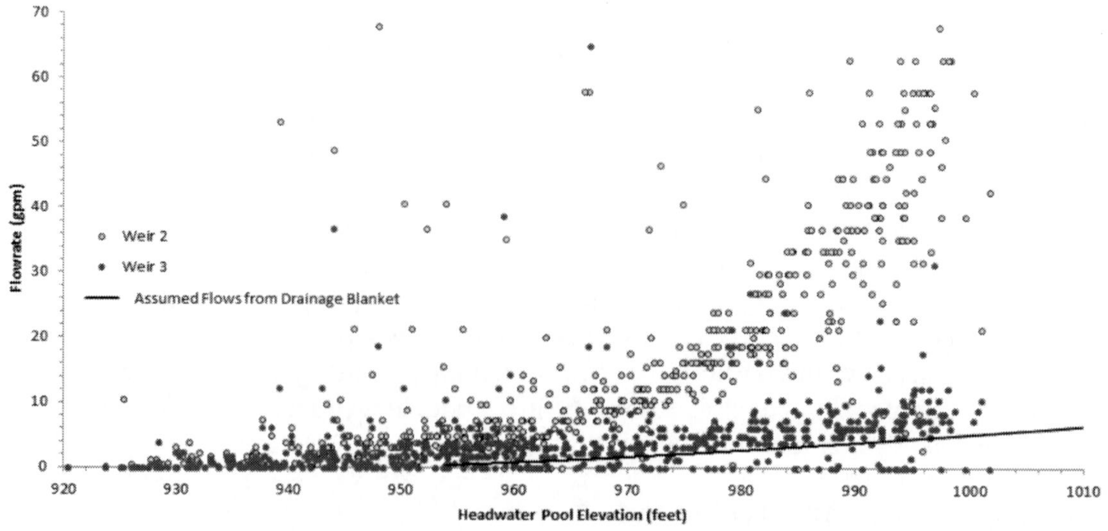

Figure 3. Flow Rates Measured from 1964 to 2014 at Weir 2 and Weir 3

During periods of wet weather, much of the measured flow is likely due to surface runoff and infiltration. Seepage through the dam and into the drainage blanket should constitute a greater portion of the observed flows during dry periods. Hence, in considering the data in Figure 3, a trend line drawn through the lower range of the data will provide a better indication of flows coming from the drainage blanket.

The data in Figure 3 shows exponentially increasing flows as the pool rises (the effect is much more pronounced at Weir 2). This behavior may be understood by considering a flownet for seepage through a cross section of the dam. The flow rate per length along the dam axis (Q/L) is given by:

$$\frac{Q}{L} = k\,\frac{n_f}{n_d}H \qquad (1)$$

where k is the hydraulic conductivity, H is the head difference, and n_f/n_d is the number of flow channels over the number of head drops in the flownet.

Water is collected by the drainage blanket over some length (L) of the dam. The full length of the drain is about 1,000 feet between the abutments; however, because the drainage blanket follows the rock surface across the valley of the dam foundation, the length actively draining seepage water changes as the pool rises. The full 1,000 feet is not active unless the pool is significantly above the ends of the blanket at both abutments. Below this elevation, a rising pool engages an increasing length of the drainage blanket, leading to increasing discharges. The active length of the blanket for a given headwater pool elevation is difficult to quantify. However, because both the head on the dam (H) and the length over which seepage is collected (L) increase with the pool elevation, there is exponential increase in flow rates from the drainage blanket (Q) with a rising pool.

The rock line and drainage blanket rise more steeply on the left side of Saddle Dam No. 1. Thus, drainage collected on the right side is more sensitive to the increasing active length of the drainage blanket. This partly explains why the flows at Weir 2 (which is connected to the drainage blanket on the flatter right abutment) increase more rapidly with higher pools, compared to the flow rates measured at Weir 3 (connected to the drainage blanket on the steeper left side). Weir 3 also captures less abutment drainage than Weir 2, which measures significant flows from a separate, shallow drainage pipe system. Given these considerations, the lower trend of the data from Weir 3 (Figure 3) provides a better indication of the seepage collected in the drainage blanket.

There are no measured flow rates for pools above about 1,001 feet. However, because the full length of the drainage blanket will be active, the flow rate should not increase exponentially for these pool levels. Linear extrapolation was used in Figure 3 to obtain flow rates expected for headwater pools above 1,001 feet.

<u>Model of Heads in the Drainage Blanket</u>

A seepage model was used to predict the piezometric conditions within the dam cross section for various pool levels. In the critical area at the downstream toe, the model is sensitive to the hydraulic head assigned to the drainage blanket. This head, representing the effectiveness of the drain pipes, was estimated for a range of headwater pools. Flow rates in the drainage system increase as the pool rises, with a concurrent increase in the gradient along the length of the drainage blanket. The head in the drainage blanket at the modeled cross section thus increases with the pool level.

Seepage along the length of the drainage blanket (along the axis of the dam) is a function of the hydraulic gradient (i_d) and can be modeled using Darcy's Law:

$$Q_d = k_d i_d A_d = k_d \frac{H_d - H_o}{L_d} A_d \qquad (2)$$

The hydraulic conductivity (k_d), drain length between the modeled cross section and the nearest lateral drain pipe (L_d), the head at the outlet (H_o), and the cross sectional area of the drainage blanket (A_d) are constants that do not change with pool level. The flow rate (Q_d) is thus proportional to the head in the drainage blanket (H_d), and the following ratio should be relatively constant for all pool levels:

$$\frac{H_d - H_o}{Q_d} = \text{constant} \qquad (3)$$

The head at the outlet (H_o) can be estimated from the physical layout of the lateral pipes that drain the blanket. The catch basin above Weir 2 has an invert elevation of 942 feet, so the lateral pipes in this area are above 942 feet. A reasonable estimate for total head in the lateral pipes nearest to the modeled dam cross section is about 943 feet.

In the seepage model, the head in the drain was set to $H_d = 950$ feet, for a pool at 991.8 feet. For this pool, the Weir 3 data (Figure 3) indicates a flow rate of about 4 gpm. Water in the drainage blanket flows from two directions toward Weir 3, so assuming about half comes from the right (the area of the modeled cross section), the appropriate flow rate is $Q_d = 2$ gpm. This leads to the following ratio for seepage in the drainage blanket, between the modeled cross section and the lateral drain pipe above Weir 3:

$$\frac{H_d - H_o}{Q_d} = \frac{950 \text{ feet} - 943 \text{ feet}}{2 \text{ gpm}} = 3.5 \text{ ft/gpm} \qquad (4)$$

Now, assuming a ratio of 3.5 ft/gpm and $H_o = 943$ feet for all pools, and obtaining the flow rate from the Weir 3 trend line (Figure 3), the head in the drain at the maximum cross section can be estimated for any pool level.

The heads in the blanket drain, plotted in Figure 4, were estimated in this manner. There is little data from the drainage blanket to verify this model. However, one CPT probe hole penetrated the drainage blanket in the winter of 2013. The measured head was 5 feet below that indicated by the relationship in Figure 4, but the pool was 20 feet above normal on that day, and pressures could have been rising as embankment seepage changed in response to the rising pool.

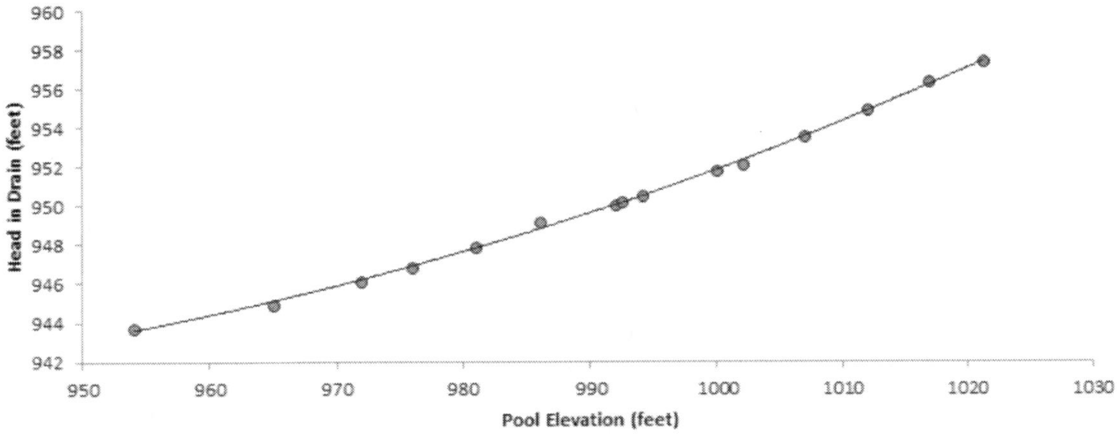

Figure 4. Estimated Head in the Blanket Drain for a Range of Pool Elevations

SEEPAGE MODEL

A two-dimensional, finite element model was developed to represent seepage through and beneath Douglas Saddle Dam No. 1, prior to installation of the relief wells. The seepage model represents the maximum section of the dam (Figure 1). The hydraulic conductivity parameters and boundary conditions were adjusted using piezometer readings measured in 2014, when the headwater pool elevation was 991.8 feet.

Boundary Conditions

The drainage blanket under the embankment was included in the seepage model. A fixed hydraulic head was assigned to one node in the drain, resulting in a flux of water out of the mesh that represents outflow from the lateral drain pipes. The heads assigned for different pool elevations, plotted in Figure 4, were determined from the historical data as previously described.

In addition to the blanket drain, the dam cross section includes a toe drain that runs the length of the dam. The toe drain consists of a concrete drainage pipe surrounded by crushed stone. Based on the elevation of the pipe and outlet, a constant head of 944 feet was assigned to the corresponding node in the seepage model.

Just downstream of the dam toe, water collected in the toe drain discharges into two, concrete-lined open channels. Groundwater in the area downstream of the dam was assumed to be hydraulically connected to the flows in the channels, where the ditch line is at approximately 941.5 feet elevation. A constant head of 942.0 feet was assigned across the area downstream of the dam toe.

Hydraulic Conductivities

Hydraulic conductivities were measured in the laboratory using retrieved soil samples. Where such data were not available, permeability parameters for vertical and horizontal seepage (k_v and k_h) were estimated based on typical values for similar materials. These values were then adjusted in the seepage model, in an attempt to better match the predicted groundwater pressures to the field piezometer readings. The resulting hydraulic conductivities for the key material zones are summarized in Table 1.

Table 1. Hydraulic Conductivity Parameters

Material	k_v (cm/s)	k_h/k_v
Embankment Fill	4.0×10^{-9}	4
Drainage Blanket	1.0×10^{-3}	2
Residual Foundation Soil	8.5×10^{-9}	5
Athens Shale	5.0×10^{-7}	10

Model Results

Predictions from the seepage model were compared to the hydraulic head measured at each piezometer, for a headwater pool of 991.8 feet. The piezometers were located beneath the dam crest, the mid-slope bench on the downstream face, at the dam toe, and in areas downstream of the toe. The instruments were located mostly toward the bottom of the dam embankment or in the upper 20 feet of the shale bedrock.

Even with an extensive effort to match the model to the piezometer readings, significant differences remained. At two locations, the model predictions were within 0.1 feet of the field measurements. Elsewhere, the predicted heads from the seepage model ranged from 8.4 feet below to 17 feet above the measurements. These discrepancies are attributed to a number of potential factors, including three-dimensional seepage, unknown seepage sources or paths, uncertainties related to the properties of the weathered shale, nonuniform seepage through the shale beds, pipes in the drainage system that may be plugged, and piezometer readings that lag changes in pool elevation.

The finite element model is helpful in understanding the seepage conditions beneath and within Saddle Dam No. 1. However, given the noted discrepancies for piezometric levels during normal pool conditions, the seepage model does not provide reliable predictions of groundwater pressures at all points within the dam cross section for PMF pool levels.

PIEZOMETRIC LEVELS AT THE DAM TOE

The potential for soil heaving at the toe of Saddle Dam No. 1 was evaluated for a range of reservoir levels. The analysis required predictions of hydraulic head (or pore water pressure) at the base of the soil column (top of rock) at the dam toe.

Piezometer Data

Pore pressures were measured in two locations at the dam toe, at piezometers P4 and P9. The profile of total head, with the pool at elevation 991.8 feet, is plotted in Figure 5. Data from P9A.2 and P9A.3 bracket the pressures in a zone of weathered shale under the soil column; an average value (949.8 feet) was used in the subsequent analyses.

The heads measured at P9A.2 and P9A.3 are also plotted in Figure 6. Unfortunately, there were no pressure measurements for higher reservoir levels. Three methods were used to extrapolate and predict the hydraulic head that is expected under the soil column at the dam toe; these predictions are plotted in Figure 6, for lake levels up to the PMF pool.

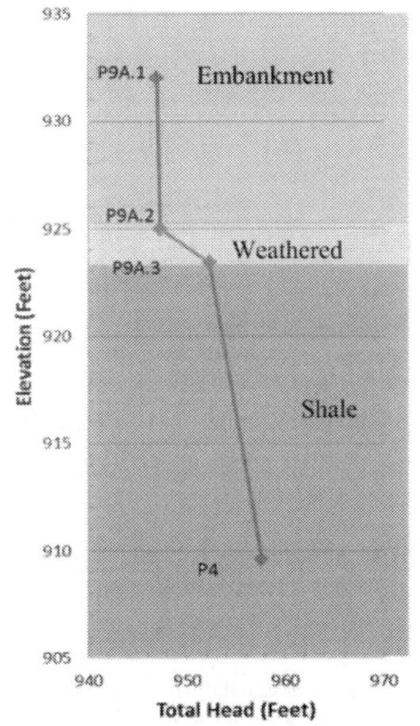

Figure 5. Head Profile at the Dam Toe, Measured when Pool at 991.8 feet

Predictions from the Seepage Model

The seepage model was used to predict heads at the top of rock for a range of lake levels. As seen in Figure 6, the seepage model predicts heads that are 5 to 10 feet higher than measured, for a headwater elevation of 991.8 feet. The seepage model thus yields a fairly conservative prediction of pressures at the dam toe.

Extrapolation Based on Confined Flownet

In the seepage model, the shale bedrock is about one hundred times more permeable than the clay embankment. Seepage through the embankment is relatively insignificant, and groundwater flow under the dam can be treated as a classic, confined aquifer problem. A confined flownet would show that the head at any location within the rock varies linearly with increases in the pool elevation. That is:

$$\frac{H_{headwater}-H}{H_{headwater}-H_{tailwater}} = \text{constant} \quad (5)$$

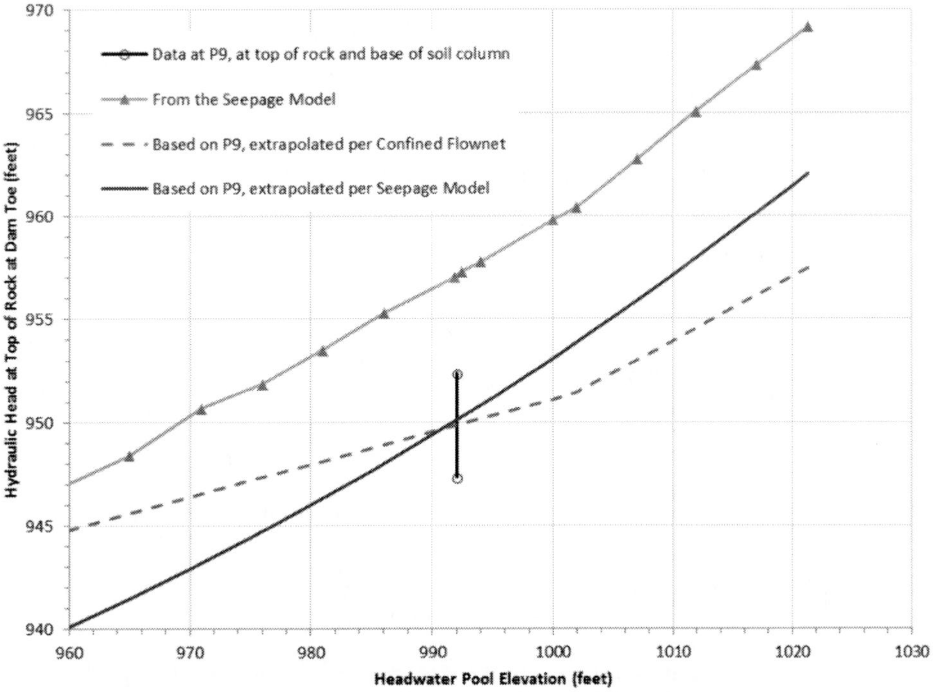

Figure 6. Predictions of Hydraulic Head at the Dam Toe

$H_{headwater}$ is the elevation of the upstream reservoir, $H_{tailwater}$ is the head at the downstream seepage exit, and H is the head at any location within the flownet.
With Equation 5, linear extrapolation can be used to predict what pressures will develop at higher or lower pool levels. The average measured head (at the top of rock at the dam toe) was H = 949.8 feet with a pool of $H_{headwater}$ = 991.8 feet. Assuming $H_{tailwater}$ = 942.0 feet, the constant in Equation 5 is 0.843. This value can then be used to estimate the head at the dam toe for other pool levels; the results are plotted in Figure 6.

During a PMF event, the downstream drainage channels are assumed to run full ($H_{tailwater}$ = 945.6 feet) due to heavy precipitation and runoff. This results in higher water pressures at the dam toe. To account for this effect, a linear adjustment was applied to the heads plotted in Figure 6, for pool elevations above 1,002 feet.

Extrapolation Based on the Seepage Model

The seepage model can also be used to extrapolate the available piezometer data, to predict pressures that will develop at the toe during higher pools.

First, the seepage model was used to predict the hydraulic head at piezometer P4, which is located about 15 feet below the top of rock at the dam toe. The resulting predictions, for a range of headwater pool levels, were fit with a quadratic equation (plotted in Figure 7). The P4 instrument was installed in 2013, so there is a lack of data for different pool elevations. A longer-term set of readings is available from observation well S-10, which is located nearby and measures average pressures in the shale over a depth interval of about 70 feet. A good model of the data from S-10 can be obtained by offsetting (by 5 feet) the trend line that was fit to the model predictions for P4 (Figure 7).

The same methodology was then used to extrapolate the total heads at the base of the soil column at the dam toe. The head predicted with the seepage model, for a range of pool levels, was fit with a quadratic trend line. This line, plotted in Figure 8, was offset 7 feet vertically to pass through the data from piezometers P9A.2 and P9A.3, for a pool of 991.8 feet (Figure 5). The vertical shift represents a correction to account for the difference between the head measured at this location and the value predicted with the

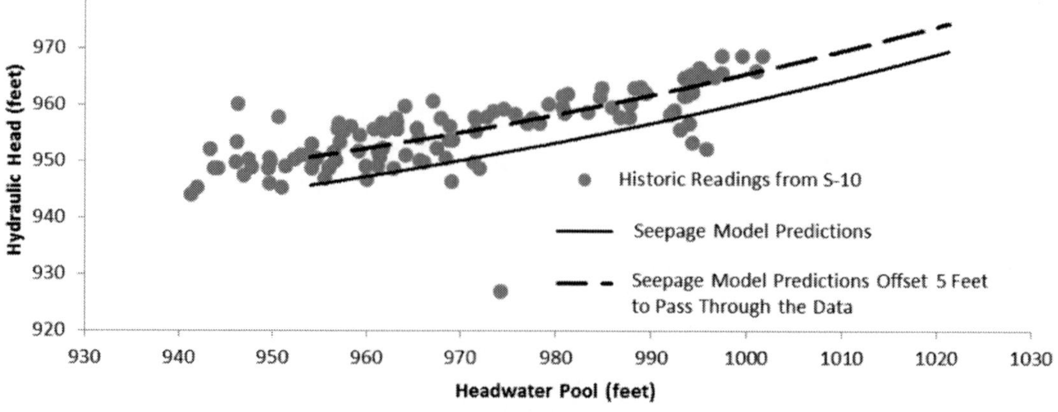

Figure 7. Seepage Model Predictions in Rock and Historic Readings from S-10

seepage model. The trend line through the data in Figure 8 is also shown on Figure 6.

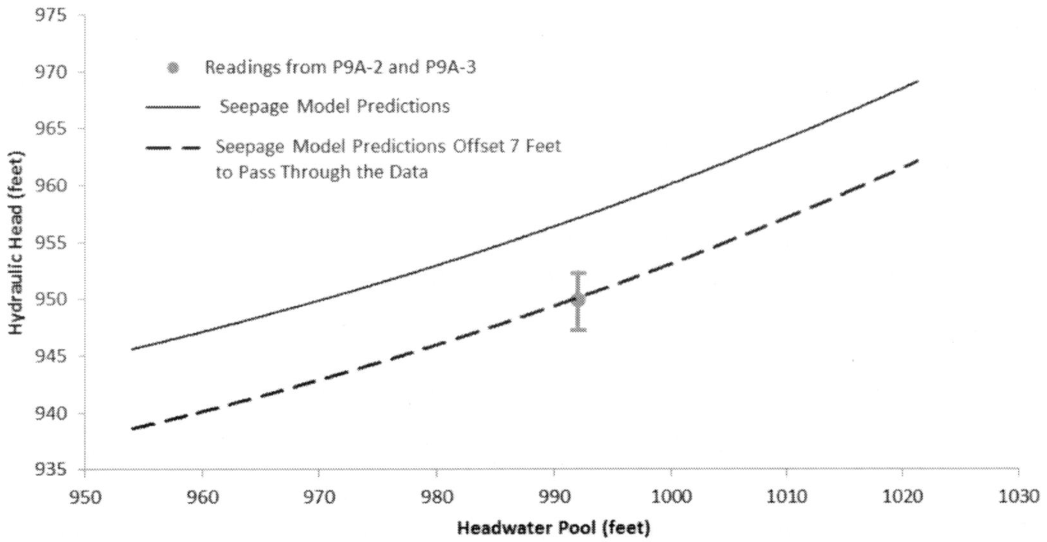

Figure 8. Seepage Model Predictions at the Base of the Soil Column at the Dam Toe

POTENTIAL FOR SOIL HEAVE

Potential Failure Mode

The assessment of PMF stability for Douglas Saddle Dam No. 1 focused on evaluating the potential to trigger soil heaving at the dam toe. Seepage is conveyed under the dam through joints, fissures, bedding planes, and laminations in the shale formation, which is hydraulically connected to the upstream reservoir. As the headwater pool rises, pore water pressures at the downstream toe of the dam increase. If high enough, groundwater pressures at the top of rock could cause heaving of the overburden soils at the dam toe. This would reduce the lateral support at the base of the dam, and could initiate sliding in the downstream face.

Assessment for the PMF Pool

Stability of the dam toe during a PMF event was evaluated in terms of the factor of safety against heaving. Two definitions for the safety factor were considered:

1. Seepage force definition, computed as the ratio of the downward gravitational pressure over the upward seepage pressure (USACE 1986).
2. Total stress definition, computed as the ratio of total vertical stress over the uplift pressure (USBR 2014).

In both cases, the factor of safety is evaluated using the groundwater pressures at the base of the soil column and the unit weight of the soil. The shear strength of the soil is conservatively neglected.

At Saddle Dam No. 1, the potential for triggering soil heave was evaluated by considering the vertical stability of a soil column at the dam toe. At the location selected for the analysis, 18.5 feet of fill material is found on top of the bedrock. The key input to this analysis is the pore pressure or hydraulic head expected at the top of the shale.

The three lines in Figure 6 represent predictions of hydraulic head at the top of rock at the dam toe, for a range of headwater pool elevations. The seepage model, which was adjusted based on piezometer readings across the dam site, provides an overly conservative indication of groundwater pressures in this critical location. The best prediction of the expected conditions was obtained by offsetting the trend line from the seepage model to pass through the measured piezometer data at the dam toe (extrapolation based on the seepage model).

The best estimate indicates that soil heaving will not occur at the dam toe during the PMF pool. However, the safety factors were less than desired for this extreme loading event. Without piezometer data for pool levels much above normal, there is significant uncertainty in the predicted water pressures in the shale bedrock. Interpreting these results with an abundance of caution, TVA initiated a dam modification project for Saddle Dam No. 1, to mitigate the risks of triggering soil heave during flood pool events.

DAM MODIFICATIONS

Design and construction of the dam modifications were accomplished on an accelerated schedule (roughly seven months). Initially, a rockfill berm was considered to stabilize the dam toe and downstream face. While this could have been built relatively quickly, the height and volume of the berm would have been excessive.

The stability concerns at Saddle Dam No. 1 are driven by seepage through the shale, so relief wells are an efficient alternative to improve the dam. A smaller toe berm was included in the design, to provide additional resistance against heaving at the toe and to improve embankment stability. The previously described seepage model was used to evaluate the effectiveness of the design. The existing drainage system was not disturbed by the modifications, but the outlet pipes were assumed to be plugged in the design analyses (the drainage blanket was assumed to have no outlet in the seepage model).

Relief Wells

A single line of 14 relief wells (Figure 9) was installed along the toe of Saddle Dam No. 1. The wells are located about 15 feet downstream of the toe drain, far enough to avoid potential interference with the existing drain pipes, but close enough to effectively lower the groundwater pressures under the downstream toe.

The wells are spaced at 25 feet on center, and aligned on an azimuth that is generally 20 degrees to the dam axis. This geometry was selected to provide access to the well heads for construction and maintenance. A drill rig can be backed up to each well, without having to maneuver between adjacent wells.

Figure 9. Relief Well System

The relief wells were inclined to intercept the fractures, fissures, and joints in the steeply dipping shale bedrock. Thirteen of the 14 wells were drilled 30 degrees from vertical. Twelve wells are inclined toward the right abutment. On the left side of the dam, the first well was oriented to penetrate under the left abutment, and the second well was installed vertically. On the right side of the dam, three wells were positioned on a natural ridge at the toe of the embankment. These wells intercept some of the seepage that appears to be coming through the shale under the right abutment. However, given the outlet elevations, these wells are not expected to produce water when the reservoir is at normal levels.

All of the wells were advanced to a tip elevation of 855 feet. With the selected inclination and spacing, the bottom tip of one well overlaps vertically with the top of the adjacent well. All planar rock fissures should be intersected by at least one well. Each well was grouted from the ground surface to a depth 2 feet below the top of competent shale (the rockline determined from rock coring along the well alignment).

Each relief well has a 65- or 85-foot long, 6-inch diameter, stainless steel well screen installed in rock. The screened intervals were set to collect seepage water from the shale rock, where there is little need to provide filtration. A coarse filter pack and a wide well screen opening were selected. The well screens and riser were installed with centralizers to maintain a uniform annular space for the filter pack.

Figure 10. Relief Well and Lateral Pipe

Given the artesian conditions in the rock, the collected seepage will drain by gravity from the relief wells (no pumps). Tee fittings were used to direct flow from each well riser to a 6-inch diameter lateral pipe (Figure 10). The laterals connect to a 24-inch collector pipe that runs parallel to the line of relief wells. Precast concrete manholes provide access for cleaning and maintenance of the drain pipes at each well. Water collected from the relief wells discharges to an open concrete channel, where flows are routed through a weir to monitor flow rates.

Toe Berm

A berm of graded sand and coarse stone aggregate was also installed on the dam toe (Figure 11). Four separate material gradations were specified: fine sand, medium sand, No. 8 coarse aggregate, and No. 3 coarse aggregate. The material gradations were checked for filter compatibility using the empirical methodology in USACE (2004).

The footprint of the toe berm was sized to protect the full width of the embankment fill from soil heave. The lateral limits of the rock fill were set to roughly cover the toe of the embankment, at the downstream extents of the dam, where the fill was originally placed on top of the exposed rock. The toe berm is approximately five feet thick and contains about 4,500 cubic yards of material.

Figure 11. Construction of Toe Berm

CONCLUSION

At TVA's Douglas Saddle Dam No. 1, instrumentation installed in 2013 recorded artesian pressures at the dam toe with a normal headwater pool. Seepage modeling was used to extrapolate this piezometric data, to predict the conditions that could develop at higher lake levels. Modifications were needed to mitigate the risks for soil heaving during a probable maximum flood event.

Relief wells and a rock berm were completed in March 2015 (Figure 12), to improve stability at the dam toe. The relief wells were inclined and arranged to intercept seepage that passes through the steeply dipping shale beds under the dam. Subsequent data has shown lowered piezometric conditions with appreciable flows from the relief well system. Project challenges included a lack of piezometric data for elevated headwater pools, development of a representative seepage model, accelerated schedules that did not permit field well tests, and installation of the wells in the shale foundation.

Figure 12. Completed Modifications to Douglas Saddle Dam No. 1

REFERENCES

Laurence, Robert A. (1943). *Geology of Douglas Dam and Reservoir.*

U.S. Army Corps of Engineers (1986). "Seep age Analysis and Control for Da ms." EM 1110-2-1901, September 30.

U.S. Army Corps of Engineers (2004). "General Design and Construction Considerations for Earth and Rock-Fill Dams." EM 1110-2-2300, July 30.

U.S. Bureau of Reclamation (2014). "Chapter 8: Seepage", *Design Standards No. 13, Embankment Dams*, DS-13(8)-4.1, January.

A DEEP SEEPAGE CUTOFF FOR THORNTON COMPOSITE RESERVOIR

Joseph R. Kovacich, PE [1]
Louis Storino, PE [2]
Raymond J. Franz, PE, D.GE [3]
Jonathan R. Bulger, PE [4]
Brian J. Kazyak, PE [5]

ABSTRACT

Chicago's Tunnel and Reservoir Plan (TARP) is one of the largest Civil Engineering projects ever constructed. As part of this megaproject, the North Lobe of the existing Thornton Quarry is designed to retain Combined Sewer Overflow (CSO) captured by the TARP Calumet Tunnel System and floodwater from Thorn Creek from a 24-hour, 20-year rainfall event. To utilize the North Lobe as a water retaining structure, the Metropolitan Water Reclamation District of Greater Chicago (District) modified the 2,700-ft long and 300-ft high Rock Dam that separates the North from the Main Lobe by physically plugging three gaps and constructing a deep seepage barrier along the length of the Rock Dam.

A 116-ft high roller compacted concrete (RCC) dam was constructed to plug the largest gap, and at lower elevations reinforced concrete tunnel plugs were constructed within two mine haul tunnels, plugging the two other gaps. To reduce seepage around these structures and through the rock mass discontinuities, state-of-the-practice grouting technology was utilized to construct a 500-ft deep, double-row grout curtain cutoff along the length of the Rock Dam and tie-in with the perimeter grout curtain along the other three sides and into a low permeable rock formation at the bottom. The physical barriers and seepage cutoff required careful design and construction.

The project was constructed beginning in 2009 and was put into service in August 2015 and experienced its first filling event in November, and by late December 2015 had been filled with approximately 70 feet of water. Instrumentation data and visual observations indicate the Rock Dam, structures, and seepage barrier are performing as intended.

[1] Principal Geological Engineer, MWH, 350 N. Orleans St. Suite 1301, Chicago IL, 60654. 312-831-3000, joseph.r.kovacich@mwhglobal.com.
[2] Principal Civil Engineer, Metropolitan Water Reclamation District of Greater Chicago, 111 E. Erie St., Chicago, IL 60611. 312-751-3167, storinol@mwrd.org.
[3] Division Manager, Hayward Baker Inc. 1350 West Lake St., Roselle, IL, 60172. 630-339-4300, RJFranz@HaywardBaker.com.
[4] Assistant Project Manager, Hayward Baker Inc. 1350 West Lake St., Roselle, IL, 60172. 630-339-4300, JRBulger@HaywardBaker.com
[5] Central Division Delivery Manager, MWH, 350 N. Orleans St. Suite 1301, Chicago IL, 60654. 312-831-3000, Brian.J.Kazyak@mwhglobal.com.

INTRODUCTION

In 1972, the District's Board of Commissioners adopted the Tunnel and Reservoir Plan (TARP) as a comprehensive pollution and flood control program for its 375 square mile combined sewer area, comprised of Chicago and 51 neighboring communities. The Deep Tunnel System (TARP Phase I) consists of 109 miles of large diameter, rock tunnels providing 2.3 billion gallons of CSO storage. Completed in 2006, these tunnels transport CSO to the three planned reservoirs around the Chicago metropolitan area. Construction of the reservoirs comprises TARP Phase II and when fully implemented will greatly increase the storage capacity of the system to approximately 18.25 billion gallons. One of the three reservoirs, the Thornton Composite Reservoir (TCR) is in service and can store nearly 8 billion gallons of CSO. The TCR is constructed in the North Lobe portion of the existing Thornton Quarry located in Thornton, IL, and is north of the I-80/294 Tollway as shown on Figure 1. The relative size and storage capacity of the TCR is shown on Figure 2.

To facilitate quarrying between the Main lobe and North lobe of the Thornton Quarry without interrupting the I-80/294 Tollway, the quarry operator excavated penetrations underneath the Tollway and through the rock rib that separates the two lobes. The main penetration measures a maximum of about 200 feet wide and 116 feet high. As quarrying progressed to deeper levels, two tunnel penetrations measuring 25-ft-high and 35-ft wide and about 450- and 600-ft long were excavated. These penetrations were used by mining equipment and haul trucks to excavate and transport quarry rock from the North Lobe to the Main Lobe crusher.

To make the North Lobe suitable to retain CSO, the rock rib, herein referred to as the Rock Dam, was modified by physically plugging the three penetrations to isolate the North Lobe from the Main Lobe. The largest penetration excavated under the I-80/294 Tollway Bridge, was plugged with a Roller Compacted Concrete (RCC) gravity dam, called the Gap Dam. The tunnels were plugged with reinforced concrete tunnel plugs each consisting of lining and bulkhead sections.

Seepage through open fractures in the Gap Dam foundation; in the rock surrounding the tunnel plugs; and through open joints and bedding planes within the Rock Dam is controlled by a continuous grout curtain constructed along the length of the Rock Dam. The 2,700-ft long seepage barrier is tied-in with the grout curtains on the east and west sides of the North Lobe and into the underlying low permeable shale layer. The intent of the grouting along the Rock Dam is to:

1. Seal interfaces between concrete and bedrock,
2. Seal and treat faults, open joints, bedding planes, and fracture zones within the rock mass, and thereby
3. Decrease the permeability of the dolomitic limestone to depths of as much as 500 feet to control seepage from the TCR into the Main Lobe.

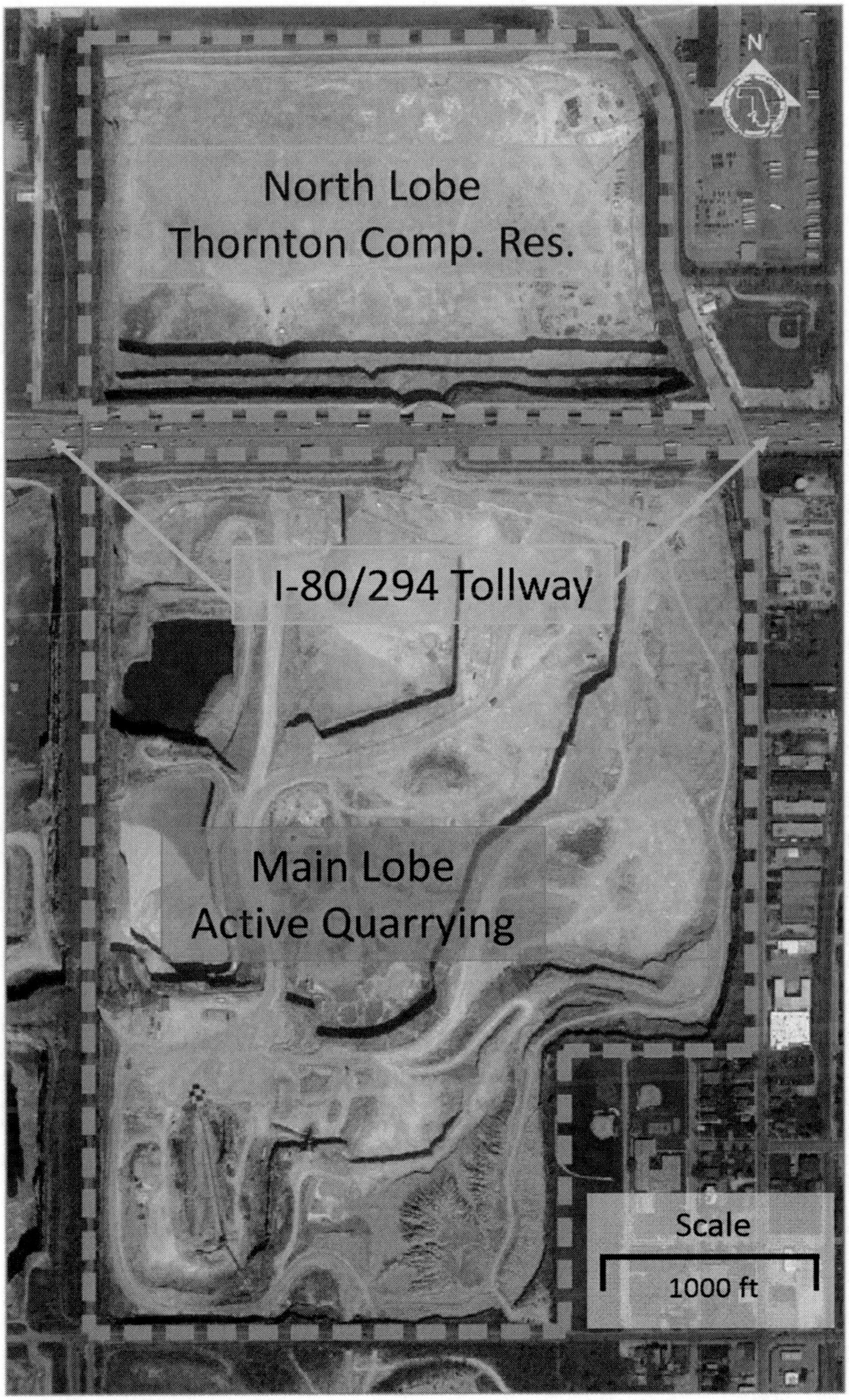

Figure 1. Aerial view of Thornton Quarry and Interstate 80/294 Tollway

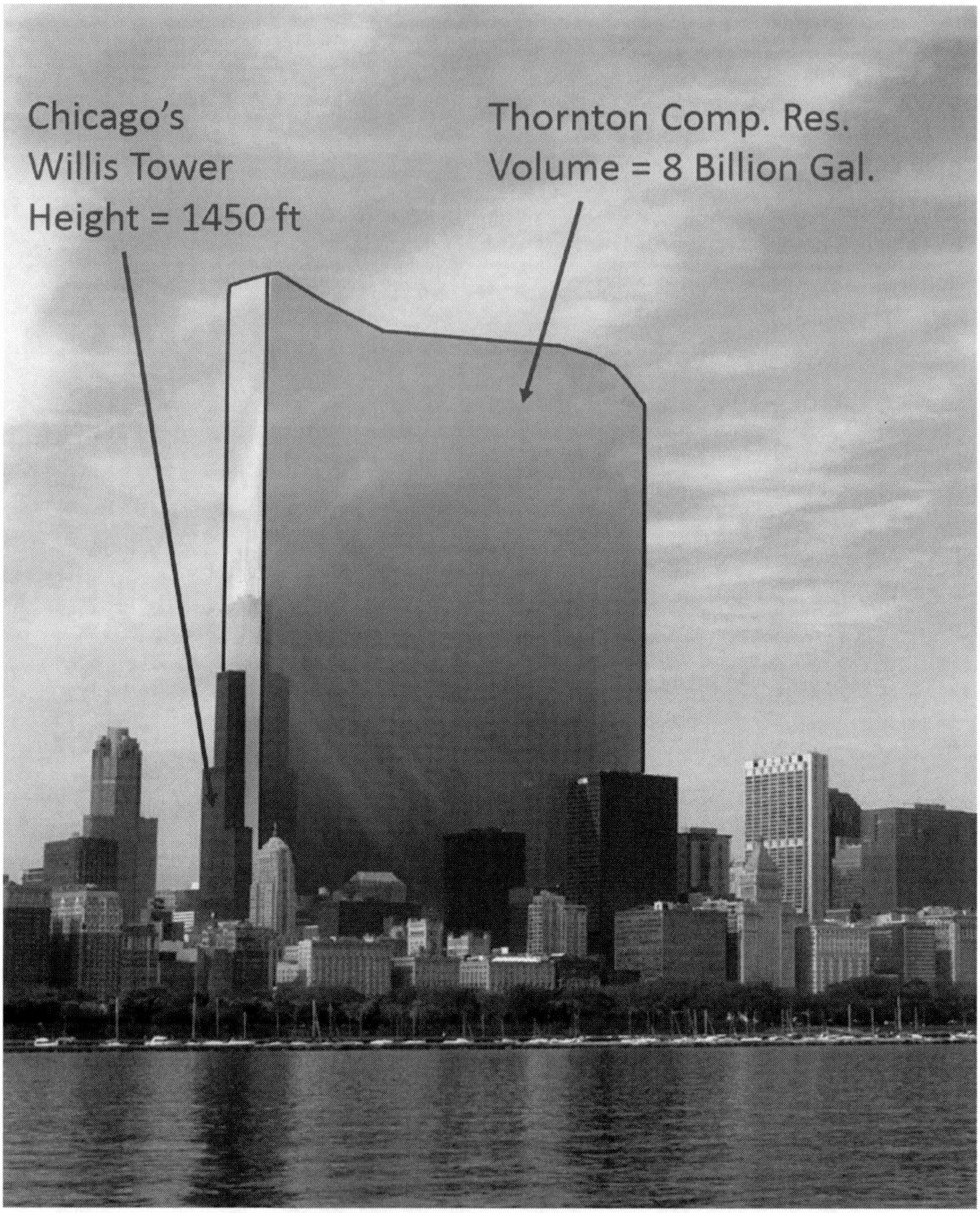

Figure 2. Relative Size and Storage Capacity of TCR

ROCK DAM GEOLOGY

Rock Dam Configuration

As shown on Figure 1, the Rock Dam is oriented in the east-west direction. The crest is approximately 330 feet wide at El. 0 ft of the City of Chicago Datum (feet, CCD) and the I-80/294 Tollway is founded on an earth embankment at approximately El. +27, as shown on Figure 3.

The north face of the Rock Dam extends from El. 0 to approximately El. -297.5. At the top of the Rock Dam (El. 0), there is a 50-foot-wide bench between the toe of the Tollway embankment and the uppermost quarry wall. Below El. 0 there are two intermediate benches: the Level 1 bench at approximately El. -109 and the Level 2 bench at approximately El. -180, each approximately 100 feet wide.

The south face of the Rock Dam currently extends from El. 0 to approximately El. -110 (east of the Tollway gap) and El. -297 (west of the Tollway gap). After construction of the TCR, quarrying in the Main Lobe will continue and ultimately extend the south face of the Rock Dam to a final elevation of approximately El. -450. The final configuration of the south face will include three benches, each approximately 30 feet wide, located at the Level 1 elevation (El. -109), the Level 2 elevation (El. -180) and the Level 3 elevation (El. -297). The two haul tunnels extend through the Rock Dam at Level 2 and Level 3, El. -180 and El. -297, respectively.

The Thorn Creek Diversion Tunnel extends the entire length of the Rock Dam. It is unlined, 22-foot-diameter, and a circular cross-section and is currently used to convey flood water from Thorn Creek (east of the TCR) to the Thornton Transitional Reservoir located in the West Lobe of the Thornton quarry (located west of the Main Lobe in Figure 1 but not shown). The diversion tunnel is parallel to the long axis (i.e., the centerline) of the Rock Dam and is approximately 35 feet south of centerline with its invert between El. -245 and El. -250. The tunnel is perpendicular to the Level 2 and Level 3 haul tunnels and crosses below the Level 2 tunnel by approximately 40 feet and above the Level 3 tunnel by approximately 30 feet. The Thorn Creek tunnel will eventually be plugged at both ends of the Rock Dam and will serve as an internal drainage gallery to collect groundwater and CSO seepage through the Rock Dam, and be pumped back into the TCR.

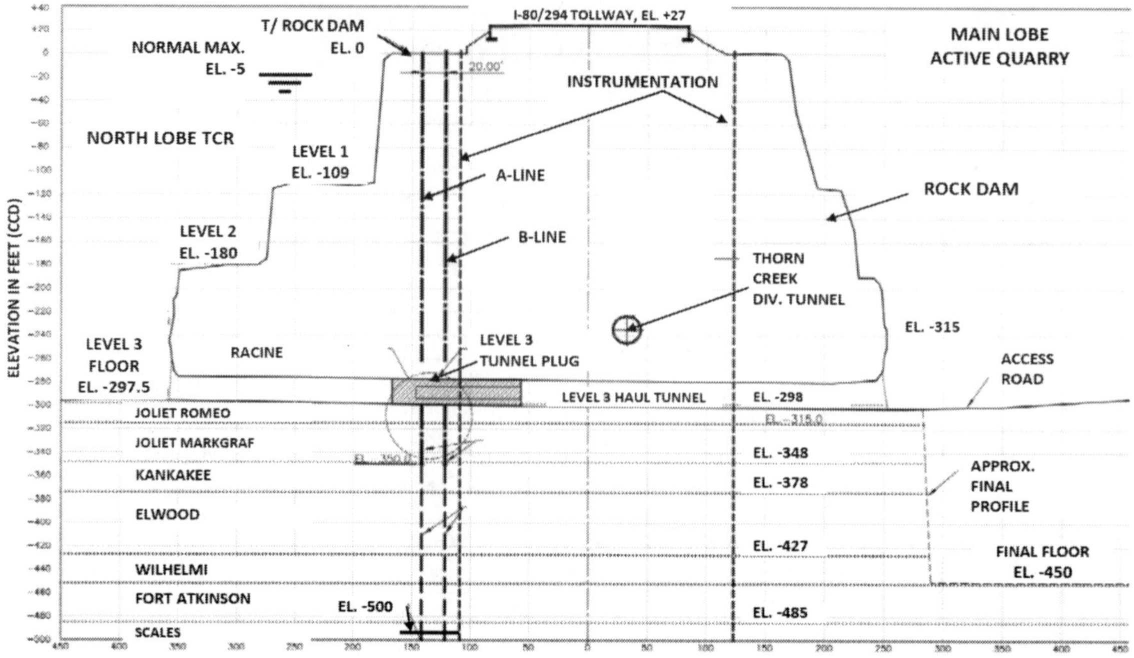

Figure 3. Typical Rock Dam Section (view to east)

Lithology

Geotechnical investigations and mapping of the project area indicated the Rock Dam is composed predominantly of the following sedimentary rock units:

- Silurian dolomitic limestone comprised of the Racine, Joliet, Kankakee, and Elwood formations extending from El. 0 to El. -427. The rock is fine to medium-grained dolomitic limestone; relatively pure to slightly argillaceous with occasional shale layers; dense to slightly porous; massive; occasional chert beds and nodules and irregularly-shaped vugs. Based on packer testing the dolomitic limestone units have a hydraulic conductivity of about 6×10^{-5} cm/s.

- Ordovician dolomitic siltstone of the Wilhelmi and Ft. Atkinson formations extending from El. -427 to El. -485. The rock is interbedded dolomitic limestone, silty claystone and shale; poorly indurated and susceptible to slaking and deterioration. Based on packer testing the dolomitic siltstones have a hydraulic conductivity of about 4×10^{-5} cm/s.

- Shale of the Scales formation extends from El. -485 to below El. -600. The rock is laminated, dolomitic shale, very uniform, susceptible to slaking and deterioration. This unit, which has a hydraulic conductivity on the order of 2×10^{-7} cm/s acts as an aquitard underlying the TCR that will restrict downward seepage.

Upon inspection of the lithology, it was determined that potential seepage could best be controlled by extending the continuous perimeter curtain to El. -500 and tying it into the underlying low permeable Scales formation beneath the reservoir.

The bedrock exposed in the walls of the Rock Dam between El. 0 to El. -297 is predominantly dolomitic limestone of the Racine Formation. The Racine Formation is subdivided into three sub-units, or "facies": the Reef, Reef-Flank, and Inter-Reef facies. Of these facies, only Reef-Flank and Inter-Reef facies are exposed in the Rock Dam. Above the Level 2 bench, much of the Rock Dam is within the Reef-Flank facies.

The Reef-Flank Facies is a transitional zone between the Reef and Inter-Reef facies and is typically comprised of relative pure to slightly argillaceous dolomitic limestone with moderately dipping beds, generally dipping between 5° and 15° to the north and northeast. The Inter-Reef facies consists of argillaceous dolomite with nearly horizontal bedding, often containing thin beds and nodules of chert and sporadic thin partings and lenses of green shale. Both the Reef-Flank and Inter-Reef facies include some relatively thin shale and siltstone layers.

Bedding planes and two prominent, nearly orthogonal sets of discontinuities (joints, faults, and shears) are prevalent throughout the Rock Dam at varying spacings, creating a well-interlocked rock mass consisting of hard, strong, prismatic blocks. Typical orientations of these features are:

- Bedding: Reef-Flank dip 5° to 15° north-northeast direction (above El. -180); Inter-Reef dip 0°to 5° north-northeast direction (below El. -180).

- Joint Set A: Near vertical, dip 80° to 90°, dip direction N40°-45°W and S40°-45°E (includes faults and shears).

- Joint Set B: Near vertical, dip 80° to 90°, dip direction N55°-60°W and S55°-60°E.

DESIGN AND CONSTRUCTION OF THE CONCRETE STRUCTURES

To utilize the North Lobe as a water retaining structure, the Rock Dam was modified by plugging three gaps and constructing a deep continuous seepage barrier. The barrier ties into the three structures as well as the grout curtain on the east and west sides of the North Lobe.

Roller Compacted Concrete Dam (Gap Dam)

Dam Design. The stability of the Gap Dam was evaluated in accordance with procedures and criteria contained in US Army Corps of Engineers Engineering Manuals following EM 1110-2-2200, Gravity Dam Design and EM 1110-2-2100, Stability Analysis of Concrete Structures. Static and pseudo-static analyses were performed for several loading cases using two-dimensional (2D) conventional methods to determine stresses on the dam and foundation as well as the sliding stability factors of safety. For determining appropriate factors of safety, the Gap Dam was considered in the Ordinary Site Information category. Assuming that failure of the dam would damage the I-80/294 Tollway bridge piers, the Gap Dam is classified as a Critical Structure.

The Gap Dam closes the gap between the North and Main Lobes at bench Level 1 (El. -109). The dam is founded on dolomitic bedrock of the Racine formation at about El. -116 and has a roughly trapezoidal shape with a width of approximately 60 feet at the bottom and approximately 210 feet at the top (El. 0). The I-80/294 Tollway spans the gap with a twin span, 367-foot-long bridge (only piers shown) downstream of the Gap Dam crest. Figure 4 and Figure 5 illustrate the configuration of the Gap Dam. The dam is a gravity structure, the arch shape is intended to facilitate construction and tie-in with the abutments.

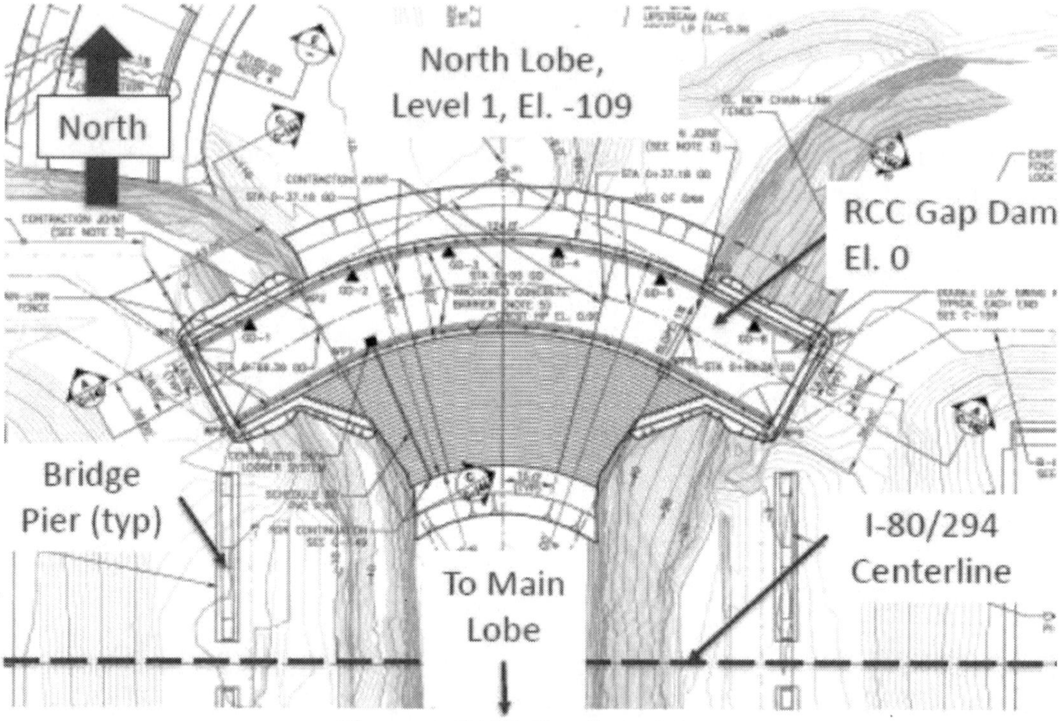

Figure 4. RCC Gap Dam Plan

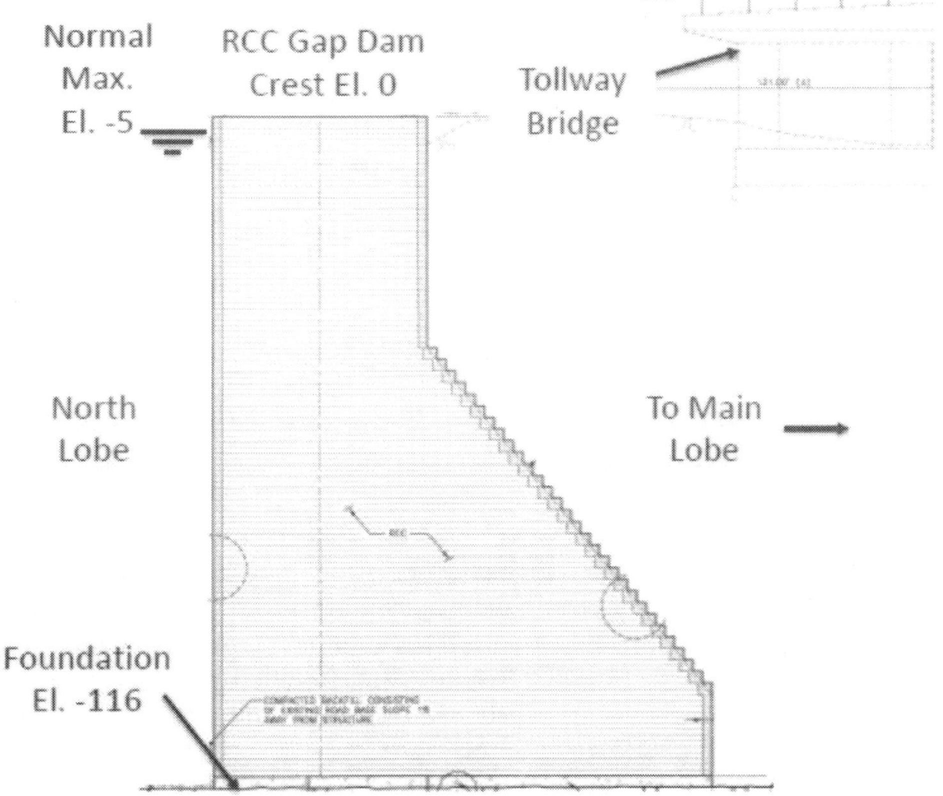

Figure 5. RCC Gap Dam Section (view to east)

<u>Foundation Treatment</u>. The Gap Dam was constructed on competent Racine formation bedrock, but foundation treatment of the base foundation and upper abutments was required. Foundation treatment consisted of surface preparation, stitch grouting, consolidation grouting, and curtain grouting. To reduce the potential for adverse impacts to the existing Tollway bridge foundations, blasting was not permitted. Instead, excavation and treatment of the foundation was completed with mechanical methods.

The abutments of the gap are vertical between El. -116 to El. -55 and then shaped to produce a relatively uniform slope suitable for the upper portion of the Gap Dam (see Figure 6). Following rough excavation using long reach excavators with a hydraulic a breakers, the slopes were power washed, and hand-scaled to remove remaining loose rock, dirt, and debris before receiving RCC.

The foundation rock within the footprint of the dam was scaled, shaped, and cleaned, and dental concrete treatment was performed along a fault/shear zone that trends northwest/southeast through the Gap Dam foundation and on the left abutment. After the fault/shear zone was cleaned of loose and fractured rock to a depth of about twice the width, the fault was backfilled with lean concrete. The trace of the fault/shear along the left vertical abutment was cleaned, formed, and filled with lean concrete.

Stitch grouting was performed along the fault/shear zone to reduce the potential for seepage along this discrete feature. The stitching extended 10 feet upstream to 10 feet downstream of the grout curtains within the Gap Dam foundation. Drilling was performed using an air-track percussion drill with a 2.5-inch diameter drill bit. Holes were inclined approximately 30 degrees from vertical to intercept the fault/shear zone at depths between 5 and 10 feet below the foundation surface. After drilling, the stitch holes were washed with pressurized water and grouted using the iGrout™ system (described below).

After completing the foundation treatment, a 2- to 3-ft thick unreinforced concrete leveling slab was placed to El. -114, with waterstops and contraction joints, to level the footprint of the Gap Dam. Curtain grouting was then performed from the leveling slab and consisted of a double-row curtain. Grouting along the A-line included 14 holes (seven vertical and seven inclined at 15 deg.) with vertical initial primary holes extending to El. -500 (i.e. into the Scales formation) and intended to work around turning points in the curtain, with the remaining holes extending to as much as El. -350. The B-Line was similarly completed with 14 holes. After closure was achieved, verification holes were performed along the centerline, in which packer testing indicated higher takes along finer cracks, which was likely resulted from quarry blasting damage and subsequent weathering and loosening of the upper part of the foundation. These findings necessitated a third row be constructed to a maximum depth of about El. -180. Ultimately, closure was achieved, geotechnical instrumentation was embedded in the foundation, and the RCC placement commenced.

RCC Construction Methodology. RCC mixing, placement and compaction followed the general guidelines and recommendations in American Concrete Institute ACI 207.5R. A construction staging area located just to the south of the Gap Dam was used to store aggregates, cement, fly ash, and admixtures, mixing equipment and the RCC conveyance system. The delivery of the RCC mix from the batching plant to the dam was completed using belt conveyors. In general, the RCC was placed in a serpentine pattern, starting in one corner of the dam and then placing material in multiple lanes parallel to the dam axis. The material was placed in approximately 12-inch thick lifts, and compacted using vibratory rollers to achieve the required density. To complete the dam as shown on Figure 7, 168 lifts comprising of 34,500 cubic yards of RCC were placed over the course of 134 working days.

To mitigate cracking, the Gap Dam was constructed with vertical transverse contraction joints. The five contraction joints were spaced approximately 40 feet apart and oriented radially and perpendicular to the upstream face. The joints extend in the upstream/downstream direction across the entire cross section and for the full height of the dam. The joints were formed by inserting a plate into the un-compacted RCC with a vibrating plate mounted on a backhoe. Plastic sheeting placed around the vertical plates was left in place as a bond breaker. Two vertical waterstops were installed at the upstream face for each contraction joint to cut off leakage through the joint. The bottom of the waterstops were embedded two feet into rock by coring the rock and then grouting the waterstops into place, and they extend to within one foot of the dam crest.

A facing system for both the upstream and downstream faces of the Gap Dam consisting of grout-enriched vibratable RCC (GEVR) was used to reduce seepage through the dam and to provide a surface more durable against freeze-thaw cycles and more resistant to the potential aggressive action of the CSO.

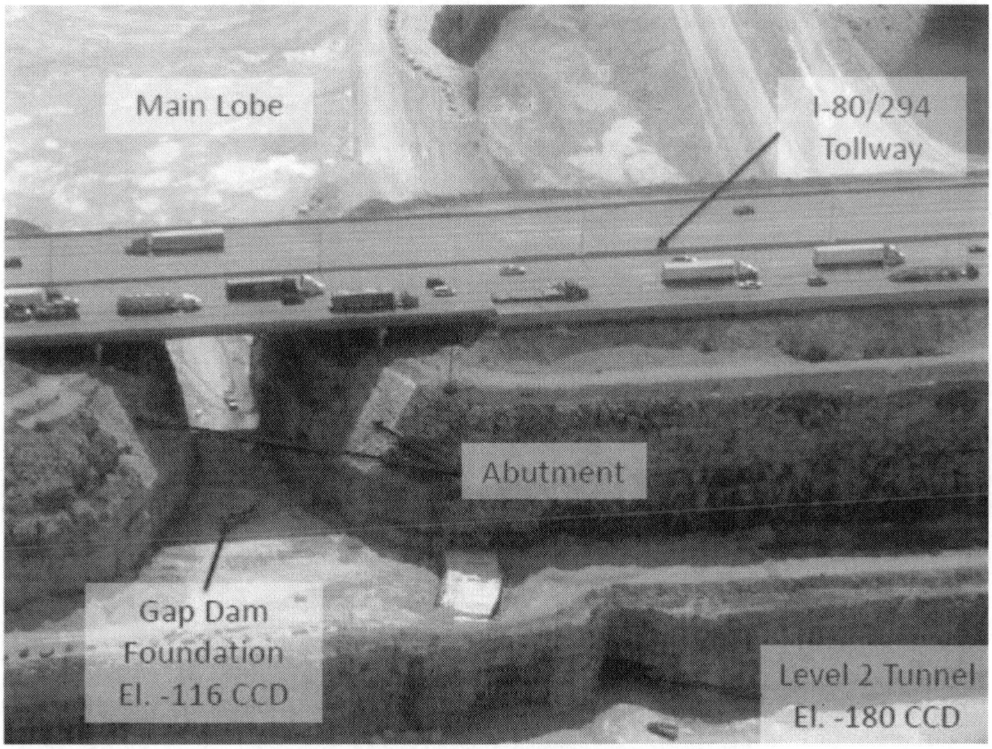

Figure 6. Aerial view of Gap Dam Foundation and Abutment Preparation (view toward the south-southwest)

Figure 7. Completed RCC Gap Dam (view to south-southwest)

Tunnel Plug Design and Construction

The two haul tunnels aligned in the north-south direction provided access through the Rock Dam at approximately El. -180 and El. -298. The geometry of the haul tunnels are as follows:

- Level Two Haul Tunnel – Approxim ately 25 feet high by 37 feet wide and 480 feet lo ng. At the upstream face of the plug the tu nnel invert is at approximately El. -183.
- Level Three Haul Tunnel – Approxim ately 24 feet high by 34 feet wide and 600 feet lo ng. At th e upstream face of the plug the tunnel inv ert is approximately El. -301.

Each plug consists of an upstream reinforced concrete bulkhead, which is designed as a two-way slab. The bulkhead is supported on the downstream side by a reinforced concrete lining, resembling a block with hollow core. The plugs are aligned with the Rock Dam grout curtain to facilitate construction of the continuous seepage barrier. After placement and curing the concrete, contact grouting was performed along the full length of the plugs in the tunnel crown and haunches walls to compensate for concrete shrinkage and to fill remaining voids not filled with concrete. Consolidation grouting was needed to improve the strength and deformation characteristics of the rock mass, which had loosened over the years and to seal potential seepage pathways near the concrete-rock interface.

In general, the construction of the plugs started with excavation of the crushed road base material and existing leveling concrete at the invert of the haul tunnels and foundation cleaning. The plug lining sections were placed in three lifts followed by the placement of the bulkheads, which were also placed in three lifts. Consolidation and contact grouting of the lining and bulkhead were completed after the lining and bulkhead had been placed and allowed to cure for a minimum of 30 days.

GROUT CURTAIN CONSTRUCTION

As a chief defense against seepage of CSO from the TCR, an approximately 9,800 LF continuous double-row grout curtain was constructed around the perimeter of the Thornton Reservoir, using state-of-the-practice drilling and grouting techniques and equipment. A particular challenge to this project involved the design and construction of the seepage barrier on the Rock Dam given that it was integrated with other structures. The construction of the south seepage barrier consisted of three parts:

1. Gap Dam foundation grouting from the leveling slab at El. -114,
2. Surface grout curtain installation from the top of the Rock Dam at approximate El. 0, and
3. Ring grouting (consolidation and contact grouting) from within the two Haul Tunnel Plugs.

Coupled seepage-stability analyses were performed for the Rock Dam to assess not only the stability of the Rock Dam, but to evaluate the effectiveness of the barrier system and estimate the leakage from the TCR. Operating at full-reservoir head, steady-state seepage analyses estimated seepage through the Rock Dam to be on the order of 200 to 500 gpm.

General Grouting Practices

The grouting program utilized a suite of balanced and stable grouts, ranging in viscosity from 34 seconds to greater than 60 seconds per liter, as measured through a Marsh Funnel. The rheological properties of the grouts were carefully selected, using varying combinations of water and cement with additives and admixtures to control bleed, cohesion, and coefficient of pressure filtration, while maximizing penetration. The grouting constituents included Type III cement, Type F flyash, bentonite, super-plasticizer, and a long-chain polymer viscosity modifier, called Diutan Gum.

Batching was performed at strategically located batch plants, using an automated colloidal mixer. Once mixed, the grout was placed into mechanically-agitated storage tanks, connected to positive-displacement progressive cavity pumps. The pumps were connected to the contractor's mobile injection units, which were capable of maneuvering to different grout hole locations. Figure 8 and Figure 9 depict the typical grout pumps and mobile injection units that were used.

Figure 8. Grout Pumps with Agitator tanks

Figure 9. Mobile Injection Grouting Cart

All grouting was performed utilizing iGrout™, the contractor's proprietary, state-of-the-practice, automated grouting control and data acquisition system. iGrout™ allowed simultaneous real-time control and monitoring for multiple grouting and water testing operations, an example of which is shown on Figure 10. Each of the contractor's mobile injection units were specially designed to communicate wirelessly with the iGrout™ system. The system was also web-based so the contractor, engineer, and owner could all monitor activities in real-time and communicate enabling field staff and expedite decision making. The system operator regulated the pump speed to reach the design injection pressure, until the refusal criterion was achieved. Refusal was generally defined as 0.1 GPM, yielding an apparent Lugeon value less than or equal to the project closure criterion of 1 Lugeon (approximately equal to $1.3x10^{-5}$ cm/s), for a minimum of 5 minutes, at the maximum grout injection pressure. Holes that exceeded the closure criterion of 3.75 gal/ft of grout hole required subsequent split-spaced grout holes on either side of the precedent hole.

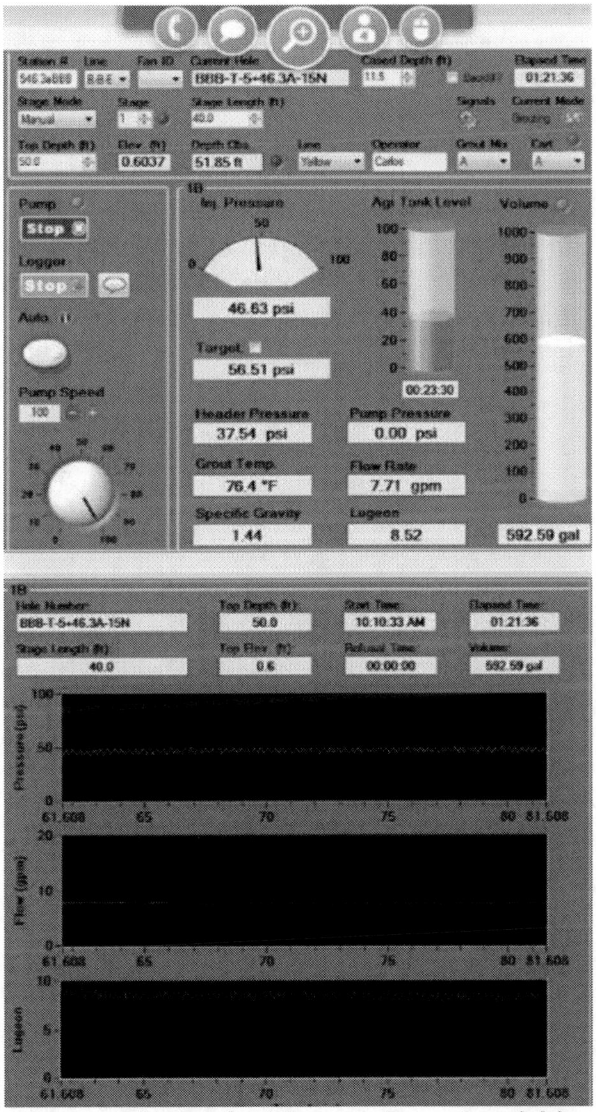

Figure 10. Computer Display for iGrout™ Data Acquisition and Control

In general, grouting was performed utilizing upstage grouting techniques. Upstage grouting is the method of grouting a hole where the packer is set in a fully-drilled hole at successively shallower depths. However, where drilling circulation was lost or holes collapsed, downstage grouting methods were used. In such cases, a packer was set above the problematic zone; the stage was grouted, and then the hole was re-drilled and grouted to the intended design depth.

Gap Dam Grouting

As previously described, prior to placing the RCC for the Gap Dam, a triple-row grout curtain was installed along the dam foundation to depths from El. -114 to as deep as El. -500, to seal gaps between the dam leveling slab and the existing bedrock, decrease the permeability of the dolomitic limestone below the dam, and treat a fault which cuts through the proposed dam location. Drilling was performed using a water-actuated down-the-hole hammer to drill 4-inch diameter holes. Grouting was monitored and recorded through iGrout™.

Once grouting was completed below the dam, the Gap Dam itself was constructed. After completion of the dam, additional holes were installed from the crest through the abutments using the water-actuated-down-the-hole hammer advanced with a purpose-built rock drill with monitoring-while-drilling capabilities. These holes were installed at angles of 0 or 15 degree from the vertical, through the dam and into the abutments and to varying lengths to maintain continuity around several turning points, as shown on Figure 11. The intention of holes inclined through the crest was to treat the RCC-rock interface and maintain grout curtain continuity with the adjacent Rock Dam grout curtain zones.

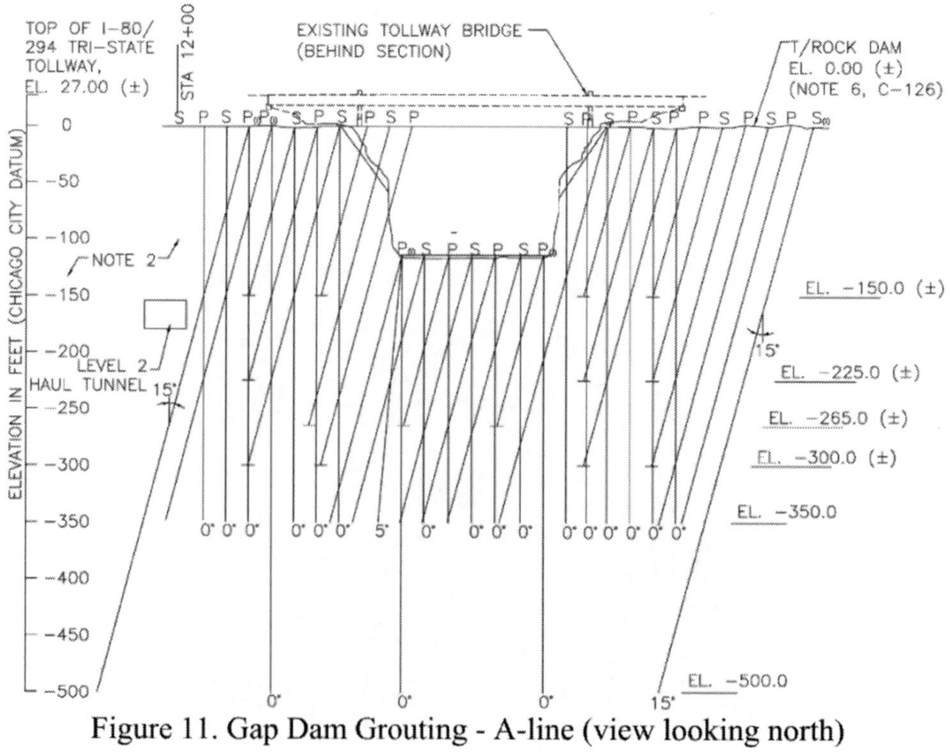

Figure 11. Gap Dam Grouting - A-line (view looking north)

Rock Dam Grouting

The grout curtain along the 2,740 lineal feet of the Rock Dam generally contained two rows (A-Line and B-Line). The project required a minimum installation of Initial Primary (IP), Primary (P), and Secondary (S) holes. When grouted holes exceeded the prescribed closure criterion, split-spacing methodology was implemented to install tertiary and higher-order holes.

The 4-inch diameter grout holes were drilled using a water-actuated down-the-hole hammer, outfitted on a purpose-built rock drill (Cubex QXW 1210). The drilling system continuously monitors drilling parameters and assists in making adjustments based on real-time data, helping to maintain the specified verticality requirements (1 foot per 30 inclined feet) for the grout holes. The drill rig is shown on Figure 12.

Figure 12. Cubex QXW 1210 on Rock Dam

The grout holes were generally inclined 15 degrees from vertical and the holes in the A-line and B-lines were in opposite directions to create a lattice to better intersect the nearly vertical fractures in the bedrock formation at multiple locations. However, at some locations, including corners and near existing utilities, fan holes ranging from 0 degrees (vertical) to 65 degrees were installed to close potential windows in the curtain. The grout holes were installed from the surface (El. 0), to elevations between El. -20 and El. -500, depending on the hole-type, and grout takes observed on adjacent holes.

To monitor verticality, the contractor generally utilized a Robertson Geologging Magnetic Survey Probe. For special instances, primarily when geologic irregularities were observed during the drilling operation, the contractor utilized the Robertson Geologging Optical Televiewer. Grout holes that exceeded the verticality requirements were redrilled. In general, the Contractor met the requirements, although several holes had to be redrilled.

After completion of drilling, the holes were pressure washed to clear any remaining fine drill cuttings from the hole. Depending on the order of the hole (i.e., initial primary, primary, etc.), pressure testing was performed to determine the relative permeability of the rock mass and select the appropriate starting grout mix. If high Lugeon values were observed during pressure testing, a more viscous starting grout mix was selected.

Stage grouting in 20-foot increments was standard throughout the Rock Dam grout curtain installation. Grouting was most commonly performed using upstage grouting but downstage grouting was used occasionally. The grouting data were recorded in iGrout™ and then output to an AutoCAD file and to PDF reports for the Engineer's review. Figure 13 shows a portion of the A-Line grouting progress drawings from the Rock Dam. The color-coding represents ranges of grout take for each stage. This system facilitated identification of low and high take zones, which could be targeted with higher order holes and verification holes.

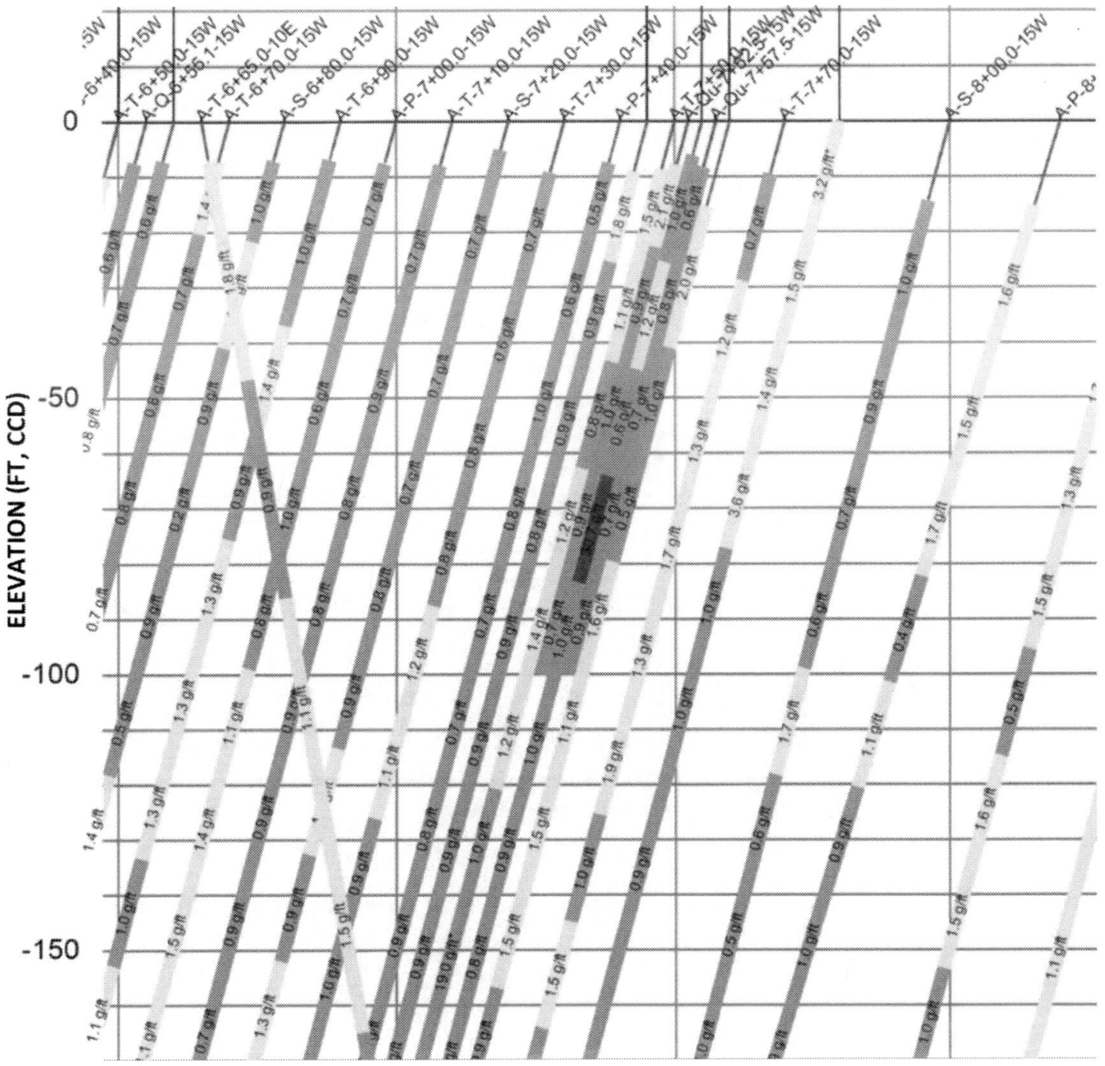

Figure 13. Typical Rock Dam Grout Curtain Record

Verification holes were installed after completion of the A-Line and B-Line grout holes. The verification holes were cored between the A-line and B-lines using a CME 75 drill with standard wireline HQ triple-tube core barrel and tooling, having an approximate core barrel diameter of 3.75-inches. Core samples were analyzed and logged by an on-site Professional Geologist, photographed, stored in core boxes, and a verification hole summary was prepared.

The verification holes were also required to be within the specified deviation requirements and have a residual hydraulic conductivity less than 1x10-5 cm/s (1 Lugeon) from water pressure testing, and not exceed the grouting closure criteria of 3.75 gal/ft. If these requirements were not satisfied, additional grout holes were installed around the verification hole until grouting closure criteria was met.

Haul Tunnel Plug Grouting

Construction of the Haul Tunnel Plugs was the final portion of the seepage barrier installed on the Rock Dam. The plugs are positioned perpendicular to the face of the Rock Dam, and contain grouting rings spaced 10 feet apart, and each consisting of 12 consolidation grout holes with 2.5-inch diameter holes as shown on Figure 14 and Figure 15. The Level 2 Tunnel Plug includes 8 rings of consolidation grout holes, and the Level 3 Tunnel Plug includes 12 rings of consolidation grout holes due to the different plug lengths.

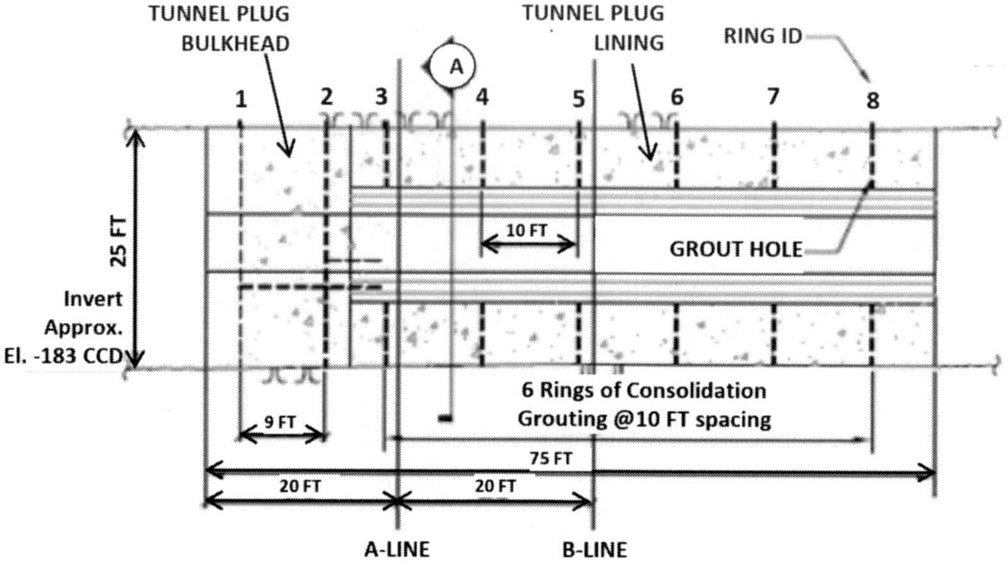

Figure 14. Typical Haul Tunnel Plug Grouting – Profile

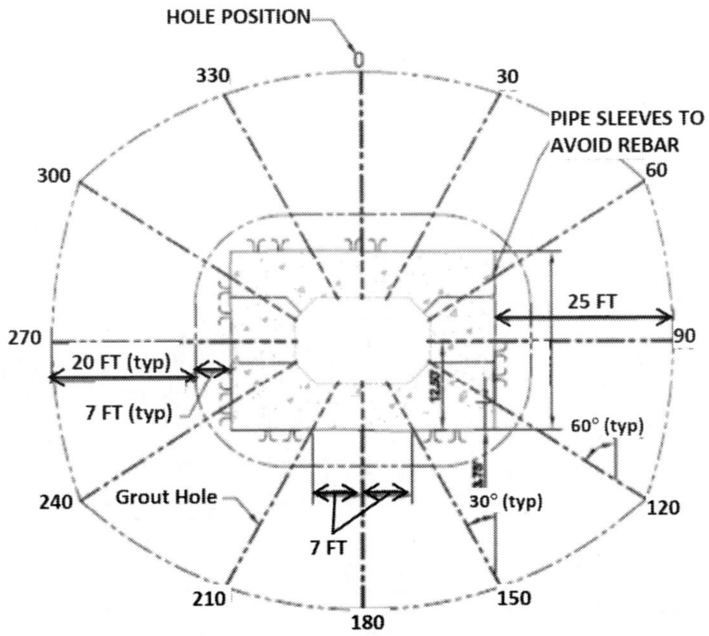

Figure 15 - Typical Haul Tunnel Plug Grouting – Lining Section

The consolidation grout holes were designed to grout voids as well as open joints and bedding planes in the rock formation around the tunnels. Consolidation grout holes ranged in length from 25 feet to 45 feet, to provide continuous coverage and overlap 20 feet with the A- and B-line surface grout curtains. The consolidation grout holes were drilled through embedded grout sleeves using a water-actuated down-the-hole hammer outfitted on an air-track percussion drill rig, as shown on Figure 16. Each hole was pressure washed after drilling, to remove cuttings.

Figure 16. Consolidation Grout Hole Drilling in Haul Tunnel Plug Lining

Consolidation grout holes below the springline were treated using traditional upstage drilling and grouting techniques. However, to grout multiple stages at and above the springline, downstage drilling and grouting methods with mechanical packers was followed. For downstage grouting, the initial stage was drilled, and the hole was grouted with the packer positioned at the concrete-rock interface. After the grout set, the mechanical packer was removed, and the remainder of the hole drilled and grouted. iGrout™ was used throughout the grouting process to provide accurate grouting records for review and decision making.

The grout takes in consolidation holes were generally low, and did not exceed the typical grout curtain closure criteria. However, because the liner was heavily reinforced on both faces, split-spacing consolidation grout holes that exceeded the closure criteria was not practical. As a result, redrilling and extending some of the consolidation grout holes and re-grouting of those holes was completed to affirm closure had been achieved.

In addition to the consolidation grouting for the ring grout curtains, contact grouting was also required to fill voids between the concrete liner and the tunnel crown. Lastly, an approximately 20-foot thick reinforced concrete bulkhead was constructed at the north end of each tunnel and was grouted in a similar fashion, albeit through pre-installed holes and injection pipes.

INSTRUMENTATION AND MONITORING

Survey monuments, optical survey targets, piezometers, and inclinometers are installed to monitor the performance of the Rock Dam and Gap Dam. Considering the variable frequency and height of reservoir filling, a manual system was installed to promote regular visual inspections of the facilities and instrumentation. Although an integrated automatic data collection and remote monitoring system would allow for continuous data collection and real-time monitoring, such systems can be less reliable and more prone to damage, malfunction, and neglect.

An array of 59 survey monuments and targets is installed on the crest of the Rock Dam and Gap Dam and along intermediate bench faces to measure potential movement of the Rock Dam, individual benches, and "key" blocks, which are prismatic rock wedges defined by mapped rock joints. The survey monuments are installed in bedrock along the Rock Dam crest and in RCC of the Gap Dam crest and are surveyed by conventional survey methods. Because no permanent access is available to the intermediate benches, survey targets mounted on the faces of key blocks are surveyed from distance.

Twenty piezometers installed along the upstream and downstream crest of the Rock Dam measure piezometric levels in the Rock Dam and its foundation. Two rows of piezometers are installed: one row is 10 to 20 feet downstream of the grout curtain, and the second row is installed from El. 0 but on the south side of the Tollway. The piezometers are spaced approximately 300 feet apart with alternating deeper- and shallower-screened intervals with those on the south side have deeper screen piezometers. The instruments take measurements at prescribed time intervals (adjustable, but currently every 4 hours) and record the data using a self-contained programmable datalogger positioned within a weather-proof enclosure affixed to the surface casing. In addition, the piezometric levels are measured by manual water-level indicators during periodic and special dam inspections.

Twelve vibrating wire piezometers are installed in the Gap Dam foundation. Piezometer cables are routed through the concrete leveling slab to the downstream side of the dam and up the back side of the dam to a centralized data logger system set on the Gap Dam crest.

Sixteen inclinometers are installed along the upstream and downstream crest of the Rock Dam to monitor Rock Dam deformations and to determine whether such deformation occurs in response to operation of the reservoir or other influences (e.g. temperature). The inclinometers are in-line with the piezometers and split-spaced between them. The casings extend from approximately El. 0.0 to about El. -455, which is below the planned bottom level of the Main Lobe. The inclinometers are measured by traversing an inclinometer probe up and down the casing to measure lateral deviations of the casing.

SUMMARY

As part of the Thornton Composite Reservoir construction project, the North Lobe of the existing Thornton Quarry is designed to retain Combined Sewer Overflow captured by the TARP Calumet Tunnel System and floodwater from Thorn Creek. The Metropolitan Water Reclamation District of Greater Chicago modified the Rock Dam by plugging three gaps and constructing a deep seepage barrier along the length of the Rock Dam to create a continuous cutoff on four sides that are tied into a low permeable layer at depth.

The Thornton Composite Reservoir was put into service in August 2015 and filled in November 2015. As of December 2015, the reservoir held as much as 70 feet of water. Instrumentation data and inspections performed during these early filling events indicate that the Gap Dam, Rock Dam, plugs, seepage barrier are operating as expected.

DESIGNING A HYDRID DAM ON A GEOLOGIC PROFILE WITH HIGHLY VARIABLE ENGINEERING PROPERTIES

Richard E. Riker, PE[1]
Jacob Esterhuizen, PE, PhD[2]
Don Deere, PE, PG[3]
Gabriel Fernandez-Delgado, PE, PhD[4]

ABSTRACT

Duck River Dam is a water supply dam in Northern Alabama. The dam, which is 2,400 feet long and nearly 140 feet high at the maximum section, is a hybrid structure comprising a 300-foot-long Roller Compacted Concrete (RCC) spillway structure along the river channel, with earth and rock fill sections at the abutments. The dam site is located on flat-lying interbedded shales and sandstones of the Pottsville Formation. Interbedded sedimentary strata present some of the most challenging foundation conditions for dams because of extreme variabilities in engineering properties of strength, deformation, cementation and permeability.

Preliminary explorations identified several features of concern, including the presence of low strength, high plasticity clay layers at the abutments, open stress relief joints in the sandstones, slickensided coal seams, occasional voids, uncemented friable and potentially erodible sandstone zones and highly variable rock mass permeability. The design approach included the bid and award of an early (Phase 1) foundation excavation contract to comply with the construction schedule; to expose, evaluate and adjust foundation levels if required; to facilitate and optimize specific supplementary exploration and testing required to reduce uncertainty and inform final design; and to reduce potential bidder's risks.

Findings from the Phase 1 foundation excavation and testing resulted in significant changes in design concepts. Measures to deal with some of the foundation challenges and project schedule constraints include (1) centering the RCC spillway section on relatively competent sandstone rock foundation, (2) constructing rock fill sections, including stability berms, on clay shale abutments, (3) excavating for the construction of a RCC cutoff in right abutment to intercept friable sandstone layers and (4) installing the grout curtain from the gallery to take grouting off the critical path and to utilize the 15-foot-thick concrete as confinement to accommodate higher injection pressures in the upper rock layers. This paper discusses the major advantages derived from this phased approach.

1 Senior Geotechnical Engineer, CH2M, Corvallis, OR 97330, rick.riker@CH2M .com
2 Senior Geotechnical Engineer, CH2M, Corvallis, OR, jacob.esterhuizen@CH2M.com
3 Principle & Chairman, Deere & Ault Consultants, Inc., Longmont, CO 80503 don.deere@deerault.com
4 Research Lecturer, University of Illinois, Urbana-Champaign, Ill, 61802-7127, gfernand@uiuc.edu

INTRODUCTION

The Cullman Utility Board (Cullman, Alabama) spent more than a decade planning and permitting a sustainable water supply solution for the City of Cullman and surrounding counties. Initial efforts were led by the U.S. Army Corps of Engineers (USACE) and included planning and feasibility studies, permitting efforts, preliminary geotechnical and geologic studies, preliminary hydrologic evaluations and conceptual design efforts. The USACE initial activities began in 1998, during which environmental challenges arose. After nearly 12 years of court hearings, the project permit application was approved. These efforts resulted in a decision to design and construct a new dam and reservoir on Duck River just 10 miles east of the City of Cullman, see Figure 1. In 2010, the Utility Board commissioned final design for a new 140-foot-high and 2,400-foot-long dam that will impound about 26,000 acre-feet. The design was completed in March of 2014, and the project was put to bid in early April. The low bidder was awarded a $51.8 MM contract in May of 2014 and construction began in earnest July of 2014. The dam construction progressed at a fast pace with substantial completion scheduled in December of 2015.

On Christmas Day of 2015, with dam construction nearly complete, the site experienced a severe storm event. More than 10 inches of rainfall fell over the watershed in a 48 hour period, (0.01 Annual Exceedance Probability (AEP)). The reservoir levels went from near empty to approximately 90 percent of the normal pool elevation in two days. This extremely rapid filling rate severely disrupted construction operations, inundated equipment and materials and has resulted in schedule delays. Efforts are underway to recover from the storm event and complete the dam. During and following this event, the dam performed quite well, with only minor damage to uncompleted components of the permanent structure. When complete, the dam and reservoir will provide the Utility Board a minimum firm yield of 32 million gallons a day, a reliable water supply for future generations, Figure 2.

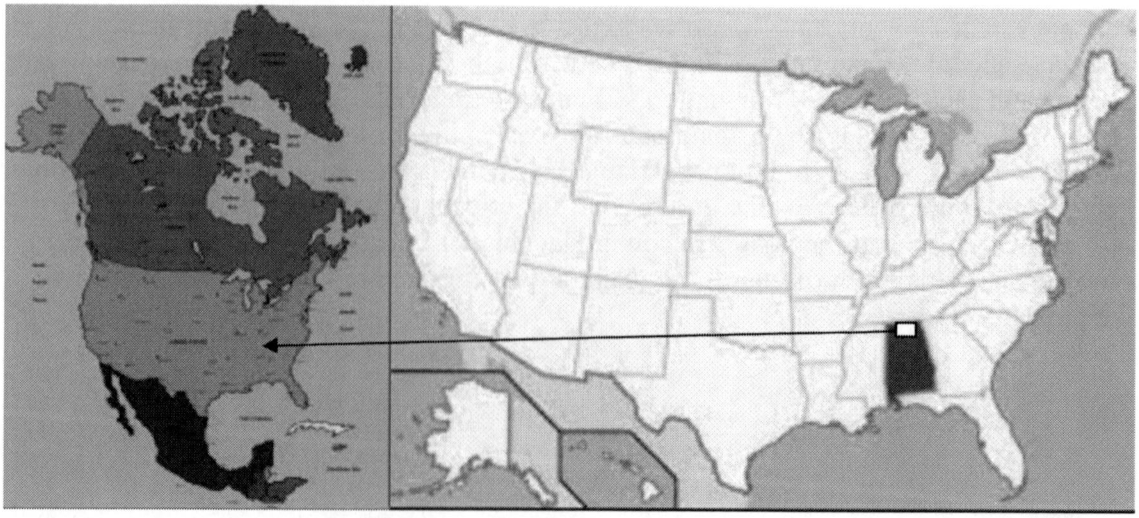

Figure 1. Project Location Map.

Figure 2. Photo of project near completion

ADAPTIVE AND PHASED APPROACH

Preliminary geotechnical studies conducted by the USACE in the late 1990's revealed a geologic profile with highly variable engineering properties. Revisions to the estimated flood magnitudes and frequencies resulted in a much larger Design Flood (PMF) than originally estimated, making the original concept of a right abutment spillway much more difficult and expensive. The updated hydrology also meant that larger floods could be expected during construction. The project design team was challenged by these technical issues, in addition to the owner's desire to accelerate the schedule and the owner's expectations regarding the construction cost. The aggressive construction schedule excluded a more conventional approach of performing detailed geotechnical site investigations and testing programs, while finalizing preliminary design concepts, followed by the preparation of final design documents, then bidding and construction. To meet these challenges, the engineer and owner selected an adaptive and phased approach.

The approach involved bidding a preliminary foundation excavation contract that included provisions for additional geotechnical investigation and testing programs. The approach was to expose, evaluate and adjust foundation levels if required; facilitate and optimize specific supplementary exploration and testing required to reduce uncertainty and inform final design; and reduce potential bidder's risks.

Based on available information, it was assumed that a competent foundation was present at the site which would be suitable for the economic development of at least one of the following dam types 1) a RCC dam, 2) an Earth Core Rock Fill dam (ECRD) or 3) a Concrete Faced Rock Fill dam (CFRD). All three dam types were initially estimated to be cost competitive. The CFRD was dropped from consideration for various reasons, including lack of recent construction experience in the U.S., which could impact competitive bidding. The RCC dam was preferable to the ECRD mainly because it would be faster to construct; construction would be less burdened by adverse weather; and there would be less risk relative to river diversion requirements, flooding and potential overtopping during construction. Given the revised hydrology, a center spillway that could be incorporated into an RCC dam became attractive. Although the RCC dam was initially preferred, the owner and design team were prepared to adopt the most appropriate dam type depending on the conditions encountered.

PHASE 1 CONTRACT

Phase 1 work was performed in the summer and fall of 2012 and included the bid and award of an early "excavation/exploration" contract that allowed construction to commence and comply with the owner's desire to accelerate the start of the project. The preliminary foundation excavation work was designed for the preferred alternative at that time, the RCC dam. At the start of the Phase 1 contract, two important and significant questions remained unanswered: 1) Were the foundation conditions suitable for constructing an RCC dam, and 2) Could an onsite quarry be developed that would allow economical production of rock fill materials and RCC aggregates? To answer these questions and others, the Phase 1work included provision for conducting additional geotechnical investigations, including vertical and angled rock coring and logging within the dam footprint, camera inspection and video logging of core holes, test pits to delineate borrow areas and quantities and laboratory testing on rock specimens and soil samples to define geotechnical parameters needed for final design. A limited quarry investigation and testing program included rock coring, test blasting, rock excavation and crushing, screening and blending of the crushed aggregates, laboratory testing of aggregates and preliminary RCC mix design testing. Given the highly variable rock mass permeability revealed by packer testing during the USACE field exploration, the contract also included provision for a limited grout testing program.

The initial excavation plan along with the locations for geotechnical explorations, quarry testing and the grout testing program are shown in Figure 3. The plan was developed utilizing the subsurface information provided in the USACE's preliminary reports. Due to the status of the 404 permit, access limitations in the river existed, such that no work could be performed within the high water boundary. The excavation plan included the entire footprint of the RCC dam alternative, except for a relatively small area in the river channel.

Geologic conditions in a river channel are often unique and different from the adjacent areas. To investigate these conditions, angle boring was drilled that extended underneath the creek. Also, a large and deep test pit was included on the left river bank and a large exploration trench was included on the right river bank. Both of these exploratory excavations were made as near as possible to the high water boundary to provide information that could be used to interpolate conditions within the river channel, where construction access was restricted. These excavations exposed rock below the expected foundation depth for visual inspection and mapping.

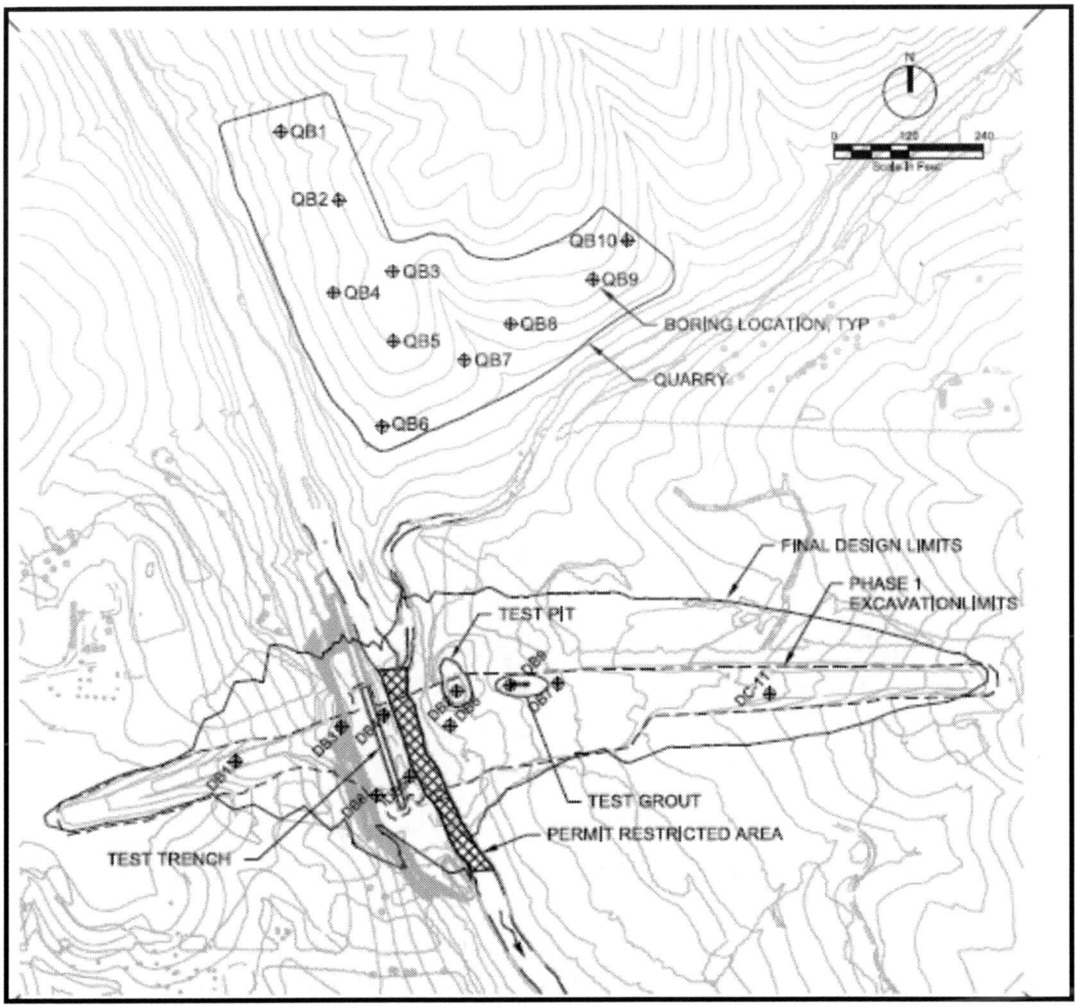

Figure 3. Excavation plan showing restricted area

FINDINGS FROM PHASE 1

Site Specific Stratigraphy and Geologic Features

The Duck River Dam site is located within the Appalachian Plateau's physiographic province and the East Warrior platform, west of the folded and faulted Valley and Ridge province. The rocks of the Plateaus are relatively flat-lying with a low regional dip of approximately 40 feet per mile to the southwest. Rock outcropping in the Plateaus province is dominantly Pennsylvania age and thickens to in excess of 5,000 feet to the southwest under the Black Warrior Basin Alabama Geological Society [ALGS], (1988). Rock at the Duck River Dam site represents the lower portion of the Pennsylvanian Pottsville Formation, which is defined as "dominated by massive, conglomeratic, orthoquartzitic sandstone interbedded with varying amounts of shale, siltstone and thin discontinuous coals." Alabama Geological Society [ALGS], (1988).

Geologic interpretations from the USACE planning efforts confirmed that the site is made up of interbedded sandstone and shales of the Pottsville Formation. The USACE reports divided the stratigraphy into six units (A through F, from the top down). Following the Phase 1 foundation excavation, the design team subdivided the upper two units as shown below in Figure 4, which represents a simplified section showing the sequence of the geologic units at the site. Detailed descriptions of the stratigraphy, structure and discontinuities for these Units are presented below.

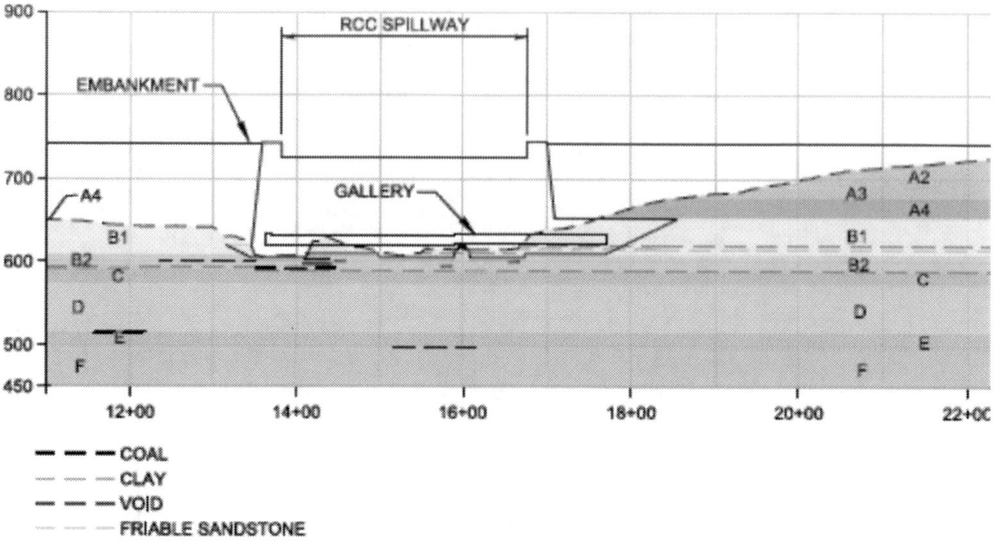

Figure 4. Simplified geologic section showing sequence of geologic units

Unit A. Unit A is predominantly shale with lesser amounts of sandstone. Most of Unit A displays some degree of weathering. Unit A can be divided into three distinct and identifiable sub-units. Sub-unit A1 represents a weathered sandstone with only a few feet of exposure at the highest elevations on the extreme left and right ends of the dam. Sub-unit A2 is dominantly shale with clay layers and seams. Sub-unit A3 is largely sandstone, but commonly interbedded with shale and clay seams with depth.

Clay layers are relatively common in Unit A. The clay layers and seams exist as either original deposited material or as a weathering product derived from the shale layers. Unit A2 has common thin inter-beds of clay within the primary shale layers, with a few thicker zones up to 5 feet thick. Several laterally persistent clay layers, generally from 2 to 6 inches thick, with occasional thicker layers up to 2 feet, were observed in Unit A3. Figure 5 shows the 5 foot clay layer encountered along the left abutment.

Figure 5. Five foot thick clay layer in Unit A2 at left abutment

Along the steeper slopes of the right abutment, Unit A contains fracture rock blocks (stress relief blocks). Weathering and decompression in the valley wall appears to allow these large blocks to separate along fractures, sliding and rotating downslope along horizontal to slightly dipping bedding planes. Vertical clay infillings, as shown in Figure 6, are present at each of these stress relief blocks and provide an indication of the amount of down valley movement and instability,

Figure 6. Vertical (12 inch thick) clay infilling at large stress relief block - right abutment

Unit B. Unit B can be divided into two distinct sub-units. Sub-unit B1 represents the structurally strongest rock encountered, Figure 7. The massive sandstone zones tend to alternate with more thinly bedded sandstone layers and with greater depth, the unit becomes interbedded with coal lenses and thin clay seams. The observed presence of clay seams or layers in the B1 sandstone is limited. Discontinuous and sporadic seams and pockets of clay are present, most of which are limited in length. A few clay seams were found to extend laterally over 10s of feet.

Figure 7. Excavation exposing Unit B sandstone in the foundation of the RCC section showing alternating bedding (massive versus more thinly bedded zones).

Relative thick layers of friable sandstone are present in the upper portion of B1 on the left abutment. Figure 8 shows a bed of friable sandstone in Unit B1 at approximate elevation 640. These weathered and friable sandstones were as much as 25 feet thick. The friable sandstone breaks down easily under the impact of a geologic hammer. The friable layers were not observed below elevation 625 where more competent quartzite is exposed.

Figure 8. Friable sandstone in Unit B1

Sub-unit B2 is differentiated by a thinning of the individual beds and an increase in the presence of interbedded shale, clay and coal seams. While Unit B2 is still predominantly sandstone, it is considered to be a transitional zone between the overlying thick sandstone of Unit B1 and the underlying shale of Unit C, and as such, can display properties of both units. Most of the clay seams observed in B2 were present for only a short distance (<10 feet), were commonly observed in association with coal in Sub-unit B2 and had undulating surfaces.

Coal seams, layers and lenses are also present in varying amounts in Sub-unit B2, but seldom in large quantities. The coal is typically only traceable as a layer over a few feet and is commonly 2 inches or less in thickness. A few coal seams over 2 feet thick were mapped in the foundation, Figure 9. The coal is commonly fractured and slickensided surfaces were frequently observed. The coal represents structurally weak material with high porosity and permeability that will allow the rapid movement of groundwater.

Figure 9. Excavation in the Unit B2 sandstone exposing coal seams

Units C and D. Units C and D were not exposed during any of the excavation work, and the descriptions are interpreted from the core borings. The two units are similar to Units A and B, respectively, but contain zones that are highly permeable. Unit C is made up of thin to medium bedded, moderately hard to hard interbedded silty shale and very fine grained sandstone. Unit C has closely spaced high angle joints. Occasional discontinuous coal seams up to 2 inches in thickness were observed in the Core. Unit D is composed of hard to very hard medium bedded very fine to fine grained sandstone with occasional zones of interbedded thin shale layers.

Jointing. Jointing and fractures within the observed rock units are abundant, particularly within Sub-units A3 and B1. The jointing data obtained from the televiewer logging were plotted on stereonets. The plots from the borings data are consistent with the field geologic mapping that most of the joints and fractures measured have a near-vertical orientation (approximately 85-degree dip), whereas bedding planes are near-horizontal. The joints and fractures are most identifiable in larger, thicker beds of sandstone and may persist for approximately 10 feet laterally and up to approximately 10 feet vertically. Jointing commonly occurs in parallel sets with spacing between 3 and 15 inches.

Although the fractures are common throughout individual bedding units, they are difficult to follow through multiple beds vertically, especially if interbedded shale or clay layers. Typical jointing and fractures for the lower portion of Sub-unit B and Unit C are shown in Figure 10. The fractures and joints frequently have weathering rinds and/or are infilled with iron oxide deposits of 1 to 3 inches wide and are indicative of water movement. An extensive grout program was anticipated based on the jointing characteristic of the rock. Grout takes were expected to be high in the bottom of B1 and in Units C and D.

Figure 10. Photo showing Rock Joint Structure in Unit B

Localized Shear Zone near River. The USACE *Engineering Studies Report* (1998) identified a possible reverse fault between borings that were drilled near the river. The presence of the fault was suspected based on apparent offset (6-feet) at the top of Unit C in boring logs. No other data from these boring indicated the presence of a fault. Because of the transitional nature of Unit B2, the exact contact with Unit C was difficult to distinguish, especially in core samples. Therefore, while the potential presence of a fault near the Duck River could not be eliminated in Phase 1, it was considered unlikely to be present.

The dipping beds shown in Figure 11 are in the river channel, approximately 10 feet above the top of Unit C. The photo was taken during construction of the dam (Phase 2). Once exposed, this area was examined and mapped by the project geologist. Observations of the rock showed evidence of riffles, imprints of fern leaves, small rounded gravel inclusions and other indications that suggest these were cross bedded fluvial deposits. No signs of shearing were observed in any of the excavations. Therefore, an offset in the top of Unit C does exist, and presence of a fault near the Duck

River was eliminated. Potential concerns such as movement and deformation along a localized shear zone were therefore considered non-issues.

Figure 11. Cross bedding below river channel

Voids. Of the 60 borings completed by the USACE within the study area, seven borings included descriptions of *possible* voids in their boring logs. The possible voids manifest as core loss, lack of rock core recovery and on occasion a rod drop and loss of drill water. None of the possible voids were confirmed and several are likely the result of soft weathered or unconsolidated material washed out during drilling by circulation of the drilling fluids. During Phase 1 investigations, the driller reported rod drops at two locations, typically only a few inches. Images of the video logged at one of these locations are shown in Figure 12. It is difficult to determine from the image if a void actually existed. In a worse case, the void is relatively small.

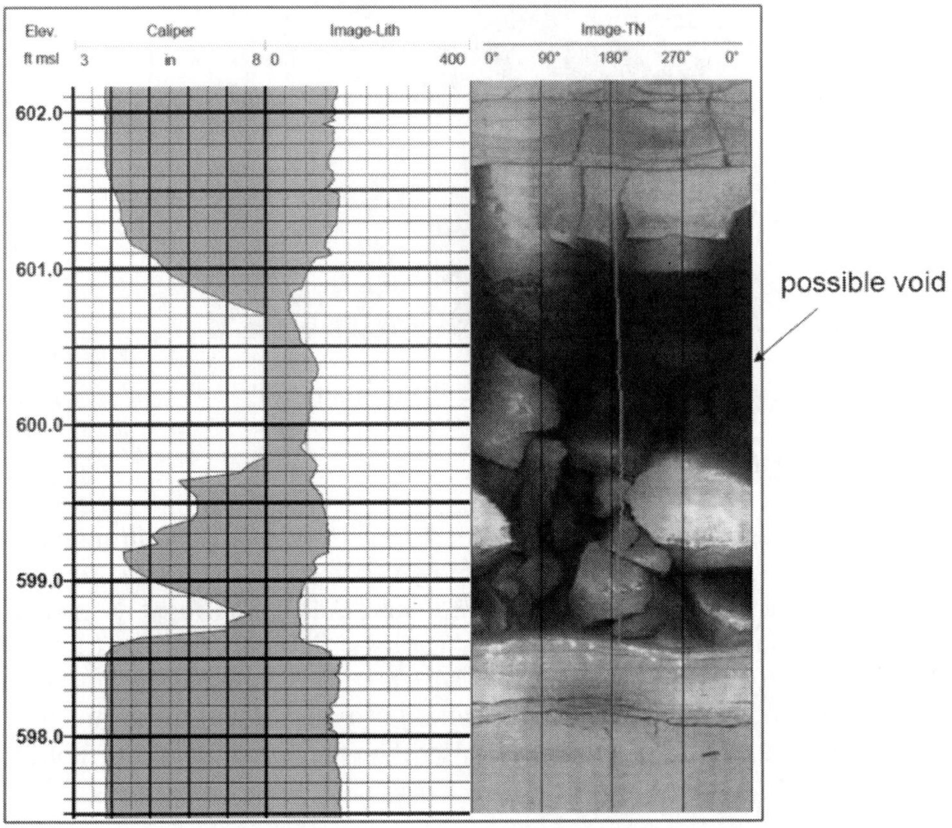

Figure 12. Televiewer log showing possible void

Most of these possible voids were in the proximity of Duck River and occurred within the lower portion of Unit B2. The variable depths of the potential void locations encountered in the borings suggest that possible voids are not interconnected. No voids were observed in any of the excavations, either Phase 1 or 2. Although significant grout takes occurred sporadically in Units B and C, these large takes were expected and were not indicative of a large void. It is possible that small voids could be related to the coal seams. The engineer concluded that the voids are likely to be small, could be related to the coal seams and do not appear to be continuous. No special treatment in the foundation was required for any voids.

Quarry Testing

Based on investigations and testing it was concluded that rock was suitable and occurred in sufficiently large quantities for producing both rock fill and RCC aggregate. The quarry developed for the project during construction is shown in Figure 13. This quarry provided high quality rock fill materials and adequate quality rock for RCC aggregates also shown on Figure 13.

Figure 13. Typical rock fill produced from quarry with a D_{50} of 12 inches

Rock Permeability and Grout Testing

Hydraulic pressure tests (packer tests) were performed while drilling through the rock foundation to estimate the hydraulic conductivity of the rock. The tests were performed in all dam footprint borings, generally at 10-foot intervals as the boring was advanced. General results are summarized in Table 1 by geologic Sub-unit. The coefficient of hydraulic conductivity should be considered a permeability value in the horizontal direction. Because of bedding and stratification, it is assumed that the vertical permeability can be as much as 10 times smaller than the horizontal permeability.

Table 1. Estimated Values of Permeability in the Dam Foundation.

Coefficient of Hydraulic Conductivity (cm/s)

Geological Unit	Best Estimate	Lower Bound	Upper Bound
Unit A	5E-4	5E-5 1E-3	
Unit B1	1E-4	1E-5 1E-3	
Unit B2	2E-3	2E-4 4E-3	
Unit C	2E-3	2E-4 4E-3	
Unit D1	2E-3	2E-4 4E-3	
Unit D2	1E-3	1E-4 1E-3	
Unit E	5E-4	5E-5 1E-3	

The results of the grout testing showed zones of low grout take in Sub-units B1 and B2. High grout takes occurred in Unit C. Moderate grout takes occurred in the upper half of Unit D with very large takes occurring in the lower half of Unit D. Correlations with grout take and permeability estimates from the packer tests were not always good and were poorest in the upper part of Sub-Unit B1 where grout takes were very small to zero.

DISCUSSION

Initial Observations and Design Challenges

Geologic conditions were expected to be locally poor given the geologic setting of interbedded sandstone and shales. Early on in Phase 1, the conditions encountered at both abutments were worse than anticipated with pervasive presence of horizontal clay seams. Both abutments (Unit A) displayed a deep zone of weathering and the rock mass was very poor. Examination of these conditions clearly indicated that an RCC gravity dam founded on Unit A was not a viable alternative.

In the right abutment, the steeper of the two abutments, the shale consisted of clay, claystone and more indurated shale. The five large distinct rock blocks were unstable. All these blocks had moved along clay/shale beds and were marked with vertical clay infillings and a clay under-bed. The pervasive layers of soft wet clay would also require removal to limit foundation deformation beneath an RCC gravity dam.

In the left abutment, although flatter and absent of stress relief blocks, clay layers were thick and more problematic regarding settlement under the heavy loads imposed by an RCC gravity dam. The complete removal of Unit A and some of Unit B1 would be required to limit deformation to establish a foundation suitable for a gravity dam.

These discoveries indicated that a gravity RCC dam founded on the Unit A shales would be difficult and expensive to construct. Complete removal of the Unit A shale under the gravity dam foundation would require a massive amount of excavation on both abutments and was not an option. Therefore, upon discovering these features, bulk excavation work on the abutments (that is, in Units A1 and A2) was discontinued. The design team had to consider alternatives to the full length RCC dam concept and adapt the dam type to these unfavorable site conditions.

A full length ECRF dam with an abutment spillway or a hybrid dam consisting of an RCC center spillway with earth and rock fill abutments were considered the next best options. Because of the large revised PMF, the ECRF dam would require a large abutment spillway. The rock conditions revealed at both abutments were poor, exposing highly weathered shale (upper Unit A) susceptible to slaking from wetting and drying cycles. The large PMF and poor rock quality would require a large, expensive, reinforced concrete lined spillway. The hybrid RCC-ECRF alternative afforded the benefit of a less expensive spillway in the river channel and in addition a benefit that the spillway could be overtopped during construction without damage. Adverse weather during construction would be less of a risk for the hybrid dam alternative.

Rock excavation (drilling and blasting) near the existing river channel in the Unit B sandstone was continued and focused on finding the top of a relatively fresh (unweathered) and more massive sandstone suitable for founding a concrete gravity dam that could serve as a spillway. Mapping of the exposed bedrock was done concurrent with the excavation progress. Unit B turned out to be highly variable and as previously discussed was divided into two Sub-units, Unit B1 and B2. The top of Unit B (designated Unit B1), was 35 feet thick and consisted primarily of highly weathered and friable sandstone judged to be susceptible to seepage and piping. The lower Unit B (designated Unit B2) was 20 feet thick and consisted of moderately weathered to fresh, hard to very hard, sandstone (quartize). Bedding, jointing and structure in the Unit B2 were highly variable and discontinuities, such as coal seams, were present.

The suitability of this lower valley bedrock as a foundation for an RCC spillway was evaluated. The primary design issues were sliding stability along discontinuities and weak bedding planes (primarily along coal seams or "dirty" bedding planes) and the potential for unacceptable deformation due to the nature of the discontinuities, zones of fractured rock and an underlying Unit C shale that was weaker than the Unit B sandstone. Seepage control and prevention of piping in Unit B2 and Unit C was also a concern due to some zones of fractured rock, friable sandstone seams at the contact and occasionally reported "voids" in the core logs.

Phase 2 Evaluations

Phase 2 consisted of evaluating the hybrid alternative for final design and preparing final design plans and specifications for bidding. At the completion of Phase 1, and to facilitate the final design, a detailed site specific geologic model was developed. This detailed map of the foundation geology included all features revealed during mapping of nearly the entire dam foundation area and proved to be as invaluable as a design tool.

Inspection of the geologic profile suggested that the RCC spillway structure could be founded on a relatively thick, quartzite layer in Unit B2 at approximate Elevation 605. This 10- to 15-foot-thick layer extends to near the interface of Unit C where inter-fingering of shale layers were observed. A discontinuous 2-inch-thick clay layer with a wavy geometry associated with a truncated coal seam, with a thin friable sandstone layer was observed at the interface of Unit B2 and C. Both the clay layers and coal seams have an irregular geometry with three dimensional waviness and corresponded with a zone of strong quartzite fingers and shale inter-bedding. Similar materials were observed in core samples taken in nearby borings. These important features of Unit B2 and C were taken into account by the stability analyses.

Key Design Considerations

Evaluation of sliding stability and deformation performance of the dam needed to be carried out during final design to determine an appropriate embankment and RCC spillway geometry. Sliding stability and deformation were investigated for the zoned earth core rock fill embankment sections using Limit Equilibrium analyses taking into account the presence of weaker layers (clay seams). In addition, FLAC (Fast Lagrangian Analysis of Continua) modeling was performed to investigate failure mechanisms associated with competent foundation layers overlying weak layers. Sliding stability and deformation were investigated for the RCC spillway structure using Limit Equilibrium methods and FLAC analyses. These analyses indicated an economical RCC spillway section could be safely constructed on these materials.

Stability of the Earth Core Rock Fill Abutments.

Based on the Phase 1 work, two important stability issues in Unit A required careful evaluation and mitigation; 1) sliding stability of the relatively thick clay layers, particularly the 5-foot-thick layer at the left abutment and 2) the large stress relief blocks on the right abutment.

Stress Relief Blocks -The large stress relief blocks on the right abutment were mapped above the top of Unit B within Sub-unit A2. The five blocks that were mapped all exhibited movement along a soft, thin (1- to 2-inch-thick), wet clay under-bed. The location of the blocks and the right abutment geometry made it feasible to excavate and remove these stress relief blocks, which was called for in the final excavation plan. The work required a series of benches excavated into the right abutment hillside, with a maximum bench height of about 45 feet.

Sliding Through Clay Layers - To address sliding/shear through the clay layers, it was necessary to establish the shear strength of these clay zones. Clay was sampled from Unit A in the dam foundation and laboratory testing was performed by Virginia Tech's geotechnical laboratory, including direct shear and torsional ring shear strength testing on reconstituted normally consolidated samples.

Fully softened shear strength were obtained from the direct shear strength test, as well as the torsional ring shear tests. Residual strength values were obtained from the torsional ring shear tests only. Fully softened friction angles obtained from the direct shear box were a few degrees higher than the values obtained using the torsional ring shear strength apparatus; fully softened friction angles ranged from 26 to 32 degrees for the direct shear strength tests, compared to fully softened friction angles of 21 to 27 degrees for the torsional ring shear test. This finding is consistent with the research performed at Virginia Tech and at the USACE's Experiment Station in Vicksburg, Mississippi.

Residual friction angles obtained from the torsional ring shear strength apparatus ranged from approximately 14 to 27 degrees.

The obtained strength values were compared with fully softened and residual friction angle correlations with liquid limit, clay-size fraction and effective stress that were developed by Stark and Eid (1994, 1997). Based on clay size fraction ranges from 25 to 45 percent, and liquid limit of about 50 percent, the fully softened and residual friction angles obtained from the correlations were approximately 28 and 18 degrees, respectively.

During a workshop held in December 2011 at Virginia Tech (Duncan et al., 2011), no consensus was made on what strengths should be used for clayey embankment foundation soils. Some were of the opinion that the full softened condition should only apply to excavated slopes in natural clay. Others thought that softening can occur at depth in foundations from changes in effective stress and saturation caused by seepage.

Another consideration is that slickensides were encountered in some locations; this would indicate the need for using even lower strengths, such as residual strengths. It is noted in the Virginia Tech workshop report that for residual strengths to be appropriate, the slickensides must align with the failure surface. "If the slickensides are randomly oriented, the residual strength should not be seen as controlling stability". At the Duck River site, it appears as if the slickensides are random and more typically associated with block movement parallel to the dam axis, and therefore, were not considered aligned with a potential failure surface.

Based on all these considerations, it was concluded that it would be appropriate to use lower bound values of the fully softened strength for the purposes of the stability analyses. The clay layers were assigned zero cohesion and an effective friction angle of 26 degrees based on the laboratory testing results.

Results for Earth Core Rock Fill Embankments - Results from the stability analyses indicated that flat slopes (i.e., flatter than typically used for rock fill dams on good foundations) or stability berms would be required where the zoned earth rock fill dams were founded in the upper hillside slopes where complete removal of the Unit A shale was not practical. However, the design included complete removal of Unit A on both abutment in the lower (bottom) reaches of the valley where Unit B was relatively shallow. Upslope of these locations, stability berms were included as part of the design, as Unit B was much deeper and excavating all of Unit A would require removal of large portions of the hillside slopes and massive excavation quantities. Typical cross sectional geometry for these two distinct dam design sections are show in Figure 14.

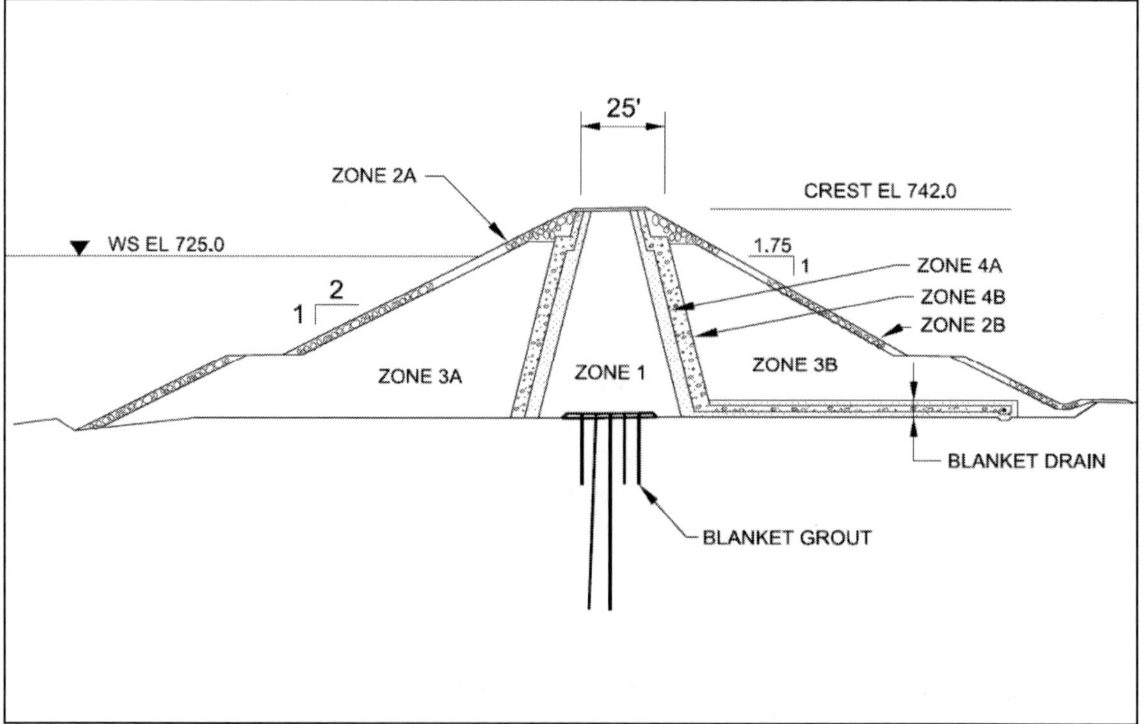

Figure 14a. Typical abutment sections with berms

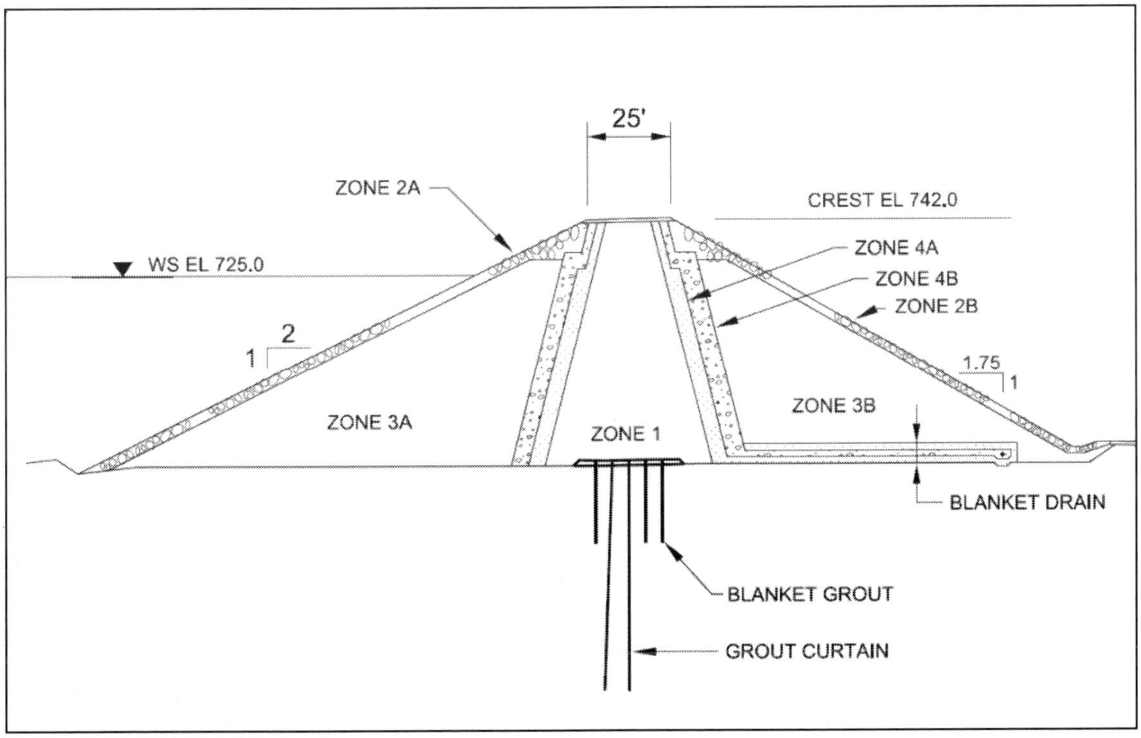

Figure 14b. Typical abutment sections without berms

Sliding Stability - RCC Gravity Spillway Section. Weak layers in the foundation under the RCC gravity spillway included coal seams with the occasional occurrence of slickensides, sandstone and shale bedding planes. The bedding planes were assumed to be mostly continuous, however, the actual lengths and extents of these coal seams were not easy to determine. To investigate the continuity of the coal seams, they were carefully mapped in the excavation exposures. Based on this mapping, rock core observations and core hole televiewer data, it was concluded that the coal seams were discontinuous. From the subsurface information, it was judged that the coal seams were likely less than one third of the foundation width, but a more conservative case was analyzed, in which it was assumed that the weak layer extends over approximately half the foundation width.

Coal Seams - The strength properties of coal seams are difficult to determine. Samples of the coal were sent to the Virginia Tech Laboratory for examination and possible direct shear testing. The quality of the coal samples did not allow for testing, and therefore, assumptions on shear strength were made based on published information provided by staff at Virginia Tech soils laboratory, ("Salamon and Munro, 1967). For shear along the horizontal seams, a friction angle value of 30 degrees and cohesion value of zero were assumed. It was also reasoned that the interface between the rock and coal is typically the critical sliding surface. Based on these considerations, a friction angle value of 26 degrees was assumed (with cohesion = 0) to account for the weaker interface between coal and rock and for any underclay below the coal.

Bedding Planes - Bedding plane shear strength was estimated for the shale and sandstone using the nonlinear failure strength criteria developed by Barton and Bandis (1990). The empirical method utilizes the results from shear tests on saw cut core samples (residual friction angle) and descriptions of surface roughness to estimate the strength along a joint.

Using this method, an equivalent joint friction angle of 32 degrees and cohesion intercept of 700 lb/ft^2 was estimated for the sandstone (Unit B) and an equivalent joint friction angle of 22 degrees and cohesion intercept of 600 lb/ft^2 for the shale (Unit C).

In addition to the joint shear strength for a small area or sample, additional strength as a result of the larger scale undulations in the rock was taken into account when considering the bedding plane shear. A waviness angle of 2 to 3 degrees was added to the joint friction angles to arrive at appropriate bedding plane shear strength values.

Results for the RCC Gravity Dam - The stability modeling incorporated the above engineering strength properties and the FLAC model allowed the design team to investigate multiple features in the rock foundation and to input various pore pressures and seepage assumptions. The FLAC analyses were performed with and without the influence of foundation drains and a grout curtain. The grout curtain was modeled as a discrete zone of low permeability material. The effects of foundation drains were modeled assuming various uplift pressures along the base of the spillway. Stability analyses were performed for different reservoir stages (full reservoir and flood stages) and corresponding tail water levels. Tail water levels determined from flood routing model results. To examine the benefits of a grout curtain, several stability runs were performed, each assuming a different permeability at the grout curtain. To examine the benefits of foundation drains various uplift pressures was assumed along the bases of the spillway. For the case of no foundation drains uplift pressure was assume to vary linear from reservoir stage at the upstream end of the spillway to tail water level at the toe of the spillway. A bi-linear uplift model was used when drains were included in the analyses assuming reductions in uplift at the drain location.

The FLAC results allow examination of outcomes such as stresses, deformation and factors of safety. The results show good agreement between the FLAC model and the LE analysis for the failure mode of the RCC spillway sliding on the rock surface. For the spillway sliding on top of rock foundation failure mode, the factor of safety values from both the FLAC and LE analyses were very high (>5.0). The factor of safety results from FLAC for the mode of sliding on a deeper bedding plane agree reasonably well with the LE result (that is, an average of approximately 2.2 from the models versus a value of 2.0 from the LE analysis). The factor of safety against sliding increases by 5 to 10 percent when a grout curtain is incorporated in the model. Similar increases in the factors of safety were noted when foundation drains were included.

The factor of safety decreases when the weak bedding plane is moved down to elevation from Sub-unit B2 to Unit C. This is a result of the bedding plane strength being lower for the deeper plane; the deeper bedding plane in Unit C is predominantly shale and significantly weaker than Unit B2.

The critical failure mechanism in the bedrock is the formation of a tension crack from the top of rock down to the bedding plane just upstream of the heel, while a horizontal sliding surface forms along the bedding plane. Results from the FLAC analyses indicated that at the downstream end, the horizontal rock layers would tend to buckle, and therefore, cross-bedding shear resistance could not be relied upon. The factor of safety results for the various models are summarized in Table 2.

Table 2. Factors of safety from gravity spillway stability analyses

		Factors of Safety against Sliding				
Case	Sliding Plane Elevation	w/o foundation drains		w foundation drains		Limit Equilibrium Model [a]
		Grout Curtain	No Grout Curtain	Grout Curtain	No Grout Curtain	
Sliding at the interface of RCC and bedrock						
Ø = 38° C = 80 psi	El 605	5.73	5.50	5.84	5.74	5.60
Sliding on a foundation bedding plane						
Ø = 32° C = 4.86 psi	El 595	2.18	1.93	2.29	2.23	2.01 [b]
Ø = 22° C = 4.17 psi	El 585	1.69	1.53	1.80	1.78	Not analyzed
Sliding on clay layer in foundation						
Ø = 26° C = 0 psi	El 595 Unit B	1.67	1.46	1.75	1.70	1.71 [b]
Ø = 26° C = 0 psi	El 585 Unit C	1.72	1.54	1.85	1.81	Not analyzed
Sliding on a bedding plane with weak coal seam beneath ½ base of dam						
El 595		1.93	1.67	2.03	1.96	Not analyzed
El 585		1.71	1.54	1.83	1.81	Not analyzed

Notes:

a) The LE model includes foundation drains (50 percent efficiency) and does not model anisotropic permeability in the foundation. It is assumed that there is not grout curtain.

b) This factor of safety is from a two-wedge LE model with the first wedge under the dam sliding on a bedding plane (or weak layer) dipping 7 degrees, while the second wedge is a passive wedge in rock with estimated Mohr Coulomb (rock mass) strength Phi = 50 degrees and C= 60 psi

The results of these evaluations dictated use of relatively flat slopes for the RCC gravity section design. The downstream slope was angled at 1 horizontal to 1 vertical (1H:1V) The upstream slope was battered at 1 horizontal to 10 vertical (1H:10V). These flatter slopes provided additional weight and normal forces to increase the frictional resistance along potential failure surfaces, and also increase the length of the potential sliding surface. In addition, the wider foundation resulted in a longer seepage path which reduced hydraulic gradients below the dam. A cross section of the final spillway configuration is shown on Figure 15.

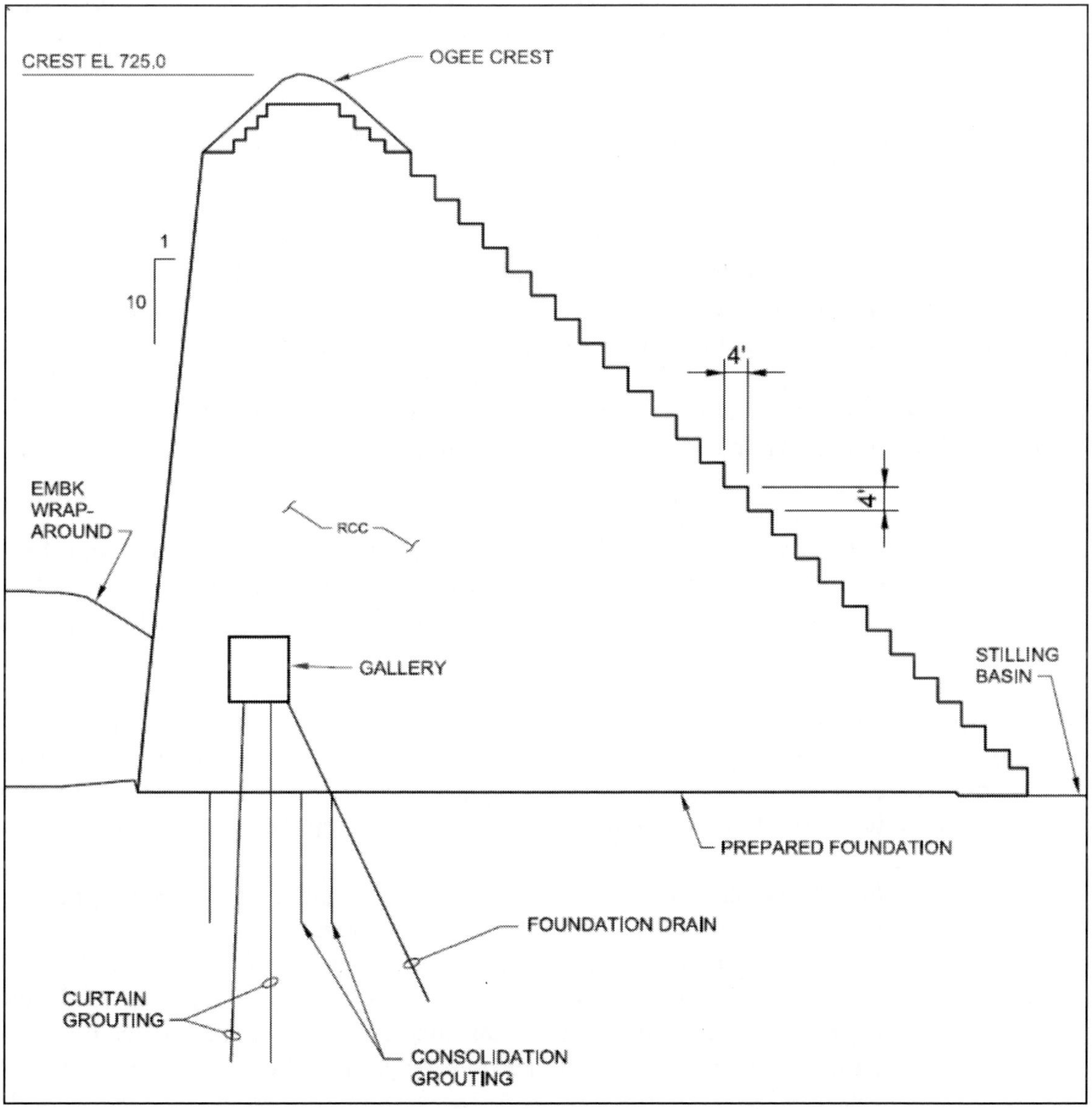

Figure 15. Final spillway geometry

Seepage and Piping Control. Isolated zones and seams of weakly cemented and friable sandstone susceptible to internal erosion and piping in upper portion of Unit B from approximately elevation 652 to 625 and near the top of Unit C and the potential for an occasional void was a concern that required mitigation. To protect against seepage and the potential for internal erosion and piping, several mitigation measures were incorporated in the design. The measures included the following:

Removal of weathered and friable sandstone in the gravity spillway section - Over excavation of weathered and friable layers was accomplished under the gravity spillway section.

Foundation Treatment in ECRF Core Section - Upon completion of the excavation, the foundation treatment of the core trench included excavation of clay infillings from joints, followed by joint treatment that included filling with bedding mortar or dental concrete. In some locations, the core trench sidewalls were treated with shotcrete.

Foundation grouting - Along nearly the entire dam, a two-row grout curtain was constructed using GIN (Grout Intensity Number) principles and methodology, but the method was modified by including some aspects of North American grouting practice such as final verification testing. The grouting program is described by Esterhuizen et al (2015). Grouting under the RCC section was performed from the gallery so that the weight of the concrete gravity dam could be utilized to allow use of higher grouting pressures in the upper foundation layers. In addition, the grouting program included blanket grouting on both the upstream and downstream sides of the curtain. Where the zoned earth rock fill dam sections were founded on Unit B1, the blanket grouting extended through the friable zones to reduce the potential for erosion of the more friable sandstone.

RCC cutoff key into right abutment - The RCC spillway was extended 100 feet into the right abutment to serve as a seepage barrier and erosion mitigation measure in a location where rock was highly fractured. This included extension of the gallery into the right abutment.

Foundation Drains under the RCC Spillway - The RCC spillway incorporated foundation drains along it entire length just downstream of the Gallery and at the toe of the spillway. These drains are effective in reducing seepage pressures that emerge from under the dam. It is important that these drains be routinely maintained to keep their effectiveness.

Filtering of drains – Foundation drains were installed in the gallery, at the toe of the spillway and the stilling basin. Where foundation rock is competent and non-erodible, such drains typically consist of a hole drilled in the rock. Because of the potentially erodible nature of the friable zones in Unit B, and potential piping of fine-grained seams present in both Units B and C, all of the drains installed on the project consisted of a slotted 2-inch diameter PVC pipe installed in a filter pack in the hole.

Blanket filters and drains for rock fill sections - Blanket (and toe) filters and drains were included for the rock fill sections, which serve to reduce the risk of internal erosion and piping and eliminate unprotected seepage exit points.

Geologic Representation and Approach to Completing Foundation Work

For the Phase 2 final design, the engineer developed an excavation plan that provided bidders the excavation contours and cross sections on 50 foot stations. Over excavation was included in the contract but only as authorized and directed by the Geotechnical Engineer. Unit price was used to establish cost for both common and rock excavation and for authorized over excavation.

Since the depth to sound rock was variable, rock excavation was defined in the contract as material that could not be ripped using a specified dozer equipped with a single ripper. To provide contractors a common basis for pricing foundation treatment, detailed geologic maps, along with all detailed notes, were included in the bidding documents. In addition, a table that listed the station, offset, elevation and detailed description of each geologic feature that was mapped within the dam footprint was provided in the contract documents. These maps clarify the foundation treatment required at the dam foundation contact. This approach proved to be effective and beneficial as there were virtually no site claims or overruns related to earthwork activities and payment items, which is unusual for a dam of this size, in highly variable site conditions.

CONCLUSIONS

Where highly variable site conditions exist in a difficult geologic setting, the adaptive and phased approach can be effective in reducing the owner's exposure to significant differing site condition claims. For this project, by excavating a large portion of the dam foundation prior to final design, the owner was able to start construction early, and the engineer was able to gather the geotechnical information required for design. The large deep excavations included in the Phase 1 investigation proved invaluable by providing visible detail on stratigraphy, rock structure and other geologic features important for evaluating geotechnical issues, such as dam stability, seepage, potential for internal piping in the foundation and seepage control measures. For the Duck River Dam Project, significant benefits to the approach included:

1) The engineer was able to change the dam type prior to the beginning of the final design to match the site-specific conditions encountered. In using this phased approach, the owner accepted significant changes in initial design concepts could be required since the final geotechnical investigation was being completed as part of the Phase 1 dam construction.

2) The engineer was able to obtain necessary information needed for final design in a timely and cost-effective manner, while allowing the owner to start construction. An early start of construction was desirable given the continued scrutiny by some environmental groups.

3) Significant changes to the dam concept were adopted after completing Phase 1 to accommodate actual foundation conditions and to provide the owner with an affordable dam.

4) The engineer was able to perform the geotechnical analyses based on accurate site-specific geologic conditions. This resulted in reducing the uncertainty in ground conditions and in the geotechnical models. Reducing the uncertainty in subsurface conditions resulted in a cost-effective design. Reducing the uncertainty in subsurface conditions resulted in efficient construction execution.

5) The most significant benefit has been the lack of claims related to the foundation work for this project. Uncertainties normally associated with the foundation work in dam construction were mitigated because a significant portion of the foundation was exposed in Phase 1. Risks were reduced, construction delays were minimized and the construction progress has kept pace with a very aggressive and optimistic construction schedule.

ACKNOWLEDGEMENTS

The authors acknowledge the invaluable contributions of the following stakeholders and individuals:

1. The Cullman Utility Board and the City of Cullman.
2. CH2M construction inspection personnel, including Jerry Gardner, Martin Murray, Eduardo Gutierrez-Pacheco, Matt Tabor.
3. CH2M employees Tom Harwell, Steve Newton and Mike Stickley, who served as Project Managers.

REFERENCES

Alabama Geological Society [ALGS]. 1988. Raymond, D.E., Osborne, W.E., Copeland, C.W. and Neathery, T.L., 1988. Alabama Stratigraphy. Geological Survey of Alabama Circular 140, 97.p.

USACE. 1998. Engineering Studies Report for Cullman/Morgan Water District Duck River Dam, Pump Station and Pipeline. Main Report, Appendix A, Hydrology and Hydraulics, and Appendix I, Cost Estimate, October 1998.

Stark, T. D. and H.T. Eid. 1994. "Drained Residual Strength of Cohesive Soils". *Journal of Geotechnical and Geoenvironmental Engineering*, Vol. 120, No. 5, pp. 856-871. ASCE.

Stark, T.D. and H.T. Eid. 1997. "Slope Stability Analyses in Stiff Fissured Clays," *Journal of Geotechnical and Geoenvironmental Engineering*, Vol. 123, No. 4, pp. 335-343. ASCE.

Barton, N. and S.C. Bandis. 1990. "Review of predictive capabilities of JRC-JCS model in engineering practice." International Symposium on Rock Joints. Loen 1990. Proceedings, pp. 603-610, 1990.

Salamon and Munro, 1967. "Evaluation of the Strength of Slender Pillars." National Institute for Occupational Safety and Health, Pittsburgh, Pennsylvania.

Duncan, J. M., T. L. Brandon, and D. R. VandenBerge. 2011. *Report of the Workshop on Shear Strength for Stability of Slopes in Highly Plastic Clays*, Report #67. Center for Geotechnical Practice and Research, Virginia Tech, Blacksburg, VA. 79.

Esterhuizen, J. J., T. Cotten, and R. E. Riker. 2015. "Application of Modified GIN Method for Grouting the Duck River Dam Foundation", Proceedings to the Annual Conference of the Association of State Dam Safety Official, New Orleans, 2015.

THINGS YOU WANTED TO KNOW ABOUT INDIVIDUAL CONSULTING, BUT WERE AFRAID TO ASK

Larry K. Nuss, P.E.[1]
Kenneth D. Hansen, P.E.[2]

ABSTRACT

Individual engineering consulting, as it relates to the dams industry, can be significantly rewarding and is usually quite enjoyable. This paper provides lessons learned and tips on starting and maintaining an individual consulting practice. Larry Nuss and Ken Hansen formed their own individual consulting businesses after their main careers ended: Larry had 36 years with the Bureau of Reclamation while Ken had 37 years with the Portland Cement Association plus nearly 12 years with Schnabel Engineering. Consulting provides an opportunity to share ones knowledge and experiences, and work with a wide range of talented engineers from many companies. Typically, individual consultants interact with their clients helping them determine appropriate paths forward and hopefully aid in making good decisions about their projects. This paper also provides guidance for some areas of consulting that may be less enjoyable and could dissuade good engineers, such as: contract clauses, payment issues, accounting, and insurance requirements. The authors share various relevant experiences during their consulting careers.

INTRODUCTION

Individual engineering consultants possess expertise in a certain field that either is not available within the client's regular staff or provides an additional level of Peer Review to the client's team. Ken specializes in RCC and soil-cement applications in dams, while Larry specializes in the design and analysis of concrete dams. This paper is written to help someone start their own individual consulting business but hopefully not to compete with us. Just joking, there is plenty of work out there for everyone. The topics in this paper are from our own experiences. As such, use the paper as a guide, but please get insurance and accounting guidance from professionals.

The individual consultant can provide a number of services:

- Planning, design, analysis, inspection, and construction
- Peer review, problem solving, Board of Consultant member
- Potential Failure Mode Analysis and Risk Analysis member or facilitator
- Mentoring, training
- Expert testimony or dispute resolution

[1] Larry K. Nuss, Structural Engineer, Nuss Engineering, LLC, 10065 Silver Maple Circle, Highlands Ranch, Colorado 80129, Larry.K.Nuss@NussEngineering.com.
[2] Kenneth D. Hansen, Consulting Engineer, 6050 Greenwood Plaza Blvd., Suite 100, Greenwood Village, Colorado, 80111, 303-695-6500, ken@ken-hansen.com.

In this paper, Consultant refers to an individual consultant, Client refers to someone that hires the consultant and may be an engineering company or owner, and Owner refers to someone that owns a dam. An owner may hire an engineering firm that in turn subcontracts to a Consultant.

According to the Institute of Electrical and Electronics Engineers (IEEE), results from a survey of individual Consultants showed that: 39% had been an individual consultant for 3 to 10 years, 67% work at home, 46% carry professional liability insurance, 26% bill 10 hours or less per week, 82% list repeat business as their main source of fees, 60 is the average consultant age, 25% have a PhD, 32% hold a Master's degree, and 72% work on an hourly fee basis.

GETTING STARTED

It is quite easy to start your own consulting business with relatively little cash outlay. When to start your consulting business is the first question to answer: early in one's career or later in retirement. Both Ken and Larry started their consulting businesses in retirement. Some advantages of starting a consulting business in retirement include: a steady retirement income while the business is getting started, established name recognition in the industry, having more friends and contacts in the industry, and developed experience. These are not necessarily the case when starting early in one's career.

Type of Business Model

The second question is the type of business model to form: Sole Proprietorships, Partnerships, and Corporations. An Accountant can be a great help in this area. One consideration is the planned level of consulting activity. A simpler business model, like a sole proprietor, may be wanted if billing less than 10 hours per week.

A sole proprietor is an individual conducting business in their own name (Kenneth D. Hansen P.E.) or under a business trade name (Hansen Consulting). The business name can be registered online with the State for a few dollars. You do not have to take any formal action to form a sole proprietorship. As long as you are the sole owner, the business status automatically comes from your business activities. Using a business name does not change the status or protections of the business. Sole proprietors still put all their personal and business assets at risk and remain liable for all the business obligations; and a sole proprietor may be taxed at a higher rate than corporations. Set up costs are low and there are no corporate reporting or filing requirements.

A Limited Liability Company (LLC) is a business entity created by state statute. The IRS will treat an LLC as a partnership (if there are 2 or more members) or a disregarded entity (if there is only one member), or as an S Corporation (if Form 2553 and 8832 are filed). The type of entity an LLC becomes is determined by you and your accountant or attorney. You tell the IRS the type of entity you wish to become when applying for the Federal ID number. State laws vary and our experience has been in Colorado. It is advisable to seek competent council from your attorney or accountant.

A corporation is considered a separate legal "person" and is owned by shareholders or members. An LLC can be set-up by you, but it is much easier and probably safer to hire an accountant or lawyer to do it correctly in a few days for a few hundred dollars. The advantages of an LLC are 1) personal assets and liabilities are separate from the business assets and liabilities, and 2) there may be tax advantages. To develop this liability separation between personal and business, the business must have separate bank accounts, credit cards, record keeping, and tax filings; must act like a separate "person"; and have member meetings. The business name must end in LLC, Corp, or Inc. to notify clients that they are working with an entity that limits liability (Nuss Engineering, LLC). These are all minor requirements considering your personal assets are protected.

The bottom line is that a sole proprietor puts their personal (home, vehicles, investments, savings, furnishings, etc.) and business assets at risk; while an LLC consultant puts only their business assets at risk. While this is true, it is rare for an individual consultant to be sued.

Starting the Business

In takes some funds to start the business. For example, Table 1 lists Larry's initial expenses to start up his business from an office in his house. Very little equipment is needed, but expenses can quickly add up. The cost of engineering software depends on the consultant's engineering specialty.

Table 1. Example of Funds Required to Start an Individual Consultant Business

Establish LLC	$400	Quickbooks Pro	$300
Laptop	$1500	Adobe Acrobat	$300
Printer (printer, copier, fax)	$350	Business credit cards	$50
Scanner	$300	Business checking account	$1000
Business cell phone	$200	Desk and chair	$650
Backup disk drive	$100	First client travel expenses	$1500
Internet connection (/month)	$30	Travel VISA	$200
Office supplies	$200	Airline International change fee	$400
Engineering software	$1000	Second client travel expenses	$1500
		TOTAL	$9,980

However as discussed below, the Consultant probably will not get paid right away. Payment might be 30 to 45 days, or longer, after a job is competed. In the interim, the Consultant has paid for airfare weeks before the job begins. Increasing expenses could cause friction with a penny-pinching spouse that is worried about you retiring too early and living off retirement income. It is a relief when the first payment is received.

To give a timeline, Nuss Engineering, LLC was established on January 20, 2012 with the accountant completing all the paperwork in about 2 weeks and obtaining a business Taxpayer ID number (similar to one's social security number but for a business). A business credit card and checking account under the Nuss Engineering, LLC name had to be set up right away. The office equipment was purchased in January and February.

Larry was contacted in February by 3 different friends to teach classes in Japan and China and be on a Review Board in Panama in May. A scanner was required to copy and return signed contracts. The business credit card was required to pay for airfare in March and travel expenses in May. The business checking account with a bank that had a Swift Code was required so the International Clients could directly deposit payments. Invoices were submitted to the Clients at the end of May and payments were made 45 days later around mid-July. So be prepared for this time lag in getting paid.

Process of Accepting Work

Individual consulting typically works this way: 1) A potential Client contacts you and asks if you would be interested in working with them on a project, shares details about the project, and suggests a Statement of Work (SOW) with tasks, 2) After the proposal is accepted, the Client's Procurement Office contacts you with a contract, 3) The contract terms (or Master Service Agreements) and fees are negotiated, agreed upon, and signed by both parties. 4) The Client's Accounting Office sends you a formal SOW with costs to be signed, a Purchase Order Number, and asks for a W9 Form, and 5) the job begins. Form W9 is a Federal Government "Request for Taxpayer – Identification Number and Certification" and lets the Client know your business structure and Taxpayer Identification Number for tax purposes.

Working for the Federal Government also requires registration on the System for Award Management (SAM). Registration is free and is online at https://governmentcontractregistration.com/. Be aware of third party vendors that have websites looking just like a Government website and they want to charge a fee. If the website asks for a credit card number you have the wrong site. Also a Dun & Bradstreet (DUNS) number is needed before registration on SAMs and can be obtained in about 10 days online at https://iupdate.dnb.com/iUpdate/.

Most of the time, the contracting process is painless, but sometimes it is daunting with a lot of forms that must be competed. Here are a few forms filled out over four years for procurement:

- Review and sign the Master Services Contract
- Submit a W9 Form
- Submit form CA587 - Nonresident Withholding Allocation Worksheet
- Submit the most recent company Audited Financial Statement
- Submit a Brief Bio
- Submit a 2-page resume of only jobs that apply to this contract
- Submit a Letter of Commitment
- Submit Standard Form 330 Part I, Sections A, B, C, D, E, F, and G
- Submit Standard Form 330 Part II - Architect-Engineer Qualifications
- Submit Schedule A - Firm-Fixed-Price Rate Schedule
- Submit Form UPM-111 - Representations, Certifications, and Other Statements
- Submit Mutual Nondisclosure Agreement
- Submit Questionnaire for Preliminary Evaluation of Conflict of Interest
- Submit Letter of Fee Justification

- Submit Minority/Women/Small Business Classification
- Register in System for Award Management (SAM)
- Register on Invoicing Processing (IPP)
- Register on Duns & Bradstreet
- Register on Ariba (an SAP Company) Network Solutions
- Submit Contract Monitoring Division Materials (CMD)
- Cyber security training

Marketing

The best marketing is word of mouth from friends and acquaintances developed over your career. During your career; establish your credibility, expertise, and name recognition by writing papers, attending conferences, serving on committees, making friends, having a mentor, working hard on projects, and sharing information. The first consulting assignment might be to complete work on a project that was started at your prior employer. However, be aware there might be ethical or conflict of interest issues doing this.

THE BUSINESS SIDE

Accounting and Invoicing

Accounting records must be kept in some way: by hand, spreadsheets, or accounting software (Quicken, Quickbooks Pro, and Intuit Online Payroll). Quickbooks Pro is an excellent and easy to learn business accounting software. It makes invoices; records income and expenses; creates balance sheets, profit and loss statements, and tracks account receivables. Intuit Online Payroll is an easy way to "pay yourself as an employee" from your business. It keeps track of all required Federal and State taxes, knows the dates when the taxes are due, and submits tax payments and forms directly to the appropriate agency. This takes a huge load off the consultant and makes it very easy. Track car mileage in a small notebook kept in the car. Mileage times $0.575 (2015) per mile is a business expense and personal non-taxable income to you. The rule of thumb is to purchase a company car if over 50% of the mileage is used for business; otherwise, take the 2015 standard $0.575 deduction per mile. Obtain and keep all receipts.

Taxes

A sole proprietor business and you are one and the same, so the business income and your personal income are the same. Form Schedule C is used to report the business income, loss, and expenses on the sole proprietor's personal tax return. The sole proprietor is responsible for self-employment tax and estimated taxes.

An S-Corp/LLC typically files a separate Schedule K-1 that reports the business income and expenses and then gets passed through on the individual member's personal 1040 tax form. An S-Corp can distribute income to the owners as 55% salary and 45% dividends (see Figure 1). Dividends are not subject to the 15% self-employment taxes. So if you make $100,000, you could save $7500 in self-employment taxes. Some advantages of incorporation are that the business can form Simple Individual Retirement Accounts for members (25% of salary), can give charitable contributions as an expense (up to 55% of gross), and pay for medical expenses of members.

Insurance

The purpose of insurance is to protect you, but there is an expense involved. Professional Liability (PL) protects you against a negligent act, error, or and omission. Commercial General Liability (CGL) protects you against bodily injury while at the client's site from property damages and bodily injury. A personal/family umbrella policy is not the same as a CGL policy. [Note: Most individual consultants do not carry PL insurance as most clients will exempt this requirement for insurance in the contracts. However, some clients require insurance and the consultant has to decide whether to obtain the insurance, have the client pay for the insurance, or reject the job.] Two advantages of liability insurance is that the insurance company has the "duty to indemnify (duty to pay claims) and also has the "duty to defend" (defend and pay for lawsuits against you). These can be huge expenses. Consultants that do not have insurance are basically "self-insured" and personally take on all liability. One argument against insurance is that a large insurance policy makes the consultant more of a target for litigation. While this may have some truth, you may not want to take the risk if you have a lot of assets to lose.

The initial draft of the contract will specify the client's desire for insurance; this may be negotiable. PL coverage over $2M is very difficult for a consultant to obtain, if not impossible. Levels of $1M are more typical. PL cost depends on the consultant's engineering discipline (structural, geotechnical, hydrology, etc.), length of time in business, any previous claims, the amount of coverage, services provided, project type, and the yearly income of the consulting business. PL coverage of $1M normally costs between $2000 and $5000 per year, but can be as high as $20,000, so shop around. CGL coverage is very reasonable at a few hundred dollars per year.

Fees, Expenses, and Invoicing

Consulting fees charged by the individual consultant are strictly up to the individual and are usually in line with other Consultants of the same stature. They are usually an hourly or daily rate. Expenses are on and above the consulting fees. If the task to be accomplished can be well defined, the fee could be a lump sum payment to include any expense involved. This provides some cost certainty to the client and eliminates having to provide receipts for every expense to the consultant. Consultant fees may be slightly lower than an equivalent Client engineer to allow your Client to markup your fee to their Client.

While the amount of funds for travel expenses incurred by the Consultant is usually considerably less than the fee, it is the detailed expenses that catch the eye of the Client's accountants.

The accountant does not want to pay more than necessary and that all charges are in accordance with the Client's policies. Examples might be:

- Requiring an invoice to be resubmitted because the receipt for dinner included an alcoholic beverage or the hotel charged an Internet Wi-Fi fee.
- Requiring receipts for small expenses such as cash tips for shuttle drivers or baggage handlers as well as for money deposited in vending machines.

One solution is to charge the client a lump daily sum for meals and incidentals. Contracts with U.S. government agencies have published per diem daily rates for meals and incidentals based on the travel destination. We have found these rates to be more than sufficient and convenient.

The Consultant should ask the Client if economy or business class airfare is desired. Business class airfare can be quite expensive and should not be assumed or expected. Considerations include cost and duration of the flight.

Consultants submit their invoices to their clients based on the terms of the contract (typically on a monthly basis). Generally, a consultant pays for their travel costs and gets reimbursed when their invoice is paid. Airfare is purchased a few weeks to a month in advance of travel; hotel, meals, and incidentals are purchased during travel. As a result these expenses can accumulate for a few months until the invoice is paid. Essentially the consultant becomes the client's banker and must have enough seed/startup money to cover this expense.

Payment of Invoices

The timely payment of invoices is usually a challenge for Consultants. The U.S. government is excellent and pays invoices within 30 days. Most, but not all, private company accountants seem to want to delay payment as long as possible with terms of 45 days (or more) or propose paid when get paid terms. Payment terms should be stated in the contract documents if the consultant wants to get paid sooner or get paid up front for travel expenses. Until the consultant gets paid, the consultant is bankrolling the client, especially for reimbursement of incurred travel expenses. This can be very burdensome for consultants just starting out. The Consultant can ask the Client to pay for airline tickets after booking, but there is no guarantee they will agree to this.

Consultants should keep track of invoices and when payment is due. If a payment is late, it is perfectly acceptable to inquire about the status of the invoice in a polite manner. Many times there is a glitch in the accounting system or the invoice is being audited and coming soon. The engineering staff the consultant is working with normally takes the lead and contacts their accounting department. They have much more leverage than the consultant. To the engineer, the consultant is a partner; to the accounting group, the consultant is just a commodity: big difference. Rarely does the consultant need to take legal action to collect fees.

CONTRACT CLAUSES

Insurance Clause

Find and read the required insurance clauses in a contract. Some Clients will waive the insurance requirements while others require substantial amounts. In reality, many times the consultant provides suggestions, and as such, is not the responsible party for the engineering work. Their input may be accepted, modified or rejected by the Client. The consultant usually does not professionally stamp drawings or documents; and usually is not the engineer of record. However, if something goes wrong with the project, the client's lawyers want to shift as much risk as they can to the individual consultant or any other party involved. Here is where the consultant must take a stand against unlimited indemnification. Contracts should have a Limited Liability clause to ensure you are not sued for millions, even if your fee is only a few thousand dollars.

Indemnity Clauses

An indemnity is an agreement by one party to bear the financial loss of another party for a specific event: essentially allocating risk or holding a party harmless. Consultants should be very careful with these clauses and probably delete them from a contract. A bad indemnity clause would make the consultant liable for the negligence of the client. So if the client is at fault, the consultant would share the liability. A good indemnity clause simply reflects that each party bears the risk of their own failures. A poorly written indemnity clause could adversely affect the coverage provided by your PL insurance. Here is an example of an acceptable Limitation of Liability Clause or a Consequential Damages Clause.

> ***Limitation of Liability*** - *In recognition of the relative risks and benefits of the project to both the Client and the Consultant, the risks have been allocated such that the Client agrees, to the fullest extent permitted by law, to limit the liability for any and all claims, losses, costs, damages of any nature whatsoever or claims expenses from any cause or causes, so that the total aggregate liability of the Consultant to all those named shall not exceed the Consultant's total fee for services rendered on this project. Such claims and causes include, but are not limited to, negligence, professional errors or omissions, strict liability, breach of contract or warranty.*
>
> ***Consequential Damages*** - *Neither party will be responsible or held liable for any consequential, special, or incidental losses or damages.*
>
> ***Third Party Beneficiaries*** – *This agreement does not and is not intended to confer any rights or remedies upon any person other than the parties.*

PAYMENT CLAUSES

Be aware of payment provisions in each contract. Contracts should state when invoices should be submitted and when payment will be made to the Consultant. Typically invoices are submitted at the end of each month or at the completion of the consultant's work. A reasonable time for payment is 30 to 45 days after the Client receives the invoice. Longer times tend to become a burden on the consultant.

Payment clauses such as "pay-if-paid" or "pay-when-paid" should be avoided. In these instances, the consultant is third in the food chain: owner-client-consultant and has limited recourse to the owner. Under a pay-if-paid clause, the consultant only gets paid if and when the owner pays the client. These clauses generally do not hold up in court. Under a pay-when-paid clause, the Consultant gets paid when the owner pays the Client or soon thereafter. If the Owner does not pay the Client, the Client must still pay the Consultant after an agreed upon time.

Some Clients permit business class airfare while others only allow economy class; some want actual daily expenses with receipts, while others allow per diem; some want hourly charges, while others would prefer lump-sum. Some contracts clearly state what is expected and what is allowed on an Invoice. At the start of a job it is highly recommended to develop an Quote specifically for the Client that contains their requirements (see Figure 2).

LEVEL OF EXPERTISE OR ACTIONS CLAUSES

The level of expertise or actions should not be over-stated in the contract. The Consultant does not have to be a super human, but the Consultant needs to display a level of expertise typical of other consultants at the same level. Words like: highest, upmost, maximum, supreme, above average, extraordinary, or topmost should be deleted from the contract; along with phrases like: any and all, warrants, guarantee, or sole option. Your insurance agent may review your contracts "for insurability issues" at no charge and make sure there are no statements that restrict their potential defense of you. A good Standard of Care clause might be:

> ***<u>Standard of Care</u>*** - *The standard of care for all professional engineering and related services performed or furnished by Engineer under this Agreement will be the care and skill ordinarily used by members of the subject profession practicing under similar circumstances at the same time and in the same locality.*

OTHER TOPICS

International Work

International work can be quite enjoyable and rewarding, but it is more challenging from a business viewpoint. The main point is to be careful when you travel, be aware of your surroundings, and be aware of the vaccines and documents. The U.S. State Department has very good websites with advice for foreign travelers. In some parts of the world traveling is not a vacation, it is an adventure. Be aware that some countries require travel VISAs and it might take time to apply and obtain them. There are companies or travel agents that specialize in obtaining VISAs quickly for a fee. Keep in mind when one travels internationally for business (outside of the US, Canada and its territories), the only insurance policy considered worldwide is the PL. Hence, one would need to consider a package policy for WC, CGL and Auto for $2,500/year (minimum premium). Here is a travel checklist:

- Backup computer files to an external device before travel
- Cash, Credit cards (no debit cards)
- Computer, USB drives, cell phone, tablets (plus all chargers), batteries
- Universal electric plugs
- Documents: Passport, Travel VISA, Letters of Invitation, work permits, Global Entry
- Health: medicines, vaccines w/ documentation (be careful), mosquito spray
- Phones, International phone $5, switch off International data, Skype, FaceTime
- Field attire: Steel toe boots, flashlight, camera, ear plugs, safety glasses, vest
- Attitude: Stay calm!! Be polite.
- Airplane comfort: Exit row, aisle seat, plus seating or business class
- Luggage: Definitely try to carry-on all your luggage
- Airplane health: Compression socks, get up every hour and move around, hydrate
- Other insurance: Airfare, foreign medical, and evacuation (SOS International is good)

Consulting Boards

Consulting Boards are quite rewarding. You get to work with some of the most knowledgeable people in the world with different backgrounds and specialties. Usually the Board reviews handout materials sent a few days before the meetings. The Board may tour the site and then listen to presentations by the team members. The Board then presents their observations and writes their review report at the meeting or after the meeting.

Experiences

Consulting as a Subconsultant to engineering firms. A consultant may be asked by an engineering firm to be part of their proposal for selection for 1) a specific project from an owner or 2) an Indefinite Delivery-Indefinite Quantity (IDIQ) contract from a government agency. Invariably, the consultant will agree to be part of the proposal. There are a few points about this process:

- Non-exclusivity. Usually there are many engineering firms vying for these contracts from Owners. The more engineering firms a Consultant can become a part of their team increases the Consultant's odds of being on a successfully selected team.
- Exclusivity. Sometimes an engineering firm will ask the Consultant to only be on their proposal. This is perfectly fine. In our view, exclusivity has a value. The value could be in the form of an agreement for a certain amount of work if the firm is selected or possibly participate in the proposal preparation or interview. In the latter case, the consultant would be paid at a reduced fee.
- Proposal preparation. Most of the time, Consultants usually spend some time in the preparation of the proposal without compensation. If the engineering firm

gets selected, the Consultant should be permitted to back charge some of this preparation time. Especially if the Consultant was a key consideration in the engineering firm being selected.

- Profitability. An issue with being a subcontractor to an engineering firm is the minimal amount of future work that the IDIQ contracts generate. The authors have not been real successful in getting work through this type of process. Sometimes the work given an engineering firm from an IDIQ contract does not match the consultant's expertise. Other times, the engineering firm retains the work in-house to keep their own employees busy or funding does not allow for consultant help.

Consulting to engineering firms or Owners. This form of contracting and work is much different than described above. In this case, an engineering firm or owner contracts directly with the consultant to perform a specific task. A statement of work is developed; a cost estimate is submitted and agreed upon; a specific contact is developed between the consultant and client; and invoicing is sent directly to the client.

Individual Consulting Glitches and Reality

The following are situations that have actually occurred to individual consultants. As can be seen, the situations are rarely with the Client's engineers, but are typically with the Client's procurement and accounting departments. The names have been deleted to protect the guilty.

1. Consultant standing at the airport ticket counters getting his boarding pass and asking the agent, "What do you mean I need a VISA to go to Brazil?" This scenario is better than one Consultant entering a foreign country and 1) being informed by the Immigration Office that unlike last year, this year a VISA is necessary, 2) spending the night in the immigration office, and 3) having to fly back to the U.S. the next day.

2. Client calls and says, "Sorry that the payment is 30 days late because the Accounting Department did not have the invoice tagged to be paid. Do you mind waiting until next month because we only cut checks once a month."

3. Consultant said to the Procurement Department, "I do not think I can get $5M in liability insurance as the insurance agent just laughed at me when I asked."

4. Consultant said to the Procurement Department, "You are a multi-million dollar company and in the contract you want me to indemnify you and pay for all your attorney fees when my consulting fee is only $6000 for this task order. Also, I am neither the engineer of record nor the responsible engineer."

5. Consultant to the County Health Department Official, "Yuk, I need a Yellow Fever shot, Hepatitis A and B shots, a Typhoid Fever Shot, Malaria medicine, and a Tetanus booster to go to this country?"

6. Consultant to Accounting Department, "So you are holding up my $12,000 invoice and I have to resubmit it because I should have not included the $10 Internet Wi-Fi charge on my Invoice?"

7. Consultant to Accounting Department, "I was on your construction job within one hour and got the project moving again in 2 days, but it will take 90 days to get paid?"

Individual Consultant Enjoyments

The following are the real joys of being an Individual Consultant:

1. The privilege of working with very talented Client engineers, alongside some of the world's best Review Board members, on challenging and diverse projects.

2. Helping solve some difficult problems by adding your experiences to solve a problem.

3. Traveling around the country and to places around the world and having lifetime memories.

4. Going out to dinner with other Consultants and Clients and enjoying their company.

5. Seeing how others approach problems.

Web Hosting

Companies like, 1and1.com or GoDaddy.com, can host your website and maintain a domain name. For instance, Larry has the domain name "NussEngineering.com" with an email Larry.K.Nuss@NussEngineering.com and Ken has "ken-hansen.com" with an email ken@ken-hansen.com. This seems a little more professional than having a Gmail or Hotmail account, but it definitely is not required. You can develop your own company website or hire a company to develop a website for you. Just purchase a book on how to develop a website and learn some simple text commands. Developing your own website gives you the advantage of being able to make small changes to the page, whereas hiring someone makes you dependent on them.

Handy Smartphone Apps

Here are some useful business related Apps:

Evernote/OneNote	- Cloud file storage available on Phone, iPad, or PC
TripIt/Trip Case	- Keeps track of all travel itineraries
United Airline	- Shows current flights, flight entertainment, and boarding passes
Trip Advisor/Yelp	- Recommends places to eat and see

Weather Channel	- Local weather
Converter +	- Units conversion and currency
Dropbox	- Backup storage to Evernote
Microsoft Office	- Word, Excel, and PowerPoint on the iPhone and iPad
Adobe Acrobat	- pdf Reader
LinkedIn	- Stay in contact and advertise
Room 77	- Shows all available hotel rooms and prices in an area
HP Calculator	- RPN Calculator

CONCLUSIONS

As stated in the Introduction, individual engineering consulting, as it relates to the dams industry, can be significantly rewarding and is quite enjoyable. Hopefully this paper will help someone start their own consulting business. Experienced individual consultants are more than willing to answer questions or give advice to younger consultants. Please feel free to get more information from Ken or Larry, if needed.

ACKNOWLEDGEMENTS

The authors would like to thank George Annandale of Golder Associates for sharing his experiences with foreign travel; to Alfred Zarlengo of USI Colorado, LLC for his review of the insurance clauses in the paper and for his help obtaining PL and CL insurance; and to Dan and Sheila Thompson at Century Accounting and Tax for helping navigate the business aspects.

REFERENCES

1] "Practical Law of Architecture, Engineering, and Geoscience, by Brian M. Samuels and Doug R. Sanders, 2011.

2] USI Colorado, LLC, 6501 S Fiddlers Green Circle, #100, Greenwood Village, CO, 80111, 800-873-8500 (Alfred Zarlengo, CIC, AAI, Professional Liability - Account Executive)

3] Century Accounting and Tax, 9894 Rosemont Ave, #101, Lone Tree, CO, 80124, 303-754-0002, tax@cbizofdenver.com (Dan and Sheila Thompson, Accountants).

4] IEEE-USA Consultants Fee Survey Report, 2014 Edition.

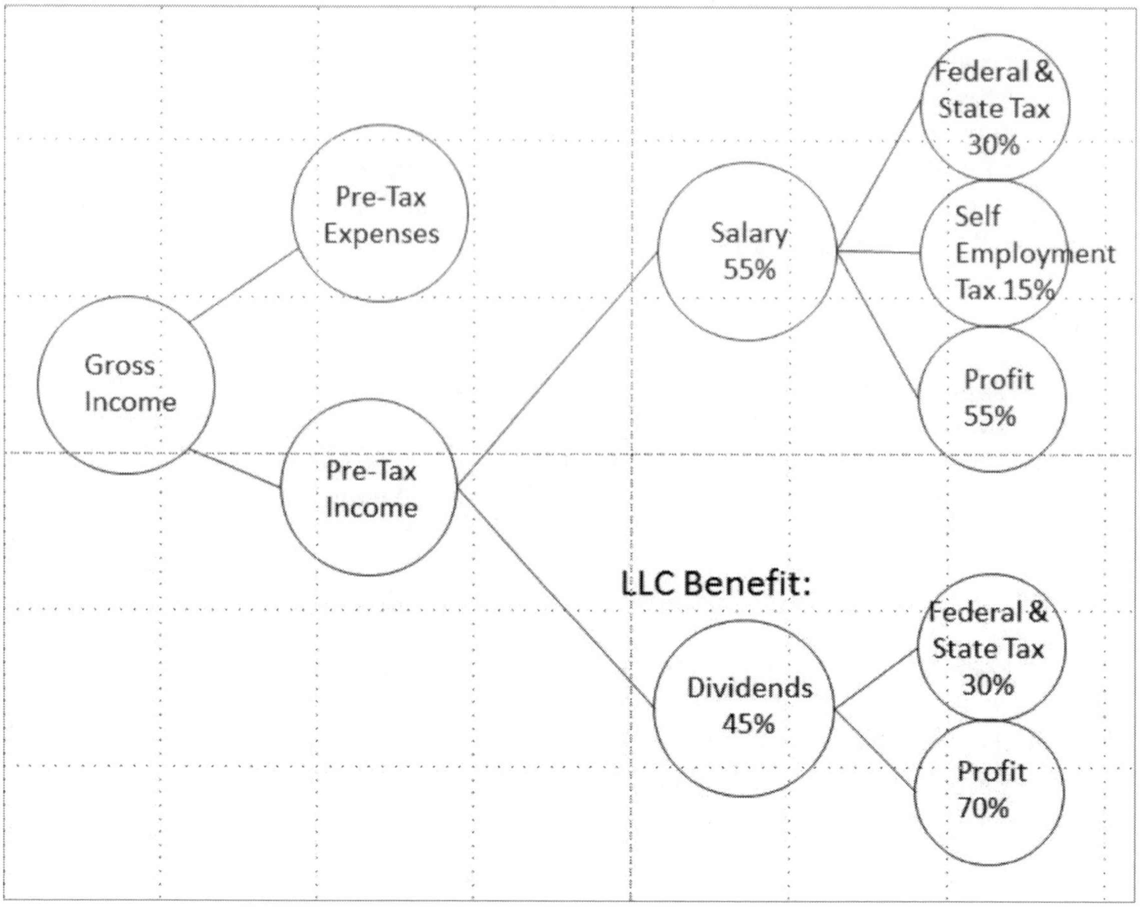

Figure 1. Distribution of income for an S-Corp/LLC

QUOTE

Nuss Engineering. LLC
Larry K. Nuss, PE
10065 Silver Maple Circle
Highlands Ranch, Colorado 80129 USA
Phone: (303) 517-8504
Email: Larry.K.Nuss@NussEngineering.com Website: http://NussEngineering.com/

Period of Performance			Quote Number:	2016-42
Start	Jan 3, 2016		Quote Date:	Jan 3, 2016
End	TDB	**Hourly and Daily Rate:**	$xxx	$xxxx

PROJECT:	**XYZ Dam and Spillway**		
Client:	**ABC Energy**		
Address	123 Main Street		
City, State, Zip			
Contract No:	TBD		
Title:	Project Manager:	Contracting Officer:	Other:
Contacts:	John Doe	Jane Doe	
Phone	xxx-xxx-xxxx	xxx-xxx-xxxx	
Email	John.Doe@abc.com	Jane.Doe@abc.com	

Estimated Amount (USD) $ -

Date	Task Description	Quantity	Unit Price	Unit	Subtotals
				per hour lump sum per day	$ -

Notes:

1. Airfare is economy class, Per Diem is approximate Federal rate, Rental car is mid-size.
2. Personally Operated Vehicle (POV) is standard Federal 2016 rate of $0.54 per mile
3. Send invoices at the end of each month to accountspayable@abc.com
4. Payment terms within 45 days of receipt of invoice.

-----END-----

Figure 2. Typical Quote Form

CONSTRUCTION OF LA ROMAINE COMPLEX IN NORTHERN QUEBEC, CANADA: BEHAVIOUR OF ASPHALT CORE ROCKFILL DAMS

Vlad Alicescu[1]
Jean Pierre Tournier[2]
Redha Kara[3]
Dan Rosculet[4]

ABSTRACT

Hydro-Québec is developing the Hydroelectric Complex of La Romaine, situated on the North shore of the St Lawrence River, in Quebec, one of the most important development projects in Canada. The 6,5 billion $CAD project consists in building 4 generating stations with a total installed power of 1550 MW and an output of 8,0 TWh. The clean and renewable electricity produced in the four power plants of the Complex will help to avoid new emissions of gas with greenhouse effect in North America with approximately 3 million tons of CO_2, if that energy would have been generated with natural gas and with about 7,5 million tons of CO_2, if coal would have been used for the same purposes. Environmental studies and measures carried out before, during and after construction until 2040 will cost over $385 million altogether. The erection of the Complex started with the $CAD 2,4 billion Romaine 2 facility, completed in 2014: the impounding of the Romaine 2 reservoir begun in May, 2014 and finished in November, when the first Megawatts reached our customers. The construction of Romaine-1 and 3 facilities is progressing very well, first kilowatts being expected by December 2015 and June 2017, respectively. Finally, the design stages for the Romaine 4 facility will be completed and the construction will start in 2016. One year after the finish of impounding the reservoir, the 6 Asphalt Core Rockfill Dams of Romaine 2 reservoir are behaving very well and a comparison with similar structures from Norway and Austria is presented in this paper.

INTRODUCTION

Hydro-Québec generates, transmits and distributes electricity and its sole shareholder is the Québec government. It uses mainly renewable generating options, in particular hydroelectricity, and supports the development of wind energy through purchases from independent power producers. It also conducts research in energy-related fields, including energy efficiency.

[1]P.Eng., MScA, MBA, Planning of Development Projects, Hydro-Québec Production, 75, Boul. René Lévesque O, Montréal, Québec, Canada H2Z 1A4, vlad@members.asce.org, Alicescu.Vlad@hydro.qc.ca
[2]P.Eng.,PhD, Expert Advisor, Hydro-Québec Equipment & SEBJ, 800, Boul. De Maisonneuve E, Montréal, Québec, Canada H2L 4M8, Tournier.Jean-Pierre@hydro.qc.ca
[3]P.Eng.,PhD, Head, Design of Hydroelectric projects, Hydro-Québec Equipment & SEBJ, 855, St. Catherine Street E, Montréal, Québec, Canada H2L 4P5, Kara.Redha@hydro.qc.ca
[4]P.Eng.,Civil engineer, Dams and civil structures, Production Manicouagan, Hydro-Québec Production, Baie Comeau, Québec, Canada G4Z 3B1, rosculet.dancristian@hydro.qc.ca

In 2014, 99% of Hydro-Québec's output comes from renewable energy sources, and the electricity sector accounts for barely 0,66% of the Quebec's GHG emissions. The *Sustainability report 2014* are showing that few electricity companies can boast as high a level of profitability and as small an environmental footprint as Hydro-Québec, who owe this performance in large part to the abundant, powerful hydraulic resources available in Québec and the vision shown by past generations who successfully tapped this potential. Hydro-Québec won the 2014 IAIR corporate award for Best Company for Leadership – Hydroelectricity, presented by the magazine IAIR – Excellence in Global Economy and Sustainability. In keeping with the Québec government's environmental policy, Hydro-Québec is working to reduce its greenhouse gas emissions by focusing on the three priorities: energy efficiency, renewable energies and technological innovation. We have to mention that in 2014, Hydro-Québec's electricity exports to the neighboring systems in Canada and U.S. avoided the emission of 9 million tons of CO_2.

The first action (initiated in the previous Action Plan) of the *Sustainable Development Action Plan 2013–2016* is "build hydropower projects" with the target: "increase installed capacity by 910 MW and annual output by 4,7 TWh by 2016" and is being accomplished by the commissioning of Romaine 1 (2015) and Romaine 2 (2014) facilities of La Romaine Hydroelectric Complex, one of the largest infrastructure project, currently under way in Canada.

LA ROMAINE HYDROELECTRIC COMPLEX

All development projects undertaken by Hydro-Québec must be successfully sustainable and therefore, they must be well integrated in their natural environment, well received by the host communities and economically attractive. We have completed the draft-design stages, in consultation with the host communities, and signed a number of partnering agreements with Native people and other local communities. Environmental studies and measures carried out before, during and after construction will cost over 385 million $CAD altogether. In addition, the project generates approximately 3,5 billion $CAD in economic spinoffs in Québec, including 1,3 billion $CAD on the North Shore region. The electricity produced in the four power plants of the Complex is helping to avoid new emissions of gas with greenhouse effect in North America with approximately 3 million tons of CO_2, if that energy would have been generated with natural gas and with about 7,5 million tons of CO_2, if coal would have been used for the same purposes.

The Complex at a glance

The project consists in building four generating stations with a total installed power of 1550 MW and an annual average output of 8,0 TWh (see Table 1 and Figures 1 & 2). Hydro-Québec conducted extensive environmental assessment studies and deposed in January 2008 the Environmental Assessment Report to the provincial and federal governments.

The formal environmental assessment process was completed by the spring of 2009 and the project started with the construction of the access road. Romaine 2 was the first facility to be developed and the first kilowatt was produced by the end of 2014. The project continues with the construction of Romaine 1 and Romaine 3 facilities and finally, Romaine 4, who must produce its first kilowatts by the summer of 2020 (see Table 1 and Figures 1 & 2). The total cost of the power generation project (without transmission grid) is estimated at 6,5 billion $CAD.

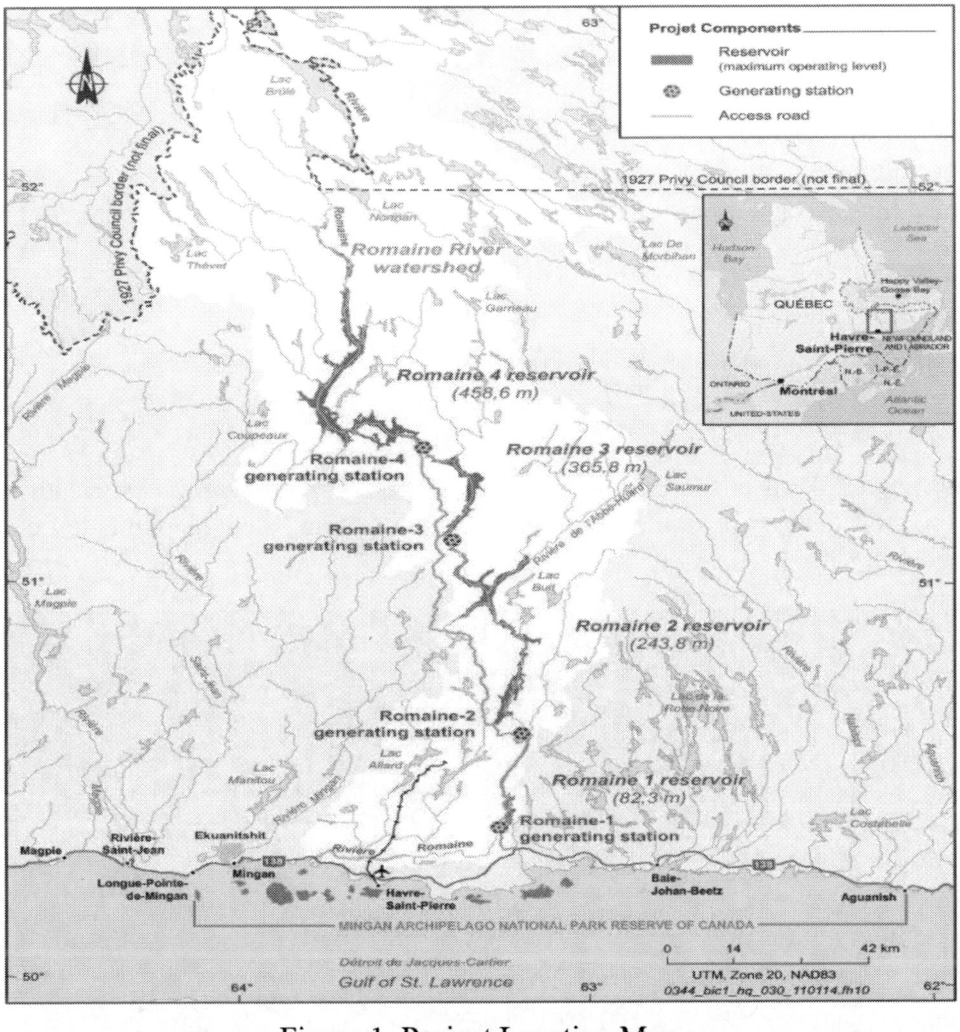

Figure 1. Project Location Map.

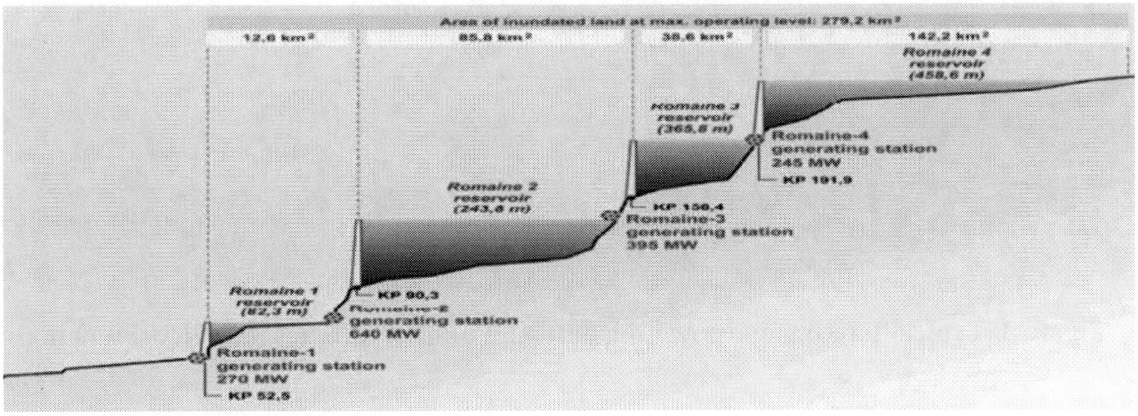

Figure 2. Scheme of La Romaine Hydroelectric Complex.

Table 1. Main characteristics of La Romaine hydroelectric Complex

Characteristics	RO-1	RO-2	RO-3	RO-4	Complex
Dam (PK)	52,5	90,3	158,4	191,9	–
Maximum flow (m^3/s)	289,4	272,4	223,2	184,5	–
Maximum operating level of the reservoir (m)	82,3	243,8	365,8	458,6	–
Area of inundated land at max. operating level (km^2)	12,6	85,8	38,6	142,2	279,2
Net hydraulic head (m)	62	158	119	88	427
Area of inundated land (km^2)	6,5	71,5	30,7	110,7	219,4
Turbines (number and type) with an UF=59%	2 Francis	2 Francis	2 Francis	2 Francis	8 Francis
Installed power (MW)	270	640	395	245	1 550
Annual output (TWh)	1,4	3,3	2,0	1,3	8,0
Dams and dykes (number)	2	7	2	1	12
Year of completion	2015	2014	2017	2020	2020

All four power plants of the Complex are or will be equipped with 2 Francis turbines and therefore their design are similar, with precast concrete and steel elements to be assembled in situ.

Figure 3. Typical La Romaine power plant with 2 Francis turbines (CATIA simulation).

Status of the construction of the Complex

Romaine 2 facility consist in a dam, 5 main dykes and a freeboard dyke (C2), a spillway with a maximum discharge capacity of 3230 (m^3/s), an intake, a 5,5 km long headrace tunnel, a surge shaft, 2 pressure shafts, a diversion tunnel, a temporary tunnel for the ecological flow outlet ($20<Q<50$ m^3/s) and a power plant with 2 Francis turbines of 320 MW each (see Table 1and Figure 2). The total cost of this project is estimated at 2,4 billion $CAN. All project works were completed by December 2014. Commercial commissioning occurred in November 21st, 2014 for the first and in December 6st, 2014 for the second one, a 640 MW contribution ready to fulfill the high electricity demand of the very cold winter of January – March 2015.

Photo 1. Romaine 2 Power plant

Romaine 1 facility consist in a diversion tunnel with a design flow of 1455 m^3/s, cofferdams, a 41 m high main dam, a spillway with a maximum discharge capacity of 3100 m^3/s, an intake structure, a 550 m long headrace channel, two 55 degree inclined concrete-lined penstocks with a reinforced concrete transition and a power house with 2 Francis turbines of 135 MW each (see Table 1 and Figure 2). The reservoir impounding was completed on October 8st, 2015. First commissioning is expected by November and the second one by December, 2015.

Photo 2. Romaine 1 Power plant, dam and spillway delivering the ecological flow

Photo 3. Romaine 1 reservoir after impounding, dam, spillway and headrace channel

Photos 4-5: Romaine 1 Power plant

Romaine 3 facility consists in a dam, a dyke, a spillway with a discharge capacity of 3320 m^3/s equipped also with small valves for the 2,2 m^3/s ecological flow, an intake, a 1675 m long headrace tunnel, a surge shaft, 2 penstocks, a diversion tunnel and a power plant with two Francis turbines of 197,5 MW each (see Table 1 and Figures 2 & 4). The project started in 2011 with the construction of the preliminary camp at Km 84 and continued with the erection of the main camp Mista, the construction of the main road to km 117 and of the secondary access roads for permanent structures. Excavation of the diversion tunnel started in May 2013 and the diversion of the river occurred in May, 2014. Powerhouse and tailrace excavation started in July 2013 and now is completed. Construction of the cofferdams, main dam and dyke B3, spillway and power intake started at the beginning of 2014 and the cofferdams are now completed. Headrace tunnel excavation commenced in October 2013 and is now 70% complete. Powerhouse concreting started in May 2015 and the equipment installation should begin in November. Finally, the reservoir impoundment must begin in October 2016 and the commissioning of the first unit must occur in June 2017.

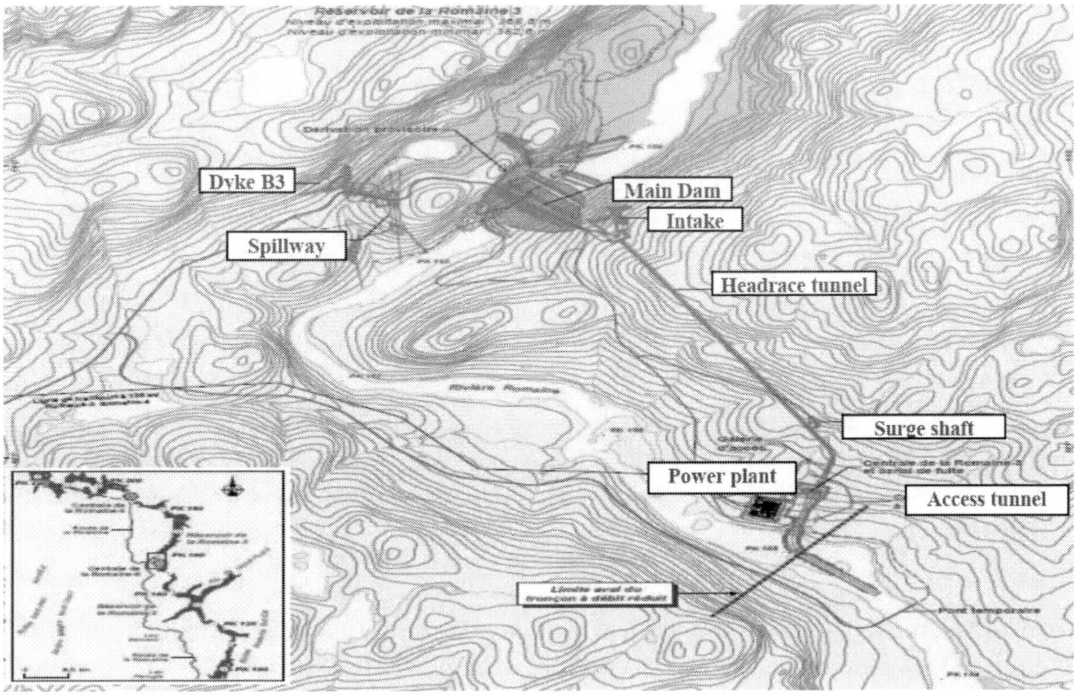

Figure 4. Plan view of Romaine 3 HEP.

Photo 6. Construction of the Main Dam, a rockfill dam with a till core

Photo 7. Construction of the Power plant

Romaine 4 Facility consist in a dam, a spillway equipped also with small valves for the ecological flow, an intake, a headrace tunnel, a surge shaft, 2 penstocks, a diversion tunnel and a power plant with 2 turbines Francis of 122,5 MW each (see Table 1 and Figure 6).

Figure 5. Romaine 4, Main Dam and spillway (CATIA Simulation)

Site investigations have been performed during the 1970s and for the environmental assessment study, in 2003-2005. Some further investigations were made in 2014 and 2015. Detailed engineering and design started in 2014 and will be completed by June 2016, when the construction on site must begin. Optimization of design and schedule of the works are in progress and the project must be completed by the summer of 2020.

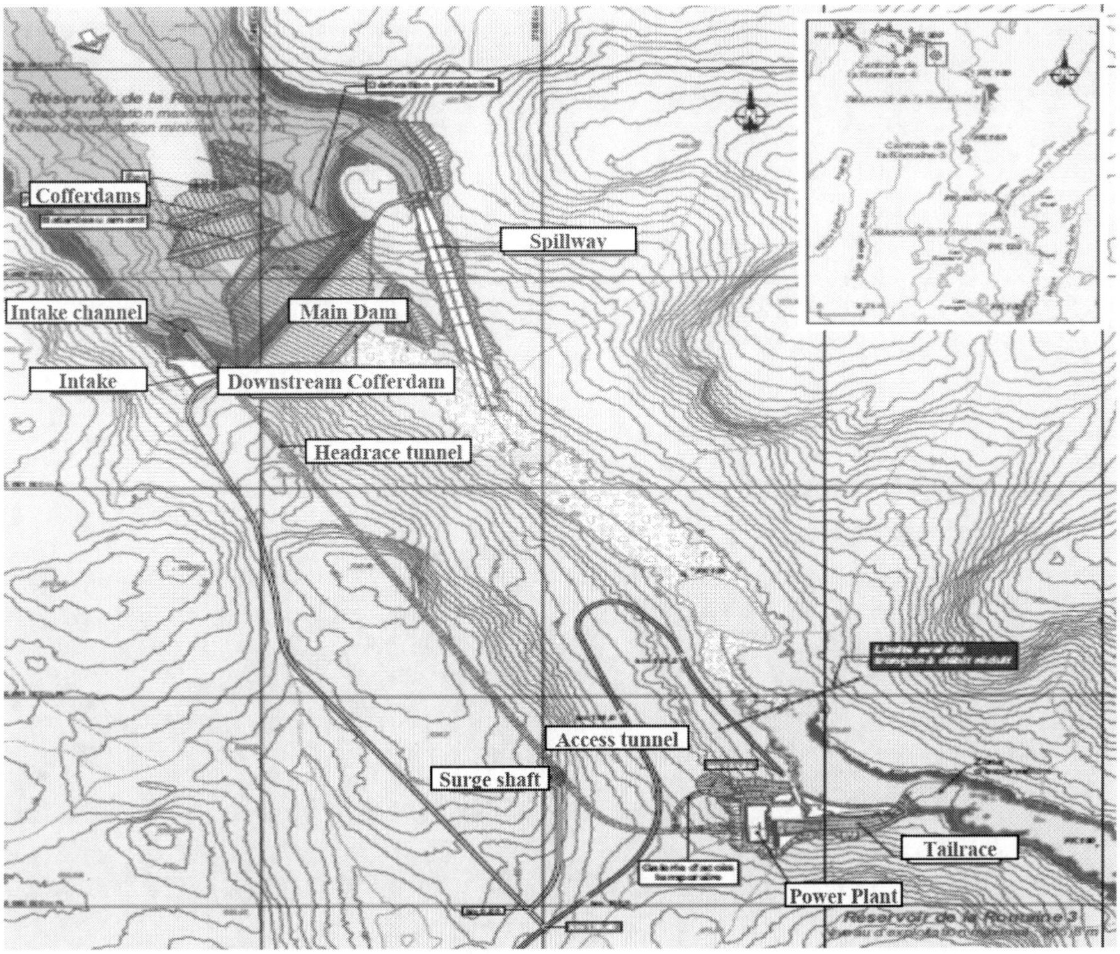

Figure 6. Plan view of Romaine 4 hydroelectric project.

Design of the retaining structures

The design of the retaining structures for the facilities of La Romaine Complex was an interesting and challenging task. After using for more than 50 years the glacial till as waterproofing material for its embankment dams, Hydro-Québec was looking forward to develop new dam concepts, mainly for the areas where natural waterproofing materials does not exist or are of poor quality. Romaine 1 and Romaine 2 sites are striking examples of particular zones with very limited quantities of natural till or silt, insufficient even for the construction of the cofferdams. For the ***Romaine 2 facility***, based on the information about the site conditions and lack of suitable earth core material, CFRD (Concrete Face Rockfill Dam) and ACRD (Asphaltic Core Rockfill Dam) options were considered for the Main Dam and dykes. The cost criteria and some other particular issues like maintenance, behavior, construction difficulties, etc. made the difference between the two variants and finally, the ACRD variant was chosen. The characteristics of the retaining structures are presented in Table 2, a typical cross section (with the Main Dam as an example), in Figure 7 and the zoning and compaction, in Table 3:

Table 2. Retaining ACRD structures of Romaine 2 facility

Structure	Type	Height (m)	Crest Lenght (m)	Asphaltic Core Volume (m^3)	Fill Volume (m^3)
Dyke A 2	ACRD	31	144	1 040	88 300
Dyke B 2	ACRD	28	115	790	73 400
Dyke D 2	ACRD	48	728	6 330	666 000
Dyke E 2	ACRD	39	407	2 470	218 000
Dyke F 2	ACRD	84	423	10 700	1 947 000
Main Dam	ACRD	131	496	18 850	4 546 000
TOTAL				40 180	7 538 700

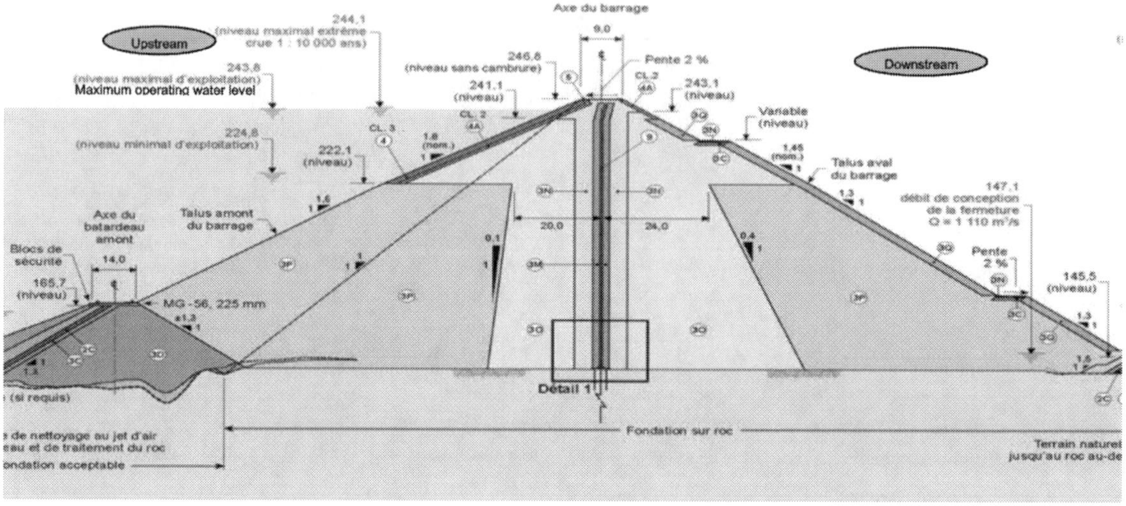

Figure 7. Typical ACRD cross section (Main Dam of Romaine 2 facility).

Table 3. Main Dam and Dykes- Zoning and compaction

Zone	Definition	Compaction
9-Core	Asphalt concrete max. 20 mm (bitumen type PG52-34)	2 passes 2 tons + 4 passes, 0,5 tons vibratory rollers (void content < 3%); layer thickness = 22,5 cm
3M -Filters	Well graded natural or crushed gravel, max. 80 mm	3-4 passes, 2,0 - 2,5 tons vibratory roller; layer thickness = 22,5 cm
3N-Transition	Well graded crushed rock max. 200 mm	4 passes, 15 tons vibratory roller; layer thickness= 45 cm
3O-Shoulder	Well graded quarried max.600 mm	8 passes, 15 tons vibratory roller; layer = 90 cm (4 passes for dykes with H<50 m)
3P-Shoulder	Well graded quarried max. 1200 mm	6 passes, 15 tons vibratory roller; layer = 180 cm (4 passes for dykes with H<50 m)
4 – Slope protection	Riprap, Min. 600 mm - max. 800 mm	Placed with a backhoe

A similar situation was noticed on ***Romaine 1 facility***: based on the information about the site conditions and lack of suitable earth core material, CFRD (Concrete Face Rockfill Dam), ACRD (Asphaltic Core Rockfill Dam) and GFRD (CARPI composite geomembrane, installed on the upstream slope of the dam) options were considered for the Main Dam. The cost criteria and some other particular issues like maintenance, behavior, etc. have made the difference between these variants and finally, the ACRD variant was chosen. The characteristics of the retaining structure are presented in Table 4, the zoning and compaction being similar to those of Romaine 2 ACRD structures.

Table 4. Retaining ACRD structures of Romaine 1 facility

Structure	Type	Height (m)	Crest Lenght (m)	Asphaltic Core Volume (m^3)	Fill Volume (m^3)
Main Dam	ACRD	41	850	5 600	608 000

Romaine 3 and Romaine 4 facilities are situated in areas with till borrow material. During Phase 1 Engineering of ***Romaine 3***, two dam types were studied: ACRD and ECRD. The cross section of the ACRD followed the design of Romaine 2 Main Dam. The cross-section of ECRD and the use of materials is the same as the typical ECRDs used at other Hydro-Québec projects (see Photo 6). Estimates of direct and indirect costs for each dam type and risk analysis comparison was undertaken, considering various uncertainties in the analysis and the conclusion was that the two types of dams were roughly equal in cost. Finally, the ECRD option was selected for the 90 m high dam of the Romaine 3 facility. The till borrow material is a non-plastic silty sand with gravel and 1-2% of clay size material. This small fraction is not clay, but finely ground rock flour. Laboratory tests indicates that the material is stable with respect to internal erosion (suffusion) and the gradation tests indicate percentages passing 0,08 mm close to 15%. Laboratory permeability tests indicate values of about 1×10^{-4} cm/s.

Table 5. Retaining structures of Romaine 3 facility

Structure	Type	Height (m)	Crest Lenght (m)	Till Core Volume (m^3)	Fill Volume (m^3)
Main Dam	ECRD	90	408	437 000	3 139 000
Dyke B3	Homogeneous	28	220	29 000	160 000
Total				466 000	3 299 000

ASPHALT MIX DESIGN

Through the design and construction of Nemiscau-1 Dam, first Asphalt Core Rockfill Dam in North-America (reference 2), Hydro-Québec wanted to gain experience with asphalt concrete core construction, before proceeding with the final design and construction of the much larger dams or dykes of La Romaine Complex. The construction of the Nemiscau-1 Dam started in June, 2008 and was completed by October, 2008. For both projects, the asphalt mix was proposed by "Entreprises Kolo Veidekke Quebec inc.", with the support of Norwegian Geotechnical Institute (NGI) for Triaxial Tests.

For Nemiscau -1 Dam, a bitumen content of 7,3 % was recommended for the asphalt mix design. Void content (less than 2 %) was calculated on laboratory samples after compaction with 30 blows on each side as allowed by the Marshall method (ASTM D6926-04). As a part of the asphalt mix design process, the stress-strain strength properties of the asphalt concrete where documented through Triaxial Tests, performed at NGI, a well-established consultant laboratory and at Laval University. The tests were performed at a temperature of 5°C which corresponds to the main temperature estimated in the core; the confining stress was equal to 0,5 MPa. The maximum deviator stress for the Nemiscau-1 Dam tests was about 2,3 MPa. The ductility (horizontal plateau) of the Nemiscau-1 Dam specimens were showing almost no strain softening and loss of resistance after the maximum level was reached, and an excellent viscous-elastic behavior. Core samples showed a more ductile behavior than laboratory samples.

For the construction of Romaine 2 and Romaine 1 retaining structures, "Entreprises Kolo Veidekke Quebec inc.", signed a contract with Hydro-Québec for core pavers leasing and technical support services. The proposed provisional mix had a bitumen content of 6,7 % of the total mix weight but adjustments to 6,9 % have been made, based on the results and experience from the field test section, build with an asphalt mix produced in the asphalt concrete batch plant. The aggregate size distribution satisfied the Fuller gradation curve with a slight increase of fine sand. An average of 13 % of filler was added for the asphalt mix of Nemiscau-1 and the mix design used for Romaine 2 retaining structures has approximately 8 % of filler content. The final mix was in accordance with the following requirements:

- Void content on laboratory made Marshall specimens (%) : < 2,0 (ASTM D3203-05)
- Permeability factor: $\leq 10^{-11}$ cm/s
- Marshall stability (N) : > 2500 (ASTM D6927-05, LC 26-060)
- Placed and compacted in up to three layers per day with a thickness of 22,5 cm after compaction and with an in situ void content less than 3 %.

Finally, for the mix design for the asphalt core of the Romaine 1 Main Dam, the bitumen content is 6,8% and the filler content approximately 9%.

Photos 8-9. Core construction at dykes A2 and D2 of Romaine 2 facility

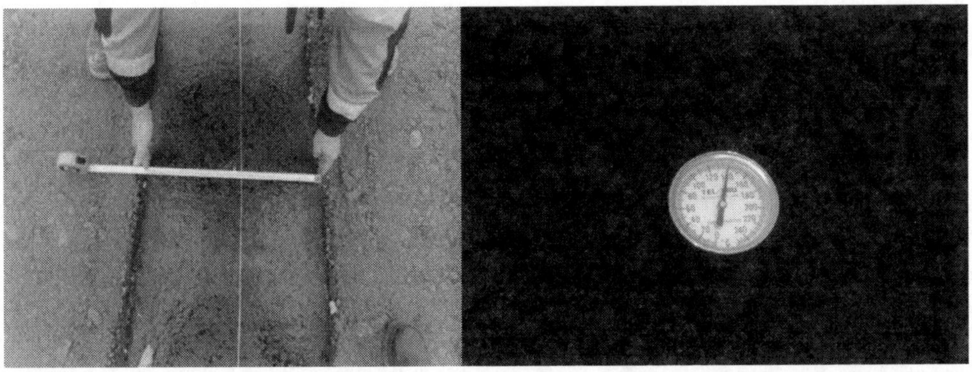

Photos 10-11. Placing quality control and compaction temperature: >130°C

Photo 12. Construction of the Dyke F2 (Romaine 2) in April 2013

BEHAVIOUR OF ROMAINE 2 ASPHALT CORE ROCKFILL DAMS

One year after the impounding of the reservoir, the 6 Asphalt Core Rockfill Dams of Romaine 2 reservoir are behaving very well and a comparison with similar structures from Norway and Austria is presented below. For this comparison, we have first chosen Storglomvatn Dam situated in Nordland, Norway (Hoeg, K and al. 1993, 2007-references 5 and 6), Finstertal Dam, situated in Austrian Alps (Pircher, W. and al 1988, Milestone projects, China 2009 – references 7 and 8) and the Romaine 2 Main Dam.

Photo 13. Finstertal Dam, Tirol, Austria

Photo 14. Storglomvatn Dam, Nordland, Norway

Photo 15. Romaine 2 Main Dam

Table 6. Comparison between Romaine 2 Main Dam, Storglomvatn Dam and Finstertal Dam

Dam	RO - 2 Main Dam	Storglomvatn Dam	Finstertal Dam
Maximum height (m)	131	125	150
Crest length (m)	496	820	652
Total fill volume (m^3)	4546000	5200000	4600000
Upstream slope	1 :1,8, 1 :1,6	1 :1,5	1 :1,5
Downstream slope	1 :1,45	1 :1,45	1 :1,45
Core height (m)	110,5	120	96
Asphalt core thickness (cm)*	85/50	95/50	70/60/50
Asphalt core volume (m^3)	18850	22500	25000
Construction period	2010-2013	1993-1997	1977-1980
Max. crown settlements upstream (cm)	22	38	25
Max. crown settlements d/stream (cm)	5	22	20
Horizontal crown displacements d/s (cm)	33	10	9
Seepage after impoundment (l/s)	1,8 (Nov. 2014)	10,0 (1997)	9,0 (1981)
Seepage in exploitation (l/s)	2,5 (May 2015)	4,5 (2007)	1,5 (2008)

* The thickness of 50 cm is recommended for the last 50 m of the core, measured from the crest below

Despite the differences in overall design, cross sections, compaction procedures, foundation characteristics and treatment, quality of the rock and other granular materials used for the construction of these dams, we consider this (qualitative) comparison very instructive. It is also an interesting indication concerning the very good behavior of these dams, well ranked between the highest Asphalt Core Rockfill Dams in the world. One may have noticed that sluicing was not used for the compaction of Romaine 2 Dam and dykes - see Table 3 – but the displacements were found to be very small as the rockfill is of high quality and well graded and the compaction was very good. As a result, the one-dimensional compression modulus for Zones 3N and 3O is as high as 170-180 MPa, which corresponds to a Young modulus of approximately 150 MPa.

In Table 7, one may see a comparison between Dyke F2 of Romaine 2 facility and the Norwegian Dams Storvatn (Arnevik, A and al. 1988 – reference 4) and Styggevatn (Hoeg, K. 1993, 2007- references 5, 6), all being impressive ACRD structures:

Photo 16. Dyke F2 of Romaine 2 facility

Photo 17. Storvatn Dam, Ulla Fore Complex, Norway

Photo 18. Styggevatn Dam, Josterdal facility, Norway

Table 7. Comparison between Dyke F2 (Romaine 2), Storvatn Dam and Styggevatn Dam

Dam	F2 Dyke, RO - 2	Storvatn Dam	Styggevatn Dam
Maximum height (m)	84	95/87 (axis)	60
Crest length (m)	423	1472	880
Total fill volume (m^3)	1947000	9500000	2500000
Upstream slope	1 :1,8, 1 :1,6	1 :1,5	1 :1,5
Downstream slope	1 :1,45	1 :1,45	1 :1,55
Core height (m)	82,5	85 (inclined)	52
Asphalt core thickness (cm)*	75/50	80/50	50
Asphalt core volume (m^3)	10700	49000	15275
Construction period	2010-2013	1981-1987	1986-1990
Max. crown settlements upstream (cm)	3,4	35	10
Max. crown settlements d/stream (cm)	1,1	18	10
Horizontal crown displacements d/s (cm)	4,7	12	7
Seepage after impoundment (l/s)	13,6 (Nov. 2014)**	10,2 (1987)	18,0 (1981)
Seepage in exploitation (l/s)	3,0 (Feb. 2015)	<5,0 (2006)	4,5 (2006)

* The thickness of 50 cm is recommended for the last 50 m of the core, measured from the crest below
** The weir is situated faraway downstream of the dyke and is collecting other water flows from both abutments

DISCUSSION

All the Asphalt Core Dams presented above are showing a very good behavior with very low seepage records, generally less than 10 l/s and very low settlements, less than 0,30 % of the height of the dam for the upstream crown and less than 0,15 % of the height of the dam for the downstream part of the crest. However, there are Asphalt Core Dams with much higher vertical displacements, like Holmvatn (H=56 m, and maximum settlements = 1,05% H) in Norway or Feistritzbach (H=88 m and maximum settlements = 1,3%H) in Austria, both with seepage flows less than 2 l/s. Thus, even with a larger deformability of the rockfill (due essentially to the poor quality of the available rock), the asphaltic core behaves very well, because of its viscoelastic-plastic and ductile properties, with a "self-healing" ability.

CONCLUSIONS

La Romaine Hydroelectric Complex is one of the most important development projects of its kind in North America and generates approximately 3,5 billion $CAD in economic spinoffs in Québec, Canada, including 1,3 billion $CAD on the North Shore of St Lawrence River region, and an average of 975 jobsite workers per year for 11 years. It is providing a reliable supply of clean and renewable energy that allow Hydro-Québec to increase its exports and is helping to avoid new emissions of gas with greenhouse effect in North America by approximately 3 million tons of CO_2, if that energy would have been generated by using natural gas and by approximately 7,5 million tons of CO_2, if coal would have been used for the same purposes. As all development projects undertaken by Hydro-Québec, La Romaine Hydroelectric Complex is successfully sustainable and therefore, well integrated in its natural environment, well received by the host communities and economically attractive. Environmental studies and attenuation measures carried out before, during and after construction will cost over 385 million $CAD.

Commercial commissioning of Romaine 2 occurred in the end of 2014, a 640 MW contribution to fulfill the high electricity peak demand in the very cold winter of 2015. Romaine 1 commissioning will follow by the end of 2015, Romaine 3 will produce its first Megawatts in 2017 and finally, Romaine 4 will be completed in 2020.

Hydro-Québec considers that technological innovation is a powerful engine for growth and performance, and new design techniques, concepts and software were used for the development of La Romaine Complex. With its 6 Asphaltic Core Rockfill Dams, Romaine 2 is the largest HEP in the world, using this type of rockfill retaining structure. The 131 m high Main Dam, and the 84 m high dyke F2 of Romaine 2 facility, are the highest Asphalt Core Rockfill Dams in the North, South and Central America.

One year after the beginning of the impounding and 8 months after its finish, the 6 Asphalt Core Rockfill Dams of Romaine 2 reservoir are behaving very well and a comparison with similar structures from Norway and Austria is showing the very good performance "on duty" of this type of dams.

ACKNOWLEDGEMENTS

Many thanks to Professor Kaare Hoeg, PhD, from Oslo University, Norway, Mrs. Anne Marit Ruud Hästein, P. Eng., MScA, from Statkraft, Norway and Mr. Dipl. Eng. Sebastian Perzlmaier, from TIWAG, Austria for their collaboration and support.

REFERENCES

Complexe de La Romaine, Étude d'impact sur l'environnement (Environmental Impact Statement), http://www.hydroquebec.com/romaine/documents/etude.html, 2008

Alicescu, V., Tournier, J.P., Vannobel, P., Moore V., 2011, Design and construction of Nemiscau-1 Dam, the first ACRD in North-America. (Proceedings of the United States Society on Dams, 2011 Annual Conference, San Diego, CA)

Alicescu, V., Tournier, J-P. and Vannobel, P. 2008. "Design and construction of Nemiscau-1 Dam, the first Asphalt Core Rockfill Dam in North-America." Canadian Dam Association Annual Conference. Winnipeg, Manitoba. October, 2008.

Arnevik, A., Kjærsli, B. and Walbø, S. 1988. "The Storvatn Dam, a rockfill dam with a central core of asphaltic concrete." *Seizième Congrès des Grands Barrages* , R. 9, Q.61, San Francisco.

Höeg, K. 1993. "Asphaltic Concrete Core for Embankment Dams." *Statkraft, Veidekke, Norwegian Geotechnical Institute*

Höeg, K., Valstad, T., Kjærsli, B. and Ruud, A.M. 2007. "Asphalt Core Embankment Dams: recent case studies and research" *Hydropower and Dams, Issue Five, 2007*

Pircher, W. and Schwab, H. 1988. "Design, construction and behaviour of the asphaltic concrete core wall of the Finstertal Dam" *Seizième Congrès des Grands Barrages*, R. 49, Q.61, San Francisco.

Milestone Projects: Finstertal Dam in Austria. 2009. *Proceedings of the 1st International Symposium on Rockfill Dams, 18th – 20th Oct. 2009*, Chengdu, China

SELECTING AN INTENSITY MEASURE FOR FRAGILITY CURVES DERIVED FROM PERFORMANCE-BASED TESTING

Spenser Anderson[1]
Kunal Menda[2]
Mo Zhao[3]
Ziyad Duron, PhD[4]

ABSTRACT

A fragility curve is a mapping between the probability of an asset becoming damaged and the intensity of the excitation or hazard loading condition applied to the asset. Before such a curve can be constructed, a means for measuring the intensity of the applied excitation must be selected. In this paper, a process is described for selecting an intensity measure for use in the construction of experimentally-derived fragility curves utilizing vibration data from Performance-Based Testing (PBT). For a seismic hazard, several examples of intensity measures compatible with PBT-based fragility curve development are presented. These intensity measures are ranked based on their correlation with the demand parameter, which in this case is the maximum displacement expected from the structure. Among the intensity measures discussed, it is found that any variable that is positively correlated with maximum structural displacement can be taken as an acceptable intensity measure when constructing a fragility curve. However, peak ground displacement is most highly correlated with maximum structural displacement, and will therefore result in the fragility curve with the steepest slope and smallest (tightest) confidence intervals. Additional considerations for selecting intensity measures are discussed, including the possibility of using ground motion predictive equations to map the chosen intensity measure onto return period – a process that enables the use of the fragility curve to assign an expected lifetime to an asset subject to hazard loading.

INTRODUCTION

Fragility curves allow an owner to visualize how the risk to their asset grows as the intensity of the loading condition rises. The curves can be used to determine the probability that their asset will fail given a hazard intensity that the owner would like to consider, or can be used to back out the hazard loading intensity that will likely produce a condition in the asset that corresponds to an unacceptable level of risk. As a result, these curves can be an invaluable tool when taking a Risk Informed Decision Making approach to asset management.

Fragility curves for large concrete dams cannot be directly measured in the field, since this would require the dam or its subsystems to actually fail. Instead, engineers use

[1] De Pietro Engineering Fellow, Harvey Mudd College, Claremont, CA91711,sanderson@hmc.edu
[2] De Pietro Engineering Fellow, Harvey Mudd College, Claremont, CA 91711, kmenda@hmc.edu
[3] De Pietro Engineering Fellow, Harvey Mudd College, Claremont, CA 91711, mzhao@hmc.edu
[4] Professor of Engineering, Harvey Mudd College, Claremont, CA 91711, duron@hmc.edu

numerical and approximate methods for estimating fragility which rely on subjective albeit experience-based judgments for building models and selecting appropriate analysis parameters. Some recent methods for fragility curve estimation, such as those developed by Ghanaat et al. (2015), estimate seismic fragility curves for dams by running many simulated earthquakes through a nonlinear finite element model of the dam and observing the fraction of the simulated earthquakes that cause the model to fail. These methods are less subjective, but are computationally expensive and use computer models that may be difficult to validate with empirical data. The methods described by Menda et al. (2016) allow the relatively fast computation of fragility curve estimates from empirical data that can be collected easily and relatively inexpensively using the Performance-Based Testing (PBT) methods described by Goldkamp et al. (2014).

The first step in the production of a fragility curve from Performance-Based Testing (PBT) data is the selection of a performance measure, as described by Zhao et al. (2016). The performance measure selected should reflect the type of damage or risk to the asset that the owner wishes to examine. The fragility curve will show the probability of this performance measure exceeding a threshold limit that indicates an unacceptable state in the dam. Therefore, the performance measure that is selected should have a clear and calculable threshold that corresponds to unacceptable risk in the eyes of the owner, where the owner is not willing to accept a structural response that maps to a performance measure beyond this threshold. This threshold does not need to correspond to actual failure of the dam – it can instead correspond to any response in the dam that the owner considers unacceptable, such as inoperable spillway gates, onset of cracking, or a stress level that is uncomfortably close to causing damage. Examples of performance measures for dams might include maximum displacement on the crest of the dam or maximum tensile stress inside the dam. If tensile stress is used as a performance measure, an example of an unacceptable threshold might be the yield stress of concrete (perhaps divided by some factor of safety).

Fragility curves must be plotted versus an intensity measure of the curve-maker's choice. It may often be appropriate to simply choose the most intuitive or widely accepted intensity measure for the fragility curve, but careful selection of the intensity measure can yield a fragility curve that provides greater predictive power than is granted by alternative choices. This paper begins with a discussion of the prerequisite and required steps that must be undertaken when constructing a fragility curve using experimental data from Performance-Based Testing. It then examines the properties of fragility curves that can make them particularly useful, and how the intensity measure can influence these properties. Examples of actual fragility curves computed with empirical data from PBT on a large concrete dam are presented. These curves are plotted against various intensity measures in order to demonstrate the process of selecting a proper intensity measure, and in order to suggest particularly effective intensity measures for similar structures and similar definitions of damage.

THE IDEAL FRAGILITY ESTIMATE — OR IS IT?

The ideal fragility curve for some applications may appear as shown in Figure 1.

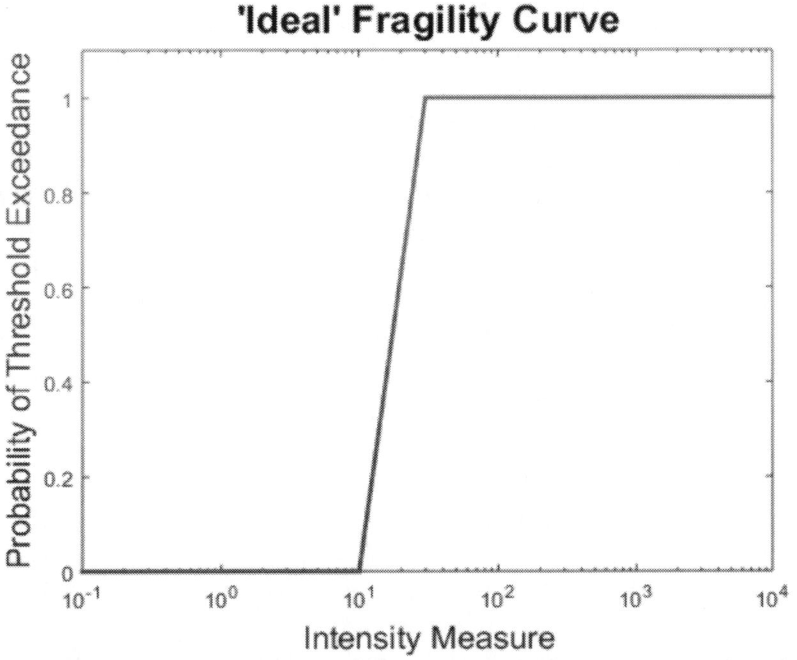

Figure 1. A sample fragility curve with a very steep slope

The fragility curve indicates the probability that a performance or engineering demand parameter has passed a predetermined threshold level as a function of intensity measure. The intensity measure is a value that is used to characterize the magnitude of the hazard loading. In the case of earthquake loading, for example, the intensity measure could be peak ground acceleration. What makes this fragility curve ideal is its steep slope, which makes it easy to identify the level of shaking (in terms of the intensity measure employed) at which the probability of threshold exceedance rises above nominal zero. In this sense, the fragility appears as a "digital pulse" with a sharp transition or rise that clearly highlights where the probability of exceedance grows. Because the slope is very large, the curve tells with a high degree of certainty whether an earthquake with a given intensity will cause exceedance of the performance threshold – the majority of this curve lies at a probability of either zero or one, and there are few intensities where we are uncertain of the dam's performance (i.e. where there is a probability between zero and one). In general, a fragility curve will have a steeper slope when the intensity measure and the performance measure are highly correlated with little scatter, as shown in the examples that follow.

However, this steep-sloped curve does not provide any information as to how the probability changes as the hazard grows in intensity. Missing is the ability to evaluate the transition region which corresponds to the lower portion of the fragility curve as the intensity measure grows. In the example shown above, the abrupt jump – or discontinuity – in the probability of exceedance as the intensity measure approaches a value of 10 does not allow an evaluation of the changing conditions in the dam that would actually lead to the abrupt change. That is, though this "ideal curve" allows perfect prediction of dam failure, it does not allow us to see how the dam gradually weakens as intensity grows. For this reason, a fragility estimate that provides a smoother transition in the lower left portion of the curve may be preferred.

In the methodology described by Menda et al. (2016), the development of the fragility estimate relies on a performance-based field test procedure that extracts a dam's characteristic behavior. This characteristic behavior is then used to develop predicted behavior of the dam subject to hazard or, as in the ensuing examples, earthquake loading. A key component of the methodology, and a factor in the eventual usefulness of the fragility estimate, is the relationship between the predicted performance or engineering demand parameter and its correlation to the intensity measure chosen for the fragility estimate. If an intensity measure can be found that is highly correlated with the performance measure, the resulting fragility curve will likely be substantially more useful.

SAMPLE SEISMIC INTENSITY MEASURES

In June 2015, Performance-Based Testing was conducted on San Clemente Dam, a concrete masonry arch dam in Carmel Valley, CA that was demolished in August 2015. The methods used were very similar to those used by Anderson et al (2015) and Goldkamp et al (2014). A Cold Gas Thruster (CGT) was mounted on the face of the dam, about twenty feet above the base of the dam and about 40 feet left of the dam's center. The CGT was used to apply a large-amplitude (greater than 10,000 lbf), short duration (less than 10 ms) impulsive load to the structure. The responses gathered by the triaxial accelerometers placed near the CGT and along the crest of the dam were used as impulse responses that were used to predict the seismic response of the structure via a convolution-based procedure described in by Menda et al (2016).

The data collected from PBT was then used to generate fragility curves for the dam in each of the 16 locations monitored along the crest (see Figure 3). The maximum predicted displacement during seismic loading at each sensor location was taken as the performance measure when generating the fragility curves. The failure threshold was taken as the displacement at that sensor location that would produce a yield stress when the dam experiences cantilever-style loading on its crest (as explained in more detail by Zhao et al (2016)). Several intensity measures were used, in order to demonstrate the impact of the chosen intensity measure on the appearance and quality of the generated fragility curve. The intensity measures chosen are explained and motivated in the following sections, and the quality of the resulting fragility curves is evaluated for each intensity measure.

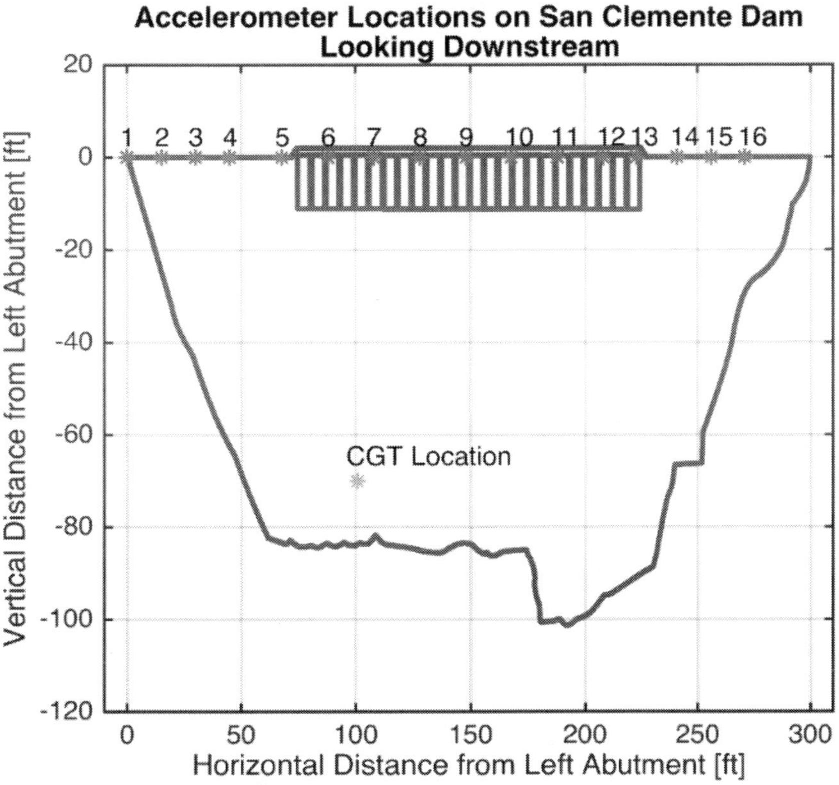

Figure 3. Instrumentation layout for the Performance-Based Testing of San Clemente Dam. The Cold Gas Thruster creates an impulse-like load, so that the responses recorded by the accelerometers can be approximated as impulse responses.

Peak Ground Acceleration

Peak ground acceleration (PGA) is a commonly accepted as an intuitive measure of the intensity of an earthquake, and so it is often used to plot fragility curves. The fragility curve for sensor location 11 on Figure 3 is plotted versus PGA in Figure 4.

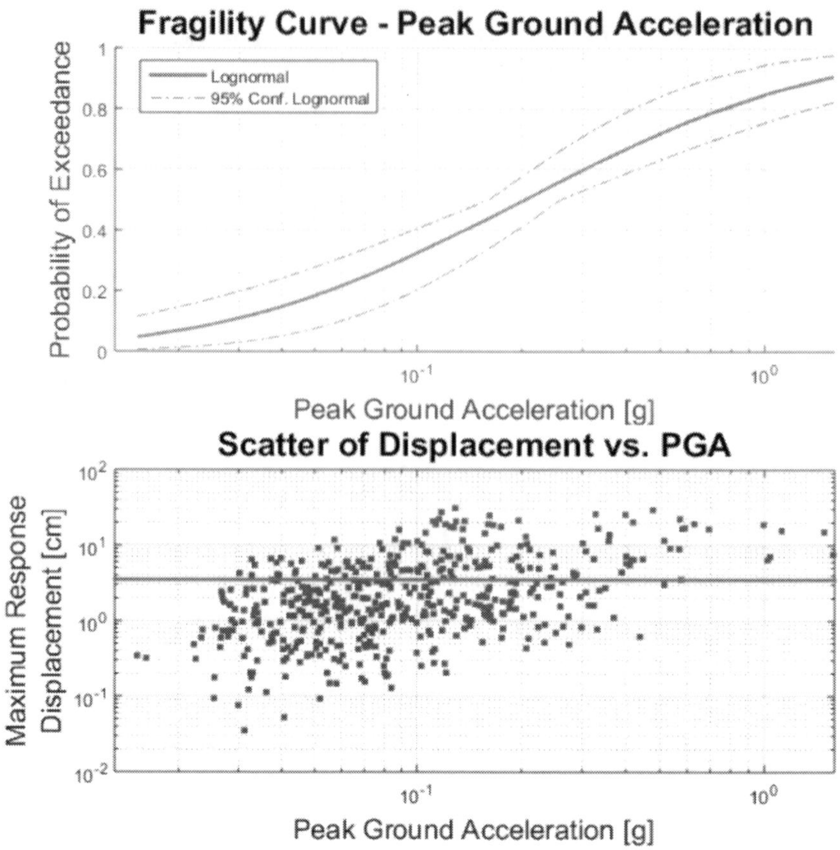

Figure 4. A fragility curve computed with peak ground acceleration as the intensity measure. Confidence intervals are denoted by the red dot-dash lines.

The relationship between the performance parameter (maximum displacement) and the peak ground acceleration intensity measure for the earthquakes used for the fragility estimate shows significant scatter. There is no strong correlation between the performance parameter and PGA as an intensity measure. The corresponding fragility estimate has large (wide) confidence intervals, is lacking the desired definition, and does not produce clear indications of how the probability of exceedance relates to PGA. This result is not considered useful for evaluating a dam's vulnerability to seismic loading.

Cumulative Absolute Velocity (CAV)

Cumulative Absolute Velocity (CAV) is the integral of the absolute value of an earthquake's acceleration time history. CAV has been found to be fairly well correlated with soil liquefaction due to earthquake ground motion, and is popular as an intensity measure when soil liquefaction is a concern. Ground motion predictive equations have recently been produced for CAV using the PEER-NGA ground motion database (Campbell and Bozorgnia 2014), extending the usefulness of CAV as an intensity measure. The fragility curve for location 11 from Figure 3 is plotted in Figure 6 against cumulative absolute velocity.

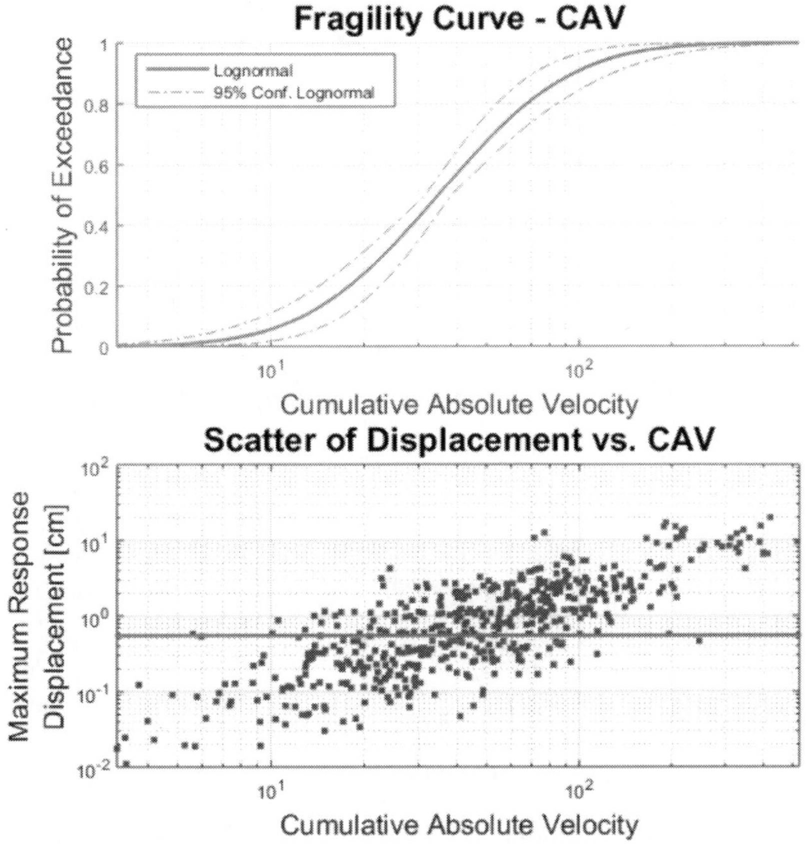

Figure 6. A fragility curve computed with cumulative absolute velocity (CAV) as the intensity measure.

In this example, there is some improved correlation between the performance parameter and the CAV intensity measure as compared to the previous correlations to PGA, but not much. The result is a fragility curve with definition that is better for evaluating a dam's vulnerabilities, but whose confidence intervals remain too large for credible use.

Arias Intensity

Arias intensity is also a commonly accepted measure of an earthquake's intensity. It is defined to be proportional to the integral of the square of the acceleration time history of an earthquake. The fragility curve for location 11 from Figure 3 is plotted in Figure 5 against integrated square acceleration.

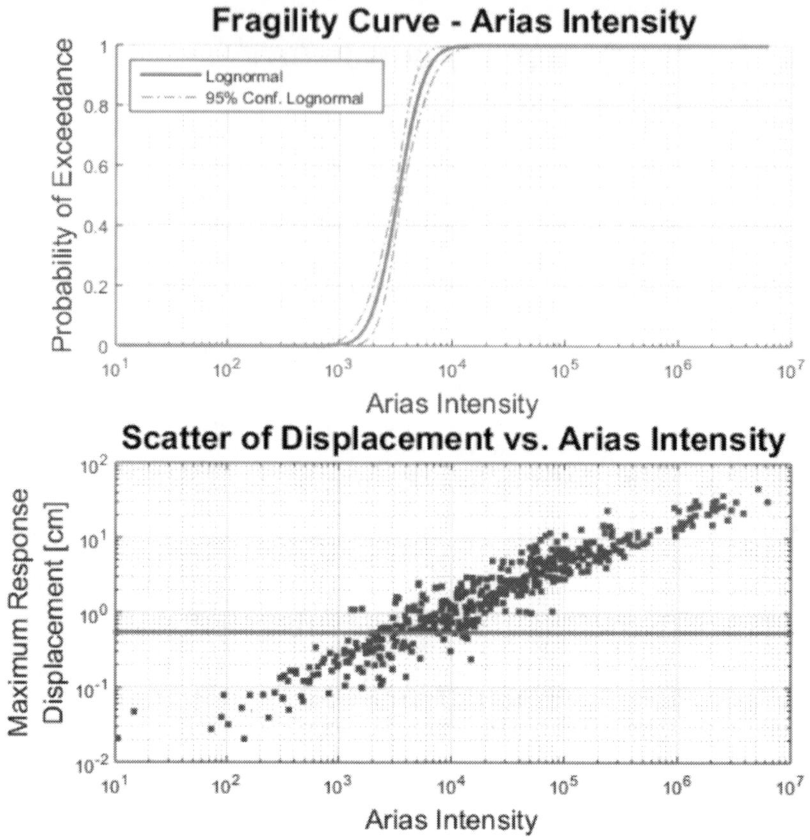

Figure 5. A fragility curve computed with an intensity measure that is proportional to Arias Intensity.

The scatter indicated in the relationship between the performance parameter and Arias Intensity is improved over the previous and weaker correlations. The result is a fragility estimate that exhibits more defined transitions, with smaller (tighter) confidence intervals. As a consequence of these smaller confidence intervals, the transition in the vicinity of 10% probability of exceedance may provide insights as to how the dam's performance changes with this intensity measure[5].

Integrated Square Velocity

The integrated square velocity (ISV) of an earthquake is defined as the integral of the squared velocities in the earthquake's time history and is computed as

$$ISV = \int v(t)^2 dt \quad (1)$$

The fragility curve location 11 from Figure 3 is plotted versus ISV in Figure 7.

[5] Of interest since most dam owners set their tolerance thresholds for probability of exceedance at or near 10%.

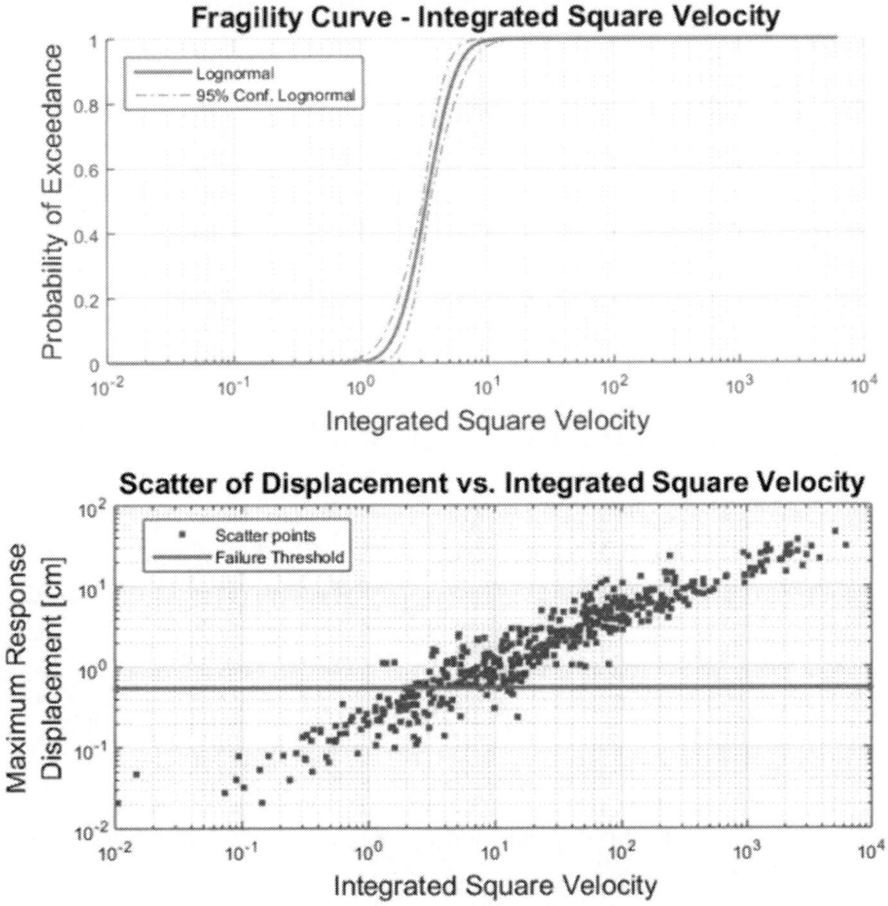

Figure 7. A fragility curve computed with integrated square velocity (ISV) as the intensity measure.

The relationship between the performance parameter and this intensity measure is about the same as the correlation with Arias Intensity. While the Integrated Square Velocity (ISV) is not a widely used intensity measure, it does provide a fairly good correlation with the performance parameter, and produces a correspondingly credible fragility curve estimate. The fragility curve contains small confidence intervals and a well-defined transition region as the probability of exceedance approaches 10%, and also offers a clearer transition into the higher probability of exceedance levels where a dam's response to extreme events can be evaluated with a higher degree of confidence.

Peak Ground Displacement

Peak ground displacement (PGD) is the maximum of the absolute value of an earthquake's displacement time history. The fragility curve for location 11 from Figure 3 is plotted versus peak ground displacement in Figure 8.

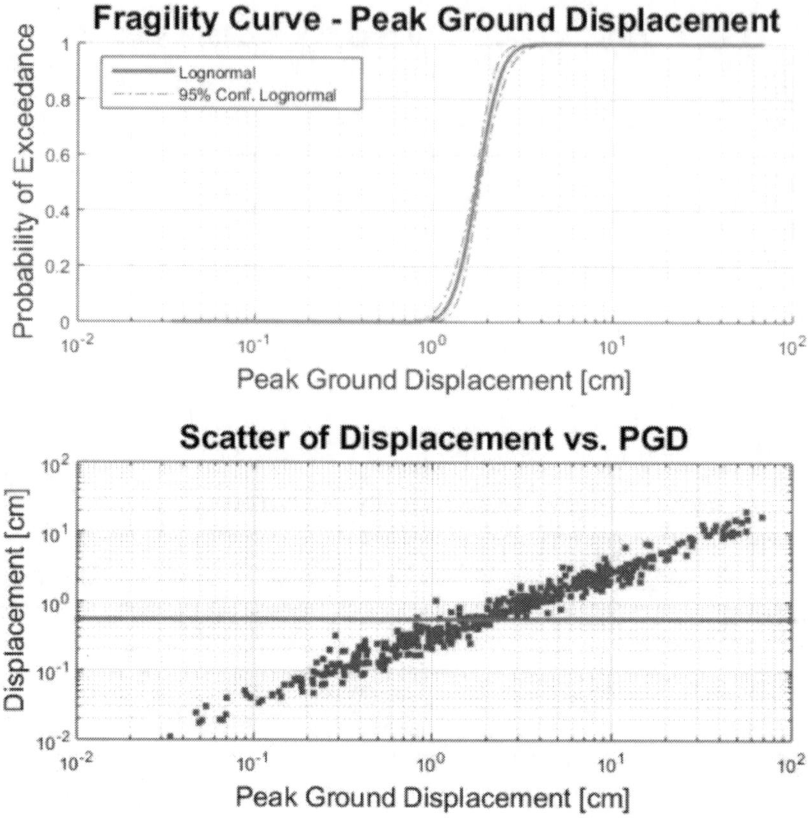

Figure 8. A fragility curve computed with peak ground displacement as the intensity measure.

This intensity measure results in perhaps the most linear correlation with the performance parameter, and contributes to a fragility estimate that offers very small confidence intervals and good resolution in the desired regions.

EXTENDING THE CURRENT APPROACH BY MAPPING ONTO RETURN PERIOD

A fragility curve is usually defined as a plot of the probability of an asset failing (for some definition of failure that is specified by the performance measure) versus an intensity measure, which is a function that maps an earthquake time history onto a scalar intensity value. This plot is useful as long as the fragility curve's user is able to produce a value of the intensity measure that represents their maximum allowable or even credible intensity. If an intensity measure has been found that is highly correlated with the chosen performance measure, this implies that the dam owner should be able to predict with a larger degree of certainty whether or not a given earthquake will cause the dam to fail (because they only need to calculate the intensity measure for that earthquake, and the curve will provide the probability of failure). However, it may not be easy for the user to determine whether or not an earthquake that is assigned a given intensity value is a credible earthquake that needs to be considered in their analysis. For example, an overly abstract intensity measure might perfectly correlate with the performance measure, but if it has become so abstract that the owner cannot look at the intensity of a seismic hazard and determine whether or not the hazard is credible, the resulting fragility curve will be of very limited practical use.

This problem is resolved in most cases if the intensity measure can be mapped onto return period, because the maximum earthquake intensity that an asset owner needs to consider is in fact typically defined in terms of the return period of a hazard earthquake. A fragility curve could simply be computed with return period as the intensity measure if it were possible to determine an earthquake's return period given its time history. Unfortunately, this is not currently possible. However, ground motion prediction equations exist for many common seismic intensity measures, and when these equations are available the intensity measure can be mapped onto return period. Ground motion prediction equations provide a means of computing the probability distribution for earthquakes occurring with various values of a selected intensity measure, given a set of parameters describing the location and site conditions of the asset. So if those site parameters can be determined, these equations provide a means of mapping the intensity measure onto annual probability of exceedance, the reciprocal of return period. The fragility curve can then be plotted against return period, which can then be used to estimate the risk of the dam for the hazard loading condition.

CONCLUSIONS

Fragility curves can be powerful tools that allow owners of dams and other assets to visualize and manage the seismic risk of their asset. Using data from Performance-Based Testing, fragility curves can be constructed that are based in empirical data rather than on subjective or assumed knowledge or on the results of numerical models, which often rely on large numbers of assumptions that have not been or cannot be experimentally validated. Furthermore, because the impulse response of the structure collected in PBT contains large amounts of information about the structure's dynamics (activating vibrational modes beyond 100 Hz), this information does not need to be computed by the numerical model. This provides a credible means for extending the linear performance of a dam into a transition region where effective decision making can take place regarding impending conditions in the dam that could lead to ultimate failure, and an uncontrolled release of water. The relatively simple nature of Performance-Based Testing, combined with the very low computational cost of the fragility-curve-generating algorithm described by Menda et al. (2016), brings the ability to generate fragility curves into the reach of most dam owners.

However, the quality of the resultant fragility curve is dependent on the intensity measure that is chosen for the curve's independent axis. A fragility curve is most useful when the intensity measure is well-correlated with the performance measure, and when resolution in the transition regions contributes to high confidence in the probabilities of exceedance approaching 10% and higher. When this is the case, the resulting fragility curve will also have a steep slope, which means that the curve's user has a greater knowledge of whether or not a given earthquake will fail the structure. It will also tend to have small uncertainty bands, meaning that the fragility curve itself can be treated as more reliable than when the intensity measure is not well correlated with the performance measure. The region that defines how the probability of exceedance varies from near zero to 10% is particularly useful in the early assessment of a dam's vulnerability to hazard loading.

When a series of fragility curves was constructed for San Clemente Dam using maximum structural displacement as a performance measure, peak ground displacement (PGD) was found to be the most highly correlated intensity measure. Integrated Square Velocity (ISV), not a widely used measure, produced a comparable fragility estimate. Although the Arias Intensity produced a credible result, other measures such as PGA and CAV produced less useful estimates.

Though fragility curves showing the probability of an asset's failure versus an intensity measure can be a valuable component of a risk assessment, in some cases, estimates may also be developed against return period. When it is possible to do this, the seismic life expectancy of the structure may be estimated from the fragility curve, providing an effective way of quantifying the seismic risk associated with a structure. Additionally, some work has been done to produce fragility curves for dams using nonlinear finite element models (Ghanaat et al. 2015). The existence of an alternate, field and performance based method for estimating fragility curves provides a valuable opportunity to utilize the strengths of each technique to test the results of the other, and should facilitate the further improvement of methods for generating fragility curves.

ACKNOWLEDGEMENTS

The authors wish to acknowledge the support from the following individuals and their organizations for making the field investigations at San Clemente Dam possible:

David Gutierrez, Chief, and Mark Schultz, Supervising Engineer from the Division of Safety of Dams Department of Water Resources, California.
Richard Svindland, VP of Operations at California American Water, dam owner.
Bill McGowan, Project Executive at Granite Construction Company.
Dylan Wade, Resident Project Engineer at San Clemente Dam for Water Systems Consulting, Inc.
Robert Carnes, consultant and logistical support to the Resident Project Engineer.

The authors are especially grateful to Mr. Wade and Mr. Carnes for their efforts in managing our support needs and safety during the performance testing at the dam.

REFERENCES

Anderson, S., P. Or me, Z. Duron, R. Nash, a nd K. Marshall. (2015). "Evaluating Relative Condition of Repeated Structural El ements with Perform ance-Based Testing." Louisville, Kentucky, 35th Annual USSD Conference.

Campbell, K. and Y. Bozorgnia. (2014). "A Ground Motion Prediction Equation for the Horizontal Component of Cumulative Absolute Velocity (CAV) Based on the PEER-NGA Strong Motion Database." *Earthquake Spectra.*

Ghanaat, Y., R. Patev, and A. Chudgar. (2015). "Seismic Fragility for Risk Assessment of Concrete Gravity Dams." Louisville, Kentucky, 35th Annual USSD Conference.

Goldkamp, M., T. Savage, J. Edelman, P. Orme, Z. Duron, and M. Knarr. (2014). "Performance-Based Analysis of a Large Concrete Dam." San Francisco, California, 34th Annual USSD Conference.

Menda, K., M. Zhao, S. Anderson, and Z. Duron. (2016). "A Performance-Based Approach to Developing Fragility Curves for Structures Subject to Seismic Loading Conditions." Denver, Colorado, 36th Annual USSD Conference.

Zhao, M., S. Anderson, K. Menda, and Z. Duron. (2016). "Selecting a Performance measure for Fragility Curves Derived from Performance-Based Testing." Denver, Colorado, 36th Annual USSD Conference.

RETROFIT AND CLOSURE OF CCR IMPOUNDMENTS TO MEET NEW REGULATIONS

Matt Bishop, EIT[1]
Brian Owens, PE[2]
Luke Williams, PE[3]

ABSTRACT

Due to recent Coal Combustion Residual (CCR) regulations, owners of coal-fired power plants have been faced with difficult decisions relating to the future of their generating facilities and accompanying ash ponds and impoundments. In many cases the requirements, costs, and challenges to either retrofit or decommission these ash ponds can be a determining factor in the future viability of the facility. In addition, regulatory guidelines in each state can demand that specific requirements be met in the designs. These requirements are typically inclusive of the hydraulic and hydrologic design event that is used in the design, the geotechnical stability of the impoundments, whether a liner system is required for the CCR materials, the manner in which the CCR impoundment must be closed, etc.

This paper presents a case study of three separate solutions that have been successfully developed for addressing these new CCR regulations: retrofitting of the ash pond and/or spillway structures, closure by capping in place, and closure by removal of the CCR materials and earthen embankments. Each solution presents unique methods of addressing the costs, construction requirements, and local regulatory requirements associated with the designs, but it is typically the local regulatory requirements for each plant location that have the greatest impact on the designs and solutions chosen. However, when equipped with the knowledge and potential challenges of each solution, owners of coal-fired power plants can make the best decision for the future of their ash ponds and impoundments.

INTRODUCTION

Owners of many coal-fired power plants have been faced with difficult decisions relating to the future of their ash ponds and impoundments due to recent Coal Combustion Residual (CCR) regulations. These decisions are largely influenced by new regulatory requirements, which may include the required hydraulic and hydrologic design event, geotechnical stability of the impoundments, whether a liner system is required for the ash pond, the manner in which the ash pond must be closed, etc.

[1] Amec Foster Wheeler, 2030 Falling Waters Road, Suite 300, Knoxville, TN 37922, matt.bishop@amecfw.com.
[2] Amec Foster Wheeler, 1070 W Main Street, Suite 5, Abingdon, VA 24210, brian.owens@amecfw.com.
[3] Amec Foster Wheeler, 2030 Falling Waters Road, Suite 300, Knoxville, TN 37922, luke.williams@amecfw.com.

A discussion of three case studies is presented in this paper, each involving a different method of addressing the regulations. Each case study also presents a coal-fired plant in a different state. The following case studies are presented in the subsequent sections:

Case Study #1

Plant status: Active
Solution: Retrofitting of the ash pond and spillway structures
State: Kentucky

Case Study #2

Plant Status: Active (to be converted to natural gas)
Solution: Closure by capping in place
State: Virginia

Case Study #3

Plant Status: Active (to be closed and replaced with a natural gas plant on plant property)
Solution: Closure by removal of CCR materials and earthen embankments
State: Undisclosed

CASE STUDY #1 — KENTUCKY COAL PLANT

Background

The first case study presents the issues and solutions implemented to address the aging infrastructure of the primary spillway system at a Kentucky ash pond. The ash pond impoundment was designed in the mid-1970s and was engineered to consist of a compacted granular fill embankment with a compacted clay core. The upstream (interior) and downstream (exterior) slopes were constructed at 2:1 slopes (horizontal:vertical). The ash pond impoundment is configured in a "U" shape with the main section parallel to a river and short sections on the east and west ends abutting into natural soils on the north side. Bottom ash material or Coal Combustion Residual (CCR) has historically been deposited within the unlined ash pond by hydraulic sluicing operation.

The original construction included a 48-inch diameter corrugated riser pipe and a 36-inch diameter outlet pipe. In 1991 a 40-inch diameter riser was slip-lined into the 48-inch riser structure. This new riser was connected to a 30-inch diameter barrel that extended into the outlet pipe. In 2013 an outside consultant investigated the principal spillway and recommended modifications or complete abandonment. In 2014 the principal spillway was reviewed by closed circuit camera, and was deemed to be in an impending failure state.

Regulatory Requirements

Upon the findings that the principal spillway pipe was unsafe, Kentucky Dam Safety was notified of the client's intent to abandon and replace the principal spillway pipe. State requirements were then followed in the design of the replacement spillway pipe. Specifically, H&H analyses were performed to determine the sizing and placement of the new spillway pipe. According to the state dam inventory, the ash pond at the site is classified as a Class "B" hazard structure. Below is the equation used to determine the design storm depth, following the Kentucky Dam Safety Regulation for a Class "B" structure.

$$P_B = P_{100} + 0.40*(PMP - P_{100}) \quad (1)$$

In which P_B denotes a 6-hour design storm rainfall, P_{100} is a 100yr – 6hr rainfall, and PMP represents a 6-hour Probable Maximum Precipitation (PMP). Therefore, design requirements were such that designing for a full PMP event was not necessary to keep the ash pond in service.

Abandonment and Replacement of the Principal Spillway

To perform abandonment and replacement of the principal spillway, the existing riser-outfall principal spillway was dewatered, and a temporary pumping system was installed to drawdown the ash pond water level. Grout abandonment of the existing principal spillway outfall pipe was then performed.

The principal spillway was then replaced with a new High Density Polyethylene (HDPE) principal spillway pipe, albeit using unconventional designs and installation methods for soil impoundments. To avoid heavy excavation work involving the impoundment, the decision was made to install a horizontal spillway through the embankment. In order to achieve this, a work pad was constructed on the downstream slope of the impoundment at the required elevation to jack and bore the new spillway pipe through the embankment, shown in Figure 1.

Extensive engineering design, analyses, and modeling were performed to prove to Kentucky Safe Dams that a horizontal spillway through the embankment was possible. Extra precautions were taken to prevent disturbances of the core materials and prevent piping along the horizontal spillway pipe. 2D and 3D seepage modeling was performed for the installation of the horizontal spillway as shown in Figure 2. The horizontal spillway was successfully installed through the embankment, and normal spillway operations were restored to the ash pond. The plant was able to remain active with the completion of the project.

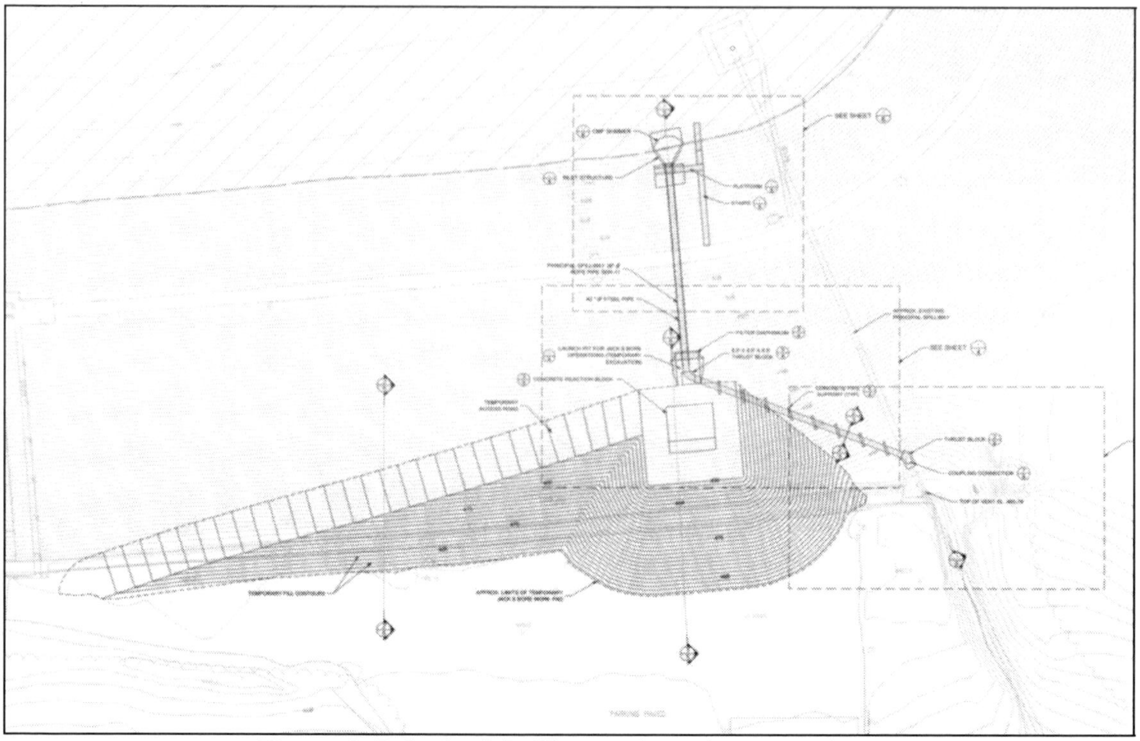

Figure 1. Principal Spillway Plan View Drawing.

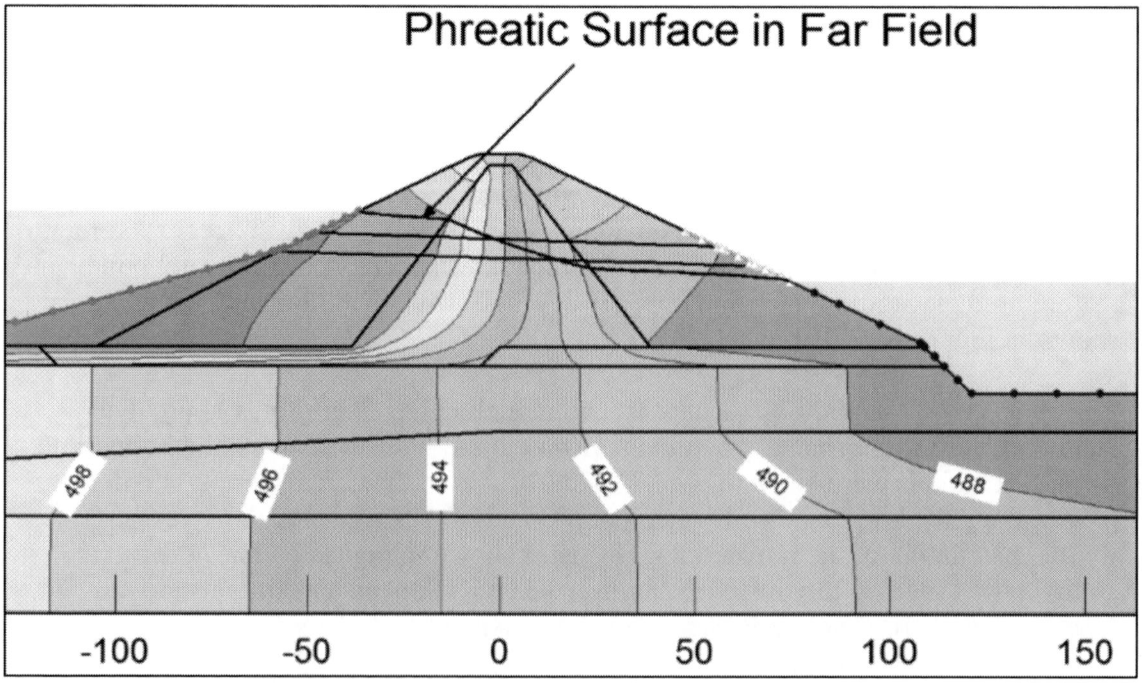

Figure 2. Seepage Model for Principal Spillway Pipe

Costs

The total cost for the installation of the horizontal spillway was approximately $10 million, with the majority of the costs incurred through Engineering Procurement Construction (EPC) contract work onsite.

CASE STUDY #2 — VIRGINIA COAL PLANT

Background

The second case study presents the issues and solutions implemented to perform an in-place closure at a Virginia ash pond. The unlined ash pond at this site is considered a side-hill impoundment and is used for sluicing and settling of ash byproducts. To facilitate settling, the sluiced ash is conveyed through serpentine channels that form the primary impounding volumes that classify the structure as an impoundment.

In the future, the client plans to close the currently active ash pond as the plant is to be converted to natural gas. As part of this conversion process, the ash pond must be closed and removed from service. To meet requirements for ash ponds and impoundments, the client has chosen to perform an in-place closure by capping the CCR materials with an impermeable liner and a soil cover.

Regulatory Requirements

The ash pond impounds water and material and is regulated as a dam by the Virginia Department of Conservation and Recreation (VDCR). In 2013, two designs were prepared for the ash pond which was, at the time, classified as a SIGNIFICANT hazard embankment. The first design provided a diversion channel to the west of the ash pond to divert water from the approximately 53-acre watershed west of the ash pond. This channel was originally designed on a 0.5 PMP event according to the current regulations at that time.

The second design provided plans and details for closure of the ash pond. Again, the 0.5 PMP and/or the 100-year precipitation events were used to design the channels that controlled runoff from the closed impoundment. These channels were designed to direct water to both the west and east of the ash pond to the creek and river located to the south of the ash pond.

Subsequent to the submittal of these plans, the pond was reclassified as a HIGH hazard embankment. HIGH hazard structures are required to safely pass stormwater runoff for the 0.9 PMP event as documented in Virginia Administrative Code Chapter 20 Impounding Structure Regulations. The 0.9 PMP storm depth for the project site is 25.9 inches for the 6 hour storm duration and 33.3 inches for the 24 hour duration, per NOAA HMR-51. The drainage features within the ash pond site are required to safely pass the stormwater runoff from the 0.9 PMP in a non-erosive manner.

Closure by Capping in Place

Additional and relatively extensive closure designs to both cap the ash pond and divert stormwater around it were developed to meet the requirements of the 0.9 PMP indicated in the previous section. A plan view of the final designs is shown on Figure 3. The closure designs eliminate impounding of water as well as convey upstream stormwater runoff safely away from the toe of the impoundment. In addition, a PVC liner will cover the entire disposal area, and two feet of clean fill will be installed above the liner. The existing toe drains will convey the contact water to the existing wastewater treatment plant, and surface runoff (i.e., non-contact water) will be conveyed to two NPDES outfalls.

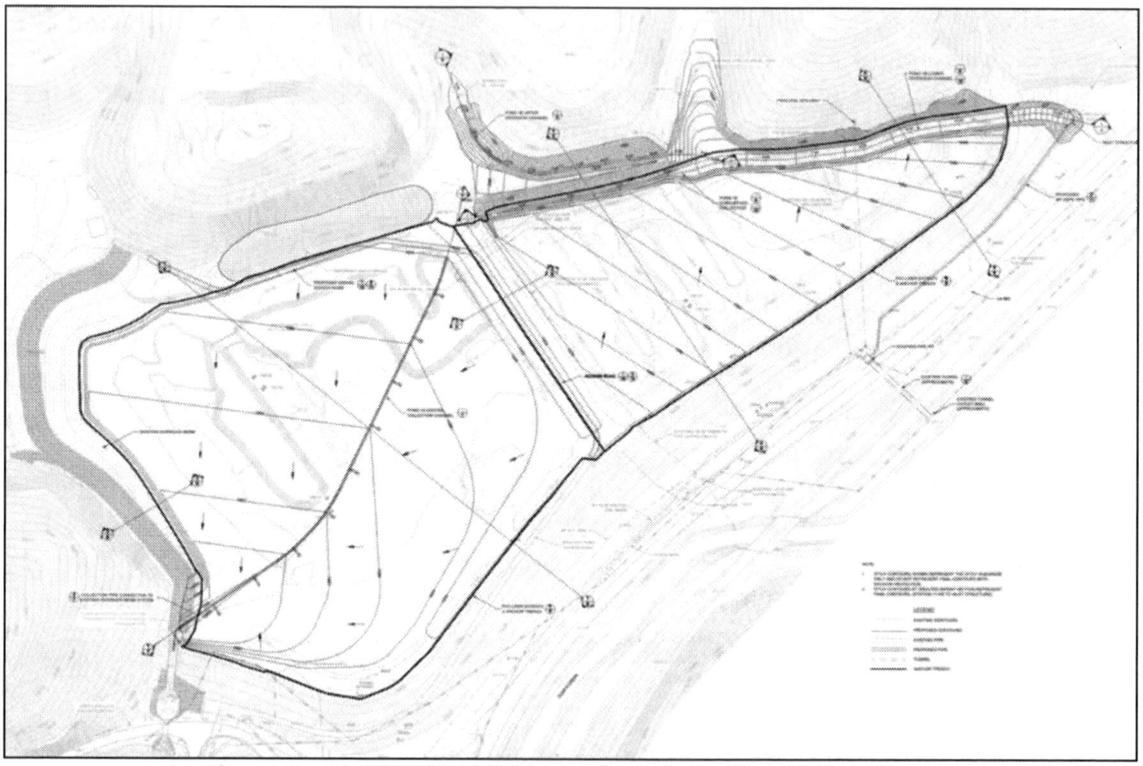

Figure 3. Closure by Capping in Place Final Grading Plan.

The ash pond site is a side-hill impoundment and, by nature, steep slopes exist. The diversion channels to the west and northeast of the ash pond each have very steep lower sections where the channels traverse the embankment slope of the dam. These steep channel sections require energy dissipation and containment of flood flows to the toe of the dam. To address these needs, a combination of several channel linings was used in the designs.

For the west diversion channel, a concrete channel was constructed for the full length, with an energy-dissipating structure at the bottom of the embankment. The northeast diversion channel consisted of several different designs. The upper portion of this channel consisted of Class 1 and Class 1A VDOT riprap. The middle portion of the channel transitioned into gabion mattresses, to help control the velocities of the stormwater. Finally, the lower and steepest portion of this diversion channel was constructed with grouted riprap to further control velocities through the steepest segment of the channel.

As noted, relatively extensive designs were undertaken to address the higher 0.9 PMP storm event. Additional diversion channels were needed to compensate for the increased stormwater flows through the site, and costs were increased for closure.

Costs

The total cost for the in-place closure of this ash pond were influenced from both the capping of the ash pond and the diversion of the 0.9 PMP stormwater. Costs associated with the capping of the pond are expected to approach approximately \$10 million, while costs associated with the diversion of the stormwater are expected to approach approximately \$5 million. Therefore, total costs incurred for this ash pond closure are expected to reach \$15 million.

CASE STUDY #3 — UNDISCLOSED COAL PLANT

Background

The third case study presents the issues and solutions implemented to perform a closure involving complete CCR removal for two ash ponds (referred to as Ash Basin 1 and Ash Basin 2). This coal-fired plant is an active coal-combustion generating facility. Historically, both fly ash and bottom ash material, or CCR, have been deposited within the two ash basins located on the plant site by hydraulic sluicing operations. The two ash basin dikes fall under the jurisdiction of State Dam Safety agency.

The two original coal-fired units are currently active. At this time, ash produced from the current operations is routed to a rim ditch, dewatered, and hauled to one of the ash basins. Ash is then being removed and hauled off-site. Current planning is to cease operation of the coal-fired units. The closure of the ash pond facilities at the station is a part of the decommissioning plan for those units.

Regulatory Requirements

According to the current state hazard-rating criteria, the dikes are considered to be large high-hazard structures, based upon potential breach impacts to downstream roads and property. The ash ponds also fall under the requirements of both the Federal CCR Rule §257.102(b)(1)(i-vi) and state regulations for closure of ash basins / surface impoundments. These requirements specify that unlined ash basins must be excavated of all CCR materials. Because of these requirements, the active and unlined ash ponds at this site are required to be excavated of all CCR materials. The CCR materials are required to be disposed of in a permitted structural fill, landfill, or similar facility.

Closure by Removal of CCR Materials and Earthen Embankments

The primary objective of the ash basin closures for the site is to remove all CCR materials from the site, breach the existing dams that form the ash basins, and to establish final grades to promote drainage, as shown on Figure 4. Ash removal from the site has been underway since 2007, with the CCR materials currently being hauled by truck to an off-site landfill facility. The total estimated volume of ash to be removed from the ash ponds is approximately 6,250,000 tons.

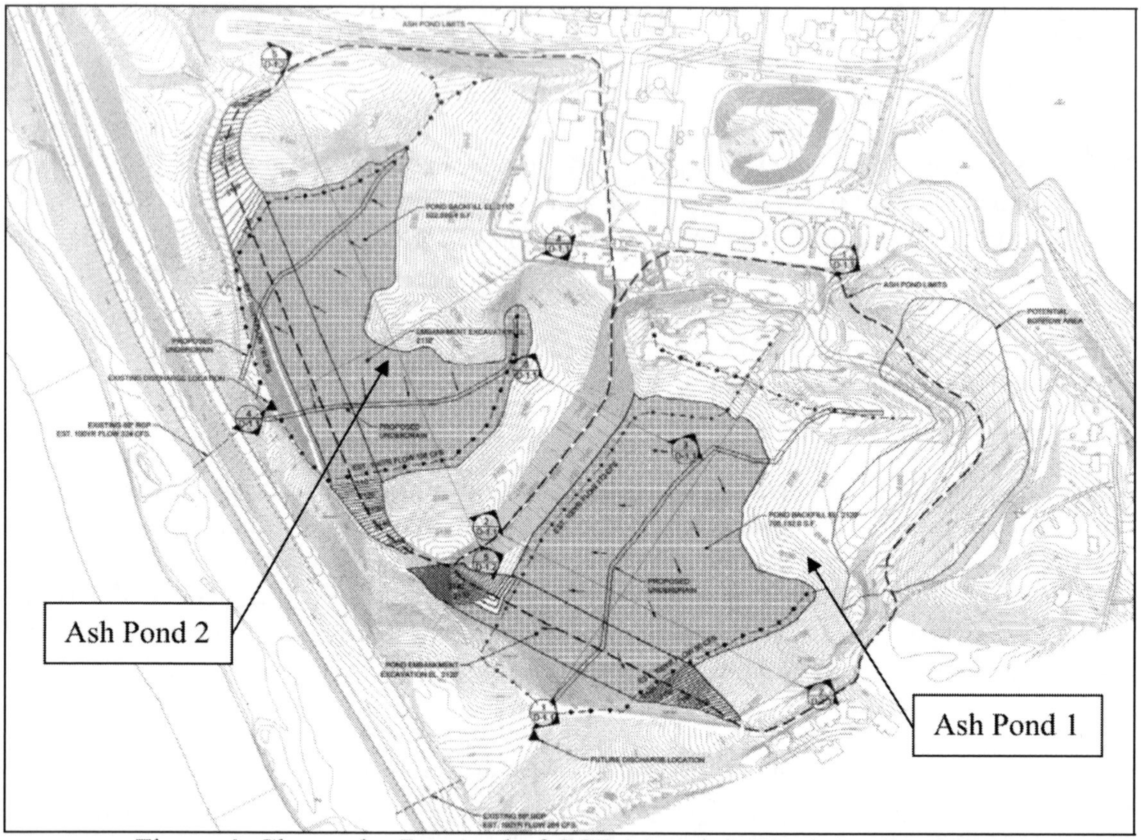

Figure 4. Closure by Removal of CCR Materials Final Grading Plan.

The closure design of the ash ponds at this site has proven to be an intricate process to coordinate between the current plant operations and future designs for the site. The first of the challenges involves the future use of Ash Pond 1. A decision was made by the client to construct a new, natural gas-fired power plant in the footprint of Ash Pond 1. Therefore, the closure design for this ash basin involves a breach of the existing impoundment and also backfilling behind the impoundment to an elevation that allows for construction of the new, proposed plant. In addition, to facilitate this decision, ash must be removed from the ash basin in a timeframe that allows construction of the new plant to begin.

Through this process, it was determined that complete ash removal from Ash Pond 1 to the off-site landfill facility could not be completed in a timeframe that allows for construction of the new plant to begin. Therefore, an engineering design was performed to temporarily stockpile the remaining CCR materials in Ash Pond 2. This process also includes the engineering design for the removal of existing engineered wetlands constructed in the footprint of Ash Pond 2, as the CCR materials will be placed in the footprint of the existing engineered wetlands. The engineering design process for this step includes the dewatering, demolition, and removal of the wetlands materials and the grading, geotechnical stability analysis, and configuration of the proposed temporary ash stacking operation.

Following the closure of Ash Pond 1 at this site, the ash removal operations will commence for Ash Pond 2. The closure design for this pond includes a "balanced breach" of the impoundment. With this design, an impoundment breach elevation was calculated that "balances" the volume of cut material from the impoundment to equal the volume of material required to backfill the existing basin up to the same elevation.

Case Study #3 has proven to be the most intricate and costly of all the case studies presented herein. Due to regulations, site constraints, and future plans for the site, a significant amount of coordination is required to address the engineering work to be performed.

Costs

Costs to close Ash Pond 1 are estimated to approach $157 million while those to close Ash Pond 2 are near $167 million. Compared to Case Studies #1 and #2, closure by removal of CCR materials has proved to be significantly more costly.

COMPARISON OF THE CASE STUDIES

Each of the case studies presents a different approach to address the future of plant operations. Case Study #1 is the only study that addressed the decision to allow the plant to continue operations indefinitely. However, it should also be noted that regulations in this state did not require as stringent of H&H requirements. Regulations in this state also did not require the unlined ash pond to be excavated. Costs incurred for the replacement of the principal spillway were less than the other case studies.

Case Study #2 addressed the client's decision to close the coal-fired generating facilities at the plant and convert it to natural gas. In this process the unlined ash pond was also not required to be excavated; however, more stringent H&H requirements were imposed by the state regulatory agencies. Because of these requirements, additional stormwater diversion design was required at the site, and costs were slightly more to address these requirements than in Case Study #1.

Case Study #3 presented the most intricate engineering design and coordination, and also the most stringent state regulations. In this state the requirements for closure dictated that complete removal of the CCR materials from the site was required. Therefore, costs were significantly higher than in Case Studies #1 and #2. Also, CCR removal requirements and future site planning resulted in additional engineering design to accommodate tight schedules and future construction of a natural gas plant at the plant site.

CONCLUSION

Owners of coal-fired power plants have begun to address recent CCR regulations in several different ways. The solutions include many different variables and approaches to addressing the issues but can be simplified by analyzing the three case studies presented herein. These case studies show that coal-fired power plants can remain active by retrofitting existing ash pond features. If a closure of the coal-fired operations is decided upon, owners can accomplish this goal and address regulatory requirements by capping in place or by removal of CCR materials. Costs associated with each option increase with the regulatory requirements and complexity of the solutions. However, each of the case studies presented effectively addresses the issues and requirements imposed upon the clients. When equipped with the knowledge and potential challenges of each solution, owners of coal-fired power plants can make the best decision for the future of their ash ponds and impoundments.

ANTECEDENT SNOW COVERED AREA ANALYSIS FOR INFLOW DESIGN FLOODS

Steven F. Daly, Ph.D., P.E., D.WRE[1]
Angela Duren, P.E., P.H., [2]

ABSTRACT

There has been an ongoing effort to assess the ability to route the Inflow Design Flood (IDF), which is also the Probable Maximum Flood (PMF) for a Success Dam in California. In 2013 a Probable Maximum Precipitation (PMP) and PMF study was done for the project using current NOAA Hydrometeorologic Report (HMR)-based procedures (HMR No. 58/59 of California) of calculating the PMP and standard USACE policy for routing the PMP to develop a PMF. Sensitivity testing reported in 2013 PMP/PMF report indicated that antecedent snow assumptions were a significant source of uncertainty. In February 2015, an antecedent snow covered area (SCA) analysis was completed by USACE Engineer Research and Development (ERDC) Cold Regions Research and Engineering Laboratory (CRREL) for the upstream watershed resulting in a change in the general assumption of the coverage of area in each elevation zone of the starting snow water equivalent (SWE). The SCA analysis included an in-depth analysis of satellite data from NASA (period of record limited to 2000-2014) and indicated that, even with the most conservative assumptions, the snow-covered area is much less than 100 percent per elevation band. After the results were incorporated into the PMF routing parameters, the peak pool elevation of the PMF was reduced by 2 feet.

INTRODUCTION

The analysis of SCA is based on the NASA-developed SCA product, MOD10A1, the MODIS/Terra Snow Cover Daily L3 Global 500 m Grid (Hall et al. 2006), which provides a daily estimate of the fractional snow covered area (fSCA) of the globe at a 500 meter pixel resolution. SCA is estimated using the normalized difference snow index (NDSI), which leverages the contrasting reflectance of snow in the visible and shortwave-infrared regions of the spectrum (Hall et al. 2006). The fSCA is estimated using an empirical relationship between fSCA and NDSI. This process may have difficulties in heavily forested areas (Salomonson and Appel 2004). Areas obscured by clouds are masked out of the MOD10A1 product based upon the MODIS cloud mask (MOD35). The MOD35 uses a host of algorithms and 19 of MODIS's 36 bands to determine if clouds are contained in the MODIS image (Ackerman et al. 2010). The large and changeable areas obscured by clouds in each daily SCA product represent a significant impediment to the use of the MOD10A1 SCA product in water resources investigations. ERDC-CRREL has developed a new technique, referred to as Time

---

[1] Research Hydraulic Engineer, U.S. Army Corps of Engineers, Engineer Research and Development Center, Cold Regions Research and Engineering Laboratory, 72 Lyme Rd, Hanover, NH 03755, Steven.F.Daly@usace.army.mil
[2] District Hydrologist, U.S. Army Corps of Engineers, Portland District333 SW 1st Ave, Portland, OR 97204, angela.m.duren@usace.army.mil

Domain Filtering (TDF), to address this issue of cloud obscuration (Morriss et al., submitted 2014). In general, mean cloud persistence, or the average time between cloud-free scenes, was found to be less than a day at Terra MODIS's temporal scale (approximately 1 day). This cloud-cover transience provides the opportunity to fill in gaps in the current day's imagery with recent data. Since multiple days' imagery is used as inputs, outputs are limited to SCA that is persistent on a multi-day scale. Each output pixel value is based upon up to 16 daily images, though only the two most recent cloud-free images influence the value of the outputs. This approach limits errors of commission in snow cover that arise from cloud and cloud shadow misclassification in the MOD10A1 detection algorithms. If SCA is present in both of the two most recent cloud-free images, the output pixel value is set to the fSCA value of the most recent image. If either pixel's value in the cloud-free pair is snow free, the output pixel is assumed to be snow free. This requirement sometimes filters transient SCA from the output product in addition to spurious SCA but creates a conservative and robust delineation of SCA extent. In the rare occurrence that a pixel does not have any cloud-free readings in the 16 input scenes, the output is cloud.

The SCA within the Success Dam watershed was determined by the fSCA for each pixel for each day by using the TDF analysis of the MOD10A snow product. The fSCA was determined for each of the 4710 MODIS pixels in the watershed for each of the 5405 days in the period of record (04 March 2000 through 20 December 2014). There were 45 days of fSCA data missing due to gaps in the MOD10A data. There were 2715 days with SCA, of which 2133 days (78.6%) had unique data; 582 days with SCA (21.4%) were duplicates caused by the presence of clouds.

The total area covered by SCA for each day of the period of record was determined by summing the fSCA of all the pixels in the watershed and multiplying the result by the area of a MODIS pixel. Figure 1 shows the time series of total watershed SCA. It is interesting to note the "spikiness" of the total SCA; the area of the watershed that is continuously snow covered each winter is relatively small (generally less than 20 square miles). Periods when much larger areas are snow covered last only a few days. The time series of total watershed SCA as a percentage of the total Success Dam watershed is presented in Figure 2, which demonstrates that the SCA generally ranges between 5% and 10% of the watershed area. The statistics of SCA for each day in the winter, including the daily average SCA and the maximum and minimum recorded SCA, is presented in Figure 3.

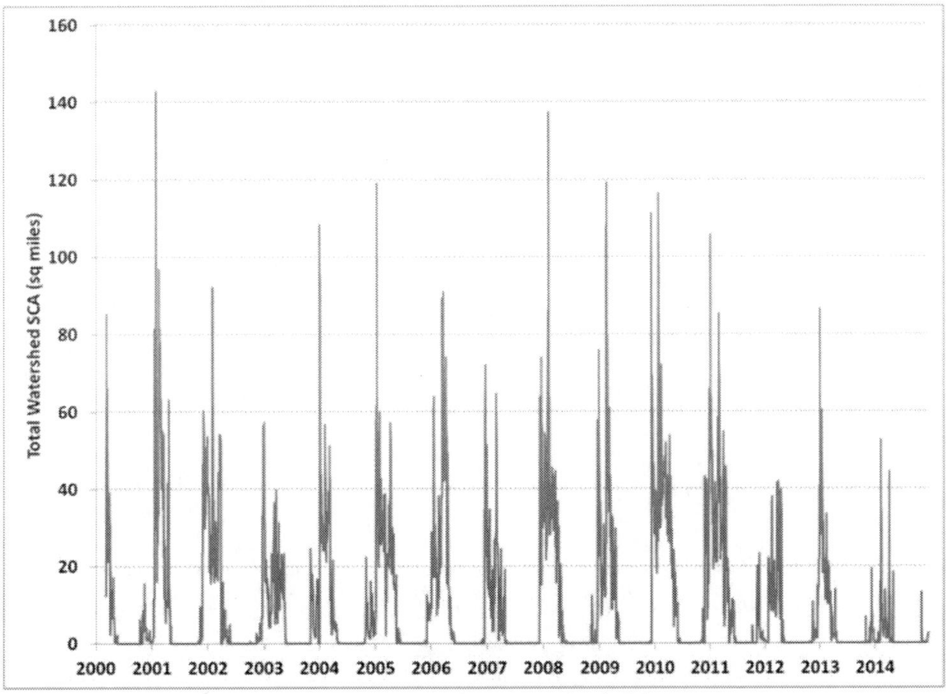

Figure 1. Total Watershed SCA

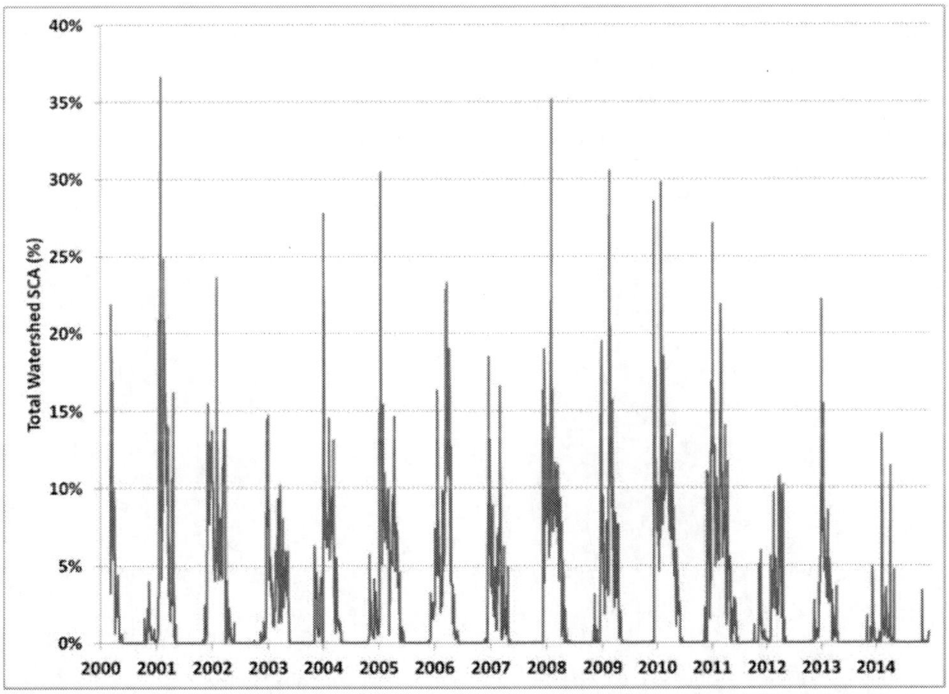

Figure 2. Total Watershed SCA as a Percentage of the Total Watershed Area

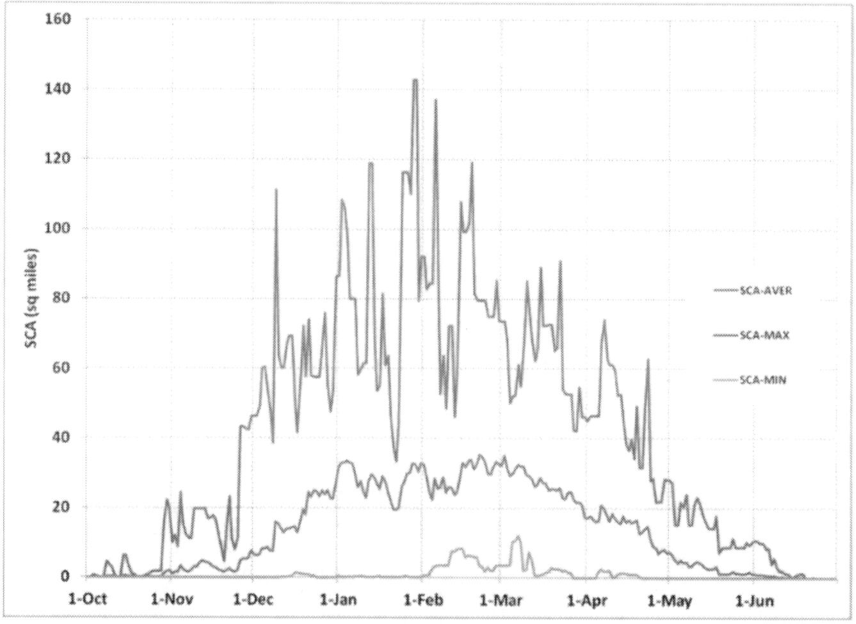

Figure 3. Average, Maximum, and Minimum Total Watershed SCA for Each Day of the Winter

DAILY SCA RESULTS

Several days in the period of record were viewed in detail: On 29 January 2001, the largest total watershed SCA over the period of record occurred (image on the left, Figure 4). The SCA map indicates that seven days later on 05 February 2001 (image on the right in Figure 4) the snow is nearly gone. The dynamic nature of the SCA can be seen in this dramatic reduction in SCA over these seven days.

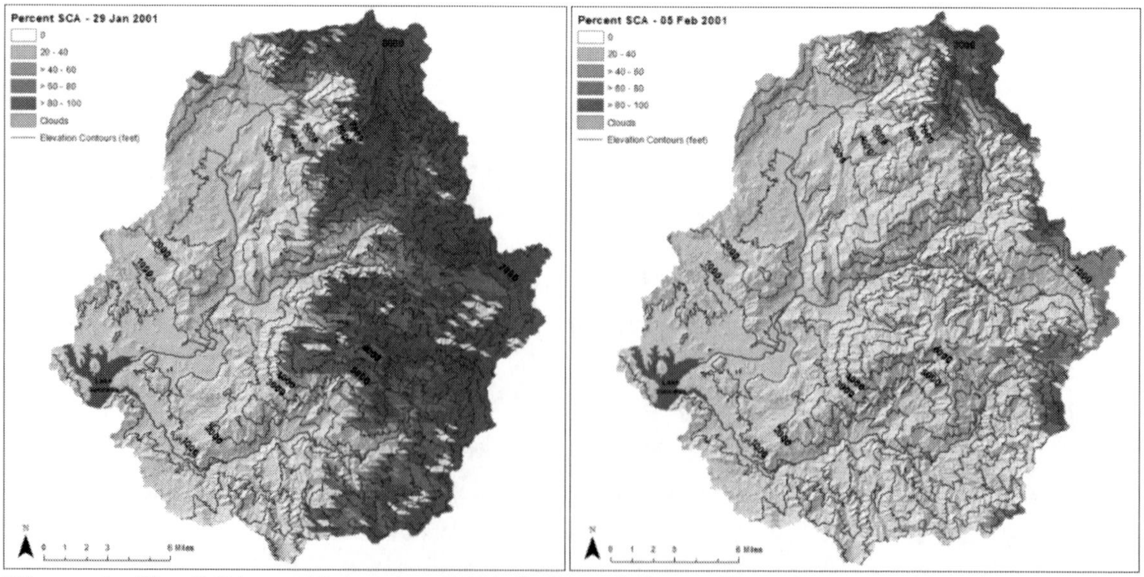

Figure 4. The SCA on 29 January 2001 (left) and Seven Days Later on 05 February 2001 (right)

The estimated SCA for 12 January 2005 is presented in Figure 5, which is one of days when the largest watershed SCA was recorded. The pixels marked as cloud covered indicate that the MODIS cloud mask interpreted the conditions over these pixels as cloud covered for 16 consecutive days or more. Note that some areas at higher elevations are snow free and that these areas match snow free areas recorded on 29 January 2001 (Figure 4).

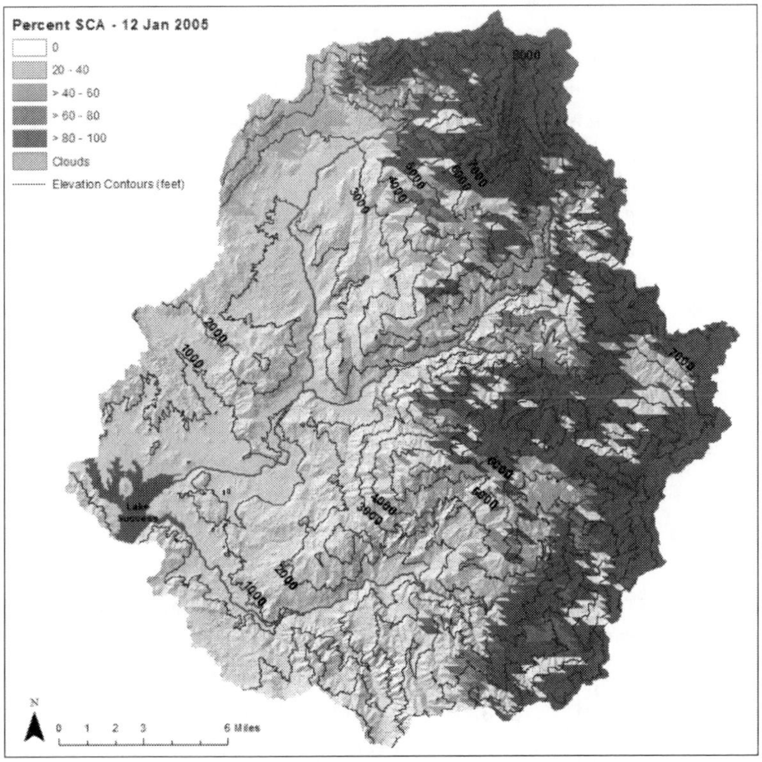

Figure 5. The SCA on 12 January 2005

The estimated SCA for 18 February 2009 is presented in Figure 6, which also had one of the largest watershed SCA recorded. In large part, the SCA follows the elevation contours, but not exclusively so. There are several instances where the SCA extends downhill in isolated salients, essentially cutting across the elevation contours. These downhill SCA projections may result from spatial variations in the distribution of snowfall that are not entirely coincidental with the contour locations, as has been shown to generally be the case for SCA in the watershed.

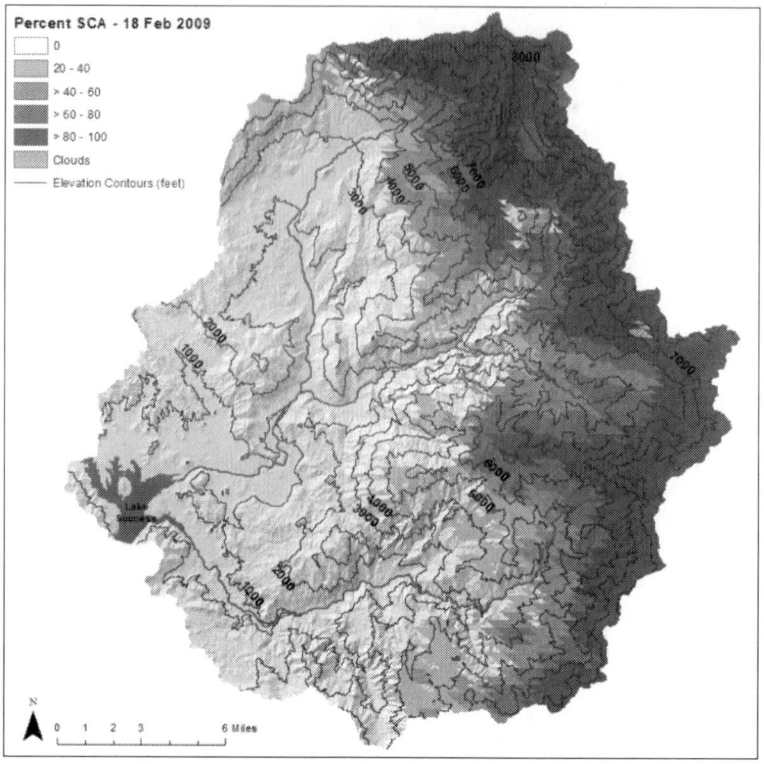

Figure 6. The SCA on 18 February 2009

GENERAL SCA DISTRIBUTION

The general distribution of SCA throughout the watershed was analyzed by calculating the longest duration each winter that each pixel was continuously covered by snow (fSCA>20%). These durations were averaged over the period of record and mapped (Figure 7), which includes the elevation contours, showing that the distribution of SCA very closely follows these contours. This close agreement of the SCA duration and the contours is taken as a partial confirmation of the TDF process. The total watershed average daily SCA (Figure 3) was used to estimate the elevation of the average effective daily snow line. Here, the effective snow line is defined by the elevation above which the area equals the total watershed SCA. The elevation information in Table 1**Error! Reference source not found.** was used to interpolate the elevation that corresponded to the area of the total watershed SCA of each day. This is a rough measure of the snow line, but the close relation of the snow line and the 32°F isotherm can be seen in Figure 8. These close results also suggest that the TDF process provided reasonable and accurate results.

was used to interpolate the elevation that corresponded to the area of the total watershed SCA of each day. This is a rough measure of the snow line, but the close relation of the snow line and the 32°F isotherm can be seen in Figure 8. These close results also suggest that the TDF process provided reasonable and accurate results.

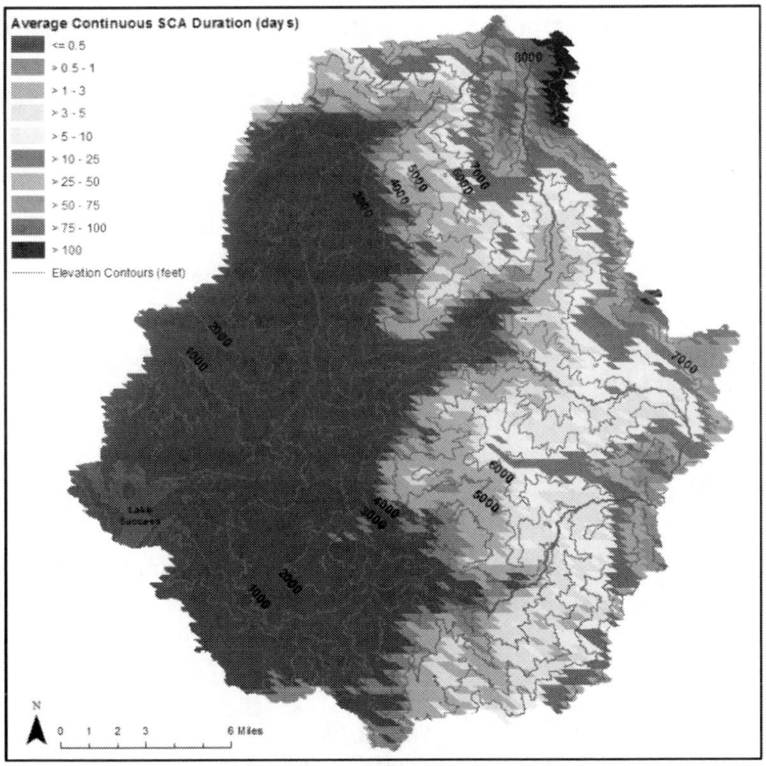

Figure 7. The Average Continuous SCA Duration Each Winter by Pixel

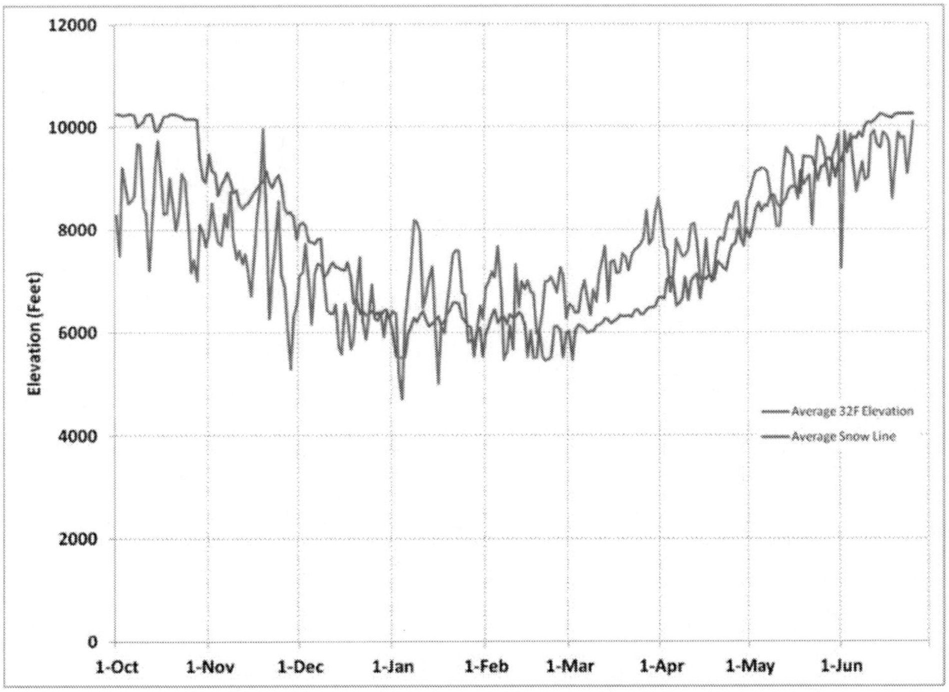

Figure 8. The Average Elevation of the 32°F isotherm and the Snow Line

The calculated the SCA for each elevation band was calculated for each day over the period of record, having determined the elevation for each pixel as described earlier. The SCA for an elevation band on day i, SCA^i_{eb}, was

$$SCA^i_{eb} = \frac{1}{n}\sum_{j=1}^{n} fSCA^i_j \tag{1}$$

where

n = the number of pixels in the elevation band and
$fSCA_j$ = the fSCA of the jth pixel.

The SCA statistics for each day of the year was computed by using the "Cyclic Analysis Statistics Math Tool" of HEC-DSSVue (USACE 2009). The daily average SCA for each elevation band (Figure 9) drops rapidly as the elevation band elevation decreases, which agrees with SCA distribution in Figure 7. The daily 95th percentile SCA for each elevation band (Figure 10) is greater than 95% of the SCA recorded on that day of the year. The 95th percentile SCA, while greater in value than the average, also drops rapidly as the band elevation decreases, especially below 7000 ft.

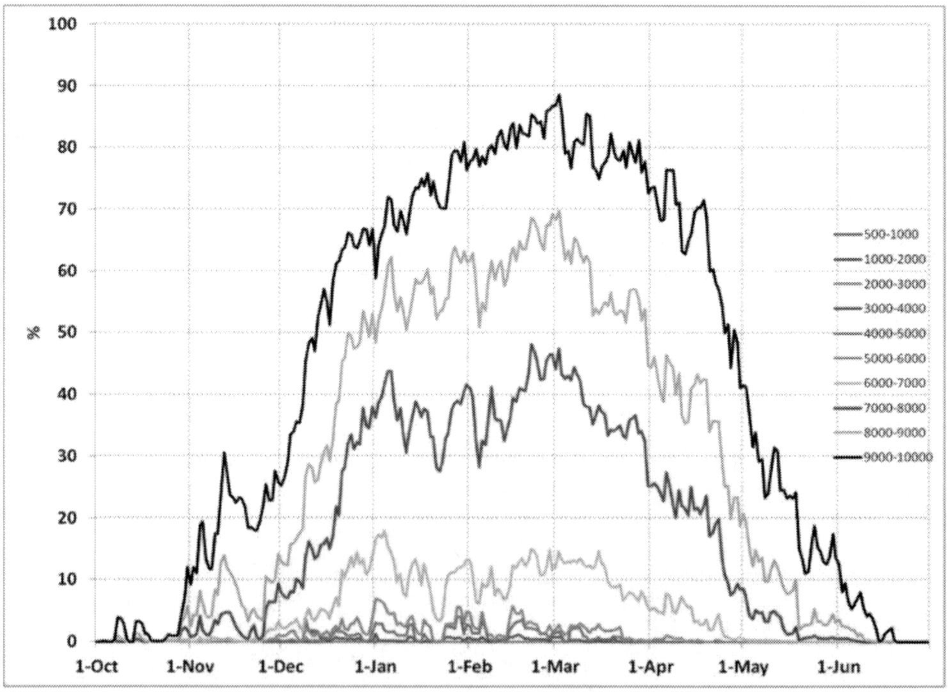

Figure 9. The Average SCA on Each Day of Winter for Each Elevation Band

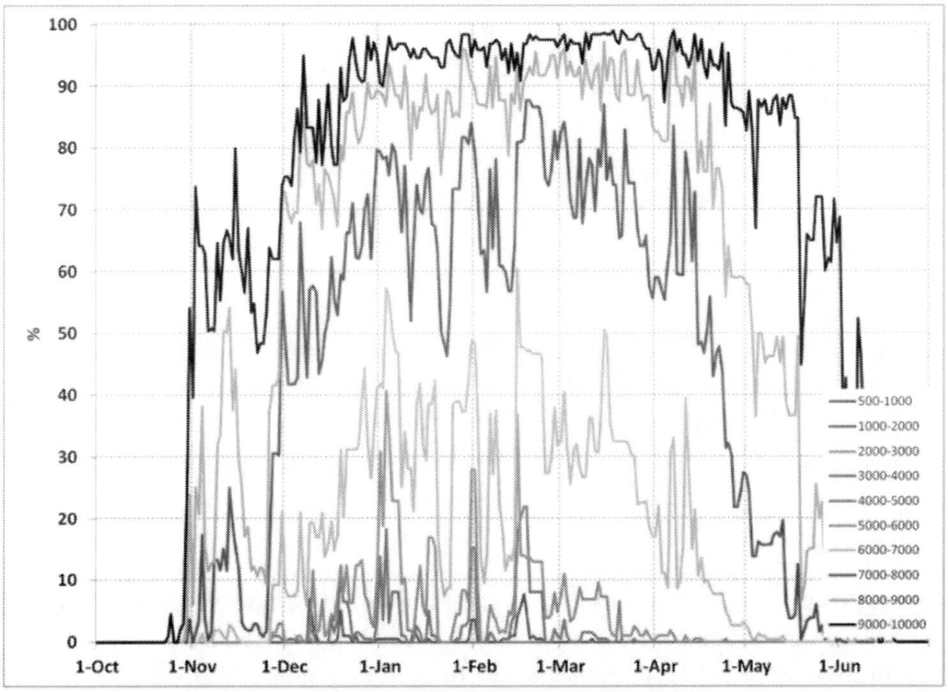

Figure 10. The 95th Percentile SCA on Each Day of the Winter for Each Elevation Band

APPLICATION OF THE SCA TO THE PMF DETERMINATION

Four possible SCA scenarios for selecting the SCA for the elevation bands for a PMF were explored (also shown in Table 1, Figure 11, and Figure 12). The 'PMF' column in Table 1 is a reference column and reflects the assumption of 100% elevation band coverage in the original PMF.

- **SCENARIO A**: This scenario is based upon average SCA in each band that was observed on the day of the maximum SWE observed at the Quaking Aspen precipitation gaging station (QUA) over the period from 2000-2014. The basis of this was that it is likely that extreme events will happen at or around the time of maximum SWE. The annual SCAs were averaged to arrive at a final value for each elevation band. These results were not particularly conservative, as the total watershed SCA of this scenario is only greater than the total watershed SCA observed on 82% of the days with a total non-zero days of SCA.

- **SCENARIO B**: This scenario also reflects the maximum SCA in each band that was observed on the day of the maximum SWE observed at the Quaking Aspen station (QUA) for the years 2000 through 2014. This scenario is similar to Scenario A except that the maximum SCA, instead of the average SCA, observed on the date of the maximum SWE was selected. These results are more conservative than Scenario A, as the total watershed SCA of this scenario is greater than the total watershed SCA observed on 96.3% of the days with a total non-zero days of SCA.

- **SCENARIO C**: This scenario reflects the average maximum 30-day SCA by taking the daily SCA for each elevation band and converted to a time step of 30 days with the value of each 30-day variable set equal to the average of the 30 days that it spanned. Then the maximum 30 day value was selected for each winter. The average over all winters was then calculated to arrive at the average maximum 30 day SCA. These results are more conservative than Scenario A, but slightly less than Scenario B. The total watershed SCA of this scenario is greater than the total watershed SCA observed on 94.1% of the days with a total non-zero days of SCA.

- **SCENARIO D**: This scenario reflects the maximum 95th percentile SCA in each elevation band (Figure 10). These results are more conservative than the other scenarios. The total watershed SCA of this scenario is greater than the total watershed SCA observed on 99.5% of the days with a total non-zero days of SCA.

Selection of SCA Scenario

The selection of the SCA scenario to use is complicated because of the highly dynamic nature of SCA in the Success Dam watershed as shown previously in Figure 1 and Figure 2. The accumulation of SWE requires time for successive snowfall events to accumulate. Highly ephemeral SCA responds to each snowfall event but quickly ablates before significant SWE can accumulate. The SCA scenario selected has to strike a balance between being sufficiently conservative in representing the largest area of the watershed that could be covered with snow when the PMF occurred and having sufficient time in place for the SWE to accumulate. The maximum 95th percentile SCA for each elevation band (Scenario D) was selected as being sufficiently conservative without representing transient events where some of the SCA had little SWE. Additional rationale behind selecting the most conservative scenario included the uncertainty associated with the limited number of years of record available for the SCA analysis (14 years), which did not cover the significant events of record in the watershed.

The total watershed SCA that results from selecting the maximum 95th percentile SCA for each elevation band is 101.6 square miles (26.03% of the watershed). This is greater than the total watershed SCA observed on 99.5% of the days with SCA greater than zero. The SWE volume that results from combining the SWE depth used in the PMF study with the 95th percentile SCA for each elevation band is 178×10^3 acre-ft. The prior distribution of SWE in the original PMF study, based upon 100% coverage in each elevation band down to 3000 ft, resulted in 232.7 square miles (59.7%) of the watershed covered with snow and a total SWE volume of 267×10^3 acre-ft.

Table 1. The SCA by Elevation and Results for Each Scenario

Elevation Band	PMF (%)	A (%)	B (%)	C (%)	D (%)
515–1000	-	0.0	0	0.0	0.0
1000–2000	-	0.0	0	0.0	0.0
2000–3000	-	0.0	0	0.0	0.8
3000–4000	100	0.0	0.5	2.9	7.7
4000–5000	100	0.1	0.5	6.8	22.0
5000–6000	100	1.6	9.7	11.2	40.7
6000–7000	100	18.0	45.4	27.4	60.6
7000–8000	100	54.0	83.0	64.9	87.6
8000–9000	100	74.8	97.4	84.8	97.7
9000–10,000	100	87.7	99.3	93.3	99.0
	PMF	A	B	C	D
Total SCA (square miles)	232.7	36.4	63.3	54.4	101.6
Total SCA (%)	59.7%	9.3%	16.2%	13.9%	26.0%
% Days Less Than or Equal to Total SCA	100%	82.0%	96.3%	94.1%	99.5%
Total SWE (acre-ft × 10^3)	266.94	91.09	144.69	115.87	177.63

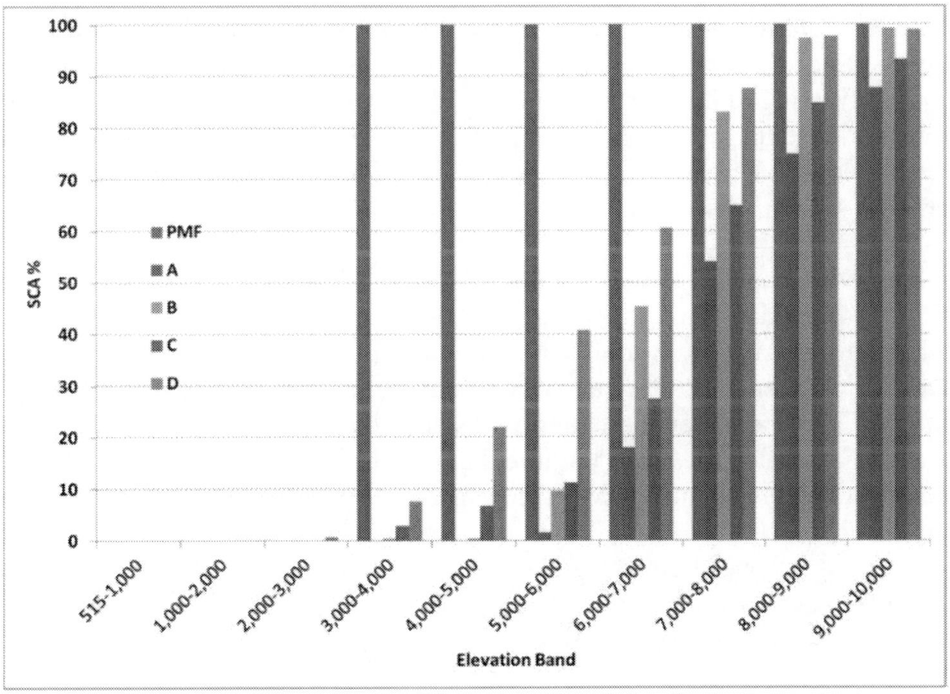

Figure 11. The SCA by Elevation Band for Each Scenario

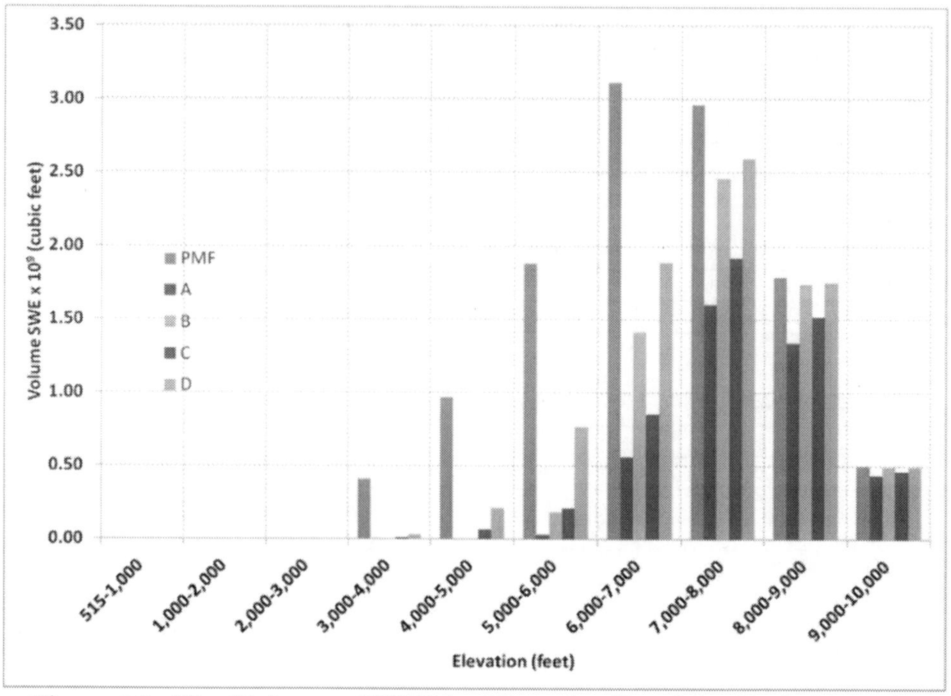

Figure 12. The Volume of SWE by Elevation Band for Each Scenario

SUMMARY AND CONCLUSIONS

ERDC-CRREL analyzed satellite imagery to determine the daily snow-covered area (SCA) of the Success Dam watershed from 2000 to 2014. This analysis was based upon time domain filtering (TDF) of NASA's daily MOD10A1 snow product. TDF results in virtually cloud-free daily snow maps displaying the fractional SCA (fSCA) of pixels not obscured by clouds. The pixel size is roughly 500 × 500 m. The Success Dam watershed is covered by 4710 pixels.

The CRREL SCA analysis found that the SCA of the Success Dam watershed is highly dynamic with large day-to-day variations. Various scenarios were considered for selecting the appropriate SCA to use in determining the initial SWE distribution and calculated the SCA statistics for each day of the winter season. The maximum 95th percentile SCA for each elevation band was selected as being sufficiently conservative without being extreme. The SWE volume that results from combining the SWE depth for each elevation band used in the PMF study with the 95th percentile SCA for each elevation band is 178×10^3 acre-ft. The peak pool elevation was reduced from the original PMF by 2 feet, which was a direct result of the reduced volume of SWE available (since the coverage of SWE in the watershed went from 100% to 59.7% and no temperature changes were made) for runoff in the final PMF routing.

REFERENCES

Ackerman, S., R. Frey, K. Strabala, Y. Lui, L. Gumley, B. Baum, and P. Menzel. 2010. *Discriminating clear-sky from cloud with MODIS algorithm theoretical basis document (MOD35), Version 6.1*. ATBD-MOD-06. Madison, WI: Cooperative Institute for Meteorological Studies, University of Wisconsin.

Anderson, E. A. 1973. *National Weather Service River Forecast System—Snow Accumulation and Ablation Model*. NOAA Technical Memorandum NWS-HYDR0-17. Washington, DC: U.S. Department of Commerce, National Oceanic and Atmospheric Administration, National Weather Service.

ESRI 2011. ArcGIS Desktop: Release 10. Redlands, CA: Environmental Systems Research Institute.

Hall D. K., G. A. Riggs, and V. V. Salomonson. 2006. *MODIS/Terra Snow Cover Daily L3 Global 500m Grid V005 (MOD10A1)*, March, 2000-June, 2014. Digital media. Boulder, Colorado USA: National Snow and Ice Data Center.

Morriss, B. F., E. Ochs, E. J. Deeb, S. D. Newman, S. F. Daly, and J. J. Gagnon. Submitted 2014. An operational snow covered area mapping algorithm for MODIS snow products. *Remote Sensing of Environment*.

Rice, R., R. C. Bales, T. H. Painter, and J. Dozier. 2011. Snow water equivalent along elevation gradients in the Merced and Tuolumne River basins of the Sierra Nevada. *Water Resources Research* 47:W08515. doi:10.1029/2010WR009278.

Salomonson, V. V., and I. Appel. 2004. Estimating fractional snow cover from MODIS using the normalized difference snow index. *Remote Sensing of Environment.* 89:351–60. http://dx.doi.org/10.1016/j.rse.2003.10.016.

U.S. Army Corps of Engineers (USACE). 2013. *Success Dam Safety Probable Maximum Precipitation (PMP)/Probable Maximum Flood (PMF) Report.* Sacramento, CA: Sacramento District, U.S. Army Corps of Engineers.

U.S. Army Corps of Engineers (USACE). 2010a. *Hydrologic Modeling System HEC-HMS User's Manual Version 3.5*. Davis, CA: Hydrologic Engineering Center.

U.S. Army Corps of Engineers (USACE). 2010b. *Statistical Software Package, HEC-SSP User's Manual Version 2.0*. Davis, CA: Hydrologic Engineering Center

U.S. Army Corps of Engineers (USACE). 2009. *HEC-DSSVue HEC Data Storage System Visual Utility Engine User's Manual Version 2.0* Davis, CA: Hydrologic Engineering Center

U.S. Army Corps of Engineers (USACE). 1998. *Tule River Basin, California, Success Dam Spillway Adequacy Study Hydrology Office Report.* Washington, DC:U.S. Army Corps of Engineers.

EVALUATION OF NEW STORAGE WITH OPTIMAL HYDROPOWER: WEISER-RIVER BASIN, IDAHO

Jeremy Giovando, PE [1]

ABSTRACT

The results of this analysis indicate there are several positive aspects to the Weiser-Galloway Project. These include increased flexibility for delivering flow for the lower Snake River; additional renewable energy sources for the Northwest; reduced flood risk for the city of Weiser; enhanced water recreation and additional water supply for the Weiser River basin. The analysis was a collaborative project completed by USACE, Idaho Department of Water Resources (IDWR), and other stakeholders, including Idaho Power Company (IPC). This study provides information about the potential operating criteria, economic benefits, and exchange of water supply associated with the Weiser-Galloway Project. For purposes of this document, only the summary results of the reservoir modeling are presented and should be considered as one example of how a new project in the Weiser River basin could be operated.

INTRODUCTION

Background

Water storage on the Weiser River at the Weiser-Galloway site has been studied for decades. USACE first received an authorizing study resolution for the Weiser River basin from the U.S. Senate Public Works Committee in 1954. In the early 1970s, Federal lands for the potential Weiser-Galloway Project were classified and withdrawn for hydropower purposes by the Federal Power Commission (now the Federal Energy Regulatory Commission). In 2008, IWRB was directed by the Idaho Legislature through HJM 8 to investigate water storage projects statewide, including the Weiser-Galloway Project.

In response to HJM 8, the IWRB initially partnered with USACE to publish the *Weiser-Galloway Gap Analysis, Economic Evaluation and Risk-Based Cost Analysis Project* (Gap Analysis), completed in March 2011 (USACE, 2011). This study addresses one of the main information gaps described in the previous report, specifically the operational details for a new storage project.

The results presented in this document are part of an ongoing investigation of potential new storage in the Weiser River basin. As such, the tables related to the reservoir modeling results only represent one option of how the new project could be operated to meet the assumed project purposes. These results do not represent final decisions on the project purposes or fully developed project benefits.

[1] Hydraulic Engineer, U.S. Army Corps of Engineers, Walla Walla District, Walla Walla, WA 99362, jeremy.j.giovando@usace.army.mil

Location

The potential dam site is located on the Weiser River near the mouth of a steep-sided canyon about 8.5 miles east of Weiser, Idaho (see Maps 1 and 2). The site is at Weiser River mile 13.5 at elevation 2200 (North American Datum 1983 [NAD83]) in Sections 25 and 26, T.11 N., R. 4 W., in Washington County, Idaho (latitude 44° 15' 18"N, longitude 116° 46' 21"W).

The topography at the potential dam site is a steep-sided, V-shaped canyon with no floodplain. Upstream of the potential dam site, the river turns northwest, and the canyon opens to a wider valley with a floodplain. Downstream of the site is a relatively wide flood plain with extensive agricultural development and the city of Weiser, Idaho.

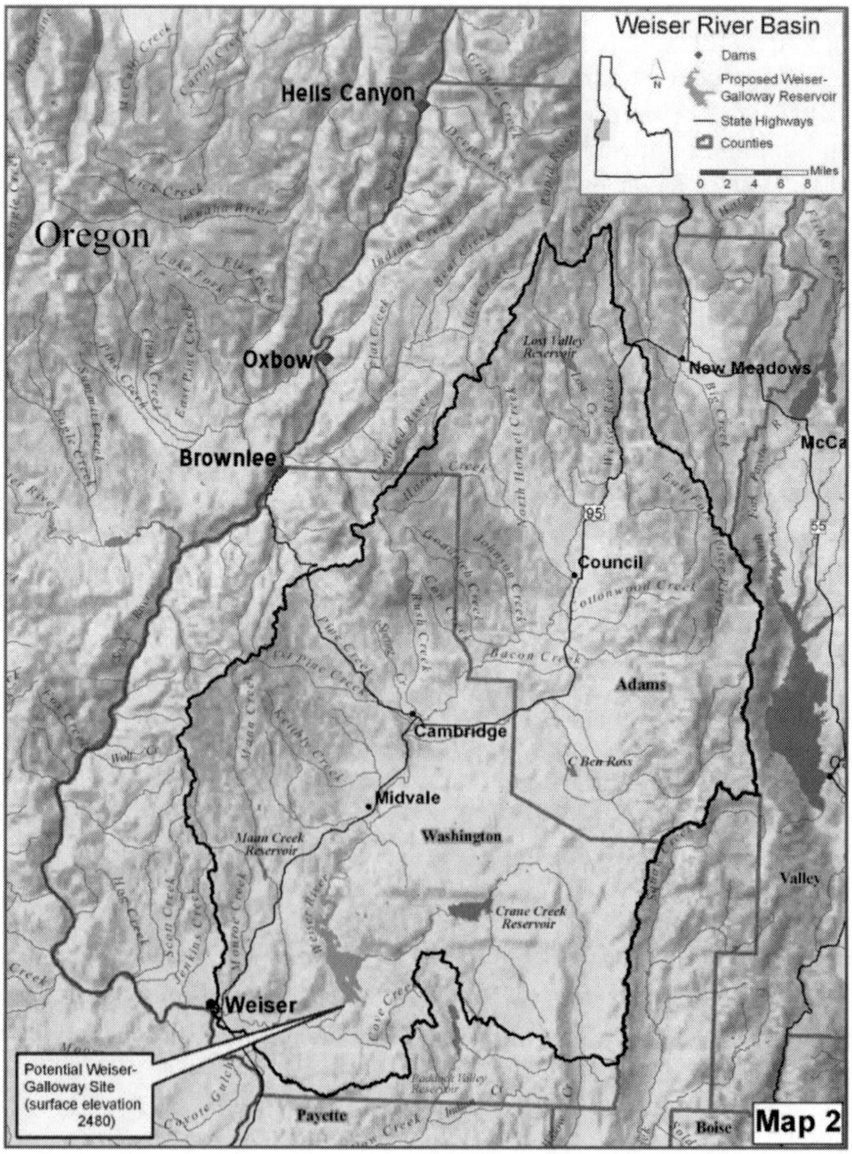

Map 1. Potential Project Location Shown Within the Weiser River Watershed

GENERAL MODEL DISCUSSION

Weiser-Snake River Optimization Modeling Process

The six operational alternatives were formulated for this study. Four of the alternatives included variations based on the level of additional water supply demand, which are termed "scenarios" for this analysis. Regardless of the alternative, the system modeling results were compared to an initial base condition, referred to as the Base Case alternative, using the same model configuration with no additional storage in the Snake River basin. This allowed for removal of any model bias and for comparison and evaluation of the expected "real" flow regime changes. A reference condition, referred to as the Reference Case alternative, was established for the Weiser River since the effects of storage from the Weiser-Galloway Project is not reflected in any observed records. These alternatives allowed comparison of other operational alternatives, relative to existing Snake River operations, as well as to potential Weiser-Galloway Project operations.

Models

RiverWare is a reservoir-simulation modeling software that is produced, developed, and supported by the University of Colorado, Center for Advanced Decision Support (CADSWES). This software is widely used by several Federal, state, and local agencies to evaluate reservoir system operations. Generally, RiverWare is utilized in one of two modes: to analyze real-time operations, or as a planning tool for new or existing projects. In this study the latter case is true.

Two state planning models were used collectively to evaluate the impact of the Weiser-Galloway Project on overall flows in the Snake River basin (together, referred to as the system model): SRPM and ESPAM. SRPM is a surface water model, and ESPAM is a groundwater model. In recent years, a combined modeling system was developed to interface between SRPM and ESPAM, allowing the two models to interact (see Section 2.3.2.1).

Hydrology

Historical hydrologic data for the Weiser River was obtained from the USGS streamflow record. For the purposes of this study, the assumed operations of upstream irrigation storage reservoirs (e.g., Crane Creek, Lost Valley, C. Ben Ross) would not significantly change after construction of the Weiser-Galloway Project. Therefore, the observed flows from the USGS record would generally reflect future depletions and operations from those projects.

Many of the reservoir model inputs datasets were not available from historical records. Any inputs that were not directly available were developed using statistical relationships. Specifically the data development was for daily flows of the Galloway and Sunnyside Canals. These datasets were devised using statistical methods to create a complete time series for the study period. For instance, the USGS canal data correlated to the observed runoff in the Weiser River, allowing development of an estimated diversion volume for each year of streamflow data. Then, using the IDWR water rights data, the statistically derived hydrographs were scaled to reflect current diversion levels. In other words, they were "normalized" to the level of diversions that would be expected currently from the Weiser River.

RESERVOIR MODELING RESULTS

The operational alternatives often had variations, based on a range of additional water supply demand, which are termed "scenarios" for this study (Table 1). Regardless of the alternative, the system modeling results were compared to an established base condition using the existing model configuration with no additional storage in the Snake River basin. For each alternative, varying Galloway Dam operations are evaluated and compared to a base condition (existing condition without Weiser-Galloway Project), or a reference condition (with Weiser-Galloway Project condition), depending on the location. Again, the Reference Case alternative is used for determining differences between alternatives for parameters at the Galloway Dam.

Although the modeling with the SRPA and ESPAM included many other projects in the Snake River basin, the results discussed in this paper will focus on the Weiser-Galloway Project results.

Table 1. Operational Alternatives

Operational Alternatives	**Scenarios**		**Plot Label**	**Description**
	Exchange Volume (kaf)	**Additional Water Supply Demands**		
Base Case	**NA**	**NA**	SRPM_brn_flow_ref	Existing Snake River Operations (without Weiser-Galloway)
Reference Case	**0**	**0**	**0_0**	Base Case, plus Weiser-Galloway
Zero Series	**0**	**25**	**0_25**	Reference Case, plus additional demands
	0	**50**	**0_50**	"
	0	**100**	**0_100**	"

200 Series	**200**	**0**	**200_0**	Zero Series, plus 200 kaf springtime water supply exchanged upstream of Milner Dam
	200	**25**	**200_25**	"
	200	**50**	**200_50**	"
	200	**100**	**200_100**	"
240 Series	**240**	**0**	**240_0**	200 Series, plus 40 kaf Boise River exchange
	240	**25**	**240_25**	"
	240	**50**	**240_50**	"
	240	**100**	**240_100**	"
200 LS Series	**200**	**0**	**200_0_LS**	200 Series, plus limited storage Apr-Sep
	200	**25**	**200_25_LS**	"
	200	**50**	**200_50_LS**	"
	200	**100**	**200_100_LS**	"

Weiser-Galloway Project Summary

The difference in seasonal discharge volume between alternatives for Galloway Dam is shown in Table 2. As discussed previously, the addition of water supply demand shifts the timing of the releases toward the current conditions. Overall, there is not a significant variation in flow regime for alternatives, which include water supply demand during the spring season. The most significant change within these alternatives is for scenarios that include 100 kaf of additional water supply. The additional water is released in December, which then results in much lower January flows, whereas releases with no additional demand result in approximately the same cumulative volume for December and January. This is primarily because the model is maximizing generation value, and December and January are relatively high generation value months.

Table 2. Weiser-Galloway Project Median Monthly Flow for Modeled Alternatives

Median Monthly Flow (cfs)												
Scenario	**Oct**	**Nov**	**Dec**	**Jan**	**Feb**	**Mar**	**Apr**	**May**	**Jun**	**Jul**	**Aug**	**Sep**
Historical	446	1030	1928	2318	2378	1144	300	238	183	158	208	317
0_0	141	50	402	648	1036	1095	338	1193	1359	2776	2358	420
0_100	135	55	850	652	981	1130	314	1213	1357	2848	2358	406
200_0	135	50	359	607	1036	1095	2246	2077	2102	816	792	222
200_100	135	50	850	81	964	1087	1773	2077	2102	1245	584	222
200_0_LS	134	76	363	538	1047	1092	2308	2603	1815	488	551	211
200_100_LS	134	50	846	68	802	1092	2308	2603	1647	1130	316	211
240_0	135	50	342	538	1036	1095	2246	2397	2266	507	321	222
240_100	135	50	850	81	964	1082	1624	2387	2210	1186	346	222

Note: Blue denotes the higher discharge values.

When considering the effects of various operating criteria for a reservoir, it is also important to evaluate the reservoir pool parameters. The results of the elevation duration shown in Table 3 suggest that providing 200 kaf of additional water demand during April through June will significantly affect the elevation at which the reservoir is at during any given year. This table was constructed using daily elevation data, and it should be expected that the reservoir would not be at or near full for long durations if maximizing hydropower generation is the underlying operational objective. Consequently, even without springtime water supply demand releases, the pool elevation would have very limited durations at full pool. With this in mind, the data demonstrates in nearly all scenarios for the majority of time (relatively small exceedance value) that the Galloway Reservoir would be less than elevation 2450 (shaded values). This indicates there may be other reservoir sizes that would provide approximately the same amount of water supply.

Table 3. Weiser-Galloway Reservoir Elevation-Duration for Modeled Alternatives

Galloway Reservoir Elevation (ft)								
	Percent Exceedance							
Percent	0_0	0_100	200_0	200_0_LS	200_100	200_100_LS	240_0	240_100
1	2470.0	2470.0	2470.0	2438.4	2470.0	2438.4	2468.4	2468.4
5	2470.0	2470.0	2449.3	2437.0	2449.3	2437.0	2445.9	2445.9
10	2467.2	2466.8	2441.0	2436.8	2440.5	2436.7	2438.4	2437.8
25	2444.3	2444.2	2433.1	2429.6	2430.7	2423.5	2431.4	2427.7
50	2422.6	2422.3	2418.0	2417.9	2417.1	2417.3	2417.4	2416.6
75	2417.1	2416.2	2416.1	2416.6	2407.7	2401.9	2415.0	2405.4
90	2415.5	2413.8	2406.3	2404.2	2393.7	2392.8	2400.6	2388.4
95	2414.9	2410.9	2399.1	2399.0	2383.2	2383.0	2397.1	2380.0
99	2401.9	2401.7	2395.2	2395.2	2380.7	2379.1	2389.7	2376.0

Note: Shaded values are less than elevation 2450.

The refill reliability shown in Table 4Table indicates that reliability is relatively high with the Zero Series alternative, but decreases substantially when spring water supply demand is added in the 200 and 240 Series alternatives. This confirms the aforementioned size consideration from the duration table.

Table 4. Weiser-Galloway Reservoir Refill Reliability for Modeled Alternatives

Galloway Reservoir Annual Maximum Pool Elevation (ft)								
	Scenario							
WY	0_0	0_100	200_0	200_100	200_0_LS	200_100_LS	240_0	240_100
1995	2470.0	2470.0	2450.1	2450.1	2437.7	2437.1	2446.1	2446.1
1996	2470.0	2470.0	2454.9	2454.9	2437.0	2437.0	2451.6	2451.6
1997	2470.0	2470.0	2449.6	2449.6	2449.6	2449.6	2449.6	2449.6
1998	2470.0	2470.0	2470.0	2470.0	2437.0	2437.0	2467.7	2467.7
1999	2470.0	2470.0	2456.1	2456.1	2437.0	2437.0	2450.4	2450.4
2000	2470.0	2470.0	2437.7	2437.7	2437.0	2437.0	2437.7	2437.7
2001	2441.2	2438.4	2437.1	2426.1	2432.7	2420.2	2437.1	2425.5
2002	2470.0	2470.0	2442.5	2437.0	2437.0	2419.7	2442.9	2437.0
2003	2470.0	2470.0	2442.9	2442.9	2434.4	2434.6	2439.0	2439.0
2004	2461.9	2462.8	2438.2	2438.2	2437.0	2437.0	2438.2	2438.2
2005	2467.4	2466.9	2440.3	2438.8	2433.1	2422.1	2437.8	2432.6
2006	2470.0	2470.0	2470.0	2470.0	2438.5	2438.5	2470.0	2470.0
2007	2435.1	2435.9	2424.3	2424.3	2424.3	2424.3	2424.3	2424.3
2008	2470.0	2470.0	2449.2	2446.9	2421.2	2418.1	2442.9	2440.1
2009	2452.6	2453.5	2418.2	2418.2	2418.1	2418.1	2418.2	2418.2
Refill Yrs	10	10	2	2	0	0	1	1
Total Yrs	15	15	15	15	15	15	15	15
% Total Yrs	66.7%	66.7%	13.3%	13.3%	0.0%	0.0%	6.7%	6.7%

Overall, the alternative that maximizes totally hydropower benefits is the Zero Series alternative. Specific scenarios in the 200 Series alternative may yield relatively high overall benefits and meet the water supply objective of the new project. There are additional potential benefits with the 240 Series; however, the complexities of quantifying the real value of water in another basin may require significantly more time and resources than just exchanging water only with the Snake River above Milner Dam.

When considering the 200 Series alternative, the amount of additional water that is available can be further refined. The initial results suggest that more than 50 kaf of additional water (beyond the release for springtime water supply) will noticeably change the pool elevation probability. This could potentially affect at-site benefits such as hydropower generation and recreation.

Weiser-Galloway Project Hydropower Generation Summary

There is significant benefit potential with the at-site hydropower.

Table 5 summarizes the average generation for the Weiser-Galloway Project for each alternative. Again, the total annual generation reflects the difference in flow regimes between alternatives. Limiting the storage season would have a significant impact on generation. Expectedly, generation in most months was reduced between the 200 and 240 kaf flow water supply volume alternatives.

Table 5. Weiser-Galloway Project Median Monthly Energy for Modeled Alternatives

Weiser-Galloway Project Median Monthly Energy (GWh)													
Scenario	**Oct**	**Nov**	**Dec**	**Jan**	**Feb**	**Mar**	**Apr**	**May**	**Jun**	**Jul**	**Aug**	**Sep**	**Annual Total (GWh)**
0_0	2.13	1.25	8.44	9.52	10.95	14.83	7.21	16.39	13.59	29.21	22.41	3.67	139.6
0_100	2.11	1.30	12.23	7.50	9.91	14.76	6.67	16.21	13.43	30.57	21.16	3.39	139.3
200_0	1.64	1.03	6.14	8.68	10.41	14.66	22.05	25.41	21.17	12.01	9.67	2.47	135.3
200_100	1.53	0.90	10.75	6.22	8.70	11.77	20.22	24.81	20.79	16.41	9.06	2.39	133.5
200_0_LS	1.40	1.21	6.00	8.48	10.06	15.22	22.92	27.77	19.61	6.04	8.33	2.23	129.3
200_100_LS	1.30	1.06	9.68	5.92	8.42	11.88	22.58	26.80	18.45	12.52	6.12	2.15	126.9
240_0	1.29	0.90	5.82	8.23	9.92	14.54	21.88	28.47	22.87	9.10	8.97	2.43	134.4
240_100	1.25	0.93	10.06	6.04	8.54	10.87	19.97	27.78	21.88	14.87	7.76	2.27	132.2

Note: Green denotes higher generation values.

CONCLUSION

Based on this investigation, the Weiser-Galloway Project may provide significant benefits for the state of Idaho and Pacific Northwest. The magnitude of benefits depends on the operational alternative implemented with the project. The results of this analysis indicate there are several positive aspects to the Weiser-Galloway Project. These include increased flexibility of delivering springtime water supply demand for the lower Snake River, hydropower benefits, reduced flood risk, and enhanced water recreation for the Weiser River basin.

The Reference Case alternative would provide the most hydropower benefits of the alternatives modeled. This is due primarily to the majority of volume released from the reservoir occurring in July and August and January through March, when there are higher generation values. The regulated flow regime shifts when water is exchanged from the Weiser River with the upper Snake River and the Boise River. This requires releases during a sub-optimal hydropower period in May and June. The relative generation value difference (compared to the Reference Case alternative) by constraining Galloway Dam to release a significant volume during springtime is approximately 10 percent on average. This becomes the tradeoff between the maximized hydropower operation and the water supply exchange operation.

Additional benefits may be realized in other parts of the Snake River basin depending on the timing and availability of water from the Weiser-Galloway Project. Further analysis will be needed to more accurately quantify these benefits.

REFERENCES

Idaho House of Representatives Resources and Conservation Committee. (2008). 59th Legislature, 2d Sess. *House Joint Memorial No. 8.*

Public Law 93-251, Section 22. (1974). *Water Resources Development Act.*

State of Idaho. (1978). Idaho Code § 42-1736B. *Water Resource Policy Actions.* Boise, Idaho.

State of Idaho. (2012). *The State Water Plan.* Boise: Idaho Water Resource Board.

University of Colorado, Boulder. (n.d.). *RiverWare.* Retrieved May 2015, from Center for Advanced Decision Support for Water and Environmental Systems: http://cadswes.colorado.edu/creative-works/riverware.

University of Idaho. (2012). *Evapotranspiration and Consumptive Irrigation Water Requirements for Idaho.* Retrieved May 2015, from ET Idaho: http://data.kimberly.uidaho.edu/ETIdaho/.

USACE. (1990). *Galloway Dam and Reservoir, Weiser River, Idaho Weiser River Basin Interim Upper Snake River and Tributaries: Technical Report.* Walla Walla: USACE, Walla Walla District.

USACE. (2010). *HEC-RAS River Analysis System User's Manual Version 4.1.*

USACE. (2011). *Weiser-Galloway Gap Analysis, Econcomic Evaluation, and Risk-Based Cost Analysis Project.* Boise: U.S. Army Corps of Engineers.

USACE. (2013). *Foundation Investigation and Evaluation, Weiser-Galloway Potential Dam Site, Weiser, Idaho.* Boise: U.S. Army Corps of Engineers.

USACE. (2013). *HEC-ResSim Reservoir System Simuation User's Manual 3.1.*

FIELD TESTS OF PLACED RIPRAP AS EROSION PROTECTION AGAINST OVERTOPPING AND LEAKAGE

Priska Helene Hiller[1]
Fredrikke Kjosavik[2]
Leif Lia[3]
Jochen Aberle[4]

ABSTRACT

Accidental overtopping or leakage can have fatal consequences for the stability of embankment dams. The downstream slopes of rockfill dams in Norway have to be protected with placed riprap as prescribed by regulations, to increase their resistance to erosion. Placed riprap consists of stones which are placed one by one in an interlocking pattern. This paper presents preliminary results from field tests carried out to investigate the resistance of placed riprap on steep slopes. Tests were also made using dumped riprap for comparison. The temporary test site was situated at the outlet channel of a spillway tunnel and water was discharged from the reservoir. For the tests, three 12-m wide and 3-m high permeable dams were specifically built and secured by riprap with a stone size of 0.4 m. The tests showed that the constructed placed ripraps could, dependent on the boundary conditions, withstand unit discharges of 8 $m^3/s/m$. A reference test with dumped riprap withstood approximately 1 $m^3/s/m$, i. e. 1/8 of the unit discharge for placed riprap. The results show that placing riprap in an interlocking pattern increases its stability considerably compared to random placement.

INTRODUCTION

Riprap is a widely used measure for erosion protection for dams and their auxiliary structures. The upper part of the upstream face of rockfill dams is usually reinforced with riprap to protect the slope against the impact of waves. In Norway, a riprap is also required on the downstream slope of rockfill dams to enhance the stability taking into account accidental loads such as overtopping or leakage. For this purpose, riprap stones must be placed in an interlocking pattern with their longest axis inclined towards the dam (OED, 2009). This is commonly defined as "placed riprap" in contrast to riprap with randomly dumped stones.

The stability of both placed and dumped riprap has been in the focus of research for a long time due to its importance for practical applications. In the following, we provide a

[1] PhD Candidate, Department of Hydraulic and Environmental Engineering, Norwegian University of Science and Technology, NTNU Trondheim, S. P. Andersens veg 5, NO-7491 Trondheim, +47 73 59 51 57, priska.hiller@ntnu.no
[2] Student, Department of Hydraulic and Environmental Engineering, Norwegian University of Science and Technology, NTNU Trondheim, S. P. Andersens veg 5, NO-7491 Trondheim, fredjoh@stud.ntnu.no
[3] Professor, Department of Hydraulic and Environmental Engineering, Norwegian University of Science and Technology, NTNU Trondheim, S. P. Andersens veg 5, NO-7491 Trondheim, leif.lia@ntnu.no
[4] Professor, Department of Hydraulic and Environmental Engineering, Norwegian University of Science and Technology, NTNU Trondheim, S. P. Andersens veg 5, NO-7491 Trondheim, jochen.aberle@ntnu.no

brief review of more recent investigations in which details on studies carried out before 2000 can be found. Abt et al. (2013) evaluated overtopping riprap design relationships on the basis of published data from 10 studies. The analysed studies were carried out with median stone diameters of 0.015 m $\leq d_{50} \leq$ 0.254 m and slopes of $0.02 \leq S \leq 0.50$. Based on their analysis, they concluded the further need for data exceeding those ranges because extrapolation of the existing relationships is limited and has to be done with care, especially for steep slopes. Furthermore, stones of $d_{50} \geq$ 0.25 m are often used in practical engineering applications such as erosion protections in rivers or auxiliary structures close to dams. Khan & Ahmad (2011) and Thornton et al. (2014) used multivariable power regression analysis to find an empirical equation based on collected data points and thus to facilitate the design of riprap. They presented an equation to determine d_{50} as a function of the riprap layer thickness, slope, coefficient of uniformity of the riprap material and the unit discharge. Thornton et al. (2014) included in addition the stone specific gravity in their equation. However, the mentioned studies were limited to data obtained with dumped riprap on gentle slopes (only 10 % of data with slopes $0.4 \leq S \leq 0.5$). Peirson et al. (2008) found that placing riprap increases its stability by 30 % compared to dumped riprap based on model tests with $0.2 \leq S \leq 0.4$. Dornack (2001) investigated placed riprap on steep slopes of up to $S = 0.67$ and described stabilizing friction forces between interconnecting riprap stones. The stability of the riprap increases because the normal forces and thus the friction in the interconnections increases with increasing slope. Riprap stability is often expressed in terms of the stone-related Froude number Fr_s

$$Fr_s = \frac{q}{\sqrt{g \cdot d^3}} \qquad (1)$$

where q denotes the unit discharge, d the stone diameter (usually $d = d_{50}$) and g = 9.81 m/s² the gravitational acceleration. Based on model tests, Dornack developed a relationship to calculate a critical stone-related Froude number $Fr_{s,cr}$, which can be used for dimensioning placed riprap for example for spillways constructed as overtoppable sections on embankment dams.

This paper presents preliminary results from a research project focusing on the development of improved design criteria for placed riprap on steep slopes with an inclination of 1:1.5 (vertical:horizontal), corresponding to a slope of $S = 0.67$. The research is based on both scale model studies and innovative large-scale experiments in the field. The latter were executed in 2012, 2013 and 2015. The experiments from 2012 and 2013 were documented in Lia et al. (2013) and Hiller & Lia (2015). The scope of the present paper is to document and report experience from the 2015 tests. They are described in this paper with focus on measuring devices. Tests were also made using dumped riprap for comparison. Preliminary results are included.

EXPERIMENTAL SETUP AND MEASURING PROCEDURE

The field site was located in a mountainous area in Southwestern Norway close to the 130 m-high rockfill dam, Svartevatn dam, owned by Sira-Kvina Power Company. The tests were run in the outlet channel from the tunnel spillway. Water was discharged from the

Svartevatn reservoir (fully supply water level FSWL = 899 m. a. s. l.) through a gate located 44 m below FSWL. The discharge could thus be estimated through the gate opening and the water level in the reservoir using the calibration curves of the gate (Vassdrags- og havnelaboratoriet ved NTH, 1973). Additionally, the discharge was validated by measurements with an Acoustic Doppler Current Profiler, ADCP (StreamPro ADCP). We note that the test site was located about 700 m downstream of the gate, so that changes in discharge at the gate became effective at the test site with a certain time lag.

Three approximately 3-m high and 12-m wide test dams (although about 9.5 m wide close to the foundations) were built as shown in Figure 1. Ideally, the dam slope with riprap exposed to overtopping should be longer, but it was limited due to the terrain at the site. Placed riprap was investigated in the first and the second test. The third test was run with dumped riprap.

Figure 1. Test dam 2 at Q = 6 m³/s. The ultrasonic sensors are marked with circles, the mounting structure of the pressure cell with an arrow, the ADCP measurement device is inside a triangle and the two video cameras are highlighted with rectangular boxes. One colored stone equipped with a pressure sensor and a Smartstone probe is scaled up on the left side. (Photo: NTNU).

The dams consisted of a permeable coarse rockfill and riprap on the dam crest and on the downstream slope (inclination of 1:1.5). Large boulders supported the upstream slope (inclination approximately 1:1). The first layers of riprap stones on the downstream slope and on the sides consisted of larger stones to create a well-defined test area and to avoid disturbance from the rough walls of the channel. Angular riprap stones with d_{50} = 0.4 m were used in the experiments. They were about half of the usually used rock size for placed riprap on the downstream slopes of rockfill dams in Norway. The grain size distribution of the riprap stones and the support fill is shown in Figure 2 and based on measurements of the stone axes and the density. The longest, a, intermediate, b, and shortest, c-axis were measured with a meter or caliper, dependent on the size of the stone, and the equivalent stone diameter d was calculated with

$$d = \sqrt{a \cdot b \cdot c} \tag{2}$$

The weight of the stone W was estimated using the relationship

$$W = d^3 \cdot \rho_s \cdot C_f \tag{3}$$

with the rock density $\rho_s = 2750$ kg/m³ and a form factor $C_f = 0.48$. The inclination angle α of the a-axis towards the dam of the riprap stones was measured with an inclination meter (Leica DISTO™ X310).

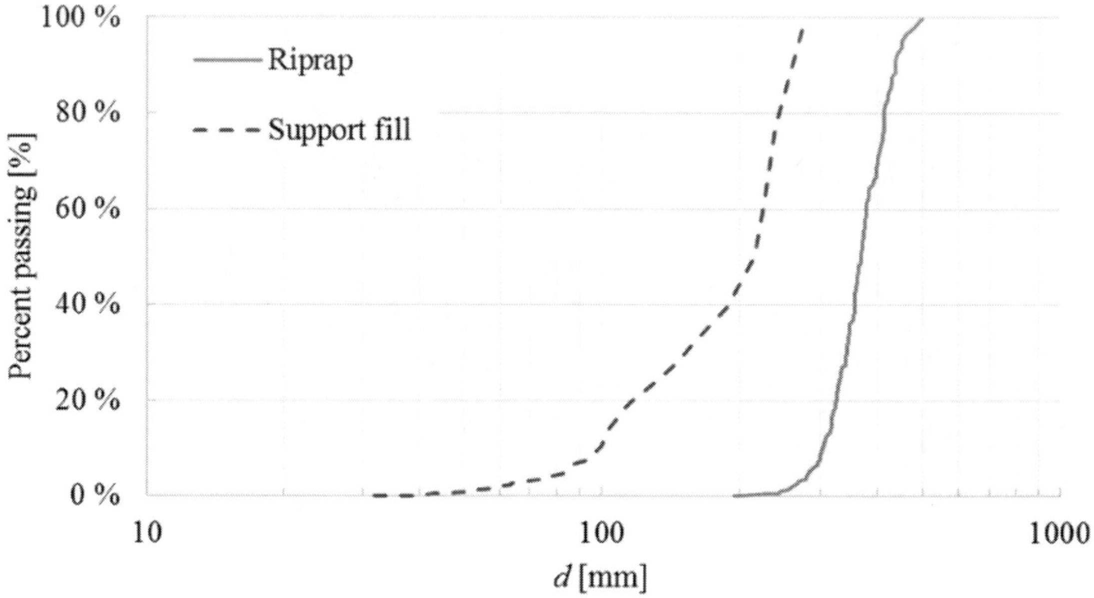

Figure 2. Grain size distribution of the riprap and the supporting fill material. The coefficient of uniformity $C_u = d_{60}/ d_{10}$ for the riprap material is $C_u = 1.24$. Note that the x-axis is in $\log_{10}$ scale.

A raster was sprayed on the test dams to facilitate visual monitoring. Two riprap stones equipped with sensors (described below) were colored to facilitate recovery after dam failure. The test dams were laser scanned before hydraulic loading. The first and the second dam were also scanned during the break between the hydraulic loadings (see description below). The coordinates of the dam corner points were measured with a GPS. All tests were monitored with three cam recorders (two of them are highlighted with rectangles in Figure 1).

The test dams with placed riprap were overtopped by a discharge of 20 m³/s (gate opening 0.4 m) for one hour. Following this initial overtopping, the discharge was stopped to inspect the dam and the riprap. After the inspection, the gate was opened 0.1 m

(6 m³/s) until stable stage was reached, followed by 15 minutes with an gate opening of 0.4 m. Afterwards, the discharge was stepwise increased in 5 minute intervals by increasing the gate opening by 0.1 m in each step (corresponding to an increase in discharge of 5 m³/s). The discharge was increased until the dam failed. Figure 3 exemplarily presents the discharge over time for test 2. In this paper, we consider failure as the state when the riprap is eroded progressively and the support fill is exposed to erosion. Hence, the erosion of a single riprap stone is not considered as dam failure.

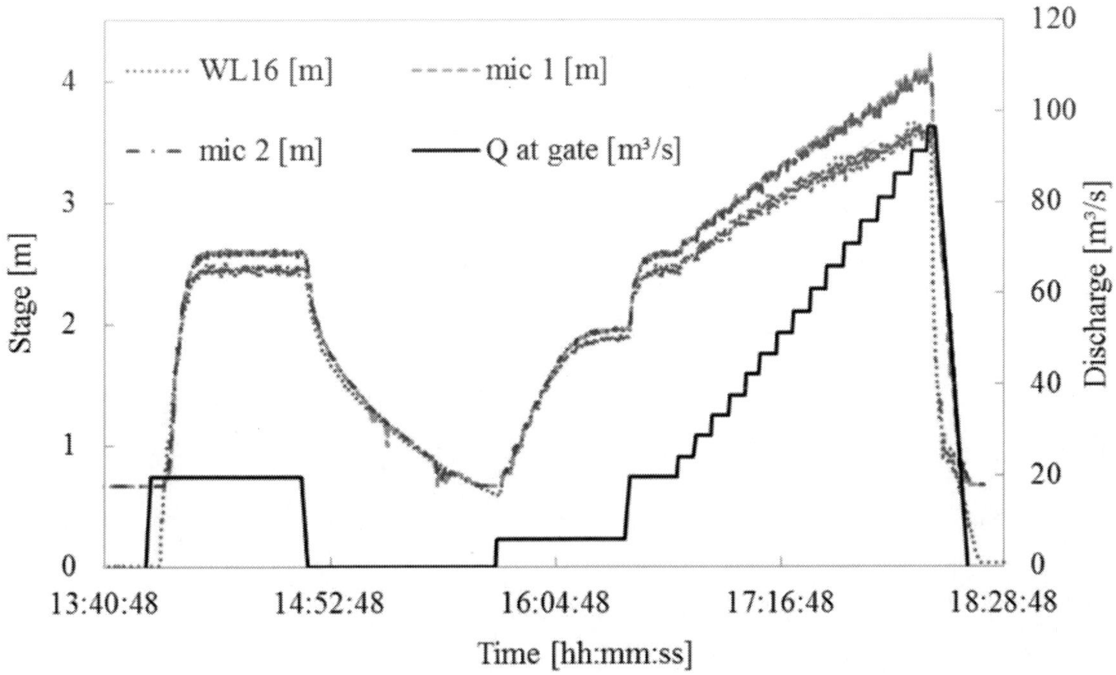

Figure 3. Measured stage (left y-axis) monitored with the pressure cell (WL16) and the ultrasonic sensors (mic 1 and mic 2) and discharge at the gate (right y-axis) for test 2. There is a time lag between the discharge at the gate and the stages measured at the test site. Waves below the ultrasonic sensors affected their measurements and resulted in an increasing deviation compared to the measurements with the pressure cell.

In test 3, dumped riprap was exposed to an initial discharge of 2.5 m³/s (0.05 m gate opening). The initial low discharge was chosen to evaluate the amount of discharge flowing through the dam. After reaching quasi stable stage with a gate opening of 0.05 m and 0.1 m, the discharge was stepwise increased until dam failure.

Stage was measured on three places upstream of the dam (highlighted in Figure 1): A pressure cell (Global Water WL16) was mounted beside the outlet where the velocity was low, determining stage every 10 seconds. Two ultrasonic sensors (microsonic mic+340) monitored the water surface above the tunnel outlet with 10 Hz frequency and stage could be determined from their data. Pressure sensors (Schlumberger Water Services, Mini-Diver, DI501), usually used in groundwater wells, were installed under the dam and in two riprap stones to monitor the stage over the downstream slope of the dam. They were mounted in drilled holes, fixed in a steal band which was bolted in the rock at the side. Three pressure sensors were placed under the dam and one immediately downstream of the dam. Two additional sensors were mounted in riprap stones, one close to the dam toe and one at the transition between dam crest and slope, respectively (see scaled up stone in Figure 1).

The experiments were also used to test the applicability of a novel sensor, a so called Smartstone, under field conditions. Hence, the riprap stone on top of the dam was in addition to the pressure sensor also equipped with a Smartstone probe. The probe can measure acceleration, orientation and rotation and uses radio communication as well as radiofrequency identification (RFID) (see Gronz et al. (2015) (article in review) for detailed description of the sensor).

PRELIMINARY RESULTS AND DISCUSSION

In the tests with placed riprap, the first stones were eroded at unit discharges between 4 and 5 $m^3/s/m$ while complete failure of the dams occurred for larger unit discharges of 6 and 8 $m^3/s/m$. In order to provide an impression of the flow conditions, Figure 4 shows test dam 2 before and during the test at the critical discharge.

Figure 4. Test dam 2 before (left) and during the test at a unit discharge of 8 $m^3/s/m$ (right). (Photo: S. R. Erland Skilnand).

Compared to the placed riprap, the dumped riprap failed when increasing the unit discharge from 0.5 to 1 $m^3/s/m$, i. e. at a significantly lower discharge. It is interesting to note that the dam failure coincided with the erosion of the first stone. Thus, the results show that placing the riprap stones in an interlocking pattern increased the stability with a factor of 6 to 8 and the factor for erosion of the first stones was approximately 4. The results are summarized in Table 1 together with the results of the field tests executed in 2012 and 2013. The results from 2015 will be compared with the results of previous field tests as well as with methods from literature after some comments about the measuring equipment and its performance during the tests.

Water levels during the experiments were accurately determined by both the pressure cell and the ultrasonic sensors. The stage fluctuated less where the pressure cell was mounted and the data from the pressure cell can therefore be considered as more representative than the data of the ultrasonic sensors. Waves in the tunnel opening affected the water stage and thus the measurements of the ultrasonic sensors. This explains why the deviation between the ultrasonic sensors and the pressure cell increased with increasing discharges as apparent in Figure 3. The pressure sensors, installed below the dam, worked well to measure the water level along the dam for all the tests. However, one sensor was damaged during dam failure in the first test and was not replaced for the subsequent tests. The measuring system for monitoring pressure below was improved compared to 2013 (described in Hiller & Lia, 2015). The pressure sensors placed in the riprap stones measured all three tests, although one had to be replaced with a new sensor because its pressure cell burst during the first test. The novel Smartstone sensor did unfortunately not measure during dam failure. The sensor needs an impact, i. e. it has to be moved a bit, to trigger switching between the configuration and the recording mode and that was not achieved. Hence, it has to be further developed before it can be used in the field in a proper way.

Table 1. Test results of the field tests in 2012, 2013 and 2015 with the corresponding stone-related Froude number. Test 2 in 2013 (in italics) failed due to instability in the dam toe and is therefore excluded from the discussion.

Test no.	Riprap size d_{50} [m]	Inclination angle α [°]	Unit failure discharge q [m³/s/m]	Stone-related Froude no. Fr_s [-]
1, 2012	0.65	random	2.1	1.3
2, 2012	0.65	37	5.1	3.1
3, 2012	0.65	63	6.3 - 8.3 *	3.8 - 5.1
4, 2012	0.65	71	> 8.3 *	> 5.1
1, 2013	0.55	61	6.5	5.1
2, 2013	*0.55*	*60*	*2.0*	*1.6*
1, 2015	0.37	53	6.2	8.9
2, 2015	0.37	55	8.0	11.4
3, 2015	0.37	random	0.6 - 1.1	0.8 - 1.5

* indicates no complete dam failure

Comparing the field tests from 2015 with the first test from 2012 using the stone-related Froude number shows that the dumped riprap tests failed both close to $Fr_s = 1$. Test 3 and 4 in 2012 did not fail completely with the available discharge capacity. Their Fr_s at failure could be in a similar range as test 1 and 2 in 2015. The reduced stone size in 2015 in combination with the same discharge capacity allowed test conditions with higher Fr_s than in 2012. Based on the results of the field tests which scatter between $3 \leq Fr_{s,\,cr} \leq 11.4$, we conclude that placed riprap can withstand $Fr_s > 3$ which is twice Fr_s for dumped riprap.

In relation to the mentioned literature, only the recommendations in Dornack (2001) are based on data of slopes up to $S = 0.67$. Using his recommendation for the critical stone-related Froude number results in $Fr_{s,\,cr} = 2.4$ for the used stone size in the field tests 2015, corresponding to an estimated failure unit discharge of 1.7 m³/s/m. This is slightly more than the unit failure discharge during the dumped riprap test and about 1/4 of the failure discharge with placed riprap. The factor compared to the erosion of the first stones is about 2. Hence, Dornack's approach seems suitable to design placed riprap on steep slopes considering that a safety factor should be added in the design. The stability gain between dumped and placed riprap in the field tests was considerably higher compared to a gain of 30 % in the results of Peirson et al. (2008). When using the proposed relationship in Khan & Ahmad (2011) for comparison, we have to keep in mind that we extrapolate out of the range the relationship is based on. The calculated failure unit discharge becomes $q = 2.3$ m³/s/m, with $d_{50} = 0.4$ m and a riprap layer thickness of the same size, and is higher than the failure unit discharge for the dumped riprap. Hence, the relationship should not be extrapolated to design dumped riprap on steep slopes with $S = 0.67$.

According to the design recommendation in the Norwegian guidelines for embankment dams (NVE, 2012), riprap with $d = 0.4$ m would withstand a unit failure discharge of 0.4 $m^3/s/m$. This is less than half of the failure discharge of the dumped riprap test. Turning the recommendation, stones with $d = 2.6$ m should be used to construct a riprap that can withstand a unit discharge of 4 $m^3/s/m$ which corresponds to the unit discharge of the erosion of the first stone in test 1 in 2015.

CONCLUSION

The results from the field tests provide data points for riprap stability on steep slopes and thus extend the range of available data points available in literature. Placing riprap in an interlocking pattern instead of random placement increases its stability with a factor of 4 compared to the erosion of the first stone and a factor of 6 to 8 compared to dam failure.

The approach to calculate the stability of placed riprap developed by Dornack (2001) results in a stone-related Froude number which is lower than the number related to the achieved failure discharge in the 2015 tests. However, his approach is evaluated as suitable considering a safety factor for the design of placed riprap. Extrapolation of the relationship developed by Khan & Ahmad (2011) to the conditions of the field tests is not recommended because it overestimates the stability of dumped riprap. Comparison of the results from the field tests with the Norwegian guidelines shows that even the dumped riprap withstood the unit discharge which is recommended for a stone size of $d = 0.4$ m.

Further analysis of the data gained during the field tests as well as physical model tests in minor scale executed in a hydraulic laboratory will complete the study about stability of placed riprap on steep slopes.

ACKNOWLEDGEMENTS

The collaboration and financial support within the research project PlaF coordinated by Energy Norway and supported by the Research Council of Norway is kindly acknowledged. Sira-Kvina Power Company provided access and water supply from the Svartevatn dam. Without this kind collaboration the field tests would not have been possible. Furthermore, we thank TT Anlegg for constructing the test dams and Bastian Dost and Geir Tesaker for the assistance with the measurements.

REFERENCES

Abt, S. R., Thornton, C. I., Scholl, B. A. & Bender, T. R. (2013). Evaluation of overtopping riprap design relationships. Journal of the American Water Resources Association, 49: 4, 923-937.

Dornack, S. (2001). Überströmbare Dämme - Beitrag zur Bemessung von Deckwerken aus Bruchsteinen. PhD Thesis, Technische Universität Dresden. Dresdener Wasserbauliche Mitteilungen, 20. (in German).

Gronz, O., Hiller, P. H., Wirtz, S., Becker, K., Iserloh, T., Seeger, M., Aberle, J., Casper, M. C. & Ries, J. (2015). Smartstones: a small 9-axis sensor implanted in stones to track their movements. In review.

Hiller, P. H. & Lia, L. 2015. Practical challenges and experience from large-scale overtopping tests with placed riprap. *In:* Toledo, M. Á., Morán, R. & Oñate, E. (eds.) *Dam Protections against Overtopping and Accidental Leakage:* 151-157. London: CRC Press/ Balkema.

Khan, D. & Ahmad, Z. (2011). Stabilization of Angular-Shaped Riprap under Overtopping Flows. World Academy of Science, Engineering and Technology, 5: 774-779.

Lia, L., Vartdal, E. A., Skoglund, M. & Campos, H. E. (2013). Rip rap protection of downstream slopes of rock fill dams - a measure to increase safety in an unpredictable future climate. In: Italian Committee on Large Dams (eds.) *Sharing Experience for Safe and Sustainable Water Storage: Electronic Proceedings of the 9th ICOLD European Club Symposium.* 10-12 April 2013, Venice, Italy. B.55.

NVE, Norwegian Water Resources and Energy Directorate (2012). Veileder for fyllingsdammer. 4/2012, NVE, Oslo. (in Norwegian).

OED, Ministry of Petroleum and Energy (2009). Forskrift om sikkerhet ved vassdragsanlegg (Damsikkerhetsforskriften/ Regulation on dam safety). FOR 2009-12-18 nr 1600. [Online] Available from: http://www.lovdata.no [Accessed 24 January 2011]. (in Norwegian).

Peirson, W. L., Figlus, J., Pells, S. E. & Cox, R. J. (2008). Placed rock as protection against erosion by flow down steep slopes. Journal of Hydraulic Engineering, 134: 9, 1370-1375.

Thornton, C., Abt, S., Scholl, B. & Bender, T. (2014). Enhanced Stone Sizing for Overtopping Flow. Journal of Hydraulic Engineering, 140: 4, 06014005.

Vassdrags- og havnelaboratoriet ved NTH (1973). Dam Svartevann, flomløp. Kalibreringskurve øvre tappeluke. (in Norwegian).

DESIGN-BUILD CONSTRUCTION OF THE PANAMA CANAL BORINQUEN DAMS 1W, 2W, AND 2E

Gabriel Martinez[1]
Wonnie Kim[2]
Carlos Zambrano[3]

ABSTRACT

The Panama Canal Expansion Project included the construction of four dams that will allow navigation access to the new Pacific Post-Panamax Locks (Pacific Locks). This paper describes the design and construction aspects of three of the dams: Borinquen Dams 1W, 2W and 2E. These Dams are part of the design-build contract for the Third Set of Locks, one of several contracts under the overall Expansion Program. Several unique challenges related to foundation conditions, high seismic hazards, and tropical environments were addressed in the design-build construction of the dams. The footprints of the dams encompass several active fault crossings and multiple geologic formations, ranging from columnar basalt to weak claystones and shales prone to slaking and degradation. In addition, the unique nature of the tropical environment implied particular challenges related to the behavior of residual soils available for the clay core as well as construction impacts due to heavy precipitation and long wet seasons.

INTRODUCTION

The Third Set of Locks Project is one of four major components of the Panama Canal Expansion Program. The Expansion Program will create a new lane of traffic for passage of larger Post-Panamax vessels, allowing more traffic and nearly doubling the existing capacity of the Canal. The Third Set of Locks Project included the design and construction of new locks and ancillary works at the Atlantic and Pacific entrances, as well as three dams (Borinquen Dams 1W, 2W and 2E) at the Pacific site. Other major components of the Expansion Program include improvements to the navigation channels and water supply.

The new Pacific Lock Complex is located adjacent to the existing Miraflores Locks. Upstream of the new Pacific Locks is the new 218-m wide Pacific Approach Channel (PAC) formed by the Borinquen Dams (See Figure 1) retaining Gatun Lake. Gatun Lake is the main waterway and water source for the operation of the Panama Canal. The Borinquen Dams will allow vessels to enter the Pacific Locks from Gatun Lake, approximately 27 m above sea level and nearly 10 m above the elevation of the entrance to the Miraflores Locks.

---

[1] M.S., Principal Geotechnical Engineer, MWH, 350 N Orleans St., Suite 1301, Chicago IL 60654, Gabriel.Martinez@mwhglobal.com
[2] P.E., M.S., Senior Geotechnical Engineer, MWH, 4832 E Kentucky Ave. Unit D, Denver, CO 80246, Wonnie.Kim@mwhglobal.com
[3] P.E., M.S., Lead Geotechnical Engineer, MWH, 350 N Orleans St., Suite 1301, Chicago IL 60654, Carlos.Zambrano@mwhglobal.com

The Borinquen Dams have a total combined length of almost 4 km, of which Dams 1W, 2W, and 2E are approximately 2.9 km. Dams 1W, 2W, and 2E, were constructed under a design-build contract awarded to the construction joint-venture Grupo Unidos por El Canal (GUPC) and the design joint-venture Consultores Internacionales LLC (CICP), of which MWH is the lead designer. Dam 1E was constructed under a separate contract. The term "Borinquen Dams" from this point forward refers to Dams 1W, 2W, and 2E.

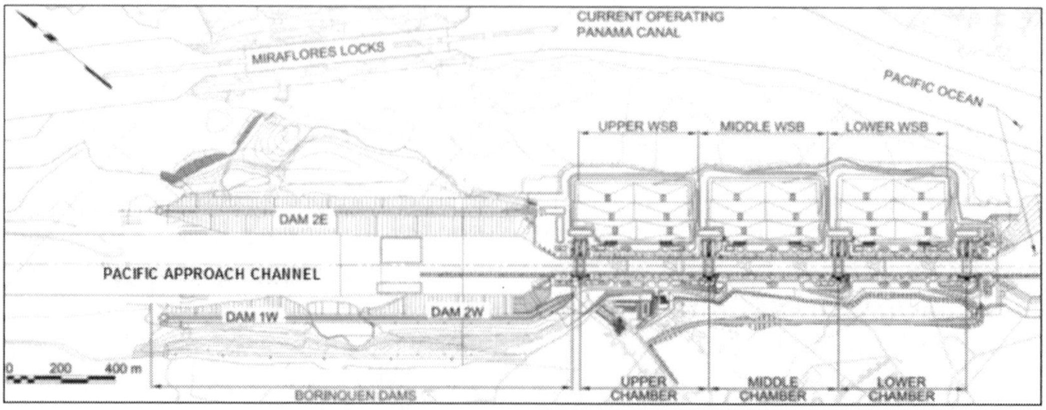

Figure 1. Third Set of Locks Project – Pacific Site Plan.

The Borinquen Dams have been designed and constructed to achieve a high degree of reliability since significant damage or failure of the dams could lead to the release of Gatun Lake and interruption to the Canal operation. The performance criteria for the dams were developed by the Owner, Autoridad del Canal de Panama (ACP), and reflect the utmost importance of reliability over the 100-year design life of the Project. Key features of the dams are summarized in Table 1 and Figure 2.

Table 1. Summary of Key Dam Features – Borinquen Dams 1W, 2W, and 2E

Dam Type	Zoned rockfill dam with central clay core
Crest Elevation (without camber)	El. 32 m PLD
Top of Core Elevation (without camber)	El. 29 m PLD
Crest Width	30 m
Max. Dam Height (above foundation)	35 m
Dam Length	0.6 km (1W), 0.9 km (2W), 1.4 km (2E)
Dam Fill Volume	7 million m^3 (approx.)
Dam/Core Freeboard (above NMOP)	4.87 m / 1.87 m
Inboard Slopes	2.87H:1V
Outboard Slopes	2.5H:1V (1W & 2W), 2.5H:1V to 3H:1V (2E)

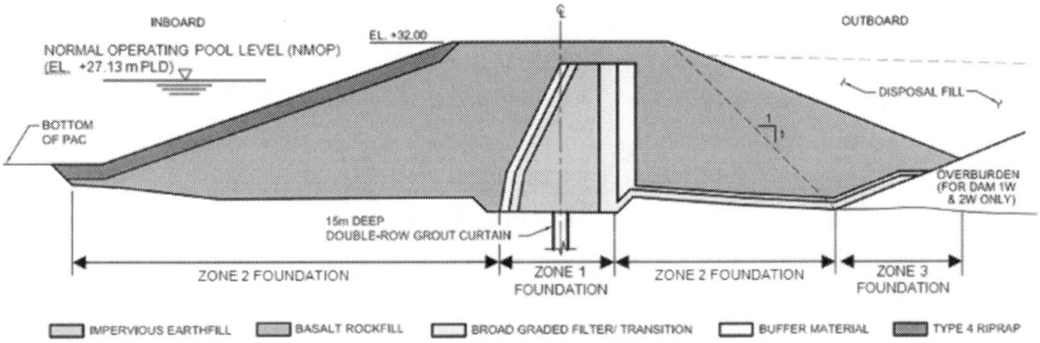

Figure 2. Typical Dam Section

FOUNDATION CONDITIONS

The Borinquen Dams' foundations are composed of several different geologic formations. The predominant formation is an intrusive igneous basalt with adjacent sedimentary and volcaniclastic formations of the La Boca and Pedro Miguel formation (see Figure 3). The La Boca formation is typically soft and weak sedimentary rock deposited in a shallow marine environment during the upper early Miocene. Sandstone, siltstone, limestone and shale make up the majority of the formation, with interlayers of tuff and other pyroclastic agglomerates from regional volcanism. Magmatic injections into the La Boca formation emplaced hard and strong plutons of gabbro, diabase, and basalt (collectively referred to as Basalt formation). The Basalt formation is typically massive with large areas of frequent fractures and columnar jointing. Along the margins between the Basalt and La Boca formations, contact metamorphism has locally altered the La Boca sedimentary units, resulting in increased strength, hardness and discoloration. This zone of altered La Boca material is informally referred to as "baked La Boca". The Pedro Miguel formation is composed of volcanics deposited in a terrestrial environment and is limited to a small portion of the Dam 2E foundation. The formation includes basalt, welded tuff and agglomerates of tuff with angular basalt blocks.

The bedrock formations are overlain by several meters of residual soils with broad areas further covered by fills and spoil from previous Canal excavations. Residual soils are typically high-plasticity silts and clays, with firm to hard consistency and high natural water contents.

The Borinquen Dams are intersected by four active faults, including the east branch of Pedro Miguel fault, the Miraflores fault and two unnamed faults. These faults are further described below.

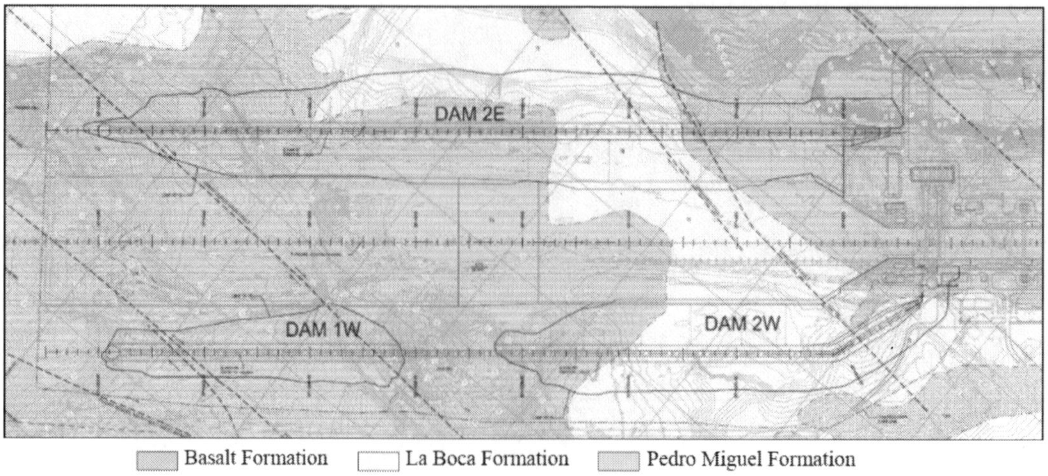

Figure 3. Borinquen Dams Geology.

Foundation Treatment

Several different foundation treatments were utilized to prepare the foundation surfaces for fill placement. Foundation treatment implemented were different based on the zone within the dam foundation (i.e., core, filter, rockfill shell), rock type, and geologic structure encountered. The Zone 1 foundation area (below the clay core and filters) required fresh, sound rock conditions with uniform surfaces prepared to facilitate placement and compaction of the dam core materials. The Zone 2 foundation areas (below the inboard shell and the major portions of the outboard shell) were excavated to weathered Basalt and sound La Boca. In the Zone 3 foundation areas (below the outboard toe of Dams 1W and 2W), all unconsolidated soils, organic material, and random fill were removed.

La Boca Foundation Treatment: Foundation treatment for the La Boca formation was largely focused on preventing rapid deterioration of the weak rock materials when exposed to air and water (i.e., slaking). Shaping the weak rock to achieve a relatively uniform foundation surface was normally achieved using excavators with straight blade buckets. Careful coordination of excavation activities, mapping, inspection, and fill/concrete placement was particularly important to limit exposure of unprotected materials.

In the Zone 1 foundation area (below the clay core and filters), lean concrete was used to protect the foundation from deterioration and provide a durable surface for foundation grouting. The final excavation was sequenced and executed in relatively small areas and coordinated with concrete placement equipment and field personnel, in addition to monitoring weather forecasts, so that the concrete protection could be placed immediately after excavation. Additional measures to prevent slaking and deterioration included final cleaning with high-pressure air from blowpipes (no water) and provisions for surface water management. After foundation treatment was completed, the clay core was overlapped the upstream and downstream edges of the concrete protection slab, allowing at least 2 m of clay core to be placed directly on the rock foundation, mitigating the potential for seepage along cracks in the concrete slab (see Figure 4).

In the Zone 2 and Zone 3 foundation areas (beneath the inboard and outboard shells), the excavation methods were similar to the core area, resulting in a nearly horizontal final excavation surface across the dam section (i.e., no core "trench"). In the shell foundation areas, the La Boca rock was left approximately 0.5 m above final grade until rockfill placement activities were ready to start. The final excavations were then executed in relatively small areas of about 50 m x 50 m, immediately followed by placement of the dam shell materials.

The excavations were typically progressed from the low-point of the dam profile to the higher elevations; with the core area typically advanced ahead of the shell area. To manage the surface water accumulation, dewatering trenches were excavated towards sumps in the shell area. As the embankment fill was placed, 300-mm diameter slotted PVC pipes were embedded vertically in filter materials and used as collection sumps for dewatering. This method has largely been successful in maintaining a dry excavation until several lifts of core and filter material were placed. After these activities were done, the PVC pipes were terminated and tremie filled with low mobility grout.

<u>Basalt Foundation Treatment</u>: The criteria for basalt foundations required the Zone 1 (core trench) excavations to follow the top of sound to slightly weathered basalt, whereas the Zone 2 (inboard and outboard shell) excavations were taken to moderately weathered basalt. The excavations in basalt were completed using mechanical methods (bulldozers with rippers, excavators, breakers, etc.) as practicable to avoid disturbance from blasting.

In the Zone 1 foundation area (core trench) and at abutments, the excavations have resulted in very irregular surfaces requiring substantial treatment to produce a relatively uniform surface for clay core compaction. Due to several fault zones and areas with a high fracture frequency and columnar jointing, the process following the top of fresh basalt produced depressions and near vertical steps up to 5 m in height. Leveling concrete and backfill concrete were used over large areas with irregular and narrow depressions. Knobs and small peaks of rock were removed, and dental concrete was placed to produce a relatively uniform surface for clay core compaction. Treatment details also included limiting the cross slopes in the core trench to a maximum slope of 5H:1V. Figure 5 shows backfill concrete placement activities in the north abutment of Dam 2E due to irregular surface conditions in columnar basalt foundation.

Figure 4. Borinquen Dam 2W - Zone 1 foundation treatment.

Figure 5. Borinquen Dam 2E - Zone 1 backfill concrete placement activities.

In addition to the basalt foundation areas, the geologic contacts between basalt and La Boca also produced a drastic change in the foundation surface, as the weathering profile for La Boca typically extends much deeper than in basalt. To avoid a sharp vertical slope/transition in the core trench, the treatment at geologic contacts included over-excavation of a portion of the basalt and reconstitution La Boca with a sloped concrete surface. These contacts also tend to be highly fractured (in both materials), so consolidation grouting was used to reduce the foundation permeability along the contacts.

In the Zone 2 (shell) foundation area, the excavations were taken down to weathered basalt and cleaned by blading with a bulldozer. The final shell foundation surface was typically 1 to 3 m above fresh/unweathered basalt.

Although the basalt does not slake or degrade when exposed to air and water, the control of water within the core trench excavation was necessary to facilitate foundation treatment and grouting operations. The control of water was primarily accomplished by diversion channels, collection sumps and dewatering pumps. Consolidation grouting was occasionally performed to reduce infiltration from prominent seams and seepage paths intersecting the core trench.

Grout Curtain: A double-row grout curtain was installed as a seepage cutoff below the dams, controlling seepage and mitigating potential piping of the foundation materials. The grout curtain typically consisted of two rows of grout holes, extending 15 m below the core foundation, with holes inclined 20° in opposite directions. The performance requirement for grout curtain was established by the Owner to achieve a residual Lugeon value of 7 or less. Primary grout holes were spaced at 6 meters, followed by split-spaced secondary holes required to achieve an effective spacing of 3 m. Water pressure tests in verification holes were used to demonstrate grout curtain closure. Tertiary and higher order holes were installed, as necessary, until the closure criteria were achieved. Several low-bleed grout mixes (< 3% bleed) were developed for the grout curtain. The mixes were composed of Type II Portland Cement, bentonite and superplasticizer: "Mix A" was a low-viscosity starter mix (<35 seconds Marsh cone); "Mix B" was a higher viscosity mix (35 to 55 seconds Marsh cone) used for high grout takes; and "Mix C" was a sanded mix used to close grout holes where limiting injection volumes cannot be achieved with either Mix A or Mix B.

A common challenge encountered in the grout curtain installation was related to excavation of the core trench at basalt abutments. Removal of weathered material resulted in steep abutments presenting difficulties for access and drilling (see Figure 6). Temporary access ramps were constructed where possible, and light drill equipment was used where required. Grout holes, normally drilled with water, were sometimes drilled with air to install the shallow inclined fan holes using light equipment.

Seismic Considerations

Panama is located in an area of high seismic hazard, predominantly controlled by crustal faults formed by the interaction of the Cocos, Nazca, Caribbean and South America tectonic plates. The Borinquen Dams are intersected by four active faults, including the east branch of the Pedro Miguel fault, the Miraflores fault and two unnamed faults (see Figure 7). These faults are capable of producing up to 3 m of horizontal displacement and 0.5 m of vertical displacement. The design earthquakes (prescribed by the Owner's Requirements) have return periods of 1,000 years and 2,500 years with corresponding horizontal peak ground accelerations of 0.72g and 0.97g, respectively. The dams and abutments were designed to withstand the 2,500-year return period earthquakes without release of Gatun Lake or overtopping, and the 1,000-year return period earthquakes without damage that requires emergency repairs or interruption of the Canal operations.

Figure 6. North Abutment of Dam 2W at Basalt – La Boca contact.

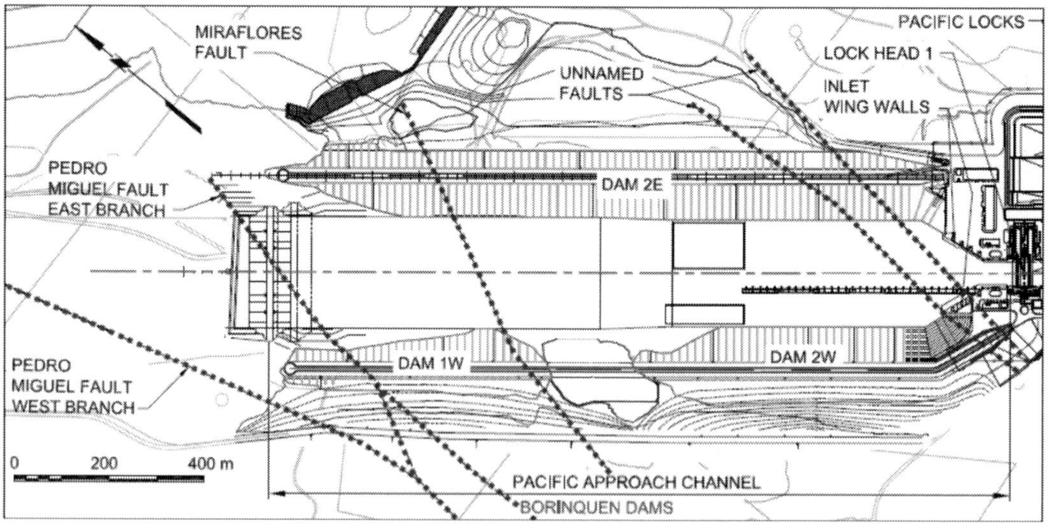

Figure 7. Borinquen Dams 1W, 2W and 2E faults.

To achieve the seismic performance requirements, the embankment design included several features to accommodate seismic deformations and fault displacements while maintaining the functionality of the core and filter zones. In particular, the thickness of the various embankment zones, and the amount of excess freeboard, were designed to encompass the anticipated seismic displacements throughout the embankment. Within 50m of the actual fault locations, the core and filter zones were also widened to accommodate the horizontal and vertical components of fault displacements.

Seismic Deformation: Two design-basis earthquakes were used for the seismic design of the dams. Each design earthquake was characterized by seven corresponding ground motion time-histories. Detailed analyses were performed to evaluate the permanent seismic deformations induced by the propagation of near-source strong ground motions. The seismic performance of the dams was evaluated with nonlinear seismic deformation analyses performed with the two-dimensional finite difference analyses (FLAC software; Itasca, 2008).

The results of the analyses indicated that the Dams 1W, 2W and 2E remain functional and retain Gatun Lake during and after the 1,000-year and 2,500-year return period earthquakes without the need for emergency repairs. Calculated permanent settlements and displacements are less than the available freeboard of 4.87 m, less than the minimum width of the core (2.9 m) and chimney filter/transition zones (2.5 m), and less than the minimum thickness of the blanket filter (1.0 m). A summary of the computed seismic displacements for the 2,500-year earthquake motions is included in Table 2. The range of crest settlements calculated using FLAC Software is in agreement with the range of settlements computed using the empirical method developed by Bureau of Reclamation (2009).

Table 2.
Calculated Permanent Seismic Displacements for 2,500-year Earthquake

Location	**Seismic Displacement** (cm)	**Displacement Direction**
30-m wide Crest El. 32.0	5 to 65	Settlement
Inboard Slope	25 to 80	Toward canal
Outboard Slope	10 to 20	Toward canal Dam 1W and Dam 2W slopes
	25 to 45	Toward the outboard side for Dam 2E slopes
Chimney Filter/Transitions	5 to 35	Toward canal
Outboard Blanket Filter	2 to 8	Settlement

Fault Treatment: The faults encountered within the dam footprints ranged from less than 5 m wide up to approximately 35 m wide and consisted of slickensided, soft to very soft silty clay gouge material and highly to moderately weathered rock. In the core trench area, the typical fault treatment consisted of removing fault gouge to a depth of 2 to 4 m. Weathered and fractured rock adjacent to fault gouge were excavated about 1 m deep and keyed approximately 1 m into sound rock on each side of the feature. The excavated fault zone was then backfilled with concrete and a third row of curtain grouting or stitch grouting was extended across the fault zone. This treatment was typically applied to faults with less than approximately 4 m width. For faults with wider shear zones, more extensive treatment measures were required, typically including a positive cutoff within the core trench, a clay blanket beneath the inboard shell, and increased thickness of the outboard blanket filter.

The Pedro Miguel fault crosses the north end of Dam 1W at an orientation of approximately 65°/315° (dip/dip direction). The fault consisted of a relatively narrow zone of slickensided soft silt and clay gouge materials, approximately 2.5 m to 4.5 m wide with adjacent zones of moderately to highly fractured but relatively sound rock. This fault zone was treated by excavating the fault gouge material and backfilling with concrete as described above.

The Miraflores fault was encountered in Dam 1W and Dam 2E and consisted of a much wider zone of fault gouge, necessitating a more extensive foundation treatment. Where the Miraflores fault crossed Dam 1W, the fault zone was up to 35 m wide and crossed the core trench at an orientation of approximately $80^{o}/290^{o}$ (dip/dip direction). Approximately 15 m of fault gouge (slickensided rock fragments in a sandy clay matrix) was bound by about 5 m of moderately weathered basalt and fractured, slightly weathered basalt on each side of the fault.

In Dam 1W, the Miraflores fault treatment included a seepage cutoff wall, an upstream clay blanket, and a downstream filter blanket. The seepage cutoff was installed parallel to the dam's axis to a minimum depth of 5 m, keying into rock on either side of the fault zone until excavation refusal, as shown in Figure 8. A grout curtain was installed through the cutoff wall and fan grout holes were drilled at the cutoff wall tie-ins to sound rock on each side of the fault zone.

For the cutoff wall, a plastic concrete mix (cement-bentonite-aggregate) was designed to provide a low permeability barrier while allowing for some deformation to occur at the dam's foundation. A perfectly rigid structure could cause cracking of the core, so the mix design targeted similar deformation characteristics as the surrounding materials. The cutoff wall was tremie placed with the following characteristics:

- Slump (ASTM C143): 18-23 cm
- Compressive Strength (ASTM D4832): 0.5 – 2.5 MPa at 28 days
- Elastic Modulus (ASTM C469): 200 – 1400 MPa at 28 days
- Design mix permeability (ASTM D5084): ≤ 1 x 10-6 cm/sec at 28 days
- In-Place Permeability: ≤ 1 x 10-5 cm/sec at 28 days

In addition to the cutoff wall, an upstream/inboard clay blanket was placed over a distance of 50 to 70 m from the clay core to lengthen the potential seepage path along the fault zone. Filter material was used to cover the clay blanket, followed by a transition material, before basalt rockfill was placed. An increased thickness of filter blanket was also placed on the downstream/outboard side of the dam to accommodate potential vertical fault displacements.

In Dam 2E, similar conditions were encountered where the Miraflores fault intersects dam axis. Although the width of the fault gouge was slightly less than in Dam 1W, the same foundation treatment measures were adopted. In particular, the fault treatment in Dam 2E included a plastic concrete seepage cutoff wall, upstream clay blanket, and downstream filter blanket (see Figures 9a and 9b).

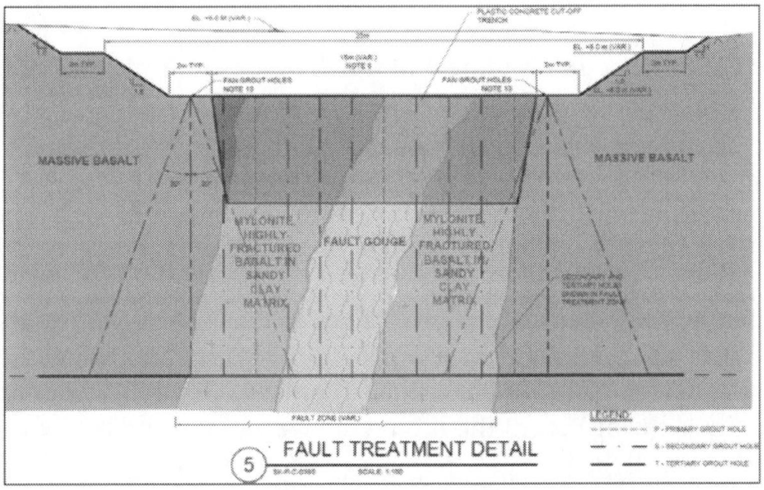

Figure 8. Miraflores fault cutoff wall detail at Dam 1W.

Figure 9a. Miraflores fault at inboard rockfill shell of Dam 2E (view towards dam axis).

Figure 9b. Miraflores fault treatment at Dam 2E, clay blanket in the inboard shell.

CONSTRUCTION IN A TROPICAL CLIMATE

The climate in the Project area is typical of low latitude environments of the inner tropics, characterized by high temperatures and humidity, abundant rainfall, and frequent thunderstorms throughout the rainy season. The rainy season extends from early May to December with average monthly precipitation of approximately 217 to 373 mm. A relatively dry season is typically experienced from December through April, with monthly rainfall averages from approximately 36 to 122 mm.

The tropical climate presented several challenges to the construction of the Borinquen Dams, in particular, the placement and compaction of clay core during the rainy season. In this regard, the design included provisions to provide flexibility in construction sequence to mitigate the effects of the prolonged wet season and heavy rainfall:

- Construction joints perpendicular to the dam axis (see Figure 10) reduce the amount of core exposed to rainfall. Slopes forming the perpendicular joints are later treated by keying-in and scarifying subsequent lifts.
- Advanced construction of rock fill shells placed on the inboard and outboard toe of the dam, independent of core and filter construction (see Figure 11). Rockfill can be placed as close as 5 m from the core/filter to protect the foundation and advance construction works during rain events.

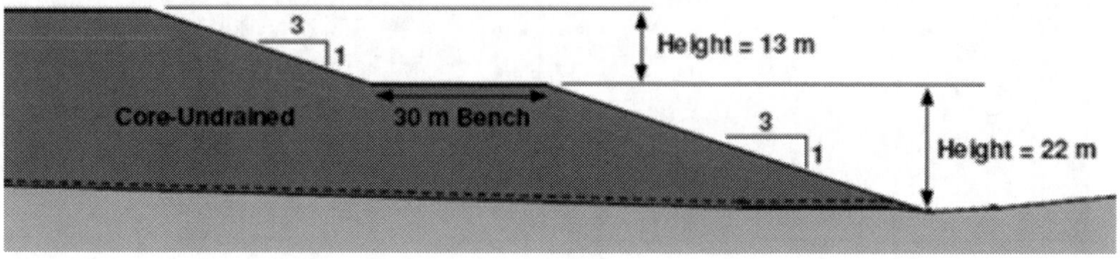

Figure 10. Construction Methods – Clay Core (profile along centerline of dam).

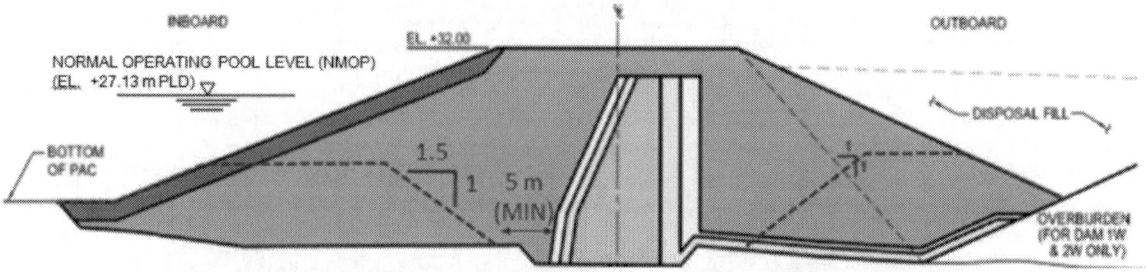

Figure 11. Construction Methods – Rockfill Shell.

Additionally, the core geometry was revised from the Owner's conceptual design to reduce the quantity of required borrow materials and excavation. The core was steepened below the controlling outboard water level without increasing maximum gradients or impacting the anticipated seepage regime through the dams. This reduction in clay core volume effectively reduced the construction and exposure time for the lower part of the core.

TROPICAL RESIDUAL SOILS FOR CLAY CORE CONSTRUCTION

Materials for the dam cores were obtained from the residual soils derived from weathering of the parent basalt rock. Borrow areas included several local hilltops and stockpiles from earlier excavations. Two of the primary sources for clay material included Sucre Hill and Cocoli Hill. Placement of these residual soils in a tropical climate necessitated an alternative approach to conventional moisture–density compaction control specifications.

The natural moisture contents of residual soils from the local borrow areas were relatively high, ranging from 30 to 45% and approximately +5 to +12% above Optimum Moisture Content (OMC) as shown in Figure 12. Despite having high natural moisture contents, the material appears dry and crumbly with clay/silt clasts.

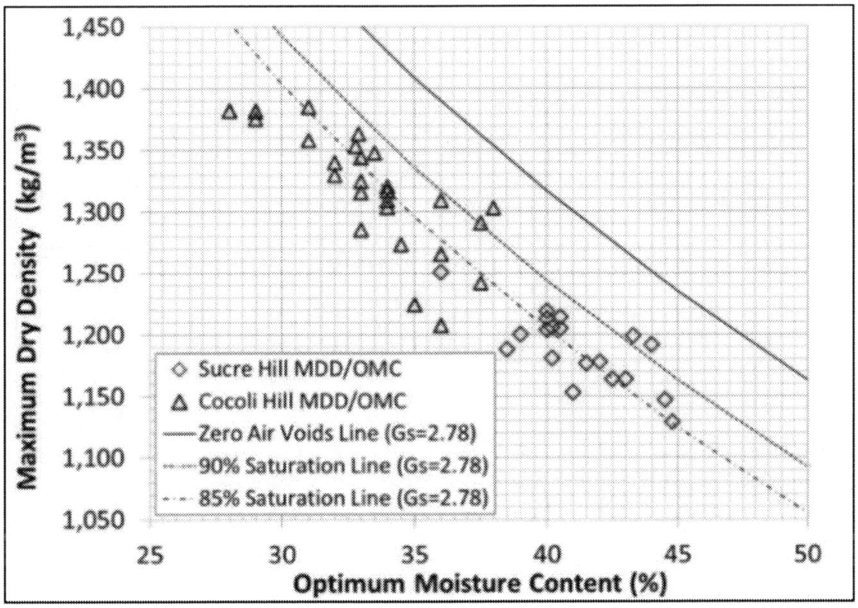

Figure 12. OMC and MDD – Sucre Hill & Cocoli Hill.

Conventional specifications for dam core materials typically require moisture contents between +1% to +3% above OMC and an average of 98% of the maximum dry density (MDD) based on Standard Proctor tests. Since the majority of borrow materials for the Borinquen Dams were found to have natural moisture contents in the range of +5 to +12% above OMC, applying conventional specifications would require a significant degree of drying and/or borrow management to reduce the natural moisture contents to within acceptable ranges. Drying and managing borrow sources was further complicated in a tropical climate with eight months of heavy rain. Therefore, due to the unique characteristics of the residual soils and the tropical climate, special placement and compaction controls were developed for materials with moisture contents greater than OMC+3%. Extensive laboratory testing and field trials (test fills) were conducted to define acceptance criteria based on Degree of Saturation (S) in addition to MDD and

OMC. Accordingly, the final compaction requirements for the Sucre Hill and Cocoli Hill source materials were developed as summarized in Table 3.

MDD and OMC were specified with Degree of Saturation to prevent compacting materials excessively wet of optimum, where strength gain is limited by oversaturation. Excessively wet materials were managed at borrow areas to avoid unsuitable materials being used in the fill placement, and observing the moisture condition at borrow areas was part of the Quality Control (QC) program. Materials outside of the specified moisture limits and/or exhibiting rutting or pumping during handling were reworked or removed. Figure 13 shows QC test results from several lifts of clay core material placed in Dam 1W. Proctor tests were completed on a regular basis from borrow area samples with materials tracked to placement areas for comparison with field densities. Two-point Proctors were regularly completed at placement areas to verify the appropriate application of source material Proctor curves.

Table 3. Sucre Hill and Cocoli Hill End-product Specification

Borrow Source	**Compaction Requirements**		**Shear Strength Requirement**
	Moisture Content	Density / Saturation Level	
Sucre Hill	+1% to +4% OMC	98% of MDD (avg.) 96% of MDD (min.)	n/a
	+4 to +10% OMC	Density equivalent to 87% (min.) and 90% (avg.) saturation	70 kPa (shear vane)
Cocoli Hill	+1% to +4% OMC	98% of MDD (average) 96% of MDD (minimum)	n/a
	+4 to +8% OMC	Density equivalent to 90% (min.) saturation	120 kPa (shear vane)

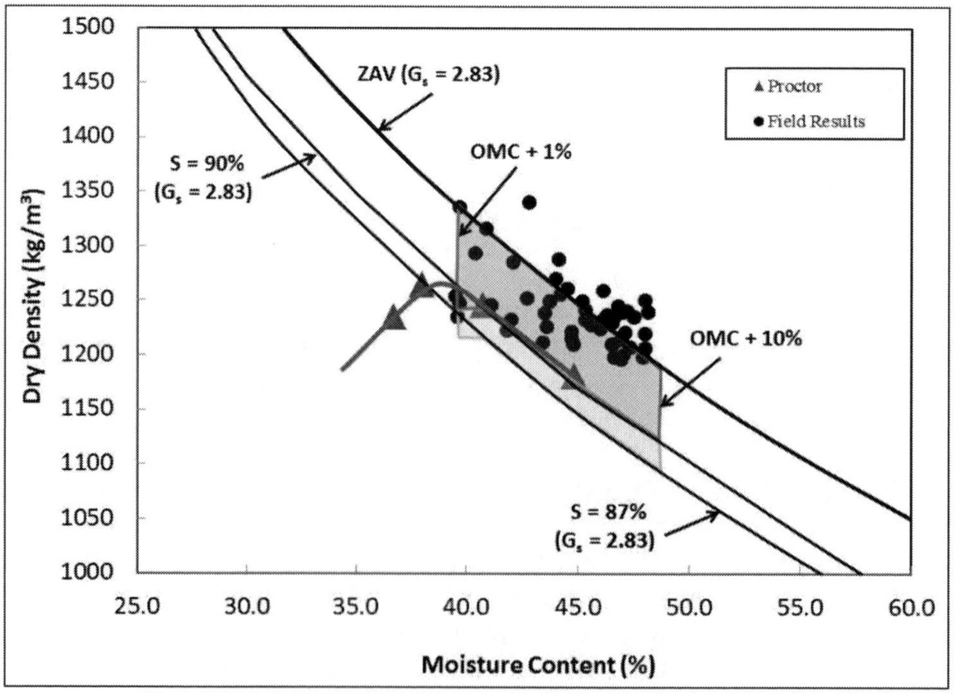

Figure 13. Dam 1W QC test data.

FILTER CONSTRUCTION

The Owner's original concept design of the dams included four different filters placed in six separate zones. However, during development of the detailed designs, an effort was made to reduce the number of filter/transition zones to mitigate several issues related to filter material availability and construction. The final design of the filter/transition zones, as shown above in Figure 2, follows U.S. Bureau of Reclamation guidelines (USBR, 2007) and contains one broadly-graded sand filter/transition zone and a graded rockfill (i.e., "buffer") adjacent to the filters/transitions.

Filter materials were primarily produced from crushed basalt rock obtained from required excavations and quarries. The basalt rock sources were also used for concrete aggregate, rockfill in the dams, and backfill around lock structures. Therefore, the crushing plant needs to keep up with the demand from several locations concurrently. Reducing the number of filter zones and the quantity of processed filter/transition materials greatly alleviated the demands on the crushing plant and also allowed lifts of the dam to be constructed quickly during intermittent "dry" working periods. Timely placement during dry weather was considered particularly important for the clay core, as it could not be placed and compacted during significant rains. Developing a simpler filter and transition zone facilitated construction and allowed more time to focus on the advancement of clay core construction.

As constructed, the dams included one broadly-graded, gravelly sand chimney filter/transition (Type 3a) at both the inboard and outboard sides of the core. A graded rockfill material was placed against the filters and consists of a coarse gravel and cobble "buffer" material obtained by screening the quarried basalt rockfill to limit the maximum particle size. A coefficient of uniformity (c_u) of less than 20 in the filter/transition zones was targeted to reduce potential for segregation within the material. Laboratory gradations and test fills were performed prior to construction to confirm placement methods would not result in significant segregation. The filter/transition, buffer and rockfill gradation limits are shown on Figure 14.

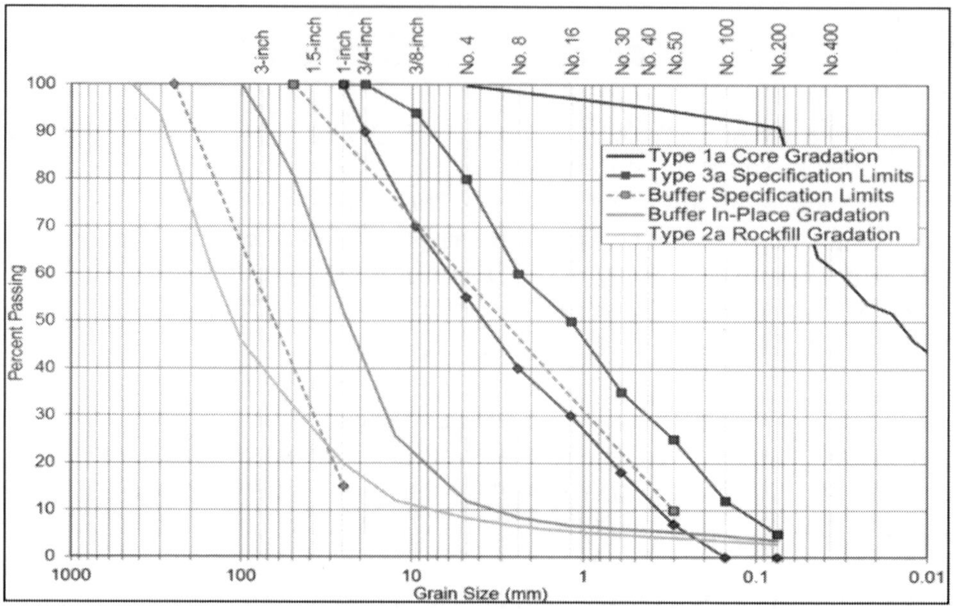

Figure 14. Filter/Transition, Buffer and Rockfill gradations.

The reduced number of filter and transition zones simplified placement and compaction of filter and core zones. Conventional excavation equipment was used to carefully place and spread filters, limiting contamination of the filters and core. Filter was placed in 450 mm lifts using a straight edged bucket and hand tools. Two passes with a 3-ton roller (without vibration) were used to meet the 70% Relative Density compaction criteria. Filter placement is shown in Figure 15.

ROCKFILL CONSTRUCTION

The shells of the dam consisted of basalt rockfill taken directly from excavations. The basalt rockfill consisted of a hard, durable, sound, and fresh to slightly weathered basalt with a maximum particle size of 800 mm and no more than 15% by weight passing the No. 4 sieve as tested after spreading and compaction. Rockfill gradations were controlled by blasting without further processing. When received at the placement area, basalt rockfill was dumped on the surface of the advancing lift and spread out to the surface of the previous lift. Lifts were spread in 900 mm thickness and compacted with 6 passes of a 20-ton smooth-drum roller with vibration (Bomag 219DH).

Figure 15. Filter placement with excavator and hand tools.

PACIFIC LOCKS COMPLEX CONNECTIONS

The south ends of Dams 2E and 2W were connected to the Inlet East (IE) Wing Wall and the small recess of Lock Head 1 (LH-1), respectively. The structural designs resulted in complex geometries and interfaces, necessitating special core and filter details to create watertight connections at the abutments. The following design measures were implemented to enhance the connections and protect the embankment clay core from internal erosion:

- Foundation preparation: Foundation treatment for structures was similar to Zone 1 of the dams with excavation and cleaning to fresh rock. Fan grout holes were added at the abutment interface to generate a complete foundation cutoff between the dam grout curtain and the grout curtains along the IE Wing Wall and across the Lock Head.
- Widening Zone 1 materials: The embankment clay core and chimney filter/transition zones were flared at the abutment. Widening Zone 1 materials enhances the watertight connection by increasing the contact area between concrete/clay and increase the level of protection against clay core piping.
- Sloped and skewed concrete surfaces:
    - o Concrete monoliths joints were built with waterstops to reduce the risk of clay core and filter/transition material erosion and piping.
    - o Concrete surfaces were shaped with 1H:2V to 1H:6V slopes to allow high-energy compaction against the concrete structure. Additionally, long term core consolidation will result in settlement against the concrete structure, promoting water tightness.
    - o Concrete surfaces at the abutments were arranged to form an 80° angle (10° from perpendicular) with the dams' axis to improve the connection. Hydrostatic

pressure applied to the core in the horizontal direction from Gatun Lake will maintain an inside-out load and ensure the connection remains in compression.

CONCLUSION

The design and construction of the Borinquen Dams 1W, 2W, and 2E faced several unique technical challenges. However, this design-build project was resourcefully implemented to address difficult conditions and provides a design that exercised flexibility in construction while meeting the strict performance requirements for a 100-year service life.

REFERENCES

Bureau, G. (2009): *Seismic Performance Review of Zipingpu Dam, China*. Proceedings of International Seminar on Earthquake and Dam Safety, March 29-April 4, Beijing China.

Bureau of Reclamation (USBR), U.S. Department of the Interior. (2007): *Design Standards No. 13 Embankment Dams*, Chapter 5, Protective Filters.

Itasca. (2008): FLAC. *Fast Lagrangian Analysis of Continua*, Itasca Consulting Group, Inc., Mill Place, 111 Third Street, Suite 450, Minneapolis, Minnesota.

A PERFORMANCE-BASED APPROACH TO DEVELOPING FRAGILITY CURVES FOR STRUCTURES SUBJECT TO SEISMIC LOADING CONDITIONS

Kunal Menda[1]
Mo Zhao[2]
Spenser Anderson[3]
Ziyad Duron, PhD[4]

ABSTRACT

Performance-Based Testing (PBT) is a new experimental technique developed to assess the performance of a structure by measuring its response to a high-amplitude, short-duration impulsive load. The response of the structure to seismic loading can obtained through convolution of the recorded impulse response and an earthquake time history. This paper outlines a method for determining experimentally derived fragility curves using information obtained through PBT. Specifically, the procedure of creating a fragility curve for a given failure criterion, demand parameter, and intensity measure is described. A structure's calculated response to an earthquake is used to determine a corresponding measurement of a demand parameter, which is then compared to a failure criterion to determine if the structure fails. By finding the responses to several earthquakes of different intensity, a probability of failure can be determined as a function of the intensity measure, thus generating the fragility curve. The ability of a finite element model to enhance the precision of these calculations is evaluated. This paper will then discuss the experimental fragility curve calculation in the context of a case study for a concrete arch dam. Finally, this paper will conclude by discussing the advantages of this approach to determining fragility curves, including its experimental basis as well as computation time.

INTRODUCTION

In the field of dam safety, modern practice relies heavily on the use of numerical models to evaluate a dam's predicted response to hazard loading conditions. The evaluation is typically based on benchmarked solutions schemes that do not typically include model validation through the use of field acquired full-scale testing techniques. A widely held view among dam safety engineers and practitioners is that computational methods have advanced to such a degree that benchmarking numerical models against known solutions is an acceptable substitute for conducting an independent verification of the model. Further, the complexity of modern hazard load cases that now include seismic events characterized by spectral content over three decades to 100 Hz as well as blast load cases whose energy can span well beyond 100 Hz, has not prompted analysts to reassess the

[1] De Pietro Engineering Fellow, Harvey Mudd College, Claremont, CA, 91711, kmenda@hmc.edu
[2] De Pietro Engineering Fellow, Harvey Mudd College, Claremont, CA, 91711, mzhao@hmc.edu
[3] De Pietro Engineering Fellow, Harvey Mudd College, Claremont, CA, 91711, sanderson@hmc.edu
[4] Professor of Engineering, Harvey Mudd College, Claremont, CA, 91711, duron@hmc.edu

use of numerical models that have not been verified beyond a dam's typical frequency range of interest. As described by Duron and Yen (2016), a true performance evaluation of a large concrete dam must encompass low, middle, and high frequency bands if the mechanisms that could lead to ultimate failure are to be truly evaluated. Duron, et al. (2013) suggest that high frequency behavior between 30Hz-100Hz can be associated with high gradient behavior, cracking mechanisms, and energy dissipation mechanisms that differ significantly from comparable behavior in the middle (15Hz-20Hz) and low (.1Hz-15H) frequency bands. As a result, models that are not built to consider the variety of mechanisms that engage as the hazard's spectral content shifts upward during loading may miss critical response predictions that could lead to better and more reliable assessments of how a dam actually handles hazard loading events.

This paper begins by reviewing the work presented by Duron, et al. (2013) on the ability to use a Cold Gas Thruster (CGT, shown in Figure 1) for measuring the impulse response of a large civil structure such as a dam, at various locations on the structure. Linear systems analysis is used to show how the response of the structure to an arbitrary loading condition may be calculated from these measured responses. The assumptions upon which the methodology is based along with a description of the approach are provided. Finally, the paper will draw on a case study for attempting to characterize the experimental fragility curve for San Clemente Dam to seismic hazard, and will evaluate the presented method for doing so.

Figure 1. Cold Gas Thruster (CGT) used for Performance Based Testing (PBT).

PERFORMANCE-BASED TESTING

Shown by Duron et al. (2013), the Cold Gas Thruster (CGT) (shown in Figure 1) is capable of delivering a broadband impulse load on the order of 20,000 lbf to a large civil structure such as a dam. The vibration response measured at various locations on the structure can be approximated as the impulse response of the structure to a point impulse load at the location of the CGT. According to linear systems theory, for a single-input single-output system, the response $y(t)$ to any input signal $f(t)$ can be calculated from the impulse response $h(t)$ as follows:

$$y(t) = f(t) * h(t) = \int_{\infty}^{\infty} f(\tau)h(t-\tau)d\tau \quad (1)$$

where $*$ denotes the convolution operator in the time domain. The response can also be calculated as:

$$Y(s) = F(s) \cdot H(s) \quad (2)$$

where $Y(s), F(s), and\ H(s)$ are the Laplace Transform s of $y(t), f(t), and\ h(t)$, respectively.

It follows that the impulse responses measured by performance-based testing should be sufficient to calculate the response of the system to any input loading signal applied as a point load *at the location* of the CGT. During loading, the impulse load propagates a broadband input through the structure that excites eigenproperties of the dam in a repeatable manner, similar in concept to an impact testing technique. From this impulse-like response, fundamental characteristics about the dam's structural performance are used to develop predicted behavior under earthquake loading.

The value of such data is that it contains the response of the structure to higher frequency loading, at frequencies well beyond the natural frequency of the first mode of vibration. While coarse Finite Element Models (FEM) may be accurate in calculating the response of a structure to static and low frequency loads, large computational resources and very well advised assumptions are required to accurately predict the behavior of the structure at the high frequencies excited by a CGT pulse, or many other potential broadband hazards.

Figure 2 shows the placement of the CGT and accelerometers in the PBT conducted at San Clemente Dam, as an example. One might express concern that the location of the CGT is somewhat arbitrary, and the response picked up by each accelerometer would vary based on its proximity to the CGT. The assumption this paper operates under is that the response picked up at any location of the dam is invariant to CGT location up to a scaling factor. This assumption is supported by Figure 3, which compares the responses at 4 locations across the top of the dam (out of the 16 shown in Figure 2) to loading by the CGT located at two different locations. The CGT was first located close to the base of the dam, as shown in Figure 2, and second located at the top of the dam. We can see that the responses stay approximately the same regardless of CGT location, up to a scaling constant (that would shift the responses up or down in the log-log plot shown). However, as expected, we find that these scaling factors are larger when locations are in closer proximity to the CGT.

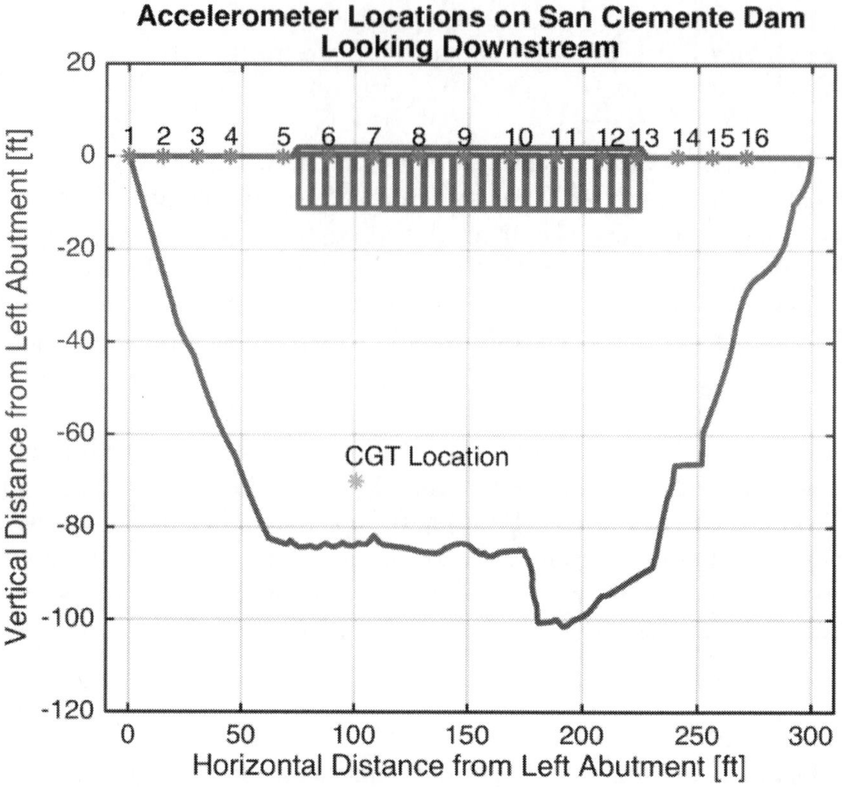

Figure 2. The locations of the accelerometers and CGT when testing San Clemente Dam, as seen looking downstream.

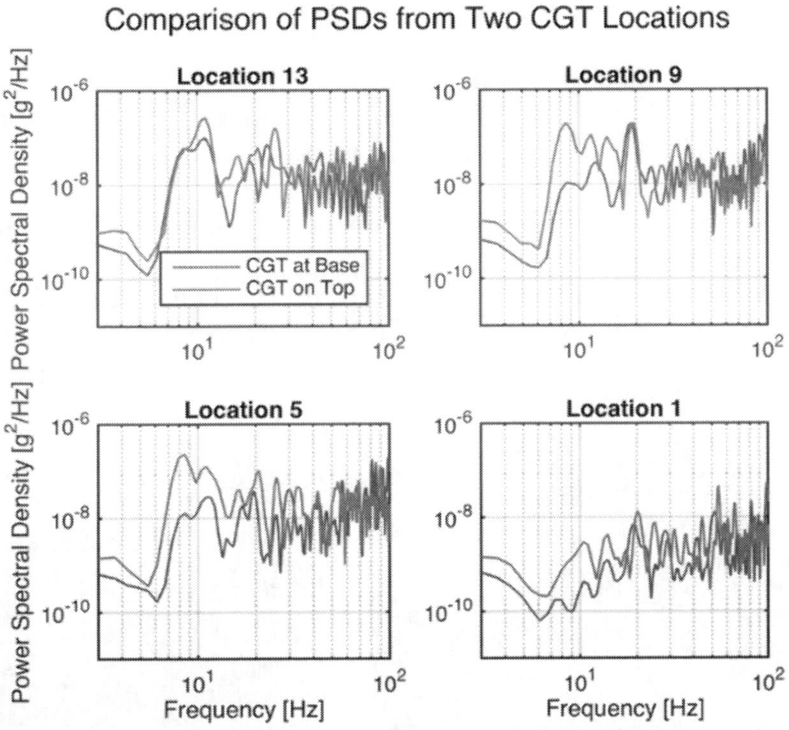

Figure 3. Comparison of responses measured at four locations on the dam to loading by the CGT at two different locations on the dam. We can see that the responses appear to be approximately the same in both tests up to a scaling constant.

When considering the 3D response to impulse ground motion for a civil structure, we not only need the impulse responses of each location as measured by CGT, but we need to scale each response such that the *relative* strength of each response *between locations* conforms to the shapes of the expected eigenmodes of the dam. Given that the relative strength of responses is *not* invariant to proximity to the CGT, we must appropriately correct for this by scaling the responses, in order to relate the responses measured to a CGT impulse load to the desired impulse response to a ground motion. The method used to do this is discussed in the next section.

RELATING CGT RESPONSES TO GROUND MOTION IMPULSE RESPONSES

The goal of the analysis is to use the data collected from PBT to calculate the response of the civil structure to an earthquake. As shown in Figure 4, an impulse load from the CGT propagates energy into the structure differently than an impulse acceleration in the foundation of the structure would. While we can expect an impulse ground motion to excite the structure uniformly, we find that locations closer the CGT experience a stronger load than the locations further away. In order to compute the response of the structure to an earthquake, we need to convolve the ground motion acceleration record of the earthquake with the impulse response to a ground motion, and not just to that of a CGT pulse. Thus, the remainder of this section will describe the method used to process the responses of the structure to a CGT impulse to yield the responses of the structure to a ground motion.

The procedure can be summarized by the following steps.

1. Remove the effects of varying CGT force pulse magnitudes by normalizing each PBT response by the response measured at the CGT attachment location.
2. Remove CGT proximity effects on each PBT response by normalizing each response in the vicinity of a selected resonance, typically the first natural frequency of the dam.
3. Impose a spatial distribution onto the set of PBT responses that reflects the actual eigenmode in the dam for the resonance selected in Step 2.

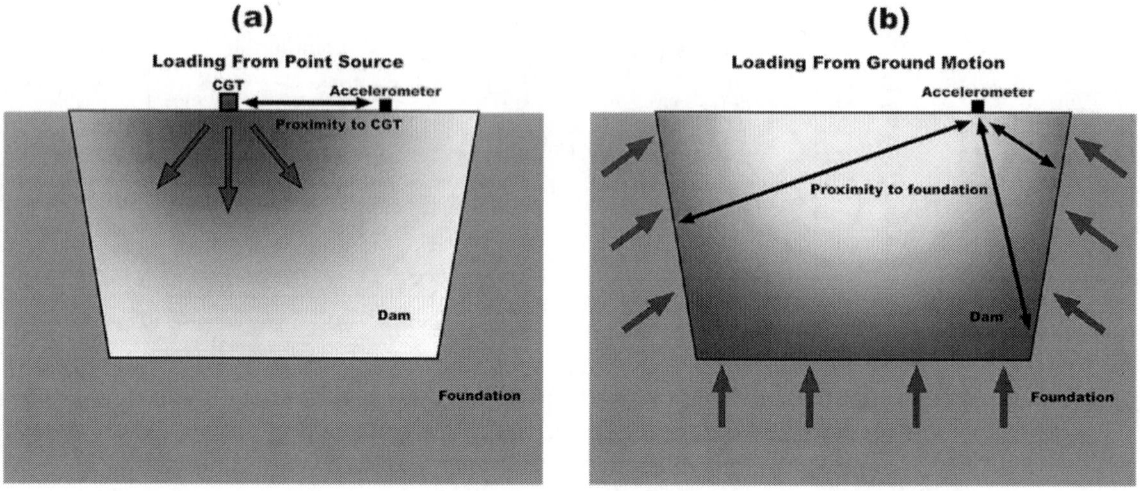

Figure 4. Energy transfer into a dam from CGT impulse (a) and impulse ground motion (b). Black arrows show the proximity of the accelerometer to the excitation source.

The first step is required since during testing, the CGT was found to have burst pressures varying from 2000-4000 psi, yielding some variation in the magnitude of the impulse delivered to the dam tested. Since the system is assumed to be linear, correcting for this is straightforward. For each response calculated, the response is normalized by the maximum acceleration recorded by the shock accelerometer placed on the CGT. Letting $a_{i,CGT}(t)$ be the acceleration response to a CGT pulse recorded at the ith location on the structure, $a_{shock}(t_0)$ be the magnitude (maximum acceleration) recorded by the shock accelerometer, we find:

$$\alpha_i(t) = \frac{a_{i,CGT}(t)}{a_{shock}(t_0)}$$

where $\alpha_i(t)$ is acceleration record for the ith location on the dam normalized for the magnitude of the CGT pulse.

The second step requires us to correct the $\alpha_i(t)$ records for the locations' proximities to the CGT. This is necessary because, while we expect each response to contain accurate information with respect to the eigenmodes excited at each location, we expect the excitation level of each location to decay with distance, as was supported by Figure 3. Thus we expect variation in amplitudes of the responses to vary not only due to physical parameters such as the relative stiffness of that location on the structure, but also the contingent parameter of proximity to the somewhat arbitrarily located CGT. This is the variation we hope to correct for. The approach used to do this is based on the assumption that a spectral estimate of each location's response can resolve the first natural frequency as a major frequency peak. The issue, then, is not that the first natural frequency has somehow been influenced by the proximity of the measured response's location to the CGT, but that the relative magnitudes between response locations that define its corresponding resonant shape has. To obtain the proper relative magnitudes across all response locations, at least for the dam's first natural resonance, the CGT induced relative amplitudes at the first natural resonance must be removed from the PBT responses. To do this, we band pass the signal $\alpha_i(t)$ in the region of the first natural frequency to observe its response to only frequencies in that neighborhood, calling this band passed signal $\alpha_i'(t)$, and then look at the amplitude of the resultant waveform as a measure of the response. We will call this factor γ_i, and calculate it as:

$$\gamma_i = \max_{t \in T} \alpha_i'(t)$$

Normalizing $\alpha_i(t)$ by γ_i corrects for the proximity to the CGT.

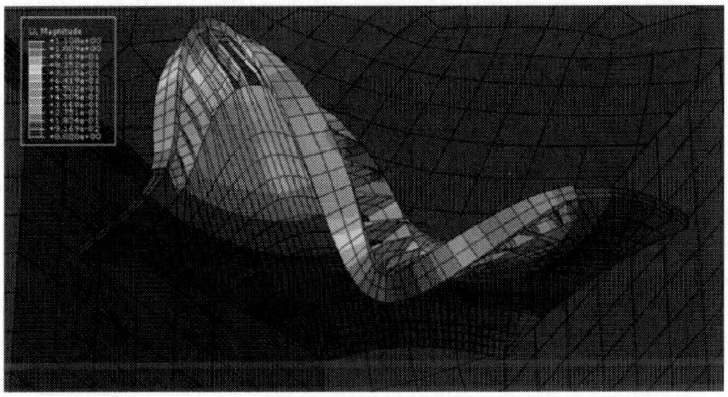

Figure 5. First eigenmode of San Clemente Dam

The last step requires us to scale all responses to achieve a spatial distribution that estimates the dam's actual response shape at the first natural frequency. This is required because the CGT excites a response at a given location that captures the local behavior over a broad frequency range, but as corrected for in the previous step, does not capture the dam's global response to a ground motion. As stated earlier, the relative magnitudes of a response at a given location, at a given natural frequency, do not necessarily conform to the shape of the corresponding eigenmode. A FEM model does this well for the lower eigenmodes – if the model is excited with a single sinusoidal frequency as a ground motion, we can expect even a course FEM model of a dam to accurately predict the first mode shapes and the magnitudes of their response. In the previous step, all PBT responses were corrected for proximity effects by removing the CGT's influence on the magnitude at the first natural frequency. Using an FEM model with a ground motion excitation at the first natural frequency, we take the oscillatory amplitude of the first eigenmode divided by the oscillatory amplitude of the base, and call this factor β_i. This factor tells us, according to the FEM, what the amplitude of the response of each location at the first natural frequency ought to be when excited by a ground motion impulse of unit magnitude, which is what we seek. Figure 5 shows the first eigenmode of San Clemente Dam, as an example. In this analysis, it was found that the β_i resemble the mode shape, as expected, where locations very close to the foundation had values close to one, implying that these locations oscillate with the same amplitude as the adjacent foundation. The locations further from the foundation had values larger than one, as high as 217, suggesting that these locations oscillate with an amplitude much larger than that of the ground motion. Using these values, the response of each location to a ground motion impulse, which we will call $h_i(t)$, can thus be calculated as:

$$h_i(t) = \alpha_i(t) \cdot \frac{\beta_i(t)}{\gamma_i(t)}$$

The steps taken to get from $a_i(t)$ to $h_i(t)$ begin with normalizing for variation in the magnitude of the CGT pulse, thus making the intermediate response *per unit CGT load*. The next step corrects for CGT proximity effects on the magnitude of the responses at their common first natural frequency. Finally, FEM is used to scale each response so that, for a ground motion oscillation at the first natural frequency, each location would respond with a magnitude appropriately conforming to the first eigenmode shape as predicted by FEM. This final response is guaranteed to the accuracy of the FEM model to predict accurate responses if the ground motion were an oscillation at the first natural frequency. Each location's response contains the information regarding its response to all other frequencies, as they are responses to the broadband impulse delivered by the CGT, thus the responses $h_i(t)$ are taken to be the responses of each location to an impulse ground motion. The next section will discuss steps taken to digitally process the $h_i(t)$ signals to efficiently convolve them with the database of earthquake records, ultimately producing the sought-after fragility curves.

FROM IMPULSE TO EARTHQUAKE RESPONSE

This section describes the steps taken to digitally convolve the ground-motion impulse responses for each location, as calculated in the previous section, with ground-motion acceleration records of earthquakes. Let $a_{i,EQ}(t)$ be the desired acceleration response of the ith location on the structure, to an earthquake with the acceleration record $EQ(t)$. Using the location's impulse response to a ground motion, we calculate $a_{i,EQ}[n]$ using the discrete time input signals and using discrete time convolution, as follows:

$$a_{i,EQ}[n] = \sum_{j=1}^{N} EQ[j]h_i[n-j]\,\Delta t$$

where $\Delta t = \frac{1}{Fs}$ is the sampling period. The convolution may also be conducted in the frequency domain as the product of the Finite Fourier Transform of the two signals. While in theory these two methods are identical, concerns exist that the output will have "edge-effects" where the output will attempt to conform to the enforced periodicity of the Fourier transform and have inaccuracies at the ends of the record. Concerns also exist that the output will have spikes due to the Gibbs phenomenon if the signal is high pass filtered. The concern regarding the Gibbs phenomenon can be ignored because an earthquake record contains more low frequency content than high frequency content, and as shown in Figure 6, the two methods of convolution are effectively identical with minimal to no edge effects.

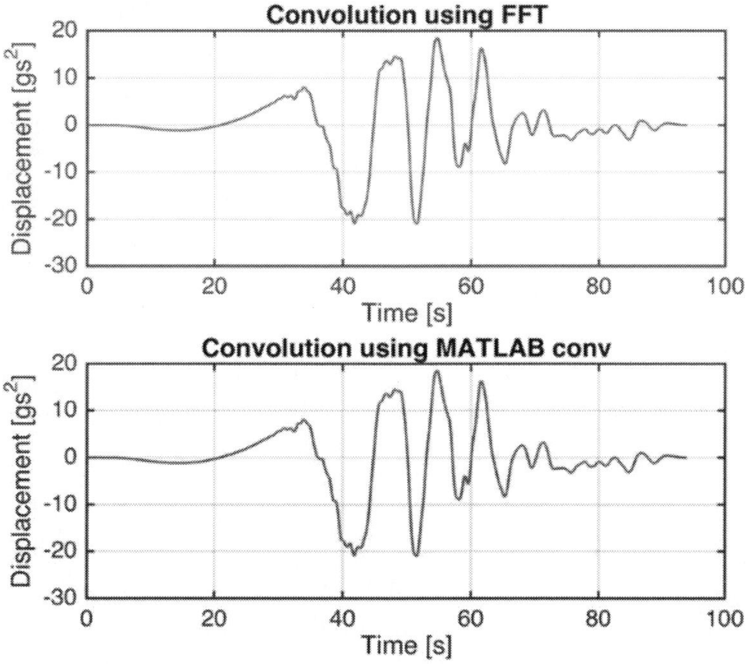

Figure 6. Convolution of input and response by FFT (top) and MATLAB's conv (bottom). The difference is negligible.

Further, there exists a concern with sampled impulse responses that it is inaccurate to integrate them for calculating velocity or displacement from acceleration. Estimating a fragility curve requires the displacement responses of the locations of the dam to be calculated, and not just accelerations. To calculate displacements from accelerations, the acceleration responses are twice integrated. The reason for concern is that if there exists unwanted DC offset anywhere in the sensor measurement of the CGT responses, then this DC offset will produce linear and quadratic trends that dominate the integrated signal. One method of dealing with this issue is to high pass the signal at the very low frequencies to remove these offset trends. Conveniently, however, convolution with earthquake time histories does the job of high pass filtering the data at very low frequencies (under 0.1 Hz) since it contains no frequency content below this (see Figure 7), and so this issue is dealt with without the introduction of additional assumptions. Having justified the method for estimating the discrete-time output record of a dam to earthquake loading, we will now describe the process of generating a fragility curve.

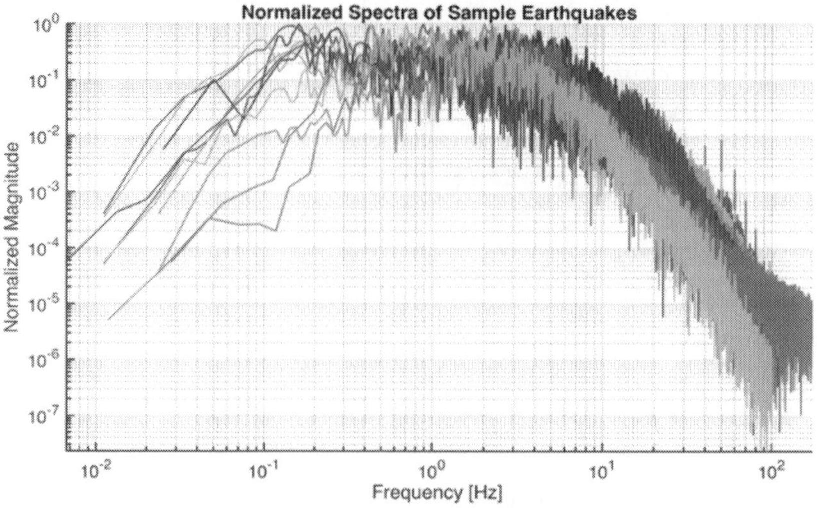

Figure 7. Normalize spectra of a sample set of earthquakes. The signal typically has bandwidth between 0.1 and 30 Hz, making it act as a bandpass filter.

MAKING A FRAGILITY CURVE

In our case study, a large number of earthquake records were acquired from the PEER Ground Motion Database. Each of these earthquakes is convolved with the $h_i(t)$ impulse response described in the previous section to produce the estimated response of the dam, at each location i. This section will outline the analysis process for each of these calculated output records in order to produce a fragility curve.

A fragility curve maps the probability of failure to the intensity of a hazard condition. Hence, in order to produce a fragility curve, we must quantify 'failure', as well as choose an appropriate intensity measure to represent the intensity of expected earthquakes. These two problems are respectively addressed in the two papers (Anderson et al. 2016, Zhao et al. 2016). For the purposes of this example analysis, we will assume that *Peak Ground Displacement (PGD)* is the intensity measure of interest, and *Peak Response Displacement (Maximum Response Displacement)* is the performance measure. However, each measure can be any function of the input earthquake signal (for intensity) and of its convolution with $h_i(t)$ (for performance). A given input earthquake signal and its corresponding output are analyzed to produce the intensity and performance measure, and the pair is plotted on a scatter plot. This is done for each earthquake in the database for a given i, and ultimately produces what is shown in Figure 8. As can be seen, PGD correlates well with Maximum Response Displacement when presented on a Log-Log scale.

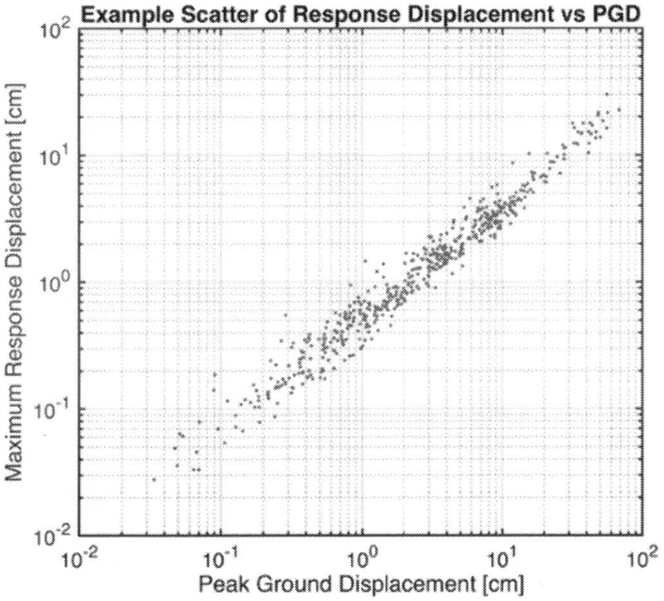

Figure 8: Scatter plot displaying the relation between PGD of the input earthquakes and the Maximum Displacement of the response at an example location on the top of a dam.

The next step in producing a fragility curve is to apply a performance limit. This is a binary condition that states that values of the performance measure above the performance limit qualify as a state of 'failure', and values below the performance measure do not. Thus, for the remainder of this paper, the 'probability of failure' is to be interpreted as the 'probability of exceeding a given performance limit'. Applying the performance limit, we can turn the scatter plot into a plot of binary values. We then conduct a probabilistic regression by fitting a functional form of:

$$p(x) = \frac{1}{2} + \frac{1}{2}\mathrm{erf}\left[\frac{ln\, x - \mu}{\sqrt{2}\,\sigma}\right] \tag{7}$$

where x is the intensity measure, $p(x)$ is the probability of failure for a given x, and μ and σ are fitting parameters. The function presented is the lognormal cumulative density function, which has a shape similar to that of a typical fragility curve. MATLAB gives us confidence limits on the fitting parameters of the regression, and these values can be used to construct the best and worst case 'confidence limits' for the fragility curves, and shown on the figures in the next section. Once this regression is performed, the resulting function is the fragility curve for location i. This process can be repeated for each location i to generate the fragility curves of the entire structure.

The next section describes the results from applying this process to the results of CGT testing San Clemente Dam and evaluating the process described to scale the impulse response to a CGT to the impulse response to a ground motion.

RESULTS FOR SAN CLEMENTE DAM CASE STUDY

This section discusses the results of applying the methods described in the entirety of this paper to the measured CGT response of San Clemente Dam. The experimental data was collected at 16 locations across the top of the dam with the CGT located at the base of the dam, as shown in Figure 2. The 16 CGT responses were then convolved with the horizontal acceleration records of 500 different earthquakes sourced from the PEER Ground Motion Database. The output acceleration records were then scaled according to the method described above. The records were analyzed using Maximum Response Displacement at the accelerometer location as the performance measure and PGD as the intensity measure of the input earthquakes. A performance limit of 5.5 cm (2.16 in.) was used, and fragility curves were produced. The fragility curves for every other location on the dam are shown in Figure 9.

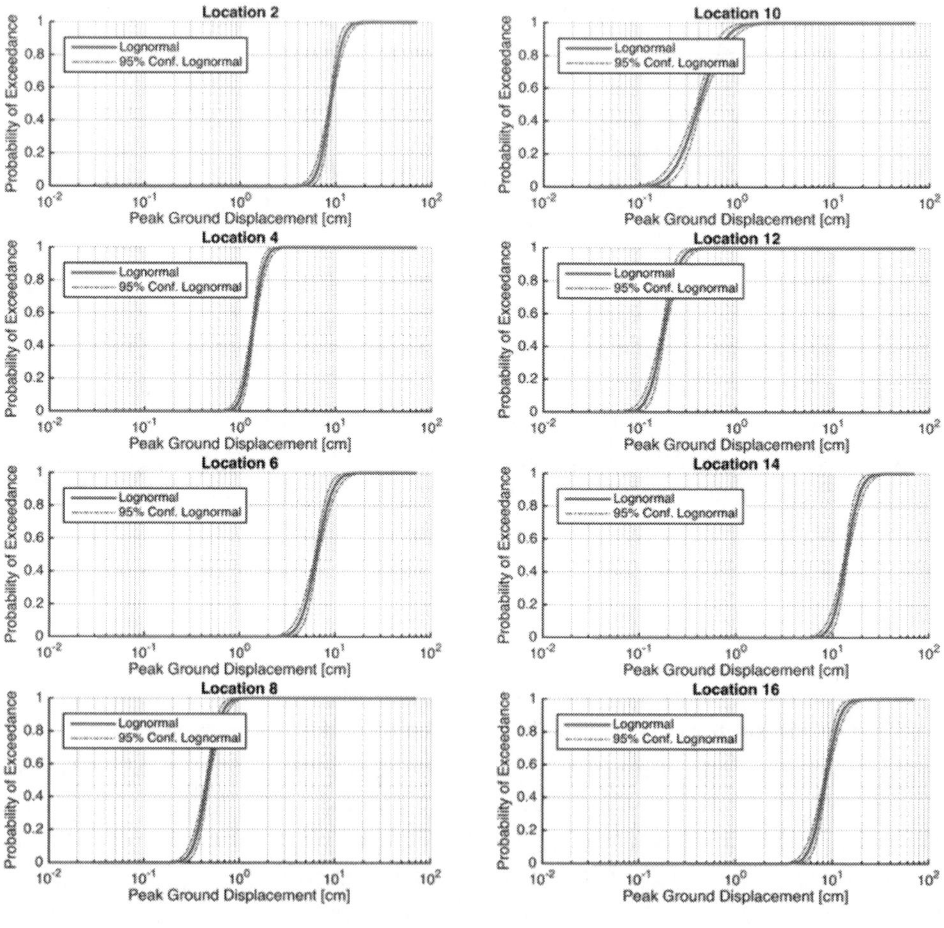

Figure 9. Fragility curves using model based method for every other location on the dam, using a 5.5 cm performance limit.

A number of observations can be made from this set of curves. The first is that the curves close to the abutments of the dam look close to identical, which is sensible given that the abutments should have motions that are quite similar to the ground motion during the earthquake, with close to a one to one correspondence between PGD and the Maximum

Displacement at these points. We would then expect to see larger Maximum Displacements further away from the abutments (closer to the middle of spillway) where the dam is presumed to be most flexible. Due to the enforcement of amplitude scaling using the first eigenmode of the dam's vibration, this *is* what we see – locations near the center with a substantially larger probability of deflecting more than 5.5 cm for the same PGD than the locations close to the abutments. This result is more clearly illustrated in Figure 10, which is a plot of range of PGDs that produce close to a 5.5 cm Maximum Response Displacements at the 16 locations of the dam. Again, we can see some conformity to the first eigenmode of the dam, where the abutments appear more rigid, and thus resilient, than the middle of the spillway. However, as seen in this plot, there isn't perfect conformity to the shape of the first eigenmode. The local deviation for this imposed constraint is due to location's specific information contained in the CGT responses. We can even highlight from Figure 10 that location 12 is likely the most susceptible to damage, which is information not given from the FEM simulations. Location 12 is shown on a picture of the dam for reference. Thus, we can see the advantages of combining FEM and performance based methods – in that the FEM analysis governs the global behavior in response across the dam, and PBT highlights the localized behavior not captured in FEM models.

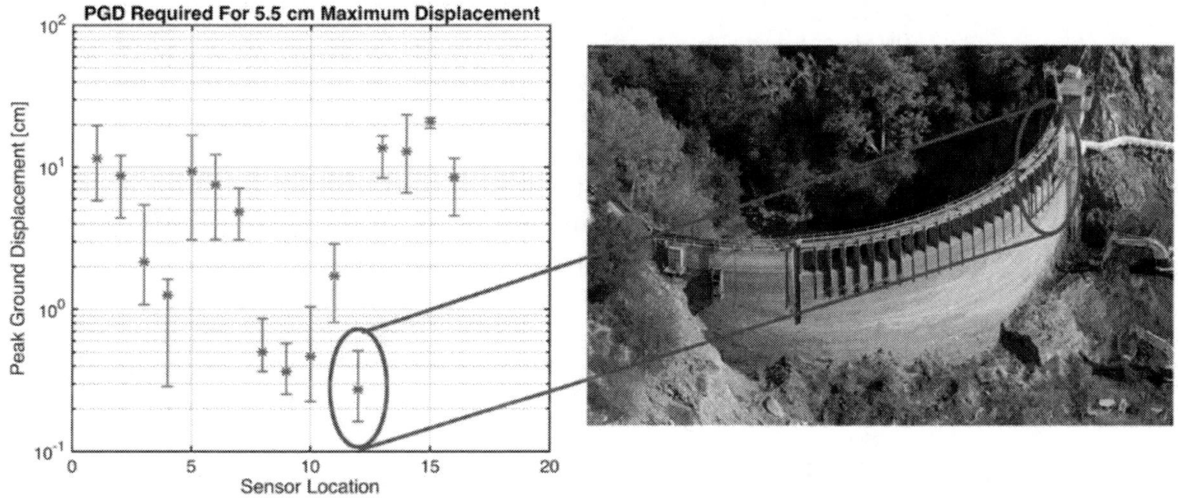

Figure 10: (Left) Figure showing the mean and range of Peak Ground Displacements that cause a Maximum Displacement of 5.5 cm. Lower PGDs imply that the location is more susceptible to damage during earthquakes. (Right) Shows the location on San Clemente Dam that looks most susceptible to damage by earthquake loading.

While it then seems like FEM is required to provide good estimates of fragility, it is important to note that the information required from FEM is only an accurate estimation of the response to sinusoidal excitation at the fundamental frequency. This is in the capacity of even a very simple model that gets the mass and geometry of the dam correct, and is likely to be an inexpensive requirement in this process.

Finally, it is of note that the process of convolution can be completed at an incredibly fast rate on an inexpensive, consumer computer. On a computer with the specifications listed the footnote[5], the convolution of a single response with the database of 500 earthquakes takes about 2.5 minutes, and thus takes about 40 minutes to complete the analysis for all 16 locations monitored at San Clemente Dam. The simulations on the FEM model of the dam were completed in about 2 hours on a computer, but this analysis can be completed prior to the conclusion of the PBT testing on the dam. Hence, the limiting factor in time required for the complete analysis is by no means the computation, which requires less than an hour of time, but rather the experimental data acquisition through PBT. The test at San Clemente Dam acquired the necessary data within three days, and an effort is underway to further reduce the time required for PBT.

The current methods for estimating fragility or the response to dynamic loading require pure reliance on FEM models, and even with band-limiting the hazards for the sake of computation, require weeks to months to perform the analysis. Hence, such a drastic improvement in computation time changes the nature of such analysis from an assessment tool to a tool that can provide real-time predictive information regarding a hazard for which short warning has been given.

CONCLUSION

This paper has outlined the method in which the impulse responses of a large civil structure, as measured by the use of a Cold Gas Thruster, can be used to estimate the structure's fragility curve for a given hazard, such as earthquakes. The paper describes the steps taken to scale the measured response of the structure to an impulse load from the Cold Gas Thruster to that of an impulse load from a ground motion. Next, the paper describes the process of digitally convolving this scaled response with a ground motion acceleration record of an earthquake. Finally, the paper describes the steps taken to generate a fragility curve of a structure from the the calculated responses to a database of earthquake records. The process is illustrated using a case study for San Clemente Dam, which confirms that the results not only match intuitive expectations, but reveal information regarding the structural health of the dam that is unlikely to easily be found through the use of Finite Element Modeling alone, showing the value of conducting Performance Based Testing.

The methods described in this paper have outlined a field-based approach for estimating a structure's fragility curve. The method offers a practical alternative to current numerical approaches that rely on randomized groupings of model parameters to address uncertainties in the dam's actual behavior, and that require significant cost and computing time to complete. The Performance Based Testing based fragility estimates can be obtained for piers, monoliths, radial gates, or any subsystem or component in a dam where vulnerability to hazard loading conditions might be suspected.

[5] Apple MacBook Pro with a 2.9 GHz Intel Core i5 processor, 512 GB Flash Hard Drive, and 16 GB RAM

ACKNOWLEDGEMENTS

The authors wish to acknowledge the support from the following individuals and their organizations for making the field investigations at San Clemente Dam possible.

David Gutierrez, Chief, and Mark Schultz, Supervising Engineer from the Division of Safety of Dams Department of Water Resources, California.
Richard Svindland, VP of Operations at California American Water, dam owner.
Bill McGowan, Project Executive at Granite Construction Company.
Dylan Wade, Resident Project Engineer at San Clemente Dam for Water Systems Consulting, Inc.
Robert Carnes, consultant and logistical support to the Resident Project Engineer.

The authors are especially grateful to Mr. Wade and Mr. Carnes for their efforts in managing our support needs and safety during the performance testing at the dam.

REFERENCES

"PEER Ground Motion Database.", Berkeley, California, Pacific Earthquake Engineering and Research Center.

Anderson, S., K. Menda, M. Zhao, and Z. Duron. (2016). "Selecting an Intensity Measure for Fragility Curves Derived from Performance-Based Testing." Denver, Colorado, 36th Annual USSD Conference.

Duron, Z., J. Yen. (2016). "A New and Advanced Dam Perform ance and Response Integration Program." Denver, Colorado, 36th Annual USSD Conference.

Duron, Z., D. Chen, M. Cuenca, S. Hanson, M. Goldkam p. (2013). "Development of a Stability Loss Monitoring Al gorithm for Burning Structur es." Gaithersburg, Maryland, Building Fire Research Laboratory, National Institute of Standards and Technology.

Duron, Z. J. Edelm an, M. Goldkamp, P. Or me, T. Savage. (2013). "Perform ance Based Testing of Large Concrete Dam s". The J ournal of Dam Safety, Association of Da m Safety Officials.

Zhao, M., S. Anderson, K. Menda, and Z. Duron. (2016). "Sel ecting a Perform ance measure for Fragility Curves Derived fr om Performance-Based Testing." Denver, Colorado, 36th Annual USSD Conference.

IMPLEMENTING LESSONS LEARNED FROM WHEELER LOCK COLLAPSE ON NEW LOCK CONSTRUCTION PROJECTS

Michael A. Morrison, PMP, PE[1]
Bernard Schulte, PE[2]

ABSTRACT

Lock walls experience hydraulic loadings and stresses very similar to concrete dams. For locks, the frequency of high and low water conditions is much greater due to chamber water levels that rise and fall quickly.

In 1961, the Tennessee Valley Authority (TVA) began construction on the Main Lock at Wheeler Dam. On June 2, 1961, the adjacent Auxiliary Lock experienced a catastrophic failure while the chamber was full, and flooded the construction site for the Main Lock. The subsequent investigation findings by independent consultants led to a change in the construction approach for TVA navigation projects from that time forward.

The new approach features additional conservatism in the analysis of existing lock walls which abut the construction sites for new locks. This conservatism can be found in the post-tensioning of lock walls and the implementation of the checkerboard excavation concept. The checkerboard excavation concept features the deliberate staging of an excavation so that pre-designed masses of rock foundation blocks are left behind adjacent to the existing lock wall during the initial construction stages of a new lock for stabilization. In this fashion, excavation work and focused blasting operations can take place while commercial river traffic passes through the existing lock, often located adjacent to the blasting area. These measures require significant investments in instrumentation and monitoring of the construction site to ensure worker safety and structure stability.

This paper will describe how the measures to maintain river traffic during construction of new locks were implemented at a new lock constructed by TVA in the 1980s and at the Kentucky Lock which is under construction at this time. Results of instrumentation and monitoring programs will be discussed as a means to document the effectiveness of the new measures.

INTRODUCTION

In 1961, the Tennessee Valley Authority (TVA) commenced construction on the Main Lock at the Wheeler Dam project. A few months after construction began, one of the monoliths from the adjacent Auxiliary Lock slid into the excavation for the Main Lock starting a progressive collapse of 436 feet of land wall, resulting in a four-day

[1]Senior Program Manager, Dam Safety, Tennessee Valley Authority, 1101 Market Street, Chattanooga, TN 37401, mamorrison0@tva.gov .
[2]Structural Engineer, Civil-Structures Section, US Army Corps of Engineers, 801 Broadway, Nashville, TN 37201, Bernard.d.schulte@usace.army.mil .

uncontrolled release of water that effectively shut down river traffic for years while repairs were made. TVA commissioned an independent investigation of the lock failure. Results and recommendations from that investigation were the beginning of a new paradigm in the design of hydraulic structures that featured staged excavations and concrete placements, post-tensioning of existing structures, and installation of instrumentation in an effort to proactively detect any movement that might impact the new structure. New TVA locks are typically designed and constructed in close proximity to existing locks. The design techniques described in this paper were implemented to provide increased safety at the construction site and to minimize interruptions of river commerce while the locks were being built.

WHEELER LOCK DISASTER

Wheeler Dam is 72-feet-high concrete dam that was constructed on the Tennessee River from 1933 to 1936 by the Tennessee Valley Authority (TVA). Like many TVA dams, Wheeler Dam was constructed to serve multiple purposes, including navigation. From a navigation perspective, Wheeler Dam is at the upstream edge of the Muscle Shoals, a barrier that blocked river transportation and commerce until construction of the Wilson Dam by the Corps of Engineers in 1919. Wheeler Dam was constructed with a 60' x 360' navigation lock. Plans for the dam included dedicating a space for a second future lock to the north (landward side) of the existing lock (See Figure 1). Initial excavation for the second lock was performed during the construction of the dam and lock.

The second lock, hereinafter referred to as the Main Lock, was designed to be larger in size, 110' wide and 600' long, so that a larger volume of river commerce could be accommodated. Subsurface exploration that preceded design of the lock began in 1960. The subsurface exploration program consisted of 24 3-inch-diameter holes drilled along the center line of each wall of the Main Lock to a depth of ten to thirty feet into the rock layer. The rock layer encountered was limestone of the Fort Payne formation. In addition to the drill holes, three 36" diameter calyx holes were drilled to depths of 19 to 25 feet in the rock layer. TVA engineers and geologists were lowered into the holes to perform direct visual inspection of the limestone and shale layers.

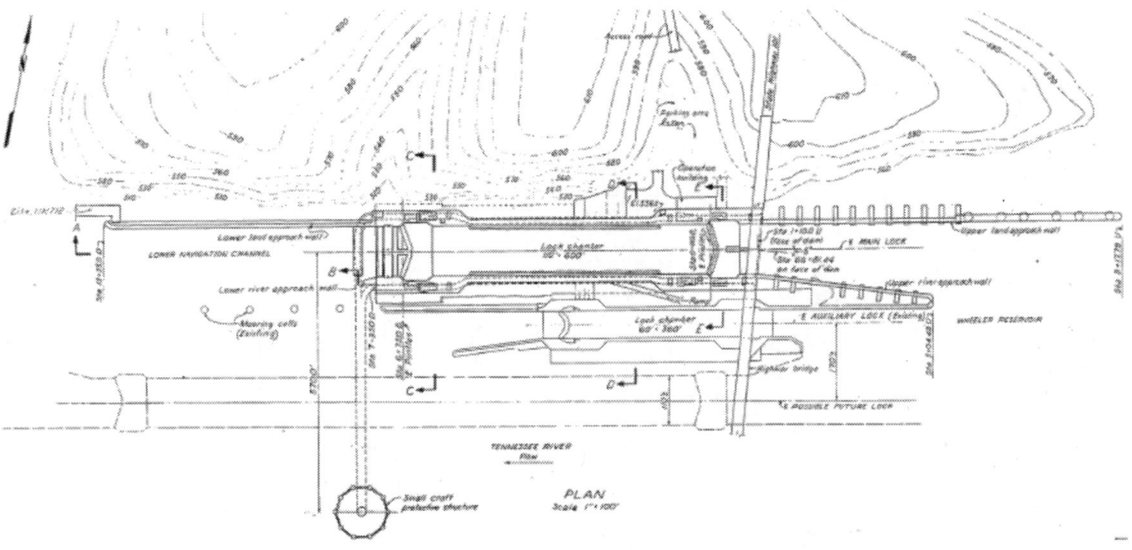

Figure 1. Locks at Wheeler Dam

The lock walls were designed to be founded on the rock layer. The rock foundation had a six-inch thick shale layer at the base of the proposed wall foundation. Lab shear tests on the shale indicated that it would have sufficient strength to support the existing wall during construction of the new lock.

Construction began on the lock in 1961. Excavation to deepen and widen the trench for the Main Lock was underway, landward of the Auxiliary Lock. On June 2, 1961, a full tow heading upstream had just cleared the lock. The downstream lock gates were closed. The upstream gates were open. The chamber was full. Unexpectedly, the land wall gate monolith began to slide into the Main Lock excavation. Water in the chamber poured into the excavation. Shortly thereafter, neighboring wall monoliths began to slide into the excavation. The failure progressed until a 436-foot-long segment of the land wall (both chamber and approach wall sections) had fallen into the new lock excavation (See Figure 2). Ground vibrations from the failure were recorded on seismographic instruments as far away as St. Louis, Missouri. The upstream-bound tow was sucked into the breach, resulting in the death of one barge sailor and one worker in the excavation. Miraculously, none of the other 40 men working in the excavation was killed. Because the upstream lock gates were open, the entirety of Wheeler Lake above the upstream gate sill was able to pour through the breach created by the failed lock wall monoliths. Four days later, the lock operators were able to erect a needle dam closure structure in the forebay of the Auxiliary Lock. Almost two weeks later, a cofferdam was constructed downstream of the disaster area and it was dewatered, allowing recovery efforts to begin in earnest.

Figure 2. Dislocated Monoliths at Wheeler Lock

In addition to the human tragedies, Wheeler Lock was shut down for a period of four years while repairs to the Auxiliary Lock were designed and constructed. Ingenious temporary methods were developed to convey cargo over the Wheeler Dam, however, none of them had the capacity of a functional Auxiliary Lock, let alone the needed capacity of the new Main Lock.

TVA commissioned an independent investigation into the causes of the Wheeler disaster, and hired three independent consultants (two engineers and a geologist) to perform a complete investigation. The initial plan for performing the investigation featured additional drill holes and in-situ testing of the shale layer. Lab testing of the shale to date had shown that the layer should have been able to withstand the compression and shear loads imposed during the construction of the Main Lock. The plan for investigations changed when a field engineer explored a calyx hole that was drilled and dewatered for the culvert outlet structure, located several hundred feet away from the Main Lock. The engineer discovered a very thin clay seam at the base of the shale layer. Subsequent calyx holes progressing toward the lock were drilled and explored. Observations within these calyx holes and at the failed wall led to the conclusion that the clay seam extended underneath the Auxiliary Lock on a nearly horizontal plane, providing an excellent, lubricated sliding surface for the land wall. The excavation for the Main Lock extended below the clay seam, thus facilitating the sliding of the wall along the seam (See Figure 3). This is how the land wall failed.

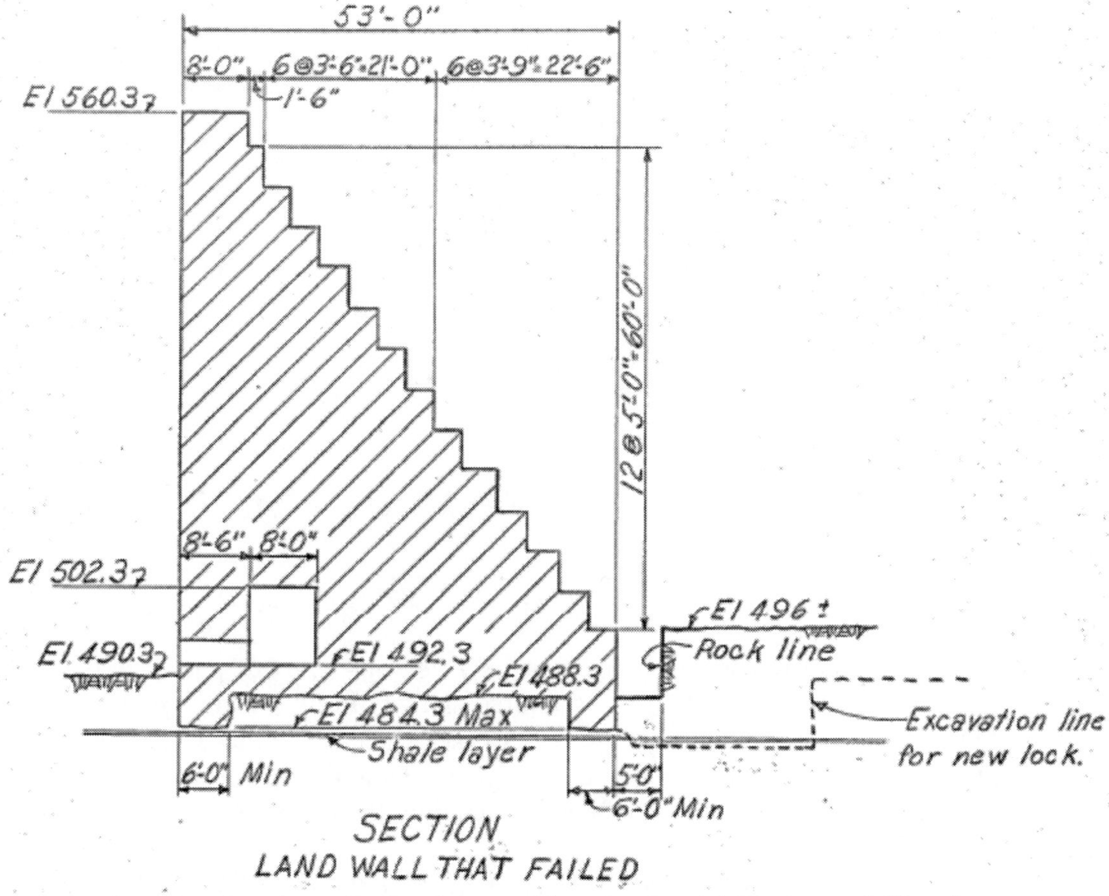

Figure 3. Section of Wheeler Auxiliary Lock with Horizontal Clay Seam

Subsequent re-examination of the pre-construction calyx holes and subsurface exploration records indicated that the clay seam may have been mistaken for drill cuttings.

The consultants' report ended with a number of recommendations for the design of the replacement Auxiliary Lock and the new Main Lock. Those recommendations can be summarized as follows:

1. All structures to be founded on rock below the shale layer.
2. Thicker wall cross sections for greater structural stability.
3. Demolition and replacement of damaged Auxiliary Lock river wall segments in accordance with the design features stated above.
4. Foundation grouting of chamber walls.
5. Post-tensioned anchors, additional foundation drains, a thicker mass concrete section, and a structural key below the shale layer for the Main Lock upstream miter gate monolith which would abut the Wheeler Dam non-overflow section.

6. Blasting associated with the removal of damaged blocks and foundation preparation would be performed with great care so as to not damage Wheeler Dam.
7. Construction of a floor slab in the Auxiliary Lock to provide greater structural stability of the empty chamber.
8. Constructing the land wall of the Main Lock against an excavated rock face on the north abutment of the dam. A substantial subsurface drain was installed above the rock layer to control hydrostatic loading on the land wall.

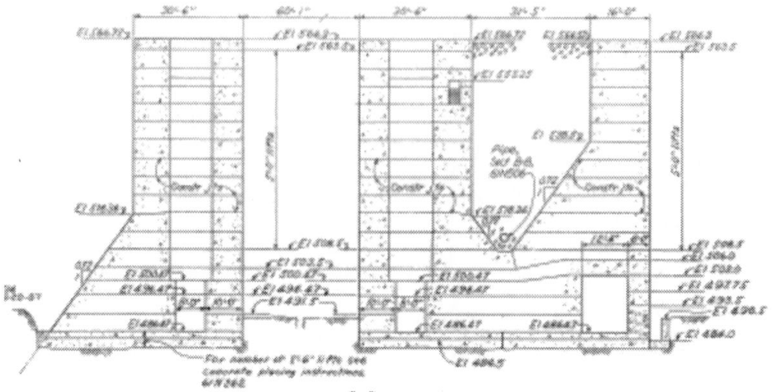

Figure 4. Section of Redesigned Wheeler Auxiliary Lock

These recommendations would form the basis for a "lessons learned" approach that would appear in the design of hydraulic structures that would be designed by or constructed for TVA. In particular, we will focus how these principles were applied to the design of Pickwick Main Lock, which was designed and funded by TVA, and Kentucky Lock, which has been designed and funded by the Corps of Engineers.

PICKWICK LOCK DESIGN AND CONSTRUCTION

Pickwick Main Lock is part of the Pickwick Dam multipurpose project located between Kentucky Dam and Wilson Dam on the Tennessee River. From a navigation perspective, Pickwick Lake is a key navigational crossroads as it forms the junction of the Tennessee-Tombigbee Waterway that provides a direct waterway connection to the Gulf port of Mobile, Alabama. The lock features a 110-foot-wide by 1000-foot-long chamber and a lift height of roughly 60-feet under summer pool conditions.

As was the case at Wheeler Main Lock, Pickwick Main Lock was designed and built several years after the dam and 110-ft by 600-ft Auxiliary Lock were constructed. The dam and Auxiliary Lock were built from 1934 to 1938. The original project design featured space for the construction of a larger additional lock on the landward side of the existing lock (See Figure 5). The Main Lock was designed and built from 1976 to 1984.

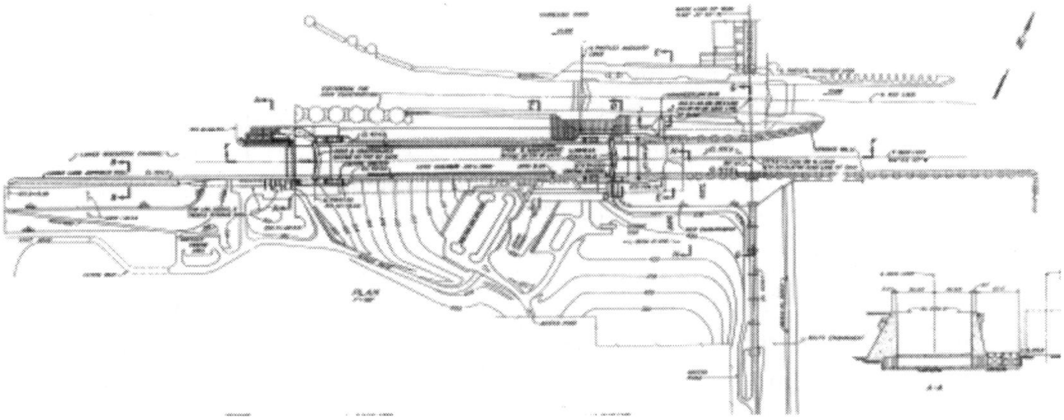

Figure 5. Layout of Pickwick Locks

Design of the Main Lock was documented in TVA's Report No. HES-92-1, *Pickwick Landing Project, Main Lock Addition, Final Design Report*. This paper will focus on design features that were implemented to prevent a failure of the Auxiliary Lock that would have been similar to the failure at Wheeler Lock.

The chamber walls of the Auxiliary and Main Locks were founded on Argillaceous limestone of the Ross Formation. Subsurface Exploration during design of the Main Lock was performed in two phases. The first phase, completed in 1964, included the boring and logging of 38 cored holes along the layout of the lock walls to a depth of approximately 77 feet into the rock layer. The second phase included 34 holes drilled in 1977 along the proposed lock wall and upper miter sill locations to depths ranging from 43 to 84 feet into the rock layer. Two 36-inch calyx holes totaling 177 feet in length were also drilled into the Auxiliary Lock land wall. During construction, additional holes, including 27 cored holes, 11 percussion holes, and two calyx holes were drilled at the upstream half of the lock in an effort to locate and evaluate horizontal weathered seams in the foundation rock. The seams were located at Elevations 334 and 318 feet. Foundation grade for most of the Main Lock wall monoliths was Elevation 331-332 feet. Foundation grade for the downstream gate and adjacent monoliths was Elevation 335. The uppermost seam was two to three inches thick.

During the design and construction of the Main Lock, several measures were taken to address concerns about the stability of the Auxiliary Lock related to new loading conditions on the Auxiliary Lock land wall:

1. Stability analyses found that routinely-occurring hydrostatic loads in the Main lock would cause unacceptable stresses in the Auxiliary Lock land wall when the Auxiliary Lock chamber was operating at routine tailwater level or when the Auxiliary Lock was being dewatered for maintenance. To address this concern over unacceptable stresses in the Auxiliary Lock land wall, post-tensioned anchors were installed to tie the monoliths to the rock foundation and mass concrete was added to the land side on six land wall monoliths.

2. The downstream cofferdam installed for the Main Lock construction connected the downstream land approach wall of the Auxiliary Lock and the left bank of the Tennessee River. Stability analyses performed on the approach wall determined that the approach wall would not be able to withstand routine hydrostatic loads imposed by normal tailwater conditions and the dewatered lock construction site. To address this deficiency, post-tensioned anchors were installed in the downstream land approach walls. The anchors were designed to be de-stressed after construction of the Main Lock was completed.
3. The five upstream wall monoliths of the Main Lock river wall were designed to match the lengths of the Auxiliary Lock land wall monoliths against which they were to be constructed. Since these walls abutted one another, it was felt that matching the lines of the monolith joints would prevent harmful stress concentrations in the Auxiliary land wall monoliths.
4. As stated previously, two horizontal seams were encountered in the vicinity of the wall monolith foundations. To prevent any potential for movement along the seams, the foundation grade for most of the Main Lock wall monoliths was set below the upper seam. Excavations and concrete placements were deliberately staged so that a substantial portion of each individual monolith on the Auxiliary Lock would be buttressed at the base by either new mass concrete or the existing rock layer. This concept would be referred to as "checkerboarding" the foundation. Checkerboarding was used for the Main Lock including the lower gate and adjacent monoliths that were founded above the upper seam.
5. Checkerboarding was also used during the excavations for the culvert discharge structures, which were situated below the base of the nearby cofferdam and below both of the known horizontal seams in the rock.
6. Additional sliding protection for the cofferdam was provided by a two-foot-thick slab anchored to the rock on the land (downstream) side of the cofferdam. Instrumentation referred to as "slip-sentries" was installed on the slabs to monitor any movements in the slabs.

Figure 6. Pickwick Auxiliary Lock Post-Tensioning

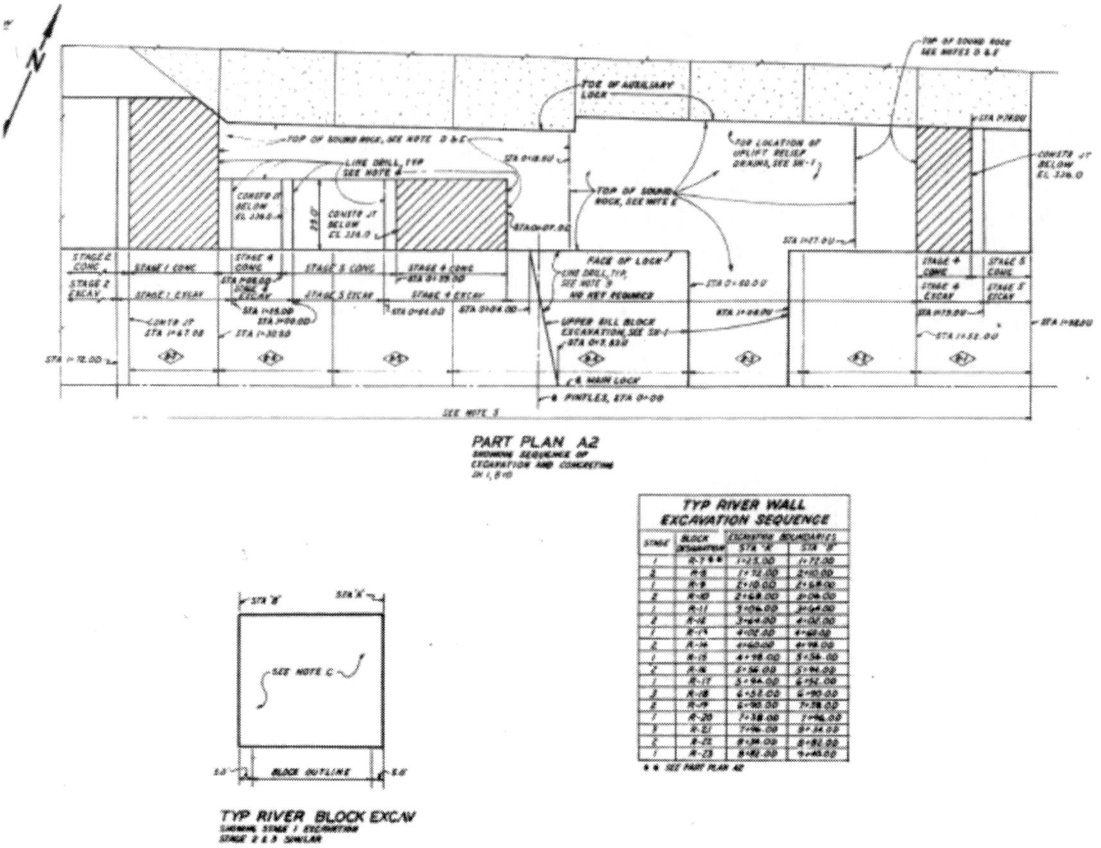

Figure 7. Layout of Checkerboard Excavations at Pickwick Locks

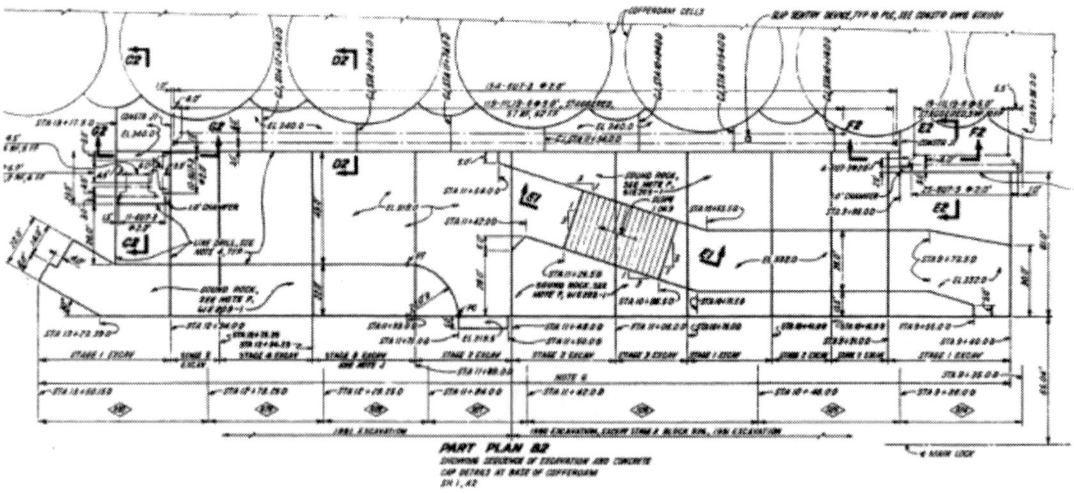

Figure 8. Slabs and Instrumentation Placed at Pickwick Locks Cofferdam for Slide Protection

KENTUCKY LOCK ADDITION DESIGN AND CONSTRUCTION

The existing Kentucky Lock is part of the Kentucky Dam multipurpose project located between Pickwick Dam and the confluence of the Tennessee River and the Ohio River. Kentucky Lock is the busiest and most heavily used lock on the Tennessee River, with over 25 million tons of material passing through its lock each year. With this large amount of traffic, shipping delays through Kentucky Lock have steadily increased each year. In order to reduce these shipping delays, a new 110-foot wide by 1200-foot long navigation lock is currently under construction landward and adjacent to the existing 110-foot by 600-ft lock.

Design and construction of the new lock structure is being completed in stages to ensure stability of the new and existing lock and to also match the variable multi-year funding stream for the project. The main stages of lock addition design and construction are:

1. Upstream Cofferdam
2. Upstream Excavation and Monoliths
3. Downstream Cofferdam
4. Downstream Excavation and Monoliths
5. Middle Wall Excavation and Monoliths.

Removal of the upstream and downstream cofferdams, construction of the upstream floating approach walls, and construction of the downstream approach walls will complete the lock addition.

Figure 9. Photograph Showing Overall Kentucky Lock Project and Upstream Excavation

The construction of the upstream monoliths is currently nearing completion. The current Upstream Monoliths contract involves the construction of two upstream u-frame monoliths, two middle wall monoliths, and five land wall monoliths. Design of the monoliths is documented in USACE's *Design Documentation Report, Phase 1, Kentucky Lock Addition, Upstream Lock Monoliths*.

The Mississippian limestone of the Warsaw and Fort Payne formations form the bedrock at the location of the existing Main Lock structure. The Fort Payne lies just below the Warsaw formation. The existing lock monoliths and the new lock addition upstream monoliths are on the Fort Payne formation. Bedding is nearly horizontal, dipping slightly in the downstream direction. Some bedding planes are tight and some are open and weathered with clayey coatings. The most notable seams, including the seam between the Warsaw and Fort Payne formations, are well above the founding elevation of the new and existing lock monoliths. For the design of the new lock structure, deep seated failure planes were not considered because the geologic data did not indicate the presence of weaker planes below the founding elevations of the monoliths.

Construction of the monoliths for the new lock addition requires the excavation of soil and rock behind the existing landward lock monoliths. The founding elevation of the new lock addition monoliths varies from 286.0 to 276.0. The founding elevation of the existing landward lock monoliths is approximately 283.0. Therefore the concrete-to-rock interface of the existing monoliths was exposed by the excavation for the new lock monoliths construction. Stability analysis of the existing monoliths considered failure planes at the horizontal planes at the interface between the foundation rock and the bottom of the monoliths, horizontal planes at the bottom of the excavation (elevation 276.0), and horizontal planes through the monoliths at changes in wall geometry.

Stability analysis of the existing lock's landward chamber monoliths found that they would either not be stable or would not meet current USACE criteria once soil was excavated from behind the walls. The worst load case involved a fully excavated site combined with headwater in the existing lock chamber. To address this concern, post-tensioned anchors were installed to tie the existing lock monoliths to the rock foundation. A total of six 30-strand anchors were installed in four of the exposed monoliths (EL-8, EL-9, EL-10, and EL-11). In addition, to add vertical weight, mass concrete was added to the back side of four monoliths (EL4, EL-5, EL-6, and EL-7).

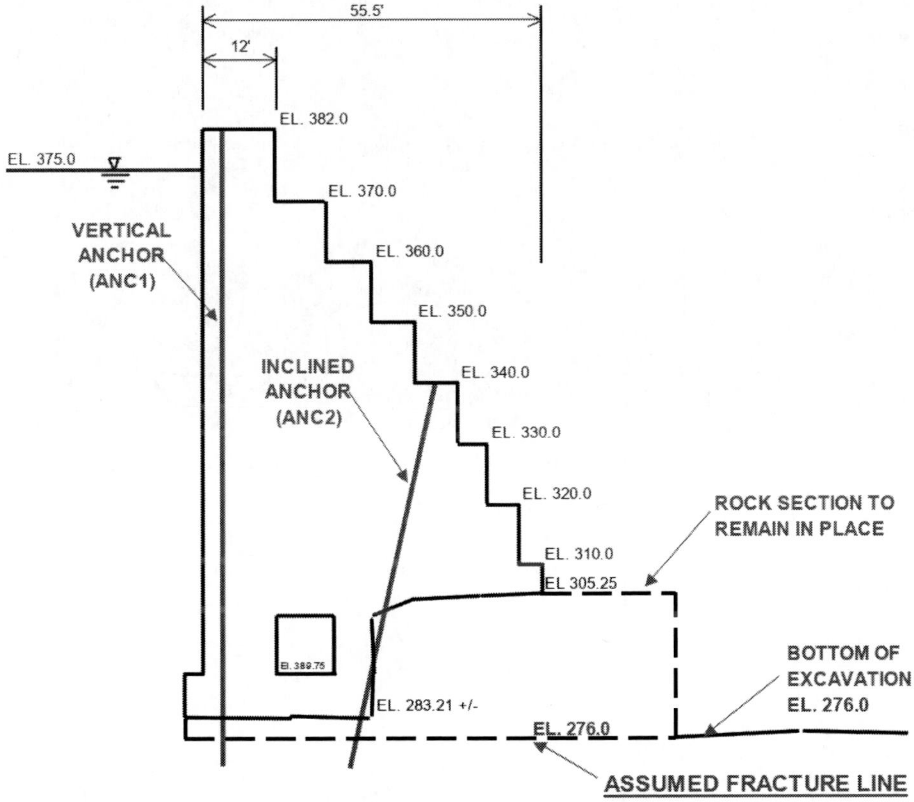

Figure 10. Kentucky Lock Addition Post-tensioning of Existing Auxiliary Lock with Checkerboard in Place

Although available geologic data did not indicate any specific horizontal seams in the vicinity of the founding elevations of the existing lock, as at Pickwick Lock, "checkerboarding" was used during excavation for the lock addition as an added level of security against potential sliding. For the checkerboarding scenario, sliding stability was conservatively checked with a cohesion value of zero to account for potential unknowns. Similar to what was done at Pickwick, excavations and concrete placements were deliberately staged so that a substantial portion of each individual monolith on the Main Lock would be buttressed at the base by either new mass concrete or the existing rock layer. Figures 11 thru 13 show the checkerboarding during excavation.

Figure 11. Partial Excavation before Checkerboard Cut in Rock at Kentucky Lock Addition

Figure 12. Checkerboard Excavation at Kentucky Lock Addition

Figure 13. Checkerboard Excavation at Kentucky Lock Addition

In addition to the physical measures to ensure the stability of the Main Lock's existing land walls, extensive instrumentation has been installed to monitor for any movement during excavation and construction of the new lock. The instrumentation includes jointmeters and relative block movement devices (RBMD) at each monolith joint to measure the relative movement between monoliths. Tiltmeters have been installed at the top of each monolith to monitor any rotation. Inclinometers were installed in each monolith down into the founding rock in order to monitor displacements at critical monolith cross-sections and at the concrete-to-rock interface.

A data logging/data acquisition system is used to collect instrumentation data. All instrumentation is read automatically by the acquisition system, except the manually read RBMD's. The system is capable of performing continuous screening of the data to detect any change in a reading from the previous readings. Readings are then compared to allowable threshold values. If readings exceed the threshold values, the system will send emails and/or voice messages to up to five (5) people. Weekly and quarterly reporting is a required part of the Upstream Monolith construction contract and the reports are used to identify long term trends in movement. At this time, other than instrument errors, such as voltage spikes, no threshold alarms have been reported for the existing lock's instrumentation.

CONCLUSION

The Tennessee Valley Authority has been able to recover from a significant structural catastrophe at Wheeler Lock that shut down river traffic for several years and use lessons learned from that catastrophe in designing new hydraulic structures. Those lessons learned include stability analyses focused on construction condition loading conditions, post-tensioning of existing structures, and phased excavation that minimizes risk to existing structures. The TVA and Corps of Engineers have implemented these lessons in construction of new structures which has resulted in improved worker safety, structure stability and maintaining river traffic with a minimum of interruptions.s Those lessons learned include stability analyses focused on construction condition loading conditions, post-tensioning of existing structures, and phased excavation that minimizes risk to existing structures.

REFERENCES

Report on Wheeler Lock Failure, Tennessee Valley Authority, Office of the Chief Engineer, Report No. 3-480, November 1961.

Pickwick Landing Project, Main Lock Addition, Final Design Report, Tennessee Valley Authority, Generation Group, Hydro Engineering Services, January 1992.

Design Documentation Report, Phase 1, Kentucky Lock Addition, Upstream Lock Monoliths, USACE, Black and Veatch Special Projects Corp., March 2003.

ASSESSING AND EVALUATING THE OPERATION OF PRESSURE CELLS IN EMBANKMENT DAMS: A CASE STUDY OF FIVE IRANIAN EMBANKMENT DAMS

Bahareh Shoghli[1]
Ali Asghar Mirghasemi[2]
Yeo Howe Lim[3]

ABSTRACT

Overtopping is one of the important dynamic factors leading up to an embankment failure. An overtopping can occur when a reservoir level reaches the crest of an embankment dam where water can either flow inside the dam body or over the downstream shells. When the reservoir level is at the near-overtopping level, the water content of permeable materials is raised, and the effective stress decreases. The probability of cracks occurring increases as shear stresses between different layers or zones of fill materials, used in embankment construction become overwhelming due to the decreasing effective stresses.

Earth pressure cell is one of the most useful instrument used for monitoring total pressures in earth-fill and embankment dams, the essential information for predicting dam failure. The cells are usually installed in clusters, and each cluster consists of three to five earth pressure cells.

Monitoring and evaluating the behavior of the dams during construction and operation time is an important role in providing the safety of dams. Accurate monitoring needs the comprehensive, accurate and truthful information. These are provided by the instruments that are installed in the important and sensitive part of the structure.

In this research, measured data from earth pressure cells in five Iranian embankment dams are used to investigate the internal consistency of the pressure cells. Depending on the type of the available data (number of pressure cells in one plane) and stage of the dam for which the data is measured, three methods are used to find the consistency of the earth pressure cells.

INTRODUCTION

To control the safety of any geotechnical structure, it is essential to install instruments to monitor the performance of structure, during construction and operation. The primary objective of geotechnical instrumentation plan is to allow an analytical assessment of the

[1] Graduate student, Department of Civil Engineering, University of North Dakota, USA, Bahareh.shoghli@und.edu.
[2] Professor, School of Civil Engineering, College of Engineering, University of Tehran, Iran, aghasemi@ut.ac.ir
[3] Associate Professor, Department of Civil Engineering, University of North Dakota, USA, howe.lim@engr.und.edu

current conditions such that the prediction of future performance, evaluation, development and verification of the future design, can be realized.

Data management and analysis are crucial for the detection of defective locations in structures and evaluating the overall performance of the structures. To have comprehensive, accurate and trustable information, the proper functioning of precise instrumentation is essential. For this objective, instruments should be installed as close as possible at the vicinity of the critical and sensitive part of the structure. Analyzing the data would be helpful in the design, maintenance and prediction of the failure of the dams. The critical parameters that ideally should be measured by the help of the instruments in the embankment dams are: external and internal displacement of dams, abutments, appurtenant structures; total soil pressure in the different part of the structure; foundation and embankment water pressure; and seepage through the embankment and its foundation.

During the time of the construction of a dam, due to the increasing weight of the materials, the body of dams will settle, because of difference in stiffness between the core and transition zones. It is obvious that different deformation and settlement will happen in the body of the dams. Regarding the nature of the clay that is the central part of the core of dams, the settling and consolidation of this material will continue many years after the construction of the dams. However, for the other zones of the dams (transition, shells), which have the coarser materials, the consolidation and settlement processes were completed during the time of construction. Effectively, the materials of the core were hang up between the materials of the transition zone in the upstream and downstream. This is the time that the arching action starts.

Arching occurs in most of the dams when the duration of the dam's construction is too short, and the materials do not have enough time for settling and consolidation. By increasing the arching in the body of dams, the pressure of the soil in the core will be less than the hydrostatics pressure of the water in the reservoir of the dams. In this period, cracks and fracturings begin to increase in the core of dam. This event is known as hydraulic fracturing of the embankment dam.

Therefore, having the information about the changes in the pressure distribution, intensity and direction of total pressure in the mass of soil is very helpful in predicting the dam failure, and this could be done with the help of pressure cells that are installed in the foundation, body and crest of the dams. Trustable information needs accurate monitoring. Among the different instruments that are installed in the dams, pressure cells have less accuracy in comparison with another instrument since the reliability of earth pressure measurement is significantly reduced by various factors such as placement configuration, size of cell, orientation of pressure cell, temperature, calibration, and thickness.

Various installation methods of earth pressure cells have been developed to minimize these compliance issues. Theroux et al. (2001), found that despite these and other developments, issues including calibration, installation, hysteresis, and arching have not been resolved regarding sensor design and installation method. Masanori Matsuu et al.

(2008) worked on the shape and embedding methods of cells for improving the measurement methods. Oossthuzian et al. (2003), and Dibiagio (2006) observed that cells installation condition affects the accuracy of cell measurement. Dibiagio (2006) studied the effects of different factors on instrument readings, including where they were placed in the embankment, and finally suggested an installation method to minimize the effect of these factors.

Dunnicliff (1993) discussed the different parameters influencing the earth pressure cells accuracy and gave suggestions to reduce these effects. Duncan (1982) performed finite element analysis on the embankment dams and compared the results with the measurement data that were obtained with the instruments. readings. They did not find the good correlation between the total pressure data. Dibiagio (2006) observed the similar difference between the measured data and calculated data, he suggests that if the installation was performed in the lower part of the dam during the dam construction (10-20 meters) of fill over the instrument, the difference between measured and calculate data is minimal. In the lower portion of an embankment dam, the stress field during the construction is essentially two- dimensional if the valley is wide and predominantly V-shaped. In this case, with the density and thickness of the fill, it is possible to calculate the pressure at each selected point.

Babbitt (2006) studied the Orovil Dam, and he also saw the discrepancy between the measured and calculate data. He figured out that the behavior of the pressure cells on this dam is questionable. According to his judgement, the principle reason for this difference is the difference in density of the soil and change of the stress field due to the installation of the cell pressure. Aufleger (2006) believes the reason for this error in the instrument was related to the mechanical interaction between the pressure cell and the soil that enveloped the instrument. Dunnicliff (1998) discusses the different influence parameters on the earth pressure accuracy and gives suggestion to minimize these effects. By considering all above, pressure cells of five different Iranian embankment dams were selected to investigate the reasons for the discrepancy in measuring data and calculating data.

PRESSURE CELLS EVALUATION METHODS

Three different methods are used for evaluations of the measured data, depending on the type (number of pressure cells in one plane), stage of the dam for which the data are measured, and finally the location of the cells.

Earth pressure cells are usually installed in five defined directions, as shown as in Figure 1. These directions are: vertical while parallel to dam axis (Figure 1-1), 45-degree to upstream (Figure 1-2), horizontal (Figure 1-3), 45-degree to downstream (Figure 1-4) and vertical while normal to dam axis (Figure 1-5).

All three methods have one thing in common. Since the pressure cells in one cluster (Figure 1) are close enough to each other, they can be considered as stresses in different directions for a single point in the embankment.The stress acting on a certain cell is calculated using Mohr circle based on the measured stresses on the other cells and then compared with the measured stress of that cell.

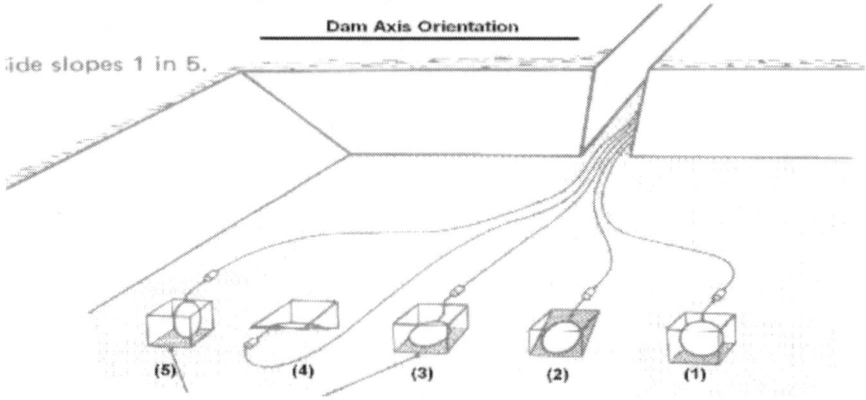

Figure 1. Cell pressure orientation in the dams

Method 1

Considering the directions of pressure cells in Figure 1, it can be observed that four cells (Situation (1) to (4)) measure stresses in one plane in different directions. Using Mohr circle the pressure acting on the "45-degree to upstream" cell (2) can be calculated using the data measured by other three cells and compared with the measured pressure by this cell. Since shear stresses cannot be measured by the earth pressure cells, this method is applicable when four measured stresses in one plane are available.

Method 2

At the end of construction and before impounding there is no horizontal force acting on the embankment due to the water reservoir pressure. In most cases the dam cross-section is almost symmetrical. Thus, at the center of the clay core the maximum principal stress direction is nearly vertical. Therefore, the vertical and horizontal stresses at the center of clay core can be considered as principal stresses

Using two principal stresses and Mohr circle third stress (oriented 45-degree to upstream) is calculated and compared with the corresponding stress measured by pressure cells. This method can be used when the cross-section of the dam is almost symmetrical, and at least three stresses in one plane are measured at the end of construction.

Method 3

In the lower portion of an embankment dam, the stress field during construction is essentially two-dimentional. If the valley is wide and not predominately V- shaped, it is possible to calculate the density and the thickness of the fill and evaluate what the measured stresses should be. This procedure is commonly used to check the performance of pressure cell installed in high embankments.

Because of the large width of the earth fill and its smaller height, the effect of the arching action is negligible and the calculated and measured stresses should be almost the same.

With the above considerations, the stress in the vertical and horizontal directions could be considered as the principal stresses.

GENERAL SPECIFICATIONS OF THE SELECTED EMBANKMENT DAMS

Five embankment dams, namely Karkheh, Maroon, Tabarakabad, Golabar and Gavooshan, have been chosen for this research. In the following sections, the brief descriptions of these dams are presented.

Karkheh Dam

Karkheh Dam is the largest dam, in terms of reservoir capacity and volume of fill placed, that has been constructed in Iran. It is a central core zoned embankment dam 127m high, 3030 m long, with an embankment volume of 32 million cubic meters. A total of 102 clusters, each of five earth pressure cells has been installed to determine the total stresses in the embankment at various locations.

Tabarakabad Dam

Tabarakabad Dam is an earth-rock fill dam with a central clay core and the maximum height of 193.5m. The instrumentation system has been arranged in three cross section of the dam. A total of 24 clusters, each consist of 5 pressure cells, are installed in the embankment.

Maroon Dam

Maroon Dam is a 170-m high earth-and rock-fill dam with an inclined clay core. It is located southwest of Iran. The crest length of this dam is 345m, and the reservoir volume is1.2 million cubic meters. For monitoring the behavior of Maroon Dam, an instrumentation system including 127 earth pressure cells is installed in five different dam cross sections.

Golabar Dam

Golabar Dam is an earth- and rock-fill dam located 50 Km southwest of Zanjan City (Northwest of Iran). The height measured from the lowest foundation is 85 m and the reservoir volume is 114.7 million cubic meters. A total of 14 clusters, each consist of 5 pressure cells, are installed in the embankment for measuring pressure at a various point in the dam.

Gavooshan Dam

Gavooshan is the dam that is located in the south of Sanandaj; the dam is an earth-rock filled dam, The height measured from the lowest foundation is 123 m and the reservoir volume is 550 million cubic meters. A total of 40 clusters, each consist of 3 pressure cells, are installed in the embankment. Moreover, they measure pressure in three directions: vertical, horizontal, and 45 degrees.

RESULTS

In this section, stresses in three embankment dams (Karkheh, Maroon, and Tabarakabad) are calculated and compared with the measured stress at the 45-degree to upstream plane.

Evaluation of Karkheh earth pressure cells

According to the describe methods, stresses at 45- degree to upstream plane were calculated, and the results were compared with corresponding measured stresses in the same direction.

Figure 2 shows the calculated stresses oriented in the 45-degree to upstream plane, which were obtained from method 1 and 2, and compared with the measured stress in the same direction. As can be seen, no consistency exists between the calculated and measured stresses.

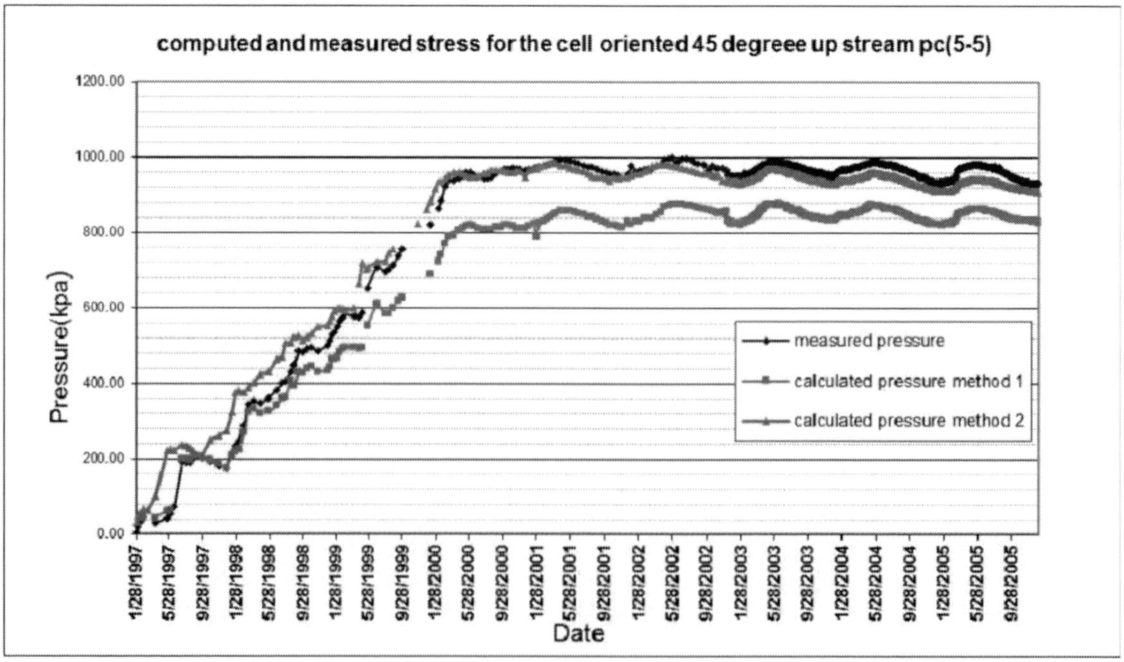

Figure 2. Calculated and measured stresses with method 1 and 2 in the same direction

Evaluation of Tabarakabad earth pressure cells

Earth pressure cells in Tabarakabad Dam are installed in three sections while the Section 9-9 is the highest one. Figure 3 shows the the calculated stresses in orientation 45-degree to upstream by method 1 and 2, and the measured stresses.

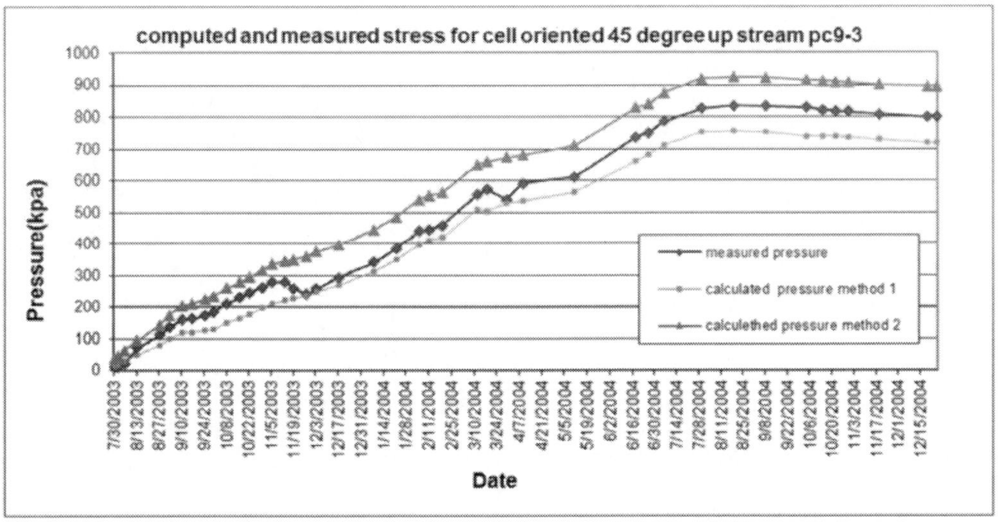

Figure 3. Calculated and measured stresses with method 1 and 2 in the same direction

Evaluation of Maroon earth pressure cells

Most of the cells used in the Maroon Dam are damaged; thus no correct data has been recorded by these cells. Out of 127 total earth pressure cells that have been installed, just 33 cells 27% are functioning.

Maroon Dam is an inclined core dam; therefore, for this dam method 2 is not applicable. As mentioned before, one of the conditions for applicability of method 3 is that the dam should be located in a wide valley. Since Maroon Dam is constructed in a "V" shaped valley, the method 3 is just applicable for section 9-9 which is located at the valley center.

Figure 4 shows the differences between calculated and measured stress in direction 45 degrees to upstream for pressure cells installed at the highest section of the dam (Section 9-9).

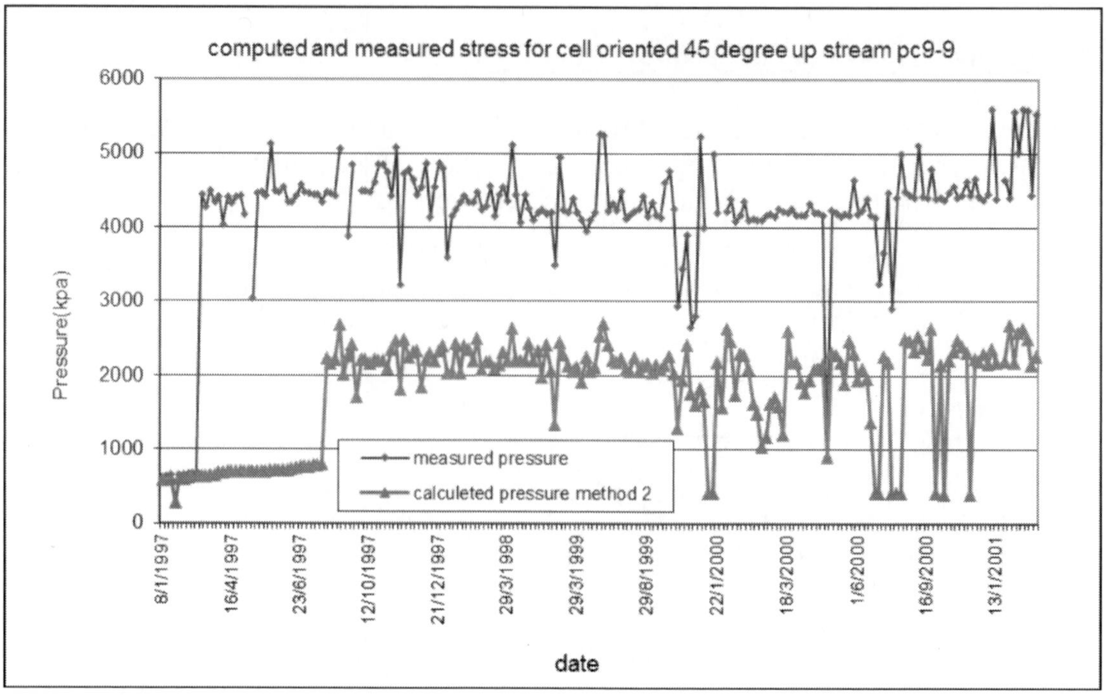

Figure 4. Calculated and measured stresses with method 1and 2 in the same direction.

Method 1 was the only method that could be used for all functioning earth pressure cells at Maroon Dam. In Table 1, the overall evaluation results are presented. As shown more than 90% of the cells have a difference greater than 10% between calculated and measured stresses in direction 45-degree to the upstream. Also, it can be seen that 45% of the cells have the error more than 70%. These differences between the calculated and the measured stresses can be accounted for by general inconsistencies found between the measured stresses by different cells installed in a cluster at different directions.

Table 1. Evaluation of earth pressure of five embankment dams with method 1

	Karkheh Dam	Tabarak abad Dam	Maroon Dam	Golaabar Dam	Gavooshan Dam
Cells have an error greater than 10 percent	89.01%	90%	91.90%	100%	-
Cells have an error greater than 20 percent	63.74%	65%	82.80%	100%	-
Cells have an error greater than 30 percent	45.05%	45%	72.73%	53.85%	-
Cells have an error greater than 40 percent	28.57%	15%	45.45%	23.08%	-
Cells have an error greater than 50 percent	23.08%	10%	45.45%	7.69%	-
Cells have an error greater than 60 percent	18.68%	5%	45.45%	0%	-
Cells have an error greater than 70 percent	12.09%	5%	45.45%	0%	-
Cells have an error greater than 80 percent	9.89%	0%	36.36%	0%	-
Cells have an error greater than 90 percent	8.79%	0%	36.36%	0%	-
Cells have an error greater than 100 percent	8.79%	0%	36.36%	0%	-

Table 2. Evaluation of earth pressure of five embankment dams with method 2

Method 2	Karkheh Dam	Tabarak Abad Dam	Maroon Dam	Golabar Dam	Gavooshan Dam
Cells have an error greater than 10 percent	74.02%	80%	-	100%	50%
Cells have an error greater than 20 percent	45.17%	60%	-	33%	-
Cells have an error greater than 30 percent	32.26%	40%	-	33.33%	-
Cells have an error greater than 40 percent	25.81%	20%	-	16.67%	-
Cells have an error greater than 50 percent	19.36%	20%	-	0%	-
Cells have an error greater than 60 percent	16.31%	0%	-	0%	-
Cells have an error greater than 70 percent	6.46%	0%	-	0%	-
Cells have an error greater than 80 percent	3.26%	0%	-	0%	-
Cells have an error greater than 90 percent	3.23%	0%	-	0%	-
Cells have an error greater than 100 percent	3.23%	0%	-	0%	-

Table 3. Evaluation of earth pressure of five embankment dams with method 3

Method 3	Karkheh Dam	Tabarak Abad Dam	Maroon Dam	Golabar Dam	Gavooshan Dam
Cells have an error greater than 10 percent	80.56%	58.5%	-	28.6%	38.10%
Cells have an error greater than 20 percent	60.48%	47.1%	-	71.4%	61.90%
Cells have an error greater than 30 percent	45.16%	41.2%	-	57.1%	33.33%
Cells have an error greater than 40 percent	33.06%	35.3%	-	57.1%	19%
Cells have an error greater than 50 percent	23.38%	29.4%	-	50.0%	4.80%
Cells have an error greater than 60 percent	18.55%	11.8%	-	35.7%	4.80%
Cells have an error greater than 70 percent	12.09%	11.8%	-	35.7%	0%
Cells have an error greater than 80 percent	7.26%	5.9%	-	28.6%	0%
Cells have an error greater than 90 percent	4.84%	5.9%	-	21.4%	0%
Cells have an error greater than 100percent	4.84%	5.9%	-	14.3%	0%

In the above tables, the pressure cells in the embankment dams are classified into ten groups for comparing the errors in each cell pressure.

These dams were selected so that none of dam has the same characteristics and each of them is located in a different region with different climatology.

Table 1 is related to the evaluation of the cells in five dams with method 1. As shown in the tables, all of the pressure cells in these dams except pressure cells in the Maroon Dam, the differences between the measured and the calculated pressures arein a range of 30 to 40%. In the Maroons Dam, there is a significant difference among the measured and the calculated data.

Gavooshan Dam doesn't have any data recorded at three pressure cells installed in the body of this dam, so we couldn't use this method.

Method two is not applicable for Maroon Dam because it has an inclined core. With this method, almost 25 percent of the cells have more than 40 percent error. In the Gavooshan

Dam, all of the errors being in the range of 0-20 % while in the Karkheh, Tabarak Abad and Golabar Dams, 74.02%, 80% and 100% of the cells, respectively, have an error greater than 10 percent. This magnitude of error is not acceptable.

Table 3 is related to the result of the method 3, there is no data for Maroon Dam, and it is because of the nature of this method and the characteristics of this dam. In this method, the valley that the dam has been installed on that should be "V" shape but the valley of the Maroon Dam is very wide, and hence this method is not applicable for this dam. Using this method, more than 80.56% of the pressure cells have errors more than 10 percent. Neglecting this error is impossible. In Tabarakabad, 58.8% of the pressure cells have an error of more than 10 percent. In Golabar Dam, 71.5% of the pressure cells registered more than 20 percent error.

CONCLUSION AND DISCUSSION

In this paper, the data measured from earth pressure cells installed in five Iranian embankment dams are used to investigate the internal consistency of the pressure cells. Three methods are employed to evaluate earth pressure measurements provided by the cells installed in a specific orientation in each cluster. It was found that there is no consistency between the data obtained by different cells in each cluster. The results of this research and other applied research in this domain show that it is usual for earth pressure cells to give results that do not meet the expectations. Many reasons contribute to the inaccuracies, but the major reasons are summarized as follow:

Installation of cell pressure

The principal reason is the procedure in which the instrument are placed it is imperative that each pressure cell installed exactly at the correct angle and plane as specified.

After the installation of the cell pressure, the backfill material should be compacted similar to the surrounding fill. The difference in compaction of backfill placed over the instrument relative to the surrounding soil will lead to local arching in the overlying fill resulting in either measured pressure that are too high or too low. So minimizing the difference in compacting of the material could improve the inaccuracy of the cell pressures.

Design of the cell pressure

Considering similar parameter in the design of the pressure cells could be very helpful in improving the results. As it mentioned above the installation procedure of cell measurement is too important for the accurate measurement. The normal stiffness of cell pressure and the surrounding backfill should be equal to the stiffness of the host medium (i.e. the embankment fill). However, this is practically impossible to achieve. It should be noted that even though having a perfect adjustment of the normal stiffness is not always helpful as the host medium itself can be subjected to gradual stiffness changes due to consolidation.

For all of these reasons, Bock (1995) suggested some criteria that must be satisfied to have pressure cells with less sensitivity to its stiffness. The key parameters in this regard are the ratio of the diameter to height (D/H). This ratio should be as high as possible to minimize the edge effects of the cells.

Calibration

An earth pressure cell is calibrated by applying a known, uniform pressure to the surface of the cell. The electrical output from the cell is recorded and plotted versus pressure. The relation between the pressure and record output should be linear. Manufacturers typically use fluid pressure for the calibration procedure. However, in reality the pressure cells embedded in embankment dams are subjected to the soil pressure that may not be uniform. A more proper calibration method may be to load the earth pressure cells through the soil to account for possible non-uniform loading and arching effects. Theroux et al. (2001) introduced a method for calibration of earth pressure cells with the uniaxial calibration device using soil pressure.

ACKNOWLEDGEMENTS

This research was funded by the Water Resources Management Co. (WRMC) under grant No: DAM&-84094

REFERENCES

Arulrajah, Arul, Bo, Myint Win, Chu, Jian, Nikraz, Hamid, Instrumentation at the Changi East land reclamation project, Singapore Proceedings of the Institution of Civil Engineers: Geotechnical Engineering, Vol. 162, No. 1 (Feb 2009), pp. 33-40

Bock, H. Some fundamental considerations on stress change monitoring with hydraulic pressure cells. – Intervals News, No. 11(1995), pp 3 -13

Bock, H."Discussions of "Karkheh Dam Instrumentation System-Some Experiences" Ali Asghar Mirghasemi" Geotechnical News, vol.24.No.1, Dec 2006.pp32-36

Dunnicliff, J., Green, Gordon E., "Geotechnical Instrumentation for Monitoring Field Performance" Willey&Sons. (1993).

Donal H.Babbitt, Discussions of "Karkheh Dam Instrumentation System-Some Experiences" Ali Asghar Mirghasemi" Geotechnical News, vol.24.No.1, March 2006.pp 32-36.

Elmo DiBiagio Discussions of "Karkheh Dam Instrumentation System-Some Experiences" Ali Asghar Mirghasemi" Geotechnical News, vol.24.No.1, March 2006.pp 32-36.

Mirghasemi, A.A. and Shoghli, B., "Evaluation of Earth Pressure Measurement in Three Iranian Embankments Dams", International Symposium on Ground Improvement Technologies and Case Histories (ISGI09), Singapore (2009)

Matsuu M., Takahashi A., Uchita Y., Ohta H, "Improvement of Earth Pressure Measuring Method in Rock-Fill Dams" International Commission on Large Dams 76th Annual Meeting Sofia, Bulgaria, June 2-6(2008)

Oosthuizen, C., Naude, P.A. & Hattingh, L.C. "Total and Pore Pressure Cells: Method in Madness." Proc 6th International Symposium on Field Measurements in Geo mechanics, Oslo, Norway, 2006

Selig, E.T,"Soil Stress Gage Calibration", Geotech Testing J.ASTM Vol.3 No.4, Dec., pp.153-158 (1980).

Shoghi, B., Lim, H., "Predictive Scheme for Failures of Embankment Dams with High Overtopping Potential: Case Studies Using Remote Sensing Data", World Environmental & Water Resources Congress 2015 (EWRCON15) Texas Austin

Theroux, B., Labuz, J.F. and Dai, S., "Field Installation of an Earth Pressure Cell" 80th Meeting of the Transportation Research Board, National Research Council (U.S) TRR, pp. 12-19 (2001).

.

US Army Corps of Engineer. Instrumentation of Embankment Dams and Levees (1984)

STABILITY OF SLOPES IN EMBANKMENT DAMS

Gopi Siddappa[1]
Santhosh G. Gowda[2]

ABSTRACT

A slope is an unsupported, inclined surface of a soil mass which is natural or manmade. These may be above ground level as embankments or below ground level as cuttings. Instability may result due to gravity, rainfall, increase in groundwater table, construction on top of the slope and changes in stress conditions. The sliding will occur if shear stress developed in soil exceeds the corresponding shear strength of the soil. The engineering solutions to slope instability problems require good understanding of analytical methods, investigative tools and stabilization measures. The computerized slope modeling is a novel method of solving geotechnical problems. In the present investigation bearing capacity, angle of friction, unit weight and cohesion of soil have been considered by varying the soil parameters and its effect on safety of slopes.

The limit equilibrium method is well known for a statically indeterminate problem and assumptions on the distributions of internal forces are required for the solution of the factor of safety. The critical failure surfaces from the limit equilibrium method and the factor of safety appear to be the primary quantity is of interest. In the present study, two slip surface options are used for finding the critical failure surface and factor of safety, for different limit equilibrium methods under different conditions. With this intension, limit equilibrium method is used. SLOPE/W, is adopted to study the causes of failure and to avoid the failures of slope as far as possible. From the parametric study, it is observed that the factor of safety increases with the increase in cohesion and friction of the soil.

INTRODUCTION

A slope is an unsupported, inclined surface of a soil mass. Slopes can be natural or manmade. These may be above ground level as embankments or below ground level as cuttings. Earth slopes are formed for railway embankments, earth dams, canal banks etc. Instability related issues in engineered as well as natural slopes are common challenges to both researchers and professionals. Instability may result due to rainfall, increase in groundwater table and changes in stress conditions. Similarly, natural slopes that have been stable for many years may suddenly fail due to changes in geometry, external forces and loss of shear strength. In addition, the long-term stability is also associated with the weathering and chemical influences that may decrease the shear strength. In such circumstances, the evaluation of slope stability conditions becomes a primary concern everywhere. When a mass of soil has an inclined surface the potential of slope to slide from higher level to lower level always exists. The sliding will occur if shear stress

[1] Professor and Head, Department of Civil Engineering, P.E.S. College of Engineering, MANDYA, Karnataka – 571401, INDIA, Ph: 0091 9448745759, E-mail: gopisiddappa@gmail.com
[2] Research Scholar, Department of Civil Engineering, P.E.S. College of Engineering, MANDYA, Karnataka – 571401, INDIA, E_mail: santhoshgg777@gmail.com

developed in the soil exceeds the corresponding shear strength of the soil. However certain practical considerations make precise stability analyses of slope difficult in practices. The engineering solutions to slope instability problems require good understanding of analytical methods, investigative tools and stabilization measures.

Several methods can be applied to analyse stability of slopes. Computer modeling and analysis of slope stability is an advanced and an accurate technique in the construction of safe slope. The computerized slope modeling is a novel method of solving geotechnical problems. From previous investigations bearing capacity, angle of friction, unit weight and cohesive of soil for computerize model have been considered The Geo-Slope software is used in identification and behavior of the earth embankment due to varying the soil parameters and its effect on safety of slopes. The limit equilibrium method has been used and it is considered as 2-D plane strain problem which satisfies equilibrium conditions and gives a more economical embankment section. The LEM is well known to be a statically indeterminate problem and assumptions on the distributions of internal forces are required for the solution of the factor of safety. The critical failure surfaces from the LEM and the FOS appear to be the primary quantity of interest. Two slip surface options are used for finding the critical failure surface and factor of safety, for different limit equilibrium methods under different conditions.

OBJECTIVES OF THE PRESENT STUDY

Present study involves analysis of the earth slopes using different methods

1) Stability analysis for a homogeneous slope by varying the soil parameter
2) Effect of geometry on slope stability
3) Comparison of factor of safety obtained independently, by slope/w and by cross checking

This study is limited to:

1) Calculation utilizes Slope/w software for limit equilibrium analysis.
2) Analysis for limit equilibrium based on ordinary or Fellenius method, Bishop's simplified method, Janbu simplified method, Spencer method, Sarma method, Corps of Engineers#1, Corps of Engineers#2, General limit equilibrium and Morgenstern-price method for limit equilibrium method.

Causes of Failure of Slopes

The important factors that cause instability in slope and lead to failure are

(1)Gravitational force. (2) Force due to seepage of water. (3) Erosion of the surface of slope due to flowing water. (4) The sudden lowering of water adjacent to the slope. (5) Forces due to earthquakes.

Types of Slope Failures

Rotational Failure. This type failure occurs by rotation along a slip surface by downward and outward movement of the soil mass.
Slope Circle Failure. In this case the failure circle intercepts the surface of the slope itself above the toe.

Toe Circle Failure. In this case the failure circle passes through the toe of the slope. This occurs in steep slopes of homogenous soils.
Base Circle Failure. In this case the failure circle passes below the toe at a depth H from top of the slope of height H. Such cases occur when slopes are flat with weak soil and a steep stratum occurs below the toe.
Translational Failure. Translational failure occurs in an infinite slope along a long failure surface parallel to the slope. The shape of the failure surface is influenced by the presence of any hard stratum at a shallow depth below the slope surface. These failures may also occur along slopes of layered materials.
Compound Failure**.** A compound failure is a combination of the rotational slips and the translational slips. A compound failure surface is curved at the two ends plane in the middle portion. A compound failure generally occurs when a hard stratum exists at considerable depth below the toe.
Wedge Failure**.** A failure along an inclined plane is known as plane failure or wedge failure or block failure. This failure may occur both in infinite and finite slope consisting of two different materials or in a homogeneous slope having cracks, fissures, joints or any other specific plane of weakness.

Limit equilibrium method is the most popular approach in slope stability analysis. This method is well known to be a statically indeterminate problem and assumptions on the interslice shear forces are required to render the problem statically determinate. Based on the assumptions on the internal forces and force and/or moment equilibrium, there are more than 10 methods developed for slope stability analysis. The famous methods include those by Fellenius (1936), Bishop (1955), Janbu (1954, 1957), Lowe and Karafiath (1960), Spencer (1967), Morgenstern-Price (1965) etc. Fellenius (1936) introduced the first method, referred to as the Ordinary or the Swedish method, for a circular slip surface. Bishop (1955) advanced the first method introducing a new relationship for the base normal force. The equation for the FOS hence became non -linear. Janbu (1954) then developed a simplified method for non-circular failure surfaces, dividing a potential sliding mass into several vertical slices (Krahn 2004).

Later, Morgenstern-Price (1965), Spencer (1967), Sarma (1973) and several others made further contributions with different assumptions for the interslice forces. A procedure of General limit equilibrium (GLE) was developed by Chugh (1986) as-an extension of the Spencer and Morgenstern-Price methods, satisfying both moment and force equilibrium conditions (Krahn 2004). These developments are reviewed in the following section, which aims to find out the key differences in the various approaches for FOS determination

Table 1. Significance of factor of safety,
Factor of safety for design (Source: in Liu and Evett,2005)

Safety Factor	Significance
Less than 1.0	Unsafe
1.0 – 1.2	Questionable safety
1.3 – 1.4	Satisfactory for cuts, fills, questionable for dams
1.5 – 1.75	Safe for dams

Ordinary or Fellenius Method

Ordinary or Fellenius Factor of safety equation in the absence of any pore water pressure:

$$F_m = \frac{\sum(c' + N\tan\phi)}{\sum W\sin\alpha} \quad (1)$$

$$N' = (W\cos\alpha - ul) \quad (2)$$

Where,
u = pore pressure,
l = slice base length and
α = inclination of slip surface at the middle of slice
In summary, OM
- satisfies moment equilibrium condition,
- neglects the interslice normal and shear forces,
- gives the most conservative FOS, and
- is useful only for demonstrations.

The forces considered in OM are shown in the sketch.

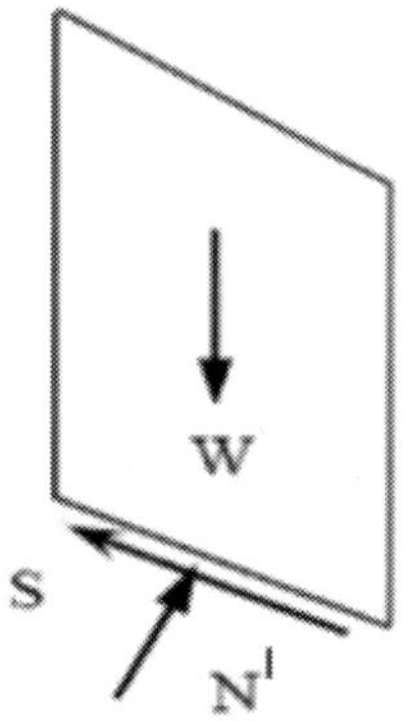

Bishop's Simplified Method

Bishop's simplified method (BSM) is very common in practice for circular shear surface (SS).This method considers the interslice normal forces but neglects the interslice shear forces. It further satisfies vertical force equilibrium to determine the effective base normal force (N'), which is given by:

$$N = \frac{1}{m_\alpha}\left(W - \frac{c'\sin\alpha}{F} - ul\cos\alpha\right) \quad (3)$$

where,

$$m_\alpha = \left(1 + \tan\alpha\frac{\tan\phi'}{F}\right) \quad (4)$$

In summary, BSM
- satisfies moment equilibrium for FOS,
- satisfies vertical force equilibrium for N,
- considers interslice normal force,
- more common in practice, and
- applies mostly for circular shear surfaces.

Janbu's Simplified Method

Janbu's simplified method Janbu's simplified method (JSM) is based on a composite SS (i.e. non-circular) and the FOS is determined by horizontal force equilibrium. As in BSM, the method considers interslice normal forces (E) but neglects the shear forces (T). The base normal force (N) is determined in the same way as in BSM and the FOS is computed by:

$$F_f = \frac{(cl(N - ul)tan\phi)sec\alpha}{\sum Wtan\alpha + \sum \Delta E} \quad (5)$$

$\Sigma\Delta E = E_2 - E_1$ = net interslice normal forces (zero if there is no horizontal force).

The derivations of the FOS based on both force and moment equilibrium. In stress terms originally presented by Janbu (1954), the above equation is written as:

$$F_o = \frac{\left\{\frac{b(c'(p-u)tan\emptyset')}{n_\alpha}\right\}}{\sum p.btan\alpha} \quad (6)$$

$$n_\alpha = cos^2\alpha\left(1 + tan\alpha\frac{tan\phi'}{F}\right) \quad (7)$$

where, p = W/b = total vertical stress and b = width of slice

In summary, JSM

- satisfies both force equilibriums,
- does not satisfy moment equilibrium,
- considers interslice normal forces, and
- is commonly used for composite shear surface.

Spencer Method

According to this method, the interslice shear force is related to:

$T = E \tan\theta$

Spencer's method is the same as Morgenstern Price method except the assumption made for interslice forces.

In summary, SM

- considers both interslice forces,
- assumes a constant interslice force function,
- satisfies both moment and force equilibrium, and
- compute FOS for force and moment equilibrium.

Morgenstern Price Method

Moment Equilibrium Condition.

$$F_m = \frac{\sum c * l + Ntan\alpha}{\sum W * sin\alpha} \quad (8)$$

Force Equilibrium Condition.

$$F_f = \frac{\sum[c * l * cos\alpha + N * tan\phi * cos\alpha]}{\sum(N * sin\alpha)} \quad (9)$$

It may be seen that moment equilibrium or force equilibrium equations contain a factor 'N' which depends on factor of safety and inter-slice forces and can be obtained by iterating process as stated in the following

Base Normal Force of the Slice. The normal force at the base of slice in vertical direction, we get the following equation

$$N=\frac{W-(T_2-T_1)-\frac{c*1*\sin\alpha}{F}}{\sum\cos\alpha+\frac{\tan\Phi*\sin\alpha}{F}} \qquad (10)$$

This is a non linear equation as factor of safety 'F' is appearing in the equation. F shall be F_m (moment factor of safety) for moment equilibrium and F_f (force factor of safety) for force equilibrium.
When inter-slice shear forces are ignored as in Bishop method the equation for 'N' becomes

$$N=\frac{W-\frac{c*1*\sin\alpha}{F}}{\sum\cos\alpha+\frac{\tan\Phi*\sin\alpha}{F}} \qquad (11)$$

Corps of Engineers Method

According to this method, the angle of the interslice resultant force can be assumed in two ways. First, it can be assumed parallel to the ground surface, i.e. $\theta=\beta$, where β is the slope angle. Two, it can be equal to the average slope an between the entry and exit points of the critical SS.
In summary, CEM:

- considers both interslice normal and shear forces,
- satisfies only force equilibrium, and
- assumes inclinations in two ways.

Sarma Method

Sarma (1973) developed a method for a non-vertical slice or for general blocks. This method satisfies both equilibrium conditions .In addition, the interslice force relationship is assumed as a linear Mohr-Coulomb expression:

$$T = ch + E\tan\varphi \qquad (12)$$

where,
c, φ = shear strength parameters, and h = slice height.

The interslice forces are adjusted until the FOS for force and moment equilibrium is satisfied.

In summary, Sarma's approach:

• considers both interslice normal and shear forces,
• satisfies both moment and force equilibrium, and
• relates the interslice forces by a quasi shear strength equation

STABILITY ANALYSIS FOR A HOMOGENEOUS SLOPE BY VARYING THE SHEAR STRENGTH PARAMETER

Case 1: Varying cohesion
Case 2: Varying Friction
Case 3: Varying Unit weight

Geometry
Fig. 1 shows a homogeneous soil slope with a slope height equal to 6 m and slope angle equal to 45°, the size of the domain is 20m in width and 10m in height is considered.

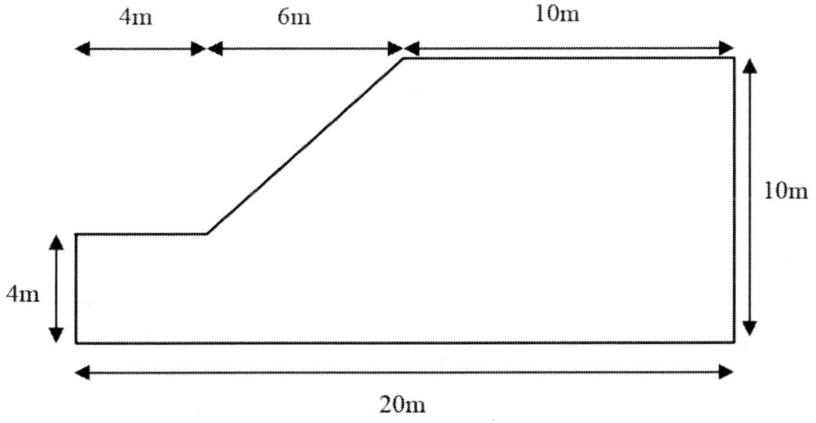

Figure 1. Homogeneous Soil Slope Geometry for "Model"

Selected methods for analysis

LE methods: the selected methods are ordinary or Fellenius method, BSM, JSM, M-PM, Spencer method, Sarma method, Corps of Engineers#1, Corps of Engineers#2 and GLE. There are two methods available in Slope/W namely, Auto Locate method and Exit and Entry method. Both the methods were used for analysis and in all the cases it is observed that the results were closely related to each other and satisfied.

Analysis and Result

Case 1. In this parametric study, the cohesive strength of the soil varies from 0, 5, 10 and 20 kPa. The density and Friction angle of the soil is kept at 20kN/m^3 and 30° respectively, in all the analysis. Figure 2(a) and 2(b) shows the compressive shear strength and factor of safety for different cohesion of soil respectively using Autolocate method and Exit and Entry method. Minimum Factor of Safety for varying cohesion using Auto Locate and Exit and Entry method is shown in Table 2(a) and 2(b) respectively.

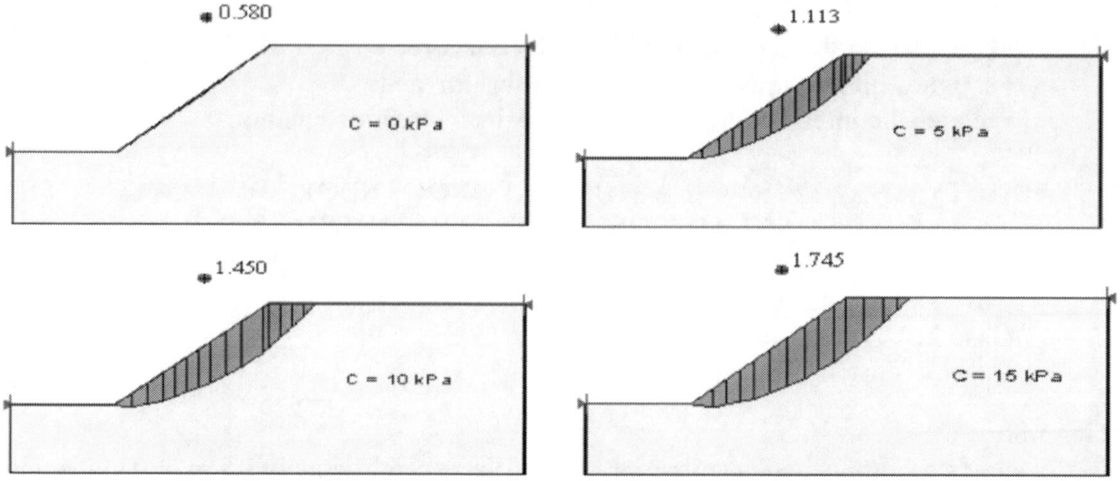

Figure 2(a). CSS and FOS by M-PM (Auto Locate)

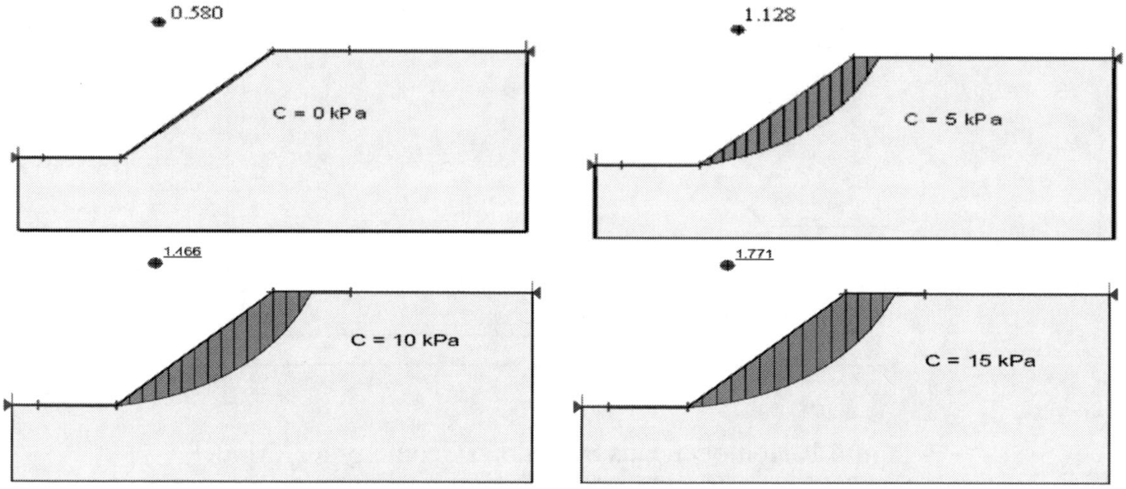

Figure 2(b). CSS and FOS by M-PM (Exit and Entry)

Table 2(a). Minimum Factor of Safety for Varying Cohesion (Auto Locate)						
No		1	2	3	4	5
γ kN/m^3		20	20	20	20	20
C (kPa)		0	5	10	15	20
φ(°)		30	30	30	30	30
M-P	Moment	0.58	1.113	1.45	1.745	2.017
	Force	0.578	1.112	1.45	1.745	2.01
Ordinary	Moment	0.578	1.052	1.388	1.664	1.929
Bishops	Moment	0.586	1.118	1.468	1.739	1.997
Janbu	Force	0.578	1.068	1.38	1.658	1.912
Spencer	Moment	0.58	1.128	1.464	1.758	2.04
	Force	0.578	1.129	1.458	1.762	2.044
GLE	Moment	0.609	1.113	1.48	1.745	2.004
	Force	0.609	1.113	1.48	1.745	2.004
Corps Engineers #1	Force	0.58	1.136	1.485	1.796	2.091
Corps Engineers #2	Force	0.579	1.143	1.531	1.822	2.13
Sarma	Moment	0.58	1.12	1.453	1.729	1.992
	Force	0.578	1.12	1.448	1.729	1.992

Table 2(b). Minimum Factor of Safety for Varying Cohesion (Entry And Exit)						
No		1	2	3	4	5
γ kN/m3		20	20	20	20	20
C (kPa)		0	5	10	15	20
φ(°)		30	30	30	30	30
M-P	Moment	0.508	1.128	1.466	1.771	2.058
	Force	0.578	1.128	1.462	1.769	2.056
Ordinary	Moment	0.578	1.076	1.408	1.71	2
Bishops	Moment	0.588	1.135	1.472	1.776	2.063
Janbu	Force	0.578	1.06	1.363	1.688	1.989
Spencer	Moment	0.58	1.129	1.466	1.772	2.059
	Force	0.58	1.129	1.468	1.774	2.062
GLE	Moment	0.626	1.128	1.465	1.771	2.058
	Force	0.626	1.128	1.465	1.771	2.058
Corps Engineers #1	Force	0.578	1.139	1.493	1.813	2.131
Corps Engineers #2	Force	0.578	1.148	1.516	1.844	2.16
Sarma	Moment	0.58	1.126	1.462	1.767	2.054
	Force	0.578	1.125	1.461	1.767	2.056

Figure 3(a) and 3(b) shows the relation between cohesion and factor of safety using Autolocate method and Exit and Entry method respectively. It is found that factor of safety increases with increase in cohesion.

Case 2. In this parametric study, the Friction angle of the soil varies from 5°, 10° 15° and 20°. The density and cohesive strength of the soil is kept at 20kN/m^3 and 10 kPa respectively, in all the analysis. Figure 4(a) and 4(b) shows the compressive shear strength and factor of safety of soil for different values of Φ respectively using Auto Locate method and Exit and Entry method. Minimum Factor of Safety for varying cohesion using Auto Locate and Exit and Entry method is shown in Table 3(a) and 3(b) respectively.

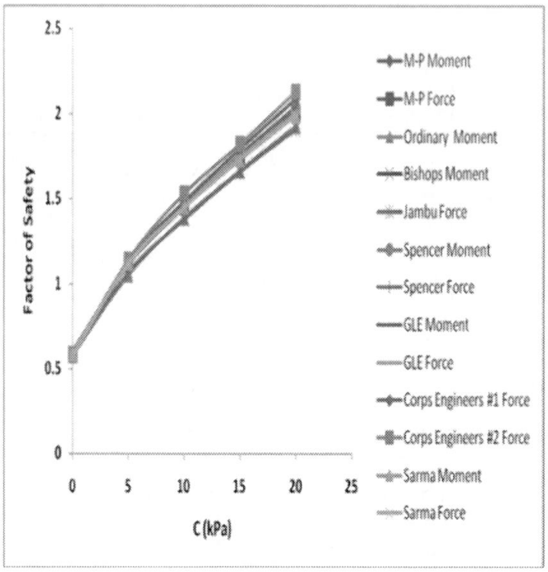

Figure 3 (a). Varying Cohesion v/s FOS (Auto locate)

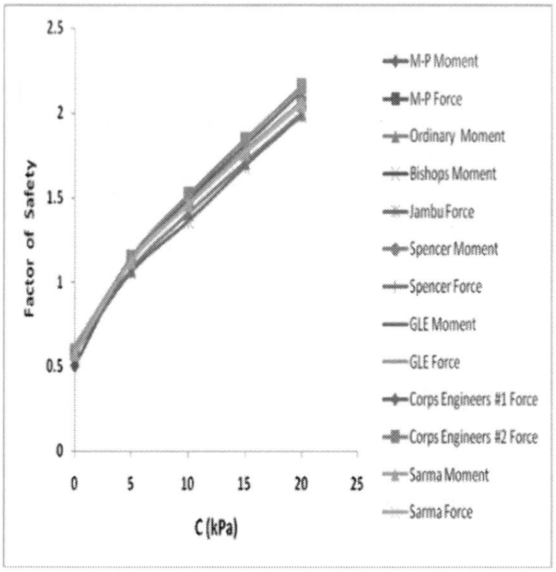

Figure 3 (b). Varying Cohesion v/s FOS (Exit and Entry)

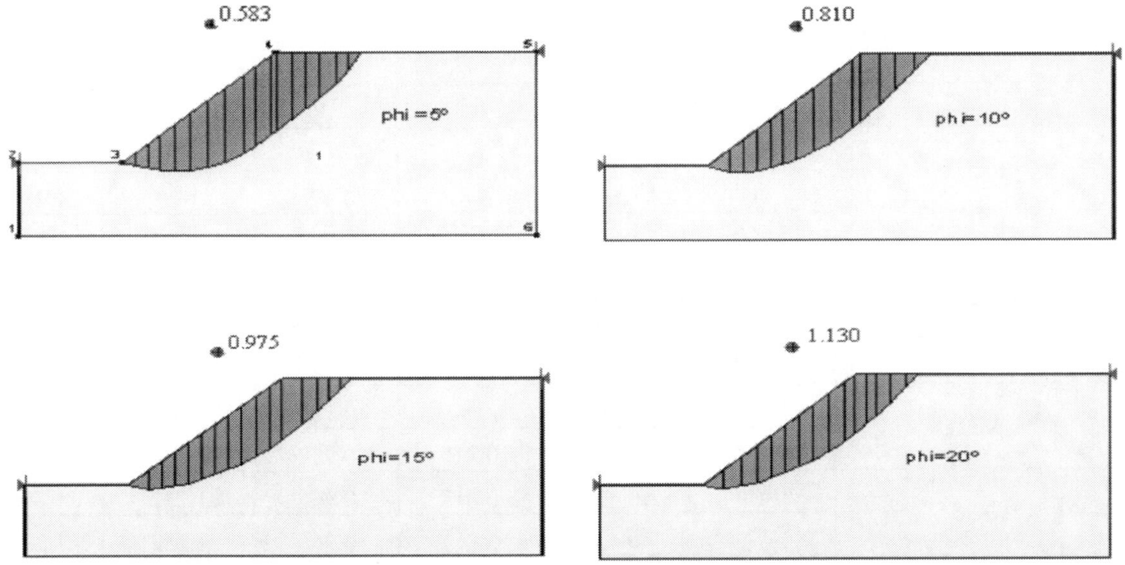

Figure 4(a). CSS and FOS by M-PM (Auto Locate)

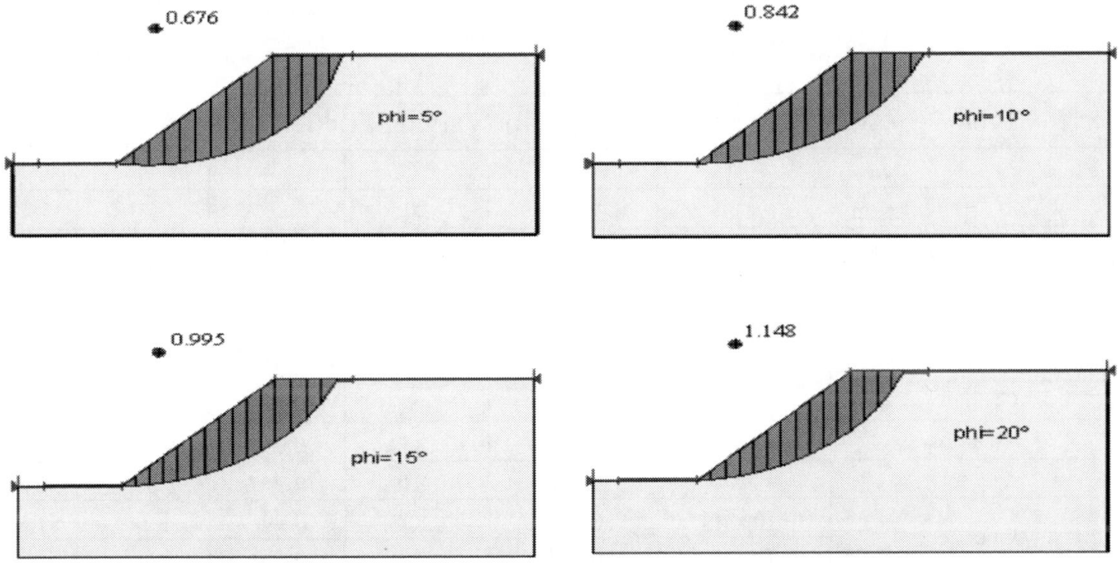

Figure 4(b). CSS and FOS by M-PM (Exit and Entry)

Table 3(a). Minimum Factor of Safety for Varying Friction (Auto Locate)						
No		1	2	3	4	5
γ kN/m^3		20	20	20	20	20
C (kPa)		10	10	10	10	10
φ(°)		5	10	15	20	25
M-P	Moment	0.583	0.81	0.975	1.13	1.285
	Force	0.593	0.8	0.974	1.132	1.287
Ordinary	Moment	0.614	0.772	0.928	1.082	1.227
Bishops	Moment	0.629	0.799	0.969	1.124	1.284
Janbu	Force	0.613	0.77	0.925	1.076	1.223
Spencer	Moment	0.652	0.82	0.986	1.14	1.299
	Force	0.654	0.811	0.987	1.143	1.303
GLE	Moment	0.42	0.817	0.968	1.132	1.285
	Force	0.42	0.817	0.968	1.132	1.285
Corps Engineers #1	Force	0.689	0.854	1.008	1.163	1.324
Corps Engineers #2	Force	0.705	0.87	1.026	1.181	1.338
Sarma	Moment	0.627	0.805	0.962	1.12	1.285
	Force	0.627	0.811	0.963	1.118	1.281

Table 3(b). Minimum Factor of Safety for Varying Friction (Entry And Exit)						
No		1	2	3	4	5
γ kN/m^3		20	20	20	20	20
C (kPa)		10	10	10	10	10
φ(°)		5	10	15	20	25
M-P	Moment	0.676	0.842	0.995	1.148	1.302
	Force	0.673	0.836	0.995	1.149	1.304
Ordinary	Moment	0.667	0.82	0.967	1.109	1.257
Bishops	Moment	0.677	0.843	0.997	1.151	1.308
Janbu	Force	0.674	0.819	0.957	1.0932	1.236
Spencer	Moment	0.678	0.843	0.996	1.149	1.303
	Force	0.681	0.834	0.996	1.15	1.305
GLE	Moment	0.677	0.839	0.995	1.148	1.302
	Force	0.677	0.839	0.995	1.148	1.302
Corps Engineers #1	Force	0.737	0.886	1.024	1.175	1.325
Corps Engineers #2	Force	0.741	0.892	1.045	1.196	1.352
Sarma	Moment	0.676	0.841	0.994	1.145	1.3
	Force	0.676	0.839	0.99	1.15	1.3

Figure 5(a) and 5(b) shows the relation between coefficient of friction and factor of safety using Autolocate method and Exit and Entry method respectively. It is found that factor of safety increase with increase in angle of friction.

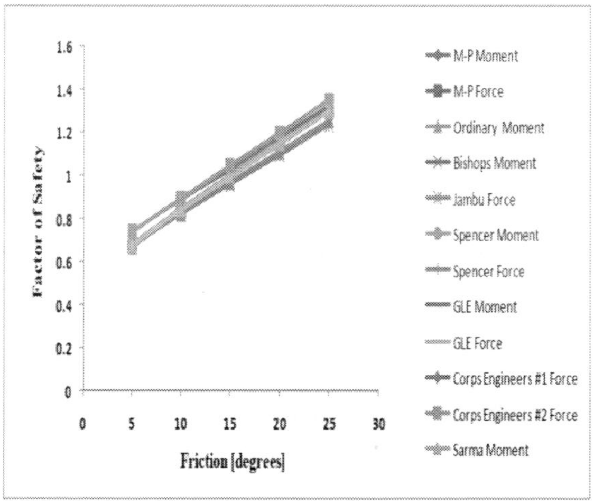

Figure 5(a). Varying Friction v/s FOS (Auto locate)

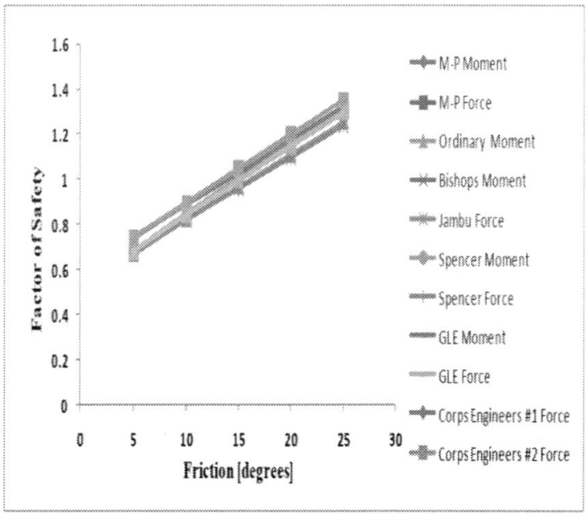

Figure 5(b). Varying Friction v/s FOS (Exit and Entry)

Case 3. In this parametric study, the density of the soil varies from 16, 17 and 18 –20 kN/m^3 . The Friction angle and cohesive strength of the soil is kept at 25° and 10 kPa respectively, in all the analysis. Figure 6(a) and 6(b) shows the compressive shear strength and factor of safety for different unit weight of soil respectively using Autolocate method and Exit and Entry method. Minimum Factor of Safety for varying cohesion using Auto Locate and Exit and Entry method is shown in Table 4(a) and 4(b) respectively.

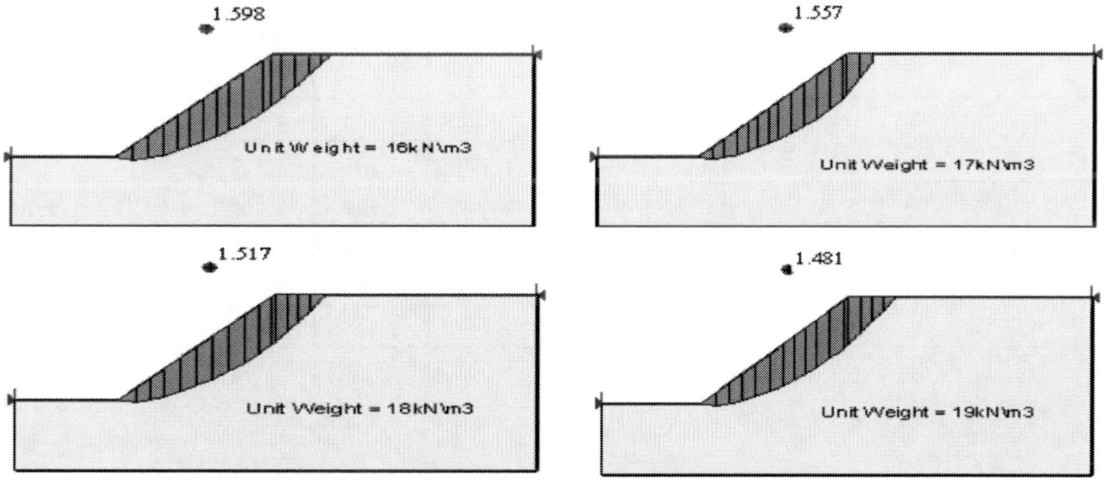

Figure 6(a). CSS and FOS by M-PM (Auto Locate)

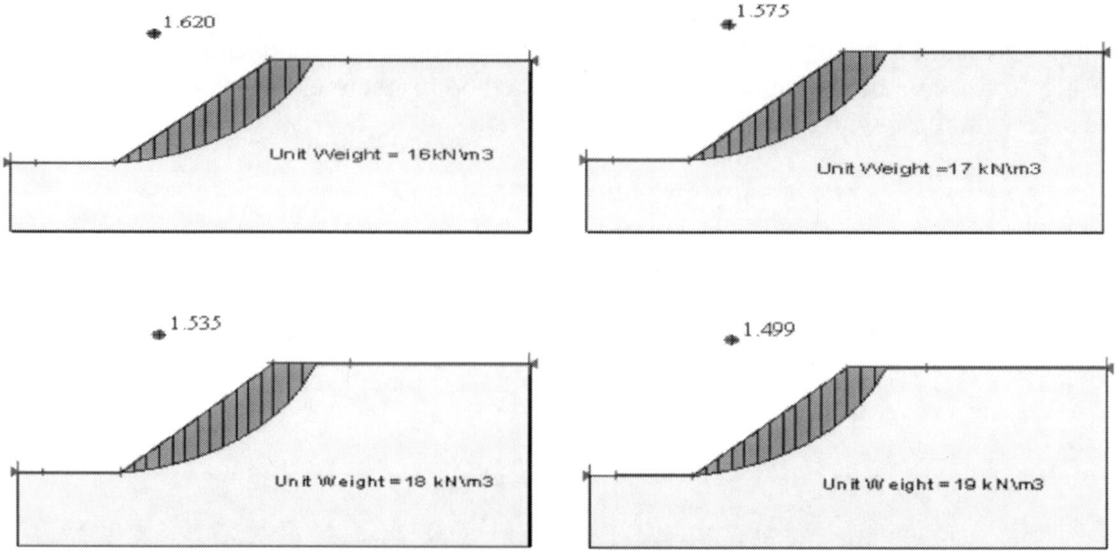

Figure 6(b). CSS and FOS by M-PM (Exit and Entry)

Table 4(a). Minimum factor of safety for varying Unit Weight (Auto locate)						
No		1	2	3	4	5
γ kN/m^3		16	17	18	19	20
C (kPa)		10	10	10	10	10
φ(°)		30	30	30	30	30
M-P	Moment	1.598	1.557	1.517	1.481	1.45
	Force	1.6	1.549	1.518	1.482	1.45
Ordinary	Moment	1.538	1.488	1.453	1.418	1.389
Bishops	Moment	1.6	1.575	1.537	1.501	1.471
Janbu	Force	1.526	1.464	1.448	1.413	1.383
Spencer	Moment	1.615	1.575	1.534	1.493	1.464
	Force	1.62	1.573	1.539	1.486	1.458
GLE	Moment	1.598	1.593	1.551	1.514	1.48
	Force	1.598	1.593	1.551	1.514	1.48
Corps Engineers #1	Force	1.643	1.599	1.564	1.521	1.485
Corps Engineers #2	Force	1.665	1.619	1.58	1.536	1.531
Sarma	Moment	1.593	1.555	1.516	1.484	1.453
	Force	1.593	1.565	1.521	1.483	1.448

Table 4(b). Minimum factor of safety for varying Unit Weight (Entry and Exit)						
No		1	2	3	4	5
γ kN/m^3		16	17	18	19	20
C (kPa)		10	10	10	10	10
$\varphi(°)$		30	30	30	30	30
M-P	Moment	1.62	1.575	1.535	1.499	1.466
	Force	1.613	1.569	1.531	1.493	1.462
Ordinary	Moment	1.564	1.519	1.478	1.441	1.408
Bishops	Moment	1.626	1.581	1.541	1.505	1.472
Janbu	Force	1.541	1.493	1.452	1.416	1.383
Spencer	Moment	1.621	1.576	1.536	1.5	1.466
	Force	1.622	1.577	1.538	1.501	1.468
GLE	Moment	1.62	1.575	1.535	1.499	1.465
	Force	1.62	1.575	1.535	1.499	1.465
Corps Engineers #1	Force	1.651	1.608	1.565	1.528	1.493
Corps Engineers #2	Force	1.682	1.633	1.592	1.552	1.516
Sarma	Moment	1.616	1.571	1.531	1.495	1.462
	Force	1.617	1.572	1.538	1.499	1.461

Figure 7(a) and 7(b) shows the relation between unit weight of soil and factor of safety using Autolocate method and Exit and Entry method respectively. It is observed that factor of safety decreases with increase in unit weight of soil.

CONCLUSION

From the parametric study, we can see that when the cohesion increases, gradually the Factor of safety also increases. When the cohesion is less than 3kPa, the factor of safety will be less than 1.Therefore, according to Liu and Evett, 2005 FOS for design slope (Table 1) will be unsafe if the FOS is less than1.

Mathematical equations for moment and force factor of safeties have been given under different conditions. FOS are computed using Microsoft excel program and also by use of software SLOPE/W. Based on the results, following broad conclusions can be drawn.

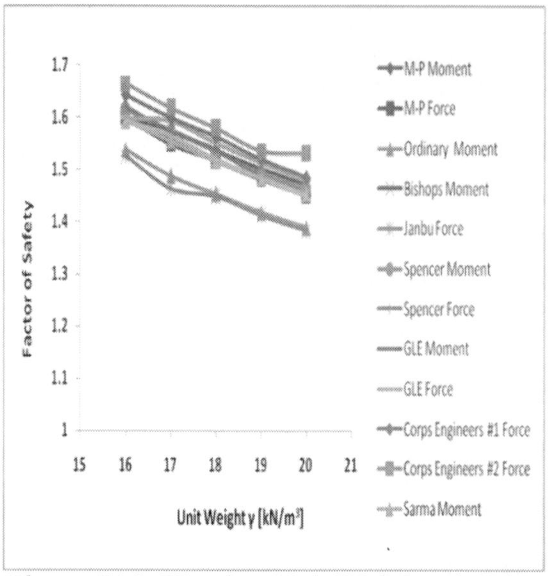

Figure 7(a). Varying Unit Weight v/s FOS (Auto locate)

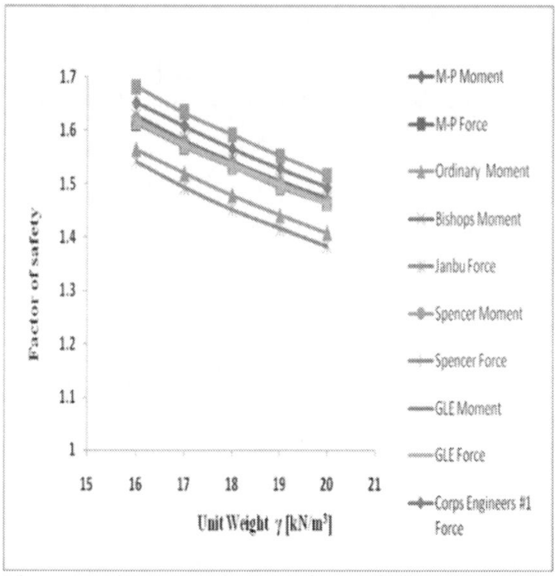

Figure 7(b). Varying Unit Weight v/s FOS (Exit and Entry)

1) The mathematical equations for obtaining factor of safety with the help of Microsoft Excel can give desired result very quickly. Convergence of iteration required for the analysis is achieved very fast. Hence use of computer is preferable to conventional graphical method of analysis.
2) The old methods (Fellenius method and Swedish Circle method) give lower factor of safety (FOS) and therefore requires a flatter slope for the specified FOS compared to the earth slope obtained on the basis of Morgenstern-Price method which takes inter-slice forces (normal and shear) into account and satisfies closed force polygon indicating equilibrium condition of the slice in a free body force diagram.
3) The Morgenstern-Price method with half-sine function, which takes into account inter-lice forces and satisfies both moment equilibrium and force equilibrium conditions, is most appropriate for the design of slope-stability.
4) GEO-SLOPE International provide a very efficient software package for analysis of slope-stability covering almost all situations. The results (output of package) tallies quite closely with the results obtained independently by solving mathematical equations using Microsoft Excel.

SCOPE OF FUTURE WORK

1) The study can be done on heterogeneous soil
2) Further the study can be carried out for a slope subjected to surcharge load and earthquake

REFERENCES

Bishop, A. W. (1955). The use of slip circles in stability analysis of slopes. Geotechnique, Vol. 5

Chugh, A. K. (1986). Variable Interslice Force Inclination in Slope Stability Analysis. Soils and Foundation, Japanese Society of SMFE, Vol 26, No. 1, pp. 115-21.

Fellunius, W. (1936). Calculations of the Stability of Earth Dams. Proceedings of the Second Congress of Large Dams. Vol. 4, pp. 445-63, Washington D. C.

Janbu, N. (1954a). Stability analysis of Slopes with Dimensionless Parameters. Thesis for the Doctor of Science in the Field of Civil Engineering, Harvard University Soil MechanicsSeries, No. 46.

Janbu, N. (1957). Earth pressure and bearing capacity calculations by generalised procedur of slices. Proceedings of the 4th International Conference, SMFE, London, 2, pp. 207-12.

Krahn, J. (2004). Stability Modelling with SLOPE/W. An Engineering Methodology, Published by GeoSlope International.61.

Lowe, J., and Karafiath, R. V. (1960). Stability of Earth Dam upon Drawdown. Proceedings of the of the first Pan American Conference on Soil Mechanics and Foundation Engineering, Maxico City, pp. 537-52.

Morgenstern, N. R. and Price, V. E. (1965). The Analysis of the Stability of General Slip Surfaces. Geotechnique, Vol. 15, No. 1 pp. 77-93.

Sarma, S. K. (1973). Stability Analysis of Embankment and Slopes. Geotechnique, Vol. 23 (3), pp.423-33.

SLOPE/W (2007). Stability Analysis. Users Guide Version 5, GeoSlope Office, Canada.

Spencer, E. (1967). A method of Analysis of the Stability of Embankments, Assuming Parallel Interslice Forces. Geotechnique, Vol. 17, pp. 11-26.

RITSCHARD DAM MONITORING:
AWARENESS LEADS TO LEARNING WHICH LEADS TO ACTION

Ray D. Tenney[1]
Don B. Meyer[2]
Michael R. May[3]

ABSTRACT

Ritschard Dam, a rockfill dam with a thin clay core, has experienced settlement of over 2 feet of its 140 foot overall height since completion in 1995. Settlement has been accompanied by 0.75 feet downstream displacement of the dam crest. Instrumentation installed in 2009 thru 2014 shows displacements are exacerbated by reservoir load. The displacement pattern is repeated in three monitored sections across the dam to varying degrees.

Inclinometer and surface monument monitoring shows the top 1/3 of the embankment is moving downstream along a 10 foot thick deformation zone where the potential for core shearing may be a concern. Internal core pore pressures are elevated above the reservoir surface and respond quickly to reservoir loading and unloading. Development of a FLAC (Fast Lagrangian Analysis of Continua) model, attributed and calibrated with laboratory and field test data and instrumentation monitoring data, has suggested that the deformations are not likely to abate in the near future. The owner is working with a design team utilizing the FLAC model to evaluate remediation needs and alternatives.

The ability to move confidently toward a recommended remedial action plan has only been possible due to the extensive monitoring program which grew over time to thoroughly observe embankment behavior. Only through extensive monitoring has it been possible to identify the causes and solutions to the ongoing deformation.

AWARENESS

Ritschard Dam is a 140 foot tall earth and rockfill dam with a narrow clay core constructed on Muddy Creek, near Kremmling, Colorado in 1995. A typical cross section is presented on Figure 1. It impounds 66,000 acre-feet of water utilized primarily for water supply. Ritschard Dam was equipped with vibrating wire piezometers in the core and chimney drain, open stand pipe piezometers in the abutment bedrock, and brass cap crest settlement monuments. The dam was constructed with 1 foot of camber and the designers expected approximately 0.5 foot of settlement[4].

---

[1] Colorado River Water Conservation District, Glenwood Springs, CO 81601, rtenney@crwcd.org
[2] Colorado River Water Conservation District, Glenwood Springs, CO 81601, dmeyer@crwcd.org
[3] AECOM, Denver, CO 80237, michael.may@aecom.com
[4] Harza Engineering Company, Design Report Embankment Dam, Wolford Mountain and Ritschard Dam Project, April 1991

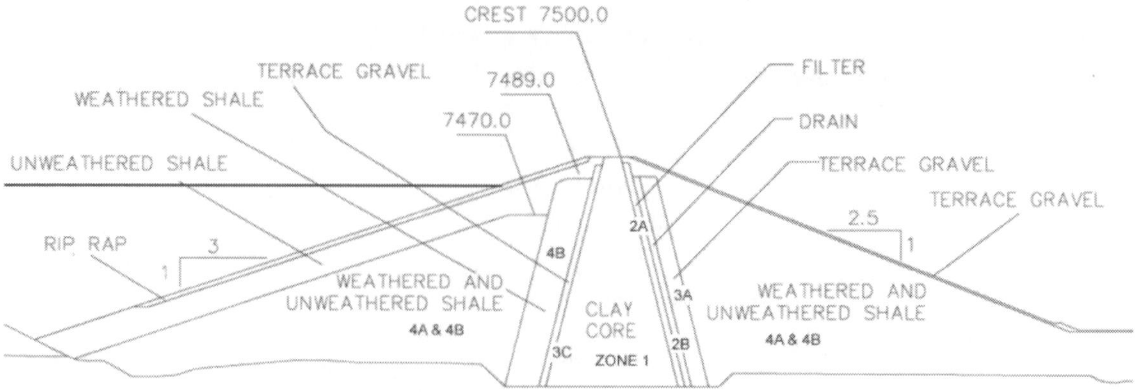

Figure 1. Ritschard Dam Cross Section

Piezometers were read regularly and crest monuments surveyed annually for 7 years after initial reservoir fill. After 7 years, crest monument surveys showed approximately 1.25 feet of settlement. During this time, vibrating wire piezometers failed one by one, perhaps because of settlement of the control building in 1998 that possibly compromised the readings by stretching the cables.

Reservoir enlargement investigations in 2006 recognized the ongoing crest settlement and recommended crest restoration by "top dressing." Annual crest monument surveys were resumed in 2007 after a three year interruption. Crest settlement had continued at a fairly uniform rate. Failure of the last functioning embankment piezometer in 2008 prompted a plan to drill new core piezometers into the embankment. At the same time, excessive opening of the outlet tower bridge joints drew attention to the horizontal component of the crest monument surveys. The center of the dam crest had displaced downstream about 0.5 feet.

Further investigation was warranted.

LEARNING

Observations of crest monument movement instilled a need to know more about the ongoing crest settlement and downstream displacement. Figures 2 and 3 show horizontal and vertical crest monument survey data. These figures illustrate crest monument deformation responses to large reservoir drawdowns in 2002, 2004, and 2012.

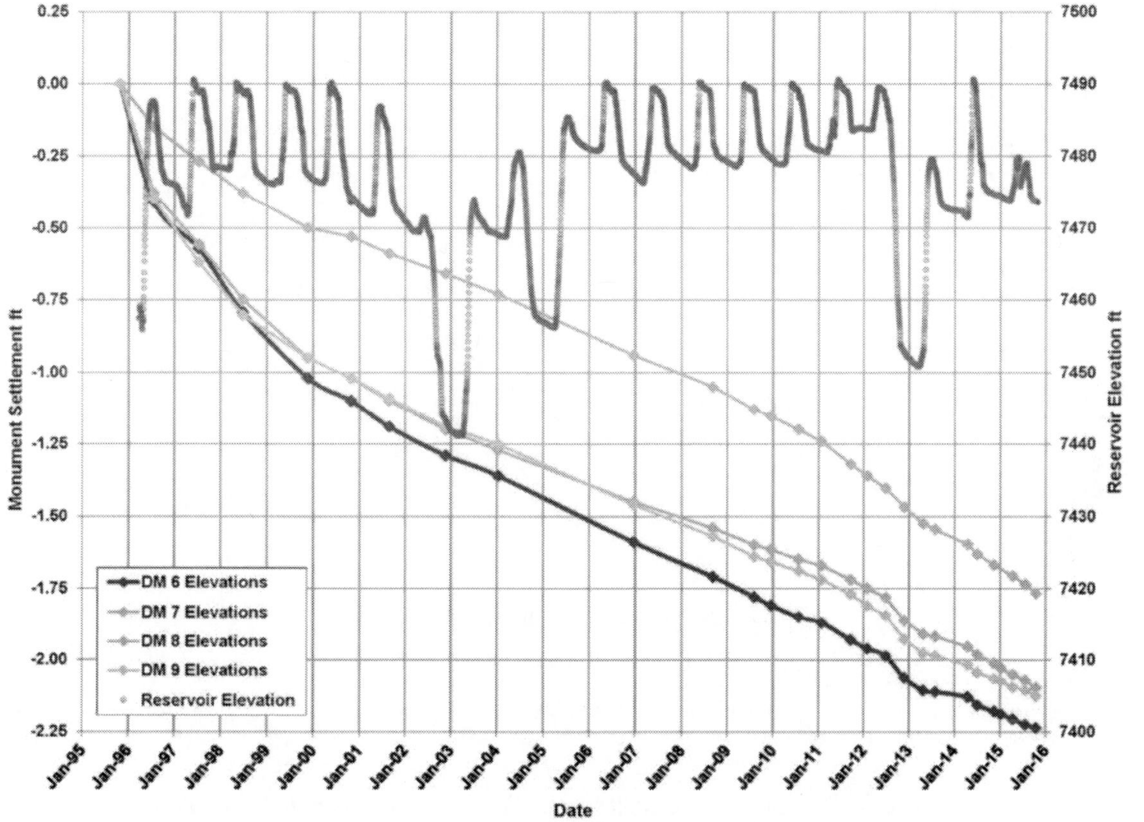

Figure 2. Vertical Crest Deformation at Ritschard Dam

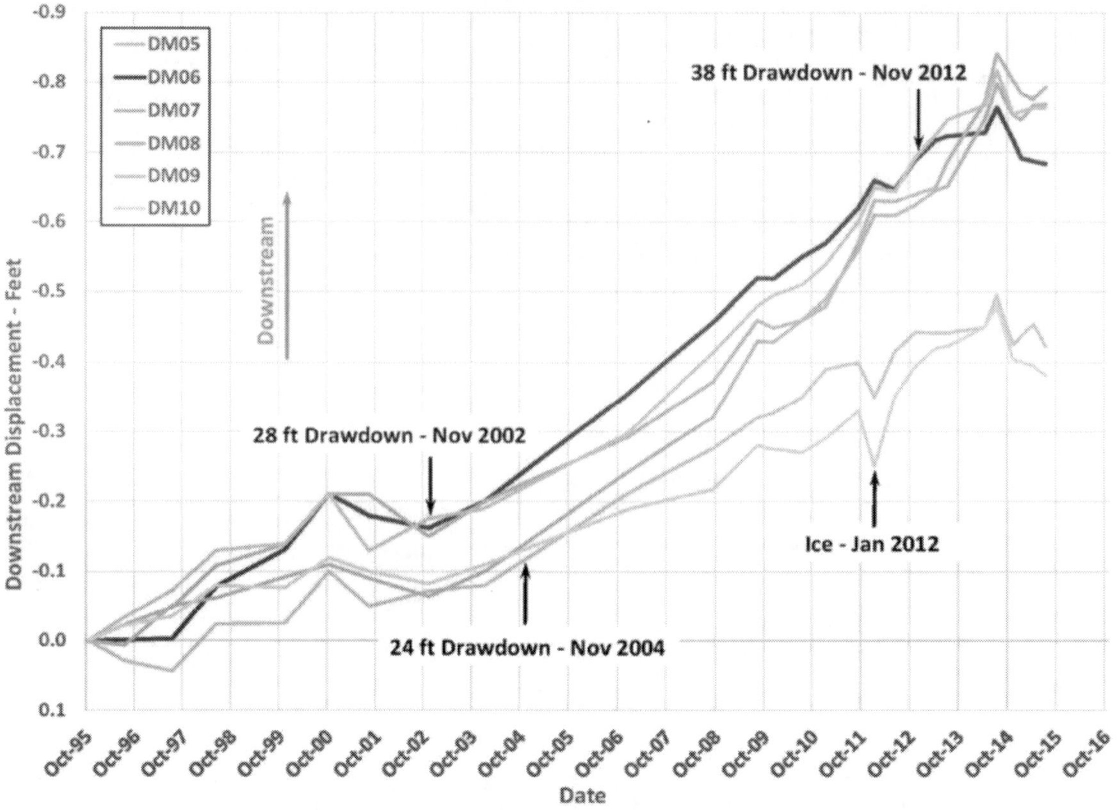

Figure 3. Horizontal Central Crest Deformation at Ritschard Dam

An instrumentation and monitoring plan developed from our initial need to replace failed embankment piezometers. In 2009 instrumentation installation included 8 core and shell piezometers, 16 settlement monuments on the downstream face of the dam, two inclinometers (with extensometers) in the embankment, and tilt meters on the outlet tower bridge piers. Inclinometer readings suggested the majority of deformation in the clay core was occurring in a narrow zone 50 feet below the crest, see Figure 4. After extensive engineering analyses and a peer review, 4 additional inclinometers and 11 piezometers in the foundation, core, and downstream shell were installed in 2011. Several of the piezometers paired open stand pipes with vibrating wires to compare performance. Additional inclinometers, extensometers, and piezometers were installed in the upstream shell in 2012. A map view of all monuments and instrumentation is shown in Figure 5.

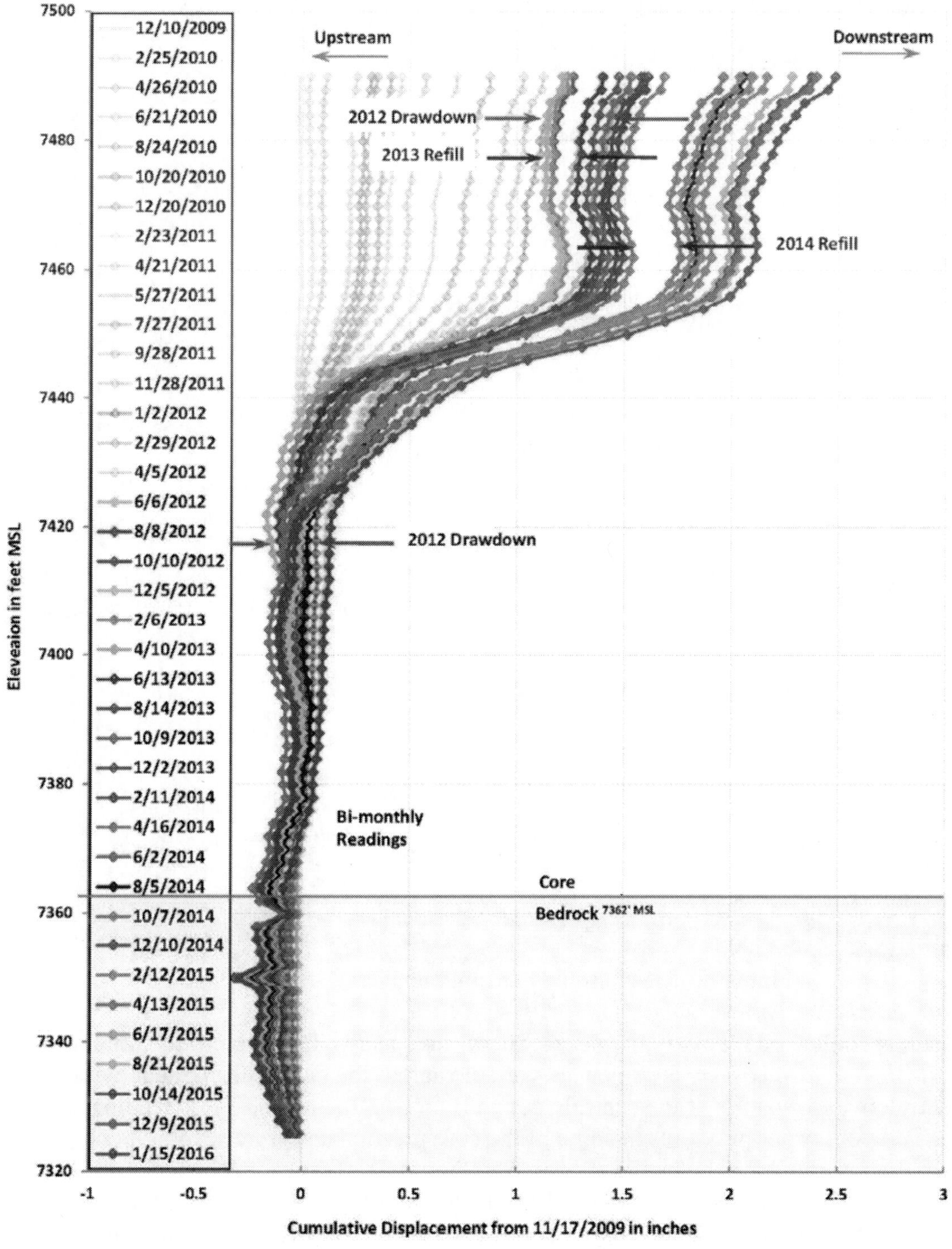

Figure 4. Ritschard Dam Core Inclinometer I-1 Displacement Pattern

Figure 5. Ritschard Dam Monuments and Instrumentation

Readings collected regularly over subsequent years showed the pattern of vertical and horizontal deformation was somewhat variable across the dam embankment, with maximum deformation at the central Section B (see Figures 6 and 11). Readings also showed pore water pressures in some of the core piezometers were several feet higher than the reservoir surface had ever been (see Figure 7). These piezometers also showed near immediate response to reservoir filling cycles, going yet higher. Increases in elevated pore water pressures (experienced with both open stand pipes and vibrating wires) were followed soon by downstream movement in core and shell inclinometers. Reservoir draw down cycles quickly reduced core pore water pressures to the previously elevated (higher than reservoir surface) values, and in the case of large drawdown, embankment retreat upstream was observed. Crest monument observations during multiple operational cycles showed that the vertical displacement of the embankment trended linearly without regard to reservoir level loading.

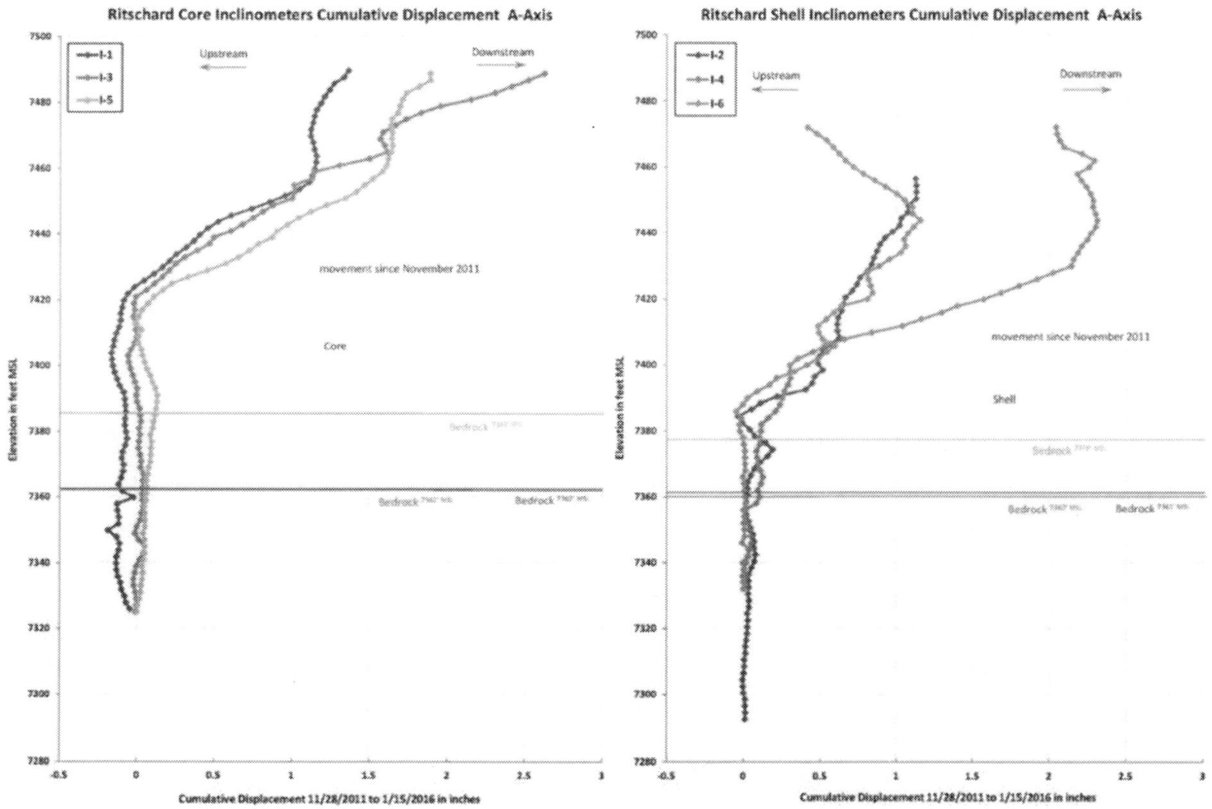

Figure 6. Ritschard Core and Shell Inclinometer Deformation Patterns across the Embankment

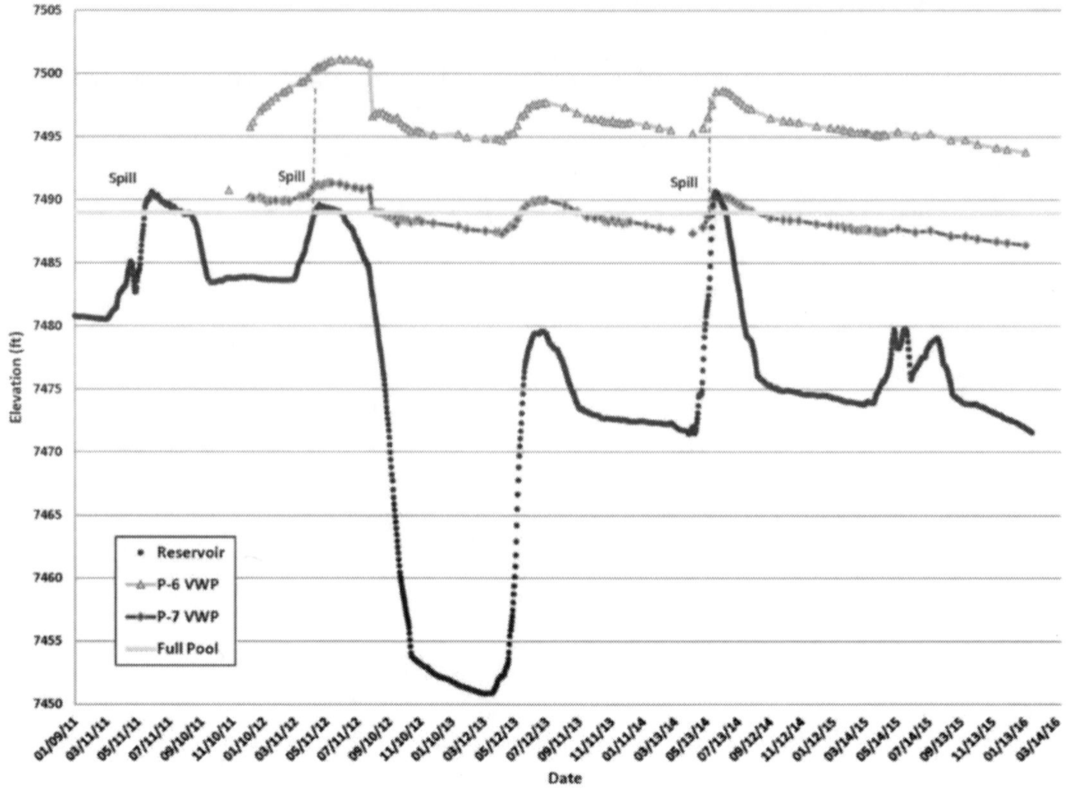

Figure 7. Selected Ritschard Dam Core Piezometer Readings

Tilt meters on the bridge piers and outlet tower brought their own lessons. The outlet tower and bridge pier 3 are founded on bedrock (see Figure 8). Two additional piers and the dam crest abutment of the bridge are founded in the upstream embankment. The abutment and pier 1 show rotation and settlement. In January 2012 routine surveying of bridge monumentation showed as much as 0.3 feet of horizontal deflection in the tower and bedrock founded pier 3. The reservoir surface was unusually high and ice pressure effects were found to be the cause. Installation of a 12 gpm well pump near the bottom of the reservoir raised warmer water under the ice sheet and melted the tower and pier 3 free of 30 inches of ice in a couple of days allowing the deflection to partially rebound. The effects of this ice event on the embankment are noticed in the crest monument surveys shown in Figure 3.

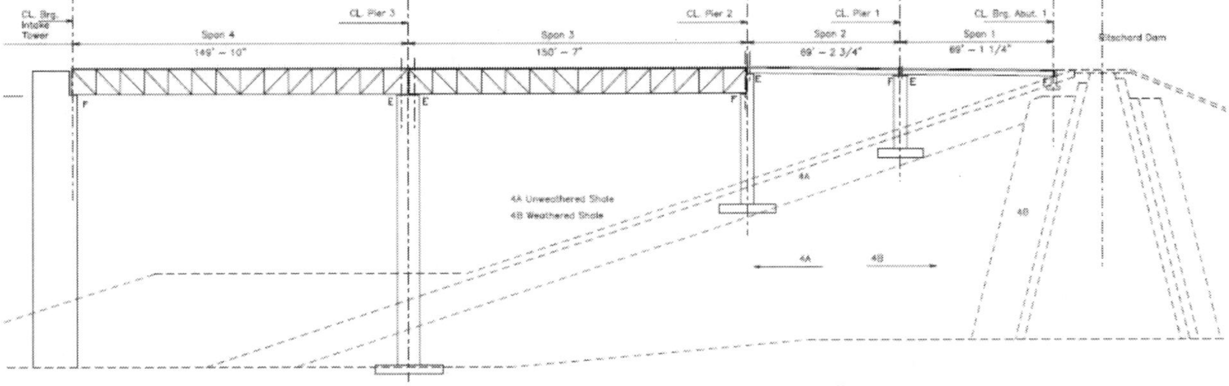

Figure 8. Ritschard Dam Intake Tower Bridge Configuration

In 2012, continued settlement of bridge pier 1 and concern for the performance of the upstream shell prompted installation of 3 inclinometers in the upstream shell. These inclinometers reflect downstream displacement during reservoir fill and upstream displacement when the reservoir is being drawn down. The 2013 reservoir operations cycle continued the patterns of core piezometer response to reservoir loading. To monitor elevated (higher than reservoir surface) core pore pressures and verify that immediate response to reservoir loading was a result of consolidation pore pressures, 6 vertical piezometer arrays were installed in 2015 and 2016. Multi-level piezometer arrays included 5 vibrating wire piezometers bracketing the zone of observed deformation, 3 upstream and 3 downstream of the center line of the core. Data collection for all core piezometers and bridge tilt meters was automated and uplinked to the internet to allow remote monitoring and more frequent observation.

The pore water pressures measured in the vertical arrays reflect the stress state in the zone of deformation, and all zones experience nearly immediate response to reservoir loading (see Figures 9 and 10). Increases in core pore water pressure appear to be followed by embankment deformation observable in the less frequently read inclinometers. Monitoring results suggest the mechanism for the unusual behavior is insufficient support of the clay core by the downstream shell.

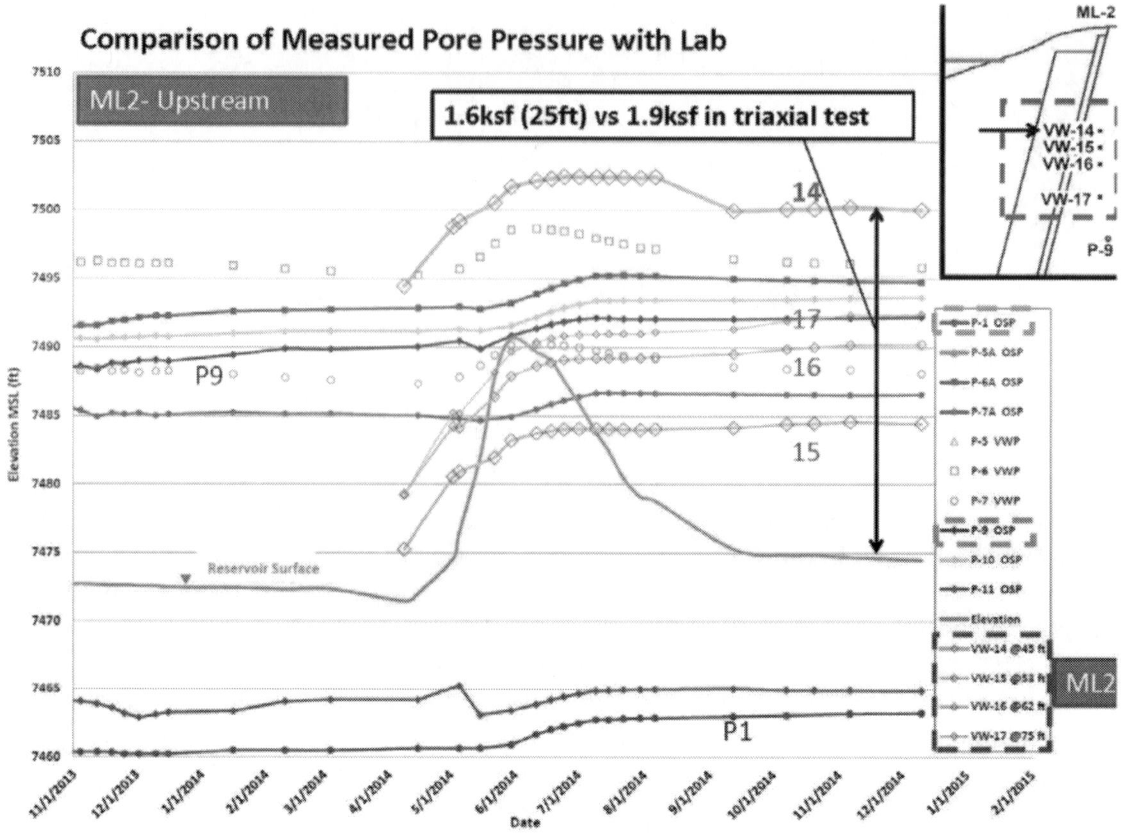

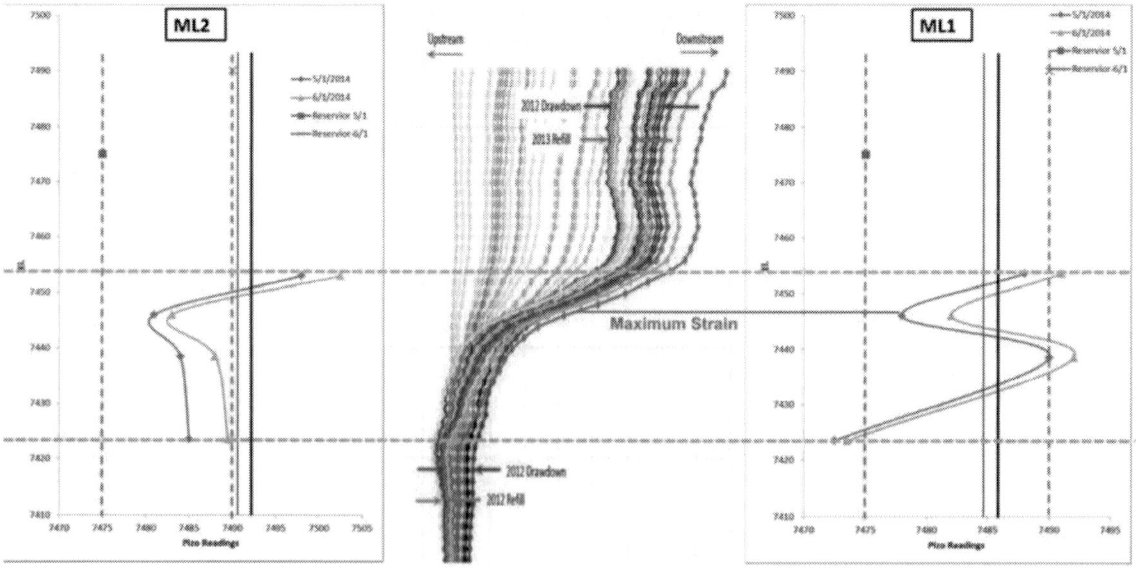

Figures 9 and 10. Ritschard Core Vertical Piezometer Array Response to Reservoir Loading

ACTION

What should be done with this dam and all the monitoring data? How could the acceptable amount of core deformation be estimated? When would the deformation stop? Is remediation necessary?

The finite element model FLAC (Fast Lagrangian Analysis of Continua) was employed to attempt to model observed embankment behavior and to predict future behavior of the embankment. Initial FLAC modeling utilizing the Mohr-Coulomb constitutive model and observations from the earliest inclinometers did not match observed behavior and was inconclusive.

Figure 11. Ritschard Dam Cross Sections Analyzed in FLAC

A second modeling effort benefitted greatly from later observations, including inclinometer and pore pressure measurements during a severe 2012 drawdown and refill cycle. Figure 11 shows three cross sections analyzed in FLAC. Development of the Ritschard FLAC model required the use of specialized constitutive models for the clay core and rockfill shells to capture the complex time-dependent behavior of these zones under reservoir cycling. To capture the permanent strain accumulation in the clay core due to annual reservoir fill and drawdown (load and unload) cycles, a custom routine was developed for the strain softening constitutive model to allow for non-linear behavior during reloading (see Figure 12). The material properties for the non-linear hyperbolic model and the degree of non-linearity during reloading were calibrated to laboratory test data from consolidation, direct simple shear, and triaxial shear tests that included load/unload cycles in the stress range of interest. To capture the time-dependent compression of the rockfill shells under self-weight and due to reservoir load and unload cycles, a custom routine was developed for the Cysoil constitutive model to induce instantaneous volumetric compression at yearly intervals, consistent with a hyperbolic settlement curve. The adjustment routine was calibrated for the various zones in the rockfill shell based on the time-dependent volumetric compression relationship required to match monitoring data for the dam, including surface monitoring points on the dam crest and downstream slope, inclinometers, and extensometers. FLAC model results for Section C suggest shear strains in the core are approaching approximately 8 percent (see Figure 13). Triaxial shear test results (see Figure 14) indicate the stress state in the core may be well advanced on the stress-strain curve, and also indicate excess pore pressures from shearing that could account for the elevated pore pressures measured by the piezometers in the clay core.

The model was calibrated to observed core pore pressures and embankment displacements (see Figure 15) and is able to simulate the complex behaviors of the dam with sufficient resolution. The model is supported by testing of recovered samples and interpretation of construction photographs, to confirm which portions of the embankment are most problematic (see Figure 16), with the red zones representing the most compressible material.

The model was used to try to predict future embankment behavior during multiple reservoir fill and draw down cycles, including possible changes in operations (lower maximum pool, see Figure 17). The FLAC modeling did not provide comfort that the dam displacements would abate anytime soon or before core integrity was possibly at risk under existing or reduced maximum pool operations. This led to a conclusion that, although no near-term risk was present, clearly a remediation plan would be required.

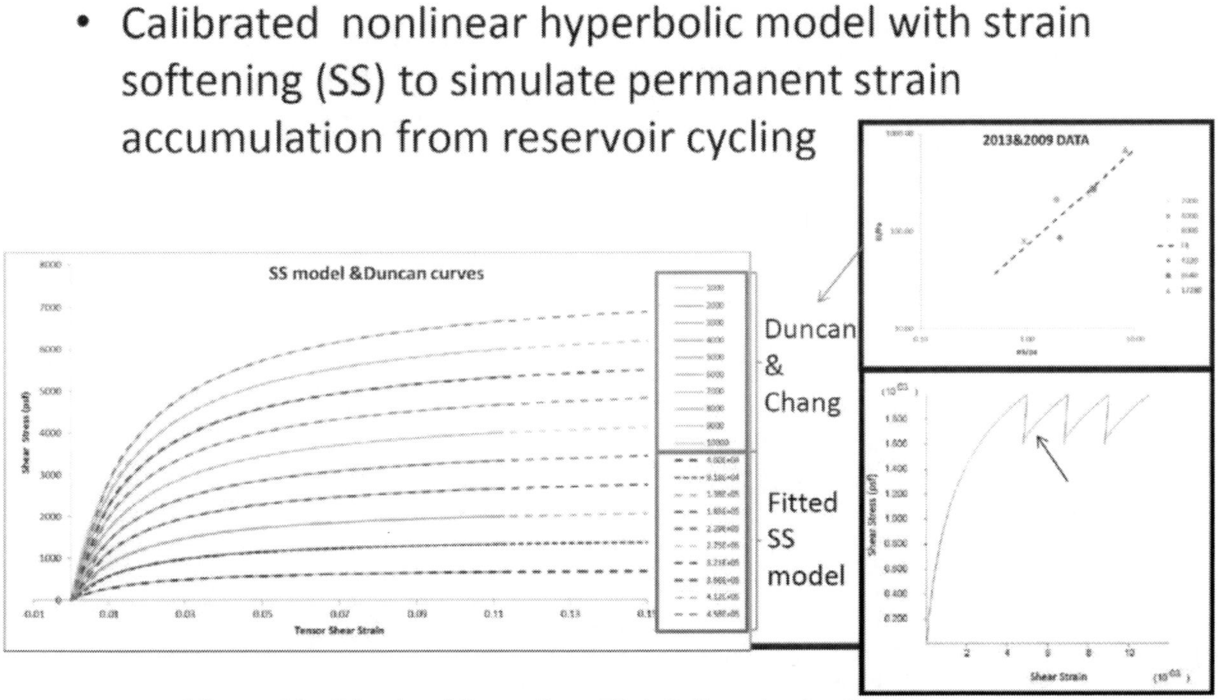

Figure 12. Ritschard Dam Core FLAC Constitutive Model

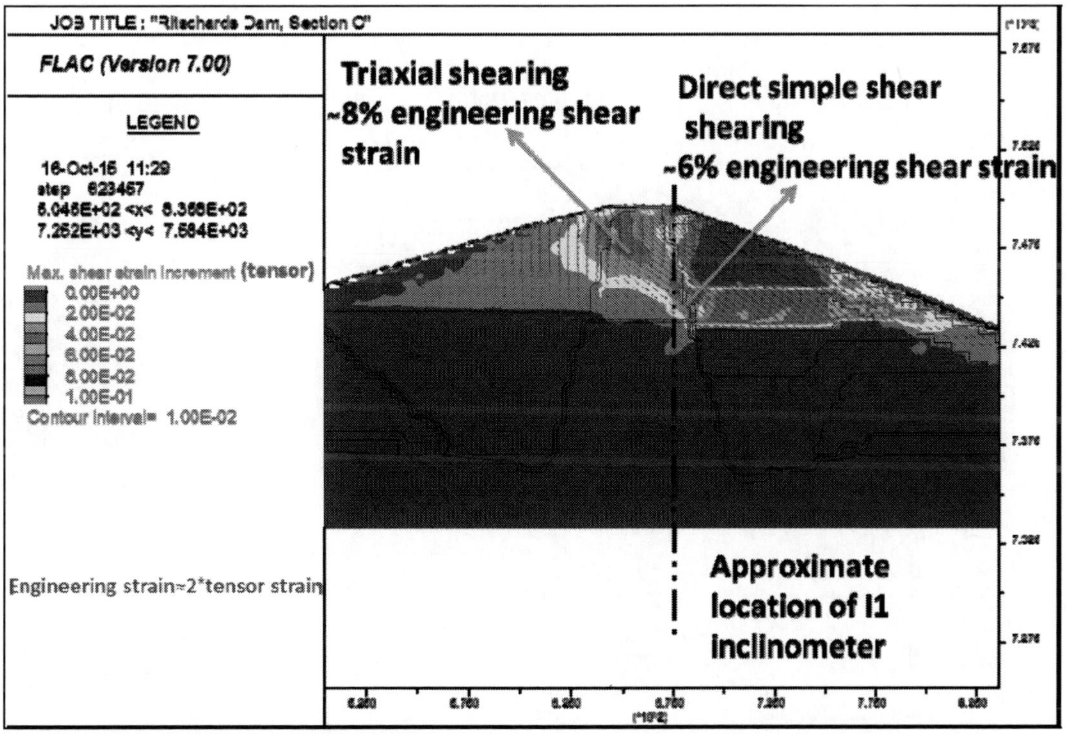

Figure 13. Ritschard Dam Core at Section C FLAC Model 2015 Shear Strain

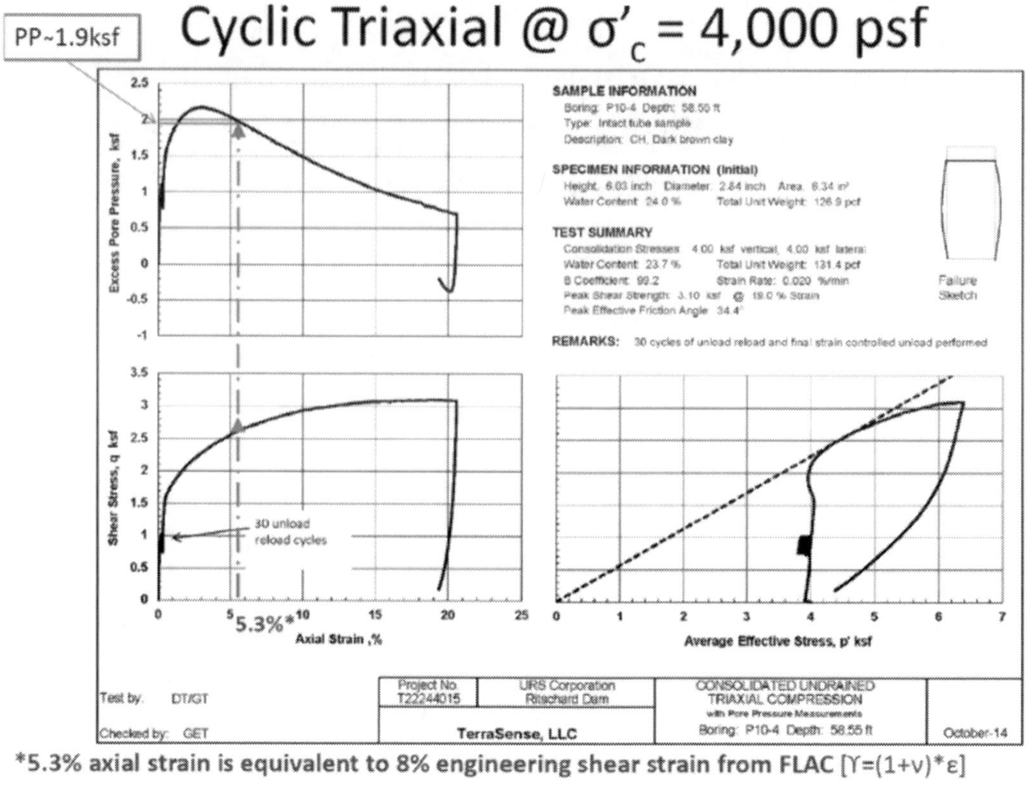

Figure 14. Ritschard Clay Core Cyclic Triaxial Testing Results

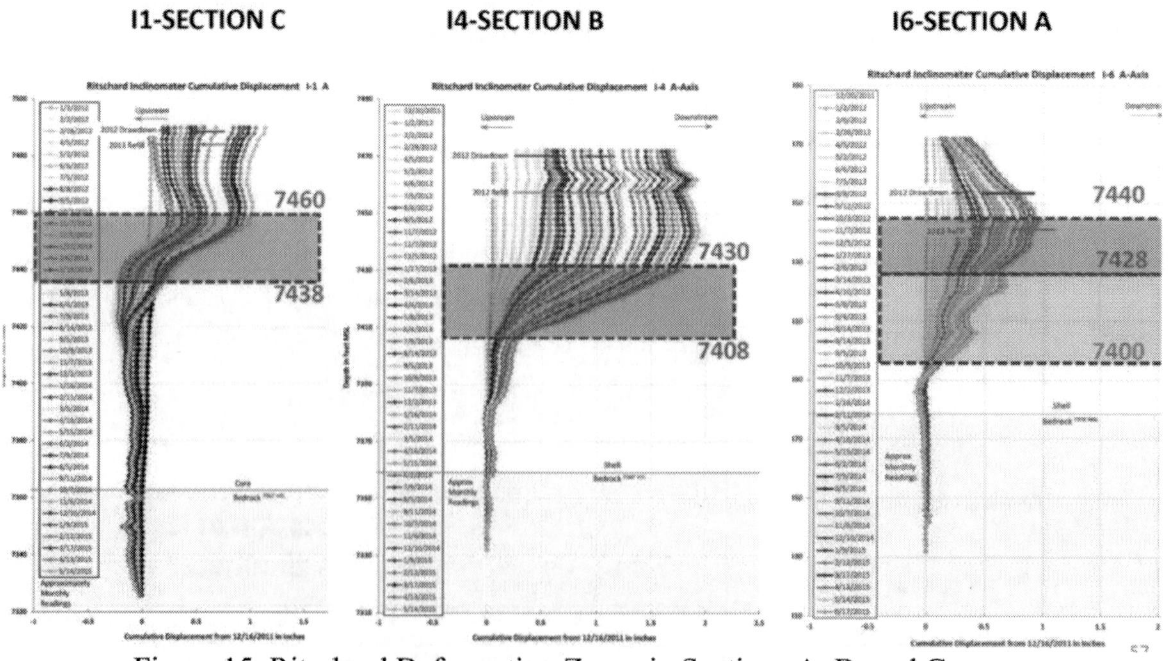

Figure 15. Ritschard Deformation Zones in Sections A, B, and C

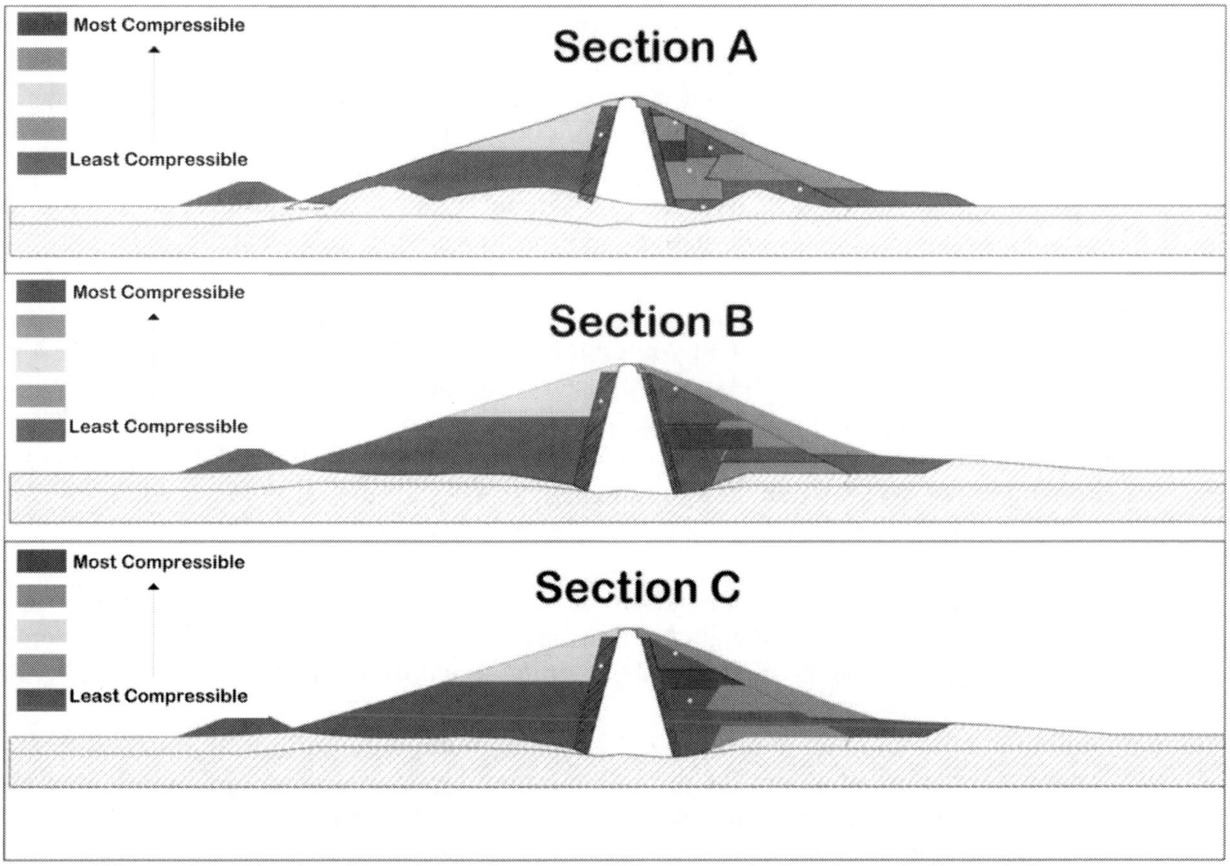

Figure 16. Ritschard Modeled Compressible Zones in Sections A, B, and C

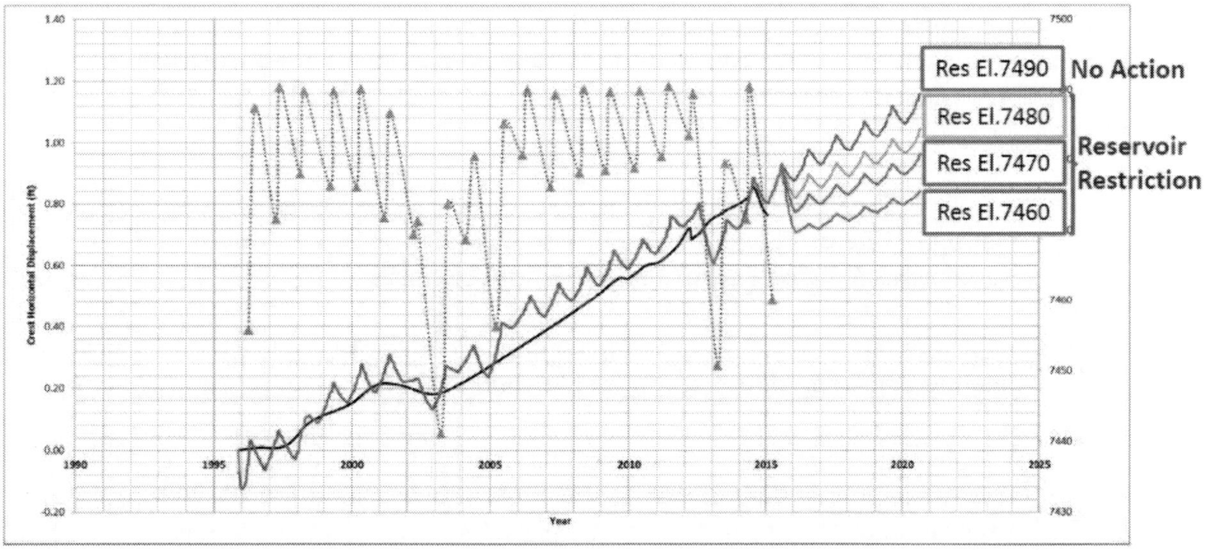

Figure 17. Ritschard Dam FLAC Model Results for No Action and Reservoir Restrictions

The FLAC model is supporting development of a remedial action plan by providing a reliable basis for evaluating and comparing alternative remediation methods. Remedial alternatives under consideration include remove and replace, and in-situ methods for shell stiffening and densification. A reservoir restriction alternative is not considered a long term solution because future demands cannot be met under the 2001 to 2004 critical hydrology. However, the reservoir is currently being operated with a 10-foot owner imposed restriction to reduce deformation. A conceptual design report is being prepared documenting the modeling, alternatives analysis, conceptual design, and estimated costs.

A review is underway using an expert panel to provide the owner assurance the monitoring data is being reasonably interpreted and utilized in the FLAC modeling effort. Remediation alternatives are being formulated with the objective of reducing the observed embankment displacements. Various in-situ solutions are being evaluated alongside a remove and replace option. Monitoring data is informing the evaluation of alternatives and will provide the assessment of success for the selected remediation alternative, once completed.

Work is targeted for a 2017 remediation effort, allowing time for formulation of the final remediation alternative, design, and permitting. Continued operation until remediation activities begin will be possible with the ongoing monitoring effort assuring the safety of the dam.

Throughout the investigation, the owner has worked with the Colorado Division of Water Resources, Dam Safety Branch and other stakeholders and interests to identify mechanisms and potential solutions.

The ability to move confidently toward a recommended remedial action plan has only been possible due to the extensive monitoring program which grew over time to thoroughly observe embankment behavior. Only through extensive monitoring has it been possible to identify the causes and solutions to the ongoing deformation.

SUMMARY

• Long term crest monument monitoring shows continued downstream movements and settlements that have not yet abated.

• Instrumentation responses to embankment behavior caused by reservoir pool loading can be used to detect possible changes in the embankment behavior.

• Multi-level vibrating wire piezometers read fast responding, elevated core pore pressures (greater than reservoir elevation) under changing reservoir pool load suggesting confined excess pore pressure from compression and displacement of the core.

• Inclinometers show localized core and shell displacements at specific depths which have not arrested, yet are predictable.

• FLAC modeling calibrated to monitoring program observations is providing a reliable basis for evaluating embankment condition and comparing alternative remediation methods.

• Operations can continue until remedial actions are implemented because the extensive embankment monitoring program allows observation of the behavior of the embankment and early detection of changes.

• Remedial actions under consideration can be carefully comparatively evaluated because monitoring data provided a sound basis for understanding and modeling the embankment behavior.

BOUNDARY DAM SPILLWAY NO. 2 MODIFICATIONS

John Werner, E.I.T.[1]
James H. Rutherford, P.E.[2]
Kim Pate, P.E.[3]
Daniel Kirschbaum, P.E.[4]
Joe Groeneveld, P.Eng.[5]
Keith Moen, P.E.[6]
Justin Arnold, P.E.[7]

ABSTRACT

The City of Seattle, City Light Department (City Light) working with Hatch Associates Consultants, Incorporated (Hatch) and Hatch's consultants at Alden Hydraulic Labs (Alden) planned, designed and provided engineering support during construction for modifications to Spillway No. 2 on the right abutment of Boundary Dam near Metaline Falls, Washington State. The modifications involved the installation of nine baffle blocks with an upstream ramp structure that provides external air to the flow over and around the blocks. The ramps and blocks create an efficient breakup of the chute jet prior to its drop into the plunge pool, thereby helping to limit overall gas transfer. The project was constructed and successfully commissioned by December 2014.

This paper presents challenges faced in planning, design and construction of the Spillway No. 2 modifications. The hydraulic design of the ramp and baffle blocks required extensive hydraulic analysis using both physical and numerical modeling. The design was informed by the installation and prototype performance experience provided by Alden. The physical model coupled with the numerical model and field studies have allowed total dissolved gas (TDG) performance to be investigated and predicted for a wide range of spill flow, spillway baffle block and ramp arrangements. Geotechnical issues included investigation of the right abutment rock structure and abutment stability in the design of 80 foot long post tensioned baffle block anchors. Three dimensional modeling of the right abutment helped to better understand the geologic setting and physical constraints for the design of the baffle block anchors.

INTRODUCTION

The Boundary Hydroelectric Project (Project) is located on the Pend Oreille River in Northeastern Washington State. Boundary Dam is the third of five dams on the Pend Oreille River (Alveni Falls, Box Canyon, Boundary, Seven Mile, Waneta) upstream of the confluence with the Columbia River. The Project operates in a load-following mode, generating power during peak-load hours and curtailing generation during off-peak hours.

[1] Civil Engineer, Hatch Associates Consultants Inc., Seattle, WA, jwerner@hatchusa.com
[2] Project Manager, Senior Civil Engineer, Hatch Associates Consultants Inc., Seattle, WA, jrutherford@hatchusa.com
[3] Chief Dam Safety Engineer, Seattle City Light, Seattle, WA, kim.pate@seattle.gov
[4] Senior Mechanical Engineer, Seattle City Light, Seattle, WA, daniel.kirschbaum@seattle.gov
[5] Engineering Department Head, Hatch Associates Consultants, Calgary, Alberta, Canada, jgroeneveld@hatch.ca
[6] NW Hydropower Engineering Manager, HDR, Seattle, WA, keith.moen@hdrinc.com
[7] Principal Engineer, Alden Laboratories Inc., Redmond, WA, jarnold@aldenlab.com

Figure 1 shows the key project features, consisting of an arch dam, reservoir, and underground power plant. Boundary Dam, constructed in the 1960s, is a variable-radius concrete arch dam with a structural height of 340 feet, a crest length of 508 feet, and a total length of 740 feet. The arch dam varies in thickness from eight feet at the crest, which is at elevation 2,004 feet (NAVD88) to 32 feet at the base at elevation 1,648 feet. The normal tailrace water surface elevation below the dam is 1,733 feet with approximately 261 feet of gross head for power purposes for a normal average upstream pool level.

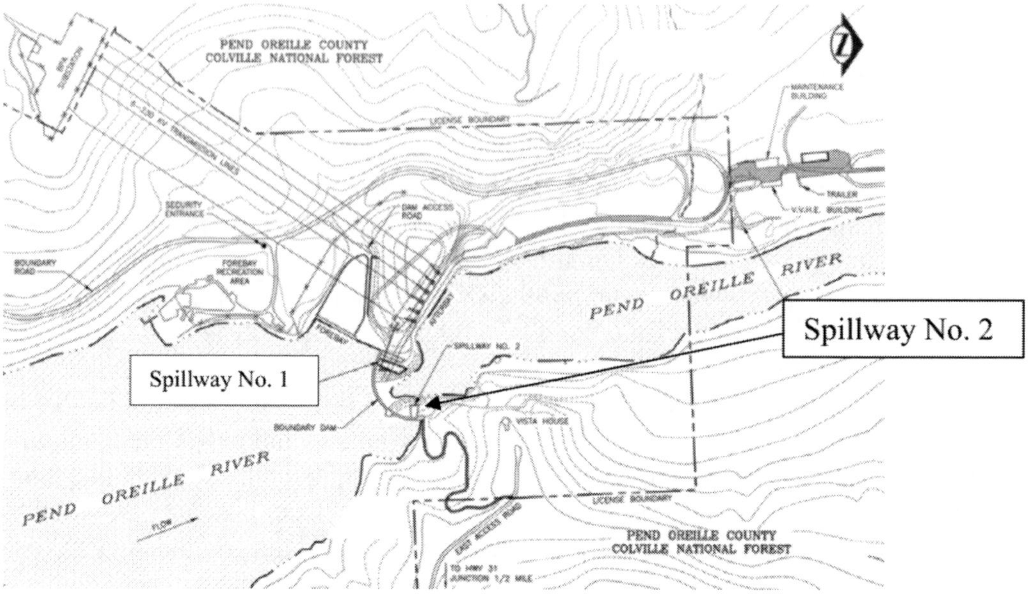

Figure 1. Boundary Dam Site Plan

Boundary Reservoir extends 17.5 miles upstream to Box Canyon Dam. Total usable storage is approximately 41,000 acre-feet from the normal maximum water surface elevation of 1,994 feet to 1,954 feet as the 40-foot operating range authorized by the current license. Figure 2 shows a downstream view of the dam looking upstream.

Since 2007 City Light (SCL) has been developing measures to reduce total dissolved gas (TDG) levels in the tailrace of the Boundary Hydroelectric Project (Project) to comply with the Pend Oreille River total maximum daily load (TMDL) for TDG. The TMDL for TDG is required to meet Project Certification, Order No. 8872, issued by the Washington Department of Ecology in November 2011 and approved as part of new license requirements issued by the Federal Energy Regulatory Commission (FERC) in March 2013.

Figure 2. Downstream View of Boundary Dam Looking Upstream

The TDG Attainment Plan that is part of the FERC license water quality certification describes the proposed process for complying with the 110 percent TDG saturation standard (Ecology 2012). City Light is required to identify all "reasonable and feasible" (Ecology 2005) improvements that could be used to meet the state of Washington's 110 percent TDG saturation standard by evaluating operational and/or structural modification alternatives at the Project.

PLANNING OF TDG UPGRADES

The SCL TDG team has convened a series of design workshops to brainstorm and develop alternatives, review Project progress, discuss technical aspects of the work, evaluate alternatives and provide direction for ongoing development of operational and structural measures to reduce TDG downstream of the Project. The initial development of alternatives occurred during a two-day workshop in 2007, attended by City Light staff and an expert panel. During this initial workshop a large number of potential mitigation measures were proposed. City Light and the expert panel selected six alternatives that they considered to be feasible and likely to be effective, and presented these in relicensing study plan reports (SCL 2007 and SCL 2009a); these are the alternatives on which the City Light TDG team has focused their subsequent design development efforts. Since 2009, the TDG mitigation design team holds an annual design workshop, and provides input to or attends periodic agency meetings. The annual team meetings are a forum to review the years' progress and significant design advances, to brainstorm ideas and discuss challenges, and to establish the goals and prioritize the studies and design development for the following year.

The TDG abatement project alternatives identified during the design workshops held in 2009 were investigated from the perspective of hydraulic performance and engineering feasibility. The three short-listed alternatives selected during this process (SCL 2009) are as follows:

- Throttle Sluice Gates:
  - Flow capacity 12,000–18,000 cfs
  - Deflector within water passage to protect gate slots
- Roughen Sluice Flow:
  - Flow capacity 16,000–20,000 cfs
  - Deflectors at exit of jet from sluiceway to flatten jet trajectory.
- Install Flow Splitters/Aerators on Spillways:
  - Flow capacity 20,000–30,000 cfs
  - Spillway deflectors, roughness elements, and/or dentated spillway

Based on both physical and numerical hydraulic analysis, as well as engineering considerations, modifications to Spillway No. 1 (SW1) and Spillway No. 2 (SW2) were selected for early implementation. SW2 modifications were scheduled first because SW2 had favorable foundation conditions that would require a simpler analysis than that of SW1. Additionally, SW2 was shorter, giving better access for construction of modifications.

The spillway modifications consist of adding Roughness Elements (REs) similar to baffle blocks to the spillway to induce turbulence, to accelerate break-up of the falling jet and to reduce jet penetration into the tailwater. The spillway REs were selected as the preferred first alternative to be implemented.

HYDRAULIC ANALYSIS AND DESIGN

Resolution of many of the hydraulic design issues for this concept relied heavily on the results of both physical and numerical hydraulic models. Both were used in complementary roles to maximize their particular strengths. Each is briefly described below.

Physical Model Studies

A 1:25 scale physical hydraulic model of the Boundary Dam spill release facilities was constructed at Alden in late 2008 and commissioned in early 2009. While the physical model cannot replicate the prototype gas transfer process and cannot, therefore, directly predict TDG performance, it can be utilized to gain an understanding of the hydrodynamics of spill releases into the tailwater that influence the gas exchange process. Direct visualization of the flow patterns generated in the tailrace and the relative depth to which entrained air bubbles are driven by the flow provide insight into how project operations and structural modifications might be developed to mitigate TDG production. The physical model has advantages over the CFD and TDG predictive modeling tools described later in this report in that, once the model has been constructed, it can be used in a relatively quick manner (hours) to compare the relative effects of operations and geometric changes to facilities, as compared to the computation time (days or weeks) required for numerical solutions to be developed.

Since commissioning, the physical hydraulic model has been used to achieve the following specific items that contributed to the design of SW2 modifications:

- Develop rating curves for all flow outlets;
- Make velocity measurements in the project tailrace (for use CFD model calibration);
- Characterize the plunge depth performance of the existing dam (Baseline Tests);
- Investigate spillway modifications including flip bucket treatments, spillway wall alterations, and roughness element placement, and;
- Optimize the configuration of roughness elements on the spillways to minimize plunge depth.

Specifically the physical model was used in the design of SW2 to evaluate the effectiveness of adding a ramp to suppress the vortices formed when flow separates from the roughness elements and its potential effect on plunge depth. Photo 1 shows vortex formation in the model prior to adding the ramp. Photo 2 shows the same location after the addition of the ramp with no evidence of vortex formation.

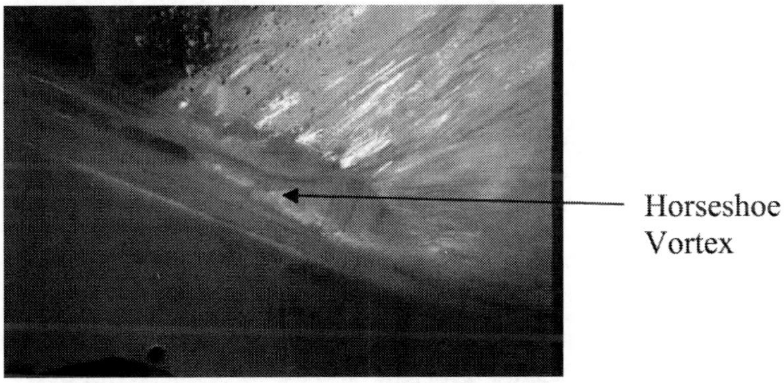

Photo 1. Vortex Formation Upstream of Elements (6,500 cfs)

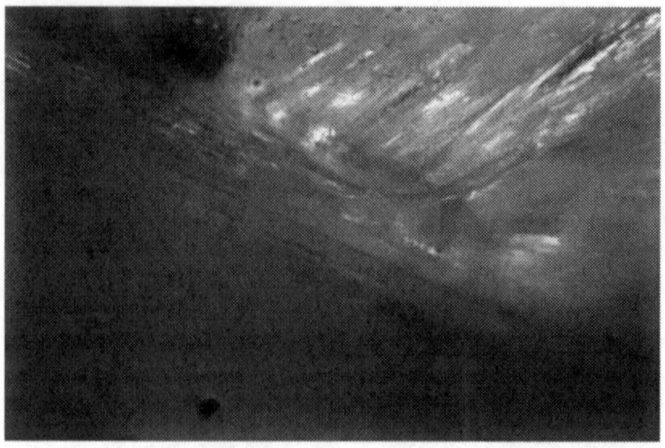

Photo 2. Ramp Installed Upstream of Elements (6,500 cfs)

Computational Fluid Dynamics (CFD) Analysis

Full three dimensional CFD models of SW1, SW2, a sluiceway, and the downstream plunge pool and river channel were developed in 2009. These models have been utilized as part of the suite of design tools used to develop an understanding of the governing hydraulic and hydrodynamic processes driving gas exchange in the tailrace of the existing Project under spill conditions, to develop the designs of structural TDG mitigation alternatives, and in combination with the TDG predictive model described in subsequent sections of this report, to predict the TDG performance of proposed TDG mitigation alternatives. The CFD models were used in conjunction with the physical model utilizing the relative strengths of the two analysis methods. The physical model was used to its greatest utility in iteratively developing physical geometry changes of the mitigation alternatives, while the CFD models were used to provide quantified information in terms of hydrodynamic loads for design purposes and flow fields for use in TDG prediction.

Results from the CFD models have been verified and validated against both physical model and field data (see Figure 3). CFD models were used extensively in the design of the SW2 modifications. Contributions have included:

- Analyze ramp and baffle block arrangements achieve reductions in TDG levels during variable spill flow;
- Estimation of hydraulic forces and pressures acting on the roughness elements;
- Estimation of pressures due to flow impingement on the left abutment under various operating scenarios;
- Investigation of the effect on tailrace flow fields associated with the addition of ramps on the spillway floor upstream of the roughness elements to suppress potential cavitation damage;
- Generation of hydrodynamic input to the TDG predictive tool for existing field data cases to validate the prediction tool; and
- Analysis to provide hydrodynamic input to the TDG predictive tool for baseline and modified operations.

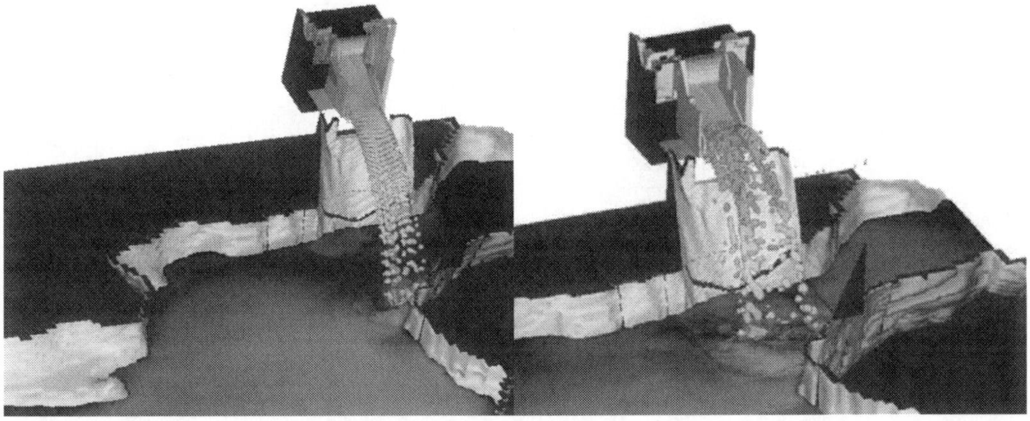

Figure 3. CFD Comparison of Unmodified and Modified Spillway No. 2 Jet Trajectories

TDG Prediction

To help evaluate the impact of the SW 2 modifications under consideration would have on downstream gas production, it was necessary to estimate expected gas transfer within the project plunge area, both for the existing or baseline condition, and with various mitigative works in place. The prediction of TDG performance is performed using the CFD model, and a particle tracking method. The particle tracking methodology first involves application of a CFD model to simulate the movement of entrained air bubbles with time in the tailwater of the project when the spillway is operating. The hydrodynamic histories (velocity and pressure) of these bubbles are then exported to and used in a spreadsheet tool, which then uses this data, in combination with gas transfer equations, to predict the equilibrium gas content of the tailrace receiving water. The methodology of acquiring and using the required bubble history information from the CFD model was initially developed and tested for the Boundary tailrace situation in 2010 and 2011. Subsequently the predictive model was applied to replicate actual spill releases at the project, and the predicted TDG levels were compared with field measurements and found to be in reasonable agreement. In subsequent years the model has been further developed and used to predict TDG performance of mitigation alternatives, including the addition of roughness elements on Spillway No. 2. As an example, Figure 4 illustrates how the time dependent average bubble plunge depth is expected to change with the addition of the ramps and roughness elements to SW2 for a discharge of 13,000 cfs.

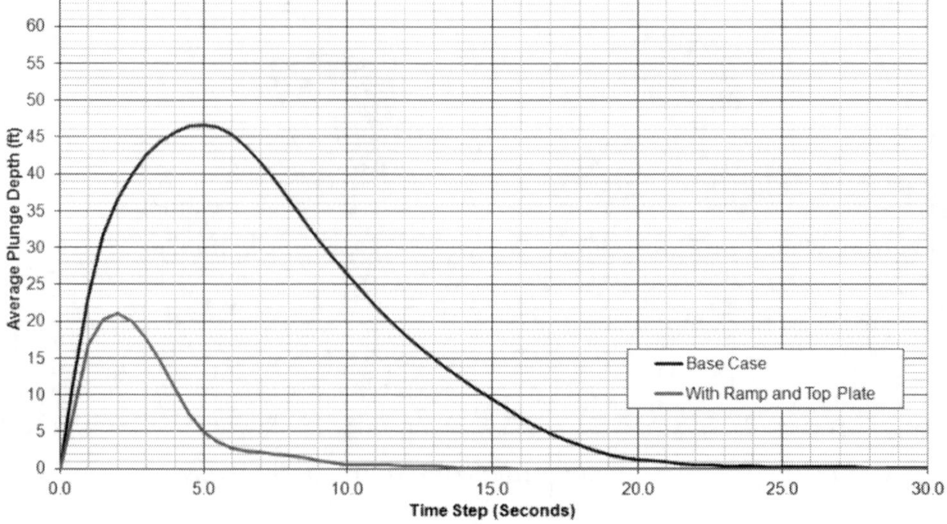

Figure 4. Time dependent air bubble plunge depth – Discharge of 13,000 cfs

STRUCTURAL ANALYSIS AND DESIGN

Detailed design of SW2 modifications was finalized in early 2014 allowing for completion of construction by the end of the year. Key structural design objectives were to; 1) ensure that the modifications are robust, and 2) avoid interference with the original design intent of the spillway.

Special consideration was given to an existing embedded drain and existing passive anchors that anchor the spillway flip bucket to the rock abutment. The new multi-strand post-tensioned (PT) anchors were designed with variable installation angles to allow the roughness elements to be fully integrated with the spillway and spillway foundation through shear friction. The variable installation angle allowed the new anchors to be installed without interfering with the existing features and to mobilize sufficient rock mass.

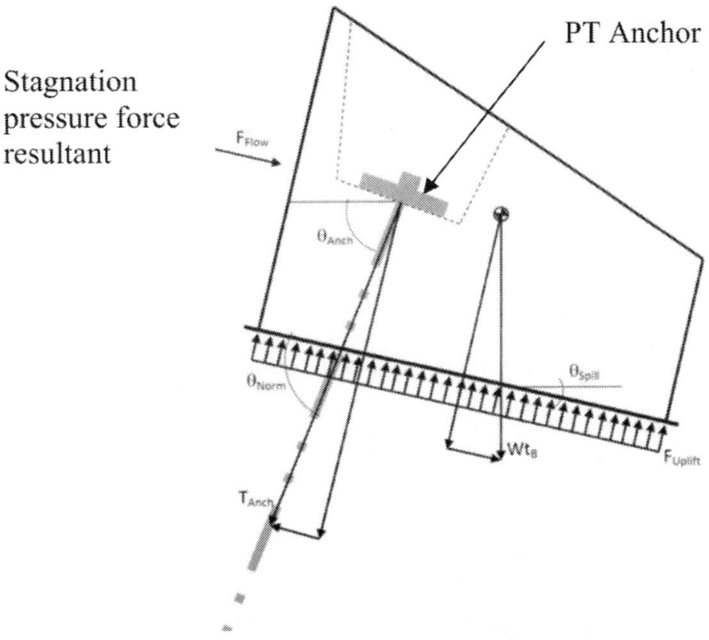

Figure 5. Roughness Element Loading Diagram

Figure 5 shows a schematic representation of the roughness element loading diagram for normal hydraulic loading. The roughness elements were designed for full stagnation pressure on the upstream face of the element. Design of the roughness elements' anchorage considered normal operating flow acting on the installed features, maximum flow due to full gate discharge conditions, fatigue due to variable hydrodynamic loads during very turbulent flow conditions, and impact resistance due to log strikes.

Loads and parameters that were considered in the design include: geologic/geotechnical parameters; hydrodynamic loads, impact load, fatigue, and cavitation. Certain load conditions were determined to be not relevant or insignificant relative to other loads that for which the modification were being designed. These include: wind load, snow load, and earthquake load

Figure 6 is a plan and section showing the SW2 modifications. Figure 7 is a 3D view of the SW2 modifications showing the roughness element anchor locations in the rock below the spillway.

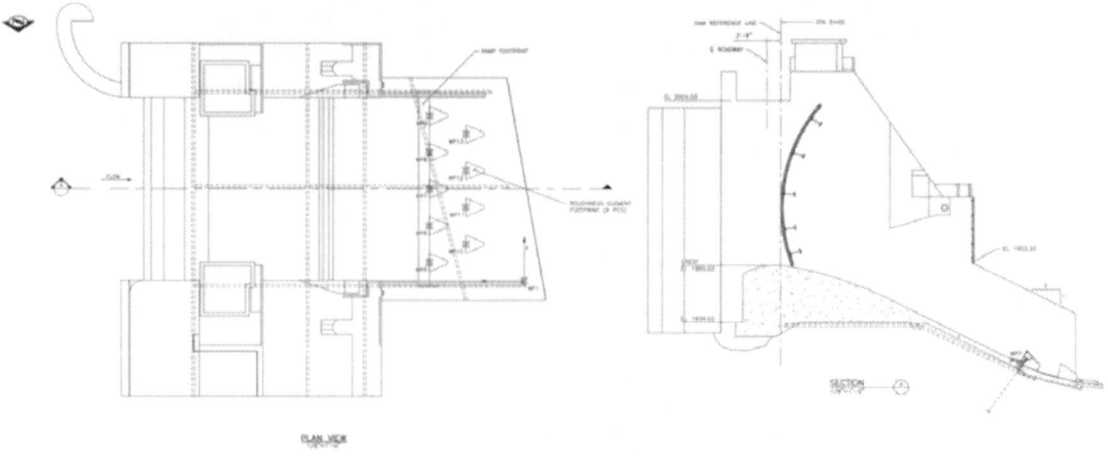

Figure 6. Spillway No. 2 TDG Modifications Plan and Section

Figure 7. 3D Representation of Spillway No. 2 Modifications

A site investigation of SW2 was performed in October of 2013 by Hatch with the help of Extreme Access, an industrial climbing company from Bend Oregon. A Schmidt Hammer was used as a non-destructive method to determine the strength of the existing spillway concrete. The rock of the right abutment, where SW2 is founded on, is considered competent rock, and presented no challenges for the design of the PT anchors.

In the case of an impact load from a log strike or other rigid body on the roughness elements during a spill event, the PT anchors are designed as a fuse. If such an event occurs, the capacity of the PT anchor tendons is the weak link and will fail first. This fuse behavior is preferred to the other failure modes (i.e., failure of bond in the bonded region or mobilization of a rock wedge), which would have potentially catastrophic effects to the integrity of Boundary Dam.

The PT anchors also considered fatigue in the design. Spill events expose the roughness elements to chaotic, violent conditions that have rapidly varying pressures. Constant cycles of load can lead to fatigue failure. The anchors were designed with an operating load large enough that the base always remains in compression even at the most extreme spill conditions (full stagnation pressure of a PMF event). Since the base always remains in compression, the anchor never experiences any additional load and is protected from fatigue.

Cavitation prevention is accomplished by the selection of robust construction materials-concrete-filled stainless steel fabrications. The cladding on the block has adequate surface durability to cavitation, and tunes the shape of the roughness element to reduce the potential for cavitation damage on the surfaces, and doubles as formwork for the placement of the concrete. A ramp is added upstream of the roughness elements to suppress vortex formation, lift the vortices that do form away from the structure surfaces, and provide a conduit for adequate air admission. Penetrations are made in the spillway walls for air admission into the hollow ramp structure.

CONSTRUCTION AND TESTING

Construction of the SW2 modifications was advertised in late March 2014. Three addendums were issued prior to opening bids in early May 2014. Six bidders responded and all were compliant. The construction contract was awarded to the low bidder, Clearwater Construction & Management of Spokane, WA, in early June 2014. Notice to proceed was given on July 7, 2014.

Major events during the 2014 construction included:

Quality Control Inspection Program	May 21
Mobilization	July 28 - 29
Concrete coring and demolition	August 8 - 15
Foundation drain reroute and spillway concrete patching	August 18 – 28
PT anchors drilling, water pressure testing, grouting	September 2 - 24
Tendon installation and bond-zone grouting	September 18 - 24
Site work temporarily suspension while steel fabrication is completed	September 29
Fabricated steel arrives at site	October 14
Roughness elements installation	October 16 - 27
Upstream ramp installation	October 22 - 31
PT anchor tensioning , testing, lock-off	November 18 - 20
Construction substantially complete	Early December
Spillway No. 2 discharge test	December 11

A combined containment and safety barrier was installed to the downstream end of the spillway prior to the work to prevent stormwater and construction debris from leaving the site. Materials were collected and pumped to holding tanks at the dam crest. A steel angle was installed on the spillway surface upstream of the work area to intercept and divert stormwater and gate leakage flows from entering the work area. After these facilities were installed, construction began with the concrete demolition and spillway drain rerouting to avoid interference with the PT anchors.

Holes for the PT anchors were drilled and then the anchors were installed and the bond zone grouted. Following their installation the construction of the roughness elements and ramp commenced. The PT anchor performance testing was conducted after the lower section of the roughness elements were filled with self-consolidating concrete (SCC) and the anchor baseplates were installed. Due to the slope of the spillway and the height of the roughness elements, a work platform was fabricated for personnel and instrumentation to perform the PT anchor performance testing. After the successful PT anchor performance testing, they were locked-off and the free stressing length was grouted. The top section of the roughness elements were filled with SCC and the final closure plates were welded closed. At each spillway wall where the ramp connects, holes were cored and bellmouth entrances were installed at the exterior of the wall to supply air to the ramp for the high air demand during spill. **Photo 3** shows the containment system in place prior PT anchor and roughness element installation.

The modified SW2 was commissioned with a full gate opening in December 2014. However low river flows during 2015 allowed only limited opportunities for field testing and evaluation of SW2 modifications on TDG performance. **Photos 4** through **8** show the construction of the modifications and the initial testing.

Photo 3. Containment and Safety Barriers Installed at Work Area

Photo 4. Post-Tension Anchors Installed

Photo 5. Roughness Elements and Ramp Installed

Photo 6. PT Anchor Performance Testing

Photo 7. Completed Spillway No. 2 Modifications Prior to Testing

Photo 8. SW2 TDG Modifications Commission Testing (December 2014)

CONCLUSION

Based on observing the performance of Spillway No. 2, it appears that the hydraulic and structural performance has met the design objectives. Aspects of this design have been applied to the design of modifications to Spillway No. 1 scheduled for modifications in summer/fall 2016. The primary difference between the right and left abutment is that the left abutment has a more complex geology and adversely oriented jointing that has required special consideration. The use of roughness elements for new and existing spillways should be considered as a tool to reduce both total dissolved gas production and energy that could cause scour.

ACKNOWLEDGEMENTS

The authors would like to acknowledge the contribution of all of the SCL TDG team with special thanks to hydraulics experts Chick Sweeney of Alden Labs and Dr. Hank Falvey, an independent consultant.

SELECTING A DAMAGE CRITERION FOR FRAGILITY CURVES DERIVED FROM PERFORMANCE-BASED TESTING

Mo Zhao[1]
Spenser Anderson[2]
Kunal Menda[3]
Ziyad Duron, Ph.D.[4]

ABSTRACT

A fragility curve is a mapping that gives the probability of an asset becoming damaged as a function of the intensity of the excitation to which the asset is exposed. In constructing such a curve, a precise definition of damage in its relation to the asset must be employed. In this paper, a process is described for quantifying damage on a large concrete dam using the calculated earthquake responses predicted by Performance-Based Testing (PBT). An advantage of the analysis procedure is that it provides a context for selecting suitable damage criteria for large concrete dams. Specifically, the maximum displacement of a point on the dam is justified as a proxy for the maximum stress endured at some point on the dam, and maximum stress in turn is used as a measure of potential damage. It is found that reasonable estimates of fragility curves can be constructed by employing simplified damage criteria based on the height of the dam alone. However, it is found that employing more complex strategies for quantifying damage that utilize finite element models enable greater specificity at the cost of additional modeling assumptions.

INTRODUCTION

Performance-Based Testing (PBT) aims to combine both experimental data with analytical techniques to evaluate the health of a dam. The PBT experiment involves using a Cold Gas Thruster (CGT) to impart a high amplitude, broadband impulse-like load onto the dam. The force delivered by the CGT is believed to be strong enough to push far into the dam's linear-elastic response without causing the concrete to yield, a necessary condition to accurately gauge the health of the dam. As a result, it is now practical to develop field testing techniques that can be used to develop performance indicators, such as fragility curves, for evaluating how a dam might respond to a large loading event. Although currently many fragility curves are made using FEA models, PBT's advantage in creating fragility curves is two-fold: it uses experimental data, and it creates fragility curves in a few hours of computation time, a sharp contrast to the weeks, and at time months it takes to compute FEA model-based fragility curves, which also use smaller earthquake data sets. PBT responses are used to develop fragility curves for earthquake hazards by convolving an experimentally determined impulse response for the dam with

[1] De Pietro Engineering Fellow, Harvey Mudd College, Claremont, CA, 91711, mzhao@hmc.edu
[2] De Pietro Engineering Fellow, Harvey Mudd College, Claremont, CA, 91711, sanderson@hmc.edu
[3] De Pietro Engineering Fellow, Harvey Mudd College, Claremont, CA, 91711, kmenda@hmc.edu
[4] Professor of Engineering, Harvey Mudd College, Claremont, CA, 91711, duron@hmc.edu

an earthquake time history to determine the dam's response to an earthquake, and then mapping that response to the an earthquake intensity measure, as outlined in Menda et al. (2016). In the typical development of a fragility curve, there is significant scrutiny over how the failure is defined and whether it is the type of failure that actually leads to an uncontrolled rapid release of water. In the current approach, instead of focusing on an exact definition of failure, fragility estimates can be developed for incremental performance limits. In this way, the dam owner is able to evaluate a variety of conditions or states in the dam that would be unacceptable for continued and safe operations.

In determining the performance limits, it is first necessary to define the engineering measure, or the metric that is representative of damage to the structure. In this paper, the primary measure used is the maximum displacement of a single point in the dam, which can be found by twice-integrating the result of the convolution described earlier in Menda et al. (2016). The reason for using displacement is because of its relationship to strain, which is in turn related to stress. Ideally, stress would be an excellent performance measure, but because it is difficult to measure, displacement is used instead.

The main focus of this paper is to discuss the influence of various damage criteria on the fragility estimate. Three damage criteria that map the displacement performance measure to a performance limit are considered. To aid in the explanation of the damage criteria, results from a PBT investigation performed on San Clemente Dam are presented as a case study. After explaining each of the damage criteria and how they are developed, this paper will discuss the shortcomings and advantages of each of the damage criteria as well as the different scenarios in which each damage criterion can be useful.

Brief Outline of the San Clemente Dam PBT Investigation

A PBT investigation was conducted on San Clemente Dam to collect data for fragility curve analysis. The investigation follows a relatively straightforward procedure: first, the CGT is placed somewhere on the dam where it can trigger a high amplitude response; then the accelerometers are placed in accessible locations of interest; and finally, the CGT is pressurized and fired to provide the impulse. The impulse was delivered several times, twice in every grouping of accelerometer locations as shown in Figure 1.

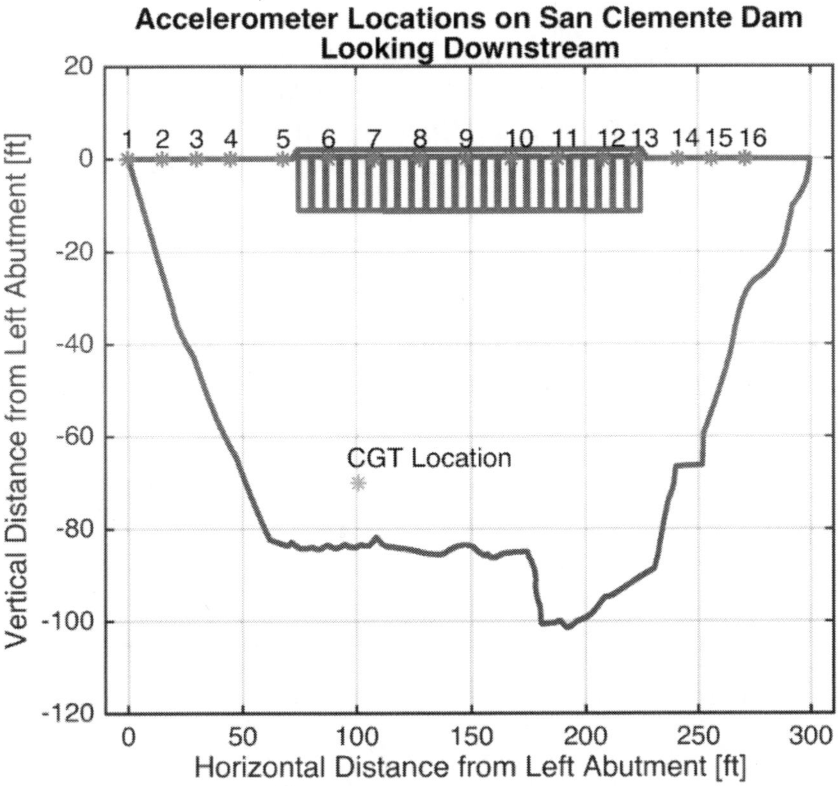

Figure 1: Instrumentation layout for the PBT experiment of San Clemente Dam.

The maximum displacements used in this paper to generate sample fragility curves are all found using the accelerometer at location 10 in Figure 1.

DAMAGE CRITERIA

The following section discusses the three damage criteria that have been found to be relevant for estimating fragility early in the consideration of a dam's vulnerabilities to hazard loading conditions. At this point, it is worth emphasizing that the development of damage criteria is completely independent of determining the dam's response to an earthquake, which is described in Menda, et al. (2016). Instead, the damage criteria attempt to map the predicted response of the dam (specifically, the maximum displacement) to the maximum stress endured by the dam.

Drift Ratio Criterion

As an initial attempt to characterize the probability of failure for concrete dams, some dam owners have put forth the following equation to determine Probability of Damage or Failure:

$$\frac{\Delta_{max}}{H_{dam}} = 0.2\% \text{ corresponds to } P(failure) = 10\% \quad (1)$$

where Δ_{max} is an estimate of the maximum deformation possible in the dam under seismic loading conditions, and H_{dam} is the height of the dam. The motivation for this equation is that similar to a high-rise building: the most vulnerable areas in a dam could be approximated as a cantilever and thereby indicating damage by the deflection normalized by its height. The practical aspect of this criterion is that it defines a performance measure that is believed to correspond to a 10% probability of failure, a probability level that few dam owners would be comfortable exceeding under any loading condition.

Using 0.2% of the height of the dam as the performance limit and 80 ft. as the height of San Clemente Dam, the fragility curve in Figure 2 was produced. That is, if the maximum displacement in the calculated displacement record exceeded 0.2% of the height of the dam, the dam would be said to have failed based on the exceedance of this criteria. The result is not an indication of ultimate dam failure; instead, this fragility curve plots the probability of exceeding the 10% probability of damage as defined by the damage criterion employed.

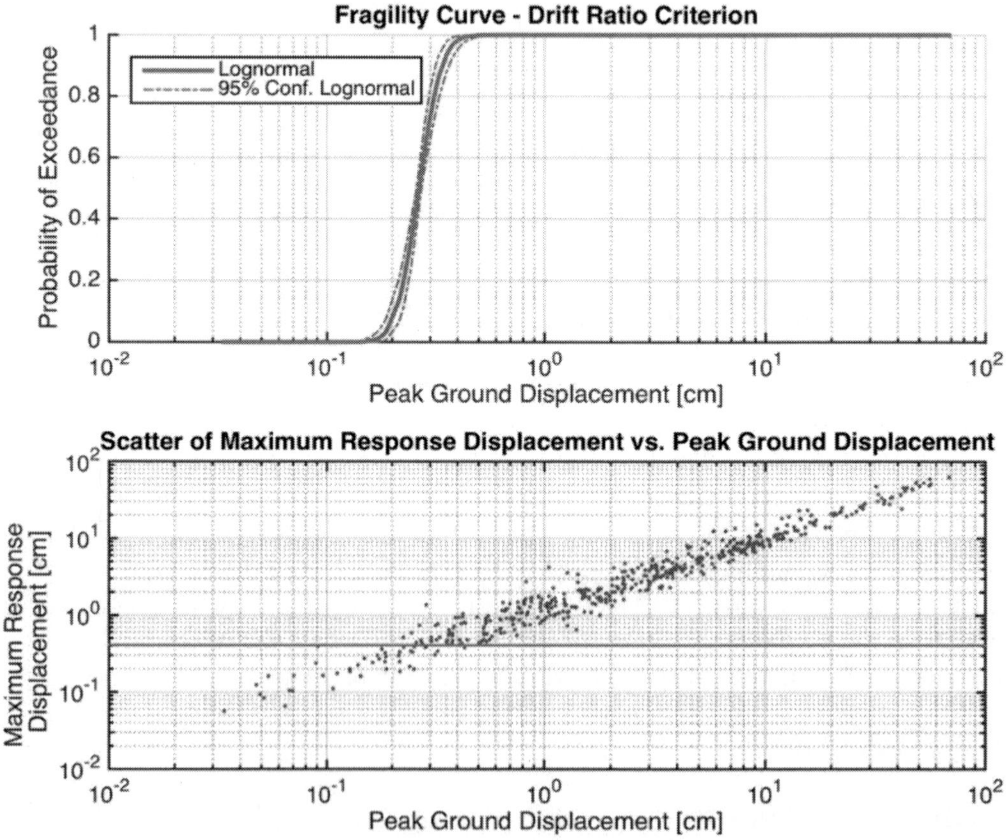

Figure 2: Fragility curve (top) made using the drift ratio criterion, which determined the performance limit to be 0.16 in., or 0.4 cm. A scatter plot (bottom) shows how the maximum displacement of the dam relates to the input earthquake's PGD.

Analytical Cantilever Criterion

A damage criterion can also be obtained by approximating the dam as a cantilever beam (see Figure 3). By modeling the dam as an Euler-Bernoulli beam, a correlation between the deflection of the dam's crest (e.g. in the stream direction and presumably at the location that will deflect the most) and the maximum stress in the dam can be made. The correlation between the cantilever model's maximum stress and displacement can then be used to evaluate the predicted displacement at which this maximum stress will occur. Using this approach, if the stress levels approach or exceed reported ultimate stress levels for the dam's concrete, an assessment of the dam's vulnerability to the hazard loading condition can be made by using the corresponding fragility curve estimates.

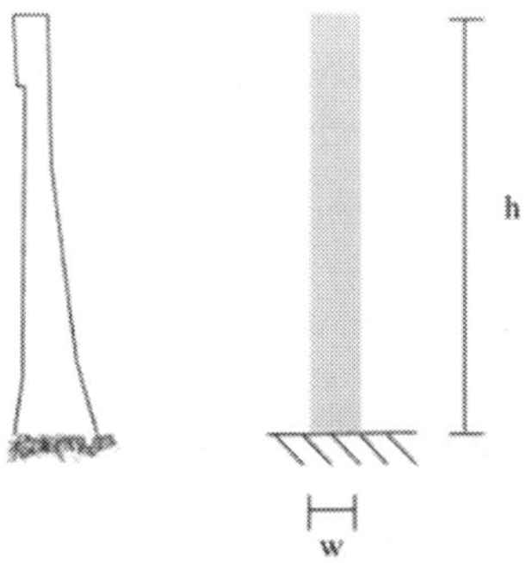

Figure 3: The original cross-section of the dam (left) and the simplified cantilever beam (right) used to model it.

As illustrated in Figure 3, the dam is modeled as a beam with rectangular cross-section of height h, length l (parallel to the cross-stream direction), and width w (parallel to the stream direction). Improved correlation may result by developing equivalent moment of inertia and section properties of the dam's actual cross section for the cantilever model. When subject to a statically applied load at the cantilever's free end, the maximum deflection occurs at the tip of the beam and is described by the equation

$$\Delta_{max} = \frac{Ph^3}{3EI} \quad (2)$$

where E is the elastic modulus, and I is the second moment of area; in particular, for a rectangular beam, the second moment of area is a constant $I = \frac{lw^3}{12}$. The corresponding maximum stress occurs at the base of the cantilever and has magnitude

$$\sigma_{max} = \frac{Ph(\frac{w}{2})}{I} \quad (3)$$

for this simplified model. Thus, in terms of the maximum stress, the maximum deflection can be described by the equation

$$\Delta_{max} = \frac{2h^2\sigma_{max}}{3Ew}. \quad (4)$$

By using the yield strength as σ_{max}, the maximum displacement at the top of the dam before yield can be approximated and thus used as the performance limit.

For the current case study, this damage criterion was applied to the data collected from the PBT experiment performed on San Clemente Dam. The width of the dam is considered to be the average of the base of the dam (approximately 14.5 ft.) and the top of the dam (approximately 4 ft.). The height of the dam is considered to be the distance from the crest of the dam to the top layer of the sediment piled on the upstream face when the test was conducted, which is approximately 80 ft. The material properties of the concrete are chosen to evaluate a range of concrete conditions to show the expected variation for both damaged and healthy concrete. Material properties include the concrete's elastic modulus and tensile yield strength which ranged from 3.5-1.5 Mpsi and 700-300 psi, respectively. The assumption is that the modulus and tensile strength values will decrease as the concrete's condition deteriorates. The tensile yield strength is used as opposed to its compressive yield strength because the potential failure modes of interest are related to concrete's ability to resist tension as opposed to compressive failure induced modes. Using this analytical cantilever criterion, the range for the performance limit was found to be 0.47-2.58 inches. The fragility curve using the average of this range (1.53 in.) as the performance limit is shown in Figure 4.

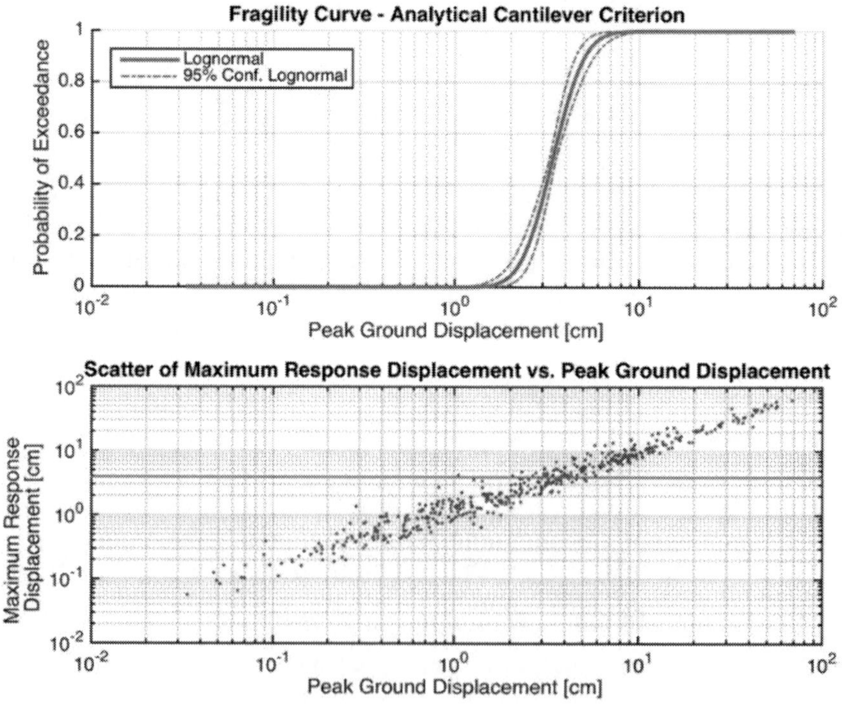

Figure 4: Fragility curve (top) made using the analytical cantilever criterion, which determined the performance limit to be 1.53 in., or 3.89 cm. A scatter plot (bottom) shows how the maximum displacement of the dam relates to the input earthquake's PGD.

Finite Element Cantilever Criterion

This criterion extends the analytical cantilever beam approximation by using a finite element model representation of the dam. Although finite element models may not always produce accurate representations of actual dynamic behavior, they do provide fairly reliable solutions to the elastic deformation of a structure under a static load. The relevant static analysis for developing the dam's fragility curve involves the static response of the model to a uniform static load applied along the entire crest of the dam.

This criterion is then applied to the fragility curve case study for San Clemente Dam. First, using exclusively the finite element model, a relationship must be established between the maximum displacement of the dam and the maximum stress experienced by the dam. Specifically, the maximum displacement is taken to be the deflection of the dam at the location of interest or the location of an accelerometer from the PBT experiment. In the top of Figure 5, the accelerometer location is chosen to be close to the center of the crest of the dam to better reflect the results of the analytical cantilever. Using the finite element model, the maximum stresses in the dam are determined. Though in most dams the maximum stress is likely to occur at the crest of the dam, in the case of San Clemente Dam, for this analysis, the maximum stress was chosen from the area in the center of the dam below the spillway in order to capture the response of the dam rather than the concentrated stresses resulting from the geometric discontinuities on the spillway piers.

Although the FEA model only represents the linear elastic behavior in the dam, the static analysis results can be used to estimate a ratio of Displacement to Principal Stress (DPS) that corresponds to presumed damaged states in the dam. This ratio can then be multiplied by the maximum tolerable yield stress of the concrete to determine the maximum tolerable deflection for the dam which is used as the performance limit, as shown in Equation 5:

$$\Delta_{max} = \sigma_{yield}(DPS) \tag{5}$$

where σ_{yield} is the estimated yield stress of concrete.

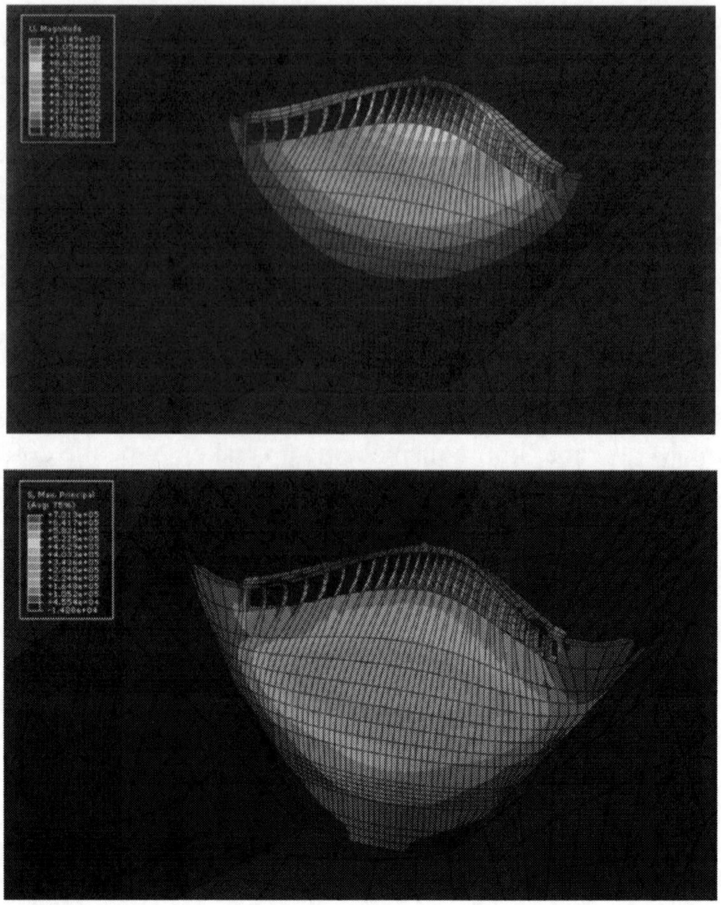

Figure 5: Abaqus model of displacement (top, in.) and maximum principal stress (bottom, psi) under uniform, static loading along the bottom of the spillway. Accelerometer location is denoted by a red dot on the crest.

Calculations were performed using an elastic modulus range of 1.5-3.5 Mpsi, which produced a corresponding DPS range of 0.314-1.43 milli-inches/psi.

To emphasize again, all the data used up to this point is determined exclusively from the finite element model; no data from the PBT experiment has been used yet. The only point where the PBT experiment is used is in the generation of the fragility curve where the maximum displacement obtained from the convolution between the PBT data and earthquake time histories is compared to the performance limit determined from the FEA model.

As with the analytical cantilever analysis, a range of performance limits can be determined for a range of conditions of concrete. Again, using a range of elastic moduli from 1.5-3.5 Mpsi and a range of yield stresses from 300-700 psi, a range of maximum tolerable deformation was computed. Using the finite element criterion, the range for the performance limit was found to be 0.4-2.2 in. The fragility curve using the average of the minimum and maximum of this range (1.3 in.) as the performance limit is shown in Figure 6.

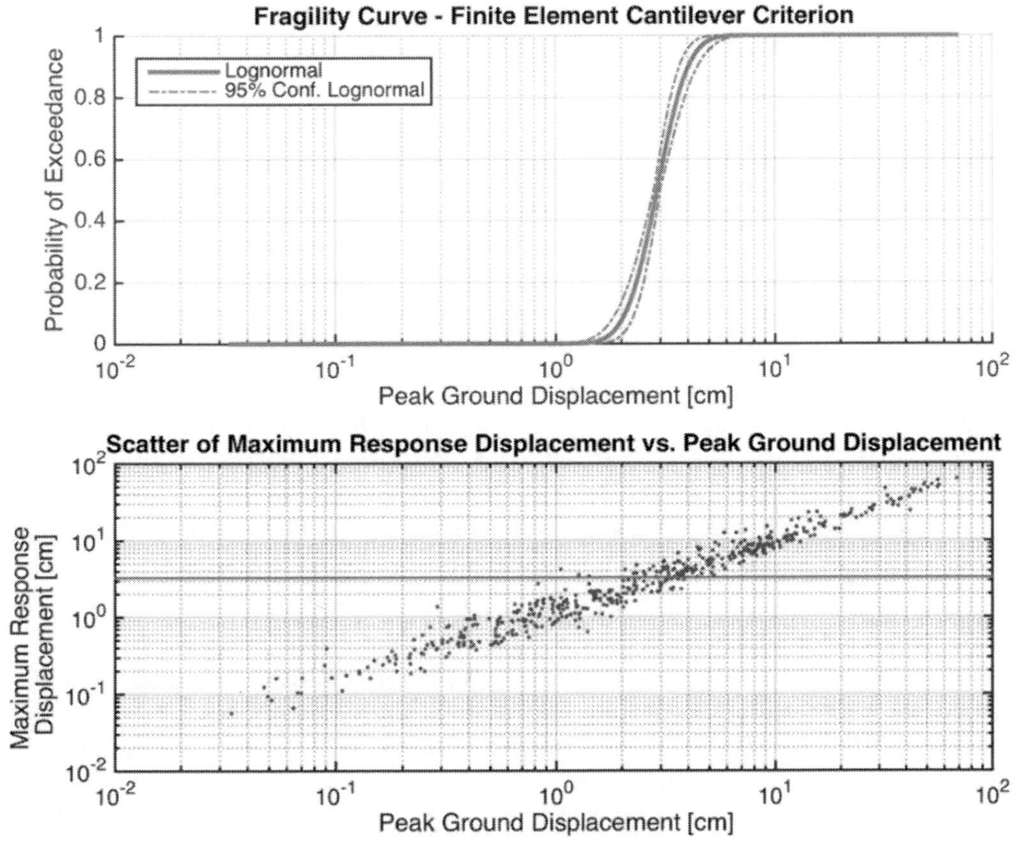

Figure 6: Fragility curve (top) made using the finite element cantilever criterion, which determined the performance limit to be 1.3 inches, or 3.3 cm. A scatter plot (bottom) shows how the maximum displacement of the dam relates to the input earthquake's PGD.

Comparison of Damage Criteria

Three damage criteria, each more complex than the one before, have been evaluated for their influence on the fragility estimate. The first criterion is based on a drift ratio that relates the maximum deflection and height of the dam to a 10% probability of failure. The advantage of using this approach as a damage criterion is that it is to some degree relevant since it is an adaptation of building story drift requirements. Nevertheless, this damage criterion is problematic primarily because it generalizes every kind of dam and uses only its height as a parameter for determining its ability to sustain damage. For example, this criterion would not encapsulate the anchoring effect of the abutments nor would it take into account the material property variation of most dams. In addition, this calculation does not allow the dam owner to define different levels or types of failure as it is completely defined by the ratio of displacement to height.

The cantilever beam model approach provides a simple, analytic, "back-of-the-envelope" calculation that can be easily, and quickly, calculated. Although this model allows the inclusion of material properties such as the yield strength and elastic modulus of concrete as well as the thickness of the dam, it is largely a two dimensional representation that does not take advantage of the known and adjacent constraints and boundary effects on the dam's center section from its neighboring monoliths and abutments.

Finally, the finite-element cantilever criterion adds another layer of complexity by introducing numerical modeling. The finite element model takes into account complex geometric conditions such as the arch in the dam and the abutment boundary, and it also provides a full three dimensional representation of the dam. The model adds versatility to the fragility estimation by allowing the analyst to focus on specific areas in the dam for different types of failures. For example, fragility may be estimated inside the spillway piers for those conditions that lead to crack or weakened piers but that do not necessary lead to uncontrolled rapid water release. And, the model can be used to focus on fragility of the dam below the piers for evaluating a failure configuration that could lead to uncontrolled rapid water release. This versatility is not provided by either of the previous damage criteria.

Although the finite element model is able to offer more geometric flexibility than the analytical cantilever, the performance limits the two criteria produce for the dam as a whole are fairly similar. In the case of San Clemente Dam, for the material concrete property combinations considered, the predicted maximum deformation ranges between 0.5 in. and 2.6 in. for the analytical cantilever, and between 0.4 in. and 2.2 in. for the finite element cantilever. If the dam owner is focused on a probability of exceedance of up to 10%, the cantilever model based fragility estimates this probability will be reached at an intensity measure (PGD) of 2.3 cm, whereas the finite element model based fragility estimate indicates the 10% probability of exceedance will be reached for an intensity measure (PGD) of about 2.1 cm. As seen in the case of San Clemente Dam, the damage threshold and the intensity measure associated with 10% probability of failure for both the analytical and finite element cantilever models are very similar. This shows that the simplified analytical cantilever model closely approximates the more complicated finite element model; moreover, the analytical model offers a slightly more conservative damage threshold. Thus, the simplified approach may be used to direct more detailed studies, should a finite element model based approach be unavailable or impractical at the time of the assessment.

CONCLUSION

Fragility curves are extremely useful tools for dam safety and decision making since they help inform dam owners about the vulnerabilities of a dam under hazard (e.g. seismic) loading conditions. A series of damage criteria can be developed that range from simple adaptations of building codes to equivalent static conditions in a dam, to a two dimensional cantilever model approach that produces a closed form damage criteria, to the use of a three dimensional finite element approach that allows for the consideration of a wider range of scenarios in which specific regions of the dam fail. These damage criteria vary in complexity, forming a tradeoff between simplicity in calculation and being able to account for more material properties, geometric factors, and unacceptable states in the dam.

In particular, the analytical and finite element cantilever models are able to provide the dam owner with a range of performance limits depending on the quality of the dam's concrete. This adds another degree of freedom in the dam owner's risk analysis: the probability of having a certain concrete condition can be combined with the probability of exceeding the corresponding performance limit (using the fragility curve) to provide an improved risk assessment of the dam. In terms of geometric factors accounted for by the damage criteria, the finite element model is able to offer a large amount of spatial information, which allows the decision-makers to determine what performance limit to pick based on the stress of a particular area of interest on the dam. However, despite the geometric flexibility available in the finite element model, finite element models are also relatively more resource intensive than the other damage criteria. To allow the dam owner to begin to evaluate the conditions in the dam that could lead to unacceptable stress, the analytical "back-of-the-envelope" calculation can be used as a first step evaluation.

ACKNOWLEDGEMENTS

The authors wish to acknowledge the support from the following individuals and their organizations for making the field investigations at San Clemente Dam possible.

David Gutierrez, Chief, and Mark Schultz, Supervising Engineer from the Division of Safety of Dams Department of Water Resources, California.
Richard Svindland, VP of Operations at California American Water, dam owner.
Bill McGowan, Project Executive at Granite Construction Company.
Dylan Wade, Resident Project Engineer at San Clemente Dam for Water Systems Consulting, Inc.
Robert Carnes, consultant and logistical support to the Resident Project Engineer.

The authors are especially grateful to Mr. Wade and Mr. Carnes for their efforts in managing our support needs and safety during the performance testing at the dam.

REFERENCES

Anderson, S. K. Menda, M. Zhao, and Z. Duron. (2016). "Selecting an Intensity Measure for Fragility Curves Derived from Performance-Based Testing." Denver, Colorado, 36th Annual USSD Conference.

Menda, K., M. Zhao, S. Anderson, and Z. Duron. (2016). "A Performance-Based Approach to Developing Fragility Curves for Structures Subject to Seismic Loading Conditions." Denver, Colorado, 36th Annual USSD Conference.

AUTHOR INDEX

Aberle, Jochen....1531
Adams, Lea G....749
Alicescu, Vlad....1463
Allerton, William....1035
Amos, Peter....829
Anderson, Spenser....1483, 1559, 1655
Annandale, George W....221
Arnold, Justin....1639
Arrocha, Jaime....1341
Auld, Bernie....1191
Ball, Andre....1365
Barrentine, Scottie....335
Barrentine, Scottie L....1379
Barrie, Richard E....711
Bartles, Michael....269
Bartojay, Katie....727
Baxter, Craig P....221
Beaty, Michael H....535
Beduhn, Robert....1
Bennett, Tony....1173
Benson, Brian....829
Bentler, David J....849, 939
Beriswill, Jeffrey A....1267
Beutel, Justin....515
Bishop, Matt....1497
Bisnett, Rachael V....1133
Blakley, Kyle R....813
Bliss, Mark....1283
Booth, Derek....1001
Bouaanani, Najib....611
Boulanger, Ross W....535
Bountry, Jennifer A....251
Boyd, M. Leslie....445
Boyer, Bradley T....165
Boyer, D....1017, 1035
Brandon, Thomas L....849, 939
Brumley, Bryan J....711
Bukhair, Layne....445
Bulger, Jonathon R....1395
Byrd, Aaron R....123
Calcagno, Frank....1161
Campbell, Dave....11
Cannon, Robert....1365
Carey, Bryan....695
Castillo-Rodríguez, Jesica Tamara....1053
Charlwood, Robin G....1119
Chowdhury, Khaled....1297
Christensen, Dan....107
Coats, J. Tyler....165
Compton, Brandon....563
Cooper, Paul J....1379
Costa, Ray....781
Craig, Matthew....581
Crookston, B. M....79, 165
Crowder, James R....165
Cui, Yantao....1001
Curtis, Dan D....675
Dalton II, Larry L....563
Daly, Steven F....1507
Darbar, Soheil Razavi....727
Davidheiser, Steven....1251
Davis, Chuck....237
De Puy, Maximiliano....1341
De Rubertis, Kim....1017, 1145
Deere, Don....1419
Depin, Nickolas J....1267
Dingrando, Jeffrey S....29, 813
Dirdal, Melinda....11
Dodson, Andrew....335
Dolen, Timothy....727
Dombeck, Elise....627
Downer, Charles W....123
Dunbar, Joseph B....799
Duren, Angela....1507
Duron, Ziyad....657, 1483, 1559, 1655
Dworak, Frank....139
Ebner, Laurie....581
Eichinger, Robert A....149
Eichler, Jason A....199
Engemoen, William....1069
Escuder-Bueno, Ignacio....1053
Esterhuizen, Jacob....1419
Feng, Lingmim....675
Fernandez-Delgado, Gabriel....1419
Fiedler, William R....1069, 1097
Fleming, Matthew....269
Fletcher, John....675
Foos, William F....1161
Forrest, Michael....969
Fortuny, William....581
Franceschi, Pastora....1341
Franz, Raymond J....1395
Freitas, Mark....395
Gagnon, J. Hawkins....1365
Ganjoo, Bal....395
Gentzler, Seth....1001
Gerlach, Michael....149
Ghanaat, Yusof....497
Gilbert, Daniel A....29, 1379
Giovando, Jeremy....1521
Goetz, Chris....1341
Goodell, Christopher....107
Gowda, Santhosh G....1603
Grant, Brian....179
Graulau-Santiago, Jaime A....123
Griebenow, Kevin....643
Groeneveld, Joe....1639
Gullo, Nathan M....199
Haberman, Ruth....97
Hamernik, David....581

AUTHOR INDEX

Hamilton, David 29
Hannes, Danielle Tripp 1201
Hansen, Kenneth D. 1447
Harder, Leslie F. 883
Harris, David W. 955
Hasan, Husein 1217
Hatton, Chris 727
Heath, Benjamin 1191
Hedien, Jason E. 303
Hess, Matt 581
Hibbs, David 415
Hiller, Priska Helene 1531
Hlepas, Georgette 315
Hou, Susan 969
Hou, Susan S. 469
Hughes, Aled 695
Hughes, David 485
Hultstrand, Doug 287
Hunt, Dina Bourliea 515
Irving, Derek 1341
Jaeger, Robert A. 865, 909
Jamieson, Stephen L. 1201
Johanson, Alan 469
Jones, Adam N. 1119
Jones, Scott 485
Kappel, Bill 287
Kara, Redha 1463
Karshenas, Afshin 563
Kay, Roger L. 547
Kazyak, Brian J. 1395
Keel, K. 79
Kees, S. 79
Keizer, Richard A. 909
Kim, Wonnie 1133, 1541
Kirschbaum, Daniel 1639
Kjosavik, Fredrikke 1531
Kleiner, David 45
Kline Jr., Robert A. 1251
Kovacich, Joseph R. 1395
Krohmer, Brian 1217
Kynett, Michael 1297
Laboon, John H. 1097
Lan, Frank 179
Laugier, Frédéric 61
Lee, Tedman 469, 969
Lia, Leif 1531
Lillycrop, Jeff 563
Lim, Yeo Howe 1591
Lin, S. Samuel 303
Lindyberg, Robert F. 563
López-Soto, Jamie 927
Lothes, Amanda 1251
Ludy, Jessica 765
Macleod, Bruce 1267
Mahnke, Steve 781
Maki, Ian P. 865
Malvick, Erik J. 985
Margo, David A. 749
Marshall, Kevin 1119
Martin, Kristen M. 985
Martinez, Gabriel 1541
Masters, Monica 445
May, Michael R. 1621
McClelland, Duane 199
McMichen, Lindsay 335
Meier, Dan 1341
Mejia, Lelio 1341
Menda, Kunal 1483, 1559, 1655
Meyer, Don B. 1621
Millet, Richard 781, 1297
Miquel, Benjamin 611
Miranda, Roberto 1341
Mirghasemi, Ali Asghar 1591
Mishalanie, David 1119
Mockler, Stephen J. 435
Moen, Keith 1639
Montalvo-Bartolomei, Axel 927
Montgomery, Donald J. 1133
Morales-Torres, Adrián 1053
Morrison, Michael A. 1575
Morse, Daniel 199
Munsey, Jeff W. 813
Muto, Matthew 1083
Neff, Keil 1191
Ng, Carman 969
Nolan, Cecily M. 1235
Nottis, Aaron 1217
Nuss, Larry K. 1447
O'Brien, Steve 485
Oestreich, Cynthia 107
Olsen, Richard S. 351
Owens, Brian 1497
Pate, Kim 1365, 1639
Paxson, Greg 11
Perlea, Mary 781, 1297
Pinchard, Thomas 61
Piotrowski, Jesse 237
Powers, Matthew 395
Pradhan, Nawa Raj 123
Queen, David 727
Quiroz, Juan E. 711
Randle, Timothy J. 251
Rauch, Alan F. 1379
Renaud, Sylvain 611
Rice, John D. 909
Riker, Richard E. 1419
Ritchie, Andy 251
Roadifer, John 969, 1001
Robbins, Bryant A. 927
Robblee, G. 79

AUTHOR INDEX

Rogers, Daniel B. 813
Rogers, Gary 1365
Rosculet, Dan....1463
Roth, Lawrence H. 765
Rutherford, James H. 1639
Saber, Robert 415
Sadden, Brian....695
Salamon, Jerzy 599
Salter, Paul 1341
Savage, Bruce 97
Sawka, Mark 581
Scheevel, Nathan 221
Schenk-McFarland, Tara 1283
Schug, David....1341
Schulte, Bernard 1575
Schweiger, Paul G. 1161
Scott, Gregg A. 1083
Serota, Tali 1173
Shewbridge, Scott E. 749
Sheydayi, Peter....1001
Shoghli, Bahareh....1591
Siddappa, Gopi....1603
Sills, George L. 883
Simpson, Mark....1267
Skahill, Brian E. 123
Snider, Frederic 1365
Snyder, David....415
Snyder, Kevin....1119
Sooch, Gurinderbir S. 675
Stanard, Tina....97
Stateler, Jay N. 379, 1017
Stephens, Isaac J. 351, 799, 927
Stiller, William B. 435
Storino, Louis 1395
Sudah, Bashar S. 711
Supranata, Y. Erwin....435
Sutley, Garrett....955
Tarbox, Glenn 695
Tenney, Ray D. 1621
Thompson, David P. 711
Thompson, Mark J. 849, 939
Toose, James....1341
Tournier, Jean Pierre 1463
Triplett, Dennis 1217
Tullis, Blake....97
Tutka, Travis....315
Urban, Robert 1341
Vandenberge, Daniel R. 849
Vargas, Boris Caro 1217
Varuso, Richard....749
Vasquez, Victor 97
Vasquez, Victor M. 445
Vermeulen, Julien 61
Von Gersdorff, Nicolas 1083
Waddell, Robert....179
Wade, Daniel....969
Wade, Daniel L. 469
Wahlin, Brian 237
Walberg, Francke....1297
Walshire, Lucas A. 799
Wan, Cheuk Y. 1235
Warms, Mikell....107
Watts, Matthew 563
Weber, Christine 515
Welshans, April L. 29
Werner, John 1639
Weston, David M. 123
Whitten, Zachary 149
Wiesner, S. 1321
Wiggins, Charles E. 563
Williams, Luke....1497
Witkowski, Doug 445
Wohletz, A. 1321
Wormer, Jeff....379
Yen, John 657, 1083
Zambrano, Carlos 1541
Zhao, Mo 1483, 1559, 1655
Zheng, Jiqin 675
Zusi, Michael 179

United States Society on Dams (USSD)
1616 Seventeenth Street #483
Denver, CO 80202
USA

ISBN 978-1-5108-2942-8

CURRAN ASSOCIATES INC.
proceedings.com

13th International Conference on Radiation Shielding (ICRS-13) & 19th Topical Meeting of the Radiation Protection and Shielding Division of the American Nuclear Society (RPSD-2016)

EPJ Web of Conferences Volume 153 (2017)

Paris, France
3 - 6 October 2016

Part 1 of 2

edp sciences